Main–Group Elements

10	11 IB	12 IIB	13 IIIA	14 IVA	15 VA	16 VIA	17 VIIA	18 VIIIA
								2 **He** 4.002602
			5 **B** 10.811	6 **C** 12.011	7 **N** 14.00674	8 **O** 15.9994	9 **F** 18.9984032	10 **Ne** 20.1797
			13 **Al** 26.981539	14 **Si** 28.0855	15 **P** 30.973762	16 **S** 32.066	17 **Cl** 35.4527	18 **Ar** 39.948
28 **Ni** 58.69	29 **Cu** 63.546	30 **Zn** 65.39	31 **Ga** 69.723	32 **Ge** 72.61	33 **As** 74.92159	34 **Se** 78.96	35 **Br** 79.904	36 **Kr** 83.80
46 **Pd** 106.42	47 **Ag** 107.8682	48 **Cd** 112.411	49 **In** 114.82	50 **Sn** 118.710	51 **Sb** 121.75	52 **Te** 127.60	53 **I** 126.90447	54 **Xe** 131.29
78 **Pt** 195.08	79 **Au** 196.96654	80 **Hg** 200.59	81 **Tl** 204.3833	82 **Pb** 207.2	83 **Bi** 208.98037	84 **Po** (209)	85 **At** (210)	86 **Rn** (222)

Inner–Transition Metals

63 **Eu** 151.965	64 **Gd** 157.25	65 **Tb** 158.92534	66 **Dy** 162.50	67 **Ho** 164.93032	68 **Er** 167.26	69 **Tm** 168.93421	70 **Yb** 173.04	71 **Lu** 174.967
95 **Am** (243)	96 **Cm** (247)	97 **Bk** (247)	98 **Cf** (251)	99 **Es** (252)	100 **Fm** (257)	101 **Md** (258)	102 **No** (259)	103 **Lr** (262)

GENERAL CHEMISTRY

THIRD · EDITION

Periodic Table of Elements

Main-Group Elements

1 H 1.00794	Atomic number Symbol Atomic weight

Transition Metals

Period	1 IA	2 IIA	3 IIIB	4 IVB	5 VB	6 VIB	7 VIIB	8	9 VIIIB
1	1 **H** 1.00794								
2	3 **Li** 6.941	4 **Be** 9.012182							
3	11 **Na** 22.989768	12 **Mg** 24.3050							
4	19 **K** 39.0983	20 **Ca** 40.078	21 **Sc** 44.955910	22 **Ti** 47.88	23 **V** 50.9415	24 **Cr** 51.9961	25 **Mn** 54.93805	26 **Fe** 55.847	27 **Co** 58.93320
5	37 **Rb** 85.4678	38 **Sr** 87.62	39 **Y** 88.90585	40 **Zr** 91.224	41 **Nb** 92.90638	42 **Mo** 95.94	43 **Tc** (98)	44 **Ru** 101.07	45 **Rh** 102.90550
6	55 **Cs** 132.90543	56 **Ba** 137.327	57 **La*** 138.9055	72 **Hf** 178.49	73 **Ta** 180.9479	74 **W** 183.85	75 **Re** 186.207	76 **Os** 190.2	77 **Ir** 192.22
7	87 **Fr** (223)	88 **Ra** (226)	89 **Ac**** (227)	104 **Unq** (261)	105 **Unp** (262)	106 **Unh** (263)	107 **Uns** (262)	108 **Uno** (265)	109 **Une** (267)

Metal

Metalloid

Nonmetal

*Lanthanides	58 **Ce** 140.115	59 **Pr** 140.90765	60 **Nd** 144.24	61 **Pm** (145)	62 **Sm** 150.36
**Actinides	90 **Th** 232.0381	91 **Pa** (231)	92 **U** 238.0289	93 **Np** (237)	94 **Pu** (244)

James F. Wilson

HOUGHTON MIFFLIN
INTERNATIONAL EDITION

GENERAL CHEMISTRY

Third Edition

Darrell D. Ebbing
Wayne State University

Consulting Editor:
Mark S. Wrighton
Massachusetts Institute of Technology

HOUGHTON MIFFLIN COMPANY
Boston London Melbourne

Warning: This text contains photographs of experiments that are potentially dangerous and harmful if undertaken without proper supervision, equipment, and safety precautions. DO NOT attempt to perform these experiments relying solely on the information presented in this text.

Credits: List of credits follows the index.

Values for atomic weights listed in the periodic table on the inside front cover of this book are from the IUPAC report ''Atomic Weights of the Elements 1988,'' *Pure and Applied Chemistry,* Vol. 60, No. 6 (1988), pp. 841–854 (© 1988 IUPAC).

Printed in the U.S.A.

Library of Congress Catalog Card Number: 89-80931

ISBN: 0-395-62301-4

ABCDEFGHIJ-VH-954321

Foreword

New editions of a good chemistry text are essential, because chemistry is a science that changes. This third edition by Darrell Ebbing builds on the strengths of the first two editions, but significant improvements have been made. Further, there are many changes that make the text timely for the 1990s. To be sure many of the principles of chemistry remain unchanged. However, illustrations of these principles can be taken from modern problem areas. In many cases we have used new illustrations of chemical principles to make the text more interesting and relevant. Other changes in a new edition must include new discoveries. One such discovery included in this edition of *General Chemistry* concerns superconductivity. High temperature superconductors discovered in the latter half of the 1980s represent a revolution in an area of science where the technological consequences range from less expensive, superior medical diagnostics to more efficient energy and transportation systems and faster electronic devices. Such new advances in science are included in *General Chemistry*.

Advances in analytical methods and instrumentation bring about improvements in understanding matter. The scanning tunneling microscope developed in the 1980s is a new tool for examining surfaces with atomic level resolution. Some results from the use of this tool are included in this edition of *General Chemistry*. Laser diagnostics tools have also been developed and the cover photo depicts experimentation directed toward probing the combustion products of coal. The international community has awakened to the possible consequences from global scale changes brought about from the use of fossil energy, and advances in combustion chemistry are needed to reduce emissions of CO_2 and chemicals responsible for acid rain.

As Consulting Editor for *General Chemistry,* I have worked closely with the author, Darrell Ebbing, and members of the Houghton Mifflin editorial, production, and art team. Our collective objective has been to produce a high-quality text that offers the best way to learn chemistry. Some of my specific activities have included reading and criticizing new text, developing responses to reviewers' comments and suggestions, contributing ideas for the cover photo and other illustrations, and writing or rewriting certain pieces for Darrell Ebbing's final review and preparation.

Learning about chemistry with *General Chemistry* will better equip an individual to deal with world-scale problems such as energy, human health, global change, and food production. In a very short period of time the users of this text will be able to participate in science. Learning about chemistry is challenging and part of the reward is being able to use it in other areas of endeavor. This edition of *General Chemistry* provides many examples of the uses of chemistry and opens many paths for future studies and careers.

Mark S. Wrighton

Preface

TO THE STUDENT:

The purpose of this textbook is to introduce you to the basic facts and principles of chemistry. Chemistry is a vibrant, growing science. It is of fundamental importance not only to all the other sciences and modern technology but to any explanation of the material things around us. Consider these diverse questions. How is it possible for the human eye to detect certain wavelengths of light? What is responsible for the red color of Io, one of Jupiter's moons? And finally, how can we see inside the brain of a patient without doing harm? Do these sound like questions involving chemistry? They are, and these are just some of the questions you will explore in your reading of this text. I hope I have piqued your curiosity. In your study of general chemistry, you will discover many things, but ultimately you will learn that there is so much more to learn and that it is exciting to discover and to question.

The challenge to any author of a general chemistry text is to present a solid understanding of the basic facts and principles of chemistry while retaining the excitement of the subject. Mark Wrighton, the consulting editor for this text, and I feel strongly that the way to do this is by constantly relating the subject matter to real substances and problems in the real world. We begin the study of chemistry with the discovery of the anticancer activity of a bright yellow substance called cisplatin. We use this discovery to illustrate the introductory ideas of the chapter and to discuss the scientific method. In Chapter 2, we start the discussion of chemical substances by looking at sodium (a soft reactive metal), chlorine (a pale green gas), and their reaction product—sodium chloride (ordinary table salt). Each of these illustrates a different type of substance. In each chapter, as we continue to study the basic principles, we keep close contact with the world of real chemical substances.

Features of the Text

Each of us learns in a different way. For that reason, we have incorporated a number of different features into the text to help you in learning the subject. We hope that by listing these features here, you will be able to tailor a study program that meets your particular needs.

CHAPTER OUTLINE Each chapter is preceded by an outline. A chapter is broken into parts, and these are divided into sections and possibly subsections. A glance at the outline will give you an overview of what you are going to read. Your instructor may refer to this outline in describing any changes in order or any omissions of material he or she plans to make. You may find this outline of value in reviewing. For example, as you review the outline, try to fill in the main details of the discussion under each heading.

CHAPTER THEME We begin each chapter with a theme, something specific that will make the topic of the chapter have relevance. As we mentioned earlier, we open Chapter 2 (Atoms, Molecules, and Ions) with a discussion of sodium, chlorine, and sodium chloride. This chapter theme then leads naturally into a series of questions (How do we explain the differences in properties of different forms of matter?), which we return to answer later in the chapter.

COLOR ILLUSTRATIONS Most of us are strongly visual in our learning. Fortunately, chemistry presents many opportunities for beautiful photography and colorful graphics. When you see something, you tend to remember it. With this in mind, we have chosen color illustrations that both clarify the discussion and help you to visualize it.

PROBLEM-SOLVING PROGRAM We only learn to the extent that we are involved with a subject, and we learn by doing. It often looks deceptively easy when your instructor explains how to solve a problem. But it is like learning to swim or to play a musical instrument. It only becomes easy with practice. To learn to solve chemistry problems, you must solve chemistry problems and solve many of them. Chemistry builds one principle on another, and fact on fact. The secret of problem-solving in chemistry (if secret is the right word) is to know what you learned earlier so well that when you approach a new problem, you know how to put the pieces together.

Recognizing the importance of problem-solving in chemistry, we felt the burden could be much reduced if we followed a consistent problem-solving program. We introduce each problem-solving skill by an *Example,* in which you are led through the reasoning that is involved in working out a particular type of problem. The skill has been selected to represent a specific category of problems encountered frequently in general chemistry. Each Example is accompanied by an *Exercise,* which is a similar problem that you can try; the answers are at the end of the book. (Some Exercises are unaccompanied by an Example because the problem-solving is not sufficiently complex to justify a formal Example; you will be able to work the Exercise by following the preceding text discussion.) At the end of the Exercise is a list of corresponding end-of-chapter *Practice Problems.* Try some of these to gain mastery of that problem-solving skill.

VOCABULARY Chemistry uses words in a precise way, and it is important to develop a vocabulary of terms in order to read and communicate effectively. Every time a new important word has been introduced in the text, we have flagged it by putting it in boldface type. The definition of that word will generally follow in the same sentence in italic type. (In any case the definition will appear in italic type close to the boldface word.) All of these words are collected at the end of the chapter in the list of *Important Terms.* They also appear, along with a few other words, in the *Glossary* at the end of the book. Whenever you are reading along and you encounter a word whose definition you do not recall, look in the Glossary.

CHECKLIST FOR REVIEW When it comes to reviewing, each of us has developed our own techniques. What we have tried to do is accommodate these differences by presenting various review possibilities. For example, you may find that the list of *Important Terms* is useful not only because it is a list of new words, but also because as you look over the words, you see the structure of the chapter. As you mentally note this structure, try to recall the ideas associated with the words. Many chapters also introduce one or more mathematical equations to be used in problem-solving. In the chapter, these are noted in blue type. Then in the Checklist for

Review, they are listed as *Key Equations*. The *Summary of Facts and Concepts* presents a verbal summary of the chapter. Study this and as you go over each statement, try to flesh out points. (Imagine you are the instructor, and try to explain the ideas and relate the facts to another student.) Finally, we present a list of *Operational Skills*. This is a summary of the problem-solving skills. Each operational skill tells you what information is needed and what is to be solved for in a given type of problem. These operational skills also refer back to the Examples that discuss that problem-solving skill.

END-OF-CHAPTER QUESTIONS AND PROBLEMS The end-of-chapter questions and problems begin with *Review Questions*. These have been designed to test your understanding of the concepts and theory. Generally, they can be answered by straightforward recall or simple extension of the chapter material. After these questions, we have listed several sections of problems to help you gain mastery of problem-solving skills. The problems are in matched pairs, with the odd-numbered ones (given in blue type) having answers at the end of the book. The problems are divided into three groups: Practice Problems, Additional Problems, and Cumulative-Skills Problems. The *Practice Problems* are keyed to a particular topic or skill by heading; the *Additional Problems* are not. The *Cumulative-Skills Problems* require you to combine several skills, often from previous chapters. By their nature, these are often challenging problems, but by working some of these, you will be building your skills.

On the inside back cover of this book is a list of *Locations of Important Information*. This lists the pages of the text or the appendix where you will find data for problem-solving

TO THE INSTRUCTOR:

In the previous edition, I noted the following objectives that we had set for the book: (1) to explain the principles as clearly as possible by always relating abstract concepts to specific real-world events; (2) to present topics in a logical, yet flexible, order; and (3) to offer an abundance of meaningful instructional aids, particularly with respect to problem-solving. The response to the second edition has been indeed gratifying. Instructors' and reviewers' comments continue to note the readability of the text and its effective approach to problem-solving. So it again seemed to me that I should not tamper with the main organization and features of the book.

In planning this third edition, however, there were three areas that we wanted to further improve. One of these was to greatly strengthen the illustration program of the book. The other was to solve the problem of presenting relevant descriptive chemistry early, while at the same time proceeding forward with the discussion of the principles that explain that chemistry. The third was to continue to show how chemistry is practiced today. This plan turned out to be an ambitious one, but one that I hope you will agree turned out very well.

Illustration Program

It would be difficult to overestimate the importance of the illustration program to a chemistry text , since it is such an important adjunct to the exposition. On the other hand, it requires constant vigilance to see that a figure conveys the proper meaning. With this in mind, each of the figures in the previous edition was scrutinized to see that it did indeed do this. As I proceeded with the revision, new photographs and line drawings were added if they seemed to clarify the exposition or make it more vivid. The result is a greatly expanded illustration program, but one that I believe adds greatly to the teachability of the text.

One of the striking features of the present edition is its new full-color design. This edition includes over 300 carefully chosen color photographs. Many of these were of chemical reactions that were specially photographed for this text. I am convinced of the usefulness of color on several grounds. First, much of chemistry involves color and while you can describe such things in writing, it is so much more vivid if you can present a good color photograph. Further, many line drawings benefit by the use of color to clarify detail. And finally, our students today are used to seeing color. If the judicious use of color can increase their interest, then it has served a purpose. While adding color, however, we are mindful that if not done well, it can detract from the exposition. I hope you find that the color here adds to the usefulness of the text.

Flexible Treatment of Descriptive Chemistry

Although there would appear to be no consensus on precisely how to present the descriptive aspects of chemistry in the general chemistry course, there does seem to be agreement that we need to present it earlier in the course. This seems reasonable to me. But a flexible approach, one that gives instructors leeway in how they want to handle the teaching of descriptive chemistry, seems called for.

I have incorporated descriptive chemistry throughout the main text, in illustrating concepts, in examples, problems, and margin notes. However, an early but planned program of descriptive chemistry also seems needed. It was for this reason that we developed the *Profile of a Chemical* series. This series, which we introduced in the second edition, is now much expanded so that one or two essays appear at the end of each chapter from Chapters 2 to 19. Each of these essays focuses on a single substance and presents some of the interesting and important properties and uses of that substance. The writing style and colorful figures were chosen to enhance student interest. Each Profile is short, and by focusing on a single substance, the Profile is easily assimilated by the student. Study questions accompany each Profile so that a student should require little or no lecture time to learn the material (although you may want to give demonstrations or show videos). The Profile series begins with two reactive elements and progresses to acids and bases, nonmetals, metals, and oxidizing and reducing agents. By the end of the series, the student will have covered a substantial amount of descriptive chemistry.

Descriptive chemistry is still treated in separate chapters in the book. Chapter 3 provides an early introduction to chemical reactions and facilitates the early treatment of descriptive chemistry. Chapter 13 further explains the acid–base concept introduced in Chapter 3 and then discusses oxidation–reduction concepts. The text ends with a block of descriptive chemistry chapters covering the main-group elements (Chapters 21 and 22), transition elements (Chapter 23), organic chemistry (Chapter 24), and biochemistry (Chapter 25).

Instrumental Methods and Related Topic Essays

In addition to the Profile of a Chemical Series, the text includes two other series of essays. One of these is the *Instrumental Methods* series. It was felt that students sometimes come away from a general chemistry course not realizing that modern chemistry is very much dependent upon sophisticated instruments. When they do realize this, they often become quite excited by the subject. Each of the essays in this series focuses on an instrumental method used by research chemists, such as infrared spectroscopy or nuclear magnetic resonance. The essays are kept short and only enough detail is given to whet the students' appetite.

The other series is the *Related Topic* essays. Each of these expands on a chemical principle (as in the essay *Enthalpy and Internal Energy*) or describes the application of a principle (as in the essay *Hemoglobin Solubility and Sickle-Cell Anemia*).

Organization of Chapters

A glance at the *Contents in Brief* will show that the chapter organization is a typical one. Chapters 1 to 6 introduce basic chemistry, including stoichiometry. The inclusion of Chapter 5 on gases allows the instructor to follow stoichiometry with problems on gas volumes. However, you can easily defer gases until just before Chapter 11 on liquids and solids. Chapters 7 to 10 discuss atomic and molecular structure, Chapters 11 and 12 discuss states of matter and solutions. Then Chapters 13 to 19 describe chemical reactions and equilibrium. The final block of chapters treats nuclear chemistry and the chemistry of the elements.

The chapter organization of the previous edition has been retained except for the following changes. The two chapters on the atmosphere and water have been deleted and the important chemistry given there has been distributed to Profiles (oxygen, nitrogen, hydrogen, ozone, carbon dioxide, and water) and to Chapter 22 on the main-group elements.

The material of the reactions chapter (Chapter 10 in the second edition) was split into two parts. The first part was rewritten and made into Chapter 3 of this edition, which now becomes an early treatment of reactions. The second part on oxidation–reduction reactions was combined with acid–base concepts to give a second reactions chapter (Chapter 13). I believe these changes add flexibility by making it possible to discuss reactions early in the course. They are also useful for the Profile series. Despite these rearrangements, if an instructor wishes to teach the reactions later, it is still possible to do so.

Within individual chapters, I have retained the division of chapter sections into chapter parts, not only because I believe that it makes the organization of the chapter transparent, but because it increases the flexibility of the chapter. For example, an instructor who wishes to treat thermodynamics before equilibrium can simply discuss the first two parts of Chapter 18 on thermodynamics before equilibrium (Chapter 15), then return to the last part of Chapter 18 that discusses free energy and equilibrium.

Chapter Revisions

I have carefully reviewed each chapter to see if I could tighten up the presentation or clarify points. The most important of these revisions are as follows. Chapter 5 on the gaseous state was condensed to emphasize the ideal gas law and the treatment of stoichiometry of gas volumes. Chapter 6 on thermochemistry was extensively rewritten to clarify the presentation and to make some points more precise. In Chapter 10, a section on molecular orbital theory of metals and semiconductors was added. In Chapter 11, the presentation of crystal lattices and unit cells was revised to clarify the discussion. Chapter 16 on acid–base equilibria now includes the material on strong acids and bases in order to unify the discussion. Chapter 19 on electrochemistry was rewritten in parts to clarify the discussion and to add new applications.

COMPLETE INSTRUCTIONAL PACKAGE

This textbook is complemented by a complete package of instructional materials:

For the Student:

Study Guide for General Chemistry, Joan I. Senyk; Larry K. Krannich, University of Alabama at Birmingham; and James R. Braun, Clayton State College. This student study guide reinforces principles and extends problem-solving presentation by offering diagnostic tests, additional worked-out examples and problems, and solutions to all in-chapter exercises.

Student Solutions Manual, George Schenk, Wayne State University. This manual provides complete worked-out solutions to all odd-numbered problems, solutions to with-in chapter exercises, and answers to review questions.

Experiments in General Chemistry, Rupert Wentworth, Indiana University. This laboratory manual includes 40 experiments, with pre-lab assignments. An instructor's manual is also available.

Qualitative Analysis and Ionic Equilibrium, George Schenk, Wayne State University. This manual covers chemical principles and laboratory procedures needed for the qualitative analysis portion of the general chemistry laboratory.

For the Instructor:

Instructor's Manual, Darrell Ebbing, Wayne State University. This manual contains additional examples, lecture demonstrations, and suggestions for alternate sequencing.

Test Bank, Ron Ragsdale, University of Utah. The printed test bank contains 2000 multiple-choice questions organized by chapter. These test items are also available on disk for the IBM-PC, Apple II, and the Macintosh. A call-in test service is also available, allowing you to order printed tests by calling Houghton Mifflin's toll-free number.

Complete Solutions Manual, George Schenk, Wayne State University. This manual contains answers to all review questions and worked-out solutions to all problems and with-in chapter exercises.

Transparencies. 200 transparencies of figures, tables, and photographs selected from the text in full color.

Video Lecture Demonstrations, John Luoma, Cleveland State University; John J. Fortman and Rubin Battino, Wright State University. Over 40 demonstrations are provided to supplement your lectures.

ACKNOWLEDGMENTS

The preparation of an introductory textbook, even a revision, is a lengthy project involving many people. The beginnings of this edition took shape soon after the completion of the second edition. By that time, I was able to see some things that could be done better. In the following months, I collected letters from instructors and notes of conversations with others. Several in-depth reviews by users as well as nonusers of the second edition were obtained. All of this became grist for the first draft of the third edition manuscript. This first draft was read in part or whole by many reviewers, whose comments were used in preparing the final manuscript. Finally at the galley proof and page proof stage, accuracy reviewers (professors teaching general chemistry) provided the final checks needed to produce a technically accurate book. I owe each of these people a debt of gratitude beyond measure. The following is a list in alphabetical order of the reviewers.

Bruce S. Ault, University of Cincinnati
Norman C. Baenziger, University of Iowa
Peter Baine, California State University–Long Beach
Jesse S. Binford, Jr., University of South Florida
John E. Davidson, Eastern Kentucky University
Michael I. Davis, The University of Texas at El Paso
Dan Decious, California State University–Sacramento
John DeKorte, Northern Arizona University
Gery Essenmacher, University of Wisconsin–Madison
Clark L. Fields, University of Northern Colorado
C. Dan Foote, Eastern Illinois University
Paula M. Getzin, Kean College
John M. Goodenow, Lawrence Technological University
Anthony V. Guzzo, University of Wyoming
Anne Harmon, Lamar University
Henry F. Henneike, Georgia State University
Floyd L. James, Miami University
Richard L. Kiefer, College of William and Mary
Robert Kren, University of Michigan–Flint
Dorothy Kurland, West Virginia Institute of Technology
John Luoma, Cleveland State University
Kenneth Magnell, Central Michigan University
David Morrissey, Michigan State University
William E. Parker, Gettysburg College
Robert Pfaff, University of Nebraska–Omaha
William Plachy, San Francisco State University
David Richardson, Riverside City College
Stephen P. Ruis, Skyline College
Jerry Sarquis, Miami University
Maurice Schwartz, University of Notre Dame
Harold S. Swofford, Jr., University of Minnesota
Wayne Tikkanen, California State University–Los Angeles
Donald Titus, Temple University
Richard S. Treptow, Chicago State University
Virginia Urbanik, Averett College
John Weyh, Western Washington University
Warren Yeakel, Henry Ford Community College
Noel S. Zaugg, Ricks College
Orville Ziebarth, Mankato State University

It has been my great fortune to have Mark Wrighton as consulting editor for this book. He has been a constant source of new ideas, and his knowledge of chemistry is prodigious. His contributions are woven into the fabric of the text, into the essays, and the illustration program. And once again he has come up with a striking cover photo. Finally, he read and commented on all of the manuscript. My special thanks to Mark.

The new setup color photographs were an important addition to this revision. I strongly felt that the text required original photographs of chemistry in action. When preparing the manuscript, I generated ideas for photos. But it is a long way from a photo idea to the final photo. I would have been lost without the help of Joe Oravec of Wayne State University. Whenever I called him with questions, he found the answer, often by doing the experiment. Then, when it was time to do one of the many photo shoots, he planned each demonstration, making certain it would work. Then off he went to M.I.T. to prepare the setups for the photography session. I am extremely grateful to him for all his work.

I also want to thank Kazi Ahmed of M.I.T. for his generous contribution of time and effort in making the initial preparations for each of the photography sessions and for his assistance during the sessions. Finally, I want to express my thanks to James Scherer, who again took all the setup photographs. The professional results I think are obvious enough, but their beauty may obscure the difficulty involved in shooting many of them.

The preparation of the answer section that appears at the end of this book required the dedication of a number of people. Professor George Schenk, of Wayne State University, again tackled the arduous task of preparing the Solutions Manual. At the same time, Lynne Hitchcock and Sharyl Majorski independently worked out the solutions and prepared an initial draft of the answer section. When they finished, they checked their answers against those of Schenk. Finally, the solutions manual was checked by John Goodenow of the Lawrence Technological University, and the galley proof of the answers was checked against his work. I want to thank each of these people for their invaluable contribution.

While Joan Senyk and Larry Krannich prepared the Study Guide to this edition, they sent me many helpful corrections and suggestions for improvement of the text. Rupert Wentworth gave me a number of suggestions for changes in the text that he thought would help the presentation of the laboratory manual, changes that I realized would also improve the text. I want to thank each of these good friends for everything they have done.

Words are wholly inadequate to express my appreciation to those in my family for all they have endured from an absorbed author. To my wife Jean, to my son Russell, to my daughters Julie and Linda and their husbands Tony and Brad, and to my grandchildren Trevor, Geneva, Warren, and Caroline, I dedicate this book.

Darrell D. Ebbing

Contents in Brief

Basics of Chemistry

Atomic and Molecular Structure

States of Matter and Solutions

Chemical Reactions and Equilibrium

Nuclear Chemistry and Chemistry of the Elements

Contents

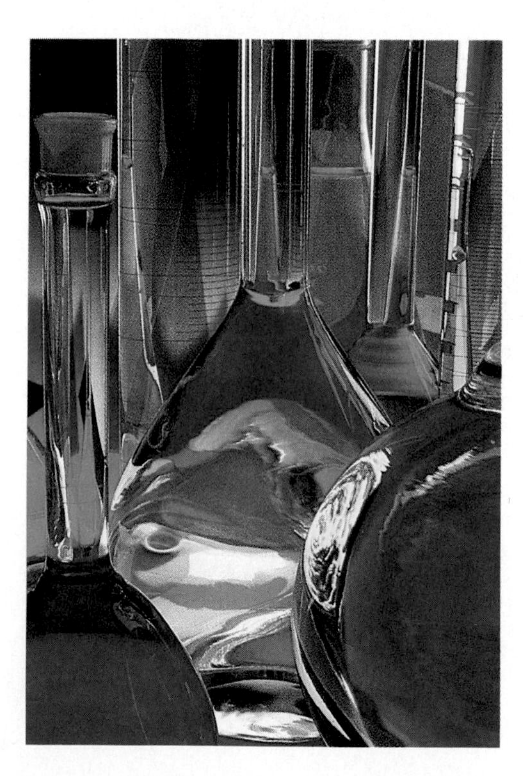

5. The Gaseous State 141

6. Thermochemistry 186

200 mL
50

List of Essays

Profile of a Chemical

Related Topic

Instrumental Methods

GENERAL CHEMISTRY

THIRD • EDITION

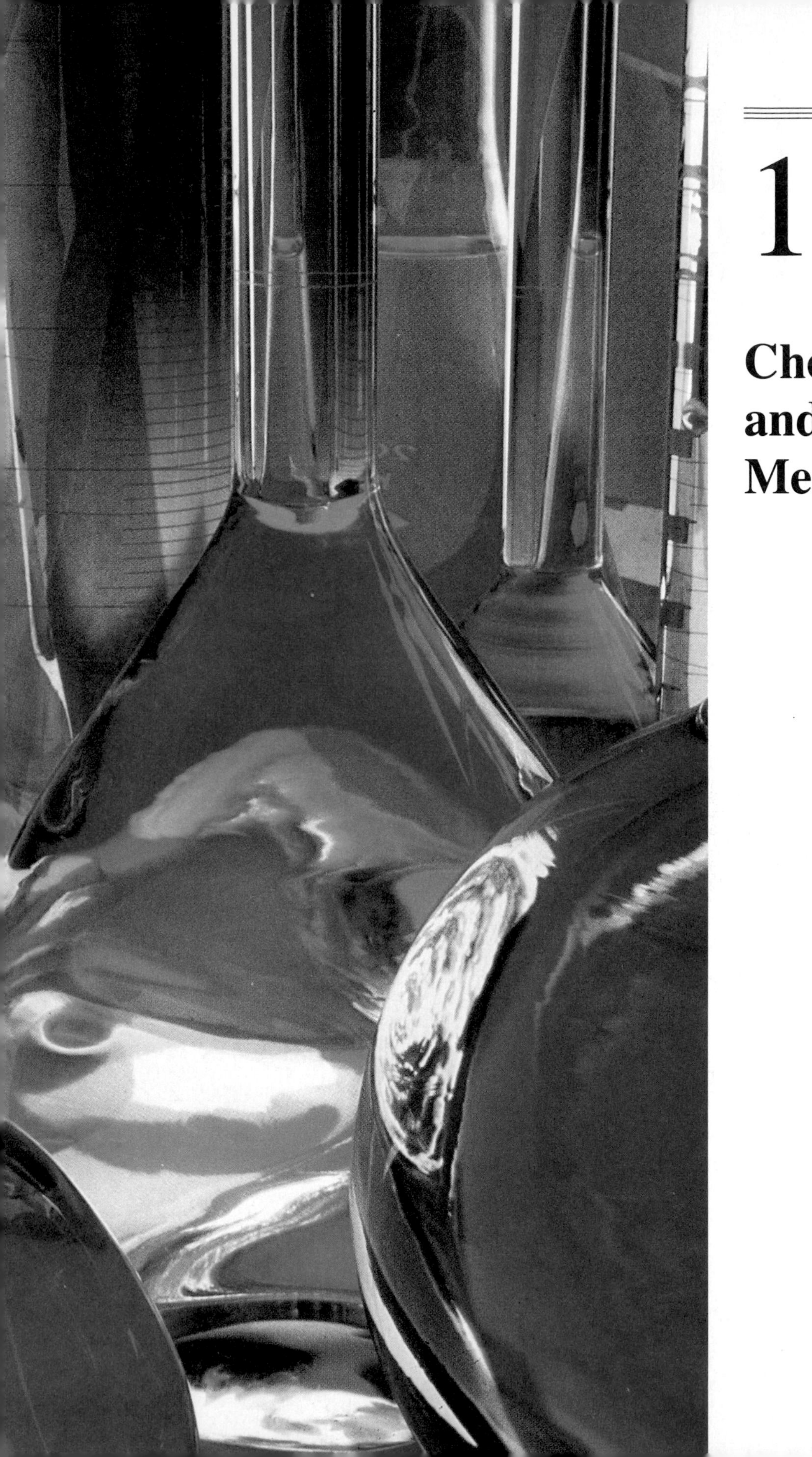

1

Chemistry and Measurement

Chapter Outline

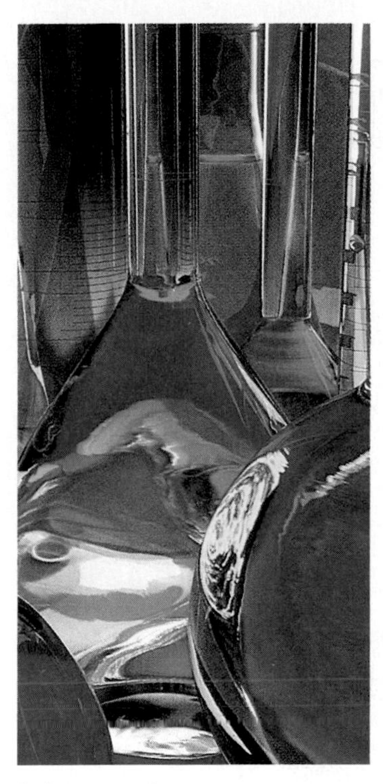

Laboratory glassware.

In 1964 Barnett Rosenberg and his coworkers at Michigan State University were studying the effects of electricity on bacterial growth. They inserted platinum electrodes (electrical connections) into a live bacterial culture. Then they allowed an electric current to pass through the culture. After 1 to 2 hours, they noted that cell division in the bacteria no longer occurred. The researchers were able to show that cell division was inhibited by a substance containing platinum, produced from the platinum electrodes by the electric current. A substance such as this one, they thought, might be useful as an anticancer drug, because cancer involves runaway cell division. Later research confirmed this view, and today the platinum-containing substance *cisplatin* is a leading anticancer drug (Figure 1.1).

This story illustrates three significant reasons to study chemistry. First, chemistry has important practical applications. The development of life-saving drugs is one, and a complete list would touch upon most areas of modern technology.

Second, chemistry is an intellectual enterprise, a way of explaining our material world. When Rosenberg and his coworkers saw that cell division in their culture had ceased, they systematically looked for the chemical substance that caused it to cease. They sought a chemical explanation for the occurrence.

Finally, chemistry figures prominently in other fields. Rosenberg's experiment began as a problem in biology; through the application of chemistry it led to an advance in medicine. Whatever your career plans, you will find your knowledge of chemistry is a useful intellectual tool for making important decisions.

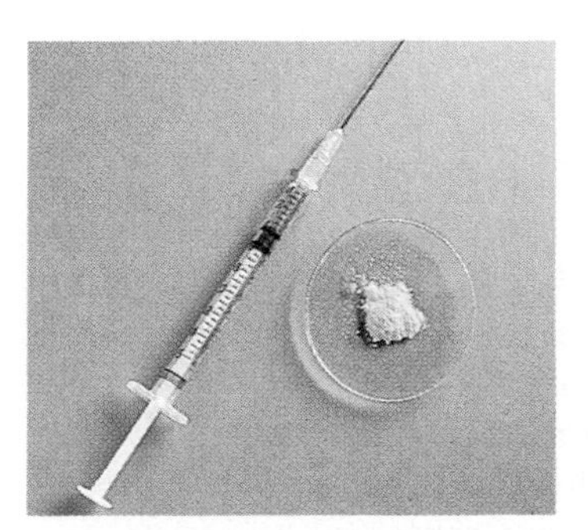

Figure 1.1
Cisplatin. Crystals of cisplatin and a hypodermic syringe containing a cisplatin solution.

Chemistry as a Quantitative Science

Chemistry is the science of the materials around us, such as air, water, rocks, and plant and animal substances. Much of chemistry involves describing these materials and the changes they undergo. However, chemistry also has a quantitative side concerned with measuring and calculating the characteristics of materials. This quantitative aspect has played, and continues to play, an important role in modern chemistry.

Figure 1.2
Modern single-pan balances.
(A) A mechanical analytical balance. *(B)* Mechanism of the analytical balance. The balance arm has a counterweight at one end which just balances the weights and the pan on the other end. When an object is placed on the pan, weights equal to the mass of the object are removed by means of a mechanism (not shown) connected to dials at the front of the balance. *(C)* A modern electronic balance with digital read-out. The mechanism is similar to the mechanical balance except that the counterweight is replaced by an electromagnetic device to balance the weight of the pan and object. Both balances obtain masses accurate to about 0.0001 gram.

A

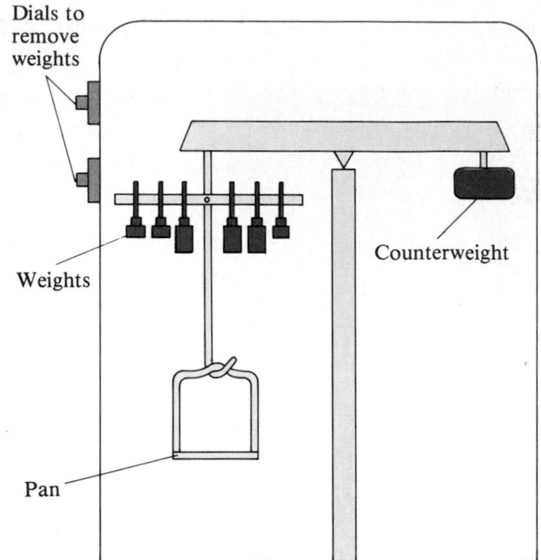

B

C

1.1 DEVELOPMENT OF MODERN CHEMISTRY

The origins of chemistry are ancient and probably began with the use of natural materials for practical purposes. Modern chemistry emerged in the eighteenth century, when the balance began to be used systematically as a tool in research. Balances measure **mass,** that is, *the quantity of matter in a material* (Figure 1.2). **Matter** is the general term for the material things around us and may be defined as *whatever occupies space and can be perceived by our senses.*

Antoine Lavoisier (1743–1794), a French chemist, insisted on the use of the balance in chemical research. His experiments demonstrated the **law of conservation of mass,** *a principle that states that the total mass remains constant during a*

Figure 1.3
The burning of magnesium.
Left: A magnesium flash bulb has a magnesium alloy wire enclosed in a bulb of oxygen gas; the wire is ignited from the heat of an electric current. *Right:* Magnesium metal reacts with oxygen in air, giving off a bright light. (Magnesium also reacts to some extent with nitrogen in the air.)

chemical change (chemical reaction).● A magnesium flash bulb gives a convenient illustration of this law. The flash from such a bulb accompanies a chemical reaction triggered by the heat of an electrical current (see Figure 1.3). A flash bulb that weighs 11.2 grams before it is flashed still weighs 11.2 grams afterward; the mass (11.2 grams) remains constant.

● Chemical reactions may gain or lose heat and other forms of energy. According to Einstein, mass and energy are equivalent. Thus, when energy is lost as heat, mass is also lost. Changes of mass in chemical reactions (billionths of a gram) are too small to detect.

In a series of experiments, Lavoisier showed that when a metal or any other substance burns, something in the air chemically combines with it. He called this component of air oxygen. For example, Lavoisier found that the liquid metal mercury was transformed in air into a red-colored substance. The substance had greater mass than the original mercury. This was due, he said, to the chemical combination of mercury with oxygen. Furthermore, the new substance, mercury(II) oxide, could be heated to recover the original mass of mercury (see Figure 1.4).● Lavoisier's explanation of the burning, or combustion, of mercury can be written

● The naming of compounds is discussed in Chapter 2.

$$\text{Mercury} + \text{oxygen} \longrightarrow \text{mercury(II) oxide}$$

Figure 1.4
Heating mercury metal in air.
Left: Mercury metal reacts with oxygen to yield mercury(II) oxide. The color of the oxide varies from red to yellow depending on particle size. *Right:* When mercury(II) oxide is heated, it decomposes to mercury and oxygen gas.

The following exercise illustrates how the law of conservation of mass can be used to investigate chemical changes such as combustion.*

Exercise 1.1

When 2.53 grams of metallic mercury are heated in air, they are converted to 2.73 grams of red-colored residue. Assuming that the chemical change is due to the reaction of the metal with oxygen from the air, and using the law of conservation of mass, determine the mass of the oxygen that has reacted. When the residue is strongly heated, it decomposes back to mercury, a silvery liquid. What is the mass of the oxygen that is lost when the residue is heated? (See Problems 1.15, 1.16, 1.17, and 1.18)

Measurements of mass before and after the combustion of various substances were necessary to convince chemists that Lavoisier's views were correct. Thereafter, measurements of mass became indispensable to chemical research.

Before we leave this section, we should note the distinction between the terms *mass* and *weight* in precise usage. The weight of an object is the force of gravity exerted on it. It is proportional to the mass of the object divided by the square of the distance between the center of mass of the object and that of the earth (or other nearby planet or satellite).● Because the earth is slightly flattened at the poles, an object weighs more at the North Pole, where it is closer to the center of the earth, than it does at the equator. The mass of an object is the same wherever it is measured.

● The force of gravity F between objects whose masses are m_1 and m_2 is Gm_1m_2/r^2, where G is the gravitational constant and r is the distance between the centers of mass of the two objects.

1.2 EXPERIMENT AND EXPLANATION

An important aspect of scientific research is careful *observation* of natural phenomena. In chemical research, observations are usually made under circumstances in which variables, such as temperature and amounts of substances, can be controlled. An **experiment** is *an observation of natural phenomena carried out in a controlled manner so that the results can be duplicated and rational conclusions obtained.* In the chapter opening it was mentioned that Rosenberg studied the effects of electricity on bacterial growth. Temperature and amounts of nutrients in a given volume of bacterial medium are important variables in such experiments. Unless these variables are controlled, the work cannot be duplicated, nor can any reasonable conclusion be drawn.

After a series of observations or experiments, a researcher may see some relationship or regularity in the results. Thus Rosenberg noted that in each experiment in which an electric current was passed through a bacterial culture by means of platinum electrodes, the bacteria ceased dividing. If the regularity or relationship is a basic one and can be simply stated, it is called a law. A **law** is *a concise statement or mathematical equation about a basic relationship or regularity of nature.* An example is the law of conservation of mass, which says that the mass remains constant during a chemical reaction.

Another aspect of scientific research is *explanation.* Explanations help us organize knowledge and predict future events. A **hypothesis** is *a tentative explanation of some regularity of nature.* Having seen that bacteria ceased to divide

*An exercise unaccompanied by a worked-out example can be solved by using the ideas just discussed in the text. The problems mentioned after the exercise appear at the end of the chapter and reinforce the skills associated with the exercise.

when an electric current from platinum electrodes passed through the culture, Rosenberg was eventually able to propose the hypothesis that platinum compounds were responsible. If a hypothesis is to be useful, it should suggest new experiments that become tests of the hypothesis. Rosenberg's hypothesis could be tested by looking for the platinum compound and testing its ability to inhibit cell division.

If a hypothesis is a basic one and successfully passes many tests, it becomes known as a theory. A **theory** is *a tested explanation of basic natural phenomena*. An example is the molecular theory of gases—the theory that all gases are composed of very small particles called molecules. This theory has withstood many tests and has been fruitful in suggesting many experiments. Note that a theory cannot be proved absolutely. It is always possible that further experiments will show the theory to be limited, or a better theory might be developed. For example, the physics of the motion of objects devised by Isaac Newton withstood experimental tests for over two centuries, until it was discovered that the equations do not hold for objects moving near the speed of light. Later it was shown that very small objects also do not follow Newton's physics. Both discoveries resulted in revolutionary developments in physics. The first led to the theory of relativity; the second to quantum mechanics, which has had an immense impact on chemistry.

The two aspects of science, observation (or experiment) and explanation, are closely related. Experiments are performed and some regularity is observed. The regularity is explained, and this explanation then leads to more experiments. The interrelationship of experiment and explanation is displayed in Figure 1.5. Thus, from his first experiments, Rosenberg explained that certain platinum compounds inhibit cell division. This explanation led him to do new experiments on the anticancer activity of these compounds.

The *general* process of advancing scientific knowledge through observation; the framing of laws, hypotheses, or theories; and the conducting of more experiments is sometimes called the *scientific method*. It is not a method for carrying out

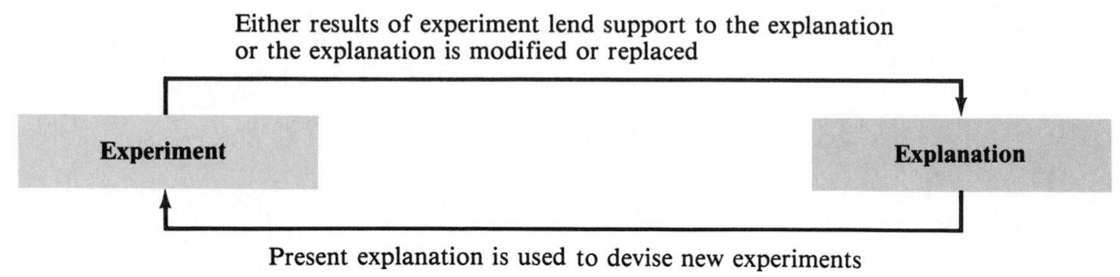
Either results of experiment lend support to the explanation
or the explanation is modified or replaced
Experiment
Explanation
Present explanation is used to devise new experiments

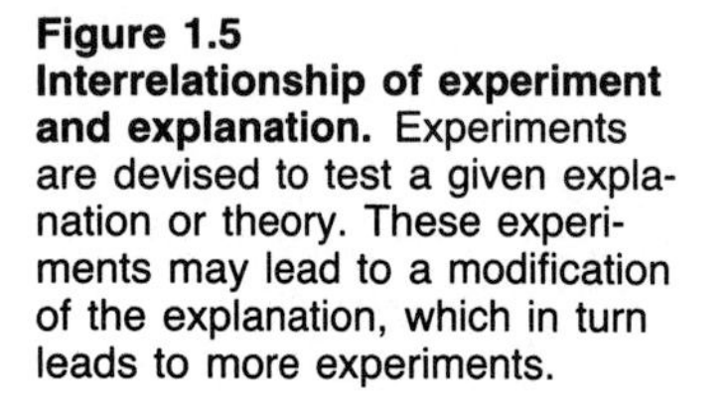

Figure 1.5 Interrelationship of experiment and explanation. Experiments are devised to test a given explanation or theory. These experiments may lead to a modification of the explanation, which in turn leads to more experiments.

a *specific* research program, because the design of experiments and their explanation draw on the creativity and individuality of a researcher.

Units of Measurement

Cisplatin, the chemical substance featured in the chapter opening, was first prepared in 1844, over a hundred years before the discovery of its anticancer activity. It is a yellow, crystalline substance known to chemists as *cis*-diamminedichloroplatinum(II).● A like substance known as *trans*-diamminedichloroplatinum(II) was discovered the same year. Although these two substances are similar, we can easily distinguish between them by testing their solubilities in water—that is, by measuring the masses of the substances that will dissolve in a given quantity of water. The solubility of cisplatin is 0.252 gram in 100 grams of water at 25°C, whereas the solubility of the second substance is 0.037 gram in 100 grams of water.

● The chemical name for cisplatin consists of parts written as one word: *cis*/di/ammine/di/chloro/platinum/(II). As we will discuss in Chapter 23, such names convey information about the composition and structure of a substance.

This is only one illustration of the many uses of measurement in chemistry. In a modern chemical laboratory, complex measurements can be made with expensive instruments, yet many experiments begin with simple measurements of mass, volume, time, and so forth. In the next few sections, we will look at units of measurement.

1.3 MEASUREMENT AND SIGNIFICANT FIGURES

Measurement is the comparison of a physical quantity to be measured with a **unit** of measurement—that is, with *a fixed standard of measurement*. On a centimeter scale, the centimeter unit is the standard of comparison.● A steel rod that measures 9.12 times the centimeter unit has a length of 9.12 centimeters. To record the measurement, we must be careful to give both the *measured number* (9.12) and the *unit* (centimeters).

● 2.54 centimeters = 1 inch

If we repeat a particular measurement, we usually do not obtain precisely the same result, because each measurement is subject to experimental error. The measured values vary slightly from one another. Suppose we perform a series of identical measurements of a quantity. The term **precision** refers to *the closeness of the set of values obtained from identical measurements of a quantity*. **Accuracy** is a related term; it refers to *the closeness of a single measurement to its true value*.● To illustrate the idea of precision, let us look at a simple measuring device, the centimeter ruler. In Figure 1.6, a steel rod has been placed next to a ruler subdivided into tenths of a centimeter. You can see that the rod measures just over 9.1 cm (cm = centimeter). With care, it is possible to estimate by eye to hundredths of a centimeter. Here we might give the measurement as 9.12 cm. Sup-

● Measurements that are of high precision are usually accurate. It is possible, however, to have a systematic error in a measurement. Suppose that, in calibrating a ruler, the first centimeter is made too small by 0.1 centimeter. Then, although the measurements of length on this ruler are still precise to 0.01 centimeter, they are accurate to only 0.1 centimeter.

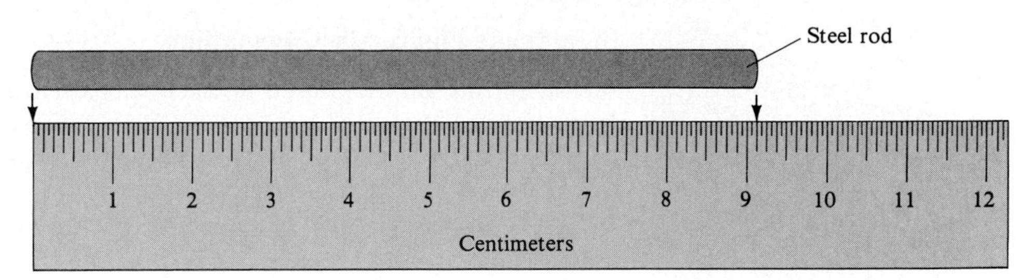

Figure 1.6
Precision of measurement with a centimeter ruler. The length of the rod is just over 9.1 centimeters. On successive measurements, we estimate the length by eye at 9.12, 9.11, and 9.13 cm. We record the length as between 9.11 cm and 9.13 cm.

pose we measure the length of this rod twice more. We find the values of these measurements to be 9.11 cm and 9 13 cm. Thus, we record the length of the rod as being somewhere between 9.11 ,m and 9.13 cm. The spread of values indicates the precision with which a measurement can be made by this centimeter ruler.

To indicate the precision of a measured number (or result of calculations on measured numbers), we often use the concept of significant figures. **Significant figures** are *those digits in a measured number (or result of a calculation with measured numbers) that include all certain digits plus a final one having some uncertainty*. When we measured the rod, we obtained the values 9.12 cm, 9.11 cm, and 9.13 cm. We could report the result as the average, 9.12 cm. The first two digits (9.1) are certain, and the next digit (2) is estimated, so has some uncertainty. It would be incorrect to write 9.120 cm for the length of the rod. This would say that the last digit (0) has some uncertainty but that the other digits (9.12) are certain, which is not true.

Number of Significant Figures

Number of significant figures refers to *the number of digits reported for the value of a measured or calculated quantity, indicating the precision of the value*. Thus, there are three significant figures in 9.12 cm, whereas 9.123 cm has four. To count the number of significant figures in a given measured quantity, we observe the following rules:

1. All digits are significant except zeros at the beginning of the number and possibly terminal zeros (one or more zeros at the end of a number). Thus, 9.12 cm, 0.912 cm, and 0.00912 cm all contain three significant figures.
2. Terminal zeros ending at the right of the decimal point are significant. Each of the following has three significant figures: 9.00 cm, 9.10 cm, 90.0 cm.
3. Terminal zeros ending to the left of the decimal point may or may not be significant. When a measurement is given as 900 cm, we do not know whether one, two, or three significant figures were intended. Any uncertainty can be removed by expressing the measurement in *scientific notation*.

Scientific notation is *the representation of a number in the form* $A \times 10^n$, *where* A *is a number with a single nonzero digit to the left of the decimal point, and* n *is an integer, or whole number*. In scientific notation, the measurement 900 cm precise to two significant figures is written 9.0×10^2 cm. If precise to three significant figures, it is written 9.00×10^2 cm. Scientific notation is also convenient for expressing very large or very small quantities. It is much easier (and it simplifies calculations) to write the speed of light as 3.00×10^8 (rather than as 300,000,000) meters per second.●

● See Appendix A for a review of scientific notation.

Figure 1.7
Significant figures and calculators. Not all of the figures that appear on a calculator are significant. In performing the calculation $0.0634 \times 100.0 \div 25.31$, the calculator shows 0.2504938. We would report the answer as 0.250, however, because the factor 0.0634 has the least number of significant figures (three).

Significant Figures in Calculations

Measurements are often used in calculations. How do we report the significant figures in a calculation? Suppose we have a substance believed to be cisplatin, and, in an effort to establish its identity, we measure its solubility (the amount that dissolves in a given quantity of water). We find that 0.0634 gram of the substance dissolves in 25.31 grams of water. The amount dissolving in 100.0 grams is

$$0.0634 \text{ gram} \times \frac{100.0 \text{ grams of water}}{25.31 \text{ grams of water}}$$

Performing the arithmetic on a pocket calculator (Figure 1.7), we get 0.2504938 for the numerical part of the answer ($0.0634 \times 100.0 \div 25.31$). But it would be incorrect to give this number as the final answer, because it is much more precise than the other numbers in the calculation. When multiplying or dividing measured quantities, we should give as many significant figures in the answer as there are in the measurement with the least number of significant figures. In the preceding calculation, 0.0634 gram has the least number of significant figures (three). Hence the answer should be reported to three significant figures—that is, 0.250 gram.

A different rule applies in the case of addition or subtraction of measured quantities. When adding or subtracting measured quantities, we should give the same number of decimal places in the answer as there are in the measurement with the least number of decimal places. Suppose we wish to add 184.2 grams and 2.324 grams. On a calculator, we find that $184.2 + 2.324 = 186.524$. But because the quantity 184.2 grams has the least number of decimal places—one, whereas 2.324 grams has three—the answer is 186.5 grams.

Exact Numbers

So far we have discussed only numbers that involve uncertainties. However, we will also encounter exact numbers. An **exact number** is *a number that arises when we count items or sometimes when we define a unit*. For example, when we say there are 9 coins in a bottle, we mean exactly 9, not 8.9 or 9.1. Also, when we say there are 12 inches to a foot, we mean exactly 12. The conventions of significant figures do not apply to exact numbers. Thus, the 12 in the expression ''12 inches to the foot'' should not be interpreted as a measured number with two significant figures. In effect, the 12 has an infinite number of significant figures, but of course it would be impossible to write out an infinite number of digits. You should make a mental note of any numbers in a calculation that are exact, because they will have no effect on the number of significant figures in a calculation. The number of significant figures in a calculation result will depend only on the numbers of significant figures in quantities having uncertainties. For example, suppose we want the total mass of 9 coins when each coin has a mass of 3.0 grams. The calculation is

$$3.0 \text{ grams} \times 9 = 27 \text{ grams}$$

The answer is reported to two significant figures because 3.0 grams has two significant figures. The number 9 is exact and does not determine the number of significant figures.

Rounding

In reporting the solubility of our substance in 100.0 grams of water as 0.250 gram, we *rounded* the number we read off the calculator (0.2504938). **Rounding** is *the procedure of dropping nonsignificant digits in a calculation result and adjusting the last digit reported.* The general procedure is as follows. Look at the leftmost digit to be dropped.

1. If this digit is greater than 5, or is 5 followed by nonzeros, add 1 to the last digit to be retained and drop all digits farther to the right. Thus, rounding 1.2151 to three significant figures gives 1.22.
2. If this digit is less than 5, simply drop it and all digits farther to the right. Rounding 1.2143 to three significant figures gives 1.21.
3. If this digit is simply 5 or 5 followed by zeros, and if the last digit to be retained is even, just drop the 5 and any zeros after it. If the last digit to be retained is odd, add 1 to it and drop the 5 and any zeros after it. For example, rounding 1.225, 1.22500, and 1.21500 to three significant figures gives 1.22 in each case.

In doing a calculation of two or more steps, it is desirable to retain additional digits for intermediate answers. This ensures that small errors from rounding do not appear in the final result. If you use a calculator, you can simply enter numbers one after the other, performing each arithmetic operation and rounding just the final answer. To keep track of the correct number of significant figures, you will need to record intermediate answers with a line under the last significant figures, as shown in the solution to part (c) of the following example.

Example 1.1 Significant Figures in Calculations

Perform the following calculations and round the answers to the correct number of significant figures (units of measurement have been omitted for clarity).

(a) $\dfrac{2.568 \times 5.8}{4.186}$ (b) $5.41 - 0.398$

(c) $4.18 - 58.16 \times (3.38 - 3.01)$

Solution

(a) The factor 5.8 has the fewest significant figures; therefore, the answer should be reported to two significant figures. Round the answer to **3.6**. (b) The number with the least number of decimal places is 5.41. Therefore, the answer is rounded to two decimal places, to **5.01**. (c) In any complicated arithmetic setup such as this, proceed step by step. Carry out operations within parentheses as a separate calculation. By convention, multiplications and divisions are performed before additions and subtractions. It is best not to round intermediate answers, but you must keep track of the rightmost digit that would be retained after rounding (shown by an underline). The steps are

$$\begin{aligned} 4.1\underline{8} - 58.1\underline{6} \times (3.3\underline{8} - 3.0\underline{1}) &= 4.1\underline{8} - 58.1\underline{6} \times 0.3\underline{7} \\ &= 4.1\underline{8} - 2\underline{1}.5 \\ &= -1\underline{7}.32 \end{aligned}$$

The final answer is **−17**.

Exercise 1.2

Give answers to the following arithmetic setups. Round to the correct number of significant figures.

(a) $\dfrac{5.61 \times 7.891}{9.1}$ (b) $8.91 - 6.435$

(c) $6.81 - 6.730$ (d) $38.91 \times (6.81 - 6.730)$

(See Problems 1.23 and 1.24.)

1.4 SI UNITS

The first measurements were probably based on the human body (the length of the foot, for example). In time, fixed standards developed, but these varied from place to place. Each country or government (and often each trade) adopted its own units. As science became more quantitative in the seventeenth and eighteenth centuries, scientists found that the lack of standard units was a problem. ● They began to seek a simple, international system of measurement. In 1791 a study committee of the French Academy of Sciences devised such a system. Called the *metric system,* it became the official system of measurement for France and was soon used by scientists throughout the world. Most nations have since adopted the metric system or, at least, have set a schedule for changing to it.

● In the system of units that Lavoisier used in the eighteenth century, there were 9216 grains to the pound (the *livre*). English chemists of the same period used a system in which there were 7000 grains in a pound—unless they were trained as apothecaries, in which case there were 5760 grains to the pound!

SI Base Units and SI Prefixes

In 1960 the General Conference on Weights and Measures adopted the **International System** of units (or **SI,** after the French *le Système International d'Unités*), which is *a particular choice of metric units*. This system has seven **SI base units,** *the SI units from which all others can be derived*. Table 1.1 lists these base units and the symbols used to represent them. In this chapter, we will discuss four base quantities: length, mass, time, and temperature. ●

● The amount of substance is discussed in Chapter 4, and the electric current (ampere) is introduced in Chapter 19. Luminous intensity will not be used.

One of the advantages of any metric system is that it is a decimal system. In SI, a larger or smaller unit for a physical quantity is indicated by an **SI prefix,** which is *a prefix used in the International System to indicate a power of ten*. For example, the base unit of length in SI is the meter (somewhat longer than a yard), and 10^{-2} meter is called a *centi*meter. Thus 2.54 centimeters equal 2.54×10^{-2} meter. The prefixes used in SI are listed in Table 1.2. Only those shown in color will be used in this book: *mega-* (10^6), *kilo-* (10^3), *deci-* (10^{-1}), *centi-* (10^{-2}), *milli-* (10^{-3}), *micro-* (10^{-6}), *nano-* (10^{-9}), and *pico-* (10^{-12}).

Length, Mass, and Time

The **meter (m)** is *the SI base unit of length*. ● By combining it with one of the SI prefixes, we can obtain a unit of appropriate size for any length measurement. For

● The meter was originally defined in terms of a standard platinum–iridium bar kept at Sèvres, France. In 1983, the meter was defined as the distance traveled by light in a vacuum in 1/299,792,458 seconds.

Table 1.1 SI Base Units

Quantity	Unit	Symbol
Length	meter	m
Mass	kilogram	kg
Time	second	s
Temperature	kelvin	K
Amount of substance	mole	mol
Electric current	ampere	A
Luminous intensity	candela	cd

Table 1.2 SI Prefixes

Multiple	Prefix	Symbol
10^{18}	exa	E
10^{15}	peta	P
10^{12}	tera	T
10^{9}	giga	G
10^{6}	mega	M
10^{3}	kilo	k
10^{2}	hecto	h
10	deka	da
10^{-1}	deci	d
10^{-2}	centi	c
10^{-3}	milli	m
10^{-6}	micro	μ*
10^{-9}	nano	n
10^{-12}	pico	p
10^{-15}	femto	f
10^{-18}	atto	a

*Greek letter pronounced "mew."

the very small lengths used in chemistry, the nanometer (1 nanometer = 10^{-9} meter) or the picometer (1 picometer = 10^{-12} meter) is an acceptable SI Unit. *A non-SI unit of length* traditionally used by chemists is the **angstrom (Å),** which equals 10^{-10} meter. (An oxygen atom, the minute particles of which the substance oxygen is composed, has a diameter of about 1.3 Å. If we could place oxygen atoms adjacent to one another, we could line up over 75 million of them in 1 centimeter.)

The **kilogram (kg)** is *the SI base unit of mass,* equal to about 2.2 pounds.● This is an unusual base unit in that it contains a prefix. In forming other SI mass units, prefixes are added to the word *gram* to give units such as the *milli*gram (mg; 1 milligram = 10^{-3} gram).

● The present standard of mass is the platinum–iridium kilogram mass kept at the International Bureau of Weights and Measures in Sèvres, France.

The **second (s)** is *the SI base unit of time* (Figure 1.8). Combining this unit with prefixes such as *milli-, micro-, nano-,* and *pico-,* we create units appropriate for measuring very rapid events. For example, the time it takes to add two 10-digit numbers on a modern high-speed computer is roughly a nanosecond. The time required for the fastest chemical processes is about a picosecond. When we measure times much longer than a few hundred seconds, we revert to *minutes* and *hours,* an obvious exception to the prefix–base format of the International System.

Exercise 1.3

Express the following quantities, using an SI prefix and a base unit. For instance, 1.6×10^{-6} m = 1.6 μm. A quantity such as 0.000168 g could be written 0.168 mg or 168 μg.

(a) 1.84×10^{-9} m (b) 5.67×10^{-12} s (c) 7.85×10^{-3} g
(d) 9.7×10^{3} m (e) 0.000732 s (f) 0.000000000154 m

(See Problems 1.27 and 1.28.)

Figure 1.8
A cesium clock. The cesium clock is the present standard for time. Cesium atoms absorb microwave radiation having a frequency of exactly 9,192,631,770 cycles per second. This absorption is used to control the time of a quartz clock accurate to one second in 150,000 years. Even more accurate clocks are being developed. Time is now known so accurately that the standard of length was defined in 1983 as the distance traveled by light in 1/299,792,458 seconds.

Temperature

Temperature is difficult to define precisely, but we all have an intuitive idea of what we mean by it. It is a measure of "hotness." A hot object placed next to a cold one becomes cooler, while the cold object becomes hotter. Heat energy passes from a hot object to a cold one, and the quantity of heat passed between the objects depends on the difference in temperature between the two. Therefore, temperature and heat are different, but related, concepts.

A thermometer is a device for measuring temperature. The common type consists of a glass capillary containing a column of liquid whose length varies with temperature. A scale alongside the capillary gives a measure of the temperature. The **Celsius scale** (formerly the centigrade scale) is *the temperature scale in general scientific use*. On this scale, the freezing point of water is 0°C and the boiling point of water at normal barometric pressure is 100°C. However, *the SI base unit of temperature* is the **kelvin (K),** a unit on an *absolute temperature* scale.● On any absolute scale, the lowest temperature that can be attained theoretically is zero. Both the Celsius and the Kelvin scales have equal-size units, but 0°C is equivalent to 273.15 K. Thus, it is easy to convert from one scale to the other, using the formula

● Note that the degree sign (°) is not used with the Kelvin scale, and the unit of temperature is simply the kelvin (not capitalized).

$$K = {}^\circ C + 273.15$$

A temperature of 20°C (about room temperature) equals 293 K.

The Fahrenheit scale is at present the common temperature scale in the United States. Figure 1.9 compares Kelvin, Celsius, and Fahrenheit scales. As the figure shows, 0°C is the same as 32°F (both exact), and 100°C corresponds to 212°F (both exact). Therefore, there are 212 − 32 = 180 Fahrenheit degrees in the range of 100 Celsius degrees. That is, there are 1.8 Fahrenheit degrees for every Celsius

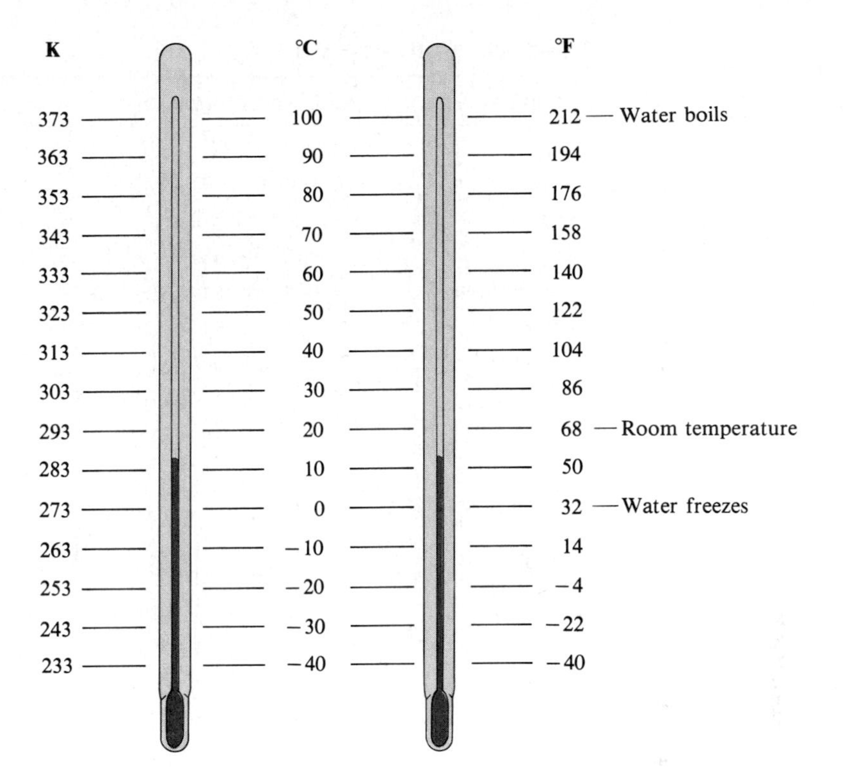

Figure 1.9 Comparison of temperature scales. Room temperature is about 20°C, 293 K, and 68°F. Water freezes at 0°C, 273 K, and 32°F. Water boils under normal pressure at 100°C, 373 K, and 212°F.

degree. Knowing this, and knowing that 0°C equals 32°F, we can derive a formula to convert degrees Celsius to degrees Fahrenheit. It is

$$°F = (1.8 \times °C) + 32$$

This formula can be rearranged to give a formula for converting degrees Fahrenheit to degrees Celsius:

$$°C = \frac{°F - 32}{1.8}$$

Example 1.2 Temperature Conversion

The hottest place on record in North America is Death Valley in California. A temperature of 134°F was reached there in 1913. What is this temperature reading in degrees Celsius? in kelvins?

Solution

Substituting, we find that

$$°C = \frac{°F - 32}{1.8} = \frac{134 - 32}{1.8} = \mathbf{56.7}$$

In kelvins,

$$K = °C + 273.15 = 56.7 + 273.15 = \mathbf{329.8}$$

Exercise 1.4

(a) A person has a fever of 102.5°F. What is this temperature in degrees Celsius? (b) A cooling mixture of dry ice and isopropyl alcohol has a temperature of −78°C. What is this temperature in kelvins? (See Problems 1.31, 1.32, 1.33, 1.34, 1.35, and 1.36.)

Table 1.3 Derived Units

Quantity	Definition of Quantity	SI Unit
Area	Length squared	m^2
Volume	Length cubed	m^3
Density	Mass per unit volume	kg/m^3
Speed	Distance traveled per unit time	m/s
Acceleration	Speed changed per unit time	m/s^2
Force	Mass times acceleration of object	$kg \cdot m/s^2$ (= newton, N)
Pressure	Force per unit area	$kg/(m \cdot s^2)$ (= pascal, Pa)
Energy	Force times distance traveled	$kg \cdot m^2/s^2$ (= joule, J)

1.5 DERIVED UNITS

Once base units have been defined for a system of measurement, we can derive other units from them. We do this by using the base units in equations that define other physical quantities. For example, *area* is defined as length times length. Therefore,

$$\text{SI unit of area} = (\text{SI unit of length}) \times (\text{SI unit of length})$$

From this, we see that the SI unit of area is meter × meter or m^2. Similarly, *speed* is defined as the rate of change of distance with time; that is, speed = distance/time. So

$$\text{SI unit of speed} = \frac{\text{SI unit of distance}}{\text{SI unit of time}}$$

The SI unit of speed is meters per second (that is, meters divided by seconds). The unit is symbolized m/s or $m \cdot s^{-1}$. The unit of speed is an example of an **SI derived unit,** which is *a unit derived by combining SI base units*. Table 1.3 defines a number of derived units. Volume and density will be discussed in this

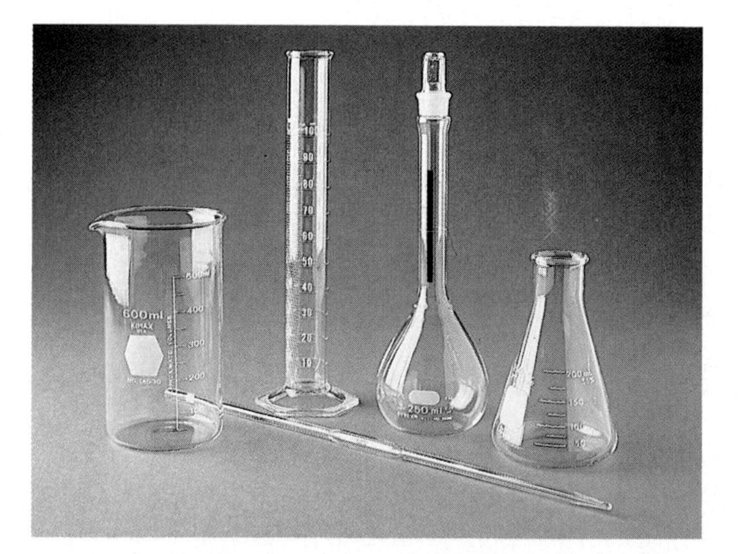

Figure 1.10
Some laboratory glassware. *Left to right:* A 600-mL beaker; a 100-mL graduated cylinder; a 250-mL volumetric flask; a 250-mL Erlenmeyer flask. *Front:* A 5-mL pipet. A graduated cylinder is used to measure approximate volumes of liquids. A pipet is calibrated to deliver a specified volume of liquid when filled to an etched line with a suction bulb. The pipet shown is accurate to about 0.01 mL. A volumetric flask is filled to an etched line and is used in making up specific volumes of solutions. The flask shown is accurate to about 0.1 mL.

section; pressure and energy will be discussed later (in Sections 5.1 and 6.1, respectively).

Volume

Volume is defined as length cubed and has the SI unit of cubic meter (m^3). This is too large a unit for normal laboratory work, so we use either cubic decimeters (dm^3) or cubic centimeters (cm^3, also written cc). Traditionally, chemists have used the **liter (L),** which is *a unit of volume equal to a cubic decimeter* (approximately one quart). In fact, most laboratory glassware (Figure 1.10) is calibrated in liters or milliliters (1000 mL = 1 L). Because 1 dm equals 10 cm, a cubic decimeter or one liter equals $(10 \text{ cm})^3 = 1000 \text{ cm}^3$. Therefore, a milliliter equals a cubic centimeter. In summary

$$1 \text{ L} = 1 \text{ dm}^3 \qquad \text{and} \qquad 1 \text{ mL} = 1 \text{ cm}^3$$

Density

The **density** of an object is its *mass per unit volume*. We can express this as

$$d = \frac{m}{V}$$

where d is the density, m is the mass, and V is the volume. Suppose an object has a mass of 15.0 g and a volume of 10.0 cm^3. Substituting, we find that

$$d = \frac{15.0 \text{ g}}{10.0 \text{ cm}^3} = 1.50 \text{ g/cm}^3$$

The density of the object is 1.50 g/cm^3 (or $1.50 \text{ g} \cdot \text{cm}^{-3}$).

Density is an important characteristic property of a material. Water, for example, has a density of 1.000 g/cm^3 at 4°C and a density of 0.998 g/cm^3 at 20°C. Lead has a density of 11.3 g/cm^3 at 20°C. (Figures 1.11 and 1.12 dramatically show the relative densities of substances.)● Oxygen gas has a density of

● The density of solid materials on earth ranges from about 1 g/cm^3 to 22.5 g/cm^3 (osmium metal). In the interior of certain stars, the density of matter is truly staggering. Black neutron stars—stars composed of neutrons, or atomic cores compressed by gravity to a superdense state—have densities of about 10^{15} g/cm^3.

Figure 1.11 ***(left)***
The relative densities of copper and mercury. Copper floats on mercury because the density of copper is less than that of mercury. (This is a copper penny; recent coins are copper-clad zinc.)

Figure 1.12 ***(right)***
Relative densities of some liquids. Shown are three liquids (dyed so that they will show up clearly): hexane, water, and methylene chloride. The water layer (blue) is less dense than methylene chloride (dark red) and floats on it. Hexane (top red layer) in turn is less dense than water and floats on it.

1.33×10^{-3} g/cm^3 at normal pressure and 20°C. (Like other gases under normal conditions, oxygen has a density that is about 1000 times smaller than liquids and solids.) Because the density is characteristic of a substance, it can be helpful in identifying it. Example 1.3 illustrates this point. Density can also be useful in determining whether a substance is pure. Consider a gold bar whose purity is questioned. Any of the metals likely to be mixed with gold, such as silver or copper, have lower densities than gold. Therefore, an adulterated (impure) gold bar can be expected to be far less dense than pure gold.

Example 1.3 Calculation of the Density

A colorless liquid, used as a solvent (a liquid that dissolves other substances), is believed to be one of the following:

SUBSTANCE	DENSITY (in g/mL)
n-butyl alcohol	0.810
ethylene glycol	1.114
isopropyl alcohol	0.785
toluene	0.866

To identify the substance, a chemist determined its density. By pouring a sample of the liquid into a graduated cylinder, she found that the volume was 35.1 mL. She also found that the sample weighed 30.5 g. What was the density of the liquid? What was the substance?

Solution

Recall that density equals mass per volume (mass divided by volume). We substitute 30.5 g for the mass and 35.1 mL for the volume.

$$d = \frac{m}{V} = \frac{30.5\ \text{g}}{35.1\ \text{mL}} = 0.869\ \text{g/mL}$$

The density of the liquid equals that of **toluene** (within experimental error).

Exercise 1.5

A piece of metal wire has a volume of 20.2 cm^3 and a mass of 159 g. What is the density of the metal? The metal is either manganese, iron, or nickel, which have densities of 7.21 g/cm^3, 7.87 g/cm^3, and 8.90 g/cm^3, respectively. From which metal is the wire made?

(See Problems 1.37, 1.38, 1.39, and 1.40.)

In addition to characterizing a substance, the density provides a useful relationship between mass and volume. For example, suppose an experiment calls for a certain mass of liquid. Rather than weigh the liquid on a balance, we might instead measure out the corresponding volume. Example 1.4 illustrates this idea.

Example 1.4 Using the Density to Relate Mass and Volume

An experiment requires 43.7 grams of isopropyl alcohol. Instead of measuring out the sample on a balance, a chemist dispenses the liquid into a graduated cylinder. The density of isopropyl alcohol is 0.785 g/mL. What volume of isopropyl alcohol should he use?

Solution

We rearrange the formula defining the density to obtain the volume.

$$V = \frac{m}{d}$$

Then we substitute into this formula:

$$V = \frac{43.7\ \text{g}}{0.785\ \text{g/mL}} = 55.7\ \text{mL}$$

Exercise 1.6

Ethanol (grain alcohol) has a density of 0.789 g/cm^3. What volume of ethanol must be poured into a graduated cylinder to equal 30.3 g?

(See Problems 1.41, 1.42, 1.43, and 1.44.)

1.6 WORKING WITH UNITS: DIMENSIONAL ANALYSIS (FACTOR-LABEL METHOD)

In performing numerical calculations with physical quantities, it is good practice to enter each quantity as a number with its associated unit. Both the numbers and the units are then carried through the indicated algebraic operations. The advantages of this are twofold:

1. The units for the answer will come out of the calculations automatically.
2. If you make an error in arranging factors in the calculation (for example, if you use the wrong formula), this will become apparent because the final units will be nonsense.

Dimensional analysis (or **factor-label method**) is *the method of calculation in which one carries along the units for quantities*. As an illustration, suppose we want to find the volume V of a cube, given s, the length of a side of the cube. Because $V = s^3$, if $s = 5.00$ cm, we find that $V = (5.00 \text{ cm})^3 = 5.00^3 \text{ cm}^3$. There is no guesswork about the unit of volume here; it is cubic centimeters (cm^3).

Suppose, however, that we wish to express the volume in liters (L), a metric unit that equals 10^3 cubic centimeters (approximately one quart). We can write this equality as

$$1 \text{ L} = 10^3 \text{ cm}^3$$

If we divide both sides of the equality by the right-hand quantity, we get

$$\frac{1 \text{ L}}{10^3 \text{ cm}^3} = \frac{\cancel{10^3 \text{ cm}^3}}{\cancel{10^3 \text{ cm}^3}} = 1$$

Observe that units are treated in the same way as algebraic quantities. Note too that the right-hand side now equals 1 and that there are no units associated with it. Because it is always possible to multiply any quantity by 1 without changing that quantity, we can multiply our previous expression for the volume by the factor $1 \text{ L}/10^3 \text{ cm}^3$ without changing the actual volume. We are changing only the way in which we express this volume:

$$V = 5.00^3 \cancel{\text{cm}^3} \times \underbrace{\frac{1 \text{ L}}{10^3 \cancel{\text{cm}^3}}}_{\text{converts cm}^3 \text{ to L}} = 125 \times 10^{-3} \text{ L} = 0.125 \text{ L}$$

The ratio $1 \text{ L}/10^3 \text{ cm}^3$ is called a **conversion factor** because it is *a factor equal to 1 that converts a quantity expressed in one unit to one expressed in another unit.* ● Note that the numbers in this conversion factor are *exact,* because one liter equals exactly one thousand cubic centimeters. Such exact conversion factors do not affect the number of significant figures in an arithmetic result. In the previous calculation, the quantity 5.00 cm (the measured length of the side) determines or limits the number of significant figures.

● It takes more room to explain conversion factors than it does to use them. With practice, you will be able to write the final conversion step without the intermediate algebraic manipulations outlined here.

The next two examples illustrate how the conversion-factor method may be used to convert one metric unit to another.

Example 1.5 Unit Conversion: Metric to Metric

Nitrogen gas is the major component of air. A sample of nitrogen gas in a glass bulb weighed 243 mg. What is this mass in SI base units of mass (kilograms)?

Solution

Do the conversion in two steps. First convert milligrams to grams; then convert grams to kilograms. To obtain the conversion factor from milligrams to grams, note that the prefix *milli-* means 10^{-3}. Therefore,

$$1 \text{ mg} = 10^{-3} \text{ g}$$

and

$$243 \cancel{\text{mg}} \times \frac{10^{-3} \text{ g}}{1 \cancel{\text{mg}}} = 2.43 \times 10^{-1} \text{ g}$$

Then, because the prefix *kilo-* means 10^3, we write

$$1 \text{ kg} = 10^3 \text{ g}$$

and

$$2.43 \times 10^{-1} \cancel{\text{g}} \times \frac{1 \text{ kg}}{10^3 \cancel{\text{g}}} = \mathbf{2.43 \times 10^{-4} \text{ kg}}$$

Note, however, that we can do the two conversions in one step, as follows:

$$243 \text{ mg} \times \underbrace{\frac{10^{-3} \text{ g}}{1 \text{ mg}}}_{\text{converts mg to g}} \times \underbrace{\frac{1 \text{ kg}}{10^3 \text{ g}}}_{\text{converts g to kg}} = 2.43 \times 10^{-4} \text{ kg}$$

Exercise 1.7

The oxygen molecule (the smallest particle of oxygen gas) consists of two oxygen atoms a distance of 121 pm apart. How many millimeters is this distance?

(See Problems 1.45, 1.46, 1.47, and 1.48.)

Example 1.6 Unit Conversion: Metric Volume to Metric Volume

The world's oceans contain approximately 1.35×10^9 km^3 of water. What is this volume in liters?

Solution

Convert cubic kilometers to cubic decimeters, which are equivalent to liters. Do this in two steps, first km^3 to m^3, then m^3 to dm^3 (the steps may be combined). For a conversion factor such as that for km^3 to m^3, note that 1 km = 10^3 m, so the factor is $(10^3 \text{ m}/1 \text{ km})^3$.

$$1.35 \times 10^9 \cancel{\text{km}^3} \times \underbrace{\left(\frac{10^3 \cancel{\text{m}}}{1 \cancel{\text{km}}}\right)^3}_{\text{converts km}^3 \text{ to m}^3} \times \underbrace{\left(\frac{1 \text{ dm}}{10^{-1} \cancel{\text{m}}}\right)^3}_{\text{converts m}^3 \text{ to dm}^3}$$

$$= 1.35 \times 10^{21} \text{ dm}^3$$

Because a cubic decimeter is equal to a liter, the volume of the oceans is $\mathbf{1.35 \times 10^{21}}$ **L**.

Exercise 1.8

A crystal is constructed by stacking small, identical pieces of crystal, much as we construct a brick wall by stacking bricks. A unit cell is the smallest such piece from which a crystal can be made. The unit cell of a crystal of gold metal has a volume of 67.6 Å^3. What is this volume in cubic decimeters?

(See Problems 1.49 and 1.50.)

The conversion-factor method can be used to convert any unit to another unit, provided a conversion equation exists between the two units. Relationships between certain U.S. units and metric units are given in Table 1.4. We can use these

Table 1.4 Relationships of Some U.S. and Metric Units

Length	Mass	Volume
1 in = 2.54 cm (exact)	1 lb = 0.4536 kg	1 qt = 0.9464 L
1 yd = 0.9144 m (exact)	1 lb = 16 oz (exact)	4 qt = 1 gal (exact)
1 mi = 1.609 km	1 oz = 28.35 g	
1 mi = 5280 ft (exact)		

to convert between U.S. and metric units. Suppose we wish to convert 0.547 lb to grams. From Table 1.4, we see that 1 lb = 453.6 g, so the conversion factor from pounds to grams is 453.6 g/1 lb. Therefore,

$$0.547 \text{ lb} \times \frac{453.6 \text{ g}}{1 \text{ lb}} = 248 \text{ g}$$

The next example illustrates a conversion requiring several steps.

Example 1.7 Unit Conversion: Any Unit to Another Unit

How many centimeters are there in 6.51 miles? Use the exact definitions 1 mi = 5280 ft, 1 ft = 12 in, and 1 in = 2.54 cm.

Solution

This problem involves several conversions, which can be done all at once. From the preceding definitions, we obtain the following conversion factors:

$$1 = \frac{5280 \text{ ft}}{1 \text{ mi}} \qquad 1 = \frac{12 \text{ in}}{1 \text{ ft}} \qquad 1 = \frac{2.54 \text{ cm}}{1 \text{ in}}$$

Then,

$$6.51 \cancel{\text{mi}} \times \underbrace{\frac{5280 \cancel{\text{ft}}}{1 \cancel{\text{mi}}}}_{\text{converts mi to ft}} \times \underbrace{\frac{12 \cancel{\text{in}}}{1 \cancel{\text{ft}}}}_{\text{converts ft to in}} \times \underbrace{\frac{2.54 \text{ cm}}{1 \cancel{\text{in}}}}_{\text{converts in to cm}} = \mathbf{1.05 \times 10^6 \text{ cm}}$$

All of the conversion factors are exact, so the number of significant figures in the result is determined by the number of significant figures in 6.51 miles.

Exercise 1.9

Using the definitions 1 in = 2.54 cm and 1 yd = 36 in (both exact), obtain the conversion factor for yards to meters. How many meters are there in 3.54 yards?

(See Problems 1.51, 1.52, 1.53, 1.54, 1.55, and 1.56.)

A Checklist for Review

IMPORTANT TERMS (The number in parentheses denotes the section in which the term is defined.)

mass (1.1)
matter (1.1)
law of conservation of mass (1.1)
experiment (1.2)
law (1.2)
hypothesis (1.2)
theory (1.2)
unit (1.3)
precision (1.3)
accuracy (1.3)
significant figures (1.3)
number of significant figures (1.3)
scientific notation (1.3)
exact number (1.3)
rounding (1.3)
International System (SI) (1.4)
SI base unit (1.4)
SI prefix (1.4)
meter (m) (1.4)
angstrom (Å) (1.4)
kilogram (kg) (1.4)

second (s) (1.4)
Celsius scale (1.4)
kelvin (K) (1.4)
SI derived unit (1.5)
liter (L) (1.5)
density (1.5)
dimensional analysis (factor-label method) (1.6)
conversion factor (1.6)

KEY EQUATIONS

$K = °C + 273.15$

$°F = (1.8 \times °C) + 32$

$d = m/V$

SUMMARY OF FACTS AND CONCEPTS

Chemistry emerged as a *quantitative science* with the work of the eighteenth-century French chemist Antoine Lavoisier. He made use of the idea that the mass remains constant during a chemical reaction *(law of conservation of mass).*

Chemistry is an experimental science in that the facts of chemistry are obtained by *experiment*. These facts are systematized and explained by *theory,* and theory suggests more experiments. The *scientific method* involves this interplay, in which the body of accepted knowledge grows as it is tested by experiment.

A quantitative science requires one to make measurements. Any measurement has limited *precision,* which we convey by writing the measured number to a certain number of *significant figures*. There are many different systems of measurement, but scientists generally use the metric system. The International System (SI) uses a particular selection of metric units. It employs seven *base units* combined with *prefixes* to obtain units of various size. Units for other quantities are derived from these. To obtain a *derived unit* in SI for a quantity such as the volume or density, we merely substitute base units into a defining equation for the quantity.

Dimensional analysis is a technique of calculating with physical quantities in which units are included and treated in the same way as numbers. Using dimensional analysis, one can obtain the *conversion factor* needed to express a quantity in new units.

OPERATIONAL SKILLS

1. **Significant figures in calculations** Given an arithmetic setup, report the answer to the correct number of significant figures and round it properly (Example 1.1).
2. **Temperature conversion** Given a temperature reading on one scale, convert it to another scale—Celsius, Kelvin, or Fahrenheit (Example 1.2).
3. **Calculation of the density** Given the mass and volume of a substance, calculate the density (Example 1.3).
4. **Using the density to relate mass and volume** Given the mass and density of a substance, calculate the volume; or given the volume and density, calculate the mass (Example 1.4).
5. **Unit conversion** Given an equation relating one unit to another (or a series of such equations), convert a measurement expressed in one unit to a new unit (Examples 1.5, 1.6, and 1.7).

Review Questions

1.1 Define the terms *matter* and *mass*. What is the difference between mass and weight?

1.2 State the law of conservation of mass. Describe how you might demonstrate this law.

1.3 A chemical reaction is often accompanied by definite changes in appearance. Thus, heat and light may be emitted, and there may be a change of color of the substances. Figures 1.3 and 1.4 show the reactions of the metals magnesium and mercury with oxygen in air. Describe the changes that occur in each case.

1.4 Define the terms *experiment* and *theory*. How are theory and experiment related? What is a hypothesis?

1.5 Illustrate the steps in the scientific method using Rosenberg's discovery of the anticancer activity of cisplatin.

1.6 What is meant by the precision of a measurement? How is it indicated?

1.7 Two rules are used to decide how to round the result of a calculation to the correct number of significant figures. Use a calculation to illustrate each rule. Explain how you obtained the number of significant figures in the answers.

1.8 Distinguish between a measured number and an exact number. Give examples of each.

1.9 How does the International System (SI) obtain units of different size from a given unit? How does the International System obtain units for all possible physical quantities from only seven base units?

1.10 What are the SI base units of length and mass? Give approximate values of U.S. equivalents of each base unit.

1.11 What is an absolute temperature scale? How are degrees Celsius related to kelvin units?

1.12 Define *density*. Describe some uses of the density.

1.13 Why should units be carried along with numbers in a calculation?

1.14 There are exactly 4 quarts in a U.S. gallon. Also, 1 quart = 0.9464 liter. What is the conversion factor from liters to U.S. gallons? Describe the intermediate steps in deriving this conversion factor from the information given.

Practice Problems

Key: These problems are for practice in applying problem-solving skills. They are divided by topic, and some are keyed to exercises (see the ends of the exercises). The problems are arranged in matched pairs; the first problem of each pair (whose number is in color) appears in the left column, and its answer is given in the back of the book.

CONSERVATION OF MASS

1.15 An 8.4-gram sample of sodium hydrogen carbonate is added to a solution of acetic acid weighing 20.0 grams. The two substances react, releasing carbon dioxide gas to the atmosphere. After reaction, the contents of the reaction vessel weigh 24.0 grams. What is the mass of carbon dioxide given off during the reaction?

1.16 Some magnesium wire weighing 2.4 grams is placed in a beaker and covered with 15.0 grams of dilute hydrochloric acid. The acid reacts with the metal and gives off hydrogen gas, which escapes into the surrounding air. After reaction, the contents of the beaker weigh 17.2 grams. What is the mass of hydrogen gas produced by the reaction?

1.17 Zinc metal reacts with yellow crystals of sulfur in a fiery reaction to produce a white powder of zinc sulfide. A chemist determines that 65.4 grams of zinc reacts with 32.1 grams of sulfur. How many grams of zinc sulfide could be produced from 20.0 grams of zinc metal?

1.18 Aluminum metal reacts with bromine, a red-brown liquid with noxious odor. The reaction is vigorous and produces aluminum bromide, a white crystalline substance. A sample of 27.0 grams of aluminum yields 266.7 grams of aluminum bromide. How many grams of bromine react with 15.0 grams of aluminum?

SIGNIFICANT FIGURES

1.19 How many significant figures are there in each of the following measurements?

(a) 73.0000 g (b) 0.0503 kg
(c) 6.300 cm (d) 0.80090 m
(e) 5.10×10^{-7} m (f) 2.001 s

1.20 How many significant figures are there in each of the following measurements?

(a) 130.0 kg (b) 0.0738 g
(c) 0.224800 m (d) 1008 s
(e) 4.380×10^{-8} m (f) 9.100×10^{4} cm

1.21 The circumference of the earth at the equator is 40,000 km. This value is precise to two significant figures. Write this in scientific notation to express correctly the number of significant figures.

1.22 The astronomical unit equals the mean distance between the earth and the sun. This distance is 150,000,000 km, which is precise to three significant figures. Express this in scientific notation to the correct number of significant figures.

1.23 Do the indicated arithmetic and give the answer to the correct number of significant figures.

(a) $\dfrac{8.71 \times 0.0301}{0.056}$
(b) $0.71 + 81.8$
(c) $934 \times 0.00435 + 107$
(d) $(847.89 - 847.73) \times 14673$

1.24 Do the indicated arithmetic and give the answer to the correct number of significant figures.

(a) $\dfrac{0.871 \times 0.23}{5.871}$
(b) $8.937 - 8.930$
(c) $8.937 + 8.930$
(d) $0.00015 \times 54.6 + 1.002$

1.25 One sphere has a radius of 5.10 cm; another has a radius of 5.00 cm. What is the difference in volume (in cubic centimeters) between the two spheres? Give the answer to the correct number of significant figures. The volume of a sphere is $(4/3)\pi r^3$, where r is the radius and $\pi = 3.1416$.

1.26 A solid cylinder of iron of circular cross section with a radius of 1.500 cm has a ruler etched along its length. What is the volume of iron contained between the marks labeled 3.10 cm and 3.50 cm? The volume of a cylinder is $\pi r^2 l$, where $\pi = 3.1416$, r is the radius, and l is the length.

SI UNITS

1.27 Write the following measurements using the most appropriate SI prefix.

(a) 5.89×10^{-12} s (b) 2.130×10^{-9} m
(c) 0.00721 g (d) 6.05×10^3 m

1.28 Write the following measurements using the most appropriate SI prefix.

(a) 4.851×10^{-6} g (b) 3.16×10^{-2} m
(c) 2.591×10^{-9} s (d) 8.93×10^{-12} g

1.29 Using the scientific notation, convert:

(a) 6.15 ps to s (b) 3.781 μm to m
(c) 1.546 Å to m (d) 9.7 mg to g

1.30 Using scientific notation, convert:

(a) 8.55 km to m (b) 1.98 ns to s
(c) 2.54 cm to m (d) 6.923 μg to g

TEMPERATURE CONVERSION

1.31 Convert:

(a) 32°F to degrees Celsius
(b) −58°F to degrees Celsius
(c) 68°F to degrees Celsius
(d) −11°F to degrees Celsius
(e) 37°C to degrees Fahrenheit
(f) −70°C to degrees Fahrenheit

1.32 Convert:

(a) 85°F to degrees Celsius
(b) 121°F to degrees Celsius
(c) 21°F to degrees Celsius
(d) −15°F to degrees Celsius
(e) −45°C to degrees Fahrenheit
(f) −65°C to degrees Fahrenheit

1.33 Salt and ice were stirred together to give a mixture to freeze ice cream. The temperature of the mixture is −21.1°C. What is this temperature in degrees Fahrenheit?

1.34 Liquid nitrogen can be used for the quick freezing of foods. The liquid boils at −196°C. What is this temperature in degrees Fahrenheit?

1.35 The manual for a computer says that the maximum operating temperature for the machine is 95°F. What is this temperature in degrees Celsius? What is it in kelvin units?

1.36 Milk can be pasteurized to kill disease-causing organisms by heating it to 161°F for 15 seconds. What is this temperature in degrees Celsius? What is it in kelvin units?

DENSITY

1.37 A certain sample of the mineral galena (lead sulfide) weighs 12.4 g and has a volume of 1.64 cm^3. What is the density of galena?

1.38 A flask contains 25.0 mL of diethyl ether weighing 17.84 g. What is the density of the ether?

1.39 A liquid with a volume of 10.7 mL has a mass of 9.42 g. The liquid is either octane, ethanol, or benzene, the densities of which are 0.702 g/cm^3, 0.789 g/cm^3, and 0.879 g/cm^3, respectively. What is the identity of the liquid?

1.40 A mineral sample has a mass of 16.3 g and a volume of 2.3 cm^3. The mineral is either sphalerite (density = 4.0 g/cm^3), cassiterite (density = 6.99 g/cm^3), or cinnabar (density = 8.10 g/cm^3). Which is it?

1.41 Platinum has a density of 22.5 g/cm^3. What is the mass of 5.9 cm^3 of this metal?

1.42 What is the mass of a 51.6-mL sample of gasoline, which has a density of 0.70 g/cm^3?

1.43 Ethyl alcohol has a density of 0.789 g/cm^3. What volume of ethyl alcohol must be poured into a graduated cylinder to give 19.8 g of alcohol?

1.44 Bromine is a red-brown liquid with a density of 3.10 g/mL. A sample of bromine weighing 88.5 g occupies what volume?

UNIT CONVERSIONS

1.45 Sodium hydrogen carbonate, known commercially as baking soda, reacts with acidic materials such as vinegar to release carbon dioxide gas. An experiment calls for 0.348 kg of sodium hydrogen carbonate. Express this mass in milligrams.

1.46 The acidic constituent in vinegar is acetic acid. A 10.0-mL sample of a certain vinegar contains 483 mg acetic acid. What is this mass of acetic acid expressed in micrograms?

1.47 The different colors of light have different wavelengths. The human eye is most sensitive to light whose wavelength is 555 nm (greenish-yellow). What is this wavelength in millimeters?

1.48 Water consists of molecules (groups of atoms). A water molecule has two hydrogen atoms, each connected to an oxygen atom. The distance between any one hydrogen atom and the oxygen atom is 0.96 Å. What is this distance in millimeters?

1.49 The total amount of fresh water on earth is estimated to be 3.73×10^8 km^3. What is this volume in cubic meters? in liters?

1.50 A submicroscopic particle suspended in a solution has a volume of 1.4 μm^3. What is this volume in liters?

1.51 An aquarium has rectangular cross section that is 47.8 in by 12.5 in; it is 19.5 in tall. How many U.S. gallons does the aquarium contain? One U.S. gallon equals exactly 231 in^3.

1.52 A spherical tank has a radius of 150.0 in. Calculate the volume of the tank in cubic inches; then convert this to Imperial gallons. The volume of a sphere is $(4/3)\pi r^3$, where r is the radius. One Imperial gallon equals 277.4 in^3.

1.53 How many grams are there in 3.58 short tons? Note that 1 g = 0.03527 oz (ounces avoirdupois), 1 lb (pound) = 16 oz, and 1 short ton = 2000 lb. (These relations are exact.)

1.54 How many liters are there in 8.46 U.S. gal? Note that 1 qt (U.S. liquid) = 946.4 cm^3 and 4 qt = 1 gal (exact).

1.55 The first measurement of sea depth was made in 1840 in the central South Atlantic, where a plummet was lowered 2425 fathoms. What is this depth in meters? Note that 1 fathom = 6 ft, 1 ft = 12 in, and 1 in = 2.54×10^{-2} m. (These relations are exact.)

1.56 The estimated amount of recoverable oil from the field at Prudhoe Bay in Alaska is 9.6×10^9 barrels. What is this amount of oil in cubic meters? One barrel = 42 gal (exact), 1 gal = 4 qt (exact), and 1 qt = 9.46×10^{-4} m^3.

Additional Problems

Key: These problems provide more practice but are not divided by topic or keyed to exercises. Each odd-numbered problem and the even-numbered problem that follows it are similar; answers to odd-numbered problems (whose numbers appear in color) are given in the back of the book.

1.57 Sodium metal reacts vigorously with water. A piece of sodium weighing 9.85 grams was added to a beaker containing 63.11 grams of water. During reaction, hydrogen gas was produced and bubbled from the solution. The solution, containing sodium hydroxide, weighed 72.53 grams. How many grams of hydrogen gas were produced?

1.58 An antacid tablet weighing 0.853 g contained calcium carbonate as the active ingredient, in addition to an inert binder. When an acid solution weighing 56.519 g was added to the tablet, carbon dioxide gas was released, producing a fizz. The resulting solution weighed 57.152 g. How many grams of carbon dioxide were produced?

1.59 When a mixture of aluminum powder and iron(III) oxide is ignited, it produces molten iron and aluminum oxide. In an experiment, 5.40 g aluminum was mixed with 18.50 g iron(III) oxide. At the end of the reaction, the mixture contained 11.17 g iron, 10.20 g aluminum oxide, and an undetermined amount of unreacted iron(III) oxide. No aluminum was left. What is the mass of the iron(III) oxide?

1.60 When chlorine gas is bubbled into a solution of sodium bromide, the sodium bromide reacts to give bromine, a red-brown liquid, and sodium chloride (ordinary table salt). A solution was made up by dissolving 20.6 g sodium bromide in 100.0 g water. After passing chlorine through the solution, investigators analyzed the mixture. It contained 16.0 g bromine and 11.7 g sodium chloride. How many grams of chlorine reacted?

1.61 A beaker weighed 53.10 g. To the beaker was added 5.348 g iron pellets and 56.1 g hydrochloric acid. What was the total mass of the beaker and the mixture (before reaction)? Express the answer to the correct number of significant figures.

1.62 A graduated cylinder weighed 68.1 g. To the cylinder was added 48.7 g water and 5.318 g sodium chloride. What was the total mass of the cylinder and the solution? Express the answer to the correct number of significant figures.

1.63 A cubic box measures 60.8 cm on an edge. What is the volume of the box in cubic centimeters? Express the answer to the correct number of significant figures.

1.64 A cylinder with circular cross section has a radius of 2.13 cm and a height of 56.32 cm. What is the volume of the cylinder? Express the answer to the correct number of significant figures.

1.65 Obtain the difference in volume between two spheres, one whose radius is 5.61 cm, the other whose radius is 5.85 cm. The volume V of a sphere is $(4/3)\pi r^3$, where r is the radius. Give the result to the correct number of significant figures.

1.66 What is the difference in surface area between two circles, one of radius 7.98 cm, the other of radius 8.50 cm? The surface area of a circle of radius r is πr^2. Obtain the result to the correct number of significant figures.

1.67 Perform the following arithmetic setups and express the answers to the correct number of significant figures.

(a) $\dfrac{56.1 - 51.1}{6.58}$ (b) $\dfrac{56.1 + 51.1}{6.58}$

(c) $(9.1 + 8.6) \times 26.91$ (d) $0.0065 \times 3.21 + 0.0911$

1.68 Perform the following arithmetic setups and report the answers to the correct number of significant figures.

(a) $\dfrac{9.345 - 9.005}{9.811}$ (b) $\dfrac{9.345 + 9.005}{9.811}$

(c) $(8.12 + 7.53) \times 3.71$ (d) $0.71 \times 0.36 + 17.36$

1.69 For each of the following, write the measurement in terms of an appropriate prefix and base unit.

(a) The mass of calcium per milliliter in a sample of blood serum is 0.0912 g.
(b) The radius of an oxygen atom is about 0.000000000066 m.
(c) A particular red blood cell measures 0.0000071 m.
(d) The wavelength of a certain ultraviolet radiation is 0.000000056 m.

1.70 For each of the following, write the measurement in terms of an appropriate prefix and base unit.

(a) The mass of magnesium per milliliter in a sample of blood serum is 0.0186 g.
(b) The radius of a carbon atom is about 0.000000000077 m.
(c) The hemoglobin molecule, a component of red blood cells, is 0.0000000065 m in diameter.
(d) The wavelength of a certain infrared radiation is 0.00000085 m.

1.71 Write each of the following in terms of the SI base unit (that is, express the prefix as the power of ten).

(a) 1.07 ps (b) 5.8 μm
(c) 319 nm (d) 15.3 ms

1.72 Write each of the following in terms of the SI base unit (that is, express the prefix as the power of ten).

(a) 7.3 mK (b) 275 pm
(c) 19.6 ms (d) 45 μm

1.73 Tungsten metal, which is used in light bulb filaments, has the highest melting point of any metal (3410°C). What is this melting point in degrees Fahrenheit?

1.74 Titanium metal is used in aerospace alloys to add strength and corrosion resistance. Titanium melts at 1677°C. What is this temperature in degrees Fahrenheit?

1.75 Calcium carbonate, a white powder used in toothpastes, antacids, and other preparations, decomposes when heated to about 825°C. What is this temperature in degrees Fahrenheit?

1.76 Sodium hydrogen carbonate (baking soda) starts to decompose to sodium carbonate (soda ash) at about 50°C. What is this temperature in degrees Fahrenheit?

1.77 Gallium metal can be melted by the heat of one's hand. Its melting point (the temperature at which it melts) is 29.8°C. What is this temperature in kelvin units? in degrees Fahrenheit?

1.78 Mercury metal is liquid at normal temperatures but freezes at −38.9°C. What is this temperature in kelvin units? in degrees Fahrenheit?

1.79 Zinc metal can be purified by distillation (transforming the liquid metal to vapor, then condensing the vapor back to the liquid). The metal boils at normal atmospheric pressure at 1666°F. What is this temperature in degrees Celsius? in kelvin units?

1.80 Iodine is a bluish-black solid. It forms a violet-colored vapor when heated. The solid melts at 236°F. What is this temperature in degrees Celsius? in kelvin units?

1.81 The density of osmium metal (a platinum-group metal) is 22.5 g/cm^3. Express this density in SI units (kg/m^3).

1.82 Vanadium metal is added to steel to impart strength. The density of vanadium is 5.96 g/cm^3. Express this in SI units (kg/m^3).

1.83 The density of quartz mineral was determined by adding a weighed piece to a graduated cylinder containing 51.2 mL water. After adding the quartz, the water level was 65.7 mL. The quartz piece weighed 38.4 g. What was the density of the quartz?

1.84 Hematite (iron ore) weighing 70.7 g was placed in a flask whose volume was 53.2 mL. The flask with hematite was then carefully filled with water and weighed. The hematite and water weighed 109.3 g. The density of the water was 0.997 g/cm^3. What was the density of the hematite?

1.85 Some bottles of colorless liquids were being labeled when the technicians accidentally mixed them up and lost track of their contents. A 15.0-mL sample withdrawn from one bottle weighed 22.3 g. The technicians knew that the liquid was either acetone, benzene, chloroform, or carbon tetrachloride (which have densities of 0.792 g/cm^3, 0.899 g/cm^3, 1.489 g/cm^3, and 1.595 g/cm^3, respectively). What was the identity of the liquid?

1.86 A solid will float on any liquid that is more dense than it is. The volume of a piece of calcite weighing 35.6 g is 12.9 cm^3. On which of the following liquids will the calcite float: carbon tetrachloride (density = 1.60 g/cm^3), methylene bromide (density = 2.50 g/cm^3), tetrabromoethane (density = 2.96 g/cm^3), methylene iodide (density = 3.33 g/cm^3)?

1.87 Platinum metal is used in jewelry; it is also used in automobile catalytic converters. What is the mass of a cube of platinum 2.20 cm on an edge? The density of platinum is 21.4 g/cm^3.

1.88 Ultrapure silicon is used to make solid-state devices, such as computer chips. What is the mass of a circular cylinder of silicon that is 12.40 cm long and has a radius of 3.50 cm? The density of silicon is 2.33 g/cm^3.

1.89 Vinegar contains acetic acid (about 5% by mass). Pure acetic acid has a strong vinegar smell but is corrosive to the skin. What volume of pure acetic acid has a mass of 35.00 g? The density of acetic acid is 1.053 g/mL.

1.90 Ethyl acetate has a characteristic fruity odor and is used as a solvent in paint lacquers and perfumes. An experiment requires 0.985 kg ethyl acetate. What volume is this (in liters)? The density of ethyl acetate is 0.902 g/mL.

1.91 Convert:
(a) 8.45 kg to milligrams
(b) 318 μs to milliseconds
(c) 93 km to nanometers
(d) 37.1 mm to centimeters

1.92 Convert:
(a) 239 Å to micrometers
(b) 19.6 kg to milligrams
(c) 24.8 cm to millimeters
(d) 4.3 ns to microseconds

1.93 Convert:
(a) 5.91 kg of chrome yellow to milligrams
(b) 753 mg of vitamin A to micrograms
(c) 90.1 MHz (megahertz), the wavelength of an FM signal, to kilohertz
(d) 498 mJ (the joule, J, is a unit of energy) to kilojoules

1.94 Convert:
(a) 7.19 μg of cyanocobalamin (vitamin B_{12}) to milligrams
(b) 104 pm, the radius of a sulfur atom, to angstroms
(c) 0.010 mm, the diameter of a typical blood capillary, to centimeters
(d) 0.0605 kPa (the pascal, Pa, is a unit of pressure) to centipascals

1.95 The largest of the Great Lakes is Lake Superior, which has a volume of 12,230 km^3. What is this volume in liters?

1.96 The average flow of the Niagara River is 0.477 km^3 of water per day. What is this volume in liters?

1.97 A room measures 10.0 ft × 12.0 ft and is 9.0 ft high. What is the volume of this room in liters?

1.98 A cylindrical settling tank is 6.0 ft deep and has circular cross section with radius of 15.0 ft. What is the volume of the tank in liters?

1.99 The Star of India sapphire weighs 563 carats. A carat equals 200 mg. What is the weight of the gemstone in grams?

1.100 Recent world production of gold was 49.6 × 10^6 troy ounces. One troy ounce equals 31.10 g. What is the world production of gold in metric tons (10^6 g)?

Cumulative-Skills Problems

Key: These problems require two or more operational skills you learned in the chapter. In later chapters, the problems under this heading will combine skills introduced in previous chapters with those given in the current one.

1.101 When 10.0 g marble chips (calcium carbonate) are treated with 50.0 mL hydrochloric acid (density 1.096 g/mL), the marble dissolves, giving a solution and releasing carbon dioxide gas. The solution weighs 60.4 g. How many liters of carbon dioxide gas are released? The density of the gas is 1.798 g/L.

1.102 Zinc ore (zinc sulfide) is treated with sulfuric acid, leaving a solution with some undissolved bits of material and releasing hydrogen sulfide gas. If 10.8 g zinc ore is treated with 50.0 mL sulfuric acid (whose density is 1.153 g/mL), 65.1 g of solution and undissolved material remains. In addition, hydrogen sulfide (density 1.393 g/L) is evolved. What is the volume (in liters) of this gas?

1.103 A steel sphere has a radius of 1.58 in. If this steel has a density of 7.88 g/cm^3, what is the mass of this steel sphere in grams?

1.104 A weather balloon filled with helium has a diameter of 3.00 ft. What is the mass in grams of the helium in the balloon at 21°C and normal pressure? The density of helium under these conditions is 0.166 g/L.

1.105 The land area of Greenland is 840,000 mi^2, with only 132,000 mi^2 free of perpetual ice. The average thickness of this ice is 5000 ft. Estimate the mass of the ice (assume two significant figures). The density of ice is 0.917 g/cm^3.

1.106 Antarctica, almost completely covered in ice, has an area of 5,500,000 mi^2 with an average height of 7500 ft. Without the ice, the height would be only 1500 ft. Estimate the mass of this ice (two significant figures). The density of ice is 0.917 g/cm^3.

1.107 A sample of an ethanol–water solution has a volume of 54.2 cm^3 and a mass of 49.6 g. What is the percentage of ethanol (by mass) in the solution? (Assume that there is no change in volume when the pure compounds are mixed.) The density of ethanol is 0.789 g/cm^3 and that of water is 0.998 g/cm^3. Alcoholic beverages are rated in *proof,* which is a measure of the relative amount of ethanol in the beverage. Pure ethanol is exactly 200 proof; a solution that is 50% ethanol by volume is exactly 100 proof. What is the proof of the given ethanol–water solution?

1.108 You have a piece of gold jewelry weighing 9.35 g. Its volume is 0.654 cm^3. Assume that the metal is an alloy (mixture) of gold and silver, which have densities of 19.3 g/cm^3 and 10.5 g/cm^3, respectively. Also assume that there is no change in volume when the pure metals are mixed. Calculate the percentage of gold (by mass) in the alloy. The relative amount of gold in an alloy is measured in *karats*. Pure gold is 24 karats; an alloy of 50% gold is 12 karats. State the proportion of gold in the jewelry in karats.

1.109 A sample of a bright red or vermillion-colored mineral was weighed in air, then weighed while suspended in water. An object is buoyed up by the mass of the fluid displaced by the object. In air, the mineral weighed 18.49 g; in water, it weighed 16.21 g. The densities of air and water are 1.205 g/L and 0.9982 g/cm^3, respectively. What is the density of the mineral?

1.110 A sample of a bright blue mineral was weighed in air, then weighed while suspended in water. An object is buoyed up by the mass of the fluid displaced by the object. In air, the mineral weighed 7.35 g; in water, it weighed 5.40 g. The densities of air and water are 1.205 g/L and 0.9982 g/cm^3, respectively. What is the density of the mineral?

2

Atoms, Molecules, and Ions

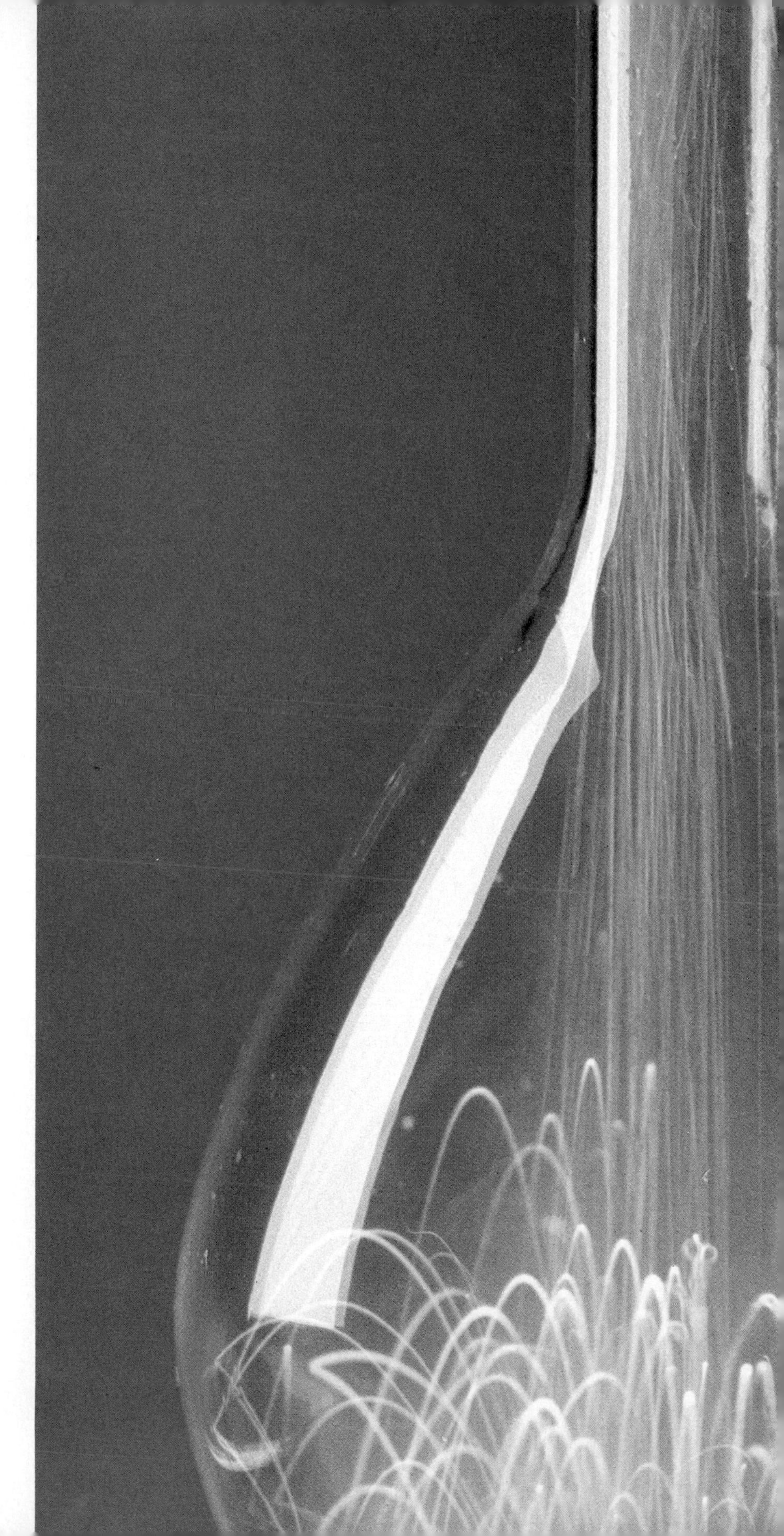

Chapter Outline

Reaction of antimony and chlorine.

Sodium is a soft, silvery metal. This metal cannot be handled with bare fingers because it reacts with any moisture on the skin, causing a burn. Chlorine is a poisonous, greenish-yellow gas with a choking odor. A dramatic chemical reaction occurs when a pea-sized piece of sodium is ignited in a flask of chlorine. The metal bursts into brilliant yellow flame and burns to a white, crystalline powder (see Figure 2.1). Watching this violent reaction of toxic substances is fascinating enough, but examining the end product provides a surprise. The white powder is sodium chloride: common table salt, an edible substance.

Figure 2.1
Reaction of sodium and chlorine. *Left:* Sodium metal and chlorine gas. *Right:* A small piece of sodium has been ignited in a flask of chlorine. The product is sodium chloride, common table salt.

Sodium metal and chlorine gas are particular forms of matter. In burning, they undergo a chemical change—a chemical reaction—in which these forms of matter change to a form of matter with very different properties. How do we explain the differences in properties of different forms of matter? And how do we explain chemical reactions such as the burning of sodium metal in chlorine gas? This chapter and the next take an introductory look at these basic questions in chemistry. In later chapters we will develop the concepts introduced here.

Matter: Description and Theory

Chemistry is both descriptive and theoretical. We just described the physical appearance of sodium and chlorine and the reaction of these two different kinds of matter. Once we have collected such descriptive information about many different kinds of matter, patterns and schemes for classifying emerge. Then the groundwork has been laid for developing a theory of our observations. The next section discusses two descriptive classifications of matter. The section following it introduces the atomic theory, which explains our observations of matter.

2.1 DESCRIPTION OF MATTER

Two principal ways of classifying matter have emerged. We classify matter (1) by its physical state as a solid, liquid, or gas, and (2) by its chemical constitution as an element, compound, or mixture. Let us look at these classification schemes.

Solids, Liquids, and Gases

Commonly, a given kind of matter exists in different physical forms under different conditions. Water, for example, exists as ice (solid water), as liquid water, and as steam (gaseous water). Sodium metal is normally solid, but it melts to a silvery liquid when heated to 98°C. Liquid sodium changes to a bluish gas if the temperature is raised to 883°C at normal barometric pressure. Similarly, chlorine, which is normally a gas, can exist as a yellow liquid or solid under the appropriate conditions.

The main identifying characteristic of solids is their rigidity: They tend to maintain their shapes when subjected to outside forces. Liquids and gases, however, are *fluids;* that is, they flow easily and change their shapes in response to slight outside forces.

What distinguishes a gas from a liquid is the characteristic of *compressibility* (and its opposite, *expansibility*). A gas is easily compressible, whereas a liquid is not. You can put more and more air into a tire, which increases only slightly in volume. In fact, a given quantity of gas can fill a container of any size. A small quantity would expand to fill the container; a larger quantity could be compressed to fill the same space. By contrast, if you were to try to force more liquid water into a glass bottle that was already full of water, it would burst.

These two characteristics, rigidity (or fluidity) and compressibility (or expansibility), can be used to frame definitions of the three states of matter:

solid *the form of matter characterized by rigidity;* a solid is relatively incompressible and has fixed shape and volume.

Table 2.1 Characteristics of the States of Matter

State of Matter	Fluidity or Rigidity	Compressibility	Relative Density
Solid	Rigid	Very low	High
Liquid	Fluid	Very low	High
Gas	Fluid	High	Low

liquid *the form of matter that is a relatively incompressible fluid;* a liquid has a fixed volume but no fixed shape.

gas *the form of matter that is an easily compressible fluid;* a given quantity of gas will fit into a container of any size and shape.

The term *vapor* is often used to refer to the gaseous state of any kind of matter that normally exists as a liquid or a solid.

Other properties, such as density, can also be used to distinguish the three states of matter. Solids and liquids have relatively high densities compared with gases. For example, the density of ice is about 0.92 g/cm^3; the density of liquid water is about 1.0 g/cm^3; and that of steam, or water vapor, is about 6×10^{-4} g/cm^3.● In most cases the liquid form of a given material is less dense than the solid form. Water is unusual in this respect. Its solid form is less dense than the liquid, which accounts for the fact that ice floats on liquid water.

● Butane is an easily liquefied gaseous fuel derived from petroleum. It is the colorless liquid in disposable cigarette lighters, where it is contained under slight pressure. When you open the valve on a lighter, the liquid turns to gas, which issues from the jet. The gas occupies a volume 240 times that of the liquid at 0°C and normal atmospheric pressure.

The three forms of matter—solid, liquid, and gas—are referred to as the **states of matter.** The characteristics of the different states of matter that we have discussed are summarized in Table 2.1.

Elements, Compounds, and Mixtures

Let us now look at the classification of matter by chemical constitution. The term *material* is used to refer to any particular kind of matter. To classify materials, we must first distinguish between physical and chemical changes and between physical and chemical properties. A **physical change** is *a change in the form of matter but not in its chemical identity*. Changes of physical state are examples of physical changes. The process of dissolving one material in another is a further example of a physical change. Thus, we can dissolve sodium chloride (table salt) in water. The result is a clear liquid, like pure water, though many of its other characteristics, including density, are different from those of pure water. The water and sodium chloride in this liquid retain their chemical identities and can be separated by some method that depends on physical changes.

Distillation is one way to separate the sodium chloride and water components of this liquid. We place the liquid in a flask to which a device called a *condenser* is attached (see Figure 2.2). The liquid in the flask is heated to bring it to a boil. (Boiling entails the formation of bubbles of the vapor in the body of the liquid.) Water vapor forms and passes from the flask into a cooled condenser, where the vapor changes back to liquid water. The liquid water is collected in another flask, called a *receiver*. The original flask now contains the solid sodium chloride. Thus, by means of physical changes (the change of liquid water to vapor and back to liquid), we have separated the sodium chloride and water which we had earlier mixed together.

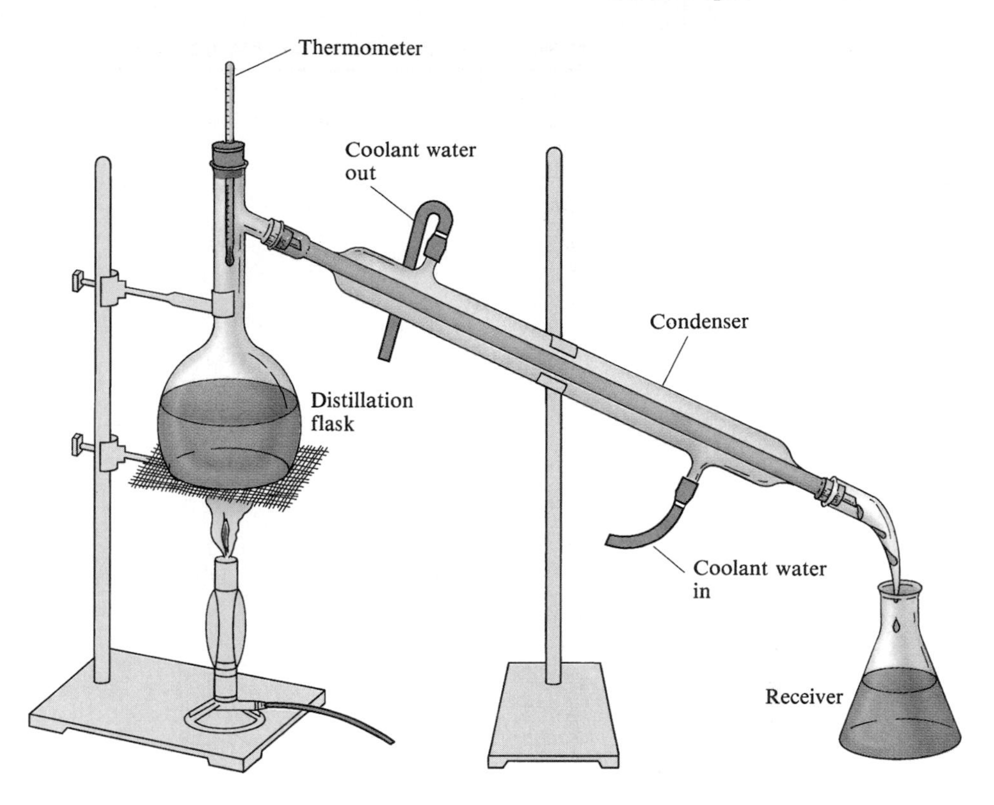

Figure 2.2
Separation by distillation. An easily vaporized liquid can be separated from another substance by distillation.

A **chemical change,** or **chemicai reaction,** is *a change in which one or more kinds of matter are transformed into a new kind of matter or several new kinds of matter*. The burning of sodium metal in chlorine gas to produce sodium chloride is a chemical change. So is the rusting of iron, during which iron combines with oxygen in air to form a new material called rust. In both of these examples, the original materials combine chemically and cannot be separated by any physical means. To recover the sodium and chlorine from sodium chloride or the iron and oxygen from rust requires some chemical change or a series of chemical changes.

Figure 2.3
Reaction of sodium with water. Sodium metal flits around the water surface as it reacts briskly, giving off hydrogen gas. The other product is sodium hydroxide, which changes a substance added to the water (phenolphthalein) from colorless to pink.

We characterize or identify a material by its various properties, which may be either physical or chemical. A **physical property** is *a characteristic that can be observed for a material without changing its chemical identity*. Examples are physical state (solid, liquid, or gas) and color. Measurable physical properties, such as mass and density, are classified as either extensive or intensive. An *extensive property* is one whose magnitude depends on the amount of material. Mass and volume are examples. An *intensive property,* however, is one whose magnitude is independent of the amount of material. It is these properties that we use to identify a material. Examples are density and melting point (the temperature at which a material melts).

A **chemical property** is *a characteristic of a material involving its chemical change*. One chemical property of sodium is its ability to react with chlorine. Another chemical property of sodium is its ability to react readily with water (see Figure 2.3).

SUBSTANCES The various materials we see around us are either substances or mixtures of substances. A **substance** is *a kind of matter that cannot be separated into other kinds of matter by any physical process*. Earlier we saw that when

Table 2.2 Some Common Elements

Name of Element	Atomic Symbol*	Physical Appearance of Element†
Aluminum	Al	Silvery-white metal
Barium	Ba	Silvery-white metal
Bromine	Br	Reddish-brown liquid
Calcium	Ca	Silvery-white metal
Carbon	C	
Graphite		Soft, black solid
Diamond		Hard, colorless crystal
Chlorine	Cl	Greenish-yellow gas
Chromium	Cr	Silvery-white metal
Cobalt	Co	Silvery-white metal
Copper	Cu (from *cuprum*)	Reddish metal
Fluorine	F	Pale yellow gas
Gold	Au (from *aurum*)	Soft, yellow metal
Helium	He	Colorless gas
Hydrogen	H	Colorless gas
Iodine	I	Bluish-black solid
Iron	Fe (from *ferrum*)	Silvery-white metal
Lead	Pb (from *plumbum*)	Bluish-white metal
Magnesium	Mg	Silvery-white metal
Manganese	Mn	Gray-white metal
Mercury	Hg (from *hydrargyrum*)	Silvery-white liquid metal
Neon	Ne	Colorless gas
Nickel	Ni	Silvery-white metal
Nitrogen	N	Colorless gas
Oxygen	O	Colorless gas
Phosphorus (white)	P	Yellowish-white, waxy solid
Potassium	K (from *kalium*)	Soft, silvery-white metal
Silicon	Si	Gray, lustrous solid
Silver	Ag (from *argentum*)	Silvery-white metal
Sodium	Na (from *natrium*)	Soft, silvery-white metal
Sulfur	S	Yellow solid
Tin	Sn (from *stannum*)	Silvery-white metal
Zinc	Zn	Bluish-white metal

*Atomic symbols are discussed in Section 2.2.

†Common form of the element under normal conditions.

sodium chloride is dissolved in water, it is possible to separate the sodium chloride from the water by the physical process of distillation. However, sodium chloride is itself a substance and cannot be separated by physical processes into new materials. Similarly, pure water is a substance.

No matter what their source, substances have definite intensive physical and chemical properties. Sodium is a solid metal having a density of 0.97 g/cm^3 and a melting point of 98°C. The metal also reacts vigorously with chlorine and with water. No matter how sodium is prepared, it always has these properties. Similarly, whether sodium chloride is obtained by burning sodium in chlorine (as described in the chapter opening) or from seawater, it is a white solid melting at 801°C.

Exercise 2.1

Potassium is a soft, silvery-colored metal melting at 64°C. The metal has a density of 0.86 g/cm^3. It reacts vigorously with water, with oxygen, and with chlorine. Identify all of the physical properties given in this description. Identify all of the chemical properties given. (See Problems 2.25, 2.26, 2.27, and 2.28.)

ELEMENTS Millions of substances have been characterized by chemists. Of these, a very small number are known as elements, from which all other substances are made. Lavoisier was the first to establish an experimentally useful definition of an element: An **element** is *a substance that cannot be decomposed by any chemical reaction into simpler substances.* In 1789 Lavoisier listed 33 substances as elements, of which over 20 are still so regarded. Today 109 elements are known. Table 2.2 lists some of the known elements (first column) and their physical appearances (third column); the symbols in this table will be explained later, after we have discussed atomic theory. Some elements are shown in Figure 2.4.

COMPOUNDS Most substances are compounds. A **compound** is *a substance composed of more than one element, chemically combined.* By the end of the eighteenth century, many compounds had been shown by Lavoisier and others to be composed of the elements in definite proportions by mass. It was Joseph Louis

Figure 2.4
Some elements. *Center:* Sulfur. *From upper left, clockwise:* mercury, bromine, iodine, chromium, and lead.

Proust (1754–1826), however, who, by his painstaking work, convinced the majority of chemists of the general validity of the **law of definite proportions** (also known as the **law of constant composition**): *A pure compound, whatever its source, always contains definite or constant proportions of the elements by mass.* For example, 1.0000 g of sodium chloride always contains 0.3934 g of sodium and 0.6066 g of chlorine, chemically combined. Sodium chloride has definite proportions of sodium and chlorine; that is, it has constant or definite composition.●

● It is now known that some compounds do not follow the law of definite proportions. These nonstoichiometric compounds, as they are called, will be described briefly in Chapter 11.

Example 2.1 Using the Law of Definite Proportions

A homogeneous material containing sodium and oxygen is obtained from several sources. Analyses of three samples gave the following data. Are these data consistent with the hypothesis that the material is a compound?

	MASS OF SAMPLE	MASS OF SODIUM	MASS OF OXYGEN
Sample A	1.020 g	0.757 g	0.263 g
Sample B	1.548 g	1.149 g	0.399 g
Sample C	1.382 g	1.025 g	0.357 g

Solution

If the material is a compound, it should exhibit definite composition. To compare the samples, we calculate the fraction of the material that is sodium. For sample A,

$$\text{Fraction of Na in sample A} = \frac{0.757 \text{ g}}{1.020 \text{ g}} = 0.742$$

Similarly,

$$\text{Fraction of O in sample A} = \frac{0.263 \text{ g}}{1.020 \text{ g}} = 0.258$$

(Multiplying the fractions by 100 would convert them to percentages.) For sample B and sample C we find the same values. Thus, all samples have the same composition, so **the data are consistent with the hypothesis that the material is a compound.**

Exercise 2.2

Potassium metal was burned in oxygen, producing a yellowish-orange material. Samples of the material gave the following data.

	MASS OF SAMPLE	MASS OF POTASSIUM	MASS OF OXYGEN
Sample A	2.502 g	1.400 g	1.102 g
Sample B	1.819 g	1.217 g	0.602 g
Sample C	2.761 g	1.832 g	0.929 g

Are these data consistent with the hypothesis that the material is a compound?

(See Problems 2.31 and 2.32.)

MIXTURES Most of the materials around us are mixtures. A **mixture** is *a material that can be separated by physical means into two or more substances.* Unlike a pure compound, a mixture has variable composition. When we dissolve sodium chloride in water, we obtain a mixture; its composition depends on the relative amount of sodium chloride dissolved. We can separate the mixture by the physical process of distillation.●

● Chromatography, another example of a physical method used to separate mixtures, is described in the Instrumental Methods at the end of this section.

Mixtures are classified into two types. A **heterogeneous mixture** is *a mixture that consists of physically distinct parts, each with different properties* (see Figure 2.5). Salt and sugar that have been stirred together constitute a heterogeneous mixture. If you look closely, you see the separate crystals of sugar and salt. A

Figure 2.5
A heterogeneous mixture. *Left:* The mixture on the watchglass consists of potassium dichromate (orange crystals) and iron filings. *Right:* A magnet separates the iron filings from the mixture.

homogeneous mixture (also known as a **solution**) is *a mixture that is uniform in its properties throughout given samples*. When sodium chloride is dissolved in water, we obtain a homogeneous mixture, or solution. Air is a gaseous solution, principally of two elementary substances, nitrogen and oxygen, which are physically mixed but not chemically combined.

A **phase** is *one of several different homogeneous materials present in the portion of matter under study*. A heterogeneous mixture composed of salt and sugar is said to be composed of two different phases; one of the phases is salt, the other is sugar. Similarly, several ice cubes in water is said to be composed of two phases; one phase is ice, the other is liquid water. Ice floating in a solution of sodium chloride in water also consists of two phases, ice and the liquid solution. Note that a phase may be either a pure substance in a particular state or a solution in a particular state (solid, liquid, or gaseous). Also, the portion of matter we are considering can consist of several phases of the same substance or several phases of different substances.

Figure 2.6 summarizes the relationship among elements, compounds, and mixtures. Materials are either substances or mixtures. Substances can be mixed by physical processes, and other physical processes can be used to separate the mixtures into substances. Substances are either elements or compounds. Elements may react chemically to yield compounds, and compounds may be decomposed by chemical reactions into elements.

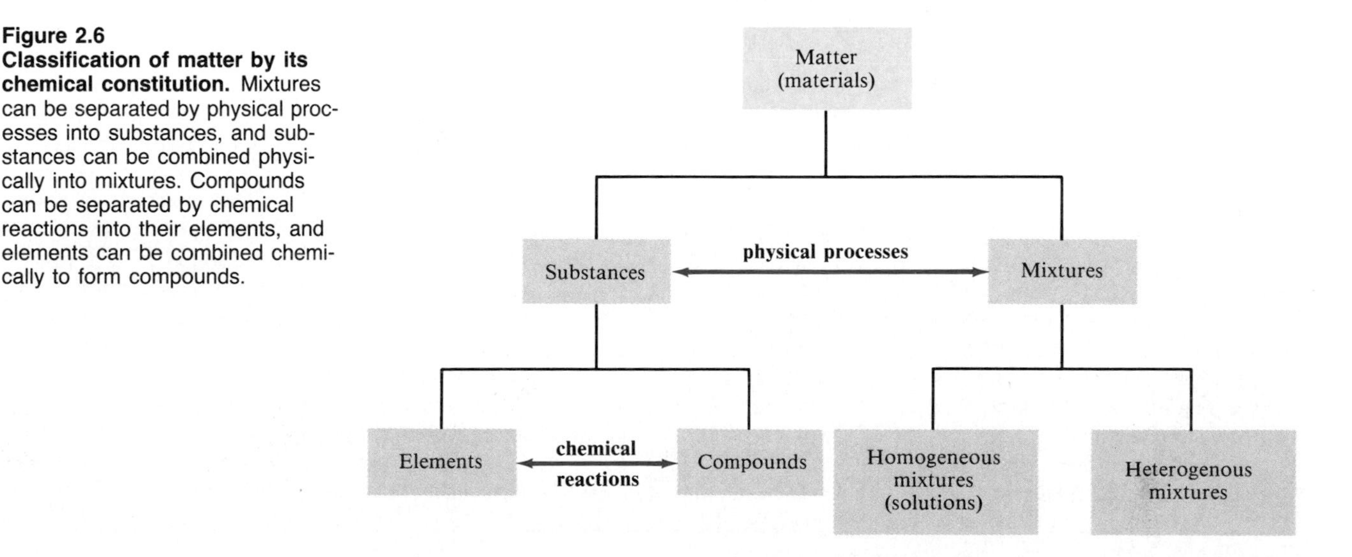

Figure 2.6
Classification of matter by its chemical constitution. Mixtures can be separated by physical processes into substances, and substances can be combined physically into mixtures. Compounds can be separated by chemical reactions into their elements, and elements can be combined chemically to form compounds.

Instrumental Methods

Figure 2.7
An illustration of paper chromatography. *A line of ink has been drawn along the lower edge of a sheet of paper. The dyes in the ink separate as a solution of methanol and water creeps upward by capillary action.*

SEPARATION OF MIXTURES BY CHROMATOGRAPHY

Chromatography is the name given to a group of similar separation techniques that depend on how fast a substance moves, in a stream of gas or liquid, past a stationary phase to which the substance may be slightly attracted. An example is provided by a simple experiment in *paper chromatography* (see Figure 2.7). In this experiment, we put a line of ink near one edge of a sheet of paper and then place the paper upright with this edge in a solution of methanol and water. As the solution creeps up the paper by capillary action, the ink moves upward, separating into a series of different-colored bands that correspond to the different dyes in the ink. The reason the ink is separated into bands containing the different dyes is that each dye is attracted to the wet paper fibers but with different strengths of attraction. As the solution moves upward, the dyes that are less strongly attracted to the paper fibers move more rapidly.

Although the phenomenon of chromatography was used in the nineteenth century by dye chemists to analyze dye solutions, the Russian botanist Mikhail Tswett was the first to understand the basis of chromatography and to apply it systematically as a method of separation. In 1906 Tswett separated pigments in plant leaves by *column chromatography*. He first dissolved the pigments from the leaves with petroleum ether, a liquid similar to gasoline. After packing a glass tube or column with powdered chalk, he poured the solution of pigments into the top of the column (see Figure 2.8). When he washed the column by pouring in more petroleum ether after the solution, it began to show distinct yellow and green bands. These bands, each containing a pure pigment, became well separated as they moved down the column, so that the pure pigments could be obtained. The name chromatography originates from this early separation of colored substances (the stem *chromato-* means "color"), though the technique is not limited to colored substances.

Gas chromatography (GC) is a more recently devised separation method. Here the moving stream is a gaseous mixture of vaporized substances plus a gas such as helium, which is called the *carrier*. The stationary material is a solid or a liquid adhering to a solid, packed in a column. As the gas passes through the column, substances in the mixture are attracted differently to the stationary column packing and thus are separated. Gas chromatography is a rapid, small-scale method of separating mixtures. It is also important in the analysis of mixtures because the time it takes for a sub-

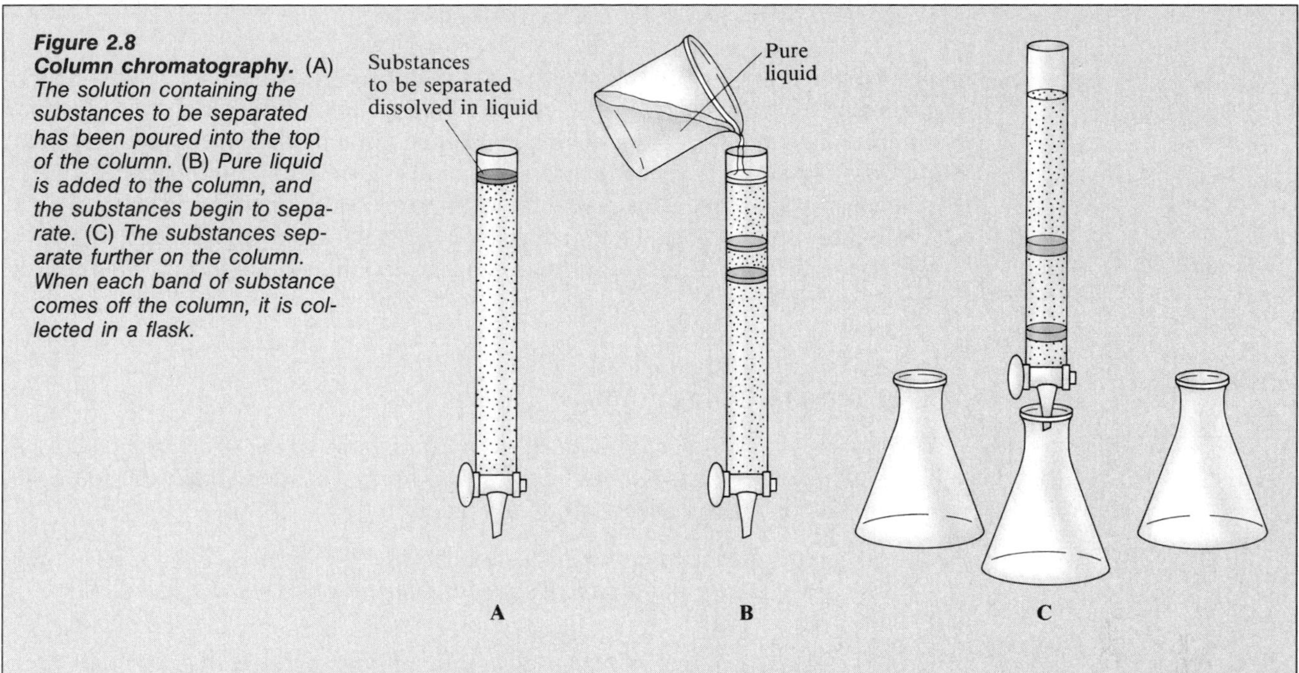

Figure 2.8
Column chromatography. (A) *The solution containing the substances to be separated has been poured into the top of the column.* (B) *Pure liquid is added to the column, and the substances begin to separate.* (C) *The substances separate further on the column. When each band of substance comes off the column, it is collected in a flask.*

stance at a given temperature to travel through the column to a detector (called the *retention time*) is fixed. Retention times can therefore be used to help identify substances. Figure 2.9 shows a gas chromatograph and a portion of a chart recording *(chromatogram)*. Each peak on the chromatogram corresponds to a specific substance. The peaks were drawn automatically by a chart recorder as the different substances in the mixture passed the detector. Complicated mixtures have been analyzed by gas chromatography. Analysis of chocolate, for example, shows that it contains over 800 flavor compounds.

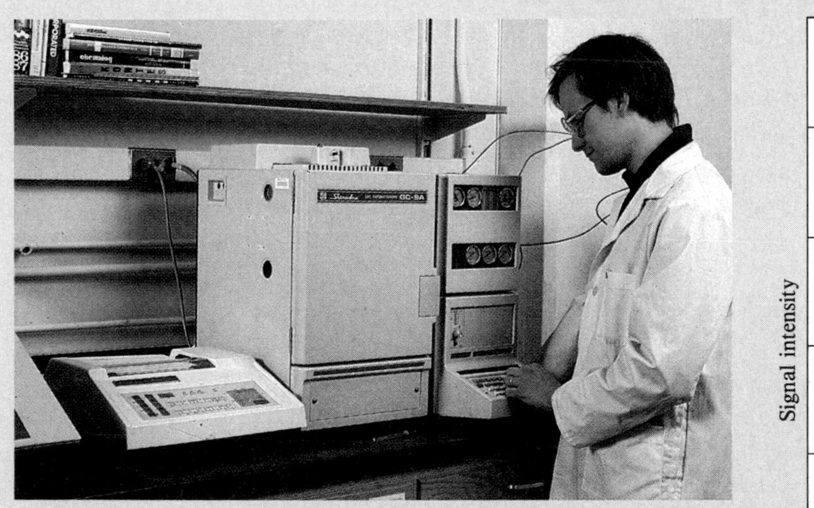

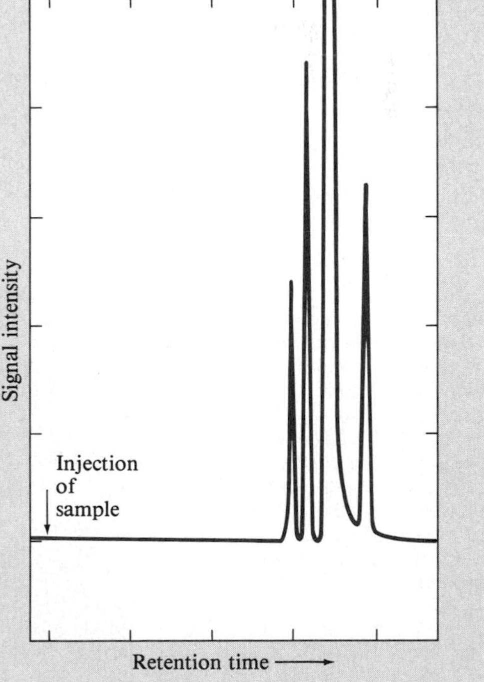

Figure 2.9
Gas chromatography. Left: *The photograph shows a gas chromatograph.* Right: *This is a chromatogram of a hexane mixture, showing its separation into four isomers (different compounds with the same molecular formula). Such hexane mixtures occur in gasoline; hexane is also used as a solvent to extract the oil from certain vegetable seeds.*

2.2 ATOMIC THEORY OF MATTER

As we noted in Chapter 1, Lavoisier laid the experimental foundation of modern chemistry. But it was the British chemist John Dalton (1766–1844) who provided the basic theory: All matter—whether element, compound, or mixture—is composed of small particles called atoms. Dalton was interested in meteorology, the science of weather, and this interest led him to speculate on the composition of air. From these speculations, he formulated a hypothesis about the structure of matter. The main features of this hypothesis, now called the atomic theory, were developed by 1803.

Postulates of Dalton's Atomic Theory

The main points of the **atomic theory,** *an explanation of the structure of matter in terms of different combinations of very small particles,* are given by the following postulates, or formal statements of the theory:

1. All matter is composed of indivisible atoms. An **atom** is *an extremely small particle of matter that retains its identity during chemical reactions.* (See Figure 2.10.)
2. An **element** is *a type of matter composed of only one kind of atom, each atom of a given kind having the same properties.* Mass is one such property. Thus, the atoms of a given element have a characteristic mass.●
3. A **compound** is *a type of matter composed of atoms of two or more elements chemically combined in fixed proportions.* The relative numbers of any two kinds of atoms in a compound occur in simple ratios. Water, for example, is a compound of the elements hydrogen and oxygen and consists of hydrogen and oxygen atoms in the ratio of 2 to 1.
4. A **chemical reaction** *consists of the rearrangement of the atoms present in the reacting substances to give new chemical combinations present in the substances formed by the reaction.* Atoms are not created, destroyed, or broken into smaller particles by any chemical reaction.●

● As we will see later in this section, it is the *average* mass of an atom that is characteristic of each element on earth.

● In Chapter 20 we will discuss nuclear reactions, in which the atom of one element can be converted into an atom of another. These are not chemical reactions, however.

Today we know that atoms are not truly indivisible; they are themselves made up of particles, as we will explain later in this section. Nevertheless, Dalton's postulates are essentially correct.

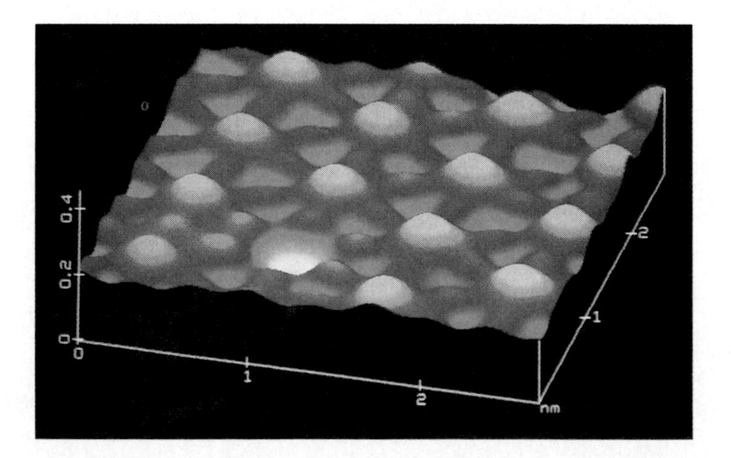

Figure 2.10
Image of iodine atoms on a platinum metal surface. This computer screen image was obtained by a scanning tunneling microscope (discussed in Chapter 7); the color was added to the image by computer. Iodine atoms are the large peaks with pink tops. Note the "vacancy" in the array of iodine atoms. A scale shows the size of the atoms in nanometers.

Deductions from Dalton's Atomic Theory

Note that atomic theory explains the difference between an element and a compound. In fact, we can use atomic theory to redefine the terms *element* and *compound*. In the foregoing postulates, these definitions appear in italics following the words in boldface type.●

● The definitions of *element* and *compound* given in the postulates can be given an experimental basis, because the atoms in a substance can be identified, for example, by the wavelengths of x rays emitted (these can be related to atomic number, as explained in Chapter 8).

Atomic theory also explains two laws we considered earlier. One of these is the law of conservation of mass, which states that the total mass remains constant during a chemical reaction. By postulate 2, every atom has a definite mass. Because a chemical reaction only rearranges the chemical combinations of atoms (postulate 4), the mass must remain constant. The other law explained by atomic theory is the law of definite proportions (constant composition). Postulate 3 defines a compound as a type of matter containing the atoms of two or more elements in definite proportions. Because the atoms have definite mass, compounds must have the elements in definite proportions by mass.

A good theory should not only explain known facts and laws but also predict new ones. The **law of multiple proportions,** deduced by Dalton from his atomic theory, is *a principle that states that when two elements form more than one compound, the masses of one element in these compounds for a fixed mass of the other element are in ratios of small whole numbers*. For example, carbon and oxygen form two compounds: carbon monoxide and carbon dioxide. Carbon monoxide contains 1.3321 g of oxygen for each 1.0000 g of carbon, whereas carbon dioxide contains 2.6642 g of oxygen for 1.0000 g of carbon. In other words, carbon dioxide contains twice the mass of oxygen as is contained in carbon monoxide ($2.6642 \text{ g} = 2 \times 1.3321 \text{ g}$) for a given mass of carbon. Atomic theory explains this by saying that carbon dioxide contains twice as many oxygen atoms for a given number of carbon atoms as does carbon monoxide. The deduction of the law of multiple proportions from atomic theory was important in convincing chemists of the validity of the theory.

Structure of the Atom; Isotopes

During the latter part of the nineteenth century and the first part of the present century, a series of important experiments showed that atoms are themselves composed of particles. An atom consists of a nucleus and one or more electrons. The *nucleus* is the central core of an atom; it has most of the mass of the atom and one or more units of positive charge. An *electron* is a very light particle that carries a unit negative electric charge and exists in the region around the positively charged nucleus. The number of electrons in an electrically neutral atom equals the positive charge on the nucleus.

The nucleus is itself composed of particles of two kinds, called protons and neutrons. A *proton* has a positive electric charge equal in magnitude, but opposite in sign, to that of the electron; its mass is 1836 times that of the electron. A *neutron* has a mass nearly identical to that of the proton but no electric charge. (It is electrically neutral, as the name indicates.)

The chemical properties of an element are determined by the number of electrons in the neutral atom, which in turn is determined by the number of protons in the nucleus. Hence this number of protons, called the *atomic number,* characterizes an element.

Though all atoms of a given element have the same atomic number and therefore the same number of protons in their nuclei, they may have different numbers of neutrons. For example, carbon has an atomic number of 6, so carbon nuclei contain 6 protons. However, in naturally occurring carbon, some nuclei contain 6 neutrons and others 7.

Because most of the mass of an atom is in the nucleus and because protons and neutrons have about the same mass, the total mass of an atom is approximately proportional to the total number of protons and neutrons in a nucleus. This total number is called the *mass number*. Thus, one form of carbon has a mass number of 12 (6 protons and 6 neutrons) and another has a mass number of 13 (6 protons and 7 neutrons). They are called carbon-12 and carbon-13, respectively. Atoms of the same element having different mass numbers, such as carbon-12 and carbon-13, are known as *isotopes* (see Figure 2.11).

A naturally occurring element consists of either a single isotope (as in the case of sodium, which contains only sodium-23) or a definite mixture of two or more isotopes. For example, all samples of naturally occurring carbon consist of 98.889% carbon-12, 1.111% carbon-13, and trace quantities of carbon-14. Such an isotopic mixture cannot be separated by usual chemical processes, because the isotopes of an element have nearly the same chemical properties—which explains why the isotopic compositions of the elements on earth have remained almost constant over time.

According to Dalton, the atoms of an element have a characteristic mass (postulate 2). In light of the existence of isotopes, we must interpret this characteristic mass as the *average* mass of the isotopes in the naturally occurring element.

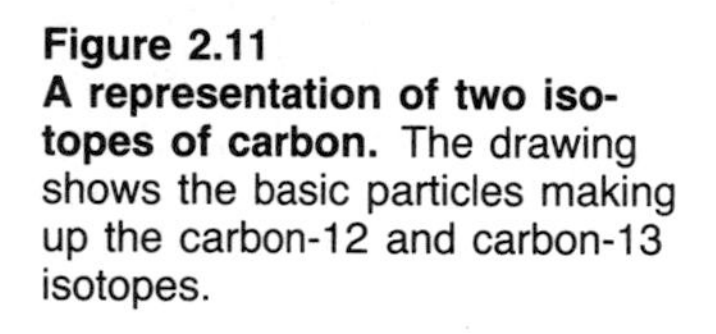

Figure 2.11
A representation of two isotopes of carbon. The drawing shows the basic particles making up the carbon-12 and carbon-13 isotopes.

Atomic Symbols

An **atomic symbol** is *a one-, two-, or three-letter notation used to represent an atom corresponding to a particular element*. Typically, the atomic symbol consists of the first letter, capitalized, from the name of the element, sometimes with an additional letter from the name in lower case. For example, chlorine has the symbol Cl. Other symbols are derived from a foreign (usually Latin) name. Sodium is given the symbol Na from its Latin name, *natrium*. Symbols of selected elements are listed in Table 2.2.●

● The names of the newly discovered elements are derived from the atomic number by putting together the numerical roots *nil* (0), *un* (1), *bi* (2), *tri* (3), *quad* (4), *pent* (5), *hex* (6), *sept* (7), *oct* (8), and *enn* (9) with the ending *-ium*. For example, the element with atomic number 106 is unnilhexium; the symbol is Unh.

2.3 ATOMIC MASS SCALE

An important feature of Dalton's atomic theory is the idea that an atom of an element has a characteristic mass, and in an early paper he showed how to obtain these characteristic masses. Because he could not weigh individual atoms, Dalton measured the relative masses of the elements required to form a compound and, from this, deduced relative atomic masses. For example, we can determine by experiment that 1.0000 g of hydrogen gas reacts with 7.9367 g of oxygen gas to give water. If we know the formula of water, we can easily determine the mass of an oxygen atom relative to that of a hydrogen atom.

Dalton did not have a way of determining the proportions of atoms of each element in water; he merely assumed the simplest possibility, that they were equal in number. From this assumption, it would follow that the oxygen atom would have a mass that was 7.9367 times that of a hydrogen atom. We now know that water contains twice as many hydrogen atoms as oxygen atoms, so the relative mass of an oxygen atom must be $2 \times 7.9367 = 15.873$ times that of a hydrogen atom.

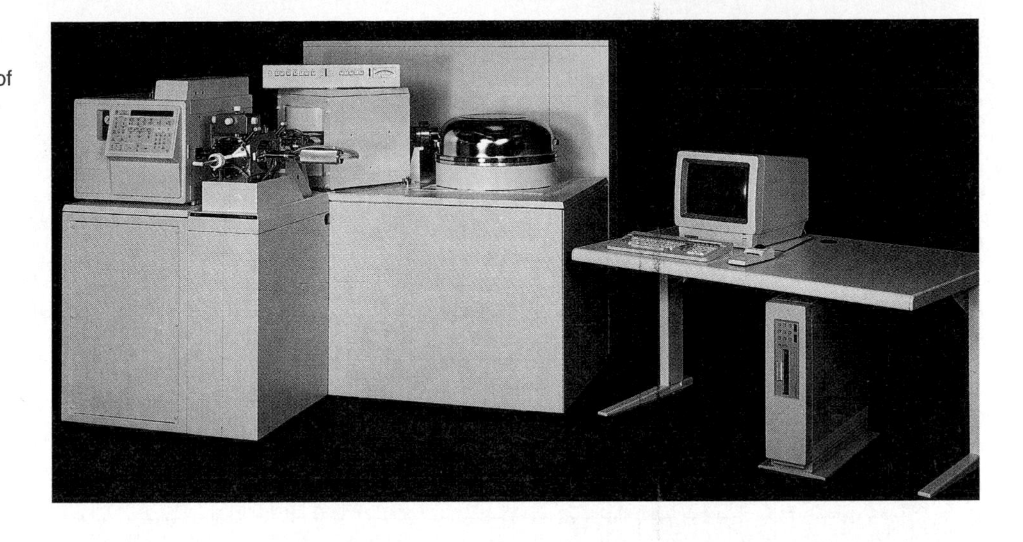

Figure 2.12
A mass spectrometer. This instrument measures the masses of atoms and molecules. The mass spectrometer is discussed in Chapter 7.

Dalton's hydrogen-based atomic mass scale was eventually replaced by a scale based on oxygen and then, in 1961, by the present carbon-12 atomic mass scale. This scale depends on measurements of atomic mass by an instrument called a *mass spectrometer* (Figure 2.12). Mass spectrometers, invented early in this century, allow us to determine the masses of atoms and molecules by removing electrons from them and observing the deflection of the resulting charged particles in electric and magnetic fields.●

● The mass spectrometer is discussed in Chapter 7.

The masses of atoms are obtained by comparison with an isotope chosen as a standard. On the *carbon-12 atomic mass scale,* the carbon-12 isotope is chosen as this standard and is arbitrarily assigned a mass of exactly 12 atomic mass units. One **atomic mass unit (amu)**, therefore, *equals exactly one-twelfth the mass of a carbon-12 atom.*

The average atomic mass for a naturally occurring element, expressed in atomic mass units, is called its **atomic weight.** (Although the word *weight* as used here, rather than *mass,* is a misnomer, it is sanctioned by long usage. We will use the term *atomic mass,* however, when referring to an individual atom rather than to an average of isotopes.) A complete table of atomic weights appears on the inside back cover of this book.

Figure 2.13
Dmitri Ivanovich Mendeleev (1834–1907). Mendeleev constructed a periodic table as part of his effort to systematize chemistry. He received many international honors for his work, but his reception in Czarist Russia was mixed. He had pushed for political reforms and made many enemies as a result.

2.4 PERIODIC TABLE OF THE ELEMENTS

In 1869 the Russian chemist Dmitri Mendeleev (1834–1907; Figure 2.13) and the German chemist J. Lothar Meyer (1830–1895), working independently, made similar discoveries. They found that when they arranged the elements in order of their atomic weights, they could place them in horizontal rows, one row under the other, so that the elements in any one vertical column had similar properties. *A tabular arrangement of elements in rows and columns, highlighting the regular repetition of properties of the elements,* is called a **periodic table.**

Eventually, more accurate determinations of atomic weights revealed discrepancies in this ordering of the elements. However, in the early part of this century, it was shown that the elements are characterized by their atomic numbers. Once the elements in the periodic table are ordered by atomic number, such discrepancies vanish.

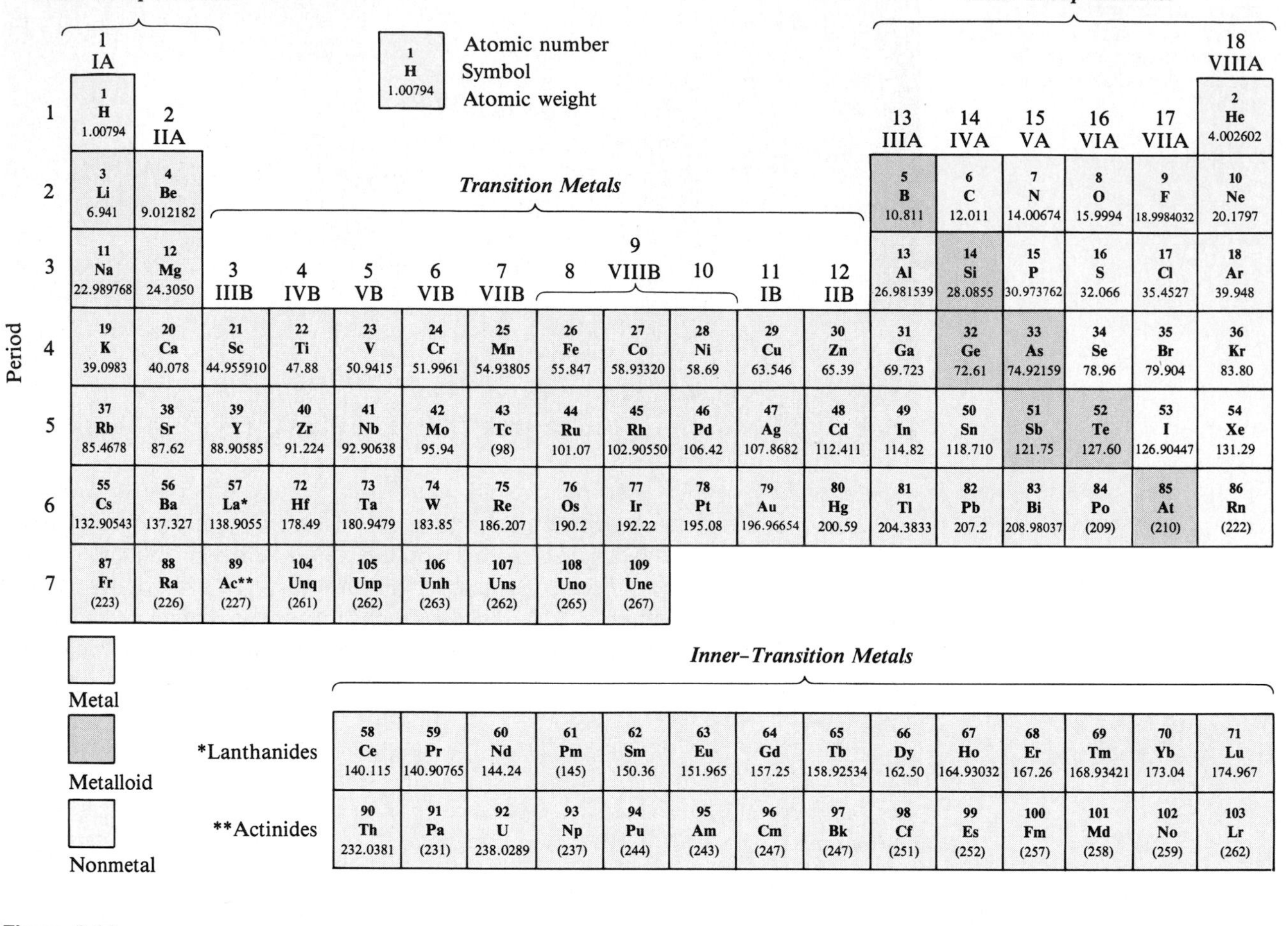

Figure 2.14
A modern form of the periodic table. This table is also given on the inside front cover of the text. The placement in the main body of the table of the two partial rows at the bottom is indicated by an asterisk (*) and double asterisk (**).

A modern version of the periodic table, with the elements arranged by atomic number, is shown in Figure 2.14 (see also inside front cover). Each entry lists the atomic number, atomic symbol, and atomic weight of an element. This is a convenient way of tabulating such information, and you should become familiar with using the periodic table for that purpose. However, as we develop the subject matter of chemistry throughout the text, you will see how the periodic table can become even more useful.

Periods and Groups

The basic structure of the periodic table is its division into rows and columns, or periods and groups. A **period** consists of *the elements in any one horizontal row of the periodic table*. A **group** consists of *the elements in any one column of the periodic table*.

The first period of elements is short, consisting of only hydrogen (H) and helium (He). The second period has 8 elements, beginning with lithium (Li) and ending with neon (Ne). There is then another period of 8 elements, and this is followed by a period having 18 elements, beginning with potassium (K) and ending with krypton (Kr). The fifth period also has 18 elements. The sixth period actually consists of 32 elements, but in order for the row to fit on a page, part of it appears at the bottom of the table. Otherwise the table would have to be expanded, with the additional elements placed after lanthanum (La, atomic number 57). The seventh period, though not complete, also has some of its elements placed as a row at the bottom of the table.

The groups are usually numbered. The numbering frequently seen in North America labels the groups with Roman numerals and A's and B's. In Europe a similar convention has been used, but some columns have the A's and B's interchanged. To eliminate this confusion, the International Union of Pure and Applied Chemistry (IUPAC) has recently suggested a convention in which the columns are numbered 1 to 18. Figure 2.14 shows the traditional North American and the IUPAC conventions. When we refer to an element by its periodic group, we will use the traditional North American convention. The A groups are called *main-group* (or *representative*) *elements;* the B groups are called *transition elements.* The two rows of elements at the bottom of the table are called *inner-transition elements* (the first row is referred to as the *lanthanides;* the second row as the *actinides*).

As we noted earlier, the elements in any one group have similar properties. For example, the elements in Group IA, often known as the *alkali metals,* are soft metals that react easily with water (hydrogen, a gas, is an exception and might better be put in a group by itself). Sodium is an alkali metal. So is potassium. The Group VIIA elements, known as *halogens,* are also reactive elements. Chlorine is a halogen. We have already noted its vigorous reaction with sodium. Bromine, which is a red-brown liquid, is another halogen. It too reacts vigorously with sodium.

Metals, Nonmetals, and Metalloids

The elements of the periodic table in Figure 2.14 are divided by a heavy "staircase" line into metals on the left and nonmetals on the right. A **metal** is *a substance or mixture that has a characteristic luster, or shine, and is generally a good conductor of heat and electricity*. Except for mercury, the metallic elements are solids at room temperature (about 20°C). They are more or less *malleable* (can be hammered into sheets) and *ductile* (can be drawn into wire).

A **nonmetal** is *an element that does not exhibit the characteristics of a metal.* Most of the nonmetals are gases (for example, chlorine and oxygen) or solids (for example, phosphorus and sulfur). The solid nonmetals are usually hard, brittle substances. Bromine is the only liquid nonmetal.

Most of the elements bordering the staircase line in the periodic table (Figure 2.14) are metalloids, or semimetals. A **metalloid,** or **semimetal,** is *an element having both metallic and nonmetallic properties*. These elements, such as silicon (Si) and germanium (Ge), are usually good *semiconductors*—elements that, when pure, are poor conductors of electricity at room temperature but become moderately good conductors at higher temperatures.●

● When these pure semiconductor elements have small amounts of certain other elements added to them (a process called *doping*), they becomc very good conductors of electricity. Semiconductors are the critical materials in solid-state electronic devices.

Exercise 2.3

By referring to the periodic table (Figure 2.14 or inside front cover), identify the group and period to which each of the following elements belongs. Then decide whether the element is a metal, nonmetal, or metalloid.

(a) Se (b) Cs (c) Fe (d) Cu (e) Br

(See Problems 2.47 and 2.48.)

Chemical Substances: Formulas and Names

Atomic theory, as we have seen, explains that the millions of known compounds are composed of only a few different kinds of atoms. Now we want to look more closely at the composition and structure of chemical substances in terms of atoms.

2.5 CHEMICAL FORMULAS; MOLECULAR AND IONIC SUBSTANCES

The **chemical formula** of a substance is *a notation that uses atomic symbols with numerical subscripts to convey the relative proportions of atoms of the different elements in the substance*. Consider the formula of aluminum oxide, Al_2O_3. This means that the compound is composed of aluminum atoms and oxygen atoms in the ratio 2:3. Consider the formula for sodium chloride, NaCl. When no subscript is written for a symbol, it is assumed to be 1. Therefore, the formula NaCl means that the compound is composed of sodium atoms and chlorine atoms in the ratio 1:1.

Additional information may be conveyed by different kinds of chemical formulas. To understand this, we need to look briefly at two main types of substances: molecular and ionic.

Molecular Substances

A **molecule** is *a definite group of atoms that are chemically bonded together—that is, tightly connected by attractive forces*. The nature of these strong forces is discussed in Chapters 9 and 10. A *molecular substance* is a substance that is composed of molecules all of which are alike. The molecules in such a substance are so small that even extremely minute samples contain tremendous numbers of them. One billionth (10^{-9}) of a drop of water, for example, contains about 2 trillion (2×10^{12}) water molecules.●

● Another way to understand the large numbers of atoms involved in normal portions of matter is to consider 1 gram of water (about one-fifth teaspoon). It contains 3.3×10^{22} water molecules. If you had a penny for every molecule in this quantity of water, your stack of pennies would be about 300 million times the distance from the earth to the sun.

A **molecular formula** is *a chemical formula that gives the exact number of different atoms of an element in a molecule*. The hydrogen peroxide molecule contains two hydrogen atoms and two oxygen atoms chemically bonded. Therefore, its molecular formula is H_2O_2. Other simple molecular substances are water, H_2O; ammonia, NH_3; carbon dioxide, CO_2; and ethanol (ethyl alcohol), C_2H_6O. In Chapter 4, we will explain how such formulas can be determined.

The atoms in a molecule are not simply piled together randomly. Rather, the atoms are chemically bonded in a definite way. A *structural formula* is a chemical formula that shows how the atoms are bonded to one another in a molecule. For example, it is known that each of the hydrogen atoms in the water molecule is bonded to the oxygen atom. Thus, the structural formula of water is H—O—H. A line joining two atomic symbols in such a formula represents the chemical bond connecting the atoms. Figure 2.15 shows some structural formulas. Structural formulas are sometimes condensed in writing. For example, the structural formula of ethanol is sometimes written CH_3CH_2OH, or C_2H_5OH, depending on the detail to be conveyed.

The atoms in a molecule are not only connected in definite ways but exhibit definite spatial arrangements as well. Molecular models can be constructed as an aid in visualizing the shapes and sizes of molecules. Figure 2.15 shows molecular

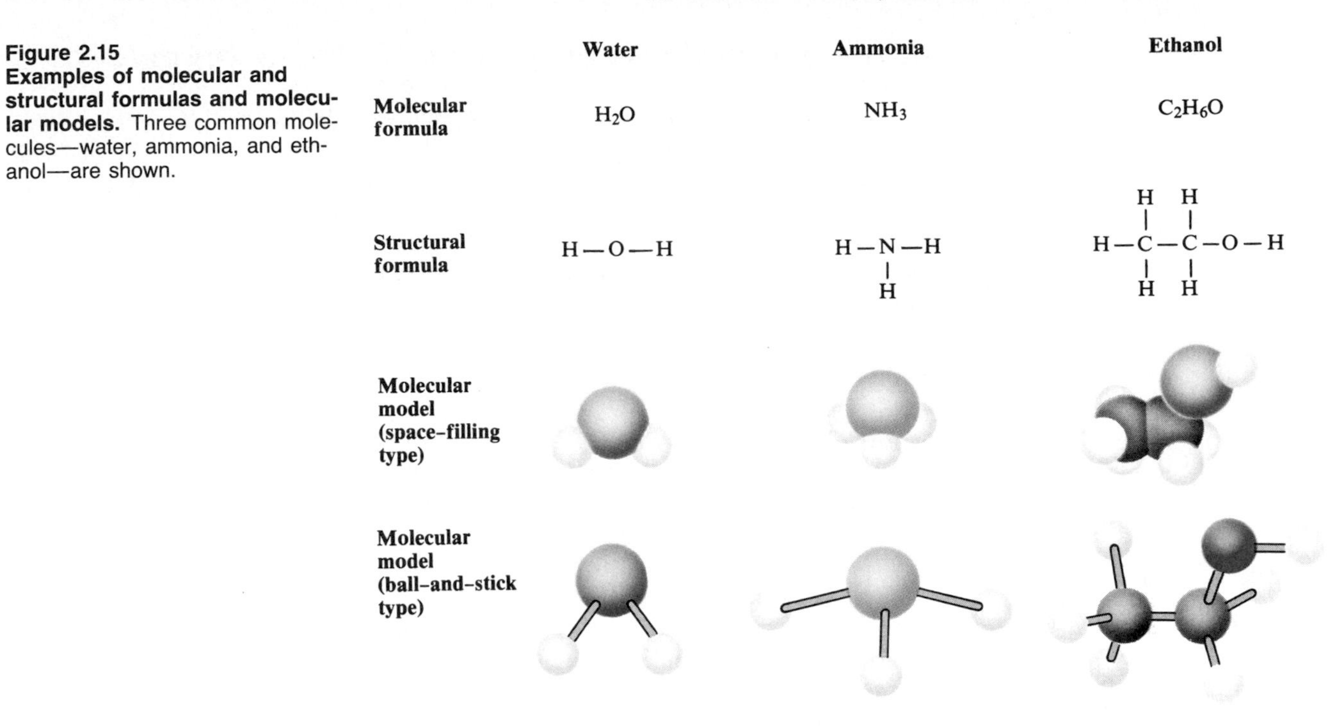

Figure 2.15
Examples of molecular and structural formulas and molecular models. Three common molecules—water, ammonia, and ethanol—are shown.

models for several compounds. The space-filling type of model gives a more realistic feeling of the space occupied by the atoms. On the other hand, the ball-and-stick type shows the bonds and bond angles more clearly.

Some elements are molecular substances and are represented by molecular formulas. Chlorine, for example, is a molecular substance and has the formula Cl_2, each molecule being composed of two chlorine atoms bonded together. Sulfur consists of molecules composed of eight atoms; its molecular formula is S_8. Helium and neon are composed of isolated atoms; their formulas are He and Ne, respectively. Other elements, such as carbon (in the form of graphite or diamond), do not have a simple molecular structure but consist of a very large, indefinite number of atoms bonded together. These elements are represented simply by their atomic symbols. Models of some elementary substances are shown in Figure 2.16.

Ionic Substances

Although many substances are molecular, others are composed of ions (pronounced "eye'-ons"). An **ion** is *an electrically charged particle obtained from an*

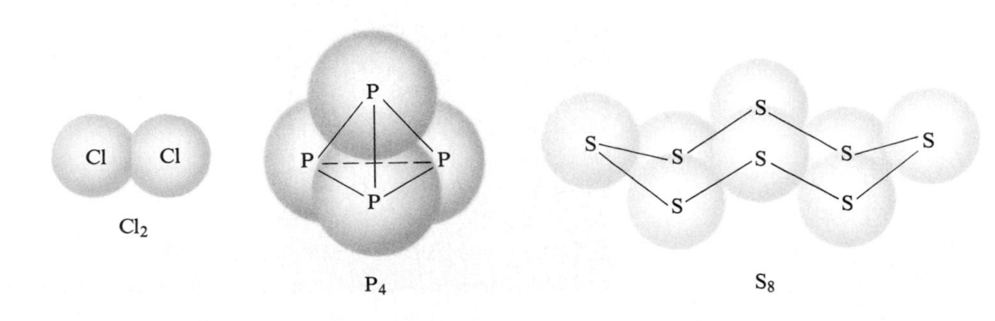

Figure 2.16
Models of some elementary substances. *Left to right:* Chlorine, Cl_2; white phosphorus, P_4; and sulfur, S_8.

atom or chemically bonded group of atoms by adding or removing electrons. Sodium chloride is a substance made up of ions.

Although atoms are normally electrically neutral and therefore contain equal numbers of positive and negative charges, they can become ions when electrons are transferred from one atom to another. An atom that picks up an extra electron becomes *a negatively charged ion,* called an **anion** (pronounced "an′-ion"). An atom that loses an electron is left as *a positively charged ion,* called a **cation** ("cat′-ion"). A sodium atom, for example, can lose an electron to form a sodium cation (denoted Na^+). A chlorine atom can gain an electron to form a chloride anion (denoted Cl^-). A calcium atom can lose two electrons to form a calcium cation, denoted Ca^{2+}. Note that the positive-two charge on the ion is indicated by a superscript 2+.

Some ions consist of two or more atoms chemically bonded but having an excess or deficiency of electrons so that the unit has an electric charge. An example is the sulfate ion, $SO_4{}^{2-}$. The superscript 2− indicates an excess of two electrons on the group of atoms.

An *ionic compound* is a compound composed of cations and anions. Sodium chloride consists of equal numbers of sodium ions, Na^+, and chloride ions, Cl^-. The strong attraction between positive and negative charges holds the ions together in a regular arrangement in space. For example, in sodium chloride, each Na^+ ion is surrounded by six Cl^- ions, and each Cl^- ion is surrounded by six Na^+ ions. Such a regular arrangement gives rise to a *crystal,* a kind of solid having a definite geometrical shape as a result of the regular arrangement of atoms, molecules, or ions making up the substance. Figure 2.17 shows sodium chloride crystals and the arrangement of ions in the crystal. Note that the size of the crystal varies with the number of ions in it.

The formula of an ionic compound is written by giving the smallest possible integer number of different ions in the substance, except that the charges on the ions are omitted so that the formulas merely indicate the atoms involved. For example, sodium chloride contains equal numbers of Na^+ and Cl^- ions. The

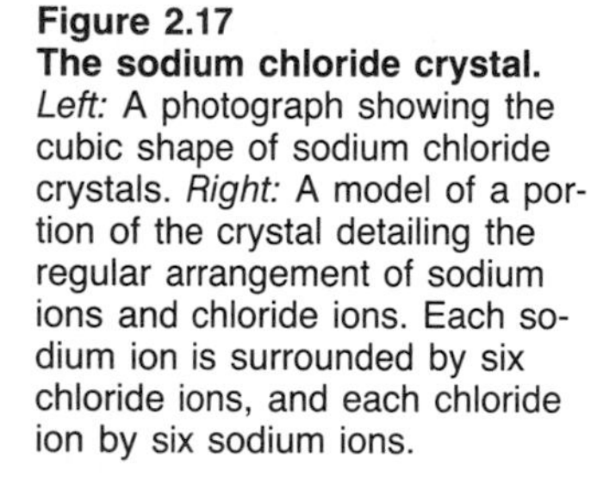

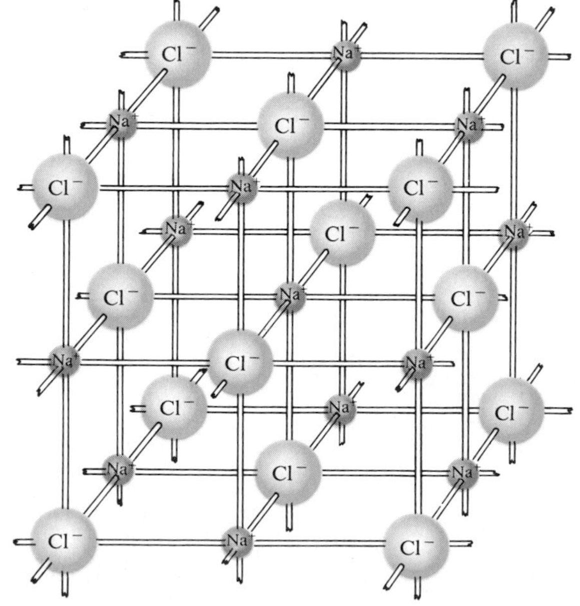

Figure 2.17
The sodium chloride crystal. *Left:* A photograph showing the cubic shape of sodium chloride crystals. *Right:* A model of a portion of the crystal detailing the regular arrangement of sodium ions and chloride ions. Each sodium ion is surrounded by six chloride ions, and each chloride ion by six sodium ions.

formula is written NaCl (not Na^+Cl^-). Iron(III) sulfate is a compound consisting of iron(III) ions, Fe^{3+}, and sulfate ions, SO_4^{2-}, in the ratio 2:3. The formula is written $Fe_2(SO_4)_3$.

Although ionic substances do not contain molecules, we can speak of the smallest unit of such a substance. The **formula unit** of a substance is *the group of atoms or ions explicitly symbolized in the formula*. For example, the formula unit of water, H_2O, is the H_2O molecule. The formula unit of iron(III) sulfate, $Fe_2(SO_4)_3$, consists of two Fe^{3+} ions and three SO_4^{2-} ions. The formula unit is the smallest unit of such substances.

All substances, including ionic compounds, are electrically neutral. We can use this fact to obtain the formula of an ionic compound, given the formulas of the ions. This is illustrated in the following example.

Example 2.2 Writing an Ionic Formula, Given the Ions

(a) Chromium(III) oxide is used as a green paint pigment (see Figure 2.18). It is a compound composed of Cr^{3+} and O^{2-} ions. What is the formula of chromium(III) oxide? (b) Strontium oxide is a compound composed of Sr^{2+} and O^{2-} ions. Write the formula of this compound.

Figure 2.18
Chromium(III) oxide. The compound is used as a green paint pigment.

Solution

(a) We can achieve electrical neutrality by taking as many cations as there are units of charge on the anion and as many anions as there are units of charge on the cation. Thus, two Cr^{3+} ions have a total charge of 6+, and three O^{2-} ions have a total charge of 6−, giving the combination a net charge of zero. The simplest ratio of Cr^{3+} to O^{2-} is 2:3, and the formula is Cr_2O_3. Note that the charge (without its sign) on one ion becomes the subscript of the other ion.

$$Cr^{③+} \;\; O^{②-} \quad \text{or} \quad Cr_2O_3$$

(b) We see that equal numbers of Sr^{2+} and O^{2-} ions will give a neutral compound. Thus, the formula is SrO. If we use the units of charge to find the subscripts, we get

$$Sr^{②+} \;\; O^{②-} \quad \text{or} \quad Sr_2O_2$$

The final formula is SrO, because this gives the simplest ratio of ions.

Exercise 2.4

Potassium chromate is an important compound of chromium (see Figure 2.19). It is composed of K^+ and CrO_4^{2-} ions. Write the formula of the compound.

(See Problems 2.63 and 2.64.)

Figure 2.19
Potassium chromate. Many compounds of chromium have bright colors, which is the origin of the name of the element. It comes from the Greek *chroma,* meaning "color."

2.6 NAMING SIMPLE COMPOUNDS

Before the structural basis of chemical substances became established, compounds were named after people, places, or particular characteristics. Examples are Glauber's salt (sodium sulfate, discovered by J. R. Glauber), sal ammoniac (ammonium chloride, named after the ancient Egyptian deity Ammon from the temple near which the substance was made), and washing soda (sodium carbonate, used for softening wash water). Today several million compounds are known and thousands of new ones are discovered every year. Without a system for naming compounds, coping with this multitude of substances would be a hopeless task. **Chemical nomenclature** is *the systematic naming of chemical compounds.*

Chemical compounds are classified as organic or inorganic. **Organic compounds** are *compounds that can be thought of as derivatives of compounds of carbon and hydrogen (hydrocarbons).* Originally, only compounds obtained from organic materials (plants or animals) were classified as organic. The present definition came into being when it was discovered that chemists could create these plant and animal compounds in the laboratory. Today most known organic compounds are obtained from chemical reactions on petroleum derivatives. We will discuss the nomenclature of these compounds later in this book. We will describe some of the simplest organic compounds, however, in our early discussions.

Inorganic compounds are *compounds composed of elements other than carbon.* A few simple compounds of carbon, including carbon monoxide, carbon dioxide, carbonates, and cyanides, are generally considered to be inorganic, however. In this section we will discuss the nomenclature of some simple inorganic compounds: ionic compounds, binary molecular compounds, acids, and hydrates.

Ionic Compounds

An ionic compound is named by naming the ions it contains, giving the cation first, then the anion. Therefore, we must look at ions and how they are named.

The simplest are monatomic ions. A **monatomic ion** is *an ion formed from a single atom.* Common monatomic ions of the main-group elements are listed in Table 2.3. Metallic elements generally form monatomic cations. The charge on a cation for a main-group element, in most cases, equals the group number (the Roman numeral). For example, aluminum, which is in Group IIIA, has an ion with charge 3+. Some exceptions are noted for elements of high atomic number (such as lead, in Group IVA, whose common ion is Pb^{2+}). Nonmetal elements generally

Table 2.3 Common Monatomic Ions of the Main-Group Elements*

	IA	IIA	IIIA	IVA	VA	VIA	VIIA
Period 1							H^-
Period 2	Li^+	Be^{2+}	B	C	N^{3-}	O^{2-}	F^-
Period 3	Na^+	Mg^{2+}	Al^{3+}	Si	P	S^{2-}	Cl^-
Period 4	K^+	Ca^{2+}	Ga^{3+}	Ge	As	Se^{2-}	Br^-
Period 5	Rb^+	Sr^{2+}	In^{3+}	Sn^{2+}	Sb	Te^{2-}	I^-
Period 6	Cs^+	Ba^{2+}	Tl^+, Tl^{3+}	Pb^{2+}	Bi^{3+}		

*Elements shown in color do not normally form compounds having monatomic ions.

Table 2.4 Common Ions of the Transition Elements

Ion	Ion Name
Cr^{3+}	Chromium(III) or chromic
Mn^{2+}	Manganese(II) or manganous
Fe^{2+}	Iron(II) or ferrous
Fe^{3+}	Iron(III) or ferric
Co^{2+}	Cobalt(II) or cobaltous
Ni^{2+}	Nickel(II) or nickel
Cu^{2+}	Copper(II) or cupric
Zn^{2+}	Zinc
Ag^{+}	Silver
Cd^{2+}	Cadmium
Hg^{2+}	Mercury(II) or mercuric

form monatomic anions. The charge on an anion equals the group number minus 8. Oxygen, in Group VIA, has an anion charge of 6 − 8 (O^{2-}).

Most transition elements have cations of several different charges (see Table 2.4). For example, iron has the cations Fe^{2+} and Fe^{3+}. Cations of the transition elements having 2+ charges are common.

A monatomic cation is given the name of the element. Thus, Al^{3+} is the aluminum ion. If there is more than one cation of that element (as in the case of iron, which has the ions Fe^{2+} and Fe^{3+}), the charge is denoted in the *Stock system* of nomenclature by a Roman numeral in parentheses following the element name. The ion Fe^{2+} is called iron(II) ion (read "iron-two ion") and Fe^{3+} is called iron(III) ion.● In an older system of nomenclature, such ions are named by adding the suffixes *-ous* and *-ic* to a stem name of the element (which may be from the

● The Roman numeral actually denotes the oxidation state, or oxidation number, of the atom in the compound. For a monatomic ion, the oxidation state equals the charge. Otherwise, the oxidation state is a hypothetical charge assigned in accordance with certain rules, which are discussed in Chapter 13.

Table 2.5 Some Common Polyatomic Ions

Name	Formula	Name	Formula
Acetate	$C_2H_3O_2^-$	Hypochlorite	ClO^-
Ammonium	NH_4^+	Hydroxide	OH^-
Carbonate	CO_3^{2-}	Mercury(I) or mercurous	Hg_2^{2+}
Chlorate	ClO_3^-	Monohydrogen phosphate	HPO_4^{2-}
Chlorite	ClO_2^-	Nitrate	NO_3^-
Chromate	CrO_4^{2-}	Nitrite	NO_2^-
Cyanide	CN^-	Oxalate	$C_2O_4^{2-}$
Dichromate	$Cr_2O_7^{2-}$	Perchlorate	ClO_4^-
Dihydrogen phosphate	$H_2PO_4^-$	Permanganate	MnO_4^-
Hydrogen carbonate (or bicarbonate)	HCO_3^-	Peroxide	O_2^{2-}
		Phosphate	PO_4^{3-}
Hydrogen sulfate (or bisulfate)	HSO_4^-	Sulfate	SO_4^{2-}
		Sulfite	SO_3^{2-}
Hydrogen sulfite (or bisulfite)	HSO_3^-	Thiosulfate	$S_2O_3^{2-}$

Latin) to indicate the ions of lower and higher charge, respectively. Thus Fe^{2+} is the ferrous ion and Fe^{3+} is the ferric ion.

The names of monatomic anions are obtained from a stem name of the element followed by the suffix *-ide*. Examples: Br^- is the bromide ion, S^{2-} is the sulfide ion, and N^{3-} is the nitride ion.

A **polyatomic ion** is *an ion consisting of two or more atoms chemically bonded together and carrying a net electric charge*. Table 2.5 lists some polyatomic ions. Many of these are *oxoanions,* which consist of oxygen with another element (called the characteristic or central element). Sulfur, for example, forms the oxoanions sulfate ion, SO_4^{2-}, and sulfite ion, SO_3^{2-}.

Example 2.3 Naming an Ionic Compound from Its Formula

Name the following: (a) Mg_3N_2, (b) $CrSO_4$.

Solution

(a) Magnesium, a Group IIA element, is expected to form only a 2+ ion (the magnesium ion). Nitrogen (Group VA) is expected to form an anion of charge equal to the group number minus 8 (N^{3-}, the nitride ion). The substance is **magnesium nitride.** (b) Chromium is a transition element and, like most such elements, has more than one monatomic ion. We can find the charge on the Cr ion if we know the formula of the anion. From Table 2.5, we see that the SO_4 in $CrSO_4$ refers to the anion SO_4^{2-} (the sulfate ion). Thus, the cation is Cr^{2+}, and the name of the compound is **chromium(II) sulfate**.

Exercise 2.5

Write the names of the following compounds: (a) CaO, (b) $PbCrO_4$.

(See Problems 2.65 and 2.66.)

Example 2.4 Writing the Formula from the Name of an Ionic Compound

Write formulas for the following compounds: (a) iron(II) phosphate, (b) titanium(IV) oxide.

Solution

(a) Iron(II) ion is Fe^{2+}; phosphate ion is PO_4^{3-}. Hence the formula is $Fe_3(PO_4)_2$. (b) The charge on oxygen is -2 and that of titanium is $+4$. Because the sum of the charges must be zero, the formula is TiO_2.

Exercise 2.6

A compound has the name thallium(III) nitrate. What is its formula?

(See Problems 2.67 and 2.68.)

Binary Molecular Compounds

A **binary compound** is *a compound composed of only two elements*. Those binary compounds composed of a metal and a nonmetal are usually ionic and are named as ionic compounds. Those composed of two nonmetals or metalloids are usually molecular. A binary molecular compound is named from the elements, following the same order in which they appear in the formula. The nonmetal or metalloid occurring first in the following sequence is written first in the formula and name:

Table 2.6 Greek Prefixes for Naming Compounds

Number	Prefix
1	mono-
2	di-
3	tri-
4	tetra-
5	penta-
6	hexa-
7	hepta-
8	octa-
9	nona-
10	deca-

Element	B	Si C	Sb As P N	H	Te Se S	I Br Cl	O	F
Group	IIIA	IVA	VA		VIA	VIIA		

The elements are arranged in order of the groups of the periodic table, listed from the bottom of the group upward, except that H is placed between Groups VA and VIA, and O is placed just before F. Examples of such formulas include NH_3, H_2S, and PF_3.

When the two elements form only one compound, we name the compound by giving the name of the first element, followed by the stem name of the second element with the suffix *-ide*. Examples:

HCl hydrogen chloride $\quad$ H_2S hydrogen sulfide

When the two elements form more than one compound, we can distinguish among these compounds by using the Greek prefixes listed in Table 2.6. The prefix *mono-* is usually omitted for the first-named element but given for the second element. Examples:

CO	carbon monoxide	CO_2	carbon dioxide
NO_2	nitrogen dioxide	N_2O_4	dinitrogen tetr(a)oxide
SF_4	sulfur tetrafluoride	SF_6	sulfur hexafluoride
ClO_2	chlorine dioxide	Cl_2O_7	dichlorine hept(a)oxide

The final vowel in a prefix is often dropped before a vowel in a stem name, as in tetroxide.

Water and some other substances of early origin, such as ammonia, are seldom given systematic names.

Example 2.5 Deciding Whether a Binary Compound Is Ionic or Molecular

Which of the following compounds are expected to be ionic and which are expected to be molecular?
(a) LiCl (b) SCl_4 (c) SbF_5

Solution

Using the periodic table (Figure 2.14), we can classify each of the elements in a compound as metallic, metalloid, or nonmetallic. Then we can decide whether the compound is expected to be ionic or molecular.
(a) Li is metallic; Cl is nonmetallic. Therefore, LiCl is expected to be **ionic**.
(b) S and Cl are nonmetals; therefore, SCl_4 is expected to be **molecular**.
(c) Sb is a metalloid; F is a nonmetal. Therefore, SbF_5 is expected to be **molecular**.

Exercise 2.7

Classify the following as ionic or molecular compounds: (a) KF, (b) P_4O_6, (c) B_2O_3.

(See Problems 2.69 and 2.70.)

Example 2.6 Naming a Binary Compound from Its Formula

Name the following compounds: (a) N_2O_3, (b) P_4O_6.

Solution

We use Greek prefixes, naming the elements in the same order as in the formulas: (a) ***di*nitrogen *tri*oxide**, (b) ***tetra*phosphorus *hex*oxide**.

Exercise 2.8

Name the following compounds: (a) NO, (b) PCl_3, (c) PCl_5.

(See Problems 2.71 and 2.72.)

Example 2.7 Writing the Formula from the Name of a Binary Compound

Give the formula of the following compounds: (a) disulfur dichloride, (b) tetraphosphorus trisulfide.

Solution

We change the names of the elements to symbols and translate the prefixes to subscripts. The formulas are (a) S_2Cl_2 and (b) P_4S_3.

Exercise 2.9

Give formulas for the following compounds: (a) carbon disulfide, (b) sulfur trioxide.

(See Problems 2.73 and 2.74.)

Acids

An **acid** is *a molecular substance that can yield one or more hydrogen ions, H^+, and an anion for each acid molecule when it is placed in aqueous (water) solution.* An example is nitric acid, HNO_3. The HNO_3 molecule yields one H^+ ion and one nitrate ion, NO_3^-, in aqueous solution.

Nitric acid, HNO_3, is an oxoacid. An **oxoacid** is *an acid containing hydrogen, oxygen, and another element.* In water the oxoacid molecule yields one or more hydrogen ions, H^+, and an oxoanion. Note that the name of the acid contains the stem name of the central atom and the suffix *-ic*. When a central element forms two oxoacids, as nitrogen does, the acids are distinguished by the suffixes *-ous* and *-ic*. The acid with the fewer oxygen atoms is given the suffix *-ous*; the acid with more oxygen atoms is given the suffix *-ic*. Nitrogen forms the oxoacids HNO_2 (nitrous acid) and HNO_3 (nitric acid). Some oxoacids are listed in Table 2.7.

When a central atom forms three or four oxoacids, they are distinguished by the prefixes *hypo-* and *per-* along with the suffixes. Consider the oxoacids of chlorine. The formulas are HClO, $HClO_2$, $HClO_3$, and $HClO_4$. The two middle acids are named chlorous acid and chloric acid, respectively. The acid HClO, having fewer oxygen atoms than chlorous acid, is called hypochlorous acid. The acid $HClO_4$, with more oxygen atoms than chloric acid, is called perchloric acid. The series of chlorine acids is listed in Table 2.7.

The names of the oxoacid and corresponding oxoanion are related. To obtain the name of the oxoanion from the oxoacid, we replace the suffix *-ic* by *-ate* and the suffix *-ous* by *-ite*, then replace the word *acid* by the word *ion*. As an example, consider carbonic acid, H_2CO_3. In water the molecule yields two H^+ ions (note the subscript 2 on H) and an oxoanion. Because two positive charges have been removed (as H^+ ions), the oxoanion must have a negative charge of 2−. The formula of the oxoanion is CO_3^{2-}. The name of the anion is carbonate ion. If you remember the names of anions, you can reverse this process to obtain the names of

Table 2.7 Some Oxoacids and Their Corresponding Oxoanions

Oxoacid		Oxoanion	
H_2CO_3	Carbonic acid	CO_3^{2-}	Carbonate ion
HNO_2	Nitrous acid	NO_2^-	Nitrite ion
HNO_3	Nitric acid	NO_3^-	Nitrate ion
H_3PO_4	Phosphoric acid	PO_4^{3-}	Phosphate ion
H_2SO_3	Sulfurous acid	SO_3^{2-}	Sulfite ion
H_2SO_4	Sulfuric acid	SO_4^{2-}	Sulfate ion
$HClO$	Hypochlorous acid	ClO^-	Hypochlorite ion
$HClO_2$	Chlorous acid	ClO_2^-	Chlorite ion
$HClO_3$	Chloric acid	ClO_3^-	Chlorate ion
$HClO_4$	Perchloric acid	ClO_4^-	Perchlorate ion

the acids. The formulas and names of the anions corresponding to the oxoacids in Table 2.7 are listed in the third and fourth columns. Note that in the case of the phosphorus and sulfur acids, a couple of letters have been dropped in forming the names of the anions.

Acid anions are anions that have H atoms they can lose as H^+ ions in water. For example, HSO_4^- has an H atom that can be removed to give H^+ and SO_4^{2-}. The anion HSO_4^- is called the hydrogen sulfate (or bisulfate) ion. Similarly, the acid anion $H_2PO_4^-$ (dihydrogen phosphate ion) has two H atoms which, when removed as H^+ ions, yield in succession HPO_4^{2-} (monohydrogen phosphate ion) and PO_4^{3-} (phosphate ion). Some acid anions are listed with other polyatomic ions in Table 2.5.

Some binary compounds of hydrogen and nonmetals yield acidic solutions when dissolved in water. These *solutions* are named like compounds by using the prefix *hydro-* and the suffix *-ic* with the stem name of the nonmetal, followed by the word *acid*. We denote the solution by the formula of the binary compound followed by (*aq*) for aqueous (water) solution. The corresponding binary compound can be distinguished from the solution by appending the state of the compound to the formula. Thus, when hydrogen chloride gas, HCl(*g*), is dissolved in water, it forms hydrochloric acid, HCl(*aq*). Here are some other examples:

BINARY COMPOUND	ACID SOLUTION
HBr(*g*), hydrogen bromide	*hydro*bromic *acid*, HBr(*aq*)
HF(*l*), hydrogen fluoride	*hydro*fluoric *acid*, HF(*aq*)

Example 2.8 Writing the Name and Formula of an Anion from the Acid

Selenium has an oxoacid H_2SeO_4, called selenic acid. What is the formula and name of the corresponding anion?

Solution

When we remove two H^+ ions from H_2SeO_4, we obtain the SeO_4^{2-} ion. We name the ion from the acid by replacing *-ic* by *-ate*. It is called the **selenate ion**.

Figure 2.20
Copper(II) sulfate. The hydrate, $CuSO_4 \cdot 5H_2O$, is blue; the anhydrous compound, $CuSO_4$, is white.

Exercise 2.10

What is the name and formula of the anion corresponding to perbromic acid, $HBrO_4$?
(See Problems 2.75 and 2.76.)

Hydrates

A **hydrate** is *a compound that contains water molecules weakly bound in its crystals*. These substances are often obtained by evaporating an aqueous solution of the compound. Consider copper(II) sulfate. When an aqueous solution of this substance is evaporated, blue crystals form in which each formula unit of copper(II) sulfate, $CuSO_4$, is associated with five molecules of water. The formula of the hydrate is written $CuSO_4 \cdot 5H_2O$, where a centered dot separates $CuSO_4$ and $5H_2O$. When the blue crystals of the hydrate are heated, the water is driven off, leaving behind white crystals of copper(II) sulfate without associated water, a substance called *anhydrous* copper(II) sulfate (see Figure 2.20).

Hydrates are named from the anhydrous compound, followed by the word *hydrate* with a prefix to indicate the number of water molecules per formula unit of the compound. For example, $CuSO_4 \cdot 5H_2O$ is known as copper(II) sulfate pentahydrate.

Example 2.9 Naming a Hydrate from Its Formula

Epsom salts has the formula $MgSO_4 \cdot 7H_2O$. What is the chemical name of the substance?

Solution

$MgSO_4$ is magnesium sulfate. $MgSO_4 \cdot 7H_2O$ is **magnesium sulfate heptahydrate**.

Exercise 2.11

Washing soda has the formula $Na_2CO_3 \cdot 10H_2O$. What is the chemical name of this substance?
(See Problems 2.77 and 2.78.)

Example 2.10 Writing the Formula from the Name of a Hydrate

The mineral gypsum has the chemical name calcium sulfate dihydrate. What is the chemical formula of this substance?

Solution

Calcium sulfate is composed of calcium ions (Ca^{2+}) and sulfate ions (SO_4^{2-}), so the formula of the anhydrous compound is $CaSO_4$. Since the mineral is a *di*hydrate, the formula of the compound is $\mathbf{CaSO_4 \cdot 2H_2O}$.

Exercise 2.12

Photographers' hypo is sodium thiosulfate pentahydrate. What is the chemical formula of this compound?

(See Problems 2.79 and 2.80.)

Profile of a Chemical

SODIUM (a Reactive Metal)

Sodium, Na, is unlike any of the metals you commonly encounter. In chemistry laboratories, it is stored in bottles where the metal is covered with a liquid such as kerosene. Sodium is stored this way to protect it from air and moisture, with which the metal reacts vigorously. Even so, the sodium in these bottles often looks nothing like a metal. Frequently it is encrusted with yellowish-brown crystals from reaction with oxygen and water, which have still managed to find their way to the metal.

If you cut through a chunk of this corrosion-covered sodium (and you can easily do that with even a very dull knife), you will see a bright, silvery metal (see Figure 2.1 at the beginning of this chapter). If you were to put a small piece in water, you would find that it floats. (But don't do it! This experiment can be dangerous—sodium metal reacts violently with water.) The density of sodium is 0.968 g/cm^3. The metal is also interesting in that it melts at 98°C, which is below the boiling point of water. Advantage is taken of this low melting point to transport the metal in special tank cars. Liquid sodium is pumped into the tanks, where it solidifies. When the metal is to be removed, it is remelted (by passing hot water through coils around the tank), then pumped out.

Sodium does not occur as the metal in nature; it is much too reactive. Early chemists, in fact, were faced with this problem: How do you separate very reactive elements from their compounds? Humphry Davy (1728–1829), a British chemist, started experimenting in 1794 with voltaic cells, or electric batteries, which were then newly invented. On October 6, 1807, he made a momentous discovery. He placed wire leads from a battery consisting of about 50 voltaic cells connected in parallel into moist crystals of caustic potash (now known as potassium hydroxide, KOH). The crystals first fused from the heat of the electric current, then began to bubble.

Figure 2.21
Indigo-dyed fabric. *The photograph shows a closeup view of indigo-dyed denim cloth (blue-jean fabric). Indigo dye is manufactured synthetically using sodium metal.*

(continued)

Davy noticed globules of a silvery metal collecting at the negative wire from his battery. He had discovered the element potassium, obtained from the *electrolysis,* or electrical decomposition, of molten potassium hydroxide. Then he performed a similar experiment using caustic soda or lye (sodium hydroxide, NaOH). On November 19 of the same year, he announced the discovery of a second new element, sodium.

Today sodium is produced in tank-car amounts by a process similar to the one Davy used but employing sodium chloride, NaCl, instead of sodium hydroxide. Until recently sodium was used to make tetraethyllead for leaded gasolines, but leaded fuels are being phased out for environmental reasons. Sodium is still important in the manufacture of many chemicals, including pharmaceuticals and dyes such as indigo. This blue dye was known to the ancients, who obtained it from certain plants. Now the dye is produced synthetically using sodium metal in one of the manufacturing steps. You know the dye well; it is used to color blue jeans (Figure 2.21).

Questions for Study

1. How is sodium metal stored in the laboratory? Explain.
2. List some physical and chemical properties of sodium metal.
3. When was the discovery of sodium announced? By whom? How did he prepare the element?
4. What is sodium used for?

Profile of a Chemical

CHLORINE (a Reactive Nonmetal)

An obvious property of chlorine, Cl_2, and the one responsible for its name, is its color (see Figure 2.1). Before 1810, chlorine was known as oxymuriatic acid, because it was thought to be a compound containing oxygen. In that year, Humphry Davy convinced other chemists that the substance was actually an element and that the name oxymuriatic acid was therefore inappropriate. He named the substance chlorine after the Greek word *chloros,* meaning "greenish-yellow."

Liquid chlorine is a golden yellow color (Figure 2.22). Its discovery in 1823 by Michael Faraday, an assistant of Davy's, led to Faraday's lengthy study of the liquefaction of gases. The initial discovery was an accident resulting from a visit by another scientist to Faraday's laboratory. The visitor noticed drops of a yellow oil inside a closed glass vessel being used in an experiment and chided Faraday on his lack of cleanliness. Faraday was puzzled about the presence of the oil, but when he broke open the vessel to examine it, there was a mild explosion and the drops vanished. Later that day Faraday found an explanation. A substance being heated in the closed vessel released chlorine gas into the small space, greatly increasing the pressure. As a result of the pressure increase, the chlorine gas liquefied. When the vessel was broken open, however, the liquid chlorine vaporized explosively. Today many gases are liquefied by increasing the pressure on them.

Figure 2.22
Liquid chlorine. *Chlorine gas was liquefied by leading it into a test tube immersed in dry ice.*

A property of chlorine with which you are no doubt familiar is its odor in low concentration. You can sometimes smell it in drinking water and in swimming pools, because chlorine is used to disinfect water. In higher concentrations, the gas has a suffocating odor and is a poison, causing the chest to constrict and the lungs to fill with liquid. Chlorine received notoriety during World War I when it was used as a war gas. Chlorine is much denser than air and so would flow along the ground and pour into troop trenches. Soon after the war, however, chlorine became the premier disinfectant, controlling deadly diseases such as typhoid fever that were being spread by polluted drinking water.

The chemical reactivity of chlorine is another notable property of the substance. In the chapter introduction, we described how sodium metal burns in chlorine. In this respect, chlorine is similar to oxygen, which is also very reactive with many other substances. Figure 2.23 shows the vigorous burning of antimony metal (actually a metalloid) powder in chlorine.

Chlorine is presently among the top ten industrial chemicals produced in the U.S. Besides its use as a disinfectant, chlorine is used as a bleach. Its greatest commercial use, however, is in the preparation of chlorinated organic compounds, including plastics such as polyvinylchloride (PVC), for bottles and for packaging film.

How is chlorine prepared? The preceding Profile on sodium described the electrolysis of molten sodium chloride, NaCl, to produce sodium metal, Na. The other product in this electrolysis is chlorine, Cl_2. However, the demand for chlorine far outstrips that for sodium. So the major industrial process for chlorine employs an aqueous solution of sodium chloride in the electrolysis. In this case, the other product is sodium hydroxide, or lye, which is in much greater demand than sodium.

Questions for Study

1. Chlorine gas was first discovered by the Swedish chemist Karl Wilhelm Scheele in 1774. However, its present name was given to the substance by Humphry Davy. Why did he rename the substance? What is the basis for the name?
2. How was the liquefaction of chlorine gas discovered? How are chlorine and many other gases liquefied?
3. List some physical and chemical properties of chlorine.
4. What are some commercial uses of chlorine?

Figure 2.23
Reaction of antimony and chlorine. *A fiery reaction occurs when powdered antimony is added to a flask of chlorine gas.*

A Checklist for Review

IMPORTANT TERMS

solid (2.1)
liquid (2.1)
gas (2.1)
states of matter (2.1)
physical change (2.1)
chemical change (chemical reaction) (2.1)
physical property (2.1)
chemical property (2.1)
substance (2.1)
element (2.1)
compound (2.1)

law of definite proportions (law of constant composition) (2.1)
mixture (2.1)
heterogeneous mixture (2.1)
homogeneous mixture (solution) (2.1)
phase (2.1)
atomic theory (2.2)
atom (2.2)
element (2.2)
compound (2.2)
chemical reaction (2.2)
law of multiple proportions (2.2)
atomic symbol (2.2)
atomic mass unit (amu) (2.3)
atomic weight (2.3)
periodic table (2.4)
period (of periodic table) (2.4)
group (of periodic table) (2.4)
metal (2.4)
nonmetal (2.4)
metalloid (semimetal) (2.4)
chemical formula (2.5)
molecule (2.5)
molecular formula (2.5)
ion (2.5)
anion (2.5)
cation (2.5)
formula unit (2.5)
chemical nomenclature (2.6)
organic compounds (2.6)
inorganic compounds (2.6)
monatomic ion (2.6)
polyatomic ion (2.6)
binary compound (2.6)
acid (2.6)
oxoacid (2.6)
hydrate (2.6)

SUMMARY OF FACTS AND CONCEPTS

Matter may be classified by its physical state as a *gas, liquid,* or *solid*. Matter may also be classified by its chemical constitution as an *element, compound,* or *mixture*. Materials are either substances or mixtures of substances. Substances are either elements or compounds, which are composed of two or more elements. Mixtures can be separated into substances by physical processes, but compounds can be separated into elements only by chemical reactions.

Atomic theory is central to chemistry. According to this theory, all matter is composed of small particles, or *atoms*. Elements are composed of the same kind of atom. Atoms are themselves composed of nuclei and electrons, and the nucleus of a given element has a definite number of protons (the *atomic number* of the element). The average mass of an atom of a given element has a definite value, called the *atomic weight* of the element.

The elements can be arranged in rows and columns by atomic number to form the *periodic table*. Elements in a given group have similar properties. Elements on the left and at the center of the table are *metals;* those on the right are *nonmetals*.

A *chemical formula* is a notation used to convey the relative proportions of the atoms of the different elements in a substance. If the substance is molecular, the formula gives the precise number of each kind of atom in the molecule. If the substance is ionic, the formula gives the relative number of different ions in the compound.

The last section of the chapter discusses chemical nomenclature, the systematic naming of compounds on the basis of their formulas or structures. Rules are given for naming *ionic compounds, binary molecular compounds, acids,* and *hydrates*.

OPERATIONAL SKILLS

1. **Using the law of definite proportions** Given the masses of the elements in several samples of a material, decide whether these data are consistent with the hypothesis that the material is a compound (Example 2.1).

2. **Writing an ionic formula, given the ions** Given the formulas of a cation and an anion, write the formula of the ionic compound of these ions (Example 2.2).

3. **Writing a name of a compound from its formula or vice versa** Given the name of a simple compound (ionic, binary molecular, acid, or hydrate) write the name (Examples 2.3, 2.6, and 2.9), or vice versa (Examples 2.4, 2.7, and 2.10).

4. **Deciding whether a binary compound is ionic or molecular** Given the elements in a binary compound, decide whether the compound is ionic or molecular (Example 2.5).

5. **Writing the name and formula of an anion from the acid** Given the name and formula of an oxoacid, write the name and formula of the oxoanion; or from the name and formula of the oxoanion, write the formula and name of the oxoacid (Example 2.8).

Review Questions

2.1 Characterize gases, liquids, and solids in terms of compressibility, fluidity, and density.

2.2 Choose a substance and give several of its physical properties and several of its chemical properties.

2.3 Give examples of an element, a compound, a heterogeneous mixture, and a homogeneous mixture.

2.4 (a) Sodium metal is partially melted. What are the two phases present? (b) A sample of sand is composed of granules of quartz (silicon dioxide) and seashells (calcium carbonate). The sand is mixed with water. What phases are present?

2.5 A material is believed to be a compound. Suppose you have several samples of this material obtained from various places around the world. Comment on what you would expect to find upon observing the melting point, density, and color for each sample. What would you expect to find upon determining the elemental composition for each sample?

2.6 Describe atomic theory and discuss how it explains the great variety of different substances. How does it explain chemical reactions?

2.7 Two compounds of iron and chlorine, A and B, contain 1.270 g and 1.904 g of chlorine, respectively, for each gram of iron. Show that these amounts are in the ratio 2:3. Is this consistent with the law of multiple proportions? Explain.

2.8 Oxygen consists of three different ________, each having eight protons but different numbers of neutrons.

2.9 Describe how Dalton obtained relative atomic masses.

2.10 Define the term *atomic weight*. Why might the values of atomic weights on a planet elsewhere in the universe be different from those on earth?

2.11 What is the name of the element in Group IVA and Period 5?

2.12 Cite some properties that are characteristic of a metal.

2.13 Ethane consists of molecules with two atoms of carbon and six atoms of hydrogen. Write the molecular formula for ethane.

2.14 What is the difference between a molecular formula and a structural formula?

2.15 Consider a mixture of 10 million O_2 molecules and 20 million H_2 molecules. In what way is this mixture similar to 20 million water molecules? In what way is it dissimilar?

2.16 Give an example of a binary compound that is ionic. Give an example of a binary compound that is molecular.

2.17 Compare the structures of the compounds H_2O and $CaCl_2$.

2.18 The compounds $CuCl$ and $CuCl_2$ were formerly called cuprous chloride and cupric chloride, respectively. What are their names using the Stock system of nomenclature? What are the advantages of the Stock system of nomenclature over the former one?

Practice Problems

SOLIDS, LIQUIDS, AND GASES

2.19 Give the normal state (solid, liquid, or gas) of each of the following.
(a) sodium hydrogen carbonate (baking soda)
(b) isopropyl alcohol (rubbing alcohol)
(c) nitrogen
(d) copper

2.20 Give the normal state (solid, liquid, or gas) of each of the following.
(a) potassium hydrogen tartrate (cream of tartar)
(b) oxygen
(c) carbon (diamond)
(d) bromine

CHEMICAL AND PHYSICAL CHANGES; PROPERTIES OF SUBSTANCES

2.21 Which of the following are physical changes and which are chemical changes?
(a) melting of sodium chloride
(b) pulverizing of rock salt
(c) burning of sulfur
(d) melting of sulfur

2.22 For each of the following, decide whether a physical or a chemical change is involved.
(a) dissolving of sugar in water
(b) rusting of iron
(c) burning of wood
(d) evaporation of alcohol

2.23 A sample of mercury(II) oxide was heated to produce mercury metal and oxygen gas. Then the liquid mercury was cooled to −40°C, where it solidified. A glowing wood splint was thrust into the oxygen, and the splint burst into flame. Identify each physical change and each chemical change.

2.24 Solid iodine, contaminated with salt, was heated until the iodine vaporized. The violet vapor of iodine was then cooled to yield the pure solid. Solid iodine and zinc metal powder were mixed and ignited to give a white powder. Identify each physical change and each chemical change.

2.25 The following are properties of substances. Decide whether each is a physical property or a chemical property.

(a) Chlorine gas liquefies at −35°C under normal pressure.
(b) Hydrogen burns in chlorine gas.
(c) Bromine is a red-brown liquid.
(d) Lithium is a soft, silvery-colored metal.
(e) Iron rusts in an atmosphere of moist air.

2.26 Decide whether each of the following is a physical property or a chemical property of the substance.

(a) Lithium metal has a density of 0.534 g/cm^3.
(b) Sodium chloride dissolves readily in water.
(c) Sodium carbonate gives off bubbles of carbon dioxide when treated with hydrochloric acid.
(d) A diamond burns in oxygen.
(e) Potassium chromate is a yellow, crystalline substance.

2.27 Iodine is a solid having somewhat lustrous, blue-black crystals. The crystals vaporize readily to a violet-colored gas. Iodine, like chlorine, combines with many metals. For example, aluminum combines with iodine to give aluminum iodide. Identify the physical and the chemical properties of iodine that are cited.

2.28 Mercury(II) oxide is an orange-red solid with a density of 11.1 g/cm^3. It decomposes when heated to give mercury and oxygen. The compound is insoluble in water (does not dissolve in water). Identify the physical and the chemical properties of mercury(II) oxide that are cited.

ELEMENTS, COMPOUNDS, AND MIXTURES

2.29 Consider the following separations of materials. State whether a physical process or a chemical reaction is involved in each separation. Refer to Table 2.2 for a list of elements.

(a) Sodium chloride is obtained from seawater by evaporation of the water.
(b) Mercury is obtained by heating the substance mercury(II) oxide; oxygen is also obtained.
(c) Pure water is obtained from ocean water by evaporating the water, then condensing it.
(d) Iron is produced from an iron ore that contains the substance iron(III) oxide.
(e) Gold is obtained from river sand by panning (allowing the heavy metal to settle in flowing water).

2.30 All of the following processes involve a separation of either a mixture into substances or a compound into elements. For each, decide whether a physical process or a chemical reaction is required. See Table 2.2 for a list of elements.

(a) Sodium metal is obtained from the substance sodium chloride.
(b) Iron filings are separated from sand by using a magnet.
(c) Sugar crystals are separated from a sugar syrup by evaporation of water.
(d) Fine crystals of silver chloride are separated from a suspension of the crystals in water.
(e) Copper is produced when zinc metal is placed in a solution of copper(II) sulfate, a compound.

2.31 Analyses of several samples of a material containing only iron and oxygen gave the following results. Could this material be a compound?

	MASS OF SAMPLE	MASS OF IRON	MASS OF OXYGEN
Sample A	1.518 g	1.094 g	0.424 g
Sample B	2.056 g	1.449 g	0.607 g
Sample C	1.873 g	1.335 g	0.538 g

2.32 A red-orange solid contains only mercury and oxygen. Analyses of three different samples gave the following results. Are the data consistent with the hypothesis that the material is a compound?

	MASS OF SAMPLE	MASS OF MERCURY	MASS OF OXYGEN
Sample A	1.0410 g	0.9641 g	0.0769 g
Sample B	1.5434 g	1.4293 g	0.1141 g
Sample C	1.2183 g	1.1283 g	0.0900 g

2.33 Label each of the following as a substance, a heterogeneous mixture, or a solution.

(a) seawater (b) sulfur
(c) fluorine (d) beach sand

2.34 Indicate whether each of the following materials is a substance, a heterogeneous mixture, or a solution.

(a) concrete (b) bromine
(c) gasoline (d) magnesium

2.35 Which of the following are pure substances and which are mixtures? For each one, list all of the different phases present.

(a) bromine liquid and its vapor
(b) a rock containing the minerals green dioptase and orange-yellow wulfenite
(c) paint, containing a liquid solution and a dispersed solid pigment
(d) partially molten iron

2.36 Which of the following are pure substances and which are mixtures? For each one, list all of the different phases present.

(a) baking powder containing sodium hydrogen carbonate and potassium hydrogen tartrate
(b) a sugar solution with sugar crystals at the bottom
(c) ink containing a liquid solution with fine particles of carbon
(d) a sand containing quartz (silicon dioxide) and calcite (calcium carbonate)

ATOMIC THEORY

2.37 Two samples of different compounds of nitrogen and oxygen have the following compositions. Show that the compounds follow the law of multiple proportions. What is the ratio of oxygen in the two compounds for a fixed amount of nitrogen?

	AMOUNT N	AMOUNT O
Compound A	1.206 g	2.755 g
Compound B	1.651 g	4.714 g

2.38 Two samples of different compounds of sulfur and oxygen have the following compositions. Show that the compounds follow the law of multiple proportions. What is the ratio of oxygen in the two compounds for a fixed amount of sulfur?

	AMOUNT S	AMOUNT O
Compound A	1.210 g	1.811 g
Compound B	1.783 g	1.779 g

2.39 The following table gives the number of protons and neutrons in the nuclei of various atoms. Which atom is the isotope of atom A? Which atom has the same mass number as atom A?

	PROTONS	NEUTRONS
Atom A	17	18
Atom B	16	19
Atom C	17	19
Atom D	18	22

2.40 The following table gives the number of protons and neutrons in the nuclei of various atoms. Which atom is the isotope of atom A? Which atom has the same mass number as atom A?

	PROTONS	NEUTRONS
Atom A	34	45
Atom B	35	44
Atom C	33	42
Atom D	34	44

2.41 What is the name of the element represented by each of the following atomic symbols?

(a) Ne
(b) Zn
(c) Ag
(d) Mg

2.42 For each atomic symbol, give the name of the element.

(a) Ca
(b) Pb
(c) Hg
(d) Sn

2.43 Give the atomic symbol for each of the following elements.

(a) potassium
(b) sulfur
(c) iron
(d) manganese

2.44 Give the atomic symbol for each of the following elements.

(a) selenium
(b) phosphorus
(c) sodium
(d) gold

2.45 Ammonia is a gas with a characteristic pungent odor. It is sold as a water solution used in household cleaning. The gas is a compound of nitrogen and hydrogen in the atomic ratio 1:3. A sample of ammonia contains 7.933 g N and 1.712 g H. What is the atomic mass of N relative to H?

2.46 Hydrogen sulfide is a gas with the odor of rotten eggs. The gas can sometimes be detected in automobile exhaust. It is a compound of hydrogen and sulfur in the atomic ratio 2:1. A sample of hydrogen sulfide contains 0.587 g H and 9.330 g S. What is the atomic mass of S relative to H?

PERIODIC TABLE

2.47 Identify the group and period for each of the following. Refer to the periodic table (Figure 2.14 or inside front cover). Label each as a metal, nonmetal, or metalloid.
(a) C (b) Po (c) Cr (d) Mg (e) B

2.48 Refer to the periodic table (Figure 2.14 or inside front cover) and obtain the group and period for each of the following elements. Also determine whether the element is a metal, nonmetal, or metalloid.
(a) Si (b) F (c) Ca (d) Co (e) Xe

2.49 Refer to the periodic table (Figure 2.14 or inside front cover) and answer the following questions.
(a) What Group VIA element is a metalloid?
(b) What is the Group IIIA element in Period 3?

2.50 Refer to the periodic table (Figure 2.14 or inside front cover) and answer the following questions.
(a) What Group VA element is a metal?
(b) What is the Group IIA element in Period 4?

2.51 Give one example (atomic symbol and name) for each of the following.
(a) a main-group (representative) element in the second period
(b) an alkali metal
(c) a transition element in the fourth period
(d) a lanthanide element

2.52 Give one example (atomic symbol and name) for each of the following.
(a) a transition element in the fifth period
(b) a halogen
(c) a main-group (representative) element in the third period
(d) an actinide element

MOLECULAR AND IONIC SUBSTANCES

2.53 The normal form of the element sulfur is a brittle, yellow solid. This is a molecular substance S_8. If this solid is vaporized, it first forms S_8 molecules; but at high temperature, S_2 molecules are formed. How do the molecules of the solid sulfur and of the hot vapor differ? How are the molecules alike?

2.54 White phosphorus is available in sticks, which have a waxy appearance. This is a molecular substance P_4. When this solid is vaporized, it first forms P_4 molecules; but at high temperature, P_2 molecules are formed. How do the molecules of white phosphorus and those of the hot vapor differ? How are the molecules alike?

2.55 A 1.50-g sample of nitrous oxide (an anesthetic, sometimes called laughing gas) contains 2.05×10^{22} N_2O molecules. How many nitrogen atoms are there in this sample? How many nitrogen atoms are there in 1.00 g of nitrous oxide?

2.56 Nitric acid is composed of HNO_3 molecules. A sample weighing 3.50 g contains 3.34×10^{22} HNO_3 molecules. How many nitrogen atoms are there in this sample? How many oxygen atoms are there in 1.50 g of nitric acid?

2.57 A sample of ammonia, NH_3, contains 3.3×10^{21} hydrogen atoms. How many NH_3 molecules are there in this sample?

2.58 A sample of ethanol (ethyl alcohol), C_2H_5OH, contains 4.2×10^{23} hydrogen atoms. How many C_2H_5OH molecules are there in this sample?

2.59 Give the molecular formula for each of the following structural formulas.

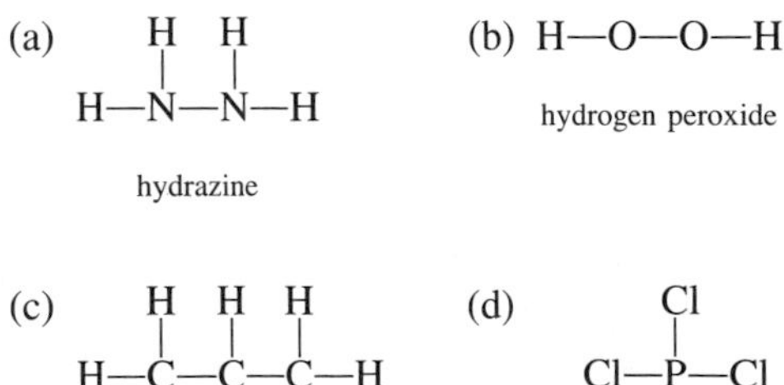

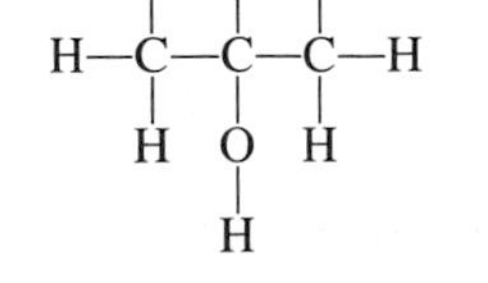

(d) Cl—P—Cl (with Cl above P)

phosphorus trichloride

2.60 What molecular formula corresponds to each of the following structural formulas?

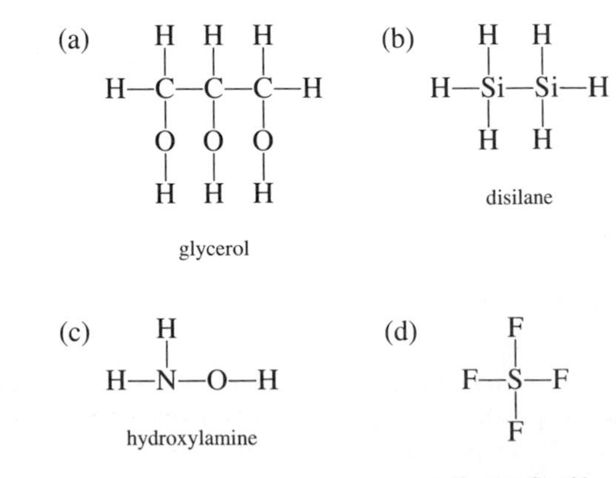

2.61 Iron(III) sulfate has the formula $Fe_2(SO_4)_3$. What is the ratio of iron atoms to oxygen atoms in this compound?

2.62 Ammonium phosphate, $(NH_4)_3PO_4$, has how many hydrogen atoms for each oxygen atom?

2.63 Write the formulas for the compounds of each of the following pairs of ions.

(a) Fe^{3+} and CN^-
(b) K^+ and SO_4^{2-}
(c) Li^+ and N^{3-}
(d) Sr^{2+} and Cl^-

2.64 For each of the following pairs of ions, write the formula of the corresponding compound.

(a) Co^{2+} and Br^-
(b) NH_4^+ and SO_4^{2-}
(c) Na^+ and PO_4^{3-}
(d) Fe^{3+} and NO_3^-

CHEMICAL NOMENCLATURE

2.65 Name the following compounds.

(a) Na_2SO_4
(b) CaO
(c) CuCl
(d) Cr_2O_3

2.66 Name the following compounds.

(a) Na_2O
(b) Mn_2O_3
(c) NH_4HCO_3
(d) $Cu(NO_3)_2$

2.67 Write the formulas of:

(a) lead(II) dichromate
(b) barium hydrogen carbonate
(c) cesium oxide
(d) iron(II) acetate

2.68 Write the formulas of:

(a) sodium thiosulfate
(b) copper(I) oxide
(c) calcium hydrogen carbonate
(d) tin(II) fluoride

2.69 For each of the following binary compounds, decide whether the compound is expected to be ionic or molecular.

(a) SeF_4
(b) LiBr
(c) SiF_4
(d) Cs_2O

2.70 For each of the following binary compounds, decide whether the compound is expected to be ionic or molecular.

(a) BaO
(b) As_4O_6
(c) ClF_3
(d) CuCl

2.71 Give systematic names to the following binary compounds.

(a) N_2O
(b) P_4O_{10}
(c) $AsCl_3$
(d) Cl_2O_7

2.72 Give systematic names to the following binary compounds.

(a) N_2F_2
(b) NCl_3
(c) N_2O_5
(d) As_4O_6

2.73 Write the formula of each of the following compounds.

(a) nitrogen tribromide
(b) xenon tetroxide
(c) oxygen difluoride
(d) dichlorine pentoxide

2.74 Write the formula of each of the following compounds.

(a) chlorine trifluoride
(b) dichlorine monoxide
(c) dinitrogen tetrafluoride
(d) bismuth pentafluoride

2.75 Give the name and formula of the anion corresponding to each of the following oxoacids.

(a) bromic acid, $HBrO_3$
(b) hyponitrous acid, $H_2N_2O_2$
(c) thiosulfuric acid, $H_2S_2O_3$
(d) arsenic acid, H_3AsO_4

2.76 Give the name and formula of the anion corresponding to each of the following oxoacids.

(a) selenious acid, H_2SeO_3
(b) bromous acid, $HBrO_2$
(c) hypoiodous acid, HIO
(d) diphosphoric acid, $H_4P_2O_7$

2.77 Glauber's salt has the formula $Na_2SO_4 \cdot 10H_2O$. What is the chemical name of this substance?

2.78 Emerald-green crystals of the substance $NiSO_4 \cdot 6H_2O$ are used in nickel plating. What is the chemical name of this compound?

2.79 Iron(II) sulfate heptahydrate is a blue-green, crystalline compound used to prepare other iron compounds. What is the formula of iron(II) sulfate heptahydrate?

2.80 Cobalt(II) chloride hexahydrate has a pink color. It loses water on heating and changes to a blue-colored compound. What is the formula of cobalt(II) chloride hexahydrate?

Additional Problems

2.81 Identify each of the following elements from its properties under normal conditions. Give the name of the element and the atomic symbol.
(a) red-brown liquid
(b) white, waxy solid
(c) soft, yellow metal
(d) soft, black solid

2.82 Identify each of the following elements from its properties under normal conditions. Give the name of the element and the atomic symbol.
(a) reddish metal
(b) brittle, yellow solid
(c) greenish-yellow gas
(d) silver-colored liquid metal

2.83 The following are results of analyses of two samples of the same or two different compounds of phosphorus and chlorine. From these results, decide whether the two samples are from the same or different compounds. If the compounds are different, obtain the ratio of Cl atoms in the compounds for a fixed amount (one gram) of P.

	AMOUNT P	AMOUNT Cl
Sample A	1.156 g	3.971 g
Sample B	1.542 g	5.297 g

2.84 The following are results of analyses of two samples of the same or two different compounds of phosphorus and oxygen. From these results, decide whether the two samples are from the same or different compounds. If the compounds are different, obtain the ratio of O atoms in the compounds for a fixed amount (one gram) of P.

	AMOUNT P	AMOUNT O
Sample A	2.581 g	3.332 g
Sample B	3.718 g	2.881 g

2.85 Rubidium, Rb, has atomic number 37. What is the charge on an ion having 36 electrons? What is the charge on an ion having 35 electrons? Write the formulas of these ions.

2.86 Astatine, At, has atomic number 85. What is the charge on an ion having 84 electrons? What is the charge on an ion having 86 electrons? Write the formulas of these ions.

2.87 Identify the following elements, giving their name and atomic symbol.
(a) a nonmetal that is normally a liquid
(b) a normally gaseous element in Group VIA
(c) a transition element in Group VB, fifth period
(d) the halogen in Period 2

2.88 Identify the following elements, giving their name and atomic symbol.
(a) a normally gaseous element in Group VIA
(b) a metal that is normally a liquid
(c) a main-group element in Group IIIA, second period
(d) the alkali metal in Period 4

2.89 Give the names of the following ions.
(a) Cr^{3+} (b) Cr^{2+}
(c) Cu^{+} (d) Cu^{2+}

2.90 Give the names of the following ions.
(a) Mn^{2+} (b) Ni^{2+}
(c) Co^{2+} (d) Co^{3+}

2.91 Write formulas for all the ionic compounds that can be formed by these ions: Na^{+}, Ni^{2+}, SO_4^{2-}, and Cl^{-}.

2.92 Write formulas for all the ionic compounds that can be formed by these ions: Ca^{2+}, Cr^{3+}, O^{2-}, and NO_3^{-}.

2.93 Name the following compounds.
(a) $Sn_3(PO_4)_2$ (b) NH_4NO_2
(c) $Mg(OH)_2$ (d) $CrSO_4$

2.94 Name the following compounds.
(a) $(NH_4)_2CO_3$ (b) $Ni(NO_3)_2$
(c) $BaSO_4$ (d) $HgCl_2$

2.95 Give the formulas for the following compounds.
(a) mercury(I) chloride
(b) copper(II) oxide
(c) ammonium dichromate
(d) zinc sulfide

2.96 Give the formulas for the following compounds.
(a) hydrogen peroxide
(b) iron(III) nitrate
(c) nickel(II) phosphate
(d) sulfur tetrafluoride

2.97 Name the following molecular compounds.
(a) $AsCl_3$ (b) SeO_2
(c) N_2O_3 (d) SiF_4

2.98 Name the following molecular compounds.
(a) ClF_4 (b) CS_2
(c) NF_3 (d) SF_6

Cumulative-Skills Problems

2.99 There are 2.619×10^{22} atoms in 1.000 g of sodium. Assume that sodium atoms are spheres of radius 1.86 Å and that they are lined up side by side. How many miles in length is the line of sodium atoms?

2.100 There are 1.699×10^{22} atoms in 1.000 g of chlorine. Assume that chlorine atoms are spheres of radius 0.99 Å and that they are lined up side by side. How many miles in length is the line of chlorine atoms?

2.101 A sample of green crystals of nickel(II) sulfate heptahydrate was heated carefully to produce the bluish-green nickel(II) sulfate hexahydrate. What are the formulas of the hydrates? If 8.753 g of the heptahydrate produces 8.192 g of the hexahydrate, how many grams of anhydrous nickel(II) sulfate could be obtained?

2.102 Cobalt(II) sulfate heptahydrate has pink-colored crystals. When heated carefully, it produces cobalt(II) sulfate monohydrate, which has red crystals. What are the formulas of these hydrates? If 3.548 g of the heptahydrate yields 2.184 g of the monohydrate, how many grams of the anhydrous cobalt(II) sulfate would be obtained?

2.103 A sample of metallic element X, weighing 3.177 g, combines with 0.6015 L of O_2 gas (at normal pressure and 20.0°C) to form the metal oxide with the formula XO. If the density of O_2 gas under these conditions is 1.330 g/L, what is the mass of this oxygen? The atomic weight of oxygen is 15.9994 amu. What is the atomic weight of X? What is the identity of X?

2.104 A sample of metallic element X, weighing 4.315 g, combines with 0.4810 L of Cl_2 gas (at normal pressure and 20.0°C) to form the metal chloride with the formula XCl. If the density of Cl_2 gas under these conditions is 2.948 g/L, what is the mass of the chlorine? The atomic weight of chlorine is 35.453 amu. What is the atomic weight of X? What is the identity of X?

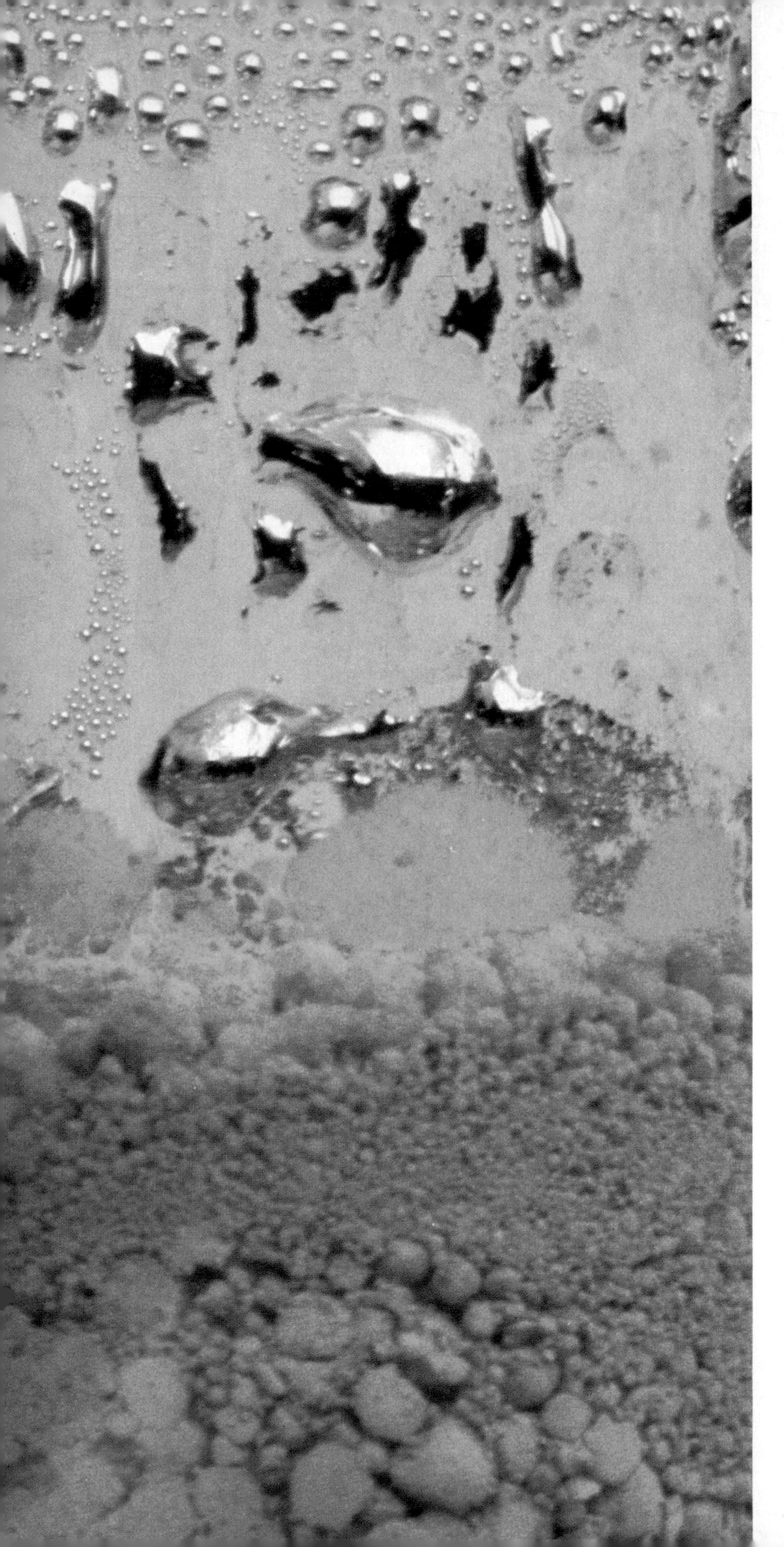

3

Chemical Reactions: An Introduction

Chapter Outline

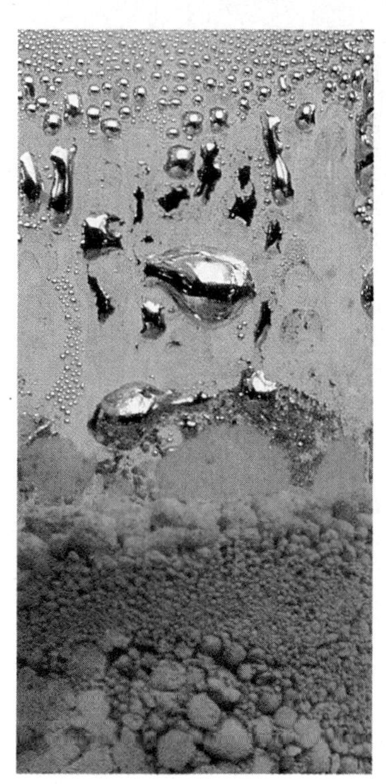

The heating of mercury(II) oxide.

A chemical reaction is accompanied by detectable changes—perhaps a change of color or the evolution of heat, but certainly a change of substances. Figure 3.1 shows the chemical reaction of hydrochloric acid, $HCl(aq)$, and sodium hydroxide, NaOH. A solution of sodium hydroxide (colorless) is added to hydrochloric acid (also colorless). To make the resulting reaction visible, we have added a substance called an indicator to the acid. The indicator we have used is colorless as long as the acid is present, but turns pink when the acid completely reacts and the sodium hydroxide is in excess. Note the formation of the pink color in Figure 3.1 (*right*), showing that sodium hydroxide has reacted with hydrochloric acid.

Chemical reactions are a key concern of chemistry. In the previous chapter, we described a number of reactions, including the burning of sodium in chlorine. Several obvious changes were seen in this reaction. A metal and a gas produced a white solid (sodium chloride, or salt). The reaction is also accompanied by a bright yellow flame. In the reaction of hydrochloric acid with sodium hydroxide, we see a color change of the indicator. But the reaction occurs even in the absence of the indicator. In this case, we add two colorless solutions and produce a colorless solution; thus, the change is less obvious. If we are careful, we can observe a temperature rise, showing that heat is released. Also, if we evaporate the final

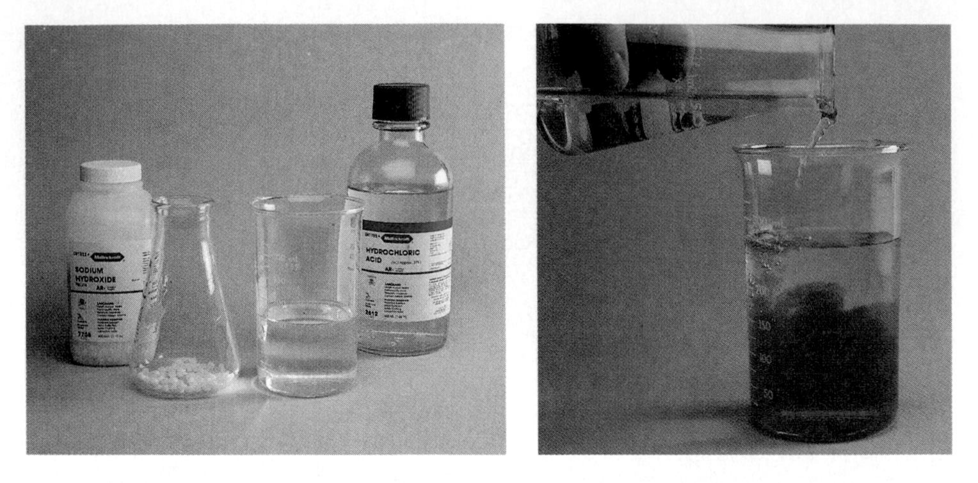

Figure 3.1
Reaction of hydrochloric acid and sodium hydroxide. *Left*: The beaker contains hydrochloric acid plus several drops of phenolphthalein indicator; the flask contains a few grams of sodium hydroxide. *Right*: Sodium hydroxide solution in the flask is being added to the hydrochloric acid; an excess of sodium hydroxide turns the indicator a deep pink color.

solution, we obtain a white, crystalline material. This proves to be sodium chloride, NaCl (salt again!), a compound different from either HCl or NaOH.

In this chapter, we begin our exploration of chemical reactions. What happens when hydrochloric acid reacts with sodium hydroxide (or sodium reacts with chlorine)? How can we represent the chemical reaction in a way that will conveniently describe what is happening? How do we classify the many possible reactions so that we can begin to understand them? These are some of the questions addressed in this chapter.

Understanding Chemical Reactions

Chemists represent chemical reactions by *chemical equations*. We will look at these equations in the first section of this chapter. Then we will look at a simple scheme for classifying reactions, which will help us to understand most of the chemical reactions we will encounter.

3.1 CHEMICAL EQUATIONS

An important feature of atomic theory is its explanation of a chemical reaction as a rearrangement of the chemical combinations of atoms. Such rearrangements of atoms are conveniently represented by chemical equations which use the chemical formulas described in the previous chapter.

Writing Chemical Equations

A **chemical equation** is *the symbolic representation of a chemical reaction in terms of chemical formulas*. For example, the burning of sodium in chlorine to produce sodium chloride is written

$$2Na + Cl_2 \longrightarrow 2NaCl$$

The formulas on the left side of an equation (before the arrow) represent the reactants; a **reactant** is *a starting substance in a chemical reaction*. The arrow means ''react to form'' or ''yield.'' The formulas on the right represent the products; a **product** is *a substance that results from a reaction*. Note the *coefficient* of 2 in front of the formula NaCl. Coefficients such as this give the relative number of molecules or formula units involved in the reaction. Coefficients of 1 are usually understood, not written.

In many cases, it is useful to indicate the states or phases of the substances in an equation. We do this by placing appropriate labels indicating the phases within parentheses following the formulas of the substances. We use the following phase labels:

$$(g) = \text{gas}, \quad (l) = \text{liquid}, \quad (s) = \text{solid}, \quad (aq) = \text{aqueous (water) solution}$$

When we use these labels, the previous equation becomes

$$2Na(s) + Cl_2(g) \longrightarrow 2NaCl(s)$$

We can also indicate in an equation the conditions under which a reaction takes place. If the reactants are heated to make the reaction go, we can indicate this by putting the symbol Δ (capital Greek delta) over the arrow. For example, the equation

$$2NaNO_3(s) \xrightarrow{\Delta} 2NaNO_2(s) + O_2(g)$$

indicates that solid sodium nitrate, $NaNO_3$, decomposes when heated to give solid sodium nitrite, $NaNO_2$, and oxygen gas, O_2.

When an aqueous solution of hydrogen peroxide, H_2O_2, comes in contact with platinum metal, Pt, the hydrogen peroxide decomposes into water and oxygen gas. The platinum acts as a *catalyst,* a substance that speeds up a reaction without undergoing any net change itself. We write the equation for this reaction as follows. Here the catalyst, Pt, is written over the arrow.●

● Platinum is a silvery-white metal used for jewelry. It is valuable as a catalyst for many reactions, including those that occur in the catalytic converters of automobiles.

$$2H_2O_2(aq) \xrightarrow{Pt} 2H_2O(l) + O_2(g)$$

Balancing Chemical Equations

When the coefficients in a chemical equation are correctly given, the numbers of atoms of each element are equal on both sides of the arrow. The equation is then said to be *balanced*. That a chemical equation should be balanced follows from atomic theory. A chemical reaction involves simply a recombination of the atoms; none are destroyed and none are created. Consider the burning of natural gas, which is composed mostly of methane, CH_4. Using atomic theory, we describe this as the chemical reaction of one molecule of methane, CH_4, with two molecules of oxygen, O_2, to form one molecule of carbon dioxide, CO_2, and two molecules of water, H_2O.

$$\underset{\text{one molecule of methane}}{CH_4} + \underset{\text{two molecules of oxygen}}{2O_2} \underset{\text{react to form}}{\longrightarrow} \underset{\text{one molecule of carbon dioxide}}{CO_2} + \underset{\text{two molecules of water}}{2H_2O}$$

Figure 3.2 represents the reaction in terms of molecular models.

Before we can write a balanced chemical equation for a reaction, we must determine *by experiment* those substances that are reactants and those that are products. We must also determine the formulas of each of these substances. Once we know these things, we can write the chemical equation.

As an example, consider the burning of propane gas (Figure 3.3). By experiment, we determine that propane reacts with oxygen in air to give carbon dioxide and water. We also determine from experiment that the formulas of propane,

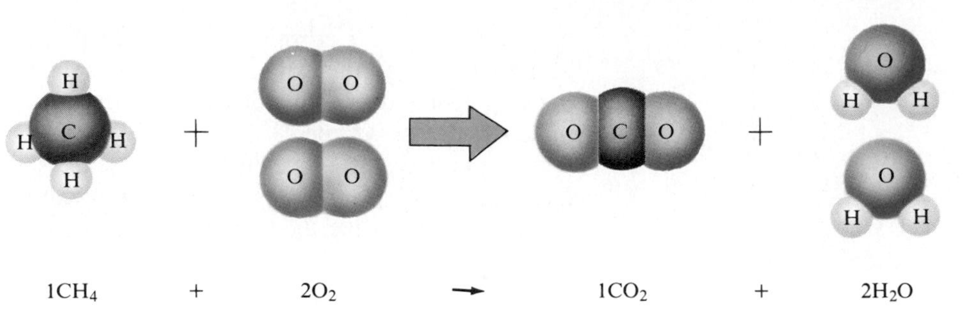

Figure 3.2
Representation of the reaction of methane with oxygen. Molecular models represent the reaction of CH_4 with O_2 to give CO_2 and H_2O.

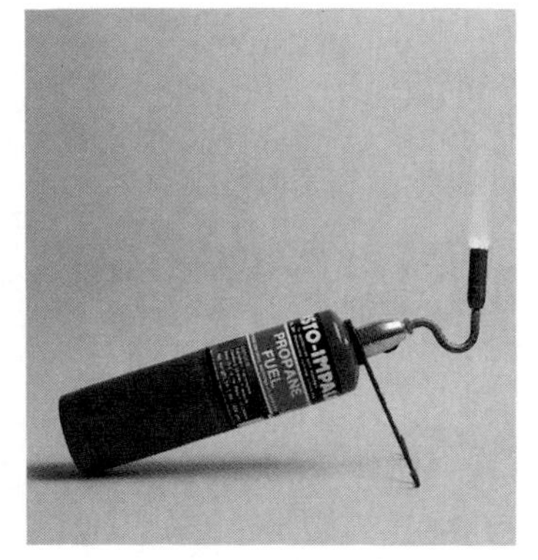

Figure 3.3
The burning of propane gas. Propane gas, C_3H_8, from the steel tank burns by reacting with oxygen, O_2, in air to give carbon dioxide, CO_2, and water, H_2O.

oxygen, carbon dioxide, and water are C_3H_8, O_2, CO_2, and H_2O, respectively. Then we can write

$$C_3H_8 + O_2 \longrightarrow CO_2 + H_2O$$

This equation is not balanced because the coefficients that give the relative number of molecules involved have not yet been determined. To balance the equation, we select coefficients that will make the numbers of atoms of each element equal on both sides of the equation.

Because there are three carbon atoms on the left side of the equation (C_3H_8), we must have three carbon atoms on the right. This is achieved by writing 3 for the coefficient of CO_2.

$$1C_3H_8 + O_2 \longrightarrow 3CO_2 + H_2O$$

(We have written the coefficient for C_3H_8 to show that it is now determined; normally when the coefficient is 1, it is omitted.) Similarly, there are eight hydrogen atoms on the left (in C_3H_8), so we write 4 for the coefficient of H_2O.

$$1C_3H_8 + O_2 \longrightarrow 3CO_2 + 4H_2O$$

The coefficients on the right are now determined, and there are ten oxygen atoms on this side of the equation: 6 from the three CO_2 molecules and 4 from the four H_2O molecules. We now write 5 for the coefficient of O_2. We drop the coefficient 1 from C_3H_8 and have the balanced equation.

$$C_3H_8 + 5O_2 \longrightarrow 3CO_2 + 4H_2O$$

It is a good idea to check your work by counting the number of atoms of each element on the left side and then on the right side of the equation.

The combustion of propane can be equally well represented by

$$2C_3H_8 + 10O_2 \longrightarrow 6CO_2 + 8H_2O$$

in which the coefficients of the previous equation have been doubled. Usually, however, it is preferable to *write the coefficients so that they are the smallest whole numbers possible*. When we refer to a balanced equation, we will assume that this is the case unless otherwise specified.

The method we have outlined—called *balancing by inspection*—is essentially a trial-and-error method.● It can be made easier, however, by observing the following rule: Balance first the atoms for elements that occur in only one substance on each side of the equation. The usefulness of this rule is illustrated in the next example. Remember that in any formula, such as $Fe_2(SO_4)_3$, a subscript to the right of the parentheses multiplies *each* subscript within the parentheses. Thus, $Fe_2(SO_4)_3$ represents 4 × 3 oxygen atoms. Remember also that you cannot change a subscript in any formula; only the coefficients can be altered to balance an equation.

● Balancing equations this way is relatively fast for all but the most complicated equations. Special methods for these equations are described in Chapter 13.

Example 3.1 Balancing Simple Equations

Balance the following equations.

(a) $H_3PO_3 \longrightarrow H_3PO_4 + PH_3$
(b) $Ca + H_2O \longrightarrow Ca(OH)_2 + H_2$
(c) $Fe_2(SO_4)_3 + NH_3 + H_2O \longrightarrow Fe(OH)_3 + (NH_4)_2SO_4$

Solution

(a) Only oxygen occurs in just one of the products (H_3PO_4). It is therefore easiest to balance O atoms first.

$$4H_3PO_3 \longrightarrow 3H_3PO_4 + PH_3$$

This equation is now also balanced in P and H atoms. Thus, the balanced equation is

$$4H_3PO_3 \longrightarrow 3H_3PO_4 + PH_3$$

(b) The equation is balanced in Ca atoms as it stands.

$$1Ca + H_2O \longrightarrow 1Ca(OH)_2 + H_2$$

O atoms occur in only one reactant and in only one product, so they are balanced next.

$$1Ca + 2H_2O \longrightarrow 1Ca(OH)_2 + H_2$$

The equation is now also balanced in H atoms. Thus, the answer is

$$Ca + 2H_2O \longrightarrow Ca(OH)_2 + H_2$$

(c) To balance the equation in Fe atoms, we write

$$1Fe_2(SO_4)_3 + NH_3 + H_2O \longrightarrow 2Fe(OH)_3 + (NH_4)_2SO_4$$

We balance the S atoms by placing the coefficient 3 for $(NH_4)_2SO_4$.

$$1Fe_2(SO_4)_3 + NH_3 + H_2O \longrightarrow 2Fe(OH)_3 + 3(NH_4)_2SO_4$$

Now we balance N atoms:

$$1Fe_2(SO_4)_3 + 6NH_3 + H_2O \longrightarrow 2Fe(OH)_3 + 3(NH_4)_2SO_4$$

Finally, the O atoms are balanced. First we count the number of O atoms on the right (18). Then we count the number of O atoms in substances on the left with known coefficients. There are 12 O's in $Fe_2(SO_4)_3$; hence, the number of remaining O's (in H_2O) must be $18 - 12 = 6$.

$$1Fe_2(SO_4)_3 + 6NH_3 + 6H_2O \longrightarrow 2Fe(OH)_3 + 3(NH_4)_2SO_4$$

The answer is

$$Fe_2(SO_4)_3 + 6NH_3 + 6H_2O \longrightarrow 2Fe(OH)_3 + 3(NH_4)_2SO_4$$

Check each of the final equations by counting the number of atoms of each element on both sides of the equations.

Exercise 3.1

Find the coefficients that balance each of the following equations.

(a) $O_2 + PCl_3 \longrightarrow POCl_3$
(b) $P_4 + N_2O \longrightarrow P_4O_6 + N_2$
(c) $As_2S_3 + O_2 \longrightarrow As_2O_3 + SO_2$
(d) $Ca_3(PO_4)_2 + H_3PO_4 \longrightarrow Ca(H_2PO_4)_2$

(See Problems 3.21, 3.22, 3.23, and 3.24.)

3.2 TYPES OF REACTIONS

In our study of general chemistry, we will describe many apparently different chemical reactions. Understanding these reactions is greatly facilitated by seeing the patterns of similarity among them. Several classification schemes have been developed to help one see these patterns. In this introductory chapter, we will look at the traditional classification based on how atoms or groups of atoms are rearranged during the reaction. This classification includes the following types of reactions:

1. Combination reactions
2. Decomposition reactions
3. Displacement reactions (single-replacement reactions)
4. Metathesis reactions (double-replacement reactions)

Figure 3.4
Reaction of copper metal and silver nitrate. When a strip of copper is dipped into silver nitrate solution, it becomes coated with crystals of silver. The blue-green color is from the copper(II) ion.

Another common type of reaction that we will include here are combustions, or reactions with oxygen.

5. Combustion reactions

We will describe each of these with examples in the remainder of this section. In studying this section, you should concentrate on recognizing the types of reactions rather than memorizing particular reactions.

Combination Reactions

A **combination reaction** is *a reaction in which two substances combine to form a third substance*. The simplest cases are those in which two elements react to form a compound of the elements. In the previous chapter, we discussed the reaction of sodium metal and chlorine gas (Figure 2.1).

$$2Na(s) + Cl_2(g) \longrightarrow 2NaCl(s)$$

Another example is the reaction of white phosphorus with chlorine. In limited quantities of chlorine, phosphorus reacts to give phosphorus trichloride, PCl_3, a colorless liquid.

$$P_4(s) + 6Cl_2(g) \longrightarrow 4PCl_3(l)$$

When chlorine is present in large amounts, however, the product is phosphorus pentachloride, PCl_5, a white, crystalline substance.

$$P_4(s) + 10Cl_2(g) \longrightarrow 4PCl_5(s)$$

Other combination reactions involve compounds as reactants. For example, phosphorus trichloride reacts with chlorine to form phosphorus pentachloride.

$$PCl_3(l) + Cl_2(g) \longrightarrow PCl_5(s)$$

Another example is the reaction of calcium oxide (quicklime) with sulfur dioxide to give calcium sulfite.

$$CaO(s) + SO_2(g) \longrightarrow CaSO_3(s)$$

This reaction can be used to remove sulfur dioxide from the stack gases produced in burning coal and other sulfur-containing materials. Sulfur dioxide has a characteristic suffocating odor (the odor of burning matches).

Decomposition Reactions

A **decomposition reaction** is *a reaction in which a single compound reacts to give two or more substances*. Often it is necessary to raise the temperature in order to decompose the compound. In Chapter 1, we described the decomposition of mercury(II) oxide into its elements when the compound is heated (Figure 1.4, *right*).

$$2HgO(s) \xrightarrow{\Delta} 2Hg(l) + O_2(g)$$

A common method of preparing oxygen in the laboratory consists of heating potassium chlorate with manganese(IV) oxide (manganese dioxide) as a catalyst.

$$2KClO_3(s) \xrightarrow[MnO_2]{\Delta} 2KCl(s) + 3O_2(s)$$

Figure 3.5
Reaction of zinc metal and hydrochloric acid. Hydrogen gas formed in the reaction bubbles from the metal surface that dips into the acid. The balloon expands as it fills with the gas.

In this reaction, a compound decomposes into a compound and an element. The industrial preparation of quicklime (calcium oxide) consists of heating limestone or seashells (both are primarily calcium carbonate).

$$CaCO_3(s) \xrightarrow{\Delta} CaO(s) + CO_2(g)$$

Here a compound decomposes into two different compounds.

Displacement Reactions

A **displacement reaction** (also called a **single-replacement reaction**) is *a reaction in which an element reacts with a compound displacing an element from it.* For example, when a copper metal strip is dipped into a solution of silver nitrate, crystals of silver metal are produced (Figure 3.4).

$$Cu(s) + 2AgNO_3(aq) \longrightarrow 2Ag(s) + Cu(NO_3)_2(aq)$$

Copper replaces silver in silver nitrate, producing copper nitrate solution and silver metal.

When a zinc metal strip is dipped into an acid solution, bubbles of hydrogen gas form on the metal and escape from the solution (Figure 3.5). In general, any active metal, such as zinc or iron, will replace hydrogen atoms in an acid. The reaction of zinc metal with hydrochloric acid is such a displacement reaction.

$$Zn(s) + 2HCl(aq) \longrightarrow H_2(g) + ZnCl_2(aq)$$

Zinc replaces hydrogen in the acid, producing zinc chloride solution and hydrogen gas.

Metathesis Reactions

A **metathesis** (me-tath′-e-sis) **reaction** (or **double-replacement reaction**) is *a reaction that appears to involve the exchange of parts of the reactants.* When the reactants are ionic compounds in solution, the parts exchanged are the cations and anions of the compounds. Figure 3.6 shows a colorless solution of potassium iodide being poured into a colorless solution of lead(II) nitrate. The ionic

Figure 3.6
Reaction of potassium iodide solution and lead(II) nitrate solution. The reactant solutions are colorless, but one of the products is lead(II) iodide, which forms as a yellow precipitate.

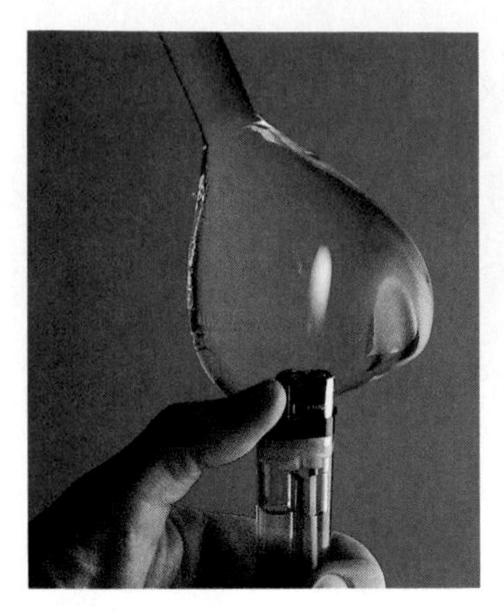

Figure 3.7
Reaction of butane and oxygen. Butane liquid in the cigarette lighter vaporizes when the valve is opened. The butane vapor burns in air to yield carbon dioxide and water vapor (note the mist of water droplets that has collected on the cool flask).

compounds in the solutions react to give a bright yellow precipitate. A **precipitate** is *a solid compound formed during a reaction in solution.*

$$2KI(aq) + Pb(NO_3)_2(aq) \longrightarrow 2KNO_3(aq) + PbI_2(s)$$

Note that iodide ions in potassium iodide exchange with nitrate ions in lead(II) nitrate. The products are colorless potassium nitrate (in solution) and a yellow precipitate of lead(II) iodide, indicated in the equation as $PbI_2(s)$.

The reaction of hydrochloric acid, $HCl(aq)$, and sodium hydroxide, $NaOH(aq)$, which we considered in the chapter opening, is another example of a metathesis reaction.

$$HCl(aq) + NaOH(aq) \longrightarrow HOH(l) + NaCl(aq)$$

Here, we have written the formula of water as HOH to emphasize the exchange.

It should be noted that these reactions actually involve ions in solution. We will discuss this aspect of metathesis reactions in the last part of the chapter.

Combustion Reactions

The reactions we have considered so far can be characterized by the type of atom rearrangement that occurs. However, it is useful to add the category of combustion reactions, which is instead characterized by the fact that one of the reactants is oxygen. A **combustion reaction** is *a reaction of a substance with oxygen, usually with the rapid release of heat to produce a flame.* Organic compounds usually burn in oxygen or air to yield carbon dioxide. If the compound contains hydrogen, water is also a product. Earlier we considered the combustion of propane (Figure

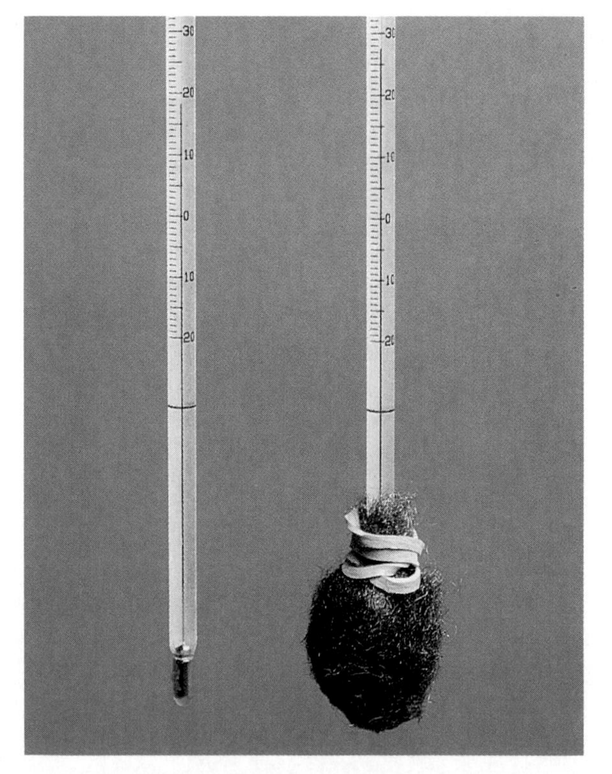

Figure 3.8
Rusting of iron wool. The temperature rises perceptibly during the rusting, showing that heat is released by the reaction just as if the iron were being burned. Note the temperature increase of the thermometer wrapped with moist iron wool.

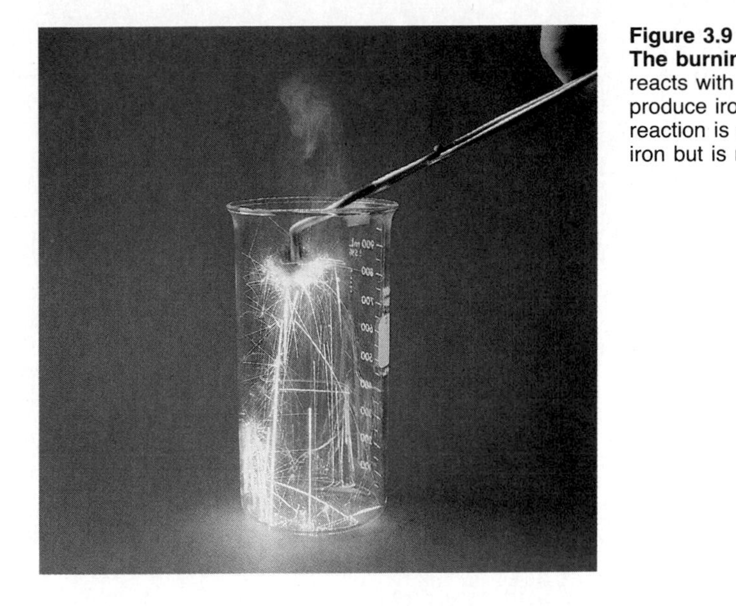

Figure 3.9
The burning of iron wool. Iron reacts with oxygen in the air to produce iron(III) oxide, Fe_2O_3. The reaction is similar to the rusting of iron but is much faster.

3.3), a hydrocarbon (binary compound of hydrogen and carbon). Similarly, butane, C_4H_{10}, burns in air to produce carbon dioxide and water vapor (Figure 3.7).

$$2C_4H_{10}(g) + 13O_2(g) \longrightarrow 8CO_2(g) + 10H_2O(g)$$

The rusting of iron, although not normally thought of as a combustion, has the essential features of such a reaction. It is the reaction of iron with oxygen, and it slowly evolves heat (Figure 3.8). The reaction is a bit complicated and involves water, but we can write the overall result of rusting approximately as follows:

$$4Fe(s) + 3O_2(g) \longrightarrow 2Fe_2O_3(s)$$

We can, in fact, burn iron wool in oxygen gas (Figure 3.9). The reaction is correctly represented by the previous equation. This could also be classified as a combination reaction, although combustions in general are not combination reactions (note the combustion of butane).

Example 3.2 Classifying Chemical Reactions

Classify each of the following reactions by its type.

(a) $2KNO_3(aq) \xrightarrow{\Delta} 2KNO_2(aq) + O_2(g)$
(b) $2H_2S(g) + 3O_2(g) \longrightarrow 2H_2O(g) + 2SO_2(g)$
(c) $HNO_3(aq) + NaOH(aq) \longrightarrow NaNO_3(aq) + H_2O(l)$
(d) $8Fe(s) + S_8(s) \xrightarrow{\Delta} 8FeS(s)$
(e) $2NaI(aq) + Cl_2(g) \longrightarrow 2NaCl(aq) + I_2(aq)$

Solution

(a) This is a **decomposition reaction**; one compound decomposes to give two other substances.
(b) This is a **combustion reaction**; hydrogen sulfide gas burns in oxygen (or air) to give water vapor and sulfur dioxide.
(c) This is a **metathesis reaction**. This can be seen more easily by writing the formula of water as the condensed structural formula HOH.

$$HNO_3(aq) + NaOH(aq) \longrightarrow NaNO_3(aq) + HOH(l)$$

Note the exchange of parts of the reactants.
(d) This is a **combination reaction**; iron combines with sulfur to give iron(II) sulfide.
(e) This is a **displacement reaction**; the iodine atom in NaI is displaced by Cl from chlorine gas that is dissolved in the solution.

Exercise 3.2

What type of reaction is represented by each of the following?

(a) $CS_2(l) + 3O_2(g) \longrightarrow CO_2(g) + 2SO_2(g)$
(b) $2Li(s) + Br_2(l) \longrightarrow 2LiBr(s)$
(c) $2H_2O_2(aq) \xrightarrow{MnO_2} 2H_2O(l) + O_2(g)$
(d) $K_2CrO_4(aq) + Pb(NO_3)_2(aq) \longrightarrow 2KNO_3(aq) + PbCrO_4(s)$
(e) $2Ga(s) + 3H_2SO_4(aq) \longrightarrow Ga_2(SO_4)_3(aq) + 3H_2(g)$

(See Problems 3.29 and 3.30.)

Before we leave this section, we should add that other ways of classifying chemical reactions exist. The traditional scheme we have just discussed is useful for simple types of reactions, but many reactions can also be placed in one of two broad categories: *acid-base reactions* and *oxidation-reduction reactions*. Combustions are examples of oxidation-reduction reactions. However, we will defer any discussion of oxidation-reduction reactions until Chapter 13, where we look at these two broad categories of reactions in some detail. In the present chapter, we will look briefly at acids and bases, and we will see that many metathesis reactions are acid-base type.

Ions in Aqueous Solution

A large number of the reactions that we will be looking at in this book are reactions that occur in aqueous solution. Later in this chapter, we will look closely at three of these kinds of reactions: precipitation, neutralization reactions, and reactions with gas formation. These reactions can usually be classified as metathesis reactions. However, they actually involve ions in solution, as we will explain. To understand this, we must first look at ions in aqueous solution. Certain compounds, called *electrolytes,* form ions when dissolved in water; we will discuss these compounds in the next section.

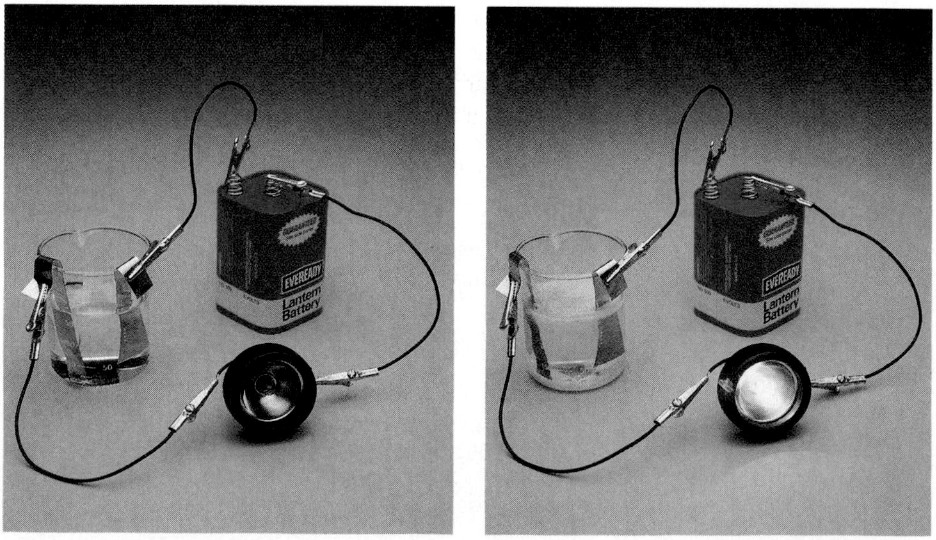

Figure 3.10
Testing the electrical conductivity of a solution. *Left:* Pure water does not conduct; therefore, the bulb does not light. *Right:* A solution of sodium chloride allows the current to pass through it, and the bulb lights.

3.3 ELECTROLYTES

Pure water is a poor conductor of electricity, but when sodium chloride crystals, NaCl, are dissolved in it, the resulting aqueous solution becomes electrically conducting. Figure 3.10 shows an apparatus used to demonstrate the conductivity of solutions. It consists of two electrodes, or electrical connections, one wired directly to a battery, the other connected to a bulb that is connected to the other side of the battery. As long as there is no electrical connection between the electrodes, the bulb cannot light; the electric circuit is not complete. The difference in electrical conductivity between pure water and a solution of NaCl becomes apparent if we first dip the electrodes into pure water. Nothing happens. But when the electrodes are dipped into a solution of NaCl, the bulb lights. Thus, the solution of NaCl is an electrical conductor and allows electricity to flow from the battery.

A substance, such as sodium chloride, that dissolves in water to give an electrically conducting solution is called an **electrolyte.** *A substance, such as sucrose, or table sugar* ($C_{12}H_{22}O_{11}$), *that dissolves in water to give nonconducting or very poorly conducting solutions* is called a **nonelectrolyte.**

Ionic Theory of Solutions

The conductivity of aqueous sodium chloride is explained by the ionic theory of solutions, proposed in 1884 by the Swedish chemist Svante Arrhenius (1859–1927).● According to this theory, an electrolyte produces ions when it dissolves in water. Sodium chloride dissolves in water as Na^+ and Cl^- ions. Suppose we dip the electrodes of the apparatus described in Figure 3.10 into this solution. One electrode is positively charged by the battery and attracts the negatively charged Cl^- ions. The other electrode is negatively charged and attracts Na^+ ions. Thus, the ions move in the solution. This movement of ions, or electric charge, is responsible for the electric current that flows in the solution. A nonelectrolyte such as sucrose dissolves in water as molecules. These molecules are not electrically charged and are not attracted to the electrodes, so no current flows.

● Arrhenius submitted his ionic theory as part of his doctoral dissertation to the faculty at Uppsala, Sweden, in 1884. It was not well received and he barely passed. In 1903, however, he was awarded the Nobel Prize in chemistry for this theory.

Most soluble ionic substances dissolve in water as ions and are therefore electrolytes. Some molecular substances also dissolve to give ions. For example, hydrogen chloride gas, HCl, reacts with water to give the ions H_3O^+ and Cl^-. Because the aqueous solution of hydrogen chloride (called hydrochloric acid) contains ions, hydrogen chloride is an electrolyte. We represent the reaction by the equation

$$HCl(g) + H_2O(l) \longrightarrow H_3O^+(aq) + Cl^-(aq)$$

Strong and Weak Electrolytes

When electrolytes dissolve in water they produce ions, but they do so to varying extents. A **strong electrolyte** is *an electrolyte that exists in solution almost entirely as ions*. When HCl dissolves in water, it reacts almost completely to give the ions H_3O^+ and Cl^-; almost no unreacted HCl molecules remain in the solution. Thus, HCl is a strong electrolyte.

Certain molecular substances dissolve in water to form only a small percentage of ions, in addition to the original molecular substance. Ammonia, NH_3, is an example. When ammonia dissolves in water, the molecules react with water to

Table 3.1 Common Acids and Bases

Name	Formula	Remarks
Acids:		
Acetic acid	$HC_2H_3O_2$	Found in vinegar
Acetylsalicylic acid	$HC_9H_7O_4$	Aspirin
Ascorbic acid	$H_2C_6H_6O_6$	Vitamin C
Citric acid	$H_3C_6H_5O_7$	Found in lemon juice
Hydrochloric acid	HCl	Found in gastric juice (digestive fluid in stomach)
Sulfuric acid	H_2SO_4	Battery acid
Bases:		
Ammonia	NH_3	Aqueous solution used as a household cleaner
Calcium hydroxide	$Ca(OH)_2$	Slaked lime (used in mortar for building construction)
Magnesium hydroxide	$Mg(OH)_2$	Milk of magnesia (antacid and laxative)
Sodium hydroxide	$NaOH$	Drain cleaners, oven cleaners

form ammonium ion, NH_4^+, and hydroxide ion, OH^-.

$$NH_3(aq) + H_2O(l) \longrightarrow NH_4^+(aq) + OH^-(aq)$$

These ions, however, react with each other to give the ammonia and water back again.

$$NH_4^+(aq) + OH^-(aq) \longrightarrow NH_3(aq) + H_2O(l)$$

Both reactions, the original one and its reverse, are going on simultaneously. Eventually the two reactions occur at the same rate, or speed, so that the numbers of different species (molecules and ions) remain constant. Such a situation is known as *chemical equilibrium*. We denote this chemical equilibrium by using a double arrow in the equation:

$$NH_3(aq) + H_2O(l) \rightleftharpoons NH_4^+(aq) + OH^-(aq)$$

The equilibrium is *dynamic;* that is, both reactions (the original one and its reverse) occur constantly, even at equilibrium. As a result of the chemical equilibrium, just a few percent of the NH_3 molecules have reacted at any given moment to form ions. Therefore, ammonia solutions are only weakly conducting.

A **weak electrolyte** is *an electrolyte that dissolves in water to give an equilibrium between a molecular substance and a relatively small quantity of ions.* Ammonia is only one example of a weak electrolyte. Many acidic substances are also weak electrolytes.

3.4 ACIDS AND BASES

Some of the most important electrolytes are acids, which we mentioned briefly in the previous chapter, and bases. These substances can be recognized by simple properties. Acids have a sour taste. Solutions of bases, on the other hand, have a bitter taste and a soapy feel. (Of course, one should never taste laboratory chemicals.) Some examples of acids are acetic acid, present in vinegar; citric acid, a

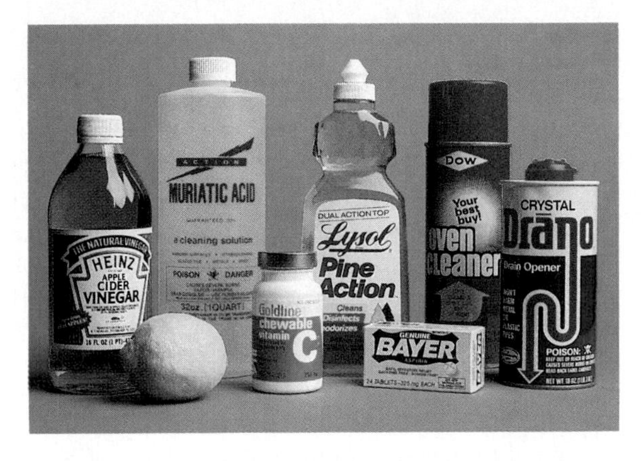

Figure 3.11
Examples of household acids and bases. Shown are vinegar (acetic acid), a lemon (citric acid), muriatic acid (hydrochloric acid), vitamin C (ascorbic acid), cleaner (contains a derivative of phenol, or carbolic acid), aspirin (acetylsalicylic acid), oven cleaner (sodium hydroxide), and drain opener (sodium hydroxide).

constituent of lemon juice; and hydrochloric acid, found in the digestive fluid of the stomach. An example of a base is aqueous ammonia, often used as a household cleaner. More examples are listed in Table 3.1; see also Figure 3.11 (p. 80).

Another property of acids and bases is their ability to cause color changes in certain dyes. An **acid-base indicator** is *a dye used to distinguish between acidic and basic solutions by means of the color changes it undergoes in these solutions*. Such dyes are common in natural substances and mixtures. The amber color of tea, for example, is lightened by the addition of lemon juice (citric acid). Red cabbage juice changes from red to green then yellow when a base is added (Figure 3.12). The green and yellow colors change back to red when an acid is added. Litmus is a common laboratory acid-base indicator. This dye, produced from certain species of lichens, turns red in acidic solution and blue in basic solution. Phenolphthalein, mentioned in the chapter opener, is another laboratory acid-base indicator. It is colorless in acidic solution and pink in basic solution.

Definitions of Acids and Bases

When Arrhenius developed his ionic theory of solutions, he also gave the classic definitions of acids and bases. To understand these definitions, we will first consider the production of ions in pure water. Although pure water is a very poor conductor of electricity, it does produce a small percentage of ions (about

Figure 3.12
Red cabbage juice as an acid-base indicator. Red cabbage juice is red in acidic solution and changes to green then yellow in basic solution. The beakers vary in acidity from highly acidic on the left to highly basic on the right.

$2 \times 10^{-7}\%$ of the molecules react to give ions). The reaction can be written

$$H_2O(l) + H_2O(l) \rightleftharpoons H_3O^+(aq) + OH^-(aq)$$

Because of this equilibrium, the hydronium ion, H_3O^+, and the hydroxide ion, OH^-, assume special significance in aqueous solution. Note that the hydronium ion, which might be thought of as H^+ bonded to H_2O, is often simply written $H^+(aq)$ and called the hydrogen ion.

In modern terms, the Arrhenius definition of an **acid** is *any substance that, when dissolved in water, increases the concentration of hydronium ion,* H_3O^+ *(hydrogen ion,* H^+*).* For example, when hydrogen chloride gas is dissolved in water, the following reaction occurs:

$$HCl(aq) + H_2O(l) \longrightarrow H_3O^+(aq) + Cl^-(aq)$$

Because the reaction produces $H_3O^+(aq)$, the solution is acidic and is called hydrochloric acid.

According to Arrhenius, a **base** is *any substance that, when dissolved in water, increases the concentration of hydroxide ion,* $OH^-(aq)$. Sodium hydroxide, NaOH, is an ionic solid that dissolves in water to give hydroxide ions. Thus, sodium hydroxide is a base.

$$NaOH(s) \xrightarrow{H_2O} Na^+(aq) + OH^-(aq)$$

Ammonia, NH_3, is a base in the Arrhenius view, because it yields hydroxide ion when it reacts with water.●

● Solutions of ammonia are sometimes called ammonium hydroxide and given the formula $NH_4OH(aq)$, based on analogy with NaOH. No NH_4OH molecule or compound has ever been found, however.

$$NH_3(aq) + H_2O(l) \rightleftharpoons NH_4^+(aq) + OH^-(aq)$$

When sodium hydroxide is added to hydrochloric acid, the properties of the base and the acid are lost, or *neutralized*. The explanation is that the hydronium ion from the acid reacts with the hydroxide ion from the base.

$$\text{(H)}_3O^+(aq) + OH^-(aq) \longrightarrow H_2O(l) + H_2O(l)$$

Note that the reaction involves the transfer of a proton (H^+) from the hydronium ion to the hydroxide ion. Similarly, hydrochloric acid reacts with the base ammonia. In this case, hydronium ion in the hydrochloric acid solution reacts with ammonia.

$$\text{(H)}_3O^+(aq) + NH_3(aq) \longrightarrow H_2O(l) + NH_4^+(aq)$$

Again, note that the reaction involves the transfer of a proton.

In 1923, Johannes N. Brønsted and Thomas M. Lowry independently realized that many reactions involve the transfer of a proton between reactants and that this proton-transfer aspect of a reaction could be used as the basis of a more general view of acid-base behavior. They defined an **acid** as *the species (molecule or ion) that donates a proton to another species in a proton-transfer reaction.* They defined a **base** as *the species (molecule or ion) that accepts a proton in a proton-transfer reaction.* Consider the reaction of hydrochloric acid with ammonia, NH_3. Hydronium ion in the hydrochloric acid solution donates a proton to ammonia. Thus, hydronium ion is the acid and ammonia is the base.

$$\underset{\text{acid}}{H_3O^+(aq)} + \underset{\text{base}}{NH_3(aq)} \longrightarrow H_2O(l) + NH_4^+(aq)$$

The Arrhenius definitions and those of Brønsted and Lowry are generally equivalent for aqueous solutions, though their points of view are different. For instance, NaOH and NH_3 are bases in the Arrhenius view because they increase the

Table 3.2 Common Strong Acids and Bases

Strong Acids	Strong Bases
$HClO_4$	LiOH
H_2SO_4	NaOH
HI	KOH
HBr	$Ca(OH)_2$
HCl	$Sr(OH)_2$
HNO_3	$Ba(OH)_2$

OH^- concentration in aqueous solution. They are bases in the Brønsted-Lowry view because they provide species (OH^- and NH_3, respectively) that can accept protons.●

● We will discuss the Brønsted-Lowry concept of acids and bases more thoroughly in Chapter 13.

Strong and Weak Acids and Bases

Acids and bases are classified as strong or weak. A **strong acid** is *an acid that ionizes completely in water* (that is, reacts completely to give ions). For example,

$$HCl(aq) + H_2O(l) \longrightarrow H_3O^+(aq) + Cl^-(aq)$$
$$HNO_3(aq) + H_2O(l) \longrightarrow H_3O^+(aq) + NO_3^-(aq)$$

Table 3.2 lists the common strong acids. Most of the other acids we will discuss are weak acids. A **weak acid** is *an acid that is only partly ionized as the result of an equilibrium reaction with water*. An example is hydrocyanic acid, HCN(*aq*).

$$HCN(aq) + H_2O(l) \rightleftharpoons H_3O^+(aq) + CN^-(aq)$$

A **strong base** is *a base that is present in aqueous solution entirely as ions, one of which is* OH^-. The ionic compound sodium hydroxide, NaOH, is an example. It dissolves in water as Na^+ and OH^-.

$$NaOH(s) \xrightarrow{H_2O} Na^+(aq) + OH^-(aq)$$

The common strong bases are hydroxides of Group IA and IIA elements, except for beryllium (see Table 3.2). A **weak base** is *a base that is only partly ionized as the result of an equilibrium reaction with water*. Ammonia, NH_3, is an example.

$$NH_3(aq) + H_2O(l) \rightleftharpoons NH_4^+(aq) + OH^-(aq)$$

Example 3.3 Classifying Acids and Bases as Strong or Weak

Label each of the following as a strong or weak acid or base: (a) LiOH, (b) HCN, (c) HBr, (d) HNO_2.

Solution

(a) As noted in Table 3.2, **LiOH is a strong base.**
(b) The strong acids are listed in Table 3.2. The others we encounter are weak. Thus, **HCN is a weak acid.**
(c) **HBr is a strong acid,** as noted in Table 3.2.
(d) HNO_2 is not one of the strong acids listed in Table 3.2; we can assume that it **is a weak acid.**

Exercise 3.3

Label each of the following as a strong or weak acid or base: (a) $HC_2H_3O_2$, (b) HClO, (c) HNO_3, (d) $Sr(OH)_2$. (See Problems 3.35 and 3.36.)

3.5 MOLECULAR AND IONIC EQUATIONS

Earlier we noted that many of the reactions that occur in aqueous solution involve ions. Precipitation reactions are simple examples. Consider the preparation of precipitated calcium carbonate, $CaCO_3$. This white, fine powdery compound is used as a paper filler (to brighten it and to retain ink), an antacid (Tums is a trade name), and in toothpastes (as a mild abrasive). One way to prepare this compound is to react calcium hydroxide with sodium carbonate.

We can write the equation for the reaction as follows:

$$Ca(OH)_2(aq) + Na_2CO_3(aq) \longrightarrow CaCO_3(s) + 2NaOH(aq)$$

We call this a **molecular equation,** *an equation in which the substances are written as if they were molecular substances, even though they may actually exist in solution as ions.* The molecular equation is useful because it is explicit about what solutions have been added and what products are obtained. Moreover, as we will see in the next chapter, molecular equations are useful when we want to calculate the masses of reactants needed and the masses of the products that can be obtained.

The molecular equation for the reaction of $Ca(OH)_2$ and Na_2CO_3 solutions, however, does not tell us that the reaction actually involves ions in solution. When solid calcium hydroxide dissolves in water, it goes into solution as Ca^{2+} and OH^- ions. Each formula unit of $Ca(OH)_2$ forms one Ca^{2+} ion and two OH^- ions. Thus, it would be more descriptive to write $Ca^{2+}(aq) + 2OH^-(aq)$ in place of $Ca(OH)_2(aq)$ in the previous equation. Similarly, we could write $2Na^+(aq) + CO_3^{2-}(aq)$ in place of $Na_2CO_3(aq)$ and write $Na^+(aq) + OH^-(aq)$ in place of $NaOH(aq)$. Then the equation becomes

$$Ca^{2+}(aq) + 2OH^-(aq) + 2Na^+(aq) + CO_3^{2-}(aq) \longrightarrow CaCO_3(s) + 2Na^+(aq) + 2OH^-(aq)$$

This is an example of an **ionic equation,** which is *a chemical equation for a reaction involving ions in solution in which soluble substances (ones that dissolve readily) are represented by the formulas of the predominant species in that solution.* The rules for converting molecular equations to ionic equations follow:

1. Ionic substances indicated in the molecular equation as dissolved in solution, such as $NaCl(aq)$, are normally written as ions.
2. Ionic substances that are insoluble (do not dissolve readily) either as reactants or products (precipitates) are represented by formulas of the compounds.
3. Molecular substances that are strong electrolytes, such as strong acids, are written as ions. Thus, $HCl(aq)$ is written as $H_3O^+(aq) + Cl^-(aq)$ or as $H^+(aq) + Cl^-(aq)$.
4. Molecular substances that are weak electrolytes or nonelectrolytes are represented by their molecular formulas.

Note that in the ionic equation we have just written, some ions appear on both sides of the arrow. These are called **spectator ions;** they are *ions in an ionic equation that do not take part in the reaction.* We can cancel them from both sides of the equation.

$$Ca^{2+}(aq) + \cancel{2OH^-(aq)} + \cancel{2Na^+(aq)} + CO_3^{2-}(aq) \longrightarrow CaCO_3(s) + \cancel{2Na^+(aq)} + \cancel{2OH^-(aq)}$$

The resulting equation is

$$Ca^{2+}(aq) + CO_3^{2-}(aq) \longrightarrow CaCO_3(s)$$

This is the **net ionic equation,** *an ionic equation from which spectator ions have been canceled*. It shows the essential reaction that occurs. In this case, calcium ions and carbonate ions come together to form solid calcium carbonate.

We can see from the net ionic equation that mixing any solution of calcium ion with any solution of carbonate ion will give this same reaction. For example, calcium nitrate, $Ca(NO_3)_2$, dissolves readily in water. Because it is an ionic compound, it goes into solution as ions, Ca^{2+} and NO_3^-. Similarly, potassium carbonate, K_2CO_3, dissolves in water to give K^+ and CO_3^{2-}. The molecular equation representing the reaction of calcium nitrate with potassium carbonate is

$$Ca(NO_3)_2(aq) + K_2CO_3(aq) \longrightarrow CaCO_3(s) + 2KNO_3(aq)$$

The ionic equation is

$$Ca^{2+}(aq) + \cancel{2NO_3^-(aq)} + \cancel{2K^+(aq)} + CO_3^{2-}(aq) \longrightarrow CaCO_3(s) + \cancel{2K^+(aq)} + \cancel{2NO_3^-(aq)}$$

and the net ionic equation is

$$Ca^{2+}(aq) + CO_3^{2-}(aq) \longrightarrow CaCO_3(s)$$

The net ionic equation is identical to the one obtained from the reaction of $Ca(OH)_2$ and Na_2CO_3. The value of the net ionic equation is in its generality. Seawater contains Ca^{2+} and CO_3^{2-} ions from various sources. But whatever the sources of these ions, we still know that sediments of calcium carbonate—and eventually limestone—can form from them.

Example 3.4 Writing Net Ionic Equations

Write a net ionic equation for each of the following molecular equations. Recall the six strong acids listed in Table 3.2.

(a) $2HClO_4(aq) + Ca(OH)_2(aq) \longrightarrow Ca(ClO_4)_2(aq) + 2H_2O(l)$

(b) $HC_2H_3O_2(aq) + NaOH(aq) \longrightarrow NaC_2H_3O_2(aq) + H_2O(l)$

Solution

(a) Perchloric acid, $HClO_4$, is a strong acid and is written in the ionic equation as $H^+(aq)$ and $ClO_4^-(aq)$. (Alternatively, we could write it as H_3O^+ and ClO_4^-, after adding an extra H_2O to the right.) Calcium hydroxide, $Ca(OH)_2$, and calcium perchlorate, $Ca(ClO_4)_2$, are dissolved ionic substances and are also written as ions. Water is a weak electrolyte, so it is written in molecular form.

$$2H^+(aq) + \cancel{2ClO_4^-(aq)} + \cancel{Ca^{2+}(aq)} + 2OH^-(aq) \longrightarrow \cancel{Ca^{2+}(aq)} + \cancel{2ClO_4^-(aq)} + 2H_2O(l)$$

After canceling spectator ions and dividing by 2, we get the following net ionic equation:

$$\mathbf{H^+(aq) + OH^-(aq) \longrightarrow H_2O(l)}$$

(b) Acetic acid, $HC_2H_3O_2$, is not one of the strong acids listed in Table 3.2. Therefore, we can assume it is a weak acid and represent it in the ionic equation by its molecular formula. Sodium hydroxide, NaOH, and sodium acetate, $NaC_2H_3O_2$, are ionic compounds and exist in solution as ions. The ionic equation is

$$HC_2H_3O_2(aq) + \cancel{Na^+(aq)} + OH^-(aq) \longrightarrow \cancel{Na^+(aq)} + C_2H_3O_2^-(aq) + H_2O(l)$$

and the net ionic equation is

$$\mathbf{HC_2H_3O_2(aq) + OH^-(aq) \longrightarrow C_2H_3O_2^-(aq) + H_2O(l)}$$

Exercise 3.4

Write ionic and net ionic equations for each of the following molecular equations.

(a) $2HNO_3(aq) + Mg(OH)_2(s) \longrightarrow 2H_2O(l) + Mg(NO_3)_2(aq)$
(b) $Pb(NO_3)_2(aq) + Na_2SO_4(aq) \longrightarrow PbSO_4(s) + 2NaNO_3(aq)$

(See Problems 3.37 and 3.38.)

Reactions in Aqueous Solution: Precipitation and Acid-Base Reactions

Among the most important reactions in aqueous solution are *precipitation reactions* and acid-base reactions (that is, reactions of acids and bases, including *neutralization reactions* and *reactions with gas formation*). Most of these reactions can be classified as metathesis reactions. In the concluding sections of this chapter, we will look at these kinds of reactions. We will learn how to predict when such reactions will occur.

3.6 WAYS TO DRIVE METATHESIS REACTIONS TO PRODUCTS

Consider once again the reaction of calcium hydroxide with sodium carbonate.

$$Ca(OH)_2(aq) + Na_2CO_3(aq) \longrightarrow CaCO_3(s) + 2NaOH(aq)$$

Note that we can obtain the products from the reactant side of the molecular equation simply by interchanging the anions (or by interchanging the cations). The reaction is therefore a metathesis reaction.

When the aqueous solutions of the reactants are first mixed, we obtain a solution of four ions. That is, if we write the left side of the molecular equation as ions, we have

$$Ca^{2+}(aq) + 2OH^-(aq) + 2Na^+(aq) + CO_3^{2-}(aq)$$

For a reaction to occur, at least two of the ions have to be removed in some way from the mixture. (Otherwise, no reaction occurs; we simply have a mixture of four ions.) In this case, calcium ions and carbonate ions come together to form calcium carbonate, which precipitates from the solution. In this way, calcium ion and carbonate ion are removed from the mixture, leaving sodium ion and hydroxide ion in solution. The net ionic equation shows this removal of ions explicitly.

$$Ca^{2+}(aq) + CO_3^{2-}(aq) \longrightarrow CaCO_3(s)$$

We can say that the metathesis reaction is driven to products by the removal of ions through *precipitation*.

Precipitation is only one way in which ions can be removed from solution. Another way is through the formation of a molecular substance that is a weak electrolyte. Consider the reaction involving hydrochloric acid and sodium hydroxide discussed in the chapter opening. Let us first write the reactants. Then, assuming a metathesis reaction, we can predict the products by interchanging the anions. We obtain

$$HCl + NaOH(aq) \longrightarrow HOH(l) + NaCl(aq)$$

This is an example of a *neutralization*—that is, a reaction of an acid with a base.

The ionic equation is

$$H^+(aq) + Cl^-(aq) + Na^+(aq) + OH^-(aq) \longrightarrow H_2O(l) + Na^+(aq) + Cl^-(aq)$$

Note that we have written HCl, NaOH, and NaCl as ions because they are strong electrolytes. Water, however, is a weak electrolyte, and we leave the formula as HOH (or H_2O). The net ionic equation is

$$H^+(aq) + OH^-(aq) \longrightarrow H_2O(l)$$

Hydrogen ion and hydroxide ion react to produce a weak electrolyte, so these ions are effectively removed from the solution as the molecule H_2O.

A third way in which ions are removed and a metathesis reaction is driven to products is through the *formation of a gaseous product*. Consider the reaction of sodium sulfide with hydrochloric acid.

$$Na_2S(aq) + 2HCl(aq) \longrightarrow 2NaCl(aq) + H_2S(g)$$

Note that this is a metathesis reaction. The ionic equation is

$$2Na^+(aq) + S^{2-}(aq) + 2H^+(aq) + 2Cl^-(aq) \longrightarrow 2Na^+(aq) + 2Cl^-(aq) + H_2S(g)$$

and the net ionic equation is

$$S^{2-}(aq) + 2H^+(aq) \longrightarrow H_2S(g)$$

Sulfide ion and hydrogen ion are removed as the gas hydrogen sulfide, which leaves the reaction vessel. (Hydrogen sulfide, H_2S, is a gas with the odor of rotten eggs.) The metathesis reaction is driven by the removal of ions through the formation of a gaseous product.

In the next sections, we will look at each of these kinds of reactions: precipitations, neutralizations, and reactions with gas formation.

3.7 PRECIPITATION REACTIONS

Precipitation reactions in aqueous solution depend on the fact that one product does not dissolve readily in water. Substances vary considerably in their ability to dissolve in water. Some compounds, such as sodium chloride, dissolve readily and are said to be *soluble*. Others, such as calcium carbonate (limestone and marble are natural forms of this compound), have very limited solubilities and are said to be *insoluble*. In order to predict whether a precipitation is likely to occur, it is necessary to know whether the potential products are soluble or insoluble. Table 3.3 lists rules for predicting the solubilities of common ionic compounds.

Let us now consider how we go about predicting whether a precipitation reaction will occur. Suppose we mix solutions of two soluble ionic compounds—for example, nickel(II) chloride and sodium phosphate. Do we expect a precipitate to form? To answer this question, let us first write the molecular equation we expect for a metathesis reaction of these compounds. To do this, we write the formulas of the reactants and then those of the products that would be obtained if the anions were interchanged. After balancing the equation, we have

$$3NiCl_2 + 2Na_3PO_4 \longrightarrow Ni_3(PO_4)_2 + 6NaCl$$

Let us verify that $NiCl_2$ and Na_3PO_4 are soluble and then check the solubilities of the products. From Table 3.3, we note that chlorides are soluble. (The listed exceptions do not include nickel(II) chloride.) Thus, nickel(II) chloride is soluble. We also see that sodium compounds are soluble. We expect Na_3PO_4 and NaCl to

Table 3.3 Empirical Rules for the Solubilities of Common Ionic Compounds

Soluble Compounds	Exceptions
Sodium, potassium, and ammonium compounds	
Acetates and nitrates	
Halides (chlorides, bromides, and iodides)	Lead(II), silver, and mercury(I) halides are insoluble
Sulfates	Calcium, strontium, barium, and lead(II) sulfates are insoluble

Insoluble Compounds	Exceptions
Carbonates and phosphates	Sodium, potassium, and ammonium compounds are soluble
Hydroxides	Sodium, potassium, and calcium compounds are soluble
Sulfides	Sodium, potassium, calcium, and ammonium compounds are soluble

be soluble. However, most metal phosphates are insoluble. Exceptions are those of sodium, potassium, and ammonium. Thus nickel(II) phosphate, $Ni_3(PO_4)_2$, is insoluble.

Now we can append the appropriate labels to the compounds in the foregoing equation.

$$3NiCl_2(aq) + 2Na_3PO_4(aq) \longrightarrow Ni_3(PO_4)_2(s) + 6NaCl(aq)$$

The reaction occurs because nickel(II) phosphate is insoluble (Figure 3.13). It forms a pale yellow-green precipitate that can be filtered from the solution of

Figure 3.13
Reaction of nickel(II) chloride and sodium phosphate. Sodium phosphate solution is added to nickel(II) chloride solution (green color). A pale yellow-green precipitate of nickel(II) phosphate forms.

● The reactants $NiCl_2$ and Na_3PO_4 must be added in correct amounts; otherwise, the excess reactant will occur along with the product NaCl.

remaining ions. If the filtrate (the solution passing through the filter) is evaporated, sodium chloride crystals form.●

To obtain the net ionic equation for this reaction, we first write the ionic equation and cancel spectator ions.

$$3Ni^{2+}(aq) + \cancel{6Cl^-(aq)} + \cancel{6Na^+(aq)} + 2PO_4^{3-}(aq) \longrightarrow Ni_3(PO_4)_2(s) + \cancel{6Na^+(aq)} + \cancel{6Cl^-(aq)}$$

The net ionic equation is

$$3Ni^{2+}(aq) + 2PO_4^{3-}(aq) \longrightarrow Ni_3(PO_4)_2(s)$$

This equation represents the essential reaction that occurs: Ni^{2+} and PO_4^{3-} ions in aqueous solution react to form the solid nickel(II) phosphate.

Example 3.5 Deciding Whether a Precipitation Will Occur

For each of the following, decide whether a precipitation reaction occurs. If it does, write the balanced molecular equation. Then write the net ionic equation. If no reaction occurs, write *NR* after the arrow

(a) $NaCl + Fe(NO_3)_2 \longrightarrow$
(b) $Al_2(SO_4)_3 + NaOH \longrightarrow$

Solution

(a) To obtain the possible metathesis reaction, we write down the reactants and interchange anions to get the products. (Remember to write the correct formulas for the compounds after making this interchange.)

$$NaCl + Fe(NO_3)_2 \longrightarrow NaNO_3 + FeCl_2 \quad \text{(not balanced)}$$

Referring to Table 3.3, we see that all the compounds in this equation are soluble. Thus, no reaction occurs.

$$NaCl(aq) + Fe(NO_3)_2(aq) \longrightarrow NR$$

(b) To obtain the metathesis equation, we write down the reactants and interchange anions to get the products.

$$Al_2(SO_4)_3 + NaOH \longrightarrow Al(OH)_3 + Na_2SO_4 \quad \text{(not balanced)}$$

From Table 3.3, we see that sulfates are soluble and hydroxides are insoluble (except as noted). Thus, $Al_2(SO_4)_3$ is soluble and $Al(OH)_3$ is insoluble, and we predict that precipitation occurs. We write the balanced equation with the labels *aq* and *s* to make it clear which substances are soluble and which precipitate. The balanced molecular equation is

$$Al_2(SO_4)_3(aq) + 6NaOH(aq) \longrightarrow 2Al(OH)_3(s) + 3Na_2SO_4(aq)$$

Check to see that the formulas of ionic compounds are correctly written.

To get the net ionic equation, we write the soluble compounds as ions in aqueous solution and cancel spectator ions.

$$2Al^{3+}(aq) + \cancel{3SO_4^{2-}(aq)} + \cancel{6Na^+(aq)} + 6OH^-(aq) \longrightarrow 2Al(OH)_3(s) + \cancel{6Na^+(aq)} + \cancel{3SO_4^{2-}(aq)}$$

The net ionic equation is

$$Al^{3+}(aq) + 3OH^-(aq) \longrightarrow Al(OH)_3(s)$$

Exercise 3.5

Aqueous solutions of sodium iodide and lead acetate are mixed. If a reaction occurs, write the molecular equation and the net ionic equation. If no reaction occurs, write *NR* after the arrow.

(See Problems 3.43, 3.44, 3.45, and 3.46.)

3.8 NEUTRALIZATION REACTIONS

A **neutralization reaction** is *a reaction of an acid and a base that results in an ionic compound and possibly water*. When a base is added to an acid solution, the acid is said to be neutralized. *The ionic compound that is a product of a neutralization reaction* is called a **salt.** Most ionic compounds other than hydroxides and oxides are salts; that is, they can be obtained from neutralization reactions.

The following are neutralization reactions:

$$\underset{\text{acid}}{2HCl(aq)} + \underset{\text{base}}{Ca(OH)_2(aq)} \longrightarrow \underset{\text{salt}}{CaCl_2(aq)} + 2H_2O(l)$$

$$\underset{\text{acid}}{HCN(aq)} + \underset{\text{base}}{KOH(aq)} \longrightarrow \underset{\text{salt}}{KCN(aq)} + H_2O(l)$$

These reactions have been written as molecular equations. Written in this form, the equation makes explicit what compounds are involved as reactants and what salt is produced. However, if we want to discuss the fundamental reaction that occurs, we need to write these as net ionic equations. The first reaction involves a strong acid, $HCl(aq)$, and a strong base, $Ca(OH)_2(aq)$. After writing $H^+(aq)$ from the strong acid as $H_3O^+(aq)$, the net ionic equation is

$$H_3O^+(aq) + OH^-(aq) \longrightarrow H_2O(l) + H_2O(l)$$

The second reaction involves a weak acid, $HCN(aq)$, and a strong base, $KOH(aq)$. The net ionic equation is

$$HCN(aq) + OH^-(aq) \longrightarrow CN^-(aq) + H_2O(l)$$

It is now clear that both reactions involve the transfer of a proton from one species to another, as we expect for acid-base reactions.

The reaction of an acid, such as sulfuric acid, with ammonia is also a neutralization reaction.

$$\underset{\text{acid}}{H_2SO_4(aq)} + \underset{\text{base}}{2NH_3(aq)} \longrightarrow \underset{\text{salt}}{(NH_4)_2SO_4(aq)}$$

The corresponding net ionic equation is

$$H^+(aq) + NH_3(aq) \longrightarrow NH_4^+(aq)$$

or

$$H_3O^+(aq) + NH_3(aq) \longrightarrow H_2O(l) + NH_4^+(aq)$$

Although the molecular equation does not have the form of a metathesis reaction, the net ionic equation shows that the reaction is nevertheless an acid-base (proton-transfer) reaction.

Example 3.6 Writing an Equation for a Neutralization

Write the molecular equation and then the net ionic equation for the neutralization of nitrous acid, HNO_2, by sodium hydroxide, NaOH, in aqueous solution.

Solution

The molecular equation for the neutralization is

$$HNO_2(aq) + NaOH(aq) \longrightarrow H_2O(l) + NaNO_2(aq)$$

To obtain the ionic equation, we note that nitrous acid is a weak acid (it is not one of the strong acids listed in Table 3.2). Both NaOH and $NaNO_2$ are ionic compounds and strong electrolytes. Therefore, the ionic equation is

$$HNO_2(aq) + \cancel{Na^+(aq)} + OH^-(aq) \longrightarrow H_2O(l) + \cancel{Na^+(aq)} + NO_2^-(aq)$$

The net ionic equation is

$$HNO_2(aq) + OH^-(aq) \longrightarrow H_2O(l) + NO_2^-(aq)$$

Exercise 3.6

Write the molecular equation and net ionic equation for the neutralization of hydrocyanic acid, HCN, by lithium hydroxide, LiOH. The salt is soluble.

(See Problems 3.47, 3.48, 3.49, and 3.50.)

Table 3.4 Ionic Compounds That Evolve Gases When Treated with Acids

Ionic Compound	Gas	Example
Carbonate (CO_3^{2-})	CO_2	$Na_2CO_3 + 2HCl \longrightarrow 2NaCl + H_2O + CO_2$
Sulfide (S^{2-})	H_2S	$Na_2S + 2HCl \longrightarrow 2NaCl + H_2S$
Sulfite (SO_3^{2-})	SO_2	$Na_2SO_3 + 2HCl \longrightarrow 2NaCl + H_2O + SO_2$

● The structural formula of acetic acid is

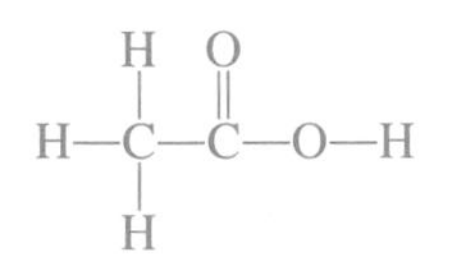

Only the H atom attached to O is acidic.

Acids such as HCl and HNO_3 that have only one acidic hydrogen atom per acid molecule are called *monoprotic* acids. Acetic acid, $HC_2H_3O_2$, is a monoprotic acid. Although the molecule has four H atoms, only one atom is acidic.● A **polyprotic acid** is *an acid that yields two or more acidic hydrogens per molecule*. An example is sulfuric acid, H_2SO_4, which is said to be *diprotic*. Phosphoric acid, H_3PO_4, is another example; it is a *triprotic* acid. By reacting a polyprotic acid with different amounts of a base, we can obtain a series of salts. Thus,

$$H_3PO_4(aq) + NaOH(aq) \longrightarrow NaH_2PO_4(aq) + H_2O(l)$$
$$H_3PO_4(aq) + 2NaOH(aq) \longrightarrow Na_2HPO_4(aq) + 2H_2O(l)$$
$$H_3PO_4(aq) + 3NaOH(aq) \longrightarrow Na_3PO_4(aq) + 3H_2O(l)$$

Salts such as NaH_2PO_4 and Na_2HPO_4 that have acidic hydrogen atoms and can undergo neutralization with bases are called *acid salts*.

Exercise 3.7

Write molecular and net ionic equations for the successive neutralizations of each acidic hydrogen of sulfuric acid with potassium hydroxide. (That is, write equations for the reaction of sulfuric acid with KOH to give the acid salt and for the reaction of the acid salt with more KOH to give potassium sulfate.) (See Problems 3.51, 3.52, 3.53, and 3.54.)

3.9 REACTIONS WITH GAS FORMATION

As we noted earlier, some reactions are driven to products by the formation of a gas. We considered the reaction of sodium sulfide with hydrochloric acid. The net ionic reaction was that of sulfide ion with hydrogen ion to produce hydrogen sulfide gas. Thus, soluble sulfides react with acids to produce hydrogen sulfide. Even insoluble sulfides may react if the acid is a strong one. Some reactions involving gas formation are listed in Table 3.4.

Note that carbonates and sulfites also react with acids to produce gaseous products. In these cases, unstable weak acids form, which decompose to give gaseous products. Consider the reaction of sodium carbonate with hydrochloric acid. We would write the following metathesis reaction:

$$Na_2CO_3(aq) + 2HCl(aq) \longrightarrow 2NaCl(aq) + H_2CO_3(aq)$$

The product carbonic acid, H_2CO_3, is unstable and decomposes to water and carbon dioxide gas.

$$H_2CO_3(aq) \longrightarrow H_2O(l) + CO_2(g)$$

The overall result of the reaction, written as a molecular equation, is

$$Na_2CO_3(aq) + 2HCl(aq) \longrightarrow 2NaCl(aq) + \underbrace{H_2O(l) + CO_2(g)}_{H_2CO_3(aq)}$$

Figure 3.14
Reaction of a carbonate with an acid. Baking soda (sodium hydrogen carbonate) reacts with acetic acid in vinegar to evolve bubbles of carbon dioxide.

The net ionic equation is

$$CO_3^{2-}(aq) + 2H^+(aq) \longrightarrow H_2O(l) + CO_2(g)$$

Thus, a carbonate compound reacts with an acid to evolve bubbles of carbon dioxide gas (Figure 3.14). This reaction is used as a simple test for carbonate minerals, such as limestone, which fizz when treated with hydrochloric acid.

Sulfite compounds react with acids in a similar fashion. The reaction gives sulfurous acid, H_2SO_3, which decomposes immediately to give water and sulfur dioxide gas. For example, when sodium sulfite is treated with hydrochloric acid, the characteristic odor of sulfur dioxide is noted.

$$Na_2SO_3(aq) + 2HCl(aq) \longrightarrow 2NaCl(aq) + \underbrace{H_2O(l) + SO_2(g)}_{H_2SO_3(aq)}$$

The net ionic equation is

$$SO_3^{2-}(aq) + 2H^+(aq) \longrightarrow H_2O(l) + SO_2(g)$$

This reaction may be used as a simple laboratory preparation for small quantities of sulfur dioxide. (Large quantities are best obtained from cylinders of liquefied SO_2.) It should be noted that this reaction and the others discussed in this section involve the transfer of a proton. Thus, they are all acid-base reactions.●

● CO_3^{2-}, S^{2-}, and SO_3^{2-} are bases. Thus, the net equations involve a basic anion with H^+, an acid.

Example 3.7 Writing an Equation for a Reaction with Gas Formation

Write the molecular equation and the net ionic equation for the reaction of zinc sulfide with hydrochloric acid.

Solution

Consider the metathesis reaction:

$$ZnS + 2HCl \longrightarrow ZnCl_2 + H_2S$$

Note from the solubility rules that ZnS is insoluble and $ZnCl_2$ is soluble. With phase labels, the molecular equation is

$$ZnS(s) + 2HCl(aq) \longrightarrow ZnCl_2(aq) + H_2S(g)$$

The ionic equation is

$$ZnS(s) + 2H^+(aq) + \cancel{2Cl^-(aq)} \longrightarrow Zn^{2+}(aq) + \cancel{2Cl^-(aq)} + H_2S(g)$$

and the net ionic equation is

$$\mathbf{ZnS(s) + 2H^+(aq) \longrightarrow Zn^{2+}(aq) + H_2S(g)}$$

Exercise 3.8

Write the molecular equation and the net ionic equation for the reaction of calcium carbonate with nitric acid. (See Problems 3.57 and 3.58.)

Profile of a Chemical

HYDROCHLORIC ACID (a Strong Acid)

Three acids are commonly found on laboratory shelves: sulfuric acid, nitric acid, and hydrochloric acid. All three are strong acids, important commercially and on the list of the top 50 industrial chemicals. All three were known to early chemists as the mineral acids, because they were prepared from various minerals.

Hydrochloric acid is an aqueous solution of hydrogen chloride, HCl, a colorless, corrosive gas. The commercially available acid is a concentrated solution that fumes when exposed to the air. The fumes form when hydrogen chloride gas escapes from the solution and attracts water molecules, giving small droplets of hydrochloric acid. Care must be taken not to breathe these fumes when opening a bottle of the concentrated acid. The acid used in general chemistry laboratories is almost always a dilute solution, which does not have this property of fuming.

Hydrochloric acid is available in hardware stores as *muriatic acid* (it may have a yellow color from iron compounds, but the pure acid is colorless). Muriatic acid is an old name that derives from the Latin *muria* (for seawater), because the original commercial method of preparation frequently used sea salt, although rock salt (which is also sodium chloride) can be used.

This preparation from salt, first described by the German chemist Johann Rudolf Glauber in 1648, consists of heating sodium chloride with a solution of sulfuric acid. At low temperatures (about 150°C), the reaction is

$$NaCl(s) + H_2SO_4(aq) \xrightarrow{\Delta} NaHSO_4(s) + HCl(g)$$

As Figure 3.15 shows, hydrogen chloride gas evolves from a mixture of sodium chloride and sulfuric acid. Note that the reaction can be classified as a metathesis reaction that is driven to products because a gas leaves the mixture. At higher temperatures, sodium chloride reacts with the sodium hydrogen sulfate produced in the previous reaction to give more hydrogen chloride.

$$NaCl(s) + NaHSO_4(s) \xrightarrow{\Delta} Na_2SO_4(s) + HCl(g)$$

Hydrogen chloride can also be prepared by the direct combination of the elements. Today, however, most hydrogen chloride is obtained as a by-product of the preparation of chlorinated hydrocarbons, including certain plastics, insecticides, and refrigerants.

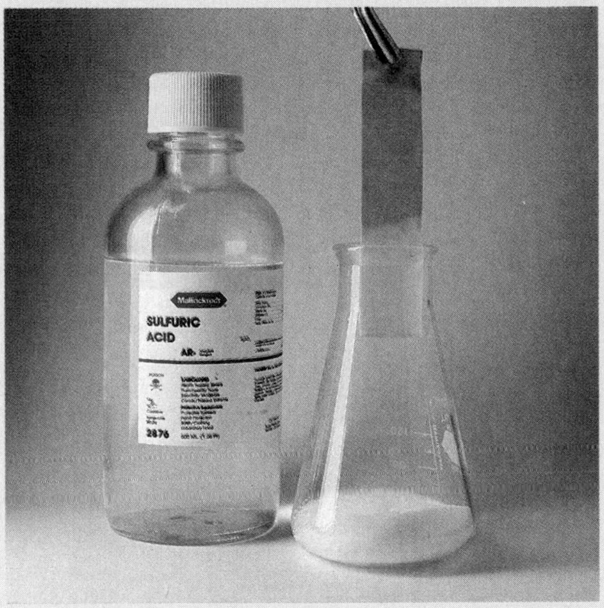

Figure 3.15
Preparation of hydrogen chloride. *Concentrated sulfuric acid was added to sodium chloride crystals in the flask, producing hydrogen chloride gas. Bromthymol-blue indicator paper changes to yellow where it is exposed to hydrogen chloride.*

To produce hydrochloric acid, hydrogen chloride gas is directed into the bottom of a tower in which water flows downward over an inert packing material. The gas dissolves in the water, and hydrochloric acid flows out the bottom of the tower.

One of the uses of the muriatic acid sold in hardware stores is to clean mortar from brick. The reactions are acid-base type. New mortar is calcium hydroxide, which reacts with hydrochloric acid.

$$2HCl(aq) + Ca(OH)_2(s) \longrightarrow CaCl_2(aq) + 2H_2O(l)$$

Mortar "sets" by reacting with carbon dioxide in the air to

(continued)

produce calcium carbonate, which also reacts with hydrochloric acid.

$$2HCl(aq) + CaCO_3(s) \longrightarrow CaCl_2(aq) + H_2O(l) + CO_2(g)$$

In either case, a solid material reacts to give a solution.

The largest single commercial use of hydrochloric acid is in the "pickling" of steel (iron alloy). Pickling consists of dipping steel into an acid bath to remove rust from its surface before processing the steel into finished goods. Rust is approximately represented as Fe_2O_3, and the reaction of rust with hydrochloric acid can be written

$$Fe_2O_3(s) + 6HCl(aq) \longrightarrow 2FeCl_3(aq) + 3H_2O(l)$$

Iron metal also reacts with hydrochloric acid but more slowly, so that the rust can be removed from steel without significant loss of iron. Iron, like most other metals, reacts with hydrochloric acid in a displacement reaction.

$$Fe(s) + 2HCl(aq) \longrightarrow FeCl_2(aq) + H_2(g)$$

The food industry uses hydrochloric acid to make corn syrup from corn starch and to make gelatin from bones. These reactions involve the breaking down of large molecules to smaller ones by the action of the acid. Interestingly, hydrochloric acid is present in the gastric juice of the stomach, where it aids in the digestion or breakdown of foods, including the changing of starches to sugars. The stomach is normally protected from the action of hydrochloric acid and other digestive substances by a thick mucous lining. If something happens to this protective lining, an ulcer can result.

Questions for Study

1. What is hydrochloric acid? List several properties of hydrochloric acid.
2. What is a common but older name for hydrochloric acid? What does this name derive from?
3. Give chemical equations for the preparation of hydrogen chloride from sodium chloride. Classify each by type of chemical reaction.
4. What is the major source of hydrogen chloride today?
5. What chemical reactions are involved in the cleaning of mortar from bricks using hydrochloric acid?
6. Write an equation for the reaction of iron with hydrochloric acid. What type of reaction is this?
7. List several uses of hydrochloric acid.

Profile of a Chemical

SODIUM HYDROXIDE (a Strong Base)

Sodium hydroxide (or lye), NaOH, can be found in drain cleaners and oven cleaners. Its use in these products is based on the fact that a strong base will break down grease and hair to soluble compounds. Drain cleaners and similar products must be handled with extreme care, because the sodium hydroxide reacts with skin just as it does with grease and hair.

Sodium hydroxide is available in the form of white crystals or as pellets. It must be stored in tightly stoppered containers; otherwise, the solid becomes wet with moisture absorbed from the atmosphere. In addition, sodium hydroxide reacts with carbon dioxide in the air, contaminating the solid with sodium hydrogen carbonate and sodium carbonate.

$$NaOH(s) + CO_2(g) \longrightarrow NaHCO_3(s)$$
$$2NaOH(s) + CO_2(g) \longrightarrow Na_2CO_3(s) + H_2O(l)$$

At one time, sodium hydroxide, known commercially as caustic soda, was prepared by a precipitation reaction using calcium hydroxide (lime) and sodium carbonate (soda ash) as reactants.

$$Ca(OH)_2(aq) + Na_2CO_3(aq) \longrightarrow CaCO_3(s) + 2NaOH(aq)$$

This lime-soda process, however, is no longer competitive with the electrolytic method. In the electrolytic method, a direct current is passed through an aqueous solution of sodium chloride. Hydrogen gas is released at the negative pole and chlorine gas is released at the positive pole. The solution at the end of the electrolysis contains sodium hydroxide (Figure 3.16). The overall reaction is

$$2NaCl(aq) + 2H_2O(l) \xrightarrow{\text{electrolysis}} H_2(g) + Cl_2(g) + 2NaOH(aq)$$

Evaporation of the solution after electrolysis yields crystals of sodium hydroxide. Chlorine is prepared commercially in the same process.

Note that some drain cleaner products contain pieces of aluminum along with the sodium hydroxide crystals. When water comes in contact with the sodium

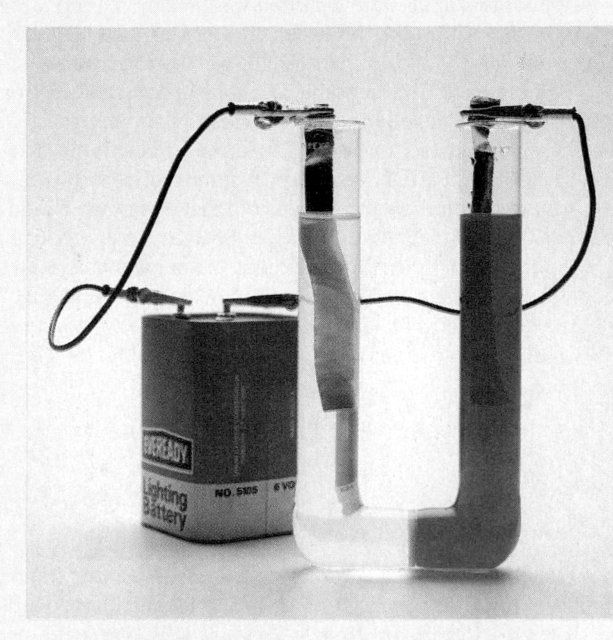

Figure 3.16
Preparation of sodium hydroxide by electrolysis. *A solution of sodium chloride is electrolyzed. Note the pink color at the negative electrode. Phenolphthalein indicator turns pink from the hydroxide ion formed at the electrode. Later, if the solution is evaporated, sodium hydroxide crystallizes out.*

hydroxide, the resulting solution reacts with the aluminum, and hydrogen gas is produced.

$$2Al(s) + 2NaOH(aq) + 6H_2O(l) \longrightarrow 2Na[Al(OH)_4](aq) + 3H_2(g)$$

In this reaction, the aluminum forms an anion, $Al(OH)_4^-$. Only metals with some nonmetallic character (those metals in the periodic table close to the staircase line that separates metals from nonmetals) react with bases in this way. The reaction produces fizzing and bubbling of the cleaner.

Major uses of sodium hydroxide include the preparation of soaps and other detergents, paper making, purification of aluminum ore, and petroleum refining.

Questions for Study

1. List some household uses for sodium hydroxide. Why are they effective?
2. Why must sodium hydroxide be kept in tightly stoppered containers?
3. Give equations for two methods of preparing sodium hydroxide.
4. Sodium hydroxide solutions should not be stored in aluminum cans. Why?
5. List some major commercial uses of sodium hydroxide.

A Checklist for Review

IMPORTANT TERMS

chemical equation (3.1)
reactant (3.1)
product (3.1)
combination reaction (3.2)
decomposition reaction (3.2)
displacement reaction (single-replacement reaction) (3.2)
metathesis reaction (double-replacement reaction) (3.2)
precipitate (3.2)
combustion reaction (3.2)
electrolyte (3.3)
nonelectrolyte (3.3)
strong electrolyte (3.3)
weak electrolyte (3.3)
acid-base indicator (3.4)
acid (3.4)
base (3.4)
strong acid (3.4)
weak acid (3.4)
strong base (3.4)
weak base (3.4)
molecular equation (3.5)
ionic equation (3.5)
spectator ions (3.5)
net ionic equation (3.5)
neutralization reaction (3.8)
salt (3.8)
polyprotic acid (3.8)

SUMMARY OF FACTS AND CONCEPTS

A chemical reaction occurs when the atoms in substances rearrange and combine into new substances. We represent a reaction by a *chemical equation,* writing a chemical formula for each reactant and product. The coefficients in the equation indicate the relative numbers of reactant and product molecules or formula units. Once the reactants and products and their formulas have been determined by experiment, we determine the coefficients by *balancing* the numbers of each kind of atom on both sides of the equation.

Most of the reactions we encounter in general chemistry can be classified into the following types: *combination reactions, decomposition reactions, displacement reactions, metathesis reactions,* and *combustion reactions*.

Substances that dissolve in water to give ions are called *electrolytes*. Electrolytes that exist in solution completely as ions are called *strong electrolytes*. Electrolytes that exist in solution in an equilibrium between molecules and ions are called *weak electrolytes*. *Acids* (substances that yield hydrogen ion in aqueous solution or donate protons) and *bases* (substances that yield hydroxide ion in aqueous solution or accept protons) are important electrolytes.

We write a *molecular equation* for a reaction as an *ionic equation* by writing the strong electrolytes as ions. By canceling the *spectator ions,* we get the *net ionic equation*.

Among the most important reactions in aqueous solution are precipitation reactions and acid-base reactions, most of which are metathesis reactions. To drive a metathesis reaction to products, certain ions must be removed from the reaction solution. There are three kinds of reactions in which ions are removed and the metathesis reaction is driven to products: *precipitation reactions* (in which ions are removed as a precipitate), *neutralization reactions* (in which H^+ and OH^- ions are removed as H_2O), and *reactions with gas formation* (in which ions are removed as a gaseous product). These reactions are discussed and illustrated.

OPERATIONAL SKILLS

1. **Balancing simple equations** Given the formulas of the reactants and products in a chemical reaction, obtain the coefficients of the balanced equation (Example 3.1).
2. **Classifying chemical reactions** Given a chemical equation, classify it by type of reaction (Example 3.2).
3. **Classifying acids and bases as strong or weak** Given the formula of an acid or base, classify it as strong or weak (Example 3.3).
4. **Writing net ionic equations** Given a molecular equation, write the corresponding net ionic equation (Example 3.4).
5. **Deciding whether a precipitation will occur** Using solubility rules, decide whether two soluble ionic compounds will react to form a precipitate. If they will, write the net ionic equation (Example 3.5).
6. **Writing an equation for a neutralization** Given an acid and a base, write the molecular equation and then the net ionic equation for the neutralization reaction (Example 3.6).
7. **Writing an equation for a reaction with gas formation** Given the reaction between a carbonate, sulfide, or sulfite and an acid, write the molecular and net ionic equations (Example 3.7).

Review Questions

3.1 To write a balanced chemical equation, what must we know from experiment and what can we do using pencil and paper?

3.2 Consider the following chemical reaction, written symbolically

$$N_2 + 3H_2 \longrightarrow 2NH_3$$

What does this mean in words? How many nitrogen atoms are there on the left side of the arrow? How many are on the right? Are there the same number on the right as on the left? Answer the same questions for hydrogen.

3.3 What is the basis of the traditional scheme for classifying chemical reactions? What are the different types of reactions?

3.4 Explain the electrical conductivity of an electrolyte solution. Use an example to illustrate.

3.5 Explain why chemical equilibrium is called a dynamic equilibrium. Use an example to illustrate.

3.6 Define the terms *strong electrolyte* and *weak electrolyte*. Give an example of each.

3.7 Define the terms *acid* and *base* and give an example of each.

3.8 What formulas are used to represent the hydronium ion in solution? Using these formulas, write equations for the ionization of water (the production of ions from water).

3.9 Explain why ammonia, NH_3, gives basic solutions in water.

3.10 What are the advantages and disadvantages of using a molecular equation to represent an ionic reaction?

3.11 What is a *spectator ion?* Illustrate with an ionic reaction.

3.12 Describe the different kinds of metathesis reactions according to the ways in which ions are removed and the reaction thereby driven to products.

3.13 Give an example of a neutralization reaction. Label the acid, base, and salt.

3.14 Give an example of a polyprotic acid and write equations for the successive neutralizations of the acidic hydrogen atoms to produce a series of salts.

3.15 The mineral magnesite fizzes when treated with hydrochloric acid. The gas released is odorless. To the extent possible, explain what reaction is occurring.

3.16 Give an example of each of the following reactions: precipitation, neutralization, and a reaction that evolves a gas.

3.17 Describe in words how you would prepare pure crystalline AgCl and $NaNO_3$ from solid $AgNO_3$ and solid NaCl.

3.18 Describe in words how you would prepare pure crystalline $CaCl_2$ starting from solid $CaCO_3$ and HCl(*aq*).

Practice Problems

CHEMICAL EQUATIONS

3.19 How many oxygen atoms are there in $As_4O_6 + 6H_2O$?

3.20 In the equation $2PbS + O_2 \longrightarrow 2PbO + 2SO_2$, how many oxygen atoms are there on the right side? Is the equation balanced as written?

3.21 Balance the following equations.
(a) $Sn + NaOH \longrightarrow Na_2SnO_2 + H_2$
(b) $Al + Fe_3O_4 \longrightarrow Al_2O_3 + Fe$
(c) $CH_3OH + O_2 \longrightarrow CO_2 + H_2O$
(d) $P_4O_{10} + H_2O \longrightarrow H_3PO_4$
(e) $PCl_5 + H_2O \longrightarrow H_3PO_4 + HCl$

3.22 Balance the following equations.
(a) $Cl_2O_7 + H_2O \longrightarrow HClO_4$
(b) $MnO_2 + HCl \longrightarrow MnCl_2 + Cl_2 + H_2O$
(c) $Na_2S_2O_3 + I_2 \longrightarrow NaI + Na_2S_4O_6$
(d) $Al_4C_3 + H_2O \longrightarrow Al(OH)_3 + CH_4$
(e) $NO_2 + H_2O \longrightarrow HNO_3 + NO$

3.23 Balance the following equations.
(a) $SbCl_5 + H_2O \longrightarrow SbOCl_3 + HCl$
(b) $Mg + SiO_2 \longrightarrow MgO + Si$
(c) $CaCl_2 + Na_2CO_3 \longrightarrow CaCO_3 + NaCl$
(d) $C_6H_6 + O_2 \longrightarrow CO_2 + H_2O$
(e) $Al_2S_3 + H_2O \longrightarrow Al(OH)_3 + H_2S$

3.24 Balance the following equations.
(a) $TiCl_4 + H_2O \longrightarrow TiO_2 + HCl$
(b) $Fe_3O_4 + H_2 \longrightarrow Fe + H_2O$
(c) $V_2O_5 + H_2 \longrightarrow V_2O_3 + H_2O$
(d) $(NH_4)_2Cr_2O_7 \longrightarrow Cr_2O_3 + H_2O + N_2$
(e) $C_4H_{10} + O_2 \longrightarrow CO_2 + H_2O$

3.25 Solid calcium phosphate and aqueous sulfuric acid solution react to give calcium sulfate, which comes out of the solution as a solid. The other product is phosphoric acid, which remains in solution. Write a balanced equation for the reaction using complete formulas for the compounds with phase labels.

3.26 Solid potassium metal reacts with water, giving a solution of potassium hydroxide and releasing hydrogen gas. Write a balanced equation for the reaction using complete formulas for the compounds with phase labels.

3.27 An aqueous solution of ammonium chloride and barium hydroxide is heated, and the compounds react to give off ammonia gas. Barium chloride solution and water are also products. Write a balanced equation for the reaction using complete formulas for the compounds with phase labels; indicate that the reactants are heated.

3.28 Lead metal is produced by heating solid lead sulfide with solid lead sulfate, resulting in liquid lead and sulfur dioxide gas. Write a balanced equation for the reaction using complete formulas for the compounds with phase labels; indicate that the reactants are heated.

TYPES OF REACTIONS

3.29 Give the type of reaction for each of the following.
(a) $2KBr(aq) + Cl_2(aq) \longrightarrow 2KCl(aq) + Br_2(aq)$
(b) $2Cu(NO_3)_2(s) \xrightarrow{\Delta} 2CuO(s) + 4NO_2(g) + O_2(g)$
(c) $2C_2H_5SH(l) + 9O_2(g) \longrightarrow 4CO_2(g) + 6H_2O(g) + 2SO_2(g)$
(d) $CuCl_2(aq) + Na_2S(aq) \longrightarrow CuS(s) + 2NaCl(aq)$
(e) $2Al(s) + 3Br_2(l) \longrightarrow Al_2Br_6(s)$

3.30 Give the type of reaction for each of the following.
(a) $NH_4Cl(s) \xrightarrow{\Delta} NH_3(g) + HCl(g)$
(b) $P_4(s) + 6Br_2(l) \longrightarrow 4PBr_3(l)$
(c) $C_2H_5OH(l) + 3O_2(g) \longrightarrow 2CO_2(g) + 3H_2O(g)$
(d) $Cd(s) + 2AgNO_3(aq) \longrightarrow Cd(NO_3)_2(aq) + 2Ag(s)$
(e) $BaCl_2(aq) + K_2CrO_4(aq) \longrightarrow BaCrO_4(s) + 2KCl(aq)$

3.31 Table sugar is the chemical substance sucrose, $C_{12}H_{22}O_{11}$. When sucrose is heated, it turns to a brown syrup which releases steam (water vapor) as bubbles. After a time the syrup turns to a black, charred mass of carbon. Write the chemical equation for the reaction. What type of reaction is this?

3.32 Metals other than the noble metals such as silver and gold react with acids to produce hydrogen gas. Thus, aluminum metal reacts with hydrochloric acid. The product other than hydrogen is aluminum chloride (in aqueous solution). Write the balanced equation for the reaction using complete formulas with phase labels. What is the type of reaction?

3.33 Milk of magnesia is the common name of magnesium hydroxide. Milk of magnesia tablets are sold as antacid tablets, which neutralize stomach acid (hydrochloric acid). The products of the reaction of solid magnesium hydroxide and hydrochloric acid are water and magnesium chloride, which is dissolved in the water. Write the chemical equation for this reaction using complete formulas with phase labels. Classify the reaction by its type.

3.34 Bar soap contains sodium compounds derived from fats. One soap compound is sodium stearate, $NaC_{18}H_{36}O_2$. It reacts with calcium compounds in hard water to give a curdy product, calcium stearate. Write the chemical equation for the reaction of an aqueous solution of sodium stearate with a solution of calcium chloride using complete formulas with phase labels. In addition to solid calcium stearate, a solution of sodium chloride is produced. What type of reaction is this?

STRONG AND WEAK ACIDS AND BASES

3.35 Classify each of the following as a strong or weak acid or base.
(a) HF (b) KOH (c) $HClO_4$ (d) HClO

3.36 Classify each of the following as a strong or weak acid or base.
(a) HBrO (b) HCNO (c) $Sr(OH)_2$ (d) HI

IONIC EQUATIONS

3.37 Write net ionic equations for the following molecular equations.
(a) $HF(aq) + KOH(aq) \longrightarrow KF(aq) + H_2O(l)$
(b) $AgNO_3(aq) + NaBr(aq) \longrightarrow AgBr(s) + NaNO_3(aq)$
(c) $CaS(aq) + 2HBr(aq) \longrightarrow CaBr_2(aq) + H_2S(g)$
(d) $NaOH(aq) + NH_4Br(aq) \longrightarrow NaBr(aq) + NH_3(g) + H_2O(l)$

3.38 Write net ionic equations for the following molecular equations.
(a) $HBr(aq) + NH_3(aq) \longrightarrow NH_4Br(aq)$
(b) $2HBr(aq) + Ba(OH)_2(aq) \longrightarrow 2H_2O(l) + BaBr_2(aq)$
(c) $Pb(NO_3)_2(aq) + 2NaBr(aq) \longrightarrow PbBr_2(s) + 2NaNO_3(aq)$
(d) $MgCO_3(s) + H_2SO_4(aq) \longrightarrow MgSO_4(aq) + H_2O(l) + CO_2(g)$

3.39 Lead(II) nitrate solution and sodium sulfate solution are mixed. Crystals of lead sulfate come out of solution, leaving a solution of sodium nitrate. Write the molecular equation and the net ionic equation for the reaction.

3.40 Potassium carbonate solution reacts with aqueous hydrobromic acid to give a solution of potassium bromide, carbon dioxide gas, and water. Write the molecular equation and the net ionic equation for the reaction.

SOLUBILITY AND PRECIPITATION

3.41 Using solubility rules, decide whether the following ionic solids are soluble or insoluble in water.
(a) AgBr (b) $Pb(NO_3)_2$
(c) $SrSO_4$ (d) Na_2CO_3

3.42 Using solubility rules, decide whether the following ionic solids are soluble or insoluble in water.
(a) $(NH_4)_2SO_4$ (b) $Ca(NO_3)_2$
(c) $BaCO_3$ (d) $PbSO_4$

3.43 Write the molecular equation and the net ionic equation for each of the following aqueous reactions. If no reaction occurs, write *NR* after the arrow.
(a) $FeSO_4 + NaCl \longrightarrow$
(b) $Na_2CO_3 + MgBr_2 \longrightarrow$
(c) $MgSO_4 + NaOH \longrightarrow$
(d) $NiCl_2 + NaBr \longrightarrow$

3.44 Write the molecular equation and the net ionic equation for each of the following aqueous reactions. If no reaction occurs, write *NR* after the arrow.
(a) $AgNO_3 + NaI \longrightarrow$
(b) $Ba(NO_3)_2 + K_2SO_4 \longrightarrow$
(c) $Mg(NO_3)_2 + K_2SO_4 \longrightarrow$
(d) $CaCl_2 + Al(NO_3)_3 \longrightarrow$

3.45 For each of the following, write molecular and net ionic equations for any precipitation reaction that occurs. If no reaction occurs, indicate this.

(a) Solutions of barium nitrate and lithium sulfate are mixed.
(b) Solutions of sodium bromide and calcium nitrate are mixed.
(c) Solutions of aluminum sulfate and sodium hydroxide are mixed.
(d) Solutions of calcium bromide and sodium phosphate are mixed.

3.46 For each of the following, write molecular and net ionic equations for any precipitation reaction that occurs. If no reaction occurs, indicate this.

(a) Zinc chloride and sodium sulfide are dissolved in water.
(b) Sodium sulfide and calcium chloride are dissolved in water.
(c) Magnesium sulfate and potassium iodide are dissolved in water.
(d) Magnesium sulfate and potassium carbonate are dissolved in water.

NEUTRALIZATION REACTIONS

3.47 Complete and balance each of the following molecular equations (in aqueous solution); include phase labels. Then, for each, write the net ionic equation.

(a) $NaOH + HNO_3 \longrightarrow$
(b) $HCl + Ba(OH)_2 \longrightarrow$
(c) $HC_2H_3O_2 + Ca(OH)_2 \longrightarrow$
(d) $NH_3 + HNO_3 \longrightarrow$

3.48 Complete and balance each of the following molecular equations (in aqueous solution); include phase labels. Then, for each, write the net ionic equation.

(a) $Al(OH)_3 + HCl \longrightarrow$
(b) $HBr + Sr(OH)_2 \longrightarrow$
(c) $Ba(OH)_2 + HC_2H_3O_2 \longrightarrow$
(d) $HNO_3 + KOH \longrightarrow$

3.49 For each of the following, write the molecular equation, including phase labels. Then write the net ionic equation. Note that the salts formed in these reactions are soluble.

(a) the neutralization of hydrobromic acid with calcium hydroxide solution
(b) the reaction of solid aluminum hydroxide with nitric acid
(c) the reaction of aqueous hydrogen cyanide with calcium hydroxide solution
(d) the neutralization of lithium hydroxide solution by aqueous hydrogen cyanide

3.50 For each of the following, write the molecular equation, including phase labels. Then write the net ionic equation. Note that the salts formed in these reactions are soluble.

(a) the neutralization of lithium hydroxide solution by aqueous perchloric acid
(b) the reaction of barium hydroxide solution and aqueous nitrous acid
(c) the reaction of sodium hydroxide solution and aqueous nitrous acid
(d) the neutralization of aqueous hydrogen cyanide by aqueous strontium hydroxide

3.51 Complete the right side of each of the following molecular equations. Then write the net ionic equations. All salts formed are soluble. Acid salts are possible.

(a) $2KOH(aq) + H_3PO_4(aq) \longrightarrow$
(b) $3H_2SO_4(aq) + 2Al(OH)_3(s) \longrightarrow$
(c) $2HC_2H_3O_2(aq) + Ca(OH)_2(aq) \longrightarrow$
(d) $H_2SO_3(aq) + NaOH(aq) \longrightarrow$

3.52 Complete the right side of each of the following molecular equations. Then write the net equations. Assume all salts formed are soluble. Acid salts are possible.

(a) $Ca(OH)_2(aq) + 2H_2SO_4(aq) \longrightarrow$
(b) $2H_3PO_4(aq) + Ca(OH)_2(aq) \longrightarrow$
(c) $NaOH(aq) + H_2SO_4(aq) \longrightarrow$
(d) $Sr(OH)_2(aq) + 2H_2CO_3(aq) \longrightarrow$

3.53 Write molecular and net ionic equations for the successive neutralizations of each acidic hydrogen of sulfurous acid by aqueous calcium hydroxide. $CaSO_3$ is insoluble; the acid salt is soluble.

3.54 Write molecular and net ionic equations for the successive neutralizations of each acidic hydrogen of phosphoric acid by calcium hydroxide solution. $Ca_3(PO_4)_2$ is insoluble; assume that the acid salts are soluble.

REACTIONS EVOLVING A GAS

3.55 The following occur in aqueous solution. Complete and balance the molecular equation using phase labels. Then write the net ionic equation.

(a) $CaS + HBr \longrightarrow$
(b) $MgCO_3(s) + HNO_3 \longrightarrow$
(c) $K_2SO_3 + H_2SO_4 \longrightarrow$

3.56 The following occur in aqueous solution. Complete and balance the molecular equation using phase labels. Then write the net ionic equation.

(a) $BaCO_3(s) + HNO_3 \longrightarrow$
(b) $K_2S + HCl \longrightarrow$
(c) $CaSO_3(s) + HI \longrightarrow$

3.57 Write the molecular equation and the net ionic equation for the reaction of solid iron(II) sulfide and hydrochloric acid. Add phase labels.

3.58 Write the molecular equation and the net ionic equation for the reaction of solid barium carbonate and hydrogen bromide in aqueous solution. Add phase labels.

Additional Problems

3.59 Balance the following equations.
(a) $C_2H_6 + O_2 \longrightarrow CO_2 + H_2O$
(b) $P_4O_6 + H_2O \longrightarrow H_3PO_3$
(c) $KClO_3 \longrightarrow KCl + KClO_4$
(d) $(NH_4)_2SO_4 + NaOH \longrightarrow NH_3 + H_2O + Na_2SO_4$
(e) $NBr_3 + NaOH \longrightarrow N_2 + NaBr + HOBr$

3.60 Balance the following equations.
(a) $NaOH + H_3PO_4 \longrightarrow Na_3PO_4 + H_2O$
(b) $SiCl_4 + H_2O \longrightarrow SiO_2 + HCl$
(c) $Ca_3(PO_4)_2 + C \longrightarrow Ca_3P_2 + CO$
(d) $H_2S + O_2 \longrightarrow SO_2 + H_2O$
(e) $N_2O_5 \longrightarrow NO_2 + O_2$

3.61 Write a sentence describing the reaction represented by the equation

$$Fe(s) + CuSO_4(aq) \longrightarrow Cu(s) + FeSO_4(aq)$$

3.62 Write a sentence describing the reaction represented by the equation

$$MgCO_3(s) \xrightarrow{\Delta} MgO(s) + CO_2(g)$$

3.63 When ammonia gas, NH_3, burns in oxygen in the presence of a platinum catalyst, nitric oxide gas, NO, and water vapor are produced. Write a balanced equation for this reaction. Include phase labels; also show the catalyst.

3.64 When solid lead sulfide, PbS, is heated in the presence of oxygen, it yields crystals of lead(II) oxide, PbO, and sulfur dioxide gas, SO_2. Write a balanced equation for this reaction. Include phase labels; also show that heat is used.

3.65 For each of the following reactions, write a balanced equation, including phase labels.
(a) When heated, ammonium dichromate crystals, $(NH_4)_2Cr_2O_7$, decompose to give nitrogen, water vapor, and solid chromium(III) oxide, Cr_2O_3.
(b) When aqueous ammonium nitrite, NH_4NO_2, is heated, it gives nitrogen and water vapor.
(c) When aqueous potassium chromate, K_2CrO_4, is added to aqueous lead(II) nitrate, $Pb(NO_3)_2$, yellow crystals of lead(II) chromate form, and a solution of potassium nitrate, KNO_3, is left behind.
(d) When gaseous ammonia, NH_3, reacts with hydrogen chloride gas, HCl, fine crystals of ammonium chloride, NH_4Cl, are formed.
(e) Aluminum added to an aqueous solution of sulfuric acid, H_2SO_4, forms a solution of aluminum sulfate, $Al_2(SO_4)_3$. Hydrogen gas is released.

3.66 For each of the following reactions, write a balanced equation, including phase labels.
(a) When solid calcium oxide, CaO, is exposed to gaseous sulfur trioxide, SO_3, solid calcium sulfate, $CaSO_4$, is formed.
(b) Solid magnesium hydroxide, $Mg(OH)_2$, reacts with aqueous nitric acid, HNO_3, producing a solution of magnesium nitrate, $Mg(NO_3)_2$, and water.
(c) Calcium metal (solid) reacts with water to produce a solution of calcium hydroxide, $Ca(OH)_2$, and hydrogen gas.
(d) When solid sodium hydrogen sulfite, $NaHSO_3$, is heated, solid sodium sulfite, Na_2SO_3; sulfur dioxide gas, SO_2; and water vapor are formed.
(e) Magnesium reacts with bromine to give magnesium bromide, $MgBr_2$.

3.67 Classify each of the reactions described in Problem 3.65 by type of reaction.

3.68 Classify each of the reactions described in Problem 3.66 by type of reaction.

3.69 Magnesium metal reacts with hydrobromic acid to produce hydrogen gas and a solution of magnesium bromide. Write the molecular equation for this reaction. Then write the corresponding net ionic equation.

3.70 Aluminum metal reacts with perchloric acid to produce hydrogen gas and a solution of aluminum perchlorate. Write the molecular equation for this reaction. Then write the corresponding net ionic equation.

3.71 Nickel(II) sulfate solution reacts with lithium hydroxide solution to produce a precipitate of nickel(II) hydroxide and a solution of lithium sulfate. Write the molecular equation for this reaction. Then write the corresponding net ionic equation.

3.72 Potassium sulfate solution reacts with barium bromide solution to produce a precipitate of barium sulfate and a solution of potassium bromide. Write the molecular equation for this reaction. Then write the corresponding net ionic equation.

3.73 Decide whether a reaction occurs for each of the following. If it does not, write *NR* after the arrow. If it does, write the balanced molecular equation; then write the net ionic equation.

(a) $LiOH + HCN \longrightarrow$
(b) $Li_2CO_3 + HNO_3 \longrightarrow$
(c) $LiCl + AgNO_3 \longrightarrow$
(d) $LiCl + MgSO_4 \longrightarrow$

3.74 Decide whether a reaction occurs for each of the following. If it does not, write *NR* after the arrow. If it does, write the balanced molecular equation; then write the net ionic equation.

(a) $Al(OH)_3 + HNO_3 \longrightarrow$
(b) $FeS + HClO_4 \longrightarrow$
(c) $CaCl_2 + NaNO_3 \longrightarrow$
(d) $MgSO_4 + Ba(NO_3)_2 \longrightarrow$

3.75 Complete and balance each of the following molecular equations, including phase labels, if a reaction occurs. Then write the net ionic equation. If no reaction occurs, write *NR* after the arrow.

(a) $Sr(OH)_2 + HC_2H_3O_2 \longrightarrow$
(b) $NH_4I + CsCl \longrightarrow$
(c) $NaNO_3 + CsCl \longrightarrow$
(d) $NH_4I + AgNO_3 \longrightarrow$

3.76 Complete and balance each of the following molecular equations, including phase labels, if a reaction occurs. Then write the net ionic equation. If no reaction occurs, write *NR* after the arrow.

(a) $HClO_4 + BaCO_3 \longrightarrow$
(b) $H_2CO_3 + Sr(OH)_2 \longrightarrow$
(c) $K_3(PO_4) + MgCl_2 \longrightarrow$
(d) $FeSO_4 + MgCl_2 \longrightarrow$

3.77 Describe in words how you would do each of the following preparations. Then give the molecular equation for each preparation.

(a) $CuCl_2(s)$ from $CuSO_4(s)$
(b) $Ca(C_2H_3O_2)_2(s)$ from $CaCO_3(s)$
(c) $NaNO_3(s)$ from $Na_2SO_3(s)$
(d) $MgCl_2(s)$ from $Mg(OH)_2(s)$

3.78 Describe in words how you would do each of the following preparations. Then give the molecular equation for each preparation.

(a) $MgCl_2(s)$ from $MgCO_3(s)$
(b) $NaNO_3(s)$ from $NaCl(s)$
(c) $Al(OH)_3(s)$ from $Al(NO_3)_3(s)$
(d) $HCl(aq)$ from $H_2SO_4(aq)$

Cumulative-Skills Problems

3.79 Lead(II) nitrate reacts with cesium sulfate in an aqueous precipitation reaction. What are the formulas of lead(II) nitrate and cesium sulfate? Write the molecular equation and net ionic equation for the reaction. What are the names of the products? Give the molecular equation for another reaction that produces the same precipitate.

3.80 Silver nitrate, $AgNO_3$, reacts with strontium chloride in an aqueous precipitation reaction. What are the formulas of silver nitrate and strontium chloride? Write the molecular equation and net ionic equation for the reaction. What are the names of the products? Give the molecular equation for another reaction that produces the same precipitate.

3.81 Elemental bromine is the source of bromine compounds. The element is produced from certain brine solutions that occur naturally. These brines are essentially solutions of calcium bromide which, when treated with chlorine gas, yield bromine in a displacement reaction. What is the molecular equation and net ionic equation for the reaction? A solution containing 40.0 g of calcium bromide requires 14.2 g of chlorine to react completely with it, and 22.2 g of calcium chloride is produced in addition to whatever bromine is obtained. How many grams of calcium bromide are required to produce 10.0 pounds of bromine?

3.82 Barium carbonate is the source of barium compounds. It is produced in an aqueous precipitation reaction from barium sulfide and sodium carbonate. (Barium sulfide is obtained by heating the mineral barite, which is barium sulfate, with carbon.) What is the molecular equation and net ionic equation for the precipitation reaction? A solution containing 33.9 g of barium sulfide requires 21.2 g of sodium carbonate to react completely with it, and 15.6 g of sodium sulfide is produced in addition to whatever barium carbonate is obtained. How many grams of barium sulfide are required to produce 10.0 tons of barium carbonate? (One ton equals 2000 pounds.)

4

Calculations with Chemical Formulas and Equations

Chapter Outline

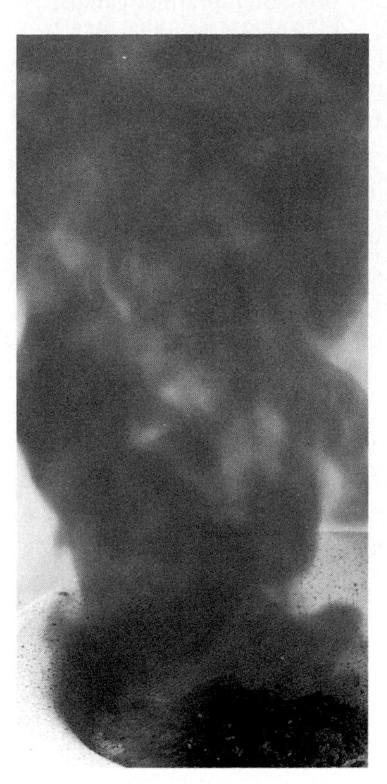

Reaction of zinc and iodine.

Acetic acid (ah-see´-tik acid) is a colorless liquid with a sharp, vinegary odor. In fact, vinegar contains about 5 percent by mass of acetic acid, which accounts for vinegar's odor and sour taste. The name vinegar derives from the French *vin aigre,* meaning "sour wine." Vinegar results from the fermentation of wine or cider by certain bacteria (Figure 4.1). These bacteria require oxygen, and the overall chemical change is the reaction of ethanol (alcohol) in wine with oxygen to give acetic acid.

An industrial preparation of acetic acid also starts from ethanol, which reacts with oxygen in two steps. First, ethanol reacts with oxygen to yield a compound called acetaldehyde, in addition to water. In the second step, the acetaldehyde reacts with more oxygen to produce acetic acid. (The body also produces acetaldehyde then acetic acid from alcohol, as it tries to eliminate alcohol from the system.)

This chapter focuses on two basic questions, which we can illustrate using these compounds: How do we determine the chemical formula of a substance such as acetic acid (or acetaldehyde)? How much acetic acid can we prepare from a given quantity of ethanol (or a given quantity of acetaldehyde)? These types of

Figure 4.1
Preparing gourmet vinegar. In the presence of air, acetic bacteria ferment wine or cider to vinegar.

questions are very important in chemistry. We must know the formulas of all the substances involved in a reaction before we can write the chemical equation, and we need the balanced chemical equation to determine the quantitative relationships among the different substances in the reaction. We begin this chapter by discussing how we relate number of atoms or molecules to grams of substance, because this is the key to answering both questions.

Mass and Moles of Substance

You buy a quantity of groceries in several ways. If you buy eggs, you buy them by the dozen—that is, by number. Eggs are easy to count out. So are oranges and lemons. Other items, though countable, are more conveniently sold by mass. A dozen peanuts is too small a number to buy, and several hundred are too difficult to count. We buy peanuts by the pound or by the kilogram—that is, by mass. Chemists also are interested in quantity—the quantity of an element or a compound, which, like grocery items, can be measured by number or by mass. But though one can easily weigh a sample of a substance, the number of atoms or molecules in it is much too large to count. (You may recall from Section 2.5 that a billionth drop of water contains 2×10^{12} H_2O molecules.) Nevertheless, chemists are interested in knowing such numbers. How many atoms of carbon are there in one molecule of acetic acid? How many molecules of acetic acid can be obtained from one molecule of ethanol? Before we look at how chemists solve this problem of measuring numbers of atoms, molecules, and ions, we must look at the concept of molecular weight and formula weight.

4.1 MOLECULAR WEIGHT AND FORMULA WEIGHT

The **molecular weight** (MW) of a substance is *the sum of the atomic weights of all of the atoms in the molecule*. It is therefore the average mass of a molecule of that substance, expressed in atomic mass units. For example, the molecular weight of water, H_2O, is 18.0 amu (2×1.0 amu from two H atoms plus 16.0 amu from one O atom). If the molecular formula for the substance is not known, we can determine the molecular weight experimentally by means of a mass spectrometer. In later chapters, we will discuss simple, inexpensive methods of molecular weight determination.

The **formula weight** (FW) of a substance is *the sum of the atomic weights of all atoms in a formula unit of the compound,* whether molecular or not. Sodium chloride, NaCl, has a formula weight of 58.44 amu (22.99 amu from Na plus 35.45 amu from Cl). NaCl is ionic, so the expression "molecular weight of NaCl" has no meaning. On the other hand, the molecular weight and the formula weight calculated from the molecular formula of a substance are identical.

Example 4.1 Calculating the Formula Weight from a Formula

Calculate the formula weight of each of the following to three significant figures, using a table of atomic weights (AW): (a) chloroform, $CHCl_3$; (b) iron(III) sulfate, $Fe_2(SO_4)_3$.

Solution

(a) The calculation is

1 × AW of C =		12.0 amu
1 × AW of H =		1.0 amu
3 × AW of Cl =	3 × 35.45 amu =	106.4 amu
FW of $CHCl_3$ =		119.4 amu

The final answer rounded to three significant figures is **119.** (b) The point to remember here is that every atomic symbol within the parentheses is multiplied by the subscript 3. The calculation is

2 × AW of Fe =	2 × 55.8 amu =	111.6 amu
3 × AW of S =	3 × 32.1 amu =	96.3 amu
3 × 4 × AW of O =	12 × 16.00 amu =	192.0 amu
FW of $Fe_2(SO_4)_3$ =		399.9 amu

The answer rounded to three significant figures is **4.00×10^2 amu.**

Exercise 4.1

Calculate the formula weights of the following compounds, using a table of atomic weights. Give the answers to three significant figures: (a) nitrogen dioxide, NO_2; (b) glucose, $C_6H_{12}O_6$; (c) sodium hydroxide, NaOH; (d) magnesium hydroxide, $Mg(OH)_2$.
(See Problems 4.15 and 4.16.)

4.2 THE MOLE CONCEPT

When we prepare a compound industrially or even study a reaction in the laboratory, we deal with tremendous numbers of molecules or ions. Suppose we wish to prepare acetic acid, starting from 10.0 g of ethanol. This small sample (less than 3 teaspoonsful) contains 1.31×10^{23} molecules, a truly staggering number. Imagine a device that counts molecules at the rate of one million per second. It would take over four billion years—nearly the age of the earth—for this device to count that many molecules! Chemists have adopted the *mole concept* as a convenient way to deal with the enormous numbers of molecules or ions in the samples they work with.

Definition of Mole

A **mole** (symbol **mol**) is defined as *the quantity of a given substance that contains as many molecules or formula units as the number of atoms in exactly 12 g of carbon-12*. One mole of ethanol, for example, contains the same number of ethanol molecules as there are carbon atoms in 12 g of carbon-12.

The number of atoms in a 12-g sample of carbon-12 is called **Avogadro's number** (to which we give the symbol N_A). Recent measurements of this number

Figure 4.2
One mole each of various substances. *Left to right:* 1-octanol ($C_8H_{17}OH$); methanol (CH_3OH); sulfur (S_8); and mercury(II) iodide (HgI_2).

give the value 6.0221367×10^{23}, which to three significant figures is 6.02×10^{23}.

A mole of a substance contains Avogadro's number (6.02×10^{23}) of molecules (or formula units). The term *mole,* like a dozen or a gross, thus refers to a particular number of things. A dozen eggs equals 12 eggs, a gross of pencils equals 144 pencils, and a mole of ethanol equals 6.02×10^{23} ethanol molecules.

In using the term *mole* for ionic substances, we mean the number of formula units of the substance. For example, a mole of sodium carbonate, Na_2CO_3, is a quantity containing 6.02×10^{23} Na_2CO_3 units. But each formula unit of Na_2CO_3 contains two Na^+ ions and one CO_3^{2-} ion. Therefore, a mole of Na_2CO_3 also contains $2 \times 6.02 \times 10^{23}$ Na^+ ions and $1 \times 6.02 \times 10^{23}$ CO_3^{2-} ions.●

● Sodium carbonate, Na_2CO_3, is a white, crystalline solid known commercially as soda ash. Large amounts of soda ash are used in the manufacture of glass. The hydrated compound, $Na_2CO_3 \cdot 10H_2O$, is known as washing soda.

When using the term *mole,* it is important to specify the formula of the unit to avoid any misunderstanding. For example, a mole of oxygen atoms (with the formula O) contains 6.02×10^{23} O atoms. A mole of oxygen molecules (formula O_2) contains 6.02×10^{23} O_2 molecules—that is, $2 \times 6.02 \times 10^{23}$ O atoms.

The **molar mass** of a substance is *the mass of one mole of the substance.* Carbon-12 has a molar mass of exactly 12 g, by definition. For all substances, the molar mass in grams is numerically equal to the formula weight in atomic mass units. (For this reason, the molar mass in grams is also called the *gram formula weight* or the *gram molecular weight.*) Ethanol, whose molecular formula is C_2H_6O (frequently written as the condensed structural formula C_2H_5OH), has a molecular weight of 46.1 amu and a molar mass of 46.1 g/mol. Figure 4.2 shows molar amounts of different substances.

Example 4.2 Calculating the Mass of an Atom or Molecule

(a) What is the mass in grams of a chlorine atom, Cl? (b) What is the mass in grams of a hydrogen chloride molecule, HCl?

Solution

(a) The atomic weight of Cl is 35.5 amu, so the molar mass of Cl is 35.5 g/mol. Because 1 mol Cl (35.5 g) contains Avogadro's number (6.02×10^{23}) of Cl atoms, dividing 35.5 g by 6.02×10^{23} gives the mass of one atom.

$$\text{Mass of a Cl atom} = \frac{35.5\ \text{g}}{6.02 \times 10^{23}} = \mathbf{5.90 \times 10^{-23}\ g}$$

(b) The molecular weight of HCl equals the AW of H plus the AW of Cl, or 1.01 amu + 35.5 amu = 36.5 amu. Therefore, 1 mol HCl equals 36.5 g, and

$$\text{Mass of an HCl molecule} = \frac{36.5\ \text{g}}{6.02 \times 10^{23}} = \mathbf{6.06 \times 10^{-23}\ g}$$

Exercise 4.2

(a) What is the mass in grams of a calcium atom, Ca? (b) What is the mass in grams of an ethanol molecule, C_2H_5OH? (See Problems 4.19, 4.20, 4.21, and 4.22.)

Mole Calculations

● Alternatively, because the molar mass is the mass per mole, we can relate mass and moles by means of the formula

Molar mass = mass/moles

Now that we know how to find the mass of one mole of substance, there are two important questions to ask. First, how much does a given number of moles of a substance weigh? Second, how many moles of a given formula unit does a given mass of substance contain? Both of these questions are easily answered using dimensional analysis, or the conversion-factor method.●

To illustrate, consider the conversion of grams of ethanol, C_2H_5OH, to moles of ethanol. The molar mass of ethanol is 46.1 g/mol, so we write

$$1 \text{ mol } C_2H_5OH = 46.1 \text{ g } C_2H_5OH$$

● If you prefer to use the formula for molar mass, the calculation is

Molar mass = mass/moles

or

Moles = mass/molar mass

Hence,

Moles = 10.0 g/46.1 g/mol = 0.217 mol

Thus, the factor converting grams of ethanol to moles of ethanol is 1 mol C_2H_5OH/46.1 g C_2H_5OH. To convert moles of ethanol to grams of ethanol, we simply invert the conversion factor (46.1 g C_2H_5OH/1 mol C_2H_5OH).

Again, suppose we are going to prepare acetic acid from 10.0 g of ethanol, C_2H_5OH. How many moles of C_2H_5OH is this?● We convert 10.0 g C_2H_5OH to moles C_2H_5OH by multiplying by the appropriate conversion factor.

$$10.0 \text{ g } \cancel{C_2H_5OH} \times \frac{1 \text{ mol } C_2H_5OH}{46.1 \text{ g } \cancel{C_2H_5OH}} = 0.217 \text{ mol } C_2H_5OH$$

Example 4.3 Converting Moles of Substance to Grams

Zinc iodide, ZnI_2, can be prepared by the direct combination of elements (Figure 4.3). A chemist determines from the amounts of elements that 0.0654 mol of ZnI_2 can form. How many grams of zinc iodide is this?

Figure 4.3
Reaction of zinc and iodine. Heat from the combination reaction of the elements causes some iodine to vaporize (violet vapor).

Solution

The molar mass of ZnI_2 is 319 g. (The formula weight is 319 amu, which is obtained by summing the atomic weights in the formula.) We can write

$$1 \text{ mol } ZnI_2 = 319 \text{ g } ZnI_2$$

Therefore,

$$0.0654 \cancel{\text{mol } ZnI_2} \times \frac{319 \text{ g } ZnI_2}{1 \cancel{\text{mol } ZnI_2}} = \mathbf{20.9 \text{ g } ZnI_2}$$

Exercise 4.3

Hydrogen peroxide, H_2O_2, is a colorless liquid. A concentrated solution of it is used as a source of oxygen for rocket propellant fuels. Dilute aqueous solutions are used as a bleach. Analysis of a solution shows that it contains 0.909 mol H_2O_2 in 1.00 L of solution. What is the mass of hydrogen peroxide in this volume of solution?

(See Problems 4.23, 4.24, 4.25, and 4.26.)

Example 4.4 Converting Grams of Substance to Moles

Lead chromate, $PbCrO_4$, is a yellow paint pigment (called chrome yellow) prepared by a precipitation reaction (Figure 4.4). In a preparation, 45.6 g of lead chromate is obtained as a precipitate. How many moles of $PbCrO_4$ is this?

Figure 4.4 Precipitation of lead chromate. When lead nitrate solution (colorless) is added to potassium chromate solution (clear yellow), a bright yellow precipitate of lead chromate forms.

Solution

The molar mass of $PbCrO_4$ is 323 g/mol. That is,

$$1 \text{ mol } PbCrO_4 = 323 \text{ g } PbCrO_4$$

Therefore,

$$45.6 \cancel{\text{g } PbCrO_4} \times \frac{1 \text{ mol } PbCrO_4}{323 \cancel{\text{g } PbCrO_4}} = \mathbf{0.141 \text{ mol } PbCrO_4}$$

Exercise 4.4

Nitric acid, HNO_3, is a colorless, corrosive liquid used in the manufacture of nitrogen fertilizers and explosives. In an experiment to develop new explosives for mining operations, a 28.5-g sample of nitric acid was poured into a beaker. How many moles of HNO_3 are there in this sample of nitric acid? (See Problems 4.27 and 4.28.)

The preceding examples show how we can relate moles and mass. Often, however, such a conversion is only part of a more complicated one. The next example shows how to put conversion factors together in a more complicated conversion. To be able to do this effectively, you must be thoroughly conversant with the individual steps.

Example 4.5 Calculating the Number of Molecules in a Given Mass

How many HCl molecules are there in 3.46 kg of hydrogen chloride?

Solution

This problem involves the conversion of mass of HCl (3.46 kg HCl) to number of HCl molecules. Though we have not discussed this particular conversion before, we have looked at all the separate steps. Recall that number of particles is related directly to moles, and moles are related directly to mass. So we begin by converting kg HCl to g HCl.

$$3.46 \text{ kg HCl} \times \frac{1 \times 10^3 \text{ g HCl}}{1 \text{ kg HCl}} = 3.46 \times 10^3 \text{ g HCl}$$

Then we convert this to moles, noting that 1 mol HCl = 36.5 g HCl.

$$3.46 \times 10^3 \text{ g HCl} \times \frac{1 \text{ mol HCl}}{36.5 \text{ g HCl}} = 94.8 \text{ mol HCl}$$

Now recall that 1 mol HCl is equivalent to 6.02×10^{23} HCl molecules (1 mol HCl = 6.02×10^{23} HCl molecules).

$$94.8 \text{ mol HCl} \times \frac{6.02 \times 10^{23} \text{ HCl molecules}}{1 \text{ mol HCl}} = 5.71 \times 10^{25} \text{ HCl molecules}$$

Once you are conversant with the individual steps, you will find that you can do them one after the other, as follows:

$$3.46 \cancel{\text{kg HCl}} \times \frac{1 \times 10^3 \cancel{\text{g HCl}}}{1 \cancel{\text{kg HCl}}} \times \frac{1 \cancel{\text{mol HCl}}}{36.5 \cancel{\text{g HCl}}} \times \frac{6.02 \times 10^{23} \text{ HCl molecules}}{1 \cancel{\text{mol HCl}}} = \mathbf{5.71 \times 10^{25} \text{ HCl molecules}}$$

Note that the units in the denominator of each conversion factor cancel the units of the previous term, finally leaving the desired units.

Exercise 4.5

Hydrogen cyanide, HCN, is a volatile, colorless liquid with the odor of certain fruit pits, such as bitter almonds, peaches, and cherries. The compound is highly poisonous. How many molecules are there in 56 mg HCN (the average toxic dose)?

(See Problems 4.31, 4.32, 4.33, and 4.34.)

Determining Chemical Formulas

As you might expect from the chapter-opening discussion, when a chemist has discovered a new compound, the first question to answer is, What is the formula? To determine this, we begin by analyzing the compound into amounts of the elements for a given amount of compound. This is conveniently expressed as **percentage composition**—that is, as *the mass percentages of each element in the compound*. We then determine the formula from this percentage composition. We must also know the molecular weight of the compound in order to determine the molecular formula.

The next section describes the calculation of mass percentages. Then, in two following sections, we describe how to determine a formula.

4.3 MASS PERCENTAGES FROM THE FORMULA

Suppose that A is a part of something—that is, part of a whole. It could be an element in a compound or one substance in a mixture. We define the **mass percentage** of A as *the parts of* A *per hundred parts of the total, by mass*. That is,

$$\text{Mass } \% \ A = \frac{\text{mass of } A \text{ in the whole}}{\text{mass of the whole}} \times 100\%$$

We can look at the mass percentage of A as the number of grams of A in 100 grams of the whole.

To get practice with the concept of mass percentage, we will start with a compound (formaldehyde, CH_2O) whose formula we are given and obtain the percentage composition.

Example 4.6 Calculating the Percentage Composition from the Formula

Formaldehyde, CH_2O, is a toxic gas with a pungent odor. Large quantities are consumed in the manufacture of plastics, and a water solution of the compound is used to preserve biological specimens. Calculate the mass percentages of each element in formaldehyde (give answers to three significant figures).

Solution

To calculate mass percentage, we need the mass of an element in a given mass of compound. We can get this information by interpreting the formula in molar terms and then converting moles to masses, using a table of atomic weights. Thus, 1 mol CH_2O has a mass of 30.0 g and contains 1 mol C (12.0 g), 2 mol H (2×1.01 g), and 1 mol O (16.0 g). Hence,

$$\% \text{ C} = \frac{12.0 \text{ g}}{30.0 \text{ g}} \times 100\% = \mathbf{40.0\%}$$

$$\% \text{ H} = \frac{2 \times 1.01 \text{ g}}{30.0 \text{ g}} \times 100\% = \mathbf{6.73\%}$$

We can calculate the percentage of O in the same way, but it can also be found by subtracting the percentages of C and H from 100%:

$$\% \text{ O} = 100\% - (40.0\% + 6.73\%) = \mathbf{53.3\%}$$

Exercise 4.6

Ammonium nitrate, NH_4NO_3, which is prepared from nitric acid, is used as a nitrogen fertilizer. Calculate the mass percentages of the elements in ammonium nitrate (to three significant figures).

(See Problems 4.43, 4.44, 4.45, and 4.46.)

Example 4.7 Calculating the Mass of an Element in a Given Mass of Compound

How many grams of carbon are there in 83.5 g of formaldehyde, CH_2O? Use the percentage composition obtained in the previous example (40.0% C, 6.73% H, 53.3% O).

Solution

CH_2O is 40.0% C, so the mass of carbon in 83.5 g CH_2O is

$$83.5 \text{ g} \times 0.400 = \mathbf{33.4 \text{ g}}$$

Exercise 4.7

How many grams of nitrogen, N, are there in a fertilizer containing 48.5 g of ammonium nitrate and no other nitrogen-containing compound? See Exercise 4.6 for the percentage composition of NH_4NO_3.

(See Problems 4.47 and 4.48.)

4.4 ELEMENTAL ANALYSIS: PERCENTAGES OF CARBON, HYDROGEN, AND OXYGEN

Suppose we have a newly discovered compound whose formula we wish to determine. The first step is to obtain its percentage composition. As an example, consider the determination of the percentages of carbon, hydrogen, and oxygen in compounds containing only these three elements. The basic idea is this: We burn a sample of the compound of known mass and get CO_2 and H_2O. Next we relate the masses of CO_2 and H_2O to the masses of carbon and hydrogen. Once we know the masses of carbon and hydrogen in the sample, we calculate the mass percentages of C and H. We then find the mass percentage of O by subtracting the mass percentages of C and H from 100.

Figure 4.5 shows an apparatus used to find the amount of carbon and hydrogen in a compound. The compound is burned in a stream of oxygen gas. The vapor of the compound and its combustion products pass over copper pellets coated with copper(II) oxide, CuO, which supplies additional oxygen and ensures that the compound is completely burned. As a result of the combustion, every mole of carbon (C) in the compound ends up as a mole of carbon dioxide (CO_2), and every mole of hydrogen (H) ends up as one-half mole of water (H_2O). The water is collected by a drying agent, a substance that has a strong affinity for water. The

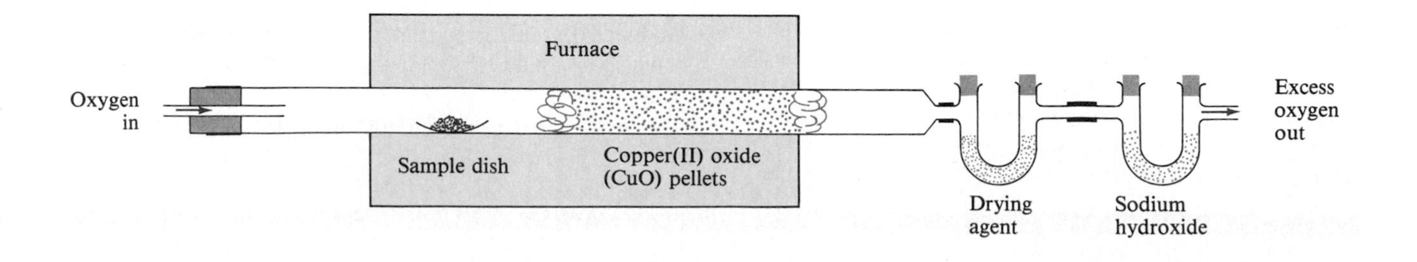

Figure 4.5
Combustion method for determining the percentages of carbon and hydrogen in a compound. The compound is placed in the sample dish and is heated by the furnace. Vapor of the compound burns in O_2 in the presence of CuO pellets, giving CO_2 and H_2O. The water vapor is collected by a drying agent, and CO_2 combines with the sodium hydroxide. Amounts of CO_2 and H_2O are obtained by weighing the U-tubes before and after combustion.

● Sodium hydroxide reacts with carbon dioxide according to the following equations:

$$NaOH + CO_2 \longrightarrow NaHCO_3$$
$$2NaOH + CO_2 \longrightarrow Na_2CO_3 + H_2O$$

carbon dioxide is collected by chemical reaction with sodium hydroxide, NaOH.● By weighing the U-tubes containing the drying agent and the sodium hydroxide before and after the combustion, it is possible to determine the masses of water and carbon dioxide produced. From these data, we can calculate the percentage composition of the compound.

The chapter opened with a discussion of acetic acid. The next example shows how to determine the percentage composition of this substance from combustion data. We will use this percentage composition in Example 4.11 to determine the formula of acetic acid.

Example 4.8 Calculating the Percentages of C and H by Combustion

Acetic acid contains only C, H, and O. A 4.24-mg sample of acetic acid is completely burned. It gives 6.21 mg of carbon dioxide and 2.54 mg of water. What is the mass percentage of each element in acetic acid?

Solution

We first convert the mass of CO_2 to moles of CO_2. Then we convert this to moles of C, noting that 1 mol $CO_2 \simeq$ 1 mol C. (The symbol $\simeq$ means "is chemically equivalent to.") Finally, we convert to mass of C. We write these conversions as follows:

$$6.21 \times 10^{-3}\ \cancel{\text{g CO}_2} \times \frac{1\ \cancel{\text{mol CO}_2}}{44.0\ \cancel{\text{g CO}_2}} \times \frac{1\ \cancel{\text{mol C}}}{1\ \cancel{\text{mol CO}_2}} \times \frac{12.0\ \text{g C}}{1\ \cancel{\text{mol C}}} = 1.69 \times 10^{-3}\ \text{g C (or 1.69 mg C)}$$

For hydrogen, noting that 1 mol $H_2O \simeq$ 2 mol H, we have the conversions

$$2.54 \times 10^{-3}\ \cancel{\text{g H}_2\text{O}} \times \frac{1\ \cancel{\text{mol H}_2\text{O}}}{18.0\ \cancel{\text{g H}_2\text{O}}} \times \frac{2\ \cancel{\text{mol H}}}{1\ \cancel{\text{mol H}_2\text{O}}} \times \frac{1.01\ \text{g H}}{1\ \cancel{\text{mol H}}} = 2.85 \times 10^{-4}\ \text{g H (or 0.285 mg H)}$$

The mass percentages of C and H in acetic acid can now be calculated.

$$\text{Mass \% C} = \frac{1.69\ \text{mg}}{4.24\ \text{mg}} \times 100\% = 39.9\%$$

$$\text{Mass \% H} = \frac{0.285\ \text{mg}}{4.24\ \text{mg}} \times 100\% = 6.72\%$$

The mass percentage of oxygen is found by subtracting the sum of these percentages from 100%:

$$\text{Mass \% O} = 100\% - (39.9\% + 6.72\%) = 53.4\%$$

Thus, the percentage composition of acetic acid is 39.9% C, 6.72% H, and 53.4% O.

Exercise 4.8

A 3.87-mg sample of ascorbic acid (vitamin C) gives 5.80 mg CO_2 and 1.58 mg H_2O on combustion. What is the percentage composition of this compound (the mass percentage of each element)? Ascorbic acid contains only C, H, and O. (See Problems 4.49 and 4.50.)

4.5 DETERMINING FORMULAS

The percentage composition of a compound leads directly to its empirical formula. An **empirical formula** (or **simplest formula**) for a compound is *the formula of a substance written with the smallest integer subscripts*. For example, hydrogen peroxide has the molecular formula H_2O_2. The molecular formula, you may recall, tells us the precise number of atoms of different elements in the substance. The empirical formula, however, merely tells us the ratio of numbers of atoms in the compound. The empirical formula of hydrogen peroxide is HO.

Compounds with different molecular formulas can have the same empirical formulas and such substances will have the same percentage composition. An example is acetylene, C_2H_2, and benzene, C_6H_6. Acetylene is a gas used as a fuel and in welding; benzene is a liquid used as a raw material for plastics, such as polystyrene. Both substances have the same empirical formula, CH, and consequently have the same percentage composition (92.3% C and 7.7% H, by mass).

In order to obtain the molecular formula of a substance, two pieces of information are needed: (1) the percentage composition, from which the empirical formula can be determined; and (2) the molecular weight. The molecular weight allows us to choose the correct multiple of the empirical formula for the molecular formula. We will illustrate these steps in the next three examples.

Empirical Formula from the Composition

The empirical formula of a compound shows the ratios of numbers of atoms in the compound. We can find this formula from the composition of the compound by converting from masses of the elements to moles. The next two examples show these calculations in detail.

Example 4.9 Determining the Empirical Formula from Percentage Composition (Binary Compound)

A compound of nitrogen and oxygen is analyzed, and a sample weighing 1.587 g is found to contain 0.483 g N and 1.104 g O. What is the empirical formula of the compound?

Solution

Convert the masses to moles, because these quantities will be proportional to the integer subscripts in the empirical formula.

$$0.483 \text{ g N} \times \frac{1 \text{ mol N}}{14.0 \text{ g N}} = 0.0345 \text{ mol N}$$

$$1.104 \text{ g O} \times \frac{1 \text{ mol O}}{16.00 \text{ g O}} = 0.06900 \text{ mol O}$$

To obtain the smallest integers from moles of elements, divide each by the smallest one. Here we need only divide 0.06900 mol by 0.0345 mol, which gives 2.00. Thus, the ratio of number of N atoms to the number of O atoms is 1 to 2. Hence, the empirical formula is $\mathbf{NO_2}$.

Exercise 4.9

A sample of compound weighing 83.5 g contains 33.4 g of sulfur. The rest is oxygen. What is the empirical formula? (See Problems 4.51 and 4.52.)

Example 4.10 Determining the Empirical Formula from Percentage Composition (General)

Chromium forms compounds of various colors (the word *chromium* comes from the Greek *khroma,* meaning ''color''); see Figure 4.6. Sodium dichromate is the most important commercial chromium compound, from which many other chromium compounds are manufactured. It is a bright orange, crystalline substance. An analysis of sodium dichromate gives the following mass percentages: 17.5% Na, 39.7% Cr, and 42.8% O. What is the empirical formula of this compound? (Sodium dichromate is ionic, so it has no molecular formula.)

Solution

Assume for the purposes of this calculation that we have 100.0 g of sodium dichromate. Of this quantity, 17.5 g is Na (the substance is 17.5% Na). Similarly, this sample contains 39.7 g Cr and 42.8 g O. We convert these amounts to moles.

$$17.5 \cancel{\text{g Na}} \times \frac{1 \text{ mol Na}}{23.0 \cancel{\text{g Na}}} = 0.761 \text{ mol Na}$$

$$39.7 \cancel{\text{g Cr}} \times \frac{1 \text{ mol Cr}}{52.0 \cancel{\text{g Cr}}} = 0.763 \text{ mol Cr}$$

$$42.8\ \cancel{g\ O} \times \frac{1\ \text{mol O}}{16.0\ \cancel{g\ O}} = 2.68\ \text{mol O}$$

Now we divide all the mole numbers by the smallest one.

$$\text{For Na:} \quad \frac{0.761\ \cancel{mol}}{0.761\ \cancel{mol}} = 1.00$$

$$\text{For Cr:} \quad \frac{0.763\ \cancel{mol}}{0.761\ \cancel{mol}} = 1.00$$

$$\text{For O:} \quad \frac{2.68\ \cancel{mol}}{0.761\ \cancel{mol}} = 3.52$$

We must decide whether these numbers are integers, within experimental error. If we round off the rightmost digit, which is subject to experimental error, we get $Na_{1.0}Cr_{1.0}O_{3.5}$. At this stage in the calculation, there are two possibilities: (1) The subscripts are all integers, in which case the empirical formula has been found. This is not the case here. (2) The subscripts are not all integers but can be made into integers by multiplying each one by some whole number. If we multiply the subscripts that we have calculated by 2, we get $Na_{2.0}Cr_{2.0}O_{7.0}$. Thus, the empirical formula is $Na_2Cr_2O_7$.

Figure 4.6
Chromium compounds of different colors. *Left to right:* Potassium chromate, K_2CrO_4; chromium (III) oxide, Cr_2O_3; potassium dichromate, $K_2Cr_2O_7$; chromium(III) chloride, $CrCl_3$; and potassium chromium(III) sulfate, $KCr(SO_4)_2$. Chromium metal is in the center.

Exercise 4.10

Benzoic acid is a white, crystalline powder used as a food preservative. The compound contains 68.8% C, 5.0% H, and 26.2% O, by mass. What is its empirical formula?

(See Problems 4.53, 4.54, 4.55, and 4.56.)

Molecular Formula from Empirical Formula

The molecular formula of a compound is a multiple of its empirical formula. Thus, the molecular formula of acetylene, C_2H_2, is equivalent to $(CH)_2$, and the molecular formula of benzene, C_6H_6, is equivalent to $(CH)_6$. Therefore, the molecular weight is some multiple of the empirical formula weight, which is obtained by summing the atomic weights from the empirical formula. For any molecular compound, we can write

$$\text{Molecular weight} = n \times \text{empirical formula weight}$$

where n is the number of empirical formula units in the molecule. We get the molecular formula by multiplying the subscripts of the empirical formula by n, which we calculate from the equation

$$n = \frac{\text{molecular weight}}{\text{empirical formula weight}}$$

Once we determine the empirical formula for a compound, we can calculate its empirical formula weight. If we have an experimental determination of its molecular weight, we can calculate n and then the molecular formula. The next example illustrates how we use percentage composition and molecular weight to determine the molecular formula of acetic acid.

Example 4.11 Determining the Molecular Formula from Percentage Composition and Molecular Weight

In Example 4.8, we found the percentage composition of acetic acid to be 39.9% C, 6.7% H, and 53.4% O. Determine the empirical formula. The molecular weight of acetic acid was determined by experiment to be 60.0 amu. What is its molecular formula?

Solution

A sample of 100.0 g of acetic acid contains 39.9 g C, 6.7 g H, and 53.4 g O. Converting these masses to moles gives 3.33 mol C, 6.6 mol H, and 3.34 mol O. Dividing these mole numbers by the smallest one gives 1.00 for C, 2.0 for H, and 1.00 for O. The empirical formula of acetic acid is CH_2O. (You may have noted that the percentage composition of acetic acid is, within experimental error, the same as that of formaldehyde—see Example 4.6—so they must have the same empirical formula.) The formula weight is 30.0 amu. Dividing the formula weight into the molecular weight gives the number by which the subscripts in CH_2O must be multiplied.

$$n = \frac{\text{MW}}{\text{FW}} = \frac{60.0 \text{ amu}}{30.0 \text{ amu}} = 2.00$$

The molecular formula of acetic acid is $(CH_2O)_2$ or $C_2H_4O_2$.

Exercise 4.11

The percentage composition of acetaldehyde is 54.5% C, 9.2% H, and 36.3% O, and its molecular weight is 44 amu. Obtain the molecular formula of acetaldehyde.

(See Problems 4.57, 4.58, 4.59, and 4.60.)

The formula of acetic acid is often written $HC_2H_3O_2$ to indicate that one of the hydrogen atoms is acidic (lost easily) while the other three are not. Now that we know the formulas of acetic acid and acetaldehyde (determined from the data in Exercise 4.11 to be C_2H_4O), we can write the equations for the industrial preparation of acetic acid described in the chapter opener.● The first step consists of reacting ethanol with oxygen to obtain acetaldehyde and water. If we write the chemical equation and balance it, we obtain

● The condensed structural formula for acetaldehyde is CH_3CHO. Its structural formula is

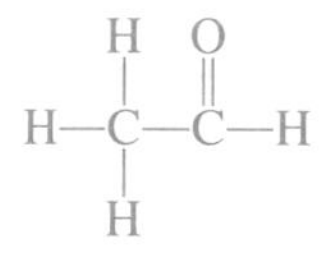

To determine this structural formula requires additional information.

$$\underset{\text{ethanol}}{2C_2H_5OH} + O_2 \longrightarrow \underset{\text{acetaldehyde}}{2C_2H_4O} + 2H_2O$$

In practice, the reaction is carried out in the gas phase at about 400°C using silver as a catalyst.

The second step consists of reacting acetaldehyde with oxygen to obtain acetic acid. Acetaldehyde liquid is mixed with a catalyst (manganese(II) acetate is used), and air is bubbled through it. The balanced equation is

$$\underset{\text{acetaldehyde}}{2C_2H_4O} + O_2 \longrightarrow \underset{\text{acetic acid}}{2HC_2H_3O_2}$$

Once we have this balanced equation, we are in position to answer quantitative questions such as, How much acetic acid could we obtain from a 10.0-g sample of acetaldehyde? We will see how to answer such questions in the next sections.

Stoichiometry: Quantitative Relations in Chemical Reactions

In Chapter 3, we described a chemical equation as a representation of what occurs when molecules (or ions) react. We will now study chemical reactions more closely to answer questions about the stoichiometry of reactions. **Stoichiometry** (pronounced "stoy-key-om′-e-tree") is *the calculation of the quantities of reactants and products involved in a chemical reaction.* It is based on the chemical equation and on the relationship between mass and moles. Such calculations are fundamental to most quantitative work in chemistry. In the next sections we will use the industrial Haber process to illustrate stoichiometric calculations.

4.6 MOLAR INTERPRETATION OF A CHEMICAL EQUATION

In the Haber process for producing ammonia, NH_3, nitrogen reacts with hydrogen at high temperature and pressure.●

● See the Profile of ammonia at the end of this chapter.

$$N_2(g) + 3H_2(g) \longrightarrow 2NH_3(g)$$

Hydrogen is usually obtained from natural gas or petroleum and so is relatively expensive. For this reason, the price of hydrogen partly determines the price of ammonia. Thus, an important question to answer is, How much hydrogen is required to give a particular quantity of ammonia? For example, how much hydrogen would be needed to produce one ton (907 kg) of ammonia? Similar kinds of questions arise throughout chemical research and industry.

To answer such quantitative questions, we must first look at the chemical equation. The equation for the preparation of ammonia by the Haber process tells us that one N_2 molecule and three H_2 molecules react to produce two NH_3 molecules. A similar statement involving multiples of these numbers of molecules is also correct. For example, 6.02×10^{23} N_2 molecules react with $3 \times 6.02 \times 10^{23}$ H_2 molecules, giving $2 \times 6.02 \times 10^{23}$ NH_3 molecules. This last statement can be put in molar terminology: One mole of N_2 reacts with three moles of H_2 to give two moles of NH_3. We conclude that we may interpret a chemical equation either in terms of numbers of molecules (or ions or formula units) or in terms of numbers of moles, depending on our needs.

Because moles can be converted to mass, we can also give a mass interpretation of a chemical equation. The molar masses of N_2, H_2, and NH_3 are 28.0, 2.02, and 17.0 g/mol, respectively. Therefore 28.0 g of N_2 reacts with 3×2.02 g of H_2 to yield 2×17.0 g of NH_3.

We summarize these three interpretations as follows:

N_2	+	$3H_2$	$\longrightarrow$	$2NH_3$	
1 molecule N_2	+	3 molecules H_2	$\longrightarrow$	2 molecules NH_3	(molecular interpretation)
1 mol N_2	+	3 mol H_2	$\longrightarrow$	2 mol NH_3	(molar interpretation)
28.0 g N_2	+	3×2.02 g H_2	$\longrightarrow$	2×17.0 g NH_3	(mass interpretation)

Suppose we ask how many grams of atmospheric nitrogen will react with 6.06 g of hydrogen. We see from the last equation that the answer is 28.0 g N_2. We formulated this question for one mole of atmospheric nitrogen. Recalling the question posed earlier, suppose we ask how much hydrogen is needed to yield 907 kg of ammonia in the Haber process. The solution to this problem depends on the fact that *the number of moles involved in a reaction is proportional to the coefficients in the balanced chemical equation.* In the next section, we will describe a procedure for solving such problems.

Exercise 4.12

In an industrial process, hydrogen chloride, HCl, is prepared by burning hydrogen gas, H_2, in an atmosphere of chlorine, Cl_2. Write the chemical equation for the reaction. Below the equation, give the molecular, molar, and mass interpretations.

(See Problems 4.61 and 4.62.)

4.7 STOICHIOMETRY OF A CHEMICAL REACTION

The chemical equation for the Haber process,

$$N_2 + 3H_2 \longrightarrow 2NH_3$$

tells us that three moles of H_2 produce two moles of NH_3. If we write this as a mathematical equation (the symbol $\simeq$ means "is chemically equivalent to"),

$$3 \text{ mol } H_2 \simeq 2 \text{ mol } NH_3$$

we can obtain a factor that converts from moles of hydrogen to moles of ammonia (2 mol NH_3/3 mol H_2) or from moles of ammonia to moles of hydrogen (3 mol H_2/2 mol NH_3).●

● These conversion factors simply express the fact that the mole ratio of NH_3 to H_2 in the reaction is 2 to 3.

Now we can see how to calculate the mass of hydrogen required to produce 907 kg ammonia (9.07×10^5 g NH_3). Because the chemical equation directly relates moles NH_3 to moles H_2, we first convert grams NH_3 to moles NH_3.

$$9.07 \times 10^5 \text{ g } NH_3 \times \frac{1 \text{ mol } NH_3}{17.0 \text{ g } NH_3} = 5.34 \times 10^4 \text{ mol } NH_3$$

Now we note from the chemical equation that 2 mol NH_3 requires 3 mol H_2. We can use this information to obtain the conversion from moles NH_3 to moles H_2.

$$5.34 \times 10^4 \text{ mol } NH_3 \times \frac{3 \text{ mol } H_2}{2 \text{ mol } NH_3} = 8.01 \times 10^4 \text{ mol } H_2$$

Finally, we convert moles H_2 to grams H_2.

$$8.01 \times 10^4 \text{ mol } H_2 \times \frac{2.02 \text{ g } H_2}{1 \text{ mol } H_2} = 1.62 \times 10^5 \text{ g } H_2$$

Normally we would do this calculation by putting all of these factors together, as follows:

$$9.07 \times 10^5 \cancel{\text{g } NH_3} \times \frac{1 \cancel{\text{mol } NH_3}}{17.0 \cancel{\text{g } NH_3}} \times \frac{3 \cancel{\text{mol } H_2}}{2 \cancel{\text{mol } NH_3}} \times \frac{2.02 \text{ g } H_2}{1 \cancel{\text{mol } H_2}} =$$
$$1.62 \times 10^5 \text{ g } H_2 \quad (\text{or } 162 \text{ kg } H_2)$$

Note that the units in the denominator of each conversion factor cancel the units in the numerator of the preceding factor. The following examples illustrate additional variations on this type of calculation.

Example 4.12 Relating the Quantity of Reactant to Quantity of Product

Hematite, Fe_2O_3, is an important ore of iron; see Figure 4.7. (An ore is a natural substance from which the metal can be profitably obtained.) The free metal is obtained by reacting hematite with carbon monoxide, CO, in a blast furnace. Carbon monoxide is formed in the furnace by partial combustion of carbon. The reaction in which the metal is produced is

$$Fe_2O_3(s) + 3CO(g) \longrightarrow 2Fe(s) + 3CO_2(g)$$

How many grams of iron can be produced from 1.00 kg Fe_2O_3?

Figure 4.7
Hematite. The name of this iron mineral stems from the Greek word for blood, which alludes to the color of certain forms of the mineral.

Solution

The calculation involves the conversion of a quantity of Fe_2O_3 to a quantity of Fe. An essential feature of this type of calculation is the conversion of moles of a given substance to moles of another substance. Therefore, we first convert the mass of Fe_2O_3 (1.00 kg Fe_2O_3 = 1.00×10^3 g Fe_2O_3) to moles Fe_2O_3. Then we convert moles Fe_2O_3 to moles Fe and to grams Fe. We use the information shown below to obtain the necessary conversion factors:

1 mol Fe_2O_3 = 160 g Fe_2O_3
(from the molar mass of Fe_2O_3)

1 mol Fe_2O_3 $\simeq$ 2 mol Fe
(from the balanced chemical equation)

1 mol Fe = 55.8 g Fe
(from the molar mass of Fe)

The calculation is as follows:

$$1.00 \times 10^3 \text{ g } Fe_2O_3 \times \frac{1 \text{ mol } Fe_2O_3}{160 \text{ g } Fe_2O_3} \times \frac{2 \text{ mol Fe}}{1 \text{ mol } Fe_2O_3} \times \frac{55.8 \text{ g Fe}}{1 \text{ mol Fe}} = \mathbf{698 \text{ g Fe}}$$

Exercise 4.13

Sodium is a soft, reactive metal that instantly reacts with water to give hydrogen gas and a solution of sodium hydroxide, NaOH. How many grams of sodium metal are needed to give 7.81 g of hydrogen by this reaction? (Remember to write the balanced equation first.) (See Problems 4.63, 4.64, 4.65, and 4.66.)

Example 4.13 Relating the Quantities of Two Reactants (or Two Products)

Today chlorine is prepared from sodium chloride by electrochemical decomposition. Formerly chlorine was produced by heating hydrochloric acid with pyrolusite (manganese dioxide, MnO_2), a common manganese ore. Small amounts of chlorine may be prepared in the laboratory by the same reaction (see Figure 4.8):

$$4HCl(aq) + MnO_2(s) \longrightarrow 2H_2O(l) + MnCl_2(aq) + Cl_2(g)$$

How many grams of HCl react with 5.00 g of manganese dioxide, according to this equation?

Figure 4.8
Preparation of chlorine. Concentrated hydrochloric acid was added to manganese dioxide in the beaker. Note the formation of yellowish-green gas (chlorine).

Solution

We write down what is given (5.00 g MnO_2) and convert this to moles, then to moles of what is desired (mol HCl). Finally we convert this to mass (g HCl). We use the following information to obtain conversion factors:

1 mol MnO_2 = 86.9 g MnO_2
(from the molar mass of MnO_2)

1 mol MnO_2 $\simeq$ 4 mol HCl
(from the balanced chemical equation)

1 mol HCl = 36.5 g HCl
(from the molar mass of HCl)

The calculations are

$$5.00 \text{ g } MnO_2 \times \frac{1 \text{ mol } MnO_2}{86.9 \text{ g } MnO_2} \times \frac{4 \text{ mol HCl}}{1 \text{ mol } MnO_2} \times \frac{36.5 \text{ g HCl}}{1 \text{ mol HCl}} = \mathbf{8.40 \text{ g HCl}}$$

Exercise 4.14

Sphalerite is a zinc sulfide (ZnS) mineral and an important commercial source of zinc metal. The first step in the processing of the ore consists of heating the sulfide with oxygen to give zinc oxide, ZnO, and sulfur dioxide, SO_2. How many kilograms of oxygen gas combine with 5.00×10^3 g of zinc sulfide in this reaction? (You must first write out the balanced chemical equation.) (See Problems 4.67 and 4.68.)

Exercise 4.15

The British chemist Joseph Priestley prepared oxygen in 1774 by heating mercury(II) oxide, HgO. Mercury metal is the other product. If 6.47 g of oxygen are collected, how many grams of mercury metal are also produced? (See Problems 4.69 and 4.70.)

4.8 LIMITING REACTANT; THEORETICAL AND PERCENTAGE YIELDS

Often reactants are added to a reaction vessel in amounts different from the molar proportions given by the chemical equation. In such cases, one of the reactants may be completely consumed at the end of the reaction, whereas some amounts of other reactants will remain unreacted. The **limiting reactant** (or **limiting reagent**) is *the reactant that is entirely consumed when a reaction goes to completion*. A reactant that is not completely consumed is often referred to as an *excess reactant*. Once one of the reactants is used up, the reaction stops. This means that the moles of product are always determined by the starting moles of limiting reactant.

To make sure you understand the concept of limiting reactant, consider a simple analogy. Suppose you are supervising the assembly of automobiles. Your plant has in stock 300 steering wheels and 900 tires, plus an excess of every other needed component. How many autos can you assemble from this stock? One way to solve this problem is to calculate the number of autos you could assemble from each component assuming everything else is in excess. From 300 steering wheels, you could assemble 300 autos; and from 900 tires, you could assemble 900 ÷ 4 = 225 autos. But by the time you have assembled 225 autos, you will have exhausted your stock of tires, and no more autos can be assembled. Tires are the *limiting* component, and they determine the number of autos you can assemble.

Now consider a chemical reaction, the burning of hydrogen in oxygen.

$$2H_2(g) + O_2(g) \longrightarrow 2H_2O(g)$$

If we add the reactants H_2 and O_2 to a vessel in the molar proportions given by the chemical equation, both will be used up when the reaction is complete. Thus, 2 mol H_2 and 1 mol O_2 react completely to give the product, H_2O. However, suppose we place only 1 mol H_2 in the reaction vessel with the same amount of O_2 (1 mol O_2). After the reaction is complete, all of the H_2 is consumed but $\frac{1}{2}$ mol O_2 remains unreacted.

We cannot obtain the amount of H_2O produced in the reaction by converting from the moles O_2 added to the reaction vessel to moles H_2O, because only part of the O_2 reacts. However, since the H_2 is entirely consumed, we can use the starting amount of this reactant (1 mol H_2) to obtain the amount of H_2O produced. From the chemical equation, we note that

$$2 \text{ mol } H_2 \simeq 2 \text{ mol } H_2O$$

This gives us the conversion factor from moles H_2 to moles H_2O. Therefore, the amount of H_2O produced is

$$1 \cancel{\text{mol } H_2} \times \frac{2 \text{ mol } H_2O}{2 \cancel{\text{mol } H_2}} = 1 \text{ mol } H_2O$$

Let us summarize the situation. Suppose we are given the amounts of reactants added to a vessel and wish to calculate the amount of product obtained when the reaction is complete. Unless we know that the reactants have been added in the molar proportions given by the chemical equation, the problem is twofold: (1) We must first identify the limiting reactant; (2) we then calculate the amount of product from the amount of limiting reactant. The next example illustrates both steps.

Example 4.14 Calculating with a Limiting Reactant (Involving Moles)

Zinc metal reacts with hydrochloric acid by the following reaction:

$$Zn(s) + 2HCl(aq) \longrightarrow ZnCl_2(aq) + H_2(g)$$

If 0.30 mol Zn are added to hydrochloric acid containing 0.52 mol HCl, how many moles H_2 are produced?

Solution

STEP 1 Which of the two reactants is the limiting reactant? To answer this, we take each reactant in turn and ask how much product (H_2) would be obtained if each were totally consumed. The reactant that gives the smaller amount of product is the limiting reactant.

$$0.30 \text{ mol Zn} \times \frac{1 \text{ mol } H_2}{1 \text{ mol Zn}} = 0.30 \text{ mol } H_2$$

$$0.52 \text{ mol HCl} \times \frac{1 \text{ mol } H_2}{2 \text{ mol HCl}} = 0.26 \text{ mol } H_2$$

We see that hydrochloric acid must be the limiting reactant and that some zinc must be left unconsumed (zinc is the excess reactant).

STEP 2 We obtain the amount of product actually obtained from the amount of limiting reactant (HCl), so the amount of H_2 produced must be **0.26 mol.**

Exercise 4.16

Aluminum chloride, $AlCl_3$, is used as a catalyst in various industrial reactions. It is prepared from hydrogen chloride gas and aluminum metal shavings.

$$2Al(s) + 6HCl(g) \longrightarrow 2AlCl_3(s) + 3H_2(g)$$

Suppose a reaction vessel contains 0.15 mol Al and 0.35 mol HCl. How many moles $AlCl_3$ can be prepared from this mixture? (See Problems 4.71 and 4.72.)

Example 4.15 Calculating with a Limiting Reactant (Involving Masses)

In an industrial process for producing acetic acid, oxygen gas is bubbled into acetaldehyde, CH_3CHO, containing manganese(II) acetate (catalyst) under pressure at 60°C.

$$2CH_3CHO(l) + O_2(g) \longrightarrow 2HC_2H_3O_2(l)$$

In a laboratory test of this reaction, 20.0 g CH_3CHO and 10.0 g O_2 were put into a reaction vessel. (a) How many grams of acetic acid can be produced by this reaction from these amounts of reactants? (b) How many grams of the excess reactant remain after the reaction is complete?

Solution

(a) This problem is similar to the preceding one, but now we must include the conversion of masses to moles.

STEP 1 We must first determine which of the reactants, CH_3CHO or O_2, is limiting. So we convert grams of reactants to moles of product $HC_2H_3O_2$.

$$20.0 \text{ } \cancel{\text{g } CH_3CHO} \times \frac{1 \text{ } \cancel{\text{mol } CH_3CHO}}{44.1 \text{ } \cancel{\text{g } CH_3CHO}} \times \frac{2 \text{ mol } HC_2H_3O_2}{2 \text{ } \cancel{\text{mol } CH_3CHO}} = 0.454 \text{ mol } HC_2H_3O_2$$

$$10.0 \text{ } \cancel{\text{g } O_2} \times \frac{1 \text{ } \cancel{\text{mol } O_2}}{32.0 \text{ } \cancel{\text{g } O_2}} \times \frac{2 \text{ mol } HC_2H_3O_2}{1 \text{ } \cancel{\text{mol } O_2}} = 0.625 \text{ mol } HC_2H_3O_2$$

Acetaldehyde, CH_3CHO, is the limiting reactant.

STEP 2 Because acetaldehyde is the limiting reactant, the grams of acetic acid produced are

$$0.454 \text{ } \cancel{\text{mol } HC_2H_3O_2} \times \frac{60.1 \text{ g } HC_2H_3O_2}{1 \text{ } \cancel{\text{mol } HC_2H_3O_2}} = 27.3 \text{ g } HC_2H_3O_2$$

(continued)

(b) To calculate the mass of the excess reactant O_2 at the completion of the reaction, we need to know how many grams have reacted (that is, how many grams of O_2 are needed to prepare 0.455 mol $HC_2H_3O_2$).

$$0.454 \cancel{\text{mol } HC_2H_3O_2} \times \frac{1 \cancel{\text{mol } O_2}}{2 \cancel{\text{mol } HC_2H_3O_2}} \times \frac{32.0 \text{ g } O_2}{1 \cancel{\text{mol } O_2}} = 7.26 \text{ g } O_2$$

To find the grams O_2 remaining, we subtract the mass consumed from the total available (10.0 g).

$$(10.0 - 7.26) \text{ g } O_2 = 2.7 \text{ g } O_2 \text{ (mass remaining)}$$

Exercise 4.17

In an experiment, 7.36 g of zinc was heated with 6.45 g of sulfur. Assume that these substances react according to the equation

$$8Zn + S_8 \longrightarrow 8ZnS$$

What amount of zinc sulfide was produced by the reaction?

(See Problems 4.73 and 4.74.)

The **theoretical yield** of a product is *the maximum amount of product that can be obtained by a reaction from given amounts of reactants*. It is the amount that we calculate from the stoichiometry based on the limiting reactant. In Example 4.15, the theoretical yield of acetic acid is 27.3 g. In practice, the *actual yield* of a product may be much less than this for several possible reasons. First, some product may be lost during the process of separating it from the final reaction mixture. Second, there may be other, competing reactions that occur simultaneously with the reaction on which the theoretical yield is based. Finally, many reactions appear to stop before they reach completion; they give mixtures of reactants and products.●

● Such reactions reach chemical equilibrium. We will discuss equilibrium quantitatively in Chapter 15.

It is important to know the actual yield from a reaction in order to make economic decisions about a method of preparation. The reactants for a given method may not be too costly per kilogram, but if the actual yield is very low, the final cost can be very high. The **percentage yield** of product is *the actual yield (experimentally determined) expressed as a percentage of the theoretical yield (calculated)*.

$$\text{Percentage yield} = \frac{\text{actual yield}}{\text{theoretical yield}} \times 100\%$$

To illustrate the calculation of percentage yield, recall that the theoretical yield of acetic acid calculated in Example 4.15 was 27.3 g. If the actual yield of acetic acid obtained in an experiment, using the amounts of reactants given in Example 4.15, is 23.8 g, then

$$\text{Percentage yield of } HC_2H_3O_2 = \frac{23.8 \cancel{\text{g}}}{27.3 \cancel{\text{g}}} \times 100\% = 87.2\%$$

Exercise 4.18

New industrial plants for acetic acid react liquid methanol with carbon monoxide in the presence of a catalyst.

$$CH_3OH(l) + CO(g) \longrightarrow HC_2H_3O_2(l)$$

In an experiment, 15.0 g of methanol and 10.0 g of carbon monoxide were placed in a reaction vessel. What is the theoretical yield of acetic acid? If the actual yield is 19.1 g, what is the percentage yield? (See Problems 4.77 and 4.78.)

Calculations Involving Solutions

Reaction between two solid reactants often proceeds very slowly or not at all. This is because the molecules or ions in a crystal tend to occupy approximately fixed positions, so that the chance of two molecules or ions coming together to react is small. For this reason, most reactions are run in liquid solutions. Reactant molecules are free to move throughout the liquid, and reaction is much faster. When we run reactions in liquid solutions, it is convenient to dispense the amounts of reactants by measuring out volumes of reactant solution. In the next sections, we will discuss calculations involving volumes of solutions.

4.9 MOLAR CONCENTRATION

When we dissolve a substance in a liquid, we call the substance the *solute* and the liquid the *solvent*. Consider ammonia solutions. Ammonia gas dissolves readily in water (Figure 4.9), and aqueous ammonia solutions are often used in the laboratory. In such solutions, ammonia gas is the solute and water is the solvent.

Concentration is the general term used in referring to the quantity of solute in a standard quantity of solution. Qualitatively, we say that a solution is *dilute* when the solute concentration is low and that the solution is *concentrated* when the solute concentration is high. Usually these terms are used in a comparative sense and do not refer to a specific concentration. We say that one solution is more dilute or less concentrated than another. However, for commercially available solutions, the term *concentrated* refers to the maximum concentration available, or approxi-

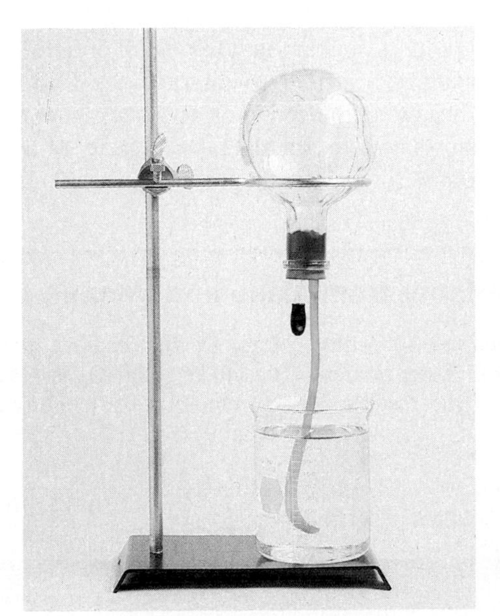

Figure 4.9
The ammonia fountain. *Left:* The flask contains ammonia gas; the beaker and dropper contain water with a small amount of phenolphthalein indicator. To start the fountain, water was squeezed from the dropper into the flask. When water enters the flask, the ammonia dissolves in it, creating a vacuum. Then, atmospheric pressure forces the water in the beaker into the flask. The ammonia turns the phenolphthalein pink. *Right:* A close-up view of the ammonia fountain.

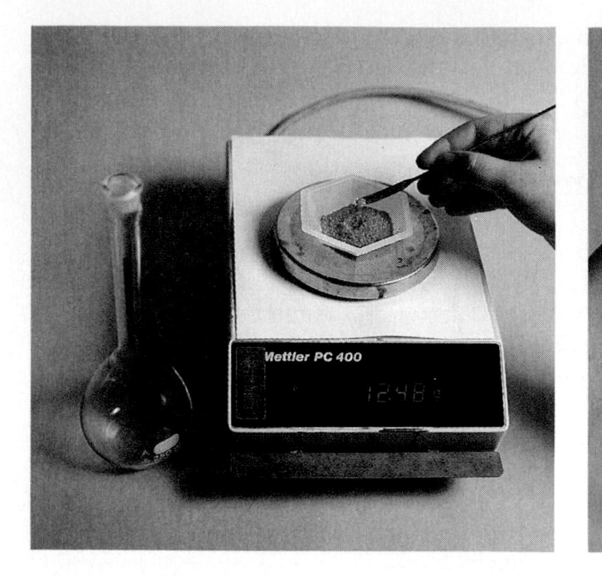

Figure 4.10
Preparing a 0.200 *M* $CuSO_4$ solution. *Left:* 0.0500 mol $CuSO_4 \cdot 5H_2O$ (12.48 g) is weighed on a platform balance. *Center:* The copper(II) sulfate pentahydrate is transferred carefully to the volumetric flask. *Right:* Water is added to bring the solution level to the mark on the neck of the 250-mL volumetric flask. The molarity is 0.0500 mol/0.250 L = 0.200 *M*.

mately this concentration. Thus, concentrated aqueous ammonia contains about 28% NH_3 by mass.

In this example, we expressed the concentration quantitatively by giving the mass percentage of solute—that is, the mass of solute in 100 grams of solution. However, our interest at the moment is in a unit of concentration that is convenient for doing reactions in solution. Molar concentration is such a unit.

Molar concentration or **molarity (*M*)** is defined as *the moles of solute dissolved in one liter (cubic decimeter) of solution.*

$$\text{Molarity } (M) = \frac{\text{moles of solute}}{\text{liters of solution}}$$

An aqueous solution that is 0.15 *M* NH_3 (read this as "0.15 molar NH_3") contains 0.15 mol NH_3 per liter of solution. If we want to prepare a solution that is, for example, 0.200 *M* $CuSO_4$, we place 0.200 mol $CuSO_4$ in a 1.000-L volumetric flask, or a proportional amount in a flask of a different size (Figure 4.10). We add a small quantity of water to dissolve the $CuSO_4$. Then we fill the flask with additional water to the mark on the neck and mix the solution. The following example shows how to calculate the molarity of a solution given the mass of solute and the volume of solution.

Example 4.16 Calculating Molarity from Mass and Volume

A sample of $NaNO_3$ weighing 0.38 g is placed in a 50.0-mL volumetric flask. The flask is then filled with water to the mark on the neck. What is the molarity of the resulting solution?

Solution

By definition, molarity is moles solute/liters solution. Therefore, we first convert grams of $NaNO_3$ to moles. (We find that 0.38 g $NaNO_3$ is 4.47×10^{-3} mol $NaNO_3$.) To obtain the molarity, we then divide the moles of solute by liters of solution (soln). Note that 50.0 mL = 50.0×10^{-3} L.

$$\text{Molarity} = \frac{4.47 \times 10^{-3} \text{ mol NaNO}_3}{50.0 \times 10^{-3} \text{ L soln}} = \mathbf{0.089}\ \boldsymbol{M}\ \mathbf{NaNO_3}$$

Exercise 4.19

A sample of sodium chloride, NaCl, weighing 0.0678 g is placed in a 25.0-mL volumetric flask. Enough water is added to dissolve the NaCl, and then the flask is filled with water and carefully shaken to mix the contents. What is the molarity of the resulting solution? (See Problems 4.79, 4.80, 4.81, and 4.82.)

The advantage of molarity as a concentration unit is that, for a solution of known molarity, the amount of solute is related to the volume of solution. Rather than having to weigh out a specified mass of substance, we can instead measure out a definite volume of solution of the substance, which is usually easier to do. As the following example illustrates, molarity can be used as a factor for converting from moles of solute to liters of solution, and vice versa.

Example 4.17 Using Molarity as a Conversion Factor

An experiment calls for the addition to a reaction vessel of 0.184 g of sodium hydroxide, NaOH, in aqueous solution. How many milliliters of 0.150 *M* NaOH should be added to the reaction vessel?

Solution

We first convert mass to moles (because molarity is a direct relation between moles and volume).

$$0.184 \text{ g NaOH} \times \frac{1 \text{ mol NaOH}}{40.0 \text{ g NaOH}} = 4.60 \times 10^{-3} \text{ mol NaOH}$$

Then we convert this amount of NaOH to volume of solution, using molarity as a conversion factor. In a 0.150 *M* NaOH solution, we have

$$1 \text{ L soln} \simeq 0.150 \text{ mol NaOH}$$

Therefore, the volume of solution equivalent to 4.60×10^{-3} mol NaOH is

$$4.60 \times 10^{-3} \text{ mol NaOH} \times \frac{1 \text{ L soln}}{0.150 \text{ mol NaOH}} = 3.07 \times 10^{-2} \text{ L soln} \quad \text{(or 30.7 mL)}$$

We need to add **30.7 mL** of 0.150 NaOH solution to the reaction vessel.

Exercise 4.20

How many milliliters of 0.163 *M* NaCl are required to give 0.0958 g of sodium chloride? (See Problems 4.83, 4.84, 4.85, and 4.86.)

Exercise 4.21

How many moles of sodium chloride should be put in a 50.0-mL volumetric flask to give a 0.15 *M* NaCl solution when the flask is filled with water? How many grams of NaCl is this? (See Problems 4.87, 4.88, 4.89, and 4.90.)

4.10 DILUTING SOLUTIONS

Commercially available aqueous ammonia (28.0% NH_3) is 14.8 *M* NH_3. Suppose, however, that we want a solution that is 1.00 *M* NH_3. We need to dilute the concentrated solution with a definite quantity of water. For this purpose, we must know the relationship between the molarity of the solution before dilution (the *initial molarity*) and that after dilution (the *final molarity*).

To obtain this relationship, first recall the equation defining molarity:

$$\text{Molarity} = \frac{\text{moles of solute}}{\text{liters of solution}}$$

This can be rearranged to give

$$\text{Moles of solute} = \text{molarity} \times \text{liters of solution}$$

The product of molarity and the volume (in liters) gives the number of moles of solute in the solution. Writing M_i for the initial molar concentration and V_i for the initial volume of solution, we get

$$\text{Moles of solute} = M_i \times V_i$$

When the solution is diluted by adding more water, the concentration and volume change to M_f (the final molar concentration) and V_f (the final volume), and the moles of solute are

$$\text{Moles of solute} = M_f \times V_f$$

Because the number of moles of solute has not changed during the dilution,

$$M_i \times V_i = M_f \times V_f$$

(Note: You can use any units for volume, but both V_i and V_f must be in the same units.) Example 4.18 shows how to use this relationship in dilution problems.

Example 4.18 Calculating the Dilution of a Solution (from One Molarity to Another)

You are given 5.00 mL of 14.8 M NH_3. What will the final volume be after this solution is diluted with water to give 1.00 M NH_3?

Solution

We write the dilution formula just given.

$$M_iV_i = M_fV_f$$

Then we rearrange it to give the final volume.

$$V_f = \frac{M_iV_i}{M_f}$$

After substituting the values given in the problem, we get

$$V_f = \frac{5.00\ \text{mL} \times 14.8\ \cancel{M}}{1.00\ \cancel{M}} = 74.0\ \text{mL}$$

Exercise 4.22

What is the final volume of a solution of sulfuric acid if you dilute 48 mL of 1.5 M H_2SO_4 to produce a solution of 0.18 M H_2SO_4? (See Problems 4.91 and 4.92.)

Exercise 4.23

You have a solution that is 14.8 M NH_3. How many milliliters of this solution are required to give 50.0 mL of 1.00 M NH_3 when diluted? (See Problems 4.93 and 4.94.)

The next example, a common laboratory problem, is somewhat more complicated than the previous one. Here again, we must emphasize the importance of being conversant with skills learned previously so that you will be able to use them with ease. This is the ''secret'' of problem solving.

Example 4.19 Calculating the Dilution of a Solution (from Mass Percentage to Molarity)

Commercially available concentrated hydrochloric acid is an aqueous solution containing 38% HCl by mass. (a) What is the molarity of this solution? The density is 1.19 g/mL. (b) How many milliliters of concentrated HCl are required to make 1.00 L of 0.10 *M* HCl?

Solution

(a) To calculate the molarity, we need to obtain the moles of HCl in some volume of the concentrated acid. We can select any convenient volume, say 1.00 L (or 1.00×10^3 mL), and calculate the mass of the solution. Knowing the mass percentage of HCl present, we can obtain the mass of HCl and then the moles of HCl (from which we obtain the molarity).

$$\begin{aligned}\text{Mass of solution} &= \text{volume} \times \text{density} \\ &= 1.00 \times 10^3 \ \cancel{\text{mL}} \times 1.19 \ \text{g}/\cancel{\text{mL}} \\ &= 1.19 \times 10^3 \ \text{g}\end{aligned}$$

Because the solution is 38% HCl by mass, the mass of HCl is

$$\begin{aligned}\text{mass HCl} &= 1.19 \times 10^3 \ \text{g} \times 0.38 \\ &= 452 \ \text{g (retaining extra digit)}\end{aligned}$$

Then $$452 \ \cancel{\text{g HCl}} \times \frac{1 \ \text{mol HCl}}{36.5 \ \cancel{\text{g HCl}}} = 12.4 \ \text{mol HCl}$$

This is the moles of HCl in 1.00 L solution, so the molarity is 12.4 *M*. Because the mass percentage of HCl was given to two significant figures, the answer should be reported as **12 *M* HCl.** We will retain the extra figure for the next calculation. (b) We wish to dilute an initial volume of acid (not known) of initial molarity (12.4 *M*) to give a final volume ($1.00 \ \text{L} = 1.00 \times 10^3$ mL) and a final molarity (0.10 *M*). We solve the dilution formula for V_i.

$$V_i = \frac{M_f V_f}{M_i}$$

Substituting, we find

$$V_i = \frac{0.10 \ \cancel{M} \times 1.00 \times 10^3 \ \text{mL}}{12.4 \ \cancel{M}} = \mathbf{8.1 \ mL}$$

Note that solving this problem requires an understanding of density, mass percentage, conversion of mass to moles, molarity, and dilution of solutions.

Exercise 4.24

Commercially available concentrated sulfuric acid is 95% H_2SO_4 by mass and has a density of 1.84 g/mL. How many milliliters of this acid are needed to give 1.0 L of 0.15 *M* H_2SO_4?

(See Problems 4.95 and 4.96.)

4.11 STOICHIOMETRY OF SOLUTION REACTIONS

Now that we see how the amount of solute can be measured out by volume of solution, let us look at the stoichiometry of solution reactions—that is, the calculation of volumes involved in a reaction. The next example illustrates the method.

Example 4.20 Calculating the Volume of Reactant Solution Needed

Consider the reaction of sulfuric acid, H_2SO_4, with sodium hydroxide, NaOH.

$$H_2SO_4(aq) + 2NaOH(aq) \longrightarrow 2H_2O(l) + Na_2SO_4(aq)$$

Suppose a beaker contains 35.0 mL of 0.175 *M* H_2SO_4. How many milliliters of 0.250 *M* NaOH must be added to just react completely with the sulfuric acid?

Solution

We convert from 35.0 mL (or 35.0×10^{-3} L) H_2SO_4 solution to moles H_2SO_4 (using the molarity of H_2SO_4), then to moles NaOH (from the chemical equation). Finally, we convert this to volume of NaOH solution (using the molarity of NaOH). The calculation is as follows:

$$\begin{aligned}&35.0 \times 10^{-3} \ \cancel{\text{L } H_2SO_4 \text{ soln}} \times \frac{0.175 \ \cancel{\text{mol } H_2SO_4}}{1 \ \cancel{\text{L } H_2SO_4 \text{ soln}}} \\ &\quad \times \frac{2 \ \cancel{\text{mol NaOH}}}{1 \ \cancel{\text{mol } H_2SO_4}} \times \frac{1 \ \text{L NaOH soln}}{0.250 \ \cancel{\text{mol NaOH}}} \\ &= 4.90 \times 10^{-2} \ \text{L NaOH soln} \quad \text{(or 49.0 mL NaOH soln)}\end{aligned}$$

Thus, 35.0 mL of 0.175 *M* sulfuric acid solution reacts with **49.0 mL** of 0.250 *M* sodium hydroxide solution.

Figure 4.11
Titration of an unknown amount of HCl with NaOH. *Left:* The flask contains HCl and a few drops of phenolphthalein indicator; the buret contains 0.101 M NaOH (the buret reading is 21.00 mL). *Center:* NaOH was added to the solution in the flask until a persistent faint pink color, which is the end point, was reached (the buret reading is 24.35 mL). The amount of HCl can be determined from the volume of NaOH used (3.35 mL); see Example 4.21. *Right:* The addition of several drops of NaOH solution beyond the end point gives a deep pink color.

Exercise 4.25

Nickel sulfate, $NiSO_4$, reacts with trisodium phosphate, Na_3PO_4, to give a pale yellow-green precipitate of nickel phosphate, $Ni_3(PO_4)_2$, and a solution of sodium sulfate, Na_2SO_4.

$$3NiSO_4(aq) + 2Na_3PO_4(aq) \longrightarrow Ni_3(PO_4)_2(s) + 3Na_2SO_4(aq)$$

How many milliliters of 0.375 M $NiSO_4$ will react with 45.7 mL of 0.265 M Na_3PO_4?

(See Problems 4.97 and 4.98.)

Exercise 4.26

Sodium hydroxide, NaOH, may be prepared by reacting calcium hydroxide, $Ca(OH)_2$, with sodium carbonate, Na_2CO_3.

$$Ca(OH)_2(aq) + Na_2CO_3(aq) \longrightarrow 2NaOH(aq) + CaCO_3(s)$$

How many milliliters of 0.150 M $Ca(OH)_2$ will react with 2.55 g Na_2CO_3?

(See Problems 4.99 and 4.100.)

An important method for determining the amount of a particular substance is based on measuring volumes of reactant solutions. Suppose a substance A reacts in solution with substance B. If we know the volume and concentration of a solution of B that just reacts with substance A in a sample, we can determine the amount of A. **Titration** is *a procedure for determining the amount of substance* A *by adding a carefully measured volume of a solution with known concentration of* B *until the reaction of* A *and* B *is just complete*.

Figure 4.11 shows a flask containing hydrochloric acid with an unknown amount of HCl being titrated with sodium hydroxide solution, NaOH, of known molarity. The reaction is

$$NaOH(aq) + HCl(aq) \longrightarrow NaCl(aq) + H_2O(l)$$

To the HCl solution are added a few drops of phenolphthalein indicator.● Phenolphthalein is colorless in the hydrochloric acid but turns pink at the completion of the reaction of NaOH with HCl. Sodium hydroxide with a concentration of 0.101 M is contained in a *buret,* a glass tube graduated to measure the volume of liquid delivered from the stopcock. The solution in the buret is added to

● An indicator is a substance that undergoes a color change when a reaction approaches completion. See Section 3.4.

the hydrochloric acid in the flask until the phenolphthalein just changes from colorless to pink. At this point, the reaction is complete and the volume of NaOH that reacts with the HCl is read from the buret. This volume of sodium hydroxide is then used to obtain the mass of HCl in the original solution.

Example 4.21 Calculating the Quantity of Substance in a Titrated Solution

A flask contains a solution with an unknown amount of HCl. This solution is titrated with 0.101 *M* NaOH. It takes 3.35 mL NaOH to complete the reaction with HCl. What is the mass of the HCl?

Solution

We convert the volume of NaOH (3.35×10^{-3} L NaOH solution) to moles NaOH (from the molarity of NaOH). Then we convert moles NaOH to moles HCl (from the chemical equation). Finally, we convert moles HCl to grams HCl. The calculation of this conversion is as follows:

$$3.35 \times 10^{-3}\ \cancel{\text{L NaOH soln}} \times \frac{0.101\ \cancel{\text{mol NaOH}}}{1\ \cancel{\text{L NaOH soln}}} \times \frac{1\ \cancel{\text{mol HCl}}}{1\ \cancel{\text{mol NaOH}}} \times \frac{36.5\ \text{g HCl}}{1\ \cancel{\text{mol HCl}}} = \mathbf{0.0123\ g\ HCl}$$

Exercise 4.27

A 5.00-g sample of vinegar is titrated with 0.108 *M* NaOH. If the vinegar requires 39.1 mL of the NaOH solution for complete reaction, what is the mass percentage of acetic acid, $HC_2H_3O_2$, in the vinegar? The reaction is

$$HC_2H_3O_2(aq) + NaOH(aq) \longrightarrow NaC_2H_3O_2(aq) + H_2O(l)$$

(See Problems 4.101 and 4.102.)

Profile of a Chemical

ACETIC ACID (a Weak Acid)

Pure acetic acid, $HC_2H_3O_2$, is often referred to as glacial acetic acid. *Glacial* means icelike. The pure acid freezes at 17°C, or 62°F, to an icelike solid (Figure 4.12), whereas dilute solutions of the acid freeze at temperatures below the freezing point of water. Pure acetic acid was first prepared about 1700 by the distillation of vinegar and became a common laboratory chemical. Early chemists were quite familiar with the solid acid, because in poorly heated laboratories pure acetic acid was frequently found frozen in the bottle. The term *glacial* came to be applied to the pure acid, whether liquid or solid. Glacial acetic acid must be handled with care, because although dilute solutions of the acid (such as vinegar) are harmless, pure acetic acid is a corrosive substance and can cause painful burns.

Interestingly, acetic acid is a nonconductor of electric current when pure but becomes conducting when water (another nonconductor) is added to it. The explanation is simple. Glacial acetic acid is a molecular substance and the molecules have no electric charge, so they cannot carry an electric current. When water is added, ions are formed, and these can carry a current.

$$HC_2H_3O_2(l) + H_2O(l) \rightleftharpoons C_2H_3O_2^-(aq) + H_3O^+(aq)$$

This reaction with water to form hydronium ion is typical of acid substances, and acetic acid solutions have properties characteristic of acids. For example, bromthymol-blue indicator turns yellow in acetic acid solution, as we expect it to in an acid. Acetic acid is a weak acid, meaning that only a portion of the molecules react with water at any given moment to form ions (note the double arrow). Thus, a 1 *M* solution of acetic acid has lower conductivity than a 1 *M* solution of a strong acid, which exists in solution completely as ions.

In the chapter opener, we noted that acetic acid is manufactured industrially from ethanol by reaction with oxygen. New plants, however, produce acetic acid from carbon monoxide and hydrogen using appropriate catalysts for the reactions:

(continued)

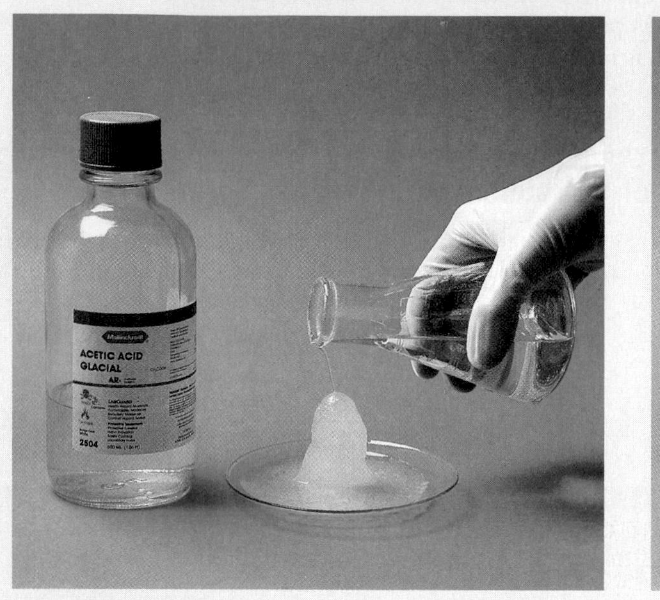

Figure 4.12
Formation of solid acetic acid. *Pure liquid acetic acid was first supercooled (cooled somewhat below its normal freezing point, 17°C) and then poured onto a piece of solid acetic acid, where it forms a pillar of the solid.*

$$CO(g) + 2H_2(g) \longrightarrow CH_3OH(g)$$
$$CO(g) + CH_3OH(l) \longrightarrow HC_2H_3O_2(l)$$

These are interesting starting materials, because in principle carbon monoxide and hydrogen can be produced from the reaction of any organic material with steam. Using coal (which is mostly carbon), the reaction is

$$C(s) + H_2O(g) \longrightarrow CO(g) + H_2(g)$$

Today most organic chemicals, including ethanol, are produced from petroleum. However, should petroleum become scarce, an organic chemicals industry based on carbon monoxide would be attractive. The use of carbon monoxide as a starting material for the manufacture of organic chemicals is currently an active area of research.

Acetic acid is an important industrial chemical. It is used primarily to prepare acetate esters, which are substances formed by reacting acetic acid with a compound whose molecules have an O—H bond in them. Ethyl acetate is a simple ester prepared from ethanol, C_2H_5OH, and acetic acid. It is a colorless liquid with fragrant odor and is used as a solvent in lacquers. Cellulose, found in cotton and wood, is a macromolecular material (a material made up of very large molecules) containing O—H bonds. It reacts with acetic acid to give cellulose acetate for film and textile fibers. Vinyl acetate, another ester of acetic acid, is used in the manufacture of polyvinyl acetate, a macromolecular material used in water-based latex paints and in glues for paper and wood.

Questions for Study

1. Why is pure acetic acid frequently called glacial acetic acid?
2. Why is it that pure acetic acid is a poor conductor of electricity while solutions of acetic acid are conductors?
3. Describe some tests which show that acetic acid solutions have acid properties.
4. Explain why 1 *M* HCl is a better electric conductor than 1 *M* $HC_2H_3O_2$.
5. New industrial plants produce acetic acid from carbon monoxide and hydrogen. What is the reaction by which CO and H_2 mixtures are prepared from coal?
6. Why is carbon monoxide being studied as a starting material for the production of organic substances?
7. Name some of the final products produced from acetic acid.

Profile of a Chemical

AMMONIA (a Weak Base)

The "ammonia" sold in grocery stores as a cleaning solution is actually an aqueous solution of ammonia gas, NH_3 (Figure 4.13). Ammonia has a characteristic penetrating odor (be careful when smelling ammonia solutions—smell only the dilute gas obtained by waving the ammonia gas from the bottle some distance from your nose), and solutions have a soapy feel typical of basic solutions. Ammonia is a weak base, reacting with water to form OH^- ion and giving an equilibrium solution in which only a small percentage of NH_3 molecules have reacted.

$$NH_3(aq) + H_2O(l) \rightleftharpoons NH_4^+(aq) + OH^-(aq)$$

Ammonia is a simple compound that figures prominently in the cycling of the element nitrogen in the biosphere, the part of the earth in which living things exist. Many important biological substances such as proteins contain nitrogen, and when animals and animal wastes decay, ammonia gas is released by bacterial processes. When this ammonia is released into the soil, it is made into nitrates by soil bacteria, and these nitrates are utilized by plants as a nitrogen fertilizer or nutrient. These plants may then be eaten by animals, and the cycle is repeated.

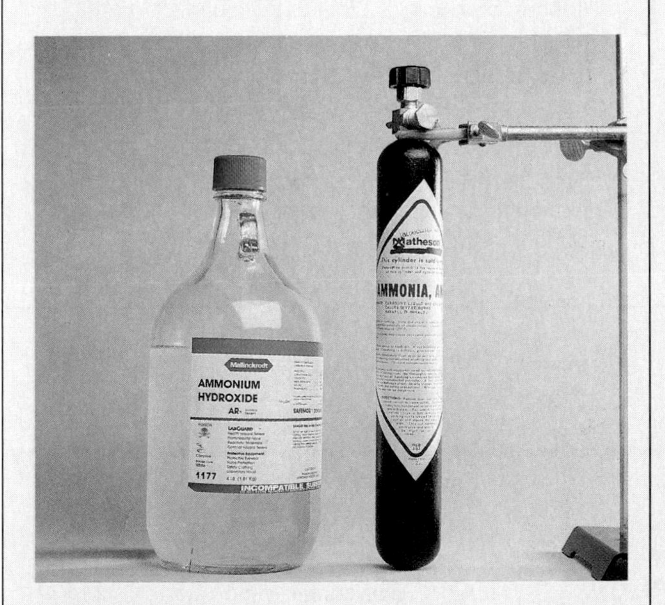

Figure 4.13
Ammonia and ammonia solutions. *Ammonia is conveniently available in steel cylinders (right), where it exists as the pure liquid under pressure. The gas dissolves readily in water to form aqueous ammonia, sometimes labeled ammonium hydroxide, although no species of NH_4OH is known.*

Figure 4.14
Preparation of ammonia from an ammonium salt. *Sodium hydroxide solution was added to ammonium chloride. Ammonia gas turns colorless phenolphthalein in the moist paper to a bright pink color.*

The name ammonia derives from the substance sal ammoniac, or the salt of Ammon (ammonium chloride). Sal ammoniac forms in deposits of animal wastes and was originally brought to medieval Europe from Egypt where it was prepared near a temple of the Egyptian god Ammon. When ammonium chloride is heated with a base, such as sodium hydroxide, it reacts to give ammonia (Figure 4.14).

$$NH_4Cl(aq) + NaOH(aq) \longrightarrow NH_3(g) + H_2O(l) + NaCl(aq)$$

Ammonia is of crucial importance to modern farming as a nitrogen fertilizer (Figure 4.15). Without commercial nitrogen fertilizers, farmers could not grow sufficient food for the world's population. Nitrogen fertilizers were important even in the nineteenth century, and by 1900 it was clear that natural deposits of nitrates from Chile would not be sufficient. Chemists therefore sought some means of converting nitrogen, N_2, from the atmosphere into nitrogen fertilizers.

Many chemists thought that the direct combination of nitrogen with hydrogen was impossible, but in 1905 the German chemist Fritz Haber (1868–1934) showed that the reaction was feasible.

(continued)

Figure 4.15
Liquid ammonia as a fertilizer. Left: *Ammonia is injected into the ground, where it dissolves in moisture in the soil.* Right: *A close-up photograph of the injection of ammonia.*

$$N_2(g) + 3H_2(g) \longrightarrow 2NH_3(g)$$

By 1913 the conditions required for an economical yield of ammonia (moderately high temperature, high pressures, and an iron catalyst) were worked out, and full-scale plants were constructed. These plants also proved to be important for Germany's war effort during World War I as ammonia was converted to nitric acid for the manufacture of explosives. Haber received the Nobel Prize in chemistry in 1918 for his work on the synthesis of ammonia.

Questions for Study

1. What is the normal form of ammonia? What is household ammonia?
2. Ammonia is said to be a weak base. What does this mean?
3. What is the origin of the word *ammonia*?
4. The name sal ammoniac is still used for a certain nitrogen compound. What is the chemical name of this compound?
5. Give the equation for the preparation of ammonia from ammonium sulfate using potassium hydroxide.
6. Give two commercial uses of ammonia.
7. Describe the Haber process for the synthesis of ammonia, including the conditions necessary to produce ammonia by this process.

A Checklist for Review

IMPORTANT TERMS

molecular weight (4.1)
formula weight (4.1)
mole (mol) (4.2)
Avogadro's number (N_A) (4.2)
molar mass (4.2)
percentage composition (p. 109)
mass percentage (4.3)
empirical (simplest) formula (4.5)
stoichiometry (p. 114)
limiting reactant (limiting reagent) (4.8)
theoretical yield (4.8)
percentage yield (4.8)
molar concentration (molarity) (M) (4.9)
titration (4.11)

KEY EQUATIONS

$$\text{Mass \% } A = \frac{\text{mass of } A \text{ in the whole}}{\text{mass of the whole}} \times 100\%$$

$$n = \frac{\text{molecular weight}}{\text{empirical formula weight}}$$

$$\text{Percentage yield} = \frac{\text{actual yield}}{\text{theoretical yield}} \times 100\%$$

$$\text{Molarity } (M) = \frac{\text{moles of solute}}{\text{liters of solution}}$$

$$M_i \times V_i = M_f \times V_f$$

SUMMARY OF FACTS AND CONCEPTS

A *formula weight* equals the sum of the atomic weights of the atoms in the formula of a compound. If the formula corresponds to that of a molecule, this sum of atomic weights equals the *molecular weight* of the compound. The mass of *Avogadro's number* (6.02×10^{23}) of formula units—that is, the mass of one *mole* of substance—equals the mass in grams that corresponds to the numerical value of the formula weight in amu. This mass is called the *molar mass.*

The *empirical formula (simplest formula)* of a compound is obtained from the *percentage composition* of the substance, which is expressed as *mass percentages* of the elements. To calculate the empirical formula, we convert mass percentages to ratios of moles, which, when expressed in smallest whole numbers, give the subscripts in the formula. A molecular formula is a multiple of the empirical formula; this multiple is determined from the experimental value of the molecular weight.

A chemical equation may be interpreted in terms of moles of reactants and products, as well as in terms of molecules. Using this *molar interpretation,* we can convert from the mass of one substance in a chemical equation to the mass of another. The maximum amount of product from a reaction is determined by the *limiting reactant,* the reactant that is completely used up; the other reactants are in excess.

Molar concentration, or *molarity,* is the moles of solute in a liter of solution. Knowing the molarity allows us to calculate the amount of solute in any volume of solution. Because the moles of solute are constant during the *dilution of a solution,* we can determine to what volume to dilute a concentrated solution to give one of desired molarity. *Titration* is a method of chemical analysis in which we measure the volume of solution of known molarity that reacts with a compound of unknown amount. We determine the amount of the compound from this volume of solution.

OPERATIONAL SKILLS

Note: A table of atomic weights is necessary for most of these skills.

1. Calculating the formula weight from a formula Given the formula of a compound and a table of atomic weights, calculate the formula weight (Example 4.1).

2. Calculating the mass of an atom or molecule Using the molar mass and Avogadro's number, calculate the mass of an atom or molecule in grams (Example 4.2).

3. Converting moles of substance to grams, and vice versa Given the moles of a compound with a known formula, calculate the mass (Example 4.3). Or, given the mass of a compound with a known formula, calculate the moles (Example 4.4).

4. Calculating the number of molecules in a given mass Given the mass of a sample of a molecular substance and its formula, calculate the number of molecules in the sample (Example 4.5).

5. Calculating the percentage composition and mass of an element from the formula Given the formula of a compound, calculate the mass percentage of elements in it (Example 4.6). From the mass percentages, calculate the mass of any element in a sample of a compound (Example 4.7).

6. Calculating the percentages of C and H by combustion Given the masses of CO_2 and H_2O obtained from the combustion of a known mass of a compound of C, H, and O, compute the mass percentages of each element (Example 4.8).

7. Determining the empirical formula from percentage composition Given the masses of elements in a known mass of compound, or given its percentage composition, obtain the empirical formula (Examples 4.9 and 4.10).

8. Determining the molecular formula from percentage composition and molecular weight Given the empirical formula and molecular weight of a substance, obtain its molecular formula (Example 4.11).

9. Relating quantities in a chemical equation Given a chemical equation and the amount of one substance, calculate the amount of another substance involved in the reaction (Examples 4.12 and 4.13).

10. Calculating with a limiting reactant Given the amounts of reactants and the chemical equation, find the limiting reactant; then calculate the amount of a product (Examples 4.14 and 4.15).

11. Calculating molarity from mass and volume Given the mass of the solute and the volume of the solution, calculate the molarity (Example 4.16).

12. Using molarity as a conversion factor Given the volume and molarity of a solution, calculate the amount of solute. Or, given the amount of solute and the molarity of a solution, calculate the volume (Example 4.17).

13. Calculating the dilution of a solution Calculate the volume of solution of known molarity required to make a specified volume of solution with different molarity (Example 4.18). Calculate this volume, given a solution whose mass percentage of solute is known (Example 4.19).

14. Calculating the volume of a reactant solution needed Given the chemical equation, calculate the volume of solution of known molarity of one substance that just reacts with a given volume of solution of another substance (Example 4.20).

15. Calculating the quantity of substance in a titrated solution Calculate the mass of one substance that reacts with a given volume of known molarity of solution of another substance (Example 4.21).

Review Questions

4.1 What is the difference between a formula weight and a molecular weight? Could a given substance have both a formula weight and a molecular weight?

4.2 Describe in words how to obtain the formula weight of a compound from the formula.

4.3 One mole of N_2 contains how many N_2 molecules? How many N atoms are there in one mole of N_2? One mole of iron(III) sulfate, $Fe_2(SO_4)_3$, contains how many moles of SO_4^{2-} ions? how many moles of O atoms?

4.4 The following compounds of nitrogen and oxygen have these percentage compositions: NO is 46.7% N, 53.3% O; NO_2 is 30.5% N, 69.5% O; and N_2O_5 is 25.9% N, 74.1% O. From this information, obtain the percentage composition of N_2O_4.

4.5 Explain what is involved in determining the composition of a compound of C, H, and O by combustion.

4.6 Explain what is involved in obtaining the empirical formula from the percentage composition.

4.7 A substance has the molecular formula $C_6H_{12}O_2$. What is its empirical formula?

4.8 Hydrogen peroxide has the empirical formula HO and an empirical formula weight of 17.0 amu. If the molecular weight is 34.0 amu, what is the molecular formula?

4.9 Describe in words the meaning of the equation

$$CH_4 + 2O_2 \longrightarrow CO_2 + 2H_2O$$

using a molecular, a molar, and then a mass interpretation.

4.10 Explain how a chemical equation can be used to relate the masses of different substances involved in a reaction.

4.11 What is a limiting reactant in a reaction mixture? Explain how it determines the amount of product.

4.12 From the definition of molar concentration, show that moles of solute equals the molar concentration times the volume of solution.

4.13 Why is the product of molar concentration and volume constant for a dilution problem?

4.14 Describe how the amount of sodium hydroxide in a mixture can be determined by titration with hydrochloric acid of known molarity.

Practice Problems

FORMULA WEIGHTS AND MOLE CALCULATIONS

4.15 Find the formula weights of the following substances to three significant figures.

(a) methanol, CH_3OH
(b) phosphorus trichloride, PCl_3
(c) potassium carbonate, K_2CO_3
(d) nickel phosphate, $Ni_3(PO_4)_2$

4.16 Find the formula weights of the following substances to three significant figures.

(a) acetic acid, $HC_2H_3O_2$
(b) phosphorus pentachloride, PCl_5
(c) potassium sulfate, K_2SO_4
(d) calcium hydroxide, $Ca(OH)_2$

4.17 Ammonium nitrate, NH_4NO_3, is used as a nitrogen fertilizer and in explosive mixtures. What is the molar mass of NH_4NO_3?

4.18 Phosphoric acid, H_3PO_4, is used to make phosphate fertilizers and detergents and is also used in carbonated beverages. What is the molar mass of H_3PO_4?

4.19 Calculate the mass (in grams) of each of the following species.
(a) Na atom (b) S atom
(c) CH_3Cl molecule (d) Na_2SO_3 formula unit

4.20 Calculate the mass (in grams) of each of the following species.
(a) Fe atom (b) F atom
(c) N_2O molecule (d) K_2CrO_4 formula unit

4.21 Diethyl ether, $(C_2H_5)_2O$, commonly known as ether, is used as an anesthetic. What is the mass in grams of a molecule of diethyl ether?

4.22 Glycerol, $C_3H_8O_3$, is used as a moistening agent for candy and is the starting material for nitroglycerine. Calculate the mass of a glycerol molecule in grams.

4.23 Calculate the mass in grams of the following.
(a) 0.15 mol Na (b) 0.594 mol S
(c) 2.78 mol CH_3Cl (d) 38 mol Na_2SO_3

4.24 Calculate the mass in grams of the following.
(a) 0.205 mol Fe (b) 0.83 mol F
(c) 5.8 mol N_2O (d) 48.1 mol K_2CrO_4

4.25 Boric acid, H_3BO_3, is a mild antiseptic and is often used as an eyewash. A sample contains 0.543 mol H_3BO_3. What is the mass of boric acid in the sample?

4.26 Carbon disulfide, CS_2, is a colorless, highly flammable liquid used in the manufacture of rayon and cellophane. A sample contains 0.0205 mol CS_2. Calculate the mass of carbon disulfide in the sample.

4.27 Obtain the moles of substance in the following.
(a) 3.43 g C (b) 7.05 g Br_2
(c) 76 g C_4H_{10} (d) 35.4 g Li_2CO_3

4.28 Obtain the moles of substance in the following.
(a) 2.57 g As (b) 7.83 g P_4
(c) 41.4 g N_2H_4 (d) 153 g $Al_2(SO_4)_3$

4.29 Calcium sulfate, $CaSO_4$, is a white, crystalline powder. Gypsum is a mineral, or natural substance, that is a hydrate of calcium sulfate. A 1.000-g sample of gypsum contains 0.791 g $CaSO_4$. How many moles of $CaSO_4$ are there in this sample? Assuming that the rest of the sample is water, how many moles of H_2O are there in the sample? Show that the result is consistent with the formula $CaSO_4 \cdot 2H_2O$.

4.30 A 1.547-g sample of blue copper(II) sulfate pentahydrate, $CuSO_4 \cdot 5H_2O$, is heated carefully to drive off the water. The white crystals of $CuSO_4$ that are left behind have a mass of 0.989 g. How many moles of H_2O were in the original sample? Show that the relative molar amounts of $CuSO_4$ and H_2O agree with the formula of the hydrate.

4.31 Calculate the following.
(a) number of atoms in 7.46 g Li
(b) number of atoms in 32.0 g Br_2
(c) number of molecules in 43 g NH_3
(d) number of formula units in 159 g $PbCrO_4$
(e) number of SO_4^{2-} ions in 14.3 g $Cr_2(SO_4)_3$

4.32 Calculate the following.
(a) number of atoms in 25.7 g Al
(b) number of atoms in 7.92 g I_2
(c) number of molecules in 38.1 g N_2O_5
(d) number of formula units in 3.31 g $NaClO_4$
(e) number of Ca^{2+} ions in 6.54 g $Ca_3(PO_4)_2$

4.33 Carbon tetrachloride, CCl_4, is a colorless liquid used in the manufacture of fluorocarbons and as an industrial solvent. How many molecules are there in 7.58 mg of carbon tetrachloride?

4.34 Chlorine trifluoride, ClF_3, is a colorless, reactive gas used in nuclear fuel reprocessing. How many molecules are there in a 5.88-mg sample of chlorine trifluoride?

MASS PERCENTAGE

4.35 A 1.836-g sample of coal contains 1.584 g C. Calculate the mass percentage of C in the coal.

4.36 A 5.69-g aqueous solution of isopropyl alcohol contains 4.87 g of isopropyl alcohol. What is the mass percentage of isopropyl alcohol in the solution?

4.37 Phosphorus oxychloride is the starting compound for preparing substances used as flame retardants for plastics. An 8.53-mg sample of phosphorus oxychloride contains 1.72 mg of phosphorus. What is the mass percentage of phosphorus in the compound?

4.38 Ethyl mercaptan is an odorous substance added to natural gas to make leaks easily detectable. A sample of ethyl mercaptan weighing 3.17 mg contains 1.64 mg of sulfur. What is the mass percentage of sulfur in the substance?

4.39 A fertilizer is advertised as containing 15.8% nitrogen (by mass). How much nitrogen is there in 4.15 kg of fertilizer?

4.40 Seawater contains 0.0065% (by mass) of bromine. How many grams of bromine are there in 1.00 L of seawater? The density of seawater is 1.025 g/cm^3.

4.41 A sample of an alloy of aluminum contains 0.0972 mol Al and 0.0381 mol Mg. What are the mass percentages of Al and Mg in the alloy?

4.42 A sample of gas mixture from a neon sign contained 0.0856 mol Ne and 0.0254 mol Kr. What are the mass percentages of Ne and Kr in the gas mixture?

CHEMICAL FORMULAS

4.43 Calculate the percentage composition for each of the following compounds (three significant figures).
(a) CO (b) CO_2
(c) $KMnO_4$ (d) $Co(NO_3)_2$

4.44 Calculate the percentage composition for each of the following compounds (three significant figures).
(a) NO (b) N_2O
(c) $KClO_4$ (d) $Mn(NO_3)_2$

4.45 Halothane, $CF_3CHBrCl$, is an inhalation anesthetic. What are the mass percentages of the elements in halothane (to three significant figures)?

4.46 Procaine hydrochloride (Novocaine), $C_{13}H_{21}ClN_2O_2$, is a local anesthetic. Calculate the mass percentages of elements in this compound (to three significant figures).

4.47 Which contains more carbon, 4.71 g of glucose, $C_6H_{12}O_6$, or 5.85 g of ethanol, C_2H_6O?

4.48 Which contains more sulfur, 40.8 g of calcium sulfate, $CaSO_4$, or 35.2 g of sodium sulfite, Na_2SO_3?

4.49 Ethylene glycol is used as an automobile antifreeze and in the manufacture of polyester fibers. The name glycol stems from the sweet taste of this poisonous compound. Combustion of 6.38 mg of ethylene glycol gives 9.06 mg of CO_2 and 5.58 mg of H_2O. The compound contains only C, H, and O. What are the mass percentages of the elements in ethylene glycol?

4.50 Phenol, commonly known as carbolic acid, was used by Joseph Lister as an antiseptic for surgery in 1865. Its principal use today is in the manufacture of phenolic resins and plastics. Combustion of 5.23 mg of phenol yields 14.67 mg CO_2 and 3.01 mg H_2O. Phenol contains only C, H, and O. What is the percentage of each element in this substance?

4.51 An oxide of osmium (symbol Os) is a pale yellow solid. If 2.89 g of the compound contains 2.16 g osmium, what is its empirical formula?

4.52 An oxide of tungsten (symbol W) is a bright yellow solid. If 5.34 g of the compound contains 4.23 g tungsten, what is its empirical formula?

4.53 Potassium manganate is a dark green, crystalline substance whose composition is 39.7% K, 27.9% Mn, and 32.5% O, by mass. What is its empirical formula?

4.54 Hydroquinone, used as a photographic developer, is 65.4% C, 5.5% H, and 29.1% O, by mass. What is the empirical formula of hydroquinone?

4.55 Acrylic acid, used in the manufacture of acrylic plastics, has the composition 50.0% C, 5.6% H, and 44.4% O. What is its empirical formula?

4.56 Malonic acid is used in the manufacture of barbiturates (sleeping pills). The composition of the acid is 34.6% C, 3.9% H, and 61.5% O. What is malonic acid's empirical formula?

4.57 Putrescine, a substance produced by decaying animals, has the empirical formula C_2H_6N. Several determinations of molecular weight give values in the range of 87 to 90 amu. Find the molecular formula of putrescine.

4.58 Compounds of boron with hydrogen are called boranes. One of these boranes has the empirical formula BH_3 and a molecular weight of 28 amu. What is its molecular formula?

4.59 Oxalic acid is a toxic substance used by laundries to remove rust stains. Its composition is 26.7% C, 2.2% H, and 71.1% O (by mass), and its molecular weight is 90 amu. What is its molecular formula?

4.60 Adipic acid is used in the manufacture of nylon. The composition of the acid is 49.3% C, 6.9% H, and 43.8% O (by mass), and the molecular weight is 146 amu. What is the molecular formula?

STOICHIOMETRY: QUANTITATIVE RELATIONS IN REACTIONS

4.61 Ethylene, C_2H_4, burns in oxygen to give carbon dioxide, CO_2, and water. Write the equation for the reaction, giving molecular, molar, and mass interpretations below the equation.

4.62 Hydrogen sulfide gas, H_2S, burns in oxygen to give sulfur dioxide, SO_2, and water. Write the equation for the reaction, giving molecular, molar, and mass interpretations below the equation.

4.63 Nitric acid, HNO_3, is manufactured by the Ostwald process, in which nitrogen dioxide, NO_2, reacts with water.

$$3NO_2 + H_2O \longrightarrow 2HNO_3 + NO$$

How many grams of nitrogen dioxide are required in this reaction to produce 5.00 g HNO_3?

4.64 White phosphorus, P_4, is prepared by fusing calcium phosphate, $Ca_3(PO_4)_2$, with carbon, C, and sand, SiO_2, in an electric furnace.

$$2Ca_3(PO_4)_2 + 6SiO_2 + 10C \longrightarrow P_4 + 6CaSiO_3 + 10CO$$

How many grams of calcium phosphate are required to give 5.00 g of phosphorus?

4.65 Tungsten metal, W, is used to make incandescent bulb filaments. The metal is produced from the yellow tungsten(VI) oxide, WO_3, by reaction with hydrogen.

$$WO_3 + 3H_2 \longrightarrow W + 3H_2O$$

How many grams of tungsten can be obtained from 4.81 kg of hydrogen with excess tungsten(VI) oxide?

4.66 Acrylonitrile, C_3H_3N, is the starting material for the production of a kind of synthetic fiber (acrylics). It can be made from propylene, C_3H_6, by reaction with nitric oxide, NO.

$$4C_3H_6 + 6NO \longrightarrow 4C_3H_3N + 6H_2O + N_2$$

How many grams of acrylonitrile are obtained from 651 kg of propylene and excess NO?

4.67 The following reaction is used to make carbon tetrachloride, CCl_4, a solvent and starting material for the manufacture of fluorocarbon refrigerants and aerosol propellants.

$$CS_2 + 3Cl_2 \longrightarrow CCl_4 + S_2Cl_2$$

Calculate the number of grams of carbon disulfide, CS_2, needed for a laboratory-scale reaction with 62.7 g of chlorine, Cl_2.

4.68 Solutions of sodium hypochlorite, NaClO, are sold as a bleach (such as Clorox). They are prepared by the reaction of chlorine with sodium hydroxide.

$$2NaOH + Cl_2 \longrightarrow NaCl + NaClO + H_2O$$

If chlorine gas, Cl_2, is bubbled into a solution containing 54.2 g NaOH, how many grams of Cl_2 will eventually react?

4.69 When dinitrogen pentoxide, N_2O_5, a white solid, is heated, it decomposes to nitrogen dioxide and oxygen.

$$2N_2O_5(s) \xrightarrow{\Delta} 4NO_2(g) + O_2(g)$$

If a sample of N_2O_5 produces 1.618 g of O_2, how many grams of NO_2 are formed?

4.70 Copper metal reacts with nitric acid. Assume that the reaction is

$$3Cu(s) + 8HNO_3(aq) \longrightarrow 3Cu(NO_3)_2(aq) + 2NO(g) + 4H_2O(l)$$

If 5.92 g $Cu(NO_3)_2$ is eventually obtained, how many grams of nitric oxide, NO, would have formed also, according to the preceding equation?

LIMITING REACTANT; THEORETICAL AND PERCENTAGE YIELDS

4.71 Potassium superoxide, KO_2, is used in rebreathing gas masks to generate oxygen.

$$4KO_2(s) + 2H_2O(l) \longrightarrow 4KOH(s) + 3O_2(g)$$

If a reaction vessel contains 0.15 mol KO_2 and 0.10 mol H_2O, what is the limiting reactant? How many moles of oxygen can be produced?

4.72 Large quantities of ammonia are burned in the presence of a platinum catalyst to give nitric oxide, as the first step in the preparation of nitric acid.

$$4NH_3(g) + 5O_2(g) \xrightarrow{Pt} 4NO(g) + 6H_2O(g)$$

Suppose a vessel contains 0.120 mol NH_3 and 0.140 mol O_2. Which is the limiting reactant? How many moles of NO could be obtained?

4.73 Methanol, CH_3OH, is prepared industrially from the gas-phase catalytic reaction

$$CO(g) + 2H_2(g) \longrightarrow CH_3OH(g)$$

In a laboratory test, a reaction vessel was filled with 35.4 g CO and 10.2 g H_2. How many grams of methanol would be produced in a complete reaction? Which reactant remains unconsumed at the end of the reaction? How many grams of it remain?

4.74 Carbon disulfide, CS_2, burns in oxygen. Complete combustion gives the reaction

$$CS_2(g) + 3O_2(g) \longrightarrow CO_2(g) + 2SO_2(g)$$

Calculate the grams of sulfur dioxide, SO_2, produced when a mixture of 15.0 g of carbon disulfide and 35.0 g of oxygen reacts. Which reactant remains unconsumed at the end of the combustion? How many grams remain?

4.75 Titanium, which is used to make airplane engines and frames, can be obtained from titanium tetrachloride, which in turn is obtained from titanium dioxide by the following process:

$$3TiO_2(s) + 4C(s) + 6Cl_2(g) \longrightarrow 3TiCl_4(g) + 2CO_2(g) + 2CO(g)$$

A vessel contains 4.15 g TiO_2, 5.67 g C, and 6.78 g Cl_2. Suppose the reaction goes to completion as written. How many grams of titanium tetrachloride can be produced?

4.76 Hydrogen cyanide, HCN, is prepared from ammonia, air, and natural gas (CH_4) by the following process:

$$2NH_3(g) + 3O_2(g) + 2CH_4(g) \xrightarrow{Pt} 2HCN(g) + 6H_2O(g)$$

Hydrogen cyanide is used to prepare sodium cyanide, which is used in part to obtain gold from gold-containing rock. If a reaction vessel contains 11.5 g NH_3, 10.0 g O_2, and 10.5 g CH_4, what is the maximum mass in grams of hydrogen cyanide that could be made, assuming the reaction goes to completion as written?

4.77 Aspirin (acetylsalicylic acid) is prepared by heating salicylic acid, $C_7H_6O_3$, with acetic anhydride, $C_4H_6O_3$. The other product is acetic acid, $C_2H_4O_2$.

$$C_7H_6O_3 + C_4H_6O_3 \longrightarrow C_9H_8O_4 + C_2H_4O_2$$

What is the theoretical yield (in grams) of aspirin, $C_9H_8O_4$, when 2.00 g of salicylic acid is heated with 4.00 g of acetic anhydride? If the actual yield of aspirin is 2.10 g, what is the percentage yield?

4.78 Methyl salicylate (oil of wintergreen) is prepared by heating salicylic acid, $C_7H_6O_3$, with methanol, CH_3OH.

$$C_7H_6O_3 + CH_3OH \longrightarrow C_8H_8O_3 + H_2O$$

In an experiment, 1.50 g of salicylic acid is reacted with 11.20 g of methanol. The yield of methyl salicylate, $C_8H_8O_3$, is 1.31 g. What is the percentage yield?

MOLARITY

4.79 A sample of 0.0341 mol iron(III) chloride, $FeCl_3$, was dissolved in water to give 25.0 mL of solution. What is the molarity of the solution?

4.80 A 50.0-mL volume of $AgNO_3$ solution contains 0.0285 mol $AgNO_3$ (silver nitrate). What is the molarity of the solution?

4.81 An aqueous solution is made from 0.798 g of potassium permanganate, $KMnO_4$. If the volume of solution is 50.0 mL, what is the molarity of $KMnO_4$ in the solution?

4.82 A sample of oxalic acid, $H_2C_2O_4$, weighing 1.274 g is placed in a 100.0-mL volumetric flask, which is then filled to the mark with water. What is the molarity of the solution?

4.83 What volume of 0.120 *M* $CuSO_4$ is required to give 0.150 mol copper(II) sulfate, $CuSO_4$?

4.84 How many milliliters of 0.126 *M* $HClO_4$ (perchloric acid) are required to give 0.00752 mol $HClO_4$?

4.85 An experiment calls for 0.0353 g of potassium hydroxide, KOH. How many milliliters of 0.0176 *M* KOH are required?

4.86 What is the volume (in milliliters) of 0.215 *M* H_2SO_4 (sulfuric acid) containing 0.949 g H_2SO_4?

4.87 Heme, obtained from red blood cells, binds oxygen, O_2. How many moles of heme are there in 25 mL of 0.0019 *M* heme solution?

4.88 Insulin is a hormone that controls the use of glucose in the body. How many moles of insulin are required to make up 28 mL of 0.0048 *M* insulin solution?

4.89 How many grams of sodium dichromate, $Na_2Cr_2O_7$, should be added to a 50.0-mL volumetric flask to prepare 0.025 *M* $Na_2Cr_2O_7$ when the flask is filled to the mark with water?

4.90 Describe how you would prepare 2.50×10^2 mL of 0.10 *M* Na_2SO_4. What mass (in grams) of sodium sulfate, Na_2SO_4, is needed?

4.91 If you dilute 34 mL of 1.32 *M* $KMnO_4$ (potassium permanganate) to 0.28 *M* $KMnO_4$, what is the final volume of the solution?

4.92 To what final volume should 25 mL of 2.4 *M* $K_2Cr_2O_7$ (potassium dichromate) be diluted to give a solution that is 0.10 *M* $K_2Cr_2O_7$?

4.93 How many milliliters of 2.00 *M* $HC_2H_3O_2$ (acetic acid) are required to make 45.0 mL of a 0.18 *M* $HC_2H_3O_2$ solution?

4.94 Describe how you would prepare 1.00 L of 0.120 *M* NaOH from a 1.20 *M* NaOH solution.

4.95 How would you make up 425 mL of 0.150 *M* HNO_3 from nitric acid that is 68.0% HNO_3? The density of 68.0% HNO_3 is 1.41 g/mL.

4.96 A solution of aqueous ammonia contains 28.0% NH_3 by mass (density 0.898 g/mL). Describe the preparation of 255 mL of 0.320 *M* NH_3 from this solution.

4.97 What volume of 0.250 *M* HNO_3 (nitric acid) reacts with 42.4 mL of 0.150 *M* Na_2CO_3 (sodium carbonate) in the following reaction?

$$2HNO_3(aq) + Na_2CO_3(aq) \longrightarrow 2NaNO_3(aq) + H_2O(l) + CO_2(g)$$

4.98 A flask contains 53.1 mL of 0.150 *M* $Ca(OH)_2$ (calcium hydroxide). How many milliliters of 0.350 *M* Na_2CO_3 (sodium carbonate) are required to just react with the calcium hydroxide in the following reaction?

$$Na_2CO_3(aq) + Ca(OH)_2(aq) \longrightarrow CaCO_3(s) + 2NaOH(aq)$$

4.99 How many milliliters of 0.150 *M* H_2SO_4 (sulfuric acid) are required to react with 2.05 g of sodium hydrogen carbonate, $NaHCO_3$, according to the following equation?

$$H_2SO_4(aq) + 2NaHCO_3(aq) \longrightarrow Na_2SO_4(aq) + 2H_2O(l) + 2CO_2(g)$$

4.100 How many milliliters of 0.238 *M* $KMnO_4$ are needed to react with 3.36 g of iron(II) sulfate, $FeSO_4$? The reaction is as follows:

$$10FeSO_4(aq) + 2KMnO_4(aq) + 8H_2SO_4(aq) \longrightarrow 5Fe_2(SO_4)_3(aq) + 2MnSO_4(aq) + K_2SO_4(aq) + 8H_2O(l)$$

4.101 A solution of hydrogen peroxide, H_2O_2, is titrated with a solution of potassium permanganate, $KMnO_4$. The reaction is

$$5H_2O_2(aq) + 2KMnO_4(aq) + 3H_2SO_4(aq) \longrightarrow 5O_2(g) + 2MnSO_4(aq) + K_2SO_4(aq) + 8H_2O(l)$$

It requires 46.9 mL of 0.145 *M* $KMnO_4$ to titrate 20.0 g of the solution of hydrogen peroxide. What is the mass percentage of H_2O_2 in the solution?

4.102 A 3.33-g sample of iron ore is transformed to a solution of iron(II) sulfate, $FeSO_4$, and this solution is titrated with 0.150 *M* $K_2Cr_2O_7$ (potassium dichromate). If it requires 41.4 mL of potassium dichromate solution to titrate the iron(II) sulfate solution, what is the percentage of iron in the ore? The reaction is

$$6FeSO_4(aq) + K_2Cr_2O_7(aq) + 7H_2SO_4(aq) \longrightarrow 3Fe_2(SO_4)_3(aq) + Cr_2(SO_4)_3(aq) + 7H_2O(l) + K_2SO_4(aq)$$

Additional Problems

4.103 Caffeine, the stimulant in coffee and tea, has the molecular formula $C_8H_{10}N_4O_2$. Calculate the mass percentage of each element in the substance. Give the answers to three significant figures.

4.104 Morphine, a narcotic substance obtained from opium, has the molecular formula $C_{17}H_{19}NO_3$. What is the mass percentage of each element in morphine (to three significant figures)?

4.105 A moth repellent, *para*-dichlorobenzene, has the composition 49.0% C, 2.7% H, and 48.2% Cl. Its molecular weight is 147 amu. What is its molecular formula?

4.106 Sorbic acid is added to food as a mold inhibitor. Its composition is 64.3% C, 7.2% H, and 28.5% O, and its molecular weight is 112 amu. What is its molecular formula?

4.107 A water-soluble compound of gold and chlorine is treated with silver nitrate to convert the chlorine completely to silver chloride, AgCl. In an experiment, 328 mg of the compound gave 464 mg of silver chloride. Calculate the percentage of Cl in the compound. What is its empirical formula?

4.108 A solution of scandium chloride was treated with silver nitrate. The chlorine in the scandium compound was converted to silver chloride, AgCl. A 58.9-mg sample of scandium chloride gave 167.4 mg of silver chloride. What are the mass percentages of Sc and Cl in scandium chloride? What is its empirical formula?

4.109 Thiophene is a liquid compound of the elements C, H, and S. A sample of thiophene weighing 7.96 mg was burned in oxygen, giving 16.65 mg CO_2. Another sample was subjected to a series of reactions that transformed all of the sulfur in the compound to barium sulfate. If 4.31 mg of thiophene gave 11.96 mg of barium sulfate, what is the empirical formula of thiophene? Its molecular weight is 84 amu. What is its molecular formula?

4.110 Aniline, a starting compound for urethane plastic foams, consists of C, H, and N. Combustion of such compounds yields CO_2, H_2O, and N_2 as products. If the combustion of 9.71 mg of aniline yields 6.63 mg H_2O and 1.46 mg N_2, what is its empirical formula? The molecular weight of aniline is 93 amu. What is its molecular formula?

4.111 Hemoglobin is the oxygen-carrying molecule of red blood cells, consisting of a protein and a nonprotein substance. The nonprotein substance is called heme. A sample of heme weighing 35.2 mg contains 3.19 mg of iron. If a heme molecule contains one atom of iron, what is the molecular weight of heme?

4.112 Penicillin V was treated chemically to convert sulfur to barium sulfate, $BaSO_4$. An 8.19-mg sample of penicillin V gave 5.46 mg $BaSO_4$. What is the percentage of sulfur in penicillin V? If there is one sulfur atom in the molecule, what is the molecular weight?

4.113 A sample of limestone (containing calcium carbonate, $CaCO_3$) weighing 413 mg is treated with oxalic acid, $H_2C_2O_4$, to give calcium oxalate, CaC_2O_4.

$$CaCO_3(s) + H_2C_2O_4(aq) \longrightarrow CaC_2O_4(s) + H_2O(l) + CO_2(g)$$

The mass of the calcium oxalate is 472 mg. What is the mass percentage of calcium carbonate in this limestone?

4.114 A titanium ore contains rutile (TiO_2) plus some iron oxide and silica. When it is heated with carbon in the presence of chlorine, titanium tetrachloride, $TiCl_4$, is formed.

$$TiO_2(s) + C(s) + 2Cl_2(g) \longrightarrow TiCl_4(g) + CO_2(g)$$

Titanium tetrachloride, a liquid, can be distilled from the mixture. If 35.4 g of titanium tetrachloride is recovered from 17.4 g of crude ore, what is the mass percentage of TiO_2 in the ore (assume all TiO_2 reacts)?

4.115 Ethylene oxide, C_2H_4O, is made by the oxidation of ethylene, C_2H_4.

$$2C_2H_4(g) + O_2(g) \longrightarrow 2C_2H_4O(g)$$

Ethylene oxide is used to make ethylene glycol for automobile antifreeze. In a pilot study, 10.6 g of ethylene gave 9.69 g of ethylene oxide. What is the percentage yield of ethylene oxide?

4.116 Nitrobenzene, $C_6H_5NO_2$, an important raw material for the dye industry, is prepared from benzene, C_6H_6, and nitric acid, HNO_3.

$$C_6H_6(l) + HNO_3(l) \longrightarrow C_6H_5NO_2(l) + H_2O(l)$$

When 20.3 g of benzene and an excess of HNO_3 are used, what is the theoretical yield of nitrobenzene? If 28.7 g of nitrobenzene is recovered, what is the percentage yield?

4.117 Zinc metal can be obtained from zinc oxide, ZnO, by reaction at high temperature with carbon monoxide, CO.

$$ZnO(s) + CO(g) \longrightarrow Zn(s) + CO_2(g)$$

The carbon monoxide is obtained from carbon.

$$2C(s) + O_2(g) \longrightarrow 2CO(g)$$

What is the maximum amount of zinc that can be obtained from 75.0 g of zinc oxide and 50.0 g of carbon?

4.118 Hydrogen cyanide, HCN, can be made by a two-step process. First, ammonia is reacted with O_2 to give nitric oxide, NO.

$$4NH_3(g) + 5O_2(g) \longrightarrow 4NO(g) + 6H_2O(g)$$

Then nitric oxide is reacted with methane, CH_4.

$$2NO(g) + 2CH_4(g) \longrightarrow 2HCN(g) + 2H_2O(g) + H_2(g)$$

When 24.2 g of ammonia and 25.1 g of methane are used, how many grams of hydrogen cyanide can be produced?

4.119 An aqueous solution contains 2.25 g of calcium chloride, $CaCl_2$, per liter. What is the molarity of $CaCl_2$? When calcium chloride dissolves in water, the calcium ions, Ca^{2+}, and chloride ions, Cl^-, in the crystal go into the solution. What is the molarity of each ion in the solution?

4.120 An aqueous solution contains 3.45 g of iron(III) sulfate, $Fe_2(SO_4)_3$, per liter. What is the molarity of $Fe_2(SO_4)_3$? When the compound dissolves in water, the Fe^{3+} ions and $SO_4{}^{2-}$ ions in the crystal go into solution. What is the molar concentration of each ion in the solution?

4.121 A stock solution of potassium dichromate, $K_2Cr_2O_7$, is made by dissolving 89.3 g of the compound in 1.00 L of solution. How many milliliters of this solution are required to prepare 1.00 L of 0.100 M $K_2Cr_2O_7$?

4.122 A 69.3-g sample of oxalic acid, $H_2C_2O_4$, was dissolved in 1.00 L of solution. How would you prepare 1.00 L of 0.150 M $H_2C_2O_4$ from this solution?

4.123 A solution contains 6.00% (by mass) NaBr (sodium bromide). The density of the solution is 1.046 g/cm^3. What is the molarity of NaBr?

4.124 An aqueous solution contains 4.50% NH_3 (ammonia) by mass. The density of the aqueous ammonia is 0.979 g/mL. What is the molarity of NH_3 in the solution?

4.125 A 0.608-g sample of fertilizer contained nitrogen as ammonium sulfate, $(NH_4)_2SO_4$. It was analyzed for nitrogen by heating with sodium hydroxide.

$$(NH_4)_2SO_4(s) + 2NaOH(aq) \longrightarrow Na_2SO_4(aq) + 2H_2O(l) + 2NH_3(g)$$

The ammonia was collected in 46.3 mL of 0.213 M HCl (hydrochloric acid), with which it reacted.

$$NH_3(g) + HCl(aq) \longrightarrow NH_4Cl(aq) + H_2O(l)$$

This solution was titrated for excess hydrochloric acid with 44.3 mL of 0.128 M NaOH.

$$NaOH(aq) + HCl(aq) \longrightarrow NaCl(aq) + H_2O(l)$$

What is the percentage of nitrogen in the fertilizer?

4.126 An antacid tablet contains sodium hydrogen carbonate, $NaHCO_3$, and inert ingredients. A 0.500-g sample of powdered tablet was mixed with 50.0 mL of 0.190 M HCl (hydrochloric acid). The mixture was allowed to stand until it reacted.

$$NaHCO_3(s) + HCl(aq) \longrightarrow NaCl(aq) + H_2O(l) + CO_2(g)$$

The excess hydrochloric acid was titrated with 47.1 mL of 0.128 M NaOH (sodium hydroxide).

$$HCl(aq) + NaOH(aq) \longrightarrow NaCl(aq) + H_2O(l)$$

What is the percentage of sodium hydrogen carbonate in the antacid?

4.127 Calcium carbide, CaC_2, used to produce acetylene, C_2H_2, is prepared by heating calcium oxide, CaO, and carbon, C, to high temperature.

$$CaO(s) + 3C(s) \longrightarrow CaC_2(s) + CO(g)$$

If a mixture contains 1.15 kg of each reactant, how many grams of calcium carbide can be prepared?

4.128 A mixture consisting of 12.8 g of calcium fluoride, CaF_2, and 13.2 g of sulfuric acid, H_2SO_4, is heated to drive off hydrogen fluoride, HF.

$$CaF_2(s) + H_2SO_4(l) \longrightarrow 2HF(g) + CaSO_4(s)$$

What is the maximum number of grams of hydrogen fluoride that can be obtained?

4.129 Alloys, or metallic mixtures, of mercury with another metal are called amalgams. Sodium in sodium amalgam reacts with water (mercury does not).

$$2Na(s) + H_2O(l) \longrightarrow 2NaOH(aq) + H_2(g)$$

If a 15.23-g sample of sodium amalgam evolves 0.108 g of hydrogen, what is the percentage of sodium in the amalgam?

4.130 A sample of sandstone consists of silica (SiO_2) and calcite ($CaCO_3$). When the sandstone is heated, calcium carbonate, $CaCO_3$, decomposes into calcium oxide, CaO, and carbon dioxide.

$$CaCO_3(s) \longrightarrow CaO(s) + CO_2(g)$$

What is the percentage of silica in the sandstone if 21.8 mg of the rock yields 4.01 mg of carbon dioxide?

Cumulative-Skills Problems

4.131 An alloy of iron (54.7%), nickel (45.0%), and manganese (0.3%) has a density of 8.17 g/cm^3. How many iron atoms are there in a block of alloy measuring 10.0 cm × 20.0 cm × 15.0 cm?

4.132 An alloy of iron (71.0%), cobalt (12.0%), and molybdenum (17.0%) has a density of 8.20 g/cm^3. How many cobalt atoms are there in a cylinder with a radius of 2.50 cm and a length of 10.0 cm?

4.133 What volume of solution of ethanol, C_2H_6O, that is 91.0% ethanol by mass contains 0.200 mol C_2H_6O? The density of the solution is 0.807 g/mL.

4.134 What volume of solution of ethylene glycol, $C_2H_6O_2$, that is 56.0% ethylene glycol by mass contains 0.350 mol $C_2H_6O_2$? The density of the solution is 1.072 g/mL.

4.135 A 10.0-mL sample of potassium iodide solution was analyzed by adding an excess of silver nitrate solution to produce silver iodide crystals, which were filtered from the solution.

$$KI(aq) + AgNO_3(aq) \longrightarrow KNO_3(aq) + AgI(s)$$

If 2.290 g of silver iodide was obtained, what was the molarity of the original KI solution?

4.136 A 25.0-mL sample of sodium sulfate solution was analyzed by adding an excess of barium chloride solution to produce barium sulfate crystals, which were filtered from the solution.

$$Na_2SO_4(aq) + BaCl_2(aq) \longrightarrow 2NaCl(aq) + BaSO_4(s)$$

If 5.483 g of barium sulfate were obtained, what was the molarity of the original Na_2SO_4 solution?

4.137 A metal, M, was converted to the sulfate $M_2(SO_4)_3$. Then a solution of the sulfate was treated with barium chloride to give barium sulfate crystals, which were filtered off.

$$M_2(SO_4)_3(aq) + 3BaCl_2(aq) \longrightarrow 2MCl_3(aq) + 3BaSO_4(s)$$

If 1.200 g of the metal gave 6.026 g of barium sulfate, what is the atomic weight of the metal? What is the metal?

4.138 A metal, M, was converted to the chloride MCl_2. Then a solution of the chloride was treated with silver nitrate to give silver chloride crystals, which were filtered from the solution.

$$MCl_2(aq) + 2AgNO_3(aq) \longrightarrow M(NO_3)_2(aq) + 2AgCl(s)$$

If 2.434 g of the metal gave 7.964 g of silver chloride, what is the atomic weight of the metal? What is the metal?

4.139 Phosphoric acid is prepared by dissolving phosphorus(V) oxide, P_4O_{10}, in water. What is the balanced equation for this reaction? How many grams of P_4O_{10} are required to make 1.50 L of aqueous solution containing 5.00% phosphoric acid by mass? The density of the solution is 1.025 g/mL.

4.140 Iron(III) chloride can be prepared by reacting iron metal with chlorine. What is the balanced equation for this reaction? How many grams of iron are required to make 2.50 L of aqueous solution containing 8.00% iron(III) chloride by mass? The density of the solution is 1.067 g/mL.

4.141 An alloy of aluminum and magnesium was treated with sodium hydroxide solution, in which only aluminum reacts.

$$2Al(s) + 2NaOH(aq) + 6H_2O(l) \longrightarrow 2NaAl(OH)_4(aq) + 3H_2(g)$$

If a sample of alloy weighing 1.118 g gave 0.1068 g of hydrogen, what is the percentage of aluminum in the alloy?

4.142 An alloy of iron and carbon was treated with sulfuric acid, in which only iron reacts.

$$2Fe(s) + 3H_2SO_4(aq) \longrightarrow Fe_2(SO_4)_3(aq) + 3H_2(g)$$

If a sample of alloy weighing 2.358 g gave 0.1228 g of hydrogen, what is the percentage of iron in the alloy?

4.143 Determine what volume of sulfuric acid solution is needed to prepare 18.7 g of aluminum sulfate, $Al_2(SO_4)_3$, by the reaction

$$2Al(s) + 3H_2SO_4(aq) \longrightarrow Al_2(SO_4)_3(aq) + 3H_2(g)$$

The sulfuric acid solution, whose density is 1.104 g/mL, contains 15.0% H_2SO_4 by mass.

4.144 Determine what volume of sodium hydroxide solution is needed to prepare 26.2 g sodium phosphate, Na_3PO_4, by the reaction

$$3NaOH(aq) + H_3PO_4(aq) \longrightarrow Na_3PO_4(aq) + 3H_2O(l)$$

The sodium hydroxide solution, whose density is 1.133 g/mL, contains 12.0% NaOH by mass.

4.145 The active ingredients of an antacid tablet contained only magnesium hydroxide and aluminum hydroxide. Complete neutralization of a sample of the active ingredients required 48.5 mL of 0.187 M hydrochloric acid. The chloride salts from this neutralization were obtained by evaporation of the filtrate from the titration; they weighed 0.4200 g. What was the percentage by mass of magnesium hydroxide in the active ingredients of the antacid tablet?

4.146 The active ingredients in an antacid tablet contained only calcium carbonate and magnesium carbonate. Complete reaction of a sample of the active ingredients required 41.33 mL of 0.08750 M hydrochloric acid. The chloride salts from the reaction were obtained by evaporation of the filtrate from this titration; they weighed 0.1900 g. What was the percentage by mass of the calcium carbonate in the active ingredients of the antacid tablet?

5

The Gaseous State

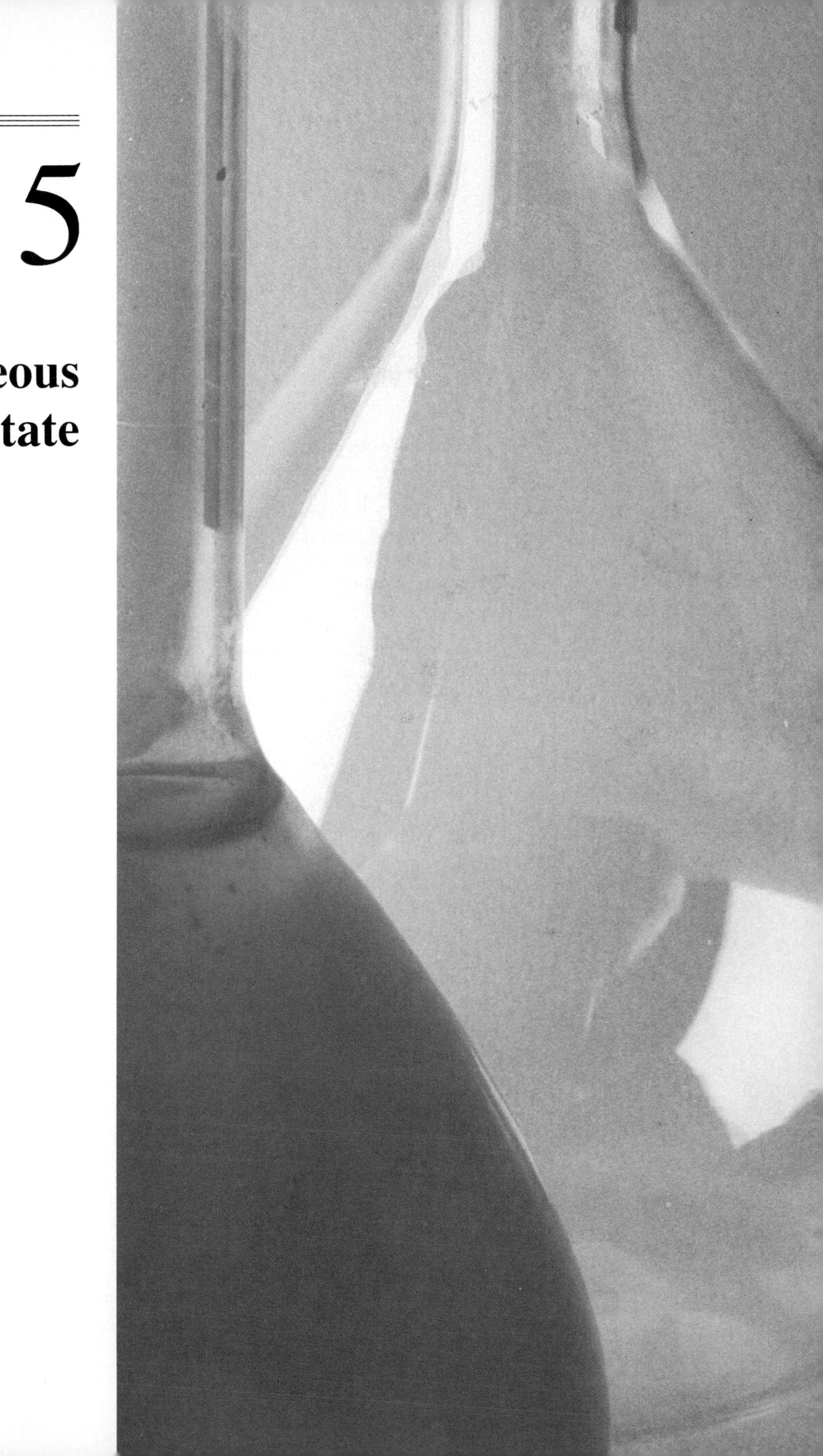

Chapter Outline

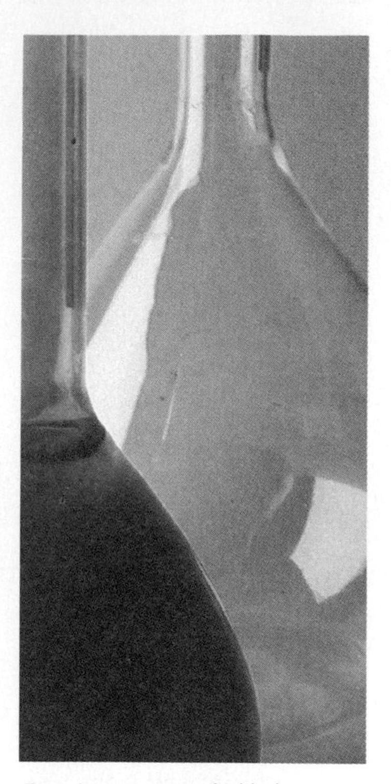

Bromine vapor and chlorine gas.

Gases are distinguished from liquids and solids by their ability to be compressed into smaller and smaller volumes. Advantage is taken of this property in transporting gases such as oxygen and nitrogen by compressing them in steel cylinders (Figure 5.1). Gases are also the simplest physical state of matter. We will see that one simple equation—the ideal gas law—relates the pressure, volume, temperature, and molar amount for all gases. We can use this equation to determine, for example, the amount of oxygen in a tank of compressed gas. To do this, we measure the pressure of the gas in the tank, the temperature, and the volume of gas (volume of the tank), and relate these properties to the amount of oxygen.

Kinetic-molecular theory describes a gas as composed of molecules in constant motion. This theory helps explain the simple relationship that exists among the pressure, volume, temperature, and amount of a gas. Kinetic theory has also enhanced our understanding of the flow of fluids, the transmission of sound, and the conduction of heat.

This chapter introduces the empirical gas laws, such as the ideal gas law, and the kinetic-molecular theory that explains these laws. After finishing the chapter, you will be able to answer questions such as, How many grams of oxygen are there in a 50.0-L tank at 21°C when the oxygen pressure is 15.7 atmospheres? (An atmosphere is a unit of pressure equal to that of the normal atmosphere.) You will also be able to calculate the average speed of an oxygen molecule in this tank (479 m/s, or 1071 mi/hr).

Table 5.1 Properties of Selected Gases

Name	Formula	Color	Odor	Toxicity
Ammonia	NH_3	Colorless	Penetrating	Toxic
Carbon dioxide	CO_2	Colorless	Odorless	Nontoxic
Carbon monoxide	CO	Colorless	Odorless	Very toxic
Chlorine	Cl_2	Pale green	Irritating	Very toxic
Helium	He	Colorless	Odorless	Nontoxic
Hydrogen	H_2	Colorless	Odorless	Nontoxic
Hydrogen chloride	HCl	Colorless	Irritating	Corrosive
Hydrogen sulfide	H_2S	Colorless	Foul	Very toxic
Methane	CH_4	Colorless	Odorless	Nontoxic
Neon	Ne	Colorless	Odorless	Nontoxic
Nitrogen	N_2	Colorless	Odorless	Nontoxic
Nitrogen dioxide	NO_2	Red-brown	Irritating	Very toxic
Oxygen	O_2	Colorless	Odorless	Nontoxic
Sulfur dioxide	SO_2	Colorless	Choking	Toxic

Gas Laws

Most substances composed of small molecules are gases under normal conditions or else are easily vaporized liquids. Table 5.1 lists some common gaseous substances, along with their formulas, color, odor, and toxicity (their harmfulness as a poison).

In the first part of this chapter, we will examine the quantitative relationships, or empirical laws, governing gases. First, however, we need to understand the concept of pressure.

Figure 5.1
Cylinders of gas. Gases such as oxygen and nitrogen can be transported as compressed gases in steel cylinders.

5.1 GAS PRESSURE AND ITS MEASUREMENT

Pressure is defined as *the force exerted per unit area of surface*. A coin resting on a table exerts a force, and therefore a pressure, downward on the table due to gravity. The air above the table exerts an additional pressure because the air is also being pulled downward by gravity.

To obtain the SI unit of pressure and a feeling for its size, let us calculate the pressure on a table from a perfectly flat coin with a radius and mass equal to that of a new penny (9.3 mm in radius and 2.5 g). The force exerted by the coin from gravity equals the mass of the coin times the constant acceleration of gravity. *Acceleration* is the change of speed per unit time, so the SI unit of acceleration is meters per second per second, which is abbreviated m/s^2. The constant acceleration of gravity is 9.81 m/s^2, and the force on the coin due to gravity is

$$\text{Force} = \text{mass} \times \text{constant acceleration of gravity}$$
$$= (2.5 \times 10^{-3}\ \text{kg}) \times (9.81\ \text{m/s}^2) = 2.5 \times 10^{-2}\ \text{kg} \cdot \text{m/s}^2$$

The cross-sectional area of the coin is $\pi \times (\text{radius})^2 = 3.14 \times (9.3 \times 10^{-3}\ \text{m})^2 = 2.7 \times 10^{-4}\ \text{m}^2$. Therefore,

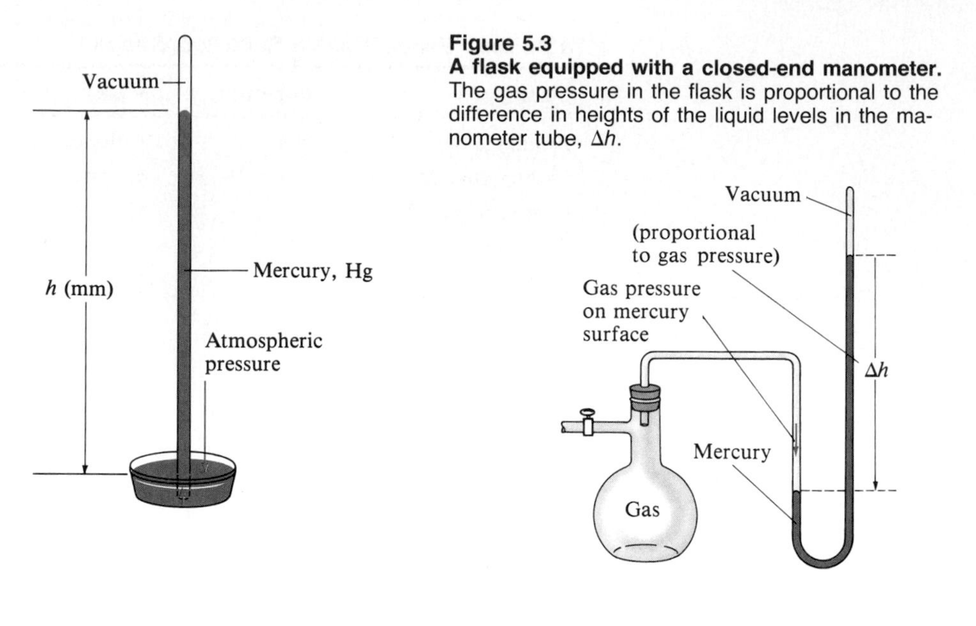

Figure 5.2
A mercury barometer. The height h is proportional to the barometric pressure. Thus, the pressure P can be given as the height of the mercury column, in units called millimeters of mercury, mmHg.

Figure 5.3
A flask equipped with a closed-end manometer. The gas pressure in the flask is proportional to the difference in heights of the liquid levels in the manometer tube, Δh.

$$\text{Pressure} = \frac{\text{force}}{\text{area}} = \frac{2.5 \times 10^{-2}\ \text{kg} \cdot \text{m/s}^2}{2.7 \times 10^{-4}\ \text{m}^2} = 93\ \text{kg/(m} \cdot \text{s}^2)$$

Note that the SI unit of pressure is kg/(m · s^2). *The SI unit of pressure, kg/(m · s^2),* is given the name **pascal (Pa),** after the French physicist Blaise Pascal (1623–1662) who studied fluid pressure. Note that the pressure exerted by a coin the size and mass of a penny is approximately 100 Pa. The pressure exerted by the atmosphere is about 1000 times larger, or about 100,000 Pa. Thus, the pascal is an extremely small unit.

Chemists have traditionally used two other units of pressure, based on the mercury barometer. A **barometer** is *a device for measuring the pressure of the atmosphere*. The mercury barometer consists of a glass tube about one meter long, filled with mercury and inverted in a dish of the same liquid metal (Figure 5.2). At sea level the mercury in the tube falls to a height of about 760 mm above the level in the dish. This height is a direct measure of the atmospheric pressure.● Air pressure downward on the surface of the mercury in the dish is transmitted through the liquid and exerts a pressure upward at the base of the mercury column, supporting it. A mercury column placed in a sealed flask, as in Figure 5.3, measures the gas pressure in the flask. It acts as a **manometer,** *a device that measures the pressure of a gas or liquid in a vessel.*

● Atmospheric, or barometric, pressure depends not only on altitude, but also on weather conditions. The "highs" and "lows" of weather reports refer to high- and low-pressure air masses. A high is associated with fair weather; a low brings unsettled weather, or storms.

The unit **millimeters of mercury (mmHg),** also called the **torr** (after Evangelista Torricelli, who invented the mercury barometer in 1643), is *a unit of pressure equal to that exerted by a column of mercury 1 mm high at 0.00°C*. The **atmosphere (atm)** is a related *unit of pressure equal to exactly 760 mmHg*.

The general relationship between the pressure P and the height h of a liquid column in a barometer or manometer is

$$P = gdh$$

Here g is the constant acceleration of gravity (9.807 m/s^2) and d stands for the density of the liquid in the manometer. If g, d, and h are in SI units, the pressure is given in pascals. Table 5.2 summarizes the relationships among the various units of pressure.

Table 5.2 Important Units of Pressure

Unit	Relationship or Definition
Pascal (Pa)	$kg/(m \cdot s^2)$ (SI unit)
Atmosphere (atm)	1 atm = 1.01325×10^5 Pa $\simeq$ 100 kPa
mmHg or torr	760 mmHg = 1 atm

Example 5.1 Relating Liquid Height and Pressure

Suppose water is used in a barometer instead of mercury. If the barometric pressure is 760.00 mmHg, what is the height of the water column in the barometer at 0.00°C? The densities of water and mercury at 0.00°C are 0.99987 g/cm^3 and 13.596 g/cm^3, respectively.

Solution

The pressure exerted by a column of water h_w, whose density is d_w, is gd_wh_w. A similar result holds for mercury, and because the pressures are equal we can equate these expressions.

$$gd_wh_w = gd_{Hg}h_{Hg}$$

Rearranging gives

$$\frac{h_w}{h_{Hg}} = \frac{d_{Hg}}{d_w}$$

This says the height of the liquid column is inversely proportional to density. Substituting, we find that

$$\frac{h_w}{760.00 \text{ mm}} = \frac{13.596 \text{ g/cm}^3}{0.99987 \text{ g/cm}^3}$$

Hence

$$h_w = 760.00 \text{ mm} \times \frac{13.596}{0.99987} = 1.0334 \times 10^4 \text{ mm}$$

(10.334 m)

Exercise 5.1

An oil whose density is 0.775 g/mL was used in a closed-end manometer to measure the pressure of a gas in a flask, as shown in Figure 5.3. If the height of the oil column was 7.68 cm, what was the pressure of the gas in the flask in mmHg?

(See Problems 5.23 and 5.24.)

5.2 EMPIRICAL GAS LAWS

All gases under moderate conditions behave quite simply with respect to pressure, temperature, volume, and molar amount. By holding any two of these physical properties constant, it is possible to show a simple relationship between the other two. The discovery of these quantitative relationships, the empirical gas laws, occurred from the mid-seventeenth to the mid-nineteenth century. We will discuss these laws in this section.

Boyle's Law

One of the characteristic properties of a gas is its *compressibility*—its ability to be squeezed into a smaller volume by the application of pressure. By comparison, liquids and solids are relatively incompressible. The compressibility of gases was first studied quantitatively by Robert Boyle in 1661.● When he poured mercury

● Robert Boyle (1627–1691) is noted for his contributions to the particle view of matter. In addition to his experiments on gases, he developed the concept of "primary" particles, a forerunner of our present concept of atoms and elements. He explained different kinds of matter in terms of the organization and motion of these primary particles.

Table 5.3 Pressure-Volume Data for 1.000 g O_2 at 0°C

P (atm)	V (L)	PV
0.2500	2.801	0.7002
0.5000	1.400	0.7000
0.7500	0.9333	0.7000
1.000	0.6998	0.6998
2.000	0.3495	0.6990
3.000	0.2328	0.6984
4.000	0.1744	0.6976
5.000	0.1394	0.6970

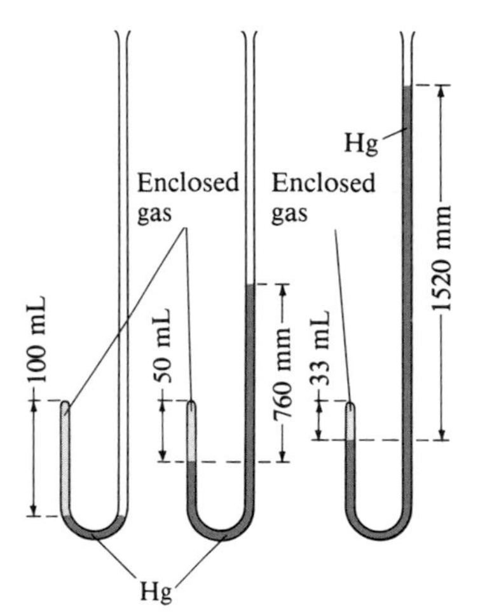

Figure 5.4
Boyle's experiment. The volume of the gas at normal atmospheric pressure (760 mmHg) is 100 mL. When the pressure is doubled by adding 760 mm of mercury, the volume is halved (to 50 mL). Tripling the pressure decreases the volume to 33 mL.

into the open end of a J-shaped tube, the volume of the enclosed gas decreased (Figure 5.4). Each addition of mercury increased the pressure on the gas, decreasing its volume. From such experiments, he formulated the law now known by his name. According to **Boyle's law,** *the volume of a sample of gas at a given temperature varies inversely with the applied pressure*. That is, $V \propto 1/P$, where V is the volume, P is the pressure, and $\propto$ means "is proportional to." Thus, if the pressure is doubled, the volume is halved.

Boyle's law can also be expressed in the form of an equation. Putting pressure and volume on the same side of the equation, we can write

Boyle's law:

$PV = \text{constant}$ (for a given amount of gas at fixed temperature)

That is, for a given amount of gas at a fixed temperature, the pressure times the volume equals a constant. Table 5.3 gives some pressure and volume data for 1.000 g O_2 at 0°C. Note that the product of the pressure and volume is nearly constant. By plotting the volume of the oxygen at different pressures (as shown in Figure 5.5), we can obtain a graph showing the inverse relationship of P and V.

We can use Boyle's law to calculate the volume occupied by a gas when the pressure changes. Consider the 50.0-L tank of oxygen mentioned in the chapter opening. The pressure of gas in the tank is 15.7 atm at 21°C. What volume of

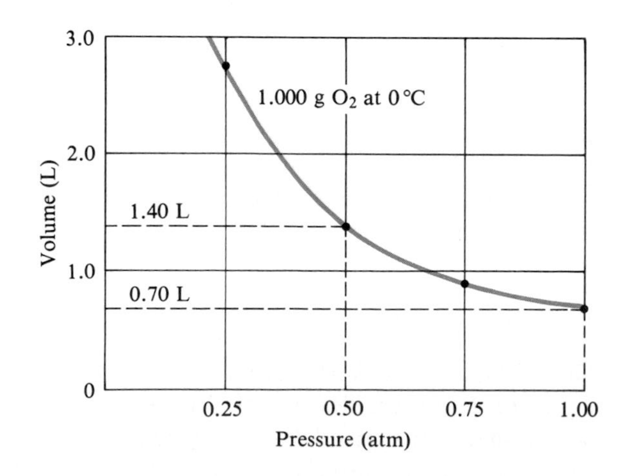

Figure 5.5
A graph of the volume of 1.000 g O_2 at 0°C for various pressures. Note how the volume decreases with increasing pressure. When the pressure is doubled (from 0.50 atm to 1.00 atm), the volume is halved.

oxygen can we get from the tank at 21°C if the atmospheric pressure is 1.00 atm? We can write P_i and V_i for the initial pressure (15.7 atm) and initial volume (50.0 L), and P_f and V_f for the final pressure (1.00 atm) and final volume (to be determined). Because the temperature does not change, the product of the pressure and volume remains constant. Thus, we can write

$$P_f V_f = P_i V_i$$

Dividing both sides of the equation by P_f gives

$$V_f = V_i \times \frac{P_i}{P_f}$$

When we substitute into this equation, we get

$$V_f = 50.0 \text{ L} \times \frac{15.7 \cancel{\text{atm}}}{1.00 \cancel{\text{atm}}} = 785 \text{ L}$$

Note that the initial volume is multiplied by a ratio of pressures. We know the oxygen gas is changing to a lower pressure and will therefore expand, so this ratio will be greater than 1. The final volume, 785 L, is that occupied by all of the gas at 1.00 atm pressure. However, because the tank holds 50.0 L, this volume of gas remains in the tank. The volume that escapes is (785 − 50.0) L = 735 L.

Example 5.2 Using Boyle's Law

A volume of air occupying 12.0 dm^3 at 98.9 kPa is compressed to a pressure of 119.0 kPa. The temperature remains constant. What is the new volume?

Solution

Putting the data for the problem in tabular form, we see what data we have and what we must find. This will suggest the method of solution.

$V_i = 12.0 \text{ dm}^3$ $P_i = 98.9 \text{ kPa}$
$V_f = ?$ $P_f = 119.0 \text{ kPa}$
} Temperature and moles remain constant.

Because P and V vary but T and n are constant, we use Boyle's law to obtain

$$V_f = V_i \times \frac{P_i}{P_f}$$

$$= 12.0 \text{ dm}^3 \times \frac{98.9 \cancel{\text{kPa}}}{119.0 \cancel{\text{kPa}}} = 9.97 \text{ dm}^3$$

Note that the pressure on the gas increases, so the gas is compressed (gives a smaller volume), and the ratio of pressures is less than 1. Thus, we could get the above result by simply multiplying the volume by the ratio of pressures, choosing the ratio so it is less than 1. (If we had expected the gas volume to increase, we would have chosen the ratio to be greater than 1.)

Exercise 5.2

A volume of carbon dioxide gas, CO_2, equal to 20.0 L was collected at 23°C and 1.00 atm pressure. What would be the volume of carbon dioxide if it were collected at 23°C and 0.830 atm? (See Problems 5.27, 5.28, 5.29, and 5.30.)

Before leaving the subject of Boyle's law, we should note that the pressure-volume product for a gas is not precisely constant. We can see this from the PV data given in Table 5.3 for oxygen. In fact, all gases follow Boyle's law at low to moderate pressures but deviate from this law at high pressures. The extent of deviation depends on the gas. We will return to this point at the end of the chapter.

Figure 5.6
The effect of temperature on the volume of a gas. A balloon immersed in liquid nitrogen shrinks because the air inside contracts in volume. When the balloon is removed from the liquid nitrogen, the air inside warms up and the balloon expands to its original size.

Charles's Law

Temperature also affects gas volume. When we immerse a balloon in liquid nitrogen (−196°C), the balloon shrinks (Figure 5.6). After we remove the balloon from the liquid nitrogen, it returns to its original size. A gas contracts when cooled and expands when heated.

One of the first quantitative observations of gases at different temperatures was made by Jacques Alexandre Charles in 1787. Charles was a French physicist and a pioneer in hot-air and hydrogen-filled balloons.● Later, John Dalton (in 1801) and Joseph Louis Gay-Lussac (in 1802) continued these kinds of experiments, which showed that a sample of gas at a fixed pressure increases in volume *linearly* with temperature, at a fixed pressure. By "linearly," we mean that if we plot the volume occupied by a given sample of gas at various temperatures, we get a straight line (Figure 5.7).

● The first ascent of a hot-air balloon carrying people was made on November 21, 1783. A few days later, Jacques Alexandre Charles made an ascent in a hydrogen-filled balloon. On landing, the balloon was attacked and torn to shreds by terrified peasants armed with pitchforks.

When we extend the straight lines in Figure 5.7 from the last experimental point toward lower temperatures—that is, when we *extrapolate* the straight lines

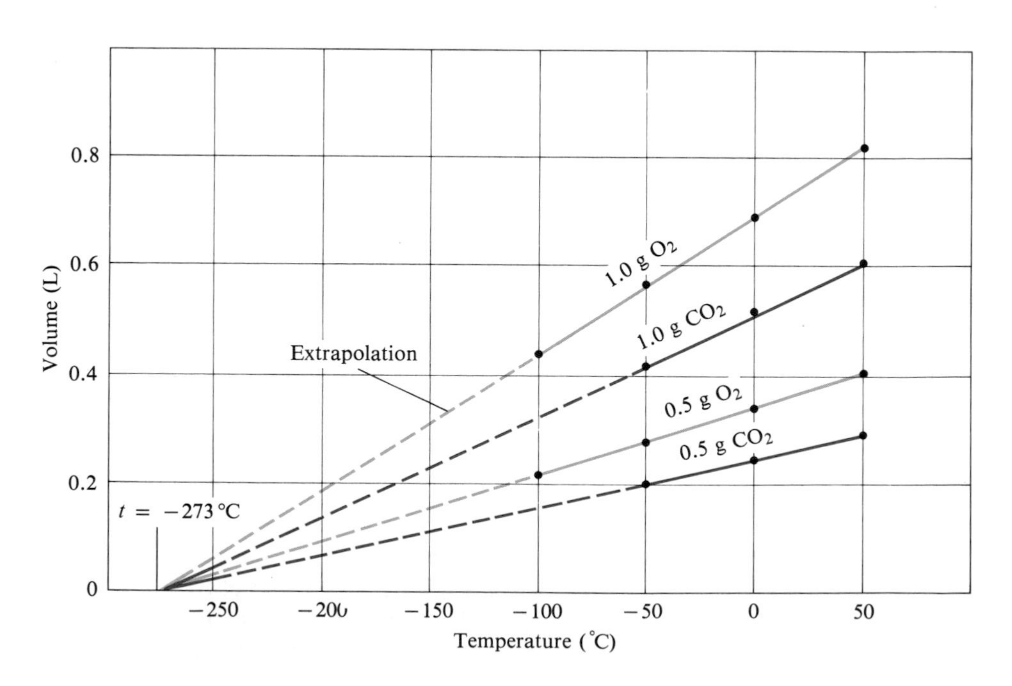

Figure 5.7
Linear relationship of gas volume and temperature at constant pressure. The graph plots gas volume versus temperature for a given mass of gas at 1.00 atm pressure. This linear relationship is independent of amount or kind of gas. Note also that all lines extrapolate to −273°C at zero volume.

backward—we find that they all intersect at a common point. This point occurs at a temperature of −273.15°C, where the graph indicates a volume of zero. This seems to say that if the substances remain gaseous, the volumes occupied will be zero at −273.15°C. This could not happen, however; all gases liquefy before they reach this temperature, and Charles's law does not apply to liquids. These extrapolations do show that we can express the volume variation of a gas with temperature more simply by choosing a different thermometer scale. Let us see how to do this.

The fact that the volume occupied by a gas varies linearly with degrees Celsius can be expressed mathematically by the following equation:

$$V = a + bt$$

where t is the temperature in degrees Celsius, and a and b are constants that determine the straight line.● We can eliminate the constant a by observing that $V = 0$ at $t = -273.15$ for any gas. Substituting into the preceding equation, we get

● The mathematical equation of a straight line is discussed in Appendix A.

$$0 = a + b(-273.15) \qquad \text{or} \qquad a = 273.15b$$

The equation for the volume can now be rewritten:

$$V = 273.15b + bt = b(t + 273.15)$$

Suppose we use a temperature scale equal to degrees Celsius plus 273.15, which you may recognize as the *Kelvin scale*.●

● The Kelvin scale was discussed in Section 1.4.

$$\text{K} = {}^\circ\text{C} + 273.15$$

If we write T for the temperature on the Kelvin scale, we obtain

$$V = bT$$

This is **Charles's law,** which we can state as follows: *The volume occupied by any sample of gas at a constant pressure is directly proportional to the absolute temperature*. Thus, doubling the absolute temperature of a gas doubles its volume.

Charles's law can be rearranged into a form that is very useful for computation.

Charles's law:

$$\frac{V}{T} = \text{constant} \qquad \text{(for a given amount of gas at a fixed pressure)}$$

This says that the volume divided by absolute temperature, for a sample of gas at a fixed pressure, remains constant.

Consider a sample of gas at a fixed pressure, and suppose the temperature changes from its initial value T_i to a final value T_f. How does the volume change? Because the volume divided by absolute temperature is constant, we can write

$$\frac{V_f}{T_f} = \frac{V_i}{T_i}$$

Or, rearranging slightly,

$$V_f = V_i \times \frac{T_f}{T_i}$$

Note that the initial volume is multiplied by a ratio of absolute temperatures. The next example illustrates a calculation involving a change in the temperature of a gas.

Example 5.3 Using Charles's Law

In the previous section, we found that the total volume of oxygen that can be obtained from a particular tank at 1.00 atm and 21°C is 785 L (including the volume remaining in the tank). What would be this volume of oxygen if the temperature had been 28°C instead?

Solution

Before we can do the gas calculation, we must express the temperatures on the Kelvin scale:

$$T_i = (21 + 273)\ \text{K} = 294\ \text{K}$$

$$T_f = (28 + 273)\ \text{K} = 301\ \text{K}$$

Then we put the data for the problem in tabular form:

$$V_i = 785\ \text{L} \quad P_i = 1.00\ \text{atm} \quad T_i = 294\ \text{K}$$

$$V_f = \ ? \quad P_f = 1.00\ \text{atm} \quad T_f = 301\ \text{K}$$

Because V and T vary but P is constant, we use Charles's law.

$$V_f = V_i \times \frac{T_f}{T_i} = 785\ \text{L} \times \frac{301\ \cancel{\text{K}}}{294\ \cancel{\text{K}}} = 804\ \text{L}$$

Note that the temperature increases, so we expect the volume to increase. This means that the ratio of absolute temperatures is greater than 1. We could get the foregoing result by simply multiplying the volume by the ratio of absolute temperatures, choosing the ratio so that it is greater than 1. (If we had expected the gas volume to decrease, we would have chosen the ratio to be less than 1.)

Exercise 5.3

If we expect a chemical reaction to produce 4.38 dm^3 of oxygen, O_2, at 19°C and 101 kPa, what would be the volume at 25°C and 101 kPa?

(See Problems 5.33, 5.34, 5.35, and 5.36.)

Although most gases follow Charles's law fairly well, they deviate from it at high pressures and low temperatures.

Combined Gas Law

Boyle's law ($V \propto 1/P$) and Charles's law ($V \propto T$) can be expressed in one statement: The volume occupied by a given amount of gas is proportional to the absolute temperature divided by the pressure ($V \propto T/P$). We can write this as an equation.

$$V = \text{constant} \times \frac{T}{P} \quad \text{or} \quad \frac{PV}{T} = \text{constant} \qquad \text{(for a given amount of gas)}$$

Consider a problem in which we wish to calculate the final volume of a gas when the pressure and temperature are changed. PV/T is constant for a given amount of gas, so we can write

$$\frac{P_f V_f}{T_f} = \frac{P_i V_i}{T_i}$$

which rearranges to

$$V_f = V_i \times \frac{P_i}{P_f} \times \frac{T_f}{T_i}$$

Thus, the final volume is obtained by multiplying the initial volume by ratios of pressures and absolute temperatures.

Example 5.4 Using the Combined Gas Law

In the Dumas method for determining the amount of nitrogen in a compound, the vapor of the substance is passed over hot copper(II) oxide. The compound is oxidized, and the nitrogen in the substance is converted to nitrogen gas (N_2). If 39.8 mg of caffeine gives 10.1 cm^3 of nitrogen gas at 23°C and 746 mmHg, what is the volume of nitrogen at 0°C and 760 mmHg?

Solution

We first express the temperatures in kelvins:

$$T_i = (23 + 273)\text{ K} = 296\text{ K}$$
$$T_f = (0 + 273)\text{ K} = 273\text{ K}$$

Now we put the data for the problem in tabular form:

$$V_i = 10.1\text{ cm}^3 \quad P_i = 746\text{ mmHg} \quad T_i = 296\text{ K}$$
$$V_f = \ ? \quad P_f = 760\text{ mmHg} \quad T_f = 273\text{ K}$$

From the data, we see that we need to use Boyle's and Charles's laws combined to find how V varies as P and T change. Note that the increase in pressure decreases the volume, making the pressure ratio less than 1. The decrease in temperature also decreases the volume, and the ratio of absolute temperatures is less than 1.

$$V_f = V_i \times \frac{P_i}{P_f} \times \frac{T_f}{T_i} = 10.1\text{ cm}^3 \times \frac{746\ \cancel{\text{mmHg}}}{760\ \cancel{\text{mmHg}}} \times \frac{273\ \cancel{\text{K}}}{296\ \cancel{\text{K}}}$$
$$= 9.14\text{ cm}^3$$

Exercise 5.4

A balloon contains 5.41 dm^3 of helium, He, at 24°C and 101.5 kPa. Suppose the gas in the balloon is heated to 35°C. If the helium pressure is now 102.8 kPa, what is its volume?

(See Problems 5.39 and 5.40.)

Avogadro's Law

In 1808 the French chemist Joseph Louis Gay-Lussac (1778–1850) concluded from experiments on gas reactions that the volumes of reactant gases at a given pressure and temperature are in ratios of small whole numbers (*the law of combining volumes*). For example, two volumes of hydrogen react with one volume of oxygen gas to produce water.

$$\underset{\text{2 volumes}}{2H_2(g)} + \underset{\text{1 volume}}{O_2(g)} \longrightarrow 2H_2O(g)$$

Three years later, the Italian chemist Amedeo Avogadro (1776–1856) interpreted the law of combining volumes in terms of what we now call **Avogadro's law:** *Equal volumes of any two gases at the same temperature and pressure contain the same number of molecules*. Thus, two volumes of hydrogen contain twice the number of molecules in one volume of oxygen, in agreement with the chemical equation for the reaction.

One mole of any gas contains the same number of molecules (Avogadro's number = 6.022×10^{23}) and by Avogadro's law must occupy the same volume at a given temperature and pressure. This *volume of one mole of gas* is called the **molar gas volume, V_m.** Volumes of gases are often compared at **standard temperature and pressure (STP),** *the reference conditions for gases chosen by convention to be 0°C and 1 atm pressure*. At STP, the molar gas volume is found to be 22.4 L/mol (Figure 5.8).

We can reexpress Avogadro's law as follows: The molar gas volume at a given temperature and pressure is a specific constant independent of the nature of the gas.

Figure 5.8
The molar volume of a gas at STP. The box occupies 22.4 L, equal to the molar volume of a gas at STP. The basketball is for comparison.

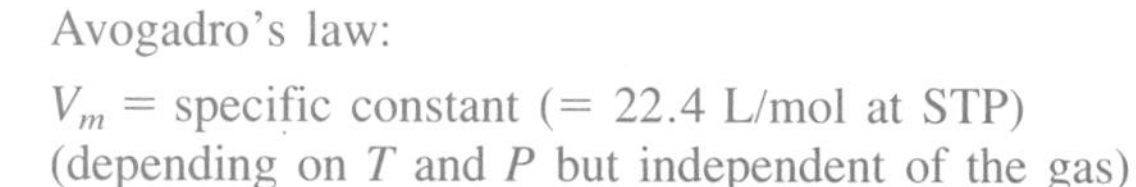

Avogadro's law:
V_m = specific constant (= 22.4 L/mol at STP)
(depending on T and P but independent of the gas)

Table 5.4 lists the molar volumes of several gases. The values agree within about two percent of the value expected from the empirical gas laws (the ''ideal gas'' value). We will discuss the reason for the deviations from this value at the end of this chapter.

5.3 THE IDEAL GAS LAW

In the previous section, we discussed the empirical gas laws. Here we will show that these laws can be combined into one equation, called the *ideal gas equation*. Earlier we combined Boyle's law and Charles's law into the equation

$$V = \text{constant} \times \frac{T}{P} \qquad \text{(for a given amount of gas)}$$

This ''constant'' is independent of the temperature and pressure but does depend on the amount of gas. For one mole, the constant will have a specific value, which we will denote as R. The molar volume, V_m, is

$$V_m = R \times \frac{T}{P}$$

According to Avogadro's law, the molar volume at a specific value of T and P is a constant independent of the nature of the gas, and this implies that R is a

Table 5.4 Molar Volumes of Several Gases

Gas	Molar Volume at 0.0°C, 1.00 atm
He	22.40 L
H_2	22.43 L
O_2	22.39 L
CO_2	22.29 L
NH_3	22.09 L
Ideal gas*	22.41 L

*An ideal gas follows the empirical gas laws.

Table 5.5 Values of the Molar Gas Constant in Various Units

Value of R
0.082058 L · atm/(K · mol)
8.3145 J/(K · mol)
8.3145 kg · m^2/(s^2 · K · mol)
8.3145 dm^3 · kPa/(K · mol)
1.9872 cal/(K · mol)

constant independent of the gas. The **molar gas constant, R,** is *the constant of proportionality relating the molar volume of a gas to* T/P. Values of R in various units are given in Table 5.5.

The preceding equation can be written for n moles of gas if we multiply both sides by n.

$$\underbrace{nV_m}_{V} = \frac{nRT}{P} \quad \text{or} \quad PV = nRT$$

Because V_m is the volume per mole, nV_m is the total volume V. *The equation* PV = nRT, *which combines all of the gas laws,* is called the **ideal gas law.**

Ideal gas law:
$PV = nRT$

The ideal gas law includes all the information contained in Boyle's, Charles's, and Avogadro's laws. In fact, starting with the ideal gas law, we can derive any of the other gas laws.

Example 5.5 Deriving Empirical Gas Laws from the Ideal Gas Law

Prove the following statement: The pressure of a given amount of gas at a fixed volume is proportional to the absolute temperature. This is sometimes called *Amontons's law*. In 1702 Guillaume Amontons constructed a thermometer based on measurement of the pressure of a fixed volume of air. The principle employed by Amontons is now used to construct gas thermometers, which are the accepted standards for calibrating other types of thermometers.

Solution

From the ideal gas law,

$$PV = nRT$$

Solving for P, we get

$$P = \left(\frac{nR}{V}\right)T$$

Note that everything in parentheses in this equation is constant. Therefore, we can write

$$P = \text{constant} \times T$$

Or, expressing this as a proportion, we get

$$P \propto T$$

Boyle's law and Charles's law follow from the ideal gas law by a similar derivation.

Exercise 5.5

Show that the moles of gas are proportional to the pressure for constant volume and temperature.

(See Problems 5.43 and 5.44.)

The limitations that apply to Boyle's, Charles's, and Avogadro's laws also apply to the ideal gas law. That is, the ideal gas law is most accurate for low to moderate pressures and for temperatures that are not too low.●

● No gas is "ideal." But the ideal gas law is very useful even though it is only an approximation. The behavior of real gases is described at the end of this chapter.

Calculations Using the Ideal Gas Law

The type of problem to which Boyle's and Charles's laws are applied involves a change in conditions (P, V, or T) of a gas. The ideal gas law allows us to solve another type of problem: Given any three of the quantities P, V, n, and T, calculate the unknown quantity. Example 5.6 illustrates such a problem.

Example 5.6 Using the Ideal Gas Law

Answer the question asked in the chapter opening: How many grams of oxygen, O_2, are there in a 50.0-L tank at 21°C when the oxygen pressure is 15.7 atm?

Solution

In asking for the mass of oxygen, we are in effect asking for moles of gas n, because mass and moles are easily related. The data given in the problem are

VARIABLE	VALUE
P	15.7 atm
V	50.0 L
T	(21 + 273) K = 294 K
n	?

From the data given, we see that we can use the ideal gas law to solve for n. The proper value to use for R depends on the units of V and P. These are in liters and atmospheres, respectively, so we use (from Table 5.5) $R = 0.0821\ \text{L} \cdot \text{atm}/(\text{K} \cdot \text{mol})$. We solve the ideal gas law for n.

$$n = \frac{PV}{RT}$$

Substituting from the table of data gives

$$n = \frac{15.7\ \cancel{\text{atm}} \times 50.0\ \cancel{\text{L}}}{0.0821\ \cancel{\text{L}} \cdot \cancel{\text{atm}}/(\cancel{\text{K}} \cdot \text{mol}) \times 294\ \cancel{\text{K}}} = 32.5\ \text{mol}$$

and converting moles to mass of oxygen yields

$$32.5\ \cancel{\text{mol O}_2} \times \frac{32.0\ \text{g O}_2}{1\ \cancel{\text{mol O}_2}} = 1.04 \times 10^3\ \text{g O}_2$$

Exercise 5.6

What is the pressure in a 50.0-L tank that contains 3.03 kg of oxygen, O_2, at 23°C?

(See Problems 5.45, 5.46, 5.47, 5.48, 5.49, and 5.50.)

Gas Density; Molecular-Weight Determination

The density of a substance depends on mass divided by volume, and because the volume of a gas varies with temperature and pressure, the density of a gas also varies with temperature and pressure (Figure 5.9). In the next example, we use the ideal gas law to calculate the density of a gas at any temperature and pressure. We do this by calculating the moles, and from this the mass, in a liter of gas at the

Figure 5.9
Hot-air ballooning. A propane gas burner on board a balloon heats the air. The heated air expands, occupying a larger volume, and therefore has a lower density than the surrounding air. The hot air and balloon rise.

given temperature and pressure; the mass per liter of gas is its density. At the end of this section, we will give an alternative method of solving this problem, using a formula relating density to temperature, pressure, and molecular weight.

Example 5.7 Calculating Gas Density

What is the density of oxygen, O_2, in grams per liter at 25°C and 0.850 atm?

Solution

Because density equals mass per unit volume, calculating the mass of 1 L of gas will give us the density of the gas. (The volume is treated as an exact quantity, so it is not used to obtain the number of significant figures in the calculation of mass.)

VARIABLE	VALUE
P	0.850 atm
V	1 L
T	(25 + 273) K = 298 K
n	?

We start from the ideal gas law and solve for n.

$$PV = nRT \quad \text{or} \quad n = \frac{PV}{RT}$$

Substituting from the table,

$$n = \frac{0.850 \cancel{\text{atm}} \times 1 \cancel{\text{L}}}{0.0821 \cancel{\text{L}} \cdot \cancel{\text{atm}}/(\cancel{\text{K}} \cdot \text{mol}) \times 298 \cancel{\text{K}}}$$
$$= 0.0347 \text{ mol}$$

Now we convert moles oxygen to grams.

$$0.0347 \cancel{\text{mol O}_2} \times \frac{32.0 \text{ g O}_2}{1 \cancel{\text{mol O}_2}} = 1.11 \text{ g O}_2$$

Therefore, the density of O_2 at 25°C and 0.850 atm is 1.11 g/L.

Suppose we wanted the density of Cl_2, instead of O_2, at this T and P. Only the conversion of moles to mass is affected. The previous calculation becomes

$$0.0347 \text{ mol Cl}_2 \times \frac{70.9 \text{ g Cl}_2}{1 \text{ mol Cl}_2} = 2.46 \text{ g Cl}_2$$

The density of Cl_2 is 2.46 g/L. Note that the density of a gas is directly proportional to its molecular weight.

Exercise 5.7

Calculate the density of helium, He, in grams per liter at 21°C and 752 mmHg. The density of air under these conditions is 1.188 g/L. What is the difference in mass between 1 liter of air and 1 liter of helium? (This mass difference is equivalent to the buoyant or lifting force of helium per liter.) (See Problems 5.51, 5.52, 5.53, and 5.54.)

Figure 5.10
A gas whose density is greater than that of air. The reddish-brown gas being poured from the beaker is bromine. Note how the gas, which is denser than air, hugs the bottom of the beaker.

From the previous example, we see that the density of a gas is directly proportional to its molecular weight. Chlorine, whose molecular weight is more than twice that of oxygen or nitrogen, the major components of air, is thus more than twice as dense as air. Figure 5.10 shows bromine gas being poured from a beaker. Note that the reddish-brown gas hugs the bottom of the beaker, because it is denser than air.

The relation between density and molecular weight of a gas suggests that we could use a measurement of gas density to determine its molecular weight. In fact, gas (vapor) density measurements provided one of the first methods of determining molecular weight. The method was worked out by the French chemist Jean-Baptiste André Dumas in 1826. It can be applied to any substance that can be vaporized without decomposing.

As an illustration, consider the determination of the molecular weight of halothane, an inhalation anesthetic. The density of halothane vapor at 71°C (344 K) and 768 mmHg (1.01 atm) is 7.05 g/L. To obtain the molecular weight, we calculate the moles of vapor in a given volume. The molar mass equals moles

● Dumas tried to obtain the atomic weight of sulfur from vapor density measurements. His results were difficult to interpret, however. Sulfur, as we now know, exists as S_8 molecules in the solid and in the liquid. In the vapor, however, the molecules are not only S_8 but S_6 and smaller species as well. Above 800°C, the vapor is mostly S_2, until about 2000°C, where sulfur exists mostly as a mon-atomic gas. These changes in molecular form can be correlated with the density measurements.

divided by the mass in the same volume. From the density, we see that one liter of vapor has a mass of 7.05 g. The moles in this volume is obtained from the ideal gas law.

$$n = \frac{PV}{RT} = \frac{1.01\ \cancel{\text{atm}} \times 1\ \cancel{\text{L}}}{0.0821\ \cancel{\text{L}} \cdot \cancel{\text{atm}}/(\cancel{\text{K}} \cdot \text{mol}) \times 344\ \cancel{\text{K}}} = 0.0358 \text{ mol}$$

Therefore, the molar mass, M_m, is

$$M_m = \frac{m}{n} = \frac{7.05 \text{ g}}{0.0358 \text{ mol}} = 197 \text{ g/mol}$$

Thus, the molecular weight is 197 amu.●

If you look at what we have done, you will see that all we need to have is the mass of vapor in *any* given volume (for a given T and P). An explicit determination of density is not necessary. We determine the moles of vapor from the volume using the ideal gas law, then determine the molar mass by dividing the mass by moles for this volume. The next example illustrates this. At the end of this section, we will derive an explicit formula for the molecular weight in terms of the density of a gas, which will give us an alternative way to solve this problem.

Example 5.8 Calculating the Molecular Weight of a Vapor

A quantity of carbon tetrachloride (a liquid), CCl_4, was put into a 200.0-mL flask. The open flask was then submerged in boiling water until all of the liquid evaporated and the vapor filled the flask (see Figure 5.11). (Any excess vapor escaped through the tube in the neck of the flask, so that the pressure of the vapor equaled the barometric pressure.) The flask was then cooled and weighed. Subtracting the mass of the empty flask gave the mass of CCl_4. The mass of the carbon tetrachloride vapor in the flask when the temperature of the boiling water was 99°C was found to be 0.970 g. The barometric pressure was 733 mmHg. What is the molecular weight of carbon tetrachloride?

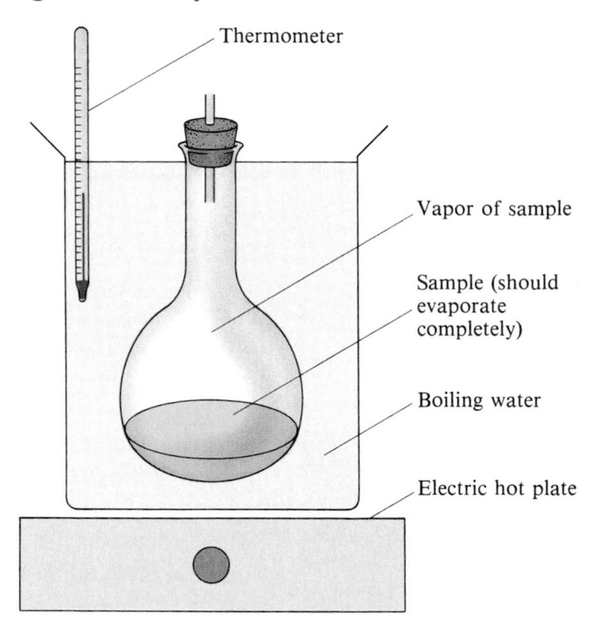

Figure 5.11
Finding the vapor density of a substance. After the flask is filled with vapor of the substance at the temperature of the boiling water, it is cooled so that the vapor condenses. The mass of the substance is found by weighing the flask and substance, then subtracting the mass of the empty flask. The pressure of the vapor equals the barometric pressure, and the temperature equals that of the boiling water.

Solution

We will calculate the moles of vapor from the ideal gas law and then calculate the molar mass by dividing mass by moles. Note that the temperature of the vapor is the same as the temperature of the boiling water and that the pressure of the vapor equals the barometric pressure. We have the following data for the vapor.

VARIABLE	VALUE
P	$733\ \cancel{\text{mmHg}} \times \frac{1 \text{ atm}}{760\ \cancel{\text{mmHg}}} = 0.964 \text{ atm}$
V	200.0 mL = 0.2000 L
T	(99 + 273) K = 372 K
n	?

From the ideal gas law, $PV = nRT$, we have

$$n = \frac{PV}{RT} = \frac{0.964\ \cancel{\text{atm}} \times 0.2000\ \cancel{\text{L}}}{0.0821\ \cancel{\text{L}}\ \cancel{\text{atm}}/(\cancel{\text{K}} \cdot \text{mol}) \times 372\ \cancel{\text{K}}}$$

$n = 0.00631 \text{ mol } CCl_4$

Dividing the mass of the vapor by moles gives the mass per mole (the molar mass).

$$\text{Molar mass} = \frac{\text{grams vapor}}{\text{moles vapor}} = \frac{0.970 \text{ g}}{0.00631 \text{ mol}} = 154 \text{ g/mol}$$

Thus, the molecular weight is **154 amu.**

Exercise 5.8

A sample of a gaseous substance at 25°C and 0.862 atm has a density of 2.26 g/L. What is the molecular weight of the substance? (See Problems 5.55, 5.56, 5.57, and 5.58.)

From the ideal gas law, we can obtain an explicit relationship between the molecular weight and density of a gas. Recall that the molar mass ($M_m = m/n$) when expressed in grams per moles is numerically equal to the molecular weight. If we substitute $n = m/M_m$ into the ideal gas law, $PV = nRT$, we obtain

$$PV = \frac{m}{M_m}RT \text{ or } PM_m = \frac{m}{V}RT$$

But m/V equals the density, d. Substituting this into the previous equation gives

$$PM_m = dRT$$

We can illustrate the use of this equation by solving Examples 5.7 and 5.8 again. In Example 5.7, we asked, "What is the density of oxygen, O_2, in grams per liter at 25°C (298 K) and 0.850 atm?" We rearrange the previous equation to give an explicit formula for the density and then substitute into this formula.

$$d = \frac{PM_m}{RT} = \frac{0.850 \cancel{\text{atm}} \times 32.0 \text{ g/}\cancel{\text{mol}}}{0.0821 \text{ L} \cdot \cancel{\text{atm}}/(\cancel{\text{K}} \cdot \cancel{\text{mol}}) \times 298 \cancel{\text{K}}} = 1.11 \text{ g/L}$$

Note that by giving M_m in g/mol and R in L · atm/(K · mol), we obtain the density in g/L.

The question in Example 5.8 is equivalent to the following. What is the molecular weight of a substance weighing 0.970 g whose vapor occupies 200.0 mL (0.200 L) at 99°C (372 K) and 733 mmHg (0.964 atm)? The density of the vapor is 0.970 g/0.200 L = 4.85 g/L. To obtain the molecular weight, we rearrange the equation $PM_m = dRT$ to give an explicit formula for M_m, then substitute into it.

$$M_m = \frac{dRT}{P} = \frac{4.85 \text{ g/}\cancel{\text{L}} \times 0.0821 \cancel{\text{L}} \cdot \cancel{\text{atm}}/(\cancel{\text{K}} \cdot \text{mol}) \times 372 \cancel{\text{K}}}{0.964 \cancel{\text{atm}}} = 154 \text{ g/mol}$$

5.4 STOICHIOMETRY PROBLEMS INVOLVING GAS VOLUMES

In Chapter 4 we learned how to find the mass of one substance in a chemical reaction from the mass of another substance in the reaction. Now that we know how to use the ideal gas law, we can extend these types of problems to include gas volumes.

Consider the following reaction, which is often used to generate small quantities of oxygen gas:

$$2KClO_3(s) \xrightarrow[MnO_2]{\Delta} 2KCl(s) + 3O_2(g)$$

Suppose we heat 0.0100 mol of potassium chlorate, $KClO_3$, in a test tube. How many liters of oxygen can we produce at 298 K and 1.02 atm?

We solve such a problem by breaking it into two problems, one involving stoichiometry and the other involving the ideal gas law. We note that 2 mol $KClO_3$ yields 3 mol O_2. Therefore,

$$0.0100 \text{ mol } KClO_3 \times \frac{3 \text{ mol } O_2}{2 \text{ mol } KClO_3} = 0.0150 \text{ mol } O_2$$

Now that we have the moles of oxygen produced, we can use the ideal gas law to calculate the volume of oxygen under the conditions given. We rearrange the ideal gas law, $PV = nRT$, and solve for the volume.

$$V = \frac{nRT}{P}$$

Then we substitute for n, T, and P. Because the pressure is given in units of atmospheres, we choose the value of R in units of L · atm/(K · mol). The answer comes out in liters.

$$V = \frac{0.0150 \text{ mol} \times 0.0821 \text{ L} \cdot \text{atm/(K} \cdot \text{mol)} \times 298 \text{ K}}{1.02 \text{ atm}} = 0.360 \text{ L}$$

The next example further illustrates this method.

Example 5.9 Solving Stoichiometry Problems Involving Gas Volumes

How many liters of chlorine gas, Cl_2, can be obtained at 40°C and 787 mmHg from 9.41 g of hydrogen chloride, HCl, according to the following equation?

$$2KMnO_4(s) + 16HCl(aq) \longrightarrow 8H_2O(l) + 2KCl(aq) + 2MnCl_2(aq) + 5Cl_2(g)$$

Solution

From the chemical equation, we find how many moles of Cl_2 can be obtained from 9.41 g HCl. Then, from the moles of Cl_2, we calculate the volume of Cl_2 under the specified conditions using the ideal gas law.

$$9.41 \text{ g HCl} \times \frac{1 \text{ mol HCl}}{36.5 \text{ g HCl}} \times \frac{5 \text{ mol } Cl_2}{16 \text{ mol HCl}} = 0.0806 \text{ mol } Cl_2$$

Before doing the gas calculation, we list the available data.

VARIABLE	VALUE
P	$787 \text{ mmHg} \times \frac{1 \text{ atm}}{760 \text{ mmHg}} = 1.036 \text{ atm}$
V	?
T	(40 + 273) K = 313 K
n	0.0806 mol

Note that we have converted the pressure in mmHg to atmospheres, and degrees Celsius to kelvins. Now we rearrange the ideal gas law and solve for the volume.

$$V = \frac{nRT}{P}$$

Substituting from our data gives

$$V = \frac{0.0806 \text{ mol} \times 0.0821 \text{ L} \cdot \text{atm/(K} \cdot \text{mol)} \times 313 \text{ K}}{1.036 \text{ atm}}$$

$$= \mathbf{2.00 \text{ L}}$$

Exercise 5.9

Lithium hydroxide, LiOH, is used in spacecraft to absorb the carbon dioxide exhaled by astronauts. The reaction is

$$2LiOH(s) + CO_2(g) \longrightarrow Li_2CO_3(s) + H_2O(l)$$

Calculate the volume of carbon dioxide at 22°C and 748 mmHg absorbed by 1.00 g of lithium hydroxide. (See Problems 5.61, 5.62, 5.63, and 5.64.)

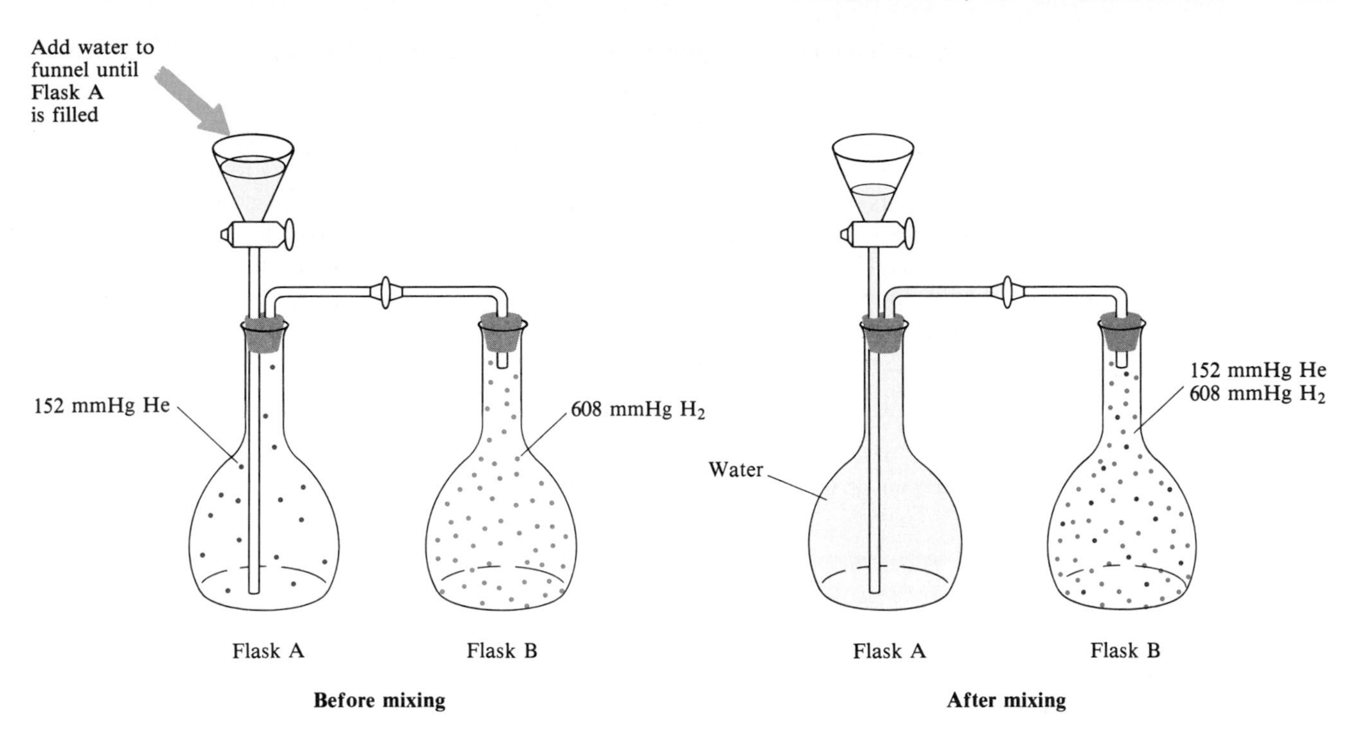

Figure 5.12
A demonstration of Dalton's law of partial pressures. Flask A is filled with water so that the helium in it is pushed into Flask B, where it mixes with hydrogen. Each gas exerts the pressure it would exert if the other were not there.

5.5 GAS MIXTURES; LAW OF PARTIAL PRESSURES

While studying the composition of air, John Dalton concluded in 1801 that each gas in a mixture of unreactive gases acts, as far as its pressure is concerned, as though it were the only gas in the mixture. To illustrate, consider two 1-L flasks, one filled with helium to a pressure of 152 mmHg at a given temperature and the other filled with hydrogen to a pressure of 608 mmHg at the same temperature. Suppose all of the helium in the one flask is put in with the hydrogen in the other flask (see Figure 5.12). After the gases are mixed in one flask, each gas occupies a volume of one liter, just as before, and has the same temperature.

According to Dalton, each gas exerts the same pressure it would exert if it were the only gas in the flask. Thus, the pressure exerted by helium in the mixture is 152 mmHg, and the pressure exerted by hydrogen in the mixture is 608 mmHg. The total pressure exerted by the gases in the mixture is 152 mmHg + 608 mmHg = 760 mmHg.

Partial Pressures and Mole Fractions

The pressure exerted by a particular gas in a mixture is the **partial pressure** of that gas. The partial pressure of helium in the preceding mixture is 152 mmHg; the partial pressure of hydrogen in the mixture is 608 mmHg. According to **Dalton's law of partial pressures,** *the sum of the partial pressures of all the different gases in a mixture is equal to the total pressure of the mixture.*

If we let P be the total pressure and P_A, P_B, P_C, . . . be the partial pressures of the component gases in a mixture, the law of partial pressures can be written as

Dalton's law of partial pressures:
$$P = P_A + P_B + P_C + \cdots$$

The individual partial pressures follow the ideal gas law. For component A,

$$P_A V = n_A RT$$

where n_A is the number of moles of component A.

The composition of a gas mixture is often described in terms of the mole fractions of component gases. The **mole fraction** of a component gas is simply *the fraction of moles of that component in the total moles of gas mixture* (or the fraction of molecules that are component molecules). Because the pressure of a gas is proportional to moles, for fixed volume and temperature ($P = nRT/V \propto n$), the mole fraction also equals the partial pressure divided by total pressure.

$$\text{Mole fraction of } A = \frac{n_A}{n} = \frac{P_A}{P}$$

Mole percent equals mole fraction $\times$ 100. Mole percent is equivalent to the percentage of the molecules that are component molecules.

Example 5.10 Calculating Partial Pressures and Mole Fractions of a Gas in a Mixture

A 1.00-L sample of dry air at 25°C and 786 mmHg contains 0.925 g N_2, plus other gases including oxygen, argon, and carbon dioxide. (a) What is the partial pressure (in mmHg) of N_2 in the air sample? (b) What is the mole fraction and mole percent of N_2 in the mixture?

Solution

(a) Each gas in a mixture follows the ideal gas law. To calculate the partial pressure of N_2, we convert 0.925 g N_2 to moles N_2.

$$0.925 \text{ g } N_2 \times \frac{1 \text{ mol } N_2}{28.0 \text{ g } N_2} = 0.330 \text{ mol } N_2$$

We substitute into the ideal gas law (noting that 25°C is 298 K).

$$P_{N_2} = \frac{n_{N_2}RT}{V}$$
$$= \frac{0.0330 \text{ mol} \times 0.0821 \text{ L} \cdot \text{atm}/(\text{K} \cdot \text{mol}) \times 298 \text{ K}}{1.00 \text{ L}}$$
$$= 0.807 \text{ atm } (= \mathbf{613\ mmHg})$$

(b) The mole fraction of N_2 in air is

$$\text{Mole fraction of } N_2 = \frac{P_{N_2}}{P} = \frac{613 \text{ mmHg}}{786 \text{ mmHg}} = 0.780$$

Air contains **78.0 mole percent** of N_2.

Exercise 5.10

A 10.0-L flask contains 1.031 g O_2 and 0.572 g CO_2 at 18°C. What are the partial pressures of oxygen and carbon dioxide? What is the total pressure? What is the mole fraction of oxygen in the mixture?

(See Problems 5.67 and 5.68.)

Collecting Gases over Water

A useful application of the law of partial pressures arises when we collect gases over water (a method used for gases that do not dissolve appreciably in water). Figure 5.13 shows how a gas, produced by chemical reaction in the flask, is collected by leading it to an inverted tube, where it displaces water. As gas bubbles through the water, it picks up molecules of water vapor that mix with it. The

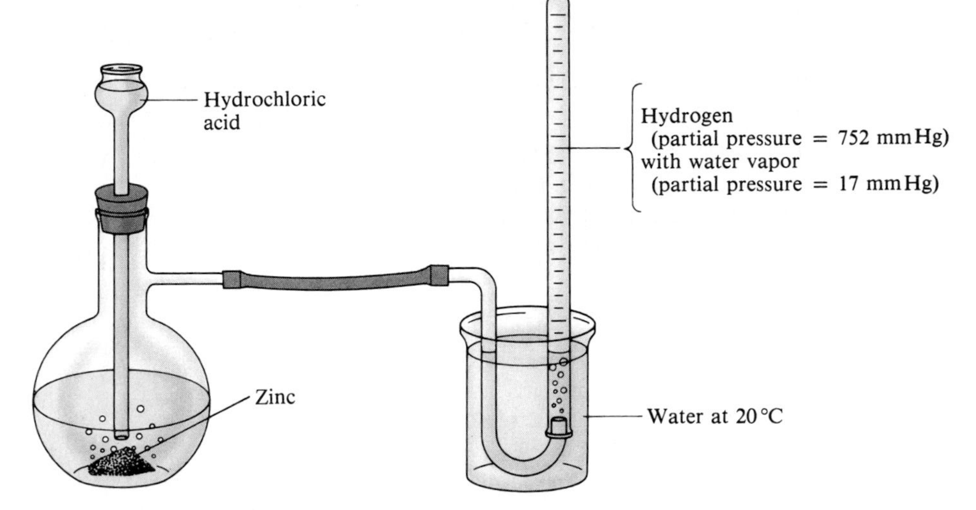

Figure 5.13
Collection of a gas over water. Hydrogen, prepared by the reaction of zinc with hydrochloric acid, is collected over water. When the gas-collection tube is adjusted so that the level in the tube is at the same height as the level in the beaker, the gas pressure in the tube equals the barometric pressure (769 mmHg). The total gas pressure equals the sum of the partial pressure of the hydrogen (752 mmHg) and the vapor pressure of water (17 mmHg).

● Vapor pressure is the maximum partial pressure of the vapor in the presence of the liquid. It is defined more precisely in Chapter 11.

partial pressure of water vapor in the gas mixture in the collection tube depends only on the temperature. This partial pressure of water vapor is called the *vapor pressure* of water.● Values of the vapor pressure of water at various temperatures are listed in Table 5.6. The following example shows how to find the partial pressure and then the mass of the collected gas.

Table 5.6 Vapor Pressure of Water at Various Temperatures

Temperature, °C	Pressure, mmHg	Temperature, °C	Pressure, mmHg
0	4.6	27	26.7
5	6.5	28	28.3
10	9.2	29	30.0
11	9.8	30	31.8
12	10.5	35	42.2
13	11.2	40	55.3
14	12.0	45	71.9
15	12.8	50	92.5
16	13.6	55	118.0
17	14.5	60	149.4
18	15.5	65	187.5
19	16.5	70	233.7
20	17.5	75	289.1
21	18.7	80	355.1
22	19.8	85	433.6
23	21.1	90	525.8
24	22.4	95	633.9
25	23.8	100	760.0
26	25.2	105	906.1

Example 5.11 Calculating the Amount of Gas Collected over Water

Hydrogen gas is produced by the reaction of hydrochloric acid, HCl, on zinc metal.

$$2HCl(aq) + Zn(s) \longrightarrow ZnCl_2(aq) + H_2(g)$$

The gas is collected over water. If 156 mL of gas is collected at 20°C (two significant figures) and 769 mmHg total pressure, what is the mass of hydrogen collected?

Solution

The gas that is collected is hydrogen mixed with water vapor. To obtain the amount of hydrogen, we must first find the partial pressure of hydrogen in the mixture, using Dalton's law (Step 1). Then we can calculate the moles of hydrogen from the ideal gas law (Step 2). Finally, we can obtain the mass of hydrogen from the moles of hydrogen (Step 3).

STEP 1 The vapor pressure of water at 20°C is 17.5 mmHg. From Dalton's law of partial pressures, we know that the total gas pressure equals the partial pressure of hydrogen, P_{H_2}, plus the partial pressure of water, P_{H_2O}.

$$P = P_{H_2} + P_{H_2O}$$

Substituting and solving for the partial pressure of hydrogen, we get

$$P_{H_2} = P - P_{H_2O} = (769 - 17.5) \text{ mmHg} = 752 \text{ mmHg}$$

STEP 2 Now we can use the ideal gas law to find the moles of hydrogen collected. The data are

VARIABLE	VALUE
P	$750 \text{ mmHg} \times \dfrac{1 \text{ atm}}{760 \text{ mmHg}} = 0.987 \text{ atm}$
V	$156 \text{ mL} = 0.156 \text{ L}$
T	$(20 + 273) \text{ K} = 293 \text{ K}$
n	?

From the ideal gas law, $PV = nRT$, we have

$$n = \frac{PV}{RT} = \frac{0.987 \cancel{\text{atm}} \times 0.156 \cancel{\text{L}}}{0.0821 \cancel{\text{L}} \cdot \cancel{\text{atm}}/(\cancel{\text{K}} \cdot \text{mol}) \times 293 \cancel{\text{K}}}$$
$$= 0.00640 \text{ mol}$$

STEP 3 We convert moles of H_2 to grams of H_2.

$$0.00640 \cancel{\text{mol } H_2} \times \frac{2.02 \text{ g } H_2}{1 \cancel{\text{mol } H_2}} = 0.0129 \text{ g } H_2$$

Exercise 5.11

Oxygen can be prepared by heating potassium chlorate, $KClO_3$, with manganese dioxide as a catalyst. The reaction is,

$$2KClO_3(s) \xrightarrow{\Delta} 2KCl(s) + 3O_2(g)$$

How many moles of O_2 would be obtained from 1.300 g of $KClO_3$? If this amount of O_2 were collected over water at 23°C and at a total pressure of 745 mmHg, what volume would it occupy?
(See Problems 5.71 and 5.72.)

Kinetic-Molecular Theory

In the following sections, we will see how interpretation of a gas in terms of the **kinetic-molecular theory** (or simply **kinetic theory**) leads to the ideal gas law. *According to this theory, a gas consists of molecules in constant random motion.* The word *kinetic* describes something in motion. Thus, kinetic energy, E_k, is the energy associated with the motion of an object of mass m. From physics,

$$E_k = \frac{1}{2}m \times (\text{speed})^2$$

We will use the concept of kinetic energy in describing the kinetic theory.

5.6 KINETIC THEORY OF AN IDEAL GAS

Our present explanation of gas pressure is that it results from the continual bombardment of the container walls by constantly moving molecules. This kinetic interpretation of gas pressure was first put forth in 1676 by Robert Hooke, who had earlier been an assistant of Boyle. Hooke did not pursue this idea, however, so the generally accepted interpretation of gas pressure remained the one given by Isaac Newton, a contemporary of Hooke.

According to Newton, the pressure of a gas was due to the mutual repulsions of the gas particles (molecules). These repulsions pushed the molecules against the walls of the gas container, much as coiled springs packed tightly in a box would push against the walls of the box. This interpretation continued to be the dominant view of gas pressure until the mid-nineteenth century.

Despite the dominance of Newton's view, some people followed the kinetic interpretation. In 1738 Daniel Bernoulli, a Swiss mathematician and physicist, gave a quantitative explanation of Boyle's law using the kinetic interpretation. He even suggested that molecules move faster at higher temperatures, in order to explain Amontons's experiments on the temperature dependence of gas volume and pressure. However, Bernoulli's paper attracted little notice. A similar kinetic interpretation of gases was submitted for publication to the Royal Society of London in 1848 by John James Waterston. His paper was rejected as "nothing but nonsense."●

● Waterston's paper, with the reviewers' comments, was discovered in the Royal Society's files by Lord Rayleigh in 1892. (Rayleigh, a physicist, codiscovered argon and received the 1904 Nobel Prize in physics.)

Soon afterward, the scientific climate for the kinetic view improved. The kinetic theory of gases was developed by a number of influential physicists, including James Clerk Maxwell (1859) and Ludwig Boltzmann (in the 1870s). Throughout the last half of the nineteenth century, research continued on the kinetic theory, making it a cornerstone of our present view of molecular substances.

Postulates of Kinetic Theory

Physical theories are often given in terms of *postulates*. These are the basic statements from which all conclusions or predictions of a theory are deduced. The postulates are accepted as long as the predictions from the theory agree with experiment. If a particular prediction did not agree with experiment, we would limit the area to which the theory applies, modify the postulates, or start over with a new theory.

The kinetic theory of an *ideal gas* (a gas that follows the ideal gas law) is based on five postulates.●

● These postulates are sometimes collected into only three or four statements, but the essential points are those given here.

POSTULATE 1 Gases are composed of molecules whose size is negligible compared to the average distance between them. Most of the volume occupied by a gas is empty space. This means that we can usually ignore the volume occupied by the molecules.

POSTULATE 2 Molecules move randomly in straight lines in all directions and at various speeds. This means that properties of a gas that depend on the motion of molecules, such as pressure, will be the same in all directions.

POSTULATE 3 The forces of attraction or repulsion between two molecules (*intermolecular forces*) in a gas are very weak or negligible, except when they collide.●

● The statements that there are no intermolecular forces and that the volume of molecules is negligible are simplifications that lead to the ideal gas law. But intermolecular forces are needed to explain how we can get the liquid state from the gaseous state, because intermolecular forces hold molecules together in the liquid state.

This means that a molecule will continue moving in a straight line with undiminished speed until it collides with another gas molecule or with the walls of the container.

POSTULATE 4 When molecules collide with one another, the collisions are *elastic.* In an elastic collision, the total kinetic energy remains constant; no kinetic energy is lost. To understand the difference between an elastic and an inelastic collision, compare the collision of two hard steel spheres with the collision of two masses of putty. The collision of steel spheres is nearly elastic (that is, the spheres bounce off each other and continue moving), but that of putty is not. Postulate 4 says that unless the kinetic energy of molecules is removed from a gas—for example, as heat—the molecules will forever move with the same average kinetic energy per molecule.

POSTULATE 5 The average kinetic energy of a molecule is proportional to the absolute temperature. This postulate establishes what we mean by temperature from a molecular point of view: the higher the temperature, the greater the molecular kinetic energy.

Qualitative Interpretation of the Gas Laws

According to the kinetic theory, the pressure of a gas results from the bombardment of container walls by molecules. Both the concentration of molecules (number per unit volume) and the average speed of the molecules are factors in determining this pressure. Molecular concentration determines the frequency of collisions with the wall. Average molecular speed determines the average force of a collision.

We can see the meaning of Boyle's law in these terms. That the temperature is constant means that the average kinetic energy of a molecule is constant (Postulate 5). Therefore, the average molecular speed and thus the average molecular force from collision remain constant. Suppose we increase the volume of a gas. This decreases the number of molecules per unit volume and so decreases the frequency of collisions per unit wall area. The pressure must decrease.

Now consider Charles's law. If we raise the temperature, we increase the average molecular speed. The average force of a collision increases. If all other factors remained fixed, the pressure would increase. For the pressure to remain constant as it does in Charles's law, it is necessary for the volume to increase so that the number of molecules per unit volume decreases, and the frequency of collisions decreases. Thus, when we raise the temperature of a gas while keeping the pressure constant, the volume must increase.

The Ideal Gas Law from Kinetic Theory

One of the most important features of kinetic theory is its explanation of the ideal gas law. To show how we can get the ideal gas law from kinetic theory, we will first find an expression for the pressure of a gas.●

● This derivation is essentially the one given by Daniel Bernoulli.

According to kinetic theory, the pressure of a gas, P, will be proportional to the frequency of molecular collisions with a surface and to the average force exerted by a molecule in collision.

$$P \propto \text{frequency of collisions} \times \text{average force}$$

The average force exerted by a molecule during a collision depends on its mass m and its average speed u—that is, on its average momentum mu. In other words, the greater the mass of the molecule and the faster it is moving, the greater the force exerted during collision. The frequency of collisions is also proportional to the average speed u, because the faster a molecule is moving, the more often it strikes the container walls. Frequency of collisions is inversely proportional to the gas volume V, because the larger the volume, the less often a given molecule strikes the container walls. Finally, the frequency of collisions is proportional to the number N of molecules in the gas volume. Putting these factors together gives

$$P \propto \left(u \times \frac{1}{V} \times N\right) \times mu$$

Bringing the volume to the left side, we get

$$PV \propto Nmu^2$$

Because the average kinetic energy of a molecule of mass m and average speed u is $\frac{1}{2}mu^2$, PV is proportional to the average kinetic energy of a molecule.● Moreover, the average kinetic energy is proportional to the absolute temperature (Postulate 5). Noting that the number of molecules, N, is proportional to the moles of molecules, n, we have

● Recall that kinetic energy is defined as $\frac{1}{2}m$ multiplied by $(\text{speed})^2$.

$$PV \propto nT$$

We can write this as an equation by inserting a constant of proportionality, R, which we identify as the molar gas constant.

$$PV = nRT$$

The next two sections give additional deductions from kinetic theory, such as the average molecular speed.

5.7 MOLECULAR SPEEDS; DIFFUSION AND EFFUSION

The principal tenet of kinetic theory is that molecules are in constant random motion. In this section, we will look at the speeds of molecules and at some conclusions of kinetic theory regarding molecular speeds.

Molecular Speeds

According to kinetic theory, the speeds of molecules in a gas vary over a range of values. The British physicist James Clerk Maxwell (1831–1879) showed theoretically—and it has since been demonstrated experimentally—that molecular speeds are distributed as shown in Figure 5.14. This distribution of speeds depends on the temperature. At any temperature, the molecular speeds vary widely, but most are close to the average speed, which is close to the speed corresponding to the maximum in the distribution curve. As the temperature increases, the average speed increases.

The **root-mean-square (rms) molecular speed,** u, is *a type of average molecular speed, equal to the speed of a molecule having the average molecular*

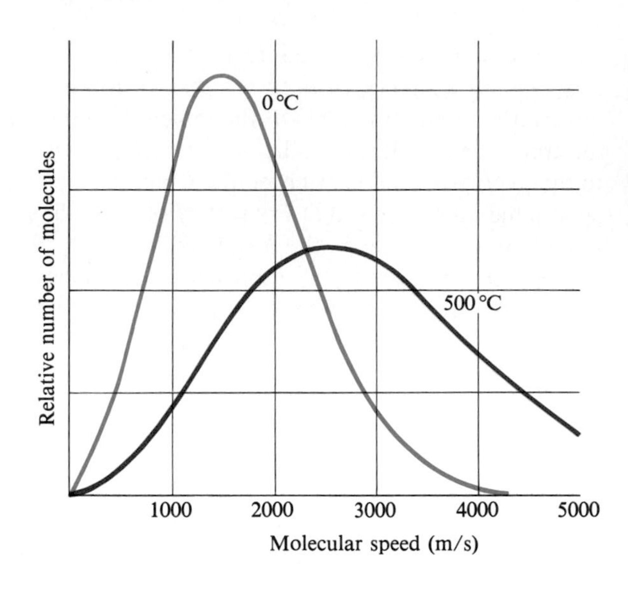

Figure 5.14
Maxwell's distribution of molecular speeds. The distributions of speeds of H_2 molecules are shown for 0°C and 500°C. Note that the speed corresponding to the maximum in the curve (this is the most probable speed) increases with temperature.

kinetic energy. It is given by the following formula:

$$u = \sqrt{\frac{3RT}{M_m}}$$

where R is the molar gas constant, T is the absolute temperature, and M_m is the molar mass for the gas. This result follows from Postulate 5 of kinetic theory.● Note that of two gases, the one with the higher molar mass will have the lower rms speed.

● According to kinetic theory, the total kinetic energy of a mole of any gas equals $\frac{3}{2}RT$. The average kinetic energy of a molecule, which by definition is $\frac{1}{2}mu^2$, is obtained by dividing $\frac{3}{2}RT$ by Avogadro's number, N_A. Therefore, the kinetic energy is $\frac{1}{2}mu^2 = 3RT/(2N_A)$. Hence $u^2 = 3RT/(N_A m)$. Or, noting that $N_A m$ equals the molar mass M_m, we get $u^2 = 3RT/M_m$, from which we get the text equation.

In applying this equation, *care must be taken to use consistent units*. If SI units are used for R (= 8.31 kg · m²/(s² · K · mol)), T(K), and M_m (kg/mol), as in the following example, the rms speed will be in meters per second. Note that in these units, H_2, whose molecular weight is 2.02 amu, has a molar mass of 2.02×10^{-3} kg/mol.

Values of the rms speed calculated from this formula indicate that molecular speeds are astonishingly high. For example, the rms speed of H_2 molecules at 20°C is 1.90×10^3 m/s (over 4000 mi/hr).

Example 5.12 Calculating the RMS Speed of a Gas Molecule

Calculate the rms speed of O_2 molecules in a tank at 21°C and 15.7 atm. See Table 5.5 for the appropriate value of R.

Solution

The rms molecular speed is independent of pressure but does depend on the absolute temperature, which is (21 + 273) K = 294 K. To calculate u, it is best to use SI units throughout. In these units, the molar mass of O_2 is 32.0×10^{-3} kg/mol, and $R = 8.31$ kg · m²/(s² · K · mol). Hence,

$$u = \left(\frac{3 \times 8.31 \text{ kg} \cdot \text{m}^2/(\text{s}^2 \cdot \text{K} \cdot \text{mol}) \times 294 \text{ K}}{32.0 \times 10^{-3} \text{ kg/mol}}\right)^{1/2}$$

$$= \mathbf{479 \text{ m/s}}$$

Exercise 5.12

What is the rms speed (in m/s) of a carbon tetrachloride molecule at 22°C?

(See Problems 5.73, 5.74, 5.75, and 5.76.)

Exercise 5.13

At what temperature do hydrogen molecules, H_2, have the same rms speed as nitrogen molecules, N_2, at 455°C? At what temperature do hydrogen molecules have the same average kinetic energy? (See Problems 5.77 and 5.78.)

Diffusion and Effusion

Gaseous **diffusion** is *the process whereby a gas spreads out through another gas to occupy the space with uniform partial pressure*. A gas or vapor having a relatively high partial pressure will spread out toward regions of lower partial pressure of that gas until the partial pressure becomes equal everywhere in the space. Thus, if we release an odorous substance at one corner of a room, minutes later it will be detected at the opposite corner. (It is well known, for example, how quickly the smell of baking apple pie draws people to the kitchen.)

When we think about our kinetic-theory calculations of molecular speed, we might ask why diffusion is not even much faster than it is. Why does it take minutes for the gas to diffuse throughout a room when the molecules are moving at perhaps a thousand miles per hour? This was, in fact, one of the first criticisms of kinetic theory. The answer is simply that a molecule never travels very far in one direction (at ordinary pressures) before it collides with another molecule and moves off in another direction. If we could trace out the path of an individual molecule, it would be a zigzagging trail. For a molecule to cross a room, it has to travel many times the straight-line distance.

Although the rate of diffusion certainly depends in part on the average molecular speed, the effect of molecular collisions makes the theoretical picture a bit complicated. *Effusion,* like diffusion, is a process involving the flow of a gas but is theoretically much simpler.

If we place a container of gas in a vacuum and then make a very small hole in the container, the gas molecules escape through the hole at the same speed they had in the container (Figure 5.15). *The process in which a gas flows through a small hole in a container* is called **effusion.** It was first studied by Thomas Graham, who discovered in 1846 that the rate of effusion of a gas is inversely proportional to the square root of its density. Today we usually state **Graham's law of effusion** in terms of molecular weight: *The rate of effusion of gas molecules from a particular hole is inversely proportional to the square root of the molecular weight of the gas at constant temperature and pressure.*

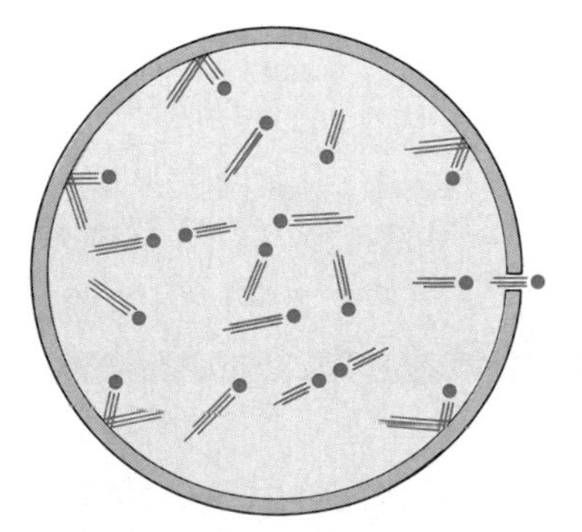

Figure 5.15
Effusion of a gas. The figure shows gas molecules effusing through a small hole in a container.

Let us consider a kinetic-theory analysis of an effusion experiment. Suppose the hole in the container is made small enough so that the gas molecules continue to move randomly (rather than moving together as they would in a wind). When a molecule happens to encounter the hole, it leaves the container. The collection of molecules leaving the container by chance encounters with the hole constitutes the effusing gas. All we have to consider for effusion is the rate at which molecules encounter the hole in the container.

The rate of effusion of molecules from a container depends on three factors: the cross-sectional area of the hole (the larger it is, the more likely molecules are to escape); the number of molecules per unit volume (the more crowded the molecules are, the more likely they are to encounter the hole); and the average molecular speed (the faster the molecules are moving, the sooner they will escape).

If we compare the effusion of different gases from the same container, at the same temperature and pressure, the first two factors will be the same. The average molecular speeds will be different, however.

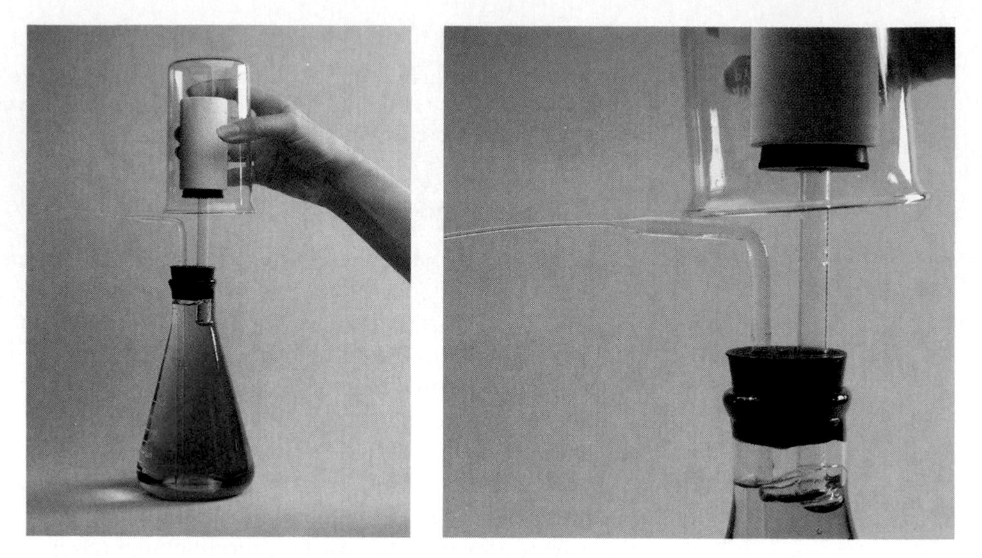

Figure 5.16
The hydrogen fountain. *Left:* A beaker containing hydrogen gas is placed over a porous clay container of air. Hydrogen effuses into the porous container faster than air effuses out. As a result, the pressure inside the porous container and the flask connected to it increases, forcing colored water out the side tube as a stream. *Right:* A close-up of the fountain.

Because the average molecular speed essentially equals $\sqrt{3RT/M_m}$, where M_m is the molar mass, the rate of effusion is proportional to $1/\sqrt{M_m}$. That is, the rate of effusion is inversely proportional to the square root of the molar mass (or molecular weight), as Graham's law states. The derivation of Graham's law from kinetic theory was considered a triumph of the theory and greatly strengthened confidence in its validity.

Graham's law of effusion:
Rate of effusion of molecules $\propto 1/\sqrt{M_m}$ (for the same container at constant T and P)

The hydrogen fountain, shown in Figure 5.16, is dependent on the differences in rates of effusion of gases.

Example 5.13 Calculating the Ratio of Effusion Rates of Gases

Calculate the ratio of effusion rates of molecules of carbon dioxide, CO_2, and sulfur dioxide, SO_2, from the same container and at the same temperature and pressure.

Solution

The two rates of effusion are inversely proportional to the square roots of their molar masses, so we can write

$$\frac{\text{Rate of effusion of } CO_2}{\text{Rate of effusion of } SO_2} = \sqrt{\frac{M_m(SO_2)}{M_m(CO_2)}}$$

where $M_m(SO_2)$ is the molar mass of SO_2 (64.1 g/mol) and $M_m(CO_2)$ is the molar mass of CO_2 (44.0 g/mol). Substituting these molar masses into the formula gives

$$\frac{\text{Rate of effusion of } CO_2}{\text{Rate of effusion of } SO_2} = \sqrt{\frac{64.1 \text{ g/mol}}{44.0 \text{ g/mol}}} = 1.21$$

In other words, carbon dioxide effuses 1.21 times faster than sulfur dioxide (because CO_2 molecules move 1.21 times faster on average than SO_2 molecules).

Exercise 5.14

If it takes 3.52 s for 10.0 mL of helium to effuse through a hole in a container at a particular temperature and pressure, how long would it take for 10.0 mL of oxygen, O_2, to effuse from the same container at the same temperature and pressure? (Note that the rate of effusion can be given in terms of volume of gas effused per second.)

(See Problems 5.79, 5.80, 5.81, and 5.82.)

Figure 5.17
A plant for separating uranium isotopes by effusion. The large metal vessels contain porous membranes through which uranium hexafluoride effuses. After many effusion stages, the gas becomes more concentrated in the uranium-235 isotope. This plant at Oak Ridge National Laboratory is generally referred to as a "gaseous diffusion" plant.

Exercise 5.15

If it takes 4.67 times as long for a particular gas to effuse as it takes hydrogen under the same conditions, what is the molecular weight of the gas? (Note that the rate of effusion is inversely proportional to the time it takes for a gas to effuse.)

(See Problems 5.83 and 5.84.)

Graham's law has practical application in the preparation of fuel rods for nuclear fission reactors. Such reactors depend on the fact that the uranium-235 nucleus undergoes fission (splits) when bombarded with neutrons. When the nucleus splits, several neutrons are emitted and a large amount of energy is liberated. These neutrons bombard more uranium-235 nuclei, and the process continues with the evolution of more energy. However, natural uranium consists of 99.27% uranium-238 (which does not undergo fission) and only 0.72% uranium-235 (which does undergo fission). A uranium fuel rod must contain about 3% uranium-235 to sustain the nuclear reaction.

To increase the percentage of uranium-235 in a sample of uranium (a process called *enrichment*), one first prepares uranium hexafluoride, UF_6, a white, crystalline solid that is easily vaporized. Uranium hexafluoride vapor is allowed to pass through a series of porous membranes. Each membrane has many small holes through which the vapor can effuse. Because the UF_6 molecules with the lighter isotope of uranium travel about 0.4% faster than the UF_6 molecules with the heavier isotope, the gas that passes through first is somewhat richer in uranium-235. When this vapor passes through another membrane, the uranium-235 vapor becomes further concentrated. It takes many effusion stages to reach the necessary enrichment; for complete separation of the isotopes, as required for bomb-grade uranium, many thousands of effusion stages are needed (Figure 5.17).

5.8 REAL GASES

In Section 5.2, we found that, contrary to Boyle's law, the pressure-volume product for O_2 at 0°C was not quite constant, particularly at high pressures (see under PV in Table 5.3). Experiments have shown that the ideal gas law describes the behavior of a real gas quite well at moderate pressures and temperatures but not so

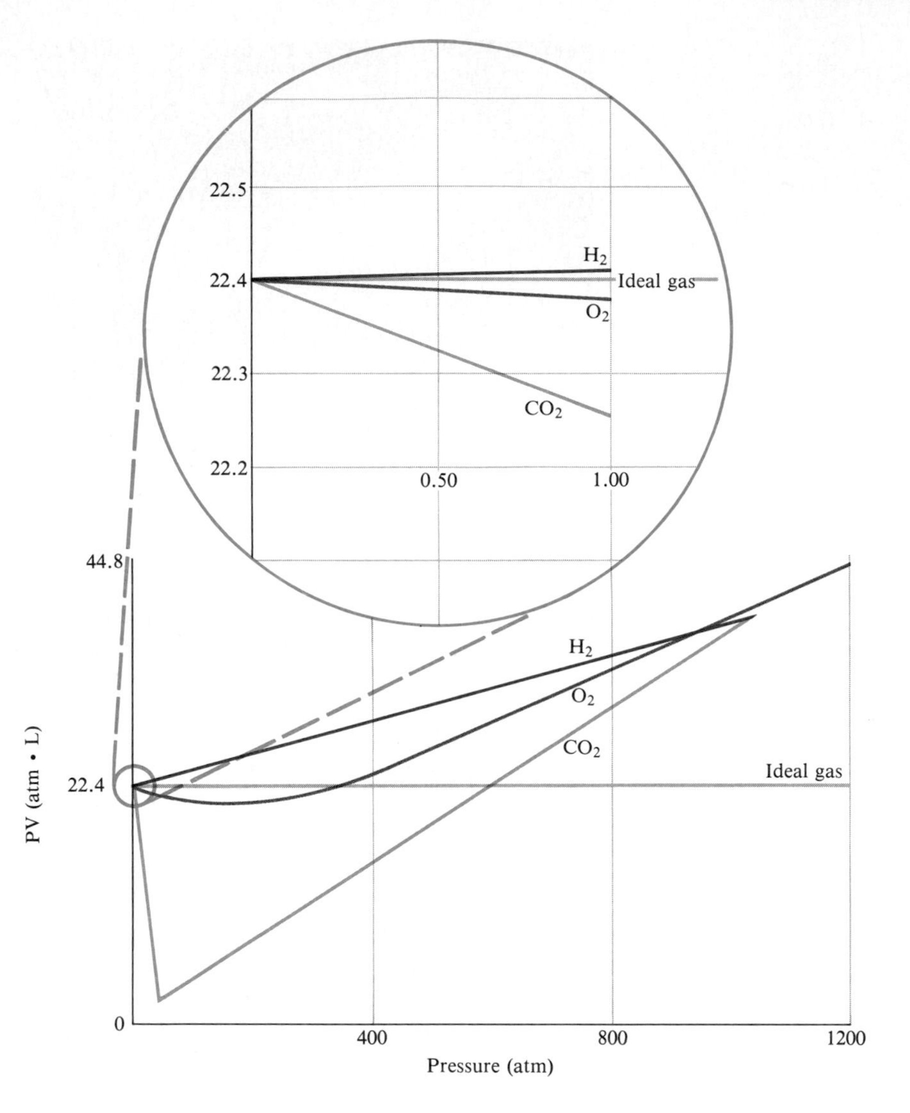

Figure 5.18 Pressure-volume product for one mole of various gases at 0°C and at different pressures. Values at low pressure are shown in the inset. For an ideal gas, the pressure-volume product is constant.

well at high pressures and low temperatures. We can see why this is so from the postulates of kinetic theory, from which the ideal gas law can be derived.

According to Postulate 1, the volume of space actually occupied by the molecules is small compared with the total volume that they occupy as a result of their motion through space. Further, according to Postulate 3, the molecules are sufficiently far apart on the average so the intermolecular forces can be disregarded. Both postulates are satisfied by a real gas when its density is low. The molecules are far enough apart to justify treating them as point particles with negligible intermolecular forces (such forces diminish as the molecules become further separated). But Postulates 1 and 3 apply less accurately as the density of a gas increases. As the molecules become more densely packed, the space occupied by molecules is no longer negligible compared to the total volume of gas. Moreover, the intermolecular forces become stronger as the molecules come closer together.

Figure 5.18 shows the behavior of the pressure-volume product at various pressures for several different gases. Note that the gases deviate from Boyle's law

Table 5.7 Van der Waals Constants of Some Gases

Gas	a $L^2 \cdot atm/mol^2$	b L/mol
Carbon dioxide, CO_2	3.592	0.04267
Ethane, C_2H_6	5.489	0.06380
Ethanol, C_2H_5OH	12.02	0.08407
Helium, He	0.03412	0.02370
Hydrogen, H_2	0.2444	0.02661
Oxygen, O_2	1.360	0.03183
Sulfur dioxide, SO_2	6.714	0.05636
Water, H_2O	5.464	0.03049

(From Weast, R. C., Ed., in *CRC Handbook of Chemistry and Physics,* 59th ed., p. D-230. Copyright CRC Press Inc., Boca Raton, Florida, 1978. With the permission of CRC Press, Inc.)

to different extents. This is because the two factors responsible for the deviation (the volume of the molecules and the magnitude of the intermolecular forces) vary with each gas.

When the ideal gas law is not sufficiently accurate for our purposes, we must use another of the many such equations available. One alternative is the **van der Waals equation,** which is *an equation that relates* P, T, V, *and* n *for nonideal gases* (but is not valid for very high pressures).

$$\left(P + \frac{n^2a}{V^2}\right)(V - nb) = nRT$$

Here a and b are constants that must be determined for each kind of gas. Table 5.7 lists van der Waals constants for several gases.

We can obtain this equation from the ideal gas law, $PV = nRT$, by replacing P by $P + n^2a/V^2$ and V by $V - nb$. To explain the change made to the volume term, we note that for the ideal gas law we assume that the molecules have negligible volume. Thus, any particular molecule can move throughout the entire volume V of the container. But if the molecules do occupy a small but finite amount of space, the volume through which any particular molecule can move is reduced. Thus, we subtract nb from V. The constant b is essentially the volume occupied by a mole of molecules.

To explain the change made to the pressure term, we need to consider the effect of intermolecular forces on the pressure. Such forces attract molecules to one another. A molecule about to collide with the wall is attracted by other molecules, and this reduces its impact with the wall. Therefore, the actual pressure is less than that predicted by the ideal gas law.

We can obtain this pressure correction by noting that the total force of attraction on any molecule about to hit the wall is proportional to the concentration of molecules, n/V. If this concentration is doubled, the total force on any molecule about to hit the wall is doubled. However, the number of molecules about to hit the wall per unit wall area is also proportional to concentration, n/V, so that the force per unit wall area, or pressure, is reduced by a factor proportional to n^2/V^2. We can write this correction factor as an^2/V^2, where a is a proportionality constant.

If we write the ideal gas equation corrected for the volume of molecules, we obtain for the pressure

$$P = \frac{RT}{V - nb}$$

Now if the pressure is reduced by the correction term an^2/V^2 to account for intermolecular forces, and we obtain

$$P = \frac{RT}{V - nb} - \frac{an^2}{V^2}$$

We can rearrange this to give the form of the van der Waals equation that we wrote earlier. The following example illustrates the use of the van der Waals equation.

Example 5.14 Using the van der Waals Equation

If sulfur dioxide were an ideal gas, the pressure at 0.0°C exerted by 1.000 mole occupying 22.41 L would be 1.000 atm (22.41 L is the molar volume of an ideal gas at STP). Use the van der Waals equation to estimate the pressure of this volume of 1.000 mol SO_2 at 0.0°C. See Table 5.7 for values of a and b.

Solution

Let us rearrange the van der Waals equation to give P in terms of the other variables.

$$P = \frac{nRT}{V - nb} - \frac{n^2a}{V^2}$$

Now we substitute $R = 0.08206\ \text{L} \cdot \text{atm/(K} \cdot \text{mol)}$, $T = 273.2\ \text{K}$, $V = 22.41\ \text{L}$, $a = 6.714\ \text{L}^2 \cdot \text{atm/mol}^2$, and $b = 0.05636\ \text{L/mol}$.

$$P = \frac{1.000\ \text{mol} \times 0.08206\ \text{L} \cdot \text{atm/K} \cdot \text{mol} \times 273.2\ \text{K}}{22.41\ \text{L} - (1.000\ \text{mol} \times 0.05636\ \text{L/mol})} - \frac{(1.000\ \text{mol})^2 \times 6.714\ \text{L}^2 \cdot \text{atm/mol}^2}{(22.41\ \text{L})^2}$$

$$= 1.003\ \text{atm} - 0.013\ \text{atm} = \mathbf{0.990\ atm}$$

Note that the first term, $nRT/(V - nb)$, gives the pressure corrected for molecular volume. This pressure, 1.003 atm, is slightly higher than the ideal value, 1.000 atm, because the volume through which a molecule is free to move, $V - nb$, is smaller than what is assumed for an ideal gas (V). Therefore, the concentration of molecules in this volume is greater, and the number of collisions per unit surface area is expected to be greater. However, the last term, n^2a/V^2, corrects for intermolecular forces, which reduce the force of the collisions, and this reduces the pressure to 0.990 atm.

Exercise 5.16

Use the van der Waals equation to calculate the pressure of 1.000 mol ethane, C_2H_6, that has a volume of 22.41 L at 0.0°C. Compare the result with the value predicted by the ideal gas law.

(See Problems 5.85 and 5.86.)

Profile of a Chemical

OXYGEN (a Component of Air)

The main rockets of a space shuttle burn hydrogen with oxygen to form water vapor. Hydrogen and oxygen, normally gases, are contained as liquids in separate compartments of the large tank attached to the space shuttle (Figure 5.19). Within $8\frac{1}{2}$ minutes of liftoff, the rocket engines consume over a half million liters of liquid oxygen.

Liquid and gaseous oxygen are produced in enormous quantities from liquid air. Air is a mixture of gases (Table 5.8), consisting mainly of nitrogen (78 mole percent), oxygen (21 mole percent), and argon (0.9 mole

Figure 5.19
The space shuttle. *The large main tank contains liquid hydrogen and liquid oxygen. Two booster rockets, burning solid fuel, are attached to each side of the main tank.*

Table 5.8 Composition of Dry Air in the Lower Atmosphere

Component	Mole Percent
Nitrogen, N_2	78.084
Oxygen, O_2	20.946
Argon, Ar	0.934
Carbon dioxide, CO_2	0.033
Neon, Ne	0.002
Helium, He	0.001

percent). It can be liquefied if it is compressed and then sufficiently cooled. The commercial equipment for doing this is called a *liquid air machine.* Figure 5.20 shows a sketch of a liquid air machine based on the principle that air, like most gases, cools when it expands. (You may have noticed the cooling effect of expanding air when it escapes from a tire.) Clean air enters the machine, where it is compressed and cooled. This cold, compressed gas then passes through a valve, where it expands. The cooling from this expansion is sufficient to liquefy the gas. Liquid air is extremely cold, boiling at about −190°C, or −310°F, at 1 atm pressure. A beaker of liquid air set on a table boils furiously.

Oxygen is obtained from liquid air by distillation. When liquid air is allowed to warm up, it begins to boil and nitrogen vaporizes. As the temperature continues to rise, argon also boils off, leaving liquid oxygen behind. Liquid oxygen boils at −183°C at a pressure of 1 atm. Although gaseous oxygen is colorless, liquid oxygen has a pale blue color (Figure 5.21).

Before the invention of the liquid air machine, oxygen was prepared in small quantities by the decomposi-

Clean air
Cooling tower
Warm water
Compressor
Expansion valve
Liquid air
Cold water

Figure 5.20
Liquid air machine. *Clean air is compressed and cooled, first in the cooling tower and then by previously cooled air. When this cold, compressed air is allowed to expand, it cools further and liquefies.*

(continued)

Figure 5.21
Liquid oxygen. *The test tube contains liquid oxygen, which has a pale blue color.*

tion of various oxygen-containing compounds. For example, when potassium chlorate, $KClO_3$, is heated, it decomposes to give oxygen.

$$2KClO_3(s) \longrightarrow 2KCl(s) + 3O_2(g)$$

The decomposition occurs at about 150°C in the presence of the catalyst manganese dioxide, MnO_2.

The principal commercial use of oxygen (over two-thirds of it) is in steel making. Iron is produced in a blast furnace where a charge of iron ore (Fe_2O_3), lime (CaO), and coke (C) is burned in a blast of air. The iron obtained from the blast furnace contains 3% to 4% carbon and is soft and brittle. To make steel, the percentage of carbon is reduced by blowing oxygen into the molten metal, thus burning the carbon to carbon dioxide.

One of the first uses of oxygen in quantity was in welding, and this is still a major usage. Any fuel burns more rapidly, and therefore hotter, in pure oxygen than it does in air. This is the basis of the classic test for pure oxygen. A wood splint glowing in air bursts into a bright white flame when it is thrust into a bottle of pure oxygen. Similarly acetylene, which burns with a yellow, sooty flame in air, burns white-hot in oxygen. The oxyacetylene torch, routinely used to weld steel, employs acetylene and oxygen to reach temperatures higher than 3000°C.

The chemical industry also uses large quantities of oxygen in the preparation of compounds such as titanium dioxide, TiO_2 (white pigment for paints, paper, and plastics), and ethylene glycol (antifreeze and polyester fibers).

Smaller amounts of oxygen are used for life-support purposes in medicine and in aviation, space, and diving systems. A diver, for example, must have a breathing mixture supplied at the pressure to which the body is subjected. As a diver goes to greater depths, this pressure increases. Compressed air can be used for breathing at moderate depths, but as the partial pressures of oxygen and nitrogen increase, this becomes hazardous. Both oxygen and nitrogen are toxic at high partial pressures. For this reason, breathing mixtures for deep diving replace nitrogen with nontoxic helium and have partial pressures of oxygen close to that in the atmosphere at normal pressure (about 0.21 atm).

Questions for Study

1. What is the reaction used in the main rockets of the space shuttle? Write the chemical equation.
2. What are the two major components of air?
3. Explain how liquid air is produced.
4. How is liquid oxygen obtained from liquid air?
5. Given an example of a reaction used to prepare small quantities of oxygen. Write the chemical equation of the reaction.
6. What is the principal commercial use of oxygen?
7. Describe the classical test for pure oxygen gas.
8. What is the oxyacetylene torch? What is its purpose?
9. Describe some end uses of compounds whose commercial production uses oxygen in their preparation.
10. What would be the approximate optimum mole fraction of oxygen in a helium-oxygen diving mixture whose total pressure is 6.0 atm?

Profile of a Chemical

NITROGEN (a Component of Air)

Almost all of the available nitrogen on earth is present as nitrogen gas, N_2. For every atom of nitrogen that exists in compound form in organic matter and in available minerals, 3500 are present as nitrogen in the atmosphere, where it exists as the colorless, odorless gas N_2.

The abundance of nitrogen in uncombined form is a result of the relative unreactivity or inertness of N_2, particularly at normal temperatures. Despite being unreactive, nitrogen forms hundreds of thousands of compounds, many of which (such as proteins and DNA) are of vital importance to life.

Natural as well as industrial processes circulate the nitrogen on earth from the atmosphere to compounds in the soil and in living organisms, then back to the atmosphere. This *nitrogen cycle* is depicted in Figure 5.22. Several processes fix atmospheric nitrogen—that is, convert it to nitrogen compounds that can be used by plants as nutrient nitrogen or fertilizer. Certain bacteria in the soil, *nitrogen-fixing bacteria,* are among the few organisms that can do this. Some nitrogen is also fixed by lightning, in which nitrogen and oxygen combine in the extremely hot air of a lightning bolt to form nitric oxide (nitrogen monoxide), NO, which is converted by reactions in air and the soil to nitrogen nutrients. Plants use these nutrients to form organic nitrogen compounds. These plants in turn are eaten by animals, which convert this nitrogen to animal protein and other substances.

Eventually, the nitrogen in these organic compounds returns to the atmosphere through the action of decay bacteria and *denitrifying bacteria.* Decay bacteria convert organic nitrogen to ammonium ion, NH_4^+, and then nitrate ion, NO_3^-. The denitrifying bacteria use nitrate ion as a source of energy and in the process convert it back to gaseous nitrogen, completing the nitrogen cycle.

The nitrogen cycle has been profoundly affected by the industrial fixation of nitrogen by the *Haber process,* which converts almost as much atmospheric nitrogen to compounds as does natural fixation. In the Haber process, nitrogen and hydrogen are combined in the presence of a catalyst to give ammonia.

$$N_2(g) + 3H_2(g) \longrightarrow 2NH_3(g)$$

This ammonia and other nitrogen fertilizer compounds produced from it, such as ammonium nitrate, have made possible the extraordinary yields of modern agriculture. Without nitrogen fertilizers, much of the world's human population would be doomed to famine.

Figure 5.22
The nitrogen cycle. *Nitrogen, N_2, is fixed by bacteria, by lightning, and by industrial synthesis of ammonia. Fixed nitrogen is used by plants and enters the food chain of animals. Later, plant and animal waste decomposes. Denitrifying bacteria complete the cycle by producing free nitrogen again.*

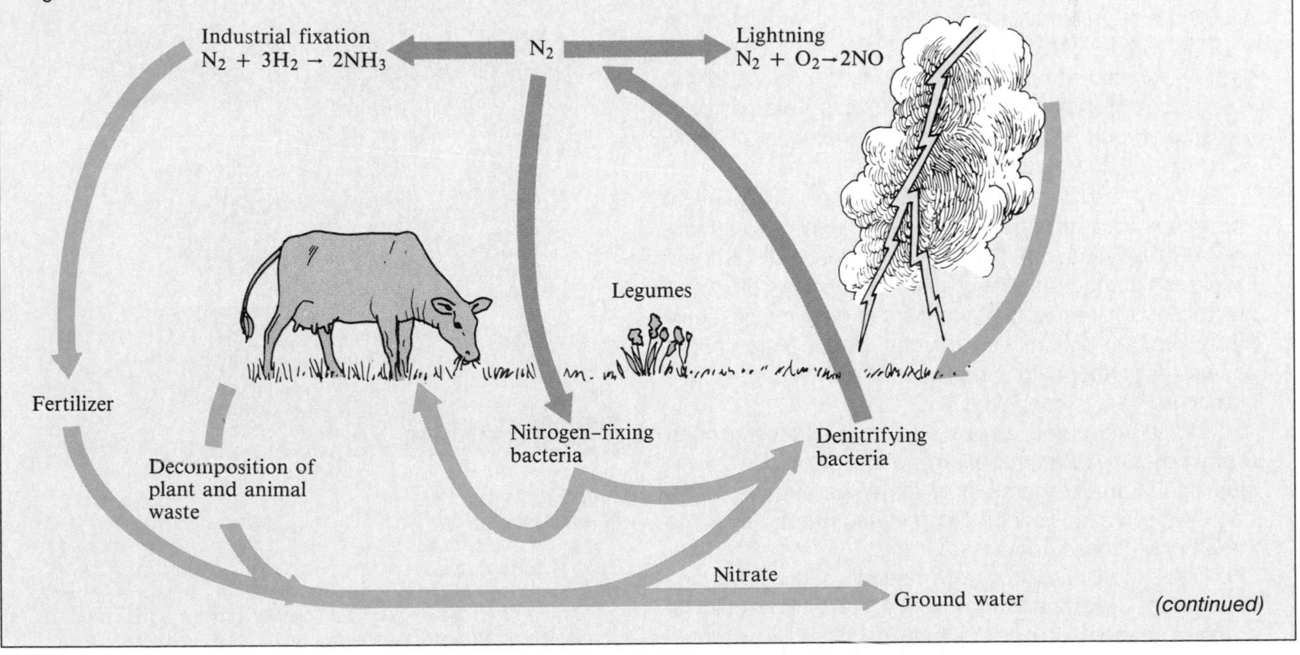

(continued)

Figure 5.23
Liquid nitrogen. *When liquid nitrogen is poured on the table (which, although at room temperature, is about 220°C above the boiling point of the liquid), it sizzles and boils away violently.*

Nitrogen in the atmosphere is the source of almost all industrial nitrogen and its compounds. Nitrogen, obtained from liquid air, ranks second (after sulfuric acid) in terms of the tons produced annually. One-third of this nitrogen is supplied as the colorless liquid, which boils at −197°C. Figure 5.23 shows liquid nitrogen being poured onto a table at room temperature, which is about 220°C above that of the boiling liquid. Just as if liquid water were being poured on a hot griddle, the liquid nitrogen sizzles and boils away violently.

Other than to produce ammonia, the main use of nitrogen is as a *protective blanket,* or protective atmosphere, which relies on the relative inertness of nitrogen. Electronic components and pharmaceuticals, for example, are often produced in a blanket of nitrogen gas, protected from reaction with oxygen in air. Figure 5.24 shows a laboratory glove box used to carry out reactions in a protective atmosphere.

Liquid nitrogen is an excellent and relatively cheap refrigerant. It is used, for example, to freeze blood samples and in freeze-grinding of materials such as beef. The beef is frozen to a brittle solid and ground easily to hamburger for fast-food restaurants.

The recent discovery of materials that are superconducting at temperatures above the boiling point of nitrogen promises to open up new, exciting uses of liquid nitrogen. A superconductor loses all resistance to an electric current when its temperature is lowered to a characteristic transition temperature. Thus, once a current is started in a superconducting magnet coil, it continues to flow indefinitely. Such magnets produce strong magnetic fields with very low power requirements. The new superconductors can be cooled below their transition temperatures with liquid nitrogen (77 K), which is much less expensive than the liquid helium (4.2 K) required by conventional superconducting materials.

Questions for Study

1. What is the predominant form of nitrogen on earth? Why is nitrogen in this form?
2. Describe the nitrogen cycle. What is nitrogen fixation?
3. What are nitrogen-fixing bacteria? What are denitrifying bacteria?
4. Describe two processes other than bacterial ones in which nitrogen is fixed. Write chemical equations for the reactions.
5. Give some physical properties of nitrogen, N_2.
6. How important is nitrogen commercially? (How does it rank in terms of tons produced?)
7. One of the uses of nitrogen is as a protective blanket. What does this mean?
8. List some commercial uses of nitrogen.

Figure 5.24
A laboratory glove box. *When a chemical reaction must be run in an inert atmosphere, it may be done in an enclosed box. Nitrogen can be led from a cylinder into the chamber at the right while the operator manipulates equipment inside by placing his or her hands in rubber gloves that attach to the holes at the front of the box.*

A Checklist for Review

IMPORTANT TERMS

pressure (5.1)
pascal (Pa) (5.1)
barometer (5.1)
manometer (5.1)
millimeters of mercury (mmHg) (or **torr**) (5.1)
atmosphere (atm) (5.1)
Boyle's law (5.2)
Charles's law (5.2)
Avogadro's law (5.2)
molar gas volume (V_m) (5.2)
standard temperature and pressure (STP) (5.2)
molar gas constant (R) (5.3)
ideal gas law (5.3)
partial pressure (5.5)
Dalton's law of partial pressures (5.5)
mole fraction (5.5)
kinetic-molecular theory of gases (kinetic theory) (p. 162)
root-mean-square (rms) molecular speed (5.7)
diffusion (5.7)
effusion (5.7)
Graham's law of effusion (5.7)
van der Waals equation (5.8)

KEY EQUATIONS

$P = gdh$

$PV = \text{constant}$ (constant n, T)

$V/T = \text{constant}$ (constant n, P)

$V_m = \text{specific constant}$ (depending on T, P; independent of gas)

$PV = nRT$

$PM_m = dRT$

$P = P_A + P_B + P_C + \cdots$

Mole fraction of $A = n_A/n = P_A/P$

$u = \sqrt{3RT/M_m}$

Rate of effusion $\propto 1/\sqrt{M_m}$ (same container at constant T and P)

$$\left(P + \frac{n^2a}{V^2}\right)(V - nb) = nRT$$

SUMMARY OF FACTS AND CONCEPTS

The *pressure* of a gas equals the force exerted per unit area of surface. It is measured by a *manometer* in units of *pascals, millimeters of mercury,* or *atmospheres*.

Gases at low to moderate pressures and moderate temperatures follow the same simple relationships, or gas laws. Thus, for a given amount of gas at constant temperature, the volume varies inversely with pressure (*Boyle's law*). Also, for a given amount of gas at constant pressure, the volume is directly proportional to the absolute temperature (*Charles's law*). These two laws, together with *Avogadro's law* (equal volumes of any two gases at the same temperature and pressure contain the same number of molecules), can be formulated as one equation, $PV = nRT$ (*ideal gas law*). This equation gives the relationship between P, V, n, and T for a gas. It also relates these quantities for each component in a gas mixture. The total gas pressure, P, equals the sum of the partial pressures of each component (*law of partial pressures*).

The ideal gas law can be explained by *kinetic-molecular theory*. We define an ideal gas as consisting of molecules with negligible volume that are in constant random motion. In the ideal gas model, there are no intermolecular forces between molecules, and the average kinetic energy of molecules is proportional to the absolute temperature. From kinetic theory, one can show that the *rms molecular speed* equals $\sqrt{3RT/M_m}$. Given this result, one can derive *Graham's law of effusion:* The rate of effusion under identical conditions is inversely proportional to the square root of the molecular weight.

Real gases deviate from the ideal gas law at high pressure and low temperature. From kinetic theory, we expect these deviations, because real molecules do have volume and intermolecular forces do exist. These two factors are partially accounted for in the *van der Waals equation*.

OPERATIONAL SKILLS

1. **Relating liquid height and pressure** Given the density of a liquid used in a barometer or manometer and the height of the column of liquid, obtain the pressure reading in mmHg (Example 5.1).

2. **Using the empirical gas laws** Given an initial volume occupied by a gas, calculate the final volume when the pressure changes at fixed temperature (Example 5.2); when the temperature changes at fixed pressure (Example 5.3); and when both pressure and temperature change (Example 5.4).

3. **Deriving empirical gas laws from the ideal gas law** Starting from the ideal gas law, derive the relationship between any two variables (Example 5.5).

4. **Using the ideal gas law** Given any three of the variables P, V, T, and n for a gas, calculate the fourth one from the ideal gas law (Example 5.6).

5. **Relating gas density and molecular weight** Given the molecular weight, calculate the density of a gas for a particular temperature and pressure (Example 5.7); or, given the gas density, calculate the molecular weight (Example 5.8).

6. **Solving stoichiometry problems involving gas volumes** Given the mass of one substance in a reaction, calculate the volume of gas produced (or used up) under given conditions (Example 5.9).

7. **Calculating partial pressures and mole fractions of a gas in a mixture** Given the masses of gases in a mixture, calculate the partial pressures and mole fractions (Example 5.10).

8. **Calculating the amount of gas collected over water** Given the volume, total pressure, and temperature of gas collected over water, calculate the mass of the dry gas (Example 5.11).

9. **Calculating the rms speed of a gas molecule** Given the molecular weight and temperature of a gas, calculate the rms molecular speed (Example 5.12).

10. **Calculating the ratio of effusion rates of gases** Given the molecular weights of two gases, calculate the ratio of rates of effusion (Example 5.13); or given the relative effusion rates of a known and an unknown gas, obtain the molecular weight of the unknown gas (as in Exercise 5.15).

11. **Using the van der Waals equation** Given n, T, V, and the van der Waals constants a and b for a gas, calculate the pressure from the van der Waals equation (Example 5.14).

Review Questions

5.1 Define *pressure*. From the definition, obtain the SI unit of pressure in terms of SI base units.

5.2 What is the purpose of a manometer? How does it work?

5.3 What are the variables that determine the height of the liquid in a manometer?

5.4 Starting with Boyle's law (stated as an equation), obtain an equation for the final volume occupied by a gas from the initial volume when the pressure is changed at constant temperature.

5.5 The volume occupied by a gas depends linearly on degrees Celsius at constant pressure, but it is not directly proportional to degrees Celsius. However, it is directly proportional to kelvins. What is the difference between a linear relationship and a direct proportion?

5.6 Explain how you would set up an absolute temperature scale based on the Fahrenheit scale, knowing that absolute zero is −273.15°C.

5.7 Starting with Charles's law (stated as an equation), obtain an equation for the final volume of a gas from its initial volume when the temperature is changed at a constant pressure.

5.8 State Avogadro's law in words. How does this law explain the law of combining volumes? Use the gas reaction $N_2 + 3H_2 \longrightarrow 2NH_3$ as an example in your explanation.

5.9 What are the standard conditions for comparing gas volumes?

5.10 What does the term *molar gas volume* mean? What is the molar gas volume (in liters) at STP for an ideal gas?

5.11 Starting from Boyle's, Charles's, and Avogadro's laws, obtain the ideal gas law, $PV = nRT$.

5.12 What variables are needed to describe a gas that obeys the ideal gas law? What are the SI units for each variable?

5.13 What is the value of R in units of $L \cdot mmHg/(K \cdot mol)$?

5.14 The ideal gas law relates the four variables. An empirical gas law relates two variables, assuming the other two are constant. How many empirical gas laws can be obtained? Give statements of each.

5.15 Give the postulates of kinetic theory and state any evidence that supports them.

5.16 Explain Boyle's law in terms of the kinetic theory.

5.17 What is the origin of gas pressure, according to kinetic theory?

5.18 How does the rms molecular speed depend on absolute temperature? on molar volume?

5.19 Explain why a gas appears to diffuse more slowly than average molecular speeds might suggest.

5.20 What is effusion? Why does a gas whose molecules have smaller mass effuse faster than one whose molecules have larger mass?

5.21 Under what conditions does the behavior of a real gas begin to differ significantly from the ideal gas law?

5.22 What is the physical meaning of the a and b constants in the van der Waals equation?

Practice Problems

Note: In these problems, the final zeros given in temperatures and pressures (for example, 20°C, 760 mmHg) are significant figures.

PRESSURE MEASUREMENTS

5.23 A closed-end manometer (as in Figure 5.3) is filled with an oil with density of 0.786 g/mL in place of mercury. What is the pressure in mmHg when height of the liquid column Δh in the manometer stands at 75.7 mm? The density of mercury at 0°C is 13.596 g/mL.

5.24 A closed-end manometer is filled with a liquid whose density is 0.786 g/mL in place of mercury. What is the pressure in atmospheres when the height of the liquid column Δh in the manometer stands at 64.8 mm? The density of mercury at 0°C is 13.596 g/mL.

5.25 A closed-end manometer was checked against a known pressure. If the manometer column is 95.6 mm high and the pressure is known to be 8.56 mmHg, what is the density of the liquid in the manometer?

5.26 The height of the liquid column Δh in a closed-end manometer is 65.7 mm. If the gas pressure is known to be 6.85 mmHg, what is the density of the liquid in the manometer?

EMPIRICAL GAS LAWS

5.27 A 6.50-L sample of oxygen gas at 20°C is confined in an inverted graduated cylinder under water at a pressure of 1.50 atm. What volume would this gas occupy if the cylinder was moved to a lower depth at 20°C where the pressure is 2.50 atm?

5.28 A balloon contains helium gas with a volume of 2.60 L at 20°C and 768 mmHg. If the balloon ascends to an altitude where the helium pressure is 614 mmHg, what is the volume of gas? Assume the temperature has not changed.

5.29 Nitrogen gas was contained in a tank at 21.6 atm pressure. If the gas were allowed to expand to a pressure equal to that of the normal atmosphere (1.00 atm) at 20°C, the total volume of gas would be 849 L. What would be the total volume of gas if it expanded from the tank to a pressure of 1.25 atm at the same temperature?

5.30 A cylinder contained carbon dioxide gas at 6.50 atm pressure. If the carbon dioxide gas were allowed to escape so that its pressure is 1.05 atm at 21°C, the total volume of CO_2 would be 24.8 L. What would be the total volume of gas if the carbon dioxide expanded so that its pressure is 1.28 atm at 21°C?

5.31 A McLeod gauge measures low gas pressures by compressing a known volume of the gas at constant temperature. If 345 cm^3 of gas is compressed to a volume of 0.0457 cm^3 under a pressure of 2.51 kPa, what was the original gas pressure?

5.32 If 456 dm^3 of krypton at 101 kPa and 21°C is compressed into a 25.0-dm^3 tank at the same temperature, what is the pressure of krypton in the tank?

5.33 A sample of nitrogen gas at 18°C and 760 mmHg has a volume of 2.67 mL. What is the volume at STP?

5.34 A mole of gas at 0°C and 760 mmHg occupies 22.41 L. What is the volume at 20°C and 760 mmHg?

5.35 Helium gas, He, at 22°C and 1.00 atm occupied a vessel whose volume was 2.54 L. What volume would this gas occupy if it was cooled to liquid nitrogen temperature (−197°C)?

5.36 An experiment called for 5.83 L of sulfur dioxide, SO_2, at 0°C and 1.00 atm. What would be the volume of this gas at 25°C and 1.00 atm?

5.37 A vessel containing 39.5 cm^3 of helium gas at 25°C and 106 kPa was inverted and placed in cold ethanol. As the gas contracted, ethanol was forced into the vessel to maintain the same pressure of helium. If this required 18.8 cm^3 of ethanol, what was the final temperature of the helium?

5.38 A sample of 58.2 cm^3 of argon gas at 18°C was contained at a pressure of 155 kPa in a J-shaped tube with mercury as in Figure 5.4. Later the temperature changed. When the mercury level was adjusted to give the same pressure of argon, the gas volume changed to 46.7 cm^3. What was the final temperature of the argon?

5.39 A bacterial culture isolated from sewage produced 41.3 mL of methane, CH_4, at 31°C and 753 mmHg. What is the volume of this methane at standard temperature and pressure (0°C, 760 mmHg)?

5.40 Pantothenic acid is a B vitamin. Using the Dumas method, we find that a sample weighing 71.6 mg gives 3.84 mL of nitrogen gas at 23°C and 785 mmHg. What is the volume of nitrogen at STP?

5.41 In the presence of a platinum catalyst, ammonia, NH_3, burns in oxygen, O_2, to give nitric oxide, NO, and water vapor. How many volumes of nitric oxide are obtained from one volume of ammonia, assuming each gas is at the same temperature and pressure?

5.42 Methanol, CH_3OH, can be produced in industrial plants by reacting carbon dioxide with hydrogen in the presence of a catalyst. Water is the other product. How many volumes of hydrogen are required for each volume of carbon dioxide when each gas is at the same temperature and pressure?

IDEAL GAS LAW

5.43 Starting from the ideal gas law, prove that the volume of a mole of gas is inversely proportional to the pressure at constant temperature (Boyle's law).

5.44 Starting from the ideal gas law, prove that the volume of a mole of gas is directly proportional to the absolute temperature at constant pressure (Charles's law).

5.45 A cylinder of neon gas, Ne, has a volume of 9.76 L. If the cylinder contains 61.2 g of neon at 23°C, what is the pressure in the cylinder?

5.46 We wish to compress 3.15 g of methane gas, CH_4, into a heavy-walled 2.00-L flask at 19°C. What must be the final pressure of methane?

5.47 An experiment calls for 2.50 mol of carbon monoxide, CO, at 55°C and 1.50 atm. What volume of carbon monoxide is this?

5.48 Calculations show that a reaction should yield 5.67 g of carbon dioxide gas, CO_2. What volume should be expected at 26°C and 782 mmHg?

5.49 The maximum safe pressure that a certain 4.00-L vessel can hold is 3.50 atm. If the vessel contains 0.410 mole of gas, what is the maximum temperature (in degrees Celsius) to which this vessel can be subjected?

5.50 A 2.50-L flask was used to collect a 5.65-g sample of propane gas, C_3H_8. After the sample was collected, the gas pressure was found to be 956 mmHg. What was the temperature of the propane in the flask?

GAS DENSITY AND MOLECULAR WEIGHT DETERMINATION

5.51 What is the density (in g/L) of methane gas, CH_4, at 125°C and 3.50 atm?

5.52 What is the density (in g/L) of carbon dioxide gas, CO_2, at 22°C and 745 mmHg?

5.53 Butane, C_4H_{10}, is an easily liquefied gaseous fuel. Calculate the density of butane gas at 1.00 atm and 25°C. Give the answer in grams per liter.

5.54 Chloroform, $CHCl_3$, is a volatile (easily vaporized) liquid solvent. Calculate the density of chloroform vapor at 99°C and 745 mmHg. Give the answer in grams per liter.

5.55 The density of the vapor of a compound at 90°C and 720 mmHg is 1.434 g/L. Calculate the molecular weight of the compound.

5.56 A liquid compound was vaporized at 100°C and 755 mmHg to give 185 mL of vapor. The vapor weighed 0.427 g. What is the molecular weight of the compound?

5.57 A 1.28-g sample of a colorless liquid was vaporized in a 250-mL flask at 121°C and 786 mmHg. What is the molecular weight of this substance?

5.58 A 2.30-g sample of white solid was vaporized in a 345-mL vessel. If the vapor has a pressure of 985 mmHg at 148°C, what is the molecular weight of the solid?

5.59 Ammonium chloride, NH_4Cl, is a white solid. When heated to 325°C, it gives a vapor that is a mixture of ammonia and hydrogen chloride.

$$NH_4Cl(s) \longrightarrow NH_3(g) + HCl(g)$$

Suppose someone contends that the vapor consists of NH_4Cl molecules rather than a mixture of NH_3 and HCl. Could you decide between these alternative views on the basis of gas-density measurements? Explain.

5.60 Phosphorus pentachloride, PCl_5, is a white solid that sublimes (vaporizes without melting) at about 100°C. At higher temperatures, the PCl_5 vapor decomposes to give phosphorus trichloride, PCl_3, and chlorine, Cl_2.

$$PCl_5(g) \longrightarrow PCl_3(g) + Cl_2(g)$$

How could gas-density measurements help to establish that PCl_5 vapor is decomposing?

STOICHIOMETRY WITH GAS VOLUMES

5.61 Small amounts of hydrogen are conveniently prepared by reacting zinc with hydrochloric acid.

$$Zn(s) + 2HCl(aq) \longrightarrow ZnCl_2(aq) + H_2(g)$$

How many grams of zinc are required to prepare 2.50 L H_2 gas at 765 mmHg and 22°C?

5.62 Carbon dioxide gas can be prepared by dropping hydrochloric acid onto marble chips ($CaCO_3$).

$$CaCO_3(s) + 2HCl(aq) \longrightarrow CaCl_2(aq) + H_2O(l) + CO_2(g)$$

How many grams of $CaCO_3$ are required to prepare 3.75 L of CO_2 at 885 mmHg and 26°C?

5.63 When carbon dioxide is bubbled into a solution of calcium hydroxide, calcium carbonate precipitates.

$$Ca(OH)_2(aq) + CO_2(g) \longrightarrow CaCO_3(s) + H_2O(l)$$

During a particular experiment, 2.35 g of calcium carbonate were formed. How many liters of carbon dioxide at 15°C and 775 mmHg must have been bubbled into a solution of calcium hydroxide to give this quantity of calcium carbonate?

5.64 In the Solvay process, ammonia and carbon dioxide are passed into a solution of sodium chloride. Sodium hydrogen carbonate, $NaHCO_3$, precipitates from solution.

$$NH_3(g) + CO_2(g) + H_2O(l) + NaCl(aq) \longrightarrow NaHCO_3(s) + NH_4Cl(aq)$$

How many liters of NH_3 at 5°C and 2.00 atm are required to produce 1.00 kg $NaHCO_3$?

5.65 Liquid oxygen was first prepared by heating potassium chlorate, $KClO_3$, in a closed vessel to obtain oxygen at high pressure. The oxygen was cooled until it liquefied.

$$2KClO_3(s) \longrightarrow 2KCl(s) + 3O_2(g)$$

If 85.0 g of potassium chlorate react in a 2.50-L vessel, which was initially evacuated, what pressure of oxygen will be attained when the temperature is finally cooled to 21°C? Use the preceding chemical equation and ignore the volume of solid product.

5.66 Raoul Pictet, the Swiss physicist who first liquefied oxygen, attempted to liquefy hydrogen. He heated potassium formate, $KCHO_2$, with KOH in a closed 2.50-L vessel.

$$KCHO_2(s) + KOH(s) \longrightarrow K_2CO_3(s) + H_2(g)$$

If 75.0 g of potassium formate react in a 2.50-L vessel, which was initially evacuated, what pressure of hydrogen will be attained when the temperature is finally cooled to 21°C? Use the preceding chemical equation and ignore the volume of solid product.

GAS MIXTURES

5.67 A 200.0-mL flask contains 1.03 mg O_2 and 0.41 mg He at 15°C. Calculate the partial pressures of oxygen and of helium in the flask. What is the total pressure?

5.68 The atmosphere in a sealed diving bell contained oxygen and helium. If the gas mixture has 0.200 atm of oxygen and a total pressure of 3.00 atm, calculate the mass of helium in 1.00 L of the gas mixture at 20°C.

5.69 The gas from a certain volcano had the following composition in mole percent (that is, mole fraction × 100): 65.0% CO_2, 25.0% H_2, 5.4% HCl, 2.8% HF, 1.7% SO_2, and 0.1% H_2S. What would be the partial pressure of each of these gases if the total pressure of volcanic gas were 760 mmHg?

5.70 In a series of experiments, the U.S. Navy developed an undersea habitat. In one experiment, the mole percent composition of the atmosphere in the undersea habitat was 79.0% He, 17.0% N_2, and 4.0% O_2. What will the partial pressure of each gas be when the habitat is 58.8 m below sea level, where the pressure is 6.91 atm?

5.71 Formic acid, $HCHO_2$, is a convenient source of small quantities of carbon monoxide. When warmed with sulfuric acid, formic acid decomposes to give CO gas.

$$HCHO_2(l) \longrightarrow H_2O(l) + CO(g)$$

If 3.85 L of carbon monoxide was collected over water at 25°C and 689 mmHg, how many grams of formic acid were consumed?

5.72 An aqueous solution of ammonium nitrite, NH_4NO_2, decomposes when heated to give off nitrogen, N_2.

$$NH_4NO_2(s) \longrightarrow 2H_2O(g) + N_2(g)$$

This reaction many be used to prepare pure nitrogen. How many grams of ammonium nitrite must have reacted if 4.16 dm^3 of nitrogen gas was collected over water at 19°C and 97.8 kPa?

MOLECULAR SPEEDS; EFFUSION

5.73 Calculate the rms speeds of N_2 molecules at 25°C and at 125°C. Sketch approximate curves of the molecular speed distributions of N_2 at 25°C and at 125°C.

5.74 Calculate the rms speed of Br_2 molecules at 23°C and 1.00 atm. What is the rms speed of Br_2 at 23°C and 1.50 atm?

5.75 Uranium hexafluoride, UF_6, is a white solid that sublimes (vaporizes without melting) at 57°C under normal atmospheric pressure. The compound is used to separate uranium isotopes by effusion. What is the rms speed (in m/s) of a uranium hexafluoride molecule at 57°C?

5.76 For a spacecraft or a molecule to leave the moon, it must reach the escape velocity (speed) of the moon, which is 2.37 km/s. The average daytime temperature of the moon's surface is 365 K. What is the rms speed (in m/s) of a hydrogen molecule at this temperature? How does this compare with the escape velocity?

5.77 At what temperature would CO_2 molecules have an rms speed equal to that of H_2 molecules at 20°C?

5.78 At what temperature does the rms speed of O_2 molecules equal 375 m/s?

5.79 What is the ratio of rates of effusion of N_2 and O_2 under the same conditions?

5.80 Obtain the ratio of rates of effusion of H_2 and H_2S under the same conditions.

5.81 If 0.10 mol I_2 vapor can effuse from an opening in a heated vessel in 52 s, how long will it take 0.10 mol H_2 to effuse under the same conditions?

5.82 If it takes 10.6 hours for 1.00 L of nitrogen, N_2, to effuse through the pores in a balloon, how long would it take for 1.00 L of helium, He, to effuse under the same conditions?

5.83 If 4.83 mL of an unknown gas effuses through a hole in a plate in the same time it takes 9.23 mL of argon, Ar, to effuse through the same hole under the same conditions, what is the molecular weight of the unknown gas?

5.84 A given volume of nitrogen, N_2, required 68.3 s to effuse from a hole in a chamber. Under the same conditions, another gas required 85.6 s for the same volume to effuse. What is the molecular weight of this gas?

VAN DER WAALS EQUATION

5.85 Calculate the pressure of ethanol vapor, $C_2H_5OH(g)$, at 82.0°C if 1.000 mol $C_2H_5OH(g)$ occupies 35.00 L. Use the van der Waals equation (see Table 5.7 for data). Compare with the result from the ideal gas law.

5.86 Calculate the pressure of water vapor at 120.0°C if 1.000 mol of water vapor occupies 32.50 L. Use the van der Waals equation (see Table 5.7 for data). Compare with the result from the ideal gas law.

5.87 Calculate the molar volume of ethane at 1.00 atm and 0°C and at 10.0 atm and 0°C, using the van der Waals equation. The van der Waals constants are given in Table 5.7. To simplify, note that the term n^2a/V^2 is small compared with P. Hence, it may be approximated with negligible error by substituting $V = nRT/P$ for this term from the ideal gas law. Then the van der Waals equation can be solved for the volume. Compare the results with the values predicted by the ideal gas law.

5.88 Calculate the molar volume of oxygen at 1.00 atm and 0°C and at 10.0 atm and 0°C, using the van der Waals equation. The van der Waals constants are given in Table 5.7. (See the note on solving the equation given in Problem 5.87.) Compare the results with the values predicted by the ideal gas law. Also compare with the values obtained from Table 5.3.

Additional Problems

5.89 A glass tumbler containing 243 cm^3 of air at 1.00×10^2 kPa (the barometric pressure) and 20°C is turned upside down and immersed in a body of water to a depth of 20.5 m. The air in the glass is compressed by the weight of water above it. Calculate the volume of air in the glass, assuming the temperature and barometric pressure have not changed.

5.90 The density of air at 20°C and 1.00 atm is 1.205 g/L. If this air were compressed at the same temperature to equal the pressure at 30.0 m below sea level, what would be its density? Assume the barometric pressure is constant at 1.00 atm. The density of seawater is 1.025 g/cm^3.

5.91 A flask contains 183 mL of argon at 21°C and 738 mmHg. What is the volume of gas, corrected to STP?

5.92 A steel bottle contains 12.0 L of a gas at 10.0 atm and 20°C. What is the volume of gas at STP?

5.93 A balloon containing 5.0 dm^3 of gas at 14°C and 100.0 kPa rises to an altitude of 2000 m, where the temperature is 20°C. The pressure of gas in the balloon is now 79.0 kPa. What is the volume of gas in the balloon at this altitude?

5.94 A volume of air is taken from the earth's surface, at 15°C and 1.00 atm, to the stratosphere, where the temperature is −20°C and the pressure is 1.00×10^{-3} atm. By what factor is the volume increased?

5.95 A radioactive metal atom decays (goes to another kind of atom) by emitting an alpha particle (He^{2+} ion). The alpha particles are collected as helium gas. A sample of helium with a volume of 12.05 mL was obtained at 765 mmHg and 23°C. How many atoms decayed during the period of the experiment?

5.96 The combustion method used to analyze for carbon and hydrogen can be adapted to give percentage N by collecting the nitrogen from combustion of the compound as N_2. A sample of a compound weighing 8.75 mg gave 1.77 mL N_2 at 25°C and 749 mmHg. What is the percentage N in the compound?

5.97 Dry air at STP has a density of 1.2929 g/L. Calculate the average molecular weight of air from the density.

5.98 A hydrocarbon gas has a density of 1.22 g/L at 20°C and 1.00 atm. An analysis gives 80.0% C and 20.0% H. What is the molecular formula?

5.99 A person exhales about 5.8×10^2 L of carbon dioxide per day (at STP). The carbon dioxide exhaled by an astronaut is absorbed from the air of a space capsule by reaction with lithium hydroxide, LiOH.

$$2LiOH(s) + CO_2(g) \longrightarrow Li_2CO_3(s) + H_2O(l)$$

How many grams of lithium hydroxide are required per astronaut per day?

5.100 Pyruvic acid, $HC_3H_3O_3$, is involved in cell metabolism. It can be assayed for (that is, the amount of it determined) by using a yeast enzyme. The enzyme makes the following reaction go to completion:

$$HC_3H_3O_3(aq) \longrightarrow C_2H_4O(aq) + CO_2(g)$$

If a sample containing pyruvic acid gives 20.3 mL of carbon dioxide gas, CO_2, at 343 mmHg and 30°C, how many grams of pyruvic acid are there in the sample?

5.101 If the rms speed of NH_3 molecules is found to be 0.510 km/s, what is the temperature (in degrees Celsius)?

5.102 If the rms speed of He atoms in the exosphere (highest region of the atmosphere) is 3.53×10^3 m/s, what is the temperature (in kelvins)?

5.103 Calculate the ratio of rates of effusion of $^{235}UF_6$ and $^{238}UF_6$, where ^{235}U and ^{238}U are isotopes of uranium. The atomic masses are ^{235}U, 235.04 amu; ^{238}U, 238.05 amu; ^{19}F (the only naturally occurring isotope), 18.998 amu. Carry five significant figures in the calculation.

5.104 Hydrogen has two stable isotopes, ^1H and ^2H, with atomic masses of 1.0078 amu and 2.0141 amu, respectively. Ordinary hydrogen gas, H_2, is a mixture consisting mostly of $^1H_2$ and $^1H^2H$. Calculate the ratio of rates of effusion of $^1H_2$ and $^1H^2H$ under the same conditions.

5.105 A 1.000-g sample of an unknown gas at 0°C gives the following data:

P (atm)	V (L)
0.2500	3.1908
0.5000	1.5928
0.7500	1.0601
1.0000	0.7930

Use these data to calculate the value of the molar mass at each of the given pressures from the ideal gas law (we will call this the "apparent molar mass" at this pressure). Plot the apparent molar masses against pressure and extrapolate to find the molar mass at zero pressure. Because the ideal gas law is most accurate at low pressures, this extrapolation will give an accurate value for the molar mass. What is the accurate molar mass?

5.106 Plot the data given in Table 5.3 for oxygen at 0°C to obtain an accurate molar mass for O_2. To do this, calculate a value of the molar mass at each of the given pressures from the ideal gas law (we will call this the "apparent molar mass" at this pressure). On a graph show the apparent molar mass versus the pressure and extrapolate to find the molar mass at zero pressure. Because the ideal gas law is most accurate at low pressures, this extrapolation will give an accurate value for the molar mass. What is the accurate molar mass?

5.107 Carbon monoxide, CO, and oxygen, O_2, react according to

$$2CO(g) + O_2(g) \longrightarrow 2CO_2(g)$$

Assuming that the reaction takes place and goes to completion, determine what substances remain and what their partial pressures are after the valve is opened in the apparatus represented in the accompanying figure. Also assume that the temperature is fixed at 300 K.

5.108 Suppose the apparatus shown in the figure accompanying Problem 5.107 contains H_2 at 0.500 atm in the left vessel separated from O_2 at 1.00 atm in the other. The valve is then opened. If H_2 and O_2 react to give H_2O when the temperature is fixed at 533 K, what substances remain and what are their partial pressures after reaction?

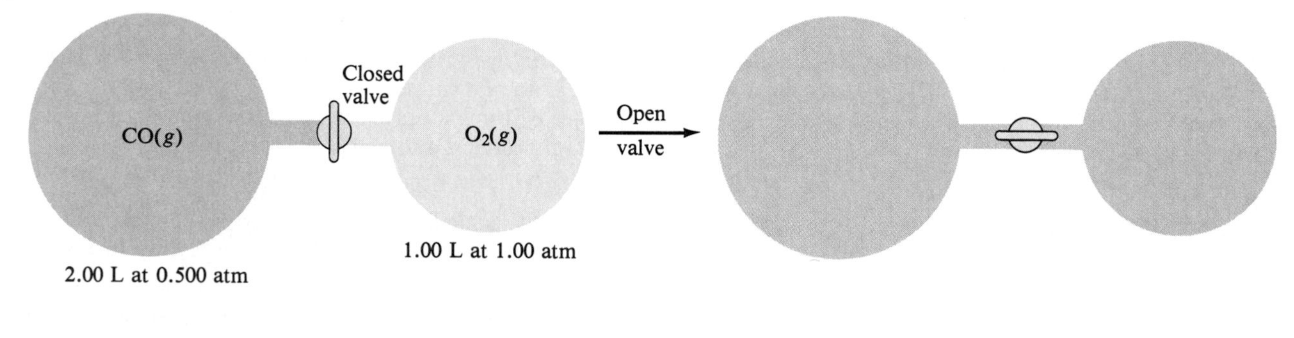

Cumulative-Skills Problems

5.109 A sample of natural gas is 85.2% methane, CH_4, and 14.8% ethane, C_2H_6, by mass. What is the density of this mixture at 18°C and 748 mmHg?

5.110 A sample of a breathing mixture for divers contained 34.3% helium, He; 51.7% nitrogen, N_2; and 14.0% oxygen, O_2 (by mass). What is the density of this mixture at 22°C and 755 mmHg?

5.111 A sample of sodium peroxide, Na_2O_2, was reacted with an excess of water.

$$2Na_2O_2(s) + 2H_2O(l) \longrightarrow 4NaOH(aq) + O_2(g)$$

All of the sodium peroxide reacted, and the oxygen was collected over water at 21°C. The barometric pressure was 771 mmHg. The apparatus was similar to that shown in Figure 5.13. However, the level of water inside the tube was 25.0 cm above the level of water outside the tube. If the volume of gas in the tube is 31.0 mL, how many grams of sodium peroxide were in the sample?

5.112 A sample of zinc metal was reacted with an excess of hydrochloric acid.

$$Zn(s) + 2HCl(aq) \longrightarrow ZnCl_2(aq) + H_2(g)$$

All of the zinc reacted, and the hydrogen gas was collected over water at 18°C; the barometric pressure was 751 mmHg. The apparatus was similar to that shown in Figure 5.13, but the level of water inside the tube was 31.0 cm above the level outside the tube. If the volume of gas in the tube is 24.0 mL, how many grams of zinc were there in the sample?

5.113 A mixture contained calcium carbonate, $CaCO_3$, and magnesium carbonate, $MgCO_3$. A sample of this mixture weighing 7.85 g was reacted with excess hydrochloric acid. The reactions are

$$CaCO_3(s) + 2HCl(aq) \longrightarrow CaCl_2(aq) + H_2O(l) + CO_2(g)$$

$$MgCO_3(s) + 2HCl(aq) \longrightarrow MgCl_2(aq) + H_2O(l) + CO_2(g)$$

If the sample reacted completely and produced 1.94 L of carbon dioxide, CO_2, at 25°C and 785 mmHg, what were the percentages of $CaCO_3$ and $MgCO_3$ in the mixture?

5.114 A mixture contained zinc sulfide, ZnS, and lead sulfide, PbS. A sample of the mixture weighing 6.12 g was reacted with an excess of hydrochloric acid. The reactions are

$$ZnS(s) + 2HCl(aq) \longrightarrow ZnCl_2(aq) + H_2S(g)$$
$$PbS(s) + 2HCl(aq) \longrightarrow PbCl_2(aq) + H_2S(g)$$

If the sample reacted completely and produced 1.049 L of hydrogen sulfide, H_2S, at 23°C and 745 mmHg, what were the percentages of ZnS and PbS in the mixture?

6

Thermochemistry

Chapter Outline

The fiery reaction of aluminum and iron(III) oxide (thermite reaction).

Nearly all chemical reactions involve either the release or the absorption of heat, a form of energy. The burning of coal and gasoline are dramatic examples of chemical reactions in which a great deal of heat is released. Such reactions are important sources of warmth and energy.

Chemical reactions that absorb heat are usually less dramatic. The reaction of barium hydroxide with an ammonium salt is an exception. If crystals of barium hydroxide octahydrate, $Ba(OH)_2 \cdot 8H_2O$, are mixed with crystals of ammonium nitrate, NH_4NO_3, in a flask, the solids form first a slush, then a liquid. Because the reaction mixture absorbs heat from the surroundings, the flask feels cool. It soon becomes so cold that if it is set in a puddle of water on a board, the water freezes; the board can then be inverted with the flask frozen to it (Figure 6.1).

In this chapter, we will be concerned with the quantity of heat released or absorbed in a chemical reaction. We want to address several questions: How do we measure the quantity of heat released or absorbed by a chemical reaction? To what extent can we relate the quantity of heat involved in a given reaction to the quantities of heat in other reactions? And how can we use this information?

Understanding Heats of Reaction

Thermodynamics is the science of the relationships between heat and other forms of energy. *Thermochemistry* is one area of thermodynamics. It concerns the study

Figure 6.1
An example of a reaction that absorbs heat. Two crystalline substances, barium hydroxide octahydrate and an ammonium salt, are mixed thoroughly in a flask. Then the flask, which feels quite cold to the touch, is set in a puddle of water on a board. In a couple of minutes, the flask and board are frozen solidly together. The board can then be inverted with the flask frozen to it.

of the quantity of heat absorbed or evolved (given off) by chemical reactions. An example of a heat-evolving reaction is the burning of fuel. There may be practical reasons why we want to know the quantity of heat evolved during the burning of a fuel: We could calculate the cost of the fuel per unit of heat energy produced; we could calculate the quantity of heat obtained per unit mass of rocket fuel; and so forth. But there are also theoretical reasons for wanting to know the quantity of heat involved in a reaction. For example, knowing such values, we are able to calculate the amount of energy needed to break a particular kind of chemical bond and so learn something about the strength of that bond. We will see later that heat measurements also provide data needed to determine whether a particular chemical reaction occurs and, if so, to what extent.●

● These questions concern chemical equilibrium and will be discussed in Chapter 18.

We have just used the terms *energy* and *heat,* assuming you have some idea of what these mean. But to proceed, we will need precise definitions of these and other terms. In the following section, we will define energy and its different forms and introduce units of energy.

6.1 ENERGY AND ITS UNITS

We can define **energy** briefly as *the potential or capacity to move matter*. According to this definition, energy is not a material thing but rather a property of matter. Energy exists in different forms that can be interconverted. We can see the relationship of a given form of energy to the motion of matter by following its interconversions into different forms.

Consider the interconversions of energy in a steam-driven electrical generator. A fuel is burned to heat water and generate steam. The steam expands against a piston, which is connected to a drive shaft that turns an electrical coil in a magnetic field. Electricity is generated in the coil. The fuel contains chemical energy, which is converted to heat. Part of the heat is then converted to motion of the drive shaft, and this motion is converted to electrical energy. The electrical energy could be used to run a motor, transforming the electrical energy back to the energy of motion. Or we could send the electricity into a light bulb, thereby converting electrical energy to heat energy and light energy. Photovoltaic cells can convert light back to electricity, which could be used to run a motor that can move matter (Figure 6.2). These examples show that energy can exist in different forms, including heat, light, and electrical energy, and these different forms can be interconverted. We also see their relationship to the energy of motion.

In this chapter, we will be especially concerned with the energy of substances, or chemical energy, and its transformation during chemical reaction into heat energy. To prepare for this, we will explore in this section the quantitative meaning of the energy of motion *(kinetic energy)*. Then we will look at the concept of *potential energy*. Finally, we will look at the concept of the *internal energy* of substances, which is defined in terms of the kinetic and potential energies of the particles making up the substance.

Kinetic Energy; Units of Energy

Kinetic energy is *the energy associated with an object by virtue of its motion*. An object of mass m and speed or velocity v has kinetic energy E_k equal to

$$E_k = \tfrac{1}{2}mv^2$$

This formula shows that the kinetic energy of an object depends on both its mass and its speed. A heavy object can move more slowly than a light object and still have the same kinetic energy.●

● In the previous chapter, we used the symbol u for *average* molecular speed. Here v is the speed of an individual object or particle.

To make sure you have a clear understanding of kinetic energy, consider the kinetic energy of a person whose mass is 59.0 kg and whose speed is 26.8 m/s. (This is equivalent to a person with a mass of 130 lbs traveling in an automobile going 60 miles per hour.) We substitute the mass and speed into the formula for the kinetic energy.

$$E_k = \tfrac{1}{2}\times(59.0\ \text{kg}) \times (26.8\ \text{m/s})^2 = 2.12 \times 10^4\ \text{kg}\cdot\text{m}^2/\text{s}^2$$

Note that the unit of energy comes out of the calculation. Because we substituted SI units of mass and speed, we obtain the SI unit of energy. *The SI unit of energy, $kg \cdot m^2/s^2$, is given the name* **joule (J)** (pronounced ''jewl'') after the English physicist James Prescott Joule (1818–1889) who studied the energy concept. We see that a person weighing 130 lbs and traveling 60 miles per hour has a kinetic energy equal to 2.12×10^4 J or 21.2 kJ (21.2 kilojoules).

The joule is an extremely small unit. To appreciate its size, note that the *watt* is a measure of the quantity of energy used per unit time and equals 1 joule per second. A 100-watt bulb, for example, uses 100 joules of energy every second. A kilowatt-hour, the unit by which electric energy is sold, equals 3600 kilowatt-seconds (because there are 3600 seconds in 1 hour), or 3.6 million joules. A household might use something like 1000 kilowatt-hours (3.6 billion joules) of electricity in a month.

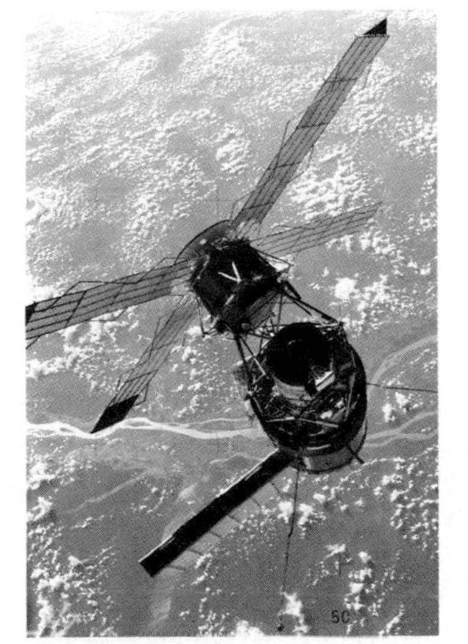

Figure 6.2
Photovoltaic cells on *Skylab*. Rectangular panels of photovoltaic cells recharge the batteries on the satellite.

The **calorie (cal)** is *a non-SI unit of energy commonly used by chemists, originally defined as the amount of energy required to raise the temperature of one gram of water by one degree Celsius*. This is only an approximate definition, however, because we now know that the energy needed to heat water depends slightly on the temperature of the water. In 1925 the calorie was defined in terms of the joule:

$$1\ \text{cal} = 4.184\ \text{J} \qquad \text{(exact definition)}$$

A person weighing 130 lbs and traveling 60 miles per hour has a kinetic energy of

$$2.12 \times 10^4\ \text{J} \times \frac{1\ \text{cal}}{4.184\ \text{J}} = 5.07 \times 10^3\ \text{cal (5.07 kcal)}$$

Example 6.1 Calculating Kinetic Energy

A regulation baseball weighs 142 g to 149 g. A baseball weighing 143 g is dropped from a window, and just before it hits the sidewalk it has a speed of 9.78 m/s. What is the kinetic energy of the baseball at this moment in joules? in calories?

Solution

We substitute into the formula $E_k = \frac{1}{2}mv^2$ using SI units.

$$E_k = \tfrac{1}{2} \times 0.143 \text{ kg} \times (9.78 \text{ m/s})^2 = \mathbf{6.84 \text{ J}}$$

To obtain the kinetic energy in calories, we multiply by the conversion factor 1 cal/4.184 J.

$$6.84 \cancel{\text{J}} \times \frac{1 \text{ cal}}{4.184 \cancel{\text{J}}} = \mathbf{1.64 \text{ cal}}$$

Exercise 6.1

An electron whose mass is 9.11×10^{-31} kg is accelerated by a positive charge to a speed of 5.0×10^6 m/s. What is the kinetic energy of the electron in joules? in calories?

(See Problems 6.29, 6.30, 6.31, and 6.32.)

Potential Energy

Potential energy is *the energy an object has by virtue of its position in a field of force*. For example, water at the top of a dam has potential energy (in addition to whatever kinetic energy it may possess), because the water is at a relatively high position in the gravitational force field of the earth. We can calculate this potential energy of the water from the formula $E_p = mgh$. Here E_p is the potential energy of a quantity of water in the dam, m is the mass of the water, g is the constant acceleration of gravity, and h is the height of the water. The height of the water, h, is measured from some standard level. The choice of this standard level is arbitrary, because only *differences* of potential energy are actually important in any physical situation. It is convenient to choose the standard level to be the surface of the earth. Then as a quantity of water falls over the dam, its potential energy decreases from mgh at the top of the dam to zero at the earth's surface.

The potential energy of the water in the dam is converted to kinetic energy when the water falls to a lower level. As the water falls, it moves faster. The potential energy decreases and the kinetic energy increases. Figure 6.3 shows the potential energy of water being converted to kinetic energy as the water falls over a dam.

Internal Energy

Consider the total energy of a quantity of water as it moves over the dam. This water has kinetic energy and potential energy as a whole. However, we know that water is itself made up of molecules, which are made up of smaller particles, electrons and nuclei. Each of these particles also has kinetic energy and potential energy. *The sum of the kinetic and potential energies of the particles making up a substance* is referred to as the **internal energy,** U, of the substance. Therefore, the total energy, E_{tot}, of a quantity of water equals the sum of its kinetic and potential energies as a whole plus its internal energy.

$$E_{tot} = E_k + E_p + U$$

Figure 6.3
Potential energy and kinetic energy. Water at the top of the dam has potential energy. As the water falls over the dam, potential energy is converted to kinetic energy.

Normally when we study a substance in the laboratory, it is at rest in a vessel. Its kinetic energy as a whole is zero. Moreover, its potential energy as a whole is constant, and we can take it to be zero. In this case, the total energy of the substance equals its internal energy, U.

Law of Conservation of Energy

We have discussed situations where one form of energy can be converted into another form. For example, when water falls over a dam, potential energy is converted into kinetic energy. Some of the kinetic energy of the water may also be converted into random molecular motion—that is, into internal energy of the water. The total energy, E_{tot}, of the water, however, remains constant, equal to the sum of the kinetic energy, E_k, the potential energy, E_p, and the internal energy, U, of the water.

This result can be stated more generally as the **law of conservation of energy:** *Energy may be converted from one form to another, but the total quantity of energy remains constant.*

6.2 HEAT OF REACTION

In the chapter opening, we mentioned chemical reactions such as the burning of coal that evolve or release heat. We also described a reaction that absorbs heat. Both types of reaction involve a *heat of reaction*. To understand this concept, we need to know what is meant by a thermodynamic *system* and its *surroundings* and we need to define the term *heat* precisely.

Suppose that we are interested in studying the change of a thermodynamic property (such as internal energy) during a physical or chemical change. *The substance or mixture of substances under study in which a change occurs* is called the **thermodynamic system** (or simply **system**). The **surroundings** are *everything in the vicinity of the thermodynamic system* (Figure 6.4).

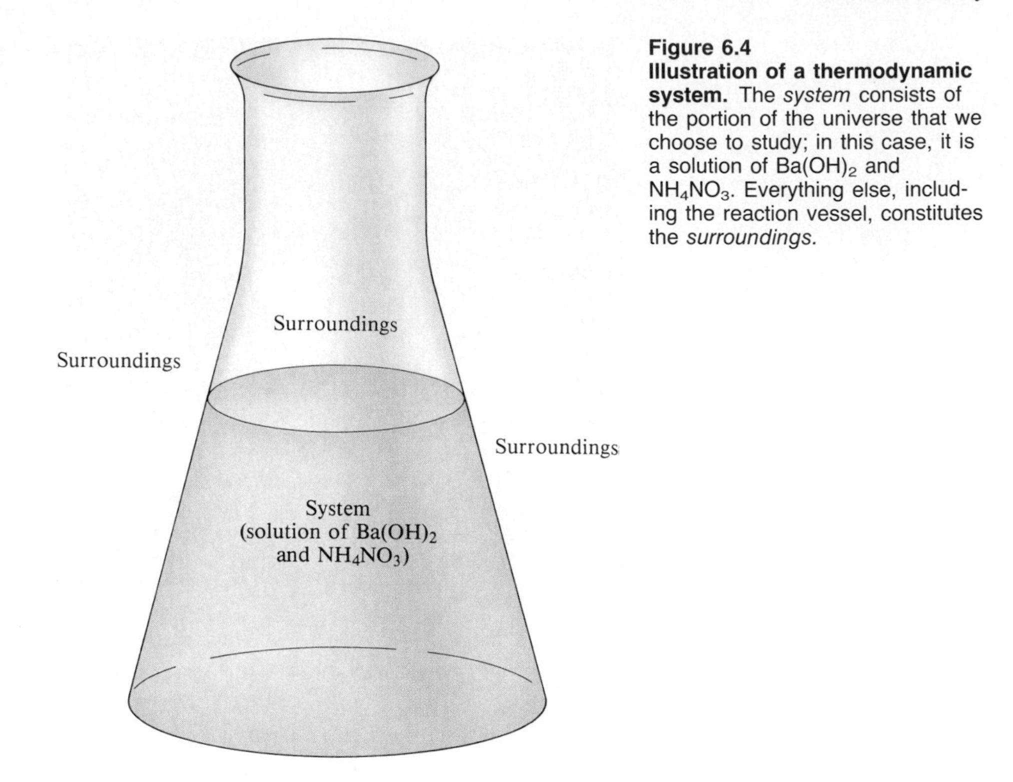

Figure 6.4
Illustration of a thermodynamic system. The *system* consists of the portion of the universe that we choose to study; in this case, it is a solution of $Ba(OH)_2$ and NH_4NO_3. Everything else, including the reaction vessel, constitutes the *surroundings*.

Definition of Heat

Heat is defined as *the energy that flows into or out of a system because of a difference in temperature between the thermodynamic system and its surroundings*. As long as a system and its surroundings are in thermal contact (that is, they are not thermally insulated from one another), energy (heat) flows between them to establish temperature equality or *thermal equilibrium*. Heat flows from a region of higher temperature to one at lower temperature; but once the temperatures become equal, heat flow stops. Note that once heat flows into a system, it appears in the system as an increase in its internal energy. We would not say that the system has heat, because heat is only energy in transport.

We can explain the transfer of energy between two regions of different temperatures in terms of kinetic-molecular theory. Imagine two vessels in contact, each containing oxygen gas and the one on the left being hotter (Figure 6.5). The

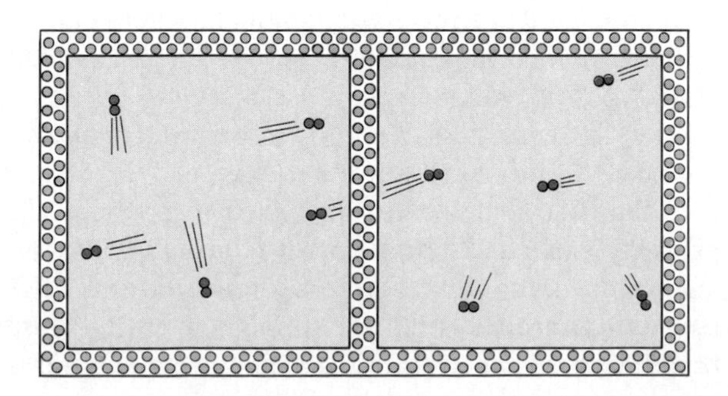

Figure 6.5
A kinetic-theory explanation of heat. The vessel on the left contains oxygen molecules at a higher temperature than the oxygen molecules on the right. Molecules collide with the vessel walls, where they lose or gain energy. The faster molecules tend to slow down, and the slower molecules tend to speed up. The net result is that energy is transferred through the vessel walls from the hot gas to the cold gas. We call this energy transfer heat.

average speed of molecules in the hotter gas is greater than that of the molecules in the colder gas. But as the molecules in their random motions collide with the vessel walls, they lose or gain energy to the walls. The faster molecules tend to slow down, while the slower molecules tend to speed up. Eventually, the average speeds of the molecules in the two vessels (and therefore the temperatures of the two gases, which are related to the average molecular speeds) become equal. The net result is that energy has transferred through the vessel walls from the hot gas to the cold gas.

Heat is denoted by the symbol q. The algebraic sign of q is chosen to be positive if heat is absorbed by the system or negative if heat is evolved. The sign of q can be remembered this way: When heat is absorbed by a system, energy is *added* to it; q is assigned a positive quantity. On the other hand, when heat is evolved by a system, energy is *subtracted* from it; q is assigned a negative number.

Heat of Reaction

Consider a system in which a chemical reaction occurs. Before the reaction, the system and surroundings are at the same temperature, say 25°C. When the reaction starts, however, the temperature changes.

Suppose the temperature falls. In that case, heat flows from the surroundings into the system. When the reaction stops, heat continues to flow until the system returns to the temperature of its surroundings (at 25°C). Heat has flowed into the system; that is, the system has absorbed heat and q is positive.

Suppose, on the other hand, that the temperature rises. In this case, heat flows from the system to the surroundings. When the reaction stops, heat continues to flow until the system returns to the temperature of its surroundings (at 25°C). In this case, heat has flowed out of the system; that is, the system has evolved heat and q is negative.

The **heat of reaction** (at a given temperature) is *the value of* q *required to return a system to the given temperature at the completion of the reaction.*

Chemical reactions or physical changes are classified as exothermic or endothermic. An **exothermic process** is *a chemical reaction or a physical change in which heat is evolved* (q is negative). An **endothermic process** is *a chemical reaction or a physical change in which heat is absorbed* (q is positive). Experimentally, we note that in the exothermic reaction, the reaction flask initially warms; in the endothermic reaction, the reaction flask initially cools. We can summarize this as follows:

TYPE OF REACTION	EXPERIMENTAL EFFECT NOTED	RESULT ON SYSTEM	SIGN OF q
Endothermic	Reaction vessel cools (heat is absorbed)	Energy added	+
Exothermic	Reaction vessel warms (heat is evolved)	Energy subtracted	−

Suppose that in an experiment, one mole of methane burns in oxygen and evolves 890 kJ of heat: $CH_4(g) + 2O_2(g) \rightarrow CO_2(g) + 2H_2O(l)$. The reaction is exothermic. Therefore, the heat of reaction, q, is -890 kJ.

Or consider the reaction described in the chapter opening, in which crystals of barium hydroxide octahydrate, $Ba(OH)_2 \cdot 8H_2O$, react with crystals of ammonium nitrate, NH_4NO_3.

$$Ba(OH)_2 \cdot 8H_2O(s) + 2NH_4NO_3(s) \longrightarrow 2NH_3(g) + 10H_2O(l) + Ba(NO_3)_2(aq)$$

When 1 mol $Ba(OH)_2 \cdot 8H_2O$ reacts with 2 mol NH_4NO_3, the reaction mixture absorbs 170.4 kJ of heat. The reaction is endothermic. Therefore the heat of reaction, q, is +170.4 kJ.

Exercise 6.2

Ammonia burns in the presence of a platinum catalyst to give nitric oxide, NO.

$$4NH_3(g) + 5O_2(g) \longrightarrow 4NO(g) + 6H_2O(l)$$

In an experiment, 4 mol NH_3 is burned and evolves 1170 kJ of heat. Is the reaction endothermic or exothermic? What is the value of q? (See Problems 6.33 and 6.34.)

6.3 ENTHALPY AND ENTHALPY CHANGE

The heat absorbed or evolved by a reaction depends on the conditions under which the reaction occurs. Usually, a reaction takes place in a vessel open to the atmosphere, where it occurs at the constant pressure of the atmosphere. We will assume that this is the case and write the heat of reaction as q_p, the subscript p indicating that the process occurs at constant pressure.

There is a property of substances called enthalpy (''en-thal'-py'') that is related to the heat of reaction q_p. **Enthalpy** (denoted H) is *an extensive property of a substance that can be used to obtain the heat absorbed or evolved in a chemical reaction*. (Recall that an extensive property is a property that depends on the amount of substance.)

Enthalpy is a state function. A **state function** is *a property of a system that depends only on its present state, which is determined by variables such as temperature and pressure, and is independent of any previous history of the system*. This means that changes in enthalpy do not depend on how the change was made, but only on the initial state and final state of the system. An analogy may help clarify the point. Suppose you are hiking in mountainous terrain. You start walking from campsite A, which is at an altitude of 1200 ft above sea level according to your pocket altimeter, and you take the curving, graded path upward to campsite B. Alternatively, you could have taken a more direct, but more difficult, route to campsite B. The distances traveled are different; the distance by the graded path is longer than that of the more direct route. However, the altitude (height above sea level) of campsite B is independent of how you got there (independent of any history). If you find the altitude of campsite B to be 4800 ft after arriving by the graded path, the altitude will be 4800 ft by any other path (assuming your altimeter is accurate, of course). Moreover, the difference in altitude of the two campsites is independent of the route you take from one to the other. Altitude is analogous to a state function, whereas distance traveled is not.

Consider a chemical reaction system. At first, the enthalpy of the system is that of the reactants. But as the reaction proceeds, the enthalpy changes and finally becomes equal to that of the products. *The change in enthalpy for a reaction at a given temperature and pressure* (called the **enthalpy of reaction**) is obtained by subtracting the enthalpy of the reactants from the enthalpy of the products. We will

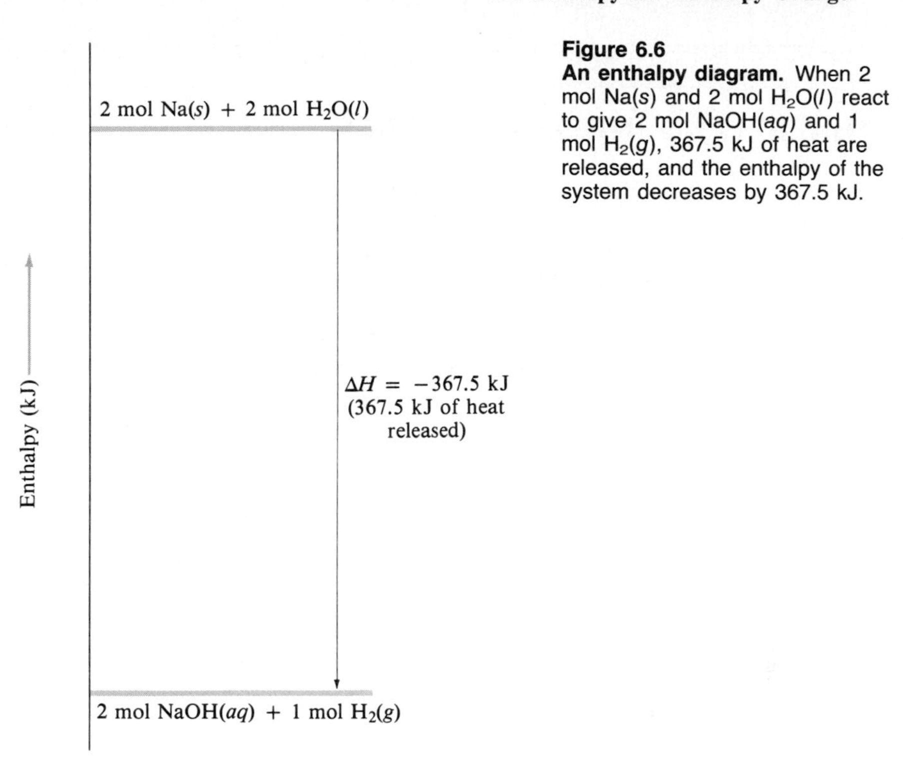

Figure 6.6
An enthalpy diagram. When 2 mol Na(*s*) and 2 mol $H_2O(l)$ react to give 2 mol NaOH(*aq*) and 1 mol $H_2(g)$, 367.5 kJ of heat are released, and the enthalpy of the system decreases by 367.5 kJ.

use the symbol Δ (meaning "change in") and write the change in enthalpy as ΔH. We apply the Δ notation by taking the final value and subtracting the initial value. Thus $\Delta H = H_{final} - H_{initial}$. Since we start from reactants and end with products, the enthalpy of reaction is

$$\Delta H = H(\text{products}) - H(\text{reactants})$$

Because H is a state function, the value of ΔH is independent of the details of the reaction. It depends only on the initial state (the reactants) and on the final state (the products).

The key relation in this chapter is that between enthalpy change and heat of reaction:

$$\Delta H = q_p$$

The enthalpy of reaction equals the heat of reaction at constant pressure.

To illustrate the concepts we have introduced, we will consider the reaction at 25°C of sodium metal and water, carried out in a beaker open to the atmosphere at 1.00 atm pressure.

$$2\text{Na}(s) + 2\text{H}_2\text{O}(l) \longrightarrow 2\text{NaOH}(aq) + \text{H}_2(g)$$

The metal and water react vigorously and heat evolves. Experiment shows that 2 moles of sodium metal react with 2 moles of water to evolve 367.5 kJ of heat. Because heat evolves, the reaction is exothermic, and we write $q_p = -367.5$ kJ. Therefore, the enthalpy of reaction, or change of enthalpy for the reaction, is $\Delta H = -367.5$ kJ. Figure 6.6 shows an *enthalpy diagram* depicting the enthalpy change for this reaction.

Related Topic

ENTHALPY AND INTERNAL ENERGY

In the preceding discussion, we noted that the enthalpy change equals the heat of reaction at constant pressure. This will be sufficient for the purpose of this chapter, which is to introduce the concepts of heat of reaction and enthalpy change. Later in the book we will look at thermodynamics in more detail. Still, it is useful at this point to note briefly the relationship of enthalpy to internal energy.

The enthalpy, H, is defined precisely as the internal energy, U, plus pressure, P, times volume, V.

$$H = U + PV$$

To obtain some understanding of this equation, consider a reaction system at constant pressure P. We will label the initial quantities (those for reactants) with a subscript i and the final quantities (those for products) with a subscript f. Then

$$\Delta H = H_f - H_i = (U_f + PV_f) - (U_i + PV_i)$$

Collecting the internal energy terms and the pressure-volume terms, we rewrite this as

$$\Delta H = (U_f - U_i) + P(V_f - V_i) = \Delta U + P\Delta V$$

where we have written ΔU for $U_f - U_i$ and ΔV for $V_f - V_i$. Finally, we rearrange this as follows:

$$\Delta U = \Delta H - P\Delta V$$

The last term in this equation ($-P\Delta V$) is the energy required by the system to change volume against the constant pressure of the atmosphere (the minus sign means that energy is required to increase the volume of the system). This required energy is called the *pressure-volume work.* Thus, the equation says that the internal energy of the system changes in two ways. It changes because energy leaves or enters the system as heat (ΔH); and it changes because the system increases or decreases in volume against the constant pressure of the atmosphere (which requires energy $-P\Delta V$).

Consider a specific reaction. When 2 moles of sodium metal and 2 moles of water react in a beaker, 1 mole of hydrogen gas forms and heat evolves. If we measure this heat, we find it to be −367.5 kJ per mole of sodium. Thus, 367.5 kJ of energy have left the system in the form of heat.

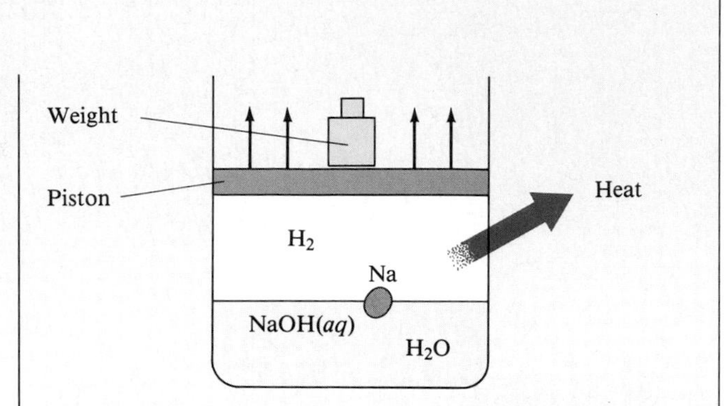

Figure 6.7
Pressure-volume work. *In this experiment, we replace the pressure of the atmosphere by a piston-and-weight assembly of equal pressure. Note that as hydrogen gas evolves, it pushes the piston and weight upward. It requires energy (the pressure-volume work) to raise the piston and weight upward in a gravitational field.*

Because hydrogen gas forms during the reaction, the volume of the system increases. To expand, the system must push back the atmosphere, and this requires energy equal to the pressure-volume work. It may be easier to see this pressure-volume work if we replace the constant pressure of the atmosphere by an equivalent pressure from a piston-and-weight assembly, as in Figure 6.7. When hydrogen gas is released during the reaction, it pushes upward on the piston and raises the weight. It requires energy to lift a weight upward in gravitational field. If we calculate this pressure-volume work at 25°C and 1.00 atm pressure, we find that it is $-P\Delta V = -2.5$ kJ.

In the sodium–water reaction, the internal energy changes by −367.5 kJ because heat evolves and by −2.5 kJ because pressure-volume work is done. The total change of internal energy is

$$\Delta U = \Delta H - P\Delta V = -367.5 \text{ kJ} - 2.5 \text{ kJ} = -370.0 \text{ kJ}$$

As you can see, ΔU does not differ a great deal from ΔH. This is the case in most reactions, so that the heat of reaction at constant pressure is approximately equal to the change of internal energy.

6.4 THERMOCHEMICAL EQUATIONS

We will often find it convenient to write the enthalpy of reaction, ΔH, with the chemical equation. A **thermochemical equation** is *the chemical equation for a* *reaction (including phase labels) in which the equation is given a molar interpretation and the enthalpy of reaction for these molar amounts is written directly after the equation.* For the reaction of sodium and water that we discussed in the previous section, we would write

$$2Na(s) + 2H_2O(l) \longrightarrow 2NaOH(aq) + H_2(g);\ \Delta H = -367.5\ \text{kJ}$$

This equation says that 2 moles of sodium react with 2 moles of water to produce 2 moles of sodium hydroxide and 1 mole of hydrogen gas, and 367.5 kJ of heat evolve.

Note that we specify that the thermochemical equation includes phase labels. This is important because the enthalpy change, ΔH, depends on the phase of the substances. Consider the reaction of hydrogen and oxygen to produce water. If the product is water vapor, 2 moles of H_2 burn to release 483.7 kJ of heat.

$$2H_2(g) + O_2(g) \longrightarrow 2H_2O(g);\ \Delta H = -483.7\ \text{kJ}$$

On the other hand, if the product is liquid water, the heat released is 571.7 kJ.

$$2H_2(g) + O_2(g) \longrightarrow 2H_2O(l);\ \Delta H = -571.7\ \text{kJ}$$

In this case, additional heat is released when water vapor condenses to liquid.●

● It takes 44.0 kJ of heat to vaporize 1 mole of liquid water at 25°C.

Example 6.2 Writing Thermochemical Equations

Aqueous sodium hydrogen carbonate solution (baking soda solution) reacts with hydrochloric acid to produce aqueous sodium chloride, water, and carbon dioxide gas. The reaction absorbs 11.8 kJ of heat at constant pressure for each mole of sodium hydrogen carbonate. Write the thermochemical equation for the reaction.

Solution

We first write the balanced chemical equation.

$$NaHCO_3(aq) + HCl(aq) \longrightarrow NaCl(aq) + H_2O(l) + CO_2(g)$$

Because the reaction absorbs heat, the enthalpy of reaction for molar amounts of this equation is 11.8 kJ. The thermochemical equation is

$$\mathbf{NaHCO_3(aq) + HCl(aq) \longrightarrow NaCl(aq) + H_2O(l) + CO_2(g);\ \Delta H = 11.8\ kJ}$$

Exercise 6.3

A propellant for rockets is obtained by mixing the liquids hydrazine, N_2H_4, and dinitrogen tetroxide, N_2O_4. These react to give gaseous nitrogen, N_2, and water vapor, evolving 1049 kJ of heat at constant pressure when 1 mol N_2O_4 reacts. Write the thermochemical equation for this reaction. (See Problems 6.35 and 6.36.)

The following are two important rules for manipulating thermochemical equations:

1. When a thermochemical equation is multiplied by any factor, the value of ΔH for the new equation is obtained by multiplying the value of ΔH in the original equation by that same factor. Consider the thermochemical equation for the synthesis of ammonia.

$$N_2(g) + 3H_2(g) \longrightarrow 2NH_3(g);\ \Delta H = -91.8\ \text{kJ}$$

Suppose we want the thermochemical equation to show what happens when twice as many moles of nitrogen and hydrogen react to produce ammonia. Because double the amounts of substances are present, the enthalpy of reaction is doubled (enthalpy is an extensive quantity). If we double the previous equation, we obtain

$$2N_2(g) + 6H_2(g) \longrightarrow 4NH_3(g);\ \Delta H = -184\ \text{kJ}$$

2. When a chemical equation is reversed, the value of ΔH is reversed in sign. Suppose we reverse the first equation we wrote for the synthesis of ammonia. Then the reaction is the dissociation of 2 moles of ammonia into its elements. The thermochemical equation is

$$2NH_3(g) \longrightarrow N_2(g) + 3H_2(g);\ \Delta H = +91.8\ \text{kJ}$$

If we want to express this in terms of 1 mole of ammonia, we simply multiply this equation by a factor of $\frac{1}{2}$.●

● If we use the molar interpretation of a chemical equation, there is nothing unreasonable about using such coefficients as $\frac{1}{2}$ and $\frac{3}{2}$.

$$NH_3(g) \longrightarrow \tfrac{1}{2}N_2(g) + \tfrac{3}{2}H_2(g);\ \Delta H = +45.9\ \text{kJ}$$

The next example further illustrates the use of these rules of thermochemical equations.

Example 6.3 Manipulating Thermochemical Equations

When 2 moles of $H_2(g)$ and 1 mole of $O_2(g)$ react to give liquid water, 572 kJ of heat evolve.

$$2H_2(g) + O_2(g) \longrightarrow 2H_2O(l);\ \Delta H = -572\ \text{kJ}$$

Write this equation for 1 mole of liquid water. Give the reverse equation, in which 1 mole of liquid water dissociates into hydrogen and oxygen.

Solution

We multiply the coefficients and ΔH in the equation by $\frac{1}{2}$:

$$H_2(g) + \tfrac{1}{2}O_2(g) \longrightarrow H_2O(l);\ \Delta H = -286\ \text{kJ}$$

If we reverse the equation, we get

$$H_2O(l) \longrightarrow H_2(g) + \tfrac{1}{2}O_2(g);\ \Delta H = +286\ \text{kJ}$$

Exercise 6.4

(a) Write the thermochemical equation for the reaction described in Exercise 6.3 for the case involving 1 mol N_2H_4. (b) Write the thermochemical equation for the reverse of the reaction described in Exercise 6.3. (See Problems 6.37, 6.38, 6.39, and 6.40.

6.5 APPLYING STOICHIOMETRY TO REACTION HEATS

As you might expect, the quantity of heat obtained from a reaction will depend on the amount of reactants. We can extend the method used to solve stoichiometry problems, described in Chapter 4, to problems involving the quantity of heat.

Consider the reaction of methane, CH_4 (the principal constituent of natural gas), burning in oxygen at constant pressure. How much heat could we obtain from 10.0 g of methane, assuming we have an excess of oxygen? We can answer this question if we know the enthalpy change for the reaction of 1 mole of methane. The thermochemical equation is

$$CH_4(g) + 2O_2(g) \longrightarrow CO_2(g) + 2H_2O(l);\ \Delta H = -890.3\ \text{kJ}$$

The calculation involves the following conversions:

$$\text{Grams of } CH_4 \longrightarrow \text{moles of } CH_4 \longrightarrow \text{kilojoules of heat}$$

The relevant information is

$$1 \text{ mol } CH_4 \simeq 16.0 \text{ g } CH_4 \quad \text{(from the molecular weight)}$$
$$1 \text{ mol } CH_4 \simeq -890.3 \text{ kJ} \quad \text{(from the thermochemical equation)}$$

Hence,

$$10.0 \text{ g } CH_4 \times \frac{1 \text{ mol } CH_4}{16.0 \text{ g } CH_4} \times \frac{-890.3 \text{ kJ}}{1 \text{ mol } CH_4} = -556 \text{ kJ}$$

That is, 10.0 g of methane burns in excess oxygen to evolve 556 kJ of heat.●

● If we wanted the result in kilocalories, we could convert this answer as follows:

$$556 \text{ kJ} \times \frac{1 \text{ kcal}}{4.184 \text{ kJ}} = 133 \text{ kcal}$$

Example 6.4 Calculating Reaction Heat from the Stoichiometry

How much heat is evolved when 907 kg of ammonia is produced according to the following equation? (Assume that the reaction occurs at constant pressure.)

$$N_2(g) + 3H_2(g) \longrightarrow 2NH_3(g); \Delta H = -91.8 \text{ kJ}$$

Solution

The calculation involves converting grams of NH_3 to moles of NH_3 and then to kilojoules of heat. Note that 907 kg equals 9.07×10^5 g. Hence,

$$9.07 \times 10^5 \text{ g } NH_3 \times \frac{1 \text{ mol } NH_3}{17.0 \text{ g } NH_3} \times \frac{-91.8 \text{ kJ}}{2 \text{ mol } NH_3} \simeq -2.45 \times 10^6 \text{ kJ}$$

Thus, 2.45×10^6 kJ of heat evolves.

Exercise 6.5

How much heat evolves when 10.0 g of hydrazine react according to the reaction described in Exercise 6.3? (See Problems 6.41, 6.42, 6.43, and 6.44.)

6.6 MEASURING HEATS OF REACTION

So far we have introduced the concept of heat of reaction, and we have related it to the enthalpy change. We also showed how to display the enthalpy change in a thermochemical equation, and from this equation we showed how to calculate the heat for the reaction of any given amount of substance. Now that we have a firm idea of what heats of reaction are, it is time to discuss how we would measure them.

First, we need to look at the heat required to raise the temperature of a substance, because a calorimetric measurement is based on the relationship between heat and temperature change. The heat required to raise the temperature of a substance is called its *heat capacity*.

Heat Capacity and Specific Heat

It requires heat to raise the temperature of a given amount of substance, and the quantity of heat depends on the temperature change. The **heat capacity** *(C)* of a sample of substance is *the quantity of heat needed to raise the temperature of the sample of substance one degree Celsius (or one kelvin).* Changing the temperature

Table 6.1 Specific Heats and Molar Heat Capacities of Some Substances

Substance	Specific Heat J/(g · °C)	Molar Heat Capacity J/(mol · °C)
Aluminum, Al	0.901	24.3
Copper, Cu	0.384	24.4
Ethanol, C_2H_5OH	2.43	112.0
Iron, Fe	0.449	25.1
Water, H_2O	4.18	75.3

of the sample from an initial temperature t_i to a final temperature t_f requires heat equal to

$$q = C\Delta t$$

where Δt is the change of temperature and equals $t_f - t_i$.●

● The heat capacity will depend on the process, whether a constant-pressure process or a constant-volume process. We will assume a constant-pressure process unless otherwise stated.

Suppose a piece of iron requires 6.70 J of heat to raise the temperature by one degree Celsius. Its heat capacity is therefore 6.70 J/°C. The quantity of heat required to raise the temperature of the piece of iron from 25.0°C to 35.0°C is

$$q = C\Delta t = (6.70 \text{ J/°C}) \times (35.0°\text{C} - 25.0°\text{C}) = 67.0 \text{ J}$$

Heat capacity is directly proportional to the amount of substance. Often heat capacities are listed for molar amounts of substances. The *molar heat capacity* of a substance is its heat capacity for one mole of substance.

Heat capacities are also compared for one-gram amounts of substances. The **specific heat capacity** (or simply **specific heat**) is *the quantity of heat required to raise the temperature of one gram of a substance by one degree Celsius (or one kelvin) at constant pressure*. To find the heat q required to raise the temperature of a sample, we multiply the specific heat of the substance, s, by the mass in grams, m, and the temperature change, Δt.

$$q = s \times m \times \Delta t$$

The specific heats and molar heat capacities of a few substances are listed in Table 6.1. Values depend somewhat on temperature, and those listed are for 25°C. Water has a specific heat of 4.18 J/(g · °C), that is, 4.18 joules per gram per degree Celsius. In terms of calories, the specific heat of water is 1.00 cal/(g · °C). Example 6.5 illustrates the use of the preceding equation.

Example 6.5 Relating Heat and Specific Heat

Calculate the heat absorbed by 15.0 g of water to raise its temperature from 20.0°C to 50.0°C (at constant pressure). The specific heat of water is 4.18 J/(g · °C).

Solution

We substitute into the equation

$$q = s \times m \times \Delta t$$

The temperature change is

$$\Delta t = t_f - t_i = 50.0°\text{C} - 20.0°\text{C} = +30.0°\text{C}$$

and therefore,

$$q = 4.18 \text{ J/(g · °C)} \times 15.0 \text{ g} \times (+30.0°\text{C}) = \mathbf{1.88 \times 10^3\ J}$$

Exercise 6.6

Iron metal has a specific heat of 0.449 J/(g · °C). How much heat is transferred to a 5.00-g piece of iron, initially at 20.0°C, when it is placed in a pot of boiling water? Assume that the temperature of the water is 100.0°C and that the water remains at this temperature, which is the final temperature of the iron. (See Problems 6.45 and 6.46.)

Measurement of Heat of Reaction

We measure the heat of reaction in a calorimeter. A **calorimeter** is *a device used to measure the heat absorbed or evolved during a physical or chemical change*. It can be as simple as the apparatus sketched in Figure 6.8, which consists of an insulated container (for example, a polystyrene coffee cup) with a thermometer in it. More elaborate calorimeters are employed when precise measurements are needed for research (Figure 6.9), although the basic idea remains the same—to measure temperature changes under controlled circumstances and relate these temperature changes to heat.

The coffee-cup calorimeter, shown in Figure 6.8, is a constant-pressure calorimeter. The heat of the reaction is calculated from the temperature change caused by the reaction, and since this is a constant-pressure process, the heat can be directly related to the enthalpy change, ΔH. Research versions of a constant-pressure calorimeter are available, and these are used whenever gases are not involved.

For reactions involving gases, a *bomb calorimeter* is generally used (Figure 6.9). Consider the heat of combustion of graphite, which is the form of carbon used in the "lead" of a pencil. To measure the heat released when graphite burns in oxygen, a sample of graphite is placed in a small cup in the calorimeter. The graphite is surrounded by oxygen, and the graphite and oxygen are sealed in a steel vessel, or bomb. An electrical circuit is activated to start the burning of the graphite. The bomb is surrounded by water in an insulated container, and the heat of

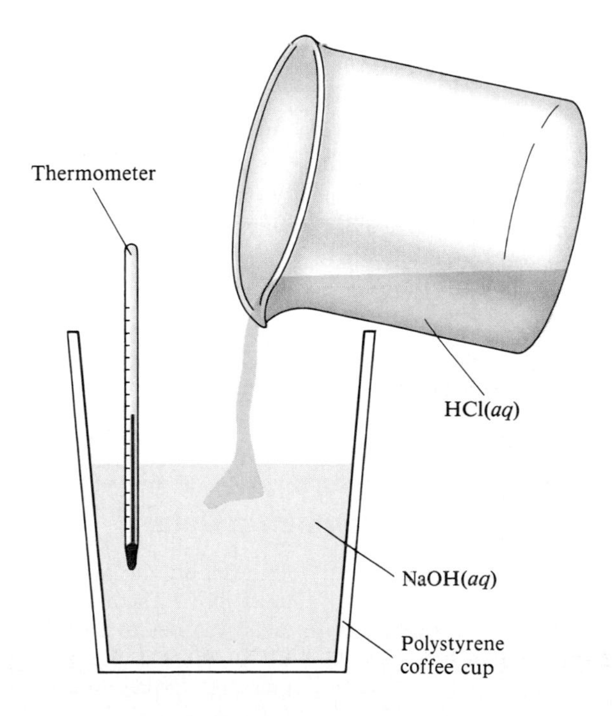

Figure 6.8
A simple coffee-cup calorimeter. After reactants are added, the cup is covered to reduce heat loss by evaporation and convection. The heat of reaction is determined by noting the temperature rise or fall.

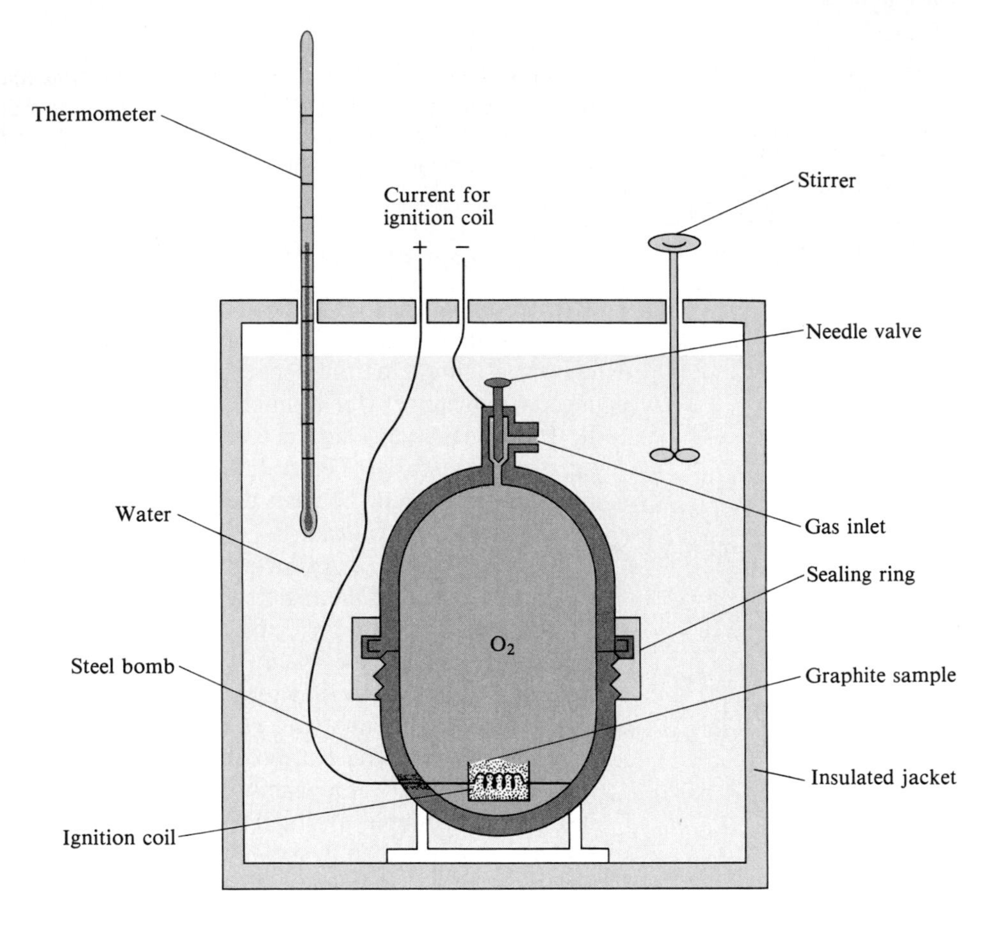

Figure 6.9
A bomb calorimeter. The heat of a reaction involving a gas (here the heat of combustion of graphite) is conveniently determined in a sealed vessel called a bomb. Reaction is started by an ignition coil running through the graphite sample.

reaction is calculated from the temperature change of the calorimeter caused by the reaction.

Because the reaction in a bomb calorimeter occurs in a closed vessel, the pressure does not remain constant. Rather the volume remains constant, and under these conditions the heat of reaction does not in general equal ΔH; a small correction is usually needed. However, this correction is negligible whenever the reaction does not involve gases or whenever the number of moles of reactant gas equals the number of moles of product gas, as in the combustion of graphite to carbon dioxide.●

● In the Related Topic at the end of Section 6.3, we noted that $\Delta H = \Delta U + P\Delta V$. The heat at constant volume equals ΔU. To obtain ΔH, we must add the correction $P\Delta V$. The change in volume is significant only if there are changes in gas volumes. From the ideal gas law, the molar volume of a gas is RT. If the moles of gas that have reacted is n_i and the moles of gas produced is n_f, then the change in volume is essentially $RT \times (n_f - n_i)$, or $RT\Delta n$, where Δn refers to the change of moles of gas. To obtain the correction in joules, we use $R = 8.31$ J/(K · mol). Thus, RT at 25°C (298 K) is 8.31 J/(K · mol) × 298 K = 2.48×10^3 J/mol. The final result is $\Delta H = \Delta U +$ (2.48 kJ) × Δn.

The next example describes the calculations needed to obtain ΔH from calorimetric measurements.

Example 6.6 Calculating ΔH from Calorimetric Data

Suppose 0.562 g of graphite is placed in a calorimeter with an excess of oxygen at 25.00°C and 1 atm pressure. Excess O_2 ensures that all carbon burns to form CO_2. The graphite is then ignited, and it burns according to the equation

$$\text{C(graphite)} + O_2(g) \longrightarrow CO_2(g)$$

On reaction, the calorimeter temperature rises from 25.00°C to 25.89°C. The heat capacity of the calorimeter and its contents was determined in a separate experiment to be 20.7 kJ/°C. What is the heat of reaction at 25.00°C and 1 atm pressure? Express the answer as a thermochemical equation.

Solution

Whatever heat is released by the reaction is absorbed by the calorimeter and its contents. Let q_{rxn} be the quantity of heat from the reaction mixture, and let C_{cal} be the heat capacity of the calorimeter and contents. The quantity of heat absorbed by the calorimeter is $C_{cal}\,\Delta t$. This will have the same magnitude as q_{rxn} but the opposite sign: $q_{rxn} = -C_{cal}\,\Delta t$. Substituting into this equation gives

$$\begin{aligned} q_{rxn} &= -C_{cal}\,\Delta t = -20.7\ \text{kJ/°C} \times (25.89\text{°C} - 25.00\text{°C}) \\ &= -20.7\ \text{kJ/°C} \times 0.89\text{°C} = -18.4\ \text{kJ} \end{aligned}$$

An extra figure has been retained in q_{rxn} for further computation. The negative sign indicates that the reaction is exothermic, as expected for a combustion.

Thus, ΔH for the combustion of 0.562 g of carbon equals -18.4 kJ. We can write

$$0.562\ \text{g C} \simeq -18.4\ \text{kJ}$$

To obtain the value for the combustion of one mole of carbon, we perform the following conversions to express one mole of carbon as kilojoules of heat:

$$\text{mol C} \longrightarrow \text{g C} \longrightarrow \text{kJ heat}$$

Thus,

$$\begin{aligned} 1\ \text{mol C} &\simeq 1\ \cancel{\text{mol C}} \times \frac{12.0\ \cancel{\text{g C}}}{1\ \cancel{\text{mol C}}} \times \frac{-18.4\ \text{kJ}}{0.562\ \cancel{\text{g C}}} \\ &\simeq -3.9 \times 10^2\ \text{kJ} \end{aligned}$$

(The final answer has been rounded to two significant figures.) When one mole of carbon burns, 3.9×10^2 kJ of heat is released. We can summarize the results by the thermochemical equation

$$\text{C(graphite)} + O_2(g) \longrightarrow CO_2(g);\ \Delta H = -3.9 \times 10^2\ \text{kJ}$$

Exercise 6.7

Suppose 33 mL of 1.20 *M* HCl is added to 42 mL of a solution containing excess sodium hydroxide, NaOH, in a coffee-cup calorimeter. The solution temperature, originally 25.0°C, rises to 31.8°C. Give the enthalpy change, ΔH, for the reaction

$$HCl(aq) + NaOH(aq) \longrightarrow NaCl(aq) + H_2O(l)$$

Express the answer as a thermochemical equation. For simplicity, assume that the heat capacity and the density of the final solution in the cup are those of water. (In more accurate work, these values must be determined.) Also assume that the total volume of the solution equals the sum of the volumes of HCl(*aq*) and NaOH(*aq*).

(See Problems 6.49, 6.50, 6.51, and 6.52.)

Using Heats of Reaction

In the first part of this chapter, we looked at the basic properties of the heat of reaction and how to measure it. Now we want to show how heats of reaction can be used. We will see that the ΔH for one reaction can be obtained from the ΔH's of other reactions. This means that we can tabulate a small number of values and use these to calculate others.

6.7 HESS'S LAW

In this section, we will show that it is possible to obtain enthalpies of reaction from other reactions, so that it is not necessary to do calorimetric measurements on all reactions. This is fortunate because calorimetric measurements are time consuming, and many reactions are not amenable to such direct measurements. A reaction may go too slowly, or it may not go ''cleanly''—that is, other reactions may occur at the same time—so that the heat obtained cannot be ascribed to a particular reaction.

This is what we would encounter if we wanted to determine the ΔH change for the reaction of graphite (carbon) and oxygen to produce carbon monoxide

$$2C(\text{graphite}) + O_2(g) \longrightarrow 2CO(g) \quad \textbf{(1)}$$

Once carbon monoxide forms, it is impossible to keep it from reacting with oxygen to give carbon dioxide.

$$2CO(g) + O_2(g) \longrightarrow 2CO_2(g) \quad \textbf{(2)}$$

At the same time, graphite may react in one step to give carbon dioxide. We described the calorimetric measurement of the enthalpy change of this reaction earlier, but in that case we could ensure a complete, clean reaction by using an excess of oxygen.

$$C(\text{graphite}) + O_2(g) \longrightarrow CO_2(g) \quad \textbf{(3)}$$

Because Reaction 1 will always be accompanied by Reactions 2 and 3, the heat evolved will be a mixture of that for all three reactions. We might try to correct this quantity of heat mathematically for Reactions 2 and 3 in order to obtain the ΔH for Reaction 1, but fortunately there is a simpler way out of this difficulty.

Recall that enthalpy is a state function. This means that the enthalpy change for a reaction is independent of the details of how we do the reaction. The enthalpy change depends only on the initial state and the final state. Suppose we can accomplish an overall reaction through a series of reaction steps, each of which has a known enthalpy change. Each reaction step involves some quantity of heat, and the heat for the overall reaction will equal the sum of the heats of the individual steps. In other words, the sum of the enthalpy changes for the steps equals the enthalpy change of the overall reaction. This means that we can calculate the enthalpy change for one reaction from that of others.

Consider the determination of ΔH for Reaction 1. Let us imagine that we do this reaction in two steps. First, we burn 2 mol graphite with 2 mol oxygen to obtain 2 mol carbon dioxide. Then we *imagine* that we can decompose this carbon dioxide to 2 mol carbon monoxide and 1 mol oxygen.

$$2C(\text{graphite}) + 2O_2(g) \longrightarrow 2CO_2(g) \quad \textbf{(first step)}$$

$$2CO_2(g) \longrightarrow 2CO(g) + O_2(g) \quad \textbf{(second step)}$$

Note that the net result of these two steps is Reaction 1. That is, 2 mol of graphite burn to 2 mol of carbon monoxide. Overall, 1 mol of oxygen reacts (2 mol of O_2 reacts in the first step, but the second step produces 1 mol O_2, so the net amount of oxygen that reacts is 1 mol).

Now we need the enthalpy changes for the two steps. Note that the first step is simply the complete combustion of graphite. Its enthalpy change can be obtained calorimetrically.

$$C(\text{graphite}) + O_2(g) \longrightarrow CO_2(g);\ \Delta H = -393.5 \text{ kJ}$$

We obtain the first step by multiplying this equation by 2.

$$2C(\text{graphite}) + 2O_2(g) \longrightarrow 2CO_2(g);\ \Delta H = 2 \times (-393.5 \text{ kJ})$$

The second step is actually the reverse of the complete combustion of carbon monoxide, whose enthalpy change can be determined calorimetrically by burning carbon monoxide in an excess of oxygen.

$$2CO(g) + O_2(g) \longrightarrow 2CO_2(g);\ \Delta H = -566.0 \text{ kJ}$$

Therefore,

$$2CO_2(g) \longrightarrow 2CO(g) + O_2(g);\ \Delta H = +566.0 \text{ kJ}$$

The enthalpy change for Reaction 1 is the sum of the enthalpy changes for these two steps; that is, $\Delta H = 2 \times (-393.5 \text{ kJ}) + (+566.0 \text{ kJ}) = -221.0 \text{ kJ}$.

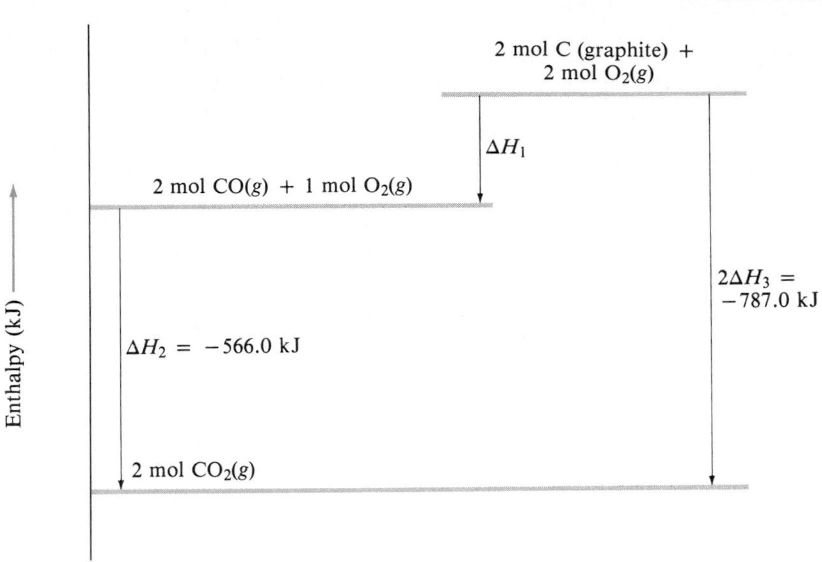

Figure 6.10
Enthalpy diagram illustrating Hess's law. The diagram shows the equality of the enthalpy change for the complete combustion of graphite to the sum of the enthalpy changes for the combustion of graphite to CO and the combustion of CO to CO_2.

$$2C(graphite) + O_2(g) \longrightarrow 2CO(g);\ \Delta H = -221.0\ kJ$$

In 1840, the Russian chemist Germain Henri Hess, a professor at the University of St. Petersburg, discovered by experiment the general result that we have been discussing. **Hess's law of heat summation** states that *if a chemical equation can be written as the sum of other equations, the ΔH of this overall equation equals a similar sum of the ΔH's for the other equations.*

Using this statement of Hess's law, we would set up the preceding problem as follows. Consider the Reactions 2 and 3, whose ΔH's are known.

$$CO(g) + 2O_2(g) \longrightarrow 2CO(g);\ \Delta H = -566.0\ kJ \qquad (2)$$

$$C(graphite) + O_2(g) \longrightarrow CO_2(g);\ \Delta H = -393.5\ kJ \qquad (3)$$

If we take Reaction 3 and multiply it by the factor 2 and add the reverse of Reaction 2, we obtain Reaction 1. (You can cancel as you would in algebraic equations.) If we add the corresponding enthalpy changes, we obtain the enthalpy change of Reaction 1. The following shows the layout of the calculation, with the enthalpy changes shown in the rightmost column.

$$\begin{array}{lll} 2C(graphite) + \cancel{2}O_2(g) \longrightarrow \cancel{2CO_2(g)} & & (-393.5\ kJ) \times (2) \\ \cancel{2CO_2(g)} \longrightarrow 2CO(g) + \cancel{O_2(g)} & & (-566.0\ kJ) \times (-1) \\ \hline 2C(graphite) + O_2(g) \longrightarrow 2CO(g) & & \Delta H = -221.0\ kJ \end{array}$$

Thus, the ΔH for Reaction 1 is −221.0 kJ. Figure 6.10 gives an enthalpy diagram that illustrates the application of Hess's law in this case. The next two examples give some practice in applying Hess's law.

Example 6.7 Applying Hess's Law (Simple Example)

Calculate the enthalpy change for the reaction

$$2S(s) + 3O_2(g) \longrightarrow 2SO_3(g)$$

given the following information:

$$S(s) + O_2(g) \longrightarrow SO_2(g);\ \Delta H = -297\ kJ \qquad (1)$$

$$2SO_3(g) \longrightarrow 2SO_2(g) + O_2(g);\ \Delta H = 198\ kJ \qquad (2)$$

Solution

We must multiply Reactions 1 and 2 by factors (perhaps re-

(continued)

versing one or both) such that, when they are added together, the desired reaction results. We can usually guess the proper factor by comparing the final equation with individual steps. In the final equation, S(s) occurs on the left with a coefficient of 2. Because Reaction 1 has S(s) on the left but with a coefficient of 1, we multiply Reaction 1 by the factor 2. Also, we note that the final equation has $2SO_3(g)$ on the right, whereas Reaction 2 has $2SO_3(g)$ on the left. Therefore, we reverse Reaction 2 (multiply it by -1). To obtain the enthalpy change for the final reaction, we must multiply ΔH's for Reactions 1 and 2 by the same factors and then add them together (in the column at the right).

$$\begin{array}{lll} 2S(s) + 2O_2(g) \longrightarrow 2SO_2(g) & & (-297\text{ kJ}) \times 2 \\ 2SO_2(g) + O_2(g) \longrightarrow 2SO_3(g) & & (198\text{ kJ}) \times (-1) \\ \hline 2S(s) + 3O_2(g) \longrightarrow 2SO_3(g) & & \Delta H = -792\text{ kJ} \end{array}$$

Exercise 6.8

Manganese metal can be obtained by reaction of manganese dioxide with aluminum.

$$4Al(s) + 3MnO_2(s) \longrightarrow 2Al_2O_3(s) + 3Mn(s)$$

What is ΔH for this reaction? Use the following data:

$$2Al(s) + \tfrac{3}{2}O_2(g) \longrightarrow Al_2O_3(s);\ \Delta H = -1676\text{ kJ}$$
$$Mn(s) + O_2(g) \longrightarrow MnO_2(s);\ \Delta H = -521\text{ kJ}$$

(See Problems 6.53 and 6.54.)

Example 6.8 Applying Hess's Law (More Complicated Case)

What is the enthalpy of reaction, ΔH, for the formation of tungsten carbide, WC, from the elements? (Tungsten carbide is very hard and is used to make cutting tools and rock drills.)

$$W(s) + C(\text{graphite}) \longrightarrow WC(s)$$

The enthalpy change for this reaction is difficult to measure directly, because the reaction occurs at 1400°C. However, the heats of combustion of the elements and of tungsten carbide can be measured:

$$2W(s) + 3O_2(g) \longrightarrow 2WO_3(s);\ \Delta H = -1680.6\text{ kJ} \quad \textbf{(1)}$$
$$C(\text{graphite}) + O_2(g) \longrightarrow CO_2(g);\ \Delta H = -393.5\text{ kJ} \quad \textbf{(2)}$$
$$2WC(s) + 5O_2(g) \longrightarrow 2WO_3(s) + 2CO_2(g);\ \Delta H = -2391.6\text{ kJ} \quad \textbf{(3)}$$

Solution

The desired reaction in this problem has W(s) on the left. Hence, we multiply Reaction 1 by $\frac{1}{2}$. (We also multiply ΔH for Reaction 1 by $\frac{1}{2}$.)

$$W(s) + \tfrac{3}{2}O_2(g) \longrightarrow WO_3(s);\quad \Delta H = \tfrac{1}{2} \times -1680.6\text{ kJ} = -840.3\text{ kJ}$$

Because the desired reaction has C(graphite) on the left side, we leave Reaction 2 as it is. The desired reaction has WC(s) on the right side. Hence, we reverse Reaction 3 and multiply it by $\frac{1}{2}$.

$$WO_3(s) + CO_2(g) \longrightarrow WC(s) + \tfrac{5}{2}O_2(g);\quad \Delta H = -\tfrac{1}{2} \times -2391.6\text{ kJ} = 1195.8\text{ kJ}$$

Note that the ΔH is obtained by multiplying the value for Reaction 3 by $-\frac{1}{2}$. Now these three reactions and the corresponding ΔH's are added together.

$$\begin{array}{lr} & \Delta H\text{, kJ} \\ W(s) + \tfrac{3}{2}O_2(g) \longrightarrow WO_3(s) & -840.3 \\ C(\text{graphite}) + O_2(g) \longrightarrow CO_2(g) & -393.5 \\ WO_3(s) + CO_2(g) \longrightarrow WC(s) + \tfrac{5}{2}O_2(g) & 1195.8 \\ \hline \end{array}$$

$$W(s) + \cancel{\tfrac{3}{2}O_2(g)} + C(\text{graphite}) + \cancel{O_2(g)} + \cancel{WO_3(s)} + \cancel{CO_2(g)} \longrightarrow \cancel{WO_3(s)} + \cancel{CO_2(g)} + WC(s) + \cancel{\tfrac{5}{2}O_2(g)}$$
$$\Delta H = -38.0\text{ kJ}$$

That is,

$$W(s) + C(\textit{graphite}) \longrightarrow WC(s);\ \mathbf{\Delta H = -38.0\ kJ}$$

Exercise 6.9

Calculate the enthalpy change for breaking the four C—H bonds in methane, CH_4.

$$CH_4(g) \longrightarrow C(g) + 4H(g)$$

Use the following information:

$$C(\text{graphite}) \longrightarrow C(g);\ \Delta H = 715.0\ \text{kJ}$$
$$H_2(g) \longrightarrow 2H(g);\ \Delta H = 436.0\ \text{kJ}$$
$$C(\text{graphite}) + O_2(g) \longrightarrow CO_2(g);\ \Delta H = -393.5\ \text{kJ}$$
$$2H_2(g) + O_2(g) \longrightarrow 2H_2O(l);\ \Delta H = -571.7\ \text{kJ}$$
$$CH_4(g) + 2O_2(g) \longrightarrow CO_2(g) + 2H_2O(l);\ \Delta H = -890.3\ \text{kJ}$$

(See Problems 6.55, 6.56, 6.57, and 6.58.)

6.8 STANDARD ENTHALPIES OF FORMATION

Because we can use Hess's law to relate the enthalpy changes of some reactions to the enthalpy changes of others, we do not need to tabulate the enthalpy changes of all possible reactions. We normally list enthalpy changes for only certain types of reactions. We also list enthalpy changes generally only for certain standard thermodynamic conditions (which are not identical to the standard conditions for gases, STP).

The term **standard state** refers to *the standard thermodynamic conditions chosen for substances when listing or comparing thermodynamic data: 1 atm pressure and the specified temperature (usually 25°C).*● These standard conditions are indicated by a superscript degree sign (°). Thus, the enthalpy change for a reaction in which reactants in their standard states yield products in their standard states is denoted $\Delta H°$ (''delta H degree,'' but often read as ''delta H zero''). The quantity $\Delta H°$ is called the *standard enthalpy of reaction.*

● The International Union of Pure and Applied Chemistry (IUPAC) recommends that the standard pressure be 1 bar (1×10^5 Pa). Thermodynamic tables are becoming available for 1 bar pressure, and in the future such tables will probably replace those for 1 atm.

As we will show in this section, it is sufficient to tabulate just the enthalpy changes for formation reactions—that is, for reactions in which compounds are formed from their elements. To specify the formation reaction precisely, however, we must specify the exact form of each element.

Some elements exist in the same physical state in two or more distinct forms. For example, oxygen in any of the physical states occurs as dioxygen (commonly called simply oxygen) with O_2 molecules, and as ozone, with O_3 molecules. Dioxygen gas is odorless; ozone gas has a characteristic pungent odor. Solid carbon has two principal crystalline forms, graphite and diamond. Graphite is a soft, black, crystalline substance; diamond is a hard, usually colorless crystal. The elements oxygen and carbon are said to exist in different allotropic forms. An **allotrope** is *one of two or more distinct forms of an element in the same physical state.*

The **reference form** of an element for the purpose of specifying the formation reaction is *the stablest form (physical state and allotrope) of the element under standard thermodynamic conditions.* The reference form of oxygen at 25°C is $O_2(g)$; the reference form of carbon at 25°C is graphite.

Table 6.2 lists standard enthalpies of formation of substances (a longer table is given in Appendix C). The **standard enthalpy of formation** (also called the **standard heat of formation**) of a substance, denoted $\Delta H_f°$, is *the enthalpy change for the formation of one mole of the substance in its standard state from its elements in their reference form and in their standard states.*

To understand this definition, consider the standard enthalpy of formation of liquid water. Note that the stablest forms of hydrogen and oxygen at 1 atm and 25°C are $H_2(g)$ and $O_2(g)$, respectively. These are therefore the reference forms of

Table 6.2 Standard Enthalpies of Formation (at 25°C)*

Formula	ΔH_f° (kJ/mol)
$e^-(g)$	0
Hydrogen	
$H^+(aq)$	0
$H(g)$	218.0
$H_2(g)$	0
Sodium	
$Na^+(g)$	609.8
$Na^+(aq)$	−239.7
$Na(g)$	107.8
$Na(s)$	0
$NaCl(s)$	−411.1
$NaHCO_3(s)$	−947.7
$Na_2CO_3(s)$	−1130.8
Calcium	
$Ca^{2+}(aq)$	−543.0
$Ca(s)$	0
$CaO(s)$	−635.1
$CaCO_3(s)$ (calcite)	−1206.9
Carbon	
$C(g)$	715.0
C(graphite)	0
C(diamond)	1.9
$CO(g)$	−110.5
$CO_2(g)$	−393.5
$HCO_3^-(aq)$	−691.1
$CH_4(g)$	−74.9
$C_2H_4(g)$	52.5
$C_2H_6(g)$	−84.7
$C_6H_6(l)$	49.0
$HCHO(g)$	−116
$CH_3OH(l)$	−238.6
$CS_2(g)$	117
$CS_2(l)$	87.9
$HCN(g)$	135
$HCN(l)$	105
$CCl_4(g)$	−96.0
$CCl_4(l)$	−139
$CH_3CHO(g)$	−166
$C_2H_5OH(l)$	−277.6
Silicon	
$Si(s)$	0
$SiO_2(s)$	−910.9
$SiF_4(g)$	−1548
Lead	
$Pb(s)$	0
$PbO(s)$	−219
$PbS(s)$	−98.3
Nitrogen	
$N(g)$	473
$N_2(g)$	0
$NH_3(g)$	−45.9
$NH_4^+(aq)$	−132.8
$NO(g)$	90.3
$NO_2(g)$	33.2
$HNO_3(aq)$	−206.6
Oxygen	
$O(g)$	249.2
$O_2(g)$	0
$O_3(g)$	143
$OH^-(aq)$	−229.9
$H_2O(g)$	−241.8
$H_2O(l)$	−285.8
Sulfur	
$S(g)$	279
$S_2(g)$	129
S_8(rhombic)	0
S_8(monoclinic)	2
$SO_2(g)$	−296.8
$H_2S(g)$	−20
Fluorine	
$F^-(g)$	−255.6
$F^-(aq)$	−329.1
$F_2(g)$	0
$HF(g)$	−273
Chlorine	
$Cl^-(aq)$	−167.5
$Cl(g)$	121.0
$Cl_2(g)$	0
$HCl(g)$	−92.3
Bromine	
$Br^-(g)$	−218.9
$Br^-(aq)$	−120.9
$Br_2(l)$	0
Iodine	
$I^-(g)$	−194.7
$I^-(aq)$	−55.9
$I_2(s)$	0
Silver	
$Ag^+(g)$	1026.4
$Ag^+(aq)$	105.9
$Ag(s)$	0
$AgF(s)$	−203
$AgCl(s)$	−127.0
$AgBr(s)$	−99.5
$AgI(s)$	−62.4

*See Appendix C for additional values.

the elements. We write the formation reaction for 1 mole of liquid water as follows:

$$H_2(g) + \frac{1}{2}O_2(g) \longrightarrow H_2O(l)$$

The standard enthalpy change for this reaction is −285.8 kJ per mole of H_2O. Therefore, the thermochemical equation is

$$H_2(g) + \frac{1}{2}O_2(g) \longrightarrow H_2O(l);\ \Delta H_f^\circ = -285.8 \text{ kJ}$$

The values of standard enthalpies of formation listed in Table 6.2 and in other tables are determined by direct measurement in some cases and by applying Hess's law in others. Oxides, such as water, can often be determined by direct calorimetric measurement of the combustion reaction. If you look back at Example 6.8, you will see an illustration of how Hess's law can be used to obtain the enthalpy of formation of tungsten carbide, WC. We will also give an example later in this section in which we obtain the standard enthalpy of formation of a substance using an equation derived from Hess's law.

Note that the standard enthalpy of formation of an element will depend on the form of the element. For example, the ΔH_f° for diamond equals the enthalpy change from the stablest form of carbon (graphite) to diamond. The thermochemical equation is

$$\text{C(graphite)} \longrightarrow \text{C(diamond)};\ \Delta H_f^\circ = 1.9 \text{ kJ}$$

On the other hand, the ΔH_f° for graphite equals zero. Note the values of ΔH_f° for the elements listed in Table 6.2; the reference forms will have zero values.

Now let us see how to use standard enthalpies of formation (listed in Table 6.2) to find the standard enthalpy change for a reaction. We will first look at this problem from the point of view of Hess's law. But when we are done, we will note a pattern in the result, which will allow us to state a simple formula for solving this type of problem.

Consider the equation

$$CH_4(g) + 4Cl_2(g) \longrightarrow CCl_4(l) + 4HCl(g);\ \Delta H^\circ = ?$$

From Table 6.2 we pick out the enthalpies of formation for $CH_4(g)$, $CCl_4(l)$, and $HCl(g)$. We can then write the following thermochemical equations:

$$\text{C(graphite)} + 2H_2(g) \longrightarrow CH_4(g);\ \Delta H_f^\circ = -74.9 \text{ kJ} \qquad \textbf{(1)}$$

$$\text{C(graphite)} + 2Cl_2(g) \longrightarrow CCl_4(l);\ \Delta H_f^\circ = -139 \text{ kJ} \qquad \textbf{(2)}$$

$$\frac{1}{2}H_2(g) + \frac{1}{2}Cl_2(g) \longrightarrow HCl(g);\ \Delta H_f^\circ = -92.3 \text{ kJ} \qquad \textbf{(3)}$$

We now apply Hess's law. Since we want CH_4 to appear on the left, and CCl_4 and 4HCl on the right, we reverse Reaction 1 and add Reaction 2 plus 4 × Reaction 3.

$CH_4(g) \longrightarrow \text{C(graphite)} + 2H_2(g)$	(−74.9 kJ) × (−1)
$\text{C(graphite)} + 2Cl_2(g) \longrightarrow CCl_4(l)$	(−139 kJ) × (1)
$2H_2(g) + 2Cl_2(g) \longrightarrow 4HCl(g)$	(−92.3 kJ) × (4)
$CH_4(g) + 4Cl_2(g) \longrightarrow CCl_4(l) + 4HCl(g)$	$\Delta H^\circ = -433$ kJ

The setup of this calculation can be greatly simplified once we closely examine what we are doing. Note that the ΔH_f° for each compound is multiplied by its coefficient in the chemical equation whose ΔH° we are calculating. Moreover, the ΔH_f° for each reactant is multiplied by a negative sign. Let us symbolize the enthalpy of formation of a substance by writing the formula in parentheses following ΔH_f°. Then our calculation can be written as follows:

$$\Delta H^\circ = [\Delta H_f^\circ(CCl_4) + 4\Delta H_f^\circ(HCl)] - [\Delta H_f^\circ(CH_4) + 4\Delta H_f^\circ(Cl_2)]$$
$$= [(-139) + 4(-92.3)] \text{ kJ} - [(-74.9) + (0)] \text{ kJ}$$
$$= -433 \text{ kJ}$$

In general, we can calculate the ΔH° for a reaction by the equation

$$\Delta H^\circ = \Sigma\, n\Delta H_f^\circ(\text{products}) - \Sigma\, m\Delta H_f^\circ(\text{reactants})$$

Here Σ is the mathematical symbol meaning "the sum of," and m and n are the coefficients of the substances in the chemical equation.

The next two examples illustrate the calculation of enthalpies of reaction from standard enthalpies of formation.

Example 6.9 Calculating the Heat of Phase Transition from Standard Enthalpies of Formation

Use values of ΔH_f° to calculate the heat of vaporization, ΔH_{vap}°, of carbon disulfide at 25°C. The vaporization process is

$$CS_2(l) \longrightarrow CS_2(g)$$

Solution

The vaporization process can be treated just like a chemical reaction. $CS_2(l)$ is the "reactant" and $CS_2(g)$ is the "product." It is convenient to read the values of ΔH_f° for substances from Table 6.2 and record them under the formulas in the equation, multiplying them by the coefficients in the equation (here, all 1's).

$$\underset{88}{CS_2(l)} \longrightarrow \underset{117}{CS_2(g)} \quad \text{(kJ)}$$

The calculation is

$$\Delta H_{vap}^\circ = \Sigma\, n\Delta H_f^\circ(\text{products}) - \Sigma\, m\Delta H_f^\circ(\text{reactants})$$
$$= \Delta H_f^\circ[CS_2(g)] - \Delta H_f^\circ[CS_2(l)]$$
$$= (117 - 88) \text{ kJ} = 29 \text{ kJ}$$

Exercise 6.10

Calculate the heat of vaporization, ΔH_{vap}°, of water, using standard enthalpies of formation (Table 6.2). (See Problems 6.59 and 6.60.)

Example 6.10 Calculating the Enthalpy of Reaction from Standard Enthalpies of Formation

Large quantities of ammonia are used to prepare nitric acid (see the Profile on nitric acid at the end of the chapter). The first step consists of the catalytic oxidation of ammonia to nitric oxide, NO.

$$4NH_3(g) + 5O_2(g) \xrightarrow{Pt} 4NO(g) + 6H_2O(g)$$

What is the standard enthalpy change for this reaction? Use Table 6.2 for data.

Solution

Here is the equation with the ΔH_f°'s recorded beneath it:

$$\underset{4(-45.9)}{4NH_3(g)} + \underset{5(0)}{5O_2(g)} \longrightarrow \underset{4(90.3)}{4NO(g)} + \underset{6(-241.8)}{6H_2O(g)} \quad \text{(kJ)}$$

We ignore the catalyst because its only effect is to speed up the reaction. The calculation is

$$\Delta H^\circ = \Sigma\, n\Delta H_f^\circ(\text{products}) - \Sigma\, m\Delta H_f^\circ(\text{reactants})$$
$$= [4\Delta H_f^\circ(NO) + 6\Delta H_f^\circ(H_2O)] - [4\Delta H_f^\circ(NH_3) + 5\Delta H_f^\circ(O_2)]$$
$$= [4(90.3) + 6(-241.8)] \text{ kJ} - [4(-45.9) + 5(0)] \text{ kJ}$$
$$= \mathbf{-906.0 \text{ kJ}}$$

Be very careful of arithmetical signs—they are a likely source of mistakes. Also pay particular attention to the state of each substance. Here, for example, we must use the ΔH_f° for $H_2O(g)$, not that for $H_2O(l)$.

Exercise 6.11

Calculate the enthalpy change for the following reaction:

$$3NO_2(g) + H_2O(l) \longrightarrow 2HNO_3(aq) + NO(g)$$

Use standard enthalpies of formation. (See Problems 6.61, 6.62, 6.63, and 6.64.)

Enthalpies of formation can also be defined for ions. In this case, because it is not possible to make thermal measurements on individual ions, we must arbitrarily define the standard enthalpy of formation of one ion as zero. Then values for all other ions can be deduced from calorimetric data. By convention, the standard enthalpy of formation of $H^+(aq)$ is taken as zero. Values of ΔH_f° for some ions are given in Table 6.2.

Exercise 6.12

Calculate the standard enthalpy change for the reaction of an aqueous solution of barium hydroxide, $Ba(OH)_2$, with an aqueous solution of ammonium nitrate, NH_4NO_3, at 25°C. (Figure 6.1 illustrated this reaction using solids instead of solutions.) The equation is

$$[Ba^{2+}(aq) + 2OH^-(aq)] + 2[NH_4^+(aq) + NO_3^-(aq)] \longrightarrow 2NH_3(g) + 2H_2O(l) + [Ba^{2+}(aq) + 2NO_3^-(aq)]$$

(See Problems 6.65 and 6.66.)

In the preceding examples, we used standard enthalpies of formation to calculate the enthalpy of reaction. We can also obtain the standard enthalpy of formation of a substance in terms of the enthalpy change for some reaction of that substance. The next example shows how this is done.

Example 6.11 Obtaining the Standard Enthalpy of Formation from an Enthalpy of Reaction

The head of a ''strike-anywhere'' match contains tetraphosphorus trisulfide, P_4S_3 (plus an oxidizer, $KClO_3$, and fillers). When tetraphosphorus trisulfide burns in excess oxygen, it evolves 3677 kJ of heat per mole of P_4S_3 at a constant pressure of 1 atm. The thermochemical equation is

$$P_4S_3(s) + 8O_2(g) \longrightarrow P_4O_{10}(s) + 3SO_2(g); \ \Delta H^\circ = -3677 \text{ kJ}$$

Calculate the standard enthalpy of formation of $P_4S_3(s)$ using this enthalpy of reaction plus ΔH_f° values for other reactants and products. The ΔH_f° for $P_4O_{10}(s)$ is -2942 kJ/mol; other values are in Table 6.2.

Solution

The enthalpy of reaction in terms of standard enthalpies of formation is

$$\Delta H^\circ = \Sigma\, n\Delta H_f^\circ(\text{products}) - \Sigma\, m\Delta H_f^\circ(\text{reactants})$$
$$= [\Delta H_f^\circ(P_4O_{10}) + 3\Delta H_f^\circ(SO_2)] - [\Delta H_f^\circ(P_4S_3) + 8\Delta H_f^\circ(O_2)]$$

Substituting data, we obtain

$$-3677 \text{ kJ} = [-2942 + 3(-296.8)] \text{ kJ} - [\Delta H_f^\circ(P_4S_3) + 8(0) \text{ kJ}]$$

Now we solve for $\Delta H_f^\circ(P_4S_3)$.

$$\Delta H_f^\circ(P_4S_3) = 3677 \text{ kJ} + [-2942 + 3(-296.8)] \text{ kJ} - [8(0)] \text{ kJ}$$
$$= \mathbf{-155 \text{ kJ}}$$

Exercise 6.13

Pyrite is a mineral whose chemical name is iron(II) disulfide, FeS_2. It is a compound of Fe^{2+} ions and disulfide ions, S_2^{2-}. A sample of pyrite was burned in oxygen in a calorimeter to $Fe_2O_3(s)$ and $SO_2(g)$, from which it was determined that the combustion of 1 mol of pyrite evolves 828 kJ of heat at a constant pressure of 1 atm. Standard enthalpies of formation of the products can be found in Appendix C. Use this information to calculate the standard enthalpy of formation of $FeS_2(s)$. (See Problems 6.67 and 6.68.)

6.9 FUELS—FOODS, COMMERCIAL FUELS, AND ROCKET FUELS

A fuel is any substance that is burned or similarly reacted to provide heat and other forms of energy. The earliest use of fuels for heat came with the control of fire, which was achieved about 750,000 years ago. This major advance allowed the human species to migrate from tropical savannas and eventually to inhabit most of the earth. Through cooking, fire also increased the variety of edible food supplies and provided some protection of the food from bacterial decay. In the mid-eighteenth century, the discovery of the steam engine, which converts the chemical energy latent in fuels to mechanical energy, ushered in the Industrial Revolution. Today fuels not only heat our homes and move our cars but are absolutely necessary for every facet of modern technology. For example, fuels generate the electricity required for our modern computing and communications technologies, and they propel the rocket engines that make possible our explorations of outer space. In this section we will look at foods as fuels, at fossil fuels (which include gas, oil, and coal), at coal gasification and liquefaction, and at rocket fuels.

Foods as Fuels

Foods fill three needs of the body: They supply substances for the growth and repair of tissue, they supply substances for the synthesis of compounds used in the regulation of body processes, and they supply energy. About 80 percent of the energy we need is for heat. The rest is used for muscular action, chemical processes, and other body processes.● The body generates energy from food by the same overall process as combustion, so the overall enthalpy change is the same as the heat of combustion, which can be determined in a calorimeter. We can get some idea of the energy available from carbohydrate foods by looking at a typical one, glucose ($C_6H_{12}O_6$). The thermochemical equation for the combustion of glucose is

● The human body requires about as much energy in a day as does a 100-watt light bulb.

$$C_6H_{12}O_6(s) + 6O_2(g) \longrightarrow 6CO_2(g) + 6H_2O(l);\ \Delta H^\circ = -2803 \text{ kJ}$$

One gram of glucose yields 15.6 kJ (3.72 kcal) of heat when burned.

A representative fat is glyceryl trimyristate, $C_{45}H_{86}O_6$. The equation for its combustion is

$$C_{45}H_{86}O_6(s) + \tfrac{127}{2}O_2(g) \longrightarrow 45CO_2(g) + 43H_2O(l);\ \Delta H^\circ = -27820 \text{ kJ}$$

One gram of this fat yields 38.5 kJ (9.20 kcal) of heat when burned. The average values quoted for carbohydrates and fats are 4.0 kcal/g and 9.0 kcal/g, respectively.● Note that fats contain more than twice the fuel value per gram as do carbohydrates. Thus, by storing its fuel as fat, the body can store more fuel for a given mass of body tissue.

● In the popular literature of nutrition, the kilocalorie is referred to as the Calorie. Thus, these values are given as 4.0 Calories and 9.0 Calories.

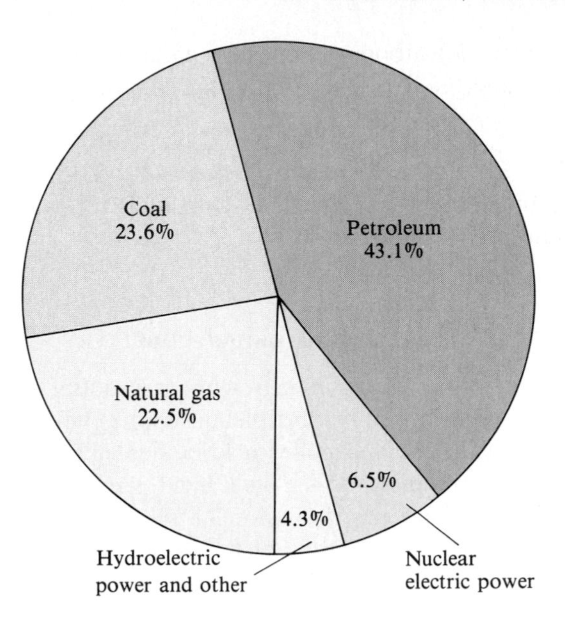

Figure 6.11
Sources of energy consumed in the United States. Data are from *Monthly Energy Review,* U.S. Department of Energy, December 1987.

Fossil Fuels

All of the fossil fuels in existence today were created millions of years ago when aquatic plants and animals were buried and compressed by layers of sediment at the bottoms of swamps and seas. Over time this organic matter was converted by bacterial decay and pressure to petroleum (oil), gas, and coal, which today are our major fuel sources. Figure 6.11 gives the percentages of the total energy consumed in the United States from various sources. Together, the fossil fuels account for nearly 90% of the total.

Anthracite, or hard coal, which is the oldest variety of coal, was laid down as much as 250 million years ago and may contain over 80% carbon. Bituminous coal, a younger variety of coal, has between 45% and 65% carbon. Fuel values of coals are rated in Btu's (British thermal units) per pound, which are essentially heats of combustion per pound of coal. A typical value is 13,200 Btu/lb. A Btu equals 1054 J, so 1 Btu/lb equals 2.32 J/g. Therefore, the combustion of coal in oxygen yields about 30.6 kJ/g. We can compare this value with the heat of combustion of pure carbon (graphite).

$$C(graphite) + O_2(g) \longrightarrow CO_2(g);\ \Delta H^\circ = -393.5 \text{ kJ}$$

The value given in the equation is for 1 mole (12.0 g) of carbon. Per gram, we get 32.8 kJ/g, which is comparable with the values obtained for coal.

Natural gas and petroleum together account for nearly three-quarters of the fossil fuels consumed per year. They are very convenient fluid fuels, being easily transportable and having no ash. Purified natural gas is primarily methane, CH_4, but it also contains small amounts of ethane, C_2H_6, propane, C_3H_8, and butane, C_4H_{10}. We would expect the fuel values of natural gas to be close to that for the heat of combustion of methane:

$$CH_4(g) + 2O_2(g) \longrightarrow CO_2(g) + 2H_2O(g);\ \Delta H^\circ = -802 \text{ kJ}$$

This value of ΔH° is equivalent to 50.1 kilojoules per gram of fuel.

Petroleum is a very complicated mixture of compounds. Gasoline, which is obtained from petroleum by chemical and physical processes, contains many different hydrocarbons (compounds of carbon and hydrogen). One such hydrocarbon is octane, C_8H_{18}. The combustion of octane evolves 5074 kJ per mole.

$$C_8H_{18}(l) + \frac{25}{2}O_2(g) \longrightarrow 8CO_2(g) + 9H_2O(g);\ \Delta H^\circ = -5074\ \text{kJ}$$

This value of ΔH° is equivalent to 44.4 kJ/g. These combustion values indicate another reason why the fluid fossil fuels are popular: They release more heat per gram than coal does.

Coal Gasification and Liquefaction

The major problem with petroleum and natural gas is their relative short supply. It has been estimated that petroleum supplies will be 80% depleted by about the year 2030. Natural gas supplies may be depleted even sooner.

Coal supplies, on the other hand, are sufficient to last several more centuries. This abundance has spurred much research into developing commercial methods for converting coal to the more easily handled liquid and gaseous fuels. Most of these methods begin by converting the coal to carbon monoxide, CO. One way involves the water–gas reaction.

$$C(s) + H_2O(g) \longrightarrow CO(g) + H_2(g)$$

In this reaction, steam is passed over hot coal. Once a mixture of carbon monoxide and hydrogen is obtained, it can be transformed by various reactions into useful products. For example, in the methanation reaction (discussed at length in Chapter 15), this mixture is reacted over a catalyst to give methane.

$$CO(g) + 3H_2(g) \longrightarrow CH_4(g) + H_2O(g)$$

Different catalysts and different reaction conditions result in liquid fuels. An added advantage of coal gasification and coal liquefaction is that sulfur, whose burning is a major source of air pollution and acid rain, can be removed during the process.

Rocket Fuels

Rockets are self-contained missiles propelled by the ejection of gases from an orifice. Usually these are hot gases propelled from the rocket by the reaction of a fuel with an oxidizer. Rockets are believed to have originated with the Chinese—perhaps before the thirteenth century, which is when they began to appear in Europe. However, it was not until the twentieth century that rocket propulsion began to be studied seriously, and since World War II rockets have become major weapons. Space exploration with satellites propelled by rocket engines began in 1957 with the Russian satellite *Sputnik I*. Today weather and communications satellites are regularly put into orbit about the earth using rocket engines.

One of the factors determining which fuel and oxidizer to use is the mass of the fuel and oxidizer required. We have already seen that natural gas and gasoline have higher fuel values per gram than does coal. The difference is caused by the higher hydrogen content of natural gas and gasoline. Hydrogen is the element of lowest density, and at the same time it reacts exothermically with oxygen to give water. You might expect hydrogen and oxygen to be an ideal fuel-oxidizer combination. The thermochemical equation for the combustion of hydrogen is

$$H_2(g) + \frac{1}{2}O_2(g) \longrightarrow H_2O(g);\ \Delta H^\circ = -242\ \text{kJ}$$

This value of $\Delta H°$ is equivalent to 120 kJ/g of fuel (H_2) compared with 50 kJ/g of methane. The second and third stages of the *Saturn V* launch vehicle that sent a three-man Apollo crew to the moon used a hydrogen/oxygen system. The launch vehicle contained liquid hydrogen (boiling at $-253°C$) and liquid oxygen, or LOX (boiling at $-183°C$). The first stage of liftoff used kerosene and oxygen, and an unbelievable 550 metric tons (550×10^3 kg) of kerosene were burned in 2.5 minutes. It is interesting to calculate the average rate of energy production in this 2.5-minute interval. Kerosene is approximately $C_{12}H_{26}$. The thermochemical equation is

$$C_{12}H_{26}(l) + \tfrac{37}{2}O_2(g) \longrightarrow 12CO_2(g) + 13H_2O(g);\ \Delta H° = -7513 \text{ kJ}$$

This value of $\Delta H°$ is equivalent to 44.1 kJ/g. Thus, 550×10^6 g of fuel generated 2.42×10^{10} kJ in 150 s (2.5 min). Each second, the average energy produced was 1.61×10^{11} J. This is equivalent to 1.61×10^{11} watts, or 216 million horsepower (1 horsepower equals 745.7 watts, or J/s).

The landing module for the Apollo mission used a fuel made of hydrazine, N_2H_4, and a derivative of hydrazine. The oxidizer was dinitrogen tetroxide, N_2O_4. These substances are normally liquids and therefore are easier to store than liquid hydrogen and oxygen. The reaction of the oxidizer with hydrazine is

$$2N_2H_4(l) + N_2O_4(l) \longrightarrow 3N_2(g) + 4H_2O(g);\ \Delta H° = -1049 \text{ kJ}$$

Solid propellants are also used as rocket fuels. The mixture used in the booster rockets of the *Columbia* space shuttle (Figure 6.12) was a fuel of polymers and rubber with aluminum metal powder. An oxidizer of ammonium perchlorate, NH_4ClO_4, was mixed with the fuel.

Figure 6.12
The launching of the *Columbia* space shuttle. The solid fuel for the booster rockets is a mixture of aluminum metal powder and polymers; ammonium perchlorate is the oxidizer. A cloud of aluminum oxide forms as the rockets burn.

Profile of a Chemical

NITRIC ACID (an Industrial Acid)

Nitric acid, HNO_3, was known to early chemists as *aqua fortis,* from the Latin meaning "strong water." The name presumably comes from the fact that nitric acid reacts with most metals, including copper and silver which are not attacked by other acids.

Nitric acid is a strong acid, ionizing completely in water.

$$HNO_3(aq) + H_2O(l) \longrightarrow H_3O^+(aq) + NO_3^-(aq)$$

The acid solution reacts with bases to produce the corresponding salts, as expected for an acid.

$$HNO_3(aq) + NH_3(aq) \longrightarrow NH_4^+(aq) + NO_3^-(aq)$$

However, the reaction of nitric acid with metals is quite different from the normal reaction of an acid with a metal. In general, acids react with active metals in a displacement reaction to produce hydrogen gas. Certain metals, such as copper, are unreactive with most acids. Nitric acid may produce some hydrogen gas with an active metal, but the main product (other than the metal ion) is one of the nitrogen oxides or the ammonium ion, depending on the metal and the acid concentration. Zinc metal with dilute nitric acid gives nitrous oxide (dinitrogen monoxide), N_2O.

$$4Zn(s) + 10H^+(aq) + 2NO_3^-(aq) \longrightarrow 4Zn^{2+}(aq) + N_2O(g) + 5H_2O(l)$$

With copper metal, the dilute acid gives nitric oxide (nitrogen monoxide), NO.

$$3Cu(s) + 8H^+(aq) + 2NO_3^-(aq) \longrightarrow 3Cu^{2+}(aq) + 2NO(g) + 4H_2O(l)$$

Nitric acid was originally prepared by heating a mixture of potassium nitrate and concentrated sulfuric acid (Figure 6.14). Nitric acid boils at 83°C under 1 atm pressure (sulfuric acid boils at a much higher temperature, about 290°C under 1 atm). Consequently, nitric acid can be distilled from the reaction vessel.

$$KNO_3(s) + H_2SO_4(l) \longrightarrow KHSO_4(s) + HNO_3(g)$$

The present industrial method of preparing nitric acid was discovered in 1902 by the German chemist Wilhelm Ostwald, who received the Nobel Prize in chemistry in 1909 for his work in catalysis. The *Ostwald process* involves the burning of ammonia in the presence of a platinum catalyst to produce nitric oxide, NO. (Figure 6.15 shows a platinum-rhodium catalyst gauze being installed in an industrial ammonia burner.) Nitric oxide reacts with more oxygen to produce nitrogen dioxide, NO_2, and this oxide reacts with water to produce

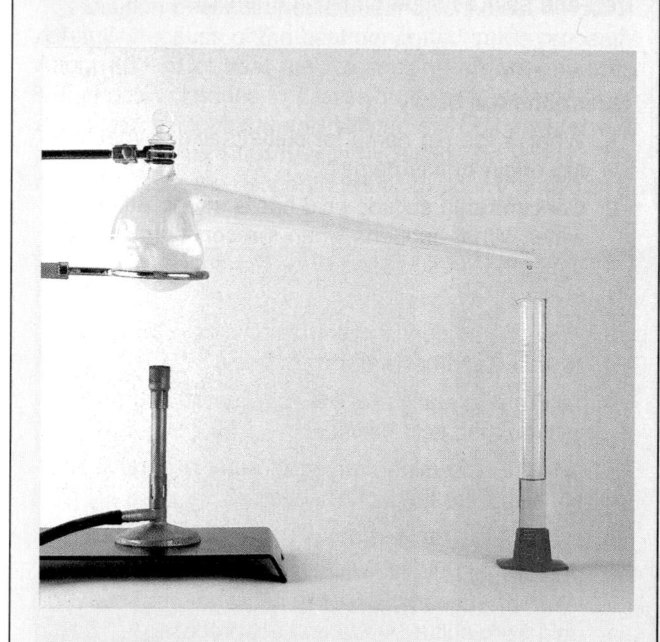

Figure 6.14
The preparation of nitric acid. *The closed vessel, called a retort, containing a mixture of potassium nitrate and concentrated sulfuric acid is heated, and nitric acid distills over. It has a brownish color because of dissolved nitrogen dioxide impurity.*

Figure 6.15
A catalyst gauze in an ammonia burner. *Workers replace a platinum-rhodium catalyst in a commercial ammonia burner, which is the heart of a nitric acid plant.*

Figure 6.16
Catalytic oxidation of ammonia (Ostwald process). *Once heated, a platinum coil continues to glow from the reaction of ammonia gas and oxygen (from air) at the surface of the platinum. The gaseous reaction catalyzed by platinum is* $4NH_3(g) + 5O_2(g) \rightarrow 4NO(g) + 6H_2O(g)$.

nitric acid, HNO_3, plus nitric oxide, which is recirculated to the second step.

$$4NH_3(g) + 5O_2(g) \xrightarrow{Pt} 4NO(g) + 6H_2O(g);\ \Delta H^\circ = -906\ \text{kJ}$$
$$2NO(g) + O_2(g) \longrightarrow 2NO_2(g);\ \Delta H^\circ = -114\ \text{kJ}$$
$$3NO_2(g) + H_2O(l) \longrightarrow 2HNO_3(aq) + NO(g);\ \Delta H^\circ = -137\ \text{kJ}$$

Note that the catalytic reaction of ammonia and oxygen is very exothermic. Figure 6.16 shows a laboratory demonstration of this reaction. A platinum coil on the end of a glass rod inserted in a rubber stopper is heated and then placed in a flask containing aqueous ammonia. Ammonia gas from the solution reacts with oxygen in the air at the surface of the platinum coil. The coil continues to glow red from the heat of reaction as long as reactants are still available.

The concentrated nitric acid available commercially is about 70% HNO_3 by mass. Nitric acid solutions are colorless when pure but decompose when exposed to light, taking on the brown color of nitrogen dioxide.

$$4HNO_3(aq) \longrightarrow 4NO_2(g) + 2H_2O(l) + O_2(g)$$

Pure nitric acid (100%) is unstable above 0°C and decomposes by the previous reaction.

Nitric acid is used to make plastics including nylon and polyurethane. The largest use of nitric acid, however, is in the production of ammonium nitrate, NH_4NO_3. Ammonium nitrate is used as a nitrogen fertilizer and in the manufacture of explosive mixtures (for example, a fuel oil-ammonium nitrate mixture is used as a mining explosive). Nitric acid is itself used to manufacture explosive compounds. Nitroglycerin (glyceryl trinitrate), $C_3H_5(NO_3)_3$, is prepared by the reaction of nitric acid and glycerol, $C_3H_5(OH)_3$, which can be made from fats or from petroleum. The reaction is carried out in the presence of sulfuric acid.

$$C_3H_5(OH)_3 + 3HNO_3 \xrightarrow{H_2SO_4} C_3H_5(NO_3)_3 + 3H_2O$$

Nitroglycerin is hazardous to handle, because it is sensitive to shock, exploding by the following reaction:

$$4C_3H_5(NO_3)_3(l) \longrightarrow 6N_2(g) + 10H_2O(g) + 12CO_2(g) + O_2(g);\ \Delta H^\circ = -5.720 \times 10^3\ \text{kJ}$$

Note the generation of heat and large volumes of gas. The gases occupy more than 1200 times the volume of the original nitroglycerin. Alfred Nobel (1833–1896), a Swedish chemist and industrialist, discovered that when nitroglycerin was mixed with clay and similar materials, it was much less shock-sensitive. The mixture is called dynamite. Much of Nobel's fortune from his explosives industry was left to establish the Nobel prizes.

Questions for Study

1. What was the old name for nitric acid? What is the origin of that name?
2. How do the reactions of nitric acid with metals differ from those of most acids? What is the main product (other than the metal ion)?
3. Write the chemical equation for the original preparation of nitric acid. How is nitric acid recovered from the reaction mixture?
4. Write the chemical equations for the preparation of nitric acid from ammonia by the Ostwald process.
5. What happens to colorless solutions of nitric acid when they are exposed to light? Write the equation for the reaction.
6. What is the largest use of nitric acid? Write the chemical equation for the reaction involved.
7. What are some other uses of nitric acid?
8. What is dynamite? Who discovered it? What was the significance of the discovery?

A Checklist for Review

IMPORTANT TERMS

energy (6.1)
kinetic energy (6.1)
joule (J) (6.1)
calorie (cal) (6.1)
potential energy (6.1)
internal energy (6.1)
law of conservation of energy (6.1)
thermodynamic system (or **system**) (6.2)
surroundings (6.2)
heat (6.2)
heat of reaction (6.2)
exothermic process (6.2)
endothermic process (6.2)
enthalpy (6.3)
state function (6.3)
enthalpy of reaction (6.3)
thermochemical equation (6.4)
heat capacity (6.6)
specific heat capacity (specific heat) (6.6)
calorimeter (6.6)
Hess's law of heat summation (6.7)
standard state (6.8)
allotrope (6.8)
reference form (6.8)
standard enthalpy of formation (standard heat of formation) (6.8)

KEY EQUATIONS

$E_k = \frac{1}{2}mv^2$

$1 \text{ cal} = 4.184 \text{ J}$

$\Delta H = q_p$

$q = C\Delta t$

$q = s \times m \times \Delta t$

$\Delta H^\circ = \Sigma\, n\Delta H_f^\circ(\text{products}) - \Sigma\, m\Delta H_f^\circ(\text{reactants})$

SUMMARY OF FACTS AND CONCEPTS

Energy exists in various forms, including *kinetic energy* and *potential energy*. The SI unit of energy is the *joule* (1 calorie = 4.184 joules). The *internal energy* of a substance is the sum of the kinetic energies and potential energies of the particles making up the substance. According to the *law of conservation of energy,* the total quantity of energy remains constant.

Reactions absorb or evolve definite quantities of heat under given conditions. At constant pressure, this *heat of reaction* is the *enthalpy of reaction,* ΔH. The chemical equation plus ΔH for molar amounts of reactants is referred to as the *thermochemical equation*. With it, we can calculate the heat for any amount of substance by *applying stoichiometry to reaction heats*. One measures the heat of reaction in a *calorimeter*. Direct calorimetric determination of the heat of reaction requires a reaction that goes to completion without other reactions occurring at the same time. Otherwise, the heat or enthalpy of reaction is determined indirectly from other enthalpies of reaction by using *Hess's law of heat summation*. Thermochemical data are conveniently tabulated as *enthalpies of formation*. If we know the values for each substance in an equation, we can easily compute the enthalpy of reaction. As an application of thermochemistry, the last section of the chapter discusses fuels.

OPERATIONAL SKILLS

1. **Calculating kinetic energy** Given the mass and speed of an object, calculate the kinetic energy (Example 6.1).

2. **Writing thermochemical equations** Given a chemical equation, states of substances, and the quantity of heat absorbed or evolved for molar amounts, write the thermochemical equation (Example 6.2).

3. **Manipulating thermochemical equations** Given a thermochemical equation, write the thermochemical equation for different multiples of the coefficients or for the reverse reaction (Example 6.3).

4. **Calculating reaction heat from the stoichiometry** Given the value of ΔH for a chemical equation, calculate the heat of reaction for a given mass of reactant or product (Example 6.4).

5. **Relating heat and specific heat** Given any three of the quantities q, s, m, and Δt, calculate the fourth one (Example 6.5).

6. **Calculating ΔH from calorimetric data** Given the amounts of reactants and the temperature change of a calorimeter of specified heat capacity, calculate the heat of reaction (Example 6.6).

7. **Applying Hess's law** Given a set of reactions with enthalpy changes, calculate ΔH for a reaction obtained from these other reactions by using Hess's law (Examples 6.7 and 6.8).

8. **Calculating the heat of phase transition from standard enthalpies of formation** Given a table of standard enthalpies of formation, calculate the heat of phase transition (Example 6.9).

9. **Calculating the enthalpy of reaction from standard enthalpies of formation** Given a table of standard enthalpies of formation, calculate the enthalpy of reaction (Example 6.10).

10. **Obtaining the standard enthalpy of formation from an enthalpy of reaction** Calculate the enthalpy of formation of a substance given an enthalpy of reaction plus enthalpies of formation for other reactants and products (Example 6.11).

Review Questions

6.1 Define *energy, kinetic energy, potential energy,* and *internal energy*.

6.2 What is the definition of the joule in terms of SI base units?

6.3 What is the original definition of the calorie? What is the present definition?

6.4 Describe the interconversions of potential and kinetic energy in a moving pendulum. A moving pendulum eventually comes to rest. Has the energy been lost? If not, what has happened to it?

6.5 Suppose heat flows into a vessel containing a gas. As the heat flows into the gas, what happens to the gas molecules? What happens to the internal energy of the gas?

6.6 Define an *exothermic* reaction and an *endothermic* reaction. Give an example of each.

6.7 The internal energy of a substance is a state function. What does this mean?

6.8 Under what condition is the enthalpy change equal to the heat of reaction?

6.9 How does the enthalpy change for an endothermic reaction occurring at constant pressure?

6.10 Why is it important to give the states of the reactants and products when giving an equation for ΔH?

6.11 If an equation for a reaction is doubled and then reversed, how is the value of ΔH changed?

6.12 Consider the reaction of methane, CH_4, with oxygen, O_2, which was discussed in Section 6.5. How would you set up the calculation if the problem had been to compute the heat if 10.0 g of H_2O were produced (instead of 10.0 g of CH_4 reacted)?

6.13 Define the heat capacity of a substance. Define the specific heat of a substance.

6.14 Describe a simple calorimeter. What are the measurements needed to determine the heat of reaction?

6.15 What property of enthalpy provides the basis of Hess's law? Explain.

6.16 The heat of fusion (also called heat of melting), ΔH_{fus}, of ice is the enthalpy change for

$$H_2O(s) \longrightarrow H_2O(l); \Delta H_{fus}$$

Similarly, the heat of vaporization, ΔH_{vap}, of liquid water is the enthalpy change for

$$H_2O(l) \longrightarrow H_2O(g); \Delta H_{vap}$$

How is the heat of sublimation, ΔH_{sub}, the enthalpy change for the reaction

$$H_2O(s) \longrightarrow H_2O(g); \Delta H_{sub}$$

related to ΔH_{fus} and ΔH_{vap}?

6.17 What is meant by the thermodynamic standard state?

6.18 What is meant by the reference form of an element? What is the standard enthalpy of formation of an element in its reference form?

6.19 What is the standard enthalpy of formation of a substance?

6.20 Write the chemical equation for the formation reaction of $H_2S(g)$.

6.21 Is the following reaction the appropriate one to use in determining the enthalpy of formation of methane, $CH_4(g)$? Why or why not?

$$C(g) + 4H(g) \longrightarrow CH_4(g)$$

6.22 What is a fuel? What are the fossil fuels?

6.23 Give chemical equations for the conversion of carbon in coal to methane, CH_4.

6.24 List some rocket fuels and corresponding oxidizers. Give thermochemical equations for the exothermic reactions of these fuels with the oxidizers.

Practice Problems

ENERGY AND ITS UNITS

6.25 The energy, E, needed to move an object a distance d by applying a force F is $E = F \times d$. What must be the SI unit of force if this equation is to be consistent with the SI unit of energy for E?

6.26 The potential energy of an object in the gravitational field of the earth is $E_p = mgh$. What must be the SI unit of g if this equation is to be consistent with the SI unit of energy for E_p?

6.27 When one mole of hydrogen gas burns in an excess of oxygen to produce liquid water, 286 kJ of heat are released. What is this heat in kilocalories?

6.28 When one mole of graphite (carbon) burns in an excess of oxygen to produce carbon dioxide gas, 394 kJ of heat are released. What is this heat in kilocalories?

6.29 What is the kinetic energy of a car of mass 4.53×10^3 lb that is moving with a speed of 45 mph? Give the answer in joules and in calories. See Table 1.4.

6.30 What is the kinetic energy of a bullet traveling 2.50×10^3 ft/s whose mass is 225 grains? A grain equals 0.0648 g. Give the answer in joules and in calories.

6.31 Calculate the kinetic energy in joules of an H_2O molecule whose speed is 648 m/s.

6.32 Calculate the kinetic energy in joules of an SO_2 molecule whose speed is 356 m/s.

HEAT OF REACTION

6.33 Hydrogen cyanide is used in the manufacture of clear plastics such as Lucite or Plexiglas. It is prepared from ammonia and natural gas (CH_4).

$$2NH_3(g) + 3O_2(g) + 2CH_4(g) \longrightarrow 2HCN(g) + 6H_2O(g)$$

The reaction evolves 939 kJ of heat per 2 mol HCN formed. Is the reaction endothermic or exothermic? What is the value of q?

6.34 Nitric acid, a source of many nitrogen compounds, is produced from nitrogen dioxide. An old process for making nitrogen dioxide employed nitrogen and oxygen.

$$N_2(g) + 2O_2(g) \longrightarrow 2NO_2(g)$$

The reaction absorbs 66.4 kJ per 2 mol of NO_2 produced. Is the reaction endothermic or exothermic? What is the value of q?

THERMOCHEMICAL EQUATIONS

6.35 When one mole of mercury(II) oxide crystals, HgO, decomposes into its elements at constant temperature and pressure, 90.8 kJ of heat is absorbed. Write an equation for this reaction, including the value of ΔH and labels for the states of reactants and products.

6.36 When one mole of zinc metal reacts with hydrochloric acid, HCl, it gives a solution of zinc chloride, $ZnCl_2$, and evolves hydrogen. At constant pressure, 152 kJ of heat is given off. Write the thermochemical equation for this reaction.

6.37 Calcium carbide, CaC_2, reacts with water to give acetylene, C_2H_2, and calcium hydroxide, $Ca(OH)_2$.

$$CaC_2(s) + 2H_2O(l) \longrightarrow C_2H_2(g) + Ca(OH)_2(s); \Delta H = -128 \text{ kJ}$$

What is ΔH for the following equation?

$$C_2H_2(g) + Ca(OH)_2(s) \longrightarrow CaC_2(s) + 2H_2O(l)$$

6.38 A laboratory preparation of oxygen consists of decomposing potassium chlorate, $KClO_3$, using a catalyst of manganese dioxide, MnO_2.

$$2KClO_3(s) \longrightarrow 2KCl(s) + 3O_2(g); \Delta H = -44.7 \text{ kJ}$$

What is ΔH for the following reaction?

$$KCl(s) + \tfrac{3}{2}O_2(g) \longrightarrow KClO_3(s)$$

6.39 Phosphoric acid, H_3PO_4, can be prepared by the reaction of phosphorus(V) oxide, P_4O_{10}, with water.

$$\tfrac{1}{4}P_4O_{10}(s) + \tfrac{3}{2}H_2O(l) \longrightarrow H_3PO_4(aq); \Delta H = -107.6 \text{ kJ}$$

What is ΔH for the reaction involving 1 mole of P_4O_{10}?

$$P_4O_{10}(s) + 6H_2O(l) \longrightarrow 4H_3PO_4(aq)$$

6.40 With a platinum catalyst, ammonia will burn in oxygen to give nitric oxide, NO.

$$4NH_3(g) + 5O_2(g) \longrightarrow 4NO(g) + 6H_2O(g); \Delta H = -906 \text{ kJ}$$

What is the enthalpy change for the following reaction?

$$NO(g) + \tfrac{3}{2}H_2O(g) \longrightarrow NH_3(g) + \tfrac{5}{4}O_2(g)$$

STOICHIOMETRY OF REACTION HEATS

6.41 White phosphorus, P_4, burns in an excess of oxygen to form tetraphosphorus decoxide, P_4O_{10}.

$$P_4(s) + 5O_2(g) \longrightarrow P_4O_{10}(s); \Delta H = -2942 \text{ kJ}$$

What is the heat evolved per gram of phosphorus burned?

6.42 Nitric oxide, NO, is formed whenever a mixture of nitrogen and oxygen gases is heated.

$$N_2(g) + O_2(g) \longrightarrow 2NO(g); \Delta H = 90.3 \text{ kJ}$$

Calculate the heat absorbed per gram of nitrogen reacted.

6.43 Ammonia burns in the presence of a copper catalyst to form nitrogen gas.

$$4NH_3(g) + 3O_2(g) \longrightarrow 2N_2(g) + 6H_2O(g); \quad \Delta H = -1267 \text{ kJ}$$

What is the enthalpy change to burn 25.6 g ammonia?

6.44 Hydrogen sulfide, H_2S, is a foul-smelling gas. It burns to form sulfur dioxide.

$$2H_2S(g) + 3O_2(g) \longrightarrow 2SO_2(g) + 2H_2O(g); \quad \Delta H = -1037 \text{ kJ}$$

Calculate the enthalpy change to burn 36.9 g hydrogen sulfide.

HEAT CAPACITY AND CALORIMETRY

6.45 You wish to heat water to make coffee. How much heat (in joules) must be used to raise the temperature of 0.180 kg of tap water (enough for one cup of coffee) from 15°C to 96°C (near the ideal brewing temperature)? Assume the specific heat is that of pure water, 4.18 J/(g · °C).

6.46 An iron skillet weighing 1.51 kg is heated on a stove to 178°C. Suppose the skillet is cooled to room temperature, 21°C. How much heat energy (in joules) must be removed to effect this cooling? The specific heat of iron is 0.450 J/(g · °C).

6.47 When steam condenses to liquid water, 2.26 kJ of heat is released per gram. The heat from 124 g of steam is used to heat a room containing 6.44×10^4 g of air (20 ft × 12 ft × 8 ft). The specific heat of air at normal pressure is 1.015 J/(g · °C). What is the change in air temperature, assuming the heat from the steam is all absorbed by air?

6.48 When ice at 0°C melts to liquid water at 0°C, it absorbs 0.334 kJ of heat per gram. Suppose the heat needed to melt 35.0 g of ice is absorbed from the water contained in a glass. If this water has a mass of 0.210 kg and a temperature of 21.0°C, what is the final temperature of the water? (Note that you will also have 35.0 g of water at 0°C from the ice.)

6.49 When 15.3 g of sodium nitrate, $NaNO_3$, was dissolved in water in a calorimeter, the temperature fell from 25.00°C to 21.56°C. If the heat capacity of the solution and the calorimeter is 1071 J/°C, what is the enthalpy change when one mole of sodium nitrate dissolves in water? The solution process is

$$NaNO_3(s) \longrightarrow Na^+(aq) + NO_3^-(aq); \Delta H = ?$$

6.50 When 23.6 g of calcium chloride, $CaCl_2$, was dissolved in water in a calorimeter, the temperature rose from 25.0°C to 38.7°C. If the heat capacity of the solution and the calorimeter is 1258 J/°C, what is the enthalpy change when one mole of calcium chloride dissolves in water? The solution process is

$$CaCl_2(s) \longrightarrow Ca^{2+}(aq) + 2Cl^-(aq); \Delta H = ?$$

6.51 A sample of ethanol, C_2H_5OH, weighing 2.84 g was burned in an excess of oxygen in a bomb calorimeter. The temperature of the calorimeter rose from 25.00°C to 33.73°C. If the heat capacity of the calorimeter and contents was 9.63 kJ/°C, what is the value of q for burning 1 mol ethanol at constant volume and 25.00°C? The reaction is

$$C_2H_5OH(l) + 3O_2(g) \longrightarrow 2CO_2(g) + 3H_2O(l)$$

6.52 A sample of benzene, C_6H_6, weighing 3.51 g was burned in an excess of oxygen in a bomb calorimeter. The temperature of the calorimeter rose from 25.00°C to 37.18°C. If the heat capacity of the calorimeter and contents was 12.05 kJ/°C, what is the value of q for burning 1 mol benzene at constant volume and 25.00°C? The reaction is

$$C_6H_6(l) + \tfrac{15}{2}O_2(g) \longrightarrow 6CO_2(g) + 3H_2O(l)$$

HESS'S LAW

6.53 Hydrogen peroxide, H_2O_2, is a colorless liquid whose solutions are used as a bleach and an antiseptic. It can be prepared in a process whose overall change is

$$H_2(g) + O_2(g) \longrightarrow H_2O_2(l)$$

Calculate the enthalpy change using the following data:

$$H_2O_2(l) \longrightarrow H_2O(l) + \tfrac{1}{2}O_2(g); \Delta H = -98.0 \text{ kJ}$$

$$2H_2(g) + O_2(g) \longrightarrow 2H_2O(l); \Delta H = -571.6 \text{ kJ}$$

6.54 Hydrazine, N_2H_4, is a colorless liquid used as a rocket fuel. What is the enthalpy change for the process in which hydrazine is formed from its elements?

$$N_2(g) + 2H_2(g) \longrightarrow N_2H_4(l)$$

Use the following reactions and enthalpy changes:

$$N_2H_4(l) + O_2(g) \longrightarrow N_2(g) + 2H_2O(l); \Delta H = -622.2 \text{ kJ}$$

$$H_2(g) + \tfrac{1}{2}O_2(g) \longrightarrow H_2O(l); \Delta H = -285.8 \text{ kJ}$$

6.55 Ammonia will burn in the presence of a platinum catalyst to produce nitric oxide, NO.

$$4NH_3(g) + 5O_2(g) \longrightarrow 4NO(g) + 6H_2O(g)$$

What is the heat of reaction at constant pressure? Use the following thermochemical equations:

$$N_2(g) + O_2(g) \longrightarrow 2NO(g);\ \Delta H = 180.6\text{ kJ}$$
$$N_2(g) + 3H_2(g) \longrightarrow 2NH_3(g);\ \Delta H = -91.8\text{ kJ}$$
$$2H_2(g) + O_2(g) \longrightarrow 2H_2O(g);\ \Delta H = -483.7\text{ kJ}$$

6.56 Hydrogen cyanide is a highly poisonous, volatile liquid. It can be prepared by the reaction

$$CH_4(g) + NH_3(g) \longrightarrow HCN(g) + 3H_2(g)$$

What is the heat of reaction at constant pressure? Use the following thermochemical equations:

$$N_2(g) + 3H_2(g) \longrightarrow 2NH_3(g);\ \Delta H = -91.8\text{ kJ}$$
$$C(\text{graphite}) + 2H_2(g) \longrightarrow CH_4(g);\ \Delta H = -74.9\text{ kJ}$$
$$H_2(g) + 2C(\text{graphite}) + N_2(g) \longrightarrow 2HCN(g);\ \Delta H = 270.3\text{ kJ}$$

6.57 Compounds with carbon–carbon double bonds, such as ethylene, C_2H_4, add hydrogen in a reaction called hydrogenation. Thus,

$$C_2H_4(g) + H_2(g) \longrightarrow C_2H_6(g)$$

Calculate the enthalpy change for this reaction, using the following combustion data:

$$C_2H_4(g) + 3O_2(g) \longrightarrow 2CO_2(g) + 2H_2O(l);\ \Delta H = -1401\text{ kJ}$$
$$C_2H_6(g) + \tfrac{7}{2}O_2(g) \longrightarrow 2CO_2(g) + 3H_2O(l);\ \Delta H = -1550\text{ kJ}$$
$$H_2(g) + \tfrac{1}{2}O_2(g) \longrightarrow H_2O(l),\ \Delta H = -286\text{ kJ}$$

6.58 Acetic acid, CH_3COOH, is contained in vinegar. Suppose acetic acid was formed from its elements, according to the following equation:

$$2C(\text{graphite}) + 2H_2(g) + O_2(g) \longrightarrow CH_3COOH(l)$$

Find the enthalpy change, ΔH, for this reaction, using the following data:

$$CH_3COOH(l) + 2O_2(g) \longrightarrow 2CO_2(g) + 2H_2O(l);\ \Delta H = -871\text{ kJ}$$
$$C(\text{graphite}) + O_2(g) \longrightarrow CO_2(g);\ \Delta H = -394\text{ kJ}$$
$$H_2(g) + \tfrac{1}{2}O_2(g) \longrightarrow H_2O(l);\ \Delta H = -286\text{ kJ}$$

STANDARD ENTHALPIES OF FORMATION

6.59 Carbon tetrachloride, CCl_4, is a liquid used as an industrial solvent and in the preparation of fluorocarbons. What is the heat of vaporization of carbon tetrachloride?

$$CCl_4(l) \longrightarrow CCl_4(g);\ \Delta H^\circ = ?$$

Use standard enthalpies of formation (Table 6.2).

6.60 The cooling effect of alcohol on the skin is due to its evaporation. Calculate the heat of vaporization of ethanol (ethyl alcohol), C_2H_5OH.

$$C_2H_5OH(l) \longrightarrow C_2H_5OH(g);\ \Delta H^\circ = ?$$

The standard enthalpy of formation of $C_2H_5OH(l)$ is -277.6 kJ/mol and that of $C_2H_5OH(g)$ is -235.4 kJ/mol.

6.61 Hydrogen sulfide gas is a poisonous gas with the odor of rotten eggs. It occurs in natural gas and is produced during the decay of organic matter, which contains sulfur. The gas burns in oxygen as follows:

$$2H_2S(g) + 3O_2(g) \longrightarrow 2H_2O(l) + 2SO_2(g)$$

Calculate the standard enthalpy change for this reaction using standard enthalpies of formation.

6.62 Carbon disulfide is a colorless liquid. When pure, it is nearly odorless, but the commercial product smells vile. Carbon disulfide is used in the manufacture of rayon and cellophane. The liquid burns as follows:

$$CS_2(l) + 3O_2(g) \longrightarrow CO_2(g) + 2SO_2(g)$$

Calculate the standard enthalpy change for this reaction using standard enthalpies of formation.

6.63 The first step in the preparation of lead from its ore (galena, PbS) consists of roasting the ore.

$$2PbS(s) + 3O_2(g) \longrightarrow 2SO_2(g) + 2PbO(s)$$

Calculate the standard enthalpy change for this reaction, using enthalpies of formation (see Appendix C).

6.64 Iron is obtained from iron ore by reduction with carbon monoxide. The overall reaction is

$$Fe_2O_3(s) + 3CO(g) \longrightarrow 2Fe(s) + 3CO_2(g)$$

Calculate the standard enthalpy change for this equation. See Appendix C for data.

6.65 Hydrogen chloride gas dissolves in water to form hydrochloric acid (an ionic solution).

$$HCl(g) \longrightarrow H^+(aq) + Cl^-(aq)$$

Find ΔH° for the above reaction. The data are given in Table 6.2.

6.66 Carbon dioxide from the atmosphere "weathers," or dissolves, limestone ($CaCO_3$) by the reaction

$$CaCO_3(s) + CO_2(g) + H_2O(l) \longrightarrow Ca^{2+}(aq) + 2HCO_3^-(aq)$$

Obtain ΔH° for this reaction. See Table 6.2 for the data.

6.67 Butane, C_4H_{10}, is a hydrocarbon fuel obtained from petroleum. It is normally a gas, but it is easily liquefied and transported in cylinders under pressure. When butane burns in an excess of oxygen at a constant pressure of 1 atm, 2855 kJ of heat evolve per mole of butane.

$$C_4H_{10}(g) + \tfrac{13}{2}O_2(g) \longrightarrow 4CO_2(g) + 5H_2O(l); \quad \Delta H^\circ = -2855 \text{ kJ}$$

Use this information and standard enthalpies of formation in Table 6.2 to calculate the standard enthalpy of formation of $C_4H_{10}(g)$.

6.68 Ethanethiol (ethyl mercaptan), C_2H_5SH, is a liquid with a particularly obnoxious odor. It is added to natural gas fuels so gas leaks can be detected. When ethanethiol burns in oxygen at a constant pressure of 1 atm, 1877 kJ of heat evolve per mole of ethanethiol.

$$C_2H_5SH(l) + \tfrac{9}{2}O_2(g) \longrightarrow 2CO_2(g) + SO_2(g) + 3H_2O(l); \quad \Delta H^\circ = -1877 \text{ kJ}$$

Use this information and standard enthalpies of formation in Table 6.2 to calculate the standard enthalpy of formation of $C_2H_5SH(l)$.

Additional Problems

6.69 Liquid hydrogen peroxide has been used as a propellant for rockets. Hydrogen peroxide decomposes into oxygen and water, giving off heat energy equal to 686 Btu per pound of propellant. What is this energy in joules per gram of hydrogen peroxide? (One Btu equals 252 cal; see also Table 1.4.)

6.70 Hydrogen is an ideal fuel in many respects; for example, the product of its combustion, water, is nonpolluting. The heat given off in burning hydrogen to gaseous water is 5.16×10^4 Btu per pound. What is this heat energy in joules per gram? (1 Btu = 252 cal; see also Table 1.4.)

6.71 Niagara Falls has a height of 167 ft (American Falls). What is the potential energy in joules of 1.00 lb of water at the top of the falls, if we take water at the bottom to have a potential energy of zero? What would be the speed of this water at the bottom of the falls, if we neglect friction during the descent of the water?

6.72 Any object, be it a space satellite or a molecule, must attain an initial upward velocity of at least 11.2 km/s in order to escape the gravitational attraction of the earth. What would be the kinetic energy in joules of a satellite weighing 2354 lbs that has the speed equal to this escape velocity of 11.2 km/s?

6.73 When calcium carbonate, $CaCO_3$ (the major constituent of limestone and seashells) is heated, it decomposes to calcium oxide (quicklime).

$$CaCO_3(s) \longrightarrow CaO(s) + CO_2(g); \; \Delta H = 178.3 \text{ kJ}$$

How much heat is required to decompose 12.0 g of calcium carbonate?

6.74 Calcium oxide (quicklime) reacts with water to produce calcium hydroxide (slaked lime).

$$CaO(s) + H_2O(l) \longrightarrow Ca(OH)_2(s); \; \Delta H = -65.2 \text{ kJ}$$

The heat released by this reaction is sufficient to ignite paper. How much heat is released when 28.4 g (1.00 oz) of calcium oxide reacts?

6.75 Formic acid, $HCHO_2$, was first discovered in ants (*formica* is Latin for "ant"). In an experiment, 5.48 g of formic acid was burned at constant pressure.

$$2HCHO_2(l) + O_2(g) \longrightarrow 2CO_2(g) + 2H_2O(l)$$

If 30.3 kJ of heat evolved, what is ΔH per mole of formic acid?

6.76 Acetic acid, $HC_2H_3O_2$, is the sour constituent of vinegar (*acetum* is Latin for "vinegar"). In an experiment, 3.58 g of acetic acid was burned.

$$HC_2H_3O_2(l) + 2O_2(g) \longrightarrow 2CO_2(g) + 2H_2O(l)$$

If 52.0 kJ of heat evolved, what is ΔH per mole of acetic acid?

6.77 A piece of lead of mass 121.6 g was heated by an electrical coil. From the resistance of the coil, the current, and the time the current flowed, it was calculated that 235 J of heat was added to the lead. The temperature of the lead rose from 20.4°C to 35.5°C. What is the specific heat of the lead?

6.78 The specific heat of copper metal was determined by putting a piece of the metal weighing 35.4 g in hot water. The quantity of heat absorbed by the metal was calculated to be 47.0 J from the temperature drop of the water. What was the specific heat of the metal if the temperature of the metal rose 3.45°C?

6.79 A 50.0-g sample of water at 100.00°C was placed in an insulated cup. Then 25.3 g of zinc metal at 25.00°C was added to the water. The temperature of the water dropped to 96.68°C. What is the specific heat of zinc?

6.80 A 19.6-g sample of a metal was heated to 61.67°C. When the metal was placed into 26.7 g of water in a calorimeter, the temperature of the water increased from 25.00°C to 30.00°C. What is the specific heat of the metal?

6.81 In a calorimetric experiment, 6.48 g of lithium hydroxide, LiOH, was dissolved in water. The temperature of the calorimeter rose from 25.00°C to 36.66°C. What is ΔH for the solution process?

$$LiOH(s) \longrightarrow Li^+(aq) + OH^-(aq)$$

The heat capacity of the calorimeter and its contents is 547 J/°C.

6.82 When 21.45 g of potassium nitrate, KNO_3, were dissolved in water in a calorimeter, the temperature fell from 25.00°C to 14.14°C. What is the ΔH for the solution process?

$$KNO_3(s) \longrightarrow K^+(aq) + NO_3^-(aq)$$

The heat capacity of the calorimeter and its contents is 682 J/°C.

6.83 A 10.00-g sample of acetic acid, $HC_2H_3O_2$, was burned in a bomb calorimeter in an excess of oxygen.

$$HC_2H_3O_2(l) + 2O_2(g) \longrightarrow 2CO_2(g) + 2H_2O(l)$$

The temperature of the calorimeter rose from 25.00°C to 35.81°C. If the heat capacity of the calorimeter and its contents is 13.43 kJ/°C, what is the enthalpy change for the reaction?

6.84 The sugar arabinose, $C_5H_{10}O_5$, is burned completely in oxygen in a calorimeter.

$$C_5H_{10}O_5(s) + 5O_2(g) \longrightarrow 5CO_2(g) + 5H_2O(l)$$

Burning a 0.548-g sample caused the temperature to rise from 20.00°C to 20.54°C. The heat capacity of the calorimeter and its contents is 15.8 kJ/°C. Calculate ΔH for the combustion reaction per mole of arabinose.

6.85 Hydrogen sulfide, H_2S, is a poisonous gas with the odor of rotten eggs. The reaction for the formation of H_2S from the elements is

$$H_2(g) + \tfrac{1}{8}S_8(rhombic) \longrightarrow H_2S(g)$$

Use Hess's law to obtain the enthalpy change for this reaction from the following enthalpy changes:

$$H_2S(g) + \tfrac{3}{2}O_2(g) \longrightarrow H_2O(g) + SO_2(g);\ \Delta H = -519\text{ kJ}$$

$$H_2(g) + \tfrac{1}{2}O_2(g) \longrightarrow H_2O(g);\ \Delta H = -242\text{ kJ}$$

$$\tfrac{1}{8}S_8(rhombic) + O_2(g) \longrightarrow SO_2(g);\ \Delta H = -297\text{ kJ}$$

6.86 Ethylene glycol, $HOCH_2CH_2OH$, is used as antifreeze. It is produced from ethylene oxide, C_2H_4O, by the reaction

$$C_2H_4O(g) + H_2O(l) \longrightarrow HOCH_2CH_2OH(l)$$

Use Hess's law to obtain the enthalpy change for this reaction from the following enthalpy changes:

$$2C_2H_4O(g) + 5O_2(g) \longrightarrow 4CO_2(g) + 4H_2O(l);\ \Delta H = -2612.2\text{ kJ}$$

$$HOCH_2CH_2OH(l) + \tfrac{5}{2}O_2(g) \longrightarrow 2CO_2(g) + 3H_2O(l);\ \Delta H = -1189.8\text{ kJ}$$

6.87 Hydrogen, H_2, is prepared by *steam reforming,* in which hydrocarbons are reacted with steam. For CH_4,

$$CH_4(g) + H_2O(g) \longrightarrow CO(g) + 3H_2(g)$$

Calculate the enthalpy change $\Delta H°$ for this reaction, using standard enthalpies of formation.

6.88 Hydrogen is prepared from natural gas (mainly methane, CH_4) by partial oxidation.

$$2CH_4(g) + O_2(g) \longrightarrow 2CO(g) + 4H_2(g)$$

Calculate the enthalpy change $\Delta H°$ for this reaction, using standard enthalpies of formation.

6.89 Calcium oxide, CaO, is prepared by heating calcium carbonate (from limestone and seashells).

$$CaCO_3(s) \longrightarrow CaO(s) + CO_2(g)$$

Calculate the standard enthalpy of reaction, using enthalpies of formation. The $\Delta H_f°$ of CaO(s) is −635 kJ/mol. Other values are given in Table 6.2.

6.90 Sodium carbonate, Na_2CO_3, is used to manufacture glass. It is obtained from sodium hydrogen carbonate, $NaHCO_3$, by heating.

$$2NaHCO_3(s) \longrightarrow Na_2CO_3(s) + H_2O(g) + CO_2(g)$$

Calculate the standard enthalpy of reaction, using enthalpies of formation (Table 6.2).

6.91 Sucrose, $C_{12}H_{22}O_{11}$, is common table sugar. The enthalpy change at 25°C and 1 atm for the complete burning of 1 mol of sucrose in oxygen to give $CO_2(g)$ and $H_2O(l)$ is −5641 kJ. From this and from data given in Table 6.2, calculate the standard enthalpy of formation of sucrose.

6.92 Acetone, CH_3COCH_3, is a liquid solvent. The enthalpy change at 25°C and 1 atm for the complete burning of 1 mol of acetone in oxygen to give $CO_2(g)$ and $H_2O(l)$ is −1791 kJ. From this and from data given in Table 6.2, calculate the standard enthalpy of formation of acetone.

Cumulative-Skills Problems

6.93 Graphite is burned in oxygen to give carbon monoxide and carbon dioxide. If the product mixture is 33% CO and 67% CO_2 by mass, what is the heat from the combustion of 1.00 g of graphite?

6.94 A sample of natural gas is 80.0% CH_4 and 20.0% C_2H_6 by mass. What is the heat from the combustion of 1.00 g of this mixture? Assume the products are $CO_2(g)$ and $H_2O(l)$.

6.95 How much heat is released when a mixture containing 10.0 g NH_3 and 20.0 g O_2 reacts by the equation

$$4NH_3(g) + 5O_2(g) \longrightarrow 4NO(g) + 6H_2O(g); \quad \Delta H° = -906 \text{ kJ}$$

6.96 How much heat is released when a mixture containing 10.0 g CS_2 and 10.0 g Cl_2 reacts by the equation

$$CS_2(g) + 3Cl_2(g) \longrightarrow S_2Cl_2(g) + CCl_4(g); \quad \Delta H° = -232 \text{ kJ}$$

6.97 Consider the Haber process:

$$N_2(g) + 3H_2(g) \longrightarrow 2NH_3(g); \ \Delta H° = -91.8 \text{ kJ}$$

The density of ammonia at 25°C and 1.00 atm is 0.696 g/L. The density of nitrogen, N_2, is 1.145 g/L, and the molar heat capacity is 29.12 J/(mol · °C). (a) How much heat is evolved in the production of 1.00 L of ammonia at 25°C and 1.00 atm? (b) What percentage of this heat is required to heat the nitrogen required for this reaction (0.500 L) from 25°C to 400°C, the temperature at which the Haber process is run?

6.98 An industrial process for manufacturing sulfuric acid, H_2SO_4, uses hydrogen sulfide, H_2S, from the purification of natural gas. In the first step of this process, the hydrogen sulfide is burned to obtain sulfur dioxide, SO_2.

$$2H_2S(g) + 3O_2(g) \longrightarrow 2H_2O(l) + 2SO_2(g); \quad \Delta H° = -1125 \text{ kJ}$$

The density of sulfur dioxide at 25°C and 1.00 atm is 2.62 g/L, and the molar heat capacity is 30.2 J/(mol · °C). (a) How much heat would be evolved in producing 1.00 L of SO_2 at 25°C and 1.00 atm? (b) Suppose heat from this reaction is used to heat 1.00 L of SO_2 from 25°C and 1.00 atm to 500°C for its use in the next step of the process. What percentage of the heat evolved is required for this?

6.99 The carbon dioxide exhaled in the breath of astronauts is often removed from the spacecraft by reaction with lithium hydroxide.

$$2LiOH(s) + CO_2(g) \longrightarrow Li_2CO_3(s) + H_2O(l)$$

Estimate the grams of lithium hydroxide required per astronaut per day. Assume that each astronaut requires 2.50×10^3 kcal of energy per day. Further assume that this energy can be equated to the heat of combustion of a quantity of glucose, $C_6H_{12}O_6$, to $CO_2(g)$ and $H_2O(l)$. From the amount of glucose required to give 2.50×10^3 kcal of heat, calculate the amount of CO_2 produced and hence the amount of LiOH required. The $\Delta H_f°$ for glucose(s) is -1273 kJ/mol.

6.100 A rebreathing gas mask contains potassium superoxide, KO_2, which reacts with moisture in the breath to give oxygen.

$$4KO_2(s) + 2H_2O(l) \longrightarrow 4KOH(s) + 3O_2(g)$$

Estimate the grams of potassium superoxide required to supply a person's oxygen needs for one hour. Assume a person requires 1.00×10^2 kcal of energy for this time period. Further assume that this energy can be equated to the heat of combustion of a quantity of glucose, $C_6H_{12}O_6$, to $CO_2(g)$ and $H_2O(l)$. From the amount of glucose required to give 1.00×10^2 kcal of heat, calculate the amount of oxygen consumed and hence the amount of KO_2 required. The $\Delta H_f°$ for glucose(s) is -1273 kJ/mol.

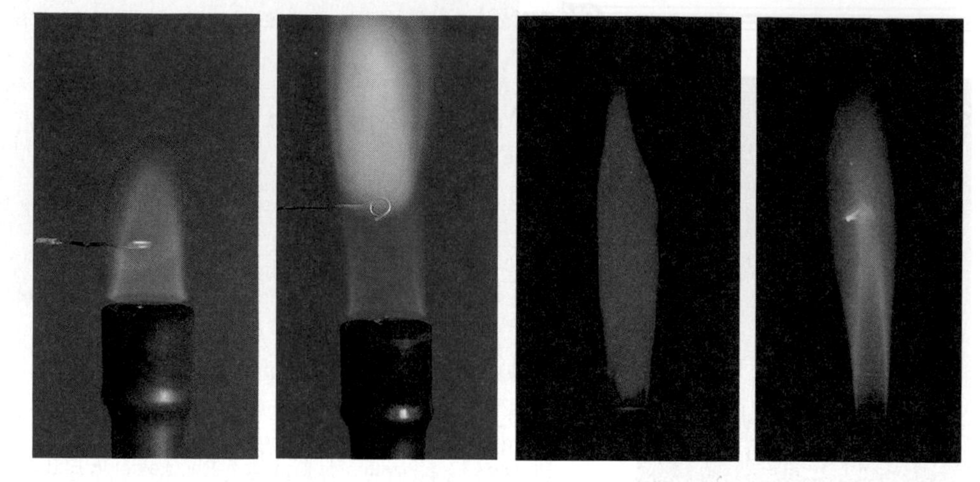

Figure 7.1
Flame tests of Groups IA and IIA elements. The flames are those of lithium (red), sodium (yellow), strontium (red), and calcium (orange).

elements. Looking at such spectra, we can ask many questions about their meaning. How is it that each atom emits particular colors of light? What does a line spectrum tell us about the structure of an atom? If we know something about the structures of atoms, can we explain the formation of ions and molecules? We will answer these questions in this and the next few chapters.

Basic Structure of Atoms

An atom consists of two basic kinds of parts, a nucleus and one or more electrons. The **nucleus** is *the central core of an atom; it has most of the mass of the atom and is positively charged*. An **electron** is *a very light, negatively charged particle that exists around a positively charged nucleus*. The atomic nucleus is itself made up of particles. A **proton** is *one of the particles found in the atomic nucleus; it is a positively charged particle with a mass over 1800 times that of the electron*. A **neutron** is *another particle found in the nucleus; it has a mass almost identical to that of the proton but has no electrical charge*.

In the next sections, we will look at some of the experiments that revealed this basic atomic structure, and we will discuss how the different atoms derive their characteristics from differences in this basic structure.

7.1 DISCOVERY OF THE ELECTRON

The neon sign with its reddish-orange glow is a familiar part of modern life. It is a tube filled with neon gas at low pressure, through which a high-voltage electric current is discharged. A neon sign is an example of a *gas discharge tube*. Gas discharge tubes have played an important part in determining the structure of the atom. In particular, the study of such tubes led to the discovery of the electron.●

● The colored glow obtained from discharge tubes is caused by the characteristic emission of light from an atom, which we mentioned in the chapter opening.

Thomson's *m/e* Experiment

When several thousand volts of electricity are discharged through a low-pressure gas contained in a glass tube, colored light characteristic of the gas filling the tube

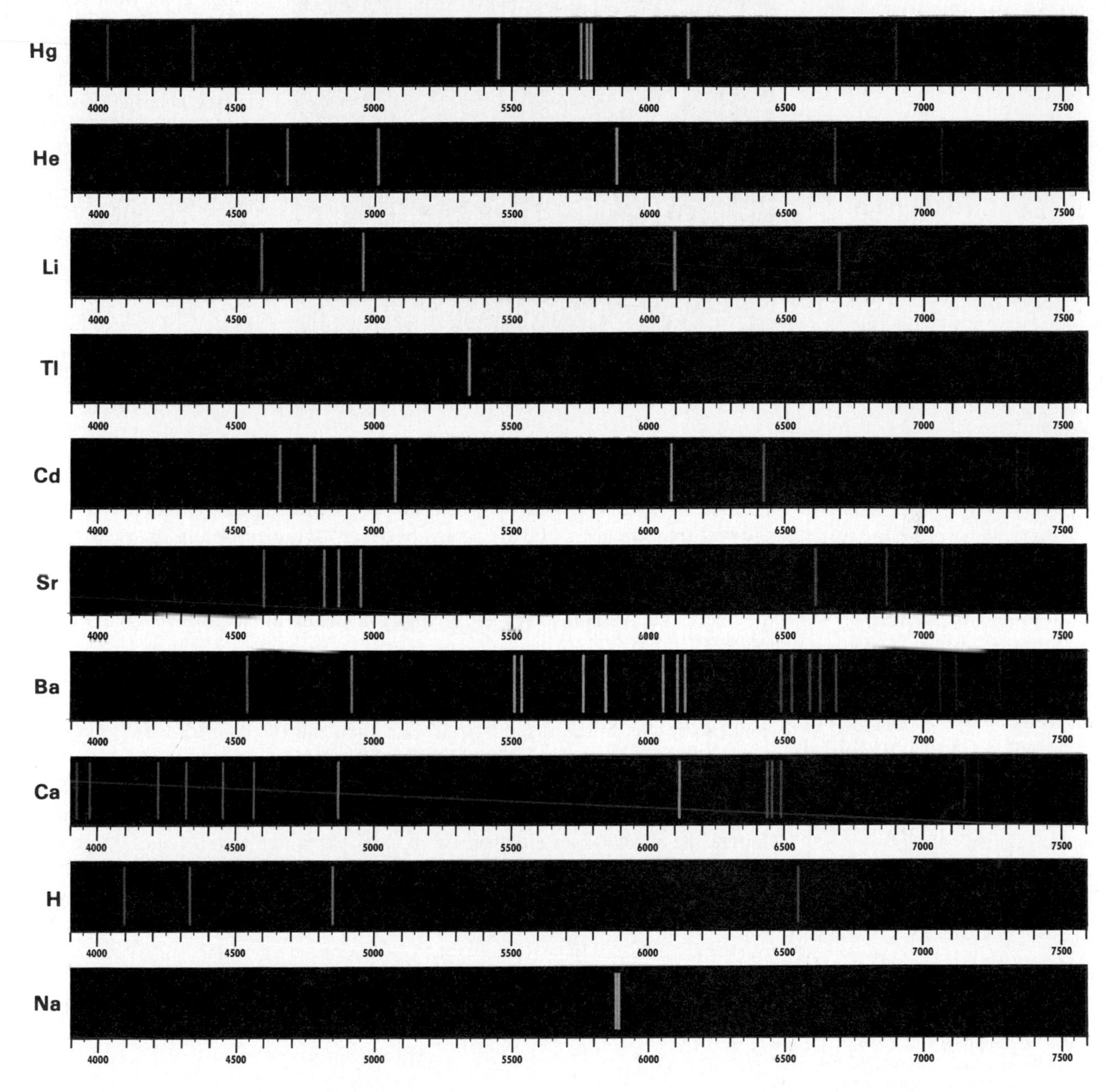

Figure 7.2
Emission (line) spectra of some elements. The lines correspond to visible light emitted by atoms.

is emitted (neon emits an orange light). If the pressure is reduced sufficiently, the colored glow vanishes, but a luminescent material can be made to glow when placed near or in the tube. Experiments with discharge tubes showed that invisible rays are emitted by the cathode (or negative electrode) in the discharge tube, but these rays can be detected because they cause certain materials (such as a zinc sulfide screen) to luminesce or glow. *The rays emitted by the cathode in a gas discharge tube* are called **cathode rays.**

In 1897, the British physicist J. J. Thomson (Figure 7.3) showed that cathode rays consist of negatively charged particles. Thus, the rays are attracted to the

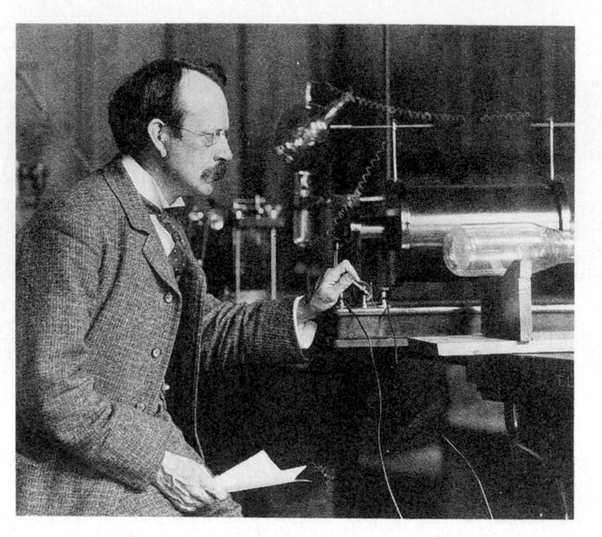

Figure 7.3
Joseph John Thomson (1856–1940). J. J. Thomson's scientific ability was recognized early with his appointment as Professor of Physics in the Cavendish Laboratory at Cambridge University when he was not quite twenty-eight years old. Soon after this appointment, Thomson began research on the discharge of electricity through gases. This work culminated in 1897 with the discovery of the electron. Thomson was awarded the Nobel Prize in physics in 1906.

anode, or positively charged electrode. If a hole is cut in the anode, some of the cathode rays will pass through the hole to form a beam (Figure 7.4). Thomson showed that when this beam passes between two electrically charged plates, it is bent toward the positive plate, showing that cathode rays are negatively charged. Cathode rays are similarly deflected by magnetic fields in a way consistent with the view that cathode rays are actually a beam of negatively charged particles (Figure 7.5).●

● A television tube works through the deflection of a cathode ray beam by electromagnetic coils. The cathode ray beam is directed toward a coated screen, where it traces a luminescent image as the beam is bent by the varying magnetism of the deflection coils.

Thomson showed that cathode rays behaved as particles of matter with a definite mass-to-charge ratio. He also showed that the cathode rays are independent of the gas in the discharge tube or the kind of material making up the electrodes. This suggested to Thomson that cathode rays are composed of particles that are constituents of all matter. These particles came to be called *electrons,* using a term coined earlier for what was then the hypothetical unit of electricity. If we denote the mass of an electron by m_e and the magnitude of its charge by e, the mass-to-charge ratio is m_e/e and has the value 5.686×10^{-12} kg/C. (The SI unit of electric charge is the *coulomb,* for which the symbol is C.)

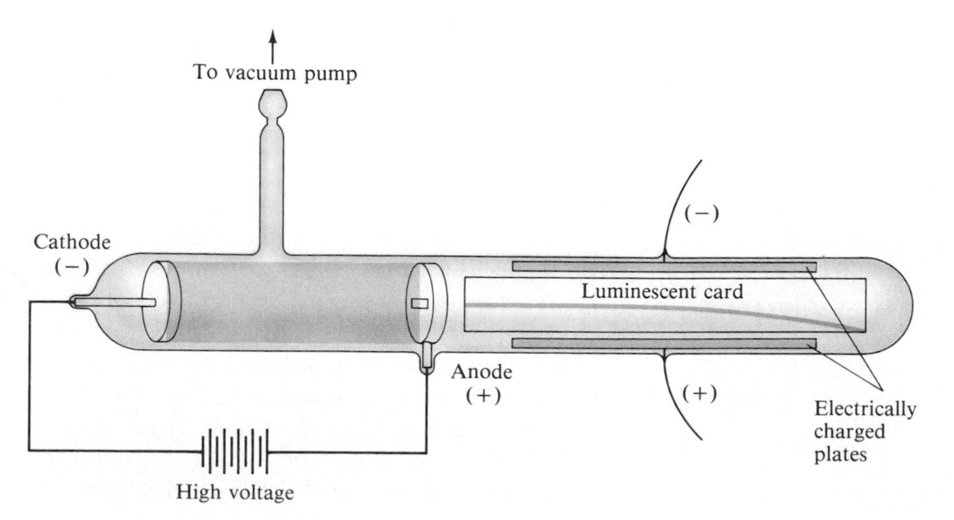

Figure 7.4
Formation of cathode rays in a discharge tube. Cathode rays leave the negative electrode, or cathode, and are accelerated toward the anode, or positive electrode. Some of the rays pass through the hole in the anode to form a beam, which then is bent by the electric plates in the tube. The beam is made visible as a green glow by a luminescent card placed in the tube.

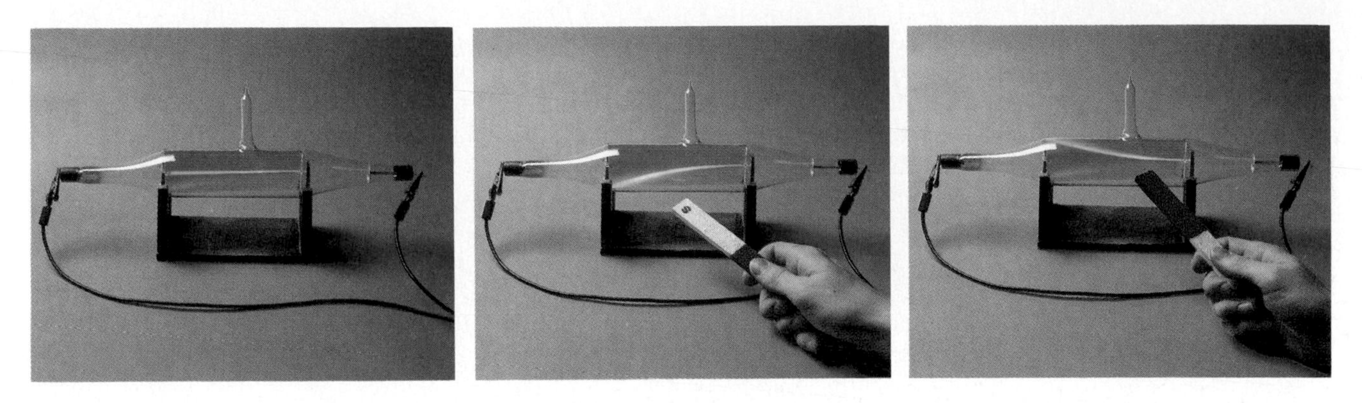

Figure 7.5
Bending of cathode rays by a magnet. *Left:* The cathode ray beam travels from right to left. *Center:* The beam bends downward as the south pole of the magnet is brought toward it, as expected for a beam of negative particles. *Right:* When the magnet is turned around, the beam bends in the opposite direction.

The experiments by Thomson on the deflection of cathode rays yielded only the mass-to-charge ratio; they did not give the mass of the electron. This crucial piece of information required another type of experiment.

Millikan's Oil Drop Experiment

In 1909 the U.S. physicist Robert Millikan (1868–1953) performed a series of experiments in which, by observing the behavior of electrically charged oil drops, he determined the charge on an electron. The experimental apparatus is shown in Figure 7.6. A spray bottle produces a fine mist of oil droplets, some of which pass through an opening into a viewing chamber, where we can observe them with a microscope.

Often these droplets have an electric charge, which is picked up from the friction of forming the mist (much as the friction of walking across a rug generates static electricity). A droplet may have one or more additional electrons on it, giving it a negative charge. If we observe the droplet while it is falling freely in air, we can determine its mass from its downward speed.●

● A falling object at first accelerates by gravity, but it soon reaches a constant speed as a result of air friction. This speed is related to the mass of the object.

As the droplet falls to the bottom of the chamber, it passes between two electrically charged plates. The droplet can be suspended between them if we adjust the voltage on the plates so that the electrical attraction upward just balances the force of gravity downward. We then use the voltage needed to establish this balance to calculate the mass-to-charge ratio for the droplet. Because we already know the mass of the droplet, we can find the charge on it.

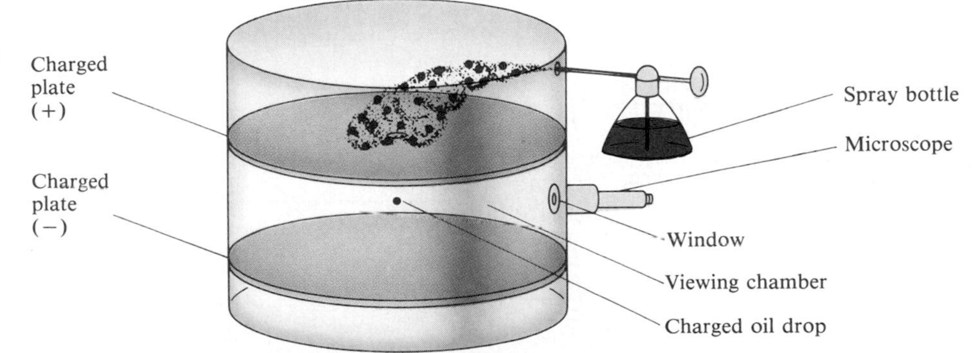

Figure 7.6
Millikan's oil drop experiment. Oil is sprayed from a bottle to produce droplets. An oil drop passes through the hole in the top plate and into the region between the charged plates, where it can be observed through a microscope. This drop, when illuminated perpendicular to the direction of view, appears in the microscope as a bright speck against a dark background.

A series of such determinations will reveal the smallest possible difference between the electric charges on two oil drops. The difference is the magnitude of the charge on a single electron.● This value (e) is 1.602×10^{-19} C. It can be used with the value of m_e/e (5.686×10^{-12} kg/C) from Thomson's experiment to calculate the mass of the electron.

● We can also induce a given droplet to change charge by irradiating the surrounding air with x rays. The smallest possible change equals the charge of one electron.

$$m_e = (m_e/e) \times e$$
$$= 5.686 \times 10^{-12} \text{ kg/C} \times 1.602 \times 10^{-19} \text{ C} = 9.109 \times 10^{-31} \text{ kg}$$

Thus, the mass of an electron is 9.109×10^{-31} kg, over 1800 times smaller than the mass of the lightest atom (hydrogen). This shows that the electron is a subatomic particle.

7.2 THE NUCLEAR MODEL OF THE ATOM

The idea that the atom has a positively charged nucleus was developed to explain experiments involving the scattering of *alpha rays* from metal foils. Alpha rays are given off by certain radioactive materials.

Discovery of Radioactivity; Alpha Rays

In 1896 the French physicist Antoine Henri Becquerel (1852–1908) found that uranium and its compounds gave off a radiation that could blacken a photographic plate even when the plate was separated from the source of radiation by thick paper. This newly discovered phenomenon of *spontaneous radiation from certain unstable elements* was called **radioactivity.**●

● Radioactivity is discussed in detail in Chapter 20.

The radiation from uranium is separable by an electric field into two electrically charged beams (Figure 7.7), one called *alpha* (α) *rays* and the other called *beta* (β) *rays*. A third type of radiation, called *gamma* (γ) *rays*, is unaffected by the field. Gamma rays are like visible light but can easily pass through a layer of matter.

Ernest Rutherford (1871–1937), a British physicist, and others showed in 1903 that alpha rays were a stream of helium ions, He^{2+}. Beta rays, on the other hand, proved to be a stream of high-speed electrons. Alpha radiation was very intriguing. It was clear evidence that uranium atoms had disintegrated to give off helium. No longer could atoms be considered unchangeable particles, as they had been viewed since the time of Dalton.

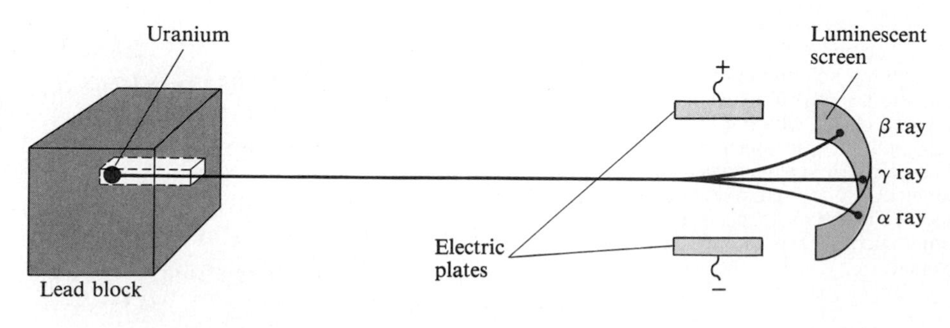

Figure 7.7
Separation of the radiation from a radioactive material (uranium). The radiation separates into alpha (α), beta (β), and gamma (γ) rays when it passes through an electric field.

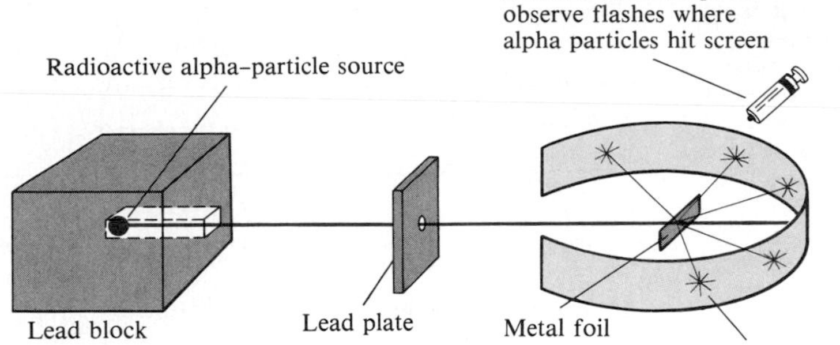

Figure 7.8
Alpha-particle scattering from metal foils. Alpha radiation is produced by a radioactive source and formed into a beam by a lead plate with a hole in it. (Lead absorbs the radiation.) Scattered alpha particles are made visible by a zinc sulfide screen, which gives tiny flashes where particles strike it. A movable microscope is used for viewing the flashes.

Alpha-Particle Scattering

After the discovery of alpha particles, Hans Geiger and Ernest Marsden, working in Rutherford's laboratory, observed the effect of bombarding thin metal foils with these positively charged atomic ions (see Figure 7.8). First they tried gold foil, and later other metals. In each case, they found that most of the alpha particles passed on through the foil as though nothing were there. But just a few of the alpha particles (about 1 in 8000) scattered at large angles and sometimes almost backwards. They behaved like bullets being shot at a wrought iron fence. Usually the bullets pass through undeflected, but occasionally one may hit an iron bar and ricochet backwards.

In 1911 Rutherford successfully explained alpha-particle scattering from foils on the basis of the *nuclear model* of the atom. According to this model, most of the mass of an atom is concentrated in a positively charged center called the *nucleus*, around which negatively charged electrons move. Although most of the mass of an atom is in its nucleus (99.95% or more), the nucleus occupies only a very small portion of the atom. Nuclei have diameters of about 10^{-15} m (10^{-5} Å), whereas atomic diameters are about 10^{-10} m (1 Å)—a hundred thousand times larger. If we were to use a golf ball to represent the nucleus, the atom would be about three miles in diameter. Atoms are mostly empty space.

Thus, when the alpha particles (helium nuclei) are directed toward a metal foil, most pass through the empty space of the atoms more or less undeflected. But when an alpha particle happens to hit a nucleus, it is scattered backwards by the repulsion of positive charges (see Figure 7.9).

7.3 NUCLEAR STRUCTURE

In 1919 Rutherford discovered that hydrogen nuclei appear to form when alpha particles collide with nitrogen atoms. Subsequently, it was shown that hydrogen nuclei form during the collision of alpha particles with other kinds of nuclei. These experiments clearly showed that atomic nuclei had structure and appeared to contain hydrogen nuclei, or *protons,* as we now call them.

Then, in 1930, the German physicists W. Bothe and H. Becker bombarded beryllium atoms with alpha particles and obtained a radiation that easily penetrated layers of matter. In 1932 the British physicist James Chadwick (1891–1974) showed that this radiation consisted of neutral particles, each with a mass approximately equal to that of a proton. Because they were neutral, these particles were

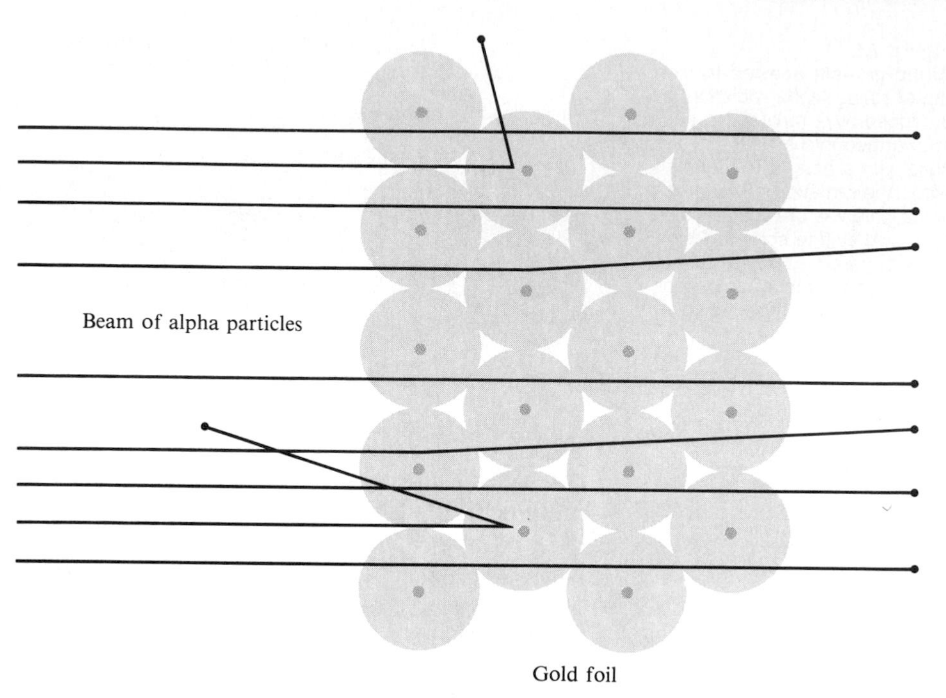

Figure 7.9
Representation of the scattering of alpha particles by a gold foil. Most of the alpha particles pass through the foil barely deflected. A few, however, collide with gold nuclei and are deflected at large angles.

called *neutrons*. Chadwick proposed that the neutron was a constituent of all nuclei. Later experiments showed that other nuclei gave off neutrons when bombarded with alpha particles.

It is now believed that atomic nuclei are composed of protons and neutrons. (The mass and charge of the proton and neutron are compared with those of the electron in Table 7.1.) Each different kind of nucleus is made up of different numbers of protons and neutrons. For example, the nucleus of a naturally occurring sodium atom contains 11 protons and 12 neutrons. The charge on a sodium nucleus is $+11e$, usually written as simply $+11$.● Similarly, the nucleus of an aluminum atom contains 13 protons and 14 neutrons; the charge on the nucleus is $+13e$, or $+13$.

● Recall that e is the magnitude of the charge on the electron. The actual charge is $-e$, because an electron has a negative charge. Therefore, the charge on a proton is $+e$.

Often we characterize a nucleus by its atomic number (symbol Z) and its mass number (A). The **atomic number (Z)** is *the number of protons in the nucleus*. The **mass number (A)** is *the total number of protons and neutrons in the nucleus*.● Thus, a sodium nucleus has atomic number 11 and mass number 23 (11 + 12); an aluminum nucleus has atomic number 13 and mass number 27 (13 + 14).

● The mass of an atom in atomic mass units is only approximately equal to the atom's mass number, partly because protons and neutrons do not have exactly unit mass (that is, neither a proton nor a neutron mass equals exactly 1 amu). Even if their masses were exactly 1, nuclear masses would not be whole numbers, because some mass is released as nuclear binding energy, according to Einstein's mass–energy equivalence. This is discussed in Chapter 20.

Any particular nucleus characterized by definite atomic number and mass number is called a **nuclide.** The notation for a nuclide consists of the symbol of the element with the atomic number written as a subscript on the left and the mass

Table 7.1 Properties of the Electron, Proton, and Neutron

Particle	Mass (kg)	Charge (C)	Mass (amu)	Charge (e)
Electron	9.10939×10^{-31}	-1.60218×10^{-19}	0.00055	-1
Proton	1.67262×10^{-27}	$+1.60218 \times 10^{-19}$	1.00728	$+1$
Neutron	1.67493×10^{-27}	0	1.00866	0

number as a superscript on the left. For the naturally occurring sodium nucleus, we have the following nuclide symbol:

$$\begin{matrix}\text{Mass number} \longrightarrow \\ \text{Atomic number} \longrightarrow\end{matrix} {}^{23}_{11}\text{Na}$$

The same information is conveyed by giving the name of the element and the mass number; ${}^{23}_{11}Na$ is also called sodium-23.

An atom is normally electrically neutral and has as many electrons moving about its nucleus as the nucleus has protons. A neutral sodium atom has a charge of +11 on its nucleus and has 11 electrons (with total charge −11) around this nucleus.

Example 7.1 Determining the Number of Electrons in an Atom or Atomic Ion

How many electrons are there in the neutral helium atom? How many electrons are there in He^+ and in He^{2+} (the alpha particle)? The atomic number of He is 2.

Solution

Helium atoms have atomic number 2 and therefore a nuclear charge of +2. **The neutral atom contains two electrons around the nucleus.** The ion He^+ is a helium atom that has lost one electron, leaving a positive charge on the atom of +1. **He^+ has one electron.** Similarly, He^{2+} is a helium atom that has lost two electrons, leaving a positive charge of +2. **He^{2+} has no electrons.** Thus, the alpha particle is a bare helium nucleus.

Exercise 7.1

The atomic number of uranium is 92. How many electrons are there in the neutral atom? How many electrons are there in the U^{2+} ion? (See Problems 7.25 and 7.26.)

Example 7.2 Writing Nuclide Symbols

What is the nuclide symbol for a nucleus that contains 38 protons and 50 neutrons?

Solution

If you look at the table of atomic weights on the inside back cover of this book, you will note that the element with atomic number 38 (the number of protons in the nucleus) is strontium, symbol Sr. The mass number is 38 + 50 = 88. **The symbol is ${}^{88}_{38}Sr$.**

Exercise 7.2

A nucleus consists of 17 protons and 18 neutrons. What is the nuclide symbol for this nucleus? (See Problems 7.27 and 7.28.)

The atomic number of an atom characterizes an element; all atoms of a given element have the same atomic number. However, the atoms of an element may have different mass numbers. **Isotopes** are *atoms whose nuclei have the same atomic number but different mass numbers;* that is, the nuclei have different numbers of neutrons.● Sodium has only one naturally occurring isotope, as does aluminum, but most elements are mixtures of isotopes. Naturally occurring oxygen, for example, contains 99.759% ${}^{16}_{8}O$, 0.037% ${}^{17}_{8}O$, and 0.204% ${}^{18}_{8}O$. (These are percentages of the total number of atoms in any given sample of the naturally occurring element.)

● Isotopes were first suspected in about 1912 when chemically identical elements with different atomic masses were found as radioactive decay products. The most convincing evidence for isotopes, however, came from the mass spectrometer (discussed in Section 7.4).

Exercise 7.3

Describe the structure of the carbon-14 atom (including the structure of the nucleus), according to the nuclear model. (See Problems 7.29 and 7.30.)

7.4 MASS SPECTROMETRY AND ATOMIC WEIGHTS

It is possible to obtain the mass-to-charge ratio for positive ions in the same way that Thomson studied cathode rays. Because the charge on an ion is equal to the magnitude of the electron charge (or to some multiple of it), these experiments can, in effect, measure the actual masses of atoms (and molecules). *An instrument, such as one based on Thomson's principles, that separates ions by mass-to-charge ratio* is called a **mass spectrometer.**

A simplified diagram of a mass spectrometer appears in Figure 7.10, showing the mass analysis of neon (a photograph of a modern instrument is shown in Figure 2.12, p. 43). Neon (which consists of neon atoms, Ne) is introduced into an *ionization chamber,* where the gas stream is crossed by an electron beam. Electrons collide with neon atoms, producing neon ions, Ne^+. Once these positive ions are formed, they are accelerated toward a negative grid and pass through slits to form a positive beam. This beam then travels through a magnetic field (the *analyzer*), which deflects the ions according to their mass-to-charge ratio. Just as a heavy car is buffeted less than a light car by a strong wind, the most massive ions are deflected the least by the magnetic field. Thus, each ion arrives in a definite place at the *detector* (charge-measuring device) according to its mass-to-charge ratio. Note that the neon ions are split into three beams by the magnetic field because neon consists of a mixture of three isotopes: $^{20}_{10}Ne$, $^{21}_{10}Ne$, and $^{22}_{10}Ne$.● The mass of each positive ion can be obtained from the amount of deflection. Moreover, the percentage of the total number of neon atoms that are neon-20 atoms, for

● Neon was the first element to be separated into isotopes. In 1913 J. J. Thomson detected a less-abundant mass at 22 amu, as well as the principal mass at 20 amu, in a sample of neon. At first he thought the mass at 22 amu might be a new element contaminating the neon sample. Later, with improved apparatus, F. W. Aston showed that most elements, including neon, are mixtures of isotopes.

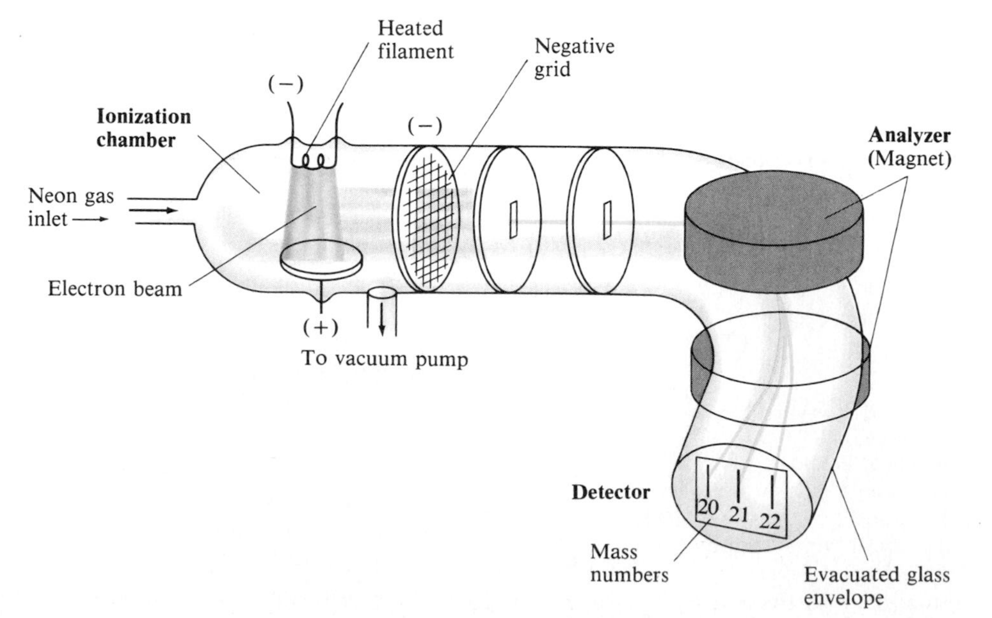

Figure 7.10
Diagram of a simple mass spectrometer, showing the separation of neon isotopes. Neon gas enters the ionization chamber, where neon atoms are ionized by a beam of electrons. Positive neon ions, Ne^+, are accelerated from the ionization chamber by the negative grid and pass to the analyzer, here shown as magnetic pole faces. The ion beam is split into three beams by the analyzer according to the mass-to-charge ratios. The three beams then travel to a detector. In most modern instruments, the detector registers an electrical signal for different positions, and these are recorded on a chart.

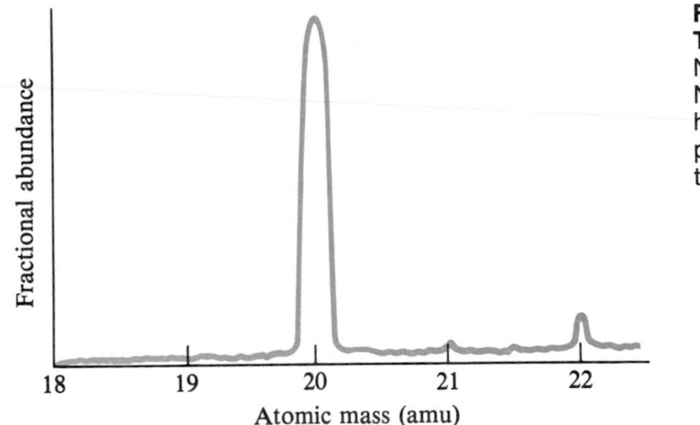

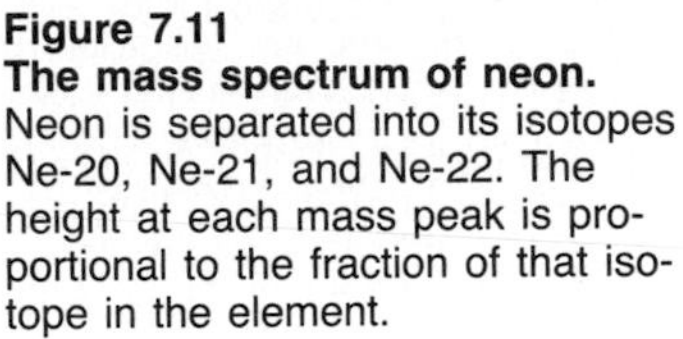
Figure 7.11
The mass spectrum of neon. Neon is separated into its isotopes Ne-20, Ne-21, and Ne-22. The height at each mass peak is proportional to the fraction of that isotope in the element.

example, can be obtained from the relative number of $^{20}_{10}Ne^+$ ions reaching the detector. The mass spectrum of neon—that is, the chart recording that shows the relative numbers of ions for various masses—is shown in Figure 7.11.

A mass analysis of a substance gives two kinds of information. It gives the masses of atoms or molecules, and it gives the relative number or relative *abundance* of each atomic or molecular ion reaching the detector. Masses are compared with those of the carbon-12 atom, which is assigned a mass of exactly 12 *atomic mass units* (amu). On this carbon-12 scale, neon isotopes have the following masses: neon-20, 19.992 amu; neon-21, 20.994 amu; and neon-22, 21.991 amu.

In a naturally occurring element, *the fraction of the total number of atoms composed of a particular isotope* is called the **fractional (isotopic) abundance.** The fractional abundances for the isotopes of neon are neon-20, 0.9051; neon-21, 0.0027; and neon-22, 0.0922. The *atomic weight* of an element is the weighted average of the masses (in atomic mass units) of the naturally occurring mixture of isotopes. For neon this is 20.179 amu. As the following example illustrates, the weighted average is found by multiplying each isotopic mass by its fractional abundance, then summing.

Example 7.3 Determining Atomic Weight from Isotopic Masses and Fractional Abundances

Chromium, Cr, has the following isotopic masses and fractional abundances:

MASS NUMBER	MASS (amu)	FRACTIONAL ABUNDANCE
50	49.9461	0.0435
52	51.9405	0.8379
53	52.9407	0.0950
54	53.9389	0.0236

What is the atomic weight of chromium?

Solution

Multiply each of the isotopic masses by its fractional abundance, then sum:

$$\begin{aligned} 49.9461 \text{ amu} \times 0.0435 &= 2.17 \text{ amu} \\ 51.9405 \text{ amu} \times 0.8379 &= 43.52 \text{ amu} \\ 52.9407 \text{ amu} \times 0.0950 &= 5.03 \text{ amu} \\ 53.9389 \text{ amu} \times 0.0236 &= \underline{1.27 \text{ amu}} \\ & 51.99 \text{ amu} \end{aligned}$$

The atomic weight of chromium is **51.99 amu.**

Exercise 7.4

Chlorine consists of the following isotopes:

ISOTOPE	MASS (amu)	FRACTIONAL ABUNDANCE
Chlorine-35	34.96885	0.75771
Chlorine-37	36.96590	0.24229

What is the atomic weight of chlorine? (See Problems 7.33, 7.34, 7.35, and 7.36.)

Instrumental Methods

MASS SPECTROMETRY AND MOLECULAR STRUCTURE

Sophisticated instruments have become indispensable in modern chemical analysis and research. One such instrument is the mass spectrometer. The mass spectrum of a compound is used in combination with other information to identify a substance or to obtain the molecular structure of a newly prepared compound.

Positive ions of molecules, like those of atoms, can be generated by bombarding the gas, or vapor of the substance, with electrons. The spectra of molecules, however, are usually much more complicated than those of atoms. One reason is that the molecular ions produced often break into fragments, giving several different kinds of positive ions. Consider the CH_2Cl_2 molecule (methylene chloride). When this molecule is struck by a high-energy electron, a positive ion $CH_2Cl_2^+$ may form.

$$CH_2Cl_2 + e^- \longrightarrow CH_2Cl_2^+ + 2e^-$$

The $CH_2Cl_2^+$ ion gains a great deal of energy from the collision of CH_2Cl_2 with the electron, and the ion frequently loses this energy by breaking into smaller pieces. One way it could do this is

$$CH_2Cl_2^+ \longrightarrow CH_2Cl^+ + Cl$$

Thus, the original molecule, even one as simple as CH_2Cl_2, can give rise to a number of ions.

The second reason for the complexity of the mass spectrum of a molecular substance is that many of the atoms in any ion can occur with different isotopic mass, so each ion often has many peaks. The mass spectrum of methylene chloride, CH_2Cl_2, is shown in Figure 7.12. Fourteen peaks are clearly visible. A larger molecule can give an even more complicated spectrum.

Because of the complexity of the mass spectrum, it can be used as a "fingerprint" in identifying a compound. Only methylene chloride has exactly the spectrum shown in Figure 7.12. Thus, by comparing the mass spectrum of an unknown substance with those in a cata-

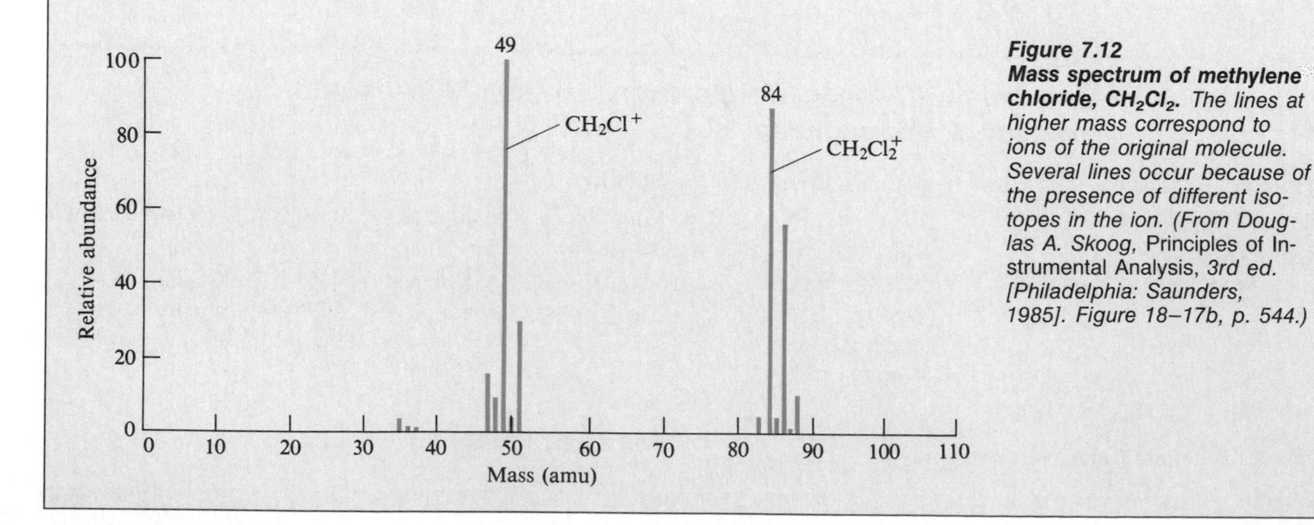

Figure 7.12
Mass spectrum of methylene chloride, CH_2Cl_2. *The lines at higher mass correspond to ions of the original molecule. Several lines occur because of the presence of different isotopes in the ion. (From Douglas A. Skoog,* Principles of Instrumental Analysis, *3rd ed. [Philadelphia: Saunders, 1985]. Figure 18–17b, p. 544.)*

log of mass spectra of known compounds, we can determine its identity. The more information we have about the compound, the shorter our search through the catalog of spectra.

The mass spectrum itself contains a wealth of information about molecular structure. Some experience is needed to analyze the spectrum of a compound, but we can get an idea of how it is done by looking at the mass spectrum shown in Figure 7.12. Suppose we do not know the identity of the compound. The most intense peaks at the greatest mass correspond to the ion from the original molecule and give us the molecular weight. Thus, we expect the peaks at mass 84 and mass 86 to be from the original molecular ion, so the molecular weight is approximately 84 to 86.

An elemental analysis is also possible. The two most intense peaks in Figure 7.12 are those at 84 amu and 49 amu. These two peaks differ by 35 amu. Perhaps the original molecule that gives the peak at 84 amu contains a chlorine-35 atom, which is lost to give the peak at 49 amu. If this is true, we should expect a weaker peak at mass 86, corresponding to the original molecular ion with chlorine-37 in place of a chlorine-35 atom. This is indeed what we see.

The relative heights of the peaks, which depend on the natural abundances of atoms, are also important because they can give us additional information about the elements present. The relative heights can also tell us how many atoms of a given element are in the original molecule. Naturally occurring chlorine is 75.8% chlorine-35 and 24.2% chlorine-37. If the original molecule contained only one Cl atom, the peaks at 84 amu and 86 amu would be in the ratio 0.758:0.242. That is, the peak at 86 amu would be about a third the height of the one at 84 amu. In fact, the relative height is twice this value. This means that the molecular ion contains two chlorine atoms, because the chance that any such ion contains one chlorine-37 atom is then twice as great. Thus, simply by comparing relative peak heights, we can both confirm the presence of particular elements and obtain the molecular formula.

Electronic Structure of Atoms

So far we have described the basic structure of atoms. To understand the formation of chemical bonds, however, we need to look at the electronic structure of atoms. Our present theory of electronic structure started with explanations for the colored light produced by atoms in hot gases and flames. Before we can discuss this, we need to describe the nature of light.

7.5 THE WAVE NATURE OF LIGHT

If we drop a stone into one end of a quiet pond, the impact of the stone with the water starts an up-and-down motion of the water surface. This up-and-down motion travels outward from where the stone hit; it is a familiar example of a wave. A *wave* is a continuously repeating change or oscillation in matter or in a physical field. Light is also a wave. It consists of oscillations in electric and magnetic fields that can travel through space. Visible light, as well as x rays and radio waves, are forms of *electromagnetic radiation*.

A wave is characterized by its wavelength and frequency. The **wavelength,** denoted by the Greek letter λ (lambda), is *the distance between any two adjacent identical points of a wave*. Thus, the wavelength is the distance between two adjacent high points (maxima). Figure 7.13 shows a cross section of a water wave at a given moment, with the wavelength identified. Radio waves have wavelengths from approximately 100 mm to several hundred meters. Visible light has much shorter wavelengths, about 10^{-6} m. Wavelengths of visible light are often given in

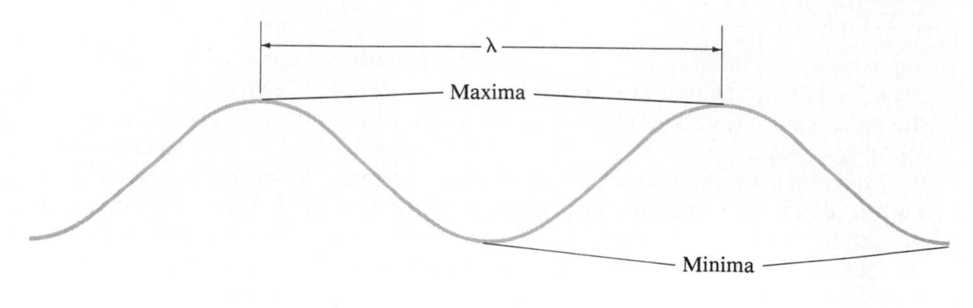

Figure 7.13
Water wave (ripple). The wavelength (λ) is the distance between two maxima (or two minima).

nanometers (1 nm = 10^{-9} m). Thus, light of wavelength 5.55×10^{-7} m, the greenish-yellow light to which the human eye is most sensitive, equals 555 nm.

The **frequency** of a wave is *the number of wavelengths of that wave that pass a fixed point in one unit of time* (usually one second). For example, imagine you are anchored in a small boat on a pond when a stone is dropped into the water. Waves travel outward from this point and move past your boat. The number of wavelengths that pass you in one second is the frequency of that wave. Frequency is denoted by the Greek letter ν (nu, pronounced "new"). The unit of frequency is /s, or s^{-1}, also called the *hertz* (Hz).

The wavelength and frequency of a wave are related to each other. With a wave of frequency ν and wavelength λ, there are ν wavelengths, each of length λ, that pass a fixed point every second. The product $\nu\lambda$ is the total length of the wave that has passed the point in one second. This length of wave per second is the speed of the wave, c. Thus,

$$c = \nu\lambda$$

The speed of light waves depends on the medium through which they are passing. In a vacuum, the speed is 3.00×10^8 m/s, and the speed in air is only slightly less. For the calculations in this section, we will assume the light is traveling in a vacuum.●

● To understand how fast the speed of light is, it might help to realize that it takes only 2.5 s for radar waves (which travel at the speed of light) to leave earth, bounce off the moon, and return—a total distance of 478,000 miles.

Example 7.4 Obtaining the Wavelength of Light from its Frequency

What is the wavelength of the yellow sodium emission, which has a frequency of 5.09×10^{14}/s?

Solution

The frequency and wavelength are related by the formula $c = \nu\lambda$. This can be rearranged to give

$$\lambda = \frac{c}{\nu}$$

in which c is the speed of light (3.00×10^8 m/s). Substituting, we have

$$\lambda = \frac{3.00 \times 10^8 \text{ m/}\cancel{s}}{5.09 \times 10^{14}/\cancel{s}} = 5.89 \times 10^{-7} \text{ m, or } \mathbf{589\ nm}$$

Exercise 7.5

The frequency of the strong red line in the spectrum of potassium is 3.91×10^{14}/s. What is the wavelength of this light in nanometers? (See Problems 7.41 and 7.42.)

Example 7.5 Obtaining the Frequency of Light from its Wavelength

What is the frequency of violet light with a wavelength of 408 nm?

Solution

The equation relating frequency and wavelength can also be rearranged to give

$$\nu = \frac{c}{\lambda}$$

Substituting for λ (408 nm = 408×10^{-9} m) gives

$$\nu = \frac{3.00 \times 10^{8}\ \cancel{m}/s}{408 \times 10^{-9}\ \cancel{m}} = \mathbf{7.35 \times 10^{14}/s}$$

Exercise 7.6

The element cesium was discovered in 1860 by Robert Bunsen and Gustav Kirchhoff, who found two bright blue lines in the spectrum of a substance isolated from a mineral water. One of these spectral lines of cesium has a wavelength of 456 nm. What is the frequency of this line? (See Problems 7.43 and 7.44.)

The range of frequencies or wavelengths of electromagnetic radiation is called the **electromagnetic spectrum,** shown in Figure 7.14. Visible light extends from the violet end of the spectrum, which has a wavelength of about 400 nm, to the red end of the spectrum, with a wavelength of less than 800 nm. Beyond these extremes, electromagnetic radiation is not visible to the human eye. Infrared radiation has wavelengths greater than 800 nm (greater than the wavelength of red light), and ultraviolet radiation has wavelengths less than 400 nm (less than the wavelength of violet light).

7.6 QUANTUM EFFECTS AND PHOTONS

Isaac Newton, who studied the properties of light in the seventeenth century, believed that light consisted of a beam of particles. In 1801, however, British

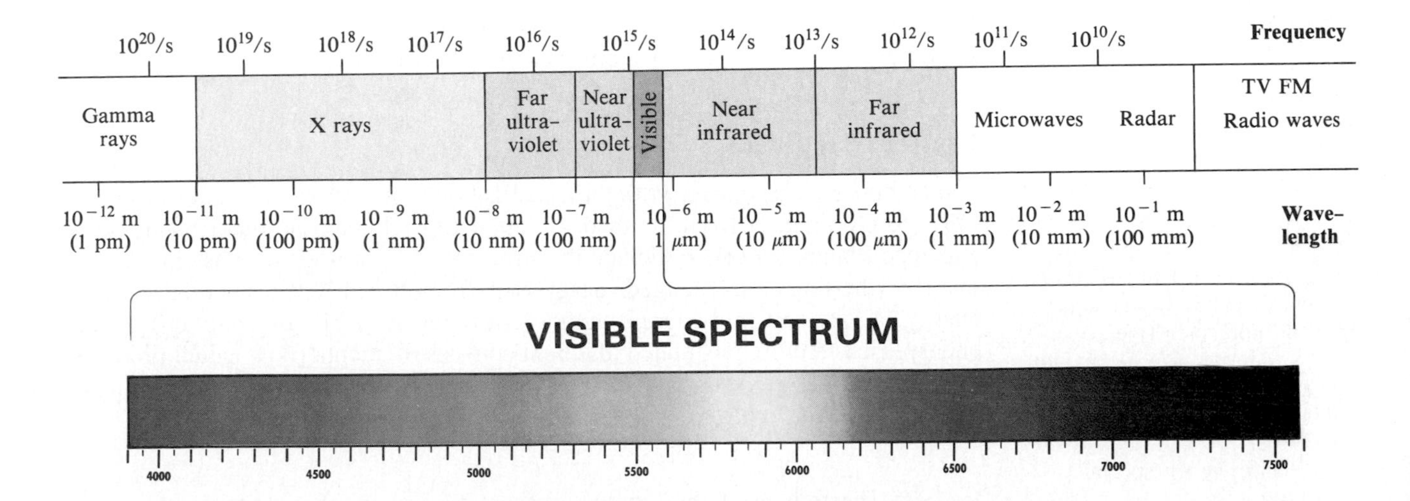

Figure 7.14
The electromagnetic spectrum. Divisions between regions are not defined precisely.

physicist Thomas Young showed that light, like waves, could be diffracted. *Diffraction* is a property of waves in which the waves spread out when they encounter an obstruction or small hole about the size of the wavelength. We can observe diffraction by viewing a light source through a hole—for example, a street light through a mesh curtain. The image of the street light is blurred by diffraction. It was further shown in the early part of this century that light has particle properties as well as wave properties. This wave–particle view of light was first postulated by the German-born U.S. physicist Albert Einstein (1879–1955), starting from the work of the German physicist Max Planck (1858–1947).

Planck's Quantization of Energy

In 1900 Max Planck found a theoretical formula that exactly describes the intensity of light of various frequencies emitted by a hot solid at different temperatures.● Earlier, others had shown experimentally that the light of maximum intensity from a hot solid varies in a definite way with temperature. A solid glows red at 750°C, then white as the temperature increases to 1200°C. At the lower temperature, chiefly red light is emitted. As the temperature increases, more yellow and blue light become mixed with the red, giving white light.●

● Max Planck was Professor of Physics at the University of Berlin when he did this research. He received the Nobel Prize in physics for it in 1918.

● Radiation emitted from the human body is mostly infrared, which is the principle behind burglar alarms that have infrared detectors. Also, observation of how the maximum intensity of infrared (which depends on temperature) varies over the body surface is used to detect breast cancer (cancerous tissue is warmer) and potential strokes (narrowed arteries reduce the surface temperature of the head).

According to Planck, the atoms of the solid oscillate, or vibrate, with a definite frequency ν. But in order to reproduce the results of experiments on glowing solids, he found it necessary to accept a strange idea: A vibrating atom could have only certain energies E, those allowed by the formula

$$E = nh\nu, \quad n = 1, 2, 3, \ldots$$

where h is a constant, now called **Planck's constant,** *a physical constant with the value* $6.63 \times 10^{-34}\ J \cdot s$. The value of n must be 1 or 2 or some other whole number. Thus, the only energies a vibrating atom can have are $h\nu$, $2h\nu$, $3h\nu$, and so forth.

The numbers symbolized by n are called *quantum numbers*. The vibrational energies of the atoms are said to be *quantized;* that is, the possible energies are limited to certain values. Quantization of energy seems contradicted by everyday experience. Imagine the same concept being applied to the energy of a moving automobile. Quantization of a car's energy would mean that only certain speeds were possible. A car could travel at, say, 10, 20, or 30 miles per hour (mi/h) but not at 12 or 25 mi/h or at any other intermediate speeds. For car energies, this seems unreasonable. But for atoms, quantization is the rule.

Photoelectric Effect

Planck himself was uneasy with the quantization assumption and tried (unsuccessfully) to eliminate it from his theory. Albert Einstein, on the other hand, boldly extended Planck's work to include the structure of light itself.● Einstein reasoned that if a vibrating atom changed energy, say, from $3h\nu$ to $2h\nu$, it would decrease in energy by $h\nu$, and this energy would be emitted as a bit (or quantum) of light energy. He therefore postulated that light consists of quanta (now called **photons**), or *particles of electromagnetic energy, with energy* E *proportional to the observed frequency of the light:*

● Albert Einstein obtained a Ph.D. from the University of Zurich in 1905. In the same year he published five papers in *Annalen der Physik,* including one on the photoelectric effect (for which he received the Nobel Prize in physics in 1921) and two on special relativity.

$$E = h\nu$$

In 1905 Einstein used this photon concept to explain the photoelectric effect.

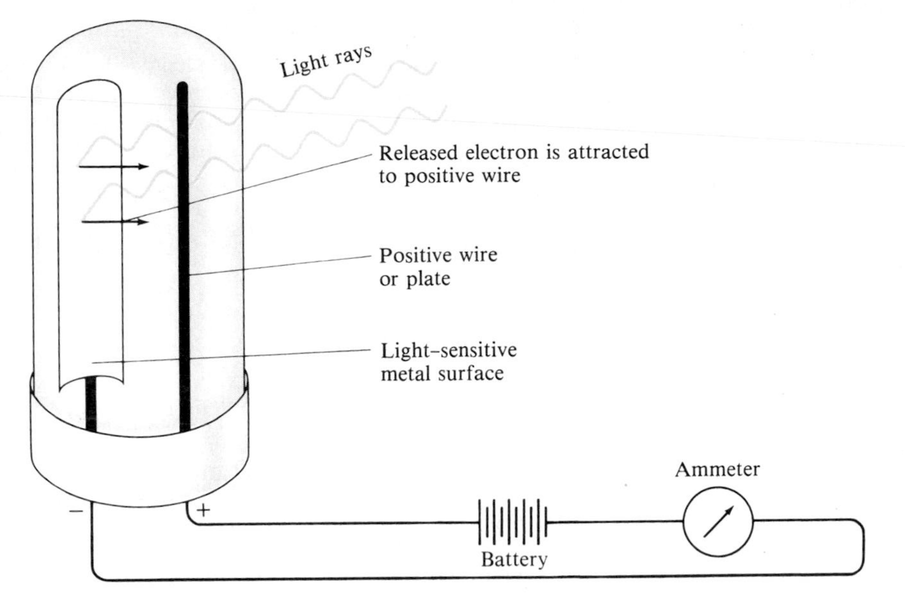

Figure 7.15
The photoelectric effect. Light shines on a metal surface, knocking out electrons. The metal surface is contained in an evacuated tube, so that the ejected electrons can be accelerated to a positively charged plate. As long as light of sufficient frequency shines on the metal, free electrons are produced, and a current flows through the tube. (The current is measured by an ammeter.) When the light is turned off, the current stops flowing.

The **photoelectric effect** is *the ejection of electrons from the surface of a metal or other material when light shines on it* (see Figure 7.15). Electrons are ejected, however, only when the frequency of light exceeds a certain *threshold value* characteristic of the particular metal. For example, although violet light will cause potassium metal to eject electrons, no amount of red light (which has a lower frequency) has any effect.

To explain this dependence of the photoelectric effect on the frequency, Einstein assumed that an electron is ejected from a metal when it is struck by a single photon. Therefore, this photon must have at least enough energy to remove the electron from the attractive forces of the metal. No matter how many photons strike the metal, if no single one has sufficient energy, an electron cannot be ejected. A photon of red light has insufficient energy to remove an electron from potassium. But a photon corresponding to the threshold frequency has just enough energy, and at higher frequencies it has more than enough energy. When the photon hits the metal, its energy $h\nu$ is taken up by the electron. The photon ceases to exist as a particle; it is said to be *absorbed*.

The wave and particle pictures of light should be regarded as complementary views of the same physical entity. This is called the *wave–particle duality* of light. Neither the wave nor the particle view alone is a complete description of light.

Example 7.6 Calculating the Energy of a Photon

The red spectral line of lithium occurs at 671 nm (6.71×10^{-7} m). Calculate the energy of one photon of this light.

Solution

The frequency of this light is

$$\nu = \frac{c}{\lambda} = \frac{3.00 \times 10^8 \cancel{\text{m}}/\text{s}}{6.71 \times 10^{-7} \cancel{\text{m}}} = 4.47 \times 10^{14}/\text{s}$$

Hence, the energy of one photon is

$$E = h\nu = 6.63 \times 10^{-34} \text{ J} \cdot \cancel{\text{s}} \times 4.47 \times 10^{14}/\cancel{\text{s}}$$
$$= \mathbf{2.96 \times 10^{-19} \text{ J}}$$

Exercise 7.7

The following are representative wavelengths in the infrared, ultraviolet, and x-ray regions of the electromagnetic spectrum, respectively: 1.0×10^{-6} m, 1.0×10^{-8} m, and 1.0×10^{-10} m. What is the energy of a photon of each radiation? Which has the greatest amount of energy per photon? Which has the least? (See Problems 7.45, 7.46, 7.47, and 7.48.)

7.7 THE BOHR THEORY OF THE HYDROGEN ATOM

According to Rutherford's nuclear model, the atom consists of a nucleus with most of the mass and a positive charge, around which move enough electrons to make the atom electrically neutral. But this model posed a dilemma. Using the then-current theory, one could show that an electrically charged particle (such as an electron) that revolves around a center would continuously lose energy as electromagnetic radiation. As an electron in an atom lost energy, it would spiral into the nucleus (in about 10^{-10} s, according to available theory). The stability of the atom could not be explained.

The solution to this theoretical dilemma was found in 1913 by Niels Bohr (1885–1962), a Danish physicist, who at the time was working with Rutherford (see Figure 7.16). Using the work of Planck and Einstein, Bohr applied a new theory to the simplest atom, hydrogen. Before we look at Bohr's theory, we need to consider the line spectra of atoms.

Atomic Line Spectra

As described in the previous section, a heated solid emits light. A heated tungsten filament in an ordinary light bulb is a typical example. We can spread out the light from a bulb with a prism to give a **continuous spectrum**—that is, *a spectrum containing light of all wavelengths,* like that of a rainbow (see Figure 7.17). The light emitted by a heated gas, however, yields different results. Rather than seeing a continuous spectrum, with all colors of the rainbow, we obtain a **line spectrum**—*a spectrum showing only certain colors or specific wavelengths of light*. When the light from a hydrogen gas discharge tube is separated into its components by a

Figure 7.16
Niels Bohr (1885–1962). After Bohr developed his quantum theory of the hydrogen atom, he used his ideas to explain the periodic behavior of the elements. Later, when the new quantum mechanics was discovered by Schrödinger and Heisenberg, Bohr spent much of his time developing its philosophical basis. He received the Nobel Prize in physics in 1922.

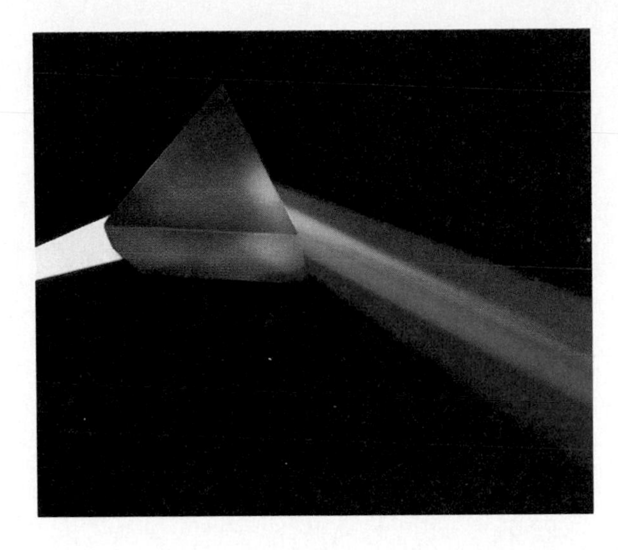

Figure 7.17
Dispersion of white light by a prism. White light, entering at the left, strikes a prism, which disperses the light into a continuous spectrum of wavelengths.

prism, it gives a spectrum of lines, each line corresponding to light of a given wavelength. The light produced in the discharge tube is emitted by hydrogen atoms. Figure 7.2 shows the line spectrum of the hydrogen atom, as well as line spectra of other atoms.

The line spectrum of the hydrogen atom is especially simple. In the visible region, it consists of only four lines (a red, a blue-green, a blue, and a violet), although others appear in the infrared and ultraviolet regions. In 1885 J. J. Balmer showed that the wavelengths λ in the visible spectrum of hydrogen could be reproduced by a simple formula:

$$\frac{1}{\lambda} = 1.097 \times 10^7/\text{m}\left(\frac{1}{2^2} - \frac{1}{n^2}\right)$$

Here n is some whole number (integer) greater than 2. By putting $n = 3$, for example, and calculating $1/\lambda$ and then λ, one finds $\lambda = 6.56 \times 10^{-7}$ m, or 656 nm, a wavelength corresponding to red light. The wavelengths of the other lines in the H-atom spectrum are obtained by successively putting $n = 4$, $n = 5$, and $n = 6$.

Bohr's Postulates

Bohr set down the following two postulates to account for (1) the stability of the hydrogen atom (that the atom exists and its electron does not continuously radiate energy and spiral into the nucleus) and (2) the line spectrum of the atom.

1. ENERGY-LEVEL POSTULATE An electron can have only *specific energy values in an atom,* which are called its **energy levels.** Thus, the atom itself can have only specific total energy values.

Bohr borrowed the idea of quantization of energy from Planck. Bohr, however, devised a rule for this quantization that could be applied to the motion of an electron in an atom. From this he derived the following formula for the energy levels of the electron in the hydrogen atom:

$$E = -\frac{R_H}{n^2} \quad n = 1, 2, 3, \ldots \infty \qquad \text{(for H atom)}$$

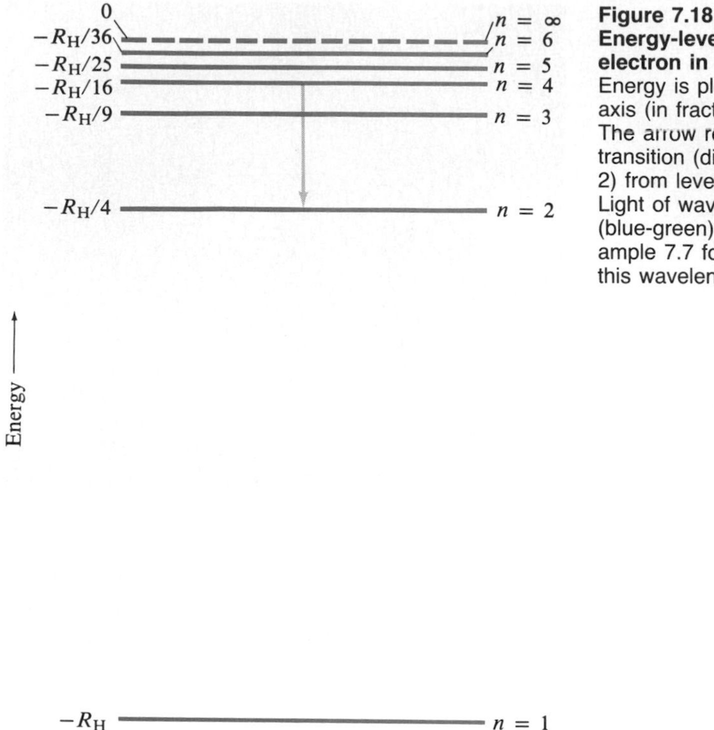

Figure 7.18
Energy-level diagram for the electron in the hydrogen atom. Energy is plotted on the vertical axis (in fractional multiples of R_H). The arrow represents an electron transition (discussed in postulate 2) from level $n = 4$ to level $n = 2$. Light of wavelength 486 nm (blue-green) is emitted. (See Example 7.7 for the calculation of this wavelength.)

where R_H is a constant (the *Rydberg constant* expressed in energy units) with the value 2.180×10^{-18} J.● Different values of the possible energies of the electron are obtained by putting in different values of n, which can have only the integral values 1, 2, 3, and so forth (up to ∞). Here n is called the *principal quantum number*. The diagram in Figure 7.18 shows the energy levels of the electron in the H atom.

● The energies have negative values because the separated nucleus and electron are taken to have zero energy. If the atom is to be stable, it must have less energy than this. Energy levels for hydrogen-like atomic ions (consisting of a nucleus of charge Z and a single electron) equal $-Z^2R_H/n^2$.

2. TRANSITIONS BETWEEN ENERGY LEVELS An electron in an atom can change energy only by going from one energy level to another energy level. By so doing, the electron undergoes a *transition*. We explain the emission of light by atoms to give a line spectrum as follows: An electron in a higher energy level (initial energy level E_i) undergoes a transition to a lower energy level (final energy level, E_f). (See Figure 7.18.) In this process, the electron loses energy, which is emitted as a photon. We obtain the energy of the photon by subtracting the lower energy of the electron, E_f, from the higher value, E_i. This energy difference (the energy of the photon) is related to the frequency of the emitted light by Einstein's equation $E = h\nu$.

$$\text{Energy of emitted photon} = E_i - E_f = h\nu$$

In this postulate, Bohr used Einstein's photon concept to explain the line spectra of atoms. By substituting values of the energy levels of the electron in the H atom, which he had derived, into the preceding equation, he was able to reproduce Balmer's formula exactly. Moreover, he was able to predict all of the lines in the spectrum of the hydrogen atom, including those in the infrared and ultraviolet regions.

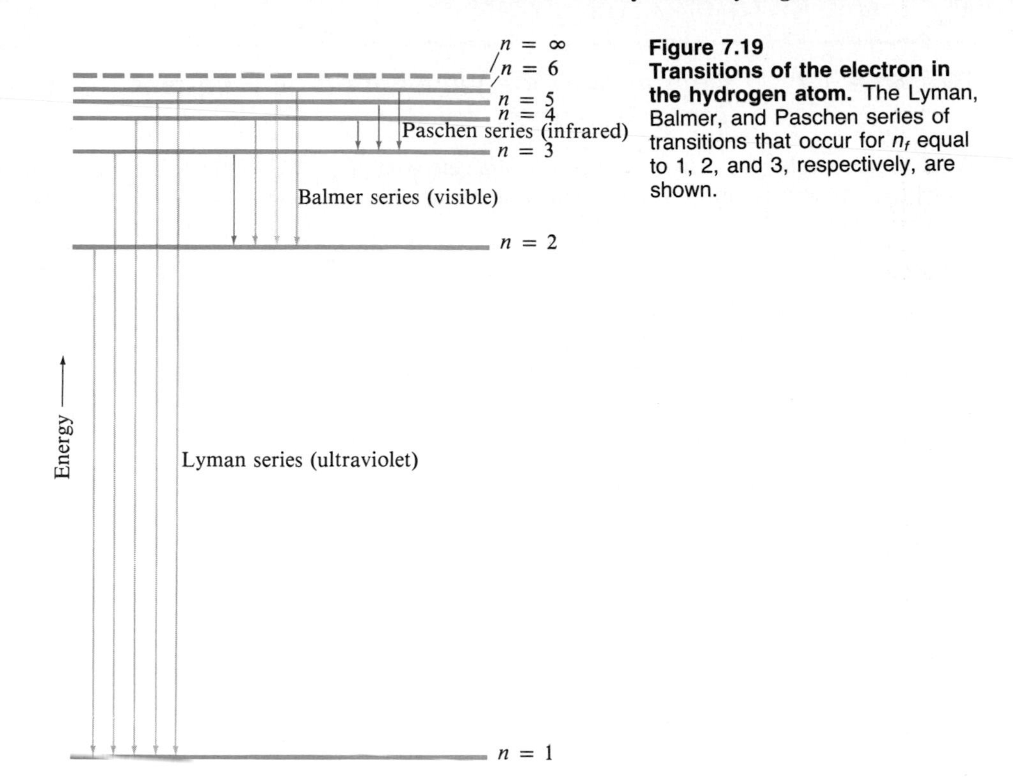

Figure 7.19
Transitions of the electron in the hydrogen atom. The Lyman, Balmer, and Paschen series of transitions that occur for n_f equal to 1, 2, and 3, respectively, are shown.

To show how Bohr obtained Balmer's formula, let us write n_i for the principal quantum number of the initial energy level, and n_f for the principal quantum number of the final energy level. Then, from postulate 1,

$$E_i = -\frac{R_H}{n_i^2} \quad \text{and} \quad E_f = -\frac{R_H}{n_f^2}$$

We substitute these into the equation in postulate 2, which states that the energy of the emitted photon, $h\nu = E_i - E_f$.

$$E_i - E_f = \left(-\frac{R_H}{n_i^2}\right) - \left(-\frac{R_H}{n_f^2}\right) = h\nu$$

That is,

$$h\nu = R_H\left(\frac{1}{n_f^2} - \frac{1}{n_i^2}\right)$$

If we now recall that $\nu = c/\lambda$, we can rewrite this as

$$\frac{1}{\lambda} = \frac{R_H}{hc}\left(\frac{1}{n_f^2} - \frac{1}{n_i^2}\right)$$

By substituting $R_H = 2.180 \times 10^{-18}$ J, $h = 6.626 \times 10^{-34}$ J · s, and $c = 2.998 \times 10^8$ m/s, we find that $R_H/hc = 1.097 \times 10^7$/m, which is the constant given in the Balmer formula. In Balmer's formula, the quantum number n_f is 2. This means that Balmer's formula gives wavelengths that occur when electrons in H atoms undergo transitions from energy levels $n_i > 2$ to level $n_f = 2$. If we change n_f to other integers, we obtain different series of lines (or wavelengths) for the spectrum of the H atom (see Figure 7.19).

Example 7.7 Determining the Wavelength or Frequency of a Hydrogen Atom Transition

What is the wavelength of light emitted when the electron in a hydrogen atom undergoes a transition from energy level $n = 4$ to level $n = 2$?

Solution

We start with the formula for the energy levels of the H atom, $E = -R_H/n^2$, and obtain the energy change for the transition. This energy difference will equal the energy of the photon, from which we can calculate frequency, then wavelength, of the emitted light. (Although we could do this problem by "plugging into" the equation for $1/\lambda$ previously derived, the method followed here requires us to remember only key formulas, shown in color.)

From the formula for the energy levels, we know that

$$E_i = \frac{-R_H}{4^2} = \frac{-R_H}{16} \quad \text{and} \quad E_f = \frac{-R_H}{2^2} = \frac{-R_H}{4}$$

We subtract the lower value from the higher value (to get a positive result). Because this equals the energy of the photon, we equate it to $h\nu$:

$$\left(\frac{-R_H}{16}\right) - \left(\frac{-R_H}{4}\right) = \frac{3R_H}{16} = h\nu$$

The frequency of the light emitted is

$$\nu = \frac{3R_H}{16\,h} = \frac{3}{16} \times \frac{2.180 \times 10^{-18}\,\cancel{J}}{6.63 \times 10^{-34}\,\cancel{J}\cdot s} = 6.17 \times 10^{14}/s$$

Since $\lambda = c/\nu$,

$$\lambda = \frac{3.00 \times 10^8\,m/\cancel{s}}{6.17 \times 10^{14}/\cancel{s}} = 4.86 \times 10^{-7}\,m\ \mathbf{(486\ nm)}$$

The color is blue-green (see Figure 7.18).

Exercise 7.8

Calculate the wavelength of light emitted from the hydrogen atom when the electron undergoes a transition from level $n = 3$ to level $n = 1$.

(See Problems 7.49, 7.50, 7.51, and 7.52.)

According to Bohr's theory, the *emission* of light from an atom occurs when an electron undergoes a transition from an upper energy level to a lower one. To complete the explanation, we need to describe how the electron gets into the upper level prior to emission. Normally, the electron in a hydrogen atom exists in its lowest, or $n = 1$, level. To get into a higher energy level, the electron must gain energy, or be *excited*. One way this can happen is through the collision of two hydrogen atoms. During this collision, some of the kinetic energy of one atom can be gained by the electron of another atom, thereby boosting or exciting the electron from the $n = 1$ level to a higher energy level.

Bohr's theory explains not only the emission but also the *absorption* of light. When an electron in the hydrogen atom undergoes a transition from $n = 3$ to $n = 2$, a photon of red light (wavelength of 656 nm) is emitted. When red light whose wavelength is 656 nm shines on a hydrogen atom in the $n = 2$ level, a photon can be absorbed. If the photon is absorbed, the energy is gained by the electron, which undergoes a transition to the $n = 3$ level. (This is the reverse of the emission process we just discussed.) Materials that have a color, such as dyed textiles and painted walls, appear colored because of the absorption of light. For example, when white light falls on a substance that absorbs red light, the color components that are not absorbed, the yellow and blue light, are reflected. The substance appears blue-green.

Postulates 1 and 2 hold for atoms other than hydrogen, except that the energy levels cannot be obtained by a simple formula. However, if we know the wavelength of the emitted light, we can relate it to ν and then to the difference in energy levels of the atom. The energy levels of atoms have been experimentally determined in this way.

Exercise 7.9

What is the difference in energy levels of the sodium atom if emitted light has a wavelength of 589 nm? (See Problems 7.55 and 7.56.)

Instrumental Methods

LASERS

Lasers are sources of intense, highly directed beams of *monochromatic* light—light of very narrow wavelength range. Figure 7.20 shows a laser being used in a research project. The word *laser* is an acronym meaning *l*ight *a*mplification by *s*timulated *e*mission of *r*adiation. Many different kinds of lasers now exist, but the general principle of a laser can be understood by looking at the ruby laser, the first type constructed (in 1960).

Ruby is aluminum oxide containing a small concentration of chromium(III) ions, Cr^{3+}, in place of some aluminum ions. The electron transitions in a ruby laser are those of Cr^{3+} ions in solid Al_2O_3. An energy-level diagram of Cr^{3+} in ruby is shown in Figure 7.21. Most of the Cr^{3+} ions are initially in the lowest energy level (level 1). If we shine light of 545 nm wavelength on a ruby crystal, the light is absorbed and Cr^{3+} ions undergo transitions from level 1 to level 3. A few of these ions in level 3 emit photons and return to level 1, but most of them undergo *radiationless transitions* to level 2. In these transitions, the ions lose energy as heat to the ruby crystal, rather than emit photons.

The Cr^{3+} ion in level 2, as any species in a level other than the lowest, is unstable and in time will undergo a transition to level 1. Thus, we can expect the Cr^{3+} ions in level 2 to undergo transitions to level 1 with the *spontaneous emission* of photons (the wavelength of this emission is 694 nm). However, this spontaneous emission of Cr^{3+} is relatively slow. If we flash a ruby rod with a bright light at 545 nm, most of the Cr^{3+} ions end up in level 2 for perhaps a fraction of a millisecond. This buildup of many excited species is crucial to the operation of a laser. If these excited ions can be triggered to emit simultaneously, or nearly simultaneously, an intense emission will be obtained.

The process of *stimulated emission* is ideal for this triggering. When a photon corresponding to 694 nm encounters a Cr^{3+} ion in level 2, it stimulates the ion to undergo the transition from level 2 to level 1. The ion emits a photon corresponding to exactly the same wavelength as the original photon. In place of just one photon, there are now two photons, the original one and the one obtained by stimulated emission. The net effect is to increase the intensity of the light at this wavelength. Thus,

(continued)

Figure 7.20
A pulse of laser light being converted to white light. *A femtosecond (10^{-15} s) pulse of red laser light is converted to white light, then dispersed with a prism to show colors of the rainbow.*

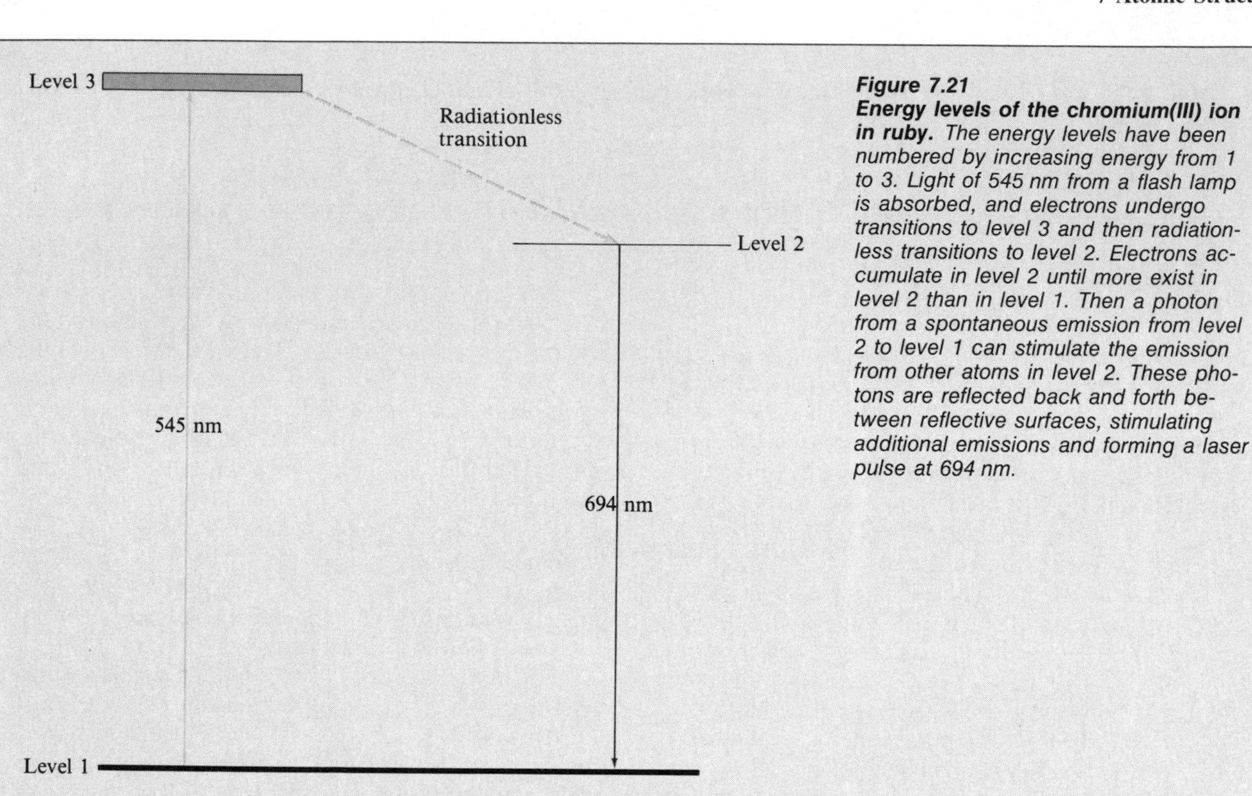

Figure 7.21
Energy levels of the chromium(III) ion in ruby. *The energy levels have been numbered by increasing energy from 1 to 3. Light of 545 nm from a flash lamp is absorbed, and electrons undergo transitions to level 3 and then radiationless transitions to level 2. Electrons accumulate in level 2 until more exist in level 2 than in level 1. Then a photon from a spontaneous emission from level 2 to level 1 can stimulate the emission from other atoms in level 2. These photons are reflected back and forth between reflective surfaces, stimulating additional emissions and forming a laser pulse at 694 nm.*

a weak light at 694 nm can be amplified by stimulated emission of the excited ruby.

We can now explain the operation of a ruby laser. A sketch of the laser is shown in Figure 7.22. It consists of a ruby rod a centimeter or less in diameter and 2 cm to 10 cm long. The ruby rod is silvered at one end to provide a completely reflective surface; the other end has a partially reflective surface. A coiled flash lamp tube is placed around the ruby rod. When the flash lamp is discharged, a bright light is emitted, and light at 545 nm (green) is absorbed by the ruby. Most of the Cr^{3+} ions end up in level 2. A few of these ions spontaneously emit photons corresponding to 694 nm (red light), and these photons stimulate other ions to emit. As the photons are reflected back and forth between the reflective surfaces, more and more ions are stimulated to emit, and the light quickly builds up in intensity until it passes through the partially reflective surface as a pulse of laser light at 694 nm.

Laser light is *coherent*. This means that the waves forming the beam are all in phase; that is, the waves

Figure 7.22
A ruby laser. *A flash lamp tube encircles a ruby rod. Light of 545 nm (green) from the flash "pumps" electrons from level 1 to level 3, then level 2, from which stimulated emission forms a laser pulse at 694 nm (red). The stimulated emission bounces back and forth between reflective surfaces at the ends of the ruby rod, building up a coherent laser beam. One end has a partially reflective surface to allow the laser beam to exit from the ruby.*

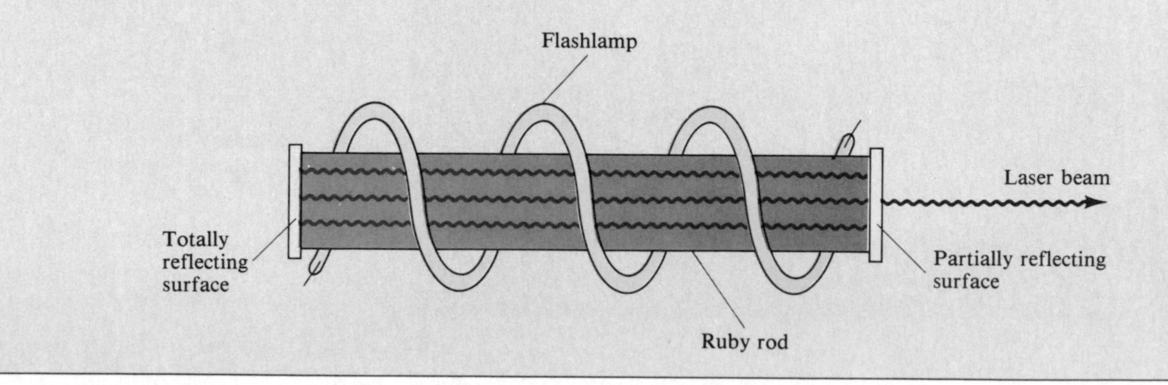

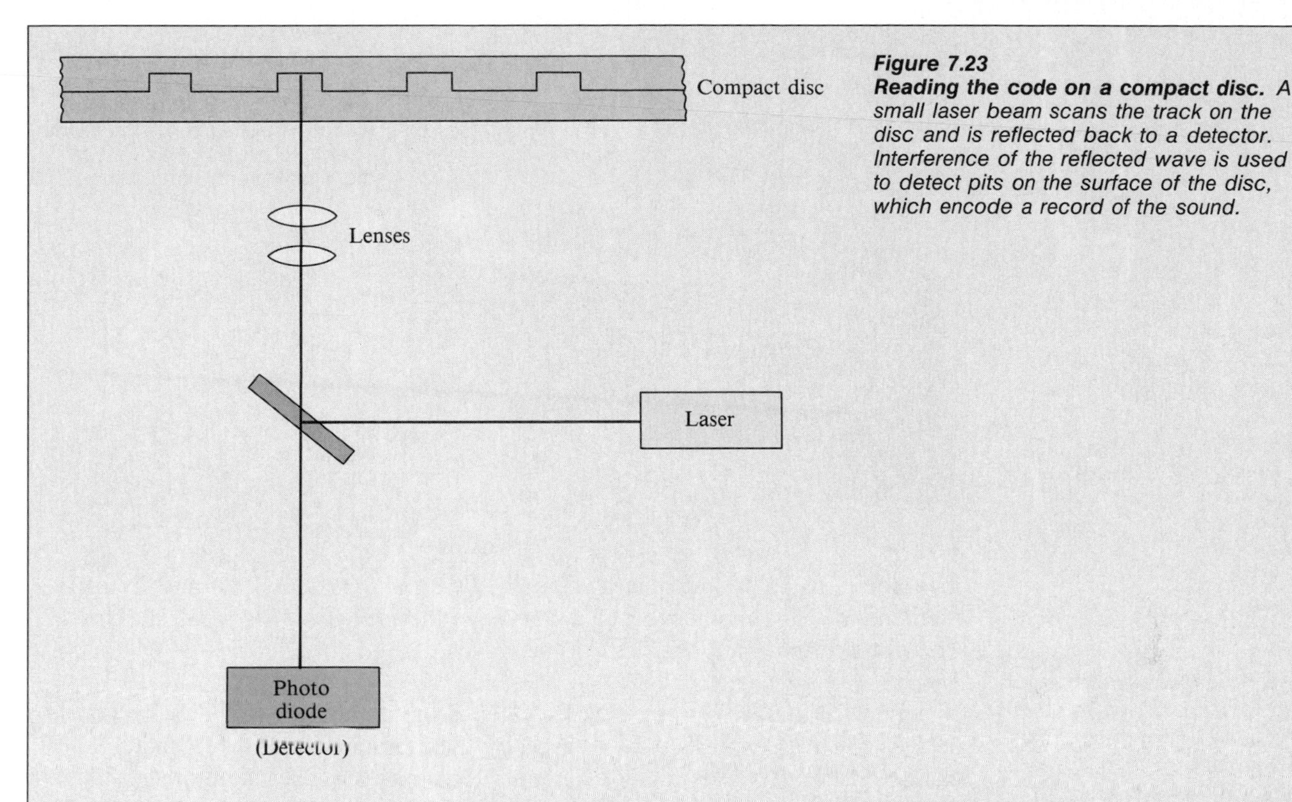

Figure 7.23
Reading the code on a compact disc. *A small laser beam scans the track on the disc and is reflected back to a detector. Interference of the reflected wave is used to detect pits on the surface of the disc, which encode a record of the sound.*

have their maxima and minima at the same points in space and time. The property of coherence of a laser beam is used in compact disc (CD) audio players. Music is encoded on the disc in the form of pits or indentations on a spiral track. When the disc is played, a small laser beam scans the track and is reflected back to a detector. Light that is reflected from an indentation is out of phase with light reflected from the main surface and interferes with it (Figure 7.23). Because of this interference, the reflected beam is diminished in intensity and gives a diminished detector signal. Fluctuations in the signal are then converted to sound.

Other properties of laser light are used in different applications. The ability to focus intense light on a spot is used in the surgical correction of a detached retina in the eye. In effect, the laser beam is used to do "spot welding" of the retina. Laser printers also employ the intensity of a laser beam. These printers follow the principle of photocopiers but use a computer to direct the laser light in a pattern of dots to form an image. In chemical research, laser beams provide intense monochromatic light for locating energy levels in molecules, to study the products of very fast chemical reactions, and to analyze samples for small amounts of substances.

7.8 QUANTUM MECHANICS

Bohr's theory firmly established the concept of atomic energy levels. It was unsuccessful, however, in accounting for the details of atomic structure and in predicting energy levels for atoms other than hydrogen. Further understanding of atomic structure required another theoretical development.

de Broglie Relation

According to Einstein, light has not only wave properties, which we characterize by frequency and wavelength, but also particle properties. Thus, a particle of light,

Figure 7.24
Scanning electron microscope. This microscope has a magnifying power up to 300,000 times. The sample is placed inside the chamber at the left and is viewed on the video screen in the center.

the photon, has a definite energy $E = h\nu$. One can also show that the photon has momentum. (The momentum of a particle is the product of its mass and speed.) This momentum, mc, is related to the wavelength of the light: $mc = h/\lambda$ or $\lambda = h/mc$.●

● A photon has a rest mass of zero, but a relativistic mass m as a result of its motion. Einstein's equation $E = mc^2$ relates this relativistic mass to the energy of the photon, which also equals $h\nu$. Therefore, $mc^2 = h\nu$ or $mc = h\nu/c = h\lambda$.

In 1923 the French physicist Louis de Broglie reasoned that if light (considered as a wave) exhibits particle aspects, then perhaps particles of matter show characteristics of waves under the proper circumstances. He therefore postulated that a particle of matter of mass m and speed v has an associated wavelength, by analogy with light:

$$\lambda = \frac{h}{mv}$$

The equation $\lambda = h/mv$ is called the **de Broglie relation.**

If matter has wave properties, why are they not commonly observed? Calculation using the de Broglie relation shows that a 1-kg mass moving at only 1 km/h has a wavelength of about 10^{-33} m, a value so incredibly small that such waves cannot be detected. On the other hand, electrons moving at experimentally accessible speeds have wavelengths the size of angstroms. Under the proper circumstances their wave character should be observable.

The wave property of electrons was first demonstrated in 1927 by C. Davisson and L. H. Germer in the United States and by George Paget Thomson (son of J. J. Thomson) in Britain. They showed that a beam of electrons, just like x rays, can be diffracted by a crystal. The German physicist Ernst Ruska used this wave property to construct the first *electron microscope* in 1933; he shared the 1986 Nobel Prize in physics for this work. A modern instrument is shown in Figure 7.24. The *resolving power,* or ability to distinguish detail, of a microscope that uses waves depends on their wavelength. To resolve detail that is the size of angstroms (several hundred picometers), we need a wavelength on the order of angstroms. X rays have wavelengths in this range, but so far no practical means have been found for focusing them. Electrons, on the other hand, are readily focused with electric and magnetic fields. Figure 7.25 shows a photograph taken with an electron microscope, showing the detail that is possible with this instrument.●

● In 1986, Thomas H. Newman collected a $1000 prize offered 26 years earlier by Richard Feynman, physics Nobel Prize winner. Newman, a student of R. Fabian Pease at Stanford University, satisfied Feynman's challenge to reduce a normal printed page to 1/25000th of its linear dimensions. Newman used a specially constructed electron-beam tool to etch the first page of *A Tale of Two Cities* onto an area 5.6 μm by 5.6 μm. The image could be read with an electron microscope, as Feynman required.

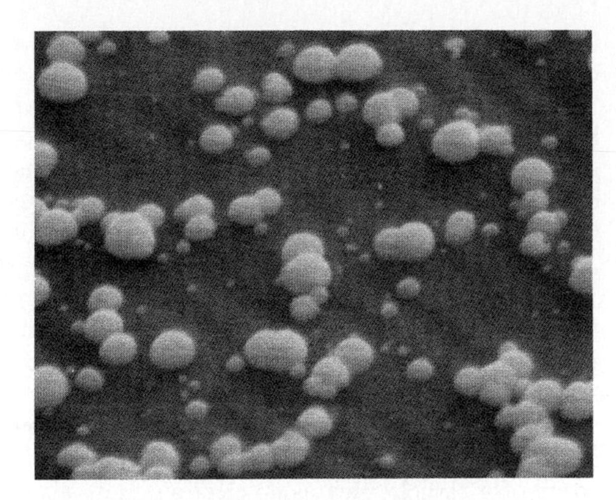

Figure 7.25
Scanning electron microscope image. Elemental platinum, Pt (bright spots), deposited on a crystal of naturally occurring molybdenite, MoS_2 (dark area). The image, magnified about 10,000 times, reveals that the surface is not uniformly coated. Instead, the platinum is found in islands.

Example 7.8 Applying the de Broglie Relation

(a) Calculate the wavelength (in meters) of the wave associated with a 1.00-kg mass moving at 1.00 km/h. (b) What is the wavelength (in angstroms) associated with an electron, whose mass is 9.11×10^{-31} kg, traveling at a speed of 4.19×10^{6} m/s? (This speed can be attained by an electron accelerated between two charged plates differing by 50.0 volts; voltages in the kilovolt range are used in electron microscopes.)

Solution

(a) A speed v of 1.00 km/h equals

$$1.00 \frac{\text{km}}{\text{h}} \times \frac{1 \text{ h}}{3600 \text{ s}} \times \frac{10^3 \text{ m}}{1 \text{ km}} = 0.278 \text{ m/s}$$

Hence, substituting quantities (all expressed in SI units for consistency), we get

$$\lambda = \frac{h}{mv} = \frac{6.63 \times 10^{-34} \text{ J} \cdot \text{s}}{1.00 \text{ kg} \times 0.278 \text{ m/s}} = \mathbf{2.38 \times 10^{-33} \text{ m}}$$

$$\text{(b) } \lambda = \frac{h}{mv} = \frac{6.63 \times 10^{-34} \text{ J} \cdot \text{s}}{9.11 \times 10^{-31} \text{ kg} \times 4.19 \times 10^{6} \text{ m/s}}$$

$$= 1.74 \times 10^{-10} \text{ m} = \mathbf{1.74 \text{ Å}}$$

Exercise 7.10

Calculate the wavelength (in angstroms) associated with an electron traveling at a speed of 2.19×10^{6} m/s.

(See Problems 7.57 and 7.58.)

● Schrödinger received the Nobel Prize in physics in 1933 for his wave formulation of quantum mechanics. Actually, Werner Heisenberg discovered quantum mechanics a few months before Schrödinger did. But Heisenberg's formulation employed matrix algebra, a mathematical discipline then seldom used by physicists. The results of Heisenberg's and Schrödinger's treatments are identical.

Wave Functions

De Broglie's relation applies quantitatively only to particles in a force-free environment. Thus, it cannot be applied directly to an electron in an atom, where the electron is subject to the attractive force of the nucleus. But in 1926 Erwin Schrödinger, guided by de Broglie's work, was able to devise a theory that could be used to find the wave properties of electrons in atoms and molecules. *The branch of physics that mathematically describes the wave properties of submicroscopic particles* is called **quantum mechanics** or **wave mechanics.**●

Without going into the mathematics of quantum mechanics here, we will discuss some of the most important conclusions of the theory. In particular, quantum mechanics alters the way we think about the motion of particles. Our usual concept of motion comes from what we see in the everyday world. We might, for instance, visually follow a ball that has been thrown. The path of the ball is given by its position at various times. We are therefore conditioned to think in terms of a continuous path for moving objects. In Bohr's theory, the electron was thought of as moving about, or orbiting, the nucleus in the way the earth orbits the sun. Quantum mechanics vastly changes this view of motion. We can no longer think of an electron as having a precise orbit in an atom. To describe such an orbit, we would have to know the exact position of the electron at various times and exactly how long it would take it to travel to a nearby position in the orbit. That is, at any moment we would have to know not only the precise position but also the precise speed of the electron.

In 1927 Werner Heisenberg showed from quantum mechanics that it is impossible to know simultaneously, with absolute precision, both the position and the speed of a particle such as an electron. Heisenberg's **uncertainty principle** is *a relation that states that the product of the uncertainty in position and the uncertainty in momentum (mass times speed) of a particle can be no smaller than Planck's constant divided by* 4π. Thus, letting Δx be the uncertainty in the x coordinate of the particle and letting Δv_x be the uncertainty in the speed in the x direction, we have

$$(\Delta x)(m\Delta v_x) \geq \frac{h}{4\pi}$$

where m is the mass of the particle. Note that if m is large enough, the uncertainties Δx and Δv_x can be quite small and still satisfy the uncertainty principle. For objects large enough to be seen, the uncertainties are small and we can easily describe their orbits. Thus, the orbit of a baseball and the orbit of the earth have meaning. But when the mass is that of an electron, the uncertainties become significant. (They become significant in just those cases where wave properties become appreciable.) The uncertainty in position of an electron in an atom is about the size of the atom. This means that it is impossible for us to describe or to know how the electron moves in an atom.

Although we cannot know how the electron moves in an atom, quantum mechanics does allow us to make *statistical* statements about the electron. For example, we can obtain the *probability* of finding an electron at a certain point in a hydrogen atom. Thus, although we cannot say that an electron will be at a particular position at a given time, we can say that the electron is likely (or not likely) to be found at this position.

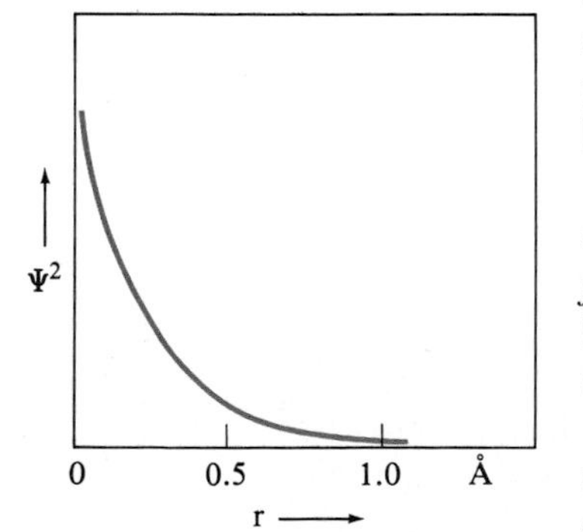

Figure 7.26
Plot of ψ^2 for the lowest energy level of the hydrogen atom. The square of the wave function is plotted along a line passing through the nucleus.

Information about a particle in a given energy level (such as an electron in an atom) is contained in a mathematical expression or function. It is called a *wave function* and is denoted by the Greek letter psi, ψ. The wave function is obtained by solving an equation of quantum mechanics (Schrödinger's equation). Its square, ψ^2, gives the probability of finding the particle within a region of space.

The wave function and its square, ψ^2, have values for all locations about a nucleus. Figure 7.26 shows values of ψ^2 for the electron in the lowest energy level of the hydrogen atom along a line through the nucleus. Note that ψ^2 is large near the nucleus, indicating that the electron is most likely to be found in this region. The value of ψ^2 decreases rapidly as the distance from the nucleus increases, but ψ^2 never goes to exactly zero, although the probability does become extremely small at large distances from the nucleus. This means that an atom does not have a definite boundary, unlike in the Bohr model of the atom.

Instrumental Methods

SCANNING TUNNELING MICROSCOPY

The 1986 Nobel Prize in physics went to three physicists for their work in developing microscopes for viewing extremely small objects. Half of the prize went to Ernst Ruska for the development of the electron microscope, described earlier in this section. The other half of the prize was awarded to Gerd Binnig and Heinrich Rohrer, at IBM's research laboratory in Zurich, Switzerland, for their invention of the *scanning tunneling microscope* in 1981. This instrument makes possible the viewing of atoms and molecules on a solid surface (Figure 7.27).

The scanning tunneling microscope is based on the concept of quantum mechanical *tunneling,* which in turn depends on the probability interpretation of quantum mechanics. Consider a hydrogen atom, which consists of an electron about a proton (which we label A), and imagine another proton (labeled B) some distance from the first one. In classical terms, if we were to move the electron from the region of proton A to the region of proton B, energy would have to be supplied to remove the electron from the attractive field of proton A. Quantum mechanics, however, gives us a different picture. The probability of the electron in the hydrogen atom being at a location far away from proton A, say near proton B, is very small but not zero. In effect, this means that the electron, which nominally belongs to proton A, may find itself near proton B without extra energy having been supplied. The electron is said to have *tunneled* from one atom to another.

Figure 7.27
Scanning tunneling microscope image of benzene molecules on a metal surface. *Benzene molecules, C_6H_6, which consist of rings of carbon atoms, each with an attached hydrogen atom, are arranged in a regular array on a rhodium metal surface. Carbon monoxide molecules are also present on the surface within holes formed by three benzene molecules but are not visible here.*

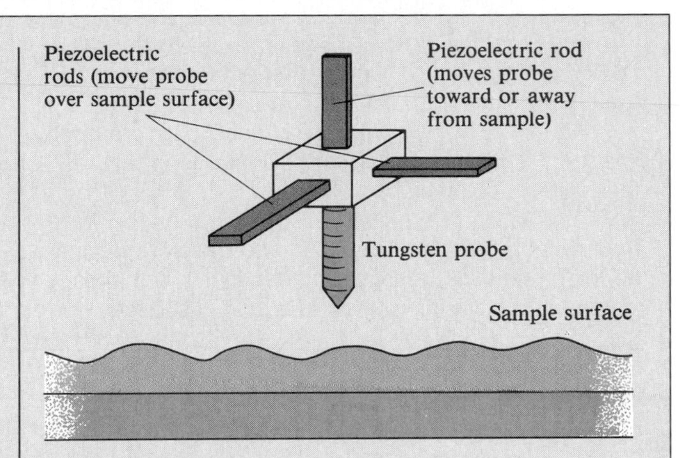

Figure 7.28
The scanning tunneling microscope. *A tunneling current is observed to flow between the probe and sample when a small voltage is impressed between them. The distance between the probe and sample (about 5 Å) is varied so that the tunneling current remains constant as the probe scans the surface of the sample. In this way, the probe essentially follows the contours of particular atoms.*

The scanning tunneling microscope consists of a tungsten metal needle with an extremely fine point (the probe) placed close to the sample to be viewed (Figure 7.28). If the probe is close enough to the sample, electrons can tunnel from the probe to the sample. The probability for this can be increased by having a small voltage applied between the probe and sample. Electrons tunneling from the probe to the sample give rise to a measurable electric current. The magnitude of this current depends on the distance between the probe and the sample (as well as on the wave function of the atom in the sample). By adjusting this distance, the current can be maintained at a fixed value. As the probe scans the sample, it moves toward or away from the sample, in effect following the contours of the sample.

The probe must move incredibly small distances, and this is done by an ingenious mechanism. Certain solids (*piezoelectric* crystals) generate a voltage when their length is changed (as when a phonograph needle rides over the grooves of a record). The reverse also occurs. Small voltage variations applied to a piezoelectric rod can generate corresponding small changes in its length. In the tunneling microscope, a variable voltage is applied to a piezoelectric rod to shorten or lengthen it. Because the probe is attached to this rod, the probe is

(continued)

moved toward or away from the sample. In this way the distance of the probe to the sample is adjusted by a voltage in order to maintain a constant tunneling current as the probe scans the sample.

Figure 7.27 shows a scanning tunneling microscope image of benzene molecules on a rhodium metal surface. (The structural formula of benzene, C_6H_6, is shown in Figure 7.29.) These images do depend on the wave functions of the material studied and therefore have to be interpreted with care, but the technique has already yielded important information about solid surfaces.

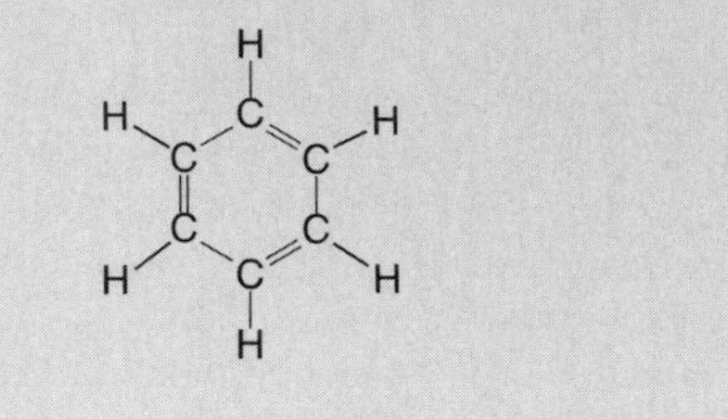

Figure 7.29
Structural formula of benzene. *The carbon atoms are bonded in a ring with a hydrogen atom attached to each carbon atom.*

7.9 QUANTUM NUMBERS AND ATOMIC ORBITALS

According to quantum mechanics, each electron in an atom is described by four different quantum numbers, three of which (n, l, and m_l) specify the wave function that gives the probability of finding the electron at various points in space.● *A wave function for an electron in an atom* is called an **atomic orbital.** An atomic orbital is pictured qualitatively by describing the region of space where there is high probability of finding the electrons. The atomic orbital so pictured has a definite shape. A fourth quantum number (m_s) refers to a magnetic property of electrons called *spin*. In this section we first look at quantum numbers, then at atomic orbitals.

● Three different quantum numbers are needed because there are three dimensions to space.

Quantum Numbers

The allowed values and general meaning of each of the four different quantum numbers of an electron in an atom are as follows:

1. PRINCIPAL QUANTUM NUMBER (*n*) *This quantum number is the one on which the energy of an electron in an atom principally depends; it can have any positive value: 1, 2, 3, and so on.* The energy of an electron in an atom depends *principally* on n. The smaller n is, the lower the energy. In the case of the hydrogen atom or any other single-electron atom (such as Li^{2+} or He^{+}), it is the only quantum number determining the energy (which is given by Bohr's formula, discussed in Section 7.7). For other atoms, the energy also depends to a slight extent on the l quantum number.

The *size* of an orbital also depends on n. The larger the value of n is, the larger the orbital. Orbitals of the same quantum state n are said to belong to the same *shell*. Shells are sometimes designated by the following letters:

Letter	*K*	*L*	*M*	*N* . . .
n	1	2	3	4 . . .

2. ANGULAR MOMENTUM QUANTUM NUMBER (*l*) (ALSO CALLED *AZIMUTHAL QUANTUM NUMBER*) *This quantum number distinguishes orbitals of given* n *having different shapes; it can have any integer value from 0 to* n − *1*. Thus, within each shell of quantum number n, there are n different kinds of orbitals, each with a distinctive shape denoted by an l quantum number. For example, if an electron has

Table 7.2 Permissible Values of Quantum Numbers for Atomic Orbitals

n	l	m_l*	Subshell Notation	Number of Orbitals in the Subshell
1	0	0	1*s*	1
2	0	0	2*s*	1
2	1	−1, 0, +1	2*p*	3
3	0	0	3*s*	1
3	1	−1, 0, +1	3*p*	3
3	2	−2, −1, 0, +1, +2	3*d*	5
4	0	0	4*s*	1
4	1	−1, 0, +1	4*p*	3
4	2	−2, −1, 0, +1, +2	4*d*	5
4	3	−3, −2, −1, 0, +1, +2, +3	4*f*	7

*Any one of the m_l quantum numbers may be associated with the n and l quantum numbers on the same line.

a principal quantum number of 3, the possible values for l are 0, 1, and 2. Thus, within the *M* shell ($n = 3$), there are three kinds of orbitals, each of which has a different shape for the region where the electron is most likely to be found. These orbital shapes will be discussed later in this section.

Orbitals of the same n, but different l, are said to belong to different *subshells* of a given shell. The different subshells are usually denoted by letters as follows:

Letter	*s*	*p*	*d*	*f*	*g* . . .
l	0	1	2	3	4 . . .

To denote a subshell within a particular shell, we write the value of the n quantum number for the shell, followed by the letter designation for the subshell. For example, 2*p* denotes a subshell with quantum numbers $n = 2$ and $l = 1$.●

● The rather odd choice of letter symbols for l quantum numbers survives from old spectroscopic terminology (describing the lines in a spectrum as *s*harp, *p*rincipal, *d*iffuse, and *f*undamental).

3. MAGNETIC QUANTUM NUMBER (m_l) *This quantum number distinguishes orbitals of given n and l—that is, of given energy and shape but having a different orientation in space; the allowed values are the integers from $-l$ to $+l$.* For $l = 0$ (*s* subshell), the allowed m_l quantum number is 0 only; there is only one orbital in the *s* subshell. For $l = 1$ (*p* subshell), $m_l = -1$, 0, and $+1$; there are three different orbitals in the *p* subshell. The orbitals have the same shape but different orientations in space. In addition, each orbital of a given subshell has the same energy. Note that there are $2l + 1$ orbitals in each subshell of quantum number l.

4. SPIN QUANTUM NUMBER (m_s) *This quantum number refers to the two possible orientations of the spin axis of an electron; possible values are $+\frac{1}{2}$ and $-\frac{1}{2}$.* An electron acts as though it were spinning on its axis like the earth. Such an electron spin would give rise to a circulating electric charge that would generate a magnetic field. Thus, an electron behaves like a small bar magnet, with a north and south pole.●

● Electron spin will be discussed further in Section 8.1.

Table 7.2 lists the permissible quantum numbers for all orbitals through the $n = 4$ shell. These values follow from the rules just given. Energies for these orbitals are shown in Figure 7.30 for the hydrogen atom. Note that all orbitals with the same principal quantum number n have the same energy. For atoms with more

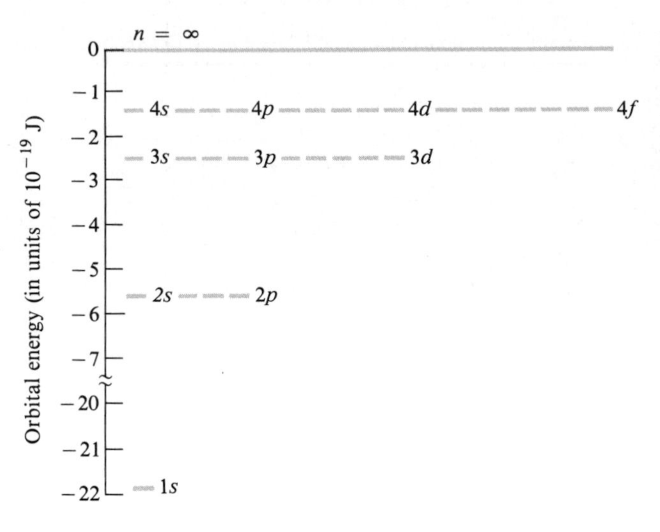

Figure 7.30
Orbital energies of the hydrogen atom. The lines for each subshell indicate the number of different orbitals of that subshell. (Note break in the energy scale.)

than one electron, however, only orbitals in the same subshell (denoted by a given n and l) have the same energy. We will have more to say about orbital energies in Chapter 8.

Example 7.9 Using the Rules for Quantum Numbers

State whether each of the following sets of quantum numbers is permissible for an electron in an atom. If a set is not permissible, explain why.

(a) $n = 1,\ l = 1,\ m_l = 0,\ m_s = +\frac{1}{2}$

(b) $n = 3,\ l = 1,\ m_l = -2,\ m_s = -\frac{1}{2}$

(c) $n = 2,\ l = 1,\ m_l = 0,\ m_s = +\frac{1}{2}$

(d) $n = 2,\ l = 0,\ m_l = 0,\ m_s = 1$

Solution

(a) **Not permitted.** The l quantum number is equal to n; it must be less than n. (b) **Not permitted.** The magnitude of the m_l quantum number (that is, the m_l value, ignoring its sign) must not be greater than l. (c) **Permitted.** (d) **Not permitted.** The m_s quantum number can only be $+\frac{1}{2}$ or $-\frac{1}{2}$.

Exercise 7.11

Explain why each of the following sets of quantum numbers is not permissible for an orbital.

(a) $n = 0,\ l = 1,\ m_l = 0,\ m_s = +\frac{1}{2}$

(b) $n = 2,\ l = 3,\ m_l = 0,\ m_s = -\frac{1}{2}$

(c) $n = 3,\ l = 2,\ m_l = 3,\ m_s = +\frac{1}{2}$

(d) $n = 3,\ l = 2,\ m_l = 2,\ m_s = 0$

(See Problems 7.67 and 7.68.)

Atomic Orbital Shapes

An s orbital has a spherical shape, though specific details of the probability distribution depend on the value of n. Figure 7.31 shows cross-sectional representations of the probability distributions of a 1s and a 2s orbital. The color shading is darker where the electron is more likely to be found. In the case of a 1s orbital (Figure

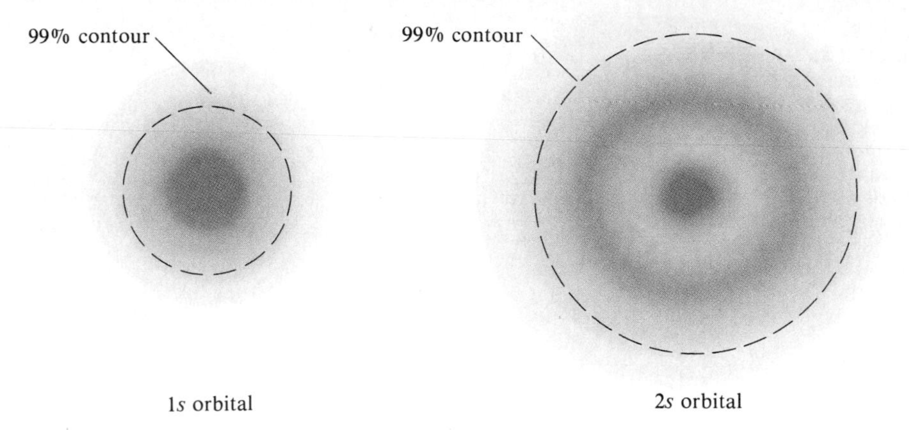

Figure 7.31
Cross-sectional representations of the probability distributions for a 1s and a 2s orbital. In a 1*s* orbital the probability distribution is largest near the nucleus. In a 2*s* orbital it is greatest in a spherical shell about the nucleus. Note the relative "size" of the orbitals, indicated by the 99% contour.

7.31, left), the electron is most likely to be found near the nucleus. The shading becomes lighter as the distance from the nucleus increases, indicating that the electron is less likely to be found far from the nucleus.

The orbital does not abruptly end at some particular distance from the nucleus. An atom, therefore, has an indefinite extension or "size." We can gauge the "size" of the orbital by means of the *99% contour*. The electron has a 99% probability of being found within the space of the 99% contour (the dashed line in the diagram).

A 2*s* orbital differs in detail from a 1*s* orbital. As shown in Figure 7.31, right, the electron in a 2*s* orbital is likely to be found in two regions, one near the nucleus and the other in a spherical shell about the nucleus (the electron is most likely to be here). The 99% contour shows that the 2*s* orbital is larger than the 1*s* orbital.

A cross-sectional diagram cannot portray the three-dimensional aspect of the 1*s* and 2*s* atomic orbitals. Figure 7.32 shows cutaway diagrams, which better illustrate this three-dimensionality.

There are three *p* orbitals in each subshell, starting with the 2*p* subshell. All *p* orbitals have the same basic shape (two lobes arranged along a straight line with the nucleus between the lobes) but differ in their orientations in space. Because the three orbitals are set at right angles to each other, we can show each one as oriented along a different coordinate axis (Figure 7.33). We denote these orbitals as $2p_x$, $2p_y$, and $2p_z$. A $2p_x$ orbital has its greatest electron probability along the

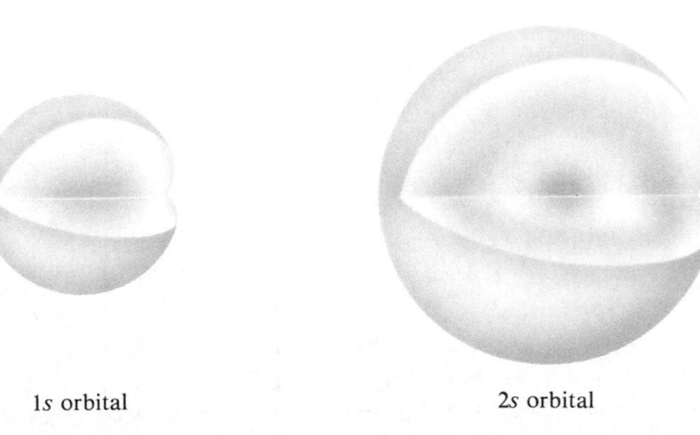

Figure 7.32
Cutaway diagrams showing the spherical shape of s orbitals. In both diagrams, the upper part of each orbital is cut away to reveal the electron distribution of the orbital.

Figure 7.33
The 2*p* orbitals. *(A)* Electron distribution in the $2p_x$ orbital. Note that it consists of two lobes oriented along the *x*-axis. *(B)* The orientation of the three 2*p* orbitals. The drawings are those usually used to depict the general shape and orientation of the orbitals, but they do not depict the detailed electron distribution as is done in *(A)*.

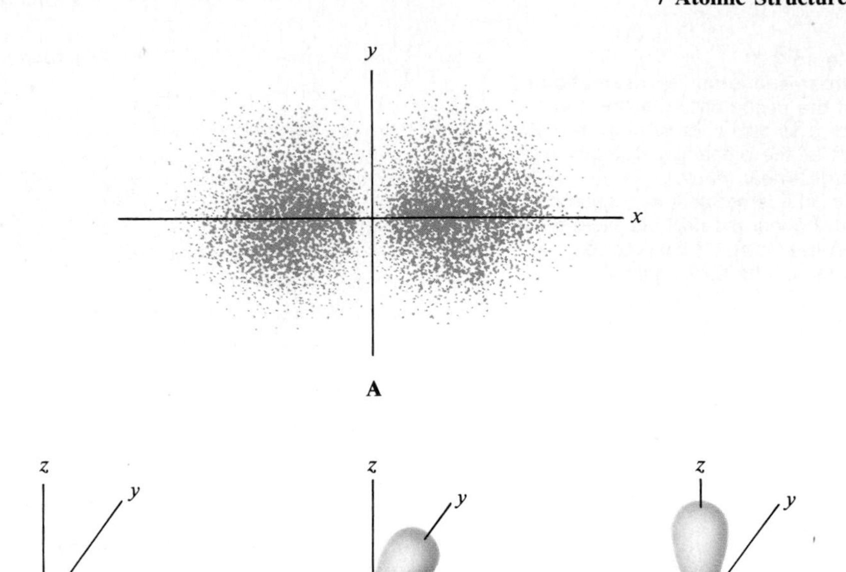

A

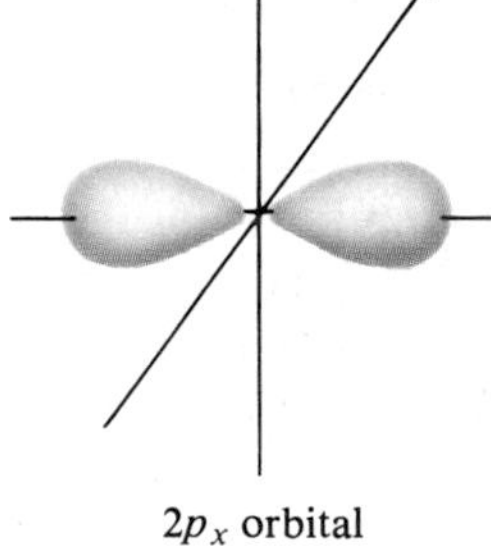

$2p_x$ orbital

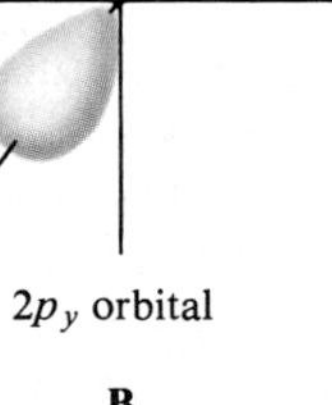

$2p_y$ orbital

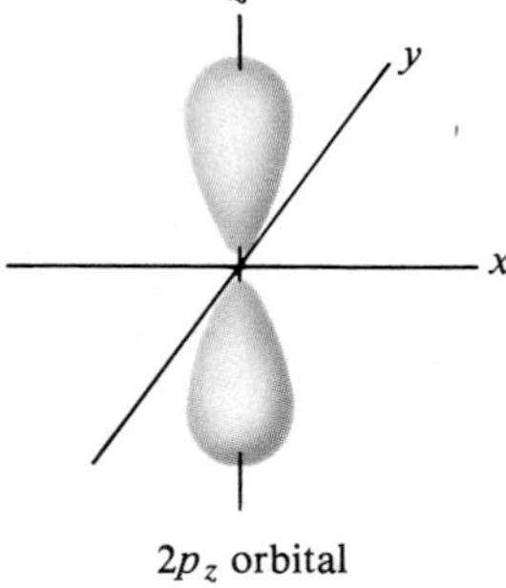

$2p_z$ orbital

B

Figure 7.34
The five 3*d* orbitals. These are labeled by subscripts, as in d_{xy}, that describe their mathematical characteristics.

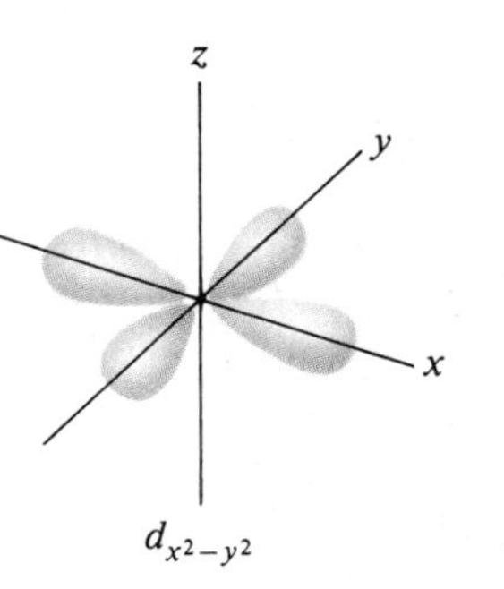

$d_{x^2-y^2}$

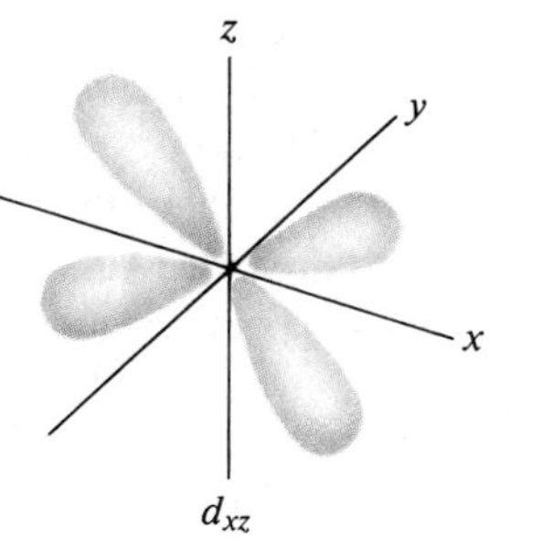

d_{xz}

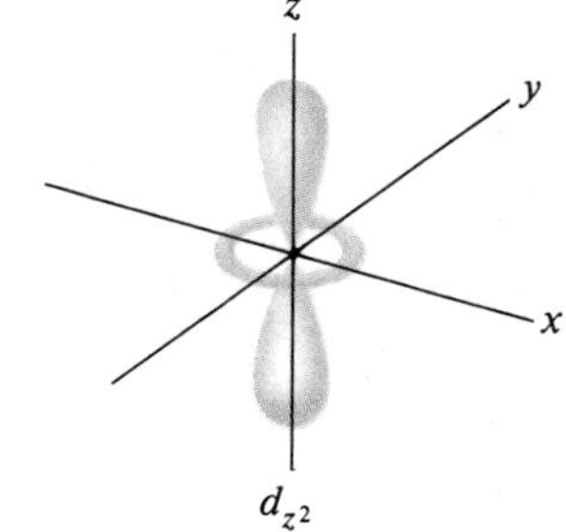

d_{z^2}

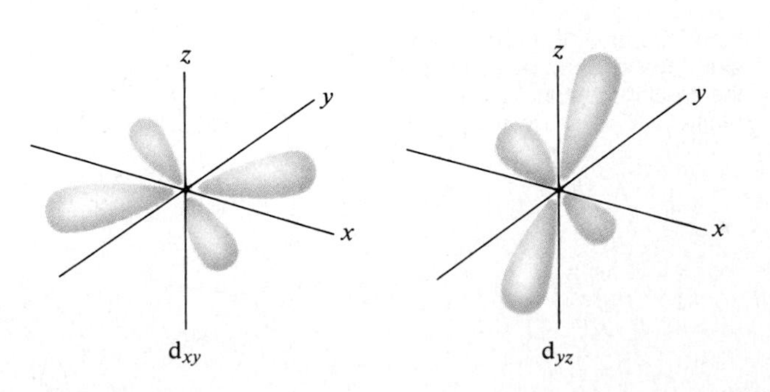

d_{xy} d_{yz}

x-axis, a $2p_y$ orbital along the y-axis, and a $2p_z$ orbital along the z-axis. Other p orbitals, such as $3p$, have this same general shape, with differences in detail depending on n. We will discuss s and p orbital shapes again in Chapter 10 in reference to chemical bonding.

There are five d orbitals, which have more complicated shapes than s and p orbitals. These are represented in Figure 7.34. We will discuss d orbitals in Chapter 23, when we look at the bonding in compounds of the transition elements.

Profile of a Chemical

HYDROGEN (Fuel of the Twenty-First Century?)

The hydrogen atom, as we have seen, has become a model for the structure of atoms. Hydrogen as a substance, however, has an equally important place in chemistry. Hydrogen is normally a colorless, odorless gas composed of H_2 molecules. Approximately 40% of the hydrogen produced commercially is used to manufacture ammonia, and about the same amount is used in petroleum refining. But the future may hold an even greater role for hydrogen as a fuel.

Liquid hydrogen, H_2, is a favorite rocket fuel (Figure 7.35). Burning it produces more heat per gram than any other fuel. In its gaseous form, hydrogen may become the favorite fuel of the twenty-first century. When hydrogen burns in air, the product is simply water. Therefore, the burning of hydrogen rather than fossil fuels (natural gas, petroleum, and coal) has important advantages.

Figure 7.35
A liquid hydrogen storage tank. *The liquid hydrogen is used as a rocket fuel.*

The burning of fossil fuels is a source of environmental pollutants. *Acid rain,* which has been shown to be injurious to the environment, is believed to be partially a result of burning coal and petroleum. These fuels contain sulfur compounds that burn to give sulfur dioxide. This in turn reacts in moist air to form sulfuric acid, a major component of acid rain. (Acid rain is discussed in the Related Topic at the end of Section 16.5.) It appears that carbon dioxide, a main product in the burning of fossil fuels, may also be a major pollutant. The percentage of carbon dioxide in the atmosphere has risen steadily since the large-scale burning of fossil fuels began in the late nineteenth century. Climatologists believe that the increase in carbon dioxide in the atmosphere is responsible for the *greenhouse effect*. The increased concentration of carbon dioxide can act like the glass on a greenhouse, retaining heat energy by absorbing and radiating infrared rays back to the surface of the earth, which might drastically affect climate. (The greenhouse effect is discussed in the Profile on carbon dioxide at the end of Chapter 11.)

Controlling carbon dioxide emissions into the atmosphere is a difficult challenge, but the answer might lie in the conversion to a *hydrogen energy economy*. In a hydrogen economy, hydrogen would become a major energy carrier. Automobiles, for example, could be modified to burn hydrogen. Hydrogen is not a primary energy source, however. It is a convenient and nonpolluting fuel, but it would have to be obtained from other energy sources.

Hydrocarbons and water are presently the main source of hydrogen. In the *steam-reforming process,* steam and hydrocarbons from natural gas or petroleum react at high temperature and pressure in the presence of a nickel catalyst. For example.

(continued)

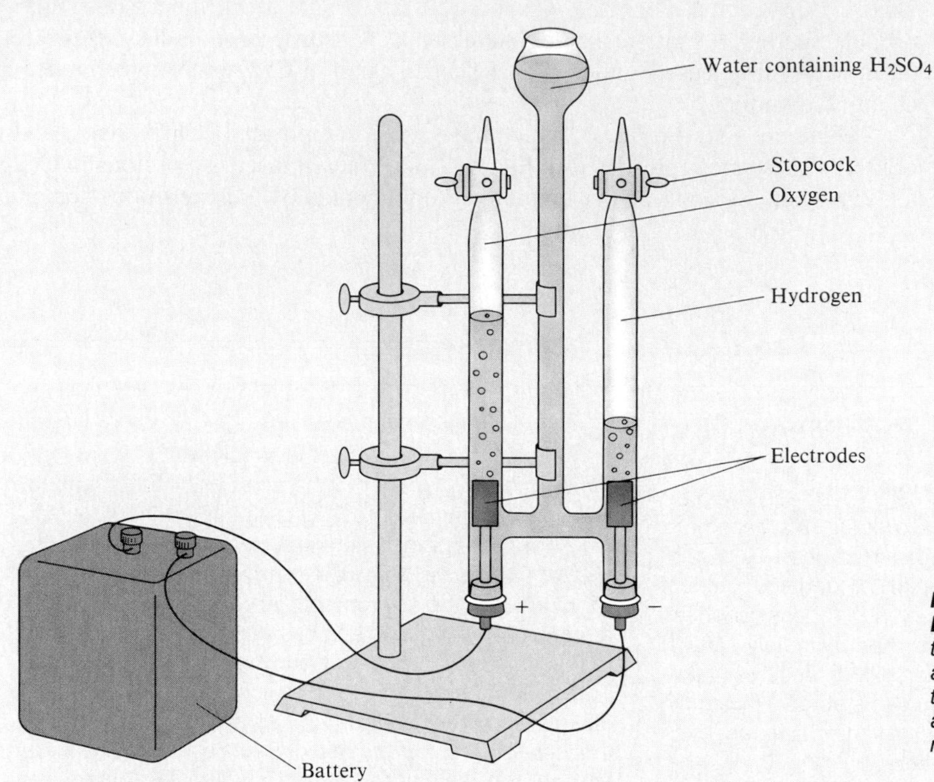

Figure 7.36
Electrolysis of water. *Note that the volume of hydrogen released at the negative electrode is twice that of the oxygen released at the other electrode, in agreement with the formula H_2O.*

$$C_3H_8(g) + 3H_2O(g) \xrightarrow{Ni} 3CO(g) + 7H_2(g)$$

The hydrocarbon, in effect, provides the energy for the production of hydrogen.

Hydrogen can also be produced from coal by the *water–gas reaction,* which is no longer used commercially but may become important again as natural gas and petroleum become more expensive. In this reaction, steam is passed over red-hot coke or coal.

$$C(s) + H_2O(g) \longrightarrow CO(g) + H_2(g)$$

These reactions produce mixtures of hydrogen with carbon monoxide. Such mixtures are used to produce various organic compounds, but to obtain pure hydrogen the carbon monoxide must be removed. First, the carbon monoxide is reacted with steam in the presence of a catalyst to give carbon dioxide and more hydrogen.

$$CO(g) + H_2O(g) \xrightarrow{\text{catalyst}} CO_2(g) + H_2(g)$$

Then the carbon dioxide is removed by dissolving it in a basic aqueous solution. When heated, this solution evolves carbon dioxide, which at present is recovered or emitted to the atmosphere. However, if the hydrogen plant is located near an ocean, the carbon dioxide can be injected deep into the seawater.

Hydrogen can be obtained directly from water that is decomposed by some form of energy. For example, electricity from nuclear power plants or from solar photovoltaic collectors can be used as a source of energy to decompose water by electrolysis (Figure 7.36). An electrolyte such as sulfuric acid or potassium hydroxide is added to the water to make it a conductor, and electrodes are inserted into the water; oxygen gas collects at the positive electrode, and hydrogen gas collects at the negative electrode. The net reaction is

$$2H_2O(l) \xrightarrow{\text{electrolysis}} 2H_2(g) + O_2(g)$$

Researchers are also looking at photochemical reactions (reactions involving light) that use solar energy to convert water directly to hydrogen and oxygen. Indeed, they have found processes that use sunlight more efficiently than green plants, which convert solar energy to chemical energy. Producing hydrogen with sunlight at competitive costs remains a challenge, however.

Questions for Study

1. Why is hydrogen useful as a rocket fuel? What environmental advantage does hydrogen have as a fuel?

2. What is acid rain? How can fossil fuels cause acid rain?
3. What is the greenhouse effect? Why do some climatologists feel that the burning of fossil fuels could cause a greenhouse effect?
4. What do we mean by a hydrogen economy?
5. The British chemist Henry Cavendish is credited with discovering hydrogen in 1766. He prepared it by reacting various metals with acids. Write the chemical equation for the preparation of hydrogen from zinc metal and hydrochloric acid.
6. Hydrogen is prepared from natural gas by reacting it with steam (steam-reforming). Natural gas is mainly methane, CH_4. Write the chemical equation for the preparation of hydrogen by steam-reforming of methane.
7. Describe how pure hydrogen could be prepared from wood charcoal (carbon). Use chemical equations where appropriate, accompanied by a verbal description.
8. How might solar energy be converted to hydrogen fuel?

A Checklist for Review

IMPORTANT TERMS

nucleus (p. 230)
electron (p. 230)
proton (p. 230)
neutron (p. 230)
cathode rays (7.1)
radioactivity (7.2)
atomic number (Z) (7.3)
mass number (A) (7.3)
nuclide (7.3)
isotopes (7.3)
mass spectrometer (7.4)
fractional (isotopic) abundance (7.4)
wavelength (λ) (7.5)
frequency (ν) (7.5)
electromagnetic spectrum (7.5)
Planck's constant (7.6)
photons (7.6)
photoelectric effect (7.6)
continuous spectrum (7.7)
line spectrum (7.7)
energy levels (7.7)
de Broglie relation (7.8)
quantum (wave) mechanics (7.8)
uncertainty principle (7.8)
atomic orbital (7.9)
principal quantum number (n) (7.9)
angular momentum quantum number (l) (7.9)
magnetic quantum number (m_l) (7.9)
spin quantum number (m_s) (7.9)

KEY EQUATIONS

$c = \nu\lambda$

$E = h\nu$

$E = -\dfrac{R_H}{n^2} \quad n = 1, 2, 3, \ldots \infty$ (for the H atom)

Energy of emitted photon $= E_i - E_f = h\nu$

$\lambda = \dfrac{h}{mv}$

SUMMARY OF FACTS AND CONCEPTS

J. J. Thomson established that the cathode rays from a gas discharge tube consist of negatively charged particles, called *electrons,* that are constituents of all atoms. Thomson measured the mass-to-charge ratio of the electron, and later Millikan measured its charge. From these measurements, the electron was found to be nearly 2000 times lighter than the lightest atom.

Rutherford proposed the nuclear model of the atom to account for the results of experiments in which alpha particles were scattered from metal foils. According to this model, the atom consists of a central core or *nucleus* around which the electrons exist. The nucleus has most of the mass of the atom and consists of *protons* (with a positive charge) and *neutrons* (with no charge). Each chemically distinct atom has a nucleus with a specific number of protons, and around the nucleus in the neutral atom are an equal number of electrons. Atoms whose nuclei have the same number of protons but a different number of neutrons are called *isotopes*.

The atomic weight can be calculated from the atomic masses and *fractional abundances* of the isotopes in a naturally occurring element. These data can be determined by a *mass spectrometer,* which separates ions according to their mass-to-charge ratios.

One way to study the electronic structure of the atom is to analyze the electromagnetic radiation that is emitted from an atom. Electromagnetic radiation is characterized by its *wavelength* λ and frequency ν, and these quantities are related to the speed of light, c ($c = \nu\lambda$).

Light was shown by Einstein to consist of particles *(photons),* each of energy $E = h\nu$, where h is Planck's constant. According to Bohr, electrons in an atom have *energy levels,* and when an electron in a higher energy level drops (or undergoes a transition) to a lower energy level, a photon is emitted. The energy of the photon equals the difference in energy between the two levels.

Electrons and other particles of matter have both particle and wave properties. For a particle of mass m and speed v, the wavelength is related to momentum mv by the *de Broglie relation:* $\lambda = h/mv$. The wave properties of a particle are described by a *wave function,* from which we can get the probability of finding the particle in different regions of space.

Each electron in an atom is characterized by four different quantum numbers. The distribution of an electron in space—its *atomic orbital*—is characterized by three of these quantum numbers: the principal quantum number, the angular momentum quantum number, and the magnetic quantum number. The fourth quantum number (spin quantum number) describes the magnetism of the electron.

OPERATIONAL SKILLS

1. **Determining the number of electrons in an atom or atomic ion.** Given an atom or atomic ion and its nuclear charge, find the number of electrons (Example 7.1).

2. **Writing nuclide symbols** Given the number of protons and neutrons in a nucleus, write its nuclide symbol (Example 7.2).

3. **Atomic weight from isotopic masses and fractional abundances** Given the isotopic masses (in atomic mass units) and fractional isotopic abundances for a naturally occurring element, calculate its atomic weight (Example 7.3).

4. **Relating wavelength and frequency of light** Given the wavelength of light, calculate the frequency, or vice versa (Examples 7.4 and 7.5).

5. **Calculating the energy of a photon** Given the frequency or wavelength of light, calculate the energy associated with one photon (Example 7.6).

6. **Determining the wavelength or frequency of a hydrogen atom transition** Given the initial and final principal quantum numbers for an electron transition in the hydrogen atom, calculate the frequency or wavelength of light emitted (Example 7.7). You need the value of R_H.

7. **Applying the de Broglie relation** Given the mass and speed of a particle, calculate the wavelength of the associated wave (Example 7.8).

8. **Using the rules for quantum numbers** Given a set of quantum numbers n, l, m_l, and m_s, state whether that set is permissible for an electron (Example 7.9).

Review Questions

7.1 Explain the operation of a cathode ray tube. Describe the deflection of cathode rays by electrically charged plates placed within the cathode ray tube. What does this imply about cathode rays?

7.2 What is the evidence that cathode rays are a part of all matter?

7.3 Explain Millikan's oil drop experiment.

7.4 Describe the different radiations emitted by radioactive nuclei. How can they be distinguished?

7.5 Describe the nuclear model of the atom. How does this model explain the results of alpha-particle scattering from metal foils?

7.6 One theory of the atom current at the time Rutherford proposed his nuclear model was that the atom consisted of a positive distribution of electricity in which electrons were embedded. This is sometimes referred to as the "plum pudding" model, in which the positive electricity is the "pudding" and the electrons are the "plums." Shooting alpha particles at this "pudding" would be like shooting at a softwood board with lead bullets. The bullets would be slowed and perhaps stopped; some might pass on through. Contrast the result expected from this model with what is actually found.

7.7 What are the different kinds of particles in the nucleus? Compare their properties with each other and with those of an electron.

7.8 Describe how protons and neutrons were discovered to be constituents of nuclei.

7.9 Briefly explain how a mass spectrometer works. What kinds of information does one obtain from the instrument?

7.10 Give a brief wave description of light. What are two characteristics of light waves?

7.11 Briefly describe the portions of the electromagnetic spectrum, starting with shortest wavelengths and going to longer wavelengths.

7.12 In your own words, explain the photoelectric effect. How does the photon concept explain this effect?

7.13 Describe the wave–particle picture of light.

7.14 Physical theory at the time Rutherford proposed his nuclear model of the atom was not able to explain how this model could give a stable atom. Explain the nature of this difficulty.

7.15 Explain the main features of Bohr's theory. Do these features solve the difficulty alluded to in Question 7.14?

7.16 Explain the process of emission of light by an atom.

7.17 Explain the process of absorption of light by an atom.

7.18 What is the evidence for electron waves? Give a practical application.

7.19 What kind of information does a wave function give about an electron in an atom?

7.20 Give the possible values of (a) the principal quantum number, (b) the angular momentum quantum number, (c) the magnetic quantum number, and (d) the spin quantum number.

7.21 What is the notation for the subshell in which $n = 4$ and $l = 3$? How many orbitals are in this subshell?

7.22 What is the general shape of an *s* orbital? of a *p* orbital?

Practice Problems

ELECTRONS, PROTONS, AND NEUTRONS

7.23 A student has determined the mass-to-charge ratio for an electron to be 5.64×10^{-12} kg/C. In another experiment, using Millikan's oil drop apparatus, he found the charge on the electron to be 1.605×10^{-19} C. What would be the mass of the electron, according to these data?

7.24 The mass-to-charge ratio for the positive ion F^+ is 1.97×10^{-7} kg/C. Using the value of 1.602×10^{-19} C for the charge on the ion, calculate the mass of the fluorine atom. (The mass of the electron is negligible compared with that of the ion, so the ion mass is essentially the atomic mass.)

7.25 Compounds of europium, Eu, are used to make color television screens. The europium nucleus has a charge of +63. How many electrons are there in the neutral atom? in the Eu^{3+} ion?

7.26 Cesium, Cs, is used in photoelectric cells ("electric eyes"). The cesium nucleus has a charge of +55. What is the number of electrons in the neutral atom? in the Cs^+ ion?

7.27 What is the nuclide symbol for the nucleus that contains 34 protons and 36 neutrons?

7.28 An atom contains five protons and six neutrons. What is the nuclide symbol for the nucleus?

7.29 Naturally occurring chlorine is a mixture of the isotopes Cl-35 and Cl-37. How many protons and how many neutrons are there in each isotope? How many electrons are there in the neutral atoms?

7.30 Naturally occurring lithium is a mixture of ${}^{6}_{3}Li$ and ${}^{7}_{3}Li$. Give the number of protons, neutrons, and electrons in the neutral atom of each isotope.

7.31 A nucleus of mass number 69 contains 38 neutrons. An atomic ion of this element has 28 electrons in it. Write the symbol for this atomic ion (give the symbol for the nucleus and give the ionic charge as a right superscript).

7.32 One isotope of a metallic element has mass number 119 and has 69 neutrons in the nucleus. An atomic ion has 48 electrons. Write the symbol for this ion (give the symbol for the nucleus and give the ionic charge as a right superscript).

ATOMIC MASSES

7.33 Calculate the atomic weight of boron, B, from the following data:

ISOTOPE	ATOMIC MASS (amu)	FRACTIONAL ABUNDANCE
B-10	10.013	0.1978
B-11	11.009	0.8022

7.34 An element has two naturally occurring isotopes with the following masses and abundances:

ATOMIC MASS (amu)	FRACTIONAL ABUNDANCE
84.9118	0.7215
86.9092	0.2785

What is the atomic weight of this element? What is the identity of the element?

7.35 Magnesium has naturally occurring isotopes with the following masses and abundances:

ISOTOPE	ATOMIC MASS (amu)	FRACTIONAL ABUNDANCE
Mg-24	23.985	0.7870
Mg-25	24.986	0.1013
Mg-26	25.983	0.1117

What is the atomic weight of magnesium, calculated from these data?

7.36 An element has naturally occurring isotopes with the following masses and abundances:

ATOMIC MASS (amu)	FRACTIONAL ABUNDANCE
27.977	0.9221
28.976	0.0470
29.974	0.0309

Calculate the atomic weight of this element. What is the identity of the element?

7.37 Silver has two naturally occurring isotopes, one of mass 106.91 amu and the other of mass 108.90 amu. Find the fractional abundances for these two isotopes. The atomic weight is 107.87 amu.

7.38 Obtain the fractional abundances for the two naturally occurring isotopes of copper. The masses of the isotopes are ${}^{63}_{29}Cu$, 62.9298 amu; ${}^{65}_{29}Cu$, 64.9278 amu. The atomic weight is 63.546 amu.

ELECTROMAGNETIC WAVES

7.39 At its closest approach, Mars is 56 million km from Earth. How long would it take to send a radio message from a space probe of Mars to Earth when the planets are at this closest distance?

7.40 The space probe *Pioneer 11* was launched April 5, 1973, and reached Jupiter in December 1974, traveling a distance of 998 million km. How long did it take an electromagnetic signal to travel to Earth from *Pioneer 11* when it was near Jupiter?

7.41 Radio waves in the AM region have frequencies in the range 550 to 1600 kilocycles per second (550 to 1600 kHz). Calculate the wavelength corresponding to a radio wave of frequency 1.255×10^6/s (that is, 1255 kHz).

7.42 Microwaves have frequencies in the range 10^9 to 10^{12}/s (cycles per second), equivalent to between 1 gigahertz and 1 terahertz. What is the wavelength of microwave radiation whose frequency is 1.145×10^{10}/s?

7.43 Light with a wavelength of 465 nm lies in the blue region of the visible spectrum. Calculate the frequency of this light.

7.44 Calculate the frequency associated with light of wavelength 656 nm. (This corresponds to one of the wavelengths of light emitted by the hydrogen atom.)

PHOTONS

7.45 What is the energy of a photon corresponding to radio waves of frequency 1.255×10^6/s?

7.46 What is the energy of a photon corresponding to microwave radiation of frequency 1.145×10^{10}/s?

7.47 The green line in the atomic spectrum of thallium has a wavelength of 535 nm. Calculate the energy of a photon of this light.

7.48 Indium compounds give a blue-violet flame test. The atomic emission responsible for this blue-violet color has a wavelength of 451 nm. Obtain the energy of a single photon of this wavelength.

BOHR THEORY

7.49 Calculate the frequency of electromagnetic radiation emitted by the hydrogen atom in the electron transition from $n = 4$ to $n = 3$.

7.50 An electron in a hydrogen atom in the level $n = 5$ undergoes a transition to level $n = 3$. What is the frequency of the emitted radiation?

7.51 The first line of the Lyman series of the hydrogen atom emission results from a transition from the $n = 2$ level to the $n = 1$ level. What is the wavelength of the emitted photon? Using Figure 7.14, describe the region of the electromagnetic spectrum in which this emission lies.

7.52 What is the wavelength of the electromagnetic radiation emitted from a hydrogen atom when the electron undergoes the transition $n = 5$ to $n = 4$? In what region of the spectrum does this line occur? See Figure 7.14.

7.53 One of the lines in the Balmer series of the hydrogen atom emission spectrum is at 397 nm. It results from a transition from an upper energy level to $n = 2$. What is the principal quantum number of the upper level?

7.54 A line of the Lyman series of the hydrogen atom spectrum has the wavelength 9.50×10^{-8} m. It results from a transition from an upper energy level to $n = 1$. What is the principal quantum number of the upper level?

7.55 What is the difference in energy between the levels that are responsible for the red emission line of the rubidium atom at 795 nm?

7.56 Calculate the difference in energy of the levels involved in the green calcium line at 554 nm.

DE BROGLIE WAVES; ATOMIC ORBITALS

7.57 What is the wavelength of a neutron traveling at a speed of 3.65 km/s? (Neutrons of these speeds are obtained from a nuclear pile.)

7.58 What is the wavelength of a proton traveling at a speed of 6.25 km/s?

7.59 At what speed must an electron travel to have a wavelength of 0.125 Å?

7.60 At what speed must a neutron travel to have a wavelength of 0.125 Å?

7.61 If the n quantum number of an atomic orbital is 4, what are the possible values of l? If the l quantum number is 3, what are the possible values of m_l?

7.62 The n quantum number of an atomic orbital is 5. What are the possible values of l? What are the possible values of m_l if the l quantum number is 4?

7.63 How many subshells are there in the M shell? How many orbitals are there in the f subshell?

7.64 How many subshells are there in the N shell? How many orbitals are there in the g subshell?

7.65 Give the notation (using letter designation for l) for the subshells denoted by the following quantum numbers:
(a) $n = 3, l = 1$ (b) $n = 4, l = 2$
(c) $n = 4, l = 0$ (d) $n = 5, l = 3$

7.66 Give the notation (using letter designations for l) for the subshells denoted by the following quantum numbers:
(a) $n = 6, l = 2$ (b) $n = 5, l = 4$
(c) $n = 4, l = 3$ (d) $n = 6, l = 1$

7.67 State which of the following sets of quantum numbers would be possible and which would be impossible for an electron in an atom.
(a) $n = 0, l = 0, m_l = 0, m_s = +\frac{1}{2}$
(b) $n = 1, l = 1, m_l = 0, m_s = +\frac{1}{2}$
(c) $n = 1, l = 0, m_l = 0, m_s = -\frac{1}{2}$
(d) $n = 2, l = 1, m_l = -2, m_s = +\frac{1}{2}$
(e) $n = 2, l = 1, m_l = -1, m_s = +\frac{1}{2}$

7.68 Explain why each of the following sets of quantum numbers would not be permissible for an electron, according to the rules for quantum numbers.
(a) $n = 1, l = 0, m_l = 0, m_s = +1$
(b) $n = 1, l = 3, m_l = 3, m_s = +\frac{1}{2}$
(c) $n = 3, l = 2, m_l = 3, m_s = -\frac{1}{2}$
(d) $n = 0, l = 1, m_l = 0, m_s = +\frac{1}{2}$
(e) $n = 2, l = 1, m_l = -1, m_s = +\frac{3}{2}$

Additional Problems

7.69 In a series of oil drop experiments, the charges measured on the oil drops were -3.20×10^{-19} C, -6.40×10^{-19} C, -9.60×10^{-19} C, and -1.12×10^{-18} C. What is the smallest difference in charge between any two drops? If this is assumed to be the charge on the electron, how many excess electrons are there on each drop?

7.70 In a hypothetical universe, an oil drop experiment gave the following measurements of charges on oil drops: -5.55×10^{-19} C, -9.25×10^{-19} C, -1.11×10^{-18} C, and -1.48×10^{-18} C. Assume that the smallest difference in charge equals the unit of negative charge in this universe. What is the value of this unit of charge? How many units of excess negative charge are there on each oil drop?

7.71 The blue line of the strontium atom emission has a wavelength of 461 nm. What is the frequency of this light? What is the energy of a photon of this light?

7.72 The barium atom has an emission with wavelength 554 nm (green). Calculate the frequency of this light and the energy of a photon of this light.

7.73 How many protons and neutrons are there in the nucleus of each of the following: $^{28}_{13}Al$, $^{41}_{20}Ca$, $^{59}_{28}Ni$?

7.74 Give the number of protons and neutrons in each of the following nuclei: $^{58}_{26}Fe$, $^{64}_{29}Cu$, $^{75}_{33}As$.

7.75 Write the nuclide symbol for the nucleus with atomic number 78 and mass number 196.

7.76 What is the nuclide symbol for the nucleus with atomic number 49 and mass number 116?

7.77 There are two naturally occurring isotopes of antimony. They have isotope masses of 120.90 amu (^{121}Sb) and 122.90 amu (^{123}Sb). What is the fractional abundance of ^{121}Sb in antimony?

7.78 Europium is a silvery-white metal. The element consists of two isotopes, ^{151}Eu and ^{153}Eu, with masses of 150.92 amu and 152.92 amu, respectively. Calculate the fractional abundance of ^{151}Eu atoms in europium.

7.79 Mass-to-charge ratios for some isotopic ions of sulfur are 3.3137×10^{-7} kg/C, 3.5205×10^{-7} kg/C, 1.6568×10^{-7} kg/C, and 1.7603×10^{-7} kg/C. What are the masses of each isotope in atomic mass units? What is the charge on each ion (in units of e)?

7.80 The main isotopic ions of potassium give mass-to-charge ratios of 4.0383×10^{-7} kg/C, 4.1430×10^{-7} kg/C, 2.0192×10^{-7} kg/C, and 2.0715×10^{-7} kg/C. Obtain the isotope masses (in amu) and the charge on each ion (in units of e).

7.81 The photoelectric work function of a metal is the minimum energy needed to eject an electron by irradiating the metal with light. For calcium, this work function equals 4.34×10^{-19} J. What is the minimum frequency of light for the photoelectric effect in calcium?

7.82 The photoelectric work function for magnesium is 5.90×10^{-19} J. (*Work function* is defined in Problem 7.81.) Calculate the minimum frequency of light required to eject electrons from magnesium.

7.83 A hydrogen-like ion has a nucleus of charge $+Ze$ and a single electron outside this nucleus. The energy levels of these ions are $-Z^2R_H/n^2$ (where Z = atomic number). Calculate the wavelength of the transition from $n = 3$ to $n = 2$ for He^+, a hydrogen-like ion. In what region of the spectrum does this emission occur?

7.84 What is the wavelength of the transition from $n = 4$ to $n = 3$ for Li^{2+}? In what region of the spectrum does this emission occur? (See Problem 7.83.)

Cumulative-Skills Problems

7.85 The energy required to dissociate the Cl_2 molecule to Cl atoms is 239 kJ/mol of Cl_2. If the dissociation of a Cl_2 molecule were accomplished by the absorption of a single photon whose energy was exactly the quantity required, what would be its wavelength (in meters)?

7.86 The energy required to dissociate the H_2 molecule to H atoms is 432 kJ/mol of H_2. If the dissociation of an H_2 molecule were accomplished by the absorption of a single photon whose energy was exactly the quantity required, what would be its wavelength (in meters)?

7.87 A microwave oven heats by radiating food with microwave radiation, which is absorbed by the food and converted to heat. Suppose an oven's radiation wavelength is 12.5 cm. A container with 0.250 L of water was placed in the oven and the temperature of the water rose from 20.0°C to 100.0°C. How many photons of this microwave radiation were required? Assume that all the energy from the radiation was used to raise the temperature of the water.

7.88 Water absorbs infrared radiation with wavelengths near 2.80 μm. Suppose this radiation is absorbed by the water and converted to heat. A 1.00-L sample of water absorbs infrared radiation, and its temperature increases from 20.0°C to 30.0°C. How many photons of this radiation are used to heat the water?

7.89 Light with a wavelength of 425 nm fell on a potassium surface, and electrons were ejected at a speed of 4.88×10^5 m/s. What energy was expended in removing the electron from the metal? Express the answer in joules (per electron) and in kilojoules per mole (of electrons).

7.90 Light with a wavelength of 405 nm fell on a strontium surface and electrons were ejected. If the speed of an ejected electron is 3.36×10^5 m/s, what energy was expended in removing the electron from the metal? Express the answer in joules (per electron) and in kilojoules per mole (of electrons).

7.91 When an electron is accelerated by a voltage difference, the kinetic energy acquired by the electron equals the voltage times the charge on the electron. Thus, 1 volt imparts a kinetic energy of 1.602×10^{-19} volt-coulombs. (The volt-coulomb equals the joule; thus, this kinetic energy is 1.602×10^{-19} J.) What is the wavelength associated with electrons accelerated by 4.00×10^3 volts?

7.92 What is the wavelength for electrons accelerated by 1.00×10^4 volts? (See Problem 7.91.)

Figure 8.1
Marie Sklodowska Curie (1867–1934). Marie Sklodowska Curie, born in Warsaw, Poland, began her doctoral work with Henri Becquerel soon after he discovered the spontaneous radiation emitted by uranium salts. She found this radiation to be an atomic property and coined the word *radioactivity* for it. In 1903 the Curies and Becquerel were awarded the Nobel Prize in physics for their discovery of radioactivity. Three years later, Pierre Curie was killed in a carriage accident. Marie Curie continued their work on radium and in 1911 was awarded the Nobel Prize in chemistry for the discovery of polonium and radium and the isolation of pure radium metal. This was the first time a scientist had received two Nobel awards (since then two others have been so honored).

compounds. We now explain this arrangement in terms of the electronic structure of atoms. In this chapter we will look at this electronic structure and its relationship to the periodic table of elements.

Electronic Structure of Atoms

In Chapter 7 we found that an electron in an atom has four quantum numbers—n, l, m_l, and m_s—associated with it. The first three quantum numbers characterize the orbital that describes the region of space where an electron is most likely to be found; we say that the electron "occupies" this orbital. The spin quantum number, m_s, describes the spin orientation of an electron. We will look at electron spin in more detail in this first section. Then we will discuss how electrons are distributed among the possible orbitals.

8.1 ELECTRON SPIN AND THE PAULI EXCLUSION PRINCIPLE

German physicists Otto Stern and Walther Gerlach first observed electron spin magnetism in 1921. They directed a beam of silver atoms into the field of a specially designed magnet. The same experiment can be done with hydrogen atoms. The beam of hydrogen atoms is split into two by the magnetic field; half of the atoms are bent in one direction and half in the other (see Figure 8.2). The fact that the atoms are affected by the laboratory magnet shows that they themselves act as magnets.

The beam of hydrogen atoms is split into two because the electron in the atom behaves as a tiny magnet with only two possible orientations. We might picture the electron as a ball of spinning charge. Like an electric current in a coil of wire, the circulating charge would create a magnetic field. Electron spin, however, is sub-

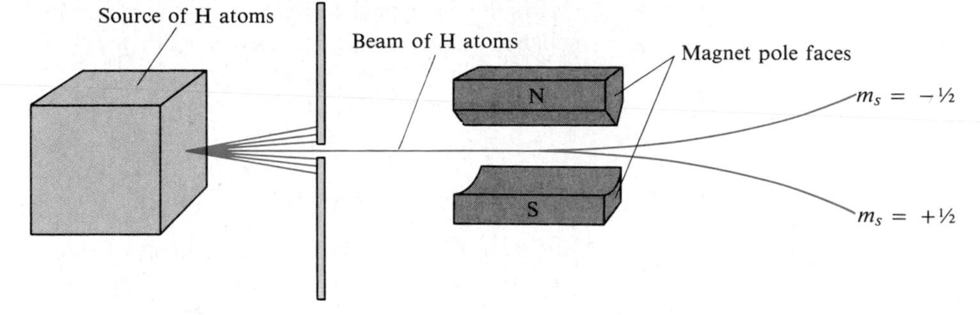

Figure 8.2
The Stern-Gerlach experiment. The diagram shows the experiment using hydrogen atoms (simpler to interpret theoretically), although the original experiment employed silver atoms. A beam of hydrogen atoms (shown in blue) is split into two by a nonuniform magnetic field. One beam consists of atoms each with an electron having $m_s = +\frac{1}{2}$; the other consists of atoms having an electron with $m_s = -\frac{1}{2}$.

ject to a quantum restriction on the possible directions of the spin axis. The resulting directions of spin magnetism, shown by models depicted in Figure 8.3, correspond to spin quantum numbers $m_s = +\frac{1}{2}$ and $m_s = -\frac{1}{2}$.●

● Protons and many nuclei also have spin. See the Instrumental Methods at the end of this section.

Electron Configurations and Orbital Diagrams

The **electron configuration** of an atom is *the particular distribution of electrons among available subshells*. It is described by a notation that lists the subshell symbols, one after the other. Each symbol has a superscript on the right giving the number of electrons in that subshell. For example, a configuration of the lithium atom (atomic number 3) with two electrons in the 1*s* subshell and one electron in the 2*s* subshell is written $1s^22s^1$.

The notation for a configuration gives the number of electrons in each subshell, but we use *a diagram to show how the orbitals of a subshell are occupied by electrons*. It is called an **orbital diagram.** In such a diagram, an orbital is represented by a circle. Each group of orbitals in a subshell is labeled by its subshell notation. An electron in an orbital is shown by an arrow; the arrow points upward when $m_s = +\frac{1}{2}$ and downward when $m_s = -\frac{1}{2}$. The orbital diagram

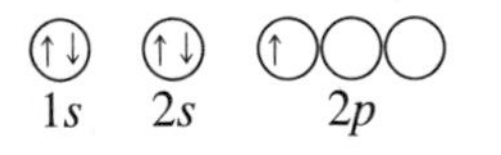

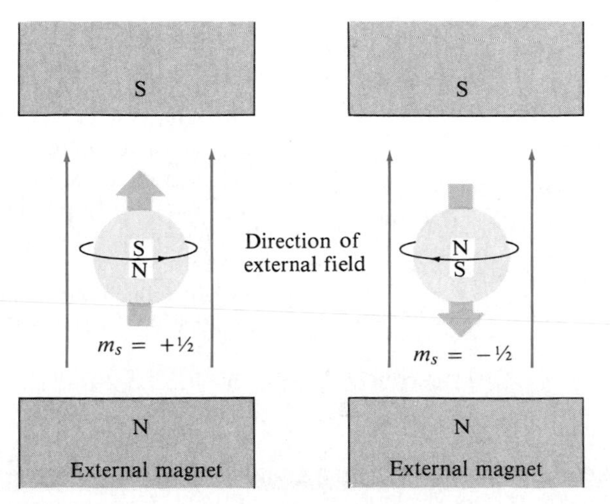

Figure 8.3
A representation of electron spin. The two possible spin orientations are indicated by the models. By convention, the spin direction is given as shown by the large arrow on the spin axis. Electrons behave as tiny bar magnets, as shown in the figure.

shows the electronic structure of an atom in which there are two electrons in the $1s$ subshell or orbital (one electron with $m_s = +\frac{1}{2}$, the other with $m_s = -\frac{1}{2}$); two electrons in the $2s$ subshell ($m_s = +\frac{1}{2}$, $m_s = -\frac{1}{2}$); and one electron in the $2p$ subshell ($m_s = +\frac{1}{2}$). The electron configuration is $1s^2 2s^2 2p^1$.

Pauli Exclusion Principle

Not all of the conceivable arrangements of electrons among the orbitals of an atom are physically possible. The **Pauli exclusion principle,** which summarizes experimental observations, is *a rule stating that no two electrons in an atom can have the same four quantum numbers*. If one electron in an atom has the quantum numbers $n = 1$, $l = 0$, $m_l = 0$, and $m_s = +\frac{1}{2}$, no other electron can have the same four quantum numbers. In other words, we cannot place two electrons with the same value of m_s in a $1s$ orbital. The orbital diagram

(↑↑)
$1s$

does not represent a possible arrangement of electrons.

Because there are only two possible values of m_s, an orbital can hold no more than two electrons, and then only if the two electrons have different spin quantum numbers. In an orbital diagram, an orbital with two electrons has to be written with arrows pointing in opposite directions. The two electrons are said to have opposite spins.

We can see that each subshell holds a maximum of twice as many electrons as the number of orbitals in the subshell. Thus, a $2p$ subshell, which has three orbitals (with $m_l = -1$, 0, and $+1$), can hold a maximum of six electrons. The maximum number of electrons in various subshells is given in the following table.

SUBSHELL	NUMBER OF ORBITALS	MAXIMUM NUMBER OF ELECTRONS
s ($l = 0$)	1	2
p ($l = 1$)	3	6
d ($l = 2$)	5	10
f ($l = 3$)	7	14

Example 8.1 Applying the Pauli Exclusion Principle

Which of the following orbital diagrams or electron configurations are possible and which are impossible, according to the Pauli exclusion principle? Explain.

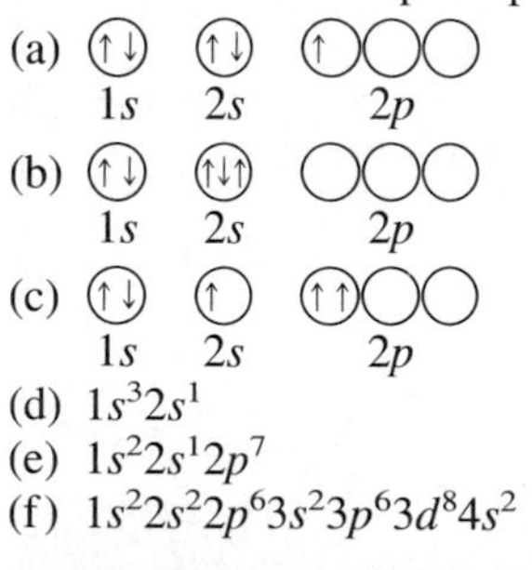

(d) $1s^3 2s^1$
(e) $1s^2 2s^1 2p^7$
(f) $1s^2 2s^2 2p^6 3s^2 3p^6 3d^8 4s^2$

Solution

(a) **Possible** orbital diagram. (b) **Impossible** orbital diagram; there are three electrons in the $2s$ orbital. (c) **Impossible** orbital diagram; there are two electrons in a $2p$ orbital with the same spin. (d) **Impossible** electron configuration; there are three electrons in the $1s$ subshell (one orbital). (e) **Impossible** electron configuration; there are seven electrons in the $2p$ subshell (which can hold only six electrons). (f) **Possible.** Note that the $3d$ subshell can hold as many as ten electrons.

Exercise 8.1

Look at the following orbital diagrams and electron configurations. Which ones are possible and which ones are not, according to the Pauli exclusion principle? Explain.

(a) (↑) (↑) ( )( )( )
   1s 2s 2p

(b) (↑) (↑) (↑↓)(↑↓)(↑↓)
   1s 2s 2p

(c) (↑↓) (↑↓) (↑↑)(↑↓)(↑↓)
   1s 2s 2p

(d) $1s^2 2s^2 2p^4$

(e) $1s^2 2s^4 2p^2$

(f) $1s^2 2s^2 2p^6 3s^2 3p^{10} 3d^{10}$

(See Problems 8.25, 8.26, 8.27, and 8.28.)

Instrumental Methods

NUCLEAR MAGNETIC RESONANCE (NMR)

We have just seen that electrons have a spin and as a result behave like tiny magnets. Protons and neutrons similarly have spins. Therefore, depending on the arrangement of protons and neutrons, a nucleus could have a spin. A nucleus with spin will act like a bar magnet, similar to the electron although many times smaller in magnitude. Examples of nuclei with spin are hydrogen-1 (proton); carbon-13 (but carbon-12, the most abundant nuclide of carbon, has no spin); and fluorine-19.

Although nuclear magnetism is much smaller than that of electrons, with the correct equipment it is easily seen and in fact forms the basis of *nuclear magnetic resonance* (NMR) spectroscopy, one of the most important methods for determining molecular structure. It also forms the basis of a new medical diagnostic tool, *magnetic resonance imaging* (Figure 8.4).

The essential features of NMR can be seen if we consider the proton. Like the electron, the proton has two spin states. In the absence of a magnetic field, these spin states have the same energy, but in the field of a

(continued)

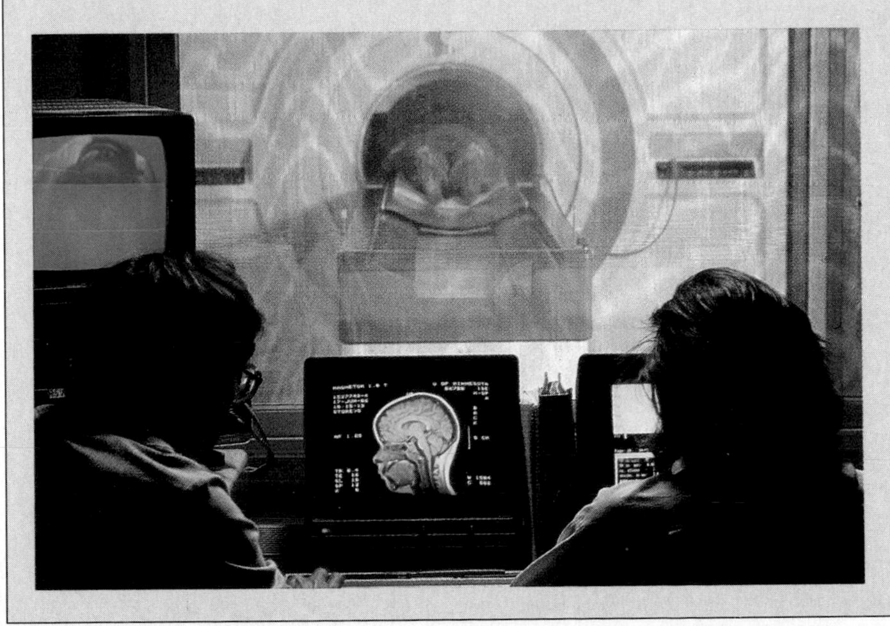

Figure 8.4
Magnetic resonance imaging. *A patient's head is placed in a large magnet and subjected to a radio pulse. Proton-spin transitions give rise to a radio wave emission that can be analyzed electronically and converted by computer to a two-dimensional image of a plane portion of the brain.*

strong magnet (external field), they have different energies. The state in which the proton magnetism is aligned with the external field, so the south pole of the proton magnet faces the north pole of the external magnet, will have lower energy. The state in which the proton magnet is turned 180°, with its south pole facing the south pole of the external magnet, will have higher energy. Now if a proton in the lower spin state is irradiated with electromagnetic waves of the proper frequency (in which the photon has energy equal to the difference in energy of the spin states), the proton will change to the higher spin state. The frequency absorbed by the proton depends on the magnitude of the magnetic field. For the magnets used in these instruments, the radiation lies in the radiofrequency range. Frequencies commonly used are 60 MHz and 100 MHz, which are in the FM region.

Figure 8.5 shows a simplified NMR spectrometer. It consists of a sample in the field of a variable electromagnet and near two coils, one a radio wave transmitter and the other a receiver coil perpendicular to the transmitter coil (so that the receiver will not pick up the signal from the transmitter). Suppose the transmitter radiates waves of 60 MHz. If the sample absorbs these radio waves, protons will undergo transitions from the lower spin state to the higher spin state. Once protons are in the higher energy state, they tend to lose energy, going back to the lower spin state and radiating 60-MHz radio waves. Thus, the sample acts like a transmitter, but one with coils in various directions, so the signal can be detected by the receiver coil.

In general, the sample will not absorb at the chosen frequency. But we can change the energy difference between spin states, and therefore the frequency that is absorbed by the sample, by increasing or decreasing the

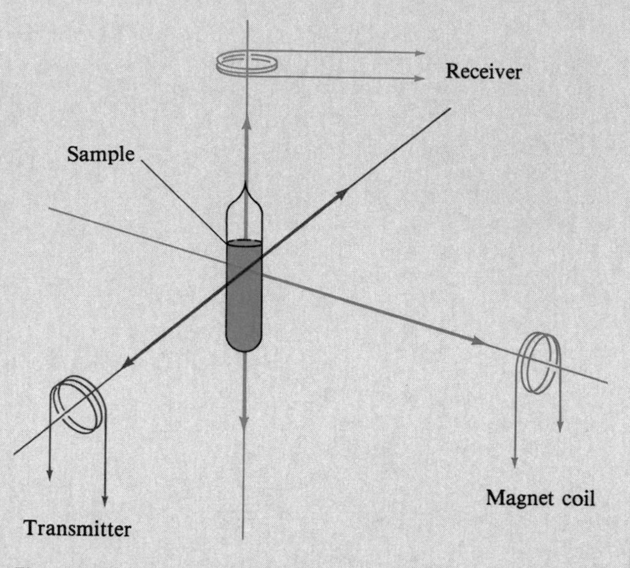

Figure 8.5
Nuclear magnetic resonance experiment. *The sample is placed in a magnetic field that can be changed by varying the electric current in a magnet coil. When the difference in energy of the nuclear spin states corresponds to the transmitter frequency, radio waves are absorbed. When the nuclear spins go back to the ground state, the radio waves are emitted in various directions and some are detected by the receiver coil. Because the receiver coil is perpendicular to the transmitter coil, it does not detect the transmitter signal but only that from the sample. (Source: John D. Roberts,* Nuclear Magnetic Resonance, *McGraw-Hill, 1959, p. 14, Figure 1–5. Reproduced with permission.)*

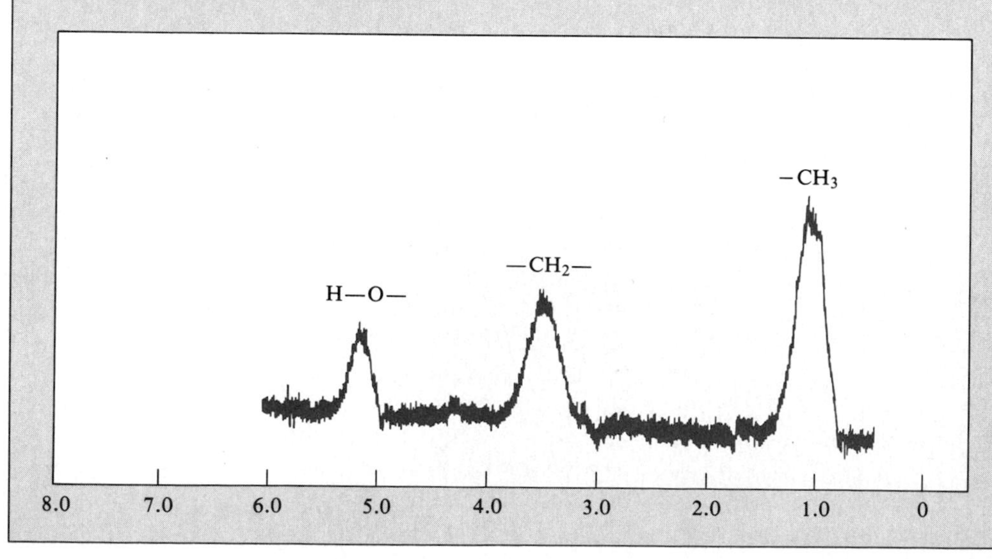

Figure 8.6
NMR spectrum of ethanol, CH_3CH_2OH (low resolution). *Note the peaks for protons in different chemical bonding environments. The area under each peak is proportional to the number of protons in each bonding environment. At the higher resolutions available with modern instruments, the center peak and right-hand peak split into several peaks because of the interactions between proton spins. These splittings offer more information about molecular structure.*

magnitude of the external magnetic field using small coils on the magnet pole faces. In so doing we can, in effect, "tune" the sample, or bring it into "resonance" with the transmitter frequency.

Because each proton in a substance is surrounded by electrons that have their own magnetic fields, the magnetic environment of a proton depends somewhat on what other atom is bonded to it. The external magnetic field needed to bring a proton spin into resonance with the 60-MHz radiation will vary with the type of chemical bond.

Figure 8.6 shows the NMR spectrum of ethanol, whose molecular structure is

```
      H  H
      |  |
H—O—C—C—H
      |  |
      H  H
```

Note that the protons in the hydrogen atoms bonded to oxygen (H—O—) absorb at lower magnetic field than do those attached to carbon atoms. Moreover, there are slight differences between the magnetic environments about the hydrogen nuclei on the $—CH_2—$ group and those on the $—CH_3$ group. Finally, the area under each absorption peak is proportional to the number of hydrogen atoms of that type. Thus, the areas corresponding to the peaks for HO—, $—CH_2—$, and $—CH_3$ are in the ratios 1:2:3. This is only part of the information that can be gleaned from the NMR spectrum of a substance, but it is enough to show the utility of NMR in structure determination.

8.2 ELECTRON CONFIGURATIONS OF ATOMS

Every atom has an infinite number of possible electron configurations. The configuration associated with the lowest energy level of the atom corresponds to a quantum-mechanical state called the *ground state*. Other configurations correspond to *excited states*, associated with energy levels other than the lowest. For example, the ground state of the sodium atom is known from experiment to have the electron configuration $1s^22s^22p^63s^1$. The electron configuration $1s^22s^22p^63p^1$ represents an excited state of the sodium atom.●

● The transition of the sodium atom from the excited state $1s^22s^22p^63p^1$ to the ground state $1s^22s^22p^63s^1$ is accompanied by the emission of yellow light at 589 nm. Thus, excited states are needed to describe the spectrum of an atom.

The chemical properties of an atom are related primarily to the electron configuration of its ground state. Table 8.1 lists the experimentally determined ground-state electron configurations of atoms Z = 1 to Z = 36. (A complete table is given in Appendix D.)

Building-up Principle (Aufbau Principle)

Most of the configurations in Table 8.1 can be explained in terms of the **building-up principle** (or **Aufbau principle**), *a scheme used to reproduce the electron configurations of the ground states of atoms by successively filling subshells with electrons in a specific order (the building-up order)*. Following this principle, we obtain the electron configuration of an atom by successively filling subshells in the following order (the *building-up order*): 1*s*, 2*s*, 2*p*, 3*s*, 3*p*, 4*s*, 3*d*, 4*p*, 5*s*, 4*d*, 5*p*, 6*s*, 4*f*, 5*d*, 6*p*, 7*s*, 5*f*. This order reproduces the experimentally determined electron configurations (with some exceptions, which we will discuss later).

The building-up order corresponds for the most part to increasing energy of the subshells. We might expect this. By filling orbitals of lowest energy first, we usually get the lowest total energy (ground state) of the atom. Recall that the energy of an orbital depends only on the quantum numbers n and l.● (The energy of the H atom, however, depends only on n.) Orbitals with the same n and l but different m_l—that is, different orbitals of the same subshell—have the same energy. The energy depends primarily on n, increasing with its value. Thus a 3*s*

● The quantum numbers and characteristics of orbitals were discussed in Section 7.9.

Table 8.1 Ground-State Electron Configurations of Atoms Z = 1 to 36*

Z	Element	Configuration	Z	Element	Configuration
1	H	$1s^1$	19	K	$1s^22s^22p^63s^23p^64s^1$
2	He	$1s^2$	20	Ca	$1s^22s^22p^63s^23p^64s^2$
3	Li	$1s^22s^1$	21	Sc	$1s^22s^22p^63s^23p^63d^14s^2$
4	Be	$1s^22s^2$	22	Ti	$1s^22s^22p^63s^23p^63d^24s^2$
5	B	$1s^22s^22p^1$	23	V	$1s^22s^22p^63s^23p^63d^34s^2$
6	C	$1s^22s^22p^2$	24	Cr	$1s^22s^22p^63s^23p^63d^54s^1$
7	N	$1s^22s^22p^3$	25	Mn	$1s^22s^22p^63s^23p^63d^54s^2$
8	O	$1s^22s^22p^4$	26	Fe	$1s^22s^22p^63s^23p^63d^64s^2$
9	F	$1s^22s^22p^5$	27	Co	$1s^22s^22p^63s^23p^63d^74s^2$
10	Ne	$1s^22s^22p^6$	28	Ni	$1s^22s^22p^63s^23p^63d^84s^2$
11	Na	$1s^22s^22p^63s^1$	29	Cu	$1s^22s^22p^63s^23p^63d^{10}4s^1$
12	Mg	$1s^22s^22p^63s^2$	30	Zn	$1s^22s^22p^63s^23p^63d^{10}4s^2$
13	Al	$1s^22s^22p^63s^23p^1$	31	Ga	$1s^22s^22p^63s^23p^63d^{10}4s^24p^1$
14	Si	$1s^22s^22p^63s^23p^2$	32	Ge	$1s^22s^22p^63s^23p^63d^{10}4s^24p^2$
15	P	$1s^22s^22p^63s^23p^3$	33	As	$1s^22s^22p^63s^23p^63d^{10}4s^24p^3$
16	S	$1s^22s^22p^63s^23p^4$	34	Se	$1s^22s^22p^63s^23p^63d^{10}4s^24p^4$
17	Cl	$1s^22s^22p^63s^23p^5$	35	Br	$1s^22s^22p^63s^23p^63d^{10}4s^24p^5$
18	Ar	$1s^22s^22p^63s^23p^6$	36	Kr	$1s^22s^22p^63s^23p^63d^{10}4s^24p^6$

*A complete table is given in Appendix D.

orbital has greater energy than a $2s$ orbital because the value of n is greater. Except for the H atom, the energies of orbitals with the same n increase with the l quantum number. A $3p$ orbital has slightly greater energy than a $3s$ orbital because l is greater. The orbital of lowest energy is $1s$; the next higher ones are $2s$ and $2p$, then $3s$ and $3p$. The $3d$ subshell, however, has an energy just below, but very close to, that of the $4s$ orbital. Figure 8.7 shows the orbital energies calculated from theory for the scandium atom ($Z = 21$). Relative values of the orbital energies for other atoms are similar. Because the $3d$ energy is very close to the $4s$ energy, the total energy of atoms (which depends on both orbital energies and the energy of repulsion between electrons) turns out to be lower for those configurations obtained by filling the $4s$ orbital before the $3d$ subshell.

Let us see how we can reproduce the electron configurations of Table 8.1 using the building-up principle. Remember that the number of electrons in a neutral atom equals the atomic number Z (the nuclear charge is $+Z$). In the case of the simplest atom, hydrogen ($Z = 1$), we obtain the ground state by placing the single electron into the $1s$ orbital, giving the configuration $1s^1$ (this is read as "one-ess-one"). Now we go to helium ($Z = 2$). The first electron goes into the $1s$ orbital, as in hydrogen, followed by the second electron, because any orbital can hold two electrons. The configuration is $1s^2$. Filling the $n = 1$ shell creates a very stable configuration and, as a result, gives rise to a chemically unreactive atom.

We continue this way through the elements, each time increasing Z by one and adding another electron. We obtain the configuration of an atom from that of the preceding element by adding an electron into the next available orbital, following the building-up order. In lithium ($Z = 3$), the first two electrons give the configuration $1s^2$, like helium, but the third electron goes into the next higher

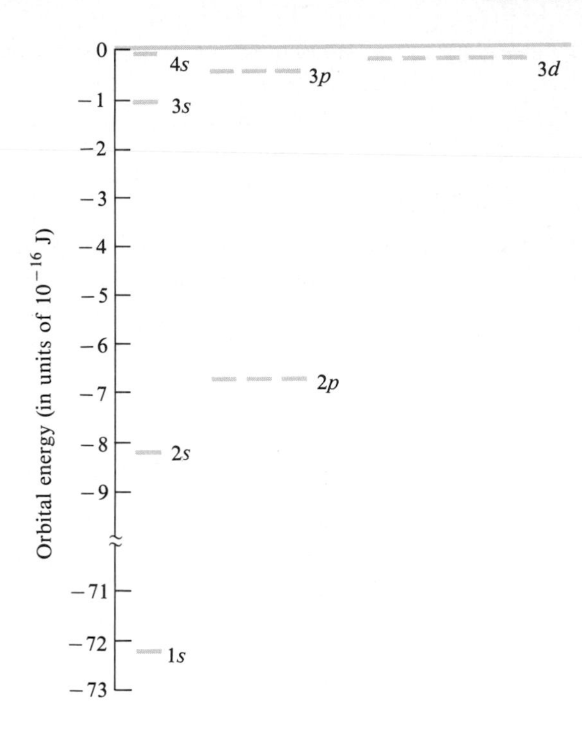

Figure 8.7
Orbital energies for the scandium atom ($Z = 21$). Note that in the scandium atom, unlike the hydrogen atom, the subshells for each n are spread apart in energy. Thus the $2p$ energy is above the $2s$. Similarly, the $n = 3$ subshells are spread to give the order $3s < 3p < 3d$. The $3d$ subshell energy is now just below the $4s$. (Values for this figure were calculated from theory by Charlotte F. Fischer, Vanderbilt University.)

orbital in the building-up order, because the $1s$ orbital is now filled. This gives the configuration $1s^22s^1$. In beryllium ($Z = 4$), the fourth electron fills the $2s$ orbital, giving the configuration $1s^22s^2$.

If we use the abbreviation [He] for $1s^2$, the configurations are

$Z = 3$	lithium	$1s^22s^1$	or	$[He]2s^1$
$Z = 4$	beryllium	$1s^22s^2$	or	$[He]2s^2$

With boron ($Z = 5$), the electrons begin filling the $2p$ subshell. We get

$Z = 5$	boron	$1s^22s^22p^1$	or	$[He]2s^22p^1$
$Z = 6$	carbon	$1s^22s^22p^2$	or	$[He]2s^22p^2$
⋮				
$Z = 10$	neon	$1s^22s^22p^6$	or	$[He]2s^22p^6$

Having filled the $2p$ subshell, we again find a particularly stable configuration. Neon is chemically unreactive as a result.

With sodium ($Z = 11$), the $3s$ orbital begins to fill. Using the abbreviation [Ne] for $1s^22s^22p^6$, we have

$Z = 11$	sodium	$1s^22s^22p^63s^1$	or	$[Ne]3s^1$
$Z = 12$	magnesium	$1s^22s^22p^63s^2$	or	$[Ne]3s^2$

Then the $3p$ subshell begins to fill.

$Z = 13$	aluminum	$1s^22s^22p^63s^23p^1$	or	$[Ne]3s^23p^1$
⋮				
$Z = 18$	argon	$1s^22s^22p^63s^23p^6$	or	$[Ne]3s^23p^6$

With the $3p$ subshell filled, a stable configuration has been attained; argon is an unreactive element.

Now the 4*s* orbital begins to fill. We get $[Ar]4s^1$ for potassium ($Z = 19$) and $[Ar]4s^2$ for calcium ($[Ar] = 1s^22s^22p^63s^23p^6$). At this point the 3*d* orbital begins to fill. We get $[Ar]3d^14s^2$ for scandium ($Z = 21$), $[Ar]3d^24s^2$ for titanium ($Z = 22$), and $[Ar]3d^34s^2$ for vanadium ($Z = 23$). Here we have written the configurations with subshells arranged in order by shells. As we will see later, this places the configuration of electrons involved in chemical reactions together at the far right. However, some prefer to arrange the subshells in the building-up order.

Let us skip to zinc ($Z = 30$). The 3*d* subshell has filled; the configuration is $[Ar]3d^{10}4s^2$. Now the 4*p* subshell begins to fill, starting with gallium ($Z = 31$), configuration $[Ar]3d^{10}4s^24p^1$, and ending with krypton ($Z = 36$), configuration $[Ar]3d^{10}4s^24p^6$.

Electron Configurations and the Periodic Table

By this time we see a pattern develop among the ground-state electron configurations of the atoms. We can see that this pattern explains the periodic table, which was briefly described in Section 2.4. Consider helium, neon, argon, and krypton, which are elements in Group VIIIA of the periodic table. Neon, argon, and krypton have configurations in which a *p* subshell has just filled (helium has a filled 1*s* subshell; no 1*p* subshell is possible).

helium	$1s^2$
neon	$1s^22s^22p^6$
argon	$1s^22s^22p^63s^23p^6$
krypton	$1s^22s^22p^63s^23p^63d^{10}4s^24p^6$

These elements are the first members of the group of elements called *noble gases* because of their relative unreactivity.

Look now at the configurations of beryllium, magnesium, and calcium, members of the group of *alkaline earth metals* (Group IIA), which are similar, moderately reactive elements.

beryllium	$1s^22s^2$	or	$[He]2s^2$
magnesium	$1s^22s^22p^63s^2$	or	$[Ne]3s^2$
calcium	$1s^22s^22p^63s^23p^64s^2$	or	$[Ar]4s^2$

Each of these configurations consists of a **noble-gas core,** that is, *an inner-shell configuration corresponding to one of the noble gases,* plus two outer electrons with an ns^2 configuration.

The elements boron, aluminum, and gallium (Group IIIA) also have similarities. Their configurations are

boron	$1s^22s^22p^1$	or	$[He]2s^22p^1$
aluminum	$1s^22s^22p^63s^23p^1$	or	$[Ne]3s^23p^1$
gallium	$1s^22s^22p^63s^23p^63d^{10}4s^24p^1$	or	$[Ar]3d^{10}4s^24p^1$

Boron and aluminum have noble-gas cores plus three electrons with the configuration ns^2np^1. Gallium has an additional filled 3*d* subshell. *The noble-gas core together with* $(n-1)d^{10}$ *electrons* is often referred to as a **pseudo-noble-gas core,** because these electrons usually are not involved in chemical reactions.

An electron in an atom outside the noble-gas or pseudo-noble-gas core is called a **valence electron.** It is such electrons that are primarily involved in chemical reaction, and similarities among the configurations of valence electrons (the

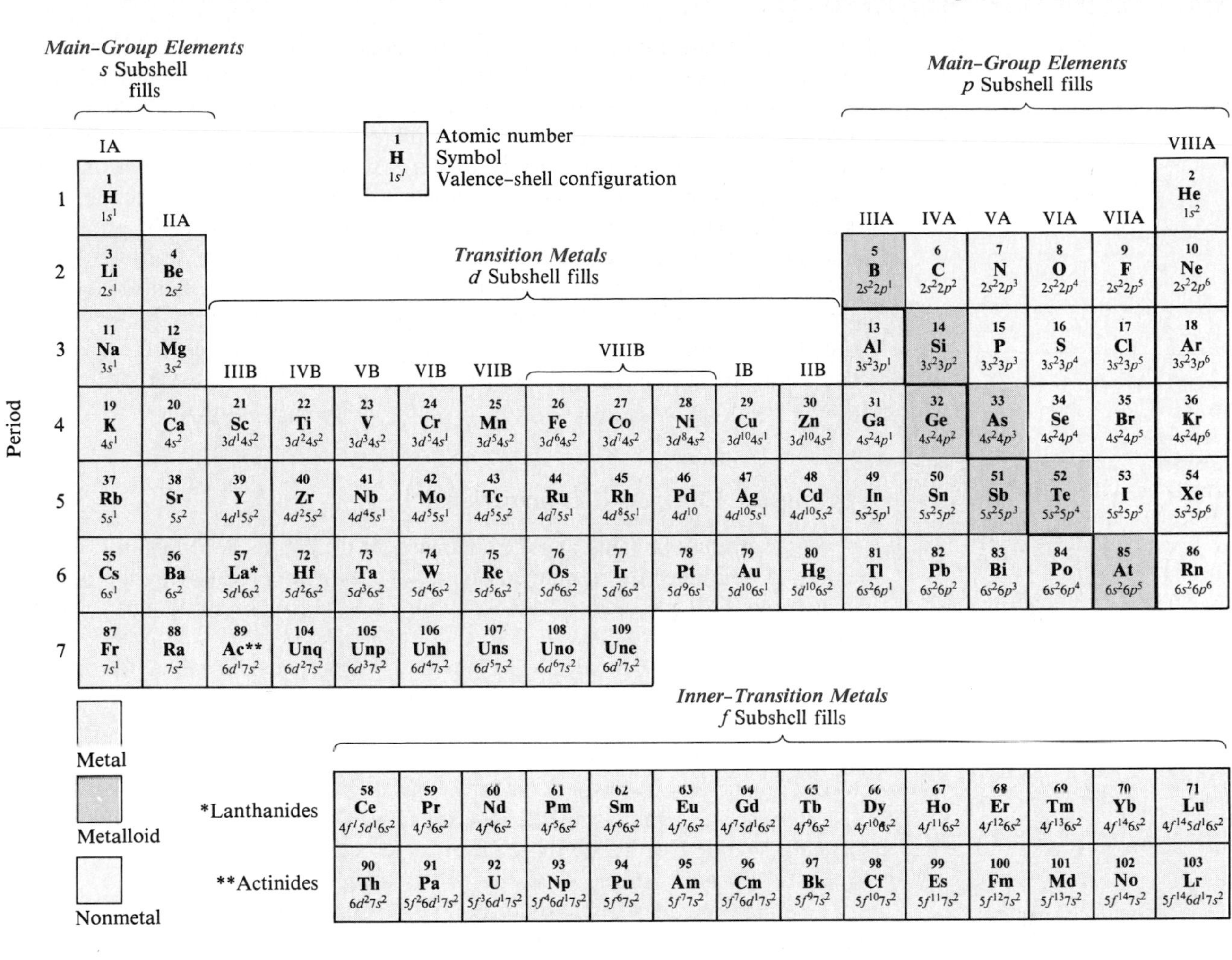

Figure 8.8
A periodic table. This table shows the valence-shell configurations of the elements.

valence-shell configurations) account for similarities of the chemical properties among groups of elements.

Figure 8.8 shows a periodic table with the valence-shell configurations included. Note the similarity in electron configuration within any group (column) of elements. This similarity explains what chemists since Mendeleev have known—the properties of elements in any group are similar.

The *main-group* (or *representative*) *elements* all have valence-shell configurations ns^anp^b, with some choice of a and b (b could be equal to 0). In other words, the outer s or p subshell is being filled. Similarly, in the *d-block transition elements* (often called simply *transition elements*), a d subshell is being filled. In the *f-block transition elements* (or *inner-transition elements*), an f subshell is being filled. (See Figure 8.8 or Appendix D for the configurations of these elements.)

Exceptions to the Building-up Principle

As we have said, the building-up principle reproduces most of the ground-state configurations correctly. There are some exceptions, however, and chromium ($Z = 24$) is the first one we encounter. The building-up principle predicts the configuration $[Ar]3d^44s^2$, though the correct one is found experimentally to be

$[Ar]3d^5 4s^1$. These two configurations are actually very close in total energy because of the closeness in energies of the $3d$ and $4s$ orbitals (note Figure 8.7 again). For that reason, small effects can influence which of the two configurations is actually lower in energy. It turns out that there is some stability in half-filled and completely filled d subshells, which explains why $[Ar]3d^5 4s^1$ is lower in energy.

Copper (Z = 29) is another exception to the building-up principle, which predicts the configuration $[Ar]3d^9 4s^2$, although experiment shows the ground state to be the configuration with a filled d subshell $[Ar]3d^{10} 4s^1$. We need not dwell on these exceptions beyond noting that they occur. The point to remember is that the configuration predicted by the building-up principle is very close in energy to the ground-state configuration (if it is not the ground state). Most of the qualitative conclusions regarding the chemistry of an element are not materially affected by arguing from the configuration given by the building-up principle.●

● More exceptions occur among the heavier transition elements, where the outer subshells are very close together. We must concede that simplicity was not the uppermost concern in the construction of the Universe!

Writing the Electron Configurations

In order to discuss bonding and the chemistry of the elements coherently, we must be able to reproduce the atomic configurations with ease, following the building-up principle. All we need is some facility in recall of the building-up order of subshells.

One approach is to recall the structure of the periodic table. Because that structure is basic, it gives us a sound way to remember the building-up order. There is a definite pattern to the order of filling of the subshells as we go through the elements in the periodic table, and from this we can write down the building-up order. Figure 8.9 shows a periodic table stressing this pattern. For example, in the blue-colored area, an ns subshell is being filled. In the green-colored area, an np subshell is being filled. The value of n is obtained from the period (row) number. In the yellow area, an $(n - 1)d$ subshell is being filled.

We read off the building-up order by starting with the first period, in which the $1s$ subshell is being filled. In the second period, we have $2s$ (blue area); then, staying in the same period but jumping across, we have $2p$ (green area). In the third period, we have $3s$ and $3p$; in the fourth period, $4s$ (blue area), $3d$ (yellow area), and then $4p$ (green area). This pattern should become clear enough to visualize with a periodic table that is not labeled with the subshells, such as the one on the inside front cover of the book. The detailed method is illustrated in the next example.

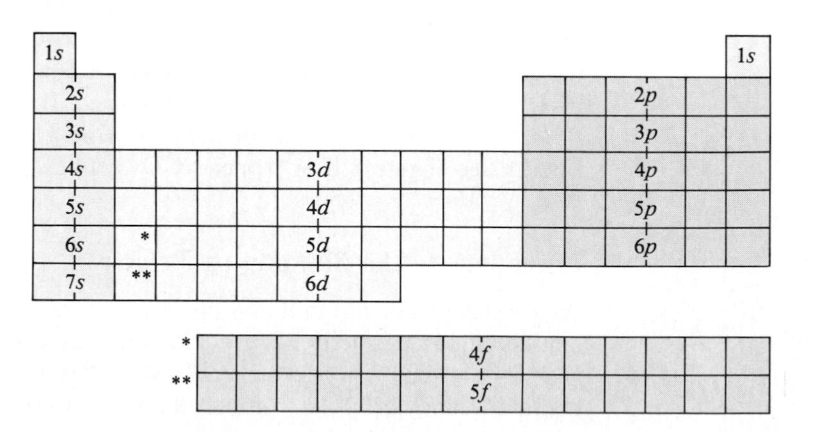

Figure 8.9
A periodic table illustrating the building-up principle. The colored areas of elements show the different subshells that are filling with those elements.

Example 8.2 Determining the Configuration of an Atom Using the Building-up Principle

Use the building-up principle to obtain the configuration for the ground state of the gallium atom (Z = 31). Give the configuration in complete form (do not abbreviate for the core).

Solution

Look at a periodic table (say the one on the inside front cover of this book). Start with hydrogen and go through the periods, writing down the subshells that are being filled; stop with gallium. We get the following order.

	$1s$	$2s\ 2p$	$3s\ 3p$	$4s\ 3d\ 4p$
Period:	first	second	third	fourth

Now we fill the subshells with electrons, remembering that we have a total of 31 electrons to distribute. We get

$$1s^22s^22p^63s^23p^64s^23d^{10}4p^1$$

Or, if we rearrange the subshells by shells, we write

$$\mathbf{1s^22s^22p^63s^23p^63d^{10}4s^24p^1}$$

Exercise 8.2

Use the building-up principle to obtain the electron configuration for the ground state of the manganese atom (Z = 25). (See Problems 8.31, 8.32, 8.33, and 8.34.)

In many cases, we need only the configuration of outer electrons. We can determine this from the position of the element in the periodic table. Recall that the valence-shell configuration of a main-group element is ns^anp^b, where n, the principal quantum number of the outer shell, also equals the period number for the element. The total number of valence electrons, which equals $a + b$, can be obtained from the group number. For example, gallium is in Period 4, so $n = 4$. It is in Group IIIA, so the number of valence electrons is 3. This gives the valence-shell configuration $4s^24p^1$. The configuration of outer shells of a transition element is obtained in a similar fashion. The next example gives the details.

Example 8.3 Determining the Configuration of an Atom Using the Period and Group Numbers

What are the configurations for the outer electrons of (a) tellurium, $Z = 52$, and (b) nickel, $Z = 28$?

Solution

(a) We locate tellurium in a periodic table and find it to be in Period 5, Group VIA. Thus, it is a main-group element, and the outer subshells are $5s$ and $5p$. These subshells contain 6 electrons, because the group is VIA. The valence-shell configuration is $\mathbf{5s^25p^4}$. (b) Nickel is a Period 4 transition element, in which the general form of the outer-shell configuration is $3d^{a-2}4s^2$. To determine a, we note that it equals the Roman numeral group number up to iron (8). After that we count Co as 9 and Ni as 10. Hence, the outer-shell configuration is $\mathbf{3d^84s^2}$.

Exercise 8.3

Using the periodic table on the inside front cover, write the valence-shell configuration of arsenic (As). (See Problems 8.35, 8.36, 8.37, and 8.38.)

Exercise 8.4

The lead atom has the ground-state configuration $[Xe]4f^{14}5d^{10}6s^26p^2$. Find the period and group for this element. From its position in the periodic table, would you classify lead as a main-group, a transition, or an inner-transition element? (See Problems 8.39 and 8.40.)

Instrumental Methods

X RAYS, ATOMIC NUMBERS, AND ORBITAL STRUCTURE (Photoelectron Spectroscopy)

In 1913 Henry G. J. Moseley, a student of Rutherford, used the technique of *x-ray spectroscopy* (just discovered by Max von Laue) to determine the atomic numbers of the elements. X rays are produced in a cathode ray tube when the electron beam (cathode ray) falls on a metal target. The explanation for the production of x rays is as follows: When an electron in the cathode ray hits a metal atom in the target, it can (if it has sufficient energy) knock an electron from an inner shell of the atom. A metal ion is produced with an electron missing from an inner orbital. Its electron configuration is unstable, and an electron from an orbital of higher energy drops into the half-filled orbital and a photon is emitted. The photon corresponds to electromagnetic radiation in the x-ray region.

The energies of the inner orbitals of an atom and the energy changes between them depend on the nuclear charge, $+Z$. Therefore, the photon energies $h\nu$ and the frequencies ν of emitted x rays depend on the atomic number Z of the metal atom in the target. Figure 8.10 shows the x-ray spectra obtained by Moseley with various metal targets.

A related technique, *x-ray photoelectron spectroscopy,* experimentally confirms our theoretical view of the orbital structure of the atom. Instead of irradiating a sample with an electron beam and analyzing the frequencies

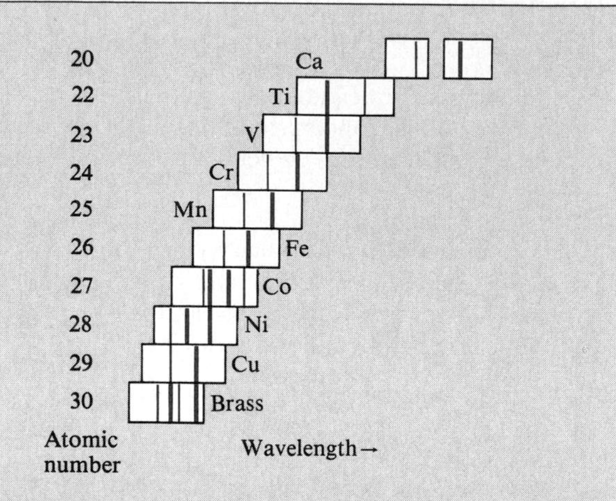

Figure 8.10
X-ray spectra of the elements calcium to zinc, obtained by Moseley. *Each line results from an emission of given wavelength. Because of the volatility of zinc, Moseley used brass (a copper-zinc alloy) to observe the spectrum of zinc. Note the copper lines in brass. Also note how the lines progress to the right (indicating increasing wavelength or decreasing energy difference) with decreasing atomic number. (From J. J. Lagowski,* The Structure of Atoms *[Boston: Houghton Mifflin, 1964], Figure 26, p. 80. Used by permission.)*

Figure 8.11
X-ray photoelectron spectrum of neon. (A) *Each peak shows the energy of the ejected electrons. The energy of each x-ray photon is 200.9×10^{-18} J.* (B) *The energy-level diagram for neon. The transitions that occur in the photoelectron spectrum are to $n = \infty$, where the energy = 0.*

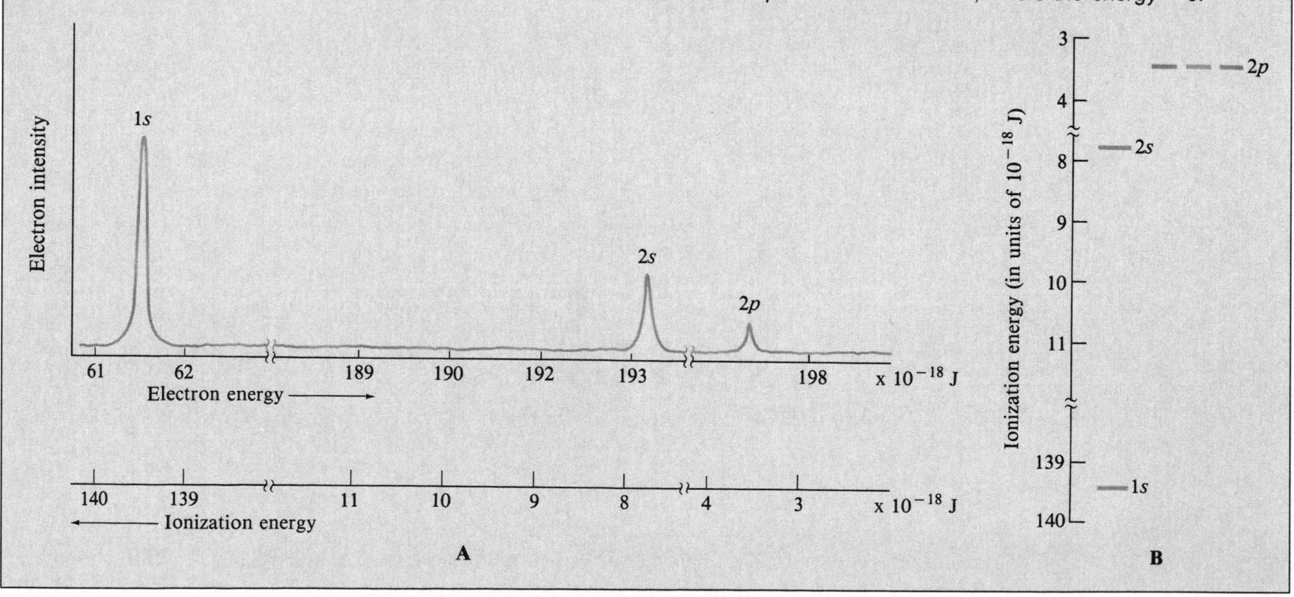

of emitted x rays, we irradiate a sample with x rays and analyze the kinetic energies of ejected electrons. In other words, we observe the *photoelectric effect* on the sample (see Section 7.6).

As an example of photoelectron spectroscopy, consider a sample of neon gas (Ne atoms). Suppose the sample is irradiated with x rays of a specific frequency great enough to remove a 1*s* electron from the neon atom. Part of the energy of the x-ray photon, $h\nu$, is used to remove the electron from the atom (this is the *ionization energy, I.E.*, for that electron). The remaining energy appears as kinetic energy, E_K, of the ejected electron. From the law of conservation of energy, we can write

$$E_K = h\nu - I.E.$$

Because $h\nu$ is fixed, E_K will be proportional to I.E., the ionization energy.

If we look at the electrons ejected from neon, we find that they have kinetic energies related to the ionization energies from all possible orbitals (1*s*, 2*s*, and 2*p*) in the atom. Thus, when we scan the various kinetic energies of ejected electrons, we see a spectrum with peaks corresponding to the different occupied orbitals (see Figure 8.11A). These ionization energies are approximately equal to the positive values of the orbital energies (Figure 8.11B), so this spectrum provides direct experimental verification of the discrete energy levels associated with the electrons of the atom.

8.3 ORBITAL DIAGRAMS OF ATOMS

In discussing the ground states of atoms, we have not yet described how the electrons are arranged in each subshell. There may be several different ways of arranging electrons in a particular configuration. Consider the carbon atom ($Z = 6$) with the ground-state configuration $1s^2 2s^2 2p^2$. Three possible arrangements are given in the following orbital diagrams.

Diagram 1: (↑↓) (↑↓) (↑)(↑)( )
1*s* 2*s* 2*p*

Diagram 2: (↑↓) (↑↓) (↑)(↓)( )
1*s* 2*s* 2*p*

Diagram 3: (↑↓) (↑↓) (↑↓)( )( )
1*s* 2*s* 2*p*

These orbital diagrams show different states of the carbon atom. Each state has a different energy and, as we will see, different magnetic characteristics.

Hund's Rule

In about 1927, the German physicist Friedrich Hund discovered an empirical rule determining the lowest energy arrangement of electrons in a subshell. **Hund's rule** states that *the lowest energy arrangement of electrons in a subshell is obtained by putting electrons into separate orbitals of the subshell with parallel spin before pairing electrons*. Let us see how this would apply to the case of the carbon atom, whose ground-state configuration is $1s^2 2s^2 2p^2$. The first four electrons go into the 1*s* and 2*s* orbitals.

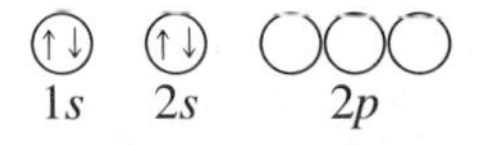

The next two electrons go into separate 2*p* orbitals, with both electrons having the same spin.

Table 8.2 Orbital Diagrams for the Ground States of Atoms from $Z = 1$ to $Z = 10$

			Orbital Diagram		
Atom	***Z***	**Configuration**	**1*s***	**2*s***	**2*p***
Hydrogen	1	$1s^1$	(↑)	( )	( )( )( )
Helium	2	$1s^2$	(↑↓)	( )	( )( )( )
Lithium	3	$1s^22s^1$	(↑↓)	(↑)	( )( )( )
Beryllium	4	$1s^22s^2$	(↑↓)	(↑↓)	( )( )( )
Boron	5	$1s^22s^22p^1$	(↑↓)	(↑↓)	(↑)( )( )
Carbon	6	$1s^22s^22p^2$	(↑↓)	(↑↓)	(↑)(↑)( )
Nitrogen	7	$1s^22s^22p^3$	(↑↓)	(↑↓)	(↑)(↑)(↑)
Oxygen	8	$1s^22s^22p^4$	(↑↓)	(↑↓)	(↑↓)(↑)(↑)
Fluorine	9	$1s^22s^22p^5$	(↑↓)	(↑↓)	(↑↓)(↑↓)(↑)
Neon	10	$1s^22s^22p^6$	(↑↓)	(↑↓)	(↑↓)(↑↓)(↑↓)

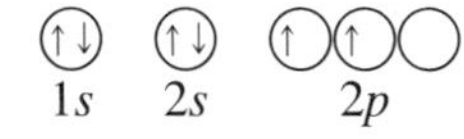

Thus, the orbital diagram corresponding to the lowest energy is the one we previously labeled as Diagram 1.

To apply Hund's rule to the oxygen atom, whose ground-state configuration is $1s^22s^22p^4$, we place the first seven electrons as follows:

(↑↓) (↑↓) (↑)(↑)(↑)
1s 2s 2p

The last electron is paired with one of the 2*p* electrons to give a doubly occupied orbital. The orbital diagram for the ground state of the oxygen atom is

(↑↓) (↑↓) (↑↓)(↑)(↑)
1s 2s 2p

In the next example, Hund's rule is used to determine the orbital diagram for the ground state of a more complicated atom. Orbital diagrams for the ground states of the first ten elements are shown in Table 8.2.

Example 8.4 Applying Hund's Rule

Write an orbital diagram for the ground state of the iron atom.

Solution

From the building-up principle, we determine that the electron configuration of the iron atom is $1s^22s^22p^63s^23p^63d^64s^2$. All the subshells except the 3*d* are filled. In placing the six electrons in the 3*d* subshell, we note that the first five go into separate 3*d* orbitals with their spin arrows in the same direction. The sixth electron must doubly occupy a 3*d* orbital. The orbital diagram is

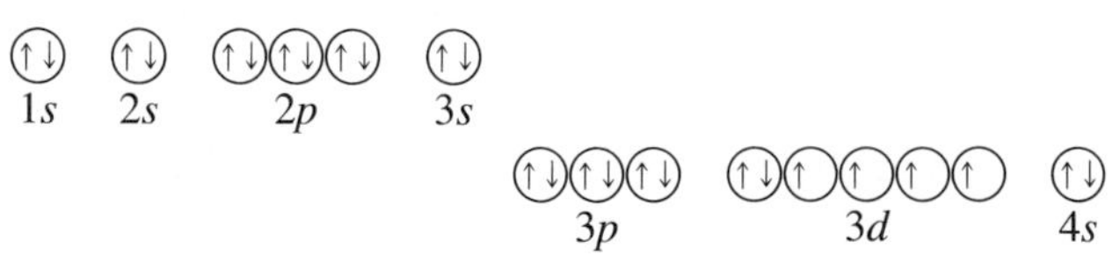

We can write this diagram in abbreviated form using [Ar] for the argon-like core of the iron atom:

[Ar] (↑↓)(↑)(↑)(↑)(↑) (↑↓)
3d 4s

Exercise 8.5

Write an orbital diagram for the ground state of the phosphorus atom ($Z = 15$). Write out all orbitals.
(See Problems 8.41 and 8.42.)

Magnetic Properties of Atoms

The magnetic properties of a substance can reveal certain information about the arrangement of electrons in an atom (or molecule). Although an electron in an atom behaves like a small magnet, the magnetic attractions from two electrons that are opposite in spin cancel each other. As a result, an atom that has only doubly occupied orbitals has no net spin magnetism. However, an atom in which there are *unpaired* electrons—that is, in which there is an excess of one kind of spin—does exhibit a net magnetism.

The magnetic properties of an atom can be observed. The most direct way is to determine whether the atomic substance is attracted to the field of a strong magnet. A **paramagnetic substance** is *a substance that is weakly attracted by a magnetic field, and this attraction is generally the result of unpaired electrons.*● For example, sodium vapor has been found experimentally to be paramagnetic. The explanation is that the vapor consists primarily of sodium atoms, each containing an unpaired electron (the configuration is $[Ne]3s^1$). A **diamagnetic substance** is *a substance that is not attracted by a magnetic field or is very slightly repelled by such a field. This property generally means that the substance has only paired electrons.* Mercury vapor is found experimentally to be diamagnetic. The explanation is that mercury vapor consists of mercury atoms, each with the electron configuration $[Xe]4f^{14}5d^{10}6s^2$, which has only paired electrons.

● The strong, permanent magnetism seen in iron objects is called *ferromagnetism* and is due to the cooperative alignment of electron spins in many iron atoms. Paramagnetism is a much weaker effect. Nevertheless, paramagnetic substances can be attracted to a strong magnet. Liquid oxygen is composed of paramagnetic O_2 molecules. When poured over a magnet, the liquid is seen to cling to the poles. (See Figure 10.20.)

We expect the orbital diagrams presented at the beginning of this section for the $1s^22s^22p^2$ configuration of the carbon atom to have different magnetic properties. Diagram 1, which is predicted by Hund's rule to be the ground state, would give a magnetic atom, whereas the other diagrams would not. If we could prepare a vapor of carbon atoms, it should be attracted to a magnet (it should be paramagnetic). It is difficult to prepare a vapor of free carbon atoms in sufficient concentration to observe a result. However, the spectrum of carbon atoms can be obtained easily from dilute vapor. From an analysis of this spectrum, it is possible to show that the ground-state atom is magnetic, which is consistent with the prediction of Hund's rule.

Periodicity of the Elements

We have seen that the periodic table that Mendeleev discovered in 1869 can be explained by the periodicity of the ground-state electron configurations of the atoms. In the next sections, we will look at various aspects of the periodicity of the elements.

8.4 MENDELEEV'S PREDICTIONS FROM THE PERIODIC TABLE

One of Mendeleev's periodic tables is reproduced in Figure 8.12. Though it is somewhat different from modern tables, it shows essentially the same arrange-

Reihen	Gruppe I. — R^2O	Gruppe II. — RO	Gruppe III. — R^2O^3	Gruppe IV. RH^4 RO^2	Gruppe V. RH^3 R^2O^5	Gruppe VI. RH^2 RO^3	Gruppe VII. RH R^2O^7	Gruppe VIII. — RO^4
1	H = 1							
2	Li = 7	Be = 9,4	B = 11	C = 12	N = 14	O = 16	F = 19	
3	Na = 23	Mg = 24	Al = 27,3	Si = 28	P = 31	S = 32	Cl = 35,5	
4	K = 39	Ca = 40	— = 44	Ti = 48	V = 51	Cr = 52	Mn = 55	Fe = 56, Co = 59, Ni = 59, Cu = 63.
5	(Cu = 63)	Zn = 65	— = 68	— = 72	As = 75	Se = 78	Br = 80	
6	Rb = 85	Sr = 87	?Yt = 88	Zr = 90	Nb = 94	Mo = 96	— = 100	Ru = 104, Rh = 104, Pd = 106, Ag = 108.
7	(Ag = 108)	Cd = 112	In = 113	Sn = 118	Sb = 122	Te = 125	J = 127	
8	Cs = 133	Ba = 137	?Di = 138	?Ce = 140	—	—	—	— — — —
9	(—)	—	—		—	—	—	
10	—	—	?Er = 178	?La = 180	Ta = 182	W = 184	—	Os = 195, Ir = 197, Pt = 198, Au = 199.
11	(Au = 199)	Hg = 200	Tl = 204	Pb = 207	Bi = 208	—	—	
12	—	—	—	Th = 231	—	U = 240	—	— — — —

Figure 8.12
Mendeleev's periodic table. This one was published in 1872.

ment. Mendeleev left spaces in his periodic table for what he felt were undiscovered elements. There are blank spaces in his row 5, for example, one directly under aluminum and another under silicon. By writing the known elements in this row with their atomic weights, he could determine approximate values (between the known ones) for the missing elements (values in parentheses).

Cu	Zn	—	—	As	Se	Br
63 amu	65 amu	(68 amu)	(72 amu)	75 amu	78 amu	80 amu

The Group III element directly under aluminum Mendeleev called eka-aluminum, with the symbol Ea. (*Eka* is the Sanskrit word meaning "first"; thus eka-aluminum is the first element under aluminum.) The known Group III elements have oxides of the form R_2O_3, so Mendeleev predicted that eka-aluminum would have an oxide with the formula Ea_2O_3.

The physical properties of this undiscovered element could be predicted by comparing values for the neighboring known elements. For eka-aluminum Mendeleev predicted a density of 5.9 g/cm^3, a low *melting point* (the temperature at which a substance melts), and a high *boiling point* (the temperature at which a substance boils).

In 1874 the French chemist Paul-Émile Lecoq de Boisbaudran found two previously unidentified lines in the atomic spectrum of a sample of sphalerite (a zinc sulfide, ZnS, mineral). Realizing he was on the verge of a discovery, Lecoq de Boisbaudran quickly prepared a large batch of the zinc mineral, from which he isolated a gram of a new element. He called this new element gallium. The properties of gallium were remarkably close to those Mendeleev predicted for eka-aluminum.

PROPERTY	PREDICTED FOR EKA-ALUMINUM	FOUND FOR GALLIUM
Atomic weight	68 amu	69.7 amu
Formula of oxide	Ea_2O_3	Ga_2O_3
Density of the element	5.9 g/cm^3	5.91 g/cm^3
Melting point of the element	Low	30.1°C
Boiling point of the element	High	1983°C

The predictive power of Mendeleev's periodic table was demonstrated again when scandium (eka-boron) was discovered in 1879 and when germanium (eka-silicon) was discovered in 1886. Both elements had properties remarkably like those predicted by Mendeleev. These early successes won acceptance for the organizational and predictive power of the periodic table.

8.5 SOME PERIODIC PROPERTIES

The electron configurations of the atoms display a periodic variation with increasing atomic number (nuclear charge). As a result, the elements show periodic variations of physical and chemical behavior. The **periodic law** is *a law stating that when the elements are arranged by atomic number, their physical and chemical properties vary periodically*. In this section, we will look at three physical properties of an atom: atomic radius, ionization energy, and electron affinity. These three quantities, especially ionization energy and electron affinity, are important in discussions of chemical bonding, as we will see in Chapter 9.

Atomic Radius

An atom does not have a definite size, because the statistical distribution of electrons does not abruptly end but merely decreases to very small values as the distance from the nucleus increases. This can be seen in the plot of the electron distribution for the argon atom, shown in Figure 8.13. Consequently, atomic size

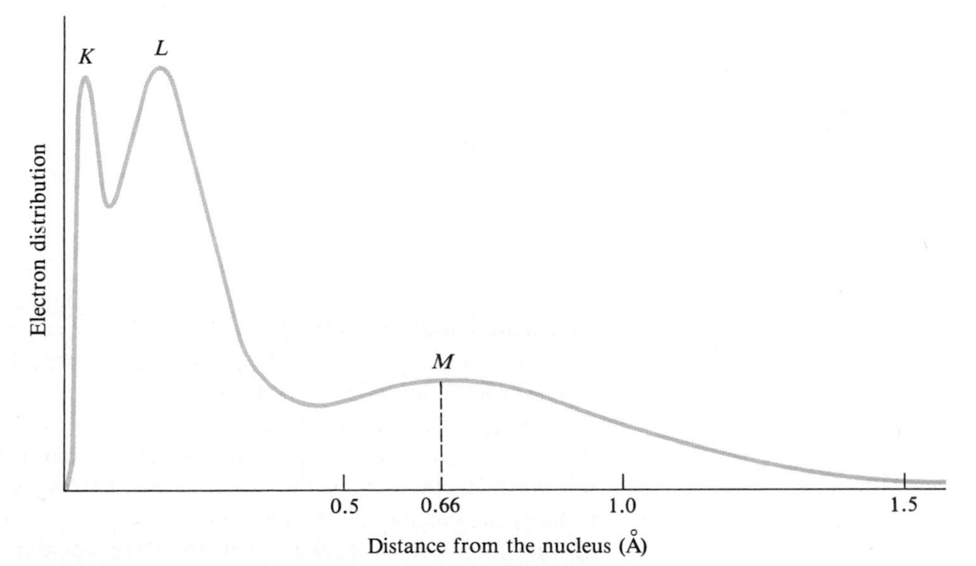

Figure 8.13
Electron distribution for the argon atom. The distribution shows three maxima, for the *K*, *L*, and *M* shells. The outermost maximum occurs at 0.66 Å (66 pm). After this maximum, the distribution falls steadily, becoming negligibly small after several angstroms.

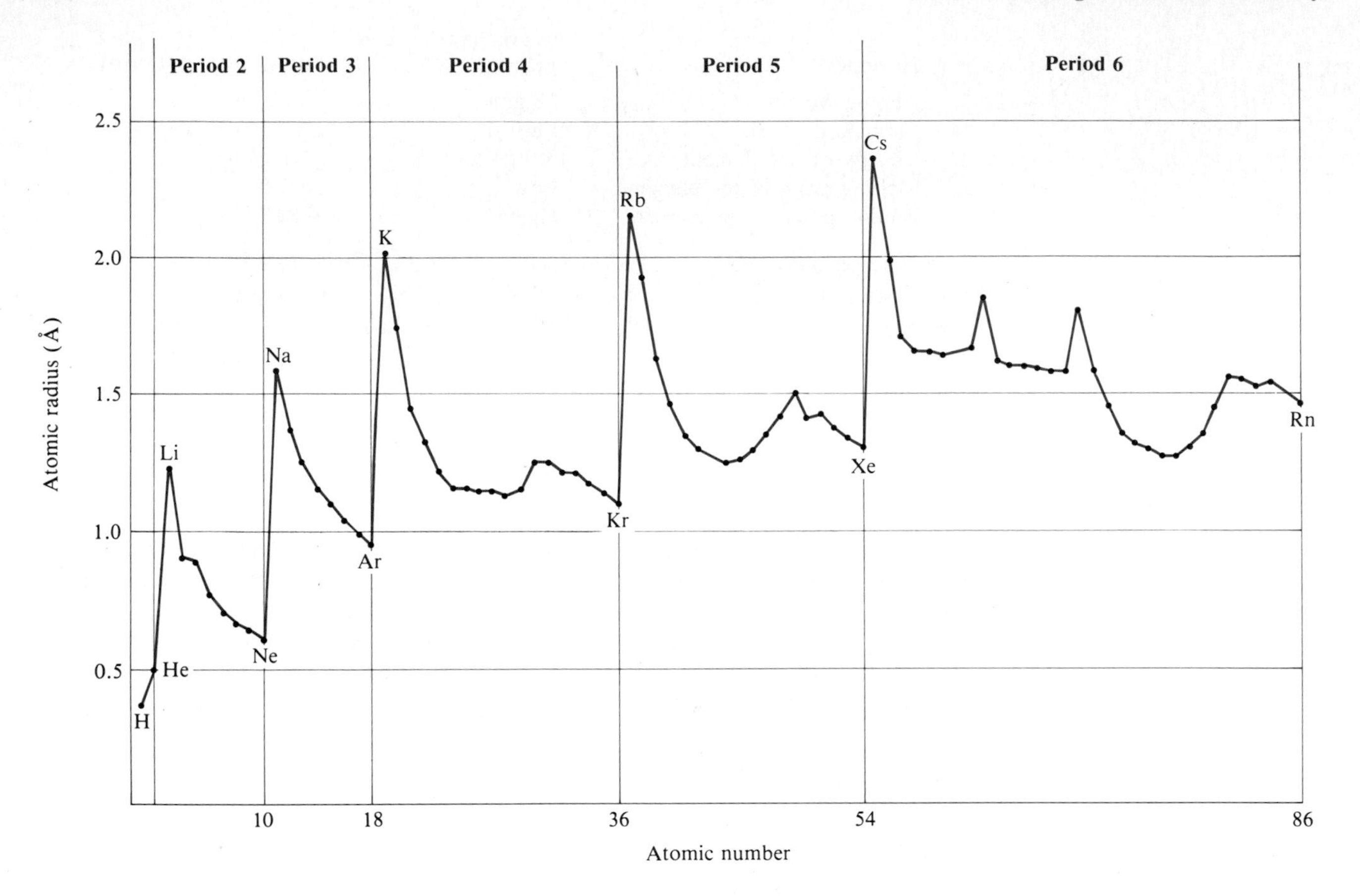

Figure 8.14
A plot of atomic radii (covalent radii) versus atomic number. Note that the curve is periodic (tends to repeat). Each period of elements begins with the Group IA atom, and the atomic radius tends to decrease until the Group VIIIA atom. (Values for He, Ne, and Ar are estimated because there are no known compounds.)

must be defined in a somewhat arbitrary manner, so various measures of atomic size exist. The atomic radii plotted in Figure 8.14 and also represented in Figure 8.15 are *covalent radii,* which are obtained from measurements of distances between atoms in the chemical bonds of molecular substances. (The determination and use of covalent radii are discussed in Chapter 9.)

Figures 8.14 and 8.15 show the following general trends in atomic radii:

1. Within each period (horizontal row), the atomic radius tends to decrease with increasing atomic number (nuclear charge). Thus, the largest atom in a period is a Group IA atom and the smallest is a noble-gas atom.

2. Within each group (vertical column), the atomic radius tends to increase with the period number.

There is a large increase in atomic radius in going from any noble-gas atom to the following Group IA atom, giving the curve in Figure 8.14 a saw-tooth appearance. A similar diagram is obtained for other measures of atomic size.

These general trends in atomic radius can be explained if we look at the two factors that primarily determine the size of the outermost orbital. One of these is the principal quantum number n of the orbital; the larger n is, the larger the size of the orbital. The other is the effective nuclear charge acting on an electron in the orbital; increasing the effective nuclear charge reduces the size of the orbital by pulling the electrons inward. The **effective nuclear charge** is *the positive charge experienced by an electron from the nucleus, equal to the nuclear charge but*

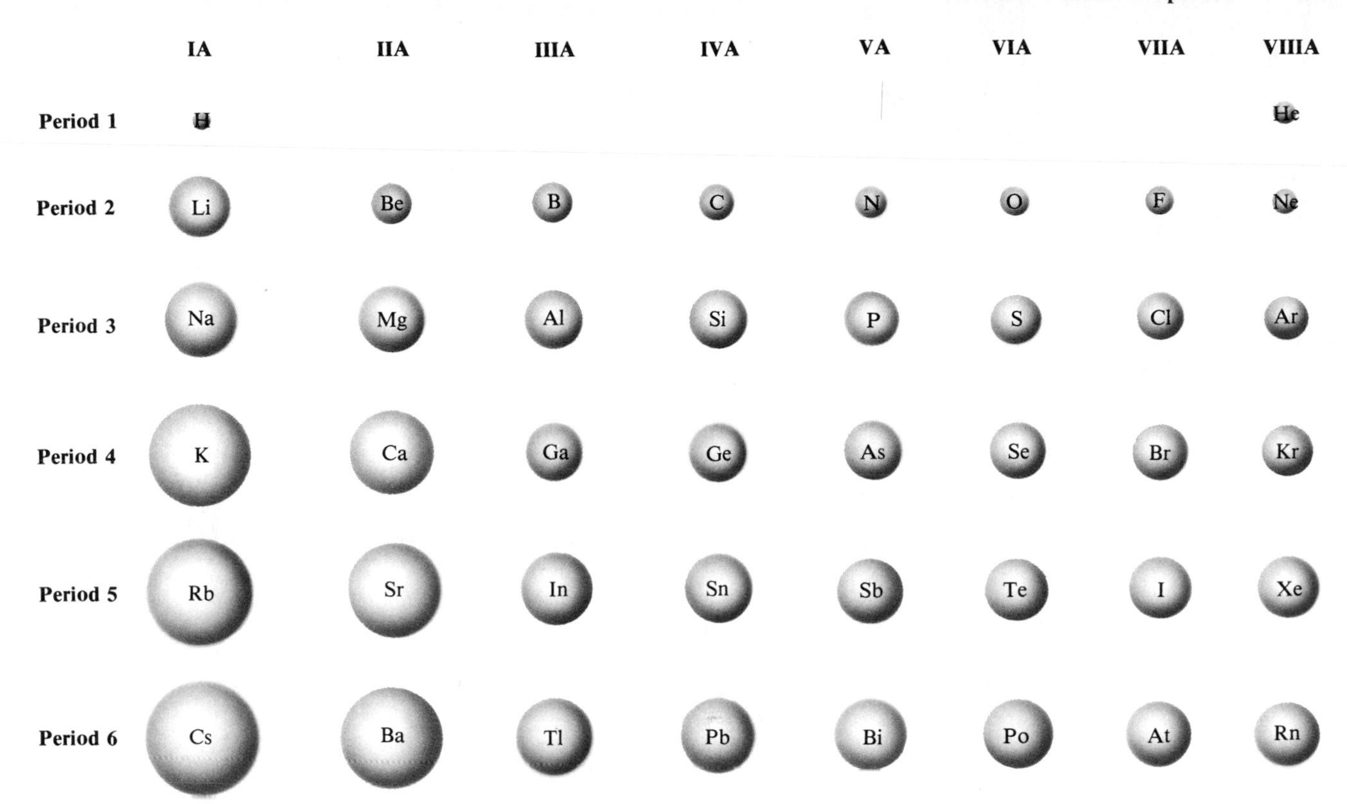

Figure 8.15 Representation of atomic radii (covalent radii) of the main-group elements. Note the trends within each period and each group.

reduced by any shielding or screening from any intervening electron distribution. Consider the effective nuclear charge on the 2*s* electron in the lithium atom (configuration $1s^22s^1$). The nuclear charge is 3*e*; but the effect of this charge on the 2*s* electron is reduced by the distribution of the two 1*s* electrons lying between the nucleus and 2*s* electron (roughly, each core electron reduces the nuclear charge by 1*e*).

Consider a given period of elements. The principal quantum number of the outer orbitals remains constant. However, the effective nuclear charge increases, because the nuclear charge increases while the number of core electrons remains constant. Consequently, the size of the outermost orbital and thus the radius of the atom decreases with increasing *Z* in any period.

Now consider a given column of elements. The effective nuclear charge remains nearly constant (approximately equal to *e* times the number of valence electrons), but *n* gets larger. Therefore, the atomic radius increases.

Example 8.5 Determining Relative Atomic Sizes from Periodic Trends

Refer to a periodic table and use the trends noted for atomic radii to arrange the following in order of increasing atomic radius: Al, C, Si.

Solution

Note that C is above Si in Group IVA. Therefore, the radius of C is smaller than that of Si (the atomic radius increases in going down a group of elements). Note that Al and Si are in the same period. Therefore the radius of Si is smaller than that of Al (radius decreases with *Z* in a period). Hence, the order of elements by increasing radius is **C, Si, Al.**

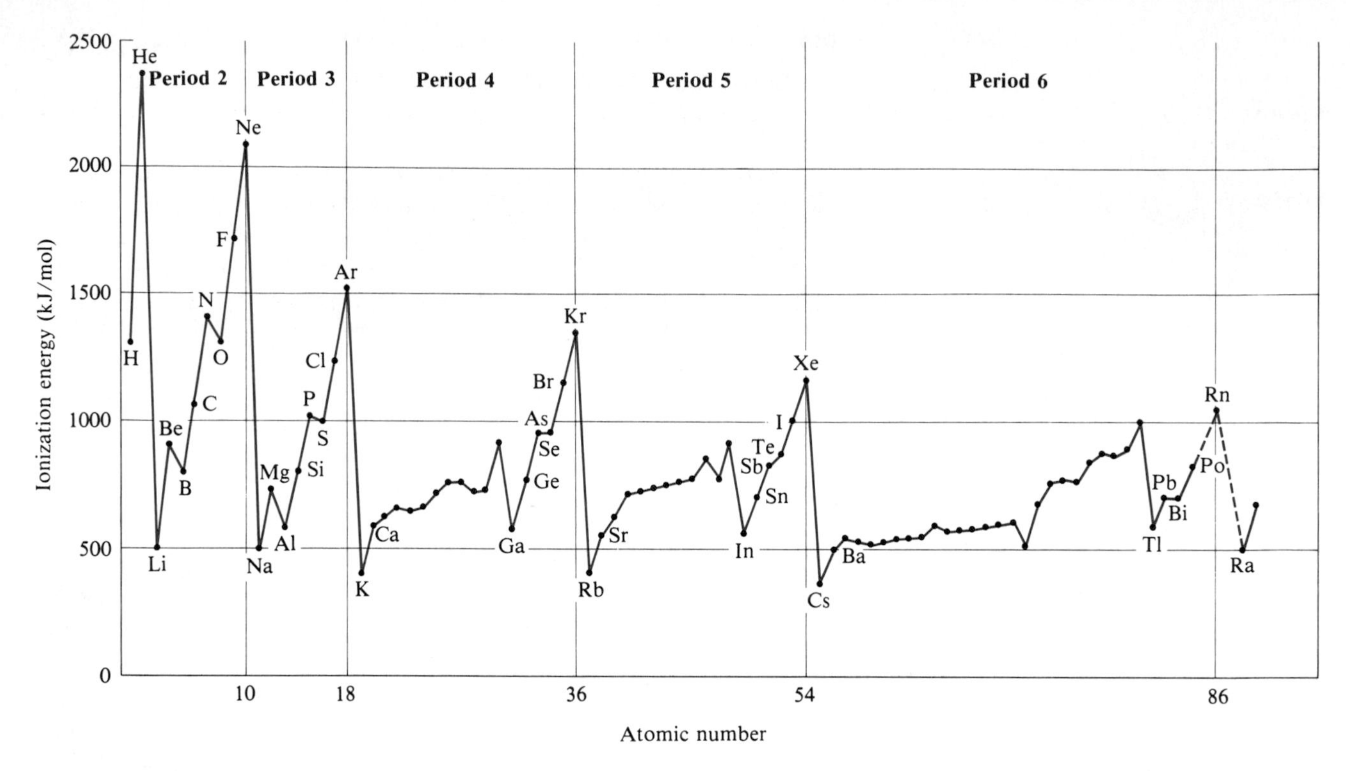

Figure 8.16
A plot of ionization energies versus atomic number. Note that the values tend to increase within each period, except for small drops in ionization energy at the Group IIIA and VA elements. Large drops occur when a new period begins.

Exercise 8.6

Using a periodic table, arrange the following in order of increasing atomic radius: Na, Be, Mg.

(See Problems 8.45 and 8.46.)

Ionization Energy

The **first ionization energy** (or **first ionization potential**) of an atom is *the minimum energy needed to remove the highest-energy (that is, the outermost) electron from the neutral atom in the gaseous state*. (When the unqualified term *ionization energy* is used, it generally means first ionization energy.) For the lithium atom, the first ionization energy is the energy needed for the following process (electron configurations are in parentheses).

$$\mathrm{Li}(1s^2 2s^1) \longrightarrow \mathrm{Li}^+(1s^2) + e^-$$

Values of this energy are usually quoted for one mole of atoms (6.02×10^{23} atoms). The ionization energy of the lithium atom is 521 kJ/mol.●

● Ionization energies are often measured in electron volts (eV). This is the amount of energy imparted to an electron when it is accelerated through an electrical potential of one volt. One electron volt is approximately 100 kJ/mol (1 eV = 96.5 kJ/mol).

Ionization energies display a periodic variation when plotted against atomic number, as Figure 8.16 shows. Within any period, values tend to increase with atomic number. Thus, the lowest values in a period are found for the Group IA elements (*alkali metals*). It is characteristic of reactive metals such as these to lose electrons easily. The largest ionization energies in any period occur for the noble-gas elements, which is a reflection of the stability of the noble-gas configuration. Because of this stability, these elements are rather unreactive.

This general trend—increasing ionization energy with atomic number in a given period—can be explained as follows: The energy needed to remove an electron from the outer shell is proportional to the effective nuclear charge divided by the average distance between electron and nucleus (this distance is inversely proportional to the effective nuclear charge). Hence, the ionization energy is proportional to the square of the effective nuclear charge and increases in going across a period.

Small deviations from this general trend occur. A IIIA element (ns^2np^1) has smaller ionization energy than the preceding IIA element (ns^2). Apparently the *np* electron of the IIIA element is more easily removed than one of the *ns* electrons of the preceding IIA element. Also note that a VIA element (ns^2np^4) has smaller ionization energy than the preceding VA element. As a result of electron repulsion, it is easier to remove an electron from the doubly occupied *np* orbital of the VIA element than from a singly occupied orbital of the preceding VA element.

Ionization energies tend to decrease in going down any column of main-group elements. This is because atomic size increases in going down the column.

Example 8.6 Determining Relative Ionization Energies from Periodic Trends

Using a periodic table only, arrange the following elements in order of increasing ionization energy: Ar, Se, S.

Solution

Note that Se is below S in Group VIA. Therefore, the ionization energy of Se should be less than that of S. Also, S and Ar are in the same period, with Z increasing from S to Ar. Therefore, the ionization energy of S should be less than that of Ar. Hence the order is Se, S, Ar.

Exercise 8.7

The first ionization energy of the chlorine atom is 1251 kJ/mol. Without looking at Figure 8.16, state which of the following values would be the more likely ionization energy for the iodine atom. Explain. (a) 1000 kJ/mol (b) 1400 kJ/mol (See Problems 8.47 and 8.48.)

The electrons of an atom can be removed successively. The energies required at each step are known as the *first* ionization energy, the *second* ionization energy, and so forth. Table 8.3 lists the successive ionization energies of the first ten elements. Note that the ionization energies for a given element increase as more electrons are removed. Thus, the first and second ionization energies of beryllium (electron configuration $1s^22s^2$) are 899 kJ/mol and 1757 kJ/mol, respectively. The first ionization energy is the energy needed to remove a 2*s* electron from the Be atom. The second ionization energy is the energy needed to remove a 2*s* electron from the positive ion Be^+. Its value is greater than that of the first ionization energy because the electron is being removed from a positive ion, which strongly attracts the electron.

Note that there is a large jump in value from the second ionization energy of Be (1757 kJ/mol) to the third ionization energy (14,848 kJ/mol). The second ionization energy corresponds to removing a valence electron (from the 2*s* orbital), which is relatively easy. The third ionization energy corresponds to removing an electron from the core of the atom (from the 1*s* orbital)—that is, from a noble-gas configuration ($1s^2$). A line in Table 8.3 separates the energies needed to remove valence electrons from those needed to remove core electrons. For each element, a large increase in ionization energy occurs when this line is crossed. The large increase results from the fact that once the valence electrons are removed, a stable noble-gas configuration is obtained. Further ionizations become much more diffi-

Questions for Study

1. What substances are believed to be responsible for the different colors of the surface of Io? Which one causes the patches of white color?
2. Describe the structure of the sulfur molecule.
3. Sulfur forms in volcanic gases from SO_2 and H_2S. Give the chemical equation for the reaction.
4. Describe the Frasch process for the recovery of sulfur from deep underground deposits.
5. Describe the Claus process for the production of sulfur from natural gas and petroleum.
6. What is the principal end product obtained from sulfur? Give the chemical equations for its production from sulfur.
7. Why is sulfur used in the production of rubber?
8. List formulas for the binary compounds of sulfur with oxygen and with carbon. If any of these compounds has a commercial use, describe it.

A Checklist for Review

IMPORTANT TERMS

electron configuration (8.1)
orbital diagram (8.1)
Pauli exclusion principle (8.1)
building-up (Aufbau) principle (8.2)
noble-gas core (8.2)
pseudo-noble-gas core (8.2)
valence electron (8.2)
Hund's rule (8.3)
paramagnetic substance (8.3)
diamagnetic substance (8.3)
periodic law (8.5)
effective nuclear charge (8.5)
first ionization energy (first ionization potential) (8.5)
electron affinity (8.5)
basic oxide (basic anhydride) (8.6)
acidic oxide (acidic anhydride) (8.6)
amphoteric substance (8.6)

SUMMARY OF FACTS AND CONCEPTS

To understand the similarities that exist among the members of a group of elements, it is necessary to know the *electron configurations* for the ground states of atoms. Only those arrangements of electrons allowed by the *Pauli exclusion principle* are possible. The ground-state configuration of an atom represents the electron arrangement that has the lowest total energy. This arrangement can be reproduced by the *building-up principle* (Aufbau principle), where electrons fill the subshells in a particular order (the building-up order) consistent with the Pauli exclusion principle. The arrangement of electrons that are in partially filled subshells is governed by *Hund's rule*.

Elements in the same group of the periodic table have similar *valence-shell configurations*. As a result, chemical and physical properties of the elements show periodic behavior. *Atomic radii,* for example, tend to decrease across any period (left to right) and increase down any group (top to bottom). *First ionization energies* tend to increase across a period and decrease down a group. *Electron affinities* of the Group VIA and VIIA elements have large negative values.

OPERATIONAL SKILLS

1. Applying the Pauli exclusion principle Given an orbital diagram or electron configuration, decide whether it is possible or not, according to the Pauli exclusion principle (Example 8.1).

2. Determining the configuration of an atom using the building-up principle Given the atomic number of an atom, write the complete electron configuration for the ground state, according to the building-up principle (Example 8.2).

3. Determining the configuration of an atom using the period and group numbers Given the period and group for an element, write the configuration of the outer electrons (Example 8.3.).

4. Applying Hund's rule Given the electron configuration for the ground state of an atom, write the orbital diagram (Example 8.4).

5. Applying periodic trends Using the known trends and referring to a periodic table, arrange a series of elements in order by atomic radius (Example 8.5) or ionization energy (Example 8.6).

Review Questions

8.1 Describe the experiment of Stern and Gerlach. How are the results for the hydrogen atom explained?

8.2 Describe the model of electron spin given in the text. What are the restrictions on electron spin?

8.3 How does the Pauli exclusion principle limit the possible configurations of an atom?

8.4 What is the maximum number of electrons that can occupy a *g* subshell ($l = 4$)?

8.5 List the orbitals in order of increasing orbital energy up to and including the $3p$ orbital.

8.6 Define each of the following: noble-gas core, pseudo-noble-gas core, valence electron.

8.7 Give two different possible orbital diagrams for the $1s^2 2s^2 2p^4$ configuration of the oxygen atom, one of which should correspond to the ground state. Label the diagram for the ground state.

8.8 Define the terms *diamagnetic substance* and *paramagnetic substance*. Does the ground-state oxygen atom give a diamagnetic or paramagnetic substance? Explain.

8.9 What kind of subshell is being filled in Groups IA and IIA? in Groups IIIA to VIIIA? in the transition elements? in the lanthanides and actinides?

8.10 How was Mendeleev able to predict the properties of gallium before it was discovered?

8.11 Describe the major trends that emerge when atomic radii are plotted versus atomic number. Describe the trends observed when first ionization energies are plotted versus atomic number.

8.12 What atom has the smallest radius among the alkaline earth elements?

8.13 What main group in the periodic table has elements with the most negative electron affinities for each period? What electron configurations of neutral atoms have only unstable negative ions?

8.14 The ions Na^+ and Mg^{2+} occur in chemical compounds, but the ions Na^{2+} and Mg^{3+} do not. Explain.

8.15 Describe the major trends in metallic character observed in the periodic table of the elements.

8.16 Distinguish between an acidic and a basic oxide. Give examples of each.

8.17 What is the name of the alkali metal atom with valence-shell configuration $5s^1$?

8.18 What would you predict for the atomic number of the halogen under astatine?

8.19 List the elements in Groups IIIA to VIA in the same order as in the periodic table. Label each element as a metal, a metalloid, or a nonmetal. Does each column of elements display the expected trend of increasing metallic characteristics?

8.20 For the list of elements you made for Question 8.19, note whether the oxides of each element are acidic, basic, or amphoteric.

8.21 Write an equation for the reaction of potassium metal with water.

8.22 From what is said in Section 8.6 about Group IIA elements, list some properties of barium.

8.23 Name and describe the two allotropes of carbon.

8.24 Match each description in the left column with the appropriate element in the right column.

(a) A waxy, white solid, normally stored under water	Sulfur
(b) A yellow solid that burns in air	Sodium
(c) A reddish-brown liquid	White phosphorus
(d) A soft, light metal that reacts vigorously with water	Bromine

Practice Problems

PAULI EXCLUSION PRINCIPLE

8.25 Which of the following orbital diagrams are allowed and which are not allowed by the Pauli exclusion principle? Explain. For those that are allowed, write the electron configuration.

(a) 1s (↑↓) 2s (↑) 2p (↑)(↑)(↑) (b) 1s (↑↓) 2s (↑↓) 2p (↑↑)(↑)(↑)

(c) 1s (↑↓) 2s (↑↓) 2p (↑↓)(↓)(↑) (d) 1s (↑↓) 2s (↑↓↑) 2p (↑)(↑)( )

8.26 Which of the following orbital diagrams are allowed by the Pauli exclusion principle? Explain how you arrived at this decision. Give the electron configuration for the allowed ones.

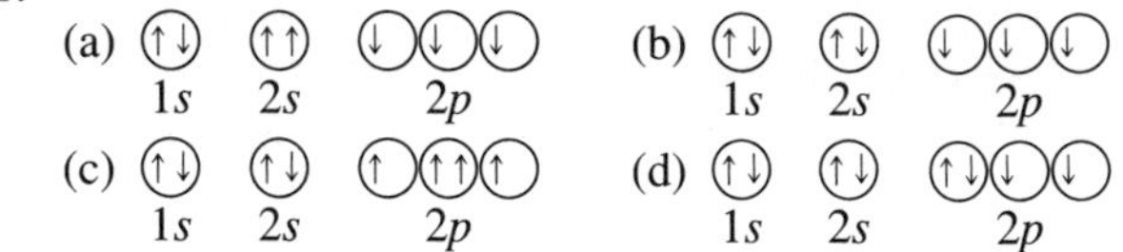

8.27 Choose the electron configurations that are possible from among the following. Explain why the others are impossible.

(a) $1s^2 2s^3 2p^6$ (b) $1s^2 2s^2 2p^4$
(c) $1s^2 2s^2 2p^8 3s^2 3p^6 3d^7$ (d) $1s^2 2s^2 2p^6 3s^1 3d^9$

8.28 Which of the following electron configurations are possible? Explain why the others are not.

(a) $1s^2 2s^1 2p^6$ (b) $1s^2 2s^2 2p^6 3s^1 3d^6$
(c) $1s^2 2s^2 2p^8$ (d) $1s^2 2s^2 2p^6 3s^2 3d^{12}$

8.29 Write out all of the possible orbital diagrams for the electron configuration $1s^2 2p^1$. (There are six different diagrams.)

8.30 Give the different orbital diagrams for the configuration $1s^1 2p^1$. (There are twelve different diagrams.)

BUILDING-UP PRINCIPLE AND HUND'S RULE

8.31 Use the building-up principle to obtain the ground-state configuration of phosphorus.

8.32 Give the electron configuration of the ground state of chlorine, using the building-up principle.

8.33 Use the building-up principle to obtain the electron configuration of the ground state of vanadium.

8.34 Give the electron configuration of the ground state of cobalt, using the building-up principle.

8.35 Bromine is a Group VIIA element in Period 4. Deduce the valence-shell configuration of bromine.

8.36 Bismuth is a Group VA element in Period 6. Write the valence-shell configuration of bismuth.

8.37 Titanium is a Group IVB element in Period 4. What would you expect to be the configuration of the outer electrons of titanium?

8.38 Cadmium is a Group IIB element in Period 5. What would you give for the valence-shell configuration of cadmium?

8.39 Thallium has a ground-state configuration $[Xe]4f^{14}5d^{10}6s^2 6p^1$. Give the group and period for this element. Classify it as a main-group, a *d*-transition, or an *f*-transition element.

8.40 The configuration for the ground state of iridium is $[Xe]4f^{14}5d^7 6s^2$. What are the group and period for this element? Is it a main-group, a *d*-transition, or an *f*-transition element?

8.41 Write the orbital diagram for the ground state of nickel. The electron configuration is $[Ar]3d^8 4s^2$.

8.42 Write the orbital diagram for the ground state of terbium. The electron configuration is $[Xe]4f^9 6s^2$.

8.43 Write an orbital diagram for the ground state of the potassium atom. Is the atomic substance diamagnetic or paramagnetic?

8.44 Write an orbital diagram for the ground state of the zinc atom. Is the atomic substance diamagnetic or paramagnetic?

PERIODIC TRENDS

8.45 From your knowledge of periodic trends, arrange the following elements in order of increasing atomic radius: F, S, Cl.

8.46 Order the following elements by increasing atomic radius according to what you expect from periodic trends: Se, S, As.

8.47 Arrange the following elements in order of increasing ionization energy: Mg, Ca, S. Do not look at Figure 8.16.

8.48 From your knowledge of periodic trends, arrange the following elements by increasing ionization energy: Ar, Na, Cl, Al.

8.49 From what you know in a general way about electron affinities, state which member of each of the following pairs has the greater negative value: (a) Cl, S (b) Se, K.

8.50 From what you know in a general way about electron affinities, state which member of each of the following pairs has the greater negative value: (a) As, Br (b) F, Li.

8.51 Write the simplest formulas expected for two oxides of tellurium.

8.52 If potassium perchlorate has the formula $KClO_4$, what formula would you expect for lithium perbromate?

Additional Problems

8.53 Write the complete ground-state electron configuration of the strontium atom, Sr, using the building-up principle.

8.54 Write the complete ground-state electron configuration of the tin atom, Sn, using the building-up principle.

8.55 Obtain the valence-shell configuration of the polonium atom, Po, using the position of this atom in the periodic table.

8.56 Obtain the valence-shell configuration of the thallium atom, Tl, using the position of this atom in the periodic table.

8.57 Write the orbital diagram for the ground state of the arsenic atom. Write out all orbitals.

8.58 Write the orbital diagram for the ground state of the germanium atom. Write out all orbitals.

8.59 For eka-lead, predict the electron configuration, whether the element is a metal or nonmetal, and the formula of an oxide.

8.60 For eka-bismuth, predict the electron configuration, whether the element is a metal or nonmetal, and the formula of an oxide.

8.61 From Figure 8.16, predict the first ionization energy of francium ($Z = 87$).

8.62 From Figure 8.16, predict the first ionization energy of astatine ($Z = 85$).

8.63 Write the orbital diagram corresponding to the ground state of Nb, whose configuration is $[Kr]4d^45s^1$.

8.64 Write the orbital diagram for the ground state of ruthenium. The configuration is $[Kr]4d^75s^1$.

8.65 Match each set of characteristics on the left with an element given in the column at the right.

(a) A reactive nonmetal; the atom has a large negative electron affinity	Sodium (Na)
(b) A soft metal; the atom has low ionization energy	Antimony (Sb)
(c) A metalloid that forms an oxide of formula R_2O_3	Argon (Ar)
(d) A chemically unreactive gas	Chlorine (Cl_2)

8.66 Match each of the elements on the right with a set of characteristics on the left.

(a) A reactive, pale yellow gas; the atom has a large negative electron affinity	Oxygen (O_2)
(b) A soft metal that reacts with water to produce hydrogen	Gallium (Ga)
(c) A metal that forms an oxide of formula R_2O_3	Barium (Ba)
(d) A colorless gas; the atom has moderately large negative electron affinity	Fluorine (F_2)

8.67 Find the electron configuration of the element with $Z = 23$. From this, give its group and period in the periodic table. Classify the element as a main-group, a *d*-block transition, or an *f*-block transition element.

8.68 Find the electron configuration of the element with $Z = 33$. From this, give its group and period in the periodic table. Is this a main-group, a *d*-block transition, or an *f*-block transition element?

Cumulative-Skills Problems

8.69 A 2.50-g sample of barium reacted completely with water. What is the equation for the reaction? How many milliliters of dry H_2 evolved at 21°C and 748 mmHg?

8.70 A sample of cesium metal reacted completely with water, evolving 48.1 mL of dry H_2 at 19°C and 768 mmHg. What is the equation for the reaction? What was the mass of cesium in the sample?

Table 9.1 Lewis Electron-Dot Symbols for Atoms of the Second and Third Periods

	IA ns^1	IIA ns^2	IIIA ns^2np^1	IVA ns^2np^2	VA ns^2np^3	VIA ns^2np^4	VIIA ns^2np^5	VIIIA ns^2np^6
Second Period	Li ·	· Be ·	· B ·	· C ·	: N ·	: O ·	: F ·	: Ne :
Third Period	Na ·	· Mg ·	· Al ·	· Si ·	: P ·	: S ·	: Cl ·	: Ar :

To understand why ionic bonding occurs, consider the transfer of a valence electron from a sodium atom (electron configuration [Ne]$3s^1$) to the valence shell of a chlorine atom ([Ne]$3s^23p^5$). We can represent the electron transfer by the following equation:

$$\text{Na}([\text{Ne}]3s^1) + \text{Cl}([\text{Ne}]3s^23p^5) \longrightarrow \text{Na}^+([\text{Ne}]) + \text{Cl}^-([\text{Ne}]3s^23p^6)$$

As a result of the electron transfer, ions are formed, each of which has a noble-gas configuration. The sodium atom has lost its 3*s* electron and has taken on the neon configuration, [Ne]. The chlorine atom has accepted the electron into its 3*p* subshell and has taken on the argon configuration, [Ne]$3s^23p^6$. Such noble-gas configurations and the corresponding ions are particularly stable. This stability of the ions accounts in part of the formation of the ionic solid NaCl. Once each cation or anion forms, it attracts ions of opposite charge. Within the sodium chloride crystal, NaCl, every Na^+ ion is surrounded by six Cl^- ions, and every Cl^- ion by six Na^+ ions. (Figure 2.17 on page 48 shows the arrangement of ions in the NaCl crystal.)

Lewis Electron-Dot Symbols

We can simplify the preceding equation for the electron transfer between Na and Cl by writing Lewis electron-dot symbols for the atoms and monatomic ions. A **Lewis electron-dot symbol** is *a symbol in which the electrons in the valence shell of an atom or ion are represented by dots placed around the letter symbol of the element.* Table 9.1 lists Lewis symbols and corresponding valence-shell electron configurations for the atoms of the second and third periods. Note that dots are placed one to each side of a letter symbol until all four sides are occupied. Then the dots are written two to a side until all valence electrons are accounted for. The exact placement of the single dots is immaterial. For example, the single dot in the Lewis symbol for chlorine can be written on any one of the four sides. (This pairing of dots does not always correspond to the pairing of electrons in the ground state. Thus, we write · Ḃ · for boron, rather than : B · , which more closely corresponds to the ground-state configuration [He]$2s^22p^1$. The first symbol better reflects boron's chemistry.)

The equation representing the transfer of an electron from the sodium atom to the chlorine atom is

$$\text{Na}\cdot + \cdot\ddot{\underset{\cdot\cdot}{\text{Cl}}}: \longrightarrow \text{Na}^+ + [:\ddot{\underset{\cdot\cdot}{\text{Cl}}}:]^-$$

The noble-gas configurations of the ions are apparent from the symbols. No dots are shown for the cation (all valence electrons have been removed, leaving the noble-gas core). There are eight dots shown in brackets for the anion (noble-gas configuration ns^2np^6).

Example 9.1 Using Lewis Symbols to Represent Ionic Bond Formation

Use Lewis electron-dot symbols to represent the transfer of electrons from magnesium to fluorine atoms to form ions with noble-gas configurations.

Solution

The Lewis symbols for the atoms are $:\ddot{\underset{..}{F}}\cdot$ and $\cdot Mg\cdot$ (see Table 9.1). The magnesium atom loses two electrons to assume a noble-gas configuration. But because a fluorine atom can accept only one electron to fill its valence shell, two fluorine atoms must take part in the electron transfer. We can represent this electron transfer as follows:

$$:\ddot{\underset{..}{F}}\cdot + \cdot Mg \cdot + \cdot\ddot{\underset{..}{F}}: \longrightarrow [:\ddot{\underset{..}{F}}:]^- + Mg^{2+} + [:\ddot{\underset{..}{F}}:]^-$$

Exercise 9.1

Represent the transfer of electrons from magnesium to oxygen atoms to assume noble-gas configurations. Use Lewis electron-dot symbols. (See Problems 9.21 and 9.22.)

Energy Involved in Ionic Bonding

We have seen in a qualitative way why a sodium atom and a chlorine atom might be expected to form an ionic bond. It is instructive, however, to look at the energy changes involved in ionic bond formation. From this analysis, we can gain further understanding of why certain atoms bond ionically and others do not.

If atoms come together and bond, there should be a net decrease in energy, because the bonded state should be more stable and therefore at a lower energy level. Consider again the formation of an ionic bond between a sodium atom and a chlorine atom. We will think of this as occurring in two steps: (1) An electron is transferred between the two separate atoms to give ions. (2) The ions then attract one another to form an ionic bond. In reality, the transfer of the electron and the formation of an ionic bond occur simultaneously, rather than in discrete steps, as the atoms approach one another. But the *net* quantity of energy involved is the same whether the steps occur one after the other or at the same time.

The first step requires removal of the 3*s* electron from the sodium atom and the addition of this electron to the valence shell of the chlorine atom. Removing the electron from the sodium atom requires energy (the first ionization energy of the sodium atom, which equals 496 kJ/mol). Adding the electron to the chlorine atom releases energy (the electron affinity of the chlorine atom, which equals −349 kJ/mol).● It requires more energy to remove an electron from the sodium atom than is gained when the electron is added to the chlorine atom. That is, the formation of ions from the atoms is not in itself energetically favorable. It requires additional energy equal to (496 − 349) kJ/mol, or 147 kJ/mol, to form ions.

● Ionization energies and electron affinities of atoms were discussed in Section 8.5.

When positive and negative ions bond, however, more than enough energy is released to supply this additional requirement. What principally determines the energy released when ions bond is the attraction of oppositely charged ions. We can estimate this energy from *Coulomb's law* if we make the simplifying assumption that the ions are spheres, just touching, with a distance between nuclei equal to that in the NaCl crystal. From experiment, this distance is 2.82 Å or 2.82×10^{-10} m. According to Coulomb's law, the energy E obtained in bringing two ions with electric charges Q_1 and Q_2 from infinite separation to a distance r apart is

$$E = \frac{kQ_1Q_2}{r}$$

Here k is a physical constant, equal to 8.99×10^9 J · m/C^2; the charge on Na^+ is $+e$ and that on Cl^- is $-e$, where e equals 1.602×10^{-19} C. Therefore,

$$E = \frac{-(8.99 \times 10^9 \text{ J} \cdot \text{m/C}^2) \times (1.602 \times 10^{-19} \text{ C})^2}{2.82 \times 10^{-10} \text{ m}} = -8.18 \times 10^{-19} \text{ J}$$

The minus sign means energy is released. This energy is for the formation of one ion pair. To express this for one mole of NaCl, we multiply by Avogadro's number, 6.02×10^{23}. We obtain -493 kJ/mol for the energy obtained when one mole of Na^+ and one mole of Cl^- come together to form NaCl ion pairs.●

● The energy value -493 kJ/mol is approximate, because of the simplifying assumption we made.

The attraction of oppositely charged ions does not stop with the bonding of pairs of ions. The maximum attraction of ions of opposite charge with the minimum repulsion of ions of the same charge is obtained with the formation of the crystalline solid. Then even more energy is released.

The energy released when a crystalline solid forms from ions is related to the lattice energy of the solid. The **lattice energy** is *the change in energy that occurs when an ionic solid is separated into isolated ions in the gas phase*. For sodium chloride, the process is

$$NaCl(s) \longrightarrow Na^+(g) + Cl^-(g)$$

The distances between ions in the crystal are continuously enlarged until the ions are very far from each other. The energy required for this process is the lattice energy of NaCl. We can calculate this energy from Coulomb's law, or we can obtain an experimental value from thermodynamic data (see the Related Topic at the end of this section). The value, $+786$ kJ/mol, is for just the opposite process of the one we were considering, in which ions come together to bond, forming the solid. When ions come together to bond, this energy is -786 kJ/mol. Consequently, the net energy obtained when gaseous Na and Cl atoms form solid NaCl is $(-786 + 147)$ kJ/mol $= -639$ kJ/mol. The negative sign shows that there has been a net decrease in energy, which we expect when stable bonding has occurred.

From this energy analysis, we can see that two atoms bond ionically if the ionization energy of one is sufficiently small and the electron affinity of the other is sufficiently large and negative. This situation exists between a reactive metal (it has low ionization energy) and a reactive nonmetal (it has large negative electron affinity). In general, bonding between a metal and a nonmetal is ionic. This energy analysis also explains why ionic bonding normally results in a solid rather than ion-pair molecules.

Related Topic

LATTICE ENERGIES FROM THE BORN-HABER CYCLE

Direct experimental determination of the lattice energy of an ionic solid is difficult. However, this quantity can be indirectly determined from experiment by means of a thermochemical "cycle" originated by Max Born and Fritz Haber in 1919 and now called the *Born-Haber cycle*. The reasoning is based on Hess's law.

To obtain the lattice energy of NaCl, we think of solid sodium chloride being formed from the elements by two different routes. These are shown in Figure 9.2. In one route, NaCl(s) is formed directly from the elements, Na(s) and $\frac{1}{2}Cl_2(g)$. The enthalpy change for this is ΔH_f°, which is given in Table 6.2 (page 208) as -411 kJ per mole of NaCl. The second route consists of five steps:

1. *Sublimation of sodium.* Metallic sodium is vaporized to a gas of sodium atoms. (*Sublimation* is the trans-

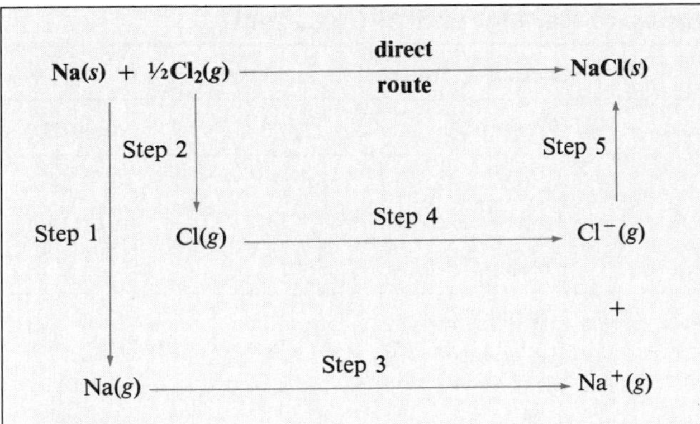

Figure 9.2
Born-Haber cycle for NaCl. *The formation of NaCl(s) from the elements is accomplished by two different routes. The direct route is the formation reaction (shown in boldface), and the enthalpy change is ΔH_f°. The indirect route occurs in five steps.*

formation of a solid to a gas.) The enthalpy change for this process has been measured experimentally and is equal to 108 kJ per mole of sodium.

2. *Dissociation of chlorine.* Chlorine molecules are dissociated to atoms. The enthalpy change for this equals the Cl—Cl bond dissociation energy, which is 240 kJ per mole of bonds or 120 kJ per mole of Cl atoms.
3. *Ionization of sodium.* Sodium atoms are ionized to Na^+ ions. The enthalpy change is essentially the ionization energy of atomic sodium, which equals 496 kJ per mole of Na.
4. *Formation of chloride ion.* The electrons from the ionization of sodium atoms are transferred to chlorine atoms. The enthalpy change for this is the electron affinity of atomic chlorine, which equals −349 kJ per mole of Cl atoms.
5. *Formation of NaCl(s) from ions.* The ions Na^+ and Cl^- formed in steps 3 and 4 combine to give solid sodium chloride. Because this process is just the reverse of the one corresponding to the lattice energy (breaking the solid into ions), the enthalpy change is the negative of the lattice energy. If we let U be the lattice energy, the enthalpy change for step 5 is $-U$.

Let us write these five steps and add them together. We also add the corresponding enthalpy changes, following Hess's law.

$$
\begin{array}{lll}
\mathrm{Na}(s) \longrightarrow \cancel{\mathrm{Na}(g)} & \Delta H_1 = & 108\ \mathrm{kJ} \\
\tfrac{1}{2}\mathrm{Cl_2}(g) \longrightarrow \cancel{\mathrm{Cl}(g)} & \Delta H_2 = & 120\ \mathrm{kJ} \\
\cancel{\mathrm{Na}(g)} \longrightarrow \cancel{\mathrm{Na^+}(g)} + \cancel{e^-(g)} & \Delta H_3 = & 496\ \mathrm{kJ} \\
\cancel{\mathrm{Cl}(g)} + \cancel{e^-(g)} \longrightarrow \cancel{\mathrm{Cl^-}(g)} & \Delta H_4 = & -349\ \mathrm{kJ} \\
\cancel{\mathrm{Na^+}(g)} + \cancel{\mathrm{Cl^-}(g)} \longrightarrow \mathrm{NaCl}(s) & \Delta H_5 = & -U \\
\hline
\mathrm{Na}(s) + \tfrac{1}{2}\mathrm{Cl_2}(g) \longrightarrow \mathrm{NaCl}(s) & \Delta H_f^\circ = & 375\ \mathrm{kJ} - U
\end{array}
$$

In summing the equations, we have canceled terms that appear on both the left and right sides of the arrows. The final equation is simply the formation reaction for NaCl(s). Adding the enthalpy changes for steps 1 to 5, we find that the enthalpy change for this formation reaction is 375 kJ − U. But the enthalpy of formation has been determined calorimetrically and equals −411 kJ. Equating these two values, we get

$$375\ \mathrm{kJ} - U = -411\ \mathrm{kJ}$$

Solving for U yields the lattice energy of NaCl:

$$U = (375 + 411)\ \mathrm{kJ} = 786\ \mathrm{kJ}$$

It is instructive to compare the lattice energy of NaCl (786 kJ) with that of MgO (determined to be 3934 kJ). Clearly MgO has much stronger bonding than NaCl. Using the ionic theory of bonding, one can calculate values for lattice energies that are close to those obtained by the Born-Haber cycle from experimental data. This result adds strength to the ionic theory of bonding. Without doing a detailed calculation, we can see that the lattice energies of NaCl and MgO are approximately as we would expect from the theory. According to Coulomb's law, the energy released in bringing a cation and an anion together is proportional to the product of the ion charges divided by the distance between the centers of the ions. For NaCl, the product of ion charges is $(+1) \times (-1) = -1$, whereas for MgO it is $(+2) \times (-2) = -4$. If the distance between ions in these solids were about the same, we would expect their lattice energies to be proportional to the negative of these products of charges. That is, we would expect the lattice energy of MgO to be four times that of NaCl. The distance between ions is not the same, of course. The distance is actually smaller in MgO, so we expect its lattice energy to be more than four times greater than in NaCl, as is the case.

9.2 ELECTRON CONFIGURATIONS OF IONS

In the previous section, we described the formation of ions from atoms. Often we can understand what monatomic ions form by looking at the electron configurations of the atoms and deciding what configurations we would expect for the ions.

Table 9.2 Ionization Energies of Na, Mg, and Al (in kJ/mol)*

	Successive Ionization Energies			
Element	First	Second	Third	Fourth
Na	496	4,562	6,912	9,543
Mg	738	1,451	7,733	10,540
Al	578	1,817	2,745	11,577

*Energies for the ionization of valence electrons lie to the left of the colored line.

Ions of the Main-Group Elements

In Chapter 2, as we discussed the naming of ionic compounds, we listed the common monatomic ions of the main-group elements (Table 2.3, p. 50). Most of the cations are obtained by removing all the valence electrons from the atoms of metallic elements. Once the atoms of these elements have lost their valence electrons, they have stable noble-gas or pseudo-noble-gas configurations. The stability of these configurations can be seen by looking at the successive ionization energies of some atoms. Table 9.2 lists the first through the fourth ionization energies of Na, Mg, and Al. The energy needed to remove the first electron from the Na atom is only 496 kJ/mol (first ionization energy). But the energy required to remove another electron (the second ionization energy) is nearly ten times greater (4562 kJ/mol). The electron in this case must be taken from Na^+, which has a neon configuration. Magnesium and aluminum atoms are similar. Their valence electrons are easily removed, but the energy needed to take an electron from either of the ions that result (Mg^{2+} and Al^{3+}) is extremely high. Thus, no compounds are found with ions having charges greater than the group number.

The loss of successive electrons from an atom requires increasingly more energy. Consequently, Group IIIA elements show less tendency to form ionic compounds than do Group IA and IIA elements, which primarily form ionic compounds. Boron, in fact, forms no compounds with B^{3+} ions. The bonding is normally covalent, a topic discussed later in this chapter. However, the tendency to form ions becomes greater going down any column of the periodic table because of decreasing ionization energy. The remaining elements of Group IIIA do form compounds containing 3+ ions.

There is also a tendency for Groups IIIA to VA elements of higher periods, particularly Period 6, to form compounds with ions having a positive charge of two less than the group number. Thus, thallium in Group IIIA has compounds with 1+ ions and compounds with 3+ ions. Ions with charge equal to the group number minus two are obtained when the *np* electrons of an atom are lost but the ns^2 electrons are retained. For example,

$$Tl([Xe]4f^{14}5d^{10}6s^2 6p^1) \longrightarrow Tl^+([Xe]4f^{14}5d^{10}6s^2) + e^-$$

Few compounds of 4+ ions are known because the energy required to form ions is so great. The first three elements of Group IVA—C, Si, and Ge—are nonmetals (or metalloids) and usually form covalent rather than ionic bonds. Tin and lead, however, commonly form compounds with 2+ ions (ionic charge equal to the group number minus two). Bismuth, in Group VA, is a metallic element that forms compounds of Bi^{3+} (ionic charge equal to the group number minus two) where only the 6*p* electrons have been lost.

Group VIA and VIIA elements, whose atoms have the highest electron affinities, would be expected to form monatomic ions by gaining electrons to give noble-gas or pseudo-noble-gas configurations. Group VIIA elements (valence-shell configuration ns^2np^5) pick up one electron to give 1− anions (ns^2np^6). Examples are F^- and Cl^-. (Hydrogen also forms compounds of the 1− ion, H^-. The hydride ion, H^-, has a $1s^2$ configuration like the noble-gas atom helium.) Group VIA elements (valence-shell configuration ns^2np^4) pick up two electrons to give 2− anions (ns^2np^6). Examples are O^{2-} and S^{2-}. Although the electron affinity of nitrogen ($2s^22p^3$) is zero, the N^{3-} ion ($2s^22p^6$) is stable in the presence of certain positive ions, including Li^+ and those of the alkaline earth elements.●

● Interestingly, lithium metal reacts with nitrogen at room temperature to form a layer of lithium nitride, Li_3N, on the metal surface.

To summarize, the common monatomic ions found in the main-group elements fall into three categories (see Table 2.3, p. 50).

1. Cations of Groups IA to IIIA having noble-gas or pseudo-noble-gas configurations. The ion charges equal the group numbers.
2. Cations of Groups IIIA to VA having ns^2 electron configurations. The ion charges equal the group numbers minus 2. Examples are Tl^+, Sn^{2+}, Pb^{2+}, and Bi^{3+}.
3. Anions of Groups VA to VIIA having noble-gas or pseudo-noble-gas configurations. The ion charges equal the group numbers minus 8.

Example 9.2 Writing the Electron Configuration and Lewis Symbol for a Main-Group Ion

Write the electron configuration and the Lewis symbol for N^{3-}.

Solution

The electron configuration of the N atom is $[He]2s^22p^3$. By gaining three electrons, the atom assumes a 3− charge and the neon configuration $[He]2s^22p^6$. The Lewis symbol is

$$[:\ddot{\underset{..}{N}}:]^{3-}$$

Exercise 9.2

Write the electron configuration and the Lewis symbol for Ca^{2+} and for S^{2-}.

(See Problems 9.23 and 9.24.)

Exercise 9.3

Write the electron configurations of Pb and Pb^{2+}.

(See Problems 9.25 and 9.26.)

Polyatomic Ions

Many ions, particularly anions, are polyatomic. Some common polyatomic ions are listed in Table 2.5 on page 51. The atoms in these ions are held together by covalent bonds, which we will discuss in Section 9.4.

Transition-Metal Ions

Most transition elements form cations of several different charges. For example, iron has the cations Fe^{2+} and Fe^{3+}. Neither of these cations has a noble-gas config-

Figure 9.3
Common transition-metal cations in aqueous solution. *Left to right:* Cr^{3+} (red-violet), Mn^{2+} (pale pink), Fe^{2+} (pale green), Fe^{3+} (pale yellow), Co^{2+} (pink), Ni^{2+} (green), Cu^{2+} (blue), Zn^{2+} (colorless).

uration; that would require the energetically impossible loss of eight electrons from the neutral atom.

In forming ions, the atoms of transition elements generally lose the *ns* electrons first; then they may lose one or more $(n-1)$ *d* electrons. The 2+ ions are common for the transition elements and are obtained by the loss of the highest-energy *s* electrons from the atom. Many transition elements also form 3+ ions by losing one $(n-1)$ *d* electron in addition to the two *ns* electrons. Table 2.4 on page 51 lists some common transition-metal ions. Many compounds of transition-metal ions are colored because of transitions involving *d* electrons, whereas the compounds of the main-group elements are usually colorless (Figure 9.3).

Example 9.3 Writing Electron Configurations of Transition-Metal Ions

Write the electron configurations of Fe^{2+} and Fe^{3+}.

Solution

The electron configuration of the Fe atom ($Z = 26$) is $[Ar]3d^64s^2$. (This can be found using the building-up principle, as in Example 8.2, or from the position of Fe in the periodic table, as in Example 8.3.) To find the ion configurations, we first remove the *ns* electrons and then the $(n-1)$ *d* electrons. The number of units of positive charge on an ion indicates the number of electrons removed. **Hence, the configuration of Fe^{2+} is $[Ar]3d^6$, and that of Fe^{3+} is $[Ar]3d^5$.**

Exercise 9.4

Write the electron configuration of Mn^{2+}. (See Problems 9.27 and 9.28.)

9.3 IONIC RADII

A monatomic ion, like an atom, is a nucleus surrounded by a distribution of electrons. The **ionic radius** is *a measure of the size of the spherical region around the nucleus of an ion within which the electrons are most likely to be found*. Like defining an atomic radius, defining an ionic radius is somewhat arbitrary, because an electron distribution never abruptly ends. However, if we imagine ions to be spheres of definite size, we can obtain their radii from known distances between nuclei in crystals. (These distances can be determined accurately by observing how crystals diffract x rays.)●

● The study of crystal structure by x-ray diffraction is discussed in Chapter 11.

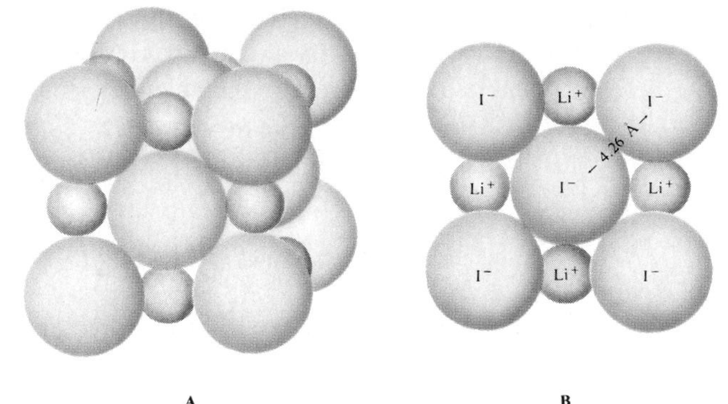

Figure 9.4
Determining the iodide ion radius in the lithium iodide (LiI) crystal. *(A)* A three-dimensional view of the crystal. *(B)* Cross section through a layer of ions. Iodide ions are assumed to be spheres in contact with one another. The distance between iodine nuclei (4.26 Å) is determined experimentally. One-half this distance (2.13 Å) equals the iodide ion radius.

To understand how to compute ionic radii, let us determine from experimental data the radius of an I^- ion in a lithium iodide (LiI) crystal. Figure 9.4 depicts a layer of ions in LiI. Note that the distance between adjacent iodine nuclei equals twice the I^- radius. From x-ray diffraction experiments, the iodine–iodine distance is found to be 4.26 Å. Therefore, the I^- radius in LiI is $1/2 \times 4.26$ Å = 2.13 Å. Other crystals give approximately the same radius for the I^- ion. Table 9.3, which lists average values of ionic radii obtained from many compounds of the main-group elements, gives 2.06 Å for the I^- radius.

That we can find values of ionic radii that agree with the known structures of many crystals is strong evidence for the existence of ions in the solid state. Moreover, these values of ionic radii compare with atomic radii in ways that we might expect. Thus, we expect a cation to be smaller and an anion to be larger than the corresponding atom (see Figure 9.5).

A cation formed when an atom loses all its valence electrons is smaller than the atom because it has one less shell of electrons. But even when only some of the valence electrons are lost from an atom, the size of the ion is smaller. With fewer electrons in the valence orbitals, the electron–electron repulsion is initially less, so

Table 9.3 Ionic Radii (in Angstroms) of Some Main-Group Elements*

	IA	IIA	IIIA	VIA	VIIA
Period 2	Li^+ 0.90	Be^{2+} 0.59		O^{2-} 1.26	F^- 1.19
Period 3	Na^+ 1.16	Mg^{2+} 0.86	Al^{3+} 0.68	S^{2-} 1.70	Cl^- 1.67
Period 4	K^+ 1.52	Ca^{2+} 1.14	Ga^{3+} 0.76	Se^{2-} 1.84	Br^- 1.82
Period 5	Rb^+ 1.66	Sr^{2+} 1.32	In^{3+} 0.94	Te^{2-} 2.07	I^- 2.06
Period 6	Cs^+ 1.81	Ba^{2+} 1.49	Tl^{3+} 1.02		

*Values are taken from R. D. Shannon, *Acta Crystallographica,* Vol. A32 (1976), p. 751.

Na $[He]2s^22p^63s^1$

Na^+ $[He]2s^22p^6$

Figure 9.5
Comparison of atomic and ionic radii. Note that the sodium atom loses its outer shell in forming the Na^+ ion. Thus, the cation is smaller than the atom. The Cl^- ion is larger than the Cl atom, because the same nuclear charge holds a greater number of electrons less strongly.

these orbitals can shrink to increase the attraction of these electrons for the nucleus. Similarly, because an anion has more electrons than the atom, the electron–electron repulsion is greater, so the valence orbitals expand. Thus, the anion radius is larger than the atomic radius.

Exercise 9.5

Which has the larger radius, S or S^{2-}? Explain. (See Problems 9.29 and 9.30.)

The ionic radii of the main-group elements shown in Table 9.3 follow a regular pattern, just as atomic radii do. *Ionic radii increase down any column because of the addition of electron shells.*

Exercise 9.6

Without looking at Table 9.3, arrange the following ions in order of increasing ionic radius: Sr^{2+}, Mg^{2+}, Ca^{2+}. (You may use a periodic table.) (See Problems 9.31 and 9.32.)

The pattern across a period becomes clear if we look first at the cations, then at the anions. For example, in the third period we have

Cation	Na^+	Mg^{2+}	Al^{3+}
Radius (Å)	1.16	0.86	0.68

All of these ions have the neon configuration $1s^22s^22p^6$, but they have different nuclear charges; that is, they are isoelectronic. **Isoelectronic** *refers to different species having the same number and configuration of electrons.* To understand the decrease in radius from Na^+ to Al^{3+}, imagine the nuclear charge (atomic number) of Na^+ to increase. With each increase of charge, the orbitals contract because of the greater attractive force of the nucleus. Thus, in any isoelectronic sequence of atomic ions, the ionic radius decreases with increasing atomic number (just as it does for the atoms).

If we look at the anions in the third-period elements, we notice that they are much larger than the cations in the same period. This abrupt increase in ionic radius is due to the fact that the anions S^{2-} and Cl^- have configurations with one more shell of electrons than the cations. And because these anions also constitute an isoelectronic sequence (with argon configuration), the ionic radius decreases with atomic number (as with the atoms):

Anion	S^{2-}	Cl^-
Radius (Å)	1.70	1.67

Thus, *in general, across a period the cations decrease in radius. When we reach the anions, there is an abrupt increase in radius, and then the radius again decreases.*

Example 9.4 Using Periodic Trends to Obtain Relative Ionic Radii

Without looking at Table 9.3, arrange the following ions in order of decreasing ionic radius: F^-, Mg^{2+}, O^{2-}. (You may use a periodic table.)

Solution

Note that F^-, Mg^{2+}, and O^{2-} are isoelectronic. If we arrange them by increasing nuclear charge, they will be in order of decreasing ionic radius. The order is O^{2-}, F^-, Mg^{2+}.

Figure 9.6
Two molecular substances: iodoform, CHI_3, and carbon tetrachloride, CCl_4. Iodoform is a low-melting, yellow solid (m.p. 120°C); carbon tetrachloride is a colorless liquid.

Exercise 9.7

Without looking at Table 9.3, arrange the following ions in order of increasing ionic radius: Cl^-, Ca^{2+}, P^{3-}. (You may use a periodic table.) (See Problems 9.33 and 9.34.)

Covalent Bond

In the preceding sections we looked at ionic substances, which are typically high-melting solids. Many substances, however, are molecular—gases, liquids, or low-melting solids consisting of molecules (Figure 9.6). A molecule is a group of atoms, frequently nonmetal atoms, strongly linked by chemical bonds. Often the forces that hold atoms together in a molecular substance cannot be understood on the basis of the attraction of oppositely charged ions. An obvious example is the molecule H_2, in which the two H atoms are held together tightly and no ions are present. In 1916 Gilbert Newton Lewis proposed that the strong attractive force between two atoms in a molecule resulted from a **covalent bond,** *a chemical bond formed by the sharing of a pair of electrons between atoms.*● In 1926 W. Heitler and F. London showed that the covalent bond in H_2 could be quantitatively explained by the newly discovered theory of quantum mechanics. We will discuss the descriptive aspects of covalent bonding in the following sections.

● G. N. Lewis (1875–1946) was Professor of Chemistry at the University of California at Berkeley. Besides his work on chemical bonding, he was noted for his research in molecular spectroscopy and thermodynamics (the study of heat involved in chemical and physical processes).

9.4 DESCRIBING COVALENT BONDS

Consider the formation of a covalent bond between two H atoms to give the H_2 molecule. As the atoms approach one another, their $1s$ orbitals begin to overlap. Each electron can then occupy the space around both atoms (Figure 9.7). In other words, the two electrons can be shared by the atoms. The electrons are attracted simultaneously by the positive charges of the two hydrogen nuclei. This attraction that bonds the electrons to both nuclei is the force holding the atoms together. Thus, while ions do not exist in H_2, the force that holds the atoms together can still be regarded as arising from the attraction of oppositely charged particles: nuclei and electrons.

It is interesting to see how the potential energy of the atoms changes as they approach and then bond. Figure 9.8 shows the potential energy of the atoms for various distances between nuclei. The potential energy of the atoms when they are some distance apart is indicated by a point on the potential-energy curve at the far

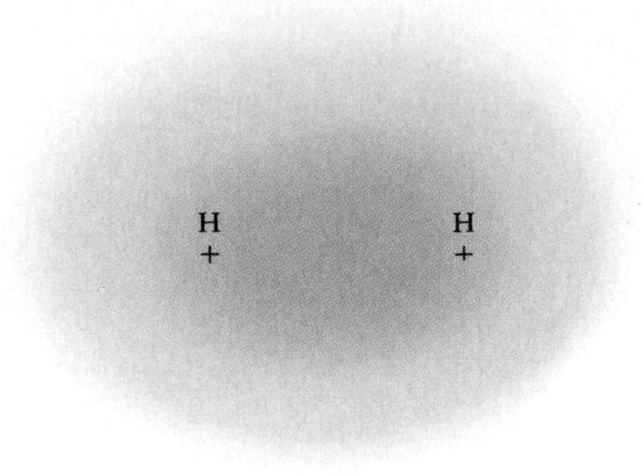

Figure 9.7
The electron probability distribution for the H_2 molecule. The electrons occupy the space around both atoms.

right. As the atoms approach (moving from the right to the left on the potential-energy curve), the potential energy gets lower and lower. The decrease in energy is a reflection of the bonding of the atoms. Eventually, as the atoms get close enough together, the repulsion of the positive charges on the nuclei becomes larger than the attraction of electrons for nuclei. In other words, the potential energy reaches a minimum value and then increases. The distance between nuclei at this minimum energy is called the *bond length* of H_2. It is the normal distance between nuclei in the molecule.

Now imagine the process to be reversed. We start with the H_2 molecule, the atoms at their normal bond length apart (at the minimum of the potential-energy curve). To separate the atoms in the molecule, energy must be added (we move along the curve to the flat portion at the right). This energy that must be added is called the *bond dissociation energy*. The larger the bond dissociation energy, the stronger the bond.

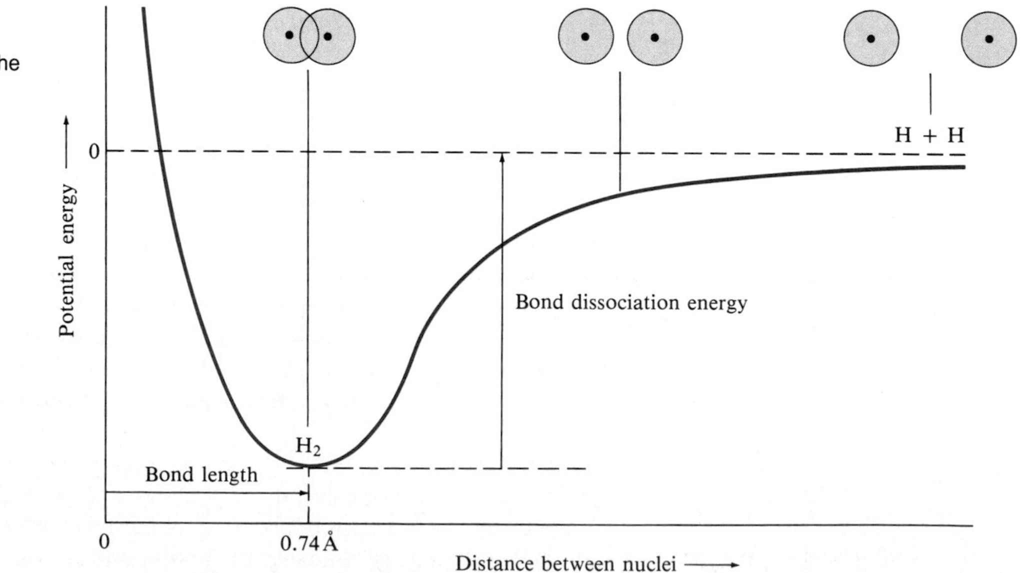

Figure 9.8
Potential-energy curve for H_2. The stable molecule occurs at the bond distance corresponding to the minimum in the potential-energy curve.

Lewis Formulas

We can represent the formation of the covalent bond in H_2 from atoms as follows:

$$H\cdot + \cdot H \longrightarrow H:H$$

Here we use the Lewis electron-dot symbol for the hydrogen atom and represent the covalent bond by a pair of dots. Recall that the two electrons from the covalent bond spend part of the time in the region of each atom. In this sense, each atom in H_2 has a helium configuration. We can draw a circle about each atom to emphasize this.

$$H:H$$

The formation of a bond between H and Cl to give an HCl molecule can be represented in a similar way.

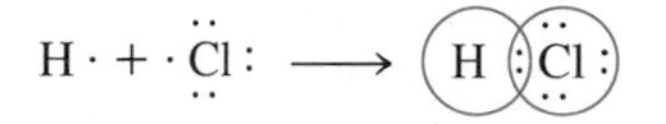

As the two atoms approach each other, unpaired electrons on each atom pair up to form a covalent bond. The pair of electrons is shared by the two atoms. Each atom then acquires a noble-gas configuration of electrons, the H atom having two electrons about it (as in He), and the Cl atom having eight valence electrons about it (as in Ar).

A formula using dots to represent valence electrons is called a **Lewis electron-dot formula.** An electron pair represented by a pair of dots in such a formula is either a **bonding pair** (*an electron pair shared between two atoms*) or a **lone,** or **nonbonding, pair** (*an electron pair that remains on one atom and is not shared*). For example,

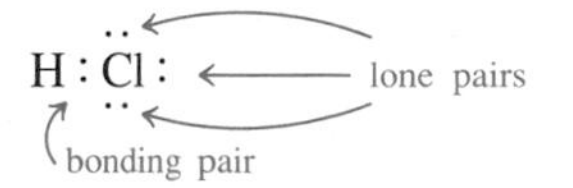

Bonding pairs are often represented by dashes rather than by pairs of dots.

Frequently, the number of covalent bonds formed by an atom equals the number of unpaired electrons shown in its Lewis symbol. Consider the formation of NH_3.

$$3H\cdot + \cdot\dot{\underset{\cdot}{N}}: \longrightarrow \begin{matrix} & H & \\ & \cdot\cdot & \\ H: & N & : \\ & \cdot\cdot & \\ & H & \end{matrix}$$

Each bond is formed between an unpaired electron on one atom and an unpaired electron on another atom. In many instances, the number of bonds formed by an atom in Groups IVA to VIIA equals the number of unpaired electrons, which is eight minus the group number. For example, a nitrogen atom (Group VA) forms $8 - 5 = 3$ covalent bonds.●

● The numbers of unpaired electrons in the Lewis symbols for the atoms of elements in Groups IA and IIIA equal the group numbers. But, except for the first elements of these groups, the atoms usually form ionic bonds.

Coordinate Covalent Bond

When bonds are formed between atoms that both donate an electron, we have

$$A\cdot + \cdot B \longrightarrow A:B$$

However, it is possible for both electrons to come from the same atom. A **coordinate covalent bond** is *a bond formed when both electrons of the bond are donated by one atom:*

$$A + :B \longrightarrow A:B$$

A coordinate covalent bond is not essentially different from other covalent bonds; it involves the sharing of a pair of electrons between two atoms. An example is the formation of the ammonium ion, in which all bonds are clearly identical.

$$H^+ + :NH_3 \longrightarrow \left[\begin{array}{c} H \\ \cdot\cdot \\ H:N:H \\ \cdot\cdot \\ H \end{array} \right]^+$$

Octet Rule

In each of the molecules we have described so far, the atoms have obtained noble-gas configurations through the sharing of electrons. Atoms other than hydrogen have obtained eight electrons in their valence shells; hydrogen atoms have obtained two. *The tendency of atoms in molecules to have eight electrons in their valence shells (two for hydrogen atoms)* is known as the **octet rule.** Many of the molecules we will discuss follow the octet rule. Some do not.

Multiple Bonds

In the molecules we have described so far, each of the bonds has been a **single bond**—that is, *a covalent bond in which a single pair of electrons is shared by two atoms*. But it is possible for atoms to share two or more electron pairs. A **double bond** is *a covalent bond in which two pairs of electrons are shared by two atoms.* A **triple bond** is *a covalent bond in which three pairs of electrons are shared by two atoms*. As examples, consider ethylene, C_2H_4, and acetylene, C_2H_2. Their Lewis formulas are

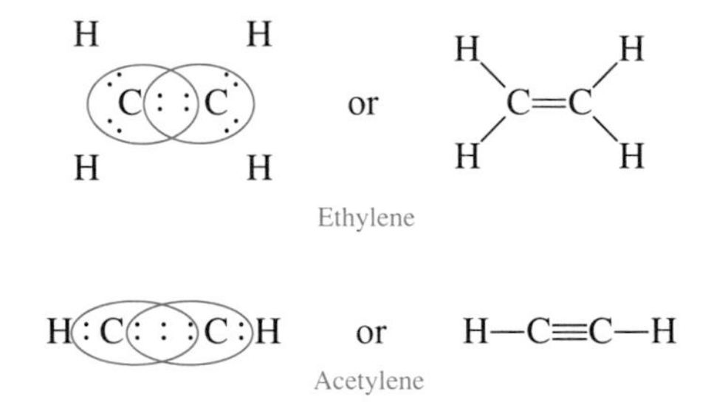

Ethylene

Acetylene

Note the octet of electrons on each C atom. Double bonds form primarily to C, N, O, and S atoms. Triple bonds form mostly to C and N atoms.

9.5 POLAR COVALENT BOND; ELECTRONEGATIVITY

A covalent bond involves the sharing of a pair of electrons between two atoms. When the atoms are alike, as in the case of the H—H bond of H_2, the bonding electrons are shared equally. That is, the electrons spend the same amount of time in the vicinity of each atom. But when the two atoms are of different elements, the

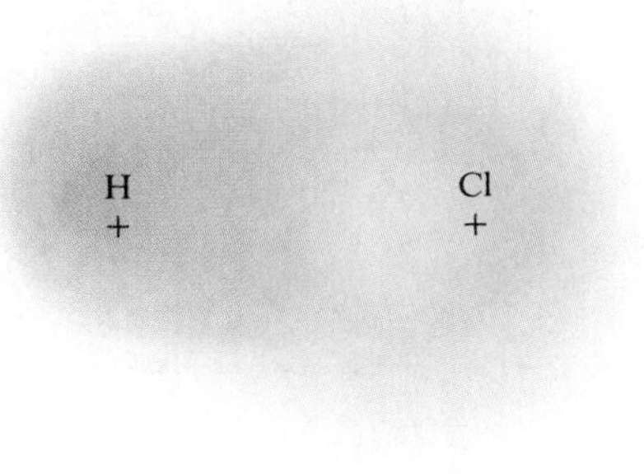

Figure 9.9
The distribution of bonding electrons in the HCl molecule. Compare this with the electron distribution for H_2 shown in Figure 9.7.

bonding electrons need not be shared equally. A **polar covalent bond** is *a covalent bond in which the bonding electrons spend more time near one atom than the other*. For example, in the case of the HCl molecule, the bonding electrons spend more time near the chlorine atom than the hydrogen atom. Figure 9.9 shows the distribution of electrons in the H—Cl bond. Thus, the H—Cl bond is polar covalent.

We can look at the polar covalent bond as intermediate between a *nonpolar* covalent bond, as in H_2, and an ionic bond, as in NaCl. From this point of view, an ionic bond is simply an extreme example of a polar covalent bond. To illustrate, we can represent the bonding in H_2, HCl, and NaCl with electron-dot formulas as follows:

H : H	H : $\ddot{\underset{..}{\text{Cl}}}$:	Na : $\ddot{\underset{..}{\text{Cl}}}$:
Nonpolar covalent	Polar covalent	Ionic

As shown, the bonding pairs of electrons are equally shared in H_2, unequally shared in HCl, and essentially not shared in NaCl. Thus, we see that it is possible to arrange different bonds to form a gradual transition from nonpolar covalent to ionic.

Electronegativity is *a measure of the ability of an atom in a molecule to draw bonding electrons to itself*. Several electronegativity scales have been proposed. In 1934 Robert S. Mulliken suggested on theoretical grounds that the electronegativity (X) of an atom be given as half its ionization energy ($I.E.$) minus electron affinity ($E.A.$).●

● Robert S. Mulliken received the Nobel Prize in chemistry in 1966 for his work on molecular orbital theory (discussed in Chapter 10).

$$X = \frac{I.E. - E.A.}{2}$$

An atom such as fluorine that tends to pick up electrons easily (large negative $E.A.$) and to hold on to them strongly (large $I.E.$) has a large electronegativity. On the other hand, an atom such as lithium or cesium that loses electrons readily (small $I.E.$) and has little tendency to gain electrons (small negative or positive $E.A.$) has small electronegativity. Until recently, only a few electron affinities had been measured. For this reason, Mulliken's scale has had limited utility. A more widely used scale was derived earlier by Linus Pauling from bond energies, which are discussed later in this chapter. Pauling's electronegativity values are given in Figure 9.10. Because electronegativities depend somewhat on the bonds formed, these values are actually average ones.●

● Linus Pauling received the Nobel Prize in chemistry in 1954 for his work on the nature of the chemical bond. In 1962 he received the Nobel Prize for peace.

Fluorine, the most electronegative element, was assigned a value of 4.0 on Pauling's scale. Lithium, at the left end of the same period, has a value of 1.0. Cesium, in the same column but below lithium, has a value of 0.7. *In general, electronegativity increases from left to right and decreases from top to bottom in*

IA	IIA	IIIB	IVB	VB	VIB	VIIB	VIIIB			IB	IIB	IIIA	IVA	VA	VIA	VIIA
								H 2.1								
Li 1.0	Be 1.5											B 2.0	C 2.5	N 3.0	O 3.5	F 4.0
Na 0.9	Mg 1.2											Al 1.5	Si 1.8	P 2.1	S 2.5	Cl 3.0
K 0.8	Ca 1.0	Sc 1.3	Ti 1.5	V 1.6	Cr 1.6	Mn 1.5	Fe 1.8	Co 1.8	Ni 1.8	Cu 1.9	Zn 1.6	Ga 1.6	Ge 1.8	As 2.0	Se 2.4	Br 2.8
Rb 0.8	Sr 1.0	Y 1.2	Zr 1.4	Nb 1.6	Mo 1.8	Tc 1.9	Ru 2.2	Rh 2.2	Pd 2.2	Ag 1.9	Cd 1.7	In 1.7	Sn 1.8	Sb 1.9	Te 2.1	I 2.5
Cs 0.7	Ba 0.9	La-Lu 1.1-1.2	Hf 1.3	Ta 1.5	W 1.7	Re 1.9	Os 2.2	Ir 2.2	Pt 2.2	Au 2.4	Hg 1.9	Tl 1.8	Pb 1.8	Bi 1.9	Po 2.0	At 2.2
Fr 0.7	Ra 0.9	Ac-No 1.1-1.7														

Figure 9.10
Electronegativities of the elements. The values given are those of Pauling.

the periodic table. Metals are the least electronegative elements (they are *electropositive*), and nonmetals are the most electronegative.

The absolute value of the difference in electronegativity of two bonded atoms gives us a rough measure of the polarity to be expected in a bond. When this difference is small, the bond is nonpolar. When it is large, the bond is polar (or, if the difference is very large, perhaps ionic). The electronegativity differences for the bonds H—H, H—Cl, and Na—Cl are 0, 0.9, and 2.1, respectively, following the expected order. Differences in electronegativity explain why ionic bonds usually form between a metal atom and a nonmetal atom; the electronegativity difference would be largest between these elements. On the other hand, covalent bonds form primarily between two nonmetals because the electronegativity differences are small.●

● The electronegativity difference in metal–metal bonding would also be small. This bonding frequently involves delocalized *metallic bonding*, briefly described in Section 9.8.

Example 9.5 Using Electronegativities to Obtain Relative Bond Polarities

Use electronegativity values (Figure 9.10) to arrange the following bonds in order of increasing polarity: P—H, H—O, C—Cl.

Solution

The absolute values of the electronegativity differences are P—H, 0.0; H—O, 1.4; C—Cl, 0.5. Hence, the order is **P—H, C—Cl, H—O.**

Exercise 9.8

Using electronegativities, decide which of the following bonds is most polar: C—O, C—S, H—Br.

(See Problems 9.41 and 9.42.)

We can use an electronegativity scale to predict the direction in which the electrons in a bond shift. Bonding electrons are pulled toward the more electronegative atom. For example, consider the H—Cl bond. Because the Cl atom (X = 3.0) is more electronegative than the H atom (X = 2.1), the bonding electrons in

H—Cl are pulled toward Cl. Because the bonding electrons spend most of their time around the Cl atom, that end of the bond acquires a partial negative charge (indicated $\delta-$). The H-atom end of the bond has a partial positive charge ($\delta+$). We can show this as follows:

$$\overset{\delta+}{\mathrm{H}}—\overset{\delta-}{\mathrm{Cl}}$$

The HCl molecule is said to be a *polar molecule*. We will say more about polar molecules when we look at molecular structures in Chapter 10.

9.6 WRITING LEWIS ELECTRON-DOT FORMULAS

The Lewis electron-dot formula of a molecule is similar to the structural formula in that it shows how atoms are bonded. Bonding electron pairs are indicated either by two dots or by a dash. In addition to the bonding electrons, however, an electron-dot formula shows the positions of lone pairs of electrons, whereas the structural formula does not. Thus, the electron-dot formula is a simple two-dimensional representation of the positions of electrons in a molecule. In the next chapter we will see how to predict the three-dimensional shape of a molecule from the two-dimensional electron-dot formula. In this section, we will discuss the steps for writing the electron-dot formula for a molecule made from atoms of the main-group elements.●

● Lewis formulas do not directly convey information about molecular shape. For example, the Lewis formula of methane, CH_4, is written as the flat (two-dimensional) formula

```
    H
    ..
H : C : H
    ..
    H
```

The actual methane molecule, however, is not flat; it has a three-dimensional structure, as explained in Chapter 10.

Skeleton Structure of a Molecule

Before we can write the Lewis formula of a molecule (or a polyatomic ion), we must know the *skeleton structure* of the molecule. The skeleton structure tells us which atoms are bonded to one another (without regard to whether the bonds are single or not). Normally this information must be found by experiment. For simple molecules, we can often predict the skeleton structure from the following four rules. There are, of course, exceptions to these rules. *The only sure way to know the correct structure of a molecule is from experiment.*

RULE 1 Many small molecules or polyatomic ions consist of a central atom around which are bonded atoms of greater electronegativity, such as F, Cl, and O. Note the following structural formulas:

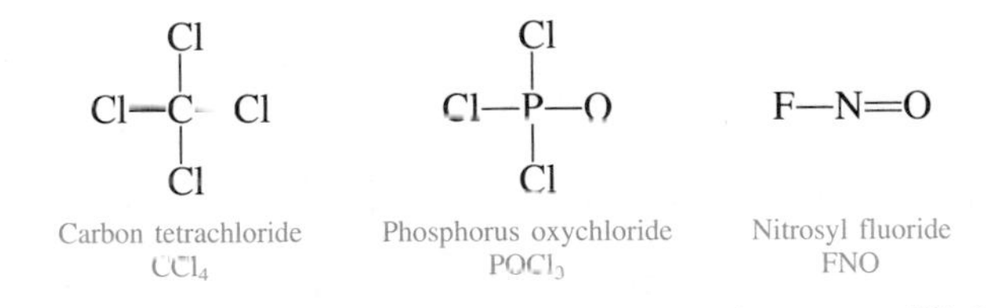

Carbon tetrachloride CCl_4 Phosphorus oxychloride $POCl_3$ Nitrosyl fluoride FNO

In some cases, H atoms surround a more electronegative atom (H_2O, NH_3, CH_4), but H cannot be a central atom because it normally forms only one bond.

RULE 2 Molecules or polyatomic ions with symmetrical formulas often have symmetrical structures. For example, disulfur dichloride, S_2Cl_2, is symmetrical, with the more electronegative Cl atoms around the S atoms: Cl—S—S—Cl.

RULE 3 *Oxoacids* are substances in which O atoms (and possibly other electronegative atoms) are bonded to a central atom, with one or more H atoms usually

bonded to O atoms. Examples are

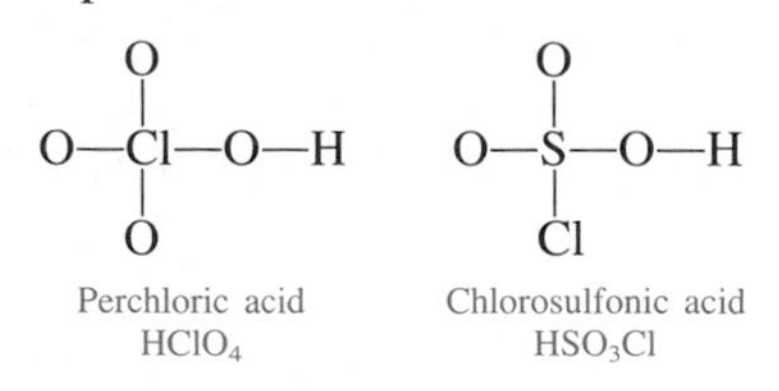

Perchloric acid $HClO_4$ Chlorosulfonic acid HSO_3Cl

RULE 4 Of several possible structural formulas, the one in which the atoms have their usual number of covalent bonds (8 − group number) is usually preferred. For example, oxygen difluoride, OF_2, has the Lewis formula on the left, not the one on the right:

:F̤̈—Ö̤—F̤̈:   :F̤̈—F̤̈—Ö̤:

Although both structures satisfy the octet rule, only the one on the left has the usual number of covalent bonds (in the one on the right, the middle F atom has two covalent bonds, instead of one). The concept of *formal charge* (discussed in the Related Topic at the end of this section) can also be used to choose among possible structures.

Steps in Writing Lewis Formulas

Once we know the skeleton structure of a molecule, we can find the Lewis formula by applying the following four steps. These steps allow us to write electron-dot formulas even for cases in which the octet rule is not followed.

STEP 1 Calculate the total number of valence electrons for the molecule by adding up the number of valence electrons (= group number) for each atom. If we are writing the Lewis formula of a polyatomic anion, we *add* the number of negative charges to this total. (Thus, for SO_4^{2-} we add 2 because the 2− charge indicates that there are 2 more electrons than are provided by the neutral atoms.) For a polyatomic cation, we *subtract* the number of positive charges from the total. (Thus, for NH_4^+ we subtract 1.)

STEP 2 Write the skeleton structure of the molecule or ion, connecting every bonded pair of atoms by a pair of dots (or a dash).

STEP 3 Distribute electrons to the atoms surrounding the central atom (or atoms) to satisfy the octet rule for these surrounding atoms.

STEP 4 Distribute the remaining electrons as pairs to the central atom (or atoms), after subtracting the number of electrons so far distributed from the total found in Step 1. If there are fewer than eight on the central atom, this suggests that a multiple bond is present. (Two electrons fewer than an octet suggests a double bond; four fewer suggests a triple bond or two double bonds.) To obtain a multiple bond, move one or two electron pairs (depending on whether the bond is to be double or triple) from a surrounding atom to the bond connecting the central atom. Atoms that often form multiple bonds are C, N, O, and S.

The next several examples illustrate how to write the Lewis electron-dot formula for a small molecule, given the molecular formula.

Example 9.6 Writing Lewis Formulas (Single Bonds Only)

Thionyl chloride, $SOCl_2$, is a liquid with a corrosive vapor. Write the Lewis formula for the molecule.

Solution

We first calculate the total number of valence electrons. The number for each kind of atom (= group number) is S = 6, O = 6, and Cl = 7. Hence, the total is $6 + 6 + (2 \times 7) = 26$. Using Rule 1, we expect the skeleton structure to consist of an S atom surrounded by O and Cl atoms. After distributing electron pairs to these surrounding atoms to satisfy the octet rule, we have

:Cl: over :Cl:S:O: or :Cl—S—O: (with :Cl: bonded above S)

This accounts for 12 pairs, or 24 electrons. Subtracting 24 from the total (26) gives 2 electrons, or 1 pair, which is placed on the S atom, giving an octet. The Lewis formula is

:Cl: over :Cl:S:O: or :Cl—S—O: (with :Cl: bonded above S)

Exercise 9.9

Dichlorodifluoromethane, CCl_2F_2, is a gas used as a refrigerant and aerosol propellant. (It is known commercially as one of the Freons.) Write the Lewis formula for CCl_2F_2.

(See Problems 9.45 and 9.46.)

Example 9.7 Writing Lewis Formulas (Including Multiple Bonds)

Carbonyl chloride, or phosgene, $COCl_2$, is a highly toxic gas used as a starting material for the preparation of polyurethane plastics. What is the electron-dot formula of $COCl_2$?

Solution

The total number of valence electrons if $4 + 6 + (2 \times 7) = 24$. From Rule 1 we expect the more electropositive atom, C, to be central, with the O and Cl atoms bonded to it. After distributing electron pairs to these surrounding atoms, we have

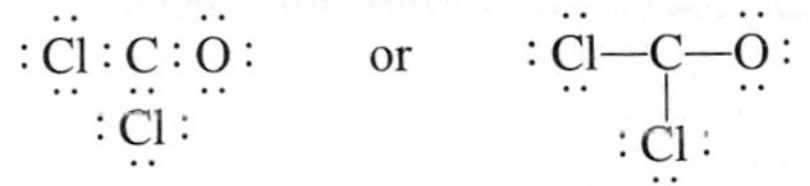

This accounts for all 24 valence electrons, leaving only 6 electrons on C. Because this is 2 fewer than an octet, a double bond is suggested. Because C and O atoms often form double bonds (see under Step 4), we move a pair of electrons on the O atom to give a carbon–oxygen double bond. These electrons still count in the octet of the O atom, but they are now shared with the C atom, giving it an octet. The electron-dot formula of $COCl_2$ is

:Cl:C::O over :Cl: or :Cl—C=O (with :Cl: bonded below C)

Check to see that each atom has an octet.

Exercise 9.10

Write the electron-dot formula of carbon dioxide, CO_2. (See Problems 9.47 and 9.48.)

Example 9.8 Writing Lewis Formulas (Ionic Species)

Obtain the electron-dot formula of the sulfite ion, SO_3^{2-}.

Solution

Both S and O atoms have 6 valence electrons. Thus, the S atom and three O atoms provide 24 valence electrons. Because the anion has a charge of 2−, it has 2 more electrons than are provided by the neutral atoms. Thus, the total number of available electrons is 24 + 2 = 26. By Rule 1 we assign S as the central atom, placing the O atoms around it. After placing electron pairs on the O atoms to satisfy the octet rule for them, we have

```
 ..               ..
:O:              :O:
 .. ..            |  ..
 S:O:      or     S—O:
 .. ..            |  ..
:O:              :O:
 ..               ..
```

So far, we have used up 12 pairs, or 24 electrons. From the total number (26), this leaves 2 electrons, or 1 pair, which is placed on the S atom. Note that the octet rule is now satisfied for the S atom. The electron-dot formula for SO_3^{2-} is

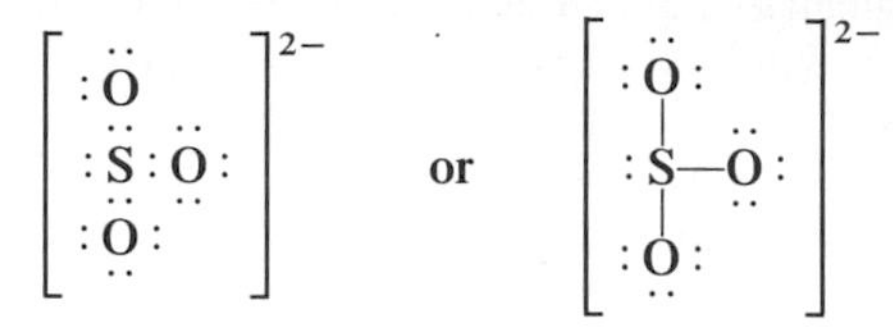

The charge on the entire ion is indicated by the superscript 2− to the square brackets enclosing the electron-dot formula.

Exercise 9.11

What is the electron-dot formula of (a) the hydronium ion, H_3O^+; (b) the chlorite ion, ClO_2^-? (See Problems 9.49 and 9.50.)

Related Topic

FORMAL CHARGE

The concept of formal charge can be used to decide among possible Lewis formulas for a molecule. The formal charge of an atom in a Lewis formula is the hypothetical charge we obtain when we assign electron dots to the different atoms in the formula in a prescribed way. The electron dots are assigned as follows:

1. Half of the dots of a bond are assigned to each atom in the bond.
2. Both dots of a lone pair are assigned to the atom to which the lone pair belongs.

The formal charge equals the number of valence electrons in the free atom (the group number) less the number of electron dots assigned to the atom by the previous rules.

As an example, consider the Lewis formula of carbonyl chloride, $COCl_2$, which was given in the solution to Example 9.7. To calculate the formal charge on a Cl atom, we first count the electron dots assigned to it: 1 from the single bond and 6 from the lone-pair electrons, for a total of 7. The formal charge equals the number of valence electrons (7) less the number of electron dots assigned (7). Thus, the formal charge on Cl atoms is zero. Similar calculations on the O atom and the C atom also yield zero. Now we discuss two rules that are useful in writing Lewis formulas.

RULE A *Whenever you have a choice of Lewis formulas for a molecule, choose the one having the lowest magnitudes of formal charges.*

As an example, consider an alternative Lewis formula for $COCl_2$:

```
 ..     ..
 Cl=C—O:
 ..  |  ..
    :Cl:
     ..
```

In the solution to Example 9.7, we noted that double bonds between C and O are common. For this reason, we did not consider the formula with a C=Cl bond. Let us see how this is handled with the concept of formal charge. We first count the electron dots assigned to the double-bonded Cl atom. We count 2 from the double bond and 4 from lone-pair electrons, for a total of 6. The formal charge is 7 − 6 = +1. For oxygen, we count 1 from the single bond and 6 from lone-pair electrons, for a total of 7. The formal charge is 6 − 7 = −1. We indicate these formal charges in the Lewis formula by inserting circled numbers near the atoms.

(+1) Cl=C—O: (−1)
|
:Cl:

Of the two formulas we have proposed for $COCl_2$, we can eliminate this one because it has formal charges. The other formula does not (and so has the "lowest magnitude of formal charges" specified in Rule A).

We can also use formal charges to help us choose the most likely skeleton structure from several possibilities. Consider thionyl chloride, $SOCl_2$, whose Lewis structure was obtained in Example 9.6. Suppose we had placed one of the Cl atoms in the center. The Lewis formulas with formal charges, this one and the one in Example 9.6, are

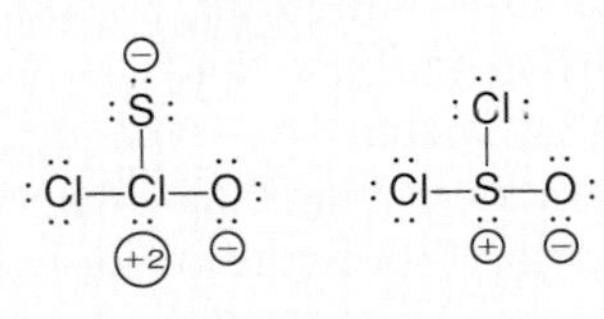

The structure on the right has the lower magnitudes of formal charges and is a more likely skeleton structure.

RULE B *When two proposed structures for a molecule have the same magnitudes of formal charges, choose the one having the negative charge on the more electronegative atom.*

Consider the following as a possible structure for thionyl chloride:

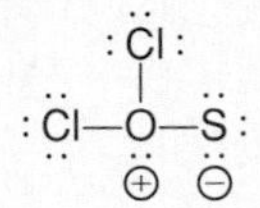

This structure differs from the preceding one (on the right) by an interchange of S and O. However, this structure associates the negative formal charge with the S atom, which is less electronegative than O. The other structure is preferred because it associates the negative formal change with the more electronegative O atom.

9.7 EXCEPTIONS TO THE OCTET RULE

Although most molecules composed of atoms of the main-group elements have electronic structures that satisfy the octet rule, a number of them do not. A few molecules, such as NO, have an odd number of electrons and so cannot satisfy the octet rule. Other exceptions to the octet rule fall into two groups. In one group are molecules with an atom having fewer than eight valence electrons around it. In the other group are molecules with an atom having more than eight valence electrons around it.

The first group consists mostly of molecules with boron atoms. For example, consider boron trifluoride, BF_3. The Lewis formula is believed to be

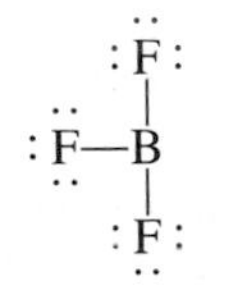

The boron atom in BF_3 has only six valence electrons around it. Many reactions of BF_3 result from this. Boron trifluoride reacts with molecules having a lone pair, such as with ammonia, NH_3, which gives the compound BF_3NH_3.

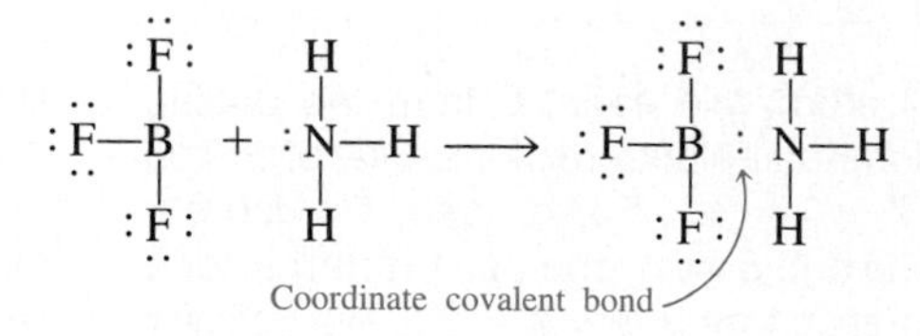

In this reaction, a coordinate covalent bond forms between the boron and nitrogen atoms, and the boron atom achieves an octet.

Although the boron atom in BF_3 does not have an octet of valence electrons around it, it does have the normal number of covalent bonds. All the valence electrons of the boron atom have been paired with unpaired electrons from fluorine atoms. The formation of BF_3 from atoms can be written

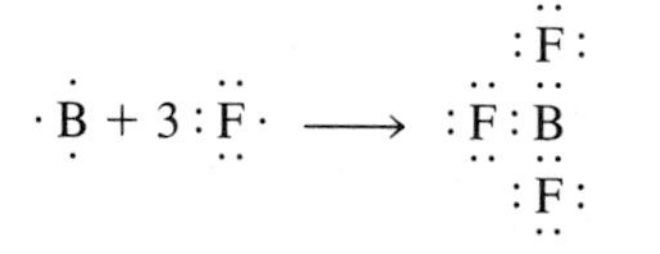

This suggests that other elements of Groups IA to IIIA might form molecules with an atom having fewer than eight electrons around it. Such molecules are uncommon, however, partly because these elements (particularly those in Groups IA and IIA) frequently form ionic bonds. Even when they form covalent bonds, the atoms tend to arrange themselves to give octets of electrons.

Aluminum chloride, $AlCl_3$, offers an interesting study in bonding. The white, crystalline substance at room temperature is an ionic solid, as might be expected for a binary compound of a metal and a nonmetal. However, the substance has a relatively low melting point (192°C) for an ionic compound. Apparently this property is due to the fact that instead of melting to a liquid of ions, as happens with most ionic solids, the compound forms Al_2Cl_6 molecules, with Lewis formula

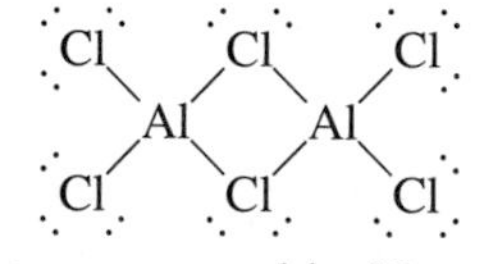

Each atom has an octet of electrons around it. Note that two of the Cl atoms are in *bridge* positions, with each Cl atom having two covalent bonds. Unlike the liquid, aluminum chloride vapor at high temperature does contain $AlCl_3$ molecules, with Lewis formula similar to that of BF_3.

Exercise 9.12

Beryllium chloride, $BeCl_2$, is a solid substance consisting of long (essentially infinite) chains of atoms with Cl atoms in bridge positions.

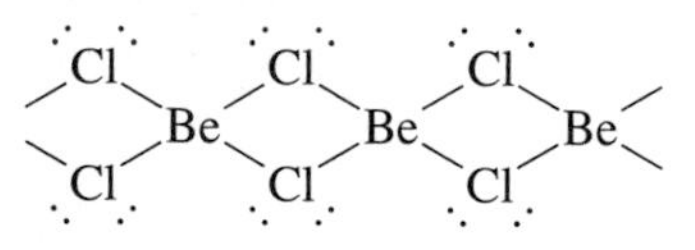

However, if the solid is heated, it forms a vapor of $BeCl_2$ molecules. Write the electron-dot formula of the $BeCl_2$ molecule. (See Problems 9.51 and 9.52.)

The second group of exceptions to the octet rule consists of a variety of molecules having a central atom with more than eight valence electrons around it. Phosphorus pentafluoride is an example. This is a colorless gas of PF_5 molecules.

Each molecule consists of a phosphorus atom surrounded by five fluorine atoms. The phosphorus atom has ten valence electrons around it.

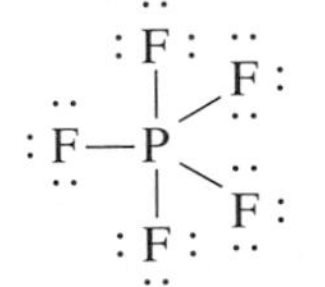

The octet rule stems from the fact that the main-group elements in most cases employ only an *ns* and three *np* valence-shell orbitals in bonding, and these orbitals hold eight electrons. Elements of the second period are restricted to these orbitals, but from the third period on, the elements also have unfilled *nd* orbitals, which may be used in bonding. For example, the valence-shell configuration of phosphorus is $3s^2 3p^3$. Using just these 3*s* and 3*p* orbitals, the phosphorus atom can accept only three additional electrons, forming three covalent bonds (as in PF_3). However, more bonds can be formed if the empty 3*d* orbitals of the atom are used. If each of the five electrons of the phosphorus atom is paired with unpaired electrons of fluorine atoms, PF_5 can be formed. Thus, phosphorus forms both the trifluoride and the pentafluoride. By contrast, nitrogen (which has no available *d* orbitals in its valence shell) forms only the trifluoride, NF_3.

Example 9.9 Writing Lewis Formulas (Exceptions to the Octet Rule)

Xenon, a noble gas, forms a number of compounds. One of these is xenon tetrafluoride, XeF_4, a white, crystalline solid first prepared in 1962. What is the electron-dot formula of the XeF_4 molecule?

Solution

Following the steps described in the previous section, we first calculate the total number of valence electrons for the molecule. There are 8 valence electrons from the Xe atom and 7 from each F atom, giving a total of 36 valence electrons. For the skeleton structure, we draw the Xe atom surrounded by the electronegative F atoms. After placing electron pairs on the F atoms to satisfy the octet rule for them, we have

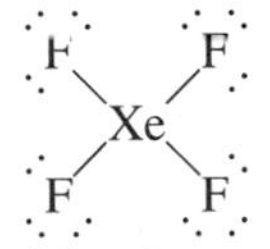

So far, this accounts for 16 pairs, or 32 electrons. A total of 36 electrons is available, so we put an additional 36 − 32 = 4 electrons (2 pairs) on the Xe atom. The Lewis formula is

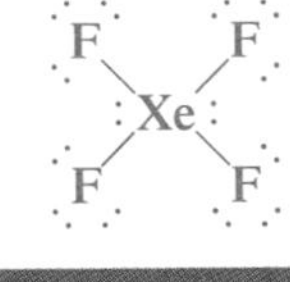

Exercise 9.13

Sulfur tetrafluoride, SF_4, is a colorless gas. Write the electron-dot formula of the SF_4 molecule.

(See Problems 9.53 and 9.54.)

9.8 DELOCALIZED BONDING; RESONANCE

We have assumed up to now that the bonding electrons are localized in the region between two atoms. There are cases, however, in which this assumption does not fit the experimental data. Suppose, for example, that we try to write an electron-dot formula for sulfur dioxide, SO_2. We find that we can write two formulas:

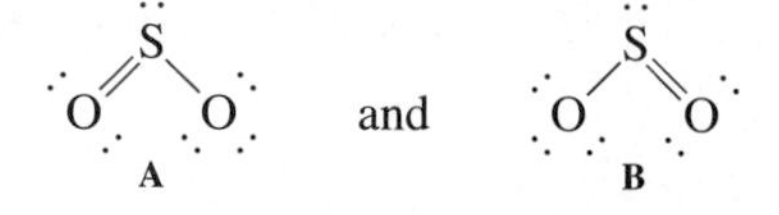

In formula A, the sulfur–oxygen bond on the left is a double bond and the sulfur–oxygen bond on the right is a single bond. In formula B, the situation is just the opposite. Experiment shows, however, that the two bonds are identical. Therefore, neither formula A nor formula B can be correct.●

● The lengths of the two sulfur–oxygen bonds (that is, the distances between the atomic nuclei) are both 1.43 Å.

According to theory, one of the bonding pairs in sulfur dioxide is spread over the region of all three atoms, rather than associated with a particular sulfur–oxygen bond. This is called **delocalized bonding,** *a type of bonding in which a bonding pair of electrons is spread over a number of atoms rather than localized between two*. We might symbolically describe the delocalized bonding in sulfur dioxide as follows:

(For clarity, only the bonding pairs are given.) The broken line indicates a bonding pair of electrons that spans three nuclei rather than only two. In effect, the sulfur–oxygen bond is neither a single bond nor a double bond but an intermediate type.

A single electron-dot formula cannot properly describe delocalized bonding. Instead, a resonance description is often used. According to the **resonance description,** *we describe the electron structure of a molecule having delocalized bonding by writing all possible electron-dot formulas*. These formulas are called the *resonance formulas* of the molecule. The actual electron distribution of the molecule is a composite of these resonance formulas.

Thus, the electron structure of sulfur dioxide can be described in terms of the two resonance formulas presented at the start of this section. By convention, we usually write all of the resonance formulas connected by double-headed arrows. For sulfur dioxide we would write

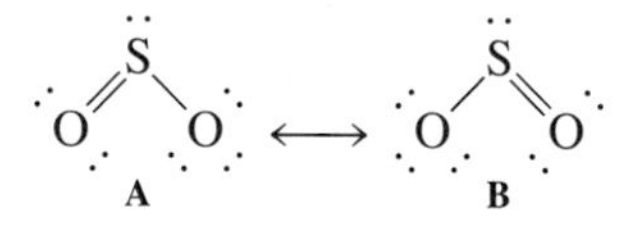

Unfortunately, this notation can be misinterpreted. It does not mean that the sulfur dioxide molecule flips back and forth between two forms. There is only one sulfur dioxide molecule. The double-headed arrow means that you should form a mental picture of the molecule by fusing the various resonance formulas. The left sulfur–oxygen bond is double in formula A and the right one is double in formula B, so you must picture an electron pair that actually encompasses both bonds.

In writing resonance formulas, it is important to realize that the nuclear positions must be the same in all electron-dot formulas. Remember that the resonance formulas describe one molecule, and every molecule has a definite nuclear arrangement. The following formula could not be a resonance formula of sulfur dioxide, because sulfur always has a central position in this molecule:

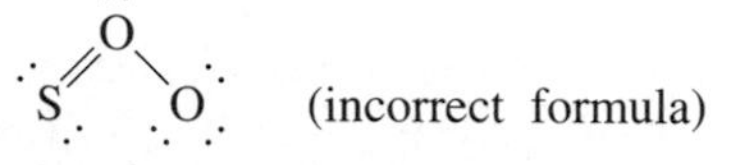

Attempting to write electron-dot formulas leads us to recognize that delocalized bonding exists in many molecules. Whenever we can write several plausible

electron-dot formulas, which usually differ merely in their allocation of single and double bonds to the same kinds of atoms (as in sulfur dioxide), we can expect delocalized bonding.

Example 9.10 Writing Resonance Formulas

Describe the electron structure of the carbonate ion, CO_3^{2-}, in terms of electron-dot formulas.

Solution

One possible electron-dot formula for the carbonate ion is

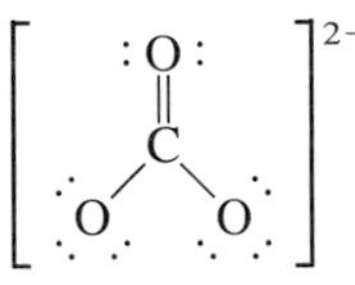

Because we expect all carbon–oxygen bonds to be equivalent, we must describe the electron structure in resonance terms.

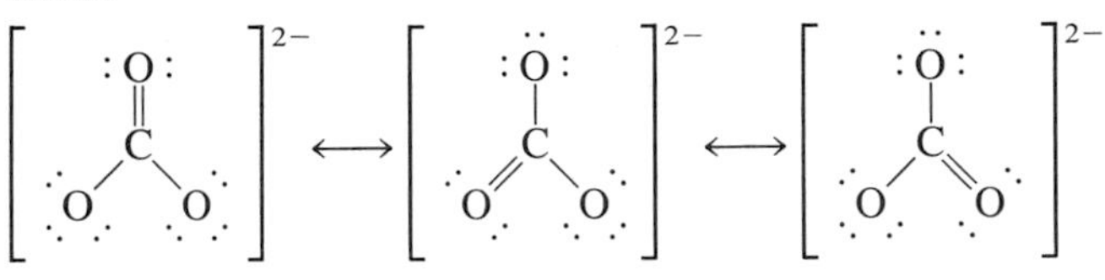

We expect that one electron pair will be delocalized over the region of all three carbon–oxygen bonds.

Exercise 9.14

Describe the bonding in NO_3^- using resonance formulas.

(See Problems 9.55, 9.56, 9.57, and 9.58.)

Metals are extreme examples of delocalized bonding. A sodium metal crystal, for example, can be regarded as an array of Na^+ ions surrounded by a ''sea'' of electrons (Figure 9.11). The valence or bonding electrons are delocalized over the entire metal crystal. The freedom of these electrons to move throughout the crystal is responsible for the electrical conductivity of a metal.

9.9 BOND LENGTH AND BOND ORDER

Bond length (or **bond distance**) is *the distance between the nuclei in a bond.* Bond lengths are determined experimentally using x-ray diffraction or the analysis

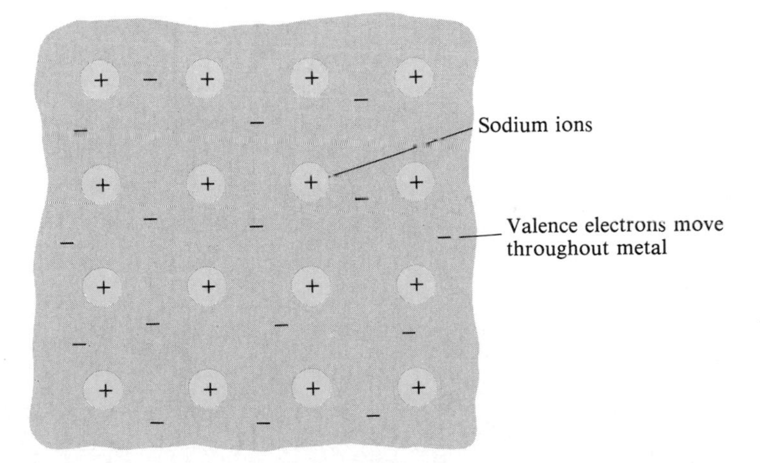

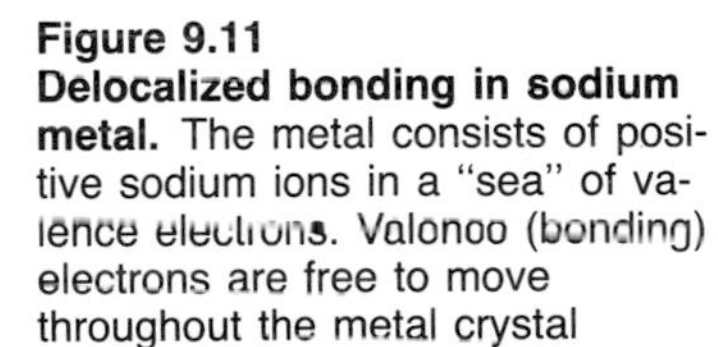

Figure 9.11
Delocalized bonding in sodium metal. The metal consists of positive sodium ions in a "sea" of valence electrons. Valence (bonding) electrons are free to move throughout the metal crystal (green area).

Table 9.4 Single-Bond Covalent Radii for Nonmetallic Elements (in Å)

				H 0.37
B 0.88	C 0.77	N 0.70	O 0.66	F 0.64
	Si 1.17	P 1.10	S 1.04	Cl 0.99
		As 1.21	Se 1.17	Br 1.14
			Te 1.37	I 1.33

of molecular spectra. Knowing the bond length in a particular molecule can sometimes give us a clue to the type of bonding present.

In many cases bond lengths for covalent single bonds in compounds can be predicted from covalent radii. **Covalent radii** are *values assigned to atoms in such a way that the sum of covalent radii of atoms A and B predicts an approximate A—B bond length*. For example, the radius of the Cl atom might be taken to be half the Cl—Cl bond length (1.98 Å). The covalent radius of Cl would be $1/2 \times 1.98$ Å $= 0.99$ Å. Table 9.4 lists single-bond covalent radii for nonmetallic elements. To predict the bond length of C—Cl, we add the covalent radii of the two atoms, C and Cl. We get $(0.77 + 0.99)$ Å $= 1.76$ Å, which compares with the experimental value of 1.78 Å found in most compounds.

● Although a double bond is stronger than a single bond, it is not necessarily less reactive than a single bond. Ethylene, $CH_2{=}CH_2$, for example, is more reactive than ethane, $CH_3{-}CH_3$, where carbon atoms are linked through a single bond. The reactivity of ethylene results from the simultaneous formation of a number of strong, single bonds. See the following section.

Exercise 9.15

Estimate the O—H bond length in H_2O from the covalent radii listed in Table 9.4.

(See Problems 9.59, 9.60, 9.61, and 9.62.)

The **bond order,** defined in terms of the Lewis formula, is *the number of pairs of electrons in a bond*. For example, in C:C the bond order is 1 (single bond); in C: :C the bond order is 2 (double bond). Bond length depends on bond order. As the bond order increases, the bond strength increases, and the nuclei are pulled inward decreasing the bond length. Look at carbon–carbon bonds. The average C—C bond length is 1.54 Å, whereas C═C is 1.34 Å long, and C≡C is 1.20 Å long. Also see Figure 9.12.●

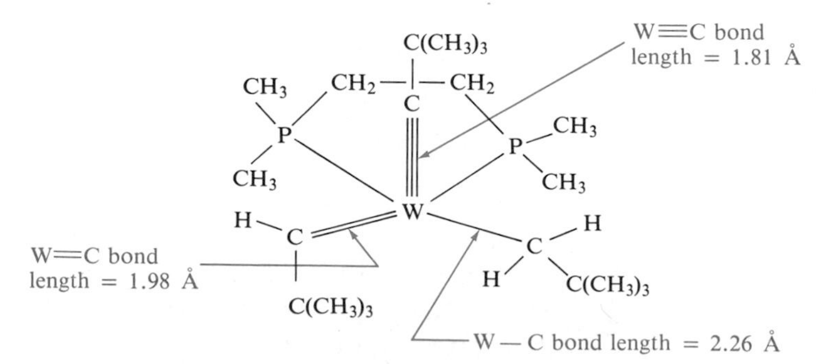

Figure 9.12
An unusual molecule containing tungsten–carbon single, double, and triple bonds. Tungsten–carbon distance decreases from single to double to triple bond (W—C > W═C > W≡C), as carbon–carbon single, double, and triple bonds do. The compound was synthesized by R. R. Schrock, MIT (the W—C bond distances were determined by M. R. Churchill, SUNY, Buffalo).

Example 9.11 Relating Bond Order and Bond Length

Consider the molecules N_2H_4, N_2, and N_2F_2. Which molecule has the shortest nitrogen–nitrogen bond? Which has the longest nitrogen–nitrogen bond?

Solution

First write the Lewis formulas

H—N̈—N̈—H (with H below each N)   :N≡N: :F̤̈—N̈=N̈—F̤̈:

N_2H_4 N_2 N_2F_2

The nitrogen–nitrogen bond should be shortest in N_2, where it is a triple bond, and longest in N_2H_4, where it is a single bond. (Experimental values for the nitrogen–nitrogen bond lengths are 1.09 Å for N_2, 1.22 Å for N_2F_2, and 1.47 Å for N_2H_4.)

Exercise 9.16

Formic acid, isolated in 1670, is the irritant in ant bites. The structure of formic acid is

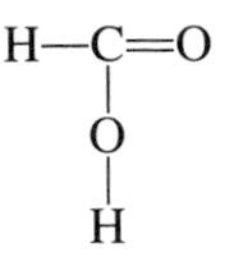

One of the carbon–oxygen bonds has a length of 1.36 Å; the other is 1.23 Å long. What is the length of the C—O bond in formic acid? (See Problems 9.63 and 9.64.)

9.10 BOND ENERGY

In Section 9.4, when we described the formation of a covalent bond, we introduced the concept of bond dissociation energy, the energy required to break a particular bond in a molecule (see Figure 9.8). The bond dissociation energy is a measure of the strength of a particular bond and is essentially the *enthalpy change* for a gas-phase reaction in which a bond breaks. The enthalpy change, ΔH, is the heat absorbed in a reaction carried out at constant pressure (heat of reaction).

Consider the experimentally determined enthalpy changes for the breaking or dissociation of a C—H bond in methane, CH_4, and in ethane, C_2H_6, in the gas phase:

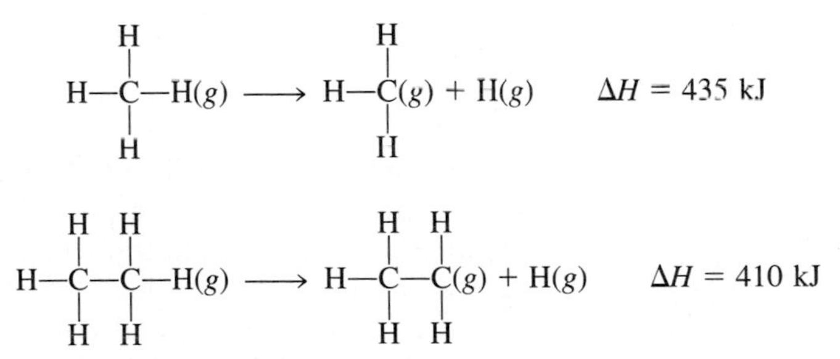

Note that the ΔH values are approximately the same in the two cases. This suggests that the enthalpy change for the dissociation of a C—H bond may be about the same in other molecules. Comparisons of this sort lead to the conclusion that we can obtain approximate values of energies of various bonds.

Table 9.5 Bond Energies (in kJ/mol)*

Single Bonds

	H	C	N	O	S	F	Cl	Br	I
H	432								
C	411	346							
N	386	305	167						
O	459	358	201	142					
S	363	272	—	—	226				
F	565	485	283	190	284	155			
Cl	428	327	313	218	255	249	240		
Br	362	285	—	201	217	249	216	190	
I	295	213	—	201	—	278	208	175	149

Multiple Bonds

C═C	602	C═N	615	C═O	799
C≡C	835	C≡N	887	C≡O	1072
N═N	418	N═O	607	S═O (in SO_2)	532
N≡N	942	O_2	494	S═O (in SO_3)	469

*Data are taken from J. E. Huheey, *Inorganic Chemistry,* 3d ed. (New York: Harper & Row, 1983), pp. A32–A40.

We define the A—B **bond energy** (denoted *BE*) as *the average enthalpy change for the breaking of an* A—B *bond in a molecule in the gas phase.* For example, to calculate a value for the C—H bond energy, or *BE*(C—H), we might look at the experimentally determined enthalpy change for the breaking of all C—H bonds in methane:

$$CH_4(g) \longrightarrow C(g) + 4H(g);\ \Delta H = 1662 \text{ kJ}$$

Because four C—H bonds are broken, we obtain an average value for one C—H bond by dividing the enthalpy change for the reaction by 4:

$$BE(\text{C—H}) = \tfrac{1}{4} \times 1662 \text{ kJ} = 416 \text{ kJ}$$

Similar calculations with other molecules, such as ethane, would yield approximately the same values for the C—H bond energy.

Table 9.5 lists values of some bond energies. Because it takes energy to break a bond, bond energies are always positive numbers. When a bond is formed, energy is released in an amount equal to the negative of the bond energy.

Bond energy is a measure of the strength of a bond: the larger the bond energy, the stronger the chemical bond. Note from Table 9.5 that the bonds C—C, C═C, and C≡C have energies of 346, 602, and 835 kJ, respectively. These numbers indicate that the triple bond is stronger than the double bond, which in turn is stronger than the single bond.

We can use this table of bond energies to estimate heats of reaction, or enthalpy changes, ΔH, for gaseous reactions. To illustrate this, let us find ΔH for the reaction

$$CH_4(g) + Cl_2(g) \longrightarrow CH_3Cl(g) + HCl(g)$$

We can *imagine* that the reaction takes place in steps involving the breaking and forming of bonds. Starting with the reactants, we suppose that one C—H bond and the Cl—Cl bond break.

$$\begin{array}{c} \text{H} \\ | \\ \text{H—C—H} \\ | \\ \text{H} \end{array} + \text{Cl—Cl} \longrightarrow \begin{array}{c} \text{H} \\ | \\ \text{H—C} \\ | \\ \text{H} \end{array} + \text{H} + \text{Cl} + \text{Cl}$$

The enthalpy change for this is BE(C—H) + BE(Cl—Cl). Now the fragments are reassembled to give the products.

$$\begin{array}{c} \text{H} \\ | \\ \text{H—C} \\ | \\ \text{H} \end{array} + \text{H} + \text{Cl} + \text{Cl} \longrightarrow \begin{array}{c} \text{H} \\ | \\ \text{H—C—Cl} \\ | \\ \text{H} \end{array} + \text{H—Cl}$$

In this case, C—Cl and H—Cl bonds are formed, and the enthalpy change equals the negative of the bond energies $-BE$(C—Cl) $-$ BE(H—Cl). Substituting bond-energy values from Table 9.5, we get the enthalpy of reaction

$$\begin{aligned} \Delta H &\simeq BE(\text{C—H}) + BE(\text{Cl—Cl}) - BE(\text{C—Cl}) - BE(\text{H—Cl}) \\ &= (411 + 240 - 327 - 428) \text{ kJ} \\ &= -104 \text{ kJ} \end{aligned}$$

The negative sign means that heat is released by the reaction. Because the bond-energy concept is only approximate, this value is only approximate (the experimental value is -101 kJ). In general, the enthalpy of reaction is (approximately) equal to the sum of the bond energies for bonds broken minus the sum of the bond energies for bonds formed.

Rather than calculating a heat of reaction from bond energies, we usually obtain the heat of reaction from thermochemical data, because these data are generally known more accurately. If the thermochemical data are not known, however, bond energies can give us an estimate of the heat of reaction.

Bond energies are perhaps of greatest value when we try to explain heats of reaction or to understand the relative stabilities of compounds. In general, a reaction is exothermic (gives off heat) if weak bonds are replaced by strong bonds (see Figure 9.13). In the reaction we just discussed, two bonds were broken and replaced by two new bonds, which overall are stronger. In the following example, a strong bond (C=C) is replaced by two weaker ones (two C—C bonds). Although the single bond is weaker than the double bond, as we would expect, two C—C bonds *together* create an energetically stabler situation than one C=C bond.

Figure 9.13
Explosion of nitrogen triiodide-ammonia complex. The red-brown complex of nitrogen triiodide and ammonia is so sensitive to explosion that it can be detonated with a feather. Nitrogen-iodine single bonds are replaced by very stable nitrogen–nitrogen triple bonds and iodine–iodine single bonds. Note the vapor of iodine, I_2, from the explosion.

Example 9.12 Estimating ΔH from Bond Energies

Polyethylene is formed by linking many ethylene molecules into long chains. Estimate the enthalpy change per mole of ethylene for this reaction (shown below), using bond energies.

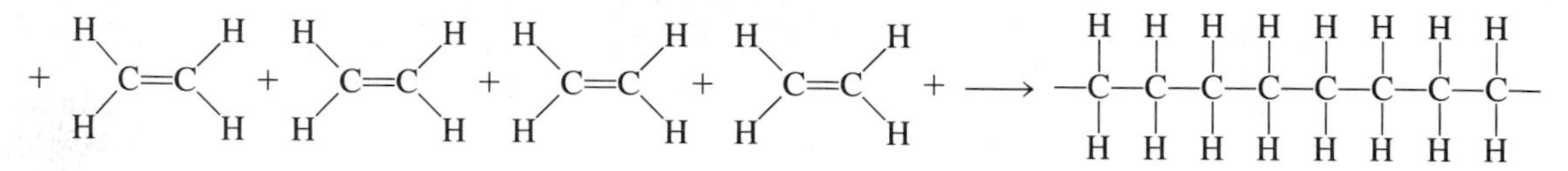

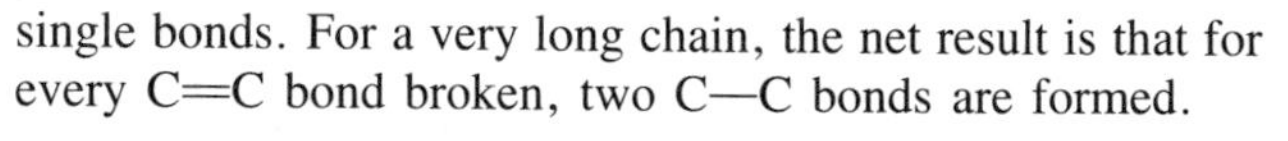

Solution

Imagine the reaction to involve the breaking of the carbon–carbon double bonds and the formation of carbon–carbon single bonds. For a very long chain, the net result is that for every C=C bond broken, two C—C bonds are formed.

$$\Delta H \simeq [602 - (2 \times 346)] \text{ kJ} = \mathbf{-90\ kJ}$$

Exercise 9.17

Use bond energies to estimate the enthalpy change for the combustion of ethylene, C_2H_4, according to the equation

$$C_2H_4(g) + 3O_2(g) \longrightarrow 2CO_2(g) + 2H_2O(g)$$

(See Problems 9.65 and 9.66.)

Instrumental Methods

INFRARED SPECTROSCOPY AND VIBRATIONS OF CHEMICAL BONDS

A chemical bond acts like a stiff spring connecting nuclei. As a result, the nuclei in a molecule vibrate, rather than maintaining fixed positions relative to each other. Nuclear vibration is depicted in Figure 9.14, which shows a spring model of HCl.

This vibration of molecules is revealed in their absorption of infrared radiation. (An instrument for observing the absorption of infrared radiation is shown in Figure 9.15.) The frequency of radiation absorbed equals the frequencies of nuclear vibrations. For example, the H—Cl bond vibrates at a frequency of 8.652×10^{13} vibrations per second. If radiation of this frequency falls on the molecule, the molecule absorbs the radiation, which is in the infrared region, and begins vibrating more strongly.

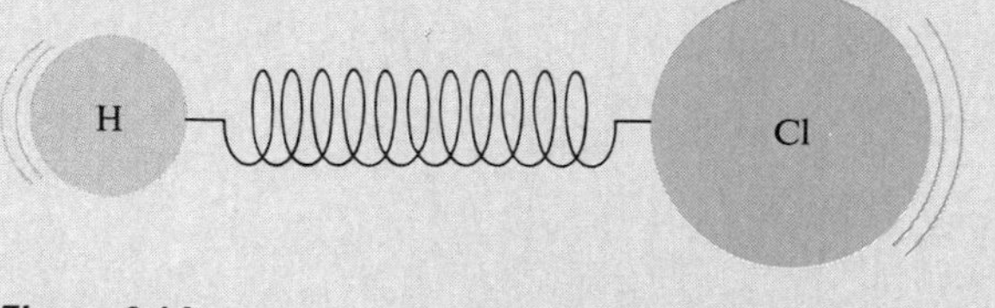

Figure 9.14
Vibration of the HCl molecule. *The vibrating molecule is represented here by a spring model. The atoms in the molecule vibrate, or move constantly back and forth.*

The infrared absorption spectrum of a molecule of even moderate size can have a rather complicated appearance. Figure 9.16 shows the infrared (IR) spectrum of ethyl butyrate, a compound present in pineapple flavor. The complicated appearance of the IR spectrum is actually an advantage: Two different compounds are very unlikely to have exactly the same IR spectrum. Consequently, the IR spectrum of a compound can act as its "fingerprint."

The IR spectrum of a compound can also yield structural information. Suppose we would like the structural formula of ethyl butyrate. The molecular formula, determined from combustion analysis (discussed in Chapter 4), is $C_6H_{12}O_2$. Important information about this structure can be obtained from Figure 9.16.

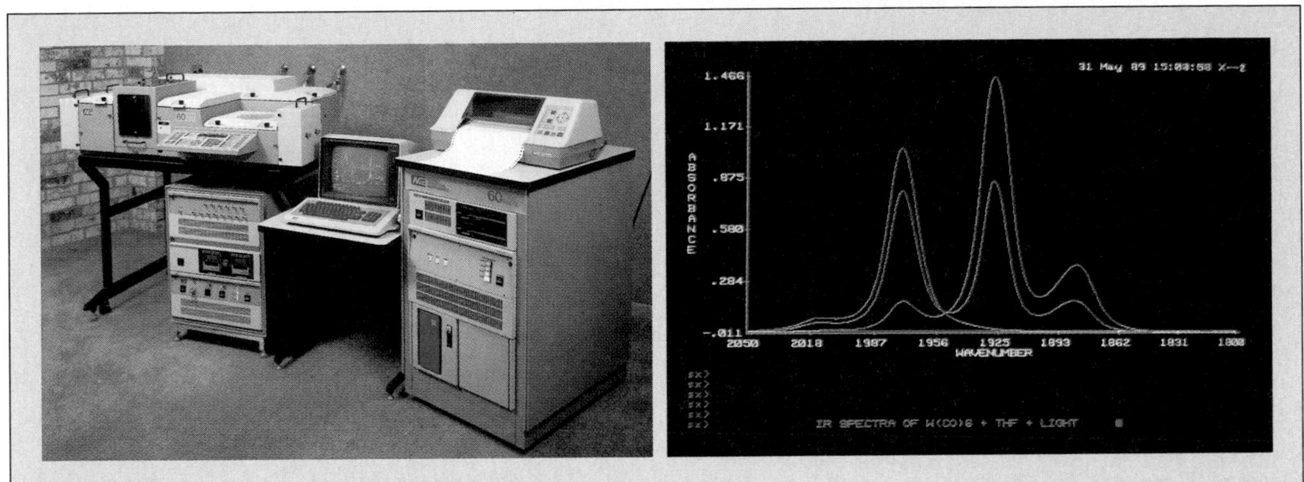

Figure 9.15
A Fourier transform infrared (FTIR) spectrometer. Left: *The instrument.* Right: *A closeup view of a cathode ray screen showing the spectrum of tungsten carbonyl,* $W(CO)_6$.

We first need to be able to read such a spectrum. Instead of plotting an IR spectrum in frequency units (the frequencies are very large), we usually give the frequencies in *wavenumbers,* which are proportional to frequency. To get the wavenumber, we divide the frequency by the speed of light expressed in centimeters per second. For example, HCl absorbs at $(8.652 \times 10^{13}\ s^{-1})/(2.998 \times 10^{10}\ cm/s) = 2886\ cm^{-1}$ (wavenumbers). Wavenumber, or sometimes wavelength, is plotted along the horizontal axis.

Percent transmittance—that is, the percent of radiation that passes through a sample—is plotted on the vertical axis. When a molecule absorbs radiation of a given frequency or wavenumber, this is seen in the spectrum as an inverted spike (peak) at that wavenumber.

(continued)

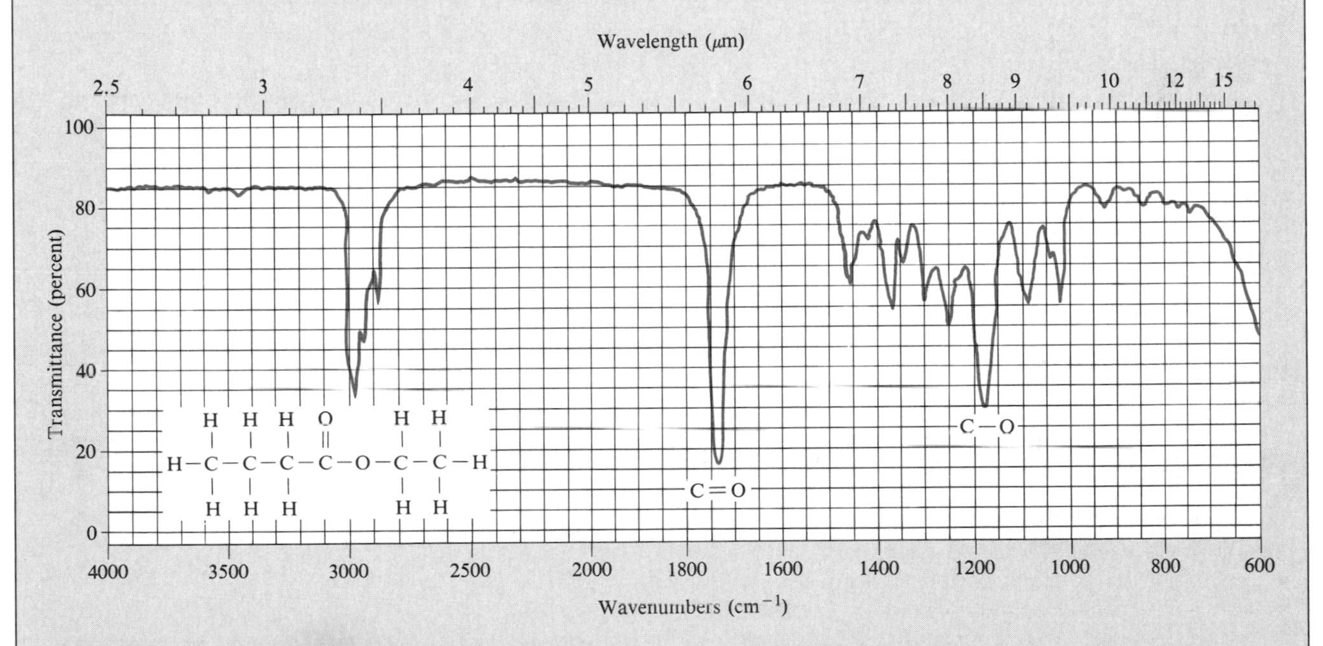

Figure 9.16
Infrared spectrum of ethyl butyrate. *Note the peak corresponding to vibrations of C═O and C—O bonds. The molecular structure is shown at the left. (From Allinger et al.,* Organic Chemistry *[New York: Worth, 1971], Figure 9.7, p. 544.)*

Certain structural features of molecules appear as absorption peaks in definite regions of the infrared spectrum. For example, the absorption peak at 1730 cm^{-1} is characteristic of the C═O bond. With some knowledge of where various bonds absorb, one can identify other peaks, including that of C—O at 1180 cm^{-1}. (Generally, the IR peak for an A—B bond occurs at lower wavenumber than that for an A═B bond.) The IR spectrum does not reveal the complete structure, but it provides important clues. Data from other instruments, such as the mass spectrometer (p. 240), give us additional clues.

Figure 9.17 shows another IR spectrum—this one of a blood sample from a person exposed to carbon monoxide, CO, a poisonous gas. It shows that the CO molecule is chemically attached to hemoglobin, a molecule present in red blood cells and responsible for carrying oxygen. Once hemoglobin bonds to CO, its oxygen-carrying function is lost.

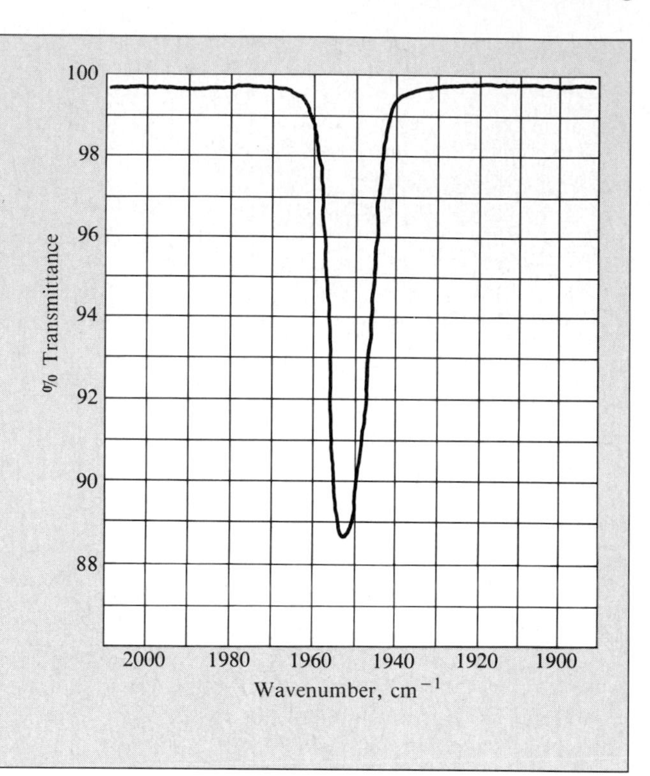

Figure 9.17
Infrared spectrum of carboxyhemoglobin (hemoglobin with CO bonded to it). *Note the peak at about 1953 cm^{-1} corresponding to the vibration of CO bonded to hemoglobin.*

Profile of a Chemical

PHOSPHORUS (a Group VA Nonmetal)

Phosphorus is the first known element to which we can attach the name of a discoverer. In 1669, the German chemist Hennig Brand, who was seeking a potion to transform base metals to gold, discovered a white, waxy solid that glowed blue-green when exposed to air. This glow was mysterious, and although Brand tried to keep his recipe secret, the preparation of the substance soon became widely known. The substance was shown later to be an element and was called white phosphorus, the word *phosphorus* deriving from the Greek meaning "bearer of light." The glow results from *chemiluminescence,* which is the emission of light from excited molecular products of a chemical reaction, in this case between white phosphorus and oxygen.

Fortunately, Brand's recipe has been much improved. He had boiled putrid urine to a paste, then strongly heated it. White phosphorus, whose molecular formula is P_4 (Figure 9.18), melts at 44°C and boils at 280°C at normal pressure. Thus, when phosphorus was formed in Brand's vessel, it distilled over. The present commercial process consists of heating calcium phosphate, $Ca_3(PO_4)_2$, with coke (carbon), C, and sand, SiO_2, in an electric furnace. The overall reaction is

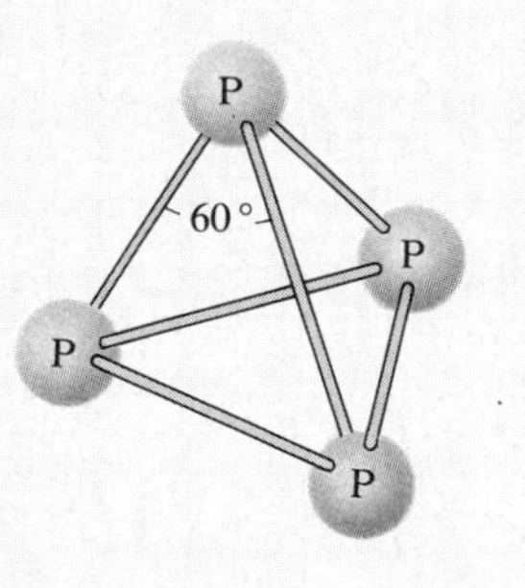

Figure 9.18
Molecular model of the P_4 molecule. *Each phosphorus atom is bonded to three other phosphorus atoms, forming a tetrahedral P_4 molecule.*

$$2Ca_3(PO_4)_2(s) + 6SiO_2(s) + 10C(s) \xrightarrow{1500^\circ C} \underset{\text{calcium silicate}}{6CaSiO_3(l)} + 10CO(g) + P_4(g)$$

Phosphorus vapor is then cooled to the liquid state. Because white phosphorus burns spontaneously in air, it is stored under water. It is shipped in tank cars as the liquid under water and in barrels as solid sticks under water.

Figure 9.19
Allotropes of phosphorus. *White phosphorus (left) and red phosphorus (right).*

At room temperature, white phosphorus is a soft, waxy solid. Its color when pure is white, but after standing in light, it takes on a yellowish cast. White phosphorus is very reactive and must be handled with tongs, because even the slight heat from your fingers will ignite it. The solid and vapor are extremely poisonous. Until the early part of this century, white phosphorus was used in matches. Then it was discovered that workers in match factories were contracting "phossy jaw," a chronic phosphorus poisoning that causes deterioration of the jawbone.

Phosphorus exists in allotropic forms other than white phosphorus (Figure 9.19). Red phosphorus is formed by heating white phosphorus in the absence of air for several hours at 240°C. It has a nonmolecular structure, and for this reason its formula is written simply P(*s*). This allotrope reacts much less readily with oxygen and does not have the blue-green chemiluminescence characteristic of white phosphorus. Red phosphorus is relatively nontoxic. It forms the active ingredient of the striking surface of a safety match, which has a head containing potassium chlorate, $KClO_3$. When the match is struck against the red phosphorus surface, a reaction of the phosphorus and potassium chlorate causes the match to ignite.

Most of the white phosphorus produced is converted to phosphoric acid, H_3PO_4. Phosphorus is burned in an excess of oxygen to form tetraphosphorus decoxide, P_4O_{10}, a white solid (see Figure 8.21, p. 302), which reacts with water to form the acid.

$$P_4(s) + 5O_2(g) \longrightarrow P_4O_{10}(s)$$
$$P_4O_{10}(s) + 6H_2O(l) \longrightarrow 4H_3PO_4(aq)$$

Phosphoric acid is used to make phosphate fertilizers and detergent preparations. The acid is also added to carbonated beverages to increase tartness.

Phosphorus forms binary compounds with many other elements, including oxygen (P_4O_6 and P_4O_{10}). The chlorides, particularly phosphorus trichloride, are especially important. Phosphorus trichloride, PCl_3, is a colorless liquid boiling at 79°C. It is prepared by bubbling chlorine gas into a suspension of liquid phosphorus in phosphorus trichloride. Phosphorus and chlorine are continuously added to the reaction vessel as phosphorus trichloride is distilled off.

$$P_4(l) + 6Cl_2(g) \longrightarrow 4PCl_3(l)$$

The quantity of chlorine must be watched because an excess would react with the phosphorus trichloride to produce the pentachloride, a yellowish solid.

$$PCl_3(l) + Cl_2(g) \longrightarrow PCl_5(s)$$

Phosphorus trichloride is made on a large scale to produce organic phosphorus compounds for plasticizers (compounds combined with plastics to keep them pliable), flame retardants, fuel additives, and insecticides.

Phosphorus pentachloride is a structurally interesting substance. Whereas the gas phase consists of PCl_5 molecules, the solid is known to have the ionic structure $[PCl_4^+][PCl_6^-]$. The Lewis formulas of the PCl_3 and PCl_5 molecules are as follows:

(continued)

Figure 9.20
"Strike-anywhere" matches. *The head of the match contains tetraphosphorus trisulfide and potassium chlorate, which react when frictional heat ignites the mixture. The products are the oxides of phosphorus and sulfur.*

Phosphorus forms several important binary compounds with sulfur. The stablest is tetraphosphorus trisulfide, P_4S_3, a yellowish-green solid melting at 173°C. Being much less toxic, it is used in place of white phosphorus in the "strike-anywhere" match (Figure 9.20). Tetraphosphorus trisulfide is prepared by heating the stoichiometric ratio of red phosphorus and sulfur above 180°C.

$$32P(s) + 3S_8(s) \longrightarrow 8P_4S_3(l)$$

Questions for Study

1. What causes the glow of white phosphorus in air?
2. What are the starting materials for the commercial production of white phosphorus?
3. Explain why white phosphorus should not be picked up with bare fingers.
4. What are the two common allotropic forms of phosphorus? List several properties of each allotrope.
5. What is the principal compound into which commercial white phosphorus is converted? Give chemical equations for the conversion.
6. List the formulas of the binary oxides, binary chlorides, and a binary sulfide of phosphorus. Give equations for the production of each.
7. What phosphorus compound is used in the head of a "strike-anywhere" match? What is used in the striking surface of a safety match?

A Checklist for Review

IMPORTANT TERMS

ionic bond (9.1)
Lewis electron-dot symbol (9.1)
lattice energy (9.1)
ionic radius (9.3)
isoelectronic (9.3)
covalent bond (p. 323)
Lewis electron-dot formula (9.4)
bonding pair (9.4)
lone (nonbonding) pair (9.4)
coordinate covalent bond (9.4)
octet rule (9.4)
single bond (9.4)
double bond (9.4)
triple bond (9.4)
polar covalent bond (9.5)
electronegativity (9.5)
delocalized bonding (9.8)
resonance description (9.8)
bond length (bond distance) (9.9)
covalent radii (9.9)
bond order (9.9)
bond energy (9.10)

SUMMARY OF FACTS AND CONCEPTS

An *ionic bond* is a strong attractive force holding ions together. An ionic bond can form between two atoms by the transfer of electrons from the valence shell of one atom to the valence shell of the other. Many similar ions attract one another to form a crystalline solid, in which positive ions are surrounded by negative ions and negative ions are surrounded by positive ions. Monatomic cations of the main-group elements have charges equal to the group number (or in some cases, the group number minus 2). Monatomic anions of the main-group elements have charges equal to the group number minus 8.

A *covalent bond* is a strong attractive force that holds two atoms together by their sharing of electrons. These bonding electrons are attracted simultaneously to both atomic nuclei, and they spend part of the time near one atom and part of the time near the other. If an electron pair is not equally shared, the bond is *polar*. This polarity results from the difference in *electronegativities* of the atoms—that is, in the unequal abilities of the atoms to draw bonding electrons to themselves.

Lewis electron-dot formulas are simple representations of the valence-shell electrons of atoms in molecules and ions and are useful for describing the covalent bonding in substances. In such formulas, the atoms usually satisfy the *octet rule;* that is, every atom has eight valence electrons about it (H has two). No single Lewis formula is sufficient to describe molecules with *delocalized bonding*. In these cases, a *resonance description* may be used. The electron distribution of the molecule is described as a composite of two or more Lewis formulas.

Bond lengths can be estimated from the covalent radii of atoms. Bond length depends on *bond order;* as the bond order increases, the bond length decreases. The A—B *bond energy* is the average enthalpy change when an A—B bond is broken. We can use values of the bond energy to estimate heats of gas reactions.

OPERATIONAL SKILLS

Note: A periodic table is needed for skills 1, 2, 3, 5, and 6.

1. **Using Lewis symbols to represent ionic bond formation** Given a metallic and a nonmetallic main-group element, use Lewis symbols to represent the transfer of electrons to form ions of noble-gas configurations (Example 9.1).
2. **Writing electron configurations of ions** Given an ion, write the electron configuration. For an ion of a main-group element, give the Lewis symbol (Examples 9.2 and 9.3).
3. **Using periodic trends to obtain relative ionic radii** Given a series of ions, arrange them in order of increasing ionic radius (Example 9.4).
4. **Using electronegativities to obtain relative bond polarities** Given the electronegativities of the atoms, arrange a series of bonds in order by polarity (Example 9.5).
5. **Writing Lewis formulas** Given the molecular formula of a simple compound or ion, write the Lewis electron-dot formula (Examples 9.6, 9.7, 9.8, and 9.9).
6. **Writing resonance formulas** Given a simple molecule with delocalized bonding, write the resonance description (Example 9.10).
7. **Relating bond order and bond length** Know the relationship between bond order and bond length (Example 9.11).
8. **Estimating ΔH from bond energies** Given a table of bond energies, estimate the heat of reaction (Example 9.12).

Review Questions

9.1 Describe the formation of a sodium chloride crystal from atoms.

9.2 Why does sodium chloride normally exist as a crystal rather than as a molecule composed of one cation and one anion?

9.3 Explain what energy terms are involved in the formation of an ionic solid from atoms. In what way should these terms change (become larger or smaller) to give the lowest energy possible for the solid?

9.4 Define lattice energy for potassium bromide.

9.5 Why do most monatomic cations of the main-group elements have a charge equal to the group number? Why do most monatomic anions of these elements have a charge equal to the group number minus 8?

9.6 The 2+ ions of transition elements are common. Explain why this might be expected.

9.7 Explain how ionic radii are obtained from known distances between nuclei in crystals.

9.8 Describe the trends shown by the radii of the monatomic ions for the main-group elements both across a period and down a column.

9.9 Describe the formation of a covalent bond in H_2 from atoms. What does it mean to say that the bonding electrons are shared by the two atoms?

9.10 Draw a potential-energy diagram for a molecule such as Cl_2. On the diagram, indicate the bond length (1.94 Å) and the bond dissociation energy (240 kJ/mol).

9.11 Give an example of a molecule that has a coordinate covalent bond.

9.12 The octet rule correctly predicts the Lewis formula of many molecules involving main-group elements. Explain why this is so.

9.13 Describe the general trends in electronegativities of the elements in the periodic table both across a period and down a column.

9.14 What is the qualitative relationship between bond polarity and electronegativity difference?

9.15 Describe the kinds of exceptions to the octet rule that we encounter in compounds of the main-group elements. Give examples.

9.16 What is a resonance description of a molecule? Why is this concept required if we wish to retain Lewis formulas as a description of the electron structure of molecules?

9.17 What is the relationship between bond order and bond length? Use an example to illustrate it.

9.18 Define bond energy. Explain how one can use bond energies to estimate the heat of reaction.

Practice Problems

IONIC BONDING

9.19 Write Lewis symbols for the following:
(a) Ba (b) Ba^{2+} (c) I (d) I^-

9.20 Write Lewis symbols for the following:
(a) In (b) In^{3+} (c) P (d) P^{3-}

9.21 Use Lewis symbols to represent the transfer of electrons between the following atoms to form ions with noble-gas configurations:
(a) Na and I (b) Na and S

9.22 Use Lewis symbols to represent the electron transfer between the following atoms to give ions with noble-gas configurations:
(a) Mg and I (b) Mg and Se

9.23 For each of the following, write the electron configuration and Lewis symbol.
(a) Mg (b) Mg^{2+} (c) Se^{2-} (d) Br^-

9.24 For each of the following, write the electron configuration and Lewis symbol:
(a) Rb (b) Rb^+ (c) I^- (d) Te^{2-}

9.25 Write the electron configurations of Sn and Sn^{2+}.

9.26 Write the electron configurations of Bi and Bi^{3+}.

9.27 Give the electron configurations of Ni^{2+} and Ni^{3+}.

9.28 Give the electron configurations of Cu^+ and Cu^{2+}.

IONIC RADII

9.29 Arrange the members of each of the following pairs in order of increasing radius and explain the order.
(a) Rb, Rb^+ (b) Se, Se^{2-}

9.30 Arrange the members of each of the following pairs in order of increasing radius and explain the order.
(a) I^-, I (b) Ca^{2+}, Ca

9.31 Without looking at Table 9.3, arrange the following by increasing ionic radius: Se^{2-}, Te^{2-}, S^{2-}. Explain how you arrived at this order. (You may use a periodic table.)

9.32 Which has the larger radius, N^{3-} or P^{3-}? Explain. (You may use a periodic table.)

9.33 Arrange the following in order of increasing ionic radius: F^-, Na^+, and N^{3-}. Explain this order. (You may use a periodic table.)

9.34 Arrange the following in order of increasing ionic radius: I^-, Cs^+, and Te^{2-}. Explain the order. (You may use a periodic table.)

COVALENT BONDING

9.35 Use Lewis symbols to show the reaction of atoms to form hydrogen sulfide. Point out which electron pairs in the Lewis formula of H_2S are bonding and which are lone pairs.

9.36 Use Lewis symbols to show the reaction of atoms to form phosphine, PH_3. Point out bonding pairs and lone pairs in the electron-dot formula of this compound.

9.37 Assuming that the atoms form the normal number of covalent bonds, give the molecular formula of the simplest compound of silicon and chlorine atoms.

9.38 Assuming that the atoms form the normal number of covalent bonds, give the molecular formula of the simplest compound of arsenic and hydrogen atoms.

POLAR COVALENT BONDS; ELECTRONEGATIVITY

9.39 With the aid of a periodic table (not Figure 9.10), arrange the following in order of increasing electronegativity.
(a) Sr, Cs, Ba (b) Ca, Ge, Ga (c) P, As, S

9.40 Using a periodic table (not Figure 9.10), arrange the following in order of increasing electronegativity.
(a) P, O, N (b) Na, Al, Mg (c) C, Al, Si

9.41 Decide which of the following bonds is least polar on the basis of electronegativities of atoms: H—Se, P—Cl, N—Cl.

9.42 Arrange the following bonds in order of increasing polarity using electronegativities of atoms: Si—O, C—Br, As—Br.

9.43 Indicate the partial charges for the bonds given in Problem 9.41, using the symbols δ^+ and δ^-.

9.44 Indicate the partial charges for the bonds given in Problem 9.42, using the symbols δ^+ and δ^-.

WRITING LEWIS FORMULAS

9.45 Write Lewis formulas for the following molecules:
(a) Br_2 (b) H_2Se (c) SOF_2 (d) H_2SO_4

9.46 Write Lewis formulas for the following molecules:
(a) FBr (b) PBr_3 (c) ClO_2F (d) H_2SO_3

9.47 Write Lewis formulas for the following molecules:
(a) P_2 (b) COSe (c) $COBr_2$ (d) HNO_2

9.48 Write Lewis formulas for the following molecules:
(a) CO (b) BrCN (c) COF_2 (d) N_2F_2

9.49 Write Lewis formulas for the following ions:
(a) ClO^- (b) ClF_2^+ (c) $SnCl_3^-$ (d) S_2^{2-}

9.50 Write Lewis formulas for the following ions:
(a) IBr_2^+ (b) BrO_3^- (c) PO_4^{3-} (d) CN^-

EXCEPTIONS TO THE OCTET RULE

9.51 Write Lewis formulas for the following:
(a) BCl_3 (b) $TlCl_2^+$ (c) $BeBr_2$

9.52 Write Lewis formulas for the following:
(a) BeF_2 (b) BeF_3^- (c) $AlBr_3$

9.53 Write Lewis formulas for the following:
(a) XeF_2 (b) SeF_4 (c) TeF_6 (d) XeF_5^+

9.54 Write Lewis formulas for the following:
(a) I_3^- (b) ClF_3 (c) IF_4^- (d) BrF_5

RESONANCE

9.55 Write a resonance description for the following:
(a) FNO_2 (b) SO_3

9.56 Write a resonance description for the following:
(a) NO_2^- (b) HNO_3

9.57 Give the resonance description of the formate ion. The skeleton structure is

$$\left[\begin{array}{c} \text{O} \\ | \\ \text{H—C—O} \end{array}\right]^-$$

9.58 Use resonance to describe the electron structure of nitromethane, CH_3NO_2. The skeleton structure is

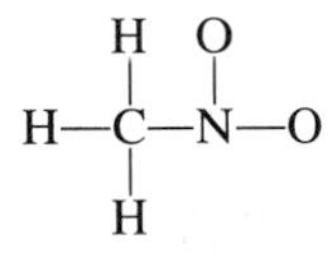

BOND LENGTH, BOND ORDER, AND BOND ENERGY

9.59 Use covalent radii (Table 9.4) to estimate the length of the P—F bond in phosphorus trifluoride, PF_3.

9.60 What do you expect for the B—Cl bond length in boron trichloride, BCl_3, on the basis of covalent radii?

9.61 Calculate the bond length for each of the following single bonds, using covalent radii (Table 9.4):
(a) C—H (b) S—Cl (c) Br—Cl (d) Si—O

9.62 Calculate the C—H and C—Cl bond lengths in chloroform, $CHCl_3$, using values for the covalent radii from Table 9.4. How do these values compare with the experimental values: C—H, 1.07 Å; C—Cl, 1.77 Å?

9.63 One of the following compounds has a carbon–nitrogen bond length of 1.16 Å; the other has a carbon–nitrogen bond length of 1.47 Å. Match a bond length with each compound.

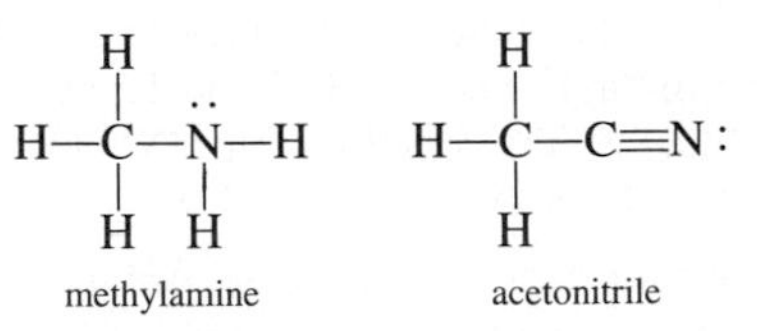

9.64 Which of the following two compounds has the shorter carbon–oxygen bond?

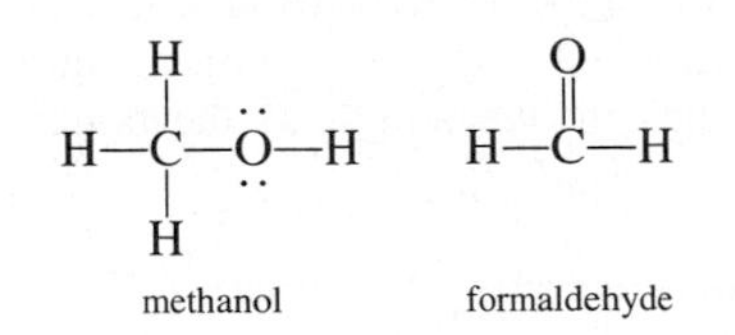

9.65 Use bond energies (Table 9.5) to estimate ΔH for the following gas-phase reaction.

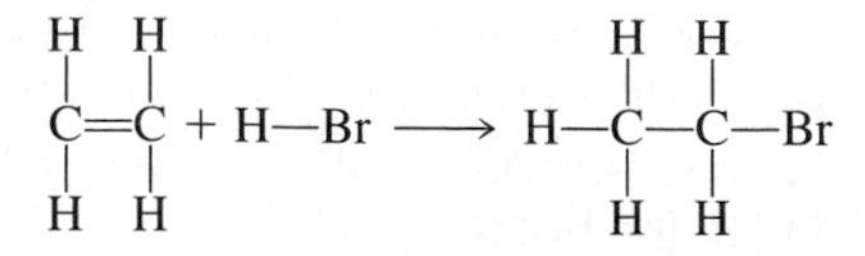

This is called an "addition" reaction, because a compound (HBr) is added across the double bond.

9.66 A commercial process for preparing ethanol (ethyl alcohol), C_2H_5OH, consists in passing ethylene gas, C_2H_4, and steam over an acid catalyst (to speed up the reaction). The gas-phase reaction is

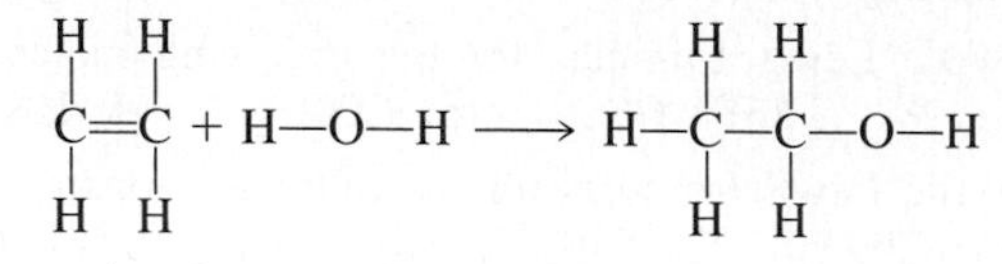

Estimate ΔH for this reaction, using bond energies (Table 9.5).

Additional Problems

9.67 For each of the following pairs of elements, state whether the binary compound formed is likely to be ionic or covalent. Give the formula and name of the compound.
(a) Sr, O (b) C, Br (c) Ga, F (d) N, Br

9.68 For each of the following pairs of elements, state whether the binary compound formed is likely to be ionic or covalent. Give the formula and name of the compound.
(a) Na, S (b) Al, F (c) Ca, Cl (d) Si, Br

9.69 Give the Lewis formula for the arsenate ion, AsO_4^{3-}. Write the formula of lead(II) arsenate.

9.70 Give the Lewis formula for the selenite ion, SeO_3^{2-}. Write the formula of aluminum selenite.

9.71 Iodic acid, HIO_3, is a colorless, crystalline compound. What is the electron-dot formula of iodic acid?

9.72 Selenic acid, H_2SeO_4, is a crystalline substance and a strong acid. Write the electron-dot formula of selenic acid.

9.73 Sodium amide, known commercially as sodamide, is used in preparing indigo, the dye used to color blue jeans. It is an ionic compound with the formula $NaNH_2$. What is the electron-dot formula of the amide anion, NH_2^-?

9.74 Lithium aluminum hydride, $LiAlH_4$, is an important reducing agent (an element or compound that generally has a strong tendency to give up electrons in its chemical reactions). Write the electron-dot formula of the AlH_4^- ion.

9.75 Nitronium perchlorate, NO_2ClO_4, is a reactive salt of the nitronium ion NO_2^+. Write the electron-dot formula of NO_2^+.

9.76 Solid phosphorus pentabromide, PBr_5, has been shown to have the ionic structure $[PBr_4^+][Br^-]$. Find the electron-dot formula of the PBr_4^+ cation.

9.77 Write electron-dot formulas for the following:
(a) $SeOCl_2$ (b) CSe_2 (c) $GaCl_4^-$ (d) C_2^{2-}

9.78 Write electron-dot formulas for the following:
(a) $POBr_3$ (b) Si_2H_6 (c) IF_2^+ (d) NO^+

9.79 Write Lewis formulas for the following:
(a) $SbCl_3$ (b) ICN (c) ICl_3 (d) IF_5

9.80 Write Lewis formulas for the following:
(a) $AlCl_4^-$ (b) AlF_6^{3-} (c) BrF_3 (d) IF_6^+

9.81 Give a resonance description for the following:
(a) SeO_2 (b) N_2O_4

9.82 Give a resonance description for the following:
(a) O_3 (b) $C_2O_4^{2-}$

9.83 The atoms in N_2O_5 are connected as follows:

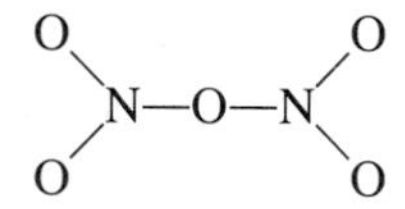

No attempt has been made to indicate whether a bond is single or double or whether there is resonance. Obtain the Lewis formula (or formulas). The N—O bond lengths are 1.18 Å and 1.36 Å. Indicate the length of the bonds in the compound.

9.84 Methyl nitrite has the structure

H—C(H)(H)—O—N—O

No attempt has been made to indicate whether a bond is single or double or whether there is resonance. Obtain the Lewis formula (or formulas). The N—O bond lengths are 1.22 Å and 1.37 Å. Indicate the lengths of the N—O bonds in the compound.

9.85 Use bond energies to estimate ΔH for the reaction

$$H_2(g) + O_2(g) \longrightarrow H_2O_2(g)$$

9.86 Use bond energies to estimate ΔH for the reaction

$$2H_2(g) + N_2(g) \longrightarrow N_2H_4(g)$$

9.87 Use bond energies to estimate ΔH for the reaction

$$N_2F_2(g) + F_2(g) \longrightarrow N_2F_4(g)$$

9.88 Use bond energies to estimate ΔH for the reaction

$$HCN(g) + 2H_2(g) \longrightarrow CH_3NH_2(g)$$

Cumulative-Skills Problems

9.89 An ionic compound has the following composition (by mass): Mg, 10.9%; Cl, 31.8%; O, 57.3%. What are the formula and name of the compound? Write the Lewis formulas for the ions.

9.90 An ionic compound has the following composition (by mass): Ca, 30.3%; N, 21.2%; O, 48.5%. What are the formula and name of the compound? Write Lewis formulas for the ions.

9.91 A gaseous compound has the following composition by mass: C, 25.0%; H, 2.1%; F, 39.6%; O, 33.3%. Its molecular weight is 48.0 amu. Write the Lewis formula for the molecule.

9.92 A liquid compound used in dry cleaning contains 14.5% C and 85.5% Cl by mass and has a molecular weight of 166 amu. Write the Lewis formula for the molecule.

9.93 A compound of tin and chlorine is a colorless liquid. The vapor has a density of 7.49 g/L at 151°C and 1.00 atm. What is the molecular weight of the compound? Why do you think the compound is molecular and not ionic? Write the Lewis formula for the molecule.

9.94 A compound of arsenic and fluorine is a gas. A sample weighing 0.100 g occupies 14.2 mL at 23°C and 765 mmHg. What is the molecular weight of the compound? Write the Lewis formula for the molecule.

9.95 Calculate the enthalpy of reaction for

$$HCN(g) \longrightarrow H(g) + C(g) + N(g)$$

from enthalpies of formation (see Appendix C). Given that the C—H bond energy is 411 kJ, obtain a value for the C≡N bond energy. Compare your result with the value given in Table 9.5.

9.96 Assume the values of the C—H and C—C bond energies given in Table 9.5. Then, using data given in Appendix C, calculate the C=O bond energy in acetaldehyde,

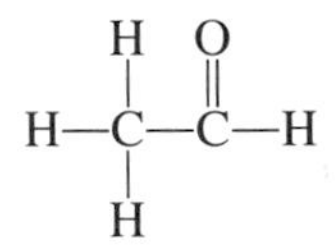

Compare your result with the value given in Table 9.5.

9.97 According to Pauling, the A—B bond energy is equal to the average of the A—A and B—B bond energies plus an energy contribution from the polar character of the bond:

$$BE(\text{A—B}) = \tfrac{1}{2}[BE(\text{A—A}) + BE(\text{B—B})] + k(X_A - X_B)^2$$

Here X_A and X_B are the electronegativities of atoms A and B, and k is a constant equal to 98.6 kJ. Assume that the electronegativity of H is 2.1. Use the formula to calculate the electronegativity of oxygen.

9.98 Because known compounds with N—I bonds tend to be unstable, there are no thermodynamic data available with which to calculate the N—I bond energy. However, we can estimate a value from Pauling's formula relating electronegativities and bond energies (see Problem 9.97). Using Pauling's electronegativities and the bond energies given in Table 9.5, calculate the N—I bond energy.

9.99 Using Mulliken's formula, calculate a value for the electronegativity of chlorine. Use values of the ionization energy from Figure 8.16 and values of the electron affinity from Table 8.4. Divide this result (in kJ/mol) by 230 to get a value comparable to Pauling's scale.

9.100 Using Mulliken's formula, calculate a value for the electronegativity of oxygen. Convert the result to a value on Pauling's scale. See Problem 9.99.

ing the F atoms has three sides). Phosphorus trifluoride is nonplanar, and the angle between any two P—F bonds is 96°. If we imagine lines connecting the fluorine atoms, these lines and those of the P—F bonds describe a pyramid with the phosphorus atom at its apex. The geometry of the PF_3 molecule is said to be *trigonal pyramidal*. How do we explain such different molecular geometries?

Subtle differences in structure are also possible. Consider the following molecular structures, which differ only in the arrangement of the atoms about the C═C bond.

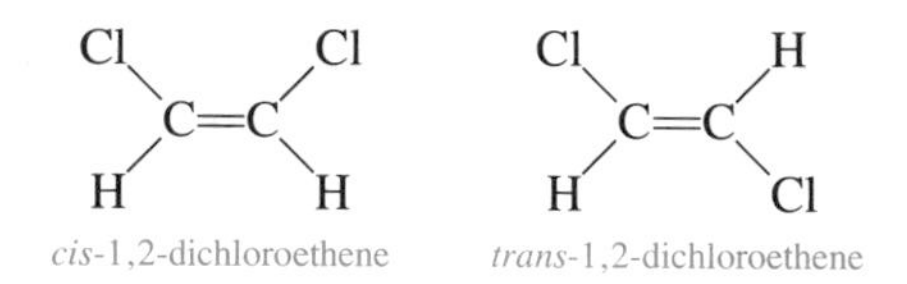

cis-1,2-dichloroethene *trans*-1,2-dichloroethene

In the *cis* compound, both H atoms are on the same side of the C═C bond; in the *trans* compound, they are on opposite sides. These structural formulas represent entirely different compounds, as their boiling points clearly demonstrate. *Cis*-1,2-dichloroethene boils at 60°C; *trans*-1,2-dichloroethene boils at 48°C. The differences between such *cis* and *trans* compounds can be quite important. The central molecular event in the detection of light by the human eye involves the transformation of a *cis* compound to its corresponding *trans* compound after the absorption of a photon. How do we explain the existence of *cis* and *trans* compounds?

In this chapter, we discuss how we explain the geometries of molecules in terms of their electronic structures. We also explore two theories of chemical bonding: valence bond theory and molecular orbital theory.

Molecular Geometry and Directional Bonding

Molecular geometry is *the general shape of a molecule, as determined by the relative positions of the atomic nuclei*. There is a simple model that allows us to predict molecular geometries or shapes from Lewis formulas. This valence-shell electron-pair model usually predicts the correct general shape of a molecule. It does not explain chemical bonding, however. For this we must look at a theory, such as valence bond theory, that is based on quantum mechanics. Valence bond theory gives us further insight into why bonds form and, at the same time, reveals that bonds have definite directions in space.

10.1 THE VALENCE-SHELL ELECTRON-PAIR REPULSION (VSEPR) MODEL

The **valence-shell electron-pair repulsion (VSEPR) model** is *a model for predicting the shapes of molecules and ions in which valence-shell electron pairs are arranged about each atom so that electron pairs are kept as far away from one another as possible, thus minimizing electron-pair repulsions.*● For example, if there are only two electron pairs in the valence shell of an atom, these pairs tend to be at opposite sides of the nucleus, so that the repulsion is minimized. This gives a *linear* arrangement of electron pairs; that is, the electron pairs mainly occupy regions of space at an angle of 180° to one another (Figure 10.2).

● The acronym VSEPR is pronounced "vesper."

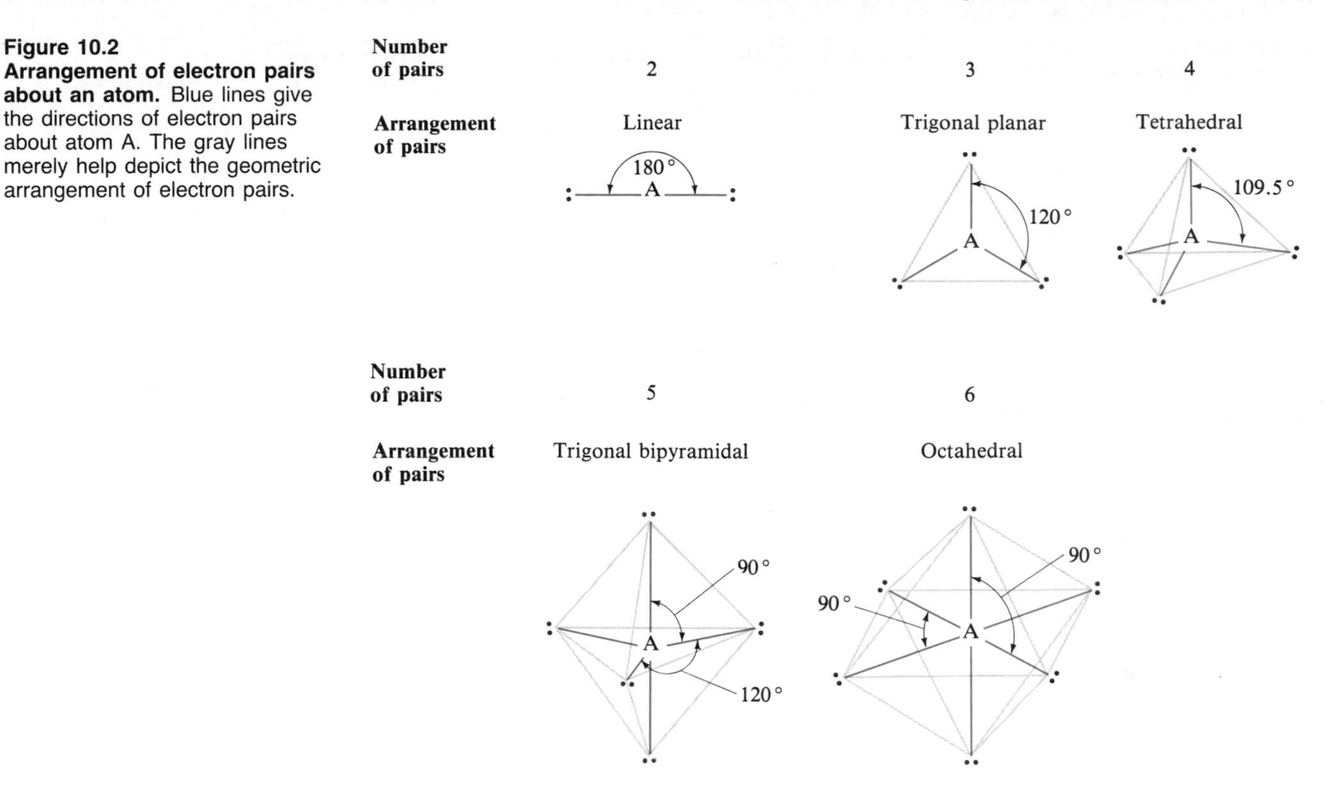

Figure 10.2
Arrangement of electron pairs about an atom. Blue lines give the directions of electron pairs about atom A. The gray lines merely help depict the geometric arrangement of electron pairs.

If three electron pairs are in the valence shell of an atom, they tend to be arranged in a plane directed toward the corners of a triangle of equal sides (equilateral triangle). This arrangement is called *trigonal planar,* in which the regions of space occupied by electron pairs are directed at 120° angles to one another (Figure 10.2).

Four electron pairs in the valence shell of an atom tend to have a *tetrahedral arrangement*. That is, if we imagine the atom at the center of a regular tetrahedron, each region of space in which an electron pair mainly lies extends toward a corner or vertex (Figure 10.2). (A regular tetrahedron is a geometrical shape with four faces, each an equilateral triangle. Thus, it has the form of a triangular pyramid.) The regions of space mainly occupied by electron pairs are directed at approximately 109.5° angles to one another.

When we determine the geometry of a molecule experimentally, we locate the positions of the atoms, not the electron pairs. To predict the relative positions of atoms around a given atom using the VSEPR model, we first note the arrangement of valence-shell electron pairs around that central atom. Some of these electron pairs are bonding pairs and some are lone pairs. *The direction in space of the bonding pairs gives us the molecular geometry*. To consider examples of the method, we will first look at molecules AX_n in which the central atom A is surrounded by two, three, or four valence-shell electron pairs. Later we will look at molecules in which A is surrounded by five or six valence-shell electron pairs.

Central Atom with Two, Three, or Four Valence-Shell Electron Pairs

Examples of geometries in which two, three, or four valence-shell electron pairs surround a central atom are shown in Figure 10.3.

Figure 10.3
Molecular geometries. Central atom with two, three, or four valence-shell electron pairs.

ELECTRON PAIRS Total	Bonding	Lone	ARRANGEMENT OF PAIRS	MOLECULAR GEOMETRY	EXAMPLE
2	2	0	Linear	Linear	BeF_2 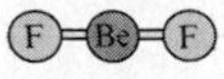
3	3	0	Trigonal planar	Trigonal planar	BF_3 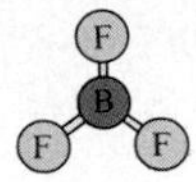
	2	1		Bent (or angular)	GeF_2 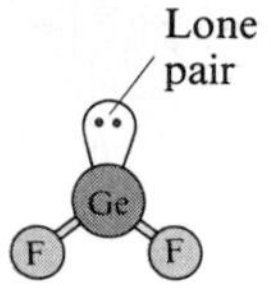
4	4	0	Tetrahedral	Tetrahedral	CH_4 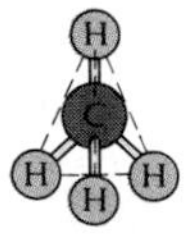
	3	1		Trigonal pyramidal	NH_3 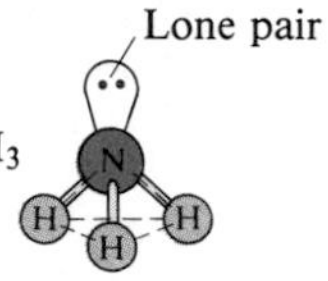
	2	2		Bent (or angular)	H_2O 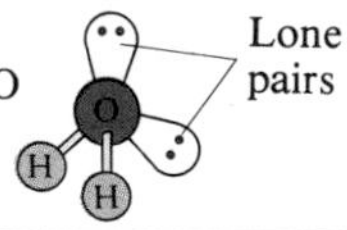

TWO ELECTRON PAIRS (LINEAR ARRANGEMENT) To find the geometry of the molecule AX_n, we first determine the number of valence-shell electron pairs around atom A. We can get this information from the electron-dot formula. For example, following the rules given in Section 9.6, we can find the electron-dot formula for the BeF_2 molecule.●

● Beryllium fluoride, BeF_2, is normally a solid. The BeF_2 molecule exists in the vapor phase at high temperature.

$$:\ddot{\underset{..}{F}}:Be:\ddot{\underset{..}{F}}:$$

There are two electron pairs in the valence shell for beryllium, and the VSEPR model predicts that they will have a linear arrangement (see Figure 10.2). Fluorine atoms are bonded in the same direction as the electron pairs. Hence, the geometry of the BeF_2 molecule is *linear*—that is, the atoms are arranged in a straight line (see Figure 10.3).

The VSEPR model can also be applied to molecules with multiple bonds. In this case, each multiple bond is treated as though it were a single electron pair (all pairs of a multiple bond are required to be in approximately the same region). To predict the geometry of carbon dioxide, for example, we first write the electron-dot formula

$$:\ddot{O}::C::\ddot{O}:$$

We have two electron groups about the carbon atom, and these are treated as though there were two pairs on carbon. Thus, according to the VSEPR model, the bonds are arranged linearly and the geometry of carbon dioxide is linear.

THREE ELECTRON PAIRS (TRIGONAL PLANAR ARRANGEMENT) Let us now predict the geometry of the boron trifluoride molecule, BF_3, introduced in the chapter opening. The electron-dot formula is

```
    ..
   :F:
 ..  ..  ..
:F : B : F:
 ..      ..
```

The three pairs on boron have a trigonal planar arrangement, and the molecular geometry, which is determined by the directions of the bonding pairs, is also *trigonal planar* (Figure 10.3).

Germanium difluoride, GeF_2, is an example of a molecule with three electron pairs, one of which is a lone pair.● The electron-dot formula is

● Germanium difluoride, GeF_2, is normally a white, crystalline solid with a polymeric structure similar to that of beryllium chloride (see Exercise 9.12, page 334). The GeF_2 molecule exists in the vapor phase.

```
 ..
:F : Ge :
 ..   ..
     :F:
      ..
```

The three electron pairs have a trigonal planar arrangement, and the molecular geometry, which is determined by the direction of bonds to germanium, is *bent* or *angular* (Figure 10.3).

To take a somewhat more complicated example, consider sulfur dioxide, SO_2. The electron-dot formula is

```
..
O: :S:
..   ..
    :O:
     ..
```

(one of two possible resonance formulas)

We need not be concerned that each bond is in fact an intermediate single–double bond. Rather, the three groups of electrons about sulfur (two bonds and one lone pair) determine the geometry. These groups have a trigonal planar arrangement, and sulfur dioxide is a bent molecule.

Note the difference between the *arrangement of electron pairs* and the *molecular geometry,* or the arrangement of nuclei. What we usually "see" by means of x-ray diffraction and similar methods are the nuclear positions—that is, the molecular geometry. Unseen, but nevertheless important, are the lone pairs, which occupy definite positions about an atom according to the VSEPR model.

FOUR ELECTRON PAIRS (TETRAHEDRAL ARRANGEMENT) The common and most important case of four electron pairs about the central atom (the octet rule) leads to three different molecular geometries, depending on the number of bonds formed. As examples, we have

```
                          H           H          H
                          ..          ..         ..
                        H:C:H        :N:H       :O:H
                          ..          ..         ..
                          H           H
```

	CH_4	NH_3	H_2O
Molecular geometry:	tetrahedral	trigonal pyramidal	bent

In each case, the electron pairs are arranged tetrahedrally, and one or more atoms are bonded in these tetrahedral directions to give the different geometries (Figure 10.3).

When all electron pairs are bonding, as in methane, CH_4, the molecular geometry is *tetrahedral*. When three of the pairs are bonding and the other is nonbonding, as in ammonia, NH_3, the molecular geometry is *trigonal pyramidal*. Note that the nitrogen atom is at the apex of the pyramid, while the three hydrogen atoms extend downward to form the triangular base of the pyramid.

Now we can answer the question raised in the chapter opening: Why is the BF_3 molecule trigonal planar and the PF_3 molecule trigonal pyramidal? This difference occurs because in BF_3 there are three pairs of electrons around boron, and in PF_3 there are four pairs of electrons around phosphorus. As in NH_3, there are three bonding pairs and one lone pair in PF_3, which give rise to the trigonal pyramidal geometry.

STEPS IN THE PREDICTION OF GEOMETRY BY THE VSEPR MODEL Let us summarize the steps to follow in order to predict the geometry of an AX_n molecule or ion by the VSEPR method (all X atoms of AX_n need not be identical).

1. Determine, from the electron-dot formula, how many electron pairs are around the central atom. Count a multiple bond as one pair.
2. Arrange the electron pairs as shown in Figure 10.2.
3. Obtain the molecular geometry from the directions of bonding pairs, as shown in Figure 10.3.

Example 10.1 Predicting Molecular Geometries (Two, Three, or Four Electron Pairs)

Predict the geometry of the following molecules, using the VSEPR method: (a) $BeCl_2$; (b) $SnCl_2$; (c) $SiCl_4$.

Solution

(a) Following the steps outlined in Section 9.6 for writing Lewis formulas, we distribute the valence electrons to the skeleton structure of $BeCl_2$ as follows:

$$:\ddot{\underset{\cdot\cdot}{Cl}}:Be:\ddot{\underset{\cdot\cdot}{Cl}}:$$

This gives fewer than an octet of electrons about Be, but because Cl atoms do not normally form multiple bonds, we leave the formula as it is. (Moreover, with the dot symbol $\cdot Be \cdot$, Be would be expected to form two single bonds.) The two pairs on Be have a linear arrangement, indicating a **linear** molecular geometry for $BeCl_2$. (See Figure 10.3 under 2 electron pairs, 0 lone pairs.) (b) Distributing the valence electrons to the skeleton structure of $SnCl_2$, we get

$$\begin{array}{l} :\ddot{\underset{\cdot\cdot}{Cl}}:\ddot{Sn}: \\ \quad\;\; :\ddot{\underset{\cdot\cdot}{Cl}}: \end{array}$$

Again, the Cl atoms do not normally form multiple bonds, so we leave this electron-dot formula as it is. The three electron pairs around the Sn atom have a trigonal planar arrangement, and thus the molecular geometry of $SnCl_2$ is **bent.** (See Figure 10.3 under 2 bonding pairs, 1 lone pair.) (c) The electron-dot formula of $SiCl_4$ is

$$\begin{array}{c} :\ddot{Cl}: \\ :\ddot{\underset{\cdot\cdot}{Cl}}:\underset{\cdot\cdot}{\ddot{Si}}:\ddot{\underset{\cdot\cdot}{Cl}}: \\ :\underset{\cdot\cdot}{Cl}: \end{array}$$

The molecular geometry is **tetrahedral.** (See Figure 10.3 under 4 bonding pairs, 0 lone pairs.)

Exercise 10.1

Use the VSEPR method to predict the geometry of the following ion and molecules: (a) ClO_3^-; (b) OF_2; (c) SiF_4. (See Problems 10.19, 10.20, 10.21, and 10.22.)

Bond Angles and the Effect of Lone Pairs

The VSEPR model allows us to predict the approximate angles between bonds in molecules. For example, this model tells us that CH_4 should have a tetrahedral geometry and that the H—C—H bond angles should be 109.5° (see Figure 10.2). Because all of the valence-shell electron pairs about the carbon atom are bonding and because all of the bonds are alike, we expect the CH_4 molecule to have an exact tetrahedral geometry. However, if one or more of the electron pairs is non-

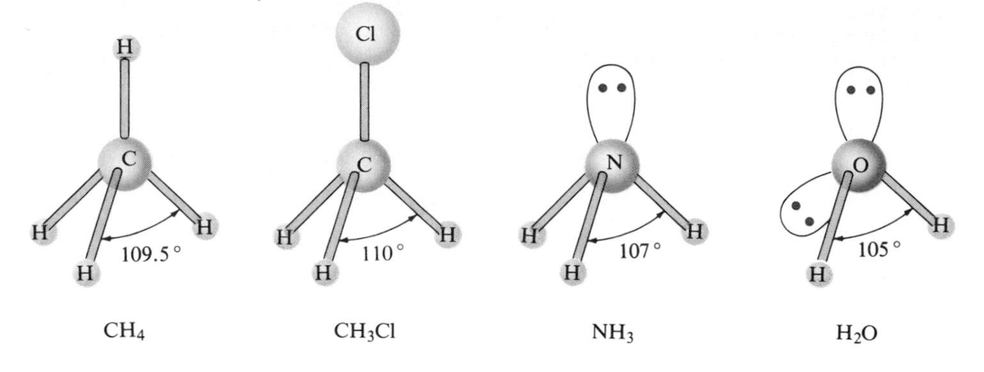

Figure 10.4
H—A—H bond angles in some molecules. Experimentally determined bond angles are shown for CH_4, CH_3Cl, NH_3, and H_2O, represented here by models.

bonding or if there are dissimilar bonds, all four valence-shell electron pairs will not be alike. Then we expect the bond angles to deviate from the 109.5° predicted by an exact tetrahedral geometry. This is indeed what we see. Experimentally determined H—A—H bond angles (A is the central atom) in CH_4, CH_3Cl, NH_3, and H_2O are shown in Figure 10.4.

The increase or decrease of bond angles from the ideal values is often predictable. *A lone pair tends to require more space than a corresponding bonding pair.* We can explain this as follows: A lone pair of electrons is attracted to only one atomic core, whereas a bonding pair is attracted to two. As a result, the lone pair is spatially more diffuse, while the bonding pair is drawn more tightly to the nuclei. Consider the trigonal pyramidal molecule NH_3. The lone pair on the nitrogen atom requires more space than the bonding pairs. Therefore, the N—H bonds are effectively pushed away from the lone pair, and the H—N—H bond angles become smaller than the tetrahedral value of 109.5°. How much smaller than the tetrahedral value, the VSEPR model cannot tell us. The experimental value of an H—N—H bond angle is a few degrees smaller (107°). But the trigonal pyramidal molecule PF_3, mentioned in the chapter opening, has an F—P—F bond angle of 96°, which is significantly smaller than the exact tetrahedral value.

Multiple bonds require more space than single bonds because of the greater number of electrons. We therefore expect the C═O bond in the formaldehyde molecule, CH_2O, to require more space than the C—H bonds. We predict that the H—C—H bond angle will be smaller than the 120° seen in an exact trigonal planar geometry, such as the F—B—F angle in BF_3. Similarly, we expect the H—C—H bond angles in the ethylene molecule, CH_2CH_2, to be smaller than the trigonal planar value. Experimental values are shown in Figure 10.5.

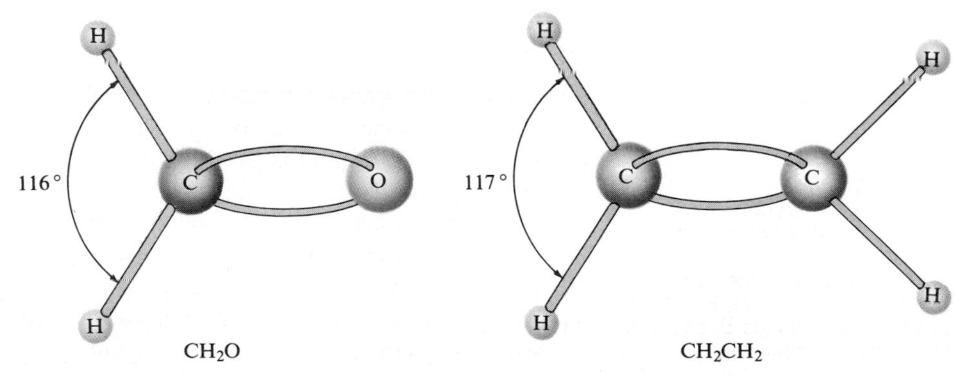

Figure 10.5
H—C—H bond angles in molecules with carbon double bond. Bond angles are shown for formaldehyde, CH_2O, and ethylene, CH_2CH_2, represented here by models.

Figure 10.6
Molecular geometries. Central atom with five or six valence-shell electron pairs.

ELECTRON PAIRS			ARRANGEMENT OF PAIRS	MOLECULAR GEOMETRY	EXAMPLE
Total	Bonding	Lone			
5	5	0	Trigonal bipyramidal	Trigonal bipyramidal	PCl_5 (Axial atom; Axial atom)
	4	1		Seesaw	SF_4 (Lone pair)
	3	2		T-Shaped	ClF_3 (Lone pairs)
	2	3		Linear	XeF_2 (Lone pairs)
6	6	0	Octahedral	Octahedral	SF_6
	5	1		Square pyramidal	IF_5 (Lone pair)
	4	2		Square planar	XeF_4 (Lone pairs)

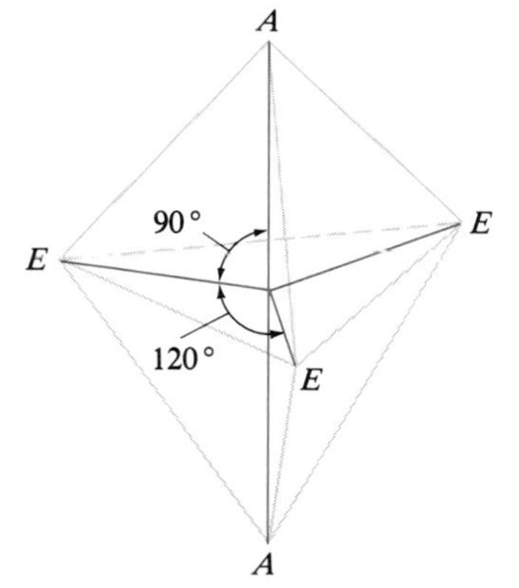

Figure 10.7
The trigonal bipyramidal arrangement of electron pairs. Electron pairs are directed along the colored lines to the vertexes of a trigonal bipyramid. Equatorial directions are labeled *E*, and axial directions are labeled *A*.

Central Atom with Five or Six Valence-Shell Electron Pairs

Five electron pairs tend to have a *trigonal bipyramidal arrangement*. The electron pairs tend to be directed to the corners of a trigonal bipyramid, a figure formed by placing the face of one tetrahedron onto the face of another tetrahedron (see Figure 10.2).

Six electron pairs tend to have an *octahedral arrangement*. The electron pairs tend to be directed to the corners of a regular octahedron, a figure that has eight triangular faces and six vertexes or corners (see Figure 10.2).

FIVE ELECTRON PAIRS (TRIGONAL BIPYRAMIDAL ARRANGEMENT) Large atoms like phosphorus can accommodate more than eight valence electrons. The phosphorus atom in phosphorus pentachloride, PCl_5, has five electron pairs in its valence shell. With five electron pairs around phosphorus, all bonding, PCl_5 should have a *trigonal bipyramidal geometry* (Figure 10.6). Note, however, that the vertexes of the trigonal bipyramid are not all equivalent (that is, the angles between the electron pairs are not all the same). Thus, the directions to which the electron pairs point are not equivalent. Two of the directions, called *axial directions,* form an axis through the central atom (see Figure 10.7). They are 180° apart. The other three directions are called *equatorial directions*. These point toward the corners of the equilateral triangle that lies on a plane through the central atom, perpendicular (at 90°) to the axial directions. The equatorial directions are 120° from each other.

Molecular geometries other than trigonal bipyramidal are possible when one or more of the five electron pairs are lone pairs. Consider the sulfur tetrafluoride molecule, SF_4.● The Lewis electron-dot formula is

● Sulfur tetrafluoride is a stable but reactive gas that is very poisonous.

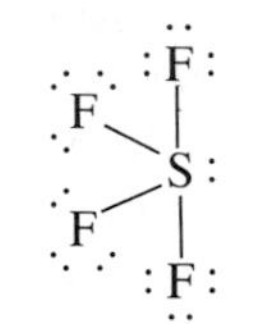

The five electron pairs about sulfur should have a trigonal bipyramidal arrangement. Because the axial and equatorial positions of the electron pairs are not equivalent, we must decide in which of these positions the lone pair appears.

The lone pair acts as though it is somewhat ''larger'' than a bonding pair. Therefore, the total repulsion between all pairs should be lower if the lone pair is in a position that puts it directly adjacent to the smallest number of other pairs.

An electron pair in an equatorial position is directly adjacent to only *two* other pairs—the two axial pairs at 90°. The other two equatorial pairs are further away at 120°. On the other hand, an electron pair in an axial position is directly adjacent to *three* other pairs—the three equatorial pairs at 90°. The other axial pair is much farther away at 180°.

Thus we expect lone pairs to occupy equatorial positions in the trigonal bipyramidal arrangement. For sulfur tetrafluoride, this gives a *seesaw* (or *distorted tetrahedral*) *geometry* (Figure 10.6).

Chlorine trifluoride, ClF_3, has the electron-dot formula

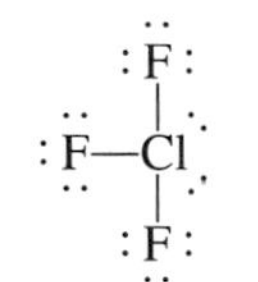

The lone pairs on chlorine occupy two of the equatorial positions of the trigonal bipyramidal arrangement, giving a *T-shaped geometry* (Figure 10.6). The four atoms of the molecule all lie in one plane, with the chlorine nucleus at the intersection of the ''T.''●

● Chlorine trifluoride is a colorless gas with a sweet, suffocating odor. It is very reactive and is used to prepare uranium hexafluoride, UF_6, for the processing of nuclear fuel.

Xenon difluoride, XeF_2, has the electron-dot formula

:F:
|
:Xe:
|
:F:

The three lone pairs on xenon occupy the equatorial positions of the trigonal bipyramidal arrangement, giving a linear molecular geometry (Figure 10.6).

SIX ELECTRON PAIRS (OCTAHEDRAL ARRANGEMENT) There are six electron pairs (all bonding pairs) about sulfur in sulfur hexafluoride, SF_6. Thus, it has an octahedral molecular geometry, with sulfur at the center of the octahedron and fluorine atoms at the vertexes (Figure 10.6).●

● Sulfur hexafluoride is relatively nonreactive and nontoxic (compare with SF_4). It is used in transformers as an insulating gas.

● The halogens form a number of molecules with one another. Except for those with only two atoms, these are mainly molecules with a chlorine, a bromine, or an iodine atom surrounded by an odd number of fluorine atoms.

Iodine pentafluoride, IF_5, has the electron-dot formula

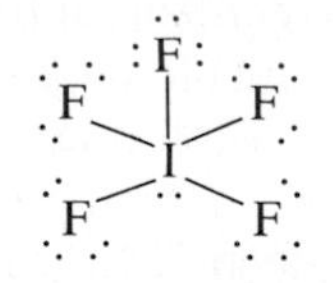

The lone pair of iodine occupies one of the six equivalent positions in the octahedral arrangement, giving a *square pyramidal geometry* (Figure 10.6). The name derives from the shape formed by drawing lines between atoms.●

Xenon tetrafluoride, XeF_4, has the electron-dot formula

● Xenon forms a series of fluorides, including XeF_2, XeF_4, and XeF_6. Xenon tetrafluoride, XeF_4, is a colorless, crystalline compound formed by the reaction of xenon, Xe, with fluorine, F_2.

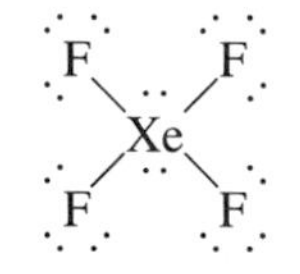

The two lone pairs on xenon occupy opposing positions in the octahedral arrangement to minimize their repulsion. The result is a *square planar geometry* (Figure 10.6).●

Example 10.2 Predicting Molecular Geometries (Five or Six Electron Pairs)

What do you expect for the geometry of tellurium tetrachloride, $TeCl_4$?

Solution

First we distribute the valence electrons to the Cl atoms to satisfy the octet rule. Then we allocate the remaining ones to the central atom, Te (following the steps outlined in Section 9.6). The electron-dot formula is

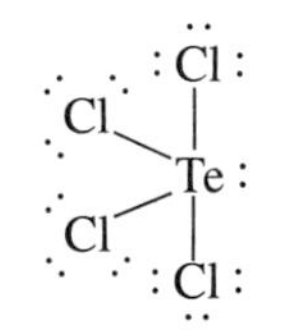

Thus, there are five electron pairs in the valence shell of Te in $TeCl_4$. Of these, four are bonding pairs and one is a lone pair. The arrangement of electron pairs is trigonal bipyramidal. We expect the lone pair to occupy an equatorial position, so $TeCl_4$ has a **seesaw molecular geometry**. (See Figure 10.6 under 4 bonding pairs, 1 lone pair.)

Exercise 10.2

According to the VSEPR method, what molecular geometry would you predict for iodine trichloride, ICl_3? (See Problems 10.23, 10.24, 10.25, and 10.26.)

10.2 DIPOLE MOMENT AND MOLECULAR GEOMETRY

The VSEPR model provides a simple procedure for predicting the geometry of a molecule. However, predictions must be verified by experiment. Information about the geometry of a molecule can sometimes be obtained from an experimental quantity called the dipole moment, which is related to the polarity of the bonds in a molecule. The **dipole moment** is *a quantitative measure of the degree of charge separation in a molecule*. The polarity of a bond, such as that in HCl, is characterized by a separation of electric charge. We can represent this in HCl by indicating partial charges, δ^+ and δ^-, on the atoms.

$$\overset{\delta^+}{\text{H}}\text{—}\overset{\delta^-}{\text{Cl}}$$

Any molecule that has a net separation of charge, as in HCl, has a dipole moment. A molecule in which the distribution of electric charge is equivalent to charges $+\delta$ and $-\delta$ separated by a distance d has a dipole moment equal to δd. Dipole moments are usually measured in units of *debyes* (D). In SI units, dipole moments are measured in coulomb-meters (C · m), and 1 D = 3.34×10^{-30} C · m.

Measurements of dipole moments are based on the fact that polar molecules (molecules having a dipole moment) can be oriented by an electric field. Figure 10.8 shows an electric field generated by charged plates. Note that the polar molecules tend to align themselves so that the negative ends of the molecules point toward the positive plate, and the positive ends point toward the negative plate. This orientation of the molecules affects the *capacitance* of the charged plates (the capacity of the plates to hold a charge). Consequently, measurements of the capacitance of plates separated by different substances can be used to obtain the dipole moments of those substances. The alignment of polar molecules (those having a dipole moment) by an electric charge is demonstrated in Figure 10.9, which shows that a polar liquid (H_2O) is attracted to an electrified rod.

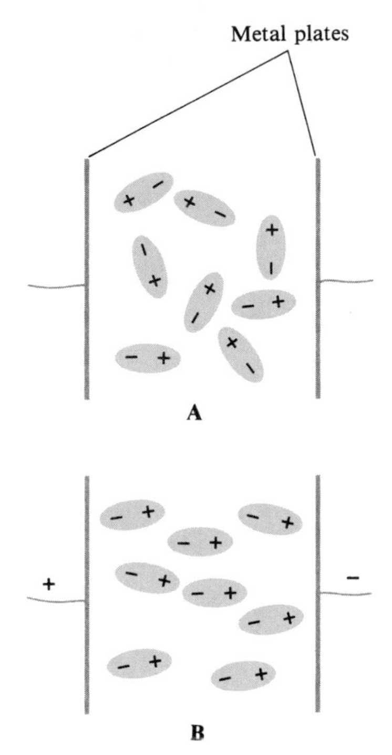

Figure 10.8
Alignment of polar molecules by an electric field. *(A)* A polar substance is placed between metal plates. *(B)* When these are connected to an electric voltage (whose direction is shown by the + and − signs), the polar molecules align so that their negative ends point toward the positive plate.

We can sometimes relate the presence or absence of a dipole moment in a molecule to its molecular geometry. For example, consider the carbon dioxide molecule. Each carbon–oxygen bond has a polarity in which the more electronegative oxygen atom has a partial negative charge.

$$\overset{\delta^-}{\text{O}}\text{=}\overset{2\delta^+}{\text{C}}\text{=}\overset{\delta^-}{\text{O}}$$

We will denote the dipole-moment contribution from each bond (the bond dipole) by an arrow (↔) with a positive sign at one end. The dipole-moment arrow points from the positive partial charge toward the negative partial charge. Thus, we can rewrite the formula for carbon dioxide as

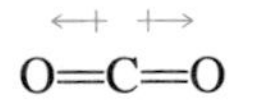

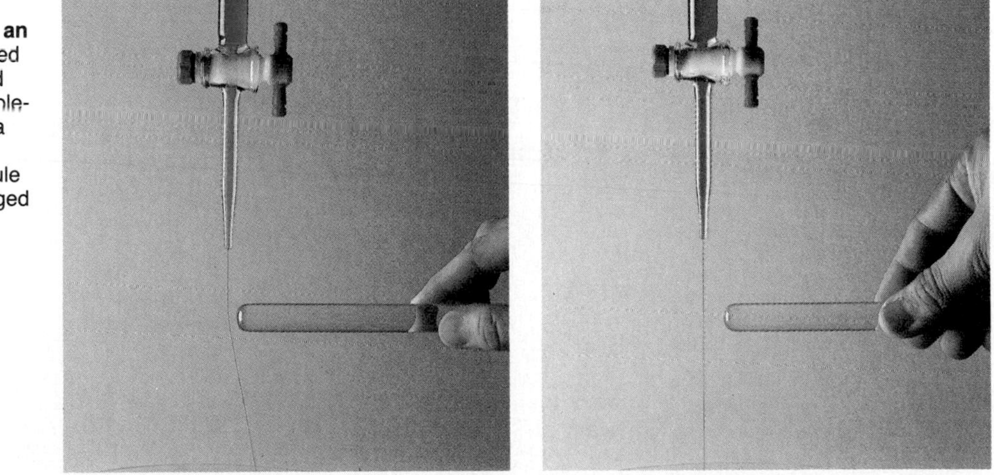

Figure 10.9
Attraction of a polar liquid to an electrified rod. *Left:* Water (dyed blue) is attracted to the charged glass rod because the water molecules align themselves to give a net attraction. *Right:* Hexane (dyed red) is a nonpolar molecule and is not attracted to the charged rod.

Each bond dipole, like a force, is a *vector* quantity; that is, it has both magnitude and direction. Like forces, two bond dipoles of equal magnitude but opposite direction cancel each other. (Think of two groups of people in a tug of war. As long as each group pulls on the rope with the same force but in the opposite direction, there is no movement—the net force is zero.) Because the two carbon–oxygen bonds in CO_2 are equal but point in opposite directions, they give a net dipole moment of zero for the molecule.

For comparison, consider the water molecule. The bond dipoles point from the hydrogen atoms toward the more electronegative oxygen.

Here, however, the two bond dipoles do not point directly toward or away from each other. As a result, they add together to give a nonzero dipole moment for the molecule.

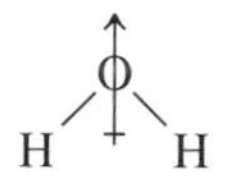

The dipole moment of H_2O has been observed to be 1.94 D.

The fact that the water molecule has a dipole moment is excellent experimental evidence for a bent geometry. If the H_2O molecule were linear, the dipole moment would be zero.

The analysis we have just made for two different geometries of AX_2 molecules can be extended to other AX_n molecules (in which all X atoms are identical). Table 10.1 summarizes the relationship between molecular geometry and dipole moment. Those geometries in which A—X bonds are directed symmetrically about the central atom (for example, linear, trigonal planar, and tetrahedral) give molecules of zero dipole moment; that is, the molecules are *nonpolar*. Those geometries in which the X atoms tend to be on one side of the molecule (for example, bent and trigonal pyramidal) can have nonzero dipole moments; that is, they can give *polar* molecules.

Table 10.1 Relationship between Molecular Geometry and Dipole Moment

Formula	Molecular Geometry	Dipole Moment*
AX	Linear	Can be nonzero
AX_2	Linear	Zero
	Bent	Can be nonzero
AX_3	Trigonal planar	Zero
	Trigonal pyramidal	Can be nonzero
	T-shaped	Can be nonzero
AX_4	Tetrahedral	Zero
	Square planar	Zero
	Seesaw	Can be nonzero
AX_5	Trigonal bipyramidal	Zero
	Square pyramidal	Can be nonzero
AX_6	Octahedral	Zero

*All X atoms are assumed to be identical.

Example 10.3 Relating Dipole Moment and Molecular Geometry

Each of the following molecules has a nonzero dipole moment. Select the molecular geometry that is consistent with this information. Explain your reasoning.

(a) SO_2 linear, bent
(b) PH_3 trigonal planar, trigonal pyramidal

Solution

(a) In the linear geometry, the S—O bond contributions to the dipole moment would cancel, giving a zero dipole moment. That would not happen in the **bent** geometry; hence, this must be the geometry for the SO_2 molecule. (b) In the trigonal planar geometry, the bond contributions to the dipole moment would cancel, giving a zero dipole moment. That would not occur in the **trigonal pyramidal** geometry; hence, this is a possible molecular geometry for PH_3.

Exercise 10.3

Bromine trifluoride, BrF_3, has a nonzero dipole moment. Indicate which of the following geometries are consistent with this information: (a) trigonal planar; (b) trigonal pyramidal; (c) T-shaped. (See Problems 10.27 and 10.28.)

Exercise 10.4

Which of the following would be expected to have a dipole moment of zero on the basis of symmetry? Explain. (a) $SOCl_2$; (b) SiF_4; (c) OF_2. (See Problems 10.29 and 10.30.)

In Example 10.3, the fact that certain molecular geometries necessarily imply a zero dipole moment allowed us to eliminate these geometries when considering a molecule with a nonzero dipole moment. The reverse argument does not follow, however. For example, suppose a molecule of the type AX_3 is found to have no measurable dipole moment. It may be that the geometry is trigonal planar. But there are two other possibilities: The geometry is trigonal pyramidal or T-shaped but the bonds are nonpolar; or the geometry is trigonal pyramidal or T-shaped but the lone pairs on the central atom offset the polarity of the bonds.

We can see this effect of lone pairs on the dipole moment of nitrogen trifluoride, NF_3. Judging by the electronegativity difference, we would expect each N—F bond to be quite polar. Yet the dipole moment of nitrogen trifluoride is only 0.2 D. (By contrast, ammonia has a dipole moment of 1.47 D.) The explanation for the small dipole moment in NF_3 appears in Figure 10.10. The lone pair on nitrogen has a dipole-moment contribution that is directed outward from the nucleus, because the electrons are offset from the nuclear center. This dipole-moment contribution thus opposes the N—F bond moments. ●

● Can you explain why NH_3 has such a large dipole moment compared with NF_3?

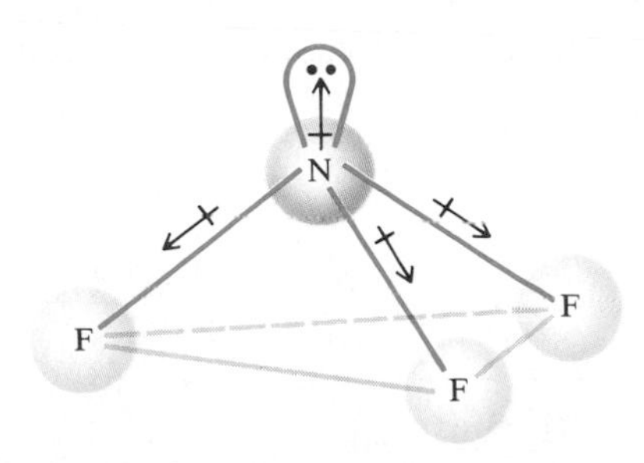

Figure 10.10
Explanation of the small dipole moment of NF_3. The lone-pair and bond contributions to the dipole moment of NF_3 tend to offset one another.

10.3 VALENCE BOND THEORY

The VSEPR model is usually a satisfactory method for predicting molecular geometries. To understand bonding and electronic structure, however, one must look to quantum mechanics. We will consider two theories stemming from quantum mechanics: valence bond theory and molecular orbital theory. Both theories use the methods of quantum mechanics but make different simplifying assumptions. In this section, we will look in a qualitative way at the basic ideas involved in **valence bond theory,** *an approximate theory to explain the electron-pair or covalent bond by quantum mechanics.*

Basic Theory

According to valence bond theory, a bond forms between two atoms when the following conditions are met:

1. An orbital on one atom comes to occupy a portion of the same region of space as an orbital on the other. The two orbitals are said to *overlap*.
2. The total number of electrons in both orbitals is no more than two.

As the orbital of one atom overlaps the orbital of another, the electrons in the orbitals begin to move about both atoms. Because they are attracted to both nuclei at once, they pull the atoms together. Strength of bonding depends on the amount of overlap; the greater the overlap, the greater the bond strength. The two orbitals cannot contain more than two electrons because a given region of space can hold only two electrons (and then only if the spins of the electrons are opposite).

For example, consider the formation of the H_2 molecule from atoms. Each atom has the electron configuration $1s^1$. As the H atoms approach each other, their $1s$ orbitals begin to overlap and a covalent bond forms (Figure 10.11). Valence bond theory also explains why two He atoms (each with electron configuration $1s^2$) do not bond. Suppose two He atoms approach one another and their $1s$ orbitals begin to overlap. Each orbital is doubly occupied, so the sharing of electrons between atoms would place the four valence electrons from the two atoms in the same region. This, of course, could not happen. As the orbitals begin to overlap, each electron pair strongly repels the other. The atoms come together, then fly apart.

Because the strength of bonding depends on orbital overlap, orbitals other than s orbitals bond only in given directions. *Orbitals bond in the directions in which they protrude or point, to obtain maximum overlap.* Consider the bonding between a hydrogen atom and a chlorine atom to give the HCl molecule. A chlorine atom has the electron configuration $[Ne]3s^23p^5$. Of the orbitals in the valence shell of the chlorine atom, three are doubly occupied by electrons and one (a $3p$

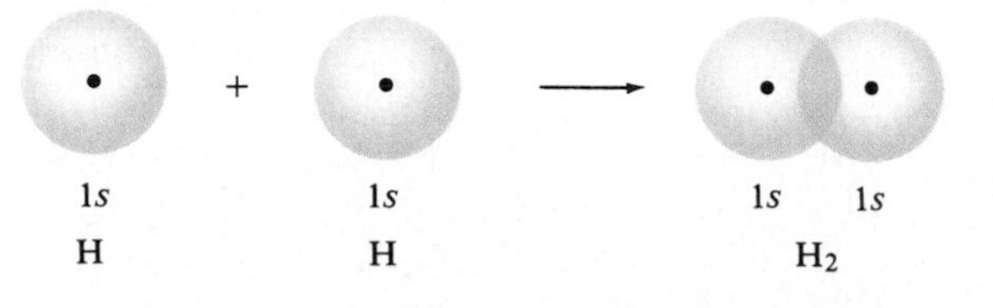

Figure 10.11
Formation of H_2. The H—H bond forms when the $1s$ orbitals, one from each atom, overlap one another.

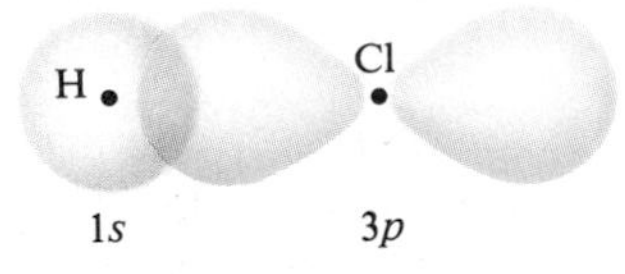

Figure 10.12
Bonding in HCl. The bond forms by the overlap of a hydrogen 1*s* orbital along the axis of a chlorine 3*p* orbital (only the main outer lobes of the 3*p* orbital are shown).

orbital) is singly occupied. The bonding of the hydrogen atom has to occur with the singly occupied 3*p* orbital of chlorine. For the strongest bonding to occur, the maximum overlap of orbitals is required. The 1*s* orbital of hydrogen must overlap along the axis of the singly occupied 3*p* orbital of chlorine (see Figure 10.12).

Hybrid Orbitals

From what has been said, one might expect the number of bonds formed by a given atom to equal the number of unpaired electrons in its valence shell. Chlorine, whose orbital diagram is

(↑↓) (↑↓) (↑↓)(↑↓)(↑↓) (↑↓) (↑↓)(↑↓)(↑)
1*s* 2*s* 2*p* 3*s* 3*p*

has one unpaired electron and forms one bond. Oxygen, whose orbital diagram is

(↑↓) (↑↓) (↑↓)(↑)(↑)
1*s* 2*s* 2*p*

has two unpaired electrons and forms two bonds, as in H_2O.

However, consider the carbon atom, whose orbital diagram is

(↑↓) (↑↓) (↑)(↑)( )
1*s* 2*s* 2*p*

We might expect this atom to bond to two hydrogen atoms to form the CH_2 molecule. Although this molecule is known to be present momentarily during some reactions, it is very reactive and cannot be isolated. But methane, CH_4, in which the carbon atom bonds to four hydrogen atoms, is well known. In fact, a carbon atom usually forms four bonds.

We might explain this as follows: Four unpaired electrons are formed when an electron from the 2*s* orbital of the carbon atom is *promoted* (excited) to the vacant 2*p* orbital.

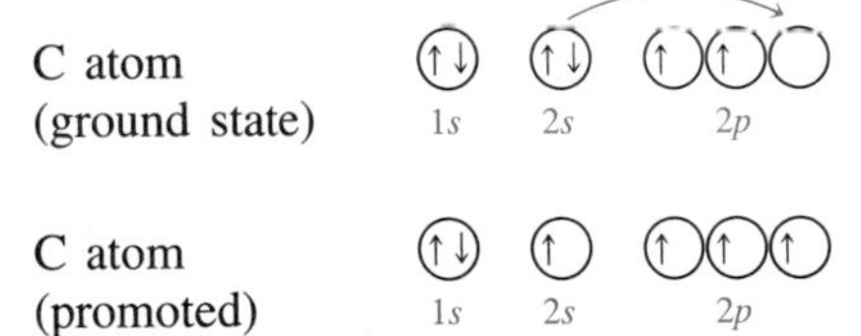

It would require energy to promote the carbon atom this way, but more than enough energy would be obtained from the formation of two additional covalent bonds. One bond would form from the overlap of the carbon 2*s* orbital with a hydrogen 1*s* orbital. Each of the other three bonds would form from a carbon 2*p* orbital and a hydrogen 1*s* orbital.

Experiment shows, however, that the four C—H bonds in methane are identical.● This implies that the carbon orbitals involved in bonding are also equivalent. For this reason, valence bond theory assumes that the four valence orbitals of the carbon atom combine during the bonding process to form four new, but equivalent, hybrid orbitals. **Hybrid orbitals** are *orbitals used to describe bonding that are obtained by taking combinations of atomic orbitals of the isolated atoms.* In this case, a set of hybrid orbitals is constructed from one *s* orbital and three *p* orbitals, so they are called sp^3 hybrid orbitals. Calculations from theory show that each sp^3 hybrid orbital has a large lobe pointing in one direction and a small lobe pointing in the opposite direction. The four sp^3 hybrid orbitals point in tetrahedral

● Nuclear magnetic resonance (page 277) and infrared spectroscopy (page 342) both show that CH_4 has four equivalent C—H bonds.

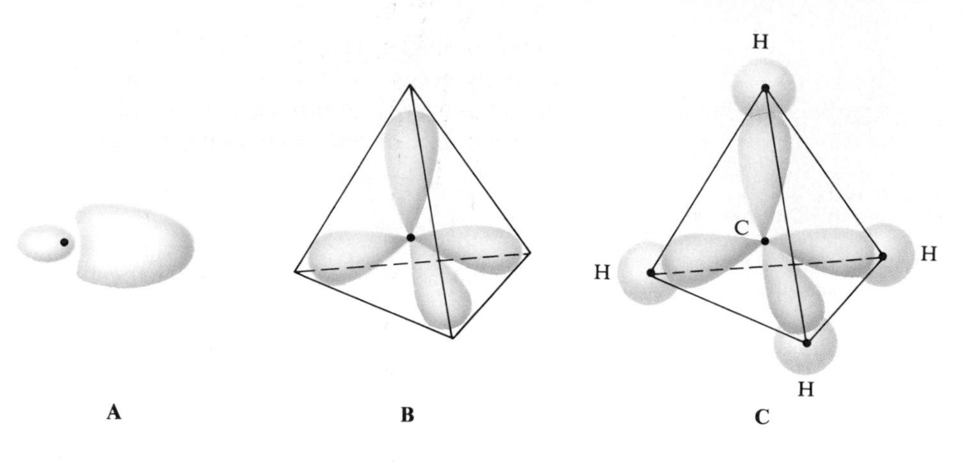

Figure 10.13
The spatial arrangement of sp^3 hybrid orbitals. *(A)* The shape of a single sp^3 hybrid orbital. *(B)* The four hybrid orbitals are arranged tetrahedrally in space (small lobes are omitted for clarity and large lobes are stylized for ease in depicting directional bonding).
(C) Bonding in CH_4. Each C—H bond is formed by the overlap of a 1*s* orbital from a hydrogen atom and an sp^3 hybrid orbital of the carbon atom.

directions. Figure 10.13 (A and B) shows the shape of a single sp^3 orbital and a set of four orbitals pointing in tetrahedral directions.

The C—H bonds in methane, CH_4, are described by valence bond theory as the overlapping of each sp^3 hybrid orbital of the carbon atom with 1*s* orbitals of hydrogen atoms (see Figure 10.13C). Thus, the bonds are arranged tetrahedrally, which is predicted by the VSEPR model. We can represent the hybridization of carbon and the bonding of hydrogen to the carbon atom in methane as follows:

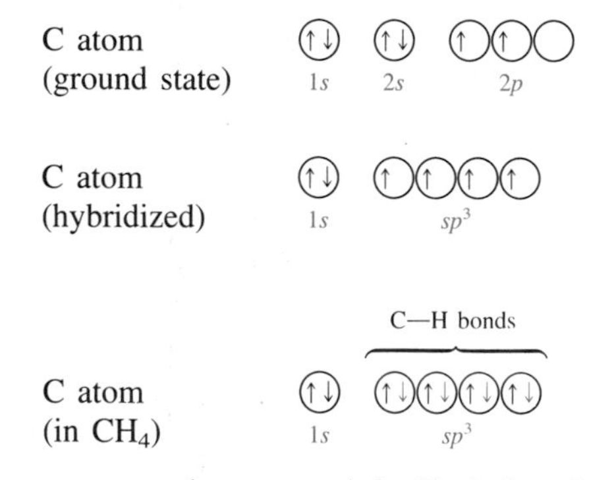

Here the blue arrows represent electrons originally belonging to hydrogen atoms.

Hybrid orbitals can be formed from various numbers of atomic orbitals. *The number of hybrid orbitals formed always equals the number of atomic orbitals used.* For example, if we combine an *s* orbital and two *p* orbitals to get a set of equivalent orbitals, we get three hybrid orbitals (they are called sp^2 hybrid orbitals). A set of hybrid orbitals always has definite directional characteristics. Thus all three sp^2 hybrid orbitals lie in a plane and are directed at 120° angles to one another. That is, they have a trigonal planar arrangement. Some possible hybrid orbitals and their geometric arrangements are listed in Table 10.2 and shown in Figure 10.14. Note that the geometric arrangements of hybrid orbitals are the same as those for electron pairs in the VSEPR model.

Only two of the three *p* orbitals are used to form sp^2 hybrid orbitals. The unhybridized *p* orbital is perpendicular to the plane of the sp^2 hybrid orbitals. Similarly, only one of the three *p* orbitals is used to form *sp* hybrid orbitals. The two unhybridized *p* orbitals are perpendicular to the axis of the *sp* hybrid orbitals

Table 10.2 Kinds of Hybrid Orbitals

Hybrid Orbitals	Geometric Arrangement	Number of Orbitals	Example
sp	Linear	2	Be in BeF_2
sp^2	Trigonal planar	3	B in BF_3
sp^3	Tetrahedral	4	C in CH_4
sp^3d	Trigonal bipyramidal	5	P in PCl_5
sp^3d^2	Octahedral	6	S in SF_6

and are perpendicular to each other. We will use these facts when we discuss multiple bonding in the next section.

Now that we know something about hybrid orbitals, let us develop a general scheme for describing the bonding about any atom (we will call this the central atom). First, notice from Table 10.2 that there is a relationship between type of hybrid orbitals on an atom and the geometric arrangement of those orbitals. If we

Figure 10.14 Diagrams of hybrid orbitals showing their spatial arrangements. Each lobe shown is one hybrid orbital (small lobes omitted for clarity).

Linear arrangement

180°

sp hybrid orbitals

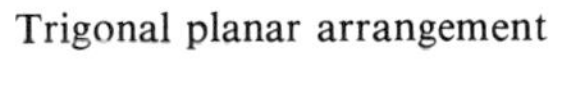

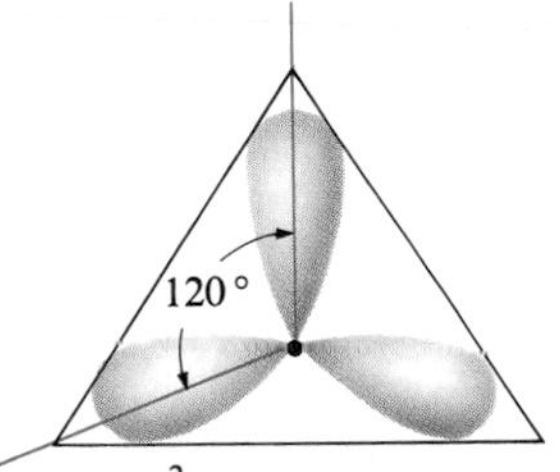

Tetrahedral arrangement

109.5°

sp^3 hybrid orbitals

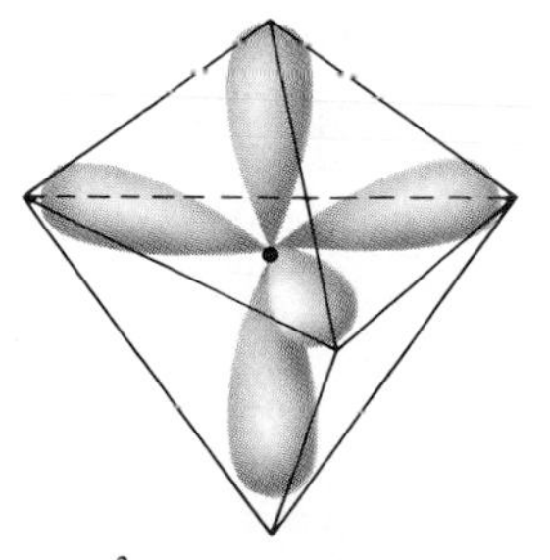

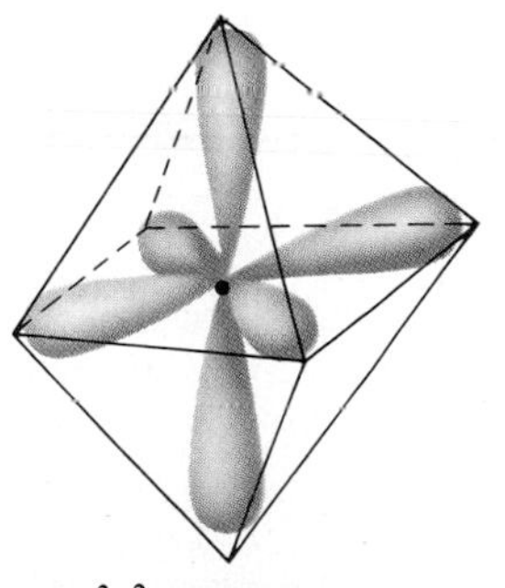

know one, we can infer the other. Thus, if we know the geometric arrangement, we know what hybrid orbitals to use in the bond description of the central atom. Our first task in describing the bonding, then, is to obtain the geometric arrangement about the central atom. In lieu of experimental information about the geometry, we can use the VSEPR model, since it proves to be a fairly good predictor of molecular geometry. To obtain the bonding description about any atom in a molecule, we proceed as follows:

1. Write the Lewis electron-dot formula of the molecule.
2. From the Lewis formula, use the VSEPR model to obtain the arrangement of electron pairs about the central atom.
3. From the geometric arrangement of the electron pairs, deduce the type of hybrid orbitals on the central atom required for the bonding description (see Table 10.2).
4. Assign electrons to the hybrid orbitals of the central atom one at a time, pairing them only when necessary.
5. Form bonds to this central atom by overlapping singly occupied orbitals of other atoms with the singly occupied hybrid orbitals of the central atom.

As an application of this scheme, let us look at the BF_3 molecule and obtain the bond description of the boron atom. Following Step 1, we write the Lewis formula of BF_3.

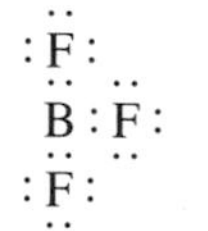

Now we apply the VSEPR model to the boron atom (Step 2). There are three electron pairs about the boron atom, so they are expected to have a planar trigonal arrangement. Looking at Table 10.2 (Step 3), we note that the three sp^2 hybrid orbitals have a planar trigonal arrangement. In Step 4, we assign the valence electrons of the boron atom to the hybrid orbitals. Finally, in Step 5, we imagine three fluorine atoms approaching the boron atom. The singly occupied $2p$ orbital on a fluorine atom overlaps one of the sp^2 hybrid orbitals on boron, forming a covalent bond. Three such B—F bonds form. Note that one of the $2p$ orbitals of boron remains unhybridized and is unoccupied by electrons. It is oriented perpendicular to the molecular plane. We can summarize these steps using orbital diagrams as follows:

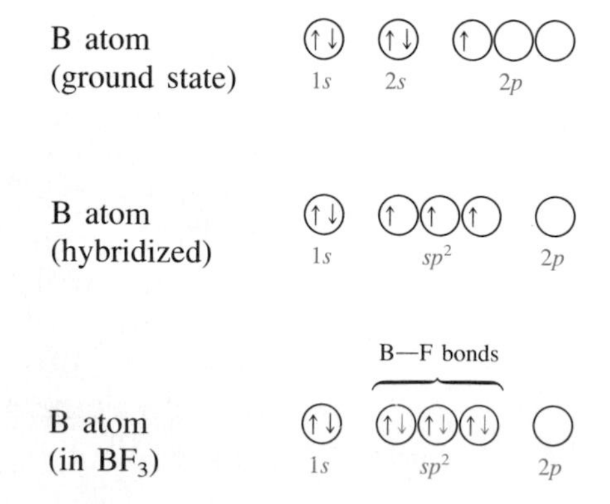

The next example further illustrates how to describe the bonding about an atom in a molecule. In order to solve problems of this type, you should memorize the geometric arrangements of the five kinds of hybrid orbitals listed in Table 10.2.

Example 10.4 Applying Valence Bond Theory (Two, Three, or Four Electron Pairs)

Describe the bonding in H_2O according to valence bond theory. Assume that the molecular geometry is the same as given by the VSEPR model.

Solution

The Lewis formula for H_2O is

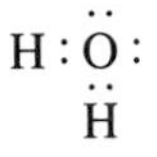

Note that there are four pairs of electrons about the oxygen atom. According to the VSEPR model, these are directed tetrahedrally, and from Table 10.2 we see that we should use sp^3 hybrid orbitals. Each O—H bond is formed by the overlap of a 1*s* orbital of a hydrogen atom with one of the singly occupied sp^3 hybrid orbitals of the oxygen atom. We can represent the bonding to the oxygen atom in H_2O as follows (see also Figure 10.15):

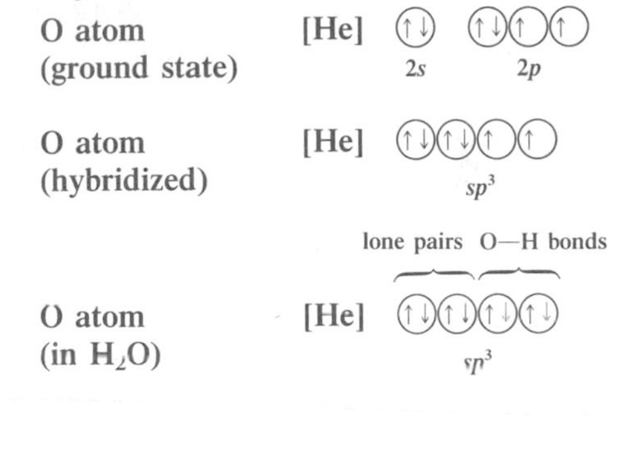

Figure 10.15
Bonding in H_2O. Orbitals on oxygen are sp^3 hybridized; bonding is tetrahedral (see Example 10.4).

Exercise 10.5

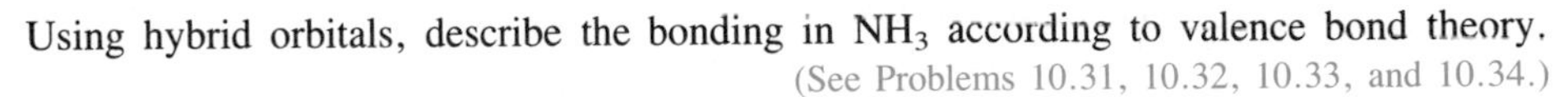

Using hybrid orbitals, describe the bonding in NH_3 according to valence bond theory.
(See Problems 10.31, 10.32, 10.33, and 10.34.)

The actual angle between O—H bonds in H_2O has been experimentally determined to be 104.5°. Because this is close to the tetrahedral angle (109.5°), the description of bonding given in Example 10.4 is essentially correct. To be more precise, we should note that the hybrid orbitals for bonds and lone pairs are not exactly equivalent. The lone pairs are somewhat larger than bonding pairs. Because they take up more space, the lone pairs push the bonding pairs closer together than they are in the exact tetrahedral case.

As the next example illustrates, hybrid orbitals are also useful for describing bonding when the central atom of a molecule is surrounded by more than eight valence electrons.

Example 10.5 Applying Valence Bond Theory (Five or Six Electron Pairs)

Describe the bonding in XeF_4 using hybrid orbitals.

Solution

The Lewis formula of XeF_4 is

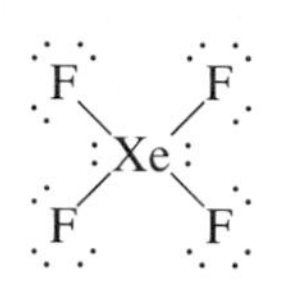

The xenon atom has four single bonds and two lone pairs. It will require six orbitals to describe the bonding. This suggests that we use sp^3d^2 hybrid orbitals on xenon (according to Table 10.2). Each fluorine atom (valence-shell configuration $2s^22p^5$) has one singly occupied $2p$ orbital, so we will assume that this orbital is used in bonding. Each Xe—F bond is formed by the overlap of a xenon sp^3d^2 hybrid orbital with a singly occupied fluorine $2p$ orbital. We can summarize this as follows:

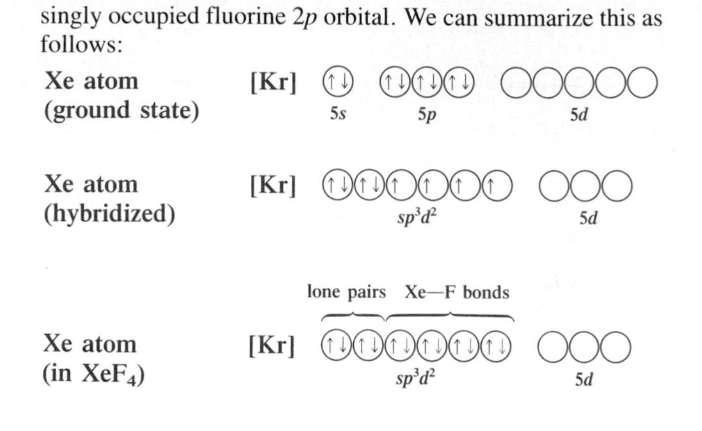

Exercise 10.6

Describe the bonding in PCl_5 using hybrid orbitals.

(See Problems 10.35, 10.36, 10.37, and 10.38.)

10.4 DESCRIPTION OF MULTIPLE BONDING

In the previous section, we described bonding as the overlap of *one* orbital from each of the bonding atoms. Now we want to consider the possibility that *more than one* orbital from each of the bonding atoms might overlap, resulting in a multiple bond.

As an example, let us look at the ethylene molecule.

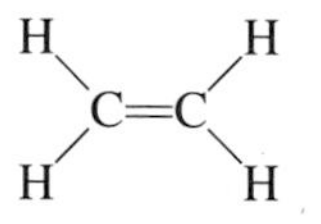

One hybrid orbital is needed for each bond (whether a single or a multiple bond) and for each lone pair. Because each carbon atom is bonded to three other atoms and there are no lone pairs, three hybrid orbitals are needed. This suggests the use of sp^2 hybrid orbitals on each carbon atom (there are three sp^2 hybrid orbitals; see Table 10.2). Thus, during bonding, the $2s$ orbital and two of the $2p$ orbitals of each carbon atom form three hybrid orbitals having trigonal planar orientation. A third $2p$ orbital on each carbon atom remains unhybridized and is perpendicular to the plane of the three sp^2 hybrid orbitals.

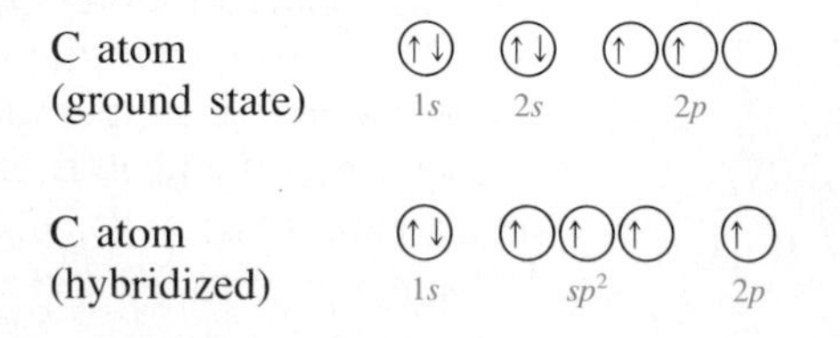

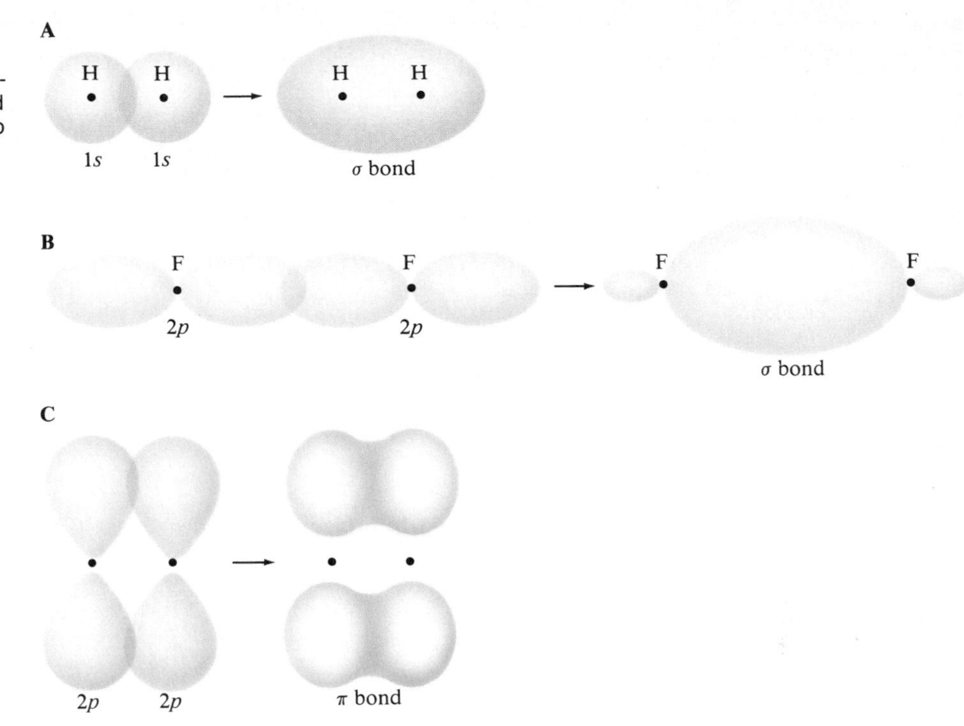

Figure 10.16
Sigma and pi bonds. *(A)* The formation of a σ bond by the overlap of two *s* orbitals. *(B)* A σ bond can also be formed by the overlap of two *p* orbitals along their axes. *(C)* When two *p* orbitals overlap sidewise, a π bond is formed.

To describe the multiple bonding in ethylene, we must distinguish between two kinds of bonds. A **σ (sigma) bond** is *a bond that has a cylindrical shape about the bond axis*. It is formed either when two *s* orbitals overlap, as in H_2 (Figure 10.16A), or when an orbital with directional character, such as a *p* orbital or a hybrid orbital, overlaps *along its axis* (Figure 10.16B). The bonds we discussed in the previous section are σ bonds.

A **π (pi) bond** is *a bond that has an electron distribution above and below the bond axis*. It is formed by the *sidewise* overlap of two parallel *p* orbitals (see Figure 10.16C). A sidewise overlap will not give so strong a bond as an along-the-axis overlap of two *p* orbitals. A π bond occurs when two parallel orbitals are still available after strong σ bonds have formed.

Now imagine that the separate atoms of ethylene move into their normal molecular positions. Each sp^2 hybrid carbon orbital overlaps a 1*s* orbital of a hydrogen atom or an sp^2 hybrid orbital of another carbon atom to form a σ bond (Figure 10.17A). Together, the σ bonds give the molecular framework of ethylene.

As we see from our orbital diagram for the hybridized C atom, a single 2*p* orbital still remains on each carbon atom. These orbitals are perpendicular to the plane of the hybrid orbitals; that is, they are perpendicular to the $—CH_2$ plane. Note that the two $—CH_2$ planes can rotate about the carbon–carbon axis without affecting the overlap of the hybrid orbitals. As these planes rotate, the 2*p* orbitals also rotate. When the $—CH_2$ planes rotate so that the 2*p* orbitals become parallel, the orbitals overlap to give a π bond (Figure 10.17B).

We therefore describe the carbon–carbon double bond as one σ bond and one π bond. Note that when the two 2*p* orbitals are parallel, the two $—CH_2$ ends of the molecule lie in the same plane. Thus, the formation of a π bond ''locks'' the two ends into a flat, rigid molecule.

We can describe the triple bonding in acetylene, $H—C{\equiv}C—H$, in similar fashion. Because each carbon atom is bonded to two other atoms and there are no

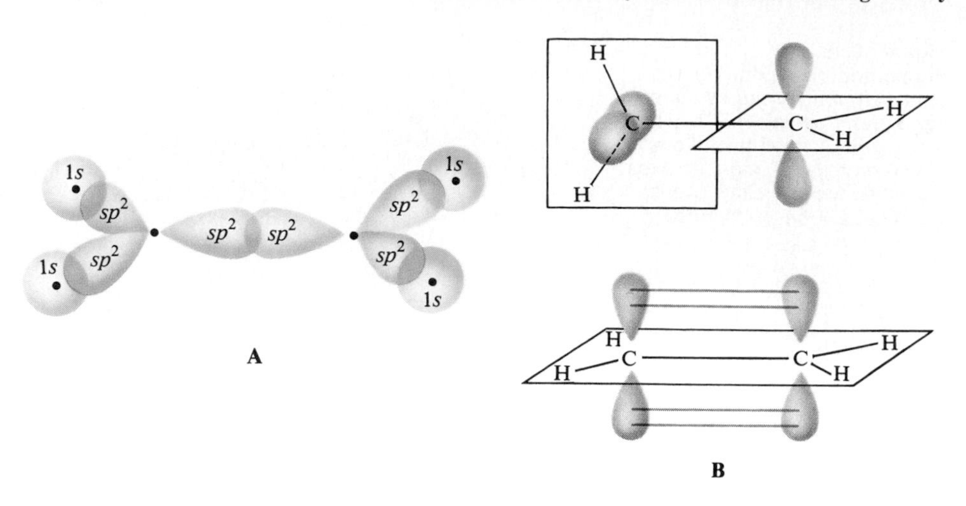

Figure 10.17
Bonding in ethylene. *(A)* The σ-bond framework in ethylene, formed by the overlap of sp^2 hybrid orbitals on C atoms and 1*s* orbitals on H atoms. *(B)* The formation of the π bonds in ethylene. When the 2*p* orbitals are perpendicular to one another, there is no overlap and no bond formation. When the two —CH_2 groups rotate so that the 2*p* orbitals are parallel, a π bond forms.

lone pairs, two hybrid orbitals are needed. This suggests *sp* hybridization (see Table 10.2).

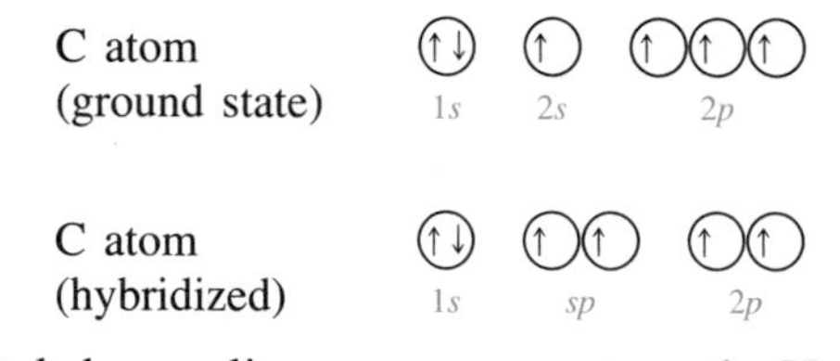

These *sp* hybrid orbitals have a linear arrangement, so the H—C—C—H geometry is linear. Bonds formed by the overlap of these hybrid orbitals are σ bonds. The two 2*p* orbitals not used to construct hybrid orbitals are perpendicular to the bond axis and to each other. They are used to form two π bonds. Thus, the carbon–carbon triple bond consists of one σ bond and two π bonds (see Figure 10.18).

Example 10.6 Applying Valence Bond Theory (Multiple Bonding)

Describe the bonding on a given N atom in dinitrogen difluoride, N_2F_2, using valence bond theory.

Solution

The electron-dot formula of N_2F_2 is

:F̤̈—N̈=N̈—F̤̈: (π marks the second N=N bond; σ marks the N—F and N=N single bonds)

Note that a double bond is described as a π bond plus a σ bond. Hybrid orbitals are needed to describe each σ bond and each lone pair (a total of three hybrid orbitals for each N atom). This suggests sp^2 hybridization (see Table 10.2). According to this description, one of the sp^2 hybrid orbitals is used to form the N—F bond, another to form the σ bond of N=N, and the third to hold the lone pair on the N atom. The 2*p* orbitals on each N atom overlap to form the π bond of N=N. Hybridization and bonding of the N atoms are shown as follows:

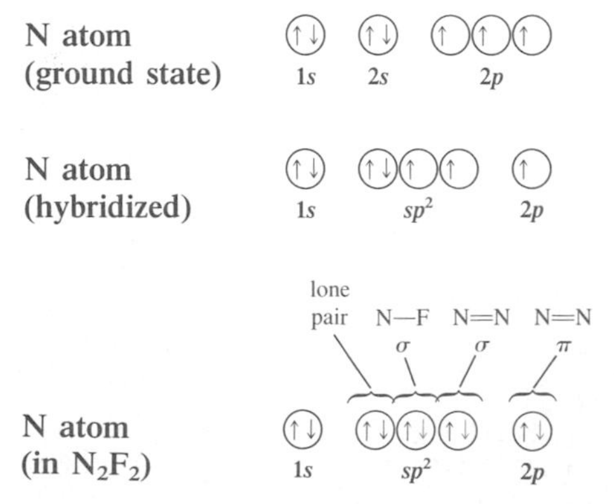

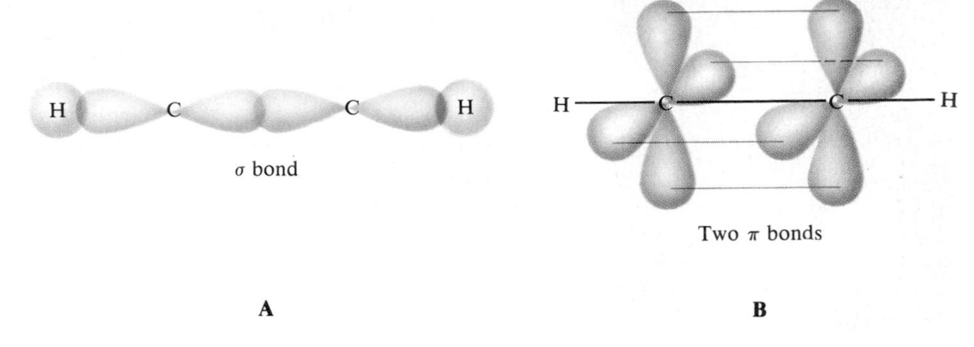

Figure 10.18
Bonding in acetylene. *(A)* The σ-bond framework. *(B)* The two π bonds.

Exercise 10.7

Describe the bonding on the carbon atom in carbon dioxide, CO_2, using valence bond theory. (See Problems 10.39 and 10.40.)

The π-bond description of the double bond agrees well with experiment. The *geometric,* or *cis–trans, isomers* of the compound 1,2-dichloroethene (described in the chapter opening) illustrate this. *Isomers* are compounds of the same molecular formula but with different arrangements of the atoms. (The numbers in the name 1,2-dichloroethene refer to the positions of the chlorine atoms. Thus, one chlorine atom is attached to carbon atom 1 and the other to carbon atom 2.) The structures of these isomers of 1,2-dichloroethene are

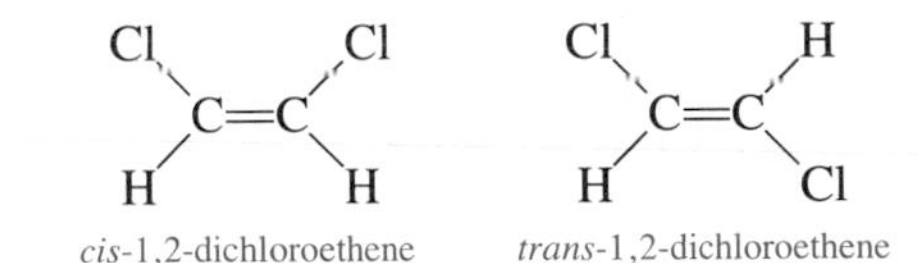

In order to transform one isomer into the other, one end of the molecule must be rotated as the other remains fixed. For this to happen, the π bond must be broken. Breaking the π bond requires considerable energy, so the *cis* and *trans* compounds are not easily interconverted. Contrast this with 1,2-dichloroethane (Figure 10.19), in which the two ends of the molecule can rotate without breaking any bonds. Here isomers corresponding to different spatial orientations of the two chlorine atoms cannot be prepared, because the two ends rotate freely with respect to one another. There is only one compound.

As we noted in the chapter opening, *cis* and *trans* isomers have different properties. The *cis* isomer of 1,2-dichloroethene boils at 60°C; the *trans* compound boils at 48°C. These isomers can also be differentiated on the basis of dipole moment. The *trans* compound has no dipole moment because it is symmetrical (the polar C—Cl bonds point in opposite directions and so cancel). However, the *cis* compound has a dipole moment of 1.85 D.

It is possible to convert one isomer to another if sufficient energy is supplied—say by chemical reaction. The role of the conversion of a *cis* isomer to its *trans* isomer in human vision is discussed in the Related Topic at the end of Section 10.7.

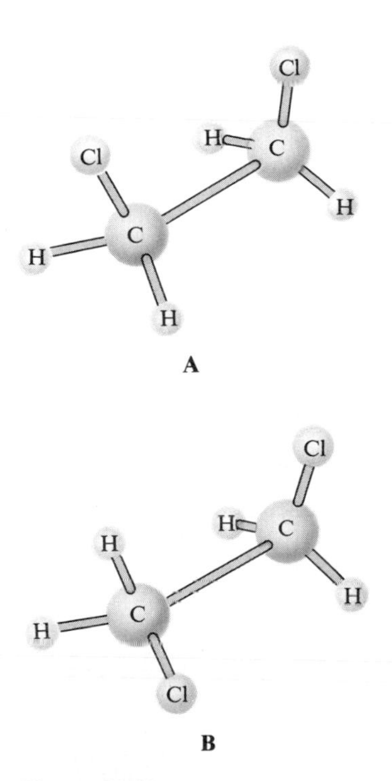

Figure 10.19
Lack of geometric isomers in 1,2-dichloroethane. Because of rotation about the carbon–carbon bond in 1,2-dichloroethane, geometric isomers are not possible. Note that the molecule pictured at the top can be twisted easily to give the molecule pictured at the bottom.

Exercise 10.8

Dinitrogen difluoride (see Example 10.6) exists as *cis* and *trans* isomers. Write structural formulas for these isomers and explain (in terms of the valence bond theory of the double bond) why they exist. (See Problems 10.41 and 10.42.)

Molecular Orbital Theory

Figure 10.20
Paramagnetism of oxygen, O_2. Liquid oxygen is poured between the poles of a strong magnet. Oxygen adheres to the poles, showing that it is paramagnetic.

In the preceding sections, we looked at a simple version of valence bond theory. Although simple valence bond theory satisfactorily describes most of the molecules we encounter, it does not apply to all of them. For example, according to this theory any molecule with an even number of electrons should be diamagnetic (not attracted to a magnet), because we assume the electrons to be paired and to have opposite spins. In fact, a few molecules with an even number of electrons are paramagnetic (attracted to a magnet), indicating that some of the electrons are not paired. The best-known example of such a paramagnetic molecule is O_2. Because of the paramagnetism of O_2, liquid oxygen sticks to a magnet when poured over it (Figure 10.20). Although valence bond theory can be extended to explain the electron structure of O_2, molecular orbital theory, an alternative bonding theory, provides a straightforward explanation of the paramagnetism of O_2.

Molecular orbital theory is *a theory of the electronic structure of molecules in terms of molecular orbitals, which may spread over several atoms or the entire molecule*. This theory views the electronic structure of molecules to be much like the electronic structure of atoms. Each molecular orbital has a definite energy. To obtain the ground state of a molecule, electrons are put into orbitals of lowest energy, consistent with the Pauli exclusion principle, just as in atoms.

10.5 PRINCIPLES OF MOLECULAR ORBITAL THEORY

We can think of a molecular orbital as being formed from a combination of atomic orbitals. As atoms approach each other and their atomic orbitals overlap, molecular orbitals are formed.

Bonding and Antibonding Orbitals

Consider the H_2 molecule. As the atoms approach to form the molecule, their 1*s* orbitals overlap. One molecular orbital is obtained by adding the two 1*s* orbitals (see Figure 10.21). Note that where the atomic orbitals overlap, their values sum to give a larger result. This means that in this molecular orbital, electrons are often

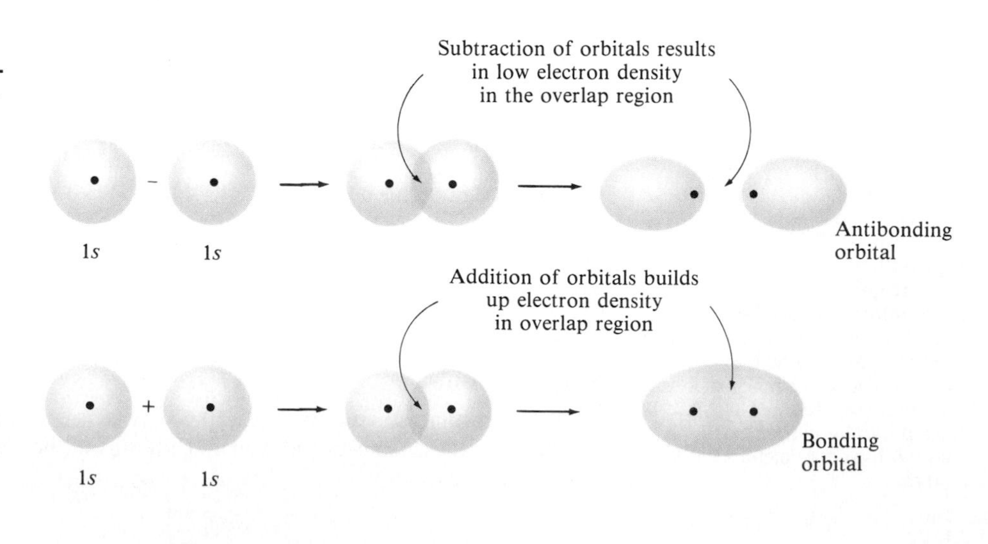
Subtraction of orbitals results in low electron density in the overlap region
−
1s
1s
Antibonding orbital
Addition of orbitals builds up electron density in overlap region
+
1s
1s
Bonding orbital

Figure 10.21
Formation of bonding and antibonding orbitals from 1s orbitals of hydrogen atoms. When the two 1*s* orbitals overlap, they can either add to give a bonding molecular orbital or subtract to give an antibonding molecular orbital.

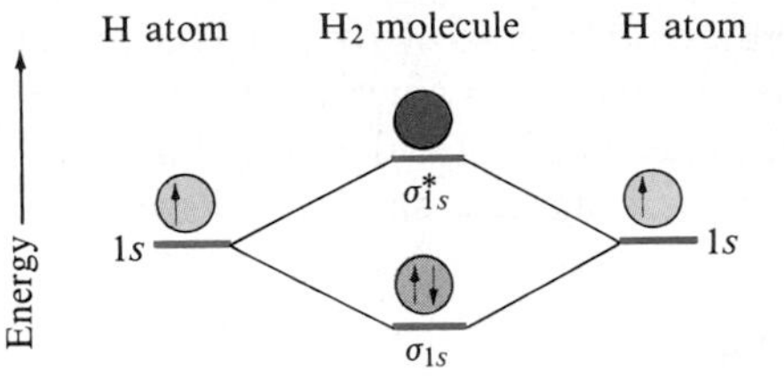

Figure 10.22
Relative energies of the 1s orbital of the H atom and the σ_{1s} and σ^*_{1s} molecular orbitals of H_2. Arrows denote occupation of the σ_{1s} orbital by electrons in the ground state of H_2.

found in the region between the two nuclei where they can hold them together. *Molecular orbitals that are concentrated in regions between nuclei* are called **bonding orbitals.** The bonding orbital in H_2, which we have just described, is denoted σ_{1s}. The σ (sigma) means that the molecular orbital has a cylindrical shape about the bond axis. The subscript 1*s* tells us that it was obtained from 1*s* atomic orbitals.

Another molecular orbital is obtained by subtracting the 1*s* orbital on one atom from the 1*s* orbital on the other (Figure 10.21). When the orbitals are subtracted, the resulting values in the region of the overlap are close to zero. This means that in this molecular orbital, the electrons spend little time between the nuclei. *Molecular orbitals that are concentrated in regions other than between two nuclei* are called **antibonding orbitals.** The antibonding orbital in H_2, which we have just described, is denoted σ^*_{1s}. The asterisk tells us that the molecular orbital is antibonding.

Figure 10.22 shows the energies of the molecular orbitals σ_{1s} and σ^*_{1s} relative to the atomic orbitals. Energies of the separate atomic orbitals are represented by heavy colored lines at the far left and right. The energies of the molecular orbitals are shown by heavy colored lines in the center. (These are connected by light gray lines to show which atomic orbitals were used to obtain the molecular orbitals.) Note that the energy of a bonding orbital is less than that of the separate atomic orbitals, whereas the energy of an antibonding orbital is higher.

We obtain the electron configuration for the ground state of H_2 by placing the two electrons (one from each atom) into the lower-energy orbital (see Figure 10.22). The orbital diagram is

(↑↓)( )
σ_{1s} σ^*_{1s}

and the electron configuration is $(\sigma_{1s})^2$. Because the energy of the two electrons is lower than their energies in the isolated atoms, the H_2 molecule is stable.

Configurations involving the σ^*_{1s} orbital describe excited states of the molecule. As an example of an excited state, we have the orbital diagram

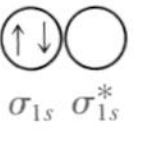

The corresponding electron configuration is $(\sigma_{1s})^1(\sigma^*_{1s})^1$.

We obtain a similar set of orbitals when we consider the approach of two helium atoms. To obtain the ground state of He_2, we note that there are four electrons, two from each atom, that would fill the molecular orbitals. Two of these electrons go into the σ_{1s} orbital and two go into the σ^*_{1s} orbital. The orbital diagram is

(↑↓)(↑↓)
σ_{1s} σ^*_{1s}

and the configuration is $(\sigma_{1s})^2(\sigma_{1s}^*)^2$. The energy decrease from the bonding electrons is offset by the energy increase from the antibonding electrons. Hence, He_2 is not a stable molecule. Molecular orbital theory therefore explains why the element helium exists as a monatomic gas, whereas hydrogen is diatomic.

Bond Order

The term *bond order* refers to the number of bonds that exist between two atoms. In molecular orbital theory, the bond order of a diatomic molecule is defined as one-half the difference between the number of electrons in bonding orbitals, n_b, and the number of electrons in antibonding orbitals, n_a.●

● For a Lewis formula, the bond order equals the number of electron pairs shared between two atoms (see Section 9.9).

$$\text{Bond order} = \tfrac{1}{2}(n_b - n_a)$$

For H_2, which has two bonding electrons, we have

$$\text{Bond order} = \tfrac{1}{2}(2 - 0) = 1$$

That is, H_2 has a single bond. For He_2, which has two bonding and two antibonding electrons, we have

$$\text{Bond order} = \tfrac{1}{2}(2 - 2) = 0$$

Bond orders need not be whole numbers; half-integral bond orders of $\tfrac{1}{2}$, $\tfrac{3}{2}$, and so forth are also possible. For example, the $H_2{}^+$ molecular ion, which is formed in mass spectrometers, has the configuration $(\sigma_{1s})^1$ and a bond order of $\tfrac{1}{2}(1 - 0) = \tfrac{1}{2}$.

Factors That Determine Orbital Interaction

The H_2 and He_2 molecules are relatively simple. A 1*s* orbital on one atom interacts with the 1*s* orbital on the other atom. But now let us consider the Li_2 molecule. Each Li atom has 1*s* and 2*s* orbitals. Which of these orbitals interact to form molecular orbitals? To find out, we need to understand the factors that determine orbital interaction.

The strength of the interaction between two atomic orbitals to form molecular orbitals is determined by two factors: (1) the energy difference between the interacting orbitals and (2) the magnitude of their overlap. *In order for the interaction to be strong, the energies of the two orbitals must be approximately equal and the overlap must be large.*

From this last statement, we see that when two Li atoms approach one another to form Li_2, only like orbitals on the two atoms interact appreciably. The 2*s* orbital of one lithium atom interacts with the 2*s* orbital of the other atom, but the 2*s* orbital from one atom does not interact with the 1*s* orbital of the other atom because their energies are quite different. Also, because the 2*s* orbitals are outer orbitals, they are able to overlap and interact strongly when the atoms approach. As in H_2, these atomic orbitals interact to give a bonding orbital (denoted σ_{2s}) and an antibonding orbital (denoted σ_{2s}^*). However, even though the 1*s* orbitals of the two atoms have the same energy, they do not overlap appreciably and so interact weakly (the difference in energy between σ_{1s} and σ_{1s}^* is very small). Figure 10.23 gives the relative energies of the orbitals.

We obtain the ground-state configuration of Li_2 by putting six electrons (three from each atom) into the molecular orbitals of lowest energy. The configuration of the diatomic molecule Li_2 is

$$Li_2 \qquad (\sigma_{1s})^2(\sigma_{1s}^*)^2(\sigma_{2s})^2$$

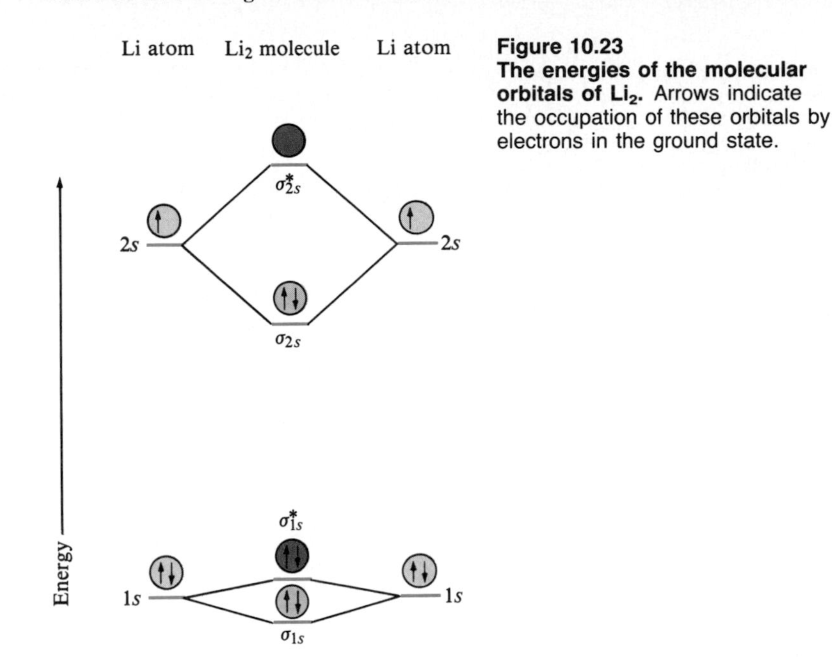

Figure 10.23
The energies of the molecular orbitals of Li_2. Arrows indicate the occupation of these orbitals by electrons in the ground state.

The $(\sigma_{1s})^2(\sigma_{1s}^*)^2$ part of the configuration is often abbreviated KK (which denotes the K shells of the two atoms). These electrons do not have a significant effect on bonding.●

● However, small shifts in the energy of the core electrons due to bonding can be measured by x-ray photoelectron spectroscopy (page 286).

$$Li_2 \qquad KK(\sigma_{2s})^2$$

In calculating bond order, we can ignore KK (it includes two bonding and two antibonding electrons). Thus, the bond order is 1, as in H_2.

In Be_2, the energy diagram is similar to that in Figure 10.23. We have eight electrons to distribute (four from each Be atom). The ground-state configuration of Be_2 is

$$Be_2 \qquad KK(\sigma_{2s})^2(\sigma_{2s}^*)^2$$

Note that the configuration has two bonding and two antibonding electrons outside the K shells. Thus, the bond order is $\frac{1}{2}(2-2) = 0$. No bond is formed, so the Be_2 molecule, like the He_2 molecule, is unstable.

10.6 ELECTRON CONFIGURATIONS OF DIATOMIC MOLECULES OF THE SECOND-PERIOD ELEMENTS

In the previous section, we looked at the electron configurations of some simple molecules: H_2, He_2, Li_2, and Be_2. These are **homonuclear diatomic molecules**—that is, *molecules composed of two like nuclei*. (**Heteronuclear diatomic molecules** are *molecules composed of two different nuclei*—for example, CO and NO.) To find the electron configurations of other homonuclear diatomic molecules, we need to have additional molecular orbitals.

We have already looked at the formation of molecular orbitals from *s* atomic orbitals. Now we need to consider the formation of molecular orbitals from *p* atomic orbitals. There are two different ways in which 2*p* atomic orbitals can

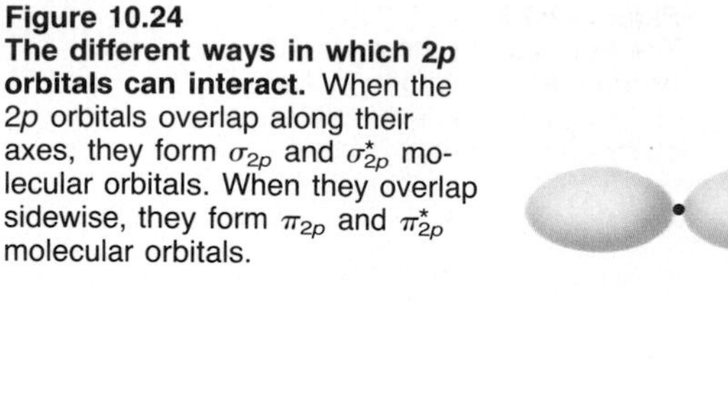

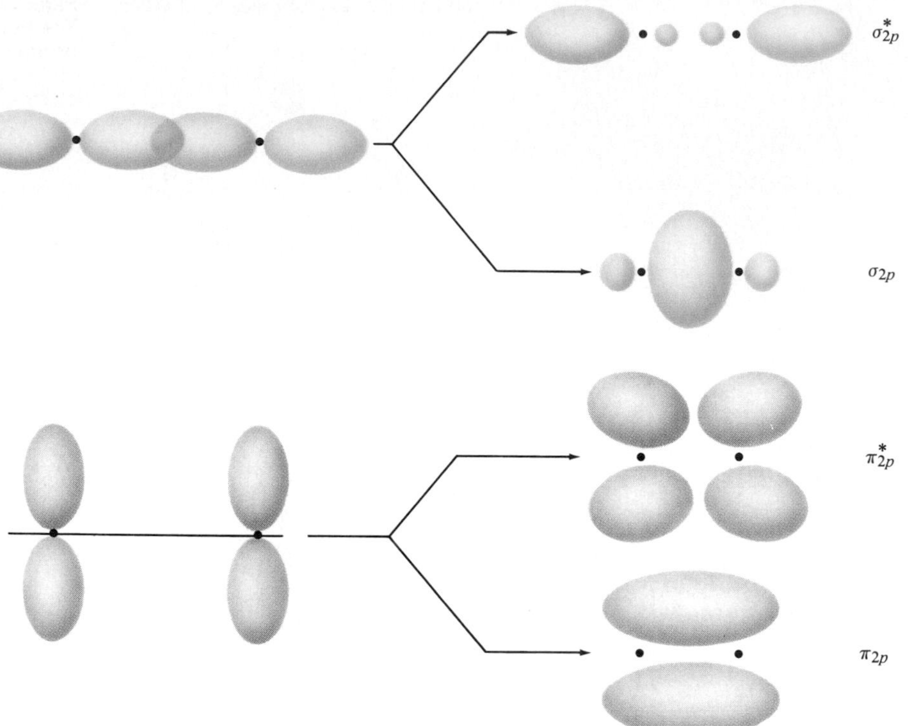

Figure 10.24
The different ways in which 2*p* orbitals can interact. When the 2*p* orbitals overlap along their axes, they form σ_{2p} and σ^*_{2p} molecular orbitals. When they overlap sidewise, they form π_{2p} and π^*_{2p} molecular orbitals.

interact. One set of 2*p* orbitals can overlap along their axes to give one bonding and one antibonding σ orbital (σ_{2p} and σ^*_{2p}). The other two sets of 2*p* orbitals then overlap sidewise to give two bonding and two antibonding π orbitals (π_{2p} and π^*_{2p}). See Figure 10.24.

Figure 10.25 shows the relative energies of the molecular orbitals obtained from 2*s* and 2*p* atomic orbitals. This order of molecular orbitals reproduces the known electron configurations of homonuclear diatomic molecules composed of

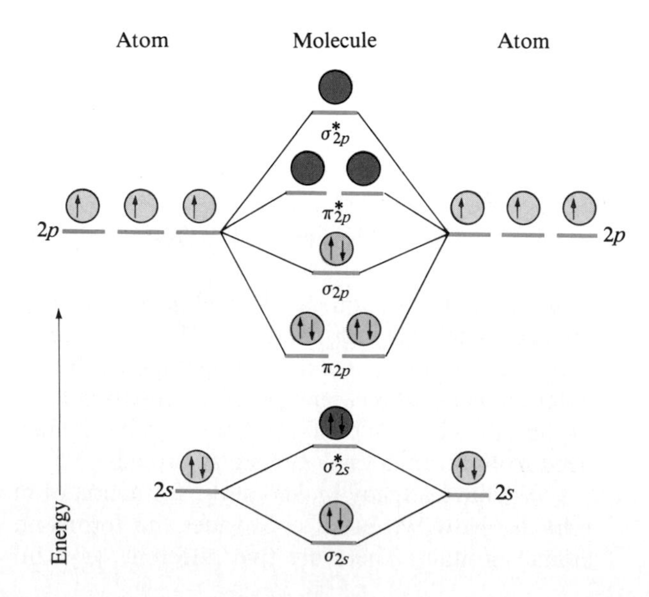

Figure 10.25
Relative energies of molecular orbitals of homonuclear diatomic molecules (excluding K shells). The arrows show the occupation of molecular orbitals by the valence electrons of N_2.

● This order gives the correct number of electrons in subshells, even though in O_2 and F_2 the energy of σ_{2p} is below π_{2p}.

elements in the second row of the periodic table.● The order of filling is

$$\sigma_{2s} \quad \sigma_{2s}^* \quad \pi_{2p} \quad \sigma_{2p} \quad \pi_{2p}^* \quad \sigma_{2p}^*$$

Note that there are two orbitals in the π subshell and two orbitals in the π^* subshell. Because each orbital can hold two electrons, a π or π^* subshell can hold four electrons. The next example shows how we can use the order of filling to obtain the orbital diagram, magnetic character, electron configuration, and bond order of a homonuclear diatomic molecule.

Example 10.7 Describing Molecular Orbital Configurations (Homonuclear Diatomic Molecules)

Give the orbital diagram of the O_2 molecule. Is the molecular substance diamagnetic or paramagnetic? What is the electron configuration? What is the bond order of O_2?

Solution

There are twelve valence electrons in O_2 (six from each atom), which occupy the molecular orbitals as shown in the following orbital diagram.

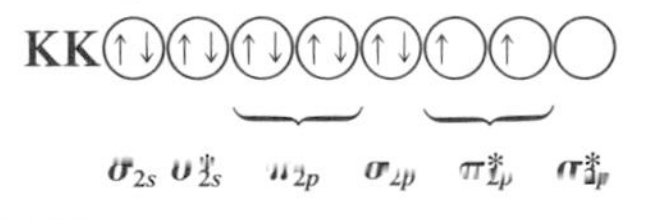

Note that the two electrons in the π_{2p}^* subshell must go into different orbitals with their spins in the same direction (Hund's rule). Because there are two unpaired electrons, the molecular substance is **paramagnetic**. The electron configuration is

$$KK(\sigma_{2s})^2(\sigma_{2s}^*)^2(\pi_{2p})^4(\sigma_{2p})^2(\pi_{2p}^*)^2$$

There are eight bonding electrons and four antibonding electrons. Therefore,

$$\text{Bond order} = \tfrac{1}{2}(8 - 4) = 2$$

Exercise 10.9

The C_2 molecule exists in the vapor phase over carbon at high temperature. Describe the molecular orbital structure of this molecule; that is, give the orbital diagram and electron configuration. Would you expect the molecular substance to be diamagnetic or paramagnetic? What is the bond order for C_2? (See Problems 10.43 and 10.44.)

Table 10.3 compares experimentally determined bond lengths, bond-dissociation energies, and magnetic character of the second-period homonuclear diatomic

Table 10.3 Theoretical Bond Orders and Experimental Data for the Second-Period Homonuclear Diatomic Molecules

Molecule	Bond Order	Bond Length (Å)	Bond-Dissociation Energy (kJ/mol)	Magnetic Character
Li_2	1	2.67	110	Diamagnetic
Be_2	0	*	*	*
B_2	1	1.59	290	Paramagnetic
C_2	2	1.24	602	Diamagnetic
N_2	3	1.10	942	Diamagnetic
O_2	2	1.21	494	Paramagnetic
F_2	1	1.42	155	Diamagnetic
Ne_2	0	*	*	*

The symbol * means no stable molecule has been observed.

molecules with the bond order calculated from molecular orbital theory, as in the preceding example. Note that as the bond order increases, bond length tends to decrease and bond-dissociation energy tends to increase. You should be able to verify that the experimentally determined magnetic character of the molecule, which is given in the last column of the table, is correctly predicted by molecular orbital theory.

When the atoms in a heteronuclear diatomic molecule are close to one another in the periodic table, the molecular orbitals have the same relative order of energies as those for homonuclear diatomic molecules. In this case we can obtain the electron configurations in the same way, as the next example illustrates.

Example 10.8 Describing Molecular Orbital Configurations (Heteronuclear Diatomic Molecules)

Write the orbital diagram for nitric oxide, NO. What is the bond order of NO?

Solution

We assume that the order of filling of orbitals is the same as for homonuclear diatomic molecules. There are eleven valence electrons in NO. Thus, the orbital diagram is

KK(↑↓)(↑↓)(↑↓)(↑↓)(↑↓)(↑)( )( )

σ_{2s} σ^*_{2s} π_{2p} σ_{2p} π^*_{2p} σ^*_{2p}

Because there are eight bonding and three antibonding electrons, we have

$$\text{Bond order} = \tfrac{1}{2}(8 - 3) = \tfrac{5}{2}$$

Exercise 10.10

Give the orbital diagram and electron configuration for the carbon monoxide molecule, CO. What is the bond order of CO? Is the molecule diamagnetic or paramagnetic?
(See Problems 10.45 and 10.46.)

When the two atoms in a heteronuclear diatomic molecule differ appreciably, we can no longer use the scheme appropriate for homonuclear diatomic molecules. The HF molecule is an example. Figure 10.26 shows the relative energies of the molecular orbitals that form. Sigma bonding and antibonding orbitals are formed by combining the 1*s* orbital on the H atom with the 2*p* orbital on the F atom that lies along the bond axis. The 2*s* orbital and the other 2*p* orbitals on fluorine remain as *nonbonding* orbitals (neither bonding nor antibonding). Note that the energy of the 2*p* subshell in fluorine is lower than the energy of the 1*s* orbital in hydrogen, because the fluorine electrons are more tightly held. As a result, the bonding orbital is made up of a greater percentage of the 2*p* fluorine orbital than of the 1*s* hydrogen orbital. This means that electrons in the bonding molecular orbital spend more time in the vicinity of the fluorine atom; that is, the H—F bond is polar, with the F atom having a small negative charge.

10.7 MOLECULAR ORBITALS AND DELOCALIZED BONDING

One of the advantages of using molecular orbital theory is the simple way in which it describes molecules with delocalized bonding. Whereas valence bond theory

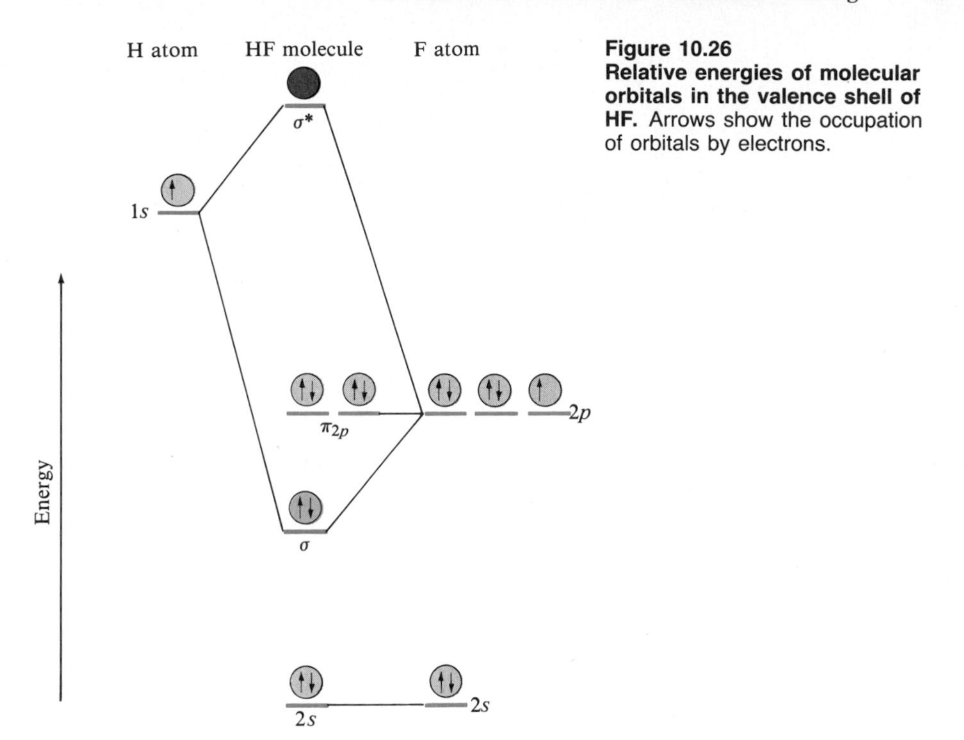

Figure 10.26
Relative energies of molecular orbitals in the valence shell of HF. Arrows show the occupation of orbitals by electrons.

requires two or more resonance formulas, molecular orbital theory describes the bonding in terms of a single electron configuration.

Consider the sulfur dioxide molecule as an example. Previously, we described this molecule in terms of resonance as

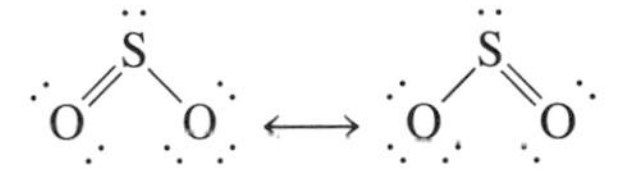

These formulas differ in the placement of the double bond and one of the lone pairs on an O atom. The part that is the same in both resonance formulas is

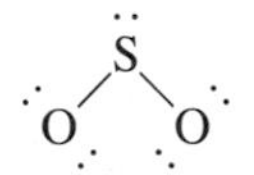

The molecular orbitals for this part of the electron structure are described as localized. The other two electron pairs, one pair of the double bond and one lone pair on an O atom, have a delocalized molecular-orbital description.

Because each atom in the localized framework has three electron pairs about it, we will assume that each atom uses sp^2 hybrid orbitals (see Table 10.2). The overlap of a hybrid orbital from sulfur with one from oxygen gives an S—O bond. Another hybrid orbital from sulfur overlaps a hybrid orbital of the oxygen atom to form the other S—O bond. This leaves one hybrid orbital on the sulfur atom and two on each oxygen to describe the lone pairs on these atoms.

After hybrid orbitals are formed, one unhybridized p orbital still remains on each atom. These three p orbitals are perpendicular to the plane of the molecule and parallel to one another. Thus, they can overlap sidewise to give three π molecular orbitals (three interacting atomic orbitals produce the same number of molecu-

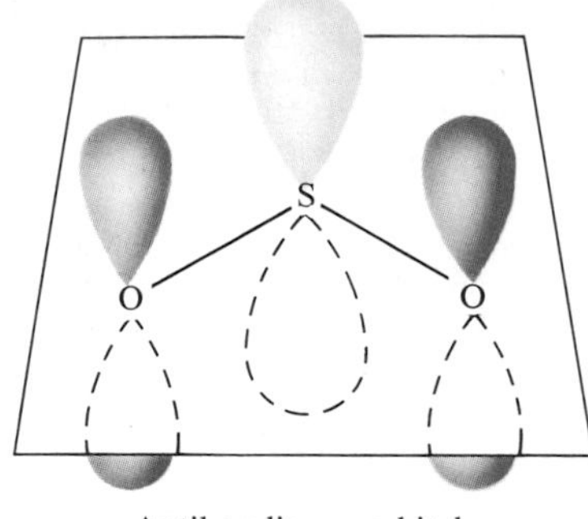

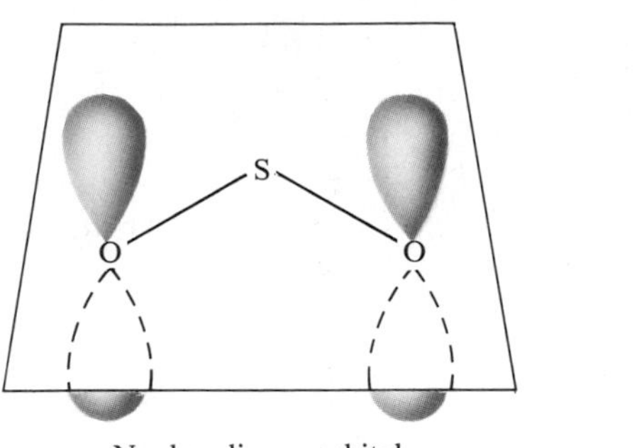

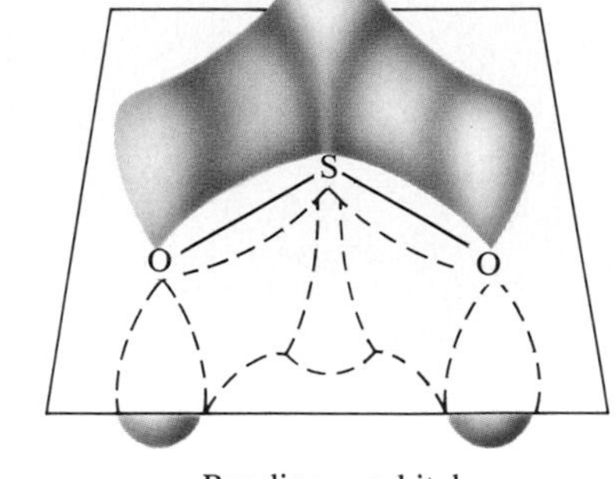

Figure 10.27
The electron distributions of the π molecular orbitals of SO_2. Note that one molecular orbital is nonbonding. It describes a lone pair of electrons that can be on either oxygen atom.

lar orbitals). All three will span the entire molecule (see Figure 10.27). One of these orbitals turns out to be bonding, one is nonbonding (that is, it has an energy equal to that of the isolated atoms), and the third is antibonding.

The bonding and nonbonding π orbitals are both doubly occupied. This agrees with the resonance description, in which one delocalized pair is bonding and the other is a lone pair on oxygen.

Related Topic

HUMAN VISION

Human vision, as well as the detection of light by other organisms, is known to involve a compound called 11-*cis*-retinal. The structural formula of this compound is shown in Figure 10.28, at the left. This compound, which is derived from vitamin A, is contained in two kinds of cells—rod cells and cone cells—located in the retina of the eye. Rod cells are responsible for vision in dim light. They give the sensation of light or dark but no sensation of color. Cone cells are responsible for color vision, but they require bright light. Color vision is possible because three types of cone cells exist: one type absorbing light in the red region of the spectrum, another absorbing in the green region, and another absorbing in the blue region. In each of these different cells, 11-*cis*-retinal is attached to a different protein molecule, which affects the region of light that is absorbed by retinal.

When a photon of light is absorbed by 11-*cis*-retinal, it is transformed to 11-*trans*-retinal (see Figure 10.28).

Figure 10.28
The cis–trans conversion of retinal on absorption of light. *Note the large movement of atoms caused by the rotation about the double bond between carbon atoms numbered 11 and 12.*

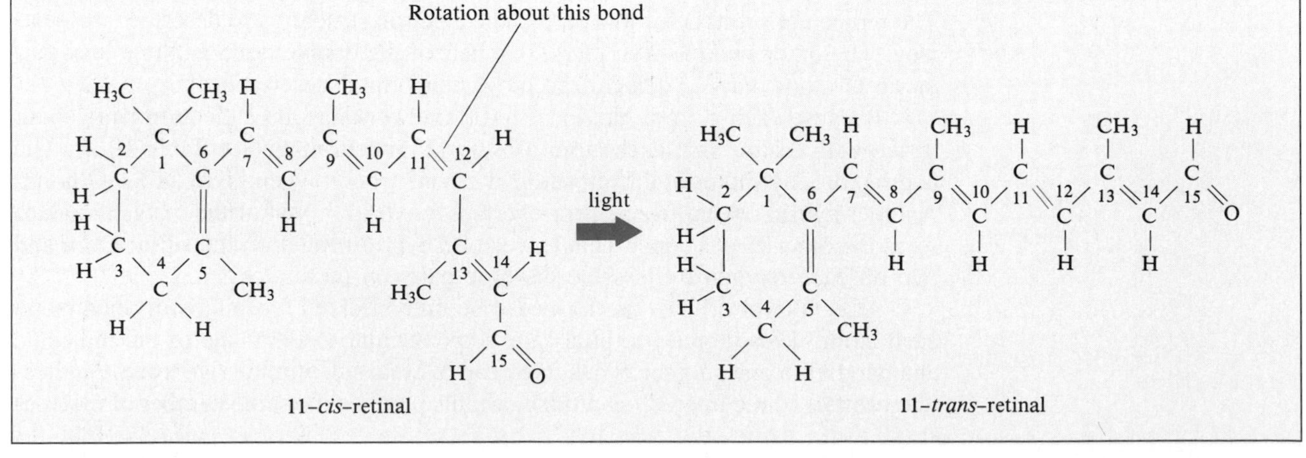

This small change at one carbon–carbon double bond results in a large movement of one end of the molecule with respect to the other. In other words, the absorption of a single photon results in a significant change in molecular geometry. This change in shape of retinal also affects the shape of the protein molecule to which it is attached, and *this* change results in a sequence of events, not completely understood, in which an electrical signal is generated.

To understand the *cis–trans* conversion when a photon is absorbed, consider the simpler molecule *cis*-1,2-dichloroethene (page 375). The double bond consists of a σ bond and a π bond. The π bond results from two electrons in a π-type molecular orbital. A corresponding antibonding orbital, π^*, has higher energy than the π orbital and is unoccupied in the ground state. When a photon of light in the ultraviolet region (near 180 nm) is absorbed by *cis*-1,2-dichloroethene, an electron in the π orbital undergoes a transition to the π^* orbital. Whereas there were previously two electrons in the bonding π orbital (contributing one bond), there is now one electron in the bonding orbital and one in the antibonding orbital (with no net bonding). As a result of the absorption of a photon, the double bond becomes a single bond and rotation about the bond is now possible. The *cis* isomer tends to rotate to the *trans* isomer, which is more stable. Then energy is lost as heat, and the electron in the π^* orbital undergoes a transition back to the π orbital; the single bond becomes a double bond again. The net result is the conversion of the *cis* isomer to the *trans* isomer.

Retinal differs from dichloroethene in having six double bonds alternating with single bonds instead of only one double bond. The alternating double and single bonds give a rigid structure, and when the middle double bond undergoes the *cis*-to-*trans* change, a large movement of atoms occurs. Another result of a long chain of alternating double and single bonds is a change in the wavelength of light absorbed in the π-to-π^* transition. Where dichloroethene absorbs in the ultraviolet region, retinal attached to its protein absorbs in the visible region (the attached protein also alters the wavelength of absorbed light). Many biological compounds that are colored are compounds consisting of long chains of alternating double and single bonds. β-Carotene, a yellow pigment in carrots, has this structure. It is broken down in the body to give vitamin A, from which retinal is derived.

10.8 MOLECULAR ORBITAL THEORY OF METALS AND SEMICONDUCTORS

In the previous section, we described the π molecular orbitals of SO_2 as delocalized, because instead of being concentrated between two atoms, they involve the entire molecule. A metal crystal may be thought of as a giant molecule in which some of the molecular orbitals encompass the entire crystal. Thus, metals are examples of extreme delocalized bonding, as we noted in Section 9.8, page 337. In a simplified way, we can think of a metal as a regular array of positive atomic cores surrounded by a ''sea'' of electrons from the valence shell. These electrons are free to move throughout the crystal, and a small voltage can attract the electrons in one direction to form an electric current. Molecular orbital theory can give us a detailed picture of this delocalized bonding. Because the energy levels of a metal, as we will see, are crowded together into ''bands'' of many levels, the molecular orbital theory of metals is often called *band theory*.

As an example of the molecular orbital description of a metal, consider a crystal of sodium. Imagine that we build the crystal by bringing sodium atoms together one at a time, and during this process we follow the formation of molecular orbitals and associated energy levels. Each isolated sodium atom has the electron configuration $1s^22s^22p^63s^1$ or $[Ne]3s^1$. When two sodium atoms approach each other, their $3s$ orbitals overlap to form two molecular orbitals (a bonding MO and an antibonding MO). Their neon cores remain essentially nonbonding. Now imagine that a third atom is brought up to this diatomic molecule. The three $3s$ orbitals overlap to form three molecular orbitals, each orbital encompassing the

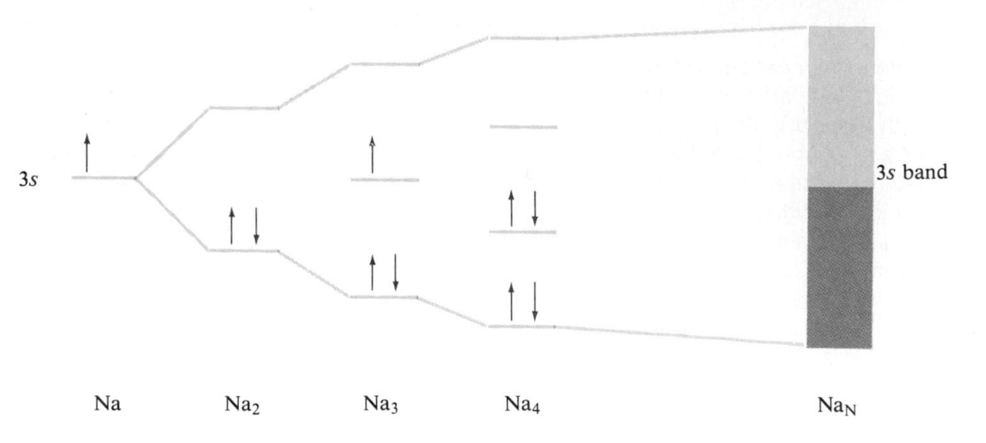

Figure 10.29
Formation of an energy band in sodium metal. Note that the number of energy levels grows until the levels merge into a continuous band of energies. The lower half of the band is occupied by electrons.

entire Na_3 molecule (much as the π orbitals of SO_2 encompass the entire molecule). When a fourth sodium atom is brought up to this assemblage, the four 3*s* orbitals of the atoms interact to form four molecular orbitals encompassing the entire molecule. When a large number N (on the order of Avogadro's number) of sodium atoms have been brought together to form a crystal, the atoms will have formed N molecular orbitals encompassing the entire crystal.

Figure 10.29 shows that at each stage the number of energy levels grows until the molecular orbital levels merge into a *band* of continuous energies. We will call this the 3*s* band of the sodium metal. A 3*s* band formed from N atoms will have N orbitals holding a maximum $2N$ electrons. Because each sodium atom has one valence electron, N atoms will supply N 3*s* electrons and will therefore half-fill the 3*s* band.

Electrons become free to move throughout a crystal when they are excited to unoccupied orbitals. In a metal, this requires very little energy, because there are unoccupied orbitals just above the occupied orbitals of highest energy. When a voltage is applied to a metal crystal, electrons are excited to the unoccupied orbitals and move toward the positive pole.

In addition to the half-filled 3*s* band in sodium, there is an unoccupied 3*p* band. Its presence does not alter our qualitative description of the sodium metal. To discuss the electrical conductivity of magnesium, however, we do need to discuss the 3*p* band. A magnesium atom has the configuration $[Ne]3s^2$. As in sodium, the 3*s* orbitals of magnesium metal overlap to form a 3*s* band. If this were the whole story, we would expect the $2N$ valence electrons of the atoms to completely fill the 3*s* band. Then when a small voltage is applied, the electrons would have no place to go. We would expect magnesium to be an insulator (a nonconducting material), but in fact it is a conductor.

The actual situation in magnesium is depicted in Figure 10.30. The 3*p* orbitals of the magnesium atoms interact, spreading the energy levels so that the bottom of the 3*p* band merges with the top of the 3*s* band. Imagine electrons filling the 3*s* band. When the electrons reach the energy where the two bands have merged, they begin to fill orbitals in both bands. As a result, the 3*s* and 3*p* bands of magnesium metal are only partially filled by the time we have accounted for all $2N$ valence electrons. Therefore, when a voltage is applied to the metal, the highest-energy electrons are easily excited into the unoccupied orbitals, giving an electrical conductor.

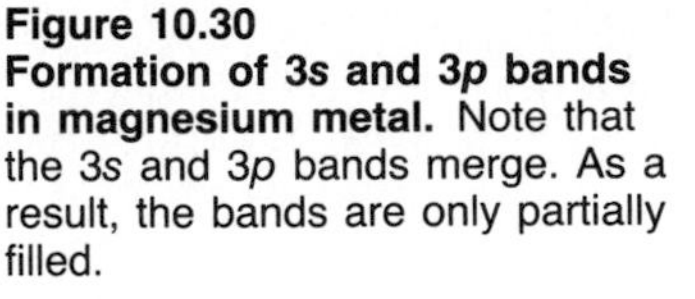
Figure 10.30
Formation of 3*s* and 3*p* bands in magnesium metal. Note that the 3*s* and 3*p* bands merge. As a result, the bands are only partially filled.

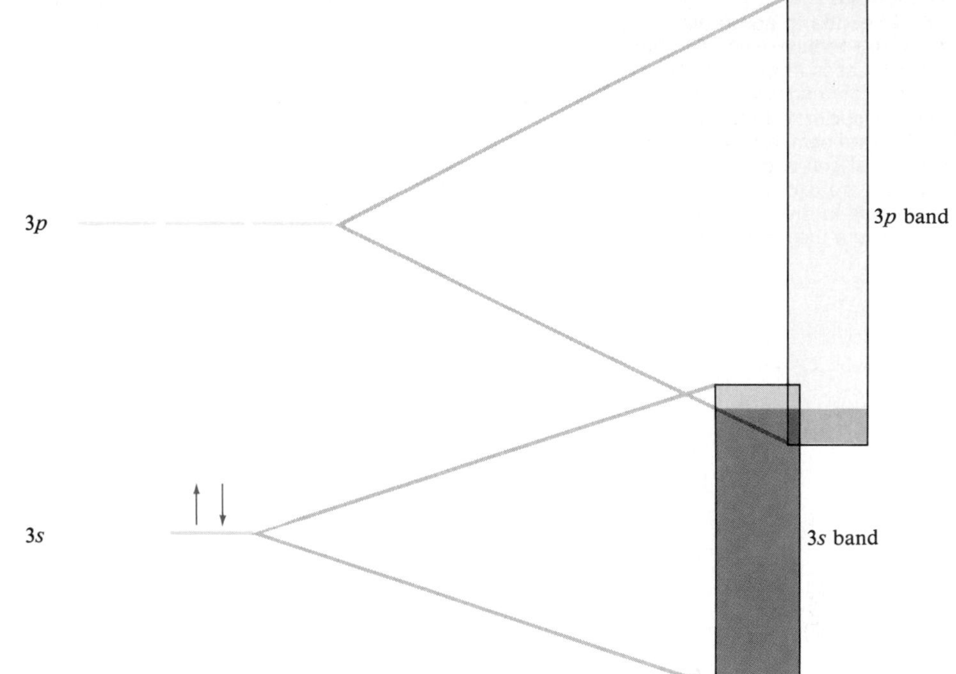

The band structure of crystals also explains the semiconducting ability of certain solids. A **semiconductor** is *a substance that is only slightly conducting at room temperature but becomes conducting at higher temperature,* although it is not normally as conducting as a metal. Metals, by contrast, become less conducting as the temperature is raised.

Consider the three solids diamond, silicon, and germanium. Diamond consists of carbon atoms, each bonded to four other atoms, with tetrahedral bonds. The other two solids have similar diamond-like structures. The band structures are also similar. In each, the highest-energy electrons occupy a completely filled band, which is separated from the lowest unoccupied band by a gap in the energy (the *band gap*). The band structures in these solids are shown in Figure 10.31.

For any of these solids to become conducting, electrons must be excited from the highest filled band (the *valence band*) to the lowest unoccupied band (the *conduction band*). If the band gap is not too large, the electrons can be excited to the conduction band by warming the solid. Silicon and germanium become significantly conducting if warmed. Diamond, however, is essentially a nonconductor, because of the very large band gap.

We can now understand the difference in resistance between a metal and a semiconductor as temperature changes. In a metal, the greater resistance to electrical conduction at higher temperatures arises from an increase in collisions of the conduction electrons with atomic cores. As the temperature increases, the atomic cores vibrate through greater amplitudes, causing more electron collisions and therefore greater electrical resistance. The same effect occurs in semiconductors, but it is normally dominated by greater conduction from the increase in conduction electrons obtained at higher temperatures.

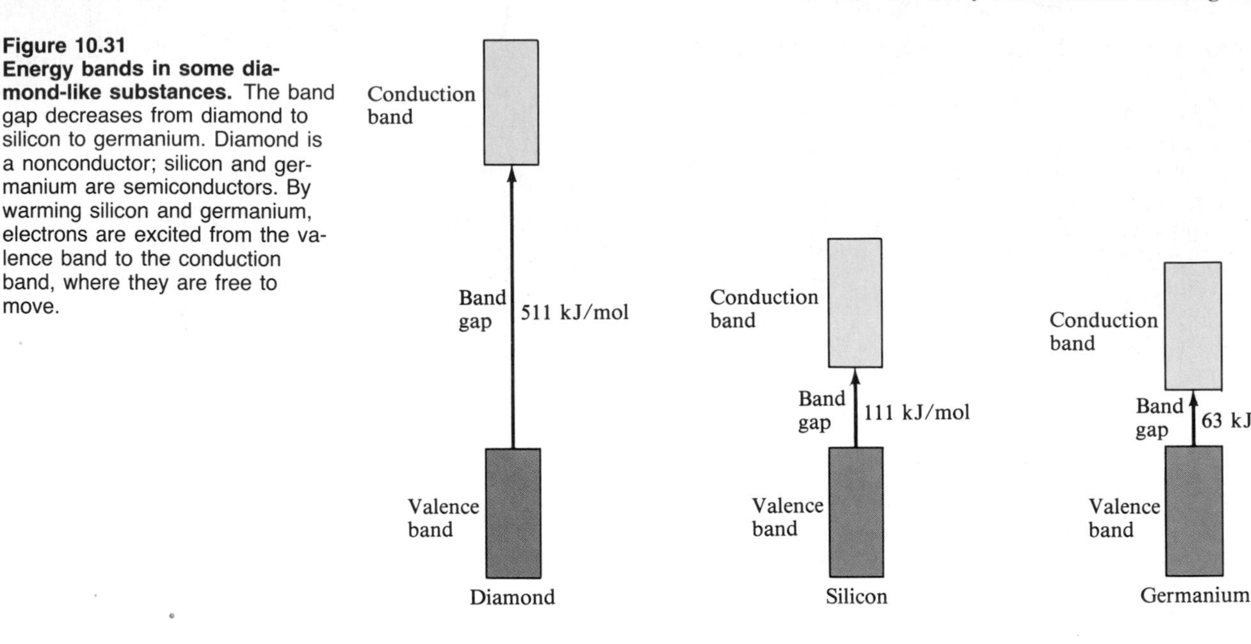

Figure 10.31
Energy bands in some diamond-like substances. The band gap decreases from diamond to silicon to germanium. Diamond is a nonconductor; silicon and germanium are semiconductors. By warming silicon and germanium, electrons are excited from the valence band to the conduction band, where they are free to move.

Related Topic

SUPERCONDUCTIVITY

A *superconductor* is a material that abruptly loses its resistance to an electric current when cooled to a definite characteristic temperature. This means that an electric current will flow in a superconductor without heat loss, unlike the current in a typical conductor. Once a current has been started in a superconducting circuit, it continues to flow indefinitely. Another intriguing property of a superconductor is its perfect diamagnetism. This means that the superconductor completely repels magnetic field lines. Figure 10.32 shows a magnet suspended in midair over one of the newly discovered ceramic superconductors. The magnet seems to levitate—it appears to hover in air in defiance of gravity. In fact, the repulsion of magnetic field lines by the superconductor holds the magnet aloft.

Superconductivity was discovered in 1911 by the Dutch physicist Heike Kamerlingh Onnes soon after he found a way to liquefy helium. By evaporating liquid helium, he could obtain temperatures near absolute zero. Kamerlingh Onnes found that mercury metal suddenly loses all resistance to an electric current when cooled to 4 K. Superconductors first became useful when a niobium metal alloy was found to become superconducting at about 23 K and to remain superconducting even when large currents flow through it (many superconducting

Figure 10.32
Levitation of a magnet by a superconductor. *The magnet (samarium-cobalt alloy) is supported above the ceramic superconductor (approximate formula is $YBa_2Cu_3O_7$). The ceramic becomes superconducting when it is cooled by liquid nitrogen.*

materials lose their superconductivity when even moderate currents flow through them). It became possible to construct superconducting magnets with high magnetic fields by starting an electric current in a superconducting circuit. Such magnets are being used in medical magnetic resonance imaging (see Instrumental Methods: Nuclear Magnetic Resonance in Chapter 8). Even though expensive liquid helium (at more than $10 per gallon) is needed to operate these magnets, they are still cheaper to use than the usual electromagnets because of their much lower power requirements.

In 1986, Johannes Georg Bednorz and Karl Alexander Müller of IBM discovered that certain copper oxide materials become superconducting at 30 K, and within months researchers had found similar materials that become superconducting at 125 K. This means that superconductors could be operated using a cheap refrigerant, such as liquid nitrogen at a few cents per gallon. Perhaps materials can even be found that are superconducting at room temperature! What remains is to find how to fabricate such superconducting materials into wires and similar objects with the ability to carry large currents. It would then be possible to construct transmission lines to carry electricity long distances without energy loss (present transmission lines dissipate between 3% and 10% of their transmitted energy as heat). Trains magnetically levitated several centimeters above the track so that they can ride at high speeds on a cushion of air become practical possibilities (Figure 10.33). The new superconductors may change radically many areas of technology.

Figure 10.33
Experimental train employing magnetic levitation. *A train in Japan uses superconducting magnets to lift it so that it rides on a cushion of air.*

Profile of a Chemical

OZONE (an Absorber of Ultraviolet Radiation in the Stratosphere)

On bright sunny days, you may notice the fresh-air smell of ozone (trioxygen, O_3) in extremely low concentrations. However, the name ozone derives from the Greek word *ozein* meaning "to smell," because the normal odor associated with ozone, at slightly higher but still low concentrations, is disagreeable. You may detect the odor around electrical equipment and faulty fluorescent lighting. Ozone is also one component of *photochemical smog,* a type of smog produced when sunlight interacts with a mixture of nitrogen oxides and hydrocarbons in polluted air.

Ozone is a faintly blue gas, produced commercially by passing ultraviolet radiation or an electrical discharge through dioxygen (normal oxygen, O_2) (Figure 10.34).

$$3O_2(g) \xrightarrow[\text{discharge}]{\text{electrical}} 2O_3(g)$$

It is used as a water disinfectant (especially in Europe) and a bleach. The gas is stable except at elevated temperatures or in the presence of a catalyst, and then it decomposes to dioxygen.

The electronic structure of ozone may be described by the resonance formulas

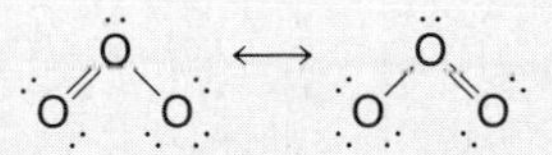

giving the oxygen–oxygen bond in O_3 partial double-bond character. This partial double-bond character is revealed when we compare the oxygen–oxygen bond lengths in hydrogen peroxide (single bond, 1.49 Å), ozone (1.28 Å), and dioxygen (double bond, 1.21 Å). The O—O—O bond angle is 117°, which is close to that predicted from the VSEPR model or from sp^2 hybridization (120°).

Ozone is an essential component of the *stratosphere,* a region of the atmosphere beginning at about 15 km (9 miles) above the surface of the earth. The lower portion of the atmosphere, near the surface of the

(continued)

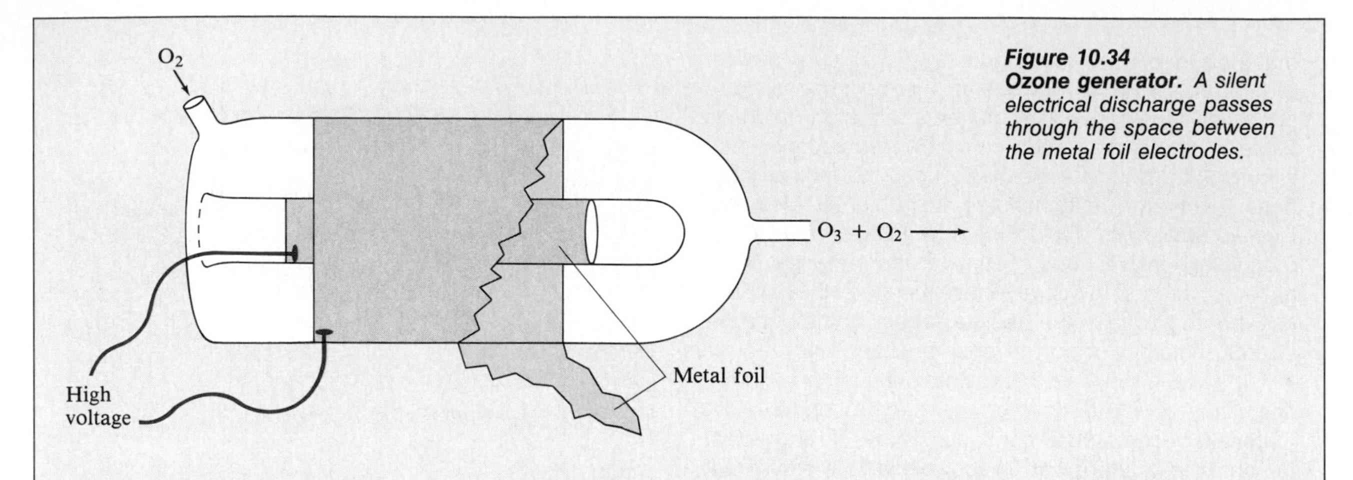

Figure 10.34
Ozone generator. *A silent electrical discharge passes through the space between the metal foil electrodes.*

earth, is called the *troposphere.* If we move upward through the troposphere, we find that the temperature generally decreases. At 10 to 15 km, however, the temperature begins to rise. This temperature increase gives rise to the stratosphere and is caused by the absorption of ultraviolet radiation between 200 nm and 300 nm by ozone, which is broken into O_2 and O by the radiation. Ozone is formed in the stratosphere by the reaction of O_2 with oxygen atoms.

$$O_2(g) + O(g) \longrightarrow O_3(g)$$

Oxygen atoms form in the stratosphere when O_2 absorbs short wavelength ultraviolet radiation (less than 200 nm).

If we were to take this ozone and compress it to 1 atm at 0°C, it would occupy a layer around the earth about 3 mm thick. Despite this seemingly small amount, ozone in the stratosphere is vitally important to us. Radiation from the sun contains ultraviolet radiation of short wavelengths, which are harmful to biological organisms. Fortunately, these harmful wavelengths are absorbed before they reach the surface of the earth. The most energetic are absorbed by O_2 in the earth's upper atmosphere. Less energetic but still harmful radiation is absorbed by the ozone in the stratosphere. Biologists believe that living organisms appeared on land only after the formation of sufficient ozone in the stratosphere about 600 million years ago. Figure 10.35 shows the

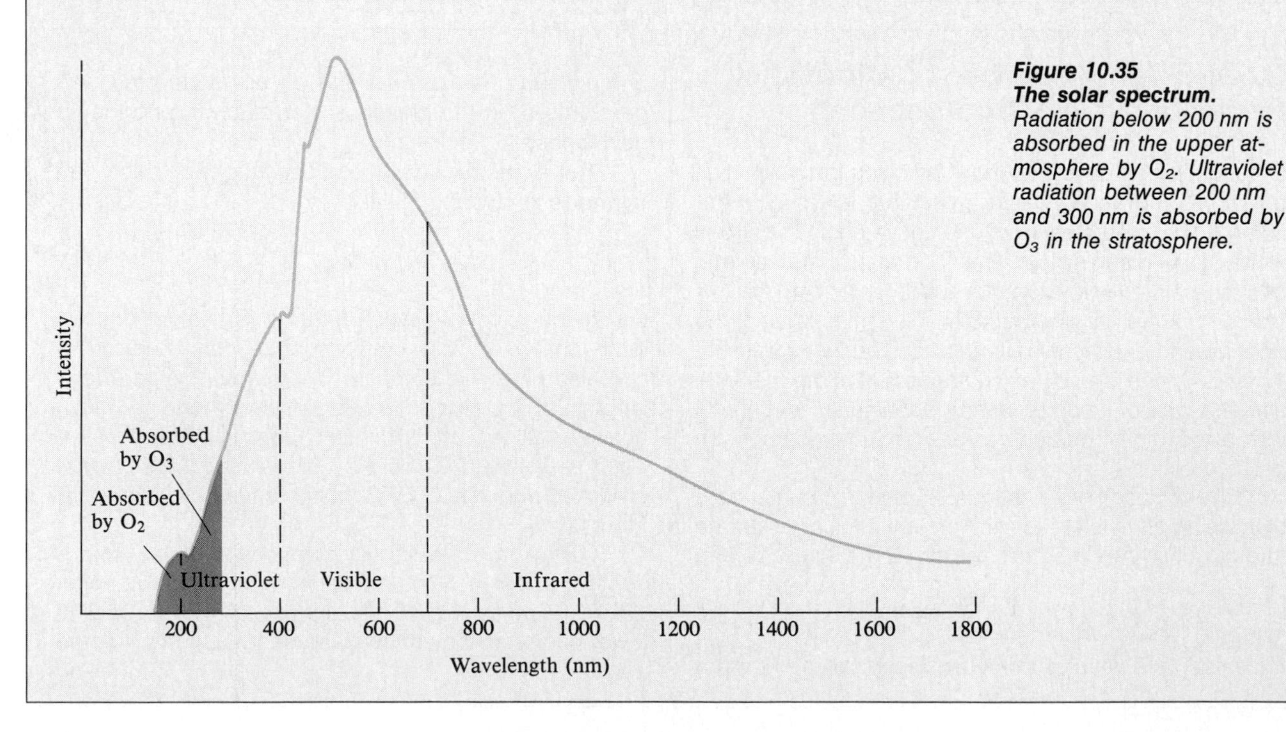

Figure 10.35
The solar spectrum. *Radiation below 200 nm is absorbed in the upper atmosphere by O_2. Ultraviolet radiation between 200 nm and 300 nm is absorbed by O_3 in the stratosphere.*

spectrum of radiation from the sun; it also shows those wavelengths absorbed in the outer atmosphere by dioxygen, O_2, and those absorbed by ozone, O_3, in the stratosphere.

In 1974, Mario J. Molina and F. Sherwood Rowland expressed concern that chlorofluorocarbons (CFCs), such as $CClF_3$ and CCl_2F_2, would be a source of chlorine atoms that could catalyze the decomposition of ozone in the stratosphere, so that it would be destroyed faster than it could be produced. The chlorofluorocarbons are relatively inert compounds used as refrigerants, spray-can propellants, and blowing agents (substances used to produce plastic foams). As a result of their inertness, however, they concentrate in the atmosphere, where they steadily rise into the stratosphere. Once in the stratosphere, ultraviolet light decomposes them to form chlorine atoms, which react with ozone to form ClO and O_2. The ClO molecules react with oxygen atoms present in the stratosphere to regenerate Cl atoms.

$$\begin{array}{l} Cl(g) + O_3(g) \longrightarrow ClO(g) + O_2(g) \\ ClO(g) + O(g) \longrightarrow Cl(g) + O_2(g) \\ \hline O_3(g) + O(g) \longrightarrow 2O_2(g) \end{array}$$

The net result is the decomposition of ozone to dioxygen. Chlorine atoms are consumed in the first step but regenerated in the second. Thus, chlorine atoms are not used up, so they function as a catalyst.

In 1985, British researchers reported that the ozone over the Antarctic had been declining precipitously each spring for several years, although it returned the following winter (Figure 10.36). Then in August 1987, an expedition was assembled to begin flights through the "ozone hole" with instruments. Researchers found that wherever ozone is depleted in the stratosphere, chlorine monoxide (ClO) appears, as expected if chlorine atoms are catalyzing the depletion (see the previous reactions). The chlorofluorocarbons are the suspected source of these chlorine atoms. In September 1987, 24 nations signed a pledge in Montreal, Canada, to freeze CFC production at 1986 levels and cut production 50% by 1999. Since then, additional nations have signed the pledge and many have agreed to an even faster phaseout.

Questions for Study

1. List some physical properties of ozone.
2. What are the commercial uses of ozone?
3. Describe the bonding in ozone. What is the geometry?
4. Describe the formation of ozone in the stratosphere.
5. Why is the presence of ozone of vital importance to us?
6. Biologists believe that life appeared on land no earlier than 600 million years ago. Why do they believe that land plants and other organisms could not have appeared before this?
7. What are the reactions responsible for the catalysis of ozone to dioxygen?
8. In 1987, airplanes flew through the Antarctic stratosphere measuring the quantity of ClO. What was the result of these measurements and what is their significance?

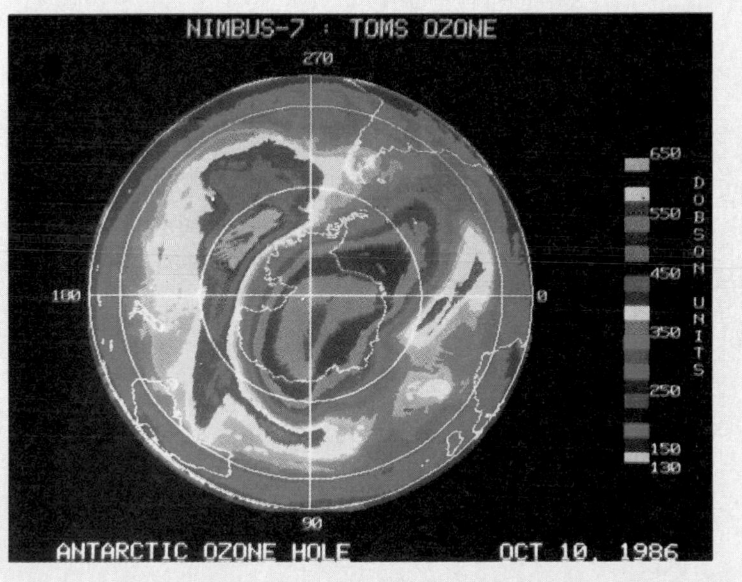

Figure 10.36
Computer map of the South Polar region showing total stratospheric ozone. *Data were obtained by the Total Ozone Mapping Spectrometer (TOMS) aboard the* Nimbus-7 *satellite in October 1986. Total ozone at any point is given in Dobson units, which equal the thickness of total ozone at 1 atm, 0°C in 10^{-2} mm. The ozone hole is shown in purple, in the region near the South Pole (Antarctica).*

Profile of a Chemical

SILICON (a Semiconducting Element)

Nearly three-fourths of the earth's crust is composed of just two elements, oxygen (45% by mass) and silicon (27% by mass). These elements form the strong Si—O bonds present in silicates, the chemical substances of the earth's rocks. One of the more common silicate minerals is quartz. It is a form of silicon dioxide or silica, SiO_2, in which each silicon atom bonds tetrahedrally to four oxygen atoms and each oxygen atom bonds to two silicon atoms, giving a strong three-dimensional structure. Pure quartz is colorless, but the mineral frequently occurs in various colors as a result of impurities. Amethyst is a purple-colored quartz used as a gem (Figure 10.37). The color is caused by Fe^{3+} ion impurities.

Pure silicon is a dark gray solid with a metallic shine or luster. Despite the prevalence of silicates, the element was not isolated until 1823 (by the Swedish chemist Jöns Jacob Berzelius). Today silicon is prepared in large quantities by heating white sand, which is fine quartz crystals, with petroleum coke (carbon) in an electric furnace to about 3000°C.

$$SiO_2(s) + 2C(s) \xrightarrow{\Delta} Si(s) + 2CO(g)$$

Commercial-grade silicon is about 98% pure and is used in various metal alloys.

Silicon is the basic material of the solid-state electronics industry. Television receivers, microcomputers, and other electronic equipment so common today employ miniature electrical circuits built on silicon chips (Figure 10.38). For this purpose, extremely pure silicon

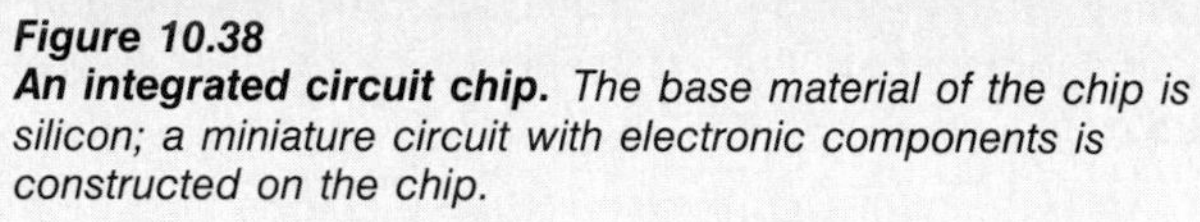

Figure 10.38
An integrated circuit chip. *The base material of the chip is silicon; a miniature circuit with electronic components is constructed on the chip.*

is required (containing about 1 impurity atom in 10^{12} silicon atoms).

The preparation of pure silicon is based on the fact that the chlorides of silicon, such as silicon tetrachloride, $SiCl_4$, or trichlorosilane, $SiHCl_3$, are volatile liquids, which can be purified by repeated distillation. Silicon tetrachloride is obtained by reacting industrial-grade silicon with chlorine.

$$Si(s) + 2Cl_2(g) \xrightarrow{\Delta} SiCl_4(g)$$

After purification, silicon tetrachloride is converted to silicon by reaction with hydrogen.

$$SiCl_4(g) + 2H_2(g) \longrightarrow Si(s) + 4HCl(g)$$

Trichlorosilane, $SiHCl_3$, is obtained as a by-product in the manufacture of silicones, which are organic compounds of silicon. It also can be converted to silicon by reaction with hydrogen.

Silicon is used in solid-state electronics devices because of its properties as a semiconductor, a material whose conductivity is intermediate between that of an insulator and a metal. Pure silicon at normal temperatures is essentially a nonconductor. Each of the atoms in silicon is bonded to four other silicon atoms through localized covalent bonds (Figure 10.39A). Because the electrons in these bonds do not move over large distances, there are no free electrons to conduct an electric current. The addition of small quantities of certain substances to pure silicon, however, greatly enhances its conductivity and makes possible the construction of

Figure 10.37
Amethyst crystals. *Amethyst is a quartz with purple color used as a gem. The purple color results from an Fe^{3+} ion impurity.*

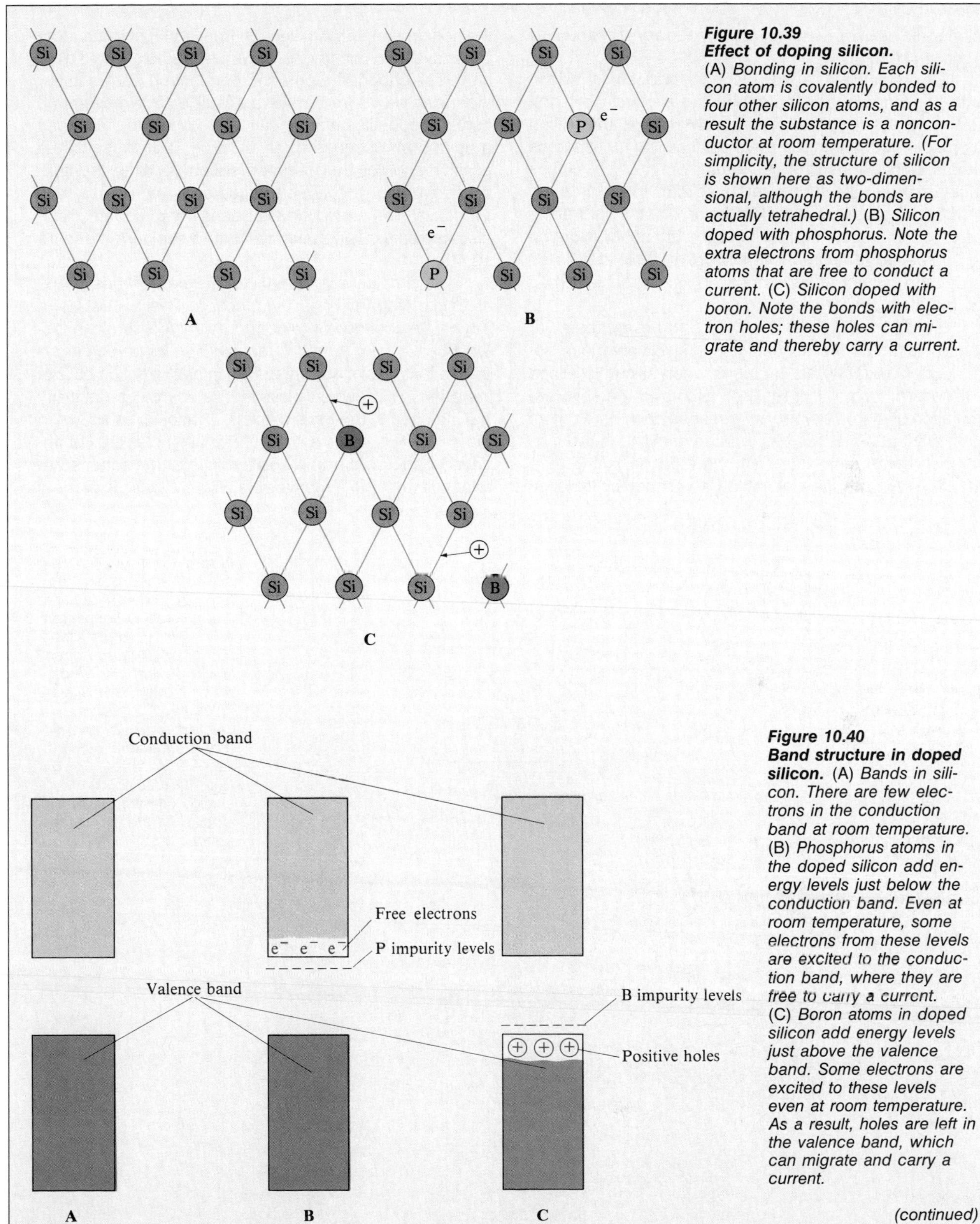

Figure 10.39
Effect of doping silicon.
(A) *Bonding in silicon. Each silicon atom is covalently bonded to four other silicon atoms, and as a result the substance is a nonconductor at room temperature. (For simplicity, the structure of silicon is shown here as two-dimensional, although the bonds are actually tetrahedral.)* (B) *Silicon doped with phosphorus. Note the extra electrons from phosphorus atoms that are free to conduct a current.* (C) *Silicon doped with boron. Note the bonds with electron holes; these holes can migrate and thereby carry a current.*

Figure 10.40
Band structure in doped silicon. (A) *Bands in silicon. There are few electrons in the conduction band at room temperature.* (B) *Phosphorus atoms in the doped silicon add energy levels just below the conduction band. Even at room temperature, some electrons from these levels are excited to the conduction band, where they are free to carry a current.* (C) *Boron atoms in doped silicon add energy levels just above the valence band. Some electrons are excited to these levels even at room temperature. As a result, holes are left in the valence band, which can migrate and carry a current.*

(continued)

electronic devices. This controlled addition of impurities is called *doping.*

Consider what happens when we dope pure silicon with phosphorus, an element having five instead of the four valence electrons of silicon. A few of the silicon atoms in the structure are replaced by phosphorus atoms. Because each phosphorus atom has five valence electrons, one electron is left over after four bonds are formed to silicon atoms. The extra electron is free to conduct an electric current, and the phosphorus-doped silicon becomes a conductor (Figure 10.39B). It is called an n-*type semiconductor,* because the current is carried by negative charges (electrons).

By doping silicon with an element having three valence electrons, the conductivity is also very much enhanced. Consider what happens when silicon is doped with boron. Some of the silicon atoms in the solid are replaced by boron atoms; but because each boron atom has only three valence electrons, one of the four bonds to each boron atom has only one electron in it (Figure 10.38C). We can think of this as a vacancy or "hole" in the bonding orbital. An electron from a neighboring atom can move in to occupy this hole. Then a hole would exist on the neighboring atom, and an electron from another atom can move into it. As a result of this movement, boron-doped silicon is an electrical conductor. Because a hole is an absence of an electron, it is essentially a positive charge. Boron-doped silicon is called a p-*type semiconductor,* because the charge is carried by positive holes. The semiconductor behavior of doped silicon also can be explained in molecular orbital terms (Figure 10.40).

The simplest semiconductor device depends on the electrical property of a *p–n junction.* This consists of a *p*-type semiconductor joined to an *n*-type semiconductor. Such a *p–n* junction can function as a *rectifier,* a device that allows current to flow in one direction but not the other. Thus, when wires from a source of electricity are attached to the ends of the *p–n* junction as shown in Figure 10.41A, a current flows. Positive holes are continuously generated at the positive wire and electrons are released from the negative wire. The two charge carriers

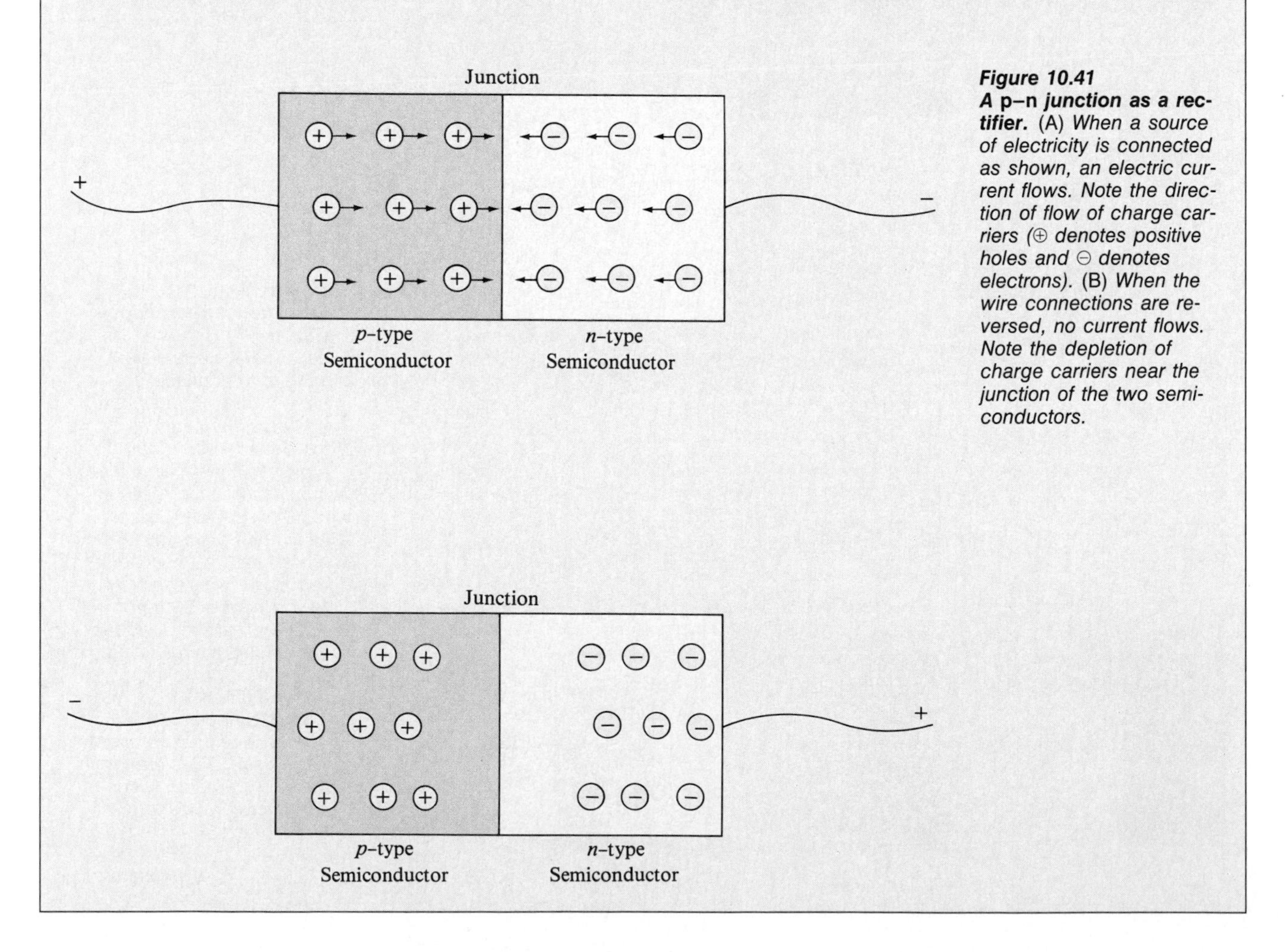

Figure 10.41
A* p–n *junction as a rectifier. (A) *When a source of electricity is connected as shown, an electric current flows. Note the direction of flow of charge carriers (⊕ denotes positive holes and ⊖ denotes electrons).* (B) *When the wire connections are reversed, no current flows. Note the depletion of charge carriers near the junction of the two semiconductors.*

(positive holes and electrons) move toward the junction where they combine. Now suppose the two wires from the electrical source are reversed with the negative wire attached to the *p*-type semiconductor and the positive wire attached to the *n*-type semiconductor (Figure 10.41B). The positive holes in the *p*-type semiconductor are attracted to the negative wire and begin to move away from the junction of the two semiconductors. At the same time, electrons in the *n*-type semiconductor are attracted to the positive wire and also begin to move away from the junction of the two semiconductors. The region of the junction becomes depleted in charge carriers (electrons and positive holes), and as a result the *p–n* junction becomes nonconducting. No current flows through the circuit.

By cleverly combining *p*-type and *n*-type semiconductors, one can construct various kinds of *transistors,* devices for controlling electric signals in complex ways. Some of the latest computer chips have as many as a million transistors per square centimeter of surface area (Figure 10.38).

Questions for Study

1. What are the two most abundant elements in the earth's crust?
2. Describe the arrangement of the atoms in quartz, SiO_2.
3. Describe the physical appearance of silicon. How is it prepared commercially?
4. List two uses of silicon.
5. Describe how ultrapure silicon is prepared from industrial-grade silicon. (What chemical reactions are used?)
6. Describe how *n*-type semiconductors and *p*-type semiconductors carry an electric current.
7. What type of semiconductor would you expect arsenic-doped germanium to be? Germanium has the same type of structure and semiconductor properties as silicon.

A Checklist for Review

IMPORTANT TERMS

molecular geometry (p. 354)
valence-shell electron-pair repulsion (VSEPR) model (10.1)
dipole moment (10.2)
valence bond theory (10.3)
hybrid orbitals (10.3)
σ (sigma) bond (10.4)
π (pi) bond (10.4)
molecular orbital theory (p. 376)
bonding orbitals (10.5)
antibonding orbitals (10.5)
homonuclear diatomic molecules (10.6)
heteronuclear diatomic molecules (10.6)
semiconductor (10.8)

SUMMARY OF FACTS AND CONCEPTS

Molecular geometry refers to the spatial arrangement of atoms in a molecule. The *valence-shell electron-pair repulsion (VSEPR) model* is a simple model for predicting molecular geometries. It is based on the idea that the valence-shell electron pairs are arranged symmetrically about an atom to minimize electron-pair repulsion. The geometry about an atom is then determined by the directions of the bonding pairs. Information about the geometry of a molecule can sometimes be obtained from the experimentally determined presence or absence of a *dipole moment*.

The bonding and geometry in a molecule can be described in terms of *valence bond theory*. In this theory, a bond is formed by the overlap of orbitals from two atoms. *Hybrid orbitals,* a set of equivalent orbitals formed by combining atomic orbitals, are often needed to describe this bond. Multiple bonds occur via the overlap of atomic orbitals to give *σ bonds* and *π bonds*. *Cis–trans* isomers result from the molecular rigidity imposed by the formation of a π bond.

Molecular orbital theory can also be used to explain bonding in molecules. According to this theory, electrons in a molecule occupy orbitals that may spread over the entire molecule. We can think of these molecular orbitals as constructed from atomic orbitals. Thus, when two atoms approach to form a diatomic molecule, the atomic orbitals interact to form *bonding* and *antibonding* molecular orbitals. The configuration of a diatomic molecule such as O_2 can be predicted from the order of filling of the molecular orbitals. From this configuration, we can predict the *bond order* and whether the molecular substance is diamagnetic or paramagnetic. Molecular orbital theory can also explain the conduction in metals and *semiconductors*.

OPERATIONAL SKILLS

1. **Predicting molecular geometries** Given the formula of a simple molecule, predict its geometry, using the VSEPR model (Examples 10.1 and 10.2).
2. **Relating dipole moment and molecular geometry** State what geometries of a molecule AX_n are consistent with the information that the molecule has a nonzero dipole moment (Example 10.3).
3. **Applying valence bond theory** Given the formula of a simple molecule, describe its bonding, using valence bond theory (Examples 10.4, 10.5, and 10.6).
4. **Describing molecular orbital configurations** Given the formula of a diatomic molecule obtained from first- or second-period elements, deduce the molecular orbital configuration, the bond order, and whether the molecular substance is diamagnetic or paramagnetic (Examples 10.7 and 10.8).

Review Questions

10.1 Describe the main features of the VSEPR model.

10.2 According to the VSEPR model, what are the arrangements of two, three, four, five, and six valence-shell electron pairs about an atom?

10.3 Why is a lone pair expected to occupy an equatorial position instead of an axial position in the trigonal bipyramidal arrangement?

10.4 Why is it possible for a molecule to have polar bonds yet have a dipole moment of zero?

10.5 Explain why nitrogen trifluoride has a small dipole moment even though it has polar bonds in a trigonal pyramidal arrangement.

10.6 Explain in terms of valence bond theory why the orbitals used on an atom give rise to a particular geometry about that atom.

10.7 What is the angle between two sp^3 hybrid orbitals?

10.8 What is the difference between a sigma bond and a pi bond?

10.9 Describe the bonding in ethylene, C_2H_4, in terms of valence bond theory.

10.10 How does the valence bond description of a carbon–carbon double bond account for *cis–trans* isomers?

10.11 What are the differences between a bonding and an antibonding molecular orbital of a diatomic molecule?

10.12 What factors determine the strength of interaction between two atomic orbitals to form a molecular orbital?

10.13 Describe the formation of bonding and antibonding molecular orbitals resulting from the interaction of two 2*s* orbitals.

10.14 Describe the formation of molecular orbitals resulting from the interaction of two 2*p* orbitals.

10.15 How does molecular orbital theory describe the bonding in the HF molecule?

10.16 Describe the bonding in SO_2, using molecular orbital theory. Compare this with the resonance description.

10.17 Explain metallic conduction in molecular orbital terms.

10.18 Explain why a semiconductor becomes more conducting as the temperature increases, whereas a metal becomes less conducting.

Practice Problems

THE VSEPR MODEL

10.19 Predict the shape or geometry of each of the following molecules, using the VSEPR model.
(a) CCl_4 (b) H_2Se (c) AsF_3 (d) $AlCl_3$

10.20 Use the electron-pair repulsion model to predict the geometry of the following molecules:
(a) CS_2 (b) SiH_4 (c) PCl_3 (d) $BeBr_2$

10.21 Predict the geometry of each of the following ions, using the electron-pair repulsion model.
(a) $TlCl_2^+$ (b) NO_2^- (c) NH_2^- (d) ClO_2^-

10.22 Use the VSEPR model to predict the geometry of the following ions:
(a) H_3O^+ (b) BeF_3^- (c) NO_2^+ (d) BrO_3^-

10.23 What geometry is expected for the following molecules, according to the VSEPR model?
(a) PF_5 (b) BrF_3 (c) BrF_5 (d) SCl_4

10.24 From the electron-pair repulsion model, predict the geometry of the following molecules:
(a) ClF_5 (b) SbF_5 (c) SeF_4 (d) TeF_6

10.25 Predict the geometries of the following ions, using the VSEPR model.
(a) $SnCl_5^-$ (b) PF_6^- (c) ClF_2^- (d) IF_4^-

10.26 Name the geometries expected for the following ions, according to the electron-pair repulsion model.
(a) BrF_6^+ (b) IF_2^- (c) ICl_4^- (d) IF_4^+

DIPOLE MOMENT AND MOLECULAR GEOMETRY

10.27 (a) The molecule AsF_3 has a dipole moment of 2.59 D. Which of the following geometries are possible: trigonal planar, trigonal pyramidal, or T-shaped? (b) The molecule H_2S has a dipole moment of 0.97 D. Is the geometry linear or bent?

10.28 (a) The molecule BrF_3 has a dipole moment of 1.19 D. Which of the following geometries are possible: trigonal planar, trigonal pyramidal, or T-shaped? (b) The molecule $TeCl_4$ has a dipole moment of 2.54 D. Is the geometry tetrahedral, distorted tetrahedral, or square planar?

10.29 Which of the following molecules would be expected to have zero dipole moments on the basis of their geometry?
(a) H_2S (b) PF_3 (c) TeF_6 (d) BeF_2

10.30 Which of the following molecules would be expected to have a dipole moment of zero because of symmetry?
(a) $AlCl_3$ (b) $SiCl_4$ (c) PCl_3 (d) ClF_5

VALENCE BOND THEORY

10.31 What hybrid orbitals would be expected for the central atom in each of the following?
(a) $AlBr_3$ (b) $BeCl_2$ (c) $SiCl_4$ (d) BeF_3^-

10.32 What hybrid orbitals would be expected for the central atom in each of the following?
(a) NCl_3 (b) BCl_3 (c) $BeBr_2$ (d) AlF_4^-

10.33 (a) Mercury(II) chloride dissolves in water to give poorly conducting solutions, indicating that the compound is largely nonionized in solution—it dissolves as $HgCl_2$ molecules. Describe the bonding of the $HgCl_2$ molecule, using valence bond theory. (b) Phosphorus trichloride, PCl_3, is a colorless liquid with a highly irritating vapor. Describe the bonding in the PCl_3 molecule, using valence bond theory. Use hybrid orbitals.

10.34 (a) Nitrogen trifluoride, NF_3, is a relatively unreactive, colorless gas. How would you describe the bonding in the NF_3 molecule in terms of valence bond theory? Use hybrid orbitals. (b) Silicon tetrafluoride, SiF_4, is a colorless gas formed when hydrofluoric acid attacks silica (SiO_2) or glass. Describe the bonding in the SiF_4 molecule, using valence bond theory.

10.35 What hybrid orbitals would be expected for the central atom in each of the following?
(a) BrF_5 (b) BrF_3 (c) $AsCl_5$ (d) ClF_4^+

10.36 What hybrid orbitals would be expected for the central atom in each of the following?
(a) PCl_5 (b) SeF_6 (c) TeF_4 (d) IF_4^-

10.37 Phosphorus pentachloride is normally a white solid. It exists in this state as the ionic compound $[PCl_4^+][PCl_6^-]$. Describe the electron structure of the PCl_6^- ion in terms of valence bond theory.

10.38 Iodine, I_2, dissolves in an aqueous solution of iodide ion, I^-, to give the triiodide ion, I_3^-. This ion consists of one iodine atom bonded to two others. Describe the bonding of I_3^- in terms of valence bond theory.

10.39 (a) Formaldehyde, H_2CO, is a colorless, pungent gas used to make plastics. Give the valence bond description of the formaldehyde molecule. (Both hydrogen atoms are attached to the carbon atom.) (b) Nitrogen, N_2, makes up about 80% of the earth's atmosphere. Give the valence bond description of this molecule.

10.40 (a) The molecule HN=NH exists as a transient species in certain reactions. Give the valence bond description of this species. (b) Hydrogen cyanide, HCN, is a very poisonous gas or liquid with the odor of bitter almonds. Give the valence bond description of HCN. (Carbon is the central atom.)

10.41 The hyponitrite ion, $^-O—N=N—O^-$, exists in solid compounds as the *trans* isomer. Using valence bond theory, explain why *cis–trans* isomers might be expected for this ion. Draw structural formulas of the two *cis–trans* isomers.

10.42 Fumaric acid, $C_4H_4O_4$, occurs in the metabolism of glucose in the cells of plants and animals. It is used commercially in beverages. The structural formula of fumaric acid is

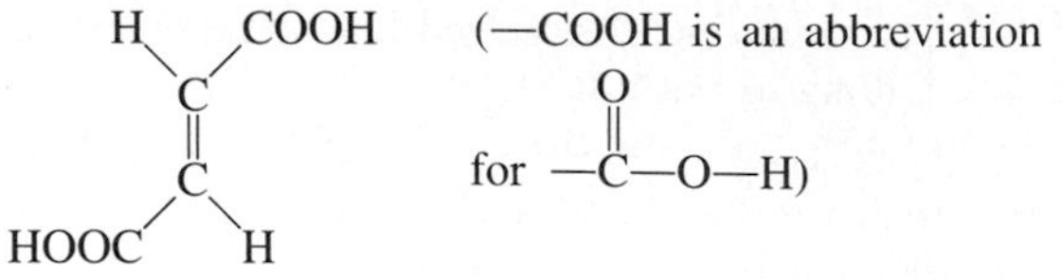

Maleic acid is the *cis* isomer of fumaric acid. Using valence bond theory, explain why these isomers are possible.

MOLECULAR ORBITAL THEORY

10.43 Describe the electronic structure of each of the following, using molecular orbital theory. Calculate the bond order of each and decide whether it should be stable. For each, state whether the substance is diamagnetic or paramagnetic.

(a) B_2 (b) B_2^+ (c) O_2^-

10.44 Use molecular orbital theory to describe the bonding in the following. For each one, find the bond order and decide whether it is stable. Is the substance diamagnetic or paramagnetic?

(a) C_2^+ (b) Ne_2 (c) C_2^-

10.45 Assume that the cyanide ion, CN^-, has molecular orbitals similar to those of a homonuclear diatomic molecule. Write the configuration and bond order of CN^-. Is a substance of the ion diamagnetic or paramagnetic?

10.46 Write the molecular orbital configuration of the diatomic molecule BN. What is the bond order of BN? Is the substance diamagnetic or paramagnetic? Use the order of energies that was given for homonuclear diatomic molecules.

Additional Problems

10.47 Predict the molecular geometry of each of the following.

(a) H_2S (b) I_3^- (c) NCl_3 (d) $HgCl_2$

10.48 Predict the molecular geometry of each of the following.

(a) $SbCl_3$ (b) PCl_4^+ (c) CO_3^{2-} (d) XeF_5^+

10.49 Which of the following molecules or ions are linear?

(a) $BeCl_2$ (b) NH_2^- (c) CS_2 (d) ICl_2^+

10.50 Which of the following molecules or ions are trigonal planar?

(a) $SnCl_3^-$ (b) BCl_3 (c) GaF_3 (d) PH_3

10.51 Describe the hybrid orbitals used by each carbon atom in the following molecules.

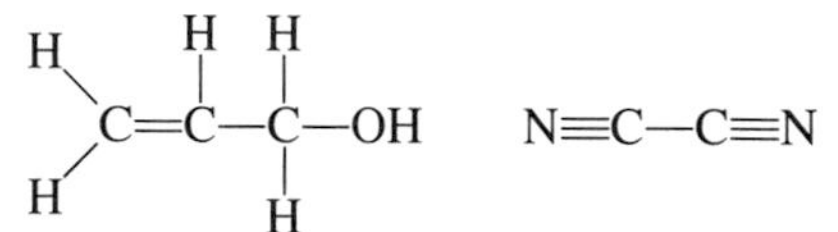

10.52 Describe the hybrid orbitals used by each nitrogen atom in the following molecules.

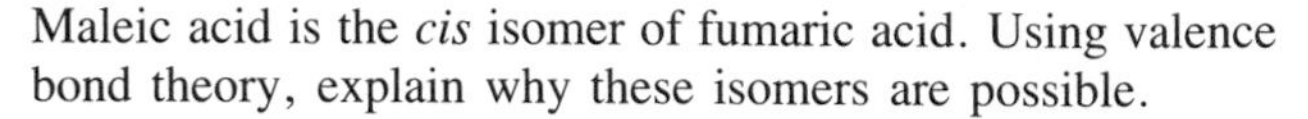

10.53 Explain how the dipole moment could be used to distinguish between the *cis* and *trans* isomers of 1,2-dibromoethene:

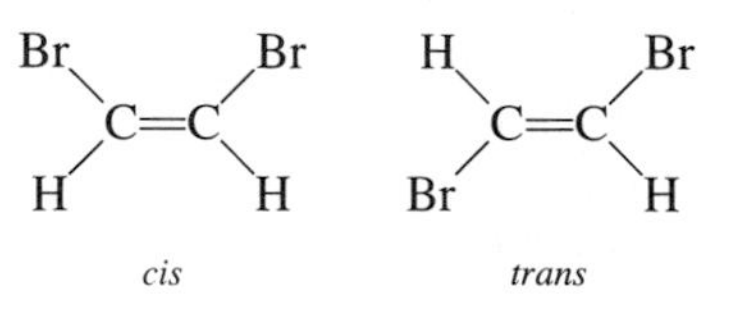

10.54 There are two compounds of the formula $Pt(NH_3)_2Cl_2$. (Compound B is *cisplatin*, mentioned in the opening to Chapter 1.) They have square planar structures. One is expected to have a dipole moment; the other is not. Which one would have a dipole moment?

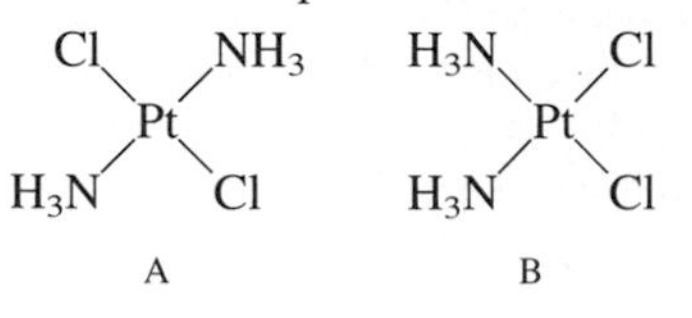

10.55 What is the molecular-orbital configuration of HeH^+? Do you expect the ion to be stable?

10.56 What is the molecular-orbital configuration of He_2^{+}? Do you expect the ion to be stable?

10.57 Calcium carbide, CaC_2, consists of Ca^{2+} and C_2^{2-} (acetylide) ions. Write the molecular-orbital configuration and bond order of the acetylide ion, C_2^{2-}.

10.58 Sodium peroxide, Na_2O_2, consists of Na^+ and O_2^{2-} (peroxide) ions. Write the molecular-orbital configuration and bond order of the peroxide ion, O_2^{2-}.

10.59 The oxygen–oxygen bond in O_2^+ is 1.12 Å and in O_2 is 1.21 Å. Explain why the bond length in O_2^+ is shorter than in O_2. Would you expect the bond length in O_2^- to be longer or shorter than that in O_2? Why?

10.60 The nitrogen–nitrogen bond distance in N_2 is 1.09 Å. On the basis of bond orders, would you expect the bond distance in N_2^+ to be less than or greater than 1.09 Å? Answer the same question for N_2^-.

Cumulative-Skills Problems

10.61 A molecular compound is composed of 60.4% Xe, 22.1% O, and 17.5% F. If the molecular weight is 217.3 amu, what is the molecular formula? What is the Lewis formula? Predict the molecular geometry using the VSEPR model. Describe the bonding, using valence bond theory.

10.62 A molecular compound is composed of 58.8% Xe, 7.2% O, and 34.0% F. If the molecular weight is 223 amu, what is the molecular formula? What is the Lewis formula? Predict the molecular geometry using the VSEPR model. Describe the bonding, using valence bond theory.

10.63 A compound of chlorine and fluorine, ClF_n, reacts at about 75°C with uranium metal to produce uranium hexafluoride, UF_6, and chlorine monofluoride, ClF. A quantity of uranium produced 3.53 g UF_6 and 343 mL ClF at 75°C and 2.50 atm. What is the formula (n) of the compound? Describe the bonding in the molecule, using valence bond theory.

10.64 Excess fluorine, $F_2(g)$, reacts at 150°C with bromine, $Br_2(g)$, to give a compound BrF_n. If 423 mL $Br_2(g)$ at 150°C and 748 mmHg produced 4.20 g BrF_n, what is n? Describe the bonding in the molecule, using valence bond theory.

10.65 Each of the following compounds has a nitrogen–nitrogen bond: N_2, N_2H_4, N_2F_2. Match each compound with one of the following bond lengths: 1.10 Å, 1.22 Å, 1.45 Å. Describe the geometry about one of the N atoms in each compound. What hybrid orbitals are needed to describe the bonding in valence bond theory?

10.66 The bond length in C_2 is 1.31 Å. Compare this with the bond lengths in C_2H_2 (1.20 Å), C_2H_4 (1.34 Å), and C_2H_6 (1.53 Å). What bond order would you predict for C_2 from its bond length? Does this agree with the molecular orbital configuration you would predict for C_2?

10.67 Draw resonance formulas of the nitric acid molecule, HNO_3. What is the geometry about the N atom? What is the hybridization on N? Use bond energies and one Lewis formula for HNO_3 to estimate ΔH_f° for $HNO_3(g)$. The actual value of ΔH_f° for $HNO_3(g)$ is -135 kJ/mol, which is lower than the estimated value because of stabilization of HNO_3 by resonance. The *resonance energy* is defined as ΔH_f°(estimated) $-$ ΔH_f°(actual). What is the resonance energy of HNO_3?

10.68 One resonance formula of benzene, C_6H_6, is

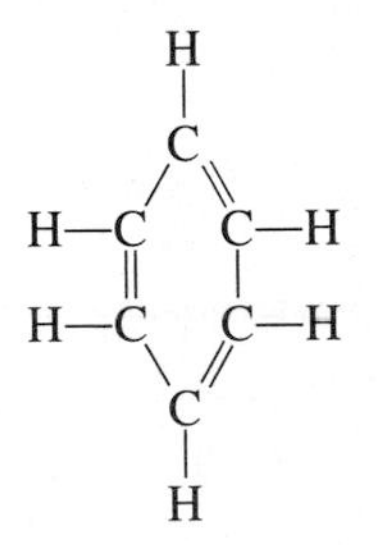

What is the other resonance formula? What is the geometry about a carbon atom? What hybridization would be used in valence bond theory to describe the bonding? The ΔH_f° for $C_6H_6(g)$ is -83 kJ/mol; ΔH_f° for C(g) is 715 kJ/mol. Obtain the resonance energy of benzene. (See Problem 10.67.)

11

States of Matter; Liquids and Solids

Chapter Outline

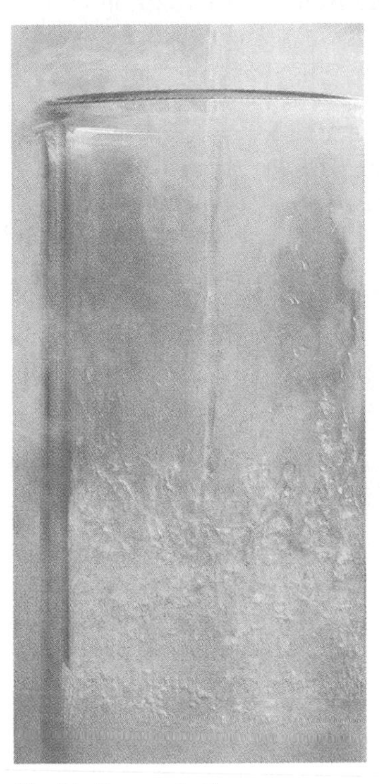

Liquid nitrogen boiling.

Dry ice is solid carbon dioxide (Figure 11.1), which has an interesting property: At normal pressures it passes directly to the gaseous state without first melting to the liquid. This property, together with the fact that the vaporizing solid is at −78°C, makes solid carbon dioxide an excellent refrigerant. It cools other objects but doesn't leave a liquid residue as ordinary ice does.

Liquid carbon dioxide exists, but only at pressures greater than 5 atm. In fact, carbon dioxide is usually transported in the liquid state either at room temperature in steel tanks at high pressure or at low temperature (about −18°C) in insulated vessels at moderate pressure. It too is an excellent refrigerant. When liquid carbon dioxide evaporates (changes to vapor), it absorbs large quantities of heat, cooling to as low as −57°C. A major use of liquid carbon dioxide is to freeze foods for grocery stores and fast-food restaurants.

Liquid carbon dioxide is obtained by compressing the gas to high pressure. If the compressed gas from the evaporating liquid is allowed to expand through a

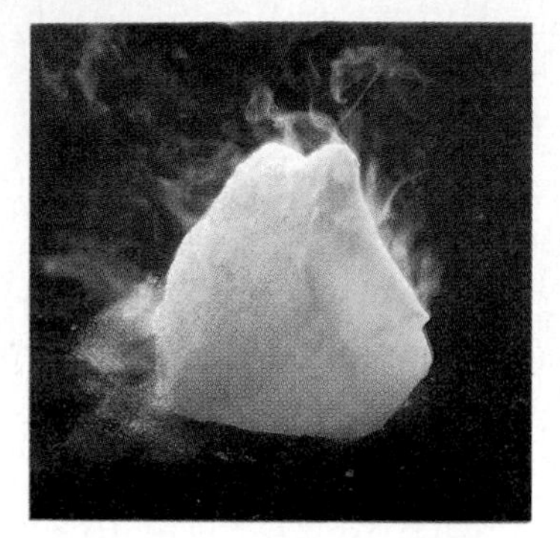

Figure 11.1
Dry ice. Dry ice is solid carbon dioxide, CO_2, whose temperature is about −78°C. It passes directly from the solid to the gaseous state. The white plumes are water ice fog, formed by condensation from the cold air.

valve, carbon dioxide ''snow'' is obtained. (Most gases cool when they expand in this manner; for example, air cools when it escapes from the valve of an inflated tire.) The carbon dioxide snow is compacted into blocks and sold as dry ice.

We see that carbon dioxide, like water and most other pure substances, exists in solid, liquid, and gaseous states and can undergo changes from one state to another. Substances change state under various temperature and pressure conditions. Can we obtain some understanding of these conditions for a change of state? Why, for example, is carbon dioxide normally a gas, whereas water is normally a liquid? What are the conditions under which carbon dioxide gas changes to a liquid or to a solid?

These are some of the questions that we will examine in this chapter. They concern certain physical properties that, as we will see, can often be related to the bonding and structure of the liquid and solid states.

11.1 COMPARISON OF GASES, LIQUIDS, AND SOLIDS

We defined the different states of matter in Section 2.1, and we discussed the gas laws and the kinetic theory of gases in Chapter 5. Here we want to recall the salient features of those discussions and then compare gases, liquids, and solids. We especially want to compare how these states are viewed in terms of kinetic-molecular theory.

Gases are compressible fluids. According to kinetic-molecular theory, gases are composed of particles called molecules that are in constant random motion throughout mostly empty space (unless the gas is highly compressed). A gas is easily compressed because the molecules can be pushed into a smaller space. It is fluid because individual molecules can move easily relative to one another.

Liquids are relatively incompressible fluids. This state of matter can also be explained by kinetic-molecular theory. According to this theory, the molecules of a liquid are in constant random motion (as in a gas) but are more tightly packed, so there is much less free space. Because the molecules can move relative to one another as in a gas, a liquid can flow (it is fluid). But the lack of free space explains why a liquid, unlike a gas, is nearly incompressible.

Solids are nearly incompressible and are rigid, not fluid. Kinetic-molecular theory explains this by saying that the particles making up a solid (which may be atoms, molecules, or ions) exist in close contact and (unlike those in a gas or liquid) do not move about but oscillate or vibrate about fixed sites. This explains the rigidity of a solid. And, of course, the compact structure explains its incompressibility. Figure 11.2 compares the gaseous, liquid, and solid states in kinetic-molecular terms.

Recall that gases normally follow closely the ideal gas law, $PV = nRT$. The simplicity of this equation is a result of the nearly negligible forces of interaction between molecules and the nearly negligible molecular size compared with the total volume of gas. Still, to be accurate we should account for both of these quantities, and the van der Waals equation, which we discussed in Section 5.8, is one attempt to do this.

$$(P + n^2a/V^2)(V - nb) = nRT$$

In this equation, the constants a and b depend on the nature of the gas. The constant a depends on the magnitude of the intermolecular forces; the constant b depends on the relative size of the molecules.

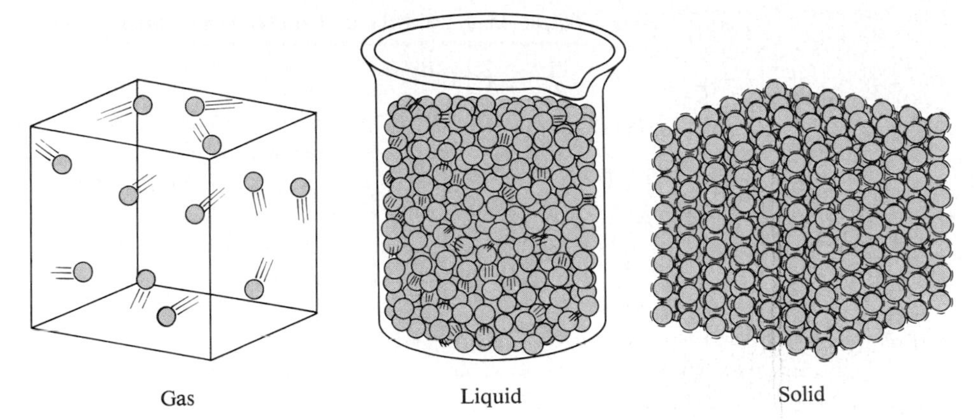

Figure 11.2
Representation of the states of matter. In a gas, the molecules are in constant random motion through largely empty space. In a liquid, the molecules are also in constant random motion, but they are more closely packed than in a gas. In a solid, the basic units (atoms, ions, or molecules) are closely packed and vibrate about fixed sites.

No such simple equations exist for the liquid and solid states. The size of the particles making up the liquid and solid states cannot be neglected relative to the total volume or accounted for in a simple way. Similarly, the forces of interaction between particles in the liquid and the solid state cannot be neglected; indeed, these forces of interaction are crucial when we describe the properties of the liquid and solid states.

Some of the most important properties of liquids and solids are concerned with changes of physical state. Therefore, we begin with this topic. In later sections, we will look specifically at the liquid and solid states and see how the properties of these states depend on the forces of attraction between particles.

Changes of State

In the chapter opener, we discussed the change of solid carbon dioxide (dry ice) directly to a gas. Such *a change of a substance from one state to another* is called a **change of state** or **phase transition.** In the next sections of this chapter, we will look at the different kinds of phase transitions and the conditions under which they occur. (A *phase* is a homogeneous portion of a system, that is, a given state of either a substance or a solution.)

11.2 PHASE TRANSITIONS

In general, each of the three states of a substance can change into either one of the other states. Table 11.1 lists the different kinds of phase transitions.

Melting is *the change of a solid to the liquid state* (melting is also referred to as *fusion*). For example,

$$\underset{\text{ice, snow}}{H_2O(s)} \longrightarrow \underset{\text{liquid water}}{H_2O(l)} \qquad \text{(melting, fusion)}$$

Freezing is *the change of a liquid to the solid state*. The freezing of liquid water to ice is a common example.

$$\underset{\text{liquid water}}{H_2O(l)} \longrightarrow \underset{\text{ice}}{H_2O(s)} \qquad \text{(freezing)}$$

The casting of a metal object involves the melting of the metal, then its freezing in a mold.

Table 11.1 Kinds of Phase Transitions

Phase Transition	Name	Examples
Solid ⟶ liquid	Melting, fusion	Melting of snow and ice
Solid ⟶ gas	Sublimation	Sublimation of dry ice, freeze-drying of coffee
Liquid ⟶ solid	Freezing	Freezing of water or a liquid metal
Liquid ⟶ gas	Vaporization	Evaporation of water or refrigerant
Gas ⟶ liquid	Condensation, liquefaction	Formation of dew, liquefaction of carbon dioxide
Gas ⟶ solid	Condensation, deposition	Formation of frost and snow

Vaporization is *the change of a solid or a liquid to the vapor*. For example,

$$\underset{\text{liquid water}}{H_2O(l)} \longrightarrow \underset{\text{water vapor}}{H_2O(g)} \qquad \text{(vaporization)}$$

The change of a solid to the vapor is specifically referred to as **sublimation.** Although the vapor pressure of many solids is quite low, some (usually molecular solids) have appreciable vapor pressure. Ice, for instance, has a vapor pressure of 4.7 mmHg at 0°C. For this reason, a pile of snow slowly disappears in winter even though the temperature is too low for it to melt. The snow is being changed directly to water vapor.

$$\underset{\text{ice, snow}}{H_2O(s)} \longrightarrow \underset{\text{water vapor}}{H_2O(g)} \qquad \text{(sublimation)}$$

Sublimation can be used to purify solids that readily vaporize. Figure 11.3 shows a simple way to purify iodine by sublimation. Impure iodine is heated in a beaker so that it vaporizes, leaving nonvolatile impurities behind. The vapor crystallizes on the bottom surface of a dish containing ice. Freeze-drying of foods is a commercial application of sublimation. Brewed coffee, for example, is frozen and placed in a vacuum to remove water vapor. The ice continues to sublime until it is all gone, leaving freeze-dried coffee. Most freeze-dried foods are easily reconstituted by adding water.

Figure 11.3
Sublimation of iodine. *Left:* The beaker contains iodine crystals, I_2; a dish of ice rests on top of the beaker. *Right:* Iodine has an appreciable vapor pressure even below its melting point (114°C); thus, when heated carefully the solid sublimes without melting. The vapor deposits as crystals on the cool underside of the dish.

Condensation is *the change of a gas to either the liquid or the solid state* (the change of vapor to solid is sometimes called *deposition*). Dew is liquid water formed by condensation of water vapor from the atmosphere. Frost is solid water formed by direct condensation of water vapor from the atmosphere without first forming liquid water. Snow is formed by a similar process in the upper atmosphere.

$$\underset{\text{water vapor}}{H_2O(g)} \longrightarrow \underset{\text{dew}}{H_2O(l)} \qquad \text{(condensation)}$$

$$\underset{\text{water vapor}}{H_2O(g)} \longrightarrow \underset{\text{frost, snow}}{H_2O(s)} \qquad \text{(condensation, deposition)}$$

When a substance that is normally a gas, such as carbon dioxide, changes to the liquid state, the phase transition is often referred to as *liquefaction*.

Vapor Pressure

Liquids, and even some solids, are constantly vaporizing. If a liquid is in a closed vessel with space above it, a partial pressure of the vapor state builds up in this space. The **vapor pressure** of a liquid is *the partial pressure of the vapor over the liquid, measured at equilibrium*. To understand what we mean by equilibrium, let us look at a simple method for measuring vapor pressure.

We introduce a few drops of water from a medicine dropper into the mercury column of a barometer (Figure 11.4A). Being less dense than mercury, the water rises in the tube to the top, where it evaporates or vaporizes.

To understand the process of vaporization, it is necessary to realize that the molecules in a liquid have a distribution of kinetic energies (Figure 11.5). Those molecules moving away from the surface will escape only if their kinetic energies are greater than a certain minimum value equal to the potential energy from the attraction of molecules in the body of the liquid.

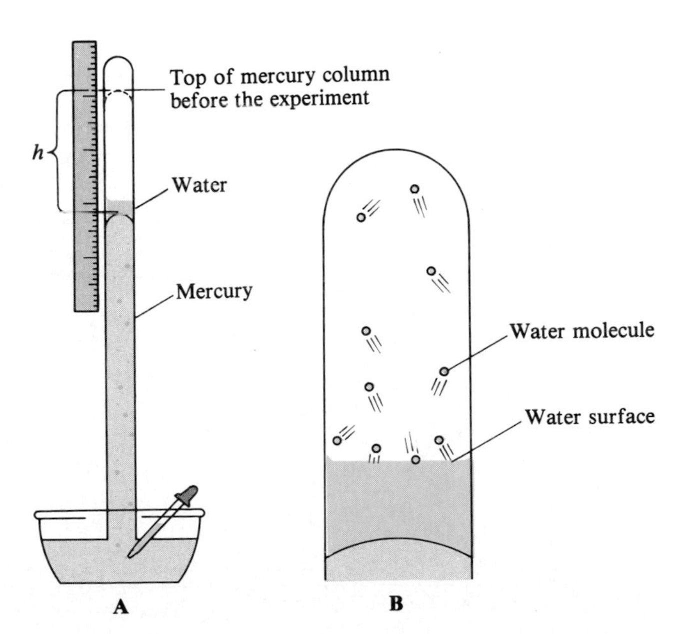

Figure 11.4
Measurement of the vapor pressure of water. *(A)* The mercury column drops a distance *h* because of the vapor pressure of water. *(B)* A detail of the upper portion of the tube, showing water molecules evaporating from and condensing on the water surface.

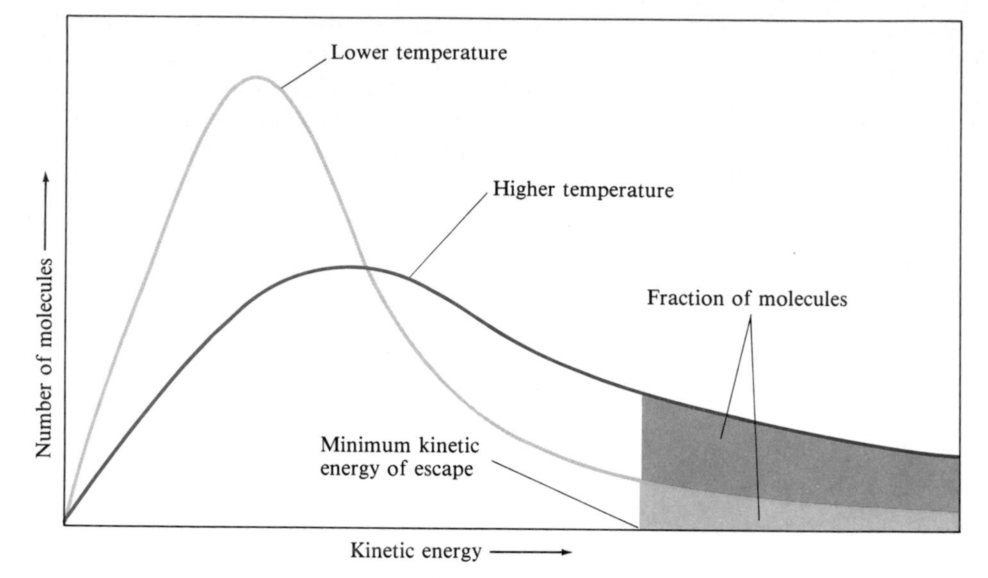

Figure 11.5
Distribution of kinetic energies of molecules in a liquid. Distributions of kinetic energies are shown for two different temperatures. The fraction of molecules having kinetic energies greater than the minimum necessary for escape is given by the colored areas (orange for the lower temperature, red plus orange area at the higher temperature). Note that the fraction of molecules having kinetic energies greater than the minimum value for vaporization increases with temperature.

As molecules at the surface gain sufficient kinetic energy (through collisions with neighboring molecules), they escape the liquid and go into the space at the top of the barometer column (see Figure 11.4B). More and more water molecules begin to fill this space. As the number of molecules in the vapor state increases, more and more gaseous molecules collide with the liquid water surface, exerting pressure on it. So, as vaporization of water proceeds, the mercury column moves downward.

Some of the molecules in the vapor collide with the liquid surface and stick; that is, the vapor condenses to liquid. The rate of condensation steadily increases as the number of molecules in the volume of vapor increases, until the rate at which molecules are condensing on the liquid equals the rate at which molecules are vaporizing (Figure 11.6).

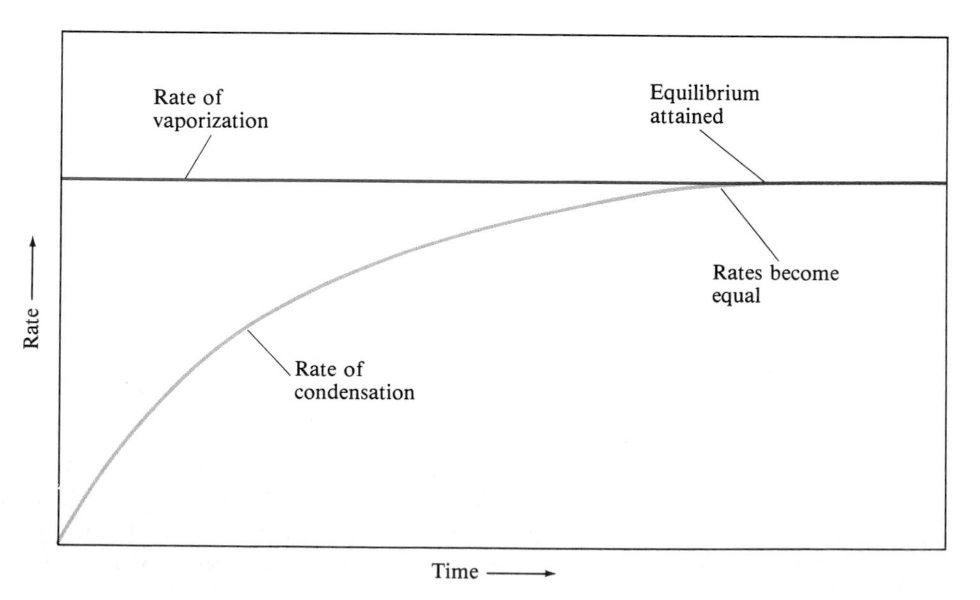

Figure 11.6
Rates of vaporization and condensation of a liquid over time. As vaporization continues at a constant rate, the rate of condensation increases. Eventually, the two rates become equal. Then the partial pressure of vapor reaches a steady value, which is the vapor pressure of the liquid. The process is said to reach equilibrium.

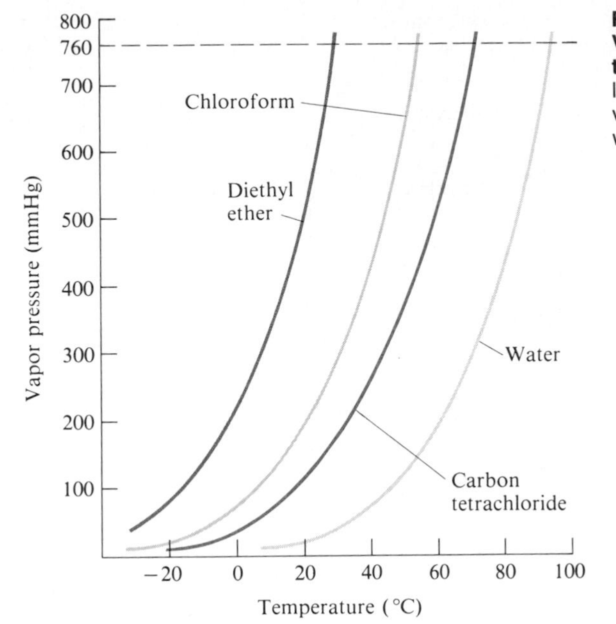

Figure 11.7
Variation of vapor pressure with temperature. Several common liquids are shown. Note that the vapor pressure increases rapidly with temperature.

When the rates of vaporization and condensation have become equal, the number of molecules in the vapor at the top of the column stops increasing and remains unchanged. The partial pressure of the vapor (as measured by the change in height of the mercury column) reaches a steady value, which is its vapor pressure. We say that the liquid and vapor are in *equilibrium*. Although the partial pressure of the vapor is unchanging, molecules are still leaving the liquid and coming back. For this reason, we speak of this as a *dynamic equilibrium*—one in which the molecular processes (in this case, vaporization and condensation) are continuously occurring. We represent this dynamic equilibrium for the vaporization and condensation of water by the equation

$$H_2O(l) \rightleftharpoons H_2O(g)$$

As Figure 11.7 shows, the vapor pressure of a substance depends on the temperature. (Appendix B gives a table of the vapor pressures of water at various temperatures.) As the temperature increases, the kinetic energy of molecular motion becomes greater, and the vapor pressure increases. Liquids and solids with relatively high vapor pressure at normal temperatures are said to be *volatile*. Chloroform and carbon tetrachloride are both volatile liquids. Naphthalene, $C_{10}H_8$, and *para*-dichlorobenzene, $C_6H_4Cl_2$, are volatile solids: They have appreciable vapor pressures at room temperature. Both are used to make moth balls.

Boiling Point and Melting Point

The temperature at which the vapor pressure of a liquid equals the atmospheric pressure is called the **boiling point** of the liquid. As the temperature of a liquid is raised, the vapor pressure increases. When the vapor pressure equals the atmospheric pressure, stable bubbles of vapor form within the body of the liquid (Figure 11.8). This process is called boiling. Once boiling starts, the temperature of the liquid remains at the boiling point (as long as sufficient heat is supplied).

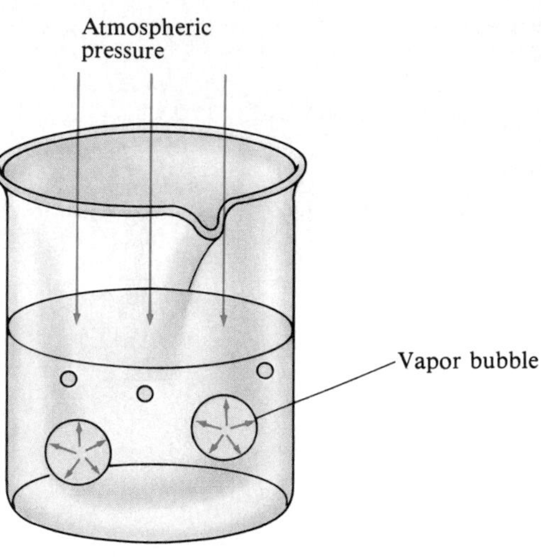

Figure 11.8
Boiling of a liquid. A liquid boils when the vapor pressure inside a bubble of vapor equals the external pressure. The temperature at which this occurs is called the boiling point.

● If you want to cook a stew in less time than is possible at atmospheric pressure, you can use a pressure cooker. This is especially valuable at high altitudes, where water boils at a lower temperature. The steam pressure inside the cooker is allowed to build up to a given value before being released by a valve. The stew then boils at a higher temperature and is cooked more quickly.

Because the atmospheric pressure can vary, the boiling point of a liquid can vary. For instance, at 1.00 atm (the average atmospheric pressure at sea level), the boiling point of water is 100°C. But at 0.83 atm (the average atmospheric pressure in Denver, Colorado, at 1609 m above sea level), the boiling point of water is 95°C. The *normal boiling point* of a liquid is the boiling point at 1 atm.●

The temperature at which a pure liquid changes to a crystalline solid, or freezes, is called the **freezing point.** *The temperature at which a crystalline solid changes to a liquid, or melts,* is called the **melting point;** it is identical to the freezing point. The melting or freezing occurs at the temperature where the liquid and solid are in dynamic equilibrium.●

● Unlike boiling points, melting points are affected only by large pressure changes. Small variations in the barometric pressure have negligible effect. The difference arises from the relative incompressibility of liquids and solids compared to gases.

$$\text{Solid} \rightleftharpoons \text{liquid}$$

Impure substances usually melt over a temperature range instead of at a definite temperature.

Both the melting point and the boiling point are characteristic properties of a substance and can be used to help identify it. Table 11.2 gives melting points and boiling points of several substances.

Heat of Phase Transition

Any change of state involves the addition or removal of energy as heat from the substance. For example, heat is needed to melt a solid or to vaporize a liquid; these are endothermic processes (processes that absorb heat).

The heat needed for the melting of a solid is called the **heat of fusion** (or *enthalpy of fusion*) and is denoted ΔH_{fus}. For ice, the heat of fusion is 6.01 kJ per mole.

$$H_2O(s) \longrightarrow H_2O(l);\ \Delta H_{fus} = 6.01 \text{ kJ/mol}$$

The heat needed for the vaporization of a liquid is called the **heat of vaporization** (or *enthalpy of vaporization*). At 100°C, the heat of vaporization of water is 40.7 kJ per mole.●

● The fact that heat is required for vaporization can be demonstrated by evaporating a dish of water inside a vessel attached to a vacuum pump. With the pressure kept low, water evaporates quickly enough to freeze the water remaining in the dish.

$$H_2O(l) \longrightarrow H_2O(g);\ \Delta H_{vap} = 40.7 \text{ kJ/mol}$$

Table 11.2 Melting Points and Boiling Points (at 1 atm) of Several Substances

Name	Type of Solid*	Melting Point, °C	Boiling Point, °C
Neon, Ne	Molecular	−249	−246
Hydrogen sulfide, H_2S	Molecular	−86	−61
Chloroform, $CHCl_3$	Molecular	−64	62
Water, H_2O	Molecular	0	100
Acetic acid, $HC_2H_3O_2$	Molecular	17	118
Mercury, Hg	Metallic	−39	357
Sodium, Na	Metallic	98	883
Tungsten, W	Metallic	3410	5660
Cesium chloride, CsCl	Ionic	645	1290
Sodium chloride, NaCl	Ionic	801	1413
Magnesium oxide, MgO	Ionic	2800	3600
Quartz, SiO_2	Covalent network	1610	2230
Diamond, C	Covalent network	3550	4827

*Types of solids are discussed in Section 11.6.

Note that much more heat is required for vaporization than for melting. Melting needs only enough energy for the molecules to escape from their sites in the solid. For vaporization, enough energy must be supplied to break most of the intermolecular attractions.

A refrigerator relies on the cooling effect accompanying vaporization. Its mechanism contains an enclosed gas that can be liquefied under pressure, such as ammonia or dichlorodifluoromethane (Freon-12), CCl_2F_2. As the liquid is allowed to evaporate, it absorbs heat and thus cools its surroundings (the interior space of the refrigerator). Gas from the evaporation is recycled to a compressor and then to a condenser, where it is liquefied again. Heat leaves the condenser, going into the surrounding air.

Example 11.1 Calculating the Heat Required for a Phase Change of a Given Mass of Substance

A particular refrigerator cools by evaporating liquefied dichlorodifluoromethane, CCl_2F_2. How many kilograms of this liquid must be evaporated to freeze a tray of water at 0°C to ice at 0°C? The mass of the water is 525 g, the heat of fusion of ice is 6.01 kJ/mol, and the heat of vaporization of dichlorodifluoromethane is 17.4 kJ/mol.

Solution

The heat that must be removed to freeze 525 g of water at 0°C is

$$525\ \cancel{\text{g H}_2\text{O}} \times \frac{1\ \cancel{\text{mol H}_2\text{O}}}{18.0\ \cancel{\text{g H}_2\text{O}}} \times \frac{-6.01\ \text{kJ}}{1\ \cancel{\text{mol H}_2\text{O}}} = -175\ \text{kJ}$$

The minus sign indicates that heat energy is taken away from the water. Consequently, the vaporization of dichlorodifluoromethane absorbs 175 kJ of heat. The mass of CCl_2F_2 that must be vaporized to absorb this quantity of heat is

$$175\ \cancel{\text{kJ}} \times \frac{1\ \cancel{\text{mol CCl}_2\text{F}_2}}{17.4\ \cancel{\text{kJ}}} \times \frac{121\ \text{g CCl}_2\text{F}_2}{1\ \cancel{\text{mol CCl}_2\text{F}_2}} = 1.22 \times 10^3\ \text{g CCl}_2\text{F}_2$$

Thus, 1.22 kg of dichlorodifluoromethane must be evaporated.

Exercise 11.1

The heat of vaporization of ammonia is 23.4 kJ/mol. How much heat is required to vaporize 1.00 kg of ammonia? How many grams of water at 0°C could be frozen to ice at 0°C by the evaporation of this amount of ammonia?

(See Problems 11.27, 11.28, 11.29, and 11.30.)

Clausius–Clapeyron Equation

We noted earlier that the vapor pressure of a substance depends on temperature. The variation of vapor pressure with temperature of some liquids was given in Figure 11.7. It can be shown that the logarithm of the vapor pressure of a liquid or solid varies with the absolute temperature according to the following approximate relation:

$$\log P = -\frac{A}{T} + B$$

Here P is the vapor pressure, and A and B are positive constants. We can confirm this relation for the liquids shown in Figure 11.7 by replotting the data. If we put $y = \log P$ and $x = 1/T$, the previous relation becomes

$$y = -Ax + B$$

This means that if we plot log P versus $1/T$, we should get a straight line with slope $-A$. The data of Figure 11.7 are replotted this way in Figure 11.9.●

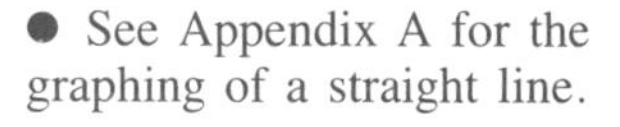

● See Appendix A for the graphing of a straight line.

The previous equation has been derived from thermodynamics, by assuming the vapor behaves like an ideal gas. The result, known as the *Clausius–Clapeyron equation,* shows that the constant A is proportional to the heat of vaporization of the liquid, ΔH_{vap}.

$$\log P = \frac{-\Delta H_{vap}}{2.303RT} + B \qquad \text{or} \qquad \ln P = \frac{-\Delta H_{vap}}{RT} + B'$$

Here ln P is the natural logarithm of P, which equals 2.303 log P. Also, $B' = 2.303\ B$.

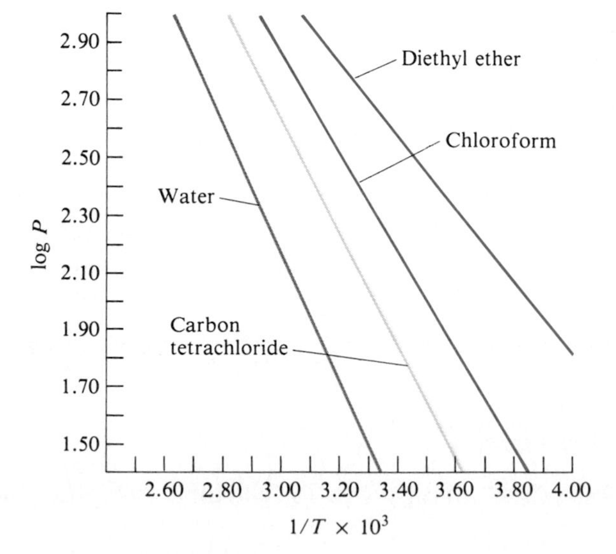

Figure 11.9
Plot of the logarithm of vapor pressure versus 1/*T*. The liquids shown in Figure 11.7 are replotted in this graph. Note the straight-line relationship.

The two-point form of the Clausius–Clapeyron equation is very useful for calculations. Let us write the previous equation for two different temperatures:

$$\log P_2 = \frac{-\Delta H_{vap}}{2.303RT_2} + B$$

$$\log P_1 = \frac{-\Delta H_{vap}}{2.303RT_1} + B$$

Here P_1 is the vapor pressure at absolute temperature T_1, and P_2 is the vapor pressure at absolute temperature T_2. If we subtract the second equation from the first, we get

$$\log P_2 - \log P_1 = \frac{-\Delta H_{vap}}{2.303RT_2} - \frac{-\Delta H_{vap}}{2.303RT_1} + B - B$$

which can be written

$$\log \frac{P_2}{P_1} = \frac{\Delta H_{vap}}{2.303R}\left(\frac{1}{T_1} - \frac{1}{T_2}\right) \quad \text{or} \quad \ln \frac{P_2}{P_1} = \frac{\Delta H_{vap}}{R}\left(\frac{1}{T_1} - \frac{1}{T_2}\right)$$

The next two examples illustrate the use of this two-point form of the Clausius–Clapeyron equation.

Example 11.2 Calculating the Vapor Pressure at a Given Temperature

Estimate the vapor pressure of water at 85°C. Note that the normal boiling point of water is 100°C and that the heat of vaporization is 40.7 kJ/mol.

Solution

The two-point form of the Clausius–Clapeyron equation relates five quantities: P_1, T_1, P_2, T_2, and ΔH_{vap}. The normal boiling point is the temperature at which the vapor pressure equals 760 mmHg. Thus, we could let P_1 equal 760 mmHg and let T_1 equal 373 K (100°C). Then P_2 would be the vapor pressure at T_2, or 358 K (85°C). The heat of vaporization, ΔH_{vap}, is 40.7×10^3 J/mol. Note that we have expressed this quantity in joules per mole to agree with the units of the molar gas constant, R, which equals 8.31 J/(K · mol). We can now substitute into the two-point equation (the equation in color just before this example).

$$\log \frac{P_2}{760 \text{ mmHg}} = \frac{40.7 \times 10^3 \text{ J/mol}}{2.303 \times 8.31 \text{ J/(K} \cdot \text{mol)}}\left(\frac{1}{373 \text{ K}} - \frac{1}{358 \text{ K}}\right)$$

$$\log \frac{P_2}{760 \text{ mmHg}} = (2127 \text{ K}) \times (-1.123 \times 10^{-4}\text{/K}) = -0.239$$

We now want to solve this equation for P_2, which is the vapor pressure of water at 85°C. To do so, we take the antilogarithm of both sides of the equation. This gives

$$\frac{P_2}{760 \text{ mmHg}} = \text{antilog}(-0.239)$$

The antilog(−0.239) is $10^{-0.239} = 0.577$. (Appendix A describes how to obtain the antilogarithm.) Thus,

$$\frac{P_2}{760 \text{ mmHg}} = 0.577$$

Therefore,

$$P_2 = 0.577 \times 760 \text{ mmHg} = \mathbf{439\ mmHg}$$

This value may be compared with the experimental value of 434 mmHg, which is given in Appendix B. Differences from the experiment stem from the fact that the Clausius–Clapeyron equation is an approximate relation.

Exercise 11.2

Carbon disulfide, CS_2, has a normal boiling point of 46°C and a heat of vaporization of 26.8 kJ/mol. What is the vapor pressure of carbon disulfide at 35°C?

(See Problems 11.33 and 11.34).

Example 11.3 Calculating the Heat of Vaporization from Vapor Pressures

Calculate the heat of vaporization of diethyl ether (often called simply "ether"), $C_4H_{10}O$, from the following vapor pressures: 400 mmHg at 18°C and 760 mmHg at 35°C (each pressure value has three significant figures).

Solution

If we let P_1 be 400 mmHg, then T_1 is 291 K (18°C), P_2 is 760 mmHg, and T_2 is 308 K (35°C). We substitute these values into the Clausius–Clapeyron equation.

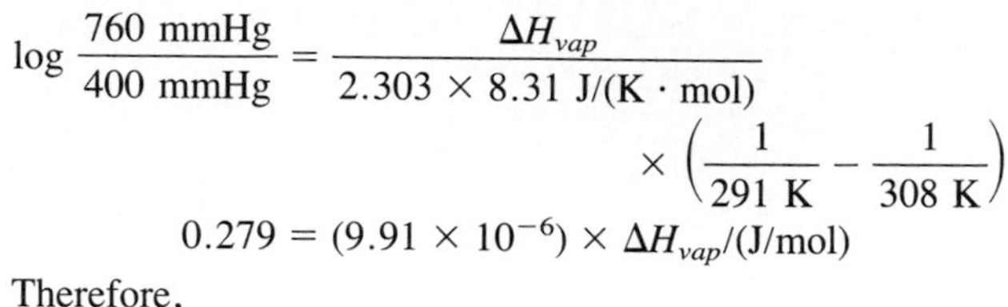

$$\log \frac{760 \text{ mmHg}}{400 \text{ mmHg}} = \frac{\Delta H_{vap}}{2.303 \times 8.31 \text{ J/(K} \cdot \text{mol)}} \times \left(\frac{1}{291 \text{ K}} - \frac{1}{308 \text{ K}}\right)$$

$$0.279 = (9.91 \times 10^{-6}) \times \Delta H_{vap}/(\text{J/mol})$$

Therefore,

$$\Delta H_{vap} = 2.82 \times 10^4 \text{ J/mol} \qquad (28.2 \text{ kJ/mol})$$

Exercise 11.3

Selenium tetrafluoride, SeF_4, is a colorless liquid. It has a vapor pressure of 757 mmHg at 105°C and 522 mmHg at 95°C. What is the heat of vaporization of selenium tetrafluoride?

(See Problems 11.35 and 11.36.)

11.3 PHASE DIAGRAMS

As we mentioned in the chapter opening, the solid, liquid, and gaseous states of carbon dioxide exist under different temperature and pressure conditions. A **phase diagram** is *a graphical way to summarize the conditions under which the different states of a substance are stable.*

Melting-Point Curve

Figure 11.10 is a phase diagram for water. It consists of three curves that divide the diagram into regions labeled "solid," "liquid," and "gas." In each given

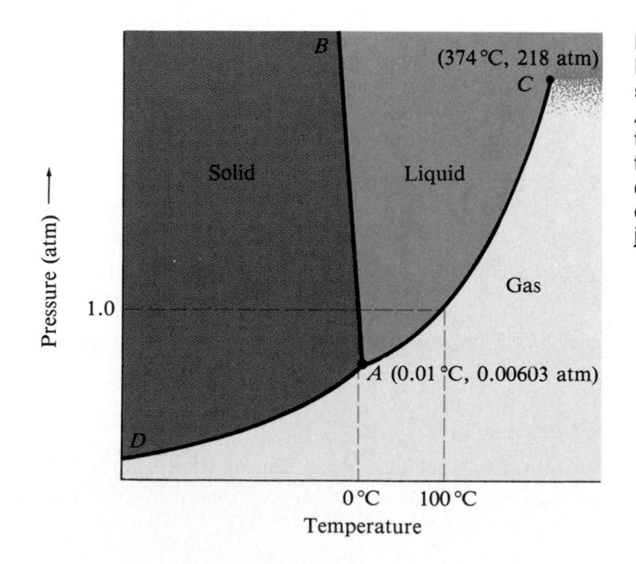

Figure 11.10
Phase diagram for water (not to scale). The curves *AB, AC,* and *AD* divide the diagram into regions that give combinations of temperature and pressure for which only one state is stable. Along any curve, the two states from the adjoining regions are in equilibrium.

region, the indicated state is stable. Every point on each of the curves indicates experimentally determined temperatures and pressures at which two states are in equilibrium. Thus, the curve *AB*, dividing the solid region from the liquid region, represents the conditions under which the solid and liquid are in equilibrium.

$$\text{Solid} \rightleftharpoons \text{liquid}$$

This curve gives the melting points of the solid at various pressures.

Usually, the melting point is only slightly affected by pressure. For this reason, the melting-point curve *AB* is nearly vertical. If a liquid is more dense than the solid, as is the case for water, the melting point decreases with pressure.● The melting-point curve in such cases leans slightly to the left. In the case of ice, the decrease is indeed slight—only 1°C for a pressure increase of 133 atm. Usually, the liquid state is less dense than the solid. In that case, the melting-point curve leans slightly to the right.

● The following is a dramatic demonstration of the effect of pressure on the melting of ice: Suspend a block of ice between two chairs. Then loop a length of wire over the block and place weights on the ends of the wire. The pressure of the wire will melt the ice under it. Liquid water will flow over the top of the wire and freeze again, because it is no longer under the pressure of the wire. As a result, the wire will cut through the ice and fall to the floor, but the block will remain as one piece.

Vapor-Pressure Curves for the Liquid and the Solid

The curve *AC* that divides the liquid region from the gaseous region gives the vapor pressures of the liquid at various temperatures. It also gives the boiling points of the liquid for various pressures. The boiling point of water at 1 atm is shown on the phase diagram (Figure 11.10).

The curve *AD* that divides the solid region from the gaseous region gives the vapor pressures of the solid at various temperatures. This curve intersects the other curves at the point *A*, the **triple point**, which is *the point on a phase diagram representing the temperature and pressure at which three phases of a substance coexist in equilibrium*. For water, a triple point occurs at 0.01°C, 0.00603 atm (4.58 mmHg), in which the solid, liquid, and vapor phases coexist.●

● Because the triple point for water occurs at a definite temperature, it is used to define the Kelvin thermometer scale. The temperature of water at its triple point is defined to be 273.16 K (0.01°C).

Suppose a solid is warmed at a pressure below the pressure at the triple point. In a phase diagram, this corresponds to moving along a horizontal line below the triple point. We can see from Figure 11.10 that such a line will intersect curve *AD*, which is the vapor-pressure curve for the solid. Thus, the solid will pass directly into the gas; that is, the solid will sublime. Freeze-drying of a food (or brewed coffee) is accomplished by placing the frozen food in a vacuum (below 0.00603 atm) so that the ice in it sublimes. Because the food can be dried at a lower temperature than if heat-dried, it retains more flavor and can often be reconstituted by adding water.

The triple point of carbon dioxide is at −57°C and 5.1 atm (Figure 11.11A). Therefore, the solid sublimes if warmed at any pressure below 5.1 atm. This is why solid carbon dioxide sublimes at normal atmospheric pressure (1 atm). Above 5.1 atm, however, the solid melts if warmed. Sulfur has a more complicated phase diagram (Figure 11.11B). It displays three triple points, one of them involving two different solid forms of sulfur (called rhombic sulfur and monoclinic sulfur), as well as the vapor.

Critical Temperature and Pressure

Imagine an experiment in which liquid and gaseous carbon dioxide are sealed into a thick-walled glass vessel at 20°C. At this temperature, the liquid is in equilibrium with its vapor at a pressure of 57 atm. We see that the liquid and vapor are separated by a well-defined boundary or meniscus. Now suppose the temperature is

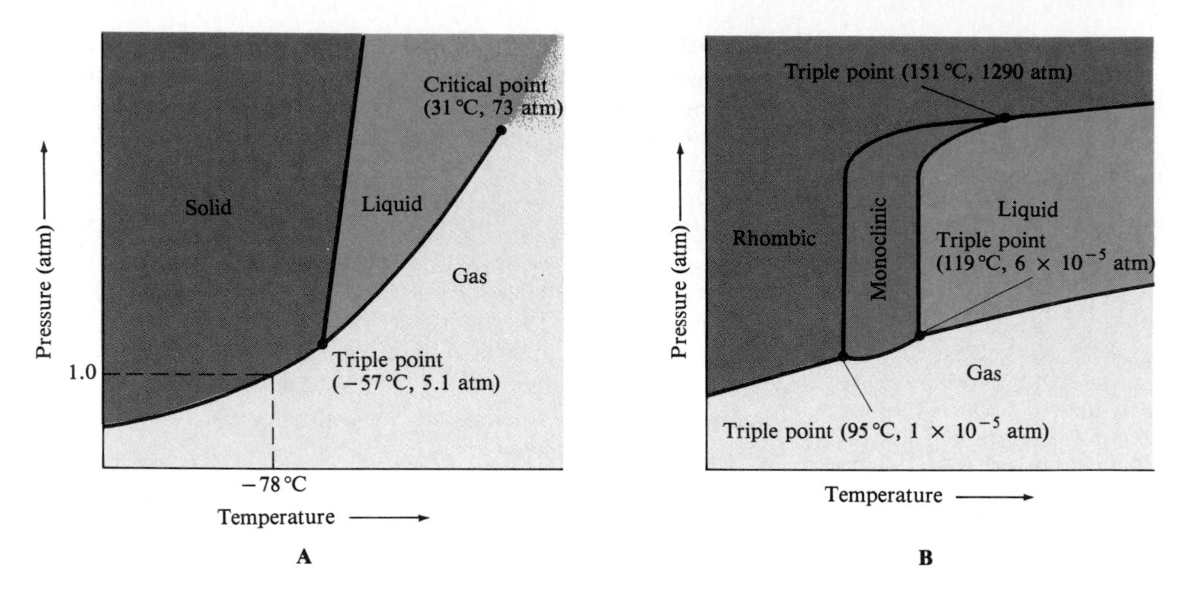

Figure 11.11
Phase diagrams for carbon dioxide and sulfur (not to scale).
(A) Carbon dioxide. At normal atmospheric pressure (1 atm), the solid sublimes when warmed.
(B) Sulfur. This substance has a complicated phase diagram with three triple points.

raised. The vapor pressure increases, and at 30°C it is 71 atm. Then, as the temperature approaches 31°C, a curious thing happens. The meniscus becomes fuzzy and less well defined. At 31°C, the meniscus disappears altogether. Above this temperature, there is only one fluid state, carbon dioxide gas (see Figure 11.12).

The temperature above which the liquid state of a substance no longer exists is called the **critical temperature.** For carbon dioxide it is 31°C. *The vapor pressure at the critical temperature* is called the **critical pressure** (73 atm for carbon dioxide). It is the minimum pressure that must be applied to a gas at the critical temperature to liquefy it.

On a phase diagram, the preceding experiment corresponds to following the vapor-pressure curve where the liquid and vapor are in equilibrium. Note that this curve in Figure 11.11A ends at a point at which the temperature and pressure have

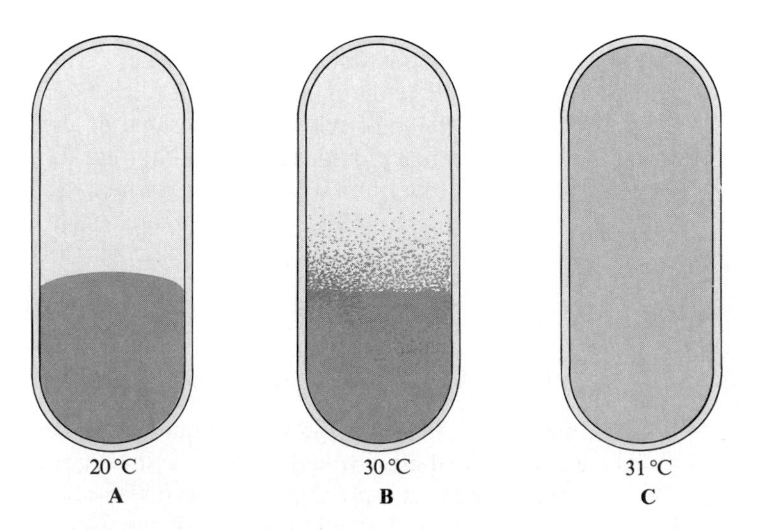

Figure 11.12
Liquid and vapor in a closed vessel at different temperatures (critical phenomenon).
(A) Carbon dioxide liquid in equilibrium with its vapor at 20°C.
(B) The same two states in equilibrium at 30°C (just below the critical point); the density of the liquid and vapor are becoming equal.
(C) Carbon dioxide at 31°C (the critical temperature); the liquid and vapor now have the same densities (in fact, the distinction between liquid and vapor has disappeared).

their critical values. This is the *critical point*. If you look at the phase diagram for water, you will see that the vapor-pressure curve for the liquid similarly ends, at point C, which is the critical point for water. In this case, the critical temperature is 374°C and the critical pressure is 218 atm.

Many important gases cannot be liquefied at room temperature. Nitrogen, for example, has a critical temperature of −147°C. This means the gas cannot be liquefied until the temperature is below −147°C.

Example 11.4 Relating the Conditions for the Liquefaction of Gases to the Critical Temperature

The critical temperatures of ammonia and nitrogen at 132°C and −147°C, respectively. Explain why ammonia can be liquefied at room temperature by merely compressing the gas to a high enough pressure, whereas the compression of nitrogen requires a low temperature as well.

Solution

The critical temperature of ammonia is well above room temperature. If ammonia gas at room temperature is compressed, it will liquefy. Nitrogen, however, has a critical temperature well below room temperature. It cannot be liquefied by compression unless the temperature is below the critical temperature.

Exercise 11.4

Describe how you could liquefy the following gases: (a) methyl chloride, CH_3Cl (critical point, 144°C, 66 atm); (b) oxygen, O_2 (critical point, −119°C, 50 atm).

(See Problems 11.39 and 11.40.)

Liquid State

In this part of the chapter, we will look at some physical properties of liquids. After that, we will explain the experimental values of these properties in terms of intermolecular forces.

11.4 PROPERTIES OF LIQUIDS: SURFACE TENSION AND VISCOSITY

We have seen that molecules tend to escape the liquid state and form a vapor. The vapor pressure is the equilibrium partial pressure of this vapor over the liquid; it increases with temperature. The boiling point is the temperature at which the vapor pressure equals the pressure applied to the liquid. Both vapor pressure and boiling point are important properties of a liquid. Values of the vapor pressure for some liquids at 20°C are listed in Table 11.3 (column 3). Two additional properties given in Table 11.3 are surface tension and viscosity, which we will discuss in this section. All of these properties, as we will see in the next section, depend on intermolecular forces, which are related to molecular structure.

Table 11.3 Properties of Some Liquids at 20°C

Substance	Molecular Weight (amu)	Vapor Pressure (mmHg)	Surface Tension (J/m^2)	Viscosity ($kg/m \cdot s$)
Water, H_2O	18	1.8×10^1	7.3×10^{-2}	1.0×10^{-3}
Carbon dioxide, CO_2	44	4.3×10^4	1.2×10^{-3}	7.1×10^{-5}
Pentane, C_5H_{12}	72	4.4×10^2	1.6×10^{-2}	2.4×10^{-4}
Glycerol, $C_3H_8O_3$	92	1.6×10^{-4}	6.3×10^{-2}	1.5×10^0
Chloroform, $CHCl_3$	119	1.7×10^2	2.7×10^{-2}	5.8×10^{-4}
Carbon tetrachloride, CCl_4	154	8.7×10^1	2.7×10^{-2}	9.7×10^{-4}
Bromoform, $CHBr_3$	253	3.9×10^0	4.2×10^{-2}	2.0×10^{-3}

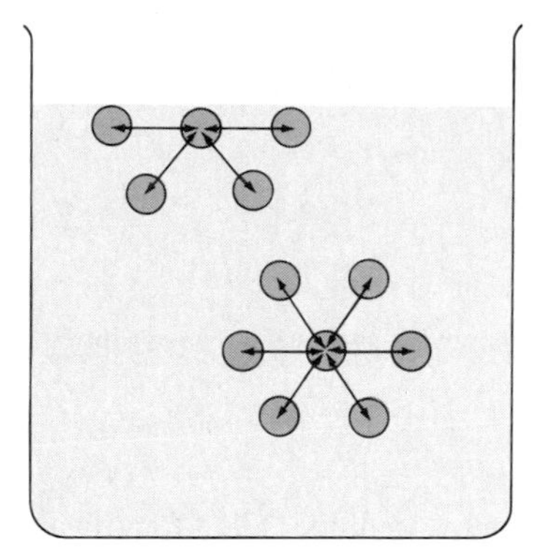

Figure 11.13
Representation of the forces between molecules in a liquid. Note that a molecule at the surface experiences a net force toward the interior of the liquid, whereas a molecule in the interior experiences no net force.

Surface Tension

As Figure 11.13 illustrates, a molecule within the body of a liquid tends to be attracted equally in all directions, so that it experiences no net force. On the other hand, a molecule at the surface of a liquid feels a net attraction by other molecules toward the interior of the liquid. As a result, there is a tendency for the surface area of a liquid to be reduced as much as possible. This explains why falling raindrops are nearly spherical. (The sphere has the smallest surface area for a given volume of any geometrical shape.)

Energy is needed to reverse the tendency toward reduction of surface area in liquids. **Surface tension** is *the energy required to increase the surface area of a liquid by a unit amount*. Column 4 in Table 11.3 lists values of the surface tension of some liquids.

The surface tension of a liquid can be affected by dissolved substances. Soaps and detergents, in particular, drastically decrease the surface tension of water. An interesting but simple experiment shows the effect of soap on surface tension. Because of surface tension, a liquid behaves as though it had a skin. Water bugs seem to skitter across this skin as if ice-skating (Figure 11.14). You can actually float a needle on water, if you carefully lay it across the surface (Figure 11.15). If you then put a drop of soap solution onto the water, the soap spreads across the surface, and the needle sinks. Water bugs also sink in soapy water.

Figure 11.14
Water bug on a water surface. The water bug seems to skate on the surface "skin" of the water, which is actually a result of surface tension.

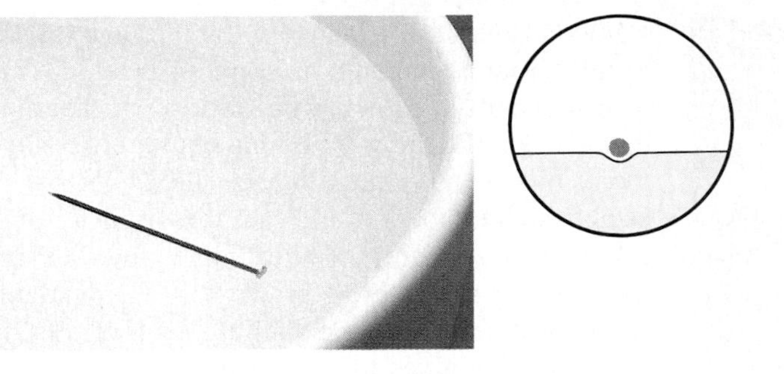

Figure 11.15
Demonstration of surface tension of water. A steel needle will float on the surface of water because of surface tension *(left)*. The water surface is depressed and slightly stretched by the needle *(right)*, but even more surface would have to be created for the needle to submerge, which takes more energy. The easiest way to float a needle is to place a square of tissue paper, with the needle on top of it, on the prongs of a fork. When the fork is lowered into the water, the tissue and needle float on the surface. The paper soon becomes water-logged and sinks, leaving the needle floating.

Capillary rise is a phenomenon that is related to surface tension. When a small-diameter glass tube, or capillary, is placed upright in water, a column of the liquid rises in the tube (Figure 11.16A). This capillary rise can be explained in the following way: Water molecules happen to be attracted to glass. Because of this attraction, a thin film of water starts to move up the inside of the glass capillary. But in order to reduce the surface area of this film, the water level begins to rise also. The final water level is a balance between the surface tension and the potential energy required to lift the water against gravity. Note that the *meniscus,* or liquid surface within the tube, has its edges curved upward; that is, it has a concave shape.

In the case of mercury, the liquid level in the capillary is lower than it is for the liquid outside (Figure 11.16B). Also, the meniscus has its edges curved downward; that is, it has a convex shape. Compare Figure 11.16B and Figure 11.16A. The difference in behavior results from the fact that the attraction between mercury atoms is greater than the attraction of mercury atoms for glass, in contrast to the situation described for water and glass.

Viscosity

Viscosity is *the resistance to flow that is exhibited by all liquids*. It can be obtained by measuring the time it takes for a given quantity of liquid to flow through a

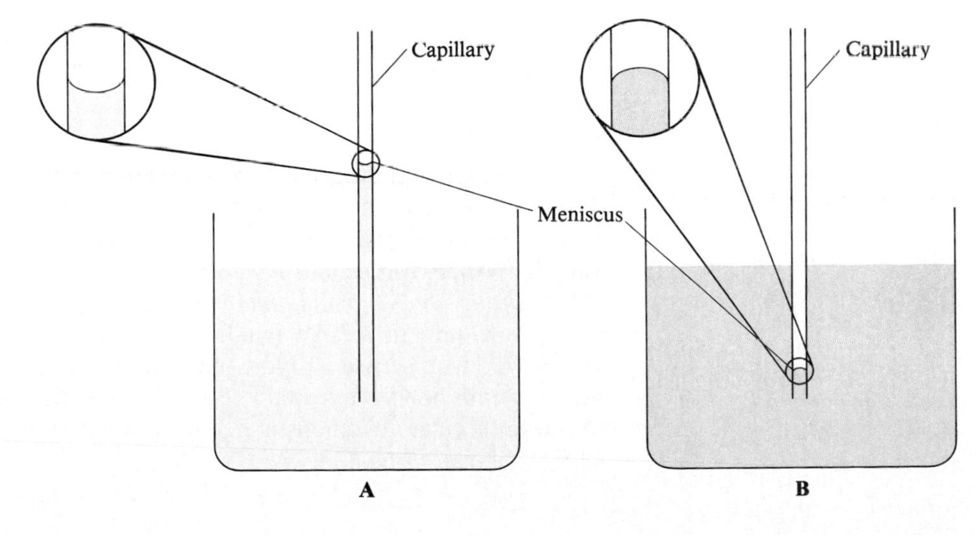

Figure 11.16
Liquid levels in capillaries. *(A)* Capillary rise, due to the attraction of water and glass. The final water level in the capillary is a balance between the force of gravity and the surface tension of water. *(B)* Depression, or lowering, of mercury level in a glass capillary. Unlike water, mercury is not attracted to glass.

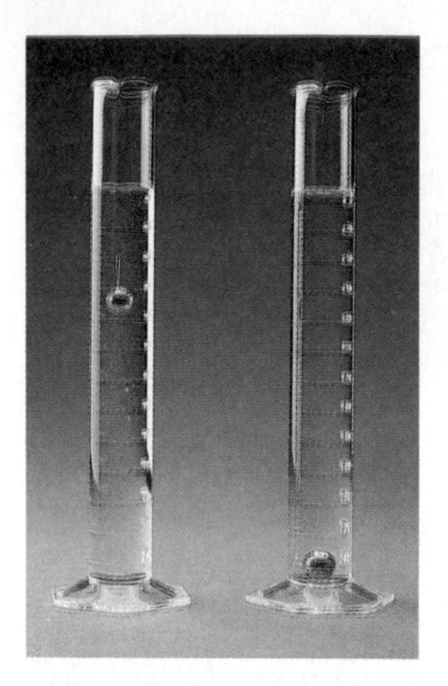

Figure 11.17
Comparison of the viscosities of two liquids. Similar steel balls were dropped simultaneously into two graduated cylinders, the right one containing water and the left one glycerol. A steel ball takes considerably longer to fall through a column of glycerol than through a similar column of water, because glycerol has a greater viscosity than does water.

capillary tube or, alternatively, by the time it takes for a steel ball of given radius to fall through a column of liquid (Figure 11.17). Even without such quantitative measurements, we know from experience that the viscosity of syrup is greater than that of water; syrup is a *viscous* liquid. Viscosity is an important characteristic of motor oils. In the United States, the Society of Automotive Engineers (SAE) has established numbers to indicate the viscosity of motor oils; higher numbers indicate greater viscosities. For example, an oil designated SAE 10W/40 has a viscosity at low temperature (0°F or −18°C, denoted by W for winter) of SAE 10, whereas at a higher temperature (210°F or 99°C), its value is SAE 40. Column 5 in Table 11.3 gives the viscosities of some liquids in SI units.

11.5 INTERMOLECULAR FORCES; EXPLAINING LIQUID PROPERTIES

In the preceding section, we described a number of properties of liquids. Now in this section we want to explain these properties. Many of the physical properties of liquids (and certain solids too) can be explained in terms of **intermolecular forces,** *the forces of interaction between molecules.* These forces are normally weakly attractive.

One of the most direct indications of the attraction between molecules is the heat of vaporization of liquids. Consider a substance like neon, which consists of molecules that are single atoms (there is no tendency for these atoms to chemically bond). Neon is normally a gas, but it liquefies when the temperature is lowered to −246°C at 1.00 atm of pressure. The heat of vaporization of the liquid at this temperature is 1.77 kJ/mol. Some of this energy is needed to push back the atmosphere when the vapor forms (0.23 kJ/mol). The remaining energy (1.54 kJ/mol) must be supplied to overcome intermolecular attractions. Because a molecule in a liquid is surrounded by several neighbor molecules, this remaining energy is some multiple of a single molecule–molecule interaction. Typically, this multiple is about 5. Thus, we expect the neon–neon interaction energy to be about 0.3 kJ/mol. Other experiments yield a similar value.● By comparison, the energy of attraction between two hydrogen atoms in the hydrogen molecule is 432 kJ/mol. Thus, the energy of attraction between neon atoms is about a thousand times smaller than that between atoms in a chemical bond.

● The potential energy of interaction between molecules can be obtained by *molecular beam* experiments, in which one beam of molecules is made to collide with another beam. The distribution of scattered molecules is measured, and the potential energy of interaction between molecules is calculated from this distribution.

Attractive intermolecular forces can be larger than those in neon. For example, chlorine, Cl_2, and bromine, Br_2, have intermolecular attractive energies of 3.0 kJ/mol and 4.3 kJ/mol, respectively. These values are still much smaller than bond energies.

Three types of attractive forces are known to exist between neutral molecules: dipole–dipole forces, London (or dispersion) forces, and hydrogen bonding forces. Each of these will be explained in the remainder of this section. The term **van der Waals forces** is *a general term for those intermolecular forces that include dipole–dipole and London forces.*● Van der Waals forces are the weak attractive forces in a large number of substances, including Ne, Cl_2, and Br_2, which we just discussed. Hydrogen bonding occurs in substances containing hydrogen atoms bonded to certain very electronegative atoms. Approximate energies of intermolecular attractions are compared with those of chemical bonds in Table 11.4.

● The Dutch physicist Johannes van der Waals (1837–1923) was the first to suggest the importance of intermolecular forces and used the concept to derive his equation for gases (see van der Waals equation, Section 5.8).

Table 11.4 Types of Intermolecular and Chemical Bonding Interactions

Type of Interaction	Approximate Energy (kJ/mol)
Intermolecular	
Van der Waals (London, dipole–dipole)	0.1 to 10
Hydrogen bonding	10 to 40
Chemical bonding	
Ionic	100 to 1000
Covalent	100 to 1000

Dipole–Dipole Forces

Polar molecules can attract one another through dipole–dipole forces. The **dipole–dipole force** is *an attractive intermolecular force resulting from the tendency of polar molecules to align themselves such that the positive end of one molecule is near the negative end of another*. Recall that a polar molecule has a dipole moment as a result of the electronic structure of the molecule.● For example, hydrogen chloride, HCl, is a polar molecule because of the difference in electronegativities of the H and Cl atoms.

● Dipole moments were discussed in Section 10.2.

$$\overset{\delta+}{\text{H}}-\overset{\delta-}{\text{Cl}}$$

Figure 11.18 shows the alignment of polar molecules in the case of HCl. Note that this alignment creates a net attraction between molecules. This attractive force is partly responsible for the fact that hydrogen chloride, which is a gas at room temperature, becomes a liquid when cooled to −85°C. At this temperature, HCl molecules have slowed enough for the intermolecular forces to hold the molecules in the liquid state.

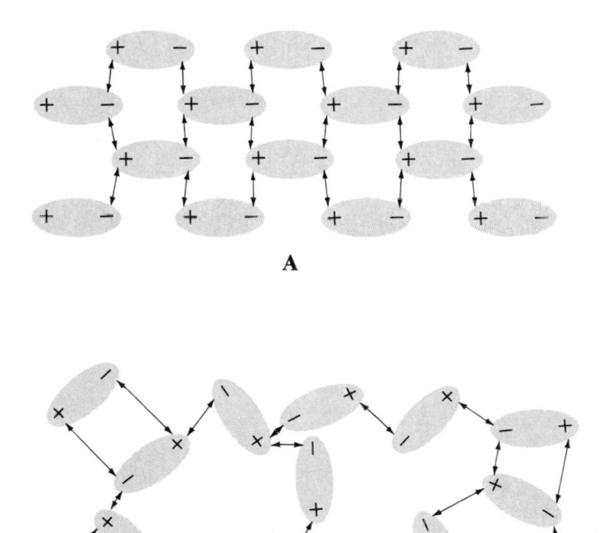

Figure 11.18
Alignment of polar molecules of HCl. *(A)* Molecules tend to line up in the solid so that positive ends point to negative ends. *(B)* The normal random motion of molecules in a liquid only partially disrupts this alignment of polar molecules.

London (Dispersion) Forces

In nonpolar molecules, there can be no dipole–dipole force. Yet such substances liquefy, so there must be another type of attractive intermolecular force. Even in hydrogen chloride, calculations show that the dipole–dipole force accounts for less than 20% of the total energy of attraction. What is the nature of this additional attractive force?

In 1930 Fritz London found he could account for a weak attraction between any two molecules. Let us use neon to illustrate his argument. The electrons of this atom move about the nucleus so that over a period of time they give a spherical distribution. The electrons spend as much time on one side of the nucleus as on the other. Experimentally, we find that the atom has no dipole moment. However, at any instant, more electrons may be on one side of the nucleus than on the other. The atom becomes a small, instantaneous dipole, with one side having a partial negative charge, δ^-, and the other side having a partial positive charge, δ^+ (see Figure 11.19A).

Now imagine that another neon atom is close to this first atom—say, next to the partial negative charge that appeared at that instant. This partial negative charge repels the electrons in the second atom, so at this instant it too becomes a small dipole (see Figure 11.19B). Note that these two dipoles are oriented with the partial negative charge of one atom next to the partial positive charge of the other atom. Therefore, an attractive force exists between the two atoms.

Electrons are in constant motion, but the motion of electrons on one atom affects the motion of electrons on the other atom. As a result, the *instantaneous dipoles* of the atoms tend to change together, maintaining a net attractive force (compare B and C of Figure 11.19). Such instantaneous changes of electron distributions occur in all molecules. For this reason, an attractive force always exists between any two molecules.

Thus, we see that **London forces** (also called **dispersion forces**) are *the weak attractive forces between molecules resulting from the small, instantaneous dipoles that occur because of the varying positions of the electrons during their motion about nuclei.*

London forces tend to increase with molecular weight. This is because molecules with larger molecular weight usually have more electrons, and London forces increase in strength with the number of electrons. Also, larger molecular weight often means larger atoms, which are more *polarizable* (more easily distorted to give instantaneous dipoles because the electrons are farther from the nuclei). The relationship of London forces to molecular weight is only approximate, however. For molecules of about the same molecular weight, the more compact one is probably less polarizable, so the London forces are smaller. Con-

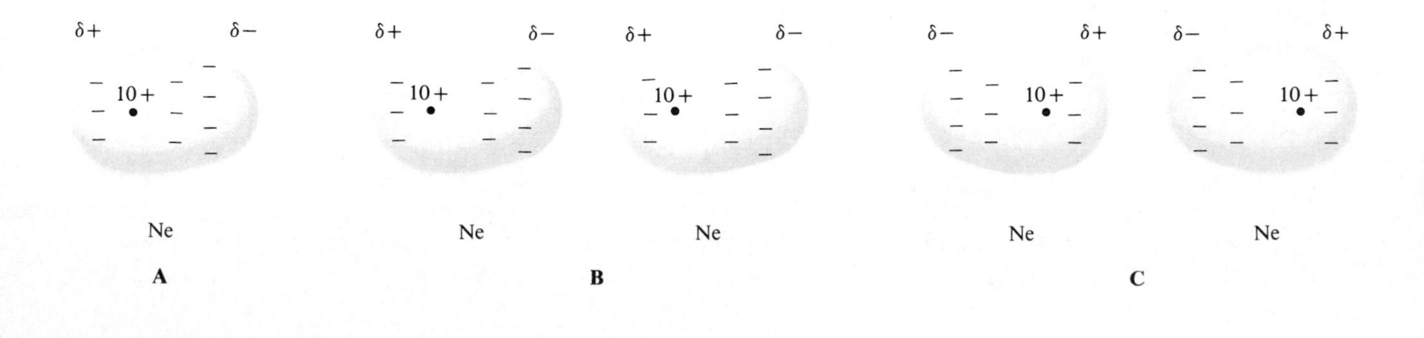

Figure 11.19
Origin of the London force.
(A) At some instant, there are more electrons on one side of a neon atom than on the other. *(B)* If this atom is near another neon atom, the electrons on that atom are repelled. The result is two instantaneous dipoles, which give an attractive force. *(C)* Later the electrons on both atoms have moved, but they tend to move together, which gives an attractive force between the atoms.

sider pentane, isopentane (2-methylbutane), and neopentane (2,2-dimethylpropane).

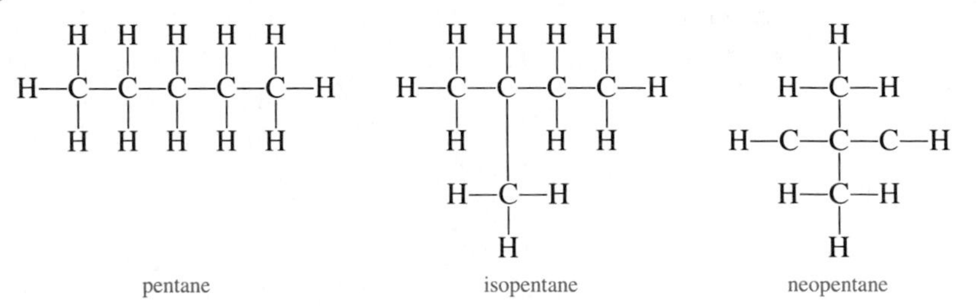

All have the same molecular formula, C_5H_{12}, and thus the same molecular weight, but they differ in the arrangement of the atoms. Pentane is a long chain of carbon atoms to which hydrogen atoms are attached. Isopentane and neopentane, however, have increasingly more compact arrangements of atoms. As a result, we expect London forces to decrease from pentane to isopentane and from isopentane to neopentane. This is what we find in the heats of vaporization. The values are 25.8 kJ/mol, 24.7 kJ/mol, and 22.8 kJ/mol for pentane, isopentane, and neopentane, respectively.

Van der Waals Forces and the Properties of Liquids

The vapor pressure of a liquid depends on intermolecular forces, because the ease or difficulty with which a molecule leaves the liquid depends on the strength of its attraction to other molecules. When the intermolecular forces in the liquid are strong, we expect the vapor pressure to be low.

Let us look again at Table 11.3, which lists vapor pressures as well as surface tensions and viscosities of liquids at 20°C. The intermolecular attractions in these liquids (except for water and glycerol) are entirely due to van der Waals forces. Of the van der Waals forces, the London force is always present and usually dominant. The dipole–dipole force is usually appreciable only in small polar molecules or in large molecules having very large dipole moments. As we mentioned earlier, the dipole–dipole attraction in the small, polar molecule HCl accounts for less than 20% of its interaction energy.

The liquids in Table 11.3 are listed in order of increasing molecular weight. Recall that London forces tend to increase with molecular weight. Therefore, London forces would be expected to increase for these liquids from the top to the bottom of the table. If London forces are the dominant attractive forces in these liquids, the vapor pressures should decrease from top to bottom in the table. And this is what we see, except for water and glycerol (their vapor pressures are relatively low, and an additional force, hydrogen bonding, is needed to explain them).

The normal boiling point of a liquid must depend on intermolecular forces, because it is related to the vapor pressure. Normal boiling points are approximately proportional to the energy of intermolecular attraction. Thus, they are lowest for liquids with the weakest intermolecular forces.

Surface tension also depends on intermolecular forces. Surface tension is the energy needed to increase the surface area of a liquid. To increase the surface area, it is necessary to pull molecules apart against the intermolecular forces of attraction. Thus, surface tension would be expected to increase with the strength of attractive forces. If London forces are the dominant attractive forces in the liquid, surface tension should increase with molecular weight—that is, it should increase from top to bottom in Table 11.3. This is what we see, except for water and glycerol, which have relatively high surface tensions.

The viscosity of a liquid depends in part on intermolecular forces, because increasing the attractive forces between molecules increases the resistance to flow. If London forces are dominant, we should expect the viscosity of the liquids in Table 11.3 to increase from the top of the column to the bottom, which it does (again with the exception of water and glycerol; these liquids have relatively high viscosities). We should note, however, that the viscosity also depends on other factors, such as the possibility of molecules tangling together. Liquids with long molecules that tangle together are expected to have high viscosities.

Hydrogen Bonding

It is interesting to compare fluoromethane, CH_3F, and methanol, CH_3OH. Both substances have about the same molecular weight (34 for CH_3F and 32 for CH_3OH) and about the same dipole moment (1.81 D for CH_3F and 1.70 D for CH_3OH). We might expect these substances to have about the same intermolecular attractive forces and therefore about the same boiling points. In fact, the boiling points are quite different. Fluoromethane boils at −78°C and is a gas under normal conditions. Methanol boils at 65°C and is normally a liquid. There is apparently an intermolecular attraction in methanol that is not present in fluoromethane.

We have already seen that the properties of water and glycerol cannot be explained in terms of van der Waals forces alone. What water, glycerol, and methanol have in common is one or more —OH groups.

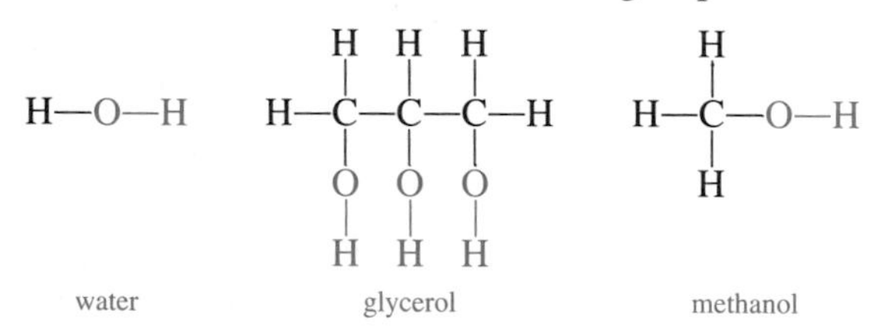

water glycerol methanol

Molecules that have this group are subject to an additional attractive force called hydrogen bonding. **Hydrogen bonding** is *a weak to moderate attractive force that exists between a hydrogen atom covalently bonded to a very electronegative atom, X, and a lone pair of electrons on another small, electronegative atom, Y.* It is represented in structural drawings by a series of dots.

$$—X—H\cdots Y—$$

Usually, hydrogen bonding is seen in cases where X and Y are the atoms F, O, and N.

Figure 11.20A shows a plot of boiling point versus molecular weight for the hydrides (binary compounds with hydrogen) of the Group VIA elements. If London forces were the only intermolecular forces present in this series of compounds, the boiling points should increase regularly with molecular weight. We do see such a regular increase in boiling point for H_2S, H_2Se, and H_2Te. On the other hand, H_2O has a much higher boiling point than we would expect if only London forces were present. This result is consistent with the view that hydrogen bonding exists in H_2O but is essentially absent in H_2S, H_2Se, and H_2Te.

Studies of the structure of H_2O in its different physical states show that hydrogen bonding is present. The hydrogen atom of one water molecule is attracted to

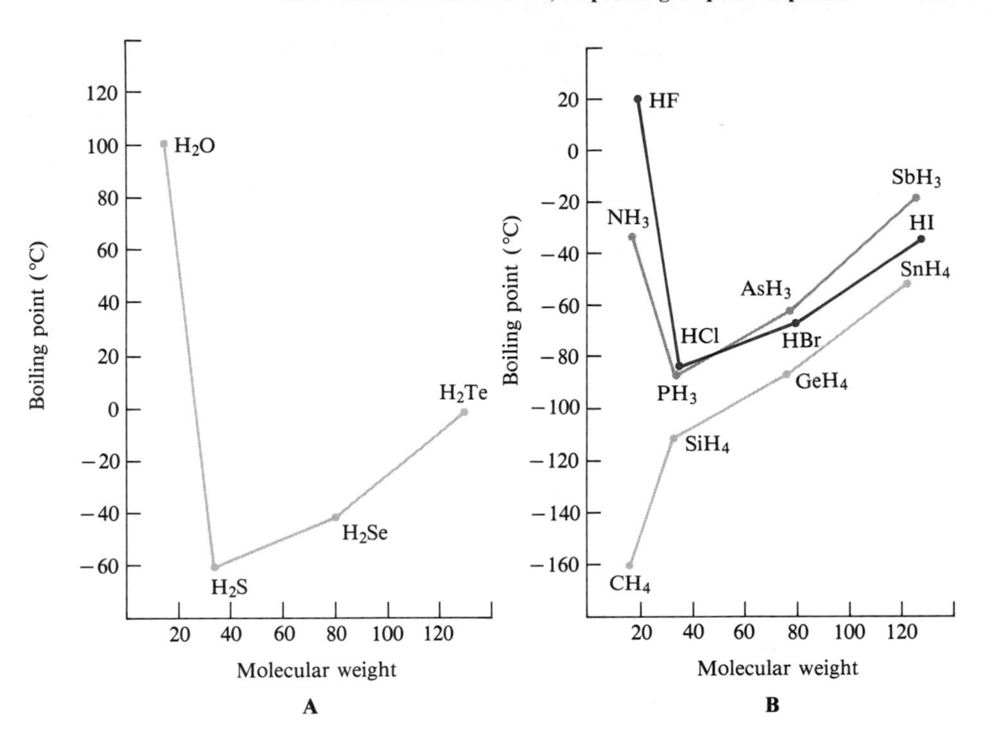

Figure 11.20
Boiling point versus molecular weight for hydrides. *(A)* Plot for the Group VIA hydrides. Note that the boiling points of H_2S, H_2Se, and H_2Te increase fairly smoothly, as we would expect if London forces are dominant. H_2O has a much higher boiling point, indicating an additional intermolecular attraction. *(B)* Boiling point versus molecular weight for the hydrides of Groups VIIA, VA, and IVA.

the electron pair of another water molecule (Figure 11.21). Many H_2O molecules can be linked this way to form clusters of molecules. Similar hydrogen bonding is seen in other molecules containing —O—H groups.

Plots of boiling point versus molecular weight for Group VIIA, VA, and IVA hydrides are shown in Figure 11.20B. Hydrogen fluoride, HF, and ammonia, NH_3, have particularly high boiling points compared with other hydrides of the same periodic group of elements. Definite structural evidence has established

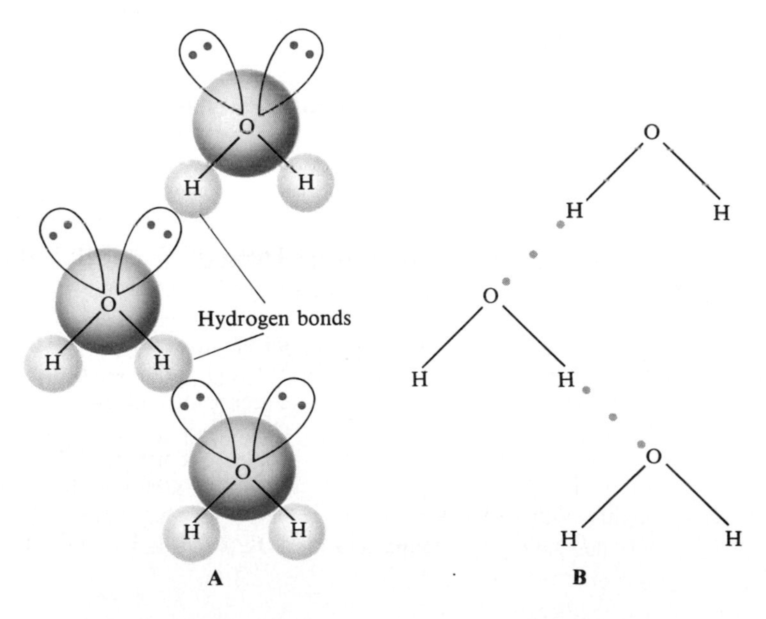

Figure 11.21
Hydrogen bonding in water. *(A)* The electrons in the O—H bonds of H_2O molecules are attracted to the oxygen atoms, leaving the positively charged protons partially exposed. A proton on one molecule is attracted to a lone pair on an oxygen atom in another water molecule. *(B)* Hydrogen bonding between water molecules is represented by a series of dots.

● Until recently, the high boiling point of ammonia was thought to be evidence for hydrogen bonding. Spectroscopic evidence now shows that although two NH_3 molecules do attract one another, the hydrogen atom of one molecule does not point toward the lone pair of the other. Thus, the intermolecular attraction appears to be something other than hydrogen bonding. See D. D. Nelson, G. T. Fraser, and W. Klemperer, *Science*, Vol. 239, p. 1670 (1987).

hydrogen bonding in HF. In solid hydrogen fluoride, the structure shows a zigzag arrangement involving hydrogen bonding.●

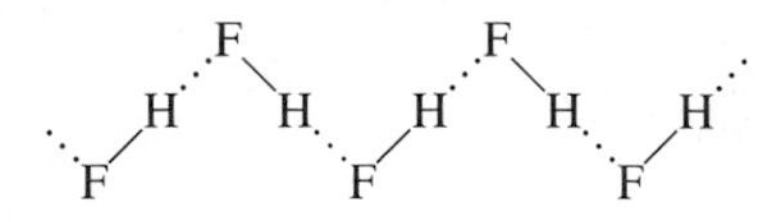

Let us look at the hydrogen-bonding attraction between two water molecules. The attractive force may be explained in part on the basis of the dipole moment of the —O—H bond.

$$\overset{\delta-}{-\text{O}}-\overset{\delta+}{\text{H}}$$

The partial positive charge on the hydrogen atom is attracted to the negative charge of a lone pair on another molecule. In addition, however, a hydrogen atom covalently bonded to an electronegative atom appears to be special. The electrons in the —O—H bond are drawn to the O atom, leaving the concentrated positive charge of the proton partially exposed. This concentrated positive charge is strongly attracted to a lone pair of electrons on another O atom.

Example 11.5 Identifying Intermolecular Forces

What kinds of intermolecular forces (London, dipole–dipole, hydrogen bonding) are expected in the following substances? (a) methane, CH_4; (b) trichloromethane (chloroform), $CHCl_3$; (c) butanol (butyl alcohol), $CH_3CH_2CH_2CH_2OH$.

Solution

(a) Methane is a nonpolar molecule. Hence, the only intermolecular attractions are **London forces.** (b) Trichloromethane is an unsymmetrical molecule with polar bonds. Thus, we expect **dipole–dipole forces,** in addition to **London forces.** (c) Butanol has a hydrogen atom attached to an oxygen atom. Therefore, we expect **hydrogen bonding.** Because the molecule is polar (from the O—H bond), we also expect **dipole–dipole forces. London forces** exist too, because such forces exist between all molecules.

Exercise 11.5

List the different intermolecular forces you would expect for each of the following compounds: (a) propanol, $CH_3CH_2CH_2OH$; (b) carbon dioxide, CO_2; (c) sulfur dioxide, SO_2.

(See Problems 11.45 and 11.46.)

Example 11.6 Determining Relative Vapor Pressure on the Basis of Intermolecular Attraction

For each of the following pairs, choose the substance you expect to have the lower vapor pressure at a given temperature: (a) carbon dioxide (CO_2) or sulfur dioxide (SO_2); (b) dimethyl ether (CH_3OCH_3) or ethanol (CH_3CH_2OH).

Solution

(a) London forces increase roughly with molecular weight. The molecular weights for SO_2 and CO_2 are 64 amu and 44 amu, respectively. Therefore, the London forces between SO_2 molecules should be greater than the London forces between CO_2 molecules. Moreover, because SO_2 is polar but CO_2 is not, there are dipole–dipole forces between SO_2 molecules but not between CO_2 molecules. **We conclude that sulfur dioxide has the lower vapor pressure.** Experimental values of the vapor pressure at 20°C are: CO_2, 56.3 atm; SO_2, 3.3 atm. (b) Dimethyl ether and ethanol have the same molecular formulas but different structural formulas. The structural formulas are

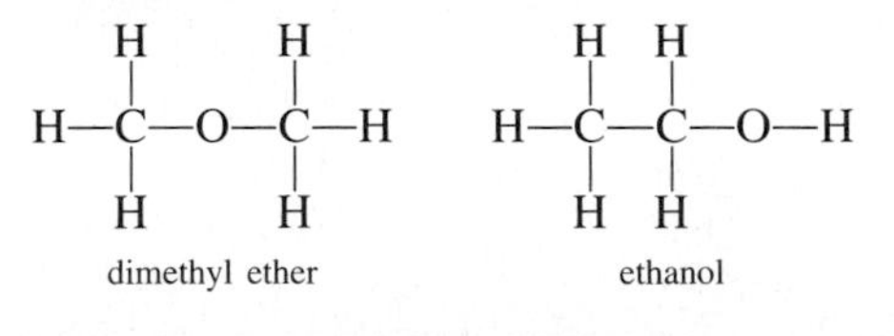

The molecular weights are equal, and therefore the London forces are approximately the same. However, there will be strong hydrogen bonding in ethanol but not in dimethyl ether. **We expect ethanol to have the lower vapor pressure.** Experimental values of vapor pressure at 20°C are: CH_3OCH_3, 4.88 atm; CH_3CH_2OH, 0.056 atm.

Exercise 11.6

Arrange the following hydrocarbons in order of increasing vapor pressure: ethane, C_2H_6; propane, C_3H_8; and butane, C_4H_{10}. Explain your answer.

(See Problems 11.49 and 11.50.)

Exercise 11.7

Methyl chloride, CH_3Cl, has a vapor pressure of 1490 mmHg, and ethanol has a vapor pressure of 42 mmHg. Explain why you might expect methyl chloride to have a higher vapor pressure than ethanol, even though methyl chloride has a somewhat larger molecular weight.

(See Problems 11.51 and 11.52.)

Solid State

A solid is a nearly incompressible state of matter with well-defined shape, because the units (atoms, molecules, or ions) making up the solid are in close contact and in fixed positions or sites. In the next section, we will look at the kinds of forces holding the units together in different types of solids. In later sections, we will look at crystalline solids and their structure.

11.6 CLASSIFICATION OF SOLIDS BY TYPE OF ATTRACTION OF UNITS

A solid consists of structural units—atoms, molecules, or ions—that are attracted to one another strongly enough to give a rigid substance. One way to classify solids is by the type of force holding the structural units together. In some cases, these forces are intermolecular and hold molecules together. In other cases, these forces are chemical bonds (metallic, ionic, or covalent) and hold atoms together. From this point of view, then, there are four different types of solids: molecular, metallic, ionic, and covalent. It should be noted, however, that a given ionic bond may have considerable covalent character or vice versa, so the distinction between ionic and covalent solids is not a precise one.

Types of Solids

A **molecular solid** is *a solid that consists of atoms or molecules held together by intermolecular forces.* Many solids are of this type. Examples: solid neon, solid water (ice), solid carbon dioxide (dry ice).

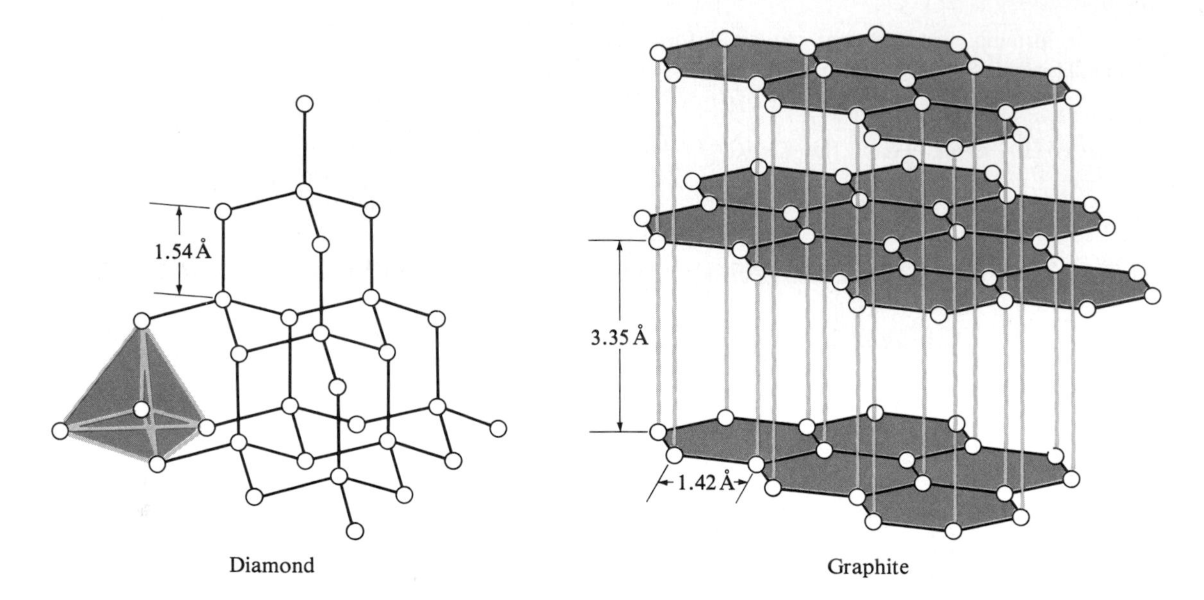

Figure 11.22
Structures of diamond and graphite. Diamond is a three-dimensional network solid; every carbon atom is covalently bonded to four others. Graphite consists of sheets of carbon atoms covalently bonded to form a hexagonal pattern of atoms; the sheets are held together by van der Waals forces.

A **metallic solid** is *a solid that consists of positive cores of atoms held together by a surrounding "sea" of electrons (metallic bonding).* In this kind of bonding, positively charged atomic cores are surrounded by delocalized electrons. Examples: iron, copper, silver.

An **ionic solid** is *a solid that consists of cations and anions held together by the electrical attraction of opposite charges (ionic bonds).* Examples: cesium chloride, sodium chloride, zinc sulfide (but ZnS has considerable covalent character).

A **covalent network solid** is *a solid that consists of atoms held together in large networks or chains by covalent bonds.* Diamond is an example of a three-dimensional network solid. Every carbon atom in diamond is covalently bonded to four others, so an entire crystal might be considered an immense molecule. It is possible to have similar two-dimensional "sheet" and one-dimensional "chain" molecules, with atoms held together by covalent bonds. Examples: diamond (three-dimensional network), graphite (sheets), asbestos (chains). Figure 11.22 shows the structures of diamond and graphite.

Table 11.5 summarizes these four types of solids.

Table 11.5 Types of Solids

Type of Solid	Structural Units	Attractive Forces Between Structural Units	Examples
Molecular	Atoms or molecules	Intermolecular forces	Ne, H_2O, CO_2
Metallic	Atoms (positive cores surrounded by electron "sea")	Metallic bonding (extreme delocalized bond)	Fe, Cu, Ag
Ionic	Ions	Ionic bonding	CsCl, NaCl, ZnS
Covalent network	Atoms	Covalent bonding	Diamond, graphite, asbestos

Example 11.7 Identifying Types of Solids

Which of the four basic types of solids would you expect the following substances to be? (a) solid ammonia, NH_3; (b) cesium, Cs; (c) cesium iodide, CsI; (d) silicon, Si.

Solution

(a) Ammonia has a molecular structure; therefore, it freezes as a molecular solid. (b) Cesium is a metal; it is a metallic solid. (c) Cesium iodide is an ionic substance; it exists as an ionic solid. (d) Silicon atoms might be expected to form covalent bonds to other silicon atoms, as carbon does in diamond. A covalent network solid would result.

Exercise 11.8

Classify each of the following solids according to the forces of attraction that exist between the structural units: (a) zinc; (b) sodium iodide, NaI; (c) silicon carbide, SiC; (d) methane, CH_4.

(See Problems 11.55, 11.56, 11.57, and 11.58.)

Physical Properties

Many physical properties of a solid can be directly related to its structure. Let us look at several of these properties.

MELTING POINT AND STRUCTURE Table 11.2 lists the melting points for various solids and also gives the type of solid. For a solid to melt, the forces holding the structural units in their sites must be overcome, at least partially. In a molecular solid, these forces are weak intermolecular attractions. Thus, molecular solids tend to have low melting points (usually below 300°C). At room temperature, many molecular substances are either liquids (such as water and ethanol) or gases (such as carbon dioxide and ammonia). Note also that the melting points of molecular solids reflect the strengths of intermolecular attractions, just as boiling points do. This can be seen in the melting points of the molecular solids listed in Table 11.2.

For ionic solids and covalent network solids to melt, chemical bonds must be broken. For that reason, the melting points of these types of solids are relatively high. The melting point of the ionic solid sodium chloride is 801°C; that of magnesium oxide is 2800°C. Melting points of covalent network solids are generally quite high. Quartz, for example, melts at 1610°C; diamond melts at 3550°C. See Table 11.2.●

● Both molecular and covalent network solids have covalent bonds, but none of the covalent bonds are broken during the melting of a molecular solid.

The difference between the melting points of sodium chloride (801°C) and magnesium oxide (2800°C) can be explained in terms of the charges on the ions. We expect the melting point to rise with lattice energy.● Lattice energies, however, depend on the product of the ionic charges. For sodium chloride (Na^+Cl^-) this product is $(+1) \times (-1) = -1$, whereas for magnesium oxide ($Mg^{2+}O^{2-}$) it is $(+2) \times (-2) = -4$. Thus, the lattice energy is greater for magnesium oxide, and this is reflected in the much higher melting point of magnesium oxide.

● Lattice energy is the energy needed to separate a crystal into isolated ions in the gaseous state. It represents the strength of attraction of ions in the solid.

Metals often have high melting points, but there is considerable variability. Mercury, which is a liquid at room temperature, melts at −39°C. Tungsten melts at 3410°C, the highest melting point of any metallic element. In general, melting points are low for the elements at the left side of the periodic table (Groups IA and IIA) but increase as we move right to the transition metals. Those elements in the center of the transition series (such as tungsten) have high melting points. Then, as we move farther to the right, the melting points decrease and are again low for Group IIB elements.

Example 11.8 Determining Relative Melting Points Based on Types of Solids

Arrange the following elements in order of increasing melting point: silicon, hydrogen, lithium. Explain your reasoning.

Solution

Hydrogen is a molecular substance (H_2) with a very low molecular weight (2 amu). Thus, we expect it to have a very low melting point. Lithium is a Group IA element and is expected to be a relatively low-melting metal. Except for mercury, however, all metals are solids below 25°C. So the melting point of lithium is well above that of hydrogen. Silicon might be expected to have a covalent network structure like carbon and to have a high melting point. Therefore, the list of these elements in order of increasing melting point is **hydrogen, lithium, and silicon.**

Exercise 11.9

Decide what type of solid is formed for each of the following substances: C_2H_5OH, CH_4, CH_3Cl, $MgSO_4$. On the basis of the type of solid and the expected magnitude of intermolecular forces (for molecular crystals), arrange these substances in order of increasing melting point. Explain your reasoning. (See Problems 11.59 and 11.60.)

HARDNESS AND STRUCTURE Hardness depends on how easily the structural units can be moved and therefore on the strength of attractive forces between the units. Thus, molecular crystals, with weak intermolecular forces, are rather soft compared with ionic crystals, in which the attractive forces are much stronger. A three-dimensional covalent network solid is usually quite hard because of the rigidity given to the structure by strong covalent bonds throughout it. Diamond and silicon carbide (SiC), which are three-dimensional covalent networks, are among the hardest substances known.

We should add that molecular and ionic crystals are generally brittle because they fracture easily along crystal planes. Metallic crystals, by contrast, are *malleable;* that is, they can be shaped by hammering (Figure 11.23).

ELECTRICAL CONDUCTIVITY AND STRUCTURE One of the characteristic properties of metals is their good electrical conductivity. The delocalized valence elec-

Table 11.6 Properties of the Different Types of Solids

Type of Solid	Melting Point of Solid	Hardness and Brittleness	Electrical Conductivity
Molecular	Low	Soft and brittle	Nonconducting
Metallic	Variable	Variable hardness; malleable	Conducting
Ionic	High to very high	Hard and brittle	Nonconducting solid (conducting liquid)
Covalent network	Very high	Very hard	Usually nonconducting

Figure 11.23
The behavior of crystals when struck. Lead is malleable; rock salt cracks along crystal planes when struck.

trons are easily moved by an electrical field and are responsible for carrying the electric current. By contrast, most covalent and ionic solids are nonconductors, because the electrons are localized to particular atoms or bonds. Ionic substances do become conducting in the liquid state, however, because the ions can move. In an ionic liquid, it is the ions that carry the electric current.

Table 11.6 summarizes the properties we discussed for the different types of solids.

11.7 CRYSTALLINE SOLIDS; CRYSTAL LATTICES AND UNIT CELLS

Solids can be crystalline or amorphous. A **crystalline solid** is *a solid composed of one or more crystals; each crystal has a well-defined ordered structure*. Sodium chloride (table salt) and sucrose (table sugar) are examples of crystalline substances. Metals are usually compact masses of crystals. An **amorphous solid** is *a solid that has a disordered structure; it lacks the well-defined arrangement of basic units (atoms, molecules, or ions) found in a crystal*. A glass is an amorphous solid obtained by cooling a liquid rapidly enough that its basic units are ''frozen'' in random positions before they can assume an ordered crystalline arrangement. Common window glass is an example, as is obsidian, a natural glass formed when molten rock from a volcanic eruption cools quickly.●

● When quartz crystals (SiO_2) are melted and then cooled rapidly, they form a glass called silica glass. Though ideal for some purposes, silica glass is difficult to work with because of its high temperature of softening. This temperature can be lowered by adding various oxides, such as Na_2O and CaO. In practice, ordinary glass is formed by melting sand (quartz crystals) with sodium carbonate and calcium carbonate. The carbonates decompose to the oxides plus carbon dioxide.

Crystal Lattices

A crystal is a three-dimensional ordered arrangement of basic units (the basic unit is either an atom, molecule, or ion, depending on the type of crystal). The ordered structure of a crystal is conveniently described in terms of a *crystal lattice*. The idea of a crystal lattice is most easily grasped by looking at two-dimensional repeating patterns such as those seen on many fabrics or wallpapers. For simplic-

Figure 11.24
A two-dimensional pattern. *(A)* A pattern of *A*'s. A point midway on the crossbar of each *A* has been selected for a lattice point. *(B)* The pattern of lattice points. *(C)* Division of the pattern into unit cells. *(D)* A single unit cell.

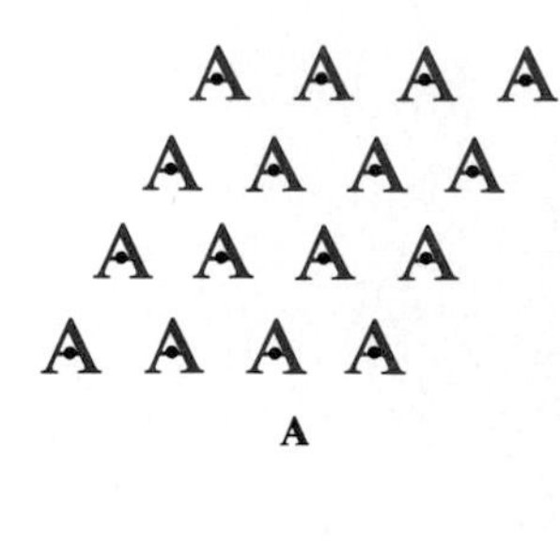

A

B

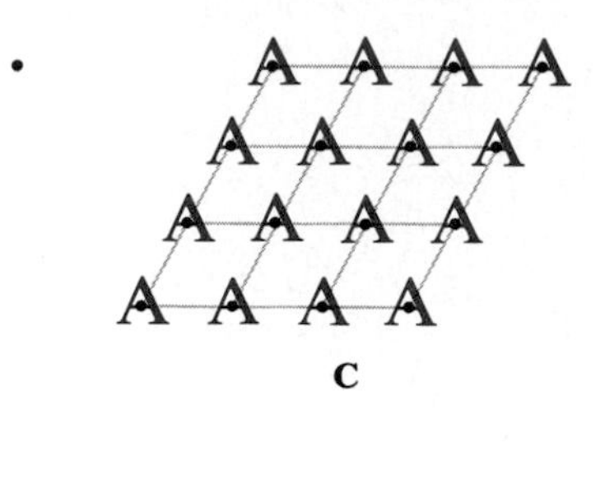

C

D

ity, consider the pattern of *A*'s shown in Figure 11.24A. This pattern consists of a repetition of basic units *A*. A lattice of points is obtained by choosing the same point in each basic unit (the choice of this point is immaterial as long as the same point is chosen in each basic unit). In the pattern of Figure 11.24, we have chosen the points to be midway on the crossbar of each *A*. The collection of these points constitutes the lattice of the pattern (Figure 11.24B). This lattice shows the essential arrangement of the basic units (*A*'s) in the pattern. We could have chosen a different location for the point in each *A*, but we would have obtained the same lattice.

Similarly, we define a **crystal lattice** as *the geometric arrangement of lattice points of a crystal, in which we choose one lattice point at the same location within each of the basic units of the crystal*. Figure 11.25 shows the crystal structure of copper metal. If we place a point at the same location in each copper atom (say at the center of the atom), we obtain the crystal lattice for copper. The crystal lattice and crystal structure are not the same. We can imagine building the crystal structure for copper from its crystal lattice by placing a copper atom at each lattice

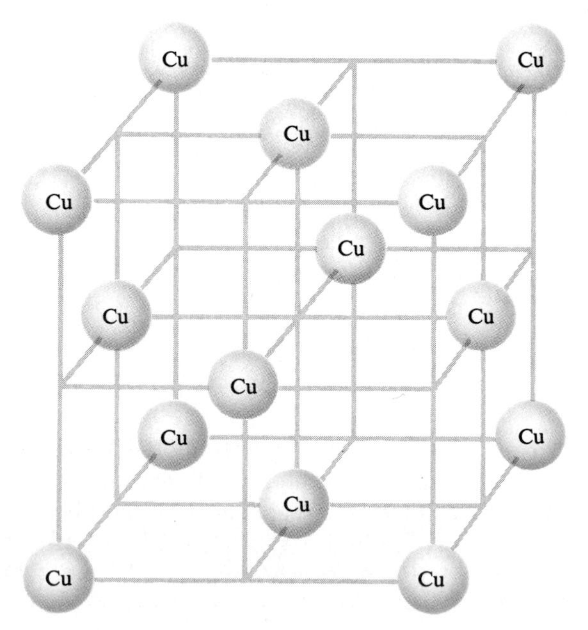

Figure 11.25
Crystal structure and crystal lattice of copper. The copper atoms have been shrunk so that the crystal structure is more visible. The crystal lattice is the geometric arrangement of lattice points, which we can take to be the centers of the atoms. Lines have been drawn to emphasize the geometry of the lattice.

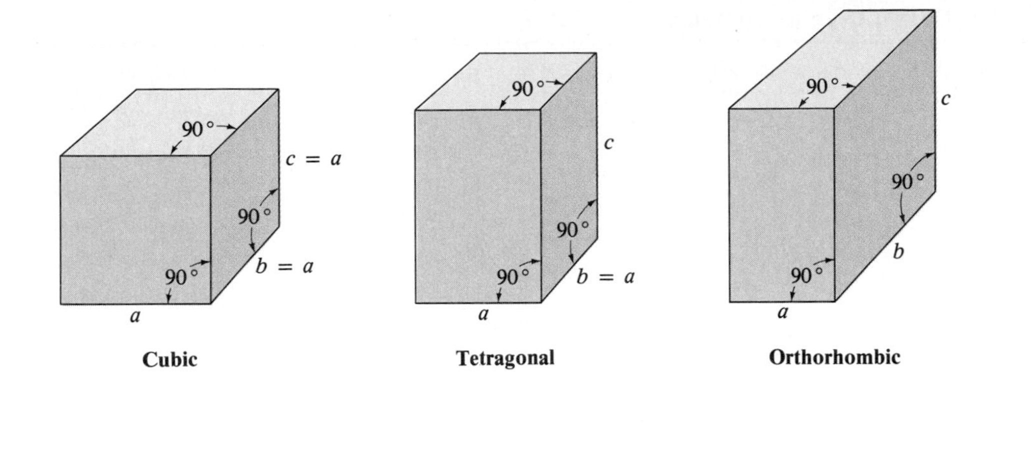

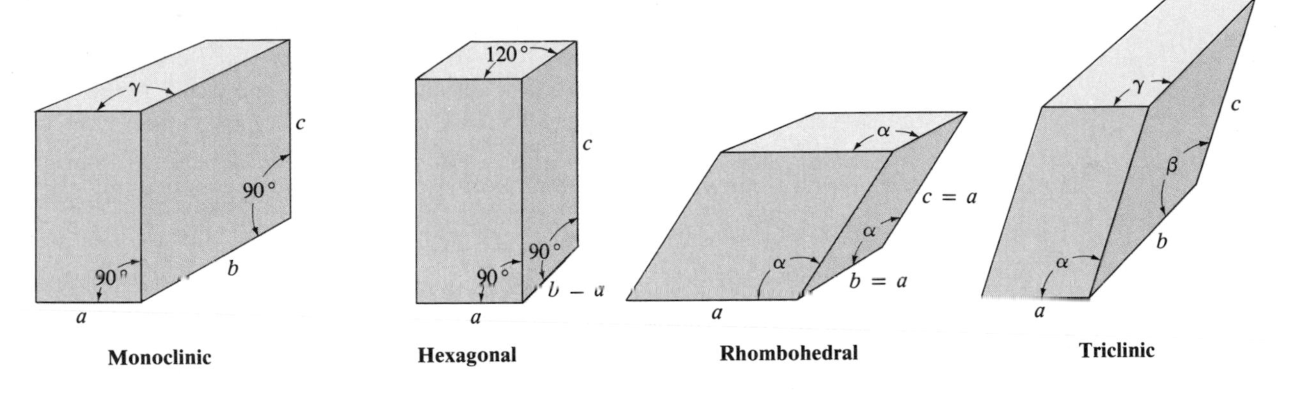

Figure 11.26
Unit-cell shapes of the different crystal systems. The unit cells of the seven different crystal systems are displayed, showing the relationships between edge lengths and angles (summarized in Table 11.7).

point. Many different crystals have the same crystal lattice. Thus, the crystal lattice for solid methane, CH_4, is the same (except for the distances between points) as that for copper. We obtain the methane crystal structure by placing CH_4 molecules at the lattice points. The crystal lattice only shows the arrangement of basic units of the crystal.

We can imagine dividing a crystal into many equivalent cells, or *unit cells,* so that these cells become the ''bricks'' from which we mentally construct the crystal. Let us look at the two-dimensional pattern in Figure 11.24C, in which we have connected lattice points by straight lines. The parallelograms formed by the straight lines divide the pattern into small cells. These are the smallest possible cells from which we could imagine constructing the pattern by stacking the individual cells—they are the unit cells of the pattern (Figure 11.24D).

In the same way, we can divide any crystal lattice into box-like cells or unit cells. The **unit cell** of a crystal is *the smallest box-like unit (each box having faces that are parallelograms) from which we can imagine constructing a crystal by stacking the units in three dimensions.*

There are seven basic shapes possible for unit cells, which give rise to the seven *crystal systems* used to classify crystals. A crystal belonging to a given crystal system has a unit cell with one of the seven shapes shown in Figure 11.26. Each unit-cell shape can be characterized by the angles between the edges of the unit cell and the relative lengths of those edges. A crystal belonging to the cubic system, for example, has 90° angles between the edges of the unit cell, and all

Table 11.7 The Seven Crystal Systems

Crystal System	Edge Length	Angles	Examples
Cubic	$a = b = c$	$\alpha = \beta = \gamma = 90°$	NaCl, Cu
Tetragonal	$a = b \neq c$	$\alpha = \beta = \gamma = 90°$	TiO_2 (rutile), Sn (white tin)
Orthorhombic	$a \neq b \neq c$	$\alpha = \beta = \gamma = 90°$	$CaCO_3$ (aragonite), $BaSO_4$
Monoclinic	$a \neq b \neq c$	$\alpha = \beta = 90°, \gamma \neq 90°$	$Na_2B_4O_7 \cdot 10H_2O$ (borax), $PbCrO_4$
Hexagonal	$a = b \neq c$	$\alpha = \beta = 90°, \gamma = 120°$	C (graphite), ZnO
Rhombohedral	$a = b = c$	$\alpha = \beta = \gamma \neq 90°$	$CaCO_3$ (calcite), HgS (cinnabar)
Triclinic	$a \neq b \neq c$	$\alpha \neq \beta \neq \gamma \neq 90°$	$K_2Cr_2O_7$, $CuSO_4 \cdot 5H_2O$

edges are of equal length. Table 11.7 summarizes the relationships between the angles and edge lengths for the unit cell of each crystal system.

Most of the crystal systems have more than one possible crystal lattice. A *simple* (or *primitive*) lattice has a unit cell in which there are lattice points only at the corners of the unit cell. Other lattices in the same crystal system have additional lattice points either within the *body* of the unit cell or on *faces* of the unit cell. As an example, we will describe the cubic system in some detail.

Cubic Unit Cells

The cubic crystal system has three possible cubic unit cells, called simple (or primitive) cubic, body-centered cubic, and face-centered cubic. The different cubic unit cells are illustrated in Figure 11.27. A **simple cubic unit cell** is *a cubic unit cell in which lattice points are situated only at the corners of the unit cell.* A **body-centered cubic unit cell** is *a cubic unit cell in which there is a lattice point at the center of the cubic cell in addition to those at the corners.* A **face-centered cubic unit cell** is *a cubic unit cell in which there are lattice points at the centers of each face of the unit cell in addition to those at the corners.*

The simplest crystal structures are those in which there is only a single atom at each lattice point. Most metals are examples. Copper metal has a face-centered cubic unit cell with one copper atom at each lattice point (see Figure 11.25). The unit cells for this and other cubic cells of simple atomic crystals are shown in Figure 11.28. Note that only the portions of the atoms actually within the unit cell are shown. For certain applications, we will need to know the number of atoms in a unit cell of such an atomic crystal. The next example shows how we count atoms in a unit cell.

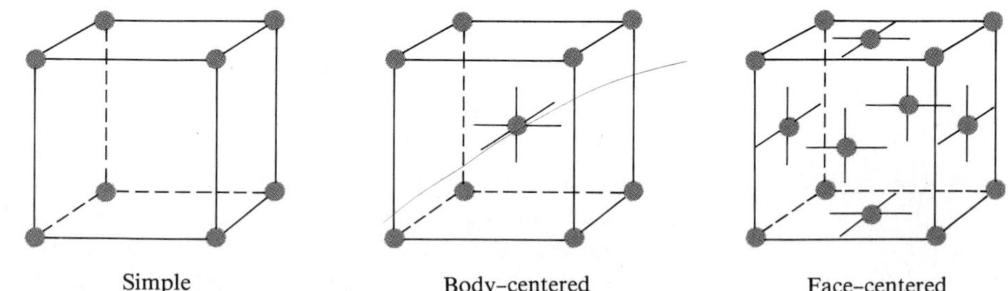

Figure 11.27
Cubic unit cells. The simple cubic unit cell has lattice points only at the corners; the body-centered cubic unit cell also has a lattice point at the center of the cell; and the face-centered cubic unit cell has lattice points at the centers of each face in addition to those at the corners.

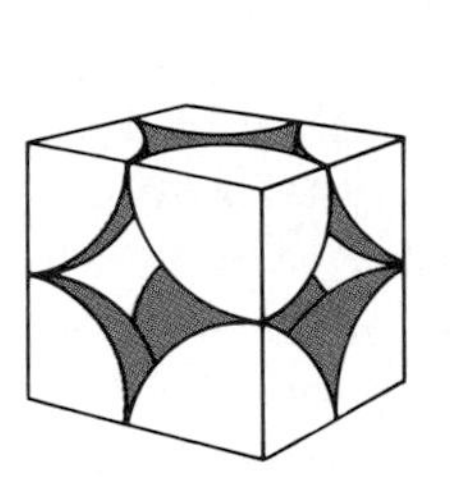

Figure 11.28
Space-filling representation of cubic unit cells. Only that portion of each atom belonging to a unit cell is shown. Note that a corner atom is shared with eight unit cells and that a face atom is shared with two.

Example 11.9 Determining the Number of Atoms per Unit Cell

How many atoms are there in the face-centered cubic unit cell of an atomic crystal having one atom at each lattice point? Remember that atoms at the corners and faces of the unit cell are shared with adjoining unit cells (see Figure 11.28).

Solution

Each atom at the center of a face is shared with one other unit cell. Hence, only one-half of the atom belongs to a particular unit cell. But there are six faces on a cube. This gives three whole atoms from the faces. Each corner atom is shared among eight unit cells. Hence, only one-eighth of a corner atom belongs to a particular unit cell. But there are eight corners on a cube. Therefore, the corners contribute one whole atom. Thus, there are a total of four atoms in a face-centered cubic unit cell. To summarize, we have

$$6 \text{ \cancel{faces}} \times \frac{1/2 \text{ atom}}{\text{\cancel{face}}} = 3 \text{ atoms}$$

$$8 \text{ \cancel{corners}} \times \frac{1/8 \text{ atom}}{\text{\cancel{corner}}} = \underline{1} \text{ atom}$$

$$\text{Total} \quad \textbf{4 atoms}$$

Exercise 11.10

Figure 11.29 shows solid dots (''atoms'') forming a two-dimensional lattice. A unit cell is marked off by the dashed line. How many ''atoms'' are there per unit cell in such a two-dimensional lattice? (See Problems 11.65 and 11.66.)

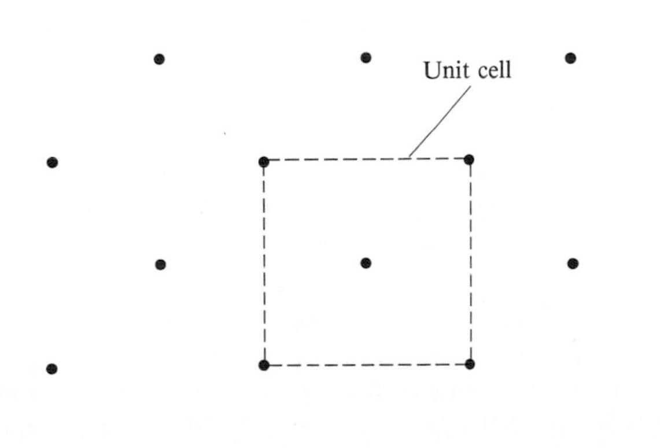

Figure 11.29
A two-dimensional lattice. See Exercise 11.10.

Crystal Defects

So far we have assumed that crystals have perfect order. In fact, real crystals have various defects or imperfections. These are principally of two kinds: chemical impurities and defects in the formation of the lattice.

Ruby is an example of a crystal with a chemical impurity. The crystal is mainly colorless aluminum oxide, Al_2O_3, but occasional aluminum ions, Al^{3+}, are replaced by chromium(III) ions, Cr^{3+}, which give a red color.

Various lattice defects occur during crystallization. Crystal planes may be misaligned, or sites in the crystal lattice may remain vacant. For example, there might be an equal number of sodium ion vacancies and chloride ion vacancies in a sodium chloride crystal. It is also possible to have an unequal number of cation and anion vacancies in an ionic crystal. For example, iron(II) oxide, FeO, usually crystallizes with some iron(II) sites left unoccupied. Enough of the remaining iron atoms have +3 charges to give an electrically balanced crystal. As a result, there are more oxygen atoms than iron atoms in the crystal. Moreover, the exact composition of the crystal can vary, so the formula FeO is only approximate. Such a compound whose composition varies slightly from its idealized formula is said to be *nonstoichiometric*. Other examples of nonstoichiometric compounds are Cu_2O and Cu_2S. Each usually has less copper than expected from the formula. The newly discovered ceramic materials having superconducting properties at relatively high temperatures (p. 389) are nonstoichiometric compounds. An example is yttrium barium copper oxide, $YBa_2Cu_3O_{7-x}$, where x is approximately 0.1. It is a compound with oxygen atom vacancies.

11.8 STRUCTURES OF SOME CRYSTALLINE SOLIDS

We have described the structure of crystals in a general way. Now we want to look in detail at the structure of several crystalline solids that represent the different types: molecular, metallic, ionic, and covalent network.

Molecular Solids; Closest Packing

The simplest molecular solids are the frozen noble gases—for example, solid neon. In this case, the molecules are single atoms and the intermolecular interactions are London forces. These forces are nondirectional (in contrast to covalent bonding, which is directional), and the maximal attraction is obtained when each atom is surrounded by the largest possible number of other atoms. The problem, then, is simply to find how identical spheres can be packed as tightly as possible into a given space.

To form a layer of close-packed spheres, you place each row of spheres in the crevices of the adjoining rows. This is illustrated in Figure 11.30A. (You may find it helpful to stack identical coins in the manner described here for spheres.) Place the next layer of spheres in the hollows in the first layer. Only half of these hollows can be filled with spheres. Once you have placed a sphere in a given hollow, it partially covers three neighboring hollows (Figure 11.30B) and completely determines the pattern of the layer (Figure 11.30C).

When we come to fill the third layer, we find that we have a choice of sites for spheres, labeled x and y in Figure 11.30C. Note that each of the x sites has a sphere in the first layer directly beneath it, but the y sites do not. When the spheres in the

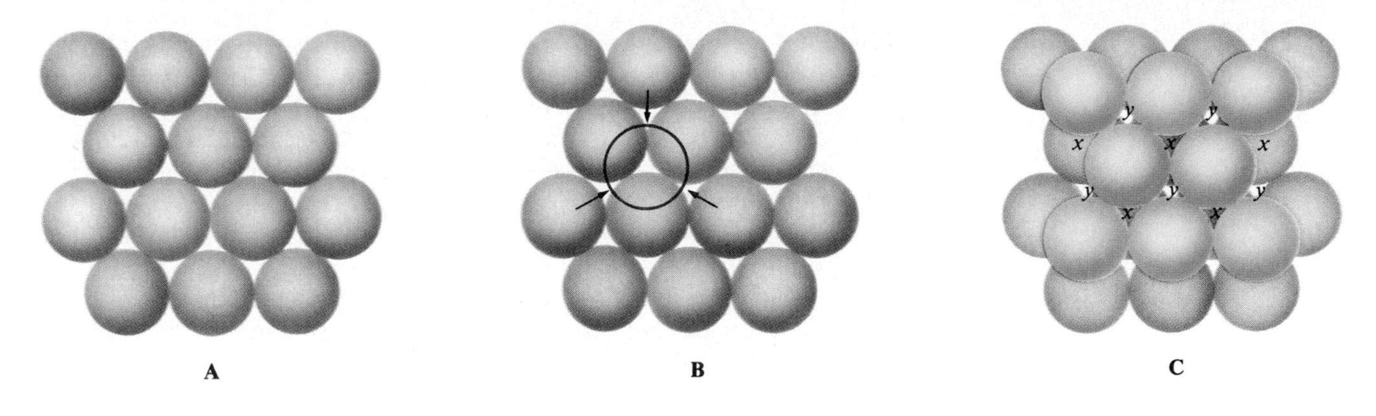

Figure 11.30
Closest packing of spheres.
(A) Each layer is formed by placing spheres in the crevices of the adjoining row. *(B)* When a sphere is placed in a hollow on top of a layer, three other hollows are partially covered. *(C)* There are two types of hollows on top of the second layer, labeled *x* and *y*. The *x* sites are directly above the spheres in the first layer; the *y* sites are not. Both are possible sites for the spheres of the third layer.

third layer are placed in the *x* sites, so that the third layer repeats the first layer, we label this stacking *ABA*. When successive layers are placed so that the spheres of each layer are directly over a layer that is one layer away, we get a stacking that we label *ABABABA* ···. (This notation indicates that the *A* layers are directly over other *A* layers, whereas the *B* layers are directly over other *B* layers.) This results in a **hexagonal close-packed structure** (hcp), *a crystal structure composed of close-packed atoms (or other units) with the stacking ABABABA ··· ; the structure has a hexagonal unit cell.*

When the spheres in the third layer are placed in *y* sites, so that the third layer is over neither the first nor the second layer, we get a stacking that we label *ABC*. The fourth layer must be over either the *A* or the *B* layer. If subsequent layers are stacked so that they are over the layer two layers below, we get the stacking *ABCABCABCA* ···, which results in a **cubic close-packed structure** (ccp), *a crystal structure composed of close-packed atoms (or other units) with the stacking ABCABCABCA ···.*●

● If we pack into a crate as many perfectly round oranges of the same size as possible, they will be close-packed. However, they need not be either hexagonal or cubic close-packed, because the *A*-, *B*-, and *C*-type layers can be placed at random (except that similar layers cannot be adjacent). A possible arrangement might be *ABCABABABC* ···. It is the forces between structural units in a close-packed crystal that determine whether one of the ordered structures, either cubic or hexagonal, will occur.

The cubic close-packed lattice is identical to the lattice having a face-centered cubic unit cell. To see this, we take portions of four layers from the cubic close-packed array (Figure 11.31, *left*). When these are placed together, they form a cube, as shown in Figure 11.31, *right*.

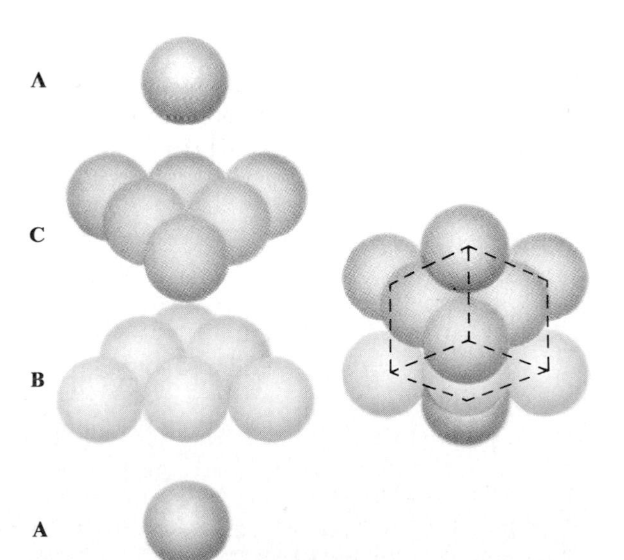

Figure 11.31
The cubic close-packed structure has a face-centered cubic lattice. *Left:* An "exploded" view of portions of four layers of the cubic close-packed structure. *Right:* These layers form a face-centered cubic unit cell.

In any close-packed arrangement, each interior atom is surrounded by twelve nearest-neighbor atoms. *The number of nearest-neighbor atoms of an atom* is called its **coordination number.** Thinking of atoms as hard spheres, one can calculate that the spheres occupy 74% of the space of the crystal. There is no way of packing identical spheres so that an atom has a coordination number greater than 12 or the spheres occupy more than 74% of the space of the crystal. All the noble-gas solids have cubic close-packed crystals except helium, which has a hexagonal close-packed crystal.

A substance composed of polyatomic molecules with approximately spherical shapes might be expected to crystallize in a close-packed structure. Methane, CH_4, is an example. Solid methane has a face-centered cubic lattice with one CH_4 molecule at each lattice point.

Metallic Solids

If we were to assume that metallic bonding is completely nondirectional, we would expect metals to crystallize in one of the close-packed structures, as the noble-gas elements do. Indeed, many of the metallic elements do have either cubic or hexagonal close-packed crystals. Copper and silver, for example, are cubic close-packed (that is, face-centered cubic with one atom at each lattice point). However, a body-centered cubic arrangement of spheres (having one sphere at each lattice point of a body-centered cubic lattice) occupies almost as great a percentage of space as the close-packed ones. In a body-centered arrangement, 68% of the space is occupied by spheres, compared with the maximal value of 74% in the close-packed arrangement. Each atom in a body-centered lattice has a coordination number of 8, rather than the 12 of each atom in a close-packed lattice.

Figure 11.32 lists the structures of the metallic elements. With few exceptions, they have either a close-packed or a body-centered structure.

Ionic Solids

The description of ionic crystals is complicated by the fact that we must give the positions in the crystal structure of both the cations and the anions. We will look at the structures of three different cubic crystals with the general formula MX (where M is the metal and X is the nonmetal): cesium chloride (CsCl), sodium chloride (NaCl), and cubic zinc sulfide or zinc blende (ZnS). Many ionic substances of the general formula MX have crystal structures that are one of these types. Once you

Figure 11.32
Crystal structures of metals. Most metals have one of the close-packed structures (hcp = hexagonal close-packed; ccp = cubic close-packed) or a body-centered cubic structure (bcc; manganese has a complicated bcc lattice, with several atoms at each lattice point). Other structures are simple cubic (sc), body-centered tetragonal (bct), orthorhombic (or), and rhombohedral (rh).

IA	IIA	IIIB	IVB	VB	VIB	VIIB	VIIIB			IB	IIB	IIIA	IVA	VA	VIA
Li bcc	Be hcp														
Na bcc	Mg hcp											Al ccp			
K bcc	Ca ccp	Sc hcp	Ti hcp	V bcc	Cr bcc	Mn bcc	Fe bcc	Co hcp	Ni ccp	Cu ccp	Zn hcp	Ga or			
Rb bcc	Sr ccp	Y hcp	Zr hcp	Nb bcc	Mo bcc	Tc hcp	Ru hcp	Rh ccp	Pd ccp	Ag ccp	Cd hcp	In bct	Sn bct		
Cs bcc	Ba bcc	La hcp	Hf hcp	Ta bcc	W bcc	Re hcp	Os hcp	Ir ccp	Pt ccp	Au ccp	Hg rh	Tl hcp	Pb ccp	Bi rh	Po sc

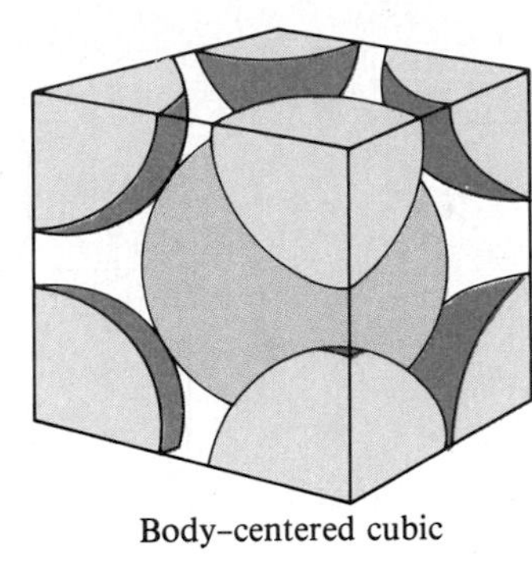

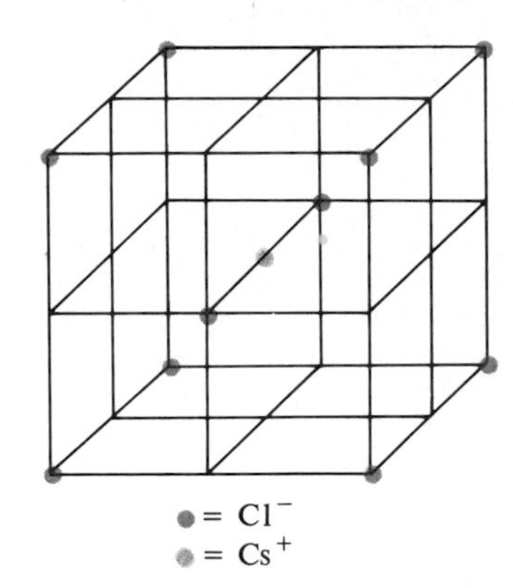

Figure 11.33
Cesium chloride unit cell. Cesium chloride has a simple cubic lattice. The figure shows a unit cell with Cl^- ions at the corners of the unit cell and a Cs^+ ion at the center (a space-filling model is on the left; a model with ions shrunk in size to emphasize the structure is on the right). An alternative unit cell would have Cs^+ ions at the corners with a Cl^- ion at the center.

have learned these structures, you can relate the structures of many other compounds to them.

Cesium chloride consists of positive ions and negative ions of about equal size arranged in a cubic array. Figure 11.33 shows the unit cell of cesium chloride. It shows both a space-filling model (ions are shown as spheres of approximately the correct relative sizes) and a model in which the ions are shrunk in size in order to display more clearly the relative positions of the ions. Note that there are chloride ions at each corner of the cube and a cesium ion at the center. Alternatively, the unit cell may be taken to have cesium ions at the corners of a cube and a chloride ion at the center. In either case, the unit cell contains one Cs^+ ion and one Cl^- ion. (Figure 11.33 has ⅛ Cl^- ion at each of eight corners plus one Cs^+ ion at the center of the unit cell.) Ammonium chloride, NH_4Cl, and thallium chloride, TlCl, are examples of other ionic compounds having this "cesium chloride" structure.

Figure 11.34A shows the unit cell of sodium chloride. It has Cl^- ions at the corners as well as at the centers of each cube face. Because the chloride ions are so much larger than the sodium ions, these ions are nearly touching and form an approximate cubic close-packed structure of Cl^- ions. The Na^+ ions are arranged in cavities of this close-packed structure. Some other compounds that crystallize in this structure are potassium chloride, KCl; calcium oxide, CaO; and silver chloride, AgCl.

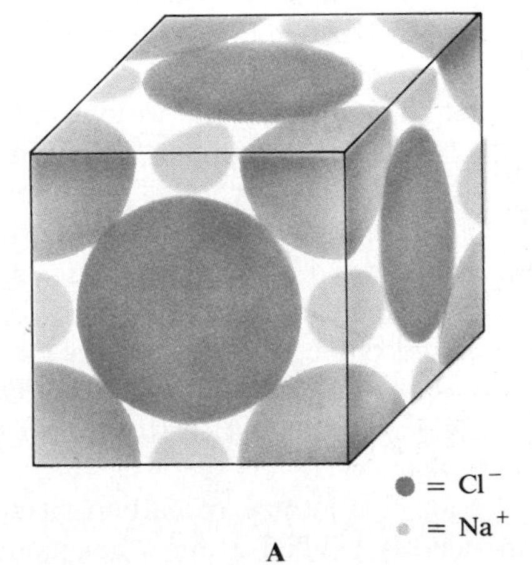

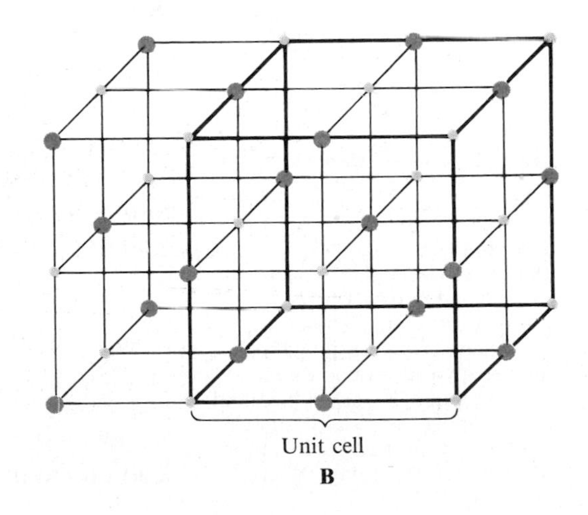

Figure 11.34
Sodium chloride unit cell. Sodium chloride has a face-centered cubic lattice. *(A)* The unit cell shown here (as a space-filling model) has Cl^- ions at the corners and at the centers of the faces of the cube. The Cl^- ions are much larger than Na^+ ions, so the Cl^- ions are nearly touching and have approximately a cubic close-packed structure; the Na^+ ions are in cavities in this structure. *(B)* The unit cell shown here has Na^+ ions at the corners and at the centers of the faces of the unit cell.

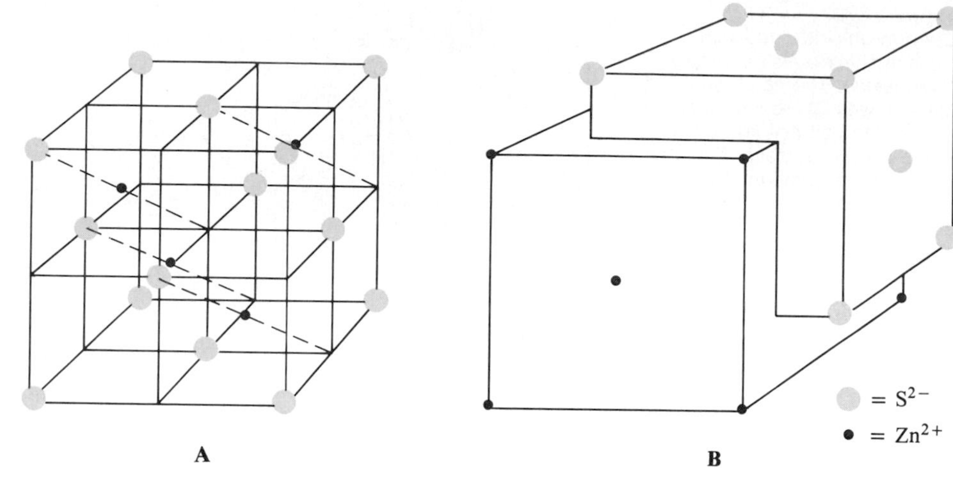

Figure 11.35
Zinc blende (cubic ZnS) unit cell. Zinc sulfide, ZnS, crystallizes in two different forms, or polymorphs: a hexagonal form (wurtzite) and a face-centered cubic form (zinc blende, or sphalerite), which is shown here. *(A)* Sulfide ions, S^{2-}, are at the corners and at the centers of the faces of the unit cell; Zn^{2+} ions are in alternate subcubes of the unit cell. *(B)* The unit cell can also be described as having Zn^{2+} ions at each corner and at the centers of each face of the unit cell, with S^{2-} ions in alternate subcubes. This diagram shows the interpenetrating face-centered cubic arrangements of the two ions.

Zinc sulfide, ZnS, crystallizes in either one of two structures. The zinc sulfide mineral having cubic structure is called zinc blende or sphalerite. Zinc sulfide also exists as hexagonal crystals (the mineral is called wurtzite). A solid substance that can occur in more than one crystal structure is said to be *polymorphic*. We will discuss only the zinc blende structure (cubic ZnS).

Figure 11.35A shows the unit cell of zinc blende. Sulfide ions are at the corners and at the centers of each face of the unit cell. The positions of the zinc ions can be described if we imagine the unit cell divided into eight cubic parts, or subcubes. Zinc ions are at the centers of alternate subcubes. The number of S^{2-} ions per unit cell is four (8 corner ions × ⅛ + 6 face ions × ½), and the number of Zn^{2+} ions is four (all are completely within the unit cell). Thus, there are equal numbers of Zn^{2+} and S^{2-} ions, as expected by the formula ZnS.

The unit cell of zinc blende can also be taken to have zinc ions at the corners and face centers, with sulfide ions in alternate subcubes. See Figure 11.35B. Zinc oxide, ZnO, and beryllium oxide, BeO, also have the "zinc blende" structure.

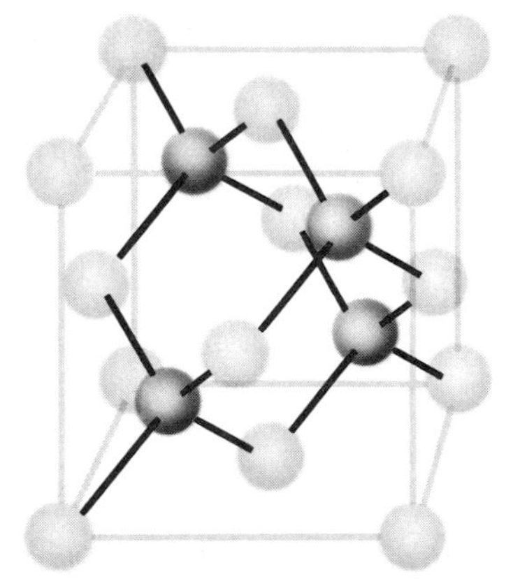

Figure 11.36
Diamond unit cell. The unit cell of diamond can be obtained from the unit cell of zinc blende by replacing Zn^{2+} ions by carbon atoms (shown here as dark spheres) and S^{2-} ions by carbon atoms (these are shown as light spheres). Note that each carbon atom is bonded to four others.

Covalent Network Solids

The structures of covalent network solids are determined primarily by the directions of covalent bonds. Diamond is a simple example. It is one allotropic form of the element carbon, in which each carbon atom is covalently bonded to four other carbon atoms in tetrahedral directions to give a three-dimensional covalent network (Figure 11.22). We can describe the bonding by assuming that every carbon atom is sp^3 hybridized. Each C—C bond forms by the overlap of one sp^3 hybrid orbital from one carbon atom with an sp^3 hybrid orbital on the other atom. The unit cell of diamond is similar to that of zinc blende (Figure 11.35A); the Zn^{2+} and S^{2-} ions are replaced by carbon atoms, as shown in Figure 11.36. The crystal lattice, which is face-centered cubic, can be obtained by placing lattice points at the centers of alternate carbon atoms (either those shown in dark gray or those shown in light gray). Silicon, germanium, and gray tin (a nonmetallic allotrope) have "diamond" structures.

Graphite is another allotrope of carbon. It consists of large, flat sheets of carbon atoms covalently bonded to form hexagonal arrays, and these sheets are stacked one on top of the other (Figure 11.22). We can describe each sheet in terms of resonance formulas such as the following:

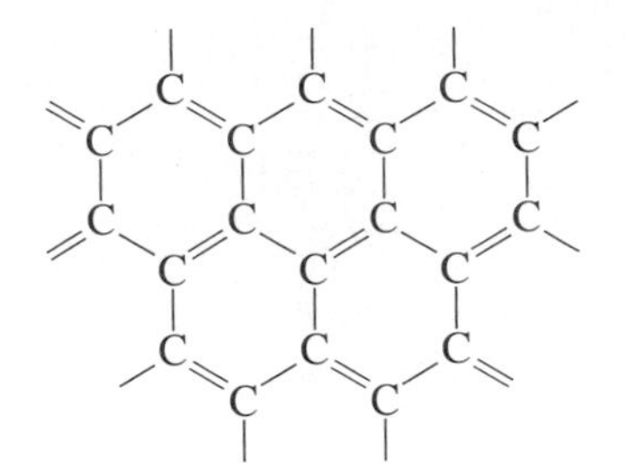

(one resonance formula)

This is just one of many possible resonance formulas, because others can be drawn by moving the double and single bonds around. This means that the π electrons are delocalized over a plane of carbon atoms. We can also describe this delocalization in molecular orbital terms. We imagine that each carbon atom is sp^2 hybridized, so each atom has three sp^2 hybrid orbitals in the molecular plane and an additional $2p$ orbital perpendicular to this plane. A σ bond forms between two carbon atoms when an sp^2 hybrid orbital on one atom overlaps an sp^2 hybrid orbital on another. The result is a planar σ-bond framework. The $2p$ orbitals of all the atoms overlap to form π orbitals encompassing the entire plane of atoms.

Each carbon–carbon distance within a graphite sheet is 1.42 Å (142 pm), which is almost midway between the length of a C—C bond (1.54 Å) and that of a C=C bond (1.34Å) and which indicates an intermediate bond order (Figure 11.22). The two-dimensional sheets or layers are stacked one on top of the other, and the attraction between sheets results from London forces. The distances between layers is 3.35 Å (335 pm), which is greater than the carbon–carbon bond length, because London forces are weaker than covalent bonding forces.

We can explain several properties of graphite on the basis of its structure. The layer structure, which we can see in electron micrographs (Figure 11.37), tends to separate easily so that sheets of graphite slide over one another. The "lead" in a

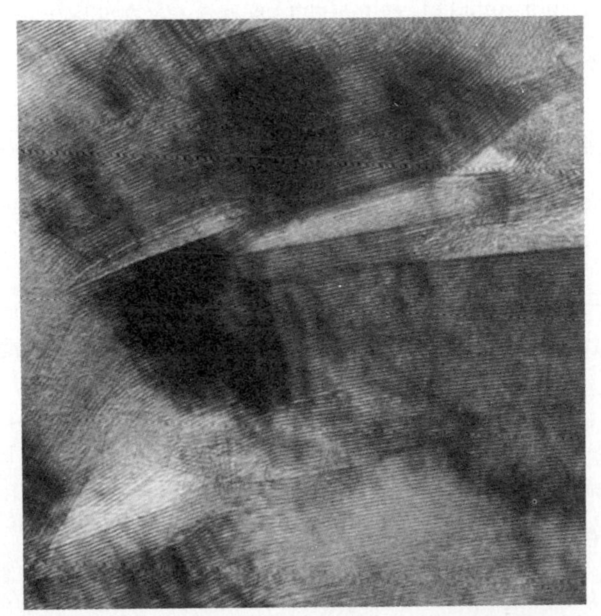

Figure 11.37
An electron micrograph of graphite. Note the layer structure of graphite.

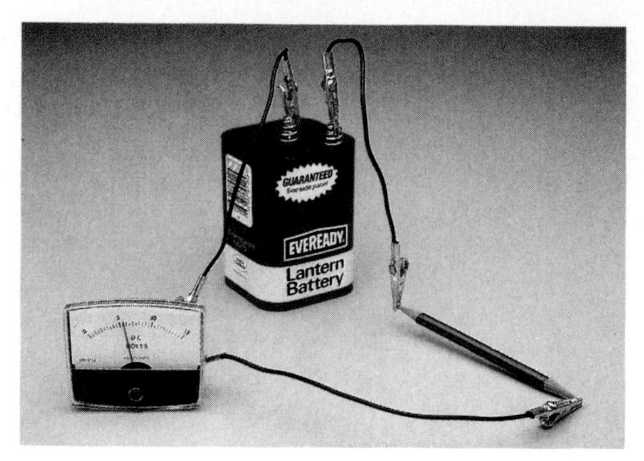

Figure 11.38
A demonstration of the electrical conductivity of graphite. The non-zero reading on the meter shows that the pencil lead (graphite) conducts electricity.

pencil is graphite. As we write with a pencil, layers of graphite rub off to form the pencil mark. Similarly, the use of graphite to lubricate locks and other closely fitting metal surfaces depends on the ability of graphite particles to move over one another as the layers slide. Another property of graphite is its electrical conductivity. Electrical conductivity is unusual in covalent network solids, because normally the electrons in such bonds are localized. Graphite is a moderately good conductor, however, because delocalization leads to mobile electrons within each carbon-atom layer. Figure 11.38 shows a demonstration of the electrical conductivity of graphite.

11.9 CALCULATIONS INVOLVING UNIT-CELL DIMENSIONS

We can determine the structure and dimensions of a unit cell by diffraction methods, which we will describe briefly in the next section. Once we know the unit-cell dimensions and the structure of a crystal, however, some interesting calculations are possible. For instance, suppose we find that a metallic solid has a face-centered cubic unit cell with all atoms at lattice points, and we determine the edge length of the unit cell. From this unit-cell dimension, we can calculate the volume of the unit cell. Then, knowing the density of the metal, we can calculate the mass of the atoms in the unit cell. Because we know that the unit cell is face-centered cubic with all atoms at lattice points and we know that such a unit cell has four atoms, we can obtain the mass of an individual atom. This determination of the mass of a single atom gave one of the first accurate determinations of Avogadro's number. The calculations are shown in the next example.

Example 11.10 Calculating Atomic Mass from Unit-Cell Dimension and Density

X-ray diffraction from crystals provides one of the most accurate ways of determining Avogadro's number. Silver crystallizes in a face-centered cubic lattice with all atoms at the lattice points. The length of an edge of the unit cell was determined by x-ray diffraction to be 4.086 Å (408.6 pm). The density of silver is 10.50 g/cm^3. Calculate the mass of a silver atom. Then, using the known value of the atomic weight, calculate Avogadro's number.

Solution

Knowing the edge length of a unit cell, we can calculate the unit-cell volume. Then, from the density, we can find the mass of the unit cell and hence the mass of a silver atom. We obtain the volume V of the unit cell by cubing the length of an edge.

$$V = (4.086 \times 10^{-10} \text{ m})^3 = 6.822 \times 10^{-29} \text{ m}^3$$

The density, d, of silver in grams per cubic meter is

$$10.50\frac{\text{g}}{\cancel{\text{cm}^3}} \times \left(\frac{1\ \cancel{\text{cm}}}{10^{-2}\ \text{m}}\right)^3 = 1.050 \times 10^7\ \text{g/m}^3$$

Density is mass per volume; hence, the mass of a unit cell equals the density times the volume of the unit cell.

$$\begin{aligned} m &= dV \\ &= 1.050 \times 10^7\ \text{g}/\cancel{\text{m}^3} \times 6.822 \times 10^{-29}\ \cancel{\text{m}^3} \\ &= 7.163 \times 10^{-22}\ \text{g} \end{aligned}$$

Because there are four atoms in a face-centered unit cell having one atom at each lattice point (see Example 11.9), the mass of a silver atom is

$$\begin{aligned} \text{Mass of 1 Ag atom} &= \tfrac{1}{4} \times 7.163 \times 10^{-22}\ \text{g} \\ &= \mathbf{1.791 \times 10^{-22}\ g} \end{aligned}$$

The known atomic weight of silver is 107.87 amu. Thus, the molar mass (Avogadro's number of atoms) is 107.87 g/mol, and Avogadro's number, N_A, is

$$N_A = \frac{107.87\ \cancel{\text{g}}/\text{mol}}{1.791 \times 10^{-22}\ \cancel{\text{g}}} = \mathbf{6.023 \times 10^{23}/mol}$$

Exercise 11.11

Lithium metal has a body-centered cubic structure with all atoms at the lattice points and a unit-cell length of 3.509 Å. Calculate Avogadro's number. The density of lithium is 0.534 g/cm^3. (See Problems 11.67 and 11.68.)

If we know or assume the structure of an atomic crystal, we can calculate the length of the unit-cell edge from the density of the substance. This is illustrated in the next example. Agreement of this value with that obtained from x-ray diffraction confirms that our view of the structure of the crystal is correct.

Example 11.11 Calculating Unit-Cell Dimension from Unit-Cell Type and Density

Platinum crystallizes in a face-centered cubic lattice with all atoms at the lattice points. It has a density of 21.45 g/cm^3 and an atomic weight of 195.08 amu. From these data, calculate the length of a unit-cell edge. Compare with the value of 3.924 Å obtained from x-ray diffraction.

Solution

The mass of an atom, and hence the mass of a unit cell, can be calculated from the atomic weight. Knowing the density and the mass of the unit cell, we can calculate the volume and the edge length of a unit cell.

We can use Avogadro's number (6.022×10^{23}/mol) to convert the molar mass of platinum (195.08 g/mol) to the mass per atom.

$$\frac{195.08\ \text{g Pt}}{1\ \cancel{\text{mol Pt}}} \times \frac{1\ \cancel{\text{mol Pt}}}{6.022 \times 10^{23}\ \text{Pt atoms}} = \frac{3.239 \times 10^{-22}\ \text{g Pt}}{1\ \text{Pt atom}}$$

Because there are four atoms per unit cell, the mass per unit cell can be calculated as follows:

$$\frac{3.239 \times 10^{-22}\ \text{g Pt}}{1\ \cancel{\text{Pt atom}}} \times \frac{4\ \cancel{\text{Pt atoms}}}{1\ \text{unit cell}} = \frac{1.296 \times 10^{-21}\ \text{g}}{1\ \text{unit cell}}$$

The volume of the unit cell is

$$V = \frac{m}{d} = \frac{1.296 \times 10^{-21}\ \cancel{\text{g}}}{21.45\ \cancel{\text{g}}/\text{cm}^3} = 6.042 \times 10^{-23}\ \text{cm}^3$$

If the edge length of the unit cell is denoted as l, the volume is $V = l^3$. Hence, the edge length is

$$\begin{aligned} l &= \sqrt[3]{V} \\ &= \sqrt[3]{6.042 \times 10^{-23}\ \text{cm}^3} \\ &= 3.924 \times 10^{-8}\ \text{cm}\ (3.924 \times 10^{-10}\ \text{m}) \end{aligned}$$

Thus, the edge length is 3.924 Å or 392.4 pm, which is in excellent agreement with the x-ray diffraction value.

Exercise 11.12

Potassium metal has a body-centered cubic structure with all atoms at the lattice points. The density of the metal is 0.856 g/cm^3. Calculate the edge length of a unit cell. (See Problems 11.69 and 11.70.)

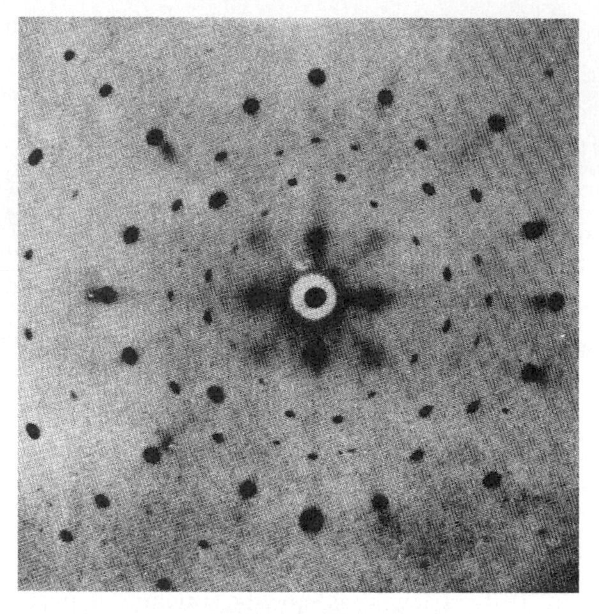

Figure 11.39
A crystal diffraction pattern. The diffraction pattern shown was obtained by diffracting x rays from a crystal of sodium chloride. The dark spots occur where x rays constructively interfere.

11.10 DETERMINING CRYSTAL STRUCTURE BY X-RAY DIFFRACTION

The determination of crystal structure by x-ray diffraction is one of the most important ways of determining the structures of molecules. Because of its ordered structure, a crystal consists of repeating planes of the same kind of atom. These planes can act as reflecting surfaces for x rays. When x rays are reflected from these planes, they show a *diffraction pattern,* which can be recorded on a photographic plate as a series of spots (see Figure 11.39). By analyzing this diffraction pattern, we can determine the positions of the atoms in the unit cell of the crystal. Once we have determined the positions of each atom in the unit cell of a molecular solid, we have also found the positions of the atoms in the molecule.●

● X-ray diffraction has been used to obtain the structure of proteins, which are large molecules essential to life. John Kendrew and Max Perutz received the Nobel Prize in chemistry in 1962 for their x-ray work in determining the structure of myoglobin (MW = 17,600 amu) and hemoglobin (MW = 66,000 amu). Hemoglobin is the oxygen-carrying protein of the red blood cells, and myoglobin is the oxygen-carrying protein in muscle tissue.

To understand how such a diffraction pattern occurs, let us look at the diffraction of two waves. Suppose two waves of the same wavelength come together in phase. By *in phase* we mean that the waves come together with their maxima at the same points and their minima at the same points (see Figure 11.40, *top*). Two waves that come together in phase form a wave with higher maxima and lower minima. We say that the waves reinforce each other, or undergo *constructive interference*. In the case of x rays, the intensity of the resultant ray is increased.

Alternatively, suppose two waves of the same wavelength come together out of phase. By *out of phase* we mean that where one wave has its maxima, the other has its minima (see Figure 11.40, *bottom*). When two waves come together out of phase, each maximum of one wave combines with a minimum of the other so that the wave is destroyed. We say that the waves undergo *destructive interference*.

Now consider the scattering, or reflection, of x rays from a crystal. Figure 11.41 shows two rays: one reflected from one plane, the other from another plane. The two rays start out in phase, but because one ray travels farther, they may end up out of phase after reflection. Only at certain angles of reflection do the two rays remain in phase. At some angles the rays constructively interfere (giving dark

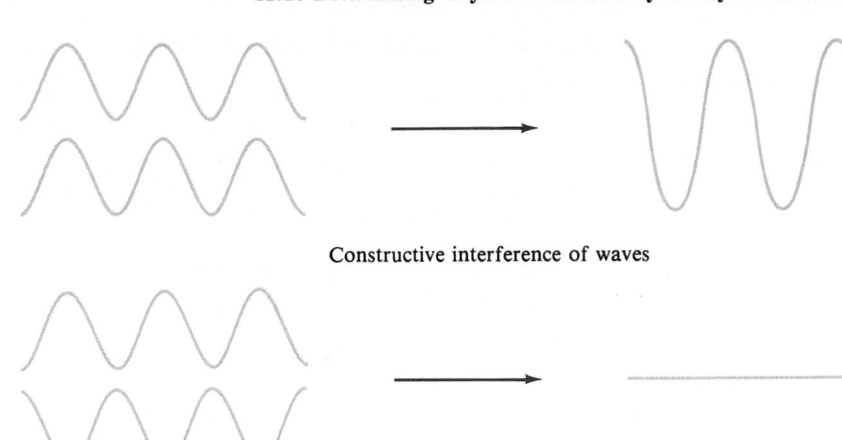

Figure 11.40
Wave interference. Two waves constructively interfere when they are in phase (that is, when their peaks match). Two waves destructively interfere when they are out of phase (that is, when the peaks of one wave match the valleys of the other).

areas on a negative photographic plate, as shown in Figure 11.39). At other angles the rays destructively interfere (giving light areas on a photographic plate).

We can relate the resulting pattern produced by the diffraction to the structure of the crystal. We can determine the type of unit cell and its size, and if the solid is composed of molecules, we can determine the position of each atom in the molecule.

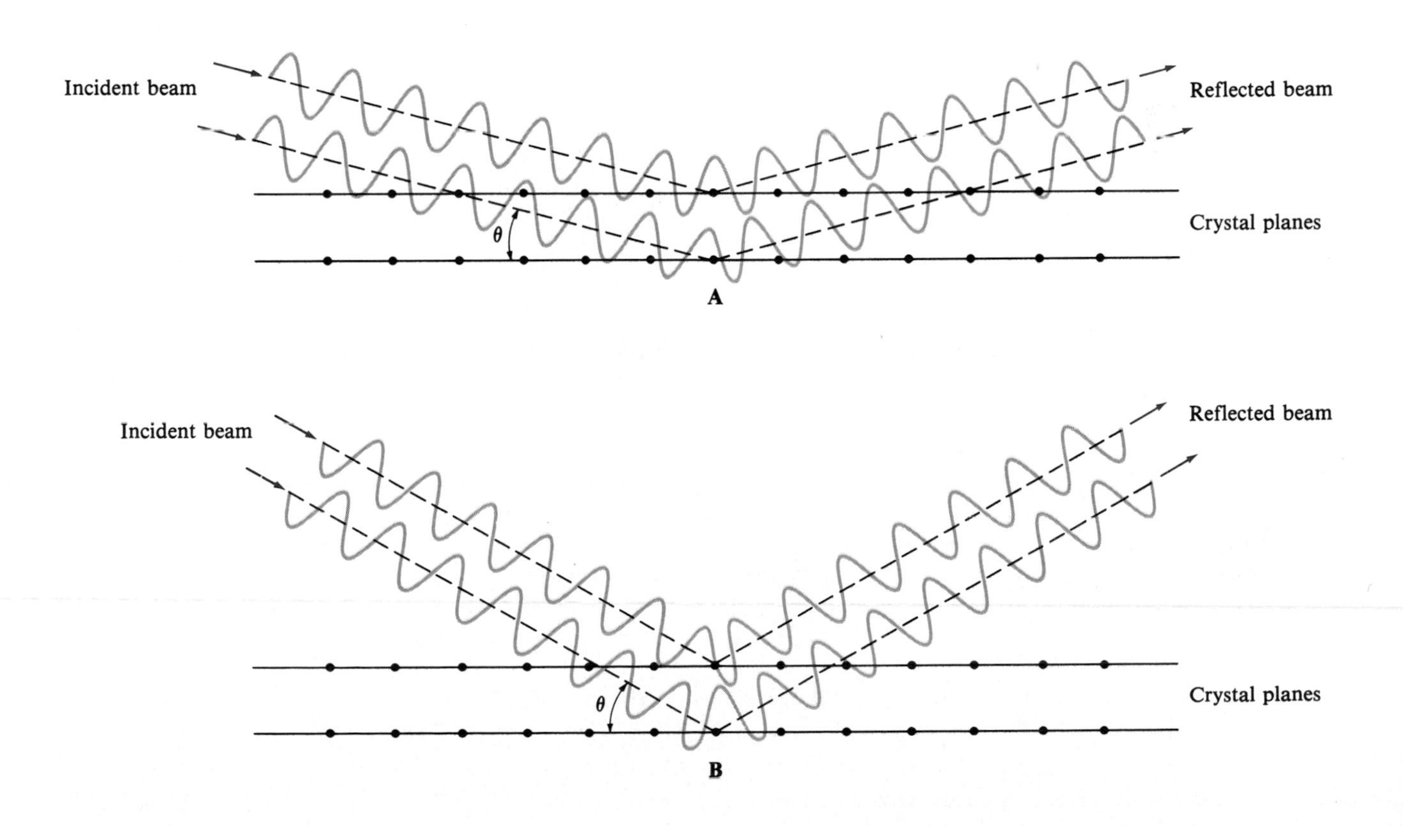

Figure 11.41
Diffraction of x rays from crystal planes. *(A)* At most angles θ, reflected waves are out of phase and destructively interfere. *(B)* At certain angles θ, however, reflected waves are in phase and constructively interfere.

Instrumental Methods

AUTOMATED X-RAY DIFFRACTOMETRY

Max von Laue, a German physicist, was the first to suggest the use of x rays for the determination of crystal structure. Soon afterward, in 1913, the British physicists William Bragg and his son Lawrence developed the method on which modern crystal-structure determination is based. They realized that the atoms in a crystal form reflecting planes for x rays, and from this idea they derived the fundamental equation of crystal-structure determination.

$$n\lambda = 2d \sin \theta, \; n = 1, 2, 3, \cdots.$$

The *Bragg equation* relates the wavelength of x rays, λ, to the distance between atomic planes, d, and the angle of reflection, θ. Note that reflections occur at several angles, corresponding to different integer values of n.

A molecular crystal has many different atomic planes, so that it reflects an x-ray beam in many different directions. By analyzing the intensities and angular directions of the reflected beams, we can determine the exact positions of all the atoms in the unit cell of the crystal and therefore obtain the structure of the molecule. The problem of obtaining the x-ray data (intensities and angular directions of the reflections) and then analyzing them, however, is not trivial. Originally, the reflected x rays were recorded on photographic plates. After taking many pictures, the scientist would pore over the negatives, measuring the densities of the spots and their positions on the plates. Then he or she would work through lengthy and laborious calculations to analyze the data. Even with early computers, the determination of a molecular structure required a year or more.

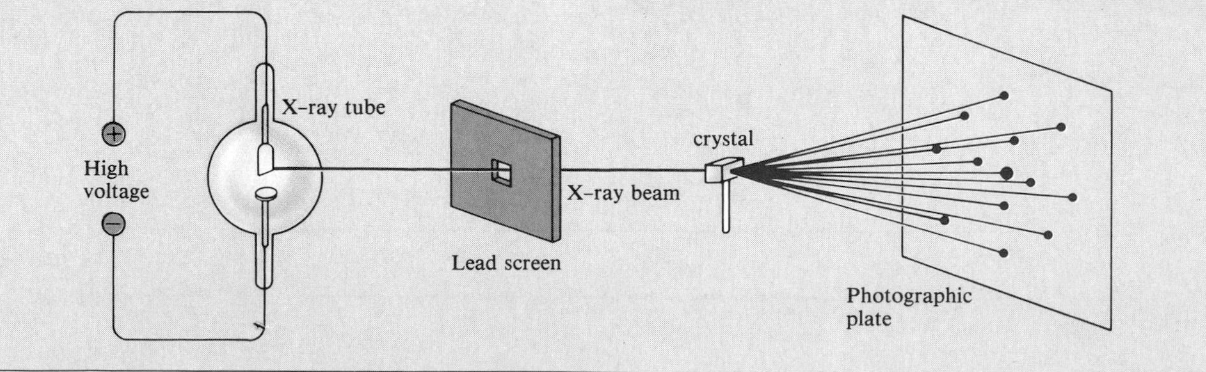

Figure 11.42
Automated x-ray diffractometer. *Top: The single-crystal specimen is mounted on a glass fiber, which is placed on a spindle within the circular assembly of the diffractometer. Note the molecular model on the cathode ray screen. Bottom: The schematic diagram shows the diffracted rays being detected by a photograph. In a modern diffractometer, the final data collection is done with an electronic detector. The data are collected and analyzed by a computer accompanying the diffractometer.*

With the development of electronic x-ray detectors and minicomputers, x-ray diffraction has become automated, so that the time and effort of determining the structure of a molecule have been substantially reduced. Now frequently the most difficult task is preparing a suitable crystal. The crystal should be several tenths of a millimeter in each dimension and without significant defects. Such crystals of protein molecules, for example, can be especially difficult to prepare.

Once a suitable crystal has been obtained, the structure of a molecule of moderate size can often be done in days. The crystal is mounted on a glass fiber (or in a glass capillary containing an inert gas, if the substance reacts with air) and placed on a pin or spindle within the circular assembly of the x-ray diffractometer (Figure 11.42). The crystal and x-ray detector (placed on the opposite side of the crystal from the x-ray tube) rotate under computer control, while the computer records the intensities and angles of thousands of x-ray reflection spots. After computer analysis of the data, the molecular structure is printed out. The cathode ray screen attached to the diffractometer in Figure 11.42 shows the structure of a molecule determined by x-ray diffraction.

Profile of a Chemical

CARBON DIOXIDE (a Commercial Chemical; the Greenhouse Effect)

In 1772, two years before he discovered oxygen, Joseph Priestley found that he could simulate certain mineral waters by dissolving carbon dioxide in ordinary drinking water. It is unlikely that he could have predicted the prodigious quantities of carbonated beverage—flavored water containing carbon dioxide under pressure—that would eventually be consumed. In the United States today, the per capita consumption of carbonated beverage rivals that of drinking water.

It is the weakly acidic character of carbonic acid that gives carbonated water its tang. When carbon dioxide dissolves in water, some of the molecules react to form carbonic acid, H_2CO_3 (Figure 11.43).

$$CO_2(aq) + H_2O(l) \rightleftharpoons H_2CO_3(aq)$$

Carbonic acid is a very weak acid, however, so carbonated beverage recipes often contain other acids, including citric and phosphoric acids, to increase their acidic character.

Carbon dioxide is a colorless gas that is nontoxic at usual concentrations. It is present in the atmosphere (about 0.03 mole percent) and in our breath, where it results from the biological oxidation of food substances. The gas does have definite physiological effects if the air breathed contains greater than a few percent of CO_2: It increases the rate of breathing, and air containing more than 5 mole percent can cause headache and unconsciousness if breathed for a sufficient length of time. Because of the density of carbon dioxide gas (about 1.5

Figure 11.43
Acidic character of carbon dioxide solutions. *When carbon dioxide gas (from dry ice) dissolves in water containing bromcresol-purple indicator, the dye changes from purple to a yellow color, showing that the solution is acidic. The acidity stems from carbonic acid, H_2CO_3, formed from carbon dioxide and water.*

(continued)

phere, temperatures fluctuate from 100°C during the lunar day to −150°C at night. The concern of climatologists is that the additional greenhouse effect from the increase in carbon dioxide in our atmosphere will drastically alter our present climate. These scientists employ computer models to predict the effect of changes of carbon dioxide, dust, and other variables on climate. At the moment, these models are subject to considerable uncertainty. To obtain good predictions, the models must include many variables and predict local climate patterns. The best guess at present is that a doubling of the carbon dioxide in the atmosphere will raise the average temperature of the earth by about 3°C (predicted to occur between 2030 and 2080). This may seem to be a small temperature increase, but local temperature variations are predicted to be much larger, so that some places will suffer more severe winters while others will experience warm, arid climates.

Questions for Study

1. What is carbonated water? What is the source of its acidity?
2. Describe some of the physical properties of carbon dioxide.
3. Why is carbon dioxide unsuitable for putting out magnesium fires?
4. Describe the production of carbon dioxide from natural gas.
5. How has the concentration of carbon dioxide in the atmosphere changed in recent years? Explain.
6. Explain what is meant by the greenhouse effect. Of what significance is the greenhouse effect on our present climate? Why is it of concern?

A Checklist for Review

IMPORTANT TERMS

change of state (phase transition) (p. 403)
melting (11.2)
freezing (11.2)
vaporization (11.2)
sublimation (11.2)
condensation (11.2)
vapor pressure (11.2)
boiling point (11.2)
freezing point (11.2)
melting point (11.2)
heat of fusion (11.2)
heat of vaporization (11.2)
phase diagram (11.3)
triple point (11.3)
critical temperature (11.3)
critical pressure (11.3)
surface tension (11.4)
viscosity (11.4)
intermolecular forces (11.5)
van der Waals forces (11.5)
dipole–dipole force (11.5)
London forces (dispersion forces) (11.5)
hydrogen bonding (11.5)
molecular solid (11.6)
metallic solid (11.6)
ionic solid (11.6)
covalent network solid (11.6)
crystalline solid (11.7)
amorphous solid (11.7)
crystal lattice (11.7)
unit cell (11.7)
simple cubic unit cell (11.7)
body-centered cubic unit cell (11.7)
face-centered cubic unit cell (11.7)
hexagonal close-packed structure (11.8)
cubic close-packed structure (11.8)
coordination number (11.8)

KEY EQUATION

$$\log \frac{P_2}{P_1} = \frac{\Delta H_{vap}}{2.303R}\left(\frac{1}{T_1} - \frac{1}{T_2}\right) \quad \text{or} \quad \ln \frac{P_2}{P_1} = \frac{\Delta H_{vap}}{R}\left(\frac{1}{T_1} - \frac{1}{T_2}\right)$$

SUMMARY OF FACTS AND CONCEPTS

Gases are composed of molecules in constant random motion throughout mostly empty space. This explains why gases are compressible fluids. Liquids are also composed of molecules in constant random motion, but the molecules are more tightly packed. Thus, liquids are incompressible fluids. Solids are composed of atoms, molecules, or ions that are in close contact and oscillate about fixed sites. Thus, solids are incompressible and rigid rather than fluid.

Any state of matter may change to another state. Melting and freezing are examples of such changes of state, or *phase transitions*. Vapor pressure, boiling point, and melting point are properties of substances that involve phase transitions. Vapor pressure increases with temperature; according to the *Clausius–Clapeyron equation*, the logarithm of the vapor pressure varies inversely with the absolute temperature. The conditions under which a given state of a substance exists are graphically summarized in a *phase diagram*.

Like vapor pressure and boiling point, *surface tension* and *viscosity* are important properties of liquids. These properties can be explained in terms of intermolecular forces. The three kinds of attractive intermolecular forces are *dipole–dipole forces*, *London forces*, and *hydrogen bonding*. London forces are weak attractive forces present in all molecules; London forces tend to increase with molecular weight. Thus, vapor pressure tends to decrease with molecular weight, whereas boiling point, surface tension, and viscosity tend to increase (unless hydrogen bonding is present). Hydrogen bonding occurs in substances containing H atoms bonded to F, O, and N atoms. When hydrogen bonding is present, vapor pressure tends to be lower than otherwise expected, whereas boiling point, surface tension, and viscosity tend to be higher.

Solids can be classified in terms of the type of force between the structural units. Thus, we have molecular, metallic, ionic, and covalent network solids. Melting point, hardness, and electrical conductivity are properties that can be related to the structure of the solid.

Solids can be crystalline or amorphous. A crystalline solid has an ordered arrangement of structural units placed at *crystal lattice* points. We may think of a crystal as constructed from *unit cells*. Cubic unit cells are of three kinds: simple cubic, body-centered cubic, and face-centered cubic. One of the most important ways of determining the structure of a crystalline solid is by *x-ray diffraction*.

OPERATIONAL SKILLS

1. **Calculating the heat required for a phase change of a given mass of substance** Given the heat of fusion (or vaporization) of a substance, calculate the amount of heat required to melt (or vaporize) a given quantity of that substance (Example 11.1).

2. **Calculating vapor pressures and heats of vaporization** Given the vapor pressure of a liquid at one temperature and its heat of vaporization, calculate the vapor pressure at another temperature (Example 11.2). Given the vapor pressures of a liquid at two temperatures, calculate the heat of vaporization (Example 11.3).

3. **Relating the conditions for the liquefaction of gases to the critical temperature** Given the critical temperature and pressure of a substance, describe the conditions necessary for liquefying the gaseous substance (Example 11.4).

4. **Identifying intermolecular forces** Given the molecular structure, state the kinds of intermolecular forces expected for a substance (Example 11.5).

5. **Determining relative vapor pressure on the basis of intermolecular attraction** Given two liquids, decide on the basis of the intermolecular forces which has the higher vapor pressure at a given temperature or which has the lower boiling point (Example 11.6).

6. **Identifying types of solids** From what you know about the bonding in a solid, classify it as molecular, metallic, ionic, or covalent network (Example 11.7).

7. **Determining the relative melting points based on types of solids** Given a list of substances, arrange them in order of increasing melting point from what you know of their structures (Example 11.8).

8. **Determining the number of atoms per unit cell** Given the description of a unit cell, find the number of atoms per cell (Example 11.9).

9. **Calculating atomic mass from unit-cell dimension and density** Given the edge length of the unit cell, crystal structure, and the density of a metal, calculate the mass of a metal atom (Example 11.10).

10. **Calculating unit-cell dimension from unit-cell type and density** Given the unit cell structure, the density, and the atomic weight for an element, calculate the edge length of the unit cell (Example 11.11).

Review Questions

11.1 List the different phase transitions that are possible and give examples of each.

11.2 Describe how you could purify iodine by sublimation.

11.3 Describe vapor pressure in molecular terms. What do we mean by saying it involves a dynamic equilibrium?

11.4 Explain why 15 g of steam at 100°C melts more ice than 15 g of liquid water at 100°C.

11.5 Why is the heat of fusion of a substance smaller than its heat of vaporization?

11.6 Explain why evaporation leads to cooling of the liquid.

11.7 Describe the behavior of a liquid and its vapor in a closed vessel as the temperature increases.

11.8 Gases that cannot be liquefied at room temperature merely by compression are called ''permanent'' gases. How could you liquefy such a gas?

11.9 The pressure in a cylinder of nitrogen continuously decreases as gas is released from it. On the other hand, a cylinder of propane maintains a constant pressure as propane is released. Explain this difference in behavior.

11.10 Why does the vapor pressure of a liquid depend on the intermolecular forces?

11.11 Explain the surface tension of a liquid in molecular terms. How does the surface tension make a liquid act as though it had a ''skin''?

11.12 Explain the origin of the London force that exists between two molecules.

11.13 Explain what is meant by hydrogen bonding. Describe the hydrogen bonding between two H_2O molecules.

11.14 Why do molecular substances have relatively low melting points?

11.15 Describe the distinguishing characteristics of a crystalline solid and an amorphous solid.

11.16 Describe the face-centered cubic unit cell.

11.17 Describe the structure of thallium(I) iodide, which has the same structure as cesium chloride.

11.18 What is the coordination number of Cs^+ in CsCl? of Na^+ in NaCl? of Zn^{2+} in ZnS?

11.19 Explain in words how Avogadro's number could be obtained from the unit-cell edge length of a cubic crystal. What other data are required?

11.20 Explain the production of an x-ray diffraction pattern by a crystal in terms of the interference of waves.

Practice Problems

PHASE TRANSITIONS

11.21 Identify the phase transition occurring in each of the following.

(a) The water level in an aquarium tank falls continuously (the tank has no leak).
(b) A mixture of scrambled eggs placed in a cold vacuum chamber slowly turns to a powdery solid.
(c) Chlorine gas is passed into a very cold test tube where it turns to a yellow liquid.
(d) When carbon dioxide gas under pressure exits from a small orifice, it turns to a white ''snow.''
(e) Molten lava from a volcano cools and turns to solid rock.

11.22 Identify the phase transition occurring in each of the following.

(a) Moth balls slowly become smaller and eventually disappear.
(b) Rubbing alcohol spilled on the palm of the hand feels cool as the volume of liquid decreases.
(c) A black deposit of tungsten metal collects on the inside of a light bulb whose filament is tungsten metal.
(d) Raindrops hit a cold metal surface, which becomes covered with ice.
(e) Candle wax turns to liquid under the heat of the candle flame.

11.23 Use Figure 11.7 to estimate the boiling point of diethyl ether, $(C_2H_5)_2O$, under an external pressure of 450 mmHg.

11.24 Use Figure 11.7 to estimate the boiling point of carbon tetrachloride, CCl_4, under an external pressure of 350 mmHg.

11.25 An electric heater coil provided heat to a 15.5-g sample of iodine, I_2, at the rate of 3.48 J/s. It took 4.54 min from the time the iodine began to melt until the iodine was completely melted. What is the heat of fusion per mole of iodine?

11.26 A 35.8-g sample of cadmium metal was melted by an electric heater providing 4.66 J/s of heat. If it took 6.92 min from the time the metal began to melt until it was completely melted, what is the heat of fusion per mole of cadmium?

11.27 Isopropyl alcohol, $CH_3CHOHCH_3$, is used in rubbing alcohol mixtures. Alcohol on the skin cools by evaporation. How much heat is absorbed by the alcohol if 10.0 g evaporates? The heat of vaporization of isopropyl alcohol is 42.1 kJ/mol.

11.28 Liquid butane, C_4H_{10}, is stored in cylinders to be used as a fuel. Suppose 31.4 g of butane gas is removed from a cylinder. How much heat must be provided to vaporize this much gas? The heat of vaporization of butane is 21.3 kJ/mol.

11.29 Water at 0°C was placed in a dish inside a vessel maintained at low pressure by a vacuum pump. After a quantity of water had evaporated, the remainder froze. If 9.31 g of ice at 0°C was obtained, how much liquid water must have evaporated? The heat of fusion of water is 6.01 kJ/mol and its heat of vaporization is 44.9 kJ/mol at 0°C.

11.30 A quantity of ice at 0.0°C was added to 36.2 g of water at 21.0°C to give water at 0.0°C. How much ice was added? The heat of fusion of water is 6.01 kJ/mol and the specific heat is 4.18 J/(g · °C).

11.31 A quantity of ice at 0°C is added to 64.3 g of water in a glass at 55°C. After the ice melted, the temperature of the water in the glass was 15°C. How much ice was added? The heat of fusion of water is 6.01 kJ/mol and the specific heat is 4.18 J/(g · °C).

11.32 Steam at 100°C was passed into a flask containing 275 g of water at 21°C, where the steam condensed. How many grams of steam must have condensed if the temperature of the water in the flask was raised to 76°C? The heat of vaporization of water at 100°C is 40.7 kJ/mol and the specific heat is 4.18 J/(g · °C).

11.33 Chloroform, $CHCl_3$, a volatile liquid, was once used as an anesthetic but has been replaced by safer compounds. It boils at 61.7°C and has a heat of vaporization of 31.4 kJ/mol. What is its vapor pressure at 25.0°C?

11.34 Methanol, CH_3OH, a colorless, volatile liquid, was formerly known as wood alcohol. It boils at 65.0°C and has a heat of vaporization of 37.4 kJ/mol. What is its vapor pressure at 22.0°C?

11.35 White phosphorus, P_4, is normally a white, waxy solid melting at 44°C to a colorless liquid. The liquid has a vapor pressure of 400.0 mmHg at 251.0°C and 760.0 mmHg at 280.0°C. What is the heat of vaporization of this substance?

11.36 Carbon disulfide, CS_2, is a volatile, flammable liquid. It has a vapor pressure of 400.0 mmHg at 28.0°C and 760.0 mmHg at 46.5°C. What is the heat of vaporization of this substance?

PHASE DIAGRAMS

11.37 Use graph paper and sketch the phase diagram of oxygen, O_2, from the following information: normal melting point, −218°C; normal boiling point, −183°C; triple point, −219°C, 1.10 mmHg; critical point, −118°C, 50.1 atm. Label each of the phase regions on the diagram.

11.38 Use graph paper and sketch the phase diagram of argon, Ar, from the following information: normal melting point, −187°C; normal boiling point, −186°C; triple point, −189°C, 0.68 atm; critical point, −122°C, 48 atm. Label each of the phase regions on the diagram.

11.39 Which of the following substances can be liquefied by applying pressure at 25°C? For those that cannot, describe the conditions under which they *can* be liquefied.

SUBSTANCE	CRITICAL TEMPERATURE	CRITICAL PRESSURE
Sulfur dioxide, SO_2	158°C	78 atm
Acetylene, C_2H_2	36°C	62 atm
Methane, CH_4	−82°C	46 atm
Carbon monoxide, CO	−140°C	35 atm

11.40 A tank of gas at 21°C has a pressure of 1.0 atm. Using the data in the table, answer the following questions. Explain your answers.

(a) If the tank contains carbon tetrafluoride, CF_4, is the liquid state also present?

(b) If the tank contains butane, C_4H_{10}, is the liquid state also present?

SUBSTANCE	BOILING POINT AT 1 ATM	CRITICAL TEMPERATURE	CRITICAL PRESSURE
CF_4	−128°C	−46°C	41 atm
C_4H_{10}	−0.5°C	152°C	38 atm

11.41 Bromine, Br_2, has a triple point at −7.3°C and 44 mmHg and a critical point at 315°C and 102 atm. The density of the solid is 3.4 g/cm^3, and the density of the liquid is 3.1 g/cm^3. Sketch a rough phase diagram of bromine, labeling all important features. Circle the correct word in each of the following sentences (and explain your answers).

(a) Bromine vapor at 40 mmHg condenses to the (liquid, solid) when cooled sufficiently.
(b) Bromine vapor at 400 mmHg condenses to the (liquid, solid) when cooled sufficiently.

11.42 Krypton, Kr, has a triple point at −169°C and 133 mmHg and a critical point at −63°C and 54 atm. The density of the solid is 2.8 g/cm^3, and the density of the liquid is 2.4 g/cm^3. Sketch a rough phase diagram of krypton. Circle the correct word in each of the following sentences (and explain your answers).

(a) Solid krypton at 130 mmHg (melts, sublimes without melting) when the temperature is raised.
(b) Solid krypton at 760 mmHg (melts, sublimes without melting) when the temperature is raised.

INTERMOLECULAR FORCES AND PROPERTIES OF LIQUIDS

11.43 The heats of vaporization of liquid Cl_2, liquid H_2, and liquid N_2 are 20.4 kJ/mol, 0.9 kJ/mol, and 5.6 kJ/mol, respectively. Are the relative values as you would expect? Explain.

11.44 The heats of vaporization of liquid O_2, liquid Ne, and liquid methanol, CH_3OH, are 6.8 kJ/mol, 1.8 kJ/mol, and 34.5 kJ/mol, respectively. Are the relative values as you would expect? Explain.

11.45 For each of the following substances, list the kinds of intermolecular forces expected.

(a) boron trifluoride, BF_3
(b) isopropyl alcohol, $CH_3CHOHCH_3$
(c) hydrogen iodide, HI
(d) krypton, Kr

11.46 Which of the following compounds would you expect to exhibit *only* London forces?

(a) carbon tetrachloride, CCl_4
(b) methyl chloride, CH_3Cl
(c) phosphorus trichloride, PCl_3
(d) phosphorus pentachloride, PCl_5

11.47 Arrange the following substances in order of increasing magnitude of the London forces: $SiCl_4$, CCl_4, $GeCl_4$.

11.48 Arrange the following substances in order of increasing magnitude of the London forces: Ar, He, Kr.

11.49 Methane, CH_4, reacts with chlorine, Cl_2, to produce a series of chlorinated hydrocarbons: methyl chloride (CH_3Cl), methylene chloride (CH_2Cl_2), chloroform ($CHCl_3$), and carbon tetrachloride (CCl_4). Which compound has the lowest vapor pressure at room temperature? Explain.

11.50 The halogens form a series of compounds with each other, which are called interhalogens. Examples are bromine chloride (BrCl), iodine bromide (IBr), bromine fluoride (BrF), and chlorine fluoride (ClF). Which compound is expected to have the lowest boiling point at any given pressure? Explain.

11.51 Predict the order of increasing vapor pressure at a given temperature for the following compounds:

(a) FCH_2CH_2F
(b) $HOCH_2CH_2OH$
(c) FCH_2CH_2OH

Explain why you chose this order.

11.52 Predict the order of increasing vapor pressure at a given temperature for the following compounds:

(a) $CH_3CH_2CH_2CH_2OH$
(b) $HOCH_2CH_2CH_2OH$,
(c) $CH_3CH_2OCH_2CH_3$

Explain why you chose this order.

11.53 List the following substances in order of increasing boiling point:

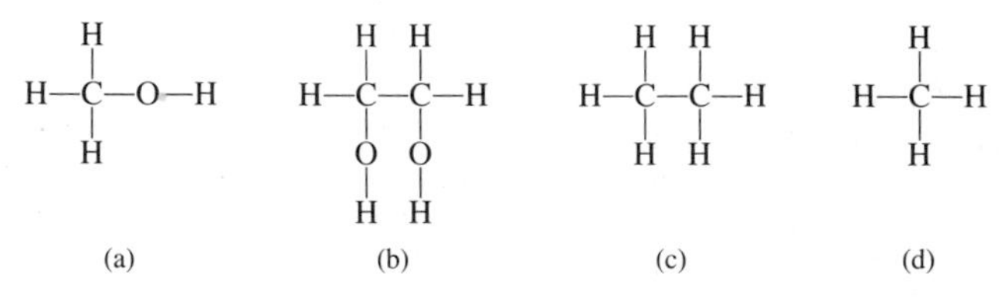

11.54 Arrange the following compounds in order of increasing boiling point.

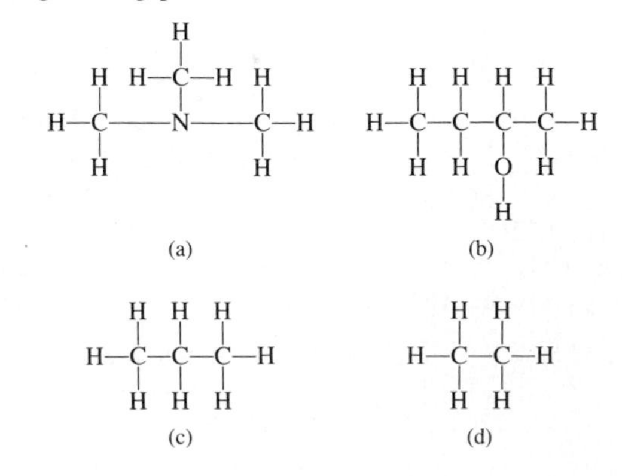

TYPES OF SOLIDS

11.55 Classify each of the following elements by the type of solid it forms: (a) Li; (b) Be; (c) B; (d) C; (e) N.

11.56 Classify each of the following by the type of solid it forms: (a) LiCl; (b) $BeCl_2$; (c) BCl_3; (d) CCl_4; (e) NCl_3.

11.57 Classify each of the following solid elements as molecular, metallic, ionic, or covalent network.

(a) tin, Sn
(b) germanium, Ge
(c) sulfur, S_8
(d) iodine, I_2

11.58 Which of the following do we expect to be molecular solids?

(a) sodium hydroxide, NaOH
(b) solid ethane, C_2H_6
(c) nickel, Ni
(d) solid silane, SiH_4

11.59 Arrange the following compounds in order of increasing melting point.

```
        H  H     H  H         H  H  H  H
        |  |     |  |         |  |  |  |
CaO  H—C—C—O—C—C—H      H—C—C—C—C—O—H     KCl
        |  |     |  |         |  |  |  |
        H  H     H  H         H  H  H  H
(a)          (b)                   (c)            (d)
```

11.60 Arrange the following substances in order of increasing melting point.

```
        H             H  H
        |             |  |
Si   H—C—O—H      H—C—C—H     NaCl
        |             |  |
        H             H  H
(a)     (b)           (c)         (d)
```

11.61 Associate each type of solid in the left-hand column with two of the properties listed in the right-hand column. Each property may be used more than once.

(a) molecular solid	low-melting
(b) ionic solid	high-melting
(c) metallic solid	brittle
(d) covalent network solid	malleable
	hard
	electrically conducting

11.62 On the basis of the description given, classify each of the following solids as molecular, metallic, ionic, or covalent network. Explain your answers.

(a) a lustrous, yellow solid that conducts electricity
(b) a hard, black solid melting at 2350°C to give a nonconducting liquid
(c) a nonconducting, pink solid melting at 650°C to give an electrically conducting liquid
(d) red crystals having a characteristic odor and melting at 171°C.

11.63 Associate each of the solids Co, LiCl, SiC, and CHI_3 with one of the following sets of properties.

(a) A white solid melting at 613°C; the liquid is electrically conducting, although the solid is not.
(b) A very hard, blackish solid subliming at 2700°C.
(c) A yellow solid with a characteristic odor having a melting point of 120°C.
(d) A gray, lustrous solid melting at 1495°C; both the solid and liquid are electrical conductors.

11.64 Associate each of the solids BN, P_4S_3, Pb, and $CaCl_2$ with one of the following sets of properties.

(a) A bluish-white, lustrous solid melting at 327°C; the solid is soft and malleable.
(b) A white solid melting at 772°C; the solid is an electrical nonconductor but dissolves in water to give a conducting solution.
(c) A yellowish-green solid melting at 172°C.
(d) A very hard, colorless substance melting at about 3000°C.

CRYSTAL STRUCTURE

11.65 How many atoms are there in a simple cubic unit cell of an atomic crystal in which all atoms are at lattice points?

11.66 How many atoms are there in a body-centered cubic unit cell of an atomic crystal in which all atoms are at lattice points?

11.67 Metallic iron has a body-centered cubic lattice with all atoms at lattice points and a unit cell whose edge length is 2.866 Å. The density of iron is 7.87 g/cm^3. What is the mass of an iron atom? Compare this value with the value you obtain from the molar mass.

11.68 Nickel has a face-centered unit cell with all atoms at lattice points and an edge length of 3.524 Å. The density of metallic nickel is 8.91 g/cm^3. What is the mass of a nickel atom? From the atomic weight, calculate Avogadro's number.

11.69 Copper metal has a face-centered cubic structure with all atoms at lattice points and a density of 8.93 g/cm^3. Its atomic weight is 63.5 amu. Calculate the edge length of the unit cell.

11.70 Barium metal has a body-centered cubic lattice with all atoms at lattice points; its density is 3.51 g/cm^3. From these data and the atomic weight, calculate the edge length of a unit cell.

11.71 Gold has cubic crystals whose unit cell has an edge length of 4.079 Å. The density of the metal is 19.3 g/cm^3. From these data and the atomic weight, calculate the number of gold atoms in a unit cell assuming all atoms are at lattice points. What type of cubic lattice does gold have?

11.72 Chromium forms cubic crystals whose unit cell has an edge length of 2.885 Å. The density of the metal is 7.20 g/cm^3. Use these data and the atomic weight to calculate the number of atoms in a unit cell assuming all atoms are at lattice points. What type of cubic lattice does chromium have?

11.73 Tungsten has a body-centered cubic lattice with all atoms at the lattice points. The edge length of the unit cell is 3.165 Å. The atomic weight of tungsten is 183.8 amu. Calculate its density.

11.74 Lead has a face-centered cubic lattice with all atoms at lattice points and a unit-cell edge length of 4.950 Å. Its atomic weight is 207.2 amu. What is the density of lead?

11.75 Metallic magnesium has a hexagonal close-packed structure and a density of 1.74 g/cm^3. Assume magnesium atoms to be spheres of radius r. Because magnesium has a close-packed structure, 74.1% of the space is occupied by atoms. Calculate the volume of each atom; then find the atomic radius, r. Note that the volume of a sphere is equal to $4\pi r^3/3$.

11.76 Metallic barium has a body-centered cubic structure (all atoms at the lattice points) and a density of 3.51 g/cm^3. Assume barium atoms to be spheres. The spheres in a body-centered array occupy 68.0% of the total space. Find the atomic radius of barium. (See Problem 11.75.)

Additional Problems

11.77 If you leave your car parked outdoors in the winter, you may find frost on the windows in the morning. If you then start the car and let the heater warm the windows, after some minutes the window will be dry. Describe all of the phase changes that have occurred.

11.78 Snow forms in the upper atmosphere in a cold air mass that is supersaturated with water vapor. When the snow later falls through a lower, warm air mass, rain forms. When this rain falls on a sunny spot, the drops evaporate. Describe all of the phase changes that have occurred.

11.79 The percent relative humidity of a sample of air is found as follows: (partial pressure of water vapor/vapor pressure of water) $\times$ 100. A sample of air at 23°C was cooled to 15°C, where moisture began to condense as dew. What was the relative humidity of the air at 23°C?

11.80 A sample of air at 21°C has a relative humidity of 58%. At what temperature will water begin to condense as dew? (See Problem 11.79.)

11.81 The vapor pressure of benzene is 100.0 mmHg at 26.1°C and 400.0 mmHg at 60.6°C. What is the boiling point of benzene at 760.0 mmHg?

11.82 The vapor pressure of water is 17.5 mmHg at 20.0°C and 355.1 mmHg at 80.0°C. Calculate the boiling point of water at 760.0 mmHg.

11.83 Describe the behavior of carbon dioxide gas when compressed at the following temperatures:

(a) 20°C (b) $-$70°C (c) 40°C

The triple point of carbon dioxide is $-$57°C and 5.1 atm, and the critical point is 31°C and 73 atm.

11.84 Describe the behavior of iodine vapor when cooled at the following pressures:

(a) 120 atm (b) 1 atm (c) 50 mmHg

The triple point of iodine is 114°C and 90.1 mmHg, and the critical point is 512°C and 116 atm.

11.85 Describe the formation of hydrogen bonds in propanol, $CH_3CH_2CH_2OH$. Represent possible hydrogen bonding structures in propanol by using structural formulas and the conventional notation for a hydrogen bond.

11.86 Describe the formation of hydrogen bonds in hydrogen peroxide, H_2O_2. Represent possible hydrogen bonding structures in hydrogen peroxide by using structural formulas and the conventional notation for a hydrogen bond.

11.87 Ethylene glycol (CH_2OHCH_2OH) is a slightly viscous liquid that boils at 198°C. Pentane (C_5H_{12}), which has approximately the same molecular weight as ethylene glycol, is a nonviscous liquid that boils at 36°C. Explain the differences in physical characteristics of these two compounds.

11.88 Pentylamine, $CH_3CH_2CH_2CH_2CH_2NH_2$, is a liquid that boils at 104°C and has viscosity of 10×10^{-4} kg/(m · s). Triethylamine, $(CH_3CH_2)_3N$, is a liquid that boils at 89°C and has a viscosity of about 4×10^{-4} kg/(m · s). Explain the differences in properties of these two compounds.

11.89 Consider the elements Al, Si, P, and S from the third row of the periodic table. In each case, identify the type of solid the element would form.

11.90 The elements in Problem 11.89 form the fluorides AlF_3, SiF_4, PF_3, and SF_4. In each case, identify the type of solid formed by the fluoride.

11.91 Decide which substance in each of the following pairs has the lower melting point. Explain how you made each choice.

(a) potassium chloride, KCl; or calcium oxide, CaO
(b) carbon tetrachloride, CCl_4; or hexachloroethane, C_2Cl_6
(c) zinc, Zn; or chromium, Cr
(d) acetic acid, CH_3COOH; or ethyl chloride, C_2H_5Cl

11.92 Decide which substance in each of the following pairs has the lower melting point. Explain how you made each choice.

(a) magnesium oxide, MgO; or hexane, C_6H_{14}
(b) 1-propanol, C_3H_7OH; or ethylene glycol, CH_2OHCH_2OH
(c) silicon, Si; or sodium, Na
(d) methane, CH_4; or silane, SiH_4

11.93 Iridium metal, Ir, crystallizes in a face-centered cubic (close-packed) structure. The edge length of the unit cell was determined by x-ray diffraction to be 3.839 Å. The density of iridium is 22.42 g/cm^3. Calculate the mass of an iridium atom. Use Avogadro's number to calculate the atomic weight of iridium.

11.94 The edge length of the unit cell of tantalum metal, Ta, is 3.306 Å; the unit cell is body-centered cubic (one atom at each lattice point). Tantalum has a density of 16.69 g/cm^3. What is the mass of a tantalum atom? Use Avogadro's number to calculate the atomic weight of tantalum.

11.95 Use your answer to Problem 11.69 to calculate the radius of the copper atom. To do this, assume that copper atoms are spheres. Then note that the spheres on any face of a unit cell touch along the diagonal.

11.96 Rubidium metal has a body-centered cubic structure (with one atom at each lattice point). The density of the metal is 1.532 g/cm^3. From this information and the atomic weight, calculate the edge length of the unit cell. Now assume that rubidium atoms are spheres. Each corner sphere of the unit cell touches the body centered sphere. Calculate the radius of a rubidium atom.

11.97 Calculate the percent of volume that is actually occupied by spheres in a body-centered cubic lattice of identical spheres. You can do this by first relating the radius of a sphere, r, to the length of an edge of a unit cell, l. (Note that the spheres do not touch along an edge, but they do touch along a diagonal passing through the body-centered sphere.) Then calculate the volume of a unit cell in terms of r. The volume occupied by spheres equals the number of spheres per unit cell times the volume of a sphere ($4\pi r^3/3$).

11.98 Calculate the percent of volume that is actually occupied by spheres in a face-centered cubic lattice of identical spheres. You can do this by first relating the radius of a sphere, r, to the length of an edge of a unit cell, l. (Note that the spheres do not touch along an edge, but they do touch along the diagonal of a face.) Then calculate the volume of a unit cell in terms of r. The volume occupied by spheres equals the number of spheres per unit cell times the volume of a sphere ($4\pi r^3/3$).

Cumulative-Skills Problems

11.99 The vapor pressure of a volatile liquid can be determined by slowly bubbling a known volume of gas through the liquid at a given temperature and pressure. In an experiment, a 5.40-L sample of nitrogen gas, N_2, at 20.0°C and 745 mmHg is bubbled through liquid isopropyl alcohol, C_3H_8O, at 20.0°C. Nitrogen containing the vapor of C_3H_8O at its vapor pressure leaves the vessel at 20.0°C and 745 mmHg. It is found that 0.6149 g C_3H_8O has evaporated. How many moles of N_2 are in the gas leaving the liquid? How many moles of alcohol are in this gaseous mixture? What is the mole fraction of alcohol vapor in the gaseous mixture? What is the partial pressure of the alcohol in the gaseous mixture? What is the vapor pressure of C_3H_8O at 20.0°C?

11.100 In an experiment, a sample of 6.35 L of nitrogen at 25.0°C and 768 mmHg is bubbled through liquid acetone, C_3H_6O. The gas plus vapor at its equilibrium partial pressure leaves the liquid at the same temperature and pressure. If 6.550 g of acetone has evaporated, what is the vapor pressure of acetone at 25.0°C? See Problem 11.99.

11.101 How much heat is needed to vaporize 10.0 mL of liquid hydrogen cyanide, HCN, at 25.0°C? The density of the liquid is 0.687 g/mL. Use standard heats of formation, which are given in Appendix C.

11.103 How much heat must be added to 25.0 g of solid white phosphorus, P_4, at 25.0°C to give the liquid at its melting point, 44.1°C? The heat capacity of solid white phosphorus is 95.4 J/(K · mol); its heat of fusion is 2.63 kJ/mol.

11.105 Acetic acid, CH_3COOH, forms stable pairs of molecules held together by two hydrogen bonds.

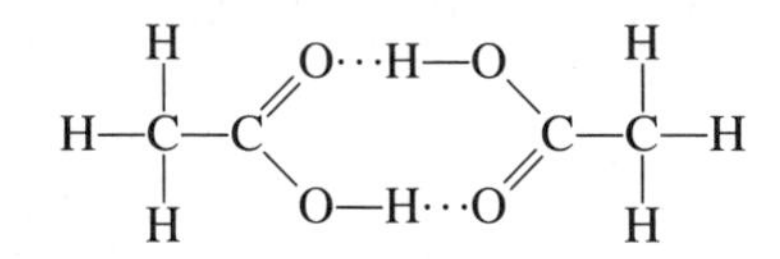

Such molecules, which are themselves formed by the association of two simpler molecules, are called dimers. The vapor over liquid acetic acid consists of a mixture of monomers (single acetic acid molecules) and dimers. At 100.6°C the total pressure of vapor over liquid acetic acid is 436 mmHg. If the vapor consists of 0.630 mole fraction of the dimer, what are the masses of monomer and dimer in 1.000 L of the vapor? What is the density of the vapor?

11.102 How much heat is needed to vaporize 10.0 mL of liquid methanol, CH_3OH, at 25.0°C? The density of the liquid is 0.787 g/mL. Use standard heats of formation, which are given in Appendix C.

11.104 How much heat must be added to 25.0 g of solid sodium, Na, at 25.0°C to give the liquid at its melting point, 97.8°C? The heat capacity of solid sodium is 28.2 J/(K · mol), and its heat of fusion is 2.60 kJ/mol.

11.106 The total pressure of vapor over liquid acetic acid at 71.3°C is 146 mmHg. If the density of the vapor is 0.702 g/L, what is the mole fraction of dimer in the vapor? See Problem 11.105.

12

Solutions

Chapter Outline

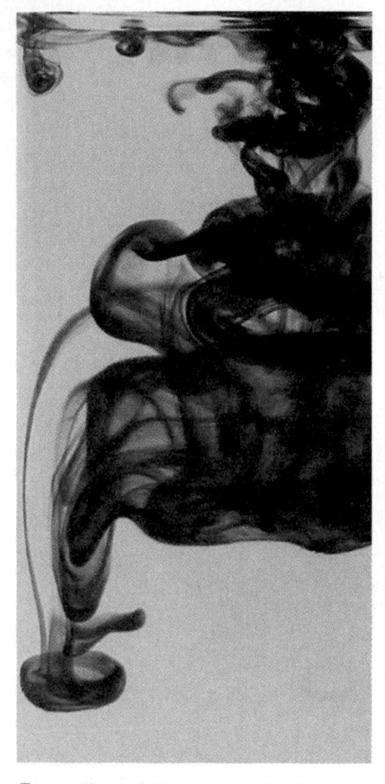

Formation of the starch-iodine complex.

There are various practical reasons for preparing solutions. For instance, most chemical reactions are run in solution. Also, solutions have particular properties that are useful. When gold is used for jewelry, it is mixed or alloyed with a small amount of silver. Gold–silver alloys are not only harder than pure gold, but they also melt at lower temperatures and are therefore easier to cast.

Solubility varies with temperature and possibly pressure. (Solubility is the equilibrium amount of one substance that dissolves in another.) The variation of solubility with pressure can be a useful property. Acetylene gas, C_2H_2, for example, is used as a fuel in welding torches. It can be transported safely under pressure in cylinders as a solution in acetone, CH_3COCH_3. Acetone is a liquid, and at 1 atm pressure 1 liter of the liquid dissolves 27 grams of acetylene. But at 12 atm, the pressure in a full cylinder, the same quantity of acetone dissolves 320 grams of acetylene, so that more can be transported. When the valve on a cylinder is opened and the pressure is reduced, acetylene gas comes out of solution.

Another useful property of solutions is their lower melting or freezing points compared with those of the major component. We have already mentioned the lowering of the melting point of gold when a small amount of silver is added. The use of ethylene glycol, CH_2OHCH_2OH, as an automobile antifreeze depends on the same property. Water containing ethylene glycol freezes at temperatures below the freezing point of pure water.

Table 12.1 Examples of Solutions

Solution	State of Matter	Description
Air	Gas	Homogeneous mixture of gases
Soda water	Liquid	Gas (CO_2) dissolved in a liquid (H_2O)
Ethanol in water	Liquid	Liquid solution of two completely miscible liquids
Brine	Liquid	Solid (NaCl) dissolved in a liquid (H_2O)
Potassium–sodium alloy	Liquid	Solution of two solids (K + Na)
Dental-filling alloy	Solid	Solution of a liquid (Hg) in a solid (Ag plus other metals)
Gold–silver alloy	Solid	Solution of two solids (Au + Ag)

What determines how much the freezing point of a solution is lowered? How does the solubility of a substance change when conditions such as pressure and temperature change? These are some of the questions we will address in this chapter.

Solution Formation

When sodium chloride dissolves in water, the resulting uniform dispersion of ions in water is called a solution. In general, a *solution* is a homogeneous mixture of two or more substances, consisting of ions or molecules. A *colloid* is similar in that it appears to be homogeneous like a solution. In fact, however, it consists of comparatively large particles of one substance dispersed throughout another substance or solution.

From the examples of solutions we mentioned in the chapter opening, we see that they may be quite varied in their characteristics. To begin our discussion, let us look at the different types of solutions we might encounter. We will discuss colloids at the end of the chapter.

12.1 TYPES OF SOLUTIONS

Solutions may exist in any of the three states of matter; that is, they may be gases, liquids, or solids. Some examples are listed in Table 12.1. The terms *solute* and *solvent* are used frequently in referring to the components of a solution. The **solute,** *in the case of a solution of a gas or solid dissolved in a liquid, is the gas or solid; in other cases, the solute is the component in smaller amount*. The **solvent,** *in a solution of a gas or solid in a liquid, is the liquid; in other cases, the solvent is the component in greater amount*. Thus, when sodium chloride is dissolved in water, sodium chloride is the solute and water is the solvent.

Gaseous Solutions

In general, nonreactive gases or vapors can mix in all proportions to give a gaseous mixture. *Fluids that mix with or dissolve in each other in all proportions* are said to be **miscible fluids.** Gases are thus miscible. (If two fluids do not mix but form two layers, they are said to be *immiscible;* see Figure 12.1.) Air, which is a

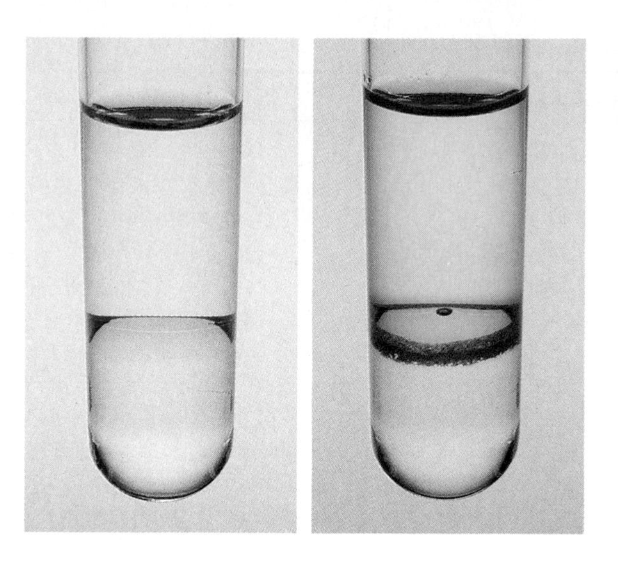

Figure 12.1
Immiscible liquids. *Left:* Water and methylene chloride are immiscible and form two layers. *Right:* Potassium chromate crystals dissolve in the water layer but float on the methylene chloride layer.

mixture of nitrogen, oxygen, and smaller amounts of other gases, is an example of a gaseous solution.

Liquid Solutions

Most liquid solutions are obtained by dissolving a gas, liquid, or solid in some liquid. Soda water, for example, consists of a solution of carbon dioxide gas in water. Ethanol, C_2H_5OH, in water is an example of a liquid–liquid solution.● Brine is water with sodium chloride (a solid) dissolved in it. Seawater contains both dissolved gases (from air) and solids (mostly sodium chloride).

It is also possible to make a liquid solution by mixing two solids together. Potassium–sodium alloy is an example of this. Both potassium and sodium are solid metals at room temperature, but a liquid solution results when the mixture contains 10% to 50% sodium.●

● Ethanol is miscible in water; that is, the two substances dissolve in each other in all proportions.

● Sodium metal melts at 98°C and potassium metal melts at 63°C. When the two metals are mixed to give a solution that is 20% sodium, for example, the melting point is lowered to −10°C. This is another example of the lowering of the freezing or melting point of solutions, which we will discuss later in the chapter. Sodium–potassium alloy is used as a heat-transfer medium in nuclear reactors.

Solid Solutions

Solid solutions are also possible. In the chapter opening, we mentioned gold–silver alloys. Dental-filling alloy is a solution of mercury (a liquid) in silver, with small amounts of other metals.

Exercise 12.1

Give an example of a solid solution prepared from a liquid and a solid.

(See Problems 12.27 and 12.28.)

12.2 SOLUBILITY AND THE SOLUTION PROCESS

The amount of substance that will dissolve in a solvent depends both on the substance and on the solvent. We describe the amount that dissolves in terms of the *solubility*.

Figure 12.2
Solubility equilibrium. The solid crystalline phase is in dynamic equilibrium with species (ions or molecules) in a saturated solution. The rate at which species leave the crystals equals the rate at which species return to the crystals.

Solubility; Saturated Solutions

To understand the concept of solubility, consider the process of dissolving sodium chloride in water. Sodium chloride is an ionic substance and it dissolves in water as Na^+ and Cl^- ions. If we could look at the dissolving of sodium chloride at the level of ions, we would see a dynamic process. Suppose we stir 40.0 grams of sodium chloride crystals into 100 mL of water at 20°C. Sodium ions and chloride ions leave the surface of the crystals and enter the solution. The ions move about at random in the solution and may by chance collide with a crystal and stick, thus returning to the crystalline state. As the sodium chloride continues to dissolve, more ions enter the solution, and the rate at which they return to the crystalline state increases (the more ions in solution, the more likely ions are to collide with the crystals and stick). Eventually, a *dynamic equilibrium* is reached in which the rate at which ions leave the crystals equals the rate at which ions return to the crystals (see Figure 12.2). We write the dynamic equilibrium this way:

$$NaCl(s) \rightleftharpoons Na^+(aq) + Cl^-(aq)$$

At equilibrium, no more sodium chloride appears to dissolve; 36.0 grams have gone into solution, leaving 4.0 grams of crystals at the bottom of the vessel. We have a **saturated solution**—that is, *a solution that is in equilibrium with respect to a given dissolved substance*. The solution is saturated with respect to NaCl, and no more NaCl can dissolve. The **solubility** of sodium chloride in water *(the amount that dissolves in a given quantity of water at a given temperature to give a saturated solution)* is 36.0 g/100 mL at 20°C. Note that if we had mixed 30.0 grams of sodium chloride with 100 mL of water, all of the crystals would have dissolved. We would have an **unsaturated solution,** *a solution not in equilibrium with respect to a given dissolved substance and in which more of the substance can dissolve*. (Saturated and unsaturated solutions are compared in Figure 12.3).

Sometimes it is possible to obtain a **supersaturated solution,** *a solution that contains more dissolved substance than a saturated solution*. For example, the solubility of sodium thiosulfate, $Na_2S_2O_3$, in water at 100°C is 231 g/100 mL. But at room temperature, the solubility is much less—about 50 g/100 mL. Suppose we prepare a solution saturated with sodium thiosulfate at 100°C. We might expect that as the water solution is cooled, sodium thiosulfate would crystallize out. In fact, if the solution is slowly cooled to room temperature, this does not occur.

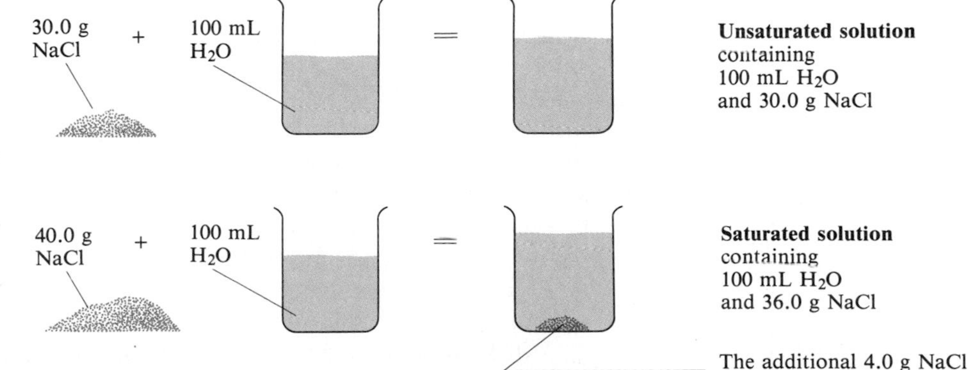

Figure 12.3
Comparison of unsaturated and saturated solutions. *Top:* The 30.0 g pile of NaCl will dissolve completely in 100 mL of water at 20°C, giving an unsaturated solution. *Bottom:* When 40.0 g NaCl is stirred into 100 mL H_2O, only 36.0 g dissolves at 20°C, leaving 4.0 g of the crystalline solid on the bottom of the beaker. This solution is saturated.

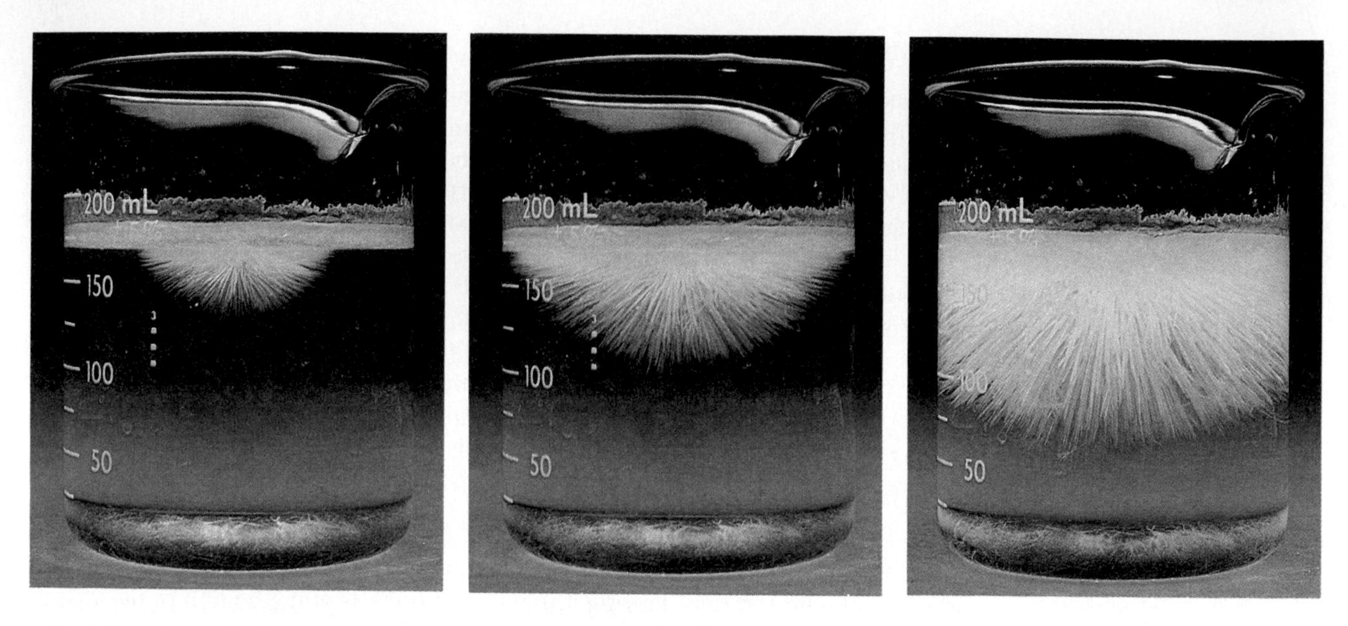

Figure 12.4
Crystallization from a supersaturated solution of sodium acetate. *Left:* Crystallization begins to occur when a small crystal of sodium acetate is added. *Center, right:* Within seconds, crystal growth spreads from the original crystal throughout the solution.

Instead the result is a solution in which 231 grams of sodium thiosulfate are dissolved in 100 mL of cold water, compared with the 50 grams we would normally expect to find dissolved.

Supersaturated solutions are not in equilibrium with the solid substance. Thus, if a small crystal of sodium thiosulfate is added to a supersaturated solution, the excess immediately crystallizes out. Crystallization is usually quite fast and dramatic (see Figure 12.4).

Factors in Explaining Solubility

The solubilities of substances in one another vary widely. We might find a substance miscible in one solvent but nearly insoluble in another. As a general rule, we find that "like dissolves like." That is, similar substances dissolve one another. Oil is miscible in gasoline. Both are mixtures of hydrocarbon substances (hydrocarbons are compounds of hydrogen and carbon only). On the other hand, oil does not mix with water. Water is a polar substance, whereas hydrocarbons are not. What we would like to know is why similar substances dissolve in one another to greater extents than do dissimilar substances. What are the factors involved in solubility?

The solubility of one substance in another can be explained in terms of two factors. One of these is the natural tendency of substances to mix. This is sometimes also referred to as the natural tendency toward disorder.● Figure 12.5A shows a vessel divided into two parts, with oxygen gas on the left and nitrogen gas on the right. If we remove the partition between the two gases, the molecules of the two gases begin to mix. Ultimately, these molecules become thoroughly mixed through their random motions (Figure 12.5B). We might expect a similar mixing of molecules or ions in other types of solutions.

● *Entropy* is a measure of disorder. Thus, the tendency toward disorder can be expressed as a tendency toward increasing entropy. The mixing of substances increases entropy. This increase is one of the primary factors leading to the spontaneous formation of a solution.

If the process of dissolving one substance in another involved nothing more than simple mixing, we would expect substances to be completely soluble in one

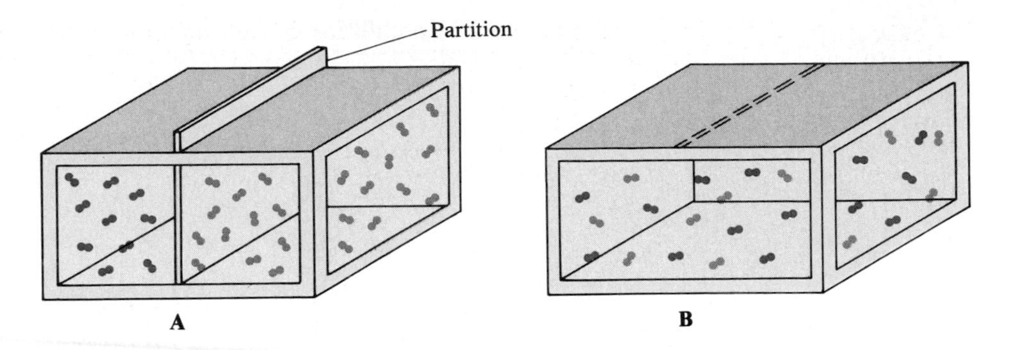

Figure 12.5
The mixing of gas molecules. *(A)* A vessel is divided by a removable partition into two parts, with oxygen gas on the left and nitrogen gas on the right. *(B)* When the partition is removed, molecules of the two gases begin to mix. Eventually, these molecules become thoroughly mixed through their random motions.

another; that is, we would expect substances to be miscible. We know this is only sometimes the case. Usually, substances have limited solubility in one another. A factor that can limit solubility is the relative forces of attraction between species (molecules or ions). Suppose there are strong interactions between solute species and strong interactions between solvent species, but weak interactions between solute and solvent species. In that case, the strongest interactions are maintained so long as the solute and solvent species do not mix. The lowest energy of the solute-solvent system is obtained then also.

The solubility of a solute in a solvent (that is, the extent of the mixing of the solute and solvent species) depends on a balance between the natural tendency for the solute and solvent species to mix and the tendency for a system to have the lowest energy possible.

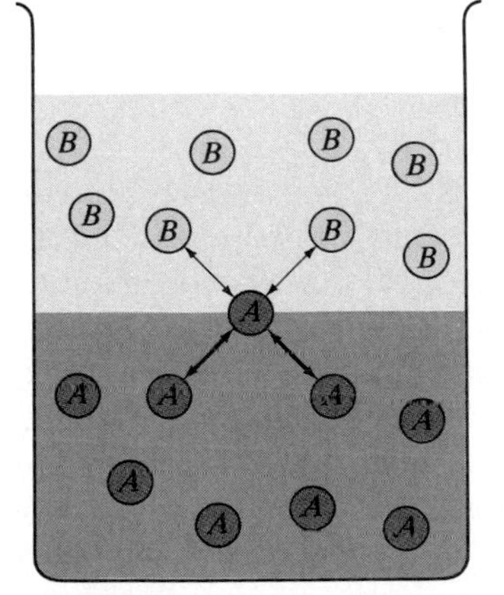

Figure 12.6
The immiscibility of liquids. Suppose an *A* molecule moves from liquid *A* into liquid *B*. If the intermolecular attraction between two *A* molecules is much stronger than the intermolecular attraction between an *A* molecule and a *B* molecule, the net force of attraction tends to pull the *A* molecule back into liquid *A*. Thus, liquid *A* will be immiscible with liquid *B*.

Molecular Solutions

The simplest example of a molecular solution is that of one gas dissolved in another. Air, which is essentially a solution of nitrogen and oxygen, is an example. The intermolecular forces in gases are weak. The only factor of importance is the natural tendency for molecules to mix (the inclination toward disorder). Gases are therefore miscible.

Substances may be miscible even when the intermolecular forces are not negligible. Consider the solution of the two similar liquid hydrocarbons heptane, C_7H_{16}, and octane, C_8H_{18}, which are components of gasoline. The intermolecular attractions are due to London forces, and those between heptane and octane molecules are nearly equal to those between octane and octane molecules and heptane and heptane molecules. The different intermolecular attractions are about the same strength, so there are no favored attractions. Octane and heptane molecules tend to move freely through one another. Therefore, the tendency of molecules to mix results in miscibility of the substances.

As a counterexample, consider the mixing of octane with water. There are strong hydrogen-bonding forces between water molecules. In order for octane to mix with water, hydrogen bonds must be broken and replaced by much weaker London forces between water and octane. In this case, the maximum forces of attraction among molecules (and therefore the lower energy) are obtained if the octane and water remain unmixed. As a result, octane and water are nearly immiscible in one another. (Also see Figure 12.6.)

Table 12.2 Solubilities of Alcohols in Water

Name	Formula	Solubility in H_2O (g/100 g H_2O at 20°C)
Methanol	CH_3OH	Miscible
Ethanol	CH_3CH_2OH	Miscible
1-Propanol	$CH_3CH_2CH_2OH$	Miscible
1-Butanol	$CH_3CH_2CH_2CH_2OH$	7.9
1-Pentanol	$CH_3CH_2CH_2CH_2CH_2OH$	2.7
1-Hexanol	$CH_3CH_2CH_2CH_2CH_2CH_2OH$	0.6

The statement ''like dissolves like'' succinctly expresses these observations. That is, substances with similar intermolecular attractions are usually soluble in one another. Thus, the two similar hydrocarbons heptane and octane are completely miscible, whereas octane and water (with dissimilar intermolecular attractions) are immiscible.

For the series of alcohols (organic, or carbon-containing, compounds with an —OH group) listed in Table 12.2, the solubility in water decreases from miscible to slightly soluble. Water and alcohols are alike in having —OH groups through which strong hydrogen-bonding attractions arise.

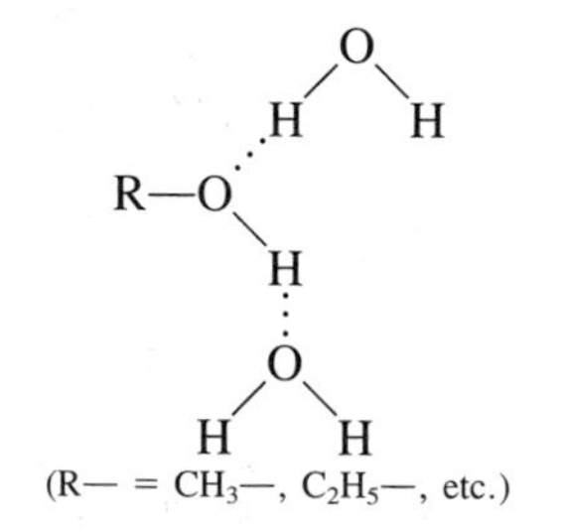

Thus, the attraction between a methanol molecule, CH_3OH, and a water molecule is nearly as strong as that between two methanol molecules or between two water molecules. Methanol and water molecules tend to mix freely. Methanol and water are miscible, as are ethanol and water and 1-propanol and water. However, as the hydrocarbon end, R—, of the alcohol becomes the more prominent portion of the molecule, the alcohol becomes less like water. Now the forces of attraction between alcohol and water molecules are weaker than those between two alcohol molecules or between two water molecules. Therefore, the solubilities of alcohols decrease with increasing length of R.

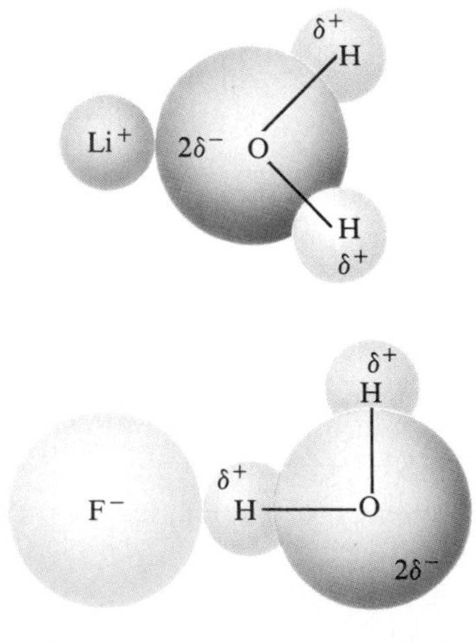

Figure 12.7
Attraction of water molecules to ions because of the ion–dipole force. The O end of H_2O orients toward the cation, whereas an H atom of H_2O orients toward the anion.

Exercise 12.2

Which of the following compounds is likely to be more soluble in water: C_4H_9OH or C_4H_9SH? Explain. (See Problems 12.29, 12.30, 12.31, and 12.32.)

Ionic Solutions

Ionic substances differ markedly in their solubilities in water. For example, sodium chloride, NaCl, has a solubility of 36 g per 100 mL of water at room temperature, whereas calcium phosphate, $Ca_3(PO_4)_2$, has a solubility of only 0.002 g per 100 mL of water. In most cases, these differences in solubility can be explained in terms of the different energies of attraction between ions in the crystal and between ions and water.

The energy of attraction between an ion and a water molecule is due to an *ion–dipole force*. Water molecules are polar, so they tend to orient with respect to nearby ions. In the case of a positive ion (Li^+, for example), water molecules orient with their oxygen atoms (the negative ends of the molecular dipoles) toward the ion. In the case of a negative ion (for instance, F^-), water molecules orient with their hydrogen atoms (the positive ends of the molecular dipoles) toward the ion (see Figure 12.7).

The attraction of ions for water molecules is called **hydration.**● Hydration of ions favors the dissolving of an ionic solid in water. Figure 12.8 illustrates the hydration of ions in the dissolving of a lithium fluoride crystal. Ions on the surface become hydrated and then move into the body of the solution as hydrated ions.

● Hydration of ions also occurs in crystalline solids. For example, copper(II) sulfate pentahydrate, $CuSO_4 \cdot 5H_2O$, contains five water molecules for each $CuSO_4$ formula unit in the crystal. Substances like this, called hydrates, were discussed in Chapter 2.

If the hydration of ions were the only factor in the solution process, we would expect all ionic solids to be soluble in water. The ions in a crystal, however, are very strongly attracted to one another. Therefore, the solubility of an ionic solid depends not only on the energy of hydration of ions (energy associated with the attraction between ions and water molecules) but also on *lattice energy*, the energy holding ions together in the crystal lattice. Lattice energy works against the solution process, so an ionic solid with relatively large lattice energy is usually insoluble.

Lattice energies depend on the charges on the ions (as well as on the distance between the centers of neighboring positive and negative ions). The greater the magnitude of ion charge, the greater the lattice energy.● For this reason, we might expect substances with singly charged ions to be comparatively soluble and those with multiply charged ions to be less soluble. This is borne out by the fact that compounds of the alkali metal ions (such as Na^+ and K^+) and ammonium ions (NH_4^+) are generally soluble, whereas those with phosphate ions (PO_4^{3-}), for example, are generally insoluble.

● According to Coulomb's law, the energy of attraction of two ions is proportional to the product of the ion charges and inversely proportional to the distance between the centers of the ions.

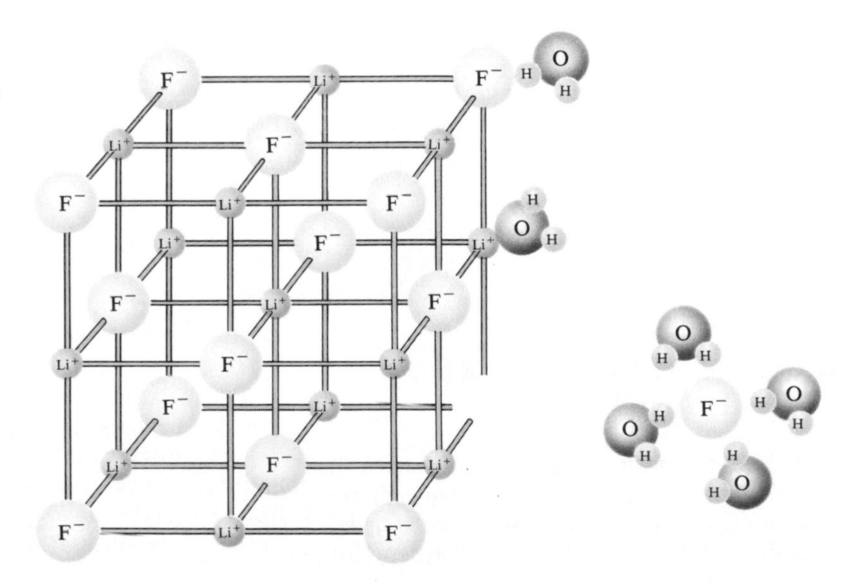

Figure 12.8
The dissolving of lithium fluoride in water. Ions on the surface of the crystal can hydrate, that is, associate with water molecules; ions at the corners are especially easy to remove because they are held by fewer lattice forces. Ions are completely hydrated in the aqueous phase and move off into the body of the liquid.

Lattice energy is also inversely proportional to the distance between neighboring ions, and this distance depends on the sum of the radii of the ions. Thus, the lattice energy of magnesium hydroxide, $Mg(OH)_2$, is inversely proportional to the sum of the radii of Mg^{2+} and OH^-. In the series of alkaline earth hydroxides—$Mg(OH)_2$, $Ca(OH)_2$, $Sr(OH)_2$, $Ba(OH)_2$—the lattice energy decreases as the radius of the alkaline earth ion increases (from Mg^{2+} to Ba^{2+}). If the lattice energy alone determines the trend in solubilities, we should expect the solubility to increase from magnesium hydroxide to barium hydroxide. In fact, this is what we find. Magnesium hydroxide is insoluble in water, and barium hydroxide is soluble. But this is not the whole story. The energy of hydration also depends on ionic radius. A small ion has a concentrated electric charge and a strong electric field that attracts water molecules. Therefore, the energy of hydration is greatest for a small ion such as Mg^{2+} and least for a large ion such as Ba^{2+}. If the energy of hydration of ions alone determined the trend in solubilities, we would expect the solubilities to decrease from magnesium hydroxide to barium hydroxide, rather than to increase. (We should also add that the energy of hydration increases with the charge on the ion. Thus, energy of hydration is greater for Mg^{2+} than for Na^+. Magnesium ion has a larger charge, as well as being a smaller ion than Na^+.)

The explanation for the observed solubility trend in the alkaline earth hydroxides is that the lattice energy decreases more rapidly in the series $Mg(OH)_2$, $Ca(OH)_2$, $Sr(OH)_2$, and $Ba(OH)_2$ than does the energy of hydration in the series of ions Mg^{2+}, Ca^{2+}, Sr^{2+}, and Ba^{2+}. For this reason, the lattice-energy factor dominates this solubility trend.

We see the opposite solubility trend when the energy of hydration decreases more rapidly, so that it dominates the trend. Consider the alkaline earth sulfates. Here the lattice energy depends on the sum of the radius of the cation and the radius of the sulfate ion. Because the sulfate ion, SO_4^{2-}, is much larger than the hydroxide ion, OH^-, the percent change in lattice energy in going through the series of sulfates from $MgSO_4$ to $BaSO_4$ is smaller than in the hydroxides. Thus, the lattice energy changes less, and the energy of hydration of the cation decreases by a greater amount. Now the energy of hydration dominates the solubility trend, and the solubility decreases from magnesium sulfate to barium sulfate. Magnesium sulfate is soluble in water, and barium sulfate is insoluble.

Exercise 12.3

Which of the following ions has the larger hydration energy, Na^+ or K^+?

(See Problems 12.33 and 12.34.)

Related Topic

HEMOGLOBIN SOLUBILITY AND SICKLE-CELL ANEMIA

Sickle-cell anemia was the first inherited disease shown to have a specific molecular basis. In people with the disease, the red blood cells tend to become elongated (sickle-shaped) when the concentration of oxygen (O_2) is low, as it is, for example, in the venous blood supply (see Figure 12.9). Once the red blood cells have sickled, they can no longer function in their normal capacity as oxygen carriers and they often break apart. Moreover, the sickled cells clog capillaries, interfering with the blood supply to vital organs.

In 1949 Linus Pauling showed that people with

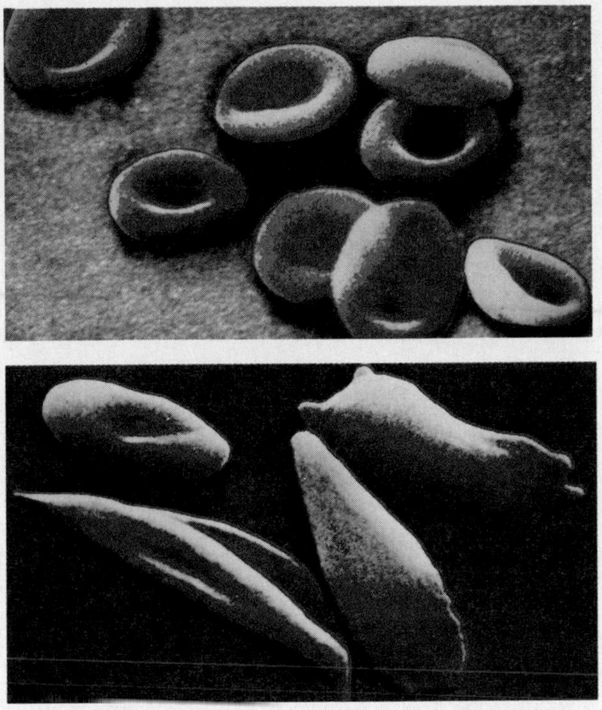

Figure 12.9
Red blood cells. *Scanning electron micrographs of normal cells* (top) *and sickled cells* (bottom). *The color is added by computer.*

sickle-cell anemia have abnormal hemoglobin. Hemoglobin is the substance in red blood cells that is responsible for carrying oxygen. It is normally present in solution within the red blood cells. But in people with sickle-cell anemia, the unoxygenated hemoglobin readily comes out of solution. It produces a fibrous precipitate that deforms the cell, giving it the characteristic sickle shape.

The normal and abnormal hemoglobins have been shown to differ very slightly. They are large molecules, with molecular weights of about 64,000 amu, and they are alike in structure except in one place. In this place, the normal hemoglobin has the group

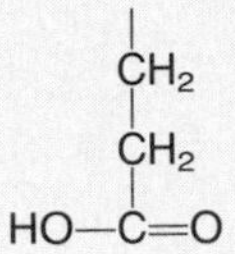

$$\begin{array}{c} | \\ CH_2 \\ | \\ CH_2 \\ | \\ HO{-}C{=}O \end{array}$$

which helps confer water solubility on the molecule because of the polarity of the group and its ability to form hydrogen bonds. The abnormal hemoglobin has the following hydrocarbon group.

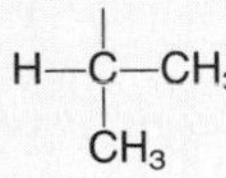

$$\begin{array}{c} | \\ H{-}C{-}CH_3 \\ | \\ CH_3 \end{array}$$

Hydrocarbon groups are nonpolar. This small change makes the molecule less water-soluble.

12.3 THE EFFECTS OF TEMPERATURE AND PRESSURE ON SOLUBILITY

In general, the solubility of a substance depends on temperature. For example, the solubility of ammonium nitrate in 100 mL of water is 118 grams at 0°C and 811 grams at 100°C. Pressure may also have an effect on solubility, as we will see.

Temperature Change

Solubilities of substances usually vary with temperature. Most gases become less soluble in water at higher temperatures. The first bubbles that appear when tap water is heated are not bubbles of water vapor. They are bubbles of air released as the increasing temperature reduces the solubility of air in water. On the other hand, the usual behavior for ionic solids is to increase in solubility in water with rising temperature.

The variations of the solubilities of the salts KNO_3, $CuSO_4$, NaCl, and $Ce_2(SeO_4)_3$ are shown in Figure 12.10. Three of these salts show the usual behavior; their solubilities increase with rising temperature. Potassium nitrate, KNO_3, changes solubility dramatically from 14 g/100 g H_2O at 0°C to 245 g/100 g H_2O at

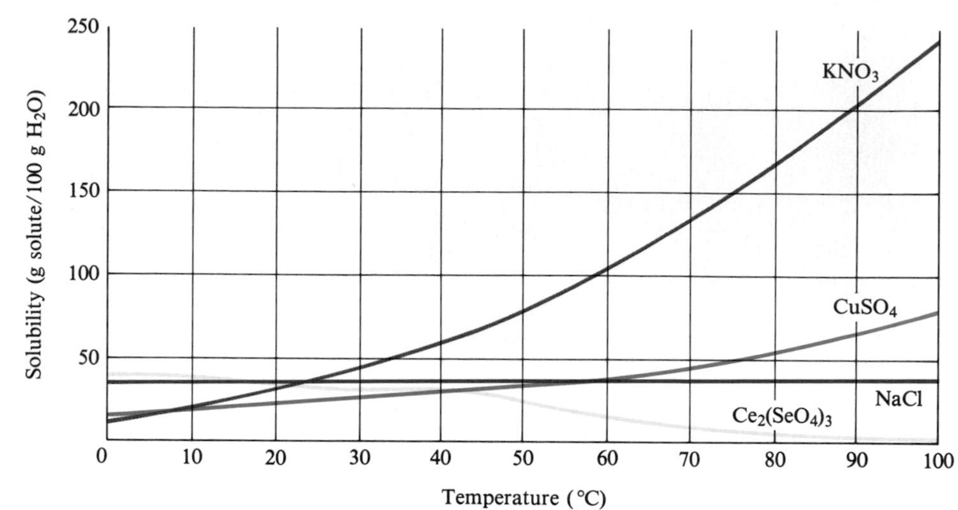

Figure 12.10
Solubility of some ionic salts at different temperatures. The solubilities of the salts NaCl, KNO_3, and $CuSO_4$ rise with increasing temperature, as is the case with most ionic solids. The solubility of $Ce_2(SeO_4)_3$, however, falls with increasing temperature.

100°C. Copper sulfate, $CuSO_4$, shows a moderate increase in solubility over this temperature interval. Sodium chloride, NaCl, increases only slightly in solubility with temperature.

A number of ionic compounds decrease in solubility with increasing temperature. Calcium sulfate, $CaSO_4$, and calcium hydroxide, $Ca(OH)_2$, are common examples. They are slightly soluble compounds that become even less soluble at higher temperatures. Cerium selenate, $Ce_2(SeO_4)_3$, whose variation in solubility with temperature is shown in Figure 12.10, is very soluble at 0°C but much less soluble at 100°C.

Heat can be released or absorbed when ionic substances are dissolved in water. In some cases, this *heat of solution* is quite noticeable. When sodium hydroxide is dissolved in water, the solution becomes hot (the solution process is exothermic). On the other hand, when ammonium nitrate is dissolved in water, the solution becomes very cold (the solution process is endothermic). This cooling effect from the dissolving of ammonium nitrate in water is the basis for instant cold packs used in hospitals and elsewhere. An instant cold pack consists of a bag of NH_4NO_3 crystals inside a bag of water (Figure 12.11). When the inner bag is

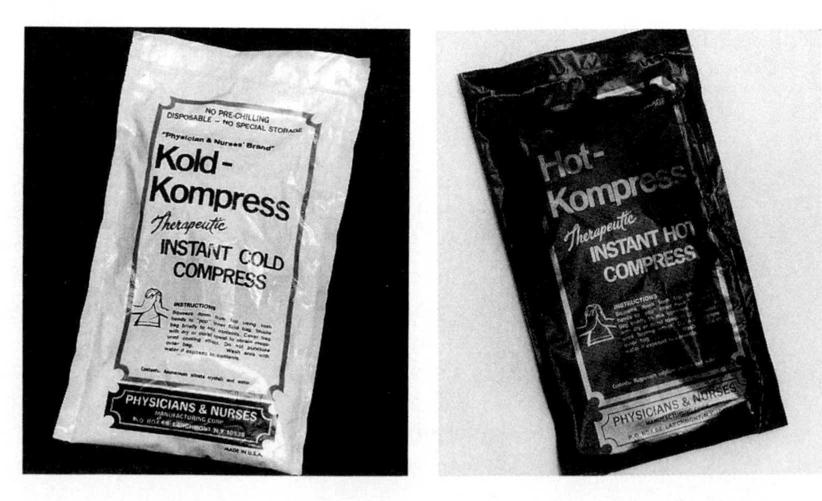

Figure 12.11
Instant cold packs and hot packs. An instant cold pack contains an inner bag of crystals of ammonium nitrate, NH_4NO_3, which when broken apart allows the crystals to dissolve in the outer bag of water. The solution process is endothermic (heat is absorbed), so the bag feels cold. An instant hot pack contains either anhydrous calcium chloride, $CaCl_2$, or anhydrous magnesium sulfate, $MgSO_4$. The solution process is exothermic (heat is released).

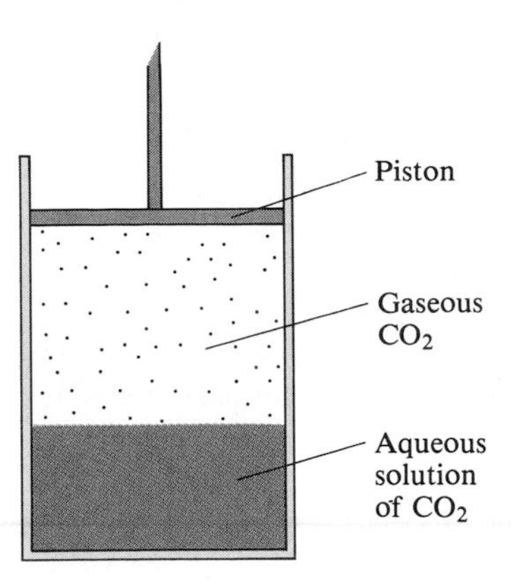

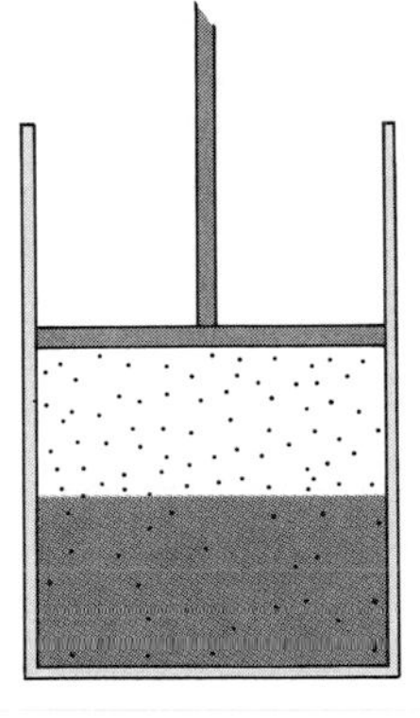

Figure 12.12
Effect of pressure on gas solubility. When the piston is pushed down, increasing the partial pressure of carbon dioxide, more gas dissolves (which tends to reduce the carbon dioxide partial pressure).

broken, NH_4NO_3 dissolves in the water. Heat is absorbed, so the bag feels cold. Similarly, hot packs are available, which contain either $CaCl_2$ or $MgSO_4$, which dissolve in water with the evolution of heat.

Pressure Change

In general, pressure change has little effect on the solubility of a liquid or solid in water. On the other hand, the solubility of a gas is very much affected by pressure. The qualitative effect of a change in pressure on the solubility of a gas can be predicted from Le Chatelier's principle. **Le Chatelier's principle** *states that when a system in equilibrium is disturbed by a change of temperature, pressure, or concentration variable, the system shifts in equilibrium composition in a way that tends to counteract this change of variable*. Let us see how Le Chatelier's principle can predict the effect on gas solubility of a change in pressure.

Imagine a cylindrical vessel that is fitted with a movable piston and contains carbon dioxide gas over its saturated water solution (Figure 12.12). The equilibrium is

$$CO_2(g) \rightleftharpoons CO_2(aq)$$

Suppose we momentarily increase the partial pressure of CO_2 gas by pushing down the piston. This change of partial pressure, according to Le Chatelier's principle, shifts the equilibrium composition in a way that tends to counteract the pressure increase. From the preceding equation, we see that the partial pressure of CO_2 gas decreases if more CO_2 dissolves.

$$CO_2(g) \longrightarrow CO_2(aq)$$

The system comes to a new equilibrium, one in which more CO_2 has dissolved.

Thus, we predict that carbon dioxide is more soluble at higher pressures. Conversely, when the partial pressure of carbon dioxide gas is reduced, its solubility is decreased. A bottle of carbonated beverage fizzes when the cap is removed: As the partial pressure of carbon dioxide is reduced, gas comes out of solution (Figure 12.13).

The argument given for carbon dioxide holds for any gas. Therefore, all gases become more soluble in a liquid at a given temperature when the partial pressure of the gas over the solution is increased.

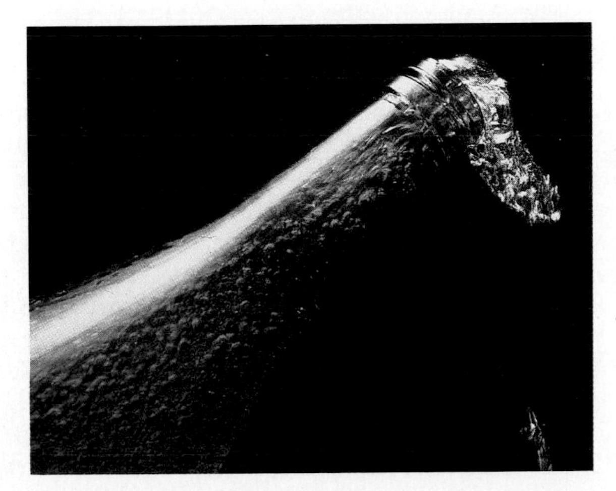

Figure 12.13
Sudden release of pressure from a carbonated beverage. A carbonated beverage is produced by dissolving carbon dioxide in beverage solution under pressure. More carbon dioxide dissolves at the higher pressure than otherwise. When this pressure is suddenly released, carbon dioxide is less soluble, and the excess bubbles from the solution.

Henry's Law

The effect of pressure on the solubility of a gas can be predicted quantitatively. According to **Henry's law,** *the solubility of a gas is directly proportional to the partial pressure of the gas above the solution*. Expressed mathematically, the law is

$$S = k_H P$$

where S is the solubility of the gas (expressed as mass of solute per unit volume of solvent), k_H is Henry's law constant for the gas, and P is the partial pressure of the gas. The next example shows how this formula is used.

Example 12.1 Applying Henry's Law

In the chapter opening, we noted that 27 g of acetylene, C_2H_2, dissolves in 1 L of acetone at 1.0 atm pressure. If the partial pressure of acetylene is increased to 12 atm, what is its solubility in acetone?

Solution

Let S_1 be the solubility of the gas at partial pressure P_1, and let S_2 be the solubility at partial pressure P_2. Then we can write Henry's law for both pressures.

$$S_1 = k_H P_1$$
$$S_2 = k_H P_2$$

When we divide the second equation by the first, we get

$$\frac{S_2}{S_1} = \frac{\cancel{k_H} P_2}{\cancel{k_H} P_1} \quad \text{or} \quad \frac{S_2}{S_1} = \frac{P_2}{P_1}$$

At 1.0 atm partial pressure of acetylene (P_1), the solubility S_1 is 27 g C_2H_2 per liter of acetone. For a partial pressure of 12 atm (P_2), we get

$$\frac{S_2}{27 \text{ g } C_2H_2/\text{L acetone}} = \frac{12 \cancel{\text{atm}}}{1.0 \cancel{\text{atm}}}$$

Hence,

$$S_2 = \frac{27 \text{ g } C_2H_2}{\text{L acetone}} \times \frac{12}{1.0} = \frac{3.2 \times 10^2 \text{ g } C_2H_2}{\text{L acetone}}$$

That is, at 12 atm partial pressure of acetylene, 1 L of acetone dissolves 3.2×10^2 g of acetylene.

Exercise 12.4

A liter of water at 25°C dissolves 0.0404 g O_2 when the partial pressure of the oxygen is 1.00 atm. What is the solubility of oxygen from air, in which the partial pressure of O_2 is 159 mmHg?

(See Problems 12.37 and 12.38.)

Colligative Properties

In the chapter opening, we mentioned that the addition of ethylene glycol, CH_2OHCH_2OH, to water lowers the freezing point of water below 0°C. For example, if 0.010 mol of ethylene glycol is added to one kilogram of water, the freezing point is lowered to −0.019°C. The magnitude of freezing-point lowering is directly proportional to the number of ethylene glycol molecules added to a quantity of water. Thus, if we add 0.020 (= 0.010 × 2) mol of ethylene glycol to one kilogram of water, the freezing point is lowered to −0.038°C (= −0.019°C × 2). Moreover, the same lowering is observed with the addition of other nonelectrolyte substances. For example, an aqueous solution of 0.020 mol of urea, $(NH_2)_2CO$, in one kilogram of water also freezes at −0.038°C.

Freezing-point lowering is a colligative property. **Colligative properties** of solutions are *properties that depend on the concentration of solute molecules or* *ions in solution but not on the chemical identity of the solute* (whether it is ethylene glycol or urea, for instance). In the remainder of this chapter, we will discuss several colligative properties, which are expressed quantitatively in terms of various concentration units. We will first look into ways of expressing the concentration of a solution.

12.4 WAYS OF EXPRESSING CONCENTRATION

The *concentration* of a solute is the amount of solute dissolved in a given quantity of solvent or solution. The quantity of solvent or solution can be expressed in terms of volume or in terms of mass or molar amount. Thus, there are several ways of expressing the concentration of a solution.

The *molarity* of a solution is the moles of solute that are in a liter of solution.●

● Molarity was discussed in Section 4.9.

$$\text{Molarity} = \frac{\text{moles of solute}}{\text{liters of solution}}$$

For example, 0.20 mol ethylene glycol dissolved in enough water to give 2.0 L of solution has a molarity of

$$\frac{0.20 \text{ mol ethylene glycol}}{2.0 \text{ L solution}} = 0.10\ M \text{ ethylene glycol}$$

The solution is 0.10 molar (denoted *M*). This unit is especially useful when we wish to dispense a given amount of solute, because using it makes the amount of solute directly related to the volume of solution.

Other concentration units are defined in terms of the mass or molar amount of solvent or solution. The most important of these are mass (or weight) percentage of solute, molality, and mole fraction.

Mass Percentage of Solute

Solution concentration is sometimes expressed in terms of the **mass percentage of solute**—that is, *the percentage by mass of solute contained in a solution.*

$$\text{Mass percentage of solute} = \frac{\text{mass of solute}}{\text{mass of solution}} \times 100\%$$

For example, an aqueous solution that is 3.5% sodium chloride by mass contains 3.5 grams of NaCl in 100.0 grams of solution. It could be prepared by dissolving 3.5 grams of NaCl in 96.5 grams of water (100.0 − 3.5 = 96.5).

Example 12.2 Calculating Mass Percentage of Solute

How would you prepare 425 g of an aqueous solution containing 2.40% by mass of sodium acetate, $NaC_2H_3O_2$?

Solution

The mass of sodium acetate in this quantity of solution is

$$\text{Mass of } NaC_2H_3O_2 = 425 \text{ g} \times 0.0240 = 10.2 \text{ g}$$

The quantity of water in the solution is

$$\text{Mass of } H_2O = \text{mass solution} - \text{mass } NaC_2H_3O_2$$
$$= 425 \text{ g} - 10.2 \text{ g} = 415 \text{ g}$$

Thus, you would prepare the solution by dissolving 10.2 g of sodium acetate in 415 g of water.

Exercise 12.5

An experiment calls for 35.0 g of hydrochloric acid that is 20.2% HCl by mass. How many grams of HCl is this? How many grams of water?

(See Problems 12.39, 12.40, 12.41, and 12.42.)

Molality

The **molality** of a solution is *the moles of solute per kilogram of solvent.*

$$\text{Molality} = \frac{\text{moles of solute}}{\text{kilograms of solvent}}$$

For example, 0.20 mol of ethylene glycol dissolved in 2.0×10^3 grams (= 2.0 kilograms) of water has a molality of

$$\frac{0.20 \text{ mol ethylene glycol}}{2.0 \text{ kg solvent}} = 0.10\ m \text{ ethylene glycol}$$

That is, the solution is 0.10 molal (denoted m). The units of molality and molarity are sometimes confused. Note that molality is defined in terms of *mass of solvent* rather than *volume of solution*.

Example 12.3 Calculating the Molality of Solute

Glucose, $C_6H_{12}O_6$, is a sugar that occurs in fruits. It is also known as "blood sugar" because it is found in blood and is the body's main source of energy (see Figure 12.14). What is the molality of a solution containing 5.67 g of glucose dissolved in 25.2 g of water?

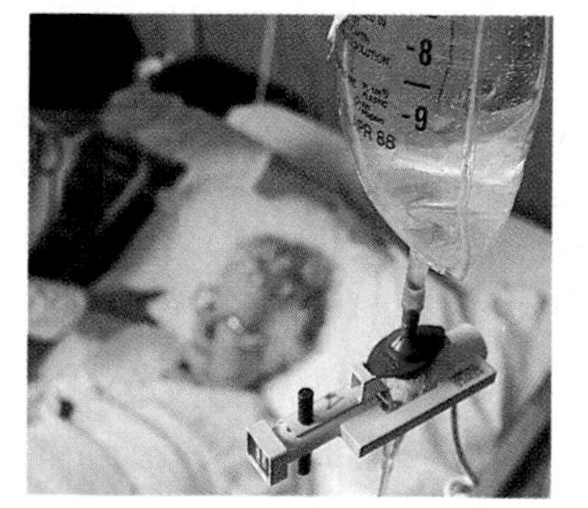

Figure 12.14 Intravenous feeding of glucose. Glucose, $C_6H_{12}O_6$, is the principal energy substance of the body and is transported to various cells by the blood. Here glucose is being fed intravenously to a patient.

Solution

The moles of glucose (MW = 180.2 amu) in 5.67 g are found as follows:

$$5.67 \cancel{\text{g } C_6H_{12}O_6} \times \frac{1 \text{ mol } C_6H_{12}O_6}{180.2 \cancel{\text{g } C_6H_{12}O_6}} = 0.0315 \text{ mol } C_6H_{12}O_6$$

We obtain the molality by dividing the moles of solute (glucose) by the mass of solvent (water) in kilograms. The mass of water is 25.2 g or 25.2×10^{-3} kg.

$$\text{Molality} = \frac{0.0315 \text{ mol } C_6H_{12}O_6}{25.2 \times 10^{-3} \text{ kg solvent}} = \mathbf{1.25\ m\ C_6H_{12}O_6}$$

Exercise 12.6

Toluene, $C_6H_5CH_3$, is a liquid compound similar to benzene, C_6H_6. It is the starting material for other substances, including trinitrotoluene (TNT). Find the molality of toluene in a solution that contains 35.6 g of toluene and 125 g of benzene.

(See Problems 12.43, 12.44, 12.45, and 12.46.)

Mole Fraction

The **mole fraction** of a component substance A (X_A) in a solution is defined as *the moles of component substance divided by the total moles of solution* (that is, moles of solute plus solvent).

$$X_A = \frac{\text{moles of substance } A}{\text{total moles of solution}}$$

For example, if a solution is made up of 1 mol of ethylene glycol and 9 mol of water, the total moles of solution are 1 mol + 9 mol = 10 mol. The mole fraction of ethylene glycol is 1/10 = 0.1, and the mole fraction of water is 9/10 = 0.9. Multiplying mole fractions by 100 gives *mole percent*. Hence, this solution is 10 mole percent ethylene glycol and 90 mole percent water. We can also say that 10% of the molecules in the solution are ethylene glycol and 90% are water. The sum of the mole fractions of all the components of a solution equals 1.

Example 12.4 Calculating the Mole Fractions of Components

What are the mole fractions of glucose and water in a solution containing 5.67 g of glucose, $C_6H_{12}O_6$, dissolved in 25.2 g of water?

Solution

This is the glucose solution described in Example 12.3. There we found that 5.67 g of glucose equals 0.0315 mol of glucose. The moles of water in the solution are

$$25.2 \cancel{\text{g } H_2O} \times \frac{1 \text{ mol } H_2O}{18.0 \cancel{\text{g } H_2O}} = 1.40 \text{ mol } H_2O$$

Hence, the total moles of solution are

$$1.40 \text{ mol} + 0.0315 \text{ mol} = 1.432 \text{ mol}$$

(We retain an extra figure in the answer for further computation.) Finally, we get

$$\text{Mole fraction glucose} = \frac{0.0315 \text{ mol}}{1.432 \text{ mol}} = \mathbf{0.0220}$$

$$\text{Mole fraction water} = \frac{1.40 \text{ mol}}{1.432 \text{ mol}} = \mathbf{0.978}$$

The sum of the mole fractions is 1.000.

Exercise 12.7

Calculate the mole fractions of toluene and benzene in the solution described in Exercise 12.6. (See Problems 12.47 and 12.48.)

Conversion of Concentration Units

It is relatively easy to interconvert concentration units when they are expressed in terms of mass or moles of solute and solvent, as the following examples show.

Example 12.5 Converting Molality to Mole Fractions

An aqueous solution is 0.120 *m* in glucose, $C_6H_{12}O_6$. What are the mole fractions of each component in the solution?

Solution

A 0.120 *m* glucose solution contains 0.120 mol of glucose in 1.00 kg of water. The moles of H_2O in 1.00 kg of water is

$$1.00 \times 10^3 \cancel{\text{g } H_2O} \times \frac{1 \text{ mol } H_2O}{18.0 \cancel{\text{g } H_2O}} = 55.6 \text{ mol } H_2O$$

Hence,

$$\text{Mole fraction glucose} = \frac{0.120 \text{ mol}}{(0.120 + 55.6) \text{ mol}} = \mathbf{0.00215}$$

$$\text{Mole fraction water} = \frac{55.6 \text{ mol}}{(0.120 + 55.6) \text{ mol}} = \mathbf{0.998}$$

Exercise 12.8

A solution is 0.120 *m* methanol dissolved in ethanol. Calculate the mole fractions of methanol, CH_3OH, and ethanol, C_2H_5OH, in the solution. (See Problems 12.49 and 12.50.)

Example 12.6 Converting Mole Fractions to Molality

A solution is 0.150 mole fraction glucose, $C_6H_{12}O_6$, and 0.850 mole fraction water. What is the molality of glucose in the solution?

Solution

One mole of solution contains 0.150 mol of glucose and 0.850 mol of water. The mass of this amount of water is

$$0.850 \; \cancel{\text{mol } H_2O} \times \frac{18.0 \text{ g } H_2O}{1 \; \cancel{\text{mol } H_2O}} = 15.3 \text{ g } H_2O$$

$$(= 0.0153 \text{ kg } H_2O)$$

Therefore the molality of glucose, $C_6H_{12}O_6$, in the solution is

$$\text{Molality of } C_6H_{12}O_6 = \frac{0.150 \text{ mol } C_6H_{12}O_6}{0.0153 \text{ kg solvent}}$$

$$= \mathbf{9.80} \; \boldsymbol{m} \; \mathbf{C_6H_{12}O_6}$$

Exercise 12.9

A solution is 0.250 mole fraction methanol, CH_3OH, and 0.750 mole fraction ethanol, C_2H_5OH. What is the molality of methanol in the solution?

(See Problems 12.51 and 12.52.)

To convert molality to molarity and vice versa, we must know the density of the solution. The calculations are described in the next two examples. As the first example shows, molality and molarity are approximately equal in dilute aqueous solutions.

Example 12.7 Converting Molality to Molarity

An aqueous solution is 0.273 *m* KCl. What is the molar concentration of potassium chloride, KCl? The density of the solution is 1.011×10^3 g/L.

Solution

There is 0.273 mol KCl per kilogram of water. To calculate the molarity, we must first find the volume of solution for a given mass of solvent. Let us consider an amount of solution containing a kilogram of water (1.000×10^3 g H_2O). The mass of potassium chloride in this quantity of solution is

$$0.273 \; \cancel{\text{mol KCl}} \times \frac{74.6 \text{ g KCl}}{1 \; \cancel{\text{mol KCl}}} = 20.4 \text{ g KCl}$$

The total mass of the solution equals the mass of water plus the mass of potassium chloride.

$$1.000 \times 10^3 \text{ g} + 20.4 \text{ g} = 1.020 \times 10^3 \text{ g}$$

The volume of solution equals the mass divided by the density of the solution.

$$\text{Volume of solution} = \frac{1.020 \times 10^3 \; \cancel{\text{g}}}{1.011 \times 10^3 \; \cancel{\text{g}}/\text{L}} = 1.009 \text{ L}$$

Hence, the molarity of the solution is

$$\frac{0.273 \text{ mol KCl}}{1.009 \text{ L solution}} = \mathbf{0.271} \; \boldsymbol{M} \; \mathbf{KCl}$$

Note that the molarity and the molality of this solution are approximately equal. This happens in cases where the solutions are dilute and the density is about 1 g/mL.

Exercise 12.10

Urea, $(NH_2)_2CO$, is used as a fertilizer. What is the molar concentration of an aqueous solution that is 3.42 *m* urea? The density of the solution is 1.045 g/mL.

(See Problems 12.53 and 12.54.)

Example 12.8 Converting Molarity to Molality

An aqueous solution is 0.907 *M* $Pb(NO_3)_2$. What is the molality of lead nitrate, $Pb(NO_3)_2$, in this solution? The density of the solution is 1.252 g/mL.

Solution

There is 0.907 mol $Pb(NO_3)_2$ per liter of solution. Let us consider 1 L (= 1.000×10^3 mL) of solution and calculate its mass. We can then calculate the mass of lead nitrate and find the mass of water by difference.

$$\begin{aligned}\text{Mass of solution} &= \text{density} \times \text{volume}\\ &= 1.252\ \text{g}/\cancel{\text{mL}} \times 1.000 \times 10^3\ \cancel{\text{mL}}\\ &= 1.252 \times 10^3\ \text{g}\end{aligned}$$

The mass of lead nitrate is

$$0.907\ \cancel{\text{mol Pb(NO}_3)_2} \times \frac{331.2\ \text{g Pb(NO}_3)_2}{1\ \cancel{\text{mol Pb(NO}_3)_2}} = 3.00 \times 10^2\ \text{g Pb(NO}_3)_2$$

The mass of the water in this solution is

$$\begin{aligned}\text{Mass of H}_2\text{O} &= \text{mass of solution} - \text{mass of Pb(NO}_3)_2\\ &= 1.252 \times 10^3\ \text{g} - 3.00 \times 10^2\ \text{g}\\ &= 9.52 \times 10^2\ \text{g}\ (= 0.952\ \text{kg})\end{aligned}$$

Hence, the molality of lead nitrate in this solution is

$$\frac{0.907\ \text{mol Pb(NO}_3)_2}{0.952\ \text{kg solvent}} = \mathbf{0.953}\ \boldsymbol{m}\ \mathbf{Pb(NO_3)_2}$$

Exercise 12.11

An aqueous solution is 2.00 *M* urea. The density of the solution is 1.029 g/mL. What is the molal concentration of urea in the solution?

(See Problems 12.55 and 12.56.)

12.5 VAPOR PRESSURE OF A SOLUTION

During the nineteenth century, chemists observed that the vapor pressure of a volatile solvent was lowered by the addition of a nonvolatile solute. **Vapor-pressure lowering** of a solvent is *a colligative property equal to the vapor pressure of the pure solvent minus the vapor pressure of the solution.* For example, water at 20°C has a vapor pressure of 17.54 mmHg. Ethylene glycol, CH_2OHCH_2OH, is a liquid whose vapor pressure at 20°C is relatively low; it can be considered to be nonvolatile compared with water. An aqueous solution containing 0.0100 mole fraction of ethylene glycol has a vapor pressure of 17.36 mmHg. Thus, the vapor-pressure lowering, ΔP, of water is

$$\Delta P = 17.54\ \text{mmHg} - 17.36\ \text{mmHg} = 0.18\ \text{mmHg}$$

Figure 12.15 shows a demonstration of vapor-pressure lowering.

In about 1886, the French chemist François Marie Raoult observed that the partial vapor pressure of solvent over a solution of a nonelectrolyte solute depends on the mole fraction of solvent in the solution. Consider a solution of volatile solvent, A, and nonelectrolyte solute, B, which may be volatile or nonvolatile. According to **Raoult's law,** *the partial pressure of solvent, P_A, over a solution equals the vapor pressure of the pure solvent, P_A°, times the mole fraction of solvent, X_A, in the solution.*

$$P_A = P_A^\circ X_A$$

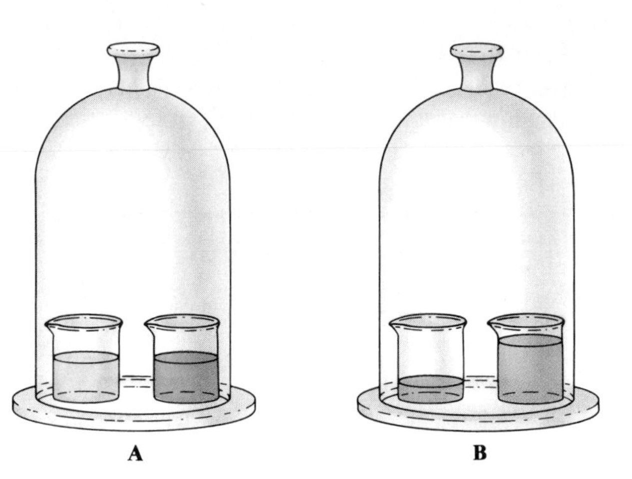

Figure 12.15
Demonstration of vapor-pressure lowering. *(A)* Two beakers, each containing water solutions of the same solute, are placed under a bell jar. The solution in the left beaker is less concentrated than that in the right beaker, so its vapor pressure is greater. The partial pressure of vapor in the bell jar is an intermediate value. It is less than the vapor pressure of the solution on the left, but more than that of the solution on the right. As a result, vapor leaves the solution on the left (which becomes more concentrated) and condenses on the solution on the right (which becomes less concentrated). *(B)* After some time, the two solutions become equal in concentration.

If the solute is nonvolatile, P_A is the total vapor pressure of the solution. Because the mole fraction of solvent in a solution is always less than 1, the vapor pressure of the solution of a nonvolatile solute is less than that for the pure solvent; the vapor pressure is lowered. In general, Raoult's law is observed to hold for dilute solutions—that is, solutions in which X_A is close to 1. If the solvent and solute are chemically similar, Raoult's law may hold for all mole fractions. Raoult's law is displayed graphically for two solutions in Figure 12.16.

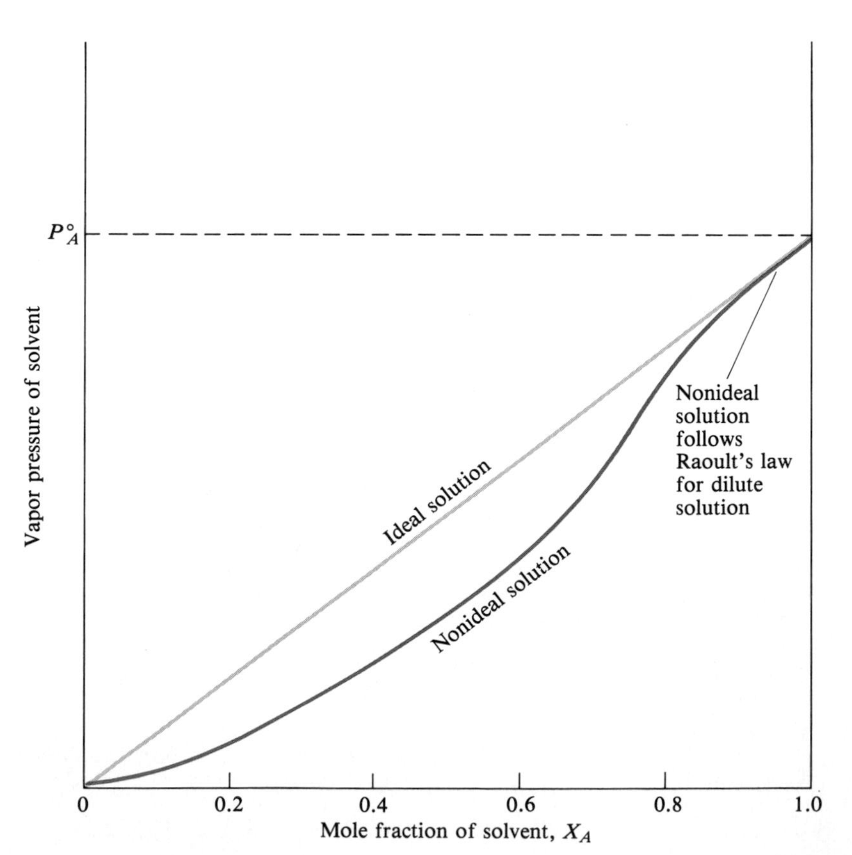

Figure 12.16
A plot of vapor pressures of solutions showing Raoult's law. Vapor pressures of two solutions have been plotted against mole fraction of solvent. In one case (labeled "ideal solution"), the vapor pressure is found to be proportional to the mole fraction of solvent for all mole fractions; it follows Raoult's law for all concentrations of solute. For the "nonideal solution," Raoult's law is followed for low solute concentrations (mole fraction of solvent near 1), but the vapor pressure deviates at other concentrations.

We can obtain an explicit expression for the vapor-pressure lowering of a solvent in a solution assuming Raoult's law holds and that the solute is a nonvolatile nonelectrolyte. The vapor-pressure lowering, ΔP, is

$$\Delta P = P_A^\circ - P_A$$

Substituting Raoult's law gives

$$\Delta P = P_A^\circ - P_A^\circ X_A = P_A^\circ(1 - X_A)$$

But the sum of the mole fractions of the components of a solution must equal 1; that is, $X_A + X_B = 1$. So $X_B = 1 - X_A$. Therefore,

$$\Delta P = P_A^\circ X_B$$

From this equation, we can see that the vapor-pressure lowering is a colligative property—one that depends on the concentration, but not on the nature, of the solute. Thus, if the mole fraction of ethylene glycol, X_B, in an aqueous solution is doubled from 0.010 to 0.020, the vapor-pressure lowering increases from 0.18 mmHg to 0.36 mmHg. Also, because the previous equation does not depend on the characteristics of the solute (other than its being nonvolatile and a nonelectrolyte), a solution that is 0.010 mole fraction urea, $(NH_2)_2CO$, has the same vapor-pressure lowering as one that is 0.010 mole fraction ethylene glycol. The next example illustrates the use of the previous equation.

Example 12.9 Calculating Vapor-Pressure Lowering

Calculate the vapor-pressure lowering of water when 5.67 g of glucose, $C_6H_{12}O_6$, is dissolved in 25.2 g of water at 25°C. The vapor pressure of water at 25°C is 23.8 mmHg. What is the vapor pressure of the solution?

Solution

This is the glucose solution described in Example 12.4. According to the calculations performed there, the solution is 0.0220 mole fraction glucose. Therefore, the vapor-pressure lowering is

$$\Delta P = P_A^\circ X_B = 23.8 \text{ mmHg} \times 0.0220 = 0.524 \text{ mmHg}$$

The vapor pressure of the solution is

$$P_A = P_A^\circ - \Delta P = (23.8 - 0.524) \text{ mmHg} = 23.3 \text{ mmHg}$$

Exercise 12.12

Naphthalene, $C_{10}H_8$, is used to make moth balls. Suppose a solution is made by dissolving 0.515 g of naphthalene in 60.8 g of chloroform, $CHCl_3$. Calculate the vapor-pressure lowering of chloroform at 20°C from the naphthalene. The vapor pressure of chloroform at 20°C is 156 mmHg. Naphthalene can be assumed to be nonvolatile compared to chloroform. What is the vapor pressure of the solution? (See Problems 12.57 and 12.58.)

An *ideal solution* of substances A and B is one in which both substances follow Raoult's law for all values of mole fractions. Such solutions occur when the substances are chemically similar so that the intermolecular forces between A and B molecules are similar to those between two A molecules or between two B molecules. The total vapor pressure over an ideal solution equals the sum of the partial vapor pressures, each of which is given by Raoult's law:

$$P = P_A^\circ X_A + P_B^\circ X_B$$

Solutions of benzene, C_6H_6, and toluene, $C_6H_5CH_3$, are ideal. Note the similarity in their structural formulas:

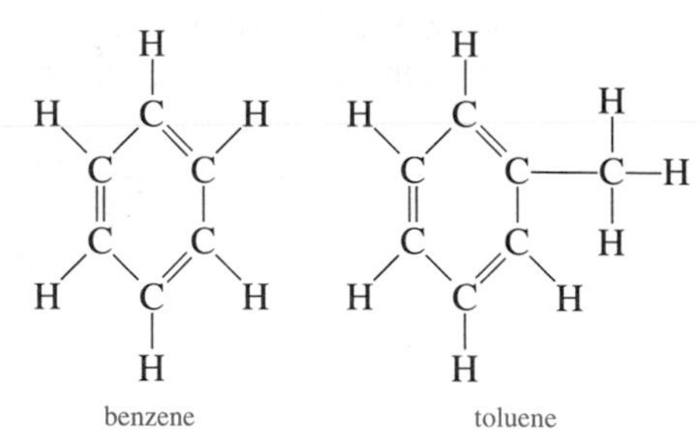

Suppose a solution is 0.70 mole fraction benzene and 0.30 mole fraction toluene. The vapor pressures of pure benzene and pure toluene are 75 mmHg and 22 mmHg, respectively. Hence, the total vapor pressure is

$$P = 75\ \text{mmHg} \times 0.70 + 22\ \text{mmHg} \times 0.30 = 59\ \text{mmHg}$$

We can easily show that the vapor over this solution is richer in the more volatile solute (benzene). The partial vapor pressure of benzene over the solution is

$$75\ \text{mmHg} \times 0.70 = 52\ \text{mmHg}$$

Because the total vapor pressure is 59 mmHg, the mole fraction of benzene in the vapor is

$$\frac{52\ \text{mmHg}}{59\ \text{mmHg}} = 0.88$$

Thus, the vapor is 0.88 mole fraction of benzene, whereas the liquid solution is 0.70 mole fraction benzene. This is a general result: The vapor over a solution is richer in the more volatile component. *Fractional distillation* employs this result to separate volatile components of a solution. If we condense the vapor that first appears over a solution, we will obtain a liquid having a greater mole fraction of

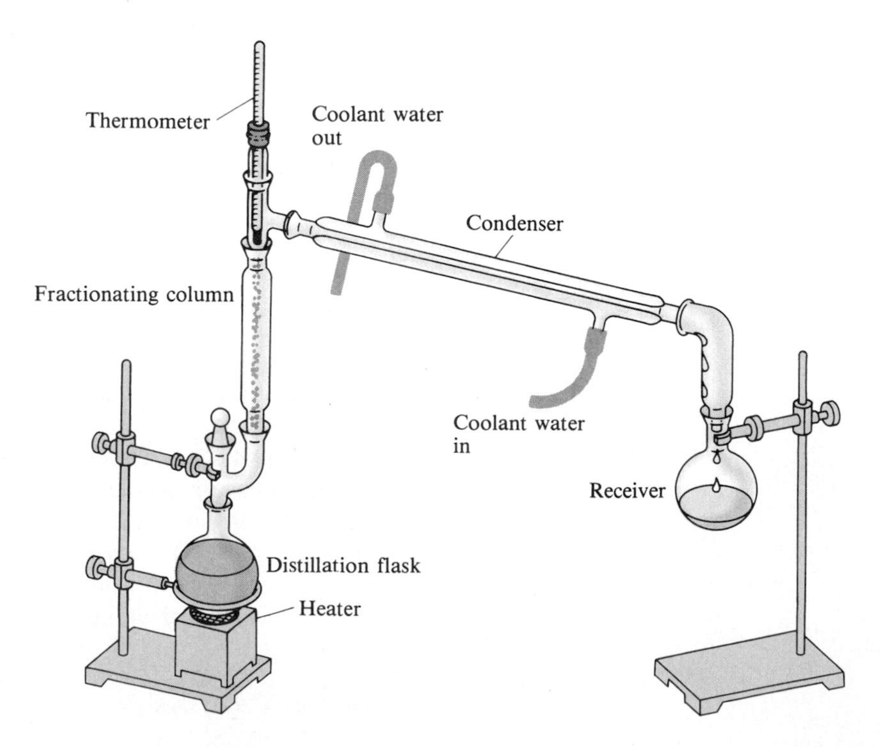

Figure 12.17
Fractional distillation. This is a typical laboratory apparatus for the fractional distillation of a liquid solution to obtain pure liquid components. Vapor boiling from the solution in the flask condenses on a packing material in cooler portions of the column. The temperature of the column changes continuously, becoming cooler and cooler going up the column. Similarly, the composition of the liquid that condenses becomes continuously richer in the more volatile component going up the column. The vapor condensing from the top of the column is that of the more volatile component. As the more volatile component is removed from the liquid in the flask, the temperature must be raised to distill over less volatile components. The same type of procedure is used to separate the components from petroleum, such as gasoline and diesel fuel.

the more volatile component. If we then vaporize a portion of this and then condense it, the liquid will be still richer in the more volatile component. Through successive distillation stages, we eventually obtain the more volatile component in pure form. In practice, the successive distillation stages are performed using a fractional distillation column, such as the one shown in Figure 12.17.

12.6 BOILING-POINT ELEVATION AND FREEZING-POINT DEPRESSION

The *normal* boiling point of a liquid is the temperature at which its vapor pressure equals 1 atm. Because the addition of a nonvolatile solute to a liquid reduces its vapor pressure, the temperature must be increased to a value greater than the normal boiling point to achieve a vapor pressure of 1 atm. Figure 12.18 shows the vapor-pressure curve of a solution. The curve is below the vapor-pressure curve of the pure liquid solvent. The **boiling-point elevation,** ΔT_b, is *a colligative property of a solution equal to the boiling point of the solution minus the boiling point of the pure solvent;* it is indicated in the diagram.

The boiling-point elevation, ΔT_b, is found to be proportional to the molal concentration, c_m, of the solution (for dilute solutions).

$$\Delta T_b = K_b c_m$$

The constant of proportionality, K_b (called the *boiling-point-elevation constant*), depends only on the solvent. Table 12.3 lists values of K_b, as well as boiling points, for some solvents. Benzene, for example, has a boiling-point-elevation

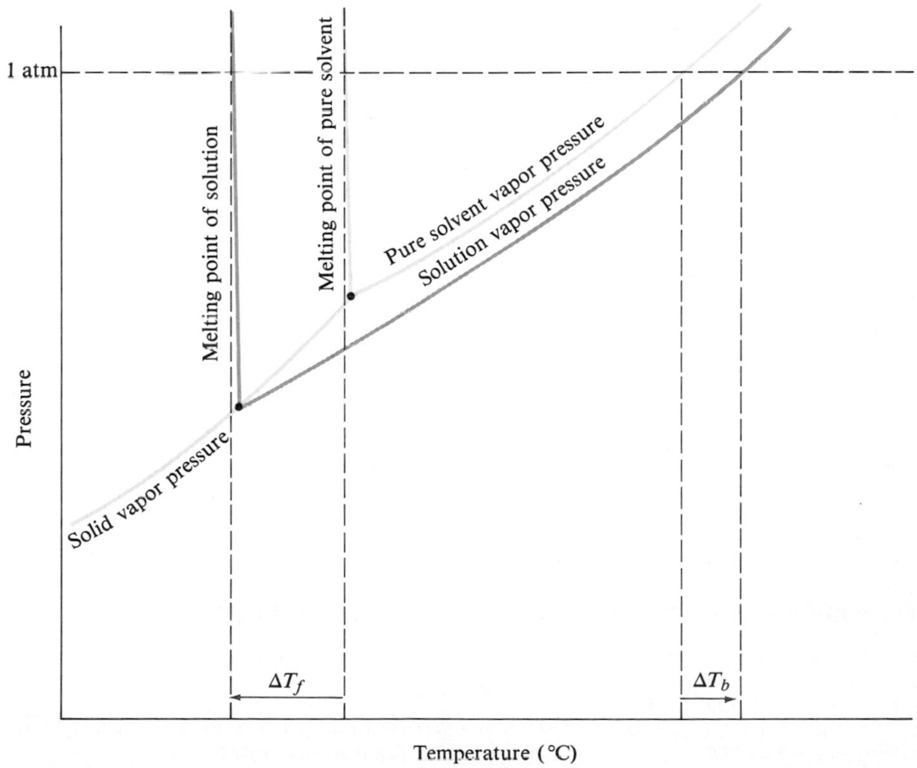

Figure 12.18
Phase diagram showing the effect of a nonvolatile solute on freezing point and boiling point. Note that the freezing point is lowered and the boiling point is elevated.

Table 12.3 Boiling-Point-Elevation and Freezing-Point-Depression Constants

Solvent	Formula	Melting Point (°C)	Boiling Point (°C)	K_f (°C/m)	K_b (°C/m)
Acetic acid	$HC_2H_3O_2$	16.60	118.5	3.59	3.08
Benzene	C_6H_6	5.455	80.2	5.065	2.61
Camphor	$C_{10}H_{16}O$	179.5	—	40	—
Carbon disulfide	CS_2	—	46.3	—	2.40
Cyclohexane	C_6H_{12}	6.55	80.74	20.0	2.79
Ethanol	C_2H_5OH	—	78.3	—	1.07
Water	H_2O	0.000	100.000	1.858	0.521

(Data are taken from Landolt-Börnstein, 6th ed., *Zahlenwerte und Functionen aus Physik, Chemie, Astronomie, Geophysik, und Technik,* Vol. II, part IIa [Heidelberg: © Springer-Verlag, 1960], pp. 844–849 and pp. 918–919.)

constant of 2.61°C/m. This means that a 0.100 m solution of a nonvolatile solute in benzene boils at 0.261°C above the boiling point of pure benzene. Pure benzene boils at 80.2°C, so a 0.100 m solution boils at 80.2°C + 0.261°C = 80.5°C.

Figure 12.18 also shows the effect of a dissolved solute on the freezing point of a solution. Usually it is the pure solvent that freezes out of solution. Sea ice, for example, is almost pure water. For that reason, the vapor-pressure curve for the solid is unchanged. Therefore, the freezing point shifts to a lower temperature. The **freezing-point depression,** ΔT_f, is *a colligative property of a solution equal to the freezing point of the pure solvent minus the freezing point of the solution.* (ΔT_f is shown in Figure 12.18.)

Freezing-point depression, ΔT_f, like boiling-point elevation, is proportional to the molal concentration, c_m (for dilute solutions).

$$\Delta T_f = K_f c_m$$

Here K_f is the *freezing-point-depression constant* and depends only on the solvent. Table 12.3 gives values of K_f for some solvents. The freezing-point-depression constant for benzene is 5.06°C/m. Thus a 0.100 m solution freezes at 0.506°C below the freezing point of pure benzene. Pure benzene freezes at 5.46°C; the freezing point of the solution is 5.46°C − 0.506°C = 4.95°C.

Example 12.10 Calculating Boiling-Point Elevation and Freezing-Point Depression

An aqueous solution is 0.0222 m glucose. What are the boiling point and the freezing point of this solution?

Solution

Table 12.3 gives K_b and K_f for water as 0.521°C/m and 1.86°C/m, respectively. Therefore,

$$\Delta T_b = K_b c_m = 0.521°\text{C}/m \times 0.0222\ m = 0.0116°\text{C}$$

$$\Delta T_f = K_f c_m = 1.86°\text{C}/m \times 0.0222\ m = 0.0413°\text{C}$$

The boiling point of the solution is 100.000°C + 0.0116°C = **100.012°C**, and the freezing point is 0.000°C − 0.0413°C = **−0.041°C**. Note that ΔT_b is added and ΔT_f is subtracted.

Exercise 12.13

How many grams of ethylene glycol, CH_2OHCH_2OH, must be added to 37.8 g of water to give a freezing point of −0.150°C? (See Problems 12.59 and 12.60.)

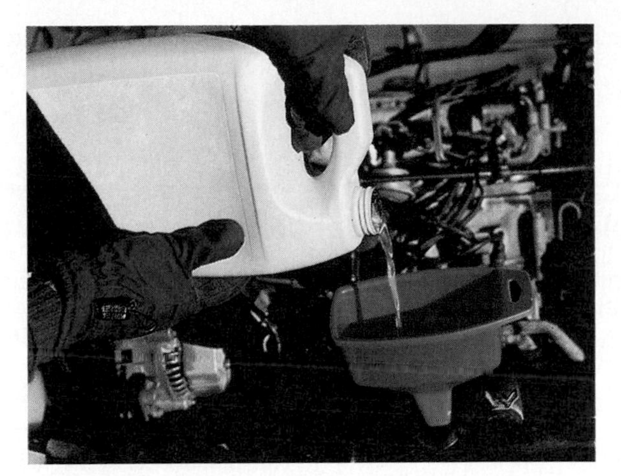

Figure 12.19
Automobile antifreeze mixtures. The main ingredient of automobile antifreeze mixtures is ethylene glycol, CH_2OHCH_2OH. This substance has low vapor pressure, so it does not easily vaporize away. It is relatively cheap, being manufactured from ethylene, C_2H_4, which is obtained from petroleum. Ethylene glycol is also used to produce polyester fibers and plastics. The substance has a sweet taste (the name glycol derives from the Greek word *glukus,* meaning "sweet") but is poisonous.

The boiling-point elevation and the freezing-point depression of solutions have a number of practical applications. We mentioned in the chapter opening that ethylene glycol is used in automobile radiators as an antifreeze because it lowers the freezing point of the coolant (Figure 12.19). The same substance also helps prevent the radiator coolant from boiling away by elevating the boiling point. Sodium chloride is spread on icy roads in the winter to lower the melting point of ice and snow below the temperature of the surrounding air.● Salt–ice mixtures are used as freezing mixtures in domestic ice cream makers. Melting ice cools the ice cream mixture to the freezing point of the solution, which is well below the freezing point of pure water.

● Calcium chloride, $CaCl_2$, is also used to melt ice on roadways. It is obtained as a by-product from the industrial production of chemicals.

Colligative properties are also used to obtain molecular weights. Although the mass spectrometer is now often used for routine determinations of the molecular weight of pure substances, colligative properties are still employed to obtain information about the species in solution. Freezing-point depression is often used because it is simple to determine a melting point or a freezing point (Figure 12.20).

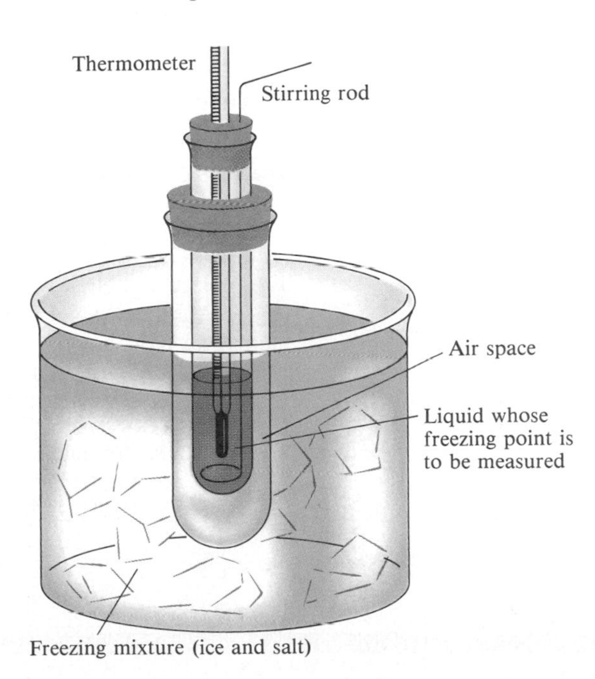

Figure 12.20
Determination of the freezing point of a liquid. The liquid is cooled by means of a freezing mixture. In order to control the rate of temperature decrease, the liquid is separated by an air space from the freezing mixture.

From the freezing-point lowering, one can calculate the molal concentration, and from the molality, one can obtain the molecular weight. The next two examples illustrate these calculations.

Example 12.11 Calculating the Molecular Weight of a Solute from Molality

A solution is prepared by dissolving 0.131 g of a substance in 25.4 g of water. The molality of the solution is determined by freezing-point depression to be 0.056 *m*. What is the molecular weight of the substance?

Solution

The molality of an aqueous solution is

$$\text{Molality} = \frac{\text{moles of substance}}{\text{kg H}_2\text{O}}$$

Hence, for the given solution,

$$0.056 \frac{\text{mol}}{\text{kg H}_2\text{O}} = \frac{\text{moles of substance}}{25.4 \times 10^{-3} \text{ kg H}_2\text{O}}$$

Or, rearranging this equation, we get

$$\text{Moles of substance} = 0.056 \frac{\text{mole}}{\cancel{\text{kg H}_2\text{O}}} \times 25.4 \times 10^{-3} \cancel{\text{kg H}_2\text{O}}$$

$$= 1.42 \times 10^{-3} \text{ mol}$$

(We have retained an extra digit for further computation.) The molar mass of the substance equals the mass of the substance divided by the number of moles.

$$\text{Molar mass} = \frac{0.131 \text{ g}}{1.42 \times 10^{-3} \text{ mol}} = 92 \text{ g/mol}$$

We round the answer to two significant figures. The molecular weight is **92 amu.**

Exercise 12.14

A 0.930-g sample of ascorbic acid (vitamin C) was dissolved in 95.0 g of water. The concentration of ascorbic acid, as determined by freezing-point depression, was 0.0555 *m*. What is the molecular weight of ascorbic acid? (See Problems 12.63 and 12.64.)

Example 12.12 Calculating the Molecular Weight from Freezing-Point Depression

Camphor is a white solid that melts at 179.5°C. It has been used to determine the molecular weights of organic compounds, because of its unusually large freezing-point-depression constant (40°C/*m*), which allows ordinary thermometers to be used. The organic substance is dissolved in melted camphor, and then the melting point of the solution is determined. (a) A 1.07-mg sample of a compound was dissolved in 78.1 mg of camphor. The solution melted at 176.0°C. What is the molecular weight of the compound? (b) If the empirical formula of the compound is CH, what is the molecular formula?

Solution

(a) The freezing-point lowering is

$$\Delta T_f = (179.5 - 176.0)°\text{C} = 3.5°\text{C}$$

so the molality of the solution is

$$\frac{\Delta T_f}{K_f} = \frac{3.5\cancel{°\text{C}}}{40\cancel{°\text{C}}/m} = 0.088\ m$$

From this we compute the moles of the compound that are dissolved in 78.1 mg of camphor (see Example 12.11).

$$0.088 \text{ mol}/\cancel{\text{kg}} \times 78.1 \times 10^{-6} \cancel{\text{kg}} = 6.9 \times 10^{-6} \text{ mol}$$

The molar mass of the compound is

$$M_m = \frac{1.07 \times 10^{-3} \text{ g}}{6.9 \times 10^{-6} \text{ mol}} = 1.6 \times 10^2 \text{ g/mol}$$

The molecular weight is **160 amu** (two significant figures). (b) The formula weight of CH is 13 amu. Therefore, the number of CH units in the molecule is

$$\frac{1.6 \times 10^2}{13} = 12$$

The molecular formula is $\mathbf{C_{12}H_{12}}$.

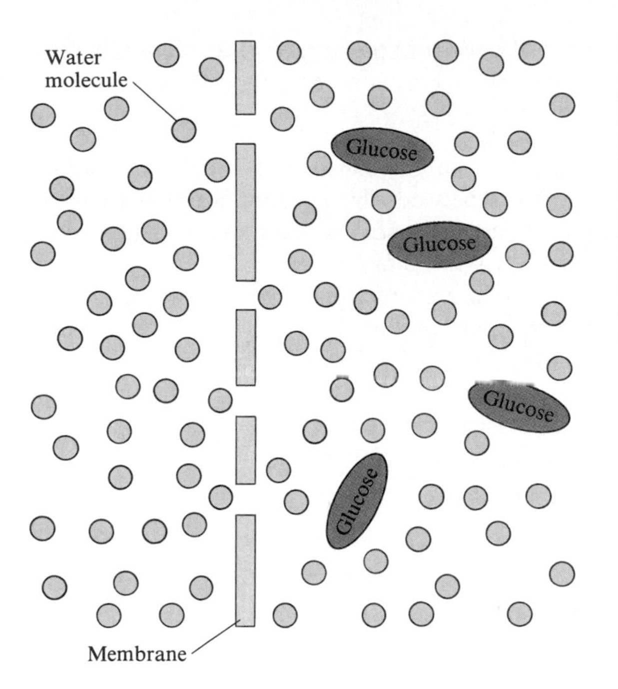

Figure 12.21
A semipermeable membrane separating water and an aqueous solution of glucose. Here the membrane is depicted as being similar to a sieve—the pores are too small to allow the glucose molecules to pass. The actual mechanism in particular cases may be more complicated. For example, water may dissolve on one side of the membrane and diffuse through it to the other side.

Exercise 12.15

A 0.205-g sample of white phosphorus was dissolved in 25.0 g of carbon disulfide, CS_2. The boiling-point elevation of the carbon disulfide solution was found to be 0.159°C. What is the molecular weight of the phosphorus in solution? What is the formula of molecular phosphorus? (See Problems 12.65 and 12.66.)

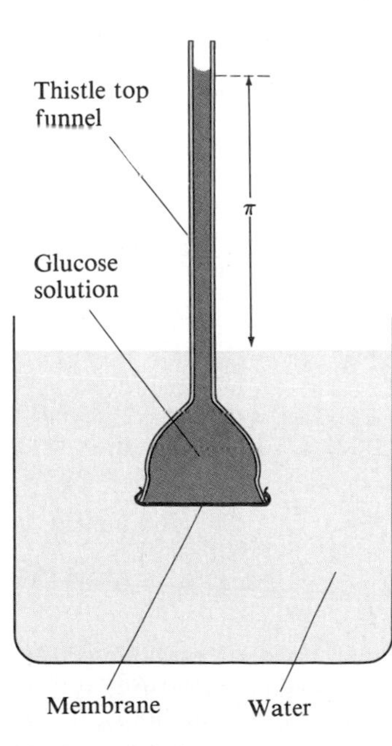

Figure 12.22
An experiment in osmosis. Water passes through the membrane into the glucose solution in the inverted funnel. The flow of water ceases when the liquid in the funnel exerts sufficient downward pressure π (the osmotic pressure).

12.7 OSMOSIS

Certain membranes will allow solvent molecules to pass through them but not solute molecules, particularly not those of large molecular weight. Such a membrane is called *semipermeable* and might be an animal bladder, a vegetable tissue, or a piece of cellophane. Figure 12.21 depicts the operation of a semipermeable membrane. **Osmosis** is *the phenomenon of solvent flow through a semipermeable membrane to equalize the solute concentrations on both sides of the membrane.* When two solutions of the same solvent are separated by a semipermeable membrane, solvent molecules migrate through the membrane from the solution of low concentration to the solution of high concentration.

Figure 12.22 shows an experiment that demonstrates osmosis. A dilute glucose solution is placed in an inverted funnel whose mouth is sealed with a semipermeable membrane. The funnel containing the glucose is then placed in a beaker of pure water. As water flows from the beaker through the membrane into the funnel, the liquid level rises in the stem of the funnel.

The glucose solution continues to rise up the funnel stem until the downward pressure exerted by the solution above the membrane eventually stops the upward flow of solvent (water). In general, the **osmotic pressure** is *a colligative property of a solution equal to the pressure that, when applied to the solution, just stops osmosis.*

The osmotic pressure π of a solution is related to the molar concentration of solute, M:

$$\pi = MRT$$

Here R is the gas constant and T is the absolute temperature. There is a formal similarity between this equation for osmotic pressure and the equation for an ideal gas. Thus, because the molar concentration M of a gas equals n/V, we know that $P = (n/V)RT = MRT$.

To see the order of magnitude of the osmotic pressure of a solution, consider the aqueous solution described in Example 12.10. There we found that a solution that is about 0.02 m has a freezing-point depression of about 0.04°C. The molarity and molality of a dilute aqueous solution are approximately equal (see Example 12.7). Therefore, this solution will also be about 0.02 M. Hence, the osmotic pressure at 25°C (298 K) is

$$\begin{aligned}\pi &= MRT \\ &= 0.02\ \cancel{\text{mol}}/\cancel{\text{L}} \times 0.082\ \cancel{\text{L}} \cdot \text{atm}/(\cancel{\text{K}} \cdot \cancel{\text{mol}}) \times 298\ \cancel{\text{K}} \\ &= 0.5\ \text{atm}\end{aligned}$$

If this pressure were to be exerted by a column of water, as in Figure 12.22, the column would have to be more than 15 feet high.

In most osmotic-pressure experiments, much more dilute solutions are employed. Often osmosis is used to determine the molecular weights of macromolecular or polymeric substances. Polymers are very large molecules generally made up from a simple, repeating unit. A typical molecule of polyethylene, for instance, might have the formula $CH_3(CH_2CH_2)_{2000}CH_3$. Although polymer solutions can be fairly concentrated in terms of grams per liter, on a mole-per-liter basis they may be quite dilute. The freezing-point depression is usually too small to measure, though the osmotic pressure may be appreciable.

Example 12.13 Calculating Osmotic Pressure

The formula for low-molecular-weight starch is $(C_6H_{10}O_5)_n$, where n averages 2.00×10^2. When 0.798 g of starch is dissolved in 100.0 mL of water solution, what is the osmotic pressure at 25°C?

Solution

The molecular weight of $(C_6H_{10}O_5)_{200}$ is 32,400 amu. Hence, the number of moles in 0.798 g of starch is

$$0.798\ \text{g starch} \times \frac{1\ \text{mol starch}}{32{,}400\ \text{g starch}} = 2.46 \times 10^{-5}\ \text{mol starch}$$

The molarity of the solution is

$$\frac{2.46 \times 10^{-5}\ \text{mol}}{0.1000\ \text{L solution}} = 2.46 \times 10^{-4}\ \text{mol/L solution}$$

and the osmotic pressure at 25°C is

$$\begin{aligned}\pi &= MRT \\ &= 2.46 \times 10^{-4}\ \text{mol/L} \times 0.0821\ \text{L} \cdot \text{atm}/(\text{K} \cdot \text{mol}) \times 298\ \text{K} \\ &= 6.02 \times 10^{-3}\ \text{atm} = 6.02 \times 10^{-3}\ \text{atm} \times \frac{760\ \text{mmHg}}{1\ \text{atm}} \\ &= \mathbf{4.58\ mmHg}\end{aligned}$$

For comparison, we can calculate the freezing-point depression. Assume that the molality is equal to the molarity (this is in effect the case for very dilute aqueous solutions). Then,

$$\begin{aligned}\Delta T_f &= 1.86°\text{C}/m \times 2.46 \times 10^{-4}\ m \\ &= 4.58 \times 10^{-4}\ °\text{C}\end{aligned}$$

which is barely detectable with generally available equipment.

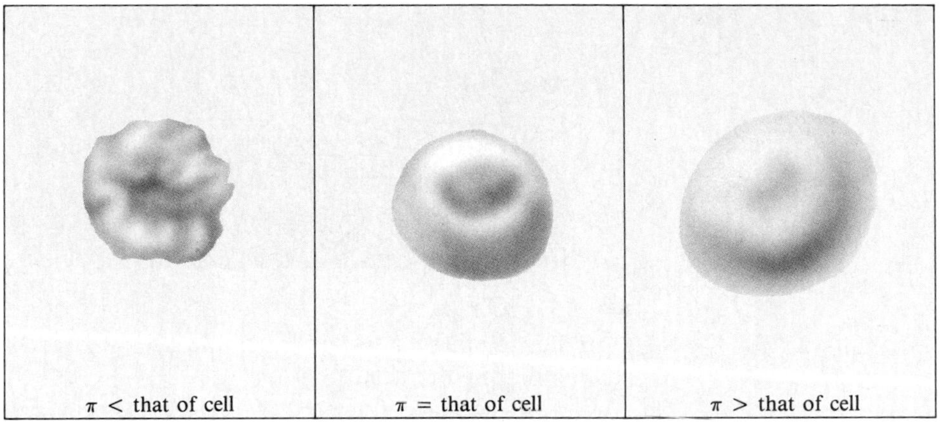

Figure 12.23
The importance of osmotic pressure to cells. Water passes through a cell membrane from one solution to another unless the osmotic pressures are equal. If a cell is placed in a solution whose osmotic pressure is too low (less than that of the cell), water passes out of the cell, which collapses. If a cell is placed in a solution whose osmotic pressure is too high (greater than that of the cell), water passes into the cell, which may burst.

π < that of cell | π = that of cell | π > that of cell

Exercise 12.16

Calculate the osmotic pressure at 20°C of an aqueous solution containing 5.0 g of sucrose, $C_{12}H_{22}O_{11}$, in 100.0 mL of solution. (See Problems 12.67 and 12.68.)

● Osmotic pressure appears to be important for the rising of sap in a tree. During the day, water evaporates from the leaves of the tree, so the aqueous solution in the leaves becomes more concentrated and the osmotic pressure increases. Sap flows upward to dilute the water solution in the leaves.

Osmosis is important in many biological processes.● A cell might be thought of (simplistically) as an aqueous solution enclosed by a semipermeable membrane. The solution surrounding the cell must have an osmotic pressure equal to that within the cell. Otherwise, water would either leave the cell, dehydrating it, or enter the cell and possibly burst the membrane. For intravenous feeding (adding a nutrient solution to the venous blood supply of a patient), it is necessary that the nutrient solution have exactly the osmotic pressure of blood plasma. If it does not, the blood cells may collapse or burst through osmosis (Figure 12.23).

The body has portions of organs that are at different osmotic pressures, and an active pumping mechanism is required to offset osmosis. For example, cells of the transparent tissue of the exterior eye, the cornea, have a more concentrated optical fluid than does the aqueous humor, a solution just behind the cornea (Figure 12.24). In order to prevent the cornea from taking up additional water from the aqueous humor, cells that pump water are located in the tissue of the cornea adjacent to the aqueous humor. Corneas that are to be stored and used for transplants must be removed from the globe of the eye soon after death. This prevents

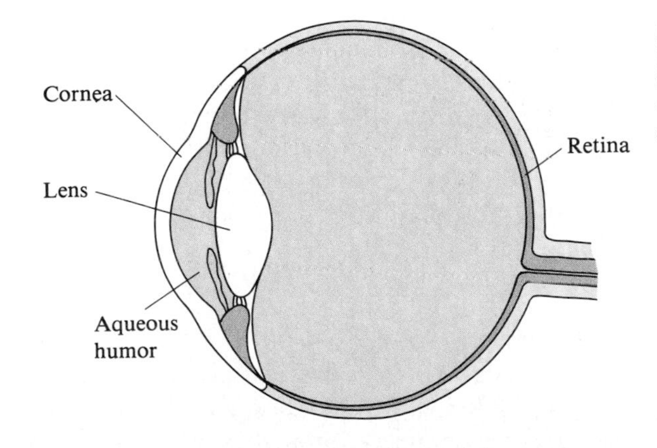

Figure 12.24
Parts of the eye. Certain cells in the cornea act as pumps to prevent osmosis of water from the aqueous humor.

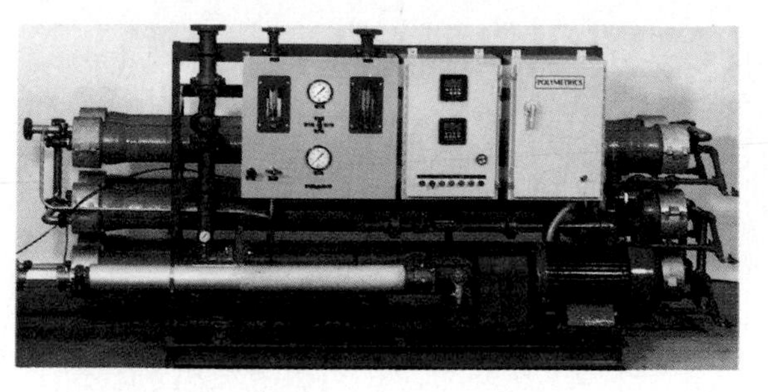

Figure 12.25
Desalination system that uses reverse osmosis. When brackish water or ocean water is subjected to a pressure that is greater than its osmotic pressure, pure water passes through a membrane, leaving a more concentrated salt solution behind.

the clouding that occurs when the pumping mechanism fails (as its energy supply is depleted at death).

The process of *reverse osmosis* has been applied to the problem of purifying water. In particular, it has been used to *desalinate* ocean water (that is, remove salts from seawater to make drinkable or industrially usable water). In normal osmosis, the solvent flows from a dilute solution through a membrane to a more concentrated solution. By applying a pressure equal to the osmotic pressure to the more concentrated solution, the process of osmosis can be stopped. By applying an even greater pressure, the osmotic process can be reversed. Then, solvent flows from the concentrated solution (which could be ocean water) through a membrane to the more dilute solution (which could be more or less pure water). Figure 12.25 shows a system using reverse osmosis to desalinate brackish water (slightly salty water) and ocean water.

12.8 COLLIGATIVE PROPERTIES OF IONIC SOLUTIONS

In order to explain the colligative properties of ionic solutions, we must realize that it is the total concentration of ions, rather than the concentration of ionic substance, that is important. For example, the freezing-point depression of 0.100 *m* sodium chloride solution is nearly twice that of 0.100 *m* glucose solution. We can explain this by saying that sodium chloride dissolves in water to form the ions Na^+ and Cl^-. Each formula unit of NaCl gives two particles. We can write the freezing-point lowering more generally as

$$\Delta T_f = iK_f c_m$$

where i is the number of ions resulting from each formula unit and c_m is the molality computed on the basis of the formula of the ionic compound. The equations for the other colligative properties must be similarly modified by the factor i.

Actually, the freezing points of ionic solutions agree with the previous equation only when the solutions are quite dilute, and i values calculated from the freezing-point depression are usually smaller than the number of ions obtained from the formula unit. For example, a 0.029 *m* aqueous solution of potassium sulfate, K_2SO_4, has a freezing point of −0.14°C. Hence,

$$i = \frac{\Delta T_f}{K_f c_m} = \frac{0.14°\text{C}}{1.86°\text{C}/m \times 0.029\ m} = 2.6$$

We might have expected a value of 3, on the basis of the fact that K_2SO_4 ionizes to give three ions.● At first this was taken as evidence that salts were not completely ionized in solution. In 1923, however, Peter Debye and Erich Hückel were able to show that the colligative properties of salt solutions could be explained by assum-

● The value of i equal to $\Delta T_f/K_f c_m$ is often called the van't Hoff factor. Thus, the van't Hoff factor for 0.029 *m* K_2SO_4 is 2.6.

ing that the salt is completely ionized in solution but that the *activities,* or effective concentrations, of the ions are less than their actual concentrations as a result of the electrical interactions of the ions in solution. The *Debye–Hückel theory* allows us to calculate these activities. When this is done, excellent agreement is obtained for dilute solutions.

Example 12.14 Determining Colligative Properties of Ionic Solutions

Estimate the freezing point of a 0.010 m aqueous solution of aluminum sulfate, $Al_2(SO_4)_3$. Assume the value of i based on the formula of the compound.

Solution

When aluminum sulfate, $Al_2(SO_4)_3$, dissolves in water, it ionizes into five ions.

$$Al_2(SO_4)_3(s) \longrightarrow 2Al^{3+}(aq) + 3SO_4^{2-}(aq)$$

Therefore, we assume $i = 5$. The freezing-point depression is

$$\Delta T_f = iK_f c_m = 5 \times 1.86°C/m \times 0.010\ m = 0.093°C$$

The freezing point of the solution is 0.000°C − 0.093°C = **−0.093°C.**

Exercise 12.17

Estimate the boiling point of a 0.050 m aqueous $MgCl_2$ solution. Assume a value of i based on the formula.

(See Problems 12.69 and 12.70.)

Colloid Formation

As we noted at the beginning of the chapter, colloids appear homogeneous like a solution, but they consist of comparatively large particles of one substance dispersed throughout another substance. We will look at colloids in this final section of the chapter.

12.9 COLLOIDS

A **colloid** is *a dispersion of particles of one substance (the dispersed phase) throughout another substance or solution (the continuous phase).* Fog is an example of a colloid: It consists of very small water droplets (dispersed phase) in air (continuous phase). A colloid differs from a true solution in that the dispersed particles are larger than normal molecules, though they are too small to be seen with a microscope. The particles are from about 10 Å to about 2000 Å in size.

Tyndall Effect

Although a colloid appears to be homogeneous because the dispersed particles are quite small, it can be distinguished from a true solution by its ability to scatter light. *The scattering of light by colloidal-sized particles* is known as the **Tyndall effect.**● For example, the atmosphere appears to be a clear gas, but a ray of sunshine against a dark background shows up many fine dust particles by light scattering. Similarly, when a beam of light is directed through clear gelatin (a colloid, not a true solution), the beam becomes visible by the scattering of light from colloidal gelatin particles. The beam appears as a ray passing through the

● Although all gases and liquids scatter light, the scattering from a pure substance or true solution is quite small and usually not detectable. However, because of the considerable depth of the atmosphere, the scattering of light by air molecules can be seen. The blue color of the sky is due to the fact that blue light is scattered more easily than red light.

Figure 12.26
A demonstration of the Tyndall effect by a colloid. A light beam is visible perpendicular to its path only if light is scattered toward the viewer. The vessel on the left contains a colloid, which scatters light. The vessel on the right contains a true solution, which scatters negligible light.

solution (Figure 12.26). When the same experiment is performed with a true solution, such as an aqueous solution of sodium chloride, the beam of light is not visible.

Types of Colloids

Colloids are characterized according to the state (solid, liquid, or gas) of the dispersed phase and the state of the continuous phase. Table 12.4 lists various types of colloids and gives some examples of each. Fog and smoke are **aerosols,** which are *liquid droplets or solid particles dispersed throughout a gas*. An **emulsion** consists of *liquid droplets dispersed throughout another liquid* (as particles of butterfat are dispersed through homogenized milk). A **sol** consists of *solid particles dispersed in a liquid.*

Hydrophilic and Hydrophobic Colloids

Colloids in which the continuous phase is water are also divided into two major classes: hydrophilic colloids and hydrophobic colloids. A **hydrophilic colloid** is *a*

Table 12.4 Types of Colloids

Continuous Phase	Dispersed Phase	Name	Example
Gas	Liquid	Aerosol	Fog, mist
Gas	Solid	Aerosol	Smoke
Liquid	Gas	Foam	Whipped cream
Liquid	Liquid	Emulsion	Mayonnaise (oil dispersed in water)
Liquid	Solid	Sol	$AgCl(s)$ dispersed in H_2O
Solid	Gas	Foam	Pumice, plastic foams
Solid	Liquid	Gel	Jelly, opal (mineral with liquid inclusions)
Solid	Solid	Solid sol	Ruby glass (glass with dispersed metal)

colloid in which there is a strong attraction between the dispersed phase and the continuous phase (water). Many such colloids consist of macromolecules (very large molecules) dispersed in water. Except for the large size of the dispersed molecules, these are like normal solutions. Protein solutions, such as gelatin in water, are hydrophilic colloids. Gelatin molecules are attracted to water molecules by London forces and hydrogen bonding.

A **hydrophobic colloid** is *a colloid in which there is a lack of attraction between the dispersed phase and the continuous phase (water).* Hydrophobic colloids are basically unstable. Given sufficient time, the dispersed phase comes out of solution by aggregating into larger particles. In this behavior, they are quite unlike true solutions and hydrophilic colloids. The time taken to separate may be extremely long, however. A colloid of gold particles in water prepared by Michael Faraday in 1857 is still preserved in the British Museum in London. This colloid is hydrophobic as well as a sol (solid particles dispersed in water).

Hydrophobic sols are often formed when a solid crystallizes rapidly from a chemical reaction or a supersaturated solution. When crystallization occurs rapidly, many centers of crystallization (called *nuclei*) are formed at once. Ions are attracted to these nuclei and very small crystals are formed. These small crystals are prevented from settling out by the random thermal motion of the solvent molecules, which continue to buffet them.

You might expect these very small crystals to aggregate into larger crystals because the aggregation would bring ions of opposite charge into contact. However, sol formation appears to happen when, for some reason, each of the small crystals gets a preponderance of one kind of charge on its surface. For example, iron(III) hydroxide forms a colloid because an excess of iron(III) ion (Fe^{3+}) is present on the surface, giving each crystal an excess of positive charge. These positively charged crystals repel one another, so aggregation to larger particles is prevented.

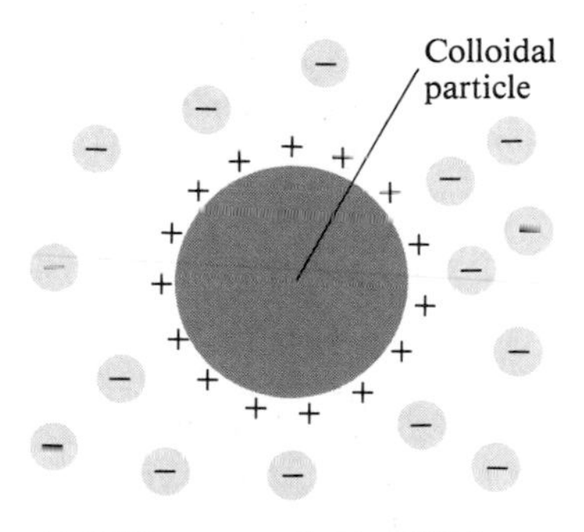

$Fe(OH)_3$ surrounded by Cl^- ions

A

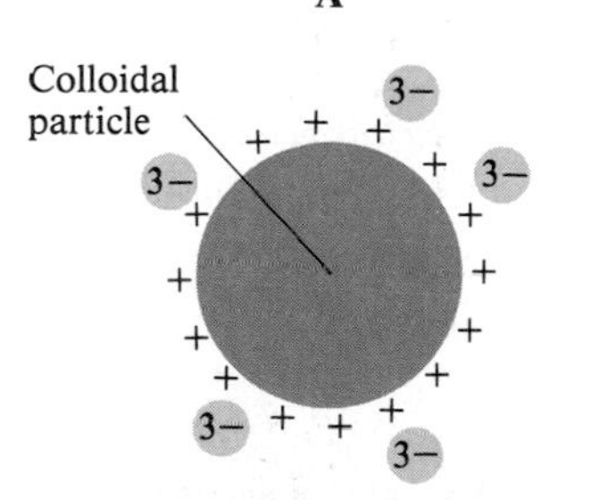

$Fe(OH)_3$ surrounded by PO_4^{3-} ions

B

Figure 12.27
Layers of ions surrounding charged colloidal particles. *(A)* A positively charged colloidal particle of iron(III) hydroxide surrounded by chloride ions. *(B)* The same colloidal particle surrounded by phosphate ions. Because these ions gather more closely to the colloidal particles, coagulation is more likely to occur.

Coagulation

An iron(III) hydroxide sol can be made to aggregate by the addition of an ionic solution, particularly if the solution contains anions with multiple charges (such as phosphate ions, PO_4^{3-}). **Coagulation** is *the process by which a colloid is made to come out of solution by aggregation.* We can picture what happens in the following way: A positively charged colloidal particle of iron(III) hydroxide gathers a layer of anions around it. The thickness of this layer is determined by the charge on the anions. The greater the magnitude of the negative charge, the more compact the layer of charge. Phosphate ions, for example, gather more closely to the positively charged colloidal particles than do chloride ions (see Figure 12.27). If the ion layer is gathered close to the colloidal particle, the overall charge is effectively neutralized. In that case, two colloidal particles can approach close enough to aggregate.

The curdling of milk when it sours is another example of coagulation. Milk is a colloidal suspension in which the particles are prevented from aggregating because they have electric charges of the same sign. The ions responsible for the coagulation (curdling) are formed when lactose (milk sugar) ferments to lactic acid. A third example is the coagulation of a colloidal suspension of soil in river water when the water meets the concentrated ionic solution of an ocean. The Mississippi Delta was formed in this way.

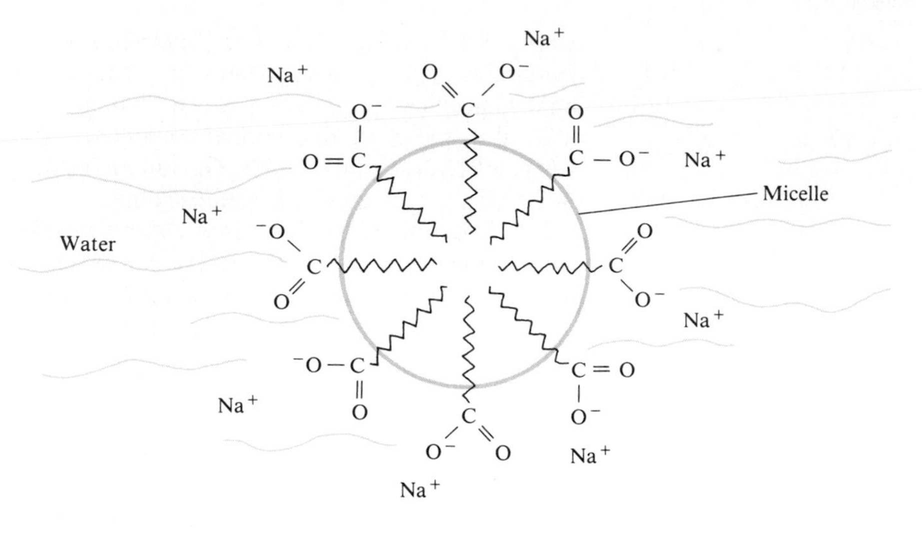

Figure 12.28
A stearate micelle in a water solution. Stearate ions associate in groups (micelles), with their hydrocarbon ends pointing inward. The ionic ends, on the outside of the micelle, point into the water solution.

Exercise 12.18

Colloidal sulfur particles are negatively charged with thiosulfate ions, $S_2O_3^{2-}$, and other ions on the surface of the sulfur. Indicate which of the following would be most effective in coagulating colloidal sulfur: NaCl, $MgCl_2$, or $AlCl_3$. (See Problems 12.75 and 12.76.)

Association Colloids

When molecules that have both a hydrophobic and a hydrophilic end are dispersed in water, they associate or aggregate to form colloidal-sized particles, or micelles. A **micelle** is *a colloidal-sized particle formed in water by the association of molecules that each have a hydrophobic end and a hydrophilic end.* The hydrophobic ends point inward toward one another, while the hydrophilic ends are on the outside of the micelle facing the water. *A colloid in which the dispersed phase consists of micelles* is called an **association colloid.**

Ordinary soap in water provides an example of an association colloid. Soap consists of compounds such as sodium stearate, $C_{17}H_{35}COONa$. The stearate ion

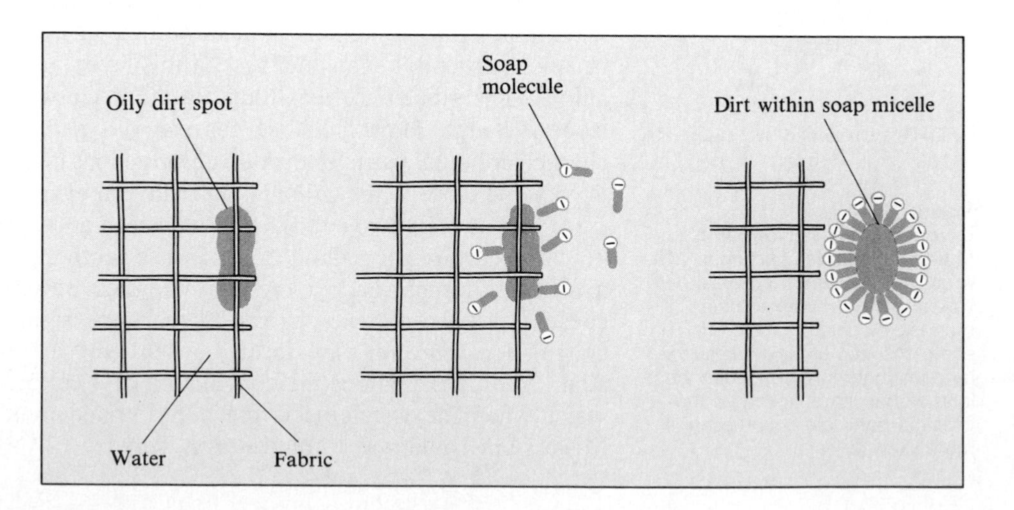

Figure 12.29
The cleansing action of soap. The hydrocarbon ends of soap molecules gather around an oil spot, forming a micelle that can be washed away in the water.

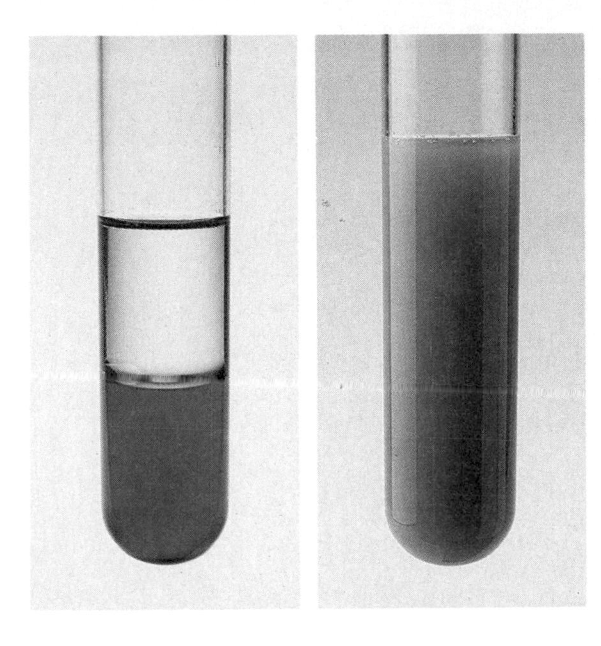

Figure 12.30
Formation of an association colloid with soap. *Left:* Vegetable oil floating on water (dyed green). *Right:* The oil is dispersed in micelles throughout the water when the mixture is shaken with soap.

has a long hydrocarbon end that is hydrophobic (because it is nonpolar) and a carboxyl group (COO^-) that is hydrophilic (because it is ionic).

$$\underbrace{CH_3CH_2CH_2CH_2CH_2CH_2CH_2CH_2CH_2CH_2CH_2CH_2CH_2CH_2CH_2CH_2CH_2}_{\text{hydrophobic end}}\underbrace{C(=O)O^-}_{\text{hydrophilic end}}$$

In water solution, the stearate ions associate into micelles in which the hydrocarbon ends point inward toward one another and away from the water, while carboxyl groups are on the outside of the micelle facing the water (Figure 12.28).

The cleaning action of soap occurs because oil and grease can be absorbed into the hydrophobic centers of soap micelles and washed away (Figures 12.29 and 12.30). Synthetic detergents are also substances that form association colloids. Sodium lauryl sulfate is a synthetic detergent present in laundry soaps, toothpastes, and shampoos.

$$CH_3CH_2CH_2CH_2CH_2CH_2CH_2CH_2CH_2CH_2CH_2CH_2OSO_3^-Na^+$$

sodium lauryl sulfate

It has a hydrophilic sulfate group ($—OSO_3^-$) and a hydrophobic dodecyl group ($C_{12}H_{25}—$, the hydrocarbon end).

The detergent molecules we have discussed so far are classified in the trade as "anionics," because they have a negative charge at the hydrophilic end. Other detergent molecules are "cationics," because they have a positive charge at the hydrophilic end. An example is

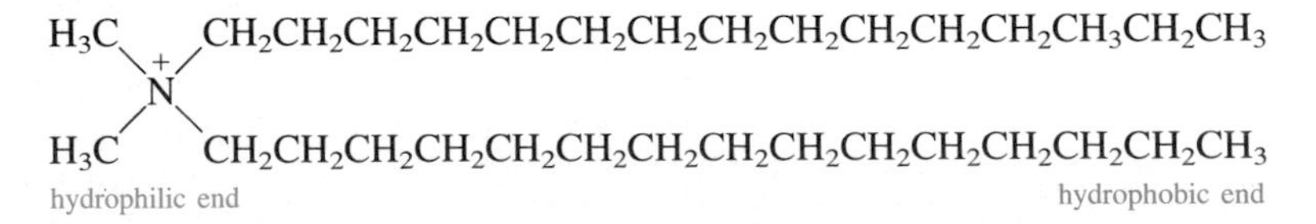

Many cationic detergents also have germicidal properties and are used in hospital disinfectants and in mouth washes (Figure 12.31).

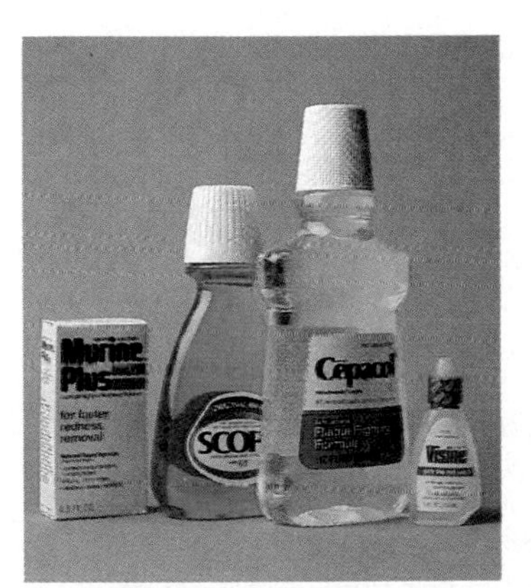

Figure 12.31
Commercial preparations containing cationic detergents. Many cationic detergents exhibit germicidal properties. For that reason, they are used in certain hospital antiseptics, mouth washes, and eye wetting solutions.

Profile of a Chemical

WATER (a Special Substance for Planet Earth)

Water is the only liquid substance (other than petroleum) to be found on earth in significant amounts. This liquid is also readily convertible under conditions on earth to the solid and gaseous forms. Perhaps it was these facts that led the ancient Greek philosopher Thales of Miletus (ca. 580 B.C.) to conceive the unitary theory in which all things are composed of water. But water has several unusual properties that set it apart from other substances. For example, its solid phase, ice, is less dense than liquid water, whereas for most substances the solid phase is more dense than the liquid. In addition, water has an unusually large heat capacity. These are some of the properties that are important in determining conditions favorable to life. In fact, it is difficult to think of life without water.

The unusual properties of water are largely linked to its ability to form hydrogen bonds. For example, ice is less dense than liquid water because ice has an open, hydrogen-bonded structure (see Figure 12.32). Each oxygen atom in the structure of ice is surrounded tetrahedrally by four hydrogen atoms: two that are close and covalently bonded to give the H_2O molecule, and two others that are farther away and held by hydrogen bonds. The tetrahedral angles give rise to a three-dimensional structure that contains open space. When ice melts, hydrogen bonds break, the ice structure partially disintegrates, and the molecules become more compactly arranged, leading to a more dense liquid.

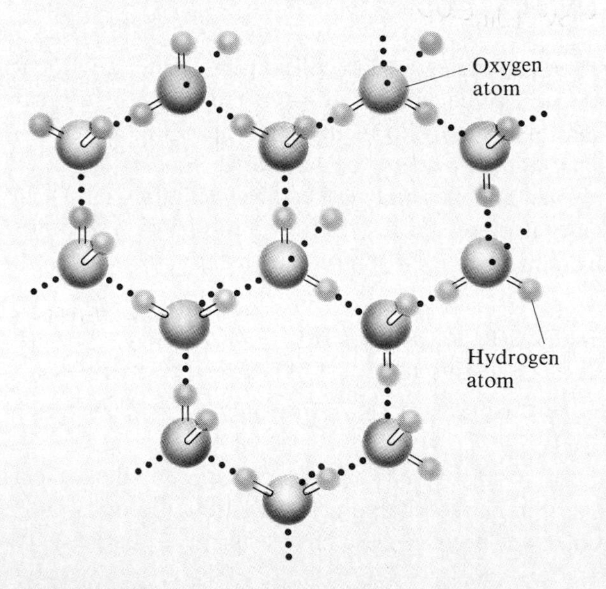

Figure 12.32
The structure of ice. *Oxygen atoms are represented by large spheres; hydrogen atoms by small spheres. Each oxygen atom is tetrahedrally surrounded by four hydrogen atoms. Two are close, giving the H_2O molecule. Two are farther away, held by hydrogen bonding (represented by three dots). The distribution of hydrogen atoms in these two types of positions is random.*

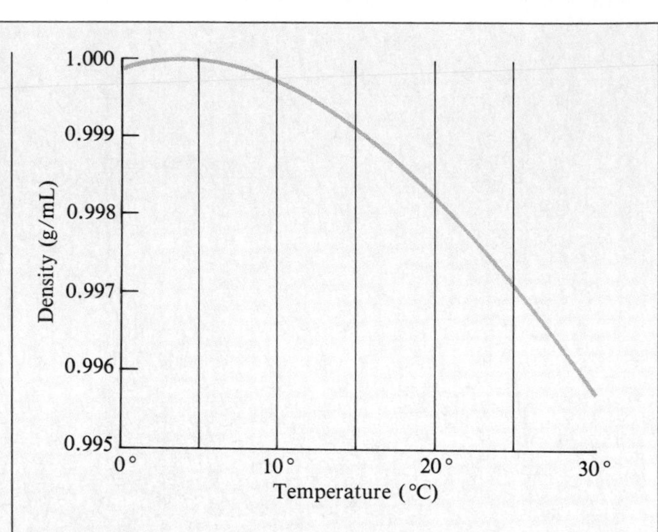

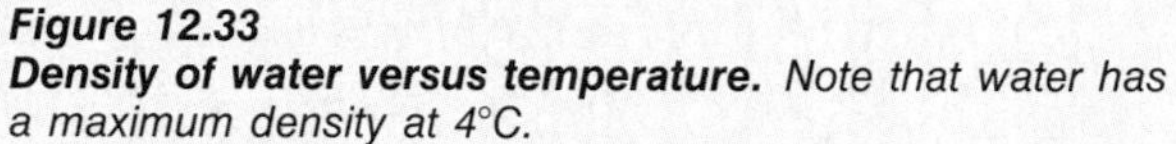

Figure 12.33
Density of water versus temperature. *Note that water has a maximum density at 4°C.*

Even liquid water has significant hydrogen bonding. Only about 15% of the hydrogen bonds are broken when ice melts. We might view the liquid as composed of ice-like clusters in which hydrogen bonds are continually breaking and forming, so clusters disappear and new ones appear. This is sometimes referred to as the "flickering cluster" model of liquid water.

As the temperature rises from 0°C, these clusters tend to break down further, giving an even more compact liquid. Thus, the density rises (see Figure 12.33). However, as is normal in any liquid when the temperature rises, the molecules begin to move and vibrate faster, and the space occupied by the average molecule increases. For most other liquids, this results in a continuous decrease in density with temperature increase. In water, this normal effect is countered by the density increase due to breaking of hydrogen bonds, but at 4°C the normal effect begins to predominate. Water shows a maximum density at 4°C and becomes less dense at higher temperatures.

The unusually large heat capacity of water is also explained by hydrogen bonding. To increase the temperature of liquid water, it is necessary to supply energy to break hydrogen bonds, in addition to the energy nor-

Table 12.5 Thermal Properties of Some Common Substances

Substance	Formula	Melting Point (°C)	Boiling Point (°C)	Heat Capacity of Liquid (J/g · °C)	Heat of Fusion (J/g)	Heat of Vaporization (J/g)
Water	H_2O	0	100	4.18	333	2257
Ammonia	NH_3	−78	−33	4.48	341	1368
Ethanol	C_2H_5OH	−117	78	2.24	104	854
Benzene	C_6H_6	6	80	1.63	127	395
Carbon tetrachloride	CCl_4	−23	77	0.83	17	194
Mercury	Hg	−39	357	0.14	12	295

mally required to increase molecular agitation. The heat of fusion and heat of vaporization are high for the same reason; hydrogen bonds must be broken to melt or to vaporize the substance. (See Table 12.5 for the thermal properties of some common substances.)

The unusually large heat of vaporization of water has an important effect on the earth's weather. Evaporation of the surface waters absorbs over 30% of the solar energy reaching the earth's surface. This energy is released when the water vapor condenses, and thunderstorms and hurricanes may result. In the process, the waters of the earth are circulated and the freshwater sources are replenished. This natural cycle of water from the oceans to freshwater sources and its return to the ocean is called the *hydrologic cycle* (Figure 12.34).

Although evaporation and condensation of water play a dominant role in our weather, other properties are important as well. The exceptionally large heat capacity of bodies of water has an important moderating effect on the surrounding temperature by warming cold air masses in winter and cooling warm air masses in summer. Worldwide, the oceans are most important, but even inland lakes have a pronounced effect. For example, the Great Lakes give Detroit a more moderate winter than cities that are somewhat farther south but have no lakes nearby.

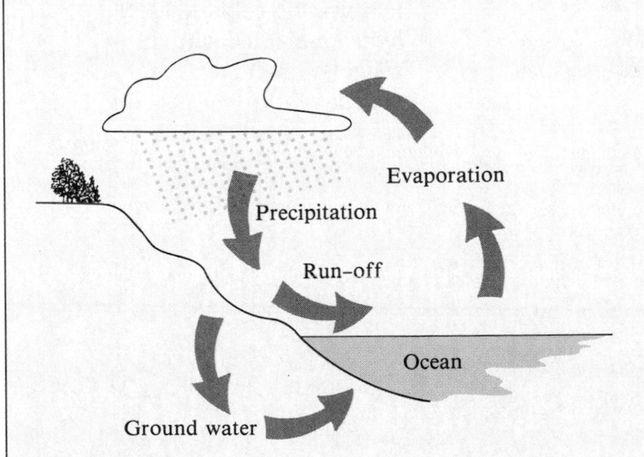

Figure 12.34
The hydrologic cycle. *Ocean water evaporates to form clouds. Then, precipitation (rain and snow) from these clouds replenishes freshwater sources. Water from these sources eventually returns to the oceans via rivers, run-off, or ground water.*

The fact that ice is less dense than water means that it forms on top of the liquid when freezing occurs. This has far-reaching effects, both for weather and for aquatic animals. When ice forms on a body of water, it insulates the underlying water from the cold air and limits further freezing. Fish depend on this for winter survival. Consider what would happen to a lake if ice were more dense than water. The ice would freeze from the bottom of the lake upward. Without the insulating effect at the surface, the lake could well freeze solid, killing the fish. Spring thaw would be prolonged, because the insulating effect of the surface water would make it take much longer for the ice at the bottom of a lake to melt.

The solvent properties of water are also unusual. Water is both a polar substance and a hydrogen-bonding molecule. As a result, water dissolves many substances, including ionic and polar compounds. These properties make water the premier solvent, biologically and industrially.

As a result of the solvent properties of water, the naturally occurring liquid always contains dissolved materials, particularly ionic substances. *Hard water* contains certain metal ions, such as Ca^{2+} and Mg^{2+}. These ions react with soaps, which are sodium salts of stearic acid and similar organic acids, to produce a curdy precipitate of calcium and magnesium salts. This precipitate adheres to clothing and bathtubs (as bathtub ring). Removing Ca^{2+} and Mg^{2+} ions from hard water is referred to as *water softening.*

(continued)

Water is often softened by *ion exchange.* Ion exchange is a process whereby a water solution is passed through a column of a material that replaces one kind of ion in solution with another kind. Home and commercial water softeners contain cation-exchange resins, which are insoluble macromolecular substances (substances consisting of giant molecules) to which negatively charged groups are chemically bonded. The negative charges are counterbalanced by ions such as Na^+. When hard water containing the Ca^{2+} ion passes through a column of this resin, the Na^+ ion in the resin is replaced by Ca^{2+} ion.

$$2NaR(s) + Ca^{2+}(aq) \longrightarrow CaR_2(s) + 2Na^+(aq)$$
(R^- = anion of exchange resin)

The water passing through the column now contains Na^+ in place of Ca^{2+} and has been softened. Once the resin has been completely converted to the calcium salt, it can be regenerated by flushing the column with a concentrated solution of NaCl to reverse the previous reaction.

Questions for Study

1. How does hydrogen bonding in water explain why ice is less dense than liquid water?
2. How does hydrogen bonding affect the heat capacity and heat of vaporization of water?
3. How does the unusually large heat capacity of water affect the Earth's weather?
4. Explain why water dissolves so many different substances.
5. What does the term *hard water* mean? Why does hard water reduce the effectiveness of soap?
6. Explain how water is softened by ion exchange.

A Checklist for Review

IMPORTANT TERMS

solute (12.1)
solvent (12.1)
miscible fluids (12.1)
saturated solution (12.2)
solubility (12.2)
unsaturated solution (12.2)
supersaturated solution (12.2)
hydration (of ions) (12.2)
Le Chatelier's principle (12.3)
Henry's law (12.3)
colligative properties (p. 470)
mass percentage of solute (12.4)
molality (12.4)
mole fraction (12.4)
vapor-pressure lowering (12.5)
Raoult's law (12.5)
boiling-point elevation (12.6)
freezing-point depression (12.6)
osmosis (12.7)
osmotic pressure (12.7)
colloid (12.9)
Tyndall effect (12.9)
aerosols (12.9)
emulsion (12.9)
sol (12.9)
hydrophilic colloid (12.9)
hydrophobic colloid (12.9)
coagulation (12.9)
micelle (12.9)
association colloid (12.9)

KEY EQUATIONS

$$S = k_H P$$

$$\text{Mass percentage of solute} = \frac{\text{mass of solute}}{\text{mass of solution}} \times 100\%$$

$$\text{Molality} = \frac{\text{moles of solute}}{\text{kilograms of solvent}}$$

$$X_A = \frac{\text{moles of substance } A}{\text{total moles of solution}}$$

$$P_A = P_A^\circ X_A$$

$$\Delta P = P_A^\circ X_B$$

$$\Delta T_b = K_b c_m$$

$$\Delta T_f = K_f c_m$$

$$\pi = MRT$$

SUMMARY OF FACTS AND CONCEPTS

Solutions are homogeneous mixtures and can be gases, liquids, or solids. Two gases, for example, will mix in all proportions to give a gaseous solution, because gases are *miscible* in one another. Often one substance will dissolve in another only to a limited extent. The maximum amount that dissolves at equilibrium is the *solubility* of that substance. Solubility is explained in terms of the natural tendency toward disorder (say by the mixing of two substances) and by the tendency to give the strongest forces of attraction between species. The dissolving of one molecular substance in another is limited when intermolecular forces strongly favor the unmixed substances. When two substances have similar types of intermolecular forces, they tend to be soluble ("like dissolves like"). Ionic substances differ greatly in their solubilities in water, because solubility depends on the relative values of *lattice energy* and *hydration energy*.

Solubilities of substances usually vary with temperature. At higher temperatures most gases will become less soluble in water, whereas most ionic solids will become more soluble. Pressure has a significant effect only on the solubility of a gas. A gas is more soluble in a liquid if the partial pressure of the gas is increased, which is in agreement with *Le Chatelier's principle*. According to *Henry's law*, the solubility of a gas is directly proportional to the partial pressure of the gas.

Colligative properties of solutions depend only on the concentration of solute particles. The concentration may be defined by the *molarity, mass percentage of solute, molality,* or *mole fraction*. One example of a colligative property is the *vapor-pressure lowering* of a volatile solvent by the addition of a nonvolatile solute. According to *Raoult's law,* the vapor pressure of the solution depends on the mole fraction of solvent in the solution. Because adding a nonvolatile solute lowers the vapor pressure, the boiling point must be raised to bring the vapor pressure back to one atmosphere *(boiling-point elevation)*. Such a solution also exhibits a *freezing-point depression*. Boiling-point elevation and freezing-point depression are colligative properties. *Osmosis* is another colligative property. In osmosis, there is a flow of solvent through a semipermeable membrane in order to equalize the concentrations of solutions on the two sides of the membrane. Colligative properties may be used to measure the concentration of solute. In this way, one can determine the molecular weight of the solute. The colligative properties of an ionic solution depend on the concentration of ions, which we can obtain from the concentration of ionic compound and its formula. Interactions of ions in solution do affect the properties significantly, however.

A *colloid* is a dispersion of particles of one substance (about 10 Å to 2000 Å in size) throughout another. Colloids can be detected by the *Tyndall effect* (the scattering of light by colloidal-sized particles). They are characterized by the state of the dispersed phase and the state of the continuous phase. An aerosol, for example, consists of liquid droplets or solid particles dispersed in a gas. Colloids in water are also classified as hydrophilic or hydrophobic. A *hydrophilic colloid* consists of a dispersed phase that is attracted to the water. Many of these colloids have macromolecules dissolved in water. In *hydrophobic colloids,* there is little attraction between the dispersed phase and the water. The colloidal particles may be electrically charged and so repel one another, preventing aggregation to larger particles. An ionic solution can neutralize this charge so that the colloid will *coagulate*—that is, aggregate. An *association colloid* consists of molecules with a hydrophobic end and a hydrophilic end dispersed in water. These molecules associate into colloidal-sized groups, or *micelles*.

OPERATIONAL SKILLS

1. Applying Henry's law Given the solubility of a gas at one pressure, find its solubility at another pressure (Example 12.1).

2. Calculating solution concentration Given the mass percent of solute, state how to prepare a given mass of solution (Example 12.2). Given the masses of solute and solvent, find the molality (Example 12.3) and mole fractions (Example 12.4).

3. Converting concentration units Given the molality of a solution, calculate the mole fractions of solute and solvent; and given the mole fractions, calculate the molality (Examples 12.5 and 12.6). Given the density, calculate the molarity from the molality, and vice versa (Examples 12.7 and 12.8).

4. Calculating vapor-pressure lowering Given the mole fraction of solute in a solution of nonvolatile solute and the vapor pressure of pure solvent, calculate the vapor-pressure lowering and vapor pressure of the solution (Example 12.9).

5. Calculating the boiling-point elevation and freezing-point depression Given the molality of a solution of nonvolatile solute, calculate the boiling-point elevation and freezing-point depression (Example 12.10).

6. Calculating molecular weights Given the masses of solvent and solute and the molality of the solution, find the molecular weight of the solute (Example 12.11). Given the masses of solvent and solute, the freezing-point depression, and K_f, find the molecular weight of the solute (Example 12.12).

7. Calculating osmotic pressure Given the molarity and the temperature of a solution, calculate its osmotic pressure (Example 12.13).

8. Determining colligative properties of ionic solutions Given the concentration of ionic compound in a solution, calculate the magnitude of a colligative property; if *i* is not given, assume the value based on the formula of the ionic compound (Example 12.14).

Review Questions

12.1 Give one example of each: a gaseous solution, a liquid solution, a solid solution.

12.2 Explain on the basis that ''like dissolves like'' why glycerol, $CH_2OHCHOHCH_2OH$, is miscible in water but benzene, C_6H_6, has very limited solubility in water.

12.3 What are the two factors needed to explain the differences in solubilities of substances?

12.4 Explain in terms of intermolecular attractions why octane is immiscible in water.

12.5 Explain why ionic substances show a wide range of solubilities in water.

12.6 Using the concept of hydration, describe the process of dissolving a sodium chloride crystal in water.

12.7 What is the usual solubility behavior of an ionic compound in water when the temperature is raised? Give an example of an exception to this behavior.

12.8 Give one example of each: a salt whose heat of solution is exothermic and a salt whose heat of solution is endothermic.

12.9 Fewer fish can be kept in an aquarium of fixed size in the summer than in the winter. Explain.

12.10 Explain why a carbonated beverage must be stored in a closed container.

12.11 Pressure has an effect on the solubility of oxygen in water but a negligible effect on the solubility of sugar in water. Why?

12.12 Four ways were discussed to express the concentration of a solute in solution. Identify them and define each concentration unit.

12.13 When two beakers containing different concentrations of a solute in water are placed in a closed cabinet for a time, one beaker gains solvent and the other loses it, so that the concentrations of solute in the two beakers become equal. Explain what is happening.

12.14 Explain the process of fractional distillation to separate a solution of two liquids into pure components.

12.15 Explain why the boiling point of a solution containing a nonvolatile solute is higher than the boiling point of a pure solvent.

12.16 It is possible to obtain drinking water from seawater by freezing. Explain the process.

12.17 List two applications of freezing-point depression.

12.18 A green, leafy salad becomes wilted if left too long in a salad dressing containing vinegar and salt. Explain what happens.

12.19 Explain the process of reverse osmosis to produce drinkable water from ocean water.

12.20 One can often see ''sunbeams'' passing through the less dense portions of clouds. What is the explanation of this?

12.21 Give an example of an aerosol, a foam, an emulsion, a sol, and a gel.

12.22 If electrodes connected to a direct current are dipped into a beaker of colloidal iron(III) hydroxide, a precipitate collects at the negative electrode. Explain.

12.23 A Cottrell precipitator consists of a column containing electrodes that are connected to a high-voltage DC source. The Cottrell precipitator is placed in smokestacks to remove smoke particles from the gas discharged from an industrial plant. Explain how you think this works.

12.24 Stearic acid can be made to form a monolayer (a layer one molecule thick) on water by placing a drop of solution of stearic acid in benzene onto the surface of the water. The solution spreads over the water and the benzene evaporates. Explain why a monolayer forms.

12.25 Explain how soap removes oil from a fabric.

12.26 Give an example of an anionic detergent and of a cationic detergent.

Practice Problems

TYPES OF SOLUTIONS

12.27 Give an example of a liquid solution prepared by dissolving a gas in a liquid.

12.28 Give an example of a solid solution prepared from two solids.

SOLUBILITY

12.29 Would boric acid, $B(OH)_3$, be more soluble in ethanol, C_2H_5OH, or in benzene, C_6H_6?

12.31 Arrange the following substances in order of increasing solubility in hexane, C_6H_{14}: CH_2OHCH_2OH, $C_{10}H_{22}$, H_2O.

12.33 Which of the following ions would be expected to have the greater energy of hydration, K^+ or Ca^{2+}?

12.35 Arrange the following alkaline-earth-metal iodates in order of increasing solubility in water; explain your reasoning: $Ba(IO_3)_2$, $Ca(IO_3)_2$, $Sr(IO_3)_2$, $Mg(IO_3)_2$. Note that IO_3^- is a large anion.

12.30 Would naphthalene, $C_{10}H_8$, be more soluble in ethanol, C_2H_5OH, or in benzene, C_6H_6?

12.32 Indicate which of the following is more soluble in ethanol, C_2H_5OH: acetic acid, CH_3COOH, or stearic acid, $C_{17}H_{35}COOH$.

12.34 Which of the following ions would be expected to have the greater energy of hydration, F^- or Cl^-?

12.36 Explain the trends in solubility (grams per 100 mL of water) of the alkali metal fluorides and permanganates.

	Li^+	Na^+	K^+	Rb^+	Cs^+
F^-	0.27	4.2	92	131	367
MnO_4^-	71	Very soluble	6.4	0.5	0.097

12.37 The solubility of carbon dioxide in water is 0.161 g CO_2 in 100 mL of water of 20°C and 1.00 atm. A soft drink is carbonated with carbon dioxide gas at 5.50 atm pressure. What is the solubility of carbon dioxide in water at this pressure?

12.38 Nitrogen, N_2, is soluble in blood and can cause intoxication at sufficient concentration. For this reason, the U.S. Navy advises divers using compressed air not to go below 125 feet. The total pressure at this depth is 4.79 atm. If the solubility of nitrogen at 1.00 atm is 1.75×10^{-3} g/100 mL of water, and the mole percent of nitrogen in air is 78.1, what is the solubility of nitrogen in water from air at 4.79 atm?

SOLUTION CONCENTRATION

12.39 How would you prepare 145 g of an aqueous solution that is 2.50% by mass of potassium iodide, KI?

12.41 What mass of solution containing 2.50% by mass of potassium iodide, KI, contains 258 mg KI?

12.43 Vanillin, $C_8H_8O_3$, occurs naturally in vanilla extract and is used as a flavoring agent. A 37.2-mg sample of vanillin was dissolved in 168.5 mg of diphenyl ether, $(C_6H_5)_2O$. What is the molality of vanillin in the solution?

12.45 Fructose, $C_6H_{12}O_6$, is a sugar occurring in honey and fruits. The sweetest sugar, it is nearly twice as sweet as sucrose (cane or beet sugar). How much water should be added to 1.75 g of fructose to give a 0.125 *m* solution?

12.40 How would you prepare 455 g of an aqueous solution that is 6.50% by mass of sodium sulfate, Na_2SO_4?

12.42 What mass of solution containing 6.50% by mass of sodium sulfate, Na_2SO_4, contains 1.50 g Na_2SO_4?

12.44 Lauryl alcohol, $C_{12}H_{25}OH$, is prepared from coconut oil; it is used to make sodium lauryl sulfate, a synthetic detergent. What is the molality of lauryl alcohol in a solution containing 15.6 g of lauryl alcohol dissolved in 148 g of ethanol, C_2H_5OH?

12.46 Caffeine, $C_8H_{10}N_4O_2$, is a stimulant found in tea and coffee. A sample of the substance was dissolved in 45.0 g of chloroform, $CHCl_3$, to give a 0.0946 *m* solution. How many grams of caffeine were in the sample?

12.47 A 100.0-g sample of a brand of rubbing alcohol contains 65.0 g of isopropyl alcohol, C_3H_7OH, and 35.0 g of water. What is the mole fraction of isopropyl alcohol in the solution? What is the mole fraction of water?

12.48 An automobile antifreeze solution contains 2.25 kg of ethylene glycol, CH_2OHCH_2OH, and 2.00 kg of water. Find the mole fraction of ethylene glycol in this solution. What is the mole fraction of water?

12.49 A bleaching solution contains sodium hypochlorite, NaClO, dissolved in water. The solution is 0.750 *m* NaClO. What is the mole fraction of sodium hypochlorite?

12.50 An antiseptic solution contains hydrogen peroxide, H_2O_2, in water. The solution is 0.655 *m* H_2O_2. What is the mole fraction of hydrogen peroxide?

12.51 Concentrated hydrochloric acid contains 1.00 mol HCl dissolved in 3.31 mol H_2O. What is the mole fraction of HCl in concentrated hydrochloric acid? What is the molal concentration of HCl?

12.52 Concentrated aqueous ammonia contains 1.00 mol NH_3 dissolved in 2.44 mol H_2O. What is the mole fraction of NH_3 in concentrated aqueous ammonia? What is the molal concentration of NH_3?

12.53 Oxalic acid, $H_2C_2O_4$, occurs as the potassium or calcium salt in many plants, including rhubarb and spinach. An aqueous solution of oxalic acid is 0.585 *m* $H_2C_2O_4$. The density of the solution is 1.022 g/mL. What is the molar concentration?

12.54 Citric acid, $H_3C_6H_5O_7$, occurs in plants. Lemons contain 5% to 8% citric acid by mass. The acid is added to beverages and candy. An aqueous solution is 0.710 *m* in citric acid. The density is 1.049 g/mL. What is the molar concentration?

12.55 A solution of vinegar is 0.763 *M* in acetic acid, $HC_2H_3O_2$. The density of the vinegar is 1.004 g/mL. What is the molal concentration of acetic acid?

12.56 A beverage contains tartaric acid, $H_2C_4H_4O_6$, a substance obtained from grapes during wine making. If the beverage is 0.271 *M* in tartaric acid, what is the molal concentration? The density of the solution is 1.016 g/mL.

COLLIGATIVE PROPERTIES

12.57 Calculate the vapor pressure at 35°C of a solution made by dissolving 20.2 g of sucrose, $C_{12}H_{22}O_{11}$, in 60.5 g of water. The vapor pressure of pure water at 35°C is 42.2 mmHg. What is the vapor-pressure lowering of the solution? (Sucrose is nonvolatile.)

12.58 What is the vapor pressure at 23°C of a solution of 1.20 g of naphthalene, $C_{10}H_8$, in 25.6 g of benzene, C_6H_6? The vapor pressure of pure benzene at 23°C is 86.0 mmHg; the vapor pressure of naphthalene can be neglected. Calculate the vapor-pressure lowering of the solution.

12.59 What is the boiling point of a solution of 0.152 g of glycerol, $C_3H_8O_3$, in 20.0 g of water? What is the freezing point?

12.60 A solution was prepared by dissolving 0.915 g of sulfur, S_8, in 100.0 g of acetic acid, $HC_2H_3O_2$. Calculate the freezing point and boiling point of the solution.

12.61 An aqueous solution of a molecular compound freezes at −0.086°C. What is the molality of the solution?

12.62 Urea, $(NH_2)_2CO$, is dissolved in 100.0 g of water. The solution freezes at −0.085°C. How many grams of urea were dissolved to make this solution?

12.63 A 0.0182-g sample of an unknown substance was dissolved in 2.135 g of benzene. The molality of this solution, determined by freezing-point depression, was 0.0698. What is the molecular weight of the unknown substance?

12.64 A solution contains 0.0653 g of a compound in 9.75 g of ethanol. The molality of the solution is 0.0368. Calculate the molecular weight of the compound.

12.65 Safrole is contained in oil of sassafras and was once used to flavor root beer. A 2.39-mg sample of safrole was dissolved in 103.0 mg of diphenyl ether. The solution had a melting point of 25.70°C. Calculate the molecular weight of safrole. The freezing point of pure diphenyl ether is 26.84°C, and the freezing-point-depression constant, K_f, is 8.00°C/*m*.

12.66 Butylated hydroxytoluene (BHT) is used as an antioxidant in processed foods (it prevents fats and oils from becoming rancid). A solution of 2.500 g of BHT in 100.0 g of benzene had a freezing point of 4.880°C. What is the molecular weight of BHT?

12.67 Dextran is a polymeric carbohydrate produced by certain bacteria. It is used as a blood plasma substitute. An aqueous solution contains 0.582 g of dextran in 106 mL of solution at 21°C. It has an osmotic pressure of 1.47 mmHg. What is the average molecular weight of the dextran?

12.68 Arginine vasopressin is a pituitary hormone. It helps regulate the amount of water in the blood by reducing the flow of urine from the kidneys. An aqueous solution containing 21.6 mg of vasopressin in 100.0 mL of solution had an osmotic pressure at 25°C of 3.70 mmHg. What is the molecular weight of the hormone?

12.69 What is the freezing point of 0.0085 *m* aqueous calcium chloride, $CaCl_2$? Use the formula of the salt to obtain *i*.

12.70 What is the freezing point of 0.0095 *m* aqueous sodium phosphate, Na_3PO_4? Use the formula of the salt to obtain *i*.

12.71 A 0.0140-g sample of an ionic compound with the formula $Cr(NH_3)_5Cl_3$ was dissolved in water to give 25.0 mL of solution at 25°C. The osmotic pressure was determined to be 119 mmHg. How many ions are obtained from each formula unit when the compound is dissolved in water?

12.72 In a mountainous location, the boiling point of pure water is found to be 95°C. How many grams of sodium chloride must be added to 1 kg of water to bring the boiling point back to 100°C? Assume that $i = 2$.

COLLOIDS

12.73 Give the type of colloid (aerosol, foam, emulsion, sol, or gel) that each of the following represents.
(a) rain cloud (b) milk of magnesia
(c) soapsuds (d) silt in water

12.74 Give the type of colloid (aerosol, foam, emulsion, sol, or gel) that each of the following represents.
(a) ocean spray (b) beaten egg white
(c) dust cloud (d) salad dressing

12.75 Arsenic(III) sulfide forms a sol with a negative charge. Which of the following ionic substances should be most effective in coagulating the sol?
(a) KCl (b) $MgCl_2$ (c) $Al_2(SO_4)_3$ (d) Na_3PO_4

12.76 Aluminum hydroxide forms a positively charged sol. Which of the following ionic substances should be most effective in coagulating the sol?
(a) NaCl (b) $CaCl_2$ (c) $Fe_2(SO_4)_3$ (d) K_3PO_4

Additional Problems

12.77 A gaseous mixture consists of 80.0 mole percent N_2 and 20.0 mole percent O_2 (the approximate composition of air). Suppose water is saturated with the gas mixture at 25°C and 1.00 atm total pressure, and then the gas is expelled from the water by heating. What is the composition in mole fractions of the gas mixture that is expelled? The solubilities of N_2 and O_2 at 25°C and 1.00 atm are 0.0175 g/L H_2O and 0.0393 g/L H_2O, respectively.

12.78 A natural gas mixture consists of 90.0 mole percent CH_4 (methane) and 10.0 mole percent C_2H_6 (ethane). Suppose water is saturated with the gas mixture at 20°C and 1.00 atm total pressure, and the gas is then expelled from the water by heating. What is the composition in mole fractions of the gas mixture that is expelled? The solubilities of CH_4 and C_2H_6 at 20°C and 1.00 atm are 0.023 g/L H_2O and 0.059 g/L H_2O, respectively.

12.79 An aqueous solution is 8.50% ammonium chloride, NH_4Cl, by mass. The density of the solution is 1.024 g/mL. What are the molality, mole fraction, and molarity of NH_4Cl in the solution?

12.80 An aqueous solution is 22.0% lithium chloride, LiCl, by mass. The density of the solution is 1.127 g/mL. What are the molality, mole fraction, and molarity of LiCl in the solution?

12.81 A 58-g sample of a gaseous fuel mixture contains 0.43 mole fraction propane, C_3H_8; the remainder of the mixture is butane, C_4H_{10}. What are the masses of propane and butane in the sample?

12.82 The diving atmosphere used by the U.S. Navy in its undersea Sea-Lab experiments consisted of 0.036 mole fraction O_2 and 0.056 mole fraction N_2, with helium (He) making up the remainder. What are the masses of nitrogen, oxygen, and helium in a 7.84-g sample of this atmosphere?

12.83 A liquid solution consists of 0.35 mole fraction ethylene dibromide, $C_2H_4Br_2$, and 0.65 mole fraction propylene dibromide, $C_3H_6Br_2$. Both ethylene dibromide and propylene dibromide are volatile liquids; their vapor pressures at 85°C are 173 mmHg and 127 mmHg, respectively. Assume that each compound follows Raoult's law in the solution. Calculate the total vapor pressure of the solution.

12.84 What is the total vapor pressure at 20°C of a liquid solution containing 0.25 mole fraction benzene, C_6H_6, and 0.75 mole fraction toluene, $C_6H_5CH_3$? Assume that Raoult's law holds for each component of the solution. The vapor pressure of pure benzene at 20°C is 75 mmHg; that of toluene at 20°C is 22 mmHg.

12.85 Urea, $(NH_2)_2CO$, has been used to melt ice from sidewalks, because the use of salt is harmful to plants. If the saturated aqueous solution contains 44% urea by mass, what is the freezing point? (The answer will be approximate because the equation in the text applies accurately only to dilute solutions.)

12.86 Calcium chloride, $CaCl_2$, has been used to melt ice from roadways. Given that the saturated solution is 32% $CaCl_2$ by mass, estimate the freezing point.

12.87 The osmotic pressure of blood at 37°C is 7.7 atm. A solution that is given intravenously must have the same osmotic pressure as the blood. What should be the molarity of a glucose solution to give an osmotic pressure of 7.7 atm at 37°C?

12.88 Maltose, $C_{12}H_{22}O_{11}$, is a sugar produced by malting (sprouting) grain. A solution of maltose at 25°C has an osmotic pressure of 5.61 atm. What is the molar concentration of maltose?

12.89 Which aqueous solution has the lower freezing point, 0.10 *m* $CaCl_2$ or 0.10 *m* glucose?

12.90 Which aqueous solution has the lower boiling point, 0.10 *m* KCl or 0.10 *m* $CaCl_2$?

Cumulative-Skills Problems

12.91 The lattice enthalpy of sodium chloride, $\Delta H°$ for $NaCl(s) \longrightarrow Na^+(g) + Cl^-(g)$, is 787 kJ/mol; the heat of solution in making up 1 *M* NaCl(*aq*) is +4.0 kJ/mol. From these data, obtain the sum of the heats of hydration of Na^+ and Cl^-. That is, obtain the sum of $\Delta H°$ values for

$$Na^+(g) \longrightarrow Na^+(aq)$$
$$Cl^-(g) \longrightarrow Cl^-(aq)$$

If the heat of hydration of Cl^- is −338 kJ/mol, what is the heat of hydration of Na^+?

12.92 The lattice enthalpy of potassium chloride is 717 kJ/mol; the heat of solution in making up 1 *M* KCl(*aq*) is +18.0 kJ/mol. Using the value for the heat of hydration of Cl^- given in Problem 12.91, obtain the heat of hydration of K^+. Compare this with the value you obtained for Na^+ in Problem 12.91. Explain the relative values of Na^+ and K^+.

12.93 A solution is made up by dissolving 15.0 g of $MgSO_4 \cdot 7H_2O$ in 100.0 g of water. What is the molality of $MgSO_4$ in this solution?

12.94 A solution is made up by dissolving 15.0 g of $Na_2CO_3 \cdot 10H_2O$ in 100.0 g of water. What is the molality of Na_2CO_3 in this solution?

12.95 An aqueous solution is 15.0% by mass of copper(II) sulfate pentahydrate, $CuSO_4 \cdot 5H_2O$. What is the molarity of $CuSO_4$ in this solution at 20°C? The density of this solution at 20°C is 1.167 g/mL.

12.96 An aqueous solution is 20.0% by mass of sodium thiosulfate pentahydrate, $Na_2S_2O_3 \cdot 5H_2O$. What is the molarity of $Na_2S_2O_3$ in this solution at 20°C? The density of this solution at 20°C is 1.174 g/mL.

12.97 The freezing point of 0.0830 *m* aqueous acetic acid is −0.159°C. Acetic acid, $HC_2H_3O_2$, is partially dissociated according to the equation

$$HC_2H_3O_2(aq) \rightleftharpoons H^+(aq) + C_2H_3O_2^-(aq)$$

Calculate the percentage of $HC_2H_3O_2$ molecules that are dissociated, assuming the equation for the freezing-point depression holds for the total concentration of molecules and ions in the solution.

12.98 The freezing point of 0.109 *m* aqueous formic acid is −0.210°C. Formic acid, $HCHO_2$, is partially dissociated according to the equation

$$HCHO_2(aq) \rightleftharpoons H^+(aq) + CHO_2^-(aq)$$

Calculate the percentage of $HCHO_2$ molecules that are dissociated, assuming the equation for the freezing-point depression holds for the total concentration of molecules and ions in the solution.

12.99 A compound of carbon, hydrogen, and oxygen was burned in oxygen, and 1.000 g of the compound produced 1.434 g CO_2 and 0.783 g H_2O. In another experiment, 0.1107 g of the compound was dissolved in 25.0 g of water. This solution had a freezing point of −0.0894°C. What is the molecular formula of the compound?

12.100 A compound of carbon, hydrogen, and oxygen was burned in oxygen, and 1.000 g of the compound produced 1.418 g CO_2 and 0.871 g H_2O. In another experiment, 0.1103 g of the compound was dissolved in 45.0 g of water. This solution had a freezing point of −0.0734°C. What is the molecular formula of the compound?

13

Chemical Reactions: Acid–Base and Oxidation–Reduction Concepts

Chapter Outline

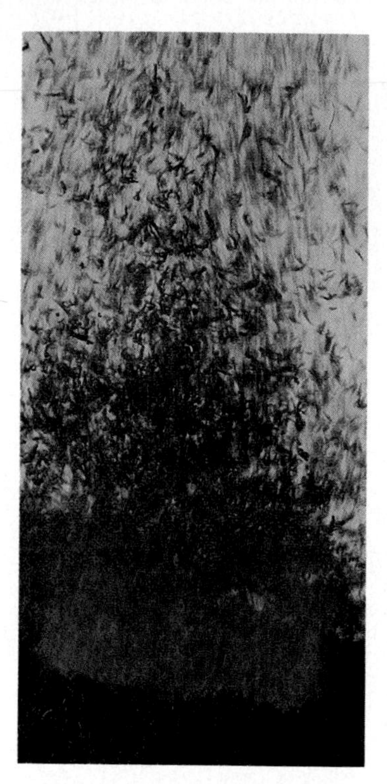

Volcano reaction (decomposition of ammonium dichromate).

Acids and bases were first recognized by simple properties, such as taste. Acids have a sour taste, whereas bases are bitter. Also, acids and bases change the color of certain dyes called indicators, such as litmus and phenolphthalein. Acids change litmus from blue to red and basic phenolphthalein from red to colorless. Bases change litmus from red to blue and phenolphthalein from colorless to pink. As we can see from these color changes, acids and bases neutralize, or reverse, the action of one another. During neutralization, acids and bases react with each other to produce ionic substances called salts. Acids react with active metals, such as magnesium and zinc, to release hydrogen.

Arrhenius framed the first successful concept of acids and bases. He defined acids and bases in terms of the effect these substances have on water. According to Arrhenius, acids are substances that increase the concentration of H^+ ion in aqueous solution, and bases increase the concentration of OH^- ion in aqueous solution. But many reactions that have characteristics of acid–base reactions in aqueous solution occur in other solvents or without a solvent. For example, hydrochloric acid reacts with aqueous ammonia, which in the Arrhenius view is a base because it increases the concentration of OH^- ion in aqueous solution. The reaction can be written

$$HCl(aq) + NH_3(aq) \longrightarrow NH_4Cl(aq)$$

Figure 13.1
Reaction of HCl(*g*) and NH_3(*g*) to form NH_4Cl(*s*). Gases from the concentrated solutions diffuse from their watch glasses (shallow dishes) and react to give a smoke of ammonium chloride.

The product is a solution of NH_4Cl—that is, a solution of NH_4^+ and Cl^- ions. A very similar reaction occurs between hydrogen chloride and ammonia dissolved in benzene, C_6H_6. The product is again NH_4Cl, which in this case precipitates from the solution.

$$HCl(benzene) + NH_3(benzene) \longrightarrow NH_4Cl(s)$$

Hydrogen chloride and ammonia even react in the gas phase. If watch glasses (shallow glass dishes) of concentrated hydrochloric acid and concentrated ammonia are placed next to each other, dense white fumes of NH_4Cl form where HCl gas and NH_3 gas come into contact (Figure 13.1).

$$HCl(g) + NH_3(g) \longrightarrow NH_4Cl(s)$$

These reactions of HCl and NH_3, in benzene and in the gas phase, are similar to the reaction in aqueous solution but cannot be explained by the Arrhenius concept. Broader acid–base concepts are needed.

In the first part of this chapter, we will discuss the Arrhenius, the Brønsted–Lowry, and the Lewis concepts of acids and bases. The Brønsted–Lowry and Lewis concepts apply to nonaqueous as well as aqueous solutions and also enlarge on the Arrhenius concept in other ways.

Acid–Base Concepts

Lavoisier was one of the first chemists to try to explain what makes a substance acidic. In 1777 he proposed that oxygen was an essential element in acids. (*Oxygen,* which he named, means "acid-former" in Greek.) But in 1808 Humphry Davy showed that hydrogen chloride, which dissolves in water to give hydrochloric acid, contains only hydrogen and chlorine. Although some chemists argued that chlorine was a compound of oxygen, chlorine was eventually proved an element. Chemists then noted that hydrogen, not oxygen, must be the essential constituent of acids. The cause of acidity and basicity was first explained in 1884 by Svante Arrhenius.

13.1 ARRHENIUS CONCEPT OF ACIDS AND BASES

A modern statement of the *Arrhenius concept* of acids and bases is as follows: *An acid is a substance that, when dissolved in water, increases the concentration of hydrogen ion, $H^+(aq)$. A base is a substance that, when dissolved in water, increases the concentration of hydroxide ion, $OH^-(aq)$.* The hydrogen ion, $H^+(aq)$, is not a bare proton but a proton chemically bonded to water—that is, $H_3O^+(aq)$. This species, called the hydronium ion, is itself associated through hydrogen bonding with a variable number of water molecules. A possibility is shown in Figure 13.2.

The special role of the hydronium ion (or hydrogen ion) and the hydroxide ion in aqueous solutions arises from the following reaction:

$$H_2O(l) + H_2O(l) \rightleftharpoons H_3O^+(aq) + OH^-(aq)$$

Thus, the addition of acids and bases alters the concentrations of these ions in water.

In Arrhenius's theory, a *strong acid* is a substance that completely ionizes in aqueous solution to give $H_3O^+(aq)$ and an anion. An example is perchloric acid, $HClO_4$.

$$HClO_4(aq) + H_2O(l) \longrightarrow H_3O^+(aq) + ClO_4^-(aq)$$

Other examples of strong acids are H_2SO_4, HI, HBr, HCl, and HNO_3. A *strong base* completely ionizes in aqueous solution to give OH^- and a cation. Sodium hydroxide is an example of a strong base.

$$NaOH(s) \xrightarrow{H_2O} Na^+(aq) + OH^-(aq)$$

The principal strong bases are the hydroxides of Group IA elements and Group IIA elements (except Be).●

● See Table 3.2 for a list of strong acids and bases.

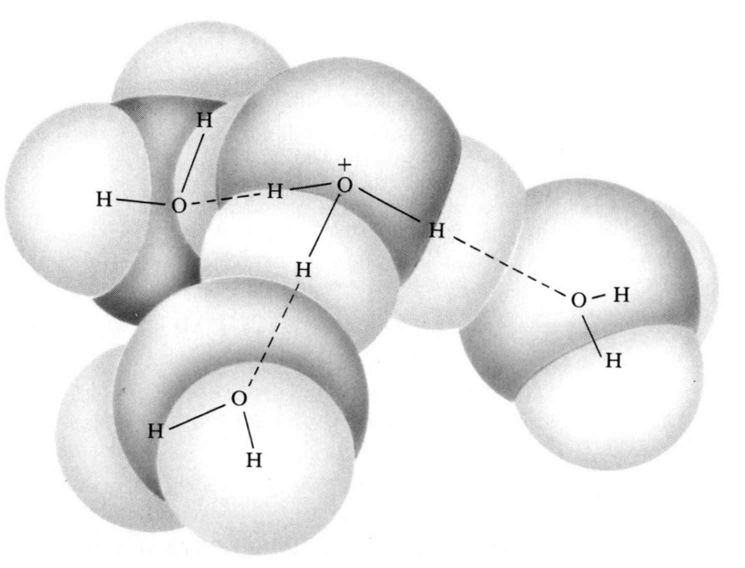

Figure 13.2
The hydronium ion, H_3O^+. It is shown here hydrogen-bonded to three water molecules.

Most of the other acids and bases we encounter are *weak*. They are not completely ionized in solution and exist in equilibrium with the corresponding ions. Consider acetic acid, $HC_2H_3O_2$. The reaction is

$$HC_2H_3O_2(aq) + H_2O(l) \rightleftharpoons H_3O^+(aq) + C_2H_3O_2^-(aq)$$

Evidence for the Arrhenius theory comes from the heat of reaction, ΔH°, for neutralization of a strong acid by a strong base. This neutralization is essentially the reaction of $H_3O^+(aq)$ with $OH^-(aq)$ and should therefore always give the same ΔH° per mole of water formed. For example, if we write the neutralization of $HClO_4$ with NaOH in ionic form, we have

$$H_3O^+(aq) + \cancel{ClO_4^-(aq)} + \cancel{Na^+(aq)} + OH^-(aq) \longrightarrow \cancel{Na^+(aq)} + \cancel{ClO_4^-(aq)} + 2H_2O(l)$$

After canceling, we get the net ionic equation

$$H_3O^+(aq) + OH^-(aq) \longrightarrow 2H_2O(l)$$

Experimentally, it is found that all neutralizations involving strong acids and bases have the same ΔH°: -55.90 kJ per mole of H^+. This indicates that the same reaction occurs in each neutralization, as Arrhenius's theory predicts.

Despite its successes, the Arrhenius concept is limited. In addition to looking at acid–base reactions only in aqueous solutions, it singles out the OH^- ion as the source of base character, when other species can play a similar role. Broader definitions of acids and bases are described in the next section.

13.2 BRØNSTED–LOWRY CONCEPT OF ACIDS AND BASES

In 1923 the Danish chemist Johannes N. Brønsted (1879–1947) and, independently, the British chemist Thomas M. Lowry (1874–1936) pointed out that acid–base reactions can be seen as proton-transfer reactions and that acids and bases can be defined in terms of this proton (H^+) transfer. According to the Brønsted–Lowry concept, an **acid** is *the species donating a proton in a proton-transfer reaction*. A **base** is *the species accepting the proton in a proton-transfer reaction*.●

● To be precise, we should say *hydrogen nucleus* instead of proton (because natural hydrogen contains some ^2H as well as ^1H). The term *proton* is conventional in this context, however.

Consider, for example, the reaction of hydrochloric acid with ammonia, which was mentioned in the chapter opening. Writing it as an ionic equation, we have

$$H_3O^+(aq) + \cancel{Cl^-(aq)} + NH_3(aq) \longrightarrow H_2O(l) + NH_4^+(aq) + \cancel{Cl^-(aq)}$$

After canceling Cl^-, we get the net ionic equation.

$$\text{(H)}_3O^+(aq) + NH_3(aq) \longrightarrow H_2O(l) + NH_4^+(aq)$$

In this reaction in aqueous solution, a proton, H^+, is transferred from the H_3O^+ ion to the NH_3 molecule, giving H_2O and NH_4^+ (see Figure 13.3). Here H_3O^+ is

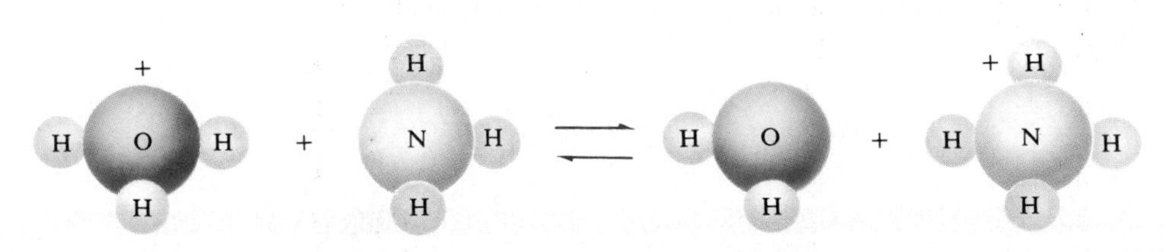

Figure 13.3
A representation of the reaction $H_3O^+ + NH_3 \rightleftharpoons H_2O + NH_4^+$. Note the transfer of a proton, H^+, from H_3O^+ to NH_3. The charges indicated for ions are *overall* charges. They are not to be associated with specific locations on the ions.

the proton donor, or acid, and NH_3 is the proton acceptor, or base. Note that in the Brønsted–Lowry concept, acids (and bases) can be ions as well as molecular substances.

The Brønsted–Lowry concept can also be applied to the reaction of HCl and NH_3 dissolved in benzene, C_6H_6, which was mentioned in the chapter opening. In benzene, HCl and NH_3 are not ionized. The equation is

$$\underset{\text{acid}}{\text{HCl}(benzene)} + \underset{\text{base}}{\text{NH}_3(benzene)} \longrightarrow \text{NH}_4\text{Cl}(s)$$

Here the HCl molecule is the proton donor, or acid, and the NH_3 molecule is the proton acceptor, or base.

In any acid–base equilibrium, both forward and reverse reactions involve proton transfers. Consider the reaction of NH_3 with H_2O.

$$\underset{\text{base}}{\text{NH}_3(aq)} + \underset{\text{acid}}{\text{H}_2\text{O}(l)} \rightleftharpoons \underset{\text{acid}}{\text{NH}_4^+(aq)} + \underset{\text{base}}{\text{OH}^-(aq)}$$

In the forward reaction, NH_3 accepts a proton from H_2O. Thus, NH_3 is a base and H_2O is an acid. In the reverse reaction, NH_4^+ donates a proton to OH^-. The NH_4^+ ion is the acid and OH^- is the base.

Note that NH_3 and NH_4^+ differ by a proton. That is, NH_3 becomes the NH_4^+ ion by gaining a proton, whereas the NH_4^+ ion becomes the NH_3 molecule by losing a proton. The species NH_4^+ and NH_3 are a conjugate acid–base pair. A **conjugate acid–base pair** consists of *two species in an acid–base reaction, one acid and one base, that differ by the gain or loss of a proton*. The acid in such a pair is called the *conjugate acid* of the base, whereas the base is the *conjugate base* of the acid. Thus, NH_4^+ is the conjugate acid of NH_3, and NH_3 is the conjugate base of NH_4^+.

Example 13.1 Identifying Acid and Base Species

In the following equations, label each of the species as an acid or a base. Show the conjugate acid–base pairs.

(a) $HCO_3^-(aq) + HF(aq) \rightleftharpoons H_2CO_3(aq) + F^-(aq)$
(b) $HCO_3^-(aq) + OH^-(aq) \rightleftharpoons CO_3^{2-}(aq) + H_2O(l)$

Solution

(a) Examine the equation to find the proton donor on each side. On the left, HF is the proton donor; on the right, H_2CO_3 is the proton donor. The proton acceptors are HCO_3^- and F^-. Once the proton donors and acceptors are identified, the acids and bases can be labeled.

$$\underset{\text{base}}{\text{HCO}_3^-(aq)} + \underset{\text{acid}}{\text{HF}(aq)} \rightleftharpoons \underset{\text{acid}}{\text{H}_2\text{CO}_3(aq)} + \underset{\text{base}}{\text{F}^-(aq)}$$

In this reaction, H_2CO_3 and HCO_3^- are a conjugate acid–base pair, as are HF and F^-.

(b) We have

$$\underset{\text{acid}}{\text{HCO}_3^-(aq)} + \underset{\text{base}}{\text{OH}^-(aq)} \rightleftharpoons \underset{\text{base}}{\text{CO}_3^{2-}(aq)} + \underset{\text{acid}}{\text{H}_2\text{O}(l)}$$

Here HCO_3^- and CO_3^{2-} are a conjugate acid–base pair, as are H_2O and OH^-. Note that although HCO_3^- functions as an acid in this reaction, it functions as a base in (a).

Exercise 13.1

For the reaction

$$H_2CO_3(aq) + CN^-(aq) \rightleftharpoons HCN(aq) + HCO_3^-(aq)$$

label each species as an acid or a base. For the base on the left, what is the conjugate acid?
(See Problems 13.23 and 13.24.)

The Brønsted–Lowry concept defines a species as an acid or base according to its function in the acid–base, or proton-transfer, reaction. As we saw in Example 13.1, some species can act either as an acid or a base. An **amphiprotic species** is *a species that can act either as an acid or a base (it can lose or gain a proton)*, depending on the other reactant.● Thus, HCO_3^- acts as an acid in the presence of OH^- but as a base in the presence of HF. Anions with ionizable hydrogens, such as HCO_3^-, and certain solvents, such as water, are amphiprotic.

● *Amphoteric* is a general term referring to a species that can act as an acid or a base. The species need not be amphiprotic, however. Thus, aluminum oxide is an amphoteric oxide, because it reacts with acids and bases. It is not amphiprotic, however, because it has no protons.

The amphiprotic characteristic of water is important in the acid–base properties of aqueous solutions. Consider, for example, the reactions of water with the base NH_3 and with the acid $HC_2H_3O_2$ (acetic acid).

$$\underset{\text{base}}{NH_3(aq)} + \underset{\text{acid}}{H_2O(l)} \rightleftharpoons \underset{\text{acid}}{NH_4^+(aq)} + \underset{\text{base}}{OH^-(aq)}$$

$$\underset{\text{acid}}{HC_2H_3O_2(aq)} + \underset{\text{base}}{H_2O(aq)} \rightleftharpoons \underset{\text{base}}{C_2H_3O_2^-(aq)} + \underset{\text{acid}}{H_3O^+(aq)}$$

In the first case, water reacts as an acid with the base NH_3. In the second case, water reacts as a base with the acid $HC_2H_3O_2$.

We have now seen several ways in which the Brønsted–Lowry concept of acids and bases has greater scope than the Arrhenius concept. In the Brønsted–Lowry concept:

1. A base is a species that accepts protons; OH^- is only one example of a base.
2. Acids and bases can be ions as well as molecular substances.
3. Acid–base reactions are not restricted to aqueous solution.
4. Some species can act as either acids or bases, depending on what the other reactant is.

13.3 RELATIVE STRENGTHS OF ACIDS AND BASES

It is useful to consider an acid–base reaction as a competition for protons. From this point of view, we can order acids and bases by their relative strengths. The stronger acids are those that lose their protons more easily than others. Similarly, the stronger bases are those that hold on to protons more strongly than others.

Recall that an acid is strong if it completely ionizes in water. In the reaction of hydrogen chloride with water, for example, water acts as a base, accepting the proton from HCl.

$$\underset{\text{acid}}{HCl(aq)} + \underset{\text{base}}{H_2O(l)} \longrightarrow \underset{\text{base}}{Cl^-(aq)} + \underset{\text{acid}}{H_3O^+(aq)}$$

The reverse reaction occurs only to an extremely small extent. Because the reaction goes almost completely to the right, we say that HCl is a strong acid. However, even though the reaction goes almost completely to products, we can consider the reverse reaction. In it, the Cl^- ion acts as the base, accepting a proton from the acid H_3O^+.

Let us look at this reaction in terms of the relative strengths of the two acids in the equation, HCl and H_3O^+. Because HCl is a strong acid, it must lose its proton readily, more readily than H_3O^+ does. We can say that HCl is a stronger acid than H_3O^+ or that, of the two, H_3O^+ is the weaker acid.

$$\underset{\text{stronger acid}}{HCl(aq)} + H_2O(l) \rightleftharpoons Cl^-(aq) + \underset{\text{weaker acid}}{H_3O^+(aq)}$$

It is important to understand that we use the terms *stronger* and *weaker* here only in a comparative sense. The H_3O^+ ion is a relatively strong acid.

An acid–base reaction normally goes in the direction of the weaker acid. We can use this fact to compare the relative strengths of any two acids, as we did in comparing the relative strengths of HCl and H_3O^+. As another example, let us look at the ionization of acetic acid, $HC_2H_3O_2$, in water.

$$HC_2H_3O_2(aq) + H_2O(l) \rightleftharpoons C_2H_3O_2^-(aq) + H_3O^+(aq)$$

Experiment reveals that in a 0.1 M acetic acid solution, only about 1% of the acetic acid molecules have ionized by this reaction. This implies that $HC_2H_3O_2$ is a weaker acid than H_3O^+. If we look at the similar situation for 0.1 M HF, we find that about 3% of the HF molecules have dissociated. Thus, HF is a weaker acid than H_3O^+ but a stronger acid than $HC_2H_3O_2$. We have already concluded that HCl is stronger than H_3O^+. Thus, we have determined that the acid strengths for these four acids are in the order $HCl > H_3O^+ > HF > HC_2H_3O_2$.

This procedure of determining the relative order of acid strengths by comparing their relative ionizations in water cannot be used to obtain the relative strengths of two strong acids such as HCl and HI. When these acids are dissolved in water, they are essentially 100% ionized. However, if we look at solutions of equal concentrations of these acids in another solvent that is less basic than water—say pure acetic acid—we do see a difference. Neither acid is completely ionized, but a greater fraction of HI molecules is found to be ionized. Thus, HI is a stronger acid than HCl. In water, the acid strengths of the strong acids appear to be the same; that is, they are "leveled out." We say that water exhibits a *leveling effect* on the strengths of the strong acids.

The first column of Table 13.1 lists acids by their strength; the strongest is at the top of the table. Note that the arrow down the left side of the table points toward the direction of the weaker acid, in the direction the reaction goes. For example, in the reaction involving HCl and H_3O^+, the arrow points from HCl toward H_3O^+, the direction in which the reaction occurs.

This same reaction can also be viewed in terms of the bases, H_2O and Cl^-. A stronger base picks up a proton more readily than does a weaker one. Water has greater base strength than the Cl^- ion; that is, H_2O picks up protons more readily than Cl^- does. In fact, Cl^- has little attraction for protons. (If it had more, HCl would not lose its proton so readily.) Because H_2O molecules compete more successfully for protons than Cl^- ions do, the reaction goes almost completely to the right. (Note that the arrow on the right in Table 13.1 points from H_2O to Cl^-.) That

Table 13.1 Relative Strengths of Acids and Bases

	Acid	Base	
Strongest acids	$HClO_4$	ClO_4^-	Weakest bases
↓	H_2SO_4	HSO_4^-	↑
	HI	I^-	
	HBr	Br^-	
	HCl	Cl^-	
	HNO_3	NO_3^-	
	H_3O^+	H_2O	
	HSO_4^-	SO_4^{2-}	
	H_2SO_3	HSO_3^-	
	H_3PO_4	$H_2PO_4^-$	
	HNO_2	NO_2^-	
	HF	F^-	
	$HC_2H_3O_2$	$C_2H_3O_2^-$	
	$Al(H_2O)_6^{3+}$	$Al(H_2O)_5OH^{2+}$	
	H_2CO_3	HCO_3^-	
	H_2S	HS^-	
	$HClO$	ClO^-	
	$HBrO$	BrO^-	
	NH_4^+	NH_3	
	HCN	CN^-	
	HCO_3^-	CO_3^{2-}	
	H_2O_2	HO_2^-	
	HS^-	S^{2-}	
Weakest acids	H_2O	OH^-	Strongest bases

is, at the completion of the reaction concentrations of product species (Cl^- and H_3O^+) are much greater than the concentrations of reactant species.

$$\underset{}{HCl(aq)} + \underset{\text{stronger base}}{H_2O(l)} \longrightarrow \underset{\text{weaker base}}{Cl^-(aq)} + H_3O^+(aq)$$

The reaction goes in the direction of the weaker base.

By comparing reactions between different pairs of bases, we can arrive at a relative order for base strengths, just as we did for acids. A definite relationship exists between acid and base strengths. When we say an acid loses its proton readily, we can also say that its conjugate base does not hold the proton very tightly. Thus, *the strongest acids have the weakest conjugate bases, and the strongest bases have the weakest conjugate acids*. This means that a list of conjugate bases of the acids in Table 13.1 will be in order of increasing base strength. The weakest bases will be at the top of the table and the strongest at the bottom.

We can use Table 13.1 to predict the direction of an acid–base reaction. The direction for an acid–base reaction always favors the weaker acid and weaker base;

that is, the normal direction of reaction is from the stronger acid and base to the weaker acid and base. For the reaction we have been discussing, we have

$$\underset{\text{stronger acid}}{HCl(aq)} + \underset{\text{stronger base}}{H_2O(l)} \longrightarrow \underset{\text{weaker base}}{Cl^-(aq)} + \underset{\text{weaker acid}}{H_3O^+(aq)}$$

The reaction follows the direction of the arrows at the left and right of Table 13.1.

Example 13.2 Deciding Whether Reactants or Products Are Favored in an Acid–Base Reaction

For the reaction

$$SO_4^{2-}(aq) + HCN(aq) \rightleftharpoons HSO_4^-(aq) + CN^-(aq)$$

use the relative strengths of acids and bases (Table 13.1) to decide which species (reactants or products) are favored at the completion of the reaction.

Solution

If we compare the relative strengths of the two acids HCN and HSO_4^-, we see that HCN is weaker. Or, comparing the bases SO_4^{2-} and CN^-, we see that SO_4^{2-} is weaker. Hence, the reaction would normally go from right to left.

$$\underset{\text{weaker base}}{SO_4^{2-}(aq)} + \underset{\text{weaker acid}}{HCN(aq)} \longleftarrow \underset{\text{stronger acid}}{HSO_4^-(aq)} + \underset{\text{stronger base}}{CN^-(aq)}$$

Thus, the reactants are favored.

Exercise 13.2

Determine the direction of the following reaction from the relative strengths of acids and bases.

$$H_2S(aq) + C_2H_3O_2^-(aq) \rightleftharpoons HC_2H_3O_2(aq) + HS^-(aq)$$

(See Problems 13.25, 13.26, 13.27, and 13.28.)

Related Topic

STRENGTHS OF ACID SOLUTIONS; pH CONCEPT

We can judge the strengths of different acids by comparing hydronium-ion concentrations of solutions having the same acid concentration. The stronger acid will have the greater H_3O^+ concentration.

Hydronium-ion concentrations in acid solutions run from high values of about 1 *M* or so down to values near 1.0×10^{-7} *M*, which is the H_3O^+ concentration present in pure water. Solutions containing 1.0×10^{-7} *M* H_3O^+ are said to be *neutral.* In a solution containing a base, the hydronium-ion concentration is even smaller. Hydronium-ion concentrations in basic solutions run from high values near 1.0×10^{-7} *M* down to values on the order of 10^{-14} *M* or so.

For convenience, pH values are often reported in place of hydronium-ion concentrations. The pH is defined as the negative logarithm (to the base 10) of the hydronium-ion concentration. If we denote the hydronium-ion concentration as $[H_3O^+]$, then

$$pH = -\log [H_3O^+]$$

Thus, a neutral solution has a pH of 7.00 ($= -\log 1.0 \times 10^{-7}$), whereas an acidic solution has a pH less than 7.00, and a basic solution has a pH greater than 7.00. Some typical pH values are as follows: for lemon juice, 2.2–2.4; vinegar, 2.4–3.4; seawater, 7.0–8.3; milk of magnesia, 10.5; and household ammonia, 11.9. Lemon juice and vinegar are acidic, of course. Seawater is neutral to slightly basic, whereas milk of magnesia and household ammonia solutions are basic.

13.4 MOLECULAR STRUCTURE AND ACID STRENGTH

The strength of an acid depends on how easily the proton, H^+, is lost or removed from an H—X bond in the acid species. By understanding the factors that determine this ease of proton loss, we will be able to predict the relative strengths of similar acids.

Two factors are important in determining relative acid strengths. One is the polarity of the bond to which the H atom is attached. The H atom should have a positive partial charge:

$$\overset{\delta+}{\text{H}}\text{—}\overset{\delta-}{\text{X}}$$

The more polarized the bond is in this direction, the easier the proton is removed and the greater the acid strength. The second factor determining acid strength is the strength of the bond—that is, how tightly the proton is held. This in turn depends on the size of atom X. The larger atom X, the weaker the bond and the greater the acid strength.

Consider a series of binary acids, HX, formed from a given column of elements of the periodic table. The acids would be compounds of these elements with hydrogen, such as the binary acids of the Group VIIA elements: HF, HCl, HBr, and HI. As we go down the column of elements, each time adding a shell of electrons to the atom, the radius increases markedly. For this reason, the size of the atom X is the dominant factor in determining the acid strength. *In going down a column of elements of the periodic table, the size of atom X increases, the H—X bond strength decreases, and the strength of the binary acid increases*. We thus predict the following order of acid strength:

$$\text{HF} < \text{HCl} < \text{HBr} < \text{HI}$$

This is the same order shown in Table 13.1 for these acids.

As we go across a row of elements of the periodic table, the atomic radius decreases slowly. For this reason, the relative strengths of the binary acids of these elements are less dependent on the sizes of atoms X. Now the polarity of the H—X bond becomes the dominant factor in determining acid strength. *Going across a row of elements of the periodic table, the electronegativity increases, the H—X bond polarity increases, and the acid strength increases*. For example, the binary acids of the last two elements of the second period are H_2O and HF. The acid strengths are

$$H_2O < HF$$

This again is the order shown in Table 13.1. Hydrogen fluoride, HF, is a weak acid, and H_2O is a very weak acid.

Now consider the oxoacids. An oxoacid has the structure

$$\text{H—O—Y—}$$

The acidic H atom is always attached to an O atom, which in turn is attached to an atom Y. Other groups, such as O atoms or O—H groups, may be attached to Y. Bond polarity appears to be the dominant factor determining relative strengths of the oxoacids. This in turn depends on the electronegativity of atom Y. If the electronegativity of atom Y is large, the H—O bond is relatively polar and the acid strength large. *For a series of oxoacids of the same structure, differing only in the atom Y, the acid strength increases with the electronegativity of Y.* Consider, for example, the acids HClO, HBrO, and HIO.● The structures are

$$\mathrm{H{-}\ddot{\underset{..}{O}}{-}\ddot{\underset{..}{Cl}}{:}} \qquad \mathrm{H{-}\ddot{\underset{..}{O}}{-}\ddot{\underset{..}{Br}}{:}} \qquad \mathrm{H{-}\ddot{\underset{..}{O}}{-}\ddot{\underset{..}{I}}{:}}$$

● The formulas of these acids may be written HXO or HOX, depending on the convention used. Formulas of oxoacids are generally written with the acidic H atoms first, followed by the characteristic element (X), then O atoms. However, the formulas of molecules composed of three atoms are often written in the order in which the atoms are bonded, which in this case is HOX.

The electronegativity of Group VIIA elements decreases going down the column of elements, so the electronegativity decreases from Cl to Br to I. Therefore, the order of acid strengths is

$$HIO < HBrO < HClO$$

For a series of oxoacids, $(HO)_mYO_n$, the acid strength increases with n, *the number of O atoms bonded to Y (excluding O atoms in OH groups).* The oxoacids of chlorine provide an example. Their structures are

$$\mathrm{H{-}\ddot{\underset{..}{O}}{-}Cl} \qquad \mathrm{H{-}\ddot{\underset{..}{O}}{-}Cl{-}\ddot{\underset{..}{O}}{:}} \qquad \mathrm{H{-}\ddot{\underset{..}{O}}{-}\overset{:\ddot{O}:}{\overset{|}{Cl}}{-}\ddot{\underset{..}{O}}{:}} \qquad \mathrm{H{-}\ddot{\underset{..}{O}}{-}\underset{:\underset{..}{O}:}{\underset{|}{\overset{:\ddot{O}:}{\overset{|}{Cl}}}}{-}\ddot{\underset{..}{O}}{:}}$$

With each additional O atom, the Cl atom becomes effectively more electronegative. As a result, the H atom becomes more acidic. Thus, the acid strengths increase in the following order:

$$HClO < HClO_2 < HClO_3 < HClO_4$$

Before we leave the subject of molecular structure and acid strength, let us look at the relative acid strengths of a polyprotic acid and its corresponding acid anions. For example, H_2SO_4 ionizes by losing a proton to give HSO_4^-, which in turn ionizes to give SO_4^{2-}. HSO_4^- can lose a proton, so it is acidic. However, because of the negative charge of the ion, which tends to attract protons, its acid strength is reduced from that of the uncharged species. That is, the acid strengths are in the order

$$HSO_4^- < H_2SO_4$$

This shows us that *the acid strength of a polyprotic acid and its anions decreases with increasing negative charge* (see Table 13.1).

Exercise 13.3

Which member of each of the following pairs is the stronger acid? (a) NH_3, PH_3 (b) HI, H_2Te (c) HSO_3^-, H_2SO_3 (d) H_3AsO_4, H_3AsO_3 (e) HSO_4^-, $HSeO_4^-$

(See Problems 13.31 and 13.32.)

13.5 LEWIS CONCEPT OF ACIDS AND BASES

Certain reactions have characteristics of acid–base reactions but do not fit the Brønsted–Lowry concept. An example is the reaction of the basic oxide Na_2O with the acidic oxide SO_3 to give the salt Na_2SO_4.

$$Na_2O(s) + SO_3(g) \longrightarrow Na_2SO_4(s)$$

G. N. Lewis, who proposed the electron-pair theory of covalent bonding, realized that the concept of acids and bases could be generalized to include reactions of acidic and basic oxides and many other reactions, as well as proton-transfer reactions. According to this concept, a **Lewis acid** is *a species that can form a covalent bond by accepting an electron pair from another species;* a **Lewis base** is *a species* *that can form a covalent bond by donating an electron pair to another species*. The Lewis and the Brønsted–Lowry concepts are simply different ways of looking at certain chemical reactions. Such different views are often helpful in devising new reactions.

Consider again the neutralization of NH_3 by HCl in aqueous solution, mentioned in the chapter opening. The reaction consists of a transfer of a proton from H_3O^+ to NH_3. The transfer of the proton to NH_3 can be written as follows:

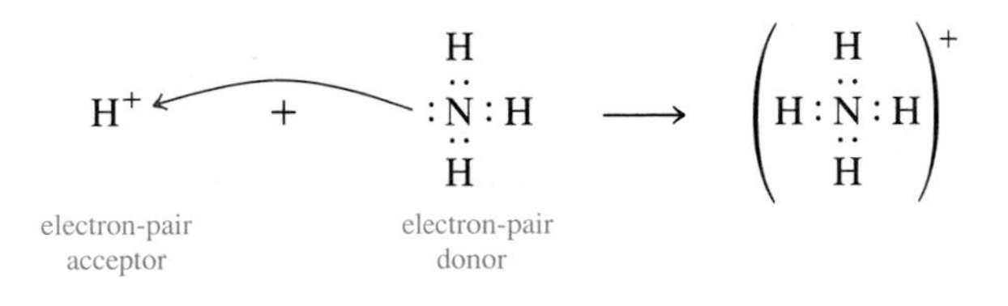

Here the red arrow shows the proton accepting an electron pair from NH_3 and an H—N bond being formed. The proton is an electron-pair acceptor, so it is a Lewis acid. Ammonia, NH_3, which has a lone pair of electrons, is an electron-pair donor and therefore a Lewis base.

Now let us look at the reaction of Na_2O with SO_3. It involves the reaction of the oxide ion, O^{2-}, from the ionic solid, Na_2O, with SO_3.

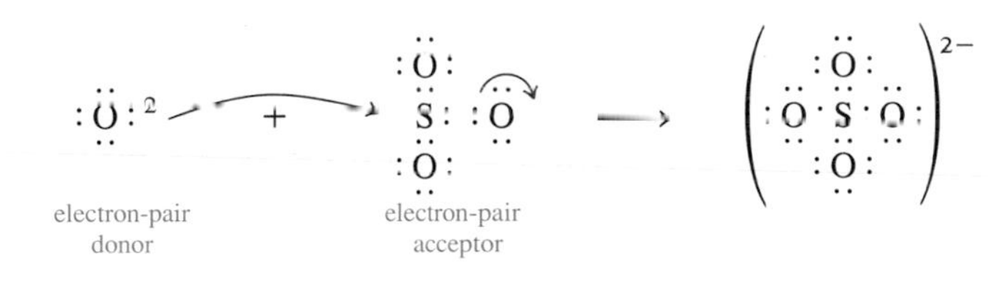

Here SO_3 accepts the electron pair from the O^{2-} ion. At the same time, an electron pair from the S═O bond moves to the O atom. Thus, O^{2-} is the Lewis base and SO_3 the Lewis acid.

The formation of *complex ions* can also be looked at as Lewis acid–base reactions. Complex ions are formed when a metal ion bonds to electron pairs from molecules such as H_2O or NH_3 or from anions such as $:C{\equiv}N:^-$. An example of a complex ion is $Al(H_2O)_6{}^{3+}$. Hydrated ions like $Al(H_2O)_6{}^{3+}$ are present in compounds (hydrates) and in aqueous solution. The formation of a hydrated metal ion, such as $Al(H_2O)_6{}^{3+}$, involves a Lewis acid–base reaction.

$$\underset{\text{Lewis acid}}{Al^{3+}} + \underset{\text{Lewis base}}{6(:\ddot{O}(H)—H)} \longrightarrow Al(:\ddot{O}(H)—H)_6{}^{3+}$$

Example 13.3 Identifying Lewis Acid and Base Species

In the following reactions, identify the Lewis acid and the Lewis base.

(a) $Ag^+ + 2NH_3 \rightleftharpoons Ag(NH_3)_2{}^+$

(b) $B(OH)_3 + H_2O \rightleftharpoons B(OH)_4{}^- + H^+$

Solution

We write out the equations using Lewis electron-dot formulas and identify the electron-pair acceptor, or Lewis acid, and electron-pair donor, or Lewis base.

(continued)

(a) The silver ion, Ag^+, forms a complex ion with two NH_3 molecules.

$$\underset{\text{Lewis acid}}{Ag^+} + \underset{\text{Lewis base}}{2\,{:}NH_3} \rightleftharpoons Ag({:}NH_3)_2{}^+$$

(b) The reaction is

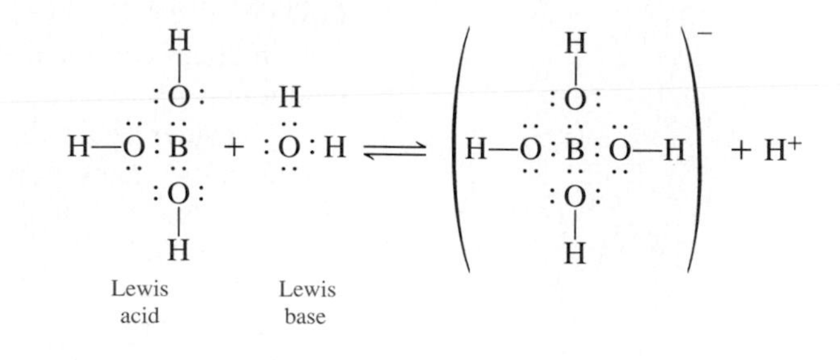

Exercise 13.4

Identify the Lewis acid and the Lewis base in each of the following reactions. Write out the chemical equations using electron-dot formulas.

(a) $BF_3 + {:}NH_3 \longrightarrow F_3B{:}NH_3$ (b) $O^{2-} + CO_2 \longrightarrow CO_3{}^{2-}$

(See Problems 13.33, 13.34, 13.35, and 13.36.)

Oxidation–Reduction Concepts

In this part of the chapter, we will look at another broad class of reactions, called *oxidation–reduction* (or *redox*) *reactions*. These either involve a transfer of electrons or else *appear* to involve a transfer of electrons. The concept of *oxidation numbers,* which we discuss in the next section, is very useful in characterizing this type of reaction.

13.6 OXIDATION NUMBERS

Oxidation number (or **oxidation state**) is defined to be *the charge an atom in a substance would have if the pairs of electrons in each bond belonged to the more electronegative atom.* This means that in a compound composed of monatomic ions, such as sodium chloride, in which the bonding pairs *do* belong to the more electronegative atom, the oxidation number equals the ionic charge. The sodium atom in NaCl has an oxidation number of +1; the chlorine atom has an oxidation number of −1.

In a covalently bonded molecule or polyatomic ion, the oxidation number represents a *hypothetical* charge. Consider the hydrogen chloride molecule, HCl. We assign both bonding electrons to chlorine, because it is more electronegative than hydrogen (3.0 for Cl, and 2.1 for H).

$$H \,\vdots\, {:}\ddot{\underset{..}{Cl}}{:}$$

Because the H atom has no electrons assigned to it, the oxidation number is +1 (one electron less than in the neutral atom). The chlorine atom has eight valence electrons, one more than in the neutral atom. Therefore, its oxidation number is

−1. If the atoms of a bond are of the same element, the two bonding electrons are assigned one to each atom. For example,

$$:\ddot{\underset{..}{Cl}}\ \vdots\ \ddot{\underset{..}{Cl}}:$$

As a result of this electron assignment, the atoms in any elementary substance have zero net charge, and their oxidation number is 0.

Usually, we do not have to analyze the Lewis formula of a species to assign oxidation numbers to the atoms; we can simply apply the following rules:

1. The oxidation number of an atom in an elementary substance is 0. Thus, the oxidation number of a chlorine atom in Cl_2 or of an O atom in O_2 is 0.
2. The oxidation number of a Group IA (alkali metal) atom in any compound is +1; the oxidation number of a Group IIA (alkaline earth) atom in any compound is +2.
3. The oxidation number of fluorine is −1 in all of its compounds.
4. The oxidation number of chlorine, bromine, and iodine is −1 in any compound containing only two elements, the halogen combined with a less electronegative element.●
5. The usual oxidation number of oxygen in a compound is −2. (The major exceptions are peroxides, such as H_2O_2 and Na_2O_2, in which the oxidation number of oxygen is −1.)
6. The oxidation number of hydrogen in most of its compounds is +1. (The exceptions are *hydrides,* compounds such as NaH in which hydrogen is bonded to metallic elements, where hydrogen has the oxidation number −1.)
7. The sum of the oxidation numbers of the atoms in a compound always equals zero. For a polyatomic ion, the oxidation numbers of the atoms add up to the charge on the ion.

● For a compound such as BrCl, we obtain +1 for the bromine and −1 for the chlorine after following the rule of assigning all of the electrons of a covalent bond to the more electronegative atom.

Example 13.4 Assigning Oxidation Numbers to Atoms in a Compound

Use the preceding rules to obtain the oxidation number of chlorine in perchloric acid, $HClO_4$.

Solution

The oxidation numbers of H and O can be assigned immediately (Rules 5 and 6). They are given above the atomic symbols.

$$\overset{+1}{H}\ \overset{?}{Cl}\ \overset{-2}{O_4}$$

Recall that the sum of the oxidation numbers for all the atoms in a compound equals zero (Rule 7):

$$x_H + x_{Cl} + 4x_O = 0$$

(Here x_H is the oxidation number of the hydrogen atoms, and so forth.) Hence,

$$+1 + x_{Cl} + 4 \times (-2) = 0$$

$$x_{Cl} = +7$$

Exercise 13.5

Find the oxidation number of chromium in potassium dichromate, $K_2Cr_2O_7$.

(See Problems 13.39 and 13.40.)

Example 13.5 Assigning Oxidation Numbers to Atoms in an Ion

What is the oxidation number of sulfur in the sulfate ion, SO_4^{2-}?

Solution

The sum of the oxidation numbers equals the ion charge (Rule 7). Hence,

$$x_S + 4x_O = -2$$
$$x_S + 4 \times (-2) = -2$$
$$x_S = +6$$

Exercise 13.6

What is the oxidation number of manganese in the permanganate ion, MnO_4^-?

(See Problems 13.41 and 13.42.)

The oxidation numbers of the elements in ionic compounds can often be most easily obtained by looking at the individual ions. Consider the compound $Fe(ClO_4)_2$. Both Fe and Cl have multiple oxidation states, so it would be difficult to determine the oxidation numbers directly from the compound. But if you remember the common polyatomic ions (Table 2.5, p. 51), you will recognize the perchlorate ion, ClO_4^-. Therefore, the ions in the compound are Fe^{2+} and ClO_4^-. We can now obtain the oxidation numbers of Fe and Cl from these ions, using the technique followed in Example 13.5. We find that Fe has oxidation number +2 and Cl has oxidation number +7.

As we will see, oxidation numbers are useful in characterizing chemical reactions. They are also useful in naming compounds. Consider a binary compound in which the first-named element has two or more oxidation states. There are then two or more binary compounds of these elements. We can distinguish them by means of the *Stock system* of nomenclature. In this system, we give the oxidation number of the first-named element as a Roman numeral within parentheses following the name.● For example,

● The Stock system was introduced in Section 2.6.

$SnCl_4$	tin(IV) chloride
$SnCl_2$	tin(II) chloride
P_4O_6	phosphorus(III) oxide
P_4O_{10}	phosphorus(V) oxide

13.7 DESCRIBING OXIDATION–REDUCTION REACTIONS

In this section, we will first discuss the terminology used to describe oxidation–reduction reactions. Then we will explore various kinds of oxidation–reduction reactions.

Terminology

When an iron metal nail is dipped into a blue solution of copper(II) sulfate, the iron becomes coated with a reddish tinge of metallic copper (Figure 13.4). The molecular equation is

$$Fe(s) + CuSO_4(aq) \longrightarrow FeSO_4(aq) + Cu(s)$$

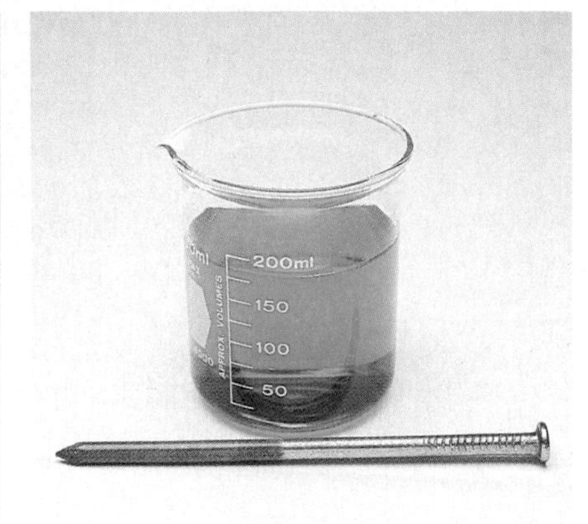

Figure 13.4
Reaction of iron with Cu^{2+}(*aq*). *Left:* Iron nail and copper(II) sulfate solution, which has a blue color. *Center:* Fe reacts with $Cu^{2+}(aq)$ to yield $Fe^{2+}(aq)$ and Cu(*s*). *Right:* The copper metal plates out on the nail.

The net ionic equation is

$$\overset{0}{\mathrm{Fe}}(s) + \overset{+2}{\mathrm{Cu}}{}^{2+}(aq) \longrightarrow \overset{+2}{\mathrm{Fe}}{}^{2+}(aq) + \overset{0}{\mathrm{Cu}}(s)$$

Oxidation numbers have been written above the atoms if the oxidation state has changed. We can see that iron metal loses electrons to form iron(II) ions, while copper(II) ions gain electrons to form copper metal. The reaction involves a transfer of electrons and is an oxidation–reduction type.

We can write the previous reaction in terms of two half-reactions. A **half-reaction** is *one of two parts of an oxidation–reduction reaction, one of which involves a loss of electrons and the other of which involves a gain of electrons*. The half-reactions are

$$\mathrm{Fe}(s) \longrightarrow \mathrm{Fe}^{2+}(aq) + 2e^- \quad \text{(electrons lost by Fe)}$$
$$\mathrm{Cu}^{2+}(aq) + 2e^- \longrightarrow \mathrm{Cu}(s) \quad \text{(electrons gained by } \mathrm{Cu}^{2+}\text{)}$$

We can also describe this reaction in terms of changes in oxidation number. The oxidation number of iron increases and the oxidation number of copper decreases.

Now consider the burning of iron wool in oxygen. The reaction is

$$4\overset{0}{\mathrm{Fe}}(s) + 3\overset{0}{\mathrm{O}}{}_2(g) \longrightarrow 2\overset{+3}{\mathrm{Fe}}{}_2\overset{-2}{\mathrm{O}}{}_3(s)$$

This reaction is also classed as an oxidation–reduction reaction, though the electron-transfer aspect is less obvious than in the previous example. Note that in this reaction the oxidation number of iron increases and the oxidation number of oxygen decreases. In effect, we have the following changes:

$$\mathrm{Fe} \longrightarrow \mathrm{Fe}^{3+} + 3e^-$$
$$\mathrm{O}_2 + 4e^- \longrightarrow 2\mathrm{O}^{2-}$$

Whether the reaction actually occurs as these half-reactions or not, the net effect appears to be a transfer of electrons from iron to oxygen. The electron-transfer aspect of the reaction is revealed by the change in oxidation numbers.

We can define an **oxidation–reduction reaction** (or **redox reaction**) more broadly as *a reaction in which electrons are transferred between species or in which atoms change oxidation number*. Such reactions consist of two parts—one called oxidation, the other called reduction. We define **oxidation** as *the part of an oxidation–reduction reaction in which there is a loss of electrons by a species or an increase in the oxidation number of an atom*. **Reduction** is *the part of an* *oxidation–reduction reaction in which there is a gain of electrons by a species or a decrease in the oxidation number of an atom.* In any reaction in which oxidation occurs, reduction must also occur. This explains why such reactions are called oxidation–reduction reactions, or redox reactions.●

● Originally, *oxidation* simply meant chemical combination with oxygen, and *reduction* meant removal of oxygen from an oxide to give the element. The terms were later broadened as the similarity among certain reactions became apparent.

A species that is *oxidized* loses electrons or contains an atom that increases in oxidation number. Similarly, a species that is *reduced* gains electrons or contains an atom that decreases in oxidation number. An **oxidizing agent** is *a species that oxidizes another species; thus, the oxidizing agent is itself reduced.* A **reducing agent** is *a species that reduces another species; it is itself oxidized.* The relationships among these terms are shown in the following diagram for the reaction of iron with copper(II) ion.

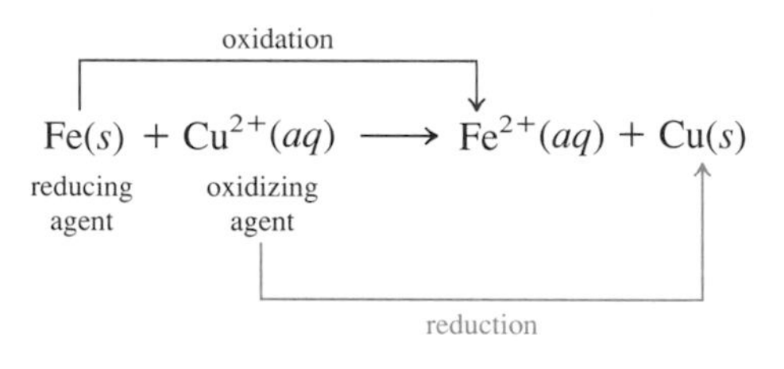

In most cases, the oxidizing and reducing agents in a reaction are separate substances. It is possible, though, for the two agents to be within the same compound. Ammonium dichromate, $(NH_4)_2Cr_2O_7$, decomposes once it is ignited to give the ''volcano'' reaction (Figure 13.5).

$$\underset{\text{orange}}{(\overset{-3}{N}H_4)_2\overset{+6}{Cr}_2O_7(s)} \longrightarrow \overset{0}{N}_2(g) + 4H_2O(g) + \underset{\text{green}}{\overset{+3}{Cr}_2O_3(s)}$$

In this exothermic reaction, the oxidizing agent is the dichromate ion, $Cr_2O_7^{2-}$, and the reducing agent is the ammonium ion, NH_4^+.

Displacement Reactions; Activity Series

The reaction of iron with copper(II) ion is an example of a simple type of oxidation–reduction reaction in which electrons are transferred between a free element

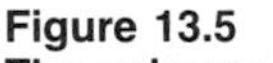

Figure 13.5
The volcano reaction. *Left:* Ammonium dichromate, $(NH_4)_2Cr_2O_7$, is an orange substance; it contains an oxidizing agent and a reducing agent in the same compound. *Right:* Once we ignite ammonium dichromate, it decomposes exothermically. Flames and chromium(III) oxide product are thrown upward by the reaction, which appears like an erupting volcano. Chromium(III) oxide is a green substance but looks black in the subdued light used to photograph the reaction.

Table 13.2 Activity Series of the Elements

Reaction with acids	Element	Reaction with water
React vigorously with acidic solutions to give H_2	Li	React vigorously with liquid water to give H_2
	K	
	Ba	
	Ca	
	Na	
React with acids to give H_2	Mg	React slowly with liquid water but readily with steam to give H_2
	Al	
	Zn	
	Cr	
	Fe	
	Cd	
	Co	
	Ni	
	Sn	
	Pb	
	H_2	
Do not react with acids to give H_2*	Cu	
	Hg	
	Ag	
	Au	

*Cu, Hg, and Ag react with HNO_3 but do not produce H_2. In these reactions, the metal is oxidized to the metal ion and NO_3^- ion is reduced to NO_2 or other nitrogen species.

and a monatomic ion. These reactions are often called *displacement reactions,* because when they are written as molecular equations, the free element appears to displace another element in a compound.

$$Fe(s) + CuSO_4(aq) \longrightarrow FeSO_4(aq) + Cu(s)$$

A further example is the reaction of zinc metal with hydrochloric acid.

$$Zn(s) + 2HCl(aq) \longrightarrow ZnCl_2(aq) + H_2(g)$$

When written as a net ionic equation, the reaction is clearly an oxidation–reduction reaction in which electrons are transferred between an element and a monatomic ion.

$$Zn(s) + 2H^+(aq) \longrightarrow Zn^{2+}(aq) + H_2(g)$$

Whether reaction will occur between a given element and a monatomic ion depends on the relative ease with which the two species gain or lose electrons. Table 13.2 shows the **activity series** of the elements, *a table that lists the elements in order of their ease of losing electrons during reactions in aqueous solution.* The metals at the top of the list are the strongest reducing agents; those at the bottom are the weakest. A free element reacts with the monatomic ion listed for another element if the free element is above that other element in the activity series. (Sodium metal and the free elements above it, however, react more readily with water than with metal ions in aqueous solution.)

For example, aluminum is above copper in the activity series. We expect aluminum to react with copper(II) ion.

$$2Al(s) + 3Cu^{2+}(aq) \longrightarrow 2Al^{3+}(aq) + 3Cu(s)$$

Aluminum loses electrons to copper(II) ion. However, we do not expect copper metal to react with aluminum ion.

$$Cu(s) + Al^{3+}(aq) \longrightarrow NR$$

Similarly, the metals above hydrogen in the activity series are expected to react with acids to give hydrogen gas. For example, magnesium metal reacts with sulfuric acid.

$$Mg(s) + H_2SO_4(aq) \longrightarrow MgSO_4(aq) + H_2(g)$$

The net ionic equation is

$$Mg(s) + 2H^+(aq) \longrightarrow Mg^{2+}(aq) + H_2(g)$$

However, copper metal does not react with sulfuric acid to give hydrogen gas. Copper is below hydrogen in the activity series. Table 13.2 also indicates the reactivities of various metals with water.

The reaction of aqueous chlorine, $Cl_2(aq)$, with potassium bromide solution, $KBr(aq)$, is an example of another type of displacement reaction. Written as a molecular equation, we have

$$Cl_2(aq) + 2KBr(aq) \longrightarrow Br_2(aq) + 2KCl(aq)$$

The net ionic equation is

$$Cl_2(aq) + 2Br^-(aq) \longrightarrow Br_2(aq) + 2Cl^-(aq)$$

In general, chlorine will oxidize bromide ion to bromine. Similarly, chlorine will oxidize iodide ion to iodine. The net ionic equation is

$$Cl_2(aq) + 2I^-(aq) \longrightarrow I_2(aq) + 2Cl^-(aq)$$

The oxidizing strengths of the halogens are in the order $Cl_2 > Br_2 > I_2$. Thus, Cl_2 will oxidize bromide ion and iodide ion, and Br_2 will oxidize iodide ion. However, I_2 will not oxidize either chloride ion or bromide ion. A typical test for confirming the presence of bromide ion or iodide ion in a solution uses these displacement reactions. Suppose we add chlorine water, $Cl_2(aq)$, to separate test tubes, one containing aqueous KBr, the other aqueous KI. Then we add methylene chloride, which forms a lower liquid layer. The halogen gives this lower layer a characteristic color—yellow-orange for Br_2 and violet for I_2.

Exercise 13.7

For each of the following, decide whether the reaction shown actually occurs.

(a) $2Ag(s) + 2H^+(aq) \longrightarrow 2Ag^+(aq) + H_2(g)$
(b) $Mg(s) + 2Ag^+(aq) \longrightarrow Mg^{2+}(aq) + 2Ag(s)$ (See Problems 13.51 and 13.52.)

Disproportionation

Disproportionation is *a reaction in which a species is both oxidized and reduced.* The following reaction in aqueous solution is an example:

$$\overset{+1}{2Cu^+}(aq) \longrightarrow \overset{0}{Cu}(s) + \overset{+2}{Cu^{2+}}(aq)$$

Note that copper(I) ion is both oxidized to copper(II) ion and reduced to copper metal. That is, copper(I) ion disproportionates in aqueous solution. Another example of a disproportionation is the reaction of chlorine with water:

$$\overset{0}{Cl_2}(aq) + H_2O(l) \longrightarrow H\overset{-1}{Cl}(aq) + H\overset{+1}{Cl}O(aq)$$

Chlorine is both reduced to HCl and oxidized to HClO.

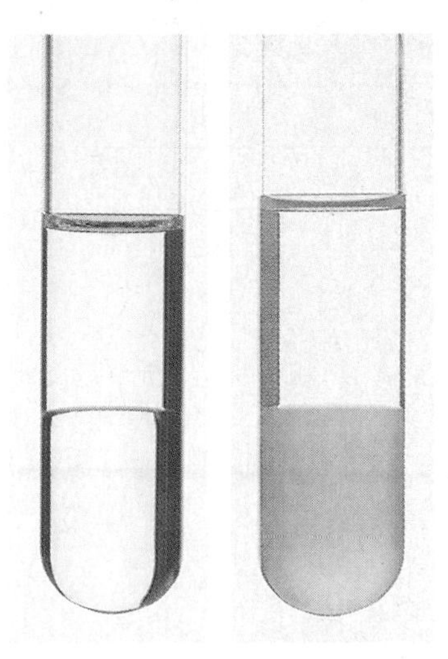

Figure 13.6
Test for bromide ion. *Left:* An aqueous solution of bromide ion above a layer of methylene chloride. *Right:* When permanganate ion and acid are added to the aqueous layer, Br^- is oxidized to Br_2, which dissolves in the methylene chloride to give an orange layer.

Oxidation–Reduction Reactions Involving Oxoanions

Not all oxidation–reduction equations are so simple in appearance as the previous reactions. Consider a standard test for bromide ion in aqueous solution. A small portion of solution suspected of containing bromide ion is treated with potassium permanganate, $KMnO_4$, in the presence of acid. If bromide ion is present, it gives bromine, Br_2. The presence of Br_2 can then be detected because Br_2 dissolves in certain liquids that are insoluble in water to give a colored layer (see Figure 13.6). The equation for the reaction is

$$10Br^-(aq) + 2MnO_4^-(aq) + 16H^+(aq) \longrightarrow 5Br_2(aq) + 2Mn^{2+}(aq) + 8H_2O(l)$$

At first glance, this equation may look a bit intimidating. It is helpful to have a way of reducing such an equation to its essentials.

To do this, we note which species contain atoms that change oxidation number. If we write down just these species, ignoring coefficients, we get a *skeleton equation*. For the previous equation, we get

$$\overset{-1}{Br^-} + \overset{+7}{MnO_4^-} \longrightarrow \overset{0}{Br_2} + \overset{+2}{Mn^{2+}}$$

The skeleton equation shows the essential part of the oxidation–reduction reaction. It shows that bromide ion is being oxidized to bromine, Br_2. The oxidizing agent is permanganate ion, MnO_4^-, and it is reduced to manganese(II) ion, Mn^{2+}. The next section demonstrates how we can obtain the balanced equation from this skeleton equation (plus the fact that the reaction occurs in acid solution). Thus, it does indeed contain the essential information of the reaction.

In understanding oxidation–reduction equations, it helps to know the usual oxidation numbers of the common elements and the species with those oxidation numbers. For bromine, we have the following data:

BROMINE SPECIES	OXIDATION NUMBER
$HBrO_4$	+7
$HBrO_3$	+5
$HBrO_2$	+3
$HBrO$	+1
Br_2	0
Br^-	−1

The most common oxidation state is −1 (shown in color) and is represented by Br^-. Like most of the nonmetals, bromine displays a series of oxidation states (except for zero all are odd numbers), from a low value equal to the group number minus eight (−1) to a high value equal to the group number (+7). Oxidation of Br^- would give a species with a higher oxidation number, which from the diagram of oxidation states would be Br_2 or perhaps an oxoacid. The actual product with MnO_4^-, we know, is Br_2.

Table 13.3 lists the oxidation states of the second and third rows of the main-group elements. In compounds, metals usually have oxidation states equal to the group number. Nonmetals, on the other hand, usually have a series of oxidation states extending from a low value equal to the group number minus eight to a high value equal to the group number. The exceptions are oxygen and fluorine. Usually, the series of oxidation states for compounds are all even or all odd (as for Br). Nitrogen is an exception; it has compounds in all oxidation states from −3 to +5.

Table 13.4 lists oxidation states of some transition elements. We see that the most stable oxidation state of manganese is +2. Manganese species in higher

Table 13.3 Oxidation States and Common Species of Some Main-Group Elements*

	Group						
	IA	IIA	IIIA	IVA	VA	VIA	VIIA
Period 2	**Li**	**Be**	**B**	**C**	**N**	**O**	**F**
	+1 Li^+	+2 Be^{2+}	+3 B_2O_3	+4 CO_2	+5 HNO_3	0 O_2	0 F_2
	0 Li	0 Be	0 B	+2 CO	+4 NO_2	−1 H_2O_2	−1 F^-
				0 C	+3 HNO_2	−2 H_2O	
				−4 CH_4	+2 NO		
					+1 N_2O		
					0 N_2		
					−1 NH_2OH		
					−2 N_2H_4		
					−3 NH_3		
Period 3	**Na**	**Mg**	**Al**	**Si**	**P**	**S**	**Cl**
	+1 Na^+	+2 Mg^{2+}	+3 Al^{3+}	+4 SiO_2	+5 H_3PO_4	+6 H_2SO_4	+7 $HClO_4$
	0 Na	0 Mg	0 Al	0 Si	+3 H_3PO_3	+4 SO_2	+5 $HClO_3$
				−4 SiH_4	0 P_4	0 S_8	+3 $HClO_2$
					−3 PH_3	−2 H_2S	+1 HClO
							0 Cl_2
							−1 Cl^-

*The most common oxidation state is shown in color.

oxidation states would function as oxidizing agents, because they would have a tendency to go to the +2 state. We see this in the reaction of MnO_4^- with Br^-. Here MnO_4^- acts as the oxidizing agent and is converted in acid solution to the stable Mn^{2+} ion. In basic solution, MnO_4^- is reduced only to MnO_2. The MnO_2 precipitates in basic solution, which concludes the reaction. (Figure 13.7 shows some compounds of manganese in various oxidation states.)

Table 13.5 lists some common oxidizing and reducing agents.

Table 13.4 Oxidation States and Common Species of Some Transition Elements*

Cr	Mn	Fe	Co	Ni
+6 CrO_4^{2-}, $Cr_2O_7^{2-}$	+7 MnO_4^-	+3 Fe^{3+}	+3 Co^{3+}	+2 Ni^{2+}
+3 Cr^{3+}	+4 MnO_2	+2 Fe^{2+}	+2 Co^{2+}	0 Ni
+2 Cr^{2+}	+2 Mn^{2+}	0 Fe	0 Co	
0 Cr	0 Mn			
Cu	**Ag**	**Zn**	**Cd**	**Hg**
+2 Cu^{2+}	+1 Ag^+	+2 Zn^{2+}	+2 Cd^{2+}	+2 Hg^{2+}
+1 Cu^+	0 Ag	0 Zn	0 Cd	+1 Hg_2^{2+}
0 Cu				0 Hg

*The most common oxidation state is shown in color.

Figure 13.7
Some compounds of manganese in various oxidation states. *From left to right:* Potassium permanganate (dark purple); manganese(IV) oxide (black); and manganese(II) chloride tetrahydrate (pink).

13.8 BALANCING OXIDATION–REDUCTION EQUATIONS

Oxidation–reduction equations are often too difficult to balance by the inspection method we used for simpler equations.● Two methods of balancing oxidation–reduction equations are described here. The first method (the *oxidation-number method*) depends on the use of oxidation numbers. In the second method (the *half-reaction method*), we break the ionic equation into what are called *half-reactions,* one half-reaction for the oxidation portion of the reaction and one for the reduction portion. Balancing such half reactions is a skill you will find useful in studying electrochemistry (Chapter 19).

● Balancing by inspection was discussed in Section 3.1.

Table 13.5 Common Oxidizing and Reducing Agents

Substance	Half-Reaction	Acid/Base
Oxidizing Agents		
Hydrogen peroxide, H_2O_2	$H_2O_2(aq) + 2H^+(aq) + 2e^- \longrightarrow 2H_2O(l)$	Acidic
Potassium permanganate, $KMnO_4$	$MnO_4^-(aq) + 8H^+(aq) + 5e^- \longrightarrow Mn^{2+}(aq) + 4H_2O(l)$	Acidic
	$MnO_4^-(aq) + 2H_2O(l) + 3e^- \longrightarrow MnO_2(s) + 4OH^-(aq)$	Basic
Chlorine, Cl_2	$Cl_2(aq) + 2e^- \longrightarrow 2Cl^-(aq)$	Acidic
Potassium dichromate, $K_2Cr_2O_7$	$Cr_2O_7^{2-}(aq) + 14H^+(aq) + 6e^- \longrightarrow 2Cr^{3+}(aq) + 7H_2O(l)$	Acidic
Nitric acid, HNO_3*	$NO_3^-(aq) + 4H^+(aq) + 3e \longrightarrow NO(g) + 2H_2O(l)$	Acidic
Sodium hypochlorite, NaClO	$ClO^-(aq) + H_2O(aq) + 2e^- \longrightarrow Cl^-(aq) + 2OH^-(aq)$	Basic
Reducing Agents		
Hydrogen peroxide, H_2O_2	$H_2O_2(aq) \longrightarrow O_2(aq) + 2H^+(aq) + 2e^-$	Acidic
Sodium iodide, NaI	$2I^-(aq) \longrightarrow I_2(s) + 2e^-$	Acidic
Tin(II) chloride, $SnCl_2$	$Sn^{2+}(aq) \longrightarrow Sn^{4+}(aq) + 2e^-$	Acidic
Zinc, Zn	$Zn(s) \longrightarrow Zn^{2+}(aq) + 2e^-$	Acidic
Sulfurous acid, H_2SO_3	$H_2SO_3(aq) + H_2O(l) \longrightarrow SO_4^{2-}(aq) + 4H^+(aq) + 2e^-$	Acidic
Sodium sulfite, Na_2SO_3	$SO_3^{2-}(aq) + 2OH^-(aq) \longrightarrow SO_4^{2-}(aq) + H_2O(l) + 2e^-$	Basic

*Product depends on reaction conditions.

Oxidation-Number Method

This method depends on the fact that the increase in oxidation number for the atoms that are oxidized must equal the absolute value (number without the sign) of the decrease in oxidation number for the atoms that are reduced.

As an example of the oxidation-number method, consider the following molecular equation:

$$HNO_3(aq) + Cu_2O(s) \longrightarrow Cu(NO_3)_2(aq) + NO(g) + H_2O(l) \qquad \text{(not balanced)}$$

In order to discover what atoms change oxidation numbers, we find the oxidation numbers of all atoms. This reveals that the N and Cu atoms change oxidation numbers.

$$H\overset{+5}{N}O_3 + \overset{+1}{Cu_2}O \longrightarrow \overset{+2}{Cu}(NO_3)_2 + \overset{+2}{N}O + H_2O$$

Note that nitric acid, HNO_3, functions as an oxidizing agent. Because the nitrogen atom is in its highest oxidation state (see Table 13.3), there are a variety of species to which it might be reduced. In practice, one finds that the reduction product depends on the concentration of HNO_3, as well as on the nature of the reducing agent. Here the product is NO. Copper(I) oxide is oxidized to Cu^{2+}, the usual ion in aqueous solution.

To balance the previous equation, we first balance each of the atoms that change oxidation number. Nitrogen atoms that change oxidation number are balanced. (N atoms that remain in oxidation state +5 are not, but we will balance them in the last step.) Copper atoms are not balanced, but we can balance them by adjusting coefficients. We get

$$H\overset{+5}{N}O_3 + \overset{+1}{Cu_2}O \longrightarrow 2\overset{+2}{Cu}(NO_3)_2 + \overset{+2}{N}O + H_2O$$

We now calculate the change in oxidation number for the copper atom. This is an increase from +1 to +2, which is an increase in oxidation number of 1. However, there are two copper atoms, so the total increase in oxidation number is 2. For the nitrogen atom, the oxidation number decreases from +5 to +2, for a total decrease of 3. We can summarize this information in the following equation:

$$\underbrace{H\overset{+5}{N}O_3 \longrightarrow \overset{+2}{N}O}_{-3} \qquad \underbrace{\overset{+1}{Cu_2}O \longrightarrow 2\overset{+2}{Cu}(NO_3)_2}_{+2}$$

$$H\overset{+5}{N}O_3 + \overset{+1}{Cu_2}O \longrightarrow 2\overset{+2}{Cu}(NO_3)_2 + \overset{+2}{N}O + H_2O$$

Now we note that the increase in oxidation number of copper must equal the absolute value of the decrease in oxidation number for nitrogen. Therefore, we multiply copper species by 3 and nitrogen species by 2. We have

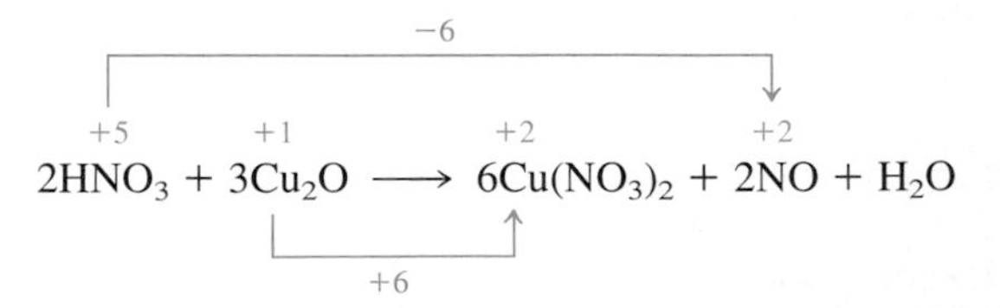

The remaining atoms are balanced by inspection. Although the nitrogen atoms that change oxidation number have been balanced, other nitrogen atoms are

present as NO_3^- ions in $6Cu(NO_3)_2$. For this reason, we add $12HNO_3$ to the $2HNO_3$ already on the left side.

$$14HNO_3 + 3Cu_2O \longrightarrow 6Cu(NO_3)_2 + 2NO + H_2O$$

Now copper and nitrogen atoms have been balanced, and all coefficients except the one for H_2O have been fixed. We can obtain the coefficient of H_2O by balancing oxygen atoms. The balanced equation (with phase labels) is

$$14HNO_3(aq) + 3Cu_2O(s) \longrightarrow 6Cu(NO_3)_2(aq) + 2NO(g) + 7H_2O(l)$$

We can now summarize the steps in the oxidation-number method for balancing a molecular equation:

1. Obtain the oxidation numbers of all atoms. From this, determine which atoms change in oxidation number. Pair up the oxidizing agent with its corresponding product; similarly, pair up the reducing agent with its product. (You may need to write a reactant or product twice. For example, if a reactant is both oxidized and reduced, write the reactant once as the oxidizing agent and once as the reducing agent.) If necessary, adjust the coefficients within a pair so that the atom that changes oxidation number is balanced.
2. Determine the changes in oxidation number in the oxidation and in the reduction portions of the equation. Make the absolute values (values without regard to sign) of these changes equal by multiplying the species involving oxidation by one factor and those involved in reduction by another factor.
3. Balance other atoms in the equation by inspection. The equation may need to be completed by adding H_2O to it; add H_2O to one side or the other to balance O atoms. Check that the equation is balanced.

The next example further illustrates these steps.

Example 13.6 Balancing Molecular Equations by the Oxidation-Number Method

Zinc metal reacts with nitric acid, HNO_3, to produce a number of products depending on the concentration of acid. In concentrated acid, zinc reduces nitrate ion to ammonium ion; zinc is oxidized to zinc ion, Zn^{2+}. The ions Zn^{2+} and NH_4^+ will have NO_3^- as corresponding negative ions, so that in a molecular equation we would write them as $Zn(NO_3)_2$ and NH_4NO_3. Writing only the species that change oxidation number, we have the skeleton equation

$$Zn(s) + HNO_3(aq) \longrightarrow Zn(NO_3)_2(aq) + NH_4NO_3(aq) \quad \text{(skeleton equation)}$$

Balance this equation by the oxidation-number method.

Solution

STEP 1 We first obtain the oxidation numbers of all atoms. From this, we find that only the Zn and N atoms change oxidation numbers (noting that N atoms in NO_3^- ions on the right have not changed oxidation state).

$$\overset{0}{Zn} + H\overset{+5}{N}O_3 \longrightarrow \overset{+2}{Zn}(NO_3)_2 + \overset{-3}{N}H_4NO_3$$

We next look to see if we must adjust coefficients so that the atoms that do change oxidation state are initially balanced. As the equation is written, there is one Zn on the left and one on the right, so no initial adjustment of coefficients of Zn species is needed. We also note that one N atom on the left is reduced to one N atom on the right, so the coefficients of these N species are initially balanced.

STEP 2 Next we determine the changes in oxidation numbers in the oxidation and in the reduction portions of the equation.

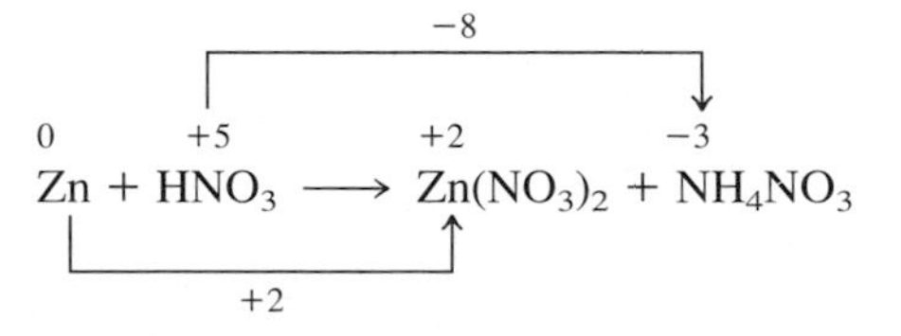

(continued)

We multiply the coefficients of the Zn species by 4 so that the changes in oxidation number are equal in absolute value.

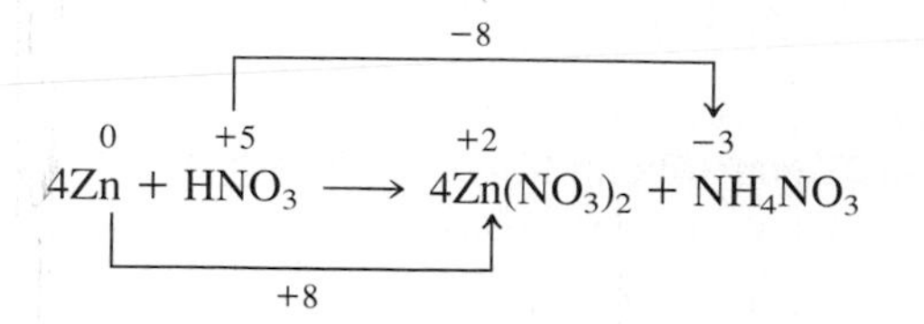

STEP 3 We now balance the remaining N atoms. There are 9 such atoms on the right, so we add $9HNO_3$ to what we already have on the left ($1HNO_3$) to obtain $10HNO_3$.

$$4Zn + 10HNO_3 \longrightarrow 4Zn(NO_3)_2 + NH_4NO_3$$

Finally, we balance the O atoms by adding H_2O's to one side or the other. There are 30 O atoms on the left and 27 O atoms on the right. Therefore, we add $3H_2O$ to the right.

$$4Zn + 10HNO_3 \longrightarrow 4Zn(NO_3)_2 + NH_4NO_3 + 3H_2O$$

The H atoms should also balance, which we can easily verify. The balanced equation with phase labels is

$$\mathbf{4Zn(\mathit{s}) + 10HNO_3(\mathit{aq}) \longrightarrow 4Zn(NO_3)_2(\mathit{aq}) + NH_4NO_3(\mathit{aq}) + 3H_2O(\mathit{l})}$$

Check that the equation is indeed balanced.

Exercise 13.8

Iodic acid, HIO_3, can be prepared by reacting iodine, I_2, with concentrated nitric acid. The skeleton equation is

$$I_2 + HNO_3 \longrightarrow HIO_3 + NO_2 \qquad \text{(skeleton equation)}$$

Balance this equation by the oxidation-number method.

(See Problems 13.55, 13.56, 13.57, and 13.58.)

The oxidation-number method also can be used to balance ionic equations of oxidation–reduction type. Consider the case of an acidic solution. In this case, we add H_2O's to balance O atoms and then add H^+ ions to balance H atoms. The charges must balance, which serves as a check on the work. The next example illustrates the use of the oxidation-number method to balance an ionic equation in acidic solution.

Example 13.7 Balancing Ionic Equations by the Oxidation-Number Method

Chlorine, Cl_2, can be prepared by oxidizing chloride ion, Cl^-, with permanganate ion, MnO_4^- in acidic solution. Permanganate ion is reduced to Mn^{2+}. Write the balanced ionic equation for this reaction.

Solution

STEP 1 The skeleton equation is

$$\overset{+7}{Mn}O_4^- + \overset{-1}{Cl^-} \longrightarrow \overset{+2}{Mn^{2+}} + \overset{0}{Cl_2}$$

Oxidation numbers have been written above the equation for the atoms that change oxidation state. We initially adjust the coefficients of Cl species so the Cl atoms balance; Mn atoms are balanced.

$$\overset{+7}{Mn}O_4^- + 2\overset{-1}{Cl^-} \longrightarrow \overset{+2}{Mn^{2+}} + \overset{0}{Cl_2}$$

STEP 2 The changes of oxidation number are

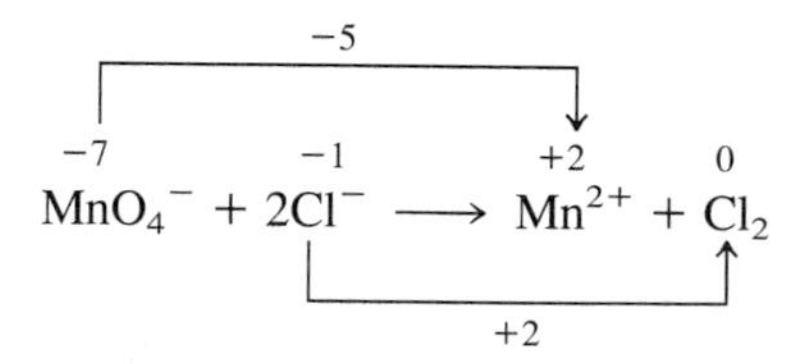

We make the absolute values of the changes in oxidation number equal by multiplying the coefficients of the Mn species by 2 and the Cl species by 5.

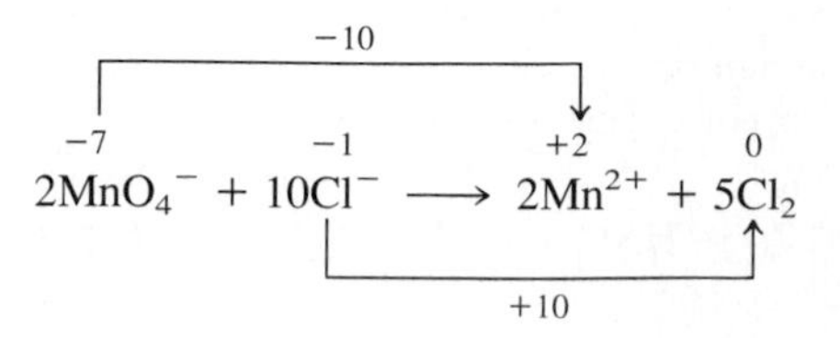

STEP 3 We balance the O atoms by adding $8H_2O$ to the right side.

$$2MnO_4^- + 10Cl^- \longrightarrow 2Mn^{2+} + 5Cl_2 + 8H_2O$$

Finally, we balance H atoms by adding $16H^+$ to the left.

$$2MnO_4^- + 10Cl^- + 16H^+ \longrightarrow 2Mn^{2+} + 5Cl_2 + 8H_2O$$

Check that the equation is indeed balanced. Note especially that the charges balance (there is a +4 charge on both sides).

Exercise 13.9

Iodide ion, I^-, is oxidized to iodate ion, IO_3^-, by permanganate ion, MnO_4^-, in acidic solution; the permanganate ion is reduced to manganese(II) ion. Obtain the balanced ionic equation using the oxidation-number method. (See Problems 13.59 and 13.60.)

To balance an equation in basic solution, we can first obtain the balanced equation as if it were in acidic solution. Then we add OH^- ions to both sides of the equation, allowing enough OH^- to react with all of the H^+ to give H_2O. Consider the oxidation of I^- to IO_3^- by MnO_4^- in basic solution. In this case, MnO_4^- is reduced to MnO_2. The skeleton equation is

$$I^- + MnO_4^- \longrightarrow IO_3^- + MnO_2$$

If we balance this as if it were in acidic solution, we obtain

$$I^- + 2MnO_4^- + 2H^+ \longrightarrow IO_3^- + 2MnO_2 + H_2O$$

Now we add $2OH^-$ to both sides of the equation. On the left side, this gives $2H_2O$.

$$I^- + 2MnO_4^- + 2H_2O \longrightarrow IO_3^- + 2MnO_2 + H_2O + 2OH^-$$

Finally, we cancel $1H_2O$ on both sides.

$$I^- + 2MnO_4^- + H_2O \longrightarrow IO_3^- + 2MnO_2 + 2OH^-$$

Exercise 13.10

Hydrogen peroxide, H_2O_2, is oxidized by permanganate ion, MnO_4^-, in basic solution to oxygen, O_2; permanganate ion is reduced to manganese dioxide, MnO_2. Use the oxidation-number method to obtain the balanced ionic equation for the reaction.

(See Problems 13.61 and 13.62.)

Half-Reaction Method

The *half-reaction method* of balancing oxidation–reduction equations is based on the separation of the equation into two half-reactions. These half-reactions are balanced and then recombined to get the balanced oxidation–reduction equation.

Consider the reaction of permanganate ion, MnO_4^-, with Fe^{2+} in acidic solution. We wish to write the skeleton equation for this reaction. Recall that the skeleton equation contains only the species that change in oxidation number (so it may contain O and H atoms on one side that do not occur on the other). The skeleton equation is

$$\overset{+7}{Mn}O_4^-(aq) + \overset{+2}{Fe}{}^{2+}(aq) \longrightarrow \overset{+2}{Mn}{}^{2+}(aq) + \overset{+3}{Fe}{}^{3+}(aq) \quad \text{(acid) skeleton equation}$$

Note that this is a reaction in which MnO_4^- (purple color) acts as an oxidizing agent in acidic solution and is reduced to Mn^{2+} (pale pink to colorless). Iron(II) ion (pale green) is oxidized to Fe^{3+} (pale yellow to colorless). From Table 13.4 we see that Fe^{2+} and Fe^{3+} are common oxidation states of iron. This reaction may be used to obtain the amount of iron in a sample of iron ore (see Figure 13.8). The iron ore is first converted to a solution of Fe^{2+}. When this solution is titrated with $KMnO_4(aq)$, the purple color of MnO_4^- ion disappears because the ion is reduced to Mn^{2+}. When all of the Fe^{2+} ion has reacted, the purple color is retained. This marks the end point of the titration.

To obtain a balanced equation for this reaction, we first split the skeleton equation into half-reactions. We write an incomplete half-reaction for manganese by copying species containing Mn from the skeleton equation.

$$MnO_4^- \longrightarrow Mn^{2+} \qquad \text{(incomplete half-reaction)}$$

Similarly, we copy the species containing Fe from the skeleton equation to get the other half-reaction.

$$Fe^{2+} \longrightarrow Fe^{3+} \qquad \text{(incomplete half-reaction)}$$

Note that the manganese half-reaction is balanced in Mn atoms. If it had not been, we would have balanced it by adjusting coefficients of reactant and product species. We balance the oxygen in this half-reaction by adding $4H_2O$ to the right side

$$MnO_4^- \longrightarrow Mn^{2+} + 4H_2O$$

Now we add $8H^+$ to the left side to balance the hydrogen. (Because the reaction occurs in acidic solution, we can assume that H^+ is available.)

$$MnO_4^- + 8H^+ \longrightarrow Mn^{2+} + 4H_2O$$

This half-reaction is not yet balanced in electric charge (there is a +7 total charge on the left and a +2 charge on the right). We add five electrons to the left side of the equation to achieve balance in charge.

$$MnO_4^- + 8H^+ + 5e^- \longrightarrow Mn^{2+} + 4H_2O \qquad \text{(reduction half-reaction)}$$

Figure 13.8
Analysis of iron ore. *Left:* Iron ore containing Fe_2O_3 is first dissolved in an acid, forming Fe^{3+}. Iron in the 3+ state is reduced to Fe^{2+} by reaction with $SnCl_2$. *Right:* The solution of Fe^{2+} is titrated with potassium permanganate, $KMnO_4$, which has an intense purple color. Permanganate ion reacts with Fe^{2+}, giving a nearly colorless solution. When all the Fe^{2+} is used up, another drop of potassium permanganate solution gives a magenta color to the contents of the flask. From the amount of potassium permanganate used, we can calculate the amount of iron in the sample. The cloudiness of the final solution is due to silicates present in the iron ore.

Check that this half-reaction is balanced in each kind of atom and that it is balanced in charge. Note that electrons are gained, as we expect for reduction.

The half-reaction for iron is balanced in iron but not in charge. Thus, we add one electron to the right side.

$$Fe^{2+} \longrightarrow Fe^{3+} + e^- \quad \text{(oxidation half-reaction)}$$

Check that this half-reaction is balanced. Note that electrons are lost, as we expect for oxidation.

The last step is to multiply each half-reaction by a factor such that when the two half-reactions are added, the electrons cancel. This is required; free electrons cannot appear in the final equation.

$$\begin{array}{lll} 1 \times (MnO_4^- + 8H^+ + 5e^- & \longrightarrow & Mn^{2+} + 4H_2O) \\ + \; 5 \times (Fe^{2+} & \longrightarrow & Fe^{3+} + e^-) \\ \hline MnO_4^- + 8H^+ + \cancel{5e^-} + 5Fe^{2+} & \longrightarrow & Mn^{2+} + 4H_2O + 5Fe^{3+} + \cancel{5e^-} \end{array}$$

The final balanced equation is

$$MnO_4^-(aq) + 5Fe^{2+}(aq) + 8H^+(aq) \longrightarrow Mn^{2+}(aq) + 5Fe^{3+}(aq) + 4H_2O(aq)$$

The half-reaction method for balancing a skeleton oxidation–reduction equation can be summarized in the following steps:

1. Split the skeleton equation into incomplete half-reactions. Usually it is obvious how to do this. If there is any difficulty, follow this procedure:
a. Obtain the oxidation numbers of the atoms in the skeleton equation in order to decide which atoms change oxidation numbers.
b. Write an incomplete half-reaction for reduction by copying only those species from the skeleton equation in which an atom decreases in oxidation number. Similarly, write an incomplete half-reaction for oxidation by copying only those species in which an atom increases in oxidation number.

2. Balance each half-reaction.
a. Balance each half-reaction in the atom being oxidized or reduced by adjusting coefficients of the reactant and product species.
b. Balance each half-reaction in O atoms by adding H_2O's to one side of the half-reaction.
c. Balance each half-reaction in H atoms by adding H^+ ions to one side of the half-reaction.
d. Balance each half-reaction in electric charge by adding electrons (e^-) to one side of the half-reaction.

3. Combine the half-reactions to get the final balanced oxidation–reduction equation.
a. Multiply each half-reaction by a factor such that when the half-reactions are added to get the balanced equation, the electrons cancel.
b. Simplify the balanced equation by canceling species that occur on both sides and reduce the coefficients to the smallest whole numbers. Check that the equation is indeed balanced.

When we balance a half-reaction in basic solution, we can assume that OH^- is available but H^+ is not. In some cases, a half-reaction in basic solution is easily balanced in O and H atoms by simply adding OH^-. For example, we can balance the incomplete half-reaction

$$Zn \longrightarrow Zn(OH)_4^{2-}$$

by adding OH^- and e^-.

$$Zn + 4OH^- \longrightarrow Zn(OH)_4^{2-} + 2e^-$$

In most cases, the balancing of O's and H's requires both OH^- and H_2O. One way to do this is to balance the half-reaction as though it were in acidic solution. Then, after Step 2d, insert the following step: Add enough OH^- to both sides of the half-reaction to combine with all H^+ ions according to the reaction $H^+ + OH^- \longrightarrow H_2O$. The effect of this can be stated as Step 2e:

2. e. If a reaction occurs in basic solution, change all H^+ ions in a half-reaction to H_2O's and add the same number of OH^- ions to the other side of the equation. Cancel any H_2O's that occur on both sides of the half-reaction.

The half-reaction method is illustrated for reactions in basic solutions in Example 13.8.

Example 13.8 Balancing Equations by the Half-Reaction Method

Permanganate ion oxidizes sulfite ion according to the following skeleton equation:

$$MnO_4^-(aq) + SO_3^{2-}(aq) \longrightarrow MnO_2(s) + SO_4^{2-}(aq) \quad \text{(basic) skeleton equation}$$

Use the half-reaction method to write the balanced equation.

Solution

Before we begin balancing the equation, it is instructive to look at the oxidation states of sulfur and manganese (Tables 13.3 and 13.4). Sulfites are salts of sulfurous acid, H_2SO_3, which is obtained by dissolving SO_2 in water. If any of these species are oxidized, we expect corresponding species in the +6 oxidation state of sulfur. The sulfites should be oxidized to sulfates (SO_4^{2-}). Permanganate ion, we know, is an oxidizing agent. In acidic solution, it goes all the way to the +2 oxidation state. But in basic solution, MnO_2 precipitates.

Following Step 1, we write the incomplete half-reactions for manganese and sulfur:

$$MnO_4^- \longrightarrow MnO_2$$

$$SO_3^{2-} \longrightarrow SO_4^{2-}$$

These are balanced in Mn and S (Step 2a).

We balance the oxygen in the manganese equation by adding H_2O's to the right (Step 2b).

$$MnO_4^- \longrightarrow MnO_2 + 2H_2O$$

Then we balance this equation in hydrogen by adding H^+ ions to the opposite side (Step 2c).

$$MnO_4^- + 4H^+ \longrightarrow MnO_2 + 2H_2O$$

Now we balance the charge by adding electrons (Step 2d).

$$MnO_4^- + 4H^+ + 3e^- \longrightarrow MnO_2 + 2H_2O$$

We can convert this half-reaction to one in basic solution by adding four OH^- ions to both sides of the equation (Step 2e).

$$MnO_4^- + 4H_2O + 3e^- \longrightarrow MnO_2 + 2H_2O + 4OH^-$$

Note that on the left we have added the four OH^- ions to four H^+ ions, giving four H_2O molecules. Canceling the extra water molecules gives

$$MnO_4^- + 2H_2O + 3e^- \longrightarrow MnO_2 + 4OH^- \quad \text{(reduction half-reaction)}$$

We can find the sulfur half-reaction in a similar manner.

$$SO_3^{2-} + 2OH^- \longrightarrow SO_4^{2-} + H_2O + 2e^- \quad \text{(oxidation half-reaction)}$$

Finally, the half-reactions are multiplied through by factors such that when they are added together, the electrons cancel (Step 3a).

$$\begin{array}{l} 2 \times (MnO_4^- + 2H_2O + 3e^- \longrightarrow MnO_2 + 4OH^-) \\ \underline{+\ 3 \times (SO_3^{2-} + 2OH^- \longrightarrow SO_4^{2-} + H_2O + 2e^-)} \end{array}$$

$$2MnO_4^- + \overset{1}{\cancel{4}}H_2O + \cancel{6e^-} + 3SO_3^{2-} + \cancel{6OH^-} \longrightarrow 2MnO_2 + \overset{2}{\cancel{8}}OH^- + 3SO_4^{2-} + \cancel{3H_2O} + \cancel{6e^-}$$

After simplification (Step 3b), the balanced equation is

$$\mathbf{2MnO_4^- + 3SO_3^{2-} + H_2O \longrightarrow 2MnO_2 + 3SO_4^{2-} + 2OH^-}$$

You should verify that the equation is balanced in charge as well as in different kinds of atoms.

Exercise 13.11

Balance the following equation using the half-reaction method.

$$Zn + NO_3^- \longrightarrow Zn^{2+} + NH_4^+ \qquad \text{(acidic)}$$

(See Problems 13.63, 13.64, 13.65, and 13.66.)

Solution Stoichiometry

In Section 4.11, we described how we could deal with stoichiometry problems using the concept of molarity. However, chemists have devised the concepts of *equivalents* and *normality* to simplify calculations in acid–base and oxidation–reduction analyses. The following section describes these concepts.

13.9 EQUIVALENTS AND NORMALITY

The definition of *equivalent* depends on whether we are discussing an acid–base reaction or an oxidation–reduction reaction. An **equivalent (eq)** *in an acid–base reaction* is *the quantity of acid that yields 1 mol H^+ ion or the quantity of base that reacts with 1 mole of H^+ ion*. In general, an equivalent is defined so that 1 equivalent of acid reacts with 1 equivalent of base. It is important to note that the definition depends on the particular chemical reaction with which we are dealing. For example, in the acid–base reaction

$$H_2SO_4(aq) + 2NaOH(aq) \longrightarrow Na_2SO_4(aq) + 2H_2O(aq)$$

each mole of H_2SO_4 supplies 2 mol H^+. Thus 1 mol H_2SO_4 supplies 2 eq H_2SO_4. Because each mole of NaOH reacts with 1 mole of H^+, 1 mol NaOH supplies 1 eq NaOH. From the equation, we see that 1 mol H_2SO_4 (= 2 eq H_2SO_4) reacts with 2 mol NaOH (= 2 eq NaOH). Thus, 1 equivalent of H_2SO_4 reacts with 1 equivalent of NaOH (which is what we expect).

Now, however, let us look at the reaction

$$H_2SO_4(aq) + NaOH(aq) \longrightarrow NaHSO_4(aq) + H_2O(l)$$

Each mole of H_2SO_4 supplies only 1 mol H^+, so an equivalent of H_2SO_4 in this reaction equals 1 mol H_2SO_4. Nonetheless, 1 equivalent of H_2SO_4 reacts with 1 equivalent of NaOH.

The **equivalent mass** of a substance is *the mass of one equivalent*. We can obtain the equivalent mass of a substance from the relationship between equivalents and moles. For example, in the first reaction between H_2SO_4 and NaOH, we have

$$1 \text{ mol } H_2SO_4 = 2 \text{ eq } H_2SO_4$$

The molar mass of H_2SO_4 is 98.1 g/mol, so

$$\text{Equivalent mass } H_2SO_4 = \frac{98.1 \text{ g } H_2SO_4}{\cancel{1 \text{ mol } H_2SO_4}} \times \frac{\cancel{1 \text{ mol } H_2SO_4}}{2 \text{ eq } H_2SO_4}$$
$$= 49.0 \text{ g } H_2SO_4/\text{eq } H_2SO_4 \text{ (49.0 g/eq)}$$

The **normality** of a solution is *the number of equivalents of a substance dissolved in a liter of solution*.

$$\text{Normality} = \frac{\text{equivalents of solute}}{\text{liters of solution}}$$

Suppose 2.00 L of a solution of sulfuric acid contains 0.30 eq of H_2SO_4. The normality of the solution is

$$\text{Normality} = \frac{0.30 \text{ eq } H_2SO_4}{2.00 \text{ L soln}} = 0.15\ N\ H_2SO_4$$

We say that the solution is 0.15 normal H_2SO_4.

From the definition of normality, note that the equivalents of solute can be obtained by multiplying the normality (N) by volume (V) of solution in liters.

$$\text{Equivalents} = N \times V$$

Example 13.9 Using Equivalents and Normality in Acid–Base Reactions

(a) A solution of 0.4203 g adipic acid was titrated with 0.1650 N NaOH solution. It required 34.87 mL of NaOH solution to obtain the end point of the titration (that is, to just react with the adipic acid). What is the equivalent mass of adipic acid?
(b) The molecular formula of adipic acid is $C_6H_{10}O_4$. How many acidic H atoms are there per molecule of adipic acid?

Solution

(a) The equivalents of NaOH used in the titration is

$$\text{Equivalents NaOH} = N \times V = 0.1650 \frac{\text{eq}}{\text{L}} \times 0.03487 \text{ L}$$
$$= 5.754 \times 10^{-3} \text{ eq}$$

By definition, the equivalents of two reactants must be equal, so

$$\text{Equivalents adipic acid} = \text{equivalents NaOH}$$
$$= 5.754 \times 10^{-3} \text{ eq}$$

The equivalent mass of adipic acid is its mass per equivalent; therefore,

$$\text{Equivalent mass of adipic acid} = \frac{0.4203 \text{ g}}{5.754 \times 10^{-3} \text{ eq}}$$
$$= \mathbf{73.04\ g/eq}$$

(b) The numerical value of the equivalents per mole equals the number of acidic H atoms in the adipic acid molecule. The molar mass, as calculated from the formula $C_6H_{10}O_4$, is 146.1 g/mol. Hence, the equivalents per mole is

$$\text{Equivalents/mole} = \frac{\text{molar mass}}{\text{equivalent mass}} = \frac{146.1 \text{ g/mol}}{73.04 \text{ g/eq}}$$
$$= 2.000 \text{ eq/mol}$$

Thus, there are **2** acidic H atoms per adipic acid molecule (that is, adipic acid is diprotic).

Exercise 13.12

A solution of 0.1916 g citric acid was titrated with 0.1156 N NaOH. It required 25.89 mL of NaOH solution to reach the end point. What is the equivalent mass of citric acid? Citric acid is triprotic. What is its molar mass? (See Problems 13.73, 13.74, 13.75, and 13.76.)

Oxidation–reduction reactions are frequently used in analysis (Figure 13.9). For analysis by titration and similar work, we find it useful to define *equivalent* for an oxidation–reduction reaction. An **equivalent (eq)** *in an oxidation–reduction reaction* is *the quantity of oxidizing or reducing agent that uses or provides one mole of electrons*. Thus, one equivalent of oxidizing agent reacts with one equivalent of reducing agent. For example, the permanganate ion, MnO_4^-, gains 5 electrons when it is reduced to Mn^{2+} in acidic solution.

$$MnO_4^-(aq) + 8H^+(aq) + 5e^- \longrightarrow Mn^{2+}(aq) + 4H_2O(l)$$

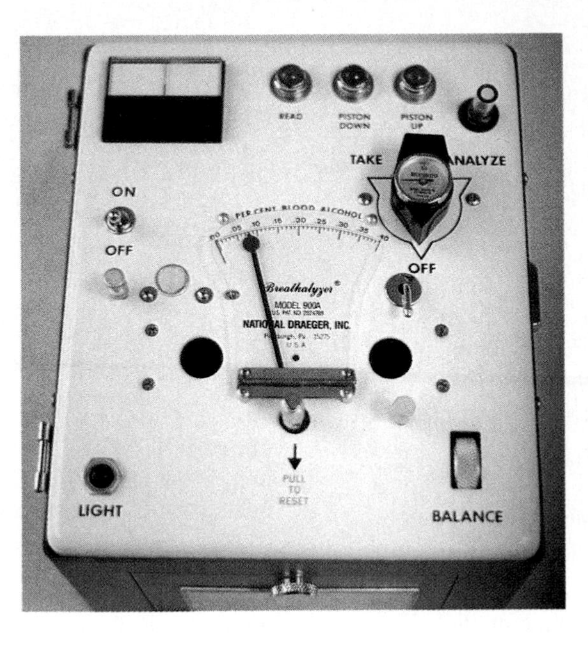

Figure 13.9
Breath analyzer for determining blood alcohol. Air in the lungs is in equilibrium with alcohol (ethanol) in the blood. A person breathes air into a cylinder, which collects a definite volume. When this sample is bubbled into a solution of potassium dichromate (orange) containing sulfuric acid, any ethanol (CH_3CH_2OH) in the breath is oxidized to acetic acid (CH_3COOH). In practice, a small quantity of silver nitrate is added as a catalyst. Ethanol reduces dichromate ion to chromium(III) ion, which is blue-green. The depth of the blue-green color is read by a photovoltaic cell, which converts the light to an electric current. The current is read on a meter scale in percent alcohol.

Therefore,

$$1 \text{ mol } MnO_4^- = 5 \text{ eq } MnO_4^-$$

We can obtain the equivalent mass of potassium permanganate, $KMnO_4$, from this result. We have

$$1 \text{ mol } KMnO_4 = 5 \text{ eq } KMnO_4$$

The molar mass of $KMnO_4$ is 158.0 g/mol, so

$$\text{Equivalent mass } KMnO_4 = \frac{158.0 \text{ g } KMnO_4}{\cancel{\text{mol } KMnO_4}} \times \frac{1 \cancel{\text{ mol } KMnO_4}}{5 \text{ eq } KMnO_4}$$
$$= 31.60 \text{ g } KMnO_4/\text{eq } KMnO_4$$

Example 13.10 Using Equivalents and Normality in Oxidation–Reduction Reactions

A 0.256-g sample of iron alloy (a mixture of iron with other elements) was dissolved in hydrochloric acid to give a solution of Fe^{2+} ion. This solution was titrated to the end point with 35.6 mL of 0.100 N $KMnO_4$, which oxidized Fe^{2+} to Fe^{3+}. What is the mass percentage of iron in the sample?

Solution

From the normality and volume of $KMnO_4$ solution, we can obtain the equivalents of $KMnO_4$ and therefore the equivalents of Fe^{2+}. We can then obtain the corresponding mass of iron and then the percentage of iron in the sample. We have

$$\text{Equivalents } Fe^{2+} = \text{equivalents } KMnO_4 = N \times V$$
$$= 0.100 \frac{\text{eq}}{\text{L}} \times 35.6 \times 10^{-3} \text{ L}$$
$$\text{Equivalents } Fe^{2+} = 3.56 \times 10^{-3} \text{ eq}$$

Because Fe^{2+} is oxidized to Fe^{3+}

$$Fe^{2+}(aq) \longrightarrow Fe^{3+}(aq) + e^-$$

one equivalent equals one mole. Therefore,

$$\text{Moles Fe} = \text{moles } Fe^{2+} = 3.56 \times 10^{-3} \text{ mol}$$

The molar mass of Fe is 55.85 g/mol, so

$$\text{Mass Fe} = 55.85 \text{ g/mol} \times 3.56 \times 10^{-3} \text{ mol} = 0.199 \text{ g}$$

The mass percentage of iron in the alloy is

$$\frac{0.199 \text{ g}}{0.256 \text{ g}} \times 100\% = 77.7\%$$

Exercise 13.13

A sample of air contains sulfur dioxide, SO_2. It was analyzed by dissolving the SO_2 in water to form $H_2SO_3(aq)$, and this was titrated with 0.0795 *N* $K_2Cr_2O_7$ (potassium dichromate). It required 28.3 mL of the $K_2Cr_2O_7$ solution to reach the end point. How many grams of SO_2 was in the air sample? The $K_2Cr_2O_7$ oxidizes H_2SO_3 to H_2SO_4 and is itself reduced to Cr^{3+}. (See Problems 13.81, 13.82, 13.83, and 13.84.)

Profile of a Chemical

POTASSIUM PERMANGANATE (an Oxidizing Agent)

Potassium permanganate, $KMnO_4$, is a deep purple, crystalline substance. Its preparation, shown in part in Figure 13.10, is interesting because it involves several oxidation states of manganese, each with distinctive color. The commercial preparation begins with manganese dioxide, MnO_2, a black substance obtained from the mineral pyrolusite. When manganese dioxide powder is mixed with potassium hydroxide and the mixture is melted in the presence of air, potassium manganate is obtained.

$$2MnO_2(s) + 4KOH(s) + O_2(g) \longrightarrow 2K_2MnO_4(s) + 2H_2O(l)$$

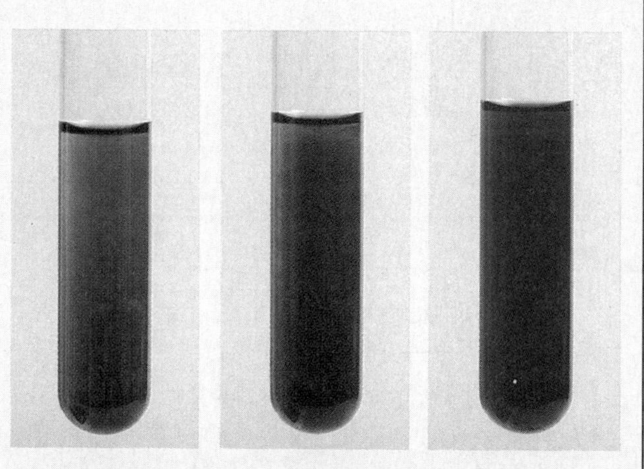

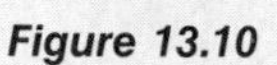

Figure 13.10
Preparation of potassium permanganate. Left: *A solution of potassium manganate, K_2MnO_4, was obtained by dissolving the solid residue obtained from heating KOH and MnO_2 to the melting point. The green color is from the MnO_4^{2-} ion; the solution contains some MnO_2 as a colloidal suspension (fine crystals).* Center: *Drops of hydrochloric acid were added to the K_2MnO_4 solution. In the acidic solution, MnO_4^{2-} ion disproportionates to MnO_4^- ion (red-purple) and MnO_2.* Right: *At the completion of the reaction, we obtain a solution of $KMnO_4$ containing a suspension of MnO_2. Some solid MnO_2 can be seen on the bottom of the test tube.*

Potassium manganate is a dark green color. Note that the anion is MnO_4^{2-} and should not be confused with permanganate ion, MnO_4^-. In this preparation, manganese in MnO_2 has an oxidation state of +4 and is oxidized by oxygen in air to MnO_4^{2-} in which manganese has an oxidation state of +6.

The solution in Figure 13.10 (left) was obtained by dissolving the melt from the preceding reaction in water; the solution has some colloidal (very small crystals) of MnO_2 in it. When the solution of potassium manganate is acidified, the manganate ion disproportionates (the ion is both oxidized and reduced).

$$\underset{\text{green}}{3MnO_4^{2-}(aq)} + 4H^+(aq) \longrightarrow \underset{\text{purple}}{2MnO_4^-(aq)} + \underset{\text{black}}{MnO_2(s)} + 2H_2O(l)$$

The change of color can be seen in Figure 13.10 (center, right), where several drops of hydrochloric acid have been added to the solution in Figure 13.10 (left). Note that manganese has changed from the +6 oxidation state to +7 (in MnO_4^-) and +4 (in MnO_2). Instead of the preceding reaction, the commercial preparation uses an electrolysis, in which electrical energy is used to accomplish the oxidation of manganate ion to permanganate ion (electrolysis is discussed in Chapter 19). The overall equation for this is

$$2MnO_4^{2-}(aq) + 2H_2O(l) \xrightarrow{\text{electrolysis}} 2MnO_4^-(aq) + 2OH^-(aq) + H_2(g)$$

Potassium permanganate is one of the most important oxidizing agents. As we noted earlier in this chapter, oxidations can be performed in acidic or basic solution. The half-reactions in acidic solution and in basic solution are as follows:

$$MnO_4^-(aq) + 8H^+(aq) + 5e^- \longrightarrow Mn^{2+}(aq) + 4H_2O(l) \quad \text{(acidic)}$$

$$MnO_4^-(aq) + 2H_2O(l) + 3e^- \longrightarrow MnO_2(s) + 4OH^-(aq) \quad \text{(basic)}$$

Potassium permanganate is used in the preparation of many organic compounds, where it is employed as an

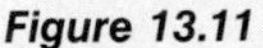

Figure 13.11
Oxidation of glycerol by potassium permanganate. *Drops of the organic compound glycerol, $C_3H_8O_3$, were added to crystals of potassium permanganate. After several seconds, the heat from the exothermic oxidation ignited the glycerol.*

oxidizing agent. (Figure 13.11 shows the highly exothermic oxidation of the organic compound glycerol, $C_3H_8O_3$). Potassium permanganate is also used commercially as a disinfectant, as an air purifier, and for waste water treatment. The compound is used in chemical analysis to determine the amount of certain reducing compounds. In this chapter, we noted its use in determining the amount of iron in a sample of iron ore by titration. Solutions of potassium permanganate for such titrations are made up approximately by mass of $KMnO_4$; then the exact concentration is determined by titration of a known quantity of oxalic acid, $H_2C_2O_4$, with the solution of $KMnO_4$. The balanced oxidation–reduction reaction is

$$2MnO_4^-(aq) + 5C_2O_4^{2-}(aq) + 16H^+(aq) \longrightarrow 2Mn^{2+}(aq) + 8H_2O(l) + 10CO_2(g)$$

Questions for Study

1. What is the name of the mineral used in the preparation of potassium permanganate?
2. Describe the colors of potassium manganate and potassium permanganate. What is the difference between the formulas of the two ions?
3. When manganate ion disproportionates in acidic solution, it forms permanganate ion and manganese dioxide. Balance this equation using either the oxidation-number method or the half-reaction method and compare with the equation given previously. Show the details of your work.
4. What are the principal uses of potassium permanganate?
5. The percentage of manganese in steel can be determined by dissolving the steel in nitric acid, converting the manganese to Mn^{2+} ion. The Mn^{2+} is then oxidized to MnO_4^- by periodate ion, IO_4^-, in acidic solution. The depth of the color of MnO_4^- is proportional to the amount of Mn originally present in the steel. Write the balanced equation for this oxidation of Mn^{2+} by IO_4^-, which is reduced to IO_3^- in acidic solution.

Profile of a Chemical

HYDROGEN PEROXIDE (an Oxidizing and Reducing Agent)

Hydrogen and oxygen form two well-known compounds, water and hydrogen peroxide, H_2O_2. Hydrogen peroxide has quite different properties from those of water. Solutions of hydrogen peroxide are commonly available in drugstores as an antiseptic and hair bleach. The hydrogen peroxide molecule has the following structural formula:

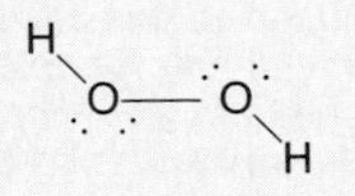

The O—O bond is characteristic of peroxide compounds.

Pure hydrogen peroxide is a syrupy liquid with a pale blue color. If very pure it is stable, but in the presence of small quantities of impurities, the liquid may decompose explosively. The decomposition is exothermic.

$$2H_2O_2(l) \longrightarrow 2H_2O(l) + O_2(g);\ \Delta H^\circ = -196.0\ \text{kJ}$$

Aqueous solutions of hydrogen peroxide similarly decompose in the presence of various catalysts, including platinum metal and ions such as Fe^{2+} and I^-.

When a 3% solution of H_2O_2 is poured on a skin wound to cleanse it, the solution fizzes violently. An en-

(continued)

Figure 13.12
Reaction of iodide ion with hydrogen peroxide. Left: *Solutions of potassium iodide, KI, and hydrogen peroxide, H_2O_2.* Right: *The experimenter pours the H_2O_2 solution into the KI solution, producing iodine, I_2. The I_2 reacts with I^- in the solution to give I_3^- ion, which appears orange-brown in dilute solution.*

zyme, or biological catalyst (called catalase) present in tissue and in blood, decomposes the hydrogen peroxide and produces bubbles of oxygen. The natural function of this enzyme is to prevent the accumulation of hydrogen peroxide in the body, since such a build-up of H_2O_2 could damage tissue. Hydrogen peroxide is formed naturally in the body during oxidation reactions involving oxygen, O_2.

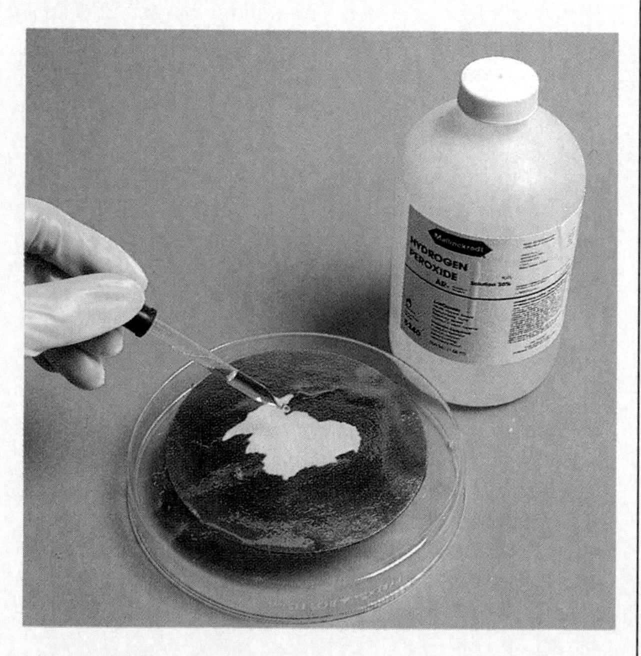

Figure 13.13
Oxidation of lead sulfide by hydrogen peroxide. *Several drops of hydrogen peroxide solution were added to lead sulfide, the brownish-black substance on the filter paper. Note the white area in the center of the filter paper. The white color is due to lead sulfate, obtained from the oxidation of lead sulfide.*

In most of its reactions, hydrogen peroxide acts as an oxidizing agent, giving water as the reduction product. For example, iodide ion in acidic solution is oxidized to iodine, I_2.

$$H_2O_2(aq) + 2H^+(aq) + 2I^-(aq) \longrightarrow 2H_2O(l) + I_2(aq)$$

In the presence of I^-, iodine forms the brown-colored ion I_3^- (Figure 13.12). The oxidizing ability of hydrogen peroxide also accounts for its use in the restoration of old paintings. The lead paint used in these paintings darkens in time as the result of the formation of black lead sulfide from small amounts of H_2S in air. Hydrogen peroxide converts PbS to $PbSO_4$, which is white (Figure 13.13). The bombardier beetle has a fascinating defense

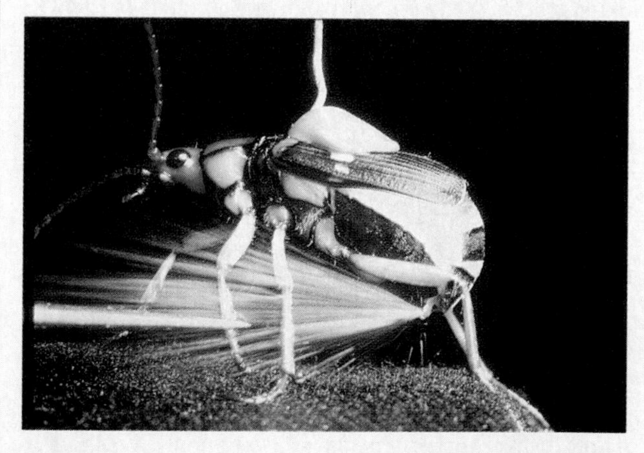

Figure 13.14
A bombardier beetle defending itself. *The beetle has an abdominal sac containing a liquid mixture of hydrogen peroxide and hydroquinone, $C_6H_6O_2$. When the beetle is agitated, enzymes are added to this mixture, initiating a chemical reaction between hydrogen peroxide and hydroquinone that generates enough heat to boil the liquid. With a popping sound, the beetle expels this steaming liquid at its foe.*

mechanism based on the oxidation of an organic compound with hydrogen peroxide (Figure 13.14). Within the beetle is a sac containing a mixture of the compound (hydroquinone) and hydrogen peroxide. An enzyme initiates the reaction, which is so exothermic that it causes the liquid to boil. The boiling liquid is then expelled at the beetle's enemies.

Hydrogen peroxide can also behave as a reducing agent in the presence of strong oxidizing agents, yielding oxygen and water. Potassium permanganate, $KMnO_4$, is one of these strong oxidizing agents. Purple-colored solutions of $KMnO_4$ are reduced to the pale pink manganese(II) ion (dilute solutions are colorless).

$$5H_2O_2(aq) + \underset{\text{purple}}{2MnO_4^-(aq)} + 6H^+(aq) \longrightarrow 8H_2O(l) + 5O_2(g) + \underset{\text{pale pink}}{2Mn^{2+}(aq)}$$

A similar reaction occurs in neutral or basic solution, except that a brownish precipitate of MnO_2 forms (Figure 13.15).

Hydrogen peroxide, as we noted earlier, is used as an antiseptic and hair bleach. However, its most important use is as a commercial bleach of textiles and paper pulp. Hydrogen peroxide is also increasingly used in treating waste water before it is discharged to natural bodies of water. It oxidizes certain pollutants to relatively harmless substances (for example, hydrogen sulfide is oxidized to sulfate ion).

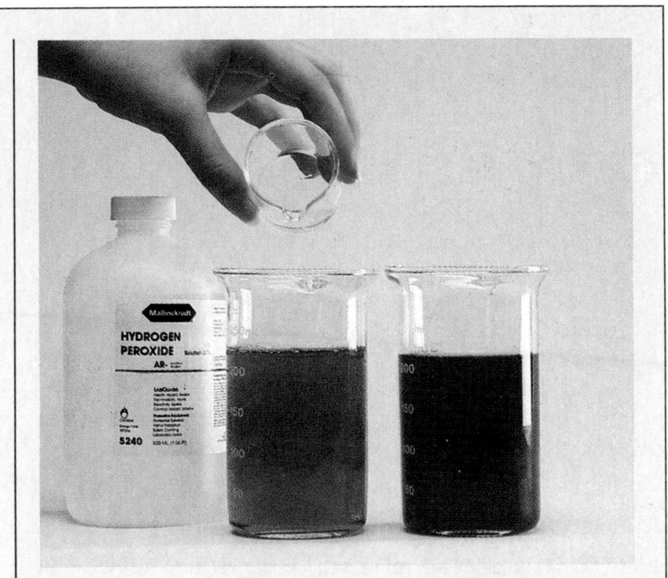

Figure 13.15
Reduction of potassium permanganate by hydrogen peroxide. *Here hydrogen peroxide solution is being added to a solution of potassium permanganate. The brown color produced is a suspension of manganese dioxide produced in the neutral mixture.*

Questions for Study

1. List some properties of hydrogen peroxide that are not possessed by water.
2. When blood is added to a solution of hydrogen peroxide, the solution bubbles furiously. What is the overall reaction that occurs?
3. Write the balanced equation for the oxidation of PbS to $PbSO_4$ by H_2O_2 in acidic solution.
4. Tetrahydroxochromate(III) ion, $Cr(OH)_4^-$, is obtained when Cr^{3+} ion is treated with a base. Write the balanced equation for the oxidation of $Cr(OH)_4^-$ ion to CrO_4^{2-} by H_2O_2 in basic solution.
5. Give an example of a reaction in which hydrogen peroxide functions as a reducing agent. Write the half-reaction for the oxidation of H_2O_2.

A Checklist for Review

IMPORTANT TERMS

acid (Brønsted–Lowry) (13.2)
base (Brønsted–Lowry) (13.2)
conjugate acid–base pair (13.2)
amphiprotic species (13.2)
Lewis acid (13.5)
Lewis base (13.5)
oxidation number (oxidation state) (13.6)
half-reaction (13.7)
oxidation–reduction (redox) reaction (13.7)
oxidation (13.7)
reduction (13.7)
oxidizing agent (13.7)
reducing agent (13.7)
activity series (13.7)
disproportionation (13.7)
equivalent (acid–base reaction) (13.9)
equivalent mass (13.9)
normality (13.9)
equivalent (oxidation–reduction reaction) (13.9)

KEY EQUATION

Equivalents = $N \times V$

SUMMARY OF FACTS AND CONCEPTS

The *Arrhenius concept* was the first successful theory of acids and bases. Then, in 1923, Brønsted and Lowry characterized acid–base reactions as proton-transfer reactions. According to the *Brønsted–Lowry concept,* an acid is a proton donor and a base is a proton acceptor.

Acid–base reactions can be viewed as a competition for protons. Equilibrium for such reactions is determined by the *relative strengths of the acids and bases*. Reaction favors the weaker acid and the weaker base. Acid strength depends on the polarity and strength of the bond involving the acidic H atom. We can therefore *relate molecular structure and acid strength*. A number of rules can be stated that allow us to predict the relative strengths of similar acids.

The *Lewis concept* is even more general than the Brønsted–Lowry concept. A Lewis acid is an electron-pair acceptor and a Lewis base is an electron-pair donor. Reactions of acidic and basic oxides and the formation of complex ions, as well as proton-transfer reactions, can be described in terms of the Lewis concept.

In an *oxidation–reduction (redox) reaction,* there is effectively a transfer of electrons from one atom to another. The atom that loses electrons and thereby increases in oxidation number is said to undergo *oxidation*. The atom that gains electrons and thereby decreases in oxidation number is said to undergo *reduction*. Oxidation and reduction must occur together in a reaction. We discuss the following oxidation–reduction reactions: displacement, disproportionation, and those involving oxoanions. *Displacement* is an oxidation–reduction reaction in which electrons are transferred from one metal to a monatomic ion of another metal. *Disproportionation* is a reaction in which a species is both oxidized and reduced. Oxidation–reduction equations can be balanced by either the *oxidation-number method* or the *half-reaction method*.

Equivalents and *normality* are defined to simplify calculations in chemical analyses. An equivalent of one reactant combines with an equivalent of another. A one-normal solution contains one equivalent per liter of solution.

OPERATIONAL SKILLS

1. **Identifying acid and base species** Given a proton-transfer reaction, label the acids and bases, and name the conjugate acid–base pairs. (Example 13.1).
2. **Deciding whether reactants or products are favored in an acid–base reaction** Given an acid–base reaction and the relative strengths of acids (or bases), decide whether reactants or products are favored (Example 13.2).
3. **Identifying Lewis acid and base species** Given a reaction involving the donation of an electron pair, identify the Lewis acid and the Lewis base (Example 13.3).
4. **Assigning oxidation numbers** Given the formula of a simple compound or ion, obtain the oxidation numbers of the atoms, using the rules given in the text (Examples 13.4 and 13.5).
5. **Balancing oxidation–reduction equations** Given the unbalanced or skeleton equation for an oxidation–reduction reaction, complete and balance it (Examples 13.6, 13.7, and 13.8).
6. **Using equivalents and normality** Given the volume and normality of a solution used to titrate a substance, obtain the equivalents and mass of the substance (Examples 13.9 and 13.10).

Review Questions

13.1 Define an acid and a base according to the Arrhenius concept. Give an example of each.

13.2 Which of the following are strong acids? Which are weak acids? (a) $HC_2H_3O_2$; (b) HClO; (c) HCl; (d) HNO_3; (e) HNO_2; (f) HCN.

13.3 Describe any thermochemical (heat of reaction) evidence for the Arrhenius concept.

13.4 Define an acid and a base according to the Brønsted–Lowry concept. Give an acid–base equation and identify each species as an acid or a base.

13.5 What is meant by the conjugate acid of a base?

13.6 Write an equation in which $H_2PO_3^-$ acts as an acid and another in which it acts as a base.

13.7 Describe four ways in which the Brønsted–Lowry concept enlarges on the Arrhenius concept.

13.8 Explain why an acid–base reaction favors the weaker acid.

13.9 Formic acid, $HCHO_2$, is a stronger acid than acetic acid, $HC_2H_3O_2$. Which is the stronger base, CHO_2^- or $C_2H_3O_2^-$?

13.10 Give two important factors that determine the strength of an acid. How does an increase in each factor affect the acid strength?

13.11 Define an acid and a base according to the Lewis concept. Give a chemical equation to illustrate.

13.12 Define *oxidation* and *reduction* in terms of electrons transferred and in terms of oxidation numbers.

13.13 Why must oxidation and reduction occur together in a reaction?

13.14 Give an example of a displacement reaction. What is the oxidizing agent? What is the reducing agent?

13.15 Define an equivalent for an acid–base reaction and for an oxidation–reduction reaction.

13.16 Barium hydroxide, $Ba(OH)_2$, is a strong base. In a neutralization, both OH^- ions from the base react. Describe how you would prepare a 0.10 *N* solution.

Practice Problems

BRØNSTED–LOWRY CONCEPT

13.17 Write the balanced equation for the ionization of HSO_3^- in water to form an acidic solution. Identify each species in the equation as either an acid or a base.

13.18 Write the balanced equation for the ionization of HPO_4^{2-} in water to form a basic solution.

13.19 Give the conjugate base to each of the following species regarded as acids.
(a) HSO_4^{2-} (b) H_2S (c) $H_2PO_4^-$ (d) NH_4^+

13.20 Give the conjugate base to each of the following species regarded as acids.
(a) $HClO$ (b) HPO_4^{2-} (c) HCN (d) $CH_3NH_3^+$

13.21 Give the conjugate acid to each of the following species regarded as bases.
(a) CN^- (b) HCO_3^- (c) SeO_4^{2-} (d) PO_4^{3-}

13.22 Give the conjugate acid to each of the following species regarded as bases.
(a) IO^- (b) SO_3^{2-} (c) HSO_3^- (d) CHO_2^-

13.23 For the following reactions, label each of the species as an acid or a base. Indicate the species that are conjugates of one another.
(a) $H_2C_2O_4 + ClO^- \rightleftharpoons HC_2O_4^- + HClO$
(b) $HPO_4^{2-} + NH_4^+ \rightleftharpoons NH_3 + H_2PO_4^-$
(c) $SO_4^{2-} + H_2O \rightleftharpoons HSO_4^- + OH^-$
(d) $Fe(H_2O)_6^{2+} + H_2O \rightleftharpoons H_3O^+ + Fe(H_2O)_5OH^+$

13.24 For the following reactions, label each of the species as an acid or a base. Indicate the species that are conjugates of one another.
(a) $HSO_4^- + HCO_3^- \rightleftharpoons SO_4^{2-} + H_2CO_3$
(b) $CN^- + H_2O \rightleftharpoons HCN + OH^-$
(c) $HCO_3^- + H_2O \rightleftharpoons OH^- + H_2CO_3$
(d) $HSO_4^- + H_2O \rightleftharpoons H_3O^+ + SO_4^{2-}$

ACID AND BASE STRENGTHS

13.25 Complete the following equation. Using Table 13.1, predict whether you would expect the reaction to occur to any significant extent or whether the reaction is more likely to occur in the opposite direction.

$$HSO_4^-(aq) + ClO^-(aq) \longrightarrow$$

13.26 Complete the following equation. Using Table 13.1, predict whether you would expect the reaction to occur to any significant extent or whether the reaction is more likely to occur in the opposite direction.

$$HCN(aq) + SO_4^{2-}(aq) \longrightarrow$$

13.27 Use Table 13.1 to decide whether the species on the left or those on the right are favored by the reaction.
(a) $NH_4^+ + H_2PO_4^- \rightleftharpoons NH_3 + H_3PO_4$
(b) $HCN + HS^- \rightleftharpoons CN^- + H_2S$
(c) $HCO_3^- + OH^- \rightleftharpoons CO_3^{2-} + H_2O$
(d) $Al(H_2O)_6^{3+} + OH^- \rightleftharpoons Al(H_2O)_5OH^{2+} + H_2O$

13.28 Use Table 13.1 to decide whether the species on the left or those on the right are favored by the reaction.
(a) $NH_4^+ + CO_3^{2-} \rightleftharpoons NH_3 + HCO_3^-$
(b) $HCO_3^- + H_2S \rightleftharpoons H_2CO_3 + HS^-$
(c) $CN^- + H_2O \rightleftharpoons HCN + OH^-$
(d) $CN^- + H_2CO_3 \rightleftharpoons HCN + HCO_3^-$

13.29 In the following reaction of trichloroacetic acid, $HC_2Cl_3O_2$, with formate ion, CHO_2^-, the formation of trichloroacetate ion, $C_2Cl_3O_2^-$, and formic acid, $HCHO_2$ are favored.

$$HC_2Cl_3O_2 + CHO_2^- \longrightarrow C_2Cl_3O_2^- + HCHO_2$$

Which is the stronger acid, trichloroacetic acid or formic acid? Explain.

13.30 In the following reaction of tetrafluoroboric acid, HBF_4, with the acetate ion, $C_2H_3O_2^-$, the formation of tetrafluoroborate ion, BF_4^-, and acetic acid are favored.

$$HBF_4 + C_2H_3O_2^- \longrightarrow BF_4^- + HC_2H_3O_2$$

Which is the weaker base, BF_4^- or acetate ion?

13.31 For each of the following pairs, give the stronger acid. Explain your answer.
(a) H_2S, HS^- (b) H_2SO_3, H_2SeO_3
(c) HBr, H_2Se (d) HIO_4, HIO_3
(e) H_2S, H_2O

13.32 Order each of the following pairs by acid strength, giving the weaker acid first. Explain your answer.
(a) HNO_3, HNO_2 (b) HCO_3^-, H_2CO_3
(c) H_2S, H_2Te (d) HCl, H_2S
(e) H_3PO_4, H_3AsO_4

LEWIS ACID–BASE CONCEPT

13.33 Complete each of the following equations. Then write the Lewis formulas of the reactants and products and identify each reactant as a Lewis acid or a Lewis base.
(a) $CO_2 + OH^- \longrightarrow$
(b) $AlCl_3 + Cl^- \longrightarrow$

13.34 Complete each of the following equations. Then write the Lewis formulas of the reactants and products and identify each reactant as a Lewis acid or a Lewis base.
(a) $CN^- + H_2S \longrightarrow$
(b) $BF_3 + PH_3 \longrightarrow$

13.35 In the following reactions, identify each of the reactants as a Lewis acid or a Lewis base.
(a) $Cr^{3+} + 6H_2O \longrightarrow Cr(H_2O)_6^{3+}$
(b) $BF_3 + (C_2H_5)_2\ddot{O}: \longrightarrow F_3B:\ddot{O}(C_2H_5)_2$

13.36 In the following reactions, label each of the reactants as a Lewis acid or a Lewis base.
(a) $BeF_2 + 2F^- \longrightarrow BeF_4^{2-}$
(b) $SnCl_4 + 2Cl^- \longrightarrow SnCl_6^{2-}$

13.37 Natural gas frequently contains hydrogen sulfide, H_2S. H_2S is removed from natural gas by passing it through aqueous ethanolamine, $HOCH_2CH_2NH_2$ (an ammonia derivative), which reacts with the hydrogen sulfide. Write the equation for the reaction. Identify each reactant as either a Lewis acid or a Lewis base. Explain how you arrived at your answer.

13.38 Coal and other fossil fuels usually contain sulfur compounds that produce sulfur dioxide, SO_2, when burned. One possible way to remove the sulfur dioxide is to pass the combustion gases into a tower packed with calcium oxide, CaO. Write the equation for the reaction. Identify each reactant as either a Lewis acid or a Lewis base. Explain how you arrived at your answer.

OXIDATION NUMBERS

13.39 Obtain the oxidation number for the element noted in each of the following.
(a) Ga in Ga_2O_3 (b) Pb in PbO_2
(c) Br in $KBrO_4$ (d) Mn in K_2MnO_4

13.40 Obtain the oxidation number for the element noted in each of the following.
(a) Cr in CrO_3 (b) Hg in Hg_2Cl_2
(c) Ga in $Ga(OH)_3$ (d) P in Na_3PO_4

13.41 Obtain the oxidation number for the element noted in each of the following.
(a) N in NH_2^- (b) I in IO_3^-
(c) Al in $Al(OH)_4^-$ (d) P in $H_2PO_4^-$

13.42 Obtain the oxidation number for the element noted in each of the following.
(a) N in NO_3^- (b) Cr in CrO_4^{2-}
(c) Zn in $Zn(OH)_4^{2-}$ (d) As in $H_2AsO_3^-$

13.43 Determine the oxidation numbers of all the elements in each of the following compounds. (*Hint:* Look at the ions present.)

(a) $Mn(ClO_3)_2$ (b) $Fe_2(CrO_4)_3$
(c) $HgCr_2O_7$ (d) $Co_3(PO_4)_2$

13.44 Determine the oxidation numbers of all the elements in each of the following compounds. (*Hint:* Look at the ions present.)

(a) $Hg_2(BrO_3)_2$ (b) $Cr_2(SO_4)_3$
(c) $CoSeO_4$ (d) Cu_2SO_3

DESCRIBING OXIDATION–REDUCTION REACTIONS

13.45 In the following reactions, label the oxidizing agent and the reducing agent.

(a) $P_4(s) + 5O_2(g) \longrightarrow P_4O_{10}(s)$
(b) $Co(s) + Cl_2(g) \longrightarrow CoCl_2(s)$

13.46 In the following reactions, label the oxidizing agent and the reducing agent.

(a) $ZnO(s) + C(g) \longrightarrow Zn(s) + CO(g)$
(b) $8Fe(s) + S_8(s) \longrightarrow 8FeS(s)$

13.47 In the following reactions, label the oxidizing agent and the reducing agent.

(a) $2Al(s) + 3F_2(g) \longrightarrow 2AlF_3(s)$
(b) $Hg^{2+}(aq) + NO_2^-(aq) + H_2O(l) \longrightarrow Hg(s) + 2H^+(aq) + NO_3^-(aq)$

13.48 In the following reactions, label the oxidizing agent and the reducing agent.

(a) $Fe_2O_3(s) + 3CO(g) \longrightarrow 2Fe(s) + 3CO_2(g)$
(b) $PbS(s) + 4H_2O_2(aq) \longrightarrow PbSO_4(s) + 4H_2O(l)$

13.49 What are the half-reactions for the following reaction?

$$Ni(s) + Cu^{2+}(aq) \longrightarrow Ni^{2+}(aq) + Cu(s)$$

Which is the oxidation half-reaction, and which is the reduction half-reaction?

13.50 What are the half-reactions for the following reaction?

$$Zn(s) + 2Ag^+(aq) \longrightarrow Zn^{2+}(aq) + 2Ag(s)$$

Which is the oxidation half-reaction, and which is the reduction half-reaction?

13.51 Decide whether each of the following reactions actually occurs.

(a) $Pb(s) + Mg^{2+}(aq) \longrightarrow Pb^{2+}(aq) + Mg(s)$
(b) $Zn(s) + 2Ag^+(aq) \longrightarrow Zn^{2+}(aq) + 2Ag(s)$
(c) $2Cr(s) + 6H^+(aq) \longrightarrow 2Cr^{3+}(aq) + 3H_2(g)$
(d) $2Au(s) + 6H^+(aq) \longrightarrow 2Au^{3+}(aq) + 3H_2(g)$

13.52 Decide whether each of the following reactions actually occurs.

(a) $Sn(s) + Zn^{2+}(aq) \longrightarrow Sn^{2+}(aq) + Zn(s)$
(b) $Hg(s) + 2H^+(aq) \longrightarrow Hg^{2+}(aq) + H_2(g)$
(c) $2H^+(aq) + Ni(s) \longrightarrow H_2(g) + Ni^{2+}(aq)$
(d) $3Ag^+(aq) + Al(s) \longrightarrow 3Ag(s) + Al^{3+}(aq)$

13.53 Identify the possible products in each of the following reactions. If there is any uncertainty about what the products might be, explain.

(a) Permanganate ion oxidizes sulfite ion in acidic solution.
(b) Nitric acid oxidizes copper metal.

13.54 Identify the possible products in each of the following reactions. If there is any uncertainty about what the products might be, explain.

(a) Dichromate ion oxidizes nitrite ion in acidic solution.
(b) Oxygen, O_2, oxidizes iodide ion in acidic solution.

BALANCING OXIDATION–REDUCTION EQUATIONS

13.55 Balance the following oxidation–reduction equations. Identify the oxidizing and reducing agents.

(a) $HI + HNO_3 \longrightarrow I_2 + NO$
(b) $Ag + H_2SO_4 \longrightarrow Ag_2SO_4 + SO_2$
(c) $MnCl_2 + KMnO_4 + KOH \longrightarrow MnO_2 + KCl$
(d) $Cr(OH)_3 + H_2O_2 + KOH \longrightarrow K_2CrO_4 + H_2O$
(e) $HClO_3 \longrightarrow HClO_4 + ClO_2$

13.56 Balance the following oxidation–reduction equations. Identify the oxidizing and reducing agents.

(a) $H_2S + HNO_3 \longrightarrow S_8 + NO$
(b) $AsH_3 + KClO_3 \longrightarrow H_3AsO_4 + KCl$
(c) $H_2SO_4 + HBr \longrightarrow SO_2 + Br_2$
(d) $P_4 + HNO_3 \longrightarrow H_3PO_4 + NO$
(e) $NO_2 \longrightarrow HNO_3 + NO$

13.57 Balance the following oxidation–reduction equations. You may have to add H_2O. Identify the oxidizing and reducing agents.
(a) $H_3AsO_4 + Zn + HNO_3 \longrightarrow AsH_3 + Zn(NO_3)_2$
(b) $SnCl_2 + O_2 + HCl \longrightarrow H_2SnCl_6$
(c) $K_2MnO_4 \longrightarrow MnO_2 + KMnO_4 + KOH$
(d) $H_2SO_4 + HI \longrightarrow H_2S + I_2$
(e) $MnSO_4 + PbO_2 + H_2SO_4 \longrightarrow HMnO_4 + PbSO_4$

13.58 Balance the following oxidation–reduction equations. You may have to add H_2O. Identify the oxidizing and reducing agents.
(a) $H_2S + H_2O_2 \longrightarrow S_8 + H_2O$
(b) $P_4 + NaOH \longrightarrow NaH_2PO_4 + PH_3$
(c) $MnO_2 + HBr \longrightarrow Br_2 + MnBr_2$
(d) $K_2S + KMnO_4 \longrightarrow S_8 + MnO_2 + KOH$
(e) $S_8 + H_2SO_4 \longrightarrow SO_2$

13.59 Balance the following oxidation–reduction equations. The reactions occur in acidic solution.
(a) $Cr_2O_7^{2-} + C_2O_4^{2-} \longrightarrow Cr^{3+} + CO_2$
(b) $Cu + NO_3^- \longrightarrow Cu^{2+} + NO$
(c) $MnO_2 + HNO_2 \longrightarrow Mn^{2+} + NO_3^-$
(d) $PbO_2 + Mn^{2+} + SO_4^{2-} \longrightarrow PbSO_4 + MnO_4^-$
(e) $HNO_2 + Cr_2O_7^{2-} \longrightarrow Cr^{3+} + NO_3^-$

13.60 Balance the following oxidation–reduction equations. The reactions occur in acidic solution.
(a) $Mn^{2+} + BiO_3^- \longrightarrow MnO_4^- + Bi^{3+}$
(b) $Cr_2O_7^{2-} + I^- \longrightarrow Cr^{3+} + IO_3^-$
(c) $MnO_4^- + H_2SO_3 \longrightarrow Mn^{2+} + SO_4^{2-}$
(d) $Cr_2O_7^{2-} + Fe^{2+} \longrightarrow Cr^{3+} + Fe^{3+}$
(e) $As + ClO_3^- \longrightarrow H_3AsO_3 + HClO$

13.61 Balance the following oxidation–reduction equations. The reactions occur in basic solution.
(a) $Mn^{2+} + H_2O_2 \longrightarrow MnO_2 + H_2O$
(b) $MnO_4^- + NO_2^- \longrightarrow MnO_2 + NO_3^-$
(c) $Mn^{2+} + ClO_3^- \longrightarrow MnO_2 + ClO_2$
(d) $MnO_4^- + NO_2 \longrightarrow MnO_2 + NO_3^-$
(e) $Cl_2 \longrightarrow Cl^- + ClO_3^-$

13.62 Balance the following oxidation–reduction equations. The reactions occur in basic solution.
(a) $Cr(OH)_4^- + H_2O_2 \longrightarrow CrO_4^{2-} + H_2O$
(b) $MnO_4^- + Br^- \longrightarrow MnO_2 + BrO_3^-$
(c) $Co^{2+} + H_2O_2 \longrightarrow Co(OH)_3 + H_2O$
(d) $Pb(OH)_4^{2-} + ClO^- \longrightarrow PbO_2 + Cl^-$
(e) $Zn + NO_3^- \longrightarrow NH_3 + Zn(OH)_4^{2-}$

13.63 Balance the following oxidation–reduction equations. The reactions occur in acidic or basic aqueous solution, as indicated.
(a) $H_2S + NO_3^- \longrightarrow NO_2 + S_8$ (acidic)
(b) $NO_3^- + Cu \longrightarrow NO + Cu^{2+}$ (acidic)
(c) $MnO_4^- + SO_2 \longrightarrow SO_4^{2-} + Mn^{2+}$ (acidic)
(d) $Bi(OH)_3 + Sn(OH)_3^- \longrightarrow Sn(OH)_6^{2-} + Bi$ (basic)

13.64 Balance the following oxidation–reduction equations. The reactions occur in acidic or basic aqueous solution, as indicated.
(a) $Hg_2^{2+} + H_2S \longrightarrow Hg + S_8$ (acidic)
(b) $S^{2-} + I_2 \longrightarrow SO_4^{2-} + I^-$ (basic)
(c) $Al + NO_3^- \longrightarrow Al(OH)_4^- + NH_3$ (basic)
(d) $MnO_4^- + C_2O_4^{2-} \longrightarrow MnO_2 + CO_2$ (basic)

13.65 Balance the following oxidation–reduction reactions. The reactions occur in acidic or basic aqueous solution, as indicated.
(a) $MnO_4^- + I^- \longrightarrow MnO_2 + IO_3^-$ (basic)
(b) $Cr_2O_7^{2-} + Cl^- \longrightarrow Cr^{3+} + Cl_2$ (acidic)
(c) $S_8 + NO_3^- \longrightarrow SO_2 + NO$ (acidic)
(d) $H_2O_2 + MnO_4^- \longrightarrow O_2 + MnO_2$ (basic)
(e) $Zn + NO_3^- \longrightarrow Zn^{2+} + N_2$ (acidic)

13.66 Balance the following oxidation–reduction reactions. The reactions occur in acidic or basic aqueous solution, as indicated.
(a) $Cr_2O_7^{2-} + H_2O_2 \longrightarrow Cr^{3+} + O_2$ (acidic)
(b) $CN^- + MnO_4^- \longrightarrow CNO^- + MnO_2$ (basic)
(c) $Cr(OH)_4^- + OCl^- \longrightarrow CrO_4^{2-} + Cl^-$ (basic)
(d) $Br_2 + SO_2 \longrightarrow Br^- + SO_4^{2-}$ (acidic)
(e) $CuS + NO_3^- \longrightarrow Cu^{2+} + 2NO + S_8$ (acidic)

EQUIVALENTS AND NORMALITY

13.67 Obtain the equivalent mass of each of the following, assuming complete neutralization in each case.
(a) H_3PO_4 (b) LiOH (c) $Mg(OH)_2$ (d) $HC_2H_3O_2$

13.68 Obtain the equivalent mass of each of the following, assuming complete neutralization in each case.
(a) KOH (b) $HClO_4$ (c) $Ca(OH)_2$ (d) H_3AsO_4

13.69 What is the normality of a solution containing 1.68 g H_2SO_4 in 145 mL of solution, assuming complete neutralization?

13.70 What is the normality of a solution containing 3.45 g H_3PO_4 in 135 mL of solution, assuming complete neutralization?

13.71 What is the normality of a solution containing 2.56 g $Ba(OH)_2$ in 135 mL of solution?

13.72 Calculate the normality of a solution containing 1.47 g $Sr(OH)_2$ in 255 mL of solution.

13.73 A 25.0-mL sample of barium hydroxide, $Ba(OH)_2$, solution was titrated with 0.150 *N* HCl. The titration required 45.3 mL of HCl. What was the normality of the barium hydroxide solution? What was the molarity?

13.74 A 35.2-mL sample of strontium hydroxide, $Sr(OH)_2$, solution was titrated with 0.250 *N* HCl. If the titration used 38.2 mL of HCl, what was the normality of the strontium hydroxide solution? What was the molarity?

13.75 An antacid tablet contains magnesium hydroxide, $Mg(OH)_2$, and an inert binding ingredient. The mass of $Mg(OH)_2$ in the tablet was determined by titration with hydrochloric acid. If one tablet required 39.1 mL of 0.2056 *N* HCl for complete neutralization, how many equivalents of magnesium hydroxide were there in the tablet? How many grams is this?

13.76 A beverage mixture contains citric acid, $H_3C_6H_5O_7$, plus sugar and flavorings. The mass of citric acid in a sample was determined by titration with sodium hydroxide solution. If a 1.500-g sample of the beverage mix required 46.65 mL of 0.1068 *N* NaOH for complete neutralization, how many equivalents of citric acid were there in the sample? How many grams is this?

13.77 Obtain the equivalent mass of the following oxidizing or reducing agents. Refer to Table 13.5 for the half-reaction.

(a) $KMnO_4$, oxidizing agent in acidic solution
(b) $K_2Cr_2O_7$, oxidizing agent in acidic solution
(c) $SnCl_2$, reducing agent in acidic solution
(d) Na_2SO_3, reducing agent in basic solution

13.78 Obtain the equivalent mass of the following oxidizing or reducing agents. Refer to Table 13.5 for the half-reaction.

(a) H_2O_2, oxidizing agent in acidic solution
(b) NaClO, oxidizing agent in basic solution
(c) $KMnO_4$, oxidizing agent in basic solution
(d) HNO_3, oxidizing agent (assume HNO_3 is reduced to NO)

13.79 What is the normality of a solution containing 6.58 g $KMnO_4$ in 125 mL of solution? Assume that $KMnO_4$ is reduced to Mn^{2+} (in acidic solution).

13.80 Calculate the normality of a solution containing 3.59 g $SnCl_2$ in 135 mL of solution. Assume that $SnCl_2$ is oxidized to Sn^{4+}.

13.81 A 25.0-mL sample of iron(II) chloride solution, $FeCl_2(aq)$, was made acidic and titrated with 0.150 *N* $KMnO_4$. If the titration required 39.2 mL of $KMnO_4$ solution, what was the normality of the $FeCl_2$ solution? What was the molarity?

13.82 A 35.0-mL sample of hydrogen peroxide solution, $H_2O_2(aq)$, was made acidic and titrated with 0.125 *N* $KMnO_4$. If the titration required 43.2 mL of $KMnO_4$ solution, what was the normality of the H_2O_2 solution? What was the molarity?

13.83 A hair bleach solution contained hydrogen peroxide, H_2O_2. The amount of H_2O_2 in a 12.5-g sample was determined by titration with potassium permanganate in acidic solution. If 39.3 mL of 0.5045 *N* $KMnO_4$ was required, how many equivalents of H_2O_2 are contained in the solution? What is the mass percentage of H_2O_2 in the bleach solution?

13.84 The amount of iron in an iron ore sample was determined by converting the iron to Fe^{2+}, then titrating with potassium dichromate in acidic solution. Iron(II) ion was oxidized to Fe^{3+}. A 1.500-g sample of ore gave an Fe^{2+} solution that required 41.6 mL of 0.1058 *N* $K_2Cr_2O_7$. How many equivalents of Fe^{2+} were in the solution? What is the mass percentage of Fe in the ore?

Additional Problems

13.85 Identify each of the following as an acid or base in terms of the Arrhenius concept. Give the chemical equation for the reaction of the substance with water, showing the origin of the acidity or basicity.

(a) BaO (b) H_2S (c) CH_3NH_2 (d) SO_2

13.86 Which of the following substances are acids in terms of the Arrhenius concept? Which are bases? Show this acid or base character by using chemical equations.

(a) P_4O_{10} (b) K_2O (c) N_2H_4 (d) H_2Se

13.87 Write a reaction for each of the following in which the species acts as a Brønsted acid. The equilibrium should favor the product side.

(a) H_2O_2 (b) HCO_3^- (c) NH_4^+ (d) $H_2PO_4^-$

13.88 Write a reaction for each of the following in which the species acts as a Brønsted base. The equilibrium should favor the product side.

(a) H_2O (b) HCO_3^- (c) NH_3 (d) $H_2PO_4^-$

13.89 Write the following compounds in order of increasing acid strength: HBr, H_2Se, H_2S.

13.90 Write the following compounds in order of increasing acid strength: $HBrO_2$, $HClO_2$, HBrO.

13.91 For each of the following, write the complete chemical equation for the acid–base reaction that occurs. Describe each using Brønsted language (if appropriate) and then using Lewis language (show electron-dot formulas).

(a) The ClO^- ion reacts with water.
(b) The reaction of NH_4^+ and NH_2^- in liquid ammonia to produce NH_3.

13.92 For each of the following, write the complete chemical equation for the acid–base reaction that occurs. Describe each using Brønsted language (if appropriate) and then using Lewis language (show electron-dot formulas).

(a) The HS^- ion reacts in water to produce H_2S.
(b) Cyanide ion, CN^-, reacts with Fe^{3+}.

13.93 Identify each of the following as an acid–base reaction or an oxidation–reduction reaction. On the reactant side, identify the acid and base or the oxidizing and reducing agent, as appropriate.
(a) $2HBr(aq) + Cl_2(aq) \longrightarrow 2HCl(aq) + Br_2(aq)$
(b) $2HBr(aq) + Ca(OH)_2(aq) \longrightarrow CaBr_2(aq) + 2H_2O(l)$
(c) $NaCN(aq) + NaHSO_4(aq) \longrightarrow HCN(aq) + Na_2SO_4(aq)$
(d) $3NaClO(aq) \longrightarrow NaClO_3(aq) + 2NaCl(aq)$

13.94 Identify each of the following as an acid–base reaction or an oxidation–reduction reaction. On the reactant side, identify the acid and base or the oxidizing and reducing agent, as appropriate.
(a) $PbO_2(s) + 4HCl(aq) \longrightarrow PbCl_2(s) + Cl_2(g) + 2H_2O(l)$
(b) $3HNO_2(aq) \longrightarrow HNO_3(aq) + 2NO(g) + H_2O(l)$
(c) $2NaNO_2(aq) + H_2SO_4(aq) \longrightarrow 2HNO_2(aq) + Na_2SO_4(aq)$
(d) $2HNO_3(aq) + Ba(OH)_2(aq) \longrightarrow Ba(NO_3)_2(aq) + 2H_2O(l)$

13.95 Balance the following skeleton equations. The reactions occur in acidic or basic aqueous solution, as indicated in parentheses.
(a) $MnO_4^- + S^{2-} \longrightarrow MnO_2 + S_8$ (basic)
(b) $IO_3^- + HSO_3^- \longrightarrow I^- + SO_4^{2-}$ (acidic)
(c) $Fe(OH)_2 + CrO_4^{2-} \longrightarrow Fe(OH)_3 + Cr(OH)_4^-$ (basic)
(d) $Cl_2 \longrightarrow Cl^- + ClO^-$ (basic)

13.96 Balance the following skeleton equations. The reactions occur in acidic or basic aqueous solution, as indicated in parentheses.
(a) $MnO_4^- + H_2S \longrightarrow Mn^{2+} + S_8$ (acidic)
(b) $Zn + NO_3^- \longrightarrow Zn^{2+} + N_2O$ (acidic)
(c) $MnO_4^{2-} \longrightarrow MnO_4^- + MnO_2$ (basic)
(d) $Br_2 \longrightarrow Br^- + BrO_3^-$ (basic)

13.97 Iron(II) hydroxide is a greenish precipitate that is formed from iron(II) ion by the addition of a base. This precipitate gradually turns to the yellowish-brown iron(III) hydroxide via oxidation by O_2 in the air. Write a balanced equation for this oxidation by O_2.

13.98 A sensitive test for bismuth(III) ion consists of shaking a solution suspected of containing the ion with a basic solution of sodium stannite, Na_2SnO_2. A positive test consists of the formation of a black precipitate of bismuth metal. Stannite ion is oxidized by bismuth(III) ion to stannate ion, SnO_3^{2-}. Write a balanced equation for the reaction just described.

13.99 The normality of a solution of iodine, I_2, can be determined by titrating a solution containing a known mass of arsenious acid, H_3AsO_3, with the solution of iodine. What is the normality of a solution of I_2 if 0.840 g H_3AsO_3 reacts with exactly 25.4 mL of $I_2(aq)$? In this reaction, H_3AsO_3 is oxidized to arsenic acid, H_3AsO_4, and I_2 is reduced to I^-.

13.100 The normality of a solution of potassium dichromate, $K_2Cr_2O_7$, can be determined by titration with potassium iodide solution. In acidic solution, $Cr_2O_7^{2-}$ is reduced to Cr^{3+} and I^- is oxidized to I_2. What is the normality of a solution of $K_2Cr_2O_7$ if 48.5 mL of the solution requires 38.3 mL of 0.0500 *N* KI for complete reaction?

Cumulative-Skills Problems

13.101 Phosphorous acid, H_3PO_3, and phosphoric acid, H_3PO_4, have approximately the same acid strengths. From this information, and noting the possibility that one or more hydrogen atoms may be directly bonded to the phosphorus atom, draw the structural formula of phosphorous acid. How many grams of sodium hydroxide would be required to completely neutralize 1.00 g of this acid?

13.102 Hypophosphorous acid, H_3PO_2, and phosphoric acid, H_3PO_4, have approximately the same acid strengths. From this information, and noting the possibility that one or more hydrogen atoms may be directly bonded to the phosphorus atom, draw the structural formula of hypophosphorous acid. How many grams of sodium hydroxide would be required to completely neutralize 1.00 g of this acid?

13.103 Boron trifluoride, BF_3, and ammonia, NH_3, react to produce $BF_3 \cdot NH_3$. A coordinate covalent bond is formed between the boron atom on BF_3 and the nitrogen atom on NH_3. Write the equation for this reaction, using Lewis electron-dot formulas. Label the Lewis acid and the Lewis base. Determine how many grams of $BF_3 \cdot NH_3$ are formed when 10.0 g of BF_3 and 10.0 g of NH_3 are placed in a reaction vessel, assuming that the reaction goes to completion.

13.104 Boron trifluoride, BF_3, and diethyl ether, $(C_2H_5)_2O$, react to produce a compound whose formula is written $BF_3 \cdot (C_2H_5)_2O$. A coordinate covalent bond is formed between the boron atom on BF_3 and the oxygen atom on $(C_2H_5)_2O$. Write the equation for this reaction, using Lewis electron-dot formulas. Label the Lewis acid and the Lewis base. Determine how many grams of $BF_3 \cdot (C_2H_5)_2O$ are formed when 10.0 g of BF_3 and 20.0 g of $(C_2H_5)_2O$ are placed in a reaction vessel, assuming that the reaction goes to completion.

Figure 14.2
Catalytic decomposition of hydrogen peroxide. The hydrogen peroxide decomposes rapidly when hydrobromic acid is added to an aqueous solution. One of the products is oxygen gas, which bubbles vigorously from the solution. In addition to this catalytic decomposition, some HBr is oxidized to Br_2, as can be seen from the red color of the liquid and vapor.

3. *Temperature at which the reaction occurs*. Usually reactions speed up when the temperature increases. It takes less time to boil an egg at sea level than it does on a mountain top, where water boils at a lower temperature. Reactions during cooking go faster at higher temperature.
4. *Surface area of a solid reactant or catalyst*. If a reaction involves a solid with a gas or liquid, the surface area of the solid affects the reaction rate. Because the reaction occurs at the surface of the solid, the rate increases with increasing surface area. A wood fire burns faster if the logs are chopped into smaller pieces. Similarly, the surface area of a solid catalyst is important to the rate of reaction. The greater the surface area per unit volume, the faster the reaction (Figure 14.3).

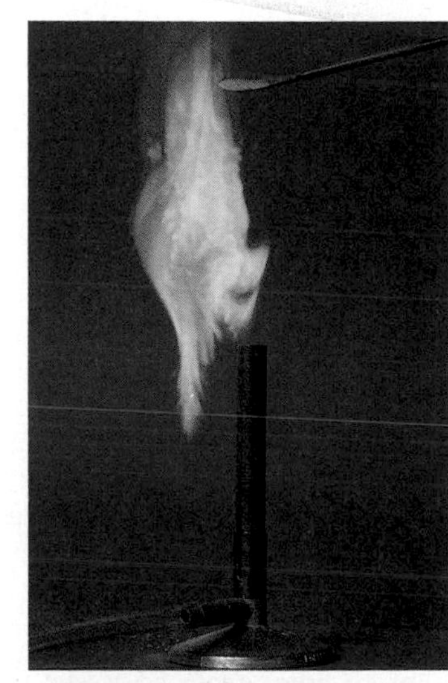

Figure 14.3
The effect of large surface area on the rate of reaction. Lycopodium powder (from the tiny spores of a club moss) ignites easily to produce a yellow flame. The powder has a large surface area per volume and burns rapidly in air.

In the first sections of this chapter, we will look primarily at reactions in the gas phase and in liquid solution where factors 1 to 3 are important. Thus, we will look at the effect of *concentrations* and *temperature* on reaction rates. Before we can explore these factors, however, we need a precise definition of reaction rate.

14.1 DEFINITION OF REACTION RATE

The rate of a reaction is the amount of product formed or the amount of reactant used up per unit of time. So that our rate calculations do not depend on the total quantity of reaction mixture used, we express the rate for a unit volume of the mixture. Therefore, the **reaction rate** is *the increase in molar concentration of product of a reaction per unit time or the decrease in molar concentration of reactant per unit time*. The usual unit of reaction rate is moles per liter per second, mol/(L · s).

Consider the gas-phase reaction we discussed in the chapter opening:

$$2N_2O_5(g) \longrightarrow 4NO_2(g) + O_2(g)$$

The rate for this reaction could be found by observing the increase in molar concentration of O_2 produced. We will denote the molar concentration of a substance by enclosing the formula of the substance in square brackets. Thus, $[O_2]$ is the molar concentration of O_2. In a given time interval Δt, the molar concentration of oxygen, $[O_2]$, in the reaction vessel increases by the amount $\Delta[O_2]$. Recall that the symbol Δ means "change in" and that we obtain the change by subtracting the

13.105 A 0.4381-g sample containing potassium iodate, KIO_3, was analyzed by reducing the iodate ion by adding an excess of potassium iodide, KI. The product of the reaction of IO_3^- and I^- is $I_2(aq)$. The iodine, I_2, was titrated with 0.1381 *M* sodium thiosulfate, $Na_2S_2O_3$, solution. In the titration, I_2 is reduced by $S_2O_3^{2-}$, which is oxidized to tetrathionate ion, $S_4O_6^{2-}$. If 38.2 mL of sodium thiosulfate was required, what was the percentage of potassium iodate in the sample?

13.106 A sample of copper ore weighing 0.4000 g was dissolved in nitric acid to produce a solution of Cu^{2+} ion. To this solution was added an excess of potassium iodide, which produces iodine and precipitates the copper(I) iodide.

$$2Cu^{2+}(aq) + 4I^-(aq) \longrightarrow 2CuI(s) + I_2(aq)$$

The iodine, I_2, was titrated with 0.1056 *M* sodium thiosulfate, $Na_2S_2O_3$, solution. In the titration, I_2 is reduced by $S_2O_3^{2-}$, which is oxidized to tetrathionate ion, $S_4O_6^{2-}$. If 24.65 mL of sodium thiosulfate solution was required, what was the percentage of copper in the ore?

13.107 The active ingredients of an antacid tablet contained only magnesium hydroxide and aluminum hydroxide. Complete neutralization of a sample of the active ingredients required 48.5 mL of 0.187 *M* hydrochloric acid. The chloride salts from this neutralization were obtained by evaporation of the filtrate from the titration; they weighed 0.420 g. What was the percentage by mass of magnesium hydroxide in the active ingredients of the antacid tablet?

13.108 The active ingredients in an antacid tablet contained only calcium carbonate and magnesium carbonate. Complete reaction of a sample of the active ingredients required 41.33 mL of 0.08750 *M* hydrochloric acid. The chloride salts were obtained by evaporation of the filtrate from this titration; they weighed 0.1900 g. What was the percentage by mass of the calcium carbonate in the active ingredients of the antacid tablet?

13.109 Calcium hydroxide solution was neutralized by hydrochloric acid. The reaction evolved 34.5 kJ of heat. How many grams of calcium hydroxide were neutralized?

13.110 A sample of calcium carbonate was mixed with an excess of hydrochloric acid. Complete reaction of the calcium carbonate evolved 15.5 kJ of heat. How many grams of calcium carbonate were in the sample?

immediately. On the other hand, the reactions that occur in a cement mixture as it hardens to concrete require years for completion.

The study of the rate, or speed, of a reaction has important applications. In the manufacture of ammonia from nitrogen and hydrogen, we may wish to know what conditions will help the reaction proceed in a commercially feasible length of time. Or we may wish to know whether the nitric oxide, NO, in the exhaust gases of supersonic transports will destroy the ozone in the stratosphere faster than the ozone is produced. Answering these questions requires knowledge of the rates of reactions.

Another reason for studying reaction rates is to understand how chemical reactions occur. By noting how the rate of a reaction is affected by changing conditions, we can sometimes learn the details of what is happening at the molecular level.

A reaction whose rate has been extensively studied under various conditions is the decomposition of dinitrogen pentoxide, N_2O_5. When this substance is heated in the gas phase, it decomposes to nitrogen dioxide and oxygen:

$$2N_2O_5(g) \longrightarrow 4NO_2(g) + O_2(g)$$

We will look at this reaction in some detail. The questions we will pose include: How is the rate of a reaction like this measured? What are the conditions that affect the rate of a reaction? How do we express the relationship of rate of a reaction to the variables that affect rate? What happens at the molecular level when N_2O_5 decomposes to NO_2 and O_2?

Reaction Rates

Chemical kinetics is the study of reaction rates, how reaction rates change under varying conditions, and what molecular events occur during the overall reaction. In the first part of this chapter, we will look at reaction rates and the variables that affect them.

What variables affect reaction rates? As we noted in the chapter opening, the rate depends on the characteristics of the reactants in a particular reaction. Some reactions are fast and others are slow, but the rate of any given reaction may be affected by the following factors:

1. *Concentrations of reactants.* Often the rate of reaction increases when the concentration of a reactant is increased. A piece of steel wool burns with some difficulty in air (20% O_2) but bursts into a dazzling white flame in pure oxygen. The rate of burning increases with the concentration of O_2. In some reactions, however, the rate is unaffected by the concentration of a particular reactant, so long as it is present at some concentration.
2. *Concentration of catalyst.* A **catalyst** is *a substance that increases the rate of reaction without being consumed in the overall reaction.* Because it is not consumed by the reaction, it does not appear in the balanced chemical equation (although its presence may be indicated by writing its formula over the arrow). A pure solution of hydrogen peroxide, H_2O_2, is stable, but when hydrobromic acid, HBr(*aq*), is added, H_2O_2 decomposes rapidly into H_2O and O_2 (Figure 14.2).

$$2H_2O_2(aq) \xrightarrow{HBr(aq)} 2H_2O(l) + O_2(g)$$

Here HBr acts as a catalyst to speed decomposition.

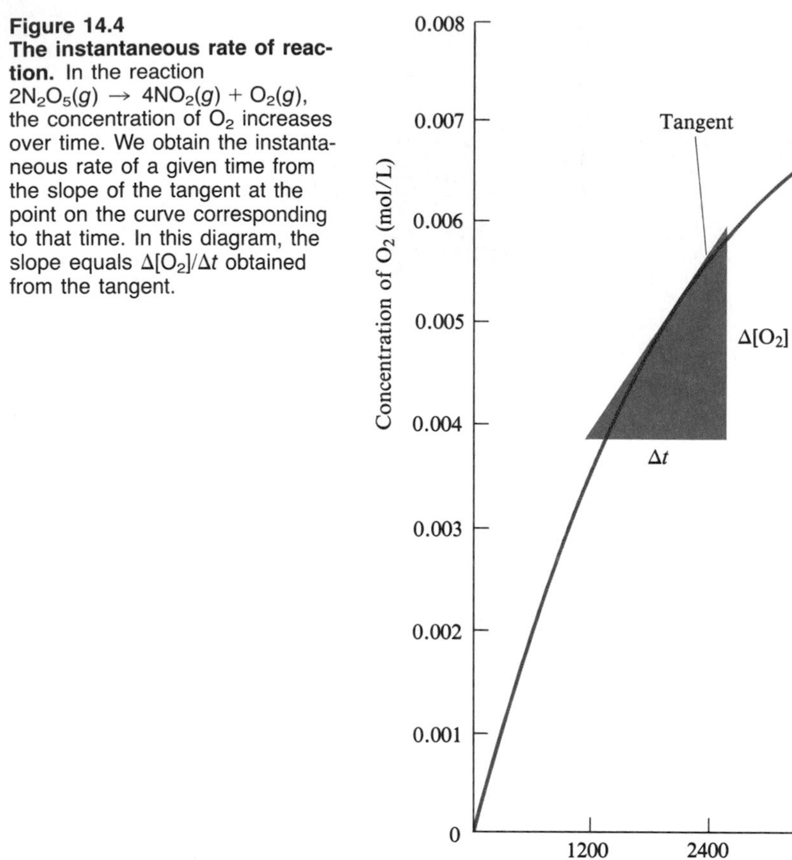

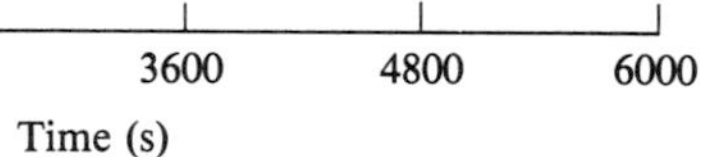

Figure 14.4
The instantaneous rate of reaction. In the reaction $2N_2O_5(g) \rightarrow 4NO_2(g) + O_2(g)$, the concentration of O_2 increases over time. We obtain the instantaneous rate of a given time from the slope of the tangent at the point on the curve corresponding to that time. In this diagram, the slope equals $\Delta[O_2]/\Delta t$ obtained from the tangent.

initial value from the final value. The rate of the reaction is given by

$$\text{Rate of formation of oxygen} = \frac{\Delta[O_2]}{\Delta t}$$

This equation gives the *average* rate over the time interval Δt. If the time interval is very short, the equation gives the *instantaneous* rate—that is, the rate at a particular instant of time. The instantaneous rate is also the value of $\Delta[O_2]/\Delta t$ for the tangent at a given instant (the straight line that just touches the curve of concentration versus time at a given point). See Figure 14.4.●

● In calculus, the rate at a given moment (the instantaneous rate) is given by the derivative $d[O_2]/dt$.

To understand the difference between average rate and instantaneous rate, it may help to think of the speed of an automobile. Speed can be defined as the rate of change of position, *x*; that is, speed equals $\Delta x/\Delta t$, where Δx is the distance traveled. If an automobile travels 84 miles in 2.0 hours, the average speed over this time interval is 84 mi/2.0 hr = 42 mi/hr. However, at any instant during this interval, the speedometer, which registers instantaneous speed, may read more or less than 42 mi/hr. At some moment on the highway, it may read 55 mi/hr, whereas on a congested city street it may read only 20 mi/hr. The quantity $\Delta x/\Delta t$ becomes more nearly an instantaneous speed as the time interval Δt is made smaller.

Figure 14.5 shows the increase in concentration of O_2 during the decomposition of N_2O_5. It shows the calculation of average rates at two positions on the curve. For example, when the time changes from 600 s to 1200 s (Δt = 600 s), the O_2 concentration increases by 0.0015 mol/L (=$\Delta[O_2]$). Therefore, the average

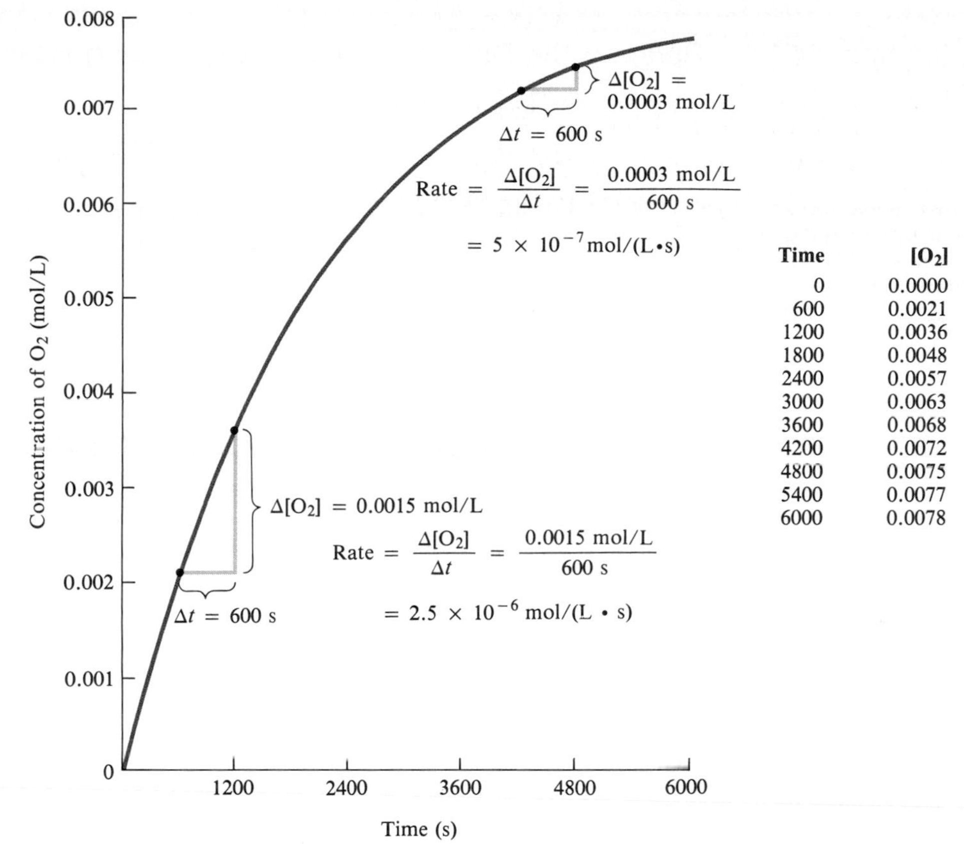

Time	$[O_2]$
0	0.0000
600	0.0021
1200	0.0036
1800	0.0048
2400	0.0057
3000	0.0063
3600	0.0068
4200	0.0072
4800	0.0075
5400	0.0077
6000	0.0078

Figure 14.5
Calculation of the average rate. The average rate of formation of O_2 during the decomposition of N_2O_5 was calculated during two different time intervals. When the time changes from 600 s to 1200 s, the average rate is 2.5×10^{-6} mol/(L · s). Later, when the time changes from 4200 s to 4800 s, the average rate has slowed to 5×10^{-7} mol/(L · s). Thus, the rate of the reaction decreases as the reaction proceeds.

rate = $\Delta[O_2]/\Delta t$ = (0.0015 mol/L)/600 s = 2.5×10^{-6} mol/(L · s). Later, during the time interval from 4200 s to 4800 s, the average rate is 5×10^{-7} mol/(L · s). Note that the rate decreases as the reaction proceeds.

Because the amounts of products and reactants are related by stoichiometry, any substance in the reaction can be used to express the rate of reaction. In the case of the decomposition of N_2O_5 to NO_2 and O_2, we gave the rate in terms of the rate of formation of oxygen, $\Delta[O_2]/\Delta t$. However, we can also express it in terms of the rate of decomposition of N_2O_5.

$$\text{Rate of decomposition of } N_2O_5 = -\frac{\Delta[N_2O_5]}{\Delta t}$$

Note the negative sign. It always occurs in a rate expression for a reactant in order to indicate a decrease in concentration and to give a positive value for the rate. Thus, because $[N_2O_5]$ decreases, $\Delta[N_2O_5]$ is negative and $-\Delta[N_2O_5]/\Delta t$ is positive.

The rate of decomposition of N_2O_5 and the rate of formation of oxygen are easily related. Two moles of N_2O_5 decompose for each mole of oxygen formed, so the rate of decomposition of N_2O_5 is twice the rate of formation of oxygen. To equate the rates, we must divide the rate of decomposition of N_2O_5 by 2 (its coefficient in the balanced chemical equation).

$$\text{Rate of formation of } O_2 = \tfrac{1}{2}(\text{rate of decomposition of } N_2O_5)$$

or

$$\frac{\Delta[O_2]}{\Delta t} = -\frac{1}{2}\frac{\Delta[N_2O_5]}{\Delta t}$$

Example 14.1 Relating the Different Ways of Expressing Reaction Rates

Consider the reaction of nitrogen dioxide with fluorine to give nitryl fluoride, NO_2F.

$$2NO_2(g) + F_2(g) \longrightarrow 2NO_2F(g)$$

How is the rate of formation of NO_2F related to the rate of reaction of fluorine?

Solution

We have

$$\text{Rate of formation of } NO_2F = \frac{\Delta[NO_2F]}{\Delta t}$$

and

$$\text{Rate of reaction of } F_2 = -\frac{\Delta[F_2]}{\Delta t}$$

We divide each rate by the coefficient of the corresponding substance in the chemical equation and then equate them:

$$\frac{1}{2}\frac{\Delta[NO_2F]}{\Delta t} = -\frac{\Delta[F_2]}{\Delta t}$$

Exercise 14.1

For the reaction given in Example 14.1, how is the rate of formation of NO_2F related to the rate of reaction of NO_2? (See Problems 14.23 and 14.24.)

Example 14.2 Calculating the Average Reaction Rate

Calculate the average rate of decomposition of N_2O_5, $-\Delta[N_2O_5]/\Delta t$, by the reaction

$$2N_2O_5(g) \longrightarrow 4NO(g) + O_2(g)$$

during the time interval from $t = 600$ s to $t = 1200$ s (regard all time figures as significant). Use the following data:

TIME	$[N_2O_5]$
600 s	$1.24 \times 10^{-2}\ M$
1200 s	$0.93 \times 10^{-2}\ M$

Solution

We calculate a Δ quantity in the rate expression by taking the final value minus the initial value.

$$\text{Average rate of decomposition of } N_2O_5 = -\frac{\Delta[N_2O_5]}{\Delta t}$$

$$= -\frac{(0.93 - 1.24) \times 10^{-2}\ M}{(1200 - 600)\ s} = -\frac{-0.31 \times 10^{-2}\ M}{600\ s}$$

$$= \mathbf{5.2 \times 10^{-6}\ M/s}$$

Note that this rate is twice the rate of formation of O_2 in the same time interval (which we calculated in the preceding text discussion to be $2.5 \times 10^{-6}\ M/s$).

Exercise 14.2

Iodide ion is oxidized by hypochlorite ion in basic solution.

$$I^-(aq) + ClO^-(aq) \rightarrow Cl^-(aq) + IO^-(aq)$$

In 1.00 M NaOH at 25°C, the iodide ion concentration (equal to the ClO^- concentration) at different times was as follows:

TIME	$[I^-]$
2.00 s	0.00169 M
8.00 s	0.00101 M

Calculate the average rate of reaction of I^- during this time interval.

(See Problems 14.27 and 14.28.)

14.2 THE EXPERIMENTAL DETERMINATION OF RATE

To obtain the rate of a reaction, we must determine the concentration of a reactant or product during the course of the reaction. One way to do this for a slow reaction is to withdraw samples from the reaction vessel at various times and analyze them. The rate of the reaction of ethyl acetate with water in acidic solution was one of the first to be determined this way.●

● Ethyl acetate is a liquid with a fruity odor that belongs to a class of organic (carbon-containing) compounds called esters. Unlike a salt, such as sodium acetate, an ester is a covalent compound.

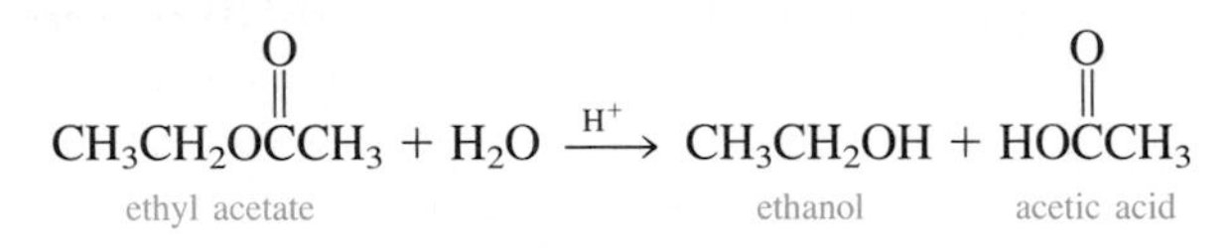

$$\underset{\text{ethyl acetate}}{CH_3CH_2O\overset{\overset{O}{\|}}{C}CH_3} + H_2O \xrightarrow{H^+} \underset{\text{ethanol}}{CH_3CH_2OH} + \underset{\text{acetic acid}}{HO\overset{\overset{O}{\|}}{C}CH_3}$$

This reaction is slow, so the acetic acid that is produced is easily obtained by titration before any significant further reaction occurs.

More convenient are techniques that can continuously follow the progress of a reaction by observing the change in some physical property of the system. These physical methods are often adaptable to fast reactions as well as to slow ones. For example, if a gas reaction involves a change in the number of molecules, the pressure of the system changes when the volume and temperature are held constant. By following the pressure change as the reaction proceeds, we can obtain the reaction rate. The decomposition of dinitrogen pentoxide in the gas phase, which we mentioned earlier, has been studied this way. Dinitrogen pentoxide crystals are sealed in a vessel equipped with a manometer (a pressure-measuring device; see Figure 14.6). The vessel is then plunged into a water bath at 45°C, at which temperature the solid vaporizes and the gas decomposes.

$$2N_2O_5(g) \longrightarrow 4NO_2(g) + O_2(g)$$

Manometer readings are taken at regular time intervals, and the pressure values are converted to partial pressures or concentrations of N_2O_5. The rates of reaction during various time intervals can be calculated as described in the previous section.

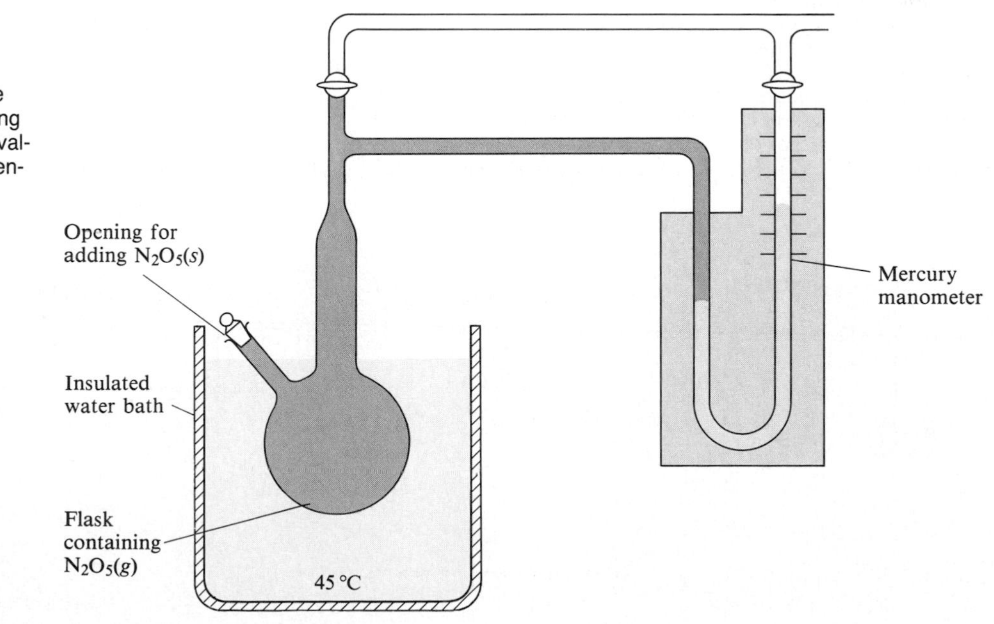

Figure 14.6
An experiment to follow the concentration of N_2O_5 as the decomposition proceeds. The total pressure is measured during the reaction at 45°C. Pressure values can be related to the concentration of N_2O_5 in the flask.

Another physical property used to follow the progress of a reaction is color, or the absorption of light by some species. Consider the reaction

$$ClO^-(aq) + I^-(aq) \longrightarrow IO^-(aq) + Cl^-(aq)$$

The hypoiodite ion, IO^-, absorbs at the blue end of the spectrum near 400 nm. The intensity of this absorption is proportional to $[IO^-]$, and we can use it to determine the reaction rate. We can also follow the decomposition of N_2O_5 from the intensity of the red-brown color of the product NO_2.

In these experiments, the intensity of absorption at a particular wavelength is measured by a spectrometer appropriate for the visible region of the spectrum. We are not limited to the use of the visible spectrometers for the determination of reaction rates, however. Depending on the reaction, other types of instruments are used, including infrared (IR) and nuclear magnetic resonance (NMR) spectrometers, which we described briefly in Related Topic boxes in earlier chapters.

14.3 DEPENDENCE OF RATE ON CONCENTRATION

Experimentally, it has been found that a reaction rate depends on the concentrations of certain reactants as well as the concentration of catalyst, if there is one (Figure 14.7). Consider the reaction of nitrogen dioxide with fluorine to give nitryl fluoride, NO_2F.

$$2NO_2(g) + F_2(g) \longrightarrow 2NO_2F(g)$$

The rate of this reaction is observed to be proportional to the concentration of nitrogen dioxide. When the concentration of nitrogen dioxide is doubled, the rate doubles. The rate is also proportional to the concentration of fluorine; doubling the concentration of fluorine also doubles the rate.

A **rate law** is *an equation that relates the rate of a reaction to the concentrations of reactants (and catalyst) raised to various powers*. The following equation is the rate law for the foregoing reaction:

$$\text{Rate} = k[NO_2][F_2]$$

Note that in this rate law both reactant concentrations have an exponent of 1. Here k, called the **rate constant,** is *a proportionality constant in the relationship between rate and concentrations*. It has a fixed value at any given temperature, but it varies with temperature. Whereas the units of rate are usually given as mol/(L · s),

A time = 0 B time = 20 seconds C time = 40 seconds

Figure 14.7
Effect of reactant concentrations on rate of reaction.
(A) Both beakers contain the same amounts of reactants: sodium thiosulfate, $Na_2S_2O_3$, in acidic solution and sodium arsenite, Na_3AsO_3. However, the beaker on the right contains more water and thus lower concentrations of reactants. *(B)* $Na_2S_2O_3$ decomposes slowly in acidic solution to yield H_2S, which reacts quickly with Na_3AsO_3 to give a bright yellow precipitate of As_2S_3. The time it takes the precipitate to form depends on the decomposition rate of $Na_2S_2O_3$. *(C)* Note that the precipitate forms more slowly in the solution of lower concentrations (the beaker on the right).

the units of k depend on the form of the rate law. For the previous rate law, we have

$$k = \frac{\text{rate}}{[NO_2][F_2]}$$

from which we get the following unit for k:

$$\frac{\text{mol/(L} \cdot \text{s)}}{(\text{mol/L})^2} = \text{L/(mol} \cdot \text{s)}$$

As a more general example, consider the reaction of substances A and B to give D and E, according to the balanced equation

$$a\text{A} + b\text{B} \xrightarrow{\text{C}} d\text{D} + e\text{E} \qquad \text{C} = \text{catalyst}$$

We could write the rate law in the form

$$\text{Rate} = k[\text{A}]^m[\text{B}]^n[\text{C}]^p$$

The exponents m, n, and p are frequently, but not always, integers. *They must be determined experimentally* and cannot be obtained simply by looking at the balanced equation. For example, note that the exponents in the equation Rate = $k[NO_2][F_2]$ have no relationship to the coefficients in the balanced equation $2NO_2 + F_2 \longrightarrow 2NO_2F$.

Once we know the rate law for a reaction and have found the value of the rate constant, we can calculate the rate of a reaction for any values of reactant concentrations. As we will see later, knowledge of the rate law is also useful in understanding how the reaction occurs at the molecular level.

Reaction Order

We can classify a reaction by its orders. The **reaction order** with respect to a given reactant species equals *the exponent of the concentration of that species in the rate law, as determined experimentally*. For the reaction of NO_2 with F_2 to give NO_2F, the reaction is first order with respect to the NO_2 because the exponent of $[NO_2]$ in the rate law is 1. Similarly, the reaction is first order with respect to F_2.

The *overall order* of a reaction equals the sum of the orders of the species in the rate law. In this example, the overall order is 2; that is, the reaction is second order overall.

Reactions display a variety of reaction orders. Some examples are listed below.

1. Cyclopropane, C_3H_6, has the molecular structure

$$\begin{array}{c} CH_2 \\ / \ \ \backslash \\ H_2C\text{—}CH_2 \end{array}$$

When heated, the carbon ring (the triangle) opens up to give propylene, $CH_2{=}CHCH_3$. Because the compounds are isomers (different compounds with the same molecular formula), the reaction is called an *isomerization*.

$$\underset{\text{cyclopropane}}{C_3H_6(g)} \longrightarrow \underset{\text{propylene}}{CH_2{=}CHCH_3(g)}$$

It has the rate law

$$\text{Rate} = k[C_3H_6]$$

The reaction is first order in cyclopropane and first order overall.

2. Nitric oxide, NO, reacts with hydrogen according to the equation

$$2NO(g) + 2H_2(g) \longrightarrow N_2(g) + 2H_2O(g)$$

The experimentally determined rate law is

$$\text{Rate} = k[NO]^2[H_2]$$

Thus, the reaction is second order in NO, first order in H_2, and third order overall.

3. Acetone, CH_3COCH_3, reacts with iodine in acidic solution.

$$CH_3COCH_3(aq) + I_2(aq) \xrightarrow{H^+} CH_3COCH_2I(aq) + HI(aq)$$

The experimentally determined rate law is

$$\text{Rate} = k[CH_3COCH_3][H^+]$$

The reaction is first order in acetone. It is zero order in iodine; that is, the rate law contains the factor $[I_2]^0 = 1$. Therefore, the rate does not depend on the concentration of I_2, as long as some concentration of I_2 is present. Note that the reaction is first order in the catalyst, H^+. Thus, the overall order is 2.

Although reaction orders frequently have whole-number values (particularly 1 or 2), they can be fractional. Zero and negative orders are also possible.

Example 14.3 Determining the Order of Reaction from the Rate Law

Bromide ion is oxidized by bromate ion in acidic solution.

$$5Br^-(aq) + BrO_3^-(aq) + 6H^+(aq) \longrightarrow 3Br_2(aq) + 3H_2O(l)$$

The empirical rate law is

$$\text{Rate} = k[Br^-][BrO_3^-][H^+]^2$$

What is the order of reaction with respect to each reactant species? What is the overall order of reaction?

Solution

The reaction is **first order with respect to Br^-** and **first order with respect to BrO_3^-**; it is **second order with respect to H^+** (an order with respect to a species equals the exponent of its concentration). **The reaction is fourth order overall (= 1 + 1 + 2).**

Exercise 14.3

What are the reaction orders with respect to each reactant species for the following reaction?

$$NO_2(g) + CO(g) \longrightarrow NO(g) + CO_2(g)$$

Assume the rate law is

$$\text{Rate} = k[NO_2]^2$$

What is the overall order? (See Problems 14.31 and 14.32.)

Determining the Rate Law

The experimental determination of the rate law for a reaction requires that we find the order of the reaction with respect to each reactant and any catalyst. The *initial-rate method* is a simple way to obtain reaction orders. It consists of doing a series of experiments in which the initial, or starting, concentrations of reactants are varied. Then the initial rates are compared, from which the reaction orders can be deduced.

To see how this method works, again consider the reaction mentioned in the chapter opening:

$$2N_2O_5(g) \longrightarrow 4NO_2(g) + O_2(g)$$

We observe this reaction in two experiments. In Experiment 2, the initial concentration of N_2O_5 is twice that in Experiment 1. We then note the initial rate of disappearance of N_2O_5 in each case. The initial concentrations and corresponding initial rates for two experiments are given in the following table.●

● In Figure 14.4, the slope of the tangent to the curve at $t = 0$ equals the initial rate of appearance of O_2, which equals one-half the initial rate of disappearance of N_2O_5.

	INITIAL N_2O_5 CONCENTRATION	INITIAL RATE OF DISAPPEARANCE OF N_2O_5
Experiment 1	1.0×10^{-2} mol/L	4.8×10^{-6} mol/(L · s)
Experiment 2	2.0×10^{-2} mol/L	9.6×10^{-6} mol/(L · s)

The rate law for this reaction will have the concentration of reactant raised to a power m.

$$\text{Rate} = k[N_2O_5]^m$$

The value of m (the reaction order) must be determined from the experimental data. Note that when the N_2O_5 concentration is doubled, we get a new rate, Rate′, given by the following equation:

$$\text{Rate}' = k(2[N_2O_5])^m = 2^m k[N_2O_5]^m$$

This rate is 2^m times the original rate.

We can now see how the rate is affected when the concentration is doubled for various choices of m. Suppose $m = 2$. We get $2^m = 2^2 = 4$. That is, when the initial concentration is doubled, the rate is multiplied by 4. We summarize the results for various choices of m as follows:

m	IF THE INITIAL CONCENTRATION IS DOUBLED, THE RATE IS MULTIPLIED BY:
−1	$\frac{1}{2}$
0	1
1	2
2	4

Let us divide the initial rate of reaction of N_2O_5 from Experiment 2 by the initial rate from Experiment 1.

$$\frac{9.6 \times 10^{-6} \text{ mol/(L} \cdot \text{s)}}{4.8 \times 10^{-6} \text{ mol/(L} \cdot \text{s)}} = 2$$

In other words, when the N_2O_5 concentration is doubled, the rate is doubled. We see that this corresponds to the result for $m = 1$. The rate law must have the form

$$\text{Rate} = k[N_2O_5]$$

We can determine the value of the rate constant k by substituting values of the rate and N_2O_5 concentrations from any one of the experiments into the rate law. Using values from Experiment 2, we get

$$9.6 \times 10^{-6} \text{ mol/(L} \cdot \text{s)} = k \times 2.0 \times 10^{-2} \text{ mol/L}$$

Hence,

$$k = \frac{9.6 \times 10^{-6}/\text{s}}{2.0 \times 10^{-2}} = 4.8 \times 10^{-4}/\text{s}$$

Example 14.4 Determining the Rate Law from Initial Rates

Iodide ion is oxidized in acidic solution to triiodide ion, I_3^-, by hydrogen peroxide (see Figure 13.13, p. 536).

$$H_2O_2(aq) + 3I^-(aq) + 2H^+(aq) \longrightarrow I_3^-(aq) + 2H_2O(l)$$

A series of four experiments was run at different concentrations, and the initial rates of I_3^- formation were determined (see table). (a) From these data, obtain the reaction orders with respect to H_2O_2, I^-, and H^+. (b) Then find the rate constant.

	INITIAL CONCENTRATIONS (MOL/L)			INITIAL RATE
	H_2O_2	I^-	H^+	(mol/(L · s))
Exp. 1	0.010	0.010	0.00050	1.15×10^{-6}
Exp. 2	0.020	0.010	0.00050	2.30×10^{-6}
Exp. 3	0.010	0.020	0.00050	2.30×10^{-6}
Exp. 4	0.010	0.010	0.00100	1.15×10^{-6}

Solution

(a) We assume that the rate law has the following form

$$\text{Rate} = k[H_2O_2]^m[I^-]^n[H^+]^p$$

and determine the reaction orders (the exponents *m*, *n*, and *p*). Comparing Experiment 1 and Experiment 2, we see that when the H_2O_2 concentration is doubled (with other concentrations constant), the rate is doubled. Therefore, $m = 1$ (the reaction is first order in H_2O_2).

To do a problem such as this in a general way, we should approach it algebraically. We write the rate law algebraically for two experiments (the subscripts denote the experiments).

$$(\text{Rate})_1 = k[H_2O_2]_1^m[I^-]_1^n[H^+]_1^p$$
$$(\text{Rate})_2 = k[H_2O_2]_2^m[I^-]_2^n[H^+]_2^p$$

Now we divide the second equation by the first.

$$\frac{(\text{Rate})_2}{(\text{Rate})_1} = \frac{k[H_2O_2]_2^m[I^-]_2^n[H^+]_2^p}{k[H_2O_2]_1^m[I^-]_1^n[H^+]_1^p}$$

The rate constant cancels. Grouping the terms, we obtain

$$\frac{(\text{Rate})_2}{(\text{Rate})_1} = \left(\frac{[H_2O_2]_2}{[H_2O_2]_1}\right)^m\left(\frac{[I^-]_2}{[I^-]_1}\right)^n\left(\frac{[H^+]_2}{[H^+]_1}\right)^p$$

Now we substitute experimental values (the units cancel in a ratio and can be omitted).

$$\frac{2.30 \cancel{\times 10^{-6}}}{1.15 \cancel{\times 10^{-6}}} = \left[\frac{0.020}{0.010}\right]^m \cancel{\left[\frac{0.010}{0.010}\right]^n} \cancel{\left[\frac{0.00050}{0.00050}\right]^p}$$

This gives $2 = 2^m$, from which we obtain $m = 1$. That is, doubling the H_2O_2 concentration doubles the rate.

We can continue in the same way. Comparing Experiment 1 and Experiment 3, we see that doubling the I^- concentration (with the other concentrations constant) doubles the rate. Therefore, $n = 1$ (the reaction is first order in I^-). Finally, comparing Experiment 1 and Experiment 4, we see that doubling the H^+ concentration (holding other concentrations constant) has no effect on the rate. Therefore, $p = 0$ (the reaction is zero order in H^+). Because $[H^+]^0 = 1$, the rate law is

$$\textbf{Rate} = k[H_2O_2][I^-]$$

Note that the orders are *not* related to the coefficients of the overall equation.

(b) We can calculate the rate constant by substituting values from any of the experiments into the rate law. Using Experiment 1, we obtain

$$1.15 \times 10^{-6} \frac{\cancel{\text{mol}}}{\cancel{\text{L}} \cdot \text{s}} = k \times 0.010 \frac{\cancel{\text{mol}}}{\cancel{\text{L}}} \times 0.010 \frac{\text{mol}}{\text{L}}$$

$$k = \frac{1.15 \times 10^{-6}/\text{s}}{0.010 \times 0.010 \times \text{mol/L}} = \mathbf{1.15 \times 10^{-2}\ L/(mol \cdot s)}$$

Exercise 14.4

The initial-rate method was applied to the decomposition of nitrogen dioxide.

$$2NO_2(g) \longrightarrow 2NO(g) + O_2(g)$$

It yielded the following results:

	INITIAL NO_2 CONCENTRATION	INITIAL RATE OF FORMATION OF O_2
Exp. 1	0.010 mol/L	7.1×10^{-5} mol/(L · s)
Exp. 2	0.020 mol/L	28×10^{-5} mol/(L · s)

Find the rate law and the value of the rate constant with respect to the O_2 formation.

(See Problems 14.35, 14.36, 14.37, and 14.38.)

14.4 CHANGE OF CONCENTRATION WITH TIME

A rate law tells us how the rate of a reaction depends on reactant concentrations at a particular moment. But often we would like to have a mathematical relationship showing how a reactant concentration changes over a period of time. Such an equation would be directly comparable to the experimental data, which are usually obtained as concentrations at various times. In addition to summarizing the experimental data, this equation would predict concentrations for all times. Using it, we could answer questions such as: How long does it take for this reaction to be 50% complete? to be 90% complete?

Moreover, as we will see, knowing exactly how the concentrations change with time for different rate laws suggests ways of plotting the experimental data on a graph. Graphical plotting provides an alternative to the initial-rate method for determining the rate law.

Using calculus, we can transform a rate law into a mathematical relationship between concentration and time. Because we will work only with the final equations, we need not go into the derivations here. We will look in some depth at first-order reactions and briefly at second-order reactions.

Concentration–Time Equations

FIRST-ORDER RATE LAW Let us first look at first-order rate laws. The decomposition of dinitrogen pentoxide has the following rate law:

$$\text{Rate} = -\frac{\Delta[N_2O_5]}{\Delta t} = k[N_2O_5]$$

Using calculus, one can show that such a first-order rate law leads to the following relationship between N_2O_5 concentration and time:●

$$\ln\frac{[N_2O_5]_t}{[N_2O_5]_0} = -kt \quad \text{or} \quad \log\frac{[N_2O_5]_t}{[N_2O_5]_0} = \frac{-kt}{2.303}$$

● The general derivation using calculus is as follows. Substituting [A] for $[N_2O_5]$, the rate law becomes

$$\frac{-d[A]}{dt} = k[A]$$

We rearrange this to give

$$\frac{-d[A]}{[A]} = k\,dt$$

Integrating from time = 0 to time = t,

$$-\int_{[A]_0}^{[A]_t}\frac{d[A]}{[A]} = k\int_0^t dt$$

gives

$$-\{\ln[A]_t - \ln[A]_0\} = k(t-0)$$

This can be rearranged to give the equation in the text.

Here $[N_2O_5]_t$ is the concentration at time t, and $[N_2O_5]_0$ is the initial concentration of N_2O_5 (that is, the concentration at $t = 0$). The symbol "ln" denotes the natural logarithm (base $e = 2.718\ldots$), and "log" denotes the logarithm to the base 10.

This equation enables us to calculate the concentration of N_2O_5 at any time, once we are given the initial concentration and the rate constant. Also, we can find the time it takes for the N_2O_5 concentration to decrease to a particular value.

More generally, let A be a substance that reacts to give products according to the equation

$$a\text{A} \longrightarrow \text{products}$$

where a is the stoichiometric coefficient of reactant A. Suppose that this reaction has a first-order rate law

$$\text{Rate} = -\frac{\Delta[A]}{\Delta t} = k[A]$$

Using calculus, we get the following equation.

$$\ln \frac{[A]_t}{[A]_0} = -kt \qquad \text{or} \qquad \log \frac{[A]_t}{[A]_0} = \frac{-kt}{2.303}$$

Here $[A]_t$ is the concentration of reactant A at time t, and $[A]_0$ is the initial concentration. The ratio $[A]_t/[A]_0$ is the fraction of reactant remaining at time t. The next example illustrates how we work with this equation.

Example 14.5 Using the Concentration–Time Equation for a First-Order Reaction

The decomposition of N_2O_5 to NO_2 and O_2 is first order, with a rate constant of 4.80×10^{-4}/s at 45°C. (a) If the initial concentration is 1.65×10^{-2} mol/L, what is the concentration after 825 s? (b) How long would it take for the concentration of N_2O_5 to decrease to 1.00×10^{-2} mol/L from its initial value, given in (a)?

Solution

(a) We need to use one of the forms of the equation relating concentration to time (the equation in color just before this example). If you are familiar with natural logarithms and your calculator has *ln* and *exp* keys, you will find the first form simple to use. We will illustrate the calculation with the form involving logarithms to the base 10. We get

$$\log \frac{[N_2O_5]_t}{1.65 \times 10^{-2}\ \text{mol/L}} = \frac{-4.80 \times 10^{-4}/\text{s} \times 825\ \text{s}}{2.303} = -0.172$$

In order to solve for $[N_2O_5]_t$, we take the antilogarithm of both sides. This removes the log from the left and gives antilog(−0.172), or $10^{-0.172}$, on the right, which equals 0.673.

$$\frac{[N_2O_5]_t}{1.65 \times 10^{-2}\ \text{mol/L}} = 0.673$$

Hence,

$$[N_2O_5]_t = 1.65 \times 10^{-2}\ \text{mol/L} \times 0.673 = \mathbf{0.0111\ mol/L}$$

(b) We substitute into one of the equations relating concentration to the time. Thus

$$\log \frac{1.00 \times 10^{-2}\ \text{mol/L}}{1.65 \times 10^{-2}\ \text{mol/L}} = \frac{-4.80 \times 10^{-4}/\text{s} \times t}{2.303}$$

The left side equals −0.217; the right side equals $-2.08 \times 10^{-4}/\text{s} \times t$. Hence,

$$0.217 = 2.08 \times 10^{-4}/\text{s} \times t$$

Or,

$$t = \frac{0.217}{2.08 \times 10^{-4}/\text{s}} = \mathbf{1.04 \times 10^3\ s\ (17.4\ min)}$$

Exercise 14.5

(a) What would be the concentration of dinitrogen pentoxide in the experiment described in Example 14.5 after 6.00×10^2 s? (b) How long would it take for the concentration of N_2O_5 to decrease to 10.0% of its initial value? (See Problems 14.41 and 14.42.)

SECOND-ORDER RATE LAW Consider the reaction

$$aA \longrightarrow \text{products}$$

and suppose it has the second-order rate law

$$\text{Rate} = -\frac{\Delta[A]}{\Delta t} = k[A]^2$$

An example is the decomposition of nitrogen dioxide at moderately high temperatures (300°C to 400°C).

$$2NO_2(g) \longrightarrow 2NO(g) + O_2(g)$$

Using calculus, we can obtain the following relationship between the concentration of A and the time.

$$\frac{1}{[A]_t} = kt + \frac{1}{[A]_0}$$

Using this equation, we can calculate the concentration of NO_2 at any time during its decomposition if we know the rate constant and the initial concentration. At 330°C, the rate constant for the decomposition of NO_2 is 0.775 L/(mol · s). Suppose the initial concentration is 0.0030 mol/L. What is the concentration of NO_2 after 645 s? By substituting into the previous equation, we get

$$\frac{1}{[NO_2]_t} = 0.775\ \text{L/(mol}\cdot\text{s)} \times 645\ \text{s} + \frac{1}{0.0030\ \text{mol/L}} = 8.3 \times 10^2\ \text{L/mol}$$

If we take the inverse of both sides of this equation, we find that $[NO_2]_t$ = 0.0012 mol/L. Thus, after 645 s, the concentration of NO_2 decreased from 0.0030 mol/L to 0.0012 mol/L.

Half-Life of a Reaction

As a reaction proceeds, the concentration of a reactant decreases, because it is being consumed. The **half-life,** $t_{1/2}$, of a reaction is *the time it takes for the reactant concentration to decrease to one-half of its initial value.*

● The concept of half-life is also used to characterize a radioactive nucleus, whose radioactive decay is a first-order process. This is discussed in Chapter 20.

For a first-order reaction, such as the decomposition of dinitrogen pentoxide, the half-life is independent of the initial concentration. To see this, let us substitute into the equation

$$\log \frac{[N_2O_5]_t}{[N_2O_5]_0} = \frac{-kt}{2.303}$$

In one half-life, the N_2O_5 concentration decreases by one-half, from its initial value, $[N_2O_5]_0$, to $[N_2O_5]_t = \frac{1}{2}[N_2O_5]_0$. After substituting, the equation becomes

$$\log \frac{\frac{1}{2}\cancel{[N_2O_5]_0}}{\cancel{[N_2O_5]_0}} = \frac{-kt_{1/2}}{2.303}$$

The expression on the left equals $\log \frac{1}{2} = -0.301$. Hence,

$$0.301 = \frac{kt_{1/2}}{2.303}$$

Solving for the half-life, $t_{1/2}$, we get

$$t_{1/2} = \frac{0.301 \times 2.303}{k} = \frac{0.693}{k}$$

Because the rate constant for the decomposition of N_2O_5 at 45°C is 4.80×10^{-4}/s, the half-life is

$$t_{1/2} = \frac{0.693}{4.80 \times 10^{-4}/\text{s}} = 1.44 \times 10^3\ \text{s}$$

Thus, the half-life is 1.44×10^3 s, or 24.0 min.

We see that the half-life of N_2O_5 does not depend on the initial concentration of N_2O_5. This means that the half-life is the same at any time during the reaction. If the initial concentration is 0.0165 mol/L, after one half-life (24.0 min) the concentration decreases to $\frac{1}{2} \times 0.0165$ mol/L = 0.0083 mol/L. After another half-life (another 24.0 min), the N_2O_5 concentration decreases to $\frac{1}{2} \times 0.0083$ mol/L = 0.0041 mol/L. Every time one half-life passes, the N_2O_5 concentration decreases by one-half again (see Figure 14.8, p. 562).

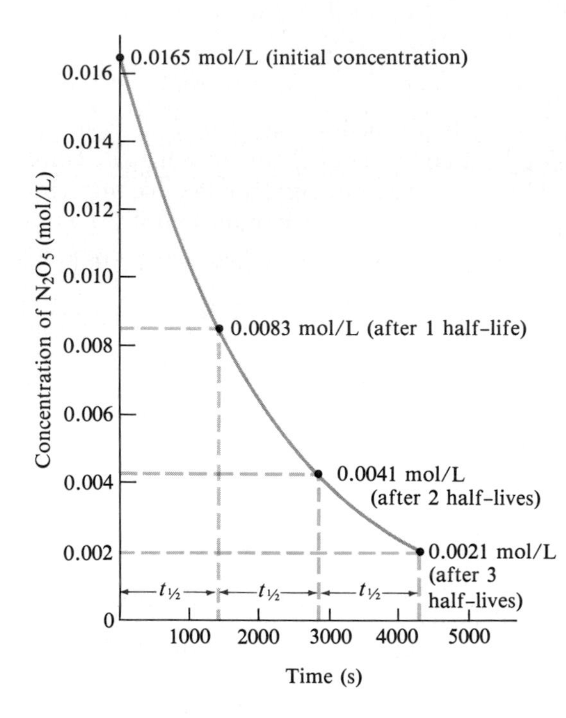

Figure 14.8
A graph illustrating that the half-life of a first-order reaction is independent of initial concentration. In one half-life (1440 s, or 24.0 min), the concentration decreases by one-half, from 0.0165 mol/L to $\frac{1}{2} \times 0.0165$ mol/L = 0.0083 mol/L. After each succeeding half-life (24.0 min), the concentration decreases by one-half again, from 0.0083 mol/L to $\frac{1}{2} \times 0.0083$ mol/L = 0.0041 mol/L, then to $\frac{1}{2} \times 0.0041$ mol/L = 0.0021 mol/L, and so forth.

The foregoing result for the half-life for the first-order decomposition of N_2O_5 is perfectly general. That is, for the general first-order rate law

$$\text{Rate} = -\frac{\Delta[\text{A}]}{\Delta t} = k[\text{A}]$$

the half-life is related to the rate constant, but it is independent of the concentration of A.

$$t_{1/2} = \frac{0.693}{k}$$

Example 14.6 Relating the Half-Life of a Reaction to the Rate Constant

Sulfuryl chloride, SO_2Cl_2, is a colorless, corrosive liquid whose vapor decomposes in a first-order reaction to sulfur dioxide and chlorine.

$$SO_2Cl_2(g) \longrightarrow SO_2(g) + Cl_2(g)$$

At 320°C, the rate constant is 2.20×10^{-5}/s. (a) What is the half-life of SO_2Cl_2 vapor at this temperature? (b) How long (in hours) would it take for 50.0% of the SO_2Cl_2 to decompose? How long would it take for 75.0% of the SO_2Cl_2 to decompose?

Solution

We substitute $k = 2.20 \times 10^{-5}$/s into the equation relating k and $t_{1/2}$.

$$t_{1/2} = \frac{0.693}{k} = \frac{0.693}{2.20 \times 10^{-5}/\text{s}} = 3.15 \times 10^4 \text{ s}$$

(b) The half-life is the time required for one-half (50.0%) of the SO_2Cl_2 to decompose. This is 3.15×10^4 s, or 8.75 h. After another half-life, one-half of the remaining SO_2Cl_2 decomposes. The total decomposed is $\frac{1}{2} + (\frac{1}{2} \times \frac{1}{2}) = \frac{3}{4}$, or 75.0%. The time required is two half-lives, or 2×8.75 h = 17.5 h.

Exercise 14.6

The isomerization of cyclopropane, C_3H_6, to propylene, $CH_2{=}CHCH_3$, is first order in cyclopropane and first order overall. At 1000°C, the rate constant is 9.2/s. What is the half-life of cyclopropane at 1000°C? How long would it take for the concentration of cyclopropane to decrease to 50% of its initial value? to 25% of its initial value?

(See Problems 14.45 and 14.46.)

It can be shown by reasoning similar to that given previously that the half-life of a second-order rate law, Rate $= k[A]^2$, is $1/(k[A]_0)$. In this case, the half-life depends on initial concentration and becomes larger as time goes on. Consider the decomposition of NO_2 at 330°C. It takes 430 s for the concentration to decrease by one-half from 0.0030 mol/L to 0.0015 mol/L. However, it takes 860 s (twice as long) for the concentration to decrease by one-half again. The fact that the half-life changes with time is evidence that the reaction is not first order.

Graphical Plotting

We saw earlier that the order of a reaction can be determined by comparing initial rates for several experiments in which different initial concentrations are used (initial-rate method). It is also possible to determine the order of a reaction by graphical plotting of the data for a particular experiment. The experimental data are plotted in several different ways, first assuming a first-order reaction, then a second-order reaction, and so forth. The order of the reaction is determined by which graph gives the best fit to the experimental data. To illustrate, we will look at how the plotting should be done for first-order and second-order reactions and then compare graphs for a specific reaction.

We have seen that the first-order rate law, $-\Delta[A]/\Delta t = k[A]$, gives the following relationship between concentration of A and time:

$$\log \frac{[A]_t}{[A]_0} = \frac{-kt}{2.303}$$

This equation can be rewritten in a slightly different form, which we can identify with the equation of a straight line. Using the property of logarithms that $\log (A/B) = \log A - \log B$, we get

$$\log [A]_t = \left(\frac{-k}{2.303}\right)t + \log [A]_0$$

A straight line has the mathematical form $y = mx + b$ when y is plotted on the vertical axis against x on the horizontal axis.● Let us now make the following identifications:

● See Appendix A for a discussion of the mathematics of a straight line.

$$\underbrace{\log [A]_t}_{y} = \underbrace{\left(\frac{-k}{2.303}\right)}_{m}\underbrace{t}_{x} + \underbrace{\log [A]_0}_{b}$$

This means that if we plot $\log [A]_t$ on the vertical axis against the time t on the horizontal axis, we will get a straight line for a first-order reaction. Figure 14.9 shows a plot of $\log [N_2O_5]$ at various times during the decomposition reaction. The fact that the points lie on a straight line is confirmation that the rate law is first order.

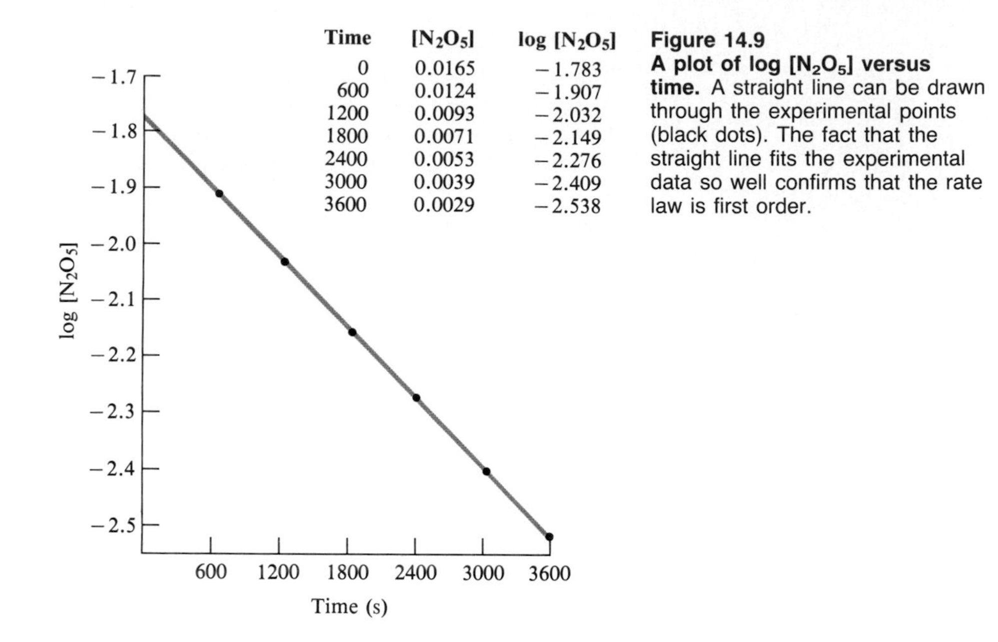

Time	$[N_2O_5]$	$\log [N_2O_5]$
0	0.0165	−1.783
600	0.0124	−1.907
1200	0.0093	−2.032
1800	0.0071	−2.149
2400	0.0053	−2.276
3000	0.0039	−2.409
3600	0.0029	−2.538

Figure 14.9
A plot of log $[N_2O_5]$ versus time. A straight line can be drawn through the experimental points (black dots). The fact that the straight line fits the experimental data so well confirms that the rate law is first order.

We can obtain the rate constant for the reaction from the slope, m, of the straight line.

$$m = \frac{-k}{2.303} \quad \text{or} \quad k = -2.303m$$

We calculate the slope of this curve in the same way we obtained the average rate of reaction from kinetic data (Example 14.2). We select two points far enough apart that when we subtract to obtain Δx and Δy for the slope, we do not lose significant figures. Using the first and last points in Figure 14.9, we get

$$m = \frac{\Delta y}{\Delta x} = \frac{(-2.538) - (-1.783)}{(3600 - 0)\text{ s}} = \frac{-0.755}{3600\text{ s}}$$
$$= -2.10 \times 10^{-4}/\text{s}$$

Therefore, $k = -2.303(-2.10 \times 10^{-4}/\text{s}) = 4.84 \times 10^{-4}/\text{s}$. (We selected two points directly from the experimental data. In precise work, we would first draw the straight line that best fits the experimental data points and then calculate the slope of this line.)

The second-order rate law, $-\Delta[\text{A}]/\Delta t = k[\text{A}]^2$, gives the following relationship between concentration of A and time:

$$\underbrace{\frac{1}{[\text{A}]_t}}_{y} = \underbrace{kt}_{mx} + \underbrace{\frac{1}{[\text{A}]_0}}_{b}$$

In this case, we get a straight line if we plot $1/[\text{A}]_t$ on the vertical axis against the time t on the horizontal axis for a second-order reaction.

As an illustration of the determination of reaction order by graphical plotting, consider the following data for the decomposition of NO_2 at 330°C.

$$2NO_2(g) \longrightarrow 2NO(g) + O_2(g)$$

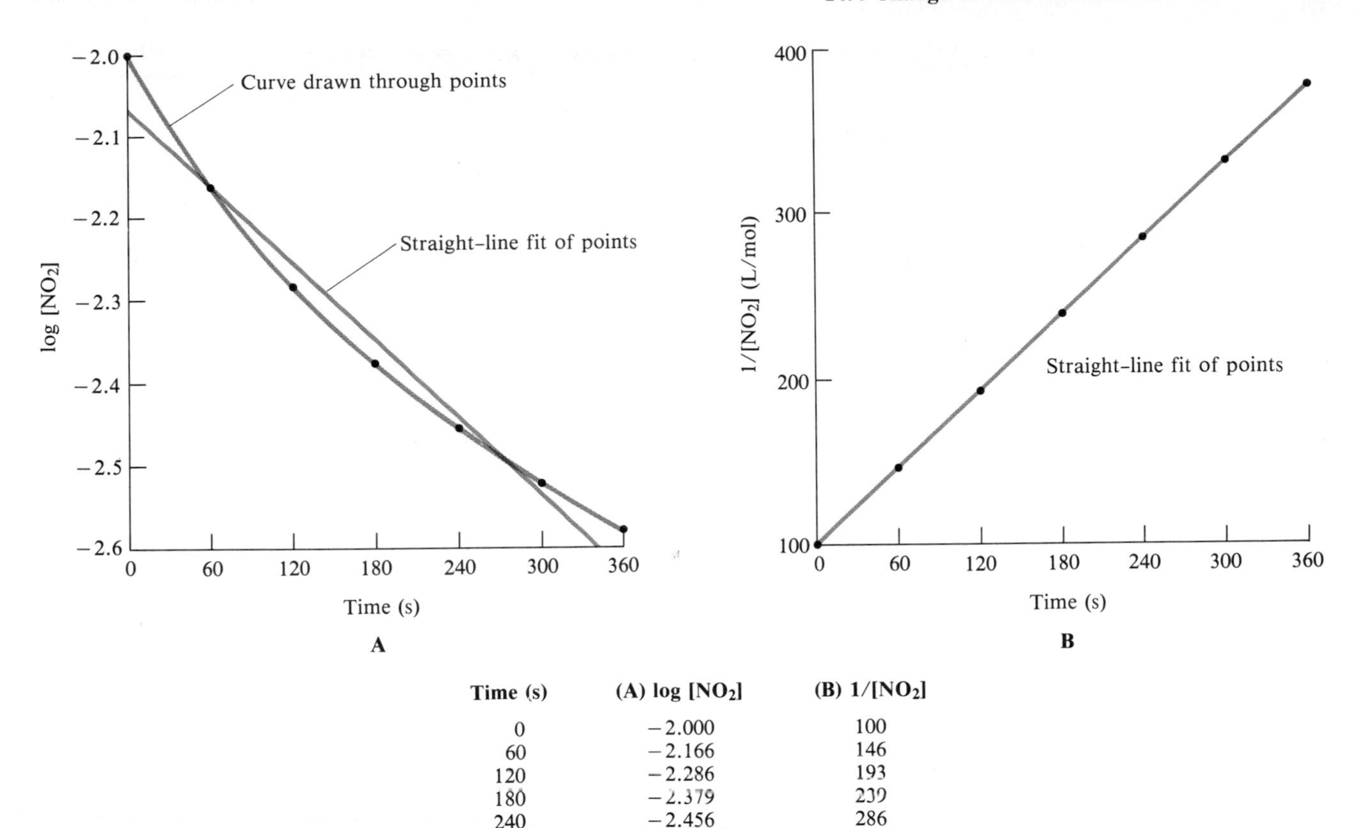

Time (s)	(A) log $[NO_2]$	(B) $1/[NO_2]$
0	−2.000	100
60	−2.166	146
120	−2.286	193
180	−2.379	239
240	−2.456	286
300	−2.521	332
360	−2.578	379

Figure 14.10
Plotting the data for the decomposition of nitrogen dioxide at 330°C. *(A)* Plot of log $[NO_2]$ against time. Note that a straight line does not fit the points well. *(B)* Plot of $1/[NO_2]$ against time. Note how closely the points follow the straight line, indicating that the decomposition is second order.

The concentrations of NO_2 for various times are

TIME, s	$[NO_2]$ IN mol/L
0	1.00×10^{-2}
60	0.683×10^{-2}
120	0.518×10^{-2}
180	0.418×10^{-2}
240	0.350×10^{-2}
300	0.301×10^{-2}
360	0.264×10^{-2}

In (A) of Figure 14.10, we have plotted log $[NO_2]$ against t, and in (B) we have plotted $1/[NO_2]$ against t. Only in (B) do the points closely follow a straight line, indicating that the rate law is second order. That is,

$$\text{Rate of disappearance of } NO_2 = -\frac{\Delta[NO_2]}{\Delta t} = k[NO_2]^2$$

We can obtain the rate constant, k, for a second-order reaction from the slope of the line, similar to the way we did for a first-order reaction. In this case, however, the slope equals k, as we can see from the equation that is plotted. Choosing the first and last points in Figure 14.10B, we get

$$k = \frac{\Delta y}{\Delta x} = \frac{(379 - 100)\ \text{L/mol}}{(360 - 0)\ \text{s}} = 0.775\ \text{L/(mol} \cdot \text{s)}$$

Table 14.1 Relationships for First-Order and Second-Order Reactions

Order	Rate Law	Concentration–Time Equation	Half-Life	Graphical Plot
1	Rate = k[A]	$\log \frac{[A]_t}{[A]_0} = \frac{-kt}{2.303}$	$0.693/k$	log [A] vs t
2	Rate = $k[A]^2$	$\frac{1}{[A]_t} = kt + \frac{1}{[A]_0}$	$1/(k[A]_0)$	$\frac{1}{[A]}$ vs t

Table 14.1 summarizes the relationships discussed in this section for first-order and second-order reactions.

14.5 TEMPERATURE AND RATE; COLLISION AND TRANSITION-STATE THEORIES

As we noted earlier, the rate of reaction depends on temperature. This shows up in the rate law through the rate constant, which is found to vary with temperature. In most cases, the rate increases with temperature (Figure 14.11). Consider the reaction of nitric oxide, NO, with chlorine, Cl_2, to give nitrosyl chloride, NOCl, and chlorine atoms.

$$NO(g) + Cl_2(g) \longrightarrow NOCl(g) + Cl(g)$$

The rate constant k for this reaction is 4.9×10^{-6} L/(mol · s) at 25°C and 1.5×10^{-5} L/(mol · s) at 35°C. Thus, in this case the rate constant and therefore the rate are more than tripled for a 10°C rise in temperature.● How do we explain this strong dependence of reaction rate on temperature? To understand it, we need to look at a simple theoretical explanation of reaction rates.

● The change in rate constant with temperature varies considerably from one reaction to another. In many cases, the rate of reaction approximately doubles for a 10°C rise, and this is often given as an approximate rule.

Figure 14.11
Effect of temperature on reaction rate. *Left:* Each test tube contains potassium permanganate, $KMnO_4$, and oxalic acid, $H_2C_2O_4$, at the same concentrations. Permanganate ion oxidizes oxalic acid to CO_2 and H_2O. One test tube was placed in a beaker of warm water (40°C); the other was kept at room temperature (20°C). *Right:* After 10 minutes, the test tube at 40°C showed noticeable reaction, whereas the other one did not.

Collision Theory

Why the rate constant depends on temperature can be explained by collision theory. **Collision theory** of reaction rates is *a theory that assumes that, in order for reaction to occur, reactant molecules must collide with an energy greater than some minimum value and with the proper orientation. The minimum energy of collision required for two molecules to react* is called the **activation energy,** E_a. The value of E_a depends on the particular reaction.

In collision theory, the rate constant for a reaction is given as a product of three factors: (1) Z, the collision frequency, (2) f, the fraction of collisions having energy greater than the activation energy, and (3) p, the fraction of collisions that occur with the reactant molecules properly oriented. Thus,

$$k = pfZ$$

We will discuss each of these factors in turn.

To have a specific reaction to relate the concepts to, we will consider the gas-phase reaction of NO with Cl_2, mentioned previously. This reaction is believed to occur in a single step. An NO molecule collides with a Cl_2 molecule. If the collision has sufficient energy and if the molecules are properly oriented, they react to produce NOCl and Cl.

Collision frequency Z, the frequency with which the reactant molecules collide, depends on temperature. As we will see, however, this dependence of collision frequency on temperature does not explain why reaction rates usually change greatly with small temperature increases. We can easily explain why the collision frequency depends on temperature. As the temperature rises, the gas molecules move faster and therefore collide more frequently. Thus, collision frequency is proportional to the root-mean-square (rms) molecular speed, which in turn is proportional to the square root of the absolute temperature, according to the kinetic theory of gases.● From kinetic theory, one can show that at 25°C, a 10°C rise in temperature increases the collision frequency by about 2%. If we were to assume that each collision of reactant molecules resulted in reaction, we would conclude that the rate would increase with temperature at the same rate as the collision frequency—that is, by 2% for a 10°C rise in temperature. This clearly does not explain the tripling of the rate (a 200% increase) that we see in the reaction of NO with Cl_2 when the temperature is raised from 25°C to 35°C.

● According to the kinetic theory of gases, the rms molecular speed equals $\sqrt{3RT/M_m}$ (see Section 5.7).

We see that the collision frequency varies only slowly with temperature. However, f, the fraction of molecular collisions having energy greater than the activation energy, changes rapidly in most reactions with even small temperature changes. It can be shown that f is related to the activation energy, E_a, this way:

$$f = e^{-E_a/RT}$$

Here $e = 2.718\ldots$, and R is the gas constant, which equals 8.31 J/(mol · K). For the reaction of NO with Cl_2, the activation energy is 8.5×10^4 J/mol. At 25°C (298 K), the fraction of collisions with sufficient energy for reaction is 1.2×10^{-15}. Thus, the number of collisions of reactant molecules that actually result in reaction is extremely small. But the frequency of collisions is very large, so the reaction rate, which depends on the product of these quantities, is not small. If the temperature is raised by 10°C to 35°C, the fraction of collisions of NO and Cl_2 molecules with sufficient energy for reaction is 3.8×10^{-15}, over three times larger than the value at 25°C! In other words, the tripling of the reaction rate when the temperature rises 10°C is explained by the temperature dependence of f.

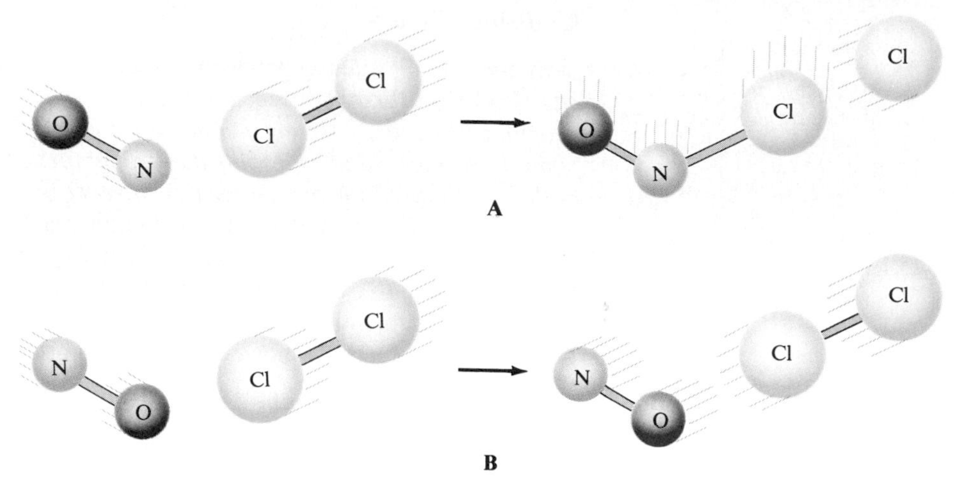

Figure 14.12
Importance of molecular orientation in the reaction of NO and Cl_2. *(A)* NO approaches with its N atom toward Cl_2, and an N—Cl bond forms. Also, the angle of approach is close to that in the product NOCl. *(B)* NO approaches with its O atom toward Cl_2. No N—Cl bond can form, so NO and Cl_2 collide and then fly apart.

From the previous equation relating f to E_a, we see that f decreases with increasing values of E_a. Because the rate constant depends on f, this means that reactions with large activation energies have small rate constants and that reactions with small activation energies have large rate constants.

We noted earlier that the reaction rate also depends on p, the proper orientation of the reactant molecules when they collide. This factor is independent of temperature changes. We can see why it is important that the reactant molecules be properly oriented by looking in some detail at the reaction of NO with Cl_2. Figure 14.12 shows two possible molecular collisions. In (A), the NO and Cl_2 molecules collide properly oriented for reaction. The NO molecule approaches with its N atom toward the Cl_2 molecule. In addition, the angle of approach is about that expected for the formation of bonds in the product molecule NOCl. In (B), an NO molecule approaches with its O atom toward the Cl_2 molecule. Because this orientation does not allow the formation of a bond between the N atom and a Cl atom, it is ineffective for reaction. The NO and Cl_2 molecules come together and then fly apart. All orientations except those close to that shown in (A) are ineffective.

Transition-State Theory

Collision theory explains some important features of a reaction, but it is limited in that it does not explain the role of activation energy. **Transition-state theory** *explains the reaction resulting from the collision of two molecules in terms of an activated complex.* An **activated complex** (transition state) is *an unstable grouping of atoms that can break up to form products.* We can represent the formation of the activated complex this way:

$$\text{O{=}N} + \text{Cl{-}Cl} \longrightarrow [\text{O{=}N{\cdots}Cl{\cdots}Cl}]$$

When the molecules come together with proper orientation, an N—Cl bond begins to form. At the same time, the kinetic energy of the collision is absorbed by the activated complex as a vibrational motion of the atoms. This energy becomes concentrated in the bonds denoted by the dashed lines and can flow between them. If, at some moment, sufficient energy becomes concentrated in one of the bonds of

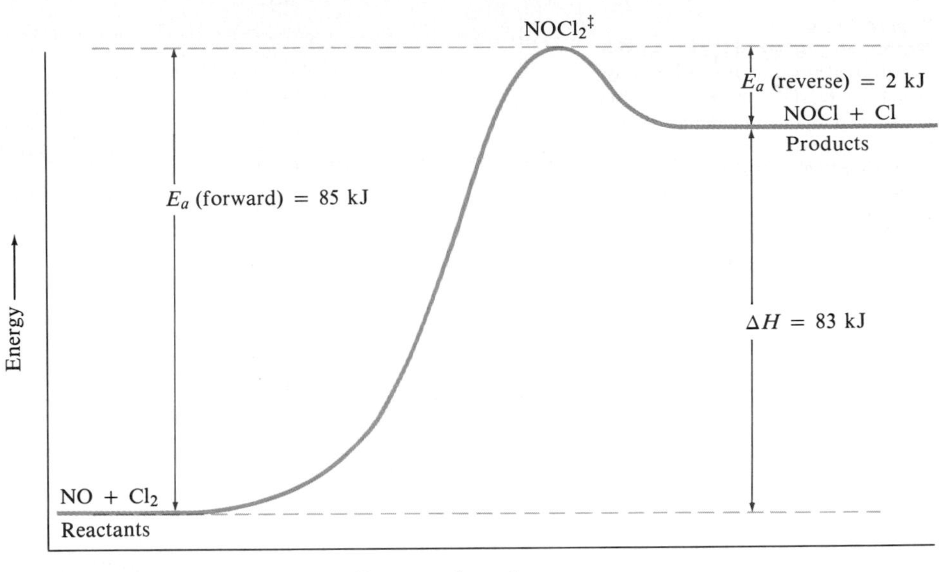

Figure 14.13
Potential-energy curve (not to scale) for the endothermic reaction $NO + Cl_2 \rightarrow NOCl + Cl$. In order for NO and Cl_2 to react, at least 85 kJ of energy must be supplied by the collision of reactant molecules. Once the activated complex forms, it may break up to products, releasing 2 kJ of energy. The difference, (85 − 2) kJ = 83 kJ, is the heat energy absorbed, ΔH.

the activated complex, that bond breaks or falls apart. Depending on whether the N----Cl or Cl----Cl bond breaks, the activated complex either reverts to the reactants or yields the products.

$$\underset{\text{reactants}}{O{=}N + Cl_2} \longleftarrow \underset{\text{activated complex}}{[O{=}N\text{---}Cl\text{---}Cl]} \longrightarrow \underset{\text{products}}{O{=}N{-}Cl + Cl}$$

Potential-Energy Diagrams for Reactions

It is instructive to consider a potential energy diagram for the reaction of NO with Cl_2. We can represent this reaction by the equation

$$NO + Cl_2 \longrightarrow NOCl_2^{\ddagger} \longrightarrow NOCl + Cl$$

Here $NOCl_2^{\ddagger}$ denotes the activated complex. Figure 14.13 shows the change in potential energy (indicated by the solid curve) that occurs during the progress of the reaction. The potential-energy curve starts at the left with the potential energy of the reactants, $NO + Cl_2$. Moving along the curve toward the right, the potential energy increases to a maximum corresponding to the activated complex. Further to the right, the potential energy decreases to that of the products, NOCl + Cl.

At the start, the NO and Cl_2 molecules have a certain quantity of kinetic energy. The total energy of these molecules equals their potential energy plus their kinetic energy, and it remains constant throughout the reaction (according to the law of conservation of energy). As the reaction progresses (going from left to right in the diagram), the reactants come together. The potential energy increases because the outer electrons of the two molecules repel. Thus, the kinetic energy decreases, and the molecules slow down. Only if the reactant molecules have sufficient kinetic energy is it possible for the potential energy to increase to the value for the activated complex. This kinetic energy must be equal to or greater than the difference in energy between the activated complex and the reactant

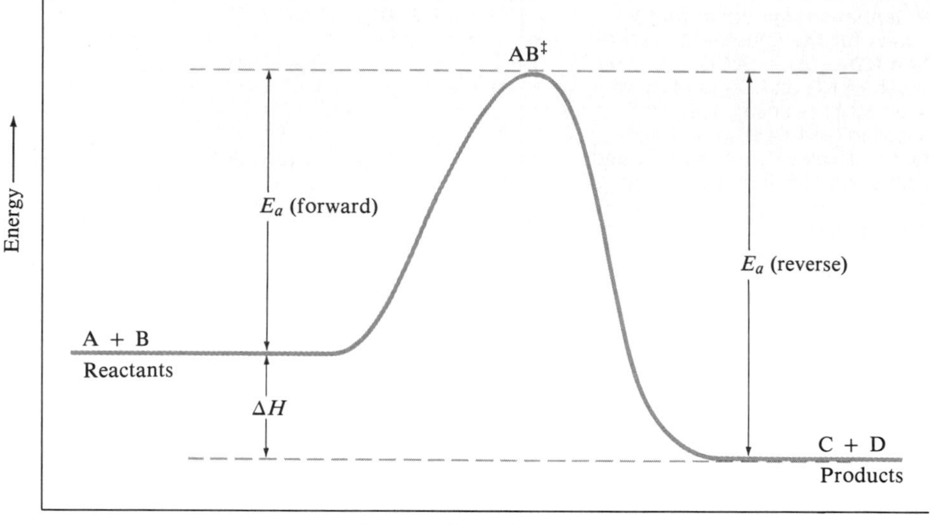

Figure 14.14
Potential-energy diagram for an exothermic reaction. The energy of the reactants is higher than that of the product, so heat energy is released when the reaction goes in the forward direction.

molecules (85 kJ/mol). The energy difference is the activation energy for the forward reaction.

At the maximum in the potential-energy curve, the reactant molecules have come together as the activated complex. When the activated complex breaks up into products, the products go to lower potential energy (by 2 kJ/mol) and gain in kinetic energy. Note that the energy of the products is higher than the energy of the reactants. The difference in energy equals the heat of reaction, ΔH. Because the energy increases, ΔH is positive and the reaction is endothermic.

Now let us look at the reverse reaction:

$$NOCl + Cl \longrightarrow NO + Cl_2$$

The activation energy for this reverse reaction is 2 kJ/mol, which is the difference in energy of the initial species, NOCl + Cl, and the activated complex. This is a smaller quantity than that for the forward reaction, so the rate constant for the reverse reaction is larger.

Figure 14.14 shows the potential-energy diagram for an exothermic reaction. In this case, the energy of the reactants is higher than that of the products, so heat energy is released when the reaction goes in the forward direction.

14.6 ARRHENIUS EQUATION

Rate constants for most chemical reactions closely follow an equation of the form

$$k = Ae^{-E_a/RT}$$

The mathematical equation $k = Ae^{-E_a/RT}$, *which expresses the dependence of the rate constant on temperature,* is called the **Arrhenius equation,** after its formulator, the Swedish chemist Svante Arrhenius.● Here e is the base of natural logarithms, 2.718 . . . ; E_a is the activation energy; R is the gas constant, 8.31 J/(K · mol); and T is the absolute temperature. *The symbol* A *in the Arrhenius equation, which is assumed to be a constant,* is called the **frequency factor.** The

● Arrhenius published this equation in 1889 and in that paper suggested that molecules must be given enough energy to become "activated" before they could react. Collision and transition-state theories, which enlarged on this concept, were developed later (1920s and 1930s, respectively).

Table 14.2 Rate Constant for the Decomposition of N_2O_5 at Various Temperatures

Temperature (°C)	k (/s)
45.0	4.8×10^{-4}
50.0	8.8×10^{-4}
55.0	1.6×10^{-3}
60.0	2.8×10^{-3}

frequency factor is related to the frequency of collisions with proper orientation (pZ). (The frequency factor does have a slight dependence on temperature, as we saw from collision theory, but usually it can be ignored.)

It is useful to recast Arrhenius's equation in logarithmic form. Taking the natural logarithm of both sides of the Arrhenius equation gives

$$\ln k = \ln A - \frac{E_a}{RT}$$

or, expressed in terms of logarithms to the base 10,

$$\log k = \log A - \frac{E_a}{2.303RT}$$

Let us make the following identification of symbols:

$$\underset{\displaystyle y}{\underset{\downarrow}{\log k}} = \underset{\displaystyle b}{\underset{\downarrow}{\log A}} + \underset{\displaystyle m}{\underset{\downarrow}{\left(\frac{-E_a}{2.303\,R}\right)}}\underset{\displaystyle x}{\underset{\downarrow}{\left(\frac{1}{T}\right)}}$$

This shows that if we plot $\log k$ against $1/T$, we should get a straight line. The slope of this line is $-E_a/(2.303\,R)$, from which we can obtain the activation energy E_a. The intercept is $\log A$. Figure 14.15 shows a plot of $\log k$ versus $1/T$ for the data given in Table 14.2. It demonstrates that the points do lie on a straight line.

We can put the previous equation into a form that is useful for computation. Let us write the equation for two different absolute temperatures T_1 and T_2. We write k_1 for the rate constant at temperature T_1 and k_2 for the rate constant at temperature T_2.

$$\log k_2 = \log A - \frac{E_a}{2.303\,RT_2}$$

$$\log k_1 = \log A - \frac{E_a}{2.303\,RT_1}$$

We eliminate A by subtracting these equations.

$$\log k_2 - \log k_1 = -\frac{E_a}{2.303\,RT_2} + \frac{E_a}{2.303\,RT_1}$$

or

$$\log \frac{k_2}{k_1} = \frac{E_a}{2.303\,R}\left(\frac{1}{T_1} - \frac{1}{T_2}\right)$$

The next example illustrates the use of this equation.

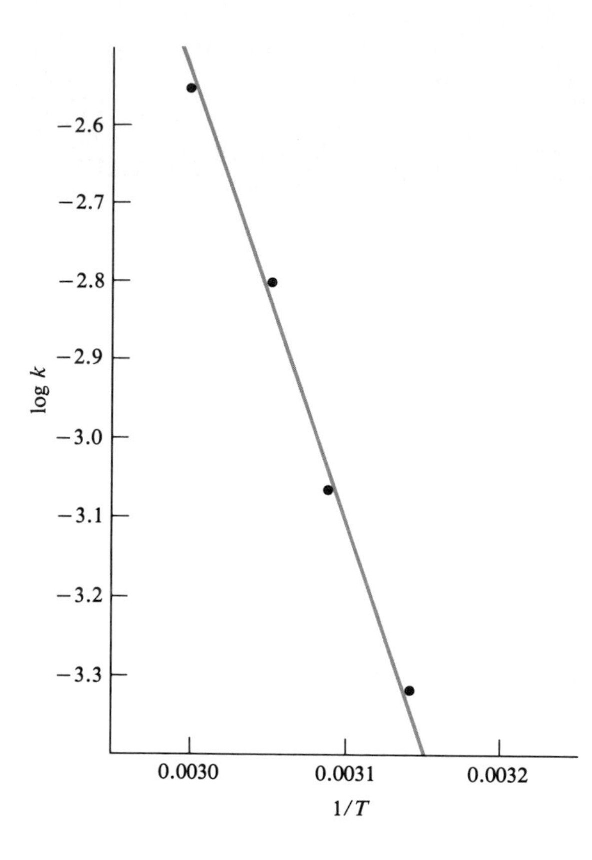

Figure 14.15
A plot of log k versus $1/T$. The logarithm of the rate constant for the decomposition of N_2O_5 (data from Table 14.2) is plotted versus $1/T$. The points are then fitted to a straight line, whose slope equals $-E_a/(2.303R)$.

Example 14.7 Using the Arrhenius Equation

The rate constant for the formation of hydrogen iodide from the elements

$$H_2(g) + I_2(g) \longrightarrow 2HI(g)$$

is 2.7×10^{-4} L/(mol · s) at 600 K and 3.5×10^{-3} L/(mol · s) at 650 K. (a) Find the activation energy E_a. (b) Then calculate the rate constant at 700 K.

Solution

(a) If we substitute the data given in the problem into the last equation, we can solve for E_a.

$$\log \frac{3.5 \times 10^{-3}}{2.7 \times 10^{-4}} = \frac{E_a}{2.303 \times 8.31 \text{ J/(mol} \cdot \not{K})} \left(\frac{1}{600 \not{K}} - \frac{1}{650 \not{K}} \right)$$

$$\log 1.30 \times 10^{1} = 1.11 = \frac{E_a}{2.303 \times 8.31 \text{ J/mol}} \times (1.28 \times 10^{-4})$$

Hence,

$$E_a = \frac{1.11 \times 2.303 \times 8.31 \text{ J/mol}}{1.28 \times 10^{-4}} = \mathbf{1.66 \times 10^5 \text{ J/mol}}$$

(b) To find the rate constant at 700 K, we write the equation obtained previously,

$$\log \frac{k_2}{k_1} = \frac{E_a}{2.303\,R} \left(\frac{1}{T_1} - \frac{1}{T_2} \right)$$

and substitute $E_a = 1.66 \times 10^5$ J/mol and

$k_1 = 2.7 \times 10^{-4}$ L/(mol · s) ($T_1 = 600$ K)
$k_2 =$ unknown as yet ($T_2 = 700$ K)

We get

$$\log \frac{k_2}{2.7 \times 10^{-4} \text{ L/(mol} \cdot \text{s)}} = \frac{1.66 \times 10^5 \text{ J/mol}}{2.303 \times 8.31 \text{ J/(mol} \cdot \text{K)}} \times \left(\frac{1}{600 \text{ K}} - \frac{1}{700 \text{ K}} \right) = 2.07$$

Taking antilogarithms, we get

$$\frac{k_2}{2.7 \times 10^{-4}\ \text{L/(mol} \cdot \text{s)}} = 10^{2.07} = 1.2 \times 10^2$$

Hence,

$$k_2 = (1.2 \times 10^2) \times (2.7 \times 10^{-4})\ \text{L/(mol} \cdot \text{s)}$$
$$= \mathbf{3.2 \times 10^{-2}\ L/(mol \cdot s)}$$

Exercise 14.7

Acetaldehyde, CH_3CHO, decomposes when heated.

$$CH_3CHO(g) \longrightarrow CH_4(g) + CO(g)$$

The rate constant for the decomposition is $1.05 \times 10^{-3}/(M^{1/2} \cdot s)$ at 759 K and $2.14 \times 10^{-2}/(M^{1/2} \cdot s)$ at 836 K. What is the activation energy for this decomposition? What is the rate constant at 865 K? (See Problems 14.55, 14.56, 14.57, and 14.58.)

Reaction Mechanisms

A balanced chemical equation is a description of the overall result of a chemical reaction. However, what actually happens at the molecular level may be more involved than is represented by this single equation. The reaction may take place in several steps. In the next sections, we will examine some reactions and see how the rate law can give us information about these steps, or *elementary reactions*.

14.7 ELEMENTARY REACTIONS

Consider the reaction of nitrogen dioxide with carbon monoxide.

$$NO_2(g) + CO(g) \longrightarrow NO(g) + CO_2(g) \quad \text{(net chemical equation)}$$

At temperatures below 500 K, this gas-phase reaction is believed to take place in two steps.

$$NO_2 + NO_2 \longrightarrow NO_3 + NO \quad \text{(elementary reaction)}$$
$$NO_3 + CO \longrightarrow NO_2 + CO_2 \quad \text{(elementary reaction)}$$

Each step, called an **elementary reaction,** is *a single molecular event, such as a collision of molecules, resulting in a reaction. The set of elementary reactions whose overall effect is given by the net chemical equation* is called the **reaction mechanism.**

According to the reaction mechanism just given, two NO_2 molecules collide and react to give the product molecule NO and the reaction intermediate NO_3. A **reaction intermediate** is *a species produced during a reaction that does not appear in the net equation because it reacts in a subsequent step in the mechanism.* Often it has a fleeting existence and cannot be isolated from the reaction mixture. The NO_3 molecule is known only from its visible light spectrum. It reacts quickly with CO to give the product molecules NO_2 and CO_2.

The overall chemical equation, which represents the net result of these two elementary reactions in the mechanism, is obtained by adding these steps together and canceling species that occur on both sides.

$$
\begin{array}{lll}
NO_2 + NO_2 \longrightarrow NO_3 + NO & & \text{(elementary reaction)} \\
NO_3 + CO \longrightarrow NO_2 + CO_2 & & \text{(elementary reaction)} \\
\hline
NO_2 + \cancel{NO_2} + \cancel{NO_3} + CO \longrightarrow \cancel{NO_3} + NO + \cancel{NO_2} + CO_2 & & \text{(overall equation)}
\end{array}
$$

Example 14.8 Writing the Overall Chemical Equation from a Mechanism

Carbon tetrachloride, CCl_4, is obtained by chlorinating methane or an incompletely chlorinated methane, such as chloroform, $CHCl_3$. The mechanism for the gas-phase chlorination of $CHCl_3$ is

$$
\begin{array}{ll}
Cl_2 \rightleftharpoons 2Cl & \text{(elementary reaction)} \\
Cl + CHCl_3 \longrightarrow HCl + CCl_3 & \text{(elementary reaction)} \\
Cl + CCl_3 \longrightarrow CCl_4 & \text{(elementary reaction)}
\end{array}
$$

Obtain the net or overall chemical equation from this mechanism.

Solution

The first step produces two Cl atoms (a reaction intermediate). One Cl atom is used in the second step and another is used in the third step. Thus, all Cl atoms cancel. Similarly, the intermediate CCl_3, produced in the second step, is used up in the third step. We can cancel all Cl and CCl_3 species.

$$
\begin{array}{ll}
Cl_2 \rightleftharpoons \cancel{2Cl} & \\
\cancel{Cl} + CHCl_3 \longrightarrow HCl + \cancel{CCl_3} & \\
\cancel{Cl} + \cancel{CCl_3} \longrightarrow CCl_4 & \\
\hline
Cl_2 + CHCl_3 \longrightarrow HCl + CCl_4 & \textbf{(overall equation)}
\end{array}
$$

Exercise 14.8

The iodide ion catalyzes the decomposition of aqueous hydrogen peroxide, H_2O_2. This decomposition is believed to occur in two steps.

$$
\begin{array}{ll}
H_2O_2 + I^- \longrightarrow H_2O + IO^- & \text{(elementary reaction)} \\
H_2O_2 + IO^- \longrightarrow H_2O + O_2 + I^- & \text{(elementary reaction)}
\end{array}
$$

What is the overall equation representing this decomposition? Note that IO^- is a reaction intermediate. The iodide ion is not an intermediate; it was added to the reaction mixture. (See Problems 14.61 and 14.62.)

Molecularity

Elementary reactions are classified according to their molecularity. The **molecularity** is *the number of molecules on the reactant side of an elementary reaction*. A **unimolecular reaction** is *an elementary reaction that involves one reactant molecule;* a **bimolecular reaction** is *an elementary reaction that involves two reactant molecules*. Bimolecular reactions are the most common. Unimolecular reactions are best illustrated by decomposition of some previously excited species. Some gas-phase reactions are thought to occur in a **termolecular reaction,** *an elementary reaction that involves three reactant molecules*. Higher molecularities are not encountered, presumably because the chance of the correct four molecules coming together at once is extremely small.

As an example of a unimolecular reaction, consider the elementary process in which an energetically excited ozone molecule (symbolized by O_3^*) spontaneously decomposes.

$$O_3^* \longrightarrow O_2 + O$$

Normally, a molecule in a sample of ozone gas is in a lower energy level. But such a molecule may be excited to a higher level if it collides with another molecule or absorbs a photon. The energy of this excited molecule is distributed among its three nuclei as vibrational energy. After a period of time, this energy becomes redistributed. If by chance most of the energy finds its way to one oxygen atom, that atom will fly off. In other words, the excited ozone molecule decomposes into an oxygen molecule with an oxygen atom.

All of the steps in the mechanism of the reaction of NO_2 with CO, given earlier in this section, are bimolecular. Consider the first step, which involves the reaction of two NO_2 molecules.

$$NO_2 + NO_2 \longrightarrow NO_3 + NO$$

When these two NO_2 molecules come together, they form an activated complex of the six atoms, $(NO_2)_2$, which immediately breaks into two new molecules, NO_3 and NO.

The overall reaction of two atoms—say, two Br atoms—to form a diatomic molecule (Br_2) is normally a termolecular process. When two bromine atoms collide, they form an excited bromine molecule, Br_2^*. This excited molecule immediately flies apart, re-forming the atoms, unless another atom or molecule is present just at the moment of molecule formation to take away the excess energy. Suppose an argon atom and the two bromine atoms all collide at the same moment.

$$Br + Br + Ar \longrightarrow Br_2 + Ar^*$$

Energy that would have been left with the bromine molecule is now picked up by the argon atom (giving the energized atom Ar^*). The bromine molecule is stabilized by being left in a lower energy level.

Example 14.9 Determining the Molecularity of an Elementary Reaction

What is the molecularity of each step in the mechanism described in Example 14.8?

$$Cl_2 \rightleftharpoons 2Cl$$
$$Cl + CHCl_3 \longrightarrow HCl + CCl_3$$
$$Cl + CCl_3 \longrightarrow CCl_4$$

Solution

The molecularity of any elementary reaction equals the number of reactant molecules. Thus, the forward part of the first step is **unimolecular**; the reverse of the first step, the second step and third step are each **bimolecular**.

Exercise 14.9

The following is an elementary reaction that occurs in the decomposition of ozone in the stratosphere by nitric oxide.

$$O_3 + NO \longrightarrow O_2 + NO_2$$

What is the molecularity of this reaction? That is, is the reaction unimolecular, bimolecular, or termolecular? (See Problems 14.63 and 14.64.)

Rate Equation for an Elementary Reaction

There is no necessarily simple relationship between the overall reaction and the rate law that we observe for it. As we stressed before, the rate law must be obtained experimentally. However, when we are dealing with an elementary reaction, the rate does have a simple, predictable form. The rate is proportional to the product of the concentration of each reactant molecule.

In order to understand this, let us look at the different possibilities. Consider the unimolecular elementary reaction.

$$A \longrightarrow B + C$$

For each A molecule there is a definite probability, or chance, that it will decompose into B and C molecules. The more A molecules there are in a given volume, the more A molecules that can decompose in that volume per unit time. In other words, the rate of reaction is proportional to the concentration of A.

$$\text{Rate} = k[A]$$

Now consider a bimolecular elementary reaction, such as

$$A + B \longrightarrow C + D$$

For the reaction to occur, the reactant molecules A and B must collide. Reaction does not occur with every collision. Nevertheless, the rate of formation of product is proportional to the frequency of molecular collisions, because a definite fraction of those collisions produce reaction. Within a given volume, the frequency of collisions is proportional to the number of A molecules, n_A, times the number of B molecules, n_B. Furthermore, the concentration of A is proportional to n_A, and the concentration of B is proportional to n_B. Therefore, the rate of this elementary reaction is proportional to [A][B].

$$\text{Rate} = k[A][B]$$

A termolecular elementary reaction has a rate equation that is obtained by similar reasoning. For the elementary reaction

$$A + B + C \longrightarrow D + E$$

the rate is proportional to the concentrations of A, B, and C.

$$\text{Rate} = k[A][B][C]$$

Any reaction we observe is likely to consist of several elementary steps, and the rate law that we find is the combined result of these steps. This is why we cannot predict the rate law by looking at the overall equation. In the next section, we look at the relationship between a reaction mechanism and the observed rate law.

Example 14.10 Writing the Rate Equation for an Elementary Reaction

Write rate equations for each of the following elementary reactions.

(a) Ozone is converted to O_2 by NO in a single step.

$$O_3 + NO \longrightarrow O_2 + NO_2$$

(b) The recombination of iodine atoms occurs as follows:

$$I + I + M \longrightarrow I_2 + M^*$$

where M is some atom or molecule that absorbs energy from the reaction.

(c) An H_2O molecule absorbs energy; some time later enough of this energy flows into one O—H bond to break it.

$$H_2O \longrightarrow H + O\text{—}H$$

Solution

The rate equation can be written directly from the elementary reaction (but *only* for an elementary reaction).

(a) **Rate = $k[O_3][NO]$**

(b) **Rate = $k[I]^2[M]$**

(c) **Rate = $k[H_2O]$**

Exercise 14.10

Write the rate equation, showing the dependence of rate on concentrations, for the elementary reaction

$$NO_2 + NO_2 \longrightarrow N_2O_4$$

(See Problems 14.65 and 14.66.)

14.8 THE RATE LAW AND THE MECHANISM

The mechanism of a reaction cannot be observed directly. A mechanism is devised to explain the experimental observations. It is like the explanation provided by a detective to explain a crime in terms of the clues found. Other explanations may be possible, and further clues may make one of these other explanations seem more plausible than the currently accepted one. So it is with reaction mechanisms. They are accepted provisionally, with the understanding that further experiments may lead us to accept another mechanism as the more probable explanation.●

● We see the scientific method in operation here, which recalls the discussion in Chapter 1. Experiments have been made from which we determine the rate law. Then a mechanism is devised to explain the rate law. This mechanism in turn suggests more experiments. These may confirm the explanation or they may disagree with it. If they disagree, a new mechanism must be devised that explains all of the experimental evidence.

An important clue in understanding the mechanism of a reaction is the rate law. The reason for its importance is that once we assume a mechanism, we can predict the rate law. If this prediction does not agree with the experimental rate law, the assumed mechanism must be wrong. Take, for example, the overall equation

$$2NO_2(g) + F_2(g) \longrightarrow 2NO_2F(g)$$

If we follow the rate of disappearance of F_2, we observe that it is directly proportional to the concentration of NO_2 and F_2.

$$\text{Rate} = k[NO_2][F_2] \quad \text{(experimental rate law)}$$

This rate law is a summary of the experimental data. Let us assume that the reaction occurs in a single elementary reaction.

$$NO_2 + NO_2 + F_2 \longrightarrow NO_2F + NO_2F \quad \text{(elementary reaction)}$$

This, then, is our assumed mechanism. Because this is an elementary reaction, we can immediately write the rate law predicted by it.

$$\text{Rate} = k[NO_2]^2[F_2] \quad \text{(predicted rate law)}$$

However, this does not agree with experiment, and our assumed mechanism must be discarded. We conclude that the reaction occurs in more than one step.

Rate-Determining Step

The reaction of NO_2 with F_2 is believed to occur in the following steps (elementary reactions).

$$NO_2 + F_2 \xrightarrow{k_1} NO_2F + F \quad \text{(slow step)}$$

$$F + NO_2 \xrightarrow{k_2} NO_2F \quad \text{(fast step)}$$

Rate constants have been written over the arrows. The second step is assumed to be much faster than the first, so that as soon as NO_2 and F_2 react, the F atom that is formed reacts with an NO_2 molecule to give another NO_2F molecule. Therefore, the rate of disappearance of F_2 is determined completely by the slow step, or rate-determining step. The **rate-determining step** is *the slowest step in the reaction mechanism*.

To better understand the significance of the rate-determining step, think of an assembly line for making "widgets." Suppose one person on the line has a much lengthier, more intricate task to perform than the others on the line. Widgets can be produced only as fast as this difficult task can be completed. This person's task is rate determining.

The rate equation for the rate-determining step of the mechanism we are discussing is

$$\text{Rate} = k_1[NO_2][F_2]$$

The mechanism agrees with the experimental rate law if we equate k_1 to k. This agreement is not absolute evidence that the mechanism is correct. However, one can perform experiments to see whether fluorine atoms react very quickly with nitrogen dioxide. Such experiments show that they do.

Example 14.11 Determining the Rate Law from a Mechanism with an Initial Slow Step

Ozone reacts with nitrogen dioxide to produce oxygen and dinitrogen pentoxide.

$$O_3(g) + 2NO_2(g) \longrightarrow O_2(g) + N_2O_5(g)$$

The proposed mechanism is

$$O_3 + NO_2 \longrightarrow NO_3 + O_2 \quad \text{(slow)}$$

$$NO_3 + NO_2 \longrightarrow N_2O_5 \quad \text{(fast)}$$

What is the rate law predicted by this mechanism?

Solution

The first step in the mechanism is rate determining, because the rate of this step is much slower than the second one. We can write the rate law directly from this step.

$$\text{Rate} = k[O_3][NO_2]$$

Exercise 14.11

The iodide ion–catalyzed decomposition of hydrogen peroxide, H_2O_2, is believed to follow the mechanism

$$H_2O_2 + I^- \xrightarrow{k_1} H_2O + IO^- \quad \text{(slow)}$$

$$H_2O_2 + IO^- \xrightarrow{k_2} H_2O + O_2 + I^- \quad \text{(fast)}$$

What rate law is predicted by this mechanism? Explain.

(See Problems 14.67 and 14.68.)

Mechanisms with an Initial Fast Step

A somewhat more complicated situation occurs when the rate-determining step follows an initial fast, equilibrium step. The decomposition of dinitrogen pentoxide,

$$2N_2O_5(g) \longrightarrow 4NO_2(g) + O_2(g) \quad \text{(overall equation)}$$

which we discussed in the chapter opening, is believed to follow this type of mechanism.

$$N_2O_5 \underset{k_{-1}}{\overset{k_1}{\rightleftharpoons}} NO_2 + NO_3 \quad \text{(fast, equilibrium)}$$

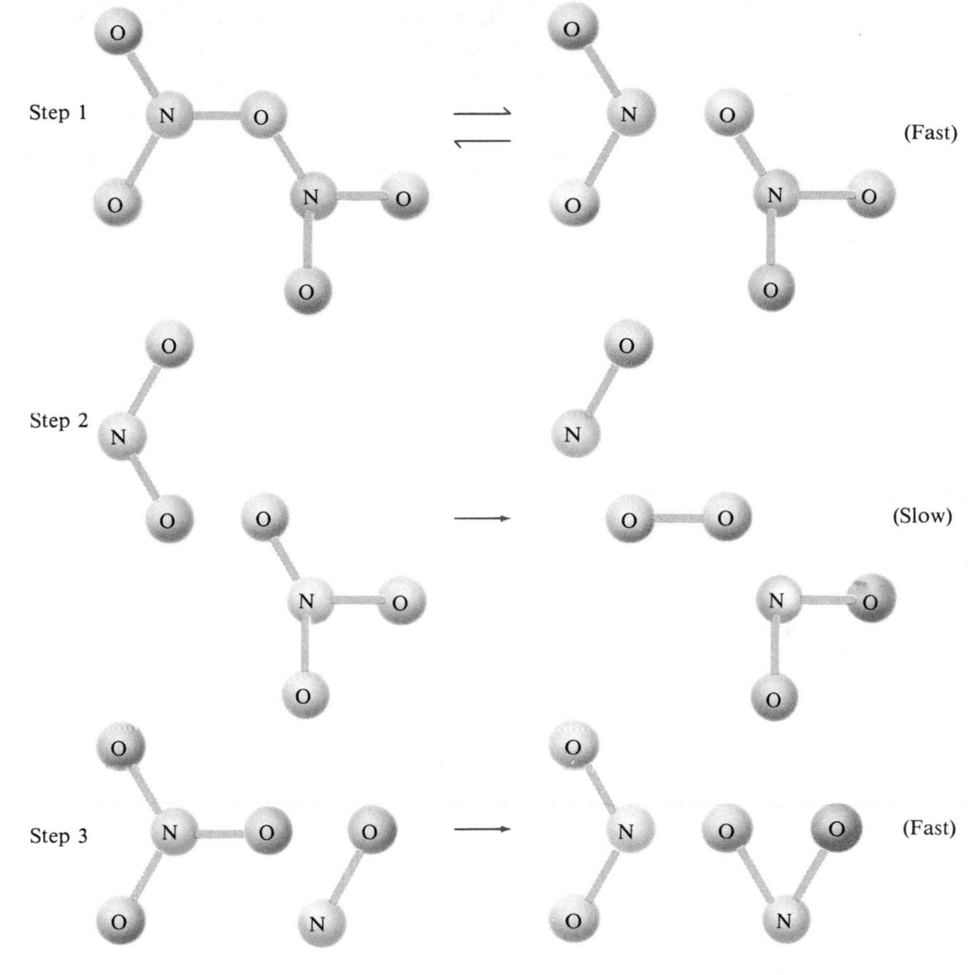

Figure 14.16 Representation of the mechanism of decomposition of N_2O_5, using molecular models. Step 1 occurs twice as Steps 2 and 3 both occur once. Note that there is little atomic rearrangement during each step.

$$NO_2 + NO_3 \xrightarrow{k_2} NO + NO_2 + O_2 \qquad \text{(slow)}$$

$$NO_3 + NO \xrightarrow{k_3} 2NO_2 \qquad \text{(fast)}$$

The first step occurs twice as the succeeding steps occur once. Figure 14.16 represents the mechanism by means of molecular models. Let us show that this mechanism is consistent with the experimentally determined rate law,

$$\text{Rate} = k[N_2O_5]$$

The second step in the mechanism is assumed to be much slower than the other steps and is therefore rate determining. Hence, the rate law predicted from this mechanism is

$$\text{Rate} = k_2[NO_2][NO_3]$$

However, this equation cannot be compared directly with experiment because it is written in terms of the reaction intermediate, NO_3. The experimental rate law will be written in terms of substances that occur in the chemical equation, not in terms of reaction intermediates. For purposes of comparison, it is necessary to re-express the rate equation, eliminating $[NO_3]$. To do this, we must look at the first step in the mechanism.

This step is fast and reversible. That is, N_2O_5 dissociates rapidly into NO_2 and NO_3, and these products in turn react to re-form N_2O_5. The rate of the forward reaction (dissociation of N_2O_5) is

$$\text{Forward rate} = k_1[N_2O_5]$$

and the rate of the reverse reaction (formation of N_2O_5 from NO_2 and NO_3) is

$$\text{Reverse rate} = k_{-1}[NO_2][NO_3]$$

When the reaction first begins, there are no NO_2 or NO_3 molecules, and the reverse rate is zero. But as N_2O_5 dissociates, the concentration of N_2O_5 decreases and the concentrations of NO_2 and NO_3 increase. Therefore, the forward rate decreases and the reverse rate increases. Soon the two rates become equal, such that N_2O_5 molecules form as often as other N_2O_5 molecules dissociate. The first step has reached *dynamic equilibrium*. Because these elementary reactions are much faster than the second step, this equilibrium is reached before any significant reaction by the second step occurs. Moreover, this equilibrium is maintained throughout the reaction.●

● Chemical equilibrium is discussed in detail in the next chapter.

At equilibrium, the forward and reverse rates are equal, so we can write

$$k_1[N_2O_5] = k_{-1}[NO_2][NO_3]$$

$$[NO_3] = \frac{k_1}{k_{-1}} \frac{[N_2O_5]}{[NO_2]}$$

Substituting into the rate law, we get

$$\text{Rate} = k_2[NO_2][NO_3] = k_2\cancel{[NO_2]} \times \frac{k_1}{k_{-1}} \frac{[N_2O_5]}{\cancel{[NO_2]}}$$

Or,

$$\text{Rate} = k_2 \frac{k_1}{k_{-1}} [N_2O_5]$$

Thus, if we identify k_1k_2/k_{-1} as k, we reproduce the experimental rate law.

Example 14.12 Determining the Rate Law from a Mechanism with an Initial Fast, Equilibrium Step

Nitric oxide can be reduced with hydrogen gas to give nitrogen and water vapor.

$$2NO(g) + 2H_2(g) \longrightarrow N_2(g) + 2H_2O(g) \quad \text{(overall equation)}$$

A proposed mechanism is

$$2NO \underset{k_{-1}}{\overset{k_1}{\rightleftharpoons}} N_2O_2 \quad \text{(fast, equilibrium)}$$

$$N_2O_2 + H_2 \xrightarrow{k_2} N_2O + H_2O \quad \text{(slow)}$$

$$N_2O + H_2 \xrightarrow{k_3} N_2 + H_2O \quad \text{(fast)}$$

What rate law is predicted by this mechanism?

Solution

According to the rate-determining step (the slow step),

$$\text{Rate} = k_2[N_2O_2][H_2]$$

However, N_2O_2 does not appear in the overall equation. Let us try to eliminate it from the rate law by looking at the first step, which is fast and reaches equilibrium. At equilibrium, the forward rate and the reverse rate are equal.

$$k_1[NO]^2 = k_{-1}[N_2O_2]$$

Therefore, $[N_2O_2] = (k_1/k_{-1})[NO]^2$, so

$$\text{Rate} = \frac{k_2k_1}{k_{-1}}[NO]^2[H_2]$$

Experimentally, we should observe the rate law

$$\textbf{Rate} = \boldsymbol{k[NO]^2[H_2]}$$

where we have replaced the constants k_2k_1/k_{-1} with k, which represents the experimentally observed rate constant.

Exercise 14.12

Nitric oxide, NO, reacts with oxygen to produce nitrogen dioxide.

$$2NO(g) + O_2(g) \longrightarrow 2NO_2(g) \quad \text{(overall equation)}$$

If the mechanism is

$$NO + O_2 \underset{k_{-1}}{\overset{k_1}{\rightleftharpoons}} NO_3 \quad \text{(fast, equilibrium)}$$

$$NO_3 + NO \xrightarrow{k_2} NO_2 + NO_2 \quad \text{(slow)}$$

what is the predicted rate law? Remember to express this in terms of substances in the chemical equation. (See Problems 14.69 and 14.70.)

14.9 CATALYSIS

Catalysis is the increase in rate of a reaction as the result of addition of a catalyst. Although a catalyst is not consumed by a reaction, it does take part in the reaction mechanism; it enters at one step and is regenerated in a later step. In the commercial preparation of sulfuric acid, its anhydride, sulfur trioxide, is obtained by oxidizing sulfur dioxide, SO_2. Sulfuric acid is then obtained from SO_3. The reaction of SO_2 with O_2 to give SO_3 is normally very slow. In an early industrial process, this reaction was carried out in the presence of nitric oxide, NO, because it is then much faster.

$$2SO_2(g) + O_2(g) \xrightarrow{NO} 2SO_3(g)$$

Nitric oxide does not appear in the overall equation, nor does it in any way affect the final composition of the reaction mixture. Thus, it acts as a catalyst.

A possible mechanism for the nitric oxide catalysis is

$$2NO + O_2 \longrightarrow 2NO_2$$
$$NO_2 + SO_2 \longrightarrow NO + SO_3$$

The last step occurs twice each time the first step occurs once. Thus, two molecules of nitric oxide are used up in the first step and are regenerated in the second step.

This catalytic oxidation of sulfur dioxide is an example of **homogeneous catalysis,** which is *the use of a catalyst in the same phase as the reacting species.* Another example occurs in the oxidation of thallium(I) to thallium(III) by cerium(IV).

$$2Ce^{4+}(aq) + Tl^{+}(aq) \longrightarrow 2Ce^{3+}(aq) + Tl^{3+}(aq)$$

The uncatalyzed reaction is very slow; presumably it involves the collision of three positive ions. The reaction can be catalyzed by manganese(II) ion, however. The mechanism is thought to be

$$Ce^{4+} + Mn^{2+} \longrightarrow Ce^{3+} + Mn^{3+}$$
$$Ce^{4+} + Mn^{3+} \longrightarrow Ce^{3+} + Mn^{4+}$$
$$Mn^{4+} + Tl^{+} \longrightarrow Tl^{3+} + Mn^{2+}$$

Each step is bimolecular.

Some of the most important industrial reactions involve **heterogeneous catalysis**—that is, *the use of a catalyst that exists in a different phase from the reacting* *species, usually a solid catalyst in contact with a gaseous or liquid solution of reactants*. Such surface or heterogeneous catalysis is thought to occur by chemical adsorption of the reactants onto the surface of the catalyst. *Adsorption* is the attraction of molecules to a surface. In *physical adsorption,* the attraction is provided by weak intermolecular forces. **Chemisorption,** by contrast, is *the binding of a species to a surface by chemical bonding forces*. It may happen that bonds in the species are broken during chemisorption, and this may provide the basis of catalytic action in certain cases.●

● Understanding the chemical processes occurring at surfaces is being advanced by new techniques, including x-ray photoelectron spectroscopy and scanning tunneling microscopy.

An example of heterogeneous catalysis involving chemisorption is provided by *catalytic hydrogenation*. This is the addition of H_2 to a compound, such as one with a carbon–carbon double bond, using a catalyst of platinum or nickel metal. Vegetable oils, which contain carbon–carbon double bonds, are changed to solid fats (shortening) when the bonds are catalytically hydrogenated. In the case of ethylene, C_2H_4, the equation is

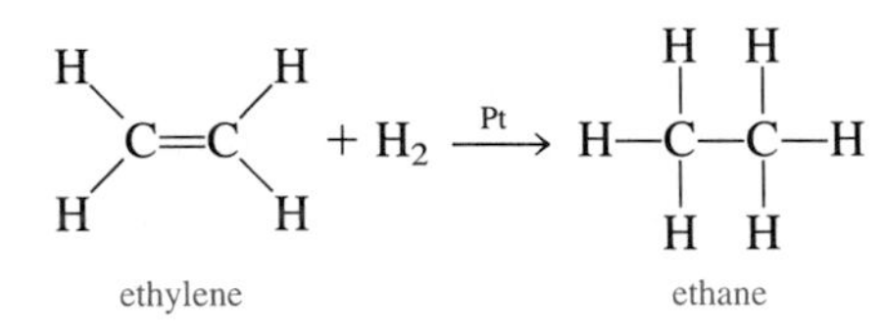

A mechanism for this reaction is represented by the four steps shown in Figure 14.17. In (A), ethylene and hydrogen molecules diffuse to the catalyst surface, where, as shown in (B), they undergo chemisorption. The pi electrons of ethylene form bonds to the metal, and hydrogen molecules break into H atoms that

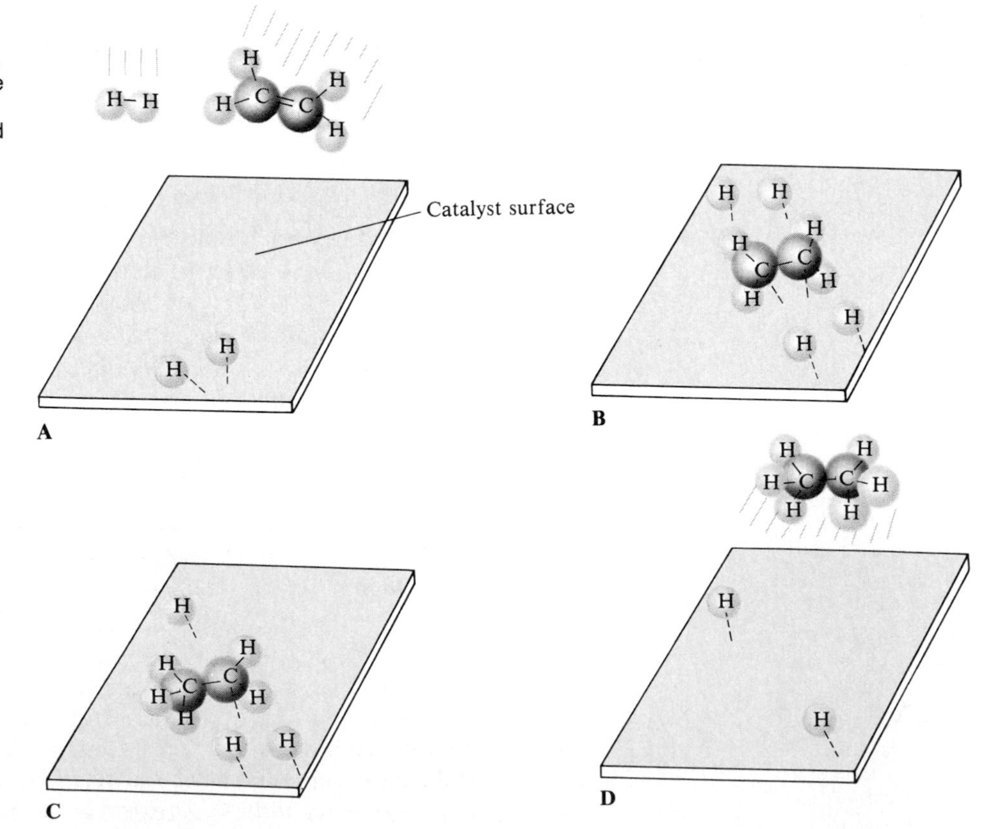

Figure 14.17
The proposed mechanism of catalytic hydrogenation of C_2H_4. *(A)* C_2H_4 and H_2 molecules diffuse to the catalyst. *(B)* The molecules form bonds (represented by dotted lines) to the catalyst surface. The H_2 molecules dissociate to atoms. *(C)* H atoms migrate to the C_2H_4 molecule, where they react to form C_2H_6. *(D)* C_2H_6 diffuses away from the catalyst.

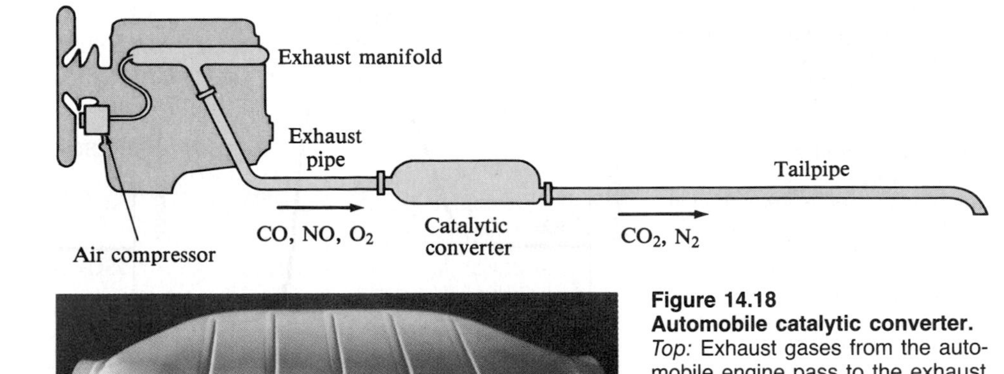

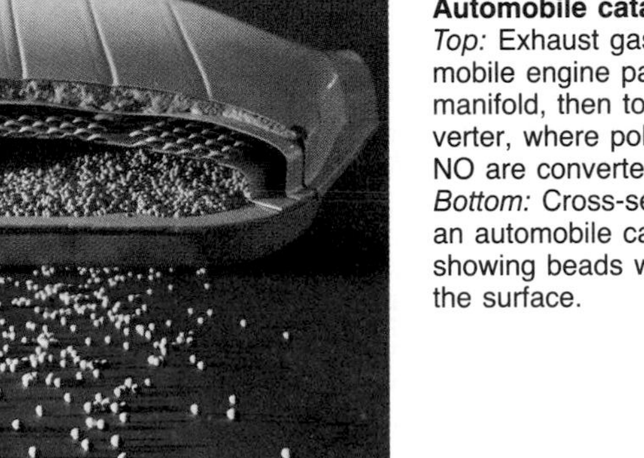

Figure 14.18
Automobile catalytic converter. *Top:* Exhaust gases from the automobile engine pass to the exhaust manifold, then to the catalytic converter, where pollutants CO and NO are converted to CO_2 and N_2. *Bottom:* Cross-sectional view of an automobile catalytic converter, showing beads with catalyst on the surface.

bond to the metal. In (C), two H atoms migrate to an ethylene molecule bonded to the catalyst, where they react to form ethane. Then, in (D), because ethane does not bond to the metal, it diffuses away. Note that the catalyst surface that was used in (B) is regenerated in (D).

At the beginning of this section, we described the catalytic oxidation of SO_2 to SO_3, the anhydride of sulfuric acid. In an early process, the homogeneous catalyst NO was used. Today, in the *contact process,* a heterogeneous catalyst, Pt or V_2O_5, is used. Surface catalysts are used in the catalytic converters of automobiles to convert substances that would be atmospheric pollutants, such as CO and NO, into harmless substances, such as CO_2 and N_2 (Figure 14.18).

Enzymes are proteins that catalyze biochemical processes. The reactant molecule, referred to as the enzyme *substrate,* attaches itself to a particular place (an *active site*) on the enzyme molecule. In this way, the substrate is held in a configuration that makes it easier for subsequent reaction to occur. The catalytic activity of enzymes is usually very specific. For example, the enzyme lysozyme, which is found in the tears of the eye, catalyzes the breakdown of the cell wall of certain bacteria. Once the cell wall of a bacterium has been removed, the cell breaks up as a result of the high osmotic pressure within. By this means, lysozyme in tears protects the surface of the eye from bacterial infection.

We have seen that a catalyst operates by providing a mechanism that facilitates the reaction. The catalyzed mechanism is faster than the uncatalyzed mechanism. We can look at this in terms of a potential-energy diagram. The activation energy for the uncatalyzed oxidation of sulfur dioxide is very large (Figure 14.19A). When a catalyst is added, however, a new mechanism is provided, in which the activation energy is lower than that for the uncatalyzed reaction in the

A Checklist for Review

IMPORTANT TERMS

catalyst (p. 548)
reaction rate (14.1)
rate law (14.3)
rate constant (14.3)
reaction order (14.3)
half-life (14.4)
collision theory (14.5)
activation energy (14.5)
transition-state theory (14.5)
activated complex (14.5)
Arrhenius equation (14.6)
frequency factor (14.6)
elementary reaction (14.7)
reaction mechanism (14.7)
reaction intermediate (14.7)
molecularity (14.7)
unimolecular reaction (14.7)
bimolecular reaction (14.7)
termolecular reaction (14.7)
rate-determining step (14.8)
homogeneous catalysis (14.9)
heterogeneous catalysis (14.9)
chemisorption (14.9)

KEY EQUATIONS

$$\ln \frac{[A]_t}{[A]_0} = -kt \quad \text{or} \quad \log \frac{[A]_t}{[A]_0} = \frac{-kt}{2.303}$$

$$\log \frac{k_2}{k_1} = \frac{E_a}{2.303R}\left(\frac{1}{T_1} - \frac{1}{T_2}\right)$$

$$t_{1/2} = \frac{0.693}{k}$$

SUMMARY OF FACTS AND CONCEPTS

The *reaction rate* is defined as the increase in moles of product per liter per second (or as the decrease in moles of reactant per liter per second). Rates of reaction are determined by following the change of concentration of a reactant or product, either by chemical analysis or by observing a physical property. It is found that reaction rates are proportional to concentrations of reactants raised to various powers (usually 1 or 2, but they can be fractional or negative). The *rate law* mathematically expresses the relationship between rates and concentrations for a chemical reaction. Although the rate law tells us how the rate of a reaction depends on concentrations at a given moment, it is possible to transform a rate law to show how concentrations change with time. The *half-life* of a reactant is the time it takes for the reactant concentration to decrease to one-half of its original concentration.

Reaction rates can often double or triple with a 10°C rise in temperature. The effect of temperature on the rate can be explained by *collision theory*. According to this theory, two molecules react after colliding only when the energy of collision is greater than the *activation energy* and when the molecules are properly oriented. It is the rapid increase in the fraction of collisions having energy greater than the activation energy that explains the large temperature dependence of reaction rates. *Transition-state theory* explains reaction rates in terms of the formation of an *activated complex* of the colliding molecules. The *Arrhenius equation* is a mathematical relationship showing the dependence of a rate constant on temperature.

A chemical equation describes the overall result of a chemical reaction that may take place in one or more steps. These steps are called *elementary reactions,* and the set of elementary reactions that describes what is happening at the molecular level in an overall reaction is called the *reaction mechanism*. In some cases, the overall reaction involves a *reaction intermediate*—a species produced in one step but used up in a subsequent one. The rate of the overall reaction depends on the rate of the slowest step (the *rate-determining step*). This rate is proportional to the product of the concentrations of each reactant molecule in that step. If this step involves a reaction intermediate, its concentration can be eliminated by using the relationship of the concentrations in the preceding fast, *equilibrium* step.

A *catalyst* is a substance that speeds up a chemical reaction without being consumed by it. The catalyst is used up in one step of the reaction mechanism but is regenerated in a later step. Catalysis is classified as *homogeneous catalysis* if the substances react within one phase and as *heterogeneous catalysis* if the substances in a gas or liquid phase react at the surface of a solid catalyst. Many industrial reactions involve heterogeneous catalysis. Catalytic activity operates by providing a mechanism for the reaction that has lower activation energy.

OPERATIONAL SKILLS

1. **Relating the different ways of expressing reaction rates** Given the balanced equation for a reaction, relate the different possible ways of defining the rate of the reaction (Example 14.1).

2. **Calculating the average reaction rate** Given the concentration of reactant or product at two different times, calculate the average rate of reaction over that time interval (Example 14.2).

3. **Determining the order of reaction from the rate law** Given an empirical rate law, obtain the orders with respect to each reactant (and catalyst, if any) and the overall order (Example 14.3).

4. **Determining the rate law from initial rates** Given initial concentrations and initial-rate data (in which the concentrations of all species are changed one at a time, holding the others constant), find the rate law for the reaction (Example 14.4).

5. **Using the concentration–time equation for a first-order reaction** Given the rate constant and initial reactant concentration for a first-order reaction, calculate the reactant concentration after a definite time, or calculate the time it takes for the concentration to decrease to a prescribed value (Example 14.5).

6. **Relating the half-life of a reaction to the rate constant** Given the rate constant for a reaction, calculate the half-life (Example 14.6).

7. **Using the Arrhenius equation** Given the values of the rate constant for two temperatures, find the activation energy and calculate the rate constant at a third temperature (Example 14.7).

8. **Writing the overall chemical equation from a mechanism** Given a mechanism for a reaction, obtain the overall chemical equation (Example 14.8).

9. **Determining the molecularity of an elementary reaction** Given an elementary reaction, state the molecularity (Example 14.9).

10. **Writing the rate equation for an elementary reaction** Given an elementary reaction, write the rate equation (Example 14.10).

11. **Determining the rate law from a mechanism** Given a mechanism with an initial slow step, obtain the rate law (Example 14.11). Given a mechanism with an initial fast, equilibrium step, obtain the rate law (Example 14.12).

Review Questions

14.1 List the four variables or factors that can affect the rate of reaction.

14.2 Define the rate of reaction of HBr in the following reaction. How is this related to the rate of formation of Br_2?

$$4HBr(g) + O_2(g) \longrightarrow 2Br_2(g) + 2H_2O(g)$$

14.3 Give at least two physical properties that might be used to determine the rate of a reaction.

14.4 A rate of reaction depends on four variables (Question 4.1). Explain by means of an example how the rate law deals with each of these variables.

14.5 The exponents in a rate law have no relationship to the coefficients in the overall balanced equation for the reaction. Give an example of a balanced equation and the rate law for a reaction that clearly demonstrates this.

14.6 The reaction

$$3I^-(aq) + H_3AsO_4(aq) + 2H^+(aq) \longrightarrow I_3^-(aq) + H_3AsO_3(aq) + H_2O(l)$$

is found to be first order with respect to each of the reactants. Write the rate law. What is the overall order?

14.7 The rate of a reaction is quadrupled when the concentration of one reactant is doubled. What is the order of the reaction with respect to this reactant?

14.8 A rate law is one-half order with respect to a reactant. What is the effect on the rate when the concentration of this reactant is doubled?

14.9 The reaction $A(g) \longrightarrow B(g) + C(g)$ is known to be first order in $A(g)$. It takes 25 s for the concentration of $A(g)$ to decrease by one-half of its initial value. How long does it take for the concentration of $A(g)$ to decrease to one-fourth of its initial value? to one-eighth of its initial value?

14.10 A reaction believed to be either first or second order has a half-life of 20 s at the beginning of an experiment but of 40 s some time later. What is the order of the reaction?

14.11 What two factors determine whether a collision between two reactant molecules will result in reaction?

14.12 Sketch a potential-energy diagram for the exothermic, elementary reaction

$$A + B \longrightarrow C + D$$

and on it denote the activation energies for the forward and reverse reactions. Also indicate the reactants, products, and activated complex.

14.13 Draw a structural formula for the activated complex in the following reaction:

$$NO_2 + NO_3 \longrightarrow NO + NO_2 + O_2$$

Refer to Figure 14.16, center, for details of this reaction. Use dashed lines for bonds about to form or break. Use a single line for all other bonds.

14.14 Rate constants for reactions often follow the Arrhenius equation. Write this equation and then identify each term in it with the corresponding factor or factors from collision theory. Give a physical interpretation of each of those factors.

14.15 By means of an example, explain what is meant by the term *reaction intermediate*.

14.16 Why is it generally impossible to predict the rate law for a reaction on the basis of the chemical equation only?

14.17 The rate law for the reaction

$$2NO_2Cl(g) \longrightarrow 2NO_2(g) + Cl_2(g) \quad \text{(overall equation)}$$

is first order in nitryl chloride, NO_2Cl.

$$\text{Rate} = k[NO_2Cl]$$

Explain why the mechanism for this reaction cannot be the single elementary reaction

$$2NO_2Cl \longrightarrow 2NO_2 + Cl_2 \quad \text{(elementary reaction)}$$

14.18 There is often one step in a reaction mechanism that is rate determining. What characteristic of such a step makes it rate determining? Explain.

14.19 The dissociation of N_2O_4 into NO_2,

$$N_2O_4(g) \rightleftharpoons 2NO_2(g)$$

is believed to occur in one step. Obtain the concentration of N_2O_4 in terms of the concentration of NO_2 and the rate constants for the forward and reverse reactions, when the reactions have come to equilibrium.

14.20 How does a catalyst speed up a reaction? How can a catalyst be involved in a reaction without being consumed by it?

14.21 Compare physical adsorption and chemisorption (chemical adsorption).

14.22 Describe the steps in the catalytic hydrogenation of ethylene.

Practice Problems

REACTION RATES

14.23 Relate the rate of decomposition of NO_2 to the rate of formation of O_2 for the following reaction:

$$2NO_2(g) \longrightarrow 2NO(g) + O_2(g)$$

14.24 For the reaction of hydrogen with iodine

$$H_2(g) + I_2(g) \longrightarrow 2HI(g)$$

relate the rate of disappearance of iodine vapor to the rate of formation of hydrogen iodide.

14.25 To obtain the rate of the reaction

$$5Br^-(aq) + BrO_3^-(aq) + 6H^+(aq) \longrightarrow 3Br_2(aq) + 3H_2O(l)$$

we might follow the Br^- concentration or instead the BrO_3^- concentration. How are the rates in terms of these species related?

14.26 To obtain the rate of the reaction

$$3I^-(aq) + H_3AsO_4(aq) + 2H^+(aq) \longrightarrow I_3^-(aq) + H_3AsO_3(aq) + H_2O(l)$$

we might follow the I^- concentration or instead the I_3^- concentration. How are the rates in terms of these species related?

14.27 Ammonium nitrite, NH_4NO_2, decomposes in solution.

$$NH_4NO_2(aq) \longrightarrow N_2(g) + 2H_2O(l)$$

The concentration of NH_4^+ ion at the beginning of an experiment was 0.500 *M*. After 3.00 hours, it was 0.432 *M*. What is the average rate of decomposition of NH_4NO_2 in this time interval?

14.28 Iron(III) chloride is reduced by tin(II) chloride.

$$2FeCl_3(aq) + SnCl_2(aq) \longrightarrow 2FeCl_2(aq) + SnCl_4(aq)$$

The concentration of Fe^{3+} ion at the beginning of an experiment was 0.03586 *M*. After 4.00 min, it was 0.02638 *M*. What is the average rate of reaction of $FeCl_3$ in this time interval?

14.29 Azomethane, CH_3NNCH_3, decomposes according to the following equation:

$$CH_3NNCH_3(g) \longrightarrow C_2H_6(g) + N_2(g)$$

The initial concentration of azomethane was 1.50×10^{-2} mol/L. After 10.0 min, this concentration decreased to 1.29×10^{-2} mol/L. Obtain the average rate of reaction during this time interval. Express the answer in units of mol/(L · s).

14.30 Nitrogen dioxide, NO_2, decomposes upon heating to form nitric oxide and oxygen according to the following equation:

$$2NO_2(g) \longrightarrow 2NO(g) + O_2(g)$$

At the beginning of an experiment, the concentration of nitrogen dioxide in a reaction vessel was 0.1103 mol/L. After 60.0 s, the concentration decreased to 0.1076 mol/L. What is the average rate of decomposition of NO_2 during this time interval, in mol/(L · s)?

RATE LAWS

14.31 Hydrogen sulfide is oxidized by chlorine in aqueous solution.

$$H_2S(aq) + Cl_2(aq) \longrightarrow S(s) + 2HCl(aq)$$

The experimental rate law is

$$\text{Rate} = k[H_2S][Cl_2]$$

What is the reaction order with respect to H_2S? with respect to Cl_2? What is the overall order?

14.32 For the reaction of nitric oxide, NO, with chlorine, Cl_2,

$$2NO(g) + Cl_2(g) \longrightarrow 2NOCl(g)$$

the observed rate law is

$$\text{Rate} = k[NO]^2[Cl_2]$$

What is the reaction order with respect to nitric oxide? with respect to Cl_2? What is the overall order?

14.33 Oxalic acid, $H_2C_2O_4$, is oxidized by permanganate ion to CO_2 and H_2O.

$$2MnO_4^-(aq) + 5H_2C_2O_4(aq) + 6H^+(aq) \longrightarrow 2Mn^{2+}(aq) + 10CO_2(g) + 8H_2O(l)$$

The rate law is

$$\text{Rate} = k[MnO_4^-][H_2C_2O_4]$$

What is the order with respect to each reactant? What is the overall order?

14.34 Iron(II) ion is iodized by hydrogen peroxide in acidic solution.

$$H_2O_2(aq) + 2Fe^{2+}(aq) + 2H^+(aq) \longrightarrow 2Fe^{3+}(aq) + 2H_2O(l)$$

The rate law is

$$\text{Rate} = k[H_2O_2][Fe^{2+}]$$

What is the order with respect to each reactant? What is the overall order?

14.35 In experiments on the decomposition of azomethane,

$$CH_3NNCH_3(g) \longrightarrow C_2H_6(g) + N_2(g)$$

the following data were obtained:

	INITIAL CONCENTRATION OF AZOMETHANE	INITIAL RATE
Exp. 1	1.13×10^{-2} M	2.8×10^{-6} M/s
Exp. 2	2.26×10^{-2} M	5.6×10^{-6} M/s

What is the rate law? What is the value of the rate constant?

14.36 Ethylene oxide, C_2H_4O, decomposes when heated to give methane and carbon monoxide.

$$C_2H_4O(g) \longrightarrow CH_4(g) + CO(g)$$

The following kinetic data were observed for the reaction at 688 K:

	INITIAL CONCENTRATION OF ETHYLENE OXIDE	INITIAL RATE
Exp. 1	0.00272 M	5.57×10^{-7} M/s
Exp. 2	0.00544 M	1.11×10^{-6} M/s

Find the rate law and the value of the rate constant for this reaction.

14.37 Nitric oxide, NO, reacts with hydrogen to give nitrous oxide, N_2O, and water.

$$2NO(g) + H_2(g) \longrightarrow N_2O(g) + H_2O(g)$$

In a series of experiments, the following initial rates of disappearance of NO were obtained:

	INITIAL CONCENTRATIONS NO	INITIAL CONCENTRATIONS H_2	INITIAL RATE OF REACTION OF NO
Exp. 1	6.4×10^{-3} *M*	2.2×10^{-3} *M*	2.6×10^{-5} *M*/s
Exp. 2	12.8×10^{-3} *M*	2.2×10^{-3} *M*	1.0×10^{-4} *M*/s
Exp. 3	6.4×10^{-3} *M*	4.5×10^{-3} *M*	5.1×10^{-5} *M*/s

Find the rate law and the value of the rate constant for the reaction of NO.

14.38 In a kinetic study of the reaction

$$2NO(g) + O_2(g) \longrightarrow 2NO_2(g)$$

the following data were obtained for the initial rates of disappearance of NO:

	INITIAL CONCENTRATIONS NO	INITIAL CONCENTRATIONS O_2	INITIAL RATE OF REACTION OF NO
Exp. 1	0.0125 *M*	0.0253 *M*	0.0281 *M*/s
Exp. 2	0.0250 *M*	0.0253 *M*	0.112 *M*/s
Exp. 3	0.0125 *M*	0.0506 *M*	0.0561 *M*/s

Obtain the rate law. What is the value of the rate constant?

14.39 Chlorine dioxide, ClO_2, is a reddish-yellow gas that is soluble in water. In basic solution it gives ClO_3^- and ClO_2^- ions.

$$2ClO_2(aq) + 2OH^-(aq) \longrightarrow ClO_3^-(aq) + ClO_2^-(aq) + H_2O$$

To obtain the rate law for this reaction, the following experiments were run and, for each, the initial rate of reaction of ClO_2 was determined. Obtain the rate law and the value of the rate constant.

	INITIAL CONCENTRATIONS (mol/L) ClO_2	INITIAL CONCENTRATIONS (mol/L) OH^-	INITIAL RATE (mol/(L · s))
Exp. 1	0.060	0.030	0.0248
Exp. 2	0.020	0.030	0.00276
Exp. 3	0.020	0.090	0.00828

14.40 Iodide ion oxidized to hypoiodite ion, IO^-, by hypochlorite ion, ClO^-, in basic solution. The equation is

$$I^-(aq) + ClO^-(aq) \xrightarrow{OH^-} IO^-(aq) + Cl^-(aq)$$

The following initial-rate experiments were run and, for each, the initial rate of formation of IO^- was determined. Find the rate law and the value of the rate constant.

	INITIAL CONCENTRATIONS (mol/L) I^-	ClO^-	OH^-	INITIAL RATE (mol/(L · s))
Exp. 1	0.010	0.020	0.010	12.2×10^{-2}
Exp. 2	0.020	0.010	0.010	12.2×10^{-2}
Exp. 3	0.010	0.010	0.010	6.1×10^{-2}
Exp. 4	0.010	0.010	0.020	3.0×10^{-2}

CHANGE OF CONCENTRATION WITH TIME; HALF-LIFE

14.41 Sulfuryl chloride, SO_2Cl_2, decomposes when heated.

$$SO_2Cl_2(g) \longrightarrow SO_2(g) + Cl_2(g)$$

In an experiment, the initial concentration of sulfuryl chloride was 0.0248 mol/L. If the rate constant is 2.2×10^{-5}/s, what is the concentration of SO_2Cl_2 after 4.5 hr? The reaction is first order.

14.42 Cyclopropane, C_3H_6, is converted to its isomer propylene, $CH_2{=}CHCH_3$, when heated. The rate law is first order in cyclopropane, and the rate constant is 6.0×10^{-4}/s at 500°C. If the initial concentration of cyclopropane is 0.0226 mol/L, what is the concentration after 955 s?

14.43 Ethyl chloride, CH_3CH_2Cl, used to produce tetraethyllead gasoline additive, decomposes when heated to give ethylene and hydrogen chloride.

$$CH_3CH_2Cl(g) \longrightarrow C_2H_4(g) + HCl(g)$$

The reaction is first order. In an experiment, the initial concentration of ethyl chloride was 0.00100 *M*. After heating at 500°C for 155 s, this was reduced to 0.00067 *M*. What was the concentration of ethyl chloride after a total of 256 s?

14.44 Cyclobutane, C_4H_8, consisting of molecules in which four carbon atoms form a ring, decomposes when heated to give ethylene.

$$C_4H_8(g) \longrightarrow 2C_2H_4(g)$$

The reaction is first order. In an experiment, the initial concentration of cyclobutane was 0.00150 *M*. After heating at 450°C for 455 s, this was reduced to 0.00119 *M*. What was the concentration of cyclobutane after a total of 968 s?

14.45 Methyl isocyanide, CH_3NC, isomerizes when heated to give acetonitrile (methyl cyanide), CH_3CN.

$$CH_3NC(g) \longrightarrow CH_3CN(g)$$

The reaction is first order. At 230°C, the rate constant for the isomerization is 6.3×10^{-4}/s. What is the half-life? How long would it take for the concentration of CH_3NC to decrease to 25% of its initial value? to 12.5% of its initial value?

14.46 Dinitrogen pentoxide, N_2O_5, decomposes when heated in carbon tetrachloride solvent.

$$N_2O_5 \longrightarrow 2NO_2 + \tfrac{1}{2}O_2(g)$$

If the rate constant for the decomposition of N_2O_5 is 6.2×10^{-4}/min, what is the half-life? (The rate law is first order in N_2O_5.) How long would it take for the concentration of N_2O_5 to decrease to 25% of its initial value? to 12.5% of its initial value?

14.47 In the presence of excess thiocyanate ion, SCN^-, the following reaction is first order in chromium(III) ion, Cr^{3+}; the rate constant is 2.0×10^{-6}/s.

$$Cr^{3+}(aq) + SCN^-(aq) \longrightarrow Cr(SCN)^{2+}(aq)$$

What is the half-life in hours? How many hours would be required for the initial concentration of Cr^{3+} to decrease to each of the following values: 25.0% left, 12.5% left, 6.25% left, 3.125% left?

14.48 In the presence of excess thiocyanate ion, SCN^-, the following reaction is first order in iron(III) ion, Fe^{3+}; the rate constant is 1.27/s.

$$Fe^{3+}(aq) + SCN^-(aq) \longrightarrow Fe(SCN)^{2+}(aq)$$

What is the half-life in seconds? How many seconds would be required for the initial concentration of Fe^{3+} to decrease to each of the following values: 25.0% left, 12.5% left, 6.25% left, 3.125% left? What is the relationship between these times and the half-life?

14.49 In the presence of excess thiocyanate ion, SCN^-, the following reaction is first order in chromium(III) ion, Cr^{3+}; the rate constant is 2.0×10^{-6}/s.

$$Cr^{3+}(aq) + SCN^-(aq) \longrightarrow Cr(SCN)^{2+}(aq)$$

If 90.0% reaction is required to obtain a noticeable color from the formation of the $Cr(SCN)^{2+}$ ion, how many hours are required?

14.50 In the presence of excess thiocyanate ion, SCN^-, the following reaction is first order in iron(III) ion, Fe^{3+}; the rate constant is 1.27/s.

$$Fe^{3+}(aq) + SCN^-(aq) \longrightarrow Fe(SCN)^{2+}(aq)$$

If 90.0% reaction is required to obtain a noticeable color from the formation of the $Fe(SCN)^{2+}$ ion, how many seconds are required?

14.51 Chlorine dioxide oxidizes iodide ion in aqueous solution to iodine; chlorine dioxide is reduced to chlorite ion.

$$2ClO_2(aq) + 2I^-(aq) \longrightarrow 2ClO_2^-(aq) + I_2(aq)$$

The order of the reaction with respect to ClO_2 was determined by starting with a large excess of I^-, so that its concentration was essentially constant. Then

$$\text{Rate} = k[ClO_2]^m[I^-]^n = k'[ClO_2]^m$$

where $k' = k[I^-]^n$. Determine the order with respect to ClO_2 and the rate constant k' by plotting the following data assuming first- and then second-order kinetics. (Data from H. Fukutomi and G. Gordon, *J. Am. Chem. Soc.*, *89*, 1362 [1967].)

TIME (s)	$[ClO_2]$ (mol/L)
0.00	4.77×10^{-4}
1.00	4.31×10^{-4}
2.00	3.91×10^{-4}
3.00	3.53×19^{-4}

14.52 Methyl acetate, CH_3COOCH_3, reacts in basic solution to give acetate ion, CH_3COO^-, and methanol, CH_3OH.

$$CH_3COOCH_3(aq) + OH^-(aq) \longrightarrow CH_3COO^-(aq) + CH_3OH(aq)$$

The overall order of the reaction was determined by starting with methyl acetate and hydroxide ion at the same concentrations, so $[CH_3COOCH_3] = [OH^-] = x$. Then

$$\text{Rate} = k[CH_3COOCH_3]^m[OH^-]^n = kx^{m+n}$$

Determine the overall order and the value of the rate constant by plotting the following data assuming first- and then second-order kinetics.

TIME (min)	$[CH_3COOCH_3]$ (mol/L)
0.00	0.01000
3.00	0.00740
4.00	0.00683
5.00	0.00634

RATE AND TEMPERATURE

14.53 Sketch a potential-energy diagram for the reaction of nitric oxide with ozone.

$$NO(g) + O_3(g) \longrightarrow NO_2(g) + O_2(g)$$

The activation energy for the forward reaction is 10 kJ; the $\Delta H°$ is -200 kJ. What is the activation energy for the reverse reaction? Label your diagram appropriately.

14.54 Sketch a potential-energy diagram for the decomposition of nitrous oxide.

$$N_2O(g) \longrightarrow N_2(g) + O(g)$$

The activation energy for the forward reaction is 251 kJ; the $\Delta H°$ is $+167$ kJ. What is the activation energy for the reverse reaction? Label your diagram appropriately.

14.77 Methyl acetate reacts in acidic solution.

$$\underset{\text{methyl acetate}}{CH_3COOCH_3} + H_2O \xrightarrow{H^+} \underset{\text{methanol}}{CH_3OH} + \underset{\text{acetic acid}}{CH_3COOH}$$

The rate law is first order in methyl acetate in acidic solution, and the rate constant at 25°C is 1.26×10^{-4}/s. How long will it take for 85% of the methyl acetate to react?

14.78 Benzene diazonium chloride, C_6H_5NNCl, decomposes by a first-order rate law.

$$C_6H_5NNCl \longrightarrow C_6H_5Cl + N_2(g)$$

If the rate constant at 20°C is 4.3×10^{-5}/s, how long will it take for 85% of the compound to decompose?

14.79 What is the half-life of methyl acetate at 25°C in the acidic solution described in Problem 14.77?

14.80 What is the half-life of benzene diazonium chloride at 20°C? See Problem 14.78 for data.

14.81 A compound decomposes by a first-order reaction. If the concentration of the compound is 0.0250 *M* after 65 s when the initial concentration was 0.0350 *M*, what is the concentration of the compound after 98 s?

14.82 A compound decomposes by a first-order reaction. The concentration of compound decreases from 0.1180 *M* to 0.0950 *M* in 5.2 min. What fraction of the compound remains after 6.7 min?

14.83 Plot the data given in Problem 14.73 to verify that the decomposition of azomethane is first order. Determine the rate constant from the slope of the straight-line plot of log $[CH_3NNCH_3]$ versus time.

14.84 The decomposition of aqueous hydrogen peroxide in a given concentration of catalyst yielded the following data:

TIME	0.0 min	5.0 min	10.0 min	15.0 min
$[H_2O_2]$	0.1000 *M*	0.0804 *M*	0.0648 *M*	0.0519 *M*

Verify that the reaction is first order. Determine the rate constant for the decomposition of H_2O_2 (in units of /s) from the slope of the straight-line plot of log $[H_2O_2]$ versus time.

14.85 The decomposition of nitrogen dioxide,

$$2NO_2(g) \longrightarrow 2NO(g) + O_2(g)$$

has a rate constant of 0.498 *M*/s at 319°C and a rate constant of 1.81 *M*/s at 354°C. What are the values of the activation energy and the frequency factor for this reaction? What is the rate constant at 383°C?

14.86 A second-order reaction has a rate constant of $8.7 \times 10^{-4}/(M \cdot s)$ at 30°C. At 40°C, the rate constant is $1.5 \times 10^{-3}/(M \cdot s)$. What are the activation energy and frequency factor for this reaction? Predict the value of the rate constant at 45°C.

14.87 At high temperature, the reaction

$$NO_2(g) + CO(g) \longrightarrow NO(g) + CO_2(g)$$

is thought to occur in a single step. What should be the rate law in that case?

14.88 Methyl chloride, CH_3Cl, reacts in basic solution to give methanol.

$$CH_3Cl + OH^- \longrightarrow CH_3OH + Cl^-$$

This reaction is believed to occur in a single step. If so, what should be the rate law?

14.89 Nitryl bromide, NO_2Br, decomposes into nitrogen dioxide and bromine.

$$2NO_2Br(g) \longrightarrow 2NO_2(g) + Br_2(g)$$

A proposed mechanism is

$$NO_2Br \longrightarrow NO_2 + Br \quad \text{(slow)}$$
$$NO_2Br + Br \longrightarrow NO_2 + Br \quad \text{(fast)}$$

Write the rate law predicted by this mechanism.

14.90 Tertiary butyl chloride reacts in basic solution according to the equation

$$(CH_3)_3CCl + OH^- \longrightarrow (CH_3)_3COH + Cl^-$$

The accepted mechanism for this reaction is

$$(CH_3)_3CCl \longrightarrow (CH_3)_3C^+ + Cl^- \quad \text{(slow)}$$
$$(CH_3)_3C^+ + OH^- \longrightarrow (CH_3)_3COH \quad \text{(fast)}$$

What should be the rate law for this reaction?

14.91 Urea, $(NH_2)_2CO$, can be prepared by heating ammonium cyanate, NH_4CNO.

$$NH_4CNO \longrightarrow (NH_2)_2CO$$

This reaction may occur by the following mechanism:

$$NH_4^+ + CNO^- \rightleftharpoons NH_3 + HCNO \quad \text{(fast, equilibrium)}$$
$$NH_3 + HCNO \longrightarrow (NH_2)_2CO \quad \text{(slow)}$$

What is the rate law predicted by this mechanism?

14.92 Acetone reacts with iodine in acidic aqueous solution to give monoiodoacetone.

$$\underset{\text{acetone}}{CH_3COCH_3} + I_2 \xrightarrow{H^+} \underset{\text{monoiodoacetone}}{CH_3COCH_2I} + HI$$

A possible mechanism for this reaction is

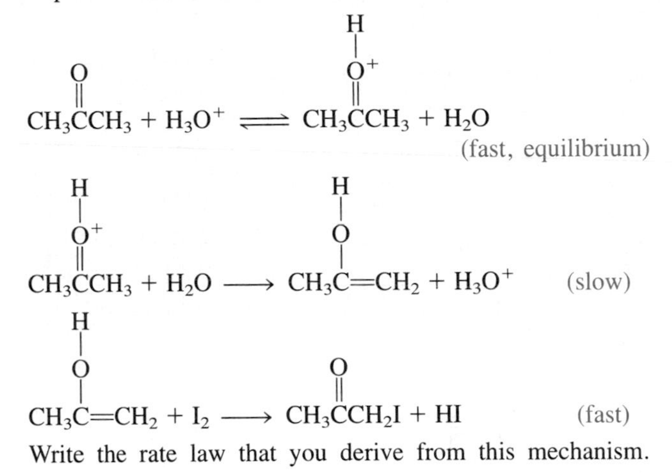

Write the rate law that you derive from this mechanism.

Cumulative-Skills Problems

14.93 Hydrogen peroxide in aqueous solution decomposes by a first-order reaction to water and oxygen. The rate constant for this decomposition is 7.40×10^{-4}/s. What quantity of heat energy is initially liberated per second from 2.00 L of solution that is 1.50 *M* H_2O_2? See Appendix C for data.

14.94 Nitrogen dioxide reacts with carbon monoxide by the overall equation

$$NO_2(g) + CO(g) \longrightarrow NO(g) + CO_2(g)$$

At a particular temperature, the reaction is second order in NO_2 and zero order in CO. The rate constant is 0.515 L/(mol · s). How much heat energy evolves per second initially from 3.50 L of reaction mixture containing 0.0250 *M* NO_2? See Appendix C for data. Assume the enthalpy change is constant with temperature.

14.95 Nitric oxide reacts with oxygen to give nitrogen dioxide.

$$2NO(g) + O_2(g) \longrightarrow 2NO_2(g)$$

The rate law is $-\Delta[NO]/\Delta t = k[NO]^2[O_2]$, where the rate constant is 1.16×10^{-5} $L^2/(mol^2 \cdot s)$ at 339°C. A vessel contains NO and O_2 at 339°C. The initial partial pressures of NO and O_2 are 155 mmHg and 345 mmHg, respectively. What is the rate of decrease of partial pressure of NO (in mmHg per second)? (*Hint:* From the ideal gas law, obtain an expression for the molar concentration of a particular gas in terms of its partial pressure.)

14.96 Nitric oxide reacts with hydrogen as follows:

$$2NO(g) + H_2(g) \longrightarrow N_2O(g) + H_2O(g)$$

The rate law is $-\Delta[H_2]/\Delta t = k[NO]^2[H_2]$, where k is 1.10×10^{-7} $L^2/(mol^2 \cdot s)$ at 826°C. A vessel contains NO and H_2 at 826°C. The partial pressures of NO and H_2 are 144 mmHg and 324 mmHg, respectively. What is the rate of decrease of partial pressure of NO? See Problem 14.95.

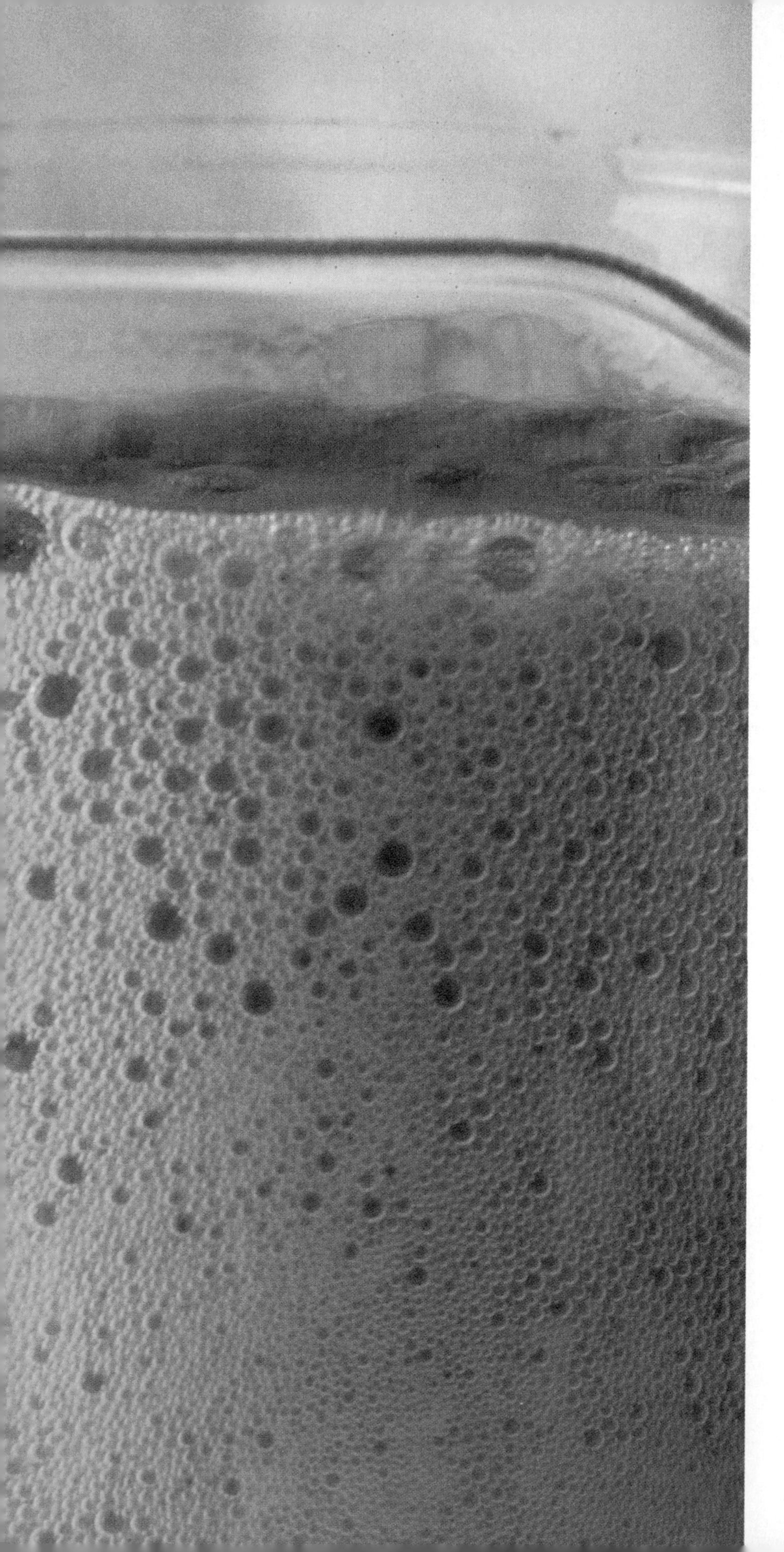

15

Chemical Equilibrium; Gaseous Reactions

Chapter Outline

Nitrogen dioxide, a red-brown gas, is in equilibrium with colorless dinitrogen tetroxide.

Chemical reactions often seem to stop before they are complete. Such reactions are *reversible*. That is, the original reactants form products, but then the products that form react with themselves to give back the original reactants. Thus, there are actually two reactions occurring, and the eventual result is a mixture of reactants and products, rather than simply a mixture of products.

Consider the gaseous reaction in which carbon monoxide and hydrogen react to produce methane and steam.

$$CO(g) + 3H_2(g) \longrightarrow \underset{\text{methane}}{CH_4(g)} + H_2O(g)$$

This reaction, which requires a catalyst to occur at a reasonable rate, is called *catalytic methanation*. Recent research on this reaction has been directed toward the economical conversion of coal to a gaseous fuel to replace natural gas, which is less plentiful than coal. Coal reacts with hot steam to form CO and H_2, and these products react by catalytic methanation to yield methane, the principal component of natural gas (Figure 15.1).

Catalytic methanation is a reversible reaction, and depending on the reaction conditions, the final reaction mixture will have varying amounts of the products methane and steam, as well as the starting substances carbon monoxide and hydrogen. It is also possible to start with methane and steam, and under the right condi-

Figure 15.1
Coal gasification plant. Gaseous fuels can be made from coal in a number of ways. One way (described in the text) is to react coal with steam to produce a mixture of CO and H_2, which then reacts by catalytic methanation to yield CH_4, the major component of natural gas.

tions form a mixture that is predominantly carbon monoxide and hydrogen. The process is called *steam reforming*.

$$CH_4(g) + H_2O(g) \longrightarrow CO(g) + 3H_2(g)$$

The product mixture of CO and H_2 (synthesis gas) is used to prepare a number of industrial chemicals.

The processes of catalytic methanation and steam reforming illustrate the reversibility of chemical reactions. Starting with CO and H_2 and using the right conditions, we can form predominantly CH_4 and H_2O. Starting with CH_4 and H_2O and using different conditions, we can obtain a reaction mixture that is predominantly CO and H_2O. An important question then is, What conditions favor the production of CH_4 and H_2O, and what conditions favor the production of CO and H_2?

As we noted earlier, certain reactions (such as catalytic methanation) appear to stop before they are complete. The reaction mixture ceases to change in any of its properties and consists of both reactants and products in definite concentrations. Such a reaction mixture is said to have reached *chemical equilibrium*. We made brief note of chemical equilibrium in Chapter 3, and we have also discussed other types of equilibria, including the equilibrium between a liquid and its vapor and the equilibrium between a solid and its saturated solution. In this chapter, we will see how to determine the composition of a reaction mixture at equilibrium and how to alter this composition by changing the conditions for the reaction.

Describing Chemical Equilibrium

Many chemical reactions are like the catalytic methanation reaction discussed in the chapter opening. Under the proper conditions, such reactions can be made to go predominantly in one direction or the other. Let us look more closely at this reversibility and see how we may characterize it quantitatively.

15.1 CHEMICAL EQUILIBRIUM—A DYNAMIC EQUILIBRIUM

When substances react, they eventually form a mixture of reactants and products in *dynamic equilibrium*.● This dynamic equilibrium consists of a forward reaction, in

● The concept of dynamic equilibrium in chemical reactions was briefly mentioned in Sections 3.3 and 14.8. Discussions on vapor pressure (Section 11.2) and solubility (Section 12.2) give detailed explanations of the role of dynamic equilibria.

which substances react to give products, and a reverse reaction, in which products react to give the original reactants. Both forward and reverse reactions occur at the same rate, or speed.

Consider the catalytic methanation reaction discussed in the chapter opening. It consists of forward and reverse reactions, which we represent by the chemical equation

$$CO(g) + 3H_2(g) \rightleftharpoons CH_4(g) + H_2O(g)$$

Suppose we put 1.000 mol CO and 3.000 mol H_2 into a 10.00-L vessel at 1200 K (927°C). The rate of the reaction of CO and H_2 depends on the concentrations of CO and H_2. At first these concentrations are large, but as the substances react their concentrations decrease (Figure 15.2). Thus, the rate of the forward reaction, which depends on reactant concentrations, is large at first but steadily decreases. On the other hand, the concentrations of CH_4 and H_2O, which are zero at first, increase with time, as Figure 15.2 shows. Therefore, the rate of the reverse reaction starts at zero and steadily increases. The forward rate decreases and the reverse rate increases until eventually they become equal. When that happens, CO and H_2 molecules are formed as fast as they react. The concentrations of reactants and products no longer change, and the reaction mixture has reached *equilibrium*. Figure 15.2 shows how the amounts of substances in the reaction mixture become constant when equilibrium is reached.

Chemical equilibrium is *the state reached by a reaction mixture when the rates of forward and reverse reactions have become equal.* If we observe the reaction mixture, we see no net change, although the forward and reverse reactions are continuing. The continuing forward and reverse reactions make the equilibrium a *dynamic* process.

Suppose we place known amounts of reactants in a vessel and let the mixture come to equilibrium. To obtain the composition of the equilibrium mixture, we need only determine the amount of one of the substances. The amounts of the others can be calculated from the amounts of substances originally placed in the

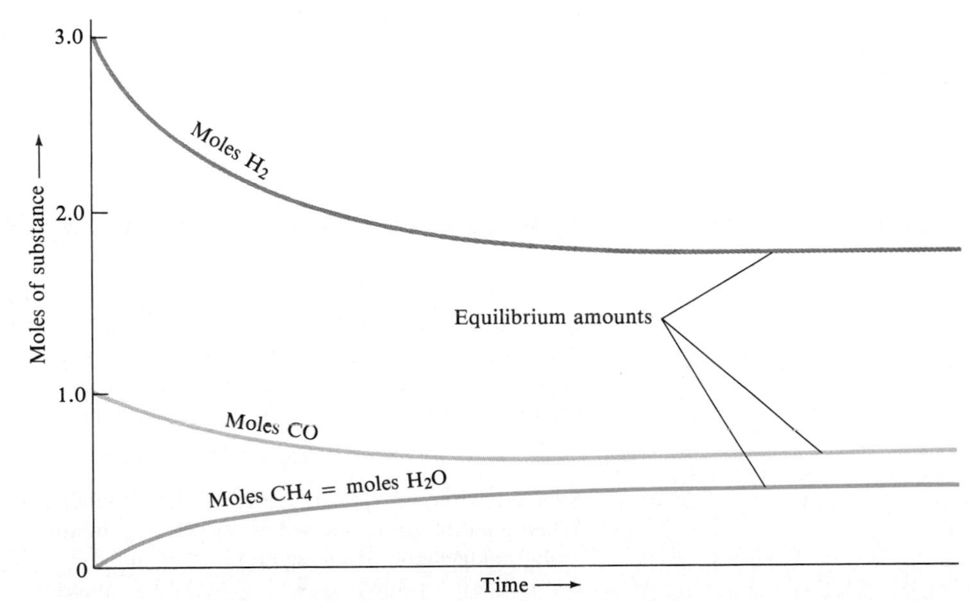

Figure 15.2
Variation in moles of substances during catalytic methanation. The experiment begins with 1.000 mol CO and 3.000 mol H_2 in a 10.00-L vessel. Note how the amounts of substances become constant at equilibrium.

vessel and the equation that represents the reaction. For example, suppose we place 1.000 mol CO and 3.000 mol H_2 in a reaction vessel at 1200 K. After equilibrium is reached, we chill the reaction mixture quickly to condense the water vapor to liquid, which we then weigh and find to be 0.387 mol. We calculate the other substances from the stoichiometry of the reaction, as shown in the following example.

Example 15.1 Applying Stoichiometry to an Equilibrium Mixture

Carbon monoxide and hydrogen react according to the following equation:

$$CO(g) + 3H_2(g) \rightleftharpoons CH_4(g) + H_2O(g)$$

When 1.000 mol CO and 3.000 mol H_2 are placed in a 10.00-L vessel at 927°C (1200 K) and allowed to come to equilibrium, the mixture is found to contain 0.387 mol H_2O. What is the molar composition of the equilibrium mixture? That is, how many moles of each substance are present?

Solution

The problem is essentially one in stoichiometry. It involves initial, or *starting,* amounts of reactants. These *change* as reaction occurs. Later, the reaction comes to *equilibrium,* and we analyze the reaction mixture for the amount of one of the reactants or products. It is very convenient to solve this problem by first setting up a table for the starting, change, and equilibrium values so that we can easily see what we have to calculate. Using the information given in the problem, we set up the table shown below.

In setting up this table, we filled in the values given in the problem. We are given the starting amounts of all substances in the equation, but we are not told explicitly what changes occur in these amounts during the reaction. Therefore, we let x be the molar change. That is, each product increases by x moles multiplied by the coefficients of the substances in the balanced equation. Reactants decrease by x moles multiplied by the corresponding coefficients. The decrease is indicated by a negative sign. Equilibrium values are equal to starting values plus the changes. For example, the starting amount of H_2 is 3.000 mol. Because this changes by $-3x$ mol, the equilibrium amount is $(3.000 - 3x)$ mol. Similarly, the starting amount of H_2O is 0, the change is x mol, and the equilibrium amount is $(0 + x)$ mol $= x$ mol. The only equilibrium amount given in the problem statement is that for H_2O (= 0.387 mol). This tells us that $x = 0.387$. We can calculate equilibrium amounts for other substances from the expressions given in the table, using this value of x. Thus,

$$\text{Equilibrium amount CO} = (1.000 - x)\text{ mol} = (1.000 - 0.387)\text{ mol} = 0.613\text{ mol}$$

$$\text{Equilibrium amount } H_2 = (3.000 - 3x)\text{ mol} = (3.000 - 3 \times 0.387)\text{ mol} = 1.839\text{ mol}$$

$$\text{Equilibrium amount } CH_4 = x\text{ mol} = 0.387\text{ mol}$$

Therefore, the amounts of substances in the equilibrium mixture are **0.613 mol CO, 1.839 mol H_2, 0.387 mol CH_4, and 0.387 mol H_2O.**

AMOUNT (MOLES)	$CO(g)$	+	$3H_2(g)$	$\rightleftharpoons$	$CH_4(g)$	+	$H_2O(g)$
Starting	1.000		3.000		0		0
Change	$-x$		$-3x$		$+x$		$+x$
Equilibrium	$1.000 - x$		$3.000 - 3x$		x		$x = 0.387$

Exercise 15.1

Synthesis gas (a mixture of CO and H_2) is increased in concentration of hydrogen by passing it with steam over a catalyst. This is the so-called water–gas shift reaction. Some of the CO is converted to CO_2, which can be removed:

$$CO(g) + H_2O(g) \rightleftharpoons CO_2(g) + H_2(g)$$

Suppose we start with a gaseous mixture containing 1.00 mol CO and 1.00 mol H_2O. When equilibrium is reached at 1000°C, the mixture contains 0.59 mol H_2. What is the molar composition of the equilibrium mixture?

(See Problems 15.15, 15.16, 15.17, and 15.18.)

15.2 THE EQUILIBRIUM CONSTANT

In the previous section, we found that when 1.000 mol CO and 3.000 mol H_2 react in a 10.00-L vessel by catalytic methanation at 1200 K, they give an equilibrium mixture containing 0.613 mol CO, 1.839 mol H_2, 0.387 mol CH_4, and 0.387 mol H_2O. Let us call this Experiment I. Now consider a similar experiment, Experiment II, in which we start with an additional mole of carbon monoxide. That is, we place 2.000 mol CO and 3.000 mol H_2 in a 10.00-L vessel at 1200 K. At equilibrium, we find that the vessel contains 1.522 mol CO, 1.566 mol H_2, 0.478 mol CH_4, and 0.478 mol H_2O. What we observe from the results of Experiments I and II is that the equilibrium composition depends on the amounts of starting substances. Nevertheless, what we will see is that all of the equilibrium compositions for a reaction at a given temperature are related by a quantity called the *equilibrium constant.*

Definition of the Equilibrium Constant K_c

Consider the general reaction

$$aA + bB \rightleftharpoons cC + dD$$

where A, B, C, and D denote reactants and products and a, b, c, and d are coefficients in the balanced chemical equation. The **equilibrium-constant expression** for a reaction is *an expression obtained by multiplying the concentrations of products together, dividing by the concentrations of reactants, and raising each concentration term to a power equal to the coefficient in the chemical equation.* The **equilibrium constant** K_c is *the value obtained for the equilibrium-constant expression when equilibrium concentrations are substituted.*● For the general reaction, we have

● In Section 15.3, we will see that we ignore concentrations of pure liquids and solids when we write the equilibrium-constant expression. The concentrations of these substances are constant and are incorporated into the value of K_c.

$$K_c = \frac{[C]^c[D]^d}{[A]^a[B]^b}$$

Here we denote the molar concentration of a substance by writing the formula of that substance in square brackets. The subscript c on the equilibrium constant means that it is defined in terms of molar concentrations. The **law of mass action** is *a relation that states that the values of the equilibrium-constant expression K_c are constant for a particular reaction at a given temperature, whatever equilibrium concentrations are substituted.*●

● "Active mass" was an early term for concentration; hence the term *mass action.*

As the following example illustrates, the equilibrium-constant expression is defined in terms of the balanced chemical equation. If the equation is rewritten with different coefficients, then the equilibrium-constant expression will be changed.

Example 15.2 Writing Equilibrium-Constant Expressions

(a) Write the equilibrium-constant expression K_c for catalytic methanation.

$$CO(g) + 3H_2(g) \rightleftharpoons CH_4(g) + H_2O(g)$$

(b) Write the equilibrium-constant expression K_c for the reverse of the reaction in (a)—that is,

$$CH_4(g) + H_2O(g) \rightleftharpoons CO(g) + 3H_2(g)$$

(c) Write the equilibrium-constant expression K_c for the synthesis of ammonia.

$$N_2(g) + 3H_2(g) \rightleftharpoons 2NH_3(g)$$

(continued)

(d) Write the equilibrium-constant expression K_c when the equation for the reaction in (c) is written

$$\tfrac{1}{2}N_2(g) + \tfrac{3}{2}H_2(g) \rightleftharpoons NH_3(g)$$

Solution

(a) The expression for the equilibrium constant is

$$K_c = \frac{[CH_4][H_2O]}{[CO][H_2]^3}$$

Note that concentrations of products are on the top and concentrations of reactants are on the bottom. Also, note that each concentration term is raised to a power equal to the coefficient of the substance in the chemical equation.

(b) When the equation is written in reverse order, the expression for K_c is inverted:

$$K_c = \frac{[CO][H_2]^3}{[CH_4][H_2O]}$$

(c) The equilibrium constant for $N_2 + 3H_2 \rightleftharpoons 2NH_3$ is

$$K_c = \frac{[NH_3]^2}{[N_2][H_2]^3}$$

(d) When the coefficients in the equation in (c) are multiplied by $\tfrac{1}{2}$ to give $\tfrac{1}{2}N_2 + \tfrac{3}{2}H_2 \rightleftharpoons NH_3$, the equilibrium-constant expression becomes

$$K_c = \frac{[NH_3]}{[N_2]^{1/2}[H_2]^{3/2}}$$

which is the square root of the previous expression.

Exercise 15.2

Write the equilibrium-constant expression K_c for the equation

$$2NO_2(g) + 7H_2(g) \rightleftharpoons 2NH_3(g) + 4H_2O(g)$$

Write the equilibrium-constant expression K_c when this reaction is written

$$NO_2(g) + \tfrac{7}{2}H_2(g) \rightleftharpoons NH_3(g) + 2H_2O(g)$$

(See Problems 15.19 and 15.20.)

Related Topic

CHEMICAL EQUILIBRIUM AND KINETICS

The law of mass action was first stated by the Norwegian chemists Cato Guldberg and Peter Waage in 1867. They were led to this law by a kinetic argument. To understand the argument, consider the decomposition of dinitrogen tetroxide, N_2O_4.

Dinitrogen tetroxide is a colorless substance that decomposes when warmed to give nitrogen dioxide, NO_2, a brown substance. Suppose we start with dinitrogen tetroxide gas and heat it above room temperature. At this higher temperature, N_2O_4 decomposes to NO_2, which can be noted by the increasing intensity of the brown color.

$$N_2O_4(g) \longrightarrow 2NO_2(g)$$

Once some NO_2 is produced, it can react to re-form N_2O_4. We will call the decomposition of N_2O_4 the forward reaction and the formation of N_2O_4 the reverse reaction. These are elementary reactions, and we can immediately write the rate equations. The rate of the forward reaction is $k_f[N_2O_4]$, where k_f is the rate constant for the forward reaction. The rate of the reverse reaction is $k_r[NO_2]^2$, where k_r is the rate constant for the reverse reaction.

At first, the reaction vessel contains mostly N_2O_4, so the concentration of NO_2 is very small. Thus, the rate of the forward reaction is relatively large, and the rate of the reverse reaction is relatively small. However, as the decomposition proceeds, the concentration of N_2O_4 decreases, so the forward rate decreases. At the same time, the concentration of NO_2 increases, so the reverse rate increases. Eventually the two rates become equal. After that, on the average, every time an N_2O_4 molecule decomposes, two NO_2 molecules recombine. Overall, the concentrations of N_2O_4 and NO_2 no longer change, although forward and reverse reactions are still occurring. The reaction has reached a dynamic equilibrium.

At equilibrium, we can write "Rate of forward reaction = rate of reverse reaction," or

$$k_f[N_2O_4] = k_r[NO_2]^2$$

Rearranging gives

$$\frac{k_f}{k_r} = \frac{[NO_2]^2}{[N_2O_4]}$$

The right side is the equilibrium-constant expression for the decomposition of dinitrogen tetroxide. The left side is the ratio of rate constants. Thus, the equilibrium-constant expression is constant for a given temperature, and we can identify the equilibrium constant K_c as the ratio of rate constants for the forward and reverse reactions.

$$K_c = \frac{k_f}{k_r}$$

If the overall reaction occurs by a multistep mechanism, one can show that the equilibrium constant equals a product of ratios of rate constants—one ratio for each step in the mechanism.

Obtaining Equilibrium Constants for Reactions

At the beginning of this section, we gave data from the results of two experiments, Experiments I and II, involving catalytic methanation. By substituting the molar concentrations from these two experiments into the equilibrium-constant expression for the reaction, we can show that we get the same value, as we expect from the law of mass action. This value equals K_c for methanation at 1200 K. Thus, besides verifying the validity of the law of mass action in this case, we will also see how an equilibrium constant can be obtained from concentration data.●

● It is also possible to determine equilibrium constants from thermochemical data, as described in Section 18.6.

EXPERIMENT I The equilibrium composition is 0.613 mol CO, 1.839 mol H_2, 0.387 mol CH_4, and 0.387 mol H_2O. The volume of the reaction vessel is 10.00 L, so the concentration of CO is

$$[CO] = \frac{0.613 \text{ mol}}{10.00 \text{ L}} = 0.0613\ M$$

Similarly, the other equilibrium concentrations are $[H_2] = 0.1839\ M$, $[CH_4] = 0.0387\ M$, and $[H_2O] = 0.0387\ M$.

Let us substitute these values into the equilibrium expression for catalytic methanation. (The expression was obtained in Example 15.2.) Although until now we have consistently carried units along with numbers in calculations, it is the usual practice to write equilibrium constants without units. We will follow that practice here.● Substitution of concentrations into the equilibrium-constant expression gives

● In thermodynamics, the equilibrium constant is defined in terms of *activities,* rather than concentrations. For an ideal mixture, the activity of a substance is the ratio of its concentration (or partial pressure if a gas) to a standard concentration of 1 M (or partial pressure of 1 atm), so that units cancel. Thus, activities have numerical values but no units.

$$K_c = \frac{[CH_4][H_2O]}{[CO][H_2]^3} = \frac{(0.0387)(0.0387)}{(0.0613)(0.1839)^3} = 3.93$$

EXPERIMENT II The equilibrium composition is 1.522 mol CO, 1.566 mol H_2, 0.478 mol CH_4, and 0.478 mol H_2O. Therefore, the concentrations, which are obtained by dividing by 10.00 L, are $[CO] = 0.1522\ M$, $[H_2] = 0.1566\ M$, $[CH_4] = 0.0478\ M$, and $[H_2O] = 0.0478\ M$. Substituting into the equilibrium-constant expression gives

$$K_c = \frac{[CH_4][H_2O]}{[CO][H_2]^3} = \frac{(0.0478)(0.0478)}{(0.1522)(0.1566)^3} = 3.91$$

Within the precision of the data, these values (3.93 and 3.91) of the equilibrium expression for different starting mixtures of these gases at 1200 K are the same. Moreover, experiment shows that when we start with CH_4 and H_2O, instead of CO and H_2, an equilibrium mixture is again reached that yields the same value

	Starting concentrations	Equilibrium concentrations	Calculated value of K_c
Experiment I	0.1000 M CO 0.3000 M H_2	0.0613 M CO 0.1839 M H_2 0.0387 M CH_4 0.0387 M H_2O	$K_c = 3.93$
Experiment II	0.2000 M CO 0.3000 M H_2	0.1522 M CO 0.1566 M H_2 0.0478 M CH_4 0.0478 M H_2O	$K_c = 3.91$
Experiment III	0.1000 M CH_4 0.1000 M H_2O	0.0613 M CO 0.1839 M H_2 0.0387 M CH_4 0.0387 M H_2O	$K_c = 3.93$

Figure 15.3
Some different equilibrium compositions for the methanation reaction. Different starting concentrations have been used in each experiment (all at 1200 K). Experiments I and II start with different concentrations of reactants CO and H_2. Experiment III starts with the products CH_4 and H_2O. All three experiments yield essentially the same value of K_c.

of K_c (see Experiment III, Figure 15.3). We can take the equilibrium constant for catalytic methanation at 1200 K to be 3.92, the average of these values.

The following exercise provides additional practice in evaluating an equilibrium constant from equilibrium compositions that are experimentally determined. Example 15.3 gives the detailed solution of a more complicated problem of this type.

Exercise 15.3

When 1.00 mol each of carbon monoxide and water reach equilibrium at 1000°C in a 10.0-L vessel, the equilibrium mixture contains 0.57 mol CO, 0.57 mol H_2O, 0.43 mol CO_2, and 0.43 mol H_2. Write the chemical equation for the equilibrium. What is the value of K_c?●

(See Problems 15.27 and 15.28.)

● The reaction described here is used industrially to adjust the ratio of H_2 to CO in synthesis gas (mixture of CO and H_2). In this process, CO reacts with H_2O to give H_2, so the ratio of H_2 to CO is increased.

Example 15.3 Obtaining an Equilibrium Constant from Reaction Composition

Hydrogen iodide, HI, decomposes at moderate temperatures according to the equation

$$2HI(g) \rightleftharpoons H_2(g) + I_2(g)$$

The amount of I_2 in the reaction mixture can be determined from the intensity of the violet color of I_2; the more intense the color, the more I_2 in the reaction vessel. When 4.00 mol HI was placed in a 5.00-L vessel at 458°C, the equilibrium mixture was found to contain 0.442 mol I_2. What is the value of K_c for the decomposition of HI at this temperature?

Solution

Note that this problem involves starting, change, and equilibrium amounts of substances. Therefore, we set up a table as we did in Example 15.1. Because we want the concentrations of substances so that we can evaluate K_c, we set this table up in terms of amounts per liter, or molar concentrations. We first calculate the concentrations of substances whose amounts were given in the problem. We divide the amounts by the volume of the reaction vessel (5.00 L).

$$\text{Starting concentration of HI} = \frac{4.00 \text{ mol}}{5.00 \text{ L}} = 0.800\ M$$

$$\text{Equilibrium concentration of } I_2 = \frac{0.442 \text{ mol}}{5.00 \text{ L}} = 0.0884\ M$$

From these values, we set up the following table:

CONCENTRATION (M)	$2HI(g)$	$\rightleftharpoons$ $H_2(g)$	+	$I_2(g)$
Starting	0.800	0		0
Change	$-2x$	x		x
Equilibrium	$0.800 - 2x$	x		$x = 0.0884$

The equilibrium concentrations of substances can be evaluated from the expressions given in the last line of this table. We know that the equilibrium concentration of I_2 is 0.0884 M and that this equals x. Therefore,

$$[HI] = (0.800 - 2x)\ M = (0.800 - 2 \times 0.0884)\ M = 0.623\ M$$

$$[H_2] = x = 0.0884\ M$$

Now we substitute into the equilibrium-constant expression for the reaction. From the chemical equation, we write

$$K_c = \frac{[H_2][I_2]}{[HI]^2}$$

Substituting, we get

$$K_c = \frac{(0.0884)(0.0884)}{(0.623)^2} = 0.0201$$

Exercise 15.4

Hydrogen sulfide, a colorless gas with a foul odor, dissociates on heating:

$$2H_2S(g) \rightleftharpoons 2H_2(g) + S_2(g)$$

When 0.100 mol H_2S was put into a 10.0-L vessel and heated to 1132°C, it gave an equilibrium mixture containing 0.0285 mol H_2. What is the value of K_c at this temperature? (See Problems 15.29, 15.30, 15.31, and 15.32.)

The Equilibrium Constant K_p

In discussing gas-phase equilibria, it is often convenient to write the equilibrium constant in terms of partial pressures of gases rather than concentrations. Note that the concentration of a gas is proportional to its partial pressure at a fixed temperature (Figure 15.4). We can see this by looking at the ideal gas law, $PV = nRT$, and solving this for n/V, which is the molar concentration of the gas. We get $n/V = P/RT$. In other words, the molar concentration of a gas equals its partial pressure divided by RT, which is constant at a given temperature.

When we express an equilibrium constant for a gaseous reaction in terms of partial pressures, we call it K_p. For catalytic methanation,

$$CO(g) + 3H_2(g) \rightleftharpoons CH_4(g) + H_2O(g)$$

the equilibrium expression in terms of partial pressures becomes

$$K_p = \frac{P_{CH_4}P_{H_2O}}{P_{CO}P_{H_2}{}^3}$$

In general, the value of K_p is different from that of K_c. From the relationship $n/V = P/RT$, one can show that

$$K_p = K_c(RT)^{\Delta n}$$

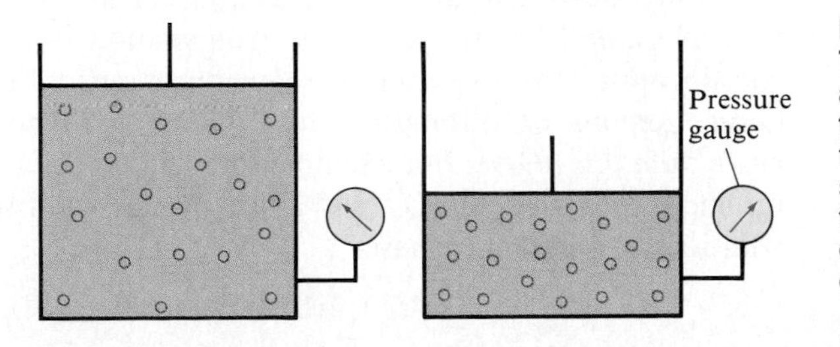

Figure 15.4
The concentration of a gas at a given temperature is proportional to the pressure. When the gas is compressed, the pressure increases (as noted from the pressure gauge) and the concentration of the gas (molecules per unit volume) increases.

where Δn is the sum of the coefficients of *gaseous* products in the chemical equation minus the sum of the coefficients of *gaseous* reactants. For the methanation reaction ($K_c = 3.92$), in which 2 mol gaseous products ($CH_4 + H_2O$) are obtained from 4 mol gaseous reactants ($CO + 3H_2$), Δn equals $2 - 4 = -2$. The usual unit of partial pressures in K_p is atmospheres; therefore, the value of R is 0.0821 L · atm/(K · mol). Hence,

$$K_p = 3.92 \times (0.0821 \times 1200)^{-2} = 4.04 \times 10^{-4}$$

Exercise 15.5

Phosphorus pentachloride dissociates on heating.

$$PCl_5(g) \rightleftharpoons PCl_3(g) + Cl_2(g)$$

If K_c equals 3.26×10^{-2} at 191°C, what is K_p at this temperature?

(See Problems 15.35, 15.36, 15.37, and 15.38.)

Equilibrium Constant for the Sum of Reactions

It is possible to determine the equilibrium constants for various chemical reactions and then use them to obtain the equilibrium constants of other reactions. A useful rule is as follows: *If a given chemical equation can be obtained by taking the sum of other equations, the equilibrium constant for the given equation equals the product of the equilibrium constants of the other equations.*

As an application of this rule, consider the following reactions at 1200 K:

(Reaction 1) $CO(g) + 3H_2(g) \rightleftharpoons CH_4(g) + H_2O(g);\ K_1 = 3.92$

(Reaction 2) $CH_4(g) + 2H_2S(g) \rightleftharpoons CS_2(g) + 4H_2(g);\ K_2 = 3.3 \times 10^4$

When we take the sum of these two equations, we get

(Reaction 3) $CO(g) + 2H_2S(g) \rightleftharpoons CS_2(g) + H_2O(g) + H_2(g)$

According to the previous rule, the equilibrium constant for this reaction, K_3, is $K_3 = K_1K_2$. The accuracy of this result is easy to verify. Substituting the expressions for K_1 and K_2 in the product K_1K_2 gives

$$K_1K_2 = \frac{\cancel{[CH_4]}[H_2O]}{[CO]\cancel{[H_2]^3}} \times \frac{[CS_2][H_2]^{\cancel{4}}}{\cancel{[CH_4]}[H_2S]^2} = \frac{[CS_2][H_2O][H_2]}{[CO][H_2S]^2}$$

which is the equilibrium-constant expression for Reaction 3. The value of the equilibrium constant is $K_1K_2 = 3.92 \times (3.3 \times 10^4) = 1.3 \times 10^5$.●

● We may use equilibrium expressions in the previous calculation in terms of concentrations or pressures.

15.3 HETEROGENEOUS EQUILIBRIA; SOLVENTS IN HOMOGENEOUS EQUILIBRIA

A **homogeneous equilibrium** is *an equilibrium that involves reactants and products in a single phase.* Catalytic methanation is an example of a homogeneous equilibrium; it involves only gaseous reactants and products. On the other hand, a **heterogeneous equilibrium** is *an equilibrium involving reactants and products in more than one phase.* For example, the reaction of iron metal filings with steam to produce iron oxide, Fe_3O_4, and hydrogen involves solid phases, Fe and Fe_3O_4, in addition to a gaseous phase.

$$3Fe(s) + 4H_2O(g) \rightleftharpoons Fe_3O_4(s) + 4H_2(g)$$

In writing the equilibrium-constant expression for a heterogeneous equilibrium, we omit concentration terms for pure solids and liquids. For the previous reaction of iron with steam, we would write

$$K_c = \frac{[H_2]^4}{[H_2O]^4}$$

Concentrations of Fe and Fe_3O_4 are omitted because, whereas the concentration of a gas can have various values, the concentration of a pure solid or a pure liquid is a constant at a given temperature.

To see why such constant concentrations are omitted in writing K_c, let us write the equilibrium-constant expression for the reaction of iron with steam, but let us include [Fe] and $[Fe_3O_4]$. We will call this K_c'.

$$K_c' = \frac{[Fe_3O_4][H_2]^4}{[Fe]^3[H_2O]^4}$$

We can rearrange the above equation, putting all of the constant factors on the left side.

$$\underbrace{\frac{[Fe]^3}{[Fe_3O_4]}}_{\text{constant factors}} \times K_c' = \underbrace{\frac{[H_2]^4}{[H_2O]^4}}_{\text{variable factors}}$$

Note that all terms on the left are constant but those on the right (the concentrations of H_2 and H_2O) are variable. The left side equals K_c. Thus, in effect, concentrations of pure solids or pure liquids are incorporated in the value of K_c. Following similar reasoning, we also omit the concentration of solvent from K_c for a homogeneous reaction, as long as the concentration of solvent remains essentially constant during reaction.●

● If we use the thermodynamic equilibrium constant, which is defined in terms of activities, we find that the activity of a pure liquid or solid is 1. Therefore, when we write the equilibrium constant, the activities of pure solids and liquids need not be given explicitly.

The fact that the concentrations of both Fe and Fe_3O_4 do not occur in the equilibrium-constant expression means that the equilibrium is not affected by the amounts of these substances, *as long as some of each is present.*

Example 15.4 Writing K_c for a Reaction with Pure Solids and Liquids

(a) Quicklime (calcium oxide, CaO) is prepared by heating a source of calcium carbonate, $CaCO_3$, such as limestone or seashells.

$$CaCO_3(s) \rightleftharpoons CaO(s) + CO_2(g)$$

Write the expression for K_c. (b) We can write the equilibrium-constant expression for a physical equilibrium, such as vaporization, as well as for chemical equilibrium. Write the expression for K_c for the vaporization of water.

$$H_2O(l) \rightleftharpoons H_2O(g)$$

Solution

In writing the equilibrium expressions, we ignore pure liquid and solid phases. (a) $\boldsymbol{K_c = [CO_2]}$; (b) $\boldsymbol{K_c = [H_2O(g)]}$.

Exercise 15.6

The Mond process for purifying nickel involves the formation of nickel carbonyl, $Ni(CO)_4$ a volatile liquid, from nickel metal and carbon monoxide.●

● Carbon monoxide is passed over impure nickel to form nickel carbonyl vapor, which when later heated, decomposes and deposits pure nickel.

$$Ni(s) + 4CO(g) \rightleftharpoons Ni(CO)_4(g)$$

Write the expression for K_c for this reaction. (See Problems 15.39 and 15.40.)

Using an Equilibrium Constant

In the preceding sections, we described how a chemical reaction reaches equilibrium and how we can characterize this equilibrium by the equilibrium constant. Now we want to see the ways in which an equilibrium constant can be used to answer important questions. We will look at the following uses:

1. *Qualitatively interpreting the equilibrium constant.* By merely looking at the magnitude of K_c, we can tell whether a particular equilibrium favors products or reactants.
2. *Predicting the direction of reaction.* Consider a reaction mixture that is not at equilibrium. By substituting the concentrations of substances that exist in a reaction mixture into an expression similar to the equilibrium constant and comparing with K_c, we can predict whether the reaction will proceed toward products or toward reactants (as defined by the way we write the chemical equation).
3. *Calculating equilibrium concentrations.* Once we know the value of K_c for a reaction, we can determine the composition at equilibrium for any set of starting concentrations.

Let us see what meaning we can attach to the value of K_c.

15.4 QUALITATIVELY INTERPRETING AN EQUILIBRIUM CONSTANT

If the value of the equilibrium constant is large, we know immediately that the products are favored at equilibrium. Consider the synthesis of ammonia from its elements.

$$N_2(g) + 3H_2(g) \rightleftharpoons 2NH_3(g)$$

At 25°C the equilibrium constant K_c equals 4.1×10^8. This means that the numerator (product concentrations) is 4.1×10^8 times larger than the denominator (reactant concentrations). In other words, at this temperature the reaction favors the formation of ammonia at equilibrium.

We can verify this by calculating one possible equilibrium composition for this reaction. Suppose the equilibrium mixture is 0.010 M in N_2 and 0.010 M in H_2. From these concentrations we can calculate the concentration of ammonia necessary to give equilibrium. Let us substitute the concentrations of N_2 and H_2 and the value of K_c into the equilibrium expression

$$K_c = \frac{[NH_3]^2}{[N_2][H_2]^3}$$

We get

$$\frac{[NH_3]^2}{(0.010)(0.010)^3} = 4.1 \times 10^8$$

Now we can solve for the concentration of ammonia.

$$[NH_3]^2 = 4.1 \times 10^8 \times (0.010)(0.010)^3 = 4.1$$

After we take the square root of both sides of this equation, we find that $[NH_3] =$ 2.0 M. The concentrations of N_2 and H_2 are both 0.010 M, so the amount of

Figure 15.5
Methane (natural gas) reacts with oxygen. The equilibrium mixture is almost entirely carbon dioxide and water. The equilibrium constant K_c for the reaction $CH_4 + 2O_2 \rightleftharpoons CO_2 + 2H_2O$ is 10^{140} at 25°C.

ammonia formed at equilibrium is 200 times that of any one reactant. Figure 15.5 shows the burning of methane, another reaction with an enormously large equilibrium constant.

If the value of the equilibrium constant is small, the reactants are favored at equilibrium. As an example, consider the reaction of nitrogen and oxygen to give nitric oxide, NO:

$$N_2(g) + O_2(g) \rightleftharpoons 2NO(g)$$

The equilibrium constant K_c equals 4.6×10^{-31} at 25°C. If we assume that the concentrations of N_2 and O_2 are 1.0 M, we find that the concentration of NO is 6.8×10^{-16} M. In this case the equilibrium constant is very small, and the concentration of product is not detectable. Reaction occurs to only a very limited extent.●

● The equilibrium constant does become large enough at higher temperatures (about 2000°C) to give appreciable amounts of nitric oxide. This is why air ($N_2 + O_2$) in flames and in auto engines becomes a source of the pollutant NO.

When the equilibrium constant is neither large nor small (around 1), neither reactants nor products are strongly favored. The equilibrium mixture contains significant amounts of all substances in the reaction. For example, in the case of the methanation reaction, the equilibrium constant K_c equals 3.92 at 1200 K. We found that if we start with 1.000 mol CO and 3.000 mol H_2 in a 10.00-L vessel, the equilibrium composition is 0.613 mol CO, 1.839 mol H_2, 0.387 mol CH_4, and 0.387 mol H_2O. Neither the reactants nor the products are predominant.

In summary, if K_c for a reaction,

$$\underbrace{aA + bB}_{\text{reactants}} \rightleftharpoons \underbrace{cC + dD}_{\text{products}}$$

is large, the equilibrium mixture is mostly products. On the other hand, if K_c is small, the equilibrium mixture is mostly reactants. When K_c is around 1, the equilibrium mixture contains appreciable amounts of both reactants and products.

Exercise 15.7

The equilibrium constant K_c for the reaction

$$2NO(g) + O_2(g) \rightleftharpoons 2NO_2(g)$$

equals 4.0×10^{13} at 25°C. Does the equilibrium mixture contain predominantly reactants

or products? If $[NO] = [O_2] = 0.50\ M$ at equilibrium, what is the equilibrium concentration of NO_2?
(See Problems 15.41, 15.42, 15.43, and 15.44.)

15.5 PREDICTING THE DIRECTION OF REACTION

Suppose a gaseous mixture from an industrial plant has the following composition at 1200 K: 0.0200 M CO, 0.0200 M H_2, 0.00100 M CH_4, and 0.00100 M H_2O. If the mixture is passed over a catalyst at 1200 K, would the reaction

$$CO(g) + 3H_2(g) \rightleftharpoons CH_4(g) + H_2O(g)$$

go toward the right or toward the left? That is, would the mixture form more CH_4 and H_2O in going toward equilibrium, or would it form more CO and H_2?

To answer this question, we substitute the concentrations of substances into the *reaction quotient* and compare its value to K_c. The **reaction quotient,** Q_c, is *an expression that has the same form as the equilibrium-constant expression, but whose concentration values are not necessarily those at equilibrium*. For catalytic methanation, the reaction quotient is

$$Q_c = \frac{[CH_4]_i[H_2O]_i}{[CO]_i[H_2]_i^3}$$

where the subscript i indicates concentrations at a particular instant i. When we substitute the concentrations of the gaseous mixture described earlier, we get

$$Q_c = \frac{(0.00100)(0.00100)}{(0.0200)(0.0200)^3} = 6.25$$

Recall that the equilibrium constant K_c for catalytic methanation is 3.92 at 1200 K. For the reaction mixture to go to equilibrium, the value of Q_c must decrease from 6.25 to 3.92. This will happen if the reaction goes to the left. In that case, the numerator of Q_c ($[CH_4]_i[H_2O]_i$) will decrease, and the denominator ($[CO]_i[H_2]_i^3$) will increase. Thus, the gaseous mixture will give more CO and H_2.

Consider the problem more generally. We are given a reaction mixture that is not yet an equilibrium. We would like to know in what direction the reaction will go as it approaches equilibrium. To answer this, we substitute the concentrations of substances from the mixture into the reaction quotient Q_c. Then, we compare Q_c to the equilibrium constant K_c.

For the general reaction $a\text{A} + b\text{B} \rightleftharpoons c\text{C} + d\text{D}$

$$Q_c = \frac{[C]_i^c[D]_i^d}{[A]_i^a[B]_i^b}$$

Then

If $Q_c > K_c$, the reaction will go to the left.
If $Q_c < K_c$, the reaction will go to the right.
If $Q_c = K_c$, the reaction mixture is at equilibrium.

Example 15.5 Using the Reaction Quotient

A 50.0-L reaction vessel contains 1.00 mol N_2, 3.00 mol H_2, and 0.500 mol NH_3. Will more ammonia, NH_3, be formed or will it dissociate when the mixture goes to equilibrium at 400°C? The equation is

$$N_2(g) + 3H_2(g) \rightleftharpoons 2NH_3(g)$$

K_c is 0.500 at 400°C.

Solution

The composition of the gas has been given in terms of moles. We convert these to molar concentrations by dividing by the volume (50.0 L). This gives 0.0200 M N_2, 0.0600 M H_2, and 0.0100 M NH_3. Substituting these concentrations into the reaction quotient gives

$$Q_c = \frac{[NH_3]_i^2}{[N_2]_i[H_2]_i^3} = \frac{(0.0100)^2}{(0.0200)(0.0600)^3} = 23.1$$

Because $Q_c = 23.1$ is greater than $K_c = 0.500$, the reaction will go to the left as it approaches equilibrium. Therefore, **ammonia will dissociate.**

Exercise 15.8

A 10.0-L vessel contains 0.0015 mol CO_2 and 0.10 mol CO. If a small amount of carbon is added to this vessel and the temperature raised to 1000°C, will more CO form? The reaction is

$$CO_2(g) + C(s) \rightleftharpoons 2CO(g)$$

The value of K_c for this reaction is 1.17 at 1000°C. Assume that the volume of gas in the vessel is 10.0 L. (See Problems 15.45 and 15.46.)

15.6 CALCULATING EQUILIBRIUM CONCENTRATIONS

Once we have determined the equilibrium constant for a reaction, we can use it to calculate the concentrations of substances in an equilibrium mixture. The next example illustrates a simple type of equilibrium problem.

Example 15.6 Obtaining One Equilibrium Concentration Given the Others

A gaseous mixture contains 0.30 mol CO, 0.10 mol H_2, and 0.020 mol H_2O, plus an unknown amount of CH_4, in each liter. This mixture is in equilibrium at 1200 K.

$$CO(g) + 3H_2(g) \rightleftharpoons CH_4(g) + H_2O(g)$$

What is the concentration of CH_4 in this mixture? The equilibrium constant K_c equals 3.92.

Solution

The equilibrium equation is

$$K_c = \frac{[CH_4][H_2O]}{[CO][H_2]^3}$$

Substituting the known concentrations and the value of K_c into this equation gives

$$3.92 = \frac{[CH_4](0.020)}{(0.30)(0.10)^3}$$

We can now solve for $[CH_4]$.

$$[CH_4] = \frac{(3.92)(0.30)(0.10)^3}{(0.020)} = 0.059$$

The concentration of CH_4 in the mixture is **0.059 mol/L.**

Exercise 15.9

Phosphorus pentachloride gives an equilibrium mixture of PCl_5, PCl_3, and Cl_2 when heated.

$$PCl_5(g) \rightleftharpoons PCl_3(g) + Cl_2(g)$$

A 1.00-L vessel contains an unknown amount of PCl_5 and 0.020 mol each of PCl_3 and Cl_2

at equilibrium at 250°C. How many moles of PCl_5 are in the vessel if K_c for this reaction is 0.0415 at 250°C?
(See Problems 15.49 and 15.50.)

Usually we begin a reaction with known starting quantities of substances and want to calculate what the quantities will be at equilibrium. The next example illustrates the steps used to solve this type of problem.

Example 15.7 Solving an Equilibrium Problem (Involving a Linear Equation in x)

The reaction

$$CO(g) + H_2O(g) \rightleftharpoons CO_2(g) + H_2(g)$$

is used to increase the ratio of hydrogen in synthesis gas (mixtures of CO and H_2). Suppose we start with 1.00 mol each of carbon monoxide and water in a 50.0-L vessel. How many moles of each substance are in the equilibrium mixture at 1000°C? The equilibrium constant K_c at this temperature is 0.58.

Solution

STEP 1 Note that we *start* with known quantities of substances, which *change* to *equilibrium* values. This suggests that we begin by setting up a table in which we list starting, change, and equilibrium quantities of substances. Because we will relate these to K_c, we should list molar amounts per liter—that is, molar concentrations. The *starting concentrations* of CO and H_2O are

$$[CO] = [H_2O] = \frac{1.00 \text{ mol}}{50.0 \text{ L}} = 0.0200 \text{ mol/L}$$

Concentrations of the products, CO_2 and H_2, are 0. The *changes in concentrations* when the mixture goes to equilibrium are not given. However, we can write them all in terms of a single unknown. If we let x be the moles of CO_2 formed per liter, then the moles of H_2 formed per liter is also x. Similarly, x moles each of CO and H_2O are consumed. We write the changes for CO and H_2O as $-x$. We obtain the *equilibrium concentrations* by adding the change in concentrations to the starting concentrations, as shown in the table below.

STEP 2 We then substitute the values for the equilibrium concentrations into the equilibrium equation,

$$K_c = \frac{[CO_2][H_2]}{[CO][H_2O]}$$

and we get

$$0.58 = \frac{(x)(x)}{(0.0200 - x)(0.0200 - x)}$$

or

$$0.58 = \frac{x^2}{(0.0200 - x)^2}$$

STEP 3 We now solve this equilibrium equation for the value of x. Note that the right-hand side is a perfect square. If we take the square root of both sides, we get

$$\pm 0.76 = \frac{x}{0.0200 - x}$$

We have written ± to indicate that we should consider both positive and negative values, because both are mathematically possible. However, we can dismiss the negative value as physically impossible (x can only be positive; it represents the concentration of CO_2 formed). Rearranging the equation gives

$$x = \frac{0.0200 \times 0.76}{1.76} = 0.0086$$

If we substitute for x in the last line of the table, the equilibrium concentrations are 0.0114 M CO, 0.0114 M H_2O, 0.0086 M CO_2, and 0.0086 M H_2. To find the moles of each substance in the 50.0-L vessel, we multiply the concentrations by the volume of the vessel. For example, the amount of CO is

$$0.0114 \text{ mol/L} \times 50.0 \text{ L} = 0.570 \text{ mol}$$

We find that the equilibrium composition of the reaction mixture is 0.570 mol CO, 0.570 mol H_2O, 0.43 mol CO_2, and 0.43 mol H_2.

CONCENTRATIONS (M)	$CO(g)$	+	$H_2O(g)$	$\rightleftharpoons$	$CO_2(g)$	+	$H_2(g)$
Starting	0.0200		0.0200		0		0
Change	$-x$		$-x$		$+x$		$+x$
Equilibrium	$0.0200 - x$		$0.0200 - x$		x		x

Exercise 15.10

What is the equilibrium composition of a reaction mixture if we start with 0.500 mol each of H_2 and I_2 in a 1.0-L vessel? The reaction is

$$H_2(g) + I_2(g) \rightleftharpoons 2HI(g) \qquad K_c = 49.7 \text{ at } 458°C$$

(See Problems 15.51 and 15.52.)

The preceding example illustrates the three steps in solving for equilibrium concentrations. To summarize, these steps are

1. Set up a table of concentrations (starting, change, and equilibrium expressions in x).
2. Substitute the expressions in x for equilibrium concentrations into the equilibrium equation.
3. Solve the equilibrium equation for the values of the equilibrium concentrations.

In the previous example, if we had not started with the same number of moles of reactants, we would not have gotten an equation with a perfect square. In that case we would have had to solve a quadratic equation. The next example illustrates how to solve such an equation.●

● A quadratic equation of the form

$$ax^2 + bx + c = 0$$

has the solutions

$$x = \frac{-b \pm \sqrt{b^2 - 4ac}}{2a}$$

This equation for x is called the *quadratic formula*.

Example 15.8 Solving an Equilibrium Problem (Involving a Quadratic Equation in x)

Hydrogen and iodine react according to the equation

$$H_2(g) + I_2(g) \rightleftharpoons 2HI(g)$$

Suppose 1.00 mol H_2 and 2.00 mol I_2 are placed in a 1.00-L vessel. How many moles of substances are in the gaseous mixture when it comes to equilibrium at 458°C? The equilibrium constant K_c at this temperature is 49.7.

Solution

As in the previous example, we note that starting concentrations change to equilibrium concentrations. We follow the same three steps: (1) Set up a table for starting, change, and equilibrium concentrations (use algebraic expressions for unknown quantities). (2) Substitute the expressions for the equilibrium concentrations into the equilibrium equation. (3) Solve the equilibrium equation for the unknown, and then work out explicit values for the equilibrium concentrations.

STEP 1 The table listing concentrations of substances is as follows:

CONCENTRATIONS (M)	$H_2(g)$	+	$I_2(g)$	$\rightleftharpoons$	$2HI(g)$
Starting	1.00		2.00		0
Change	$-x$		$-x$		$2x$
Equilibrium	$1.00 - x$		$2.00 - x$		$2x$

Note that the changes in concentrations equal x multiplied by the coefficient of that substance in the balanced chemical equation. The change is negative for a reactant and positive for a product. Equilibrium concentrations equal starting concentrations plus the changes in concentrations.

STEP 2 Substituting into the equilibrium equation,

$$K_c = \frac{[HI]^2}{[H_2][I_2]}$$

we get

$$49.7 = \frac{(2x)^2}{(1.00 - x)(2.00 - x)}$$

STEP 3 Because the right-hand side is not a perfect square, we must use the quadratic formula to solve for x. The previous equation rearranges to give

$$(1.00 - x)(2.00 - x) = (2x)^2/49.7 = 0.0805x^2$$

or

$$0.920x^2 - 3.00x + 2.00 = 0$$

Hence,

$$x = \frac{3.00 \pm \sqrt{9.00 - 7.36}}{1.84} = 1.63 \pm 0.70$$

There are two mathematical solutions to a quadratic equation. We obtain one by taking the upper (positive) sign in $\pm$ and the other by taking the lower (negative) sign. We get

$$x = 2.33 \qquad \text{and} \qquad x = 0.93$$

(continued)

However, $x = 2.33$ gives a negative value to $1.00 - x$ (the equilibrium concentration of H_2), which is physically impossible. Thus, we are left with $x = 0.93$. We substitute this value of x into the last line of the table in Step 1 to get the equilibrium concentrations and then multiply these by the volume of the vessel (1.00 L) to get the amounts of substances. The last line of the table is rewritten as shown below. **The equilibrium composition is 0.07 mol H_2, 1.07 mol I_2, and 1.86 mol HI.**

CONCENTRATIONS (*M*)	$H_2(g)$	+	$I_2(g)$	$\rightleftharpoons$	$2HI(g)$
Equilibrium	$1.00 - x = 0.07$		$2.00 - x = 1.07$		$2x = 1.86$

Exercise 15.11

Phosphorus pentachloride, PCl_5, decomposes when heated.

$$PCl_5(g) \rightleftharpoons PCl_3(g) + Cl_2(g)$$

If the initial concentration of PCl_5 is 1.00 mol/L, what is the equilibrium composition of the gaseous mixture at 160°C? The equilibrium constant K_c at 160°C is 0.0211.

(See Problems 15.53 and 15.54.)

Changing the Reaction Conditions; Le Chatelier's Principle

Getting the maximum amount of product from a reaction depends on the proper selection of reaction conditions. By changing these conditions, we can increase or decrease the yield of product. There are three ways in which we can alter the equilibrium composition of a gaseous reaction mixture and possibly increase the yield of product. We might change the yield by

1. Changing concentrations by removing products or adding reactants to the reaction vessel.
2. Changing the partial pressure of gaseous reactants and products.
3. Changing the temperature.

It should be noted that a catalyst cannot alter equilibrium composition, although it can change the rate at which a product is formed.

15.7 REMOVING PRODUCTS OR ADDING REACTANTS

One way to increase the yield of a desired product is to change concentrations in a reaction mixture by removing a product or adding a reactant. Consider the methanation reaction,

$$CO(g) + 3H_2(g) \rightleftharpoons CH_4(g) + H_2O(g)$$

If we place 1.000 mol CO and 3.000 mol H_2 in a 10.00-L reaction vessel, the equilibrium composition at 1200 K is 0.613 mol CO, 1.839 mol H_2, 0.387 mol CH_4, and 0.387 mol H_2O. Can we alter this composition by removing or adding one of the substances to improve the yield of methane?

Table 15.1 The Effect of Removing Water Vapor from a Methanation Mixture

Stage of Process	Mol CO	Mol H_2	Mol CH_4	Mol H_2O
Original reaction mixture	0.613	1.839	0.387	0.387
After removing water (before equilibrium)	0.613	1.839	0.387	0
When equilibrium is re-established	0.491	1.473	0.509	0.122

● Le Chatelier's principle was introduced in Section 12.3, where it was used to determine the effect of pressure on solubility.

To answer this question, we can apply **Le Chatelier's principle,** which *states that when a system in chemical equilibrium is disturbed by a change of temperature, pressure, or a concentration, the system shifts in equilibrium composition in a way that tends to counteract this change of variable.*● Suppose we remove a substance from or add one to the equilibrium mixture in order to alter the concentration of the substance. Chemical reaction then occurs to partially restore the initial concentration of the removed or added substance. (However, if the concentration of substance cannot be changed, as in the case of a pure solid or liquid reactant or product, changes in amount will have no effect on the equilibrium.)

For example, suppose that water vapor is removed from the reaction vessel containing the equilibrium mixture for methanation. Le Chatelier's principle predicts that net chemical change will occur to partially reinstate the original concentration of water vapor. This means that the methanation reaction momentarily goes in the forward direction,

$$CO(g) + 3H_2(g) \longrightarrow CH_4(g) + H_2O(g)$$

until equilibrium is re-established. Going in the forward direction, the concentrations of both water vapor and methane increase.

A practical way to remove water vapor in this reaction might be to cool the reaction mixture quickly to condense the water. Liquid water could be removed and the gases reheated until equilibrium was again established. The concentration of water vapor would build up again as the concentration of methane increased. Table 15.1 lists the amounts of each substance at each stage of this process. Note how the yield of methane has been improved.

It is often useful to add an excess of a cheap reactant in order to force the reaction toward more products. In this way, the more expensive reactant is made to react to a greater extent that it would otherwise.

Consider the ammonia synthesis

$$N_2(g) + 3H_2(g) \rightleftharpoons 2NH_3(g)$$

If we wished to convert as much hydrogen to ammonia as possible, we might increase the concentration of nitrogen. To understand the effect of this, first suppose that a mixture of nitrogen, hydrogen, and ammonia is at equilibrium. If nitrogen is now added to this mixture, the equilibrium is disturbed. According to Le Chatelier's principle, the reaction will now go in the direction that will use up some of the added nitrogen.

$$N_2(g) + 3H_2(g) \longrightarrow 2NH_3(g)$$

Consequently, adding more nitrogen has the effect of converting a greater quantity of hydrogen to ammonia. Figure 15.6 illustrates the effect on another chemical equilibrium of adding or removing a reactant.

A B C

Figure 15.6
Effect on a chemical equilibrium of changing the concentration of a substance. The concentration is changed by adding or removing a substance. *(A)* The center beaker contains an orange precipitate of HgI_2 in equilibrium with I^- (colorless) and HgI_4^{2-} (colorless): $HgI_2(s) + 2I^-(aq) \rightleftharpoons HgI_4^{2-}(aq)$. *(B)* The solution of NaI (right beaker) is added to the equilibrium mixture in the center beaker. By adding I^-, the reaction is forced to the right: $HgI_2(s) + 2I^-(aq) \longrightarrow HgI_4^{2-}(aq)$. The orange precipitate of HgI_2 disappears. *(C)* A solution of NaClO (left beaker) is now added to the reaction mixture; the NaClO removes I^- by oxidizing it to IO^-. By removing I^-, the reaction shifts to the left: $HgI_2(s) + 2I^-(aq) \longleftarrow HgI_4^{2-}(aq)$. An orange precipitate of HgI_2 appears.

We can look at these situations in terms of the reaction quotient, Q_c. Consider the methanation reaction, in which

$$Q_c = \frac{[CH_4]_i[H_2O]_i}{[CO]_i[H_2]_i^3}$$

If the reaction mixture is at equilibrium, $Q_c = K_c$. Suppose we remove some H_2O from this equilibrium mixture. Now $Q_c < K_c$, and from what we said in Section 15.5, we know the reaction will proceed in the forward direction to restore equilibrium.

We can now summarize the conclusions from this section:

When more reactant is added to, or some product is removed from, an equilibrium mixture, thereby changing the concentration of reactant or product, net reaction occurs left to right (that is, in the forward direction) to give a new equilibrium, and more products are produced.

When more product is added to, or some reactant is removed from, an equilibrium mixture, thereby changing the concentration of reactant or product, net reaction occurs right to left (that is, in the reverse direction) to give a new equilibrium, and more reactants are produced.

Example 15.9 Applying Le Chatelier's Principle When a Concentration Is Altered

Predict the direction of reaction when H_2 is removed from a mixture (lowering its concentration) in which the following equilibrium is established:

$$H_2(g) + I_2(g) \rightleftharpoons 2HI(g)$$

Solution

When H_2 is removed from the reaction mixture, lowering its concentration, the reaction goes in the **reverse direction** (more HI dissociates to H_2 and I_2) to partially restore the H_2 that was removed.

$$H_2(g) + I_2(g) \longleftarrow 2HI(g)$$

Exercise 15.12

Consider each of the following equilibria that are disturbed in the manner indicated. Predict the direction of reaction.

(a) The equilibrium

$$CaCO_3(s) \rightleftharpoons CaO(s) + CO_2(g)$$

is disturbed by increasing the pressure (that is, concentration) of carbon dioxide.
(b) The equilibrium

$$2Fe(s) + 3H_2O(g) \rightleftharpoons Fe_2O_3(s) + 3H_2(g)$$

is disturbed by increasing the concentration of hydrogen.

(See Problems 15.57 and 15.58.)

A reaction whose equilibrium constant is extremely small remains almost completely as reactants and cannot be shifted to products by adding an excess of one reactant. For example, the reaction

$$CO_2(g) + 2H_2O(g) \rightleftharpoons CH_4(g) + 2O_2(g)$$

has an equilibrium constant K_c equal to 10^{-140}. This value is so small that the equilibrium mixture practically consists only of carbon dioxide and water. Adding more carbon dioxide to the reaction vessel has no appreciable effect. The reaction is essentially irreversible.

15.8 CHANGING THE PRESSURE AND TEMPERATURE

The optimum conditions for catalytic methanation involve moderately elevated temperatures and normal to moderately high pressures.

$$CO(g) + 3H_2(g) \xrightleftharpoons[\text{1 atm–100 atm}]{\text{230°C–450°C}} CH_4(g) + H_2O(g)$$

Let us see whether we can gain insight into why these might be the optimum conditions for the reaction.

Effect of Pressure Change

A pressure change *obtained by changing the volume* can effect the yield of product in a gaseous reaction if the reaction involves a change in total moles of gas. The methanation reaction, $CO + 3H_2 \rightleftharpoons CH_4 + H_2O$, is an example of a change in moles of gas. When the reaction goes in the forward direction, four moles of reactant gas ($CO + 3H_2$) become two moles of product gas ($CH_4 + H_2O$).

To see the effect of such a pressure change, consider what happens when an equilibrium mixture from the methanation reaction is compressed to one-half of its original volume at a fixed temperature (see Figure 15.7). The total pressure is doubled (PV = constant at a fixed temperature, according to Boyle's law, so halving V requires that P double). Because the partial pressures and therefore the concentrations of reactants and products have changed, the mixture is no longer at equilibrium. The direction in which the reaction goes to re-establish equilibrium can be predicted by applying Le Chatelier's principle. Reaction should go in the forward direction, because then the moles of gas decrease, and the pressure (which is proportional to moles of gas) decreases. In this way, the initial pressure increase is partially reduced.

We find the same result by looking at the reaction quotient Q_c. Let [CO], $[H_2]$, $[CH_4]$, and $[H_2O]$ be the molar concentrations at equilibrium for the metha-

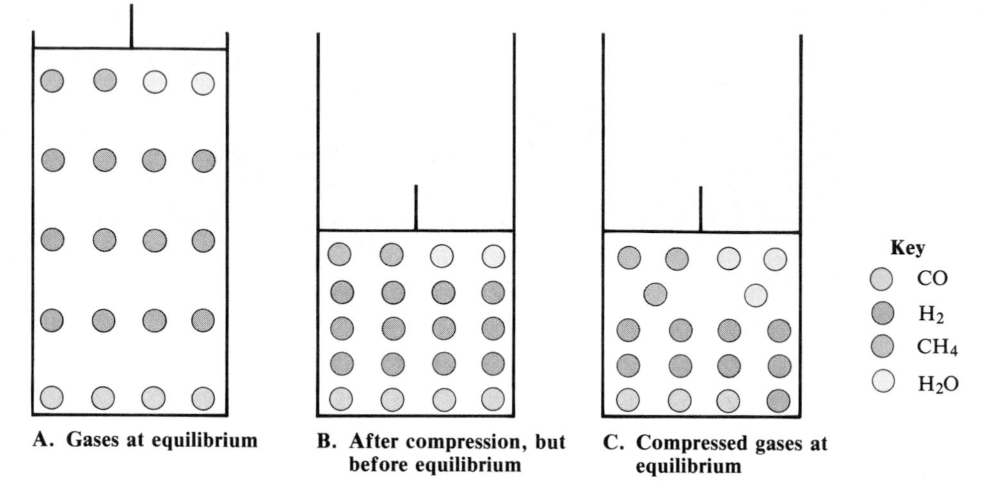

Figure 15.7
Effect on chemical equilibrium of changing the pressure. The effect on the methanation reaction; the approximate composition is represented by the proportion of different colored circles (see key at right). *(A)* The original equilibrium mixture of CO, H_2, CH_4, and H_2O molecules. *(B)* The gases are compressed to one-half their initial volume, increasing their partial pressures, so that the mixture is no longer at equilibrium. *(C)* Equilibrium is re-established when the reaction goes in the forward direction: $CO + 3H_2 \longrightarrow CH_4 + H_2O$. In this way, the total number of molecules is reduced, which reduces the initial pressure increase.

nation reaction. When the volume of an equilibrium mixture is halved, the partial pressures and therefore the concentrations are doubled. We obtain the reaction quotient at that moment by replacing each equilibrium concentration by double its value.

$$Q_c = \frac{(2[CH_4])(2[H_2O])}{(2[CO])(2[H_2])^3} = \frac{K_c}{4}$$

Because $Q_c < K_c$, the reaction proceeds in the forward direction.

We can see the quantitative effect of the pressure change by solving the equilibrium problem. Recall that if we put 1.000 mol CO and 3.000 mol H_2 into a 10.00-L vessel, the equilibrium composition at 1200 K, where K_c equals 3.92, is 0.613 mol CO, 1.839 mol H_2, 0.387 mol CH_4, and 0.387 mol H_2O. Suppose the volume of the reaction gases is halved so that the initial concentrations are doubled. The temperature remains at 1200 K, so K_c is still 3.92. But if we solve for the new equilibrium composition (following the method of Example 15.8), we find 0.495 mol CO, 1.485 mol H_2, 0.505 mol CH_4, and 0.505 mol H_2O. Note that the amount of CH_4 has increased from 0.387 mol to 0.505 mol. We conclude that high pressure of reaction gases favors high yields of methane.

In order to decide the direction of reaction when the pressure of the reaction mixture is increased (say by decreasing the volume), we ignore liquids and solids. These are not much affected by pressure changes because they are nearly incompressible. Consider the reaction

$$C(s) + CO_2(g) \rightleftharpoons 2CO(g)$$

The moles *of gas* increase when the reaction goes in the forward direction (1 mol CO_2 goes to 2 mol CO). Therefore, when we increase the pressure of the reaction mixture by decreasing its volume, the reaction goes in the reverse direction. The moles of gas decrease, and the initial increase of pressure is partially reduced, as Le Chatelier's principle leads us to expect.

It is important to note that an increase or a decrease in pressure of a gaseous reaction must result in changes of partial pressures of substances in the chemical equation if it is to have an effect on the equilibrium composition. Only changes in partial pressures, or concentrations, of these substances can affect the reaction quotient. Consider the effect of increasing the pressure in the methanation reaction by adding helium gas. Although the total pressure increases, the partial pressures of CO, H_2, CH_4, and H_2O do not change. Thus, the equilibrium composition is not affected. However, changing the pressure by changing the volume of this system changes the partial pressures of all gases, so the equilibrium composition is affected.

We can summarize these conclusions: If the pressure is increased by decreasing the volume of a reaction mixture, the reaction shifts in the direction of fewer moles of gas. The following example illustrates applications of this statement.

Example 15.10 Applying Le Chatelier's Principle When the Pressure Is Altered

Look at each of the following equations and decide whether an increase of pressure obtained by decreasing the volume will increase, decrease, or have no effect on the amounts of products.

(a) $CO(g) + Cl_2(g) \rightleftharpoons COCl_2(g)$
(b) $2H_2S(g) \rightleftharpoons 2H_2(g) + S_2(g)$
(c) $C(graphite) + S_2(g) \rightleftharpoons CS_2(g)$

Solution

(a) Reaction decreases the number of molecules of gas (from two to one). According to Le Chatelier's principle, an increase of pressure **increases** the amount of product. (b) Reaction increases the number of molecules of gas (from two to three); hence, an increase of pressure **decreases** the amounts of products. (c) Reaction does not change the number of molecules of gas. (We ignore the change in volume due to consumption of solid carbon because the change in volume of a solid is insignificant. Look only at gas volumes when deciding the effect of pressure change on equilibrium composition.) Pressure change has **no effect.**

Exercise 15.13

Can you increase the amount of product in each of the following equations by increasing the pressure? Explain.

(a) $CO_2(g) + H_2(g) \rightleftharpoons CO(g) + H_2O(g)$
(b) $4CuO(s) \rightleftharpoons 2Cu_2O(s) + O_2(g)$
(c) $2SO_2(g) + O_2(g) \rightleftharpoons 2SO_3(g)$

(See Problems 15.59 and 15.60.)

Effect of Temperature Change

Temperature has a profound effect on most reactions. In the first place, reaction rates usually increase with an increase in temperature, meaning that equilibrium is reached sooner. Many gaseous reactions are sluggish or have imperceptible rates at room temperature but speed up enough at higher temperature to become commercially feasible processes.

Second, equilibrium constants vary with temperature. Table 15.2 gives values of K_c for methanation at various temperatures. Note that K_c equals 4.9×10^{27} at 298 K. Thus, an equilibrium mixture at room temperature is mostly methane and water.

Table 15.2 Equilibrium Constant for Methanation at Different Temperatures

Temperature (K)	K_c
298	4.9×10^{27}
800	1.38×10^{5}
1000	2.54×10^{2}
1200	3.92

Whether we should raise or lower the temperature of a reaction mixture to increase the equilibrium amount of product can be shown by Le Chatelier's principle. Again, consider the methanation reaction,

$$CO(g) + 3H_2(g) \rightleftharpoons CH_4(g) + H_2O(g); \Delta H° = -206.2 \text{ kJ}$$

The value of $\Delta H°$ shows this reaction to be quite exothermic. Thus, as products are formed, considerable heat is released. According to Le Chatelier's principle, as the temperature is raised, the reaction shifts to form more reactants, thereby absorbing

Figure 15.8
Effect on chemical equilibrium of changing the temperature. The reaction shown is endothermic: $Co(H_2O)_6^{2+}(aq) + 4Cl^-(aq) \rightleftharpoons CoCl_4^{2-}(aq) + 6H_2O(l)$. At room temperature, the equilibrium mixture is blue, from $CoCl_4^{2-}$. When cooled by the ice bath, the equilibrium mixture turns pink, from $Co(H_2O)_6^{2+}$.

heat and attempting to counter the increase in temperature.

$$CO(g) + 3H_2(g) \longleftarrow CH_4(g) + H_2O(g) + \text{heat}$$

Thus, we would predict the equilibrium constant to be smaller for higher temperatures, in agreement with the values of K_c given in Table 15.2. Figure 15.8 illustrates the effect of temperature on another chemical equilibrium.●

● The quantitative effect of temperature on the equilibrium constant is discussed in Chapter 18.

The conclusions from Le Chatelier's principle regarding temperature effects on an equilibrium can be summarized this way: For an endothermic reaction (ΔH positive), the amounts of products are increased at equilibrium by an increase in temperature (K_c is larger at higher T). For an exothermic reaction (ΔH negative), the amounts of products are increased at equilibrium by a decrease in temperature (K_c is larger at lower T).

Example 15.11 Applying Le Chatelier's Principle When the Temperature Is Altered

Carbon monoxide is formed when carbon dioxide reacts with solid carbon (graphite).

$$CO_2(g) + C(\text{graphite}) \rightleftharpoons 2CO(g);\ \Delta H^\circ = 172.5\ \text{kJ}$$

Is a high or low temperature more favorable to the formation of carbon monoxide?

Solution

The reaction absorbs heat in the forward direction.

$$\text{Heat} + CO_2(g) + C(\text{graphite}) \longrightarrow 2CO(g)$$

As the temperature is raised, reaction occurs in the forward direction, using heat and thereby tending to lower the temperature. Thus, **high temperature** is more favorable to the formation of carbon monoxide. This is one reason why combustions of carbon and organic materials can produce significant amounts of carbon monoxide.

Exercise 15.14

Consider the possibility of converting carbon dioxide to carbon monoxide by the endothermic reaction

$$CO_2(g) + H_2(g) \rightleftharpoons CO(g) + H_2O(g)$$

Is a high or a low temperature more favorable to the production of carbon monoxide? Explain.

(See Problems 15.61 and 15.62.)

Choosing the Optimum Conditions for Reaction

We are now in a position to understand the optimum conditions for the methanation reaction. Because the reaction is exothermic, low temperatures should favor high yields of methane; that is, the equilibrium constant is large for low temperature. However, gaseous reactions are often very slow at room temperature. In practice, the methanation reaction is run at moderately elevated temperatures (230°C–450°C) in the presence of a nickel catalyst, where the rate of reaction is sufficiently fast but the equilibrium constant is not too small. Because the methanation reaction involves a decrease in moles of gas, the yield of methane should increase as the pressure increases. However, the equilibrium constant is large at the usual operating temperatures, so very high pressures are not needed to obtain economical yields of methane. Pressures of 1 atm–100 atm are usual for this reaction.

As another example, consider the Haber process for the synthesis of ammonia.

$$N_2(g) + 3H_2(g) \xrightleftharpoons{\text{Fe catalyst}} 2NH_3(g); \Delta H^\circ = -91.8 \text{ kJ}$$

Because the reaction is exothermic, the equilibrium constant is larger for lower temperatures. But the reaction proceeds too slowly at room temperature to be practical, even in the presence of the best available catalysts.● The optimum choice of temperature, found experimentally to be about 450°C, is a compromise between an increased rate of reaction at higher temperature and an increased yield of ammonia at lower temperature. Because the formation of ammonia decreases the moles of gases, the yield of product is improved by high pressures. The equilibrium constant K_c is only 0.159 at 450°C, so higher pressures (up to 600 atm) are required for an economical yield of ammonia. Ammonia from the Haber reactor is removed from the reaction mixture by cooling the compressed gases until NH_3 liquefies. Unreacted N_2 and H_2 circulate back to the reactor.

● This technological problem has stimulated a great deal of basic research into understanding how certain bacteria "fix" nitrogen at atmospheric pressure to make NH_3. An enzyme called nitrogenase is responsible for N_2 fixation in these bacteria. This enzyme contains Fe and Mo, which may play a role in the catalysis.

Exercise 15.15

Consider the reaction

$$2CO_2(g) \rightleftharpoons 2CO(g) + O_2(g); \Delta H^\circ = 566 \text{ kJ}$$

Discuss the temperature and pressure conditions that would give the best yield of carbon monoxide. (See Problems 15.65 and 15.66.)

15.9 EFFECT OF A CATALYST

A *catalyst* is a substance that increases the rate of a reaction but is not consumed by it. The significance of a catalyst can be seen in the reaction of sulfur dioxide with oxygen to give sulfur trioxide.

$$2SO_2(g) + O_2(g) \rightleftharpoons 2SO_3(g)$$

The equilibrium constant K_c for this reaction is 1.7×10^{26}, which indicates that for all practical purposes the reaction should go almost completely to products. Yet, when sulfur is burned in air or oxygen, it forms predominantly SO_2 and very little SO_3. Oxidation of SO_2 to SO_3 is simply too slow to give a significant amount of product. However, the rate of the reaction is appreciable in the presence of a platinum or divanadium pentoxide catalyst. The oxidation of SO_2 in the presence of a catalyst is the main step in the *contact process* for the industrial production of sulfuric acid, H_2SO_4. Sulfur trioxide reacts with water to form sulfuric acid. (In the industrial process, SO_3 is dissolved in concentrated H_2SO_4, which is then diluted.)●

● Sulfur dioxide from the combustion of coal and from other sources appears to be a major cause of the marked increase in acidity of rain in the eastern United States in the last few decades. This *acid rain* has been shown to contain sulfuric and nitric acids. The SO_2 is oxidized in moist, polluted air to H_2SO_4. Acid rain is discussed in the Related Topic at the end of Section 16.5.

It is important to understand that *a catalyst has no effect on the equilibrium composition of a reaction mixture. A catalyst merely speeds up the attainment of equilibrium*. For example, suppose we mix 2.00 mol SO_2 and 1.00 mol O_2 in a 100.0-L vessel. In the absence of a catalyst, these substances appear unreactive. Much later, if we analyze the mixture, we find essentially the same amounts of SO_2 and O_2. But when a catalyst is added, the rates of both forward and reverse reactions are very much increased. As a result, the reaction mixture comes to equilibrium in a short time. The amounts of SO_2, O_2, and SO_3 can be calculated from the equilibrium constant. We find that the mixture is mostly SO_3 (2.00 mol), with only 1.7×10^{-8} mol SO_2 and 8.4×10^{-9} mol O_2.

A catalyst is useful for a reaction, such as $2SO_2 + O_2 \rightleftharpoons 2SO_3$, that is normally slow but has a large equilibrium constant. However, if the reaction has an exceedingly small equilibrium constant, a catalyst is of little help. The reaction

$$N_2(g) + O_2(g) \rightleftharpoons 2NO(g)$$

has been considered for the industrial production of nitric acid (NO reacts with O_2 and H_2O to give nitric acid). At 25°C, however, the equilibrium constant K_c equals 4.6×10^{-31}. An equilibrium mixture would contain an extremely small concentration of NO. We cannot expect a catalyst to give a significant yield at this temperature; a catalyst merely speeds up the attainment of equilibrium. The equilibrium constant increases as the temperature is raised, so that at 2000°C, air (which is a mixture of N_2 and O_2) forms about 0.4% NO at equilibrium. An industrial plant was set up in Norway in 1905 to prepare nitrate fertilizers using this reaction. The plant was eventually made obsolete by the *Ostwald process* for making nitric acid (discussed in the Profile on nitric acid at the end of Chapter 6), in which NO is prepared by the oxidation of ammonia. This latter reaction is more economical than the direct reaction of N_2 and O_2, in part because the equilibrium constant is larger at moderate temperatures.

Although a catalyst cannot affect the composition at true equilibrium, in some cases it can affect the product in a reaction because it affects the rate of one reaction out of several possible reactions. The importance of rate for determining the product in a reaction is illustrated by a simple example in Figure 15.9. Mercury(II) ion reacts with iodide ion to form a precipitate of mercury(II) iodide.

$$Hg^{2+}(aq) + 2I^-(aq) \longrightarrow HgI_2(s)$$

In concentrated solutions, orange tetragonal crystals of HgI_2 form. In dilute solution, however, yellow rhombic crystals form faster and are the initial product. Later, when true equilibrium is attained, the orange crystals appear.

The Ostwald process presents an interesting example of the effect of a catalyst in determining a product when several possibilities exist. Two reactions of ammonia with oxygen are possible. The reaction used in the Ostwald process is

$$4NH_3(g) + 5O_2(g) \rightleftharpoons 4NO(g) + 6H_2O(g)$$

However, nitric oxide dissociates to its elements,

$$2NO(g) \rightleftharpoons N_2(g) + O_2(g)$$

and the equilibrium constant for this is quite large (2.2×10^{30} at 25°C), so at true

time = 0 time = 45 minutes

Figure 15.9
An example of how rates of reaction can affect the kind of product. *Left:* At time = 0, a bright orange precipitate of tetragonal crystals of HgI_2 forms from a concentrated solution of $HgCl_2$ and KI. In the dilute solution in the right beaker, however, the HgI_2 precipitates as rhombic crystals with a yellow color. Although less stable, the yellow crystals form faster in the dilute solution. *Right:* In 45 minutes the yellow crystals change to the more stable orange crystals.

equilibrium the products of the reaction of ammonia with oxygen might be expected to be nitrogen and water.

$$4NH_3(g) + 3O_2(g) \rightleftharpoons 2N_2(g) + 6H_2O(g)$$

Ammonia can be made to burn in oxygen, and these are the products. (The reaction occurs most readily in the presence of a copper catalyst; see Figure 15.10.) However, what Ostwald discovered was that the first reaction, to form NO from NH_3 and O_2, was catalyzed by platinum. Therefore, by using this catalyst at moderate temperatures, nitric oxide could be selectively formed. The dissociation of nitric oxide to its elements is normally too slow at moderate temperatures to be significant. Many other examples could be cited in which the products of given reactants depend on the catalyst.

Figure 15.10
The oxidation of ammonia using a copper catalyst. The products are N_2 and H_2O, whereas a platinum catalyst results in NO and H_2O as products (see Figure 6.16, p. 219).

Profile of a Chemical

CARBON MONOXIDE (Starting Material for Organic Compounds)

Carbon has two well-known oxides: carbon dioxide, CO_2, and carbon monoxide, CO. We discussed the properties of carbon dioxide in the Profile in Chapter 11. Carbon monoxide, CO, is a colorless, odorless gas that is very toxic. It poisons by combining with the hemoglobin present in the red blood cells. The normal function of hemoglobin is to combine with oxygen from air that is breathed into the lungs and to release it later to the cells of the body. When carbon monoxide combines with hemoglobin, it binds so strongly that it is not easily released; therefore, the hemoglobin is not available to combine with oxygen. As a result, the blood becomes less able to carry oxygen and eventually body cells die.

We have seen that carbon dioxide is the acid anhydride of carbonic acid (that is, it reacts with water to form carbonic acid). Carbon monoxide does not react directly with water under normal conditions, but in a formal sense we can regard it as the acid anhydride of formic acid, $HCHO_2$. Concentrated sulfuric acid removes the elements of water from formic acid. We can write the following equation for the reaction (see Figure 15.11).

$$HCHO_2(l) \xrightarrow{H_2SO_4} H_2O(l) + CO(g)$$

Figure 15.11
Laboratory preparation of carbon monoxide.
Concentrated sulfuric acid is added to formic acid, $HCHO_2$. The concentrated sulfuric acid removes the elements of water from the formic acid forming carbon monoxide, which burns in air with a blue flame.

Carbon monoxide is a very reactive compound and use is made of its reactivity to prepare a number of organic compounds. Commercially, carbon monoxide is prepared along with hydrogen either by reacting natural gas (mostly CH_4) with steam or by partially oxidizing natural gas with air.

$$CH_4(g) + H_2O(g) \xrightarrow{Ni} CO(g) + 3H_2(g)$$
$$2CH_4(g) + O_2(g) \longrightarrow 2CO(g) + 4H_2(g)$$

Carbon monoxide can also be prepared by the *water-gas reaction,* in which steam is passed over red-hot carbon (coal or coke) to give a gaseous mixture of carbon monoxide and hydrogen.

$$C(s) + H_2O(g) \longrightarrow CO(g) + H_2(g)$$

Mixtures of carbon monoxide and hydrogen (called *synthesis gas*) yield a number of products depending on the temperature, pressure, and catalyst. Methanol (methyl alcohol), which is used as a solvent and in the manufacture of formaldehyde for plastics, is prepared using a zinc oxide–chromium(III) oxide catalyst.

$$CO(g) + 2H_2(g) \xrightarrow{ZnO-Cr_2O_3} CH_3OH(g)$$

Acetic acid is prepared by bubbling carbon monoxide into liquid methanol at 215°C and 140 atm using a nickel(II) iodide catalyst.

$$CO(g) + CH_3OH(l) \xrightarrow{NiI_2} CH_3COOH(l)$$

Many other reactions of carbon monoxide have been studied, and in principle most organic compounds can be prepared from it through a series of reactions.

At the present time, natural gas and petroleum are the principal sources of organic compounds, in addition to functioning as fuels. However, if these materials should become in short supply, carbon monoxide could replace them as a source of organic compounds and fuels. In that case, carbon monoxide would be prepared from the partial oxidation of some carbon-containing material (such as coal, wood, or organic waste). Hydrogen for synthesis gas mixtures could be obtained by reacting carbon monoxide with steam. The reaction is

$$CO(g) + H_2O(g) \xrightarrow{\text{ZnO–CuO}} CO_2(g) + H_2(g)$$

Alternately, synthesis gas can be obtained from the water-gas reaction, which was described previously.

As we noted in the chapter opening, mixtures of carbon monoxide and hydrogen can be converted by catalytic methanation to methane for use as a fuel.

$$CO(g) + 3H_2(g) \xrightarrow{\text{Ni}} CH_4(g) + H_2O(l)$$

It is also possible to convert synthesis gas mixtures to synthetic gasolines by using the appropriate catalyst and reaction conditions. These liquid fuels contain various hydrocarbons, such as octane, C_8H_{18}. The equation for the preparation of octane from CO is

$$8CO(g) + 17H_2(g) \xrightarrow{\text{Fe–Co}} C_8H_{18}(g) + 8H_2O(g)$$

A commercial plant for producing liquid fuels from coal is presently operating in South Africa.

Questions for Study

1. Carbon monoxide is a poisonous gas. Explain why this is so.
2. A simple laboratory preparation of carbon monoxide consists of adding concentrated sulfuric acid to formic acid. What is the chemical equation for the overall reaction?
3. How is carbon monoxide prepared commercially?
4. List two organic compounds that can be produced from carbon monoxide. Give equations for the preparations.
5. Describe the steps in producing synthetic gasoline by one method from coal and water.

A Checklist for Review

IMPORTANT TERMS

chemical equilibrium (15.1)
equilibrium-constant expression (15.2)
equilibrium constant (15.2)
law of mass action (15.2)
homogeneous equilibrium (15.3)
heterogeneous equilibrium (15.3)
reaction quotient (15.5)
Le Chatelier's principle (15.7)

KEY EQUATIONS

$$K_c = \frac{[C]^c[D]^d}{[A]^a[B]^b} \qquad Q_c = \frac{[C]_i^c[D]_i^d}{[A]_i^a[B]_i^b}$$

SUMMARY OF FACTS AND CONCEPTS

Chemical equilibrium can be characterized by the *equilibrium constant* K_c. The expression for K_c has the concentration of products in the numerator and the concentration of reactants in the denominator. Pure liquids and solids are ignored in writing the equilibrium-constant expression. When K_c is very large, the equilibrium mixture is mostly products, and when K_c is very small, the equilibrium mixture is mostly reactants. The reaction quotient Q_c takes the form of the equilibrium-constant expression. If we substitute the concentrations of substances in a reaction mixture into Q_c, we can predict the direction the reaction must go to attain equilibrium. We can use K_c to calculate the composition of the reaction mixture at equilibrium, starting from various initial compositions.

The *choice of conditions,* including *catalysts,* can be very important to the success of a reaction. Removing a product from the reaction mixture, for example, shifts the equilibrium composition to give more product. Changing the pressure and temperature can also affect the product yield. *Le Chatelier's principle* is useful in predicting the effect of such changes.

OPERATIONAL SKILLS

1. Applying stoichiometry to an equilibrium mixture Given the starting amounts of reactants and the amount of one substance at equilibrium, find the equilibrium composition (Example 15.1).

2. Writing equilibrium-constant expressions Given the chemical equation, write the equilibrium-constant expression (Examples 15.2 and 15.4).

3. Obtaining an equilibrium constant from reaction composition Given the equilibrium composition, find K_c (Example 15.3).

4. Using the reaction quotient Given the concentrations of substances in a reaction mixture, predict the direction of reaction (Example 15.5).

5. Obtaining one equilibrium concentration given the others Given K_c and all concentrations of substances but one in an equilibrium mixture, calculate the concentration of this one substance (Example 15.6).

6. Solving equilibrium problems Given the starting composition and K_c of a reaction mixture, calculate the equilibrium composition (Examples 15.7 and 15.8).

7. Applying Le Chatelier's principle Given a reaction, use Le Chatelier's principle to decide the effect of adding or removing a substance (Example 15.9), changing the pressure (Example 15.10), or changing the temperature (Example 15.11).

Review Questions

15.1 Consider the reaction $N_2O_4(g) \rightleftharpoons 2NO_2(g)$. Draw a graph illustrating the changes of concentrations of N_2O_4 and NO_2 as equilibrium is approached when one starts with pure N_2O_4. Describe how the rates of forward and reverse reactions change as the mixture approaches dynamic equilibrium. Why is this called a *dynamic* equilibrium?

15.2 When 1.0 mol each of $H_2(g)$ and $I_2(g)$ are mixed at a certain high temperature, they react to give a final mixture consisting of 0.5 mol each of $H_2(g)$ and $I_2(g)$ and 1.0 mol $HI(g)$. Why is it that you obtain the same final mixture when you bring 2.0 mol $HI(g)$ to the same temperature?

15.3 Explain why the equilibrium constant for a gaseous reaction can be written in terms of partial pressures instead of concentrations.

15.4 Obtain the equilibrium constant for the reaction

$$HCN(aq) \rightleftharpoons H^+(aq) + CN^-(aq)$$

from the following

$$HCN(aq) + OH^-(aq) \rightleftharpoons CN^-(aq) + H_2O(l); \quad K_1 = 4.9 \times 10^4$$

$$H_2O(l) \rightleftharpoons H^+(aq) + OH^-(aq); \; K_2 = 1.0 \times 10^{-14}$$

15.5 Which of the following reactions involve homogeneous equilibria and which involve heterogeneous equilibria? Explain the difference.

(a) $2NO(g) + O_2(g) \rightleftharpoons 2NO_2(g)$
(b) $2Cu(NO_3)_2(s) \rightleftharpoons 2CuO(s) + 4NO_2(g) + O_2(g)$
(c) $2N_2O(g) \rightleftharpoons 2N_2(g) + O_2(g)$
(d) $2NH_3(g) + 3CuO(s) \rightleftharpoons 3H_2O(g) + N_2(g) + 3Cu(s)$

15.6 Explain why pure liquids and solids can be ignored when writing the equilibrium expression.

15.7 What qualitative information can one get from the magnitude of the equilibrium constant?

15.8 What is the reaction quotient? How is it useful?

15.9 In Example 15.7, we found that starting with 1.00 mol each of CO and H_2O in a 50.0-L vessel at 1000°C gave an equilibrium concentration of H_2 equal to 0.0086 *M* for the reaction

$$CO(g) + H_2O(g) \rightleftharpoons CO_2(g) + H_2(g)$$

What would be the equilibrium concentration of H_2 if we had started with 1.00 mol each of CO_2 and H_2?

15.10 List the possible ways in which one can alter the equilibrium composition of a reaction mixture.

15.11 Two moles of H_2 are mixed with 1 mol O_2 at 25°C. No observable reaction takes place, although K_c for the reaction to form water is very large at this temperature. When a piece of platinum is added, however, the gases react rapidly. Explain the role of platinum in this reaction. How does it affect the equilibrium composition of the reaction mixture?

15.12 How is it possible for a catalyst to give products from a reaction mixture that are different from those that are obtained when no catalyst or a different catalyst is used? Give an example.

15.13 When a stream of hydrogen gas is passed over magnetic iron oxide, Fe_3O_4, metallic iron and water are formed. On the other hand, when a stream of water vapor is passed over metallic iron, Fe_3O_4 and hydrogen are formed. Explain by means of Le Chatelier's principle why the reaction goes in one direction in one case but in the reverse direction in the other.

15.14 List four ways in which the yield of ammonia in the reaction

$$N_2(g) + 3H_2(g) \rightleftharpoons 2NH_3(g);\ \Delta H° < 0$$

can be improved for a given amount of H_2. Explain the principle behind each way.

Practice Problems

REACTION STOICHIOMETRY

15.15 A 1.500-mol sample of phosphorus pentachloride, PCl_5, dissociates at 160°C and 1 atm to give 0.203 mol phosphorus trichloride, PCl_3, at equilibrium.

$$PCl_5(g) \rightleftharpoons PCl_3(g) + Cl_2(g)$$

What is the composition of the final reaction mixture?

15.16 Nitric oxide, NO, reacts with bromine, Br_2, to give nitrosyl bromide, NOBr.

$$2NO(g) + Br_2(g) \rightleftharpoons 2NOBr(g)$$

A sample of 0.0873 mol NO with 0.0437 mol Br_2 gives an equilibrium mixture containing 0.0518 mol NOBr. What is the composition of the equilibrium mixture?

15.17 Methanol, CH_3OH, formerly known as wood alcohol, is manufactured commercially by the following reaction:

$$CO(g) + 2H_2(g) \rightleftharpoons CH_3OH(g)$$

A 1.500-L vessel was filled with 0.1500 mol CO and 0.3000 mol H_2. When this mixture came to equilibrium at 500 K, the vessel contained 0.1187 mol CO. How many moles of each substance were in the vessel?

15.18 In the contact process, sulfuric acid is manufactured by first oxidizing SO_2 to SO_3, which is then reacted with water. The reaction of SO_2 with O_2 is

$$2SO_2(g) + O_2(g) \rightleftharpoons 2SO_3(g)$$

A 2.000-L flask was filled with 0.0400 mol SO_2 and 0.0200 mol O_2. At equilibrium at 900 K, the flask contained 0.0296 mol SO_3. How many moles of each substance were in the flask at equilibrium?

THE EQUILIBRIUM CONSTANT AND ITS EVALUATION

15.19 Write equilibrium-constant expressions K_c for each of the following reactions.

(a) $PCl_3(g) + Cl_2(g) \rightleftharpoons PCl_5(g)$
(b) $3O_2(g) \rightleftharpoons 2O_3(g)$
(c) $2NOCl(g) \rightleftharpoons 2NO(g) + Cl_2(g)$
(d) $4NH_3(g) + 3O_2(g) \rightleftharpoons 2N_2(g) + 6H_2O(g)$

15.20 Write equilibrium-constant expressions K_c for each of the following reactions.

(a) $POCl_3(g) \rightleftharpoons POCl(g) + Cl_2(g)$
(b) $2H_2(g) + O_2(g) \rightleftharpoons 2H_2O(g)$
(c) $2C_2H_4(g) + O_2(g) \rightleftharpoons 2CH_3CHO(g)$
(d) $2H_2S(g) + 3O_2(g) \rightleftharpoons 2H_2O(g) + 2SO_2(g)$

15.21 The equilibrium-constant expression for a gas reaction is

$$K_c = \frac{[CS_2][H_2]^4}{[CH_4][H_2S]^2}$$

Write the balanced chemical equation corresponding to this expression.

15.22 The equilibrium-constant expression for a gas reaction is

$$K_c = \frac{[NH_3]^4[O_2]^5}{[NO]^4[H_2O]^6}$$

Write the balanced chemical equation corresponding to this expression.

15.23 The equilibrium-constant expression for a reaction is

$$K_c = \frac{[N_2][H_2O]^2}{[NO]^2[H_2]^2}$$

What is the equilibrium-constant expression when the equation for this reaction is halved and then reversed?

15.24 The equilibrium-constant expression for a reaction is

$$K_c = \frac{[H_2O]^2[Cl_2]^2}{[HCl]^4[O_2]}$$

What is the equilibrium-constant expression when the equation for this reaction is halved and then reversed?

15.25 The equilibrium constant K_c for the equation

$$2HI(g) \rightleftharpoons H_2(g) + I_2(g)$$

at 425°C is 1.84. What is the value of K_c for the following equation?

$$H_2(g) + I_2(g) \rightleftharpoons 2HI(g)$$

15.26 The equilibrium constant K_c for the equation

$$CS_2(g) + 4H_2(g) \rightleftharpoons CH_4(g) + 2H_2S(g)$$

at 900°C is 27.8. What is the value of K_c for the following equation?

$$\tfrac{1}{2}CS_2(g) + 2H_2(g) \rightleftharpoons \tfrac{1}{2}CH_4(g) + H_2S(g)$$

15.27 A 5.00-L vessel contained 0.0185 mol phosphorus trichloride, 0.0158 mol phosphorus pentachloride, and 0.0870 mol chlorine at 230°C in an equilibrium mixture. Calculate the value of K_c for the reaction

$$PCl_3(g) + Cl_2(g) \rightleftharpoons PCl_5(g)$$

15.28 An 8.00-L reaction vessel at 491°C contained 0.650 mol H_2, 0.275 mol I_2, and 3.000 mol HI. Assuming that the substances are at equilibrium, find the value of K_c at 491°C for the reaction of hydrogen and iodine to give hydrogen iodide. The equation is

$$H_2(g) + I_2(g) \rightleftharpoons 2HI(g)$$

15.29 Obtain the value of K_c for the following reaction at 500 K:

$$CO(g) + 2H_2(g) \rightleftharpoons CH_3OH(g)$$

Use the data given in Problem 15.17.

15.30 Obtain the value of K_c for the following reaction at 900 K:

$$2SO_2(g) + O_2(g) \rightleftharpoons 2SO_3(g)$$

Use the data given in Problem 15.18.

15.31 At 77°C, 2.00 mol of nitrosyl bromide, NOBr, placed in a 1.00-L flask dissociates to the extent of 9.4%; that is, for each mole of NOBr before reaction, (1.000 − 0.094) mol NOBr remains after dissociation. Calculate the value of K_c for the dissociation reaction

$$2NOBr(g) \rightleftharpoons 2NO(g) + Br_2(g)$$

15.32 A 2.00-mol sample of nitrogen dioxide was placed in an 80.0-L vessel. At 200°C, the nitrogen dioxide was 6.0% decomposed according to the equation

$$2NO_2(g) \rightleftharpoons 2NO(g) + O_2(g)$$

Calculate the value of K_c for this reaction at 200°C. (See Problem 15.31.)

15.33 Write equilibrium-constant expressions K_p for each of the following reactions.

(a) $H_2(g) + Br_2(g) \rightleftharpoons 2HBr(g)$
(b) $CS_2(g) + 4H_2(g) \rightleftharpoons CH_4(g) + 2H_2S(g)$
(c) $4HCl(g) + O_2(g) \rightleftharpoons 2H_2O(g) + 2Cl_2(g)$
(d) $CO(g) + 2H_2(g) \rightleftharpoons CH_3OH(g)$

15.34 Write equilibrium-constant expressions K_p for each of the following reactions.

(a) $N_2O_4(g) \rightleftharpoons 2NO_2(g)$
(b) $2NO(g) + Br_2(g) \rightleftharpoons 2NOBr(g)$
(c) $2SO_2(g) + O_2(g) \rightleftharpoons 2SO_3(g)$
(d) $4NH_3(g) + 5O_2(g) \rightleftharpoons 4NO(g) + 6H_2O(g)$

15.35 The value of K_c for the following reaction at 900°C is 0.28.

$$CS_2(g) + 4H_2(g) \rightleftharpoons CH_4(g) + 2H_2S(g)$$

What is K_p at this temperature?

15.36 The equilibrium constant K_c equals 10.5 for the following reaction at 227°C.

$$CO(g) + 2H_2(g) \rightleftharpoons CH_3OH(g)$$

What is the value of K_p at this temperature?

15.37 The reaction

$$SO_2(g) + \tfrac{1}{2}O_2(g) \rightleftharpoons SO_3(g)$$

has K_p equal to 6.55 at 627°C. What is the value of K_c at this temperature?

15.38 Fluorine, F_2, dissociates into atoms on heating.

$$\tfrac{1}{2}F_2(g) \rightleftharpoons F(g)$$

The value of K_p at 842°C is 7.55×10^{-2}. What is the value of K_c at this temperature?

15.39 Write the expression for the equilibrium constant K_c for each of the following equations.

(a) $C(s) + CO_2(g) \rightleftharpoons 2CO(g)$
(b) $FeO(s) + CO(g) \rightleftharpoons Fe(s) + CO_2(g)$
(c) $Na_2CO_3(s) + SO_2(g) + \tfrac{1}{2}O_2(g) \rightleftharpoons Na_2SO_4(s) + CO_2(g)$
(d) $PbI_2(s) \rightleftharpoons Pb^{2+}(aq) + 2I^-(aq)$

15.40 For each of the following equations, give the expression for the equilibrium constant K_c.

(a) $NH_4Cl(s) \rightleftharpoons NH_3(g) + HCl(g)$
(b) $C(s) + 2N_2O(g) \rightleftharpoons CO_2(g) + 2N_2(g)$
(c) $2NaHCO_3(s) \rightleftharpoons Na_2CO_3(s) + H_2O(g) + CO_2(g)$
(d) $Fe^{3+}(aq) + 3OH^-(aq) \rightleftharpoons Fe(OH)_3(s)$

USING THE EQUILIBRIUM CONSTANT

15.41 On the basis of the value of K_c, decide whether or not you expect nearly complete reaction at equilibrium for each of the following:

(a) $2H_2(g) + O_2(g) \rightleftharpoons 2H_2O(g)$; $K_c = 3 \times 10^{81}$
(b) $2HF(g) \rightleftharpoons H_2(g) + F_2(g)$; $K_c = 1 \times 10^{-95}$

15.42 Would either of the following reactions go almost completely to product at equilibrium?

(a) $N_2(g) + 2O_2(g) \rightleftharpoons 2NO_2(g)$; $K_c = 3 \times 10^{-17}$
(b) $2SO_2(g) + O_2(g) \rightleftharpoons 2SO_3(g)$; $K_c = 8 \times 10^{25}$

15.43 Hydrogen fluoride decomposes according to the equation

$$2HF(g) \rightleftharpoons H_2(g) + F_2(g)$$

The value of K_c at room temperature is 1.0×10^{-95}. From the magnitude of K_c, do you think the decomposition occurs to any great extent at room temperature? If an equilibrium mixture in a 1.0-L vessel contains 1.0 mol HF, what is the amount of H_2 formed? Does this result agree with what you expect from the magnitude of K_c?

15.44 Suppose sulfur dioxide reacts with oxygen at 25°C.

$$2SO_2(g) + O_2(g) \rightleftharpoons 2SO_3(g)$$

The equilibrium constant K_c equals 8.0×10^{35} at this temperature. From the magnitude of K_c, do you think this reaction occurs to any great extent when equilibrium is reached at room temperature? If an equilibrium mixture is 1.0 M SO_3 and has equal concentrations of SO_2 and O_2, what is the concentration of SO_2 in the mixture? Does this result agree with what you expect from the magnitude of K_c?

15.45 The following reaction has an equilibrium constant K_c equal to 3.59 at 900°C.

$$CH_4(g) + 2H_2S(g) \rightleftharpoons CS_2(g) + 4H_2(g)$$

For each of the following compositions, decide whether the reaction mixture is at equilibrium. If it is not, decide which direction the reaction should go.

(a) $[CH_4] = 1.15\ M$, $[H_2S] = 1.20\ M$, $[CS_2] = 1.51\ M$, $[H_2] = 1.08\ M$
(b) $[CH_4] = 1.07\ M$, $[H_2S] = 1.20\ M$, $[CS_2] = 0.90\ M$, $[H_2] = 1.78\ M$
(c) $[CH_4] = 1.10\ M$, $[H_2S] = 1.49\ M$, $[CS_2] = 1.10\ M$, $[H_2] = 1.68\ M$
(d) $[CH_4] = 1.45\ M$, $[H_2S] = 1.29\ M$, $[CS_2] = 1.25\ M$, $[H_2] = 1.75\ M$

15.46 The following reaction has an equilibrium constant K_c equal to 3.07×10^{-4} at 24°C.

$$2NOBr(g) \rightleftharpoons 2NO(g) + Br_2(g)$$

For each of the following compositions, decide whether the reaction mixture is at equilibrium. If it is not, decide which direction the reaction should go.

(a) $[NOBr] = 0.0610\ M$, $[NO] = 0.0151\ M$, $[Br_2] = 0.0108\ M$
(b) $[NOBr] = 0.115\ M$, $[NO] = 0.0169\ M$, $[Br_2] = 0.0142\ M$
(c) $[NOBr] = 0.181\ M$, $[NO] = 0.0123\ M$, $[Br_2] = 0.0201\ M$

15.47 Methanol, CH_3OH, is manufactured industrially by the reaction

$$CO(g) + 2H_2(g) \rightleftharpoons CH_3OH(g)$$

A gaseous mixture at 500 K is 0.020 M CH_3OH, 0.10 M CO, and 0.10 M H_2. What will be the direction of reaction if this mixture goes to equilibrium? The equilibrium constant K_c equals 10.5 at 500 K.

15.48 Sulfur trioxide, used to manufacture sulfuric acid, is obtained commercially from sulfur dioxide.

$$2SO_2(g) + O_2(g) \rightleftharpoons 2SO_3(g)$$

The equilibrium constant K_c for this reaction is 4.17×10^{-2} at 727°C. What is the direction of reaction when a mixture that is 0.20 M SO_2, 0.10 M O_2, and 0.40 M SO_3 approaches equilibrium?

15.49 Phosgene, $COCl_2$, used in the manufacture of polyurethane plastics, is prepared from CO and Cl_2.

$$CO(g) + Cl_2(g) \rightleftharpoons COCl_2(g)$$

An equilibrium mixture at 395°C contains 0.012 mol CO and 0.025 mol Cl_2 per liter, as well as $COCl_2$. If K_c at 395°C is 1.23×10^3, what is the concentration of $COCl_2$?

15.50 Nitric oxide, NO, is formed in automobile exhaust by the reaction of N_2 and O_2 (from air).

$$N_2(g) + O_2(g) \rightleftharpoons 2NO(g)$$

The equilibrium constant K_c is 0.0025 at 2127°C. If an equilibrium mixture at this temperature contains 0.023 mol N_2 and 0.031 mol O_2 per liter, what is the concentration of NO?

15.51 Iodine and bromine react to give iodine monobromide, IBr.

$$I_2(g) + Br_2(g) \rightleftharpoons 2IBr(g)$$

What is the equilibrium composition of a mixture at 150°C that initially contained 0.0015 mol each of iodine and bromine in a 5.0-L vessel? The equilibrium constant K_c for this reaction at 150°C is 1.2×10^2.

15.52 Initially a mixture contains 0.850 mol each of N_2 and O_2 in an 8.00-L vessel. Find the composition of the mixture when equilibrium is reached at 3900°C. The reaction is

$$N_2(g) + O_2(g) \rightleftharpoons 2NO(g)$$

and $K_c = 0.0123$ at 3900°C.

15.53 The equilibrium constant K_c for the reaction

$$PCl_3(g) + Cl_2(g) \rightleftharpoons PCl_5(g)$$

equals 49 at 230°C. If 0.500 mol each of phosphorus trichloride and chlorine are added to a 5.0-L reaction vessel, what is the equilibrium composition of the mixture at 230°C?

15.54 Calculate the composition of the gaseous mixture obtained when 1.00 mol of carbon dioxide is exposed to hot carbon at 800°C in a 1.00-L vessel. The equilibrium constant K_c at 800°C is 14.0 for the reaction

$$CO_2(g) + C(s) \rightleftharpoons 2CO(g)$$

15.55 Suppose 1.000 mol CO and 3.000 mol H_2 are put in a 10.00-L vessel at 1200 K. The equilibrium constant K_c for

$$CO(g) + 3H_2(g) \rightleftharpoons CH_4(g) + H_2O(g)$$

equals 3.92. Find the equilibrium composition of the reaction mixture.

15.56 The equilibrium constant K_c for the reaction

$$N_2(g) + 3H_2(g) \rightleftharpoons 2NH_3(g)$$

at 450°C is 0.159. Calculate the equilibrium composition when 1.00 mol N_2 is mixed with 3.00 mol H_2 in a 2.00-L vessel.

Le CHATELIER'S PRINCIPLE

15.57 (a) Predict the direction of reaction when chlorine gas is added to an equilibrium mixture of PCl_3, PCl_5, and Cl_2. The reaction is

$$PCl_3(g) + Cl_2(g) \rightleftharpoons PCl_5(g)$$

(b) What is the direction of reaction when chlorine gas is removed from an equilibrium mixture of these gases?

15.58 Consider the equilibrium

$$FeO(s) + CO(g) \rightleftharpoons Fe(s) + CO_2(g)$$

When carbon dioxide is removed from the equilibrium mixture (say by passing the gases through water to absorb CO_2), what is the direction of net reaction as the new equilibrium is achieved?

15.59 What would you expect to be the effect of an increase of pressure on each of the following reactions? Would the pressure change cause reaction to go to the right or left?

(a) $CH_4(g) + 2S_2(g) \rightleftharpoons CS_2(g) + 2H_2S(g)$
(b) $H_2(g) + Br_2(g) \rightleftharpoons 2HBr(g)$
(c) $CO_2(g) + C(s) \rightleftharpoons 2CO(g)$

15.60 Indicate whether either an increase or a decrease of pressure obtained by changing the volume would increase the amount of product in each of the following reactions.

(a) $CO(g) + 2H_2(g) \rightleftharpoons CH_3OH(g)$
(b) $2SO_2(g) + O_2(g) \rightleftharpoons 2SO_3(g)$
(c) $N_2O_4(g) \rightleftharpoons 2NO_2(g)$

15.61 Methanol is prepared industrially from synthesis gas (CO and H_2).

$$CO(g) + 2H_2(g) \rightleftharpoons CH_3OH(g); \ \Delta H° = -21.7 \text{ kcal}$$

Would the fraction of methanol obtained at equilibrium be increased by raising the temperature? Explain.

15.62 One way of preparing hydrogen is by the decomposition of water.

$$2H_2O(g) \rightleftharpoons 2H_2(g) + O_2(g); \ \Delta H° = 484 \text{ kJ}$$

Would you expect the decomposition to be favorable at high or low temperature? Explain.

15.63 Use thermochemical data (Appendix C) to decide whether the equilibrium constant for the following reaction will increase or decrease with temperature.

$$2NO_2(g) + 7H_2(g) \rightleftharpoons 2NH_3(g) + 4H_2O(g)$$

15.64 Use thermochemical data (Appendix C) to decide whether the equilibrium constant for the following reaction will increase or decrease with temperature.

$$CH_4(g) + 2H_2S(g) \rightleftharpoons CS_2(g) + 4H_2(g)$$

15.65 What would you expect to be the general temperature and pressure conditions for an optimum yield of nitric oxide, NO, by the oxidation of ammonia?

$$4NH_3(g) + 5O_2(g) \rightleftharpoons 4NO(g) + 6H_2O(g); \ \Delta H° < 0$$

15.66 Predict the general temperature and pressure conditions for the optimum conversion of ethylene (C_2H_4) to ethane (C_2H_6).

$$C_2H_4(g) + H_2(g) \rightleftharpoons C_2H_6(g); \ \Delta H° < 0$$

Additional Problems

15.67 A mixture of carbon monoxide, hydrogen, and methanol, CH_3OH, is at equilibrium according to the equation

$$CO(g) + 2H_2(g) \rightleftharpoons CH_3OH(g)$$

At 250°C, the mixture is 0.096 *M* CO, 0.191 *M* H_2, and 0.015 *M* CH_3OH. What is K_c for this reaction at 250°C?

15.68 An equilibrium mixture of SO_3, SO_2, and O_2 at 727°C is 0.0160 *M* SO_3, 0.0056 *M* SO_2, and 0.0021 *M* O_2. What is the value of K_c for the following reaction?

$$SO_2(g) + \tfrac{1}{2}O_2(g) \rightleftharpoons SO_3(g)$$

15.69 At 850°C and 1.000 atm pressure, a gaseous mixture of carbon monoxide and carbon dioxide in equilibrium with solid carbon is 90.55% CO by mass.

$$C(s) + CO_2(g) \rightleftharpoons 2CO(g)$$

Calculate K_c for this reaction at 850°C.

15.70 An equilibrium mixture of dinitrogen tetroxide, N_2O_4, and nitrogen dioxide, NO_2, is 65.8% NO_2 by mass at 1.00 atm pressure and 25°C. Calculate K_c at 25°C for the reaction

$$N_2O_4(g) \rightleftharpoons 2NO_2(g)$$

15.71 A 2.00-L vessel contains 1.00 mol N_2, 1.00 mol H_2, and 2.00 mol NH_3. What is the direction of reaction (forward or reverse) needed to attain equilibrium at 400°C? The equilibrium constant K_c for the reaction

$$N_2(g) + 3H_2(g) \rightleftharpoons 2NH_3(g)$$

is 0.51 at 400°C.

15.72 A vessel originally contained 0.200 mol iodine monobromide (IBr), 0.0010 mol I_2, and 0.0010 mol Br_2. The equilibrium constant K_c for the reaction

$$I_2(g) + Br_2(g) \rightleftharpoons 2IBr(g)$$

is 1.2×10^2 at 150°C. What is the direction (forward or reverse) needed to attain equilibrium at 150°C?

15.73 A gaseous mixture containing 1.00 mol each of CO, H_2O, CO_2, and H_2 is exposed to a zinc oxide–copper oxide catalyst at 1000°C. The reaction is

$$CO(g) + H_2O(g) \rightleftharpoons CO_2(g) + H_2(g)$$

and the equilibrium constant K_c is 0.58 at 1000°C. What is the direction of reaction (forward or reverse) as the mixture attains equilibrium?

15.74 A 2.0-L reaction flask initially contains 0.10 mol CO, 0.20 mol H_2, and 0.50 mol CH_3OH (methanol). If this mixture is brought in contact with a zinc oxide–chromium(III) oxide catalyst, the equilibrium

$$CO(g) + 2H_2(g) \rightleftharpoons CH_3OH(g)$$

is attained. The equilibrium constant K_c for this reaction at 300°C is 1.1×10^{-2}. What is the direction of reaction (forward or reverse) as the mixture attains equilibrium?

15.75 Hydrogen bromide dissociates when heated according to the equation

$$2HBr(g) \rightleftharpoons H_2(g) + Br_2(g)$$

The equilibrium constant K_c equals 1.6×10^{-2} at 200°C. What are the moles of substances in the equilibrium mixture at 200°C if we start with 0.010 mol HBr in a 1.0-L vessel?

15.76 Iodine monobromide, IBr, occurs as brownish-black crystals that vaporize with decomposition:

$$2IBr(g) \rightleftharpoons I_2(g) + Br_2(g)$$

The equilibrium constant K_c at 100°C is 0.026. If 0.010 mol IBr is placed in a 1.0-L vessel at 100°C, what are the moles of substances at equilibrium in the vapor?

15.77 Phosgene, $COCl_2$, is a toxic gas used in the manufacture of urethane plastics. The gas dissociates at high temperature.

$$COCl_2(g) \rightleftharpoons CO(g) + Cl_2(g)$$

At 400°C, the equilibrium constant K_c is 8.05×10^{-4}. Find the percentage of phosgene that dissociates at this temperature when 1.00 mol phosgene is placed in a 25.0-L vessel.

15.78 Dinitrogen tetroxide, N_2O_4, is a colorless gas (boiling point, 21°C), which dissociates to give nitrogen dioxide, NO_2, a reddish-brown gas.

$$N_2O_4(g) \rightleftharpoons 2NO_2(g)$$

The equilibrium constant K_c at 25°C is 0.125. What percentage of dinitrogen tetroxide is dissociated when 0.0300 mol N_2O_4 is placed in a 1.00-L flask at 25°C?

15.79 Suppose one starts with a mixture of 1.00 mol CO and 4.00 mol H_2 in a 10.00-L vessel. Find the moles of substances present at equilibrium at 1200 K for the reaction

$$CO(g) + 3H_2(g) \rightleftharpoons CH_4(g) + H_2O(g);\ K_c = 3.92$$

You will get an equation of the form

$$f(x) = 3.92$$

where $f(x)$ is an expression in the unknown x (the amount of CH_4). Solve this equation by guessing values of x, then computing values of $f(x)$. Find values of x such that the values of $f(x)$ bracket 3.92. Then choose values of x to get a smaller bracket around 3.92. Obtain x to two significant figures.

15.80 What are the moles of substances present at equilibrium at 450°C if 1.00 mol N_2 and 4.00 mol H_2 in a 10.0-L vessel react according to the following equation:

$$N_2(g) + 3H_2(g) \rightleftharpoons 2NH_3(g)$$

The equilibrium constant K_c is 0.153 at 450°C. Use the numerical procedure described in Problem 15.79.

15.81 The amount of nitrogen dioxide formed by dissociation of dinitrogen tetroxide,

$$N_2O_4(g) \rightleftharpoons 2NO_2(g)$$

increases as the temperature rises. Is the dissociation of N_2O_4 endothermic or exothermic?

15.82 The equilibrium constant K_c for the synthesis of methanol, CH_3OH,

$$CO(g) + 2H_2(g) \rightleftharpoons CH_3OH(g)$$

is 4.3 at 250°C and 1.8 at 275°C. Is this reaction endothermic or exothermic?

Chapter Outline

Limestone caves are the result of certain acid–base equilibria.

Many well-known substances are weak acids or bases. The following are weak acids: aspirin (acetylsalicylic acid, a headache remedy); phenobarbital (a sedative); saccharin (a sweetener); and niacin (nicotinic acid, a B vitamin). Because these are weak acids, their reactions with water do not go to completion. To discuss such acid–base reactions, we need to look at the equilibria involved and be able to calculate the concentrations of species in a reaction mixture.

Consider, for example, how we could answer the following questions: What is the hydrogen-ion concentration of 0.10 *M* niacin (nicotinic acid)? What is the hydrogen-ion concentration of the solution obtained by dissolving one 5.00-grain tablet of aspirin (acetylsalicylic acid) in 0.500 L of water? If these were solutions of strong acids, the calculations would be simple; 0.10 *M* monoprotic acid would yield 0.10 *M* H^+ ion. However, because niacin is a weak monoprotic acid, the H^+ ion concentration is less than 0.10 *M*. To find the concentration, we will need the equilibrium constant for the reaction involved, and we will need to solve an equilibrium problem.

A similar process is involved in finding the hydrogen-ion concentration of 0.10 *M* sodium nicotinate (sodium salt of niacin). In this case, we will need to look at the acid–base equilibrium of the nicotinate ion. Let us see how we answer questions of this sort.

Self-Ionization of Water; Solutions of a Strong Acid or Base

Because we will be looking at acid–base equilibria in aqueous solution, we will need first to discuss the self-ionization of water—that is, the production of ions from water. This is an equilibrium that occurs in all aqueous solutions. Before we can look at the ionization of weak acids and bases, we will need to explore the simpler case of solutions of a strong acid or base.

16.1 SELF-IONIZATION OF WATER

Although pure water is often considered a nonelectrolyte (nonconductor of electricity), precise measurements do show a very small conduction. This conduction results from **self-ionization** (or **autoionization**), *a reaction in which two like molecules react to give ions*. In the case of water, a proton from one H_2O molecule is transferred to another H_2O molecule, leaving behind an OH^- ion and forming a hydronium ion, $H_3O^+(aq)$.

$$H_2O(l) + H_2O(l) \rightleftharpoons H_3O^+(aq) + OH^-(aq)$$

We can see the slight extent to which the self-ionization of water occurs by noting the small value of its equilibrium constant K_c.

$$K_c = \frac{[H_3O^+][OH^-]}{[H_2O]^2}$$

The value of K_c at room temperature is 3.2×10^{-18}. Because the concentration of ions formed is very small, the concentration of H_2O remains essentially constant, about 56 M at 25°C. If we arrange this equation, placing $[H_2O]^2$ with K_c, the ion product $[H_3O^+][OH^-]$ equals a constant.

$$\underbrace{[H_2O]^2K_c}_{\text{constant}} = [H_3O^+][OH^-]$$

We call *the equilibrium value of the ion product* $[H_3O^+][OH^-]$ the **ion-product constant for water,** which is written $\boldsymbol{K_w}$.● At 25°C, the value of K_w is 1.0×10^{-14}. Like any equilibrium constant, K_w varies with temperature. At body temperature (37°C), K_w equals 2.5×10^{-14}. Because we often write $H^+(aq)$ for $H_3O^+(aq)$, the ion-product for water can be written

● The thermodynamic equilibrium constant is defined in terms of activities. The activity of water is essentially 1, so we do not include water explicitly in the equilibrium expression.

$$K_w = [H^+][OH^-] = 1.0 \times 10^{-14} \text{ at } 25°\text{C}$$

Using K_w, we can calculate the concentrations of H^+ and OH^- ions in pure water. These ions are produced in equal numbers in pure water, so their concentrations are equal. Let $x = [H^+] = [OH^-]$. Then, substituting into the equation for the ion-product constant,

$$K_w = [H^+][OH^-]$$

we get, at 25°C,

$$1.0 \times 10^{-14} = x^2$$

Hence, x equals 1.0×10^{-7}. Thus, the concentrations of H^+ and OH^- are both $1.0 \times 10^{-7}\ M$ in pure water.

If we add an acid or a base to water, the concentrations of H^+ and OH^- will no longer be equal. The equilibrium equation $K_w = [H^+][OH^-]$ will still hold, however, as we will discuss in the next section.

16.2 SOLUTIONS OF A STRONG ACID OR BASE

Let us now consider an aqueous solution of a strong acid or base. Some common strong acids and bases are listed in Table 16.1. Suppose we dissolve 0.10 mol HCl in 1.0 L of aqueous solution, giving 0.10 *M* HCl. We would like to know the concentration of H^+ ion in this solution. In addition to the self-ionization of water, which we can write

$$H_2O(l) \rightleftharpoons H^+(aq) + OH^-(aq)$$

we have the reaction of HCl with water, which also produces H^+ ion. A strong acid such as hydrochloric acid, HCl(*aq*), essentially reacts completely with water.

$$HCl(aq) + H_2O(l) \longrightarrow H_3O^+(aq) + Cl^-(aq)$$

We can write this more simply as

$$HCl(aq) \longrightarrow H^+(aq) + Cl^-(aq)$$

Because we started with 0.10 mol HCl in 1.0 L of solution, the reaction will produce 0.10 mol H^+, so the concentration of H^+ ion from HCl is 0.10 *M*.

Now consider the product of H^+ ion by the self-ionization of water. In pure water, the concentration of H^+ produced is $1.0 \times 10^{-7}\ M$; in an acid solution, the contribution of H^+ from water will be even smaller. We can see this by applying Le Chatelier's principle to the self-ionization reaction. When we add H^+ to water by adding an acid, the self-ionization of water reverses until a new equilibrium is obtained.

$$H_2O(l) \longleftarrow H^+(aq) + OH^-(aq)$$

Consequently, the concentration of H^+ produced by the self-ionization of water ($< 1 \times 10^{-7}\ M$) is negligible in comparison with that produced from HCl (0.10 *M*). So, 0.10 *M* HCl has a concentration of H^+ ion equal to 0.10 *M*.

Table 16.1 Common Strong Acids and Bases

Strong Acids	Strong Bases*
$HClO_4$	LiOH
H_2SO_4	NaOH
HI	KOH
HBr	$Ca(OH)_2$
HCl	$Sr(OH)_2$
HNO_3	$Ba(OH)_2$

*In general, the Group IA and IIA hydroxides (except beryllium hydroxide) are strong bases. $Mg(OH)_2$ is a strong base in the sense that it exists in solution completely as the ions Mg^{2+} and OH^-. However, $Mg(OH)_2(s)$ is not very soluble in water, so aqueous solutions of $Mg(OH)_2$ do not contain high concentrations of OH^-.

In a solution of a strong acid, we can normally ignore the self-ionization of water as a source of H^+. The H^+ concentration is usually determined by the strong acid concentration. (This is not true when the acid solution is extremely dilute, however. In a solution that is 1.0×10^{-7} *M* HCl, the self-ionization of water produces an amount of H^+ comparable to that produced by HCl.)

Although we normally ignore the self-ionization of water in calculating the H^+ ion concentration in a solution of a strong acid, the self-ionization equilibrium still exists and is responsible for the presence of a small concentration of OH^- ion. We can use the ion-product constant for water to calculate this concentration. As an example, let us calculate the concentration of OH^- ion in 0.10 *M* HCl. We substitute $[H^+] = 0.10\ M$ into the equilibrium equation for K_w (for 25°C).

$$K_w = [H^+][OH^-]$$
$$1.0 \times 10^{-14} = 0.10 \times [OH^-]$$

We solve this for $[OH^-]$.

$$[OH^-] = \frac{1.0 \times 10^{-14}}{0.10} = 1.0 \times 10^{-13}$$

Thus, the OH^- concentration is $1.0 \times 10^{-13}\ M$.

Now let us consider a solution of a strong base, such as 0.010 *M* NaOH. What are the OH^- and H^+ concentrations in this solution? Because NaOH is a strong base, all of the NaOH is present in the solution as ions. One mole of NaOH dissolves in water as one mole of Na^+ and one mole of OH^-. Therefore, the concentration of OH^- obtained from NaOH in 0.010 *M* NaOH solution is 0.010 *M*. The concentration of OH^- produced from the self-ionization of water in this solution ($< 1 \times 10^{-7}\ M$) is negligible and can be ignored. Thus, the concentration of OH^- ion in the solution is 0.010 *M*. Hydrogen ion, H^+, is produced by the self-ionization of water. To obtain its concentration, we substitute into the equilibrium equation for K_w (for 25°C).

$$K_w = [H^+][OH^-]$$
$$1.0 \times 10^{-14} = [H^+] \times 0.010$$

Solving for H^+ concentration, we obtain

$$[H^+] = \frac{1.0 \times 10^{-14}}{0.010} = 1.0 \times 10^{-12}$$

Thus, the H^+ concentration is $1.0 \times 10^{-12}\ M$.

The following example further illustrates the calculation of H^+ and OH^- concentrations in solutions of a strong acid or base.

Example 16.1 Calculating Concentrations of H^+ and OH^- in Solutions of a Strong Acid or Base

Calculate the concentrations of hydrogen ion and hydroxide ion at 25°C in (a) 0.15 *M* HNO_3 and (b) 0.010 *M* $Ca(OH)_2$.

Solution

(a) Nitric acid is a strong acid and is present in solution completely as ions. Every mole of HNO_3 contributes one mole of H^+ ion. Consequently, the H^+ concentration is equal to that of the acid; that is, the H^+ concentration is 0.15 *M*. (We can ignore any contribution to the H^+ concentration from the self-ionization of water.) The OH^- concentration is obtained from the equation for K_w.

$$K_w = [H^+][OH^-]$$
$$1.0 \times 10^{-14} = 0.15 \times [OH^-]$$

(continued)

Example 16.4 Determining K_a from the Solution pH

Nicotinic acid (niacin) is a monoprotic acid with the formula $HC_6H_4NO_2$. A solution that is 0.012 *M* in nicotinic acid has a pH of 3.39 at 25°C. What is the acid-ionization constant, K_a, for this acid at 25°C? What is the degree of ionization of nicotinic acid in this solution?

Solution

It is important to realize that when we say the solution is 0.012 *M* nicotinic acid, this refers to *how the solution is prepared*. The solution is made up by adding 0.012 mol of nicotinic acid to enough water to give a liter of solution. However, once the solution is prepared, some of the nicotinic acid molecules ionize. Hence, the actual concentration of nicotinic acid in solution is somewhat less than 0.012 *M*. To solve for K_a, we follow the three steps outlined in Chapter 15 for solving equilibrium problems.

STEP 1 Let us abbreviate the formula for nicotinic acid as HNic. Then 1 L of solution contains 0.012 mol HNic and 0 mol Nic^-, the acid anion, before ionization. The H^+ concentration at the start is that from the self-ionization of water. It is usually much smaller than that obtained from the acid (unless the solution is extremely dilute or K_a is quite small). So we write $[H^+] = \sim 0$ (meaning approximately zero). If x mol HNic ionizes, x mol each of H^+ and Nic^- is formed, leaving $(0.012 - x)$ mol HNic in solution. We can summarize the situation as follows:

CONCENTRATION (*M*)	HNic(*aq*)	⇌	$H^+(aq)$ +	$Nic^-(aq)$
Starting	0.012		~ 0	0
Change	$-x$		$+x$	$+x$
Equilibrium	$0.012 - x$		x	x

Thus, the molar concentrations of HNic, H^+, and Nic^- at equilibrium are $(0.012 - x)$, x, and x, respectively.

STEP 2 The equilibrium equation is

$$K_a = \frac{[H^+][Nic^-]}{[HNic]}$$

When we substitute the expressions for the equilibrium concentrations, we get

$$K_a = \frac{x^2}{(0.012 - x)}$$

STEP 3 The value of x equals the numerical value of the molar hydrogen-ion concentration, and this can be obtained from the pH of the solution.

$$x = [H^+] = \text{antilog}(-\text{pH})$$
$$= \text{antilog}(-3.39) = 4.1 \times 10^{-4} = 0.00041$$

We can substitute this value of x into the equation obtained in Step 2. Note first, however, that

$$0.012 - x = 0.012 - 0.00041 = 0.01159$$
$$\simeq 0.012 \text{ (to two significant figures)}$$

This means that the concentration of undissolved acid is equal to the original concentration of the acid within the precision of the data. (We will make use of this type of observation in later problem solving.) Therefore,

$$K_a = \frac{x^2}{(0.012 - x)} \simeq \frac{x^2}{0.012}$$
$$\simeq \frac{(0.00041)^2}{0.012} = 1.4 \times 10^{-5}$$

To obtain the degree of ionization, note that x mol out of 0.012 mol of nicotinic acid ionizes. Hence,

$$\text{Degree of ionization} = \frac{x}{0.012} = \frac{0.00041}{0.012} = 0.034$$

The percent ionization is obtained by multiplying this by 100; we get 3.4%.

Exercise 16.7

Lactic acid, $HC_3H_5O_3$, is found in sour milk, where it is produced by the action of lactobacilli on lactose, or milk sugar.● A 0.025 *M* solution of lactic acid has a pH of 2.75. What is the ionization constant K_a for this acid? What is the degree of ionization?

(See Problems 16.51 and 16.52.)

● Lactic acid has the structural formula

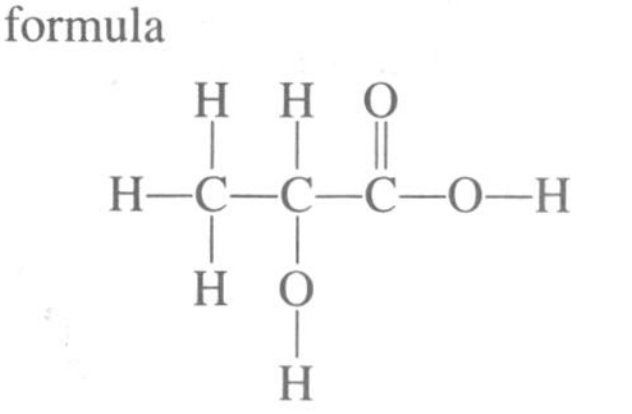

Table 16.2 lists acid-ionization constants for various weak acids. The weakest acids have the smallest values.

Table 16.2 Acid-Ionization Constants at 25°C*

Substance	Formula	K_a
Acetic acid	$HC_2H_3O_2$	1.7×10^{-5}
Benzoic acid	$HC_7H_5O_2$	6.3×10^{-5}
Boric acid	H_3BO_3	5.9×10^{-10}
Carbonic acid	H_2CO_3	4.3×10^{-7}
	HCO_3^-	4.8×10^{-11}
Cyanic acid	$HCNO$	3.5×10^{-4}
Formic acid	$HCHO_2$	1.7×10^{-4}
Hydrocyanic acid	HCN	4.9×10^{-10}
Hydrofluoric acid	HF	6.8×10^{-4}
Hydrogen sulfate ion	HSO_4^-	1.1×10^{-2}
Hydrogen sulfide	H_2S	8.9×10^{-8}
	HS^-	1.2×10^{-13}†
Hypochlorous acid	$HClO$	3.5×10^{-8}
Nitrous acid	HNO_2	4.5×10^{-4}
Oxalic acid	$H_2C_2O_4$	5.6×10^{-2}
	$HC_2O_4^-$	5.1×10^{-5}
Phosphoric acid	H_3PO_4	6.9×10^{-3}
	$H_2PO_4^-$	6.2×10^{-8}
	HPO_4^{2-}	4.8×10^{-13}
Phosphorous acid	H_2PHO_3	1.6×10^{-2}
	$HPHO_3^-$	7×10^{-7}
Propionic acid	$HC_3H_5O_2$	1.3×10^{-5}
Pyruvic acid	$HC_3H_3O_3$	1.4×10^{-4}
Sulfurous acid	H_2SO_3	1.3×10^{-2}
	HSO_3^-	6.3×10^{-8}

*The ionization constants for polyprotic acids are for successive ionizations. Thus, for H_3PO_4, the equilibrium is $H_3PO_4 \rightleftharpoons H^+ + H_2PO_4^-$. For $H_2PO_4^-$, the equilibrium is $H_2PO_4^- \rightleftharpoons H^+ + HPO_4^{2-}$.

†This value is in doubt. Recent evidence suggests that it is about 10^{-19}. See R. J. Myers, *J. Chem. Educ.*, **63**, 687 (1986).

Calculations with K_a

Once we know the value of K_a for an acid HA, we can calculate the equilibrium concentrations of species HA, A^-, and H^+ for solutions of different molarities. The general method for doing this was discussed in Chapter 15. Here we illustrate the use of a simplifying approximation that can often be used for weak acids. As an example, we will look at the first question posed in the chapter opening.

Example 16.5 Calculating Concentrations of Species in a Weak Acid Solution Using K_a (Approximation Method)

What are the concentrations of nicotinic acid, hydrogen ion, and nicotinate ion in a solution of 0.10 *M* nicotinic acid, $HC_6H_4NO_2$, at 25°C? What is the pH of the solution? What is the degree of ionization of nicotinic acid? The acid ionization constant, K_a, was determined in the previous example to be 1.4×10^{-5}.

(continued)

Solution

STEP 1 At the start (before ionization), the concentration of nicotinic acid, HNic, is 0.10 *M* and that of its conjugate base, Nic^-, is 0. The concentration of H^+ is essentially zero (~ 0), assuming that the contribution from the self-ionization of water can be neglected. Let us assume that we have one liter of solution. Then the nicotinic acid in this volume of solution ionizes to give x mol H^+ and x mol Nic^-, leaving $(0.10 - x)$ mol of nicotinic acid. These data are summarized in the following table:

CONCENTRATION (*M*)	$HNic(aq)$	$\rightleftharpoons$	$H^+(aq)$ +	$Nic^-(aq)$
Starting	0.10		~ 0	0
Change	$-x$		$+x$	$+x$
Equilibrium	$0.10 - x$		x	x

The equilibrium concentrations of HNic, H^+, and Nic^- are $(0.10 - x)$, x, and x, respectively.

STEP 2 We now substitute these concentrations and the value of K_a into the equilibrium equation for acid ionization:

$$\frac{[H^+][Nic^-]}{[HNic]} = K_a$$

We get

$$\frac{x^2}{(0.10 - x)} = 1.4 \times 10^{-5}$$

STEP 3 Now we solve this equation for the value of x. This is actually a quadratic equation, but it can be simplified so that the value of x is easily found. Because the acid-ionization constant is small, the value of x is small. Let us assume that x is much smaller than 0.10, so that

$$0.10 - x \simeq 0.10$$

We will need to check that this assumption is valid after we obtain a value for x. The equilibrium equation becomes

$$\frac{x^2}{0.10} \simeq 1.4 \times 10^{-5}$$

Or

$$x^2 \simeq 1.4 \times 10^{-5} \times 0.10 = 1.4 \times 10^{-6}$$

Hence,

$$x \simeq 1.2 \times 10^{-3} = 0.0012$$

At this point, we should check to make sure our assumption that $0.10 - x \simeq 0.10$ is valid. We substitute the value obtained for x into $0.10 - x$.

$$0.10 - x = 0.10 - 0.0012$$
$$= 0.10 \text{ (to two significant figures)}$$

The assumption is indeed valid.

Now we can substitute the value of x into the last line of the table we wrote in Step 1 to find the concentrations of species. Thus, the concentrations of nicotinic acid, hydrogen ion, and nicotinate ion are **0.10 *M*, 0.0012 *M*, and 0.0012 *M***, respectively.

The pH of the solution is

$$pH = -\log [H^+] = -\log 0.0012 = \mathbf{2.92}$$

The degree of ionization equals the amount per liter of nicotinic acid that ionizes ($x = 0.0012$) divided by the total amount per liter of nicotinic acid initially present (0.10). Thus, the degree of ionization is $0.0012/0.10 = \mathbf{0.012}$.

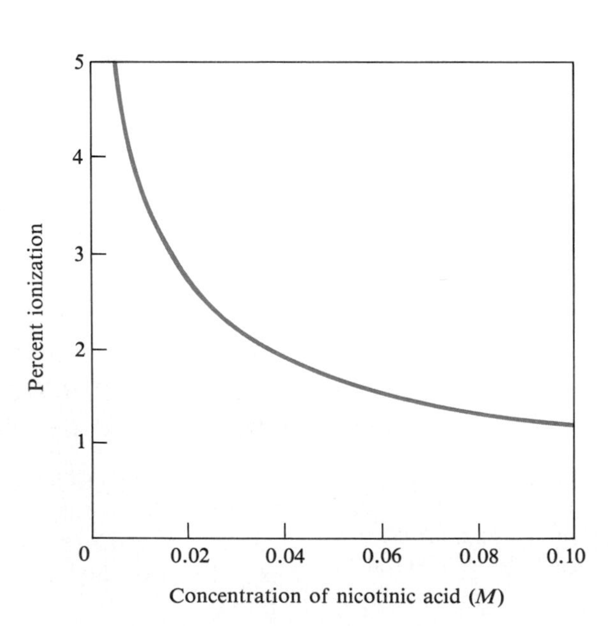

Figure 16.7
Variation of percent ionization of a weak acid with concentration. Shown here is nicotinic acid. Note that the percent ionization is greatest for the most dilute solutions.

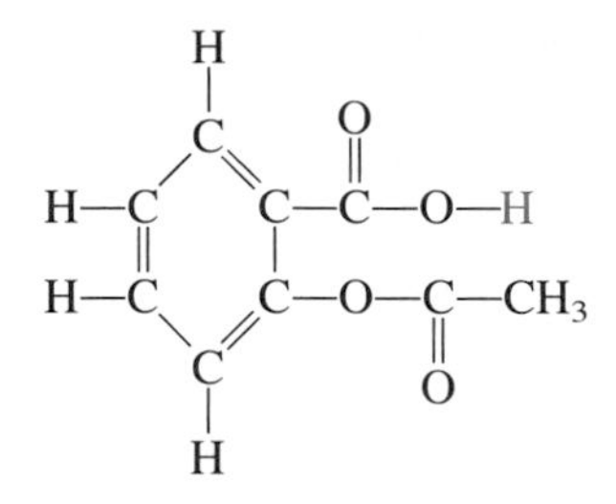

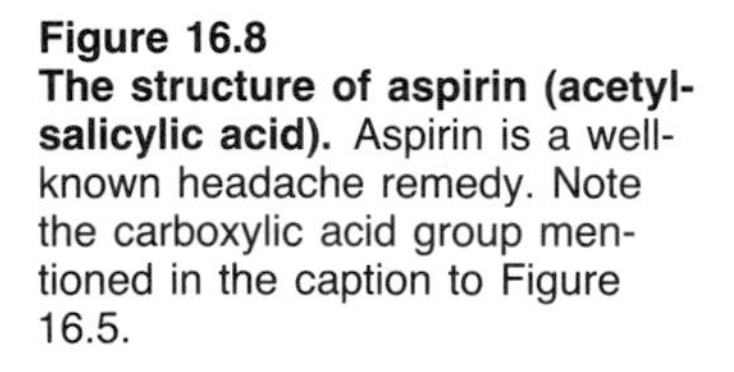

Figure 16.8
The structure of aspirin (acetylsalicylic acid). Aspirin is a well-known headache remedy. Note the carboxylic acid group mentioned in the caption to Figure 16.5.

Exercise 16.8

What are the concentrations of hydrogen ion and acetate ion in a solution of 0.10 M acetic acid, $HC_2H_3O_2$? What is the pH of the solution? What is the degree of ionization? See Table 16.2 for the value of K_a. (See Problems 16.53 and 16.54)

In Example 16.5, the degree of ionization of nicotinic acid (0.012) is relatively small in 0.10 M solution. That is, only 1.2% of the molecules ionize. It is the small value of the degree of ionization that allows us to neglect x in the term $0.10 - x$ and thereby simplify the calculation.

The degree of ionization of a weak acid depends on both K_a and the concentration of the acid solution. For a given concentration, the larger K_a, the greater the degree of ionization. For a given value of K_a, however, the more dilute the solution, the greater the degree of ionization. Figure 16.7 shows how the percent ionization (degree of ionization × 100) varies with the concentration of solution.

How can we know when we can use the simplifying assumption of Example 16.5, where we neglected x in the denominator of the equilibrium equation? *It can be shown that this simplifying assumption gives an error of less than 3% if* K_a *divided by the concentration is* 10^{-3} *or less.*● Thus, for K_a equal to 10^{-5} and a concentration of 10^{-2} M, the assumption will be valid. But it will not be if the concentration of the same acid is 0.001 M.

● The percent error is $100x/C$, where C is the concentration of acid. But $K_a = x^2/(C - x) \simeq x^2/C$, so $x^2/C^2 \simeq K_a/C$ and the percent error is $100\sqrt{K_a/C}$. Suppose the percent error is 3%. Then $100\sqrt{K_a/C} = 3$, or $K_a/C \simeq 1 \times 10^{-3}$.

If the simplifying assumption of Example 16.5 is not valid, we can solve the equilibrium equation exactly by using the *quadratic formula*. We illustrate this in the next example with a solution of aspirin, $HC_9H_7O_4$, a common headache remedy (Figure 16.8 shows the structural formula).

Example 16.6 Calculating Concentrations of Species in a Weak Acid Solution Using K_a (Quadratic Formula)

What is the pH at 25°C of the solution obtained by dissolving a 5.00-grain tablet of aspirin (acetylsalicylic acid) in 0.500 L of water? The tablet contains 5.00 grains, or 0.325 g, of acetylsalicylic acid, $HC_9H_7O_4$. The acid is monoprotic, and K_a equals 3.3×10^{-4} at 25°C.

Solution

The molar mass of $HC_9H_7O_4$ is 180.2 g. From this we find that an aspirin tablet contains 0.00180 mol of the acid. Hence, the concentration of the solution is 0.00180 mol/0.500 L, or 0.0036 M. (We retain two significant figures, the number of significant figures in K_a.)

STEP 1 We will abbreviate the formula for acetylsalicylic acid as HAcs and let x be the amount of H^+ ion formed in one liter of solution. The amount of acetylsalicylate ion formed is also x mol, and the amount of un-ionized acetylsalicylic acid is $(0.0036 - x)$ mol. These data are summarized in the following table:

CONCENTRATION (M)	HAcs(aq)	⇌	H^+(aq) +	Acs^-(aq)
Starting	0.0036		~0	0
Change	$-x$		$+x$	$+x$
Equilibrium	$0.0036 - x$		x	x

STEP 2 If we substitute the equilibrium concentrations and the value of K_a into the equilibrium equation, we get

$$\frac{[H^+][Acs^-]}{[HAcs]} = K_a$$

$$\frac{x^2}{0.0036 - x} = 3.3 \times 10^{-4}$$

STEP 3 Note that K_a divided by the concentration is $3.3 \times 10^{-4}/0.0036 = 9.2 \times 10^{-2}$. Thus, we can expect that x cannot be neglected compared with 0.0036. (You should try the approximation method but then show that x cannot be neglected in $0.0036 - x$.) We can solve the equation exactly using the quadratic formula. Let us rearrange the preceding

(continued)

equation to put it in the form $ax^2 + bx + c = 0$. We get

$$x^2 = (0.0036 - x) \times 3.3 \times 10^{-4}$$
$$= 1.2 \times 10^{-6} - 3.3 \times 10^{-4}\,x$$

Or

$$x^2 + 3.3 \times 10^{-4}\,x - 1.2 \times 10^{-6} = 0$$

We now substitute into the quadratic formula.

$$x = \frac{-b \pm \sqrt{b^2 - 4ac}}{2a}$$
$$= \frac{-3.3 \times 10^{-4} \pm \sqrt{(3.3 \times 10^{-4})^2 + 4(1.2 \times 10^{-6})}}{2}$$
$$= \frac{-3.3 \times 10^{-4} \pm 2.2 \times 10^{-3}}{2}$$

If we take the lower sign in $\pm$, we will get a negative value for x. But x equals the concentration of H^+ ion, so it must be positive. Therefore we ignore the negative solution. Taking the upper sign, we get

$$x = [H^+] = 9.4 \times 10^{-4}$$

Now we can calculate the pH.

$$pH = -\log [H^+] = -\log 9.4 \times 10^{-4} = 3.03$$

Exercise 16.9

What is the pH of an aqueous solution that is 0.0030 *M* pyruvic acid, $HC_3H_3O_3$?●

(See Problems 16.59 and 16.60.)

● Pyruvic acid is an important biochemical intermediate formed during the breakdown of glucose by a cell. In muscle tissue, glucose is broken down to pyruvic acid, which is eventually oxidized to carbon dioxide and water. The structure of pyruvic acid is

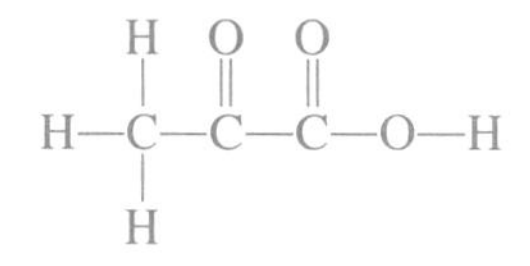

16.5 POLYPROTIC ACIDS

In the preceding section, we dealt only with acids releasing one H^+ ion or proton. Some acids, however, have two or more such protons (these are called *polyprotic acids*). Sulfuric acid, for example, can lose two protons in aqueous solution. The first one is lost completely (sulfuric acid is a strong acid).

$$H_2SO_4(aq) \longrightarrow H^+(aq) + HSO_4^-(aq)$$

The hydrogen sulfate ion, HSO_4^-, that is left may then lose the second proton. In this case, an equilibrium exists.

$$HSO_4^-(aq) \rightleftharpoons H^+(aq) + SO_4^{2-}(aq)$$

For a weak diprotic acid like carbonic acid, H_2CO_3, there are two simultaneous equilibria to consider.

$$H_2CO_3(aq) \rightleftharpoons H^+(aq) + HCO_3^-(aq)$$
$$HCO_3^-(aq) \rightleftharpoons H^+(aq) + CO_3^{2-}(aq)$$

Carbonic acid is in equilibrium with the hydrogen carbonate ion, HCO_3^-, which is in turn in equilibrium with the carbonate ion, CO_3^{2-}. Each equilibrium has an associated acid-ionization constant. For the loss of the first proton, we have

$$K_{a1} = \frac{[H^+][HCO_3^-]}{[H_2CO_3]} = 4.3 \times 10^{-7}$$

and for the loss of the second proton we have

$$K_{a2} = \frac{[H^+][CO_3^{2-}]}{[HCO_3^-]} = 4.8 \times 10^{-11}$$

Note that K_{a1} for carbonic acid is much larger than K_{a2} (by a factor of about 1×10^4). This indicates that carbonic acid loses the first proton more easily than it does the second one, because the first proton separates from an ion of single negative charge, whereas the second proton separates from an ion of double negative charge. This double negative charge strongly attracts the proton back to it. In general, the second ionization constant K_{a2}, of a polyprotic acid is much smaller than the first ionization constant, K_{a1}. In the case of a triprotic acid, the third ionization constant, K_{a3}, is much smaller than the second one, K_{a2}. See the values for phosphoric acid, H_3PO_4, in Table 16.2.

Calculating the concentrations of various species in a solution of a polyprotic acid might appear to be complicated because several equilibria occur at once. However, reasonable assumptions can be made that simplify the calculation, as we show in the next example.

Example 16.7 Calculating Concentrations of Species in a Solution of a Diprotic Acid

Ascorbic acid (vitamin C) is a diprotic acid, $H_2C_6H_6O_6$ (Figure 16.9). What is the pH of a 0.10 *M* solution? What is the concentration of ascorbate ion, $C_6H_6O_6^{2-}$? The acid ionization constants are $K_{a1} = 7.9 \times 10^{-5}$ and $K_{a2} = 1.6 \times 10^{-12}$.

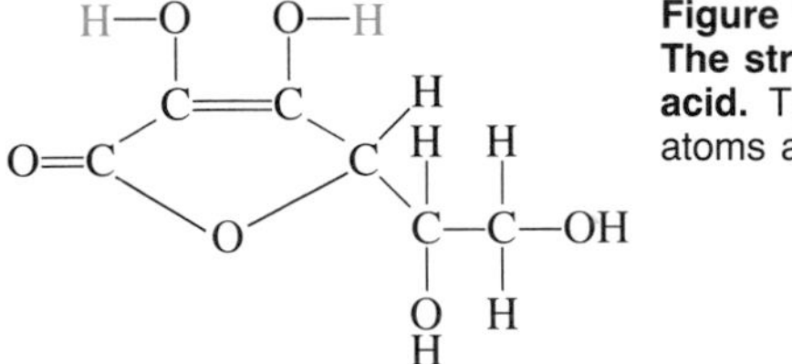

Figure 16.9
The structure of ascorbic acid. The acidic hydrogen atoms are shown in color.

Solution

CALCULATION OF pH Let us abbreviate the formula for ascorbic acid as H_2Asc. Hydrogen ions are produced by two successive acid ionizations.

$$H_2Asc(aq) \rightleftharpoons H^+(aq) + HAsc^-(aq);\ K_{a1} = 7.9 \times 10^{-5}$$

$$HAsc^-(aq) \rightleftharpoons H^+(aq) + Asc^{2-}(aq);\ K_{a2} = 1.6 \times 10^{-12}$$

To be exact, we should account for both reactions to obtain the pH. However, K_{a2} is so much smaller than K_{a1} that the amount of hydrogen ion produced in the second reaction can be neglected in comparison with that produced in the first. Therefore, we can find the pH by considering only the first reaction.

If we let x be the amount of H^+ formed, we get the following results:

CONCENTRATION (*M*)	$H_2Asc(aq)$	$\rightleftharpoons$	$H^+(aq)$ +	$HAsc^-(aq)$
Starting	0.10		~ 0	0
Change	$-x$		$+x$	$+x$
Equilibrium	$0.10 - x$		x	x

We now substitute into the equilibrium equation for the first ionization.

$$\frac{[H^+][HAsc^-]}{[H_2Asc]} = K_{a1}$$

$$\frac{x^2}{0.10 - x} = 7.9 \times 10^{-5}$$

Assuming x to be much smaller than 0.10, we get

$$\frac{x^2}{0.10} \simeq 7.9 \times 10^{-5}$$

Or

$$x^2 \simeq 7.9 \times 10^{-5} \times 0.10$$

$$x \simeq 2.8 \times 10^{-3} = 0.0028$$

(Note that $0.10 - x = 0.10 - 0.0028 = 0.10$, correct to two significant figures. Thus, the assumption that $0.10 - x \simeq 0.10$ is correct.)

The hydrogen-ion concentration is 0.0028 *M*, so

$$pH = -\log [H^+] = -\log(0.0028) = 2.55$$

ASCORBATE-ION CONCENTRATION Ascorbate ion, Asc^{2-}, is produced only in the second reaction. We assume the starting concentrations of H^+ and $HAsc^-$ for this reaction to be those from the first equilibrium. Let us write y for the concentration of ascorbate ion produced. The amounts of each species in one liter of solution are as follows:

CONCENTRATION (*M*)	$HAsc^-(aq)$	$\rightleftharpoons$	$H^+(aq)$	+ $Asc^{2-}(aq)$
Starting	0.0028		0.0028	0
Change	$-y$		$+y$	$+y$
Equilibrium	$0.0028 - y$		$0.0028 + y$	y

We now substitute into the equilibrium equation for the second ionization.

$$\frac{[H^+][Asc^{2-}]}{[HAsc^-]} = K_{a2}$$

(continued)

$$\frac{(0.0028 + y)y}{0.0028 - y} = 1.6 \times 10^{-12}$$

This equation can be simplified if we assume that y is much smaller than 0.0028. (Again, this assumes that the reaction occurs to only a small extent, as we expect from the magnitude of the equilibrium constant.) That is,

$$0.0028 + y \simeq 0.0028$$
$$0.0028 - y \simeq 0.0028$$

Then the equilibrium equation reads

$$\frac{(0.0028)y}{0.0028} \simeq 1.6 \times 10^{-12}$$

Hence,

$$y \simeq 1.6 \times 10^{-12}$$

(Note that 1.6×10^{-12} is indeed much smaller than 0.0028, as we assumed.) The concentration of ascorbate ion equals K_{a2}, or **1.6×10^{-12} *M*.**

Exercise 16.10

Sulfurous acid, H_2SO_3, is a diprotic acid with $K_{a1} = 1.3 \times 10^{-2}$ and $K_{a2} = 6.3 \times 10^{-8}$. The acid forms when sulfur dioxide (a gas with a suffocating odor) dissolves in water. What is the pH of a 0.25 *M* solution of sulfurous acid? What is the concentration of sulfite ion, SO_3^{2-}, in the solution? Note that K_{a1} is relatively large.

(See Problems 16.63 and 16.64.)

Example 16.7 shows that $[H^+]$ and $[HA^-]$ in a solution of a diprotic acid H_2A can be calculated from the first ionization constant, K_{a1}. The concentration of the ion A^{2-}, on the other hand, equals the second ionization constant, K_{a2}.

Related Topic

ACID RAIN

Acid rain is a term referring to rain having a pH lower than that of natural rain (whose pH is 5.6). Natural rain dissolves carbon dioxide from the atmosphere to give a slightly acidic solution, which accounts for the fact that rainwater has a pH somewhat lower than that of pure water (volcanic gases and other sources contribute to the lower pH of natural rain). The pH of rain in eastern North America and western Europe, however, is approximately 4 and sometimes lower. This acidity is primarily the result of the dissolving in rainwater of sulfur oxides and nitrogen oxides from human activities. In the northeastern United States, the strong acid components in acid rain are about 62% sulfuric acid, 32% nitric acid, and 6% hydrochloric acid.

The sulfuric acid in acid rain has been traced to the burning of fossil fuels and to the burning of sulfide ores in the production of metals, such as zinc and copper. Coal, for example, contains some sulfur mainly as pyrite or iron(II) disulfide, FeS_2. When this burns in air, it produces sulfur dioxide.

$$4FeS_2(s) + 11O_2(g) \longrightarrow 2Fe_2O_3(s) + 8SO_2(g)$$

In the presence of dust particles and other substances in polluted air, the sulfur dioxide oxidizes further to give sulfur trioxide, which reacts with water to form sulfuric acid.

$$2SO_2(g) + O_2(g) \longrightarrow 2SO_3(g)$$
$$SO_3(g) + H_2O(l) \longrightarrow H_2SO_4(aq)$$

Almost all of the sulfuric acid dissociates into the ions H^+, HSO_4^-, and SO_4^{2-}. The first ionization of H_2SO_4 is complete, whereas the second ionization (that of HSO_4^-) is partial ($K_{a2} = 1.1 \times 10^{-2}$).

Acid rain can be harmful to some plants, to fish (by changing the pH of lake water), and to structural materials and monuments (Figure 16.10). Marble, for example, is composed of calcium carbonate, $CaCO_3$, which dissolves in water of low pH.

$$H^+(aq) + CaCO_3(s) \longrightarrow Ca^{2+}(aq) + HCO_3^-(aq)$$

Figure 16.10
Effect of acid rain on marble. *These marble columns on a New York building have been damaged by acid rain.*

It is instructive to calculate the pH of natural rain, that is, water saturated with carbon dioxide from the atmosphere. For this, we need to solve two equilibrium problems. First, we need the equilibrium for the solution of CO_2 in water.

$$CO_2(g) + H_2O(l) \rightleftharpoons H_2CO_3(aq)$$

$$K = \frac{[H_2CO_3]}{P_{CO_2}} = 3.5 \times 10^{-2}$$

Here $[H_2CO_3]$ represents the concentration of molecular species formed when CO_2 dissolves in water. The partial pressure of CO_2 in the atmosphere is 0.00033 atm. If we substitute this for P_{CO_2} and solve for $[H_2CO_3]$, we obtain 1.2×10^{-5} *M*. (Note that the equilibrium equation is essentially Henry's law, discussed in Section 12.3.) Now we need to account for the acid ionization of H_2CO_3. We look only at the first ionization, which produces almost all of the H^+.

$$H_2CO_3(aq) \rightleftharpoons H^+(aq) + HCO_3^-(aq)$$

$$K_{a1} = \frac{[H^+][HCO_3^-]}{[H_2CO_3]} = 4.3 \times 10^{-7}$$

Denoting $[H^+] = [HCO_3^-]$ by x, we can write the equilibrium equation

$$\frac{x^2}{1.2 \times 10^{-5} - x} = 4.3 \times 10^{-7}$$

Making our usual assumption about neglecting x in the denominator, we obtain $x = [H^+] = 2.3 \times 10^{-6}$. This assumption and the neglect of the self-ionization of water affect the answer in the second significant figure. Retaining one significant figure, the pH $= -\log(2 \times 10^{-6}) = 5.7$, which is essentially what we gave earlier for the pH of natural rain.

16.6 BASE-IONIZATION EQUILIBRIA

Equilibria involving weak bases are treated similarly to those for weak acids. Ammonia, for example, ionizes in water as follows:

$$NH_3(aq) + H_2O(l) \rightleftharpoons NH_4^+(aq) + OH^-(aq)$$

The corresponding equilibrium constant is

$$K_c = \frac{[NH_4^+][OH^-]}{[NH_3][H_2O]}$$

Because the concentration of H_2O is nearly constant, we can rearrange this equation as we did for acid ionization.●

● In terms of the thermodynamic equilibrium constant, the activity of H_2O is nearly constant and essentially 1, so it does not appear explicitly in the equilibrium constant.

$$K_b = [H_2O]K_c = \frac{[NH_4^+][OH^-]}{[NH_3]}$$

Exercise 16.12

What is the hydrogen-ion concentration of a 0.20 *M* solution of ammonia in water? See Table 16.3 for K_b. (See Problems 16.69 and 16.70.)

16.7 ACID–BASE PROPERTIES OF SALT SOLUTIONS; HYDROLYSIS

A salt may be regarded as an ionic compound obtained by a neutralization reaction in aqueous solution. The resulting salt solution may be neutral, but often it is acidic or basic (Figure 16.12). One of the successes of the Brønsted–Lowry concept of acids and bases was in pointing out that some ions can act as acids or bases. Thus, the acidity or basicity of a salt solution is explained in terms of the acidity or basicity of individual ions in the solution.

Consider a solution of sodium cyanide, NaCN. A 0.1 *M* solution has a pH of 11.1 and is therefore fairly basic. Sodium cyanide dissolves in water to give Na^+ and CN^- ions.

$$NaCN(s) \xrightarrow{H_2O} Na^+(aq) + CN^-(aq)$$

Sodium ion, Na^+, is unreactive with water, but the cyanide ion, CN^-, does react.

$$CN^-(aq) + H_2O(l) \rightleftharpoons HCN(aq) + OH^-(aq)$$

From the Brønsted–Lowry point of view, the CN^- ion acts as a base, because it accepts a proton from H_2O. We can also see, however, that OH^- ion is a product, so we expect the solution to have a basic pH. This explains why solutions of NaCN are basic.

The reaction of the CN^- ion with water is referred to as the *hydrolysis* of CN^-. The **hydrolysis** of an ion is *the reaction of an ion with water to produce the conjugate acid and hydroxide ion or the conjugate base and hydrogen ion*. We saw that the CN^- ion hydrolyzes to give the conjugate acid and OH^- ion. As another example, consider the ammonium ion, NH_4^+, which hydrolyzes as follows.

$$NH_4^+(aq) + H_2O(l) \rightleftharpoons NH_3(aq) + H_3O^+(aq)$$

The ammonium ion acts as an acid, donating a proton to H_2O and forming NH_3. Note that this equation has the form of an acid ionization, so we could write the K_a expression for it. Similarly, the hydrolysis reaction for the CN^- ion has the form of a base ionization, so we could write the K_b expression for it.

Figure 16.12
Salt solutions are acidic, neutral, or basic. A pH meter measures the pH of solutions of the salts NH_4Cl, NaCl, and Na_2CO_3.

Two questions become apparent. First, how can we predict whether a particular salt solution will be acidic, basic, or neutral? Then, how can we calculate the concentrations of H^+ ion and OH^- ion in the salt solution (or equivalently, how do we predict the pH of the salt solution)? We will now look at the first of these questions.

Prediction of Whether a Salt Solution Is Acidic, Basic, or Neutral

When the CN^- ion hydrolyzes, it produces the conjugate acid, HCN. Hydrogen cyanide, HCN, is a weak acid. That means that it tends to hold on to the proton strongly (does not ionize readily). In other words, the cyanide ion, CN^-, tends to pick up a proton easily, so it acts as a base. This argument can be generalized: *The anions of weak acids are basic. On the other hand, the anions of strong acids have hardly any basic character; that is, these ions do not hydrolyze.* For example, the Cl^- ion, which is conjugate to the strong acid HCl, shows no appreciable reaction with water.

$$Cl^-(aq) + H_2O(l) \longrightarrow \text{no reaction}$$

Now consider a cation conjugate to a weak base. The simplest example is the NH_4^+ ion, which we just discussed. We see that it behaves like an acid. *The cations of weak bases are acidic. On the other hand, the cations of strong bases (metal ions of Groups IA and IIA elements—except Be) have hardly any acidic character, that is, these ions do not hydrolyze.* For example,

$$Na^+(aq) + H_2O(l) \longrightarrow \text{no reaction}$$

Aqueous metal ions, other than the cations of the strong bases, usually hydrolyze by acting as acids. These ions usually form hydrated metal ions. For example, the aluminum ion Al^{3+} forms the hydrated ion $Al(H_2O)_6{}^{3+}$. The bare Al^{3+} ion acts as a Lewis acid, forming bonds to the electron pairs on O atoms of H_2O molecules. Because the electrons are drawn away from the O atoms by the Al atom, the O atoms in turn tend to draw electrons from the O—H bonds, making them highly polar. As a result, the H_2O molecules in $Al(H_2O)_6{}^{3+}$ are acidic. The $Al(H_2O)_6{}^{3+}$ ion hydrolyzes in aqueous solution by donating a proton (from one of the acidic H_2O molecules on the ion) to water molecules in the solvent. After the proton leaves the hydrated metal ion, the ion charge is reduced by 1 and an H_2O molecule on the ion becomes a hydroxide ion bonded to the metal. Thus, the formula of the ion becomes $Al(H_2O)_5(OH)^{2+}$.

$$Al(H_2O)_6{}^{3+}(aq) + H_2O(l) \rightleftharpoons Al(H_2O)_5(OH)^{2+}(aq) + H_3O^+(aq)$$

To predict the acidity or basicity of a salt solution, we need to examine the acidity or basicity of the ions composing the salt. Consider potassium acetate, $KC_2H_3O_2$. The ions are K^+ and $C_2H_3O_2^-$ (acetate ion). Potassium is a Group IA element, so K^+ does not hydrolyze. However, the acetate ion is the conjugate of acetic acid, a weak acid. Therefore, the acetate ion is basic.

$$K^+(aq) + H_2O(l) \longrightarrow \text{no reaction}$$

$$C_2H_3O_2^-(aq) + H_2O(l) \rightleftharpoons HC_2H_3O_2(aq) + OH^-(aq)$$

A solution of potassium acetate is predicted to be basic.

From this discussion, we can derive a set of rules for deciding whether a salt solution will be neutral, acidic, or basic. These rules apply to normal salts (those in which the anion has no acidic hydrogen atoms).

1. **A salt of a strong base and a strong acid.** The salt has no hydrolyzable ions and so gives a neutral aqueous solution. An example is NaCl.
2. **A salt of a strong base and a weak acid.** The anion of the salt is the conjugate of the weak acid. It hydrolyzes to give a basic solution. An example is NaCN.
3. **A salt of a weak base and a strong acid.** The cation of the salt is the conjugate of the weak base. It hydrolyzes to give an acidic solution. An example is NH_4Cl.
4. **A salt of a weak base and a weak acid.** Both ions hydrolyze. Whether the solution is acidic or basic depends on the relative acid–base strengths of the two ions. To determine this, we need to compare the K_a of the cation with the K_b of the anion. If the K_a is larger, the solution is acidic. If the K_b is larger, the solution is basic. Consider solutions of ammonium formate, NH_4CHO_2. These solutions are slightly acidic because the K_a for NH_4^+ ($= 5.6 \times 10^{-10}$) is somewhat larger than the K_b for formate ion, CHO_2^- ($= 5.9 \times 10^{-11}$).

The next example illustrates the application of these rules.

Example 16.9 Predicting Whether a Salt Solution Is Acidic, Basic, or Neutral

Decide whether aqueous solutions of the following salts are acidic, basic, or neutral: (a) KCl (b) NaF (c) $Zn(NO_3)_2$ (d) NH_4CN

Solution

(a) KCl is a salt of a strong base (KOH) and a strong acid (HCl), so none of the ions hydrolyze, and a solution of KCl is **neutral.** (b) NaF is a salt of a strong base (NaOH) and a weak acid (HF), so a solution of NaF is **basic** because of the hydrolysis of F^-. (We assume HF is a weak acid, because it is not one of the common strong acids listed in Table 16.1.) (c) $Zn(NO_3)_2$ is the salt of a weak base ($Zn(OH)_2$) and a strong acid (HNO_3), so a solution of $Zn(NO_3)_2$ is **acidic.** Note also that Zn^{2+} is not an ion of a Group IA or Group IIA element, so it would be expected to form a metal hydrate ion that hydrolyzes to give an acidic solution. (d) NH_4CN is a salt of a weak base (NH_3) and a weak acid (HCN). Therefore, to answer the question, we would need the K_a for NH_4^+ and the K_b for CN^-. Although these are listed in some tables, more frequently they are not. However, as we will see later in this section, values are easily calculated from the molecular substances that are conjugates (NH_3 and HCN). The K_a for NH_4^+ is 5.6×10^{-10}, and the K_b for CN^- is 2.0×10^{-5}. From these values, we conclude that a solution of NH_4CN is **basic.**

Exercise 16.13

Consider solutions of the following salts: (a) NH_4NO_3; (b) KNO_3; (c) $Al(NO_3)_3$. Which solution is acidic? Which is basic? Which is neutral? (See Problems 16.75 and 16.76.)

The pH of a Salt Solution

Suppose we would like to calculate the pH of 0.10 *M* NaCN. As we have seen, the solution is basic because of the hydrolysis of the CN^- ion. The calculation is essentially the same as that for any weak base (Example 16.8). We would require the base-ionization constant of the cyanide ion. However, frequently the ionization constants for ions are not listed directly in tables, because the values are easily related to those for the conjugate molecular species. Thus, the K_b for CN^- is related to the K_a for HCN.

To see this relationship between K_a and K_b for conjugate acid–base pairs, consider the acid ionization of HCN and the base ionization of CN^-. When these two reactions are added, we get the ionization of water.

$$\begin{array}{lll} HCN(aq) & \rightleftharpoons H^+(aq) + CN^-(aq) & K_a \\ CN^-(aq) + H_2O(l) & \rightleftharpoons HCN(aq) + OH^-(aq) & K_b \\ \hline H_2O(l) & \rightleftharpoons H^+(aq) + OH^-(aq) & K_w \end{array}$$

● Another way to obtain this result is by multiplying the expression for K_b by $[H^+]/[H^+]$ and rearranging.

$$K_b = \frac{[HCN][OH^-]}{[CN^-]} \times \frac{[H^+]}{[H^+]} = [H^+][OH^-] \Big/ \frac{[H^+][CN^-]}{[HCN]}$$

which is equivalent to $K_b = K_w/K_a$ or $K_aK_b = K_w$.

The equilibrium constants for the reactions are shown symbolically at the right. In Section 15.2, it was shown that when two reactions are added, their equilibrium constants are multiplied.●

$$K_aK_b = K_w$$

This relationship is general and shows that the product of acid- and base-ionization constants in aqueous solution for conjugate acid–base pairs equals the ion-product constant for water, K_w.

Example 16.10 Obtaining K_a from K_b or K_b from K_a

Use Tables 16.2 and 16.3 to obtain the following at 25°C: (a) K_b for CN^-; (b) K_a for NH_4^+.

Solution

(a) The conjugate acid of CN^- is HCN, whose K_a is 4.9×10^{-10} (Table 16.2). Hence,

$$K_b = \frac{K_w}{K_a} = \frac{1.0 \times 10^{-14}}{4.9 \times 10^{-10}} = 2.0 \times 10^{-5}$$

Note that K_b is approximately equal to K_b for ammonia (which equals 1.8×10^{-5}). Thus, the base strength of CN^- is comparable to that of NH_3. (b) The conjugate base of NH_4^+ is NH_3, whose K_b is 1.8×10^{-5} (Table 16.3). Hence,

$$K_a = \frac{K_w}{K_b} = \frac{1.0 \times 10^{-14}}{1.8 \times 10^{-5}} = 5.6 \times 10^{-10}$$

NH_4^+ is a relatively weak acid. Acetic acid, by way of comparison, has K_a equal to 1.7×10^{-5}.

Exercise 16.14

Calculate the following, using Tables 16.2 and 16.3: (a) K_b for F^-; (b) K_a for $C_6H_5NH_3^+$ (conjugate acid of aniline, $C_6H_5NH_2$). (See Problems 16.79 and 16.80.)

If we have a solution of a salt in which only one of the ions hydrolyzes, the calculation of the concentrations of species present follows that for solutions of weak acids or bases. The only difference is that we must first obtain K_a or K_b for the ion that hydrolyzes. The next example illustrates the reasoning and calculations involved.

Example 16.11 Calculating Concentrations of Species in a Salt Solution

What is the pH of 0.10 *M* sodium nicotinate at 25°C (a problem posed in the chapter opening)? The K_a for nicotinic acid was determined in Example 16.4 to be 1.4×10^{-5} at 25°C.

Solution

Sodium nicotinate gives Na^+ and nicotinate ions in solution. Only the nicotinate ion hydrolyzes. Let us write HNic for nicotinic acid and Nic^- for the nicotinate ion. The hydrolysis of nicotinate ion is

$$Nic^-(aq) + H_2O(l) \rightleftharpoons HNic(aq) + OH^-(aq)$$

Thus, nicotinate ion acts as a base, and we can calculate the concentration of species in solution as in Example 16.8.

(continued)

First, however, we need K_b for the nicotinate ion. This is related to K_a for nicotinic acid by the equation $K_aK_b = K_w$. Substituting, we get

$$K_b = \frac{K_w}{K_a} = \frac{1.0 \times 10^{-14}}{1.4 \times 10^{-5}} = 7.1 \times 10^{-10}$$

Now we can proceed with the equilibrium calculation. We will only sketch this calculation, because it is similar to that in Example 16.8. We let x = [HNic] = [OH^-], then substitute into the equilibrium equation

$$\frac{[\text{HNic}][\text{OH}^-]}{[\text{Nic}^-]} = K_b$$

This gives

$$\frac{x^2}{0.10 - x} = 7.1 \times 10^{-10}$$

Solving this equation, we find that $x = [\text{OH}^-] = 8.4 \times 10^{-6}$. Hence,

$$\text{pH} = 14.00 - \text{pOH} = 14.00 + \log(8.4 \times 10^{-6}) = \mathbf{8.92}$$

As expected, the solution has a pH greater than 7.00.

Exercise 16.15

Benzoic acid, $HC_7H_5O_2$, and its salts are used as food preservatives. What is the concentration of benzoic acid in an aqueous solution of 0.015 *M* sodium benzoate? What is the pH of the solution? K_a for benzoic acid is 6.3×10^{-5}.

(See Problems 16.81, 16.82, 16.83, and 16.84.)

Solutions of a Weak Acid or Base with Another Solute

In the preceding sections, we looked at solutions that contained either a weak acid, a weak base, or a salt of a weak acid or base. In the remaining sections of this chapter, we will look at the effect of adding another solute to a solution of a weak acid or base. The solutes we will look at are those that significantly affect acid or base ionization—that is, strong acids or bases and salts that contain an ion that is produced in the acid or base ionization. These solutes affect the equilibrium through the *common-ion effect,* which we discuss in the next section.

16.8 COMMON-ION EFFECT

The **common-ion effect** is *the shift in an ionic equilibrium caused by the addition of a solute that provides an ion that takes part in the equilibrium.* To understand this definition, consider a solution of acetic acid, $HC_2H_3O_2$, in which we have the following acid-ionization equilibrium.

$$HC_2H_3O_2(aq) + H_2O(l) \rightleftharpoons C_2H_3O_2^-(aq) + H_3O^+(aq)$$

Suppose we add HCl(aq) to this solution. What is the effect of this addition on the acid-ionization equilibrium? Because HCl(aq) is a strong acid, it provides H_3O^+ ion, which is present on the right side of the equation for acetic acid ionization. According to Le Chatelier's principle, the equilibrium composition should shift to the left. ●

● Le Chatelier's principle was applied in Section 15.7 to the problem of adding substances to an equilibrium mixture.

$$HC_2H_3O_2(aq) + H_2O(l) \longleftarrow C_2H_3O_2^-(aq) + \underset{\text{Added}}{H_3O^+(aq)}$$

Thus, the degree of ionization of acetic acid is decreased by the addition of a strong acid. This repression of the ionization of acetic acid by $HCl(aq)$ is an example of the common-ion effect. The next example illustrates quantitatively this repression of the ionization.

Example 16.12 Calculating the Common-Ion Effect on Acid Ionization (Effect of a Strong Acid)

The degree of ionization of acetic acid, $HC_2H_3O_2$, in a 0.10 *M* aqueous solution at 25°C is 0.013. K_a at this temperature is 1.7×10^{-5}. Calculate the degree of ionization of $HC_2H_3O_2$ in a 0.10 *M* solution at 25°C to which sufficient HCl is added to make it 0.010 *M* HCl. How is the degree of ionization affected?

Solution

STEP 1 Starting concentrations are $[HC_2H_3O_2] = 0.10\ M$, $[H^+] = 0.010\ M$ (from HCl), and $[C_2H_3O_2^-] = 0$. The acetic acid ionizes to give an additional x mol/L of H^+ and x mol/L of $C_2H_3O_2^-$. The table is

CONCENTRATION (*M*)	$HC_2H_3O_2(aq)$ $\rightleftharpoons$	$H^+(aq)$ +	$C_2H_3O_2^-(aq)$
Starting	0.10	0.010	0
Change	$-x$	$+x$	$+x$
Equilibrium	$0.10 - x$	$0.010 + x$	x

STEP 2 We substitute into the equilibrium equation

$$\frac{[H^+][C_2H_3O_2^-]}{[HC_2H_3O_2]} = K_a$$

obtaining

$$\frac{(0.010 + x)x}{0.10 - x} = 1.7 \times 10^{-5}$$

STEP 3 To solve this equation, let us assume that x is small compared to 0.010. Then

$$0.010 + x \simeq 0.010$$

$$0.10 - x \simeq 0.10$$

The equation becomes

$$\frac{0.010x}{0.10} \simeq 1.7 \times 10^{-5}$$

Solving for x, we get

$$x = 1.7 \times 10^{-5} \times \frac{0.10}{0.010} = 1.7 \times 10^{-4}$$

(Check that x can indeed be neglected in $0.010 + x$ and $0.10 - x$.) The degree of ionization of $HC_2H_3O_2$ is $x/0.10 = 0.0017$. This is much smaller than the value for 0.10 *M* $HC_2H_3O_2$ (0.013), because **the addition of HCl represses the ionization of $HC_2H_3O_2$.**

Exercise 16.16

What is the concentration of formate ion, CHO_2^-, in a solution at 25°C that is 0.10 *M* $HCHO_2$ and 0.20 *M* HCl? What is the degree of ionization of formic acid, $HCHO_2$?

(See Problems 16.85 and 16.86.)

As we will see in the following section, solutions that contain a weak acid or a weak base and a corresponding salt are especially important. Therefore, we will need to be able to calculate the concentrations of species present in such solutions. A solution of acetic acid and sodium acetate is an example. Note that the acetate ion represses the ionization of acetic acid by the common-ion effect, just as the addition of H^+ ion did. Therefore, the pH of an acetic acid solution is raised by adding sodium acetate. (We can also look at this as the addition of a base, $NaC_2H_3O_2$, to an acid, $HC_2H_3O_2$, which of course raises the pH of the acid solution.) The next example shows how we can calculate the concentrations of species in such a solution.

Example 16.13 Calculating the Common-Ion Effect on Acid Ionization (Effect of a Conjugate Base)

A solution is prepared to be 0.10 *M* acetic acid, $HC_2H_3O_2$, and 0.20 *M* sodium acetate, $NaC_2H_3O_2$. What is the pH of this solution at 25°C? K_a for acetic acid at 25°C is 1.7×10^{-5} *M*.

Solution

STEP 1 We consider the equilibrium

$$HC_2H_3O_2(aq) \rightleftharpoons H^+(aq) + C_2H_3O_2^-(aq)$$

Initially, 1 L of solution contains 0.10 mol of acetic acid. Sodium acetate is a strong electrolyte, so 1 L of solution contains 0.20 mol of acetate ion. When the acetic acid ionizes, it gives *x* mol of hydrogen ion and *x* mol of acetate ion. This is summarized in the table below.

STEP 2 The equilibrium equation is

$$\frac{[H^+][C_2H_3O_2^-]}{[HC_2H_3O_2]} = K_a$$

Substituting into this equation gives

$$\frac{x(0.20 + x)}{0.10 - x} = 1.7 \times 10^{-5}$$

STEP 3 To solve the equation, let us assume that *x* is small compared with 0.10 and 0.20. Then

$$0.20 + x \simeq 0.20$$
$$0.10 - x \simeq 0.10$$

The equilibrium equation becomes

$$\frac{x(0.20)}{0.10} \simeq 1.7 \times 10^{-5}$$

Hence,

$$x \simeq 1.7 \times 10^{-5} \times \frac{0.10}{0.20} = 8.5 \times 10^{-6}$$

(Note that *x* is indeed much smaller than 0.10 or 0.20.) Thus, the hydrogen-ion concentration is 8.5×10^{-6} *M*, and

$$pH = -\log [H^+] = -\log(8.5 \times 10^{-6}) = \mathbf{5.07}$$

For comparison, the pH of 0.10 *M* acetic acid is 2.88.

CONCENTRATION (*M*)	$HC_2H_3O_2(aq)$ $\rightleftharpoons$	$H^+(aq)$ +	$C_2H_3O_2^-(aq)$
Starting	0.10	~0	0.20
Change	$-x$	$+x$	$+x$
Equilibrium	$0.10 - x$	x	$0.20 + x$

Exercise 16.17

One liter of solution was prepared by dissolving 0.025 mol of formic acid, $HCHO_2$, and 0.018 mol of sodium formate, $NaCHO_2$, in water. What was the pH of the solution? K_a for formic acid is 1.7×10^{-4}. (See Problems 16.87 and 16.88.)

The previous examples considered the common-ion effect in solutions of weak acids. We also encounter the common-ion effect in solutions of weak bases. For example, the base ionization of NH_3 is

$$NH_3(aq) + H_2O(l) \rightleftharpoons NH_4^+(aq) + OH^-(aq)$$

This suggests that we could repress, or decrease the effect, of the base ionization by adding an ammonium salt (providing NH_4^+ ion) or a strong base (providing OH^- ion). The calculations are similar to those for acid ionization, except that we use the appropriate base ionization.

16.9 BUFFERS

A **buffer** is *a solution characterized by the ability to resist changes in pH when limited amounts of acid or base are added to it.* If 0.01 mol of hydrochloric acid is

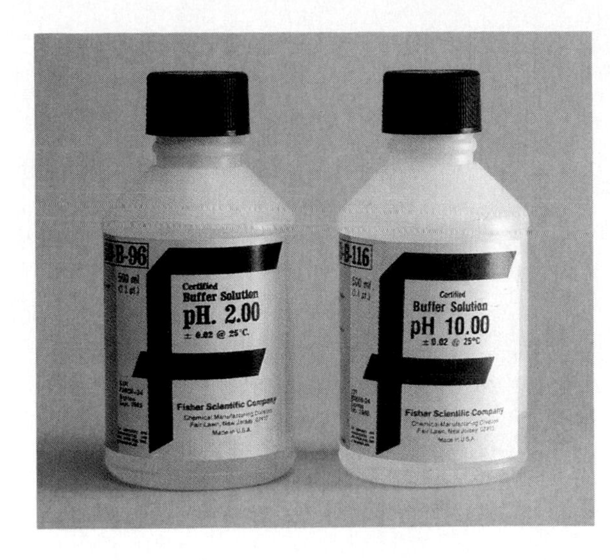

Figure 16.13
Some laboratory buffers. These are commercially prepared buffer solutions.

added to 1 L of pure water, the pH changes from 7.0 to 2.0—a pH change of 5.0 units. By contrast, the addition of this amount of hydrochloric acid to 1 L of buffered solution might change the pH by only 0.1 unit. Biological fluids, such as blood, are usually buffer solutions because the control of pH is vital to their proper functioning. The oxygen-carrying function of blood depends on its being maintained very near a pH of 7.4. If the pH were to change by a tenth of a unit, the capacity of the blood to carry oxygen would be lost.

Buffers contain either a weak acid and its conjugate base or a weak base and its conjugate acid. Thus, a buffer solution contains both an acid species and a base species in equilibrium. To understand the action of a buffer, consider one that contains approximately equal molar amounts of a weak acid HA and its conjugate base A^-. When a strong acid is added to the buffer, it supplies hydrogen ions that react with the base A^-.

$$H^+(aq) + A^-(aq) \longrightarrow HA(aq)$$

On the other hand, when a strong base is added to the buffer, it supplies hydroxide ions. These react with the acid HA.

$$OH^-(aq) + HA(aq) \longrightarrow H_2O(l) + A^-(aq)$$

Thus, a buffer solution resists changes in pH through its ability to combine with both H^+ and OH^- ions.

Blood, as we mentioned, is a buffer solution. It contains H_2CO_3 and HCO_3^-, as well as other conjugate acid–base pairs. A buffer frequently used in the laboratory contains the conjugate acid–base pair $H_2PO_4^-$ and HPO_4^{2-} (Figure 16.13).

Buffers also have commercial applications. For example, the label on a package of artificial fruit-juice mix says that it contains "citric acid to provide tartness and sodium citrate to regulate tartness." A solution of citric acid and its base conjugate, citrate ion (provided by sodium citrate), functions as an acid–base buffer, which is what "to regulate tartness" means. The pH of the buffer is in the acid range.

Two important characteristics of a buffer are its pH and its *buffer capacity,* which is the amount of acid or base the buffer can react with before giving a significant pH change. Buffer capacity depends on the amount of acid and conju-

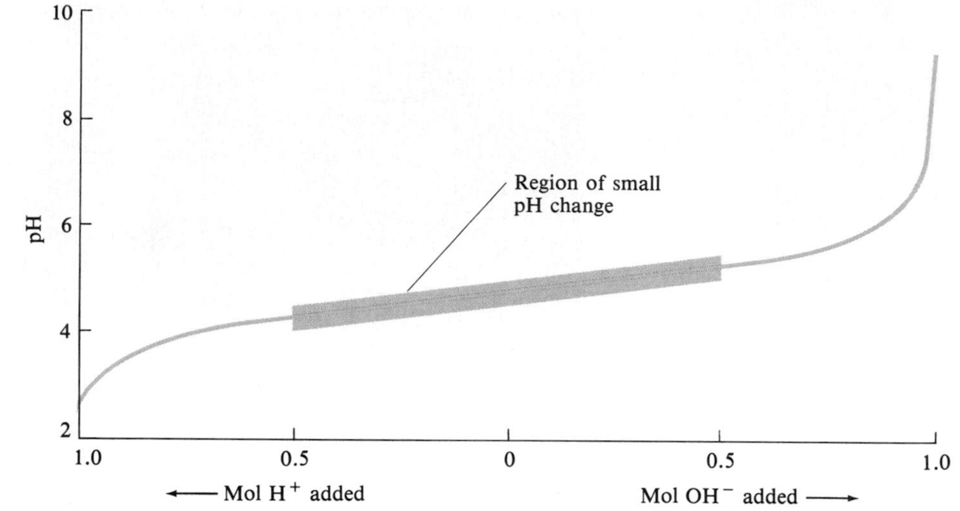

Figure 16.14
Effect of added acid or base on a buffer solution. The buffer contains 1.0 mol acetic acid and 1.0 mol acetate ion in 1.00 L of solution. Note that the addition of 0.5 mol or less of strong acid or base gives only a small change of pH.

gate base in the solution. Figure 16.14 illustrates the change in pH of a buffer solution containing 1.0 mol acetic acid and 1.0 mol acetate ion to which varying amounts of H^+ and OH^- are added. This buffer changes less than 0.5 pH units as long as no more than 0.5 mol of H^+ or OH^- ion is added. Note that this is one-half (or less than one-half) the amounts of acid and conjugate base in the solution. The ratio of amounts of acid and conjugate base is also important. Unless this ratio is close to 1 (between 1:10 and 10:1), the buffer capacity will be too low to be useful.

The other important characteristic of a buffer is its pH. Let us now look at how to calculate the pH of a buffer.

The pH of a Buffer

The solution described in Example 16.13 in the previous section is a buffer, because it is a solution of a weak acid (0.10 *M* acetic acid) and its conjugate base (0.20 *M* acetate ion from sodium acetate). This example described how to calculate the pH of such a buffer (we obtained a pH of 5.07). The next example is similar, except now you are given a recipe for making up a buffer from volumes of solutions and you are asked to calculate the pH. In this example, the buffer consists of a molecular base (NH_3) and its conjugate acid (NH_4^+).

Example 16.14 Calculating the pH of a Buffer from Given Volumes of Solution

Instructions for making up a buffer say to mix 60.0 mL of 0.100 *M* NH_3 with 40.0 mL of 0.100 *M* NH_4Cl. What is the pH of this buffer?

Solution

The buffer contains a base and its conjugate acid in equilibrium. The equation is

$$NH_3(aq) + H_2O(l) \rightleftharpoons NH_4^+(aq) + OH^-(aq)$$

The problem is to obtain the concentration of H^+ or OH^- in this equilibrium mixture. To do the equilibrium calculation, we need the starting concentrations in the solution obtained by mixing the NH_3 and NH_4Cl solutions. For this, we calculate the moles of NH_3 and moles of NH_4^+ added to the buffer solution, then divide by the total volume of buffer.

How many moles of NH_3 are added? Recall that

$$\text{Molarity } NH_3 = \frac{\text{moles } NH_3}{\text{liters } NH_3 \text{ solution}}$$

Note that the instructions say to add 60.0 mL (or 0.060 L) of 0.100 M NH_3. So

$$\text{Moles } NH_3 = \text{molarity } NH_3 \times \text{liters } NH_3 \text{ solution}$$

$$= 0.100 \frac{\text{mol}}{\text{L}} \times 0.060 \text{ L} = 0.0060 \text{ mol } NH_3$$

In the same way, we find that we have added 0.0040 mol NH_4^+ (from NH_4Cl). We assume that the total volume of buffer equals the sum of the volumes of the two solutions.

$$\text{Total volume buffer} = 60.0 \text{ mL} + 40.0 \text{ mL}$$

$$= 100.0 \text{ mL } (0.1000 \text{ L})$$

Therefore, the concentrations of base and conjugate acid are

$$[NH_3] = \frac{0.0060 \text{ mol}}{0.1000 \text{ L}} = 0.060\ M$$

$$[NH_4^+] = \frac{0.0040 \text{ mol}}{0.1000 \text{ L}} = 0.040\ M$$

Now we are ready to do the equilibrium calculation.

STEP 1 We fill in the concentration table for the acid–base equilibrium (base ionization of NH_3) as shown below.

STEP 2 We substitute the equilibrium concentrations into the equilibrium equation.

$$\frac{[NH_4^+][OH^-]}{[NH_3]} = K_b$$

$$\frac{(0.040 + x)x}{0.060 - x} = 1.8 \times 10^{-5}$$

STEP 3 To solve this equation, we assume that x is small compared with 0.040 and 0.060. So this equation becomes

$$\frac{0.040x}{0.060} = 1.8 \times 10^{-5}$$

Therefore, $x = (0.060/0.040) \times (1.8 \times 10^{-5}) = 2.7 \times 10^{-5}$. (Check that x can be neglected in $0.040 + x$ and $0.060 - x$.) Thus, the hydroxide-ion concentration is $2.7 \times 10^{-5}\ M$. The pH of the buffer is

$$\text{pH} = 14.00 - \text{pOH} = 14.00 + \log(2.7 \times 10^{-5}) = \mathbf{9.43}$$

(If you have 2.7×10^{-5} in your calculator from the previous calculation, you obtain the pH by pressing the *log* button, then adding 14.)

CONCENTRATION (M)	$NH_3(aq)$ + $H_2O(l)$ $\rightleftharpoons$	$NH_4^+(aq)$ +	$OH^-(aq)$
Starting	0.060	0.040	0
Change	$-x$	$+x$	$+x$
Equilibrium	$0.060 - x$	$0.040 + x$	x

Exercise 16.18

What is the pH of a buffer prepared by adding 30.0 mL of 0.15 M $HC_2H_3O_2$ (acetic acid) to 70.0 mL of 0.20 M $NaC_2H_3O_2$ (sodium acetate)? (See Problems 16.91 and 16.92.)

Adding an Acid or a Base to a Buffer

We noted earlier that a buffer resists changes in pH, and we also described qualitatively how this occurs. What we want to do here is to show quantitatively that a buffer does indeed tend to resist pH change when a small quantity of an acid or a base is added to it.

In all of the previous acid–base examples, we dealt with situations in which an acid and its conjugate base were close to equilibrium. In the following example, however, we add a strong acid to an acetic acid/acetate ion buffer. Hydrogen ion reacts with the conjugate base (acetate ion) in the buffer, so that the concentrations of species in solution change markedly from their initial value.

Therefore, we do the problem in two parts. First, we assume that the H^+ ion from the strong acid and the conjugate base from the buffer react completely. This is a stoichiometric calculation. Actually, the H^+ ion and the base from the buffer reach equilibrium just before complete reaction. So we now solve the equilibrium problem using concentrations from the stoichiometric calculation. Because these concentrations are not far from equilibrium, we can use our usual simplifying assumption about x. The next example illustrates this type of problem.

Example 16.15 Calculating the pH of a Buffer When a Strong Acid or Strong Base Is Added

Calculate the pH of 75 mL of the buffer solution described in Example 16.13 (0.10 M $HC_2H_3O_2$ and 0.20 M $NaC_2H_3O_2$) to which 9.5 mL of 0.10 M hydrochloric acid is added. Compare the pH change with what would occur if this amount of acid were added to pure water.

Solution

When hydrogen ion (from hydrochloric acid) is added to the buffer, it reacts with acetate ion.

$$H^+(aq) + C_2H_3O_2^- \longrightarrow HC_2H_3O_2(aq)$$

Because acetic acid is a weak acid, we can assume as a first approximation that the reaction goes to completion. This part of the problem is simply a stoichiometric calculation. Then we assume that the acetic acid ionizes slightly. This part of the problem involves an acid-ionization equilibrium.

STOICHIOMETRIC CALCULATION We must first calculate the amounts of hydrogen ion, acetate ion, and acetic acid present in the solution before reaction. The molar amount of hydrogen ion will equal the molar amount of hydrochloric acid added, which we obtain by converting the volume of hydrochloric acid, HCl, to moles of HCl. To do this, we note for 0.10 M HCl that

$$1 \text{ L HCl} \simeq 0.10 \text{ mol HCl}$$

Hence, to convert 9.5 mL ($= 9.5 \times 10^{-3}$ L) of hydrochloric acid, we have

$$9.5 \times 10^{-3} \cancel{\text{L HCl}} \times \frac{0.10 \text{ mol HCl}}{1 \cancel{\text{L HCl}}} \simeq 0.00095 \text{ mol HCl}$$

Hydrochloric acid is a strong acid, so it exists in solution as the ions H^+ and Cl^-. Thus, the amount of H^+ added is 0.00095 mol.

The amounts of acetate ion and acetic acid are found in a similar way. The buffer in Example 16.13 contains 0.20 mol acetate ion and 0.10 mol acetic acid in 1 L of solution. The amounts in 75 mL ($= 0.075$ L) of solution are obtained by converting to moles.

$$0.075 \cancel{\text{L soln}} \times \frac{0.20 \text{ mol } C_2H_3O_2^-}{1 \cancel{\text{L soln}}} \simeq 0.015 \text{ mol } C_2H_3O_2^-$$

$$0.075 \cancel{\text{L soln}} \times \frac{0.10 \text{ mol } HC_2H_3O_2}{1 \cancel{\text{L soln}}} \simeq 0.0075 \text{ mol } HC_2H_3O_2$$

We now assume that all of the hydrogen ion added (0.00095 mol) reacts with acetate ion. Therefore, 0.00095 mol of acetic acid is produced and 0.00095 mol of acetate ion is used up. Hence, after reaction we have

$$\text{Moles of acetate ion} = (0.015 - 0.00095) \text{ mol } C_2H_3O_2^- = 0.014 \text{ mol } C_2H_3O_2^-$$

$$\text{Moles of acetic acid} = (0.0075 + 0.00095) \text{ mol } HC_2H_3O_2 = 0.0084 \text{ mol } HC_2H_3O_2$$

EQUILIBRIUM CALCULATION We first calculate the concentrations of $HC_2H_3O_2$ and $C_2H_3O_2^-$ present in the solution before we consider the acid ionization equilibrium. Note that the total volume of solution (buffer plus hydrochloric acid) is 75 mL + 9.5 mL, or 84 mL (0.084 L). Hence, the starting concentrations are

$$[HC_2H_3O_2] = \frac{0.0084 \text{ mol}}{0.084 \text{ L}} = 0.10\ M$$

$$[C_2H_3O_2^-] = \frac{0.014 \text{ mol}}{0.084 \text{ L}} = 0.17\ M$$

From this, we construct the following table:

CONCENTRATION (M)	$HC_2H_3O_2(aq)$ $\rightleftharpoons$	$H^+(aq)$ +	$C_2H_3O_2^-(aq)$
Starting	0.10	~0	0.17
Change	$-x$	$+x$	$+x$
Equilibrium	$0.10 - x$	x	$0.17 + x$

The equilibrium equation is

$$\frac{[H^+][C_2H_3O_2^-]}{[HC_2H_3O_2]} = K_a$$

Substituting, we get

$$\frac{x(0.17 + x)}{0.10 - x} = 1.7 \times 10^{-5}$$

If we assume that x is small enough that $0.17 + x \simeq 0.17$ and $0.10 - x \simeq 0.10$, this equation becomes

$$\frac{x(0.17)}{0.10} = 1.7 \times 10^{-5}$$

or

$$x = 1.7 \times 10^{-5} \times \frac{0.10}{0.17} = 1.0 \times 10^{-5}$$

Note that x is indeed small, so the assumptions we made earlier are correct. Thus, the H^+ ion concentration is 1.0×10^{-5} M. The pH is

$$\text{pH} = -\log [H^+] = -\log(1.0 \times 10^{-5}) = 5.00$$

Because the pH of the buffer was 5.07 (see Example 16.13), the pH has changed by $5.07 - 5.00 = 0.07$ units.

ADDING HCl TO PURE WATER If 9.5 mL of 0.10 M hydrochloric acid were added to 75 mL of pure water, the hydrogen-ion concentration would change to

$$[H^+] = \frac{\text{amount of } H^+ \text{ added}}{\text{total volume of solution}}$$

$$[H^+] = \frac{0.00095 \text{ mol } H^+}{0.084 \text{ L solution}} = 0.011\ M$$

(The total volume is 75 mL of water plus 9.5 mL HCl, assuming no change of volume on mixing.) The pH is

$$pH = -\log [H^+] = -\log 0.011 = 1.96$$

The pH of pure water is 7.00, so the change in pH is 7.00 − 1.96 = 5.04 units, compared with 0.07 units for the buffered solution.

Exercise 16.19

What is the pH of the solution described in Exercise 16.17 if 50.0 mL of 0.10 M sodium hydroxide is added to 1 L of solution? (See Problems 16.93 and 16.94.)

Suppose we had added a strong base to the buffer, instead of a strong acid. The stoichiometric part of the calculation would depend on the reaction

$$OH^-(aq) + HC_2H_3O_2(aq) \longrightarrow H_2O(l) + C_2H_3O_2^-(aq)$$

The results of this calculation would give the concentrations of $HC_2H_3O_2$ and $C_2H_3O_2^-$, which we would then use in an equilibrium calculation, following the same procedure as in the previous example.

Henderson–Hasselbalch Equation

How do we prepare a buffer of given pH? We can show that the buffer must be prepared from a conjugate acid–base pair in which the acid ionization constant is approximately equal to the desired H^+ ion concentration. To illustrate, consider a buffer made up of a weak acid HA and its conjugate base A^-. The acid-ionization equilibrium is

$$HA(aq) \rightleftharpoons H^+(aq) + A^-(aq)$$

and the acid-ionization constant is

$$K_a = \frac{[H^+][A^-]}{[HA]}$$

By rearranging this, we can get an equation for the H^+ ion concentration.

$$[H^+] = K_a \times \frac{[HA]}{[A^-]}$$

This equation expresses the H^+ ion concentration in terms of K_a for the acid and the ratio of concentrations of HA and A^-. This equation was derived from the equilibrium constant, so the concentrations of HA and A^- should be equilibrium values. But because the presence of A^- represses the ionization of HA, these concentrations do not differ significantly from the values used to prepare the buffer. If [HA] and [A^-] are approximately equal, the hydrogen-ion concentration of the buffer is approximately equal to K_a.

We can use the preceding equation to derive an equation for the pH of a buffer. Let us take the negative logarithm of both sides of the equation. That is,

$$-\log [H^+] = -\log \left(K_a \times \frac{[HA]}{[A^-]}\right) = -\log K_a - \log \frac{[HA]}{[A^-]}$$

The left side equals the pH. We can also simplify the right side. The pK_a of a weak acid is defined in a manner similar to pH and pOH.●

● Acid- and base-ionization constants are often listed as pK_a and $pK_b(= -\log K_b)$.

$$pK_a = -\log K_a$$

The previous equation can be written

$$pH = pK_a - \log \frac{[HA]}{[A^-]} = pK_a + \log \frac{[A^-]}{[HA]}$$

More generally, we can write

$$pH = pK_a + \log \frac{[base]}{[acid]}$$

This is *an equation relating the pH of a buffer for different concentrations of conjugate acid and base;* it is known as the **Henderson–Hasselbalch equation.** By substituting the value of pK_a for the conjugate acid and the ratio [base]/[acid], we obtain the pH of the buffer.

The question we asked earlier was how to prepare a buffer of a given pH—for example, pH 4.90. We can see that we need to find a conjugate acid–base pair in which the pK_a of the acid is close to the desired pH. Thus, K_a for acetic acid is 1.7×10^{-5}, and its pK_a is $-\log(1.7 \times 10^{-5}) = 4.77$. We can get a pH somewhat higher by increasing the ratio [base]/[acid].

Consider the calculation of the pH of a buffer containing 0.10 *M* NH_3 and 0.20 *M* NH_4Cl. The conjugate acid is NH_4^+, whose K_a we can calculate from K_b for NH_3 ($= 1.8 \times 10^{-5}$). K_a for NH_4^+ is 5.6×10^{-10}, and the pK_a is $-\log(5.6 \times 10^{-10}) = 9.25$. Hence,

$$pH = 9.25 + \log \frac{0.10}{0.20} = 8.95$$

The solution is basic, as we should have guessed, because the buffer contains the weak base NH_3 and the very weak acid NH_4^+.

16.10 ACID–BASE TITRATION CURVES

An acid–base titration is a procedure for determining the amount of acid (or base) in a solution by determining the volume of base (or acid) of known concentration that will completely react with it.● An **acid–base titration curve** is *a plot of the pH of a solution of acid (or base) against the volume of added base (or acid).* Such curves are used to gain insight into the titration process. We can use the titration curve to choose an indicator that will show when the titration is complete.

● The technique of titration was discussed in Section 4.11.

Titration of a Strong Acid by a Strong Base

Figure 16.15 shows a curve for the titration of 25.0 mL of 0.100 *M* HCl by 0.100 *M* NaOH. Note that the pH changes slowly at first until the molar amount of base added nearly equals that of the acid—that is, until the titration is near the equivalence point. The **equivalence point** is *the point in a titration when a stoichiometric amount of reactant has been added.* At the equivalence point, the pH of this solution of NaOH and HCl is 7.0 because it contains a salt, NaCl, that does not hydrolyze. However, the pH changes rapidly near the equivalence point, from a pH of about 3 to a pH of about 11. To detect the equivalence point, we add an

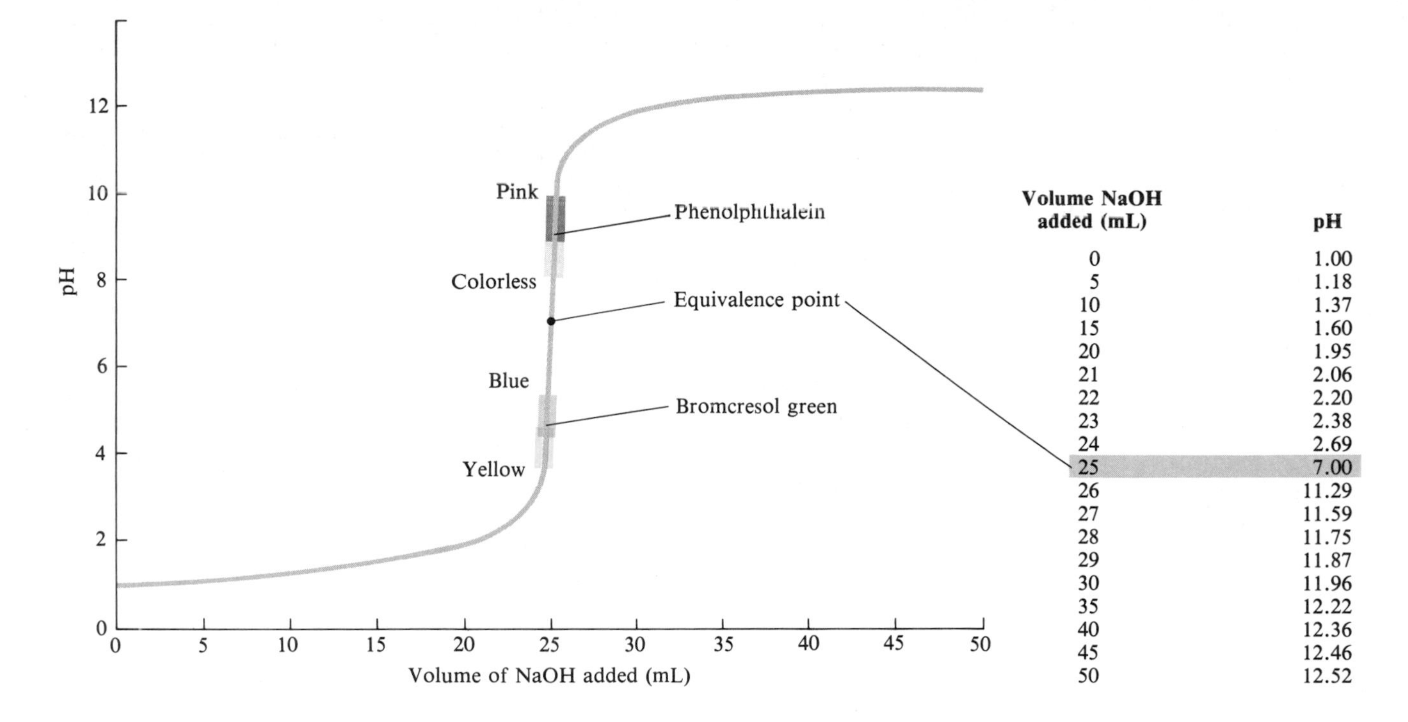

Volume NaOH added (mL)	pH
0	1.00
5	1.18
10	1.37
15	1.60
20	1.95
21	2.06
22	2.20
23	2.38
24	2.69
25	7.00
26	11.29
27	11.59
28	11.75
29	11.87
30	11.96
35	12.22
40	12.36
45	12.46
50	12.52

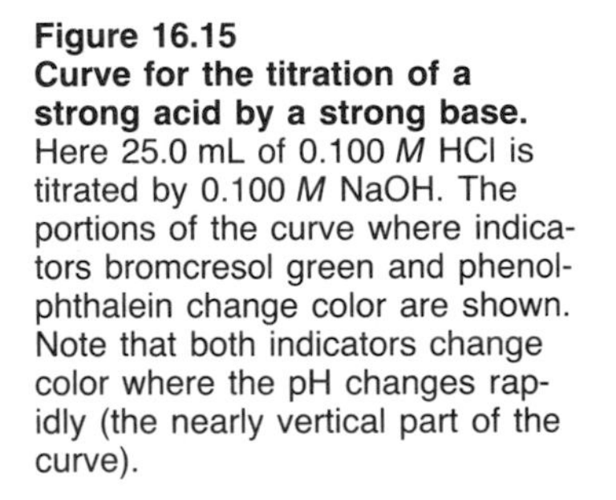

Figure 16.15
Curve for the titration of a strong acid by a strong base. Here 25.0 mL of 0.100 *M* HCl is titrated by 0.100 *M* NaOH. The portions of the curve where indicators bromcresol green and phenolphthalein change color are shown. Note that both indicators change color where the pH changes rapidly (the nearly vertical part of the curve).

indicator that changes color within the pH range 3–11. Phenolphthalein can be used, because it changes from colorless to pink in the pH range 8.2–10.0. (Figure 16.3 shows the pH ranges for the color changes of indicators.) Even though this color change occurs on the basic side, only a fraction of a drop of base is required to change the pH several units when the titration is near the equivalence point. The indicator bromcresol green, whose color changes in the pH range 3.8 to 5.4, would also work. Because the pH change is so large, many other indicators could be used.

The following example shows how to calculate a point on the titration curve of a strong acid and a strong base.

Example 16.16 Calculating the pH of a Solution of a Strong Acid and a Strong Base

Calculate the pH of a solution in which 10.0 mL of 0.100 *M* NaOH is added to 25.0 mL of 0.100 *M* HCl.

Solution

Because the reactants are strong acid and strong base, the reaction is essentially complete. The equation is

$$H^+(aq) + OH^-(aq) \longrightarrow H_2O(l)$$

We get the amounts of reactants by multiplying the volume (in liters) of each solution by its molar concentration.

$$\text{Mol } H^+ = 0.0250 \text{ L} \times 0.100 \text{ mol/L} = 0.00250 \text{ mol}$$

$$\text{Mol } OH^- = 0.0100 \text{ L} \times 0.100 \text{ mol/L} = 0.00100 \text{ mol}$$

All of the OH^- reacts, leaving an excess of H^+.

$$\text{Excess } H^+ = (0.00250 - 0.00100) \text{ mol} = 0.00150 \text{ mol } H^+$$

We obtain the concentration of H^+ by dividing this amount of H^+ by the total volume of solution (= 0.0250 L + 0.0100 L = 0.0350 L).

$$[H^+] = \frac{0.00150 \text{ mol}}{0.0350 \text{ L}} = 0.0429$$

Hence, pH = $-\log [H^+] = -\log(0.0429) =$ **1.368**

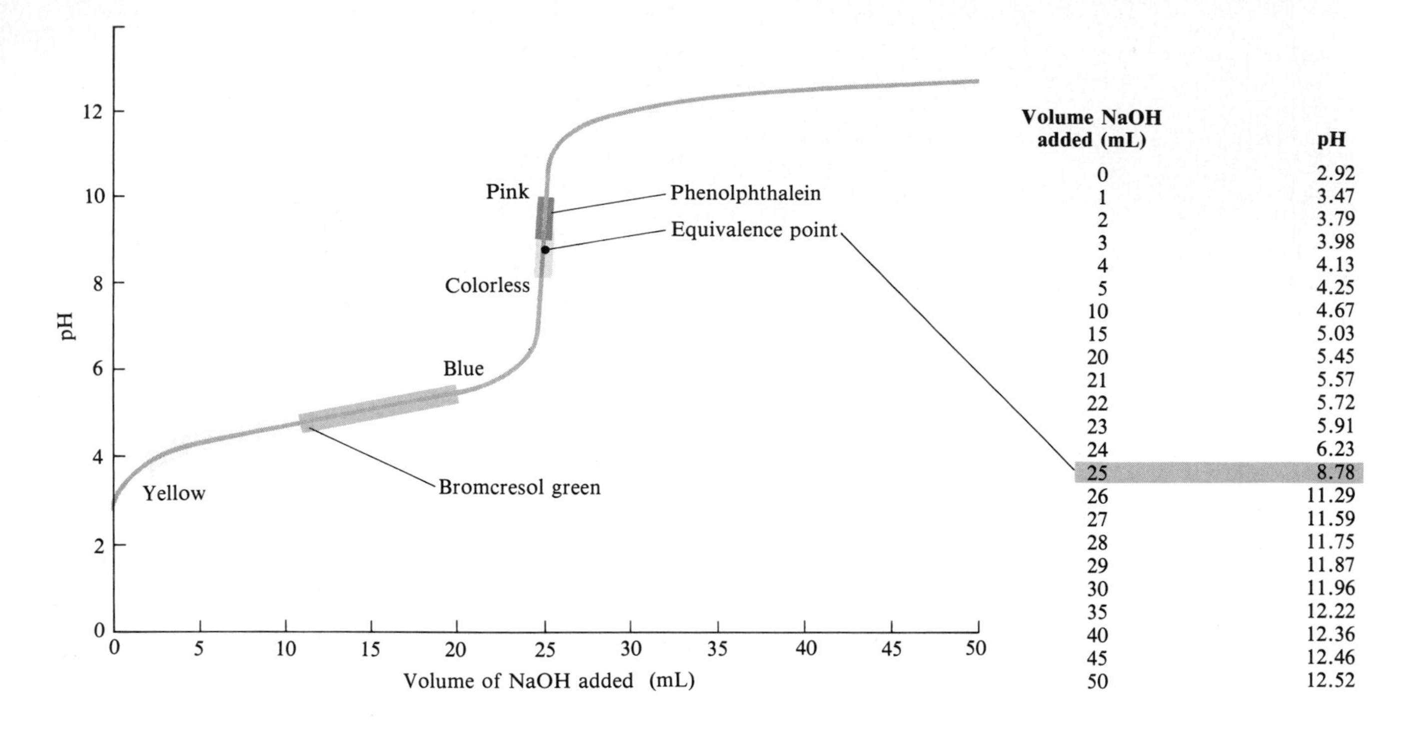

Volume NaOH added (mL)	pH
0	2.92
1	3.47
2	3.79
3	3.98
4	4.13
5	4.25
10	4.67
15	5.03
20	5.45
21	5.57
22	5.72
23	5.91
24	6.23
25	8.78
26	11.29
27	11.59
28	11.75
29	11.87
30	11.96
35	12.22
40	12.36
45	12.46
50	12.52

Figure 16.16
Curve for the titration of a weak acid by a strong base. Here 25.0 mL of 0.100 *M* nicotinic acid, a weak acid, is titrated by 0.100 *M* NaOH. Note that bromcresol green changes color during the early part of the titration, well before the equivalence point. Phenolphthalein changes color where the pH changes rapidly (near the equivalence point). Thus, phenolphthalein could be used as an indicator for the titration, whereas bromcresol green could not.

Exercise 16.20

What is the pH of a solution in which 15 mL of 0.10 *M* NaOH has been added to 25 mL of 0.10 *M* HCl? (See Problems 16.101 and 16.102.)

Titration of a Weak Acid by a Strong Base

The titration of a weak acid by a strong base gives a somewhat different curve. Figure 16.16 shows the curve for the titration of 25.0 mL of 0.100 *M* nicotinic acid, $HC_6H_4NO_2$, by 0.100 *M* NaOH. The titration starts at a higher pH than the titration of HCl because nicotinic acid is a weak acid. As before, the pH changes slowly at first, then rapidly near the equivalence point. The pH range in which the rapid change is seen occurs from about pH 7 to pH 11. Note that the pH range is shorter than that for the titration of a strong acid by a strong base. This means that the choice of an indicator is more critical. Phenolphthalein would work; it changes color in the range 8.2–10.0. Bromcresol green would not work. This indicator changes color in the range 3.8–5.4, which occurs before the titration curve rises steeply.

Note also that the equivalence point for the titration curve of nicotinic acid occurs on the basic side. This happens because at the equivalence point, the solution is that of the salt, sodium nicotinate, which is basic from the hydrolysis of the nicotinate ion. The optimum choice of indicator would be one that changes color over a range that includes the pH of the equivalence point.

The following example shows how to calculate the pH at the equivalence point in the titration of a weak acid and a strong base.

Example 16.17 Calculating the pH at the Equivalence Point in the Titration of a Weak Acid by a Strong Base

Calculate the pH of the solution at the equivalence point when 25 mL of 0.10 M nicotinic acid is titrated by 0.10 M sodium hydroxide. K_a for nicotinic acid equals 1.4×10^{-5}.

Solution

At the equivalence point, equal molar amounts of nicotinic acid and sodium hydroxide react to give a solution of sodium nicotinate. We first calculate the concentration of nicotinate ion (stoichiometry problem). Then we find the pH of this solution (hydrolysis problem).

CONCENTRATION OF NICOTINATE ION We assume that the reaction of the base with the acid is complete. In this case, 25 mL of 0.10 M sodium hydroxide is needed to react with 25 mL of 0.10 M nicotinic acid. The molar amount of nicotinate ion formed equals the initial molar amount of nicotinic acid.

$$25 \times 10^{-3}\ \cancel{\text{L soln}} \times \frac{0.10 \text{ mol nicotinate ion}}{1\ \cancel{\text{L soln}}} = 2.5 \times 10^{-3} \text{ mol nicotinate ion}$$

The total volume of solution is 50 mL (25 mL NaOH solution plus 25 mL of nicotinic acid solution, assuming that there is no change in volume on mixing). Dividing the molar amount of nicotinate ion by the volume of solution in liters gives the molar concentration of nicotinate ion.

$$\text{Molar concentration} = \frac{2.5 \times 10^{-3} \text{ mol}}{50 \times 10^{-3} \text{ L}} = 0.050\ M$$

HYDROLYSIS OF NICOTINATE ION This portion of the calculation follows the method given in Example 16.11. We find that K_b for nicotinate ion is 7.1×10^{-10} and that the concentration of hydroxide ion is $6.0 \times 10^{-6}\ M$. The pH is **8.78**. Try working through the calculation to verify these numbers.

Exercise 16.21

What is the pH at the equivalence point when 25 mL of 0.10 M HF is titrated by 0.15 M NaOH?

(See Problems 16.103 and 16.104.)

Figure 16.17
Curve for the titration of a weak base by a strong acid. Here 25.0 mL of 0.100 M NH_3 is titrated by 0.100 M HCl. Methyl red can be used as an indicator for the titration.

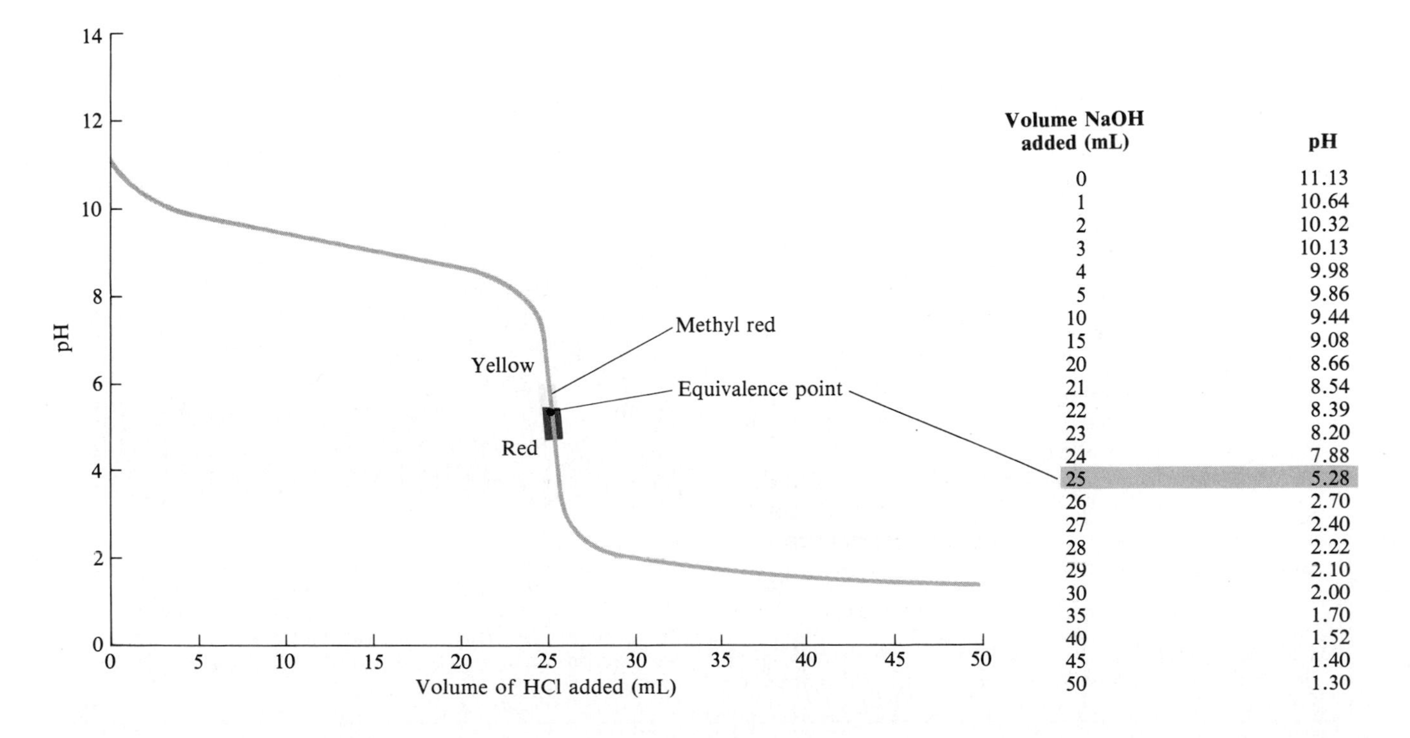

Volume NaOH added (mL)	pH
0	11.13
1	10.64
2	10.32
3	10.13
4	9.98
5	9.86
10	9.44
15	9.08
20	8.66
21	8.54
22	8.39
23	8.20
24	7.88
25	5.28
26	2.70
27	2.40
28	2.22
29	2.10
30	2.00
35	1.70
40	1.52
45	1.40
50	1.30

Titration of a Weak Base by a Strong Acid

When we titrate a weak base by a strong acid, we get a titration curve similar to that obtained when a weak acid is titrated by a strong base. Figure 16.17 shows the pH changes during the titration of 25.0 mL of 0.100 *M* NH_3 by 0.100 *M* HCl. In this case, the pH declines slowly at first, then falls abruptly from about pH 7 to pH 3. Methyl red, which changes color from yellow at pH 6.0 to red at pH 4.8, is a possible indicator for this titration. Note that phenolphthalein could not be used to find the equivalence point.

Exercise 16.22

What is the pH at the equivalence point when 35 mL of 0.20 *M* ammonia is titrated by 0.12 *M* hydrochloric acid? K_b for ammonia is 1.8×10^{-5}.

(See Problems 16.105 and 16.106.)

Profile of a Chemical

ALUMINUM (an Amphoteric Metal)

Aluminum is such a common metal today that it may be difficult to imagine people replacing their luxury tableware by aluminum ware, as Napoleon III did soon after the metal became commercially available in 1855. But at that time aluminum was an exotic metal, despite the fact that aluminum is the third most abundant element on earth, present in clay and many rocks. The original commercial process for the preparation of aluminum involved the reduction of aluminum chloride with sodium, then an expensive metal itself.

$$AlCl_3(l) + 3Na(l) \longrightarrow Al(s) + 3NaCl(s)$$

The price of aluminum remained high until just after 1886, when Charles M. Hall in the United States and Paul L. T. Héroult in France independently discovered that they could obtain aluminum by electrolyzing aluminum oxide in a molten bath of cryolite, a mineral of aluminum, $NaAlF_6$. (Both Hall and Héroult were 23 years old at the time.) This is essentially the process used today. It starts with bauxite, a hydrated aluminum oxide ore. So far no one has succeeded in economically producing aluminum from clay.

Aluminum is a very reactive metal. When a mixture of aluminum powder and iron(III) oxide (called *thermite*) is ignited, it reacts in a spectacular incandescent shower, producing molten iron (Figure 16.18).

$$2Al(s) + Fe_2O_3(s) \longrightarrow 2Fe(l) + Al_2O_3(s)$$

The molten iron from the reaction of thermite has been used for welding. Aluminum also reacts with many ele-

Figure 16.18
Thermite reaction. *A mixture of aluminum powder and iron(III) oxide (thermite mixture) is ignited with burning magnesium. Once started, the thermite mixture burns violently giving off an incandescent shower. Molten iron is formed by the reaction, which has been used as a method of welding railroad tracks.*

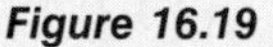

Figure 16.19
Reaction of aluminum and bromine. *When bromine liquid is poured over aluminum foil, a vigorous reaction occurs; aluminum bromide is the product.*

ments, including the halogens. Figure 16.19 shows the vigorous reaction of aluminum foil with bromine to form aluminum bromide.

A newly exposed aluminum surface quickly dulls as the metal reacts with oxygen in air to form the oxide. The oxide coating adheres tightly to the surface and prevents further reaction of the metal. This is an obvious advantage in many applications, since corroding is not a problem, as it is with iron. But it is also a source of an environmental problem: Aluminum cans do not disintegrate by corrosion the way coated steel cans do, so they accumulate along roadways and in land fills. Recycling of aluminum is an attractive solution to this problem and also helps save on the electrical energy that would otherwise be needed for the preparation of new metal from its ore.

Aluminum is an *amphoteric* metal; that is, the metal reacts with both acids and bases. As with other reactive metals, it displaces hydrogen from acids. The equation is:

$$2Al(s) + 6H^+(aq) \longrightarrow 2Al^{3+}(aq) + 3H_2(g)$$

Reaction with a base involves the formation of the tetrahydroxoaluminate ion, $Al(OH)_4^-(aq)$.

$$2Al(s) + 2OH^-(aq) + 6H_2O(l) \longrightarrow 2[Al(OH)_4^-(aq)] + 3H_2(g)$$

Commercial drain cleaners are principally sodium hydroxide, and some contain small pieces of aluminum, which fizz when the cleaner comes in contact with water. The fizzing results from the formation of hydrogen bubbles as the aluminum metal reacts with hydroxide ion in aqueous solution (Figure 16.20).

Questions for Study

1. Give the chemical equation for the original commercial preparation of aluminum.
2. How is aluminum prepared today? What is the ore used in this process?
3. Aluminum is a reactive metal. Give chemical equations for the reaction of aluminum with (a) Fe_2O_3; (b) Cl_2; (c) Br_2; (d) O_2.
4. Iron is also a reactive metal, and if unprotected it oxidizes in moist air to form rust, eventually crumbling to powder. Although aluminum is even more reactive with oxygen, aluminum bars and sheets do not appear to corrode or oxidize so easily. Explain why.
5. Aluminum is an amphoteric metal. Explain what that means. Then give equations for appropriate reactions that illustrate this property of aluminum.

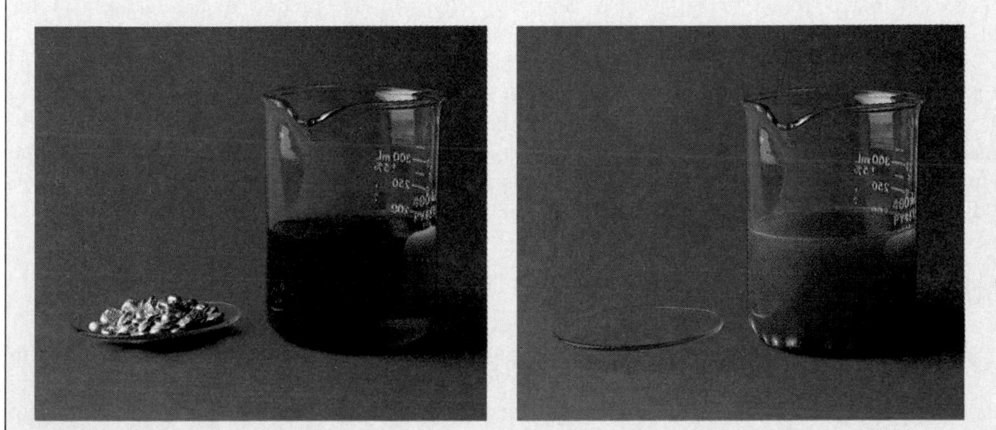

Figure 16.20
Reaction of aluminum with sodium hydroxide solution. *Aluminum metal reacts with aqueous strong base to form the tetrahydroxoaluminate ion, $Al(OH)_4^-$, and releasing bubbles of hydrogen gas, H_2. The solution contains phenolphthalein indicator, which is pink in basic solution.*

A Checklist for Review

IMPORTANT TERMS

self-ionization (autoionization) (16.1)
ion-product constant for water (K_w) (16.1)
pH (16.3)
acid-ionization constant (16.4)
degree of ionization (16.4)
base-ionization constant (16.6)
hydrolysis (16.7)
common-ion effect (16.8)
buffer (16.9)
Henderson–Hasselbalch equation (16.9)
acid–base titration curve (16.10)
equivalence point (16.10)

KEY EQUATIONS

$$K_w = [H^+][OH^-] = 1.00 \times 10^{-14} \text{ at } 25°C$$

$$pH = -\log [H^+]$$

$$pH + pOH = 14.00$$

$$K_a = \frac{[H^+][A^-]}{[HA]}$$

$$K_b = \frac{[HB^+][OH^-]}{[B]}$$

$$K_aK_b = K_w$$

$$pH = pK_a + \log \frac{[\text{base}]}{[\text{acid}]}$$

SUMMARY OF FACTS AND CONCEPTS

Water ionizes to give hydrogen ion and hydroxide ion. The concentrations of these ions in aqueous solution are related by the *ion-product constant for water (K_w)*. Thus, we can describe the acidity or basicity of a solution by the hydrogen-ion concentration. We often use the pH, which is the negative logarithm of the hydrogen-ion concentration, as a measure of acidity.

When a weak acid dissolves in water, it ionizes to give hydrogen ions and the conjugate base ions. The equilibrium constant for this *acid ionization* is $K_a = [H^+][A^-]/[HA]$, where HA is the general formula for the acid. The constant K_a can be determined from the pH of an acid solution of known concentration. Once obtained, the acid-ionization constant can be used to find the concentrations of species in any solution of the acid. In the case of a *diprotic* acid, H_2A, the concentration of H^+ and HA are calculated from K_{a1}, and the concentration of A^{2-} equals K_{a2}. Similar considerations apply to *base ionizations*.

Solutions of salts may be acidic or basic because of *hydrolysis* of the ions. The equilibrium constant for hydrolysis equals K_a for a cation or K_b for an anion. Calculation of the pH of a solution of a salt in which one ion hydrolyzes is fundamentally the same as calculation of the pH of a solution of an acid or base. However, K_a or K_b for an ion is usually obtained from the conjugate base or acid by using the equation $K_aK_b = K_w$.

A *buffer* is a solution that can resist changes in pH when small amounts of acid or base are added to it. A buffer contains either a weak acid and its conjugate base or a weak base and its conjugate acid. The concentrations of acid and base conjugates are approximately equal. Different pH ranges are possible for acid–base conjugates, depending on their ionization constants.

An *acid–base titration curve* is a plot of the pH of the solution against the volume of reactant added. During the titration of a strong acid by a strong base, the pH changes slowly at first. Then, as the amount of base nears the stoichiometric value—that is, nears the *equivalence point*—the pH rises abruptly, changing by several units. The pH at the equivalence point is 7.0. A similar curve is obtained when a weak acid is titrated by a strong base. However, the pH changes less at the equivalence point. Moreover, the pH at the equivalence point is greater than 7.0 because of hydrolysis of the salt produced. An indicator must be chosen that changes color within a pH range near the equivalence point, where the pH changes rapidly.

OPERATIONAL SKILLS

1. **Calculating concentrations of H^+ and OH^- in solutions of a strong acid or base** Given the concentration of a strong acid or base, calculate the hydrogen-ion and hydroxide-ion concentrations (Example 16.1).

2. **Calculating the pH from the hydrogen-ion concentration, or vice versa** Given the hydrogen-ion concentration, calculate the pH (Example 16.2); or given the pH, calculate the hydrogen-ion concentration (Example 16.3).

3. **Determining K_a (or K_b) from the solution pH** Given the molarity and pH of a solution of a weak acid, calculate the acid-ionization constant K_a (Example 16.4). The K_b for a base can be determined in a similar way (see Exercise 16.11).

4. **Calculating concentrations of species in a weak acid solution using K_a** Given K_a, calculate the hydrogen-ion concentration and pH of a solution of a weak acid of known molarity (Examples 16.5 and 16.6). Given K_{a1}, K_{a2}, and the molarity of a diprotic acid solution, calculate the pH and the concentrations of H^+, HA^-, and A^{2-} (Example 16.7).

5. **Calculating concentrations of species in a weak base solution** Given K_b, calculate the hydrogen-ion concentration and pH of a solution of a weak base of known molarity (Example 16.8).

6. **Predicting whether a salt solution is acidic, basic, or neutral** Decide whether an aqueous solution of a given salt is acidic, basic, or neutral (Example 16.9).

7. **Obtaining K_a from K_b or K_b from K_a** Calculate K_a for a cation or K_b for an anion from the ionization constant of the conjugate base or acid (Example 16.10).

8. **Calculating concentrations of species in a salt solution** Given the concentration of a solution of a salt in which one ion hydrolyzes, and given the ionization constant of the conjugate acid or base of this ion, calculate the H^+ ion concentration (Example 16.11).

9. **Calculating the common-ion effect on acid ionization** Given K_a and the concentrations of weak acid and strong acid in a solution, calculate the degree of ionization of the weak acid (Example 16.12). Given K_a and the concentrations of weak acid and its salt in a solution, calculate the pH (Example 16.13).

10. **Calculating the pH of a buffer from given volumes of solution** Given concentrations and volumes of acid and conjugate base from which a buffer is prepared, calculate the buffer pH (Example 16.14).

11. **Calculating the pH of a buffer when a strong acid or strong base is added** Calculate the pH of a given volume of buffer solution (given the concentrations of conjugate acid and base in the buffer) to which a specified amount of strong acid or strong base is added (Example 16.15).

12. **Calculating the pH of a solution of a strong acid and a strong base** Calculate the pH during the titration of a strong acid by a strong base, given the volumes and concentrations of the acid and base (Example 16.16).

13. **Calculating the pH at the equivalence point in a titration** Calculate the pH at the equivalence point for the titration of a weak acid by a strong base (Example 16.17). Be able to do the same type of calculation for the titration of a weak base by a strong acid.

Review Questions

16.1 What is meant by the self-ionization of water? Write the expression for K_w. What is its value at 25°C?

16.2 What is meant by the pH of a solution? Describe two ways of measuring pH.

16.3 What is the pH of a neutral solution at 37°C, where K_w equals 2.5×10^{-14}?

16.4 What is the sum of the pH and the pOH for a solution at 37°C, where K_w equals 2.5×10^{-14}?

16.5 Write an equation for the ionization of hydrogen cyanide, HCN, in aqueous solution. What is the equilibrium expression K_a for this acid ionization?

16.6 Which one of the following is the weakest acid: $HClO_4$, HCN, or $HC_2H_3O_2$? See Table 16.1 and Table 16.2.

16.7 Briefly describe two methods for determining K_a for a weak acid.

16.8 Describe how the degree of ionization of a weak acid changes as the concentration increases.

16.9 Consider a solution of 0.010 *M* HF ($K_a = 6.8 \times 10^{-4}$). In solving for the concentrations of species in this solution, could you use the simplifying assumption in which you neglect x in the denominator of the equilibrium equation? Explain.

16.10 Phosphorous acid, H_2PHO_3, is a diprotic acid. Write equations for the acid ionizations. Write the expressions for K_{a1} and K_{a2}.

16.11 What is the concentration of oxalate ion, $C_2O_4{}^{2-}$, in 0.10 *M* oxalic acid, $H_2C_2O_4$? K_{a1} is 5.6×10^{-2}, and K_{a2} is 5.1×10^{-5}.

16.12 Write the equation for the ionization of aniline, $C_6H_5NH_2$, in aqueous solution. Write the expression for K_b.

16.13 Which of the following is the strongest base: NH_3, $C_6H_5NH_2$, or CH_3NH_2? See Table 16.3.

16.14 Do you expect a solution of anilinium chloride (aniline hydrochloride), $C_6H_5NH_3Cl$, to be acidic or basic? (Anilinium chloride is the salt of aniline and hydrochloric acid.) Write the equation for the reaction involved. What is the equilibrium expression? How would you obtain the value for the equilibrium constant from Table 16.3?

16.15 What is meant by the common-ion effect? Give an example.

16.16 The pH of 0.10 *M* CH_3NH_2 (methylamine) is 11.8. When the chloride salt of methylamine, CH_3NH_3Cl, is added to this solution, does the pH increase or decrease? Explain, using Le Chatelier's principle and the common-ion effect.

16.17 Define the term *buffer*. Give an example of one.

16.18 What is meant by the capacity of a buffer? Describe a buffer with low capacity and the same buffer with greater capacity.

16.19 Describe the pH changes that occur during the titration of a weak base by a strong acid. What is meant by the term *equivalence point?*

16.20 If the pH is 8.0 at the equivalence point for the titration of a certain weak acid with sodium hydroxide, what indicator might you use? See Figure 16.3. Explain your choice.

Practice Problems

SOLUTIONS OF A STRONG ACID OR BASE

16.21 What are the concentrations of H^+ and OH^- in each of the following?
(a) 0.25 *M* HCl (b) 1.25 *M* NaOH
(c) 0.0035 *M* $Ca(OH)_2$ (d) 2.5 *M* HNO_3

16.22 What are the concentrations of H^+ and OH^- in each of the following?
(a) 1.50 *M* KOH (b) 0.015 *M* $Ba(OH)_2$
(c) 0.015 *M* $HClO_4$ (d) 0.150 *M* HCl

16.23 A solution of hydrochloric acid is 0.050 *M* HCl. What is the hydrogen-ion concentration at 25°C? What is the hydroxide-ion concentration at 25°C?

16.24 A solution is 0.020 *M* HNO_3 (nitric acid). What is the hydrogen-ion concentration at 25°C? What is the hydroxide-ion concentration at 25°C?

16.25 What are the hydrogen-ion and the hydroxide-ion concentrations of a solution at 25°C that is 0.0050 *M* strontium hydroxide, $Sr(OH)_2$?

16.26 A saturated solution of magnesium hydroxide is 3.2×10^{-4} *M* $Mg(OH)_2$. What are the hydrogen-ion and hydroxide-ion concentrations in the solution at 25°C?

16.27 The following are solution concentrations. Indicate whether each solution is acidic, basic, or neutral.
(a) 5×10^{-9} *M* H^+ (b) 5×10^{-9} *M* OH^-
(c) 1×10^{-7} *M* OH^- (d) 2×10^{-6} *M* H^+

16.28 The following are solution concentrations. Indicate whether each solution is acidic, basic, or neutral.
(a) 2×10^{-4} *M* OH^- (b) 2×10^{-6} *M* H^+
(c) 6×10^{-10} *M* OH^- (d) 6×10^{-10} *M* H^+

16.29 A shampoo solution at 25°C has a hydroxide-ion concentration of 1.5×10^{-9} *M*. Is the solution acidic, neutral, or basic?

16.30 An antiseptic solution at 25°C has a hydroxide-ion concentration of 8.4×10^{-5} *M*. Is the solution acidic, neutral, or basic?

CALCULATIONS INVOLVING pH

16.31 Which of the following pH values indicate an acidic solution at 25°C? Which are basic and which are neutral?
(a) 11.2 (b) 7.0 (c) 1.2 (d) 6.1

16.32 Which of the following pH values indicate an acidic solution at 25°C? Which are basic and which are neutral?
(a) 13.1 (b) 2.3 (c) 5.8 (d) 8.5

16.33 For each of the following, state whether the solution at 25°C is acidic, neutral, or basic: (a) A beverage solution has a pH of 3.5. (b) A 0.50 *M* solution of potassium bromide, KBr, has a pH of 7.0. (c) A 0.050 *M* solution of pyridine, C_5H_5N, has a pH of 9.0. (d) A solution of iron(III) chloride has a pH of 5.5.

16.34 For each of the following, state whether the solution at 25°C is acidic, neutral, or basic: (a) A 0.1 *M* solution of trisodium phosphate, Na_3PO_4, has a pH of 12.0. (b) A 0.1 *M* solution of calcium chloride, $CaCl_2$, has a pH of 7.0. (c) A 0.2 *M* solution of copper(II) sulfate, $CuSO_4$, has a pH of 4.0. (d) A sample of rainwater has a pH of 5.7.

16.35 Obtain the pH corresponding to the following hydrogen-ion concentrations.
(a) 1.0×10^{-8} *M* (b) 5.0×10^{-12} *M*
(c) 7.5×10^{-3} *M* (d) 6.35×10^{-9} *M*

16.36 Obtain the pH corresponding to the following hydrogen-ion concentrations.
(a) 1.0×10^{-4} *M* (b) 3.2×10^{-10} *M*
(c) 2.3×10^{-5} *M* (d) 2.91×10^{-11} *M*

16.37 A sample of vinegar has a hydrogen-ion concentration of 7.5×10^{-3} *M*. What is the pH of the vinegar?

16.38 Some lemon juice has a hydrogen-ion concentration of 5.0×10^{-3} *M*. What is the pH of the lemon juice?

16.39 A solution of washing soda (sodium carbonate, Na_2CO_3) has a hydroxide-ion concentration of 0.0040 *M*. What is the pH at 25°C?

16.40 A solution of lye (sodium hydroxide, NaOH) has a hydroxide-ion concentration of 0.050 *M*. What is the pH at 25°C?

16.41 The pH of a cup of coffee (at 25°C) was found to be 5.12. What is the hydrogen-ion concentration?

16.42 A wine was tested for acidity, and its pH was found to be 3.85 at 25°C. What is the hydrogen-ion concentration?

16.43 A detergent solution has a pH of 11.63 at 25°C. What is the hydroxide-ion concentration?

16.44 Morphine is a narcotic that is used to relieve pain. A solution of morphine has a pH of 9.61 at 25°C. What is the hydroxide-ion concentration?

16.45 A 1.00-L aqueous solution contained 5.80 g of sodium hydroxide, NaOH. What was the pH of the solution at 25°C?

16.46 A 1.00-L aqueous solution contained 6.78 g of barium hydroxide, $Ba(OH)_2$. What was the pH of the solution at 25°C?

16.47 A certain sample of rainwater gives a yellow color with methyl red and a yellow color with bromthymol blue. What is the approximate pH of the water? Is the rainwater acidic, neutral, or basic? (See Figure 16.3.)

16.48 A drop of thymol blue gave a yellow color with a solution of aspirin. A sample of the same aspirin solution gave a yellow color with bromphenol blue. What was the pH of the solution? Was the solution acidic, neutral, or basic? (See Figure 16.3.)

Note: Values of K_a and K_b that are not given in the following problems can be obtained from Tables 16.2 and 16.3.

ACID IONIZATION

16.49 Write the chemical equations for the acid ionizations of each of the following weak acids. (First write these in terms of H_3O^+, then in terms of H^+.) Then write the K_a expressions for each acid in terms of H^+.

(a) HNO_2 (b) $HClO$ (c) HCN (d) $HCHO_2$

16.50 Write the chemical equations for the acid ionizations of each of the following weak acids. (First write these in terms of H_3O^+, then in terms of H^+.) Then write the K_a expressions for each acid in terms of H^+.

(a) HF (b) $HCNO$ (c) $HBrO$ (d) $HC_7H_5O_2$

16.51 A solution of 0.20 *M* hydrazoic acid, HN_3, has a pH of 3.21. What is the K_a for hydrazoic acid?

16.52 Calculate K_a for propionic acid, $HC_3H_5O_2$. The pH of a 0.055 *M* aqueous solution is 3.06.

16.53 Boric acid, $B(OH)_3$, is used as a mild antiseptic. What is the pH of a 0.025 *M* aqueous solution of boric acid? What is the degree of ionization of boric acid in this solution? The hydrogen ion arises principally from the reaction

$$B(OH)_3(aq) + H_2O(l) \rightleftharpoons B(OH)_4^-(aq) + H^+(aq)$$

The equilibrium constant for this reaction is 5.9×10^{-10}.

16.54 Formic acid, $HCHO_2$, is used to make methyl formate (a fumigant for dried fruit) and ethyl formate (an artificial rum flavor). What is the pH of a 0.12 *M* solution of formic acid? What is the degree of ionization of $HCHO_2$ in this solution? See Table 16.2 for K_a.

16.55 $C_6H_4NH_2COOH$, *para*-aminobenzoic acid (PABA), is used in some sunscreen agents. Calculate the concentrations of hydrogen ion and *para*-aminobenzoate ion, $C_6H_4NH_2COO^-$, in a 0.050 *M* solution of the acid. The value of K_a is 2.2×10^{-5}.

16.56 Barbituric acid, $HC_4H_3N_2O_3$, is used to prepare various barbiturate drugs (used as sedatives). Calculate the concentrations of hydrogen ion and barbiturate ion in a 0.20 *M* solution of the acid. The value of K_a is 9.8×10^{-5}.

16.57 A solution of acetic acid, $HC_2H_3O_2$, on a laboratory shelf was of undetermined concentration. If the pH of the solution was found to be 2.68, what was the concentration of the acetic acid?

16.58 A chemist wanted to determine the concentration of a solution of lactic acid, $HC_3H_5O_3$. She found that the pH of the solution was 2.51. What was the concentration of the solution? The K_a of lactic acid is 1.4×10^{-4}.

16.59 Hydrofluoric acid, HF, unlike hydrochloric acid, is a weak electrolyte. What is the hydrogen-ion concentration and the pH of a 0.040 *M* aqueous solution of HF?

16.60 Chloroacetic acid, $HC_2H_2ClO_2$, has a greater acid strength than acetic acid, because the electronegative chlorine atom pulls electrons away from the O—H bond and thus weakens it. Calculate the hydrogen-ion concentration and the pH of a 0.015 *M* solution of chloroacetic acid. K_a is 1.3×10^{-3}.

16.61 What is the hydrogen-ion concentration of a 2.00 *M* solution of 2,6-dinitrobenzoic acid, $(NO_2)_2C_6H_3COOH$, for which $K_a = 7.94 \times 10^{-2}$?

16.62 What is the hydrogen-ion concentration of a 2.00×10^{-4} *M* solution of *p*-bromobenzoic acid, BrC_6H_4COOH, for which $K_a = 1.00 \times 10^{-4}$?

16.63 Phthalic acid, $H_2C_8H_4O_4$, is a diprotic acid used in the synthesis of phenolphthalein indicator. $K_{a1} = 1.2 \times 10^{-3}$, and $K_{a2} = 3.9 \times 10^{-6}$. (a) Calculate the hydrogen-ion concentration of a 0.015 *M* solution. (b) What is the concentration of the $C_8H_4O_4^{2-}$ ion in the solution?

16.64 Carbonic acid, H_2CO_3, can be found in a wide variety of body fluids (from dissolved CO_2). (a) Calculate the hydrogen-ion concentration of a 5.45×10^{-4} *M* H_2CO_3 solution. (b) What is the concentration of CO_3^{2-}?

BASE IONIZATION

16.65 Write the chemical equation for the base ionization of ethylamine, $C_2H_5NH_2$. Write the K_b expression for ethylamine.

16.66 Write the chemical equation for the base ionization of pyridine, C_5H_5N. Write the K_b expression for pyridine.

16.67 Ethanolamine, $HOC_2H_4NH_2$, is a viscous liquid with an ammonia-like odor; it is used to remove hydrogen sulfide from natural gas. A 0.15 *M* aqueous solution of ethanolamine has a pH of 11.34. What is K_b for ethanolamine?

16.68 Trimethylamine, $(CH_3)_3N$, is a gas with a fishy, ammonia-like odor. An aqueous solution that is 0.25 *M* trimethylamine has a pH of 11.63. What is K_b for trimethylamine?

16.69 What is the concentration of hydroxide ion in a 0.080 *M* aqueous solution of methylamine, CH_3NH_2? What is the pH?

16.70 What is the concentration of hydroxide ion in a 0.15 *M* aqueous solution of hydroxylamine, NH_2OH? What is the pH?

ACID–BASE PROPERTIES OF SALT SOLUTIONS; HYDROLYSIS

16.71 Note whether hydrolysis occurs for each of the following ions. If hydrolysis does occur, write the chemical equation for it. Then write the equilibrium expression for the acid or base ionization (whichever occurs).

(a) I^- (b) CHO_2^- (c) $CH_3NH_3^+$ (d) IO^-

16.72 Note whether hydrolysis occurs for each of the following ions. If hydrolysis does occur, write the chemical equation for it. Then write the equilibrium expression for the acid or base ionization (whichever occurs).

(a) NO_2^- (b) Br^- (c) NO_3^- (d) $NH_2NH_3^+$

16.73 Write the equation for the acid ionization of the $Zn(H_2O)_6^{2+}$ ion.

16.74 Write the equation for the acid ionization of the $Cu(H_2O)_6^{2+}$ ion.

16.75 For each of the following salts, indicate whether the aqueous solution will be acidic, basic, or neutral.

(a) $Fe(NO_3)_3$ (b) Na_2CO_3
(c) $Ca(CN)_2$ (d) NH_4ClO_4

16.76 Note whether aqueous solutions of each of the following salts will be acidic, basic, or neutral.

(a) Na_2S (b) $Cu(NO_3)_2$
(c) $KClO_4$ (d) CH_3NH_3Cl

16.77 Decide whether solutions of the following salts are acidic, neutral, or basic.

(a) ammonium acetate (b) anilinium acetate

16.78 Decide whether solutions of the following salts are acidic, neutral, or basic.

(a) ammonium cyanate (b) pyridinium cyanate

16.79 Obtain (a) the K_b value for NO_2^-; (b) the K_a value for $C_5H_5NH^+$ (pyridinium ion).

16.80 Obtain (a) the K_b value for ClO^-; (b) the K_a value for NH_3OH^+ (hydroxylammonium ion).

16.81 What is the pH of a 0.025 *M* aqueous solution of sodium propionate, $NaC_3H_5O_2$? What is the concentration of propionic acid in the solution?

16.82 Calculate the OH^- concentration and pH of a 0.010 *M* aqueous solution of sodium cyanide, NaCN. Finally, obtain the hydrogen-ion, CN^-, and HCN concentrations.

16.83 Calculate the concentration of pyridine, C_5H_5N, in a solution that is 0.15 *M* pyridinium bromide, C_5H_5NHBr. What is the pH of the solution?

16.84 What is the pH of a 0.35 *M* solution of methylammonium chloride, CH_3NH_3Cl? What is the concentration of methylamine in the solution?

COMMON-ION EFFECT

16.85 Calculate the degree of ionization of
(a) 0.80 *M* HF
(b) the same solution that is also 0.10 *M* HCl

16.86 Calculate the degree of ionization of
(a) 0.20 *M* $HCHO_2$ (formic acid)
(b) the same solution that is also 0.10 *M* HCl

16.87 What is the pH of a solution that is 0.10 *M* KNO_2 and 0.15 *M* HNO_2 (nitrous acid)?

16.88 What is the pH of a solution that is 0.20 *M* KCNO and 0.10 *M* HCNO (cyanic acid)?

16.89 What is the pH of a solution that is 0.10 *M* CH_3NH_2 (methylamine) and 0.15 *M* CH_3NH_3Cl (methylammonium chloride)?

16.90 What is the pH of a solution that is 0.15 *M* $C_2H_5NH_2$ (ethylamine) and 0.10 *M* $C_2H_5NH_3Br$ (ethylammonium bromide)?

BUFFERS

16.91 A buffer is prepared by adding 45.0 mL of 0.15 *M* NaF to 35.0 mL of 0.10 *M* HF. What is the pH of the final solution?

16.92 A buffer is prepared by adding 115 mL of 0.30 *M* NH_3 to 145 mL of 0.15 *M* NH_4NO_3. What is the pH of the final solution?

16.93 What is the pH of a buffer solution that is 0.10 *M* NH_3 and 0.10 *M* NH_4^+? What is the pH if 12 mL of 0.20 *M* hydrochloric acid is added to 125 mL of buffer?

16.94 A buffer is prepared by mixing 525 mL of 0.50 *M* formic acid, $HCHO_2$, and 475 mL of 0.50 *M* sodium formate, $NaCHO_2$. Calculate the pH. What would be the pH of 85 mL of the buffer to which 8.5 mL of 0.15 *M* hydrochloric acid has been added?

16.95 What is the pH of a buffer solution that is 0.10 *M* chloroacetic acid and 0.15 *M* sodium chloroacetate? $K_a = 1.4 \times 10^{-3}$.

16.96 What is the pH of a buffer solution that is 0.20 *M* propionic acid and 0.10 *M* sodium propionate?

16.97 What is the pH of a buffer solution that is 0.15 *M* pyridine and 0.10 *M* pyridinium bromide?

16.98 What is the pH of a buffer solution that is 0.20 *M* methylamine and 0.15 *M* methylammonium chloride?

16.99 How many moles of sodium acetate must be added to 2.0 L of 0.10 *M* acetic acid to give a solution that has a pH equal to 5.00? Ignore the volume change due to the addition of sodium acetate.

16.100 How many moles of hydrofluoric acid, HF, must be added to 500.0 mL of 0.30 *M* sodium fluoride to give a buffer of pH 3.50? Ignore the volume change due to the addition of hydrofluoric acid.

TITRATION CURVES

16.101 What is the pH of a solution in which 20 mL of 0.10 *M* NaOH is added to 25 mL of 0.10 *M* HCl?

16.102 What is the pH of a solution in which 35 mL of 0.10 *M* NaOH is added to 25 mL of 0.10 *M* HCl?

16.103 A 1.24-g sample of benzoic acid was dissolved in water to give 50.0 mL of solution. This solution was titrated with 0.180 *M* NaOH. What was the pH of the solution when the equivalence point was reached?

16.104 A 0.400-g sample of propionic acid was dissolved in water to give 50.0 mL of solution. This solution was titrated with 0.150 *M* NaOH. What was the pH of the solution when the equivalence point was reached?

16.105 Find the pH of the solution obtained when 32 mL of 0.087 *M* ethylamine is titrated to the equivalence point with 0.15 *M* HCl.

16.106 What is the pH at the equivalence point when 22 mL of 0.20 *M* hydroxylamine is titrated with 0.15 *M* HCl?

16.107 Calculate the pH of a solution obtained by mixing 500.0 mL of 0.10 *M* NH_3 with 200.0 mL of 0.15 *M* HCl.

16.108 Calculate the pH of a solution obtained by mixing 35.0 mL of 0.15 *M* acetic acid with 25.0 mL of 0.10 *M* sodium acetate.

Additional Problems

16.109 Salicylic acid, $C_6H_4OHCOOH$, is used in the manufacture of acetylsalicylic acid (aspirin) and methyl salicylate (wintergreen flavor). A saturated solution of salicylic acid contains 2.2 g of the acid per liter of solution and has a pH of 2.43. What is the value of K_a?

16.110 Cyanoacetic acid, $CH_2CNCOOH$, is used in the manufacture of barbiturate drugs. An aqueous solution containing 5.0 g in a liter of solution has a pH of 1.89. What is the value of K_a?

16.111 A 0.050 *M* aqueous solution of sodium hydrogen sulfate, $NaHSO_4$, has a pH of 1.73. Calculate K_{a2} for sulfuric acid. Sulfuric acid is a strong electrolyte, so you can ignore hydrolysis of the HSO_4^- ion.

16.112 A 0.10 *M* aqueous solution of sodium dihydrogen phosphate, NaH_2PO_4, has a pH of 4.10. Calculate K_{a2} for phosphoric acid. You can ignore hydrolysis of the $H_2PO_4^-$ ion.

16.113 Compare the acid-ionization constant of HCO_3^- with its base-ionization (hydrolysis) constant. Which is more extensive for HCO_3^-, acid ionization or hydrolysis? What is the pH of a 0.10 *M* solution of sodium hydrogen carbonate, $NaHCO_3$? Use the more extensive equilibrium only.

16.114 Compare the acid-ionization constant of HPO_4^{2-} with its base-ionization (hydrolysis) constant. Which is more extensive for HPO_4^{2-}, acid ionization or hydrolysis? What is the pH of a 0.10 *M* solution of disodium hydrogen phosphate, Na_2HPO_4? Use the more extensive equilibrium only.

16.115 Calculate the base-ionization constants for CN^- and CO_3^{2-}. Which ion is the stronger base?

16.116 Calculate the base-ionization constants for PO_4^{3-} and SO_4^{2-}. Which ion is the stronger base?

16.117 Calculate the pH of a 0.15 *M* aqueous solution of aluminum chloride, $AlCl_3$. The acid ionization of hydrated aluminum ion is

$$Al(H_2O)_6^{3+}(aq) + H_2O(l) \rightleftharpoons Al(H_2O)_5OH^{2+}(aq) + H_3O^+(aq)$$

and K_a is 1.4×10^{-5}.

16.118 Calculate the pH of a 0.15 *M* aqueous solution of zinc chloride, $ZnCl_2$. The acid ionization of hydrated zinc ion is

$$Zn(H_2O)_6^{2+}(aq) + H_2O(l) \rightleftharpoons Zn(H_2O)_5OH^{+}(aq) + H_3O^+(aq)$$

and K_a is 2.5×10^{-10}.

16.119 An artificial fruit beverage contains 11.0 g of tartaric acid, $H_2C_4H_4O_6$, and 20.0 g of its salt, potassium hydrogen tartrate, per liter. What is the pH of the beverage? $K_{a1} = 1.0 \times 10^{-3}$.

16.120 A buffer is made by dissolving 13.0 g of sodium dihydrogen phosphate, NaH_2PO_4, and 15.0 g of disodium hydrogen phosphate, Na_2HPO_4, in a liter of solution. What is the pH of the buffer?

16.121 Blood contains several acid–base systems that tend to keep its pH constant at about 7.4. One of the most important buffer systems involves carbonic acid and hydrogen carbonate ion. What must be the ratio of $[HCO_3^-]$ to $[H_2CO_3]$ in the blood if the pH is 7.40?

16.122 Codeine, $C_{18}H_{21}NO_3$, is an alkaloid ($K_b = 6.2 \times 10^{-9}$) used as a painkiller and cough suppressant. A solution of codeine is acidified with hydrochloric acid to pH 4.50. What is the ratio of the concentration of the conjugate acid of codeine to that of the base codeine?

16.123 Calculate the pH of a solution that is obtained by mixing 456 mL of 0.10 *M* hydrochloric acid with 285 mL of 0.15 *M* sodium hydroxide. Assume the combined volume is the sum of the two original volumes.

16.124 Calculate the pH of a solution that is made up from 2.0 g of potassium hydroxide dissolved in 115 mL of 0.19 *M* perchloric acid. Assume the change in volume due to adding potassium hydroxide is negligible.

16.125 Find the pH of the solution obtained when 25 mL of 0.065 *M* benzylamine, $C_7H_7NH_2$, is titrated to the equivalence point with 0.050 *M* hydrochloric acid. K_b for benzylamine is 4.7×10^{-10}.

16.126 What is the pH of the solution obtained by titrating 1.24 g of sodium hydrogen sulfate, $NaHSO_4$, dissolved in 50.0 mL of water with 0.180 *M* sodium hydroxide until the equivalence point is reached? Assume that any volume change due to adding the sodium hydrogen sulfate or to mixing the solutions is negligible.

16.127 Ionization of the first proton from H_2SO_4 is complete (H_2SO_4 is a strong acid); the acid-ionization constant for the second proton is 1.1×10^{-2}. (a) What would be the approximate hydrogen-ion concentration in 0.100 *M* H_2SO_4 if ionization of the second proton were ignored? (b) The ionization of the second proton must be considered for a more exact answer, however. Calculate the hydrogen-ion concentration in 0.100 *M* H_2SO_4, accounting for the ionization of both protons.

16.128 Ionization of the first proton from H_2SeO_4 is complete (H_2SeO_4 is a strong acid); the acid-ionization constant for the second proton is 1.2×10^{-2}. (a) What would be the approximate hydrogen-ion concentration in 0.150 *M* H_2SeO_4 if ionization of the second proton were ignored? (b) The ionization of the second proton must be considered for a more exact answer, however. Calculate the hydrogen-ion concentration in 0.150 *M* H_2SeO_4, accounting for the ionization of both protons.

Cumulative-Skills Problems

16.129 Consider the reaction of SO_4^{2-} and HCN to yield HSO_4^{2-} and CN^-. Calculate the value of the equilibrium constant for this reaction. *Hint:* Use the rule (given in Section 15.2) that the equilibrium constant of a reaction equal to the sum of two reactions is the product of the equilibrium constants of the summed reactions.

16.130 Consider the reaction of H_2S and $C_2H_3O_2^-$ to yield $HC_2H_3O_2$ and HS^-. Calculate the value of the equilibrium constant for this reaction. (See the hint in Problem 16.129.)

16.131 The pH of a white vinegar solution is 2.45. This vinegar is an aqueous solution of acetic acid with a density of 1.09 g/mL. What is the mass percent of acetic acid in the solution?

16.132 The pH of a household cleaning solution is 11.87. This cleanser is an aqueous solution of ammonia with a density of 1.00 g/mL. What is the mass percent of ammonia in the solution?

16.133 What is the freezing point of 0.92 *M* aqueous acetic acid? The density of this solution is 1.008 g/cm^3.

16.134 What is the freezing point of 0.87 *M* aqueous ammonia? The density of this solution is 0.992 g/cm^3.

16.135 A chemist needs a buffer with pH 4.35. How many milliliters of pure acetic acid (density = 1.049 g/mL) must be added to 465 mL of 0.0941 *M* NaOH solution to obtain such a buffer?

16.136 A chemist needs a buffer with pH 3.50. How many milliliters of pure formic acid (density = 1.220 g/mL) must be added to 325 mL of 0.0857 *M* NaOH solution to obtain such a buffer?

17

Solubility and Complex-Ion Equilibria

Chapter Outline

Formation of lead(II) iodide precipitate.

Many natural processes depend on the precipitation or the dissolving of a slightly soluble salt. For example, caves form in limestone (calcium carbonate) over thousands of years as ground water seeps through cracks, dissolving out cavities in the rock. Kidney stones form when salts such as calcium phosphate or calcium oxalate slowly precipitate in these organs (Figure 17.1). Poisoning by oxalic acid is also explained by the precipitation of the calcium salt. If oxalic acid is accidentally ingested, the oxalate-ion concentration in the blood may increase sufficiently to precipitate calcium oxalate. Calcium ion, which is needed for proper muscle control, is then removed from the blood, and muscle tissues go into spasm.

To understand such phenomena quantitatively, we must be able to solve problems in solubility equilibria. Calcium oxalate kidney stones form when the concentrations of calcium ion and oxalate ion are sufficiently great. What is the relationship between the concentrations of ions and the solubility of a salt? What is the minimum concentration of oxalate ion that gives a precipitate of the calcium salt from a 0.0025 *M* solution of Ca^{2+} (the approximate concentration of calcium ion in blood plasma)? What is the effect of pH on the solubility of this salt? We will look at questions such as these in this chapter.

Figure 17.1
A kidney stone. Kidney stones are usually calcium phosphate or calcium oxalate that precipitates as a crystalline mass.

Solubility Equilibria

To deal quantitatively with an equilibrium, we require the equilibrium constant. In the next section, we will look at the equilibria of slightly soluble, or nearly insoluble, ionic compounds and show how we can determine their equilibrium constants.

Once we find values for various ionic compounds, we can use them to answer solubility or precipitation questions.

17.1 THE SOLUBILITY PRODUCT CONSTANT

When an ionic compound is dissolved in water, it usually goes into solution as the ions. When an excess of a slightly soluble ionic compound is mixed with water, an equilibrium occurs between the solid compound and the ions in the saturated solution. For the salt calcium oxalate, CaC_2O_4, we have the following equilibrium:

$$CaC_2O_4(s) \xrightleftharpoons{H_2O} Ca^{2+}(aq) + C_2O_4{}^{2-}(aq)$$

The equilibrium constant for this solubility process, called the *solubility product constant* of CaC_2O_4, is written

$$K_{sp} = [Ca^{2+}][C_2O_4{}^{2-}]$$

In general, the **solubility product constant (K_{sp})** is *the equilibrium constant for the solubility equilibrium of a slightly soluble (or nearly insoluble) ionic compound*. It equals the product of the equilibrium concentrations of the ions in the compound, each concentration raised to a power equal to the number of such ions in the formula of the compound. Like any equilibrium constant, K_{sp} depends on the temperature, but at a given temperature it has a constant value for various concentrations of the ions.●

● An ionic equilibrium is affected to a small extent by the presence of ions not directly involved in the equilibrium. We will ignore this effect here.

Lead(II) iodide, PbI_2, is another example of a slightly soluble salt. The equilibrium in water is

$$PbI_2(s) \rightleftharpoons Pb^{2+}(aq) + 2I^-(aq)$$

and the expression for the equilibrium constant or solubility product constant is

$$K_{sp} = [Pb^{2+}][I^-]^2$$

Example 17.1 Writing Solubility Product Expressions

Write the solubility product expressions for the following salts: (a) AgCl; (b) Hg_2Cl_2; (c) $Pb_3(AsO_4)_2$.

Solution

The equilibria and solubility product expressions are

(a) $AgCl(s) \rightleftharpoons Ag^+(aq) + Cl^-(aq)$; $K_{sp} = [Ag^+][Cl^-]$

(b) $Hg_2Cl_2(s) \rightleftharpoons Hg_2{}^{2+}(aq) + 2Cl^-(aq)$;
$K_{sp} = [Hg_2{}^{2+}][Cl^-]^2$

Note that the mercury(I) ion is $Hg_2{}^{2+}$. (Mercury also has salts with the mercury(II) ion, Hg^{2+}.)

(c) $Pb_3(AsO_4)_2(s) \rightleftharpoons 3Pb^{2+}(aq) + 2AsO_4{}^{3-}(aq)$;
$K_{sp} = [Pb^{2+}]^3[AsO_4{}^{3-}]^2$

Exercise 17.1

Give solubility product expressions for the following: (a) barium sulfate; (b) iron(III) hydroxide; (c) calcium phosphate. (See Problems 17.13 and 17.14.)

The solubility product constant, K_{sp}, of a slightly soluble ionic compound is expressed in terms of the molar concentrations of ions in the saturated solution. These ion concentrations are in turn related to the *molar solubility* of the ionic compound, which is the moles of compound that dissolve to give a liter of saturated solution. The next two examples show how to determine the solubility product constant from the solubility of a slightly soluble ionic compound.

Example 17.2 Calculating K_{sp} from the Solubility (Simple Example)

A liter of a solution saturated at 25°C with calcium oxalate, CaC_2O_4, is evaporated to dryness, giving a 0.0061-g residue of CaC_2O_4. Calculate the solubility product constant for this salt at 25°C.

Solution

The solubility of calcium oxalate is 0.0061 g/L *of solution*. We can also express this as a molar solubility—that is, as the number of moles of salt that dissolve per liter of solution. We convert grams per liter to moles per liter (the formula weight of CaC_2O_4 is 128 amu).

Molar solubility of CaC_2O_4 =

$$0.0061\ \cancel{g\ CaC_2O_4}/L \times \frac{1\ mol\ CaC_2O_4}{128\ \cancel{g\ CaC_2O_4}}$$

Molar solubility of $CaC_2O_4 = 4.8 \times 10^{-5}$ mol CaC_2O_4/L

Let us look at the equilibrium problem.

STEP 1 Suppose we mix solid CaC_2O_4 in a liter of solution. Of this solid, 4.8×10^{-5} mol will dissolve to form 4.8×10^{-5} mol of each ion. The results are summarized in the table below. (Because the concentration of the solid does not appear in K_{sp}, we do not include it in the table.)

STEP 2 We now substitute into the equilibrium equation:

$$K_{sp} = [Ca^{2+}][C_2O_4{}^{2-}] = (4.8 \times 10^{-5})(4.8 \times 10^{-5})$$
$$= \mathbf{2.3 \times 10^{-9}}$$

CONCENTRATION (*M*)	$CaC_2O_4(s)$	$\rightleftharpoons$	$Ca^{2+}(aq)$	+	$C_2O_4{}^{2-}(aq)$
Starting			0		0
Change			$+4.8 \times 10^{-5}$		$+4.8 \times 10^{-5}$
Equilibrium			4.8×10^{-5}		4.8×10^{-5}

Exercise 17.2

Silver ion may be recovered from used photographic fixing solution by precipitating it as silver chloride. The solubility of silver chloride is 1.9×10^{-3} g/L. Calculate K_{sp}.

(See Problems 17.15 and 17.16.)

The next example is similar, except that the salt produces unequal numbers of cations and anions.

Example 17.3 Calculating K_{sp} from Solubility (More Complicated Example)

By experiment, it is found that 1.2×10^{-3} mol of lead(II) iodide, PbI_2, dissolves in 1 L of aqueous solution at 25°C. What is the solubility product constant at this temperature?

Solution

STEP 1 Suppose the solid lead(II) iodide is mixed into 1 L of solution. We find that 1.2×10^{-3} mol dissolves to form 1.2×10^{-3} mol Pb^{2+} and $2 \times (1.2 \times 10^{-3})$ mol I^-, as summarized in the table below.

STEP 2 We substitute into the equilibrium equation.

$$K_{sp} = [Pb^{2+}][I^-]^2 = (1.2 \times 10^{-3}) \times (2 \times 1.2 \times 10^{-3})^2$$
$$= \mathbf{6.9 \times 10^{-9}}$$

CONCENTRATION (*M*)	$PbI_2(s)$	$\rightleftharpoons$	$Pb^{2+}(aq)$	+	$2I^-(aq)$
Starting			0		0
Change			$+1.2 \times 10^{-3}$		$+2 \times (1.2 \times 10^{-3})$
Equilibrium			1.2×10^{-3}		$2 \times (1.2 \times 10^{-3})$

Exercise 17.3

Lead(II) arsenate, $Pb_3(AsO_4)_2$, has been used as an insecticide. It is only slightly soluble in water. If the solubility is 3.0×10^{-5} g/L, what is the solubility product constant? Assume that the solubility equilibrium is the only important one.

(See Problems 17.17 and 17.18.)

Table 17.1 gives a list of solubility product constants for various ionic compounds. If the solubility product constant is known, the solubility of the compound can be calculated. The problem is the reverse of finding K_{sp} from the solubility.

Example 17.4 Calculating the Solubility from K_{sp}

The mineral fluorite is calcium fluoride, CaF_2. Calculate the solubility (in grams per liter) of calcium fluoride in water from the known solubility product constant (3.4×10^{-11}).

Solution

STEP 1 Let x be the molar solubility of CaF_2. When solid CaF_2 is mixed into a liter of solution, x mol dissolves, forming x mol Ca^{2+} and $2x$ mol F^-.

CONCENTRATION (M)	$CaF_2(s) \rightleftharpoons$ $Ca^{2+}(aq)$	+ $2F^-(aq)$
Starting	0	0
Change	$+x$	$+2x$
Equilibrium	x	$2x$

STEP 2 We substitute into the equilibrium equation.

$$[Ca^{2+}][F^-]^2 = K_{sp}$$
$$(x) \times (2x)^2 = 3.4 \times 10^{-11}$$
$$4x^3 = 3.4 \times 10^{-11}$$

STEP 3 We now solve for x.

$$x = \sqrt[3]{\frac{3.4 \times 10^{-11}}{4}} = 2.0 \times 10^{-4}$$

(With minor variations, you obtain this answer on a scientific calculator as follows: While the result of dividing 3.4×10^{-11} by 4 is in the calculator, press INV, then y^x; enter *3* and press =.) Thus, the molar solubility is 2.0×10^{-4} mol CaF_2 per liter. To get the solubility in grams per liter, we convert, using the molar mass of CaF_2 (78.1 g/mol).

$$\text{Solubility} = 2.0 \times 10^{-4}\ \cancel{\text{mol CaF}_2}/\text{L} \times \frac{78.1\ \text{g CaF}_2}{1\ \cancel{\text{mol CaF}_2}}$$
$$= \mathbf{1.6 \times 10^{-2}\ g\ CaF_2/L}$$

Exercise 17.4

Anhydrite is a calcium sulfate mineral deposited when seawater evaporates. What is the solubility of calcium sulfate, in grams per liter? Table 17.1 gives the solubility product for calcium sulfate. (See Problems 17.21, 17.22, 17.23, and 17.24.)

These examples illustrate the relationship between the solubility of a slightly soluble ionic compound in pure water and its solubility product constant. In the next sections, we will see how the solubility product constant can be used to calculate the solubility in the presence of other ions. It is also useful in deciding whether to expect precipitation under given conditions.

17.2 SOLUBILITY AND THE COMMON-ION EFFECT

The importance of the solubility product constant becomes apparent when we consider the solubility of one salt in the solution of another having the same cation or anion. For example, suppose we wish to know the solubility of calcium oxalate

Table 17.1 Solubility Product Constants, K_{sp}

Substance	Formula	K_{sp}
Aluminum hydroxide	$Al(OH)_3$	4.6×10^{-33}
Barium chromate	$BaCrO_4$	1.2×10^{-10}
Barium fluoride	BaF_2	1.0×10^{-6}
Barium sulfate	$BaSO_4$	1.1×10^{-10}
Cadmium oxalate	CdC_2O_4	1.5×10^{-8}
Cadmium sulfide	CdS	8×10^{-27}
Calcium carbonate	$CaCO_3$	3.8×10^{-9}
Calcium fluoride	CaF_2	3.4×10^{-11}
Calcium oxalate	CaC_2O_4	2.3×10^{-9}
Calcium phosphate	$Ca_3(PO_4)_2$	1×10^{-26}
Calcium sulfate	$CaSO_4$	2.4×10^{-5}
Cobalt(II) sulfide	CoS	4×10^{-21}
Copper(II) hydroxide	$Cu(OH)_2$	2.6×10^{-19}
Copper(II) sulfide	CuS	6×10^{-36}
Iron(II) hydroxide	$Fe(OH)_2$	8×10^{-16}
Iron(II) sulfide	FeS	6×10^{-18}
Iron(III) hydroxide	$Fe(OH)_3$	2.5×10^{-39}
Lead(II) arsenate	$Pb_3(AsO_4)_2$	4×10^{-36}
Lead(II) chloride	$PbCl_2$	1.6×10^{-5}
Lead(II) chromate	$PbCrO_4$	1.8×10^{-14}
Lead(II) iodide	PbI_2	6.5×10^{-9}
Lead(II) sulfate	$PbSO_4$	1.7×10^{-8}
Lead(II) sulfide	PbS	2.5×10^{-27}
Magnesium arsenate	$Mg_3(AsO_4)_2$	2×10^{-20}
Magnesium carbonate	$MgCO_3$	1.0×10^{-5}
Magnesium hydroxide	$Mg(OH)_2$	1.8×10^{-11}
Magnesium oxalate	MgC_2O_4	8.5×10^{-5}
Manganese(II) sulfide	MnS	2.5×10^{-10}
Mercury(I) chloride	Hg_2Cl_2	1.3×10^{-18}
Mercury(II) sulfide	HgS	1.6×10^{-52}
Nickel(II) hydroxide	$Ni(OH)_2$	2.0×10^{-15}
Nickel(II) sulfide	NiS	3×10^{-19}
Silver acetate	$AgC_2H_3O_2$	2.0×10^{-3}
Silver bromide	$AgBr$	5.0×10^{-13}
Silver chloride	$AgCl$	1.8×10^{-10}
Silver chromate	Ag_2CrO_4	1.1×10^{-12}
Silver iodide	AgI	8.3×10^{-17}
Silver sulfide	Ag_2S	6×10^{-50}
Strontium carbonate	$SrCO_3$	9.3×10^{-10}
Strontium chromate	$SrCrO_4$	3.5×10^{-5}
Strontium sulfate	$SrSO_4$	2.5×10^{-7}
Zinc hydroxide	$Zn(OH)_2$	2.1×10^{-16}
Zinc sulfide	ZnS	1.1×10^{-21}

Figure 17.2
Demonstration of the common-ion effect. When the experimenter adds lead(II) nitrate solution (colorless) from the dropper to the saturated solution of lead(II) chromate (pale yellow), a yellow precipitate of lead chromate forms.

in a solution of calcium chloride. Each salt contributes the same cation (Ca^{2+}). The effect of the calcium ion provided by the calcium chloride is to make calcium oxalate less soluble than it would be in pure water.

We can explain this decrease in solubility in terms of Le Chatelier's principle. Suppose we first mix crystals of calcium oxalate in a quantity of pure water to establish the equilibrium

$$CaC_2O_4(s) \rightleftharpoons Ca^{2+}(aq) + C_2O_4^{2-}(aq)$$

Now imagine that we add some calcium chloride. Calcium chloride is a soluble salt, so it dissolves to give an increase in calcium-ion concentration. We can regard this increase as a stress on the original equilibrium. According to Le Chatelier's principle, the ions will react to remove some of the added calcium ion.

$$CaC_2O_4(s) \longleftarrow Ca^{2+}(aq) + C_2O_4^{2-}(aq)$$

In other words, some calcium oxalate precipitates from the solution. The solution now contains less calcium oxalate. We conclude that calcium oxalate is less soluble in a solution of calcium chloride than it is in pure water.

A solution of soluble calcium chloride has an ion *in common* with the slightly soluble calcium oxalate. The decrease in solubility of calcium oxalate in a solution of calcium chloride is an example of the *common-ion effect* (first discussed in Section 16.8). In general, any ionic equilibrium is affected by a substance producing an ion involved in the equilibrium, as we would predict from Le Chatelier's principle. Figure 17.2 illustrates the principle of the common-ion effect for lead(II) chromate, $PbCrO_4$, which is only slightly soluble in water at 25°C. When the very soluble $Pb(NO_3)_2$ is added to a saturated solution of $PbCrO_4$, the concentration of common ion, Pb^{2+}, increases and $PbCrO_4$ precipitates. The equilibrium is

$$PbCrO_4(s) \rightleftharpoons Pb^{2+}(aq) + CrO_4^{2-}(aq)$$

The next example shows how we can calculate the solubility of a slightly soluble salt in a solution containing a substance with a common ion.

Example 17.5 Calculating the Solubility of a Slightly Soluble Salt in a Solution of a Common Ion

What is the molar solubility of calcium oxalate in 0.15 *M* calcium chloride? Compare this molar solubility with that found earlier (see Example 17.2) for CaC_2O_4 in pure water (4.8×10^{-5} *M*). The solubility product constant for calcium oxalate is 2.3×10^{-9}.

Solution

As we noted earlier, the two salts have a common ion (Ca^{2+}). One of the salts is soluble ($CaCl_2$) and provides Ca^{2+} ion that suppresses the solubility of the slightly soluble salt by the common-ion effect.

STEP 1 Suppose solid CaC_2O_4 is mixed into 1 L of 0.15 *M* $CaCl_2$ and that the molar solubility of the CaC_2O_4 is *x M*. Thus, at the start (before CaC_2O_4 dissolves), there is 0.15 mol Ca^{2+} in the solution. Of the solid CaC_2O_4, *x* mol dissolves to give *x* mol of additional Ca^{2+} and *x* mol $C_2O_4^{2-}$. The following table summarizes these results:

CONCENTRATION (*M*)	$CaC_2O_4(s) \rightleftharpoons$	$Ca^{2+}(aq)$	+ $C_2O_4^{2-}(aq)$
Starting		0.15	0
Change		$+x$	$+x$
Equilibrium		$0.15 + x$	x

STEP 2 We substitute into the equilibrium equation.

$$[Ca^{2+}][C_2O_4^{2-}] = K_{sp}$$

$$(0.15 + x)x = 2.3 \times 10^{-9}$$

STEP 3 Let us rearrange this equation in the form

$$x = \frac{2.3 \times 10^{-9}}{0.15 + x}$$

Because calcium oxalate is only slightly soluble, we might expect x to be negligible compared to 0.15. In that case,

$$0.15 + x \simeq 0.15$$

and the previous equation becomes

$$x \simeq \frac{2.3 \times 10^{-9}}{0.15} = 1.5 \times 10^{-8}$$

Note that x is indeed much smaller than 0.15, so our assumption was correct. Therefore, the molar solubility of calcium oxalate in 0.15 M $CaCl_2$ is **1.5×10^{-8} M**. In pure water, the molar solubility is 4.8×10^{-5} M, which is over **3000 times greater.**

Exercise 17.5

(a) Calculate the molar solubility of barium fluoride, BaF_2, in water at 25°C. The solubility product constant for BaF_2 at this temperature is 1.0×10^{-6}.
(b) What is the molar solubility of barium fluoride in 0.15 M NaF at 25°C? Compare the solubility in this case with that of BaF_2 in pure water.

(See Problems 17.25, 17.26, 17.27, and 17.28.)

17.3 PRECIPITATION CALCULATIONS

In the chapter opening, we mentioned that calcium oxalate precipitates to form kidney stones. The same salt would precipitate in the body if oxalic acid (a poison) were to be accidentally ingested, because Ca^{2+} ion is present in the blood. To understand processes such as these, we must understand the conditions under which precipitation occurs. Precipitation is merely another way of looking at a solubility equilibrium. Rather than ask how much of a substance will dissolve in a solution, we ask, Will precipitation occur for given starting ion concentrations?

Criterion for Precipitation

The question just asked can be stated more generally: Given the concentrations of substances in a reaction mixture, will the reaction go in the forward or the reverse direction? To answer this, we evaluate the *reaction quotient* Q_c and compare it with the equilibrium constant K_c.● The reaction quotient has the same form as the equilibrium-constant expression, but the concentrations of substances are not necessarily equilibrium values. Rather, they are concentrations at the start of a reaction. To predict the direction of reaction, we compare Q_c with K_c.

● The reaction quotient was discussed in Section 15.5.

If $Q_c < K_c$, the reaction should go in the forward direction.
If $Q_c = K_c$, the reaction mixture is at equilibrium.
If $Q_c > K_c$, the reaction should go in the reverse direction.

Let us apply this test to a precipitation reaction. Suppose lead(II) nitrate, $Pb(NO_3)_2$, and sodium chloride, NaCl, are added to water to give a solution that is 0.050 M Pb^{2+} and 0.10 M Cl^-. Will lead(II) chloride, $PbCl_2$, precipitate? To answer this, we first write the solubility equilibrium.

$$PbCl_2(s) \xrightleftharpoons{H_2O} Pb^{2+}(aq) + 2Cl^-(aq)$$

The reaction quotient has the form of the equilibrium-constant expression, which in this case is the K_{sp} expression, though the concentrations of the products are starting values, denoted by the subscript i.

$$Q_c = [Pb^{2+}]_i[Cl^-]_i^2$$

Here Q_c for a solubility reaction is often called the **ion product** (rather than reaction quotient), because it is *the product of ion concentrations in a solution, each concentration raised to a power equal to the number of ions in the formula of the ionic compound.*

To evaluate the ion product Q_c, we substitute the concentrations of Pb^{2+} and Cl^- ions that are in the solution at the start of the reaction. We have 0.050 *M* Pb^{2+} and 0.10 *M* Cl^-. Substituting, we find that

$$Q_c = (0.050)(0.10)^2 = 5.0 \times 10^{-4}$$

Because K_{sp} for $PbCl_2$ is 1.6×10^{-5} (Table 17.1), we see that Q_c is greater than K_{sp}. Therefore the reaction goes in the reverse direction. That is,

$$PbCl_2(s) \longleftarrow Pb^{2+}(aq) + 2Cl^-(aq)$$

In other words, Pb^{2+} and Cl^- react to precipitate $PbCl_2$. As precipitation occurs, the ion concentrations, and hence the ion product, decrease. Precipitation ceases when the ion product equals K_{sp}. The reaction mixture is then at equilibrium.

We can summarize our conclusions in terms of the following criterion for precipitation. Precipitation is expected to occur if the ion product for a solubility reaction is greater than K_{sp}.● If the ion product is less than K_{sp}, precipitation will not occur (the solution is unsaturated with respect to the ionic compound). If the ion product equals K_{sp}, the reaction is at equilibrium (the solution is saturated with the ionic compound). The next example is a simple application of this criterion for precipitation.

● Precipitation may not occur even though the ion product has been exceeded. In such a case, the solution is supersaturated. Usually a small crystal forms after a time, and then precipitation occurs rapidly.

Example 17.6 Predicting Whether Precipitation Will Occur (Given Ion Concentrations)

The concentration of calcium ion in blood plasma is 0.0025 *M*. If the concentration of oxalate ion is 1.0×10^{-7} *M*, do you expect calcium oxalate to precipitate? K_{sp} for calcium oxalate is 2.3×10^{-9}.

Solution

The ion product for calcium oxalate is

$$\begin{aligned}\text{Ion product} &= [Ca^{2+}]_i[C_2O_4^{2-}]_i \\ &= (0.0025) \times (1.0 \times 10^{-7}) \\ &= 2.5 \times 10^{-10}\end{aligned}$$

This value is smaller than the solubility product constant, so we do not expect precipitation to occur.

Exercise 17.6

Anhydrite is a mineral whose chemical composition is $CaSO_4$ (calcium sulfate). An inland lake has Ca^{2+} and SO_4^{2-} concentrations of 0.0052 *M* and 0.0041 *M*, respectively. If these concentrations were doubled by evaporation, would you expect calcium sulfate to precipitate? (See Problems 17.31 and 17.32.)

The following example from quantitative analysis is a typical industrial or laboratory problem in precipitation. We are given the volumes and concentrations of two solutions and asked whether a precipitate will form when the solutions are mixed.

Example 17.7 Predicting Whether Precipitation Will Occur (Given Solution Volumes)

Sulfate ion, SO_4^{2-}, in solution is often determined quantitatively by precipitating it as barium sulfate, $BaSO_4$. The sulfate ion may have been formed from a sulfur compound. Analysis for the amount of sulfate ion then indicates the percentage of sulfur in the compound. Is a precipitate expected to form at equilibrium when 50.0 mL of 0.0010 *M* $BaCl_2$ is added to 50.0 mL of 0.00010 *M* Na_2SO_4? The solubility product constant for barium sulfate is 1.1×10^{-10}. Assume that the total volume of solution, after mixing, equals the sum of the volumes of the separate solutions.

Solution

Let us first calculate the concentrations of Ba^{2+} and SO_4^{2-}, assuming that no precipitate of barium sulfate has formed. The molar amount of Ba^{2+} present in 50.0 mL (= 0.0500 L) of 0.0010 *M* $BaCl_2$ is

$$\text{Amount of Ba}^{2+} = \frac{0.0010 \text{ mol Ba}^{2+}}{1 \cancel{\text{L soln}}} \times 0.050 \cancel{\text{L soln}}$$

$$\text{Amount of Ba}^{2+} = 5.0 \times 10^{-5} \text{ mol Ba}^{2+}$$

The molar concentration of Ba^{2+} in the total solution equals the molar amount of Ba^{2+} divided by the total volume (0.0500 L $BaCl_2$ + 0.0500 L Na_2SO_4 = 0.1000 L).

$$[\text{Ba}^{2+}] = \frac{5.0 \times 10^{-5} \text{ mol}}{0.1000 \text{ L soln}} = 5.0 \times 10^{-4}\ M$$

Similarly, we find

$$[\text{SO}_4^{2-}] = 5.0 \times 10^{-5}\ M$$

(Try the calculations for SO_4^{2-}.) The ion product is

$$Q_c = [\text{Ba}^{2+}]_i[\text{SO}_4^{2-}]_i = (5.0 \times 10^{-4}) \times (5.0 \times 10^{-5})$$
$$= 2.5 \times 10^{-8}$$

Because the ion product is greater than the solubility product constant (1.1×10^{-10}), **we expect barium sulfate to precipitate.**

Exercise 17.7

A solution of 0.00016 *M* lead(II) nitrate, $Pb(NO_3)_2$, was poured into 456 mL of 0.00023 *M* sodium sulfate, Na_2SO_4. Would a precipitate of lead(II) sulfate, $PbSO_4$, be expected to form if 255 mL of the lead nitrate solution were added?

(See Problems 17.33 and 17.34.)

Completeness of Precipitation

For various reasons, we may want to know how nearly completely an ion precipitates from a solution. We may want to know whether more than 99% of the Mg^{2+} ion in seawater can be removed by precipitating the ion as $Mg(OH)_2$ with a specified concentration of OH^- (Figure 17.3). Or we may want to know whether a

Figure 17.3
Vats for the precipitation of magnesium hydroxide from seawater. Seawater contains magnesium ion (in addition to other ions). When base is added to it, magnesium hydroxide precipitates. This is the source of magnesium metal.

precipitation is adequate for the quantitative determination of an ion. As indicated in Example 17.7, we can determine the amount of SO_4^{2-} ion in solution by precipitating it as $BaSO_4$. The precipitate is filtered from the solution, dried, and then weighed. If this determination of SO_4^{2-} ion is to be useful, most of the ion must be precipitated, and very little (less than 0.1%) should remain in solution. The next example shows how to calculate the concentration of an ion remaining after precipitation and the percentage of the ion that was not precipitated.

Example 17.8 Determining the Completeness of Precipitation

In Example 17.7, solutions of $BaCl_2$ and Na_2SO_4 were mixed and a precipitate of $BaSO_4$ formed. *Before precipitation,* we have a solution containing 0.00050 mol Ba^{2+} per liter and 0.000050 mol SO_4^{2-} per liter. Calculate the molar concentration of SO_4^{2-} ion left *after $BaSO_4$ precipitates*. What percentage of SO_4^{2-} is not precipitated?

Solution

Before we try to set up the calculation, let us think about the reaction (a good thing to do in any case). Because the solubility product constant for $BaSO_4$ is very small (1.1×10^{-10}, according to Table 17.1), the Ba^{2+} and SO_4^{2-} ions will react until almost all of the ion that is the limiting reactant is consumed. The reaction is $Ba^{2+} + SO_4^{2-} \longrightarrow BaSO_4$, so equal molar amounts of the ions react. The solution initially contains 0.00050 mol Ba^{2+} and 0.000050 mol SO_4^{2-} per liter. Therefore, the SO_4^{2-} ion is the limiting reactant. If the reaction were to go to completion, we would get a precipitate of $BaSO_4$ in a solution of unconsumed Ba^{2+} ion. However, we would expect $BaSO_4$ to dissolve to a slight extent in the solution of the common ion, Ba^{2+}, and this would produce some SO_4^{2-} ion.

This suggests that we solve the problem in two parts. First, we assume the reaction goes to completion. We do a *stoichiometry calculation* to find the concentration of Ba^{2+}. Then, to find the SO_4^{2-} concentration, we do an *equilibrium calculation* of the solubility of $BaSO_4$ in a solution of the common ion Ba^{2+} (as in Example 17.5).

STOICHIOMETRY CALCULATION The precipitation equation, with initial and final amounts of substance, assuming the reaction were to go to completion, is shown in the table below.

MOLES OF SUBSTANCE	$Ba^{2+}(aq)$	+	$SO_4^{2-}(aq)$	→	$BaSO_4(s)$
Initial moles	0.00050		0.000050		0
Final moles	(0.00050 − 0.000050) = 0.00045		0		0.000050

EQUILIBRIUM CALCULATION The concentration table for the solubility equation follows.

CONCENTRATION (*M*)	$BaSO_4(s)$	$\rightleftharpoons$	$Ba^{2+}(aq)$	+	$SO_4^{2-}(aq)$
Starting			0.00045		0
Change			$+x$		$+x$
Equilibrium			$0.00045 + x$		x

Because K_{sp} for $BaSO_4$ is 1.1×10^{-10}, the equilibrium equation is

$$[Ba^{2+}][SO_4^{2-}] = K_{sp}$$

$$(0.00045 + x)x = 1.1 \times 10^{-10}$$

If we assume that x can be neglected in comparison to 0.00045, this equation is easily solved for x.

$$x \simeq \frac{1.1 \times 10^{-10}}{0.00045} = 2.4 \times 10^{-7}$$

Note that x can be neglected compared to 0.00045. **Thus, the SO_4^{2-} concentration is 2.4×10^{-7} *M*.** The concentration of the ion before precipitation of $BaSO_4$ was 5.0×10^{-5} *M*, so the percentage of SO_4^{2-} remaining in solution is

$$\frac{2.4 \times 10^{-7}}{5.0 \times 10^{-5}} \times 100\% = \mathbf{0.48\%}$$

This is somewhat higher than the 0.1% required for a precise analytical procedure. However, we could reduce the concentration of SO_4^{2-} that remains by increasing the concentration of Ba^{2+}.

Exercise 17.8

Lead chromate, $PbCrO_4$, is a yellow pigment used in paints. Suppose 0.50 L of 1.0×10^{-5} *M* $Pb(C_2H_3O_2)_2$ and 0.50 L of 1.0×10^{-3} *M* K_2CrO_4 are mixed. Calculate the equilibrium concentration of Pb^{2+} ion remaining in solution after $PbCrO_4$ precipitates. What is the percentage of Pb^{2+} remaining in solution after precipitation of $PbCrO_4$?

(See Problems 17.39 and 17.40.)

Fractional Precipitation

Fractional precipitation is *the technique of separating two or more ions from a solution by adding a reactant that precipitates first one ion, then another, and so forth*. For example, suppose a solution is 0.10 M Ba^{2+} and 0.10 M Sr^{2+}. As we will show in the following paragraphs, when we slowly add a concentrated solution of potassium chromate, K_2CrO_4, to the solution of Ba^{2+} and Sr^{2+} ions, barium chromate precipitates first. After most of the Ba^{2+} ion has precipitated, strontium chromate begins to come out of solution. It is therefore possible to separate Ba^{2+} and Sr^{2+} ions from a solution by fractional precipitations using K_2CrO_4.

To understand why Ba^{2+} and Sr^{2+} ions can be separated this way, let us calculate (1) the concentration of CrO_4^{2-} necessary to just begin the precipitation of $BaCrO_4$ and (2) that concentration necessary to just begin the precipitation of $SrCrO_4$. We will ignore any volume change in the solution of Ba^{2+} and Sr^{2+} ions resulting from the addition of the concentrated K_2CrO_4 solution. To calculate the CrO_4^{2-} concentration when $BaCrO_4$ begins to precipitate, we substitute the initial Ba^{2+} concentration into the solubility product equation. K_{sp} for $BaCrO_4$ is 1.2×10^{-10}.

$$[Ba^{2+}][CrO_4^{2-}] = K_{sp} \text{ (for } BaCrO_4)$$
$$(0.10)[CrO_4^{2-}] = 1.2 \times 10^{-10}$$
$$[CrO_4^{2-}] = \frac{1.2 \times 10^{-10}}{0.10} = 1.2 \times 10^{-9}\ M$$

In the same way, we can calculate the CrO_4^{2-} concentration when $SrCrO_4$ begins to precipitate. K_{sp} for $SrCrO_4$ is 3.5×10^{-5}.

$$[Sr^{2+}][CrO_4^{2-}] = K_{sp} \text{ (for } SrCrO_4)$$
$$(0.10)[CrO_4^{2-}] = 3.5 \times 10^{-5}$$
$$[CrO_4^{2-}] = \frac{3.5 \times 10^{-5}}{0.10} = 3.5 \times 10^{-4}\ M$$

Note that $BaCrO_4$ precipitates first because the CrO_4^{2-} concentration necessary to form the $BaCrO_4$ precipitate is smaller.

These results reveal that as the solution of K_2CrO_4 is slowly added to the solution of Ba^{2+} and Sr^{2+}, barium chromate begins to precipitate when the CrO_4^{2-} concentration reaches $1.2 \times 10^{-9}\ M$. Barium chromate continues to precipitate as K_2CrO_4 is added. When the concentration of CrO_4^{2-} reaches $3.5 \times 10^{-4}\ M$, strontium chromate begins to precipitate.

What is the percentage of Ba^{2+} ion remaining just as $SrCrO_4$ begins to precipitate? Let us first calculate the concentration of Ba^{2+} at this point. We write the solubility product equation and substitute $[CrO_4^{2-}] = 3.5 \times 10^{-4}$, which is the concentration of chromate ion when $SrCrO_4$ begins to precipitate.

$$[Ba^{2+}][CrO_4^{2-}] = K_{sp} \text{ (for } BaCrO_4)$$
$$[Ba^{2+}](3.5 \times 10^{-4}) = 1.2 \times 10^{-10}$$
$$[Ba^{2+}] = \frac{1.2 \times 10^{-10}}{3.5 \times 10^{-4}} = 3.4 \times 10^{-7}\ M$$

Figure 17.4
Titration of chloride ion by silver nitrate using potassium chromate as an indicator. *Left:* A small amount of K_2CrO_4 (yellow) is added to a solution containing an unknown amount of Cl^- ion. *Center:* The solution is titrated by $AgNO_3$ solution, giving a white precipitate of AgCl. *Right:* When nearly all of the Cl^- ion has precipitated as AgCl, silver chromate begins to precipitate. Silver chromate, Ag_2CrO_4, has a red-brown color, and the appearance of this color signals the end of the titration. An excess of Ag^+ was added to show the color of Ag_2CrO_4 more clearly.

To calculate the percentage of Ba^{2+} ion remaining, we divide this concentration of Ba^{2+} by the initial concentration (0.10 *M*), and multiply by 100%.

$$\frac{3.4 \times 10^{-7}}{0.10} \times 100\% = 0.00034\%$$

The percentage of Ba^{2+} ion remaining in solution is quite low, so most of the Ba^{2+} ion has precipitated by the time $SrCrO_4$ begins to precipitate. We conclude that Ba^{2+} and Sr^{2+} can indeed be separated by fractional precipitation. (Another application of fractional precipitation is shown in Figure 17.4.)

17.4 EFFECT OF pH ON SOLUBILITY

In discussing solubility in the previous sections, we assumed that the only equilibrium of interest was the one between the solid ionic compound and its ions in solution. Sometimes, however, it is necessary to account for other reactions the ions might undergo. For example, if the anion is the conjugate base of a weak acid, it reacts with H^+ ion. We should expect the solubility to be affected by pH. We will look into this possibility in this section.

Qualitative Effect of pH

Consider the equilibrium between solid calcium oxalate, CaC_2O_4, and its ions in aqueous solution:

$$CaC_2O_4(s) \rightleftharpoons Ca^{2+}(aq) + C_2O_4{}^{2-}(aq)$$

Because the oxalate ion is conjugate to a weak acid (hydrogen oxalate ion, $HC_2O_4{}^-$), we would expect it to react with any H^+ ion that is added—say, from a strong acid:

$$C_2O_4{}^{2-}(aq) + H^+(aq) \rightleftharpoons HC_2O_4{}^-(aq)$$

According to Le Chatelier's principle, as $C_2O_4{}^{2-}$ ion is removed by reaction with H^+ ion, more calcium oxalate dissolves to replenish some of the $C_2O_4{}^{2-}$ ion.

$$CaC_2O_4(s) \longrightarrow Ca^{2+}(aq) + C_2O_4{}^{2-}(aq)$$

Therefore, we expect calcium oxalate to be more soluble in acidic solution (low pH) than it is in pure water.

In general, salts of weak acids should be expected to be more soluble in acidic solutions. As an example, consider the process of tooth decay. Bacteria on the teeth produce an acidic medium as a result of the metabolism of sugar. Teeth are normally composed of a calcium phosphate mineral hydroxyapatite, which we can denote as either $Ca_5(PO_4)_3OH$ or $3Ca_3(PO_4)_2 \cdot Ca(OH)_2$. This mineral salt of the weak acid H_2O (conjugate to OH^-) dissolves in the presence of the acid medium, producing cavities in the teeth. Fluoride toothpastes provide F^- ion, which gradually replaces the OH^- ion in the tooth to produce fluorapatite, $Ca_5(PO_4)_3F$ or $3Ca_3(PO_4)_2 \cdot CaF_2$, which is much less soluble than hydroxyapatite.

Example 17.9 Determining the Qualitative Effect of pH on Solubility

Consider two slightly soluble salts, calcium carbonate and calcium sulfate. Which of these would have its solubility more affected by the addition of strong acid? Would the solubility of the salt increase or decrease?

Solution

Calcium carbonate gives the solubility equilibrium

$$CaCO_3(s) \rightleftharpoons Ca^{2+}(aq) + CO_3^{2-}(aq)$$

When a strong acid is added, the hydrogen ion reacts with carbonate ion, because it is conjugate to a weak acid (HCO_3^-).

$$H^+(aq) + CO_3^{2-}(aq) \rightleftharpoons HCO_3^-(aq)$$

As carbonate ion is removed, calcium carbonate dissolves. Moreover, the hydrogen carbonate ion itself is removed in further reaction.

$$H^+(aq) + HCO_3^-(aq) \longrightarrow H_2CO_3(aq) \longrightarrow H_2O(l) + CO_2(g)$$

Bubbles of carbon dioxide gas appear as more calcium carbonate dissolves.

In the case of calcium sulfate, the corresponding equilibria are

$$CaSO_4(s) \rightleftharpoons Ca^{2+}(aq) + SO_4^{2-}(aq)$$
$$H^+(aq) + SO_4^{2-}(aq) \rightleftharpoons HSO_4^-(aq)$$

Again, we see that the anion of the insoluble salt is removed by reaction with hydrogen ion. We would expect calcium sulfate to become more soluble in strong acid (except in sulfuric acid, which would supply sulfate ion and therefore precipitate calcium sulfate by the common-ion effect). However, HSO_4^- is a much stronger acid than HCO_3^-, as we can see by comparing acid-ionization constants. (The values K_{a2} for H_2CO_3 and K_a for HSO_4^- in Table 16.2 are 4.8×10^{-11} and 1.1×10^{-2}, respectively.) **Thus, calcium carbonate becomes much more soluble in acidic solution, whereas the solubility of calcium sulfate is less affected.**

Exercise 17.9

Which of the following salts would have its solubility more affected by changes in pH: silver chloride or silver cyanide? (See Problems 17.45 and 17.46.)

Separation of Metal Ions by Sulfide Precipitation

Many metal sulfides are insoluble in water but dissolve in acidic solution. The change in solubility with pH, or hydrogen-ion concentration, can be used to separate a mixture of metal ions.● As an example, consider an aqueous solution containing 0.10 *M* zinc ion, Zn^{2+}, and 0.10 *M* lead(II) ion, Pb^{2+}. We would like to separate these ions from one another. Let us first look at the possibility of precipitating the metal sulfides by dissolving H_2S gas in the aqueous solution.

● A separation scheme based on this variation of solubility is discussed in Section 17.7.

When H_2S dissolves in water, it ionizes as a diprotic acid.

$$H_2S(aq) \rightleftharpoons H^+(aq) + HS^-(aq)$$
$$HS^-(aq) \rightleftharpoons H^+(aq) + S^{2-}(aq)$$

Sulfide ion, S^{2-}, produced in this way may react with metal ions to precipitate the metal sulfides. For example,

$$Zn^{2+}(aq) + S^{2-}(aq) \rightleftharpoons ZnS(s)$$

To predict whether we will get a precipitate of ZnS, we calculate the ion product of ZnS and compare it with its K_{sp}. We can similarly predict whether PbS will precipitate.

We know the metal-ion concentrations, but to calculate the ion products for ZnS and PbS we also need the S^{2-} ion concentration. Recall that the concentration of the anion from the second ionization of a diprotic acid equals K_{a2}.● Thus, the second ionization of H_2S gives the anion S^{2-}, whose concentration should equal K_{a2} for H_2S. From Table 16.2, we see that K_{a2} for H_2S is 1.2×10^{-13}. Hence $[S^{2-}] = 1.2 \times 10^{-13}$ M.

● This result was obtained in Section 16.5.

The ion product of ZnS is

$$[Zn^{2+}]_i[S^{2-}]_i = (0.10)(1.2 \times 10^{-13}) = 1.2 \times 10^{-14}$$

This is much larger than K_{sp} for ZnS (1.1×10^{-21}), so we conclude that zinc sulfide should precipitate. The ion product of PbS also equals 1.2×10^{-14}. This is larger than K_{sp} for PbS (2.5×10^{-27}), so lead(II) sulfide is also expected to precipitate. Because ZnS and PbS precipitate together, we have not obtained a separation of the metal ions.

Now consider the effect of adding a strong acid to the solution of the Zn^{2+} and Pb^{2+} ions before saturating it with H_2S gas. The H^+ ion from the strong acid will repress the ionization of H_2S, giving a lower concentration of S^{2-} ion. By adjusting the H^+ ion concentration, we can control the S^{2-} concentration and therefore the ion products for ZnS and PbS. As we will show in the next example, we can adjust the H^+ ion concentration so that only PbS precipitates when the solution is saturated with H_2S. The precipitate of PbS can be filtered from the solution of Zn^{2+} ion. In this way, we can separate a mixture of the metal ions.

In order to obtain the H^+ ion concentration necessary to keep some ions from precipitating while others form the sulfides, let us look at the net, or overall, equation for the acid ionizations of H_2S. We obtain this net equation by taking the sum of the acid ionizations.

$$\begin{array}{l} H_2S(aq) \rightleftharpoons H^+(aq) + HS^-(aq) \\ \underline{HS^-(aq) \rightleftharpoons H^+(aq) + S^{2-}(aq)} \\ H_2S(aq) \rightleftharpoons 2H^+(aq) + S^{2-}(aq) \end{array}$$

The equilibrium constant for an equation obtained by taking the sum of two or more equations is obtained by multiplying the equilibrium constants of the equations that were summed.● Thus, the equilibrium constant for the equation obtained by taking the sum of the acid ionizations of H_2S is $K_{a1}K_{a2}$, where K_{a1} and K_{a2} are the equilibrium constants for the successive ionizations of H_2S. We can verify this by substituting the expressions for K_{a1} and K_{a2} into $K_{a1}K_{a2}$.

● See the end of Section 15.2.

$$K_{a1}K_{a2} = \frac{[H^+][\cancel{HS^-}]}{[H_2S]} \times \frac{[H^+][S^{2-}]}{[\cancel{HS^-}]} = \frac{[H^+]^2]S^{2-}]}{[H_2S]}$$

From Table 16.2, we see that K_{a1} equals 8.9×10^{-8} and that K_{a2} equals 1.2×10^{-13}. Hence, $K_{a1}K_{a2}$ equals $(8.9 \times 10^{-8}) \times (1.2 \times 10^{-13})$, or 1.1×10^{-20}.

A saturated solution of hydrogen sulfide is 0.10 M H_2S, and this is not significantly altered by changes in H^+ ion concentration. We substitute this concentration of H_2S into the equilibrium equation.

$$\frac{[H^+]^2[S^{2-}]}{[H_2S]} = K_{a1}K_{a2}$$

This gives

$$\frac{[H^+]^2[S^{2-}]}{0.10} = 1.1 \times 10^{-20}$$

Note that as $[H^+]$ increases, $[S^{2-}]$ decreases. Thus, by adjusting the H^+ ion concentration, we can obtain any desired S^{2-} concentration.

Example 17.10 Separating Metal Ions by Sulfide Precipitation

Consider a solution of 0.10 *M* Zn^{2+} and 0.10 *M* Pb^{2+} saturated with H_2S. What range of H^+ ion concentration will give a precipitate of one of the metal sulfides, leaving the other metal ion in solution? The concentration of H_2S in a saturated solution is 0.10 *M*.

Solution

We first calculate the minimum S^{2-} ion concentration that must be present before each metal sulfide can precipitate. Once we have the S^{2-} ion concentrations necessary to precipitate ZnS and PbS, we can calculate the corresponding H^+ ion concentrations.

STEP 1 We find the minimum S^{2-} ion concentration necessary to give a precipitate of ZnS by substituting the Zn^{2+} ion concentration into the equilibrium equation

$$[Zn^{2+}][S^{2-}] = K_{sp}$$

to give

$$(0.10)[S^{2-}] = 1.1 \times 10^{-21}$$

Hence, the minimum S^{2-} ion concentration needed to give a precipitate of ZnS is

$$[S^{2-}] = \frac{1.1 \times 10^{-21}}{0.10} = 1.1 \times 10^{-20}$$

Similarly, the minimum S^{2-} ion concentration needed to precipitate PbS is

$$[S^{2-}] = \frac{K_{sp} \text{ (for PbS)}}{[Pb^{2+}]} = \frac{2.5 \times 10^{-27}}{0.10} = 2.5 \times 10^{-26}$$

STEP 2 If we substitute these minimum S^{2-} ion concentrations into the equilibrium equation for the double ionization of H_2S,

$$\frac{[H^+]^2[S^{2-}]}{[H_2S]} = 1.1 \times 10^{-20}$$

we can obtain the corresponding H^+ ion concentrations. For ZnS, we get

$$[H^+]^2 = 1.1 \times 10^{-20} \times \frac{[H_2S]}{[S^{2-}]}$$

$$= 1.1 \times 10^{-20} \times \frac{0.10}{1.1 \times 10^{-20}} = 0.10$$

Or

$$[H^+] = 0.32$$

At any concentration of H^+ ion equal to or less than 0.32 *M*, the S^{2-} ion concentration will be equal to or greater than that necessary to give a precipitate of ZnS. When the H^+ ion concentration is greater than 0.32 *M*, the Zn^{2+} ion will remain in solution.

For PbS, we get

$$[H^+]^2 = 1.1 \times 10^{-20} \times \frac{[H_2S]}{[S^{2-}]}$$

$$= 1.1 \times 10^{-20} \times \frac{0.10}{2.5 \times 10^{-26}} = 4.4 \times 10^4$$

$$[H^+] = 2.1 \times 10^2$$

At any H^+ ion concentration equal to or less than 2.1×10^2 *M*, PbS will precipitate. (It is not possible to get an H^+ ion concentration this large, so we can expect PbS to precipitate in any solution.)

When saturated with H_2S, any solution of 0.10 *M* Zn^{2+} and 0.10 *M* Pb^{2+} in which the H^+ ion concentration is greater than 0.32 *M* will give a precipitate of PbS but not of ZnS.

Exercise 17.10

Find the range of pH that will allow only one of the metal ions in a solution that is 0.050 *M* Cu^{2+}, 0.050 *M* Fe^{2+}, and saturated with H_2S to precipitate as a sulfide.

(See Problems 17.47 and 17.48.)

dilute hydrochloric acid. These ions are precipitated as the chlorides AgCl, Hg_2Cl_2, and $PbCl_2$, which are removed by filtering.

The other metal ions remain in the filtrate, the solution that passes through the filter. Many of these ions can be separated by precipitating them as metal sulfides with H_2S. This separation takes advantage of the fact that only the least soluble metal sulfides precipitate in acidic solution. After these are removed and the solution is made basic, other metal sulfides precipitate.●

● The solubility of the metal sulfides at different pH was discussed in Section 17.4.

Analytical Group II consists of metal ions precipitated as metal sulfides from a solution that is 0.3 *M* H^+ and is saturated with H_2S.● The ions are As^{3+}, Bi^{3+}, Cd^{2+}, Cu^{2+}, Hg^{2+}, Pb^{2+}, Sb^{3+}, and Sn^{4+}. Lead(II) ion can appear in Analytical Group II as well as in Group I, because $PbCl_2$ is somewhat soluble and therefore Pb^{2+} may not be completely precipitated by HCl.

● Hydrogen sulfide is often prepared in the solution by adding thioacetamide, CH_3CSNH_2, and warming.

$$CH_3CSNH_2 + H_2O \longrightarrow CH_3CONH_2 + H_2S$$

Analytical Group III consists of the metal ions in the filtrate from Group II that are precipitated in weakly basic solution with H_2S. The ions Co^{2+}, Fe^{2+}, Mn^{2+}, Ni^{2+}, and Zn^{2+} are precipitated as the sulfides. The ions Al^{3+} and Cr^{3+} are precipitated as hydroxides.

The filtrate obtained after the Group III metal ions have precipitated contains the alkali metal and the alkaline earth ions. The Analytical Group IV ions, Ba^{2+}, Ca^{2+}, Mg^{2+}, and Sr^{2+}, are precipitated as carbonates or phosphates by adding $(NH_4)_2CO_3$ or $(NH_4)_2HPO_4$. The filtrate from this separation contains the Analytical Group V ions, K^+ and Na^+.

To illustrate the separation scheme, suppose we have a solution containing the metal cations Ag^+, Cu^{2+}, Zn^{2+}, Ca^{2+}, and Na^+. When we add dilute HCl(*aq*) to this solution, Ag^+ precipitates as AgCl.

$$Ag^+(aq) + Cl^-(aq) \longrightarrow \underset{\text{white}}{AgCl(s)}$$

We filter off the precipitate of AgCl. The filtrate contains the remaining ions.

When we make the filtrate 0.3 *M* H^+ and then saturate it with H_2S, the Cu^{2+} ion precipitates as CuS.

$$Cu^{2+}(aq) + S^{2-}(aq) \longrightarrow \underset{\text{black}}{CuS(s)}$$

Zinc(II) ion also forms a sulfide, but it is soluble in acidic solution. After removing the precipitate of CuS, the solution is made weakly basic with $NH_3(s)$. This increases the S^{2-} concentration sufficiently to precipitate ZnS.

$$Zn^{2+}(aq) + S^{2-}(aq) \longrightarrow \underset{\text{white}}{ZnS(s)}$$

The filtrate from this separation contains Ca^{2+} and Na^+. Calcium ion can be precipitated as the phosphate or the carbonate. For example, by adding $(NH_4)_2CO_3$, we get the reaction

$$Ca^{2+}(aq) + CO_3^{2-}(aq) \longrightarrow \underset{\text{white}}{CaCO_3(s)}$$

The filtrate from this solution contains Na^+ ion. Now the five ions have been separated.

In the complete analysis, once the ions are separated into analytical groups, these are further separated into the individual ions. We can illustrate this with Analytical Group I, which is precipitated as a mixture of the chlorides AgCl, Hg_2Cl_2, and $PbCl_2$. Of these chlorides, only $PbCl_2$ is significantly soluble in hot water. Thus, if the Analytical Group I precipitate is mixed with hot water, lead chloride dissolves and the silver and mercury(I) chlorides can be filtered off. Any lead(II) ion is in the filtrate, and can be revealed by adding potassium chromate. Lead(II) ion gives a bright yellow precipitate of lead(II) chromate, $PbCrO_4$.

$$Pb^{2+}(aq) + CrO_4^{2-}(aq) \longrightarrow \underset{\text{yellow}}{PbCrO_4(s)}$$

If there is precipitate remaining after the extraction with hot water, it consists of AgCl, Hg_2Cl_2, or both. However, only silver ion forms a stable ammonia complex ion. Thus, silver chloride dissolves in ammonia.

$$AgCl(s) + 2NH_3(aq) \longrightarrow Ag(NH_3)_2^+(aq) + Cl^-(aq)$$

Mercury(I) chloride, however, is simultaneously oxidized and reduced in ammonia solution, giving a precipitate of mercury(II) amido chloride, $HgNH_2Cl$, and mercury metal, which appears black because of its finely divided state.

$$Hg_2Cl_2(s) + 2NH_3(aq) \longrightarrow \underbrace{HgNH_2Cl(s) + Hg(l)}_{\text{black or gray}} + NH_4^+(aq) + Cl^-(aq)$$

The presence of silver ion in the filtrate is revealed by adding hydrochloric acid, which combines with the NH_3, releasing $Ag^+(aq)$ ion. The $Ag^+(aq)$ ion reacts with $Cl^-(aq)$ to give a white precipitate of AgCl.

$$Ag(NH_3)_2^+(aq) + 2H^+(aq) + Cl^-(aq) \longrightarrow AgCl(s) + 2NH_4^+(aq)$$

The other analytical groups are separated in similar fashion. Once a given ion has been separated, its presence is usually confirmed by a particular reactant—perhaps one that gives a distinctive precipitate.

Profile of a Chemical

LEAD (a Main-Group Metal)

Lead was one of the seven metals known to the ancients (the others were gold, silver, mercury, copper, tin, and iron). It is very easy to reduce galena (PbS), a common lead ore, to the metal. Moderate heating in air with charcoal (carbon) produces molten lead (the melting point of lead is only 327°C). The metal is also easily worked and can be fashioned into implements. The early Romans used lead to make water pipes, and today many pipes (especially those used in the chemical industry) are still made of lead. The word *plumbing* derives from the Latin word *plumbum* for "lead."

Lead remains one of the most useful metals and is fifth in order by tons produced. It is used to make storage battery plates, tetraethyllead for gasoline antiknock compounds, paint pigments, and ammunition. The use of tetraethyllead in gasoline and the use of lead compounds in paints are being phased out because of the metal's toxicity. Lead compounds poison by combining with enzymes (biological catalysts).

Lead is obtained today from galena by a process similar to that used by the ancients. The ore is first roasted, a process in which the galena is burned in air to form lead(II) oxide.

$$2PbS(s) + 3O_2(g) \longrightarrow 2PbO(s) + 2SO_2(g)$$

Then the lead(II) oxide is reduced with coke (carbon) in a blast furnace. The reactions are

$$PbO(s) + C(s) \longrightarrow Pb(l) + CO(g)$$
$$PbO(s) + CO(g) \longrightarrow Pb(l) + CO_2(g)$$

Most lead compounds are prepared from lead(II) oxide (known commercially as litharge), a yellow compound produced by burning lead in air. Other oxides are trilead tetroxide, Pb_3O_4, and lead(IV) oxide or lead dioxide, PbO_2 (Figure 17.8). Trilead tetroxide, also called red lead, is a red-orange compound used as a coating for steel beams to prevent rusting. It is prepared by carefully heating lead(II) oxide. Lead(IV) oxide has a dark brown color and is the principal constituent of one of the plates

(continued)

Figure 17.8
Lead and its oxides. *Lead metal is in the center and the oxides arranged around it: lead(II) oxide (yellow), trilead tetroxide, Pb_3O_4 (red-orange), and lead(IV) oxide (dark brown).*

of a storage battery. It forms in the battery when it is first charged.

Most lead compounds are insoluble. Lead(II) nitrate and lead(II) acetate are two common soluble compounds of lead from which many other lead compounds are prepared by precipitation. Figure 17.9 shows precipitates of lead(II) sulfate (white), lead(II) chromate (yellow), and lead(II) sulfide (black). The solubility products of these compounds are 1.7×10^{-8}, 1.8×10^{-14}, and 2.5×10^{-27}, respectively. Lead(II) chromate is used as a yellow pigment in certain paints where health hazards from lead poisoning is not a problem.

Figure 17.9
Preparation of some lead compounds. *Many lead compounds are prepared by precipitation starting with a soluble lead compound such as lead(II) nitrate. All beakers were initially filled with lead(II) nitrate solution. To the second, third, and fourth beakers, were added respectively: sodium sulfide, potassium chromate, and sodium sulfate. The precipitates that form are lead(II) sulfide (black), lead(II) chromate (yellow), and lead(II) sulfate (white).*

Questions for Study

1. What metals other than lead were known to the ancients? Why was lead discovered so early?
2. What are some of the principal uses of lead?
3. Give the chemical equations involved in the preparation of lead from its ore galena.
4. List the formulas and colors of three oxides of lead.
5. Which oxide is used commercially to prepare other lead compounds? How could you prepare lead nitrate from it?
6. How could you prepare lead(II) chromate?

A Checklist for Review

IMPORTANT TERMS

solubility product constant (K_{sp}) (17.1)
ion product (17.3)
fractional precipitation (17.3)
complex ions (p. 696)
ligand (p. 696)
formation (stability) constant (K_f) (17.5)
dissociation constant (of a complex) (K_d) (17.5)
amphoteric hydroxide (17.5)
qualitative analysis (p. 701)

SUMMARY OF FACTS AND CONCEPTS

The equilibrium constant for the equilibrium between a slightly soluble ionic solid and its ions in solution is called the *solubility product constant,* K_{sp}. Its value can be determined from the solubility of the solid. Conversely, when the solubility product constant is known, the solubility of the solid can be calculated. The solubility is decreased by the addition of a soluble salt that supplies a common ion. Qualitatively, this can be seen to follow from Le Chatelier's principle. Quantitatively, the *common-ion effect* on solubility can be obtained from the solubility product constant.

Rather than look at the solubility process as the dissolving of a solid in a solution, we can look at it as the *precipitation* of the solid from the solution. We can decide whether precipitation will occur by computing the *ion product*. Precipitation occurs when the ion product is greater than K_{sp}.

Solubility is affected by the pH if the compound supplies an anion conjugate to a weak acid. As the pH decreases (H^+ ion concentration increases), the anion concentration decreases because the anion forms the weak acid. Therefore, the ion product decreases and the solubility increases.

The concentration of a metal ion in solution is decreased by *complex-ion formation*. The equilibrium constant for the formation of the complex ion from the aqueous metal ion and the ligands is called the *formation constant* (or stability constant), K_f. Because complex-ion formation reduces the concentration of aqueous metal ion, an ionic compound of the metal is more soluble in a solution of the ligand.

The sulfide scheme of *qualitative analysis* separates a mixture of metal ions by using precipitation reactions. Variation of solubility with pH and with complex-ion formation is used to aid in the separation.

OPERATIONAL SKILLS

1. **Writing solubility product expressions** Write the solubility product expression for a given ionic compound (Example 17.1).

2. **Calculating K_{sp} from the solubility, or vice versa** Given the solubility of a slightly soluble ionic compound, calculate K_{sp} (Examples 17.2 and 17.3). Given K_{sp}, calculate the solubility of an ionic compound (Example 17.4).

3. **Calculating the solubility of a slightly soluble salt in a solution of a common ion** Given the solubility product constant, calculate the molar solubility of a slightly soluble ionic compound in a solution that contains a common ion (Example 17.5).

4. **Predicting whether precipitation will occur** Given the concentrations of ions originally in solution, determine whether a precipitate is expected to form (Example 17.6). Determine whether a precipitate is expected to form when two solutions of known volume and molarity are mixed (Example 17.7). For both problems, you will need the solubility product constant.

5. **Determining the completeness of precipitation** Calculate the concentration and percentage of an ion remaining after the corresponding ionic compound precipitates from a solution of known concentrations of ions (Example 17.8). K_{sp} is required.

6. **Determining the qualitative effect of pH on solubility** Decide whether the solubility of a salt will be greatly increased by decreasing the pH (Example 17.9).

7. **Separating metal ions by sulfide precipitation** Given the metal-ion concentrations in solution, calculate the range of pH required to separate a mixture of two metal ions by precipitating one as the metal sulfide (Example 17.10). K_{sp} values are required.

8. **Calculating the concentration of a metal ion in equilibrium with a complex ion** Calculate the concentration of aqueous metal ion in equilibrium with the complex ion, given the original metal-ion and ligand concentrations (Example 17.11). The formation constant K_f of the complex ion is required.

9. **Predicting whether a precipitate will form in the presence of the complex ion** Predict whether an ionic compound will precipitate from a solution of known concentrations of cation, anion, and ligand that complexes with the cation (Example 17.12). K_f and K_{sp} are required.

10. **Calculating the solubility of an ionic compound in a solution of the complex ion** Calculate the molar solubility of a slightly soluble ionic compound in a solution of known concentration of a ligand that complexes with the cation (Example 17.13). K_{sp} and K_f are required.

Review Questions

17.1 Suppose the molar solubility of nickel hydroxide, $Ni(OH)_2$, is x M. Show that K_{sp} for nickel hydroxide equals $4x^3$.

17.2 In your own words, explain why calcium sulfate is less soluble in sodium sulfate solution than in pure water.

17.3 What must be the concentration of silver ion in a solution that is in equilibrium with solid silver chloride and that is 0.10 M in Cl^-?

17.4 Discuss briefly how you could predict whether a precipitate will form when solutions of lead nitrate and potassium iodide are mixed. What information do you need to have?

17.5 Explain why barium fluoride dissolves in dilute hydrochloric acid but is insoluble in water.

17.6 Explain how metal ions such as Pb^{2+} and Zn^{2+} are separated by precipitation with hydrogen sulfide.

17.7 Lead chloride at first precipitates when sodium chloride is added to a solution of lead nitrate. Later, when the solution is made more concentrated in chloride ion, the precipitate dissolves. Explain what is happening. What equilibria are involved? Note that lead ion forms the complex ion $PbCl_4^{2-}$.

17.8 A precipitate forms when a small amount of sodium hydroxide is added to a solution of aluminum sulfate. This precipitate dissolves when more sodium hydroxide is added. Explain what is happening.

17.9 Describe how you would separate the metal ions in a solution containing silver ion, copper(II) ion, and nickel(II) ion, using the sulfide scheme of qualitative analysis.

17.10 A solution containing calcium ion and magnesium ion is buffered with ammonia–ammonium chloride. When carbonate ion is added to the solution, calcium carbonate precipitates but magnesium carbonate does not. Explain.

Practice Problems

SOLUBILITY AND K_{sp}

17.11 Use the solubility rules (p. 88) to decide which of the following are expected to be soluble and which insoluble.
(a) Na_2CO_3 (b) $MgCO_3$ (c) $PbSO_4$ (d) NH_4Br

17.12 Use the solubility rules (p. 88) to decide which of the following are expected to be soluble and which insoluble.
(a) $PbBr_2$ (b) KBr (c) $BaBr_2$ (d) $BaSO_4$

17.13 Write solubility product expressions for the following.
(a) $BaCrO_4$ (b) $Fe(OH)_2$
(c) $Pb_3(AsO_4)_2$ (d) Ag_2CrO_4

17.14 Write solubility product expressions for the following.
(a) $CaCO_3$ (b) BaF_2
(c) $Mg_3(AsO_4)_2$ (d) Ag_2S

17.15 The solubility of silver bromate, $AgBrO_3$, in water is 0.0072 g/L. Calculate K_{sp}.

17.16 The solubility of magnesium oxalate, MgC_2O_4, in water is 0.0093 mol/L. Calculate K_{sp}.

17.17 Calculate the solubility product constant for copper(II) iodate, $Cu(IO_3)_2$. The solubility of copper(II) iodate in water is 0.13 g/100 mL.

17.18 The solubility of silver chromate, Ag_2CrO_4, in water is 0.022 g/L. Calculate K_{sp}.

17.19 The pH of a saturated solution of magnesium hydroxide (''milk of magnesia'') was found to be 10.52. From this, find K_{sp} for magnesium hydroxide.

17.20 A solution saturated in calcium hydroxide (''limewater'') has a pH of 12.35. What is K_{sp} for calcium hydroxide?

17.21 Celestite (strontium sulfate) is an important mineral of strontium. Calculate the solubility of strontium sulfate, $SrSO_4$, from the solubility product constant (see Table 17.1).

17.22 Barite (barium sulfate, $BaSO_4$) is a common barium mineral. From the solubility product constant (Table 17.1), find the solubility of barium sulfate in grams per liter of water.

17.23 What is the solubility of PbF_2 in water? The K_{sp} for PbF_2 is 2.7×10^{-8}.

17.24 What is the solubility of MgF_2 in water? The K_{sp} for MgF_2 is 7.1×10^{-9}.

COMMON-ION EFFECT

17.25 What is the solubility (in grams per liter) of strontium sulfate, $SrSO_4$, in 0.15 M sodium sulfate, Na_2SO_4? See Table 17.1.

17.26 What is the solubility (in grams per liter) of lead(II) chromate, $PbCrO_4$, in 0.20 M potassium chromate, K_2CrO_4? See Table 17.1.

17.27 The solubility of magnesium fluoride, MgF_2, in water is 0.0076 g/L. What is the solubility (in grams per liter) of magnesium fluoride in 0.020 M sodium fluoride, NaF?

17.28 The solubility of silver sulfate, Ag_2SO_4, in water has been determined to be 8.0 g/L. What is the solubility in 0.75 M sodium sulfate, Na_2SO_4?

17.29 What is the solubility (in grams per liter) of magnesium oxalate, MgC_2O_4, in 0.020 M sodium oxalate, $Na_2C_2O_4$? Solve the equation exactly. See Table 17.1 for K_{sp}.

17.30 Calculate the molar solubility of strontium sulfate, $SrSO_4$, in 0.0015 M sodium sulfate, Na_2SO_4. Solve the equation exactly. See Table 17.1 for K_{sp}.

PRECIPITATION

17.31 Lead(II) chromate, $PbCrO_4$, is used as a yellow paint pigment ("chrome yellow"). When a solution is prepared that is 5.0×10^{-4} M in lead ion, Pb^{2+}, and 5.0×10^{-5} M in chromate ion, CrO_4^{2-}, would you expect some of the lead(II) chromate to precipitate? See Table 17.1.

17.32 Lead sulfate, $PbSO_4$, is used as a white paint pigment. When a solution is prepared that is 5.0×10^{-4} M in lead ion, Pb^{2+}, and 1.0×10^{-5} M in sulfate ion, SO_4^{2-}, would you expect some of the lead sulfate to precipitate? See Table 17.1.

17.33 The following solutions are mixed: 1.0 L of 0.00010 M NaOH and 1.0 L of 0.0020 M $MgSO_4$. Is a precipitate expected? Explain.

17.34 A 45-mL sample of 0.015 M calcium chloride, $CaCl_2$, is added to 55 mL of 0.010 M sodium sulfate, Na_2SO_4. Is a precipitate expected? Explain.

17.35 A 45.0-mL sample of 0.0015 M $BaCl_2$ was added to a beaker containing 75.0 mL of 0.0025 M KF. Will a precipitate form?

17.36 A 65.0-mL sample of 0.010 M $Pb(NO_3)_2$ was added to a beaker containing 40.0 mL of 0.035 M KCl. Will a precipitate form?

17.37 How many moles of calcium chloride, $CaCl_2$, can be added to 1.5 L of 0.020 M potassium sulfate, K_2SO_4, before a precipitate is expected? Assume that the volume of the solution is not changed significantly by the addition of calcium chloride.

17.38 Magnesium sulfate, $MgSO_4$, is added to 456 mL of 0.040 M sodium hydroxide, NaOH, until a precipitate just forms. How many grams of magnesium sulfate were added? Assume that the volume of the solution is not changed significantly by the addition of magnesium sulfate.

17.39 What are the concentration and the percentage of Ag^+ ion remaining after Ag_2CrO_4 precipitates when 25.0 mL of 0.10 M $AgNO_3$ is added to 25.0 mL of 0.10 M K_2CrO_4?

17.40 What are the concentration and the percentage of Ca^{2+} ion remaining after $CaCO_3$ precipitates when 25.0 mL of 0.10 M $CaCl_2$ is added to 25.0 mL of 0.10 M Na_2CO_3?

17.41 What is the I^- concentration just as AgCl begins to precipitate when 1.0 M $AgNO_3$ is slowly added to a solution containing 0.015 M Cl^- and 0.015 M I^-?

17.42 What is the Cl^- concentration just as Ag_2CrO_4 begins to precipitate when 1.0 M $AgNO_3$ is slowly added to a solution containing 0.015 M Cl^- and 0.015 M CrO_4^{2-}?

EFFECT OF pH ON SOLUBILITY

17.43 Write the net ionic equation in which the slightly soluble salt magnesium oxalate, MgC_2O_4, dissolves in dilute hydrochloric acid.

17.44 Write the net ionic equation in which the slightly soluble salt calcium fluoride, CaF_2, dissolves in dilute hydrochloric acid.

17.45 Which of the following salts would you expect to dissolve readily in acidic solution: barium sulfate or barium fluoride? Explain.

17.46 Which of the following salts would you expect to dissolve readily in acidic solution: calcium phosphate, $Ca_3(PO_4)_2$, or calcium sulfate, $CaSO_4$? Explain.

17.47 A solution is 0.10 M Co^{2+} and 0.10 M Hg^{2+}. Calculate the range of pH in which only one of the metal sulfides precipitates when the solution is saturated in H_2S.

17.48 A solution is 0.10 M Ni^{2+} and 0.10 M Cd^{2+}. Calculate the range of pH in which only one of the metal sulfides precipitates when the solution is saturated in H_2S.

COMPLEX IONS

17.49 Write the chemical equation for the formation of the $Ag(CN)_2^-$ ion. Write the K_f expression.

17.51 Sufficient sodium cyanide, NaCN, was added to 0.015 *M* silver nitrate, $AgNO_3$, to give a solution that was initially 0.100 *M* in cyanide ion, CN^-. What is the concentration of silver ion, Ag^+, in this solution after $Ag(CN)_2^-$ forms? The formation constant K_f for the complex ion $Ag(CN)_2^-$ is 5.6×10^{18}.

17.53 Predict whether cadmium oxalate, CdC_2O_4, will precipitate from a solution that is 0.0020 *M* $Cd(NO_3)_2$, 0.010 *M* $Na_2C_2O_4$, and 0.10 *M* NH_3. Note that cadmium ion forms the $Cd(NH_3)_4^{2+}$ complex ion.

17.55 What is the molar solubility of CdC_2O_4 in 0.10 *M* NH_3?

17.50 Write the chemical equation for the formation of the $Ag(S_2O_3)_2^{3-}$ ion. Write the K_f expression.

17.52 The formation constant K_f for the complex ion $Zn(OH)_4^{2-}$ is 2.8×10^{15}. What is the concentration of zinc ion, Zn^{2+}, in a solution that is initially 0.20 *M* in $Zn(OH)_4^{2-}$?

17.54 Predict whether nickel(II) hydroxide, $Ni(OH)_2$, will precipitate from a solution that is 0.0020 *M* $NiSO_4$, 0.010 *M* NaOH, and 0.10 *M* NH_3. Note that nickel(II) ion forms the $Ni(NH_3)_6^{2+}$ complex ion.

17.56 What is the molar solubility of NiS in 0.10 *M* NH_3?

QUALITATIVE ANALYSIS

17.57 Describe how you could separate the following mixture of metal ions: Cd^{2+}, Pb^{2+}, and Sr^{2+}.

17.59 A student dissolved a compound in water, then added hydrochloric acid. No precipitate formed. Then she bubbled H_2S into this solution, but again no precipitate formed. However, when she made the solution basic with ammonia and bubbled in H_2S, a precipitate formed. Which of the following are possible as the cation in the compound?
(a) Ag^+ (b) Ca^{2+} (c) Mn^{2+} (d) Cd^{2+}

17.58 Describe how you could separate the following mixture of metal ions: Na^+, Hg^{2+}, and Ca^{2+}.

17.60 A student was asked to identify a compound. In an effort to do so, he first dissolved the compound in water. He found that no precipitate formed when hydrochloric acid was added, but when H_2S was bubbled into this acidic solution, a precipitate formed. Which one of the following could be the precipitate?
(a) $PbCrO_4$ (b) CdS (c) MnS (d) Ag_2S

Additional Problems

17.61 Lead(II) sulfate is often used as a test for lead(II) ion in qualitative analysis. Using the solubility product constant (Table 17.1), calculate the molar solubility of lead(II) sulfate in water.

17.63 Mercury(I) chloride, Hg_2Cl_2, is an unusual salt in that it dissolves to form Hg_2^{2+} and $2Cl^-$. Use the solubility product constant (Table 17.1) to calculate the following: (a) the molar solubility of Hg_2Cl_2 and (b) the solubility of Hg_2Cl_2 in grams per liter.

17.65 For cerium(III) hydroxide, $Ce(OH)_3$, K_{sp} equals 2.0×10^{-20}. Calculate (a) its molar solubility (recall that taking the square root twice gives the fourth root) and (b) the pOH of the saturated solution.

17.67 What is the solubility of magnesium hydroxide in a solution buffered at pH 8.80?

17.69 What is the molar solubility of $Mg(OH)_2$ in a solution containing 1.0×10^{-1} *M* NaOH? See Table 17.1 for K_{sp}.

17.62 Mercury(II) ion is often precipitated as mercury(II) sulfide in qualitative analysis. Using the solubility product constant (Table 17.1), calculate the molar solubility of mercury(II) sulfide in water, assuming that no other reactions occur.

17.64 Magnesium ammonium phosphate is an unusual salt in that it dissolves to form Mg^{2+}, NH_4^+, and PO_4^{3-} ions. K_{sp} for magnesium ammonium phosphate equals 2.5×10^{-13}. Calculate (a) its molar solubility and (b) its solubility in grams per liter.

17.66 Copper(II) ferrocyanide, $Cu_2Fe(CN)_6$, dissolves to give Cu^{2+} and $[Fe(CN)_6]^{4-}$ ions; K_{sp} for $Cu_2Fe(CN)_6$ equals 1.3×10^{-16}. Calculate: (a) the molar solubility and (b) the solubility in grams per liter of copper(II) ferrocyanide.

17.68 What is the solubility of silver oxide, Ag_2O, in a solution buffered at pH 10.50? The equilibrium is $Ag_2O(s) + H_2O(l) \rightleftharpoons 2Ag^+(aq) + 2OH^-(aq)$; $K_c = 2.0 \times 10^{-8}$.

17.70 What is the molar solubility of $Al(OH)_3$ in a solution containing 1.0×10^{-3} *M* NaOH? See Table 17.1 for K_{sp}.

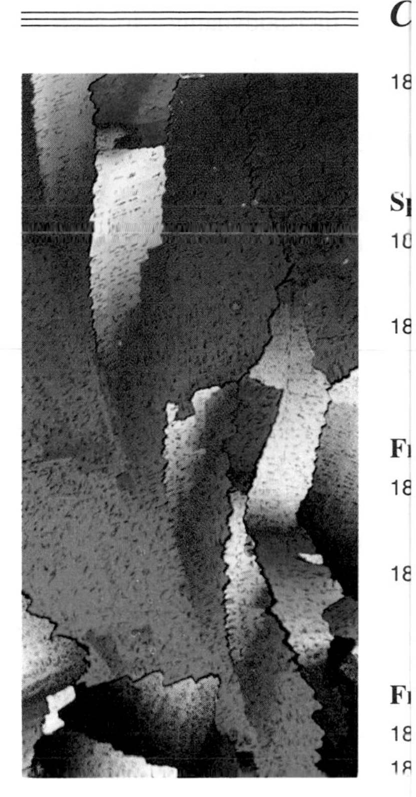

Photomicrograph of urea crystals under polarized light.

Figure 18.1
Urea. Urea is used as a plant fertilizer because it slowly decomposes in the soil to provide ammonia.

17.71 What must be the concentration of sulfate ion in order to precipitate calcium sulfate, $CaSO_4$, from a solution that is 0.0030 *M* Ca^{2+}?

17.72 What must be the concentration of chromate ion in order to precipitate strontium chromate, $SrCrO_4$, from a solution that is 0.0025 *M* Sr^{2+}? K_{sp} for strontium chromate is 5.7×10^{-5}.

17.73 A 3.20-L solution of 1.25×10^{-3} *M* $Pb(NO_3)_2$ is mixed with a 0.80-L solution of 5.0×10^{-1} *M* NaCl. Calculate Q_c for the dissolution of $PbCl_2$. No precipitate has formed. Is the solution supersaturated, saturated, or not saturated?

17.74 A 0.150-L solution of 2.4×10^{-5} *M* $MgCl_2$ is mixed with 0.050 L of 4.0×10^{-3} *M* NaOH. Calculate Q_c for the dissolution of $Mg(OH)_2$. No precipitate has formed. Is the solution supersaturated, saturated, or not saturated?

17.75 How many grams of sodium chloride can be added to 785 mL of 0.0015 *M* silver nitrate before a precipitate forms?

17.76 How many grams of sodium sulfate can be added to 435 mL of 0.0028 *M* barium chloride before a precipitate forms?

17.77 A solution is 0.10 *M* in sodium sulfate, Na_2SO_4. When 50.0 mL of 0.10 *M* barium nitrate, $Ba(NO_3)_2$, is added to 50.0 mL of this solution, what fraction of the sulfate ion is not precipitated?

17.78 A solution is 0.10 *M* in sodium chloride. When 50.0 mL of 0.10 *M* silver nitrate is added to 50.0 mL of this solution, what fraction of the chloride ion is not precipitated?

17.79 Solid KSCN was added to a 2.00 *M* Fe^{3+} solution so that it was also initially 2.00 *M* SCN^-. These ions then react to give the complex ion $Fe(SCN)^{2+}$, whose formation constant was 9.0×10^2. What is the concentration of $Fe^{3+}(aq)$ at equilibrium? Be sure to check any simplifying assumption you make.

17.80 Solid KSCN was added to a 2.00 *M* Co^{2+} solution so that it was also initially 2.00 *M* SCN^-. These ions then react to give the complex ion $Co(SCN)^+$, whose formation constant was 1.0×10^2. What is the concentration of $Co^{2+}(aq)$ at equilibrium? Be sure to check any simplifying assumption you make.

17.81 Calculate the molar solubility of silver bromide, AgBr, in 5.0 *M* NH_3.

17.82 Calculate the molar solubility of silver iodide, AgI, in 2.0 *M* NH_3.

17.83 The solubility of zinc oxalate, ZnC_2O_4, in 0.0150 *M* ammonia is 3.6×10^{-4} mol/L. What is the oxalate-ion concentration in the saturated solution? If the solubility product constant for zinc oxalate is 1.5×10^{-9}, what must be the zinc-ion concentration in the solution? Now calculate the formation constant for the complex ion $Zn(NH_3)_4{}^{2+}$.

17.84 The solubility of cadmium oxalate, CdC_2O_4, in 0.150 *M* ammonia is 6.1×10^{-3} mol/L. What is the oxalate-ion concentration in the saturated solution? If the solubility product constant for cadmium oxalate is 1.5×10^{-8}, what must be the cadmium-ion concentration in the solution? Now calculate the formation constant for the complex ion $Cd(NH_3)_4{}^{2+}$.

Cumulative-Skills Problems

17.85 A solution is 1.5×10^{-4} *M* Zn^{2+} and 0.20 *M* HSO_4^-. The solution also contains Na_2SO_4. What should be the minimum molarity of Na_2SO_4 to prevent the precipitation of zinc sulfide when the solution is saturated with hydrogen sulfide (0.10 *M* H_2S)?

17.86 A solution is 1.8×10^{-4} *M* Co^{2+} and 0.20 *M* HSO_4^-. The solution also contains Na_2SO_4. What should be the minimum molarity of Na_2SO_4 to prevent the precipitation of cobalt(II) sulfide when the solution is saturated with hydrogen sulfide (0.10 *M* H_2S)?

17.87 What is the solubility of calcium fluoride in a buffer solution containing 0.45 *M* $HCHO_2$ (formic acid) and 0.20 *M* $NaCHO_2$? (*Hint:* Consider the equation $CaF_2(s) + 2H^+(aq) \rightleftharpoons Ca^{2+}(aq) + 2HF(aq)$ and solve the equilibrium problem.)

17.88 What is the solubility of magnesium fluoride in a buffer solution containing 0.45 *M* $HC_2H_3O_2$ (acetic acid) and 0.20 *M* $NaC_2H_3O_2$? The K_{sp} for magnesium fluoride is 6.5×10^{-9}. (See the hint for Problem 17.87.)

17.89 A 67.0-mL sample of 0.350 *M* of $MgSO_4$ is added to 45.0 mL of 0.250 *M* $Ba(OH)_2$. What is the net ionic equation for the reaction that occurs? What are the concentrations of ions in the mixture at equilibrium?

17.90 A 50.0-mL sample of 0.0150 *M* of Ag_2SO_4 is added to 25.0 mL of 0.0100 *M* $PbCl_2$. What is the net ionic equation for the reaction that occurs? What are the concentrations of ions in the mixture at equilibrium?

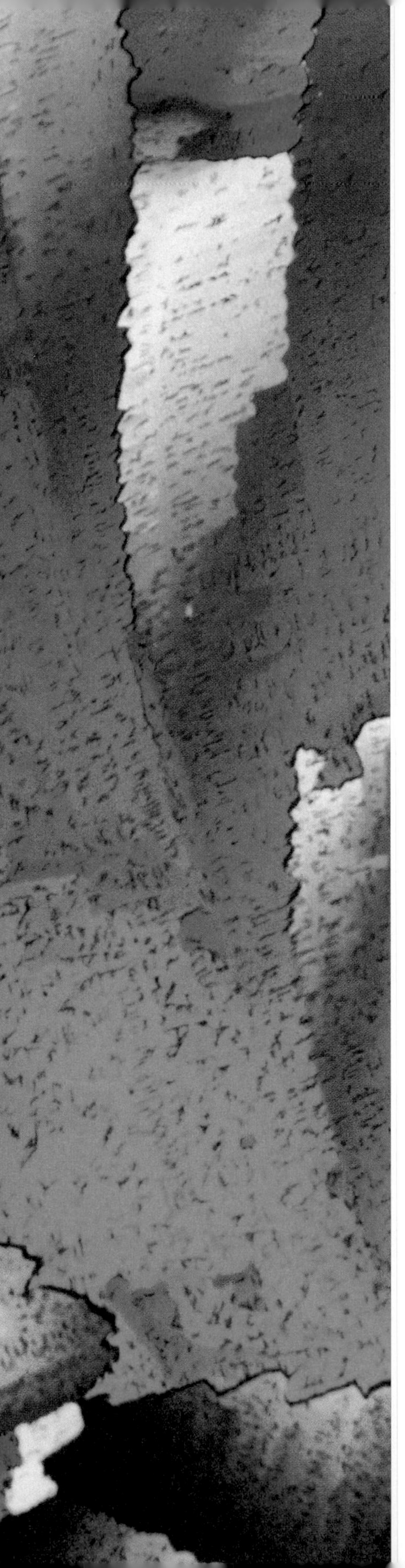

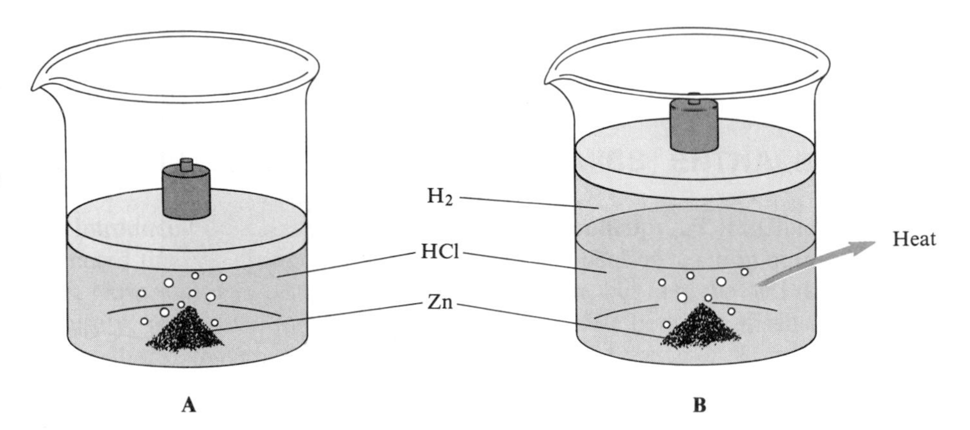

Figure 18.3 Reaction of zinc metal with hydrochloric acid at constant pressure. (*A*) The beginning of the reaction. The constant pressure of the atmosphere has been replaced by a piston and weight to give an equivalent pressure. (*B*) The reaction produces hydrogen gas. This increases the volume of the system, so that the piston and weight are lifted upward. Work is done by the system on the piston and weight.

When this reaction is carried out in a beaker open to the atmosphere, the reaction is exothermic, evolving 152.4 kJ of heat per mole of zinc consumed. We write $q_p = -152.4$ kJ, where the subscript p indicates that the process occurs at constant pressure (the pressure of the atmosphere).

The hydrogen gas that is produced increases the volume of the system. As hydrogen is evolved, work must be done by the system to push back the atmosphere. How can we calculate this work? Let us imagine for the moment that the atmosphere is replaced by a piston and weights, whose downward force from gravity F creates a pressure on the gas equivalent to that of the atmosphere. The pressure P equals F divided by the cross-sectional area of the piston, A. (See Figure 18.3.)

Suppose the increase in volume of the system due to the production of hydrogen is ΔV. Because the volume of a cylinder equals its height h times its cross-sectional area A, the change in volume is $\Delta V = A \times h$. Thus, $h = \Delta V/A$. The work done by the system in expanding equals the force of gravity times the distance the piston moves.

$$w = -F \times h = -F \times \frac{\Delta V}{A} = -\frac{F}{A} \times \Delta V$$

The negative sign is given because w is work done by the system and represents energy lost by it. Note that F/A is the pressure, P, which equals that of the atmosphere. Therefore,

$$w = -P\Delta V$$

This formula tells us that we can calculate the work done by a chemical reaction carried out in an open vessel by multiplying the atmospheric pressure P by the change in volume of the chemical system, ΔV. For example, when 1.00 mol Zn reacts with excess hydrochloric acid, 1.00 mol H_2 is produced. At 25°C and 1.00 atm ($= 1.01 \times 10^5$ Pa), this amount of H_2 occupies 24.5 L ($= 24.5 \times 10^{-3}$ m^3). The work done by the chemical system in pushing back the atmosphere is

$$w = P\Delta V = -(1.01 \times 10^5 \text{ Pa}) \times (24.5 \times 10^{-3} \text{ m}^3)$$
$$= -2.47 \times 10^3 \text{ J, or } -2.47 \text{ kJ}$$

If we apply the first law to this chemical system, we can relate the change in internal energy of the system to the heat of reaction. We have

$$\Delta U = q_p + w = q_p - P\Delta V$$

For the reaction of Zn with HCl, we have $q_p = -152.4$ kJ and $w = -P\Delta V = -2.47$ kJ, so

$$\Delta U = -152.4 \text{ kJ} - 2.47 \text{ kJ} = -154.9 \text{ kJ}$$

We can now summarize what happens when 1.00 mol Zn reacts with excess hydrochloric acid. When the reaction occurs, the internal energy changes as the kinetic and potential energies change in going from reactants to products. This energy change, ΔU, equals -154.9 kJ. Energy leaves the system mostly as heat ($q_p = -152.4$ kJ) but partly as expansion work ($w = -2.47$ kJ).

Exercise 18.2

Consider the combustion (burning) of methane, CH_4, in oxygen.

$$CH_4(g) + 2O_2(g) \longrightarrow CO_2(g) + 2H_2O(l)$$

The heat of reaction at 25°C and 1.00 atm is -890.2 kJ. What is the change in volume when 1.00 mol CH_4 reacts with 2.00 mol O_2? (You can ignore the volume of liquid water, which is insignificant compared with volumes of gases.) What is w for this change? Calculate ΔU for the change indicated by the foregoing chemical equation.

(See Problems 18.19 and 18.20.)

Enthalpy and Enthalpy Change

In Chapter 6, we defined enthalpy tentatively in terms of the relationship of ΔH to the heat of reaction at constant pressure, q_p. We now define **enthalpy,** H, precisely as *the quantity* U + PV. Because U, P, and V are state functions, H is also a state function. This means that for a given temperature and pressure, a given amount of a substance has a definite enthalpy. Thus, if we know the enthalpies of substances, we can calculate the change of enthalpy, ΔH, for a reaction.

Let us now show the relationship of ΔH to the heat of reaction q_p. We note that ΔH is the final enthalpy H_f minus the initial enthalpy H_i.

$$\Delta H = H_f - H_i$$

Now, we substitute defining expressions for H_f and H_i, using the subscripts f and i to indicate *final* and *initial,* respectively. Note, however, that the pressure is constant.

$$\begin{aligned}\Delta H &= (U_f + PV_f) - (U_i + PV_i) = (U_f - U_i) + P(V_f - V_i)\\ &= \Delta U + P\Delta V\end{aligned}$$

Earlier we showed that $\Delta U = q_p - P\Delta V$. Hence,

$$\Delta H = (q_p - P\Delta V) + P\Delta V = q_p$$

This shows that $\Delta H = q_p$, as we noted.

In practice, what we do is measure certain heats of reaction and use them to tabulate enthalpies of formation, ΔH_f°. The standard enthalpy change for a reaction is

$$\Delta H^\circ = \Sigma\ n\Delta H_f^\circ(\text{products}) - \Sigma\ m\Delta H_f^\circ(\text{reactants})$$

Standard enthalpies of formation for selected substances and ions are listed in Table 6.2 (p. 208) and in Appendix C.

Suppose we would like to know how much heat is absorbed or evolved in the reaction given in the chapter opening. The following are standard enthalpies of formation at 25°C for the substances indicated (in kJ/mol): $NH_3(g)$, −45.9; $CO_2(g)$, −393.5; $NH_2CONH_2(aq)$, −319.2; $H_2O(l)$, −285.8. Substituting into the previous equation yields

$$\Delta H^\circ = [(-319.2 - 285.8) - (-2 \times 45.9 - 393.5)]\text{ kJ}$$
$$= -119.7\text{ kJ}$$

From the minus sign in the value of ΔH°, we conclude that heat is evolved. The reaction is exothermic.

Spontaneous Processes and Entropy

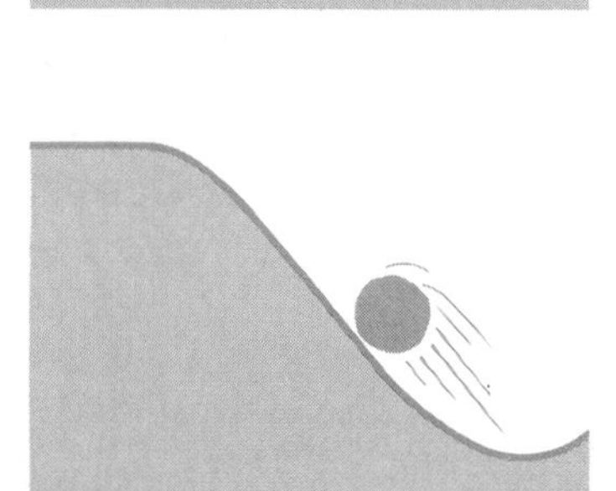

Figure 18.4
Examples of a spontaneous and a nonspontaneous process. *Top*: The rolling of a rock downhill is a spontaneous process. The rock eventually comes to equilibrium at the bottom of the hill. *Bottom*: The rolling of a rock uphill by itself is a nonspontaneous process.

Why does a chemical reaction go naturally in a particular direction? To answer this question, we need to look at spontaneous processes. A **spontaneous process** is *a physical or chemical change that occurs by itself.* It requires no continuing outside agency to make it happen. A rock at the top of a hill rolls down (Figure 18.4, *top*). Heat flows from a hot object to a cold one. An iron object rusts in moist air (Figure 18.5). These processes occur spontaneously, or naturally, without requiring an outside force or agency. They continue until equilibrium is reached. If these processes were to go in the opposite direction, they would be *nonspontaneous*. The rolling of a rock uphill by itself is not a natural process; it is nonspontaneous (see Figure 18.4, *bottom*). The rock could be moved to the top of the hill, but work would have to be expended. Heat can be made to flow from a cold to a hot object, but a heat pump or refrigerator is needed. Rust can be converted to iron, but the process requires chemical reactions used in the manufacture of iron from its ore (iron oxide).

Figure 18.5
The rusting of iron in moist air is a spontaneous reaction. This sculpture in Police Plaza in New York was allowed to rust to give a pleasing effect.

18.2 ENTROPY AND THE SECOND LAW OF THERMODYNAMICS

When we ask whether a chemical reaction goes in the direction in which it is written, as we did in the chapter opening, we are asking whether the reaction is spontaneous in this direction. The first law of thermodynamics cannot help us answer such a question. It does help us keep track of the various forms of energy in a chemical change, using the constancy of total energy (conservation of energy). But although at one time it was thought that spontaneous reactions must be exothermic ($\Delta H < 0$), many spontaneous reactions are now known to be endothermic ($\Delta H > 0$); recall the reaction shown in Figure 6.1, p. 188.

The second law of thermodynamics, which we will discuss in this section, gives us a way to answer questions about the spontaneity of a reaction. The second law is expressed in terms of a quantity called *entropy*.

Entropy

Entropy, S, is *a thermodynamic quantity that is a measure of the randomness or disorder in a system*. As we will discuss later, the SI unit of entropy is joules per kelvin (J/K). Entropy, like enthalpy, is a state function. That is, the quantity of entropy in a given amount of substance depends only on variables such as temperature and pressure that determine the state of the substance. If these variables are fixed, the quantity of entropy is fixed. Thus, 1 mol of ice at 0°C and 1 atm pressure has an entropy that has been determined experimentally to be 41 J/K. One mole of liquid water at 0°C and 1 atm has an entropy of 63 J/K. (In the next section, we will describe how entropy values are measured.)

We would expect the entropy (disorder) to increase when ice melts to the liquid. Ice has an ordered crystalline structure (see Figure 12.32, p. 492). When ice melts to liquid water, the crystalline structure breaks down, and a less-ordered liquid structure results. In ice, the H_2O molecules occupy regular, fixed positions in a crystal lattice; that is, ice has a relatively low entropy. But in the liquid, the molecules move freely about, giving a disordered structure—one having greater entropy. We can see this in the entropy values for these states.

We calculate the entropy change, ΔS, for a process similarly to the way we calculate ΔH. If S_i is the initial entropy and S_f is the final entropy, the change in entropy is

$$\Delta S = S_f - S_i$$

For the melting of ice to liquid water,

$$H_2O(s) \longrightarrow H_2O(l)$$
$$\Delta S = (63 - 41) \text{ J/K} = 22 \text{ J/K}$$

Thus, when 1 mol of ice melts at 0°C, the water increases in entropy by 22 J/K. The entropy increases, as we expect, because a substance becomes more disordered when it melts.

Second Law of Thermodynamics

A process occurs naturally as a result of an *overall* increase in disorder. There is a natural tendency for things to mix and to break down, events that represent increasing disorder. Where we see order or structure being built, it results from the use of greater order elsewhere. Order in one place is used to build order in another. Thus, a house is built (and order produced) at the expense of order in the food molecules of workers and order in substances needed to construct building materials. Moreover, during the overall process of building the house, some order is degraded to disorder. The *net* result, as in any natural process, is to increase disorder.

We can put this more precisely in terms of the **second law of thermodynamics,** *a law stating that the total entropy of a system and its surroundings always increases for a spontaneous process.* Note that entropy is quite different from energy. Energy can be neither created nor destroyed during chemical change (law of conservation of energy). But entropy is created during a spontaneous, or natural, process.

For a spontaneous process carried out at a given temperature, the second law can be restated in a form that refers only to the system. To arrive at that form, consider the changes in entropy that occur *in the system.* As the process takes place, entropy is created. At the same time, heat flows into or out of the system, and entropy accompanies that heat flow. Thus, when heat flows into the system, entropy flows into the system. In general, the entropy change associated with heat q at absolute temperature T can be shown to equal q/T. The net change in entropy of the system, ΔS, equals the sum of the entropy created during the spontaneous process and the change in entropy associated with the heat flow.

$$\Delta S = \text{entropy created} + \frac{q}{T} \qquad \text{(spontaneous process)}$$

Normally the quantity of entropy created during a spontaneous process cannot be directly measured. Because it is a positive quantity, however, if we delete it from the right side of the equation, we can conclude that ΔS must be greater than q/T for a spontaneous process.

$$\Delta S > \frac{q}{T} \qquad \text{(spontaneous process)}$$

The restatement of the second law is as follows: *For a spontaneous process at a given temperature, the change in entropy of the system is greater than the heat divided by the absolute temperature.*

Entropy Change for a Phase Transition

Certain processes occur at equilibrium or, more precisely, very close to equilibrium. For example, ice at 0°C is in equilibrium with liquid water at 0°C. If heat is slowly absorbed by the system, it remains very near equilibrium, but the ice melts. Under these essentially equilibrium conditions, no significant amount of entropy is created. The entropy change results entirely from the absorption of heat. Therefore,

$$\Delta S = \frac{q}{T} \qquad \text{(equilibrium process)}$$

Other phase changes, such as the vaporization of a liquid, also occur under equilibrium conditions.●

● When a system is at equilibrium, a small change in a condition can make the process go in one direction or the other. Thus, the process is said to be *reversible*. Think of a two-pan balance in which the weight on each pan has been adjusted to make the balance beam level—that is, so the beam is at equilibrium. A small weight on either pan will tip the scale one way or the other.

We can use the previous equation to obtain the entropy change for a phase change. Consider the melting of ice. The heat absorbed is the heat of fusion,

ΔH_{fus}, which is known from experiment to be 6.0 kJ for 1 mol of ice. We get the entropy change for melting by dividing ΔH_{fus} by the absolute temperature of the phase transition, 273 K (0°C). Because entropy changes are usually expressed in joules per kelvin, we convert ΔH_{fus} to 6.0×10^3 J.

$$\Delta S = \frac{\Delta H_{fus}}{T} = \frac{6.0 \times 10^3 \text{ J}}{273 \text{ K}} = 22 \text{ J/K}$$

Note that this is the ΔS value we obtained earlier for the conversion of ice to water.

Example 18.1 Calculating the Entropy Change for a Phase Transition

The heat of vaporization, ΔH_{vap}, of carbon tetrachloride, CCl_4, at 25°C is 43 kJ/mol.

$$CCl_4(l) \longrightarrow CCl_4(g);\ \Delta H_{vap} = 43.0 \text{ kJ/mol}$$

If 1 mol of liquid carbon tetrachloride at 25°C has an entropy of 214 J/K, what is the entropy of 1 mol of the vapor in equilibrium with the liquid at this temperature?

Solution

When the liquid evaporates, it absorbs heat: ΔH_{vap} = 43 kJ/mol (43×10^3 J/mol) at 25°C, or 298 K. The entropy change, ΔS, is

$$\Delta S = \frac{\Delta H_{vap}}{T} = \frac{43.0 \times 10^3 \text{ J/mol}}{298 \text{ K}} = 144 \text{ J/(mol} \cdot \text{K)}$$

In other words, 1 mol of carbon tetrachloride increases in entropy by 144 J/K when it vaporizes. The entropy of 1 mol of the vapor equals the entropy of 1 mol of liquid (214 J/K) plus 144 J/K.

$$\begin{aligned}\text{Entropy of vapor} &= (214 + 144) \text{ J/(mol} \cdot \text{K)} \\ &= \mathbf{358\ J/(mol \cdot K)}\end{aligned}$$

Exercise 18.3

Liquid ethanol, $C_2H_5OH(l)$, at 25°C has an entropy of 161 J/(mol · K). If the heat of vaporization, ΔH_{vap}, at 25°C is 42.3 kJ/mol, what is the entropy of the vapor in equilibrium with the liquid at 25°C? (See Problems 18.21 and 18.22.)

Now we can see how thermodynamics is applied to the question of whether a reaction is spontaneous. Suppose that we propose a chemical reaction for the preparation of a substance. We will also assume that the reaction occurs at constant temperature and pressure. For urea, we propose the reaction

$$2NH_3(g) + CO_2(g) \longrightarrow NH_2CONH_2(aq) + H_2O(l)$$

Is this a spontaneous reaction? That is, does it go left to right as written? We can use the second law in the form $\Delta S > q/T$ to answer this question if we know both ΔH and ΔS for the reaction.

Recall that the heat of reaction at constant pressure, q_p, equals the enthalpy change ΔH. The second law for a spontaneous reaction at constant temperature and pressure becomes

$$\Delta S > \frac{q_p}{T} = \frac{\Delta H}{T} \quad \text{(spontaneous reaction, constant } T \text{ and } P\text{)}$$

Thus, ΔS is greater than $\Delta H/T$ for a spontaneous reaction at constant temperature and pressure. Therefore, if we subtract ΔS from $\Delta H/T$, we will get a negative quantity for such a process. That is,

$$\frac{\Delta H}{T} - \Delta S < 0 \quad \text{(spontaneous reaction, constant } T \text{ and } P\text{)}$$

Multiplying each term of this inequality by the positive quantity T, we get

$$\Delta H - T\Delta S < 0 \qquad \text{(spontaneous reaction, constant } T \text{ and } P\text{)}$$

This inequality is important. If we had a table of entropies of substances, we could calculate ΔS for the proposed reaction. From Table 6.2, we could also calculate ΔH. If $\Delta H - T\Delta S$ were negative for the reaction, we would predict that it is spontaneous left to right, as written. However, if $\Delta H - T\Delta S$ were positive, we would predict that the reaction is nonspontaneous in the direction written but spontaneous in the opposite direction. If $\Delta H - T\Delta S$ is zero, the reaction is at equilibrium.

In the next section, we will see how to obtain the entropies of substances and then the entropy change of a reaction.

18.3 STANDARD ENTROPIES AND THE THIRD LAW OF THERMODYNAMICS

To determine experimentally the entropy of a substance, we first measure the heat absorbed by the substance by warming it at various temperatures. That is, we find the heat capacity at different temperatures.● We then calculate the entropy as we will describe. This determination of the entropy is based on the third law of thermodynamics.

● Heat capacity was discussed in Section 6.6.

Third Law of Thermodynamics

The **third law of thermodynamics** is *a law stating that a substance that is perfectly crystalline at 0 K has an entropy of zero*. This seems reasonable. A perfectly crystalline substance at 0 K should have perfect order. When the temperature is raised, however, the substance becomes more disordered as it absorbs heat.

We can determine the entropy of a substance at a temperature other than 0 K—say, at 298 K (25°C)—by slowly heating the substance from near 0 K to 298 K. Recall that the entropy change ΔS that occurs when heat is absorbed at a temperature T is q/T. Suppose we heat the substance from near 0.0 K to 2.0 K, and the heat absorbed is 0.19 J. We find the entropy change by dividing the heat

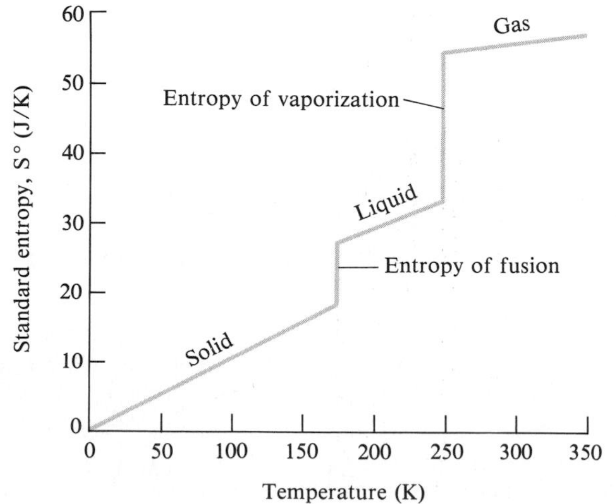

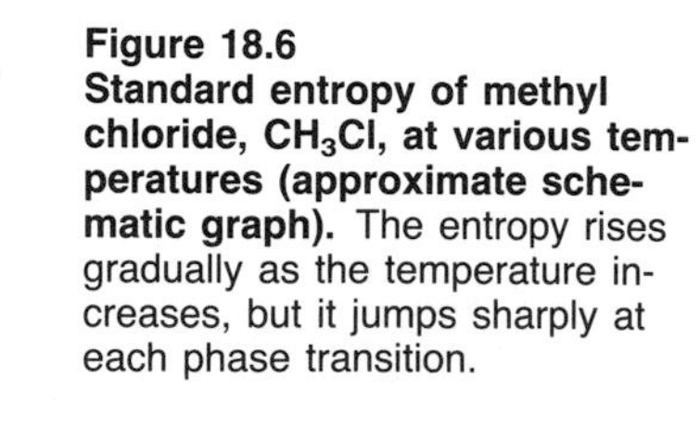
Figure 18.6 Standard entropy of methyl chloride, CH_3Cl, at various temperatures (approximate schematic graph). The entropy rises gradually as the temperature increases, but it jumps sharply at each phase transition.

Table 18.1 Standard Entropies (at 25°C)

Formula	$S°$, J/(mol · K)	Formula	$S°$, J/(mol · K)	Formula	$S°$, J/(mol · K)
Hydrogen		***Carbon (continued)***		***Sulfur***	
$H^+(aq)$	0	$CS_2(l)$	151.0	$S_2(g)$	228.1
$H_2(g)$	130.6	$HCN(g)$	201.7	S(rhombic)	31.9
Sodium		$HCN(l)$	112.8	S(monoclinic)	32.6
$Na^+(aq)$	60.2	$CCl_4(g)$	309.7	$SO_2(g)$	248.1
$Na(s)$	51.4	$CCl_4(l)$	214.4	$H_2S(g)$	205.6
$NaCl(s)$	72.1	$CH_3CHO(g)$	266	***Fluorine***	
$NaHCO_3(s)$	102	$C_2H_5OH(l)$	161	$F^-(aq)$	−9.6
$Na_2CO_3(s)$	139	***Silicon***		$F_2(g)$	202.7
Calcium		$Si(s)$	18.0	$HF(g)$	173.7
$Ca^{2+}(aq)$	−55.2	$SiO_2(s)$	41.5	***Chlorine***	
$Ca(s)$	41.6	$SiF_4(g)$	285	$Cl^-(aq)$	55.1
$CaO(s)$	38.2	***Lead***		$Cl_2(g)$	223.0
$CaCO_3(s)$	92.9	$Pb(s)$	64.8	$HCl(g)$	186.8
Carbon		$PbO(s)$	66.3	***Bromine***	
C(graphite)	5.7	$PbS(s)$	91.3	$Br^-(aq)$	80.7
C(diamond)	2.4	***Nitrogen***		$Br_2(l)$	152.2
$CO(g)$	197.5	$N_2(g)$	191.5	***Iodine***	
$CO_2(g)$	213.7	$NH_3(g)$	193	$I^-(aq)$	109.4
$HCO_3^-(aq)$	95.0	$NO(g)$	210.6	$I_2(s)$	116.1
$CH_4(g)$	186.1	$NO_2(g)$	239.9	***Silver***	
$C_2H_4(g)$	219.2	$HNO_3(aq)$	146	$Ag^+(aq)$	73.9
$C_2H_6(g)$	229.5	***Oxygen***		$Ag(s)$	42.7
$C_6H_6(l)$	172.8	$O_2(g)$	205.0	$AgF(s)$	84
$HCHO(g)$	219	$O_3(g)$	238.8	$AgCl(s)$	96.1
$CH_3OH(l)$	127	$OH^-(aq)$	−10.5	$AgBr(s)$	107.1
$CS_2(g)$	237.8	$H_2O(g)$	188.7	$AgI(s)$	114
		$H_2O(l)$	69.9		

absorbed by the average absolute temperature [$\frac{1}{2}(0.0 + 2.0)\text{K} = 1.0$ K]. Hence, ΔS equals 0.19 J/1.0 K = 0.19 J/K. This gives us the entropy of the substance at 2.0 K. Now we heat the substance from 2.0 K to 4.0 K, and this time 0.88 J of heat is absorbed. The average temperature is $\frac{1}{2}(2.0 + 4.0)\text{K} = 3.0$ K, and the entropy change is 0.88 J/3.0 K = 0.29 J/K. The entropy of the substance at 4.0 K is (0.19 + 0.29) J/K = 0.48 J/K. Proceeding this way, we can eventually get the entropy at 298 K.●

● The process just described is essentially the numerical evaluation of an integral, which we can write as follows: The heat absorbed for temperature change dT is $C_p(T)dT$, and the entropy change is $C_p(T)dT/T$. The standard entropy at temperature T is

$$\int_0^T \frac{C_p(T)dT}{T}$$

Figure 18.6 shows how the entropy of a substance changes with temperature. Note that the entropy increases gradually as the temperature increases. But when there is a phase change (for example, from solid to liquid), the entropy increases sharply. The entropy change for the phase transition is calculated from the enthalpy of the phase transition, as described earlier (see Example 18.1). The **standard entropy** of a substance or ion (Table 18.1), also called its **absolute entropy,**

$S°$, is *the entropy value for the standard state of the species* (indicated by the superscript degree sign). For a substance, the standard state is the pure substance at 1 atm pressure. For a species in solution, the standard state is the 1 *M* solution. Table 18.1 gives standard entropies of various substances at 25°C and 1 atm. Note that the elements have nonzero values, unlike standard enthalpies of formation, $\Delta H_f°$, which by convention are zero. The symbol $S°$, rather than $\Delta S°$, is chosen for standard entropies to emphasize that they originate from the third law.●

● Entropies of substances must be positive. Those for ions, however, can be negative because they are derived by arbitrarily setting $S°$ for $H^+(aq)$ equal to zero.

Entropy Change for a Reaction

Once we determine the standard entropies of all substances in a reaction, we can calculate the change of entropy, $\Delta S°$, for the reaction. A sample calculation is described in Example 18.3. Even without knowing values for the entropies of substances, one can sometimes predict the sign of $\Delta S°$ for a reaction. The entropy usually increases in the following situations:

1. A reaction in which a molecule is broken into two or more smaller molecules.
2. A reaction in which there is an increase in moles of gas. (This may result from a molecule breaking up, in which case Rules 1 and 2 are related.)
3. A process in which a solid changes to a liquid or gas or a liquid changes to a gas.

Example 18.2 illustrates how we can apply these rules (especially Rule 2) to find the sign of $\Delta S°$ for certain reactions involving gases.

Example 18.2 Predicting the Sign of the Entropy Change of a Reaction

(a) Is $\Delta S°$ positive or negative for the following reaction?

$$\underset{\text{glucose}}{C_6H_{12}O_6(s)} \longrightarrow \underset{\text{ethanol}}{2C_2H_5OH(l)} + 2CO_2(g)$$

Explain. The equation represents the essential change that takes place during the fermentation of glucose (grape sugar) to ethanol (ethyl alcohol). (b) Do you expect the entropy to increase or decrease in the preparation of urea from NH_3 and CO_2,

$$2NH_3(g) + CO_2(g) \longrightarrow NH_2CONH_2(aq) + H_2O(l)$$

as described in the chapter opening? Explain. (c) What is the sign of $\Delta S°$ for the following reaction?

$$CO(g) + H_2O(g) \longrightarrow CO_2(g) + H_2(g)$$

Solution

(a) A molecule (glucose) breaks into smaller molecules (C_2H_5OH) and (CO_2). Moreover, this results in a gas being released. We predict that $\Delta S°$ for this reaction is **positive**. That is, the entropy increases. (b) In this reaction, the moles of gas decrease (by 3 mol), which would decrease the entropy. We predict that the entropy should **decrease**. That is, $\Delta S°$ is negative. (c) Because there is no change in the number of moles of gas, **we cannot predict the sign of $\Delta S°$ from the rules given.**

Exercise 18.4

Predict the sign of $\Delta S°$ for each of the following reactions.

(a) $CaCO_3(s) \longrightarrow CaO(s) + CO_2(g)$
(b) $CS_2(l) \longrightarrow CS_2(g)$
(c) $2Hg(l) + O_2(g) \longrightarrow 2HgO(s)$
(d) $2Na_2O_2(s) + 2H_2O(l) \longrightarrow 4NaOH(aq) + O_2(g)$

(See Problems 18.25 and 18.26.)

It is useful to be able to predict the sign of $\Delta S°$. We gain some understanding of the reaction, and we can use the prediction for qualitative work. For quantitative work, however, we need to find the value of $\Delta S°$. We can find the standard change of entropy, $\Delta S°$, for a reaction by subtracting the standard entropies of reactants from the standard entropies of products, similar to the way we obtained $\Delta H°$.

$$\Delta S° = \Sigma\, nS°(\text{products}) - \Sigma\, mS°(\text{reactants})$$

The next example illustrates the calculations.

Example 18.3 Calculating $\Delta S°$ for a Reaction

Calculate the change of entropy, $\Delta S°$, at 25°C for the reaction in which urea is formed from NH_3 and CO_2.

$$2NH_3(g) + CO_2(g) \longrightarrow NH_2CONH_2(aq) + H_2O(l)$$

The standard entropy of $NH_2CONH_2(aq)$ is 174 J/(mol · K). See Table 18.1 for other values.

Solution

The calculation is similar to that used to obtain $\Delta H°$ from $\Delta H_f°$ values. We put the standard entropy values multiplied by stoichiometric coefficients below the formulas in the balanced equation.

$$\begin{array}{lcccc} & 2NH_3(g) & + \; CO_2(g) & \longrightarrow \; NH_2CONH_2(aq) & + \; H_2O(l) \\ S°: & 2 \times 193 & 214 & 174 & 70 \end{array}$$

Then we calculate the entropy change by subtracting the entropy of the reactants from the entropy of the products.

$$\Delta S° = \Sigma\, nS°(\text{products}) - \Sigma\, mS°(\text{reactants})$$
$$= [(174 + 70) - (2 \times 193 + 214)]\ \text{J/K} = -356\ \text{J/K}$$

Note that the sign of $\Delta S°$ agrees with the solution of Example 18.2(b).

Exercise 18.5

Calculate the change of entropy, $\Delta S°$, for the reaction given in Example 18.2(a). The standard entropy of glucose, $C_6H_{12}O_6(s)$, is 212 J/(mol · K). See Table 18.1 for other values. (See Problems 18.27 and 18.28.)

Free-Energy Concept

At the end of Section 18.2, we saw that the quantity $\Delta H - T\Delta S$ can serve as a criterion of spontaneity of a reaction at constant temperature and pressure. If the value of this quantity is negative, the reaction is spontaneous. If it is positive, the reaction is nonspontaneous. If it equals zero, the reaction is at equilibrium.

As an application of this criterion, consider the reaction described in the chapter opening, in which urea is prepared from NH_3 and CO_2. The heat of reaction $\Delta H°$ was calculated from enthalpies of formation in Section 18.1, where we obtained −119.7 kJ. Then, in Example 18.3, we calculated the entropy of reaction $\Delta S°$ and found a value of −356 J/K, or −0.356 kJ/K. Let us substitute these values, and T = 298 K (25°C), into the expression $\Delta H° - T\Delta S°$.

$$\Delta H° - T\Delta S° = (-119.7\ \text{kJ}) - (298\ \text{K})(-0.356\ \text{kJ/K})$$
$$= -13.6\ \text{kJ}$$

We see that $\Delta H° - T\Delta S°$ is a negative quantity and so conclude that the reaction is spontaneous under standard conditions.

18.4 FREE ENERGY AND SPONTANEITY

It is very convenient to define a new thermodynamic quantity in terms of H and S that will be directly useful as a criterion of spontaneity. For this purpose, the American physicist J. Willard Gibbs (1839–1903) introduced the concept of **free energy,** G, which is *a thermodynamic quantity defined by the equation* $G = H - TS$. This quantity gives a direct criterion of spontaneity of reaction.

As a reaction proceeds at a given temperature and pressure, reactants form products and the enthalpy H and entropy S change. These changes in H and S, denoted ΔH and ΔS, result in a change in free energy, ΔG, given by the equation

$$\Delta G = \Delta H - T\Delta S$$

Note that the change in free energy, ΔG, equals the quantity $\Delta H - T\Delta S$ that we just saw serves as a criterion of spontaneity of a reaction. Thus, if we can show that ΔG for a reaction at a given temperature and pressure is negative, we can predict that the reaction will be spontaneous.

Standard Free-Energy Change

Recall that for purposes of tabulating thermodynamic data, certain *standard states* are chosen, which are indicated by a superscript degree sign on the symbol of the quantity. The standard states are as follows: for pure liquids and solids, 1 atm pressure; for gases, 1 atm partial pressure; for solutions, 1 M concentration. The temperature is the temperature of interest, usually 25°C (298 K).

The standard free-energy change, $\Delta G°$, is the free-energy change that occurs when reactants in their standard states are converted to products in their standard states. Example 18.4 illustrates the calculation of the standard free-energy change, $\Delta G°$, from $\Delta H°$ and $\Delta S°$.

$$\Delta G° = \Delta H° - T\Delta S°$$

Example 18.4 Calculating $\Delta G°$ from $\Delta H°$ and $\Delta S°$

What is the standard free-energy change, $\Delta G°$, for the following reaction at 25°C?

$$N_2(g) + 3H_2(g) \longrightarrow 2NH_3(g)$$

Use values of $\Delta H_f°$ and $S°$ from Tables 6.2 and 18.1.

Solution

Let us write the balanced equation and add, below each formula, values of $\Delta H_f°$ and $S°$ multiplied by stoichiometric coefficients.

	$N_2(g)$	+ $3H_2(g)$	$\longrightarrow$ $2NH_3(g)$
$\Delta H_f°$:	0	0	2 × (−45.9) kJ
$S°$:	191.5	3 × 130.6	2 × 193 J/K

We calculate $\Delta H°$ and $\Delta S°$ by taking values for products and subtracting those for reactants.

$$\Delta H° = \Sigma\, n\Delta H_f°(\text{products}) - \Sigma\, m\Delta H_f°(\text{reactants})$$
$$= [2 \times (-45.9) - 0]\text{ kJ} = -91.8\text{ kJ}$$
$$\Delta S° = \Sigma\, nS°(\text{products}) - \Sigma\, mS°(\text{reactants})$$
$$= [2 \times 193 - (191.5 + 3 \times 130.6)]\text{ J/K} = -197\text{ J/K}$$

We now substitute into the equation for $\Delta G°$ in terms of $\Delta H°$ and $\Delta S°$. Note that we substitute $\Delta S°$ in units of kJ/K.

$$\Delta G° = \Delta H° - T\Delta S°$$
$$= -91.8\text{ kJ} - (298\text{ K})(-0.197\text{ kJ/K}) = \mathbf{-33.1\text{ kJ}}$$

Exercise 18.6

Calculate $\Delta G°$ for the following reaction at 25°C. Use data given in Tables 6.2 and 18.1.

$$CH_4(g) + 2O_2(g) \longrightarrow CO_2(g) + 2H_2O(g)$$

(See Problems 18.31 and 18.32.)

Standard Free Energies of Formation

The **standard free energy of formation,** $\Delta G_f°$, of a substance is defined similarly to the standard enthalpy of formation. That is, $\Delta G_f°$ is *the free-energy change that occurs when 1 mol of substance is formed from its elements in their stablest states at 1 atm and at a specified temperature* (usually 25°C). For example, the standard free energy of formation of $NH_3(g)$ is the free-energy change for the reaction

$$\tfrac{1}{2}N_2(g) + \tfrac{3}{2}H_2(g) \longrightarrow NH_3(g)$$

The reactants, N_2 and H_2, each at 1 atm, are converted to the product, NH_3, at 1 atm pressure. In Example 18.4, we found $\Delta G°$ for the formation of 2 mol NH_3 from its elements to be -33.1 kJ. Hence $\Delta G_f°$ (NH_3) = -33.1 kJ/2 mol = -16.6 kJ/mol.

As in the case of standard enthalpies of formation, the standard free energies of formation of elements in their stablest states are assigned the value zero. By tabulating $\Delta G_f°$ for substances, as in Table 18.2 (p. 726), we can easily calculate $\Delta G°$ for any reaction involving those substances. We simply subtract the standard free energies of reactants from the standard free energies of products:

$$\Delta G° = \Sigma\, n\Delta G_f°(\text{products}) - \Sigma\, m\Delta G_f°(\text{reactants})$$

The next example illustrates the calculations.

Example 18.5 Calculating $\Delta G°$ from Standard Free Energies of Formation

Calculate $\Delta G°$ for the combustion of 1 mol of ethanol, C_2H_5OH, at 25°C.

$$C_2H_5OH(l) + 3O_2(g) \longrightarrow 2CO_2(g) + 3H_2O(g)$$

Use the standard free energies of formation given in Table 18.2.

Solution

We write the balanced equation with values of $\Delta G_f°$ multiplied by stoichiometric coefficients below each formula.

	$C_2H_5OH(l)$ +	$3O_2(g)$ $\longrightarrow$	$2CO_2(g)$ +	$3H_2O(g)$
$\Delta G_f°$:	-174.8	0	$2(-394.4)$	$3(-228.6)$ kJ

The calculation is

$$\begin{aligned}\Delta G° &= \Sigma\, n\Delta G_f°(\text{products}) - \Sigma\, m\Delta G_f°(\text{reactants})\\ &= [2(-394.4) + 3(-228.6) - (-174.8)]\ \text{kJ}\\ &= \mathbf{-1299.8\ kJ}\end{aligned}$$

Exercise 18.7

Calculate $\Delta G°$ at 25°C for the following reaction, using values of $\Delta G_f°$.

$$CaCO_3(s) \longrightarrow CaO(s) + CO_2(g)$$

(See Problems 18.35 and 18.36.)

Table 18.2 Standard Free Energies of Formation (at 25°C)

Formula	ΔG_f°, kJ/mol	Formula	ΔG_f°, kJ/mol	Formula	ΔG_f°, kJ/mol
Hydrogen		***Carbon (continued)***		***Sulfur***	
$H^+(aq)$	0	$CS_2(l)$	63.6	$S_2(g)$	80.1
$H_2(g)$	0	$HCN(g)$	125	S(rhombic)	0
Sodium		$HCN(l)$	121	S(monoclinic)	0.10
$Na^+(aq)$	−261.9	$CCl_4(g)$	−53.7	$SO_2(g)$	−300.2
$Na(s)$	0	$CCl_4(l)$	−68.6	$H_2S(g)$	−33
$NaCl(s)$	−348.0	$CH_3CHO(g)$	−133.7	***Fluorine***	
$NaHCO_3(s)$	−851.9	$C_2H_5OH(l)$	−174.8	$F^-(aq)$	−276.5
$Na_2CO_3(s)$	−1048.1	***Silicon***		$F_2(g)$	0
Calcium		$Si(s)$	0	$HF(g)$	−275
$Ca^{2+}(aq)$	−553.0	$SiO_2(s)$	−856.5	***Chlorine***	
$Ca(s)$	0	$SiF_4(g)$	−1506	$Cl^-(aq)$	−131.2
$CaO(s)$	−603.5	***Lead***		$Cl_2(g)$	0
$CaCO_3(s)$	−1128.8	$Pb(s)$	0	$HCl(g)$	−95.3
Carbon		$PbO(s)$	−189	***Bromine***	
C(graphite)	0	$PbS(s)$	−96.7	$Br^-(aq)$	−102.8
C(diamond)	2.9	***Nitrogen***		$Br_2(l)$	0
$CO(g)$	−137.2	$N_2(g)$	0	***Iodine***	
$CO_2(g)$	−394.4	$NH_3(g)$	−16	$I^-(aq)$	−51.7
$HCO_3^-(aq)$	−587.1	$NO(g)$	86.60	$I_2(s)$	0
$CH_4(g)$	−50.8	$NO_2(g)$	51	***Silver***	
$C_2H_4(g)$	68.4	$HNO_3(aq)$	−110.5	$Ag^+(aq)$	77.1
$C_2H_6(g)$	−32.9	***Oxygen***		$Ag(s)$	0
$C_6H_6(l)$	124.5	$O_2(g)$	0	$AgF(s)$	−185
$HCHO(g)$	−110	$O_3(g)$	163	$AgCl(s)$	−109.7
$CH_3OH(l)$	−166.2	$OH^-(aq)$	−157.3	$AgBr(s)$	−95.9
$CS_2(g)$	66.9	$H_2O(g)$	−228.6	$AgI(s)$	−66.3
		$H_2O(l)$	−237.2		

ΔG° as a Criterion for Spontaneity

We have already seen that the quantity $\Delta G = \Delta H - T\Delta S$ can be used as a criterion for the spontaneity of a reaction. The change of free energy, ΔG, should be calculated for the conditions at which the reaction occurs. If the reactants are at standard conditions and these give products at standard conditions, the free-energy change we need to look at is ΔG°. The calculation is simple, as shown in Example 18.5. For other conditions, we should look at the appropriate ΔG value. This would be a more complicated calculation. Nevertheless, the standard free-energy change ΔG° is still a useful guide to the spontaneity of reaction in these cases. The following rules are useful in judging the spontaneity of a reaction:

1. When ΔG° is a large negative number (more negative than about −10 kJ), the reaction is spontaneous as written, and reactants transform almost entirely to products when equilibrium is reached.

2. When $\Delta G°$ is a large positive number (larger than about 10 kJ), the reaction is nonspontaneous as written, and reactants do not give significant amounts of products at equilibrium.
3. When $\Delta G°$ has a small negative or positive value (less than about 10 kJ), the reaction gives an equilibrium mixture with significant amounts of both reactants and products.

Example 18.6 Interpreting the Sign of $\Delta G°$

Calculate $\Delta H°$ and $\Delta G°$ for the reaction

$$2KClO_3(s) \longrightarrow 2KCl(s) + 3O_2(g)$$

Interpret the signs obtained for $\Delta H°$ and $\Delta G°$. Values of $\Delta H_f°$ (in kJ/mol) are as follows: $KClO_3(s)$, −391.2; $KCl(s)$, −436.7. Similarly, values of $\Delta G_f°$ (in kJ/mol) are as follows: $KClO_3(s)$, −289.9; $KCl(s)$, −408.8. Note that $O_2(g)$ is the reference form of the element, so $\Delta H_f° = \Delta G_f° = 0$ for it.

Solution

The problem is set up as follows:

	$2KClO_3(s)$	$\longrightarrow$	$2KCl(s)$	+	$3O_2(g)$
$\Delta H_f°$:	$2 \times (-391.2)$		$2 \times (-436.7)$		0 kJ
$\Delta G_f°$:	$2 \times (-289.9)$		$2 \times (-408.8)$		0 kJ

Then,

$$\Delta H° = [2 \times (-436.7) - 2 \times (-391.2)] \text{ kJ} = \mathbf{-91.0 \text{ kJ}}$$

$$\Delta G° = [2 \times (-408.8) - 2 \times (-289.9)] \text{ kJ} = \mathbf{-237.8 \text{ kJ}}$$

The reaction is exothermic, liberating 91.0 kJ of heat. The large negative value for $\Delta G°$ indicates that the equilibrium composition is mostly potassium chloride and oxygen.

Exercise 18.8

Which of the following reactions are spontaneous in the direction written? See Table 18.2 for data.

(a) $C(\text{graphite}) + 2H_2(g) \longrightarrow CH_4(g)$
(b) $2H_2(g) + O_2(g) \longrightarrow 2H_2O(l)$
(c) $4HCN(g) + 5O_2(g) \longrightarrow 2H_2O(l) + 4CO_2(g) + 2N_2(g)$
(d) $Ag^+(aq) + I^-(aq) \longrightarrow AgI(s)$

(See Problems 18.39 and 18.40.)

18.5 INTERPRETATION OF FREE ENERGY

We have seen that the free-energy change serves as a criterion of spontaneity of a chemical reaction. This gives us some idea of what free energy is and how we interpret it. In this section, we will look more closely at the meaning of free energy.

Maximum Work

Theoretically, spontaneous reactions can be used to obtain useful work. By useful work, we mean energy that can be used directly to move objects of normal size. We use the combustion of gasoline to move an automobile, and we use a reaction in a battery to generate electricity to drive a motor. Similarly, biochemical reactions in muscle tissue occur in such a way as to contract muscle fibers and lift a weight.

Often reactions are not carried out in a way that does useful work. The reactants are simply poured together in a reaction vessel, and products are separated from the mixture. As the reaction occurs, the free energy of the system decreases and entropy is produced. No useful work is obtained.

In principle, if a reaction is carried out to obtain the maximum useful work, no entropy is produced. It can be shown that the maximum useful work, w_{max}, for a spontaneous reaction is ΔG.

$$w_{max} = \Delta G$$

The term *free energy* comes from this result. *The free-energy change is the maximum energy available, or free, to do useful work.* As a reaction occurs in such a way as to give the maximum useful work, the free energy decreases, and a corresponding quantity of useful work is obtained.●

● It is possible to obtain the maximum work, and therefore ΔG, for some reactions from electrochemical cells (batteries), as we will show in Chapter 19.

The concept of maximum work from a chemical reaction is an idealization. In any real situation, less than this maximum work is obtained and some entropy is created. When this work is eventually expended, it appears in the environment as additional entropy.

Coupling of Reactions

One kind of useful work is work that is needed to effect a nonspontaneous chemical change. Consider the direct decomposition of iron(III) oxide—essentially rust—to iron.

$$2Fe_2O_3(s) \longrightarrow 4Fe(s) + 3O_2(g);\ \Delta G^\circ = +1487\ \text{kJ}$$

The reaction is nonspontaneous, because ΔG° is a large positive quantity. This is in agreement with common knowledge. Iron tends to rust in air. We do not expect a rusty wrench to turn spontaneously into shiny iron and oxygen. But this does not mean we cannot change iron(III) oxide to iron. It only means that we will have to do work on the iron(III) oxide to reduce it. We must find a way to couple this reaction with one having a more negative ΔG°.●

● In addition to coupling, free energy can be supplied to chemical reactions by electrical work. The process is called electrolysis (see Chapter 19).

Consider the reaction

$$2CO(g) + O_2(g) \longrightarrow 2CO_2(g);\ \Delta G^\circ = -514.4\ \text{kJ}$$

For 3 mol O_2, ΔG° is -1543 kJ, which is more negative than that for the direct decomposition of 2 mol Fe_2O_3 to its elements. Let us add the two reactions:

$$\begin{array}{llll}
2Fe_2O_3(s) & \longrightarrow 4Fe(s) + \cancel{3O_2(g)}; & \Delta G^\circ = & 1487\ \text{kJ} \\
6CO(g) + \cancel{3O_2(g)} & \longrightarrow 6CO_2(g); & \Delta G^\circ = & -1543\ \text{kJ} \\
\hline
2Fe_2O_3(s) + 6CO(g) & \longrightarrow 4Fe(s) + 6CO_2(g); & \Delta G^\circ = & -56\ \text{kJ}
\end{array}$$

Thus, iron(III) oxide can be reduced spontaneously to free iron with carbon monoxide. In fact, this is the reaction that occurs in a blast furnace, where iron ore (mainly Fe_2O_3) is reduced to iron.

The concept of coupling two chemical reactions, one that is spontaneous and one that is nonspontaneous, to give a spontaneous change is a very useful one in biochemistry.● Adenosine triphosphate, or ATP, is a large molecule containing phosphate groups. It plays a central role in the transfer of energy in living systems. ATP can react with water in the presence of an enzyme (biochemical catalyst) to give adenosine diphosphate, ADP, and a phosphate ion.

● Coupling of reactions will be discussed again in Chapter 25.

$$\text{ATP} + H_2O \longrightarrow \text{ADP} + \text{phosphate ion};\ \Delta G^\circ = -31\ \text{kJ}$$

Figure 18.7
The chemiluminescent reaction of a firefly. *Left*: The glow of a firefly results from the chemiluminescent reaction of a compound called luciferin, an organic peroxide, whose free energy is provided by ATP molecules in the organism. *Right*: Prof. Paul Schaap demonstrates the chemiluminescence of a similar compound, which he and his students prepared at Wayne State University. The chemiluminescence can be triggered by enzymes (biological catalysts) and is being used as an extremely sensitive test of certain biological substances, including antibodies in blood.

ATP is first synthesized in a living organism; the energy used is obtained from food. The spontaneous reaction of ATP to give ADP is then coupled to various nonspontaneous reactions to accomplish the necessary reactions of the cell (see Figure 18.7).

Free-Energy Change During Reaction

We have seen that the free-energy change is related to the work done during a chemical reaction. Consider the combustion of gasoline in O_2. The reaction is spontaneous, so the free-energy change is negative. That is, the free energy of the system changes to a lower value as the reactants are converted to products. Figure 18.8 shows the free-energy change that occurs during this reaction.

When the gasoline is burned in a gasoline stove, the decrease in free energy shows up as an increase in entropy of the system and the surroundings. However, when the gasoline is burned in an automobile engine, some of this decrease in free energy shows up as work done. Theoretically, all of the free-energy decrease can

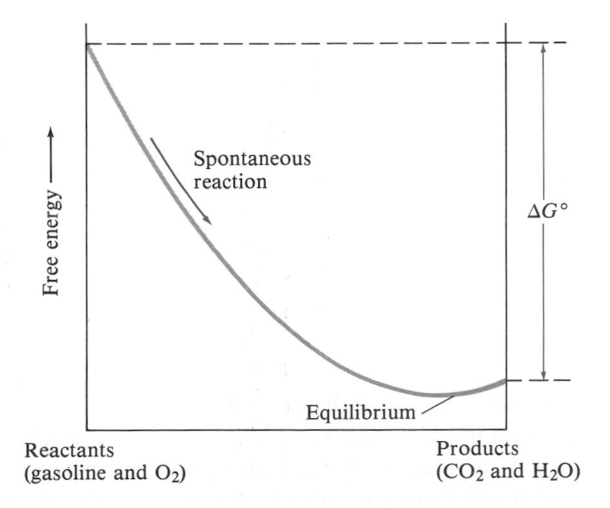

Figure 18.8
Free-energy change during a spontaneous reaction (combustion of gasoline). Note that the free energy decreases as the reaction proceeds. At equilibrium, the free energy is a minimum, and the equilibrium mixture is mostly products.

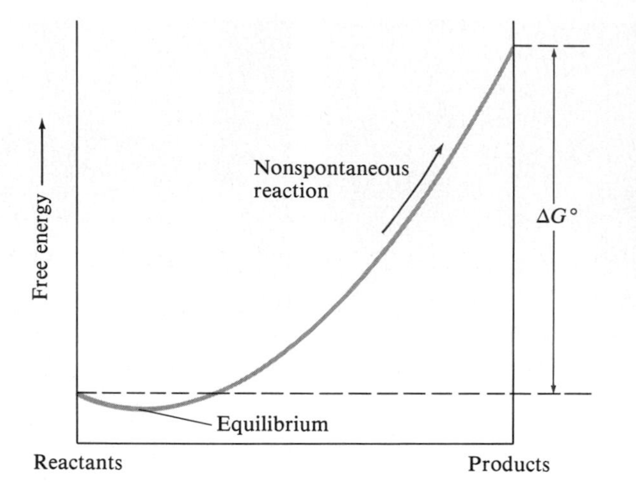

Figure 18.9
Free-energy change during a nonspontaneous reaction. The free energy decreases until equilibrium, at which the minimum value is reached. There is very little reaction, however, because the equilibrium mixture is mostly reactants. To change reactants to mostly products is to undergo a nonspontaneous reaction.

be used to do work. This gives the maximum work. In practice, less work is obtained, and the difference appears as an increase of entropy. Ultimately, the work is itself used up—that is, changed to entropy.

Let us look at Figure 18.8 again. At the start of the reaction, the system contains only reactants: gasoline and O_2. The free energy has a relatively high value. This decreases as the reaction proceeds. The decrease in free energy appears either as an increase in entropy or as work done. Eventually, the free energy of the system reaches its minimum value. Then the net reaction stops; it has come to equilibrium.

Before we leave this section, let us consider a reaction in which $\Delta G°$ is positive. We would predict that the reaction is nonspontaneous. Figure 18.9 shows the free-energy change as the reaction proceeds. Note that there is a small decrease in free energy as the system goes to equilibrium. Some reaction occurs in order to give the equilibrium mixture. But this mixture consists primarily of reactants, because the reaction does not go very far before coming to equilibrium. To change reactants completely to products is a nonspontaneous reaction (shown by the arrow along the curve in Figure 18.9).

Free Energy and Equilibrium Constants

In the previous section, we saw that for any spontaneous reaction, the total free energy of the substances in the reaction mixture decreases as the reaction proceeds (that is, the free-energy change is negative). Earlier, we saw that the standard free-energy change, $\Delta G°$, can be used as a criterion of the spontaneity of a reaction. When $\Delta G°$ is negative, the reaction is spontaneous. We will now see that the standard free-energy change is related to the equilibrium constant.

18.6 RELATING $\Delta G°$ TO THE EQUILIBRIUM CONSTANT

One of the most important results of chemical thermodynamics is an equation relating the standard free-energy change for a reaction to the equilibrium constant. Before we look at this equation, we must discuss the thermodynamic form of the equilibrium constant which occurs in that equation.

● To be precise, K is defined in terms of activities, which are dimensionless quantities numerically equal to "effective" concentrations and pressures.

The **thermodynamic equilibrium constant, K,** is *the equilibrium constant in which the concentrations of gases are expressed in partial pressures in atmospheres, whereas the concentrations of solutes in liquid solutions are expressed in molarities*. For a reaction involving only solutes in liquid solution, K is identical to K_c; for reactions involving only gases, K equals K_p.●

Example 18.7 Writing the Expression for a Thermodynamic Equilibrium Constant

Write expressions for the thermodynamic equilibrium constants for each of the following reactions:

(a) The reaction given in the chapter opening,

$$2NH_3(g) + CO_2(g) \rightleftharpoons NH_2CONH_2(aq) + H_2O(l)$$

(b) The solubility process

$$AgCl(s) \rightleftharpoons Ag^+(aq) + Cl^-(aq)$$

Solution

(a) Note that H_2O is a solvent, so it does not appear explicitly in K. The gases appear in K as partial pressures, and the solute appears as a molar concentration.

$$K = \frac{[NH_2CONH_2]}{P_{CO_2}P^2_{NH_3}}$$

(b) Note that AgCl is a solid, so it does not appear explicitly in K. The solutes appear as molar concentrations.

$$K = [Ag^+][Cl^-]$$

This is identical to K_{sp}.

Exercise 18.9

Give the expression for K for each of the following reactions.

(a) $CaCO_3(s) \rightleftharpoons CaO(s) + CO_2(g)$

(b) $PbI_2(s) \rightleftharpoons Pb^{2+}(aq) + 2I^-(aq)$

(c) $H^+(aq) + HCO_3^-(aq) \rightleftharpoons H_2O(l) + CO_2(g)$ (See Problems 18.45 and 18.46.)

The standard free energy change, $\Delta G°$, for a reaction can be calculated using data from thermodynamic tables. If we wish the free-energy change when reactants in nonstandard states are changed to products in nonstandard states (ΔG), we can obtain it from the standard free-energy change $\Delta G°$ using the following equation.

$$\Delta G = \Delta G° + RT \ln Q$$

Here Q is the thermodynamic form of the reaction quotient. It has the same general appearance as the thermodynamic equilibrium constant K, but the concentrations and partial pressures are those for a mixture at some instant, perhaps at the beginning of a reaction. We obtain ΔG from $\Delta G°$ by adding $RT \ln Q$, where $\ln Q$ is the natural logarithm of Q ($\ln Q = 2.303 \log Q$).

We can derive the equation relating $\Delta G°$ to the equilibrium constant K from the preceding equation. In the previous section, we saw that as a chemical reaction approaches equilibrium, the free energy decreases and continues to decrease until equilibrium is reached. At equilibrium, the free energy ceases to change; then $\Delta G = 0$. Also, the reaction quotient Q becomes equal to the equilibrium constant K. If we substitute $\Delta G = 0$ and $Q = K$ into the preceding equation, we obtain

$$0 = \Delta G° + RT \ln K$$

This result easily rearranges to give the basic equation relating the standard free-energy change to the equilibrium constant.

$$\Delta G° = -RT \ln K \quad \text{or} \quad \Delta G° = -2.303\, RT \log K$$

The next example illustrates the calculation of a thermodynamic equilibrium constant from the standard free-energy change, ΔG°.

Example 18.8 Calculating *K* from the Standard Free-Energy Change (Molecular Equation)

Find the value of the equilibrium constant K at 25°C (298 K) for the reaction

$$2NH_3(g) + CO_2(g) \rightleftharpoons NH_2CONH_2(aq) + H_2O(l)$$

The standard free-energy change, ΔG°, at 25°C equals −13.6 kJ. (We calculated this value of $\Delta G^\circ = \Delta H^\circ - T\Delta S^\circ$ just before Section 18.4.)

Solution

We rearrange the equation $\Delta G^\circ = -2.303RT \log K$ to give

$$\log K = \frac{\Delta G^\circ}{-2.303RT}$$

Note that ΔG° and R must be in compatible units. We normally express ΔG° in joules and set R equal to 8.31 J/(K · mol). Substituting numerical values into this equation, we get

$$\log K = \frac{-13.6 \times 10^3}{-2.303 \times 8.31 \times 298} = 2.38$$

Hence

$$K = \text{antilog } 2.38 = 10^{2.38} = \mathbf{2.4 \times 10^2}$$

Note that although the value of K indicates that products predominate at equilibrium, K is only moderately large. We would expect that the composition could be easily shifted toward reactants if we could remove either NH_3 or CO_2 (according to Le Chatelier's principle). This is what happens when urea is used as a fertilizer. As NH_3 is used up, more NH_3 is produced by the decomposition of urea.

Exercise 18.10

Use data from Table 18.2 to obtain the equilibrium constant K_p at 25°C for the reaction

$$CaCO_3(s) \rightleftharpoons CaO(s) + CO_2(g)$$

Note that values of ΔG_f° are needed for $CaCO_3$ and CaO, even though the substances do not appear in $K_p = P_{CO_2}$.
(See Problems 18.47, 18.48, 18.49, and 18.50.)

The method of calculating K from ΔG° also works for net ionic equations. In Example 18.9, we obtain a solubility product constant from thermodynamic data.

Example 18.9 Calculating *K* from the Standard Free-Energy Change (Net Ionic Equation)

Calculate the equilibrium constant K_{sp} at 25°C for the reaction

$$AgCl(s) \rightleftharpoons Ag^+(aq) + Cl^-(aq)$$

Use ΔG_f° values from Table 18.2.

Solution

Note that K_{sp} equals K. We first calculate ΔG°. Writing ΔG_f° values below the formulas in the equation gives

$$\begin{array}{lccc} & AgCl(s) \rightleftharpoons & Ag^+(aq) \;+ & Cl^-(aq) \\ \Delta G_f^\circ: & -109.7 & 77.1 & -131.2 \text{ kJ} \end{array}$$

Hence,

$$\Delta G^\circ = [(-131.2 + 77.1) - (-109.7)] \text{ kJ}$$

$$\Delta G^\circ = 55.6 \text{ kJ (or } 55.6 \times 10^3 \text{ J)}$$

We now substitute numerical values into the equation relating $\log K$ and ΔG°.

$$\log K = \frac{\Delta G^\circ}{-2.303\,RT} = \frac{55.6 \times 10^3}{-2.303 \times 8.31 \times 298} = -9.75$$

Therefore,

$$K = \text{antilog }(-9.75) = 10^{-9.75} = \mathbf{1.8 \times 10^{-10}}$$

Exercise 18.11

Calculate the solubility product constant for $Mg(OH)_2$ at 25°C. The ΔG_f° values (in kJ/mol) are as follows: $Mg^{2+}(aq)$, −456.0; $OH^-(aq)$, −157.3; $Mg(OH)_2(s)$, −933.9.

(See Problems 18.51 and 18.52.)

We can understand the use of ΔG° as a criterion of spontaneity in terms of its relationship to the equilibrium constant, $\Delta G^\circ = -2.303RT \log K$. When the equilibrium constant is greater than 1, log K is positive and ΔG° is negative. Similarly, when the equilibrium constant is less than 1, log K is negative and ΔG° is positive. This agrees with the first two rules listed at the end of Section 18.4. We can get the third rule by substituting $\Delta G^\circ = \pm 10 \times 10^3$ J into $\Delta G^\circ = -2.303\ RT \log K$ and solving for K. We find that K is between 0.018 and 57. In this range, the equilibrium mixture contains significant amounts of reactants as well as products.

18.7 CHANGE OF FREE ENERGY WITH TEMPERATURE

In the previous sections, we obtained the free-energy change and equilibrium constant for a reaction at 25°C, the temperature at which the thermodynamic data were given. How do we find ΔG° or K at another temperature? Precise calculations are possible but are rather involved. Instead, we will look at a simple method that gives approximate results.

In this method, we assume that ΔH° and ΔS° are essentially constant with respect to temperature. (This is only approximately true.) We get the value of ΔG° at any temperature by substituting values of ΔH° and ΔS° at 25°C (obtained from appropriate tables) into the following equation.●

● This approximation is most accurate for temperatures not too different from the temperature for which the ΔH° and ΔS° values are obtained. Much different temperatures give greater error.

$$\Delta G^\circ = \Delta H^\circ - T\Delta S^\circ$$

Remember that the superscript degree sign (°) refers to substances in standard states, which are substances at 1 atm and at the *specified temperature*. Although until now this was 25°C (298 K), we now consider other temperatures. Note that in general, ΔG° depends strongly on temperature.

Spontaneity and Temperature Change

All of the four possible choices of signs for ΔH° and ΔS°, listed in Table 18.3, give different temperature behaviors for ΔG°. Consider the case in which ΔH° is

Table 18.3 Effect of Temperature on the Spontaneity of Reactions

ΔH°	ΔS°	ΔG°	Description*	Example
−	+	−	Spontaneous at all T	$C_6H_{12}O_6(s) \longrightarrow 2C_2H_5OH(l) + 2CO_2(g)$
+	−	+	Nonspontaneous at all T	$3O_2(g) \longrightarrow 2O_3(g)$
−	−	+ or −	Spontaneous at low T; nonspontaneous at high T	$2NH_3(g) + CO_2(g) \longrightarrow NH_2CONH_2(aq) + H_2O(l)$
+	+	+ or −	Nonspontaneous at low T; spontaneous at high T	$Ba(OH)_2 \cdot 8H_2O(s) + 2NH_4NO_3(s) \longrightarrow Ba(NO_3)_2(aq) + 2NH_3(g) + 10H_2O(l)$

*The terms *low temperature* and *high temperature* are relative. For a particular reaction, high temperature could mean room temperature.

negative and $\Delta S°$ is positive. An example is the reaction

$$\underset{\text{glucose}}{C_6H_{12}O_6(s)} \longrightarrow \underset{\text{ethanol}}{2C_2H_5OH(l)} + 2CO_2(g)$$

This represents the overall change of glucose (grape sugar) to ethanol (ethyl alcohol). The signs of $\Delta H°$ and $\Delta S°$ are easily explained. The formation of more stable bonds, such as occur in CO_2, releases energy as heat. Thus, the reaction is exothermic and $\Delta H°$ is negative. As explained in Example 18.2, the breaking up of a molecule ($C_6H_{12}O_6$) into smaller ones and the formation of a gas are expected to increase the entropy, so $\Delta S°$ is positive.

When $\Delta H°$ is negative and $\Delta S°$ is positive, both terms in $\Delta G°$ (that is, $\Delta H°$ and $-T\Delta S°$) are negative. Therefore, $\Delta G°$ is always negative and the reaction is spontaneous whatever the temperature.

If the signs of $\Delta H°$ and $\Delta S°$ are reversed (that is, if $\Delta H°$ is positive, or endothermic, and $\Delta S°$ is negative), $\Delta G°$ is always positive. Thus, the reaction is nonspontaneous at all temperatures. An example is the reaction in which oxygen gas is converted to ozone.

$$3O_2(g) \longrightarrow 2O_3(g)$$

To accomplish this conversion, oxygen is passed through a tube in which an electrical discharge occurs. The electrical discharge supplies the necessary free energy for this otherwise nonspontaneous change.

The reaction described in the chapter opening,

$$2NH_3(g) + CO_2(g) \longrightarrow NH_2CONH_2(aq) + H_2O(l)$$

is one in which both $\Delta H°$ and $\Delta S°$ are negative. In this case, the sign of $\Delta G°$ depends on the relative magnitudes of the terms $\Delta H°$ and $-T\Delta S°$, which have opposite signs. At some temperature, these terms just cancel and $\Delta G°$ equals zero. Below this temperature, $\Delta G°$ is negative. Above it, $\Delta G°$ is positive. Therefore, this reaction is spontaneous at low temperatures but becomes nonspontaneous at sufficiently high temperatures. This particular reaction is spontaneous at 25°C but becomes nonspontaneous at about 60°C.

A reaction in which both $\Delta H°$ and $\Delta S°$ are positive is the one described in the opening to Chapter 6.●

● The reaction mixture spontaneously cools enough to freeze water. See Figure 6.1, p. 188.

$$Ba(OH)_2 \cdot 8H_2O(s) + 2NH_4NO_3(s) \longrightarrow Ba(NO_3)_2(aq) + 2NH_3(g) + 10H_2O(l)$$

The reaction is endothermic, and because crystalline solids change to a solution and a gas, the entropy increases. Again, the sign of $\Delta G°$ depends on the relative magnitudes of the terms $\Delta H°$ and $-T\Delta S°$. The reaction is spontaneous at room temperature, but it would be nonspontaneous at a sufficiently low temperature. Table 18.3 summarizes this discussion.

Calculation of $\Delta G°$ at Various Temperatures

As an application of the method of calculating $\Delta G°$ at various temperatures, assuming $\Delta H°$ and $\Delta S°$ are constant, we will look at the following reaction:

$$CaCO_3(s) \longrightarrow CaO(s) + CO_2(g)$$

At 25°C, $\Delta G°$ equals +130.9 kJ, and the equilibrium partial pressure of CO_2 calculated from this is 1.1×10^{-23} atm. The very small value of this partial pressure shows that $CaCO_3$ is quite stable at room temperature. In the next example, we will see how $\Delta G°$ and K_p for this reaction change at higher temperature.

Example 18.10 Calculating $\Delta G°$ and K at Various Temperatures

(a) What is $\Delta G°$ at 1000°C for the following reaction?

$$CaCO_3(s) \rightleftharpoons CaO(s) + CO_2(g)$$

Is this reaction spontaneous at 1000°C and 1 atm? (b) What is the value of K_p at 1000°C for this reaction? What is the partial pressure of CO_2?

Solution

(a) From Tables 6.2 and 18.1, we have

	$CaCO_3(s)$ $\rightleftharpoons$	$CaO(s)$ +	$CO_2(g)$
$\Delta H_f°$:	−1206.9	−635.1	−393.5 kJ
$S°$:	92.9	38.2	213.7 J/K

We calculate $\Delta H°$ and $\Delta S°$ from these values.

$$\Delta H° = [(-635.1 - 393.5) - (-1206.9)] \text{ kJ} = 178.3 \text{ kJ}$$

$$\Delta S° = [(38.2 + 213.7) - (92.9)] \text{ J/K} = 159.0 \text{ J/K}$$

Now we substitute $\Delta H°$, $\Delta S°$ (= 0.1590 kJ/K), and T (= 1273 K) into the equation for $\Delta G°$.

$$\Delta G° = \Delta H° - T\Delta S° = 178.3 \text{ kJ} - (1273 \text{ K})(0.1590 \text{ kJ/K})$$

$$= \mathbf{-24.1 \text{ kJ}}$$

Because $\Delta G°$ is negative, the reaction should be **spontaneous** at 1000°C and 1 atm. (b) We substitute the value of $\Delta G°$ at 1273 K, which equals -24.1×10^3 J, into the equation relating log K and $\Delta G°$.

$$\log K = \frac{\Delta G°}{-RT\ 2.303} = \frac{-24.1 \times 10^3}{-8.31 \times 1273 \times 2.303}$$

$$= 0.989$$

$$K = K_p = \text{antilog } 0.989 = \mathbf{9.75}$$

$K_p = P_{CO_2}$, so the partial pressure of CO_2 is **9.75 atm.**

Exercise 18.12

The thermodynamic equilibrium constant for the vaporization of water,

$$H_2O(l) \rightleftharpoons H_2O(g)$$

is $K_p = P_{H_2O}$. Use thermodynamic data to calculate the vapor pressure of water at 45°C. Compare your answer with the value given in Table 5.6, p. 161.

(See Problems 18.57 and 18.58.)

We can easily use the method described in Example 18.10 to find the temperature at which a reaction such as the decomposition of $CaCO_3$ changes from nonspontaneous to spontaneous under standard conditions (1 atm for reactants and products). At this temperature, $\Delta G°$ equals zero.

$$\Delta G° = 0 = \Delta H° - T\Delta S°$$

Solving for T gives

$$T = \frac{\Delta H°}{\Delta S°}$$

For the decomposition of $CaCO_3$, using values obtained in Example 18.10, we get

$$T = \frac{178.3 \text{ kJ}}{0.1590 \text{ kJ/K}} = 1121 \text{ K (848°C)}$$

Thus, $CaCO_3$ should be stable to thermal decomposition to CaO and CO_2 at 1 atm until heated to 848°C. This is only approximate, of course, because we assumed that $\Delta H°$ and $\Delta S°$ are constant with temperature.

Exercise 18.13

To what temperature must magnesium carbonate be heated to decompose it to MgO and CO_2 at 1 atm? Is this higher or lower than the temperature required to decompose $CaCO_3$?

Values of ΔH_f° (in kJ/mol) are as follows: MgO(s), −601.2; $MgCO_3$(s), −1112. Values of S° (in J/K) are as follows: MgO(s), 26.9; $MgCO_3$(s), 65.9. Data for CO_2 are given in Tables 6.2 and 18.1.
(See Problems 18.59 and 18.60.)

Profile of a Chemical

IRON (a Transition Metal)

The process for preparing iron from its ores was discovered very early, certainly before the thirteenth century B.C. Perhaps the metal was found in the ashes of a fire built on an outcropping of iron ore, such as hematite (Fe_2O_3), magnetite (Fe_3O_4), or siderite ($FeCO_3$). The ore would have been reduced by charcoal (carbon) at the high temperatures of the fire to yield iron.

Today, iron is produced in a blast furnace by a similar reduction. A mixture of iron ore, coke (carbon produced by heating coal), and limestone is added at the top of the furnace (Figure 18.10), and a blast of heated air enters at the bottom. Near the bottom of the furnace, the coke burns to carbon dioxide. As the carbon dioxide rises through the heated coke, it is reduced to carbon monoxide. The carbon monoxide then reduces the iron oxides in the iron ore to metallic iron. The overall reduction of Fe_2O_3 can be written

$$Fe_2O_3(s) + 3CO(g) \longrightarrow 2Fe(l) + 3CO_2(g)$$

Molten iron flows to the bottom of the blast furnace, where it is drawn off. Impurities in the iron ore react with calcium oxide from the limestone to produce a glassy material called *slag*. Molten slag collects in a layer floating on the molten iron and is drawn off periodically.

Steels are alloys containing over 50% iron and up to about 1.5% carbon. The iron obtained from a blast furnace (called pig iron) contains a number of impurities, including 3% to 4% carbon, that make it brittle. To produce steel, we must remove these impurities and reduce the carbon content of the iron. The *basic oxygen process* is a method of making steel by blowing oxygen into the molten iron to oxidize impurities and decrease the amount of carbon present. Other metals may be added to the steel to give it desired properties. Stainless steels, for example, contain 12% to 18% chromium and 8% nickel.

Pure iron is a soft, reactive metal. It reacts with acids, giving off hydrogen gas and producing the pale green iron(II) ion, $Fe^{2+}(aq)$, also called the ferrous ion.

$$Fe(s) + 2H^+(aq) \longrightarrow Fe^{2+}(aq) + H_2(g)$$

The iron(II) ion is a mild reducing agent and is slowly oxidized in aqueous solution by molecular oxygen.

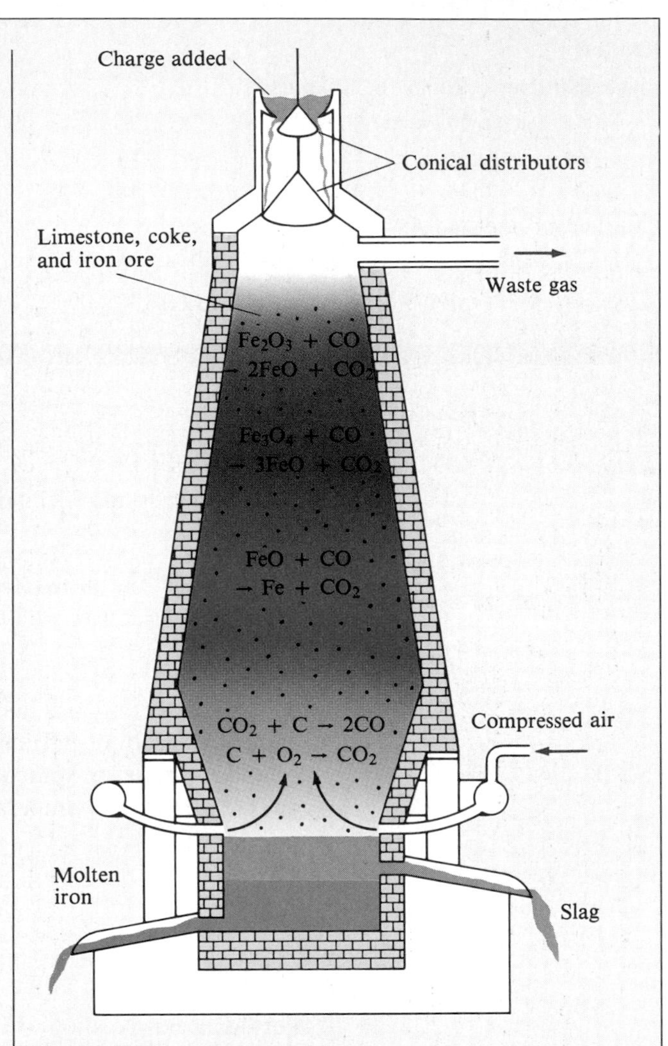

Figure 18.10
Blast furnace for the reduction of iron ore to iron metal. *Iron ore, coke, and limestone are added at the top of the blast furnace, and air enters at the bottom, where it burns the coke. The carbon monoxide from the burning reduces the iron ore to metallic iron.*

$$4Fe^{2+}(aq) + O_2(aq) + 4H^+(aq) \longrightarrow 4Fe^{3+}(aq) + 2H_2O(l)$$

The iron(III) ion, Fe^{3+}, exists in aqueous solution as the hydrated ion $Fe(H_2O)_6^{3+}(aq)$ and hydroxo species, such as $Fe(H_2O)_5(OH)^{2+}$. The hydroxo species have a yellow

Figure 18.11
Preparation of some iron(III) compounds. *An aqueous solution of $FeCl_3$ (pale yellow) was added to four beakers. Then the following solutions were added to the second, third, and fourth beakers, respectively: NaOH(aq), KSCN(aq), and $K_4[Fe(CN)_6](aq)$. A red-brown precipitate of $Fe(OH)_3$ formed in the second beaker, the dark red complex ion $Fe(H_2O)_5(SCN)^{2+}$ formed in the third beaker, and a dark blue precipitate of Prussian blue formed in the fourth beaker.*

color, and aqueous solutions of iron(III) salts (also called ferric salts) usually have a yellow color; see Figure 18.11.

When solutions of iron(II) salts are made basic in the absence of oxygen, a white precipitate of iron(II) hydroxide appears.

$$Fe^{2+}(aq) + 2OH^-(aq) \longrightarrow Fe(OH)_2(s)$$

In the presence of air, however, this precipitate is quickly oxidized to a red-brown precipitate, which we can represent as $Fe(OH)_3$ (see Figure 18.11).

$$4Fe(OH)_2(s) + O_2(aq) + 2H_2O(l) \longrightarrow 4Fe(OH)_3(s)$$

The substance is perhaps better represented as the hydrated oxide $Fe_2O_3 \cdot xH_2O$. When heated, it forms iron(III) oxide. Rust is the hydrated oxide.

A number of complex ions of iron(III) are well known. When thiocyanate ion is added to a solution of aqueous iron(III) ion, a blood-red thiocyanate complex forms. The ion has been used in the analysis of iron.

$$Fe(H_2O)_6{}^{3+}(aq) + SCN^-(aq) \longrightarrow \underset{\text{blood-red}}{Fe(H_2O)_5(SCN)^{2+}(aq)} + H_2O(l)$$

The addition of potassium ferrocyanide, $K_4[Fe(CN)_6]$, to an aqueous solution of Fe^{3+} ion yields an intense dark blue precipitate of Prussian blue (see Figure 18.11). The formula is not definitely established but may be $Fe_4[Fe(CN)_6]_3$. Prussian blue is used as a pigment in paints and inks.

Questions for Study

1. Describe how iron ore is reduced in a blast furnace. What is the chemical equation for the overall reduction of Fe_3O_4 by CO?
2. What is the purpose of the basic oxygen process?
3. What are the products of the reaction of iron and hydrochloric acid?
4. Iron(II) ion is a mild reducing agent. Write the balanced equation for (a) the reduction of Sn^{4+} to Sn^{2+} by Fe^{2+} and (b) the oxidation of Fe^{2+} by $Cr_2O_7{}^{2-}$ in acidic solution.
5. Calculate $\Delta G°$ for the following reaction:
$$Fe_2O_3(s) + 3H_2(g) \longrightarrow 2Fe(s) + 3H_2O(g)$$
What is the value of K at 25°C?
6. How is Prussian blue prepared? What is it used for?

A Checklist for Review

IMPORTANT TERMS

internal energy, U (18.1)
state function (18.1)
work (18.1)
first law of thermodynamics (18.1)
enthalpy, H (18.1)
spontaneous process (p. 716)
entropy, S (18.2)
second law of thermodynamics (18.2)
third law of thermodynamics (18.3)
standard (absolute) entropy (18.3)
free energy, G (18.4)
standard free energy of formation, ΔG_f° (18.4)
thermodynamic equilibrium constant, K (18.6)

KEY EQUATIONS

$\Delta U = q + w$

$w = -P\Delta V$

$\Delta S > \frac{q}{T}$ (spontaneous process)

$\Delta S = \frac{q}{T}$ (equilibrium process)

$\Delta S^\circ = \Sigma\, nS^\circ(\text{products}) - \Sigma\, mS^\circ(\text{reactants})$

$\Delta G^\circ = \Delta H^\circ - T\Delta S^\circ$

$\Delta G^\circ = \Sigma\, n\Delta G_f^\circ(\text{products}) - \Sigma\, m\Delta G_f^\circ(\text{reactants})$

$\Delta G = \Delta G^\circ + RT \ln Q$ or $\Delta G = \Delta G^\circ + 2.303\, RT \log Q$

$\Delta G^\circ = -RT \ln K$ or $\Delta G^\circ = -2.303\, RT \log K$

SUMMARY OF FACTS AND CONCEPTS

The *first law of thermodynamics* states that $\Delta U = q + w$. For a chemical reaction at constant pressure, this becomes $\Delta U = q_p - P\Delta V$. If we define *enthalpy, H,* as $U + PV$, we can show that for a reaction at constant pressure, $\Delta H = \Delta U + P\Delta V = q_p$. *Entropy, S,* is a thermodynamic measure of randomness or disorder in a system. According to the *second law of thermodynamics,* the total entropy of a system and its surroundings increases for a *spontaneous process,* one that occurs of its own accord. For a spontaneous reaction at constant T and P, one can show that the entropy change ΔS is greater than $\Delta H/T$. For an equilibrium process, such as a phase transition, $\Delta S = \Delta H/T$. The standard entropy S° of a substance is determined by measuring the heat absorbed in warming it at increasing temperatures at 1 atm. The method depends on the *third law of thermodynamics,* which states that perfectly crystalline substances at 0 K have zero entropy.

Free energy, G, is defined as $H - TS$. The change in free energy, ΔG, for a reaction at constant T and P equals $\Delta H - T\Delta S$, which is negative for a spontaneous change. *Standard free-energy changes,* ΔG°, can be calculated from ΔH_f° and S° values or from *standard free energies of formation,* ΔG_f°. ΔG_f° for a substance is the standard free-energy change for the formation of the substance from the elements in their stablest states. The standard free-energy change, ΔG°, serves as a *criterion of spontaneity* of a reaction. A negative value means the reaction is spontaneous. A value of zero means the reaction is at equilibrium. A positive value means the reaction is nonspontaneous.

The free-energy change, ΔG, is related to the *maximum work* that can be done by a reaction. This maximum work equals ΔG. A nonspontaneous reaction can be made to go by coupling it with a reaction that has a negative ΔG, so that ΔG for the overall result is negative. The reaction with the negative ΔG does work on the nonspontaneous reaction.

The *thermodynamic equilibrium constant, K,* can be calculated from the standard free-energy change by the relationship $\Delta G^\circ = -2.303\, RT \log K$. ΔG° at various temperatures can be obtained by using the approximation that ΔH° and ΔS° are constant. Therefore ΔG°, which equals $\Delta H^\circ - T\Delta S^\circ$, can be easily calculated for any T.

OPERATIONAL SKILLS

1. **Calculating the entropy change for a phase transition** Given the heat of phase transition and the temperature of the transition, calculate the entropy change of the system, ΔS (Example 18.1).

2. **Predicting the sign of the entropy change of a reaction** Predict the sign of ΔS° for a reaction to which the rules listed in the text can be clearly applied (Example 18.2).

3. **Calculating ΔS° for a reaction** Given the standard entropies of reactants and products, calculate the entropy of reaction, ΔS° (Example 18.3).

4. **Calculating ΔG° from ΔH° and ΔS°** Given enthalpies of formation and standard entropies of reactants and products, calculate the standard free-energy change ΔG° for a reaction (Example 18.4).

5. **Calculating ΔG° from standard free energies of formation** Given the free energies of formation of reactants and products, calculate the standard free-energy change ΔG° for a reaction (Example 18.5).

6. **Interpreting the sign of ΔG°** Use the standard free-energy change to determine the spontaneity of a reaction (Example 18.6).

7. **Writing the expression for a thermodynamic equilibrium constant** For any balanced chemical equation, write the expression for the thermodynamic equilibrium constant (Example 18.7).

8. **Calculating K from the standard free-energy change** Given the standard free-energy change for a reaction, calculate the thermodynamic equilibrium constant (Examples 18.8 and 18.9).

9. **Calculating ΔG° and K at various temperatures** Given ΔH° and ΔS° at 25°C, calculate ΔG° and K for a reaction at a temperature other than 25°C (Example 18.10).

Review Questions

18.1 What is a spontaneous process? Give three examples of spontaneous processes. Give three examples of nonspontaneous processes.

18.2 Which contains greater entropy, a quantity of frozen benzene or the same quantity of liquid benzene at the same temperature? Explain in terms of the degree of order of the substance.

18.3 State the second law of thermodynamics.

18.4 The entropy change ΔS for a phase transition equals $\Delta H/T$, where ΔH is the enthalpy change. Why is it that the entropy change for a system in which a chemical reaction occurs spontaneously does not equal $\Delta H/T$?

18.5 Describe how the standard entropy of hydrogen gas at 25°C can be obtained from heat measurements.

18.6 Describe what you would look for in a reaction involving gases in order to predict the sign of $\Delta S°$. Explain.

18.7 Define the free energy G. How is ΔG related to ΔH and ΔS?

18.8 What is meant by the standard free-energy change $\Delta G°$ for a reaction? What is meant by the standard free energy of formation $\Delta G_f°$ of a substance?

18.9 Explain how $\Delta G°$ can be used to decide whether a chemical equation is spontaneous in the direction written.

18.10 What is the useful work obtained in the ideal situation in which a chemical reaction with free-energy change ΔG is run so that it produces no entropy?

18.11 Give an example of a chemical reaction used to obtain useful work.

18.12 How is the concept of coupling of reactions useful in explaining how a nonspontaneous change could be made to occur?

18.13 Explain how the free energy changes as a spontaneous reaction occurs. Show by means of a diagram how G changes with the extent of reaction.

18.14 Explain how an equilibrium constant can be obtained from thermal data alone (that is, from measurements of heat only).

18.15 Discuss the different sign combinations of $\Delta H°$ and $\Delta S°$ that are possible for a process carried out at constant temperature and pressure. For each combination, state whether the process must be spontaneous or not, or whether both situations are possible. Explain.

18.16 Consider a reaction in which $\Delta H°$ and $\Delta S°$ are positive. Suppose the reaction is nonspontaneous at room temperature. How would you estimate the temperature at which the reaction becomes spontaneous?

Practice Problems

FIRST LAW OF THERMODYNAMICS

18.17 A gas is cooled and loses 65 J of heat. The gas contracts as it cools, and work done on the system equal to 22 J is exchanged with the surroundings. What are q, w, and ΔU?

18.18 An ideal gas is cooled isothermally (at constant temperature). The internal energy of an ideal gas remains constant during an isothermal change. If q is -65 J, what are ΔU and w?

18.19 What is ΔU when 1.00 mol of liquid water vaporizes at 100°C? The heat of vaporization, $\Delta H°_{vap}$, of water at 100°C is 40.66 kJ/mol.

18.20 What is ΔU for the following reaction at 25°C?

$$N_2(g) + 3H_2(g) \longrightarrow 2NH_3(g);\ \Delta H° = -91.8 \text{ kJ}$$

ENTROPY CHANGES

18.21 Chloroform, $CHCl_3$, is a solvent and has been used as an anesthetic. The heat of vaporization of chloroform at its boiling pont (61.2°C) is 29.6 kJ/mol. What is the entropy change when 1 mol $CHCl_3$ vaporizes at its boiling point?

18.22 Diethyl ether (known simply as ether), $(C_2H_5)_2O$, is a solvent and anesthetic. The heat of vaporization of diethyl ether at its boiling point (35.6°C) is 26.7 kJ/mol. What is the entropy change when 1 mol $(C_2H_5)_2O$ vaporizes at its boiling point?

18.23 The enthalpy change when liquid methanol, CH_3OH, vaporizes at 25°C is 37.4 kJ/mol. What is the entropy change when 1 mol of vapor in equilibrium with the liquid condenses to liquid at 25°C? The entropy of this vapor at 25°C is 252 J/(mol · K). What is the entropy of the liquid at this temperature?

18.24 The heat of vaporization of carbon disulfide, CS_2, at 25°C is 29 kJ/mol. What is the entropy change when 1 mol of carbon disulfide vapor in equilibrium with liquid condenses to liquid at 25°C? The entropy of this vapor at 25°C is 248 J/(mol · K). What is the entropy of the liquid at this temperature?

18.25 Predict the sign of $\Delta S°$, if possible, for each of the following reactions. If you cannot predict the sign for any reaction, state why.

(a) $C_2H_2(g) + 2H_2(g) \longrightarrow C_2H_6(g)$
(b) $N_2(g) + O_2(g) \longrightarrow 2NO(g)$
(c) $2C_2H_2(g) + 3O_2(g) \longrightarrow 4CO(g) + 2H_2O(g)$
(d) $2C(s) + O_2(g) \longrightarrow 2CO(g)$

18.26 Predict the sign of $\Delta S°$, if possible, for each of the following reactions. If you cannot predict the sign for any reaction, state why.

(a) $2SO_3(g) \longrightarrow 2SO_2(g) + O_2(g)$
(b) $C_2H_5OH(l) + 3O_2(g) \longrightarrow 2CO_2(g) + 3H_2O(l)$
(c) $P_4(g) \longrightarrow P_4(s)$
(d) $2NaHCO_3(s) \longrightarrow Na_2CO_3(s) + H_2O(g) + CO_2(g)$

18.27 Calculate $\Delta S°$ for the following reactions, using standard entropy values.

(a) $2Na(s) + Cl_2(g) \longrightarrow 2NaCl(s)$
(b) $NaCl(s) \longrightarrow NaCl(aq)$
(c) $CS_2(l) + 3O_2(g) \longrightarrow CO_2(g) + 2SO_2(g)$
(d) $2CH_3OH(l) + 3O_2(g) \longrightarrow 2CO_2(g) + 4H_2O(g)$

18.28 Calculate $\Delta S°$ for the following reactions, using standard entropy values.

(a) $2Ca(s) + O_2(g) \longrightarrow 2CaO(s)$
(b) $CaCl_2(s) \longrightarrow CaCl_2(aq)$
(c) $CS_2(g) + 4H_2(g) \longrightarrow CH_4(g) + 2H_2S(g)$
(d) $C_2H_4(g) + 3O_2(g) \longrightarrow 2CO_2(g) + 2H_2O(g)$

18.29 Calculate $\Delta S°$ for the reaction

$$CH_4(g) + 2O_2(g) \longrightarrow CO_2(g) + 2H_2O(l)$$

See Table 18.1 for values of standard entropies. Does the entropy of the chemical system increase or decrease as you expect? Explain.

18.30 What is the change in entropy, $\Delta S°$, for the reaction

$$CaCO_3(s) + 2H^+(aq) \longrightarrow Ca^{2+}(aq) + H_2O(l) + CO_2(g)$$

See Table 18.1 for values of standard entropies. Does the entropy of the chemical system increase or decrease as you expect? Explain.

FREE-ENERGY CHANGE AND SPONTANEITY

18.31 Using enthalpies of formation (Appendix C), calculate $\Delta H°$ for the following reaction at 25°C. Also calculate $\Delta S°$ for this reaction from standard entropies at 25°C. Use these values to calculate $\Delta G°$ for the reaction at this temperature.

$$2CH_3OH(l) + 3O_2(g) \longrightarrow 2CO_2(g) + 4H_2O(l)$$

18.32 Using enthalpies of formation (Appendix C), calculate $\Delta H°$ for the following reaction at 25°C. Also calculate $\Delta S°$ for this reaction from standard entropies at 25°C. Use these values to calculate $\Delta G°$ for the reaction at this temperature.

$$4HCN(l) + 5O_2(g) \longrightarrow 2H_2O(g) + 4CO_2(g) + 2N_2(g)$$

18.33 The free energy of formation of one mole of compound refers to a particular chemical equation. For each of the following, write that equation.

(a) $NaCl(s)$ (b) $HCN(l)$
(c) $SO_2(g)$ (d) $PH_3(g)$

18.34 The free energy of formation of one mole of compound refers to a particular chemical equation. For each of the following, write that equation.

(a) $CaO(s)$ (b) $CH_3NH_2(g)$
(c) $CS_2(l)$ (d) $P_4O_{10}(s)$

18.35 Calculate the standard free energy of the following reactions at 25°C, using standard free energies of formation.

(a) $CH_4(g) + 2O_2(g) \longrightarrow CO_2(g) + 2H_2O(g)$
(b) $CaCO_3(s) + 2H^+(aq) \longrightarrow Ca^{2+}(aq) + H_2O(l) + CO_2(g)$

18.36 Calculate the standard free energy of the following reactions at 25°C, using standard free energies of formation.

(a) $C_2H_4(g) + 3O_2(g) \longrightarrow 2CO_2(g) + 2H_2O(g)$
(b) $Na_2CO_3(s) + H^+(aq) \longrightarrow 2Na^+(aq) + HCO_3^-(aq)$

18.37 On the basis of $\Delta G°$ for each of the following reactions, decide whether the reaction is spontaneous or nonspontaneous as written. Or, if you expect an equilibrium mixture with significant amounts of both reactants and products, say so.

(a) $SO_2(g) + 2H_2S(g) \longrightarrow 3S(s) + 2H_2O(g)$; $\Delta G° = -91$ kJ

(b) $2H_2O_2(aq) \longrightarrow O_2(g) + 2H_2O(l)$; $\Delta G° = -211$ kJ

(c) $HCOOH(l) \longrightarrow CO_2(g) + H_2(g)$; $\Delta G° = 119$ kJ

(d) $I_2(s) + Br_2(l) \longrightarrow 2IBr(g)$; $\Delta G° = 7.5$ kJ

(e) $NH_4Cl(s) \longrightarrow NH_3(g) + HCl(g)$; $\Delta G° = 92$ kJ

18.38 For each of the following reactions, state whether the reaction is spontaneous or nonspontaneous as written or is easily reversible (that is, is a mixture with significant amounts of reactants and products).

(a) $HCN(g) + 2H_2(g) \longrightarrow CH_3NH_2(g)$; $\Delta G° = -92$ kJ

(b) $N_2(g) + O_2(g) \longrightarrow 2NO(g)$; $\Delta G° = 173$ kJ

(c) $2NO(g) + 3H_2O(g) \longrightarrow 2NH_3(g) + \frac{5}{2}O_2(g)$; $\Delta G° = 479$ kJ

(d) $H_2(g) + Cl_2(g) \longrightarrow 2HCl(g)$; $\Delta G° = -191$ kJ

(e) $H_2(g) + I_2(s) \longrightarrow 2HI(g)$; $\Delta G° = 2.6$ kJ

18.39 Calculate $\Delta H°$ and $\Delta G°$ for the following reactions at 25°C, using thermodynamic data from Appendix C; interpret the signs of $\Delta H°$ and $\Delta G°$.

(a) $Al_2O_3(s) + 2Fe(s) \longrightarrow Fe_2O_3(s) + 2Al(s)$

(b) $COCl_2(g) + H_2O(l) \longrightarrow CO_2(g) + 2HCl(g)$

18.40 Calculate $\Delta H°$ and $\Delta G°$ for the following reactions at 25°C, using thermodynamic data from Appendix C; interpret the signs of $\Delta H°$ and $\Delta G°$.

(a) $2PbO(s) + N_2(g) \longrightarrow 2Pb(s) + 2NO(g)$

(b) $CS_2(l) + 2H_2O(l) \longrightarrow CO_2(g) + 2H_2S(g)$

MAXIMUM WORK

18.41 Consider the reaction of 2 mol $H_2(g)$ at 25°C and 1 atm with 1 mol $O_2(g)$ at the same temperature and pressure to produce liquid water at these same conditions. If this reaction is run in a controlled way to generate work, what is the maximum useful work that can be obtained? How much entropy is produced in this case?

18.42 Consider the reaction of 1 mol $H_2(g)$ at 25°C and 1 atm with 1 mol $Cl_2(g)$ at the same temperature and pressure to produce gaseous HCl at these same conditions. If this reaction is run in a controlled way to generate work, what is the maximum useful work that can be obtained? How much entropy is produced in this case?

18.43 What is the maximum work that could be obtained from 5.00 g of zinc metal in the following reaction at 25°C?

$$Zn(s) + Cu^{2+}(aq) \longrightarrow Zn^{2+}(aq) + Cu(s)$$

18.44 What is the maximum work that could be obtained from 5.00 g of zinc metal in the following reaction at 25°C?

$$Zn(s) + 2H^+(aq) \longrightarrow Zn^{2+}(aq) + H_2(g)$$

CALCULATION OF EQUILIBRIUM CONSTANTS

18.45 Give the expression for the thermodynamic equilibrium constant for each of the following reactions.

(a) $CO(g) + H_2O(g) \rightleftharpoons CO_2(g) + H_2(g)$

(b) $Mg(OH)_2(s) \rightleftharpoons Mg^{2+}(aq) + 2OH^-(aq)$

(c) $2Li(s) + 2H_2O(l) \rightleftharpoons 2Li^+(aq) + 2OH^-(aq) + H_2(g)$

18.46 Write the expression for the thermodynamic equilibrium constant for each of the following reactions.

(a) $CO(g) + 2H_2(g) \rightleftharpoons CH_3OH(g)$

(b) $2Ag^+(aq) + CrO_4^{2-}(aq) \rightleftharpoons Ag_2CrO_4(s)$

(c) $CaCO_3(s) + 2H^+(aq) \rightleftharpoons Ca^{2+}(aq) + H_2O(l) + CO_2(g)$

18.47 What is the standard free-energy change $\Delta G°$ at 25°C for the following reaction? Obtain necessary information from Table 18.2.

$$H_2(g) + Cl_2(g) \longrightarrow 2HCl(g)$$

What is the value of the thermodynamic equilibrium constant K?

18.48 What is the standard free-energy change $\Delta G°$ at 25°C for the following reaction? See Table 18.2 for data.

$$C(graphite) + O_2(g) \longrightarrow CO_2(g)$$

Calculate the value of the thermodynamic equilibrium constant K.

18.49 Calculate the standard free-energy change and the equilibrium constant K_p for the following reaction at 25°C. See Table 18.2 for data.

$$CO(g) + 3H_2(g) \rightleftharpoons CH_4(g) + H_2O(g)$$

18.50 Calculate the standard free-energy change and the equilibrium constant K_p for the following reaction at 25°C. See Appendix C for data.

$$CO(g) + 2H_2(g) \rightleftharpoons CH_3OH(g)$$

18.51 Obtain the equilibrium constant K_c at 25°C from the free-energy change for the reaction

$$Mg(s) + Cu^{2+}(aq) \rightleftharpoons Mg^{2+}(aq) + Cu(s)$$

See Appendix C for data.

18.52 Calculate the equilibrium constant K_c at 25°C from the free-energy change for the following reaction:

$$Zn(s) + Cu^{2+}(aq) \rightleftharpoons Zn^{2+}(aq) + Cu(s)$$

See Appendix C for data.

FREE ENERGY AND TEMPERATURE CHANGE

18.53 Find the sign of $\Delta S°$ for the reaction

$$2N_2O_5(s) \longrightarrow 4NO_2(g) + O_2(g)$$

The reaction is endothermic and spontaneous at 25°C. Explain the spontaneity of the reaction in terms of enthalpy and entropy changes.

18.54 The combustion of acetylene, C_2H_2, is a spontaneous reaction given by the equation

$$2C_2H_2(g) + 5O_2(g) \longrightarrow 4CO_2(g) + 2H_2O(l)$$

As expected for a combustion, the reaction is exothermic. What is the sign of $\Delta H°$? What do you expect for the sign of $\Delta S°$? Explain the spontaneity of the reaction in terms of the enthalpy and entropy changes.

18.55 Estimate the value of $\Delta H°$ for the following reaction from bond energies (Table 9.5).

$$H_2(g) + Cl_2(g) \longrightarrow 2HCl(g)$$

Is the reaction exothermic or endothermic? Note that the reaction involves the breaking of symmetrical molecules (H_2 and Cl_2) and the formation of a less symmetrical product (HCl). From this, would you expect $\Delta S°$ to be positive or negative? Comment on the spontaneity of the reaction in terms of the changes in enthalpy and entropy.

18.56 Compare the energies of the bonds broken and those formed (see Table 9.5) for the reaction

$$HCN(g) + 2H_2(g) \longrightarrow CH_3NH_2(g)$$

From this, conclude whether the reaction is exothermic or endothermic. What is the sign of $\Delta S°$? Explain. The reaction is spontaneous at 25°C. Explain this in terms of the enthalpy and entropy changes.

18.57 Use data given in Tables 6.2 and 18.1 to obtain the value of K_p at 1000°C for the reaction

$$C(\text{graphite}) + CO_2(g) \rightleftharpoons 2CO(g)$$

Carbon monoxide is known to form during combustion of carbon at high temperatures. Do the data agree with this? Explain.

18.58 Use data given in Tables 6.2 and 18.1 to obtain the value of K_p at 2000°C for the reaction

$$N_2(g) + O_2(g) \rightleftharpoons 2NO(g)$$

Nitric oxide is known to form in hot flames in air, which is a mixture of N_2 and O_2. It is present in auto exhaust from this reaction. Are the data in agreement with this result? Explain.

18.59 Sodium carbonate, Na_2CO_3, can be prepared by heating sodium hydrogen carbonate, $NaHCO_3$.

$$2NaHCO_3(s) \longrightarrow Na_2CO_3(s) + H_2O(g) + CO_2(g)$$

Estimate the temperature at which $NaHCO_3$ decomposes to products at 1 atm. See Appendix C for data.

18.60 Oxygen was first prepared by heating mercury(II) oxide, HgO.

$$2HgO(s) \longrightarrow 2Hg(g) + O_2(g)$$

Estimate the temperature at which HgO decomposes to O_2 at 1 atm. See Appendix C for data.

Additional Problems

18.61 Acetic acid, CH_3COOH, freezes at 16.6°C. The heat of fusion, ΔH_{fus}, is 69.0 J/g. What is the change of entropy, ΔS, when 1 mol of liquid acetic acid freezes to the solid?

18.62 Acetone, CH_3COCH_3, boils at 56°C. The heat of vaporization of acetone at this temperature is 29.1 kJ/mol. What is the entropy change when 1 mol of liquid acetone vaporizes at 56°C?

18.63 Without doing any calculations, decide what the sign of ΔS° will be for each of the following reactions.

(a) $2LiOH(aq) + CO_2(g) \longrightarrow Li_2CO_3(aq) + H_2O(l)$

(b) $(NH_4)_2Cr_2O_7(s) \longrightarrow N_2(g) + 4H_2O(g) + Cr_2O_3(s)$

(c) $2N_2O_5(g) \longrightarrow 4NO_2(g) + O_2(g)$

(d) $O_2(g) + 2F_2(g) \longrightarrow 2OF_2(g)$

18.64 For each of the following reactions, decide whether there is an increase or a decrease in entropy. Why do you think so? (No calculations are needed.)

(a) $N_2(g) + 3H_2(g) \longrightarrow 2NH_3(g)$

(b) $NH_4Cl(s) \longrightarrow NH_3(g) + HCl(g)$

(c) $CO(g) + 2H_2(g) \longrightarrow CH_3OH(l)$

(d) $Li_3N(s) + 3H_2O(l) \longrightarrow 3LiOH(aq) + NH_3(g)$

18.65 The following equation shows how nitrogen dioxide reacts with water to produce nitric acid:

$$3NO_2(g) + H_2O(l) \longrightarrow 2HNO_3(l) + NO(g)$$

Predict the sign of ΔS° for this reaction.

18.66 Ethanol burns in air or oxygen according to the equation

$$C_2H_5OH(l) + 3O_2(g) \longrightarrow 2CO_2(g) + 3H_2O(g)$$

Predict the sign of ΔS° for this reaction.

18.67 Acetic acid in vinegar results from the bacterial oxidation of ethanol.

$$C_2H_5OH(l) + O_2(g) \longrightarrow CH_3COOH(l) + H_2O(l)$$

What is ΔS° is this reaction? Use standard entropy values. (See Appendix C for data.)

18.68 Methanol is produced commercially from carbon monoxide and hydrogen.

$$CO(g) + 2H_2(g) \longrightarrow CH_3OH(l)$$

What is ΔS° for this reaction? Use standard entropy values.

18.69 Is the following reaction spontaneous as written? Explain. Do whatever calculation is needed to answer the question.

$$SO_2(g) + H_2(g) \longrightarrow H_2S(g) + O_2(g)$$

18.70 Is the following reaction spontaneous as written? Explain. Do whatever calculation is needed to answer the question.

$$CH_4(g) + N_2(g) \longrightarrow HCN(g) + NH_3(g)$$

18.71 The reaction

$$CO_2(g) + H_2(g) \longrightarrow CO(g) + H_2O(g)$$

is nonspontaneous at room temperature but becomes spontaneous at a much higher temperature. What can you conclude from this about the signs of ΔH° and ΔS°, assuming that the enthalpy and entropy changes are not greatly affected by the temperature change? Explain your reasoning.

18.72 The reaction

$$N_2(g) + 3H_2(g) \longrightarrow 2NH_3(g)$$

is spontaneous at room temperature but becomes nonspontaneous at a much higher temperature. From this fact alone, obtain the signs of ΔH° and ΔS°, assuming that ΔH° and ΔS° do not change much with temperature. Explain your reasoning.

18.73 Calculate ΔG° at 25°C for the reaction

$$CaF_2(s) \rightleftharpoons Ca^{2+}(aq) + 2F^-(aq)$$

The value of ΔG_f° at 25°C for $CaF_2(s)$ is −1162 kJ/mol. See Table 18.2 for other values. What is the value of the solubility product constant, K_{sp}, for this reaction at 25°C?

18.74 Calculate ΔG° at 25°C for the reaction

$$BaSO_4(s) \rightleftharpoons Ba^{2+}(aq) + SO_4^{2-}(aq)$$

The values of ΔG_f° at 25°C are (in kJ/mol): $BaSO_4(s)$, −1353; $Ba^{2+}(aq)$, −561; $SO_4^{2-}(aq)$, −742. What is the value of the solubility product constant, K_{sp}, for this reaction at 25°C?

18.75 Consider the decomposition of phosgene, $COCl_2$.

$$COCl_2(g) \longrightarrow CO(g) + Cl_2(g)$$

Calculate ΔH° and ΔS° at 25°C for this reaction. See Appendix C for data. What is ΔG° at 25°C? Assume that ΔH° and ΔS° are constant with respect to a change of temperature. Now calculate ΔG° at 800°C. Compare the two values of ΔG°. Briefly discuss the spontaneity of the reaction at 25°C and at 800°C.

18.76 Consider the reaction

$$CS_2(g) + 4H_2(g) \rightleftharpoons CH_4(g) + 2H_2S(g)$$

Calculate ΔH°, ΔS°, and ΔG° at 25°C for this reaction. Assume ΔH° and ΔS° are constant with respect to a change of temperature. Now calculate ΔG° at 650°C. Compare the two values of ΔG°. Briefly discuss the spontaneity of the reaction at 25°C and at 650°C.

Cumulative-Skills Problems

18.77 Hydrogen bromide dissociates into its gaseous elements, H_2 and Br_2, at elevated temperatures. Calculate the percent dissociation at 375°C and 1.00 atm. What would be the percent dissociation at 375°C and 10.0 atm? Use data from Appendix C and make any reasonable approximation to obtain K.

18.78 Hydrogen gas and iodine gas react to form hydrogen iodide. If 0.500 mol H_2 and 1.00 mol I_2 are placed in a closed, 10.0-L vessel, what is the mole fraction of HI in the mixture when equilibrium is reached at 205°C? Use data from Appendix C and make any reasonable approximations to obtain K.

18.79 A 20.0-L vessel is filled with 1.00 mol of ammonia, NH_3. What percent of ammonia dissociates to the elements if equilibrium is reached at 345°C? Use data from Appendix C and make any reasonable approximation to obtain K.

18.80 A 25.0-L vessel is filled with 0.0100 mol CO and 0.0300 mol H_2. How many moles of CH_4 and how many moles of H_2O are produced when equilibrium is reached at 785°C? Use data from Appendix C and make any reasonable approximation to obtain K.

18.81 K_a for acetic acid at 25.0°C is 1.754×10^{-5}. At 50.0°C, K_a is 1.633×10^{-5}. What are $\Delta H°$ and $\Delta S°$ for the ionization of acetic acid?

18.82 K_{sp} for silver chloride at 25.0°C is 1.782×10^{-10}. At 35.0°C, K_{sp} is 4.159×10^{-10}. What are $\Delta H°$ and $\Delta S°$ for the reaction?

19

Electrochemistry

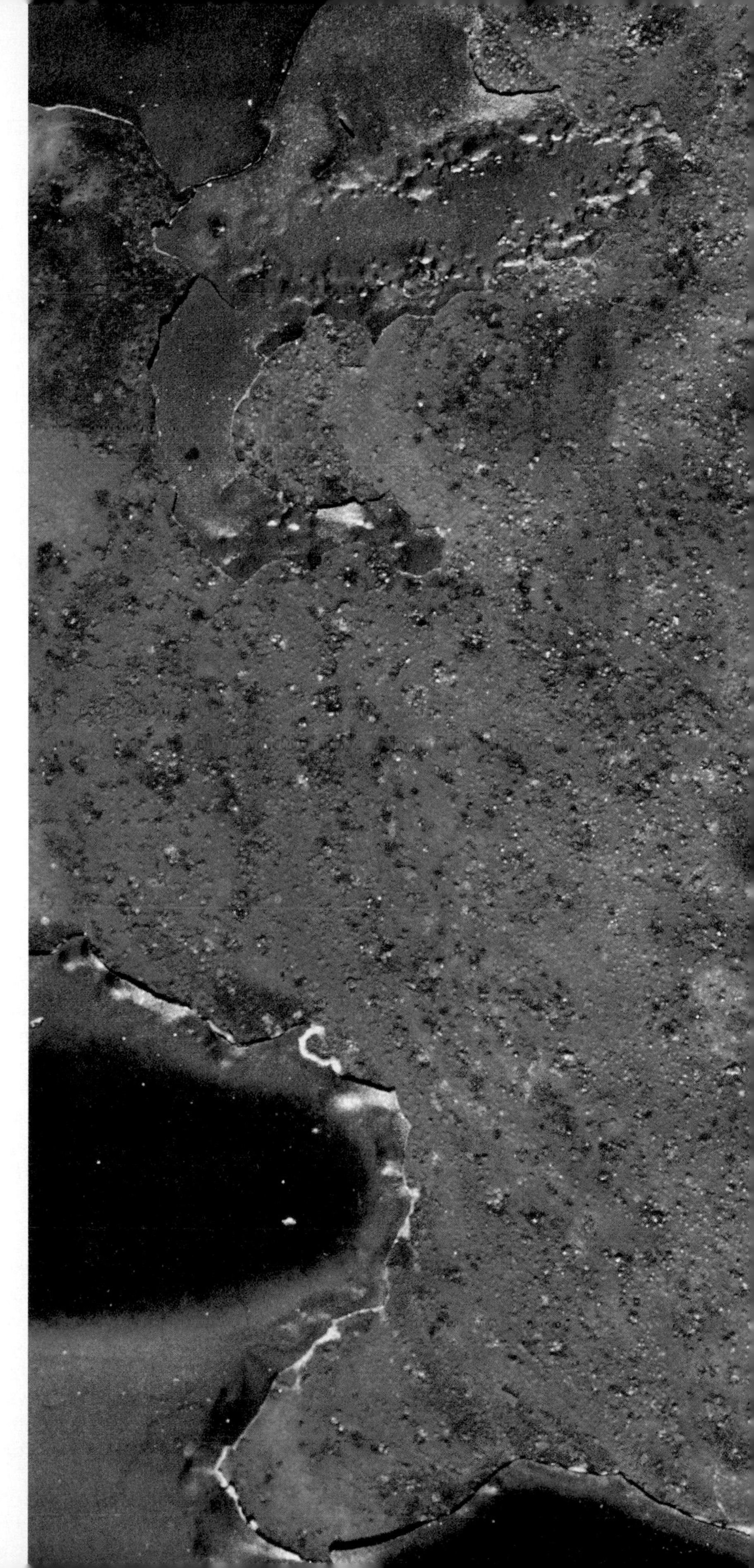

Chapter Outline

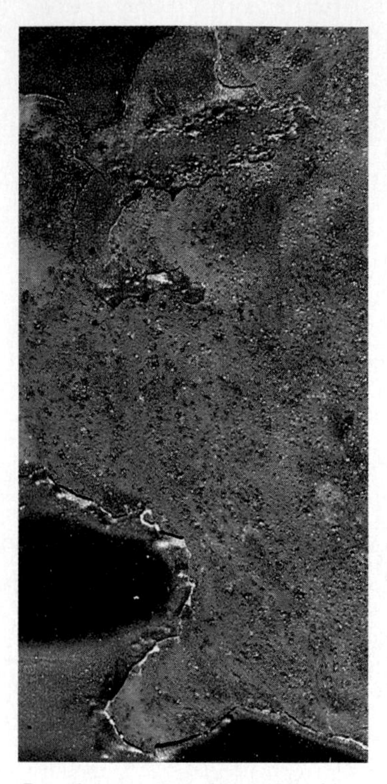

Rust forms in an electrochemical process.

The first battery was invented by Alessandro Volta about 1800. He assembled a pile consisting of pairs of zinc and silver disks separated by paper disks soaked in salt water. With a tall pile, he could detect a weak electric shock when he touched the two ends of the pile. Later Volta showed that any two different metals could be used to make such a voltaic pile (Figure 19.1).

A battery cell that became popular during the nineteenth century was constructed in 1836 by the English chemist John Frederick Daniell. It used zinc and copper. The basic principle was that of Volta's battery pile, but each metal was surrounded by a solution of the metal ion, and the solutions were kept separate by a porous ceramic barrier. Each metal with its solution was a half-cell; a zinc half-cell and a copper half-cell made up one voltaic cell. This construction became the standard form of such cells, which exploit the spontaneous chemical reaction

$$Zn(s) + Cu^{2+}(aq) \longrightarrow Zn^{2+}(aq) + Cu(s)$$

to generate electric energy.

Today many different kinds of batteries have been developed. They include miniature button batteries for pocket calculators and watches, and larger batteries for storing the energy from the solar collectors of communications satellites. But all such batteries operate on the same general principles of Volta's pile or Daniell's cell. In this chapter, we will look at the general principles involved in setting up a chemical reaction as a battery. We will answer such questions as, What voltage can we expect from a particular battery? and, How can we relate the battery voltage to the equilibrium constant for the reaction?

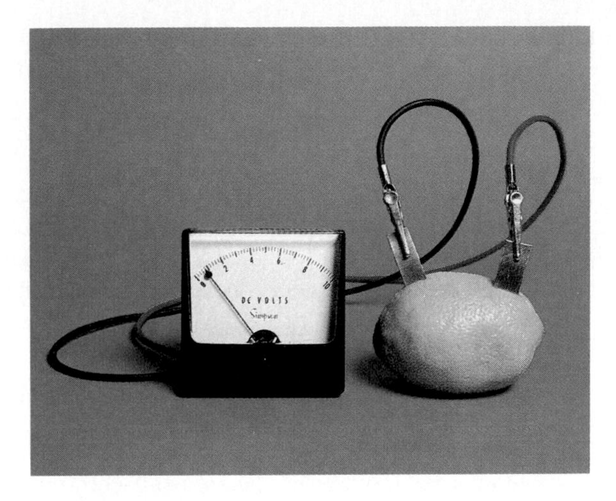

Figure 19.1
Lemon battery. Zinc and copper strips in a lemon generate a voltage, as this photograph shows.

Voltaic Cells

In discussing oxidation–reduction reactions in Chapter 13, we described the separation of such reactions into half-reactions. The separation into half-reactions was done only on paper, as a way to balance chemical equations. It is possible in some cases, however, to physically separate a reaction into half-reactions. In that event, the useful energy or work available from a spontaneous reaction may be used to drive electrons from the oxidation half-reaction through an external circuit to the reduction half-reaction. This is what happens in the operation of a battery.

A battery is a kind of electrochemical cell. An **electrochemical cell** is *a system consisting of electrodes that dip into an electrolyte and in which a chemical reaction either uses or generates an electric current*. A **voltaic,** or **galvanic, cell** is *an electrochemical cell in which a spontaneous reaction generates an electric current*. An **electrolytic cell** is *an electrochemical cell in which an electric current drives an otherwise nonspontaneous reaction*. In the next sections, we will discuss the basic principles behind voltaic cells and then explore some commercial uses of these cells.

19.1 CONSTRUCTION OF VOLTAIC CELLS

A voltaic cell consists of two *half-cells* that are electrically connected. Each **half-cell** is *the portion of an electrochemical cell in which a half-reaction takes place*. A simple half-cell can be made from a metal strip that dips into a solution of its metal ion. An example is the zinc–zinc ion half-cell (often called simply a zinc electrode), which consists of a zinc metal strip dipping into a solution of a zinc salt. Another simple half-cell consists of a copper metal strip dipping into a solution of a copper salt (copper electrode).

In a voltaic cell, two half-cells are connected in such a way that electrons flow from one metal electrode to another through an external circuit, while ions flow from one half-cell to another through an internal cell connection. Consider a voltaic cell consisting of a zinc electrode and a copper electrode. Because zinc tends to lose electrons more readily than copper, the zinc electrode takes on a negative charge relative to the copper electrode. When the two electrodes are connected through an external metallic wire, electrons flow from the zinc through this external circuit to the copper. The following half-reactions occur:

$$Zn(s) \longrightarrow Zn^{2+}(aq) + 2e^-$$
$$Cu^{2+}(aq) + 2e^- \longrightarrow Cu(s)$$

The first half-reaction shows zinc metal forming zinc ions, which go into the solution. The electrons from this half-reaction travel through the zinc metal to the external circuit and then to the copper electrode. The negative charge on the copper metal attracts positive copper ions to it, which react with the electrons to deposit copper metal on the electrode.

The two half-cells need to be connected internally to allow ions to flow between them. Thus, as zinc ions continue to be produced, the zinc ion solution begins to build up a positive charge. Similarly, as copper ions plate out as copper, the solution builds up a negative charge. The half-cell reactions will stop unless positive ions can move from the zinc half-cell to the copper half-cell, and negative ions from the copper half-cell can move to the zinc half-cell. It is necessary that these ion flows occur without mixing of the zinc ion and copper ion solutions. If copper ion were to come in contact with the zinc metal, for example, direct reaction would occur without an electric current being generated. The voltage would drop, and the battery would run down quickly.

$$Zn(s) + Cu^{2+}(aq) \longrightarrow Zn^{2+}(aq) + Cu(s)$$

Figure 19.2A shows the two half-cells of a voltaic cell connected by a salt bridge. A **salt bridge** is *a tube of an electrolyte in a gel that is connected to the two half-cells of a voltaic cell; it allows the flow of ions but prevents the mixing of the different solutions that would allow direct reaction of the cell reactants.* In Figure 19.2B, the half-cells are connected externally so that an electric current flows. Figure 19.2C shows an actual setup of the zinc–copper cell, but in this case the salt bridge is replaced by a porous glass plate, which accomplishes a similar purpose.

● The porous barrier in the original Daniell cell serves the same purpose as the salt bridge.

The two half-cell reactions, as we noted earlier, are

$$Zn(s) \longrightarrow Zn^{2+}(aq) + 2e^- \quad \text{(oxidation half-reaction)}$$
$$Cu^{2+}(aq) + 2e^- \longrightarrow Cu(s) \quad \text{(reduction half-reaction)}$$

The first half-reaction, which loses electrons, is the oxidation half-reaction. *The electrode at which oxidation occurs* is called the **anode.** The second half-reaction, which gains electrons, is the reduction half-reaction. *The electrode at which reduction occurs* is called the **cathode.** These definitions of anode and cathode hold for all electrochemical cells, including electrolytic cells, as we will see when we discuss them.

Note that the sum of the two half-reactions

$$Zn(s) + Cu^{2+}(aq) \longrightarrow Zn^{2+}(aq) + Cu(s)$$

is *the net reaction that occurs in the voltaic cell;* it is called the **cell reaction.**

Once we know which electrode is the anode and which is the cathode, we can easily determine the direction of electron flow in the external portion of the circuit. Electrons are given up by the anode (from the oxidation half-reaction) and thus flow from it, whereas electrons are used up by the cathode (by the reduction half-reaction) and so flow into this electrode. The anode has a negative sign, because electrons flow from it. The cathode has a positive sign. Look again at Figure 19.2B and note the labeling of the electrodes as anode and cathode; also note the direction of electron flow in the external circuit and the signs of the

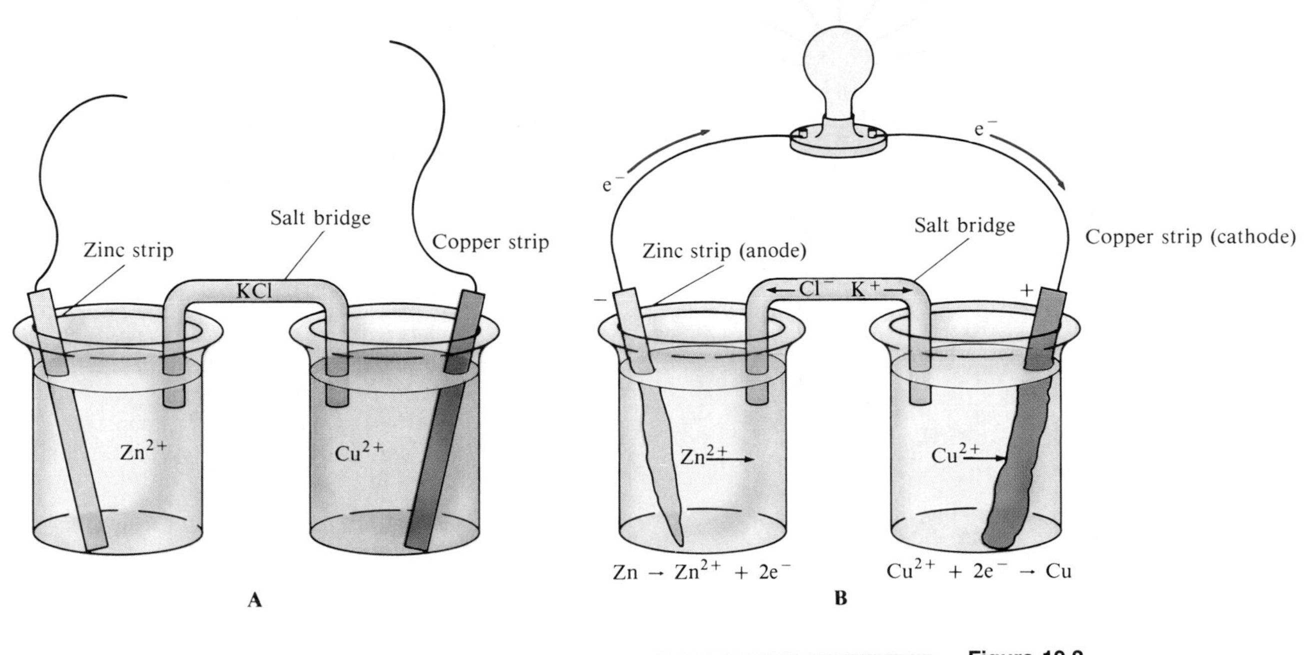

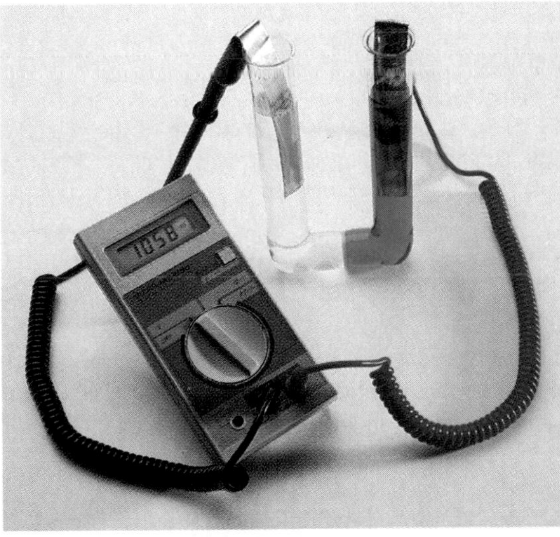

Figure 19.2
A zinc–copper voltaic cell.
(A) A zinc electrode and a copper electrode, without an external circuit; there is no cell reaction. *(B)* When the two electrodes are connected by an external circuit (with a light bulb), chemical reaction occurs. *(C)* A similar cell, with the salt bridge replaced by a porous glass plate and the bulb replaced by a voltmeter. Note the blue color of $Cu^{2+}(aq)$.

electrodes. Note too the migration of ions in the solutions. The following example further illustrates these points about a voltaic cell.

Example 19.1 Sketching and Labeling a Voltaic Cell

A voltaic cell is constructed from a half-cell in which a cadmium rod dips into a solution of cadmium nitrate, $Cd(NO_3)_2$, and another half-cell in which a silver rod dips into a solution of silver nitrate, $AgNO_3$. The two half-cells are connected by a salt bridge. Silver ion is reduced during operation of the voltaic cell. Draw a sketch of the cell. Label the anode and cathode, showing the corresponding half-reactions at these

(continued)

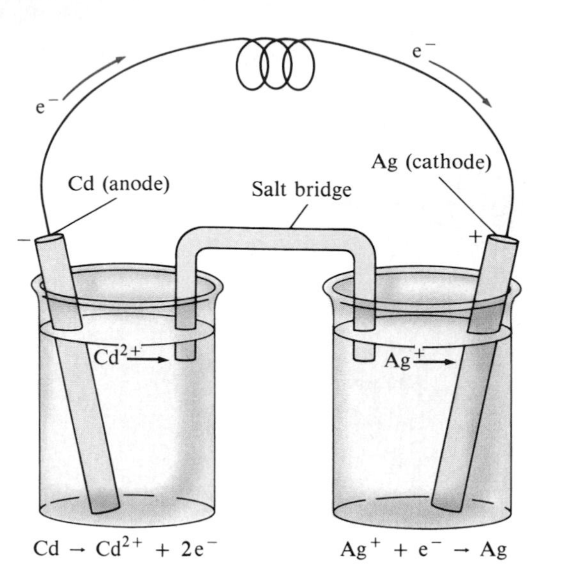

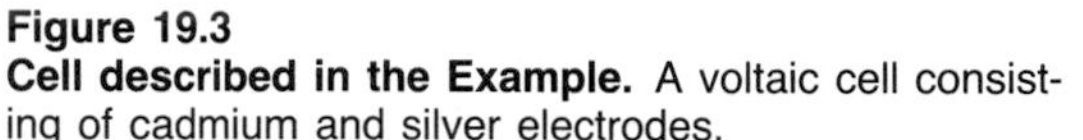

Figure 19.3
Cell described in the Example. A voltaic cell consisting of cadmium and silver electrodes.

electrodes. Indicate the electron flow in the external circuit, the signs of the electrodes, and the direction of cation migration in the half-cells.

Solution

Because silver ion is reduced at the silver electrode, the silver electrode is the cathode. The half-reaction is

$$Ag^{+}(aq) + e^{-} \longrightarrow Ag(s)$$

The cadmium electrode must be the anode (electrode at which oxidation occurs); the half-reaction is

$$Cd(s) \longrightarrow Cd^{2+}(aq) + 2e^{-}$$

Now that we have labeled the electrodes, we can see that the electron flow in the external circuit is from the cadmium electrode (anode) to the silver electrode (cathode). Positive ions will flow in the solution portion of the circuit opposite to the direction of the electrons. The sketch for this cell, including the labeling, is given in Figure 19.3.

Exercise 19.1

A voltaic cell consists of a silver–silver ion half-cell and a nickel–nickel(II) ion half-cell. Silver ion is reduced during operation of the cell. Sketch the cell, labeling the anode and cathode and indicating the corresponding electrode reactions. Show the direction of electron flow in the external circuit and the direction of cation movement in the half-cells.

(See Problems 19.19 and 19.20.)

19.2 NOTATION FOR VOLTAIC CELLS

It is convenient to have a shorthand way of designating particular voltaic cells. The cell described earlier, consisting of a zinc metal–zinc ion half-cell and a copper metal–copper ion half-cell, is written

$$Zn(s)|Zn^{2+}(aq)\|Cu^{2+}(aq)|Cu(s)$$

In this notation, the anode, or oxidation half-cell, is always written on the left; the cathode, or reduction half-cell, is written on the right. The two electrodes are electrically connected by means of a salt bridge, denoted by two vertical bars.

$$\underset{\text{anode}}{Zn(s)|Zn^{2+}(aq)} \quad \underset{\text{salt bridge}}{\|} \quad \underset{\text{cathode}}{Cu^{2+}(aq)|Cu(s)}$$

The cell terminals are at the extreme ends in this cell notation, and a single vertical bar indicates a phase boundary—say, between a solid terminal and the electrode solution. For the anode of the same cell, we have

$$\underset{\text{anode terminal}}{Zn(s)} \quad \underset{\text{phase boundary}}{|} \quad \underset{\text{anode electrolyte}}{Zn^{2+}(aq)}$$

When the half-reaction involves a gas, an inert material such as platinum serves as a terminal and as an electrode surface on which the half-reaction occurs. The platinum catalyzes the half-reaction but otherwise is not involved in it. Figure

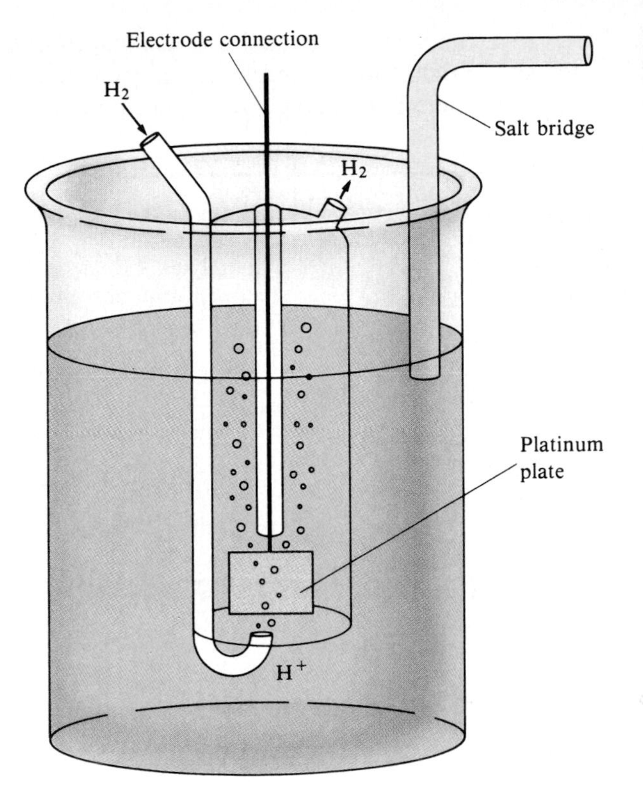

Figure 19.4
A hydrogen electrode. Hydrogen gas bubbles over a platinum surface, where the half-reaction $2H^+(aq) + 2e^- \rightleftharpoons H_2(g)$ occurs.

19.4 shows a hydrogen electrode; hydrogen bubbles over a platinum plate that is immersed in an acidic solution. The cathode half-reaction is

$$2H^+(aq) + 2e^- \rightleftharpoons H_2(g)$$

The notation for the hydrogen electrode, written as a cathode, is

$$H^+(aq)|H_2(g)|Pt$$

To write such an electrode as an anode, we simply reverse the notation:

$$Pt|H_2(g)|H^+(aq)$$

Here are several additional examples of this notation for electrodes (written as cathodes). A comma separates ions present in the same solution. We will write the cathode with oxidized species before reduced species, in the same order as in the half-reaction.●

● Gas electrodes are frequently written with the gas next to the metal electrode, however. In that case, the chlorine electrode is written $Cl^-|Cl_2|Pt$.

CATHODE	CATHODE REACTION
$Cl_2(g)\|Cl^-(aq)\|Pt$	$Cl_2(aq) + 2e^- \rightleftharpoons 2Cl^-(aq)$
$Fe^{3+}(aq)$, $Fe^{2+}(aq)\|Pt$	$Fe^{3+}(aq) + e^- \rightleftharpoons Fe^{2+}(aq)$
$Cd^{2+}(aq)\|Cd(s)$	$Cd^{2+}(aq) + 2e^- \rightleftharpoons Cd(s)$

Exercise 19.2

Write the notation for a cell in which the electrode reactions are

$$2H^+(aq) + 2e^- \longrightarrow H_2(g)$$
$$Zn(s) \longrightarrow Zn^{2+}(aq) + 2e^-$$

(See Problems 19.25 and 19.26.)

We can write the overall cell reaction from the cell notation by first writing the appropriate half-cell reactions and then summing these in such a way that the electrons cancel. Example 19.2 shows how this is done.

Example 19.2 Writing the Cell Reaction from the Cell Notation

(a) Write the cell reaction for the voltaic cell

$$Tl(s)|Tl^{+}(aq)||Sn^{2+}(aq)|Sn(s)$$

(b) Write the cell reaction for the voltaic cell

$$Zn(s)|Zn^{2+}(aq)||Fe^{3+}(aq), Fe^{2+}(aq)|Pt$$

Solution

(a) The half-cell reactions are

$$Tl(s) \longrightarrow Tl^{+}(aq) + e^{-}$$
$$Sn^{2+}(aq) + 2e^{-} \longrightarrow Sn(s)$$

Multiplying the anode reaction by 2 and then summing the half-cell reactions gives

$$2Tl(s) + Sn^{2+}(aq) \longrightarrow 2Tl^{+}(aq) + Sn(s)$$

(b) The half-cell reactions are

$$Zn(s) \longrightarrow Zn^{2+}(aq) + 2e^{-}$$
$$Fe^{3+}(aq) + e^{-} \longrightarrow Fe^{2+}(aq)$$

and the cell reaction is

$$Zn(s) + 2Fe^{3+}(aq) \longrightarrow Zn^{2+}(aq) + 2Fe^{2+}(aq)$$

Exercise 19.3

Give the overall cell reaction for the voltaic cell

$$Cd(s)|Cd^{2+}(aq)||H^{+}(aq)|H_2(g)|Pt$$

(See Problems 19.29 and 19.30.)

To fully specify a voltaic cell, it is necessary to give the concentrations of solutions or ions and the pressure of gases. In the cell notation, these are written within parentheses for each species. For example,

$$Zn(s)|Zn^{2+}(1.0\ M)||H^{+}(1.0\ M)|H_2(1.0\ \text{atm})|Pt$$

19.3 ELECTROMOTIVE FORCE

Work is needed to move electrons in a wire or to move ions through a solution to an electrode. The situation is analogous to pumping water from one point to another. Work must be expended to pump the water. The analogy may be extended. Water moves from a point at high pressure to a point at low pressure. Thus, a pressure difference is required. The work expended in moving water through a pipe depends on the volume of water and the pressure difference. The situation with electricity is similar. An electric charge moves from a point at high electric potential (high electrical pressure) to a point at low electric potential (low electrical pressure). The work needed to move an electric charge through a conductor depends on the total charge moved and the potential difference. **Potential difference** is *the difference in electric potential (electrical pressure) between two points*. We measure this quantity in volts. The **volt, V,** is *the SI unit of potential difference*. The electrical work expended in moving a charge through a conductor is

$$\text{Electrical work} = \text{charge} \times \text{potential difference}$$

Corresponding SI units for the terms in this equation are

$$\text{Joules} = \text{coulombs} \times \text{volts}$$

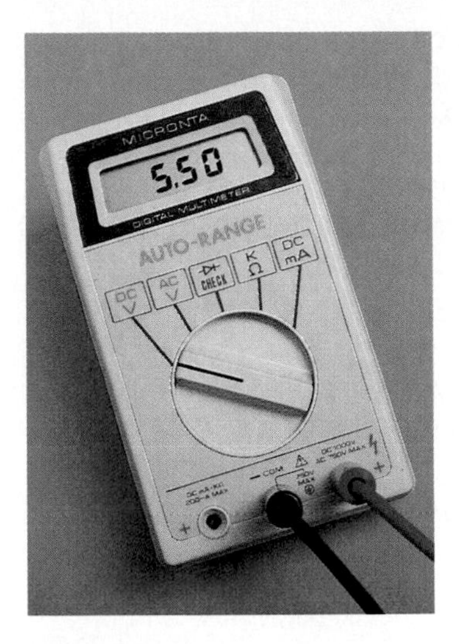

Figure 19.5
A digital voltmeter. A digital voltmeter draws negligible current, so it can be used to measure cell emf's.

The **faraday constant, *F*,** is *the magnitude of charge on one mole of electrons; it equals 9.65×10^4 C* (96,500 coulombs). In moving one faraday of charge from one electrode to another, the work done by a voltaic cell is the product of the faraday constant F times the potential difference between the electrodes. The work w is the negative of this, because the voltaic cell loses energy as it does work on the surroundings.

$$w = -F \times \text{potential difference}$$

In the normal operation of a voltaic cell, the potential difference (voltage) across the electrodes is less than the maximum possible voltage of the cell. One reason for this is that it takes energy or work to drive a current through the cell itself. The decrease in cell voltage as current is drawn reflects this energy expenditure within the cell; and the greater the current, the lower the voltage. Thus, the cell voltage has its maximum value only when no current flows. The situation is analogous to measuring the pressure difference between that of water in a faucet and that of the outside atmosphere. We expect this difference to be higher if we measure it by a device that does not require us to run water from the faucet. When we do run water, the pressure difference drops.

The maximum potential difference between the electrodes of a voltaic cell is referred to as the **electromotive force (emf)** of the cell, denoted E_{cell}. It is measured by an electronic digital voltmeter (Figure 19.5), which draws negligible current.

We can also determine which electrode of a cell is the anode and which is the cathode from this measurement. The digital voltmeter gives the reading with a sign attached, from which we can deduce the sign, or polarity, of each electrode. The anode of a voltaic cell has negative polarity, whereas the cathode has positive polarity.

We can now write an expression for the maximum work obtainable from a voltaic cell. Let n be the number of electrons transferred in the overall cell equation. The maximum electrical work of a voltaic cell for molar amounts of reactants (according to the cell equation *as written*) is

$$w_{max} = -nFE_{cell}$$

Here E_{cell} is the cell emf, and F is the faraday constant, 9.65×10^4 C.

Example 19.3 Calculating the Quantity of Work from a Given Amount of Cell Reactant

The emf of a particular voltaic cell with the cell reaction

$$Hg_2^{2+}(aq) + H_2(g) \rightleftharpoons 2Hg(l) + 2H^+(aq)$$

is 0.650 V. Calculate the maximum electrical work of this cell when 0.500 g H_2 is consumed.

Solution

To get n, it is necessary to obtain the half-reactions, which we find by using the method described in Section 13.8 for balancing oxidation–reduction equations by the half-reaction method. The half-reactions are

$$Hg_2^{2+}(aq) + 2e^- \rightleftharpoons 2Hg(l)$$

$$H_2(g) \rightleftharpoons 2H^+(aq) + 2e^-$$

Thus, n equals 2, and the maximum work for the reaction as written is

$$\begin{aligned} w_{max} = -nFE_{cell} &= -2 \times 9.65 \times 10^4 \text{ C} \times 0.650 \text{ V} \\ &= -1.25 \times 10^5 \text{ V} \cdot \text{C} \\ &= -1.25 \times 10^5 \text{ J} \end{aligned}$$

(Remember that a joule is equal to a volt-coulomb. Also, note that the negative sign means work is done by the cell—that is, energy is lost by the cell.) For 0.500 g H_2, the maxi-

(continued)

mum work is

$$0.500\ \cancel{g\ H_2} \times \frac{1\ mol\ \cancel{H_2}}{2.02\ \cancel{g\ H_2}} \times \frac{-1.25 \times 10^5\ J}{1\ mol\ \cancel{H_2}} = -3.09 \times 10^4\ J$$

Note that the conversion factor -1.25×10^5 J/1 mol H_2 is determined by the chemical equation as written. In this equation, 1 mol H_2 reacts and the maximum work produced is -1.25×10^5 J.

Exercise 19.4

What is the maximum electrical work that can be obtained from 6.54 g of zinc metal that reacts in a Daniell cell, described in the chapter opening, whose emf is 1.10 V? The overall cell reaction is

$$Zn(s) + Cu^{2+}(aq) \longrightarrow Zn^{2+}(aq) + Cu(s)$$

(See Problems 19.33, 19.34, 19.35, and 19.36.)

19.4 STANDARD CELL emf's AND STANDARD ELECTRODE POTENTIALS

A cell emf is a measure of the driving force of the cell reaction. This reaction occurs in the cell as separate half-reactions, an oxidation half-reaction and a reduction half-reaction. The general form of these half-reactions are

$$\text{Reduced species} \longrightarrow \text{oxidized species} + ne^- \quad \text{(oxidation/anode)}$$

$$\text{Oxidized species} + ne^- \longrightarrow \text{reduced species} \quad \text{(reduction/cathode)}$$

We can imagine that the cell emf is composed of a contribution from the anode (whose value depends on the ability of the oxidation half-reaction to lose electrons) and a contribution from the cathode (the value depending on the ability of the reduction half-reaction to gain electrons). We will call these contributions the *oxidation potential* and the *reduction potential,* respectively. Then

$$E_{cell} = \text{oxidation potential} + \text{reduction potential}$$

A reduction potential is a measure of the tendency for an oxidized species to gain electrons in the reduction half-reaction. If we can construct a table of reduction potentials, we will have a list of strengths of oxidizing agents, in addition to having a way of calculating cell emf's.

We can look at an oxidation half-reaction as the reverse of a corresponding reduction half-reaction. The oxidation potential for an oxidation half-reaction equals the negative of the reduction potential for the reverse half-reaction (which is a reduction).

$$\text{Oxidation potential for a half-reaction} = -\text{reduction potential for the reverse half-reaction}$$

This means that in practice we need to tabulate only oxidation potentials or reduction potentials. The choice, by convention, is to tabulate reduction potentials. We call these *electrode potentials,* and we denote them by the symbol E.

To illustrate the ideas we have just presented, consider the zinc–copper cell we described earlier.

$$Zn(s)|Zn^{2+}(aq)\|Cu^{2+}(aq)|Cu(s)$$

The half-reactions are

$$Zn(s) \longrightarrow Zn^{2+}(aq) + 2e^-$$
$$Cu^{2+}(aq) + 2e^- \longrightarrow Cu(s)$$

The first half-reaction is an oxidation. If we write E_{Zn} for the electrode potential corresponding to the reduction half-reaction $Zn^{2+}(aq) + 2e^- \rightarrow Zn(s)$, then $-E_{Zn}$ is the potential for the oxidation half-reaction $Zn(s) \rightarrow Zn^{2+}(aq) + 2e^-$. The copper half-reaction is a reduction. We will write E_{Cu} for the electrode potential.

The cell emf is the sum of the potentials for the reduction and oxidation half-reactions that occur. For the cell we have been describing, this is the sum of the reduction potential (electrode potential) for the copper half-cell and the oxidation potential (negative of the electrode potential) for the zinc half-cell.

$$E_{cell} = E_{Cu} + (-E_{Zn}) = E_{Cu} - E_{Zn}$$

Note that the cell emf equals the difference in the two electrode potentials. Thus, we can think of the electrode potential as the electric potential on the electrode, and we obtain the cell emf as a potential difference, in which we subtract the anode potential from the cathode potential.

$$E_{cell} = E_{cathode} - E_{anode}$$

The electrode potential is an intensive property. This means that its value is independent of the amount of species in the reaction. Thus, the electrode potential for the half-reaction

$$2Cu^{2+}(aq) + 4e^- \longrightarrow 2Cu(s)$$

is the same as for

$$Cu^{2+}(aq) + 2e^- \longrightarrow Cu(s)$$

Tabulating Standard Electrode Potentials

The emf of a voltaic cell depends on the concentrations of substances and the temperature of the cell. For purposes of tabulating electrochemical data, it is usual to choose thermodynamic standard-state conditions for voltaic cells. The **standard emf,** E°_{cell}, is *the emf of a voltaic cell operating under standard-state conditions (solute concentrations are each 1 M, gas pressures are each 1 atm, and the temperature has a specified value—usually 25°C).* Note the superscript degree sign (°), which signifies standard-state conditions.●

● To be precise, the standard-state conditions are those in which the activities (ideal concentrations) of solutes and gases equal 1. This is equivalent to 1 *M* only for ideal solutions. Because of the strong attractions between ions, substantial deviations from ideal conditions exist in electrolyte solutions, except when very dilute. We will ignore these deviations here.

If we can derive a table of electrode potentials, we can calculate cell emf's from them. This gives us a great advantage over tabulating cell emf's. From a small table of electrode potentials, we can obtain the emf's of all the cells that we could construct from pairs of electrodes. For instance, a table of 40 electrode potentials would give the emf's of nearly 800 voltaic cells.

However, it is not possible to measure the potential of a single electrode; only emf's of cells can be measured. What we can do, though, is measure the emf's of cells constructed from various electrodes connected in turn to one particular electrode, which we choose as a reference. We arbitrarily assign this reference electrode a potential equal to zero and obtain the potentials for the other electrodes by measuring the emf's. By convention, the reference chosen for comparing electrode potentials is the *standard hydrogen electrode*. This is an electrode like that pictured in Figure 19.4 operating under standard-state conditions.

The **standard electrode potential,** E°, is *the electrode potential when the concentrations of solutes are 1 M, the gas pressures are 1 atm, and the temperature has a specified value (usually 25°C).* The superscript degree sign (°) signifies standard-state conditions.

To understand how standard electrode potentials are obtained, let us look at

Table 19.1 Standard Electrode (Reduction) Potentials in Aqueous Solution at 25°C*

Cathode (Reduction) Half-Reaction	Standard Potential, E° (Volts)
$Li^+(aq) + e^- \rightleftharpoons Li(s)$	−3.04
$Na^+(aq) + e^- \rightleftharpoons Na(s)$	−2.71
$Mg^{2+}(aq) + 2e^- \rightleftharpoons Mg(s)$	−2.38
$Al^{3+}(aq) + 3e^- \rightleftharpoons Al(s)$	−1.66
$2H_2O(l) + 2e^- \rightleftharpoons H_2(g) + 2OH^-(aq)$	−0.83
$Zn^{2+}(aq) + 2e^- \rightleftharpoons Zn(s)$	−0.76
$Cr^{3+}(aq) + 3e^- \rightleftharpoons Cr(s)$	−0.74
$Fe^{2+}(aq) + 2e^- \rightleftharpoons Fe(s)$	−0.41
$Cd^{2+}(aq) + 2e^- \rightleftharpoons Cd(s)$	−0.40
$Ni^{2+}(aq) + 2e^- \rightleftharpoons Ni(s)$	−0.23
$Sn^{2+}(aq) + 2e^- \rightleftharpoons Sn(s)$	−0.14
$Pb^{2+}(aq) + 2e^- \rightleftharpoons Pb(s)$	−0.13
$Fe^{3+}(aq) + 3e^- \rightleftharpoons Fe(s)$	−0.04
$2H^+(aq) + 2e^- \rightleftharpoons H_2(g)$	0.00
$Sn^{4+}(aq) + 2e^- \rightleftharpoons Sn^{2+}(aq)$	0.15
$Cu^{2+}(aq) + e^- \rightleftharpoons Cu^+(aq)$	0.16
$Cu^{2+}(aq) + 2e^- \rightleftharpoons Cu(s)$	0.34
$IO^-(aq) + H_2O(l) + 2e^- \rightleftharpoons I^-(aq) + 2OH^-(aq)$	0.49
$Cu^+(aq) + e^- \rightleftharpoons Cu(s)$	0.52
$I_2(s) + 2e^- \rightleftharpoons 2I^-(aq)$	0.54
$Fe^{3+}(aq) + e^- \rightleftharpoons Fe^{2+}(aq)$	0.77
$Hg_2^{2+}(aq) + 2e^- \rightleftharpoons 2Hg(l)$	0.80
$Ag^+(aq) + e^- \rightleftharpoons Ag(s)$	0.80
$Hg^{2+}(aq) + 2e^- \rightleftharpoons Hg(l)$	0.85
$ClO^-(aq) + H_2O(l) + 2e^- \rightleftharpoons Cl^-(aq) + 2OH^-(aq)$	0.90
$2Hg^{2+}(aq) + 2e^- \rightleftharpoons Hg_2^{2+}(aq)$	0.90
$NO_3^-(aq) + 4H^+(aq) + 3e^- \rightleftharpoons NO(g) + 2H_2O(l)$	0.96
$Br_2(l) + 2e^- \rightleftharpoons 2Br^-(aq)$	1.07
$O_2(g) + 4H^+(aq) + 4e^- \rightleftharpoons 2H_2O(l)$	1.23
$Cr_2O_7^{2-}(aq) + 14H^+(aq) + 6e^- \rightleftharpoons 2Cr^{3+}(aq) + 7H_2O(l)$	1.33
$Cl_2(g) + 2e^- \rightleftharpoons 2Cl^-(aq)$	1.36
$MnO_4^-(aq) + 8H^+(aq) + 5e^- \rightleftharpoons Mn^{2+}(aq) + 4H_2O(l)$	1.49
$H_2O_2(aq) + 2H^+(aq) + 2e^- \rightleftharpoons 2H_2O(l)$	1.78
$S_2O_8^{2-}(aq) + 2e^- \rightleftharpoons 2SO_4^{2-}(aq)$	2.01
$F_2(g) + 2e^- \rightleftharpoons 2F^-(aq)$	2.87

*See Appendix I for a more extensive table.

how we would find the standard electrode potential, $E°$, for the zinc electrode. We connect a standard zinc electrode to a standard hydrogen electrode. When we measure the emf of the cell with a voltmeter, we obtain 0.76 V, with the zinc electrode acting as the anode.

Let us write the cell emf in terms of the electrode potentials. The cell is

$$Zn(s)|Zn^{2+}(aq)||H^{+}(aq)|H_2(g)|Pt$$

and the half-reactions with corresponding half-cell potentials (oxidation or reduction potentials) are

$$Zn(s) \longrightarrow Zn^{2+}(aq) + 2e^{-};\ -E°_{Zn}$$

$$2H^{+}(aq) + 2e^{-} \longrightarrow H_2(g);\ E°_{H_2}$$

The cell emf is the sum of the half-cell potentials.

$$E_{cell} = E°_{H_2} + (-E°_{Zn})$$

We now substitute 0.76 V for the cell emf and 0.00 V for the standard hydrogen electrode potential. This gives $E°_{Zn} = -0.76$ V.

Proceeding in this way, we can obtain the electrode potential for a series of half-cell reactions. Table 19.1 lists standard electrode potentials for selected half-cells at 25° C.

Strengths of Oxidizing and Reducing Agents

Standard electrode potentials are useful in determining the strengths of oxidizing and reducing agents under standard-state conditions. Because electrode potentials are reduction potentials, those reduction half-reactions in Table 19.1 with the larger (that is, more positive) electrode potentials have the greater tendency to go left to right as written. A reduction half-reaction has the general form

$$\text{Oxidized species} + ne^{-} \longrightarrow \text{reduced species}$$

The oxidized species acts as an oxidizing agent. Consequently, *the strongest oxidizing agents in a table of standard electrode potentials are the oxidized species corresponding to half-reactions with the largest (most positive) $E°$ values.*

Those reduction half-reactions with lower (that is, more negative) electrode potentials have a greater tendency to go right to left. That is,

$$\text{Reduced species} \longrightarrow \text{oxidized species} + ne^{-}$$

The reduced species acts as a reducing agent. Consequently, *the strongest reducing agents in a table of standard electrode potentials are the reduced species corresponding to half-reactions with the smallest (most negative) $E°$ values.*

The first two and last two entries in Table 19.1 are as follows:

$$Li^{+}(aq) + e^{-} \longrightarrow Li(s)$$

$$Na^{+}(aq) + e^{-} \longrightarrow Na(s)$$

. . .

. . .

$$S_2O_8^{2-}(aq) + 2e^{-} \longrightarrow 2SO_4^{2-}(aq)$$

$$F_2(g) + 2e^{-} \longrightarrow 2F^{-}(aq)$$

The strongest oxidizing agents are the species at the lower left in the table (given here in red). The strongest reducing agents are the species at the upper right in the table (given here in blue).

Example 19.4 Determining the Relative Strengths of Oxidizing and Reducing Agents

(a) Order the following oxidizing agents by increasing strength under standard-state conditions: $Cl_2(g)$, $H_2O_2(aq)$, $Fe^{3+}(aq)$. (b) Order the following reducing agents by increasing strength under standard-state conditions: $H_2(g)$, $Al(s)$, $Cu(s)$.

Solution

(a) The simplest procedure is to read down the table of electrode potentials picking out the reduction half-reactions in which the species of interest occur as reactants. The order from *top to bottom* in which these species occur in the table is the order of increasing oxidizing power of the species. The half-reactions and corresponding electrode potentials are as follows:

$$Fe^{3+}(aq) + e^- \longrightarrow Fe^{2+}(aq) \qquad 0.77\ V$$
$$Cl_2(g) + 2e^- \longrightarrow 2Cl^-(aq) \qquad 1.36\ V$$
$$H_2O_2(aq) + 2H^+(aq) + 2e^- \longrightarrow 2H_2O(l) \qquad 1.78\ V$$

The order by increasing oxidizing strength is $Fe^{3+}(aq)$, $Cl_2(g)$, and $H_2O_2(aq)$

(b) We pick out the half-reactions in which the species of interest occur as products. The order from *bottom to top* in which these species occur in the table is the order of increasing reducing strength of the species. The half-reactions and corresponding electrode potentials are

$$Al^{3+}(aq) + 3e^- \longrightarrow Al(s) \qquad -1.66\ V$$
$$2H^+(aq) + 2e^- \longrightarrow H_2(g) \qquad 0.00\ V$$
$$Cu^{2+}(aq) + 2e^- \longrightarrow Cu(s) \qquad 0.34\ V$$

The order by increasing reducing strength is $Cu(s)$, $H_2(g)$, $Al(s)$.

Exercise 19.5

Which is the stronger oxidizing agent, $NO_3^-(aq)$ in acidic solution (to NO) or $Ag^+(aq)$?

(See Problems 19.37, 19.38, 19.39, and 19.40.)

We can use a table of electrode potentials to predict the direction of spontaneity of an oxidation–reduction reaction. We need only note the relative strengths of the oxidizing agents on the left and right sides of the equation. The stronger oxidizing agent will be on the reactant side of the equation when it is written as a spontaneous reaction. (Alternatively, we can look at the reducing agents; the stronger reducing agent will be on the reactant side of the spontaneous reaction.)

Example 19.5 Determining the Direction of Spontaneity from Electrode Potentials

Consider the reaction

$$Zn^{2+}(aq) + 2Fe^{2+}(aq) \longrightarrow Zn(s) + 2Fe^{3+}(aq)$$

Does the reaction go spontaneously in the direction indicated, under standard conditions?

Solution

We can look at the oxidizing agents. In this reaction, Zn^{2+} is the oxidizing agent on the left; Fe^{3+} is the oxidizing agent on the right. The corresponding standard electrode potentials are

$$Zn^{2+}(aq) + 2e^- \longrightarrow Zn(s);\ E^\circ = -0.76\ V$$
$$Fe^{3+}(aq) + e^- \longrightarrow Fe^{2+}(aq);\ E^\circ = 0.77\ V$$

The stronger oxidizing agent is the one involved in the half-reaction with the more positive standard electrode potential, so Fe^{3+} is the stronger oxidizing agent. The reaction is nonspontaneous as written.

We could instead look at the reducing agents (Fe^{2+} on the left and Zn on the right). The stronger reducing agent is the one involved in the half-reaction with the more negative standard electrode potential. Zn is the stronger reducing agent. We reach the same conclusion: The reaction is nonspontaneous as written.

Exercise 19.6

Does the following reaction occur spontaneously in the direction indicated, under standard conditions?

$$Cu^{2+}(aq) + 2I^-(aq) \longrightarrow Cu(s) + I_2(s)$$

(See Problems 19.41, 19.42, 19.43, and 19.44.)

Calculating Cell emf's from Standard Potentials

The emf of a voltaic cell constructed from standard electrodes is easily calculated using a table of electrode potentials. As an example, consider the cell constructed from a cadmium electrode and a silver electrode having the following reduction half-reactions and corresponding standard electrode potentials (reduction potentials):

$$Cd^{2+}(aq) + 2e^- \longrightarrow Cd(s) \qquad E^\circ_{Cd} = -0.40\ V$$
$$Ag^+(aq) + e^- \longrightarrow Ag(s) \qquad E^\circ_{Ag} = 0.80\ V$$

We will need to reverse one of these half-reactions to obtain the oxidation part of the cell reaction. The cell reaction is spontaneous with the stronger reducing agent on the left. This will be Cd, because it is the reactant in the half-reaction with the more negative electrode potential. Therefore, we reverse the first half-reaction, as well as the sign of the half-cell potential (to obtain the oxidation potential, which equals $-E^\circ_{Cd}$).

$$Cd(s) \longrightarrow Cd^{2+}(aq) + 2e^- \qquad -E^\circ_{Cd} = 0.40\ V$$
$$Ag^+(aq) + e^- \longrightarrow Ag(s) \qquad E^\circ_{Ag} = 0.80\ V$$

To obtain the cell reaction, we must multiply the half-reactions by factors so that when the half-reactions are added together, the electrons cancel. This does not affect the half-cell potentials, because they are intensive quantities and so do not depend on amount of substance. We multiply the second half-reaction by 2, then add.

$$Cd(s) \longrightarrow Cd^{2+}(aq) + 2e^- \qquad -E^\circ_{Cd} = 0.40\ V$$
$$2Ag^+(aq) + 2e^- \longrightarrow 2Ag(s) \qquad E^\circ_{Ag} = 0.80\ V$$
$$\overline{Cd(s) + 2Ag^+(aq) \longrightarrow Cd^{2+}(aq) + 2Ag(s) \qquad E^\circ_{Ag} - E^\circ_{Cd} = 1.20\ V}$$

The cell emf, E°_{cell}, is 1.20 V. The cell diagram is

$$Cd(s)|Cd^{2+}(aq)||Ag^+(aq)|Ag(s)$$

Suppose that at the beginning of this problem, we had mistakenly reversed the silver half-reaction, instead of correctly reversing the cadmium half-reaction. Then our work would have looked like this:

$$Cd^{2+}(aq) + 2e^- \longrightarrow Cd(s) \qquad E^\circ_{Cd} = -0.40\ V$$
$$2Ag(s) \longrightarrow 2Ag^+(aq) + 2e^- \qquad -E^\circ_{Ag} = -0.80\ V$$
$$\overline{Cd^{2+}(aq) + 2Ag(s) \longrightarrow Cd(s) + 2Ag^+(aq) \qquad E^\circ_{Cd} - E^\circ_{Ag} = -1.20\ V}$$

The corresponding cell notation would have been

$$Ag(s)|Ag^+(aq)||Cd^{2+}(aq)|Cd(s)$$

Everything is simply the reverse of the correct result. At this point, we realize our mistake, because the cell emf is found to be negative. A negative emf merely indicates that the cell reaction is nonspontaneous as written. To obtain the spontaneous reaction and a positive emf, we simply reverse both half-reactions and corresponding half-cell potentials (we would also reverse the cell notation). This changes the sign of the emf.

Note that the emf of the cell equals the standard electrode potential of the cathode minus the standard electrode potential of the anode. So, alternatively, we can calculate cell emf's from the equation

$$E^\circ_{cell} = E^\circ_{cathode} - E^\circ_{anode}$$

We obtain

$$E^\circ_{cell} = E^\circ_{Ag} - E^\circ_{Cd} = 0.80\ V - (-0.40\ V) = 1.20\ V$$

Example 19.6 Calculating the emf from Standard Potentials

Calculate the standard emf of the following cell at 25°C using standard electrode potentials.

$$Al(s)|Al^{3+}(aq)||Fe^{2+}(aq)|Fe(s)$$

What is the cell reaction?

Solution

The reduction half-reactions and standard electrode potentials are

$$Al^{3+}(aq) + 3e^- \longrightarrow Al(s) \qquad E^\circ_{Al} = -1.66\ V$$

$$Fe^{2+}(aq) + 2e^- \longrightarrow Fe(s) \qquad E^\circ_{Fe} = -0.41\ V$$

Note that the aluminum half-cell is assumed to be the anode, according to the cell notation. Because by convention the anode is written on the left, the aluminum half-reaction should be written as an oxidation. Therefore, we reverse the first half-reaction and its half-cell potential to obtain

$$Al(s) \longrightarrow Al^{3+}(aq) + 3e^- \qquad -E^\circ_{Al} = 1.66\ V$$

$$Fe^{2+}(aq) + 2e^- \longrightarrow Fe(s) \qquad E^\circ_{Fe} = -0.41\ V$$

We obtain the cell emf by adding the half-cell potentials. Because we also want the cell reaction, we multiply the first half-reaction by 2 and the second half-reaction by 3, so that when the half-reactions are added, the electrons cancel. Note that we do not multiply the half-cell potentials by these factors; half-cell potentials are intensive quantities, so they do not depend on amount of substance. The addition of half-reactions is displayed below. The cell emf is 1.25 V.

Note that the cell emf is the standard electrode potential for the cathode minus the standard electrode potential for the anode. We could have calculated the cell emf from the formula

$$E^\circ_{cell} = E^\circ_{cathode} - E^\circ_{anode}$$

We obtain

$$E^\circ_{cell} = E^\circ_{Fe} - E^\circ_{Al} = -0.41\ V - (-1.66\ V) = 1.25\ V$$

$$2Al(s) \longrightarrow 2Al^{3+}(aq) + 6e^- \qquad -E^\circ_{Al} = 1.66\ V$$

$$3Fe^{2+}(aq) + 6e^- \longrightarrow 3Fe(s) \qquad E^\circ_{Fe} = -0.41\ V$$

$$2Al(s) + 3Fe^{2+}(aq) \longrightarrow 2Al^{3+}(aq) + 3Fe(s) \qquad E^\circ_{Fe} - E^\circ_{Al} = 1.25\ V$$

Exercise 19.7

Using standard electrode potentials, calculate E°_{cell} at 25°C for the following cell.

$$Zn(s)|Zn^{2+}(aq)||Cu^{2+}(aq)|Cu(s)$$

(See Problems 19.45 and 19.46.)

19.5 EQUILIBRIUM CONSTANTS FROM emf's

Some of the most important results of electrochemistry are the relationships among cell emf, free-energy change, and equilibrium constant. Recall that the free-energy change ΔG for a reaction equals the maximum useful work of the reaction (Section 18.5).

$$\Delta G = w_{max}$$

For a voltaic cell, this work is the electrical work, $-nFE_{cell}$ (where n is the number of electrons transferred in a reaction), so when the reactants and products are in their standard states, we have

$$\Delta G^\circ = -nFE^\circ_{cell}$$

With this equation, emf measurements become an important source of thermodynamic information. Alternatively, thermodynamic data can be used to calculate cell emf's. These calculations are shown in the following examples.

Example 19.7 Calculating the Free-Energy Change from Electrode Potentials

Using standard electrode potentials, calculate the standard free-energy change at 25°C for the reaction

$$Zn(s) + 2Ag^+(aq) \longrightarrow Zn^{2+}(aq) + 2Ag(s)$$

Solutions

The half-reactions, corresponding half-cell potentials, and their sums are displayed below. Note that each half-reaction involves two electrons; hence, $n = 2$. Also, $E^\circ_{cell} = 1.56$ V, and the faraday constant, F, is 9.65×10^4 C. Therefore,

$$\Delta G^\circ = -nFE^\circ_{cell} = -2 \times 9.65 \times 10^4 \text{ C} \times 1.56 \text{ V}$$
$$= -3.01 \times 10^5 \text{ J}$$

Recall that (coulombs) × (volts) = joules. Thus, the standard free-energy change is **−301 kJ**.

$$Zn(s) \longrightarrow Zn^{2+}(aq) + 2e^- \qquad -E^\circ = 0.76 \text{ V}$$
$$2Ag^+(aq) + 2e^- \longrightarrow 2Ag(s) \qquad E^\circ = 0.80 \text{ V}$$
$$Zn(s) + 2Ag^+(aq) \longrightarrow Zn^{2+}(aq) + 2Ag(s) \qquad E^\circ_{cell} = 1.56 \text{ V}$$

Exercise 19.8

What is ΔG° at 25°C for the reaction

$$Sn^{2+}(aq) + 2Hg^{2+}(aq) \longrightarrow Sn^{4+}(aq) + Hg_2^{2+}(aq)$$

For data, see Table 19.1. (See Problems 19.49 and 19.50.)

Example 19.8 Calculating the Cell emf from Free-Energy Change

Suppose the reaction of zinc metal and chlorine gas is utilized in a cell in which zinc ions and chloride ions are formed in aqueous solution.

$$Zn(s) + Cl_2(g) \xrightarrow{H_2O} Zn^{2+}(aq) + 2Cl^-(aq)$$

Calculate the standard emf for this cell at 25°C from standard free energies of formation (see Appendix C).

Solution

We write the equation with ΔG°_f's beneath.

$$\begin{array}{lcccc} & Zn(s) + & Cl_2(g) \longrightarrow & Zn^{2+}(aq) + & 2Cl^-(aq) \\ \Delta G^\circ_f: & 0 & 0 & -147 & 2 \times (-131) \text{ kJ} \end{array}$$

Hence,

$$\Delta G^\circ = \Sigma\, n\Delta G^\circ_f(\text{products}) - \Sigma\, m\Delta G^\circ_f(\text{reactants})$$
$$= [-147 + 2 \times (-131)] \text{ kJ}$$
$$= -409 \text{ kJ} = -4.09 \times 10^5 \text{ J}$$

We obtain n by splitting the reaction into half-reactions.

$$Zn(s) \longrightarrow Zn^{2+}(aq) + 2e^-$$
$$Cl_2(g) + 2e^- \longrightarrow 2Cl^-(aq)$$

Note that each half-reaction involves two electrons, so $n = 2$.

(continued)

Now we substitute into

$$\Delta G^\circ = -nFE^\circ_{cell}$$

$$-4.09 \times 10^5 \text{ J} = -2 \times 9.65 \times 10^4 \text{ C} \times E^\circ_{cell}$$

Solving for E°_{cell}, we get

$$E^\circ_{cell} = 2.12 \text{ V}$$

Exercise 19.9

Use standard free energies of formation (Appendix C) to obtain the standard emf of a cell at 25°C with the reaction

$$Mg(s) + Cu^{2+}(aq) \longrightarrow Mg^{2+}(aq) + Cu(s)$$

(See Problems 19.53 and 19.54.)

The measurement of cell emf's gives us yet another way to obtain equilibrium constants. Combining the previous equation, $\Delta G^\circ = -nFE^\circ_{cell}$, with the equation $\Delta G^\circ = -RT \ln K$ from Section 18.6, we get

$$nFE^\circ_{cell} = RT \ln K$$

or

$$E^\circ_{cell} = \frac{RT}{nF} \ln K = \frac{2.303\ RT}{nF} \log K$$

Substituting values for the constants R and F at 25°C gives the equation

$$E^\circ_{cell} = \frac{0.0592}{n} \log K \quad \text{(values in volts at 25°C)}$$

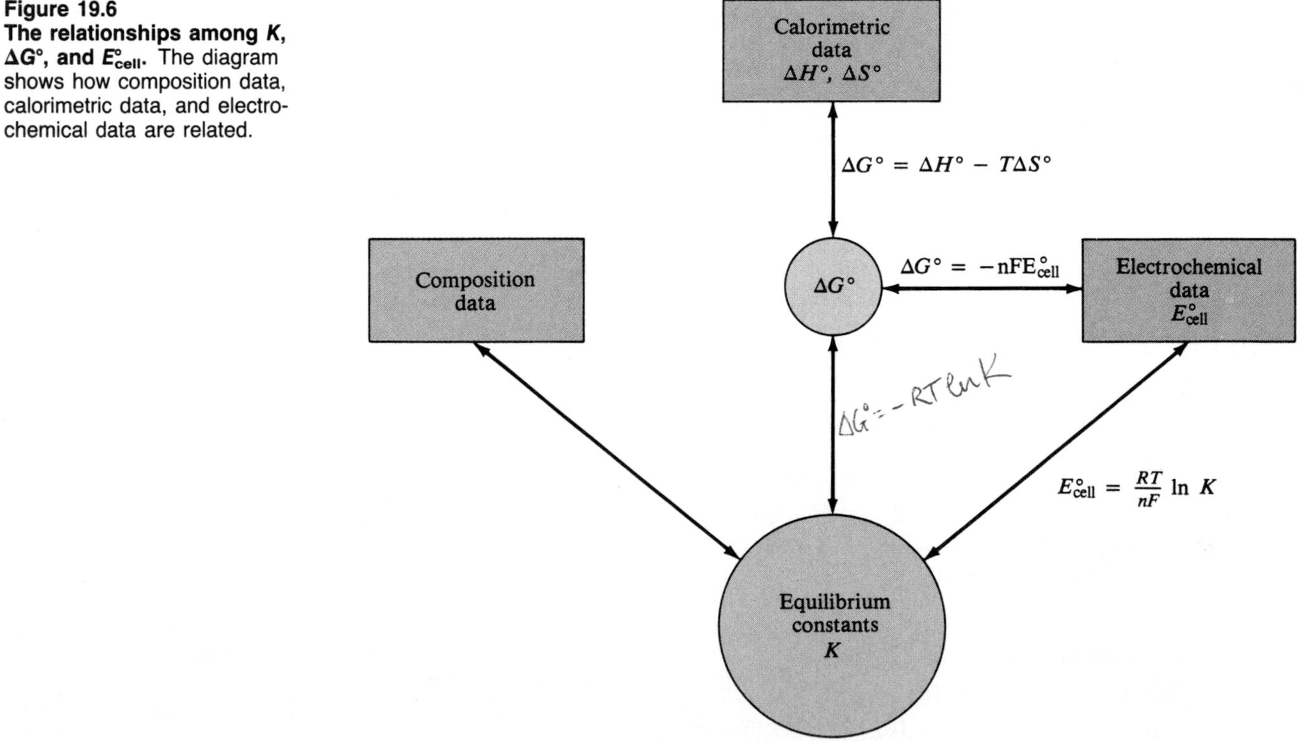

Figure 19.6
The relationships among K, ΔG°, and E°_{cell}. The diagram shows how composition data, calorimetric data, and electrochemical data are related.

Figure 19.6 summarizes the various relationships among K, ΔG°, and E°_{cell}.

The following example illustrates calculation of the thermodynamic equilibrium constant from the cell emf.

Example 19.9 Calculating the Equilibrium Constant from Cell emf

The standard emf for the following voltaic cell is 1.10 V.

$$Zn(s)|Zn^{2+}(aq)\|Cu^{2+}(aq)|Cu(s)$$

Calculate the equilibrium constant K_c for the reaction

$$Zn(s) + Cu^{2+}(aq) \rightleftharpoons Zn^{2+}(aq) + Cu(s)$$

Solution

The reaction corresponds to the one for the voltaic cell. By splitting the reaction into half-reactions, we note that $n = 2$. Note also that $K = K_c$. Substituting into the equation relating E°_{cell} and K gives

$$1.10 = \frac{0.0592}{2} \log K_c$$

Solving for $\log K_c$, we find

$$\log K_c = 37.2$$

Thus,

$$K_c = 2 \times 10^{37}$$

Note that the antilog of 37.2 gives 1.6×10^{37}, which, when rounded to the correct number of significant figures (one), is 2×10^{37}.

Exercise 19.10

Calculate the equilibrium constant K_c for the following reaction from standard electrode potentials.

$$Fe(s) + Sn^{4+}(aq) \rightleftharpoons Fe^{2+}(aq) + Sn^{2+}(aq)$$

(See Problems 19.57 and 19.58.)

19.6 DEPENDENCE OF emf ON CONCENTRATION

The emf of a cell depends on the concentrations of ions and on gas pressures. For that reason, cell emf's provide a way to measure ion concentrations. The pH meter, for example, depends on the variation of cell emf with hydrogen-ion concentration. We can relate cell emf's for various concentrations of ions and various gas pressures to standard electrode potentials by means of an equation first derived by the German chemist Walther Nernst (1864–1941).●

● Nernst also formulated the third law of thermodynamics, for which he received the Nobel Prize in 1920.

Nernst Equation

Recall that the free-energy change, ΔG, is related to the standard free-energy change, ΔG°, by the following equation (Section 18.6):

$$\Delta G = \Delta G^\circ + RT \ln Q$$

Here Q is the thermodynamic reaction quotient. The reaction quotient has the form of the equilibrium constant, except that the concentrations and gas pressures are those that exist in a reaction mixture at a given instant. We can apply this equation to a voltaic cell. In that case, the concentrations and gas pressures are those that exist in the cell at a particular instant. If we substitute $\Delta G = -nFE_{cell}$ and $\Delta G^\circ = -nFE^\circ_{cell}$ into this equation, we obtain

$$-nFE_{cell} = -nFE^\circ_{cell} + RT \ln Q$$

This result rearranges to give the **Nernst equation,** *an equation relating the cell emf to its standard emf and the reaction quotient.*

$$E_{cell} = E^\circ_{cell} - \frac{RT}{nF} \ln Q \qquad \text{or} \qquad E_{cell} = E^\circ_{cell} - \frac{2.303\ RT}{nF} \log Q$$

If we substitute 298 K (25°C) for the temperature in the Nernst equation and put in values for R and F, we get (using common logarithms)

$$E_{cell} = E^\circ_{cell} - \frac{0.0592}{n} \log Q \qquad \text{(values in volts at 25°C)}$$

We can show from the Nernst equation that the cell emf E_{cell} decreases as the cell reaction proceeds. As the reaction occurs in the voltaic cell, the concentrations of products increase and the concentrations of reactants decrease. Therefore, Q and $\log Q$ increase. The second term in the Nernst equation, $(0.0592/n) \log Q$, increases, so that the difference $E^\circ_{cell} - (0.0592/n) \log Q$ decreases. Thus, the cell emf E_{cell} becomes smaller. Eventually the cell emf goes to zero, and the cell reaction comes to equilibrium.

As an example of the computation of the reaction quotient, consider the following voltaic cell:

$$Cd(s)|Cd^{2+}(0.0100\ M)||H^+(1.00\ M)|H_2(1.00\ atm)|Pt$$

The cell reaction is

$$Cd(s) + 2H^+(aq) \rightleftharpoons Cd^{2+}(aq) + H_2(g)$$

and the expression for the equilibrium constant is

$$K = \frac{[Cd^{2+}]P_{H_2}}{[H^+]^2}$$

Note that the hydrogen-gas concentration is given here in terms of the pressure (in atmospheres). The expression for the reaction quotient has the same form as K, except that the values for the ion concentrations and hydrogen-gas pressures are those that exist in the cell.● Hence,

● We have omitted subscript i's on the concentrations and pressures in Q in order to simplify the notation.

$$Q = \frac{[Cd^{2+}]P_{H_2}}{[H^+]^2} = \frac{0.0100 \times 1.00}{(1.00)^2} = 0.0100$$

The next example illustrates a complete calculation of emf from the ion concentrations in a voltaic cell.

Example 19.10 Calculating the Cell emf for Nonstandard Conditions

What is the emf of the following cell at 25°C?

$$Zn(s)|Zn^{2+}(1.00 \times 10^{-5}\ M)||Cu^{2+}(0.100\ M)|Cu(s)$$

The standard emf of this cell is 1.10 V.

Solution

The cell reaction is

$$Zn(s) + Cu^{2+}(aq) \rightleftharpoons Zn^{2+}(aq) + Cu(s)$$

The number of electrons transferred is two; hence, $n = 2$, and the reaction quotient is

$$Q = \frac{[Zn^{2+}]}{[Cu^{2+}]} = \frac{1.00 \times 10^{-5}}{0.100} = 1.00 \times 10^{-4}$$

The standard emf is 1.10 V, so the Nernst equation becomes

$$E_{cell} = E^\circ_{cell} - \frac{0.0592}{n} \log Q$$

$$E_{cell} = 1.10 - \frac{0.0592}{2}\log(1.0 \times 10^{-4})$$
$$= 1.10 - (-0.12) = 1.22$$

Thus, the cell emf is 1.22 V. This result is qualitatively what we would expect. Because the concentration of product (Zn^{2+}) is much less than the standard value (1 *M*), whereas the concentration of reactant (Cu^{2+}) is 0.100 *M*, the spontaneity of the reaction as measured by E_{cell} is greater than the standard value.

Exercise 19.11

What is the emf of the following cell at 25°C?

$$Zn(s)|Zn^{2+}(0.200\ M)\|Ag^{+}(0.00200\ M)|Ag(s)$$

(See Problems 19.61 and 19.62.)

Electrode Potentials for Nonstandard Conditions

In addition to using the Nernst equation to obtain cell emf's, we can use it to calculate the potential of an electrode when the concentration of an ion is other than 1 molar. In other words, we can use this equation to calculate the potential of an electrode under nonstandard conditions.

Consider this problem: What is the potential of the zinc electrode $Zn^{2+}(0.100\ M)|Zn(s)$ at 25°C? The standard potential of the hydrogen electrode is defined to be zero, so the emf of the cell

$$Pt|H_2(1\ atm)|H^{+}(1\ M)\|Zn^{2+}(0.100\ M)|Zn(s)$$

equals

$$E_{cell} = E(Zn^{2+}|Zn) - E^\circ(H^{+}|H_2)$$
$$= E(Zn^{2+}|Zn)$$

Thus, the Nernst equation for the cell at 25°C becomes

$$E(Zn^{2+}|Zn) = E^\circ(Zn^{2+}|Zn) - \frac{0.0592}{n}\log Q$$

Values for n and Q follow from the overall cell reaction.

$$Zn^{2+}(aq) + H_2(g) \rightleftharpoons Zn(s) + 2H^{+}(aq)$$

We see that the number of electrons transferred is two; hence $n = 2$. Also, the reaction quotient Q is

$$Q = \frac{[H^{+}]^2}{[Zn^{2+}]P_{H_2}} = \frac{1^2}{[Zn^{2+}] \times 1} = \frac{1}{[Zn^{2+}]}$$

Or, because $[H^{+}]$ and P_{H_2} have numerical values of 1, Q equals $1/[Zn^{2+}]$.

In effect, the Nernst equation can be applied to the zinc half-reaction

$$Zn^{2+}(aq) + 2e^{-} \rightleftharpoons Zn(s)$$

The value of n follows from the half-reaction, and the expression required to find Q is obtained by writing a reaction quotient for the half-reaction (ignoring e^{-}). We get $E(Zn^{2+}|Zn) = -0.76 + (0.0592/2)\log 0.100 = -0.79$ V.

Example 19.11 Calculating the Electrode Potential for Nonstandard Conditions

What is the potential of the hydrogen electrode $H^+(0.100\ M)|H_2(1\ atm)|Pt$ at 25°C (298 K)?

Solution

The half-reaction is

$$2H^+(aq) + 2e^- \rightleftharpoons H_2(g)$$

For this equation, $n = 2$ and

$$Q = \frac{P_{H_2}}{[H^+]^2} = \frac{1}{(0.100)^2}$$

Because $E° = 0$ for the standard hydrogen electrode, the Nernst equation for the nonstandard hydrogen electrode is

$$E = E° - \frac{0.0592}{n} \log Q$$

$$= 0 - \frac{0.0592}{2} \log \frac{1}{(0.100)^2}$$

$$= \mathbf{-0.0592\ V}$$

Exercise 19.12

What is the potential of the copper electrode $Cu^{2+}(0.0350\ M)|Cu(s)$ at 25°C?

(See Problems 19.65 and 19.66.)

Determination of pH

The determination of pH from the measurement of emf is especially useful. To measure the pH of any test solution, we can dip a hydrogen electrode into the solution and connect this to a standard hydrogen electrode. The cell is

$$Pt|H_2(1\ atm)|H^+(\text{test solution})\|H^+(1\ M)|H_2(1\ atm)|Pt$$

The emf of this cell equals the potential developed by the test solution half-cell, whose half-reaction is

$$\tfrac{1}{2}H_2(1\ atm) \rightleftharpoons H^+(\text{test solution}) + e^-$$

Then, according to the Nernst equation for this electrode at 25°C,

$$E_{cell} = -0.0592 \log [H^+]$$

where $[H^+]$ is the hydrogen-ion concentration of the test solution. This equation shows that there is a direct relationship between the cell emf E_{cell} that we measure and the hydrogen-ion concentration of the test solution.

If we wish, we can rewrite the preceding equation in terms of the pH.

$$pH = -\log [H^+]$$

We get

$$E_{cell} = 0.0592\ pH$$

Or, at 25°C,

$$pH = \frac{E_{cell}}{0.0592}$$

The standard hydrogen electrode, which requires hydrogen gas at standard pressure, is too cumbersome for routine laboratory use. Also, the platinum foil is easily fouled by the presence of other substances in solution. The hydrogen electrode is often replaced by a *glass electrode*. This compact electrode (see Figure 19.7) consists of a silver wire coated with silver chloride immersed in a solution of dilute hydrochloric acid. (The silver–silver chloride electrode serves as an internal reference.) The electrode solution is separated from the test solution by a thin glass membrane, which develops a potential across it depending on the hydrogen-ion

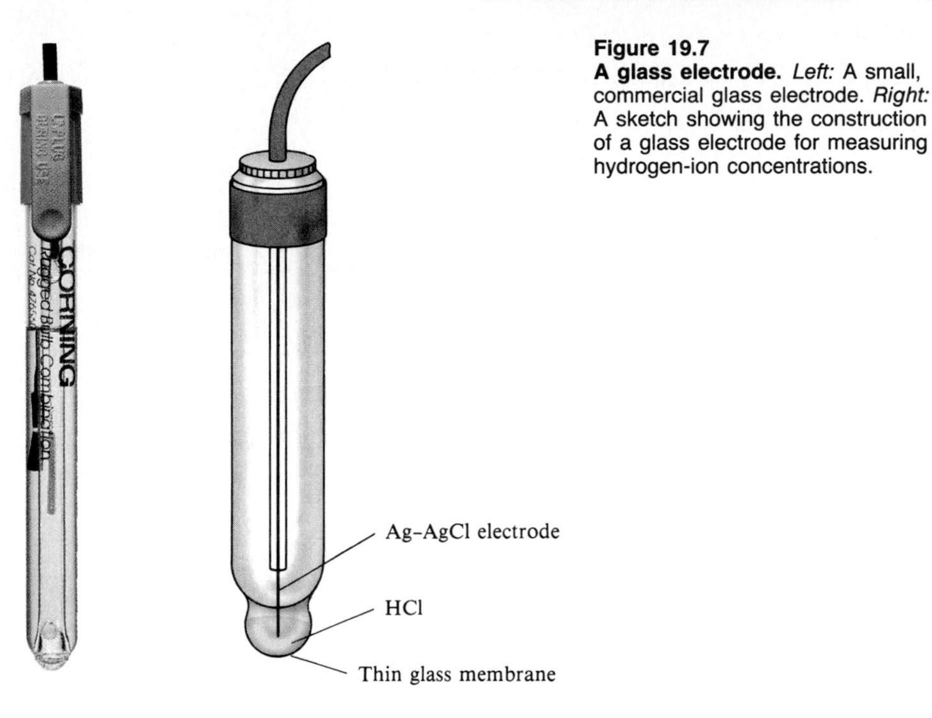

Figure 19.7
A glass electrode. *Left:* A small, commercial glass electrode. *Right:* A sketch showing the construction of a glass electrode for measuring hydrogen-ion concentrations.

concentrations on its inner and outer surfaces. A mercury–mercury(I) chloride (calomel) electrode is often used as the other electrode. The emf of the cell depends linearly on the pH. In a popular arrangement, the emf is measured with a voltmeter that reads pH directly (see Figure 16.2).

The glass electrode is an example of an *ion-selective electrode*. Many electrodes have been developed recently that are sensitive to a particular ion, such as K^+, NH_4^+, Ca^{2+}, or Mg^{2+}. They can be used to monitor solutions of that ion. It is even possible to measure the concentration of a nonelectrolyte. To measure urea, NH_2CONH_2, in solution, one uses an electrode selective to NH_4^+ that is coated with a gel containing the enzyme urease. (An enzyme is a biochemical catalyst.) The gel is held in place around the electrode by means of a nylon net. Urease catalyzes the decomposition of urea to ammonium ion, whose concentration is measured.

$$NH_2CONH_2(aq) + 2H_2O(l) + H^+(aq) \xrightarrow{\text{urease}} 2NH_4^+(aq) + HCO_3^-(aq)$$

Exercise 19.13

What is the nickel(II)-ion concentration in the cell

$$Zn(s)|Zn^{2+}(1.00\ M)||Ni^{2+}(aq)|Ni(s)$$

if the emf is 0.34 V at 25°C? (See Problems 19.67 and 19.68.)

19.7 SOME COMMERCIAL VOLTAIC CELLS

We commonly use voltaic cells as convenient, portable sources of energy. Flashlights and radios are examples of devices that are often powered by the **zinc–carbon,** or **Leclanché, dry cell** (Figure 19.8). *This voltaic cell has a zinc can as*

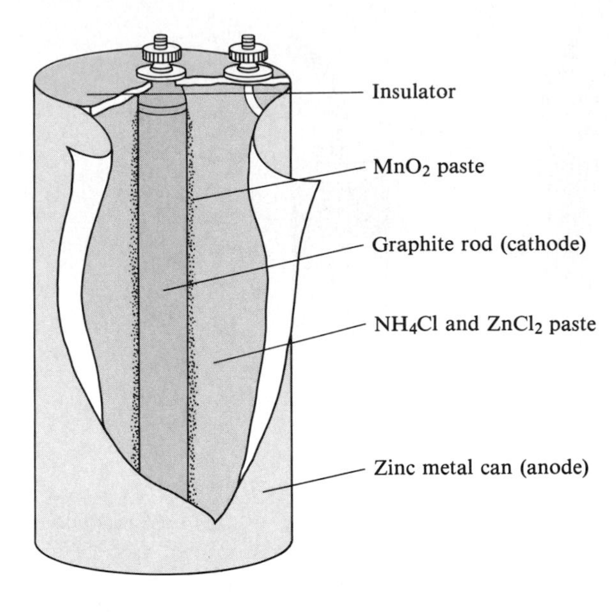

Figure 19.8
Leclanché dry cell. The cell has a zinc anode and a carbon rod with a paste of MnO_2 as the cathode. The electrolyte consists of a paste of NH_4Cl and $ZnCl_2$.

the anode; a graphite rod in the center, immediately surrounded by a paste of manganese dioxide and carbon black, is the cathode. Around this is another paste, this one containing ammonium and zinc chlorides. The electrode reactions are complicated but are approximately these:

$$Zn(s) \longrightarrow Zn^{2+}(aq) + 2e^- \qquad \text{(anode)}$$
$$2NH_4^+(aq) + 2MnO_2(s) + 2e^- \longrightarrow Mn_2O_3(s) + H_2O(l) + 2NH_3(aq) \qquad \text{(cathode)}$$

The voltage of this dry cell is initially about 1.5 V, but it decreases as current is drawn off. The voltage also deteriorates rapidly in cold weather.

An **alkaline dry cell** (Figure 19.9) is *similar to the Leclanché cell, but it has potassium hydroxide in place of ammonium chloride.* This cell performs better under current drain and in cold weather. The half-reactions are

$$Zn(s) + 2OH^-(aq) \longrightarrow Zn(OH)_2(s) + 2e^- \qquad \text{(anode)}$$
$$2MnO_2(s) + H_2O(l) + 2e^- \longrightarrow Mn_2O_3(s) + 2OH^-(aq) \qquad \text{(cathode)}$$

A dry cell is not truly "dry," because the electrolyte is an aqueous paste. Solid-state batteries have been developed, however. One of these is a **lithium–iodine battery,** *a voltaic cell in which the anode is lithium metal and the cathode*

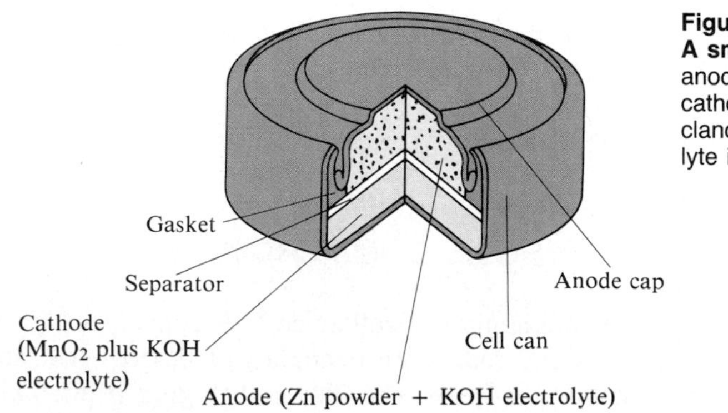

Figure 19.9
A small alkaline dry cell. The anode is zinc powder and the cathode is MnO_2, as in the Leclanché cell; however, the electrolyte is KOH.

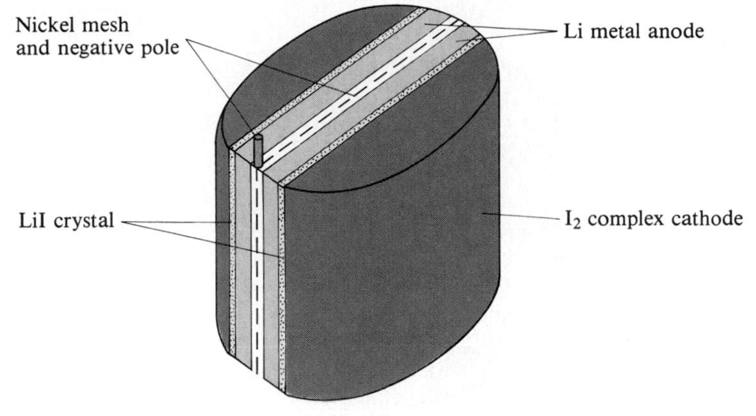

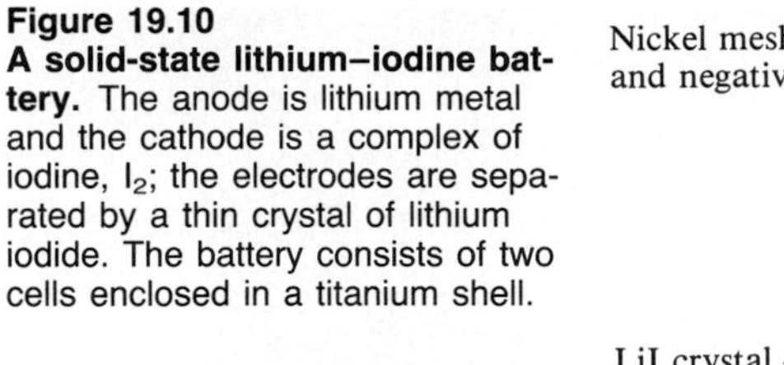
Figure 19.10
A solid-state lithium–iodine battery. The anode is lithium metal and the cathode is a complex of iodine, I_2; the electrodes are separated by a thin crystal of lithium iodide. The battery consists of two cells enclosed in a titanium shell.

is an I_2 complex. These solid-state electrodes are separated by a thin crystal of lithium iodide (Figure 19.10). Current is carried through the crystal by diffusion of Li^+ ions. Although the cell has high resistance and therefore low current, the battery is very reliable and is used to power heart pacemakers. The battery is implanted within the chest and lasts about ten years before it has to be replaced.

Once a dry cell is completely discharged (has come to equilibrium), the cell cannot be reversed, or recharged, and is discarded. Some types of cell are rechargeable after use, however. The best known of these is the **lead storage cell.** *This voltaic cell consists of electrodes of lead alloy grids; one electrode is packed with a spongy lead to form the anode, and the other is packed with lead dioxide to form the cathode* (see Figure 19.11). Both are bathed in an aqueous solution of sulfuric acid, H_2SO_4. The half-cell reactions during discharge are

$$Pb(s) + HSO_4^-(aq) \longrightarrow PbSO_4(s) + H^+(aq) + 2e^- \quad \text{(anode)}$$

$$PbO_2(s) + 3H^+(aq) + HSO_4^-(aq) + 2e^- \longrightarrow PbSO_4(s) + 2H_2O(l) \quad \text{(cathode)}$$

White lead(II) sulfate coats each electrode during discharge, and sulfuric acid is consumed. Each cell delivers about 2 V, and a battery consisting of six cells in series gives about 12 V.

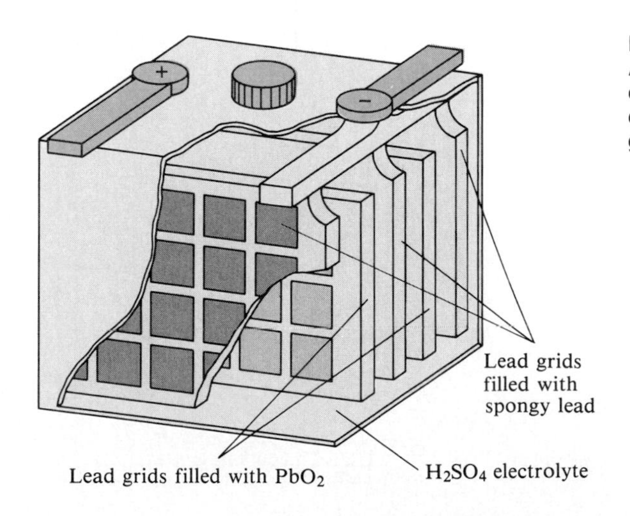

Figure 19.11
A lead storage cell. Each cell delivers about 2 V, and a battery consisting of six cells in series gives about 12 V.

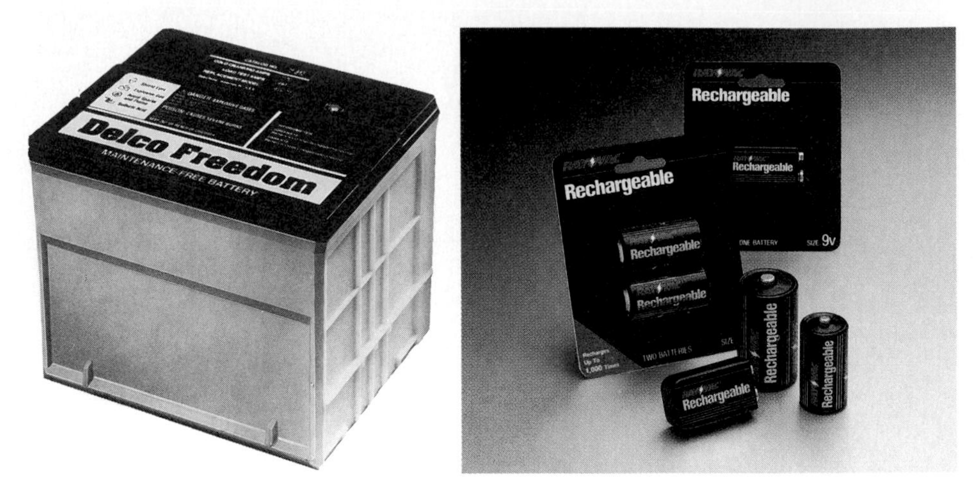

Figure 19.12 ***(left)***
A maintenance-free battery. This lead storage battery contains calcium–lead alloy grids for electrodes. This alloy resists the decomposition of water, so the battery can be sealed.

Figure 19.13 ***(right)***
A nicad storage battery. This rechargeable cell has a cadmium anode and a hydrated nickel oxide cathode.

After the lead storage battery is discharged, it is recharged from an external electric current. The previous half-reactions are reversed. Some water is decomposed into hydrogen and oxygen gas during this recharging, so more water may have to be added at intervals. However, newer batteries use lead electrodes containing some calcium metal; the calcium-lead alloy resists the decomposition of water. These *maintenance-free* batteries are sealed (Figure 19.12).

The **nickel–cadmium cell** (nicad cell) is a common storage battery. It is *a voltaic cell consisting of an anode of cadmium and a cathode of hydrated nickel oxide (approximately NiOOH) on nickel; the electrolyte is potassium hydroxide.* Nicad batteries are used in calculators, portable power tools, shavers, and toothbrushes. The half-cell reactions during discharge are

$$Cd(s) + 2OH^-(aq) \longrightarrow Cd(OH)_2(s) + 2e^- \quad \text{(anode)}$$

$$NiOOH(s) + H_2O(l) + e^- \longrightarrow Ni(OH)_2(s) + OH^-(aq) \quad \text{(cathode)}$$

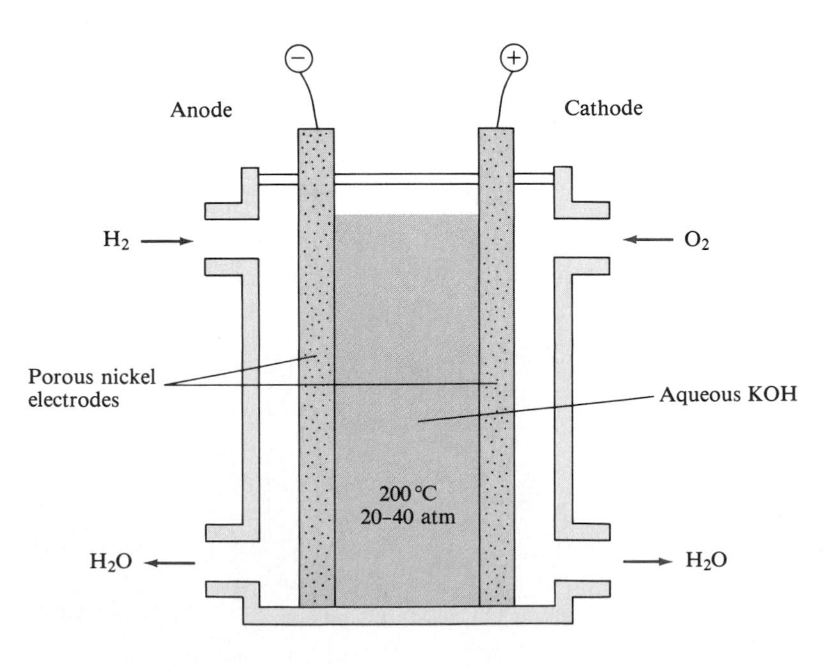

Figure 19.14
A hydrogen–oxygen fuel cell. Hydrogen gas passes into a chamber where the gas is in contact with a porous material in contact with a hot aqueous solution of potassium hydroxide. This forms the anode at which hydrogen is oxidized to water. Oxygen gas enters a similar electrode and is reduced to hydroxide ion. The net chemical change is the reaction of hydrogen with oxygen to form water.

Figure 19.15
A power plant for a space shuttle orbiter. It contains hydrogen–oxygen fuel cells.

These half-reactions are reversed when the cell is recharged. Nicad batteries can be recharged and discharged many times (see Figure 19.13).

A **fuel cell** is *essentially a battery, but it differs by operating with a continuous supply of energetic reactants or fuel.* Figure 19.14 shows a fuel cell that uses hydrogen and oxygen.● At one electrode, oxygen passes through a porous material that catalyzes the following reaction:

● The first fuel cell was invented in 1839 by William Groves. He obtained the cell by electrolyzing water and collecting the hydrogen and oxygen in small, inverted test tubes. After removing the battery current from the electrolysis cell, he found that a current would flow from the cell. In effect, the electrolysis charged the cell, after which it operated as a fuel cell.

$$O_2(g) + 2H_2O(l) + 4e^- \longrightarrow 4OH^-(aq) \quad \text{(cathode)}$$

At the other electrode, hydrogen reacts.

$$2H_2(g) + 4OH^-(aq) \longrightarrow 4H_2O(l) + 4e^- \quad \text{(anode)}$$

The sum of these half-cell reactions is

$$2H_2(g) + O_2(g) \longrightarrow 2H_2O(l)$$

which is the net reaction in the fuel cell. Such cells are used in the space shuttle orbiters to supply electric energy (Figure 19.15). Other types of cells employing hydrocarbon fuels have been constructed.

Another use of voltaic cells is to control the corrosion of underground pipelines and tanks. Such pipelines and tanks are usually made of steel, an alloy of iron, and their corrosion or rusting is an electrochemical process.

Consider the rusting that occurs when a drop of water is in contact with iron. The edge of the water drop exposed to the air becomes one pole of a voltaic cell (see Figure 19.16). At this edge, molecular oxygen from air is reduced to hydroxide ion in solution.

$$O_2(g) + 2H_2O(l) + 4e^- \longrightarrow 4OH^-(aq)$$

The electrons for this reduction are supplied by the oxidation of metallic iron at the

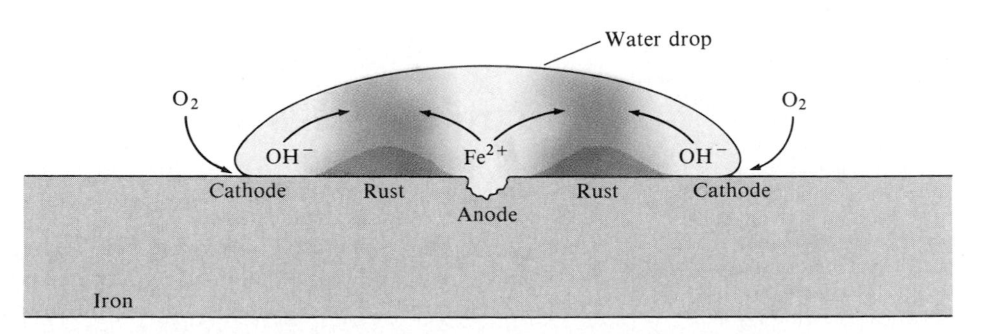

Figure 19.16
The electrochemical process involved in the rusting of iron. Here a single drop of water containing ions forms a voltaic cell in which iron is oxidized to iron(II) ion at the center of the drop (this is the anode). Oxygen gas from air is reduced to hydroxide ion at the periphery of the drop. Hydroxide ions and iron(II) ions migrate together and react to form iron(II) hydroxide. This is oxidized to iron(III) hydroxide by more O_2 that dissolves at the surface of the drop. Iron(III) hydroxide precipitates, and this settles to form rust on the surface of the iron.

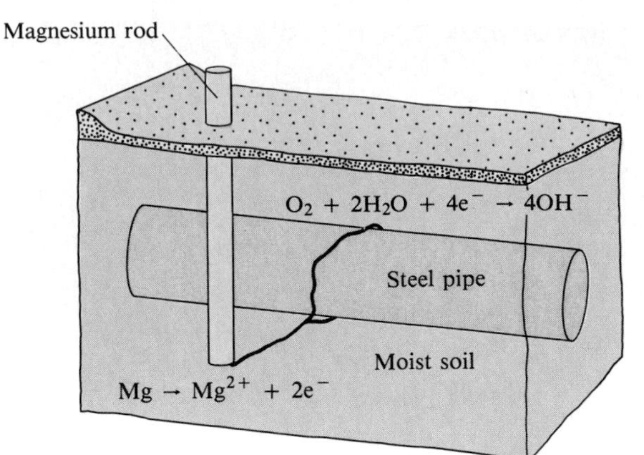

Figure 19.17
Cathodic protection of a buried steel pipe. Iron in the steel becomes the cathode in an iron–magnesium voltaic cell. Magnesium is then oxidized in preference to iron.

center of the drop, which acts as the other pole of the voltaic cell.

$$Fe(s) \longrightarrow Fe^{2+}(aq) + 2e^-$$

These electrons flow from the center of the drop through the metallic iron to the edge of the drop. The metallic iron functions as the external circuit between the cell poles.

Ions move within the water drop, completing the electric circuit. Iron(II) ions move outward from the center of the drop, and hydroxide ions move inward from the edge. The two ions meet in a doughnut-shaped region, where they react to precipitate iron(II) hydroxide.

$$Fe^{2+}(aq) + 2OH^-(aq) \longrightarrow Fe(OH)_2(s)$$

This precipitate is quickly oxidized by oxygen to rust (approximated by the formula $Fe_2O_3 \cdot H_2O$).

$$4Fe(OH)_2(s) + O_2(g) \longrightarrow 2Fe_2O_3 \cdot H_2O(s) + 2H_2O(l)$$

If a buried steel pipeline (Figure 19.17) is connected to an active metal (that is, a highly electropositive substance) such as magnesium, then a voltaic cell is formed; the active metal is the anode and iron becomes the cathode. Wet soil forms the electrolyte, and the electrode reactions are

$$Mg(s) \longrightarrow Mg^{2+}(aq) + 2e^- \qquad \text{(anode)}$$
$$O_2(g) + 2H_2O(l) + 4e^- \longrightarrow 4OH^-(aq) \qquad \text{(cathode)}$$

As the cathode, the iron-containing steel pipe is protected from oxidation. Of course, the magnesium rod is eventually consumed and must be replaced, but this is cheaper than digging up the pipeline. This use of an active metal to protect iron from corrosion is called *cathodic protection*. See Figure 19.18 for a laboratory demonstration of cathodic protection.

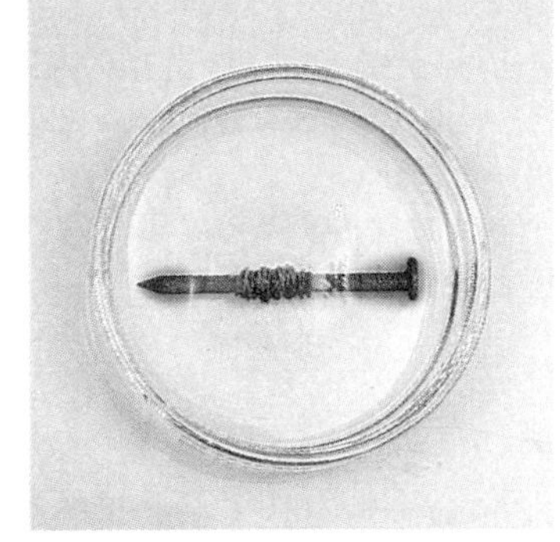

Figure 19.18
A demonstration of cathodic protection. The nails are in a gel containing phenolphthalein indicator and potassium ferricyanide. Iron corrosion yields Fe^{2+}, which reacts with ferricyanide ion to give a blue precipitate. Where OH^- forms, phenolphthalein appears pink. The unprotected nail is on the top. The nail on the bottom has magnesium wrapped around the center. Note that no corrosion (blue regions) appears on it.

Electrolytic Cells

An *electrolytic cell,* you may recall, is an electrochemical cell in which an electric current drives an otherwise nonspontaneous reaction. *The process of producing a chemical change in an electrolytic cell* is called **electrolysis.** Many important substances, including aluminum and chlorine, are produced commercially by electrolysis. We will begin by looking at the electrolysis of molten salts.

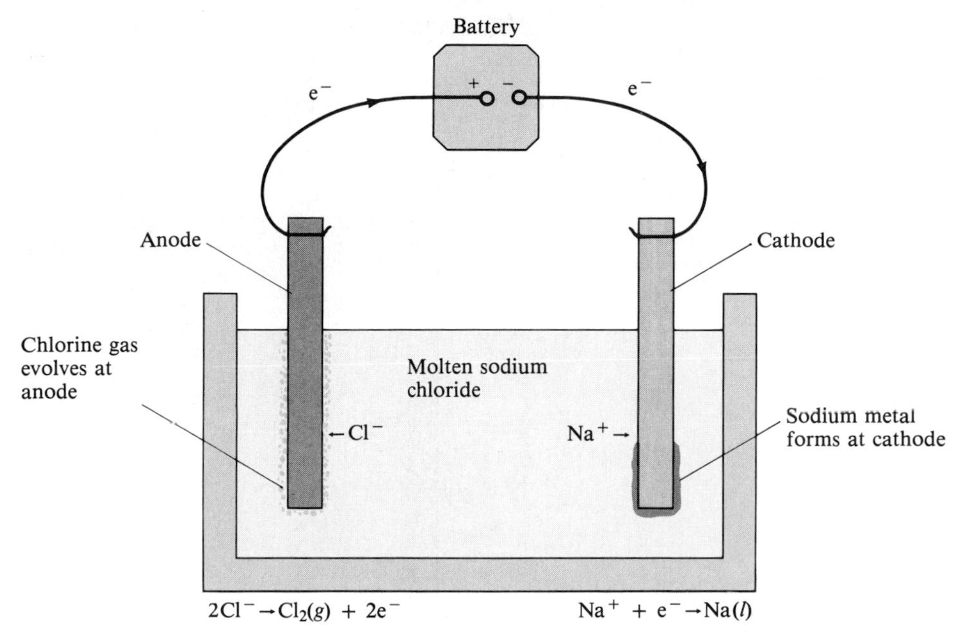

Figure 19.19
Electrolysis of molten sodium chloride. Sodium metal forms at the cathode from the reduction of Na^+ ion; chlorine gas forms at the anode from the oxidation of Cl^- ion. Sodium metal is produced commercially this way, although the commercial cell must be designed to collect the products and to keep them away from one another.

19.8 ELECTROLYSIS OF MOLTEN SALTS

Figure 19.19 shows a simple electrolytic cell. Wires from a battery are connected to electrodes that dip into molten sodium chloride (NaCl melts at 801°C). At the electrode connected to the negative pole of the battery, globules of sodium metal form; chlorine gas evolves from the other electrode. The half-reactions are

$$Na^+(l) + e^- \longrightarrow Na(l)$$
$$Cl^-(l) \longrightarrow \tfrac{1}{2}Cl_2(g) + e^-$$

As we noted earlier (Section 19.1), the *anode* is the electrode at which oxidation occurs, and the *cathode* is the electrode at which reduction occurs (these definitions hold for electrolytic cells as well as for voltaic cells). Thus, during the electrolysis of molten NaCl, the reduction of Na^+ to Na occurs at the cathode, and the oxidation of Cl^- to Cl_2 occurs at the anode (note the labeling of electrodes in Figure 19.19).

This electrolysis is used commercially to obtain sodium metal from sodium chloride. A **Downs cell** is *a commercial electrochemical cell used to obtain sodium metal by the electrolysis of molten sodium chloride* (Figure 19.20). The cell is constructed to keep the products of the electrolysis separate, because they would otherwise react. Calcium chloride is added to the sodium chloride to lower the melting point from 801°C for NaCl to about 580°C for the mixture (remember that the melting point, or freezing point, of a substance is lowered by the addition of a solute). We obtain the cell reaction by adding the previous half-reactions.

$$\begin{array}{r} Na^+(l) + e^- \longrightarrow Na(l) \\ Cl^-(l) \longrightarrow \tfrac{1}{2}Cl_2(g) + e^- \\ \hline Na^+(l) + Cl^-(l) \longrightarrow Na(l) + \tfrac{1}{2}Cl_2(g) \end{array}$$

A number of other reactive metals are obtained by the electrolysis of a molten salt or ionic compound. Lithium, magnesium, and calcium metals are all obtained

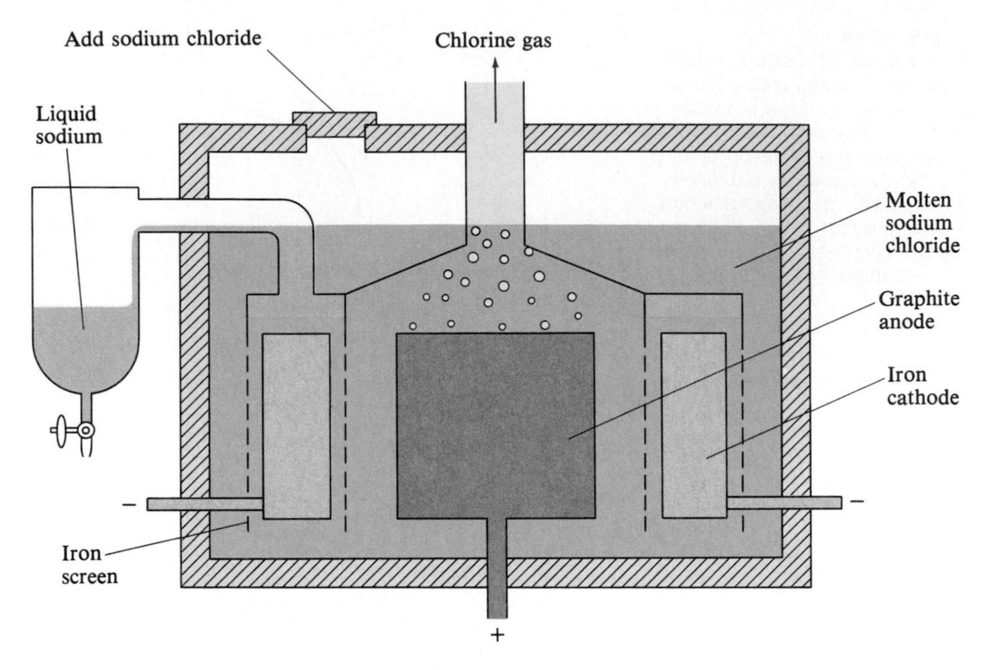

Figure 19.20
A Downs cell for the preparation of sodium metal. In this commercial cell, sodium is produced by the electrolysis of molten sodium chloride. The salt contains calcium chloride, which is added to lower the melting point of the mixture. Liquid sodium forms at the cathode, where it rises to the top of the molten salt and collects in a tank. Chlorine gas is a by-product.

by the electrolysis of the chlorides. The first commercial preparation of sodium metal adapted the method used by Humphry Davy when he discovered the element in 1807. Davy electrolyzed molten sodium hydroxide, NaOH, whose melting point (318°C) is relatively low for an ionic compound. The half-reactions are

$$Na^+(l) + e^- \longrightarrow Na(l) \qquad \text{(cathode)}$$

$$4OH^-(l) \longrightarrow O_2(g) + 2H_2O(g) + 4e^- \qquad \text{(anode)}$$

Many of the commercial uses of electrolysis involve aqueous solutions. We will look at electrolysis in aqueous solution in the next section.

Exercise 19.14

Write the half-reactions for the electrolysis of the following molten compounds: (a) KCl; (b) KOH.

(See Problems 19.69 and 19.70.)

19.9 AQUEOUS ELECTROLYSIS

In the electrolysis of a molten salt, the possible half-reactions are usually limited to those involving ions from the salt. When we electrolyze an aqueous solution of an ionic compound, however, we must consider the possibility that water is involved at one or both electrodes. Let us look at the possible half-reactions involving water.

Water can be reduced or oxidized in half-reactions, and we can easily obtain these half-reactions. To do this, we first note the species that are likely to be involved. In addition to H_2O, these are H_2, O_2, H^+, and OH^-. Only H_2 and O_2 involve a change of oxidation state. Hydrogen in H_2 has a lower oxidation state (0) than that in H_2O (in which it is +1), whereas the oxygen in O_2 has a higher oxidation state (0) than that in H_2O (in which it is -2). Thus, we can reduce water to H_2 or oxidize it to O_2.

Consider the reduction half-reaction. It must involve the reduction of H_2O to H_2. We will need to balance the half-reaction by putting an oxygen-containing species on the right. The only species in which there are no changes in oxidation states from those in H_2O is OH^-. Therefore, the balanced half-reaction is

$$2H_2O(l) + 2e^- \longrightarrow H_2(g) + 2OH^-(aq)$$

We can obtain the oxidation half-reaction for water in a similar way. It involves the oxidation of H_2O to O_2. We will need to put a hydrogen-containing species on the right side of the equation to balance it. The only species in which there is no change in oxidation state is H^+. Therefore, the balanced half-reaction is

$$2H_2O(l) \longrightarrow O_2(g) + 4H^+(aq) + 4e^-$$

Now let us consider the electrolysis of different aqueous solutions and try to decide what the half-reactions might be. Once we have the half-reactions, we can obtain the overall chemical change due to the electrolysis.

Electrolysis of Sulfuric Acid Solutions

To decide what is likely to happen during the electrolysis of a solution of sulfuric acid, H_2SO_4, we must consider the half-reactions involving ionic species from H_2SO_4, in addition to those involving water. Sulfuric acid is a strong acid and ionizes completely into H^+ and HSO_4^-. The HSO_4^- ion is relatively strong and for the most part ionizes into H^+ and SO_4^{2-}. Therefore, the species we have to look at are H^+, SO_4^{2-}, and H_2O.

At the cathode, the possible reduction half-reactions are

$$2H^+(aq) + 2e^- \longrightarrow H_2(g)$$
$$2H_2O(l) + 2e^- \longrightarrow H_2(g) + 2OH^-(aq)$$

At the anode, the possible oxidation half-reactions are

$$2SO_4^{2-}(aq) \longrightarrow S_2O_8^{2-}(aq) + 2e^-$$
$$2H_2O(l) \longrightarrow O_2(g) + 4H^+(aq) + 4e^-$$

In the first half-reaction, the sulfate ion, SO_4^{2-} is oxidized to the peroxydisulfate ion, $S_2O_8^{2-}$. The $S_2O_8^{2-}$ ion contains the peroxy group —O—O— and therefore has O in the -1 oxidation state. Thus, oxygen in this half-reaction is oxidized from oxidation state -2 to -1.

Consider the cathode reaction. As we can easily show, the two reduction half-reactions (reduction of H^+ and the reduction of H_2O) in acidic solution are essentially equivalent, although this is not immediately obvious. Consider the reduction of water to give H_2 and OH^-. In acidic solution, OH^- reacts with H^+ to give H_2O. The net result is obtained by adding the two reactions.

$$2H_2O(l) + 2e^- \longrightarrow H_2(g) + 2OH^-(aq)$$
$$2H^+(aq) + 2OH^-(aq) \longrightarrow 2H_2O(l)$$
$$\overline{2H^+(aq) + 2e^- \longrightarrow H_2(g)}$$

We may consider the cathode reaction to be the reduction of H^+.

For the anode, we consider the oxidation half-reactions and their oxidation potentials (which equal the electrode potentials with signs reversed).

$$2SO_4^{2-}(aq) \longrightarrow S_2O_8^{2-}(aq) + 2e^-; \ -E° = -2.01 \text{ V}$$
$$2H_2O(l) \longrightarrow O_2(g) + 4H^+(aq) + 4e^-; \ -E° = -1.23 \text{ V}$$

The species whose oxidation half-reaction has the larger (less negative) oxidation potential is the more easily oxidized. Thus, under standard conditions, we expect H_2O to be oxidized in preference to SO_4^{2-}.

A note of caution must be added here. Electrode potentials are measured under conditions in which the half-reactions are at or very near equilibrium. In electrolysis, in which the half-reactions can be far from equilibrium, we may have to supply a larger voltage than predicted from electrode potentials. This additional voltage, called *overvoltage,* can be fairly large (several tenths of a volt), particularly for forming a gas. This means that in trying to predict which of two half-reactions is the one that actually occurs at an electrode, we must be especially careful if the electrode potentials are within several tenths of a volt of one another. For example, the standard oxidation potential for O_2 is -1.23 V, but because of overvoltage, the effective oxidation potential may be several tenths of a volt lower (more negative). The sulfate ion is so difficult to oxidize ($-E° = -2.01$ V), however, that our previous conclusion about the anode half-reaction in electrolyzing aqueous sulfuric acid is unaltered.

We obtain the cell reaction by adding the half-reactions that occur at the electrodes.

$$2[2H^+(aq) + 2e^- \longrightarrow H_2(g)] \qquad \text{(cathode)}$$
$$2H_2O(l) \longrightarrow O_2(g) + 4H^+(aq) + 4e^- \qquad \text{(anode)}$$
$$\overline{2H_2O(l) \longrightarrow 2H_2(g) + O_2(g)}$$

The net cell reaction is simply the *electrolysis of water*. Where electricity is cheap, hydrogen can be prepared commercially by the electrolysis of water. Here we used H_2SO_4 as an electrolyte; however, various electrolytes are used, including NaOH.

Electrolysis of Sodium Chloride Solutions

When we electrolyze an aqueous solution of sodium chloride, NaCl, the possible species involved in half-reactions are Na^+, Cl^-, and H_2O. The possible cathode half-reactions are

$$Na^+(aq) + e^- \longrightarrow Na(s); \ E° = -2.71 \text{ V}$$
$$2H_2O(l) + 2e^- \longrightarrow H_2(g) + 2OH^-(aq); \ E° = -0.83 \text{ V}$$

Under standard conditions, we expect H_2O to be reduced in preference to Na^+, which agrees with what one actually observes. Hydrogen gas evolves at the cathode.●

● The OH^- concentration is actually 1×10^{-7} *M*. From Nernst's equation, *E* for the reduction of H_2O is -0.41 V; we still conclude that H_2O is reduced in preference to Na^+.

The possible half-reactions at the anode are

$$2Cl^-(aq) \longrightarrow Cl_2(g) + 2e^-; \ -E° = -1.36 \text{ V}$$
$$2H_2O(l) \longrightarrow O_2(g) + 4H^+(aq) + 4e^-; \ -E° = -1.23 \text{ V}$$

Under standard-state conditions, we might expect H_2O to be oxidized in preference to Cl^-. However, the potentials are close and overvoltages at the electrodes could alter this conclusion.

It is possible nevertheless to give a general statement about the product expected at the anode. Electrode potentials, as we have seen, depend on concentrations. It turns out that when the solution is concentrated enough in Cl^-, Cl_2 is the

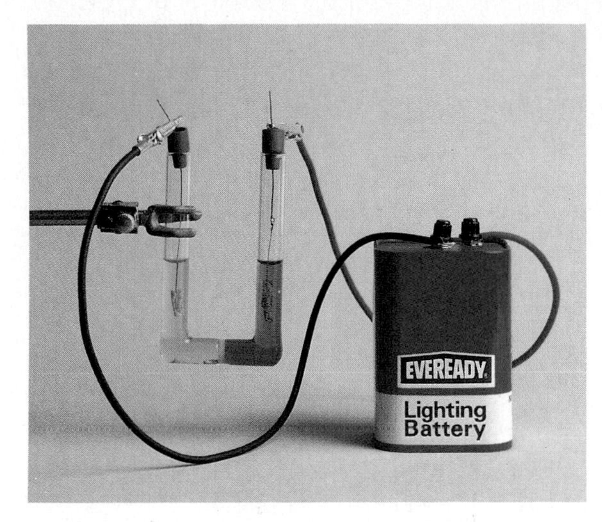

Figure 19.21
Electrolysis of aqueous potassium iodide. Note the formation of I_2 at the anode (brown color). Water is reduced to hydrogen gas and hydroxide ion; the solution contains phenolphthalein, which gives a pink color with OH^- at the cathode.

product; but in dilute solution, O_2 is the product. To see this, we would simply apply the Nernst equation to the $Cl^-|Cl_2$ half-reaction.

$$2Cl^-(aq) \longrightarrow Cl_2(g) + 2e^-$$

Starting with very dilute NaCl solutions, we would find that the oxidation potential of Cl^- is very negative, so H_2O is reduced in preference to Cl^-. But as we increase the NaCl concentration, we would find that the oxidation potential of Cl^- increases until eventually Cl^- is oxidized in preference to H_2O. The product changes from O_2 to Cl_2.

The half-reactions and cell reaction for the electrolysis of aqueous sodium chloride to chlorine and hydroxide ion are as follows:

$$2H_2O(l) + 2e^- \longrightarrow H_2(g) + 2OH^-(aq) \quad \text{(cathode)}$$
$$2Cl^-(aq) \longrightarrow Cl_2(g) + 2e^- \quad \text{(anode)}$$
$$\overline{2H_2O(l) + 2Cl^-(aq) \longrightarrow H_2(g) + Cl_2(g) + 2OH^-(aq)}$$

Because the electrolysis started with sodium chloride, the cation in the electrolyte solution is Na^+. When we evaporate the electrolyte solution at the cathode, we obtain sodium hydroxide, NaOH. Figure 19.21 shows the similar electrolysis of aqueous potassium iodide, KI.

The electrolysis of aqueous sodium chloride is the basis of the *chlor-alkali industry,* which is the major commercial source of chlorine and sodium hydroxide. Commercial cells are of several types, but in each the main problem is to keep the products separate, because chlorine reacts with aqueous sodium hydroxide.

Figure 19.22 shows a modern **chlor-alkali membrane cell,** *a cell for the electrolysis of aqueous sodium chloride in which the anode and cathode compartments are separated by a special plastic membrane that allows only cations to pass through it.* Sodium chloride solution is added to the anode compartment, where chloride ion is oxidized to chlorine. The sodium ions carry the current from the anode to the cathode by passing through the membrane. Water is added to the top of the cathode compartment, where it is reduced to hydroxide ion, and sodium hydroxide solution is removed at the bottom of the cathode compartment.

The older **chlor-alkali mercury cell** is *a cell for the electrolysis of aqueous sodium chloride in which mercury metal is used as the cathode* (Figure 19.23). At

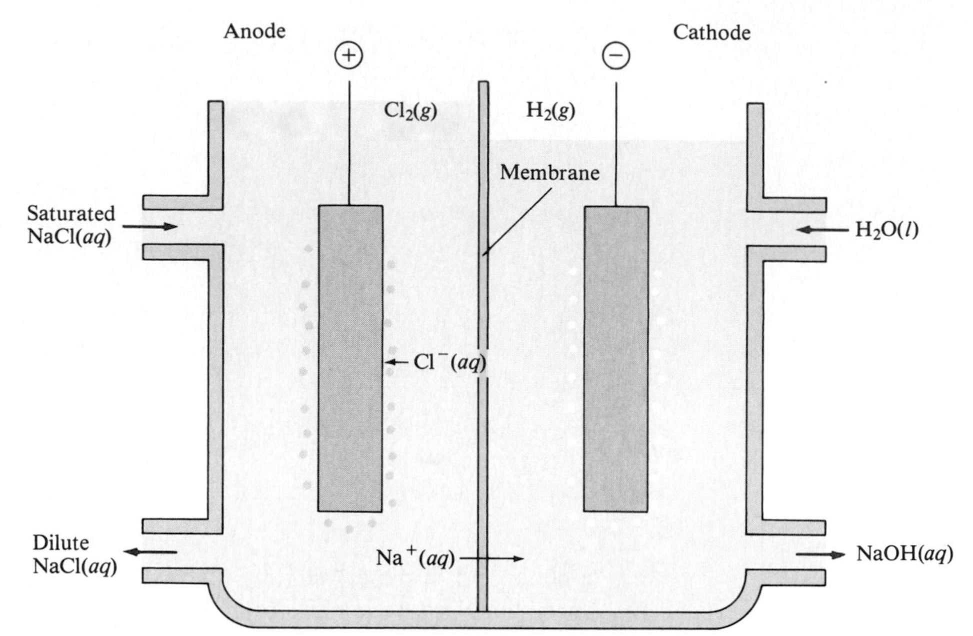

Figure 19.22
A chlor-alkali membrane cell. Sodium chloride solution enters the anode compartment, where Cl^- ion is oxidized to Cl_2. Sodium ion migrates from the anode compartment through the membrane to the cathode compartment. Here water is reduced to hydrogen and OH^- ion. Sodium hydroxide solution is removed at the bottom of the cathode compartment.

the mercury cathode, sodium ion is reduced in preference to water. Sodium ion is reduced to sodium to form a liquid sodium–mercury alloy called sodium amalgam. (An *amalgam* is an alloy of mercury with any of various other metals.)

$$Na^+(aq) + e^- \longrightarrow Na(\text{amalgam})$$

Sodium amalgam circulates from the electrolytic cell to the amalgam decomposer. The amalgam and graphite particles in the decomposer form the electrodes of many voltaic cells, in which sodium reacts with water to give sodium hydroxide solution and hydrogen gas. Loss of mercury from these cells has been a source of

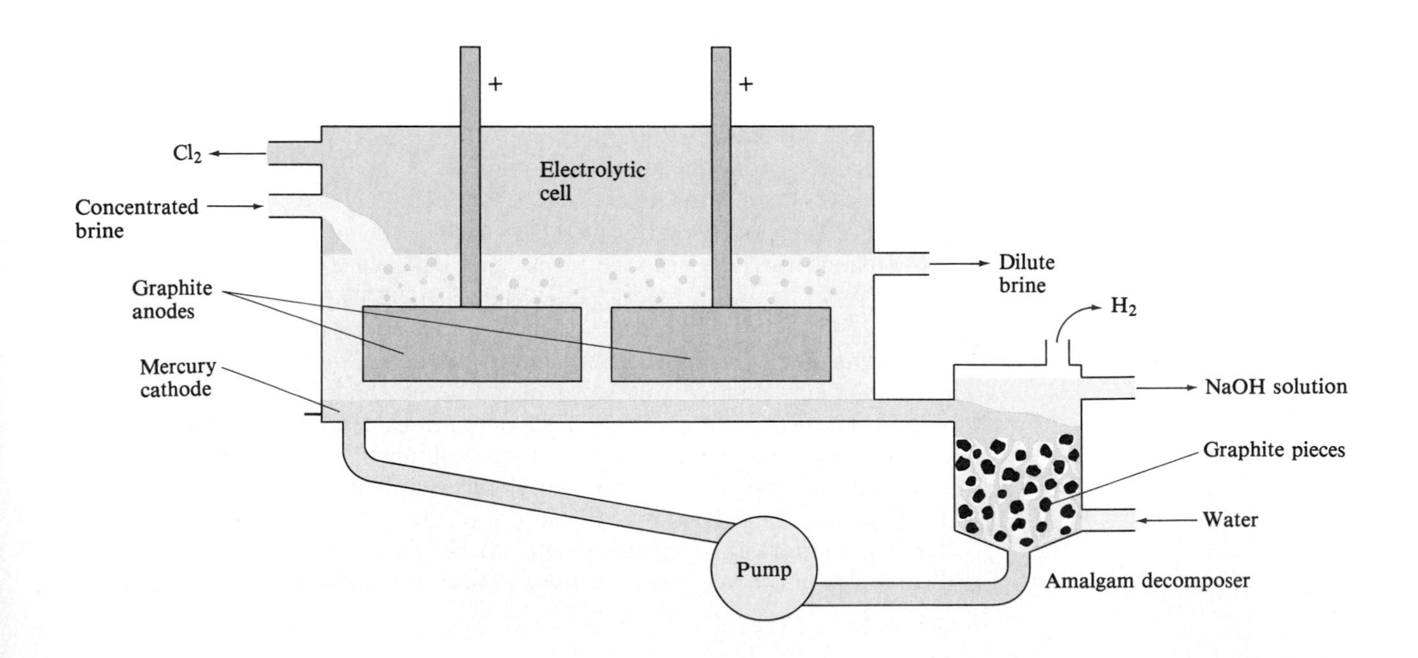

Figure 19.23
A chlor-alkali mercury cell. Chloride ion oxidizes to chlorine gas at the graphite anodes (+ electrodes), and sodium deposits at the mercury cathode. Sodium amalgam (a liquid sodium–mercury alloy) circulates to the amalgam decomposer, where sodium reacts with water to form $NaOH(aq)$ and H_2.

mercury pollution of waterways. Under environmental regulations, filters and coolers have been added to reduce the mercury losses.

Electroplating of Metals

Many metals are protected from corrosion by plating them with other metals. Zinc coatings are often used to protect steel, because the coat protects the steel by cathodic protection even when the zinc coat is scratched (see Section 19.7). A thin zinc coating can be applied to steel by *electrogalvanizing* or zinc electroplating (galvanized steel has a thick zinc coating obtained by dipping the object in molten zinc). The steel object is placed in a bath of zinc salts and made the cathode in an electrolytic cell. The cathode half-reaction is

$$Zn^{2+}(aq) + 2e^- \longrightarrow Zn(s)$$

Electrolysis is also used to purify some metals. For example, copper for electrical use, which must be very pure, is purified by electrolysis. Slabs of impure copper serve as anodes, with pure copper sheets serving as cathodes; the electrolyte bath is copper(II) sulfate, $CuSO_4$ (Figure 19.24). During the electrolysis, copper(II) ions leave the anode slabs and plate out on the cathode sheets. Less reactive metals, such as gold, silver, and platinum, form a valuable mud that collects on the bottom of the electrolytic cell. Metals more reactive than copper

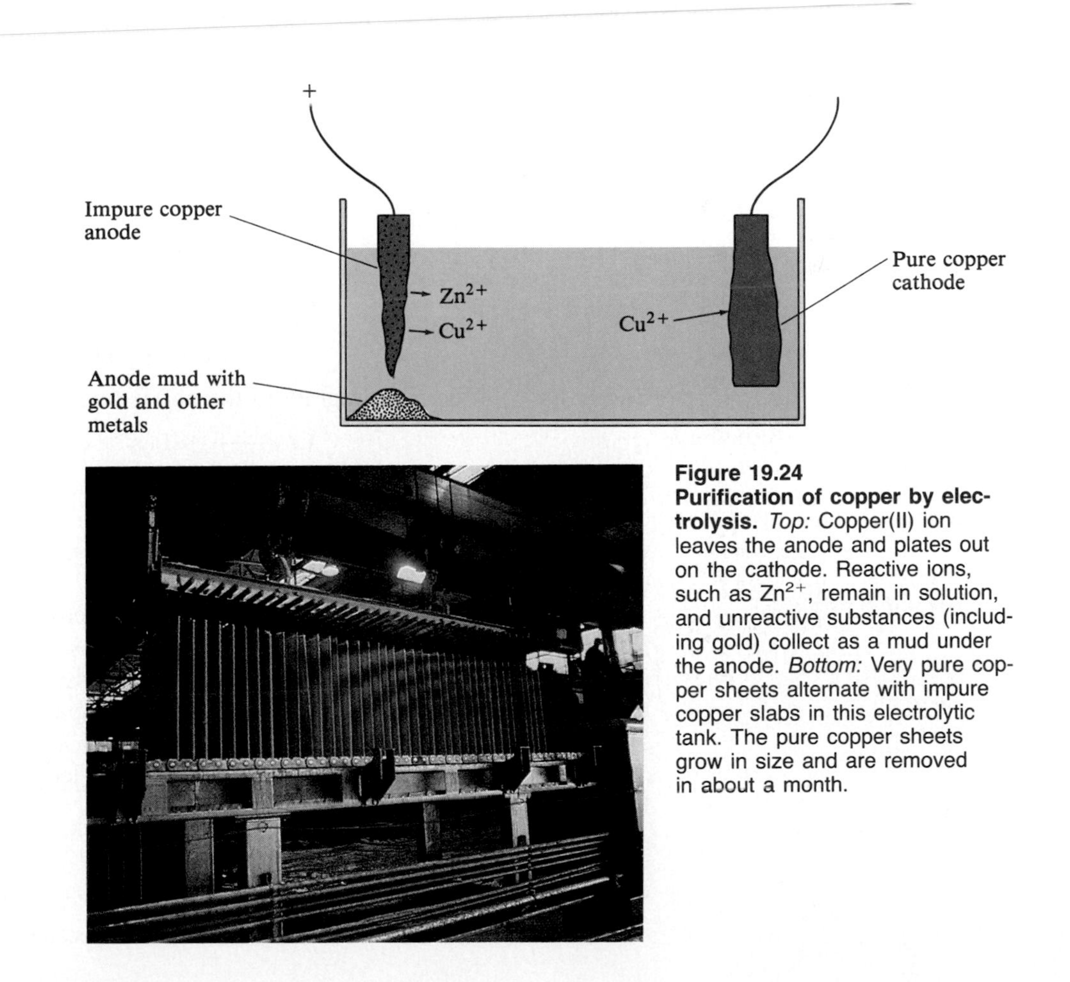

Figure 19.24
Purification of copper by electrolysis. *Top:* Copper(II) ion leaves the anode and plates out on the cathode. Reactive ions, such as Zn^{2+}, remain in solution, and unreactive substances (including gold) collect as a mud under the anode. *Bottom:* Very pure copper sheets alternate with impure copper slabs in this electrolytic tank. The pure copper sheets grow in size and are removed in about a month.

remain as ions in the electrolytic bath. After about a month in the electrolytic cell, the pure copper cathodes are much enlarged and are removed from the cell bath.

Example 19.12 Predicting the Half-Reactions in an Aqueous Electrolysis

What do you expect to be the half-reactions in the electrolysis of aqueous copper(II) sulfate?

Solution

The species we should consider for half-reactions are $Cu^{2+}(aq)$, $SO_4^{2-}(aq)$, and H_2O. Possible cathode half-reactions are

$$Cu^{2+}(aq) + 2e^- \longrightarrow Cu(s);\ E^\circ = 0.34\ V$$
$$2H_2O(l) + 2e^- \longrightarrow H_2(g) + 2OH^-(aq);\ E^\circ = -0.83\ V$$

Because the copper electrode potential is much larger than the reduction potential of water, we expect it to be reduced.

The possible anode half-reactions are

$$2SO_4^{2-}(aq) \longrightarrow S_2O_8^{2-}(aq) + 2e^-;\ -E^\circ = -2.01\ V$$
$$2H_2O(l) \longrightarrow O_2(g) + 4H^+(aq) + 4e^-;\ -E^\circ = -1.23\ V$$

We expect H_2O to be oxidized.

The expected half-reactions are

$$\mathbf{Cu^{2+}(aq) + 2e^- \longrightarrow Cu(s)}$$
$$\mathbf{2H_2O(l) \longrightarrow O_2(g) + 4H^+(aq) + 4e^-}$$

Exercise 19.15

Give the half-reactions that occur when aqueous silver nitrate is electrolyzed. Nitrate ion is not oxidized during the electrolysis. (See Problems 19.71 and 19.72.)

19.10 STOICHIOMETRY OF ELECTROLYSIS

In 1831 and 1832, the British chemist and physicist Michael Faraday showed that one can relate the amounts of substances released at the electrodes during electrolysis to the total charge that has flowed in the electric circuit.● If we look at the electrode reactions, we see that the relationship is a stoichiometric one.

● Faraday's results were summarized in two laws: (1) The quantity of a substance liberated at an electrode is directly proportional to the quantity of electric charge that has flowed in the circuit. (2) For a given quantity of electric charge, the amount of any metal that is deposited is proportional to its equivalent weight (atomic weight divided by the charge on the metal ion). These laws follow directly from the stoichiometry of electrolysis.

When molten sodium chloride is electrolyzed, sodium ions migrate to the cathode, where they react with electrons.

$$Na^+ + e^- \longrightarrow Na(l)$$

Similarly, chloride ions migrate to the anode and release electrons.

$$Cl^- \longrightarrow \tfrac{1}{2}Cl_2(g) + e^-$$

Therefore, when one mole of electrons reacts with sodium ions, one faraday of charge (the magnitude of charge on one mole of electrons) passes through the circuit. One mole of sodium metal is deposited at one electrode, and one-half mole of chlorine gas evolves at the other.

What is new in this type of stoichiometric problem is the measurement of numbers of electrons. We do not weigh them out as we do substances. Rather, we measure the quantity of electric charge that has passed through the circuit. We use the fact that one faraday (9.65×10^4 C) is equivalent to the charge on one mole of electrons.

If we know the current in a circuit and the length of time that it has been flowing, we can calculate the electric charge.

$$\text{Electric charge} = \text{electric current} \times \text{time lapse}$$

Corresponding units are

$$\text{Coulombs} = \text{amperes} \times \text{seconds}$$

The **ampere (A)** is *the base unit of current in the International System (SI)*. The coulomb (C), which is the SI unit of electric charge, is equivalent to an ampere-second. So a current of 0.50 amperes flowing for 84 seconds gives a charge of $0.50\ A \times 84\ s = 42\ A \cdot s$, or 42 C.

If we are given the current and the time of electrolysis, we can calculate the amount of substance produced at an electrode. On the other hand, if we are given the amount of substance produced at an electrode and the length of time of electrolysis, we can determine the current. The next two examples illustrate these calculations.

Example 19.13 Calculating the Amount of Charge from the Amount of Product in an Electrolysis

When an aqueous solution of copper(II) sulfate, $CuSO_4$, is electrolyzed, copper metal is deposited.

$$Cu^{2+}(aq) + 2e^- \longrightarrow Cu(s)$$

(The other electrode reaction gives oxygen: $2H_2O \longrightarrow O_2 + 4H^+ + 4e^-$.) If a constant current was passed for 5.00 hr and 404 mg of copper metal was deposited, what must have been the current?

Solution

From the electrode equation for copper, we can write

$$1\ \text{mol Cu} \simeq 2\ \text{mol e}^-$$

Because 1 mol e^- is equivalent to 9.65×10^4 C (1 faraday), the charge equivalent to 404 mg of copper is

$$0.404\ \cancel{\text{g Cu}} \times \frac{1\ \cancel{\text{mol Cu}}}{63.6\ \cancel{\text{g Cu}}} \times \frac{2\ \cancel{\text{mol e}^-}}{1\ \cancel{\text{mol Cu}}} \times \frac{9.65 \times 10^4\ \text{C}}{1\ \cancel{\text{mol e}^-}} \simeq 1.2\underline{2}6 \times 10^3\ \text{C}$$

The time lapse, 5.00 hr, equals 1.80×10^4 s. Thus,

$$\text{Current} = \frac{\text{charge}}{\text{time}} = \frac{1.226 \times 10^3\ \text{C}}{1.80 \times 10^4\ \text{s}} = \mathbf{6.81 \times 10^{-2}\ A}$$

Exercise 19.16

A constant electric current deposits 365 mg of silver in 216 min from an aqueous silver nitrate solution. What is the current? (See Problems 19.73 and 19.74.)

Example 19.14 Calculating the Amount of Product from the Amount of Charge in an Electrolysis

When an aqueous solution of potassium iodide is electrolyzed using platinum electrodes, the half-reactions are

$$2I^-(aq) \longrightarrow I_2(aq) + 2e^-$$

$$2H_2O(l) + 2e^- \longrightarrow H_2(g) + 2OH^-(aq)$$

How many grams of iodine are produced when a current of 8.52 mA flows through the cell for 10.0 min?

Solution

When the current flows for 6.00×10^2 s (10.0 min), the amount of charge is

$$8.52 \times 10^{-3}\ \text{A} \times 6.00 \times 10^2\ \text{s} = 5.11\ \text{C}$$

Note that two moles of electrons are equivalent to one mole of I_2. Hence,

$$5.11\ \cancel{\text{C}} \times \frac{1\ \cancel{\text{mol e}^-}}{9.65 \times 10^4\ \cancel{\text{C}}} \times \frac{1\ \cancel{\text{mol I}_2}}{2\ \cancel{\text{mol e}^-}} \times \frac{254\ \text{g I}_2}{1\ \cancel{\text{mol I}_2}} \simeq \mathbf{6.73 \times 10^{-3}\ g\ I_2}$$

Exercise 19.17

How many grams of oxygen are liberated by the electrolysis of water after 185 s with a current of 0.0565 A? (See Problems 19.75 and 19.76.)

Profile of a Chemical

ZINC (a Metal for Batteries)

Metal workers in ancient Rome prepared copper, lead, and iron by reducing the respective ores with carbon at high temperature (much as we do today). They probably unknowingly prepared zinc this way as well; but because the element has such a low boiling point (907°C), the zinc would have vaporized without being collected. Not until the thirteenth century in India was the process of preparing zinc metal discovered. The process involved heating zinc ores with carbon and condensing the resulting zinc vapor. This method for producing zinc metal later spread to China and then to Europe.

Today zinc is prepared by a similar method. The most common zinc ore is sphalerite, ZnS. The ore is first roasted in air, where it forms the oxide.

$$2ZnS(s) + 3O_2(g) \longrightarrow 2ZnO(s) + 2SO_2(g)$$

The zinc oxide is heated with coke (carbon) to reduce it to the metal vapor, which is condensed to the liquid from which the solid metal freezes.

$$ZnO(s) + C(s) \longrightarrow Zn(g) + CO(g)$$

Zinc is a reactive metal. It reacts with acids to produce zinc ion, Zn^{2+}, and hydrogen gas. It also reduces metal ions that have higher reduction potentials than Zn^{2+}. For example, the standard reduction potential for Pb^{2+} to Pb is -0.13 V, whereas the standard reduction potential for Zn^{2+} to Zn is -0.76 V. Therefore, lead(II) ion is reduced by zinc metal to lead metal.

$$Pb^{2+}(aq) + Zn(s) \longrightarrow Pb(s) + Zn^{2+}(aq)$$

This reaction is shown in Figure 19.25. A zinc rod is placed in a solution of lead(II) nitrate, and the rod becomes covered with metallic crystals of lead.

***Figure 19.25* Displacement reaction of zinc and lead(II) ion.** *A zinc rod was placed in an aqueous solution of lead(II) nitrate, $Pb(NO_3)_2$. After some time, lead crystals covered the submerged surface of the rod.*

The most important use of zinc is in providing a protective coating for other metals. Although zinc is an active metal, it forms an adherent oxide coat that protects the zinc from further oxidation by air. Galvanized steel is made by either dipping steel sheets in molten zinc or by electrolytic deposition of zinc on steel. Paints containing zinc powder are also used. The zinc protects the steel directly by keeping the oxygen away. But even if the zinc coating is broken, the steel remains uncorroded because of cathodic protection. The zinc metal coating becomes the anode and oxidizes in preference to the exposed steel, which becomes the cathode. Zinc is also used as the anode in batteries, including the zinc–carbon dry cell, the alkaline dry cell, and the mercury(II) oxide cell.

Brass is an alloy of copper with 20%–50% zinc. Its discovery predates by many centuries the discovery of zinc. The early Romans prepared brass by heating together zinc ore, charcoal, and copper. Today brass is produced by melting a mixture of copper and zinc metals. Brass and similar alloys are important for making castings.

The oxidation state of zinc in its compounds is +2. The most important compound of zinc is zinc oxide, ZnO, which is prepared by burning zinc vapor in air.

$$2Zn(g) + O_2(g) \longrightarrow 2ZnO(s)$$

Zinc oxide is used in making rubber, where it imparts desirable characteristics to the material. Zinc oxide is also used as a pigment in white paint and in enamels and glazes.

Zinc oxide (as well as certain other materials) is also used as a photoconductive surface in photocopiers. A photoconductive material becomes electrically conducting when exposed to light. The zinc oxide surface in the copier is electrically charged, then exposed to a light image from a printed document. Lighted areas of the photoconducting surface become electrically conducting, so that the electrical charge in these areas is drained away. Dark areas remain electrically charged, however. When a black powder, called the toner, is spread over the zinc oxide surface, it sticks to the electrically charged areas, producing an image of the document. This image is then transferred with heat to a sheet of paper.

Questions for Study

1. Explain why zinc was discovered long after brass, which is an alloy of zinc.
2. Describe how zinc is prepared from sphalerite, ZnS.
3. Will zinc metal displace nickel from its salts? If so, write the net ionic equation for the reaction. If it does not, give an example of a metal salt from which zinc will displace the metal, and then write the net ionic equation.
4. Describe some important commercial uses of zinc metal.
5. Describe the function of zinc oxide in a photocopier.

A Checklist for Review

IMPORTANT TERMS

electrochemical cell (p. 747)
voltaic (galvanic) cell (p. 747)
electrolytic cell (p. 747)
half-cell (19.1)
salt bridge (19.1)
anode (19.1)
cathode (19.1)
cell reaction (19.1)
potential difference (19.3)
volt (V) (19.3)
faraday constant (*F*) (19.3)
electromotive force (emf) (19.3)
standard emf (19.4)
standard electrode potential (19.4)
Nernst equation (19.6)
zinc–carbon (Leclanché) dry cell (19.7)
alkaline dry cell (19.7)
lithium–iodine battery (19.7)
lead storage cell (19.7)
nickel–cadmium cell (19.7)
fuel cell (19.7)
electrolysis (p. 772)
Downs cell (19.8)
chlor-alkali membrane cell (19.9)
chlor-alkali mercury cell (19.9)
ampere (A) (19.10)

KEY EQUATIONS

$$w_{max} = -nFE_{cell}$$

$$E^{\circ}_{cell} = E^{\circ}_{cathode} - E^{\circ}_{anode}$$

$$\Delta G^{\circ} = -nFE^{\circ}_{cell}$$

$$E^{\circ}_{cell} = \frac{0.0592}{n} \log K \quad \text{(values in volts at 25°C)}$$

$$E_{cell} = E^{\circ}_{cell} - \frac{0.0592}{n} \log Q \quad \text{(values in volts at 25°C)}$$

SUMMARY OF FACTS AND CONCEPTS

Electrochemical cells are of two types: voltaic and electrolytic. A *voltaic cell* uses a spontaneous chemical reaction to generate an electric current. It does this by physically separating the reaction into its oxidation and reduction half-reactions. These half-reactions take place in *half-cells.* The half-cell in which reduction occurs is called the *cathode;* the half-cell in which oxidation occurs is called the *anode*. Electrons flow in the external circuit from the anode to the cathode.

The *electromotive force (emf)* is the maximum voltage of a voltaic cell. It can be directly related to the maximum work that can be done by the cell. A *standard electrode potential,* or reduction potential, refers to the potential of an electrode in which molar concentrations and gas pressures (in atmospheres) have unit values. A table of standard electrode potentials is useful for establishing the direction of spontaneity of an oxidation–reduction reaction and for calculating the *standard emf* of a cell.

The standard free-energy change, standard emf, and equilibrium constant are all related. Knowing one quantity, we can calculate the others. Electrochemical measurements can therefore give us equilibrium or thermodynamic information.

An electrode potential depends on concentrations of the electrode substances, according to the *Nernst equation*. Because of this relationship, cell emf's can be used to measure ion concentrations. This is the basic principle of a pH meter, a device that measures the hydrogen-ion concentration.

Voltaic cells are used commercially as portable energy sources (batteries). In addition, the basic principle of the voltaic cell is employed in the *cathodic protection* of buried pipelines and tanks.

Electrolytic cells represent the other type of electrochemical cell. They use an external voltage source to push a reaction in a nonspontaneous direction. The electrolysis of an aqueous solution often involves the oxidation or reduction of water at the electrodes. Electrolysis of concentrated sodium chloride solution, for example, gives hydrogen at the cathode. The amounts of substances released at an electrode are related to the amount of charge passed through the cell. This relationship is stoichiometric and follows from the electrode reactions.

OPERATIONAL SKILLS

1. **Sketching and labeling a voltaic cell** Given a verbal description of a voltaic cell, sketch the cell, labeling the anode and cathode, and give the directions of electron flow and ion migration (Example 19.1).

2. **Writing the cell reaction from the cell notation** Given the notation for a voltaic cell, write the overall cell reaction (Example 19.2). Alternatively, given the cell reaction, write the cell notation.

3. **Calculating the quantity of work from a given amount of cell reactants** Given the emf and overall reaction for a voltaic cell, calculate the maximum work that can be obtained from a given amount of reactant (Example 19.3).

4. **Determining the relative strengths of oxidizing and reducing agents** Given a table of standard electrode potentials, list oxidizing or reducing agents by increasing strength (Example 19.4).

5. **Determining the direction of spontaneity from electrode potentials** Given standard electrode potentials, decide the direction of spontaneity for an oxidation–reduction reaction under standard conditions (Example 19.5).

6. **Calculating the emf from standard potentials** Given standard electrode potentials, calculate the standard emf of a voltaic cell (Example 19.6).

7. **Calculating the free-energy change from electrode potentials** Given standard electrode potentials, calculate the standard free-energy change for an oxidation–reduction reaction (Example 19.7).

8. **Calculating the cell emf from free-energy change** Given a table of standard free energies of formation, calculate the standard emf of a voltaic cell (Example 19.8).

9. **Calculating the equilibrium constant from cell emf** Given standard potentials (or standard emf), calculate the equilibrium constant of an oxidation–reduction reaction (Example 19.9).

10. **Calculating the cell emf and electrode potential for nonstandard conditions** Given standard electrode potentials and the concentrations of substances in a voltaic cell, calculate the cell emf (Example 19.10). Given the standard electrode potential, calculate the electrode potential for nonstandard conditions (Example 19.11).

11. **Predicting the half-reactions in an aqueous electrolysis** Using values of electrode potentials, decide which electrode reactions actually occur in the electrolysis of an aqueous solution (Example 19.12).

12. **Relating the amounts of charge and product in an electrolysis** Given the amount of product obtained by electrolysis, calculate the amount of charge that flowed (Example 19.13). Given the amount of charge that flowed, calculate the amount of product obtained by electrolysis (Example 19.14).

Review Questions

19.1 Describe the difference between a voltaic cell and an electrolytic cell.

19.2 Define *cathode* and *anode*. Describe the migration of cations and anions in a cell.

19.3 What is the SI unit of electrical potential?

19.4 Define the *faraday*.

19.5 Why is it necessary to measure the voltage of a voltaic cell when no current is flowing in order to obtain the cell emf?

19.6 How are standard electrode potentials defined?

19.7 Express the SI unit of energy as the product of two electrical units.

19.8 Give the mathematical relationships between the members of each possible pair of the three quantities ΔG°, E°_{cell}, and K.

19.9 Using the Nernst equation, explain how the corrosion of iron would be affected by the pH of a water drop.

19.10 Describe the zinc–carbon, or Leclanché, dry cell and the lead storage battery.

19.11 What is a fuel cell? Describe an example.

19.12 Explain the electrochemistry of rusting.

19.13 Iron may be protected by coating with tin (tin cans) or with zinc (galvanized iron). Galvanized iron does not corrode as long as zinc is present. By contrast, when a tin can is scratched, the exposed iron underneath corrodes rapidly. Explain the difference between zinc and tin as protective coatings against iron corrosion.

19.14 The electrolysis of water is often done by passing a current through a dilute solution of sulfuric acid. What is the function of the sulfuric acid?

19.15 Describe a method for the preparation of sodium metal from sodium chloride.

19.16 Potassium was discovered by the British chemist Humphry Davy when he electrolyzed molten potassium hydroxide. What would be the anode reaction?

19.17 Briefly explain why different products are obtained from the electrolysis of molten NaCl and the electrolysis of dilute aqueous solution of NaCl.

19.18 Write the Nernst equation for the electrode reaction $2Cl^-(aq) \longrightarrow Cl_2(g) + 2e^-$. With this equation, explain why the electrolysis of concentrated sodium chloride solution might be expected to release chlorine gas rather than oxygen gas at the anode.

Practice Problems

ELECTROCHEMICAL CELLS

19.19 A voltaic cell is constructed from the following half-cells: a zinc electrode in zinc sulfate solution and a nickel electrode in nickel sulfate solution. The half-reactions are

$$Zn(s) \longrightarrow Zn^{2+}(aq) + 2e^-$$
$$Ni^{2+}(aq) + 2e^- \longrightarrow Ni(s)$$

Sketch the cell, labeling the anode and cathode (with the electrode reactions), and show the direction of electron flow and the movement of cations.

19.20 Half-cells were made from a nickel rod dipping in a nickel sulfate solution and a copper rod dipping in a copper sulfate solution. The half-reactions that occurred in a voltaic cell using these half-cells were

$$Cu^{2+}(aq) + 2e^- \longrightarrow Cu(s)$$
$$Ni(s) \longrightarrow Ni^{2+}(aq) + 2e^-$$

Sketch the cell and label the anode and cathode, showing the corresponding electrode reactions. Give the direction of electron flow and the movement of cations.

19.21 Zinc reacts spontaneously with silver ion.

$$Zn(s) + 2Ag^+(aq) \longrightarrow Zn^{2+}(aq) + 2Ag(s)$$

Describe a voltaic cell using this reaction. What are the half-reactions?

19.22 Cadmium reacts spontaneously with copper(II) ion.

$$Cd(s) + Cu^{2+}(aq) \longrightarrow Cd^{2+}(aq) + Cu(s)$$

Obtain half-reactions for this, then describe a voltaic cell using these half-reactions.

19.23 A silver oxide–zinc cell maintains a fairly constant voltage during discharge (1.60 V). The button form of this cell is used in watches, hearing aids, and other electronic devices. The half-reactions are

$$Zn(s) + 2OH^-(aq) \longrightarrow Zn(OH)_2(s) + 2e^-$$
$$Ag_2O(s) + H_2O(l) + 2e^- \longrightarrow 2Ag(s) + 2OH^-(aq)$$

Identify each as the anode or cathode reaction. What is the overall reaction in the voltaic cell?

19.24 A mercury battery, used for hearing aids and electric watches, delivers a constant voltage (1.35 V) for long periods. The half-reactions are

$$HgO(s) + H_2O(l) + 2e^- \longrightarrow Hg(l) + 2OH^-(aq)$$
$$Zn(s) + 2OH^-(aq) \longrightarrow Zn(OH)_2(s) + 2e^-$$

Which half-reaction occurs at the anode and which occurs at the cathode? What is the overall cell reaction?

VOLTAIC CELL NOTATION

19.25 Write the cell notation for a voltaic cell with the following half-reactions.

$$Cd(s) \longrightarrow Cd^{2+}(aq) + 2e^-$$
$$Pb^{2+}(aq) + 2e^- \longrightarrow Pb(s)$$

19.26 Write the cell notation for a voltaic cell with the following half-reactions.

$$Al(s) \longrightarrow Al^{3+}(aq) + 3e^-$$
$$2H^+(aq) + 2e^- \longrightarrow H_2(g)$$

19.27 Give the notation for a voltaic cell constructed from a hydrogen electrode (cathode) in 1.0 *M* HCl and a nickel electrode (anode) in 1.0 *M* $NiSO_4$ solution. The electrodes are connected by a salt bridge.

19.28 A voltaic cell has an iron rod in 0.30 *M* iron(III) chloride solution for the cathode, and a zinc rod in 0.20 *M* zinc sulfate solution for the anode. The half-cells are connected by a salt bridge. Write the notation for this cell.

19.29 Write the overall cell reaction for the following voltaic cell.

$$Fe(s)|Fe^{2+}(aq)||Ag^+(aq)|Ag(s)$$

19.30 Write the overall cell reaction for the following voltaic cell.

$$Pt|H_2(g)|H^+(aq)||Br_2(l)|Br^-(aq)|Pt$$

19.31 Consider the voltaic cell

$$Cd(s)|Cd^{2+}(aq)||Ni^{2+}(aq)|Ni(s)$$

Write the half-cell reactions and the overall cell reaction. Make a sketch of this cell and label it. Include labels showing the anode, cathode, and direction of electron flow.

19.32 Consider the voltaic cell

$$Zn(s)|Zn^{2+}(aq)||Ag^+(aq)|Ag(s)$$

Write the half-cell reactions and the overall cell reaction. Make a sketch of this cell and label it. Include labels showing the anode, cathode, and direction of electron flow.

ELECTRODE POTENTIALS AND CELL emf's

19.33 A voltaic cell whose cell reaction is

$$2Fe^{3+}(aq) + Zn(s) \longrightarrow 2Fe^{2+}(aq) + Zn^{2+}(aq)$$

has an emf of 0.72 V. What is the maximum electrical work that can be obtained from this cell per mole of iron(III) ion?

19.34 A particular voltaic cell operates on the reaction

$$Zn(s) + Cl_2(g) \longrightarrow Zn^{2+}(aq) + 2Cl^-(aq)$$

giving an emf of 0.853 V. Calculate the maximum electrical work generated when 20.0 g of zinc metal is consumed.

19.35 What is the maximum work that you can obtain from 15.0 g of nickel in the following cell when the emf is 0.97 V?

$$Ni(s)|Ni^{2+}(aq)||Ag^+(aq)|Ag(s)$$

19.36 Calculate the maximum work available from 25.0 g of aluminum in the following cell when the emf is 1.15 V.

$$Al(s)|Al^{3+}(aq)||H^+(aq)|O_2(g)|Pt$$

Note that O_2 is reduced to H_2O.

19.37 Order the following oxidizing agents by increasing strength under standard-state conditions: $O_2(g)$; $H_2O_2(aq)$; $NO_3^-(aq)$ (in acidic solution).

19.38 Order the following oxidizing agents by increasing strength under standard-state conditions: $Ag^+(aq)$; $I_2(aq)$; $MnO_4^-(aq)$ (in acidic solution).

19.39 Consider the reducing agents $Cu^+(aq)$, $Zn(s)$, and $Fe(s)$. Which is strongest? Which is weakest?

19.40 Consider the reducing agents $Sn^{2+}(aq)$, $Al(s)$, and $I^-(aq)$. Which is strongest? Which is weakest?

19.41 Consider the following reactions. Are they spontaneous in the direction written, under standard conditions at 25°C?

(a) $Sn^{4+}(aq) + 2Fe^{2+}(aq) \longrightarrow Sn^{2+}(aq) + 2Fe^{3+}(aq)$

(b) $4MnO_4^-(aq) + 12H^+(aq) \longrightarrow 4Mn^{2+}(aq) + 5O_2(g) + 6H_2O(l)$

19.42 Answer the following questions by referring to standard electrode potentials at 25°C.

(a) Will dichromate ion, $Cr_2O_7^{2-}$, oxidize iron(II) ion in acidic solution under standard conditions?

(b) Will copper metal reduce 1.0 *M* $Ni^{2+}(aq)$ to metallic nickel?

19.43 What would you expect to happen when chlorine gas, Cl_2, at 1 atm pressure is bubbled into a solution containing 1.0 *M* F^- and 1.0 *M* Br^- at 25°C? Write a balanced equation for the reaction that occurs.

19.44 Dichromate ion, $Cr_2O_7^{2-}$, is added to an acidic solution containing Br^- and Mn^{2+}. Write a balanced equation for any reaction that occurs. Assume standard conditions at 25°C.

19.45 Calculate the standard emf of the following cell at 25°C.

$$Cr(s)|Cr^{3+}(aq)||Hg_2^{2+}(aq)|Hg(l)$$

19.46 Calculate the standard emf of the following cell at 25°C.

$$Sn(s)|Sn^{2+}(aq)||Cu^{2+}(aq)|Cu(s)$$

19.47 What is the standard emf you would obtain from a cell at 25°C using an electrode in which $I^-(aq)$ is in contact with $I_2(s)$ and another electrode in which a chromium strip dips into a solution of $Cr^{3+}(aq)$?

19.48 What is the standard emf you would obtain from a cell at 25°C using an electrode in which $Hg_2^{2+}(aq)$ is in contact with mercury metal and another electrode in which an aluminum strip dips into a solution of $Al^{3+}(aq)$?

RELATIONSHIPS AMONG $E°_{cell}$, $\Delta G°$, AND K

19.49 Calculate the standard free-energy change at 25°C for the following reaction.

$$3Cu(s) + 2NO_3^-(aq) + 8H^+(aq) \longrightarrow 3Cu^{2+}(aq) + 2NO(g) + 4H_2O(l)$$

Use standard electrode potentials.

19.50 Calculate the standard free-energy change at 25°C for the following reaction.

$$4Al(s) + 3O_2(g) + 12H^+(g) \longrightarrow 4Al^{3+}(aq) + 6H_2O(l)$$

Use standard electrode potentials.

19.51 What is $\Delta G°$ for the following reaction?

$$2I^-(aq) + Cl_2(g) \longrightarrow I_2(s) + 2Cl^-(aq)$$

Use data given in Table 19.1.

19.52 Using electrode potentials, calculate the standard free-energy change for the reaction

$$Na(s) + \tfrac{1}{2}Cl_2(g) \xrightarrow{aq} Na^+(aq) + Cl^-(aq)$$

19.53 Calculate the standard emf at 25°C for the following cell reaction from standard free energies of formation (Appendix C).

$$Al(s) + 3Ag^+(aq) \longrightarrow Al^{3+}(aq) + 3Ag(s)$$

19.54 Calculate the standard emf at 25°C for the following cell reaction from standard free energies of formation (Appendix C).

$$2Al(s) + 3Cu^{2+}(aq) \longrightarrow 2Al^{3+}(aq) + 3Cu(s)$$

19.55 Calculate the standard emf of the lead storage cell whose overall reaction is

$$PbO_2(s) + 2HSO_4^-(aq) + 2H^+(aq) + Pb(s) \longrightarrow 2PbSO_4(s) + 2H_2O(l)$$

See Appendix C for free energies of formation.

19.56 Calculate the standard emf of the cell corresponding to the oxidation of oxalic acid, $H_2C_2O_4$, by permanganate ion, MnO_4^-.

$$5H_2C_2O_4(aq) + 2MnO_4^-(aq) + 6H^+(aq) \longrightarrow 10CO_2(g) + 2Mn^{2+}(aq) + 8H_2O(l)$$

See Appendix C for free energies of formation; $\Delta G_f°$ for $H_2C_2O_4(aq)$ is -698 kJ.

19.57 Calculate the equilibrium constant K for the following reaction at 25°C from standard electrode potentials.

$$Fe^{3+}(aq) + Sn^{2+}(aq) \longrightarrow Fe^{2+}(aq) + Sn^{4+}(aq)$$

19.58 Calculate the equilibrium constant K for the following reaction at 25°C from standard electrode potentials.

$$Fe^{3+}(aq) + 2Hg(l) \longrightarrow Fe^{2+}(aq) + Hg_2^{2+}(aq)$$

19.59 Copper(I) ion can act as both an oxidizing agent and a reducing agent. Hence, it can react with itself.

$$2Cu^+(aq) \longrightarrow Cu(s) + Cu^{2+}(aq)$$

Calculate the equilibrium constant at 25°C for this reaction, using appropriate values of electrode potentials.

19.60 Use electrode potentials to calculate the equilibrium constant at 25°C for the reaction

$$2ClO_3^-(aq) \rightleftharpoons ClO_4^-(aq) + ClO_2^-(aq)$$

See Appendix I for data.

NERNST EQUATION

19.61 Calculate the emf of the following cell at 25°C.

$$Cr(s)|Cr^{3+}(1.0 \times 10^{-2}\ M)||Ni^{2+}(2.0\ M)|Ni(s)$$

19.62 What is the emf of the following cell at 25°C?

$$Ni(s)|Ni^{2+}(1.0\ M)||Sn^{2+}(1.0 \times 10^{-4}\ M)|Sn(s)$$

19.63 Calculate the emf of a cell operating with the following reaction at 25°C, in which $[MnO_4^-] = 0.010\ M$, $[Br^-] = 0.010\ M$, $[Mn^{2+}] = 0.15\ M$, and $[H^+] = 1.0\ M$.

$$2MnO_4^-(aq) + 10Br^-(aq) + 16H^+(aq) \longrightarrow 2Mn^{2+}(aq) + 5Br_2(l) + 8H_2O(l)$$

19.64 Calculate the emf of a cell operating with the following reaction at 25°C, in which $[Cr_2O_7^{2-}] = 0.020\ M$, $[I^-] = 0.015\ M$, $[Cr^{3+}] = 0.20\ M$, and $[H^+] = 1.0\ M$.

$$Cr_2O_7^{2-}(aq) + 6I^-(aq) + 14H^+(aq) \longrightarrow 2Cr^{3+}(aq) + 3I_2(s) + 7H_2O(l)$$

19.65 What is the potential of the following electrode at 25°C?

$$Zn^{2+}(2.0 \times 10^{-3}\ M)|Zn(s)$$

19.66 What is the potential of the following electrode at 25°C?

$$Ag^+(1.0 \times 10^{-7}\ M)|Ag(s)$$

19.67 The voltaic cell

$$Cd(s)|Cd^{2+}(aq)||Ni^{2+}(1.0\ M)|Ni(s)$$

has an electromotive force of 0.240 V at 25°C. What is the concentration of cadmium ion? ($E^\circ_{cell} = 0.170$ V.)

19.68 The emf of the following cell at 25°C is 0.131 V.

$$Pt|H_2(1.00\ atm)|H^+(\text{test solution})||H^+(1.00\ M)|H_2(1.00\ atm)|Pt$$

What is the pH of the test solution?

ELECTROLYSIS

19.69 What are the half-reactions in the electrolysis of (a) $CaCl_2(l)$; (b) $CsOH(l)$?

19.70 What are the half-reactions in the electrolysis of (a) $MgBr_2(l)$; (b) $Ca(OH)_2(l)$?

19.71 Describe what you expect to happen when the following solutions are electrolyzed: (a) aqueous Na_2SO_4; (b) aqueous KBr. That is, what are the electrode reactions? What is the overall reaction?

19.72 Describe what you expect to happen when the following solutions are electrolyzed: (a) aqueous $CuCl_2$; (b) aqueous $Cu(NO_3)_2$. That is, what are the electrode reactions? What is the overall reaction? Nitrate ion is not oxidized.

19.73 In the commercial preparation of aluminum, aluminum oxide, Al_2O_3, is electrolyzed at 1000°C. (The mineral cryolite is added as a solvent.) Assume that the cathode reaction is

$$Al^{3+} + 3e^- \longrightarrow Al$$

How many coulombs of electricity are required to give 5.12 kg of aluminum?

19.74 Chlorine, Cl_2, is produced commercially by the electrolysis of aqueous sodium chloride. The anode reaction is

$$2Cl^-(aq) \longrightarrow Cl_2(g) + 2e^-$$

How long will it take to produce 1.18 kg of chlorine if the current is 5.00×10^2 A?

19.75 When molten lithium chloride, LiCl, is electrolyzed, lithium metal is liberated at the cathode. How many grams of lithium are liberated when 5.00×10^3 C of charge passes through the cell?

19.76 How many grams of cadmium are deposited from an aqueous solution of cadmium sulfate, $CdSO_4$, when an electric current of 1.51 A flows through the solution for 156 min?

Additional Problems

19.77 Give the notation for a voltaic cell that uses the reaction

$$Mg(s) + Cl_2(g) \longrightarrow Mg^{2+}(aq) + 2Cl^-(aq)$$

What is the half-cell reaction for the anode? for the cathode? What is the standard emf of the cell?

19.78 Give the notation for a voltaic cell whose overall cell reaction is

$$Mg(s) + 2Ag^+(aq) \longrightarrow Mg^{2+}(aq) + 2Ag(s)$$

What are the half-cell reactions? Label them as anode or cathode reactions. What is the standard emf of this cell?

19.79 Use electrode potentials to answer the following questions. (a) Is the oxidation of nickel by iron(III) ion a spontaneous reaction under standard conditions?

$$Ni(s) + 2Fe^{3+}(aq) \longrightarrow Ni^{2+}(aq) + 2Fe^{2+}(aq)$$

(b) Will iron(III) ion oxidize tin(II) ion to tin(IV) ion under standard conditions?

$$2Fe^{3+}(aq) + Sn^{2+}(aq) \longrightarrow 2Fe^{2+}(aq) + Sn^{4+}(aq)$$

19.80 Use electrode potentials to answer the following questions, assuming standard conditions. (a) Do you expect permanganate ion (MnO_4^-) to oxidize chloride ion to chlorine gas in acidic solution? (b) Will dichromate ion ($Cr_2O_7^{2-}$) oxidize chloride ion to chlorine gas in acidic solution?

19.81 Determine the emf of the following cell.

$$Pb|PbSO_4(s),\ SO_4^{2-}(1.0\ M)||H^+(1.0\ M)|H_2(1.0\ atm)|Pt$$

The anode is essentially a lead electrode, $Pb|Pb^{2+}(aq)$. However, the anode solution is saturated with lead sulfate, so that the lead(II)-ion concentration is determined by the solubility product of $PbSO_4$ ($= 1.7 \times 10^{-8}$).

19.82 Determine the emf of the following cell.

$$Pt|H_2(1.0\ atm)|H^+(1.0\ M)||Cl^-(1.0\ M),\ AgCl(s)|Ag$$

The cathode is essentially a silver electrode, $Ag^+(aq)|Ag$. However, the cathode solution is saturated with silver chloride, so that the silver-ion concentration is determined by the solubility product of AgCl ($= 1.8 \times 10^{-10}$).

19.83 (a) Calculate the equilibrium constant for the following reaction at 25°C.

$$Sn(s) + Pb^{2+}(aq) \rightleftharpoons Sn^{2+}(aq) + Pb(s)$$

The standard emf of the corresponding voltaic cell is 0.010 V. (b) If an excess of tin metal is added to 1.0 M Pb^{2+}, what is the concentration of Pb^{2+} at equilibrium?

19.84 (a) Calculate the equilibrium constant for the following reaction at 25°C.

$$Ag^{+}(aq) + Fe^{2+}(aq) \rightleftharpoons Ag(s) + Fe^{3+}(aq)$$

The standard emf of the corresponding voltaic cell is 0.030 V. (b) When equal volumes of 1.0 M solutions of Ag^{+} and Fe^{2+} are mixed, what is the equilibrium concentration of Fe^{2+}?

19.85 How many faradays are required for each of the following processes? How many coulombs are required?

(a) Reduction of 1.0 mol Na^{+} to Na
(b) Reduction of 1.0 mol Cu^{2+} to Cu
(c) Oxidation of 1.0 g H_2O to O_2
(d) Oxidation of 1.0 g Cl^{-} to Cl_2

19.86 How many faradays are required for each of the following processes? How many coulombs are required?

(a) Reduction of 1.0 mol Fe^{3+} to Fe^{2+}
(b) Reduction of 1.0 mol Fe^{3+} to Fe
(c) Oxidation of 1.0 g Sn^{2+} to Sn^{4+}
(d) Reduction of 1.0 g Au^{3+} to Au

19.87 In an analytical determination of arsenic, a solution containing arsenious acid, H_3AsO_3, potassium iodide, and a small amount of starch is electrolyzed. The electrolysis produces free iodine from iodide ion, and the iodine immediately oxidizes the arsenious acid to hydrogen arsenate ion, $HAsO_4^{2-}$.

$$I_2(aq) + H_3AsO_3(aq) + H_2O(l) \longrightarrow 2I^{-}(aq) + HAsO_4^{2-}(aq) + 4H^{+}(aq)$$

When the oxidation of arsenic is complete, the free iodine combines with the starch to give a deep blue color. If, during a particular run, it takes 65.3 s for a current of 10.5 mA to give an end point (indicated by the blue color), how many grams of arsenic are present in the solution?

19.88 The amount of lactic acid, $HC_3H_5O_3$, produced in a sample of muscle tissue was analyzed by reaction with hydroxide ion. Hydroxide ion was produced in the sample mixture by electrolysis. The cathode reaction was

$$2H_2O(l) + 2e^{-} \longrightarrow H_2(g) + 2OH^{-}(aq)$$

Hydroxide ion reacts with lactic acid as soon as it is produced. The end point of the reaction is detected with an acid–base indicator. It required 115 s for a current of 15.6 mA to reach the end point. How many grams of lactic acid (a monoprotic acid) were present in the sample?

Cumulative-Skills Problems

19.89 Under standard conditions for all concentrations, the following reaction is spontaneous at 25°C.

$$O_2(g) + 4H^{+}(aq) + 4Br^{-}(aq) \longrightarrow 2H_2O(l) + 2Br_2(l)$$

If $[H^{+}]$ is decreased so that the pH = 3.60, what value will E_{cell} have, and will the reaction be spontaneous at this $[H^{+}]$?

19.90 Under standard conditions for all concentrations, the following reaction is spontaneous at 25°C.

$$O_3(g) + 2H^{+}(aq) + 2Co^{2+}(aq) \longrightarrow O_2(g) + H_2O(l) + 2Co^{3+}(l)$$

If $[H^{+}]$ is decreased so that the pH = 8.40, what value will E_{cell} have, and will the reaction be spontaneous at this $[H^{+}]$?

19.91 Under standard conditions for all concentrations, the following reaction is spontaneous at 25°C.

$$O_2(g) + 4H^{+}(aq) + 4Br^{-}(aq) \longrightarrow 2H_2O(l) + 2Br_2(l)$$

If $[H^{+}]$ is adjusted by adding a buffer of 0.10 M NaCNO and 0.10 M HCNO ($K_a = 3.5 \times 10^{-4}$), what value will E_{cell} have, and will the reaction be spontaneous at this $[H^{+}]$?

19.92 Under standard conditions for all concentrations, the following reaction is spontaneous at 25°C.

$$O_3(g) + 2H^{+}(aq) + 2Co^{2+}(aq) \longrightarrow O_2(g) + H_2O(l) + 2Co^{3+}(aq)$$

If $[H^{+}]$ is adjusted by adding a buffer of 0.10 M NaClO and 0.10 M HClO ($K_a = 3.5 \times 10^{-8}$), what value will E_{cell} have, and will the reaction be spontaneous at this $[H^{+}]$?

19.93 An electrode is prepared by dipping a silver strip into a solution saturated with silver thiocyanate, AgSCN, and containing 0.10 M SCN^{-}. The emf of the voltaic cell constructed by connecting this, as the cathode, to the standard hydrogen half-cell as the anode is 0.45 V. What is the solubility product of silver thiocyanate?

19.94 An electrode is prepared from liquid mercury in contact with a saturated solution of mercury(I) chloride, Hg_2Cl_2, containing 0.10 M Cl^{-}. The emf of the voltaic cell constructed by connecting this, as the anode, to the standard hydrogen half-cell as the cathode is 0.28 V. What is the solubility product of mercury(I) chloride?

20

Nuclear Chemistry

Chapter Outline

Uranium fuel rods in water.

Technetium is an unusual element. Although a *d*-transition element (under manganese in Group VIIB) with a small atomic number ($Z = 43$), it has no stable isotopes. The nucleus of every technetium isotope is radioactive and decays, or disintegrates, to give an isotope of another element. Many of the technetium isotopes decay by emitting an electron from the nucleus.

Because of its nuclear instability, technetium is not found naturally on earth. Nevertheless, it is produced commercially in kilogram quantities from other elements by nuclear reactions, processes in which nuclei are transformed into different nuclei. Technetium (from the Greek *tekhnetos,* meaning "artificial") was the first new element produced in the laboratory from another element. It was discovered in 1937 by Carlo Perrier and Emilio Segrè when the element molybdenum was bombarded with deuterons (nuclei of hydrogen, each having one proton and one neutron). Later, technetium was found to be a product of the fission, or splitting, of uranium nuclei. Thus, technetium is produced in nuclear fission reactors used to generate electricity.

Technetium is one of the principal isotopes used in medical diagnostics based on radioactivity. A compound of technetium is injected into a vein where it is attracted to certain body organs. The energy emitted by technetium nuclei is detected by special equipment and gives an image of these body organs. Figure 20.1 shows the image of a person's skeleton obtained from technetium that was administered in this manner. The technetium is eliminated by the body after several hours.

In this chapter, we will look at nuclear processes such as those that we have described for technetium. We will answer such questions as the following: How do we describe the radioactive decay of technetium? How do we describe the transformation of a molybdenum nucleus into technetium? How is technetium produced from uranium by nuclear fission, or splitting? What are some practical applications of nuclear processes?

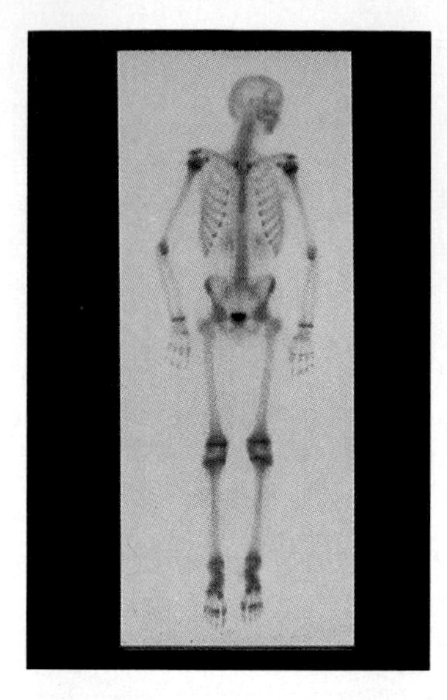

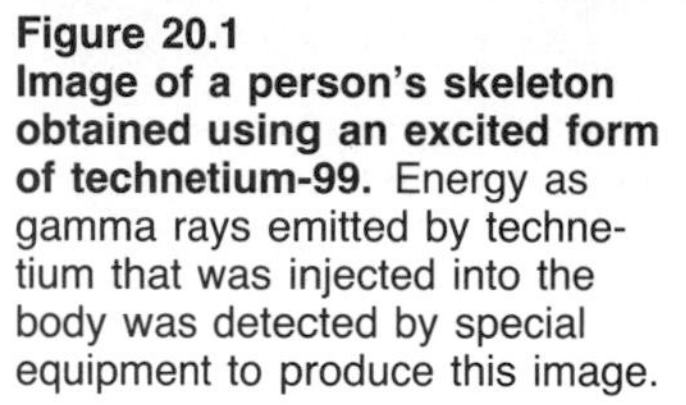

Figure 20.1
Image of a person's skeleton obtained using an excited form of technetium-99. Energy as gamma rays emitted by technetium that was injected into the body was detected by special equipment to produce this image.

Radioactivity and Nuclear Bombardment Reactions

In chemical reactions, only the outer electrons of the atoms are disturbed. The nuclei of the atoms are not affected. In nuclear reactions, however, nuclear changes occur. The compound in which the atoms are found is immaterial.

We will look at two types of nuclear reactions. One type is **radioactive decay,** *the process in which a nucleus spontaneously disintegrates, giving off radiation*. Natural radioactive elements emit, or give off, three kinds of radiation, called alpha, beta, and gamma radiation. *Alpha rays* consist of helium-4 nuclei (with two protons and two neutrons). *Beta rays* consist of electrons. *Gamma rays* are electromagnetic radiation, like light, with very short wavelengths (about 0.01 Å, or 1×10^{-12} m).

The second type of nuclear reaction is a **nuclear bombardment reaction,** *a nuclear reaction in which a nucleus is bombarded, or struck, by another nucleus or by a nuclear particle*. If there is sufficient energy in this collision, the nuclear particles of the reactants rearrange to give a product nucleus or nuclei. In the next section, we will look at radioactive decay.

20.1 RADIOACTIVITY

The phenomenon of radioactivity was discovered by Antoine Henri Becquerel in 1896. He discovered that photographic plates developed bright spots when exposed to uranium minerals, and he concluded that the mineral was giving off some sort of radiation. It was later shown that uranium nuclei emit alpha particles and thereby decay, or disintegrate, to thorium nuclei.

A sample of uranium-238 decays, or disintegrates, spontaneously over a period of billions of years. After about 30 billion years, the sample would be nearly gone. Strontium-90, formed by nuclear reactions that occur in nuclear weapons testing and nuclear power reactors, decays more rapidly. A sample of strontium-90 would be nearly gone after a couple of hundred years. In either case, it is impossible to know when a particular nucleus will decay, although as we will see in Section 20.4, precise information can be given about the rate of decay of any radioactive sample.

Nuclear Equations

We can write an equation for the nuclear reaction corresponding to the decay of uranium-238 much as we would write an equation for a chemical reaction. We represent the uranium-238 nucleus by the *nuclide symbol* ${}^{238}_{92}\text{U}$.● The radioactive decay of ${}^{238}_{92}\text{U}$ by alpha-particle emission (loss of a ${}^{4}_{2}\text{He}$ nucleus) is written

● Nuclide symbols were introduced in Section 7.3. For uranium-238, we have

$$\begin{array}{l} \text{mass number} \longrightarrow {}^{238} \\ \text{atomic number} \longrightarrow {}_{92}\text{U} \end{array}$$

$$^{238}_{92}\text{U} \longrightarrow {}^{234}_{90}\text{Th} + {}^{4}_{2}\text{He}$$

The product, in addition to helium-4, is thorium-234. This is an example of a **nuclear equation,** which is *a symbolic representation of a nuclear reaction*. Normally, only the nuclei are represented. It is not necessary to indicate the chemical compound or the electron charges for any ions involved, because the chemical environment has no effect on nuclear processes.

Reactant and product nuclei are represented in nuclear equations by their nuclide symbols. Other particles are given the following symbols, in which the subscript equals the charge and the superscript equals the total number of protons and neutrons in the particle (mass number).

Proton	${}^{1}_{1}H$	or	${}^{1}_{1}p$
Neutron	${}^{1}_{0}n$		
Electron	${}^{0}_{-1}e$	or	${}^{0}_{-1}\beta$
Positron	${}^{0}_{1}e$	or	${}^{0}_{1}\beta$
Gamma photon	${}^{0}_{0}\gamma$		

The decay of a nucleus with the emission of an electron, ${}^{0}_{-1}e$, is usually called beta emission, and the emitted electron is sometimes labeled ${}^{0}_{-1}\beta$. A **positron** is *a particle similar to an electron, with the same mass but a positive charge.* A **gamma photon** is *a particle of electromagnetic radiation of short wavelength (about 0.01 Å, or 10^{-12} m) and high energy.*

Example 20.1 Writing a Nuclear Equation

Write the nuclear equation for the radioactive decay of radium-226 by alpha decay to give radon-222. A radium-226 nucleus emits one alpha particle, leaving behind a radon-222 nucleus.

Solution

Looking at a list of elements (inside back cover), we find that the atomic number of radium is 88. Hence, the nuclide symbol is ${}^{226}_{88}Ra$. Similarly, the nuclide symbols for radon-222 and the alpha particle are ${}^{222}_{86}Rn$ and ${}^{4}_{2}He$. Therefore, the equation is

$$ {}^{226}_{88}Ra \longrightarrow {}^{222}_{86}Rn + {}^{4}_{2}He $$

Exercise 20.1

Potassium-40 is a naturally occurring radioactive isotope. It decays to calcium-40 by beta emission. When a potassium-40 nucleus decays by beta emission, it emits one beta particle and gives a calcium-40 nucleus. Write the nuclear equation for this decay.

(See Problems 20.19, 20.20, 20.21, and 20.22.)

The total charge is conserved, or remains constant, during a nuclear reaction. This means that the sum of the subscripts (number of protons, or positive charges, in the nuclei) for the products must equal the sum of the subscripts for the reactants. For the equation in Example 20.1, the subscript for the reactant ${}^{226}_{88}Ra$ is 88. For the products, the sum of the subscripts is $86 + 2 = 88$.

Similarly, the total number of *nucleons* (protons and neutrons) is conserved, or remains constant, during a nuclear reaction. This means that the sum of the superscripts (the mass numbers) for the reactants equals the sum of the superscripts for the products. For the equation in Example 20.1, the superscript for the reactant nucleus is 226. For the products, the sum of the superscripts is $222 + 4 = 226$.

Note that if all reactants and products but one are known in a nuclear equation, the identity of that one nucleus or particle can be easily obtained. This is illustrated in the next example.

Example 20.2 Deducing a Product or Reactant in a Nuclear Equation

Technetium-99 is a long-lived radioactive isotope of technetium. Each nucleus decays by emitting one beta particle. What is the product nucleus?

Solution

Let us write the nuclear equation for the decay of technetium-99. Looking at a list of elements, we see that technetium has atomic number 43. Thus, the nuclide symbol is ${}^{99}_{43}\text{Tc}$. A beta particle is an electron, whose symbol is ${}^{0}_{-1}\text{e}$. For the unknown product nucleus, we write ${}^{A}_{Z}\text{X}$, where A is the mass number and Z is the atomic number. The nuclear equation is

$${}^{99}_{43}\text{Tc} \longrightarrow {}^{A}_{Z}\text{X} + {}^{0}_{-1}\text{e}$$

From the superscripts, we can write

$$99 = A + 0, \text{ or } A = 99$$

Similarly, from the subscripts, we get

$$43 = Z - 1, \text{ or } Z = 43 + 1 = 44$$

Hence $A = 99$ and $Z = 44$, so the product is ${}^{99}_{44}\text{X}$. Because element 44 is ruthenium, symbol Ru, we write the product nucleus as ${}^{99}_{44}\textbf{Ru}$.

Exercise 20.2

Plutonium-239 decays by alpha emission, with each nucleus emitting one alpha particle. What is the other product of this decay? (See Problems 20.23, 20.24, 20.25, and 20.26.)

Nuclear Stability

At first glance, the existence of several protons in the small space of a nucleus is puzzling. Why wouldn't the protons be strongly repelled by their like electric charges? The existence of stable nuclei with more than one proton is due to the nuclear force. The **nuclear force** is *a strong force of attraction between nucleons that acts only at very short distances (about 10^{-15} m)*. Beyond nuclear distances, these nuclear forces become negligible. Therefore, two protons that are much farther apart than 10^{-15} m repel one another by their like electric charges. Inside the nucleus, however, two protons are close enough together for the nuclear force between them to be effective. This force in a nucleus can more than compensate for the repulsion of electric charges and thereby give a stable nucleus.

The protons and neutrons in a nucleus appear to have energy levels much as the electrons in an atom have energy levels. The **shell model of the nucleus** is *a nuclear model in which protons and neutrons exist in levels, or shells, analogous to the shell structure that exists for electrons in an atom*. Recall that, in an atom, filled shells of electrons are associated with the special stability of the noble gases. The total numbers of electrons for these stable atoms are 2 (for He), 10 (for Ne), 18 (for Ar), and so forth. Experimentally, it is noted that nuclei with certain numbers of protons or neutrons appear to be very stable. These numbers, called *magic numbers* and associated with specially stable nuclei, were later explained by the shell model. According to this theory, a **magic number** is *the number of nuclear particles in a completed shell of protons or neutrons*. Because nuclear forces differ from electrical forces, these numbers are not the same as those for electrons in atoms. For protons, the magic numbers are 2, 8, 20, 28, 50, and 82. Neutrons have these same magic numbers, as well as the magic number 126. For protons, calculations show that 114 should also be a magic number.

Some of the evidence for these magic numbers, and therefore for the shell model, is as follows. Many radioactive nuclei decay by emitting alpha particles, or

Table 20.1 Number of Stable Isotopes with Even and Odd Numbers of Protons and Neutrons

	Number of Stable Isotopes			
	157	52	50	5
Number of protons	Even	Even	Odd	Odd
Number of neutrons	Even	Odd	Even	Odd

$^{4}_{2}He$ nuclei. There appears to be special stability in the $^{4}_{2}He$ nucleus. It contains 2 protons and 2 neutrons; that is, it contains a magic number of protons (2) and a magic number of neutrons (also 2).

Another piece of evidence is seen in the final products obtained in natural radioactive decay. For example, uranium-238 decays to thorium-234, which in turn decays to protactinium-234, and so forth. Each product is radioactive and decays to another nucleus until the final product, $^{206}_{82}Pb$, is reached. This nucleus is stable. Note that it contains 82 protons, which is a magic number. Other radioactive decay series end at $^{207}_{82}Pb$ and $^{208}_{82}Pb$, both of which have a magic number of protons. Note that $^{208}_{82}Pb$ also has a magic number of neutrons ($208 - 82 = 126$).

Evidence also points to the special stability of pairs of protons and pairs of neutrons, analogous to the stability of pairs of electrons in molecules. Table 20.1 lists the number of stable isotopes that have an even number of protons and an even number of neutrons (157). By comparison, there are only 5 stable isotopes having an odd number of protons and an odd number of neutrons.

Finally, when we plot each stable nuclide on a graph with the number of protons (Z) on the horizontal axis and the number of neutrons (N) on the vertical

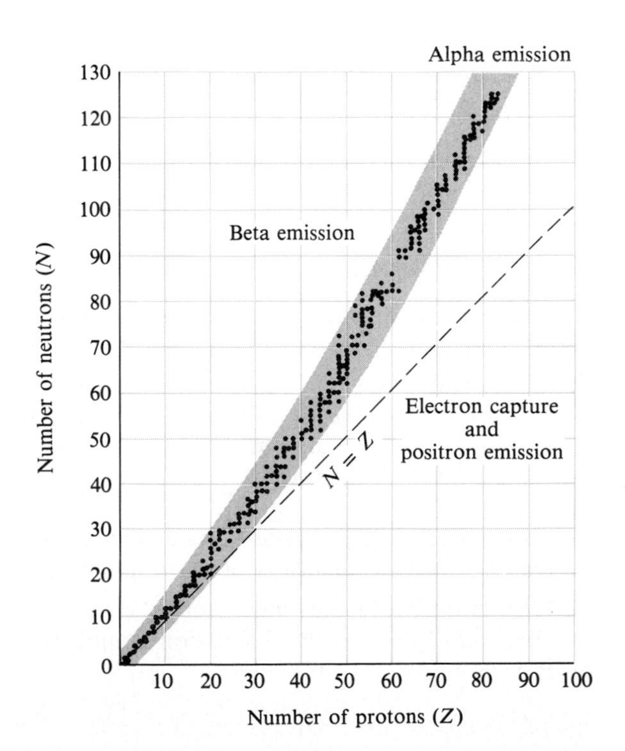

Figure 20.2
Band of stability. The stable nuclides, indicated by black dots, cluster in a band. Nuclides to the left of the band of stability usually decay by beta emission, whereas those to the right usually decay by positron emission or electron capture. Nuclides of $Z > 83$ often decay by alpha emission.

axis, these nuclides fall in an area, or band. The **band of stability** is *the region in which stable nuclides lie in a plot of number of protons against number of neutrons*. Figure 20.2 (p. 795) shows the band of stability; the rest of the figure is explained later in this section. For nuclides up to $Z = 20$, the ratio of neutrons to protons is about 1.0 to 1.1. As Z increases, however, the neutron-to-proton ratio increases to about 1.5. This increase in neutron-to-proton ratio with Z is believed to result from the increasing repulsions of protons from their electric charges. More neutrons are required to give attractive nuclear forces to offset these repulsions.

It appears that when the number of protons becomes very large, the proton–proton repulsions become so great that stable nuclides are impossible. Thus, no stable nuclides are known with atomic numbers greater than 83. On the other hand, all elements with Z equal to 83 or less have one or more stable nuclides, with the exception of technetium ($Z = 43$), as we noted in the chapter opening, and promethium ($Z = 61$).

Example 20.3 Predicting the Relative Stabilities of Nuclides

One of the nuclides in each of the following pairs is radioactive; the other is stable. Which is radioactive and which is stable? Explain.

(a) $^{208}_{84}Po$, $^{209}_{83}Bi$ (b) $^{39}_{19}K$, $^{40}_{19}K$ (c) $^{71}_{31}Ga$, $^{76}_{31}Ga$

Solution

Note that the problem states that one nucleus is radioactive and the other stable. We must decide which is more likely to be radioactive or which is more likely to be stable. (a) Polonium has an atomic number greater than 83, so **$^{208}_{84}Po$ is radioactive.** Bismuth-209 has 126 neutrons (a magic number), so **$^{209}_{83}Bi$ is expected to be stable.** (b) Of these two isotopes, $^{39}_{19}K$ has a magic number of neutrons (20), so **$^{39}_{19}K$ is expected to be stable.** The isotope $^{40}_{19}K$ has an odd number of protrons (19) and an odd number of neutrons (21). Because stable odd–odd nuclei are rare, we might expect **$^{40}_{19}K$ to be radioactive.** (c) Of the two isotopes, $^{76}_{31}Ga$ lies farther from the center of the band of stability, so it is more likely to be radioactive. For this reason, **we expect $^{76}_{31}Ga$ to be radioactive and $^{71}_{31}Ga$ to be stable.**

Exercise 20.3

Of the following nuclides, two are radioactive. Which are radioactive and which is stable? Explain. (a) $^{118}_{50}Sn$; (b) $^{76}_{33}As$; (c) $^{227}_{89}Ac$. (See Problems 20.27 and 20.28.)

Types of Radioactive Decay

There are five common types of radioactive decay. These are listed in Table 20.2 and discussed here.

1. **Alpha emission** (abbreviated α): *emission of a $^{4}_{2}He$ nucleus, or alpha particle, from an unstable nucleus*. An example is the radioactive decay of radium-226, written as

$$^{226}_{88}Ra \longrightarrow ^{222}_{86}Rn + ^{4}_{2}He$$

The product nucleus has an atomic number that is two less, and a mass number that is four less, than that of the original nucleus.

2. **Beta emission** (abbreviated β or β^-): *emission of a high-speed electron from an unstable nucleus*. Beta emission is equivalent to the conversion of a neutron to a proton.●

$$^{1}_{0}n \longrightarrow ^{1}_{1}p + ^{0}_{-1}e$$

● Outside the nucleus, the neutron is unstable and decays in a few minutes to a proton and an electron.

Table 20.2 Types of Radioactive Decay

Type of Decay	Radiation	Equivalent Process	Nuclear Change: Atomic Number	Nuclear Change: Mass Number	Usual Nuclear Condition
Alpha emission (α)	${}^{4}_{2}He$	—	-2	-4	$Z > 83$
Beta emission (β)	${}^{0}_{-1}e$	${}^{1}_{0}n \longrightarrow {}^{1}_{1}p + {}^{0}_{-1}e$	$+1$	0	N/Z too large
Positron emission (β^+)	${}^{0}_{1}e$	${}^{1}_{1}p \longrightarrow {}^{1}_{0}n + {}^{0}_{1}e$	-1	0	N/Z too small
Electron capture (EC)	x rays	${}^{1}_{1}p + {}^{0}_{-1}e \longrightarrow {}^{1}_{0}n$	-1	0	N/Z too small
Gamma emission (γ)	${}^{0}_{0}\gamma$	—	0	0	Excited nucleus

An example of beta emission is the radioactive decay of carbon-14.

$${}^{14}_{6}C \longrightarrow {}^{14}_{7}N + {}^{0}_{-1}e$$

The product nucleus has an atomic number that is one more than that of the original nucleus. The mass number remains the same.

3. **Positron emission** (abbreviated β^+): *emission of a positron from an unstable nucleus*. A positron, denoted in nuclear equations as ${}^{0}_{1}e$, is a particle identical to an electron in mass but having a positive charge instead of a negative one. Positron emission is equivalent to the conversion of a proton to a neutron.●

● Positrons are annihilated as soon as they encounter electrons. When a positron and an electron collide, both particles vanish with the emission of two gamma photons that carry away the energy.

$${}^{0}_{1}e + {}^{0}_{-1}e \longrightarrow 2{}^{0}_{0}\gamma$$

$${}^{1}_{1}p \longrightarrow {}^{1}_{0}n + {}^{0}_{1}e$$

An example of positron emission is

$${}^{95}_{43}Tc \longrightarrow {}^{95}_{42}Mo + {}^{0}_{1}e$$

The product nucleus has an atomic number that is one less than that of the original nucleus. The mass number remains the same.

4. **Electron capture** (abbreviated EC): *the decay of an unstable nucleus by capturing, or picking up, an electron from an inner orbital of an atom*. In effect, a proton is changed to a neutron, as in positron emission.

$${}^{1}_{1}p + {}^{0}_{-1}e \longrightarrow {}^{1}_{0}n$$

An example is given by potassium-40, which has a natural abundance of 0.012%.● It can decay by electron capture, as well as by beta and positron emissions. The equation for electron capture in potassium-40 is

● Most of the argon in the atmosphere is believed to have resulted from the radioactive decay of ${}^{40}_{19}K$.

$${}^{40}_{19}K + {}^{0}_{-1}e \longrightarrow {}^{40}_{18}Ar$$

The product nucleus has an atomic number that is one less than that of the original nucleus. The mass number remains the same. When another orbital electron fills the vacancy in the inner-shell orbital created by electron capture, an x-ray photon is emitted.

5. **Gamma emission** (abbreviated γ): *emission from an excited nucleus of a gamma photon, corresponding to radiation with a wavelength of about 10^{-12} m*. In many cases, radioactive decay results in a product nucleus that is in an excited state. As in the case of atoms, the excited state is unstable and goes to a lower-energy state with the emission of electromagnetic radiation. For nuclei, this radiation is in the gamma-ray region of the spectrum.

   Often gamma emission occurs very quickly after radioactive decay. In some cases, however, an excited state has significant lifetime before it emits a gamma photon. A **metastable nucleus** is *a nucleus in an excited state with a lifetime of at least one nanosecond (10^{-9} s).* In time, the metastable nucleus decays by gamma emission. An example is metastable technetium-99, denoted ${}^{99m}_{43}Tc$, which is used in medical diagnosis, as discussed in Section 20.5.

$$ {}^{99m}_{43}Tc \longrightarrow {}^{99}_{43}Tc + {}^{0}_{0}\gamma $$

The product nucleus is simply a lower-energy state of the original nucleus, so there is no change of atomic number or mass number.

Nuclei outside the band of stability are generally radioactive. Those to the left of the band of stability have a neutron-to-proton ratio (N/Z) larger than that needed for stability. This ratio is reduced by beta emission, because in this process a neutron is changed to a proton. Nuclei to the right of the band of stability have an N/Z ratio that is smaller than that needed for stability. These nuclei tend to decay by either positron emission or electron capture. Both processes convert a proton to a neutron.

As examples of radioactive isotopes, consider carbon-14 and phosphorus-30. Looking at Figure 20.2, we see that carbon-14 lies to the left of the band of stability. Thus, we expect it to decay by beta emission, which is what is observed. On the other hand, phosphorus-30 lies to the right of the band of stability. We expect it to decay by either positron emission or electron capture. Positron emission is observed.

Heavier nuclei, especially those with Z greater than 83, often decay by alpha emission. Uranium-238 decays by alpha emission, as we noted earlier. The isotopes ${}^{226}_{88}Ra$ and ${}^{232}_{90}Th$ are other examples of alpha emitters.

Example 20.4 Predicting the Type of Radioactive Decay

Using Figure 20.2, predict the possible type of radioactive decay for each of the following radioactive nuclides: (a) ${}^{47}_{20}Ca$; (b) ${}^{25}_{13}Al$.

Solution

(a) The nucleus of calcium-47 has 20 protons and $47 - 20 = 27$ neutrons. This places the nuclide to the left of the band of stability. Thus, it is expected to decay by **beta emission,** which is observed. (b) Aluminum-25 (${}^{25}_{13}Al$) has 13 protons and 12 neutrons, placing it to the right of the band of stability. We expect it to decay either by **positron emission or by electron capture.** Positron emission is actually observed.

Exercise 20.4

Predict the type of decay expected for each of the following radioactive nuclides: (a) ${}^{13}_{7}N$; (b) ${}^{26}_{11}Na$. Refer to Figure 20.2. (See Problems 20.29 and 20.30.)

Radioactive Decay Series

All nuclides with atomic number greater than $Z = 83$ are radioactive, as we have noted. Many of these nuclides decay by alpha emission. Alpha particles, or ${}^{4}_{2}He$ nuclei, are especially stable and are formed in the radioactive nucleus at the moment of decay. By emitting an alpha particle, the nucleus reduces its atomic number, becoming more stable. However, if the nucleus has a very large Z, the product nucleus is also radioactive. Natural radioactive elements, such as uranium-238,

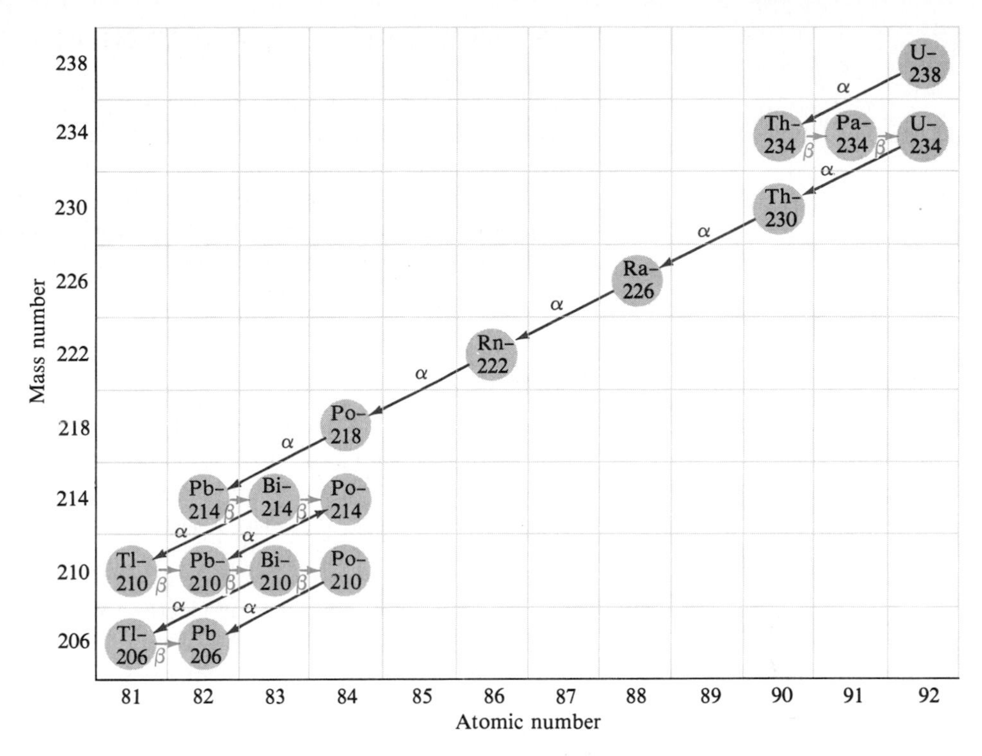

Figure 20.3
Uranium-238 radioactive decay series. Each nuclide occupies a position on the graph determined by its atomic number and mass number. Alpha decay is shown by a diagonal line. Beta decay is shown by a short horizontal line.

give a **radioactive decay series,** *a sequence in which one radioactive nucleus decays to a second, which then decays to a third, and so forth*. Eventually, a stable nucleus is reached. For the natural radioactive decay series, this stable nucleus is an isotope of lead.

There are three radioactive decay series found naturally. One of these begins with uranium-238. Figure 20.3 shows the sequence of nuclear decay processes. In the first step, uranium-238 decays by alpha emission to thorium-234.

$$^{238}_{92}U \longrightarrow ^{234}_{90}Th + ^{4}_{2}He$$

Note that this step is represented in Figure 20.3 by a long diagonal line labeled α. Each alpha decay reduces the atomic number by 2 and the mass number by 4. The atomic number and mass number of each nuclide are given by its horizontal and vertical position on the graph. Thorium-234 in turn decays by beta emission to protactinium-234, which decays by beta emission to uranium-234.

$$^{234}_{90}Th \longrightarrow ^{234}_{91}Pa + ^{0}_{-1}e$$
$$^{234}_{91}Pa \longrightarrow ^{234}_{92}U + ^{0}_{-1}e$$

Note that a beta emission is represented in Figure 20.3 by a short horizontal line. Each beta decay increases the atomic number by one but has no effect on the mass number. After the decay of protactinium-234 and the formation of uranium-234, there are a number of alpha-decay steps. The final product of the series is lead-206.

Natural uranium is 99.28% $^{238}_{92}U$, which decays as we have described. However, the natural element also contains 0.72% $^{235}_{92}U$. This isotope starts a second radioactive decay series, which consists of a sequence of alpha and beta decays, ending with lead-207. The third naturally occurring radioactive decay series begins with thorium-232 and ends with lead-208.●

● Thorium is thought to be three times more abundant than uranium and may become a major source of nuclear power (see Section 20.7).

20.2 NUCLEAR BOMBARDMENT REACTIONS

The nuclear reactions discussed in the previous section are radioactive decay reactions, in which a nucleus spontaneously decays to another nucleus and emits a particle, such as an alpha or beta particle. In 1919, Ernest Rutherford discovered that it was possible to change the nucleus of one element into the nucleus of another by processes that could be controlled in the laboratory. **Transmutation** is *the change of one element to another by bombarding the nucleus of the element with nuclear particles or nuclei.*

Transmutation

In his experiments, Rutherford used a radioactive element as a source of alpha particles and allowed these particles to collide with nitrogen nuclei. He discovered that protons were ejected in the process. The equation for the nuclear reaction is

$$^{14}_{7}N + ^{4}_{2}He \longrightarrow ^{17}_{8}O + ^{1}_{1}H$$

These experiments were repeated on other light nuclei, most of which were transmuted to other elements with the ejection of a proton. These experiments yielded two significant results. First, they strengthened the view that all nuclei contain protons. Second, they showed for the first time that it was possible to change one element into another under laboratory control.

When beryllium was bombarded with alpha particles, a penetrating radiation was given off that was not deflected by electric or magnetic fields. Therefore, the radiation did not consist of charged particles. The British physicist James Chadwick (1891–1974) suggested in 1932 that the radiation from beryllium consisted of neutral particles, each with a mass approximately that of a proton. The particles were called neutrons. The reaction that resulted in the discovery of the neutron is

$$^{9}_{4}Be + ^{4}_{2}He \longrightarrow ^{12}_{6}C + ^{1}_{0}n$$

In 1933, a nuclear bombardment reaction was used to produce the first artificial radioactive isotope. Irène and Frédéric Joliot-Curie found that aluminum bombarded with alpha particles produced phosphorus-30, which decayed by emitting positrons. The reactions are

$$^{27}_{13}Al + ^{4}_{2}He \longrightarrow ^{30}_{15}P + ^{1}_{0}n$$

$$^{30}_{15}P \longrightarrow ^{30}_{14}Si + ^{0}_{1}e$$

Phosphorus-30 was the first radioactive nucleus produced in the laboratory. Since then over a thousand radioactive isotopes have been made.●

● Some of the uses of radioactive isotopes are discussed in Section 20.5.

Nuclear bombardment reactions are often referred to by an abbreviated notation. For example, the reaction

$$^{14}_{7}N + ^{4}_{2}He \longrightarrow ^{17}_{8}O + ^{1}_{1}H$$

is abbreviated $^{14}_{7}N(\alpha, p)^{17}_{8}O$. In this notation, one first writes the nuclide symbol for the original nucleus (target), then in parentheses the symbol for the projectile particle (incoming particle), followed by a comma and the symbol for the ejected particle. After the last parenthesis, the nuclide symbol for the product nucleus is given. The following symbols are used for particles:

Neutron	n
Proton	p
Deuteron, $^{2}_{1}H$	d
Alpha, $^{4}_{2}He$	α

Example 20.5 Using the Notation for a Bombardment Reaction

(a) Write the abbreviated notation for the following bombardment reaction, in which neutrons were first discovered.

$$^{9}_{4}Be + ^{4}_{2}He \longrightarrow ^{12}_{6}C + ^{1}_{0}n$$

(b) Write the nuclear equation for the bombardment reaction denoted $^{27}_{13}Al(p, d)^{26}_{13}Al$.

Solution

(a) The notation is $^{9}_{4}Be(\alpha, n)^{12}_{6}C$. (b) The nuclear equation is

$$^{27}_{13}Al + ^{1}_{1}H \longrightarrow ^{26}_{13}Al + ^{2}_{1}H$$

Exercise 20.5

(a) Write the abbreviated notation for the reaction

$$^{40}_{20}Ca + ^{2}_{1}H \longrightarrow ^{41}_{20}Ca + ^{1}_{1}H$$

(b) Write the nuclear equation for the bombardment reaction $^{12}_{6}C(d, p)^{13}_{6}C$.

(See Problems 20.33, 20.34, 20.35, and 20.36.)

Elements of large atomic number merely scatter, or deflect, alpha particles from natural sources, rather than giving a transmutation reaction. These elements have nuclei of large positive charge, and the alpha particle must be traveling very fast in order to penetrate the nucleus and react. Alpha particles from natural sources do not have sufficient kinetic energy. In order to shoot charged particles into heavy nuclei, it is necessary to accelerate the charged particles.

A **particle accelerator** is *a device used to accelerate electrons, protons, and alpha particles and other ions to very high speeds*. The basis of a particle accelerator is the fact that a charged particle will accelerate toward a plate having a charge opposite in sign to that of the particle. It is customary to measure the kinetic energies of these particles in units of electron volts. An **electron volt** (eV) is *the quantity of energy that would need to be imparted to an electron (whose charge is 1.602×10^{-19} C) to accelerate it by one volt potential difference.*

$$1\ \text{eV} = (1.602 \times 10^{-19}\ \text{C}) \times (1\ \text{V}) = 1.602 \times 10^{-19}\ \text{J}$$

Typically, particle accelerators give charged particles energies of millions of electron volts (MeV). In order to keep the accelerated particles from colliding with molecules of gas, the apparatus is enclosed and evacuated to low pressures, about 10^{-6} mmHg or less.

A charged particle can be accelerated in stages to very high kinetic energy. Figure 20.4 shows a diagram of a **cyclotron,** *a type of particle accelerator consisting of two hollow, semicircular metal electrodes, called dees (because the shape of a dee resembles the letter D) in which charged particles are accelerated by stages to higher and higher kinetic energies.* Ions introduced at the center of the cyclotron are accelerated in the space between the two dees. Magnet poles (not shown in the figure) above and below the dees keep the ions moving in an enlarging spiral path. The dees are connected to a high-frequency electric current that changes their polarity so that each time the ion moves into the space between the dees, it is accelerated. Thus, the ion is continually accelerated until it finally leaves the cyclotron at high speed. Outside the cyclotron, the ions are directed toward a target element in order that investigators may study nuclear reactions or prepare isotopes.

Technetium, whose discovery was mentioned in the chapter opening, was first prepared by directing **deuterons,** or *nuclei of hydrogen-2 atoms,* from a

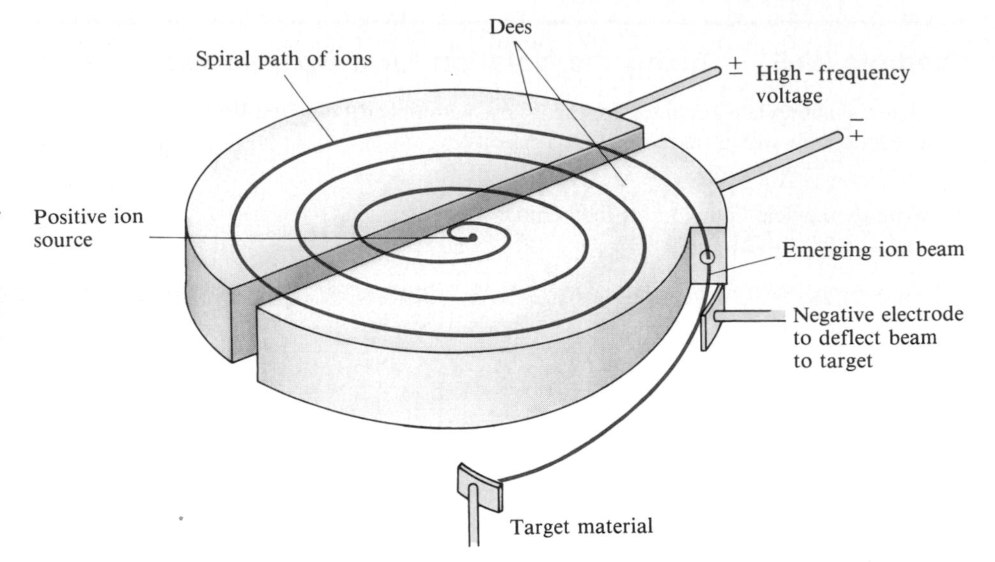

Figure 20.4
A cyclotron. Positive ions are introduced at the center of the cyclotron. A fraction of the ions will be crossing the gap between the dees when the electric polarities are just right to accelerate them (the positive ions are accelerated toward the negative dee and away from the positive dee). The dees alternate in polarities, so that these ions continue to be accelerated each time they pass between the gap in the dees. Magnet poles above and below the dees produce a magnetic field that keeps the ions moving in a spiral path within the dees. At the end of their path, the ions encounter a negative electrode that deflects them to a target material.

● The hydrogen-2 atom is often called deuterium and given the symbol D. It is a stable isotope with a natural abundance of 0.015%. Deuterium was discovered in 1931 by Harold Urey and coworkers and was prepared in pure form by G. N. Lewis.

cyclotron at a molybdenum target.● The nuclear reaction is ${}^{96}_{42}Mo(d, n){}^{97}_{43}Tc$, or

$$ {}^{96}_{42}Mo + {}^{2}_{1}H \longrightarrow {}^{97}_{43}Tc + {}^{1}_{0}n $$

The energies of particles obtained from a cyclotron are limited (to about 20 MeV for deuterons, for example). For higher energies, more sophisticated accelerators are required. Figure 20.5 shows the accelerator at Fermilab in Batavia, Illinois. Larger and larger particle accelerators are being built to study the basic constituents of matter. The proposed Superconducting Supercollider (SSC) would generate particle energies of about 40,000 GeV (1 GeV = 10^9 eV).

Transuranium Elements

The **transuranium elements** are *those elements with atomic numbers greater than that of uranium (Z = 92), the naturally occurring element of greatest* Z. In 1940, E. M. McMillan and P. H. Abelson, at the University of California at Berkeley, discovered the first transuranium element. They produced an isotope of element

Figure 20.5
The Fermilab accelerator in Batavia, Illinois. The main-ring accelerator. Protons are accelerated in the upper ring of red and blue magnets to 400 billion electron volts. In the lower ring of yellow and red superconducting magnets, protons are accelerated to almost a trillion electron volts.

93, which they named neptunium, by bombarding uranium-238 with neutrons. This gave uranium-239, by the capture of a neutron, and this nucleus decayed in a few days by beta emission to neptunium-239.

$$^{238}_{92}\text{U} + ^{1}_{0}\text{n} \longrightarrow ^{239}_{92}\text{U}$$
$$^{239}_{92}\text{U} \longrightarrow ^{239}_{93}\text{Np} + ^{0}_{-1}\text{e}$$

The next transuranium element to be discovered was plutonium ($Z = 94$). Deuterons, the positively charged nuclei of hydrogen-2, were accelerated by a cyclotron and directed at a uranium target to give neptunium-238, which decayed to plutonium-238.

$$^{238}_{92}\text{U} + ^{2}_{1}\text{H} \longrightarrow ^{238}_{93}\text{Np} + 2^{1}_{0}\text{n}$$
$$^{238}_{93}\text{Np} \longrightarrow ^{238}_{94}\text{Pu} + ^{0}_{-1}\text{e}$$

Another isotope of plutonium, plutonium-239, is now produced in large quantity in nuclear reactors, as described in Section 20.7. Plutonium-239 is used for nuclear weapons.

The discovery of the next two transuranium elements, americium ($Z = 95$) and curium ($Z = 96$), depended on an understanding of the correct positions in the periodic table of the elements beyond actinium ($Z = 89$). It had been thought that these elements should be placed after actinium under the *d*-transition elements. Thus, uranium was placed in Group VIIB under tungsten. However, Glenn T. Seaborg at the University of California, Berkeley, postulated a second series of elements to be placed at the bottom of the periodic table, under the lanthanides, as shown in modern tables (see inside front cover). These elements, the actinides, would be expected to have chemical properties similar to those of the lanthanides.

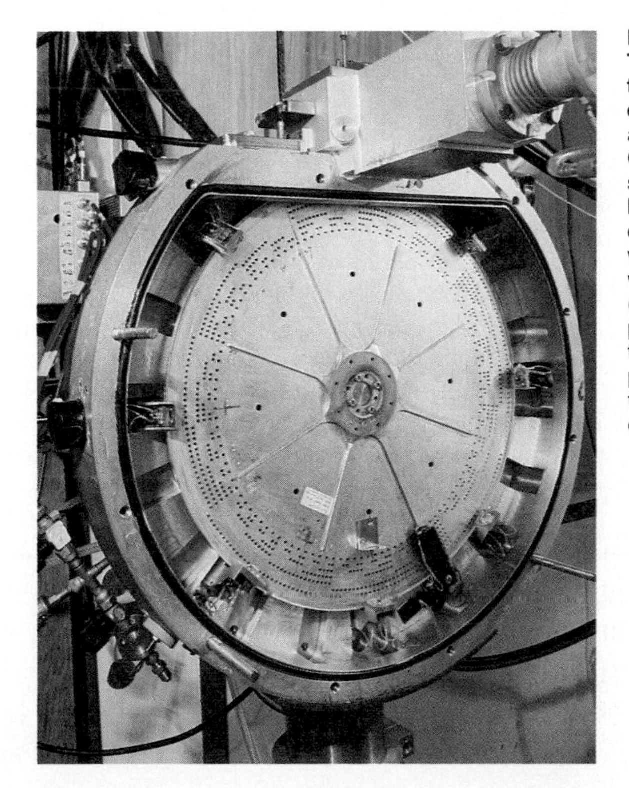

Figure 20.6
The vertical wheel, an apparatus used in the discovery of elements 104, 105, and 106. The apparatus was designed by Albert Ghiorso and colleagues. They synthesized element 106 by bombarding californium-249 with oxygen-18. The products were whisked by gas jets to the vertical wheel, where they were deposited (at the top). Detectors around the periphery of the wheel monitored the alpha activity to identify the products. Binary-coded holes label the positions around the wheel for computer collection of data.

Figure 20.7 The High Flux Isotope Reactor at Oak Ridge National Laboratory. The reactor in which transuranium isotopes are produced lies beneath a protecting pool of water. Visible light is emitted by high-energy particles moving through the water.

Once they understood this, Seaborg and others soon identified curium and americium. Elements up to $Z = 109$ have been reported. (Figure 20.6, p. 803, shows the apparatus used in the discovery of elements 104, 105, and 106.)●

● A West German group reported the detection of one atom of element 109 (with a half-life of about 0.005 s) in 1982. More recently, Soviet physicists claimed the creation of element 110, though this is still unconfirmed. Physicists hope to test the shell model by trying to make elements with atomic numbers near 114 (a magic number of protons) and approximately 184 neutrons (also a magic number). If the theory is correct, such elements should have special stability.

The transuranium elements have a number of commercial uses. Plutonium-238 emits only alpha radiation, which is easily stopped by shielding. The isotope has been used as a power source for space satellites, navigation buoys, and heart pacemakers. Americium-241 is both an alpha-ray and a gamma-ray emitter. The gamma rays are used in devices that measure the thickness of materials such as metal sheets. Americium-241 is also used in home smoke detectors, in which the alpha radiation ionizes the air in a chamber within the detector and renders it electrically conducting. Smoke reduces the conductivity of the air, and this reduced conductivity is detected by an alarm circuit.

Figure 20.7 shows the High Flux Isotope Reactor at Oak Ridge National Laboratory, where technicians produce transuranium elements by starting with plutonium-239 and bombarding it with neutrons. Neutrons are produced by nuclear fission, a process discussed in Section 20.7.

Example 20.6 Determining the Product Nucleus in a Nuclear Bombardment Reaction

Plutonium-239 was bombarded by alpha particles. Each $^{239}_{94}Pu$ nucleus was struck by one alpha particle and emitted one neutron. What was the product nucleus?

Solution

We can write the nuclear equation as follows:

$$^{239}_{94}Pu + ^{4}_{2}He \longrightarrow ^{A}_{Z}X + ^{1}_{0}n$$

To balance this equation in charge (subscripts) and mass number (superscripts), we write the equations

$$239 + 4 = A + 1 \quad \text{(from superscripts)}$$
$$94 + 2 = Z + 0 \quad \text{(from subscripts)}$$

Hence,

$$A = 239 + 4 - 1 = 242$$
$$Z = 94 + 2 = 96$$

The product is $^{242}_{96}Cm$.

Exercise 20.6

Carbon-14 is produced in the upper atmosphere when a particular nucleus is bombarded with neutrons. A proton is ejected for each nucleus that reacts. What is the identity of the nucleus that produces carbon-14 by this reaction?

(See Problems 20.39, 20.40, 20.41, and 20.42.)

20.3 RADIATIONS AND MATTER: DETECTION AND BIOLOGICAL EFFECTS

Radiations from nuclear processes affect matter in part by dissipating energy in it. An alpha, beta, or gamma particle traveling through matter dissipates energy by ionizing atoms or molecules, producing positive ions and electrons. In some cases, these radiations may also excite electrons in matter. When these electrons undergo transitions back to their ground states, light is emitted. The ions, free electrons, and light produced in matter can be used to detect nuclear radiations. Because nuclear radiations can ionize molecules and break chemical bonds, they adversely affect biological organisms. We will first look at the detection of nuclear radiations and then briefly discuss biological effects and radiation dosage in humans.

Radiation Counters

Two types of devices—*ionization counters* and *scintillation counters*—are used to count particles emitted from radioactive nuclei and other nuclear processes. Ionization counters depend on the production of ions in matter. Scintillation counters detect the production of scintillations, or flashes of light.

A **Geiger counter** (Figure 20.8), *a kind of ionization counter used to count particles emitted by radioactive nuclei, consists of a metal tube filled with gas, such as argon*. The tube is fitted with a thin glass or plastic window through which radiation enters. A wire runs down the center of the tube and is insulated from the tube. The tube and wire are connected to a high-voltage source so that the tube becomes the negative electrode and the wire the positive electrode. Normally the gas in the tube is an insulator and no current flows through it. However, when radiation, such as an alpha particle, $^{4}_{2}He^{2+}$, passes through the window of the tube and into the gas, atoms are ionized. Free electrons are quickly accelerated to the wire. As they are accelerated to the wire, additional atoms may be ionized from collisions with these electrons and more electrons set free. An avalanche of electrons is created, and this gives a pulse of current that is detected by electronic equipment. The amplified pulse activates a digital counter or gives an audible "click."

Alpha and beta particles can be detected directly by a Geiger counter.● To detect neutrons, boron trifluoride is added to the gas in the tube. Neutrons react

● Gamma and low-energy alpha particles are better detected by scintillation counters. These counters are filled with solids or liquids, which are more likely to stop gamma rays. Also, low-energy alpha particles may be absorbed by the window of the Geiger counter and go undetected.

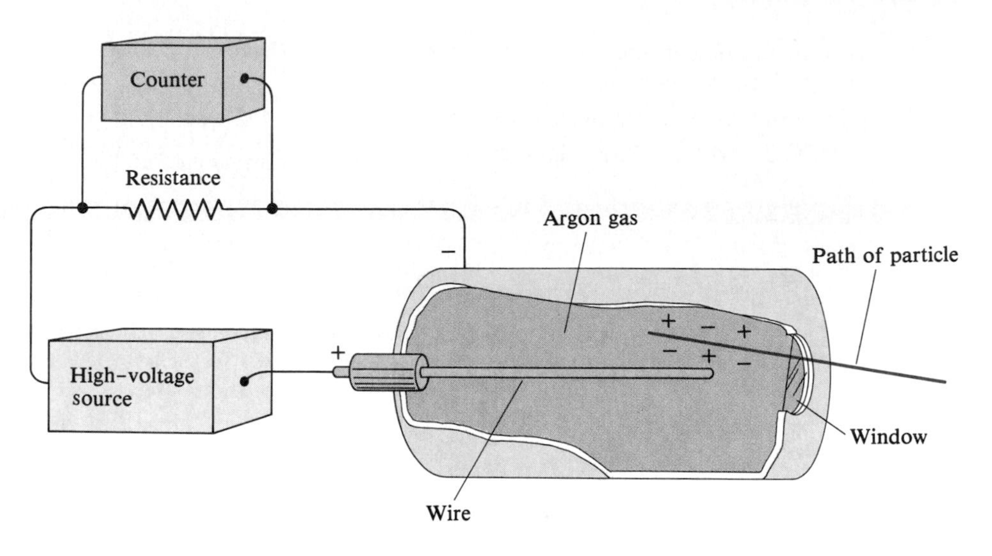

Figure 20.8
A Geiger counter. A particle of radiation enters the thin window and passes into the gas. Energy from the particle ionizes gas molecules, giving positive ions and electrons, which are accelerated to the electrodes. The electrons, which move faster, strike the wire anode and create a pulse of current. Pulses from radiation particles are counted.

with boron-10 nuclei to produce alpha particles, which can then be detected.

$$^{1}_{0}n + ^{10}_{5}B \longrightarrow ^{7}_{3}Li + ^{4}_{2}He$$

A **scintillation counter** (Figure 20.9) is *a device that detects nuclear radiations from flashes of light generated in a material by the radiation.* A *phosphor* is a substance that emits flashes of light when struck by radiation. Rutherford had used zinc sulfide as a phosphor to detect alpha particles. A sodium iodide crystal containing thallium(I) iodide is used as a phosphor for gamma radiation. (Excited technetium-99 emits gamma rays and is used for medical diagnostics, as mentioned in the chapter opening. The gamma rays are detected by a scintillation counter.)

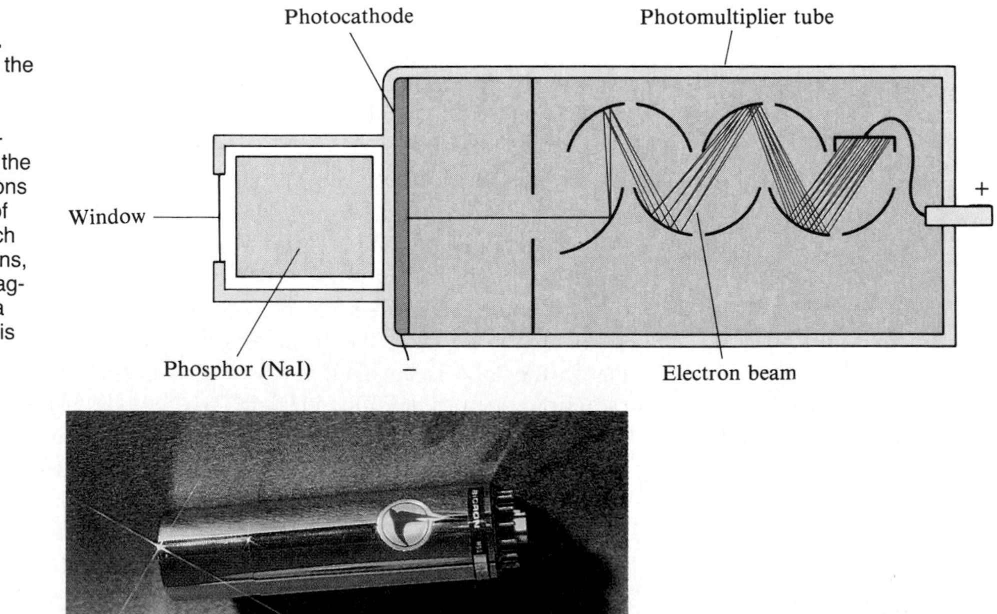

Figure 20.9
A scintillation counter probe. *Top:* Radiation passes through the window into a phosphor (here, NaI). Flashes of light from the phosphor fall on the photocathode, which ejects electrons by the photoelectric effect. The electrons are accelerated to electrodes of increasing positive voltage, each electrode ejecting more electrons, so that the original signal is magnified. *Bottom:* Photograph of a scintillation probe (the window is on the right).

The flashes of light from the phosphor are detected by a *photomultiplier*. A photon of light from the phosphor hits a photoelectric-sensitive surface (the photocathode).● This emits an electron, which is accelerated by a positive voltage to another electrode, from which several electrons are emitted. These electrons are accelerated by a higher voltage to the next electrode, from which more electrons are emitted, and so forth. The result is that a single electron may produce a million electrons and therefore a detectable pulse of electric current.

● The photoelectric effect was discussed in Section 7.6.

A radiation counter can be used to measure the rate of nuclear distintegrations in a radioactive material. The **activity of a radioactive source** is *the number of nuclear disintegrations per unit time occurring in a radioactive material.* A **curie (Ci)** is *a unit of activity equal to 3.700×10^{10} disintegrations per second.*● Thus, a sample of technetium having an activity of 1.0×10^{-2} Ci is decaying at the rate of $(1.0 \times 10^{-2}) \times (3.7 \times 10^{10}) = 3.7 \times 10^{8}$ nuclei per second.

● The curie was originally defined as the number of disintegrations per second from 1.0 g of radium-226.

Biological Effects and Radiation Dosage

Although the quantity of energy dissipated in a biological organism from a radiation dosage might be small, the effects can be quite adverse because important chemical bonds may be broken. DNA in the chromosomes of the cell is especially affected, which interferes with cell division. Cells that divide the fastest, such as those in the blood-forming tissue in bone marrow, are most affected by nuclear radiations.

To monitor the effect of nuclear radiations on biological tissue, it is necessary to have a measure of radiation dosage. The **rad** (from *r*adiation *a*bsorbed *d*ose) is *the dosage of radiation that deposits 1×10^{-2} J of energy per kilogram of tissue.* However, the biological effect of radiation depends not only on the energy deposited in the tissue but also on the type of radiation. For example, neutrons are more destructive than gamma rays of the same radiation dosage measured in rads. A **rem** is *a unit of radiation dosage used to relate various kinds of radiation in terms of biological destruction. It equals the rad times a factor for the type of radiation, called the relative biological effectiveness* (RBE).

$$\text{rems} = \text{rads} \times \text{RBE}$$

Beta and gamma radiations have an RBE of about 1, whereas neutron radiation has an RBE of about 5 and alpha radiation an RBE of about 10.●

● Sources of alpha radiation outside the body are relatively harmless, because the radiation is absorbed by the skin. Internal sources, however, are very destructive.

The effects of radiation on a person depend not only on the dosage but also on the length of time in which the dose was received. A series of small doses has less overall effect than these dosages given all at once. A single dose of about 500 rems is fatal to most people, and survival from a much smaller dose can be uncertain or leave the person chronically ill. Detectable effects are seen with doses as low as 30 rems. Continuous exposure to such low levels of radiation may result in cancer or leukemia. At even lower levels, the answer to whether the radiation dose is safe depends on the possible genetic effects of the radiation. Because radiation can cause chromosome damage, heritable defects are possible.

Safe limits for radiations are much debated. Although there may not be a strictly safe limit, it is important to have in mind the magnitude of the radiations humans may be subjected to. The background radiation that we all receive results from cosmic rays (radiations from space) and natural radioactivity. This averages about 0.1 rem per year but varies considerably with location. Uranium and its decay products in the soil are an important source of this background radiation.● Another source is potassium-40, a radioactive isotope with a natural abundance of

● Soils contain varying amounts of uranium-238, which decays in several steps to radium-226, then to radon-222, a gas (see Figure 20.3). Some homes situated in areas of high uranium content have been found to accumulate radon gas. Radon has a half-life of 3.8 days, decaying by alpha emission to radioactive lead, bismuth, and polonium. These decay products can remain in the lungs and may lead to lung cancer.

Although we might be able to obtain the half-life of a radioactive nucleus by direct observation in some cases, this is impossible for many nuclei because they decay too quickly or too slowly. Uranium-238, for example, has a half-life of 4.51 billion years, which is much too long to be directly observed. The usual method of determining the half-life is by measuring decay rates and relating them to half-lives.

We relate the half-life for radioactive decay, $t_{1/2}$, to the decay constant k by the equation

$$t_{1/2} = \frac{0.693}{k}$$

The next example illustrates the use of this relation.

Example 20.8 Calculating the Half-Life from the Decay Constant

The decay constant for the beta decay of $^{99}_{43}Tc$ was obtained in Example 20.7. There we found that k equals 1.0×10^{-13}/s. What is the half-life of this isotope in years?

Solution

We substitute the value of k into the preceding equation.

$$t_{1/2} = \frac{0.693}{k} = \frac{0.693}{1.0 \times 10^{-13}/\text{s}} = 6.9 \times 10^{12}\ \text{s}$$

Then we convert this half-life in seconds to years.

$$6.9 \times 10^{12}\ \text{s} \times \frac{1\ \text{min}}{60\ \text{s}} \times \frac{1\ \text{h}}{60\ \text{min}} \times \frac{1\ \text{d}}{24\ \text{h}} \times \frac{1\ \text{y}}{365\ \text{d}} = 2.2 \times 10^{5}\ \text{y}$$

Exercise 20.8

Cobalt-60, used in cancer therapy, decays by beta and gamma emission. The decay constant is 4.18×10^{-9}/s. What is the half-life in years? (See Problems 20.47 and 20.48.)

Tables of radioactive nuclei often list the half-life. When we want the decay constant or the activity of a sample, we can calculate them from the half-life. This calculation is illustrated in the next example.

Example 20.9 Calculating the Decay Constant and Activity from the Half-Life

Tritium, $^{3}_{1}H$, is a radioactive nucleus of hydrogen. It is used in luminous watch dials. Tritium decays by beta emission with a half-life of 12.3 years. What is the decay constant (in /s)? What is the activity (in curies) of a sample containing 2.5 μg of tritium? The atomic mass of $^{3}_{1}H$ is 3.02 amu.

Solution

Let us convert the half-life to its value in seconds and then calculate k. After that we can use the rate equation to find the rate of decay of the sample (in nuclei/s) and finally the activity.

The conversion of the half-life to seconds gives

$$12.3\ \text{y} \times \frac{365\ \text{d}}{1\ \text{y}} \times \frac{24\ \text{h}}{1\ \text{d}} \times \frac{60\ \text{min}}{1\ \text{h}} \times \frac{60\ \text{s}}{1\ \text{min}} = 3.88 \times 10^{8}\ \text{s}$$

Because $t_{1/2} = 0.693/k$, we solve this for k and substitute the half-life in seconds.

$$k = \frac{0.693}{t_{1/2}} = \frac{0.693}{3.88 \times 10^{8}\ \text{s}} = 1.79 \times 10^{-9}/\text{s}$$

Before substituting into the rate equation, we need to know the number of tritium nuclei in a sample containing

2.5×10^{-6} g of tritium. We get

$$2.5 \times 10^{-6} \text{ g H-3} \times \frac{1 \text{ mol H-3}}{3.02 \text{ g H-3}} \times \frac{6.02 \times 10^{23} \text{ H-3 nuclei}}{1 \text{ mol H-3}} = 5.0 \times 10^{17} \text{ H-3 nuclei}$$

Now we substitute into the rate equation.

$$\text{Rate} = kN_t = 1.79 \times 10^{-9}\text{/s} \times 5.0 \times 10^{17} \text{ nuclei} = 9.0 \times 10^{8} \text{ nuclei/s}$$

We obtain the activity of the sample by dividing the rate in disintegrations of nuclei per second by 3.70×10^{10} disintegrations of nuclei per second per curie.

$$\text{Activity} = \frac{9.0 \times 10^{8} \text{ nuclei/s}}{3.70 \times 10^{10} \text{ nuclei/(s} \cdot \text{Ci)}} = \mathbf{0.024 \text{ Ci}}$$

Exercise 20.9

Strontium-90, $^{90}_{38}Sr$, is a radioactive decay product of nuclear fallout from nuclear weapons testing. Because of its chemical similarity to calcium, it is incorporated into the bones if present in food. The half-life of strontium-90 is 28.1 y. What is the decay constant of this isotope? What is the activity of a sample containing 5.2 ng (5.2×10^{-9} g) of strontium-90?

(See Problems 20.49, 20.50, 20.51, and 20.52.)

Once we know the decay constant for a radioactive isotope, we can calculate the fraction of the radioactive nuclei that remains after a given period of time by the following equation.●

$$\ln \frac{N_t}{N_0} = -kt \qquad \text{or} \qquad \log \frac{N_t}{N_0} = \frac{-kt}{2.303}$$

Here N_0 is the number of nuclei in the original sample ($t = 0$). After a period of time t, the number of nuclei decay to the number N_t. The fraction of nuclei remaining after time t is N_t/N_0. The next example illustrates the use of this equation.

● A similar equation is given in Chapter 14 for the reactant concentration at time t, $[A]_t$:

$$\log \frac{[A]_t}{[A]_0} = \frac{-kt}{2.303}$$

where $[A]_0$ is the concentration of A at $t = 0$.

Example 20.10 Determining the Fraction of Nuclei Remaining after a Specified Time

Phosphorus-32 is a radioactive isotope with a half-life of 14.3 d. A biochemist has a vial containing a compound of phosphorus-32. If the compound is used in an experiment 5.5 d after the compound was prepared, what fraction of the radioactive isotope originally present remains? Suppose the sample in the vial originally contained 0.28 g of phosphorus-32. How many grams remain after 5.5 d?

Solution

If N_0 is the original number of P-32 nuclei in the vial, and N_t is the number after 5.5 d, the fraction remaining is N_t/N_0. We can obtain this fraction from the equation

$$\log \frac{N_t}{N_0} = \frac{-kt}{2.303}$$

We substitute $k = 0.693/t_{1/2}$.

$$\log \frac{N_t}{N_0} = \frac{-0.693t}{2.303t_{1/2}}$$

Because $t = 5.5$ d and $t_{1/2} = 14.3$ d, we obtain

$$\log \frac{N_t}{N_0} = \frac{-0.693 \times 5.5 \text{ d}}{2.303 \times 14.3 \text{ d}} = -0.116$$

(We have retained an additional digit for further calculation.) Hence,

$$\text{Fraction nuclei remaining} = \frac{N_t}{N_0} = 10^{-0.116} = \mathbf{0.77}$$

We have rounded 0.766 to two significant figures. Thus, 77% of the nuclei remain. The mass of $^{32}_{15}P$ in the vial after 5.5 d is

$$0.28 \text{ g} \times 0.766 = \mathbf{0.21 \text{ g}}$$

Exercise 20.10

A nuclear power plant emits into the atmosphere a very small amount of krypton-85, a radioactive isotope with a half-life of 10.76 y. What fraction of this krypton-85 remains after 25.0 y? (See Problems 20.55 and 20.56.)

Radioactive Dating

Fixing the dates of relics and stone implements or pieces of charcoal from ancient campsites is an application based on radioactive decay rates. Because the rate of radioactive decay of a nuclide is constant, this rate can serve as a clock for dating very old rocks and human implements. Dating wood and similar carbon-containing objects that are several thousand to fifty thousand years old can be done with radioactive carbon, carbon-14, which has a half-life of 5730 y.

Carbon-14 is present in the atmosphere as a result of cosmic-ray bombardment of earth. Cosmic rays are radiations from space that consist of protons and alpha particles, as well as heavier ions. These radiations produce other kinds of particles, including neutrons, as they bombard the upper atmosphere. The collision of a neutron with a nitrogen-14 nucleus (the most abundant nitrogen nuclide) can produce a carbon-14 nucleus.

$$^{14}_{7}\text{N} + ^{1}_{0}\text{n} \longrightarrow ^{14}_{6}\text{C} + ^{1}_{1}\text{H}$$

Carbon dioxide containing carbon-14 mixes with the lower atmosphere. Because of the constant production of $^{14}_{6}C$ and its radioactive decay, a small, constant fractional abundance of carbon-14 is maintained in the atmosphere.

Living plants, which continuously use atmospheric carbon dioxide, also maintain a constant abundance of carbon-14. Similarly, living animals, by feeding on plants, have a constant fractional abundance of carbon-14. But once an organism dies, it is no longer in chemical equilibrium with atmospheric CO_2. The ratio of carbon-14 to carbon-12 begins to decrease by radioactive decay. Thus, this ratio of carbon isotopes becomes a clock measuring the time since the death of the organism.

If we assume that the ratio of carbon isotopes in the lower atmosphere has remained at the present level for the last 50,000 years (presently 1 out of 10^{12} carbon atoms is carbon-14), we can deduce the age of any dead organic object by measuring the level of beta emissions that arise from the radioactive decay of carbon-14.●

● Analyses of tree rings have shown that this assumption is not quite valid. Before 1000 B.C., the levels of carbon-14 were somewhat higher than they are today. Moreover, recent human activities (burning of fossil fuels and atmospheric nuclear testing) have changed the fraction of carbon-14 in atmospheric CO_2.

$$^{14}_{6}\text{C} \longrightarrow ^{14}_{7}\text{N} + ^{0}_{-1}\text{e}$$

In this way, bits of campfire charcoal, parchment, jaw bones, and the like have been dated.

Example 20.11 Applying the Carbon-14 Dating Method

A piece of charcoal from a tree killed by the volcanic eruption that formed the crater in Crater Lake (in Oregon) gave 7.0 disintegrations of carbon-14 nuclei per minute per gram of total carbon. Present-day carbon (in living matter) gives 15.3 disintegrations per minute per gram of total carbon. Determine the date of the volcanic eruption.

Solution

We substitute $k = 0.693/t_{1/2}$ into the equation for the number of nuclei in a sample after time t.

$$\log \frac{N_t}{N_0} = \frac{-kt}{2.303} = \frac{-0.693t}{2.303t_{1/2}}$$

Hence,

$$t = \frac{2.303 t_{1/2}}{0.693} \log \frac{N_0}{N_t}$$

To get N_0/N_t, we assume that the ratio of $^{14}_{6}C$ to $^{12}_{6}C$ in the atmosphere has remained constant. Then we can say that 1.00 gram of total carbon from the living tree gave 15.3 disintegrations per minute. The ratio of the number of $^{14}_{6}C$ nuclei originally present to the number that existed at the time of dating equals the ratio of rates of disintegration. That is,

$$\frac{N_0}{N_t} = \frac{15.3}{7.0} = 2.2$$

Therefore, substituting this value of N_0/N_t and $t_{1/2} = 5730$ y into the previous equation gives

$$t = \frac{2.303 t_{1/2}}{0.693} \log \frac{N_0}{N_t} = \frac{2.303 \times 5730 \text{ y}}{0.693} \log 2.2$$
$$= 6.5 \times 10^3 \text{ y}$$

Thus, the date of the eruption was about **4500** B.C.

Exercise 20.11

A jawbone from the archaeological site at Folsom, New Mexico, was dated by analysis of its radioactive carbon. The activity of the carbon from the jawbone was 4.5 disintegrations per minute per gram of total carbon. What was the age of the jawbone? Carbon from living material gives 15.3 disintegrations per minute per gram of carbon.

(See Problems 20.61 and 20.62.)

For the age of rocks and meteorites, other similar methods of dating have been used. One method depends on the radioactivity of naturally occurring potassium-40, which decays by positron emission and electron capture (as well as by beta emission).

$$^{40}_{19}K \longrightarrow {}^{40}_{18}Ar + {}^{0}_{1}e$$
$$^{40}_{19}K + {}^{0}_{-1}e \longrightarrow {}^{40}_{18}Ar$$

Potassium occurs in many rocks. Once such a rock forms by solidification of molten material, the argon from the decay of potassium-40 is trapped. To obtain the age of a rock, one first determines the number of $^{40}_{19}K$ atoms and the number of $^{40}_{18}Ar$ atoms in a sample. The number of $^{40}_{19}K$ atoms equals N_t. The number originally present, N_0, equals N_t plus the number of argon atoms, because each argon atom resulted from the decay of a $^{40}_{19}K$ nucleus. One then calculates the age of the rock from the ratio N_0/N_t.

The oldest rocks on earth have been dated at 3.8×10^9 y. The rocks at the earth's surface have been subjected to extensive weathering, so even older rocks may have existed. This age, 3.8×10^9 y, therefore represents the minimum possible age of the earth—the time since the solid crust first formed. Ages of meteorites, which are assumed to have solidified at the same time as other solid objects in the solar system, including earth, have been determined to be 4.4×10^9 y to 4.6×10^9 y. It is now believed from this and other evidence that the age of the earth is 4.6×10^9 y.

20.5 APPLICATIONS OF RADIOACTIVE ISOTOPES

We have already described two applications of nuclear chemistry. One was the preparation of elements not available naturally. We noted that the discovery of the transuranium elements clarified the position of the heavy elements in the periodic table. In the section just completed, we discussed the use of radioactivity in dating objects. We will discuss practical uses of nuclear energy in the last section of the chapter. Here we will look at the applications of radioactive isotopes to chemical analysis and to medicine.

Chemical Analysis

A **radioactive tracer** is *a radioactive isotope added to a chemical, biological, or physical system to study the system*. The advantage of a radioactive tracer is that it behaves chemically just as a nonradioactive isotope does, but it can be detected in exceedingly small amounts by measuring the radiations emitted.

As an illustration of the use of radioactive tracers, consider the problem of establishing that chemical equilibrium is a dynamic process. Let us look at the equilibrium of solid lead(II) iodide and its saturated solution, containing $Pb^{2+}(aq)$ and $I^-(aq)$. The equilibrium is

$$PbI_2(s) \rightleftharpoons Pb^{2+}(aq) + 2I^-(aq)$$

In two separate beakers, we prepare saturated solutions of PbI_2 in contact with the solid. One beaker contains only natural iodine atoms with nonradioactive isotopes. The other beaker contains radioactive iodide ion, $^{131}I^-$. Some of the solution, but no solid, containing the radioactive iodide ion is now added to the other beaker containing nonradioactive iodide ion. Both solutions are saturated, so the amount of solid in this beaker remains constant. Yet after a time the solid lead iodide, which was originally nonradioactive, becomes radioactive. This is evidence for a dynamic equilibrium, in which radioactive iodide ions in the solution substitute for nonradioactive iodide ions in the solid.

With only naturally occurring iodine available, it would have been impossible to detect the dynamic equilibrium. By using ^{131}I as a radioactive tracer, we can easily follow the substitution of radioactive iodine into the solid by measuring its radioactivity.

A series of experiments using tracers was carried out in the late 1950s by Melvin Calvin at the University of California, Berkeley, in order to discover the mechanism of photosynthesis in plants.● The overall process of photosynthesis involves the reaction of CO_2 and H_2O to give glucose, $C_6H_{12}O_6$, and O_2. Energy for photosynthesis comes from the sun.

● Melvin Calvin received the Nobel Prize in chemistry in 1961 for his work on photosynthesis.

$$6CO_2(g) + 6H_2O(l) \xrightarrow{\text{sunlight}} C_6H_{12}O_6(aq) + 6O_2(g)$$

This equation represents only the net result of photosynthesis. As Calvin was able to show, the actual process consists of many separate steps. In several experiments, algae (single-celled plants) were exposed to carbon dioxide containing much more radioactive carbon-14 than occurs naturally. Then the algae were extracted with a solution of alcohol and water. The various compounds in this solution were separated by chromatography and identified.● Those compounds that contained radioactive carbon were produced in the different steps of photosynthesis. Eventually, Calvin was able to use tracers to show the main steps in photosynthesis.

● Chromatography was discussed in the Instrumental Methods at the end of Section 2.1.

Another example of the use of radioactive tracers in chemistry is **isotope dilution,** *a technique designed to determine the quantity of a substance in a mixture or the total volume of solution by adding a known amount of an isotope to it.* After removal of a portion of the mixture, the fraction by which the isotope has been diluted provides a way of determining the quantity of substance or volume of solution.

As an example, suppose we wish to obtain the volume of water in a tank but are unable to drain the tank. We add 100 mL of water containing a radioactive isotope. After allowing this to mix completely with the water in the tank, we withdraw 100 mL of solution from the tank. We find that the activity of this solution in curies is 1/1000 that of the original solution. The isotope has been diluted by a factor of 1000, so the volume of the tank is 1000×100 mL = 100,000 mL (100 L).

A typical chemical example of isotope dilution is the determination of the amount of vitamin B_{12}, a cobalt-containing substance, in a sample of food. Although part of the vitamin in food can be obtained in pure form, not all of the vitamin can be separated. Therefore, we cannot precisely determine the quantity of vitamin B_{12} in a sample of food by separating the pure vitamin and weighing it. But we can determine the amount of vitamin B_{12} by isotope dilution. Suppose we add 2.0×10^{-7} g of vitamin B_{12} containing radioactive cobalt-60 to 125 g of food and mix well. We then separate 5.4×10^{-7} g of pure vitamin B_{12} from the food and find that the activity in curies of this quantity of the vitamin contains 5.6% of the activity added from the radioactive cobalt. The mass of vitamin B_{12} in the food, including the amount added (2.0×10^{-7} g), is

$$5.4 \times 10^{-7} \text{ g} \times \frac{100}{5.6} = 9.6 \times 10^{-6} \text{ g}$$

Subtracting the amount added in the analysis gives

$$9.6 \times 10^{-6} \text{ g} - 2.0 \times 10^{-7} \text{ g} = 9.4 \times 10^{-6} \text{ g}$$

Neutron activation analysis is *an analysis of elements in a sample based on the conversion of stable isotopes to radioactive isotopes by bombarding a sample with neutrons*. Human hair contains trace amounts of many elements. By determining the exact amounts and the position of the elements in the hair shaft, we can identify whom the hair comes from (assuming we have a sample known to be that person's hair). Consider the analysis of human hair for arsenic, for example. When the natural isotope ${}^{75}_{33}As$ is bombarded with neutrons, a metastable nucleus ${}^{76m}_{33}As$ is obtained.

$${}^{75}_{33}\text{As} + {}^{1}_{0}\text{n} \longrightarrow {}^{76m}_{33}\text{As}$$

A metastable nucleus is in an excited state. It decays, or undergoes a transition, to a lower state by emitting gamma rays. The frequencies, or energies, of the gamma rays emitted are characteristic of the element and serve to identify it. Also, the intensities of the gamma rays emitted are proportional to the amount of the element present. The method is very sensitive; it can identify as little as 10^{-9} g of arsenic.●

● Neutron activation analysis has been used to authenticate oil paintings by giving an exact analysis of pigments used. Pigment compositions have changed, so it is possible to detect fraudulent paintings done with more modern pigments. The analysis can be done without affecting the painting.

Medical Therapy and Diagnosis

The use of radioactive isotopes has had a profound effect on the practice of medicine. Radioisotopes were first used in medicine in the treatment of cancer. This treatment is based on the fact that rapidly dividing cells, such as those in cancer, are more adversely affected by radiation from radioactive substances than are those cells that divide more slowly. Radium-226 and its decay product radon-222 were used for cancer therapy a few years after the discovery of radioactivity. Today gamma radiation from cobalt-60 is more commonly used.

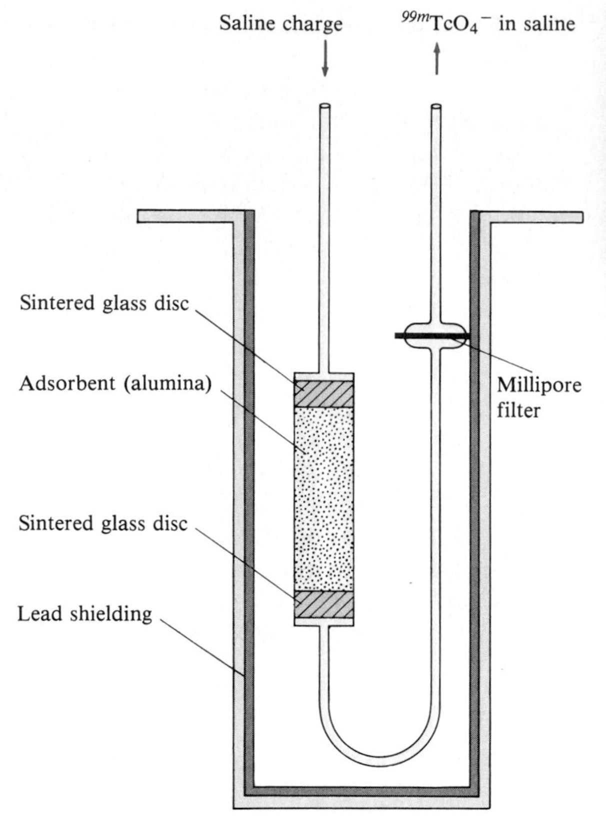

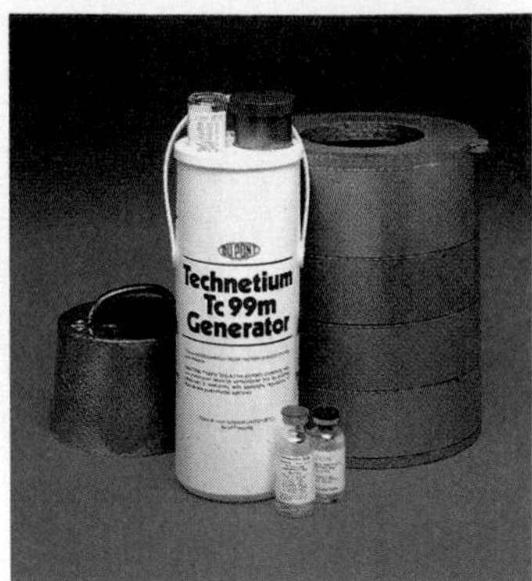

Figure 20.11
A technetium-99*m* generator.
Left: A schematic view of the generator. Molybdenum-99, in the form of MoO_4^{2-} ion adsorbed on alumina, decays to technetium-99*m*. The technetium is leached from the generator with a salt solution (saline charge) as TcO_4^-. *Above:* Photograph of a technetium-99*m* generator, with lead case and vials of saline charge.

Cancer therapy, however, is only one of the ways in which radioactive isotopes are used in medicine. The greatest advances in the use of radioactive isotopes have been in the diagnosis of disease. Radioactive isotopes are used for diagnosis in two ways. They are used to develop images of internal body organs, so that their functioning can be examined. And they are used as tracers in the analysis of minute amounts of substances, such as a growth hormone in blood, in order to deduce possible disease conditions.

Technetium-99*m* is the radioactive isotope used most often to develop pictures or images of internal body organs. It has a half-life of 6.02 h, decaying by gamma emission to technetium-99 in its nuclear ground state. The image is prepared by scanning the body for gamma rays with a scintillation detector. Figure 20.1, described in the chapter opening, shows the image of a person's skeleton obtained with technetium-99*m*. The technetium is soon excreted by the body, and the gamma radioactivity decays to negligible levels within hours.

In a hospital, the technetium isotope is produced in a special container, or generator, shown in Figure 20.11. The generator contains radioactive molybdate ion, MoO_4^{2-}, adsorbed on alumina granules. Radioactive molybdenum-99 is produced at a nuclear reactor facility by bombarding the natural, nonradioactive, isotope molybdenum-98 with neutrons.

$$^{98}_{42}Mo + ^{1}_{0}n \longrightarrow ^{99}_{42}Mo$$

This radioactive molybdenum, adsorbed on alumina, is placed in the generator and sent to the hospital. Pertechnetate ion is obtained when the molybdenum-99 nucleus in MoO_4^{2-} decays. The nuclear equation is

$$^{99}_{42}\text{Mo} \longrightarrow {}^{99m}_{43}\text{Tc} + {}^{0}_{-1}\text{e}$$

Each day pertechnetate ion, TcO_4^-, is leached from the generator with a salt solution whose osmotic pressure is the same as that of blood. Pertechnetate ion can be used to obtain images of the brain. However, other compounds of technetium are prepared to obtain images of other organs. Certain complex compounds of technetium bind to damaged heart tissues. They can be used to diagnose heart attacks. An active area of research is the synthesis of compounds of radioactive isotopes that may make it possible to see the functioning of the various organs of the body.●

● Over a hundred radioactive isotopes have been used in medicine. Some examples are iodine-131, used to measure thyroid gland activity; phosphorus-32, used to locate tumors; and iron-59, used to measure the rate of formation of red blood cells.

Radioimmunoassay is a newly developed technique for analyzing blood and other body fluids for very small quantities of biologically active substances. The technique depends on the reversible binding of the substance to an antibody. Antibodies are produced in animals as protection against foreign substances. They protect by binding to the substance and countering its biological activity. Consider, for example, the analysis for insulin in a sample of blood from a patient. Before the analysis, a solution of insulin-binding antibodies has been prepared from laboratory animals. This solution is combined with insulin containing a radioactive isotope, in which the antibodies bind with radioactive insulin. Now the sample containing an unknown amount of insulin is added to the antibody–radioactive insulin mixture. The nonradioactive insulin replaces some of the radioactive insulin bound to the antibody. As a result, the antibody loses some of its radioactivity. The loss in radioactivity can be related to the amount of insulin in the blood sample.

Related Topic

POSITRON EMISSION TOMOGRAPHY (PET)

Positron emission tomography (PET) is a new technique for following biochemical processes within the organs (brain, heart, and so forth) of the human body. Like magnetic resonance imaging (see Instrumental Methods at the end of Section 8.1), a PET scan produces an image of a two-dimensional slice through a body organ of a patient. The image shows the distribution of some positron-emitting isotope present in a compound that was administered earlier to the patient by injection. By comparing the PET scan of the patient with that of a healthy subject, a physician can diagnose the presence or absence of disease (Figure 20.12). The PET scan of the brain of an Alzheimer's patient differs markedly from that of a healthy subject. Similarly, the PET scan of a patient with a heart damaged by a coronary attack clearly shows the damaged area. Moreover, the PET scan may help the physician assess the likelihood of success of bypass surgery.

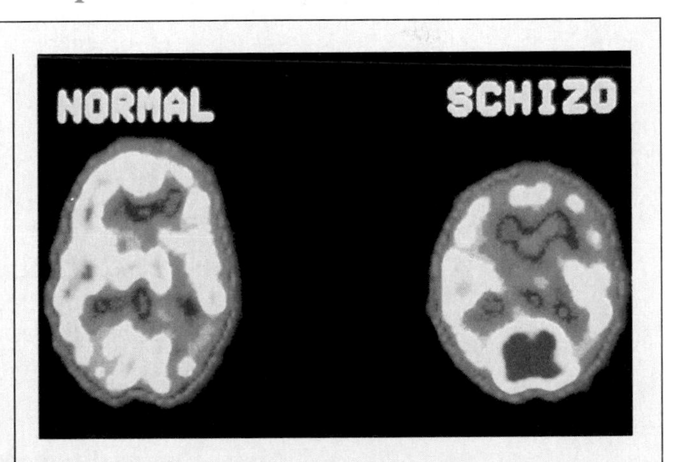

Figure 20.12
PET scans of normal and diseased patients. *The scans are of the brains of a normal patient and a schizophrenic patient. The scans show differences in brain activities measured by glucose usages—yellow to red colors show high values; green to blue are low values.*

(continued)

Some isotopes used in PET scans are carbon-11, nitrogen-13, oxygen-15, and fluorine-18. All have short half-lives, so the radiation dosage to the patient is minimal. However, because of the short half-life, a chemist must prepare the diagnostic compound containing the radioactive nucleus shortly before the physician administers it. The preparation of this compound requires a cyclotron, whose cost (several million dollars) is a major deterrent to the general use of PET scans.

Figure 20.13 shows a patient undergoing a PET scan of the brain. The instrument actually detects gamma radiation. When a nucleus emits a positron within the body, the positron travels only a few millimeters before it reacts with an electron. This reaction is an example of the annihilation of matter (an electron) by antimatter (a positron). Both the electron and the positron disappear and produce two gamma photons. The gamma photons easily pass through human tissue, so they can be recorded by scintillation detectors placed around the body. You can see the circular bank of detectors in Figure 20.13. The detectors record the distribution of gamma radiation, and from this information a computer constructs a cathode-ray image.

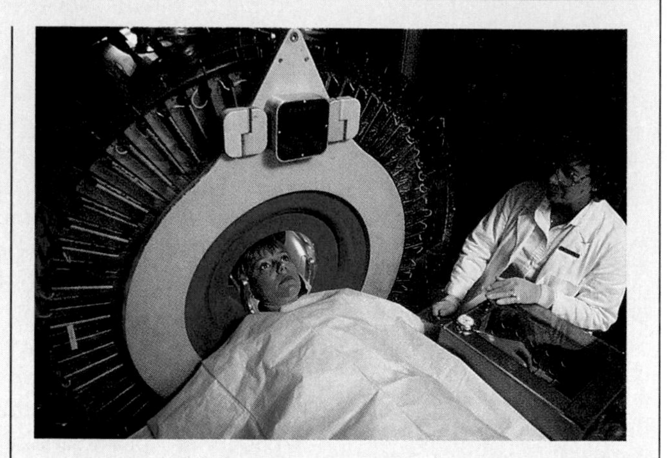

Figure 20.13
A patient undergoing a PET scan of the brain. *Note the circular bank of gamma-ray detectors.*

Energy of Nuclear Reactions

Nuclear reactions, like chemical reactions, involve changes of energy. However, the changes of energy in nuclear reactions are enormous by comparison with chemical reactions. The energy released by certain nuclear reactions is used in nuclear power reactors.

20.6 MASS–ENERGY CALCULATIONS

When nuclei decay, they form products of lower energy. The change of energy is related to changes of mass, according to the mass–energy equivalence relation derived by Albert Einstein in 1905. Using this relation, we can obtain the energies of nuclear reactions from mass changes.

Mass–Energy Equivalence

One of the conclusions from Einstein's theory of special relativity is that the mass of a particle changes with its speed: the greater the speed, the greater the mass. Or, because kinetic energy depends on speed, the greater the kinetic energy of a particle, the greater its mass. This result, according to Einstein, is even more general. Energy and mass are equivalent and are related by the equation

$$E = mc^2$$

Here c is the speed of light, 3.00×10^8 m/s.

The meaning of this equation is that to any mass there is an associated energy, or to any energy there is an associated mass. Thus, if any system loses energy, it must also lose mass. For example, when carbon burns in oxygen, it releases heat energy.

$$C(graphite) + O_2(g) \longrightarrow CO_2(g),\ \Delta H = -393.5\ kJ$$

Because this chemical system loses energy, it should also lose mass. In principle, we could obtain ΔH for the reaction by measuring the change in mass and relating this by Einstein's equation to the change in energy.● However, weight measurements are of no practical value in determining heats of reaction. The changes in mass are simply too small to detect.

● The change in mass gives the change in internal energy, ΔU. The enthalpy change, ΔH, equals $\Delta U + P\Delta V$. For the reaction given in the text, $P\Delta V$ is essentially zero.

Calculation of the mass change for a typical chemical reaction, the burning of carbon in oxygen, will show just how small this quantity is. When the energy changes by an amount ΔE, the mass changes by an amount Δm. We can write Einstein's equation in the form

$$\Delta E = (\Delta m)c^2$$

The change in energy when 1 mol of carbon reacts is -3.935×10^5 J, or $-3.935 \times 10^5\ kg \cdot m^2/s^2$. Hence,

$$\Delta m = \frac{\Delta E}{c^2} = \frac{-3.935 \times 10^5\ kg \cdot \cancel{m^2/s^2}}{(3.00 \times 10^8\ \cancel{m/s})^2} = -4.37 \times 10^{-12}\ kg$$

For comparison, a good analytical balance can detect a mass as small as 1×10^{-7} kg, but this is ten thousand times greater than the mass change caused by the release of heat during the combustion of 1 mol of carbon.

Table 20.3 Masses of Some Nuclei and Other Atomic Particles*

Symbol	Z	A	Mass (amu)	Symbol	Z	A	Mass (amu)
e^-	−1	0	0.000549	Co	27	59	58.9184
n	0	1	1.00867	Ni	28	58	57.9199
H or p	1	1	1.00728	Pb	82	206	205.9295
	1	2	2.01345		82	207	206.9309
	1	3	3.01550		82	208	207.9316
He	2	3	3.01493	Po	84	210	209.9368
	2	4	4.00150		84	218	217.9628
Li	3	6	6.01347	Rn	86	222	221.9703
	3	7	7.01435	Ra	88	226	225.9771
Be	4	9	9.00999	Th	90	230	229.9837
B	5	10	10.0102		90	234	233.9942
	5	11	11.0066	Pa	91	234	233.9931
C	6	12	11.9967	U	92	233	232.9890
	6	13	13.0001		92	234	233.9904
O	8	16	15.9905		92	235	234.9934
Cr	24	52	51.9273		92	238	238.0003
Fe	26	56	55.9206	Pu	94	239	239.0006

*The mass of an atom is obtained by adding the masses of the electrons to the nuclear mass given in this table. For example, the mass of the $^{12}_{6}C$ atom is 11.9967 + 6(0.000549) = 12.0000. (From R. C. Weast, ed., *CRC Handbook of Chemistry and Physics*, 59th ed. [Boca Raton, Fla.: CRC Press, Inc., 1978]. With permission of CRC Press, Inc.)

By contrast, the mass changes in nuclear reactions are approximately a million times larger per mole of reactant than those in chemical reactions. Consider the alpha decay of uranium-238 to thorium-234. The nuclear equation is

$$\underset{238.0003}{{}^{238}_{92}\text{U}} \longrightarrow \underset{233.9942}{{}^{234}_{90}\text{Th}} + \underset{4.00150 \text{ amu}}{{}^{4}_{2}\text{He}}$$

Here we have written the nuclear mass (in amu) beneath each nuclide symbol. (Table 20.3, p. 819, lists masses of some nuclei and other atomic particles.) The change in mass for this nuclear reaction, starting with molar amounts, is

$$\Delta m = (233.9942 + 4.00150 - 238.0003) \text{ g} = -0.0046 \text{ g}$$

As in calculating ΔH and similar quantities, we subtract the value for the reactant from the sum of the values for the products. The minus sign indicates a loss of mass. This loss of mass is clearly large enough to detect.

From a table of nuclear masses, such as Table 20.3, we can use Einstein's equation to calculate the energy change for a nuclear reaction. This is illustrated in the next example. Recall from Section 20.2 that 1 eV = 1.602×10^{-19} J. Therefore, 1 MeV equals 1.602×10^{-13} J.●

● From Einstein's equation, one can show that 1 amu is equivalent to 931 MeV. This number can be used to obtain energies of nuclear reactions. Thus, for the reaction given in Example 20.12, the mass change is −0.01960 amu. Multiplying this by 931 MeV gives the energy change, −18.2 MeV, which agrees with the answer given to three significant figures.

Example 20.12 Calculating the Energy Change for a Nuclear Reaction

(a) Calculate the energy change in joules (four significant figures) for the following nuclear reaction per mole of ${}^{2}_{1}\text{H}$:

$${}^{2}_{1}\text{H} + {}^{3}_{2}\text{He} \longrightarrow {}^{4}_{2}\text{He} + {}^{1}_{1}\text{H}$$

Nuclear masses are given in Table 20.3. (b) What is the energy change in MeV for one ${}^{2}_{1}\text{H}$ nucleus?

Solution

We write the nuclear masses below each nuclide symbol and then calculate Δm. Once we have Δm, we can obtain ΔE.

$$\underset{2.01345}{{}^{2}_{1}\text{H}} + \underset{3.01493}{{}^{3}_{2}\text{He}} \longrightarrow \underset{4.00150}{{}^{4}_{2}\text{H}} + \underset{1.00728 \text{ amu}}{{}^{1}_{1}\text{H}}$$

Hence,

$$\Delta m = (4.00150 + 1.00728 - 2.01345 - 3.01493) \text{ amu}$$
$$= -0.01960 \text{ amu}$$

(a) To obtain the energy change for molar amounts, we note that the molar mass of a nucleus in grams is numerically equal to the mass of a single nucleus in amu. Therefore, the mass change for molar amounts in this nuclear reaction is −0.01960 g, or -1.960×10^{-5} kg. The energy change is

$$\Delta E = (\Delta m)c^2 = (-1.960 \times 10^{-5} \text{ kg})(2.998 \times 10^8 \text{ m/s})^2$$
$$\Delta E = -1.762 \times 10^{12} \text{ kg} \cdot \text{m}^2/\text{s}^2, \text{ or } \mathbf{-1.762 \times 10^{12} \text{ J}}$$

(b) The mass change for the reaction of one ${}^{2}_{1}\text{H}$ nucleus is −0.01960 amu. Let us change this to grams. Recall that 1 amu equals 1/12 the mass of a ${}^{12}_{6}\text{C}$ atom, whose mass is 12 g/6.022×10^{23}. Thus, 1 amu = 1 g/6.022×10^{23}. Hence, the mass change in grams is

$$\Delta m = -0.01960 \text{ amu} \times \frac{1 \text{ g}}{1 \text{ amu} \times 6.022 \times 10^{23}}$$
$$= -3.255 \times 10^{-26} \text{ g (or } -3.255 \times 10^{-29} \text{ kg)}$$

Then,

$$\Delta E = (\Delta m)c^2 = (-3.255 \times 10^{-29} \text{ kg})(2.998 \times 10^8 \text{ m/s})^2$$
$$= -2.926 \times 10^{-12} \text{ J}$$

We now convert this to MeV.

$$\Delta E = -2.926 \times 10^{-12} \text{ J} \times \frac{1 \text{ MeV}}{1.602 \times 10^{-13} \text{ J}} = \mathbf{-18.26 \text{ MeV}}$$

Exercise 20.12

(a) Calculate the energy change in joules when 1.00 g ${}^{234}_{90}\text{Th}$ decays to ${}^{234}_{91}\text{Pa}$ by beta emission. (b) What is the energy change in MeV when one ${}^{234}_{90}\text{Th}$ nucleus decays? Use Table 20.3 for these calculations.

(See Problems 20.67 and 20.68.)

Nuclear Binding Energy

The equivalence of mass and energy explains the otherwise puzzling fact that the mass of a nucleus is always less than the sum of the masses of its constituent nucleons. For example, the helium-4 nucleus consists of two protons and two neutrons, giving the following sum:

$$\begin{aligned} \text{Mass of 2 protons} &= 2 \times 1.00728 \text{ amu} = 2.01456 \text{ amu} \\ \text{Mass of 2 neutrons} &= 2 \times 1.00867 \text{ amu} = \underline{2.01734} \text{ amu} \\ \text{Total mass of nucleons} &= \phantom{2 \times 1.00867 \text{ amu} =} 4.03190 \text{ amu} \end{aligned}$$

The mass of the helium-4 nucleus is 4.00150 amu (see Table 20.3), so the mass difference is

$$\Delta m = (4.00150 - 4.03190) \text{ amu} = -0.03040 \text{ amu}$$

This mass difference is explained as follows. When the nucleons come together to form a nucleus, energy is released. (The nucleus has lower energy and is therefore more stable than the separate nucleons.) According to Einstein's equation, there must be an equivalent decrease in mass.

The **binding energy** of a nucleus is *the energy needed to break a nucleus into its individual protons and neutrons*. Thus, the binding energy of the helium-4 nucleus is the energy change for the reaction

$${}^{4}_{2}\text{He} \longrightarrow 2{}^{1}_{1}\text{p} + 2{}^{1}_{0}\text{n}$$

The **mass defect** of a nucleus is *the total nucleon mass minus the nuclear mass*. In the case of helium-4, the mass defect is 4.03190 amu − 4.00150 amu = 0.03040 amu (this is the positive value of the mass difference we calculated earlier). Both the binding energy and the corresponding mass defect are reflections of the stability of the nucleus.

To compare the stabilities of various nuclei, it is useful to compare binding energies per nucleon (proton or neutron). Figure 20.14 shows values of this quantity (in MeV per nucleon) plotted against the mass number for various nuclides. Most of the points lie near the smooth curve drawn on the graph.

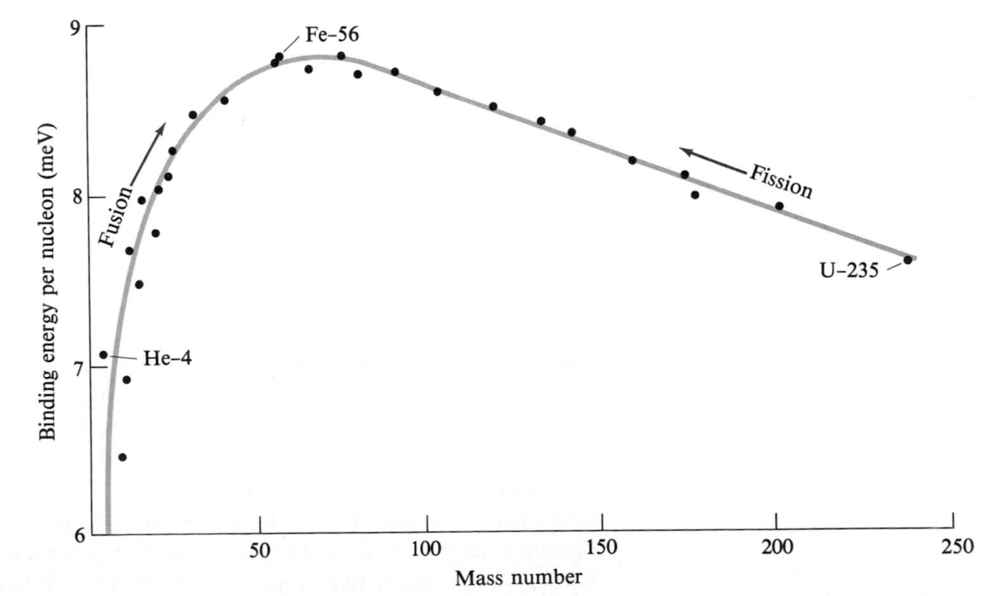

Figure 20.14
Plot of nuclear binding energy per nucleon versus mass number. The binding energy of each nuclide is divided by the number of nucleons (total of protons and neutrons), then plotted at the mass number of the nuclide. Note that nuclides near mass number 50 have the largest binding energies per nucleon. Thus, heavy nuclei are expected to undergo fission to approach this mass number, whereas light nuclei are expected to undergo fusion.

Note that nuclides near mass number 50 have the largest binding energies per nucleon. This means that a group of nucleons would tend to form those nuclides, because they would thus form nuclei of the lowest energy. For this reason, heavy nuclei might be expected to split to give lighter nuclei, and light nuclei might be expected to react, or combine, to form heavier nuclei.

Nuclear fission is *a nuclear reaction in which a heavy nucleus splits into lighter nuclei and energy is released.* For example, californium-252 decays both by alpha emission (97%) and by spontaneous fission (3%). During spontaneous fission, the nucleus splits into two stabler, lighter nuclei plus several neutrons. There are many possible ways in which the nucleus can split. One way is represented by the following equation:

$$^{252}_{98}Cf \longrightarrow {}^{142}_{56}Ba + {}^{106}_{42}Mo + 4\,{}^{1}_{0}n$$

In some cases, a nucleus can be induced to undergo fission by bombarding it with neutrons. An example is the nuclear fission of uranium-235. When a neutron strikes the $^{235}_{92}U$ nucleus, the nucleus splits into roughly equal parts, giving off several neutrons. Three possible splittings are shown in the following equations:

$$^{1}_{0}n + {}^{235}_{92}U \longrightarrow {}^{142}_{54}Xe + {}^{90}_{38}Sr + 4\,{}^{1}_{0}n$$
$$^{1}_{0}n + {}^{235}_{92}U \longrightarrow {}^{139}_{56}Ba + {}^{94}_{36}Kr + 3\,{}^{1}_{0}n$$
$$^{1}_{0}n + {}^{235}_{92}U \longrightarrow {}^{144}_{55}Cs + {}^{90}_{37}Rb + 2\,{}^{1}_{0}n$$

Nuclear fission is discussed further in Section 20.7.

Nuclear fusion is *a nuclear reaction in which light nuclei combine to give a stabler, heavier nucleus plus possibly several neutrons, and energy is released.* An example of nuclear fusion is

$$^{2}_{1}H + {}^{3}_{1}H \longrightarrow {}^{4}_{2}He + {}^{1}_{0}n$$

Even though a nuclear reaction is energetically favorable, the reaction may be imperceptibly slow unless the correct conditions are present (see Section 20.7).

20.7 NUCLEAR FISSION AND NUCLEAR FUSION

We have seen that the stablest nuclei are those of intermediate size (with mass numbers around 50). Nuclear fission and nuclear fusion are reactions in which nuclei attain sizes closer to this intermediate range. In doing so, these reactions release tremendous amounts of energy. Nuclear fission of uranium-235 is employed in nuclear power plants to generate electricity. Nuclear fusion may supply us with energy in the future.

Nuclear Fission; Nuclear Reactors

Nuclear fission was discovered as a result of experiments to produce transuranium elements. Soon after the neutron was discovered in 1932, experimenters realized that this particle, being electrically neutral, should easily penetrate heavy nuclei. They began using neutrons in bombardment reactions, hoping to produce isotopes that would decay to new elements. In 1938, Otto Hahn, Lise Meitner, and Fritz Strassmann in Berlin identified barium in uranium samples that had been bombarded with neutrons. Soon afterward, the presence of barium was explained as a result of fission of the uranium-235 nucleus. When this nucleus is struck by a

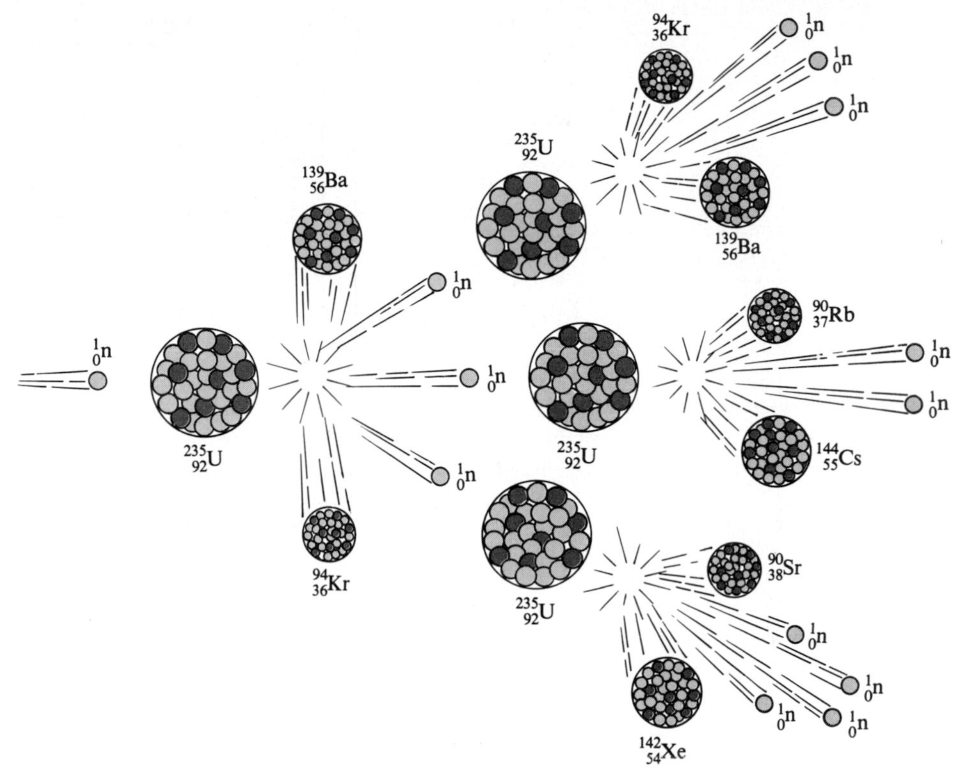

Figure 20.15
Representation of a chain reaction of nuclear fissions. Each nuclear fission produces two or more neutrons, which can in turn cause more nuclear fissions. At each stage, a greater number of neutrons are produced, so that the number of nuclear fissions multiplies quickly. Such a chain reaction is the basis of nuclear power and nuclear weapons. Note that the fission products vary, although the original nucleus splits roughly into halves.

neutron, it splits into two nuclei. Fissions of uranium nuclei produce approximately 30 different elements of intermediate mass, including barium.

When the uranium-235 nucleus splits, approximately 2 to 3 neutrons are released. If the neutrons from each nuclear fission are absorbed by other uranium-235 nuclei, these nuclei split and release even more neutrons. Thus, a chain reaction can occur. A nuclear **chain reaction** is *a self-sustaining series of nuclear fissions caused by the absorption of neutrons released from previous nuclear fissions*. The number of nuclei that fission quickly multiply as a result of the absorption of neutrons released from previous nuclear fissions. Figure 20.15 shows how such a chain reaction occurs. The chain reaction of nuclear fissions is the basis of nuclear power and nuclear weapons.

In order to sustain a chain reaction in a sample of fissionable material, a nucleus that splits must give an average of one neutron that results in the fission of another nucleus, and so on. If the sample is too small, many of the neutrons leave the sample before they have a chance to be absorbed. There is thus a **critical mass** for a particular fissionable material, which is *the smallest mass of fissionable material in which a chain reaction can be sustained*. If the mass is much larger than this (a *supercritical* mass), the number of nuclei that split multiply rapidly. An atomic bomb is detonated with a small amount of chemical explosive that pushes together two or more masses of fissionable material to get a supercritical mass (Figure 20.16). A rapid chain reaction results in the splitting of most of the fissionable nuclei—and in the explosive release of an enormous amount of energy.

A **nuclear fission reactor** is *a device that permits a controlled chain reaction of nuclear fissions*. In power plants, a nuclear reactor is used to produce heat, which in turn is used to produce steam to drive an electric generator. A nuclear reactor consists of fuel rods alternating with control rods contained within a vessel.

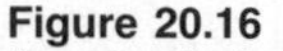

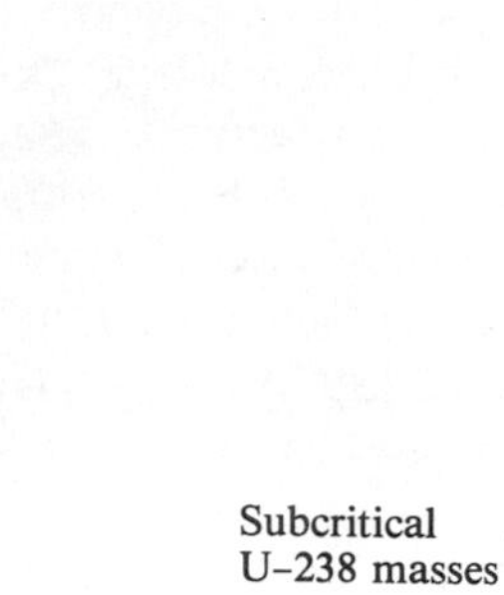

Figure 20.16
An atomic bomb. In the gun type bomb, one piece of uranium-235 of subcritical mass is hurled into another piece by a chemical explosive. The two pieces together give a supercritical mass, and a nuclear explosion results. Bombs using plutonium-239 require an implosion technique, in which wedges of plutonium arranged on a spherical surface are pushed into the center of the sphere by a chemical explosive, where a supercritical mass of plutonium results in a nuclear explosion.

The **fuel rods** are *the cylinders that contain fissionable material*. In the light-water (ordinary water) reactors commonly used in the United States (see Figure 20.17), these fuel rods contain uranium dioxide pellets in a zirconium alloy tube. Natural uranium contains only 0.72% uranium-235, which is the isotope that undergoes fission. The uranium used for fuel in these reactors is ''enriched'' so that it contains about 3% of the uranium-235 isotope. **Control rods** are *cylinders composed of substances that absorb neutrons, such as boron and cadmium, and can therefore slow the chain reaction*. By varying the depth of the control rods within the fuel-rod assembly (reactor core), one can increase or decrease the absorption of neutrons. If necessary, these rods can be dropped all the way into the fuel-rod assembly to stop the chain reaction.

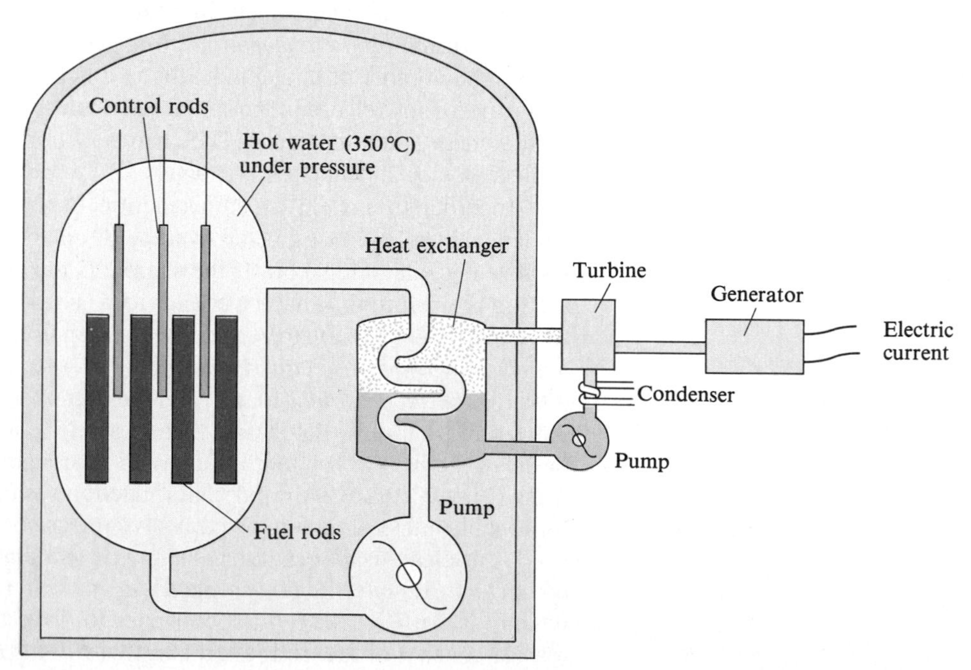

Figure 20.17
Light-water nuclear reactor (pressurized-water design). The fuel rods heat the water that is circulated to a heat exchanger. Steam is produced in the heat exchanger and passes to the turbine that drives a generator. A confinement shell surrounds the components that contain radioactive material so that leaks of radioactive material are confined.

A **moderator,** which is *a substance that slows down neutrons,* is required if uranium-235 is the fuel and this isotope is present as a small fraction of the total fuel. The neutrons that are released by the splitting of uranium-235 nuclei are absorbed more readily by uranium-238 than by other uranium-235 nuclei. However, when the neutrons are slowed down by a moderator, they are more readily absorbed by uranium-235, so it is possible to sustain a chain reaction with low fractional abundance of this isotope. Commonly used moderators are heavy water ($^2_1H_2O$), light water, and graphite.

In the light-water reactor, ordinary water acts as both a moderator and a coolant. Figure 20.17 shows a pressurized-water design of this type of reactor. Water in the reactor is maintained at about 350°C under high pressure (150 atm) so it does not boil. The hot water is circulated to a heat exchanger, where the heat is used to produce steam to run a turbine and generate electricity.

After a period of time, fission products that absorb neutrons accumulate in the fuel rods. This interferes with the chain reaction, so eventually the fuel rods must be replaced. Originally, the intention was to send these fuel rods to *reprocessing plants,* where fuel material could be chemically separated from the radioactive wastes. Opposition to constructing these plants has been intense, however. Plutonium-239 would be one of the fuel materials separated from the spent fuel rods. This isotope is produced during the operation of the reactor when uranium-238 is bombarded with neutrons. It is fissionable and can be used to construct atomic bombs. For this reason, many people believe that the availability of this element in large quantities would increase the chance that many countries and terrorist groups could divert enough plutonium to produce atomic bombs.

Whether or not the spent fuel rods are reprocessed, a pressing problem facing the nuclear power industry is how to dispose of radioactive wastes safely. One of many proposals is to encase the waste in a ceramic material and store the solids deep in the earth, perhaps in salt mines.

Breeder Reactors

Only about 1% of the energy available in natural uranium is actually released in a conventional nuclear reactor. This energy is released primarily from the splitting of uranium-235, which makes up only 0.72% of natural uranium. An additional small quantity of energy is obtained from the splitting of plutonium-239, produced when uranium-238 absorbs neutrons. In principle, if more of the uranium-238, the most abundant isotope of uranium, could be converted to plutonium, more energy could be obtained from a given quantity of uranium. This is the basic motivation for developing breeder reactors.

A **breeder reactor** is *a nuclear reactor designed to produce more fissionable material than it consumes*. In one possible design, the fuel would consist of plutonium-239 and uranium-235 oxides, which would be surrounded by uranium-238. No moderator would be used, but the high concentration of nuclear fuel would allow a chain reaction to occur. Because no moderator would be used, many more of the neutrons would be absorbed by uranium-238, so more of this isotope would be converted to plutonium-239. The net result of the breeder reactor would be to produce energy and to convert uranium-238 to plutonium-239, which could be used for additional fuel.

Critics of this design say that because reprocessing is needed to obtain the plutonium fuel, plutonium could be diverted to the manufacture of atomic bombs.

For this reason, breeder reactors have been proposed that use a molten salt bath containing dissolved nuclear fuels. The molten salt would circulate from the reactor to a reprocessing plant, then back to the reactor. This enclosed reprocessing operation is designed to reduce the risk of theft of bomb materials.

Up to this point we have discussed the possibility of using breeder reactors only to extend supplies of uranium. However, it is possible to convert thorium-232, which is the most abundant isotope of thorium, to fissionable uranium-233 using a breeder reactor. The reserves of thorium are more plentiful than those of uranium.

Nuclear Fusion

As we noted in Section 20.6, energy can be obtained by combining light nuclei into a heavier nucleus by nuclear fusion. Such fusion reactions have been observed in the laboratory by means of bombardment using particle accelerators. Deuterons (^2_1H nuclei), for example, can be accelerated toward targets containing deuterium (^2_1H atoms) or tritium (^3_1H atoms). The reactions are

$$^2_1H + ^2_1H \longrightarrow ^3_2He + ^1_0n$$
$$^2_1H + ^3_1H \longrightarrow ^4_2He + ^1_0n$$

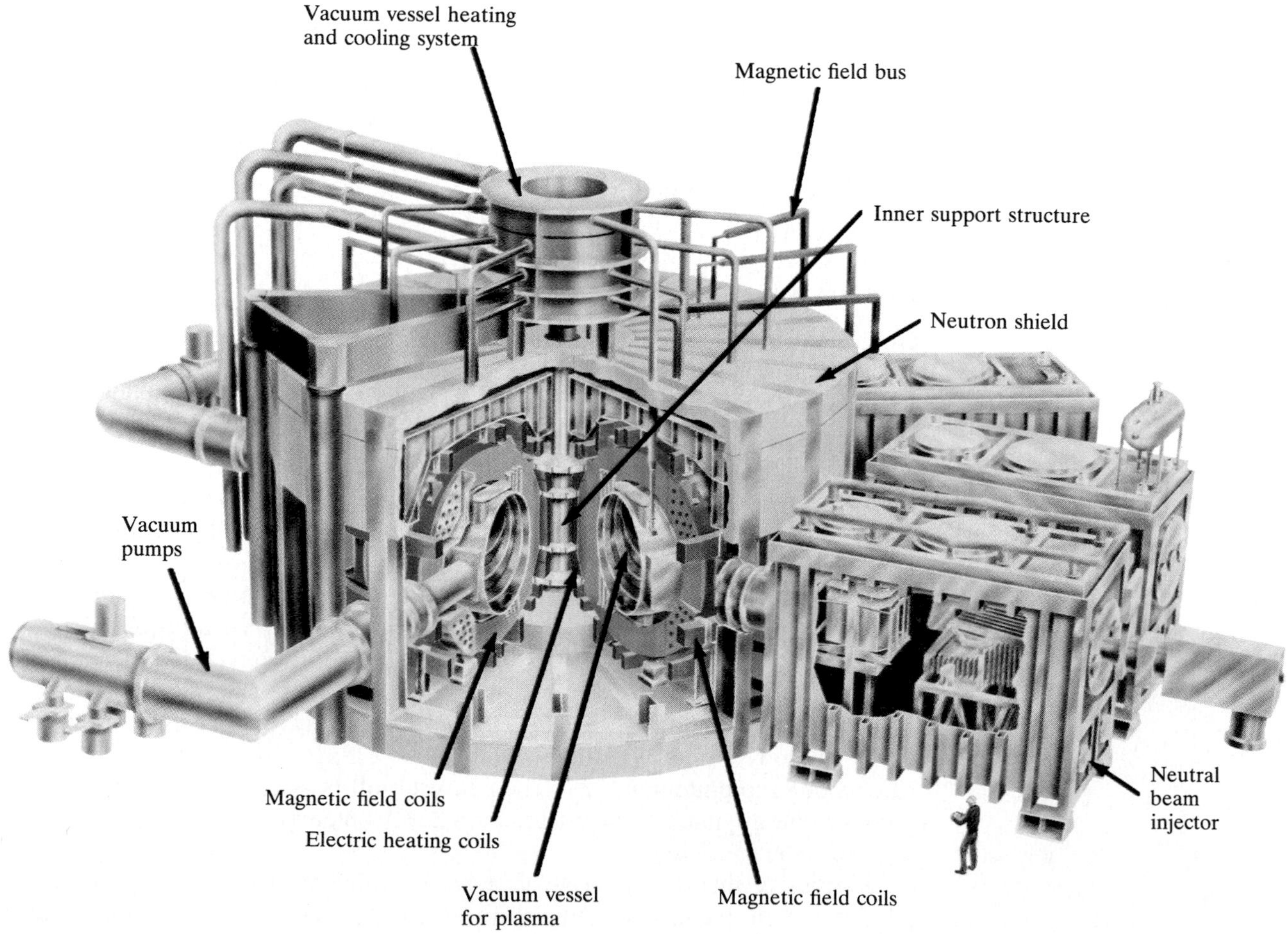

Figure 20.18
Tokamak nuclear fusion test reactor. This reactor at Princeton University's Plasma Physics Laboratory will test the feasibility of nuclear fusion power. A doughnut-shaped plasma is contained by a magnetic field. Initially, the plasma will be heated by a combination of methods, including electric heating and injection of a heated neutral atom beam into the plasma. Once heated to about 100 million °C, the plasma will be maintained at this temperature by nuclear fusion.

To get the nuclei to react, the bombarding nucleus must have enough kinetic energy to overcome the repulsion of electric charges of the nuclei. The first reaction uses only deuterium, which is present in ordinary water.● It is therefore very attractive as a source of energy. But, as we will discuss, the second reaction is more likely to be used first.

● Natural hydrogen contains 0.015% deuterium.

Energy cannot be obtained in a practical way using particle accelerators. Another way to give nuclei sufficient kinetic energy to react is by heating the nuclear materials to a sufficiently high temperature. The reaction of deuterium, $^{2}_{1}H$, and tritium, $^{3}_{1}H$, turns out to require the lowest temperature of any fusion reaction. For this reason it is likely to be the first fusion reaction developed as an energy source. For practical purposes, this temperature will need to be about 100 million °C. At this temperature, all atoms have been stripped of their electrons so that a plasma results. A **plasma** is *an electrically neutral gas of ions and electrons*. At 100 million °C, the plasma is essentially separate nuclei and electrons. Thus, the development of nuclear fusion requires study of the properties of plasmas at high temperatures.

It is now believed that the energy of stars, including our sun, where extremely high temperatures exist, derives from nuclear fusion. The hydrogen bomb also employs nuclear fusion for its destructive power. High temperature is first attained by a fission bomb. This then ignites fusion reactions in surrounding material of deuterium and tritium.

The main problem in developing controlled nuclear fusion is how to heat a plasma to high temperature and maintain those temperatures. When a plasma touches any material whatever, heat is quickly conducted away, and the plasma temperature quickly falls.● A *tokamak nuclear fusion reactor* uses a doughnut-shaped magnetic field to hold the plasma away from any material (see Figure 20.18). A *laser fusion reactor* employs a bank of lasers aimed at a single point. Pellets containing deuterium and tritium would drop into the reactor where they would be heated to 100 million °C by bursts of laser light.

● Even though the plasma is nearly 100 million °C above the melting point of any material, the total quantity of heat that could be transferred from the plasma is very small because its concentration is extremely low. Therefore, if the plasma were to touch the walls of the reactor, the plasma would cool but the walls would not melt.

Related Topic

THE CHERNOBYL NUCLEAR ACCIDENT

The nuclear accident that occurred at the Chernobyl reactor, north of Kiev in the U.S.S.R., on April 26, 1986, renewed fears in some about the safely of nuclear reactors. It is important to understand the nature of the accident at Chernobyl. It is not possible for a nuclear reactor using normal fuel elements to become an atomic bomb. However, without proper design and safeguards, it is possible for a malfunction of a reactor to disperse dangerous radioactivity over a populated area. This is in fact what occurred at Chernobyl.

The Chernobyl reactor had serious design problems. Unlike U.S. reactors, the Chernobyl reactor had no confinement shell surrounding it that would contain radioactive spills. Another problem was the design of the cooling system. The fuel core had a graphite moderator with water cooling in which some of the liquid water changed to steam. Liquid water is a good absorber of neutrons, whereas steam is not. This meant that as the reactor ran hotter, producing a greater percentage of steam, more neutrons became available for nuclear fission. Unless these extra neutrons were absorbed, the reactor would become even hotter and then would run out of control. Normally, as the reactor became hotter, control rods were automatically pushed in to absorb these extra neutrons. In a U.S. light-water reactor, the water coolant is always under pressure to maintain the liquid phase. As long as coolant water is present, the reactor is under control.

On the day of the Chernobyl accident, operators disabled the safety system on the reactor while they proceeded with an experimental test of the reactor. During the test, the reactor cooled excessively and threatened

(continued)

to shut down. If this happened, the reactor could not be restarted for a long period. Therefore, the operators removed most of the control rods (counter to safety instructions). Then the reactor began to overheat. With the safety system disabled, the operators were unable to work fast enough to correct the overheating, and the reactor went out of control.

The fuel rods melted and spilled their hot contents into the superheated water, which flashed into steam. The sudden pressure increase blew the top off the reactor and the roof off the building, spewing radioactive material skyward. Hot steam reacted with the zirconium shell of the fuel rods and with the graphite moderator to produce hydrogen gas, which ignited. The graphite moderator burned for a long period, spreading more radioactivity.

The cost of the Chernobyl accident was enormous. Thirty-one people died, and several hundred were hospitalized. Thousands of people had to be evacuated and resettled. The accident was the direct result of a faulty reactor coupled with a disregard of reactor safety procedures.

A Checklist for Review

IMPORTANT TERMS

radioactive decay (p. 792)
nuclear bombardment reaction (p. 792)
nuclear equation (20.1)
positron (20.1)
gamma photon (20.1)
nuclear force (20.1)
shell model of the nucleus (20.1)
magic number (20.1)
band of stability (20.1)
alpha emission (20.1)
beta emission (20.1)
positron emission (20.1)
electron capture (20.1)
gamma emission (20.1)
metastable nucleus (20.1)
radioactive decay series (20.1)
transmutation (20.2)
particle accelerator (20.2)
electron volt (20.2)
cyclotron (20.2)
deuterons (20.2)
transuranium elements (20.2)
Geiger counter (20.3)
scintillation counter (20.3)
activity of a radioactive source (20.3)
curie (Ci) (20.3)
rad (20.3)
rem (20.3)
radioactive decay constant (20.4)
half-life (20.4)
radioactive tracer (20.5)
isotope dilution (20.5)
neutron activation analysis (20.5)
binding energy (20.6)
mass defect (20.6)
nuclear fission (20.6)
nuclear fusion (20.6)
chain reaction (20.7)
critical mass (20.7)
nuclear fission reactor (20.7)
fuel rods (20.7)
control rods (20.7)
moderator (20.7)
breeder reactor (20.7)
plasma (20.7)

KEY EQUATIONS

$$\text{Rate} = kN_t \qquad t_{1/2} = \frac{0.693}{k} \qquad \ln\frac{N_t}{N_0} = -kt \quad \text{or} \quad \log\frac{N_t}{N_0} = \frac{-kt}{2.303} \qquad E = mc^2$$

SUMMARY OF FACTS AND CONCEPTS

Nuclear reactions are of two types, *radioactive decay* and *nuclear bombardment reactions*. Such reactions are represented by nuclear equations, each nucleus being denoted by a nuclide symbol. These equations must be balanced in charge (subscripts) and in nucleons (superscripts).

According to the *nuclear shell model,* the nucleons are arranged in shells. *Magic numbers* are the numbers of nucleons in a completed shell of protons or neutrons. Nuclei with magic numbers of protons or neutrons are especially stable. Pairs of protons and pairs of neutrons are also especially stable. When placed on a plot of N versus Z, stable nuclei fall in a *band of stability*. Those radioactive nuclides that fall to the left of the band of stability in this plot usually decay by beta emission. Those radioactive nuclides that fall to the right of the band of stability usually decay by positron emission or electron capture. However, nuclides with $Z > 83$ often decay by alpha emission. Uranium-238 forms a *radioactive decay series*. In such series, one element decays to another, which decays to another, and so forth, until a stable isotope is reached (lead-206, in the case of the uranium-238 series).

Transmutation of elements has been carried out in the laboratory by bombarding nuclei with various atomic particles. Alpha particles from natural sources can be used as reactants with light nuclei. For heavier nuclei, positive ions such as alpha particles must first be accelerated in a *particle accelerator*. Many of the *transuranium elements* have been obtained by bombardment of elements with accelerated particles. For example, plutonium was first made by bombarding uranium-238 with deuterons ($^{2}_{1}H$ nuclei) from a *cyclotron*, a type of particle accelerator.

Particles of radiation from nuclear processes can be counted by Geiger counters or scintillation counters. In a *Geiger counter*, the particle ionizes a gas, which then conducts a pulse of electricity between two electrodes. In a *scintillation counter*, the particle hits a phosphor, and this emits a flash of light that is detected by a photomultiplier tube. The activity of a radioactive source, or the number of nuclear disintegrations per unit time, is measured in units of *curies* (3.700×10^{10} disintegrations per second).

Radiation affects biological organisms by breaking chemical bonds. The *rad* is the measure of radiation dosage that deposits 1×10^{-2} J of energy per kilogram of tissue. A *rem* equals the number of rads times a factor to account for the relative biological effectiveness of the radiation.

Radioactive decay is a first-order rate process. The rate is characterized by the *decay constant, k,* or by the *half-life*, $t_{1/2}$. The quantities k and $t_{1/2}$ are related. Knowing one or the other, we can calculate how long it will take for a given radioactive sample to decay by a certain fraction. Methods of *radioactive dating* depend on determining the fraction of a radioactive isotope that has decayed and, from this, the time that has elapsed.

Radioactive isotopes are used as *radioactive tracers* in chemical analysis and medicine. *Isotope dilution* is one application of radioactive tracers in which the dilution of the tracer can be related to the original quantity of nonradioactive isotope. *Neutron activation analysis* is a method of analysis that depends on the conversion of elements to radioactive isotopes by neutron bombardment.

According to Einstein's *mass–energy equivalence*, mass is related to energy by the equation $E = mc^2$. A nucleus has less mass than the sum of the masses of the separate nucleons. The positive value of this mass difference is called the *mass defect;* it is equivalent to the *binding energy* of the nucleus. Nuclides having mass numbers near 50 have the largest binding energies per nucleon. It follows that heavy nuclei should tend to split, a process called *nuclear fission*, and that light nuclei should tend to combine, a process called *nuclear fusion*. Tremendous amounts of energy are released in both processes. Nuclear fission is used in conventional nuclear power reactors. Nuclear fusion reactors are in the experimental stage.

OPERATIONAL SKILLS

1. **Writing a nuclear equation** Given a word description of a radioactive decay process, write the nuclear equation (Example 20.1).

2. **Deducing a product or reactant in a nuclear equation** Given all but one of the reactants and products in a nuclear reaction, find that one nuclide (Examples 20.2 and 20.6).

3. **Predicting the relative stabilities of nuclides** Given a number of nuclides, determine which are most likely to be radioactive and which are most likely to be stable (Example 20.3).

4. **Predicting the type of radioactive decay** Predict the type of radioactive decay that is most likely for given nuclides (Example 20.4).

5. **Using the notation for a bombardment reaction** Given an equation for a nuclear bombardment reaction, write the abbreviated notation, or vice versa (Example 20.5).

6. **Calculating the decay constant from the activity** Given the activity (disintegrations per second) of a radioactive isotope, obtain the decay constant (Example 20.7).

7. **Relating the decay constant, half-life, and activity** Given the decay constant of a radioactive isotope, obtain the half-life (Example 20.8), or vice versa (Example 20.9). Given the decay constant and mass of a radioactive isotope, calculate the activity of the sample (Example 20.9).

8. **Determining the fraction of nuclei remaining after a specified time** Given the half-life of a radioactive isotope, calculate the fraction remaining after a specified time (Example 20.10).

9. **Applying the carbon-14 dating method** Given the disintegrations of carbon-14 nuclei per gram of carbon in a dead organic object, calculate the age of the object—that is, the time since its death (Example 20.11).

10. **Calculating the energy change for a nuclear reaction** Given nuclear masses, calculate the energy change for a nuclear reaction (Example 20.12). Obtain the answer in joules per mole or MeV per particle.

Review Questions

20.1 What are the two types of nuclear reaction? Give an example of a nuclear equation for each type.

20.2 What are *magic numbers?* Give several examples of nuclei with magic numbers of protons.

20.3 List characteristics to look for in a nucleus in order to predict whether it is stable.

20.4 What are the five common types of radioactive decay? What is the condition that usually leads to each type of decay?

20.5 What are the isotopes that begin each of the naturally occurring radioactive decay series?

20.6 Give equations for (a) the first transmutation of an element obtained in the laboratory by nuclear bombardment, (b) the reaction that produced the first artificial, radioactive isotope.

20.7 What is a particle accelerator, and how does one operate? Why are they required for certain nuclear reactions?

20.8 In what major way has the discovery of transuranium elements affected the form of modern periodic tables?

20.9 Describe how a Geiger counter works. How does a scintillation counter work?

20.10 Define the units *curie, rad,* and *rem.*

20.11 The half-life of cesium-137 is 30.2 y. How long will it take for a sample of cesium-137 to decay to 1/8 its original mass?

20.12 What is the age of a rock that contains equal numbers of $^{40}_{19}K$ and $^{40}_{18}Ar$ nuclei? The half-life of $^{40}_{19}K$ is 1.28×10^9 y.

20.13 What is a radioactive tracer? Give an example of the use of such a tracer in chemistry.

20.14 Isotope dilution has been used to obtain the volume of blood supply in a living animal. Explain how this could be done.

20.15 Briefly describe neutron activation analysis.

20.16 The deuteron, $^{2}_{1}H$, has a mass that is smaller than the sum of the masses of its constituents, the proton plus the neutron. Explain why this is so.

20.17 Certain stars obtain their energy from nuclear reactions such as

$$^{12}_{6}C + ^{12}_{6}C \longrightarrow ^{23}_{11}Na + ^{1}_{1}H$$

Explain in a sentence or two why this reaction might be expected to release energy.

20.18 Briefly describe how each of the following operates: (a) nuclear fission reactor; (b) breeder reactor; (c) tokamak fusion reactor.

Practice Problems

RADIOACTIVITY

20.19 Rubidium-87, which forms about 28% of natural rubidium, is radioactive, decaying by the emission of a single beta particle to strontium-87. Write the nuclear equation for this decay of rubidium-87.

20.20 Write the nuclear equation for the decay of phosphorus-32 to sulfur-32 by beta emission. A phosphorus-32 nucleus emits a beta particle and gives a sulfur-32 nucleus.

20.21 Thorium is a naturally occurring radioactive element. Thorium-232 decays by emitting a single alpha particle to produce radium-228. Write the nuclear equation for this decay of thorium-232.

20.22 Radon is a radioactive noble gas formed in soil containing radium. Radium-226 decays by emitting a single alpha particle to produce radon-222. Write the nuclear equation for this decay of radium-226.

20.23 Fluorine-18 is an artificially produced radioactive isotope. It decays by emitting a single positron. Write the nuclear equation for this decay.

20.24 Sodium-22 is an artificially produced radioactive isotope. It decays by emitting a single positron. Write the nuclear equation for this decay.

20.25 Polonium was discovered in uranium ores by Marie and Pierre Curie. Polonium-210 decays by emitting a single alpha particle. Write the nuclear equation for this decay.

20.26 Actinium was discovered in uranium ore residues by André-Louis Debierne. Actinium-227 decays by emitting a single alpha particle. Write the nuclear equation for this decay.

20.27 From each of the following pairs, choose the nuclide that is radioactive (one is known to be radioactive, the other stable). Explain your choice.
(a) $^{122}_{51}Sb$, $^{136}_{54}Xe$ (b) $^{204}_{82}Pb$, $^{204}_{85}At$ (c) $^{87}_{37}Rb$, $^{80}_{37}Rb$

20.28 From each of the following pairs, choose the nuclide that is radioactive (one is known to be radioactive, the other stable). Explain your choice.
(a) $^{102}_{47}Ag$, $^{109}_{47}Ag$ (b) $^{25}_{12}Mg$, $^{24}_{10}Ne$ (c) $^{203}_{81}Tl$, $^{223}_{90}Th$

20.29 Predict the type of radioactive decay process that is likely for each of the following nuclides. See Figure 20.2.
(a) $^{228}_{92}U$ (b) $^{8}_{5}B$ (c) $^{68}_{29}Cu$

20.30 Predict the type of radioactive decay process that is likely for each of the following nuclides. See Figure 20.2.
(a) $^{60}_{30}Zn$ (b) $^{6}_{2}He$ (c) $^{241}_{93}Np$

20.31 Four radioactive decay series are known—three naturally occurring, and one beginning with the synthetic isotope $^{241}_{94}Pu$. To which of these decay series does the isotope $^{219}_{86}Rn$ belong? To which series does $^{220}_{86}Rn$ belong? Note that each isotope in these series decays by either α emission or β emission. How do these decay processes affect the mass number?

20.32 Four radioactive decay series are known—three naturally occurring, and one beginning with the synthetic isotope $^{241}_{94}Pu$. To which of these decay series does the isotope $^{227}_{89}Ac$ belong? To which series does $^{225}_{89}Ac$ belong? Note that each isotope in these series decays by either α emission or β emission. How do these decay processes affect the mass number?

NUCLEAR BOMBARDMENT REACTIONS

20.33 Write the abbreviated notations for the following bombardment reactions.
(a) $^{26}_{12}Mg + ^{2}_{1}H \longrightarrow ^{24}_{11}Na + ^{4}_{2}He$
(b) $^{16}_{8}O + ^{1}_{0}n \longrightarrow ^{16}_{7}N + ^{1}_{1}p$

20.34 Write the abbreviated notations for the following bombardment reactions.
(a) $^{14}_{7}N + ^{1}_{1}p \longrightarrow ^{11}_{6}C + ^{4}_{2}He$
(b) $^{63}_{29}Cu + ^{4}_{2}He \longrightarrow ^{66}_{31}Ga + ^{1}_{0}n$

20.35 Write out the nuclear equations for the following bombardment reactions.
(a) $^{27}_{13}Al(d, \alpha)^{25}_{12}Mg$ (b) $^{10}_{5}B(\alpha, p)^{13}_{6}C$

20.36 Write out the nuclear equations for the following bombardment reactions.
(a) $^{45}_{21}Sc(n, \alpha)^{42}_{19}K$ (b) $^{63}_{29}Cu(p, n)^{63}_{30}Zn$

20.37 A proton is accelerated to 13.8 MeV per particle. What is this energy in kJ/mol?

20.38 An alpha particle is accelerated to 23.1 MeV per particle. What is this energy in kJ/mol?

20.39 Fill in the missing spaces in the following reactions.
(a) $^{6}_{3}Li + ^{1}_{0}n \longrightarrow ? + ^{3}_{1}H$
(b) $^{232}_{90}Th(?, n)^{235}_{92}U$

20.40 Fill in the missing spaces in the following reactions.
(a) $^{27}_{13}Al + ^{3}_{1}H \longrightarrow ^{27}_{12}Mg + ?$
(b) $^{12}_{6}C(^{3}H, ?)^{14}_{6}C$

20.41 Curium was first synthesized by bombarding an element with alpha particles. The products were curium-242 and a neutron. What was the target element?

20.42 Californium was first synthesized by bombarding an element with alpha particles. The products were californium-245 and a neutron. What was the target element?

RATE OF RADIOACTIVE DECAY

20.43 Tritium, or hydrogen-3, is prepared by bombarding lithium-6 with neutrons. A 0.250-mg sample of tritium decays at the rate of 8.94×10^{10} disintegrations per second. What is the decay constant (in /s) of tritium, whose atomic mass is 3.02 amu?

20.44 The first isotope of plutonium discovered was plutonium-238. It is used to power batteries for heart pacemakers. A sample of plutonium-238 weighing 2.8×10^{-6} g decays at the rate of 1.8×10^{6} disintegrations per second. What is the decay constant of plutonium-238 in reciprocal seconds (/s)?

20.45 Sulfur-35 is a radioactive isotope used in chemical and medical research. A 0.48-mg sample of sulfur-35 has an activity of 20.4 Ci. What is the decay constant of sulfur-35 (in /s)?

20.46 Sodium-24 is used in medicine to study the circulatory system. A sample weighing 5.2×10^{-6} g has an activity of 45.3 Ci. What is the decay constant of sodium-24 (in /s)?

20.47 Rubidium-87 is a radioactive isotope occurring in natural rubidium. The decay constant is 4.6×10^{-19}/s. What is the half-life in years?

20.48 Neptunium-237 was the first isotope of a transuranium element to be discovered. The decay constant is 1.03×10^{-14}/s. What is the half-life in years?

20.49 Carbon-14 has been used to study the mechanisms of reactions that involve organic compounds. The half-life of carbon-14 is 5.73×10^{3} y. What is the decay constant (in /s)?

20.50 Promethium-147 has been used in luminous paint for dials. The half-life of this isotope is 2.5 y. What is the decay constant (in /s)?

20.51 Gold-198 has a half-life of 2.69 d. What is the activity (in curies) of a 0.43-mg sample?

20.52 Cesium-134 has a half-life of 2.05 y. What is the activity (in curies) of a 0.75-mg sample?

20.53 A sample of a phosphorus compound contains phosphorus-32. This sample of radioactive isotope is decaying at the rate of 6.0×10^{12} disintegrations per second. How many grams of ^{32}P are in the sample? The half-life of ^{32}P is 14.3 d.

20.54 A sample of sodium thiosulfate, $Na_2S_2O_3$, contains sulfur-35. Determine the mass of ^{35}S in the sample from the decay rate, which was determined to be 7.7×10^{11} disintegrations per second. The half-life of ^{35}S is 88 d.

20.55 A sample of sodium-24 was administered to a patient to test for faulty blood circulation by comparing the radioactivity reaching various parts of the body. What fraction of the sodium-24 nuclei would remain undecayed after 24.0 h? The half-life is 15.0 h. If a sample contains 5.0 μg of ^{24}Na, how many micrograms remain after 24.0 h?

20.56 A solution of sodium iodide containing iodine-131 was given to a patient to test for malfunctioning of the thyroid gland. What fraction of the iodine-131 nuclei would remain undecayed after 7.0 d? If a sample contains 5.0 μg of ^{131}I, how many micrograms remain after 7.0 d? The half-life of I-131 is 8.07 d.

20.57 If 28.0% of a sample of silver-112 decays in 1.52 h, what is the half-life of this isotope (in hours)?

20.58 If 18.0% of a sample of zinc-65 decays in 69.9 d, what is the half-life of this isotope (in days)?

20.59 A sample of iron-59 initially registers 125 counts per second on a radiation counter. After 10.0 d, the sample registers 107 counts per second. What is the half-life (in days) of iron-59?

20.60 A sample of copper-64 gives a reading of 88 counts per second on a radiation counter. After 9.5 h, the sample gives a reading of 53 counts per second. What is the half-life (in hours) of copper-64?

20.61 Carbon from a cypress beam obtained from the tomb of Sneferu, a king of ancient Egypt, gave 8.1 disintegrations of ^{14}C per minute per gram of carbon. How old is the cypress beam? Carbon from living material gives 15.3 disintegrations of ^{14}C per minute per gram of carbon.

20.62 Carbon from the Dead Sea Scrolls, very old manuscripts found in Israel, gave 12.1 disintegrations of ^{14}C per minute per gram of carbon. How old are the manuscripts? Carbon from living material gives 15.3 disintegrations of ^{14}C per minute per gram of carbon.

20.63 Several hundred pair of sandals found in a cave in Oregon were found by carbon-14 dating to be 9.0×10^3 years old. What must have been the activity of the carbon-14 in the sandals in disintegrations per minute per gram? Assume the original activity was 15.3 disintegrations per minute per gram.

20.64 Carbon in some mammoth bones found in Arizona were found by carbon-14 dating to be 1.13×10^4 years old. What must have been the activity of the carbon-14 in the bones in disintegrations per minute per gram? Assume the original activity was 15.3 disintegrations per minute per gram.

MASS–ENERGY EQUIVALENCE

20.65 Find the change of mass (in grams) resulting from the release of heat when 1 mol H_2 reacts with 1 mol Cl_2.

$$H_2(g) + Cl_2(g) \longrightarrow 2HCl(g),\ \Delta H = -185 \text{ kJ}$$

20.66 Find the change of mass (in grams) resulting from the release of heat when 1 mol SO_2 is formed from the elements.

$$S(s) + O_2(g) \longrightarrow SO_2(g),\ \Delta H = -297 \text{ kJ}$$

20.67 Calculate the energy change for the following nuclear reaction (in joules per mole of ^2_1H).

$$^2_1H + ^3_1H \longrightarrow ^4_2He + ^1_0n$$

Give the energy change in MeV per ^2_1H nucleus. See Table 20.3.

20.68 Calculate the change in energy, in joules per mole of ^1_1H, for the following nuclear reaction.

$$^1_1H + ^1_1H \longrightarrow ^2_1H + ^0_1e$$

Give the energy change in MeV per ^1_1H nucleus. See Table 20.3.

20.69 Obtain the mass defect (in amu) and binding energy (in MeV) for the ^6_3Li nucleus. What is the binding energy (in MeV) per nucleon? See Table 20.3.

20.70 Obtain the mass defect (in amu) and binding energy (in MeV) for the $^{56}_{26}Fe$ nucleus. What is the binding energy (in MeV) per nucleon? See Table 20.3.

Additional Problems

20.71 Sodium-23 is the only stable isotope of sodium. Predict how sodium-20 will decay and how sodium-26 will decay.

20.72 Aluminum-27 is the only stable isotope of aluminum. Predict how aluminum-24 will decay and how aluminum-30 will decay.

20.73 A uranium-235 nucleus decays by a series of alpha and beta emissions until it reaches lead-207. How many alpha emissions and how many beta emissions occur in this series of decays?

20.74 A thorium-232 nucleus decays by a series of alpha and beta emissions until it reaches lead-208. How many alpha emissions and how many beta emissions occur in this series of decays?

20.75 A bismuth-209 nucleus reacts with an alpha particle to produce an astatine nucleus and two neutrons. Write the complete nuclear equation for this reaction.

20.76 A bismuth-209 nucleus reacts with a deuteron to produce a polonium nucleus and a neutron. Write the complete nuclear equation for this reaction.

20.77 Complete the following equation by filling in the blank.

$$^{238}_{92}U + ^{12}_{6}C \longrightarrow \text{———} + 4^{1}_{0}n$$

20.78 Complete the following equation by filling in the blank.

$$^{246}_{96}Cm + ^{12}_{6}C \longrightarrow \text{———} + 4^{1}_{0}n$$

20.79 Tritium, or hydrogen-3, is formed in the upper atmosphere by cosmic rays, similar to the formation of carbon-14. Tritium has been used to determine the age of wines. A certain wine that has been aged in a bottle has a tritium content only 78% of that in a similar wine of the same mass that has just been bottled. How long has the aged wine been in the bottle? The half-life of tritium is 12.3 y.

20.80 The naturally occurring isotope rubidium-87 decays by beta emission to strontium-87. This decay is the basis of a method for determining the ages of rocks. A sample of rock contains 102.1 μg ^{87}Rb and 5.3 μg ^{87}Sr. What is the age of the rock? The half-life of rubidium-87 is 4.8×10^{10} y.

20.81 When a positron and an electron collide, they are annihilated and two gamma photons of equal energy are emitted. Calculate the wavelength corresponding to this gamma emission.

20.82 When technetium-99*m* decays to technetium-99, a gamma photon corresponding to an energy of 0.143 MeV is emitted. What is the wavelength of this gamma emission? What is the difference in mass between Tc-99*m* and Tc-99?

20.83 Calculate the energy released when 1.00 kg of uranium-235 undergoes the following fission process.

$$^{1}_{0}n + ^{235}_{92}U \longrightarrow ^{136}_{53}I + ^{96}_{39}Y + 4^{1}_{0}n$$

The masses of $^{136}_{53}I$ and $^{96}_{39}Y$ nuclei are 135.8401 amu and 95.8629 amu, respectively. Other masses are given in Table 20.3. Compare this energy with the heat released when 1.00 kg C(graphite) burns to $CO_2(g)$.

20.84 Calculate the energy released when 1.00 kg of hydrogen-1 undergoes fusion to helium-4, according to the following reaction.

$$4^{1}_{1}H \longrightarrow ^{4}_{2}He + 2^{0}_{1}e$$

This reaction is one of the principal sources of energy from the sun. See Table 20.3 for data. Compare the energy released by 1.00 kg of $^{1}_{1}H$ in this reaction to the heat released when 1.00 kg of C(graphite) burns to $CO_2(g)$.

Cumulative-Skills Problems

20.85 A sample of sodium phosphate, Na_3PO_4, weighing 54.5 mg contains radioactive phosphorus-32 (with mass 32.0 amu). If 15.6% of the phosphorus atoms in the compound is phosphorus-32 (the remainder is naturally occurring phosphorus), how many disintegrations of this nucleus occur per second in this sample? Phosphorus-32 has a half-life of 14.3 d.

20.86 A sample of sodium thiosulfate, $Na_2S_2O_3$, weighing 38.1 mg contains radioactive sulfur-35 (with mass 35.0 amu). If 22.3% of the sulfur atoms in the compound is sulfur-35 (the remainder is naturally occurring sulfur), how many disintegrations of this nucleus occur per second in this sample? Sulfur-35 has a half-life of 87.9 d.

20.87 Polonium-210 has a half-life of 138.4 days, decaying by alpha emission. Suppose the helium gas originating from the alpha particles in this decay were collected. What volume of helium at 25°C and 735 mmHg could be obtained from 1.0000 g of polonium dioxide, PoO_2, in a period of 48.0 hrs?

20.88 Radium-226 decays by alpha emission to radon-222, a noble gas. What volume of pure radon-222 at 23°C and 785 mmHg could be obtained from 543.0 mg of radium bromide, $RaBr_2$, in a period of 20.2 yrs? The half-life of radium-226 is 1602 years.

20.89 What is the energy (in joules) evolved when 1 mol of helium-4 nuclei is produced from protons and neutrons? How many liters of ethane, C_2H_6, at 25°C and 725 mmHg are needed to evolve the same quantity of energy when the ethane is burned in oxygen to CO_2 and H_2O? See Table 20.3 and Appendix C for data.

20.90 Plutonium-239 has been used as a power source for heart pacemakers. What is the energy obtained from the following decay of 215 mg of plutonium-239?

$$^{239}_{94}Pu \longrightarrow ^{4}_{2}He + ^{235}_{92}U$$

Suppose the electric energy produced from this amount of plutonium-239 is 25.0% of this value. What is the minimum grams of zinc that would be needed for the standard voltaic cell $Zn|Zn^{2+}||Cu^{2+}|Cu$ to obtain the same electric energy?

21

The Main-Group Elements: Groups IA to IIIA; Metallurgy

Chapter Outline

Gallium arsenide rod from which chips are made for electronic devices.

The main-group or representative metals—those elements in the A columns of the periodic table—have valence-shell configurations $ns^a np^b$. Within a particular column or family of elements, the valence-shell configurations are similar. As a result, the elements in a column often show similar physical and chemical properties. The alkali metals, for example, are very much alike both physically and chemically. So too are the alkaline earth metals.

Perhaps even more noteworthy are the systematic trends we see when we read down a given column or across a given period. Ionization energies, electron affinities, electronegativities, and other properties change from one element to the next in a fairly definite way. Indeed, this is what allowed Mendeleev to predict the properties of germanium 15 years before the element was discovered.

In this chapter and the next, we will examine the similarities within a family of elements and the trends in a given column or period.

21.1 GENERAL OBSERVATIONS ABOUT THE MAIN-GROUP ELEMENTS

We pointed out in the chapter opening that we can observe trends in the behavior of the elements across a row or down a column of the periodic table. As you study the chemical and physical properties of the main-group elements, you should have these trends firmly in mind. Let us first look at the variation in metallic–nonmetallic character of these elements.

Metallic–Nonmetallic Character

Table 21.1 compares properties of the metallic and nonmetallic elements. A metal has a characteristic shine or luster and is a conductor of heat and electricity. Most metallic elements are malleable, ductile solids. The nonmetallic elements are nonlustrous. Most are diatomic gases (O_2, N_2, Cl_2) or hard, brittle solids (C, S_8, I_2) at

Table 21.1 Comparison of Metallic and Nonmetallic Elements

Metals	Nonmetals
Lustrous	Nonlustrous
Solids at 20°C (except Hg, which is a liquid)	Solids or gases at 20°C (except Br_2, which is a liquid)
Solids are malleable and ductile	Solids are usually hard and brittle
Conductors of heat and electricity	Nonconductors of heat and electricity (except graphite, an allotrope of C)
Low ionization energies	Moderate to high ionization energies
Low electronegativities	Moderate to high electronegativities
Form cations	Form monatomic anions or oxoanions
Oxides are basic (unless metal is in a high oxidation state)	Oxides are acidic

20°C and 1 atm. Figure 21.1 shows some main-group elements, displaying a range of metallic–nonmetallic character.

The metallic elements have low ionization energies and low electronegativities compared with the nonmetallic elements. As a result, the metals often form cations in compounds or in aqueous solution (Na^+, Ca^{2+}, Al^{3+}). Nonmetals, on the other hand, form monatomic anions (O^{2-}, Cl^-) and oxoanions (NO_3^-, SO_4^{2-}).

Oxides of the metallic elements are usually basic. The oxides of the most active metals react with water to give basic solutions. For example,

$$Na_2O(s) + H_2O(l) \longrightarrow 2Na^+(aq) + 2OH^-(aq)$$
$$CaO(s) + H_2O(l) \longrightarrow Ca^{2+}(aq) + 2OH^-(aq)$$

As the bonding in a metal oxide becomes less ionic (as it does in higher oxidation states), the oxide becomes less basic.● Aluminum oxide, Al_2O_3, is an example of a metal oxide that is amphoteric. An **amphoteric substance** is *a substance that has both acidic and basic properties*. Aluminum oxide dissolves in acids to produce the cation, as expected for a metal oxide.

● Some transition-metal oxides in high oxidation states are acidic. For example, chromium trioxide, CrO_3, is an acidic oxide. The corresponding acid is chromic acid, H_2CrO_4, which forms chromate salts.

$$Al_2O_3(s) + 6H^+(aq) \longrightarrow 2Al^{3+}(aq) + 3H_2O(l)$$

But the oxide also dissolves in strong base.

$$Al_2O_3(s) + 3H_2O(l) + 2OH^-(aq) \longrightarrow 2Al(OH)_4^-(aq)$$

Figure 21.1
Some main-group elements. *Left to right:* Silicon (metalloid); rhombic sulfur (nonmetal); red phosphorus (nonmetal); aluminum (metal).

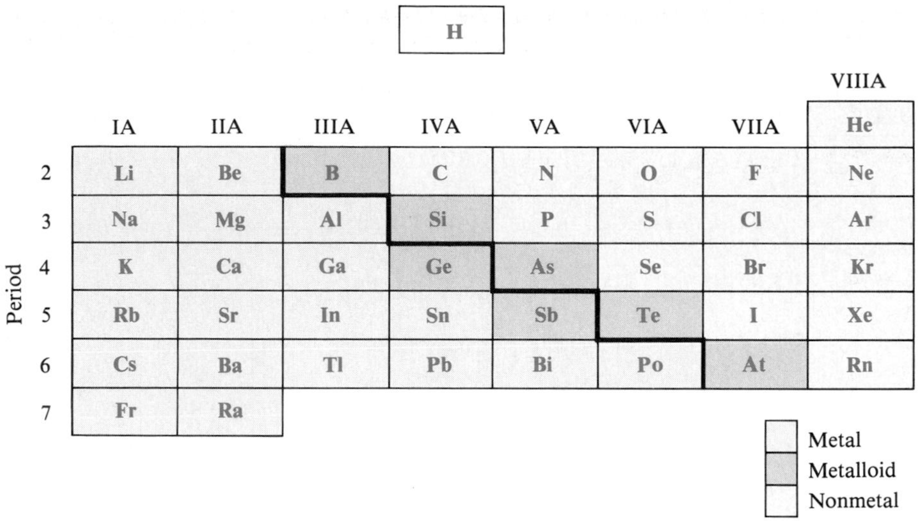

Figure 21.2
An abridged periodic table showing the main-group elements. Elements to the left of the heavy staircase line are largely metallic in character; those to the right are largely nonmetallic.

				H				
Period	IA	IIA	IIIA	IVA	VA	VIA	VIIA	VIIIA
								He
2	Li	Be	B	C	N	O	F	Ne
3	Na	Mg	Al	Si	P	S	Cl	Ar
4	K	Ca	Ga	Ge	As	Se	Br	Kr
5	Rb	Sr	In	Sn	Sb	Te	I	Xe
6	Cs	Ba	Tl	Pb	Bi	Po	At	Rn
7	Fr	Ra						

Metal
Metalloid
Nonmetal

In this case the aluminate anion, $Al(OH)_4^-$, is formed. Note that Al—O bonds, which have considerable covalent character, are retained in going from Al_2O_3 to the ion $Al(OH)_4^-$.

Nonmetal oxides are acidic. In this case, covalent bonds between oxygen and the element are retained in reacting with water or base by forming oxoanions. For example,

$$SO_2(g) + H_2O(l) \rightleftharpoons H^+(aq) + HSO_3^-(aq)$$
$$SO_2(g) + 2OH^-(aq) \longrightarrow H_2O(l) + SO_3^{2-}(aq)$$

Figure 21.2 shows a periodic table in which only the main-group elements appear. Those elements to the left of the heavy "staircase" line are largely metallic in behavior; those to the right are largely nonmetallic. Elements along the line are metalloids, or semimetals. These elements exhibit characteristics of both metals and nonmetals. Though nonmetallic in much of their chemical behavior, these elements have allotropes that have a luster and conduct electricity, though not so well as metals. They are *semiconductors*.

The metallic characteristics of the elements in the periodic table decrease in going across a period from left to right. Figure 21.2 illustrates this in a broad way. In any period, the elements on the left are metals and those on the right are nonmetals. Also, note the second-period elements shown in Figure 21.1. When placed in the order of the periodic table, they are: aluminum (a metal), silicon (metalloid), phosphorus (nonmetal), and sulfur (nonmetal). Table 21.2, which lists the oxides of the main-group elements, more clearly shows the gradual change from metallic to nonmetallic characteristics in a given period. Sodium oxide, Na_2O, on the far left, is a strongly basic oxide (characteristic of a metal). Proceeding to the right, we encounter magnesium oxide, MgO (strongly basic); aluminum oxide, Al_2O_3 (amphoteric); silicon dioxide, SiO_2 (weakly acidic); phosphorus(V) oxide, P_4O_{10} (moderately acidic); sulfur trioxide, SO_3 (strongly acidic); and dichlorine heptoxide, Cl_2O_7 (strongly acidic).

The metallic characteristics of the elements in the periodic table become more important going down any column (group). This trend is most pronounced in Groups IIIA to VA. For example, in Group IVA, carbon (second period) is a

Table 21.2 Acid–Base Behavior of the Oxides of the Main-Group Elements*

Period	Group						
	IA	IIA	IIIA	IVA	VA	VIA	VIIA
2	Li_2O (s.b.)	BeO (amph.)	B_2O_3 (w.a.)	CO_2 (w.a.)	N_2O_5 (s.a.) N_2O_3 (w.a.)	—	—
3	Na_2O (s.b.)	MgO (s.b.)	Al_2O_3 (amph.)	SiO_2 (w.a.)	P_4O_{10} (a.) P_4O_6 (w.a.)	SO_3 (s.a.) SO_2 (w.a.)	Cl_2O_7 (s.a.) Cl_2O (w.a.)
4	K_2O (s.b.)	CaO (s.b.)	Ga_2O_3 (amph.)	GeO_2 (w.a.)	As_2O_5 (w.a.) As_4O_6 (amph.)	SeO_3 (s.a.) SeO_2 (w.a.)	Br_2O (w.a.)
5	Rb_2O (s.b.)	SrO (s.b.)	In_2O_3 (w.b.)	SnO_2 (amph.) SnO (amph.)	Sb_2O_5 (w.a.) Sb_4O_6 (amph.)	TeO_3 (w.a.) TeO_2 (amph.)	I_2O_5 (a.)
6	Cs_2O (s.b.)	BaO (s.b.)	Tl_2O_3 (w.b.) Tl_2O (s.b.)	PbO_2 (amph.) PbO (amph.)	Bi_2O_5 (w.a.) Bi_2O_3 (w.b.)	PoO_2 (amph.) PoO (w.b.)	—

*The acid–base behavior is indicated as follows: s.b. = strongly basic; w.b. = weakly basic; amph. = amphoteric; w.a. = weakly acidic; a. = moderately acidic; s.a. = strongly acidic.

nonmetal, germanium (fourth period) is a metalloid, and lead (sixth period) is a metal. Note the changes in acid–base behavior of the oxides in each column of elements.

Example 21.1 Predicting the Relative Metallic–Nonmetallic Character of Elements

Choose the more metallic element in each of the following pairs: (a) Li or Be; (b) Be or Mg; (c) Al or K. Explain your answers in terms of the elements' positions in the periodic table.

Solution

(a) Li and Be are in the same period, and Li falls to the left of Be. Therefore, **Li is the more metallic.**

(b) Be and Mg are in the same column, and Mg falls below Be. Therefore, **Mg is the more metallic.**

(c) Al is in the same period as Na, which is to its left and therefore is more metallic. K is below Na and more metallic than it. Hence, **K is more metallic than Al.**

Exercise 21.1

For each of the following pairs, select the element that is more nonmetallic: (a) Ca or Ga; (b) Ga or B; (c) Be or Cs. Explain your answers. (See Problems 21.41 and 21.42.)

Oxidation States

Table 21.3 shows the oxidation states displayed by compounds of the main-group elements. From this, we can draw a number of conclusions.

1. The elements of Groups IA to IIIA have a common oxidation state equal to the group number (the number of valence electrons). This is the only common oxidation state found in compounds of Groups IA and IIA metals, and it is the only important one in compounds of Group IIIA elements except for thallium. In the case of thallium, the +1 oxidation state is most common.

Table 21.3 Oxidation States in Compounds of the Main-Group Elements*

	Group						
Period	IA	IIA	IIIA	IVA	VA	VIA	VIIA
2	Li +1	Be +2	B +3	C +4 +2 −4	N +5 +4 +3 +2 +1 −3	O −1 −2	F −1
3	Na +1	Mg +2	Al +3	Si +4 −4	P +5 +3 −3	S +6 +4 +2 −2	Cl +7 +5 +3 +1 −1
4	K +1	Ca +2	Ga +3	Ge +4 +2	As +5 +3 −3	Se +6 +4 −2	Br +7 +5 +1 −1
5	Rb +1	Sr +2	In +3 +1	Sn +4 +2	Sb +5 +3 −3	Te +6 +4 −2	I +7 +5 +1 1
6	Cs +1	Ba +2	Tl +3 +1	Pb +4 +2	Bi +5 +3	Po +4 +2	At +5 −1

*The most common oxidation state is shown in color. Some uncommon oxidation states are not shown.

2. The more electronegative nonmetals have a most common oxidation state equal to the group number minus 8. The absolute value of this difference equals the number of electrons gained or shared in bonding. Thus, for oxygen (Group VIA), the most common oxidation state is $6 - 8 = -2$. For fluorine (Group VIIA), the oxidation state $7 - 8 = -1$ is the only one exhibited by compounds.

3. Except for oxygen and fluorine, the nonmetals and metalloids of Groups IVA to VIIA have a variety of oxidation states in compounds, extending from the most positive (equal to the group number) to the most negative (equal to the group number minus 8). With the exception of nitrogen, the common oxidation states for any one of these elements are either all even or all odd. For example, chlorine (Group VIIA) has common oxidation states of +7, +5, +3, +1, and −1 in its compounds.

4. For Periods 2 and 3 of Groups IIIA to VIA, the most common positive oxidation state equals the group number (except for oxygen, which has no common positive oxidation states). But as one progresses down a column of these elements, the oxidation state equal to the group number minus 2 assumes greater importance. In the heavier metals (Periods 5 and 6), this oxidation state is an important one, being the most common in the Period 6 metals (Tl, Pb, Bi, Po). It appears that the *s* electrons of the valence shell are less likely to be involved

● This noninvolvement of the *ns* electrons in the heavier metals is somctimes called the "inert pair" effect. The explanation depends on the fact that bond strengths decrease with increasing atomic size. To form more bonds in the higher oxidation state requires promotional energy ($ns^2np^2 \rightarrow ns^1np^3$ in Group IVA elements), but less energy is available from bond formation in the heavier elements.

in bonding in these elements.● Consider the Group VA elements. The most common positive oxidation state of nitrogen is +5, the group number. But in arsenic, the +3 (= 5 − 2) oxidation state becomes important, and in antimony and bismuth this is the most common oxidation state.

Elements that display multiple oxidation states can have several oxides. As Table 21.2 shows, the oxide in which the element is in the lower oxidation state is more basic than the oxide with the element in a higher oxidation state. For example, arsenic(V) oxide is weakly acidic, but arsenic(III) oxide is amphoteric. The reason for this general behavior is that, as the oxidation state increases, the bonding becomes more covalent. As we saw earlier, the more covalent the oxide, the more acidic it is. Greater covalent character of the higher oxidation state can be seen in other compounds also. For example, chlorides, bromides, and iodides show a noticeable increase in covalent character with an increase in oxidation state. Thus, tin(II) chloride is a white, crystalline solid melting at 246°C, but tin(IV) chloride is a colorless liquid that freezes at −33°C. That tin(IV) chloride has such a low freezing point is an indication of its molecular nature and the covalent character of the Sn—Cl bond.

Example 21.2 Predicting Covalent Character from Oxidation State

Lead forms two chlorides, $PbCl_2$ and $PbCl_4$. Which compound do you expect to be less stable? In which compound is the bonding more covalent?

Solution

Lead is a metal of the sixth period, so the oxidation number corresponding to the group number is expected to give less stable compounds than the oxidation number equal to the group number minus 2 (4 − 2 = +2). **We expect $PbCl_2$ to be more stable than $PbCl_4$. Moreover, we expect the bonding in $PbCl_4$ to be more covalent than that in $PbCl_2$.** [Lead(IV) chloride is a yellow, oily liquid that explodes near its boiling point of 105°C. Lead(II) chloride is a white, crystalline solid that is slightly soluble in water, dissolving to give Pb^{2+} ions. Its melting point is 501°C.]

Exercise 21.2

Thallium forms two chlorides, TlCl and $TlCl_3$. One chloride is a crystalline substance that melts at 25°C and decomposes on heating. The other is a crystalline substance melting at 430°C. Identify the formulas of the chlorides with the properties given.

(See Problems 21.43 and 21.44.)

Differences in Behavior of the Second-Row Elements

The chemical and physical properties of a second-row element are often rather different from those of the other elements in the same group. This difference in behavior is due in part to the relatively small atom in these elements. Thus, the second-row elements of Groups IA to IIIA give relatively small, polarizing cations. These elements, therefore, display greater covalent character in their compounds. Beryllium compounds are often covalent, whereas compounds of other Group IIA elements (Mg, Ca, Sr, Ba) are primarily ionic. Similarly, boron compounds are covalent, but aluminum and other Group IIIA elements (Ga, In, Tl) have many ionic compounds.

The relatively small atom in a second-row nonmetal gives rise to relatively high electronegativity (electron-withdrawing power). For example, nitrogen has an electronegativity of 3.0, but other Group VA elements have electronegativities between 1.9 and 2.1. Similarly, oxygen has an electronegativity of 3.5; other Group VIA elements have electronegativities between 2.0 and 2.5.

Another reason for the difference in behavior between a second-row element and the other elements of the same group has to do with the fact that bonding in the second-row elements involves only *s* and *p* orbitals, whereas the other elements may use *d* orbitals. This places a limit on the types of compound formed by the second-row elements. For example, although nitrogen forms only the trihalides (such as NCl_3), phosphorus has both trihalides (PCl_3) and pentahalides (PCl_5), which it forms by using 3*d* orbitals.

Unlike other elements of the same group, the second-row nonmetals often display strong multiple bonding, in which π orbitals are formed by the overlap of *p* orbitals. For example, carbon has many compounds with multiple bonds; silicon has very few. The structures of the oxides CO_2 and SiO_2 are quite different. Carbon dioxide is molecular (O=C=O), but silicon dioxide is macromolecular and has silicon–oxygen single bonds. Elementary nitrogen, N_2, has triply bonded atoms (N≡N); phosphorus consists of P_4 molecules with phosphorus–phosphorus single bonds. The explanation for this difference is that π bonds form between two sidewise-approaching *p* orbitals only if they can get close enough for good overlap. The overlap is largest for the small atoms of the second-row elements. It should be noted, however, that the elements beyond the second row (such as phosphorus and sulfur) do show some multiple bonding involving *d* orbitals.

Although the elements of the second row differ in some respects from others in the same group, each of the first elements shows a resemblance to the element to its right in the third row. Lithium, for example, shows some similarities to magnesium. They are said to exhibit a **diagonal relationship,** which refers to *the resemblance of each of the first three elements of the second row of the periodic table to the element located diagonally to the right of it in the third row.*

Li Be B C

Na Mg Al Si

Thus, when lithium is heated in nitrogen, it forms a nitride.

$$6Li(s) + N_2(g) \longrightarrow 2Li_3N(s)$$

No other alkali metal combines directly with nitrogen, but magnesium does.

$$3Mg(s) + N_2(g) \longrightarrow Mg_3N_2(s)$$

Table 21.4 compares some properties of lithium with those of the other alkali metals and magnesium. Note the similarity of lithium and magnesium. As we will

Table 21.4 Comparison of Some Properties of Lithium with Those of Other Alkali Metals and Magnesium

Other Alkali Metals	Lithium	Magnesium
Burn in air to form peroxides or superoxides but no nitrides	Burns in air to form normal oxide and nitride	Burns in air to form normal oxide and nitride
Carbonates decompose when strongly heated	Carbonate decomposes when moderately heated	Carbonate decomposes when moderately heated
Fluorides, carbonates, and phosphates are soluble	Fluoride, carbonate, and phosphate are slightly soluble	Fluoride, carbonate, and phosphate are insoluble

see, the diagonal relationships between beryllium and aluminum and between boron and silicon are even more striking.

21.2 GROUP IA: THE ALKALI METALS

Group IA elements, also known as the alkali metals, are the most reactive metallic elements, readily losing the ns^1 valence electron to form compounds in the +1 oxidation state. The majority of these compounds are ionic and water-soluble.

Because of their reactivity, Group IA elements always occur in nature as compounds, never as free metals. Sodium and potassium are abundant in the rocks of the earth's crust. Most of these rocks are composed of insoluble aluminosilicate minerals—substances containing silicon, aluminum, and oxygen with positive ions such as Na^+ and K^+.● (A **mineral** is *a naturally occurring solid substance or solid solution with definite crystalline form*.) Normal weathering of these insoluble minerals releases the alkali metal ions as soluble salts, which eventually find their way to the oceans. Seawater is principally a water solution of the salts of the alkali and alkaline earth (Group IIA) metals, with sodium chloride as the main constituent.

● The structure of silicates and aluminosilicates is discussed in Section 22.1.

When seawater evaporates, the dissolved salts crystallize out, often as pure substances. The deposits of soluble minerals of sodium and potassium found in many parts of the world, including halite or rock salt (NaCl), sylvite (KCl), and carnallite ($KCl \cdot MgCl_2 \cdot 6H_2O$), probably originated when enclosed bodies of seawater slowly evaporated. These minerals are now important commercial sources of sodium chloride and potassium chloride (Figure 21.3). Sodium chloride is also obtained commercially from seawater by solar evaporation.

The other alkali metals (except francium) are obtained mainly from aluminosilicate minerals. Lithium is present in spodumene, $LiAl(SiO_3)_2$, an aluminosilicate mineral that occurs in several areas of the world. Cesium is available from the relatively rare aluminosilicate mineral pollucite, $CsAl(SiO_3)_2 \cdot H_2O$. And rubidium occurs as an impurity in many minerals, including pollucite; however, there are no known minerals in which rubidium is a major constituent.

Francium is a very rare, radioactive element. It was discovered in 1939 by Marguerite Perey, a French chemist. She found that about 1% of actinium-227

Figure 21.3
A salt mine. Underground sodium chloride beds are often several hundred feet thick.

Table 21.5 Properties of Group IA Elements

Property	Lithium	Sodium	Potassium	Rubidium	Cesium
Electron configuration	[He]$2s^1$	[Ne]$3s^1$	[Ar]$4s^1$	[Kr]$5s^1$	[Xe]$6s^1$
Melting point, °C	181	97.8	63.6	38.9	28.4
Boiling point, °C	1347	883	774	688	678
Density, g/cm^3	0.53	0.97	0.86	1.53	1.88
Ionization energy (first), kJ/mol	520	496	419	403	376
(second), kJ/mol	7298	4562	3051	2632	2420
Electronegativity (Pauling scale)	1.0	0.9	0.8	0.8	0.7
Standard potential (volts), $M^+ + e^- \rightleftharpoons M$	−3.04	−2.71	−2.92	−2.92	−2.92
Covalent radius, Å	1.23	1.57	2.02	2.16	2.35
Ionic radius, Å	0.90	1.16	1.52	1.66	1.81

decays by alpha emission to francium-223; the remainder decays by beta emission to thorium-227.

$$^{227}_{89}\text{Ac} \longrightarrow \,^{223}_{87}\text{Fr} + \,^{4}_{2}\text{He}$$

Francium-223, the longest-lived isotope of the element, has a half-life of 21 minutes. It has been estimated that there is less than 25 grams of francium on earth. The element has no commercial uses, and very little of its chemistry is known.

Properties of the Elements

All Group IA elements are silvery-white, metallic solids, although cesium, which melts at 28°C, would be liquid on a warm day. The metals are soft; sodium can be easily cut with a knife. Table 21.5 lists melting points and other physical properties of the alkali metals. Note that lithium, sodium, and potassium have densities less than 1.00 g/mL and would float on water. (But the three metals react vigorously with water!)

The softness of the metals and their low melting points are indications of weak metal bonding. This weak bonding is due to the relatively large size of the alkali metallic atoms, compared with those of the elements that follow in the same period, and to the fact that there is only one valence electron. Note how the melting points decrease from lithium to cesium. This is expected because the atomic radii increase, reducing the strength of the metal bonding.

As we noted earlier, the alkali metals are all chemically very reactive. They are strong reducing agents, as their standard potentials reveal (see Table 21.5 for physical data). And considering their low electronegativities, we expect them to form ionic compounds. This is also apparent from their low first ionization energies. Second ionization energies are quite high, however, so that only +1 ions are formed in compounds.

All of the alkali metals react with water. The reactions involve the displacement of hydrogen from water by the metal. For example,

$$2\text{Na}(s) + 2\text{H}_2\text{O}(l) \longrightarrow 2\text{NaOH}(aq) + \text{H}_2(g)$$

Because the alkali metal hydroxides are strong bases, the formula NaOH(aq) should be interpreted as a solution of $Na^+(aq)$ and $OH^-(aq)$ ions.

Figure 21.4
Oxygen compounds of alkali metals. Sodium peroxide, Na_2O_2 (yellowish-white); and potassium superoxide, KO_2 (orange-yellow).

The alkali metals become increasingly reactive as we progress down the column of elements. A piece of lithium metal added to water moves around the surface releasing hydrogen gas. Sodium spins around, vigorously hissing as hydrogen evolves. The hydrogen may catch fire from the heat of reaction (see Figure 2.3, p. 33). Potassium reacts violently and bursts into flame, colored purple from the emission spectrum of potassium vapor (Figure 8.17, p. 299).

The alkali metals burn in air and in oxygen. All of them form oxides (compounds of O^{2-}), but other compounds of oxygen are possible and may be the predominant product (Figure 21.4). Lithium forms lithium oxide, Li_2O, a white solid. Although sodium forms the oxide, the predominant product when it is burned in air is sodium peroxide, Na_2O_2, a yellowish-white compound. This is an ionic compound of sodium ion and peroxide ion, $O_2{}^{2-}$. When potassium burns in air, the predominant product is potassium superoxide, KO_2, an orange-yellow compound. This is an ionic compound of potassium ion and the superoxide ion, $O_2{}^{-}$. The oxidation states of oxygen in the ions O^{2-}, $O_2{}^{2-}$, and $O_2{}^{-}$ are -2, -1, and $-\frac{1}{2}$, respectively. (The molecular orbital structures of the ions $O_2{}^{-}$ and $O_2{}^{2-}$ are those of oxygen but with one and with two additional antibonding electrons, so that the bond orders are $\frac{3}{2}$ and 1, respectively.)

The alkali metals also react easily with the halogens (F_2, Cl_2, Br_2, I_2) to give the halides and at elevated temperatures with hydrogen to give ionic hydrides (such as Li^+H^-). Lithium, as we discussed earlier, is the only alkali metal that combines directly with nitrogen.

Because of their reactivity with air and moisture, the alkali metals must be stored under an inert liquid such as kerosene or in an inert atmosphere. They must not be picked up with the bare fingers because they react with any moisture, producing a burn (from the alkali metal hydroxide and heat of reaction).

Preparation of the Elements

Sodium and lithium metals are most easily prepared by electrolysis of their fused salts. Sodium was first isolated in 1807 by Humphry Davy, who electrolyzed molten sodium hydroxide.● This was a major industrial method of preparation. Today most sodium is produced by the electrolysis of molten sodium chloride–calcium chloride mixtures in a Downs cell (see Section 19.8); calcium chloride is added to lower the melting point to about 580°C. Lithium is obtained from the electrolysis of molten lithium chloride–potassium chloride mixtures.

● The electrode reactions are

$$Na^+ + e^- \longrightarrow Na$$
$$4OH^- \longrightarrow 2H_2O + O_2 + 4e^-$$

Potassium is more easily prepared by chemical reduction rather than by electrolysis of the chloride. In the commercial process, molten potassium chloride reacts with sodium metal at 870°C.

$$Na(l) + KCl(l) \longrightarrow NaCl(l) + K(g)$$

The reaction goes in the direction written because potassium vapor leaves the reaction chamber and is condensed.

The other alkali metals, rubidium and cesium, are also prepared by chemical reduction of their salts. For example, when molten cesium chloride is heated at 700°C to 800°C with calcium metal at low pressure, cesium vapor distills over.

$$2CsCl(l) + Ca(l) \longrightarrow CaCl_2(l) + 2Cs(g)$$

Uses of the Elements

Sodium and lithium are the most important of the alkali metals. The largest single use of sodium metal has been in the production of tetraethyllead, $(C_2H_5)_4Pb$, a gasoline additive that controls engine knocking. Sodium–lead alloy reacts with ethyl chloride, C_2H_5Cl, to give tetraethyllead. This use of sodium is declining as leaded gasolines are being phased out. Large quantities of sodium are used in the preparation of certain sodium compounds, such as sodium peroxide, Na_2O_2, and sodium amide, $NaNH_2$. And because sodium is a strong reducing agent, the metal is useful in the production of other metals (titanium, for example) and in the preparation of organic compounds. Small amounts of the metal are used in sodium vapor lamps, those bright yellow lamps along roadways. Sodium is also being increasingly used as a heat-transfer agent in, for example, nuclear reactors.

The use of lithium metal has greatly expanded with the development of strong aluminum–lithium alloys. Because these alloys are light (the density of lithium is about one-fifth that of aluminum), they are used in aircraft construction. Another growing use of lithium metal is as a battery anode. Because lithium has low density and large negative standard potential, it is a concentrated source of electric energy (see Figure 21.5).

Small amounts of potassium are prepared, primarily to make potassium superoxide, KO_2, for use in rebreathing gas masks. These gas masks consist of a

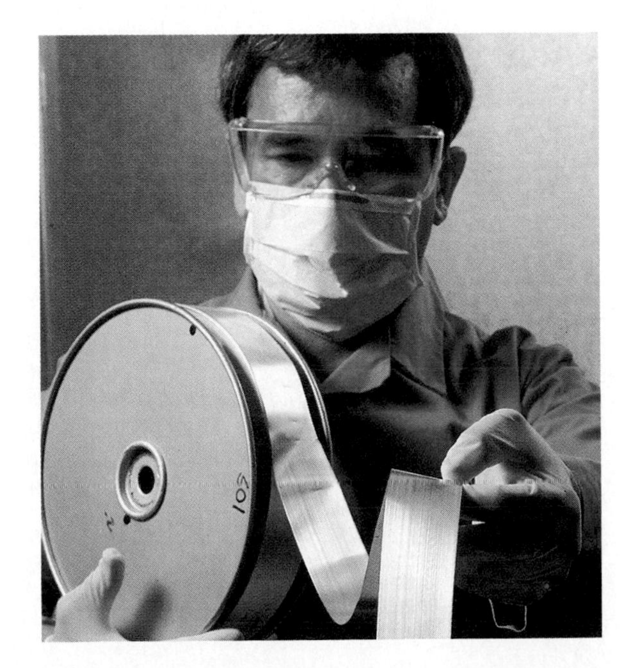

Figure 21.5
A roll of lithium metal for batteries. The metal must be handled in humidity-free rooms to prevent corrosion.

closed system in which air is circulated through a canister of KO_2. Oxygen is released when moisture in the breath attacks the superoxide.

$$4KO_2(s) + 2H_2O(l) \longrightarrow 4KOH(s) + 3O_2(g)$$

The potassium hydroxide produced in this reaction removes carbon dioxide from the exhaled air.

$$KOH(s) + CO_2(g) \longrightarrow KHCO_3(s)$$

Uses of the Compounds

Table 21.6 lists some of the most important compounds of the alkali metals. Lithium carbonate is a slightly soluble salt obtained from the processing of lithium ores. It is used to make lithium hydroxide. Calcium hydroxide (lime) reacts with solutions of lithium carbonate to precipitate calcium carbonate, leaving a solution of LiOH.

$$[Ca^{2+}(aq) + 2OH^-(aq)] + [2Li^+(aq) + CO_3^{2-}(aq)] \longrightarrow CaCO_3(s) + 2[Li^+(aq) + OH^-(aq)]$$

Lithium hydroxide is used to manufacture lithium soaps for lubricating greases.

Sodium chloride is the most important compound of sodium, because it is the source of the metal and other sodium compounds. Another important sodium com-

Table 21.6 Uses of Alkali Metal Compounds

Compound	Use
Li_2CO_3	Aluminum production (added to the molten electrolyte) Preparation of LiOH
LiOH	Manufacture of lithium soaps for lubricating greases
LiH	Reducing agent in organic syntheses
$LiNH_2$	Preparation of antihistamines and other pharmaceuticals
NaCl	Source of sodium and sodium compounds Condiment and food preservative Soap manufacture (precipitates soap from reaction mixture)
NaOH	Pulp and paper industry Extraction of aluminum oxide from ore Manufacture of viscose rayon Petroleum refining Manufacture of soap
Na_2CO_3	Manufacture of glass Used in detergents and water softeners
Na_2O_2	Textile bleach
$NaNH_2$	Preparation of indigo dye for denim (blue jeans)
KCl	Fertilizer Source of other potassium compounds
KOH	Manufacture of soft soap Manufacture of other potassium compounds
K_2CO_3	Manufacture of glass
KNO_3	Fertilizers Explosives and fireworks

Figure 21.6
A laboratory demonstration of the Solvay process. *Left:* A concentrated solution of ammonia is saturated with sodium chloride. When pieces of dry ice (solid CO_2) are added, a water cloud forms. *Right:* Sodium hydrogen carbonate precipitates from the reaction mixture.

pound is sodium hydroxide, NaOH. It is produced by the electrolysis of aqueous sodium chloride. The overall electrolysis reaction is

$$2NaCl(aq) + 2H_2O(l) \longrightarrow 2NaOH(aq) + H_2(g) + Cl_2(g)$$

Chlorine is also a major product of this electrolysis.●

● The electrolysis of aqueous NaCl was discussed in Section 19.9.

Sodium hydroxide is a strong base that has many important applications in chemical processing. Large quantities are used to make paper, to separate aluminum oxide from its ore, and to refine petroleum.

Sodium carbonate is another important compound of sodium. The anhydrous compound, Na_2CO_3, is called soda ash. Large quantities of soda ash are consumed in making glass. The decahydrate, $Na_2CO_3 \cdot 10H_2O$, called washing soda, is used as a water softener and is added to detergent preparations. Most of the soda ash used in the United States is obtained from the mineral trona ($Na_2CO_3 \cdot NaHCO_3 \cdot H_2O$), which is mined from deposits in southwestern Wyoming.

Sodium carbonate is also prepared by the **Solvay process,** *an industrial method for obtaining sodium carbonate from sodium chloride, ammonia, and carbon dioxide*. In this process, ammonia is first dissolved in a saturated solution of sodium chloride. Then carbon dioxide is bubbled into the solution, which results in a precipitate of sodium hydrogen carbonate (baking soda). Figure 21.6 shows a laboratory demonstration of the Solvay process.

$$NH_3(g) + \underbrace{H_2O(l) + CO_2(g)}_{H_2CO_3(aq)} + [Na^+(aq) + Cl^-(aq)] \longrightarrow NaHCO_3(s) + [NH_4^+(aq) + Cl^-(aq)]$$

We can think of this as a metathesis reaction of ammonium hydrogen carbonate ($NH_3 + H_2CO_3$) and NaCl. The sodium hydrogen carbonate is filtered from the solution of ammonium chloride and washed. When heated to 175°C, it decomposes to sodium carbonate.●

● In the complete commercial process, CO_2 from this step is recycled to produce more $NaHCO_3$. Additional CO_2 is obtained by heating limestone ($CaCO_3$).

$$CaCO_3(s) \xrightarrow{\Delta} CaO(s) + CO_2(g)$$

The CaO from this step is used to recover NH_3 from the NH_4Cl solution.

$$CaO(s) + 2NH_4Cl(aq) \xrightarrow{\Delta} CaCl_2(aq) + 2NH_3(g) + H_2O(l)$$

Overall, the net result is

$$CaCO_3 + 2NaCl \longrightarrow Na_2CO_3 + CaCl_2$$

$$2NaHCO_3(s) \xrightarrow{\Delta} Na_2CO_3(s) + CO_2(g) + H_2O(g)$$

Potassium chloride is the most important compound of potassium. Over 90% of KCl is used as an agricultural fertilizer, because potassium ion is an important plant nutrient. The remainder is used to prepare potassium and potassium compounds. Potassium hydroxide, from which many potassium compounds are produced, is obtained by the electrolysis of aqueous KCl.

Example 21.3 Using Chemical Reactions of the Group IA Elements and Compounds

Show by equations how potassium nitrate, KNO_3, could be prepared from potassium chloride, KCl, and nitric acid (in two steps).

Solution

Potassium hydroxide is produced commercially from potassium chloride by electrolysis of aqueous solutions.

$$2[K^+(aq) + Cl^-(aq)] + 2H_2O(l) \xrightarrow{\text{electrolysis}} 2[K^+(aq) + OH^-(aq)] + H_2(g) + Cl_2(g)$$

When the aqueous potassium hydroxide from this electrolysis is neutralized with nitric acid, the salt potassium nitrate is obtained.

$$[K^+(aq) + OH^-(aq)] + [H^+(aq) + NO_3^-(aq)] \longrightarrow [K^+(aq) + NO_3^-(aq)] + H_2O(l)$$

Exercise 21.3

Starting from potassium chloride, show by means of equations how potassium superoxide is prepared (in two steps). (See Problems 21.53 and 21.54.)

21.3 GROUP IIA: THE ALKALINE EARTH METALS

Group IIA elements, or alkaline earth metals, are chemically reactive, though less so than the alkali metals. During reaction, the alkaline earth elements use the ns^2 valence electrons to form compounds in the +2 oxidation state. In the case of calcium, strontium, barium, and radium, these compounds are nearly always ionic and contain the +2 metal ion. However, bonding in magnesium often shows some covalent character, and in beryllium it is predominantly covalent.

Like the alkali metals, Group IIA elements always occur in nature as compounds. Magnesium and calcium are very abundant in the rocks of the earth's crust. This outer portion of the earth was originally in the form of silicates (compounds of silicon and oxygen with cations) and aluminosilicates. Magnesium and calcium, with sodium and potassium, are present in these rocks as cations. Weathering of the rocks produces soluble compounds of these cations, which eventually reach the seas. Calcium ion in seawater is used by shellfish to form their outer shell of calcium carbonate. Shells from dead animals have accumulated over geological time to form limestone deposits. Magnesium ion in seawater has reacted with these calcium carbonate sediments to form dolomite, $CaCO_3 \cdot MgCO_3$. Most of the magnesium ion has remained in the oceans, however.

The chief commercial sources of magnesium are seawater, underground brines, and the minerals dolomite and magnesite, $MgCO_3$. Calcium compounds are obtained from seashells and limestone. Gypsum, $CaSO_4 \cdot 2H_2O$, is also an important mineral. Other alkaline earth elements are much less common than magnesium and calcium. The principal ore of beryllium is the aluminosilicate mineral beryl, $Be_3Al_2(SiO_3)_6$. (An **ore** is *a rock or mineral from which a metal can be economically produced*.) Gem-quality forms of beryl are aquamarine (light blue) and emerald (dark green). Strontium is found in celestite, $SrSO_4$, and strontianite, $SrCO_3$. Barium is found in barite, $BaSO_4$, and witherite, $BaCO_3$. Radium occurs in small amounts in uranium ores, as the radioactive isotope of mass number 226, which has a half-life of 1620 years and decays by alpha emission.

$$^{226}_{88}Ra \longrightarrow {}^{222}_{86}Rn + {}^{4}_{2}He$$

All other isotopes of radium are also radioactive. Some minerals of the alkaline earth elements are shown in Figure 21.7.

Figure 21.7
Some minerals of the alkaline earth elements. *Left to right:* Beryl, $Be_3Al_2(SiO_3)_6$ (blue) in quartz; dolomite, $CaCO_3 \cdot MgCO_3$; celestite, $SrSO_4$.

Properties of the Elements

Beryllium is a gray metal almost as hard as iron and hard enough to scratch glass. The other alkaline earth elements are silvery metals much softer than beryllium but still harder than the alkali metals. Table 21.7 lists some of their properties. Note that the melting points of Group IIA metals are well above those of Group IA metals (see Table 21.5). The higher melting points and greater hardness of Group IIA metals are due to the increased strength of bonding from two valence electrons.

The first two ionization energies of Group IIA elements are relatively low (see Table 21.7) and, as expected, the elements give +2 cations. Thus, the most active elements (Ca, Sr, Ba, Ra) react vigorously with water, producing the metal ions.

$$Ca(s) + 2H_2O(l) \longrightarrow [Ca^{2+}(aq) + 2OH^-(aq)] + H_2(g)$$

Magnesium reacts slowly with water at normal temperatures but rapidly with steam. Beryllium is rather unreactive with water. Beryllium is also unlike the other Group IIA elements in forming complex anions, such as $Be(OH)_4^{2-}$. This is an

Table 21.7 Properties of Group IIA Elements

Property	Beryllium	Magnesium	Calcium	Strontium	Barium
Electron configuration	$[He]2s^2$	$[Ne]3s^2$	$[Ar]4s^2$	$[Kr]5s^2$	$[Xe]6s^2$
Melting point, °C	1278	649	839	769	725
Boiling point, °C	2970	1090	1484	1384	1640
Density, g/cm^3	1.85	1.74	1.54	2.6	3.51
Ionization energy (first), kJ/mol	899	738	590	549	503
(second), kJ/mol	1757	1451	1145	1064	965
(third), kJ/mol	14848	7733	4912	4210	3430
Electronegativity (Pauling scale)	1.5	1.2	1.0	1.0	0.9
Standard potential (volts), $M^{2+} + 2e^- \rightleftharpoons M$	−1.70	−2.38	−2.76	−2.89	−2.90
Covalent radius, Å	0.89	1.36	1.74	1.92	1.98
Ionic radius, Å	0.59	0.86	1.14	1.32	1.49

indication of its partial nonmetallic character. Thus, beryllium not only reacts with acids but also dissolves in strong base (as does aluminum, with which beryllium has a diagonal relationship).

$$Be(s) + 2[Na^+(aq) + OH^-(aq)] + 2H_2O(l) \longrightarrow \underset{\text{sodium beryllate}}{[2Na^+(aq) + Be(OH)_4{}^{2-}(aq)]} + H_2(g)$$

Magnesium metal is unreactive with basic solutions. Of course, Group IIA metals below magnesium react with the water in such solutions.

The alkaline earth metals burn in oxygen to form the oxides or, in the case of barium, the peroxide. Barium peroxide forms at low temperatures but begins to decompose to the oxide at 700°C. Calcium, strontium, and barium react exothermically with hydrogen to form the ionic hydrides. For example,

$$Ca(s) + H_2(g) \longrightarrow CaH_2(s);\ \Delta H^\circ = -183\text{ kJ}$$

Magnesium reacts with hydrogen only at high pressure and in the presence of a catalyst (MgI_2). All Group IIA elements react directly with the halogens to form halides and with nitrogen, on heating, to give the nitrides.

Preparation of the Elements

The alkaline earth metals are prepared by electrolysis of the molten halides (usually the chlorides) or by chemical reduction of either the halides or the oxides. Magnesium, which is commercially the most important alkaline earth metal, is produced by the electrolysis of fused magnesium chloride. Seawater provides an inexhaustible source of magnesium ion, and seashells, which are mostly calcium carbonate, are a source of needed base to isolate the magnesium ion. The process for doing this, developed by the Dow Chemical Company, is shown in Figure 21.8. When oyster shells are heated, the calcium carbonate decomposes to the oxide.●

● Dolomite, $CaCO_3 \cdot MgCO_3$, is sometimes used in place of calcium carbonate. When heated, it yields MgO as well as CaO. The magnesium oxide provides magnesium ion, in addition to that from seawater.

$$CaCO_3(s) \xrightarrow{\Delta} CaO(s) + CO_2(g)$$

Addition of calcium oxide (a basic oxide) to seawater precipitates magnesium hydroxide.

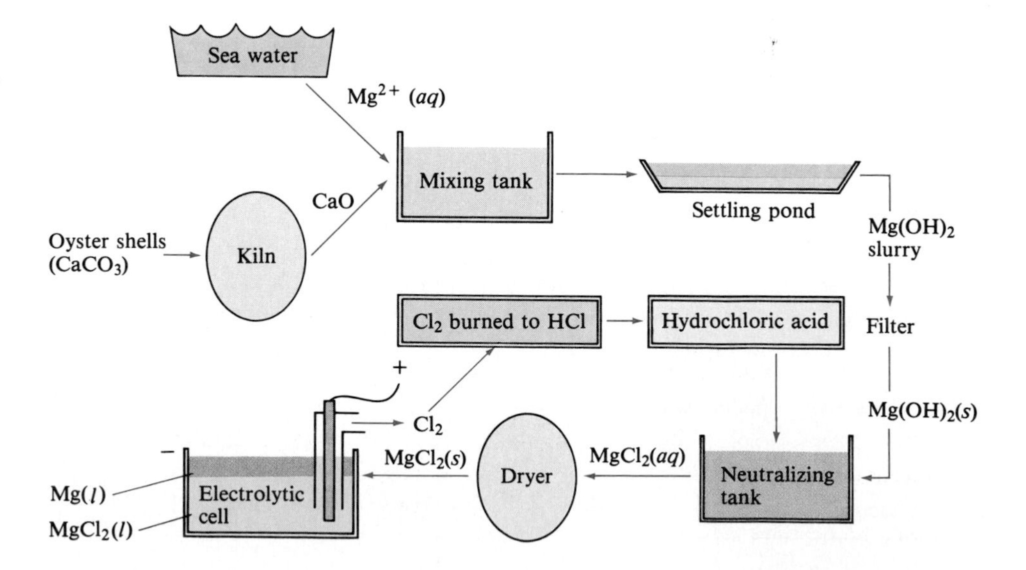

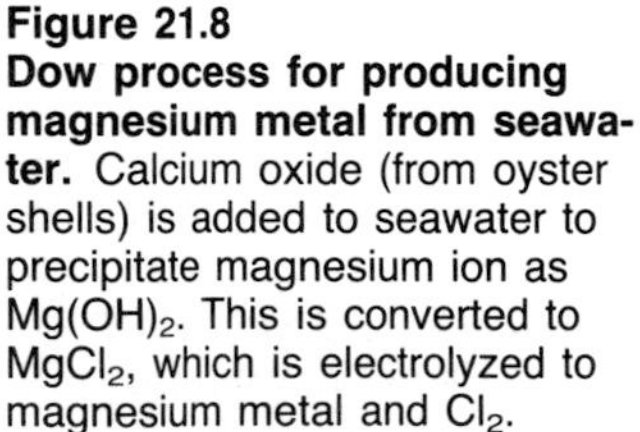

Figure 21.8
Dow process for producing magnesium metal from seawater. Calcium oxide (from oyster shells) is added to seawater to precipitate magnesium ion as $Mg(OH)_2$. This is converted to $MgCl_2$, which is electrolyzed to magnesium metal and Cl_2.

$$Mg^{2+}(aq) + CaO(s) + H_2O(l) \longrightarrow Mg(OH)_2(s) + Ca^{2+}(aq)$$

The magnesium hydroxide is filtered off and treated with hydrochloric acid to convert it to the chloride.

$$Mg(OH)_2(s) + 2[H^+(aq) + Cl^-(aq)] \longrightarrow [Mg^{2+}(aq) + 2Cl^-(aq)] + 2H_2O(l)$$

The dry salt, obtained by evaporation of the solution, is melted and electrolyzed at 700°C (Figure 21.9).

$$MgCl_2(l) \xrightarrow{\text{electrolysis}} Mg(l) + Cl_2(g)$$

The by-product chlorine can be sold or burned with methane (natural gas) to provide hydrochloric acid for the process.

$$2CH_4(g) + O_2(g) + 4Cl_2(g) \longrightarrow 8HCl(g) + 2CO(g)$$

Magnesium is also obtained from magnesite or dolomite by decomposing it to MgO and then reducing the oxide with ferrosilicon, an alloy of silicon and iron.

Other alkaline earth metals are manufactured in smaller amounts. Beryllium can be obtained by electrolysis of beryllium chloride, $BeCl_2$, to which sodium chloride is added to increase the conductivity of the molten salt. However, most beryllium is prepared by chemical reduction of the fluoride with magnesium.

$$BeF_2(l) + Mg(l) \xrightarrow[950°C]{\Delta} MgF_2(l) + Be(s)$$

Calcium is prepared by electrolysis of molten calcium chloride and by reduction of calcium oxide by aluminum in a vacuum, where the calcium produced distills off.●

● Calcium oxide (a basic oxide) reacts with aluminum oxide (an amphoteric oxide) to give tricalcium aluminate.

$$3CaO + Al_2O_3 \longrightarrow Ca_3Al_2O_6$$

Therefore, the overall reaction of CaO with Al can be written

$$6CaO + 2Al \longrightarrow 3Ca + Ca_3Al_2O_6$$

$$3CaO(s) + 2Al(l) \xrightarrow[1200°C]{\Delta} 3Ca(g) + Al_2O_3(s)$$

Barium is also produced by reduction of the oxide by aluminum, and although very little strontium is used commercially, it can be produced by a similar process.

Uses of the Elements

Commercially, the most important Group IIA metal is magnesium. It is used in large quantities to make aluminum alloy, to which it imparts hardness as well as corrosion resistance. Similarly, magnesium alloys often contain small quantities of

Figure 21.9
Production of magnesium. Molten magnesium chloride is electrolyzed to molten magnesium and chlorine gas.

aluminum. These alloys are used to make aircraft parts, auto parts, and ladders. Smaller quantities of magnesium are used as a reducing agent to prepare other metals, such as uranium and beryllium. Some photographic flash lamps use a fine magnesium wire in oxygen and give off a brilliant white light when the metal burns.

Beryllium is an expensive metal, but it has some special characteristics that make it useful. When added in small amounts to copper, it gives an alloy as hard as steel. The alloy is used to make nonsparking electrical contacts. Because beryllium is transparent to x rays, the metal is used to make windows for x-ray tubes.● Beryllium is also a moderator of neutrons (that is, it slows them down) and is used in nuclear reactors and weapons. Special precautions must be taken when working with beryllium and its alloys; the metallic dust is very toxic. (Beryllium compounds are also very toxic.)

● X rays are scattered by the electrons in atoms. For this reason, atoms of high atomic number (which have many electrons) scatter x rays more effectively than atoms of low atomic number. Therefore, elements of low atomic number, such as Be, are transparent to x rays; elements such as lead are opaque.

Calcium is used as a scavenger (agent to remove impurities in materials) in producing certain metals and preparing various alloys. When added to lead, for example, it produces a hard metal for storage battery grids (electrodes). The electrodes resist the electrolysis of water during charging, so the batteries can be sealed to give *maintenance-free* storage batteries. Calcium is also used as a reducing agent in preparing some of the less common metals, such as thorium.

$$ThO_2(s) + 2Ca(l) \xrightarrow[1000°C]{\Delta} Th(s) + 2CaO(s)$$

Barium is used in small amounts in the manufacture of television and vacuum tubes to remove traces of air. Strontium has few commercial uses.

Uses of the Compounds

Some of the most important compounds of Group IIA elements are formed by calcium. Calcium carbonate minerals and seashells are the chief commercial sources of these compounds. When heated to 900°C, the carbonate decomposes, releasing carbon dioxide.

$$CaCO_3(s) \xrightarrow[900°C]{\Delta} CaO(s) + CO_2(g)$$

The product is calcium oxide, known commercially as lime or quicklime. It is one of the most important industrial chemicals, second only to sulfuric acid in tons produced. (See Table 21.8 for a list of alkaline earth compounds and their uses.) Most of the calcium oxide is consumed in steel making. Added to molten iron containing silicates or silicon dioxide, it combines to give a slag (a glassy waste material) that floats to the top of the metal. This is essentially an acid–base reaction.

$$\underset{\text{basic oxide}}{CaO(s)} + \underset{\text{acidic oxide}}{SiO_2(s)} \longrightarrow \underset{\text{calcium silicate slag}}{CaSiO_3(l)}$$

Calcium oxide reacts exothermically with water to produce the hydroxide, known commercially as slaked lime.

$$CaO(s) + H_2O(l) \longrightarrow Ca(OH)_2(s);\ \Delta H° = -65.7\text{ kJ}$$

Because the heat released is sufficient to ignite paper, care must be taken in storing the oxide.

Table 21.8 Uses of Alkaline Earth Compounds

Compound	Use
MgO	Refractory bricks (for furnaces) Animal feeds
$Mg(OH)_2$	Source of magnesium for the metal and compounds Milk of magnesia (antacid and laxative)
$MgSO_4 \cdot 7H_2O$	Fertilizer Medicinal uses (laxative and analgesic) Mordant (used in dyeing fabrics)
CaO and $Ca(OH)_2$	Manufacture of steel Neutralizer for chemical processing Water treatment Mortar Stack-gas scrubber (to remove H_2S and SO_2)
$CaCO_3$	Paper coating and filler Antacids, dentifrices
$CaSO_4$	Plaster, wallboard Portland cement
$Ca(H_2PO_4)_2$	Soluble phosphate fertilizer
$BaSO_4$	Oil-well drilling mud Gastrointestinal x-ray photography Paint pigment (lithopone)

Table 21.9 gives the solubilities of some Group IIA compounds. Aqueous solutions of calcium hydroxide react with carbon dioxide to give a white precipitate of calcium carbonate.

$$[Ca^{2+}(aq) + 2OH^-(aq)] + CO_2(g) \longrightarrow CaCO_3(s) + H_2O(l)$$

The reaction is the basis of a test for carbon dioxide (Fig. 21.10 (left) and (center)). It is also used to prepare a pure form of finely divided calcium carbonate as a filler in making paper, and for toothpowders, antacids, and other purposes (Figure 21.11). Mortar, used in bricklaying, is made by mixing slaked lime with sand and water. The mortar hardens as the mixture dries and calcium hydroxide crystal-

Table 21.9 Solubilities of Some Group IIA Compounds at 25°C

	Hydroxides	Carbonates	Sulfates
Be	Insoluble $K_{sp} = 2.0 \times 10^{-18}$	—*	Soluble
Mg	Insoluble $K_{sp} = 1.8 \times 10^{-11}$	Slightly soluble $K_{sp} = 1.0 \times 10^{-5}$	Soluble
Ca	Slightly soluble $K_{sp} = 5.5 \times 10^{-6}$	Insoluble $K_{sp} = 3.8 \times 10^{-9}$	Slightly soluble $K_{sp} = 2.4 \times 10^{-5}$
Sr	Soluble	Insoluble $K_{sp} = 9.3 \times 10^{-10}$	Insoluble $K_{sp} = 2.5 \times 10^{-7}$
Ba	Soluble	Insoluble $K_{sp} = 4.9 \times 10^{-9}$	Insoluble $K_{sp} = 1.1 \times 10^{-10}$

*Only an insoluble hydroxide carbonate, $Be(OH)_2 \cdot BeCO_3$, is known.

Figure 21.10
Reaction of carbon dioxide with calcium hydroxide solution. *Left:* Calcium hydroxide solution with bromothymol-blue indicator. *Center:* Carbon dioxide from dry ice reacts with calcium ion to precipitate calcium carbonate. Bromothymol-blue changes to yellow in acidic solution. *Right:* In an excess of carbon dioxide, the calcium carbonate dissolves to form a solution of calcium ion and hydrogen carbonate ion, HCO_3^-.

lizes. Later it slowly sets to a harder solid as the hydroxide reacts with carbon dioxide in the air to form calcium carbonate.

Interestingly, large amounts of calcium oxide (or hydroxide) are used to remove calcium ion from hard water. Water that has passed through limestone deposits usually contains soluble calcium hydrogen carbonate, $Ca(HCO_3)_2$. Although calcium carbonate is insoluble in pure water, it does dissolve in water containing acidic substances, such as carbon dioxide (contained in ground water exposed to air). (See Figures 21.10 (right) and 21.12.) We can write the reaction as

$$\underset{\text{limestone}}{CaCO_3(s)} + CO_2(g) + H_2O(l) \longrightarrow \underset{\text{calcium hydrogen carbonate}}{[Ca^{2+}(aq) + 2HCO_3^-(aq)]}$$

When hard water containing $Ca(HCO_3)_2$ is treated with the correct amount of calcium hydroxide, nearly all the calcium ion is precipitated as calcium carbonate.

$$[Ca^{2+}(aq) + 2HCO_3^-(aq)] + [Ca^{2+}(aq) + 2OH^-(aq)] \longrightarrow 2CaCO_3(s) + 2H_2O(l)$$

Magnesium carbonate, like calcium carbonate, decomposes when heated.

$$MgCO_3(s) \xrightarrow{\Delta} MgO(s) + CO_2(g)$$

If the carbonate is heated strongly (above 1400°C), the magnesium oxide that results is chemically rather inert. In this form, it is used to make refractory bricks

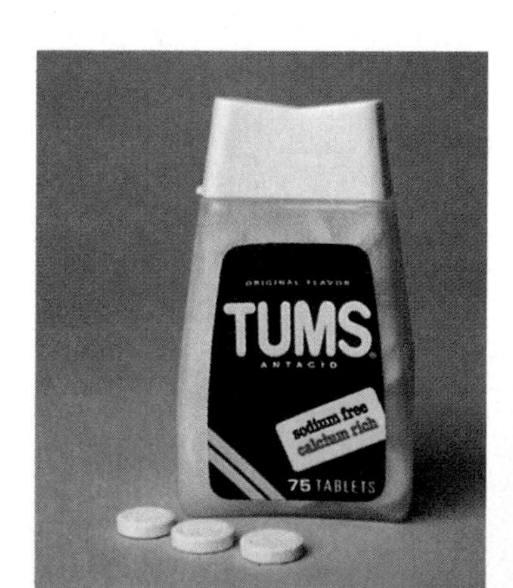

Figure 21.11
Some antacids contain calcium carbonate. The antacid preparation consists of purified calcium carbonate with flavorings and a binder. Such antacids are sometimes prescribed as a nutrient calcium supplement.

Figure 21.12
A limestone cave (Harrison's Caves, Barbados). Such caves are formed by the action of acidic ground water on limestone deposits. The iciclelike formations (stalagmites and stalactites) in the caves are caused by reprecipitation of calcium carbonate.

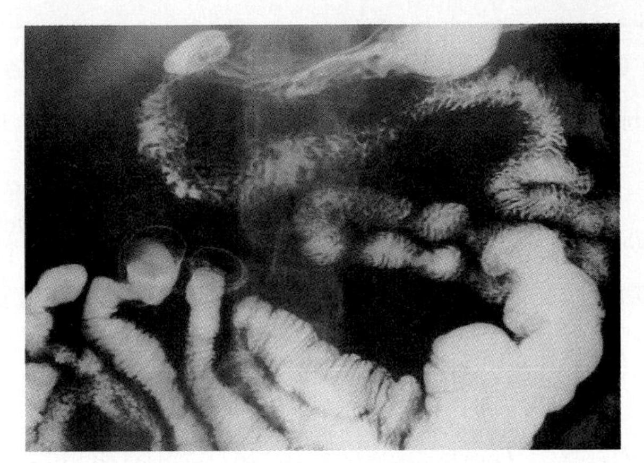

Figure 21.13
An x-ray photograph of the gastrointestinal tract. Barium sulfate is administered to the patient, because it is opaque to x rays and renders the gastrointestinal tract visible.

(to line high-temperature furnaces). When the oxide is prepared at lower temperatures (about 700°C), however, it is obtained as a powder that dissolves easily in acids. It is used in animal feeds to provide magnesium ion as a nutrient.

The most important compound of barium is the sulfate, which occurs as the mineral barite. Barium sulfate is a most insoluble compound. It is used in oil drilling in the form of a mud, which makes a seal around the drilling bit. Barium, like all elements with a high atomic number, is opaque to x rays. For this reason, barium sulfate is given orally or as an enema to obtain diagnostic x-ray photographs of the stomach and intestines. Soluble barium compounds cannot be used because they are toxic, but suspensions of barium sulfate provide negligible amounts of barium ion (Figure 21.13).

Beryllium is unlike the other alkaline earth elements in several ways. As we saw earlier, it forms the beryllate ion, $Be(OH)_4^{2-}$. Consequently, beryllium hydroxide, $Be(OH)_2$, is amphoteric, in contrast to the other Group IIA hydroxides. It dissolves in acids to form $Be^{2+}(aq)$, but it also dissolves easily in strong base to form $Be(OH)_4^{2-}$. For example,

$$2[Na^+(aq) + OH^-(aq)] + Be(OH)_2(s) \longrightarrow [2Na^+(aq) + Be(OH)_4^{2-}(aq)]$$

In this reaction, beryllium hydroxide is acting as a Lewis acid, accepting electron pairs from OH^- ions to give an octet of electrons about the Be atom. As we will see later, a similar type of reaction occurs with aluminum hydroxide, which is also amphoteric, thus showing the diagonal relationship between Be and Al.

The tendency of Be^{2+} to accept electrons is due to its small size. As a result, bonding in its compounds is highly covalent. The melting point of beryllium chloride (450°C), for example, is low compared with that of the other chlorides of Group IIA elements ($MgCl_2$ melts at 714°C). Also, the molten salt has a low electrical conductivity, so electrolysis of $BeCl_2$ requires the addition of a salt such as NaCl. In the solid phase, beryllium chloride has a polymeric structure (that is, the substance is covalent with an infinitely repeating unit), in which each chlorine atom "bridges" two beryllium atoms.

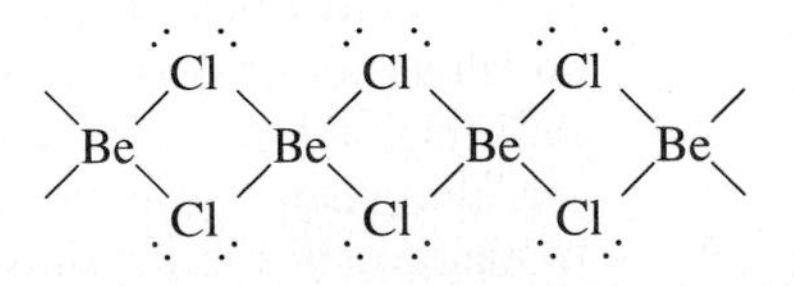

Some aluminum halides form molecular substances with a somewhat similar structure.

Example 21.4 Using Chemical Tests to Identify Group IIA Compounds

By means of a chemical test, distinguish between $MgCl_2$ and $BaCl_2$.

Solution

From Table 21.9, we note that $MgSO_4$ is soluble in water but $BaSO_4$ is insoluble. Therefore, if we add sodium sulfate to an aqueous solution of each compound, $BaCl_2$ will give a precipitate of the sulfate, and $MgCl_2$ will not.

$$Na_2SO_4(aq) + BaCl_2(aq) \longrightarrow 2NaCl(aq) + BaSO_4(s)$$

Exercise 21.4

Use a chemical test to distinguish between $BeCl_2$ and $MgCl_2$.

(See Problems 21.57, 21.58, 21.59, and 21.60.)

Example 21.5 Using Chemical Reactions of the Group IIA Elements and Compounds

Show how you could prepare CaH_2 from limestone ($CaCO_3$), aluminum, and hydrogen in three steps.

Solution

The steps are

$$CaCO_3(s) \xrightarrow{\Delta} CaO(s) + CO_2(g)$$

$$3CaO(s) + 2Al(l) \xrightarrow{\Delta} 3Ca(g) + Al_2O_3(s)$$

$$Ca(s) + H_2(g) \longrightarrow CaH_2(s)$$

Exercise 21.5

By means of equations, show how you could prepare magnesium sulfate from limestone, seawater, and sulfuric acid in three steps.

(See Problems 21.61 and 21.62.)

21.4 GROUP IIIA: BORON AND THE ALUMINUM FAMILY METALS

As we noted earlier, the metallic character of the elements increases in moving down a column of the periodic table. This behavior is quite evident in the Group IIIA elements. The first element, boron, is a semiconducting solid, and its compounds display distinctly nonmetallic properties. For example, the oxide B_2O_3 is acidic (see Table 21.2). The rest of the elements in this group are metals, but the oxides and hydroxides change from amphoteric for aluminum and gallium to basic for indium and thallium. Bonding to boron is covalent; the atom shares its $2s^22p^1$ valence electrons to give the +3 oxidation state. Although aluminum has many covalent compounds, it also has definitely ionic ones, such as AlF_3. The cation $Al^{3+}(aq)$ is present in aqueous solutions of aluminum salts. Gallium, indium, and thallium also have ionic compounds and give cations in aqueous solution. The +3 oxidation state is the only important one for aluminum, but some +1 compounds are known for gallium and indium, and in the case of thallium, both +1 and +3 states are important.

Figure 21.14
Gem-quality corundum. Shown here is sapphire, whose color is due to iron and titanium ions as impurities.

Aluminum is the third most abundant element in the earth's crust (after oxygen and silicon). It occurs primarily in aluminosilicate minerals found in the original rocks of this outer portion of the earth. Weathering of these rocks has formed aluminum-containing clays, which are an essential part of most soils. Further weathering of clays has given deposits of bauxite, an aluminum ore containing $AlO(OH)$ and $Al(OH)_3$ in various proportions. Corundum is a hard mineral of aluminum oxide, Al_2O_3. The pure oxide is colorless, but the presence of impurities can give various colors to it. Sapphire (usually blue) and ruby (deep red) are gem-quality corundum (Figure 21.14).

Boron, although not an abundant element, is well known because it occurs concentrated in accessible minerals, such as borax ($Na_2B_4O_7 \cdot 10H_2O$), kernite ($Na_2B_4O_7 \cdot 4H_2O$), and colemanite ($Ca_2B_6O_{11} \cdot 5H_2O$). These minerals were formed by the evaporation of hot springs and hot lakes. The major world supply of boron comes from southern California. The other elements of Group IIIA—gallium, indium, and thallium—are relatively rare and are obtained as by-products from the processing of other metals. For example, gallium is obtained from bauxite during aluminum production.

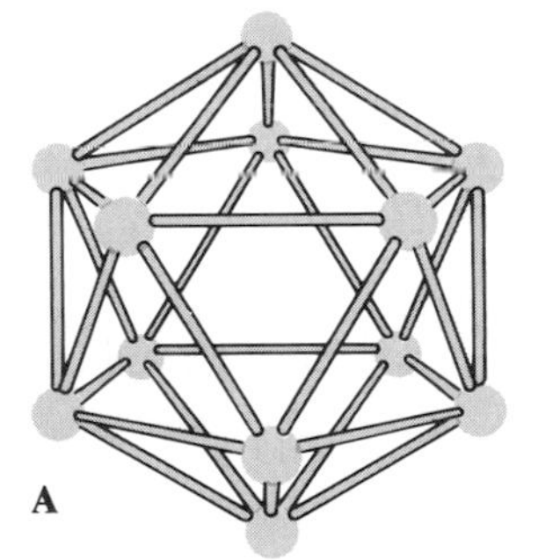

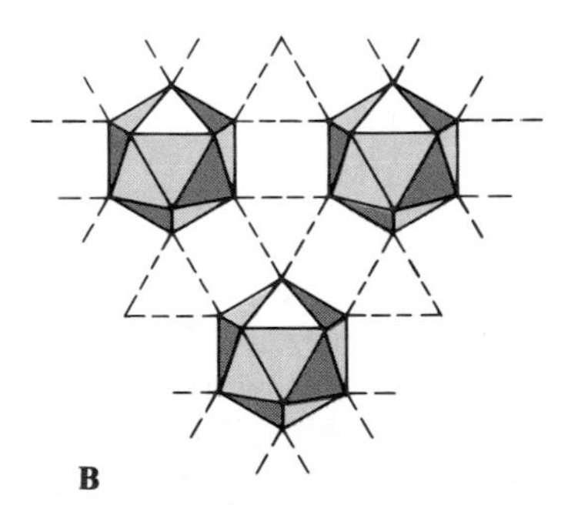

Figure 21.15
Structure of boron. *(A)* The repeating unit in several allotropes of boron consists of 12 boron atoms arranged at the corners of a regular icosahedron. *(B)* The structure of α-rhombohedral boron. Icosahedra of boron atoms are linked by chemical bonds to form a solid.

Properties of the Elements

Boron has a number of allotropic forms. Three of these are well characterized, and all consist of repeating units of twelve boron atoms arranged in regular icosahedra (20-sided figures, as shown in Figure 21.15). These icosahedra of boron atoms are covalently bonded to one another to form covalent network solids; the allotropes differ in the exact arrangement of these icosahedra. The allotropes are semiconductors, and the most stable one is a hard, black solid. It has a high melting point (2300°C), as expected for a covalent network solid. Table 21.10 lists properties of boron and other Group IIIA elements. Except at very high temperatures, crystalline boron is inert to chemical attack.

The other Group IIIA elements are metals. Aluminum has a moderately high melting point (660°C); the others melt at lower temperatures. However, there is no simple pattern as we go down the column of elements. Gallium melts easily when held in the hand, but indium and thallium have higher melting points. Except for thallium, the metals form an adherent oxide coating that gives them some protection against further oxidation. The elements form the +3 metal oxides such as Al_2O_3. Thallium also forms the +1 metal oxide, Tl_2O. Similarly, aluminum,

Table 21.10 Properties of Group IIIA Elements

Property	Boron	Aluminum	Gallium	Indium	Thallium
Electron configuration	$[He]2s^22p^1$	$[Ne]3s^23p^1$	$[Ar]3d^{10}4s^24p^1$	$[Kr]4d^{10}5s^25p^1$	$[Xe]4f^{14}5d^{10}6s^26p^1$
Melting point, °C	2300	660	30	157	304
Boiling point, °C	2550	2467	2403	2080	1457
Density, g/cm^3	2.34	2.70	5.90	7.30	11.85
Ionization energy (first), kJ/mol	801	578	579	558	589
(second), kJ/mol	2427	1817	1979	1821	1971
(third), kJ/mol	3660	2745	2963	2704	2878
(fourth), kJ/mol	25025	11577	6200	5200	4890
Electronegativity (Pauling scale)	2.0	1.5	1.6	1.7	1.8
Standard potential (volts), $M^{3+} + 3e^- \rightleftharpoons M$	—	−1.66	−0.56	−0.34	0.72
Covalent radius, Å	0.88	1.25	1.25	1.50	1.55
Ionic radius, Å	—	0.68	0.76	0.94	1.02

gallium, and indium combine directly with the halogens to give the metal(III) halides. Thallium forms both thallium(I) halides and thallium(III) halides [except for thallium(III) iodide, which is unknown●].

● The iodide ion presumably reduces thallium(III) to thallium(I). A compound TlI_3 is known, but it consists of $Tl^+I_3^-$, not $Tl^{3+}(I^-)_3$. It is therefore thallium(I) triiodide.

As expected for active metals, Group IIIA elements react with acids to release hydrogen.

$$2Al(s) + 6[H^+(aq) + Cl^-(aq)] \longrightarrow 2[Al^{3+}(aq) + 3Cl^-(aq)] + 3H_2(g)$$

(Nitric acid, however, renders aluminum unreactive, or *passive,* because a thick, adhering oxide layer forms.) Aluminum and gallium also dissolve in strong base to form the hydroxo anions and hydrogen.

$$2Al(s) + 2[Na^+(aq) + OH^-(aq)] + 6H_2O(l) \longrightarrow 2[Na^+(aq) + \underbrace{Al(OH)_4^-(aq)]}_{\text{sodium aluminate}} + 3H_2(g)$$

$$2Ga(s) + 2[Na^+(aq) + OH^-(aq)] + 6H_2O(l) \longrightarrow 2[Na^+(aq) + \underbrace{Ga(OH)_4^-(aq)]}_{\text{sodium gallate(III)}} + 3H_2(g)$$

Preparation of the Elements

Aluminum is the only element in Group IIIA produced in large quantity. It is made by the **Hall–Héroult process,** *a commercial method of producing aluminum by electrolysis of aluminum oxide dissolved in molten cryolite,* Na_3AlF_6. The cryolite mixture is electrolyzed at about 950°C.● An electrolysis cell is shown in Figure 21.16. Although the electrolysis is complicated and not completely understood, the final result can be represented by the following simplified electrode reactions:

● Cryolite, Na_3AlF_6, is an ore of aluminum found in Greenland. For use in aluminum production, it is now usually prepared from aluminum hydroxide obtained from bauxite.

$$12HF + 2Al(OH)_3 + 6NaOH \longrightarrow 2Na_3AlF_6 + 12H_2O$$

$$4Al^{3+} + 12e^- \longrightarrow 4Al(l) \quad \text{(cathode)}$$

$$6O^{2-} + 3C(s) \longrightarrow 3CO_2(g) + 12e^- \quad \text{(anode)}$$

$$\underbrace{2[2Al^{3+} + 3O^{2-}]}_{Al_2O_3} + 3C(s) \xrightarrow{\text{electrolysis}} 4Al(l) + 3CO_2(g) \quad \text{(overall)}$$

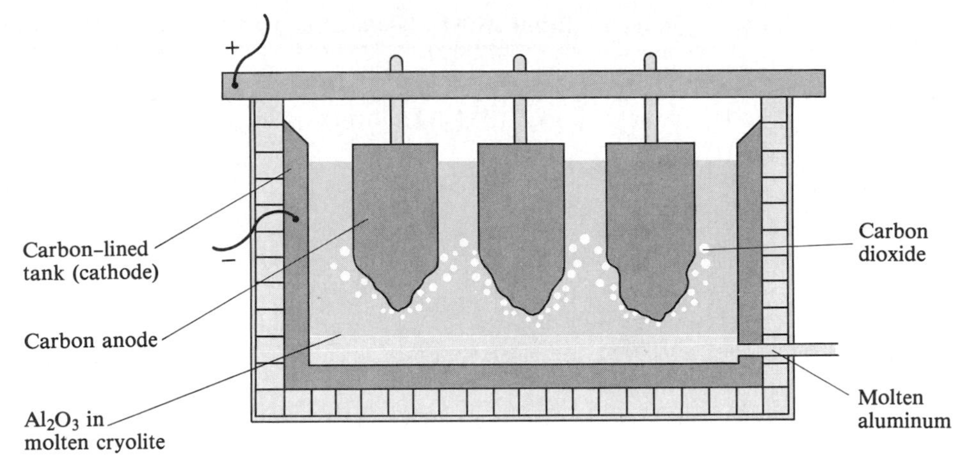

Figure 21.16
Hall–Héroult cell for the production of aluminum. Aluminum oxide is electrolyzed in molten cryolite (the electrolyte). Molten aluminum forms at the cathode (tank lining) and collects at the bottom of the cell, where it is periodically withdrawn.

The carbon anodes are made from carbonized petroleum and must be replaced continuously.

Boron may be prepared by reducing boric oxide, B_2O_3, with magnesium.

$$B_2O_3(l) + 3Mg(l) \xrightarrow{\Delta} 2B(s) + 3MgO(s)$$

After the magnesium oxide has been leached out with hydrochloric acid, the product is a brown, amorphous powder of rather impure boron. Pure boron can be prepared by reducing boron trichloride vapor with hydrogen on a hot tungsten wire.

$$2BCl_3(g) + 3H_2(g) \xrightarrow[1500°C]{W} 2B(s) + 6HCl(g)$$

Lustrous, black crystals are obtained.

Figure 21.17
Preparation of gallium arsenide, GaAs, semiconductor material. Gallium arsenide is used to make semiconductor diode lasers for use in fiber optic communications, compact disc players, and laser printers. Also, computer chips are being developed using gallium arsenide, because it conducts an electrical signal five times faster than presently used silicon.

Uses of the Elements

Commercially, the most important Group IIIA element is aluminum, which is made in large quantities for alloys. Although the pure metal is soft and corrodes easily, the addition of small quantities of other metals, such as copper, magnesium, and manganese, gives hard, corrosion-resistant alloys. Some aluminum is used to produce other metals. Chromium is obtained by the **Goldschmidt process,** *a method of preparing a metal by reduction of its oxide with powdered aluminum.* The reaction of Cr_2O_3 with Al is exothermic.

$$Cr_2O_3(s) + 2Al(l) \longrightarrow Al_2O_3(l) + 2Cr(l); \Delta H° = -536 \text{ kJ}$$

A similar reaction produces iron for welding from a mixture of powdered aluminum and iron(III) oxide. Once the mixture (called *thermite*) is ignited, the reaction is self-sustaining and gives a spectacular incandescent shower. The reaction is also the basis of incendiary bombs.

The other Group IIIA elements are becoming increasingly useful. Boron has been used to make filaments to reinforce plastic and metal parts for special applications. Gallium and indium have been used to make semiconductors for solid-state electronics (Figure 21.17).

Table 21.11 Uses of Boron and Aluminum Compounds

Compound	Use
$Na_2B_4O_7 \cdot 10H_2O$ (borax)	Source of boron and its compounds Detergent formulations
$B(OH)_3$	Preparation of boric oxide Antiseptic and preservative
B_2O_3	Fiber glass and borosilicate glass Enamels and glazes
Al_2O_3	Source of aluminum and its compounds Abrasive Refractory bricks and furnace linings Synthetic sapphires and rubies
$Al_2(SO_4)_3 \cdot 18H_2O$	Making of paper Water purification Mordant
$AlCl_3$	Preparation of aluminum Catalyst in organic reactions
$AlCl_3 \cdot 6H_2O$	Antiperspirant

Uses of the Compounds

Compounds of both boron and aluminum are important (see Table 21.11). Boron resembles silicon in forming a large number of compounds and polymeric materials containing bonds between the element and oxygen.● Boron and oxygen atoms are often arranged in rings and more complicated structures involving trigonal planar BO_3 units and tetrahedral BO_4 units. Figure 21.18 shows the structure of the anion in the mineral borax, also known as sodium tetraborate decahydrate. Note the alternating trigonal planar and tetrahedral boron atoms in the structure. The chemical formula of borax is usually written $Na_2B_4O_7 \cdot 10H_2O$, but the compound consists of an anion whose formula is better written as $B_4O_5(OH)_4^{2-}$. Thus, borax is $Na_2[B_4O_5(OH)_4] \cdot 8H_2O$.

● This characteristic is indicative of the diagonal relationship between boron and silicon.

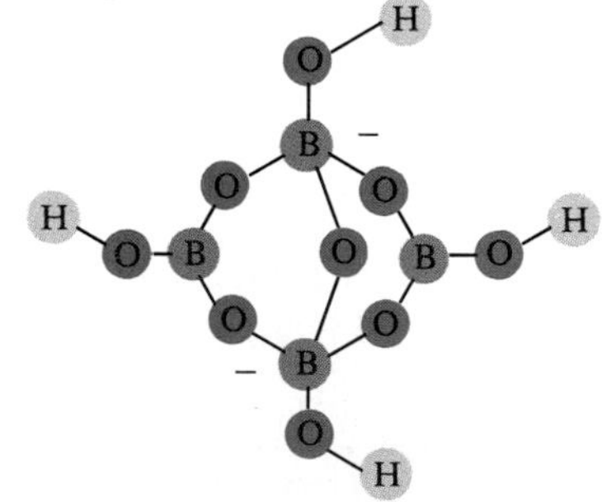

Figure 21.18
Structure of the tetraborate ion, $B_4O_5(OH)_4^{2-}$. Two of the boron atoms are part of the tetrahedral BO_4 units (top and bottom B atoms in the drawing); the other two boron atoms are part of the trigonal planar BO_3 units.

Borax is used in soaps and detergents. But much of the borax produced is converted to boric acid, then to boric oxide. Boric acid, $B(OH)_3$, is prepared by adding sulfuric acid to borax solutions.

$$\underbrace{[2Na^+(aq) + B_4O_5(OH)_4^{2-}(aq)]}_{\text{borax}} + [2H^+(aq) + SO_4^{2-}(aq)] + 3H_2O(l) \longrightarrow [2Na^+(aq) + SO_4^{2-}(aq)] + \underbrace{4B(OH)_3(s)}_{\text{boric acid}}$$

The boric acid crystallizes from the solution. Although the formula is often written H_3BO_3, the acid is monoprotic and functions as a Lewis acid by accepting an electron pair of OH^- (from water).

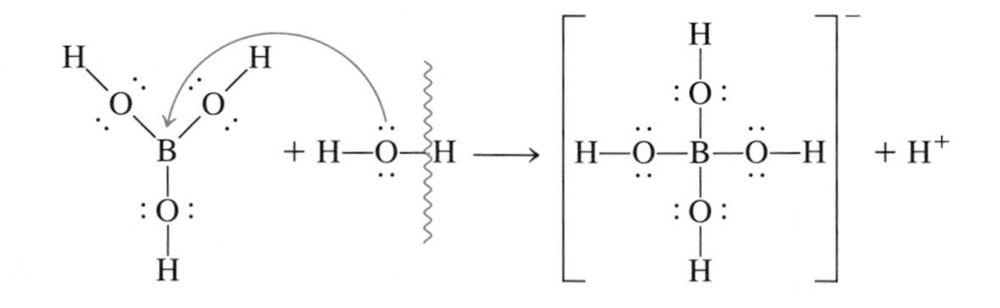

Boric acid is toxic to bacteria and fungi and is used as an antiseptic and as a preservative for wood and leather. It is also used as a fire retardant for cellulose products (paper and cloth).

Most of the boric acid produced is converted to boric oxide by heating.

$$2B(OH)_3(s) \xrightarrow{\Delta} B_2O_3(s) + 3H_2O(g)$$

Large amounts of boric oxide are used to make fiber glass and borosilicate glass, which has a very small coefficient of thermal expansion (Pyrex is a brand name). Boric oxide is also used in porcelain enamels and ceramic glazes.

Boron forms a series of **boranes,** *binary compounds of boron and hydrogen,* which are of interest because of their unique electronic structure. The simplest member is diborane, B_2H_6, a gas that is spontaneously flammable in moist air. Its structure may be written

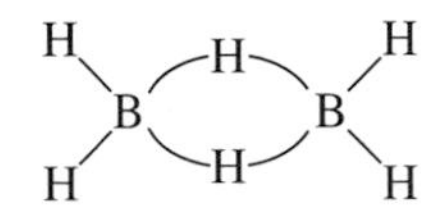

The end B—H bonds are the normal covalent sort. Each interior B—H—B bond, however, is a **three-center bond,** *a bond in which three atoms are held together by two electrons*. To get a more detailed picture of the bonding, let us assume that the bonding orbitals on the boron atoms are sp^3 hybrids. Two of the sp^3 hybrid orbitals on each boron atom are involved in bonding to the end hydrogen atoms. The other sp^3 hybrid orbitals are used in bonding to the "bridge" hydrogens. Each three center bond is obtained by the overlapping of an sp^3 hybrid orbital from each boron with the 1*s* of hydrogen, giving a three-center molecular orbital (see Figure 21.19). Note that the plane containing the end hydrogens must be perpendicular to the plane containing the bridge hydrogens.

The most important compound of aluminum is alumina, or aluminum oxide, Al_2O_3. It is prepared by heating aluminum hydroxide, which is obtained from bauxite.

$$2Al(OH)_3(s) \xrightarrow{\Delta} Al_2O_3(s) + 3H_2O(g)$$

Prepared at low temperature (550°C), it is a porous or powdery white solid. It is used as a carrier for catalysts and for other purposes, though most alumina is used in the production of aluminum metal. When aluminum oxide is fused at high temperature (2045°C), it forms corundum and can be used as an abrasive or refractory. When it is fused with certain metal impurities, synthetic sapphires and rubies

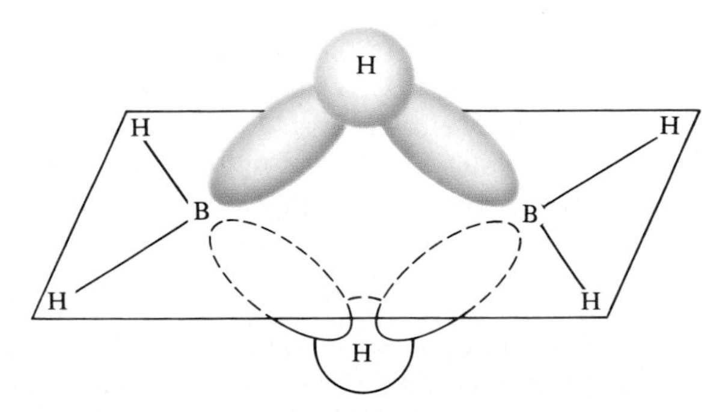

Figure 21.19
Bonding in diborane, B_2H_6. The three-center B—H—B bonds (one shown in color) are perpendicular to the plane of the rest of the molecule.

Figure 21.20
Deterioration of the paper in books. Aluminum sulfate and similar acidic compounds have been used as the sizing in many kinds of paper, including those used in books. However, acids decompose the cellulose fibers in paper, so that such paper quickly deteriorates. The Library of Congress estimates that 25% of its collection is brittle from deteriorating paper. Calcium carbonate, a basic compound, has been suggested as an alternative for paper sizing.

are obtained. Thus, synthetic rubies contain about 2.5% chromium oxide, Cr_2O_3. They are used primarily in watches and instruments as ''jewel'' bearings.

Aluminum sulfate octadecahydrate, $Al_2(SO_4)_3 \cdot 18H_2O$, is the commonest soluble salt of aluminum. It is prepared by dissolving bauxite in sulfuric acid. The salt is acidic in water solution. In water, aluminum ion forms a strong hydration complex, $Al(H_2O)_6{}^{3+}$, and this ion in turn hydrolyzes.

$$Al(H_2O)_6{}^{3+}(aq) + H_2O(l) \rightleftharpoons Al(H_2O)_5OH^{2+}(aq) + H_3O^+(aq)$$

Aluminum sulfate is used in large amounts by the paper industry during the *sizing* of paper. In this process, substances such as clay are added to the cellulose fibers to give a less-porous material with a smooth finish. Colloidal clay is coagulated onto the paper fibers by aluminum ion (Figure 21.20).● Aluminum sulfate is also used to treat waste water in paper pulping, as well as to purify municipal water. When a base is added to an aqueous solution of an aluminum salt, a gelatinous precipitate of aluminum hydroxide forms.

● Coagulation of colloids by multiply charged ions was discussed in Section 12.9.

$$Al^{3+}(aq) + 3OH^-(aq) \longrightarrow Al(OH)_3(s)$$

The aluminum ion coagulates colloidal suspensions, which are then adsorbed on the aluminum hydroxide precipitate and can be filtered off. Aluminum hydroxide is often precipitated onto the fibers in fabrics, where it adsorbs certain dyes. It is said to act as a *mordant* (fixative).

Aluminum hydroxide dissolves in acid to form the Al^{3+} ion and, in excess base, to form the aluminate ion, $Al(OH)_4{}^-$.

$$Al(OH)_3(s) + 3H^+(aq) \longrightarrow Al^{3+}(aq) + 3H_2O(l)$$
$$Al(OH)_3(s) + OH^-(aq) \longrightarrow Al(OH)_4{}^-(aq)$$

Thus, aluminum hydroxide is amphoteric.

Aluminum chloride hexahydrate, $AlCl_3 \cdot 6H_2O$, can be prepared by dissolving alumina in hydrochloric acid. It is used as an antiperspirant and disinfectant. The anhydrous salt, $AlCl_3$, cannot be prepared by heating the hydrate, because it decomposes to give HCl. Anhydrous aluminum chloride is obtained by reacting scrap aluminum with chlorine gas.

$$2Al(s) + 3Cl_2(g) \longrightarrow 2AlCl_3(s)$$

The solid is believed to be ionic, but when the substance is heated at near its melting point, the molecular substance Al_2Cl_6 is formed. Aluminum chloride sublimes at its melting point (180°C), and Al_2Cl_6 molecules are present in the vapor.

The structure of Al_2Cl_6 has chlorine ''bridges'' as in solid $BeCl_2$.

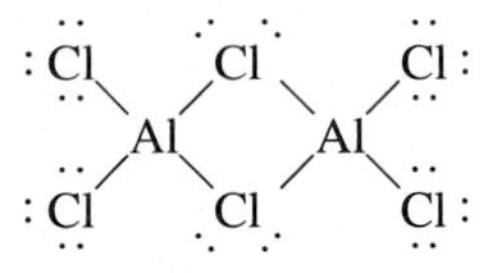

At higher temperatures, the Al_2Cl_6 molecules in the vapor dissociate to trigonal planar $AlCl_3$ molecules. Anhydrous aluminum chloride is used as a catalyst, for example, in the preparation of ethylbenzene. Ethylbenzene is used to prepare styrene, the starting material for polystyrene plastic and foam (Styrofoam).

Example 21.6 Using Chemical Tests to Identify Groups IA to IIIA Compounds

A solution is either $NaCl(aq)$, $MgCl_2(aq)$, or $AlCl_3(aq)$. Devise a chemical test to differentiate among these solutions. You may use only one other substance.

Solution

Add NaOH(*s*) slowly to each solution. When NaOH(*s*) is slowly added to $AlCl_3(aq)$, a white precipitate of $Al(OH)_3$ forms. This later dissolves as more NaOH is added to give the soluble aluminate, $Na[Al(OH)_4]$. When NaOH is added to a solution of $MgCl_2$, a white precipitate of $Mg(OH)_2$ forms. This does not dissolve when more base is added. Sodium chloride solution does not give a precipitate with NaOH.

Exercise 21.6

The following white solids have been placed in separate unlabeled beakers so that you do not know which is which: KOH; $Al_2(SO_4)_3 \cdot 18H_2O$; $BaCl_2 \cdot 2H_2O$. Devise a series of chemical tests, using only water in addition to the substances in these beakers, to establish the identity of each solid. (See Problems 21.73 and 21.74.)

21.5 METALLURGY

With the exception of boron, the elements we have studied in this chapter are metals. Many of them have commercial importance. In fact, the world production of aluminum exceeds that of all metals but iron. **Metallurgy** is *the scientific study of the production of metals from their ores and the making of alloys*. In this section, we will look at the basic steps in the production of a metal from its ore.

1. **Preliminary treatment.** Usually, the ore must first be concentrated in the metallic element. The result may be a relatively pure compound of the metal. Often this compound must then be converted by chemical reaction to one that can be reduced to the metal.
2. **Reduction.** A compound of the metal is reduced to the free element by electrolysis or chemical reduction.
3. **Refining.** The metal obtained from the reduction may need to be refined, or purified.

Preliminary Treatment

Most metals are obtained from ores that contain varying amounts of impurities along with the mineral of the metal. *The impurities in an ore* are called **gangue.**

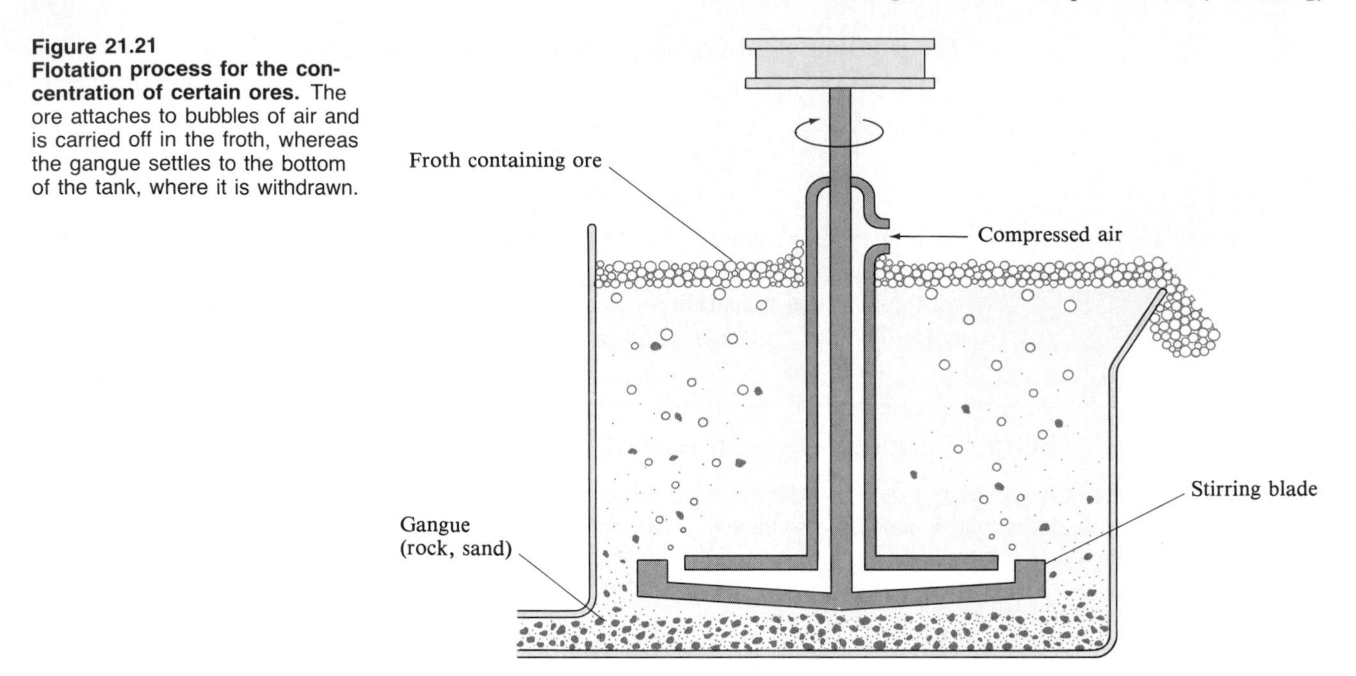

Figure 21.21
Flotation process for the concentration of certain ores. The ore attaches to bubbles of air and is carried off in the froth, whereas the gangue settles to the bottom of the tank, where it is withdrawn.

Ores may be concentrated in the metallic element by separating off the gangue, using physical or chemical methods. Panning for gold is a simple physical separation. It depends on the difference in density between gold and sand. Water easily flushes sand and dirt aside, leaving grains of gold at the bottom of the pan. **Flotation** is *a physical method of separating a mineral from the gangue that depends on differences in their wettability*. The ore is crushed to a fine powder, then mixed into a tank of water (usually containing a flotation agent) where a fast stream of air forms a froth (see Figure 21.21). Particles that are wetted by water, such as the gangue, sink to the bottom of the tank, whereas particles that are not wetted become attached to air bubbles and float to the top of the tank. A flotation agent in the water gives the mineral particles a hydrophobic coating (a coating that is not wetted by water). Each molecule of the flotation agent has a polar end that is attracted to the surface of the mineral and a nonpolar end that is hydrophobic. These molecules align themselves on the mineral surface with their nonpolar ends pointing outward, giving the mineral particle a hydrophobic surface. Sulfide ores, such as those of copper, lead, and zinc, are concentrated this way by flotation.

The Bayer process is an example of a chemical method of concentrating an ore. The **Bayer process** is *a process in which purified aluminum oxide is obtained from bauxite*. It depends on the fact that aluminum hydroxide is amphoteric. Bauxite is an ore containing aluminum hydroxide, $Al(OH)_3$, or the oxide hydroxide, $AlO(OH)$. When bauxite is mixed with hot, aqueous sodium hydroxide solution, the aluminum minerals dissolve to give the aluminate ion, $Al(OH)_4^-$.

$$Al(OH)_3(s) + OH^-(aq) \longrightarrow Al(OH)_4^-(aq)$$
$$AlO(OH)(s) + OH^-(aq) + H_2O(l) \longrightarrow Al(OH)_4^-(aq)$$

Silicate and iron oxide impurities remain undissolved and are filtered off. As the hot solution of sodium aluminate cools, aluminum hydroxide precipitates. In practice, the solution is seeded with aluminum hydroxide to start the precipitation.

Figure 21.22
Precipitation of aluminum hydroxide by acidifying an aluminate ion solution. *Left:* Carbon dioxide gas is bubbled into a solution of aluminate ion, $Al(OH)_4^-$, containing aluminon dye. *Right:* The CO_2 precipitates aluminum hydroxide, $Al(OH)_3$, from solution. Aluminum hydroxide is normally white, but adsorbs the dye to form a pink-colored precipitate (called a *lake*). A standard test for aluminum ion is its precipitation as the hydroxide in the presence of aluminon dye to form this lake.

$$Al(OH)_4^-(aq) \longrightarrow Al(OH)_3(s) + OH^-(aq)$$

Aluminum hydroxide can also be precipitated by slightly acidifying the solution (Figure 21.22).● The precipitate is filtered off and heated to convert it to a white powder of anhydrous aluminum oxide.

● Gallium can be separated from bauxite at this stage of the process. Gallate ion is more stable than aluminate ion and tends to stay in solution, whereas aluminate ion reverts to the hydroxide on cooling or with mild acidification with CO_2. Gallium hydroxide precipitates only after further acidification.

$$2Al(OH)_3(s) \xrightarrow{\Delta} Al_2O_3(s) + 3H_2O(g)$$

Once an ore is concentrated, it may be necessary to convert the mineral to a compound suitable for reduction. **Roasting** is *the process of heating a mineral in air to obtain the oxide*. Thus, zinc ore (ZnS) is converted to zinc oxide by roasting in air.

$$2ZnS(s) + 3O_2(g) \longrightarrow 2ZnO(s) + 2SO_2(g)$$

The roasting is exothermic and, once it is started, does not require additional heating.

Reduction

The free metal is obtained by reduction of a compound, using either electrolysis or a chemical reducing agent. In the commercial production of zinc, zinc oxide is reduced by heating it with some form of carbon, such as anthracite coal or coke.

$$ZnO(s) + C(s) \xrightarrow{\Delta} Zn(g) + CO(g)$$

Zinc metal vapor leaves the reaction chamber and condenses to the liquid, which then solidifies. Hydrogen and active metals, such as sodium, magnesium, and aluminum, are also used as reducing agents when carbon is unsuitable. Tungsten, for example, combines chemically with carbon. To produce the pure metal, hydrogen is used instead.

$$WO_3(s) + 3H_2(g) \xrightarrow{\Delta} W(s) + 3H_2O(g)$$

Several examples of the production of a metal using an active metal as reducing agent were given in earlier sections.

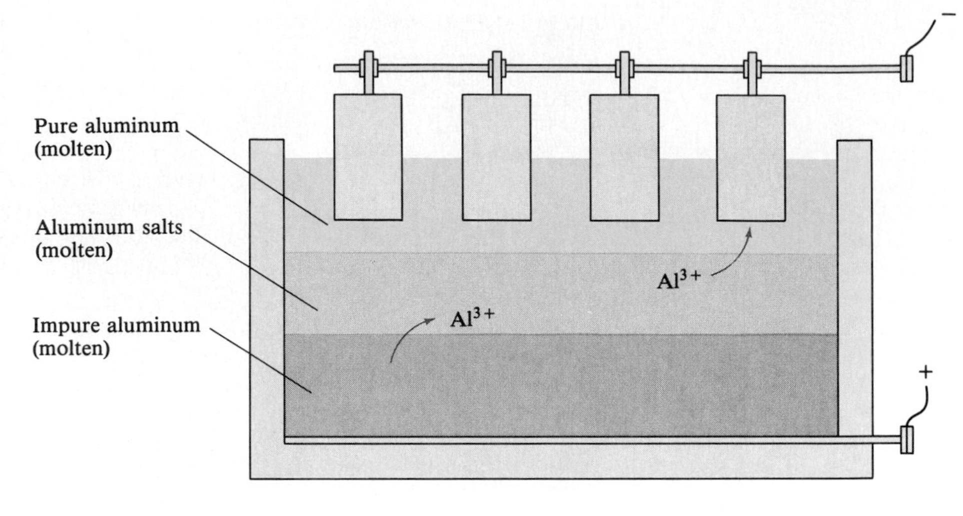

Figure 21.23
Hoopes cell for the refining of aluminum. Aluminum is refined by electrolysis. Aluminum ions move from the anode (impure aluminum) to the cathode (pure aluminum). During operation of the cell, impure aluminum is continually added to the bottom and pure aluminum is drawn off the top.

Refining of a Metal

Often the metal obtained from a reduction process contains impurities and must be purified or refined before it can be used. Zinc, for example, is contaminated by lead, cadmium, and iron. It is refined by fractional distillation. Some metals, such as copper, nickel, and aluminum, are refined electrolytically. The **Hoopes process** is *a process for the electrolytic refining of aluminum*. Impure aluminum forms the anode and pure aluminum forms the cathode of the Hoopes cell, which contains three liquid layers (Figure 21.23). The bottom layer is molten, impure aluminum; the middle is a fused salt layer containing aluminum fluoride; and the top layer is pure aluminum. At the anode (bottom layer), aluminum passes into solution as aluminum ions, Al^{3+}, and at the cathode (top layer), these ions are reduced to the pure metal. In operation, impure, molten metal is added to the bottom of the cell and pure aluminum is drawn off the top.

Exercise 21.7

(a) Write equations for the metallurgy of magnesium that illustrate the concentration of magnesium ion from seawater and the conversion of the resulting compound to one that can be electrolytically reduced.
(b) Write the equation for the electrolytic reduction of this compound to give metallic magnesium. (See Problems 21.75 and 21.76.)

A Checklist for Review

IMPORTANT TERMS

amphoteric substance (21.1)
diagonal relationship (21.1)
mineral (21.2)
Solvay process (21.2)
ore (21.3)
Hall–Héroult process (21.4)
Goldschmidt process (21.4)
boranes (21.4)
three-center bond (21.4)
metallurgy (21.5)
gangue (21.5)
flotation (21.5)
Bayer process (21.5)
roasting (21.5)
Hoopes process (21.5)

SUMMARY OF FACTS AND CONCEPTS

Before examining the properties of the main-group elements, we can make several *general observations*. First, the metallic characteristics of the elements decrease in going across a period and increase in going down a column. Also, the oxidation states of these elements vary predictably. Finally, the elements of the second row differ somewhat from the other elements in the same column; in fact, lithium, beryllium, and boron are similar in many ways to the elements located diagonally to the right of them *(diagonal relationships)*.

Group IA elements (alkali metals) occur as ions in the aluminosilicate rocks of the earth's crust. Commercial sources of sodium are halite (NaCl) and seawater; sources of potassium are sylvite (KCl) and carnallite ($KCl \cdot MgCl_2 \cdot 6H_2O$). The alkali metals are soft and chemically very reactive. All react vigorously with water, for example. Sodium and lithium are prepared by electrolysis of the molten chlorides. Other Group IA elements are prepared by reduction of the molten chlorides with sodium. Sodium is the most important alkali metal. Important compounds of sodium are sodium hydroxide and sodium carbonate. Sodium hydroxide, obtained by electrolysis of aqueous NaCl, is used to make paper and to extract aluminum oxide from bauxite. Sodium carbonate is used to make glass. Potassium chloride is important as a fertilizer.

Group IIA elements (alkaline earth metals) also occur as ions in aluminosilicate rocks. The main commercial source of magnesium ion is seawater, and calcium compounds are prepared from seashells and limestone. The most active alkaline earth metals are calcium and the metals below it; they react easily with water. Magnesium is less reactive with water but dissolves readily in acids. Beryllium reacts with both acids and bases. Magnesium is prepared by electrolysis of the chlorides; the other Group IIA elements are usually prepared by chemical reduction. Magnesium is an important metal used in lightweight alloys. Calcium oxide, obtained by heating calcium carbonate, is produced in large quantities for the manufacture of steel, for water treatment, and for various chemical processes.

Of the Group IIIA elements, aluminum is the most abundant (the third most abundant element on earth). Boron is much less abundant than aluminum, but it occurs in accessible minerals, such as borax ($Na_2B_4O_7 \cdot 10H_2O$). Elemental boron is a semiconductor but is nonmetallic in its chemical properties. Aluminum is a reactive metal dissolving in both acids and bases. The metal is prepared in large quantity for alloys by the electrolytic reduction of aluminum oxide *(Hall–Héroult process)* obtained from the ore bauxite. Boron compounds are used to make detergents (borax) and glass (boric oxide). *Boranes* are a class of compounds, including diborane (B_2H_6), that are interesting because of their unique bonding: They contain *three-center bonds* (B—H—B). The most important aluminum compound is the sulfate, $Al_2(SO_4)_3 \cdot 18H_2O$. This compound is used by the paper industry and for water purification. Aluminum chloride is an ionic compound that vaporizes as Al_2Cl_6 molecules, which have a bridge chlorine structure.

The last section discusses the basic steps in *metallurgy*. These steps are *preliminary treatment* of the ore or other natural source (concentration and conversion to a reducible compound); *reduction* to the metal; and *metal refining*, or purification. In the production of aluminum, pure aluminum oxide is prepared from bauxite by treating it with hot sodium hydroxide solution *(Bayer process)*. The aluminum minerals dissolve, leaving impurities to be filtered off. Aluminum hydroxide precipitates from the cool solution and is converted to the oxide by heating. The oxide is reduced to the metal *(Hall–Héroult process)* and refined electrolytically *(Hoopes process)*. In a Hoopes cell, impure aluminum forms the anode and pure aluminum the cathode. During electrolysis, aluminum leaves the anode as the ion and deposits as the pure metal at the cathode.

OPERATIONAL SKILLS

1. **Predicting the relative metallic–nonmetallic character of elements** Given two elements, decide which has the greater metallic character by knowing their location in the periodic table (Example 21.1).

2. **Predicting covalent character from oxidation state** Given two binary compounds of the same two elements, but differing in the oxidation states of one element, predict which compound has greater covalent character in its bonds (Example 21.2).

3. **Using chemical reactions of the Group IA elements and compounds** Devise a method of preparing a Group IA substance from given starting substances by recalling important chemical reactions of the group (Example 21.3).

4. **Using chemical tests to identify Group IIA compounds** Devise chemical tests for distinguishing Group IIA compounds by recalling important chemical reactions of the group (Example 21.4).

5. **Using chemical reactions of the Group IIA elements and compounds** Devise a method of preparing a Group IIA substance from given starting substances by recalling important chemical reactions of the group (Example 21.5).

6. **Using chemical tests to identify Groups IA to IIIA compounds** Devise chemical tests for distinguishing Groups IA to IIIA compounds by recalling important chemical reactions of the group (Example 21.6).

Review Questions

21.1 Give four characteristics of a metal.

21.2 Describe the different ways in which the metallic characteristics of the elements vary in going through the periodic table.

21.3 Consider the Group VIA elements. For each element, state the rule that applies for predicting the most common oxidation state.

21.4 Without looking at Table 21.2, choose the oxide from each of the following pairs that you would expect to be more acidic. Explain your choices. (a) As_4O_6, As_2O_5; (b) Tl_2O, Tl_2O_3; (c) Sb_2O_5, Sb_4O_6; (d) SO_2, SO_3; (e) SnO, SnO_2.

21.5 What are some reasons why the behavior of a second-row element differs from that of the remaining elements in the same column?

21.6 Explain what is meant by a diagonal relationship between elements. Illustrate this explanation for some pair of elements.

21.7 Give a commercial source (ore or other natural source) of each of the following elements: Na, K, Mg, Ca, B, Al.

21.8 Francium was discovered as a radioactive-decay product of actinium-227. Write the equation for the formation of francium from actinium-227 by alpha emission.

21.9 Write balanced equations for the reactions of lithium with oxygen, with water, and with nitrogen. Do the same for sodium. Compare. (If no reaction occurs, indicate this by writing NR.)

21.10 Ethanol, C_2H_5OH, reacts with sodium metal because the hydrogen atom attached to oxygen is slightly acidic. Write a balanced equation for the reaction of sodium with ethanol to give the salt sodium ethylate, $NaOC_2H_5$.

21.11. (a) Write electrode reactions for the electrolysis of fused lithium chloride. (b) Do the same for fused lithium hydroxide. The hydroxide ion is oxidized to oxygen and water.

21.12 Write the equation for the preparation of rubidium by the reduction of rubidium chloride with calcium. Be sure to show the phase for each substance. How is the reaction forced in the direction of the product (Rb)?

21.13 What is the principal use of sodium metal? of potassium?

21.14 How is sodium hydroxide manufactured? What is another important product in this process?

21.15 How is lithium hydroxide produced from lithium carbonate?

21.16 One of the alkaline earth elements has only radioactive isotopes. What is the name of this element?

21.17 Write equations for the reactions of calcium metal with the following substances: (a) water, (b) oxygen, (c) hydrogen, (d) nitrogen.

21.18 Starting from seawater, limestone, and hydrochloric acid, write equations for the preparation of magnesium metal.

21.19. Strontium can be produced by chemical reduction of strontium oxide. Write a chemical equation showing how this can be done.

21.20 Calcium carbonate is used in some antacid preparations to neutralize HCl in the stomach. Write the equation for the antacid reaction.

21.21 Write the equation that represents the setting of mortar.

21.22 Barium sulfate is given orally to obtain x-ray photographs of the stomach. Barium carbonate is a water-insoluble compound. Why could the carbonate not be used in place of the sulfate?

21.23 Write equations in which (a) $Be(OH)_2$ reacts with $HCl(aq)$, and (b) $Be(OH)_2$ reacts with KOH. Compare the behavior of $Be(OH)_2$ in these reactions with that of $Mg(OH)_2$ with the same substances.

21.24 What is the principal chemical constituent in ruby?

21.25 Aluminum reacts with both acids and bases. Write balanced equations illustrating this.

21.26 Nitric acid may be shipped in aluminum drums. Explain how this is possible even though aluminum is an active metal.

21.27 Aluminum can be produced by electrolysis. Write the electrode reactions.

21.28 Describe a method for preparing pure boron from boron trichloride.

21.29 Aluminum oxide has a large negative enthalpy of formation. How is this characteristic useful for welding with thermite?

21.30 How is boric acid obtained from borax? Write the chemical equation for the reaction.

21.31 Explain the acidity of boric acid. How does its acidity differ from that of an acid like HCN?

21.32 Describe the bonding in tetraborane, B_4H_{10}.

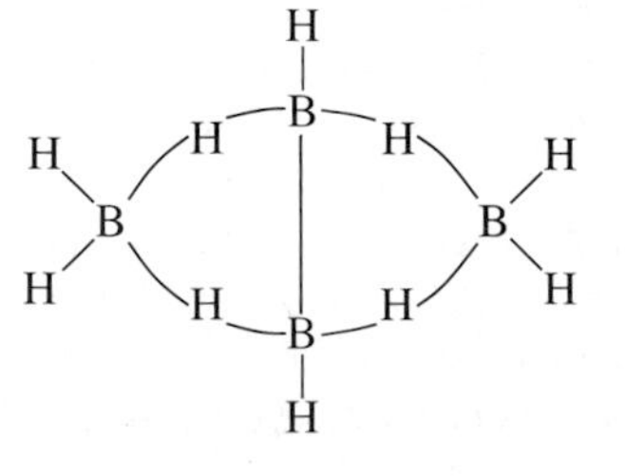

21.33 Aluminum nitrate dissolves in water to give acidic solutions. Explain why this is so.

21.34 Compare the structures of solid $BeCl_2$ and the molecule Al_2Cl_6. How are they similar to each other? How are they different?

21.35 What are the basic steps in the production of a pure metal from a natural source? Illustrate each step with the production of aluminum.

21.36 How does flotation separate an ore mineral from impurities?

21.37 What is the purpose of the roasting of zinc ore?

21.38 Carbon is often used as a cheap reducing agent, but it is sometimes unsuitable. Give an example in which another reducing agent is preferred.

Practice Problems

GENERAL OBSERVATIONS ABOUT THE MAIN-GROUP ELEMENTS

21.39 Decide whether each of the following main-group elements is a metal, a metalloid, or a nonmetal. List all the evidence that points to this conclusion. (a) Element X is a nonlustrous, yellow solid that burns in air, producing an oxide XO_2. The oxide dissolves in water to give a solution that turns blue litmus red. When NaOH is added to this solution, the ion XO_3^{2-} is formed. (b) Element X is a lustrous solid that conducts electricity. The oxide reacts with $HCl(aq)$, forming a solution of the salt XCl_3. It reacts with $NaOH(aq)$ to form a solution of the salt $NaX(OH)_4$.

21.40 Decide whether each of the following main-group elements is a metal, a metalloid, or a nonmetal. List all the evidence that points to this conclusion. (a) Element X is a lustrous, black solid that is a poor conductor of electricity at room temperature but becomes a moderately good conductor at elevated temperatures. The oxide of the element forms an acid H_3XO_3. (b) Element X is a soft, lustrous solid that reacts with $HCl(aq)$, forming a solution of the salt XCl_2. The element is a good conductor of electricity at room temperature.

21.41 Identify the more metallic element in each of the following pairs: (a) Na or Mg; (b) Na or Be; (c) Ga or Cs. Explain your answers.

21.42 Identify the more metallic element in each of the following pairs: (a) K or Ga; (b) K or Ca; (c) Rb or Al. Explain your answers.

21.43 Lead forms the oxides PbO and PbO_2. One of these oxides is a yellow solid that melts at 888°C; the other is a dark brown powder that decomposes at 290°C. Match the formula of each oxide with its properties. Explain.

21.44 Antimony has two fluorides. One is a colorless, oily liquid that freezes at 7°C. The other is a colorless solid that melts at 292°C. What is the probable formula of each of these compounds? Explain.

21.45 After studying Sections 21.2 to 21.4, make a list of ways in which the second-row elements Li, Be, and B are similar to other members of the same group. Then make a list of ways in which they differ.

21.46 After studying Sections 21.3 and 21.4, make a list of the properties that beryllium and aluminum share.

THE GROUP IA ELEMENTS

21.47 Francium was discovered as a minor decay product of actinium-227. The major decay of actinium-227 is by beta emission to thorium. Write the nuclear equation for the decay of actinium-227 by beta emission.

21.48 Francium-223 is a radioactive alkali metal that decays by beta emission. Write the nuclear equation for this decay process.

21.49 Titanium metal is produced from one of its ores [ilmenite ($FeTiO_3$) and rutile (TiO_2)] by chlorination to give titanium tetrachloride. For example,

$$3TiO_2(s) + 4C(s) + 6Cl_2(g) \longrightarrow 3TiCl_4(g) + 2CO(g) + 2CO_2(g)$$

The titanium tetrachloride vapor is reduced by passing it into liquid sodium or magnesium. Write the balanced equation for the reduction of $TiCl_4$ with sodium.

21.50 Liquid ammonia has many chemical properties similar to those of water. Sodium metal, for example, reacts with liquid ammonia, producing sodium amide ($NaNH_2$) and hydrogen. (The reaction is slow, however, unless a catalyst such as Fe^{3+} is present.) Write the balanced equation for this reaction.

21.51 Potassium hydroxide may be prepared by the aqueous reaction of potassium sulfate with barium hydroxide (a strong, soluble base). Write the balanced equation for this reaction, including phases for the species. Why does the reaction go in this direction?

21.52 Pure sodium hydrogen carbonate (baking soda) is prepared by dissolving carbon dioxide gas in a saturated solution of sodium carbonate. Sodium hydrogen carbonate precipitates from the solution. Write the balanced equation, indicating phases for each species.

21.53 Caustic soda, NaOH, has been manufactured from sodium carbonate in a manner similar to the manufacture of lithium hydroxide. Write balanced equations (in three steps) for the preparation of NaOH starting from lime ($Ca(OH)_2$), salt (NaCl), carbon dioxide, and ammonia.

21.54 Sodium cyanide, NaCN, is used in electroplating. It is made from hydrogen cyanide, HCN, which is prepared from ammonia, air, and natural gas (CH_4).

$$2NH_3(g) + 3O_2(g) + 2CH_4(g) \xrightarrow[1030°C]{Pt} 2HCN(g) + 6H_2O(g)$$

Write balanced equations for the preparation of NaCN from salt (NaCl), ammonia, air, and natural gas.

THE GROUP IIA ELEMENTS

21.55 Radium is found in uranium minerals. It results from the alpha decay of thorium-230, which is the decay product of uranium-234. Write the nuclear equation representing the decay of thorium-230 to radium.

21.56 A short-lived isotope of radium decays by alpha emission to radon-219. Write the nuclear equation for this decay process.

21.57 Describe a chemical test to distinguish between beryllium and magnesium metals.

21.58 Barium hydroxide and sodium hydroxide are strong bases. What simple test could you use to tell a solution of one from a solution of the other?

21.59 Devise a chemical method for separating a solution of $BeCl_2$, $MgCl_2$, and $BaCl_2$ into three solutions or compounds each containing a different one of the elements Be, Mg, and Ba.

21.60 Describe a chemical method for separating a solution of NaCl, $MgCl_2$, and $SrCl_2$ into three solutions or compounds each containing a different one of the elements Na, Mg, and Sr.

21.61 How could you prepare beryllium metal from beryllium fluoride, magnesium hydroxide, and hydrochloric acid in three steps? Write the balanced equations.

21.62 How could you prepare calcium metal from gypsum ($CaSO_4 \cdot 2H_2O$), soda ash, and aluminum? Write the balanced equations for the preparation (in three steps).

21.63 When magnesium metal burns in air and the resulting ash is treated with water, the odor of ammonia can be detected. Write the balanced equation for the formation of the compound during combustion that subsequently gives the ammonia. Write the equation representing the reaction of the compound with water.

21.64 Finely divided magnesium burns in air. Neither water nor carbon dioxide can be used to extinguish magnesium fires, because in both cases magnesium reacts to produce the oxide. Water is reduced to hydrogen, and carbon dioxide is reduced to carbon. Write balanced equations for these reactions.

THE GROUP IIIA ELEMENTS

21.65 Write the balanced equation for the reaction that occurs when thermite (Al and Fe_2O_3) is ignited.

21.66 Manganese metal can be obtained from Mn_3O_4 by the Goldschmidt process. Write the balanced equation for this.

21.67 Hans Christian Oersted, a Danish physicist and chemist, first prepared aluminum by the reduction of aluminum chloride with potassium (actually potassium amalgam, which is a potassium–mercury alloy). He obtained the aluminum chloride by reaction of alumina (Al_2O_3) and charcoal with chlorine. Write the balanced equation for this preparation of aluminum, starting with alumina.

21.68 Boron was discovered by the French chemists Joseph Louis Gay-Lussac and Louis Jacques Thenard. They isolated boron by heating boric acid with potassium. During this heating, boric acid is first transformed to boric oxide, which is then reduced to boron. Write balanced equations for these reactions.

21.69 What are the reactions that occur during the neutralization of boric acid by aqueous ammonia, $NH_3(aq)$? Write the equations in terms of electron-dot formulas.

21.70 Hydrofluoric acid, HF, reacts with boron trifluoride to give fluoroboric acid, HBF_4. Write the equation for this reaction in terms of electron-dot formulas.

21.71 Baking powders contain sodium hydrogen carbonate (or potassium hydrogen carbonate) and an acidic substance. When water is added to a baking powder, carbon dioxide is released. One kind of baking powder contains $NaHCO_3$ and sodium aluminum sulfate, $NaAl(SO_4)_2 \cdot 12H_2O$. Write the net ionic equation for the reaction that occurs in water solution.

21.72 When aluminum sulfate is dissolved in water, it produces an acidic solution. Suppose the pH of this solution is raised by the dropwise addition of aqueous sodium hydroxide. (a) Describe what you would observe as the pH continues to rise. (b) Write balanced equations for any reactions that occur.

21.73 The following solid substances are in separate but unlabeled test tubes: $AlCl_3 \cdot 6H_2O$, $BaCl_2 \cdot 2H_2O$, $BeSO_4 \cdot 4H_2O$, and KOH. Describe how they can be identified by chemical tests, using only these substances and water.

21.74 Unlabeled test tubes contain solid $AlCl_3 \cdot 6H_2O$ in one, $Ba(OH)_2 \cdot 8H_2O$ in another, and $MgSO_4 \cdot 7H_2O$ in another. How could you find out what is in each test tube, using chemical tests that involve only the compounds in these test tubes? Water is available as a solvent.

METALLURGY

21.75 Galena, PbS, is an important lead ore. (a) Write the balanced equation for the roasting of galena. Assume that the mineral is converted to lead(II) oxide. (b) Write a chemical equation for the reduction of lead(II) oxide by carbon to give metallic lead. (c) Describe what happens at each electrode in the electrolytic refining of lead. Electrodes, one of impure lead and the other of pure lead, are suspended in a bath of fluorosilicic acid, H_2SiF_6.

21.76 Nickel ore, NiS, is roasted to nickel(II) oxide and then converted with hydrogen to the metal. Nickel is refined by the Mond process, in which nickel combines with carbon monoxide to form volatile nickel carbonyl, $Ni(CO)_4$. When nickel carbonyl is heated to 200°C, it decomposes to free nickel and carbon monoxide gas. Write balanced equations for these reactions. Label each reaction according to whether it involves preliminary treatment, reduction, or refining.

Additional Problems

21.77 Aluminum chloride vaporizes as Al_2Cl_6 molecules. At still higher temperatures, these molecules dissociate to $AlCl_3$ molecules. What is the geometry of Cl atoms about aluminum in the $AlCl_3$ molecule? What is the geometry of Cl atoms about aluminum in the Al_2Cl_6 molecule?

21.78 Beryllium chloride has a polymeric bridge Cl structure in the solid phase, but it consists of $BeCl_2$ molecules in the gas phase. What is the geometry of Cl atoms about beryllium in solid $BeCl_2$? What is the geometry of Cl atoms about beryllium in the $BeCl_2$ molecule?

21.79 Natural sources of potassium are 0.0118% radioactive $^{40}_{19}K$. This isotope has a half-life of 1.28×10^9 years. What is the value of the radioactive decay constant for this isotope? What fraction of a sample of this isotope would remain after 100 million years (1.00×10^8 years)?

21.80 Natural rubidium is 27.85% radioactive $^{87}_{37}Rb$, which decays by beta emission to $^{87}_{38}Sr$. This decay of rubidium to strontium has been used to date mineralogical samples. It is assumed that when a rubidium mineral first crystallized, the ratio of strontium atoms to rubidium atoms was zero. What would be the value of this ratio 5×10^9 years (approximately the age of the earth) after such a mineral crystallized? The half-life of $^{87}_{37}Rb$ is 5×10^{11} years.

21.81 Calculate the enthalpy change, $\Delta H°$, for the production of 1 mol Fe by the following reaction at 25°C. (The value of $\Delta H°$ changes rather slowly with temperature, so the value at 25°C is an approximate value for higher temperatures.)

$$\tfrac{1}{2}Fe_2O_3(s) + Al(s) \longrightarrow Fe(s) + \tfrac{1}{2}Al_2O_3(s)$$

Is the reaction exothermic or endothermic? See Appendix C for data.

21.82 Calculate the enthalpy change, $\Delta H°$, for the production of 1 mol Ca by the following reaction at 25°C.

$$CaO(s) + \tfrac{2}{3}Al(s) \longrightarrow Ca(s) + \tfrac{1}{3}Al_2O_3(s)$$

Is the reaction exothermic or endothermic? See Appendix C for data.

21.83 Estimate the temperature at which strontium carbonate decomposes to strontium oxide and CO_2 at 1 atm.

$$SrCO_3(s) \longrightarrow SrO(s) + CO_2(g)$$

To do this, calculate $\Delta H°$ and $\Delta S°$ at 25°C. Assume that these values do not change appreciably with temperature. Then calculate the temperature at which $\Delta G° = 0$. See Appendix C for data.

21.84 Estimate the temperature at which barium carbonate decomposes to barium oxide and CO_2 at 1 atm.

$$BaCO_3(s) \longrightarrow BaO(s) + CO_2(g)$$

See Problem 21.83.

21.85 Calculate $E°$ for the disproportionation of $In^+(aq)$.

$$3In^+(aq) \rightleftharpoons 2In(s) + In^{3+}(aq)$$

(*Disproportionation* is a reaction in which a species undergoes both oxidation and reduction.) Use the following standard potentials:

$$In^+(aq) + e^- \rightleftharpoons In(s) \qquad E° = -0.21\ V$$
$$In^{3+}(aq) + 2e^- \rightleftharpoons In^+(aq) \qquad E° = -0.40\ V$$

From $E°$, calculate $\Delta G°$ for the reaction (in kilojoules). Does the disproportionation occur spontaneously?

21.86 Calculate $E°$ for the disproportionation of $Tl^+(aq)$,

$$3Tl^+(aq) \rightleftharpoons 2Tl(s) + Tl^{3+}(aq)$$

from the following standard potentials:

$$Tl^+(aq) + e^- \rightleftharpoons Tl(s) \qquad E° = -0.34\ V$$
$$Tl^{3+}(aq) + 2e^- \rightleftharpoons Tl^+(aq) \qquad E° = 1.25\ V$$

From $E°$, calculate $\Delta G°$ for the reaction (in kilojoules). Does the disproportionation occur spontaneously? See Problem 21.85.

21.87 Metallic sodium has a body-centered cubic lattice. Assume that sodium atoms in the metal consist of spheres packed as closely as allowed by this type of lattice. From the known density of the metal (see Table 21.5), calculate the radius of a sodium atom (in angstroms). (Note that spheres that are packed body-centered cubic touch one another along the diagonal running between opposite corners of a unit cell.)

21.88 Metallic lithium has a body-centered cubic lattice. Assume that lithium atoms in the metal consist of spheres packed as closely as allowed by this type of lattice. From the known density of the metal (see Table 21.5), calculate the radius of a lithium atom (in angstroms). See Problem 21.87.

21.89 Seawater contains 1272 g of magnesium ion per metric ton (1 metric ton = 1 megagram). What is the minimum amount of slaked lime, $Ca(OH)_2$, that must be added to 1.00 metric ton of seawater to precipitate the magnesium ion?

21.90 A bauxite sample is approximately 52% Al_2O_3. How many grams of aluminum can be obtained from 1.00 metric ton (1.00×10^6 g) of the bauxite?

21.91 Lithium hydroxide has been used in spaceships to absorb carbon dioxide exhaled by astronauts. Assuming that the product is lithium carbonate, determine what mass of lithium hydroxide is needed to absorb the carbon dioxide from 1.00 L of air containing 30.0 mmHg partial pressure of CO_2 at 25°C.

21.92 Potassium chlorate, $KClO_3$, is used in fireworks and explosives. It can be prepared by bubbling chlorine into hot potassium hydroxide.

$$6KOH(aq) + 3Cl_2(g) \longrightarrow KClO_3(s) + 5KCl(aq) + 3H_2O(l)$$

How many grams of $KClO_3$ can be obtained from 156 L of Cl_2 whose pressure is 784 mmHg at 25°C?

22

The Main-Group Elements: Groups IVA to VIIIA

Chapter Outline

Natural sulfur crystals.

In studying the elements of Groups IA to IIIA, we saw the emergence of the two major trends in metallic and nonmetallic character. First, as we move across a period from left to right, the elements become less metallic and more nonmetallic. This trend is especially pronounced in the second period. Thus, lithium, in Group IA, is an active metal that reacts in most cases to give ionic compounds; the oxide is basic. Beryllium, in Group IIA, is a metal that shows a tendency toward covalent bonding; the oxide is amphoteric. Boron, in Group IIIA, is a metalloid with definite nonmetallic characteristics. The element has a covalent network structure, and the oxide B_2O_3 is acidic.

The second major trend we saw in the elements of Groups IA to IIIA was an increase in metallic character as we move down a column. Thus, the first element in Group IIIA, boron, is a metalloid, but the other elements are metals. The oxides progress from acidic (B_2O_3) to amphoteric (Al_2O_3, Ga_2O_3) to weakly basic (In_2O_3, Tl_2O_3).

As we study the elements of Groups IVA to VIIIA, we see these major trends with even greater clarity. And whereas the elements of Groups IA to IIIA were metals (except for boron), the elements of these groups are primarily nonmetals. Only those elements at the bottom of Groups IVA to VIA are metals, and even they show some nonmetallic characteristics.

22.1 GROUP IVA: THE CARBON FAMILY

The elements of Group IVA show in a more striking fashion than the earlier groups the normal trend to greater metallic character in progressing down a column. The first element in the group, carbon, is distinctly nonmetallic. Silicon and germanium are metalloids; tin and lead are metals. All Group IVA elements have compounds in the +4 oxidation state, though these compounds are essentially covalent. Carbon and silicon also have compounds in the −4 oxidation state. In some of the metal carbides, such as Al_4C_3, the negative ion C^{4-} is present. Most of the compounds in the −4 state are covalent. The tetravalent state (that is, the +4 or −4 oxidation state) is expected from the ns^2np^2 valence configuration of the atoms, and covalent bonding is usually tetrahedral, involving sp^3 hybridization. However, all of these elements, with the exception of carbon, can also form six bonds by utilizing sp^3d^2 hybridization. Some examples are the complex ions SiF_6^{2-}, $GeCl_6^{2-}$, $Sn(OH)_6^{2-}$, and $Pb(OH)_6^{2-}$. The divalent (+2) state is important for tin and lead, and +2 cations of these elements are found in compounds and aqueous solutions.

Silicon, the most abundant element of Group IVA, occurs in silica (SiO_2) and silicate minerals. The next most abundant element is carbon, which is found in all living matter as well as in the fossil fuels. The largest sources of carbon, however, are the carbonate minerals limestone and dolomite. Carbon also occurs as CO_2 in the atmosphere and as graphite and diamond, which are allotropic forms of the free element. The other Group IVA elements are much less abundant. Germanium is rather rare and is found in several minerals, such as argyrodite, a mixed silver–germanium sulfide ore ($4Ag_2S \cdot GeS_2$). The most important commercial source of this element is flue dust from zinc ore roasting and from coal combustion. The principal tin ore is cassiterite (SnO_2). Lead is obtained mainly from the mineral galena (PbS).

Physical Properties of the Elements

CARBON The crystalline allotropes of carbon—graphite and diamond—are covalent network solids. Graphite is a soft, black solid with a slippery feel. It has a luster and conducts electricity. The solid has a layer structure.● Within a layer, each carbon atom is covalently bonded to three others so that a hexagonal, or "chicken-wire," pattern is formed. The bonding involves delocalized π electrons, which are responsible for the conductivity of the solid. Two layers are attracted to each other by London forces, so that the layers easily slide over one another. This accounts for the softness and slippery feel of graphite. Amorphous forms of carbon, such as charcoal and carbon black, have a layer structure like that of graphite. The difference lies in the stacking of layers. In graphite, the stacking is ordered; in amorphous carbon, the stacking is more or less disordered.

● The structures of graphite and diamond were discussed in detail in Section 11.8.

Diamond is a colorless, transparent solid.● It is very hard (the hardest substance known) and an electrical insulator (nonconductor). The difference between its properties and those of graphite is explained by the difference in structure. Diamond has a three-dimensional network structure. Each carbon atom is covalently bonded to four other atoms. The bonding is tetrahedral, involving sp^3 hybridization of the atoms. Diamond requires the breaking of many strong C—C bonds in order to move one portion of the crystal relative to another. Once these strong directional bonds are broken, they are unlikely to reform. Thus, diamond is both hard and brittle. As expected for a covalent network solid, the melting point of diamond is high (about 3550°C; see Table 22.1).

● Impurities and crystal imperfections can give diamond various colors. A trace of boron, for example, gives a blue diamond.

SILICON AND GERMANIUM The graphite-type structure is unique to carbon, but the diamond-type structure is not.● Silicon, germanium, and one allotrope of tin have a crystal structure like that of diamond. Unlike diamond, however, these substances are semiconductors. They are slightly conducting at room temperature but become good conductors at higher temperatures. Silicon is a hard, lustrous, gray solid that melts at 1410°C. Germanium has a silvery-gray metallic luster, but unlike a metal it shatters as easily as glass.

● The $p\pi$ bonding in a graphite-type structure would be expected to become weaker as the atomic size increases and the orbital overlap decreases.

Table 22.1 Properties of Group IVA Elements

Property	Carbon	Silicon	Germanium	Tin	Lead
Electron configuration	$[He]2s^22p^2$	$[Ne]3s^23p^2$	$[Ar]3d^{10}4s^24p^2$	$[Kr]4d^{10}5s^25p^2$	$[Xe]4f^{14}5d^{10}6s^26p^2$
Melting point, °C	3550 (diamond)	1410	937	232	328
Boiling point, °C	4827	2355	2830	2260	1740
Density, g/cm^3	3.51 (diamond)	2.33	5.35	7.28 (white Sn)	11.3
Ionization energy (first), kJ/mol	1086	786	762	709	716
Electron affinity, kJ/mol	−122	−134	−120	−121	−110
Electronegativity (Pauling scale)	2.5	1.8	1.8	1.8	1.8
Standard potential (volts), $M^{2+} + 2e^- \rightleftharpoons M$	—	—	—	−0.14	−0.13
Covalent radius, Å	0.77	1.17	1.22	1.40	1.54
Ionic radius (for M^{2+}), Å	—	—	—	1.06	1.33

TIN AND LEAD Tin exists in two allotropic forms. The common form, white tin, is metallic and malleable. It is a metallic conductor of electricity; that is, it conducts moderately well at room temperature but becomes less conducting at elevated temperatures. White tin undergoes a transition at 13°C to gray tin, a brittle semiconductor. It is this allotrope that has the diamondlike structure.

$$\text{Gray tin} \overset{13°C}{\rightleftharpoons} \text{white tin}$$

The transition between these allotropes is normally slow, and white tin can be cooled below 13°C without changing to gray tin. However, a tin pipe (for example) kept for some time at below 13°C begins to crumble to gray tin. The crystallization begins at isolated points that form centers for further accelerated crystallization. Crumbly spots start to grow all over the pipe—an effect called tin disease or tin pest.

Lead occurs only in a metallic form with a cubic closest-packed structure. When the metal is freshly cut, it has a bright silvery luster with bluish cast. It soon acquires a dull gray appearance as an adherent oxide coating forms.

Chemical Properties of the Elements

At room temperature, the Group IVA elements are relatively unreactive, particularly if they are in massive crystalline form. They do all burn in air once the temperature has been raised. Carbon burns in an excess of oxygen with a flameless glow to give carbon dioxide. In an inadequate supply of oxygen, it burns to a mixture of CO and CO_2. Silicon, germanium, and tin all react with oxygen to yield the dioxides. Tin, for example, burns in air with a white flame, giving tin(IV) oxide, SnO_2. Lead, however, usually gives lead(II) oxide, PbO, unless the temperature is kept below 500°C. In this case, it forms the red oxide Pb_3O_4, with lead in +2 and +4 oxidation states.

Group IVA elements are relatively unreactive with water, although carbon and silicon do react at red heat with steam.

$$C(s) + H_2O(g) \xrightarrow{\Delta} CO(g) + H_2(g)$$

$$Si(s) + 2H_2O(g) \xrightarrow{\Delta} SiO_2(s) + 2H_2(g)$$

Group IVA elements, except for the metals tin and lead, are also unreactive with dilute acids. Even the metallic elements react only very slowly at room temperature, but upon heating they react more vigorously with acids to release hydrogen gas.

$$Sn(s) + 2[H^+(aq) + Cl^-(aq)] \longrightarrow [Sn^{2+}(aq) + 2Cl^-(aq)] + H_2(g)$$
$$Pb(s) + 2[H^+(aq) + Cl^-(aq)] \longrightarrow [Pb^{2+}(aq) + 2Cl^-(aq)] + H_2(g)$$

Tin and lead are also attacked by hot solutions of alkali metal hydroxides, giving hydrogen and the hydroxo ions, $Sn(OH)_3^-$ or $Pb(OH)_3^-$.● For example,

$$Sn(s) + [Na^+(aq) + OH^-(aq)] + 2H_2O(l) \longrightarrow [Na^+(aq) + Sn(OH)_3^-(aq)] + H_2(g)$$

● Hydroxo ions such as $Sn(OH)_3^-$ result from the acidic character of the aqueous metal ion. We can write $Sn^{2+}(aq)$ as $Sn(H_2O)_6^{2+}(aq)$ to indicate explicitly those water molecules directly bonded to the metal atom. In basic solution, we have

$$Sn(H_2O)_6^{2+}(aq) + OH^-(aq) \rightleftharpoons Sn(H_2O)_5(OH)^+(aq) + H_2O(l)$$

The transfer of three protons gives the anion $Sn(H_2O)_3(OH)_3^-$, also written $Sn(OH)_3^-(aq)$.

Except for diamond and silicon, Group IVA elements can be oxidized by concentrated nitric acid. Germanium and tin give the dioxides—for example,

$$3Ge(s) + 4HNO_3(aq) \longrightarrow 3GeO_2(s) + 4NO(g) + 2H_2O(l)$$

whereas lead gives Pb^{2+}.

$$3Pb(s) + 8HNO_3(aq) \longrightarrow 3[Pb^{2+}(aq) + 2NO_3^-(aq)] + 2NO(g) + 4H_2O(l)$$

The Group IVA elements Si, Ge, Sn, and Pb combine with the halogens at elevated temperatures. The tetrahalide is the usual product, except for lead, where the lead(II) halide is usually formed. Here again we see the stability of the +2 oxidation state of lead.

Preparation and Uses of the Elements

CARBON Although carbon occurs in the free state as diamond and graphite and in amorphous form as coal, commercial quantities of the various allotropes of carbon are also manufactured. Carbon in the form of carbon black, for example, is a major chemical product. It is prepared by burning natural gas (CH_4) or petroleum hydrocarbons in a limited supply of air so that heat ''cracks'' or breaks bonds in the hydrocarbon.

$$CH_4(g) \xrightarrow{\Delta} \underset{\text{carbon black}}{C(s)} + 2H_2$$

Carbon black is used in the manufacture of rubber tires (to increase wear) and as a pigment in black printing inks. Coke is an amorphous carbon obtained by heating coal or petroleum in the absence of air. Petroleum coke is used wherever a pure form of carbon is needed.

Although natural graphite is essential for certain uses, such as the making of pencil ''leads,'' synthetic graphite must be manufactured to meet the demands of the electrochemical industry for electrodes. Graphite is prepared by heating petroleum coke in an electric furnace to 3500°C, using sand and iron as catalysts. Amorphous carbon and graphite fibers, used for reinforcing high-temperature materials, are made by heating textile fibers to high temperature.

The positive free energy of formation of diamond at 1 atm shows that it is thermodynamically unstable and should change to graphite.

$$C(\text{graphite}) \longrightarrow C(\text{diamond});\ \Delta G_f^\circ = +2.9\ \text{kJ/mol}$$

Figure 22.1
Synthetic diamonds. These diamonds *(right)* were made by heating graphite *(left)* under high pressure with a catalyst. Pencil lead is graphite mixed with clay.

Fortunately, the change of diamond to graphite is imperceptibly slow at normal temperatures. When diamond is heated in an inert atmosphere at 1000°C, however, the change is rapid. Of course, the possibility of changing graphite to diamond is much more interesting. Because graphite is less dense than diamond, we might expect it to change to diamond at sufficiently high pressure.● High temperature would also be needed so that the change would occur in a reasonable period of time. Although people have tried to synthesize diamond ever since these conditions for the transformation were understood, the first successful experiments were performed only in 1955 by scientists at the General Electric Company. The process, which produces cheap industrial diamonds for grinding and cutting tools, requires temperatures of about 3000°C and pressures of 100,000 atm (1×10^{10} Pa or 10 GPa), in addition to a transition-metal catalyst. Gem-quality diamonds have been made, but the cost is too high for them to compete with natural diamonds. Some synthetic diamonds are shown in Figure 22.1.●

● According to Le Chatelier's principle, high pressure would be expected to shift the composition of the equilibrium

$$\text{Graphite} \rightleftharpoons \text{diamond}$$

toward the solid that has the smaller volume—that is, to the more dense diamond phase.

● Diamond films have recently been produced under moderate laboratory conditions. A mixture of CH_4 and H_2 is decomposed in a high electric field, and carbon deposits as a diamond film on surfaces at 500°C to 1000°C. Possible uses are diamond-coated tools and diamond lasers.

SILICON AND GERMANIUM The production of other Group IVA elements involves at some stage the reduction of an oxide. Silicon, for example, is prepared by reducing quartz sand (SiO_2) with coke (C) in an electric furnace at 3000°C.

$$SiO_2(l) + 2C(s) \xrightarrow{\Delta} Si(l) + 2CO(g)$$

When used for metallurgical purposes, silicon is prepared as an alloy with iron, called ferrosilicon. This alloy is obtained by reducing iron(III) oxide along with sand.

Ultrapure silicon is required for solid-state semiconductor devices, such as transistors, solar cells, and microcomputer chips. To prepare it, impure silicon (from the reduction of sand) is heated with chlorine to give silicon tetrachloride, $SiCl_4$, a low-boiling liquid (b.p. 58°C) purified by fractional distillation.

$$Si(s) + 2Cl_2(g) \xrightarrow{\Delta} SiCl_4(l)$$

Then a mixture of silicon tetrachloride vapor and hydrogen is passed through a hot tube, where very pure silicon crystallizes on the surface of a pure silicon rod.

$$SiCl_4(g) + 2H_2(g) \xrightarrow{\Delta} Si(s) + 4HCl(g)$$

Silicon prepared this way has only $10^{-8}\%$ impurities (Figure 22.2).

Germanium is prepared from its ores or from flue dusts obtained in the production of zinc. The material is treated with concentrated hydrochloric acid to

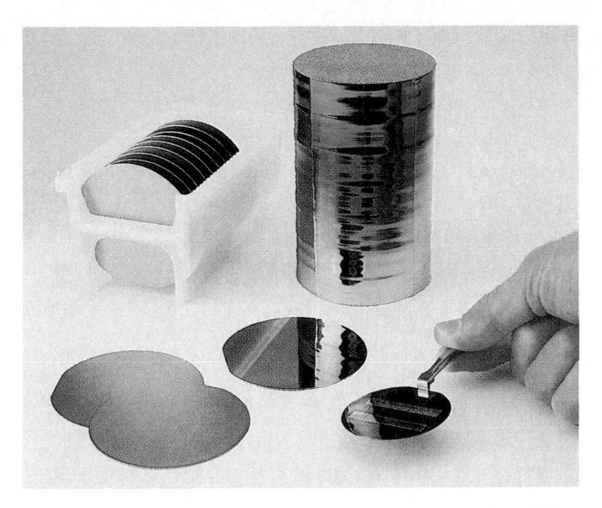

Figure 22.2
Ultrapure silicon rod and wafers cut from a similar rod. Silicon wafers form the base material for integrated circuit chips used in solid-state electronic devices.

produce germanium tetrachloride, $GeCl_4$, which is distilled from the mixture. After purifying by fractional distillation, the germanium tetrachloride is reacted with water (hydrolyzed) to give the dioxide.

$$GeCl_4(l) + 2H_2O(l) \longrightarrow GeO_2(s) + 4HCl(aq)$$

Germanium dioxide is reduced to the free element by heating with hydrogen or carbon. The element is purified for semiconductor applications by **zone refining,** *a purification process in which a solid rod of a substance is melted in a small, moving band or zone that carries the impurities out of the rod.* The process depends on the greater solubility of the impurities in the liquid phase than in the solid phase.● (See Figure 22.3.) In addition to its use in semiconductor devices, germanium is used to make infrared-transmitting optics.

● For details, see the Related Topic on zone melting and refining.

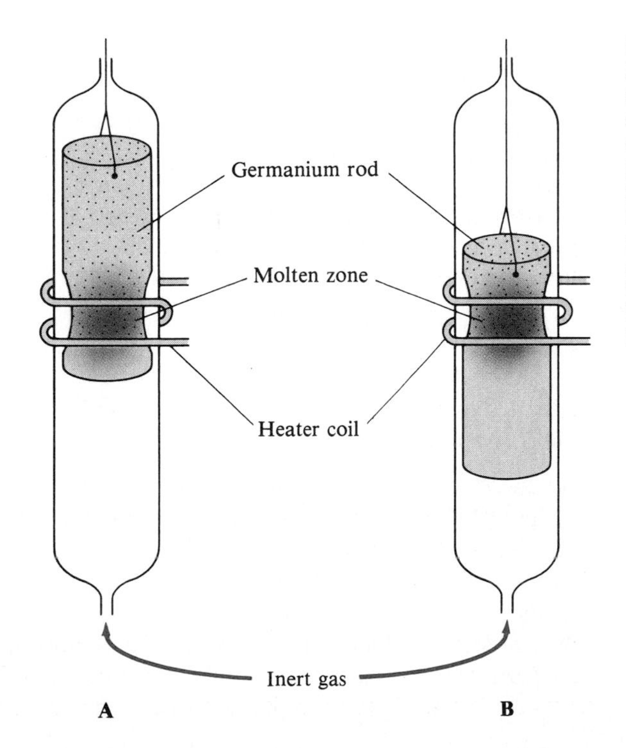

Figure 22.3
Zone refining. In the method shown here, a germanium rod is slowly lowered through a heater coil, which melts the rod. The impurities concentrate in the molten zone that moves upward. Surface tension holds the two parts of the rod together at the molten zone. After cooling, the previously molten zone at the end of the rod can be cut off, removing all the impurities.

TIN AND LEAD Tin is obtained from purified tin(IV) oxide, SnO_2, by reducing it with carbon in a furnace.

$$SnO_2(s) + 2C(s) \xrightarrow{\Delta} Sn(l) + 2CO(g)$$

The metal is used in tin-plating of steel, as a protective coating. Some tin is also used for alloys such as bronze and solder.

Lead is produced by roasting galena, a lead(II) sulfide ore. In roasting, the ore is burned in air, where the sulfide is converted to the oxide.

$$2PbS(s) + 3O_2(g) \longrightarrow 2PbO(s) + 2SO_2(g)$$

Some free metal is also produced during the roasting.

$$PbS(s) + 2PbO(s) \longrightarrow 3Pb(s) + SO_2(g)$$

The fused mass from the roasting is broken up and fed into blast furnaces. Here the lead(II) oxide is reduced with carbon monoxide produced by the partial oxidation of coke.

$$PbO(s) + CO(g) \longrightarrow Pb(l) + CO_2(g)$$

Large quantities of the metal are used to make electrodes for storage batteries. It is also used in alloys for cable coverings and for solder (lead with tin). Important compounds made from lead include tetraethyllead, $(C_2H_5)_4Pb$ (gasoline additive); litharge, PbO; and red lead, Pb_3O_4.

Related Topic

ZONE MELTING AND ZONE REFINING

Zone melting is a technique in which a rod of some material is melted in a narrow band or zone that travels the length of the rod. One of the more important uses of zone melting is in the purification of certain substances. The process is called zone refining. Consider the zone refining of germanium. In this process, a germanium rod is melted at one end by moving either the germanium rod or the heater (Figure 22.3). Because a solution containing dissolved solute (impurities in the germanium solvent) normally has a lower melting point than pure solvent, impurities concentrate in the melted zone, which moves from one end of the rod to the other. The process can be repeated until most of the impurities have been removed.

Zone melting is also used to make single crystals of silicon for semiconductor applications. When a silicon rod is first produced, it consists of a tightly packed mass of small crystals, unsuitable for semiconductor devices. To form this into a single crystal, the rod is placed on top of a wafer of single-crystal silicon. This part of the rod is melted, and as the melted zone is moved upward, the single crystal of silicon grows until it becomes the entire rod (see Figure 22.2).

Important Compounds

CATENATION Carbon is unique in that it forms stable compounds containing long chains of atoms of the element covalently bonded to one another. **Catenation** is *the ability of an atom to bond covalently to like atoms*. Because the strengths of carbon–carbon bonds (C—C, C=C, C≡C) are comparable to those of carbon bonds with hydrogen and with oxygen, a wide variety of carbon compounds have been prepared. Over a million such compounds (called organic compounds) are known.

Table 22.2 Hydrides of the Group IVA Elements

Element	Hydride with Longest Chain of the Element
Carbon	Unlimited length
Silicon	Si_8H_{18}
Germanium	Ge_5H_{12}
Tin	Sn_2H_6
Lead	PbH_4

Other Group IVA elements also display limited catenation in the hydrides (binary compounds with hydrogen); Table 22.2 shows them. Silicon, for example, forms hydrides called *silanes* that can contain chains of single-bonded Si atoms.

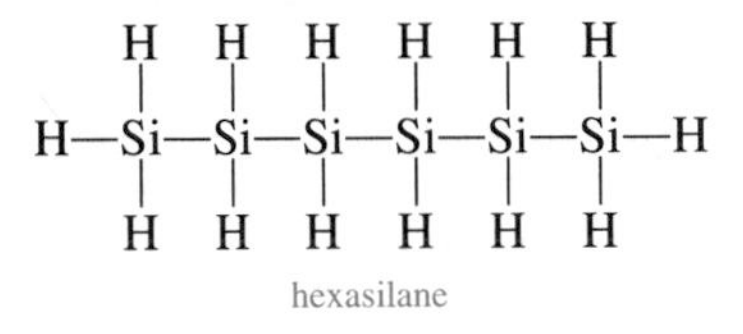

hexasilane

Compounds containing C=C or C≡C bonds are common, and a few compounds with Si=Si bonds have been discovered recently. Similar compounds of the other Group IVA elements are unknown.

CARBON OXIDES The dioxides of all Group IVA elements are known. Carbon dioxide is the usual product of the combustion of organic compounds and elementary carbon when there is an excess of oxygen. However, an equilibrium exists among carbon, CO, and CO_2 that favors carbon monoxide at temperatures above about 700°C.

$$CO_2(g) + C(s) \rightleftharpoons 2CO(g)$$

For this reason, CO is always a product of the combustion of carbon unless excess oxygen is present. Commercially, carbon monoxide is prepared from natural gas (CH_4) or petroleum hydrocarbons. For example,

$$CH_4(g) + H_2O(g) \xrightarrow{\Delta} CO(g) + 3H_2(g)$$

The resulting mixture is used to prepare methanol, CH_3OH. (See Table 22.3 for a list of uses of Group IVA compounds.) Many other organic compounds could be made from carbon monoxide if supplies of petroleum, the present source of most organic compounds, become scarce. In that case, carbon monoxide could be made from any source of carbon by the water–gas reaction.

$$C(s) + H_2O(g) \xrightarrow{\Delta} CO(g) + H_2(g)$$

Although carbon monoxide is not soluble or reactive with water under normal conditions, it does react with hot, aqueous sodium hydroxide at high pressure to give sodium formate.

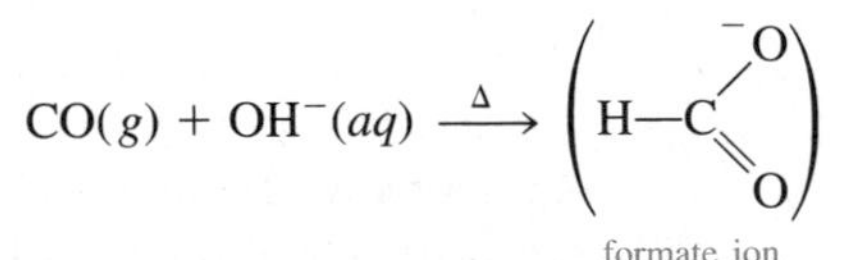

formate ion

Table 22.3 Uses of Group IVA Compounds

Compound	Use
CO	Fuel; reducing agent Synthesis of methanol, CH_3OH
CO_2	Refrigerant Carbonation of beverages
SiO_2	Source of silicon and its compounds Abrasives Glass
SnO_2	Source of tin and its compounds
PbO	Lead storage batteries Ceramic glazes and glass
PbO_2	Cathode in lead storage batteries
Pb_3O_4	Pigment for painting structural steel

● Concentrated sulfuric acid has a strong attraction for water.

Formic acid, HCOOH, can be dehydrated with concentrated sulfuric acid to give carbon monoxide.● This is a convenient laboratory preparation for producing small quantities of CO. Thus, in a formal way, CO is the anhydride of formic acid, HCOOH.

Carbon dioxide is the anhydride of unstable carbonic acid, H_2CO_3. The gas dissolves in water to give the equilibrium

$$CO_2(aq) + H_2O(l) \rightleftharpoons H_2CO_3(aq)$$

Although carbonic acid is unstable and cannot be isolated from an aqueous solution, many of its salts—the carbonates and hydrogen carbonates—are well known. Carbon dioxide is liquefied by pressure (about 57 atm at 20°C), and the liquid is used in large quantities to refrigerate foods. Cooling is caused by vaporization of the liquid. Large quantities of the gas are also used to carbonate beverages.

SILICON DIOXIDE AND SILICATES The principal oxide of silicon is silica or silicon dioxide, SiO_2. It has a quite different structure from carbon dioxide, which is a linear molecule. Silica is a covalent network solid in which each Si atom is covalently bonded to four O atoms that are in turn bonded to other Si atoms (see Figure 22.4). Quartz is one of several different crystalline forms of silica and is a constituent of many rocks. Weathering of these rocks releases quartz particles, which are a major component of most sands. Quartz melts at 1610°C to give a viscous liquid that cools to form a glass (an amorphous solid) called silica glass.

Silica reacts with hot, aqueous alkali metal hydroxide or fused alkali metal carbonates to form soluble silicates. **Silicates** are *compounds of silicon and oxygen with various metals*. They may be thought of as derived from silicic acid, H_4SiO_4 or $Si(OH)_4$, although this substance has never been isolated from solution. Solutions of sodium silicate appear to contain the ion

```
          O⁻
          |
H—O—Si—O—H
          |
          O₋
```

but undoubtedly they also contain ions with two or more silicon atoms. Such ions can form from simple silicate ions by condensation reactions. A **condensation**

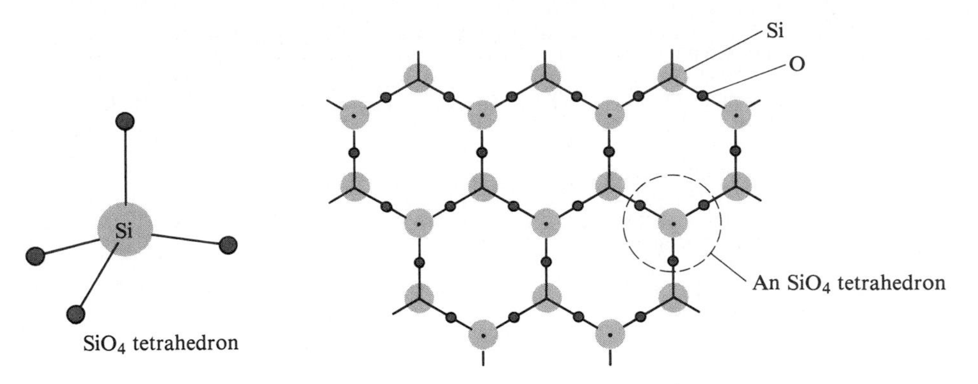

Figure 22.4
Structure of silica (SiO_2). Silica consists of SiO_4 tetrahedra linked to one another by their oxygen atoms. A portion of a plane of atoms within silica is shown. Alternate Si atoms have bonds to O atoms pointing upward; the other Si atoms have bonds to O atoms pointing downward. These O atoms, in turn, have bonds to layers of Si atoms above and below this plane. Thus, silica has a three-dimensional network structure.

reaction is *a reaction in which two molecules or ions are joined by the elimination of a small molecule such as H_2O.*

$$\text{H—O—}\underset{\text{O}^-}{\overset{\text{O}^-}{\text{Si}}}\text{—O—H} + \text{H—O—}\underset{\text{O}^-}{\overset{\text{O}^-}{\text{Si}}}\text{—O—H} \longrightarrow \text{H—O—}\underset{\text{O}^-}{\overset{\text{O}^-}{\text{Si}}}\text{—O—}\underset{\text{O}^-}{\overset{\text{O}^-}{\text{Si}}}\text{—O—H} + H_2O$$

Ordinary glass, which is obtained by melting silica with lime (CaO) and sodium carbonate, is a polymeric silicate with a network of silicon–oxygen bonds. Silicon–oxygen chains and networks are a dominant feature of silicon chemistry.

An enormous variety of silicate minerals exists (Figure 22.5). The structure of each can be understood in terms of SiO_4 tetrahedra as basic units. A few minerals contain $SiO_4{}^{4-}$ as a discrete ion. Zircon, $ZrSiO_4$, is an example. It has a very high melting point (2550°C), as would be expected from the high charge on the ions. Bonding between ions certainly has considerable covalent character.

Silicate minerals usually consist of more complicated structures in which SiO_4 tetrahedra are linked through a common oxygen—for example,

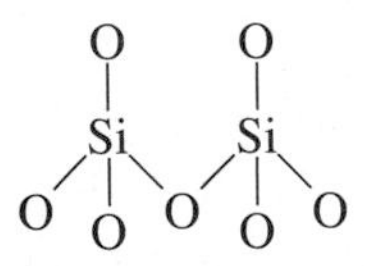

Figure 22.5
Some silicate minerals. *Left:* Asbestos; note the fibers on its surface. *Center:* Mica; this mineral cleaves easily into sheets. *Right:* Orthoclase, a feldspar mineral; this is a three-dimensional network silicate.

Figure 22.6
Structures of some silicate minerals. A number of silicate minerals consist of finite silicate anions, such as SiO_4^{4-} and $Si_2O_7^{6-}$. Beryl contains the cyclic anion $Si_6O_{18}^{12-}$. Other silicate minerals have anions with very long chains or double chains, as in the asbestos mineral tremolite. The chains are linked to one another by cations. In asbestos, this linkage between chains is weak, and the mineral frays easily into fibers.

represents the SiO_4 tetrahedron.

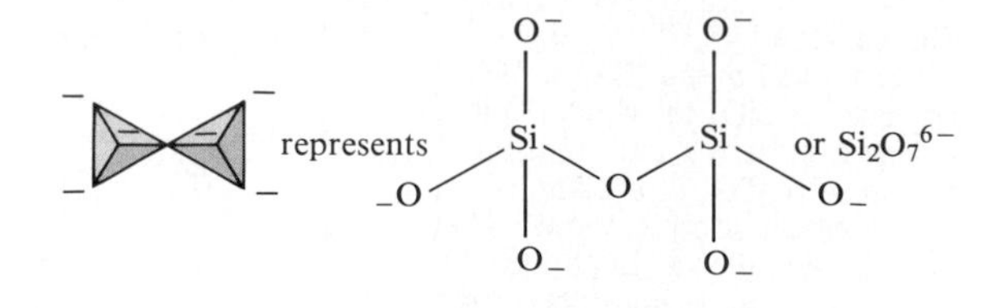

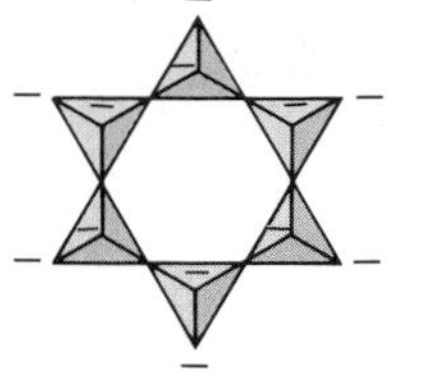

Here SiO_4 tetrahedra are linked into a ring to give the $Si_6O_{18}^{12-}$ ion. Beryl, $Be_3Al_2Si_6O_{18}$, is the best-known mineral with this ion. Emerald is a green variety of beryl.

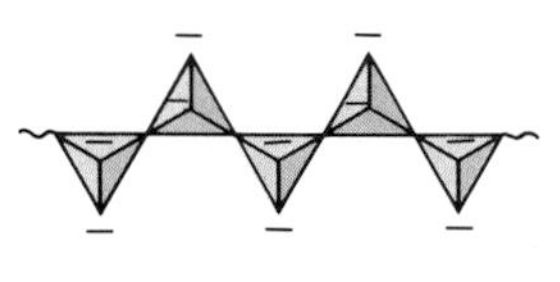

An infinite chain of SiO_4 tetrahedra, giving the empirical formula SiO_3^{2-}. Diopside, $CaMg(SiO_3)_2$, is an example of a mineral with this structure.

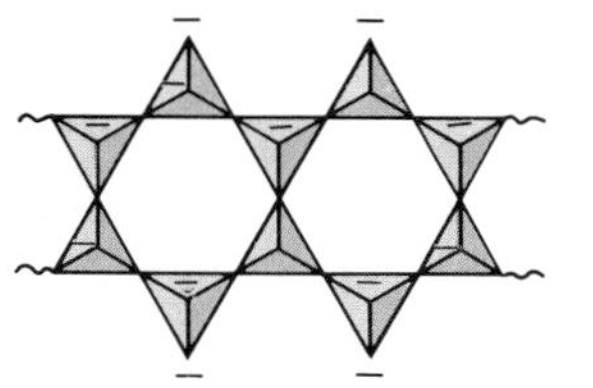

An infinite double chain of SiO_4 tetrahedra, giving the empirical formula $Si_4O_{11}^{6-}$. The asbestos mineral tremolite, $Ca_2Mg_5(OH)_4(Si_4O_{11})_2$, is an example.

In this way, it is possible to form chain or ring anions or polymeric chains of indefinite length (as, for example, in the fibrous asbestos minerals). Silicate anion chains are linked to one another through cations. In the micas, there is a two-dimensional crosslinking of chains to form silicate sheets. These anionic sheets are held together through bonds to cations. This type of mineral cleaves easily along the silicate sheets. The structures of different silicate minerals are shown in Figure 22.6.

When SiO_4 tetrahedra are linked in a three-dimensional network, silica results. The **aluminosilicate minerals,** such as the feldspars, are *a type of silicate consisting of three-dimensional networks in which some of the* SiO_4 *tetrahedra are replaced by* AlO_4 *tetrahedra*. When Si is replaced by Al, additional positive charge is needed; it is usually supplied by cations of Groups IA and IIA elements.

TIN AND LEAD COMPOUNDS Tin(II) oxide, a black solid, is obtained by heating the white precipitate of $Sn(OH)_2$ formed by reacting Sn^{2+} ions with hydroxide ions.

$$Sn^{2+}(aq) + 2OH^-(aq) \longrightarrow Sn(OH)_2(s)$$

$$Sn(OH)_2(s) \xrightarrow{\Delta} SnO(s) + H_2O(g)$$

The oxide and hydroxide are amphoteric. They dissolve in acids to form Sn^{2+} and in base to form the stannite ion, $Sn(OH)_3^-$. For example,

$$Sn(OH)_2(s) + [Na^+(aq) + OH^-(aq)] \longrightarrow Na^+(aq) + \underbrace{Sn(OH)_3^-(aq)}_{\text{stannite ion}}$$

Tin(IV) oxide, SnO_2, is also amphoteric. It dissolves in sulfuric acid to give a solution of tin(IV) sulfate, $Sn(SO_4)_2$, and in aqueous sodium hydroxide to give a solution of stannate ion, $Sn(OH)_6^{2-}$.

$$SnO_2(s) + 2[Na^+(aq) + OH^-(aq)] + 2H_2O(l) \longrightarrow 2Na^+(aq) + \underset{\text{stannate ion}}{Sn(OH)_6^{2-}(aq)}$$

The stannite ion is a strong reducing agent and is oxidized to the stannate ion, for example by H_2O_2.

Lead(II) oxide, PbO, is prepared commercially by exposing molten lead to air. This oxide, which is a reddish-yellow solid at normal temperatures, is called litharge in commercial trade. Its largest use is in making lead storage batteries (a paste of litharge and sulfuric acid is packed into the electrode grids). It is also used in glasses and enamels. Lead(II) oxide and lead(II) hydroxide are amphoteric, as are the corresponding tin compounds. For example, normally insoluble lead(II) hydroxide dissolves in acids to give Pb^{2+} and in bases to give the plumbite ion, $Pb(OH)_3^-$. Lead(IV) oxide, PbO_2, can be prepared by oxidizing plumbite ion with hypochlorite ion, OCl^-.

$$Pb(OH)_3^-(aq) + OCl^-(aq) \longrightarrow PbO_2(s) + OH^-(aq) + H_2O(l) + Cl^-(aq)$$

It is a dark brown solid and a strong oxidizing agent in acid solution. It is formed in the cathodes of lead storage batteries during charging.● Lead(IV) oxide reacts with strong base to form the plumbate ion, $Pb(OH)_6^{2-}$. The oxide PbO_2 is thermally unstable and decomposes on heating to PbO and Pb_3O_4. Trilead tetroxide, Pb_3O_4, also called red lead, is a bright red powder used as a pigment in a protective paint for steel. It contains both lead(II) and lead(IV).

● The lead storage cell was described in Section 19.7.

Example 22.1 Using Chemical Reactions of the Group IVA Elements and Compounds

Sodium stannate, $Na_2Sn(OH)_6$, is a white, crystalline solid, solutions of which are used in the electrolytic plating of tin. Show by means of balanced equations how you could prepare sodium stannate from tin metal in two steps.

Solution

First prepare a solution of sodium stannite.

$$Sn(s) + NaOH(aq) + 2H_2O(l) \longrightarrow NaSn(OH)_3(aq) + H_2(g)$$

Then oxidize the stannite ion to stannate ion—say, with H_2O_2. Hydrogen peroxide is reduced to H_2O. Note that the reaction must be in basic solution, because the stannite ion would decompose to Sn^{2+} in acid solution. The balanced equation is

$$Sn(OH)_3^-(aq) + H_2O_2(aq) + OH^-(aq) \longrightarrow Sn(OH)_6^{2-}(aq)$$

or

$$NaSn(OH)_3(aq) + H_2O_2(aq) + NaOH(aq) \longrightarrow Na_2Sn(OH)_6(aq)$$

Exercise 22.1

Use balanced equations to describe how lead(IV) oxide can be prepared from lead(II) oxide (in two steps). (See Problems 22.53 and 22.54.)

Exercise 22.2

Solutions of sodium stannite, $NaSn(OH)_3$, are unstable and decompose to Sn and $Sn(OH)_6^{2-}$. Write the balanced equation for this reaction.

(See Problems 22.55 and 22.56.)

22.2 GROUP VA: NITROGEN AND THE PHOSPHORUS FAMILY

Like the carbon family, the Group VA elements show the distinct trend from nonmetallic to metallic. The first members, nitrogen and phosphorus, are definitely nonmetals; arsenic and antimony are metalloids; bismuth is a metal. Nitrogen, however, shows only slight resemblance to the other members. We see this, for example, in the formulas of the elements and compounds. Elementary nitrogen is N_2; white phosphorus is P_4. Similarly, the common +5 oxoacid of nitrogen is HNO_3; that of phosphorus is H_3PO_4.

Except for bismuth, Group VA elements have stable compounds in the +5 oxidation state. In the case of nitrogen, these compounds are oxidizing agents. Thus, nitric acid, HNO_3, is reduced to NO_2 (oxidation state +4), NO (+2), N_2 (0), and NH_3 (−3). The +5 state of phosphorus is quite stable, however, and phosphoric acid, H_3PO_4, is nonoxidizing. For the remaining elements, the +3 state becomes progressively more stable.

Phosphorus is the most abundant Group VA element and occurs in phosphate minerals, such as fluorapatite, $Ca_5(PO_4)_3F$—also written $3Ca_3(PO_4)_2 \cdot CaF_2$ to emphasize the presence of calcium phosphate. The other elements, except for nitrogen, are much less abundant and occur as oxide and sulfide ores. Bismuth also occurs as the free element.

Physical Properties of the Elements

Nitrogen exists as a colorless, odorless gas consisting of N_2 molecules (the structure of the molecule is $:N{\equiv}N:$). Nitrogen gas can be liquefied if cooled below its critical temperature (−147°C), then compressed (to 33.5 atm at the critical temperature). The colorless liquid boils at −196°C at 1 atm pressure.

Phosphorus has two common allotropes: white phosphorus and red phosphorus (Figure 22.7). White phosphorus is a waxy, white solid. It is highly toxic and very reactive. Because of its reactivity with oxygen, white phosphorus is stored under water, in which it is insoluble. As we might expect from its low melting point (44°C; see Table 22.4), white phosphorus is a molecular solid (P_4). The phosphorus atoms in the P_4 molecule are arranged at the corners of a regular tetrahedron so that each atom is single-bonded to the other three (Figure 22.8). The P—P—P bond angle is 60° and thus is much smaller than the normal bond angle

Figure 22.7
Allotropes of phosphorus. *Left:* White phosphorus. *Right:* Red phosphorus.

Table 22.4 Properties of Group VA Elements

Property	Nitrogen	Phosphorus	Arsenic	Antimony	Bismuth
Electron configuration	[He]$2s^2 2p^3$	[Ne]$3s^2 3p^3$	[Ar]$3d^{10} 4s^2 4p^3$	[Kr]$4d^{10} 5s^2 5p^3$	[Xe]$4f^{14} 5d^{10} 6s^2 6p^3$
Melting point, °C	−210	44 (white P)	613 (sublimes)	631	271
Boiling point, °C	−196	280		1750	1560
Density, g/cm³	1.25×10^{-3}	1.82	5.73	6.68	9.80
Ionization energy (first), kJ/mol	1402	1012	947	834	703
Electron affinity, kJ/mol	≥0	−72	−77	−101	−110
Electronegativity (Pauling scale)	3.0	2.1	2.0	1.9	1.9
Covalent radius, Å	0.70	1.10	1.21	1.41	1.52
Ionic radius, Å	1.32 (N^{3-})	1.85 (P^{3-})	0.72 (As^{3+})	0.90 (Sb^{3+})	1.17 (Bi^{3+})

for *p* bonding (90°). This gives a weaker P—P bond than otherwise (because there is smaller overlap of the *p* orbitals), and it accounts for the great reactivity seen in this phosphorus allotrope. Red phosphorus is a rather unreactive network solid.

Arsenic is normally a brittle, gray solid of metallic luster. Gray arsenic sublimes at 615°C. If the vapor is rapidly cooled, a yellow, nonmetallic solid crystallizes. Yellow arsenic is believed to be a molecular solid, As_4, analogous to white phosphorus. It is unstable at room temperature and reverts to gray arsenic. Antimony is a silvery, lustrous solid. A yellow, nonmetallic form is known but is stable only at very low temperatures. Bismuth is a white metal with a pinkish tinge.

Chemical Properties of the Elements

Nitrogen is relatively unreactive at normal temperatures, in part because of the strength of the N≡N bond. It does react with a number of elements at elevated temperatures, however. With oxygen, it gives nitric oxide, NO.

$$N_2(g) + O_2(g) \longrightarrow 2NO(g)$$

Burning in air generally results in the production of some nitric oxide. Nitrogen, N_2, and oxygen, O_2, are the major constituents of air, and they react at the high temperatures in a flame. The nitrogen oxide air pollutants found in automobile exhaust are produced in this way. Nitrogen also combines with hydrogen at elevated temperatures (in the presence of a catalyst) to form ammonia.

$$N_2(g) + 3H_2(g) \longrightarrow 2NH_3(g)$$

This is an important industrial preparation (the Haber process) and the commercial source of nitrogen compounds.

White phosphorus ignites spontaneously in air because of its reactivity and forms a dense white cloud of phosphorus oxides. In an excess of oxygen, the element burns to phosphorus(V) oxide, P_4O_{10}.

$$P_4(s) + 5O_2(g) \longrightarrow P_4O_{10}(s)$$

In a deficient supply of oxygen, phosphorus(III) oxide, P_4O_6, forms. Arsenic, antimony, and bismuth burn when heated in air. Arsenic forms arsenic(III) oxide,

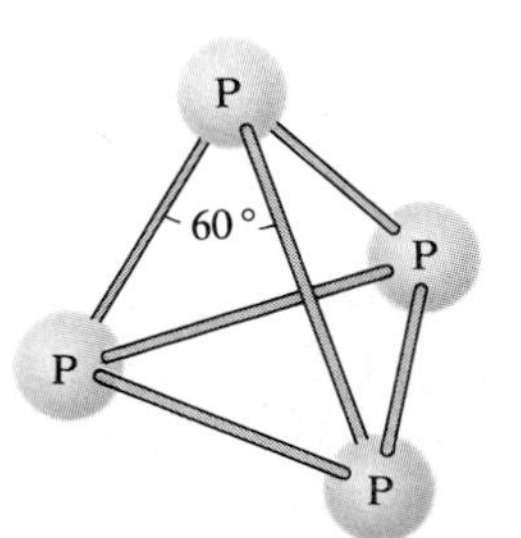

Figure 22.8
Structure of the P_4 molecule. The reactivity of white phosphorus results from the small P—P—P angle (60°).

As_4O_6. Antimony forms antimony(III) oxide, Sb_4O_6, and diantimony tetroxide, Sb_2O_4, with antimony in +3 and +5 oxidation states. Bismuth forms bismuth(III) oxide, Bi_2O_3.

Phosphorus and the heavier elements (As, Sb, and Bi) react directly with the halogens. Phosphorus gives the pentahalides (PF_5, PCl_5, PBr_5, but not PI_5) as well as the trihalides (PF_3, PCl_3, PBr_3, and PI_3). The other elements give primarily the trihalides, although SbF_5, $SbCl_5$, and AsF_5 can form.

Preparation and Uses of the Elements

NITROGEN Most of the available nitrogen on earth is present as nitrogen gas in the atmosphere, which consists of 78.1% N_2 by mass. Air is the major commercial source of the element. The components in air are separated by liquefaction, followed by distillation. Nitrogen is the most volatile component in liquid air, so it is the first to distill off, leaving behind a liquid that is primarily oxygen with a small amount of noble gases (mostly argon).

Liquid nitrogen is used as a refrigerant to freeze foods, to freeze soft or rubbery materials prior to grinding them, and to freeze biological materials. However, most nitrogen is used as a *blanketing gas,* whose purpose is to protect a material from oxygen during processing or storage. Thus, electronic components are often made under a nitrogen atmosphere.

PHOSPHORUS White phosphorus, a major industrial chemical, is prepared by heating phosphate rock (fluorapatite, $3Ca_3(PO_4)_2 \cdot CaF_2$) with coke (C) and sand (SiO_2) in an electric furnace. The reaction can be written

$$2Ca_3(PO_4)_2(s) + 6SiO_2(s) + 10C(s) \xrightarrow{1500^\circ C} \underset{\text{calcium silicate}}{6CaSiO_3(l)} + 10CO(g) + P_4(g)$$

The gases from the furnace are cooled to condense phosphorus vapor to the liquid, which is stored under water until pumped into tank cars. Slag, consisting of calcium silicate and calcium fluoride (from fluorapatite), is periodically drained from the furnace. Most of the white phosphorus produced is used to manufacture phosphoric acid, H_3PO_4. For this, phosphorus is burned in excess air, and the oxide mist is sprayed with water. Some white phosphorus is converted to red phosphorus, for use in making matches, by heating at 240°C in an inert atmosphere.

ARSENIC, ANTIMONY, AND BISMUTH Arsenic may be obtained from various ores, such as the sulfide, As_4S_6, from which it is prepared by roasting in air, followed by reduction of the oxide with coke.

$$As_4S_6(s) + 9O_2(g) \longrightarrow As_4O_6(s) + 6SO_2(g)$$

$$As_4O_6(s) + 6C(s) \xrightarrow{\Delta} As_4(g) + 6CO(g)$$

Arsenic(III) oxide, present in flue gases from the roasting of copper ores, is also used as a source of arsenic. Antimony is obtained from stibnite, Sb_4S_6, by roasting to the oxide, followed by reduction with coke. Bismuth is obtained as a by-product in the electrolytic refining of copper. It is present in the mud that collects near the anode.●

● The electrolytic refining of copper was discussed in Section 19.9.

Arsenic, antimony, and bismuth are used in alloys. Arsenic is used to harden lead for lead shot; antimony is alloyed with lead for storage battery plates; and bismuth is mixed with tin and lead in low-melting alloys for automatic fire sprinklers and other uses.

Table 22.5 Some Compounds of Nitrogen in Various Oxidation States

Oxidation State	Compound	Formula
−3	Ammonia	NH_3
	Lithium nitride	Li_3N
−2	Hydrazine	N_2H_4
−1	Hydroxylamine	NH_2OH
0	Nitrogen	N_2
+1	Nitrous oxide	N_2O
+2	Nitric oxide	NO
+3	Dinitrogen trioxide	N_2O_3
	Nitrous acid	HNO_2
+4	Nitrogen dioxide	NO_2
	Dinitrogen tetroxide	N_2O_4
+5	Dinitrogen pentoxide	N_2O_5
	Nitric acid	HNO_3

Important Compounds

NITROGEN COMPOUNDS Nitrogen forms compounds in all oxidation states from −3 to +5. Table 22.5 lists some simple compounds.

Ammonia, NH_3, is the most important commercial compound of nitrogen. It is a colorless gas with a characteristic irritating or pungent odor. Ammonia is prepared commercially by the Haber process from N_2 and H_2, as described earlier. Figure 22.9 shows a flow diagram of the industrial preparation of ammonia from natural gas, steam, and air. Small laboratory amounts can be obtained by the reaction of an ammonium salt with a strong base, such as NaOH or $Ca(OH)_2$.

$$NH_4^+(aq) + OH^-(aq) \longrightarrow NH_3(g) + H_2O(l)$$

Ammonia is easily liquefied, and the liquid is used as a nitrogen fertilizer. Ammonium salts, such as the sulfate and nitrate, are also sold as fertilizers. Large quantities of ammonia are converted to urea, NH_2CONH_2, which is used as a fertilizer, livestock feed supplement, and in the manufacture of urea–formaldehyde plastics.

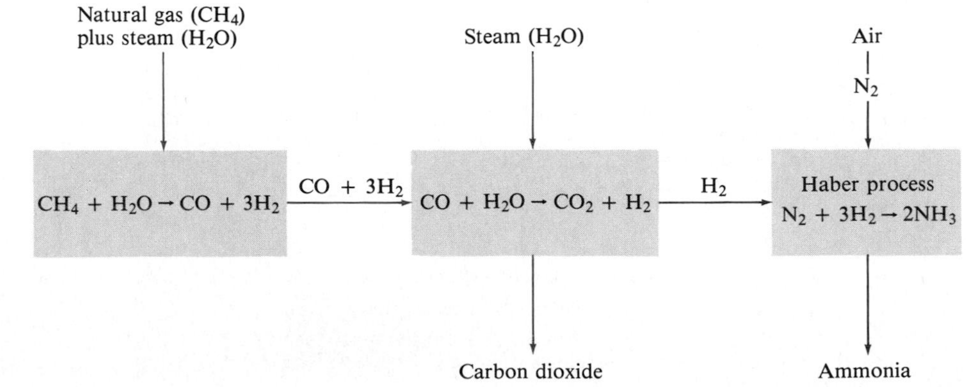

Figure 22.9
Flow diagram of the industrial preparation of ammonia. The raw materials are natural gas, water, and air. Hydrogen for the Haber process is obtained by reacting natural gas with steam to give carbon monoxide and hydrogen. In the next step, carbon monoxide is reacted with steam to give carbon dioxide and additional hydrogen. The carbon dioxide is removed by dissolving it in water solution.

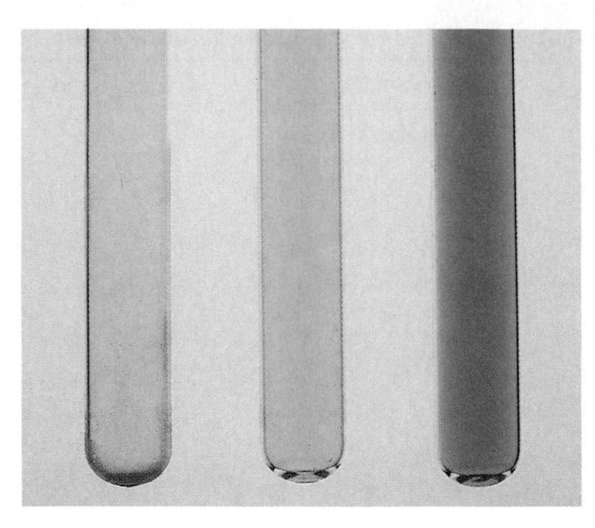

Figure 22.10
Effect of temperature on the equilibrium of NO_2 and N_2O_4. The temperature increases in the test tubes from left to right. N_2O_4 (colorless) predominates at cooler temperatures, whereas NO_2 (red-brown) predominates at higher temperatures.

Nitrous oxide, N_2O, is a colorless gas with a sweet odor. It can be prepared by careful heating of molten ammonium nitrate. *(Strong heating may cause an explosion.)*

$$NH_4NO_3(s) \xrightarrow{\Delta} N_2O(g) + 2H_2O(g)$$

Nitrous oxide, or laughing gas, is used as a dental anesthetic. It is also useful as a propellant in whipped-cream dispensers. The gas dissolves in cream under pressure. When the cream is dispensed, the gas bubbles out, forming a foam.

Nitric oxide, NO, is a colorless gas. Although it can be prepared by the direct combination of the elements at elevated temperatures, large amounts are prepared from ammonia as the first step in the commercial preparation of nitric acid. The ammonia is oxidized in the presence of a platinum catalyst.

$$4NH_3(g) + 5O_2(g) \xrightarrow{Pt} 4NO(g) + 6H_2O(g)$$

Nitric oxide reacts rapidly with oxygen to give nitrogen dioxide. Nitrogen dioxide, NO_2, is a reddish-brown gas or liquid (its boiling point is 21°C). It exists

Figure 22.11
Reaction of copper metal with dilute nitric acid. *Left:* Copper metal is oxidized by nitric acid to $Cu^{2+}(aq)$ ion, which has a blue color. The principal reduction product from dilute nitric acid is nitric oxide, NO, a colorless gas. *Right:* When the stopper in the flask is removed, air enters, and the NO reacts with O_2 to produce NO_2, a red-brown gas.

Figure 22.12
Reaction of copper metal with concentrated nitric acid. The principal reduction product from concentrated nitric acid is nitrogen dioxide, NO_2 (red-brown gas).

in equilibrium with the colorless compound dinitrogen tetroxide, N_2O_4.

$$2NO_2(g) \rightleftharpoons N_2O_4(g)$$

This reaction is exothermic, so dinitrogen tetroxide predominates at lower temperatures. Above 140°C, the mixture is mostly nitrogen dioxide (Figure 22.10).

Nitric acid, HNO_3, is an important industrial acid and is used to prepare explosives, nylon, and polyurethane plastics. It is produced commercially by the **Ostwald process,** which is *an industrial preparation of nitric acid by the catalytic oxidation of ammonia*. In this process, ammonia is burned in the presence of a platinum catalyst to give NO, which then reacts with oxygen to form NO_2. The nitrogen dioxide is dissolved in water where it reacts to form nitric acid and nitric oxide.

$$4NH_3(g) + 5O_2(g) \xrightarrow{Pt} 4NO(g) + 6H_2O(g)$$
$$2NO(g) + O_2(g) \longrightarrow 2NO_2(g)$$
$$3NO_2(g) + H_2O(l) \longrightarrow 2HNO_3(aq) + NO(g)$$

The nitric oxide produced in the last step is recycled for use in the second step.

Nitric acid is a strong oxidizing agent. Although copper metal is unreactive to most acids, it is oxidized by nitric acid. In dilute acid, nitric oxide is a principal reduction product (Figure 22.11).

$$3Cu(s) + 8H^+(aq) + 2NO_3^-(aq) \longrightarrow 3Cu^{2+}(aq) + 2NO(g) + 4H_2O(l)$$

With concentrated nitric acid, nitrogen dioxide is obtained (Figure 22.12).

$$Cu(s) + 4H^+(aq) + 2NO_3^-(aq) \longrightarrow Cu^{2+}(aq) + 2NO_2(g) + 2H_2O(l)$$

PHOSPHORUS OXIDES AND OXOACIDS The phosphorus oxides P_4O_6 and P_4O_{10} have related structures (Figure 22.13). Phosphorus(III) oxide, P_4O_6, has a tetrahedron of phosphorus atoms with oxygen atoms between each pair of phosphorus atoms to give P—O—P bonds. Phosphorus(V) oxide, P_4O_{10}, is similar but has an additional oxygen atom bonded to each phosphorus atom. These phosphorus–oxygen bonds are much shorter than the other P—O bonds (1.39 Å versus 1.62 Å); hence, they have considerable double-bond character.

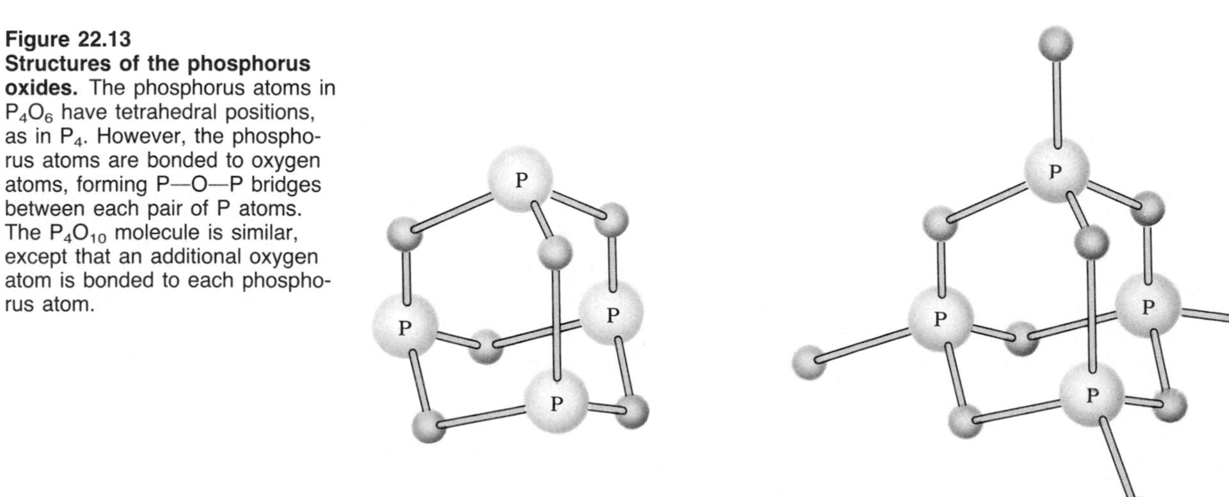

Figure 22.13
Structures of the phosphorus oxides. The phosphorus atoms in P_4O_6 have tetrahedral positions, as in P_4. However, the phosphorus atoms are bonded to oxygen atoms, forming P—O—P bridges between each pair of P atoms. The P_4O_{10} molecule is similar, except that an additional oxygen atom is bonded to each phosphorus atom.

Phosphorus(III) oxide is a low-melting solid (m.p. 23°C) and is the anhydride of phosphor*ous* acid, H_3PO_3.●

● Note the suffix *-ous*, indicating the lower oxidation state (+3) of phosphorus.

```
      O
      |
H—O—P—H
      |
      O
      |
      H
```

Note that one of the hydrogen atoms is directly attached to phosphorus. It is not acidic, so phosphorus acid is diprotic rather than triprotic as the formula H_3PO_3 would suggest.

Phosphorus(V) oxide is a white solid that sublimes at 360°C. It combines vigorously with water, making it useful in the laboratory as a drying agent. The substance is produced in large quantities by burning white phosphorus in excess air. Most of the P_4O_{10} produced is not isolated but is reacted immediately with excess water to obtain orthophosphoric acid, H_3PO_4.

$$P_4O_{10}(s) + 6H_2O(l) \longrightarrow 4H_3PO_4(aq)$$

Orthophosphoric acid (often called simply phosphoric acid) is a colorless solid that melts at 42°C when pure. It is usually sold as an aqueous solution. Orthophosphoric acid is triprotic and has the structure

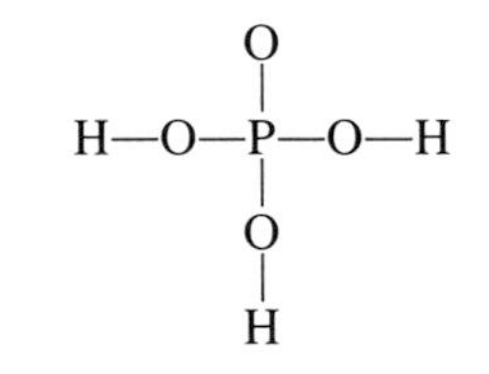

The possible sodium salts of phosphoric acid are sodium dihydrogen phosphate (NaH_2PO_4), disodium hydrogen phosphate (Na_2HPO_4), and trisodium phosphate (Na_3PO_4). Phosphoric acid produced from phosphorus, as described in the previous paragraph, is relatively pure and is used primarily in the detergent and food and beverage industries. An impure acid, produced in large quantities for the manufacture of fertilizers, is obtained by treating phosphate rock (fluorapatite) with sulfuric acid.●

● Because phosphate rock contains CaF_2, hydrofluoric acid, HF, can also form. When HF reacts with silica (SiO_2) in the phosphate rock, it forms hexafluorosilicic acid, H_2SiF_6. This by-product is used to make AlF_3 and synthetic cryolite for the aluminum industry.

$$Ca_3(PO_4)_2(s) + 3H_2SO_4(aq) \longrightarrow 3CaSO_4(s) + 2H_3PO_4(aq)$$

When phosphate rock is treated with orthophosphoric acid, it dissolves to give a solution of calcium dihydrogen phosphate, $Ca(H_2PO_4)_2$.

$$Ca_3(PO_4)_2(s) + 4H_3PO_4(aq) \longrightarrow 3Ca(H_2PO_4)_2(aq)$$

By this process, insoluble phosphate rock is converted to a soluble phosphate fertilizer. In the trade, this fertilizer is called triple superphosphate. Uses of some phosphorus compounds are listed in Table 22.6.

Phosphorus(V) oxide is not only the anhydride of orthophosphoric acid but is also the anhydride of two series of acids obtained by condensation reactions from orthophosphoric acid.● For example,

● Condensation reactions were described in Section 22.1.

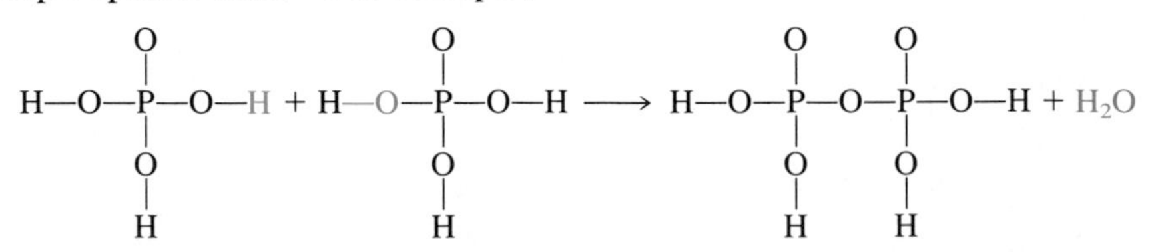

Table 22.6 Uses of Some Phosphorus Compounds

Compound	Use
$Ca(H_2PO_4)_2 \cdot H_2O$	Phosphate fertilizer Baking powder
$CaHPO_4 \cdot 2H_2O$	Animal feed additive Toothpowder
H_3PO_4	Manufacture of phosphate fertilizers
PCl_3	Manufacture of $POCl_3$ Manufacture of pesticides
$POCl_3$	Manufacture of plasticizers (substances that keep plastics pliable) Manufacture of flame retardants
P_4S_{10}	Manufacture of lubricant additives and pesticides
$Na_5P_3O_{10}$	Detergent additive

One series consists of the linear **polyphosphoric acids,** *acids with the general formula* $H_{n+2}P_nO_{3n+1}$, *which are formed from chains of* P—O *bonds.*

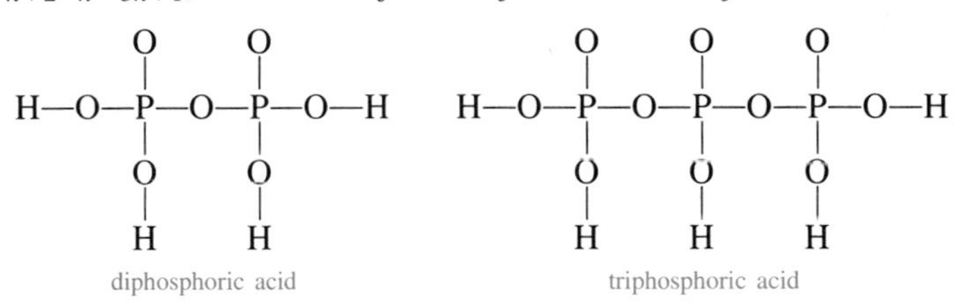

The other series consists of the **metaphosphoric acids,** which are *acids with the general formula* $(HPO_3)_n$. Figure 22.14 shows the structure of a cyclic, or ring, metaphosphoric acid. When a linear polyphosphoric acid chain is very long, the formula becomes $(HPO_3)_n$, with n very large, and the acid is called a polymetaphosphoric acid.

The polyphosphates and metaphosphates are used in detergents, where they act as water softeners by complexing with metal ions in the water. Sodium triphosphate ($Na_5P_3O_{10}$), one of the most commonly used polyphosphates, is manufac-

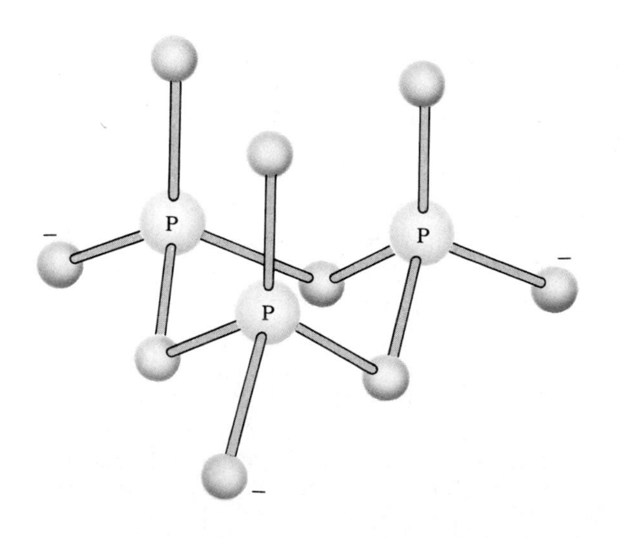

Figure 22.14
The structure of trimetaphosphate ion, $P_3O_9^{3-}$. This is a cyclic metaphosphate ion; the general formula of the metaphosphate ions is $(PO_3^-)_n$.

tured by adding sufficient sodium carbonate to phosphoric acid to give a solution of the salts NaH_2PO_4 and Na_2HPO_4. When this solution is sprayed into a hot kiln, the phosphate ions condense to sodium triphosphate. The use of phosphates in detergents has been criticized for contributing to the overfertilization of algae in lakes. Such lakes become oxygen-deficient from decomposing algae, and the fish die.

ARSENIC, ANTIMONY, AND BISMUTH COMPOUNDS Compounds of the elements As, Sb, and Bi are produced in small quantities. Arsenic(III) oxide is a white powder and is slightly soluble in water, producing a solution of the very weak acid H_3AsO_3 (arsenious acid). Metal arsenites, with the ion AsO_3^{3-}, are known. Arsenic acid, H_3AsO_4, is obtained when arsenic(III) oxide is oxidized with nitric acid. Arsenates, such as Na_3AsO_4, are well-known compounds. Arsenic(V) oxide, As_2O_5, whose structure is unknown, is obtained by heating arsenic acid. Arsenic compounds are poisonous to many animals and have been used as insecticides.

Antimony(III) oxide, Sb_4O_6, is amphoteric and forms salts of the cation Sb^{3+}, such as $Sb_2(SO_4)_3$, as well as antimonites—salts of the anion SbO_2^-. Antimony(V) oxide, Sb_2O_5, whose structure is unknown, is prepared by oxidation of Sb_4O_6 with nitric acid. It is weakly acidic and reacts with strong base, such as KOH, to form the antimonate anion, $Sb(OH)_6^-$. Sodium antimonate, $Na[Sb(OH)_6]$, is one of the few insoluble sodium salts.

Bismuth(III) oxide, Bi_2O_3, which is an ionic oxide, is weakly basic and forms salts, such as $Bi(NO_3)_3$. A white, gelatinous precipitate of $Bi(OH)_3$ is obtained when NaOH is added to a solution of bismuth salt.

Example 22.2 Using Chemical Reactions of the Group VA Elements and Compounds

With balanced equations, describe how you might prepare sodium antimonate, $Na[Sb(OH)_6]$, from antimony(III) oxide.

Solution

We can oxidize Sb_4O_6 to Sb_2O_5 with nitric acid. Antimony(III) oxide is amphoteric and first dissolves in nitric acid to form Sb^{3+}.

$$Sb_4O_6(s) + 12H^+(aq) \longrightarrow 4Sb^{3+}(aq) + 6H_2O(l)$$

The Sb^{3+} ion is oxidized to Sb_2O_5 by nitrate ion. If we assume the NO_3^- is reduced to NO, we get

$$6Sb^{3+}(aq) + 4NO_3^-(aq) + 7H_2O(l) \longrightarrow 3Sb_2O_5(s) + 14H^+(aq) + 4NO(g)$$

The +5 oxide can be dissolved in a strong base, such as KOH,

$$Sb_2O_5(s) + 2OH^-(aq) + 5H_2O(l) \longrightarrow 2Sb(OH)_6^-(aq)$$

and sodium antimonate is precipitated from the solution by addition of a sodium salt.

$$Na^+(aq) + Sb(OH)_6^-(aq) \longrightarrow Na[Sb(OH)_6](s)$$

Exercise 22.3

With balanced equations, describe how you can prepare trisodium phosphate, Na_3PO_4, from white phosphorus.

(See Problems 22.65 and 22.66.)

Exercise 22.4

Arsenic compounds can be reduced to arsine, AsH_3, by zinc in acidic solution. This forms the basis of the Marsh test for arsenic. Arsine is a gas. When passed through a heated tube, it decomposes, leaving a black deposit of arsenic. Write the balanced equation for the reduction of arsenious acid by zinc in acidic solution.

(See Problems 22.67, 22.68, 22.69, and 22.70.)

22.3 GROUP VIA: OXYGEN AND THE SULFUR FAMILY

Group VIA elements, like those of Groups IVA and VA, show the trend from nonmetallic to metallic. Oxygen and sulfur are strictly nonmetals. Although the chemistries of selenium and tellurium are predominantly those of nonmetals, they do have semiconducting allotropes, which is expected of metalloids. Polonium is metallic.

Oxygen has rather different properties from the other members of Group VIA. It is a very electronegative element and bonding involves only *s* and *p* orbitals. For the other members of the group, *d* orbitals become a factor in bonding. Oxygen has compounds mainly in the -2 oxidation state. The other Group VIA elements have compounds in this state also, but the $+4$ and $+6$ states are common. Selenium and tellurium in the $+6$ state are strong oxidizing agents, showing the greater stability of the $+4$ state in the heavier Group VIA elements.

Oxygen is the most abundant element on earth, making up 48% by mass of the outer portion (crust, atmosphere, and surface waters). Sulfur is also an abundant element. It occurs in sulfate minerals, such as gypsum ($CaSO_4 \cdot 2H_2O$), and in sulfide minerals, which are important metal ores. Sulfur is present in coal and petroleum as organic sulfur compounds and in natural gas as hydrogen sulfide. Free sulfur occurs in some volcanic areas, perhaps formed by the reaction of hydrogen sulfide and sulfur dioxide, which are present in volcanic gases.

$$16H_2S(g) + 8SO_2(g) \longrightarrow 16H_2O(l) + 3S_8(s)$$

Commercial deposits of the free element also occur in salt domes, which are massive columns of salt embedded in rock a hundred meters or more below the earth's surface.● None of the other Group VIA elements is abundant. Selenium and tellurium occur mixed with sulfide ores, and polonium-210 occurs in thorium and uranium ores. Polonium-210 has a half-life of 138 days, decaying by alpha emission.

● Such sulfur deposits are found in the United States along the shore of the Gulf of Mexico.

Physical Properties of the Elements

OXYGEN The common form of oxygen is dioxygen, O_2 (the element also exists as the allotrope ozone, O_3). Dioxygen (usually simply called oxygen) is a colorless, odorless gas under standard conditions. The critical temperature is $-118°C$. Thus, oxygen can be liquefied if the gas is first cooled below $-118°C$ and then compressed. Both the liquid and solid O_2 have a pale blue color. The melting point of the solid is $-218°C$, and the boiling point of the liquid at 1 atm is $-183°C$.

The structure of molecular oxygen is not readily described by Lewis dot formulas. Neither of the following Lewis formulas is adequate:

$$\underset{(a)}{:\ddot{O}{=}\ddot{O}:} \qquad \underset{(b)}{:\dot{\ddot{O}}{-}\dot{\ddot{O}}:}$$

Formula (a) describes oxygen as double-bonded and diamagnetic. Molecular oxygen, however, is known to be paramagnetic. Formula (b) describes oxygen as paramagnetic but single-bonded, which is not in agreement with the known O_2 bond length (1.21 Å). By comparison, the O—O single-bond length in hydrogen peroxide, HOOH, is 1.49 Å. The much smaller value of the O_2 bond length suggests that O_2 is multiple-bonded. Molecular orbital theory correctly describes the bonding. The paramagnetism, according to this theory, is due to two unpaired

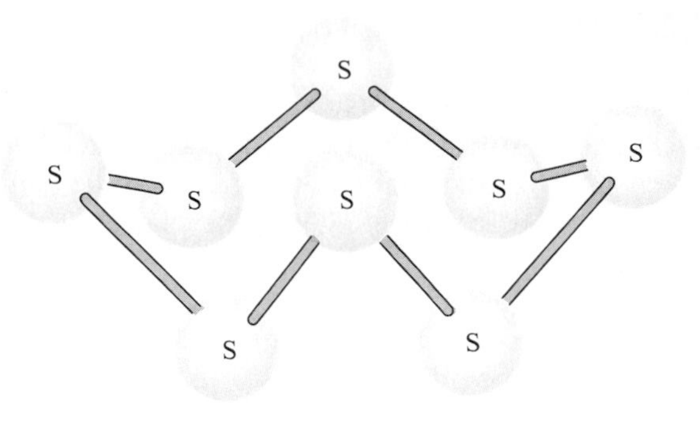

Figure 22.15
The structure of the S_8 molecule. Each molecule consists of eight S atoms arranged in a ring (in the shape of a crown).

electrons in antibonding π orbitals. Molecular orbital theory also predicts a double bond (bond order of 2).●

● The molecular-orbital description of O_2 was given in Section 10.6.

SULFUR, SELENIUM, AND TELLURIUM The stable form of sulfur, called rhombic sulfur, is a yellow, crystalline solid with a lattice of crown-shaped S_8 molecules (Figure 22.15). The catenation evident here also shows up in some sulfur compounds, such as the metal polysulfides, which have linear chain ions like S_6^{2-}. Rhombic sulfur melts at 113°C (see Table 22.7) to give an orange-colored liquid. Upon continued heating, this changes to a dark reddish-brown, viscous liquid. The original melt consists of S_8 molecules, but these open up, and the fragments join to give long spiral chains of sulfur atoms. The viscosity increases as compact S_8 molecules are replaced by long molecular chains that can intertwine.● At temperatures greater than 200°C, the chains begin to break apart, and the viscosity decreases. Sulfur boils at 445°C, giving a vapor of S_8, S_6, S_4, and S_2 molecules. Figure 22.16 shows the appearance of sulfur at various temperatures.

● If this viscous liquid is cooled quickly, it gives a rubbery form of sulfur. The rubberlike properties are due to the long spiral molecular chains that can be stretched along their length.

Selenium also exists in several allotropic modifications. An amorphous, red selenium, Se_8, precipitates when solutions of selenious acid, H_2SeO_3, are reduced. When this is heated below its melting point, it changes to a crystalline, red selenium. The stable, gray form crystallizes from molten selenium. Gray selenium is a **photoconductor,** *a material that is normally a poor conductor of electricity but becomes a good conductor when light falls on it.*

This property of selenium is used in *xerography,* a photocopying technique. The photocopier in Figure 22.17 contains a belt coated with selenium. During the photocopying process, this belt is electrostatically charged and then exposed to a light and dark image from a printed page. Electric charge drains away from the

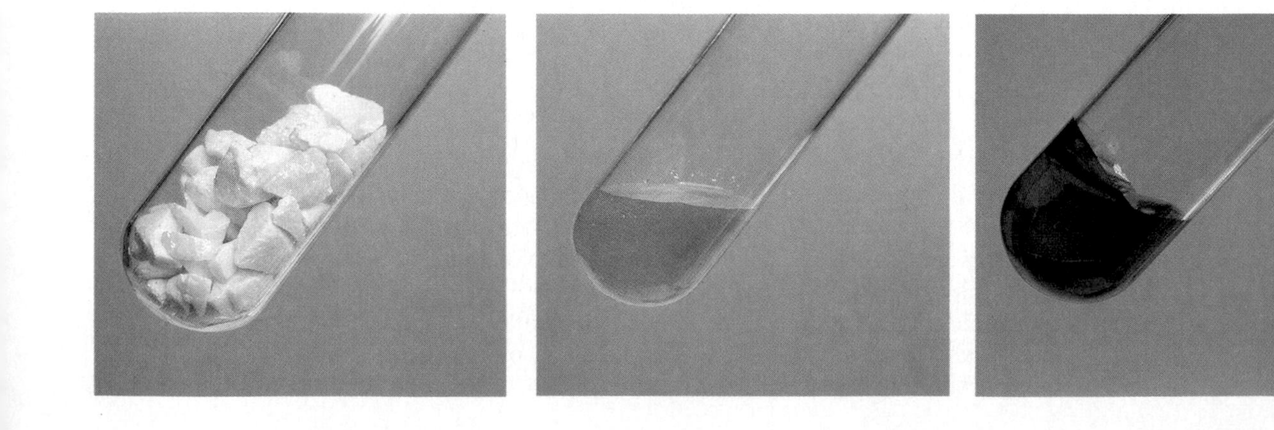

Figure 22.16
Sulfur at various temperatures. *Left:* Solid rhombic sulfur at 20°C. *Center:* Orange-colored liquid sulfur at 120°C. *Right:* Viscous liquid sulfur at 200°C.

Table 22.7 Properties of Group VIA Elements

Property	Oxygen	Sulfur	Selenium	Tellurium	Polonium
Electron configuration	$[He]2s^22p^4$	$[Ne]3s^23p^4$	$[Ar]3d^{10}4s^24p^4$	$[Kr]4d^{10}5s^25p^4$	$[Xe]4f^{14}5d^{10}6s^26p^4$
Melting point, °C	−218	113	217 (gray Se)	452	254
Boiling point, °C	−183	445	685	1390	962
Density, g/cm³	1.43×10^{-3}	2.07	4.81 (gray Se)	6.25	9.32
Ionization energy (first), kJ/mol	1314	1000	941	869	812
Electron affinity, kJ/mol	−141	−200	−195	−190	−180
Electronegativity (Pauling scale)	3.5	2.5	2.4	2.1	2.0
Standard potential (volts), $X + 2e^- \rightleftharpoons X^{2-}$	—	−0.51	−0.78	−0.92	—
Covalent radius, Å	0.66	1.04	1.17	1.37	1.53
Ionic radius (for X^{2-}), Å	1.26	1.70	1.84	2.07	—

areas made conductive by exposure to light. The belt now has an electrostatic image of the printed page on its surface. The image is developed with a dry powder that is attracted to the charged areas of the belt. In a final step, the powder on the belt is transferred to a sheet of plain paper, which is heated to fuse the powder to the paper.

Tellurium exists only as a silvery solid with metallic luster. Polonium is a metal—the only one with a simple cubic lattice.

Chemical Properties of the Elements

OXYGEN Molecular oxygen is a very reactive gas and combines directly with many substances. The products are usually oxides. Oxides are binary compounds with oxygen in the −2 oxidation state.

Figure 22.17 The selenium-coated belt of the Xerox 9900 photocopying machine. After the belt is electrostatically charged, it is exposed to a light and dark image, and charge drains away from those areas exposed to light. When the belt is dusted with a dry powder (toner), the powder adheres to the charged areas. The resulting image from the toner is then transferred from the belt to paper.

Most metals react readily with oxygen to form oxides. For example, magnesium wire and iron wool burn brightly in air to yield the oxides.

$$Mg(s) + O_2(g) \longrightarrow 2MgO(s)$$
$$4Fe(s) + 3O_2(g) \longrightarrow 2Fe_2O_3(s)$$

Oxides of metals in low oxidation state such as MgO and Fe_2O_3 are basic oxides. Those of metals in high oxidation state may be acidic.

Lithium reacts vigorously with oxygen to give the basic oxide, Li_2O, but the other alkali metals form predominantly peroxides and superoxides. **Peroxides** are *compounds with oxygen in the −1 oxidation state* (they contain either the O_2^{2-} ion or else the covalently bonded group —O—O—). **Superoxides** are *binary compounds with oxygen in the* $-\frac{1}{2}$ *oxidation state*; they contain the superoxide ion, O_2^-. Sodium metal burns in air to give mainly the peroxide.

$$2Na(s) + O_2(g) \longrightarrow Na_2O_2(s)$$

Potassium, rubidium, and cesium form mainly the superoxides.

$$K(s) + O_2(g) \longrightarrow KO_2(g)$$

Nonmetals react to form covalent oxides, most of which are acidic. For example, carbon (such as charcoal or graphite) burns in an excess of oxygen to give carbon dioxide.

$$C(s) + O_2(g) \longrightarrow CO_2(g) \qquad \text{(excess } O_2\text{)}$$

In a limited supply of oxygen or at high temperature, carbon monoxide is the usual product.

$$2C(s) + O_2(g) \longrightarrow 2CO(g) \qquad \text{(limited } O_2\text{)}$$

Sulfur, S_8, burns in oxygen to give sulfur dioxide, SO_2.

$$S_8(s) + 8O_2(g) \longrightarrow 8SO_2(g)$$

Sulfur forms another oxide, sulfur trioxide, SO_3, but only small amounts are obtained during the burning of sulfur in air.

Compounds in which at least one element is in a reduced state are oxidized by oxygen, giving compounds that would be expected to form when the individual elements are burned in oxygen. For example, a hydrocarbon such as octane, C_8H_{18}, burns to give carbon dioxide and water.

$$2C_8H_{18}(l) + 25O_2(g) \longrightarrow 16CO_2(g) + 18H_2O(g)$$

Some other examples are given in the following equations.

$$2H_2S(g) + 3O_2(g) \longrightarrow 2H_2O(g) + 2SO_2(g)$$
$$CS_2(l) + 3O_2(g) \longrightarrow CO_2(g) + 2SO_2(g)$$
$$2ZnS(s) + 3O_2(g) \longrightarrow 2ZnO(s) + 2SO_2(g)$$

SULFUR, SELENIUM, AND TELLURIUM Sulfur reacts with nearly all elements. It burns in air with a characteristic blue flame, giving off sulfur dioxide gas, which is recognizable by its sharp, choking odor. Although sulfur also has the trioxide, SO_3, very little of it forms even when sulfur burns in excess air. Like sulfur, the elements selenium and tellurium burn in air to form the dioxides, which in these cases are solids.

Sulfur, selenium, and tellurium also react directly with the halogens. For example, they react vigorously with fluorine to give the hexafluorides.

$$S(s) + 3F_2(g) \longrightarrow SF_6(g)$$

(We will henceforth follow convention and write the formula of sulfur with the symbol S, rather than the strictly correct S_8, in order to simplify the coefficients in equations.) Group VIA elements react with most metals. Thus, sulfur gives sulfides (containing S^{2-}) and, in some cases, both sulfides and disulfides (with S_2^{2-}):●

● Many of the metal sulfides (FeS and CuS, for example) are nonstoichiometric compounds—that is, solid substances that deviate from their idealized formulas and can exhibit variable composition. See Section 11.7.

$$Hg(l) + S(s) \xrightarrow[\text{temperature}]{\text{room}} HgS(s)$$

$$Fe(s) + S(s) \xrightarrow{\Delta} \underset{\text{iron(II) sulfide}}{FeS(s)}$$

$$Fe(s) + 2S(s) \xrightarrow{\Delta} \underset{\text{iron(II) disulfide}}{FeS_2(s)}$$

Sulfur reacts with hot, concentrated nitric acid to give H_2SO_4, an acid in the +6 oxidation state.

$$S(s) + 6HNO_3(aq) \longrightarrow H_2SO_4(aq) + 6NO_2(g) + 2H_2O(l)$$

Selenium and tellurium are oxidized only to the +4 acids. For example,

$$Se(s) + 4HNO_3(aq) \longrightarrow H_2SeO_3(aq) + 4NO_2(g) + H_2O(l)$$

Preparation and Uses of the Elements

OXYGEN Oxygen is produced in enormous quantities from air. As described in the discussion of nitrogen, the air is first liquefied, then distilled. Nitrogen and argon are more volatile components of air and distill off, leaving liquid oxygen behind.

Oxygen can be prepared in small quantities by decomposing certain oxygen-containing compounds. Both the Swedish chemist Karl Wilhelm Scheele and the British chemist Joseph Priestley are credited with the discovery of oxygen. Priestley obtained the gas in 1774 by heating mercury(II) oxide.

$$2HgO(s) \xrightarrow{\Delta} 2Hg(l) + O_2(g)$$

A reaction used for the laboratory preparation of oxygen consists of the moderate heating of potassium chlorate, $KClO_3$, with manganese dioxide, MnO_2, as a catalyst.

$$2KClO_3(s) \xrightarrow[MnO_2]{\Delta} 2KCl(s) + 3O_2(g)$$

Over two-thirds of the oxygen produced is used in the making of steel, where its purpose is to oxidize impurities in the steel. Oxygen is also the oxidizing agent in many chemical processes and in the treatment of waste water. Small amounts of oxygen are used in welding, for medical purposes, and for rocket propulsion.

SULFUR, SELENIUM, AND TELLURIUM Free sulfur is mined by the **Frasch process,** *a process in which underground deposits of solid sulfur are melted in place with superheated water and the molten sulfur is forced upward as a froth, using air under pressure* (see Figure 22.18). The sulfur obtained this way is 99.6% pure. It is used primarily in the manufacture of sulfuric acid. Selenium is obtained from the flue dusts from the roasting of sulfide ores and from the anode mud formed by the electrolytic refining of copper. When these materials are leached with various

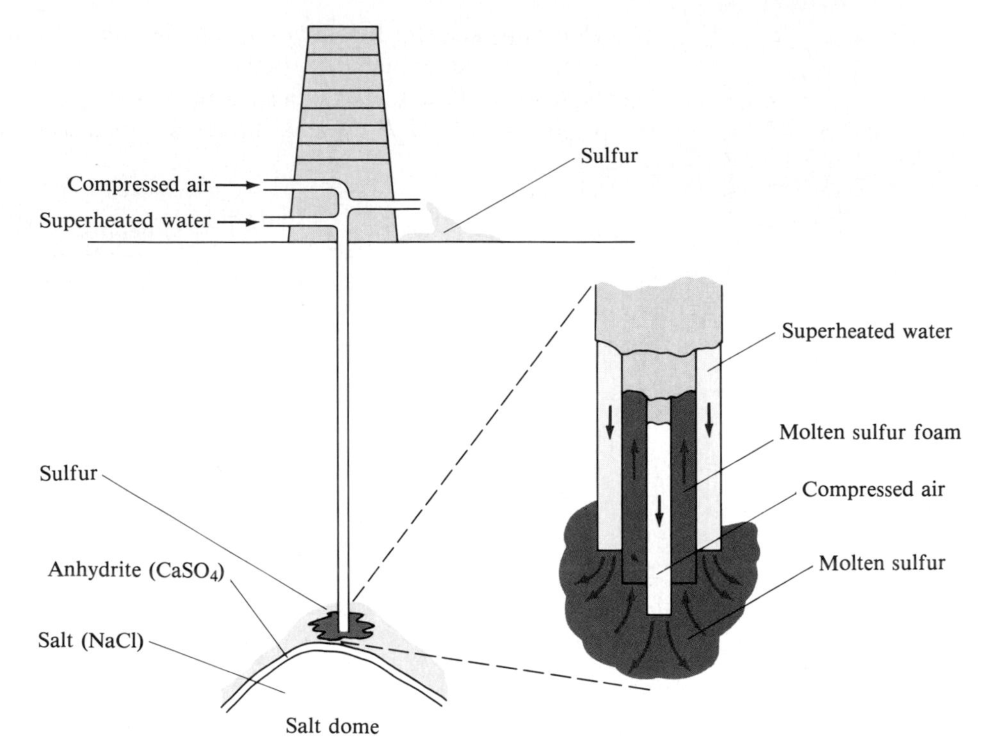

Figure 22.18
The Frasch process for mining sulfur. The well consists of concentric pipes. Superheated water passing down the outer jacket exits into the sulfur deposit, melting it. Compressed air from the inner pipe pushes the molten sulfur up the middle jacket. Molten sulfur flows from the top of the well onto the ground to cool.

oxidizing agents, H_2SeO_3 and H_2SeO_4 are obtained. These are reduced to elemental selenium with sulfur dioxide. For example,

$$H_2SeO_3(aq) + 2SO_2(g) + H_2O(l) \longrightarrow \underset{\text{red selenium}}{Se(s)} + 2H_2SO_4(aq)$$

Selenium has been used to make photoelectric cells and rectifiers for electronic equipment. These uses have been largely displaced by cheaper silicon semiconductor devices. Red selenium is also used to color red glass and enamels. Tellurium is useful in alloys.

Important Compounds

HYDROGEN SULFIDE, SELENIDE, AND TELLURIDE Hydrogen sulfide, H_2S, is a colorless gas with the strong odor of rotten eggs. It is quite poisonous. Hydrogen sulfide is a very weak, diprotic acid ($K_1 = 8.9 \times 10^{-8}$; $K_2 = 1.2 \times 10^{-13}$). It forms hydrogen sulfide salts (such as NaHS) and sulfide salts (Na_2S). In acid solution, hydrogen sulfide is a reducing agent and with mild oxidizing agents yields sulfur.

$$2Fe^{3+}(aq) + H_2S(aq) \longrightarrow 2Fe^{2+}(aq) + 2H^+(aq) + S(s)$$

Stronger oxidizing agents give sulfate ion.

Hydrogen sulfide is used in qualitative analysis to separate metal ions. The separation is based on the different solubilities of the metal sulfides formed from the metal ions with H_2S. The gas can be prepared by the reaction of an acid on a metal sulfide.

$$2[H^+(aq) + Cl^-(aq)] + ZnS(s) \longrightarrow [Zn^{2+}(aq) + 2Cl^-(aq)] + H_2S(g)$$

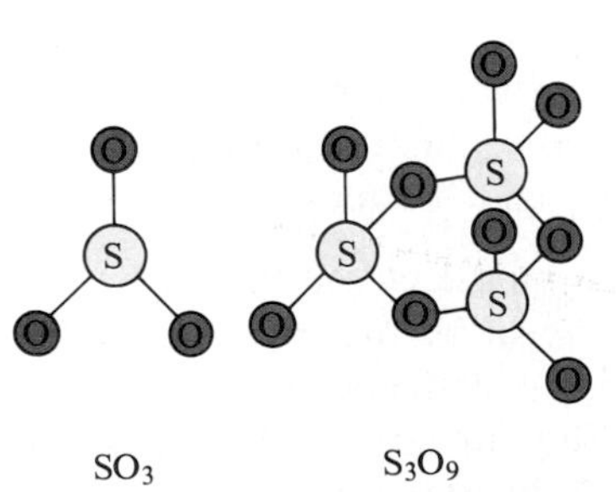

Figure 22.19
The structures of SO_3 and S_3O_9. These molecules are in equilibrium in liquid sulfur trioxide.

Table 22.8 Uses of Some Sulfur Compounds

Compound	Use
CS_2	Manufacture of rayon and cellophane Manufacture of CCl_4
SO_2	Manufacture of H_2SO_4 Food preservative Textile bleach
H_2SO_4	Manufacture of phosphate fertilizers Petroleum refining Manufacture of various chemicals
$Na_2S_2O_3$	Photographic fixer

An aqueous solution of H_2S is conveniently prepared in the laboratory by warming a solution of thioacetamide, CH_3CSNH_2.

$$\underset{\text{thioacetamide}}{CH_3CSNH_2(aq)} + 2H_2O(l) \xrightarrow{\Delta} \underset{\text{ammonium acetate}}{NH_4{}^+(aq) + CH_3COO^-(aq)} + H_2S(aq)$$

Hydrogen selenide, H_2Se, and hydrogen telluride, H_2Te, are also poisonous gases. They are weak acids but are stronger than H_2S. Both are strong reducing agents and give the elements as products.

SULFUR OXIDES AND OXOACIDS Sulfur dioxide, SO_2, is a colorless gas with a suffocating odor. It is obtained by burning sulfur or sulfides. The gas dissolves in water to give an acidic solution of sulfurous acid, H_2SO_3. These solutions contain only a small fraction of H_2SO_3 and are primarily $SO_2(aq)$.

$$SO_2(aq) + H_2O(l) \rightleftharpoons H_2SO_3(aq)$$

Sulfurous acid, which is stable only in aqueous solution, is diprotic, the first dissociation being moderately strong ($K_1 = 1.3 \times 10^{-2}$), the second rather weak ($K_2 = 6.3 \times 10^{-8}$). Two series of salts are known: the hydrogen sulfites (also called bisulfites) and the sulfites. Acidified solutions of these salts release SO_2.

$$2H^+(aq) + SO_3{}^{2-}(aq) \longrightarrow H_2O(l) + SO_2(g)$$

Sulfur dioxide solutions and sulfites are reducing agents, being oxidized to sulfates. Calcium hydrogen sulfite, made from SO_2 and $Ca(OH)_2$, is used to manufacture paper pulp by dissolving the natural cement (lignin) that holds the cellulose fibers together in wood. Sulfur dioxide gas is used to preserve dried fruit by inhibiting the growth of fungi. (Table 22.8 lists uses of sulfur compounds.)

Sulfur trioxide is unusual; it can exist in three different forms at room temperature. One form is a volatile liquid containing SO_3 and S_3O_9 molecules in equilibrium (Figure 22.19). Two solid forms of more complicated structure (not shown) are also possible. Sulfur trioxide is produced in only small amounts when sulfur burns, although thermodynamically it should be favored. It is prepared commercially as part of the contact process. The **contact process** is *an industrial method for the manufacture of sulfuric acid. It consists of the oxidation of sulfur dioxide to sulfur trioxide with a solid catalyst (platinum or divanadium pentoxide), followed by the reaction of sulfur trioxide with water.*●

● Sulfur dioxide for the contact process is obtained from several sources. Traditionally, sulfur mined by the Frasch process has been used. However, antipollution laws have made it necessary to recover SO_2 from plants that roast sulfide ores. Hydrogen sulfide from natural gas is another source of SO_2 (by burning the H_2S).

$$2SO_2(g) + O_2(g) \xrightarrow[V_2O_5]{\Delta} 2SO_3(g)$$

$$SO_3(g) + H_2O(l) \longrightarrow H_2SO_4(aq)$$

Figure 22.20
Environmental effects of acid rain. This bronze statue at St. Mark's Basilica in Venice shows corrosion and pitting from acid rain.

The sulfur trioxide mists that result from the contact process are difficult to dissolve completely in water. Therefore, in the industrial preparation of sulfuric acid, the trioxide is first dissolved in concentrated H_2SO_4. The major species in this solution is $H_2S_2O_7$ (pyrosulfuric acid). The solution is diluted with water to give concentrated sulfuric acid.

Sulfuric acid is a component of acid rain and forms in air by reactions similar to those involved in the commercial preparation of this acid. The burning of fossil fuels and sulfide ores produces sulfur dioxide. After residing in the atmosphere for some time, sulfur dioxide oxidizes to sulfur trioxide, which dissolves in rain to give $H_2SO_4(aq)$. Figure 22.20 shows the effect of acid rain on a bronze sculpture.

Sulfuric acid is diprotic; the first dissociation in water is strong, the second moderately strong ($K_2 = 1.1 \times 10^{-2}$). Both sulfate and hydrogen sulfate (bisulfate) salts are known. The sulfate anion, which has a tetrahedral geometry, might be written

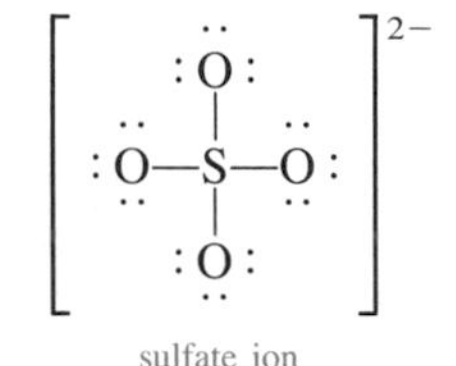

sulfate ion

though the S—O bond length (1.51 Å), when compared with the sum of the sulfur and oxygen single-bonded covalent radii (1.70 Å), indicates substantial double-bond character.● Partial double bonding can be achieved by overlapping of $2p$ orbitals on the oxygen atoms with $3d$ orbitals on the sulfur atom. One may represent this by resonance structures of the following sort:

● The covalent radii of Group VIA elements are given in Table 22.7.

$$\left[\begin{array}{c} :\ddot{O}: \\ \| \\ :\ddot{O}-S-\ddot{O}: \\ \| \\ :\ddot{O}: \end{array}\right]^{2-}$$

Concentrated sulfuric acid is a viscous liquid and a powerful dehydrating agent (see Figure 6.13, p. 216). The concentrated acid is also an oxidizing agent. Copper is not dissolved by most acids ($E°$ for $Cu^{2+}|Cu$ is 0.34 V), but it is dissolved by concentrated sulfuric acid, which is reduced to sulfur dioxide.

$$Cu(s) + 2H_2SO_4(l) \longrightarrow Cu^{2+}(aq) + SO_4^{2-}(aq) + SO_2(g) + 2H_2O(l)$$

More sulfuric acid is manufactured than any other chemical. Most of the acid is used to make soluble phosphate and ammonium sulfate fertilizers. Sulfuric acid is also used in petroleum refining and in the manufacture of many chemicals.

Thiosulfate ion, $S_2O_3^{2-}$, is related structurally to the sulfate ion, with sulfur replacing an oxygen atom. (The prefix *thio-* means that sulfur has replaced oxygen in the compound or ion whose name follows the prefix.)

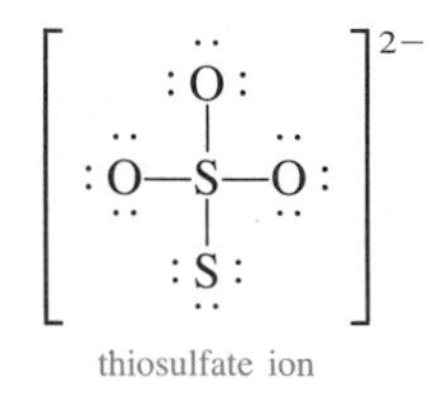

thiosulfate ion

Sodium thiosulfate is prepared by heating a suspension of sulfur in sodium sulfite solution.

$$[2Na^+(aq) + SO_3^{2-}(aq)] + S(s) \xrightarrow{\Delta} [2Na^+(aq) + S_2O_3^{2-}(aq)]$$

Thiosulfate ion decomposes in acidic solution to give sulfur dioxide and a precipitate of sulfur, essentially reversing the reaction just given.

Sodium thiosulfate pentahydrate, $Na_2S_2O_3 \cdot 5H_2O$, is known as photographer's "hypo." Photographic film has a layer of silver halide embedded in gelatin. When exposed to light, the silver halide decomposes to very small grains of metallic silver, forming a "latent" image. A *developer* enlarges these grains and brings out the image by reducing nearby crystals of silver halide. Then a *fixer* of sodium thiosulfate solution dissolves unexposed silver halide from the film by forming the complex ion $Ag(S_2O_3)_2^{3-}$.

SELENIUM AND TELLURIUM OXIDES AND OXOACIDS Selenium dioxide is a colorless, volatile solid prepared by burning selenium in air. The vapor phase contains SeO_2 molecules. The solid dissolves in water to give a solution of selenious acid, from which crystals of selenious acid may be isolated. Tellurium dioxide, prepared by burning tellurium in air, is a colorless solid. It is barely soluble in water, and no aqueous acid is known. Tellurium dioxide does dissolve readily in solutions of alkali metal hydroxides, giving tellurite salts such as Na_2TeO_3. The dioxide also dissolves in acids, showing that it is amphoteric.

Selenium trioxide is a white solid; tellurium trioxide is an orange solid. Both are prepared by heating the corresponding acid. For example,

$$\underset{\text{selenic acid}}{H_2SeO_4(s)} \longrightarrow H_2O(g) + SeO_3(s)$$

Selenic acid is prepared by strongly oxidizing selenious acid—for example, by heating it with 30% hydrogen peroxide.

$$H_2SeO_3(aq) + H_2O_2(aq) \longrightarrow [H^+(aq) + HSeO_4^-(aq)] + H_2O(l)$$

It is a strong, diprotic acid. Both SeO_3 and H_2SeO_4 are strong oxidizing agents. Telluric acid may be prepared similarly to selenic acid, by oxidizing TeO_2 with $H_2O_2(aq)$. Surprisingly, the acid has the formula H_6TeO_6 or $Te(OH)_6$, rather than the formula we would expect by analogy with sulfuric and selenic acids. Telluric acid is a weak, diprotic acid and a strong oxidizing agent. The weak acidity is an indication of the greater metallic character of tellurium.

Example 22.3 Using Chemical Reactions of the Group VIA Elements and Compounds

Prepare sulfuric acid from $H_2S(g)$, air, and water (plus any required catalysts).

Solution

Hydrogen sulfide burns in air to give H_2O and SO_2.

$$2H_2S(g) + 3O_2(g) \longrightarrow 2H_2O(g) + 2SO_2(g)$$

Then, using heat and a catalyst, we obtain

$$2SO_2(g) + O_2(g) \xrightarrow[\text{catalyst}]{\Delta} 2SO_3(g)$$

Finally, we dissolve SO_3 in water.

$$SO_3(g) + H_2O(l) \longrightarrow H^+(aq) + HSO_4^-(aq)$$

Exercise 22.5

Prepare sodium thiosulfate from sulfur, air, and sodium carbonate solution.

(See Problems 22.75 and 22.76.)

Exercise 22.6

Thiosulfate ion is used to titrate iodine by reducing it to I^-. Thiosulfate ion is oxidized to tetrathionate ion, $S_4O_6^{2-}$, which has the structure

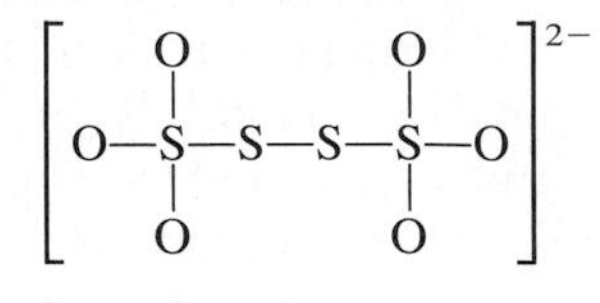

Write the balanced equation for the reaction. (See Problems 22.77 and 22.78.)

22.4 GROUP VIIA: THE HALOGENS

For the most part, the Group VIIA elements, or halogens, have very similar properties—or at least properties that change smoothly in progressing down the column. All are reactive nonmetals, except perhaps for astatine, whose chemistry is not well known. Fluorine, as a second-period element, does exhibit some differences from the other elements of Group VIIA, although these are not so pronounced as we have seen for the second-period elements in Groups IIIA to VIA. The solubilities of the fluorides in water, for example, are often quite different from those of the other halides. Calcium chloride, bromide, and iodide are very soluble in water. Calcium fluoride, however, is insoluble. Silver chloride, bromide, and iodide are insoluble, but silver fluoride is soluble.

All of the halogens form stable compounds in the -1 oxidation state. For fluorine compounds, this is the only oxidation state. Chlorine, bromine, and iodine also have compounds in which the halogen is in one of the positive oxidation states: $+1$, $+3$, $+5$, or $+7$.

Fluorine and chlorine are abundant elements. Fluorine is widely available in fluorapatite, $3Ca_3(PO_4)_2 \cdot CaF_2$, and in fluorite, CaF_2. Chlorine is abundant in the oceans and in salt deposits as NaCl. Bromine is much less abundant, but it occurs as Br^- in recoverable concentrations in certain brines and ocean water. Iodine occurs in very low concentration as I^- in seawater, but this is assimilated by kelp (seaweed), a commercial source of the element. Iodine also occurs as sodium iodide in certain oil-well brines and as sodium iodate, $NaIO_3$, in Chilean nitrate deposits.

Properties of the Elements

All of the halogens exist as diatomic molecular substances; they are F_2, Cl_2, Br_2, and I_2. Fluorine and chlorine are gases; fluorine is pale yellow, and chlorine is greenish-yellow. Bromine is a volatile, reddish-brown liquid with reddish-brown vapor. Iodine is a shiny, black solid that sublimes readily, giving a violet vapor. Melting and boiling points and other properties are listed in Table 22.9. The increasing melting points and boiling points from F_2 to I_2 are explained by the argument that the London forces between molecules increase as the halogen molecules increase in size.

The halogens are all oxidizing agents, though the oxidizing power decreases from F_2 to I_2. We can see this readily by considering the oxidation of water by the halogens. The emf of these reactions can be calculated from the standard potentials for the electrodes $X_2|X^-$, given in Table 22.9, and that for the following half-

Table 22.9 Properties of Group VIIA Elements

Property	Fluorine	Chlorine	Bromine	Iodine	Astatine
Electron configuration	$[He]2s^22p^5$	$[Ne]3s^23p^5$	$[Ar]3d^{10}4s^24p^5$	$[Kr]4d^{10}5s^25p^5$	$[Xe]4f^{14}5d^{10}6s^26p^5$
Melting point, °C	−220	−101	−7	114	—
Boiling point, °C	−188	−35	59	184	—
Density, g/cm^3	1.69×10^{-3}	3.21×10^{-3}	3.12	4.93	—
Ionization energy (first), kJ/mol	1681	1251	1140	1008	—
Electron affinity, kJ/mol	−328	−349	−325	−295	−270
Electronegativity (Pauling scale)	4.0	3.0	2.8	2.5	2.2
Standard potential (volts), $X_2 + 2e^- \rightleftharpoons 2X^-$	2.87	1.36	1.07	0.54	0.3
Covalent radius, Å	0.64	0.99	1.14	1.33	1.48
Ionic radius (for X^-), Å	1.19	1.67	1.82	2.06	—

● In neutral water solution, $[H^+] = 1.0 \times 10^{-7}\ M$. The concentrations of all other species have standard values.

reaction in neutral water:●

$$4H^+(aq) + O_2(g) + 4e^- \longrightarrow 2H_2O(l);\ E = 0.82\ V$$

For the fluorine reaction, we get

$$2F_2(g) + 2H_2O(l) \longrightarrow 4H^+(aq) + 4F^-(aq) + O_2(g);\ E^\circ_{cell} = 2.05\ V$$

Thus, fluorine reacts vigorously with water. For the reaction of water with Cl_2, Br_2, and I_2, we get 0.54 V, 0.24 V, and −0.28 V, respectively. The emf values show that chlorine and bromine should react to some extent with water to give oxygen, whereas the reverse reaction occurs for iodine: I^- is oxidized by molecular oxygen to I_2. In the case of chlorine and bromine, however, the following reactions are faster and occur more readily than the reactions to form O_2, particularly in basic solution:

$$Cl_2(g) + H_2O(l) \rightleftharpoons HClO(aq) + [H^+(aq) + Cl^-(aq)]$$
$$Br_2(aq) + H_2O(l) \rightleftharpoons HBrO(aq) + [H^+(aq) + Br^-(aq)]$$

The relative oxidizing powers of Cl_2, Br_2, and I_2 can be seen by observing which of the halogens react with the halide ion of another. When chlorine water, $Cl_2(aq)$, is added to dilute solutions of bromide or iodide ion, the elements are released.

$$Cl_2(aq) + 2Br^-(aq) \longrightarrow 2Cl^-(aq) + Br_2(aq)$$
$$Cl_2(aq) + 2I^-(aq) \longrightarrow 2Cl^-(aq) + I_2(aq)$$

Although bromine is not strong enough as an oxidizing agent to oxidize Cl^-, it easily oxidizes I^-.

$$Br_2(aq) + 2I^-(aq) \longrightarrow 2Br^-(aq) + I_2(aq)$$

Iodine, of course, is not strong enough to oxidize either Cl^- or Br^-. The first two reactions have been used as tests of Br^- and I^-. When methylene chloride, CH_2Cl_2, is poured into a test tube of Br^- or I^- to which chlorine water has been added, it forms a colored layer of the halogen in CH_2Cl_2 at the bottom of the tube. Bromine gives an orange layer; iodine gives a violet layer.

The halogens react directly with most other elements. They react with themselves to form **interhalogens**—*binary compounds of one halogen with another,* such as ClF, BrF, IBr, ClF_3, ClF_5, and IF_7.

Preparation and Uses of the Elements

Fluorine is such a strong oxidizing agent that it has been prepared only by electrolysis. The cell electrolyte is potassium fluoride dissolved in liquid hydrogen fluoride. Electrolysis produces hydrogen at the cathode and fluorine at the anode. Fluorine is produced in commercial quantities for the manufacture of uranium nuclear fuel rods. Uranium metal is reacted with excess fluorine to produce uranium hexafluoride, UF_6, a volatile, white solid. The vapor of this compound is separated by diffusion to give mixtures containing more of the fissionable uranium-235 isotope than is present in the natural source. Nuclear fuel rods are 3%–4% uranium-235; it constitutes 0.72% of natural uranium.

Chlorine, Cl_2, is a major industrial chemical. It is prepared by the electrolysis of aqueous sodium chloride.● The major use of chlorine is in the manufacture of various chlorinated hydrocarbons, such as vinyl chloride, $CH_2{=}CHCl$ (for plastics); carbon tetrachloride, CCl_4 (for fluorocarbons); and methyl chloride, CH_3Cl (for silicones and tetramethyllead). Large quantities of chlorine are also used to disinfect water supplies and to bleach paper pulp and textiles.

● The electrolysis of NaCl(*aq*) was described in Section 19.9.

Bromine, Br_2, can be obtained from seawater or brine by oxidation of the bromide ion in solution with chlorine. Most of the bromine produced in the United States is obtained from brine wells in Arkansas and Michigan. These brines contain such high concentrations of bromide ion that the bromine can be distilled from the oxidized solution. In the commercial process, hot brine is led into the top of a tower, while steam and chlorine are fed in at the bottom. Bromine and water vapor leaving from the top of the tower are condensed, giving a distillate of separate layers: bromine on the bottom, water on top. The bromine layer is drawn off and purified by further distillation.

Bromine is used to manufacture bromine compounds, including methyl bromide, CH_3Br (as a pesticide); silver bromide (for photographic film); and alkali metal bromides (for sedatives).

Iodine, I_2, is produced from natural brines by oxidizing I^- with chlorine.

$$2I^-(aq) + Cl_2(g) \longrightarrow I_2(s) + 2Cl^-(aq)$$

The precipitated iodine is filtered from the solution. The element is also produced from sodium iodate, an impurity in Chilean saltpeter, $NaNO_3$, by reducing iodate ion with sodium hydrogen sulfite, $NaHSO_3$. Iodine is used to make compounds such as silver iodide for photographic film and potassium iodide as a human nutritional and animal-feed supplement.

Important Compounds

The most important inorganic compounds of the halogens are the *hydrogen halides,* the *halogen oxoacids,* and their salts. Table 22.10 lists uses of some halogen compounds.

HYDROGEN HALIDES Each of the halogens forms a binary compound with hydrogen: HF, HCl, HBr, and HI. All are colorless gases with sharp, penetrating odors. Hydrogen fluoride, whose boiling point (20°C) is near room temperature, is easily

Table 22.10 Uses of Some Halogen Compounds

Compound	Use
AgBr, AgI	Photographic film
CCl_4	Manufacture of fluorocarbons
CH_3Br	Pesticide
$C_2H_4Br_2$	Lead scavenger in leaded gasolines
$C_2H_4Cl_2$	Manufacture of vinyl chloride (plastics) Lead scavenger in leaded gasolines
C_2H_5Cl	Manufacture of tetraethyllead for gasoline
HCl	Metal treating Food processing
NaClO	Household bleach Manufacture of hydrazine for rocket fuel
$NaClO_3$	Paper pulp bleaching (with ClO_2)
KI	Human nutritional and animal-feed supplement

liquefied. Its boiling point is high compared with those of the other hydrogen halides (HCl, −85°C; HBr, −67°C; HI, −35°C), whose boiling points increase as expected for molecular substances having little or no hydrogen bonding. The much higher boiling point of hydrogen fluoride is good evidence for strong hydrogen bonding in this substance.

The hydrogen halides dissolve in water to give acidic solutions called *hydrohalic acids*. For example, hydrogen chloride gas dissolves in water to give hydrochloric acid (see Figure 22.21). Hydrochloric, hydrobromic, and hydriodic acids are strong acids, the molecular species HX ionizing completely in solution.

Figure 22.21
The hydrogen chloride fountain. *Left:* The flask contains hydrogen chloride gas. When water is added to the flask from the dropper, the hydrogen chloride dissolves in it, reducing the pressure in the flask. Atmospheric pressure pushes the water in the beaker into the flask, forming a stream or fountain. Water in the beaker has bromthymol blue indicator in it. The indicator solution in the flask is yellow because it is acidic. *Right:* A close-up view of the HCl fountain.

$$HCl(g) \xrightarrow{H_2O} H^+(aq) + Cl^-(aq)$$

Hydrofluoric acid, by contrast, is a weak acid. An equilibrium exists in aqueous solution between molecular HF and the ions H^+ and F^-.

$$HF(aq) \rightleftharpoons H^+(aq) + F^-(aq)$$

The hydrogen halides can be formed by direct combination of the elements. Fluorine and hydrogen react violently. The reaction has no commercial importance because fluorine is normally prepared from hydrogen fluoride. However, chlorine is burned in an excess of hydrogen to produce hydrogen chloride for industrial use.

$$H_2(g) + Cl_2(g) \longrightarrow 2HCl(g)$$

Hydrogen bromide and hydrogen iodide are prepared in a similar way. In these cases, the elements are heated with a platinum catalyst—for example,

$$H_2(g) + Br_2(g) \xrightarrow[Pt]{\Delta} 2HBr(g)$$

The hydrogen halides can be prepared by heating the salts with a nonvolatile acid. The commercial method of preparing hydrogen fluoride uses concentrated sulfuric acid.

$$CaF_2(s) + H_2SO_4(l) \xrightarrow{\Delta} CaSO_4(s) + 2HF(g)$$

The reaction is driven to the right by removing gaseous hydrogen fluoride, the other substances being relatively nonvolatile. Hydrogen chloride can be prepared in a similar reaction from sodium chloride and sulfuric acid.

$$NaCl(s) + H_2SO_4(l) \xrightarrow{\Delta} NaHSO_4(s) + HCl(g)$$

On stronger heating, the reaction goes to sodium sulfate.

$$NaCl(s) + NaHSO_4(s) \xrightarrow{\Delta} Na_2SO_4(s) + HCl(g)$$

Neither HBr nor HI can be prepared with H_2SO_4, because the hot concentrated acid oxidizes Br^- and I^- to the elements (see Figure 22.22). For these halides, phosphoric acid can be used because it is nonoxidizing as well as nonvolatile.

$$NaBr(s) + H_3PO_4(l) \xrightarrow{\Delta} HBr(g) + NaH_2PO_4(s)$$

Figure 22.22
The action of concentrated sulfuric acid on halide salts. Concentrated sulfuric acid was added to watch glasses containing, from left to right, NaCl, NaBr, and NaI. Note the formation of Br_2 in the center watch glass and the formation of I_2 in the watch glass on the right. These halogens form when concentrated sulfuric acid oxidizes the corresponding halide ions. Chloride ion is not oxidized by H_2SO_4.

Although some hydrogen chloride is prepared commercially from sodium chloride and sulfuric acid, as well as by direct combination of the elements, today most HCl is obtained as a by-product of the chlorination of organic compounds. For example, when methane, CH_4, reacts with chlorine, it produces various chloromethanes, such as methyl chloride (monochloromethane), and HCl.

$$CH_4(g) + Cl_2(g) \longrightarrow \underset{\text{methyl chloride}}{CH_3Cl(g)} + HCl(g)$$

The quantity of chlorinated hydrocarbons produced for various purposes is so large that by-product HCl satisfies most of the commercial demand for the acid.

A major use of hydrogen fluoride is in the preparation of organic fluorine compounds. For example, trichlorofluoromethane, CCl_3F, which is used as a refrigerant and aerosol propellant, is prepared from carbon tetrachloride and HF, with antimony pentafluoride as a catalyst.

$$CCl_4(l) + HF(g) \xrightarrow{SbF_5} CCl_3F(l) + HCl(g)$$

Polytetrafluoroethylene plastic (Teflon) is also an organic fluorine material requiring HF in its manufacture. Another major use of HF is in the preparation of the electrolyte for the production of aluminum.● The reactions of hydrofluoric acid with silica (SiO_2) and glass are unique among the hydrohalic acids.

● The production of aluminum was discussed in Section 21.4.

$$6HF(aq) + \underset{\text{silica}}{SiO_2(s)} \longrightarrow H_2SiF_6(aq) + 2H_2O(l)$$

$$\underset{\text{glass}}{CaSiO_3(s)} + 8HF(aq) \longrightarrow \underset{\text{hexafluorosilicic acid}}{H_2SiF_6(aq)} + CaF_2(aq) + 3H_2O(l)$$

Small quantities of hydrofluoric acid are used to etch glass by this reaction.

Hydrochloric acid is the fourth most important industrial acid (after sulfuric, phosphoric, and nitric acids). It is used to clean metal surfaces of oxides (a process called "pickling") and to extract certain metal ores, such as those of tungsten. Its use in the preparation of magnesium was mentioned earlier.●

● See Section 21.3.

HALOGEN OXOACIDS The halogens form a variety of oxoacids (Table 22.11). Figure 22.23 shows the structures of the oxoacids of chlorine. The acidic character of these acids increases with the number of oxygen atoms bonded to the halogen atom; that is, acid strength increases from HClO to $HClO_4$. Perchloric acid is the strongest of the common acids. All of the halogen oxoacids are oxidizing agents.

The chemistry of the chlorine oxoacids can be understood in part by referring to the standard-potential diagram shown in Figure 22.24. A **standard-potential**

Table 22.11 Halogen Oxoacids

Oxidation State	Fluorine Oxoacids	Chlorine Oxoacids	Bromine Oxoacids	Iodine Oxoacids	General Name
+1	HFO*	HClO†	HBrO†	HIO†	Hypohalous acid
+3	—	$HClO_2$†	$HBrO_2$†	—	Halous acid
+5	—	$HClO_3$†	$HBrO_3$†	HIO_3	Halic acid
+7	—	$HClO_4$	$HBrO_4$†	HIO_4 H_5IO_6	Perhalic acid

*The oxidation state of F in HFO is −1.

†These acids are known only in aqueous solution.

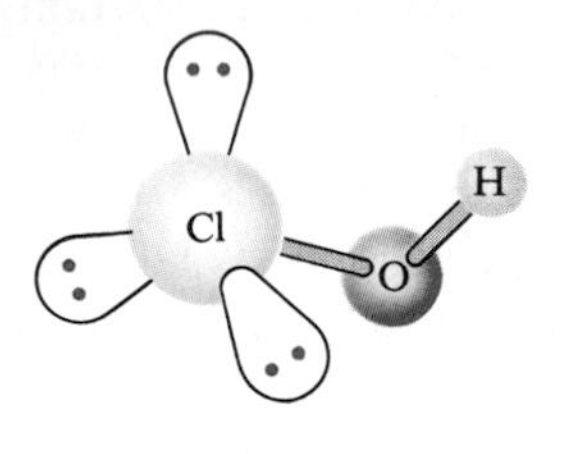

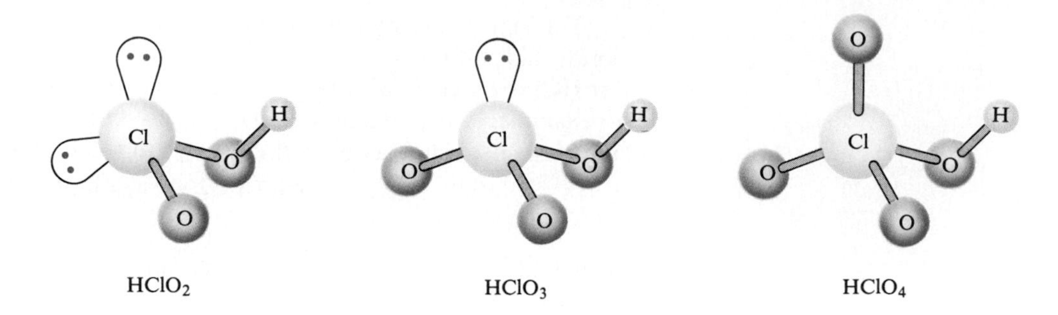

Figure 22.23
Structures of the chlorine oxoacids. The models of the oxoacids also include lone pairs on the Cl atoms.

diagram is *a convenient graphic presentation of the standard potentials of an element*. Species of different oxidation states are connected in the diagram by an arrow, and the standard potential (in volts) for the half-reaction between them is written over the arrow. For example,

$$HClO \xrightarrow{+1.63} Cl_2 \qquad \text{(acidic solution)}$$

represents the half-reaction for the reduction of HClO to Cl_2 in acid solution. We can also write this out in full:●

● The method of obtaining the half-reaction was described in Section 13.8.

$$2HClO(aq) + 2H^+(aq) + 2e^- \longrightarrow Cl_2(g) + 2H_2O(l); E° = 1.63 \text{ V}$$

The feasibility of the disproportionation (simultaneous oxidation and reduction) of Cl_2 in basic solution can be seen from a glance at the diagram. The relevant portion is

$$ClO^- \xrightarrow{+0.44} Cl_2 \xrightarrow{+1.36} Cl^- \qquad \text{(basic solution)}$$

If Cl_2 is to disproportionate in basic solution, it must be reduced to Cl^- and oxidized to ClO^-. The emf for this reaction is $+1.36 \text{ V} - (+0.44 \text{ V}) = +0.92 \text{ V}$. Because this is positive, the reaction is spontaneous. In general, a species disproportionates if the standard potential on the left in the diagram is less than that on the right.

The disproportionation of chlorine in basic solution is used to prepare sodium hypochlorite, NaClO. Chlorine, released by the electrolysis of aqueous sodium chloride, is allowed to mix with the cold sodium hydroxide solution also obtained

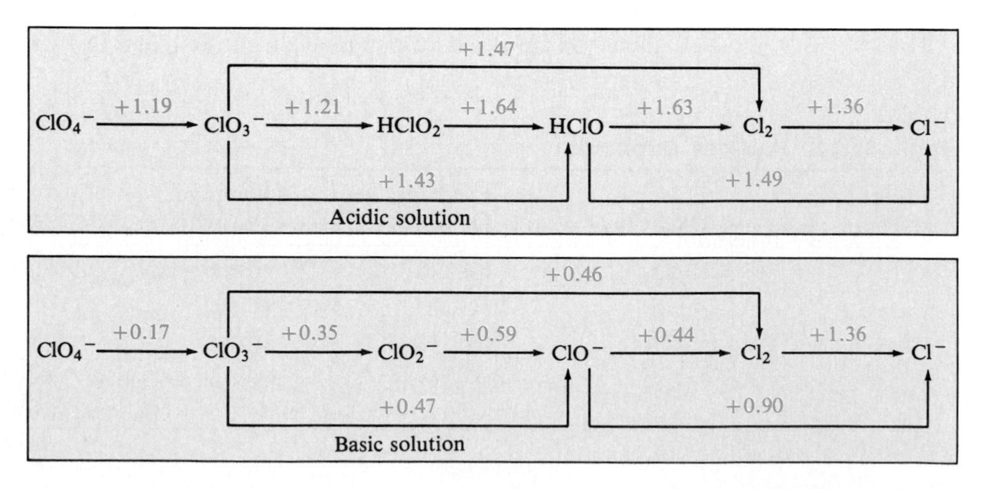

Figure 22.24
Standard-potential diagram for chlorine and its compounds. The upper part of the diagram gives electrode potentials for reactions in 1 *M* H^+ (acidic solution); the lower part gives electrode potentials for 1 *M* OH^- (basic solution).

in the electrolysis. The reaction is

$$Cl_2(g) + 2[Na^+(aq) + OH^-(aq)] \longrightarrow [Na^+(aq) + ClO^-(aq)] + [Na^+(aq) + Cl^-(aq)] + H_2O(l)$$

Solutions of sodium hypochlorite are sold as a bleach. The corresponding acid, HClO, is unstable except in dilute aqueous solution. Like the other hypohalous acids, it is a weak acid.

Hypochlorite ion is itself unstable, disproportionating into chlorate ion, ClO_3^-, and chloride ion.

$$3ClO^-(aq) \longrightarrow ClO_3^-(aq) + 2Cl^-(aq);\ E^\circ_{cell} = +0.43\ V \quad \text{(basic)}$$

The reaction in cold basic solution is very slow, but it is fast in hot solution. Therefore, when chlorine reacts with hot sodium hydroxide solution, sodium chlorate is a product instead of NaClO.

$$3Cl_2(g) + 6[Na^+(aq) + OH^-(aq)] \longrightarrow [Na^+(aq) + ClO_3^-(aq)] + 5[Na^+(aq) + Cl^-(aq)] + 3H_2O(l)$$

The sodium chlorate is obtained from the solution by partial evaporation. Sodium chloride crystallizes from the hot solution. The filtrate is then evaporated to recover $NaClO_3$. Chloric acid is stable only in aqueous solution.

Sodium perchlorate and potassium perchlorate are produced commercially by electrolysis of a saturated solution of the chlorate. The anode reaction is

$$ClO_3^-(aq) + H_2O(l) \longrightarrow ClO_4^-(aq) + 2H^+(aq) + 2e^-$$

Hydrogen is evolved at the cathode. Perchloric acid, $HClO_4$, can be prepared by treating a perchlorate salt with sulfuric acid.

$$KClO_4(s) + H_2SO_4(l) \longrightarrow KHSO_4(s) + HClO_4(l)$$

The perchloric acid is distilled from the mixture at reduced pressure (to keep the temperature below 92°C, where perchloric acid decomposes explosively).

Example 22.4 Using Chemical Reactions of the Group VIIA Elements and Compounds

Sodium hexafluorosilicate, Na_2SiF_6, is used for the fluoridation of water supplies. (The fluorine becomes incorporated into tooth enamel, which is then less susceptible to decay.) Describe with balanced equations how this salt can be obtained from calcium fluoride, sand (SiO_2), sulfuric acid, sodium hydroxide, and water.

Solution

The reactions are

$$CaF_2(s) + H_2SO_4(l) \longrightarrow CaSO_4(s) + 2HF(g)$$

$$HF(g) \xrightarrow{H_2O} HF(aq)$$

$$6HF(aq) + SiO_2(s) \longrightarrow H_2SiF_6(aq) + 2H_2O(l)$$

$$H_2SiF_6(aq) + 2NaOH(aq) \longrightarrow Na_2SiF_6(aq) + 2H_2O(l)$$

Exercise 22.7

Chlorine dioxide, ClO_2, is a reddish-yellow gas used to bleach paper pulp. It may be prepared by reducing aqueous sodium chlorate with sulfur dioxide in acid solution. (This gas is safely prepared only in low concentrations; otherwise it may detonate.) Describe with balanced equations how chlorine dioxide could be made from aqueous sodium chloride, sulfur dioxide, and sulfuric acid. (See Problems 22.85 and 22.86.)

Exercise 22.8

Perbromic acid was discovered in 1968. The sodium salt may be prepared by oxidation of sodium bromate by fluorine in basic solution. Write the balanced equation for this reaction.

(See Problems 22.87, 22.88, 22.89, and 22.90.)

Exercise 22.9

Show by calculating the standard emf that the disproportionation of Cl_2 in acid solution is not spontaneous under standard conditions (see Figure 22.24 for data).

(See Problems 22.91 and 22.92.)

22.5 GROUP VIIIA: THE NOBLE GASES

The noble gases were unknown until argon was discovered in 1894. Soon afterward, the other noble gases were discovered. In our discussions of bonding, we pointed out the relative stability of the noble-gas configuration. For many years, it was thought that because the atoms of these elements had completed octets, the noble gases would be completely unreactive. Consequently, these elements were known as *inert gases*. Then, in 1962, the first compound of xenon was discovered.

Discovery of the Noble Gases

In 1892, the English physicist Lord Rayleigh discovered that the density of molecular nitrogen obtained from air (1.2561 g/L at STP) was noticeably greater than the density of nitrogen obtained by decomposition of nitrogen compounds (1.2498 g/L at STP). He concluded that one of these two nitrogen sources was contaminated with another substance.

Soon after this, Rayleigh began collaborating with the Scottish chemist and physicist William Ramsay. Ramsay passed atmospheric nitrogen over hot magnesium to remove the nitrogen as magnesium nitride, Mg_3N_2, and obtained a nonreactive residual gas. He placed this gas in a sealed glass tube and subjected it to a high-voltage electrical discharge to study the emission of light. Analysis of the emission showed a spectrum of red and green lines unlike that of any known element. In 1894 Ramsay and Rayleigh concluded that they had discovered a new element, which they called *argon* from the Greek word *argos*, meaning "lazy."● They also surmised that argon was a member of a new column of elements in the periodic table, lying between the halogens and the alkali metals.

● *Lazy* here refers to argon's lack of chemical reactivity.

Other noble gases were discovered soon after argon. Emission lines of helium (from the Greek *helios*, meaning "sun") had actually been observed in the sun's spectrum before helium was discovered on earth in the uranium ore clevite. Neon, krypton, and xenon were all obtained by fractional distillation of liquid air. Radon was discovered as a gaseous decay product of radium. All known isotopes of radon are radioactive.

Preparation and Uses of the Noble Gases

Commercially, all the noble gases except helium and radon are obtained by the distillation of liquid air. The principal sources of helium are certain natural gas wells located in the western United States. Helium has the lowest boiling point

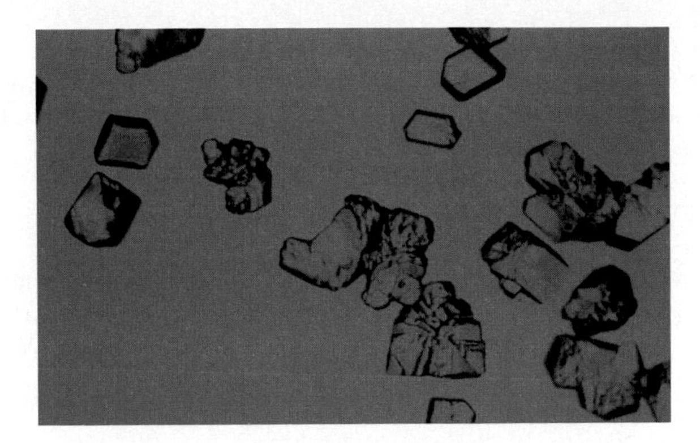

Figure 22.25
Crystals of xenon tetrafloride. These crystals were obtained in the experiment that first produced a binary compound of xenon. The crystals are enlarged 100 times.

(−268.9°C) of any substance and is very important in low-temperature research. The major use of argon is as a blanketing gas (inert gas) in metallurgical processes. It is also used as a mixture with nitrogen to fill incandescent light bulbs. Here the gas mixture conducts heat away from the hot tungsten filament. All the noble gases are used in gas discharge tubes. Neon gives a highly visible red-orange emission and has long been used in advertising signs. The noble gases are also used in a number of lasers. The helium–neon laser was the first continuously operating gas laser. It emits red light at a wavelength of 632.8 nm.

Compounds of the Noble Gases

Neil Bartlett, while working at the University of British Columbia, prepared the first noble-gas compound after he discovered that molecular oxygen reacts with platinum hexafluoride, PtF_6, to form the ionic solid $[O_2^+][PtF_6^-]$. Because the ionization energy of xenon (1.17×10^3 kJ/mol) is close to that of molecular oxygen (1.21×10^3 kJ/mol), Bartlett reasoned that xenon should react with platinum hexafluoride also. In 1962, he reported the synthesis of an orange-yellow compound with the approximate formula $XePtF_6$.● Later in the same year, chemists from Argonne National Laboratory near Chicago reported that xenon reacts directly with fluorine at 400°C to give the tetrafluoride.

● The product actually has variable composition and can be represented by the formula $Xe(PtF_6)_n$, where n is between 1 and 2.

$$Xe(g) + 2F_2(g) \longrightarrow XeF_4(s)$$

The product is a volatile, colorless solid (Figure 22.25). Since then a number of noble-gas compounds have been prepared, all involving bonds to the highly electronegative elements fluorine and oxygen. Most are compounds of xenon (see Table 22.12), but a few are compounds of krypton and radon.

Table 22.12 Some Compounds of Xenon

Compound	Formula	Description
Xenon difluoride	XeF_2	Colorless crystals
Xenon tetrafluoride	XeF_4	Colorless crystals
Xenon hexafluoride	XeF_6	Colorless crystals
Xenon trioxide	XeO_3	Colorless crystals, explosive
Xenon tetroxide	XeO_4	Colorless gas, explosive

Exercise 22.10

Give the Lewis formula for KrF_2. What hybridization is expected for the krypton atom in this molecule? Using the VSEPR model, deduce the molecular geometry of KrF_2.

(See Problems 22.95 and 22.96.)

Exercise 22.11

Xenon difluoride is a powerful oxidizing agent. In acidic solution it is reduced to xenon. Hydrogen fluoride is the other product. Write a half-reaction for this reduction. Hydrochloric acid is oxidized to Cl_2 by xenon difluoride. Write a balanced equation for this reaction.

(See Problems 22.97 and 22.98.)

A Checklist for Review

IMPORTANT TERMS

zone refining (22.1)
catenation (22.1)
silicates (22.1)
condensation reaction (22.1)
aluminosilicate minerals (22.1)
Ostwald process (22.2)
polyphosphoric acids (22.2)
metaphosphoric acids (22.2)
photoconductor (22.3)
peroxides (22.3)
superoxides (22.3)
Frasch process (22.3)
contact process (22.3)
interhalogens (22.4)
standard-potential diagram (22.4)

SUMMARY OF FACTS AND CONCEPTS

Group IVA elements show a definite progression from nonmetallic to metallic as we read down the column in the periodic table. Carbon occurs as its nonmetallic allotropes, graphite and diamond. Silicon and germanium have diamondlike structures, but whereas diamond is an electrical insulator, silicon and germanium are semiconductors. Although one allotrope of tin has the diamondlike structure, the common form is metallic. Lead has only a metallic form. The +4 oxidation state is common for carbon, silicon, and germanium, but the +2 state occurs in tin and is the most important oxidation state for lead. We see this in the reaction of the elements. All except lead burn in excess oxygen to form the dioxides; lead gives PbO. Germanium and tin are oxidized to the dioxides by concentrated nitric acid, but lead gives Pb^{2+}.

Silicon is the second most abundant element in the earth's crust, where it occurs as silica (SiO_2) and *silicates*. These minerals consist of SiO_4 tetrahedra linked through a common oxygen. Silica is acidic and reacts with basic oxides, such as CaO, to form glass. Tin(II) and lead(II) oxides are amphoteric, forming the cations Sn^{2+} and Pb^{2+} in acid and the anions $Sn(OH)_3^-$ and $Pb(OH)_3^-$ in base.

Nitrogen, in Group VA, exists as the molecular substance N_2, which is normally a gas. Phosphorus has two common allotropes. White phosphorus is very reactive; it has a molecular structure with the formula P_4. Red phosphorus is less reactive. The common forms of arsenic and antimony are metallike; bismuth is definitely metallic.

Nitrogen, the first member of Group VA, forms compounds in all oxidation states from −3 to +5. Ammonia, NH_3, is the most important nitrogen compound. It is used as a nitrogen fertilizer and in the preparation of nitric acid, HNO_3, by the *Ostwald process*. In this process, NH_3 is oxidized in the presence of a catalyst to NO, then to NO_2, which reacts with water to give nitric acid.

The other group VA elements have compounds in +5 and +3 oxidation states, though +3 is the only stable state of bismuth. Phosphorus burns in air, forming P_4O_6 and P_4O_{10}. Phosphorus and its oxoacid H_3PO_4 are major industrial chemicals, and the soluble phosphates are important agricultural fertilizers. Orthophosphoric acid, H_3PO_4, undergoes condensation reactions to form the polyphosphoric and metaphosphoric acids. (In a *condensation reaction*, two molecules are joined by eliminating a small molecule, in this case H_2O.) Salts of these condensed phosphates are used in detergent formulations.

Oxygen, in Group VIA, is the most abundant element on earth, making up 48% by mass of the outer portion (crust, atmosphere, and surface waters). In the atmosphere, oxygen is present as the molecular substance O_2, from which it is separated by liquefaction. Sulfur, selenium, and tellurium, in Group VIA, are similar in their chemical properties. Sulfur is strictly nonmetallic, whereas selenium and tellurium are metalloids. Thus, sulfur is a molecular solid, S_8. Selenium has both a molecular form, Se_8, and a gray, semiconducting allotrope. Tellurium has only a metallike form. The oxidation states +6, +4, and −2 are all important in these elements, although the +6 state is more strongly oxidizing in selenium and tellurium. Free sulfur is mined by the *Frasch process* for use in manufacturing sulfuric acid by the *contact process*. In this process, SO_2 is catalytically oxidized with O_2 to SO_3, which is then dissolved in concentrated sulfuric acid. When this solution is diluted, it gives the concentrated acid.

Group VIIA elements, or halogens, have similar properties. They are reactive nonmetals with the molecular formula X_2. Fluorine and chlorine are abundant and important elements. Fluorine, F_2, is very reactive; it oxidizes water to O_2. Though less reactive, chlorine gas is nevertheless an active nonmetal. It is obtained by the electrolysis of aqueous NaCl. The gas dissolves readily in cold basic solution, disproportionating to Cl^- and OCl^-. In hot solution, the hypochlorite ion, OCl^-, disproportionates to Cl^- and ClO_3^-. The chlorate ion can be electrolytically oxidized to perchlorate ion, ClO_4^-. Thus, a series of oxoacid salts can be made readily.

The first Group VIIIA element or noble gas was discovered in 1894, when argon was isolated from air. All of the noble gases, except helium and radon, are now prepared by distillation of liquid air. Although the noble gases are relatively unreactive as a group, compounds have been prepared for the heavier elements.

OPERATIONAL SKILLS

1. **Using chemical reactions of the Group IVA elements and compounds** Devise a method of preparing a Group IVA substance from given starting substances by recalling important chemical reactions of the group (Example 22.1).

2. **Using chemical reactions of the Group VA elements and compounds** Devise a method of preparing a Group VA substance from given starting substances by recalling important chemical reactions of the group (Example 22.2).

3. **Using chemical reactions of the Group VIA elements and compounds** Devise a method of preparing a Group VIA substance from given starting substances by recalling important chemical reactions of the group (Example 22.3).

4. **Using chemical reactions of the Group VIIA elements and compounds** Devise a method of preparing a Group VIIA substance from given starting substances by recalling important chemical reactions of the group (Example 22.4).

Review Questions

22.1 Why is it that silicon forms the ion SiF_6^{2-} but carbon has no similar ion?

22.2 How is carbon black similar to graphite? How does it differ?

22.3 Which Group IVA elements have allotropes with diamondlike structures?

22.4 Why would high pressure be an expected condition for the transformation of graphite to more dense diamond?

22.5 Describe the steps in preparing ultrapure silicon from quartz sand.

22.6 What is meant by the term *catenation?* Give an example of a compound that displays catenation.

22.7 Through the use of equations, show how an ion that has three silicon atoms could form from $Si(OH)_2O_2^{2-}$ ions.

22.8 Give reactions involving the dioxides of carbon, silicon, tin, and lead that show their acid–base behavior.

22.9 Write equations for the reactions of silicon, tin, and lead with (a) Br_2, (b) O_2, (c) $HCl(aq)$, and (d) $HNO_3(aq)$. Write NR if no reaction occurs.

22.10 Which of the following ions is the better reducing agent: $Pb(OH)_3^-$ or $Sn(OH)_3^-$? Explain why this is expected.

22.11 What is the Haber process? Why is it important?

22.12 List the different nitrogen oxides. What is the oxidation number of nitrogen in each?

22.13 Describe the steps in the Ostwald process for making nitric acid from ammonia.

22.14 Describe the structure of white phosphorus. How does the structure account for its chemical reactivity?

22.15 What are the products when each of the Group VA elements—phosphorus, arsenic, antimony, and bismuth—reacts with O_2?

22.16 Write the equations for the production of antimony from stibnite, Sb_4S_6.

22.17 What acids correspond to the following anhydrides: P_4O_6, P_4O_{10}, As_4O_6, As_2O_5?

22.18 Hypophosphorus acid, H_3PO_2, has the following structure:

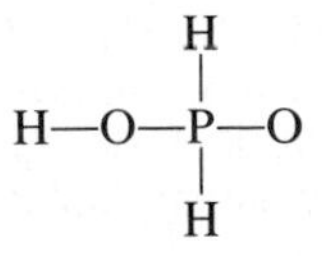

Do you expect this acid to be monoprotic, diprotic, or triprotic? Explain.

22.19 Describe two different methods used to manufacture phosphoric acid, H_3PO_4, starting from $Ca_3(PO_4)_2$.

22.20 By means of an equation, show how triphosphoric acid could be formed from phosphoric acid and diphosphoric acid.

22.21 What is the purpose of adding polyphosphates to a detergent?

22.22 What reaction was used by Priestley in preparing pure oxygen?

22.23 What is the most important commercial means of producing oxygen?

22.24 Describe the molecular-orbital configurations of O_2, O_2^-, and O_2^{2-}. State which species give paramagnetic substances and which give diamagnetic substances. What is the bond order of each?

22.25 What is the difference between an oxide, a peroxide, and a superoxide? Give an example of each kind of compound.

22.26 List three natural sources of sulfur or sulfur compounds.

22.27 What is the structure of the stable form of sulfur?

22.28 Describe the changes that occur as sulfur melts and the temperature of the liquid rises. Explain what happens.

22.29 Sulfur hexafluoride is used as a gaseous insulator in electrical transformers. How could you prepare this compound?

22.30 Describe the Frasch process for mining sulfur.

22.31 Each of the following substances often reacts as an oxidizing or reducing agent. Fill in the table, noting whether the substance listed usually acts as an oxidizing or reducing agent. Then give the usual product formed.

SUBSTANCE	OXIDIZING OR REDUCING AGENT?	USUAL PRODUCT
H_2S		
$Na_2SO_3(aq)$		
Hot, conc. H_2SO_4		
$Na_2S_2O_3(aq)$		

22.32 Give equations for two different methods of preparing each of the following: (a) H_2S, (b) SO_2.

22.33 Give the equations for the steps in the contact process for manufacturing sulfuric acid from sulfur.

22.34 Pyrosulfuric acid molecules are in equilibrium with H_2SO_4 in concentrated sulfuric acid. The reaction is a condensation, similar to that in which pyrophosphoric (diphosphoric) acid is formed from phosphoric acid. Write the equation for the equilibrium between H_2SO_4 and $H_2S_2O_7$, using structural formulas for the species involved.

22.35 Describe the preparation of sodium thiosulfate.

22.36 Thiosulfuric acid cannot be prepared. Explain.

22.37 Explain how sodium thiosulfate is used in photography. What would happen to the negative if it were not fixed?

22.38 Fill in the following table, giving the formula of the acid corresponding to each anhydride listed. Then state whether the acid is strong or weak. If the corresponding acid is unknown, give this information in the "Acid" column.

ANHYDRIDE	ACID	STRONG OR WEAK?
SO_2		
SeO_2		
TeO_2		
SO_3		
SeO_3		
TeO_3		

22.39 What is the difference in behavior of F_2 and Cl_2 with water? Give equations for the reactions.

22.40 Complete and balance the following equations. Write NR if no reaction occurs.

(a) $I_2(aq) + Cl^-(aq) \longrightarrow$
(b) $Cl_2(aq) + Br^-(aq) \longrightarrow$
(c) $Br_2(aq) + I^-(aq) \longrightarrow$
(d) $Br_2(aq) + Cl^-(aq) \longrightarrow$

22.41 A test tube contains a solution of one of the following salts: NaCl, NaBr, NaI. Describe a single test that can distinguish among these possibilities.

22.42 What is an interhalogen? Give an example.

22.43 Give a natural source for each of the halogens (except astatine). Describe how the element is obtained from that source.

22.44 Hydrogen fluoride has a boiling point near room temperature, but hydrogen chloride boils at −85°C. Explain why HF has the higher boiling point, though we might have expected otherwise on the basis of the molecular weights of HF and HCl.

22.45 Hydrogen chloride can be prepared by heating NaCl with concentrated sulfuric acid. Why is substituting NaBr for sodium chloride not a satisfactory way to prepare HBr?

22.46 Phosphate rock is primarily fluorapatite, a calcium fluoride phosphate mineral. In the preparation of phosphate fertilizer, hydrogen fluoride is produced, and at one time it was vented into the surrounding atmosphere. Today it is converted into compounds such as hexafluorosilicic acid. Write the equation for the reaction of HF(aq) with silica (SiO_2).

22.47 What is the standard potential for the reduction of ClO_3^- to $HClO_2$? Write the half-reaction.

22.48 How is sodium hypochlorite prepared? Give the balanced equation.

22.49 Do you expect an aqueous solution of sodium hypochlorite to be acidic, neutral, or basic? What about an aqueous solution of sodium perchlorate?

22.50 What was the argument used by Bartlett that led him to the first synthesis of a noble-gas compound?

Practice Problems

THE GROUP IVA ELEMENTS

22.51 Formic acid, HCOOH, is used to produce certain artificial flavorings. The acid is produced from sodium formate. By means of balanced equations, show how sodium formate can be made in two steps, starting from methane, CH_4.

22.52 One of the most important salts of carbonic acid is sodium carbonate, Na_2CO_3. Use balanced equations to show how sodium carbonate could be made in two steps, starting with carbon.

22.53 Show how you could prepare a solution of tin(II) chloride from tin(IV) oxide. Give balanced equations.

22.54 Write balanced equations for the preparation of $Pb(OH)_2(s)$ from lead metal.

22.55 A test for bismuth ion, Bi^{3+}, consists of precipitating it as $Bi(OH)_3$ (white) and then reducing this with stannite ion to finely divided bismuth metal (black). Write the balanced equation for the reduction of $Bi(OH)_3$ to Bi by sodium stannite solution.

22.56 Lead(IV) oxide will oxidize hydrochloric acid to chlorine, Cl_2. Write the balanced equation for this reaction.

22.57 The $Sn^{2+}(aq)$ ion can be written in more detail as $Sn(H_2O)_6{}^{2+}$. This ion is acidic by hydrolysis. Write a possible equation for this hydrolysis.

22.58 Lead(II) nitrate, one of the few soluble lead salts, gives a solution with a pH of about 3 to 4. Write an equation for a possible reaction to explain why the solution is not neutral.

22.59 Solid tin(II) chloride dihydrate, $SnCl_2 \cdot 2H_2O$, consists of molecules of $SnCl_2(H_2O)$, in which H_2O is directly attached to the tin atom by a coordinate bond from oxygen. The other H_2O molecule is not bonded to tin. What is the expected geometry of the $SnCl_2(H_2O)$ molecule?

22.60 Ethylene dichloride, $C_2H_4Cl_2$, and ethylene dibromide, $C_2H_4Br_2$, are added to leaded gasolines to produce volatile compounds, such as PbBrCl, in the engine (to keep lead from depositing). What is the expected geometry of the PbBrCl molecule?

THE GROUP VA ELEMENTS

22.61 You have the following substances: NH_3, O_2, Pt, and P_4O_{10} (plus H_2O). Write equations for the steps in the preparation of N_2O from these substances.

22.62 Give equations for the preparation of N_2O. You can use NaOH, $NaNO_3$, H_2SO_4, and $(NH_4)_2SO_4$ (plus H_2O). Several steps may be required.

22.63 Give the formulas of three compounds for each of the following oxidation states of phosphorus: (a) +3, (b) +5.

22.64 Give the formulas of three compounds for each of the following oxidation states of arsenic: (a) +3, (b) +5.

22.65 Using balanced equations, describe how arsenic(V) oxide can be prepared from arsenic(III) oxide.

22.66 Starting from bismuth metal, show how to prepare $Bi(OH)_3$. Show the steps with balanced equations.

22.67 Phosphorous acid is oxidized to phosphoric acid by hot, concentrated sulfuric acid, which is reduced to SO_2. Write the balanced equation for this reaction.

22.68 Phosphorous acid is oxidized to phosphoric acid by nitric acid, which is reduced to NO. Write a balanced equation for this reaction.

22.69 Arsenate ion is reduced to arsenite ion in aqueous solution by hydrazinium chloride, $(N_2H_5)Cl$. Note that aqueous solutions of hydrazinium chloride are acidic (why?). Hydrazinium chloride is oxidized to N_2. Write the balanced equation for this reaction.

22.70 Arsenic(III) sulfide is oxidized to arsenate ion by hydrogen peroxide in basic solution. Write the balanced equation for this reaction.

22.71 Although phosphorus pentabromide exists as PBr_5 molecules in the vapor, the solid is ionic, with the structure $[PBr_4{}^+]Br^-$. What is the expected geometry of $PBr_4{}^+$? Describe the bonding to phosphorus.

22.72 Antimony(V) oxide dissolves in strong base to give the $Sb(OH)_6{}^-$ ion. What is the expected geometry of this ion? Describe the bonding to antimony in this ion.

THE GROUP VIA ELEMENTS

22.73 Write an equation for each of the following:
(a) burning of lithium in oxygen
(b) burning of methylamine, CH_3NH_2, in excess oxygen (assume the nitrogen in methylamine ends up as N_2)
(c) the preparation of phosphoric acid, H_3PO_4, from phosphorus, P_4, in two steps

22.74 Write an equation for each of the following:
(a) burning of calcium metal in air
(b) burning of phosphine, PH_3, in excess oxygen
(c) preparation of calcium carbonate, $CaCO_3$, from carbon, C, and an aqueous solution of calcium hydroxide, $Ca(OH)_2$, in two steps

22.75 With equations, show how sulfur, S_8, could be prepared from only $H_2S(g)$ and air.

22.76 With equations, describe how selenic acid could be prepared from selenium, air, H_2O_2, and water.

22.77 Selenium dioxide can be detected in qualitative analysis by converting it to selenious acid and reducing this with H_2S to selenium. A yellow precipitate of sulfur and selenium forms. Write the balanced equation for the reaction of selenious acid and H_2S.

22.78 Concentrated sulfuric acid oxidizes iodide ion to iodine, I_2. Write the balanced equation for this reaction.

22.79 Describe the bonding in each of the following: (a) H_2Se, (b) SeF_4. What are the expected geometries?

22.80 Describe the bonding in each of the following: (a) SO_3^{2-}, (b) SeF_6. What are the expected geometries?

22.81 What are the oxidation numbers of sulfur in each of the following: (a) SF_6, (b) SO_3, (c) H_2S, (d) $CaSO_3$?

22.82 What are the oxidation numbers of sulfur in each of the following: (a) S_8, (b) CaS, (c) $CaSO_4$, (d) SCl_4?

THE GROUP VIIA ELEMENTS

22.83 When silica or glass reacts with hydrogen fluoride gas, the product is silicon tetrafluoride, $SiF_4(g)$. Write the balanced equation for the reaction of silica with $HF(g)$ to form SiF_4 and H_2O.

22.84 A solution of chloric acid may be prepared by reacting a solution of barium chlorate with sulfuric acid. Barium sulfate precipitates. Write the balanced equation for the reaction.

22.85 Uranium hexafluoride is a volatile solid and is used in the gaseous-diffusion method of separating uranium isotopes. With balanced equations, show how this compound can be prepared from uranium metal, calcium fluoride, sulfuric acid, and potassium fluoride.

22.86 Silver iodide is a light-sensitive compound used in photographic film. With balanced equations, show how this compound could be prepared, starting from a solution of sodium iodate, silver nitrate, sodium hydrogen sulfite, and hydrogen.

22.87 Chlorine may be prepared in the laboratory by oxidizing chloride ion with a sufficiently strong oxidizing agent. In one method, hydrochloric acid is heated with potassium dichromate, $K_2Cr_2O_7$. Write the balanced equation for the reaction. Dichromate ion is reduced to Cr^{3+}.

22.88 Iodic acid can be prepared by oxidizing elemental iodine with concentrated nitric acid, which is reduced to $NO_2(g)$. Write the balanced equation for this reaction.

22.89 Iodine is prepared from sodium iodate with sodium hydrogen sulfite. Write the balanced equation for the reaction.

22.90 When solid potassium chlorate is carefully heated, it disproportionates to give potassium chloride and potassium perchlorate. Write the balanced equation for this reaction.

22.91 By calculating the standard emf, decide whether sodium hypochlorite will oxidize $Fe^{2+}(aq)$ to $Fe^{3+}(aq)$ in acidic solution under standard conditions. See Figure 22.24 and Table 19.1 for data.

22.92 Using the data from Figure 22.24, calculate E°_{cell} for the disproportionation of chlorous acid to hypochlorous acid and chlorate ion. Is the reaction spontaneous under standard conditions?

22.93 Discuss the bonding in the following molecules or ions. What is the expected geometry? (a) Cl_2O; (b) BrO_3^-; (c) BrF_3.

22.94 Discuss the bonding in the following molecules or ions. What is the expected geometry? (a) HFO; (b) $SiCl_4$; (c) SiF_6^{2-}.

NOBLE GASES

22.95 Xenon tetrafluoride, XeF_4, is a colorless solid. Give the Lewis formula for the XeF_4 molecule. What is the hybridization of the xenon atom in this compound? What geometry is predicted by the VSEPR model for this molecule?

22.96 Xenon tetroxide, XeO_4, is a colorless, unstable gas. Give the Lewis formula for the XeO_4 molecule. What is the hybridization of the xenon atom in this compound? What geometry would you expect for this molecule?

22.97 Xenon difluoride, XeF_2, is hydrolyzed (broken up by water) in basic aqueous solution to give xenon, fluoride ion, and molecular oxygen as products. Write a balanced chemical equation for the reaction.

22.98 Xenon trioxide, XeO_3, is reduced to xenon in acidic solution by iodide ion. Iodide ion is oxidized to iodine, I_2. Write a balanced equation for the reaction.

Additional Problems

22.99 Identify each of the following substances from the description.

(a) a yellow solid that burns with a blue flame, giving off a gas with a choking odor
(b) a white solid that reacts with water to give phosphoric acid
(c) a reddish-brown liquid that reacts vigorously with sodium metal to give a white solid
(d) a metal that crumbles to a powder when exposed to temperatures below 13°C for a period of time

22.100 Identify each of the following substances from the description.

(a) a white, waxy solid, normally stored under water because it spontaneously bursts into flames when exposed to air
(b) a viscous liquid that reacts with table sugar, giving a charred mass
(c) an acid that etches glass
(d) a pale green gas that dissolves in aqueous sodium hydroxide to give a solution used as a bleach

22.101 You are given three unlabeled test tubes. One test tube contains a solution of sodium sulfate, one contains a solution of sodium hydrogen sulfate, and the third contains a solution of sodium hydrogen sulfite. Describe how you could identify the solutions.

22.102 You are given three unlabeled test tubes. One contains a solution of sodium fluoride, the second contains a solution of sodium chloride, and the third contains a solution of sodium iodide. Describe how you could determine the identity of each solution.

22.103 Arsenious acid is a very weak acid.

$$H_3AsO_3(aq) \rightleftharpoons H^+(aq) + H_2AsO_3^-(aq); \quad K_a = 6 \times 10^{-10}$$

What is the pH of 0.050 M sodium dihydrogen arsenite, NaH_2AsO_3?

22.104 Calculate the pH of a 0.015 M solution of arsenic acid. K_a for the ionization of the first hydrogen is 6.0×10^{-3}. Neglect any further ionizations of the acid.

22.105 Chlorine gas can be prepared by adding dilute $HCl(aq)$ dropwise onto potassium permanganate crystals, $KMnO_4$. The $KMnO_4$ is reduced to $Mn^{2+}(aq)$. What volume (in liters) of 1.50 M $HCl(aq)$ is required to react with 12.0 g $KMnO_4$?

22.106 Iodic acid can be prepared by oxidizing iodine with concentrated nitric acid. What volume (in liters) of 15.8 M HNO_3 is required to produce 15.0 g of iodic acid? Assume that the nitric acid is reduced to NO_2.

22.107 The main ingredient in many phosphate fertilizers is calcium dihydrogen phosphate monohydrate, $Ca(H_2PO_4)_2 \cdot H_2O$. What is the mass percentage of phosphorus in this salt? If a fertilizer is 15.5 mass percent P, and all of this phosphorus is present in the fertilizer as $Ca(H_2PO_4)_2 \cdot H_2O$, what is the mass percentage of this salt in the fertilizer?

22.108 A fertilizer contains phosphorus in two compounds, $Ca(H_2PO_4)_2 \cdot H_2O$ and $CaHPO_4$. The fertilizer contains 30.0 mass percent $Ca(H_2PO_4)_2 \cdot H_2O$ and 10.0 mass percent $CaHPO_4$. What is the mass percentage of phosphorus in the fertilizer?

22.109 Sodium hypochlorite solution is produced by the electrolysis of cold sodium chloride solution. The electrolysis is carried out so as to thoroughly mix the products (NaOH and Cl_2). How long must a cell operate to produce 1.00×10^3 L of 5.25% solution of NaClO if the cell current is 2.50×10^3 A? Assume that the density of the solution is 1.00 g/mL.

22.110 Sodium perchlorate is produced by electrolysis of sodium chlorate. If a current of 2.50×10^3 A passes through an electrolytic cell, how many kilograms of sodium perchlorate are produced per hour?

22.111 The amount of sodium hypochlorite in a bleach solution can be determined by using a given volume of bleach to oxidize excess iodide ion to iodine, because the reaction goes to completion. The amount of iodine produced is then determined by titration with sodium thiosulfate, which is oxidized to sodium tetrathionate, $Na_2S_4O_6$. Potassium iodide was added in excess to 5.00 mL of bleach (density = 1.00 g/mL). This solution, containing the iodine released in the reaction, was titrated with 0.100 *M* $Na_2S_2O_3$. If 34.6 mL of sodium thiosulfate was required to reach the end point (detected by disappearance of the blue color of starch–iodine complex), what was the mass percent of NaClO in the bleach?

22.112 Ascorbic acid (vitamin C), $C_6H_8O_6$, is a reducing agent. It can be determined quantitatively by a titration procedure involving iodine, I_2.

$$\underset{\text{ascorbic acid}}{C_6H_8O_6} + I_2 \longrightarrow \underset{\text{dehydroascorbic acid}}{C_6H_6O_6} + 2[H^+ + I^-]$$

A 30.0-g sample of an orange-flavored beverage mix was placed in a flask to which 10.00 mL of 0.0500 *M* KIO_3 and excess KI were added. The IO_3^- and I^- ions react in acid solution to give I_2, which then reacts with ascorbic acid. Excess iodine is titrated with sodium thiosulfate (see Problem 22.111). If 29.5 mL of 0.0300 *M* $Na_2S_2O_3$ is required to titrate the excess I_2, how many grams of ascorbic acid are there in 100.0 g of beverage mix?

22.113 Consider the following standard potentials:

$$Sn(OH)_3^-(aq) + 2e^- \rightleftharpoons Sn(s) + 3OH^-(aq); \quad E° = -0.79 \text{ V}$$

$$Sn(OH)_6^{2-}(aq) + 2e^- \rightleftharpoons Sn(OH)_3^-(aq) + 3OH^-(aq); \quad E° = -0.96 \text{ V}$$

Would you expect stannite ion to disproportionate in basic solution to give metallic tin and stannate ion? Explain.

22.114 Consider the following standard potentials:

$$H_3PO_3(aq) + 2H^+(aq) + 2e^- \rightleftharpoons H_3PO_2(aq) + H_2O(l); \quad E° = -0.50 \text{ V}$$

$$H_3PO_4(aq) + 2H^+(aq) + 2e^- \rightleftharpoons H_3PO_3(aq) + H_2O(l); \quad E° = -0.28 \text{ V}$$

Would you expect phosphorous acid to disproportionate into phosphoric acid and hypophosphorous acid, H_3PO_2? Explain.

22.115 Complete the following standard-potential diagram by calculating the standard potential for the change indicated by the question mark.

$$H_3PO_4 \xrightarrow{-0.28} H_3PO_3 \xrightarrow{-0.50} H_3PO_2 \xrightarrow{-0.51} P_4 \xrightarrow{-0.04} PH_3$$

(upper bracket: $H_3PO_2 \rightarrow PH_3$, ?; lower bracket: $H_3PO_3 \rightarrow P_4$, −0.50)

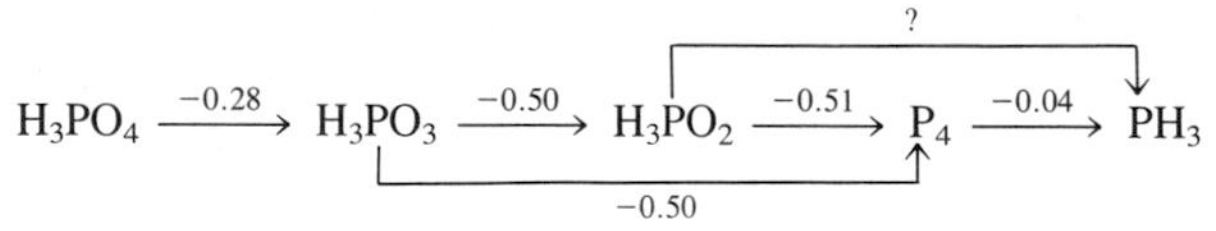

Hint: The standard potentials are related to the standard free-energy changes for the reaction involving the indicated reduction and the oxidation of H_2 to H^+. Also, the free-energy changes for reactions can be added, as in Hess's law, to obtain other values.

22.116 Complete the following standard-potential diagram by calculating the standard potential for the change indicated by the question mark. (See Problem 22.115.)

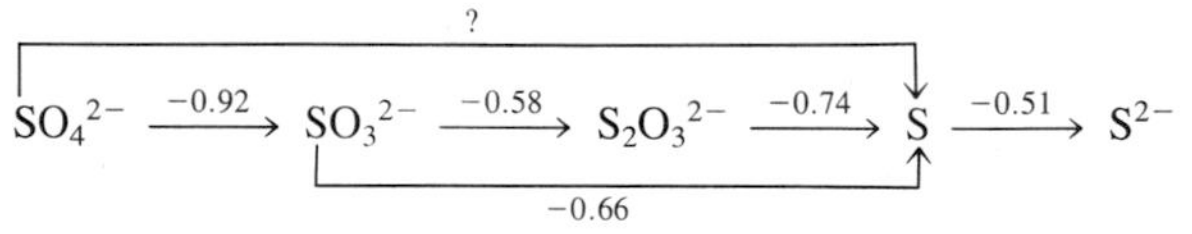

$$SO_4^{2-} \xrightarrow{-0.92} SO_3^{2-} \xrightarrow{-0.58} S_2O_3^{2-} \xrightarrow{-0.74} S \xrightarrow{-0.51} S^{2-}$$

(upper bracket: $SO_4^{2-} \rightarrow S$, ?; lower bracket: $SO_3^{2-} \rightarrow S$, −0.66)

23

The Transition Elements

Chapter Outline

Copper tubing.

In the previous chapters, we studied the main-group elements—the A groups in the periodic table. Between columns IIA and IIIA are ten columns of the transition elements (the B groups). Among these elements are metals with familiar commercial applications: iron tools, copper wire, silver jewelry and coins. Many catalysts for important industrial reactions involve transition elements. Examples are found in petroleum refining and in the synthesis of ammonia from N_2 and H_2.

In addition to their commercial usefulness, many transition elements have biological importance. Iron compounds, for example, are found throughout the plant and animal kingdoms. Iron is present in hemoglobin, the molecule in red blood cells that is responsible for the transport of oxygen, O_2, from the lungs to other body tissue. Myoglobin, in muscle, is a very similar molecule containing iron. It takes oxygen from hemoglobin, holding it until it is required by the muscle cells. Cytochromes are iron-containing compounds within each cell and are involved in the oxidation of food molecules. In these cases, the transition element is central to the structure and function of the biological molecule. Hemoglobin and myoglobin are examples of *metal complexes* or *coordination compounds,* in which the metal atom is surrounded by other atoms bonded to it by the electron pairs these atoms donate. In hemoglobin and myoglobin, the O_2 molecule bonds to the iron atom.

Properties of the Transition Elements

The transition elements are strictly defined as those elements having a partially filled *d* or *f* subshell in any common oxidation state (including the 0 oxidation state). For example, copper, which has the configuration [Ar]$3d^{10}4s^1$ in the 0 oxidation state, has the configuration [Ar]$3d^9$ in the common +2 oxidation state. Thus, copper(II) has a partially filled *d* subshell, so copper is a transition element. Figure 23.1 shows the various divisions of the transition elements.

The ***d*-block transition elements** are *those transition elements with an unfilled* d *subshell in common oxidation states.* These elements are frequently referred to simply as "the transition elements." Although zinc, cadmium, and mercury, in Group IIB, are not *d*-block transition elements by the strict definition, they are often included with them because of their similar properties. We will accept this broader definition and concentrate on these elements in the center of the periodic table.

The elements with a partially filled f *subshell in common oxidation states* are known as ***f*-block transition elements** or **inner transition elements.** The *f*-block transition elements are the two rows of elements at the bottom of the periodic table. The elements in the first row are called the *lanthanides* or *rare earths.*● The elements in the second row are called *actinides.* All of the actinides are radioactive and, from neptunium (Element 93) on, are synthetic. They have been produced in nuclear reactors or by using particle accelerators.

● The lowest-energy configuration of the lutetium atom is [Xe]$4f^{14}5d^16s^2$, and in its only common oxidation state (+3) the configuration is [Xe]$4f^{14}$. Although both configurations have a filled *f* subshell, lutetium is usually considered a lanthanide element.

23.1 PERIODIC TRENDS IN THE TRANSITION ELEMENTS

The transition elements have a number of characteristics that set them apart from the main-group elements.

Figure 23.1
Classification of the transition elements. The classification into *d*-block or *f*-block element depends on which subshell is filling.

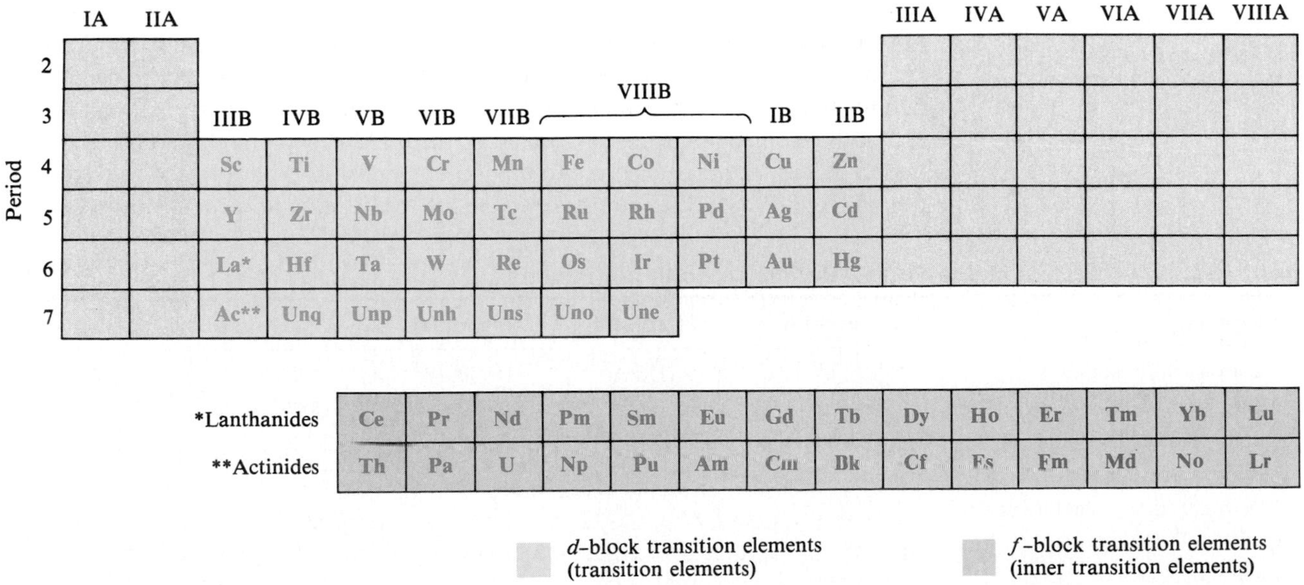

Period	IA	IIA	IIIB	IVB	VB	VIB	VIIB	VIIIB			IB	IIB	IIIA	IVA	VA	VIA	VIIA	VIIIA
2																		
3																		
4			Sc	Ti	V	Cr	Mn	Fe	Co	Ni	Cu	Zn						
5			Y	Zr	Nb	Mo	Tc	Ru	Rh	Pd	Ag	Cd						
6			La*	Hf	Ta	W	Re	Os	Ir	Pt	Au	Hg						
7			Ac**	Unq	Unp	Unh	Uns	Uno	Une									

*Lanthanides	Ce	Pr	Nd	Pm	Sm	Eu	Gd	Tb	Dy	Ho	Er	Tm	Yb	Lu
**Actinides	Th	Pa	U	Np	Pu	Am	Cm	Bk	Cf	Es	Fm	Md	No	Lr

d-block transition elements (transition elements)

f-block transition elements (inner transition elements)

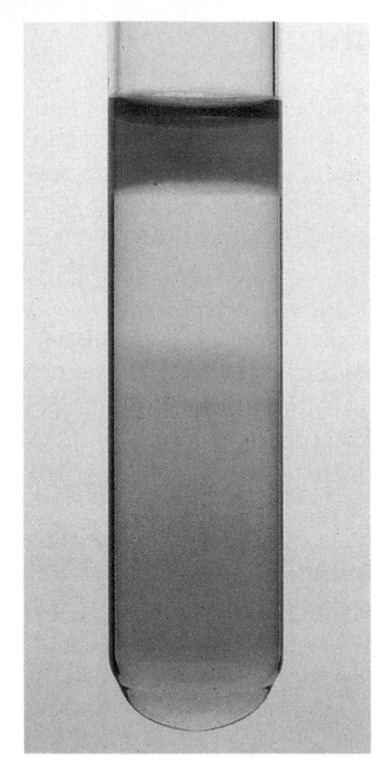

Figure 23.2
Oxidation states of vanadium. Potassium permanganate solution is poured to form a layer over a vanadium(II) ion solution. In a few hours the vanadium solution develops layers, each containing a different vanadium ion species. The ion species are (*from bottom to top*): V^{2+} (violet), V^{3+} (green), VO^{2+} (blue), and VO_2^+ (pale yellow). (The brown and pink layers at the top contain MnO_2 and MnO_4^-, respectively.)

1. All of the transition elements are metals and, except for the IIB elements, have high melting points and high boiling points, and are hard solids. Thus, in the fourth-period elements from scandium to copper, the lowest-melting metal is copper (1083°C) and the highest-melting metal is vanadium (1890°C). In the main-group metals, only beryllium melts above 1000°C; the rest melt at appreciably lower temperatures.
2. With the exception of the IIIB and IIB elements, each transition element has several oxidation states. Vanadium, for example, exists as aqueous ions in all oxidation states from +2 to +5 (Figure 23.2). Of the main-group metals, only the heavier ones display several oxidation states. Because of their multiplicity of oxidation states, the transition elements are often involved in oxidation–reduction reactions.
3. Transition-metal compounds are often colored, and many are paramagnetic. Most of the compounds of the main-group metals are colorless and diamagnetic.

We will look at points 1 and 2 in some detail in this section. In particular, we will examine trends in melting points, boiling points, hardness, and oxidation states. We will also look at trends in covalent radii and ionization energies, which we can relate to chemical properties. We will discuss the color and paramagnetism of transition-metal complexes later in the chapter.

Electron Configurations

Electronic structure is central to any discussion of the transition elements. Table 23.1 lists the electron configurations of the fourth-period transition elements. For the most part, these configurations are those predicted by the building-up principle. Following the building-up principle, the $3d$ subshell begins to fill after calcium (configuration $[Ar]4s^2$). Thus, scandium has the configuration $[Ar]3d^14s^2$, and, as we go across the period, additional electrons go into the $3d$ subshell. We get the configuration $[Ar]3d^24s^2$ for titanium, and $[Ar]3d^34s^2$ for vanadium. Then in chromium the configuration predicted by the building-up principle is $[Ar]3d^44s^2$, but the actual configuration is $[Ar]3d^54s^1$. This is usually explained as due to the special stability of a half-filled d subshell. The configurations for the rest of the elements are those predicted by the building-up principle, until we get to copper. Here, the predicted configuration is $[Ar]3d^94s^2$, but the actual configuration is $[Ar]3d^{10}4s^1$. This is explained as due to the stability of a filled d subshell.

Table 23.1 Properties of the Fourth-Period Transition Elements

Property	Scandium	Titanium	Vanadium	Chromium	Manganese
Electron configuration	$[Ar]3d^14s^2$	$[Ar]3d^24s^2$	$[Ar]3d^34s^2$	$[Ar]3d^54s^1$	$[Ar]3d^54s^2$
Melting point, °C	1541	1660	1890	1857	1244
Boiling point, °C	2831	3287	3380	2672	1962
Density, g/cm^3	3.0	4.5	6.0	7.2	7.2
Electronegativity (Pauling scale)	1.3	1.5	1.6	1.6	1.5
Covalent radius, Å	1.44	1.32	1.22	1.18	1.17
Ionic radius (for M^{2+}), Å	—	1.00	0.93	0.87	0.81

Melting Points and Boiling Points

Table 23.1 reveals that the melting points of the transition metals increase from 1541°C for scandium to 1890°C for vanadium and 1857°C for chromium, then decrease to 1083°C for copper and 420°C for zinc. The same pattern is observed in the fifth-period and sixth-period elements. As we read across a row of transition elements, the melting points increase, reaching a maximum at the Group VB or VIB elements, after which the melting points decrease. Thus, tungsten in Group VIB has the highest melting point (3410°C) of any metallic element, and mercury in Group IIB has the lowest (−39°C). Similar trends can be seen in the boiling points of the metals. These properties depend on the strengths of metal bonding, which in turn depend roughly on the number of unpaired electrons in the metal atoms. At the beginning of a period of transition elements, there is one unpaired *d* electron. The number of unpaired *d* electrons increases across a period until Group VIB, after which the electrons begin to pair.

Atomic Radii

Trends in atomic radii are of concern because chemical properties are determined in part by atomic size. Looking at the covalent radii (one measure of atomic size) in Table 23.1, we see that they decrease quickly from scandium (1.44 Å) to titanium (1.32 Å) and vanadium (1.22 Å). This decrease in atomic size across a row is the behavior we see in the main-group elements. It is due to an increase in *effective nuclear charge* that acts on the outer electrons and pulls them in more strongly. The effective nuclear charge is the positive charge "felt" by an electron; it equals the nuclear charge minus the shielding or screening of the positive charge by intervening electrons. After vanadium, the covalent radii decrease slowly from 1.18 Å for chromium to 1.15 Å for nickel. Then the covalent radii increase slightly to 1.17 Å for copper and 1.25 Å for zinc.● The relative constancy in covalent radii for the later elements is partly responsible for the similarity in properties of the Group VIIIB elements iron, cobalt, and nickel.

● The small increase in covalent radius for copper and zinc has no simple explanation.

Figure 23.3 compares the covalent radii of the transition elements. The atomic radii increase in going from a fourth-period to a fifth-period element within any column. For example, reading down the Group IIIB elements, we find that the covalent radius of scandium is 1.44 Å and the radius of yttrium is 1.62 Å. We expect an increase of radius from scandium to yttrium because of the addition of a shell of electrons. Continuing down the column of IIIB elements, we find a small

Table 23.1 (continued)

Property	Iron	Cobalt	Nickel	Copper	Zinc
Electron configuration	$[Ar]3d^6 4s^2$	$[Ar]3d^7 4s^2$	$[Ar]3d^8 4s^2$	$[Ar]3d^{10} 4s^1$	$[Ar]3d^{10} 4s^2$
Melting point, °C	1535	1495	1453	1083	420
Boiling point, °C	2750	2870	2732	2567	907
Density, g/cm^3	7.9	8.9	8.9	8.9	7.1
Electronegativity (Pauling scale)	1.8	1.8	1.8	1.9	1.6
Covalent radius, Å	1.17	1.16	1.15	1.17	1.25
Ionic radius (for M^{2+}), Å	0.75	0.79	0.83	0.87	0.88

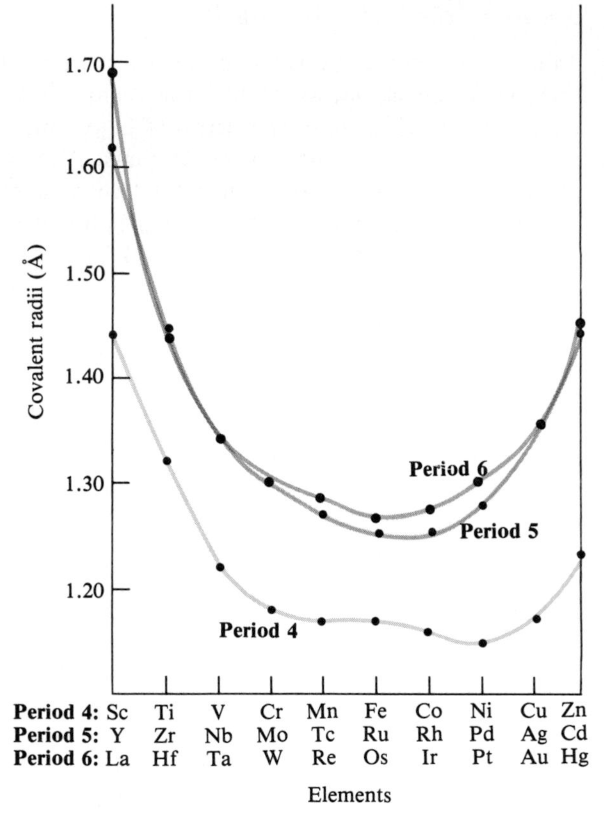

Figure 23.3 Comparison of the covalent radii for transition elements of different periods. Note that elements in the same column of the fifth and sixth periods have approximately the same radii.

increase of covalent radius to 1.69 Å in lanthanum. But all of the remaining elements of the sixth period have nearly the same covalent radii as the corresponding elements in the fifth period.

This similarity of radii in the fifth- and sixth-period transition elements can be explained in terms of the *lanthanide contraction*. Between lanthanum and hafnium in the sixth period are 14 lanthanide elements (cerium to lutetium), in which the $4f$ subshell fills. The covalent radius decreases slowly from cerium (1.65 Å) to lutetium (1.56 Å), but the total decrease is substantial. By the time the $4f$ subshell is complete, the covalent radii of the transition elements from hafnium on are similar to those of the elements in the preceding row of the periodic table. Therefore, the covalent radius of hafnium (1.46 Å) is approximately the same as that of zirconium (1.45 Å).

Table 23.2 First Ionization Energies of the Transition Elements (in kJ/mol)

	IIIB	IVB	VB	VIB	VIIB	VIIIB			IB	IIB
Fourth period	Sc 631	Ti 658	V 650	Cr 652	Mn 717	Fe 759	Co 758	Ni 737	Cu 745	Zn 906
Fifth period	Y 616	Zr 660	Nb 664	Mo 685	Tc 702	Ru 711	Rh 720	Pd 805	Ag 731	Cd 868
Sixth period	La 538	Hf 680	Ta 761	W 770	Re 760	Os 840	Ir 880	Pt 870	Au 890	Hg 1007

The chemical properties of the transition elements parallel the pattern seen in covalent radii. That is, the fourth-period elements have substantially different properties from the elements in the same group in the fifth and sixth periods. However, elements in the same group of the fifth period and the sixth period are very much alike. Hafnium, for example, is so much like zirconium that it remained undiscovered until it was identified in 1923 in zirconium ores from its x-ray spectrum.● Zirconium (mixed with hafnium as an impurity) was discovered over a hundred years earlier, in 1787.

● The atomic number of an atom can be obtained from its x-ray spectrum, which therefore gives unequivocal identification of an element. See the Instrumental Methods at the end of Section 8.2.

Ionization Energies

Looking at the first ionization energies of the fourth-period elements (Table 23.2), we see that although they vary somewhat irregularly, they tend to increase from left to right. The other rows of transition elements behave similarly. Most noteworthy, however, is that all of the sixth-period elements after lanthanum have ionization energies higher than the fourth-period and fifth-period transition elements. This behavior is opposite to what we found in the main-group elements, where the ionization energies decrease regularly down a column. The high ionization energies of the sixth-period elements from osmium to mercury are no doubt one determinant of the relative unreactivity of these elements.

Oxidation States

Table 23.3 gives oxidation states in compounds of the fourth-period transition elements. Scandium occurs as the 3+ ion, and zinc occurs as the 2+ ion. The other elements, however, exhibit several oxidation states. This multiplicity of oxidation states is due to the varying involvement of the *d* electrons in bonding.

Most of the transition elements have a doubly filled *ns* orbital. Because the *ns* electrons ionize before the $(n - 1)d$ electrons, we might expect the +2 oxidation state to be common. This oxidation state is in fact seen in all of the fourth-period elements except scandium, where the +3 ion with an Ar configuration is especially stable.

The maximum oxidation state possible equals the number of *s* and *d* electrons in the valence shell (which equals the group number for the elements up to iron).

Table 23.3 Oxidation States of the Fourth-Period Transition Elements

IIIB	IVB	VB	VIB	VIIB	VIIIB			IB	IIB
Sc	Ti	V	Cr	Mn	Fe	Co	Ni	Cu	Zn
							+1	**+1**	
	+2	+2	**+2**	**+2**	**+2**	**+2**	**+2**	**+2**	**+2**
+3	+3	**+3**	**+3**	+3	**+3**	**+3**	+3	+3	
	+4	**+4**	+4	**+4**	+4	+4	+4		
		+5	+5	+5	+5				
			+6	+6	+6				
				+7					

Key: Common oxidation states are in boldface. Additional oxidation states, particularly zero and negative values, may be observed in complexes with CO and with organic compounds.

● The maximum oxidation state is generally found in compounds of the transition element with very electronegative elements, such as F and O (in oxides and oxoanions).

● Presumably, Fe(VIII) would oxidize any element to which it would bond.

● As one might expect, osmium tetroxide is a strong oxidizing agent and can be reduced to lower oxides or to the metal, both of which are black. It is used to stain biological specimens for microscopic examination; the specimen reduces OsO_4 to give a black stain. The substance is hazardous, because its vapors can similarly stain the eyeball.

Thus, titanium in Group IVB has a maximum oxidation state of +4. Similarly, vanadium (Group VB), chromium (Group VIB), and manganese (Group VIIB) exhibit maximum oxidation states of +5, +6, and +7, respectively.●

The total number of oxidation states actually observed for the elements increases from scandium (+3) to manganese (all states from +2 to +7). From iron on, however, the maximum oxidation state is not attained, so the number of observed states decreases. The highest oxidation state seen in iron is +6.● Thereafter, the highest observed oxidation number decreases until, for zinc compounds, we find only the +2 oxidation state. The Zn^{2+} ion has a stable $[Ar]3d^{10}$ configuration.

In the fifth and sixth periods, there is also a multiplicity of oxidation states, particularly in the middle of the periods. The higher oxidation states of these elements tend to be more stable than those states of corresponding fourth-period elements. As an example, we can compare Cr and W in Group VIB. The CrO_4^{2-} ion, with chromium(VI), is a strong oxidizing agent (that is, it is unstable) and is reduced to Cr^{3+} (which is stable). On the other hand, the WO_4^{2-} ion, with tungsten(VI), is not oxidizing. Thus, the +6 state in tungsten is very stable. The stability of the higher oxidation states in the heavier elements is also evident in their greater range of observed oxidation numbers. Osmium, in the sixth period of Group VIIIB, has oxidation numbers from +2 to +8, the +8 state being observed in the tetroxide, OsO_4.● However, iron in the fourth period of the same group, as we said, has +6 as its highest observed state.

As in the main-group elements, the transition elements become less characteristically metallic in higher oxidation states. Thus, the halides become more covalent, and the oxides more acidic, as the metal atom goes from a lower to a higher oxidation state. For example, titanium dichloride, $TiCl_2$, is a high-melting solid (m.p. 1035°C), whereas titanium tetrachloride, $TiCl_4$, is a liquid (m.p. −24°C). We conclude that titanium dichloride is ionic and that titanium tetrachloride is a covalent, molecular substance. As for the oxides, titanium monoxide, TiO, is weakly basic, and titanium dioxide, TiO_2, is amphoteric (acidic and basic).

Exercise 23.1

Compound A is a colorless liquid melting at 20°C. Compound B is a greenish-yellow powder melting at 1406°C. One of these compounds is VF_3; the other is VF_5. Which is Compound A? Why?

(See Problems 23.29 and 23.30.)

As in the main-group elements, the transition elements become more metallic going down a column. Thus, the oxides of the fifth- and sixth-period elements are generally more basic than the oxides of the corresponding elements in the fourth period. For example, chromium(VI) oxide, CrO_3, is strongly acidic. Molybdenum(VI) and tungsten(VI) oxides, MoO_3 and WO_3, are weakly acidic.

Example 23.1 Predicting Relative Oxidizing Strengths of Transition-Metal Compounds

Consider the +3 oxidation state of cobalt and rhodium. In one of these elements, the 3+ metal ion, $M^{3+}(aq)$, is stable in water, and 3+ salts can be crystallized from aqueous solution. In the other, the $M^{3+}(aq)$ ion oxidizes water and is therefore unstable in aqueous solution. Which metal ion is unstable in aqueous solution?

Solution

Of two transition elements in the same column and the same oxidation state, the one in the fourth period is expected to be the more oxidizing. Thus, we expect $Co^{3+}(aq)$ to be more oxidizing than $Rh^{3+}(aq)$. Because the problem states that one of these ions oxidizes water, this ion must be $\mathbf{Co^{3+}(aq)}$, which would therefore be unstable in water.

Exercise 23.2

Consider the compounds $KMnO_4$ and $KReO_4$. Which would you expect to be the stronger oxidizing agent? Explain. (See Problems 23.31 and 23.32.)

23.2 THE CHEMISTRY OF TWO TRANSITION METALS

Iron and titanium are the most abundant of the transition elements. In fact, iron is the fourth most abundant element in the earth's crust (after oxygen, silicon, and aluminum). However, many transition elements, particularly some in the sixth period, are rare. Surprisingly, technetium (Element 43) has only radioactive isotopes and does not occur naturally.● Promethium (Element 61), a lanthanide, is the only other element with atomic number less than 83 that has no stable isotopes.

● Technetium-99 is used in medical diagnosis; see Section 20.5.

The transition elements occur in most cases as oxide or sulfide ores. Table 23.4 lists some of these ores. The least reactive elements, such as gold, silver, copper, and platinum, are found as uncombined metals. Many of the transition metals are used commercially in alloys. The most important of these are the steels, which are alloys of iron with small amounts of carbon and often other transition

Table 23.4 Some Ores of the Transition Metals

Transition Metal	Ore
Titanium, Ti	Ilmenite, $FeTiO_3$ Rutile, TiO_2
Vanadium, V	Carnotite, $K_2(UO_2)_2(VO_4)_2 \cdot 3H_2O$ Vanadinite, $Pb_5(VO_4)_3Cl$
Chromium, Cr	Chromite, $FeCr_2O_4$
Molybdenum, Mo	Molybdenite, MoS_2
Tungsten, W	Scheelite, $CaWO_4$ Wolframite, $FeWO_4$-$MnWO_4$
Manganese, Mn	Pyrolusite, MnO_2
Iron, Fe	Hematite, Fe_2O_3 Magnetite, Fe_3O_4 Siderite, $FeCO_3$
Cobalt, Co	Cobaltite, CoAsS-FeAsS Skutterudite, $CoAs_3$-$NiAs_3$
Nickel, Ni	Pentlandite, NiS-FeS
Copper, Cu	Native copper, Cu Chalcocite, Cu_2S Cuprite, Cu_2O
Zinc, Zn	Sphalerite (zinc blende), ZnS

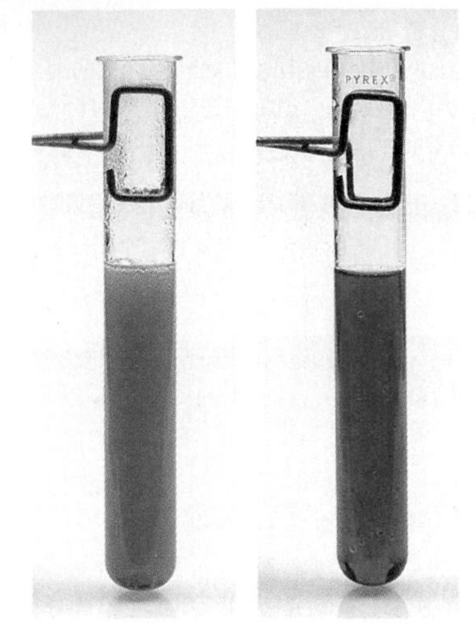

Figure 23.4
Aqueous chromium ion. *Left:* Chromium metal reacts with hydrochloric acid to produce $Cr(H_2O)_6^{2+}$, which is blue. *Right:* $Cr(H_2O)_6^{2+}$ oxidizes with air to give a chromium(III) species, which is green.

elements such as chromium, manganese, and nickel. In this section we will look at the chemistry of two transition elements: chromium (Group VIB) and copper (Group IB).

Chromium

Chromium metal reacts with acids, releasing hydrogen gas and giving a bright blue solution of the chromium(II) or chromous ion, $Cr(H_2O)_6^{2+}(aq)$ (Figure 23.4, *left*).

$$Cr(s) + 2[H^+(aq) + Cl^-(aq)] + 6H_2O(l) \longrightarrow [Cr(H_2O)_6^{2+}(aq) + 2Cl^-(aq)] + H_2(g)$$

The chromium(II) ion is easily oxidized by O_2 to chromium(III) ion species; so to prepare chromium(II) salts, we must carry out this reaction in the absence of air (Figure 23.4, *right*). Chromium(II) ion is even oxidized by hydrogen ion, but the reaction is slow.

The most stable oxidation state of chromium is +3. Chromium metal burns in oxygen to give chromium(III) oxide, Cr_2O_3 (also called chromic oxide). This oxide is a hard, green solid and is amphoteric. In acid, the oxide dissolves, forming the violet-colored ion $Cr(H_2O)_6^{3+}(aq)$. (Hydrochloric acid gives green-colored complex ions.) Addition of base to a solution of this ion precipitates chromium(III) hydroxide.

$$Cr(H_2O)_6^{3+}(aq) + 3OH^-(aq) \longrightarrow Cr(H_2O)_3(OH)_3(s) + 3H_2O(l)$$

When further base is added, the precipitate dissolves, giving a green solution of the hydroxo ion, the formula of which may be $Cr(H_2O)_2(OH)_4^-$.

$$Cr(H_2O)_3(OH)_3(s) + OH^-(aq) \longrightarrow Cr(H_2O)_2(OH_4)^-(aq) + H_2O(l)$$

Chromium(III) oxide also reacts with strong base to give hydroxo ions. The above reactions are sometimes written more simply without showing the water molecules bonded to the metal atom—that is,

$$Cr^{3+}(aq) + 3OH^-(aq) \longrightarrow Cr(OH)_3(s)$$
$$Cr(OH)_3(s) + OH^-(aq) \longrightarrow Cr(OH)_4^-(aq)$$

Chromium(III) can be oxidized to chromium(VI) species, especially in basic solution. (Transition-metal ions are often more easily oxidized to oxoanions in basic solution.) For example, hydrogen peroxide oxidizes chromium(III) hydroxide to the chromate ion, CrO_4^{2-} in basic solution.

$$2Cr(OH)_3(s) + 3HO_2^-(aq) + OH^-(aq) \longrightarrow 2CrO_4^{2-}(aq) + 5H_2O(l)$$

(We have written hydrogen peroxide as HO_2^- because it is a weak acid and exists in basic solution as the anion.) The violet solution of Cr^{3+} turns to the bright yellow of the CrO_4^{2-} ion.

When the yellow solution of a chromate salt is acidified, it turns orange. The color change is due to the formation of the dichromate ion, $Cr_2O_7^{2-}(aq)$ (see Figure 23.5).

$$\underset{\text{yellow}}{2CrO_4^{2-}(aq)} + 2H^+(aq) \rightleftharpoons \underset{\text{orange}}{Cr_2O_7^{2-}(aq)} + H_2O(l)$$

The chromate and dichromate ions are in equilibrium, which is sensitive to pH changes. Both chromate and dichromate salts are well known. The dichromate ion is a strong oxidizing agent in acid solution.

$$Cr_2O_7^{2-}(aq) + 14H^+(aq) + 6e^- \longrightarrow 2Cr^{3+}(aq) + 7H_2O(l); E^\circ = 1.33 \text{ V}$$

Figure 23.5
Chromate–dichromate equilibrium. The beaker on the left contains CrO_4^{2-} (yellow). When the experimenter adds sulfuric acid to a similar solution in the beaker on the right, CrO_4^{2-} is converted to $Cr_2O_7^{2-}$ (orange).

Chromate ion in basic solution, however, is much less oxidizing.

$$CrO_4^{2-}(aq) + 4H_2O(l) + 3e^- \longrightarrow Cr(OH)_3(s) + 5OH^-(aq);\ E° = -0.12\ V$$

Chromium(VI) oxide (chromium trioxide, CrO_3) is a red, crystalline compound. It precipitates when concentrated sulfuric acid is added to concentrated solutions of a dichromate salt.●

● Chromium trioxide is a strong oxidizing agent. Vapors of ethanol, C_2H_5OH, spontaneously ignite in the presence of the solid oxide.

$$[2K^+(aq) + Cr_2O_7^{2-}(aq)] + 2H_2SO_4(l) \longrightarrow [2K^+(aq) + 2HSO_4^-(aq)] + 2CrO_3(s) + H_2O(l)$$

Chromium trioxide is an acidic oxide. It dissolves in water, forming acidic solutions of chromate and dichromate ions. When the oxide is heated above its melting point of 196°C, it decomposes by losing oxygen to give chromium(III) oxide.

$$4CrO_3(s) \xrightarrow{\Delta} 2Cr_2O_3(s) + 3O_2(g)$$

The chief commercial source of chromium is the ore chromite, $FeCr_2O_4$, which is an iron(II) chromium(III) oxide. This is reduced with carbon in an electric furnace to obtain an alloy of iron and chromium.

$$FeCr_2O_4(s) + 4C(s) \xrightarrow{\Delta} Fe(l) + 2Cr(l) + 4CO(g)$$

This alloy, called ferrochrome, is used to make steels. Pure chromium is prepared by the exothermic reaction of chromium(III) oxide with aluminum (the Goldschmidt process). Once the mixture is ignited, the heat of reaction produces molten chromium.

$$Cr_2O_3(s) + 2Al(s) \longrightarrow 2Cr(l) + Al_2O_3(l)$$

Chromium(III) oxide and other chromium compounds are obtained from sodium chromate. This salt is prepared by strongly heating chromite ore with sodium carbonate in air.

$$4FeCr_2O_4(s) + 8Na_2CO_3(s) + 7O_2(g) \xrightarrow{1100°C} 8Na_2CrO_4(s) + 2Fe_2O_3(s) + 8CO_2(g)$$

Sodium chromate is soluble and is leached from the mixture with water. It can be converted to sodium dichromate with acid. Sodium dichromate is reduced to chromium(III) oxide by heating with carbon.

$$Na_2Cr_2O_7(s) + 2C(s) \longrightarrow Cr_2O_3(s) + Na_2CO_3(s) + CO(g)$$

Copper

Copper and the other Group IB elements silver and gold can be found as the free elements. This is a reflection of the stability of the zero oxidation states of these elements. Copper is not attacked by most acids. It does react with concentrated sulfuric acid and with nitric acid. In these cases, the anion of the acid acts as the oxidizing agent (rather than H^+, which is the usual oxidizing agent in acids). Sulfuric acid is reduced to SO_2.

$$Cu(s) + 2H_2SO_4(l) \longrightarrow [Cu^{2+}(aq) + SO_4^{2-}(aq)] + SO_2(g) + 2H_2O(l)$$

Dilute nitric acid is reduced to NO, concentrated acid to NO_2.

$$3Cu(s) + 8[H^+(aq) + NO_3^-(aq)] \longrightarrow 3[Cu^{2+}(aq) + 2NO_3^-(aq)] + 2NO(g) + 4H_2O(l)$$

$$Cu(s) + 4[H^+(aq) + NO_3^-(aq)] \longrightarrow [Cu^{2+}(aq) + 2NO_3^-(aq)] + 2NO_2(g) + 2H_2O(l)$$

The oxidation product of copper in the foregoing reactions is the copper(II) ion, $Cu(H_2O)_6^{2+}(aq)$, also called the cupric ion. It has a bright blue color. Hydrated copper(II) salts, such as copper(II) sulfate pentahydrate, $CuSO_4 \cdot 5H_2O$, also have a blue color. Four of the water molecules of copper(II) sulfate pentahydrate are associated with Cu^{2+}, and the fifth is hydrogen-bonded to the sulfate ion as well as to the water molecules on the copper ion. We can write the formula as $[Cu(H_2O)_4]SO_4 \cdot H_2O$ to better represent its structure. When this salt is heated, it loses its water of hydration in stages. The hydrate $CuSO_4 \cdot H_2O$ and the anhydrous salt $CuSO_4$ are colorless substances, which show that the complexing of water molecules to the copper ion is responsible for the blue color.

When solutions of copper(II) ion are made basic, a turquoise-blue precipitate of copper(II) hydroxide, $Cu(OH)_2$, forms (Figure 23.6, *left*). Heating the solution converts the hydroxide to copper(II) oxide, CuO, which is black. The oxide and hydroxide are amphoteric and dissolve in strong base to form deep blue solutions, perhaps of the anion $Cu(OH)_4^{2-}$ (Figure 23.6, *right*).

Although most of the aqueous chemistry of copper involves the +2 oxidation state, there are a number of important binary compounds and complexes of copper(I). When copper is heated in oxygen below 1000°C, it forms the black copper(II) oxide, CuO. But above this temperature, it forms the red copper(I) oxide, Cu_2O. This oxide is found naturally as the mineral cuprite.

In water, the copper(I) ion, $Cu(H_2O)_6^+$, disproportionates (that is, it reduces and oxidizes itself).

$$2Cu(H_2O)_6^+(aq) \longrightarrow Cu(s) + Cu(H_2O)_6^{2+}(aq) + 6H_2O(l)$$

For this reason, copper(I) compounds that might be expected to give the $Cu(H_2O)_6^+$ ion are unstable in aqueous solution. However, the preceding reaction can be shifted to the left if an insoluble copper(I) compound is formed. Thus, insoluble copper(I) chloride can be prepared by boiling an acidic mixture of copper(II) chloride and copper metal.

$$[Cu^{2+}(aq) + 2Cl^-(aq)] + Cu(s) \longrightarrow \underset{\text{white precipitate}}{2CuCl(s)}$$

Figure 23.6
Amphoteric character of copper(II) ion. *Left:* A turquoise-blue precipitate of $Cu(OH)_2$ forms when the experimenter adds NaOH solution to $CuSO_4$ solution. *Right:* When more NaOH solution is added, the precipitate dissolves to form a deep blue solution of the hydroxo ion $Cu(OH)_4^{2-}$.

Similarly, the reduction of Cu^{2+} to Cu^+ is easily accomplished when the concentration of Cu^+ is kept low by the formation of a sufficiently insoluble copper(I) salt. Consider the reduction of Cu^{2+} to copper(I) by the iodide ion to form a white precipitate of copper(I) iodide.

$$2Cu^{2+}(aq) + 4I^-(aq) \longrightarrow 2CuI(s) + I_2(aq)$$

The reaction goes to the right, even though the iodide ion is a very mild reducing agent, because of the formation of very insoluble copper(I) iodide.

The principal commercial use of copper is as an electrical conductor. Common ores of copper are native copper (the free metal), copper oxides, and copper sulfides. Most copper is presently obtained by open-pit mining of low-grade rock containing only a few percent copper as copper sulfides. The ore is concentrated in copper by flotation.● In this process, a slurry of the crushed ore is agitated with air, and the copper sulfides are carried away in the froth. This concentrated ore is then treated in several steps that result in the production of molten copper(I) sulfide, Cu_2S. This molten material, called *matte,* is reduced to copper by blowing air through it.

● The general steps in the production of a metal from its ore, including flotation, are discussed in Section 21.5.

$$Cu_2S(l) + O_2(g) \longrightarrow 2Cu(l) + SO_2(g)$$

The metal produced in this step is called *blister copper* and is about 99% pure. For electrical use, the copper must be further purified or refined by electrolysis (see Figure 19.24, p. 779).

Example 23.2 Preparing Compounds of Chromium or Copper

Write molecular equations describing the preparation of chromium trioxide from chromium(III) sulfate.

Solution

Chromium(III) ion is oxidized by H_2O_2 in basic solution to chromate ion. When this solution is acidified, it forms the dichromate ion. If the solution is concentrated, and concentrated sulfuric acid added, chromium trioxide precipitates.

The molecular equations are

$$Cr_2(SO_4)_3(aq) + 3H_2O_2(aq) + 10KOH(aq) \longrightarrow 2K_2CrO_4(aq) + 3K_2SO_4(aq) + 8H_2O(l)$$

$$2K_2CrO_4(aq) + 2H_2SO_4(aq) \longrightarrow K_2Cr_2O_7(aq) + 2KHSO_4(aq) + H_2O(l)$$

$$K_2Cr_2O_7(aq) + 2H_2SO_4(aq) \longrightarrow 2CrO_3(s) + 2KHSO_4(aq) + H_2O(l)$$

Exercise 23.3

Write molecular equations describing the preparation of copper(I) iodide from copper metal. (See Problems 23.33 and 23.34.)

Exercise 23.4

What is the oxidation number of tungsten in scheelite, $CaWO_4$? (See Problems 23.35 and 23.36.)

Exercise 23.5

Write a balanced equation for the reaction of dichromate ion and iodide ion in acidic solution. (See Problems 23.37 and 23.38.)

Complex Ions and Coordination Compounds

As shown in the previous section, ions of the transition elements exist in aqueous solution as *complex ions*. Iron(II) ion, for example, exists in water as $Fe(H_2O)_6^{2+}$. The water molecules in this ion are arranged about the iron atom with their oxygen atoms bonded to the metal by donating electron pairs to it. Replacing the H_2O molecules by six CN^- ions gives the $Fe(CN)_6^{4-}$ ion. The biological activity of the transition elements and their role in human nutrition (Table 23.5) depend in most cases on the formation of *complexes,* or *coordination compounds,* which exhibit the type of bonding that occurs in $Fe(H_2O)_6^{2+}$ and $Fe(CN)_6^{4-}$. In the chapter opening, for example, we saw that hemoglobin, a complex of iron, is vital to the transport of oxygen by the red blood cells.

23.3 FORMATION AND STRUCTURE OF COMPLEXES

A metal atom, particularly a transition-metal atom, often functions in chemical reactions as a Lewis acid, accepting electron pairs from molecules or ions.● Thus, Fe^{2+} and H_2O can bond to one another in a Lewis acid–base reaction.

● Lewis acid–base reactions are discussed in Section 13.5.

Table 23.5 Fourth-Period Transition Elements Essential to Human Nutrition

Element	Some Biochemical Substances	Function
Chromium	Glucose tolerance factor	Utilization of glucose
Manganese	Isocitrate dehydrogenase	Cell energetics
Iron	Hemoglobin and myoglobin	Transport and storage of oxygen
	Cytochrome *c*	Cell energetics
	Catalase	Decomposition of hydrogen peroxide
Cobalt	Cobalamin (vitamin B_{12})	Development of red blood cells
Copper	Ceruloplasmin	Synthesis of hemoglobin
	Cytochrome oxidase	Cell energetics
Zinc	Carbonic anhydrase	Elimination of carbon dioxide
	Carboxypeptidase A (pancreatic juice)	Protein digestion
	Alcohol dehydrogenase	Oxidation of ethanol

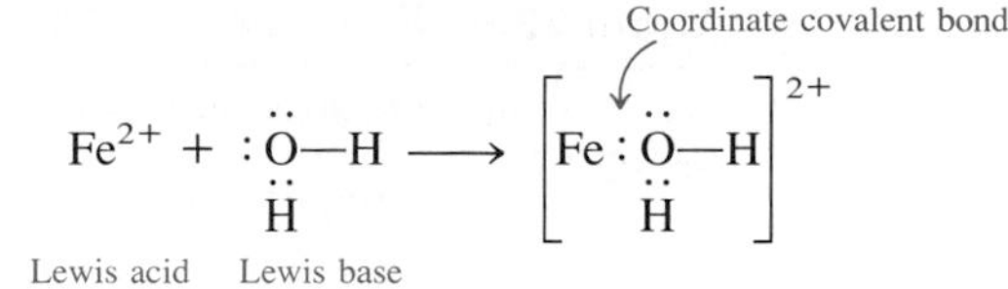

A pair of electrons on the oxygen atom of H_2O forms a coordinate covalent bond to Fe^{2+}. In water, the Fe^{2+} ion ultimately bonds to six H_2O molecules to give the $Fe(H_2O)_6{}^{2+}$ ion.

The Fe^{2+} ion also undergoes a similar Lewis acid–base reaction with cyanide ions, CN^-. In this case, the Fe^{2+} ion bonds to the electron pair on the carbon atom of CN^-.

$$Fe^{2+} + :C{\equiv}N:^- \longrightarrow (Fe:C{\equiv}N:)^+$$

Finally, a very stable ion $Fe(CN)_6{}^{4-}$ is obtained with six cyanide ions.● Note that the charge on the $Fe(CN)_6{}^{4-}$ ion equals the sum of the charges on the ions from which it is formed: $+2 + 6(-1) = -4$.

● Although the ion $Fe(CN)_6{}^{4-}$ is a complex of Fe^{2+} and CN^- ions, a solution of $Fe(CN)_6{}^{4-}$ contains negligible concentration of CN^-. Thus, a substance such as $K_4[Fe(CN)_6]$ is relatively nontoxic, even though the free cyanide ion is a violent poison.

In some cases, a neutral species is produced from a metal ion with anions. Cisplatin, the anticancer drug discussed in the opening of Chapter 1, has the structure

$$\begin{array}{c} :\ddot{Cl}: \\ H_3N:\ddot{Pt}:\ddot{Cl}: \\ NH_3 \end{array}$$

It consists of Pt^{2+} with two NH_3 molecules (neutral) and two Cl^- ions, giving a neutral species. Iron pentacarbonyl, $Fe(CO)_5$, is an example of a neutral species formed from a neutral iron atom with CO molecules.

Basic Definitions

A **complex ion** is *a metal atom or ion with Lewis bases attached to it through coordinate covalent bonds.* A **complex** (or **coordination compound**) is *a compound consisting either of complex ions with other ions of opposite charge* (for example, the compound $K_4[Fe(CN)_6]$ of the complex ion $Fe(CN)_6{}^{4-}$ with four K^+ ions) *or of a neutral complex species* (such as cisplatin).

Ligands are *the Lewis bases attached to the metal atom in a complex.* They are electron-pair donors, so ligands may be neutral molecules (such as H_2O or NH_3) or anions (such as CN^- or Cl^-) that have at least one atom with a lone pair of electrons. Cations only rarely function as ligands. We might expect this, because an electron pair on a cation is held securely by the positive charge, so it would not be involved in coordinate bonding.●

● A cation in which the positive charge is far removed from an electron pair that could be donated can function as a ligand. An example is the pyrazinium ion:

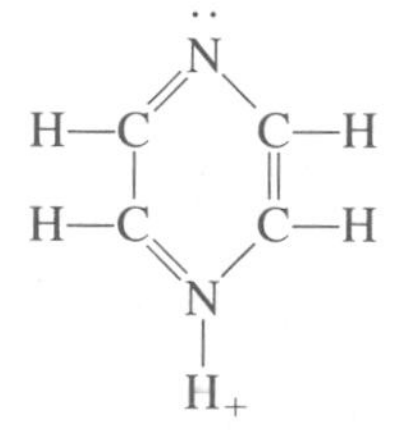

The **coordination number** of a metal atom in a complex is *the total number of bonds the metal atom forms with ligands.* In $Fe(H_2O)_6{}^{2+}$, the iron atom bonds to each of the oxygen atoms in the six water molecules. Therefore, the coordination number of iron in this ion is 6, by far the most common coordination number. Coordination number 4 is also well known, and many examples of number 5 have been recently discovered. Table 23.6 gives some examples of complexes for the coordination numbers 2 to 8. The coordination number for an atom depends on several factors, but size of the metal atom is important. Thus, coordination numbers 7 and 8 are seen primarily in fifth- and sixth-period elements.●

● Very high coordination numbers (9 to 12) are known for some complex ions of the lanthanide elements.

Table 23.6 Examples of Complexes of Various Coordination Numbers

Complex	Coordination Number
$Ag(NH_3)_2^+$	2
HgI_3^-	3
$PtCl_4^{2-}$, $Ni(CO)_4$	4
$Fe(CO)_5$, $Co(CN)_5^{3-}$	5
$Co(NH_3)_6^{3+}$, $W(CO)_6$	6
$Mo(CN)_7^{3-}$	7
$W(CN)_8^{4-}$	8

Polydentate Ligands

The ligands we have discussed so far bond to the metal atom through one atom of the ligand. Thus, ammonia bonds through the nitrogen atom. It is called a **monodentate ligand** (meaning ''one-toothed'' ligand)—that is, *a ligand that bonds to a metal atom through one atom of the ligand.* A **bidentate ligand** (''two-toothed'' ligand) is *a ligand that bonds to a metal atom through two atoms of the ligand.* Ethylenediamine is an example of such a ligand.

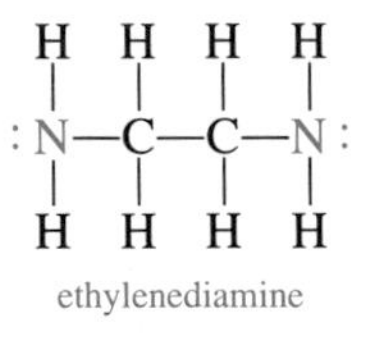

ethylenediamine

Nitrogen atoms at the ends of the molecule have lone pairs of electrons that can form coordinate covalent bonds. In forming a complex, the ethylenediamine molecule bends around so that both nitrogen atoms coordinate to the metal atom, M.

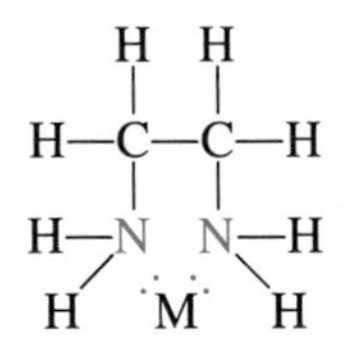

Because ethylenediamine is a common bidentate ligand, it is frequently abbreviated in formulas as ''en.'' Figure 23.7 shows the structure of the stable ion $Co(en)_3^{3+}$. The oxalate ion, $C_2O_4^{2-}$, is another common bidentate ligand.

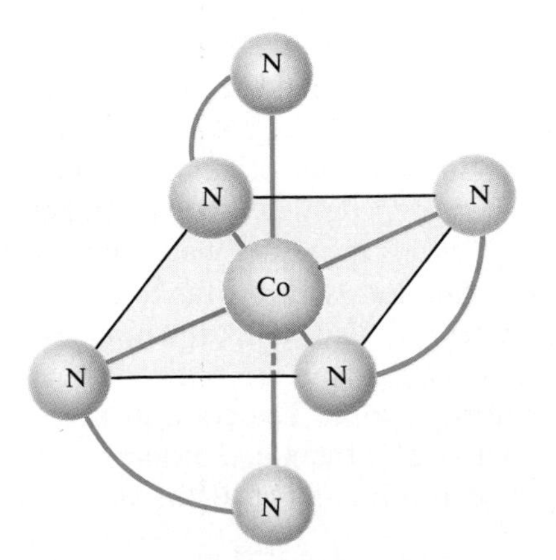

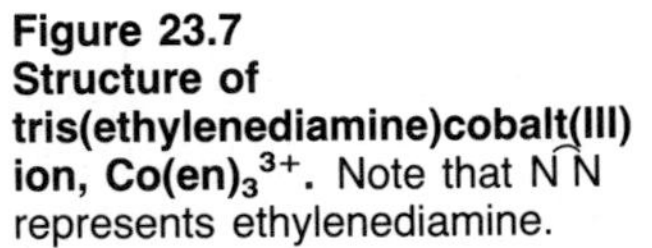

Figure 23.7 Structure of tris(ethylenediamine)cobalt(III) ion, $Co(en)_3^{3+}$. Note that N͡N represents ethylenediamine.

The hemoglobin molecule in red blood cells is an example of a complex with a *quadridentate ligand*—one that bonds to the metal atom through four ligand atoms. Hemoglobin consists of a protein *globin* chemically bonded to *heme,* whose structure is shown in Figure 23.8 (*left*). Heme is a planar molecule consisting of iron(II) to which a quadridentate ligand is bonded through its four nitrogen atoms.

Ethylenediaminetetraacetate ion (EDTA) is a ligand that bonds through six of its atoms.

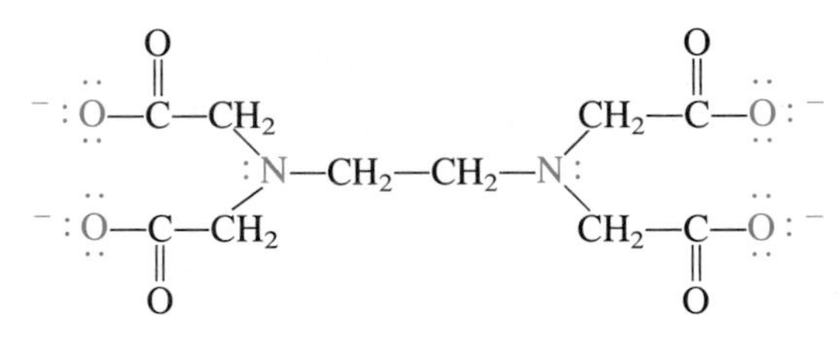

It can completely envelop a metal atom, simultaneously occupying all six positions in an octahedral geometry (Figure 23.8, *right*).

A **polydentate ligand** ("having many teeth") is *a ligand that can bond with two or more atoms to a metal atom. A complex formed by polydentate ligands* is frequently quite stable and is called a **chelate.**● Because of the stability of chelates, polydentate ligands (also called *chelating agents*) are often used to remove metal ions from a chemical system. EDTA, for example, is added to certain canned foods to remove transition-metal ions that can catalyze the deterioration of the food. The same chelating agent has been used to treat lead poisoning because it binds Pb^{2+} ions as the chelate, which can then be excreted by the kidneys.

● The term *chelate* is derived from the Greek *chele* for "claw," because a polydentate ligand appears to attach itself to the metal atom like crab claws to some object.

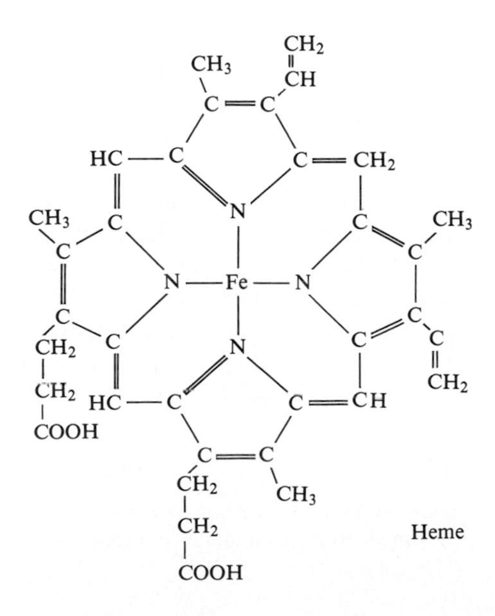

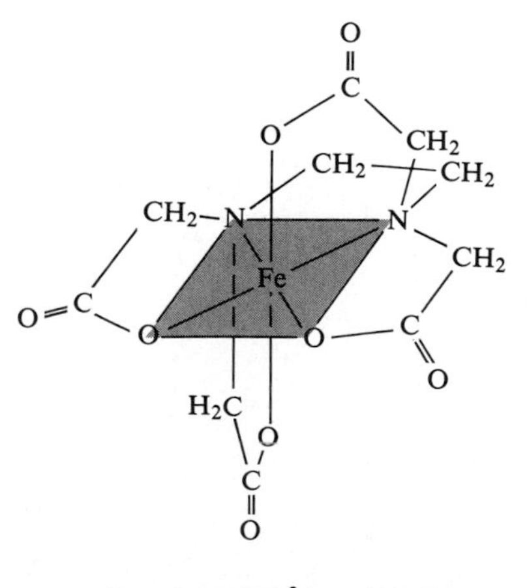

Figure 23.8
Some complexes with ligands that bond with more than one atom to the metal atom (polydentate ligands). *Left:* The structure of heme. *Right:* Complex of Fe^{2+} and ethylenediaminetetra-acetate ion (EDTA). Note how the EDTA ion envelops the metal ion.

Table 23.7 Some Platinum(IV) Complexes Studied by Werner

Old Formula	Modern Formula	Number of Ions	Number of Free Cl^- Ions
$PtCl_4 \cdot 6NH_3$	$[Pt(NH_3)_6]Cl_4$	5	4
$PtCl_4 \cdot 4NH_3$	$[Pt(NH_3)_4Cl_2]Cl_2$	3	2
$PtCl_4 \cdot 3NH_3$	$[Pt(NH_3)_3Cl_3]Cl$	2	1
$PtCl_4 \cdot 2NH_3$	$[Pt(NH_3)_2Cl_4]$	0	0

Discovery of Complexes; Formula of a Complex

In 1798 B. M. Tassaert found that a solution of cobalt(II) chloride in aqueous ammonia exposed to air (an oxidizing agent) deposits orange-yellow crystals. He assigned the formula $CoCl_3 \cdot 6NH_3$ to these crystals. Later, a similar compound of platinum was assigned the formula $PtCl_4 \cdot 6NH_3$. Because these formulas suggested that the substances were somehow composed of two stable compounds, platinum(IV) chloride and ammonia in the latter case, they were called complex compounds or simply complexes.

The basic explanation for the structure of these complexes was given by the Swiss chemist Alfred Werner in 1893. According to Werner, a metal atom exhibits two kinds of valences, a primary valence and a secondary valence. The primary valence is what we now call the oxidation number of the metal. The secondary valence corresponds to what we now call the coordination number, which is often 6. In Werner's view, the substance previously represented by the formula $PtCl_4 \cdot 6NH_3$ is composed of the ion $Pt(NH_3)_6{}^{4+}$, with six ammonia molecules directly attached to the platinum atom. The charge of this ion is balanced by four Cl^- ions, giving a neutral compound with the structural formula $[Pt(NH_3)_6]Cl_4$.

Table 23.7 lists a series of platinum(IV) complexes studied by Werner. It gives both the old and Werner's modern formulas. (Note that square brackets are used in modern formulas to separate the metal and its associated groups from the rest of the formula.) According to Werner's theory, these complexes should dissolve to give different numbers of ions per formula unit. For example, $[Pt(NH_3)_4Cl_2]Cl_2$ dissolves to give three ions: $Pt(NH_3)_4Cl_2{}^{2+}$ and two Cl^- ions. Werner was able to show that the electrical conductances of solutions of these complexes were equal to what was expected for the number of ions predicted by his formulas. He also demonstrated that the chloride ions in the platinum complexes were of two kinds: those that could be precipitated from solution as AgCl using silver nitrate and those that could not. He explained that the chloride ions within the platinum complex ion are securely attached to the metal atom and that only those outside the complex ion can be precipitated with silver nitrate. The number of free Cl^- ions (those not attached to platinum) determined this way agreed exactly with his formulas. Over a period of twenty years, Werner carried out many experiments on complexes, all of them in basic agreement with his theory. He received the Nobel Prize for his work in 1913.

Exercise 23.6

Another complex studied by Werner had a composition corresponding to the formula $PtCl_4 \cdot 2KCl$. From electrical-conductance measurements, he determined that each formula unit contained three ions. He also found that silver nitrate did not give a precipitate of AgCl with this complex. Write a formula for this complex that agrees with this information. (See Problems 23.39 and 23.40.)

Related Topic

ON THE STABILITY OF CHELATES

The special stability of chelates stems from the additional entropy obtained when they are formed. This leads to a large negative $\Delta G°$, which is equivalent to a large equilibrium constant for the formation of the chelate. To understand how this happens, consider the formation of the chelate $Co(en)_3{}^{3+}$ from the complex ion $Co(NH_3)_6{}^{3+}$, with monodentate ligands NH_3.

$$Co(NH_3)_6{}^{3+} + 3en \rightleftharpoons Co(en)_3{}^{3+} + 6NH_3$$

Each en molecule replaces two NH_3 molecules. Therefore, the number of particles in the reaction mixture is increased when the reaction goes to the right. In most cases, an increase in number of particles increases the possibilities for randomness or disorder. Thus, when the reaction goes to the right, there is an increase in entropy; that is, $\Delta S°$ is positive.

At the same time, the reaction involves very little change of internal energy ($\Delta U°$) or enthalpy ($\Delta H° = \Delta U° + P\Delta V$), because the bonds are similar (all consist of a nitrogen atom coordinated to a cobalt atom). Six nitrogen-metal bonds in $Co(NH_3)_6{}^{3+}$ are broken, and six new nitrogen-metal bonds of about the same energy are formed in $Co(en)_3{}^{3+}$. Thus, $\Delta H° \simeq 0$.

The spontaneity of a reaction depends on $\Delta G°$, which equals $\Delta H° - T\,\Delta S°$. But because $\Delta H° \simeq 0$,

$$\Delta G° = \Delta H° - T\Delta S°$$
$$= -T\Delta S°$$

The entropy change is positive, so $\Delta G°$ is negative, and the reaction is spontaneous from left to right.

The fact that the equilibrium for the foregoing reaction favors the chelate $Co(en)_3{}^{3+}$ means that the chelate has thermodynamic stability. A similar argument could be made for any equilibrium involving the replacement of monodentate ligands by polydentate ligands. Reaction tends to favor the chelate.

23.4 NAMING COORDINATION COMPOUNDS

Thousands of coordination compounds are now known. A systematic method of naming such compounds, or *nomenclature,* should give us basic information about the structure of a coordination compound. What is the metal in the complex? Does the metal atom occur in the cation or the anion? What is the oxidation state of the metal? What are the ligands? We can answer these questions by following the rules of nomenclature agreed upon by the International Union of Pure and Applied Chemistry (IUPAC). These rules are essentially an extension of those originally given by Werner.

1. When we are naming a salt, the name of the cation precedes the name of the anion. (This is a familiar rule.) For example,

$K_4[Fe(CN)_6]$ is named $\underbrace{\text{potassium}}_{\text{cation}}\ \underbrace{\text{hexacyanoferrate(II)}}_{\text{anion}}$

and

$[Co(NH_3)_6]Cl_3$ is named $\underbrace{\text{hexaamminecobalt(III)}}_{\text{cation}}\ \underbrace{\text{chloride}}_{\text{anion}}$

2. The name of the complex—whether anion, cation, or neutral species—consists of two parts written together as one word. Ligands are named first, and the metal atom is named second. For example,

$Fe(CN)_6^{4-}$ is named hexacyanoferrate(II) ion

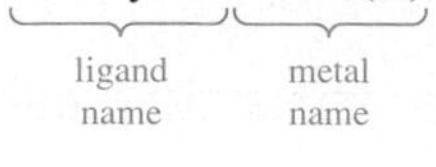

$Co(NH_3)_6^{3+}$ is named hexaamminecobalt(III) ion

ligand name — metal name

3. The complete ligand name consists of a Greek prefix denoting the number of ligands, followed by the specific name of the ligand. When there are two or more ligands, the ligands are written in alphabetical order (disregarding Greek prefixes).

a. Anionic ligands end in *-o*. Some examples are given in the following table.

ANION NAME	LIGAND NAME
Bromide, Br^-	Bromo
Carbonate, CO_3^{2-}	Carbonato
Chloride, Cl^-	Chloro
Cyanide, CN^-	Cyano
Fluoride, F^-	Fluoro
Hydroxide, OH^-	Hydroxo
Oxalate, $C_2O_4^{2-}$	Oxalato
Oxide, O^{2-}	Oxo
Sulfate, SO_4^{2-}	Sulfato

b. Neutral ligands are usually given the name of the molecule. There are, however, several important exceptions, which are given in the following table.

MOLECULE	LIGAND NAME
Ammonia, NH_3	Ammine
Carbon monoxide, CO	Carbonyl
Water, H_2O	Aqua ●

● *Aquo* is used in the earlier literature.

c. The prefixes used to denote the number of ligands are *mono-* (1) (usually omitted); *di-* (2); *tri-* (3); *tetra-* (4); *penta-* (5); *hexa-* (6); and so forth. To see how the ligand name is formed, consider the complex ions

$Fe(CN)_6^{4-}$ or hexacyanoferrate(II) ion

6 — CN^- ligands

$Co(NH_3)_6^{3+}$ or hexaamminecobalt(III) ion

6 — NH_3 ligands

d. When the name of the ligand also has a Greek prefix, the number of ligands is denoted with *bis* (2), *tris* (3), *tetrakis* (4), and so forth. The name of the ligand follows in parentheses. For example, the complex $[Co(en)_3]Cl_3$ is named as follows:

tris(ethylenediamine)cobalt(III) chloride

3 — ligand name

4. The complete metal name consists of the name of the metal, followed by *-ate* if the complex is an anion, which in turn is followed by the oxidation number of the metal, indicated by Roman numerals in parentheses. (An oxidation state of zero is indicated by 0 in parentheses.) When there is a Latin name for the metal, it is used to name the anion (except for mercury). These names are given in the following table.

ENGLISH NAME	LATIN NAME	ANION NAME
Copper	Cuprum	Cuprate
Gold	Aurum	Aurate
Iron	Ferrum	Ferrate
Lead	Plumbum	Plumbate
Silver	Argentum	Argentate
Tin	Stannum	Stannate

Examples are

hexacyanoferrate(II) — ferr = ferrum = iron; -ate indicates an anion; (II) oxidation number 2

hexaamminecobalt(III) — cobalt: metal name; (III) oxidation number 3

Example 23.3 Writing the IUPAC Name Given the Structural Formula of a Compound

Give the IUPAC name of each of the following coordination compounds: (a) $[Pt(NH_3)_4Cl_2]Cl_2$; (b) $[Pt(NH_3)_2Cl_2]$; (c) $K_2[PtCl_6]$.

Solution

(a) The cation is listed first in the formula.

$$\underbrace{[Pt(NH_3)_4Cl_2]}_{\text{cation}}\underbrace{Cl_2}_{\text{anions}}$$

There are two Cl^- anions, so the charge on the cation is +2: $Pt(NH_3)_4Cl_2^{2+}$. The oxidation number of platinum plus the sum of the charges on the ligands (−2) equals the cation charge +2. Therefore, the oxidation number of Pt is +4. Hence, the name of the compound is **tetraamminedichloroplatinum(IV) chloride.** Note that the ligands are listed in alphabetical order (that is, ammine before chloro). (b) This is a neutral complex species. The oxidation number of platinum must balance that of the two chloride ions. The name of the compound is **diamminedichloroplatinum(II).** (c) The complex anion is $PtCl_6^{2-}$. The oxidation number of platinum is +4, and the name of the compound is **potassium hexachloroplatinate(IV).**

Exercise 23.7

Give the IUPAC names of (a) $[Co(NH_3)_5Cl]Cl_2$; (b) $K_2[Co(H_2O)(CN)_5]$; (c) $Fe(H_2O)_5(OH)^{2+}$.

(See Problems 23.47, 23.48, 23.49, and 23.50.)

Example 23.4 Writing the Structural Formula Given the IUPAC Name of a Compound

Write the structural formula corresponding to each of the following IUPAC names: (a) hexaaquairon(II) chloride; (b) potassium tetrafluoroargentate(III); (c) pentachlorotitanate(II) ion.

(continued)

Solution

(a) The complex cation hexaaquairon(II) is Fe^{2+} with six H_2O ligands: $Fe(H_2O)_6^{2+}$. The formula of the compound is $[Fe(H_2O)_6]Cl_2$. (Remember to enclose the formula of the complex ion in parentheses.) (b) The compound contains the complex anion tetrafluoroargentate(III)—that is, Ag^{3+} with four F^- ligands. The formula of the ion is AgF_4^-. Hence, the formula of the compound is **$K[AgF_4]$**. (c) The ion contains Ti^{2+} and five Cl^- ligands. Thus, the charge on the ion is $2 + 5(-1) = -3$. The formula of the complex ion is $TiCl_5^{3-}$.

Exercise 23.8

Write structural formulas for each of the following: (a) potassium hexacyanoferrate(II), (b) tetraamminedichlorocobalt(III) chloride, (c) tetrachloroplatinate(II) ion.

(See Problems 23.51 and 23.52.)

23.5 STRUCTURE AND ISOMERISM IN COORDINATION COMPOUNDS

Although we described the formation of a complex as a Lewis acid–base reaction (Section 23.3), we did not go into any details of structure. We did not look at the geometry of complex ions, nor did we inquire about the precise nature of the bonding. There are three properties of complexes that have proved pivotal in determining these details.

1. *Isomerism* Isomers are compounds with the same molecular formula (or the same simplest formula, in the case of ionic compounds) but with different arrangements of atoms. Because their atoms are differently arranged, isomers have different properties. There are many possibilities for isomerism in coordination compounds. The study of isomerism can lead to information about atomic arrangement in coordination compounds. Werner's research on the isomerism of coordination compounds finally convinced others that his views were essentially correct.
2. *Paramagnetism* Paramagnetic substances are attracted to a strong magnetic field. Paramagnetism is due to unpaired electrons in a substance. (*Ferromagnetism* in solid iron is also due to unpaired electrons, but in this case the magnetism of many iron atoms is aligned, giving a magnetic effect that is perhaps a million times stronger than that seen in paramagnetic substances.) Many complex compounds are paramagnetic. The magnitude of this paramagnetism can be measured with a *Gouy balance,* in which the force of magnetic attraction is balanced with weights (Figure 23.9). Because paramagnetism is related to the electron configuration of the complex, these measurements can give information about the bonding.
3. *Color* A substance is colored because it absorbs light in the visible region of the spectrum. The absorption of visible light is due to a transition between closely spaced electronic energy levels. As we have seen, many coordination compounds are highly colored. This color is related to the electronic structure of the compounds.●

● Recall that the wavelength of light absorbed by a substance is related to the difference between two energy levels in the substance. See Section 7.7.

In this section, we will investigate the relationship between structure and isomerism. We will look into explanations of the paramagnetism and color of coordination compounds in the following sections.

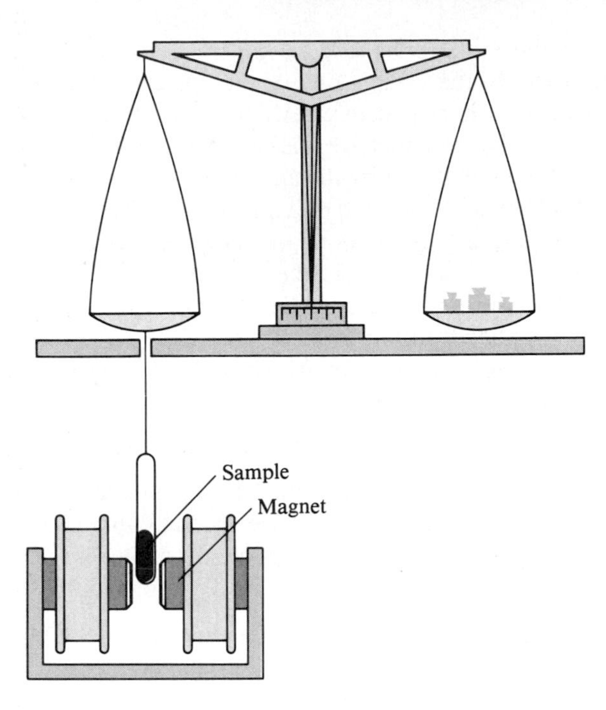

Figure 23.9
Gouy balance for measuring the paramagnetism of a substance. If the sample is attracted into the field of the magnet, the left pan of the balance will be subjected to a downward force. This force is balanced by weights added to the right pan.

Structural Isomerism

There are two major kinds of isomers. **Structural isomers** are *isomers that differ in how the atoms are joined together*—that is, in the order in which the atoms are bonded to one another. (For example, H—N═C═O and N≡C—O—H are structural isomers because the H atom is bonded to N in one case and to O in the other.) **Stereoisomers,** on the other hand, are *isomers in which the atoms are bonded to each other in the same order but that differ in the precise arrangement of these atoms in space*.

The simplest structural isomers of complexes are ionization isomers. **Ionization isomers** are *isomers of a complex that differ in the anion that is coordinated to the metal atom*. The following cobalt complexes provide an example.

$[Co(NH_3)_5(SO_4)]Br$	red compound
$[Co(NH_3)_5Br]SO_4$	violet compound

These isomers differ in whether the sulfate ion or the bromide ion is attached to the cobalt atom. The red compound gives a precipitate of AgBr when mixed with silver nitrate solution, but it gives no precipitate of $BaSO_4$ when mixed with barium chloride solution. This implies that the sulfate ion is strongly attached to the metal. The violet compound, however, gives a precipitate of $BaSO_4$ when treated with barium chloride, but it gives no precipitate of AgBr when treated with silver nitrate.

Hydrate isomers are *isomers of a complex that differ in the placement of water molecules in the complex*. A well-studied example is that of the three chromium complexes with composition $CrCl_3 \cdot 6H_2O$.

$[Cr(H_2O)_6]Cl_3$	violet compound
$[Cr(H_2O)_5Cl]Cl_2 \cdot H_2O$	light green compound
$[Cr(H_2O)_4Cl_2]Cl \cdot 2H_2O$	dark green compound

In the violet complex, all the water molecules are coordinated to the chromium atom. In the others, one or two water molecules are not directly attached to the metal atom but occur elsewhere in the crystal lattice. The isomers can be differentiated by conductance measurements (which determine the number of ions per formula unit) and by the amount of AgCl precipitated by excess silver nitrate (Cl^- bonded to Cr is not precipitated). Moreover, the water molecules not directly coordinated to a Cr atom are easily lost. Thus, the isomers behave differently when exposed to a dry atmosphere in contact with concentrated sulfuric acid. (Concentrated sulfuric acid is a dehydrating agent.)

Coordination isomers are *isomers consisting of complex cations and complex anions that differ in the way the ligands are distributed between the metal atoms*. An example is

$$[Cu(NH_3)_4][PtCl_4] \quad \text{and} \quad [Pt(NH_3)_4][CuCl_4]$$

In the first compound, the NH_3 ligands are associated with copper and the Cl^- ligands with platinum. In the second compound, the ligands are transposed. It is also possible for ligands to be distributed in different ways between two metal atoms of the same element. An example of this sort of coordination isomerism is

$$[Pt(NH_3)_4][PtCl_4] \quad \text{and} \quad [Pt(NH_3)_3Cl][Pt(NH_3)Cl_3]$$

Linkage isomers are *isomers of a complex that differ in the atom of a ligand that is bonded to the metal atom*. A ligand such as the nitrite ion, NO_2^-, can coordinate to a metal atom either with the electron pair on the nitrogen atom or with an electron pair on an oxygen atom. A complex with an O-bonded NO_2 ligand has the structure $[Co(NH_3)_5(ONO)]Cl_2$, where the nitrite ligand is written ONO to emphasize its O-bonding to cobalt. The compound is red and slowly changes to the yellow-brown isomer $[Co(NH_3)_5(NO_2)]Cl_2$, in which the nitrite group is N-bonded. Ligands such as NO_2^- that can bond through one atom or another are called *ambidentate*. Another ambidentate ligand is the thiocyanate ion, SCN^-, which can bond through the sulfur atom or the nitrogen atom.

Exercise 23.9

Give the name describing the type of structural isomerism displayed by each of the following pairs.
(a) $[Co(en)_3][Cr(CN)_6]$ and $[Cr(en)_3][Co(CN)_6]$
(b) $[Mn(CO)_5(SCN)]$ and $[Mn(CO)_5(NCS)]$
(c) $[Co(NH_3)_5(NO_3)]SO_4$ and $[Co(NH_3)_5(SO_4)]NO_3$
(d) $[Co(NH_3)_4(H_2O)Cl]Cl_2$ and $[Co(NH_3)_4Cl_2]Cl \cdot H_2O$

(See Problems 23.53 and 23.54.)

Exercise 23.10

A complex has the composition $Co(NH_3)_4(H_2O)Cl_3$. Conductance measurements show that there are three ions per formula unit, and precipitation of AgCl with silver nitrate shows that there are two Cl^- ions not coordinated to cobalt. What is the structural formula of the compound? Write the structural formula of an isomer. (See Problems 23.55 and 23.56.)

Stereoisomerism

The existence of various types of structural isomers is strong evidence for the view that complexes consist of groups directly bonded to a central metal atom. The

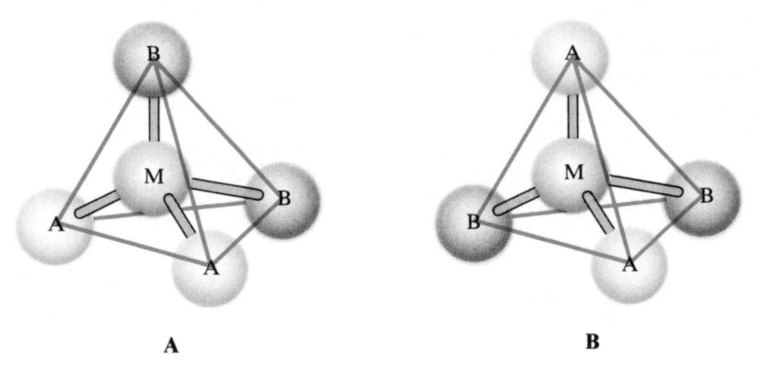

Figure 23.10
The tetrahedral complex MA_2B_2 has no geometric isomers. Molecule A can be rotated to look like molecule B.

existence of stereoisomers not only strengthens this view but also helps explain how these groups are arranged about the central atom.

Geometric isomers are *isomers in which the atoms are joined to one another in the same way but differ because some atoms occupy different relative positions in space*. For an example, consider the complex diamminedichloroplatinum(II), $[Pt(NH_3)_2Cl_2]$. Two compounds of this composition are known. One is an orange-yellow compound with solubility at 25°C of 0.252 g per 100 g of water. The other compound is pale yellow and is much less soluble (0.037 g per 100 g of water at 25°C).● How do we explain the occurrence of these two isomers?

● The isomers of $[Pt(NH_3)_2Cl_2]$ also differ in their biological properties. The orange-yellow compound acts as an anticancer drug, but the other isomer has little anticancer activity.

There are two symmetrical geometries of complexes of four ligands: tetrahedral and square planar. Let us write MA_2B_2 for a complex with ligands A and B about the metal atom M. As Figure 23.10 shows, a tetrahedral geometry for MA_2B_2 allows only one arrangement of ligands. The square planar geometry for MA_2B_2, however, gives two different arrangements. (These two arrangements for $[Pt(NH_3)_2Cl_2]$ are shown in Figure 23.11.) One, labeled *cis,* has the two A ligands on one side of the square and the two B ligands on the other side. The other arrangement, labeled *trans,* has A and B ligands across the square from one another. The *cis* and *trans* arrangements of MA_2B_2 are examples of geometric isomers.

That there are two isomers of $[Pt(NH_3)_2Cl_2]$ is evidence for the square planar geometry in this complex. But how do we identify the substances and their properties with the *cis* and *trans* arrangements? We can distinguish between these arrangements by predicting their polarity. The *trans* arrangement, being completely symmetrical, is nonpolar. However, the *cis* arrangement, with electronegative Cl atoms on one side of the platinum atom, is polar. This difference between *cis* and

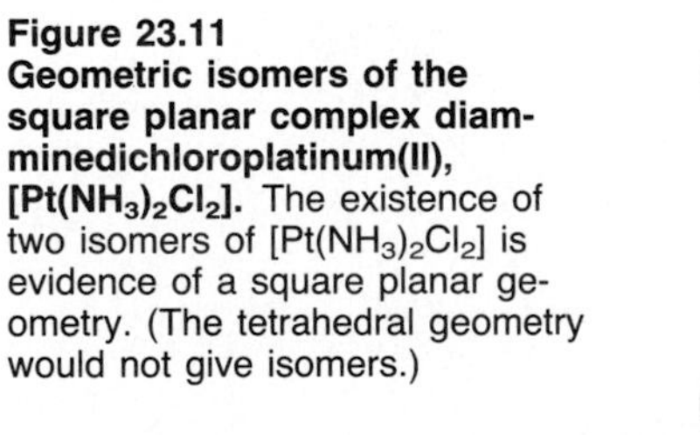

Orange-yellow compound
Solubility = 0.252 g/100 g H_2O

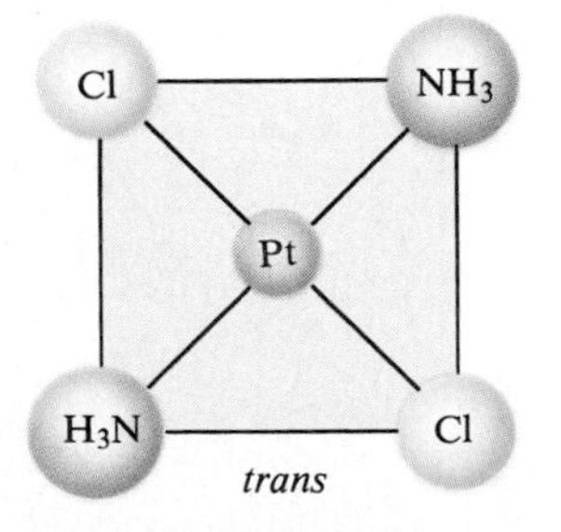

Pale yellow compound
Solubility = 0.037 g/100 g H_2O

Figure 23.11
Geometric isomers of the square planar complex diamminedichloroplatinum(II), $[Pt(NH_3)_2Cl_2]$. The existence of two isomers of $[Pt(NH_3)_2Cl_2]$ is evidence of a square planar geometry. (The tetrahedral geometry would not give isomers.)

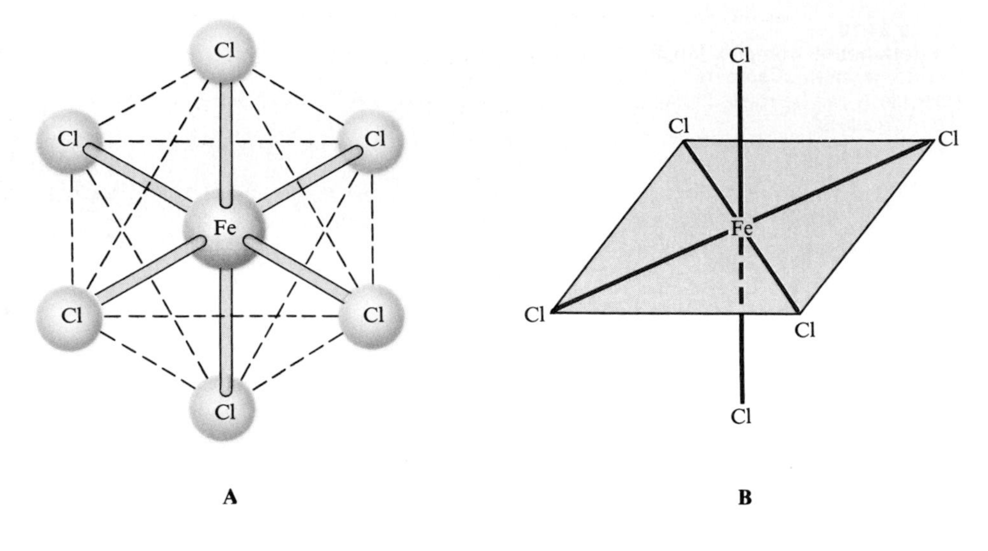

Figure 23.12
The octahedral geometry. (*A*) Each position in this geometry is equivalent. (*B*) The octahedral geometry is often shown like this.

trans isomers should be reflected in their solubilities in water, because polar substances are more soluble in water (itself a polar substance) than nonpolar substances. Thus, we expect the more soluble isomer to have the *cis* arrangement.

Although the difference in solubility is revealing, the most direct evidence of polarity in a molecule comes from measurements of dipole moment. These show that the less soluble platinum(II) complex has no dipole moment and must be *trans* and that the other isomer does have a dipole moment and must be *cis*. Figure 23.11 shows the two isomers and their properties.

Six-coordinate complexes have only one symmetrical geometry: octahedral (Figure 23.12). Geometric isomers are possible for this geometry also. Consider the complex MA_4B_2 in which two of the A ligands occupy positions just opposite one another. The other four ligands have a square planar arrangement in which *cis–trans* isomers are possible. Tetraamminedichlorocobalt(III) chloride, $[Co(NH_3)_4Cl_2]Cl$, is an example of a complex with such geometric isomers. The *trans* compound is green; the *cis* compound is purple. See Figure 23.13.

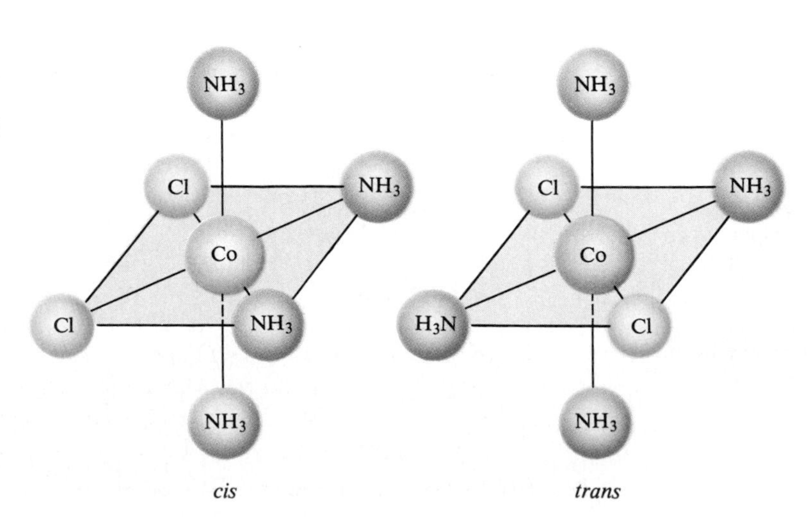

Figure 23.13
Geometric isomers of tetraamminedichlorocobalt(III) ion, $Co(NH_3)_4Cl_2^+$. The *cis* compound is purple; the *trans* compound is green.

Example 23.5 Deciding Whether Geometric Isomers Are Possible

Are there any geometric isomers of the stable octahedral complex $[Co(NH_3)_3(NO_2)_3]$? If so, draw them.

Solution

Yes, there are geometric isomers. This is easily seen if we first draw an NH_3 ligand and an NO_2^- ligand opposite one another in an octahedral geometry. Then, the other ligands can have *cis* or *trans* arrangements on the square perpendicular to the axis of the first NH_3 and NO_2^- ligands. **See Figure 23.14.**

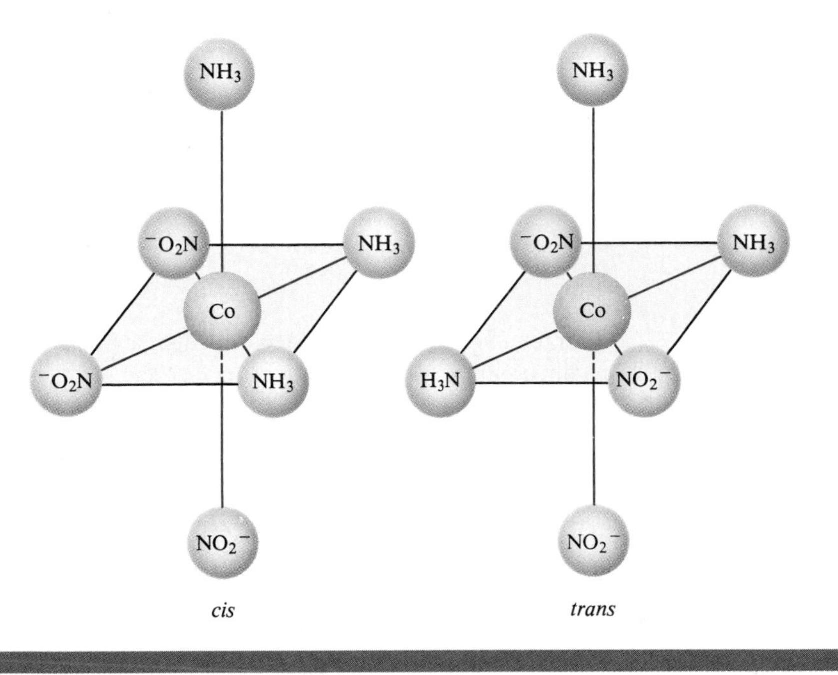

Figure 23.14
***Cis–trans* isomers of $[Co(NH_3)_3(NO_2)_3]$.** Note the arrangements of groups on the colored squares.

Exercise 23.11

Do any of the following stable octahedral complexes have geometric isomers? If so, draw them.

(a) $Co(NH_3)_5Cl^{2+}$ (b) $Co(NH_3)_4(H_2O)_2^{3+}$
(c) $[Cr(NH_3)_3(SCN)_3]$ (d) $Co(NH_3)_6^{3+}$

(See Problems 23.57 and 23.58.)

Bidentate ligands increase the potential for isomerism in octahedral complexes. Figure 23.15 shows the isomers of the dichlorobis(ethylenediamine)-cobalt(III) ion, $CoCl_2(en)_2^+$. Isomer A is the *trans* isomer (it has a green color). Note that both B and C are *cis* isomers (both have a violet color). Yet isomers B and C are not identical molecules. They are **enantiomers,** or **optical isomers;** that is, they are *isomers that are nonsuperimposable mirror images of one another.*

To better understand the nature of enantiomers, note that the two *cis* isomers have the same relationship to each other as your left hand and right hand. The mirror image of a left hand looks like the right hand, and vice versa (Figure 23.16). But neither the right hand nor the mirror image of the left hand can be turned in any way to look exactly like the left hand; that is, the right and left hands cannot be *superimposed* on one another. (Remember that a left-handed glove does not fit a right hand.) Any physical object *possessing the quality of handedness*—whose mirror image is not identical with itself—is said to be **chiral** (from the

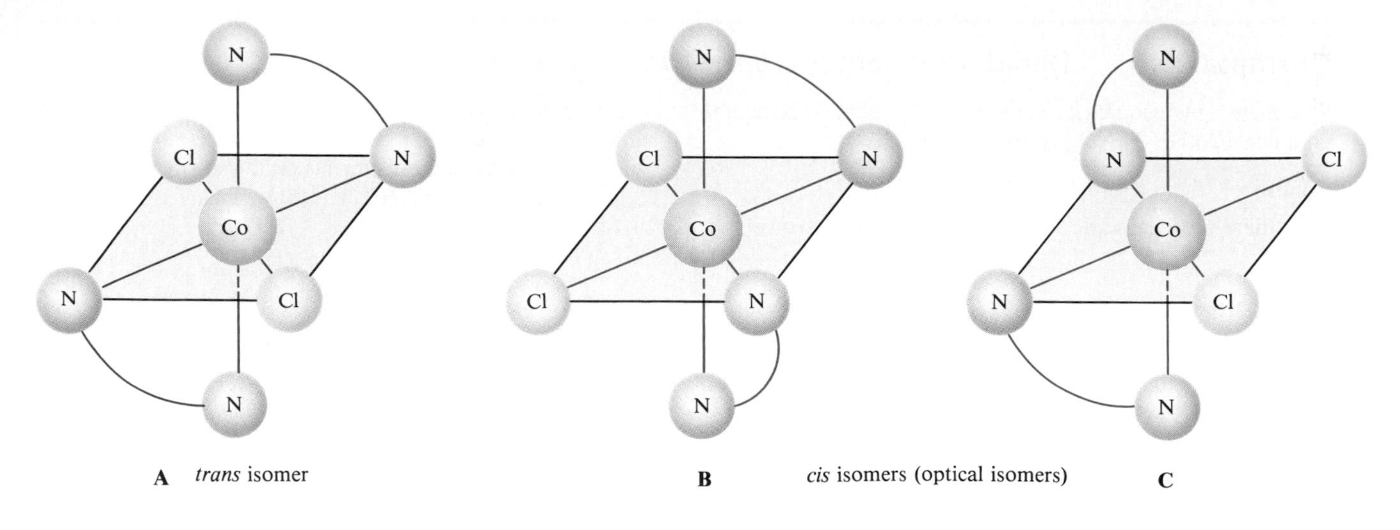

Figure 23.15
Isomers of dichlorobis(ethylenediamine)cobalt(III) ion, $CoCl_2(en)_2^+$. (*A*) The *trans* isomer. (*B* and *C*) Optical isomers of the *cis* form. Note that N͡N represents ethylenediamine.

Greek *cheir,* meaning ''hand''). Thus, a glove is chiral. However a pencil, whose mirror image looks identical to the real pencil, is *achiral.* Isomer A in Figure 23.15 is achiral; its mirror image is identical to itself. However, isomer B is chiral; its mirror image, isomer C, is not superimposable on B.

Enantiomers (optical isomers) have identical properties in a symmetrical environment. They have identical melting points and the same solubilities and colors. Enantiomers are usually differentiated by the manner in which they affect *plane-polarized* light. Normally, a light beam consists of electromagnetic waves vibrating in all possible planes about the direction of the beam (Figure 23.17). One of these planes may be selected out by passing a beam of light through a *polarizer* (say, through a Polaroid lens). When this plane-polarized light is passed through a solution containing an enantiomer, such as one isomer of *cis*-$[CoCl_2(en)_2]Cl$, the plane of the light wave is twisted. One of the enantiomers twists the plane to the right (as the light comes out toward you); the other isomer twists the plane to the left by the same angle. Because of *the ability to rotate the plane of light waves, either as pure substances or in solution,* enantiomers are said to be **optically active.** (This is also the origin of the term *optical isomer*.) Figure 23.18 shows a sketch of a *polarimeter,* an instrument that determines the angular change in the plane of a light wave made by an optically active compound.

A compound whose solution rotates the plane of polarized light to the right

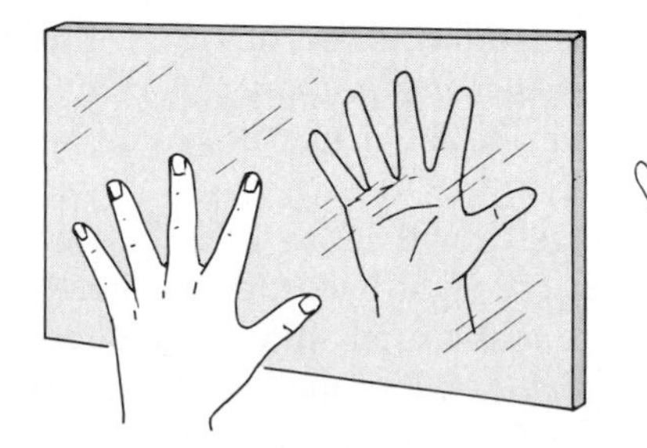

Figure 23.16
Nonsuperimposable mirror images. The mirror image of the left hand looks like the right hand, but the left hand itself cannot be superimposed on the right hand.

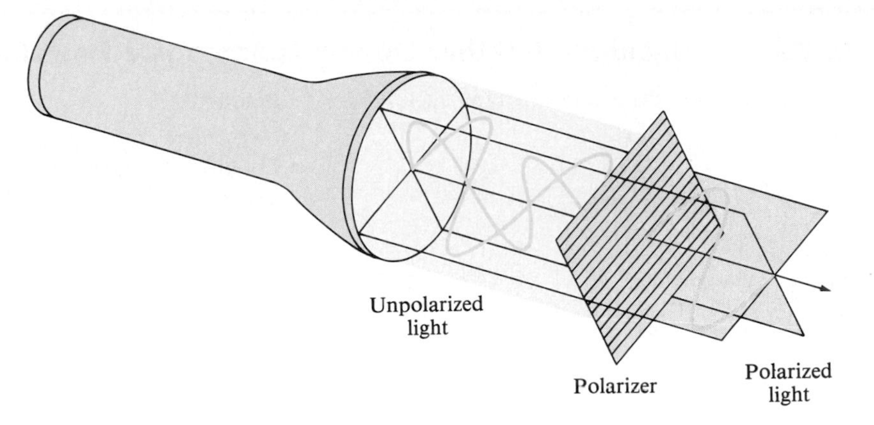

Figure 23.17
Polarization of light. Light from the source consists of waves vibrating in various planes along any axis. The polarizer filters out all waves except those vibrating in a particular plane.

(*when looking toward the source of light*) is called **dextrorotatory** and is labeled *d*. *A compound whose solution rotates the plane of polarized light to the left* (*when looking toward the source of light*) is called **levorotatory** and is labeled *l*. Thus, the dextrorotatory isomer of *cis*-$[CoCl_2(en)_2]^+$ is *d-cis*-dichlorobis(ethylenediamine)cobalt(III).

A chemical reaction normally produces *a mixture of equal amounts of optical isomers,* called a **racemic mixture.** Because the two isomers rotate the plane of polarized light in equal but opposite directions, a racemic mixture has no net effect on polarized light.

In order to show that optical isomers exist, a racemic mixture must be separated into its *d* and *l* isomers; that is, the racemic mixture must be *resolved*. One way to resolve a mixture containing *d* and *l* complex ions is to prepare a salt with an optically active ion of opposite charge. For example, the tartaric acid, $H_2C_4H_4O_6$, prepared from the white substance in wine vats is the optically active isomer *d*-tartaric acid. When the racemic mixture of *cis*-$[CoCl_2(en)_2]Cl$ is treated with *d*-tartaric acid, the *d*-tartrate salts of *d*- and *l-cis*-$[CoCl_2(en)_2]^+$ may be crystallized. These salts will no longer be optical isomers of one another and will have different solubilities.

Figure 23.18
A sketch of a polarimeter. [From Hart, *Organic Chemistry: A Short Course,* 6th ed. (Boston: Houghton Mifflin, 1983), p. 124. *Used by permission*.]

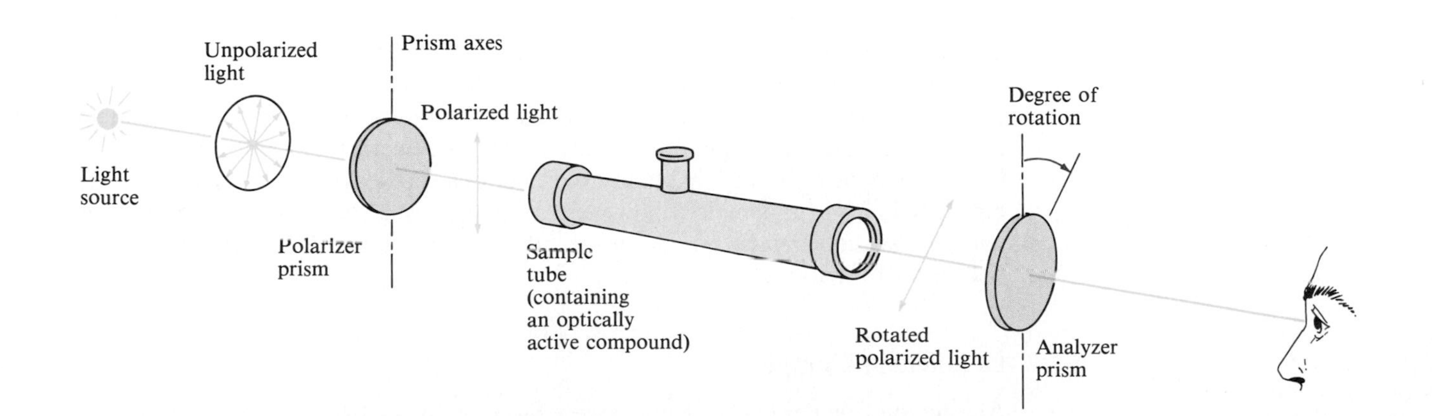

Example 23.6 Deciding Whether Optical Isomers Are Possible

Are there optical isomers of the complex $[Co(en)_3]Cl_3$? If so, draw them.

Solution

Yes, there are optical isomers, because the complex $Co(en)_3^{3+}$ has nonsuperimposable mirror images. **See Figure 23.19.**

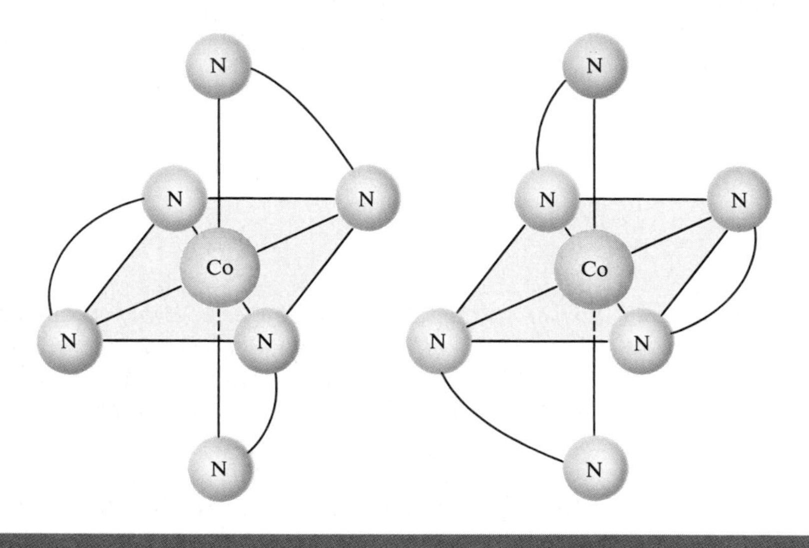

Figure 23.19
Optical isomers of the complex ion $Co(en)_3^{3+}$. Note that the isomers are nonsuperimposable mirror images of one another.

Exercise 23.12

Do any of the following have optical isomers? If so, draw them.
(a) *trans*-$Co(en)_2(NO_2)_2^+$ (b) *cis*-$Co(en)_2(NO_2)_2^+$
(c) $Ir(en)_3^{3+}$ (d) *cis*-$[Ir(H_2O)_3Cl_3]$ (See Problems 23.59 and 23.60.)

23.6 VALENCE BOND THEORY OF COMPLEXES

A complex is formed, as we noted earlier, when electron pairs from ligands are donated to a metal ion. This does not explain how the metal ion can accept electron pairs. Nor does it explain the paramagnetism often observed in complexes. *Valence bond theory* provided the first detailed explanation of the electronic structure of complexes. This explanation is essentially an extension of the view of covalent bonding we described earlier.● According to this view, a covalent bond is formed by the overlap of two orbitals, one from each bonding atom. In the usual covalent bond formation, each orbital originally holds one electron, and after the orbitals overlap, a bond is formed that holds two electrons. In the formation of a coordinate covalent bond in a complex, however, a ligand orbital containing two electrons overlaps an unoccupied orbital on the metal atom. Figure 23.20 diagrams these two bond formations.

● See Section 10.3.

Octahedral Complexes

Among the first complexes studied were those of chromium(III). There are a great number of these, including the violet complex hexaaquachromium(III) chloride,

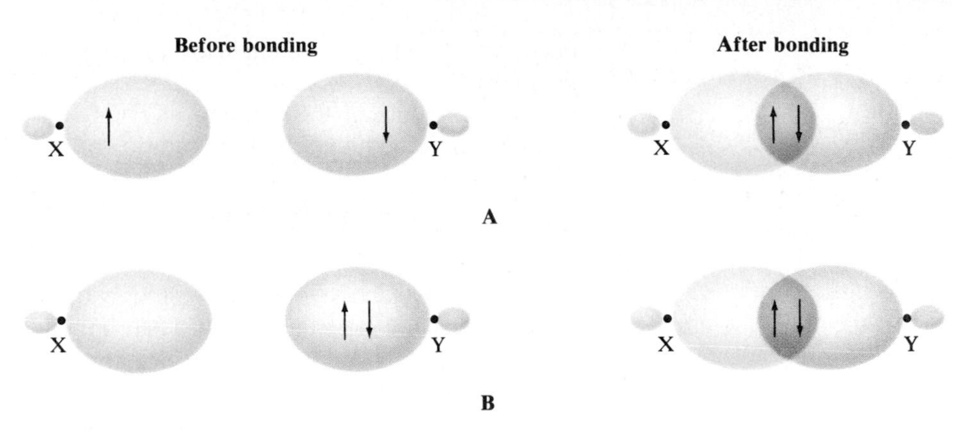

Figure 23.20
Covalent bond formation between atoms X and Y. (*A*) In the usual case, each of the overlapping orbitals contains one electron. (*B*) When a coordinate covalent bond forms, one orbital containing a lone pair of electrons overlaps an empty orbital.

$[Cr(H_2O)_6]Cl_3$, and the yellow complex hexaamminechromium(III) chloride, $[Cr(NH_3)_6]Cl_3$. Almost all chromium(III) complexes have the coordination number 6 and thus are octahedral complexes. All are paramagnetic, displaying a magnetism equivalent to that of three unpaired electrons. Let us see how valence bond theory explains the bonding and paramagnetism of the complex ions of chromium(III), such as $Cr(NH_3)_6{}^{3+}$.

The electron configuration of the chromium atom is $[Ar]3d^54s^1$. In the formation of a transition-metal ion, the outer *s* electrons ($4s^1$, here) are lost first, then the outer *d* electrons. Thus, Cr^{3+} has the configuration $[Ar]3d^3$. The orbital diagram is

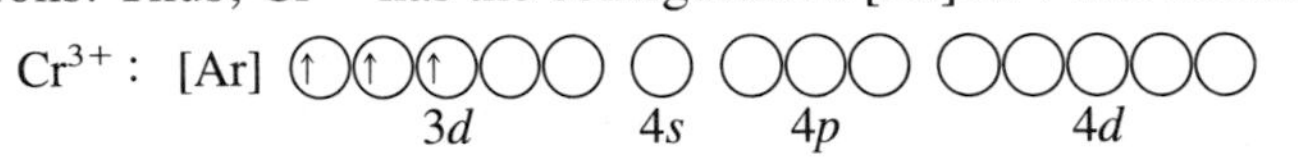

Note that the $3d^3$ electrons are placed in separate orbitals with their spins in the same direction, following Hund's rule.

If electron pairs from six $:NH_3$ ligands are to bond to Cr^{3+}, forming six equivalent bonds, hybrid orbitals will be required. Hybrid orbitals with an octahedral arrangement can be formed with two *d* orbitals, the 4*s* orbital, and the three 4*p* orbitals. The *d* orbitals could be either 3*d* or 4*d*; the two available 3*d* orbitals are used because they have lower energy. We will call these d^2sp^3 hybrid orbitals (rather than sp^3d^2) to emphasize that the *d* orbitals have a principal quantum number of 1 less than that of the *s* and *p* orbitals. We can now write the orbital diagram for the metal atom in the complex.

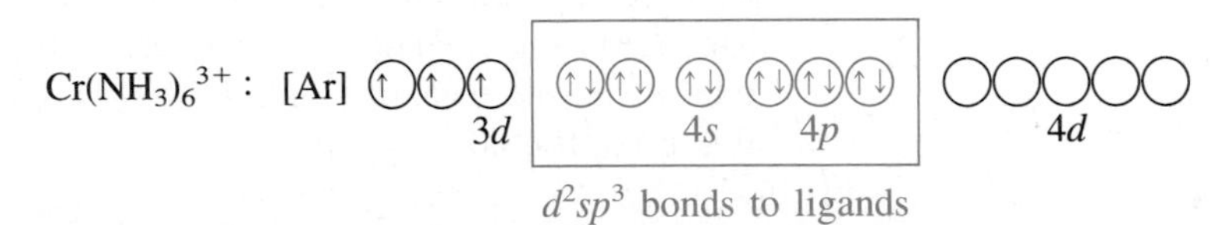

Electron pairs donated from ligands are shown in color. Note that there are three unpaired electrons in 3*d* orbitals on the chromium atom, which explains the paramagnetism of this complex ion. The bonding in other octahedral complexes of chromium(III) is essentially the same.

The bonding in complexes of iron(II) is more diverse. Most of the complexes are octahedral and paramagnetic, and the magnitude of this paramagnetism indicates four unpaired electrons. An example is the pale green hexaaquairon(II) ion, $Fe(H_2O)_6{}^{2+}$. However, the yellow hexacyanoferrate(II) ion, $Fe(CN)_6{}^{4-}$, is an example of a diamagnetic iron(II) complex ion. (Iron in oxidation state +2 also occurs occasionally in complexes of coordination number 4, with tetrahedral geometry.)

Let us consider the bonding in $Fe(H_2O)_6^{2+}$. The configuration of the iron atom is $[Ar]3d^64s^2$, and the configuration of Fe^{2+} is $[Ar]3d^6$. The orbital diagram of the ion is

Fe^{2+} : [Ar] 3*d* 4*s* 4*p* 4*d*

According to Hund's rule, the first five of the $3d^6$ electrons go into separate orbitals; the sixth electron must then pair up with one of the others. Because the 3*d* orbitals have electrons in them, they cannot be used for bonding to ligand orbitals unless some of the electrons are moved. Suppose we use two of the empty 4*d* orbitals instead. We will call the hybrid orbitals formed from them sp^3d^2 to emphasize that the *d* orbitals have the same principal quantum number as that of the *s* and *p* orbitals. Then the orbital diagram of the complex ion would be

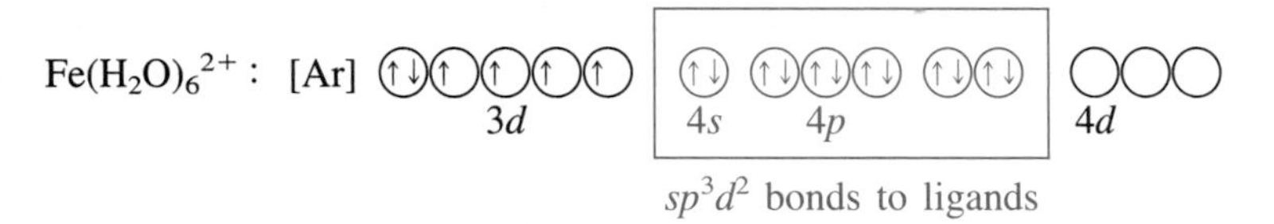

This bonding picture correctly explains the paramagnetism of the complex ion and shows that it should correspond to that of four unpaired electrons, in agreement with experiment.

Suppose, however, that in forming a complex ion of Fe^{2+}, two of the 3*d* electrons pair up so that two 3*d* orbitals are unoccupied and can be used to form d^2sp^3 hybrid orbitals. The configuration of this excited state of the Fe^{2+} ion is

Fe^{2+}(excited) : [Ar] 3*d* 4*s* 4*p* 4*d*

available for d^2sp^3 bonds

The 3*d* orbitals are lower in energy than the 4*d* orbitals, so they will be preferred for bonding if available. However, pairing of the electrons to make the two 3*d* orbitals available for d^2sp^3 bonding requires energy. We would expect this type of bonding to occur only if the bonding were sufficiently strong to provide the energy needed for electron pairing. This is apparently the case for the hexacyanoferrate(II) ion, $Fe(CN)_6^{4-}$. (The CN^- ion tends to give strong bonding to metal ions, in contrast to the weaker bonding of H_2O.) The orbital diagram for the iron atom in this complex is

$Fe(CN)_6^{4-}$: [Ar] 3*d* 4*s* 4*p* 4*d*

d^2sp^3 bonds to ligands

This diagram correctly describes the ion as diamagnetic (no unpaired electrons).

If we similarly examine any transition-metal ion that has configurations d^4, d^5, d^6, or d^7, we find two bonding possibilities. In one case, the bonding is sp^3d^2. In the other, some electrons are paired and the bonding is d^2sp^3. The number of unpaired electrons in the latter type of complex is lower than that with sp^3d^2

bonding. A **low-spin complex ion** is *a complex ion in which there is maximum pairing of electrons in the orbitals of the metal atom.* A **high-spin complex ion** is *a complex ion in which there is minimum pairing of electrons in the orbitals of the metal atom.* Because low-spin complexes require energy for pairing of electrons, they are expected to occur with ligands that bond relatively strongly. Weakly bonding ligands form high-spin complexes.

Example 23.7 Describing the Bonding in an Octahedral Complex Ion (Valence Bond Theory)

Cobalt(II) has both high-spin and low-spin octahedral complex ions. Most of the octahedral complex ions are high-spin, such as the pink ion $Co(H_2O)_6^{2+}$. There are also a few low-spin complex ions such as $Co(CN)_6^{4-}$. Describe the bonding in both $Co(H_2O)_6^{2+}$ and $Co(CN)_6^{4-}$, using valence bond theory. How many unpaired electrons are there in each complex ion?

Solution

The electron configuration of cobalt is [Ar]$3d^74s^2$, and that of the cobalt(II) ion is [Ar]$3d^7$. A high-spin complex ion such as $Co(H_2O)_6^{2+}$ would be obtained from sp^3d^2 hybrid orbitals. The complex would be paramagnetic, with three unpaired electrons.

$Co(H_2O)_6^{2+}$:

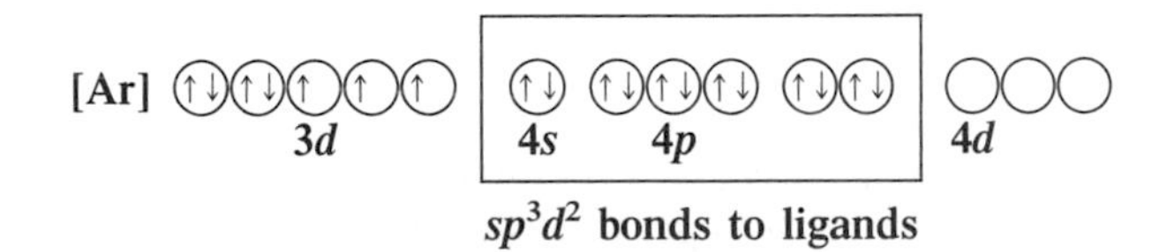

Bonding of the cobalt(II) ion using d^2sp^3 hybrid orbitals would require that two unpaired electrons in the $3d$ subshell be moved. One could pair up with another electron in a $3d$ orbital, but the other would have to be promoted to a higher unoccupied orbital ($4d$). The orbital diagram for a low-spin complex ion, such as $Co(CN)_6^{4-}$, follows. The complex is paramagnetic, with one unpaired electron.

$Co(CN)_6^{4-}$:

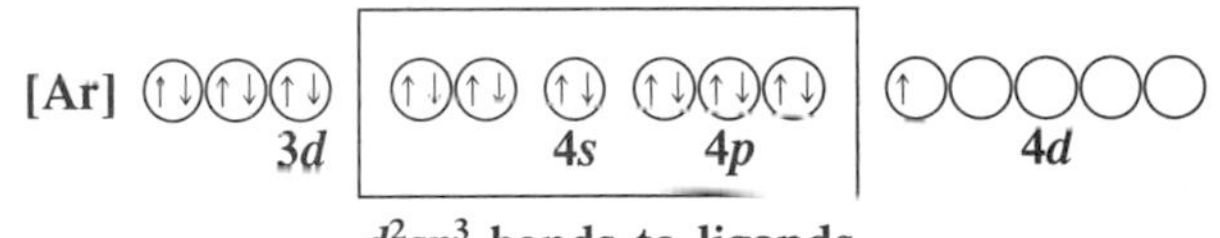

Exercise 23.13

Cobalt(III) forms many stable complex ions, including $Co(NH_3)_6^{3+}$. Most of these are octahedral and diamagnetic. The complex CoF_6^{3-}, however, is paramagnetic. Describe the bonding in $Co(NH_3)_6^{3+}$ and CoF_6^{3-}, using valence bond theory. How many unpaired electrons are there in each complex ion? (See Problems 23.61 and 23.62.)

If we examine metal ions with configurations d^8 and d^9, we again find only one spin type of octahedral complex possible. (Recall that the d^3 configuration, as in Cr^{3+}, gives only one type octahedral complex; d^1 and d^2 are similar.) An example of a d^8 complex is the green ion $Ni(H_2O)_6^{2+}$. The nickel atom has the configuration [Ar]$3d^84s^2$, and the Ni^{2+} ion has the configuration [Ar]$3d^8$. The orbital diagram for this ion is

Ni^{2+}: [Ar] (↑↓)(↑↓)(↑↓)(↑)(↑) ( ) ( )( )( ) ( )( )( )( )( )
3d 4s 4p 4d

Because three of the $3d$ orbitals are doubly occupied, two empty orbitals cannot be created by pairing of electrons. However, we could obtain octahedral hybrid

Table 23.8 Hybrid Orbitals for Various Coordination Numbers and Geometries

Coordination Number	Geometry	Hybrid Orbital
2	Linear	sp
4	Tetrahedral	sp^3
	Square planar	dsp^2
6	Octahedral	d^2sp^3 or sp^3d^2

orbitals using the 4*d* orbitals. Thus,

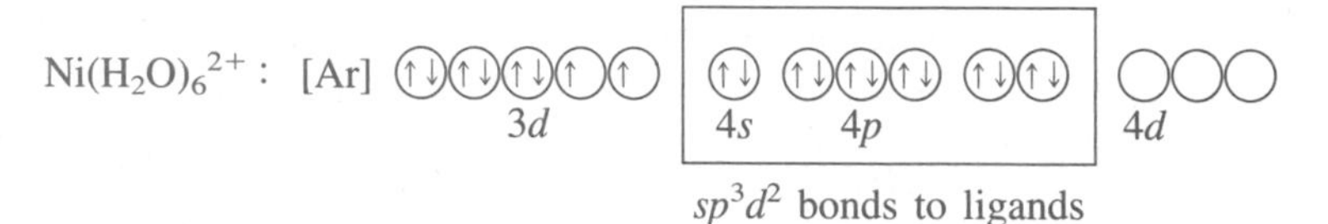

This would predict a paramagnetic complex ion with two unpaired electrons, which is what is observed experimentally.

Tetrahedral and Square Planar Complexes

Although coordination number 6 (octahedral complexes) is very common, coordination number 4 is nearly as common. In this case, the geometries are either tetrahedral or square planar. Table 23.8 lists the hybrid orbitals used to describe various geometries. The tetrahedral geometry is described by sp^3 hybrid orbitals. The square planar geometry uses dsp^2 orbitals.

Nickel(II) complex ions with four ligands are common. Most of these, such as $Ni(CN)_4^{2-}$, are diamagnetic. That $Ni(CN)_4^{2-}$ is diamagnetic is evidence for a square planar geometry.● To see this, recall that the electron configuration of the Ni^{2+} ion is $[Ar]3d^8$, with three doubly occupied 3*d* orbitals and two singly occupied 3*d* orbitals. To form dsp^2 hybrid orbitals, the two unpaired 3*d* electrons must first be paired, giving an unoccupied 3*d* orbital. The orbital diagram for the complex ion is

● Care must be used when predicting geometry from the magnetic characteristics of a complex, however. For example, a complex of nickel(II) with the bidentate ligand acetylacetonate (= acac) has the simplest formula $[Ni(acac)_2]$ and is paramagnetic. It appears from this formula that the Ni atom is four-coordinate and, because of its paramagnetism, has a tetrahedral geometry (see Example 23.8). In fact, the molecular formula is $[Ni(acac)_2]_3$, in which the Ni atom has an octahedral geometry. This geometry is expected to be paramagnetic.

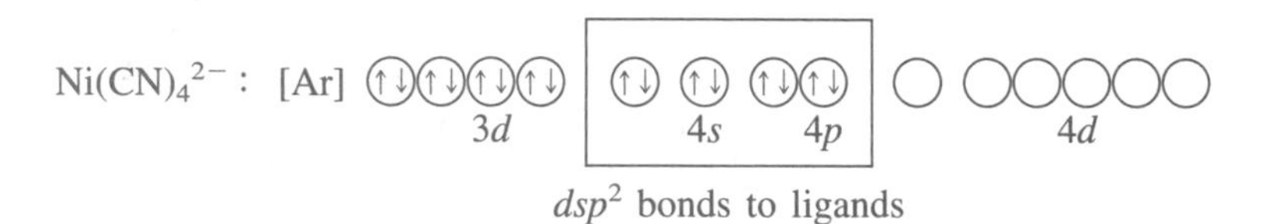

Example 23.8 Describing the Bonding in a Four-Coordinate Complex Ion (Valence Bond Theory)

Nickel(II) forms some tetrahedral complex ions such as $Ni(NH_3)_4^{2+}$. Discuss the bonding in this complex ion and describe its magnetic characteristics.

Solution

The orbital diagram for Ni^{2+}, whose electron configuration is $[Ar]3d^8$, is as follows:

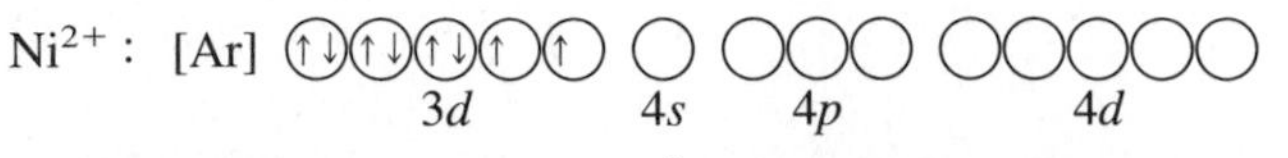

A tetrahedral geometry uses sp^3 hybrid orbitals. Therefore, none of the 3*d* electrons in Ni^{2+} need to be paired up or

promoted. The orbital diagram of $Ni(NH_3)_4^{2+}$ follows:

$Ni(NH_3)_4^{2+}$:

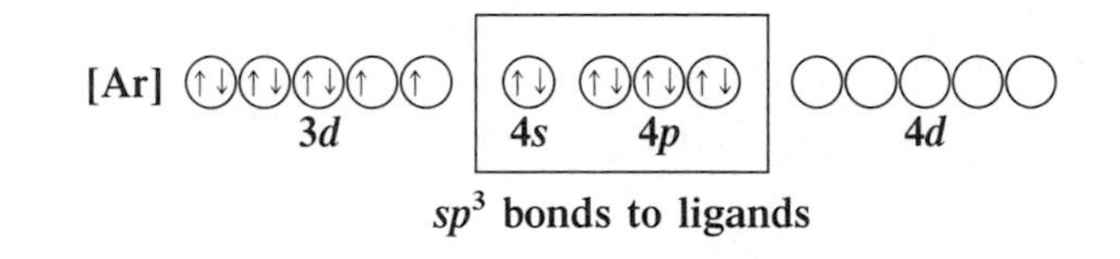

Because there are two unpaired electrons, the tetrahedral complex ions of nickel(II) are **paramagnetic,** in contrast to the square planar complex ions, which are diamagnetic.

Exercise 23.14

The complex ion $CoCl_4^{2-}$ is paramagnetic, with a magnetism corresponding to three unpaired electrons. If the complex ion indeed has four ligands, as suggested by the formula, what geometry is indicated? (See Problems 23.63 and 23.64.)

23.7 CRYSTAL FIELD THEORY

Although valence bond theory explains the bonding and magnetic properties of complexes in straightforward fashion, it is limited in two important ways. First, the theory cannot simply explain the color of complexes. Second, the theory is difficult to extend quantitatively. Consequently, another theory—crystal field theory—has emerged as the prevailing view of transition-metal complexes.

Crystal field theory is *a model of the electronic structure of transition-metal complexes that considers how the energies of the* d *orbitals of a metal ion are affected by the electric field of the ligands.* According to this theory, the ligands in a transition-metal complex are treated as point charges. Thus, a ligand anion becomes simply a point of negative charge. A neutral molecule, with its electron pair that it donates to the metal atom, is replaced by a partial negative charge, representing the negative end of the molecular dipole. In the electric field of these negative charges, the five *d* orbitals of the metal atom no longer have exactly the same energy. The result, as we will see, explains both the paramagnetism and the color observed in certain complexes.

The simplifications used in crystal field theory are drastic. Treating the ligands as point charges is essentially the same as treating the bonding as ionic. However, it turns out that the theory can be extended to include covalent character in the bonding. Simple extension is usually referred to as *ligand field theory,* but after including several levels of refinements, the theory becomes equivalent to molecular orbital theory.

Effect of an Octahedral Field on the *d* Orbitals

All five *d* orbitals of an isolated metal atom have the same energy. But if the atom is brought into the electric field of several point charges, these *d* orbitals may be affected in different ways and therefore may have different energies. To understand how this can happen, we must first see what these *d* orbitals look like. We will then be able to picture what happens to them in the crystal field theory of an octahedral complex.

Figure 23.21 shows the shapes of the five *d* orbitals. The orbital labeled d_{z^2} has a dumbbell shape along the *z*-axis, with a collar in the *x–y* plane surrounding

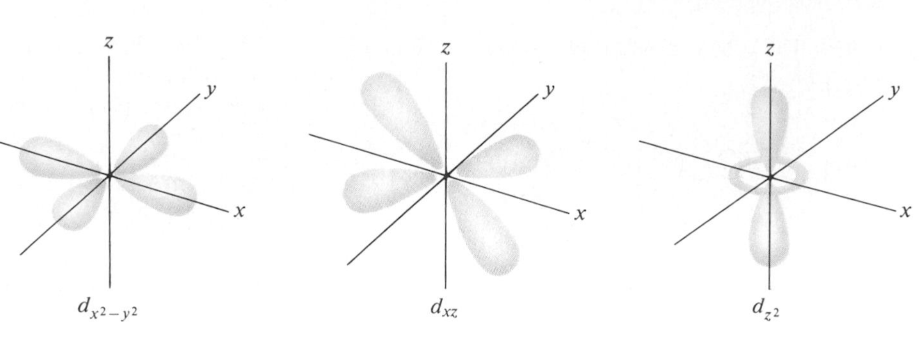

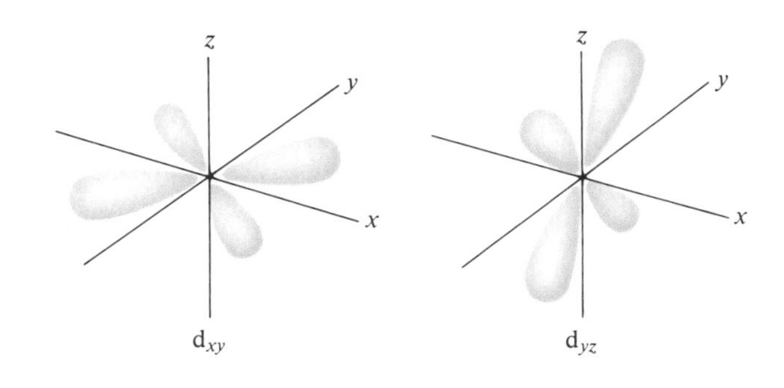

Figure 23.21
The five *d* orbitals. In an isolated atom, these orbitals have the same energy; but when the atom is placed in the electric field similar to that in an octahedral complex ion, the orbitals split into two sets with different energies.

this dumbbell. Remember that this shape represents the volume most likely to be occupied by an electron in this orbital. The other four *d* orbitals have "cloverleaf" shapes, each differing from one another only in the orientation of the orbitals in space. Thus, the "cloverleaf" orbital $d_{x^2-y^2}$ has its lobes along the *x*-axis and the *y*-axis. Orbitals d_{xy}, d_{yz}, and d_{xz} have their lobes directed between the two sets of axes designated in the orbital label. Orbital d_{xy}, for example, has its lobes lying between the *x*- and *y*-axes.

A complex ion with six ligands will have those ligands arranged about the metal atom to reduce their mutual repulsions. Normally, an octahedral arrangement achieves this. Now imagine that the ligands are replaced by negative charges. If the ligands are anions, they are replaced by the anion charge. If the ligands are neutral molecules, they are replaced by the partial negative charge from the molecular dipole. These six charges are placed at equal distances from the metal atom, one charge on each of the positive and negative sides of the *x*-, *y*-, and *z*-axes (see Figure 23.22).

Fundamentally, the bonding in this model of a complex is due to the attraction of the positive metal ion for the negative charges of the ligands. However, an electron in a *d* orbital of the metal atom is repelled by the negative charge of the ligands. This repulsion alters the energy of the *d* orbital depending on whether it is directed *toward* ligands or *between* ligands. For example, consider the difference in the repulsive effect of ligands on metal-ion electrons in the d_{z^2} and the d_{xy} orbitals. Because the d_{z^2} orbital is directed at the two ligands on the *z*-axis (one on the $-z$ side and the other on the $+z$ side), an electron in the orbital is rather strongly repelled by them. Thus, the energy of the d_{z^2} orbital becomes greater.

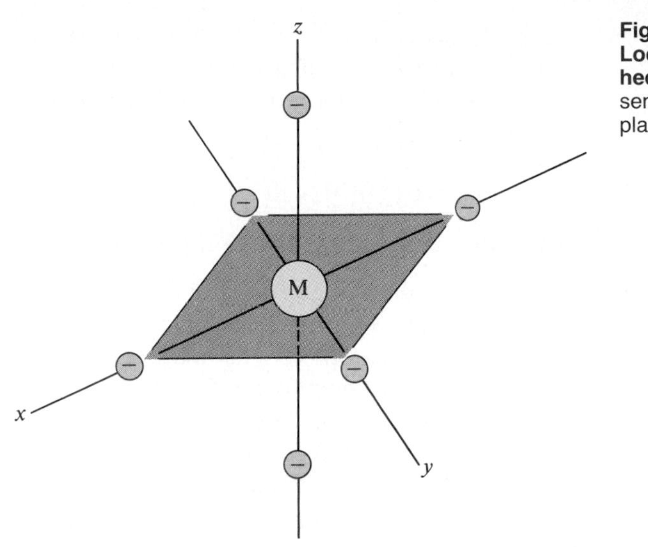

Figure 23.22
Location of ligands in an octahedral complex. Ligands, represented by negative charges, are placed on the *x*-, *y*-, and *z*-axes.

Similarly, an electron in the d_{xy} orbital is repelled by the negative charge of the ligands, but because the orbital is not pointed directly at the ligands, this repulsive effect is smaller. Its energy is raised, but less than the energy of the d_{z^2} orbital is raised.

If we look at the five *d* orbitals in an octahedral field (electric field of octahedrally arranged charges), we see that we can divide them into two sets. Orbitals d_{z^2} and $d_{x^2-y^2}$ are both directed toward ligands, and orbitals d_{xy}, d_{yz}, and d_{xz} are directed between ligands. The orbitals in the first set (d_{z^2} and $d_{x^2-y^2}$) have higher energy than those in the second set (d_{xy}, d_{yz}, and d_{xz}). Figure 23.23 shows the

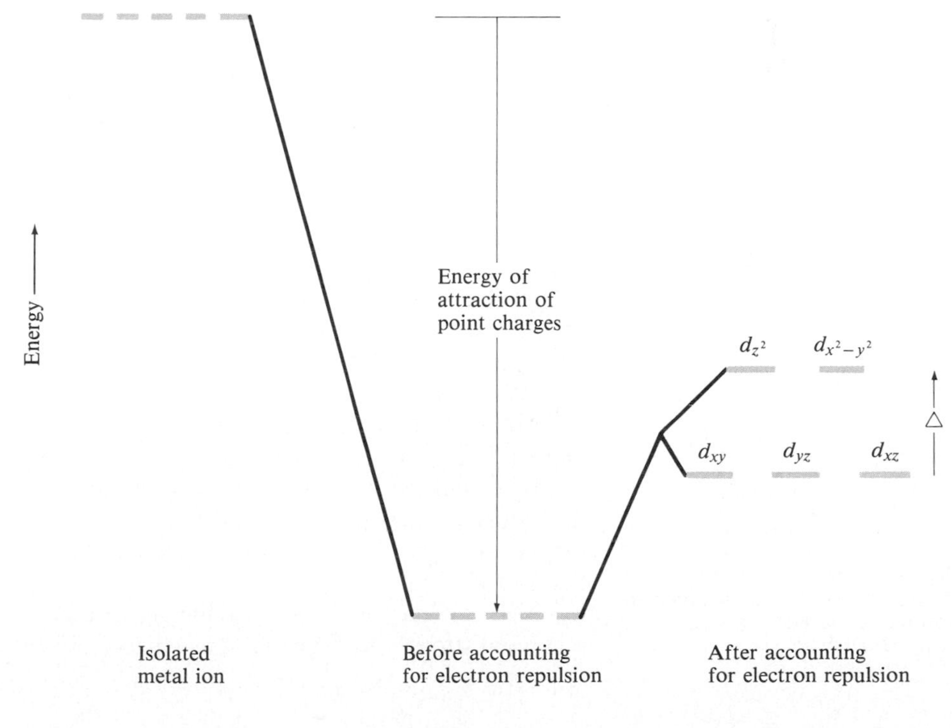

Figure 23.23
Energy levels of *d* orbitals in an octahedral field. The positive metal ion is attracted to the negative charges (ligands), but electrons in the *d* orbitals are repelled by them. Thus, although there is an overall attraction, the *d* orbitals no longer have the same energy.

energy levels of the d orbitals in an octahedral field. *The difference in energy between the two sets of* d *orbitals on a central metal ion that arises from the interaction of the orbitals with the electric field of the ligands* is called the **crystal field splitting, Δ.**

Figure 23.24
Occupation of the 3*d* orbitals in an octahedral complex of Cr^{3+}. Note that the electrons occupy different orbitals with the same spin (Hund's rule).

High-Spin and Low-Spin Complexes

Once we have the energy levels for the d orbitals in an octahedral complex, we can decide how the d electrons of the metal ion are distributed in them. Knowing this distribution, we can predict the magnetic characteristics of the complex.

Consider the complex ion $Cr(NH_3)_6{}^{3+}$. According to crystal field theory, it consists of the Cr^{3+} ion surrounded by NH_3 molecules treated as partial negative charges. The effect of these charges is to split the d orbitals of Cr^{3+} into two sets as shown in Figure 23.23. We now ask how the d electrons are distributed among the d orbitals of the Cr^{3+} ion. Because the electron configuration of Cr^{3+} is $[Ar]3d^3$, there are three d electrons to distribute. These electrons are placed in the d orbitals of lower energy, following Hund's rule (Figure 23.24). We see that the complex ion $Cr(NH_3)_6{}^{3+}$ has three unpaired electrons and is therefore paramagnetic.

As another example, consider the complex ion $Fe(H_2O)_6{}^{2+}$. What are its magnetic characteristics? Remember, we need only look at the d electrons of the metal ion, Fe^{2+}. The electron configuration of the ion is $[Ar]3d^6$. We distribute six electrons among the d orbitals of the complex in such a way as to get the lowest total energy. If we place all six electrons into the lower three d orbitals, we get the distribution shown by the energy-level diagram in Figure 23.25A. All of the electrons are paired, so we would predict that this distribution gives a diamagnetic complex. The $Fe(H_2O)_6{}^{2+}$ ion, however, is paramagnetic. Thus, the distribution we gave in Figure 23.25A is not correct. What was our mistake?

The mistake we made above was to ignore the **pairing energy, *P*,** *the energy required to put two electrons into the same orbital*. When an orbital is already occupied by an electron, it requires energy to put another electron into that orbital because of their mutual repulsion. Suppose that this pairing energy is greater than the crystal field splitting; that is, suppose $P > \Delta$. In that case, once the first three electrons have singly occupied the three lower-energy d orbitals, the fourth electron will go into one of the higher-energy d orbitals. It will take less energy to do that than to pair up with an electron in one of the lower-energy orbitals. Similarly, the fifth electron will go into the last empty d orbital. The sixth electron must pair up, so it goes into one of the lower-energy orbitals. Figure 23.25B shows this electron distribution. In this case, there are four unpaired electrons and the complex is paramagnetic.

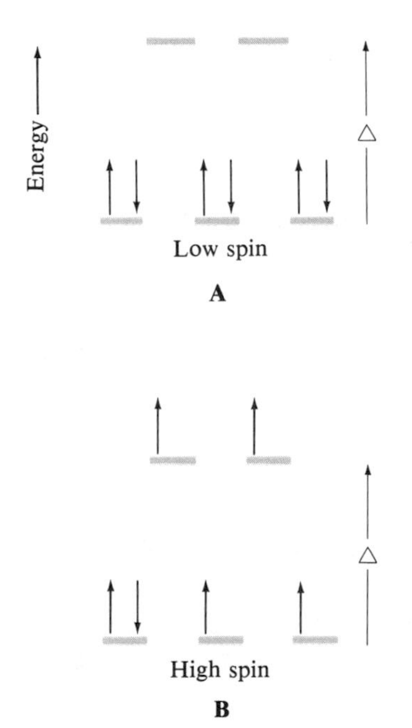

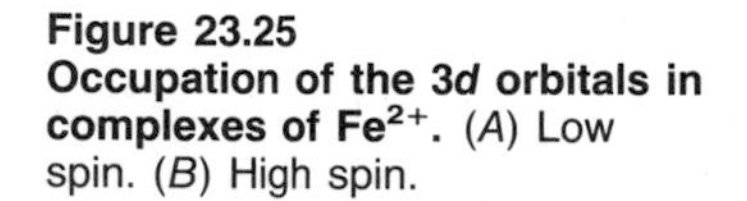

Figure 23.25
Occupation of the 3*d* orbitals in complexes of Fe^{2+}. (*A*) Low spin. (*B*) High spin.

We see that crystal field theory predicts two possibilities: a *low-spin complex* when $P < \Delta$ and a *high-spin complex* when $P > \Delta$. The value of Δ, as we will explain later, can be obtained from the spectrum of a complex, and the value of P can be calculated theoretically. Even in the absence of these numbers, however, the theory predicts that a paramagnetic octahedral complex of Fe^{2+} should have a magnetism equal to that of four unpaired electrons. This is what we find for the $Fe(H_2O)_6{}^{2+}$ ion.

We would expect low-spin diamagnetic Fe^{2+} complexes to occur for ligands that bond strongly to the metal ion—that is, for those giving large Δ. Ligands that might give a low-spin complex are suggested by a look at the spectrochemical series. The **spectrochemical series** is *an arrangement of ligands according to the relative magnitudes of the crystal field splittings they induce in the* d *orbitals of a metal ion*. The following is a short version of the spectrochemical series.

Weak-bonding ligands — Strong-bonding ligands

$$I^- < Br^- < Cl^- < F^- < OH^- < H_2O < NH_3 < en < NO_2^- < CN^- < CO$$

Increasing Δ ⟶

From this series, we see that the CN^- ion bonds more strongly than H_2O, which explains why $Fe(CN)_6^{4-}$ is a low-spin complex ion and $Fe(H_2O)_6^{2+}$ is a high-spin complex. We can also see why carbon monoxide might be expected to be poisonous. We know that O_2 bonds reversibly to the Fe(II) atom of hemoglobin, so the bonding is only moderately strong. According to the spectrochemical series, however, carbon monoxide, CO, forms a strong bond. The bonding in this case is irreversible (or practically so). It results in a very stable complex of CO and hemoglobin, which cannot function then as a transporter of O_2.

Example 23.9 Describing the Bonding in an Octahedral Complex Ion (Crystal Field Theory)

Describe the distribution of *d* electrons in the complex ion $Co(H_2O)_6^{2+}$, using crystal field theory. The hexaaquacobalt(II) ion is a high-spin complex ion. What would be the distribution of *d* electrons in an octahedral cobalt(II) complex ion that is low spin? How many unpaired electrons are there in each ion?

Solution

The electron configuration of Co^{2+} is $[Ar]3d^7$. The high-spin and low-spin distributions in the *d* orbitals are

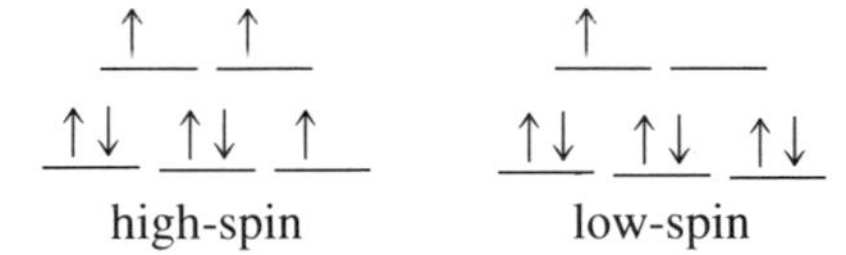

Thus, $Co(H_2O)_6^{2+}$, a high-spin complex, has three unpaired electrons. A low-spin complex would have one unpaired electron.

Exercise 23.15

Describe the distribution of *d* electrons in $Ni(H_2O)_6^{2+}$, using crystal field theory. How many unpaired electrons are there in this ion? (See Problems 23.65 and 23.66.)

Tetrahedral and Square Planar Complexes

When a metal ion bonds with tetrahedrally arranged ligands, the *d* orbitals of the ion split to give two *d* orbitals at lower energy and three *d* orbitals at higher energy (just the opposite of what we found for an octahedral field). See Figure 23.26A. On the other hand, in the field of ligands in a square planar arrangement, the *d* orbitals split as shown in Figure 23.26B.

The observed splittings in a tetrahedral field are approximately one-half the size of those seen in comparable octahedral complexes. As a result, only high-spin complexes are observed. In the square planar case, only low-spin complexes have been found.

Figure 23.26
Energy splittings of the *d* orbitals in a complex with four ligands. (*A*) A tetrahedral field. The crystal field splitting, Δ, is smaller than in a comparable octahedral complex. (*B*) A square planar field.

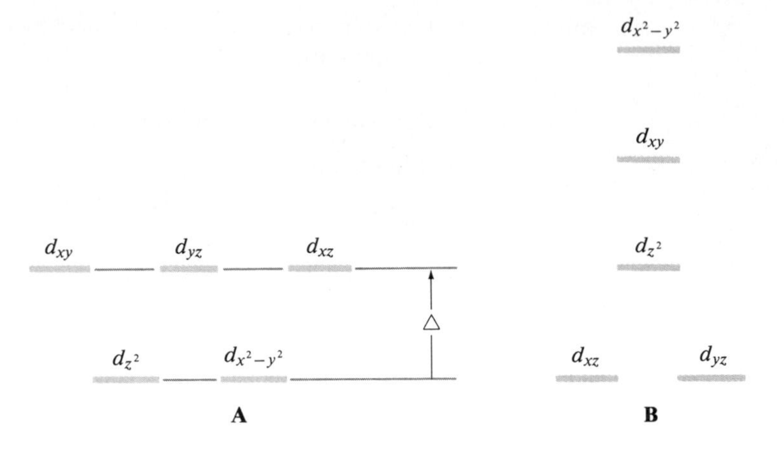

Example 23.10 Describing the Bonding in a Four-Coordinate Complex Ion (Crystal Field Theory)

Describe the *d*-electron distributions of the complexes $Ni(NH_3)_4{}^{2+}$ and $Ni(CN)_4{}^{2-}$, according to crystal field theory. The tetraamminenickel(II) ion is paramagnetic, and the tetracyanonickelate(II) ion is diamagnetic.

Solution

We expect the tetrahedral field to give high-spin complexes and the square planar field to give low-spin complexes. Thus, the geometry of the $Ni(NH_3)_4{}^{2+}$ ion, which is paramagnetic, is probably tetrahedral. The distribution of *d* electrons in the Ni^{2+} ion (configuration d^8) is

↑↓ ↑ ↑

↑↓ ↑↓

$Ni(NH_3)_4{}^{2+}$ (tetrahedral)

The geometry of $Ni(CN)_4{}^{2-}$, which is diamagnetic, is probably square planar; the distribution of *d* electrons is

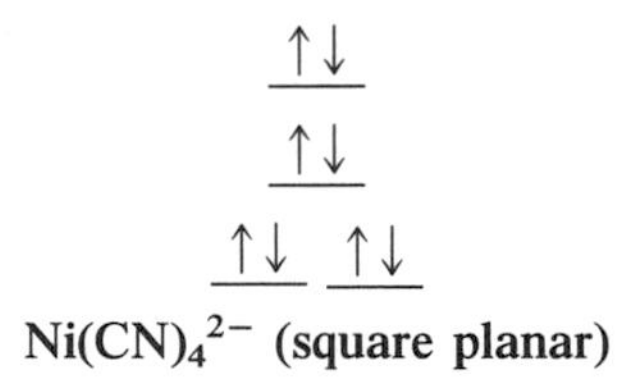

$Ni(CN)_4{}^{2-}$ (square planar)

Exercise 23.16

Describe the distribution of *d* electrons in the $CoCl_4{}^{2-}$ ion. The ion has a tetrahedral geometry. Assume a high-spin complex. (See Problems 23.67 and 23.68.)

Visible Spectra of Transition-Metal Complexes

Frequently, substances absorb light only in regions outside the visible spectrum and reflect, or pass on (transmit), all of the visible wavelengths. As a result, these substances appear white or colorless (white light is a mixture of all visible wavelengths). However, some substances absorb certain wavelengths in the visible spectrum and transmit the remaining ones. Thus, they appear colored. Many transition-metal complexes, as we have noted, are colored substances. The color

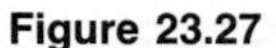

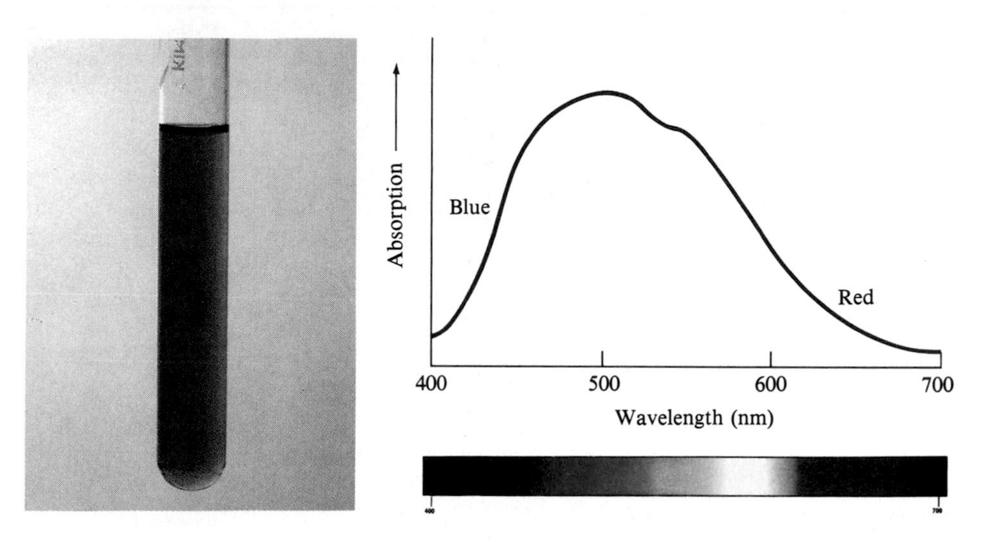

Figure 23.27
Color and visible spectrum of $Ti(H_2O)_6{}^{3+}$. *Left:* A test tube containing a solution of $Ti(H_2O)_6{}^{3+}$ solution. *Right:* Visible spectrum of $Ti(H_2O)_6{}^{3+}$. Unlike atomic spectra, which show absorption lines, spectra of ions and molecules in solution yield broad bands resulting from changes in nuclear motion that accompany the electronic transitions.

results from electron jumps, or transitions, between the two closely spaced *d* orbital energy levels that come from the crystal field splitting.●

● The colors of many gemstones are due to transition-metal-ion impurities in the mineral. Ruby has Cr^{3+} in alumina, Al_2O_3, and emerald has Cr^{3+} in beryl, $Be_3Al_2(SiO_3)_6$.

The spectrum of a d^1 configuration complex is particularly simple. Hexaaquatitanium(III) ion, $Ti(H_2O)_6{}^{3+}$, is an example. Titanium has the configuration $[Ar]3d^24s^2$, and Ti^{3+} has the configuration $[Ar]3d^1$. According to crystal field theory, the *d* electron occupies one of the lower-energy *d* orbitals of the octahedral complex. Figure 23.27 shows the visible spectrum of $Ti(H_2O)_6{}^{3+}$. It results from a transition, or jump, of the *d* electron from a lower-energy *d* orbital to a higher-energy *d* orbital, as shown in Figure 23.28. Note that the energy change equals the crystal field splitting, Δ. Consequently, the wavelength, λ, of light that is absorbed is related to Δ.●

● Recall from Section 7.7 that the energy change during a transition equals $h\nu$.

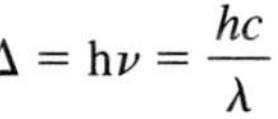

$$\Delta = h\nu = \frac{hc}{\lambda}$$

Or,

$$\lambda = \frac{hc}{\Delta}$$

When white light, which contains all visible wavelengths (from 400 nm to 750 nm), falls on a solution containing $Ti(H_2O)_6{}^{3+}$, blue-green light is absorbed. (The maximum light absorption is observed at 500 nm, which is blue-green light. See Table 23.9.) The other wavelengths of visible light, including red and some blue light, pass through the solution, giving it a red-purple color.

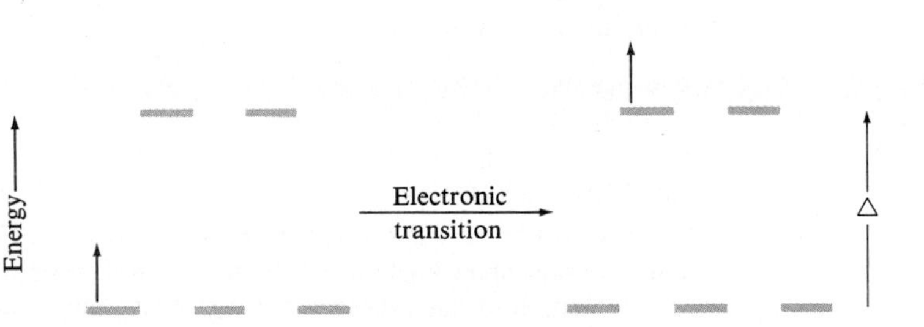

Figure 23.28
The electronic transition responsible for the visible absorption in $Ti(H_2O)_6{}^{3+}$. An electron undergoes a transition from a lower-energy *d* orbital to a higher-energy *d* orbital. The energy change equals the crystal field splitting, Δ.

Table 23.9 Color Observed for Given Absorption of Light by an Object

Wavelength Absorbed (nm)	Color Absorbed	Approximate Color Observed*
410	Violet	Green-yellow
430	Violet-blue	Yellow
480	Blue	Orange
500	Blue-green	Red
530	Green	Purple
560	Green-yellow	Violet
580	Yellow	Violet-blue
610	Orange	Blue
680	Red	Blue-green
720	Purple-red	Green

*The exact color depends on the relative intensities of various wavelengths coming from the object and on the response of the eye to those wavelengths.

When the ligands in the Ti^{3+} complex are changed, it changes Δ and therefore the color of the complex. For example, replacing H_2O ligands by weaker F^- ligands should give a smaller crystal field splitting and therefore an absorption at longer wavelengths. The absorption of $TiF_6{}^{3-}$ is at 590 nm, in the yellow rather than the blue-green, and the color observed is violet-blue.

From this discussion, we see that the visible spectrum can be related to the crystal field splitting, and values of Δ can therefore be obtained by spectroscopic analysis. However, when there is more than one *d* electron, several excited states can be formed. Consequently, the spectrum generally consists of several lines, and the analysis is more complicated than that for Ti^{3+}.

Example 23.11 Predicting the Relative Wavelengths of Absorption of Complex Ions

When water ligands in $Ti(H_2O)_6{}^{3+}$ are replaced by CN^- ligands to give $Ti(CN)_6{}^{3-}$, the maximum absorption shifts from 500 nm to 450 nm. Is this shift in the expected direction? Explain. What color do you expect to observe for this ion?

Solution

According to the spectrochemical series, CN^- is a more strongly bonding ligand than H_2O. Consequently, Δ should increase, and the wavelength of the absorption ($\lambda = hc/\Delta$) should decrease. **Thus, the shift of the absorption is in the expected direction.** Because the absorbed light is between blue and violet-blue (see Table 23.9), the observed color is **orange-yellow** (this is the *complementary color* of the color between blue and violet-blue).

Exercise 23.17

The $Fe(H_2O)_6{}^{3+}$ ion has a pale purple color, and the $Fe(CN)_6{}^{3-}$ ion has a ruby-red color. What are the approximate wavelengths of the maximum absorption for each ion? Is the shift of wavelength in the expected direction? Explain.

(See Problems 23.69, 23.70, 23.71, and 23.72.)

Related Topic

THE COOPERATIVE RELEASE OF OXYGEN FROM OXYHEMOGLOBIN

Hemoglobin is an iron-containing substance in red blood cells responsible for the transport of O_2 from the lungs to various parts of the body. Myoglobin is a similar substance in muscle tissue. It acts as a reservoir for the storage of O_2 and as a transporter of O_2 within muscle cells. The explanation for the different actions of these two substances involves some fascinating transition-metal chemistry.

Myoglobin consists of heme—a complex of Fe(II) bonded to a quadridentate ligand (Figure 23.8)—and globin. Globin, a protein, is attached through a nitrogen atom to one of the octahedral positions of Fe(II). The sixth position is vacant in free myoglobin, but is occupied by O_2 in oxymyoglobin. Hemoglobin is essentially a four-unit structure of myoglobinlike units—that is, a *tetramer* of myoglobin.

Myoglobin and hemoglobin exist in equilibrium with the oxygenated forms oxymyoglobin and oxyhemoglobin, respectively. For example, hemoglobin (Hb) and O_2 are in equilibrium with oxyhemoglobin (HbO_2).

$$Hb + O_2 \rightleftharpoons HbO_2$$

Although hemoglobin is a tetramer of myoglobin, it does not function simply as four independent units of myoglobin. For it to function efficiently as a transporter of O_2 from the lungs and then be able to release that O_2 easily to myoglobin, hemoglobin must be less strongly attached to O_2 in the vicinity of a muscle cell than is myoglobin. In hemoglobin, the release of O_2 from one heme group triggers the release of O_2 from another heme group of the same molecule. In other words, there is a *cooperative release* of O_2 from hemoglobin that makes it possible for it to give up its O_2 to myoglobin.

The mechanism that has been postulated for this cooperative release of O_2 depends on a change of iron(II) from a low-spin form to a high-spin form, with a corresponding change in radius of the iron atom. In oxyhemoglobin, iron(II) exists in the low-spin form. When O_2 leaves, the iron atom goes to a high-spin form with two electrons in the higher-energy *d* orbital. These higher-energy orbitals are antibonding molecular orbitals and are somewhat larger in size than the lower-energy *d* orbitals, which are nonbonding.

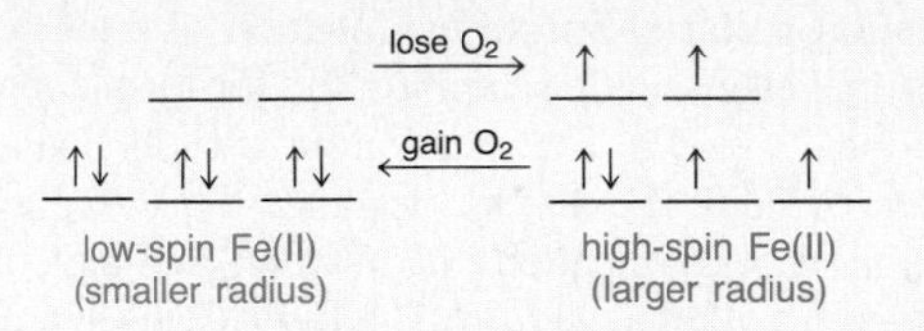

When an O_2 molecule leaves a heme group, the radius of the iron atom increases, and the atom pops out of the heme plane by about 0.7 Å. In hemoglobin, this change triggers the cooperative release of another O_2 molecule. As the iron atom moves, the attached globin group moves with it. This motion of one globin group causes an adjacent globin group in the tetramer to alter its shape, which in turn makes possible the easy release of an O_2 molecule from its heme unit.

A Checklist for Review

IMPORTANT TERMS

***d*-block transition elements** (p. 923)
***f*-block (inner) transition elements** (p. 923)
complex ion (23.3)
complex (coordination compound) (23.3)
ligands (23.3)
coordination number (23.3)
monodentate ligand (23.3)
bidentate ligand (23.3)
polydentate ligand (23.3)
chelate (23.3)
structural isomers (23.5)
stereoisomers (23.5)
ionization isomers (23.5)
hydrate isomers (23.5)
coordination isomers (23.5)
linkage isomers (23.5)
geometric isomers (23.5)
enantiomers (optical isomers) (23.5)
chiral (23.5)
optically active (23.5)
dextrorotatory (23.5)
levorotatory (23.5)
racemic mixture (23.5)
low-spin complex ion (23.6)
high-spin complex ion (23.6)
crystal field theory (23.7)
crystal field splitting, Δ (23.7)
pairing energy, *P* (23.7)
spectrochemical series (23.7)

SUMMARY OF FACTS AND CONCEPTS

Transition elements are defined as those elements having a partially filled *d* subshell in any common oxidation state. They have a number of characteristics, including high melting points and a multiplicity of oxidation states. Compounds of transition elements are frequently colored and many are paramagnetic. These properties are due to the participation of *d* orbitals in bonding. We described the chemical properties of two transition elements: Cr and Cu. Chromium metal reacts with acids to give Cr^{2+} ion, which is readily oxidized to Cr^{3+}. Although the +3 oxidation state is stable, Cr^{3+} can be oxidized in basic solution to the chromium(VI) species CrO_4^{2-}. Copper metal reacts only with acids having strongly oxidizing anions, such as HNO_3; it gives Cu^{2+} ion. Copper(I) ion disproportionates in aqueous solution to Cu and Cu^{2+}, but insoluble compounds of copper(I) can be prepared in water solutions.

Transition-metal atoms often function as Lewis acids, reacting with groups called *ligands* by forming coordinate covalent bonds to them. The metal atom with its ligands is a *complex ion* or neutral *complex*. Ligands that bond through more than one atom are called *polydentate,* and the complex that is formed is called a *chelate*. The IUPAC has agreed on a nomenclature of complexes that gives basic structural information about the species. The presence of isomers in coordination compounds is evidence for particular geometries. For example, $[Pt(NH_3)_2Cl_2]$ has two isomers, which is evidence for a square planar geometry having *cis–trans isomers*. Octahedral complexes with bidentate ligands may have *optical isomers*—that is, isomers that are mirror images of one another.

Valence bond theory gave the earliest theoretical description of the electronic structure of a complex. According to this theory, a complex forms when doubly occupied ligand orbitals overlap unoccupied orbitals of the metal atom. Octahedral complexes with metal-ion configuration d^4, d^5, d^6, or d^7 give rise to two bonding types. In the case of strong-bonding ligands, *d* electrons are paired so that d^2sp^3 hybrid orbitals can be used, and a *low-spin complex* forms. Otherwise, the *d* electrons of the metal ion remain unpaired, and a *high-spin complex* forms using sp^3d^2 hybrid orbitals.

Crystal field theory also predicts high-spin and low-spin complexes in these configurations of the metal ion. Following this theory, the ligands are treated as electric charge points that affect the energy of the *d* orbitals of the metal ion. In an octahedral complex, two of the *d* orbitals have higher energy than the other three. A high-spin complex forms when the *pairing energy* is greater than the *crystal field splitting,* so that electrons would "prefer" to occupy a higher-energy *d* orbital rather than pair up with an electron in a lower-energy orbital. When the pairing energy is smaller than the crystal field splitting, the *d* orbitals are occupied in the normal fashion, giving a low-spin complex. Color in transition-metal complexes is explained as due to a jump of an electron from the lower-energy to the higher-energy *d* orbitals. The crystal field splitting can be obtained experimentally from the visible spectrum of a complex.

OPERATIONAL SKILLS

1. Predicting relative oxidizing strengths of transition-metal compounds Given similar compounds of two transition elements in the same column having the same oxidation state, decide which is the stronger oxidizing agent (Example 23.1).

2. Preparing compounds of chromium or copper Write balanced equations describing the preparation of a compound of Cr or Cu (Example 23.2).

3. Writing the IUPAC name given the structural formula of a compound, and vice versa Given the structural formulas of coordination compounds, write the IUPAC names (Example 23.3); given the IUPAC names of complexes, write the structural formulas (Example 23.4).

4. Deciding whether isomers are possible Given the formula of a complex, decide whether geometric isomers are possible and, if so, draw them (Example 23.5). Given the structural formula of a complex, decide whether enantiomers (optical isomers) are possible and, if so, draw them (Example 23.6).

5. Describing the bonding in a complex ion Given a transition-metal complex ion, describe the bonding types (high-spin and low-spin, if both exist), using valence bond theory for octahedral and four-coordinate complexes. Give the number of unpaired electrons in the complex (Examples 23.7 and 23.8). Do the same using crystal field theory (Examples 23.9 and 23.10).

6. Predicting the relative wavelengths of absorption of complex ions Given two complexes that differ only in the ligands, predict, on the basis of the spectrochemical series, which complex absorbs at higher wavelength. Given the absorption maxima, predict the colors of the complexes (Example 23.11).

Review Questions

23.1 What characteristics of the transition elements set them apart from the main-group elements?

23.2 According to the building-up principle, what is the electron configuration of the ground state of the technetium atom (atomic number 43)?

23.3 The highest melting point for metals in the fifth period occurs for molybdenum. Explain why this is expected.

23.4 Iron, cobalt, and nickel are similar in properties and are sometimes studied together as the "iron triad." For example, each is a fairly active metal that reacts with acids to give hydrogen and the +2 ions. In addition to the +2 ions, the +3 ions of the metals also figure prominently in the chemistries of the elements. Explain why these elements are similar.

23.5 Palladium and platinum are very similar to one another. Thus, they are unreactive toward most acids. However, nickel, which is in the same column of the periodic table, is an active metal. Explain why this difference exists.

23.6 Write balanced equations for the reactions of Cr and Cu metals with HCl(*aq*). If no reaction occurs, write NR.

23.7 Write balanced equations for the reactions of chromium(III) oxide with a strong acid and with a strong base.

23.8 Describe the structure of copper(II) sulfate pentahydrate. What color change occurs when the salt is heated? What causes the color change?

23.9 Copper(I) sulfate dissolves in water to give copper(II) sulfate. What is the other product? Write the balanced equation for the reaction.

23.10 What evidence did Werner obtain to show that the platinum complex $PtCl_4 \cdot 4NH_3$ has the structural formula $[Pt(NH_3)_4Cl_2]Cl_2$?

23.11 Define the terms *complex ion, ligand,* and *coordination number*. Use an example to illustrate the use of these terms.

23.12 Define the term *bidentate ligand*. Give two examples.

23.13 Rust spots on clothes can be removed by dissolving them in oxalic acid. The oxalate ion forms a stable complex with Fe^{3+}. Using an electron dot formula, indicate how an oxalate ion bonds to the metal ion.

23.14 Describe step by step how the name *potassium hexacyanoferrate(II)* leads one to the structural formula $K_4[Fe(CN)_6]$.

23.15 What three properties of coordination compounds have been important in determining the details of their structure and bonding?

23.16 Define each of the following and give an example of each: (a) ionization isomerism, (b) hydrate isomerism, (c) coordination isomerism, (d) linkage isomerism.

23.17 Define *geometric isomerism* and *optical isomerism* and give an example of each.

23.18 Compounds A and B are known to be stereoisomers of one another. Compound A has a violet color; compound B has a green color. Are compounds A and B geometric or optical isomers?

23.19 Explain the difference in behavior of *d* and *l* isomers with respect to polarized light.

23.20 What is a racemic mixture? Describe one method of resolving a racemic mixture.

23.21 Describe the formation of a coordinate covalent bond between a metal-ion orbital and a ligand orbital.

23.22 (a) Describe the steps in the formation of a high-spin octahedral complex of Fe^{2+} in valence bond terms. (b) Do the same for a low-spin complex.

23.23 Explain why *d* orbitals of a transition-metal atom may have different energies in the octahedral field of six negative charges. Describe how each of the *d* orbitals is affected by the octahedral field.

23.24 What is meant by the term *crystal field splitting?* How is it determined experimentally?

23.25 What is meant by the term *pairing energy?* How do the relative values of pairing energy and crystal field splitting determine whether a complex is low-spin or high-spin?

23.26 (a) Use crystal field theory to describe a high-spin octahedral complex of Fe^{2+}. (b) Do the same for a low-spin complex.

23.27 What is the spectrochemical series? Use the ligands CN^-, H_2O, Cl^-, and NH_3 to illustrate the term. Then arrange them in order, describing the meaning of this order.

23.28 A complex absorbs red light from a single electron transition. What color is this complex?

Practice Problems

PROPERTIES OF THE TRANSITION ELEMENTS

23.29 Chromium forms several oxides. One of these is a dark red, crystalline solid melting at 197°C. It dissolves readily in water, giving an acidic solution. Another oxide is a dark green solid melting at 2435°C. It is insoluble in water but dissolves in acids and bases. If one of these oxides is CrO_3 and the other is Cr_2O_3, which is which? Explain your answer.

23.30 Iron forms iron(II) chloride, $FeCl_2$, and iron(III) chloride, $FeCl_3$. One of these chlorides is a dark brown solid melting at 306°C; the other is a white, crystalline solid with greenish tint that melts at 674°C. Which description fits iron(II) chloride? Why do you think so?

23.31 Of the elements chromium and tungsten, only one is known to form a hexachloride. Which element would that be? Explain.

23.32 Manganese and rhenium form oxides in the +7 oxidation state. One of these metal(VII) oxides is a bright yellow solid melting at 300°C. The other is a red-brown oily liquid that explodes on warming. Which of these is Mn_2O_7? Why do you think so?

23.33 Give balanced equations, using complete formulas, for the preparation of $Cr(OH)_3$ from chromium metal.

23.34 Give balanced equations, with complete formulas, for the preparation of chromium(III) sulfate from chromium metal.

23.35 Find the oxidation numbers of the transition metal in each of the following compounds: (a) $FeCO_3$, (b) MnO_2, (c) CuCl, (d) CrO_2Cl_2.

23.36 Find the oxidation numbers of the transition metal in each of the following compounds: (a) $CoSO_4$, (b) Ta_2O_5, (c) $Cu_2(OH)_3Cl$.

23.37 Write the balanced equation for the reaction of iron(II) ion with nitrate ion in acidic solution. Nitrate ion is reduced to NO.

23.38 Write the balanced equation for the reaction of sulfurous acid with dichromate ion.

STRUCTURAL FORMULAS AND NAMING OF COMPLEXES

23.39 A cobalt complex whose composition corresponded to the formula $Co(NO_2)_2Cl \cdot 4NH_3$ gave an electrical conductance equivalent to two ions per formula unit. Excess silver nitrate solution immediately precipitates 1 mol AgCl per formula unit. Write a structural formula consistent with these results.

23.40 A cobalt complex has a composition corresponding to the formula $Co(NO_3)Cl_2 \cdot 4NH_3$. From electrical-conductance measurements, it was determined that there are two ions per formula unit. Silver nitrate solution gave no immediate precipitate. Write a structural formula consistent with this information.

23.41 Give the coordination number of the transition-metal atom in each of the following complexes.
(a) $Au(CN)_4^-$ (b) $[Co(NH_3)_4(H_2O)_2]Cl_3$
(c) $[Au(en)_2]Cl_3$ (d) $Cr(en)_2(C_2O_4)^+$

23.42 Give the coordination number of the transition element in each of the following complexes.
(a) $[Ni(NH_3)_6](ClO_3)_2$ (b) $[Cu(NH_3)_4]SO_4$
(c) $[Cr(en)_3]Cl_3$ (d) $K_2[Ni(CN)_4]$

23.43 Determine the oxidation number of the transition element in each of the following complexes.
(a) $K_2[Ni(CN)_4]$ (b) $Mo(en)_3^{3+}$
(c) $Cr(C_2O_4)_3^{3-}$ (d) $[Co(NH_3)_5(NO_2)]Cl_2$

23.44 For each of the following complexes, determine the oxidation state of the transition-metal atom.
(a) $[CoCl(en)_2(NO_2)]NO_2$ (b) $PtCl_4^{2-}$
(c) $K_3[Cr(CN)_6]$ (d) $Fe(H_2O)_5(OH)^{2+}$

23.45 Consider the complex ion $Cr(NH_3)_2Cl_2(C_2O_4)^-$.
(a) What is the oxidation state of the metal atom?
(b) Give the formula and name of each ligand in the ion.
(c) What is the coordination number of the metal atom?
(d) What would be the charge on the complex if all ligands were chloride ions?

23.46 Consider the complex ion $Mn(NH_3)_2(H_2O)_3(OH)^{2+}$.
(a) What is the oxidation state of the metal atom?
(b) Give the formula and name of each ligand in the ion.
(c) What is the coordination number of the metal atom?
(d) What would be the charge on the complex if all ligands were chloride ions?

23.47 Write the IUPAC name for each of the following coordination compounds.
(a) $K_3[FeF_6]$ (b) $Cu(NH_3)_2(H_2O)_2^{2+}$
(c) $(NH_4)_2[Fe(H_2O)F_5]$ (d) $Ag(CN)_2^-$

23.48 Name the following complexes, using IUPAC rules.
(a) $K_4[Mo(CN)_8]$ (b) CrF_6^{3-}
(c) $V(C_2O_4)_3^{2-}$ (d) $K_2[FeCl_4]$

23.49 Give the IUPAC name for each of the following.
(a) $Fe(CO)_5$ (b) $Rh(CN)_2(en)_2^+$
(c) $Cr(NH_3)_4SO_4]Cl$ (d) MnO_4^-

23.50 Give the IUPAC name for each of the following.
(a) $W(CO)_8$ (b) $[Co(H_2O)_2(en)_2](SO_4)_3$
(c) $K[Mo(CN)_8]$ (d) CrO_4^{2-}

23.51 Write the structural formula for each of the following compounds.
(a) potassium hexacyanomanganate(III)
(b) sodium tetracyanozincate(II)
(c) tetraamminedichlorocobalt(III) nitrate
(d) hexaamminechromium(III) tetrachlorocuprate(II)

23.52 Give the structural formula for each of the following complexes.
(a) diaquadicyanocopper(II)
(b) potassium hexachloroplatinate(IV)
(c) tetraamminenickel(II) perchlorate
(d) tetraammineplatinum(II) tetrachlorocuprate(II)

ISOMERISM

23.53 Give the type of structural isomerism shown by each of the following.
(a) $Co(CN)_5(NCS)^{3-}$ and $Co(CN)_5(SCN)^{3-}$
(b) $[Co(NH_3)_6][Cr(C_2O_4)_3]$ and $[Cr(NH_3)_6][Co(C_2O_4)_3]$
(c) $[Co(NH_3)_3(H_2O)_2Cl]Br_2$ and $[Co(NH_3)_3(H_2O)ClBr]Br \cdot H_2O$
(d) $[Co(NH_3)_4Cl(NO_2)]Cl$ and $[Co(NH_3)_4Cl_2]NO_2$

23.54 Give the type of structural isomerism shown by each of the following.
(a) $[CoCl(en)_2(NO_2)]NO_2$ and $[Co(en)_2(NO_2)_2]Cl$
(b) $[Co(NH_3)_6][Cr(NO_2)_6]$ and $[Cr(NH_3)_6][Co(NO_2)_6]$
(c) $[Co(en)_2(ONO)_2Cl$ and $[Co(en)_2(NO_2)_2]Cl$
(d) $[Cr(H_2O)_2Cl_2(py)_2]Cl$ and $[Cr(H_2O)Cl_3(py)_2] \cdot H_2O$ (py = pyridine, C_5H_5N)

23.55 A complex has a composition corresponding to the formula $CoBr_2Cl \cdot 4NH_3$. What is the structural formula if conductance measurements show two ions per formula unit? Silver nitrate solution gives an immediate precipitate of AgCl but no AgBr. Write the structural formula of an isomer.

23.56 Studies of a complex gave a composition corresponding to the formula $CoBr(C_2O_4) \cdot 4NH_3$. Conductance measurements indicate that there are two ions per formula unit. If calcium nitrate gives no immediate precipitate of calcium oxalate, but silver nitrate precipitates silver bromide, what is the structural formula of the complex? Write the structural formula of an isomer.

23.57 Draw *cis–trans* structures of any of the following square planar or octahedral complexes that exhibit geometric isomerism. Label the drawings *cis* or *trans*.
(a) $[Pd(NH_3)_2Cl_2]$ (b) $Pd(NH_3)_3Cl^+$
(c) $Pd(NH_3)_4^{2+}$ (d) $Ru(NH_3)_4Br_2^+$

23.58 If any of the following octahedral complexes display geometric isomerism, draw the structures and label them as *cis* or *trans*.
(a) $Co(NO_2)_4(NH_3)_2^-$ (b) $Co(NH_3)_5(NO_2)^{2+}$
(c) $Pt(NH_3)_3Br_3^+$ (d) $Cr(NH_3)_5Cl^{2+}$

23.59 Determine whether there are optical isomers of any of the following. If so, sketch the isomers.
(a) *cis*-$Co(NH_3)_2(en)_2^{3+}$ (b) *trans*-$IrCl_2(C_2O_4)_2^{3-}$

23.60 Sketch mirror images of each of the following. From these sketches, determine whether optical isomers exist and note this fact on the drawings.
(a) $Rh(en)_3^{3+}$ (b) *cis*-$Cr(NH_3)_2(SCN)_4^-$

VALENCE BOND AND CRYSTAL FIELD THEORIES

23.61 For each of the following, first draw orbital diagrams for the isolated metal atom and metal ion. Then, using valence bond theory, draw the orbital diagram for the metal atom in the octahedral complex.
(a) $V(H_2O)_6^{3+}$ (b) $Fe(en)_3^{3+}$ (high spin)
(c) $Rh(CN)_6^{3-}$ (low spin)

23.62 For each of the following, first draw orbital diagrams for the isolated metal atom and metal ion. Then, using valence bond theory, draw the orbital diagram for the metal atom in the octahedral complex.
(a) $V(H_2O)_6^{2+}$ (b) FeF_6^{3-} (high spin)
(c) $Co(en)_3^{3+}$ (low spin)

Nomenclature of Alkanes

As we noted in the chapter opening, several million organic compounds are known. A nomenclature for these substances has developed over the years as a way of understanding and classifying their structures. This nomenclature is now formulated in rules agreed upon by the International Union of Pure and Applied Chemistry (IUPAC).

The first four straight-chain alkanes (methane, ethane, propane, and butane) have long-established names. Higher members of the series are named from the Greek words indicating the number of carbon atoms in the molecule, with the suffix *-ane* added. For example, the straight-chain alkane with the formula C_5H_{12} is named pentane. Table 24.1 gives the names of the first ten straight-chain alkanes.

The following four IUPAC rules are applied in naming the branched-chain alkanes:

1. Determine the longest continuous (not necessarily straight) chain of carbon atoms in the molecule. The base name of the branched-chain alkane is that of the straight-chain alkane (Table 24.1) corresponding to the number of carbon atoms in this longest chain. For example, in

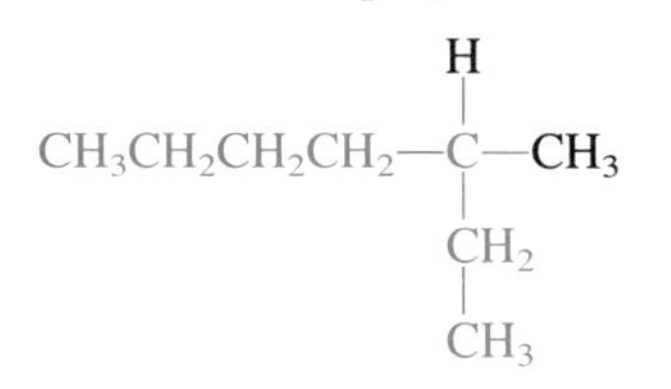

   the longest continuous carbon chain, shown in color, has seven carbon atoms, giving the base name heptane. The full name for the alkane includes the names of any branched chains. These names are placed in front of the base name, as described in the remaining rules.

2. Any chain branching off the longest chain is named as an alkyl group. An **alkyl group** is *an alkane less one hydrogen atom.* (Table 24.2 lists some alkyl groups.) When a hydrogen atom is removed from an end carbon atom of a straight-chain alkane, the resulting alkyl group is named by changing the suffix *-ane* of the alkane to *-yl*. Thus, removing a hydrogen atom from methane gives the methyl group, —CH_3. The structure shown in rule 1 has a methyl group as a branch on the heptane chain.

Table 24.2 Important Alkyl Groups

Original Alkane	Structure of Alkyl Group	Name of Alkyl Group
Methane, CH_4	CH_3—	Methyl
Ethane, CH_3CH_3	CH_3CH_2—	Ethyl
Propane, $CH_3CH_2CH_3$	$CH_3CH_2CH_2$—	Propyl
Propane, $CH_3CH_2CH_3$	CH_3CHCH_3 |	Isopropyl
Butane, $CH_3CH_2CH_2CH_3$	$CH_3CH_2CH_2CH_2$—	Butyl
Isobutane, CH_3CHCH_3 | CH_3	| CH_3CCH_3 | CH_3	*Tertiary*-butyl (*t*-butyl)

3. The complete name of a branch requires a number that locates that branch on the longest chain. For this purpose, we number each carbon atom on the longest chain in whichever direction gives the smaller numbers for the locations of all branches. The structural formula in rule 1 is numbered as shown in the following structure:

$$\overset{7}{C}H_3\overset{6}{C}H_2\overset{5}{C}H_2\overset{4}{C}H_2-\overset{3}{C}H(-CH_3)-{}^{2}CH_2-{}^{1}CH_3 \quad \left(\text{not } \overset{1}{C}H_3\overset{2}{C}H_2\overset{3}{C}H_2\overset{4}{C}H_2-\overset{5}{C}H(-CH_3)-{}^{6}CH_2-{}^{7}CH_3\right)$$

Thus, the methyl branch is located at carbon 3 of the heptane chain (not carbon 5). The complete name of the branch is 3-methyl, and the compound is named 3-methylheptane. Note that the branch name and the base name are written as a single word, with a hyphen following the number.

4. When there are more than one alkyl branch of the same kind (say, two methyl groups), this number is indicated by a Greek prefix, such as *di-*, *tri-*, or *tetra-*, used with the name of the alkyl group. The position of each group on the longest chain is given by numbers. For example,

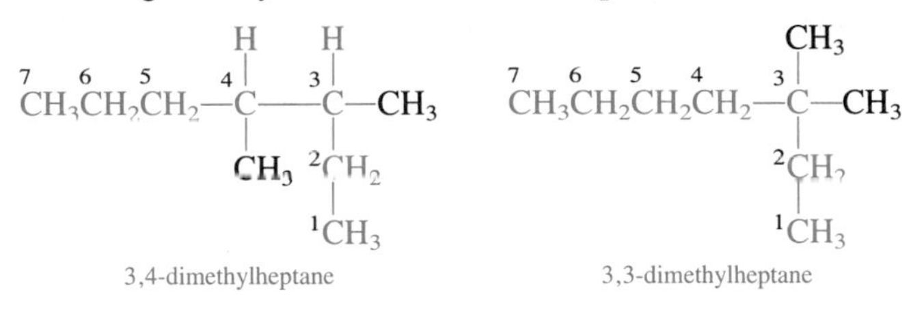
3,4-dimethylheptane   3,3-dimethylheptane

Note that the position numbers are separated by a comma and are followed by a hyphen.

When there are two or more different alkyl branches, the name of each branch, with its position number, precedes the base name. The branch names are placed in alphabetical order. For example,

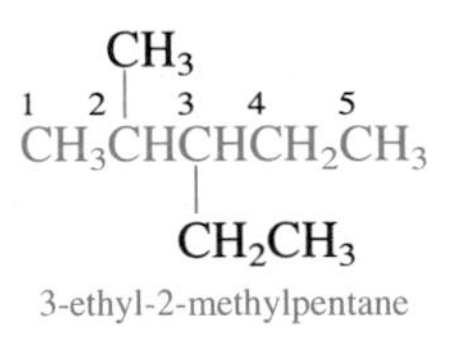
3-ethyl-2-methylpentane

Note the use of hyphens.

Example 24.1 Writing the IUPAC Name of an Alkane Given the Structural Formula

Give the IUPAC name for each of the following compounds.

(a) $(CH_3CH_2CH_2)(CH_3CH_2CH_2CH_2)CHCH_2CH_2CH_3$

(b) $CH_3-C(CH_3)_2-CH_2-CH_2-CH_2-CH_3$

(continued)

Solution

(a) The longest continuous chain is numbered as follows:

$$
\begin{array}{l}
\qquad CH_3CH_2CH_2\diagdown \;{}^{4\quad 3\quad 2\quad 1} \\
\qquad\qquad\qquad\qquad CHCH_2CH_2CH_3 \\
{}^{8\quad 7\quad 6\quad 5}\qquad\qquad \diagup \\
CH_3CH_2CH_2CH_2
\end{array}
$$

The name of the compound is **4-propyloctane.** If the longest chain had been numbered in the opposite direction, we would have the name 5-propyloctane. But because 5 is larger than 4, this name is unacceptable. (b) The numbering of the longest chain is

$$
\begin{array}{c}
{}^{1}CH_3 \\
| \\
CH_3\!-\!{}^{2}C\!-\!CH_3 \\
| \\
{}^{3}CH_2 \\
| \\
{}^{4}CH_2 \\
| \\
{}^{5}CH_2 \\
| \\
{}^{6}CH_3
\end{array}
$$

Any one of the methyl carbon atoms could be given the number 1, and the other methyl groups branch off carbon atom 2. Hence, the name is **2,2-dimethylhexane.**

Exercise 24.1

What is the IUPAC name for each of the following hydrocarbons?

(a)
$$
\begin{array}{c}
\quad CH_3 \\
\quad | \\
CH_3CHCHCH_3 \\
| \\
CH_3
\end{array}
$$

(b)
$$
\begin{array}{c}
\qquad CH_2CH_2CH_3 \\
| \\
CH_3CHCHCH_2CH_3 \\
| \\
CH_3
\end{array}
$$

(See Problems 24.21 and 24.22.)

Example 24.2 Writing the Structural Formula of an Alkane Given the IUPAC Name

Write the condensed structural formula of 4-ethyl-3-methylheptane.

Solution

First write out the carbon skeleton for heptane.

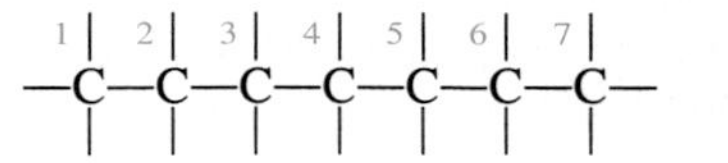

Then attach the alkyl groups.

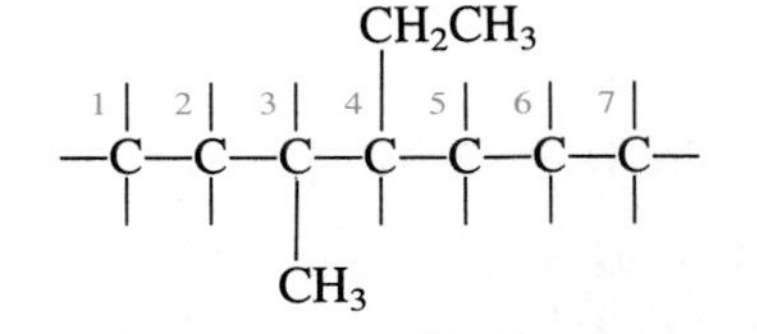

After filling out the structure with H atoms, we have

$$
\begin{array}{c}
\qquad CH_2CH_3 \\
| \\
CH_3CH_2CHCHCH_2CH_2CH_3 \\
| \\
CH_3
\end{array}
$$

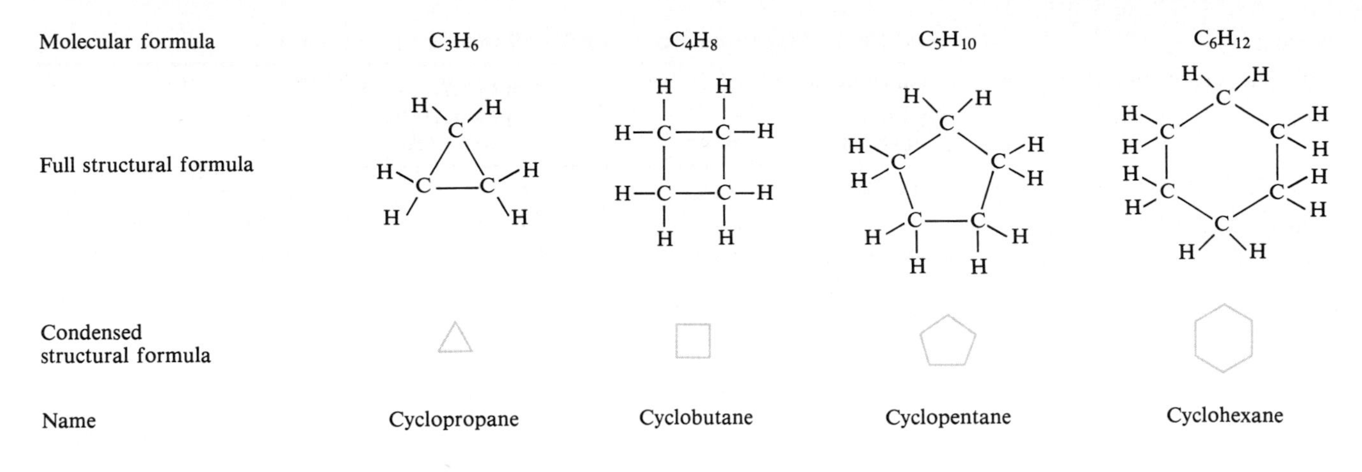

Molecular formula
C_3H_6
C_4H_8
C_5H_{10}
C_6H_{12}
Full structural formula
Condensed structural formula
Name
Cyclopropane
Cyclobutane
Cyclopentane
Cyclohexane

Figure 24.4
First four members of the cycloalkane series. These are saturated aliphatic hydrocarbons characterized by carbon-atom rings.

Exercise 24.2

Write the condensed structural formula of 3,3-dimethyloctane.

(See Problems 24.23 and 24.24.)

Cycloalkanes

Cycloalkanes are *saturated hydrocarbons in which the carbon atoms form a ring; the general formula is* C_nH_{2n}. Figure 24.4 gives the names and structural formulas of the first four members of the cycloalkane series. In the condensed structural formulas, a carbon atom and its attached hydrogen atoms are assumed to be at each corner.

Sources of Alkanes and Cycloalkanes

Fossil fuels (natural gas, petroleum, and coal) are the principal sources of hydrocarbons. Natural gas is mainly methane with smaller amounts of the other gaseous alkanes (ethane, propane, and butane). Petroleum is a mixture of alkanes and cycloalkanes with smaller amounts of aromatic hydrocarbons. These hydrocarbons are separated by distillation into fractions such as gasoline and kerosene (see Table 24.3). These fractions are usually processed further—for example, to obtain a greater quantity of gasoline with the desired fuel characteristics. This processing is called *petroleum refining*.

24.2 ALKENES AND ALKYNES

Unsaturated hydrocarbons are *hydrocarbons that do not contain the maximum number of hydrogen atoms for a given carbon-atom framework*. Such compounds have carbon–carbon multiple bonds and, under the proper conditions, add molecular hydrogen to give a saturated compound. For example, ethylene adds hydrogen to give ethane.

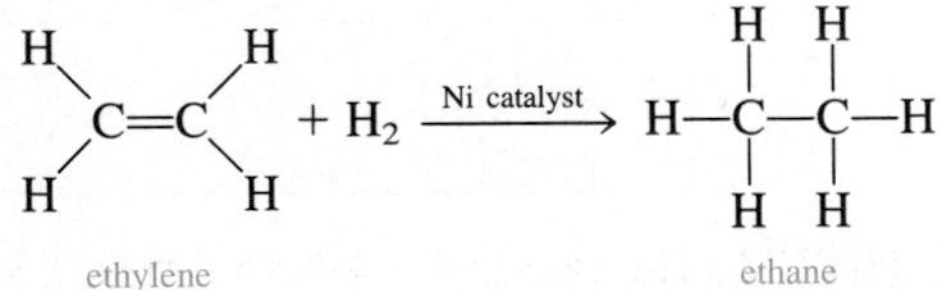

Table 24.3 Fractions from the Distillation of Petroleum

Boiling Range, °C	Name	Range of Carbon Atoms per Molecule	Use
Below 20	Gases	C_1 to C_4	Heating, cooking, and chemical raw material
20–200	Naphtha; straight-run gasoline	C_5 to C_{12}	Fuel; lighter fractions (such as petroleum ether, b.p. 30°C–60°C) are also used as laboratory solvents
200–300	Kerosene	C_{12} to C_{15}	Fuel
300–400	Fuel oil	C_{15} to C_{18}	Heating homes, diesel fuel
Over 400		Over C_{18}	Lubricating oil, greases, paraffin waxes, asphalt

(From Harold Hart, *Organic Chemistry: A Short Course,* 7th Ed. [Boston: Houghton Mifflin, 1987], p. 102.)

Alkenes

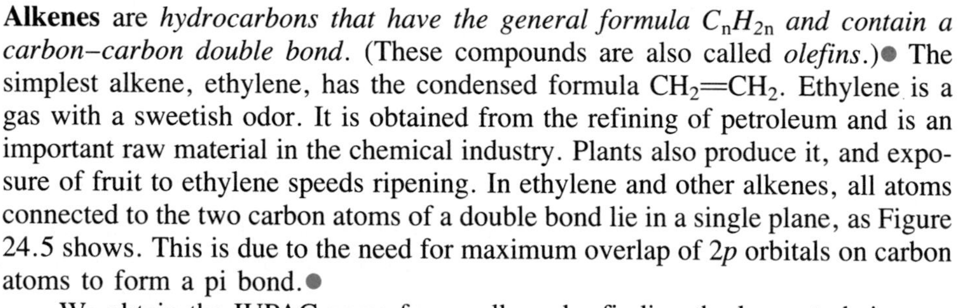

● Olefin means "oil-forming." Many alkenes react with Cl_2 to form oily compounds.

● Bonding in ethylene was discussed in detail in Section 10.4.

Alkenes are *hydrocarbons that have the general formula C_nH_{2n} and contain a carbon–carbon double bond.* (These compounds are also called *olefins.*)● The simplest alkene, ethylene, has the condensed formula $CH_2{=}CH_2$. Ethylene is a gas with a sweetish odor. It is obtained from the refining of petroleum and is an important raw material in the chemical industry. Plants also produce it, and exposure of fruit to ethylene speeds ripening. In ethylene and other alkenes, all atoms connected to the two carbon atoms of a double bond lie in a single plane, as Figure 24.5 shows. This is due to the need for maximum overlap of 2*p* orbitals on carbon atoms to form a pi bond.●

We obtain the IUPAC name for an alkene by finding the longest chain containing the double bond. As with the alkanes, this longest chain gives us the stem name, but the suffix is *-ene* rather than *-ane*. The carbon atoms of the longest chain are numbered from the end nearest the carbon–carbon double bond, and the position of the double bond is given the number of the first carbon atom of that bond. This number is written in front of the stem name of the alkene. Branch chains are named as in the alkanes. The simplest alkene, $CH_2{=}CH_2$, is called ethene, although the common name is ethylene.

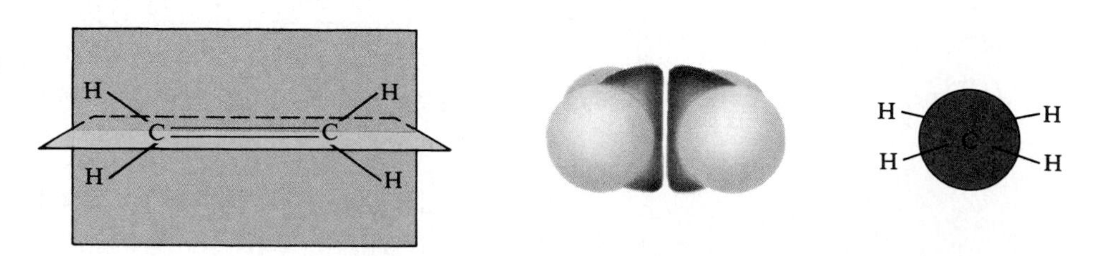

Figure 24.5 Representations of the ethylene molecule showing its planar geometry. In the drawing on the right, the view is down the carbon–carbon axis.

Example 24.3 Writing the IUPAC Name of an Alkene Given the Structural Formula

What is the IUPAC name of each of the following alkenes?

(a) $CH_2{=}CHCHCH_2CH_3$ (with CH_3 attached to the third carbon)

(b) $CH_3CH_2CH_2CH_2CHCH{=}CHCH_3$ (with $-CH_2-CH_2-CH_2-CH_2-CH_3$ attached to the CH carbon)

Solution

(a) The numbering of the carbon chain is

$$\overset{1}{C}H_2{=}\overset{2}{C}H\overset{3}{C}H\overset{4}{C}H_2\overset{5}{C}H_3 \quad (CH_3 \text{ on carbon } 3)$$

Because the longest chain containing a double bond has five carbon atoms, this is a pentene. Moreover, it is a 1-pentene, because the double bond is between carbon atoms 1 and 2. The name of the compound is **3-methyl-1-pentene.** Note the placement of hyphens. If the numbering had been in the opposite direction, we would have named the compound as a 4-pentene. But this is unacceptable, because 4 is greater than 1. (b) The numbering of the longest chain containing the double bond is

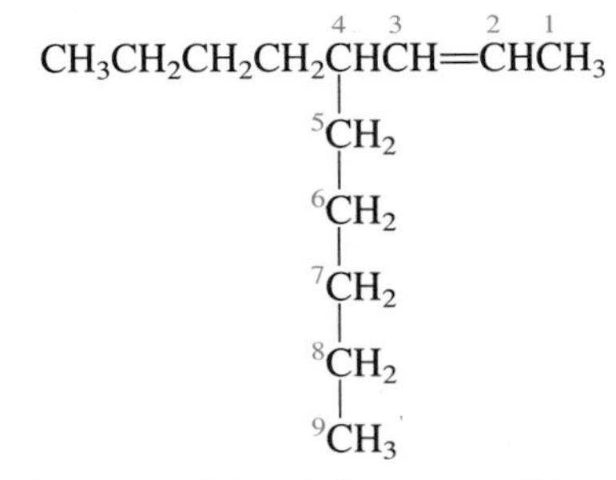

This gives the name **4-butyl-2-nonene.** There is a longer chain (of ten carbon atoms), but it does not contain the double bond.

Exercise 24.3

Give the IUPAC name for each of the following compounds.

(a) $CH_3C{=}CHCHCH_3$ (with CH_3 on C-2 and CH_2CH_3 on C-4)

(b) $CH_3CH_2CH_2CHCH_2CH_2CH_3$ (with $CH{=}CH_2$ on C-4)

(See Problems 24.25 and 24.26.)

Exercise 24.4

Write the condensed structural formula of 2,5-dimethyl-2-heptene.

(See Problems 24.27 and 24.28.)

Rotation about a carbon–carbon double bond cannot occur without breaking the pi bond. This requires energies comparable to those in chemical reactions, so rotation normally does not occur. This lack of rotation about the double bond gives rise to isomers in certain alkenes. For example, there are two isomers of 2-butene.

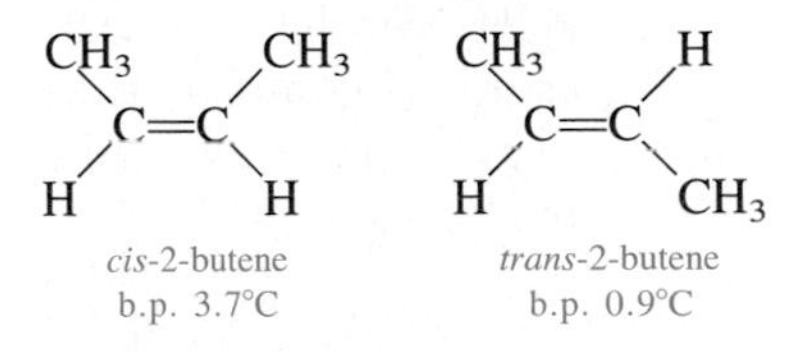

cis-2-butene b.p. 3.7°C   *trans*-2-butene b.p. 0.9°C

The different boiling points indicate that these are indeed different compounds.

Cis-2-butene and *trans*-2-butene are geometric isomers. **Geometric isomers** are *isomers in which the atoms are joined to one another in the same way but that differ because some atoms occupy different relative positions in space*. The *cis* isomer has identical groups (in the case of the butenes, alkyl groups) attached to the same side of the double bond, whereas the *trans* isomer has them on opposite sides. An alkene with the general formula

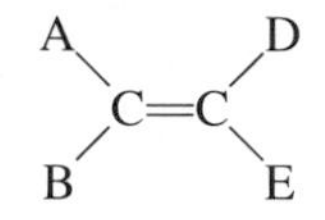

has a geometric isomer only if groups A and B are different and groups D and E are different. Thus, there is no isomer of propene, $CH_2{=}CHCH_3$.

Example 24.4 Predicting *Cis–Trans* Isomers

For each of the following alkenes, decide whether *cis–trans* isomers are possible. If so, draw structural formulas of the isomers and give their IUPAC names (labeling *cis* or *trans*).

(a) $CH_3CH_2CH{=}C(CH_3)_2$ (b) $CH_3CH{=}CHCH_2CH_3$

Solution

(a) If we write out the structure, we have

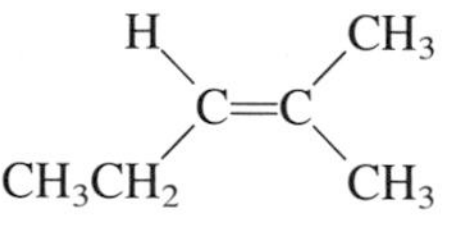

Because two methyl groups are attached to the second carbon atom of the double bond, **geometric isomers are not possible.** (b) **Geometric isomers are possible.** They are

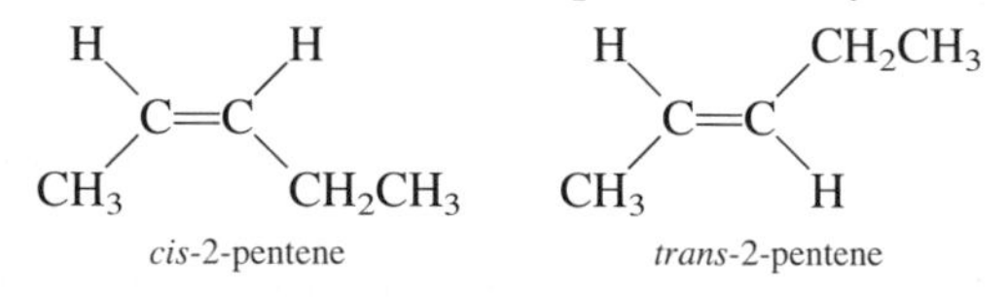

Exercise 24.5

Decide whether *cis–trans* isomers are possible for each of the following compounds. If any have isomers, draw the structural formulas and give the IUPAC names, labeling *cis* or *trans*.

(a) $CH_3CH{=}CHCH_2CH_2CH_3$ (b) $CH_3CH_2CH{=}CH_2$

(See Problems 24.29 and 24.30.)

Alkynes

Alkynes are *unsaturated hydrocarbons containing a carbon–carbon triple bond*. The general formula is C_nH_{2n-2}. The simplest alkyne is acetylene (ethyne).●

● The *-ene* ending in the common name *acetylene* does not follow IUPAC rules.

$$H{-}C{\equiv}C{-}H$$

Acetylene has a linear molecule. This very reactive gas is used to form other chemical compounds and plastics. It burns with oxygen in the oxyacetylene torch to give a very hot flame (about 3000°C). Acetylene is produced commercially from methane.

$$2CH_4 \xrightarrow{1600^\circ C} CH{\equiv}CH + 3H_2$$

Acetylene is also prepared from calcium carbide, CaC_2, which is obtained by

Figure 24.6
Preparation of acetylene gas. Here acetylene is prepared by the reaction of water on calcium carbide. The acetylene burns with a sooty flame.

heating calcium oxide and coke (carbon) in an electric furnace:

$$CaO(s) + 3C(s) \xrightarrow{2000°C} CaC_2(l) + CO(g)$$

The calcium carbide is cooled until it solidifies. The carbide ion, C_2^{2-}, is strongly basic and reacts with water to produce acetylene (Figure 24.6).

$$CaC_2(s) + 2H_2O(l) \longrightarrow Ca(OH)_2(aq) + C_2H_2(g)$$

The alkynes are named by IUPAC rules in the same way as the alkenes, except that the stem name is determined from the longest chain containing the carbon–carbon triple bond. The suffix for this stem name is *-yne*.

Exercise 24.6

Give the IUPAC name for each of the following alkynes.

(a) $CH_3C{\equiv}CH$ (b) $CH{\equiv}CCHCH_3$ with CH_2CH_3 on the third carbon

(See Problems 24.31 and 24.32.)

24.3 AROMATIC HYDROCARBONS

Aromatic hydrocarbons contain benzene rings, or six-membered rings of carbon atoms with alternating single and double carbon–carbon bonds. The electronic structure of these rings may be represented by resonance formulas. For benzene, we have

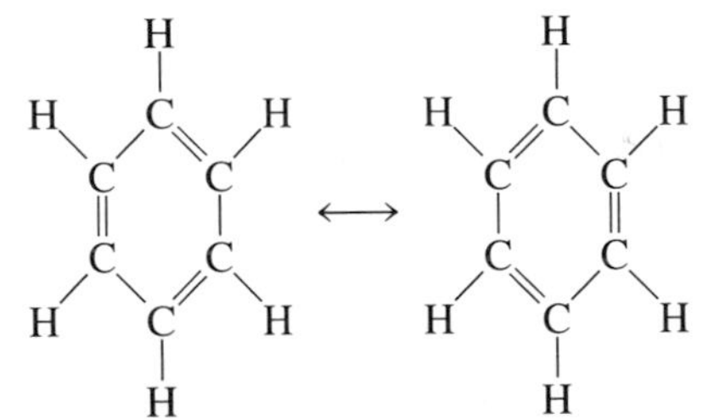

This electronic structure can also be described in molecular-orbital terms (Figure 24.7). In this description, pi molecular orbitals encompass the entire carbon-atom

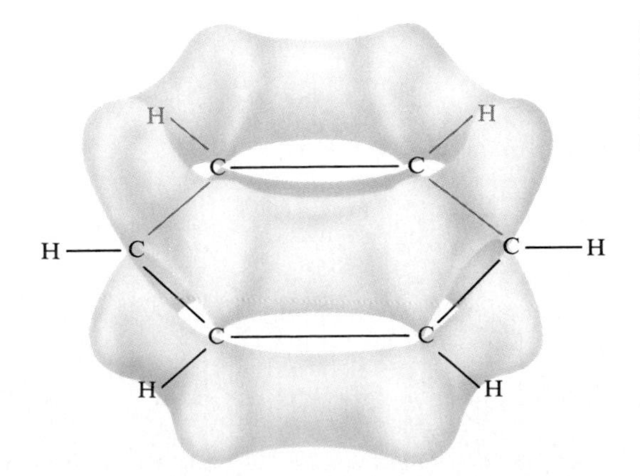

Figure 24.7
Lowest-energy π molecular orbital of benzene. Note how the orbital encompasses the entire ring of carbon atoms.

ring. The pi electrons are said to be delocalized. Benzene is given the condensed formula

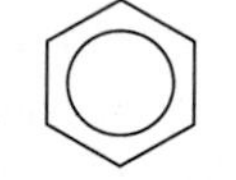

where the circle represents the delocalization of pi electrons.

Delocalization of the pi electrons means that the "double" bonds in benzene do not behave as isolated double bonds. When we discuss the chemical reactions of benzene, we will see that it behaves quite differently from an alkene. Figure 24.8 gives formulas of some *fused-ring,* or *polynuclear,* aromatic hydrocarbons. In the fused-ring hydrocarbons, two or more rings share carbon atoms.

Alkyl groups may replace one or more hydrogen atoms of benzene, giving hydrocarbon derivatives of benzene. For example, toluene (methylbenzene) has the structural formula

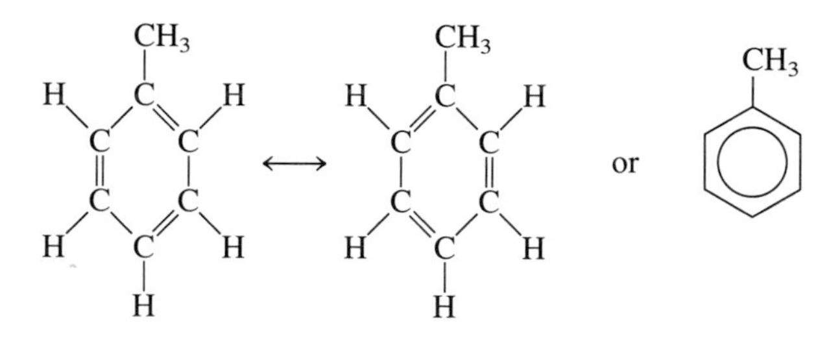

When two groups are on the benzene ring, three isomers are possible. They may be distinguished by the prefixes *ortho-* (*o*), *meta-* (*m*), and *para-* (*p*). For example,

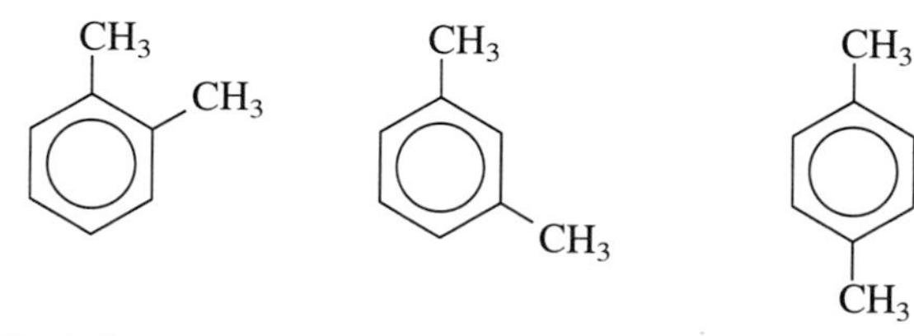

A numbering system is also used to show the positions of two or more groups.

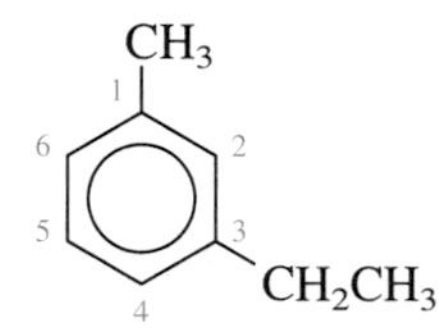

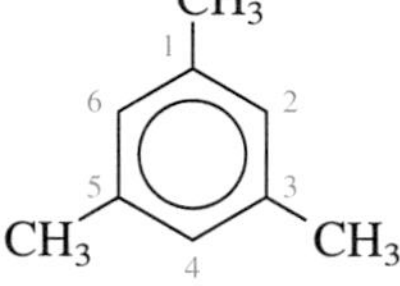

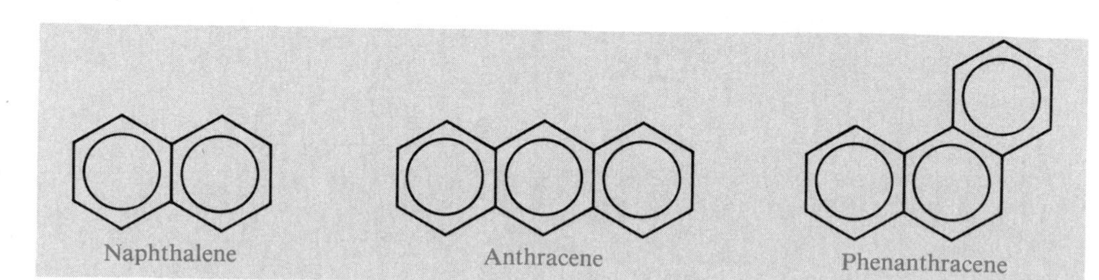

Figure 24.8
Formulas of some fused-ring aromatic hydrocarbons.
Naphthalene is the simplest member of this series. It is a white, crystalline substance used in manufacturing plastics and plasticizers (to keep plastics pliable). Small amounts are used for moth balls.

It is sometimes preferable to name a compound containing a benzene ring by regarding the ring as a group in the same manner as alkyl groups. Pulling a hydrogen atom from benzene leaves the phenyl group, C_6H_5—. For example, we name the following compound diphenylmethane by using methane as the stem name.

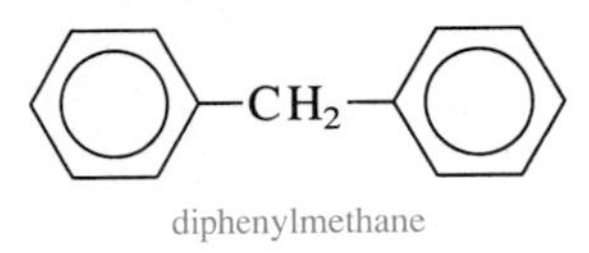

diphenylmethane

Exercise 24.7

Write the structural formula of (a) ethylbenzene; (b) 1,2-diphenylethane.

(See Problems 24.33 and 24.34.)

24.4 REACTIONS OF HYDROCARBONS

Natural gas and petroleum, which are mixtures of hydrocarbons, are the major sources of organic chemicals. By various reactions they are converted to final products—solvents, plastics, textile fibers, and so forth. The saturated hydrocarbons, forming the bulk of petroleum, are not readily reactive at normal temperatures, except with a few reagents. At higher temperatures with the proper catalysts, however, they can be made to break up or rearrange, giving unsaturated hydrocarbons. These unsaturated hydrocarbons are quite reactive.

Oxidation

All hydrocarbons burn in an excess of O_2 to give carbon dioxide and water.

$$\underset{\text{ethane}}{C_2H_6(g)} + \tfrac{7}{2}O_2(g) \longrightarrow 2CO_2(g) + 3H_2O(l);\ \Delta H^\circ = -1560\ \text{kJ/mol}$$

$$\underset{\text{benzene}}{C_6H_6(l)} + \tfrac{15}{2}O_2(g) \longrightarrow 6CO_2(g) + 3H_2O(l);\ \Delta H^\circ = -3267\ \text{kJ/mol}$$

The large negative ΔH's explain why the hydrocarbons are useful as fuels.

Unsaturated hydrocarbons are oxidized under milder conditions than saturated hydrocarbons. For example, when aqueous potassium permanganate, $KMnO_4(aq)$, is added to an alkene or an alkyne, the purple color of $KMnO_4$ fades and a precipitate of brown manganese dioxide forms.

$$\underset{\text{1-hexene}}{3C_4H_9CH{=}CH_2} + 2MnO_4^-(aq) + 4H_2O \longrightarrow$$

$$3C_4H_9\overset{\displaystyle H}{\underset{\displaystyle \underset{\displaystyle H}{O}}{C}}-\overset{\displaystyle H}{\underset{\displaystyle \underset{\displaystyle H}{O}}{C}}-H + 2MnO_2(s) + 2OH^-(aq)$$

Saturated hydrocarbons are unreactive with $KMnO_4(aq)$, so this reagent can be used to distinguish them from unsaturated hydrocarbons (this is called the *Bayer test* of unsaturation).●

● A positive Bayer test indicates only that the compound is easily oxidized. Other tests are required to definitely identify a compound as an alkene.

Substitution Reactions of Alkanes

Alkanes react with the halogens F_2, Cl_2, and Br_2. Reaction with Cl_2 requires sunlight (indicated *hv*) or heat.

$$\text{H}-\underset{\underset{\text{H}}{|}}{\overset{\overset{\text{H}}{|}}{\text{C}}}-\text{H} + \text{Cl}-\text{Cl} \xrightarrow{hv} \text{H}-\underset{\underset{\text{H}}{|}}{\overset{\overset{\text{H}}{|}}{\text{C}}}-\text{Cl} + \text{H}-\text{Cl}$$

This is an example of a substitution reaction. A **substitution reaction** is *a reaction in which a part of the reagent molecule is substituted for an H atom on a hydrocarbon or hydrocarbon group*. All of the H atoms of an alkane may undergo substitution, leading to a mixture of products.

$$CH_3Cl + Cl_2 \xrightarrow{hv} CH_2Cl_2 + HCl$$

$$CH_2Cl_2 + Cl_2 \xrightarrow{hv} CHCl_3 + HCl$$

$$CHCl_3 + Cl_2 \xrightarrow{hv} CCl_4 + HCl$$

Fluorine is very reactive with saturated hydrocarbons and usually gives complete substitution. Bromine is less reactive than Cl_2 and often requires elevated temperatures for substitution.

Addition Reactions of Alkenes

Alkenes are more reactive than alkanes because of the presence of the double bond. Many reagents *add* to the double bond. A simple example is the addition of a halogen, such as Br_2, to propene.

$$CH_3CH{=}CH_2 + Br_2 \longrightarrow CH_3\underset{\underset{\text{Br}}{|}}{C}H-\underset{\underset{\text{Br}}{|}}{C}H_2$$

An **addition reaction** is *a reaction in which parts of a reagent are added to each carbon atom of a carbon–carbon multiple bond, which then becomes a C—C single bond*. The addition of Br_2 to an alkene is fast. In fact, it occurs so readily that bromine dissolved in carbon tetrachloride, CCl_4, is a useful reagent to test for unsaturation. When a few drops of the solution are added to an alkene, the red-brown color of the Br_2 is immediately lost. There is no reaction with alkanes, and the solution retains the red-brown color of the Br_2.

Unsymmetrical reagents, such as HCl and HBr, add to unsymmetrical alkenes to give two products that are isomers of one another. For example,

$$CH_3CH{=}CH_2 + HBr \rightarrow \overset{3}{C}H_3\overset{2}{\underset{\underset{\text{Br}}{|}}{C}}H-\overset{1}{\underset{\underset{\text{H}}{|}}{C}}H_2$$

2-bromopropane

$$CH_3CH{=}CH_2 + HBr \rightarrow \overset{3}{C}H_3\overset{2}{\underset{\underset{\text{H}}{|}}{C}}H-\overset{1}{\underset{\underset{\text{Br}}{|}}{C}}H_2$$

1-bromopropane

In one case, the hydrogen atom of HBr adds to carbon atom 1, giving 2-bromopropane; in the other case, the hydrogen atom of HBr adds to carbon atom 2, giving 1-bromopropane. (The name 1-bromopropane means that a bromine atom is substituted for a hydrogen atom at carbon atom 1.)

The two products are not formed in equal amounts; one is more likely to form. **Markownikoff's rule** is *a generalization stating that the major product formed by the addition of an unsymmetrical reagent such as H—Cl, H—Br, or H—OH is the one obtained when the H atom of the reagent adds to the carbon atom of the multiple bond that already has the more hydrogen atoms attached to it.* In the preceding example, the H atom of HBr should add preferentially to carbon atom 1, which has two hydrogen atoms attached to it. The major product then is 2-bromopropane.

Example 24.5 Predicting the Major Product in an Addition Reaction

What is the major product of the following reaction?

$$\underset{1}{CH_3}-\underset{2}{\overset{\overset{\displaystyle CH_3}{|}}{C}}=\underset{3}{\overset{\overset{\displaystyle H}{|}}{C}}-\underset{4}{CH_3} + HCl \longrightarrow$$

Solution

There is one H atom attached to carbon 3 and none attached to carbon 2. Thus, the major product is

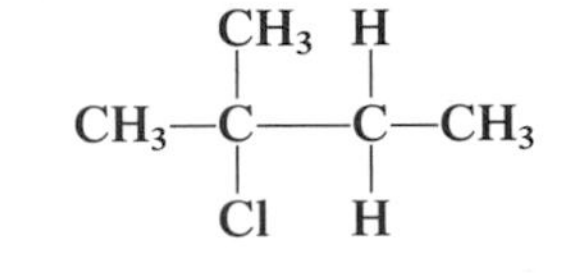

Exercise 24.8

Predict the main product when HBr is added to 1-butene.

(See Problems 24.37 and 24.38.)

The alkynes also undergo addition reactions, usually adding two molecules of the reagent for each C≡C bond. The major product is the isomer predicted by Markownikoff's rule. Thus,

$$CH_3-C\equiv C-H + 2HCl \longrightarrow CH_3-\overset{\overset{\displaystyle Cl}{|}}{\underset{\underset{\displaystyle Cl}{|}}{C}}-\overset{\overset{\displaystyle H}{|}}{\underset{\underset{\displaystyle H}{|}}{C}}-H$$

Substitution Reactions of Aromatic Hydrocarbons

Although benzene, C_6H_6, is an unsaturated hydrocarbon, it does not usually undergo addition reactions. The delocalized π electrons of benzene are more stable than localized π electrons. Therefore, benzene does not react readily with Br_2 in carbon tetrachloride, as an alkene would. The usual reactions of benzene are substitution reactions. In the presence of iron(III) bromide as a catalyst, a bromine atom of Br_2 substitutes for a hydrogen atom on the benzene ring.

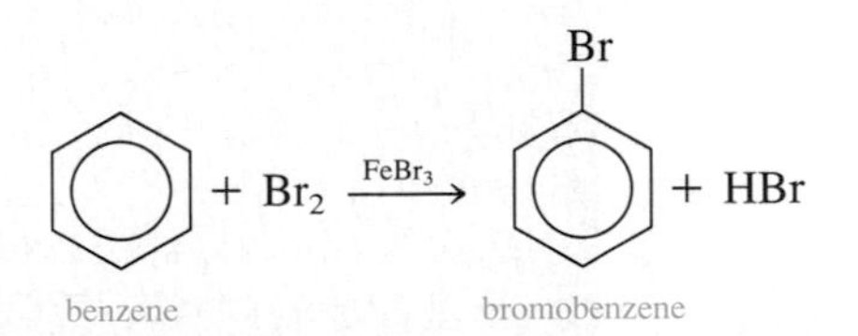

Similarly, benzene undergoes substitution with nitric acid in the presence of sulfuric acid to give nitrobenzene.

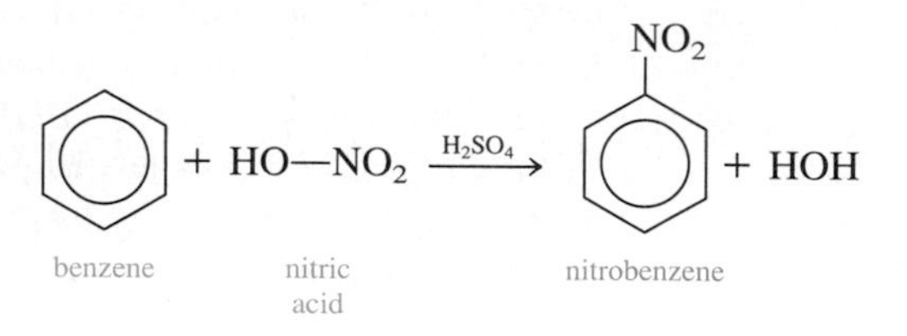

Petroleum Refining

Petroleum, as we noted earlier, is a mixture of hydrocarbons, principally alkanes and cycloalkanes. The object of *petroleum refining* is to obtain various hydrocarbon products from this mixture. In part, this is accomplished by fractional distillation of the petroleum, but the demand for certain products, particularly gasoline, is greater than what can be supplied by distillation. For that reason, petroleum refiners resort to various chemical processes to increase the amount of desired products.

One of these chemical processes is *catalytic cracking* (Figure 24.9). Although alkanes are usually stable compounds, when the hydrocarbon vapor is passed over a heated catalyst of alumina (Al_2O_3) and silica (SiO_2), the hydrocarbon molecule breaks up, or "cracks," to give hydrocarbons of lower molecular weight.

In order for a gasoline to function properly in an engine, it should not begin to burn before it is ignited by the spark plug. If it does, it gives engine "knock." The antiknock characteristics of a gasoline are rated by the *octane number* scale. This scale is based on heptane, which is given an octane number of 0, and on 2,2,4-trimethylpentane (an octane isomer), which is given an octane number of 100. The higher the octane number, the better the antiknock characteristics.

Substances, such as tetraethyllead, may be added to gasoline to raise the

Figure 24.9
Catalytic cracking unit of a petroleum refinery. In catalytic cracking, petroleum hydrocarbons are broken into smaller molecules to increase the yield of gasoline.

octane number, but the octane rating can also be improved by changing straight-chain hydrocarbons to branched-chain isomers, which can be accomplished by heating the hydrocarbon vapor in the presence of a catalyst.

Derivatives of Hydrocarbons

Certain groups of atoms in many organic molecules are particularly reactive and have characteristic chemical properties. A **functional group** is *a reactive portion of a molecule that undergoes predictable reactions*. The C=C bond in a compound, for example, reacts readily with reagents Br_2 and HBr in addition reactions. For this reason, the C=C bond is a functional group. Many functional groups contain an atom other than carbon and have lone pairs of electrons on it. These lone pairs contribute to the reactivity of the functional group. Others, such as C=O, have multiple bonds that are reactive. Table 24.4 lists some common functional groups.

In the previous sections of the chapter, we discussed the hydrocarbons and their reactions. All other organic compounds can be considered as derivatives of hydrocarbons. In these compounds, one or more hydrogen atoms of a hydrocarbon have been replaced by atoms other than carbon to give a functional group.

Table 24.4 Some Organic Functional Groups

Structure of General Compound* (Functional Group in Color)	Name of Functional Group
R—Cl: R—Br:	Organic halide
R—O—H	Alcohol
R—O—R′	Ether
R—C(=O)—H	Aldehyde
R—C(=O)—R′	Ketone
R—C(=O)—O—H	Carboxylic acid
R—C(=O)—O—R′	Ester
R—NH—H (N bonded to H); R—N(R′)—H; R—N(R′)—R″	Amine
R—C(=O)—N(H)—R′	Amide

*R, R′, and R″ are general hydrocarbon groups.

24.5 ORGANIC COMPOUNDS CONTAINING OXYGEN

Many of the important functional groups in organic compounds contain oxygen. Examples are alcohols, ethers, aldehydes, ketones, carboxylic acids, and esters. We will look at characteristics of these compounds in this section.

Alcohols and Ethers

Structurally, we may think of an **alcohol** as *a compound obtained by substituting a hydroxyl group (—OH) for an —H atom on an aliphatic hydrocarbon*. Some examples are

CH_3OH — methanol (methyl alcohol)

CH_3CH_2OH — ethanol (ethyl alcohol)

$CH_3CH(OH)CH_3$ — 2-propanol (isopropyl alcohol)

Alcohols are named by IUPAC rules similar to those used for naming the hydrocarbons, except that the stem name is determined from the longest chain containing the carbon atom to which the —OH group is attached. The suffix for this stem name is *-ol*. The position of the —OH group is indicated by a number preceding the stem name. (The number is omitted if it is unnecessary, as in ethanol.)

Alcohols are usually classified by the number of carbon atoms attached to the carbon atom to which the —OH group is bonded. A *primary alcohol* has one such carbon atom, a *secondary alcohol* has two, and a *tertiary alcohol* has three. The following are examples:

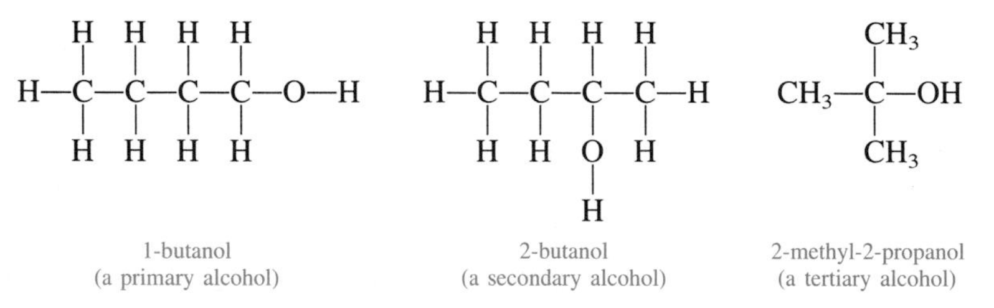

1-butanol (a primary alcohol) 2-butanol (a secondary alcohol) 2-methyl-2-propanol (a tertiary alcohol)

Methanol, ethanol, ethylene glycol, and glycerol are some common alcohols. Methanol, CH_3OH, was at one time separated from the liquid distilled from sawdust—hence the common name wood alcohol. It is a toxic liquid prepared by reacting carbon monoxide with hydrogen at high pressure in the presence of a catalyst. It is used as a solvent and as the starting material for the preparation of formaldehyde. Ethanol is manufactured by the fermentation of glucose (a sugar) or by the addition of water to the double bond of ethylene. The latter reaction is carried out by heating ethylene with water in the presence of sulfuric acid.

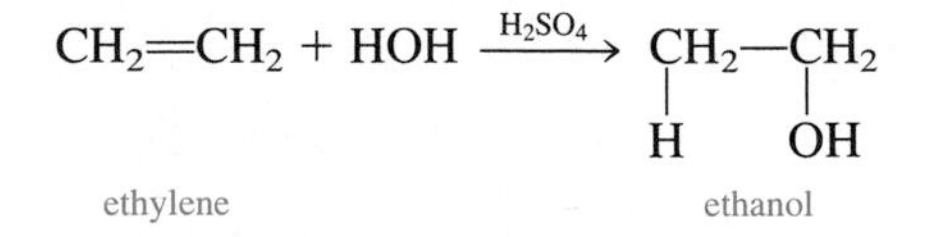

ethylene ethanol

Alcoholic beverages contain ethanol. Ethanol is also a solvent and a starting material for many organic compounds. It is mixed with gasoline and sold as gasohol, an automotive fuel. Ethylene glycol (IUPAC name: 1,2-ethanediol) and glycerol (1,2,3-propanetriol) are alcohols containing more than one hydroxyl group.

$$\underset{\text{ethylene glycol}}{\begin{array}{cc} CH_2 - & CH_2 \\ | & | \\ OH & OH \end{array}} \qquad \underset{\text{glycerol}}{\begin{array}{ccc} CH_2 - & CH - & CH_2 \\ | & | & | \\ OH & OH & OH \end{array}}$$

Ethylene glycol is a liquid prepared from ethylene and is used as an antifreeze agent. It is also used in the manufacture of polyester plastics and fibers. ● Glycerol is a nontoxic, sweet-tasting liquid obtained from fats during the making of soap. It is used in foods and candies to keep them soft and moist.

● Polyesters are discussed in Section 24.8.

Formally, an alcohol may be thought of as a derivative of water in which one H atom of H_2O has been replaced by a hydrocarbon group, R. Similarly, an **ether** is *a compound formally obtained by replacing both H atoms of H_2O by hydrocarbon groups R and R′.*

$$\underset{\text{water}}{H-O-H} \qquad \underset{\text{an alcohol}}{R-O-H} \qquad \underset{\text{an ether}}{R-O-R'}$$

Common names for ethers are formed by naming the hydrocarbon groups and adding the word *ether*. Thus, $CH_3OCH_2CH_2CH_3$ is called methyl propyl ether. By IUPAC rules, the ethers are named as derivatives of the longest hydrocarbon chain. For example, $CH_3OCH_2CH_2CH_3$ is 1-methoxypropane; the methoxy group is $CH_3O—$. The best-known ether is diethyl ether, $CH_3CH_2OCH_2CH_3$, a volatile liquid used as a solvent and as an anesthetic.

Exercise 24.9

Give the IUPAC name of the following compound.

$$\begin{array}{c} OH \\ | \\ CH_3CH_2CH_2CCH_2CH_3 \\ | \\ CH_2CH_3 \end{array}$$

(See Problems 24.43 and 24.44.)

Exercise 24.10

Give the common name of each of the following compounds: (a) CH_3OCH_3; (b) $CH_3OCH_2CH_3$. (See Problems 24.47 and 24.48.)

Aldehydes and Ketones

Aldehydes and ketones are compounds containing a *carbonyl group*.

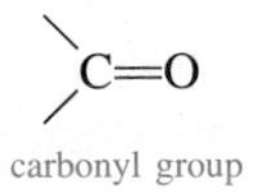

carbonyl group

An **aldehyde** is *a compound containing a carbonyl group with at least one H atom attached to it.*

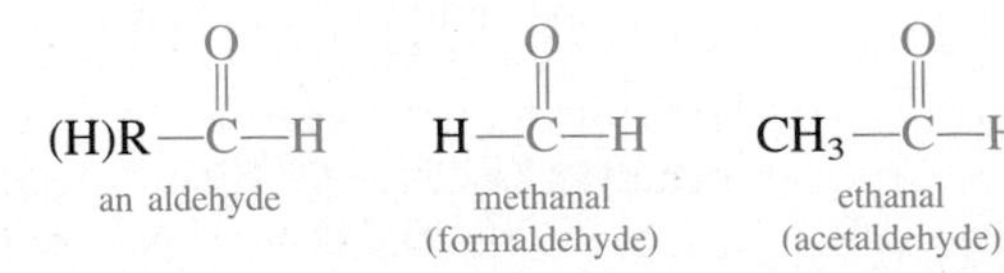

an aldehyde methanal (formaldehyde) ethanal (acetaldehyde)

Figure 24.10
Some aldehydes of aromatic hydrocarbons. The name *aromatic* in aromatic hydrocarbons derives from the aromatic odors of many derivatives of these hydrocarbons, including the aldehydes shown here.

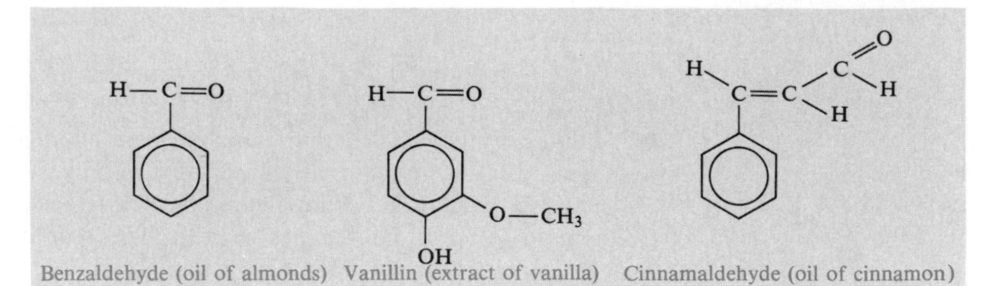

Here (H)R indicates a hydrocarbon group or H atom. The aldehyde function is usually abbreviated —CHO, and the structural formula of acetaldehyde is written CH_3CHO.

A **ketone** is *a compound containing a carbonyl group with two hydrocarbon groups attached to it.*

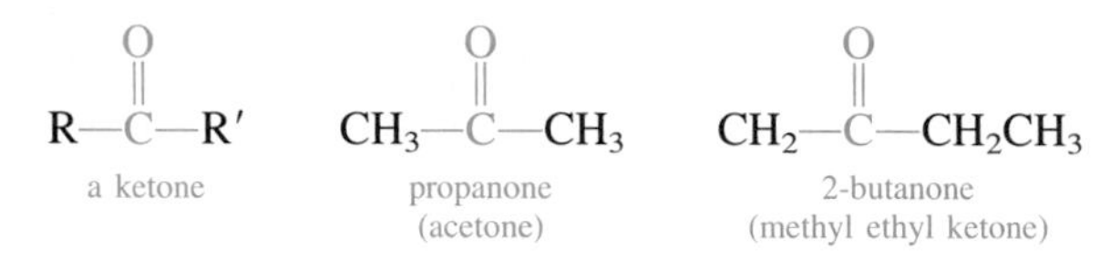

The ketone functional group is abbreviated —CO—; thus, acetone is written CH_3COCH_3.

Aldehydes and ketones are named according to IUPAC rules similar to those for naming alcohols. We first locate the longest carbon chain containing the carbonyl group to get the stem hydrocarbon name. Then we change the *-e* ending of the hydrocarbon to *-al* for aldehydes and *-one* for ketones. In the case of aldehydes, the carbon atom of the —CHO group is always the number-1 carbon. In ketones, however, the carbonyl group may occur in various nonequivalent positions on the carbon chain. When that happens, the position of the carbonyl group is indicated by a number before the stem name, just as the position of the hydroxyl group is indicated in alcohols. The carbon chain is numbered to give the smaller number for the position of the carbonyl group.

The aldehydes of lower molecular weight have sharp, penetrating odors. Formaldehyde (methanal), HCHO, and acetaldehyde (ethanal), CH_3CHO, are examples. With increasing molecular weight, the aldehydes become more fragrant. Some aldehydes of aromatic hydrocarbons have especially pleasant odors (see Figure 24.10). Formaldehyde is a gas produced by the oxidation of methanol. The gas is very soluble in water, and a 37% aqueous solution called Formalin is marketed as a disinfectant and as a preservative of biological specimens. The main use of formaldehyde is in the manufacture of plastics and resins. Acetone, CH_3COCH_3, is the simplest ketone. It is a liquid with a fragrant odor. The liquid is an important solvent for lacquers, paint removers, and nail polish remover.

Exercise 24.11

Name the following compounds by IUPAC rules:

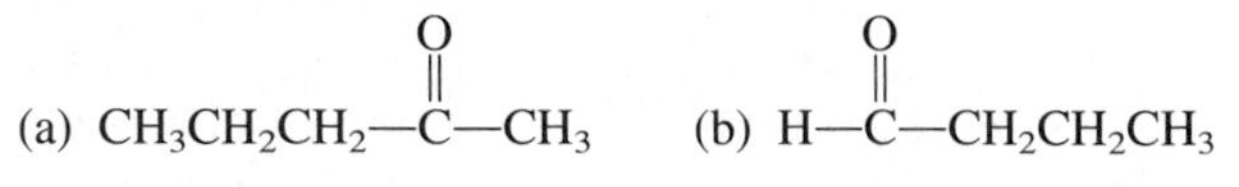

(See Problems 24.49 and 24.50.)

Table 24.5 Common Carboxylic Acids

Carbon Atoms	Formula	Source	Common Name	IUPAC Name
1	$HCOOH$	Ants (Latin, *formica*)	Formic acid	Methanoic acid
2	CH_3COOH	Vinegar (Latin, *acetum*)	Acetic acid	Ethanoic acid
3	CH_3CH_2COOH	Milk (Greek, *protos pion,* "first fat")	Propionic acid	Propanoic acid
4	$CH_3(CH_2)_2COOH$	Butter (Latin, *butyrum*)	Butyric acid	Butanoic acid
5	$CH_3(CH_2)_3COOH$	Valerian root (Latin, *valere,* "to be strong")	Valeric acid	Pentanoic acid
6	$CH_3(CH_2)_4COOH$	Goats (Latin, *caper*)	Caproic acid	Hexanoic acid
7	$CH_3(CH_2)_5COOH$	Vine blossom (Greek, *oenanthe*)	Enanthic acid	Heptanoic acid
8	$CH_3(CH_2)_6COOH$	Goats (Latin, *caper*)	Caprylic acid	Octanoic acid
9	$CH_3(CH_2)_7COOH$	Pelargonium (an herb with stork-shaped capsules; Greek, *pelargos,* "stork")	Pelargonic acid	Nonanoic acid
10	$CH_3(CH_2)_8COOH$	Goats (Latin, *caper*)	Capric acid	Decanoic acid

(From Harold Hart, *Organic Chemistry: A Short Course,* 7th Ed. [Boston: Houghton Mifflin, 1987], p. 260.)

Carboxylic Acids and Esters

A **carboxylic acid** is *a compound containing the carboxyl group, —COOH.*

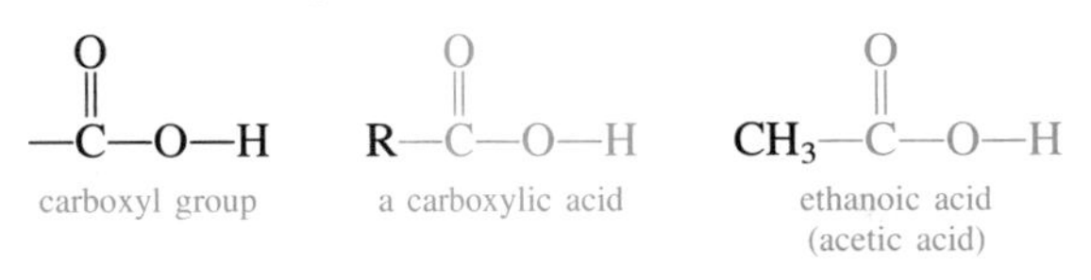

These compounds are named by IUPAC rules like those for the aldehydes, except that the ending on the stem name is *-oic* followed by the word *acid*. Many carboxylic acids have been known for a long time and are usually referred to by common names (see Table 24.5). The carboxylic acids are weak acids because of the acidity of the H atom on the carboxyl group. Acid ionization constants are about 10^{-5}.

An **ester** is *a compound formed from a carboxylic acid, RCOOH, and an alcohol, R'OH.* The general structure is

$$\underset{\text{RC—O—R'}}{\overset{\text{:O:}}{\|}}$$

We will discuss these compounds in the next section, which deals with reactions of oxygen-containing organic compounds.

24.6 REACTIONS OF OXYGEN-CONTAINING ORGANIC COMPOUNDS

Alcohols, aldehydes, ketones, and carboxylic acids are especially reactive compounds. (Ethers are rather unreactive.) In this section, we will discuss the oxidation and reduction of these oxygen-containing organic compounds. We will also discuss the reaction of alcohols with carboxylic acids to form esters.

Oxidation–Reduction Reactions

When discussing organic reactions, it is convenient to define *oxidation* as the addition of oxygen atoms to, or the removal of hydrogen atoms from, an organic compound. *Reduction* is defined as the addition of hydrogen atoms to, or the removal of oxygen atoms from, an organic compound. A general oxidizing agent is written as (O), and a general reducing agent as (H). For example, the oxidation of ethanol, CH_3CH_2OH, to acetaldehyde, CH_3CHO, is written

$$\underset{\text{ethanol}}{CH_3CH_2OH} + \underset{\text{an oxidizing agent}}{(O)} \longrightarrow \underset{\text{acetaldehyde (ethanal)}}{CH_3CHO} + H_2O$$

Note that two hydrogen atoms have been taken away from ethanol to give acetaldehyde. Acetaldehyde may be oxidized further:

$$\underset{\text{acetaldehyde (ethanal)}}{CH_3CHO} + \underset{\text{an oxidizing agent}}{(O)} \longrightarrow \underset{\text{acetic acid}}{CH_3COOH}$$

In this case, an oxygen atom has been added to acetaldehyde to give acetic acid.

Example 24.6 Balancing Oxidation–Reduction Equations Involving Organic Compounds

2-Propanol can be oxidized to acetone (2-propanone).

$$\underset{\displaystyle\overset{|}{OH}}{CH_3CHCH_3} + (O) \longrightarrow \underset{\displaystyle\overset{\|}{O}}{CH_3CCH_3} + H_2O$$

Write the balanced equation for the oxidation of 2-propanol to acetone by dichromate ion, $Cr_2O_7{}^{2-}$, in acidic solution. (The balancing of oxidation–reduction equations was discussed in Section 13.8.)

Solution

The oxidation half-reaction is

$$\underset{\displaystyle\overset{|}{OH}}{CH_3CHCH_3} \longrightarrow \underset{\displaystyle\overset{\|}{O}}{CH_3CCH_3} + 2H^+ + 2e^-$$

Because dichromate ion is reduced to chromium(III) ion, the reduction half-reaction is

$$\underset{\text{orange}}{Cr_2O_7{}^{2-}} + 14H^+ + 6e^- \longrightarrow \underset{\text{green}}{2Cr^{3+}} + 7H_2O$$

(A change from the orange color of $Cr_2O_7{}^{2-}$ to the green color of a Cr^{3+} complex species is seen as the reaction occurs.) The balanced equation is

$$3\underset{\displaystyle\overset{|}{OH}}{CH_3CHCH_3} + Cr_2O_7{}^{2-} + 8H^+ \longrightarrow 3\underset{\displaystyle\overset{\|}{O}}{CH_3CCH_3} + 2Cr^{3+} + 7H_2O$$

Exercise 24.12

Write the balanced equation for the oxidation of ethanol to acetaldehyde by permanganate ion in acidic solution. Permanganate is reduced to Mn^{2+} in acidic solution.

(See Problems 24.51 and 24.52.)

Primary and secondary alcohols are easily oxidized. In both, hydrogen atoms are bonded to the carbon atom carrying the hydroxyl group, and these hydrogen atoms are removed in the oxidation to give a carbonyl group. In general, for oxidizing agents such as dichromate or permanganate ions in acidic solution, we get the following reactions:

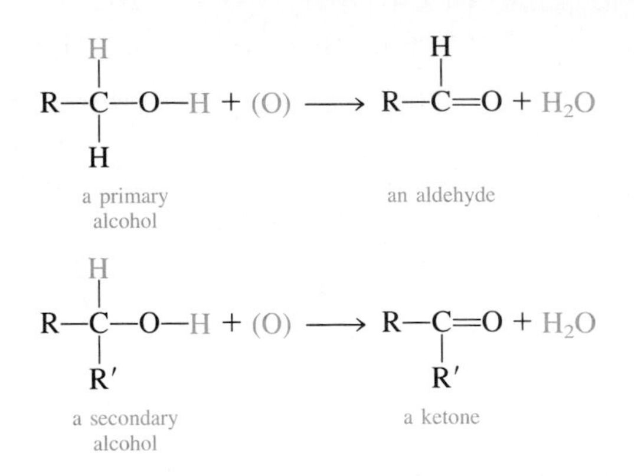

An aldehyde is easily oxidized to a carboxylic acid. Therefore, unless the aldehyde is removed from the reaction vessel, a primary alcohol gives the carboxylic acid upon oxidation. Because aldehydes boil at temperatures below those for the corresponding primary alcohols, however, aldehydes are easily distilled from the reaction vessel. Tertiary alcohols are unreactive with these oxidizing agents.

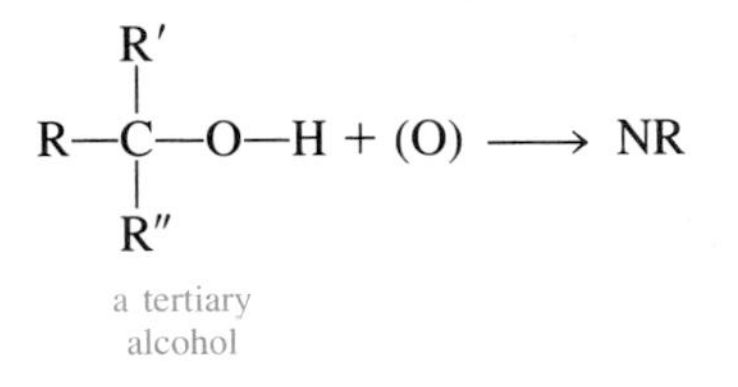

Of course, they can be oxidized to H_2O and CO_2 with very strong oxidizing agents (such as O_2), as can most organic compounds.

As we mentioned, aldehydes are easily oxidized to the corresponding carboxylic acid. Ketones, however, are resistant to oxidation, except under severe conditions where the carbon chain is broken up. With oxidizing agents under moderate conditions, we get

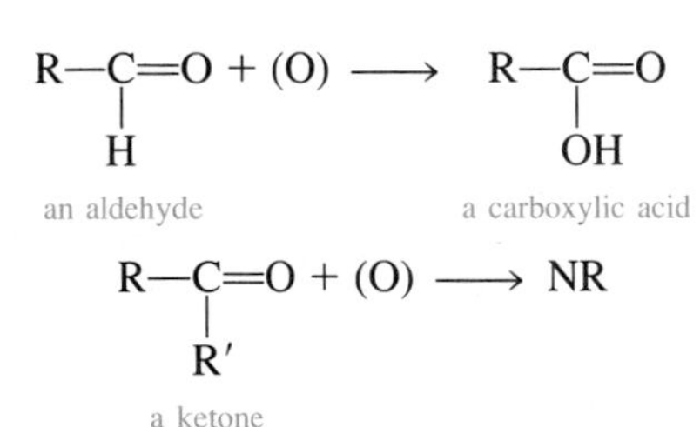

Figure 24.11
Tollen's test for aldehydes. Here a flask was coated with silver by reducing silver ion with formaldehyde.

Note that, as with an alcohol, the presence of an H atom on the carbon atom carrying the functional group determines the reactivity with oxidizing agents.

Tollen's test for distinguishing between aldehydes and ketones relies on the difference in ease of oxidation of these two classes of compounds. In this test, silver ion in ammonia solution functions as a mild oxidizing agent. Silver(I) is reduced to silver metal by aldehydes but not by ketones. To test a compound, we put a sample of it with Tollen's reagent in a clean test tube or flask. If the compound is an aldehyde, silver metal will deposit in a few minutes on the inside of the glass vessel as a reflective coating (Figure 24.11). (The same reaction can be used to produce mirrors.)

Aldehydes, ketones, and carboxylic acids can all be reduced to the corresponding alcohols. Lithium aluminum hydride, $LiAlH_4$, is a reducing agent commonly used for these reactions.

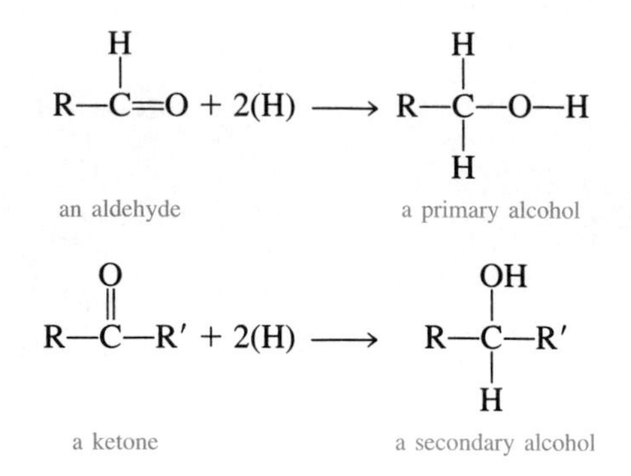

Exercise 24.13

Complete each of the following equations. If no reaction occurs, write NR. Name any organic products.

(a) $(CH_3)_3COH + (O) \longrightarrow$

(b) $CH_3CH(CHO)CH_2CH_3 + (O) \longrightarrow$

(c) $CH_3CH(CHO)CH_2CH_3 + (H) \longrightarrow$

(d) $CH_3CH(OH)CH_2CH_3 + (O) \longrightarrow$

(See Problems 24.53 and 24.54.)

Esterification and Saponification

Carboxylic acids react with alcohols when these substances are heated together in the presence of an inorganic acid, such as H_2SO_4, to produce an ester.

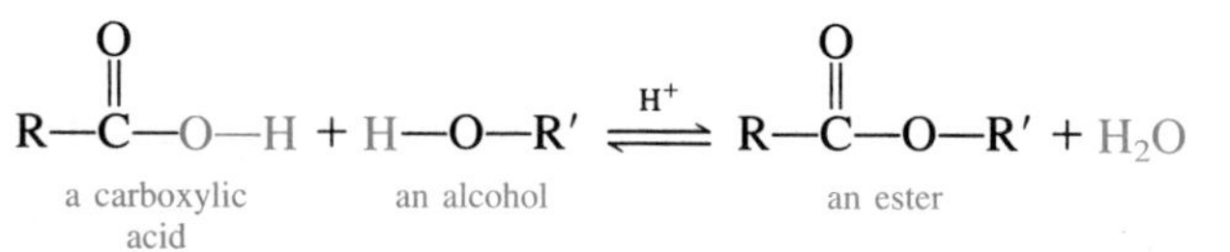

This is a type of **condensation reaction,** *a reaction in which two molecules are joined by the elimination of a small molecule such as water.*

Table 24.6 Odors of Esters

Name	Formula	Odor*
Ethyl formate	$HCOOCH_2CH_3$	Rum
Pentyl acetate	$CH_3COOCH_2CH_2CH_2CH_2CH_3$	Banana
Octyl acetate	$CH_3COOCH_2CH_2CH_2CH_2CH_2CH_2CH_2CH_3$	Orange
Methyl butyrate	$CH_3CH_2CH_2COOCH_3$	Apple
Ethyl butyrate	$CH_3CH_2CH_2COOCH_2CH_3$	Pineapple
Pentyl butyrate	$CH_3CH_2CH_2COOCH_2CH_2CH_2CH_2CH_3$	Apricot
Methyl salicylate	*o*-$C_6H_4(OH)COOCH_3$	Wintergreen

*Natural flavors are generally complex mixtures of esters and other constituents.

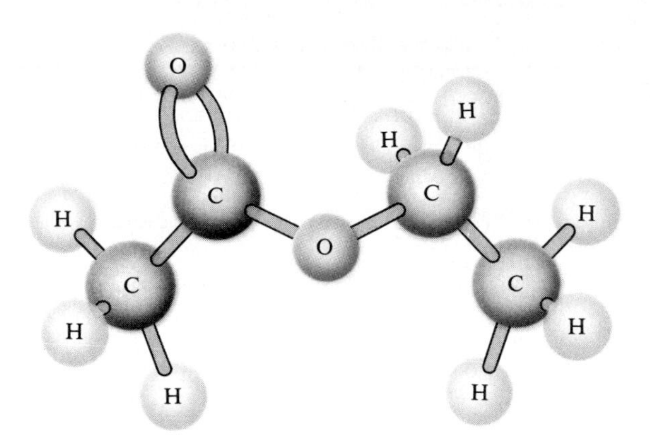

Figure 24.12
Model of the ethyl acetate molecule, $CH_3COOCH_2CH_3$. Ethyl acetate is used as a solvent in lacquers and similar protective coatings.

Esters are intriguing substances because many of them have very pleasant odors. They are present in natural flavors and are used to make artificial flavorings. Table 24.6 lists some esters and the odors associated with them. Esters are named by giving the name of the hydrocarbon group corresponding to the alcohol part of the molecule followed by the name of the carboxylic acid, with *-oic* (or *-ic*) changed to *-ate* (similar to the naming of salts). Thus, ethanol and acetic acid give the ester ethyl acetate (Figure 24.12).

The reaction of a carboxylic acid and an alcohol is reversible. Thus, an ester reacts with water in the presence of hydrogen ion to give the carboxylic acid and the alcohol. Such a reaction of a compound with water is called a *hydrolysis*. The hydrolysis goes to completion in the presence of a base, and in this case the products are the alcohol and the salt of the carboxylic acid.

Saponification is *the hydrolysis of an ester in the presence of a base*. The term, which comes from the Latin for ''soap'' *(sapon)*, originated from the soap-making process. In this process, an animal fat or a vegetable oil is boiled with a strong base, usually NaOH. Animal fats and vegetable oils are esters of glycerol with various long-chain carboxylic acids called *fatty acids*.● An equation for the saponification of a fat that is an ester of glycerol with stearic acid, $CH_3(CH_2)_{16}COOH$ or $C_{17}H_{35}COOH$, follows. In general, the three fatty-acid groups in a fat or an oil need not be identical.

● Fatty acids will be discussed further in the next chapter. See Table 25.3 for a list of some important fatty acids.

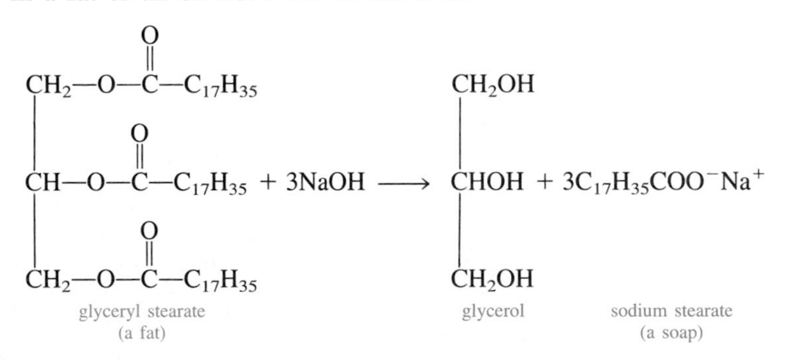

$$\begin{array}{l} CH_2{-}O{-}\overset{\overset{O}{\|}}{C}{-}C_{17}H_{35} \\ | \\ CH{-}O{-}\overset{\overset{O}{\|}}{C}{-}C_{17}H_{35} \\ | \\ CH_2{-}O{-}\overset{\overset{O}{\|}}{C}{-}C_{17}H_{35} \end{array} + 3NaOH \longrightarrow \begin{array}{l} CH_2OH \\ | \\ CHOH \\ | \\ CH_2OH \end{array} + 3C_{17}H_{35}COO^-Na^+$$

glyceryl stearate (a fat) — glycerol — sodium stearate (a soap)

Exercise 24.14

Write an equation for the preparation of methyl propionate. Note any catalyst used.

(See Problems 24.55 and 24.56.)

Table 24.7 Some Common Amines

Name	Formula	Boiling Point, °C
Methylamine	CH_3-NH_2	−6.5
Dimethylamine	$CH_3-\underset{\underset{H}{\vert}}{N}-CH_3$	7.4
Trimethylamine	$CH_3-\underset{\underset{CH_3}{\vert}}{N}-CH_3$	3.5
Ethylamine	$CH_3CH_2-NH_2$	16.6
Piperidine	$N-H$	106
Aniline	$-NH_2$	184

(From Robert D. Whitaker et al., *Concepts of General, Organic, and Biological Chemistry* [Boston: Houghton Mifflin, 1981], p. 343.)

24.7 ORGANIC COMPOUNDS CONTAINING NITROGEN

Alcohols and ethers, you may recall, can be considered derivatives of H_2O, where one or both H atoms are replaced by hydrocarbon groups, R. Thus, the general formula of an alcohol is ROH, and that of an ether is ROR′. Another important class of organic compounds is obtained by similarly substituting R groups for the H atoms of ammonia, NH_3.

Most organic bases are **amines,** which are *compounds that are structurally derived by replacing one or more hydrogen atoms of ammonia with hydrocarbon groups.*

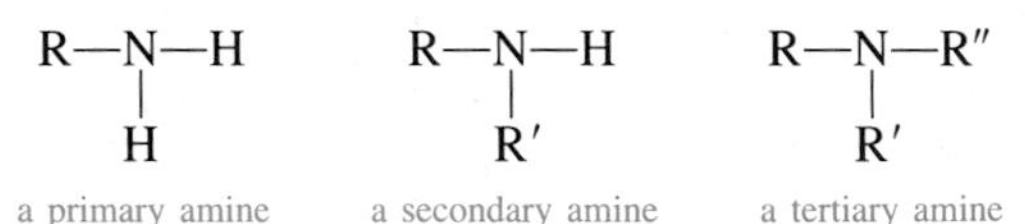

Table 24.7 lists some common amines.

Amines are bases, because the nitrogen atom has an unshared electron pair that can accept a proton to form a substituted ammonium ion. For example,

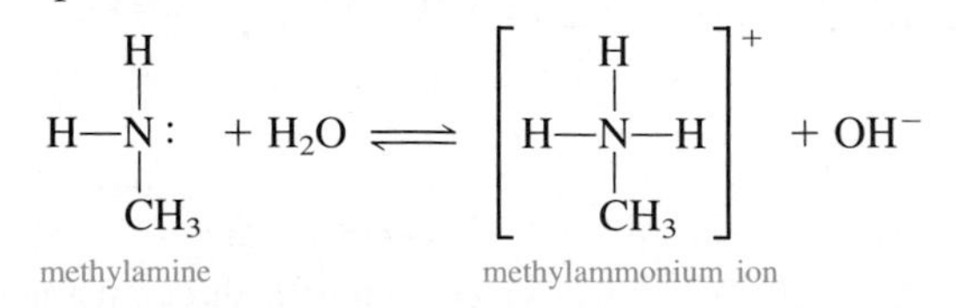

Like ammonia, amines are weak bases. Table 16.3 gives base ionization constants of some amines.

Amides are *compounds derived from the reaction of ammonia, or a primary or secondary amine, with a carboxylic acid.* For example, when ammonia is strongly heated with acetic acid, they react to give the amide acetamide.●

● The condensation to give an amide occurs under milder conditions if the ammonia or amine is reacted with the acid chloride, a derivative of the carboxylic acid obtained by replacing

$$\overset{\large O}{\overset{\|}{—COH}} \text{ by } \overset{\large O}{\overset{\|}{—CCl}}.$$

In this reaction, HCl is a product.

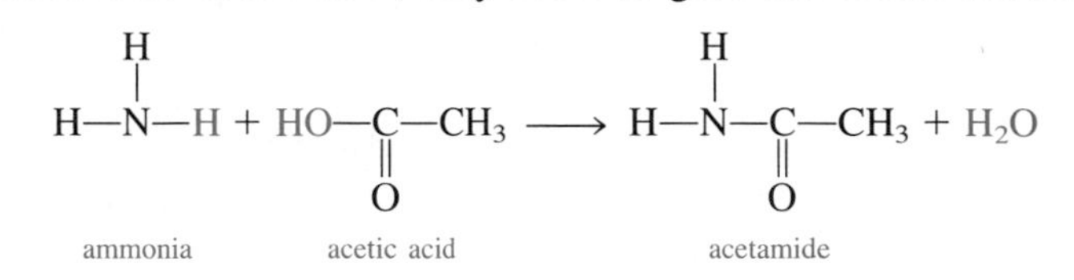

ammonia   acetic acid   acetamide

With methylamine and acetic acid, we get N-methylacetamide.

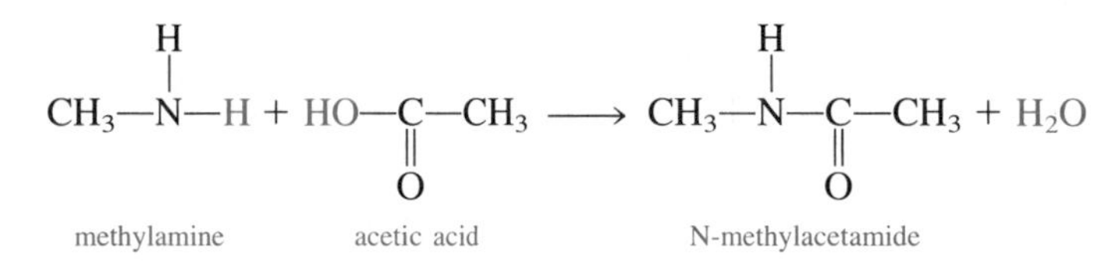

methylamine   acetic acid   N-methylacetamide

24.8 ORGANIC POLYMERS

A **polymer** is *a chemical species of very high molecular weight that is made up from many repeating units of low molecular weight. A compound used to make a polymer (and from which the polymer's repeating unit arises)* is called a **monomer.** A simple example is polyethylene, which consists of many ethylene units, $—CH_2CH_2—$, chemically bonded to one another to form long-chain molecules consisting of thousands of ethylene units. The name polyethylene stems from the name of the monomer (ethylene) together with the prefix *poly-,* meaning "many."

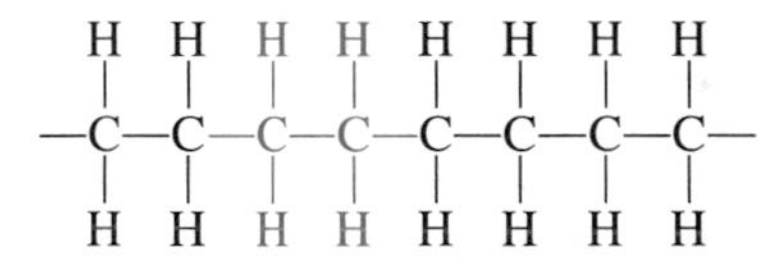

One ethylene unit is shown in color.

The first synthetic polymer was discovered in 1909 by Leo Baekeland, a Belgian-born U.S. chemist. The polymer, marketed under the trademark Bakelite, was prepared from phenol and formaldehyde, and became an important plastic and resin for adhesives and paints. The study of polymers developed rapidly once the structures of some natural polymers, including cellulose and rubber, were determined. Chemists have continued to prepare new polymer materials with remarkably diverse properties. For example, organic polymers having the high electrical conductivity of copper have been discovered recently.

We will describe natural polymers in the next chapter. In this section, we will look at several synthetic polymers. Polymers are classified as *addition polymers* or *condensation polymers,* depending on the type of reaction used in forming them.

Addition Polymers

An **addition polymer** is *a polymer formed by linking together many molecules by addition reactions.* The monomers must have multiple bonds that would undergo addition reactions. For example, when propylene (whose IUPAC name is propene) is heated under pressure with a catalyst, it forms polypropylene.

$$\ldots + \overset{\overset{\displaystyle CH_3}{|}}{CH}{=}CH_2 + \overset{\overset{\displaystyle CH_3}{|}}{CH}{=}CH_2 + \overset{\overset{\displaystyle CH_3}{|}}{CH}{=}CH_2 + \ldots \longrightarrow$$

$$\underset{\text{polypropylene}}{-\overset{\overset{\displaystyle CH_3}{|}}{CH}-CH_2-\overset{\overset{\displaystyle CH_3}{|}}{CH}-CH_2-\overset{\overset{\displaystyle CH_3}{|}}{CH}-CH_2-}$$

Note that the π electrons in the double bonds of the propylene molecules form new σ bonds that unite the monomers.

The preparation of an addition polymer is often induced by an *initiator,* a compound that produces free radicals (species having an unpaired electron). Organic peroxides (organic compounds with the —O—O— group) are frequently used as initiators. If we write R for an organic group, then the general formula of such a peroxide is R—O—O—R. When heated, the —O—O— bond in the organic peroxide breaks and free radicals form (the unpaired electron is symbolized by a dot).

$$R{-}O{-}O{-}R \xrightarrow{\Delta} R{-}O\cdot + \cdot O{-}R$$

The free radical reacts with the alkene monomer to produce a new free-radical species, which in turn reacts with another monomer molecule. For example, the polymerization of propylene occurs as follows:

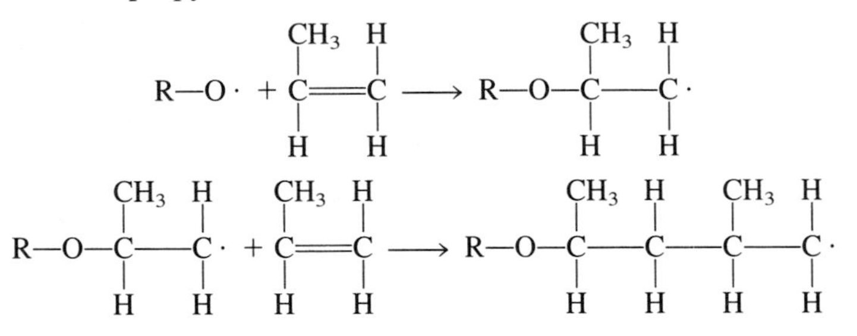

At each addition step, the unpaired electron from the free radical pairs with one of the π electrons of the alkene monomer to form a σ bond. One end of the chain retains an unpaired electron and can add another monomer molecule. The reaction

Figure 24.13
Nylon being pulled from a reaction mixture. It is formed by reacting hexamethylene diamine (bottom water layer) with adipic acid or, as in this experiment, with adipoyl chloride, $ClCO(CH_2)_4COCl$ (top hexane layer). Nylon, formed at the interface of the two layers, is continuously removed.

Table 24.8 Some Addition Polymers

Polymer	Monomer	Uses
Polyethylene	$CH_2{=}CH_2$	Bottles, plastic tubing
Polypropylene	$CH_2{=}\overset{\overset{\displaystyle CH_3}{|}}{CH}$	Bottles, textiles
Polytetrafluoroethylene (Teflon®)	$CF_2{=}CF_2$	Nonstick surface for frying pans
Polyvinyl chloride (PVC)	$CH_2{=}\overset{\overset{\displaystyle Cl}{|}}{CH}$	Plastic pipes
Polyacrylonitrile (Orlon®, Acrilan®)	$CH_2{=}\overset{\overset{\displaystyle CN}{|}}{CH}$	Carpets, textiles

terminates when the free electron at the end of the chain reacts with another free radical. Some addition polymers are listed in Table 24.8.

Exercise 24.15

An addition polymer is prepared from vinylidene chloride, $CH_2{=}CCl_2$. Write the structure of the addition polymer.
(See Problems 24.59 and 24.60.)

Condensation Polymers

A **condensation polymer** is *a polymer formed by linking together many molecules by condensation reactions*. Wallace Carothers (1896–1937), a chemist at E. I. Du Pont de Nemours and Company, realized that polymers could be prepared by condensation reactions. He and his coworkers produced a number of polymers this way, including polyamides (nylon) and polyesters. The first synthetic fibers, produced from nylon, were announced by Du Pont in 1939.

POLYESTERS A substance with two alcohol groups reacts with one containing two carboxylic acid groups to form a *polyester,* a polymer whose repeating units are joined by ester groups. The reactant molecules join as links of a chain to form a very long molecule. The polyester Dacron, used as a textile fiber, is prepared from ethylene glycol and terephthalic acid.

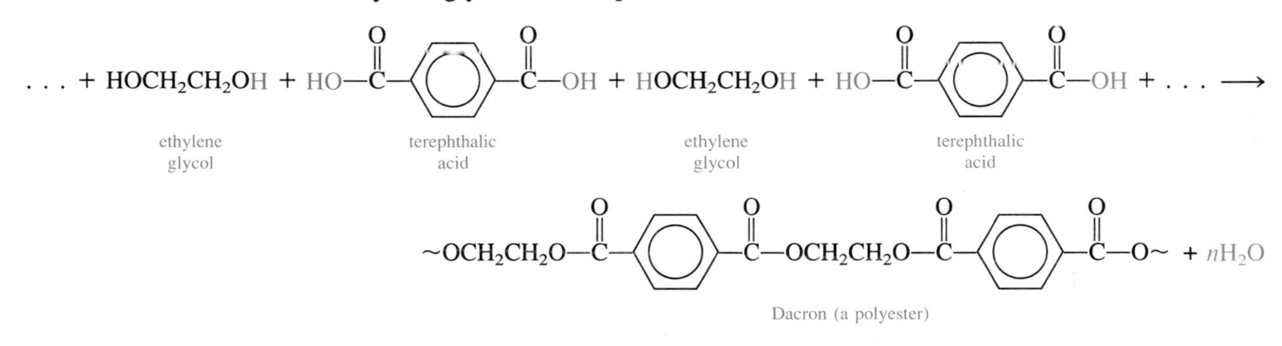

A water molecule is split out during the formation of each ester linkage.

POLYAMIDES When a compound containing two amine groups reacts with one containing two carboxylic acid groups, a condensation polymer called a *polyamide* is formed (Figure 24.13). Nylon-66 is an example. It is prepared by heating hexamethylene diamine (1,6-diaminohexane) and adipic acid (hexanedioic acid).

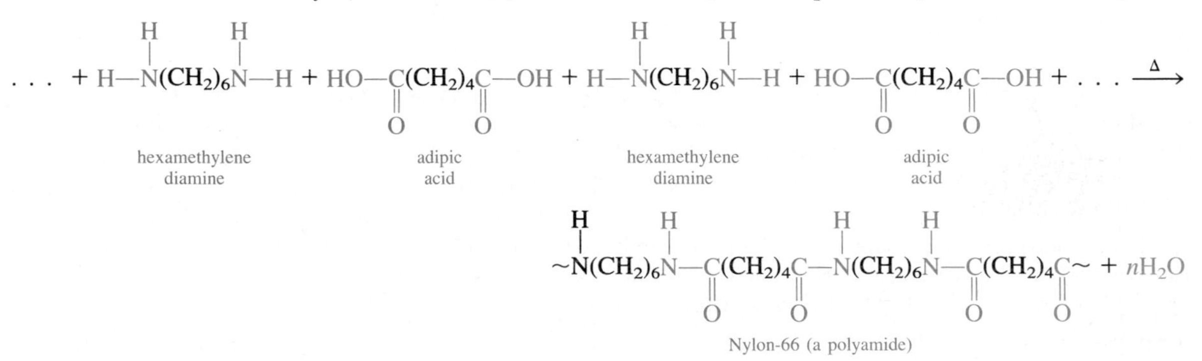

A Checklist for Review

IMPORTANT TERMS

hydrocarbons (p. 971)
aromatic hydrocarbons (24.1)
aliphatic hydrocarbons (24.1)
saturated hydrocarbons (24.1)
alkanes (24.1)
homologous series (24.1)
alkyl group (24.1)
cycloalkanes (24.1)
unsaturated hydrocarbons (24.2)
alkenes (24.2)
geometric isomers (24.2)
alkynes (24.2)
substitution reaction (24.4)
addition reaction (24.4)
Markownikoff's rule (24.4)
functional group (p. 987)
alcohol (24.5)
ether (24.5)
aldehyde (24.5)
ketone (24.5)
carboxylic acid (24.5)
ester (24.5)
condensation reaction (24.6)
saponification (24.6)
amines (24.7)
amides (24.7)
polymer (24.8)
monomer (24.8)
addition polymer (24.8)
condensation polymer (24.8)

SUMMARY OF FACTS AND CONCEPTS

Organic compounds are hydrocarbons or derivatives of hydrocarbons. The two main types of hydrocarbons are *aromatic* (containing benzene rings) or *aliphatic*. The simplest aliphatic hydrocarbons are *alkanes,* compounds with the general formula C_nH_{2n+2}. These are *saturated hydrocarbons*—that is, hydrocarbons in which the carbon atoms are bonded to four other atoms (either hydrogens or other carbons). Another *homologous series* of saturated hydrocarbons are the *cycloalkanes,* compounds of the general formula C_nH_{2n} in which the carbon atoms are joined in a ring. The *alkenes* and *alkynes* are unsaturated hydrocarbons—that is, hydrocarbons not containing the maximum number of H atoms but containing double or triple carbon–carbon bonds. An *aromatic hydrocarbon* is one containing benzene rings—that is, six-membered rings of carbon atoms with alternating single and double bonds in resonance formulas.

The alkanes and aromatic hydrocarbons usually undergo *substitution reactions*. Alkenes and alkynes undergo *addition reactions*. *Markownikoff's rule* predicts the major product in the addition of an unsymmetrical reagent to an unsymmetrical alkene.

A *functional group* is a portion of an organic molecule that reacts readily in predictable ways. Important functional groups containing oxygen are *alcohols* (ROH), *aldehydes* (RCHO), *ketones* (RCOR′), and *carboxylic acids* (RCOOH). Alcohols (except tertiary alcohols) and aldehydes are readily oxidized. Primary alcohols are oxidized to aldehydes, which are then oxidized to carboxylic acids. Secondary alcohols are oxidized to ketones. Aldehydes, ketones, and carboxylic acids can be reduced to the corresponding alcohols. Carboxylic acids and alcohols react to produce *esters*. In basic solution, an ester can be *saponified* to give the alcohol and carboxylate salt. *Amines* are organic derivatives of ammonia. They react with carboxylic acids to give *amides*.

Polymers are species of very high molecular weight consisting of many repeating units *(monomers)*. They are classified as *addition polymers* or *condensation polymers,* depending on the type of reaction used in forming them.

OPERATIONAL SKILLS

1. Writing the IUPAC name of a hydrocarbon given the structural formula, and vice versa Given the structure of a hydrocarbon, state the IUPAC name (Examples 24.1 and 24.3). Given the IUPAC name of a hydrocarbon, write the structural formula (Example 24.2).

2. Predicting *cis–trans* isomers Given a condensed structural formula of an alkene, decide whether *cis* and *trans* isomers are possible, and then draw the structural formulas (Example 24.4).

3. Predicting the major product in an addition reaction Predict the major product in the addition of an unsymmetrical reagent to an unsymmetrical alkene (Example 24.5).

4. Balancing oxidation–reduction equations involving organic compounds Write a complete balanced equation for the oxidation or reduction of an organic compound (Example 24.6).

Review Questions

24.1 Give the molecular formula of an alkane with 30 carbon atoms.

24.2 Why would you expect the melting points of the alkanes to increase in the series methane, ethane, propane, and so on?

24.3 Draw structural formulas of the five isomers of C_6H_{14}.

24.4 Draw structural formulas of an alkane, a cycloalkane, an alkene, and an aromatic hydrocarbon, each with seven carbon atoms.

24.5 Explain why there are two isomers of 2-butene. Draw their structural formulas and name the isomers.

24.6 Draw structural formulas for the isomers of ethylmethylbenzene.

24.7 Fill in the following table.

HYDROCARBON	SOURCE	USE
Methane		
Octane		
Ethylene		
Acetylene		

24.8 Give condensed structural formulas of all possible substitution products of ethane and Cl_2.

24.9 Define the terms *substitution reaction* and *addition reaction*. Give examples of each.

24.10 What would you expect to be the major product when two molecules of HCl add successively to acetylene? Explain.

24.11 What is the purpose of the octane number scale? How is this scale defined?

24.12 What is a functional group? Give an example of one and explain how it fits this definition.

24.13 An aldehyde contains the carbonyl group. Ketones, carboxylic acids, and esters also contain the carbonyl group. What distinguishes these latter compounds from an aldehyde?

24.14 Fill in the following table.

COMPOUND	SOURCE	USE
Methanol		
Ethanol		
Ethylene glycol		
Glycerol		
Formaldehyde		

24.15 Identify and name the functional group in each of the following.

(a) CH_3COCH_3 (b) $CH_3OCH_2CH_3$
(c) $CH_3CH{=}CH_2$ (d) CH_3CH_2COOH
(e) CH_3CH_2CHO (f) $CH_3CH_2CH_2OH$

24.16 What are the products, if any, of oxidation by an inorganic reagent (such as acidic $K_2Cr_2O_7$) of a primary alcohol, a secondary alcohol, a tertiary alcohol, an aldehyde, and a ketone? Write an equation for the oxidation of a particular compound of each type. Use (O) for the oxidizing agent. If no reaction occurs, write NR.

24.17 Consider the following formulas of two esters:

$$CH_3CH_2{-}\overset{\overset{\large O}{\|}}{C}{-}O{-}CH_3 \qquad CH_3CH_2{-}O{-}\overset{\overset{\large O}{\|}}{C}{-}CH_3$$

One of these is ethyl ethanoate (ethyl acetate) and one is methyl propanoate (methyl propionate). Which is which?

24.18 Write the equation for the saponification of ethyl acetate.

24.19 What is the source of basicity of an amine? Illustrate by writing the equation for the reaction of triethylamine, $(CH_3CH_2)_3N$, with acetic acid.

24.20 What is the difference between an addition polymer and a condensation polymer? Give an example of each, writing the equation for its formation.

Practice Problems

NAMING HYDROCARBONS

24.21 Give the IUPAC name for each of the following hydrocarbons.

(a) $CH_3CHCH_2CHCH_3$
$\quad CH_3 \quad CH_3$

(b) $\quad CH_3 \qquad CH_3$
$CH_3CCH_2CH_2CH_2CCH_3$
$\quad CH_3 \qquad CH_3$

(c) $CH_3CH_2CHCH_2CH_2CH_2CH_3$
$\qquad CH_2CH_2CH_3$

(d) $\quad CH_3$
$CH_3CHCHCH_2CH_2CH_2$
$\quad CH_2CH_3 \qquad CH_3$

24.22 What is the IUPAC name of each of the following compounds?

(a) $\qquad CH_3$
$CH_3CH_2CHCHCH_3$
$\qquad CH_3$

(b) $\qquad CH_2CH_3$
$CH_3CH_2CH_2CHCH_2CH_3$

(c) $CH_3CH_2CH_2CHCH_2CH_2CH_3$
$\qquad CH_3CHCH_3$

(d) $\qquad CH_3$
$CH_3CHCH_2CH_2CHCH_2CH_3$
$\quad CH_2CH_3$

24.23 Write the condensed structural formula for each of the following compounds.

(a) 2,3-dimethylhexane
(b) 3-ethylhexane
(c) 2-methyl-4-isopropylheptane
(d) 2,2,3,3-tetramethylpentane

24.24 Write the condensed structural formula for each of the following compounds.

(a) 2,2-dimethylbutane
(b) 3-isopropylhexane
(c) 3-ethyl-4-methyloctane
(d) 3,4,4,5-tetramethylheptane

24.25 Give the IUPAC name of each of the following.

(a) $CH_2{=}CHCH_2CH_2CH_3$

(b) $CH_3C{=}CHCH_2CHCH_3$
$\quad CH_3 \qquad CH_3$

24.26 For each of the following, write the IUPAC name.

(a) CH_3CH_2
$\qquad C{=}CHCH_2CH_3$
CH_3CH_2

(b) $CH_3CH_2CCH_2CH_2CH_3$
$\qquad \| $
$\qquad CH_2$

24.27 Give the condensed structural formula for each of the following compounds.

(a) 3-ethyl-2-pentene
(b) 4-ethyl-2-methyl-2-hexene

24.28 Write the condensed structural formula for each of the following compounds.

(a) 2,3-dimethyl-2-pentene
(b) 2-methyl-4-propyl-3-heptene

24.29 If there are geometric isomers for the following, draw structural formulas showing the isomers. Label the isomers with their IUPAC names, including *cis* and *trans* designations.

(a) $CH_3CH_2CH{=}CHCH_2CH_3$

(b) $CH_3C{=}CHCH_2CH_3$
$\quad CH_2CH_3$

24.30 One or both of the following have geometric isomers. Draw the structures of any geometric isomers and label the isomers with IUPAC names, including the prefix *cis* or *trans*.

(a) $CH_3CHCH{=}CHCH_3$
$\quad CH_3$

(b) $CH_3C{=}CHCH_2CH_3$
$\quad CH_3$

24.31 Give the IUPAC name of each of the following compounds.

(a) $CH_3C{\equiv}CCH_3$

(b) $CH{\equiv}CCHCH_3$
$\qquad CH_3$

24.32 Write the IUPAC name of each of the following hydrocarbons.

(a) $CH_3CHC{\equiv}CH$
$\quad CH_3$

(b) $CH_3C{\equiv}CCH_2CH_3$

24.33 Write structural formulas for (a) 1,1,1-triphenylethane, (b) *o*-ethylmethylbenzene.

24.34 Write structural formulas for (a) 1,2,3-trimethylbenzene, (b) *p*-diethylbenzene.

REACTIONS OF HYDROCARBONS

24.35 Complete and balance the following equations. Note any catalyst used.

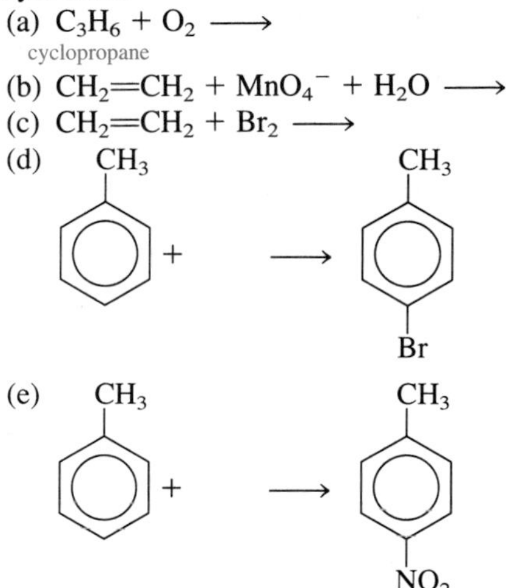

24.36 Complete and balance the following equations. Note any catalyst used.

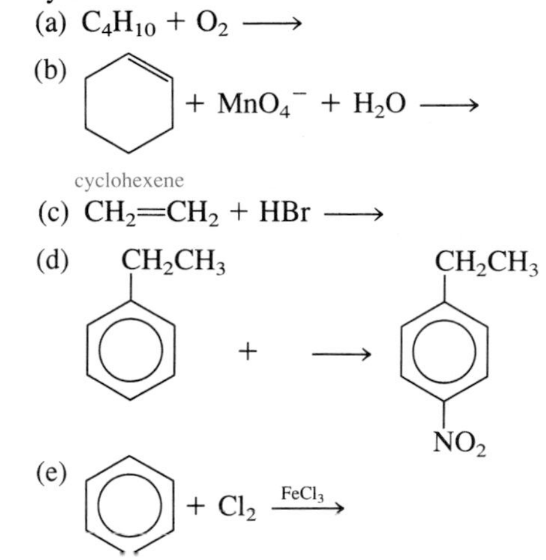

24.37 What is the major product when HBr is added to methylpropene?

24.38 Complete the following equation, giving only the main product.

$$CH_2{=}CHCH_3 + H{-}OH \xrightarrow[\text{catalyst}]{H_2SO_4}$$

24.39 Write two equations for the cracking of pentane into different products.

24.40 Write two equations for the cracking of hexane into different products.

NAMING OXYGEN-CONTAINING ORGANIC COMPOUNDS

24.41 Circle and name the functional group in each compound.

(a) $CH_3{-}\overset{\overset{\displaystyle O}{\|}}{C}{-}CH_2CH_2CH_3$

(b) $CH_3{-}\underset{\underset{\displaystyle H}{|}}{\overset{\overset{\displaystyle OH}{|}}{C}}{-}CH_2CH_3$

(c) $HO{-}\overset{\overset{\displaystyle O}{\|}}{C}{-}CH_2CH_3$

(d) $H{-}\overset{\overset{\displaystyle O}{\|}}{C}{-}CH_2CH_3$

24.42 Circle and name the functional group in each compound.

(a) $CH_2{=}CHCH_3$

(b) $O{=}\underset{\underset{\displaystyle CH_3CHCH_3}{|}}{C}{-}OH$

(c) $HO{-}CH_2\underset{\underset{\displaystyle CH_3}{|}}{\overset{\overset{\displaystyle CH_3}{|}}{CH}}$

(d) $O{=}\underset{\underset{\displaystyle CH_3CHCH_3}{|}}{C}{-}CH_3$

24.43 Give the IUPAC name for each of the following.

(a) $HOCH_2CH_2CH_2CH_2CH_3$

(b) $CH_3CHCH_2CH_2CH_3$ (with OH on the second carbon)

(c) $CH_3CH_2CH_2CHCH_2CH_2CH_3$

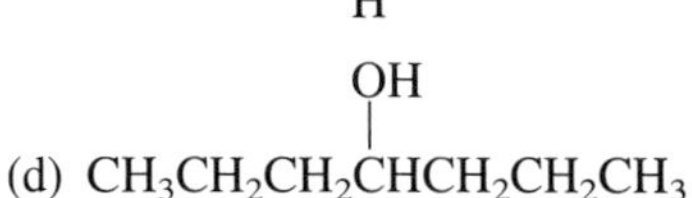

(d) $CH_3CH_2CH_2CHCH_2CH_2CH_3$ (with OH on the fourth carbon)

24.44 Write the IUPAC name for each of the following.

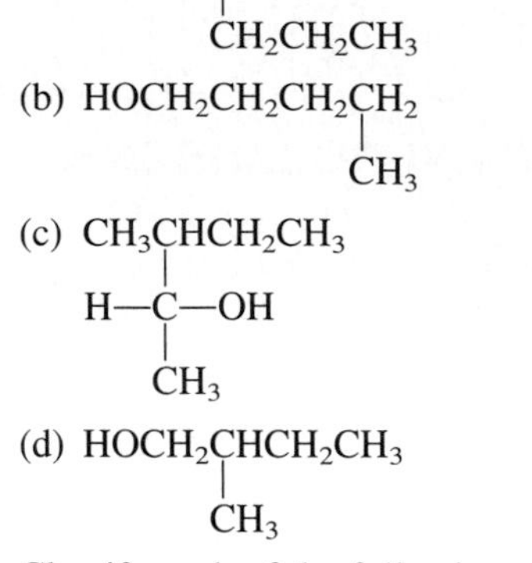

(a) $HOCH_2CHCH_2CH_3$ / $CH_2CH_2CH_3$

(b) $HOCH_2CH_2CH_2CH_2$ / CH_3

(c) $CH_3CHCH_2CH_3$ / H—C—OH / CH_3

(d) $HOCH_2CHCH_2CH_3$ / CH_3

24.45 State whether each of the following alcohols is primary, secondary, or tertiary.

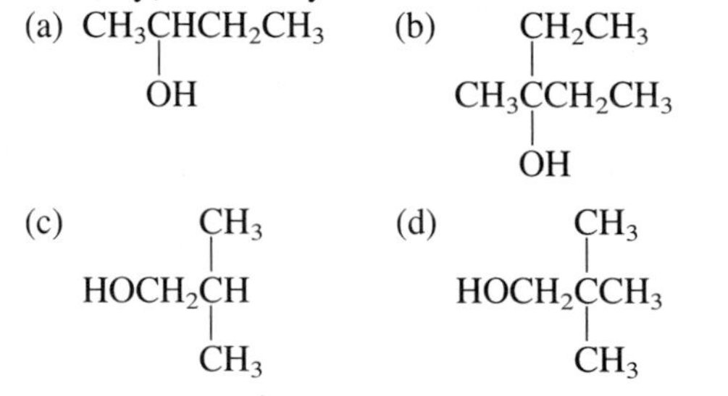

24.46 Classify each of the following as a primary, secondary, or tertiary alcohol.

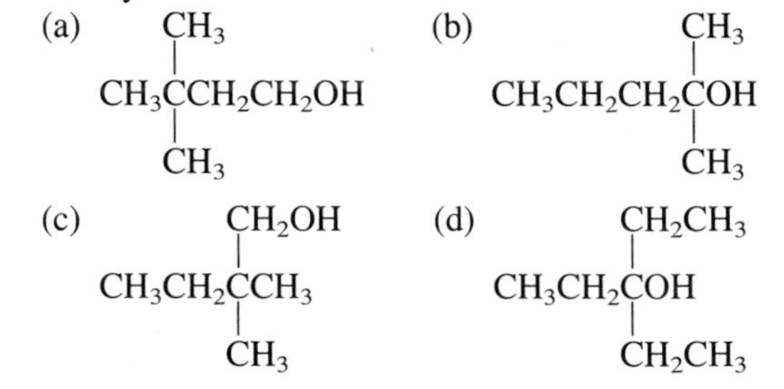

24.47 What is the common name of each of the following compounds?

(a) $CH_3CH_2OCH_2CH_2CH_3$

(b)

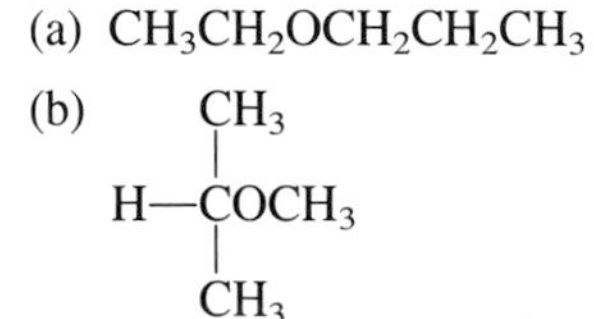

24.48 What is the common name of each of the following compounds?

(a)

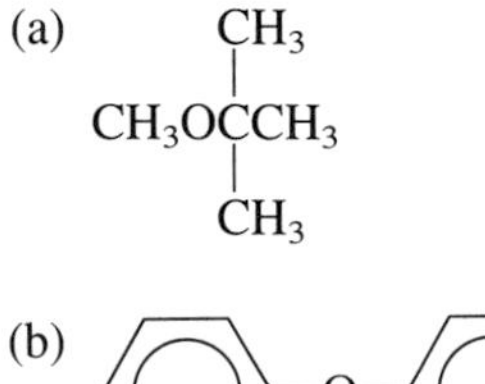

(b) C₆H₅—O—C₆H₅ (two benzene rings joined by O)

24.49 According to IUPAC rules, what is the name of each of the following compounds?

(a) $CH_3COCH_2CH_3$ (b) $CH_3CH_2CH_2CHO$

(c) $H{-}\overset{O}{\overset{\|}{C}}CH_2CH_2\overset{CH_3}{\overset{|}{C}}CH_3$ with a second CH_3 below the quaternary carbon

(d) $CH_3\underset{CH_2CH_3}{\underset{|}{C}H}\overset{O}{\overset{\|}{C}}CH_3$

24.50 Write the IUPAC name of each of the following compounds.

(a) $CH_3\underset{CHO}{\underset{|}{C}H}CH_3$ (b) $CH_3\underset{COCH_3}{\underset{|}{C}H}CH_3$

(c) CH_3CH_2 / $CH_2C{-}H$ (with =O on the C) (d) CH_3CHCH_3 / $CH_2{-}\underset{O}{\underset{\|}{C}}{-}CH_2CH_3$

REACTIONS OF OXYGEN-CONTAINING ORGANIC COMPOUNDS

24.51 Benzaldehyde, C_6H_5CHO, is oxidized by potassium permanganate, $KMnO_4$, in basic solution to benzoic acid, C_6H_5COOH. Permanganate ion is reduced to manganese dioxide. Write a balanced equation for the reaction.

24.52 Cyclohexanol, $C_6H_{11}OH$, is oxidized by chromium trioxide, CrO_3, in acidic solution to cyclohexanone, $C_6H_{10}O$. Chromium trioxide is reduced to Cr^{3+}. Write a balanced equation for the reaction.

24.53 Complete the following equations. If no reaction occurs, write NR. Name any organic products.

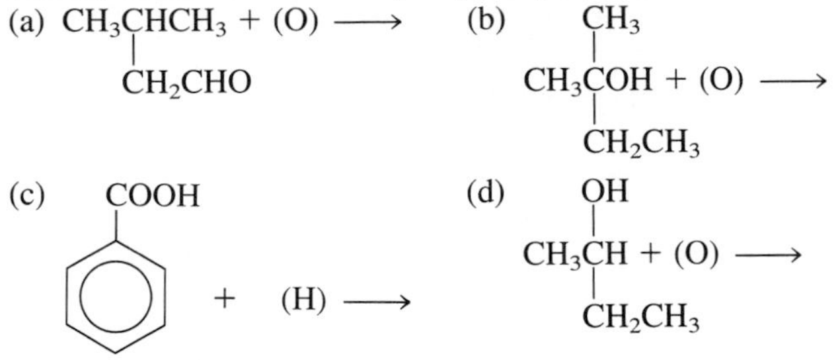

(a) CH_3CHCH_3 (with CH_2CHO on the middle C) + (O) ⟶

(b) CH_3COH (with CH_3 above and CH_2CH_3 below the central C) + (O) ⟶

(c) COOH on a benzene ring + (H) ⟶

(d) CH_3CH (with OH above and CH_2CH_3 below the central C) + (O) ⟶

24.54 Give the structural formula of and write the IUPAC name for the organic product, if any, of the following equations. Note any cases where no reaction occurs.

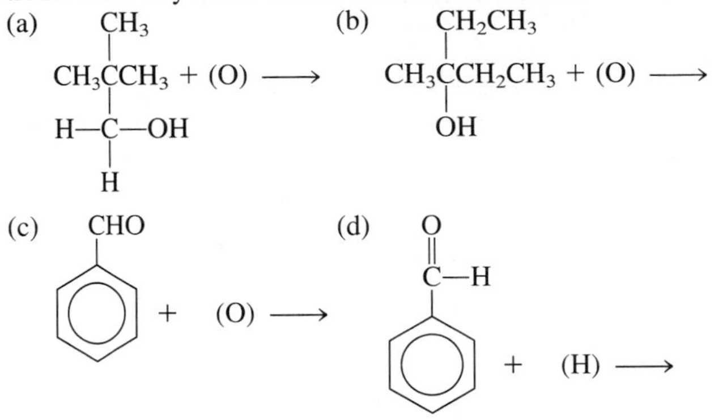

(a) CH_3CCH_3 (with CH_3 above and H—C—OH, H below the central C) + (O) ⟶

(b) $CH_3CCH_2CH_3$ (with CH_2CH_3 above and OH below the central C) + (O) ⟶

(c) CHO on a benzene ring + (O) ⟶

(d) C(=O)—H on a benzene ring + (H) ⟶

24.55 Write equations for the following. Note any catalyst used.

(a) Preparation of ethyl butyrate
(b) Saponification of methyl formate

24.56 Write equations for the following. Note any catalyst used.

(a) Preparation of isopropyl acetate
(b) Saponification of a fat that is an ester of glycerol with stearic, palmitic, and oleic acids. (*Note:* The structures of these acids are given in the next chapter in Table 25.3.)

ORGANIC COMPOUNDS CONTAINING NITROGEN

24.57 Identify each of the following compounds as a primary, secondary, or tertiary amine or amide.

(a) NH_2 on a benzene ring

(b) $CH_3CH_2NHCH_2CH_3$

24.58 Identify each of the following compounds as a primary, secondary, or tertiary amine or amide.

(a) 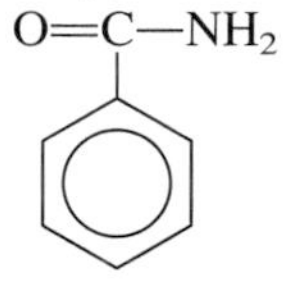 O=C—NH_2 on a benzene ring (b) $CH_3CH_2CH_2NH_2$

ORGANIC POLYMERS

24.59 Teflon is an addition polymer of 1,1,2,2-tetrafluoroethene. Write the equation for the formation of the polymer.

24.60 Polyvinyl chloride (PVC) is an addition polymer of vinyl chloride, $CH_2{=}CHCl$. Write the equation for the formation of the polymer.

Additional Problems

24.61 Give the IUPAC name of each of the following compounds.

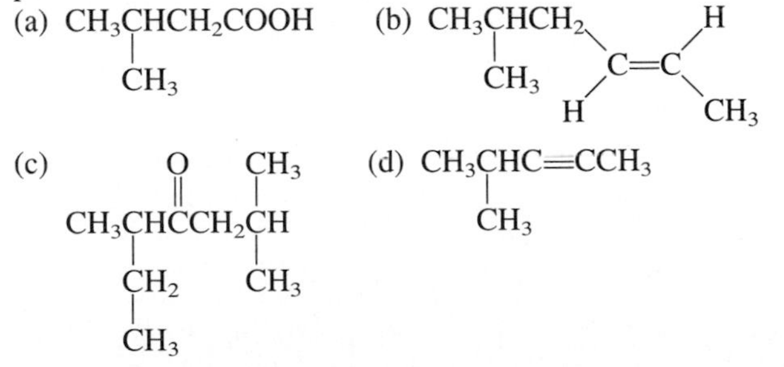

(a) CH_3CHCH_2COOH (with CH_3 on C-2)

(b) CH_3CHCH_2 (with CH_3 on C-2) attached to C=C, with H and H on opposite sides and CH_3 on the other carbon

(c) CH_3CHCCH_2CH (with O double-bonded on the carbonyl C, CH_2CH_3 on the second C, and CH_3 above and CH_3 below the last CH)

(d) $CH_3CHC{\equiv}CCH_3$ (with CH_3 on C-2)

24.62 Give the IUPAC name of each of the following compounds.

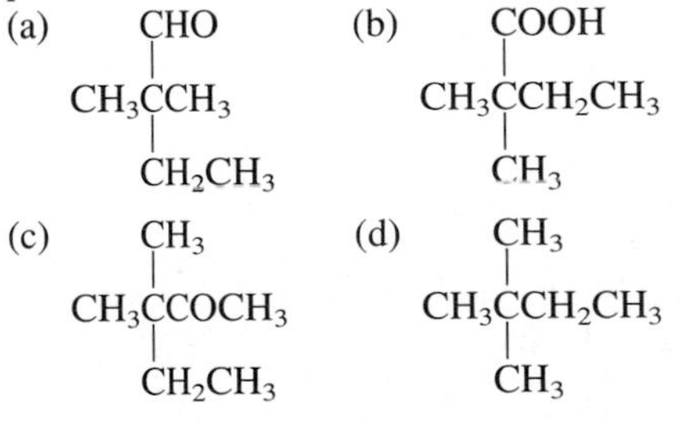

(a) CH_3CCH_3 (with CHO above and CH_2CH_3 below the central C)

(b) $CH_3CCH_2CH_3$ (with COOH above and CH_3 below the central C)

(c) CH_3CCOCH_3 (with CH_3 above and CH_2CH_3 below the central C)

(d) $CH_3CCH_2CH_3$ (with CH_3 above and CH_3 below the central C)

24.63 Write the structural formula for each of the following compounds.

(a) isopropyl propionate
(b) *t*-butylamine
(c) 2,2-dimethylhexanoic acid
(d) *cis*-3-hexene

24.65 Describe chemical tests that could distinguish between:

(a) propionaldehyde (propanal) and acetone (propanone)
(b) $CH_2{=}CH{-}C{\equiv}C{-}CH{=}CH_2$ and benzene

24.67 Identify each of the following compounds from the description given.

(a) A gas with a sweetish odor that promotes the ripening of green fruit.
(b) An unsaturated compound of the formula C_7H_8 that gives a negative test with bromine in carbon tetrachloride.
(c) A compound with an ammonialike odor that acts as a base; its molecular formula is CH_5N.
(d) An alcohol used as a starting material for the manufacture of formaldehyde.

24.69 A compound that is 85.6% C and 14.4% H and has a molecular weight of 56.1 amu reacts with water and sulfuric acid to produce a compound that reacts with acidic potassium dichromate solution to produce a ketone. What is the name of the original hydrocarbon?

24.64 Write the structural formula for each of the following compounds.

(a) 3-ethyl-1-pentene
(b) 1,1,2,2-tetraphenylethane
(c) 1-phenyl-2-butanone
(d) cyclopentanone

24.66 Describe chemical tests that could distinguish between:

(a) acetic acid and acetaldehyde (ethanal)
(b) toluene (methylbenzene) and 2-methylcyclohexene

24.68 Identify each of the following compounds from the description given.

(a) An acidic compound that also has properties of an aldehyde; its molecular formula is CH_2O_2.
(b) A compound used as a preservative for biological specimens and as a raw material for plastics.
(c) A saturated hydrocarbon boiling at 0°C, which is liquefied and sold in cylinders as a fuel.
(d) A saturated hydrocarbon that is the main constituent of natural gas.

24.70 A compound with a fragrant odor reacts with dilute acid to give two organic compounds, A and B. Compound A is identified as an alcohol with a molecular weight of 32.0 amu. Compound B is identified as an acid. It can be reduced to give a compound whose composition is 60.0% C, 13.4% H, and 26.6% O and whose molecular weight is 60.1 amu. What is the name of the original compound?

25

Biochemistry

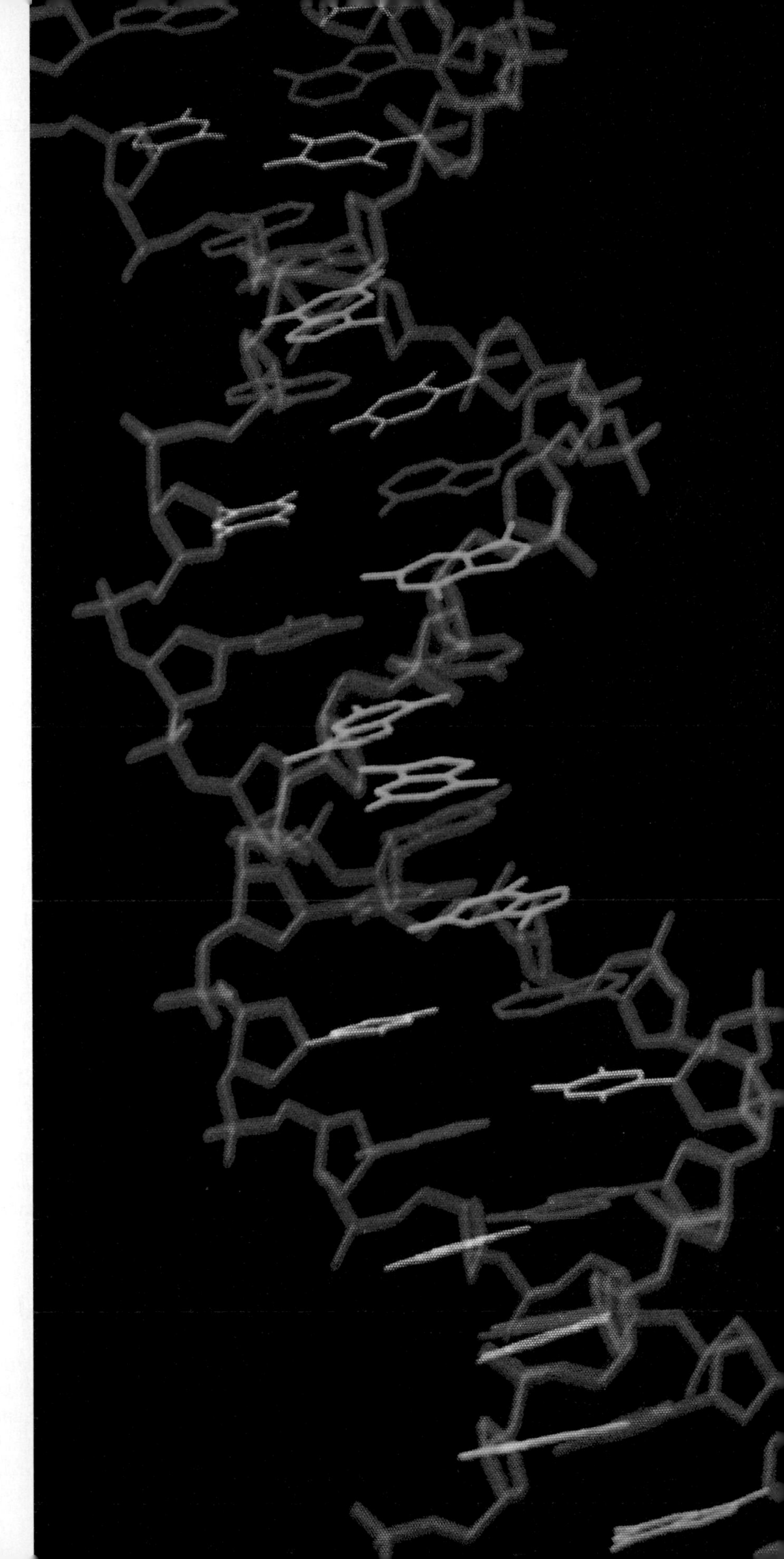

Chapter Outline

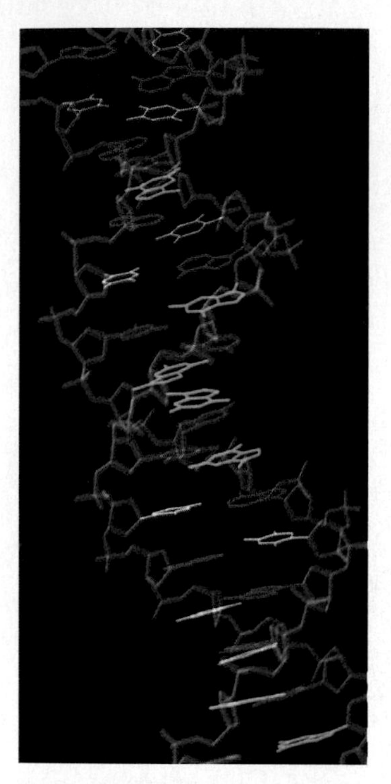
Representation of a DNA molecule generated by a computer.

In 1856, the chemist Louis Pasteur, who was a professor at Lille, France, was asked by a local industrialist to determine why his factory was having difficulty manufacturing alcohol from fermenting sugar beets. (In this process, the beet sugar is broken down to alcohol and carbon dioxide.) Examining droplets of the liquor under the microscope, Pasteur discovered that the good and bad samples contained different microorganisms. He concluded that particular yeasts (previously identified by others as living organisms) were necessary for the desired production of alcohol. He postulated that the beet sugar provided food for the yeast and that the alcohol was the residue of the food breakdown. Later, Pasteur was asked to investigate the diseases of wines, and he discovered that good wines resulted only when wild yeasts normally present in the bloom of the grape were present to ferment the grape sugar. Wines of poor quality resulted when other microorganisms contaminated the fermenting grape juice and formed products that caused bad flavors. Pasteur also showed that certain bacteria that ferment milk sugar to lactic acid were responsible for the souring of milk, and he made many other important contributions to the then-infant sciences of bacteriology and biological chemistry.

Yeasts and bacteria used in the production of foods and beverages provide an everyday example of the ability of organisms to transform matter and energy. These biological processes are major concerns of biochemistry, which we introduce in this chapter.

Introduction to Biological Systems

Living systems contain less than 30 chemical elements, but have features not shared by nonliving systems. They are able to exchange matter and energy with their surroundings and to respond to changes in those surroundings. They can transform energy and matter into different forms according to their needs. They can grow and reproduce. All these properties are due to the highly organized state of living systems.

The elaborate organization of matter in even the most complex living organisms begins with water and small organic molecules. Many of these small molecules are combined into large polymers, including proteins, complex carbohydrates, and nucleic acids. These larger molecules aggregate, forming fibers and membranes, which in turn are organized into cells. Cells, then tissues and body organs, and finally organisms represent the culmination of this hierarchy of structure.

25.1 THE CELL: UNIT OF BIOLOGICAL STRUCTURE

The cell is the smallest structure that shows all the attributes of life—ability to grow, to transform matter and energy, to respond to external stimuli, and to reproduce.● Most of the plants and animals large enough for us to see with the naked eye are multicellular organisms. In multicellular organisms, most cells possess specialized functions. There are muscle cells, blood cells, nerve cells, and so forth. The red blood cell, for example, is little more than a bag of hemoglobin. Hemoglobin is a protein that captures O_2 molecules when O_2 concentrations (partial pressures) are high (in the lungs) and releases them when O_2 concentrations are low (in other tissues).

● Most cells range in radius between 0.5 and 20 μm. Most covalently bonded atoms are between 0.05 and 0.15 nm in radius. The lower limit on cell size is imposed because any living cell must have a certain minimum molecular content.

Cells are constructed from organic molecules, such as proteins and carbohydrates. The organic molecules and complex aggregates of molecules that make up cells are manufactured by the cells themselves. Cells usually employ as raw materials organic molecules taken in as foods, although green plants and the photosynthetic bacteria and algae are able to construct complex organic molecules from simple inorganic species (CO_2, NH_3 or NO_3^-, SO_4^{2-}, PO_4^{3-}) and water.

Metabolism is *the process of building up and breaking down organic molecules in cells*. It involves a great variety of organic reactions catalyzed by many different enzymes. An **enzyme** is *a protein that catalyzes a biochemical reaction*. Much of the protein content of a cell is in its many enzymes.● A sequence of interconnected enzyme-catalyzed reactions is called a *metabolic pathway*. The reactions in this sequence are subject to *regulation,* that is, to being slowed down or speeded up by controlling the enzyme activity. In this way the cell can exert control over the rates of synthesis and breakdown of biological molecules. We will further consider enzymes in Section 25.3.

● Enzymes often increase reaction rates a billionfold or more.

25.2 ENERGY AND THE BIOLOGICAL SYSTEM

A continuing input of energy is necessary to maintain life. The living system represents a highly organized state, thus a very low-entropy state. Recall from the discussion of thermodynamics in Chapter 18 that for any reaction or process to proceed spontaneously, it must have a negative free-energy change associated with it. The free-energy change, ΔG, is related to the changes in enthalpy and entropy that occur in the process:

$$\Delta G = \Delta H - T\Delta S$$

If a process yields a more highly organized state, as do most processes in the formation and maintenance of biological systems, the entropy change will be negative and ($-T\Delta S$) will be a positive number. Unless the ΔH term is sufficiently large and negative to override the entropy term (which is not usually the case), the ΔG will be positive, and the reaction or process will not occur spontaneously unless energy is made available to it.

Say, for example, a cell needs to synthesize a quantity of protein, making low-entropy complex molecules out of high-entropy simpler molecules. This process requires not only raw materials (the simpler molecules) but also a source of free energy. This free energy can be supplied through the *coupling* of chemical reactions. If a reaction requiring free energy, such as protein synthesis, and a reaction releasing free energy are coupled, that is, caused to proceed together, the overall free-energy change for the two coupled reactions is the sum of their ΔG's, and the unfavorable reaction can occur.●

● Synthesis of one gram of protein would require ΔG equal to about 17 kJ.

What is the source of energy that powers cell metabolism, the mechanical work of muscle contraction, and the electrical work of nerve-impulse transmission? The original source of energy for all purposes is the sun. However, most cells are unable to use solar energy directly. Only photosynthetic cells can absorb radiant energy and transform it into the chemical energy of biological molecules.● Through photosynthesis, green plants use solar energy to convert simple inorganic substances into the complex organic chemical compounds that all other organisms oxidize for energy. Thus, all animals, most bacteria, and nonphotosynthetic plants such as fungi depend on the food-making abilities of photosynthetic organisms.

● Plant cells containing one of the green pigments called *chlorophylls* trap radiant energy for conversion to chemical energy.

Biological Molecules

We classify biological molecules, or biomolecules, into four groups: proteins, carbohydrates, nucleic acids, and lipids. Many of these biomolecules are macromolecules. **Macromolecules** are *very large molecules having molecular weights that may be several millions of atomic mass units*. There is an underlying simplicity in the structures of these biological macromolecules, however. They are usually linear chains of small, similar molecules covalently bonded together. Each is a type of biological polymer built up by *condensation reactions* between small units or building blocks.● In this section we will look at the structure of the four types of biomolecules and how they contribute to cell structure and function (metabolism).

● In a condensation reaction, water is lost when a new covalent bond is formed (see Section 24.8 for discussion of condensation polymers).

25.3 PROTEINS

Proteins are *biological polymers of small molecules called amino acids*. Their molecular weights range from 6000 to many millions, so they can be very large molecules. Many proteins also contain non-amino acid components such as metal ions (e.g., Fe^{2+}, Zn^{2+}, Cu^{+}, Mg^{2+}) or certain complex organic molecules that are usually derived from vitamins. (Vitamins are organic molecules necessary in small quantities for normal cell structure and metabolism.)

Proteins are very important molecules in cells and organisms, playing both structural and functional roles. The proteins of bone and connective tissue, for example, are of major structural importance. Some proteins are enzymes, catalyzing specific metabolic reactions; some transport materials in the bloodstream or across biological membranes; and some (hormones) carry chemical messages to coordinate the body's activities. *Insulin* and *glucagon,* for example, are protein hormones made in the pancreas and secreted to regulate the body's blood-sugar level.

Amino Acids

Amino acids are *molecules containing a protonated amino group (NH_3^+) and an ionized carboxyl group (COO^-).* The building blocks of protein are alpha-amino acids (α-amino acids). An α-amino acid has the general structure

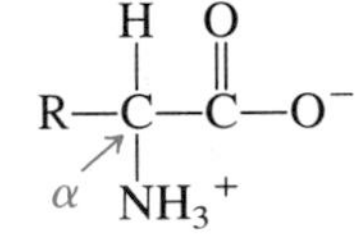

Here, the carbon atom next to the carboxyl carbon is labeled α and is the one bearing the amino group. Note that the amino acid is shown in doubly ionized form, called the *zwitterion,* since this is the form that predominates in the near-neutral pH of biological systems. The carboxyl group is a fairly strong acid, so its proton dissociates to a large extent. Conversely, the amino group is a fairly strong base, so it holds on to protons quite firmly at neutral pH. If the amino acid were in acidic solution, the $—COO^-$ group would tend to bind hydrogen ions,

$$—COO^- + H^+ \rightleftharpoons —COOH$$

to an extent dictated by its acid dissociation constant (K_a). Similarly, in basic solution the $—NH_3^+$ group would tend to lose protons,

$$—NH_3^+ \rightleftharpoons —NH_2 + H^+$$

depending on the base dissociation constant (K_b). Note that the total charge (net charge, + or −) of the amino-acid molecule will therefore depend on the pH of its solution. ●

● Various laboratory methods for separating mixtures of compounds are based on differences in net charges. Ion exchange chromatography is an example.

Example 25.1 Drawing the Zwitterion of an Amino Acid

The structure of the amino acid azaserine (an antibiotic and anticancer agent produced by some microorganisms but not found in proteins) is

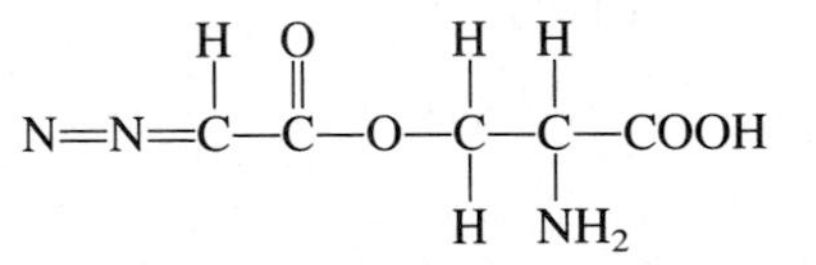

At neutral pH, is the carboxyl group negatively or positively charged? Draw the zwitterion of azaserine.

Solution

Since the carboxyl group is a fairly strong acid, at neutral pH most of these carboxyl groups will be ionized, and the resulting groups are **negatively charged.**

The zwitterion has both carboxyl and amino groups in ionized form:

$$N{=}N{=}\overset{H}{\overset{|}{C}}—\overset{O}{\overset{\|}{C}}—O—CH_2—\overset{NH_3^+}{\overset{|}{C}H}—COO^-$$

Exercise 25.1

A common amino acid in the body is ornithine. It is involved in the excretion of excess nitrogen into the urine. The structural formula of ornithine is

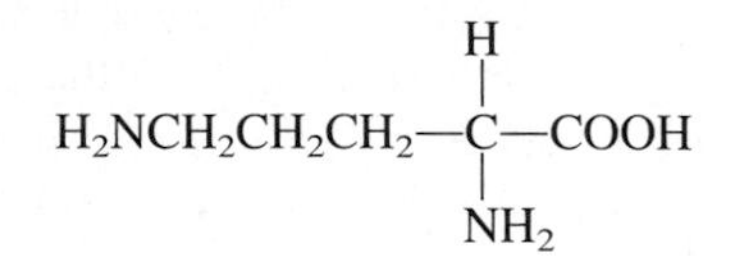

Write the fully ionized form of the molecule. (Note that there are two ionizable amino groups.) (See Problems 25.21 and 25.22.)

An organic compound with four different substituents on any one carbon is *chiral* (Section 23.5), capable of existing as isomers that are nonsuperimposable mirror images of one another. These isomers are called *enantiomers,* or D- and *L-isomers*. Amino acids are chiral (except for glycine, CH_2NH_2COOH), and thus each can exist as the D-isomer or as the L-isomer. All the amino acids of known naturally occurring proteins are L-amino acids, however.

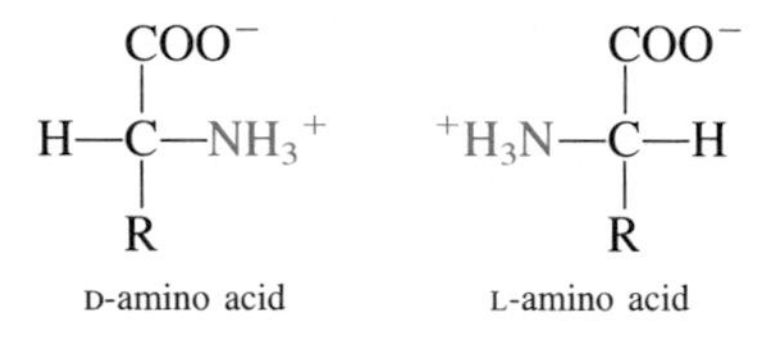

D-amino acid L-amino acid

To get the three-dimensional, tetrahedral arrangement about a carbon atom from these flat formulas, proceed as follows. The groups that are attached along the horizontal bonds bend toward you. Those that are attached along the vertical bonds bend away from you. Thus, in the D-amino acid the H atom and NH_3^+ group are toward you, and the COO^- and R groups are away from you. The L-amino acid is similar, except that it is the mirror image of the D-amino acid.

In a protein, amino acids are linked together by **peptide** (or **amide**) **bonds,** which are *the C—N bonds resulting from a condensation reaction between the carboxyl group of one amino acid and the amino group of a second amino acid:*

$$^{+}H_3N-\underset{H}{\overset{R_1}{C}}-\overset{O}{\overset{\|}{C}}-O^- + {}^{+}H_3N-\underset{H}{\overset{R_2}{C}}-\overset{O}{\overset{\|}{C}}-O^- \longrightarrow {}^{+}H_3N-\underset{H}{\overset{R_1}{C}}-\overset{O}{\overset{\|}{C}}-\underset{H}{N}-\underset{H}{\overset{R_2}{C}}-\overset{O}{\overset{\|}{C}}-O^- + H_2O$$

Peptide bond

This is an example of a *dipeptide,* a molecule formed by linking together two amino acids. Similarly, a *tripeptide* is formed from three amino acids. A **polypeptide** is *a polymer formed by the linking of many amino acids by peptide bonds.*●

● A polypeptide is *any* amino acid polymer, which may or may not have a biological function. The term *protein* describes a polypeptide that has a biological function.

Example 25.2 Writing the Structural Formula of a Dipeptide

Two common amino acids are

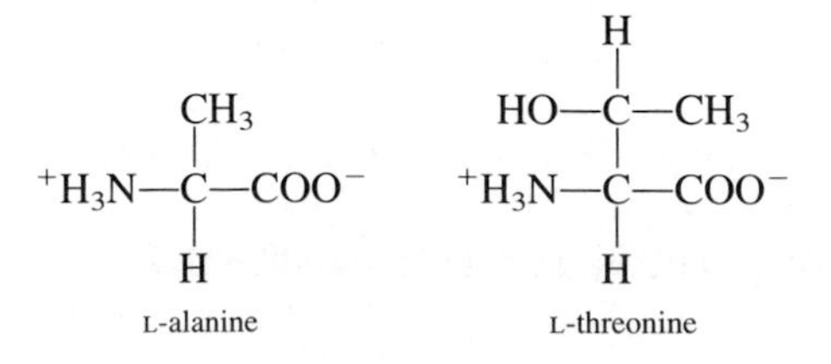

L-alanine L-threonine

Write the structural formulas of the two dipeptides that they could form.

Solution

The carboxyl group of either of the amino acids could be peptide bonded to the amino group of the other. The structural formulas of these dipeptides are:

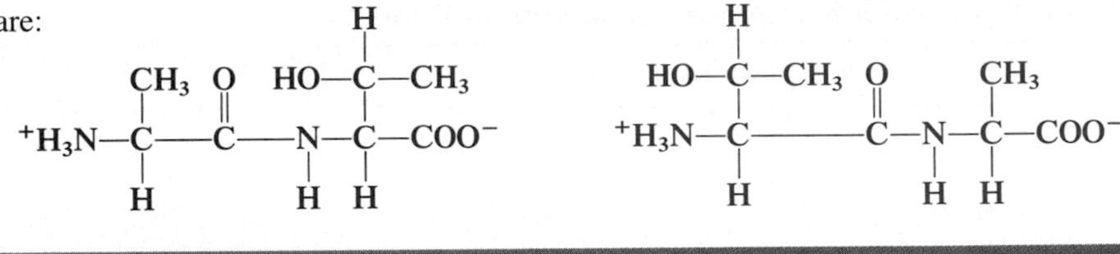

Exercise 25.2

Write the structural formulas of all possible tripeptides with the composition of two glycines and one serine. (See the structural formulas in Table 25.1.)

(See Problems 25.23 and 25.24.)

Of the many known amino acids, 20 different kinds are found in most proteins. Each has a different R group, or *side chain*. The side chain determines the properties of a protein. Nine of these amino acids have nonpolar, or hydrocarbon, side chains. The others have polar side chains, capable of ionizing or forming hydrogen bonds with other amino acids or with water. Table 25.1 lists these amino acids. The side chains are shown in color. The commonly used three-letter abbreviation of each amino acid is given below the full name.

Example 25.3 Forming a Hydrogen Bond Between Two Amino Acids

Show by sketching structural formulas of two amino acid molecules how their side chains might form a hydrogen bond between them.

Solution

Because a hydrogen bond only requires bringing together an oxygen or nitrogen with an attached hydrogen atom and another group containing an oxygen or nitrogen, there are several pairs of polar amino acids we could choose. One such pair is L-serine and L-aspartate (shown below):

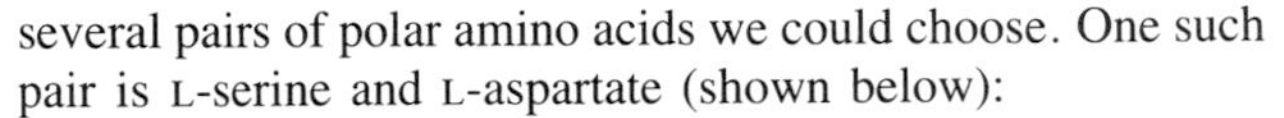

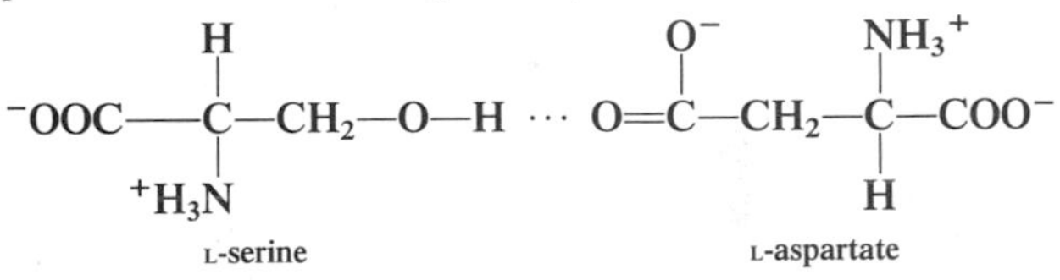

Exercise 25.3

Show how the side chains of L-threonine and L-glutamate could form a hydrogen bond between them.

(See Problems 25.25 and 25.26.)

Protein Primary Structure

The **primary structure** of a protein refers to *the order or sequence of the amino-acid units in the protein*. This order of amino acids is conveniently shown by denoting the amino acids using their three-letter codes, each amino-acid code in the sequence being separated by a dash. In this notation, it is understood that the amino group is on the left and the carboxyl group is on the right. The dipeptides in Example 25.2 would be written ala–thr (the first dipeptide given in the solution) and thr–ala. Consider a polypeptide containing the following five amino-acid units: glycine, alanine, valine, histidine, and serine. These five units can be

Table 25.1 The Amino Acids Found in Most Proteins

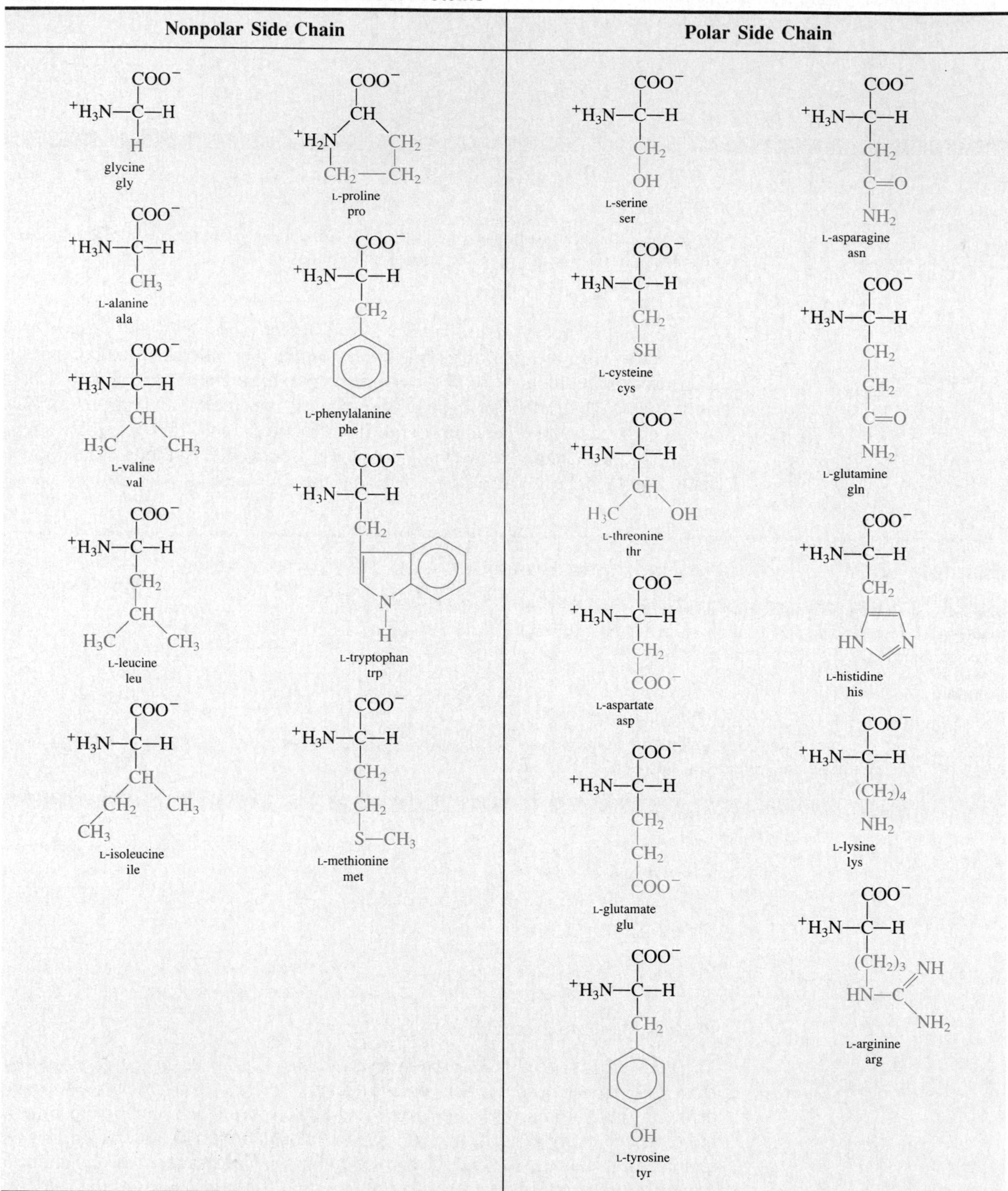

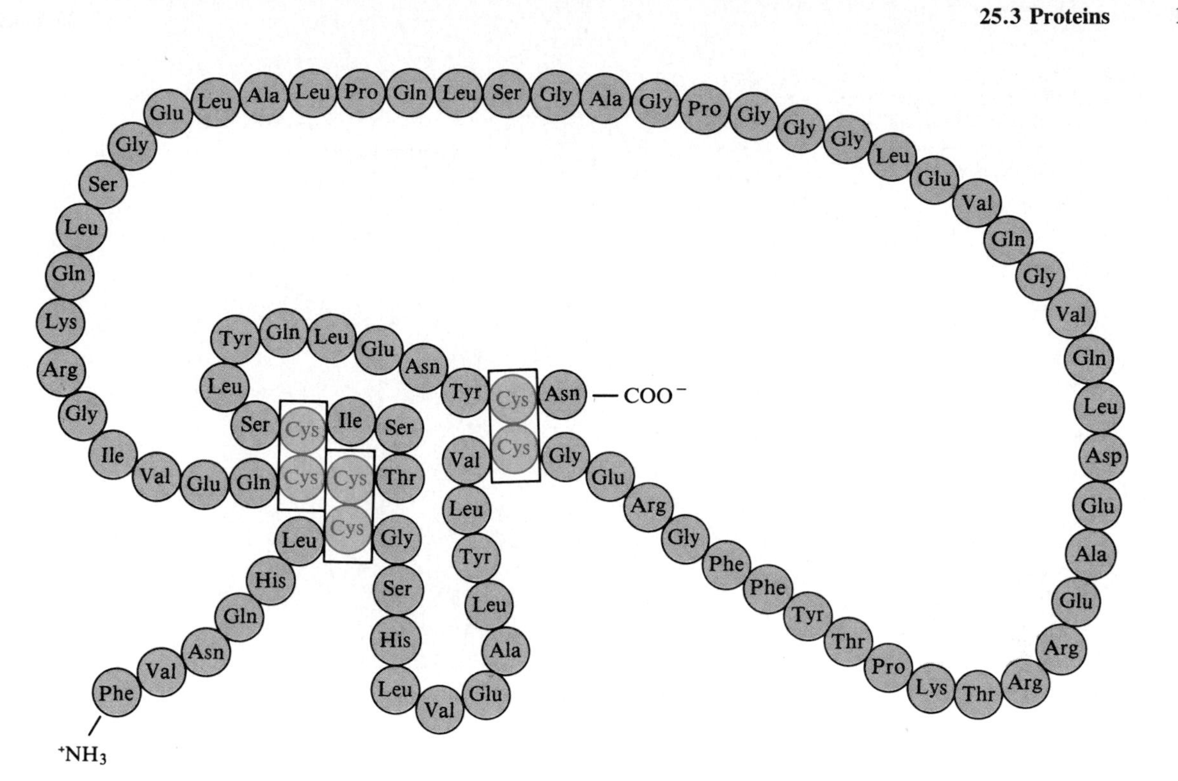

Figure 25.1
Primary structure of human proinsulin. The formula is arranged to show the position of the disulfide linkages (in Cys–Cys) and is not an indication of three-dimensional shape.

ordered in many different ways to make a polypeptide. Three possible primary structures of the polypeptide are

gly–ser–ala–val–his

his–ala–ser–val–gly

ser–ala–gly–his–val

Each of the possible sequences would produce a different peptide with different properties.

In any protein, which may contain as few as 50 or more than 1000 amino-acid units, the amino acids are arranged in one unique sequence, the primary structure of that protein molecule. Figure 25.1 diagrams the primary structure of one form of the hormone insulin.

Shapes of Proteins

A protein molecule of a unique amino-acid sequence spontaneously folds and coils into a characteristic three-dimensional conformation in aqueous solution (Figure 25.2). The side chains of the amino-acid units, with their different chemical properties—nonpolar or polar—determine the characteristic shape of the protein. For an energetically stable shape, those parts of the chain with nonpolar amino-acid side chains are buried within the structure away from water, because nonpolar groups are hydrophobic (not attracted to water). Conversely, most polar groups are stablest on the surface, where they can hydrogen bond with water or with other polar side chains. Occasionally, side chains form ionic bonds.

One type of covalent linkage is important to protein shape. The amino acid cysteine, with its thiol (—SH) side chain, is able to react with a second cysteine in

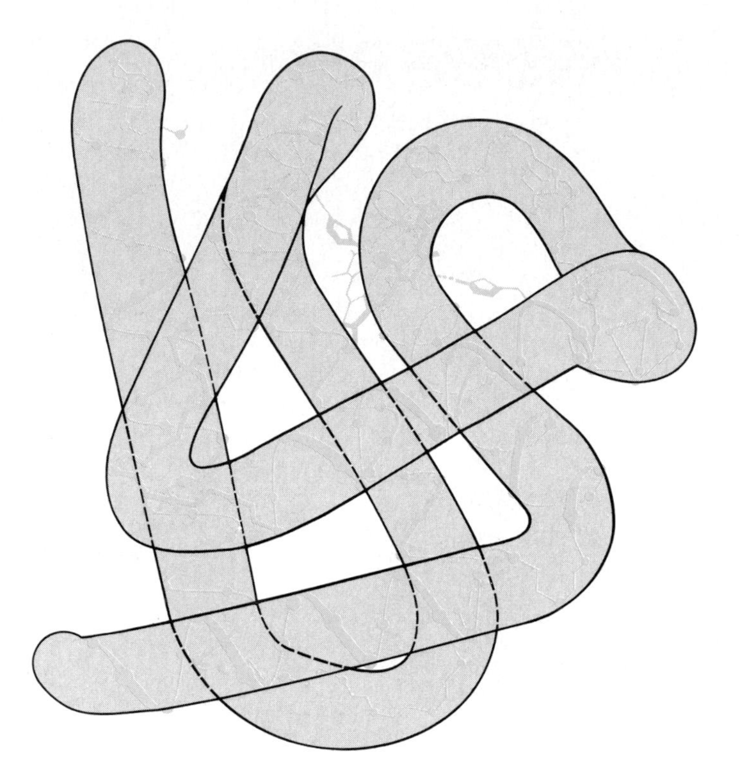

Figure 25.2
Three-dimensional structure of the protein myoglobin. Note the regions of helical structure. This structure was deduced from x-ray diffraction data.

the presence of an oxidizing agent, as follows:

$$\text{cys—SH} + \text{HS—cys} + (\text{O}) \longrightarrow \text{cys—S—S—cys} + H_2O$$

The resulting group (—S—S—) is a *disulfide*. When two cysteine side chains in a polypeptide are brought close together by folding of the molecule, then oxidized, they form a *disulfide cross link,* which helps anchor the folded chain into position (Figure 25.1).●

● Hair can be permanent waved by the use of disulfide cross links. Hair proteins have many disulfides that can be chemically reduced and re-formed again while the hair is rolled on curlers. The new disulfides hold the hair shafts in their new wavy conformation.

Proteins may be classed as fibrous proteins or globular proteins on the basis of their arrangements in space. **Fibrous proteins** are *proteins that form long coils or align themselves in parallel to form long water-insoluble fibers. The relatively simple coiled or parallel arrangement of a protein molecule* is called its **secondary structure. Globular proteins** are *proteins in which long coils fold into compact, roughly spherical shapes*. Thus, in addition to their secondary structure (long coil), they have a **tertiary structure,** or *structure associated with the way the protein coil is folded*. Most globular proteins are water soluble because they are relatively small in dimension and have hydrophilic (''water-loving'') surfaces that bind water molecules.

The characteristic coiling or folding pattern of protein molecules produces unique surface features, such as grooves or indentations, and hydrophobic or charged areas (Figure 25.3). These surface features are fundamental to protein function. They allow protein molecules to interact with other molecules in specific ways. For example, certain protein molecules characteristically aggregate because of complementary surface features, forming tubules or filaments or other organized structures. Particular surface features are also responsible for specific binding and catalysis by enzyme proteins.

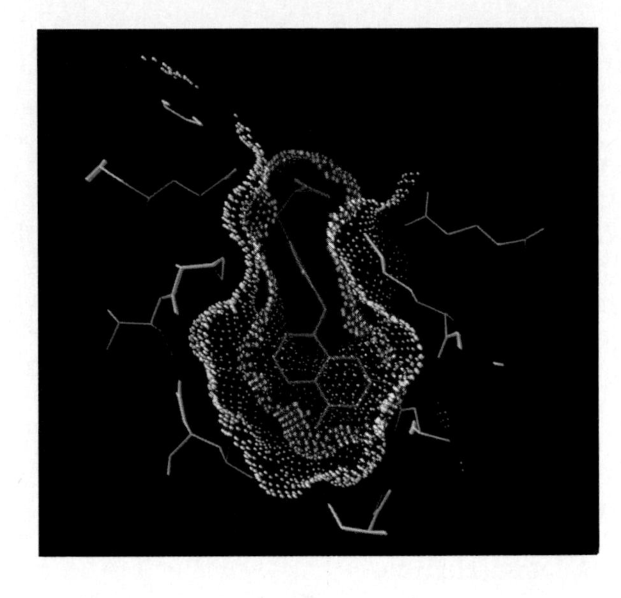

Figure 25.3
Computer-generated image of a derivative of thyroxine (thyroid hormone) bound to a site on the protein prealbumin. Thyroxine is carried in the blood by prealbumin. The surface of prealbumin to which the hormone binds is shown by blue dots; the surface of thyroxine is shown by orange dots. Note how thyroxine "fits" the binding site on albumin.

Enzymes

Enzymes are marvelously efficient and specific catalysts of biochemical reactions. They are usually globular proteins.● An enzyme possesses a characteristic *active site* at which a reactant molecule binds and catalysis takes place. *The reactant substance whose reaction the enzyme catalyzes* is called the **substrate.** Figure 25.4 shows the substrate molecule fitting into the active site as a key fits in a lock (similar to the bonding of thyroxine to prealbumin shown in Figure 25.3). Some flexibility is probably possible, however, in the shapes of both the active site and the substrate.

● Until recently, all enzymes were thought to be proteins. Now Thomas R. Cech at the University of Colorado and Sidney Altman at Yale University have discovered that some ribonucleic acids (RNA) behave as enzymes. They shared the 1989 Nobel Prize in Chemistry for their discovery. (RNA is discussed in Section 25.5.)

The rate of an enzyme-catalyzed reaction does not show a linear dependence on the substrate concentration, because the product forms only after the substrate has been bound by the enzyme to form an *enzyme–substrate complex*. It is the concentration of this enzyme–substrate complex on which the rate depends.

Many enzymes are capable of increasing reaction rates by a factor of 10^9 or more. Reactions that would be so slow as to be insignificant may proceed extremely rapidly in the presence of the appropriate enzyme. Living organisms could not exist if they had to depend on uncatalyzed reaction rates for oxidation of foods or synthesis of cellular structures.

In addition to speed, a second feature of enzyme function is *specificity*. Because of the special properties of its active site, an enzyme is said to be specific for only certain substrates or certain types of reaction.

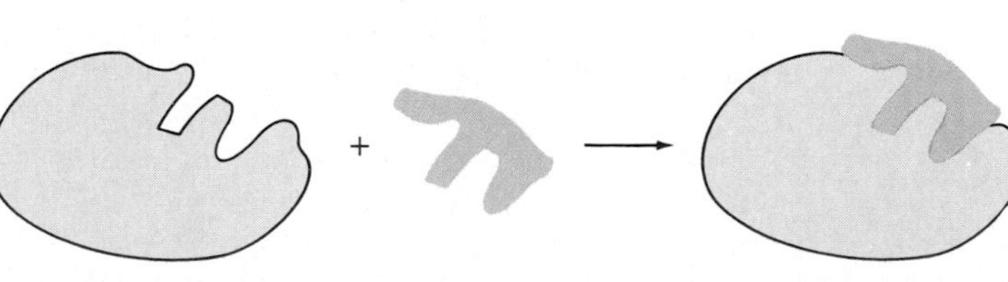

Figure 25.4
"Lock-and-key" model of the interaction of enzyme and substrate. Only specific substrates will fit into the active sites on the enzyme.

25.4 CARBOHYDRATES

Carbohydrates are either *polyhydroxy aldehydes or ketones or they are substances that yield such compounds when they react with water.*● Some carbohydrates, such as cellulose and chitin, are structural elements in plants and animals; others, such as glucose, provide energy and raw materials for cell activities. Beet sugar, grape sugar, and milk sugar, introduced in the chapter opening as important compounds in Pasteur's work, are carbohydrates.

● Many carbohydrates have empirical formulas $C_m(H_2O)_n$; hence the name carbohydrate.

Carbohydrates may be divided into three categories: monosaccharides, oligosaccharides, and polysaccharides. **Monosaccharides** are *simple sugars, each containing three to nine carbon atoms, all but one of which bear a hydroxyl group. The remaining carbon atom is part of a carbonyl group.* **Oligosaccharides** are *short polymers of two to ten simple sugar units,* and **polysaccharides** are *polymers consisting of more than ten simple sugar units.*

Monosaccharides

Monosaccharides exist as D- and L-isomers, because there is at least one carbon atom in each molecule with four different groups attached. Because of their polar structure, monosaccharides are highly soluble in water.

Only a few of the many monosaccharides are of major biological importance. All are D-sugars, and most are 5- or 6-carbon sugars (pentoses or hexoses).● D-glucose, D-fructose, D-ribose, and 2-deoxy-D-ribose are by far the most important sugars. D-glucose is common blood sugar, an important energy source for cell function. D-fructose is the common sugar in fruits (such as grapes) and a food source. D-ribose and 2-deoxy-D-ribose are parts of nucleic acids, which we will study later.

● NAME OF SUGAR	NUMBER OF CARBON ATOMS
Triose	3
Tetrose	4
Pentose	5
Hexose	6
Heptose	7
Octose	8
Nonose	9

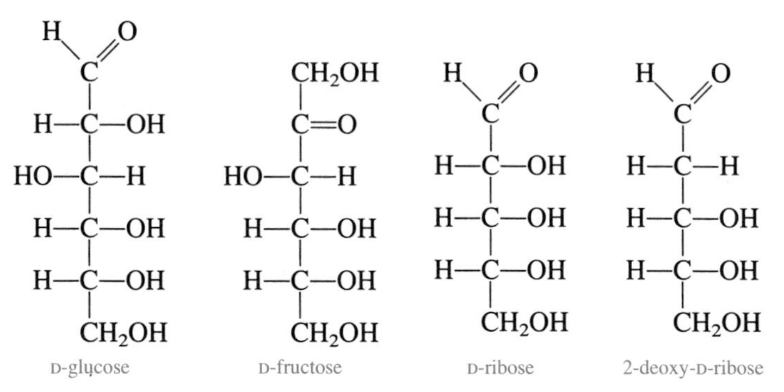

Although we often draw the simple sugars as straight-chain molecules, they do not exist predominantly in that form. The chain curls up because carbonyl groups react readily with alcohol groups. With aldehydes, alcohols give *hemiacetals,* or compounds in which an —OH group, an —OR group, and an H atom are attached to the same carbon atom.

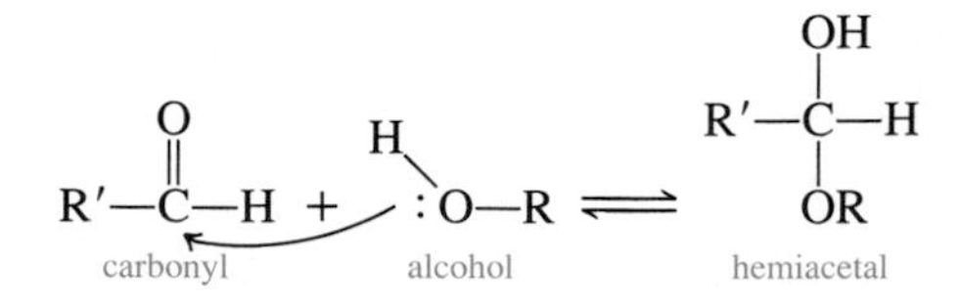

A ketone and an alcohol give a *hemiketal,* which is similar to a hemiacetal, except the H atom is replaced by an R″ group.

The straight-chain form of D-glucose, whose formula we gave earlier, is a flexible molecule that can bend around to give a cyclic hemiacetal.

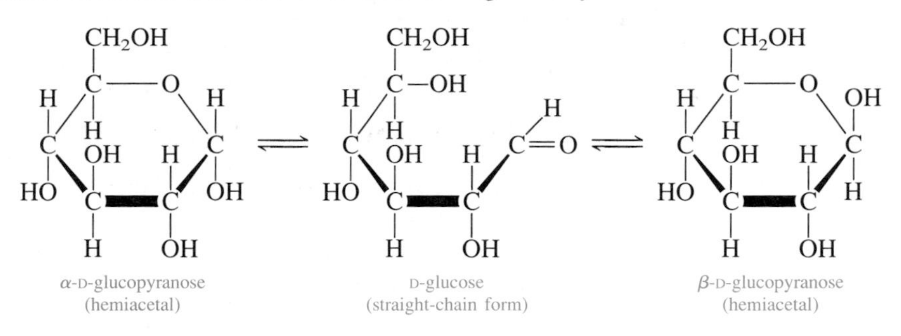

The straight-chain form, in the center, is shown bent around so that an —OH group and the —CHO group can react to give a cyclic hemiacetal. A new —OH group is obtained (on the right in each hemiacetal) and can point either down or up. Thus, there are two isomers of the hemiacetal—one labeled α (—OH down) and the other labeled β (—OH up). Note that these hemiacetals have six-membered rings. Monosaccharides with six-membered rings are called *pyranoses;* those with five-membered rings are called *furanoses*. In the case of glucose, the hemiacetals are called α-D-gluco*pyranose* and β-D-gluco*pyranose* to indicate that they are six-membered rings.

Example 25.4 Predicting the Cyclic Forms of a Monosaccharide

Write the reactions showing the equilibria between the straight-chain form of D-ribose and its furanose, hemiacetal forms. The formula of D-ribose was given earlier.

Solution

We write the straight-chain form of D-ribose bent so that the —CHO group can react with that —OH group that gives a five-membered ring.

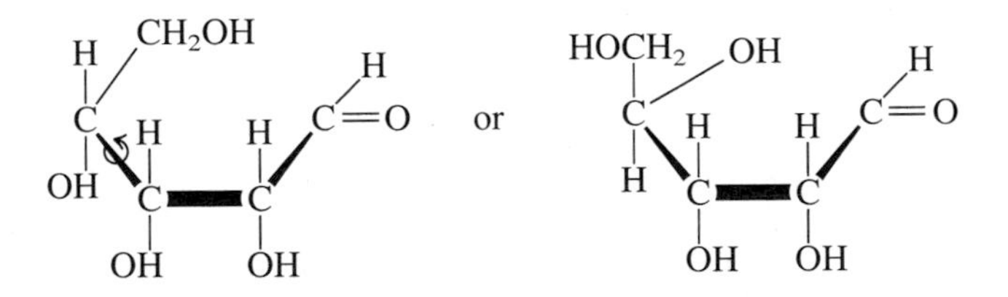

Note that the formula on the left is simply the formula given earlier but laid on its side and bent around. Thus, all —OH groups that were on the right are now pointing down; those on the left are now pointing up. The —CH_2OH, the —H, and —OH groups can be rotated to get the formula on the right. (Remember, we can twist a portion of a molecule about a C—C bond.) In that position, the —OH and —CHO groups can react to give a five-membered ring. The equilibria are as follows:

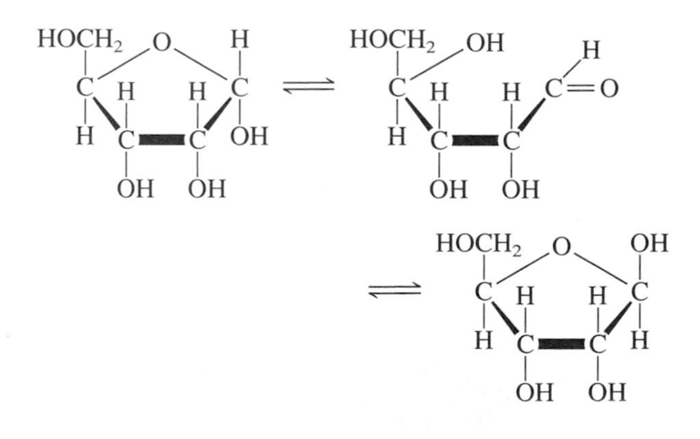

Exercise 25.4

Write the equilibria for the reaction between the straight-chain form of D-fructose and each of its furanose, hemiketal forms. The formula of D-fructose was given earlier.

(See Problems 25.27 and 25.28.)

Oligosaccharides and Polysaccharides

Oligosaccharides and polysaccharides are formed from monosaccharide building blocks. The hemiacetal carbon atom of one monosaccharide is attached to the alcohol oxygen atom of another monosaccharide by a condensation reaction. The most common oligosaccharide is sucrose (a disaccharide), found in sugar beets and sugar cane. In sucrose, an O atom from carbon 1 of glucose is joined to carbon 2 of fructose. (The carbon atoms of each monosaccharide are numbered from the end closest to the carbonyl group, or what was the carbonyl group.)

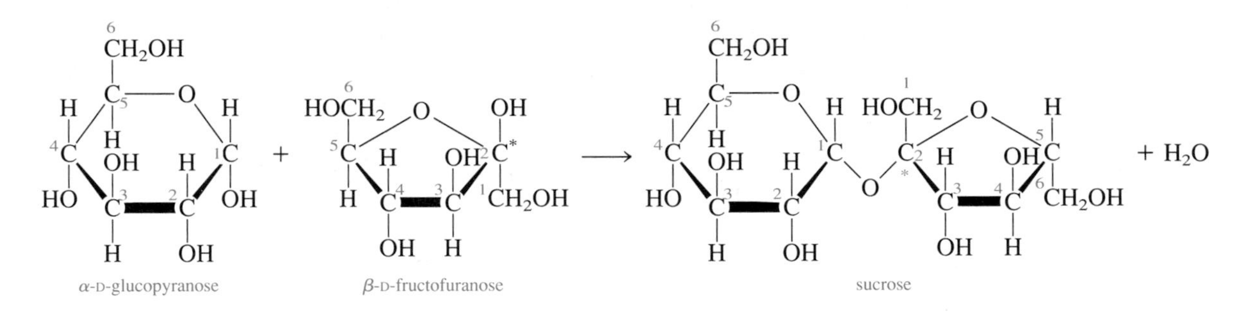

α-D-glucopyranose β-D-fructofuranose sucrose

Note that the fructose is flipped upside down and rotated 180° to form the bond (compare the positions of the starred carbon atom in β-D-fructofuranose and sucrose). *Maltose* and *lactose* are also common disaccharides with properties similar to sucrose. Maltose is composed of two α-D-glucose units.

Example 25.5 Predicting the Disaccharide Given the Monosaccharides

Lactose (milk sugar), introduced in the chapter opening, is composed of β-D-galactopyranose and α-D-glucopyranose, joined by an O atom from carbon 1 of galactose to carbon 4 of glucose. Diagram the reaction of its formation from the monosaccharides. D-galactose has the straight-chain form

H O
C
H—C—OH
HO—C—H
HO—C—H
H—C—OH
CH_2OH

Solution

We first write the hemiacetal structures, then link them by a condensation reaction. This reaction is shown at the top of the following page.

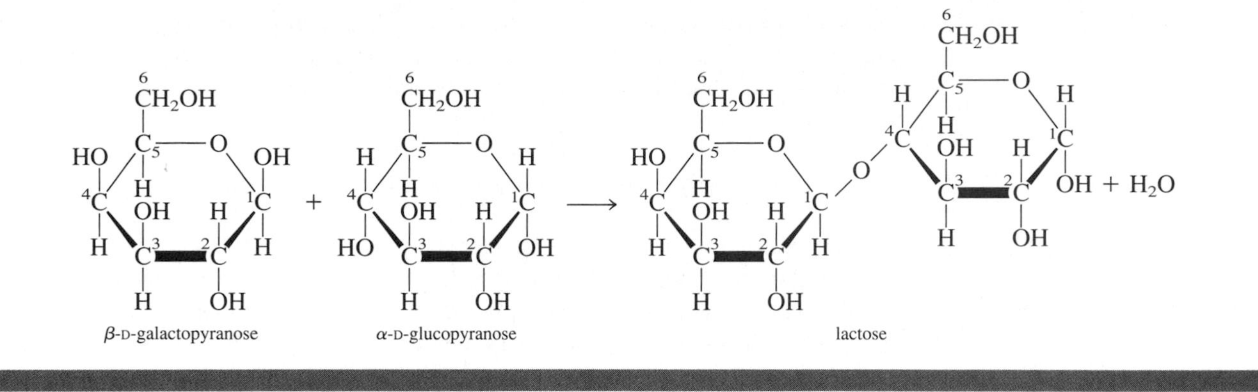

β-D-galactopyranose α-D-glucopyranose lactose

Exercise 25.5

Write the structural formula for maltose, which consists of two α-D-glucopyranose units linked by an O atom from carbon 1 of one unit to carbon 4 of the other.

(See Problems 25.29 and 25.30.)

Polysaccharides usually contain only one type of building block, or sometimes two alternating building-block units. Most polysaccharides are linear polymers—one main chain with no branching units.

Polysaccharides are structural molecules and energy-storage polymers. *Cellulose,* a polysaccharide, is a structural carbohydrate abundant among plant species. Plant cell walls contain fibers of cellulose that preserve the structural integrity of the cells they surround. Cellulose makes possible the *turgor* (water pressure in cells) that gives nonwoody plants their shapes.

Cellulose is a linear polymer of β-D-glucopyranose units:

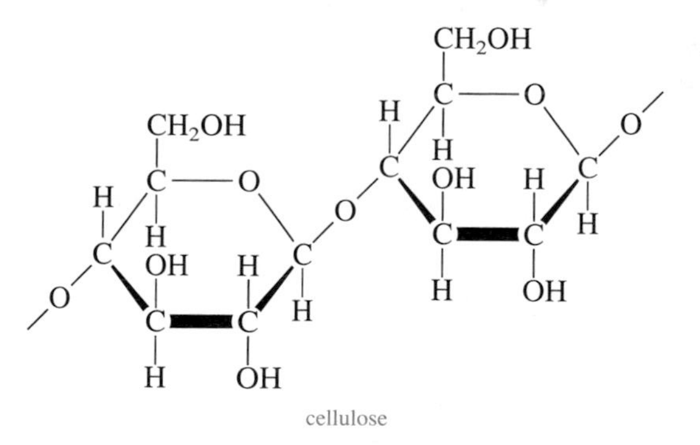

cellulose

Molecular weights of cellulose molecules are in the millions.

The major energy-storage polysaccharides in plants are starches. One kind of starch, *amylose,* contains long, unbranched chains of α-D-glucopyranose units:

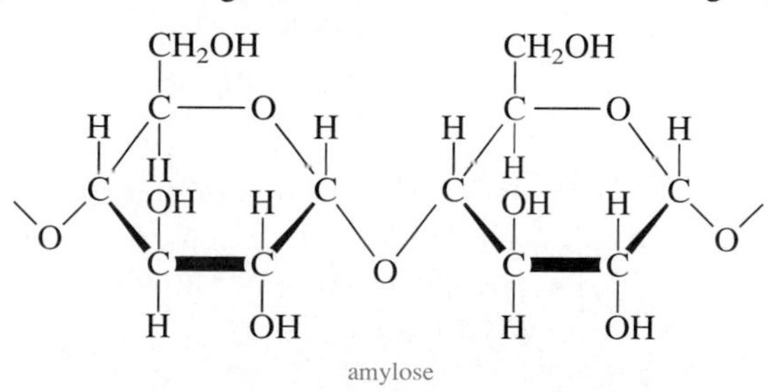

amylose

A second kind of starch, *amylopectin,* is a branched polysaccharide. The structure is like that of amylose, except that a branch occurs about every 18 to 20 units along the main chain. Starch molecules vary considerably in size, but molecular weights of several million are probably common.

Human beings cannot use cellulose for food because their starch-digesting enzyme attacks only α-glucose polymers, not β-glucose polymers such as cellulose.

Animals store sugars in the form of the polysaccharide *glycogen*. Glycogen is very similar in structure to amylopectin except that branches occur more frequently. The molecular weight of glycogen obtained from animal tissues is similarly very high. Plant and animal polysaccharides represent a rich source of chemical energy, and they are stored along with fats (discussed in Section 25.6) within cells to provide an energy reservoir to be drawn from as needed.● Branched polysaccharides are thought to be more quickly added to (or broken down) when the cell's needs dictate, because there are many more "free ends" in such a polymer than in a single linear chain.

● Runners who do *carbohydrate loading* are trying to increase the amount of glycogen stored in their muscle cells so they will have a larger energy reserve for a race. They deplete their muscle glycogen by a long run, minimize the carbohydrate in their diets for a few days, then load with carbohydrate by eating a starchy food, such as spaghetti, for a day or so before the race.

25.5 NUCLEIC ACIDS

Nucleic acids are vital to the life cycles of cells because they are the carriers of species inheritance.● There are two types of nucleic acids: *deoxyribonucleic acid (DNA)* and *ribonucleic acid (RNA)*. Both are polymers of *nucleotide* building blocks. Their structures are examined in detail in this section.

● *Genetic engineering* refers to the manipulation of nucleic acids to change the characteristics of organisms (for example, to correct genetic flaws).

Nucleotides

Nucleotides are *the building blocks of nucleic acids*. They are of two types: *ribonucleotides* and *deoxyribonucleotides*. The general structure of nucleotides consists of an organic base (described below) linked to carbon 1 of a pentose (5-carbon sugar), which is linked in turn at carbon 5 to a phosphate group. The base plus the sugar group is called a *nucleoside*. The only difference between the two types of nucleotides is in the sugar. Ribonucleotides contain β-D-ribose; deoxyribonucleotides contain 2-deoxy-β-D-ribose. Both sugar groups have furanose rings.

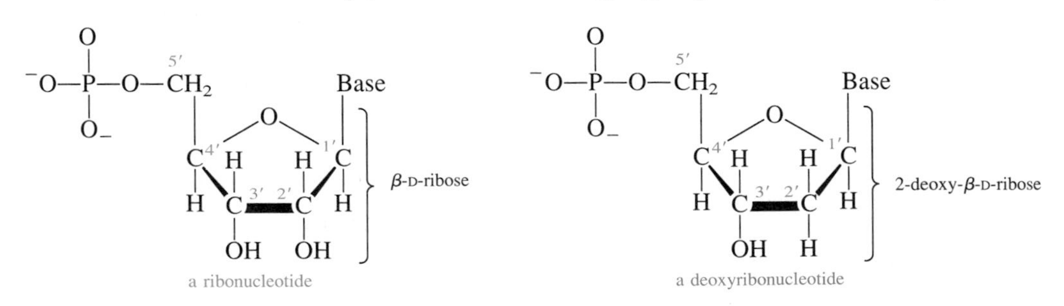

a ribonucleotide a deoxyribonucleotide

The carbon atoms of ribose are labeled with numbers 1′, 2′, . . . 5′. (Numbers without primes are used for carbon atoms in the organic base.) Note that the phosphate group is shown in ionized form, as this form predominates near neutral pH in cells.

Five organic bases (amines) are most often found in nucleotides. They are *adenine, guanine, cytosine, uracil,* and *thymine*. They link to the sugars through the indicated nitrogen (in color) in each of the following structural formulas:

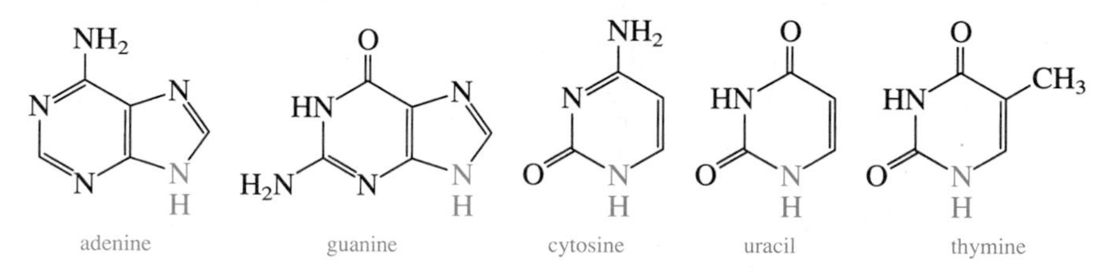

Nucleosides are named from the bases. For example, the nucleoside composed of adenine with β-D-ribose is called adenosine. The nucleoside composed of adenine with 2-deoxy-β-D-ribose is called deoxyadenosine. A nucleotide is named by adding monophosphate (or diphosphate or triphosphate) after the nucleoside name. A number with a prime indicates the position of the phosphate group on the ribose ring. Thus, adenosine-5′-monophosphate is a nucleotide composed of adenine, β-D-ribose, and a phosphate group at the 5′ position of β-D-ribose.

Example 25.6 Drawing the Structure of a Nucleotide

Draw the structural formula of the nucleotide containing guanine, 2-deoxy-β-D-ribose, and a phosphate group at the 5′ position. Name the compound.

Solution

The structural formula is shown below. The name of the compound is **deoxyguanosine monophosphate**.

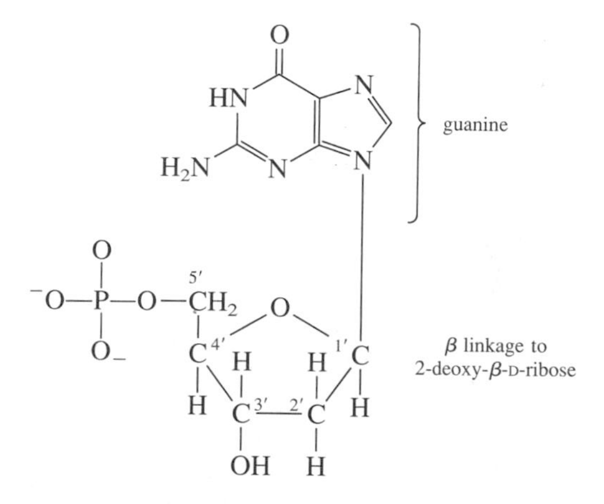

Exercise 25.6

Write the structural formula for cytidine-5′-monophosphate.

(See Problems 25.33 and 25.34.)

Polynucleotides and Their Conformations

A **polynucleotide** is *a linear polymer of nucleotide units* linked from the hydroxyl group at the 3′ carbon of the pentose of one nucleotide to the phosphate group of

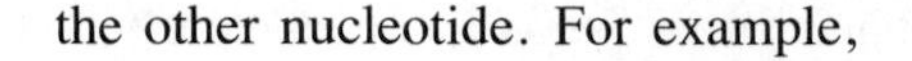
the other nucleotide. For example,

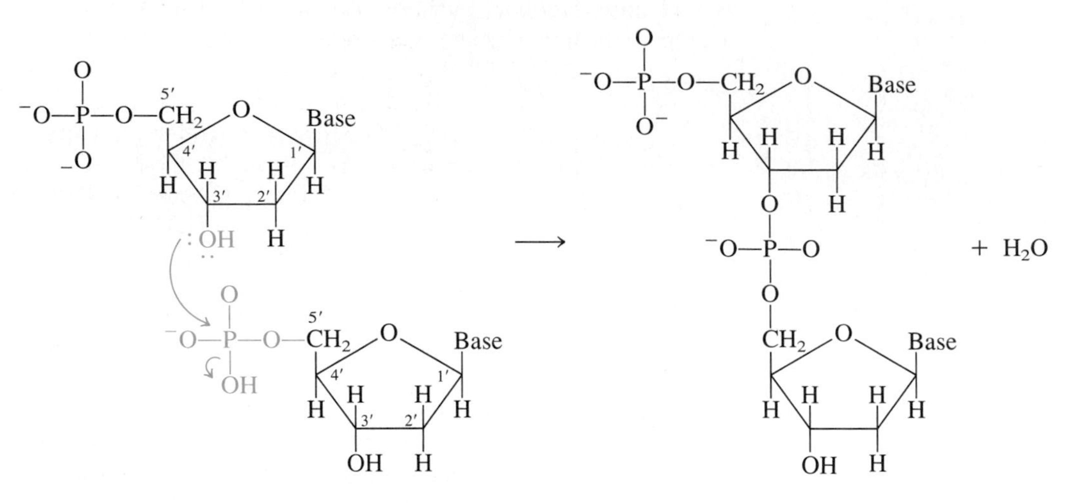

Just as the unique sequence of amino acids in a protein determines the protein's nature, so the sequence of nucleotides determines the particular properties and functions of a polynucleotide.

Nucleic acids are *polynucleotides folded or coiled into specific three-dimensional shapes.* **Complementary bases** are *nucleotide bases that form strong hydrogen bonds with one another.* Adenine and thymine are complementary bases, as are adenine and uracil, and guanine and cytosine. Hydrogen bonding of complementary bases, called *base pairing,* is the key to nucleic acid structure and function (see Figure 25.5).

Deoxyribonucleic acid (DNA) is *the hereditary constituent of cells and consists of two polymer strands of deoxyribonucleotide units.* The two strands coil

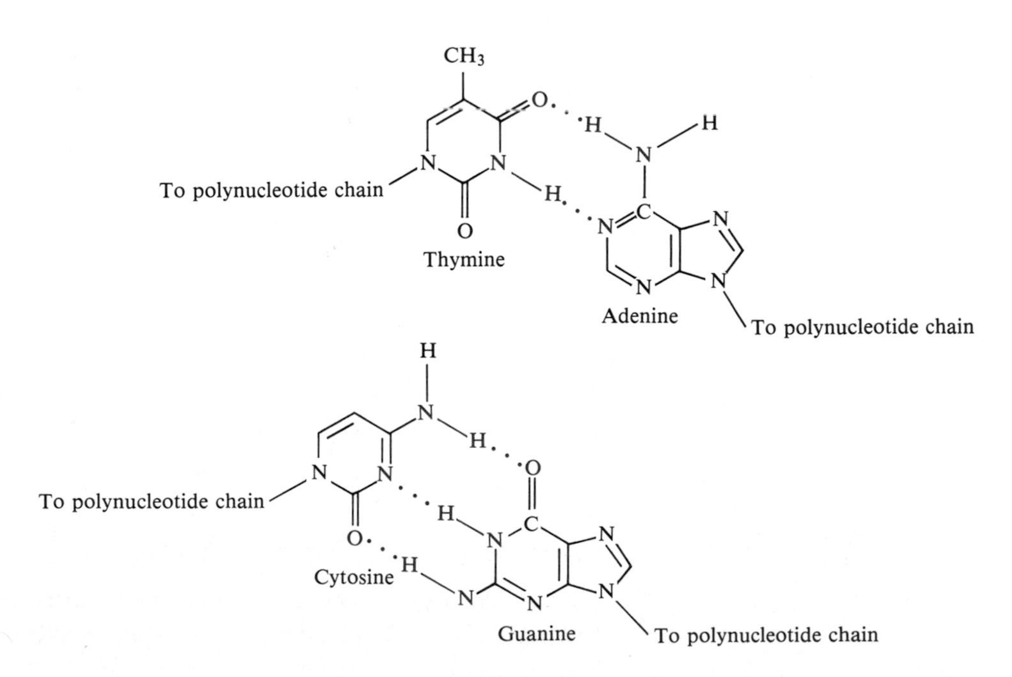

Figure 25.5
Hydrogen-bonded complementary bases. Each base hydrogen bonds strongly to only one other base, its complementary base.

about each other in a double helix, with base pairing along the entire lengths of the strands (Figure 25.6). **Ribonucleic acid (RNA)** is *a constituent of cells used to manufacture proteins from genetic information. It is a polymer of ribonucleotide units.* Both DNA and RNA have nucleotides with the bases adenine, guanine, and cytosine. However, DNA contains thymine but not uracil, and RNA contains uracil but not thymine. The following example discusses how to write complementary sequences for DNA. We will find it convenient in writing the bases to use the abbreviations A = adenine, C = cytosine, G = guanine, T = thymine, and U = uracil.

Example 25.7 Writing the Complementary Sequence for a DNA Sequence

Noting the three complementary base pairs and which bases are found in DNA or RNA, write the DNA sequence complementary to the following sequence:

ATGCTACGGATTCAA

Solution

Since DNA does not contain uracil, the complementary base for adenine is thymine. Thus, the proper sequence of base pairs is

ATGCTACGGATTCAA
TACGATGCCTAAGTT

Exercise 25.7

Write the RNA sequence complementary to the sequence given in Example 25.7 above. (See Problems 25.37 and 25.38.)

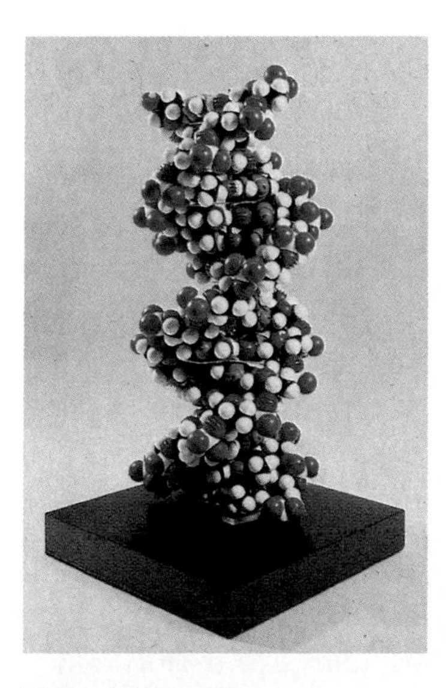

Figure 25.6
Double helix of the DNA molecule. The structure was first deduced by James Watson, an American scientist, and Francis Crick, a British scientist, in the early 1950s.

DNA and the Nature of the Genetic Code

Photomicrographs of dividing cells show structures, called chromosomes, as dense, thick rods (Figure 25.7). *Chromosomes* are cell structures that contain DNA and proteins; the DNA contains the genetic inheritance of the cell and organism. Before cell division, the cell synthesizes a new and identical set of chromosomes, or more particularly a new and identical set of DNA molecules—the genetic information—to be transmitted to the new cell. Thus, the new cell will have all the necessary instructions for normal structure and function.

What is the nature of this genetic information? The genetic information is coded into the linear sequence of nucleotides in the DNA molecules. Each DNA molecule is composed of hundreds of genes. A **gene** is *a sequence of nucleotides in a DNA molecule that codes for a given protein.* The nucleotides in a gene are grouped in sets of three, or triplets. Each triplet codes for one amino acid in a protein.

Occasionally, an error is made during the synthesis of new DNA. Such a change in the genetic information, or genetic error, is called a *mutation.* One possible consequence could be the synthesis of a faulty or inactive protein. We know of various genetic diseases that result from a single error leading to a change in amino-acid sequence that has profound effects on the protein's activity and thus the life of the organism.

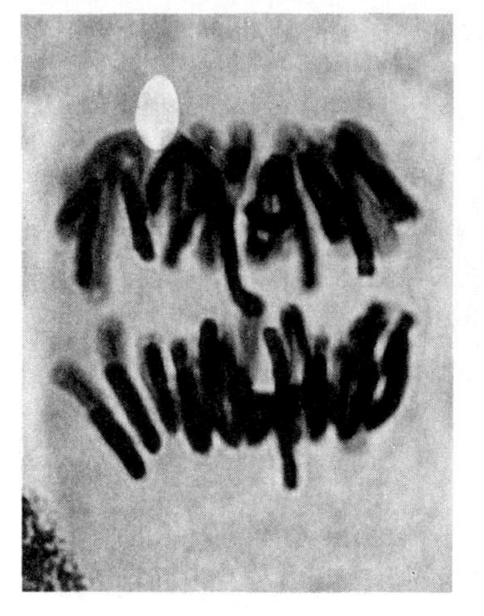

Figure 25.7
Chromosomes in a dividing cell. The chromosomes appear as dense, thick rods.

RNA and the Transmission of the Genetic Code

RNA is used to translate the genetic information stored in DNA into protein structure. There are three classes of RNA: ribosomal RNA, messenger RNA, and transfer RNA.

Ribosomes are *tiny cellular particles on which protein synthesis takes place.* They are constructed of numerous proteins plus three or four RNA molecules. *The RNA in a ribosome* is called **ribosomal RNA.** Ribosomes provide a surface on which to organize the process of protein synthesis, and they also contain enzymes that catalyze the process.

A **messenger RNA** molecule is *a relatively small RNA molecule that can diffuse about the cell and attach itself to a ribosome, where it serves as a pattern for protein synthesis.* The first step in protein synthesis, called *transcription,* is synthesis of a messenger RNA molecule that has a sequence of bases complementary to that of a gene. *A sequence of three bases in a messenger RNA molecule that serves as the code for a particular amino acid* is called a **codon.**

When the messenger RNA molecule attaches itself to a ribosome, its codons provide the sequence of amino acids in a protein that will be synthesized. *Translation* is the synthesis of protein using messenger RNA codons. Table 25.2, the complete messenger RNA code-word dictionary, shows the specific amino acids coded for by messenger RNA codons. There are 64 possible arrangements of the

Table 25.2 Genetic Code Dictionary*

	U	C	A	G
U	UUU Phe	UCU Ser	UAU Tyr	UGU Cys
	UUC Phe	UCC Ser	UAC Tyr	UGC Cys
	UUA Leu	UCA Ser	UAA End	UGA End
	UUG Leu	UCG Ser	UAG End	UGG Trp
C	CUU Leu	CCU Pro	CAU His	CGU Arg
	CUC Leu	CCC Pro	CAC His	CGC Arg
	CUA Leu	CCA Pro	CAA Gln	CGA Arg
	CUG Leu	CCG Pro	CAG Gln	CGG Arg
A	AUU Ile	ACU Thr	AAU Asn	AGU Ser
	AUC Ile	ACC Thr	AAC Asn	AGC Ser
	AUA Ile	ACA Thr	AAA Lys	AGA Arg
	AUG Met	ACG Thr	AAG Lys	AGG Arg
G	GUU Val	GCU Ala	GAU Asp	GGU Gly
	GUC Val	GCC Ala	GAC Asp	GGC Gly
	GUA Val	GCA Ala	GAA Glu	GGA Gly
	GUG Val	GCG Ala	GAG Glu	GGG Gly

*The four nucleotide bases of RNA—U, C, A, and G—are arranged along the left side of the table and the top. Combining these two, and then adding a third (again U, C, A, or G) gives the 64 three-nucleotide codons shown in capital letters in the table. Next to each codon appears the abbreviation for the amino acid it codes for during protein synthesis. In three cases, the word *End* appears because each of these codons serves as a signal to terminate protein synthesis. (From Robert D. Whitaker et al., *Concepts of General, Organic, and Biological Chemistry* [Boston: Houghton Mifflin Co., 1981], p. 697.)

four RNA bases; so there are 64 codons. Three of the codons do not signify amino acids but signify the end of a message and thus are called *termination codons*. The remaining 61 codons signify particular amino acids. Because only 20 different amino acids exist in proteins, there are a number of instances in which 2, 3, 4, or even 6 codons translate to the same amino acid.

Example 25.8 Writing the Amino-Acid Sequence Corresponding to an RNA Sequence

What amino-acid sequence would result if the following messenger-RNA sequence were translated from left to right?

AGAGUCCGAGACUUGACGUGA

Solution

We mark the message off into triplets, beginning at the left, and consult the codon dictionary in Table 25.2 to obtain

AGA	GUC	CGA	GAC	UUG	ACG	UGA
arg	val	arg	asp	leu	thr	end

Exercise 25.8

Give one of the nucleotide sequences that would translate to the peptide lys–pro–ala–phe–trp–glu–his–gly. (See Problems 25.43 and 25.44.)

A **transfer RNA** molecule is *the smallest RNA molecule*. It bonds to *a particular* amino acid and carries it to a ribosome; then it attaches itself (through base pairing) to a messenger-RNA codon.

To picture this process, imagine that we have a ribosome, with a messenger RNA attached in the proper way for translation, and the first codon is in position to be read:

AUG GGA CCG ACG UGC GAG CUC . . . (messenger-RNA codons)

The first messenger-RNA codon is AUG, the codon for methionine. Methionine is carried to the ribosome by a transfer RNA that has a triplet sequence, called an *anticodon,* complementary to the codon AUG. The messenger-RNA codon and the transfer-RNA anticodon with methionine pair up. The next codon in the message is GGA, which specifies glycine. A transfer RNA with the anticodon complementary to GGA carries glycine into position to be peptide bonded to methionine. Each succeeding codon in the message is handled in a similar way until a termination codon appears to signal the end of the polypeptide chain. Then the finished product is released from the ribosome.

25.6 LIPIDS

Lipids are *biological substances that are soluble in nonpolar organic solvents, such as chloroform and carbon tetrachloride*. They include the familiar food fats and oils, the lipids of biological membranes, steroid hormones, and many more unusual and exotic compounds. In this section we will discuss fats and oils and the membrane lipids. Lipids differ from the other classes of biological molecules in that there are no lipid polymers analogous to polymers of amino acids, sugars, or nucleotides. We will see, however, that lipid structures formed with noncovalent bonds are very important.

Fats and Oils

At normal temperatures, *fats* are solids and *oils* are liquids. Nevertheless, fats and oils have the same basic structure. Both are **triacylglycerols** (commonly called triglycerides), which are *esters formed from glycerol (a trihydroxy alcohol) and three fatty acids (long-chain carboxylic acids):*●

● Esters were discussed in Section 24.6.

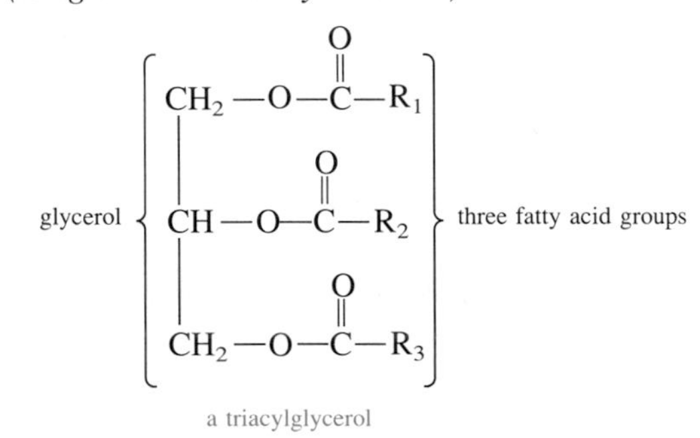

a triacylglycerol

Hydrolysis of a triacylglycerol with strong base (a process called saponification) yields fatty acid salts (soaps) and free glycerol.

Usually, in naturally occurring fats and oils, fatty acids (called ''fatty'' because of their long nonpolar ''tails'') are long-chain molecules of 12, 14, 16, 18, 20, 22, or 24 carbons. They may be saturated (containing no double bonds) or unsaturated (containing one or more double bonds). A fat or oil molecule typically is formed from two or three different fatty acids. Table 25.3 summarizes structures

Table 25.3 Some Naturally Occurring Fatty Acids

Structure	Common Name	m.p., °C
Saturated fatty acids		
$CH_3(CH_2)_{10}CO_2H$	Lauric acid	44.2
$CH_3(CH_2)_{12}CO_2H$	Myristic acid	53.9
$CH_3(CH_2)_{14}CO_2H$	Palmitic acid	63.1
$CH_3(CH_2)_{16}CO_2H$	Stearic acid	69.6
$CH_3(CH_2)_{18}CO_2H$	Arachidic acid	76.5
$CH_3(CH_2)_{22}CO_2H$	Lignoceric acid	86.0
Unsaturated fatty acids		
$CH_3(CH_2)_5CH{=}CH(CH_2)_7CO_2H$	Palmitoleic acid	−0.5
$CH_3(CH_2)_7CH{=}CH(CH_2)_7CO_2H$	Oleic acid	13.4
$CH_3(CH_2)_4CH{=}CHCH_2CH{=}CH(CH_2)_7CO_2H$	Linoleic acid	−5
$CH_3CH_2CH{=}CHCH_2CH{=}CHCH_2CH{=}CH(CH_2)_7CO_2H$	Linolenic acid	−11
$CH_3(CH_2)_4(CH{=}CHCH_2)_3CH{=}CH(CH_2)_3CO_2H$	Arachidonic acid	−49.5

(From Robert D. Whitaker et al., *Concepts of General, Organic, and Biological Chemistry* [Boston: Houghton Mifflin Co., 1981], p. 507.)

of the most common fatty acids. Vegetable oils, which contain unsaturated fatty acids, are liquids at room temperature. This is because fatty acids with double bonds do not pack together well. Conversely, animal fats, composed of the more easily aligned saturated fatty acids, are usually solids at room temperature.

In most organisms, triacylglycerols are a long-term form of energy storage. Excess food taken in is stored as fat, whether it originated as carbohydrate, protein, or fat itself. Fats are a rich source of chemical energy. Stored in compact, nonhydrated droplets in cells, they are an ideal energy reservoir.● Fat stored in animal *adipose tissue,* specialized fat-storage tissue, also serves a structural purpose, providing insulation against cold and padding to protect delicate body organs. The higher-melting saturated fats of animals are better suited to these purposes than liquid oils would be.

● Fats yield 38 kJ/g compared with 17 kJ/g for oxidation of carbohydrates and proteins.

Example 25.9 Writing the Structure of a Triacylglycerol

Write a structural formula for a triacylglycerol containing two 16-carbon saturated fatty acids and one 18-carbon unsaturated fatty acid (see Table 25.3).

Solution

A triacylglycerol contains glycerol in ester linkage with three fatty acids. We chose two palmitic acids and an oleic acid to construct this triacylglycerol:

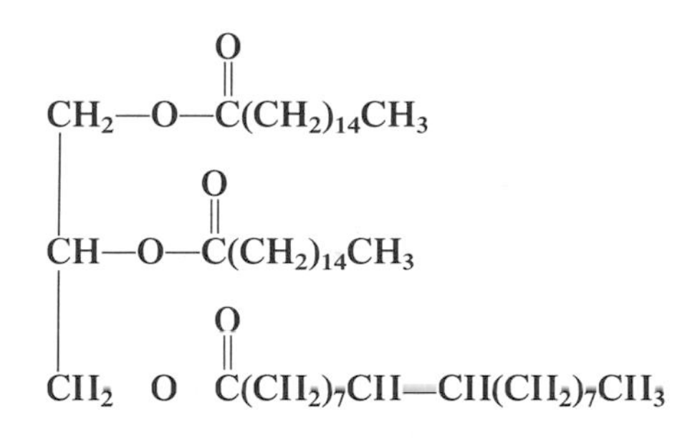

Exercise 25.9

Write a structural formula for a triacylglycerol containing three different unsaturated fatty acids. (See Problems 25.49 and 25.50.)

Biological Membranes

Membranes are basic for life processes. The membrane that surrounds a cell is the boundary between cell and "not cell." The properties of the membrane largely dictate what can get into or out of the cell. Most cells also contain extensive systems of internal membranes, which organize and regulate cell function.

A biological membrane (Figure 25.8) is composed of proteins inserted into a phospholipid matrix. A **phospholipid** *resembles a triacylglycerol, but only two fatty acids are present; the third glycerol —OH is bonded to a phosphate group that is bonded in turn to an alcohol.* Phospholipids have both hydrophobic and hydrophilic properties. They are hydrophobic at the end with the hydrocarbon chains of fatty acids, and they are hydrophilic at the end with the phosphate group and the alcohol (called the *polar head*). Because of these properties, phospholipids tend to aggregate spontaneously in water to form an extensive sheet called a *phospholipid bilayer,* two molecules in thickness. The interior of the lipid bilayer is hydrophobic, consisting, as shown in Figure 25.8, of the hydrocarbon chains of the fatty acids of phospholipids. It presents an effective barrier to charged or polar

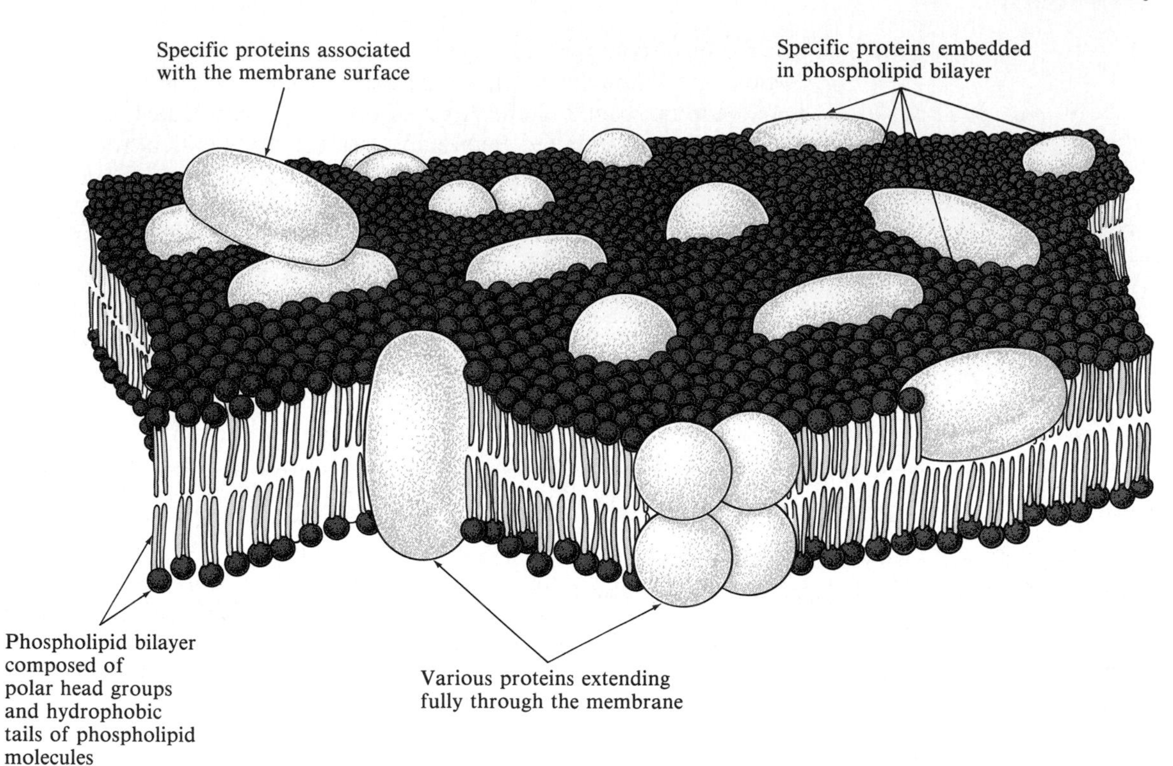

Figure 25.8
A model of membrane structure. The membrane consists of a phospholipid bilayer with protein molecules inserted in it.

substances, which cannot spontaneously enter or pass through a hydrophobic environment. Usually, most substances found in living organisms do not readily pass through membranes because they are charged, polar, and/or relatively large in size. The lipid bilayer does, however, provide a supporting framework for proteins that catalyze transport of particular substances across the membrane.

A Checklist for Review

IMPORTANT TERMS

metabolism (25.1)
enzymes (25.1)
macromolecules (p. 1010)
proteins (25.3)
amino acids (25.3)
peptide (amide) bonds (25.3)
polypeptide (25.3)
primary structure (of a protein) (25.3)
fibrous proteins (25.3)
secondary structure (of a protein) (25.3)
globular proteins (25.3)
tertiary structure (of a protein) (25.3)
substrate (25.3)
carbohydrates (25.4)
monosaccharides (25.4)
oligosaccharides (25.4)
polysaccharides (25.4)
nucleotides (25.5)
polynucleotide (25.5)
nucleic acids (25.5)
complementary bases (25.5)
deoxyribonucleic acid (DNA) (25.5)
ribonucleic acid (RNA) (25.5)
gene (25.5)
ribosomes (25.5)
ribosomal RNA (25.5)
messenger RNA (25.5)
codon (25.5)
transfer RNA (25.5)
lipids (25.6)
triacylglycerols (25.6)
phospholipid (25.6)

SUMMARY OF FACTS AND CONCEPTS

The cell is the smallest organizational unit possessing all the attributes of life. Cells are constructed of *proteins, carbohydrates, nucleic acids,* and *lipids.*

Living organisms transform matter and energy in sequences of *enzyme*-catalyzed reactions called *metabolic pathways,* which together make up *metabolism.* Life requires input of free energy. Most reactions responsible for the highly organized living system have positive free energies and must occur coupled to reactions releasing free energy.

The *macromolecules* of life—proteins, polysaccharides, and nucleic acids—are composed of simple units. Proteins are polymers of *α-amino acids* linked by *peptide bonds.* The 20 common amino acids of protein have different *side chains,* which may be *polar* or *nonpolar.* The unique *primary structure* of a polypeptide (the amino-acid sequence) is responsible for spontaneous folding into a unique shape maintained by interaction between side chains and disulfide bonds. This folding produces surface features responsible for the protein's function. An enzyme has an *active site* where a *substrate* binds and catalysis takes place. The size, shape, and polarity of the active site determine specificity of the enzyme.

Carbohydrates include *monosaccharides, oligosaccharides,* and *polysaccharides.* A monosaccharide is a water-soluble polyhydroxyl carbonyl compound of 3 to 9 carbons. Straight-chain monosaccharides exist in equilibrium with cyclic *hemiacetals* having α and β isomers. D-glucose, D-ribose, 2-deoxy-D-ribose, D-fructose, and their derivatives are the most common monosaccharides. Oligosaccharides contain 2 to 10 monosaccharides; polysaccharides are longer polymers. Structural polysaccharides include cellulose and chitin, and energy-storage polysaccharides include amylose, amylopectin, and glycogen.

Nucleic acids are polymers of *nucleotides.* Nucleotides are base–sugar–phosphate compounds. Conformations of nucleic acids are the result of hydrogen bonding, called base pairing, between *complementary bases. DNA* is the genetic material of chromosomes. DNA consists of two complementary polynucleotides coiled into a *double helix;* each has all the information necessary to direct synthesis of a new complementary strand. The genetic code is the relationship between nucleotide sequence and protein amino-acid sequence. Sixty-four *codons* (3-nucleotide sequences) are the basis of the code. *Transcription* is the synthesis of a *messenger RNA* molecule, which represents a copy of the information in a DNA *gene.* The messenger RNA binds to a *ribosome,* where *translation* of the message into an amino-acid sequence occurs. *Transfer RNA* molecules attach to amino acids and bind to the messenger RNA by *codon–anticodon* pairing, bringing each amino acid in turn into the position specified by the codon sequence.

Lipids are classified on the basis of their solubility in organic solvents and poor solubility in water. Fats and oils are *triacylglycerols,* which are energy-storage lipids. *Phospholipids* are structural lipids and are the main components of the *phospholipid bilayer* of membranes. Membranes also contain proteins inserted into or bound to the surfaces of the bilayer to transport the water-soluble materials into the cell and to catalyze many other important reactions.

OPERATIONAL SKILLS

1. **Drawing the zwitterion of an amino acid** Given the structural formula of an amino acid with a nonpolar side chain, draw the zwitterion (Example 25.1).

2. **Writing the structural formula of a dipeptide** Given the structural formulas of two α-amino acids, draw the structure of the two dipeptides they could form (Example 25.2).

3. **Forming a hydrogen bond between two amino acids** Given the structural formulas of amino acids with polar side chains, show how two amino-acid side chains might form a hydrogen bond between them (Example 25.3).

4. **Predicting the cyclic forms of a monosaccharide** Given the straight-chain structural formula of a 6-carbon monosaccharide, write the reactions for the equilibria existing between that form and the cyclic forms. Or, given the cyclic form of the sugar, draw the straight-chain form (Example 25.4).

5. **Predicting the disaccharide given the monosaccharides** Given the structural formulas for the cyclic forms of two 6-carbon monosaccharides, diagram the reaction to give a disaccharide that could form between them, linked from carbon 1 of the monosaccharide to carbon 4 of the other (Example 25.5).

6. **Drawing the structure of a nucleotide** Given a pentose, an organic base, and a phosphate group, draw the structural formula of the nucleotide they could form and name it (Example 25.6).

7. **Writing the complementary sequence for a DNA sequence** Given a DNA sequence denoted by the bases, write a complementary sequence (Example 25.7).

8. Writing the amino-acid sequence corresponding to an RNA sequence Given a messenger-RNA sequence denoted by the bases, write the amino-acid sequence that would result. Or, given the amino-acid sequence, write the messenger-RNA sequence (Example 25.8). See Table 25.2.

9. Writing the structure of a triacylglycerol Given the structures of three fatty acids, draw the structural formula for the triacylglycerol they would give. Or, given the structural formula of a triacylglycerol, identify each fatty-acid unit (Example 25.9). See Table 25.3.

Review Questions

25.1 Most reactions concerned with formation and maintenance of biological systems have positive free-energy changes. How are such reactions caused to proceed?

25.2 Describe the primary structure of protein. What makes one protein different from another protein of the same size? What is the basis of the unique conformation of a protein?

25.3 Distinguish between secondary and tertiary structures of protein.

25.4 What is an enzyme? What physical features does this molecule have that explain enzyme specificity? Describe the "lock-and-key" model of catalysis.

25.5 Define the terms *monosaccharide, oligosaccharide,* and *polysaccharide*. Name the four most important monosaccharides and their biological functions.

25.6 What is the difference in structure between cellulose and amylose?

25.7 What are the structural forms of D-glucose that are present in the blood?

25.8 Name the complementary base pairs. Describe the DNA double helix.

25.9 How do ribonucleotides and deoxyribonucleotides differ in structure? Do they form polymers in the same way?

25.10 Explain the nature of the genetic code.

25.11 Define *codon* and *anticodon*. How do they interact?

25.12 Show mathematically why there are 64 possible triplet codons.

25.13 Outline how the genetic message in a gene is translated to produce a polypeptide. Include the roles of the gene, messenger RNA, ribosomes, and transfer RNA.

25.14 Distinguish between a fat and an oil.

25.15 Describe the structure of a triacylglycerol.

25.16 Outline the structure of the biological membrane. Relate the structure of the membrane and its lipid components to its tendency to allow or bar passage to hydrophobic or hydrophilic molecules.

Practice Problems

ENERGY AND METABOLISM

25.17 What is the free-energy change associated with a pair of coupled reactions if the ΔG of one is $+7.32$ kcal/mol and ΔG of the second is -10.07 kcal/mol?

25.18 The overall free-energy change for a pair of coupled reactions is -1.5 kcal/mol. What is the ΔG of the energy-requiring reaction if the energy-releasing reaction has $\Delta G = -5.2$ kcal/mol?

25.19 A metabolic pathway catalyzes the breakdown of fat into CO_2 and H_2O. Would you expect such a pathway to release free energy or to require the input of free energy? Explain.

25.20 Some 12 enzymes work together to catalyze the formation of glucose (a 6-carbon sugar) from simpler raw materials (3- and 4-carbon compounds) in the animal liver. Would you expect the overall change to release free energy or to require free energy? Explain.

AMINO ACIDS AND PRIMARY STRUCTURE

25.21 Alanine has the structure

$$CH_3-\underset{NH_2}{\overset{H}{C}}-COOH$$

Draw the zwitterion that would exist at neutral pH.

25.22 Valine has the structure

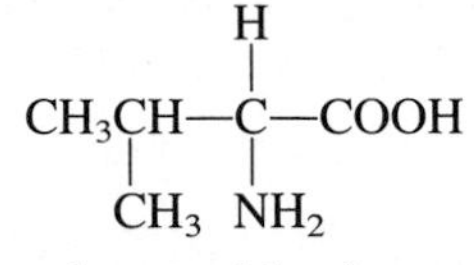

Draw the zwitterion that would exist at neutral pH.

25.23 Write the structural formula of a dipeptide formed from the reaction of L-alanine and L-histidine. How many dipeptides are possible?

25.24 Write the structural formulas of two tripeptides formed from the reaction of L-tryptophan, L-glutamate, and L-tyrosine. How many tripeptides are possible?

25.25 Draw the following amino acids with hydrogen-bonding between their side chains:

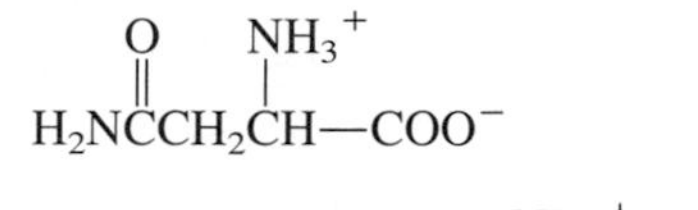

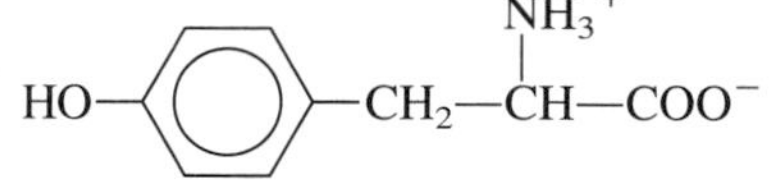

25.26 Which two of these three amino acids have side chains that could undergo hydrophobic interaction?

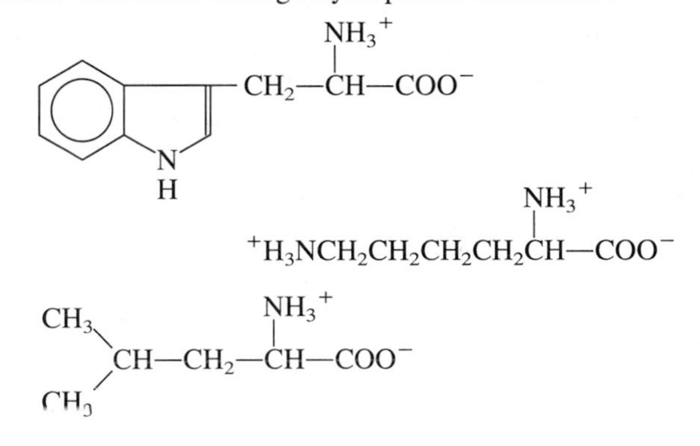

MONOSACCHARIDES, OLIGOSACCHARIDES, AND POLYSACCHARIDES

25.27 Write the reactions for the equilibria existing in a solution of D-galactose:

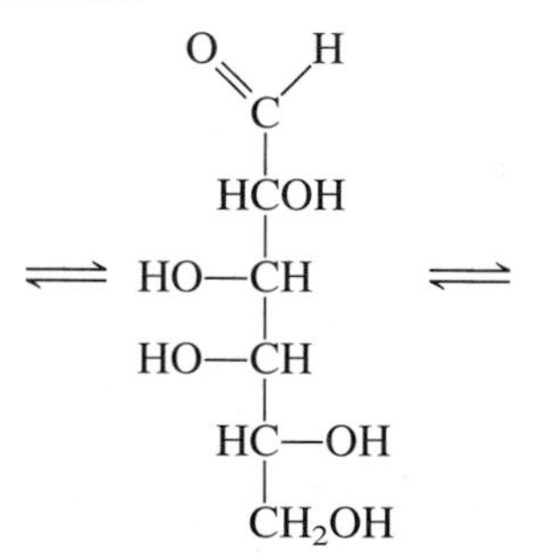

25.28 Write the reactions for the equilibria existing in a solution of D-sorbose:

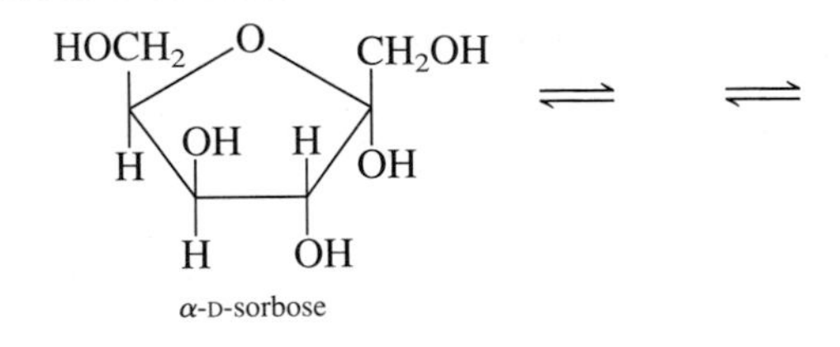

25.29 Write reactions describing the formation of two different disaccharides involving α-D-glucopyranose and β-D-galactopyranose linked between carbon 1 of one molecule and carbon 4 of the other. Is one of your products lactose?

25.30 Write a reaction to show formation of a trisaccharide with two α-D-glucopyranose molecules and one β-D-galactopyranose, linked from carbon 1 of one molecule to carbon 4 of another. How many trisaccharides are possible?

NUCLEOTIDES AND POLYNUCLEOTIDES

25.31 If adenine, thymine, guanine, and cytosine were all analyzed separately in a sample of DNA, what molar ratios of A:T and G:C would you expect to find?

25.32 If a sample of DNA isolated from a microorganism culture were analyzed and found to contain 1.5 moles of cytosine nucleotides and 0.5 moles of adenosine nucleotides, what would be the amounts of guanine and thymine nucleotides in the sample?

25.33 Write a structural formula for the nucleotide adenosine-5′-monophosphate.

25.34 Write a structural formula for the nucleotide deoxyadenosine-5′-monophosphate.

25.35 How many hydrogen bonds link a guanine–cytosine base pair? an adenine–uracil base pair? Would you expect any difference between the strength of guanine–cytosine bonding and adenine–uracil bonding? Explain.

25.36 How many hydrogen bonds link an adenine–thymine base pair? Would there be any difference in strength between adenine–thymine bonding and adenine–uracil bonding? between adenine–thymine and cytosine–guanine bonding? Explain.

25.37 Write the DNA sequence complementary to the sequence ACTGACGCAATTGACCGC.

25.38 Write the sequence of an RNA strand that could base pair with the sequence UAGCUUUACGAAGUGGA.

DNA, RNA, AND PROTEIN SYNTHESIS

25.39 If the codon were two nucleotides, how many codons would be possible? Would this be a workable code for the purpose of protein synthesis?

25.40 If the codon were four nucleotides, how many codons would be possible? Would this be workable as an amino-acid code?

25.41 Nucleic acids can be denatured by heat as proteins can. What bonds are broken when a DNA molecule is denatured? Would DNA of greater percent composition of guanine and cytosine denature more or less readily than DNA of greater percent composition of adenine and thymine?

25.42 Would RNA of a certain percent composition of guanine and cytosine denature more or less readily than RNA with a lower percent of guanine and cytosine? Why?

25.43 Consulting Table 25.2, write the amino-acid sequence resulting from left-to-right translation of the mRNA sequence

GGAUCCCGCUUUGGGCUGAAAUAG

25.44 Write the amino-acid sequence obtained from left-to-right translation of the mRNA sequence

AUUGGCGCGAGAUCGAAUGAGCCCAGU

See Table 25.2.

25.45 List the codons to which the following anticodons would form base pairs:

Anticodon: GAC UGA GGG ACC

Codon:

25.46 List the anticodons to which the following codons would form base pairs:

Codon: UUG CAC ACU GAA

Anticodon:

25.47 Give one of the nucleotide sequences that would translate to

leu–ala–val–glu–asp–cys–met–trp–lys

25.48 Write one nucleotide sequence that would translate to

tyr–ile–pro–his–leu–his–thr–ser–phe–met

LIPIDS AND MEMBRANES

25.49 Write a reaction for the formation of a triacylglycerol from a mole of glycerol and three moles of stearic acid (see Table 25.3).

25.50 Write a reaction for the formation of a triacylglycerol from a mole of glycerol, two moles of oleic acid, and one mole of myristic acid (see Table 25.3).

25.51 Write a reaction for the saponification (base hydrolysis) of the product of the reaction in Problem 25.49.

25.52 Write a reaction for the base hydrolysis of the product of the reaction in Problem 25.50.

25.53 Which of the following molecules or ions would you expect to diffuse slowly across a cell membrane? Which would diffuse more rapidly?

$CH_3CH_2OH \quad Cl^- \quad H_2O \quad Na^+$

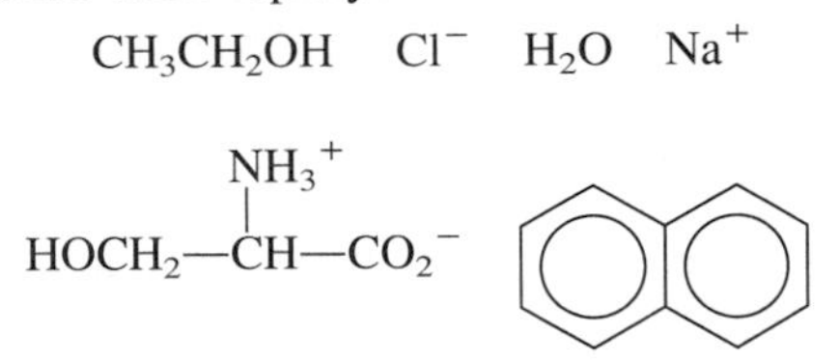

25.54 In general, would ionic or neutral compounds more readily diffuse across cell membranes?

Additional Problems

25.55 A portion of a DNA gene has the nucleotide sequence AGTCGACCGTTAAT. Write a complementary sequence.

25.56 An RNA transcribed from a DNA sequence has the nucleotide sequence UAGCACGGGACUUGG. Write the complementary DNA sequence.

25.57 A peptide contains six amino acids: L-arginine, L-proline, L-glutamate, L-glycine, L-asparagine, and L-glutamine. How many different peptides of this amino-acid composition are possible?

25.58 Using abbreviations for the amino acids, give three possible sequences of the peptide described in Problem 25.57.

25.59 Name each of the three major constituents of the nucleotide adenosine-5′-monophosphate.

25.60 Write the structural formula of deoxyguanosine-5′-monophosphate.

25.61 Which of the following amino acids has a nonpolar side chain?

$CH_3SCH_2CH_2CH(NH_3^+){-}CO_2^- \qquad HSCH_2CH(NH_3^+){-}CO_2^-$

25.62 Which of the following amino acids has a polar side chain?

$^+H_3NCH_2CH_2CH_2CH_2CH(NH_3^+){-}CO_2^-$

$CH_3{-}CH(CH_3){-}CH_2{-}CH(NH_3^+){-}CO_2^-$

25.63 Write the structural formula of one of the two dipeptides that could be formed from the amino acids in Problem 25.61.

25.64 Write the structural formula of one of the two peptides that could be formed by the amino acids in Problem 25.62.

25.65 Draw the structure of the ionic form expected for L-valine at high pH.

25.66 Draw the structure of the ionic form expected for L-valine at low pH. Write an equation for the acid dissociation reaction that occurs when the pH is raised to neutral.

25.67 A common triacylglycerol is

$CH_2{-}O{-}C({=}O){-}(CH_2)_{14}CH_3$

$CH{-}O{-}C({=}O){-}(CH_2)_7CH{=}CH(CH_2)_7CH_3$

$CH_2{-}O{-}C({=}O){-}(CH_2)_7CH{=}CH(CH_2)_5CH_3$

Write structural formulas for each molecule from which the triacylglycerol is constructed.

25.68 Write a balanced equation to describe the saponification of the triacylglycerol shown in Problem 25.67.

Appendix A
Mathematical Skills

Only a few basic mathematical skills are required for the study of general chemistry. But in order to concentrate your attention on the concepts of chemistry, you will find it necessary to have a firm grasp on these basic mathematical skills. In this appendix, we will review scientific (or exponential) notation, logarithms, simple algebraic operations, the solution of quadratic equations, and the plotting of straight-line graphs.

A.1 SCIENTIFIC (EXPONENTIAL) NOTATION

In chemistry, we frequently encounter very large and very small numbers. Thus, the number of molecules in a liter of air at 20°C and normal barometric pressure is 25,000,000,000,000,000,000,000. Similarly, the distance between two hydrogen atoms in a hydrogen molecule is 0.000,000,000,074 meters. In these forms, such numbers are both inconvenient to write and difficult to read. For this reason, we normally express them in scientific, or exponential, notation. Scientific calculators also use this notation.

In scientific notation, a number is written in the form $A \times 10^n$. A is a number greater than or equal to 1 and less than 10, and the exponent n (the nth power of ten) is a positive or negative integer. For example, 4853 would be written in scientific notation as 4.853×10^3, which is 4.853 multiplied by three factors of 10:

$$4.853 \times 10^3 = 4.853 \times 10 \times 10 \times 10 = 4853$$

The number 0.0568 would be written in scientific notation as 5.68×10^{-2}, which is 5.68 divided by two factors of 10:

$$5.68 \times 10^{-2} = \frac{5.68}{10 \times 10} = 0.0568$$

Any number can be conveniently transformed to scientific notation by moving the decimal point in the number to obtain a number, A, greater than or equal to 1 and less than 10. If the decimal point is moved to the left, we multiply A by 10^n, where n equals the number of places moved. If the decimal point is moved to the right, we multiply A by 10^{-n}. Consider the number 0.00731. We must move the decimal point to the right three places. Therefore, 0.00731 equals 7.31×10^{-3}. To transform a number written in scientific notation to one in usual form, the process is reversed. If the exponent is positive, the decimal point is shifted right. If the exponent is negative, the decimal point is shifted left.

Example 1 Expressing Numbers in Scientific Notation

Express the following numbers in scientific notation:

(a) 843.4 (b) 0.00421 (c) 1.54

Solution

Shift the decimal point to get a number between 1 and 10; count the number of positions shifted.

(a) 8 4 3 . 4 = $\mathbf{8.434 \times 10^2}$
(b) 0 . 0 0 4 2 1 = $\mathbf{4.21 \times 10^{-3}}$
(c) **leave as is or write** $\mathbf{1.54 \times 10^0}$

Exercise 1

Express the following numbers in scientific notation:

(a) 4.38 (b) 4380 (c) 0.000483

Example 2 Converting Numbers in Scientific Notation to Usual Form

Convert the following numbers in scientific notation to usual form:

(a) 6.39×10^{-4} (b) 3.275×10^2

Solution

(a) 0 0 0 0 6 . 3 9 $\times 10^{-4}$ = **0.000639**
(b) 3 . 2 7 5 $\times 10^2$ = **327.5**

Exercise 2

Convert the following numbers in scientific notation to usual form:

(a) 7.025×10^3 (b) 8.97×10^{-4}

Addition and Subtraction

Before adding or subtracting two numbers written in scientific notation, it is necessary to express both to the same power of 10. After adding or subtracting, it may be necessary to shift the decimal point to express the result in scientific notation.

Example 3 Adding and Subtracting in Scientific Notation

Carry out the following arithmetic; give the result in scientific notation.

$$(9.42 \times 10^{-2}) + (7.6 \times 10^{-3})$$

Solution

We can shift the decimal point in either number in order to obtain both to the same power of 10. For example, to get both numbers to 10^{-2} we shift the decimal point one place to the left and add 1 to the exponent in the expression 7.6×10^{-3}.

$$7.6 \times 10^{-3} = 0.76 \times 10^{-2}$$

Now we can add the two numbers.

$$(9.42 \times 10^{-2}) + (0.76 \times 10^{-2}) = (9.42 + 0.76) \times 10^{-2} = 10.18 \times 10^{-2}$$

Since 10.18 is not between 1 and 10, we shift the decimal point to express the final result in scientific notation.

$$10.18 \times 10^{-2} = \mathbf{1.018 \times 10^{-1}}$$

Exercise 3

Add the following and express the sum in scientific notation:

$$(3.142 \times 10^{-4}) + (2.8 \times 10^{-6})$$

Multiplication and Division

To multiply two numbers in scientific notation, first we multiply the two powers of 10 by adding their exponents. Then we multiply the remaining factors. Division is handled similarly. We first move any power of 10 in the denominator to the numerator, changing the sign of the exponent. After multiplying the two powers of 10, we carry out the indicated division.

Example 4 Multiplying and Dividing in Scientific Notation

Do the following arithmetic, and express the answers in scientific notation.

(a) $(6.3 \times 10^2) \times (2.4 \times 10^5)$

(b) $\dfrac{6.4 \times 10^2}{2.0 \times 10^5}$

Solution

(a) $(6.3 \times 10^2) \times (2.4 \times 10^5) = (6.3 \times 2.4) \times 10^7$
$= 15.12 \times 10^7$
$= 1.512 \times 10^8$

(b) $\dfrac{6.4 \times 10^2}{2.0 \times 10^5} = \dfrac{6.4}{2.0} \times 10^2 \times 10^{-5} = \dfrac{6.4}{2.0} \times 10^{-3}$
$= 3.2 \times 10^{-3}$

Exercise 4

Perform the following operations, expressing the answers in scientific notation:

(a) $(5.4 \times 10^{-7}) \times (1.8 \times 10^8)$ (b) $\dfrac{5.4 \times 10^{-7}}{6.0 \times 10^{-5}}$

Powers and Roots

A number $A \times 10^n$ raised to a power p is evaluated by raising A to the power p and multiplying the exponent in the power of 10 by p:

$$(A \times 10^n)^p = A^p \times 10^{n \times p}$$

We extract the rth root of a number $A \times 10^n$ by first moving the decimal point in A so that the exponent in the power of 10 is exactly divisible by r. Suppose this has been done, so that n in the number $A \times 10^n$ is exactly divisible by r. Then

$$\sqrt[r]{A \times 10^n} = \sqrt[r]{A} \times 10^{n/r}$$

Example 5 Finding Powers and Roots in Scientific Notation

Simplify the following expressions:

(a) $(5.29 \times 19^2)^3$

(b) $\sqrt{2.31 \times 10^7}$

Solution

(a) $(5.29 \times 10^2)^3 = (5.29)^3 \times 10^6 = 148 \times 10^6$
$= 1.48 \times 10^8$

(b) $\sqrt{2.31 \times 10^7} = \sqrt{23.1 \times 10^6} = \sqrt{23.1} \times 10^{6/2}$
$= 4.81 \times 10^3$

Exercise 5

Obtain the values of the following, and express them in scientific notation:
(a) $(3.56 \times 10^3)^4$ (b) $\sqrt[3]{4.81 \times 10^2}$

Electronic Calculators

Scientific calculators will perform all of the arithmetic operations we have just described (as well as those discussed in the next section). The basic operations of addition, subtraction, multiplication, and division are usually similar and straightforward on most calculators. However, more variation exists in raising a number to a power and extracting a root. If you have the instructions, by all means read them. Otherwise, the following information may help.

Squares and square roots are usually obtained with special keys, perhaps labeled x^2 and $\sqrt{x}$. Thus, to obtain $(5.15)^2$, you enter 5.15 and press x^2. To obtain $\sqrt{5.15}$, you enter 5.15 and press $\sqrt{x}$ (or perhaps INV, for inverse, and x^2). Other powers and roots require a y^x (or a^x) key. The answer to Example 5(a), which is $(5.29 \times 10^2)^3$, would be obtained by a sequence of steps such as the following. Enter 5.29×10^2, press the y^x key, enter 3, and press the = key.

The same sequence can be used to extract a root. Suppose we want $\sqrt[5]{2.18 \times 10^6}$. This is equivalent to $(2.18 \times 10^6)^{1/5}$ or $(2.18 \times 10^6)^{0.2}$. If the calculator has a $1/x$ key, the sequence would be as follows. Enter 2.18×10^6, press the y^x key, enter 5, press the $1/x$ key, then press the = key. Some calculators have a $\sqrt[x]{y}$ key, so this can be used to extract the xth root of y, using a sequence of steps similar to that for y^x.

A.2 LOGARITHMS

The *logarithm* to the *base a* of a number x, denoted $\log_a x$, is the exponent of the constant a needed to equal the number x. For example, suppose a is 10, and we would like the logarithm of 1000, that is, we would like the value of $\log_{10}1000$. This is the exponent, y, of 10 such that 10^y equals 1000. The value of y is 3. Thus, $\log_{10}1000 = 3$.

Common logarithms are logarithms in which the base is 10. The common logarithm of a number x is often denoted simply as $\log x$. It is easy to see how to obtain the common logarithms of 10, 100, 1000, and so forth. But logarithms are defined for all positive numbers, not just the powers of 10. In general, the exponents or values of the logarithm will be decimal numbers. To understand the meaning of a decimal exponent, consider $10^{0.400}$. This is equivalent to $10^{400/1000} = 10^{2/5} = \sqrt[5]{10^2} = 2.51$. Therefore, $\log 2.51 = 0.400$. Any decimal exponent is essentially a fraction, p/r, so by evaluating the expressions $10^{p/r} = \sqrt[r]{10^p}$ one could construct a table of logarithms. In practice, power series or other methods are used. A table of common logarithms is given in Appendix B.

The following are fundamental properties of all logarithms.

$$\log_a 1 = 0 \tag{1}$$

$$\log_a(A \times B) = \log_a A + \log_a B \tag{2}$$

$$\log_a \frac{A}{B} = \log_a A - \log_a B \tag{3}$$

$$\log_a A^p = p \log_a A \tag{4}$$

$$\log_a \sqrt[r]{A} = \frac{1}{r}\log_a A \tag{5}$$

These properties are very useful in working with logarithms.

Electronic calculators that evaluate logarithms are now available for about \$20 or so. Their simplicity of operation makes them well worth the price. To obtain the logarithm of a number, you enter the number and press the LOG key.

Exercise 6

Find the values of (a) log 0.00582 (b) log 689

Antilogarithm

The antilogarithm (abbreviated antilog) is the inverse of the common logarithm. Thus, antilog x is simply 10^x. If your electronic calculator has a 10^x key, you obtain the antilogarithm of a number by entering the number and pressing the 10^x key. (It may be necessary to press an inverse key before pressing a 10^x/LOG key.) If your calculator has a y^x (or a^x) key, you enter 10, press y^x, enter x, then press the = key.

Exercise 7

Evaluate (a) antilog 5.728 (b) antilog (5.728)

Natural Logarithms

The mathematical constant $e = 2.71828\ldots$, like π, occurs in many scientific and engineering problems. It is frequently seen in the natural exponential function $y = e^x$. The inverse function is called the *natural logarithm*, $x = \ln y$, where $\ln y$ is simplified notation for $\log_e y$.

It is possible to express the natural logarithm in terms of the logarithm to the base 10, or the common logarithm. Let us take the common logarithm of both sides of the equation $y = e^x$. Using property 4, given earlier, we get

$$\log y = \log e^x = x \log e$$

Since x equals $\ln y$, and $\log e$ is 0.4343, we can write this equation as

$$\log y = \ln y \log e = 0.4343 \ln y$$

Finally, solving for $\ln y$

$$\ln y = \frac{1}{0.4343}\log y = 2.303 \log y$$

If your calculator has an LN key, you can obtain the natural logarithm of a number by entering the number and pressing the LN key.

A.3 ALGEBRAIC OPERATIONS AND GRAPHING

Often we are given an algebraic formula that we would like to rearrange in order to solve for a particular quantity. As an example, suppose we would like to solve the following equation for V.

$$PV = nRT$$

We can eliminate P from the left-hand side by dividing by P. But to maintain the equality, we must perform the same operation on both sides of the equation:

$$\frac{PV}{P} = \frac{nRT}{P}$$

Or,

$$V = \frac{nRT}{P}$$

Quadratic Formula

A quadratic equation is one involving only powers of x in which the highest power is two. The general form of the equation can be written

$$ax^2 + bx + c = 0$$

where a, b, and c are constants. For given values of these constants, only certain values of x are possible (in general, there will be two values). These values of x are said to be the solutions of the equation.

These solutions are given by the *quadratic formula:*

$$x = \frac{-b \pm \sqrt{b^2 - 4ac}}{2a}$$

In this formula, the symbol $\pm$ means that there are two possible values of x, one obtained by taking the positive sign, the other by taking the negative sign.

Example 6 Obtaining the Solutions of a Quadratic Equation

Obtain the solutions of the following quadratic equation:

$$2.00x^2 - 1.72x - 2.86 = 0$$

Solution

Using the quadratic formula, we substitute $a = 200$, $b = -1.72$, and $c = -2.86$. We get

$$x = \frac{1.72 \pm \sqrt{(-1.72)^2 - 4 \times 2.00 \times (-2.86)}}{2 \times 2.00}$$

$$= \frac{1.72 \pm 5.08}{4.00} = -0.84 \text{ and } +1.70$$

Although mathematically there are two solutions, in any real problem one may not be allowed. For example, if the solution is some physical quantity that can have only positive values, a negative solution must be rejected.

Exercise 8

Find the positive solution (or solutions) to the following equation:

$$1.80x^2 + 0.850x - 9.50 = 0$$

The Straight-Line Graph

A graph is a visual means of representing a mathematical relationship or physical data. Consider the following data in which values of y from some experiment are given for four values of x.

x	y
1	−1
2	1
3	3
4	5

By plotting these x, y points on a graph (Figure A.1), we can see that they fall on a straight line. This suggests (but does not prove) that other points from this type of experiment might fall on the same line. It would be useful to have the mathematical equation for this line.

The general form of a straight line is

$$y = mx + b$$

The constant m is called the *slope* of the straight line. It is obtained by dividing the vertical distance between any two points on the line by the horizontal distance. If the two points are (x_1, y_1) and (x_2, y_2), the slope is given by the following formula.

$$\text{slope} = \frac{y_2 - y_1}{x_2 - x_1}$$

Suppose we choose the points (2, 1) and (4, 5) from our data. Then

$$\text{slope} = \frac{5 - 1}{4 - 2} = 2$$

Thus, $m = 2$ for the straight line in Figure A.1.

The constant b is called the *intercept*. It is the value of y at $x = 0$. From Figure A.1, we see that the intercept is −3. Therefore, $b = -3$. Hence, the equation of the straight line is

$$y = 2x - 3$$

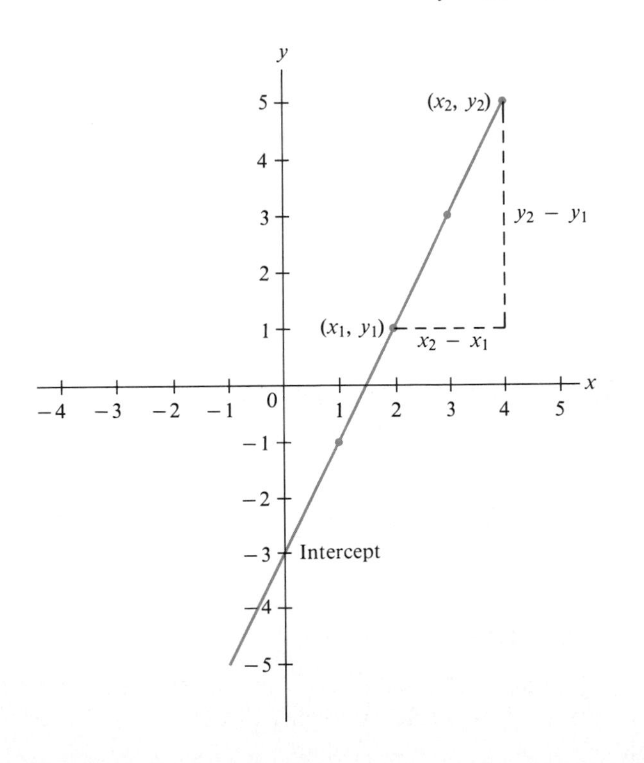

Figure A.1
A straight-line plot of some data.

Appendix B
Vapor Pressure of Water at Various Temperatures

Temperature, °C	Pressure, mmHg	Temperature, °C	Pressure, mmHg
0	4.6	27	26.7
5	6.5	28	28.3
10	9.2	29	30.0
11	9.8	30	31.8
12	10.5	35	42.2
13	11.2	40	55.3
14	12.0	45	71.9
15	12.8	50	92.5
16	13.6	55	118.0
17	14.5	60	149.4
18	15.5	65	187.5
19	16.5	70	233.7
20	17.5	75	289.1
21	18.7	80	355.1
22	19.8	85	433.6
23	21.1	90	525.8
24	22.4	95	633.9
25	23.8	100	760.0
26	25.2	105	906.1

Appendix C Thermodynamic Quantities for Substances and Ions at 25°C

Substance or Ion	ΔH_f° (kJ/mol)	ΔG_f° (kJ/mol)	S° (J/K · mol)
$e^-(g)$	0	0	20.87
$H^+(g)$	1536.3	1517.1	108.83
$H^+(aq)$	0	0	0
$H(g)$	218.0	203.30	114.60
$H_2(g)$	0	0	130.6
Group IA			
$Li^+(g)$	687.163	649.989	132.91
$Li^+(aq)$	−278.46	−293.8	14
$Li(g)$	161	128	138.67
$Li(s)$	0	0	29.10
$LiF(s)$	−616.9	−588.7	35.66
$LiCl(s)$	−408	−384	59.30
$LiBr(s)$	−351	−342	74.1
$LiI(s)$	−270	−270	85.8
$Na^+(g)$	609.839	574.877	147.85
$Na^+(aq)$	−239.66	−261.87	60.2
$Na(g)$	107.76	77.299	153.61
$Na(s)$	0	0	51.446
$NaF(s)$	−575.4	−545.1	51.21
$NaCl(s)$	−411.1	−384.0	72.12
$NaBr(s)$	−361	−349	86.82
$NaI(s)$	−288	−285	98.5
$NaHCO_3(s)$	−947.7	−851.9	102
$Na_2CO_3(s)$	−1130.8	−1048.1	139
$K^+(g)$	514.197	481.202	154.47
$K^+(aq)$	−251.2	−282.28	103
$K(g)$	89.2	60.7	160.23
$K(s)$	0	0	64.672
$KF(s)$	−568.6	−538.9	66.55
$KCl(s)$	−436.68	−408.8	82.55
$KBr(s)$	−394	−380	95.94
$KI(s)$	−328	−323	106.39
$Rb^+(g)$	495.04		
$Rb^+(aq)$	−246	−282.2	124
$Rb(g)$	85.81	55.86	169.99
$Rb(s)$	0	0	69.5

Substance or Ion	ΔH_f° (kJ/mol)	ΔG_f° (kJ/mol)	S° (J/K · mol)
$RbF(s)$	−549.28		
$RbCl(s)$	−430.58		
$RbBr(s)$	−389.2	−378.1	108.3
$RbI(s)$	−328	−326	118.0
$Cs^+(g)$	458.5	427.1	169.72
$Cs^+(aq)$	−248	−282.0	133
$Cs(g)$	76.7	49.7	175.5
$Cs(s)$	0	0	85.15
$CsF(s)$	−554.7	−525.4	88
$CsCl(s)$	−442.8	−414	101.18
$CsBr(s)$	−395	−383	121
$CsI(s)$	−337	−333	130
Group IIA			
$Mg^{2+}(g)$	2351		
$Mg^{2+}(aq)$	−461.96	−456.01	−118
$Mg^+(g)$	894.1		
$Mg(g)$	150	115	148.55
$Mg(s)$	0	0	32.69
$MgCl_2(s)$	−641.6	−592.1	89.630
$MgO(s)$	−601.2	−569.0	26.9
$Mg_3N_2(s)$	−461	−401	88
$MgCO_3(s)$	−1112	−1028	65.86
$Ca^{2+}(g)$	1934.1		
$Ca^{2+}(aq)$	−542.96	−553.04	−55.2
$Ca^+(g)$	788.6		
$Ca(g)$	192.6	158.9	154.78
$Ca(s)$	0	0	41.6
$CaF_2(s)$	−1215	−1162	68.87
$CaCl_2(s)$	−795.0	−750.2	114
$CaO(s)$	−635.1	−603.5	38.2
$CaCO_3(s)$	−1206.9	−1128.8	92.9
$CaSO_4(s)$	−1432.7	−1320.3	107
$Ca_3(PO_4)_2(s)$	−4138	−3899	263
$Sr^{2+}(g)$	1784		
$Sr^{2+}(aq)$	−545.51	−557.3	−39
$Sr^+(g)$	719.6		
$Sr(g)$	164	110	164.54
$Sr(s)$	0	0	54.4
$SrCl_2(s)$	−828.4	−781.2	117
$SrO(s)$	−592.0	−562.4	55.5
$SrCO_3(s)$	−1218	−1138	97.1
$SrSO_4(s)$	−1445	−1334	122
$Ba^{2+}(g)$	1649.9		
$Ba^{2+}(aq)$	−538.36	−560.7	13
$Ba^+(g)$	684.6		
$Ba(g)$	175.6	144.8	170.28
$Ba(s)$	0	0	62.5
$BaCl_2(s)$	−806.06	−810.9	126
$BaO(s)$	−548.1	−520.4	72.07
$BaCO_3(s)$	−1219	−1139	112
$BaSO_4(s)$	−1465	−1353	132
Group IIIA			
B(β-rhombohedral)	0	0	5.87
$B_2O_3(s)$	−1272	−1193	53.8

Substance or Ion	ΔH_f° (kJ/mol)	ΔG_f° (kJ/mol)	S° (J/K · mol)
Al(*s*)	0	0	28.3
Al^{3+}(*aq*)	−524.7	−481.2	−313
Al_2O_3(*s*)	−1676	−1582	50.94
Group IVA			
C(*g*)	715.0	669.6	158.0
C(graphite)	0	0	5.686
C(diamond)	1.896	2.866	2.439
CO(*g*)	−110.5	−137.2	197.5
CO_2(*g*)	−393.5	−394.4	213.7
CO_2(*aq*)	−412.9	−386.2	121
CO_3^{2-}(*aq*)	−676.26	−528.10	−53.1
HCO_3^-(*aq*)	−691.11	−587.06	95.0
H_2CO_3(*aq*)	−698.7	−623.42	191
CH_4(*g*)	−74.87	−50.81	186.1
C_2H_2(*g*)	227	209	200.85
C_2H_4(*g*)	52.47	68.36	219.22
C_2H_6(*g*)	−84.667	−32.89	229.5
C_6H_6(*l*)	49.0	124.5	172.8
CH_3OH(*g*)	−201.2	−161.9	238
CH_3OH(*l*)	−238.6	−166.2	127
HCHO(*g*)	−116	−110	219
$HCOO^-$(*aq*)	−410	−335	91.6
HCOOH(*l*)	−409	−346	129.0
HCOOH(*aq*)	−410	−356	164
C_2H_5OH(*l*)	−277.63	−174.8	161
CH_3CHO(*g*)	−166	−133.7	266
CH_3COOH(*l*)	−487.0	−392	160
CN^-(*aq*)	151	166	118
HCN(*g*)	135	125	201.7
HCN(*l*)	105	121	112.8
HCN(*aq*)	105	112	129
CS_2(*g*)	117	66.9	237.79
CS_2(*l*)	87.9	63.6	151.0
CH_3Cl(*g*)	−83.7	−60.2	234
CH_2Cl_2(*l*)	−117	−63.2	179
$CHCl_3$(*l*)	−132	−71.5	203
CCl_4(*g*)	−96.0	−53.7	309.7
CCl_4(*l*)	−139	−68.6	214.4
$COCl_2$(*g*)	−220	−206	283.74
Si(*s*)	0	0	18.0
SiO_2(*s*)	−910.9	−856.5	41.5
Sn(gray)	3	4.6	44.8
Sn(white)	0	0	51.5
$SnCl_4$(*l*)	−545.2	−474.0	259
Pb^{2+}(*aq*)	1.6	−24.3	21
Pb(*s*)	0	0	64.785
PbO(*s*)	−218	−198	68.70
PbO_2(*s*)	−276.6	−219.0	76.6
PbS(*s*)	−98.3	−96.7	91.3
$PbCl_2$(*s*)	−359	−314	136
$PbSO_4$(*s*)	−918.39	−811.24	147
Group VA			
N(*g*)	473	456	153.2
N_2(*g*)	0	0	191.5
NO(*g*)	90.29	86.60	210.65
NO_2(*g*)	33.2	51	239.9

Substance or Ion	ΔH_f° (kJ/mol)	ΔG_f° (kJ/mol)	S° (J/K · mol)
$N_2O_4(g)$	9.16	97.7	304.3
$N_2O_5(g)$	11	118	346
$NH_3(g)$	−45.9	−16	193
$NH_3(aq)$	−80.83	26.7	110
$NO_3^-(aq)$	−206.57	−110.5	146
$HNO_3(l)$	−173.23	−79.914	155.6
$HNO_3(aq)$	−206.57	−110.5	146
$P(g)$	333.9	292.0	163.1
P(red)	0	0	22.8
P_4(white)	68	48	164
$P_2(g)$	179	127	218
$P_4(g)$	129	72.5	280
$PCl_3(g)$	−271	−258	312
$PCl_5(g)$	−382	−313	353
$P_4O_{10}(s)$	−2942	−2675	229
$PO_4^{3-}(aq)$	−1266	−1013	−218
$HPO_4^{2-}(aq)$	−1281	−1082	−36
$H_2PO_4^-(aq)$	−1285	−1135	89.1
$H_3PO_4(aq)$	−1277		
Group VIA			
$O(g)$	249.2	231.7	160.95
$O_2(g)$	0	0	205.0
$O_3(g)$	143	163	238.82
$OH^-(aq)$	−229.94	−157.30	−10.54
$H_2O(g)$	−241.826	−228.60	188.72
$H_2O(l)$	−285.840	−237.192	69.940
$H_2O_2(l)$	−187.8	−120.4	110
$H_2O_2(aq)$	−191.2	−134.1	144
$S(g)$	279	239	168
$S_2(g)$	129	80.1	228.1
$S_8(g)$	101	49.1	430.211
S(rhombic)	0	0	31.9
S(monoclinic)	0.30	0.096	32.6
$S^{2-}(aq)$	41.8	83.7	22
$HS^-(aq)$	−17.7	12.6	61.1
$H_2S(g)$	−20.2	−33	205.6
$H_2S(aq)$	−39	−27.4	122
$SO_2(g)$	−296.8	−300.2	248.1
$SO_3(g)$	−396	−371	256.66
$SO_4^{2-}(aq)$	−907.51	−741.99	17
$HSO_4^-(aq)$	−885.75	−752.87	126.9
$H_2SO_4(l)$	−813.989	−690.059	156.90
$H_2SO_4(aq)$	−907.51	−741.99	17
Group VIIA			
$F(g)$	78.9	61.8	158.64
$F^-(g)$	−255.6	−262.5	145.47
$F^-(aq)$	−329.1	−276.5	−9.6
$F_2(g)$	0	0	202.7
$HF(g)$	−273	−275	173.67
$Cl(g)$	121.0	105.0	165.1
$Cl^-(g)$	−234	−240	153.25
$Cl^-(aq)$	−167.46	−131.17	55.10
$Cl_2(g)$	0	0	223.0
$HCl(g)$	−92.31	−95.30	186.79
$HCl(aq)$	−167.46	−131.17	55.06

Substance or Ion	ΔH_f° (kJ/mol)	ΔG_f° (kJ/mol)	S° (J/K · mol)
Br(*g*)	111.9	82.40	174.90
Br^-(*g*)	−218.9		
Br^-(*aq*)	−120.9	−102.82	80.71
Br_2(*g*)	30.91	3.13	245.38
Br_2(*l*)	0	0	152.23
HBr(*g*)	−36	−53.5	198.59
I(*g*)	106.8	70.21	180.67
I^-(*g*)	−194.7		
I^-(*aq*)	−55.94	−51.67	109.4
I_2(*g*)	62.442	19.38	260.58
I_2(*s*)	0	0	116.14
HI(*g*)	25.9	1.3	206.33
Group IB			
Cu^+(*aq*)	51.9	50.2	−26
Cu^{2+}(*aq*)	64.39	64.98	−98.7
Cu(*g*)	341.1	301.4	166.29
Cu(*s*)	0	0	33.1
Ag^+(*aq*)	105.9	77.111	73.93
Ag(*g*)	289.2	250.4	172.892
Ag(*s*)	0	0	42.702
AgF(*s*)	−203	−185	84
AgCl(*s*)	−127.03	−109.72	96.11
AgBr(*s*)	−99.50	−95.939	107.1
AgI(*s*, II)	−62.38	−66.32	114
Ag_2S(*s*)	31.8	40.3	146
Group IIB			
Zn^{2+}(*aq*)	−152.4	−147.21	−106.5
Zn(*g*)	130.5	94.93	160.9
Zn(*s*)	0	0	41.6
ZnO(*s*)	−348.0	−318.2	43.9
ZnS(*s*, zinc blende)	−203	−198	57.7
Cd^{2+}(*aq*)	−72.38	−77.74	−61.1
Cd(*g*)	112.8	78.20	167.64
Cd(*s*)	0	0	51.5
CdS(*s*)	−144	−141	71
Hg^{2+}(*aq*)		164.8	
$Hg_2{}^+$(*aq*)		153.9	
Hg(*g*)	61.30	31.8	174.87
Hg(*l*)	0	0	76.027
$HgCl_2$(*s*)	−230	−184	144
Hg_2Cl_2(*s*)	−264.9	−210.66	196
HgO(*s*)	−90.79	−58.50	70.27
Group VIB			
$[Cr(H_2O)_6]^{3+}$(*aq*)	−1971		
Cr(*s*)	0	0	23.8
$CrO_4{}^{2-}$(*aq*)	−863.2	−706.3	38
$Cr_2O_7{}^{2-}$(*aq*)	−1461	−1257	214
Group VIIB			
Mn^{2+}(*aq*)	−219	−223	−84
Mn(*s*, α)	0	0	31.8
MnO_2(*s*)	−520.9	−466.1	53.1
$MnO_4{}^-$(*aq*)	−518.4	−425.1	190

Substance or Ion	ΔH_f° (kJ/mol)	ΔG_f° (kJ/mol)	S° (J/K · mol)
Group VIIIB			
$Fe^{3+}(aq)$	−47.7	−10.5	−293
$Fe^{2+}(aq)$	−87.9	−84.94	113
$Fe(s)$	0	0	27.3
$FeO(s)$	−272.0	−251.4	60.75
$Fe_2O_3(s)$	−825.5	−743.6	87.400
$Fe_3O_4(s)$	−1121	−1018	145.3
$Co^{2+}(aq)$	−67.4	−51.5	−155
$Co(s)$	0	0	30
$Ni^{2+}(aq)$	−64.0	−46.4	−159
$Ni(s)$	0	0	30.1

Appendix D
Electron Configurations of Atoms in the Ground State

Z	Element	Configuration	Z	Element	Configuration
1	H	$1s^1$	41	Nb	$[Kr]4d^4 5s^1$
2	He	$1s^2$	42	Mo	$[Kr]4d^5 5s^1$
3	Li	$[He]2s^1$	43	Tc	$[Kr]4d^5 5s^2$
4	Be	$[He]2s^2$	44	Ru	$[Kr]4d^7 5s^1$
5	B	$[He]2s^2 2p^1$	45	Rh	$[Kr]4d^8 5s^1$
6	C	$[He]2s^2 2p^2$	46	Pd	$[Kr]4d^{10}$
7	N	$[He]2s^2 2p^3$	47	Ag	$[Kr]4d^{10} 5s^1$
8	O	$[He]2s^2 2p^4$	48	Cd	$[Kr]4d^{10} 5s^2$
9	F	$[He]2s^2 2p^5$	49	In	$[Kr]4d^{10} 5s^2 5p^1$
10	Ne	$[He]2s^2 2p^6$	50	Sn	$[Kr]4d^{10} 5s^2 5p^2$
11	Na	$[Ne]3s^1$	51	Sb	$[Kr]4d^{10} 5s^2 5p^3$
12	Mg	$[Ne]3s^2$	52	Te	$[Kr]4d^{10} 5s^2 5p^4$
13	Al	$[Ne]3s^2 3p^1$	53	I	$[Kr]4d^{10} 5s^2 5p^5$
14	Si	$[Ne]3s^2 3p^2$	54	Xe	$[Kr]4d^{10} 5s^2 5p^6$
15	P	$[Ne]3s^2 3p^3$	55	Cs	$[Xe]6s^1$
16	S	$[Ne]3s^2 3p^4$	56	Ba	$[Xe]6s^2$
17	Cl	$[Ne]3s^2 3p^5$	57	La	$[Xe]5d^1 6s^2$
18	Ar	$[Ne]3s^2 3p^6$	58	Ce	$[Xe]4f^1 5d^1 6s^2$
19	K	$[Ar]4s^1$	59	Pr	$[Xe]4f^3 6s^2$
20	Ca	$[Ar]4s^2$	60	Nd	$[Xe]4f^4 6s^2$
21	Sc	$[Ar]3d^1 4s^2$	61	Pm	$[Xe]4f^5 6s^2$
22	Ti	$[Ar]3d^2 4s^2$	62	Sm	$[Xe]4f^6 6s^2$
23	V	$[Ar]3d^3 4s^2$	63	Eu	$[Xe]4f^7 6s^2$
24	Cr	$[Ar]3d^5 4s^1$	64	Gd	$[Xe]4f^7 5d^1 6s^2$
25	Mn	$[Ar]3d^5 4s^2$	65	Tb	$[Xe]4f^9 6s^2$
26	Fe	$[Ar]3d^6 4s^2$	66	Dy	$[Xe]4f^{10} 6s^2$
27	Co	$[Ar]3d^7 4s^2$	67	Ho	$[Xe]4f^{11} 6s^2$
28	Ni	$[Ar]3d^8 4s^2$	68	Er	$[Xe]4f^{12} 6s^2$
29	Cu	$[Ar]3d^{10} 4s^1$	69	Tm	$[Xe]4f^{13} 6s^2$
30	Zn	$[Ar]3d^{10} 4s^2$	70	Yb	$[Xe]4f^{14} 6s^2$
31	Ga	$[Ar]3d^{10} 4s^2 4p^1$	71	Lu	$[Xe]4f^{14} 5d^1 6s^2$
32	Ge	$[Ar]3d^{10} 4s^2 4p^2$	72	Hf	$[Xe]4f^{14} 5d^2 6s^2$
33	As	$[Ar]3d^{10} 4s^2 4p^3$	73	Ta	$[Xe]4f^{14} 5d^3 6s^2$
34	Se	$[Ar]3d^{10} 4s^2 4p^4$	74	W	$[Xe]4f^{14} 5d^4 6s^2$
35	Br	$[Ar]3d^{10} 4s^2 4p^5$	75	Re	$[Xe]4f^{14} 5d^5 6s^2$
36	Kr	$[Ar]3d^{10} 4s^2 4p^6$	76	Os	$[Xe]4f^{14} 5d^6 6s^2$
37	Rb	$[Kr]5s^1$	77	Ir	$[Xe]4f^{14} 5d^7 6s^2$
38	Sr	$[Kr]5s^2$	78	Pt	$[Xe]4f^{14} 5d^9 6s^1$
39	Y	$[Kr]4d^1 5s^2$	79	Au	$[Xe]4f^{14} 5d^{10} 6s^1$
40	Zr	$[Kr]4d^2 5s^2$	80	Hg	$[Xe]4f^{14} 5d^{10} 6s^2$

Z	Element	Configuration	Z	Element	Configuration
81	Tl	$[Xe]4f^{14}5d^{10}6s^{2}6p^{1}$	96	Cm	$[Rn](5f^{7}6d^{1}7s^{2})$
82	Pb	$[Xe]4f^{14}5d^{10}6s^{2}6p^{2}$	97	Bk	$[Rn](5f^{9}7s^{2})$
83	Bi	$[Xe]4f^{14}5d^{10}6s^{2}6p^{3}$	98	Cf	$[Rn](5f^{10}7s^{2})$
84	Po	$[Xe]4f^{14}5d^{10}6s^{2}6p^{4}$	99	Es	$[Rn](5f^{11}7s^{2})$
85	At	$[Xe](4f^{14}5d^{10}6s^{2}6p^{5})$	100	Fm	$[Rn](5f^{12}7s^{2})$
86	Rn	$[Xe]4f^{14}5d^{10}6s^{2}6p^{6}$	101	Md	$[Rn](5f^{13}7s^{2})$
87	Fr	$[Rn](7s^{1})$	102	No	$[Rn](5f^{14}7s^{2})$
88	Ra	$[Rn]7s^{2}$	103	Lr	$[Rn](5f^{14}6d^{1}7s^{2})$
89	Ac	$[Rn]6d^{1}6s^{2}$	104	Unq	$[Rn](5f^{14}5d^{2}7s^{2})$
90	Th	$[Rn]6d^{2}7s^{2}$	105	Unp	$[Rn](5f^{14}6d^{3}7s^{2})$
91	Pa	$[Rn](5f^{2}6d^{1}7s^{2})$	106	Unh	$[Rn](5f^{14}6d^{4}7s^{2})$
92	U	$[Rn](5f^{3}6d^{1}7s^{2})$	107	Uns	$[Rn](5f^{14}6d^{5}7s^{2})$
93	Np	$[Rn](5f^{4}6d^{1}7s^{2})$	108	Uno	$[Rn](5f^{14}6d^{6}7s^{2})$
94	Pu	$[Rn]5f^{6}7s^{2}$	109	Une	$[Rn](5f^{14}6d^{7}7s^{2})$
95	Am	$[Rn]5f^{7}7s^{2}$			

Configurations in this table, except those in parentheses, are taken from Charlotte E. Moore, *National Standard Reference Data Series,* National Bureau of Standards, 34 (September 1970). Those in parentheses were obtained on the basis of their assumed position in the periodic table.

Appendix E
Acid Ionization Constants at 25°C

Substance	Formula	K_a
Acetic acid	$HC_2H_3O_2$	1.7×10^{-5}
Arsenic acid*	H_3AsO_4	6.5×10^{-3}
	$H_2AsO_4^-$	1.2×10^{-7}
	$HAsO_4^{2-}$	3.2×10^{-12}
Ascorbic acid*	$H_2C_6H_6O_6$	6.8×10^{-5}
	$HC_6H_6O_6^-$	2.8×10^{-12}
Benzoic acid	$HC_7H_5O_2$	6.3×10^{-5}
Boric acid	H_3BO_3	5.9×10^{-10}
Carbonic acid*	H_2CO_3	4.3×10^{-7}
	HCO_3^-	4.8×10^{-11}
Chromic acid*	H_2CrO_4	1.5×10^{-1}
	$HCrO_4^-$	3.2×10^{-7}
Cyanic acid	$HCNO$	3.5×10^{-4}
Formic acid	$HCHO_2$	1.7×10^{-4}
Hydrocyanic acid	HCN	4.9×10^{-10}
Hydrofluoric acid	HF	6.8×10^{-4}
Hydrogen peroxide	H_2O_2	1.8×10^{-12}
Hydrogen sulfate ion	HSO_4^-	1.1×10^{-2}
Hydrogen sulfide*	H_2S	8.9×10^{-8}
	HS^-	1.2×10^{-13}
Hypochlorous acid	$HClO$	3.5×10^{-8}
Lactic acid	$HC_3H_5O_3$	1.3×10^{-4}
Nitrous acid	HNO_2	4.5×10^{-4}
Oxalic acid*	$H_2C_2O_4$	5.6×10^{-2}
	$HC_2O_4^-$	5.1×10^{-5}
Phenol	C_6H_5OH	1.1×10^{-10}
Phosphoric acid*	H_3PO_4	6.9×10^{-3}
	$H_2PO_4^-$	6.2×10^{-8}
	HPO_4^{2-}	4.8×10^{-13}
Phosphorous acid*	H_2PHO_3	1.6×10^{-2}
	$HPHO_3^-$	7×10^{-7}
Propionic acid	$HC_3H_5O_2$	1.3×10^{-5}
Pyruvic acid	$HC_3H_3O_3$	1.4×10^{-4}
Sulfurous acid*	H_2SO_3	1.3×10^{-2}
	HSO_3^-	6.3×10^{-8}

*The ionization constants for polyprotic acids are for successive ionizations. Thus for H_3PO_4, the equilibrium is $H_3PO_4 \rightleftharpoons H^+ + H_2PO_4^-$. For $H_2PO_4^-$, the equilibrium is $H_2PO_4^- \rightleftharpoons H^+ + HPO_4^{2-}$.

Appendix F
Base Ionization Constants at 25°C

Substance	Formula	K_b
Ammonia	NH_3	1.8×10^{-5}
Aniline	$C_6H_5NH_2$	4.2×10^{-10}
Dimethylamine	$(CH_3)_2NH$	5.1×10^{-4}
Ethylamine	$C_2H_5NH_2$	4.7×10^{-4}
Ethylene diamine	$NH_2CH_2CH_2NH_2$	5.2×10^{-4}
Hydrazine	N_2H_4	1.7×10^{-6}
Hydroxylamine	NH_2OH	1.1×10^{-8}
Methylamine	CH_3NH_2	4.4×10^{-4}
Pyridine	C_5H_5N	1.4×10^{-9}
Trimethylamine	$(CH_3)_3N$	6.5×10^{-5}
Urea	NH_2CONH_2	1.5×10^{-14}

Appendix G Solubility Product Constants at 25°C

Substance	Formula	K_{sp}
Aluminum hydroxide	$Al(OH)_3$	4.6×10^{-33}
Barium chromate	$BaCrO_4$	1.2×10^{-10}
Barium fluoride	BaF_2	1.0×10^{-6}
Barium sulfate	$BaSO_4$	1.1×10^{-10}
Cadmium oxalate	CdC_2O_4	1.5×10^{-8}
Cadmium sulfide	CdS	8×10^{-27}
Calcium carbonate	$CaCO_3$	3.8×10^{-9}
Calcium fluoride	CaF_2	3.4×10^{-11}
Calcium oxalate	CaC_2O_4	2.3×10^{-9}
Calcium phosphate	$Ca_3(PO_4)_2$	1×10^{-26}
Calcium sulfate	$CaSO_4$	2.4×10^{-5}
Cobalt(II) sulfide	CoS	4×10^{-21}
Copper(II) hydroxide	$Cu(OH)_2$	2.6×10^{-19}
Copper(II) sulfide	CuS	6×10^{-36}
Iron(II) hydroxide	$Fe(OH)_2$	8×10^{-16}
Iron(II) sulfide	FeS	6×10^{-18}
Iron(III) hydroxide	$Fe(OH)_3$	2.5×10^{-39}
Lead(II) arsenate	$Pb_3(AsO_4)_2$	4×10^{-36}
Lead(II) chloride	$PbCl_2$	1.6×10^{-5}
Lead(II) chromate	$PbCrO_4$	1.8×10^{-14}
Lead(II) iodide	PbI_2	6.5×10^{-9}
Lead(II) sulfate	$PbSO_4$	1.7×10^{-8}
Lead(II) sulfide	PbS	2.5×10^{-27}
Magnesium arsenate	$Mg_3(AsO_4)_2$	2×10^{-20}
Magnesium carbonate	$MgCO_3$	1.0×10^{-5}
Magnesium hydroxide	$Mg(OH)_2$	1.8×10^{-11}
Magnesium oxalate	MgC_2O_4	8.5×10^{-5}
Manganese(II) sulfide	MnS	2.5×10^{-10}
Mercury(I) chloride	Hg_2Cl_2	1.3×10^{-18}
Mercury(II) sulfide	HgS	1.6×10^{-52}
Nickel(II) hydroxide	$Ni(OH)_2$	2.0×10^{-15}
Nickel(II) sulfide	NiS	3×10^{-19}

Substance	Formula	K_{sp}
Silver acetate	$AgC_2H_3O_2$	2.0×10^{-3}
Silver bromide	$AgBr$	5.0×10^{-13}
Silver chloride	$AgCl$	1.8×10^{-10}
Silver chromate	Ag_2CrO_4	1.1×10^{-12}
Silver iodide	AgI	8.3×10^{-17}
Silver sulfide	Ag_2S	6×10^{-50}
Strontium carbonate	$SrCO_3$	9.3×10^{-10}
Strontium chromate	$SrCrO_4$	3.5×10^{-5}
Strontium sulfate	$SrSO_4$	2.5×10^{-7}
Zinc hydroxide	$Zn(OH)_2$	2.1×10^{-16}
Zinc sulfide	ZnS	1.1×10^{-21}

Appendix H
Formation Constants of Complex Ions at 25°C

Complex Ion	K_f
$Ag(CN)_2^-$	5.6×10^{18}
$Ag(NH_3)_2^+$	1.7×10^7
$Ag(S_2O_3)_2^{3-}$	2.9×10^{13}
$Cd(NH_3)_4^{2+}$	1.0×10^7
$Cu(CN)_2^-$	1.0×10^{16}
$Cu(NH_3)_4^{2+}$	4.8×10^{12}
$Fe(CN)_6^{4-}$	1.0×10^{35}
$Fe(CN)_6^{3-}$	9.1×10^{41}
$Ni(CN)_4^{2-}$	1.0×10^{31}
$Ni(NH_3)_6^{2+}$	5.6×10^8
$Zn(NH_3)_4^{2+}$	2.9×10^9
$Zn(OH)_4^{2-}$	2.8×10^{15}

Appendix I Standard Electrode (Reduction) Potentials in Aqueous Solution at 25°C

Cathode (Reduction) Half-Reaction	Standard Potential, $E°$ (Volts)
$Li^+(aq) + e^- \rightleftharpoons Li(s)$	−3.04
$K^+(aq) + e^- \rightleftharpoons K(s)$	−2.92
$Ca^{2+}(aq) + 2e^- \rightleftharpoons Ca(s)$	−2.76
$Na^+(aq) + e^- \rightleftharpoons Na(s)$	−2.71
$Mg^{2+}(aq) + 2e^- \rightleftharpoons Mg(s)$	−2.38
$Al^{3+}(aq) + 3e^- \rightleftharpoons Al(s)$	−1.66
$2H_2O(l) + 2e^- \rightleftharpoons H_2(g) + 2OH^-(aq)$	−0.83
$Zn^{2+}(aq) + 2e^- \rightleftharpoons Zn(s)$	−0.76
$Cr^{3+}(aq) + 3e^- \rightleftharpoons Cr(s)$	−0.74
$Fe^{2+}(aq) + 2e^- \rightleftharpoons Fe(s)$	−0.41
$Cd^{2+}(aq) + 2e^- \rightleftharpoons Cd(s)$	−0.40
$Ni^{2+}(aq) + 2e^- \rightleftharpoons Ni(s)$	−0.23
$Sn^{2+}(aq) + 2e^- \rightleftharpoons Sn(s)$	−0.14
$Pb^{2+}(aq) + 2e^- \rightleftharpoons Pb(s)$	−0.13
$Fe^{3+}(aq) + 3e^- \rightleftharpoons Fe(s)$	−0.04
$2H^+(aq) + 2e^- \rightleftharpoons H_2(g)$	0.00
$Sn^{4+}(aq) + 2e^- \rightleftharpoons Sn^{2+}(aq)$	0.15
$Cu^{2+}(aq) + e^- \rightleftharpoons Cu^+(aq)$	0.16
$ClO_4^-(aq) + H_2O(l) + 2e^- \rightleftharpoons ClO_3^-(aq) + 2OH^-(aq)$	0.17
$AgCl(s) + e^- \rightleftharpoons Ag(s) + Cl^-(aq)$	0.22
$Cu^{2+}(aq) + 2e^- \rightleftharpoons Cu(s)$	0.34
$ClO_3^-(aq) + H_2O(l) + 2e^- \rightleftharpoons ClO_2^-(aq) + 2OH^-(aq)$	0.35
$IO^-(aq) + H_2O(l) + 2e^- \rightleftharpoons I^-(aq) + 2OH^-(aq)$	0.49
$Cu^+(aq) + e^- \rightleftharpoons Cu(s)$	0.52
$I_2(s) + 2e^- \rightleftharpoons 2I^-(aq)$	0.54
$ClO_2^-(aq) + H_2O(l) + 2e^- \rightleftharpoons ClO^-(aq) + 2OH^-(aq)$	0.59
$Fe^{3+}(aq) + e^- \rightleftharpoons Fe^{2+}(aq)$	0.77
$Hg_2^{2+}(aq) + 2e^- \rightleftharpoons 2Hg(l)$	0.80
$Ag^+(aq) + e^- \rightleftharpoons Ag(s)$	0.80
$Hg^{2+}(aq) + 2e^- \rightleftharpoons Hg(l)$	0.85

Cathode (Reduction) Half-Reaction	Standard Potential, $E°$ (Volts)
$ClO^-(aq) + H_2O(l) + 2e^- \rightleftharpoons Cl^-(aq) + 2OH^-(aq)$	0.90
$2Hg^{2+}(aq) + 2e^- \rightleftharpoons Hg_2^{2+}(aq)$	0.90
$NO_3^-(aq) + 4H^+(aq) + 3e^- \rightleftharpoons NO(g) + 2H_2O(l)$	0.96
$Br_2(l) + 2e^- \rightleftharpoons 2Br^-(aq)$	1.07
$O_2(g) + 4H^+(aq) + 4e^- \rightleftharpoons 2H_2O(l)$	1.23
$Cr_2O_7^{2-}(aq) + 14H^+(aq) + 6e^- \rightleftharpoons 2Cr^{3+}(aq) + 7H_2O(l)$	1.33
$Cl_2(g) + 2e^- \rightleftharpoons 2Cl^-(aq)$	1.36
$Ce^{4+}(aq) + e^- \rightleftharpoons Ce^{3+}(aq)$	1.44
$MnO_4^-(aq) + 8H^+(aq) + 5e^- \rightleftharpoons Mn^{2+}(aq) + 4H_2O(l)$	1.49
$H_2O_2(aq) + 2H^+(aq) + 2e^- \rightleftharpoons 2H_2O(l)$	1.78
$Co^{3+}(aq) + e^- \rightleftharpoons Co^{2+}(aq)$	1.82
$S_2O_8^{2-}(aq) + 2e^- \rightleftharpoons 2SO_4^{2-}(aq)$	2.01
$O_3(g) + 2H^+(aq) + 2e^- \rightleftharpoons O_2(g) + H_2O(l)$	2.07
$F_2(g) + 2e^- \rightleftharpoons 2F^-(aq)$	2.87

Answers to Exercises

Chapter 1

1.1 0.20 g, 0.20 g **1.2** (a) 4.9 (b) 2.48 (c) 0.08 (d) 3 **1.3** (a) 1.84 nm (b) 5.67 ps (c) 7.85 mg (d) 9.7 km (e) 0.732 ms or 732 μs (f) 0.154 nm or 154 pm **1.4** (a) 39.2°C (b) 195 K **1.5** 7.87 g/cm^3; the object is made of iron **1.6** 38.4 cm^3 **1.7** 1.21×10^{-7} mm **1.8** 6.76×10^{-26} dm^3 **1.9** 3.24 m

Chapter 2

2.1 Physical properties: soft; silver color; melts at 64°C; density is 0.86 g/cm^3. Chemical properties: metal, reacts with water; reacts with oxygen; reacts with chlorine. **2.2** Fraction of potassium in sample A = 0.5596; in sample B = 0.6690; in sample C = 0.6635. The material is not a compound. **2.3** (a) Se: Group VIA, Period 4; nonmetal (b) Cs: Group IA, Period 6; metal (c) Fe: Group VIIIB, Period 4; metal (d) Cu: Group IB, Period 4; metal (e) Br: Group VIIA, Period 4; nonmetal **2.4** K_2CrO_4 **2.5** (a) calcium oxide (b) lead(II) chromate **2.6** $Tl(NO_3)_3$ **2.7** (a) ionic (b) molecular (c) molecular **2.8** (a) nitrogen monoxide (b) phosphorus trichloride (c) phosphorus pentachloride **2.9** (a) CS_2 (b) SO_3 **2.10** perbromate ion, BrO_4^- **2.11** sodium carbonate decahydrate **2.12** $Na_2S_2O_3 \cdot 5H_2O$

Chapter 3

3.1 (a) $O_2 + 2PCl_3 \longrightarrow 2POCl_3$
(b) $P_4 + 6N_2O \longrightarrow P_4O_6 + 6N_2$
(c) $2As_2S_3 + 9O_2 \longrightarrow 2As_2O_3 + 6SO_2$
(d) $Ca_3(PO_4)_2 + 4H_3PO_4 \longrightarrow 3Ca(H_2PO_4)_2$
3.2 (a) combustion reaction (b) combination reaction (c) decomposition reaction (d) metathesis reaction (e) displacement reaction
3.3 (a) weak acid
(b) weak acid
(c) strong acid
(d) strong base
3.4 (a) $2H^+(aq) + 2NO_3^-(aq) + Mg(OH)_2(s) \longrightarrow 2H_2O(l) + Mg^{2+}(aq) + 2NO_3^-(aq)$
$2H^+(aq) + Mg(OH)_2(s) \longrightarrow 2H_2O(l) + Mg^{2+}(aq)$
(b) $Pb^{2+}(aq) + 2NO_3^-(aq) + 2Na^+(aq) + SO_4^{2-}(aq) \longrightarrow PbSO_4(s) + 2Na^+(aq) + 2NO_3^-(aq)$
$Pb^{2+}(aq) + SO_4^{2-}(aq) \longrightarrow PbSO_4(s)$
3.5 $2NaI(aq) + Pb(C_2H_3O_2)_2(aq) \longrightarrow PbI_2(s) + 2NaC_2H_3O_2(aq)$
$2I^-(aq) + Pb^{2+}(aq) \longrightarrow PbI_2(s)$
3.6 $HCN(aq) + LiOH(aq) \longrightarrow H_2O(l) + LiCN(aq)$
$HCN(aq) + OH^-(aq) \longrightarrow H_2O(l) + CN^-(aq)$
3.7 $H_2SO_4(aq) + KOH(aq) \longrightarrow H_2O(l) + KHSO_4(aq)$
$KHSO_4(aq) + KOH(aq) \longrightarrow H_2O(l) + K_2SO_4(aq)$
$H^+(aq) + OH^-(aq) \longrightarrow H_2O(l)$
$HSO_4^-(aq) + OH^-(aq) \longrightarrow H_2O(l) + SO_4^{2-}(aq)$
3.8 $CaCO_3(s) + 2HNO_3(aq) \longrightarrow Ca(NO_3)_2(aq) + CO_2(g) + H_2O(l)$
$CaCO_3(s) + 2H^+(aq) \longrightarrow Ca^{2+}(aq) + CO_2(g) + H_2O(l)$

Chapter 4

4.1 (a) 46.0 amu (b) 180 amu (c) 40.0 amu (d) 58.3 amu **4.2** (a) 6.66×10^{-23} g/atom (b) 7.65×10^{-23} g/molecule **4.3** 30.9 g H_2O_2 **4.4** 0.452 mol HNO_3 **4.5** 1.2×10^{21} HCN molecules **4.6** 35.0% N, 60.0% O, 5.04% H **4.7** 17.0 g N **4.8** 40.9% C, 4.57% H, 54.5% O **4.9** SO_3 **4.10** $C_7H_6O_2$ **4.11** C_2H_4O
4.12

H_2	+	Cl_2	→	2HCl
1 molecule H_2	+	1 molecule Cl_2	→	2 molecules HCl
1 mol H_2	+	1 mol Cl_2	→	2 mol HCl
2.02 g H_2	+	70.9 g Cl_2	→	2 × 36.5 g HCl

4.13 178 g Na **4.14** 2.46 kg O_2 **4.15** 81.1 g Hg **4.16** 0.12 mol $AlCl_3$ **4.17** 11.0 g ZnS **4.18** 21.4 g; 89.3% **4.19** 0.0464 *M* NaCl **4.20** 10.1 mL **4.21** 0.0075 mol NaCl; 0.44 g NaCl **4.22** 4.0×10^2 mL **4.23** 3.38 mL **4.24** 8.4 mL **4.25** 48.4 mL $NiSO_4$ **4.26** 1.60×10^2 mL **4.27** 5.07%

Chapter 5

5.1 4.38 mm Hg **5.2** 24.1 L **4.3** 4.47 dm^3 **5.4** 5.54 dm^3 **5.5** Use $PV = nRT$. Solve for n: $n = PV/RT = P(V/RT)$. Everything within parentheses is constant; therefore, n = constant × P (or $n \propto P$). **5.6** 46.0 atm **5.7** density of He = 0.164 g/L; difference in mass

between 1 L of air and 1 L of He = 1.024 g
5.8 64.2 amu **5.9** 0.515 L CO_2 **5.10** P_{O_2} = 0.0769 atm; P_{CO_2} = 0.0311 atm; P = 0.1080 atm; X_{O_2} = 0.712 **5.11** 0.01591 mol O_2; V = 0.406 L
5.12 219 m/s **5.13** T = 52.4 K, 728 K **5.14** 9.96 s
5.15 44.0 amu **5.16** P_{vdW} = 0.992 atm; P_{ideal} = 1.000 atm (larger)

Chapter 6

6.1 1.1×10^{-17} J, 2.7×10^{-18} cal
6.2 exothermic; $q = -1170$ kJ
6.3 $2N_2H_4(l) + N_2O_4(l) \longrightarrow 3N_2(g) + 4H_2O(g)$; $\Delta H = -1049$ kJ
6.4 (a) $N_2H_4(l) + \frac{1}{2}N_2O_4(l) \longrightarrow \frac{3}{2}N_2(g) + 2H_2O(g)$; $\Delta H = -5.245 \times 10^2$ kJ
(b) $4H_2O(g) + 3N_2(g) \longrightarrow N_2O_4(l) + 2N_2H_4(l)$; $\Delta H = 1049$ kJ
6.5 164 kJ **6.6** 1.80×10^2 J **6.7** -54 kJ; $HCl(aq) + NaOH(aq) \longrightarrow NaCl(aq) + H_2O(l)$; $\Delta H = -54$ kJ
6.8 $\Delta H = -1789$ kJ **6.9** $\Delta H = 1661.9$ kJ
6.10 $\Delta H^\circ_{vap} = 44.0$ kJ **6.11** $\Delta H^\circ = -136.7$ kJ
6.12 $\Delta H^\circ = +62.0$ kJ **6.13** -178 kJ

Chapter 7

7.1 The neutral atom has 92 electrons. The U^{+2} ion has 90 electrons. **7.2** ${}^{35}_{17}Cl$ **7.3** Carbon-14 is an isotope of carbon. It has a nucleus of 6 protons and 8 neutrons; there are 6 electrons around the nucleus. **7.4** 35.453 amu
7.5 767 nm **7.6** 6.58×10^{14}/s **7.7** 2.0×10^{-19} J, 2.0×10^{-17} J, and 2.0×10^{-15} J; greatest energy: x-ray region; least energy: infrared region **7.8** 103 nm
7.9 3.38×10^{-19} J **7.10** $\lambda = 3.32$ Å **7.11** (a) The value of n must be a positive whole number. (b) The values of l range from 0 to $n - 1$. Here, l has a value greater than n. (c) The values for m_l range from $-l$ to $+l$. Here, m_l has a value greater than that of l. (d) The values for m_s are $+\frac{1}{2}$ or $-\frac{1}{2}$, not 0.

Chapter 8

8.1 (a) possible (b) possible (c) impossible. There are two electrons in a $2p$ orbital with the same spin.
(d) possible (e) impossible. Only two electrons are allowed in an s subshell. (f) impossible. Only six electrons are allowed in a p subshell.
8.2 $1s^22s^22p^63s^23p^63d^54s^2$ **8.3** $4s^24p^3$ **8.4** Group IVA, Period 6; main-group element
8.5 (↑↓) (↑↓) (↑↓)(↑↓)(↑↓) (↑↓) (↑)(↑)(↑)
1s 2s 2p 3s 3p

8.6 In order of increasing radius: Be, Mg, Na **8.7** It is more likely that 1000 kJ/mol is the ionization energy for iodine because ionization energies tend to decrease with atomic number in a column **8.8** F **8.9** (a) H_2Se (b) $CaSeO_4$

Chapter 9

9.1 $\cdot Mg \cdot + :\dot{\underset{\cdot\cdot}{O}}\cdot \longrightarrow Mg^{2+} + \left[:\ddot{\underset{\cdot\cdot}{O}}:\right]^{2-}$ **9.2** The electron configuration for Ca^{2+} is [Ar] and its Lewis symbol is Ca^{2+}. The electron configuration for S^{2-} is $[Ne]3s^23p^6$ and its Lewis symbol is $\left[:\ddot{\underset{\cdot\cdot}{S}}:\right]^{2-}$.
9.3 The electron configurations of Pb and Pb^{2+} are $[Xe]4f^{14}5d^{10}6s^26p^2$ and $[Xe]4f^{14}5d^{10}6s^2$, respectively.
9.4 $[Ar]3d^5$ **9.5** One would expect S^{2-}, because it has two additional electrons. **9.6** In order of increasing ionic radius: Mg^{2+}, Ca^{2+}, Sr^{2+} **9.7** In order of increasing ionic radius: Ca^{2+}, Cl^-, P^{3-} **9.8** C—O is the most polar bond. **9.9**

:Cl:
:F:C:F:
:Cl:

9.10 O::C::O or O=C=O

9.11 (a) [H:O:H with H above O]$^+$ (b) [:O:Cl:O:]$^-$

9.12 :Cl:Be:Cl: **9.13** S with four F atoms (F, F above; F, F below)

9.14 [:O:N:O: with :O: above N]$^-$ ⟷ [O::N:O: with :O: above N]$^-$ ⟷ [:O:N::O with :O: above N]$^-$

9.15 1.03 Å **9.16** 1.23 Å **9.17** -1304 kJ

Chapter 10

10.1 (a) trigonal pyramidal (b) bent (c) tetrahedral
10.2 T-shaped **10.3** (b), (c) **10.4** (b) **10.5** For the N atom:
N (ground state) = [He] (↑↓) (↑)(↑)(↑) and
2s 2p

N (hybridized) = [He] (↑↓)(↑)(↑)(↑)
sp^3

Each N—H bond is formed by the overlap of a $1s$ orbital of a hydrogen atom with one of the sp^3 hybrid orbitals of the nitrogen atom. **10.6** Each P—Cl bond is formed by the overlap of a phosphorus sp^3d hybrid orbital with a singly occupied $3p$ chlorine orbital. **10.7** The C atom is sp hybridized. A carbon–oxygen bond is double and is composed of a σ bond and a π bond. The σ bond is formed by the overlap of a hybrid C orbital with a $2p$ orbital of O that lies along the axis. The π bond is formed by the sidewise overlap of a $2p$ on C with a $2p$ on O. **10.8** The structural formulas for the isomers are as follows:

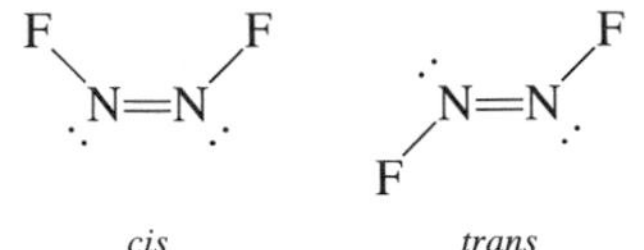

cis *trans*

These compounds exist as separate isomers. If they were interchangeable, one end of the molecule would have to remain fixed as the other rotated. This would require breaking the π bond and a considerable expenditure of energy.

10.9 KK(↑↓) (↑↓) (↑↓)(↑↓) ( ) ( )( ) ( );
σ_{2s} $\sigma^{\star}_{2s}$ π_{2p} σ_{2p} $\pi^{\star}_{2p}$ $\sigma^{\star}_{2p}$

$KK(\sigma_{2s})^2(\sigma^{\star}_{2s})^2(\pi_{2p})^4$; diamagnetic; bond order = 2

10.10 KK(↑↓) (↑↓) (↑↓)(↑↓) (↑↓) (↑)( ) ( );
σ_{2s} $\sigma^{\star}_{2s}$ π_{2p} σ_{2p} $\pi^{\star}_{2p}$ $\sigma^{\star}_{2p}$

$KK(\sigma_{2s})^2(\sigma^{\star}_{2s})^2(\pi_{2p})^4(\sigma_{2p})^2$; bond order = 3; diamagnetic

Chapter 11

11.1 1.37×10^3 kJ, 4.12×10^3 g **11.2** $P_2 = 5.30 \times 10^2$ mm Hg **11.3** $\Delta H_{vap} = 43.0$ kJ/mol
11.4 (a) Methyl chloride can be liquefied by sufficiently increasing the pressure as long as the temperature is below 144°C. (b) Oxygen can be liquefied by compression as long as the temperature is below −119°C.
11.5 (a) hydrogen bonding, dipole-dipole forces, and London forces (b) London forces (c) dipole-dipole forces and London forces **11.6** In order of increasing vapor pressure we have: butane (C_4H_{10}), propane (C_3H_8), and ethane (C_2H_6). London forces increase with increasing molecular weight. Therefore, one would expect the lowest vapor pressure to correspond to the highest molecular weight. **11.7** Hydrogen bonding is negligible in methyl chloride but strong in ethanol, which explains the lower vapor pressure of ethanol. **11.8** (a) metallic solid (b) ionic solid (c) covalent network solid (d) molecular solid **11.9** C_2H_5OH, molecular; CH_4, molecular; CH_3Cl, molecular; $MgSO_4$, ionic. In order of increasing melting point: CH_4, CH_3Cl, C_2H_5OH, and $MgSO_4$.
11.10 2 atoms **11.11** $N_A = 6.02 \times 10^{23}$/mol
11.12 5.33 Å

Chapter 12

12.1 A dental filling, made up of liquid mercury and solid silver, is a solid solution. **12.2** C_4H_9OH

12.3 Na^+ **12.4** 8.45×10^{-3} g O_2/L **12.5** 7.07 g HCl; 27.9 g H_2O **12.6** 3.09 *m* $C_6H_5CH_3$ **12.7** mole fraction toluene = 0.195; mole fraction benzene = 0.805
12.8 mole fraction methanol = 0.00550; mole fraction ethanol = 0.995 **12.9** 7.24 *m* CH_3OH **12.10** 2.96 *M* $(NH_2)_2CO$ **12.11** 2.20 *m* $(NH_2)_2CO$ **12.12** ΔP = 1.22 mm Hg; P = 155 mm Hg **12.13** 1.89×10^{-1} g CH_2OHCH_2OH **12.14** 176 amu **12.15** 124 g/mol; P_4
12.16 $\pi = 3.5$ atm **12.17** 100.078°C **12.18** $AlCl_3$

Chapter 13

13.1 $H_2CO_3(aq) + CN^-(aq) \rightleftharpoons HCN(aq) + HCO_3^-(aq)$
acid base acid base
HCN is the conjugate acid of CN^-.
13.2 The reactants are favored. **13.3** (a) PH_3 (b) HI (c) H_2SO_3 (d) H_3AsO_4 (e) HSO_4^-

13.4

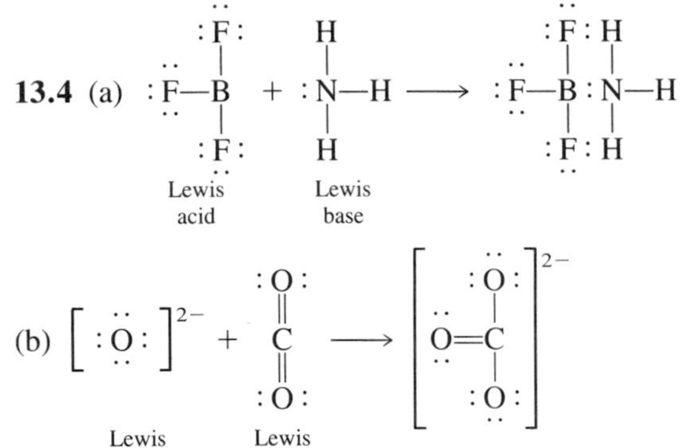

13.5 +6 **13.6** +7 **13.7** (a) will not react (b) will react
13.8 $I_2 + 10\,HNO_3 \longrightarrow 2HIO_3 + 10NO_2 + 4H_2O$
13.9 $18H^+ + 5I^- + 6MnO_4^- \longrightarrow 5IO_3^- + 6Mn^{2+} + 9H_2O$
13.10 $3H_2O_2 + 2MnO_4^- \longrightarrow 3O_2 + 2MnO_2 + 2H_2O + 2OH^-$
13.11 $4Zn + NO_3^- + 10H^+ \longrightarrow 4Zn^{2+} + NH_4^+ + 3H_2O$
13.12 Equivalent mass = 64.02 g/equiv
Molar mass = 192.0 g/mole **13.13** 0.0721g SO_2

Chapter 14

14.1 $\dfrac{\Delta[NO_2F]}{\Delta t} = -\dfrac{\Delta[NO_2]}{\Delta t}$
14.2 1.1×10^{-4} M/s **14.3** zero order in [CO]; second order in $[NO_2]$; second order overall **14.4** rate = $k[NO_2]^2$; $k = 0.70$ L/(mol · s) **14.5** (a) $[N_2O_5]_t = 1.2 \times 10^{-2}$ mol/L (b) $t = 4.80 \times 10^3$ s **14.6** half-life = 0.075 s; decreases to 50% of initial concentration in 0.075 s, to 25% in 0.150 s **14.7** $E_a = 2.08 \times 10^5$ J; k =

$5.79 \times 10^{-2}/(M^{1/2} \cdot s)$ **14.8** $2H_2O_2 \longrightarrow 2H_2O + O_2$ **14.9** bimolecular **14.10** rate = $k[NO_2]^2$ **14.11** rate = $k_1[H_2O_2][I^-]$ **14.12** rate = $k_2(k_1/k_{-1})[NO]^2[O_2]$

Chapter 15

15.1 0.41 mol CO, 0.41 mol H_2O, 0.59 mol CO_2, and 0.59 mol H_2
15.2 $K_c = \frac{[NH_3]^2[H_2O]^4}{[NO_2]^2[H_2]^7}$; $K_c = \frac{[NH_3][H_2O]^2}{[NO_2][H_2]^{7/2}}$
15.3 $CO(g) + H_2O(g) \rightleftharpoons H_2(g) + CO_2(g)$; 0.57
15.4 $K_c = 2.3 \times 10^{-4}$ **15.5** $K_p = 1.24$
15.6 $K_c = \frac{[Ni(CO)_4]}{[CO]^4}$ **15.7** products; 2.2×10^6 *M*
15.8 $Q_c = 0.67$; more CO will form **15.9** 0.0096 moles of PCl_5 **15.10** 0.11 mol H_2, 0.11 mol I_2, 0.78 mol HI **15.11** 0.86 mol PCl_5, 0.135 mol PCl_3, 0.135 mol Cl_2 **15.12** (a) in the reverse direction (b) in the reverse direction **15.13** (a) Pressure has no effect. (b) no (c) yes **15.14** high temperature **15.15** High temperatures and low pressure would give the best yield of product.

Chapter 16

16.1 Hydrogen ion concentration is 4.0×10^{-14} *M*; hydroxide ion concentration is 0.250 *M*. **16.2** basic **16.3** 1.35 **16.4** 12.40 **16.5** 6.9×10^{-4} *M*
16.6 4×10^{-4} *M* **16.7** 1.4×10^{-4}, 0.071
16.8 $[H^+] = [C_2H_3O_2^-] = 1.3 \times 10^{-3}$ *M*; pH = 2.89; 0.013 **16.9** 3.24 **16.10** pH = 1.29; $[SO_3^{2-}] = 6.3 \times 10^{-8}$ *M* **16.11** 3.3×10^{-6} **16.12** 5.3×10^{-12} *M*
16.13 (a) acidic (b) neutral (c) acidic
16.14 (a) 1.5×10^{-11} (b) 2.4×10^{-5} **16.15** 1.5×10^{-6}; 8.19 **16.16** 8.5×10^{-5} *M*, 8.5×10^{-4}
16.17 3.63 **16.18** 5.26 **16.19** 3.83 **16.20** 1.60
16.21 7.97 **16.22** 5.19

Chapter 17

17.1 (a) $K_{sp} = [Ba^{2+}][SO_4^{2-}]$ (b) $K_{sp} = [Fe^{3+}][OH^-]^3$ (c) $K_{sp} = [Ca^{2+}]^3[PO_4^{3-}]^2$ **17.2** 1.8×10^{-10}
17.3 4.5×10^{-36} **17.4** 0.68 g/L **17.5** (a) 6.3×10^{-3} mol/L (b) 4.4×10^{-5} mol/L **17.6** Precipitation is expected. **17.7** No precipitate will form. **17.8** 3.6×10^{-11} *M*; 7.2×10^{-4} % **17.9** silver cyanide **17.10** pH below 2.50 will precipitate only CuS. **17.11** 1.2×10^{-9} *M* **17.12** No precipitate will form.
17.13 0.44 mol/L

Chapter 18

18.1 3.89 J **18.2** −48.92 L; 4.94 kJ; −885.3 kJ
18.3 303 J/(mol · K)
18.4 (a) positive (b) positive (c) negative (d) positive
18.5 537 J/K **18.6** −800.7 kJ **18.7** 130.9 kJ
18.8 All four reactions are spontaneous in the direction written.
18.9 (a) $K = P_{CO_2}$ (b) $K = [Pb^{2+}][I^-]^2$ (c) $K = \frac{P_{CO_2}}{[H^+][HCO_3^-]}$ **18.10** 1.1×10^{-23} **18.11** 2.4×10^{-29}
18.12 1×10^{-1} atm **18.13** T = 671.4 K, which is lower than that for $CaCO_3$.

Chapter 19

19.1 (a) anode: $Ni(s) \longrightarrow Ni^{2+}(aq) + 2e^-$; cathode: $Ag^+(aq) + e^- \longrightarrow Ag(s)$; electrons flow from anode to cathode; Ag^+ flows to Ag, Ni^{2+} flows away from Ni.
19.2 $Zn(s)|Zn^{2+}(aq)||H^+(aq)|H_2(g)|Pt$
19.3 $Cd(s) + 2H^+(aq) \longrightarrow Cd^{2+}(aq) + H_2(g)$
19.4 -2.12×10^4 J **19.5** NO_3^- **19.6** No
19.7 1.10 V **19.8** -1.4×10^2 kJ **19.9** 2.699 V
19.10 log K_c = 18.95, $K_c = 9 \times 10^{18}$ (no significant figures) **19.11** 1.42 V **19.12** +0.30 V
19.13 4×10^{-7} *M* Ni^{2+}
19.14 (a) $K^+(l) + e^- \longrightarrow K(l)$
$Cl^-(l) \longrightarrow \frac{1}{2}Cl_2(g) + e^-$
(b) $K^+(l) + e^- \longrightarrow K(l)$
$4OH^-(l) \longrightarrow O_2(g) + 2H_2O(g) + 4e^-$
19.15 $Ag^+(aq) + e^- \longrightarrow Ag(s)$ (cathode)
$2H_2O(l) \longrightarrow O_2(g) + 4H^+(aq) + 4e^-$ (anode)
19.16 2.52×10^{-2} A **19.17** 8.66×10^{-4} g

Chapter 20

20.1 ${}^{40}_{19}K \longrightarrow {}^{40}_{20}Ca + {}^{0}_{-1}e$ **20.2** ${}^{235}_{92}U$ **20.3** (a) is stable; (b) and (c) are radioactive. **20.4** (a) positron emission (b) beta emission **20.5** (a) ${}^{40}_{20}Ca(d,p){}^{41}_{20}Ca$ (b) ${}^{12}_{6}C + {}^{2}_{1}H \longrightarrow {}^{13}_{6}C + {}^{1}_{1}H$ **20.6** ${}^{14}_{7}N$ **20.7** 3.2×10^{-5}/s **20.8** 5.26 years **20.9** 7.82×10^{-10}/s; 7.4×10^{-7} Ci **20.10** 0.200 **20.11** 1.0×10^4 years
20.12 -2.12×10^8 J; −0.514 MeV

Chapter 21

21.1 (a) Ga (b) B (c) Be **21.2** TlCl melts at 430°C; $TlCl_3$ melts at 25°C.
21.3 $Na(l) + KCl \longrightarrow NaCl(l) + K(g)$
$K(s) + O_2(g) \longrightarrow KO_2(s)$

21.4 Add NaOH; hydroxides precipitate, but $Be(OH)_2$ then dissolves to give $Be(OH)_4^{2-}$.

21.5 $CaCO_3(s) \xrightarrow{\Delta} CaO(s) + CO_2(g)$
$CaO(s) + Mg^{2+}(aq) + H_2O(l) \longrightarrow Mg(OH)_2(s) + Ca^{2+}(aq)$
$Mg(OH)_2(s) + H_2SO_4(aq) \longrightarrow MgSO_4(aq) + 2H_2O(l)$
21.6 Make solutions of the compounds. If any two are mixed and give no precipitate, one is KOH and the other is $BaCl_2$. The one that gives a precipitate with an excess of the third solution, which is $Al_2(SO_4)_3$, must be $BaCl_2$.
21.7 (a) $Mg^{2+}(aq) + 2OH^-(aq) \longrightarrow Mg(OH)_2(s)$; $Mg(OH)_2(s) + 2HCl(aq) \longrightarrow MgCl_2(aq) + 2H_2O(l)$
(b) $MgCl_2(s) \longrightarrow Mg(l) + Cl_2(g)$

Chapter 22

22.1 $PbO(s) + OH^-(aq) + H_2O(l) \longrightarrow Pb(OH)_3^-$
$Pb(OH)_3^-(aq) + OCl^-(aq) \longrightarrow PbO_2(s) + OH^-(aq) + H_2O(l) + Cl^-(aq)$
22.2 $2NaSn(OH)_3(aq) \longrightarrow Sn(s) + Na_2Sn(OH)_6(aq)$
22.3 $P_4(s) + 5O_2(g) \longrightarrow P_4O_{10}(s)$;
$P_4O_{10}(s) + 6H_2O(l) \longrightarrow 4H_3PO_4(aq)$;
$H_3PO_4(aq) + 3NaOH(aq) \longrightarrow Na_3PO_4(aq) + 3H_2O(l)$
22.4 $H_3AsO_3(aq) + 3Zn(s) + 6H^+(aq) \longrightarrow AsH_3(g) + 3Zn^{2+}(aq) + 3H_2O(l)$
22.5 $S(s) + O_2(g) \longrightarrow SO_2(g)$
$SO_2(g) + Na_2CO_3(aq) \longrightarrow Na_2SO_3(aq) + CO_2(g)$
$Na_2SO_3(aq) + S(s) \xrightarrow{\Delta} Na_2S_2O_3(aq)$
22.6 $I_2(aq) + 2S_2O_3^{2-}(aq) \longrightarrow 2I^-(aq) + S_4O_6^{2-}(aq)$
22.7 $2NaCl(aq) + 2H_2O(l) \xrightarrow{\text{electrolysis}} 2NaOH(aq) + H_2(g) + Cl_2(g)$
$3Cl_2(g) + 6NaOH(aq) \longrightarrow NaClO_3(aq) + 5NaCl(aq) + 3H_2O(l)$
$2NaClO_3(aq) + SO_2(g) + H_2SO_4(aq) \longrightarrow 2ClO_2(g) + 2NaHSO_4(aq)$
22.8 $NaBrO_3(aq) + F_2(g) + 2NaOH(aq) \longrightarrow NaBrO_4(aq) + 2NaF(aq) + H_2O(l)$
22.9 $E^\circ_{cell} = -0.27$ V **22.10** $:\ddot{F}:\ddot{Kr}:\ddot{F}:$, sp³d, linear
22.11 $XeF_2 + 2H^+ + 2e^- \longrightarrow Xe + 2HF$
$2HCl + XeF_2 \longrightarrow Cl_2 + Xe + 2HF$

Chapter 23

23.1 VF_5 **23.2** $KMnO_4$
23.3 $Cu(s) + 2H_2SO_4(l) \longrightarrow [Cu^{2+}(aq) + SO_4^{2-}(aq)] + SO_2(g) + 2H_2O(l)$
$2Cu^{2+}(aq) + 4I^-(aq) \longrightarrow 2CuI(s) + I_2(aq)$
23.4 +6
23.5 $Cr_2O_7^{2-}(aq) + 14H^+(aq) + 6I^-(aq) \longrightarrow 2Cr^{3+}(aq) + 3I_2 + 7H_2O$

23.6 $K_2[PtCl_6]$ **23.7** (a) pentaamminechlorocobalt(III) chloride (b) potassium aquapentacyanocobaltate(III) (c) pentaaquahydroxoiron(III) ion **23.8** (a) $K_4[Fe(CN)_6]$ (b) $[Co(NH_3)_4Cl_2]Cl$ (c) $PtCl_4^{2-}$ **23.9** (a) coordination isomers (b) linkage isomers (c) ionization isomers (d) hydrate isomers **23.10** $[Co(NH_3)_4(H_2O)Cl]Cl_2$; $[Co(NH_3)_4Cl_2]Cl \cdot H_2O$ is a possibility. **23.11** Geometric isomers are possible for (b) and (c). **23.12** Only (b) and (c); (b) is similar to Figure 23.15; (c) is similar to Figure 23.19. **23.13** $Co(NH_3)_6^{3+}$ has d^2sp^3 bonding and 0 unpaired electrons; CoF_6^{3-} has sp^3d^2 bonding and 4 unpaired electrons. **23.14** tetrahedral

23.15 ↑↓ ↑↓ ↑↓ (lower); ↑ ↑ (upper)

2 unpaired electrons **23.16** ↑↓ ↑↓ (lower); ↑ ↑ ↑ (upper)

23.17 530 nm, 500 nm, yes

Chapter 24

24.1 (a) 2,3-dimethylbutane (b) 3-ethyl-2-methylhexane
24.2 $CH_3CH_2-C(CH_3)_2-CH_2CH_2CH_2CH_2CH_3$
24.3 (a) 2,4-dimethyl-2-hexene (b) 3-propyl-1-hexene
24.4 $CH_3-C(CH_3)=CH-CH_2-CH(CH_3)-CH_2CH_3$
24.5 (a) *cis*: H(CH₃)C=C(H)(CH₂CH₂CH₃), H atoms on the same side; *trans*: H(CH₃)C=C(CH₂CH₂CH₃)(H), H atoms on opposite sides
(b) none **24.6** (a) propyne (b) 3-methyl-1-pentyne
24.7 (a) benzene ring–CH_2CH_3 (b) two benzene rings joined through H_2C-CH_2
24.8 2-bromobutane **24.9** 3-ethyl-3-hexanol
24.10 (a) dimethyl ether (b) methyl ethyl ether
24.11 (a) 2-pentanone (b) butanal
24.12 $5CH_3CH_2OH + 2MnO_4^- + 6H^+ \longrightarrow 5CH_3CH(=O) + 2Mn^{2+} + 8H_2O$
24.13 (a) NR (b) $CH_3CH(COOH)CH_2CH_3$

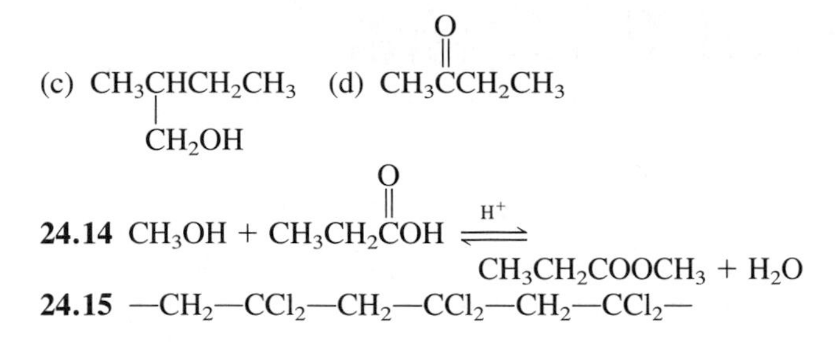

(c) $CH_3CHCH_2CH_3$ with CH_2OH (d) $CH_3C(=O)CH_2CH_3$

24.14 $CH_3OH + CH_3CH_2COH \overset{H^+}{\rightleftharpoons} CH_3CH_2COOCH_3 + H_2O$

24.15 —CH_2—CCl_2—CH_2—CCl_2—CH_2—CCl_2—

Chapter 25

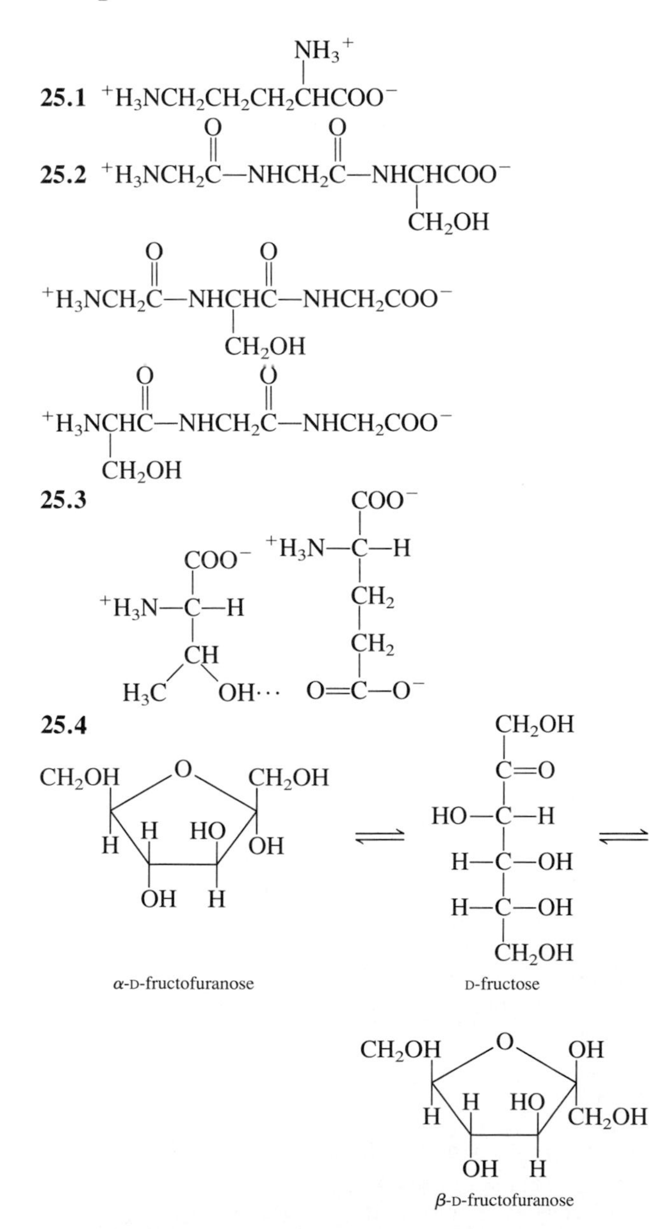

25.1 $^+H_3NCH_2CH_2CH_2CH(NH_3^+)COO^-$

25.2 $^+H_3NCH_2C(=O)$—$NHCH_2C(=O)$—$NHCH(CH_2OH)COO^-$

$^+H_3NCH_2C(=O)$—$NHCH(CH_2OH)C(=O)$—$NHCH_2COO^-$

$^+H_3NCH(CH_2OH)C(=O)$—$NHCH_2C(=O)$—$NHCH_2COO^-$

25.3

25.4

α-D-fructofuranose

D-fructose

β-D-fructofuranose

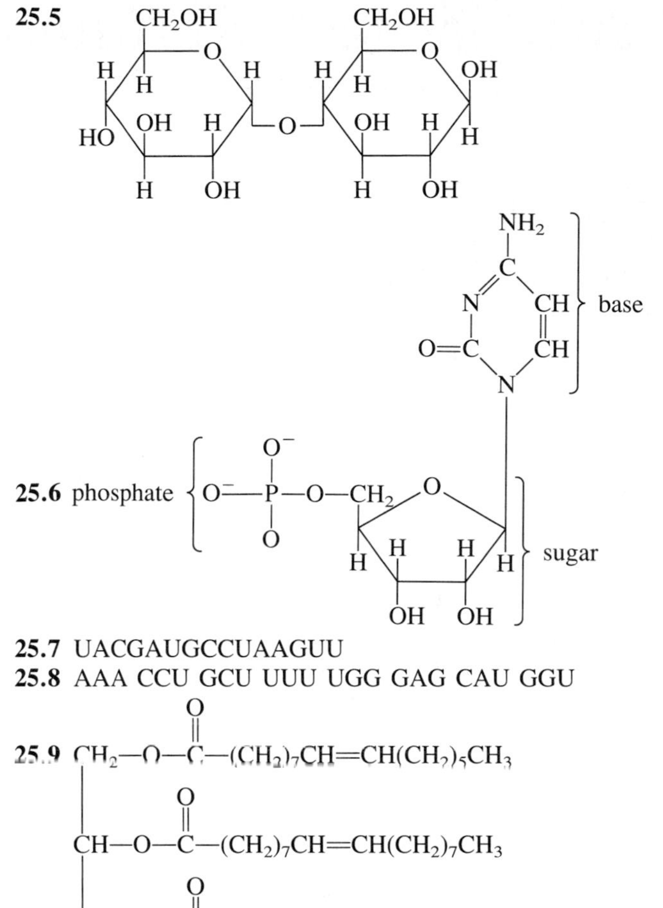

25.5

25.6

25.7 UACGAUGCCUAAGUU

25.8 AAA CCU GCU UUU UGG GAG CAU GGU

25.9 CH_2—O—$C(=O)$—$(CH_2)_7CH{=}CH(CH_2)_5CH_3$

CH—O—$C(=O)$—$(CH_2)_7CH{=}CH(CH_2)_7CH_3$

CH_2—O—$C(=O)$—$(CH_2)_7CH{=}CHCH_2CH{=}CH(CH_2)_4CH_3$

Answers to Odd-Numbered Problems

Chapter 1

1.15 4.4 g **1.17** 29.8 g **1.19** (a) 6 (b) 3 (c) 4 (d) 5 (e) 3 (f) 4 **1.21** 4.0×10^4 km **1.23** (a) 4.7 (b) 82.5 (c) 111 (d) 2.3×10^3 **1.25** 32 cm^3 **1.27** (a) 5.89 ps (b) 2.130 nm (c) 7.21 mg (d) 6.05 km **1.29** (a) 6.15×10^{-12} s (b) 3.781×10^{-6} m (c) 1.546×10^{-10} m (d) 9.7×10^{-3} g **1.31** (a) 0°C (b) −50°C (c) 20°C (d) −24°C (e) 99°F (f) −94°F **1.33** −6.0°F **1.35** 35°C, 308 K **1.37** 7.56 g/cm^3 **1.39** benzene **1.41** 1.3×10^2 g **1.43** 25.1 mL **1.45** 3.48×10^5 mg **1.47** 5.55×10^{-4} mm **1.49** 3.73×10^{17} m^3; 3.73×10^{20} L **1.51** 50.4 gallons **1.53** 3.25×10^6 g **1.55** 4.435×10^3 m **1.57** 0.43 g **1.59** 2.53 g **1.61** 114.5 g **1.63** 2.25×10^5 cm^3 **1.65** 9.9×10^1 cm^3 **1.67** (a) 7.6×10^{-1} (b) 1.63×10^1 (c) 4.76×10^2 (d) 1.12×10^{-1} **1.69** (a) 9.12 cg/mL (b) 66 pm (c) 7.1 μm (d) 56 nm **1.71** (a) 1.07×10^{-12} s (b) 5.8×10^{-6} m (c) 3.19×10^{-7} m (d) 1.53×10^{-2} s **1.73** 6170°F **1.75** 1.52×10^{3}°F **1.77** 303.0 K, 85.6°F **1.79** 907.8°C, 1181.0 K **1.81** 2.25×10^4 kg/m^3 **1.83** 2.65 g/mL **1.85** chloroform **1.87** 2.28×10^2 g **1.89** 33.24 mL **1.91** (a) 8.45×10^6 mg (b) 3.18×10^{-1} ms (c) 9.3×10^{13} nm (d) 3.71 cm **1.93** (a) 5.91×10^6 mg (b) 7.53×10^5 μg (c) 9.01×10^4 kHz (d) 4.98×10^{-4} kJ **1.95** 1.2230×10^{16} L **1.97** 3.1×10^4 L **1.99** 1.13×10^2 g **1.101** 2.4 L **1.103** 2.13×10^3 g **1.105** 2.6×10^{21} g **1.107** 34.1%, 79.2 proof **1.109** 8.10 g/cm^3

Chapter 2

2.19 (a) solid (b) liquid (c) gas (d) solid **2.21** (a) physical (b) physical (c) chemical (d) physical **2.23** heating of mercury(II) oxide = chemical change; solidification of mercury = physical change; burning of wood = chemical change **2.25** (a) physical (b) chemical (c) physical (d) physical (e) chemical **2.27** Physical properties: lustrous; blue-black crystals that vaporize readily to a violet gas. Chemical properties: combines with many metals, including aluminum. **2.29** (a) physical (b) chemical (c) physical (d) chemical (e) physical **2.31** no **2.33** (a) solution (b) substance (c) substance (d) heterogeneous mixture **2.35** (a) pure substance; bromine liquid, bromine gas (b) mixture; dioptase, wulfenite (c) mixture; liquid solution and solid (d) pure substance; nonmolten portion(solid), liquid **2.37** The ratio of O masses is 4/5 (A to B). **2.39** C, B **2.41** (a) neon (b) zinc (c) silver (d) magnesium **2.43** (a) K (b) S (c) Fe (d) Mn **2.45** 13.90 **2.47** (a) Group IVA, period 2, nonmetal (b) Group VIA, period 6, metal (c) Group VIB, period 4, metal (d) Group IIA, period 3, metal (e) Group IIIA, period 2, metalloid **2.49** (a) Te (b) Al **2.51** (a) O, oxygen (b) Li, lithium (c) Cr, chromium (d) Nd, neodymium **2.53** Differences: as a solid, each molecule contains 8 sulfur atoms, whereas gas molecules contain various numbers of sulfur atoms. Alike: molecules of solid and gas contain sulfur atoms. **2.55** 4.10×10^{22} N atoms, 2.73×10^{22} N atoms **2.57** 1.1×10^{21} NH_3 molecules **2.59** (a) N_2H_4 (b) H_2O_2 (c) C_3H_8O (d) PCl_3 **2.61** 1 Fe to 6 O **2.63** (a) $Fe(CN)_3$ (b) K_2SO_4 (c) Li_3N (d) $SrCl_2$ **2.65** (a) sodium sulfate (b) calcium oxide (c) copper(I) chloride (d) chromium(III) oxide **2.67** (a) $PbCr_2O_7$ (b) $Ba(HCO_3)_2$ (c) Cs_2O (d) $Fe(C_2H_3O_2)_2$ **2.69** (a) molecular (b) ionic (c) molecular (d) ionic **2.71** (a) dinitrogen monoxide (b) tetraphosphorus decoxide (c) arsenic trichloride (d) dichlorine heptoxide **2.73** (a) NBr_3 (b) XeO_4 (c) OF_2 (d) Cl_2O_5 **2.75** (a) BrO_3^-, bromate (b) $N_2O_2^{2-}$, hyponitrite (c) $S_2O_3^{2-}$, thiosulfate (d) AsO_4^{3-}, arsenate **2.77** sodium sulfate decahydrate **2.79** $FeSO_4 \cdot 7H_2O$ **2.81** (a) bromine, Br (b) phosphorus, P (c) gold, Au (d) carbon, C **2.83** same compound **2.85** 1+, 2+, Rb^+, Rb^{2+} **2.87** (a) bromine, Br (b) oxygen, O (c) niobium, Nb (d) fluorine, F **2.89** (a) chromium(III) ion (b) chromium(II) ion (c) copper(I) ion (d) copper(II) ion **2.91** Na_2SO_4, NaCl, $NiSO_4$, $NiCl_2$ **2.93** (a) tin(II) phosphate (b) ammonium nitrite (c) magnesium hydroxide (d) chromium(II) sulfate **2.95** (a) Hg_2Cl_2 (b) CuO (c) $(NH_4)_2Cr_2O_7$ (d) ZnS **2.97** (a) arsenic trichloride (b) selenium dioxide (c) dinitrogen trioxide (d) silicon tetrafluoride **2.99** 6.06×10^9 miles **2.101** $NiSO_4 \cdot 7H_2O$, $NiSO_4 \cdot 6H_2O$, 4.826 g **2.103** 0.8000 g, 63.54 amu, Cu

Chapter 3

3.19 12 **3.21** (a) $Sn + 2NaOH \longrightarrow Na_2SnO_2 + H_2$
(b) $8Al + 3Fe_3O_4 \longrightarrow 4Al_2O_3 + 9Fe$
(c) $2CH_3OH + 3O_2 \longrightarrow 2CO_2 + 4H_2O$
(d) $P_4O_{10} + 6H_2O \longrightarrow 4H_3PO_4$

(e) $PCl_5 + 4H_2O \longrightarrow H_3PO_4 + 5HCl$
3.23 (a) $SbCl_5 + H_2O \longrightarrow SbOCl_3 + 2HCl$
(b) $2Mg + SiO_2 \longrightarrow 2MgO + Si$
(c) $CaCl_2 + Na_2CO_3 \longrightarrow CaCO_3 + 2NaCl$
(d) $2C_2H_6 + 15O_2 \longrightarrow 12CO_2 + 6H_2O$
(e) $Al_2S_3 + 6H_2O \longrightarrow 2Al(OH)_3 + 3H_2S$
3.25 $Ca_3(PO_4)_2(s) + 3H_2SO_4(aq) \longrightarrow 2H_3PO_4(aq) + 3CaSO_4(s)$
3.27 $2NH_4Cl(aq) + Ba(OH)_2(aq) \xrightarrow{\Delta} 2NH_3(g) + BaCl_2(aq) + 2H_2O(l)$
3.29 (a) displacement (b) decomposition
(c) combustion (d) metathesis (e) combination
3.31 $C_{12}H_{22}O_{11}(s) \xrightarrow{\Delta} 11H_2O(g) + 12C(s)$, decomposition
3.33 $Mg(OH)_2(s) + 2HCl(aq) \longrightarrow 2H_2O(l) + MgCl_2(aq)$, metathesis
3.35 (a) weak acid (b) strong base (c) strong acid
(d) weak acid
3.37 (a) $HF(aq) + OH^-(aq) \longrightarrow F^-(aq) + H_2O(l)$
(b) $Ag^+(aq) + Br^-(aq) \longrightarrow AgBr(s)$
(c) $S^{2-}(aq) + 2H^+(aq) \longrightarrow H_2S(g)$
(d) $OH^-(aq) + NH_4^+(aq) \longrightarrow NH_3(g) + H_2O(l)$
3.39 $Pb(NO_3)_2(aq) + Na_2SO_4(aq) \longrightarrow PbSO_4(s) + 2NaNO_3(aq)$;
$Pb^{2+}(aq) + SO_4^{2-}(aq) \longrightarrow PbSO_4(s)$
3.41 (a) insoluble (b) soluble (c) insoluble
(d) soluble
3.43 (a) $FeSO_4(aq) + NaCl(aq) \longrightarrow$ *NR*
(b) $Na_2CO_3(aq) + MgBr_2(aq) \longrightarrow MgCO_3(s) + 2NaBr(aq)$;
$CO_3^{2-}(aq) + Mg^{2+}(aq) \longrightarrow MgCO_3(s)$
(c) $MgSO_4(aq) + 2NaOH(aq) \longrightarrow Mg(OH)_2(s) + Na_2SO_4(aq)$;
$Mg^{2+}(aq) + 2OH^-(aq) \longrightarrow Mg(OH)_2(s)$
(d) $NiCl_2(aq) + NaBr(aq) \longrightarrow$ *NR*
3.45 (a) $Ba(NO_3)_2(aq) + Li_2SO_4(aq) \longrightarrow BaSO_4(s) + 2LiNO_3(aq)$;
$Ba^{2+}(aq) + SO_4^{2-}(aq) \longrightarrow BaSO_4(s)$
(b) $NaBr(aq) + Ca(NO_3)_2(aq) \longrightarrow$ *NR*
(c) $Al_2(SO_4)_3(aq) + 6NaOH(aq) \longrightarrow 2Al(OH)_3(s) + 3Na_2SO_4(aq)$;
$2Al^{3+}(aq) + 6OH^-(aq) \longrightarrow 2Al(OH)_3(s)$
(d) $3CaBr_2(aq) + 2Na_3PO_4(aq) \longrightarrow Ca_3(PO_4)_2(s) + 6NaBr(aq)$;
$3Ca^{2+}(aq) + 2PO_4^{3-}(aq) \longrightarrow Ca_3(PO_4)_2(s)$
3.47 (a) $NaOH(aq) + HNO_3(aq) \longrightarrow NaNO_3(aq) + H_2O(l)$;
$H^+(aq) + OH^-(aq) \longrightarrow H_2O(l)$
(b) $2HCl(aq) + Ba(OH)_2(aq) \longrightarrow 2H_2O(l) + BaCl_2(aq)$;
$H^+(aq) + OH^-(aq) \longrightarrow H_2O(l)$
(c) $2HC_2H_3O_2(aq) + Ca(OH)_2(s) \longrightarrow Ca(C_2H_3O_2)_2(aq) + 2H_2O(l)$;
$2HC_2H_3O_2(aq) + Ca(OH)_2(s) \longrightarrow 2H_2O(l) + Ca^{2+}(aq) + 2C_2H_3O_2^-(aq)$
(d) $NH_3(aq) + HNO_3(aq) \longrightarrow NH_4NO_3(aq)$;
$H^+(aq) + NH_3(aq) \longrightarrow NH_4^+(aq)$

3.49 (a) $2HBr(aq) + Ca(OH)_2(aq) \longrightarrow CaBr_2(aq) + 2H_2O(l)$;
$H^+(aq) + OH^-(aq) \longrightarrow H_2O(l)$
(b) $Al(OH)_3(s) + 3HNO_3(aq) \longrightarrow Al(NO_3)_3(aq) + 3H_2O(l)$;
$3H^+(aq) + Al(OH)_3(s) \longrightarrow 3H_2O(l) + Al^{3+}(aq)$
(c) $2HCN(aq) + Ca(OH)_2(aq) \longrightarrow Ca(CN)_2(aq) + 2H_2O(l)$
$HCN(aq) + OH^-(aq) \longrightarrow H_2O(l) + CN^-(aq)$
(d) $LiOH(aq) + HCN(aq) \longrightarrow H_2O(l) + LiCN(aq)$
$HCN(aq) + OH^-(aq) \longrightarrow H_2O(l) + CN^-(aq)$
3.51 (a) $2KOH(aq) + H_3PO_4(aq) \longrightarrow 2H_2O(l) + K_2HPO_4(aq)$
$H_3PO_4(aq) + 2OH^-(aq) \longrightarrow 2H_2O(l) + HPO_4^{2-}(aq)$
(b) $3H_2SO_4(aq) + 2Al(OH)_3(s) \longrightarrow Al_2(SO_4)_3(aq) + 6H_2O(l)$;
$3H^+(aq) + Al(OH)_3(s) \longrightarrow Al^{3+}(aq) + 3H_2O(l)$
(c) $2HC_2H_3O_2(aq) + Ca(OH)_2(aq) \longrightarrow 2H_2O(l) + Ca(C_2H_3O_2)_2(aq)$;
$HC_2H_3O_2(aq) + OH^-(aq) \longrightarrow H_2O(l) + C_2H_3O_2^-(aq)$
(d) $H_2SO_3(aq) + NaOH(aq) \longrightarrow NaHSO_3(aq) + H_2O(l)$
$H_2SO_3(aq) + OH^-(aq) \longrightarrow HSO_3^-(aq) + H_2O(l)$
3.53 $2H_2SO_3(aq) + Ca(OH)_2(aq) \longrightarrow Ca(HSO_3)_2(aq) + 2H_2O(l)$;
$Ca(HSO_3)_2(aq) + Ca(OH)_2(aq) \longrightarrow 2CaSO_3(s) + 2H_2O(l)$;
$H_2SO_3(aq) + OH^-(aq) \longrightarrow H_2O(l) + HSO_3^-(aq)$;
$2Ca^{2+}(aq) + 2HSO_3^-(aq) + 2OH^-(aq) \longrightarrow 2CaSO_3(s) + 2H_2O(l)$
3.55 (a) $CaS(aq) + 2HBr(aq) \longrightarrow H_2S(g) + CaBr_2(aq)$;
$2H^+(aq) + S^{2-}(aq) \longrightarrow H_2S(g)$
(b) $MgCO_3(s) + 2HNO_3(aq) \longrightarrow CO_2(aq) + H_2O(l) + Mg(NO_3)_2(aq)$;
$2H^+(aq) + MgCO_3(s) \longrightarrow CO_2(aq) + H_2O(l) + Mg^{2+}(aq)$
(c) $K_2SO_3(aq) + H_2SO_4(aq) \longrightarrow SO_2(g) + H_2O(l) + K_2SO_4(aq)$;
$2H^+(aq) + SO_3^{2-}(aq) \longrightarrow SO_2(g) + H_2O(l)$
3.57 $FeS(s) + 2HCl(aq) \longrightarrow H_2S(g) + FeCl_2(aq)$;
$FeS(s) + 2H^+(aq) \longrightarrow Fe^{2+}(aq) + H_2S(g)$
3.59 (a) $2C_2H_6 + 7O_2 \longrightarrow 4CO_2 + 6H_2O$
(b) $P_4O_6 + 6H_2O \longrightarrow 4H_3PO_3$
(c) $4KClO_3 \longrightarrow 3KClO_4 + KCl$
(d) $(NH_4)_2SO_4 + 2NaOH \longrightarrow 2NH_3 + 2H_2O + Na_2SO_4$
(e) $2NBr_3 + 3NaOH \longrightarrow N_2 + 3NaBr + 3HOBr$
3.61 When metallic iron is added to a solution of copper(II) sulfate, insoluble copper metal and a solution of iron(II) sulfate are produced.
3.63 $4NH_3(g) + 5O_2(g) \xrightarrow{Pt} 4NO(g) + 6H_2O(g)$
3.65 (a) $(NH_4)_2Cr_2O_7(s) \xrightarrow{\Delta} N_2(g) + 4H_2O(g) + Cr_2O_3(s)$
(b) $NH_4NO_2(aq) \xrightarrow{\Delta} N_2(g) + 2H_2O(g)$
(c) $K_2CrO_4(aq) + Pb(NO_3)_2(aq) \longrightarrow PbCrO_4(s) + 2KNO_3(aq)$

(d) $NH_3(g) + HCl(g) \longrightarrow NH_4Cl(s)$
(e) $2Al(s) + 3H_2SO_4(aq) \longrightarrow Al_2(SO_4)_3(aq) + 3H_2(g)$
3.67 (a) decomposition (b) decomposition
(c) metathesis (d) combination (e) displacement
3.69 $Mg(s) + 2HBr(aq) \longrightarrow H_2(g) + MgBr_2(aq)$;
$Mg(s) + 2H^+(aq) \longrightarrow H_2(g) + Mg^{2+}(aq)$
3.71 $NiSO_4(aq) + 2LiOH(aq) \longrightarrow Ni(OH)_2(s) + Li_2SO_4(aq)$;
$Ni^{2+}(aq) + 2OH^-(aq) \longrightarrow Ni(OH)_2(s)$
3.73 (a) $LiOH(aq) + HCN(aq) \longrightarrow H_2O(l) + LiCN(aq)$;
$OH^-(aq) + HCN(aq) \longrightarrow H_2O(l) + CN^-(aq)$
(b) $Li_2CO_3(aq) + 2HNO_3(aq) \longrightarrow 2LiNO_3(aq) + H_2O(l) + CO_2(g)$
$CO_3^{2-}(aq) + 2H^+(aq) \longrightarrow H_2O(l) + CO_2(g)$
(c) $LiCl(aq) + AgNO_3(aq) \longrightarrow LiNO_3(aq) + AgCl(s)$
$Cl^-(aq) + Ag^+(aq) \longrightarrow AgCl(s)$
(d) $LiCl(aq) + MgSO_4(aq) \longrightarrow$ *NR*
3.75 (a) $Sr(OH)_2(aq) + 2HC_2H_3O_2(aq) \longrightarrow Sr(C_2H_3O_2)_2(aq) + 2H_2O(l)$;
$OH^-(aq) + HC_2H_3O_2(aq) \longrightarrow H_2O(l) + C_2H_3O_2^-(aq)$
(b) $NH_4I(aq) + CsCl(aq) \longrightarrow$ *NR*
(c) $NaNO_3(aq) + CsCl(aq) \longrightarrow$ *NR*
(d) $NH_4I(aq) + AgNO_3(aq) \longrightarrow AgI(s) + NH_4NO_3(aq)$
$I^-(aq) + Ag^+(aq) \longrightarrow AgI(s)$
3.77 (a) $BaCl_2(aq) + CuSO_4(aq) \longrightarrow BaSO_4(s) + CuCl_2(aq)$;
$BaSO_4$ is filtered off, and the solution evaporated to give solid $CuCl_2$
(b) $CaCO_3(s) + 2HC_2H_3O_2(aq) \longrightarrow CO_2(g) + H_2O(l) + Ca(C_2H_3O_2)_2(aq)$;
evaporate H_2O to leave $Ca(C_2H_3O_2)_2(s)$
(c) $Na_2SO_3(s) + 2HNO_3(aq) \longrightarrow SO_2(g) + H_2O(l) + 2NaNO_3(aq)$;
evaporate H_2O to leave $NaNO_3(s)$
(d) $Mg(OH)_2(s) + 2HCl(aq) \longrightarrow 2H_2O(l) + MgCl_2(aq)$;
evaporate H_2O to leave $MgCl_2(s)$
3.79 $Pb(NO_3)_2$; Cs_2SO_4;
$Pb(NO_3)_2(aq) + Cs_2SO_4(aq) \longrightarrow PbSO_4(s) + 2CsNO_3(aq)$;
$Pb^{2+}(aq) + SO_4^{2-}(aq) \longrightarrow PbSO_4(s)$;
lead(II) sulfate; cesium nitrate;
$Pb(NO_2)_2(aq) + Na_2SO_4(aq) \longrightarrow PbSO_4(s) + 2NaNO_2(aq)$
3.81 $CaBr_2(aq) + Cl_2(g) \longrightarrow Br_2(l) + CaCl_2(aq)$;
$2Br^-(aq) + Cl_2(g) \longrightarrow Br_2(l) + 2Cl^-(aq)$; 5.67×10^3 g

Chapter 4

4.15 (a) 32.0 amu (b) 137 amu (c) 138 amu (d) 366 amu **4.17** 80.17 g/mol **4.19** (a) 3.82×10^{-23} g/atom (b) 5.33×10^{-23} g/atom (c) 8.38×10^{-23} g/molecule (d) 2.09×10^{-22} g/unit **4.21** 1.23×10^{-22} g/molecule **4.23** (a) 3.4 g Na (b) 19.0 g S (c) 1.40×10^2 g CH_3Cl (d) 4.8×10^3 g Na_2SO_3
4.25 33.6 g **4.27** (a) 0.286 mol (b) 0.0441 mol (c) 1.3 mol (d) 0.479 mol **4.29** 5.81×10^{-3} mol $CaSO_4$; 1.16×10^{-2} mol H_2O; for every mole of $CaSO_4$ there are two moles of H_2O. **4.31** (a) 6.47×10^{23} atoms (b) 2.41×10^{23} atoms (c) 1.5×10^{24} molecules (d) 2.96×10^{23} units (e) 6.58×10^{22} SO_4^{2-} ions **4.33** 2.97×10^{19} molecules **4.35** 86.27%
4.37 20.2% **4.39** 6.56×10^{-1} kg **4.41** 73.9% Al, 26.1% Mg
4.43 (a) 42.9% C, 57.1% O
(b) 27.3% C, 72.7% O
(c) 24.7% K, 34.8% Mn, 40.5% O
(d) 32.2% Co, 15.3% N, 52.5% O
4.45 12.2% C, 28.9% F, 0.511% H, 40.5% Br, 18.0% Cl
4.47 Ethanol **4.49** 38.8% C, 9.79% H, 51.5% O
4.51 OsO_4 **4.53** K_2MnO_4 **4.55** $C_3H_4O_2$
4.57 $C_4H_{12}N_2$ **4.59** $C_2H_2O_4$
4.61

C_2H_4	+	$3O_2$	$\longrightarrow$	$2CO_2$	+	$2H_2O$
1 molecule C_2H_4	+	3 molecules O_2	$\longrightarrow$	2 molecules CO_2	+	2 molecules H_2O
1 mole C_2H_4	+	3 moles O_2	$\longrightarrow$	2 moles CO_2	+	2 moles H_2O
28.052 g C_2H_4	+	3×32.00 g O_2	$\longrightarrow$	2×44.01 g CO_2	+	2×18.016 g H_2O

4.63 5.48 g NO_2 **4.65** 1.46×10^5 g W **4.67** 22.4 g CS_2 **4.69** 9.306 g NO_2 **4.71** KO_2 is the limiting reactant; 0.11 moles of oxygen **4.73** 40.5 g CH_3OH; H_2 remains; 5.1 g H_2 **4.75** 9.07 g **4.77** 2.61 g is the theoretical yield; 80.5% **4.79** 1.36 M **4.81** 0.101 M
4.83 1.25 L **4.85** 35.8 mL **4.87** 4.8×10^{-5} moles
4.89 0.33 g **4.91** 1.6×10^2 mL **4.93** 4.0 mL
4.95 4.19 mL; Take 4.19 mL of the 68.0% HNO_3 solution and dilute to 425 mL. **4.97** 50.9 mL nitric acid
4.99 81.3 mL **4.101** 2.89% **4.103** 49.5% C, 5.19% H, 28.9% N, 16.5% O **4.105** $C_6H_4Cl_2$
4.107 35.0% Cl, $AuCl_3$ **4.109** C_4H_4S, C_4H_4S
4.111 616 amu **4.113** 89.3% **4.115** 58.2%
4.117 60.3 g **4.119** 0.0203 M $CaCl_2$, 0.0203 M Ca^{2+}, 0.0406 M Cl^- **4.121** 329 mL **4.123** 0.610 M
4.125 9.66% **4.127** 1.31×10^3 g CaC_2 **4.129** 16.2%
4.131 1.45×10^{26} Fe atoms **4.133** 0.0122 L
4.135 0.975 mol/L **4.137** 69.71 g/mol, gallium
4.139 $P_4O_{10} + 6H_2O \longrightarrow 4H_3PO_4$; 55.7 g P_4O_{10}
4.141 85.23% **4.143** 0.0971 L **4.145** 61.7%

Chapter 5

5.23 4.38 mmHg **5.25** 1.22 g/mL **5.27** 3.90 L
5.29 679 L **5.31** 3.32×10^{-4} kPa **5.33** 2.50 mL
5.35 0.65 L **5.37** 156 K **5.39** 36.7 mL **5.41** 1 volume **5.43** PV = nRT; V = nRT/P; if temperature and moles remain constant, V = constant × 1/P
5.45 7.55 atm **5.47** 44.9 L **5.49** 143°C **5.51** 1.72 g/L **5.53** 2.38 g/L **5.55** 45.1 amu **5.57** 160 g/mol

5.59 Gas density depends on molecular weight or average molecular weight of a mixture. Thus, the density of a gas of NH_4Cl would be greater than that of a mixture of NH_3 and HCl, since NH_3 and HCl have lower molecular weights than NH_4Cl. **5.61** 6.80 g Zn **5.63** 0.544 L CO_2 **5.65** 10.0 atm **5.67** $P(O_2) = 3.80 \times 10^{-3}$ atm; P(He) = 0.012 atm; P(total) = 0.016 atm **5.69** $P(CO_2) =$ 494 mmHg; $P(H_2) =$ 190 mmHg; P(HCl) = 41 mmHg; P(HF) = 21 mmHg; $P(SO_2) =$ 13 mmHg; $P(H_2S) =$ 0.8 mmHg **5.71** 6.34 g **5.73** At $T = 298$ K, $u = 5.15 \times 10^2$ m/s; at $T = 398$ K, $u = 5.95 \times 10^2$ m/s; graph as in Figure 5.14. **5.75** 1.53×10^2 m/s **5.77** 6.40×10^3 K (6.13×10^{3}°C) **5.79** rate N_2/rate O_2 = 1.069 **5.81** 4.6 s **5.83** 146 amu **5.85** 0.8250 atm (0.8328 atm from ideal gas law) **5.87** At 1.00 atm: 22.4 L from ideal gas law, 22.2 L from van der Waals equation; at 10.0 atm: 2.24 L from ideal gas law, 2.08 L from van der Waals equation. **5.89** 80.7 cm^3 **5.91** 165 mL **5.93** 6.5 dm^3 **5.95** 3.01×10^{20} atoms **5.97** 28.979 amu **5.99** 1.2×10^3 g LiOH **5.101** −95°C **5.103** 1.0043 **5.105** 28.07 g/mol **5.107** $P(O_2) = 0.167$ atm; P(CO) = 0.333 atm **5.109** 0.710 g/L **5.111** 0.194 g Na_2O_2 **5.113** 76% $CaCO_3$, 24% $MgCO_3$

Chapter 6

6.25 kg · m/s² **6.27** 68.4 kcal **6.29** 4.2×10^5 J, 9.9×10^4 cal **6.31** 6.28×10^{-21} J **6.33** exothermic, −939 kJ **6.35** $HgO(s) \longrightarrow Hg(l) + \frac{1}{2}O_2(g); \Delta H = 90.8$ kJ **6.37** +128 kJ **6.39** −430.4 kJ **6.41** −23.75 kJ **6.43** −476 kJ **6.45** 6.1×10^4 J **6.47** 4.29°C **6.49** 20.5 kJ **6.51** -1.36×10^3 kJ **6.53** −187.8 kJ **6.55** −906.3 kJ **6.57** −137 kJ **6.59** 43 kJ **6.61** −1125 kJ **6.63** −835 kJ **6.65** −75.2 kJ **6.67** −148 kJ/mole **6.69** 1.59×10^3 J/g **6.71** 226 J, 31.6 m/s **6.73** 21.4 kJ **6.75** −255 kJ/mol **6.77** 0.128 J/g°C **6.79** 0.383 J/g°C **6.81** −23.6 kJ/mol **6.83** −871.7 kJ/mol **6.85** −20 kJ **6.87** 206.2 kJ **6.89** 178 kJ **6.91** −2225 kJ/mol **6.93** −22 kJ **6.95** −113 kJ **6.97** (a) −1.88 kJ (b) 11.9% **6.99** 1.07×10^3 g LiOH

Chapter 7

7.23 9.05×10^{-3} kg **7.25** 63, 60 **7.27** $^{70}_{34}Se$ **7.29** Cl-35: 18 neutrons, 17 protons, 17 electrons; Cl-37: 20 neutrons, 17 protons, 17 electrons **7.31** $^{69}_{31}Ga^{3+}$ **7.33** 10.812 amu **7.35** 24.31 amu **7.37** Ag-106: 0.518; Ag-108: 0.582 **7.39** 1.9×10^2 s **7.41** 238.9 m **7.43** 6.45×10^{14}/s **7.45** 8.316×10^{-28} J **7.47** 3.72×10^{-19} J **7.49** 1.60×10^{14}/s **7.51** 1.22×10^{-7} m **7.53** 7 **7.55** 2.50×10^{-19} J **7.57** 1.08 Å **7.59** 5.82×10^7 m/s **7.61** $l = 0, 1, 2, 3$; $m_l = -3, -2, -1, 0, 1, 2, 3$ **7.63** 3, 7 **7.65** (a) 3*p* (b) 4*d* (c) 4*s* (d) 5*f* **7.67** (a) impossible (b) impossible (c) possible (d) impossible (e) possible **7.69** -1.6×10^{-19} C; 2, 4, 6, 7 **7.71** 6.51×10^{14}/s, 4.31×10^{-19} J **7.73** 13 protons, 15 neutrons; 20 protons, 21 neutrons; 28 protons, 31 neutrons **7.75** $^{196}_{78}Pt$ **7.77** 0.575 **7.79** 31.972 amu, 33.968 amu, 31.971 amu, 33.968 amu; 1, 1, 2, 2 **7.81** 6.55×10^{14}/s **7.83** 1.640×10^{-7} m; near UV **7.85** 5.01×10^{-7} m **7.87** 5.26×10^{28} **7.89** 3.59×10^{-19} J, 2.16×10^2 kJ/mol **7.91** 1.94×10^{-11} m

Chapter 8

8.25 (a) allowed; $1s^22s^12p^3$ (b) not allowed; electron spins must be in opposite directions in the same orbital (c) allowed; $1s^22s^22p^4$ (d) not allowed; 2*s* subshell can hold no more than 2 electrons **8.27** (a) impossible, 2*s* subshell can hold no more than 2 electrons
(b) possible
(c) impossible, 2*p* subshell can hold no more than 6 electrons
(d) possible

8.29
(↑↓) (↓)( )( )
(↑↓) (↑)( )( )
(↑↓) ( )(↓)( )
(↑↓) ( )(↑)( )
(↑↓) ( )( )(↓)
(↑↓) ( )( )(↑)
1*s* 2*p*

8.31 $1s^22s^22p^63s^23p^3$ **8.33** $1s^22s^22p^63s^23p^63d^34s^2$ **8.35** $4s^24p^5$ **8.37** $3d^24s^2$ **8.39** Group IIIA, period = 6, main-group element

8.41 [Ar] (↑↓)(↑↓)(↑↓)(↑)(↑) (↑↓)
3*d* 4*s*

8.43 (↑↓) (↑↓) (↑↓)(↑↓)(↑↓) (↑↓) (↑↓)(↑↓)(↑↓) (↑)
1*s* 2*s* 2*p* 3*s* 3*p* 4*s*
Paramagnetic

8.45 F, Cl, S **8.47** Ca, Mg, S **8.49** (a) Cl (b) Se **8.51** TeO_2, TeO_3 **8.53** $1s^22s^22p^63s^23p^63d^{10}4s^24p^65s^2$ **8.55** $6s^26p^4$

8.57 (↑↓) (↑↓) (↑↓)(↑↓)(↑↓) (↑↓) (↑↓)(↑↓)(↑↓) (↑↓)(↑↓)(↑↓)(↑↓)(↑↓)
1*s* 2*s* 2*p* 3*s* 3*p* 3*d*
(↑↓) (↑)(↑)( )
4*s* 4*p*

8.59 [Rn] $5f^{14}6d^{10}7s^27p^2$, metal, eka-PbO or eka-PbO_2
8.61 370 kJ/mol
8.63 [Kr] (↑)(↑)(↑)(↑)( ) (↑) — $4d$, $5s$
8.65 (a) Cl_2 (b) Na (c) Sb (d) Ar
8.67 $1s^22s^22p^63s^23p^63d^34s^2$, Group VB, Period 4, *d*-block transition element
8.69 $Ba(s) + 2H_2O(l) \longrightarrow H_2(g) + Ba(OH)_2(aq)$; 447 mL **8.71** RaO, 93.4% Ra **8.73** 108 J
8.75 1.313×10^3 kJ/mol H **8.77** −639 kJ/mol

Chapter 9

9.19 (a) ·Ba· (b) Ba^{2+} (c) :Ï· (d) [:Ï:]⁻
9.21 (a) Na· + ·I: ⟶ Na^+ + [:I:]⁻
(b) Na· + ·S· + ·Na ⟶ $2Na^+$ + [:S:]²⁻
9.23 (a) $1s^22s^22p^63s^2$, ·Mg·
(b) $1s^22s^22p^6$, Mg^{2+}
(c) $1s^22s^22p^63s^23p^63d^{10}4s^24p^6$, [:Se:]²⁻
(d) $1s^22s^22p^63s^23p^63d^{10}4s^24p^6$, [:Br:]⁻
9.25 (a) [Kr] $4d^{10}5s^25p^2$ (b) [Kr] $4d^{10}5s^2$
9.27 (a) [Ar] $3d^8$ (b) [Ar] $3d^7$
9.29 (a) Rb^+, Rb. The rubidium ion has one less shell of electrons, so it is smaller. (b) Se, Se^{2-}. The electron–electron repulsion is greater in the anion, so the valence orbitals expand to give a larger radius. **9.31** S^{2-}, Se^{2-}, Te^{2-}. The radius increases with increasing number of filled shells. **9.33** Na^+, F^-, N^{3-}. The ions are isoelectronic, so the atomic radius increases with decreasing nuclear charge.

9.35 2H· + :S· ⟶ :S:H (with H below) — Lone (nonbonding) electron pairs; Bonding electron pairs

9.37 $SiCl_4$ **9.39** (a) Cs, Ba, Sr (b) Ca, Ga, Ge (c) As, P, S **9.41** least polar is N—Cl bond
9.43 (a) $\overset{\delta+}{H}$—$\overset{\delta-}{Se}$ (b) $\overset{\delta+}{P}$—$\overset{\delta-}{Cl}$ (c) nonpolar
9.45 (a) :Br—Br: (b) H—Se—H
(c) :F—S(—O:)—F: (d) H—O—S(:O:)(:O:)—O—H
9.47 (a) :P≡P: (b) :O=C=Se:
(c) :O=C(—Br:)—Br: (d) H—O—N=O:

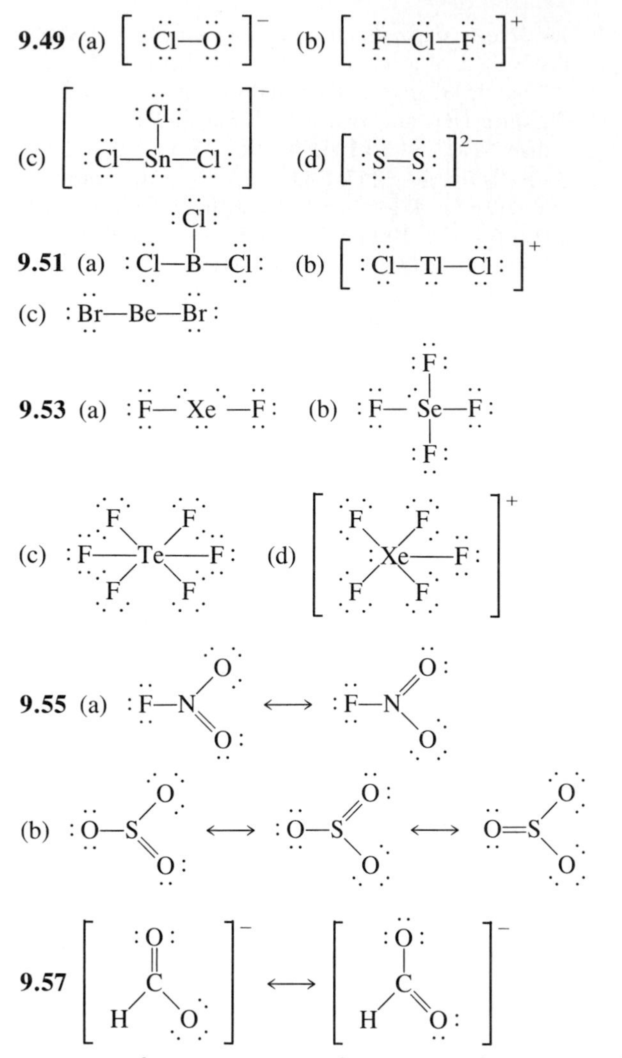

9.59 1.74 Å **9.61** (a) 1.14 Å (b) 2.03 Å (c) 2.13 Å (d) 1.83 Å **9.63** 1.47 Å, methylamine; 1.16 Å, acetonitrile **9.65** −78 kJ **9.67** (a) ionic, SrO, strontium oxide (b) covalent, CBr_4, carbon tetrabromide (c) ionic, GaF_3, gallium(III) fluoride (d) covalent, NBr_3, nitrogen tribromide

9.69 [:O—As(:O:)(:O:)—O:]³⁻, $Pb_3(AsO_4)_2$
9.71 H—O—I(—O:)—O: **9.73** [H—N—H]⁻
9.75 [:O=N=O:]⁺ **9.77** (a) :Cl—Se(:O:)—Cl:
(b) :Se=C=Se:

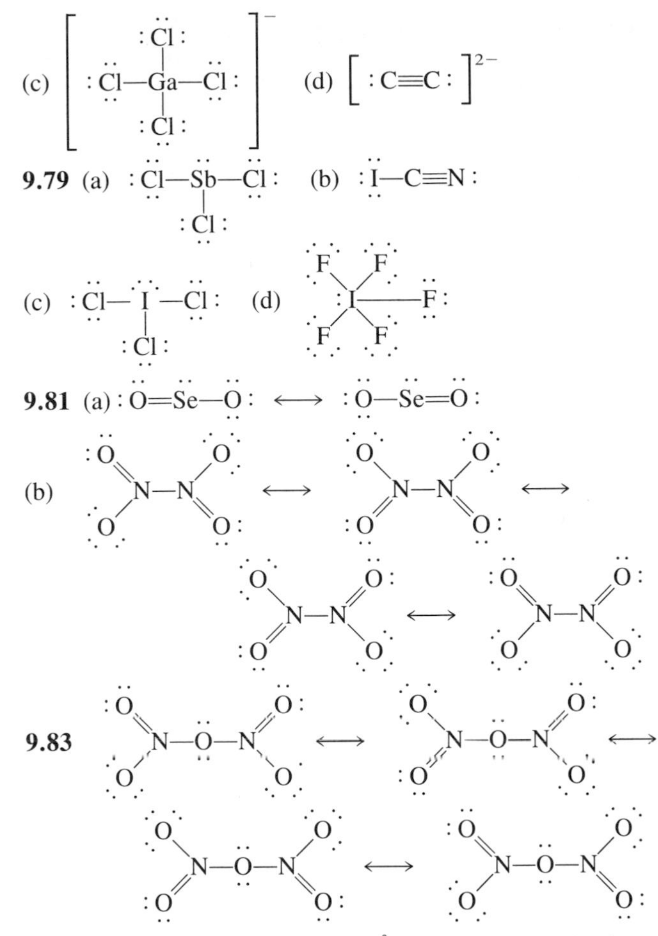

The outer N—O bonds are 1.18 Å, and the inner N—O bonds are 1.36 Å.

9.85 −134 kJ **9.87** −160 kJ

9.89 $Mg(ClO_4)_2$, magnesium perchlorate,

Mg^{2+} and $\left[\begin{array}{c} :\ddot{O}: \\ | \\ :\ddot{O}-\ddot{Cl}-\ddot{O}: \\ | \\ :\ddot{O}: \end{array}\right]^-$ **9.91** $\begin{array}{l} H \\ \quad \backslash \\ \quad\; C{=}\ddot{O}: \\ \quad / \\ :\ddot{F}: \end{array}$

9.93 261 amu; the compound is volatile, so is molecular;

$$\begin{array}{c} :\ddot{Cl}: \\ | \\ :\ddot{Cl}-Sn-\ddot{Cl}: \\ | \\ :\ddot{Cl}: \end{array}$$

9.95 860 kJ/mol compared to table value of 887 kJ/mol

9.97 3.4 **9.99** 3.47

Chapter 10

10.19 (a) tetrahedral (b) bent (c) trigonal pyramidal (d) trigonal planar **10.21** (a) linear (b) bent (c) bent (d) bent **10.23** (a) trigonal bipyramidal (b) T-shaped (c) square pyramidal (d) seesaw **10.25** (a) trigonal bipyramidal (b) octahedral (c) linear (d) square planar **10.27** (a) trigonal pyramidal (b) bent **10.29** (a) has dipole (b) has dipole (c) zero dipole (octahedral) (d) zero dipole (linear) **10.31** (a) sp^2 (b) sp (c) sp^3 (d) sp^2 **10.33** (a) Two single bonds and no lone pair on Hg suggests sp hybridization. An Hg—Cl bond is formed by overlapping an Hg hybrid orbital with a $3p$ orbital of Cl. (b) The presence of three single bonds and one lone pair on P suggests sp^3 hybridization. Each hybrid orbital overlaps a $3p$ orbital of a Cl atom to form a P—Cl bond. The fourth hybrid orbital contains a lone pair. **10.35** (a) sp^3d^2 (b) sp^3d (c) sp^3d (d) sp^3d **10.37** The P atom in PCl_6^- has 6 single bonds around it and no lone pairs. This suggests sp^3d^2 hybridization. Each bond is a σ bond formed by overlap of an sp^3d^2 hybrid orbital on P with a $3p$ orbital on Cl. **10.39** (a) The C atom is bonded to three other atoms; therefore, we assume sp^2 hybrid orbitals. One $2p$ orbital remains unhybridized. Each carbon–hydrogen bond is a σ bond formed by the overlap of an sp^2 hybrid orbital on C with a $1s$ orbital on H. The remaining sp^2 hybrid orbital on C overlaps with a $2p$ orbital on O to form a σ bond. The unhybridized $2p$ orbital on C overlaps with a parallel $2p$ orbital on O to form a π bond. Together, the σ and π bonds describe the C=O bond. (b) The nitrogen atoms are sp hybridized, and a σ bond is formed by the overlap of an sp hybrid orbital from each N. The remaining sp hybrid orbitals contain lone pairs of electrons. The two unhybridized $2p$ orbitals on one N overlap with the unhybridized $2p$ orbitals on the other N to form two π bonds.

10.41

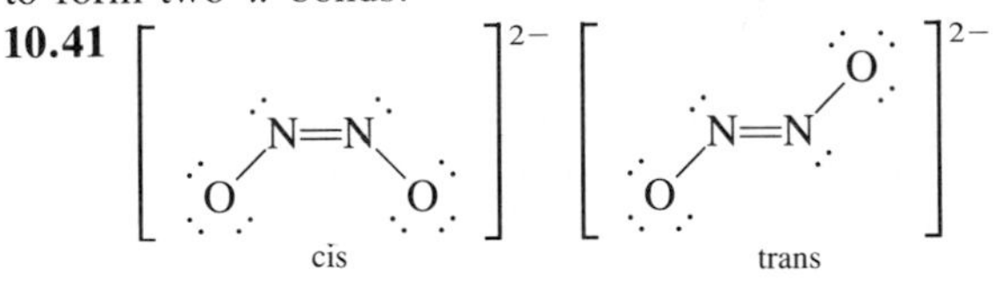

Each of the N atoms has a lone pair of electrons and is bonded to two atoms. The N atoms are sp^2 hybridized. The two possible arrangements of the oxygen atoms relative to one another are shown above. Since the π bond between the nitrogen atoms must be broken to interconvert these two forms, it is expected that the hyponitrite ion will exhibit *cis-trans* isomerization.

10.43 (a) $KK(\sigma_{2s})^2(\sigma_{2s}^*)^2(\pi_{2p})^2$, bond order = 1, stable, paramagnetic (b) $KK(\sigma_{2s})^2(\sigma_{2s}^*)^2(\pi_{2p})^1$, bond order = 1/2, stable, paramagnetic (c) $KK(\sigma_{2s})^2(\sigma_{2s}^*)^2(\pi_{2p})^4(\sigma_{2p})^2(\pi_{2p}^*)^3$, bond order = 3/2, stable, paramagnetic

10.45 $KK(\sigma_{2s})^2(\sigma_{2s}^*)^2(\pi_{2p})^4(\sigma_{2p})^2$, bond order = 3, diamagnetic **10.47** (a) bent (b) linear (c) trigonal pyramidal (d) linear **10.49** (a) linear (b) bent (c) linear (d) bent **10.51** Left: Both carbon atoms with the double bonds are sp^2 hybridized. The carbon atom with the OH group is sp^3 hybridized. Right: Both carbon

atoms are *sp* hybridized. **10.53** The *trans* isomer is expected to have a zero dipole moment, whereas the *cis* isomer is not. **10.55** $(\sigma_{1s})^2$, stable
10.57 $KK(\sigma_{2s})^2(\sigma_{2s}^*)^2(\pi_{2p})^4(\sigma_{2p})^2$; 3 **10.59** O_2^+ has one less antibonding electron than O_2; O_2^- has a longer bond because it has one more antibonding electron.

10.61 XeO_3F_2; 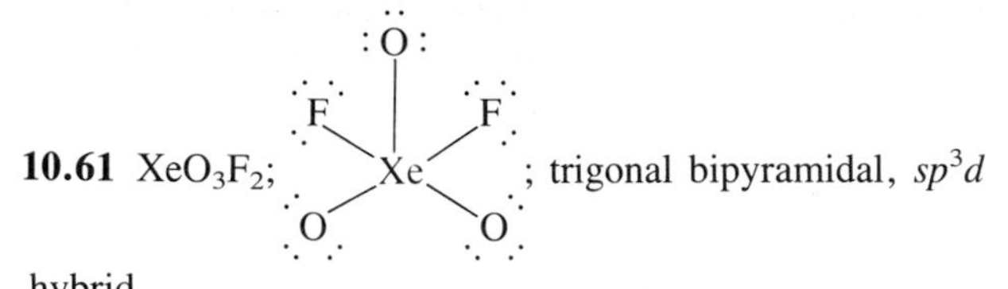 ; trigonal bipyramidal, sp^3d hybrid
10.63 ClF_3; the Cl atom has sp^3d hybrid orbitals, giving three Cl—F bonds and two lone pairs. **10.65** (a) N_2: 1.10 Å, linear geometry, *sp* hybrids. (b) N_2F_2: 1.22 Å, trigonal planar geometry, sp^2 hybrids. (c) N_2H_4: 1.45 Å, tetrahedral geometry, sp^3 hybrids.

10.67 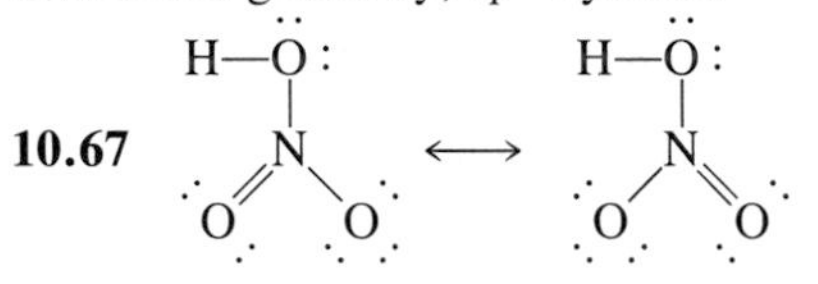

trigonal planar geometry, sp^2 hybrids, $\Delta H_f^\circ = -40$ kJ/mol, resonance energy = 95 kJ

Chapter 11

11.21 (a) vaporization (b) sublimation (c) condensation (d) condensation (deposition) (e) freezing **11.23** 18°C **11.25** 15.5 kJ/mol **11.27** 7.00 kJ **11.29** 1.25 g **11.31** 27.1 g **11.33** 1.90×10^2 mmHg **11.35** 53.3 kJ/mol **11.39** SO_2 and C_2H_2 can be liquefied at 25°C; to liquefy CH_4, reduce temperature below −82°C, then apply pressure greater than 46 atm; to liquefy CO, reduce temperature below −140°C, then apply pressure greater than 35 atm. **11.41** (a) solid (b) liquid
11.43 Values increase with molecular weight as expected. **11.45** (a) London forces (b) dipole–dipole forces, London forces, hydrogen bonding (c) dipole–dipole forces, London forces (d) London forces **11.47** CCl_4, $SiCl_4$, $GeCl_4$ **11.49** CCl_4; highest molecular weight and greatest London forces. **11.51** $HOCH_2CH_2OH$, FCH_2CH_2OH, FCH_2CH_2F; hydrogen bonding decreases in magnitude from left to right. **11.53** CH_4, C_2H_6, CH_3OH, CH_2OHCH_2OH **11.55** (a) metallic (b) metallic (c) molecular (d) covalent network (e) molecular **11.57** (a) metallic (b) covalent network (c) molecular (d) molecular **11.59** $C_2H_5OC_2H_5$, C_4H_9OH, KCl, CaO **11.61** (a) low-melting, brittle (b) high-melting, brittle (c) malleable, electrically conducting (d) high-melting, hard **11.63** (a) LiCl (b) SiC (c) CHI_3 (d) Co **11.65** 1 **11.67** 9.26×10^{-23} g; 9.274×10^{-23} g from AW **11.69** 3.61 Å
11.71 4, fcc **11.73** 19.25 g/cm^3 **11.75** 1.72×10^{-23} cm^3, 1.60 Å **11.77** Water vapor deposits as frost (deposition), frost melts (melting), liquid water evaporates (vaporization) **11.79** 60.7% **11.81** 80°C
11.83 (a) condenses to liquid (and to solid at high pressure) (b) condenses to solid (c) remains a gas
11.85 Hydrogen bonds form between H atom of OH group on one molecule and O atom on another molecule.
11.87 Hydrogen bonding between ethylene glycol molecules leads to higher boiling point and greater viscosity. **11.89** Al, metallic; Si, covalent network; P, molecular; S, molecular. **11.91** (a) KCl; smaller charges on ions (b) CCl_4; lower molecular weight and smaller London forces (c) Zn; Group IIB metal is lower melting (d) C_2H_5Cl; no hydrogen bonding. **11.93** 3.171×10^{-22} g, 191.0 amu **11.95** 1.28 Å **11.97** 68.0%
11.99 0.220 mol N_2, 0.01023 mol C_3H_8O, 0.0444, 33.1 mmHg, 33.1 mmHg **11.101** 7.6 kJ **11.103** 898 J
11.105 0.42 g monomer, 1.42 g dimer, 1.84 g/L

Chapter 12

12.27 Aqueous ammonia **12.29** Ethanol **12.31** H_2O, CH_2OHCH_2OH, $C_{10}H_{22}$ **12.33** Ca^{2+} **12.35** $Ba(IO_3)_2$, $Sr(IO_3)_2$, $Ca(IO_3)_2$, $Mg(IO_3)_2$. Because IO_3^- is large, the change in lattice energy in these iodates is small. Therefore, the hydration energy of the cation dominates the solubility trend. **12.37** 0.886 g/100 mL
12.39 Dissolve 3.63 g KI in 141 g H_2O **12.41** 10.3 g
12.43 1.45 *m* **12.45** 77.7 g **12.47** 0.358, 0.642
12.49 0.0133 **12.51** 0.232, 16.8 *m* **12.53** 0.568 *M*
12.55 0.796 *m* **12.57** 41.5 mmHg, 0.729 mmHg
12.59 100.043°C, −0.153°C **12.61** 4.6×10^{-2} *m*
12.63 122 amu **12.65** 163 amu **12.67** 6.85×10^4 amu
12.69 −0.047°C **12.71** 2.78 ≃ 3 **12.73** (a) aerosol (b) sol (c) foam (d) sol **12.75** $Al_2(SO_4)_3$
12.77 0.671, 0.329 **12.79** 1.74 *m*, 0.0304, 1.63 *M*
12.81 21 g C_3H_8, 37 g C_4H_{10} **12.83** 143 mmHg
12.85 −24°C **12.87** 0.30 *M* **12.89** 0.10 *m* $CaCl_2$
12.91 −783 kJ/mol, −445 kJ/mol **12.93** 0.565 *m*
12.95 0.701 *M* **12.97** 3.1% **12.99** $C_3H_8O_3$

Chapter 13

13.17 $\underset{\text{acid}}{HSO_3^-} + \underset{\text{base}}{H_2O} \rightleftharpoons \underset{\text{acid}}{H_3O^+} + \underset{\text{base}}{SO_3^{2-}}$
13.19 (a) SO_4^{2-} (b) HS^- (c) HPO_4^{2-} (d) NH_3
13.21 (a) HCN (b) H_2CO_3 (c) $HSeO_4^-$ (d) HPO_4^{2-}
13.23 (a) $\underset{\text{acid}}{H_2C_2O_4} + \underset{\text{base}}{ClO^-} \rightleftharpoons \underset{\text{base}}{HC_2O_4^-} + \underset{\text{acid}}{HClO}$
$H_2C_2O_4$ and $HC_2O_4^-$ are conjugates, and ClO^- and HClO are conjugates.

(b) $HPO_4^{2-} + NH_4^+ \rightleftharpoons NH_3 + H_2PO_4^-$
base acid base acid
HPO_4^{2-} and $H_2PO_4^-$ are conjugates, and NH_4^+ and NH_3 are conjugates.
(c) $SO_4^{2-} + H_2O \rightleftharpoons HSO_4^- + OH^-$
base acid acid base
SO_4^{2-} and HSO_4^- are conjugates, and H_2O and OH^- are conjugates.
(d) $Fe(H_2O)_6^{2+} + H_2O \rightleftharpoons H_3O^+ + Fe(H_2O)_5OH^+$
acid base acid base
$Fe(H_2O)_6^{2+}$ and $Fe(H_2O)_5OH^+$ are conjugates, and H_2O and H_3O^+ are conjugates.
13.25 $HSO_4^-(aq) + ClO^-(aq) \longrightarrow SO_4^{2-}(aq) + HClO(aq)$
Reaction occurs as written. **13.27** (a) left (b) left (c) right (d) right **13.29** trichloroacetic acid
13.31 (a) H_2S; acid strength decreases with negative anion charge
(b) H_2SO_3; acid strength increases with electronegativity
(c) HBr; acid strength increases with electronegativity
(d) HIO_4; acid strength increases with number of O atoms
(e) H_2S; acid strength increases with size of central atom
13.33 (a) $CO_2 + OH^- \longrightarrow HCO_3^-$

$$:O{=}C{=}O: + [:\ddot{O}{-}H]^- \longrightarrow [C(:O:)(:O:)(:O{-}H)]^-$$
Lewis acid, Lewis base

(b) $AlCl_3 + [:\ddot{Cl}:]^- \longrightarrow [AlCl_4]^-$
Lewis acid, Lewis base

13.35 (a) $Cr^{3+} + 6H_2O \longrightarrow Cr(H_2O)_6^{3+}$
Lewis acid, Lewis base
(b) $BF_3 + (C_2H_5)_2\ddot{O}: \longrightarrow F_3B:\ddot{O}(C_2H_5)_2$
Lewis acid, Lewis base
13.37 $H_2S + HOCH_2CH_2NH_2 \longrightarrow HS^- + HOCH_2CH_2NH_3^+$
Lewis acid, Lewis base
The H^+ ion from H_2S accepts a pair of electrons from the N atom in $HOCH_2CH_2NH_2$. **13.39** (a) +3 (b) +4 (c) +7 (d) +6 **13.41** (a) −3 (b) +5 (c) +3 (d) +5 **13.43** (a) Mn, +2; Cl, +5; O, −2 (b) Fe, +3; Cr, +6; O, −2 (c) Hg, +2; Cr, +6; O, −2 (d) Co, +2; P, +5; O, −2
13.45 (a) $P_4(s) + 5O_2(g) \longrightarrow P_4O_{10}(s)$
reducing agent, oxidizing agent
(b) $Co(s) + Cl_2(g) \longrightarrow CoCl_2(s)$
reducing agent, oxidizing agent

13.47 (a) reducing agent: Al; oxidizing agent: F_2
(b) oxidizing agent: Hg^{2+}; reducing agent: NO_2^-
13.49 $Ni(s) \longrightarrow Ni^{2+}(aq) + 2e^-$ (oxidation); $Cu^{2+}(aq) + 2e^- \longrightarrow Cu(s)$ (reduction) **13.51** (a) *NR* (b) reaction occurs (c) reaction occurs (d) *NR*
13.53 (a) Sulfite ion is oxidized to sulfate ion; permanganate ion is reduced in acid solution to Mn^{2+}.
(b) Copper metal can be oxidized to Cu^+ or Cu^{2+} (usually Cu^{2+}); nitrate ion can be reduced to several products, including NO_2 and NO.
13.55 (a) $6HI + 2HNO_3 \longrightarrow 3I_2 + 2NO + 4H_2O$
reducing agent, oxidizing agent
(b) $2Ag + 2H_2SO_4 \longrightarrow Ag_2SO_4 + SO_2 + 2H_2O$
reducing agent, oxidizing agent
(c) $3MnCl_2 + 2KMnO_4 + 4KOH \longrightarrow 5MnO_2 + 6KCl + 2H_2O$
reducing agent, oxidizing agent
(d) $2Cr(OH)_3 + 3H_2O_2 + 4KOH \longrightarrow 2K_2CrO_4 + 8H_2O$
reducing agent, oxidizing agent
(e) $3HClO_3 \longrightarrow HClO_4 + 2ClO_2 + H_2O$
oxidizing and reducing agent
13.57 (a) $H_3AsO_4 + 4Zn + 8HNO_3 \longrightarrow AsH_3 + 4Zn(NO_3)_2 + 4H_2O$
oxidizing agent, reducing agent
(b) $2SnCl_2 + O_2 + 8HCl \longrightarrow 2H_2SnCl_6 + 2H_2O$
reducing agent, oxidizing agent
(c) $3K_2MnO_4 + 2H_2O \longrightarrow MnO_2 + 2KMnO_4 + 4KOH$
oxidizing and reducing agent
(d) $H_2SO_4 + 8HI \longrightarrow H_2S + 4I_2 + 4H_2O$
oxidizing agent, reducing agent
(e) $2MnSO_4 + 5PbO_2 + 3H_2SO_4 \longrightarrow 2HMnO_4 + 5PbSO_4 + 2H_2O$
reducing agent, oxidizing agent
13.59 (a) $Cr_2O_7^{2-} + 3C_2O_4^{2-} + 14H^+ \longrightarrow 2Cr^{3+} + 6CO_2 + 7H_2O$
(b) $3Cu + 2NO_3^- + 8H^+ \longrightarrow 3Cu^{2+} + 2NO + 4H_2O$
(c) $MnO_2 + HNO_2 + H^+ \longrightarrow Mn^{2+} + NO_3^- + H_2O$
(d) $5PbO_2 + 2Mn^{2+} + 5SO_4^{2-} + 4H^+ \longrightarrow 5PbSO_4 + 2MnO_4^- + 2H_2O$
(e) $3HNO_2 + Cr_2O_7^{2-} + 5H^+ \longrightarrow 2Cr^{3+} + 3NO_3^- + 4H_2O$
13.61 (a) $Mn^{2+} + H_2O_2 + 2OH^- \longrightarrow MnO_2 + 2H_2O$
(b) $2MnO_4^- + 3NO_2^- + H_2O \longrightarrow 2MnO_2 + 3NO_3^- + 2OH^-$
(c) $Mn^{2+} + 2ClO_3^- \longrightarrow MnO_2 + 2ClO_2$
(d) $MnO_4^- + 3NO_2 + 2OH^- \longrightarrow MnO_2 + 3NO_3^- + H_2O$
(e) $3Cl_2 + 6OH^- \longrightarrow 5Cl^- + ClO_3^- + 3H_2O$
13.63 (a) $8H_2S + 16NO_3^- + 16H^+ \longrightarrow 16NO_2 + S_8 + 16H_2O$

(b) $2NO_3^- + 3Cu + 8H^+ \longrightarrow 2NO + 3Cu^{2+} + 4H_2O$
(c) $2MnO_4^- + 5SO_2 + 2H_2O \longrightarrow 2Mn^{2+} + 5SO_4^{2-} + 4H^+$
(d) $2Bi(OH)_3 + 3Sn(OH)_3^- + 3OH^- \longrightarrow 2Bi + 3Sn(OH)_6^{2-}$
13.65 (a) $2MnO_4^- + I^- + H_2O \longrightarrow 2MnO_2 + IO_3^- + 2OH^-$
(b) $Cr_2O_7^{2-} + 6Cl^- + 14H^+ \longrightarrow 2Cr^{3+} + 3Cl_2 + 7H_2O$
(c) $3S_8 + 32NO_3^- + 32H^+ \longrightarrow 24SO_2 + 32NO + 16H_2O$
(d) $3H_2O_2 + 2MnO_4^- \longrightarrow 3O_2 + 2MnO_2 + 2OH^- + 2H_2O$
(e) $5Zn + 2NO_3^- + 12H^+ \longrightarrow 5Zn^{2+} + N_2 + 6H_2O$
13.67 (a) 32.67 g/eq (b) 23.95 g/eq (c) 29.17 g/eq (d) 60.05 g/eq **13.69** 0.236 *N* **13.71** 0.221 *N* **13.73** 0.272 *N*, 0.136 *M* **13.75** 0.00804 eq, 0.234 g **13.77** (a) 31.60 g/eq (b) 49.03 g/eq (c) 94.80 (d) 63.02 **13.79** 1.67 *N* **13.81** 0.235 *N*, 0.235 *M* **13.83** 0.0198 eq, 2.70% **13.85** (a) BaO is a base; $BaO + H_2O \longrightarrow Ba^{2+} + 2OH^-$ (b) H_2S is an acid; $H_2S + H_2O \longrightarrow H_3O^+ + HS^-$ (c) CH_3NH_2 is a base; $CH_3NH_2 + H_2O \longrightarrow CH_3NH_3^+ + OH^-$ (d) SO_2 is an acid; $SO_2 + 2H_2O \longrightarrow H_3O^+ + HSO_3^-$
13.87 (a) $H_2O_2(aq) + S^{2-}(aq) \longrightarrow HO_2^-(aq) + HS^-(aq)$
(b) $HCO_3^-(aq) + OH^-(aq) \longrightarrow CO_3^{2-}(aq) + H_2O(l)$
(c) $NH_4^+(aq) + CN^-(aq) \longrightarrow NH_3(aq) + HCN(aq)$
(d) $H_2PO_4^-(aq) + OH^-(aq) \longrightarrow HPO_4^{2-}(aq) + H_2O(aq)$
13.89 H_2S, H_2Se, HBr
13.91 (a) $ClO^-(aq) + H_2O(l) \rightleftharpoons HClO(aq) + OH^-(aq)$

Brønsted base, Brønsted acid

$[:\ddot{Cl}-\ddot{O}:]^- + H-\ddot{O}:(-H) \rightleftharpoons :\ddot{Cl}-\ddot{O}:H + [:\ddot{O}-H]^-$

Lewis base, Lewis acid

(b) $NH_4^+ + NH_2^- \rightleftharpoons 2NH_3$

Brønsted acid, Brønsted base

$[H-N(H)(H)-H]^+ + [:\ddot{N}(H)-H]^- \rightleftharpoons 2H-\ddot{N}(H)-H$

Lewis acid, Lewis base

13.93 (a) oxidation–reduction reaction; Cl_2 is the oxidizing agent; HBr is the reducing agent. (b) acid–base reaction; HBr is the acid; $Ca(OH)_2$ is the base. (c) acid–base reaction; NaCN is the base; $NaHSO_4$ is the acid. (d) oxidation–reduction reaction; NaClO is both the oxidizing agent and the reducing agent.
13.95 (a) $16MnO_4^- + 24S^{2-} + 32H_2O \longrightarrow 16MnO_2 + 3S_8 + 64OH^-$
(b) $IO_3^- + 3HSO_3^- \longrightarrow I^- + 3SO_4^{2-} + 3H^+$
(c) $3Fe(OH)_2 + CrO_4^{2-} + 4H_2O \longrightarrow 3Fe(OH)_3 + Cr(OH)_4^- + OH^-$
(d) $Cl_2 + 2OH^- \longrightarrow Cl^- + ClO^- + H_2O$
13.97 $4Fe(OH)_2 + O_2 + 2H_2O \longrightarrow 4Fe(OH)_3$
13.99 0.525 *N*
13.101 HO—P(OH)(H)—O; 0.976 g NaOH
13.103 $BF_3 + :NH_3 \longrightarrow F_3B:NH_3$; BF_3 is the Lewis acid, and NH_3 is the Lewis base; 12.5 g **13.105** 43.0% **13.107** 61.7% **13.109** 22.9

Chapter 14

14.23 $-\frac{1}{2}\Delta[NO_2]/\Delta t = \Delta[O_2]/\Delta t$ **14.25** $\frac{1}{5}\Delta[Br^-]/\Delta t = \Delta[BrO_3^-]/\Delta t$ **14.27** 2.3×10^{-2} *M*/hr **14.29** 3.5×10^{-6} mol/(L · s) **14.31** 1, 1, 2 **14.33** Orders with respect to MnO_4^-, $H_2C_2O_4$, and H^+ are 1, 1, 0, respectively. Overall order is 2. **14.35** rate = $k[CH_3NNCH_3]$, $k = 2.5 \times 10^{-4}$/s **14.37** rate = $k[NO]^2[H_2]$, $k = 2.9 \times 10^2/(M^2 \cdot s)$ **14.39** rate = $k[ClO_2]^2[OH^-]$, $k = 2.3 \times 10^2/(M^2 \cdot s)$ **14.41** 0.017 *M* **14.43** 5.16×10^{-4} *M* **14.45** 1.1×10^3 s, 2.2×10^3 s, 3.3×10^3 s **14.47** 96 hr, 1.9×10^2 hr, 2.9×10^2 hr, 3.8×10^2 hr, 4.8×10^2 hr **14.49** 3.2×10^2 hr
14.51 First order, $k = 0.101$/s **14.53** $E_a = 210$ kJ
14.55 1.0×10^5 J/mol, 1.7×10^{-3}/s **14.57** 83.8 kJ/mol
14.59 114 kJ **14.61** $NOCl_2$; $2NO + Cl_2 \longrightarrow 2NOCl$
14.63 (a) bimolecular (b) bimolecular (c) unimolecular (d) termolecular **14.65** (a) Rate = $k[O_3]$ (b) Rate = $k[NOCl_2][NO]$ **14.67** Rate = $k[C_3H_6]^2$ **14.69** Rate = $(k_2k_1/k_{-1})[H_2][I_2] = k[H_2][I_2]$ **14.71** Cl_2; $2N_2O \longrightarrow 2N_2 + O_2$ **14.73** 3.5×10^{-6} *M*/s, 3.2×10^{-6} *M*/s, 2.5×10^{-6} *M*/s **14.75** Average $k = 2.5 \times 10^{-4}$/s **14.77** 1.51×10^4 s **14.79** 5.50×10^3 s
14.81 0.0211 *M* **14.83** 2.5×10^{-4}/s
14.85 1.14×10^5 J/mol, 5×10^9, 5 *M*/s
14.87 Rate = $k[NO_2][CO]$ **14.89** Rate = $k_1[NO_2Br]$
14.91 Rate = $(k_2k_1/k_{-1}) \times [NH_4^+][CNO^-] = k[NH_4^+][CNO^-]$ **14.93** 0.218 kJ/s
14.95 6.60×10^{-8} mmHg/s

Chapter 15

15.15 1.297 mol PCl_5, 0.203 mol PCl_3, 0.203 mol Cl_2
15.17 0.1187 mol CO, 0.2374 mol H_2, 0.0313 mol CH_3OH **15.19** (a) $K_c = \frac{[PCl_5]}{[PCl_3][Cl_2]}$ (b) $K_c = \frac{[O_3]^2}{[O_2]^3}$
(c) $K_c = \frac{[NO]^2[Cl]}{[NOCl]^2}$ (d) $K_c = \frac{[N_2]^2[H_2O]^6}{[NH_3]^4[O_2]^3}$
15.21 $CH_4(g) + 2H_2S(g) \rightleftharpoons CS_2(g) + 4H_2(g)$

15.23 $K_c = \frac{[NO][H_2]}{[N_2]^{1/2}[H_2O]}$ **15.25** $K_c = 0.543$
15.27 49.1 **15.29** 10.6 L^2/mol^2 **15.31** 1.0×10^{-3}
15.33 (a) $K_p = \frac{P_{HBr}^2}{P_{H_2}P_{Br_2}}$ (b) $K_p = \frac{P_{CH_4}P_{H_2S}^2}{P_{CS_2}P_{H_2}^4}$ (c) $K_p = \frac{P_{H_2O}^2 P_{Cl_2}^2}{P_{HCl}^4 P_{O_2}}$ (d) $K_p = \frac{P_{CH_3OH}}{P_{CO}P_{H_2}^2}$ **15.35** 3.0×10^{-5}
15.37 56.3 **15.39** (a) $K_c = \frac{[CO]^2}{[CO_2]}$ (b) $K_c = \frac{[CO_2]}{[CO]}$
(c) $K_c = \frac{[CO_2]}{[SO_2][O_2]^{1/2}}$ (d) $K_c = [Pb^{2+}][I^-]^2$
15.41 (a) nearly complete (b) not complete **15.43** no, 3.2×10^{-48}, yes **15.45** (a) goes to right (b) goes to left (c) equilibrium (d) goes to left **15.47** goes to left **15.49** 0.37 *M* **15.51** $[I_2] = [Br_2] = 4.7 \times 10^{-5}$ *M*, $[IBr] = 5.1 \times 10^{-4}$ *M* **15.53** 0.18 mol PCl_3, 0.18 mol Cl_2, 0.32 mol PCl_5 **15.55** [CO] = 0.0613 *M*, $[H_2]$ = 0.1839 *M*, $[CH_4]$ = 0.0387 *M*, $[H_2O]$ = 0.0387 *M*
15.57 (a) forward (b) reverse **15.59** (a) no effect (b) no effect (c) goes to left **15.61** No, the fraction of methanol would decrease. **15.63** decrease **15.65** low temperature and low pressure **15.67** 4.3 **15.69** 0.153
15.71 reverse **15.73** reverse **15.75** 0.008 mol HBr, 0.0010 mol H_2, 0.0010 mol Br_2 **15.77** 13.2%
15.79 0.48 mol CO, 2.44 mol H_2, 0.52 mol CH_4, 0.52 mol H_2O **15.81** endothermic **15.83** $K_p = \frac{P_{NH_3}^2}{P_{N_2}P_{H_2}^3} = \frac{[NH_3]^2(RT)^2}{[N_2](RT)[H_2]^3(RT)^3} = K_c/(RT)^2$, so $K_c = K_p(RT)^2$ **15.85** 0.430 **15.87** 0.408 atm

Chapter 16

16.21 (a) $[H^+] = 0.25$ *M*, $[OH^-] = 4.0 \times 10^{-14}$ *M*
(b) $[H^+] = 8.0 \times 10^{-15}$ *M*, $[OH^-] = 1.25$ *M*
(c) $[H^+] = 1.4 \times 10^{-12}$ *M*, $[OH^-] = 0.0070$ *M*
(d) $[H^+] = 2.5$ *M*, $[OH^-] = 4.0 \times 10^{-15}$ *M*
16.23 $[H^+] = 0.050$ *M*, $[OH^-] = 2.0 \times 10^{-13}$ *M*
16.25 $[H^+] = 1.0 \times 10^{-12}$ *M*, $[OH^-] = 0.010$ *M*
16.27 (a) basic (b) acidic (c) neutral (d) acidic
16.29 acidic **16.31** (a) basic (b) neutral (c) acidic (d) acidic **16.33** (a) acidic (b) neutral (c) basic (d) acidic **16.35** (a) 8.00 (b) 11.30 (c) 2.12 (d) 8.197 **16.37** 2.12 **16.39** 11.60 **16.41** 7.6×10^{-6} *M* **16.43** 4.3×10^{-3} *M* **16.45** 13.16 **16.47** 5.5 to 6.5, acidic
16.49 (a) $HNO_2(aq) + H_2O(l) \rightleftharpoons H_3O^+(aq) + NO_2^-(aq)$
$HNO_2(aq) \rightleftharpoons H^+(aq) + NO_2^-(aq)$
$K_a = \frac{[H^+][NO_2^-]}{[HNO_2]}$
(b) $HClO(aq) + H_2O(l) \rightleftharpoons H_3O^+(aq) + ClO^-(aq)$
$HClO(aq) \rightleftharpoons H^+(aq) + ClO^-(aq)$
$K_a = \frac{[H^+][ClO^-]}{[HClO]}$
(c) $HCN(aq) + H_2O(l) \rightleftharpoons H_3O^+(aq) + CN^-(aq)$
$HCN(aq) \rightleftharpoons H^+(aq) + CN^-(aq)$
$K_a = \frac{[H^+][CN^-]}{[HCN]}$
(d) $HCHO_2(aq) + H_2O(l) \rightleftharpoons H_3O^+(aq) + CHO_2^-(aq)$
$HCHO_2(aq) \rightleftharpoons H^+(aq) + CHO_2^-(aq)$
$K_a = \frac{[H^+][CHO_2^-]}{[HCHO_2]}$
16.51 1.9×10^{-6} **16.53** 5.42, 1.5×10^{-4}
16.55 $[H^+] = [C_6H_4NH_2COO^-] = 1.0 \times 10^{-3}$ *M*
16.57 0.26 *M* **16.59** 4.9×10^{-3} *M*, 2.31
16.61 0.361 *M* **16.63** (a) 3.7×10^{-3} *M* (b) 3.9×10^{-6} *M*
16.65 $C_2H_5NH_2(aq) + H_2O(l) \rightleftharpoons C_2H_5NH_3^+(aq) + OH^-(aq)$
$K_b = \frac{[C_2H_5NH_3^+][OH^-]}{[C_2H_5NH_2]}$
16.67 3.2×10^{-5} **16.69** 5.7×10^{-3} *M*, 11.76
16.71 (a) No hydrolysis
(b) $CHO_2^-(aq) + H_2O(l) \rightleftharpoons HCHO_2(aq) + OH^-(aq)$
$K_b = \frac{[HCHO_2][OH^-]}{[CHO_2^-]}$
(c) $CH_3NH_3^+(aq) + H_2O(l) \rightleftharpoons H_3O^+(aq) + CH_3NH_2(aq)$
$K_a = \frac{[H_3O^+][CH_3NH_2]}{[CH_3NH_3^+]}$
(d) $IO^-(aq) + H_2O(l) \rightleftharpoons HIO(aq) + OH^-(aq)$
$K_b = \frac{[HIO][OH^-]}{[IO^-]}$
16.73 $Zn(H_2O)_6^{2+}(aq) + H_2O(l) \rightleftharpoons Zn(H_2O)_5(OH)^+(aq) + H_3O^+(aq)$
16.75 (a) acidic (b) basic (c) basic (d) acidic
16.77 (a) nearly neutral (b) acidic
16.79 (a) 2.2×10^{-11} (b) 7.1×10^{-6} **16.81** 8.64, 4.4×10^{-6} *M* **16.83** 1.0×10^{-3} *M*, 2.99
16.85 (a) 0.029 (b) 0.0064 **16.87** 3.17 **16.89** 10.47
16.91 3.45 **16.93** 9.25, 9.09 **16.95** 3.03 **16.97** 5.32
16.99 0.34 mol **16.101** 2.0 **16.103** 8.59 **16.105** 5.97
16.107 9.08 **16.109** 1.1×10^{-3} **16.111** 1.1×10^{-2}
16.113 hydrolysis, 9.68 **16.115** $K_b(CN^-) = 2.0 \times 10^{-5}$; $K_b(CO_3^{2-}) = 2.1 \times 10^{-4}$; CO_3^{2-} **16.117** 2.84
16.119 3.16 **16.121** 11/1 **16.123** 2.4 **16.125** 3.11
16.127 (a) 0.100 *M* (b) 0.109 *M* **16.129** 4.5×10^{-8}
16.131 4.1% **16.133** −1.8°C **16.135** 9.1 mL

Chapter 17

17.11 (a) soluble (b) insoluble (c) insoluble (d) soluble **17.13** (a) $[Ba^{2+}][CrO_4^{2-}]$
(b) $[Fe^{2+}][OH^-]^2$ (c) $[Pb^{2+}]^3[AsO_4^{3-}]^2$
(d) $[Ag^+]^2[CrO_4^{2-}]$ **17.15** 9.3×10^{-10}

17.17 1.3×10^{-7} **17.19** 1.8×10^{-11} **17.21** 0.092 g/L **17.23** 1.9×10^{-3} *M* **17.25** 3.1×10^{-4} g/L **17.27** 1.1×10^{-6} g/L **17.29** 0.4 g/L **17.31** Lead chromate will precipitate. **17.33** no precipitate **17.35** No precipitate **17.37** 0.0018 mol **17.39** 6.6×10^{-6} *M*, 0.013% **17.41** 6.9×10^{-9} *M*
17.43 $MgC_2O_4(s) + H^+(aq) \longrightarrow Mg^{2+}(aq) + HC_2O_4^-(aq)$
17.45 BaF_2; F^- is the conjugate base of the weak acid HF. **17.47** pH less than 0.8
17.49 $Ag^+(aq) + 2CN^-(aq) \longrightarrow Ag(CN)_2^-(aq)$
$K_f = \frac{[Ag(CN)_2^-]}{[Ag^+][CN^-]^2}$
17.51 5.5×10^{-19} *M* **17.53** will precipitate **17.55** 3.0×10^{-3} *M* **17.57** Add HCl; Pb^{2+} will be precipitated as the chloride. Filter off $PbCl_2$; add H_2S in 0.3 *M* H^+. Cd^{2+} will precipitate as the sulfide; filter off CdS. The filtrate contains Sr^{2+}. **17.59** Ca^{2+} and Mn^{2+} **17.61** 1.3×10^{-4} *M* **17.63** (a) 6.9×10^{-7} *M*; (b) 3.2×10^{-4} g/L **17.65** (a) 5.2×10^{-6} *M*; (b) pOH = 4.81 **17.67** 26 g/L **17.69** 1.8×10^{-9} *M* **17.71** 8.0×10^{-3} *M* **17.73** 1.0×10^{-5}; not saturated **17.75** 5.5×10^{-6} g **17.77** 2.1×10^{-4} **17.79** 4.7×10^{-2} *M* **17.81** 1.4×10^{-2} *M* **17.83** 3.6×10^{-4} *M*, 4.2×10^{-6} *M*, 2.5×10^9 **17.85** 0.18 *M* **17.87** 2.7×10^{-4} *M* **17.89** $[SO_4^{2-}] = 0.109$ *M*; $[Ba^{2+}] = 1.0 \times 10^{-9}$ *M*; $[Mg^{2+}] = 0.109$ *M*; $[OH^-] = 1.3 \times 10^{-5}$ *M*

Chapter 18

18.17 −65 J, 22 J, −43 J **18.19** 37.57 kJ **18.21** 88.5 J/(mol · K) **18.23** −125 J/K, 127 J/K **18.25** (a) negative (b) not predictable (c) positive (d) positive **18.27** (a) −181.6 J/K (b) 43.2 J/K (c) −56.1 J/K (d) 313 J/K **18.29** −242.6 J/K **18.31** −1453.2 kJ, −162 J/K, 1404.9 kJ **18.33** (a) $Na(s) + \frac{1}{2}Cl_2(g) \longrightarrow NaCl(s)$
(b) $\frac{1}{2}H_2(g) + C(graphite) + \frac{1}{2}N_2(g) \longrightarrow HCN(l)$
(c) $S(rhombic) + O_2(g) \longrightarrow SO_2(g)$
(d) $P(red) + \frac{3}{2}H_2(g) \longrightarrow PH_3(g)$
18.35 (a) −800 kJ (b) −55.8 kJ
18.37 (a) spontaneous (b) spontaneous (c) nonspontaneous (d) equilibrium mixture with significant amounts of reactants and products (e) nonspontaneous **18.39** (a) 850 kJ; 838 kJ; endothermic reaction with mainly reactants at equilibrium. (b) −72 kJ; −142 kJ; exothermic reaction with mainly products at equilibrium. **18.41** −474.4 kJ; 0
18.43 −16.2 kJ **18.45** (a) $K = \frac{P_{CO_2}P_{H_2}}{P_{CO}P_{H_2O}}$ (b) $K = [Mg^{2+}][OH^-]^2$ (c) $K = [Li^+]^2[OH^-]^2P_{H_2}$
18.47 −190.6 kJ; 2×10^{33} **18.49** −142.2 kJ; 8.0×10^{24} **18.51** −520.99 kJ; 2.03×10^{91} **18.53** Reaction is spontaneous because $\Delta S°$ is large and positive.
18.55 −184 kJ; exothermic; positive; both ΔH and $-T\Delta S$ are negative, so reaction is spontaneous at all *T*. **18.57** 1.2×10^2; yes **18.59** 383 K **18.61** −14.3 J/K **18.63** (a) negative (b) positive (c) positive (d) negative **18.65** negative **18.67** −136 J/K **18.69** no, $\Delta G°$ is +267 kJ **18.71** $\Delta H°$ is positive; $\Delta S°$ is positive **18.73** 56 kJ; 2×10^{-10} **18.75** 1.10×10^2 kJ; 136 J/K; 69 kJ; −37 kJ; nonspontaneous at 25°C, but spontaneous at 800°C **18.77** 0.0040%, same; 4.0×10^{-10} **18.79** 85% **18.81** −2.29 kJ/mol; −98.74 J/(mol · K)

Chapter 19

19.19 Anode: $Zn \longrightarrow Zn^{2+} + 2e^-$
Cathode: $Ni^{2+} + 2e^- \longrightarrow Ni$; Electron flow is in circuit is from Zn to Ni.
19.21 $Zn(s) \longrightarrow Zn^{2+}(aq) + 2e^-$
$Ag^+(aq) + e^- \longrightarrow Ag(s)$
19.23 Anode: $Zn(s) + 2OH^-(aq) \longrightarrow Zn(OH)_2(s) + 2e^-$
Cathode: $Ag_2O(s) + H_2O(l) + 2e^- \longrightarrow 2Ag(s) + 2OH^-(aq)$
Overall: $Zn(s) + Ag_2O(s) + H_2O(l) \longrightarrow 2Ag(s) + Zn(OH)_2(s)$
19.25 $Cd(s)|Cd^{2+}(aq)||Pb^{2+}(aq)|Pb(s)$
19.27 $Ni(s)|Ni^{2+}(1\ M)||H^+(1\ M)|H_2(g)|Pt$
19.29 $2Ag^+(aq) + Fe(s) \longrightarrow Fe^{2+}(aq) + 2Ag(s)$
19.31 Anode: $Cd(s) \longrightarrow Cd^{2+}(aq) + 2e^-$
Cathode: $Ni^{2+}(aq) + 2e^- \longrightarrow Ni(s)$
Overall: $Cd(s) + Ni^{2+}(aq) \longrightarrow Cd^{2+}(aq) + Ni(s)$
19.33 -7.0×10^1 kJ **19.35** −48 kJ **19.37** $NO_3^-(aq)$, $O_2(g)$, $H_2O_2(aq)$ **19.39** strongest is Zn(*s*); weakest is $Cu^+(aq)$ **19.41** (a) not spontaneous (b) spontaneous **19.43** Cl_2 will oxidize Br^-; $Cl_2(s) + 2Br^-(aq) \longrightarrow Br_2(l) + 2Cl^-(aq)$
19.45 1.54 V **19.47** 1.28 V **19.49** -3.6×10^5 J **19.51** -1.6×10^5 J **19.53** 2.46 V **19.55** 1.92 V **19.57** 1×10^{21} **19.59** 1×10^6 **19.61** 0.56 V **19.63** 0.29 V **19.65** −0.84 V **19.67** 4×10^{-3} *M*
19.69 (a) $Ca^{2+}(l) + 2e^- \longrightarrow Ca(l)$
$Cl^-(l) \longrightarrow (1/2)Cl_2(g) + e^-$
(b) $Cs^+(l) + e^- \longrightarrow Cs(l)$
$4OH^-(l) \longrightarrow O_2(g) + 2H_2O(g) + 4e^-$
19.71 (a) Anode: $2H_2O(l) \longrightarrow O_2(g) + 4H^+(aq) + 4e^-$; cathode: $2H_2O(l) + 2e^- \longrightarrow H_2(g) + 2OH^-(aq)$; overall: $2H_2O(l) \longrightarrow 2H_2(g) + O_2(g)$ (b) Anode: $2Br^-(aq) \longrightarrow Br_2(l) + 2e^-$; cathode: $2H_2O(l) + 2e^- \longrightarrow H_2(g) + 2OH^-(aq)$; overall: $2Br^-(aq) + 2H_2O(l) \longrightarrow Br_2(l) + H_2(g) + 2OH^-(aq)$
19.73 5.49×10^7 C **19.75** 0.360 g **19.77** 3.74 V **19.79** (a) spontaneous (b) Fe^{3+} will oxidize Sn^{2+} to Sn^{4+}. **19.81** 0.36 V **19.83** (a) 2.2 (b) 0.3 *M*
19.85 (a) 1.0 F, 9.6×10^4 C (b) 2.0 F, 1.9×10^5 C (c) 0.11 F, 1.1×10^4 C (d) 2.8×10^{-2} F, 2.7×10^3 C

19.87 2.66×10^{-4} g **19.89** -0.05 V; not spontaneous **19.91** -0.04 V; not spontaneous **19.93** $K_{sp} = 1 \times 10^{-7}$

Chapter 20

20.19 ${}^{87}_{37}\text{Rb} \longrightarrow {}^{87}_{38}\text{Sr} + {}^{0}_{-1}\text{e}$

20.21 ${}^{232}_{90}\text{Th} \longrightarrow {}^{4}_{2}\text{He} + {}^{228}_{88}\text{Ra}$ **20.23** ${}^{18}_{9}\text{F} \longrightarrow {}^{18}_{8}\text{O} + {}^{0}_{1}\text{e}$ **20.25** ${}^{210}_{84}\text{Po} \longrightarrow {}^{4}_{2}\text{He} + {}^{206}_{82}\text{Pb}$ **20.27** (a) ${}^{122}_{51}\text{Sb}$ (b) ${}^{204}_{85}\text{At}$ (c) ${}^{80}_{37}\text{Rb}$ **20.29** (a) α emission (b) positron emission or electron capture (c) β emission **20.31** ${}^{235}_{92}\text{U}$ decay series; ${}^{232}_{90}\text{Th}$ decay series. Alpha emission decreases the mass number by 4. Beta emission has no effect upon the mass number.

20.33 (a) ${}^{26}_{12}\text{Mg}(d, \alpha){}^{24}_{11}\text{Na}$ (b) ${}^{16}_{8}\text{O}(n, p){}^{14}_{7}\text{N}$

20.35 (a) ${}^{27}_{13}\text{Al} + {}^{2}_{1}\text{H} \longrightarrow {}^{25}_{12}\text{Mg} + {}^{4}_{2}\text{He}$

(b) ${}^{10}_{5}\text{B} + {}^{4}_{2}\text{He} \longrightarrow {}^{1}_{1}\text{p} + {}^{13}_{6}\text{C}$

20.37 1.33×10^{9} kJ/mol **20.39** (a) ${}^{4}_{2}\text{He}$ (b) α

20.41 plutonium–239 **20.43** $1.80 \times 10^{-9}\ \text{s}^{-1}$

20.45 $9.1 \times 10^{-8}\ \text{s}^{-1}$ **20.47** 4.8×10^{10} years

20.49 3.84×10^{-12}/s **20.51** 1.1×10^{2} Ci **20.53** 5.7×10^{-4} g **20.55** 0.330, 1.7 μg **20.57** 3.21 h

20.59 44.6 d **20.61** 5.3×10^{3} y **20.63** 5.2 disintegrations/(min·g) **20.65** -2.06×10^{-9} g

20.67 1.688×10^{12} J; -17.50 MeV

20.69 0.03438 amu; 32.02 MeV; 5.338 MeV

20.71 Na-20 is expected to decay by electron capture or positron emission. Na-26 is expected to decay by beta emission. **20.73** 7 alpha emissions and 4 beta emissions

20.75 ${}^{209}_{83}\text{Bi} + {}^{4}_{2}\text{He} \longrightarrow {}^{211}_{85}\text{At} + 2{}^{1}_{0}n$ **20.77** ${}^{246}_{98}\text{Cf}$

20.79 4.4 y **20.81** 2.42 pm **20.83** -1.01×10^{11} kJ; -3.28×10^{4} kJ **20.85** 1.75×10^{13} nuclei/s

20.87 0.001 L **20.89** -2.732×10^{12} J; 4.91×10^{7} L

Chapter 21

21.39 (a) nonmetal (b) metal **21.41** (a) Na is more metallic. (b) Na is more metallic. (c) Cs is more metallic. **21.43** Brown solid with lower melting point is lead(IV) oxide, PbO_2. **21.45** See chapter for discussion.

21.47 ${}^{227}_{89}\text{Ac} \longrightarrow {}^{227}_{90}\text{Th} + {}^{0}_{-1}\text{e}$

21.49 $TiCl_4(g) + 4Na(l) \longrightarrow Ti(s) + 4NaCl(s)$

21.51 $K_2SO_4(aq) + Ba(OH)_2(aq) \longrightarrow 2KOH(aq) + BaSO_4(s)$

$BaSO_4(s)$ precipitates out of the solution to drive the reaction to the right.

21.53 $CO_2(g) + NH_3(g) + NaCl(aq) + H_2O(l) \longrightarrow NaHCO_3(s) + NH_4Cl(aq)$

$2NaHCO_3(s) \xrightarrow{\Delta} Na_2CO_3(s) + CO_2(g) + H_2O(g)$

$Ca(OH)_2(aq) + Na_2CO_3(aq) \longrightarrow 2NaOH(aq) + CaCO_3(s)$

21.55 ${}^{230}_{90}\text{Th} \longrightarrow {}^{226}_{88}\text{Ra} + {}^{4}_{2}\text{He}$ **21.57** Place a small amount of metal in strong base. If it is Be, it will react with evolution of H_2 gas. Mg gives no reaction.

21.59 Add Na_2SO_4; $BaSO_4$ precipitates and is filtered off. Add NaOH to filtrate; $Mg(OH)_2$ precipitates, leaving $Be(OH)_4{}^{2-}$.

21.61 $Mg(OH)_2(s) + 2HCl(aq) \longrightarrow MgCl_2(aq) + 2H_2O(l)$

$MgCl_2(l) \xrightarrow{\text{electrolysis}} Mg(l) + Cl_2(g)$

$BeF_2(l) + Mg(l) \xrightarrow[950°C]{\Delta} Be(s) + MgF_2(l)$

21.63 $3Mg(s) + N_2(g) \longrightarrow Mg_3N_2(s)$

$Mg_3N_2(s) + 6H_2O(l) \longrightarrow 3Mg(OH)_2(aq) + 2NH_3(aq)$

21.65 $Fe_2O_3(s) + 2Al(l) \longrightarrow Al_2O_3(l) + 2Fe(l)$

21.67 $2Al_2O_3(s) + 3C(s) + 6Cl_2(g) \xrightarrow{\Delta} 4AlCl_3(g) + 3CO_2(g)$

$AlCl_3(s) + 3K(\text{amalgam}) \longrightarrow 3KCl(s) + Al(s)$

21.69

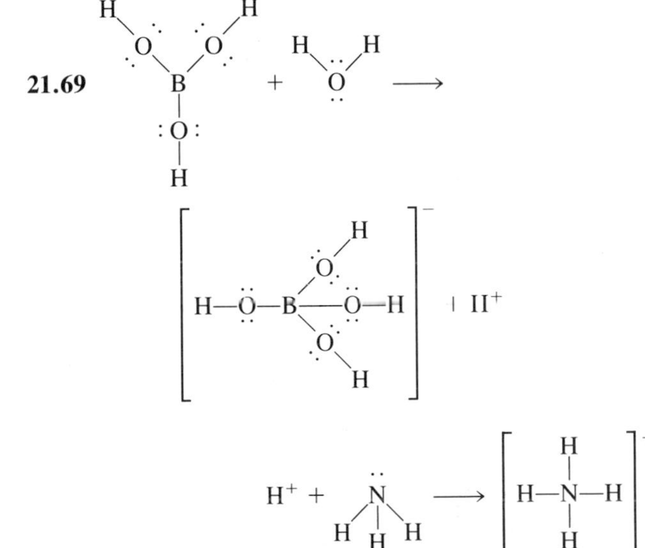

21.71 $Al(H_2O)_6{}^{3+}(aq) + HCO_3{}^{-}(aq) \longrightarrow Al(H_2O)_5OH^{2+}(aq) + H_2O(l) + CO_2(g)$

21.73 Test portions of solutions of each compound with the others; the results can differentiate the compounds. $AlCl_3$ reacts only with KOH, giving a precipitate that dissolves in excess of KOH. $BaCl_2$ reacts only with $BeSO_4$, giving a precipitate. $BeSO_4$ also gives a precipitate with KOH that dissolves in excess KOH.

21.75 (a) $2PbS(s) + 3O_2(g) \longrightarrow 2PbO(s) + 2SO_2(g)$

(b) $PbO(s) + C(s) \xrightarrow{\Delta} Pb(s) + CO(g)$

(c) At the anode the impure lead is oxidized and goes into solution as Pb^{2+}. At the cathode the Pb^{2+} ions are reduced to Pb and deposited on the cathode, leaving the impurities in solution or at the impure anode.

21.77 $AlCl_3$ has a trigonal planar geometry; the geometry of Cl atoms about Al in Al_2C_6 is tetrahedral.

21.79 5.41×10^{-10}/y; 0.947 **21.81** -425 kJ; exothermic **21.83** 1.35×10^{3} K **21.85** $E° = 0.19$ V; -37 kJ; yes **21.87** 1.9 Å **21.89** 3878 g

21.91 7.73×10^{-2} g

Chapter 22

22.51 $CH_4(g) + H_2O(g) \xrightarrow{\Delta} CO(g) + 3H_2(g)$

$CO(g) + NaOH(aq) \xrightarrow{\Delta} NaCOOH(aq)$

22.53 $SnO_2(s) + 2C(s) \xrightarrow{\Delta} Sn(l) + 2CO(g)$
$Sn(s) + 2HCl(aq) \longrightarrow SnCl_2(aq) + H_2(g)$
22.55 $3NaOH + 3NaSn(OH)_3 + 2Bi(OH)_3 \longrightarrow 3Na_2Sn(OH)_6 + 2Bi$
22.57 $Sn(H_2O)_6^{2+}(aq) + H_2O(l) \longrightarrow Sn(H_2O)_5(OH)^+(aq) + H_3O^+(aq)$
22.59 trigonal pyramidal

22.61 $4NH_3(g) + 5O_2(g) \xrightarrow{Pt} 4NO(g) + 6H_2O(g)$
$2NO(g) + O_2(g) \longrightarrow 2NO_2(g)$
$3NO_2(g) + H_2O(l) \longrightarrow 2HNO_3(aq) + NO(g)$
$HNO_3(g) + NH_3(aq) \longrightarrow NH_4NO_3(aq)$
$NH_4NO_3(aq) \longrightarrow NH_4NO_3(s)$

$NH_4NO_3(s) \xrightarrow{\Delta} N_2O(g) + 2H_2O(g)$
22.63 (a) P_4O_6, H_3PO_3, PF_3 (b) P_4O_{10}, H_3PO_4, PF_5
22.65 $3As_4O_6(s) + 8HNO_3(aq) + 14H_2O(l) \longrightarrow 12H_3AsO_4(aq) + 8NO(g)$

$2H_3AsO_4(s) \xrightarrow{\Delta} As_2O_5(s) + 3H_2O(g)$
22.67 $H_3PO_3 + H_2SO_4 \longrightarrow H_3PO_4 + SO_2 + H_2O$
22.69 $2AsO_4^{3-} + N_2H_5^+ \longrightarrow 2AsO_3^{3-} + N_2 + 2H_2O + H^+$

N_2H_5 hydrolyzes
22.71 Tetrahedral. Each P—Br bond is formed by the overlap of a $3p$ orbital on Br and an sp^3 orbital on P.
22.73 (a) $4Li(s) + O_2(g) \longrightarrow 2Li_2O(s)$
(b) $4CH_3NH_2(g) + 9O_2(g) \longrightarrow 4CO_2(g) + 10H_2O(g) + 2N_2(g)$
(c) $P_4(s) + 5O_2(g) \longrightarrow P_4O_{10}(s)$
$P_4O_{10}(s) + 6H_2O(l) \longrightarrow 4H_3PO_4(aq)$
22.75 $2H_2S(g) + 3O_2(g) \longrightarrow 2H_2O(g) + 2SO_2(g)$
$16H_2S(g) + 8SO_2(g) \longrightarrow 16H_2O(l) + 3S_8(s)$
22.77 $H_2SeO_3 + 2H_2S \longrightarrow Se + 2S + 3H_2O$
22.79 (a) sp^3; bent (b) sp^3d hybrid; seesaw
22.81 (a) +6 (b) +6 (c) −2 (d) +4
22.83 $SiO_2(s) + 4HF(g) \longrightarrow 2H_2O(l) + SiF_4(g)$
22.85 $CaF_2(s) + H_2SO_4(l) \longrightarrow CaSO_4(s) + 2HF(g)$

$2HF(l) \xrightarrow[KF]{electrolysis} H_2(g) + F_2(g)$

$U(s) + 3F_2(g) \longrightarrow UF_6(s)$
22.87 $8H^+ + 6HCl + K_2Cr_2O_7 \longrightarrow 3Cl_2 + 2Cr^{3+} + 7H_2O + 2K^+$
22.89 $2IO_3^- + 5HSO_3^- \longrightarrow I_2 + 5SO_4^{2-} + 3H^+ + H_2O$

or

$2NaIO_3 + 5NaHSO_3 \longrightarrow I_2 + 2Na_2SO_4 + 3NaHSO_4 + H_2O$
22.91 $E^\circ_{cell} = 0.86$ V; ClO^- will oxidize Fe^{2+} to Fe^{3+}.
22.93 (a) sp^3; bent
(b) sp^3, trigonal pyramidal
(c) sp^3d; T-shaped

22.95 :F̤ F̤: Xe :F̤ F̤: sp^3d^2, square planar

22.97 $2XeF_2 + 4OH^- \longrightarrow 2Xe + 4F^- + O_2 + 2H_2O$
22.99 (a) S_8 (b) P_4O_{10} (c) Br_2 (d) Sn
22.101 Test the pH. $Na_2SO_4(aq)$ is neutral; other solutions are acidic. Add HCl to acidic solutions. $NaHSO_3(aq)$ will evolve SO_2 (characteristic odor). **22.103** 11.0
22.105 0.405 L **22.107** 24.6% P, 63.1% $Ca(H_2PO_4)_2 \cdot H_2O$ **22.109** 15.1 hr **22.111** 2.58% NaOCl
22.113 Disproportionation is expected. **22.115** −0.16 V

Chapter 23

23.29 CrO_3 is dark red and low-melting. **23.31** W
23.33 $Cr(s) + 2HCl(aq) \longrightarrow CrCl_2(aq) + H_2(g)$
$4CrCl_2(aq) + O_2(g) + 4HCl(aq) \longrightarrow 4CrCl_3(aq) + 2H_2O(l)$
$CrCl_3(aq) + 3NaOH(aq) \longrightarrow Cr(OH)_3(s) + 3NaCl(aq)$
23.35 (a) +2 (b) +4 (c) +1 (d) +6
23.37 $3Fe^{2+} + NO_3^- + 4H^+ \longrightarrow 3Fe^{3+} + NO + 2H_2O$
23.39 $[Co(NH_3)_4(NO_2)_2]Cl$ **23.41** (a) 4 (b) 6 (c) 4 (d) 6 **23.43** (a) +2 (b) +3 (c) +3 (d) +3
23.45 (a) +3 (b) (c) 6 (d) −3

Name	*Formula*
ammine	NH_3
chloro	Cl^-
oxalato	$C_2O_4^{2-}$

23.47 (a) potassium hexafluoroferrate(III)
(b) diamminediaquacopper(II) ion
(c) ammonium aquapentafluoroferrate(III)
(d) dicyanoargentate(I) ion
23.49 (a) pentacarbonyliron(0)
(b) dicyanobis(ethylenediamine)rhodium(III) ion
(c) tetraaminesulfatochromium(III) chloride
(d) tetraoxomanganate(VII) ion
23.51 (a) $K_3[Mn(CN)_6]$ (b) $Na_2[Zn(CN)_4]$
(c) $[Co(NH_3)_4Cl_2]NO_3$ (d) $[Cr(NH_3)_6]_2[CuCl_4]_3$
23.53 (a) linkage (b) coordination (c) hydrate (d) ionization **23.55** $[Co(NH_3)_4Br_2]Cl$; an ionization isomer is $[Co(NH_3)_4BrCl]Br$

23.57 (a)

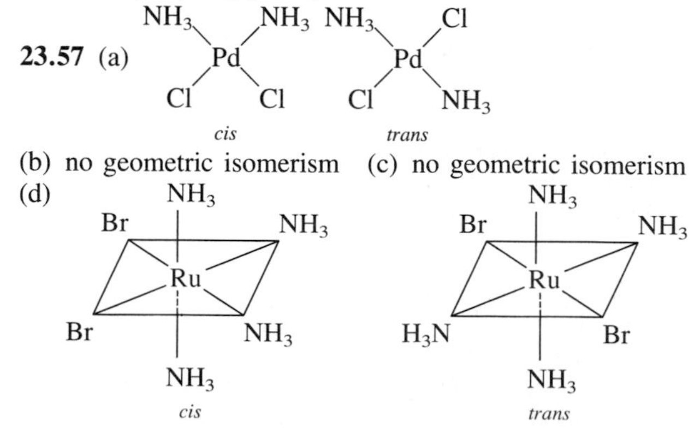

23.59 (a) Has optical isomers, similar to those in Figure 23.15B. (b) No
23.61 (a) V: [Ar](↑)(↑)(↑)( )( ) (↑↓) ( )( )( )
3d 4s 4p

V^{3+}: [Ar](↑)(↑)( )( )( ) ( ) ( )( )( )
3d 4s 4p

$V(H_2O)_6{}^{3+}$: [Ar](↑)(↑)( ) [(↑↓)(↑↓) (↑↓) (↑↓)(↑↓)(↑↓)]
3d 4s 4p
d^2sp^3 bonds to ligands

(b) Fe: [Ar](↑↓)(↑)(↑)(↑)(↑) (↑↓) ( )( )( )
3d 4s 4p
( )( )( )( )( )
4d

Fe^{3+}: [Ar](↑)(↑)(↑)(↑)(↑) ( ) ( )( )( ) ( )( )( )( )( )

$Fe(en)_3{}^{3+}$: [Ar](↑)(↑)(↑)(↑)(↑)
[(↑↓) (↑↓)(↑↓)(↑↓) (↑↓)(↑↓)] ( )( )( )
sp^3d^2 bonds to ligands

(c) Rh: [Kr](↑↓)(↑↓)(↑)(↑)(↑) (↑↓) ( )( )( )
4d 5s 5p

Rh^{3+}: [Kr](↑↓)(↑)(↑)(↑)(↑) ( ) ( )( )( )

$Rh(CN)_6{}^{3-}$: [Kr](↑↓)(↑↓)(↑↓) [(↑↓)(↑↓) (↑↓) (↑↓)(↑↓)(↑↓)]
d^2sp^3 bonds to ligands

23.63 (a) square planar (b) square planar geometry (c) tetrahedral (d) tetrahedral **23.65** (a) 2 unpaired electrons (b) 3 unpaired electrons (c) 2 unpaired electrons **23.67** (a) 8 electrons in square planar field (low spin) (b) 7 electrons in square planar field (low spin) (c) 5 electrons in tetrahedral field (high spin) (d) 7 electrons in tetrahedral field (high spin)
23.69 purple **23.71** Yes, H_2O is a weaker bonding ligand. **23.73** 239 kJ/mol **23.75** no d electrons
23.77 3 **23.79** $[Cu^{2+}] = 6.1 \times 10^{-4}\ M$; $[NH_3] = 2.4 \times 10^{-3}\ M$; $[Cu(NH_3)_4{}^{2+}] = 0.10\ M$

Chapter 24

24.21 (a) 2,4-dimethylpentane
(b) 2,2,6,6-tetramethylheptane (c) 4-ethyloctane
(d) 3,4-dimethyloctane

24.23 (a) $CH_3CH(CH_3)CH(CH_3)CH_2CH_2CH_3$

(b) $CH_3CH_2CH(CH_2CH_3)CH_2CH_2CH_3$

(c) $CH_3CH(CH_3)CH_2CH(CH(CH_3)CH_3)CH_2CH_2CH_3$

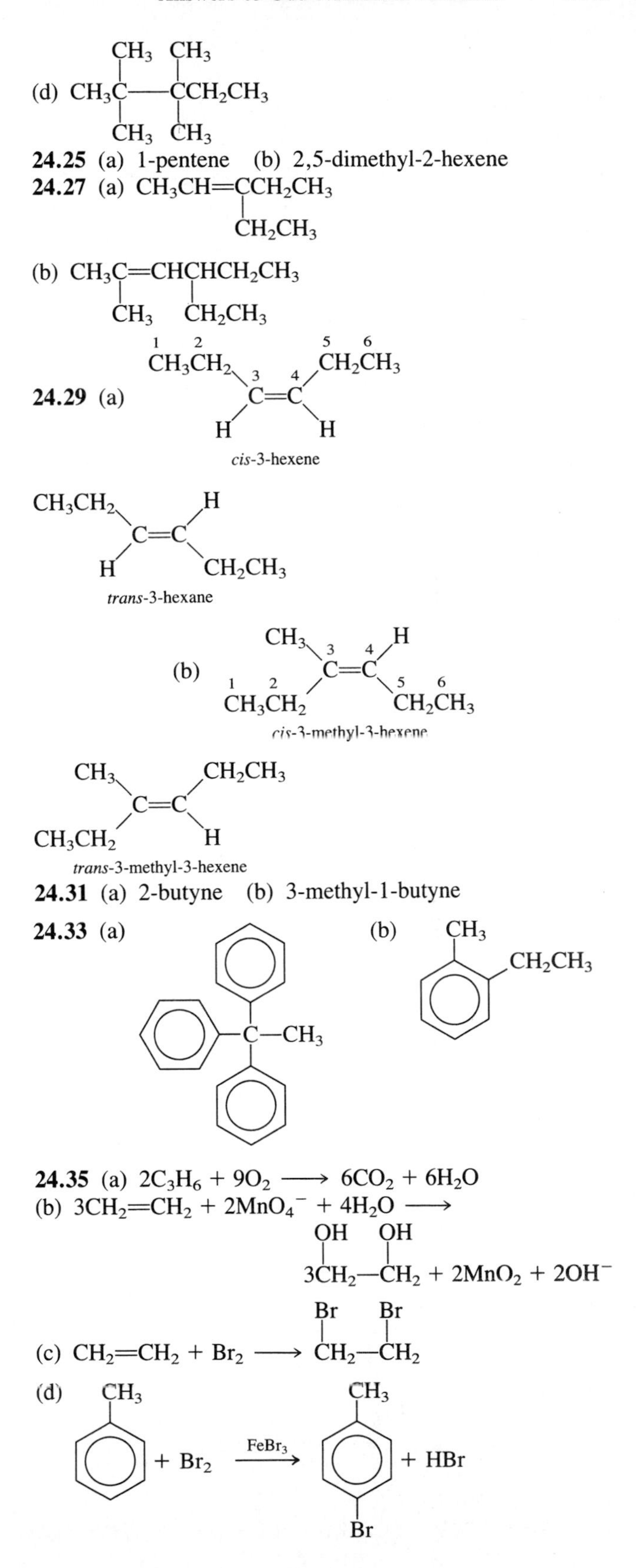

(e) 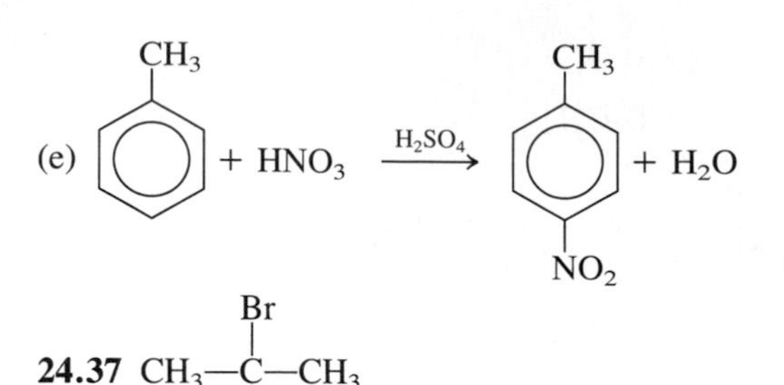

24.37 $CH_3—C(Br)(CH_3)—CH_3$

24.39 $CH_3CH_2CH_2CH_2CH_3 \xrightarrow[\Delta]{Al_2O_3—SiO_2} CH_4 + CH_2{=}CHCH_2CH_3$

$CH_3CH_2CH_2CH_2CH_3 \xrightarrow[\Delta]{Al_2O_3—SiO_2} CH_2{=}CH_2 + CH_3CH_2CH_3$

24.41 (a) $CH_3—\overset{O}{\overset{\|}{C}}—CH_2CH_2CH_3$ ketone (b) $CH_3—\overset{OH}{\overset{|}{C}}HCH_2CH_3$ alcohol

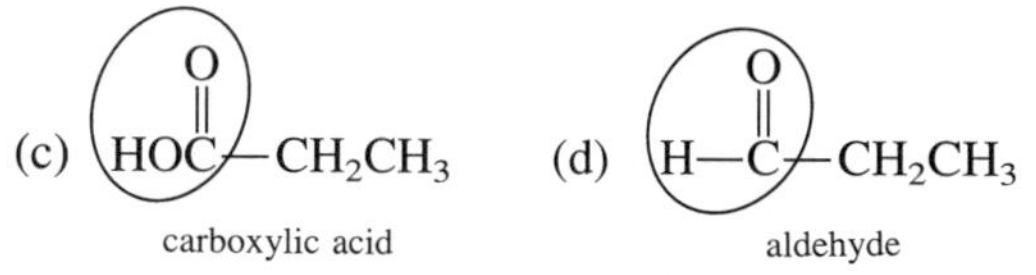

(c) $HO\overset{O}{\overset{\|}{C}}—CH_2CH_3$ carboxylic acid (d) $H—\overset{O}{\overset{\|}{C}}—CH_2CH_3$ aldehyde

24.43 (a) 1-pentanol (b) 2-pentanol (c) 2-propyl-1-pentanol (d) 4-heptanol **24.45** (a) secondary alcohol (b) tertiary alcohol (c) primary alcohol (d) primary alcohol **24.47** (a) ethyl propyl ether (b) methyl isopropyl ether **24.49** (a) 2-butanone (b) butanal (c) 4,4-dimethylpentanal (d) 3-methyl-2-pentanone

24.51 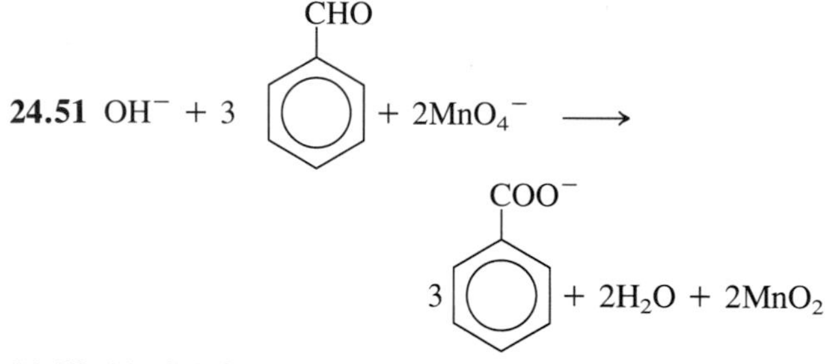

$OH^- + 3\,C_6H_5CHO + 2MnO_4^- \longrightarrow 3\,C_6H_5COO^- + 2H_2O + 2MnO_2$

24.53 (a) CH_3CHCH_3 with CH_2COOH (b) no reaction

3-methylbutanoic acid

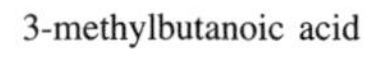

(c) 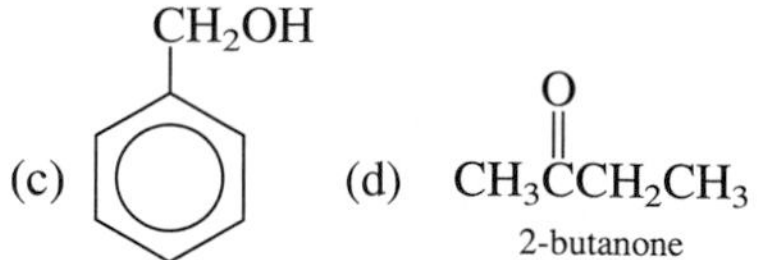

phenylmethanol

(d) $CH_3\overset{O}{\overset{\|}{C}}CH_2CH_3$

2-butanone

24.55 (a) $CH_3CH_2CH_2\overset{O}{\overset{\|}{C}}—OH + HO—CH_2CH_3 \xrightarrow{H^+} CH_3CH_2CH_2\overset{O}{\overset{\|}{C}}—O—CH_2CH_3 + H_2O$

(b) $H\overset{O}{\overset{\|}{C}}OCH_3 + NaOH \longrightarrow H\overset{O}{\overset{\|}{C}}O^-Na^+ + CH_3OH$

24.57 (a) primary amine (b) secondary amine

24.59 $nCF_2{=}CF_2 \longrightarrow —CF_2—CF_2—CF_2—CF_2—CF_2—CF_2—$

24.61 (a) 3-methylbutanoic acid (b) *trans*-5-methyl-2-hexene (c) 2,5-dimethyl-4-heptanone (d) 4-methyl-2-pentyne

24.63 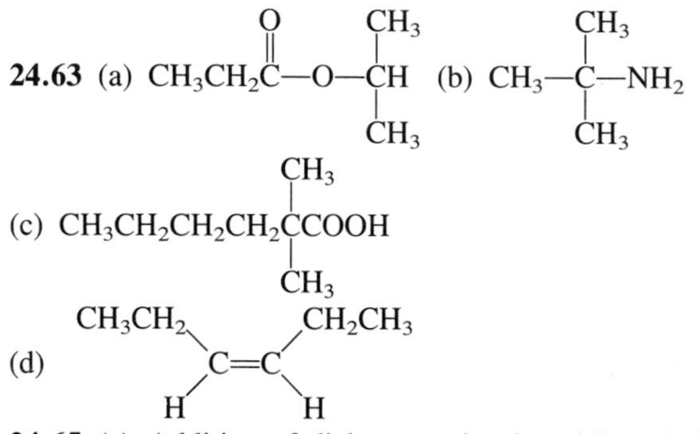

(a) $CH_3CH_2\overset{O}{\overset{\|}{C}}—O—CH(CH_3)_2$ (b) $CH_3—C(CH_3)_2—NH_2$

(c) $CH_3CH_2CH_2CH_2C(CH_3)_2COOH$

(d) $CH_3CH_2(H)C{=}C(H)CH_2CH_3$

24.65 (a) Addition of dichromate ion in acidic solution to propionaldehyde will cause the reagent to change from orange to green as the aldehyde is oxidized. Under similar conditions, acetone would not react. (b) Add Br_2 in CCl_4 to each compound. First compound reacts; the color of Br_2 fades. Benzene does not react. **24.67** (a) ethylene (b) toluene (c) methylamine (d) methanol

24.69 1-butene or 2-butene

Chapter 25

25.17 −2.75 kcal/mol **25.19** releases free energy

25.21 $CH_3—C(H)(\overset{+}{N}H_3)—COO^-$ **25.23** 2 dipeptides are possible

25.25

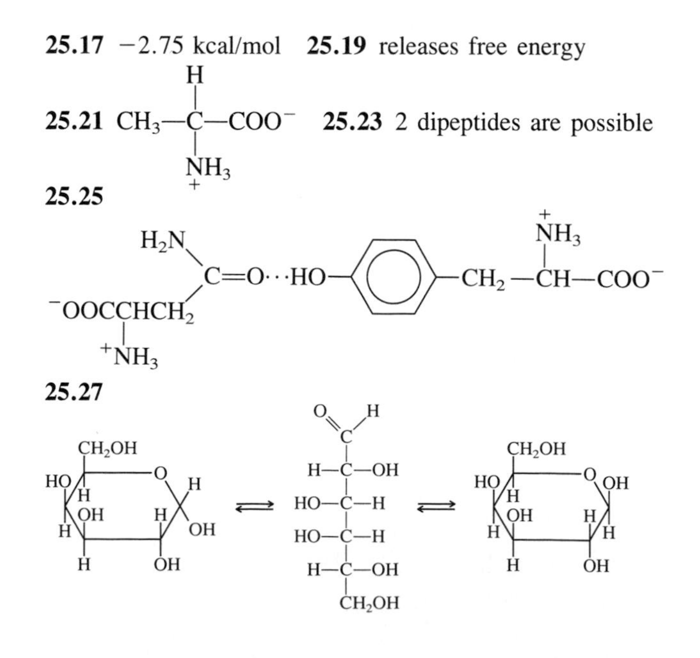

25.27

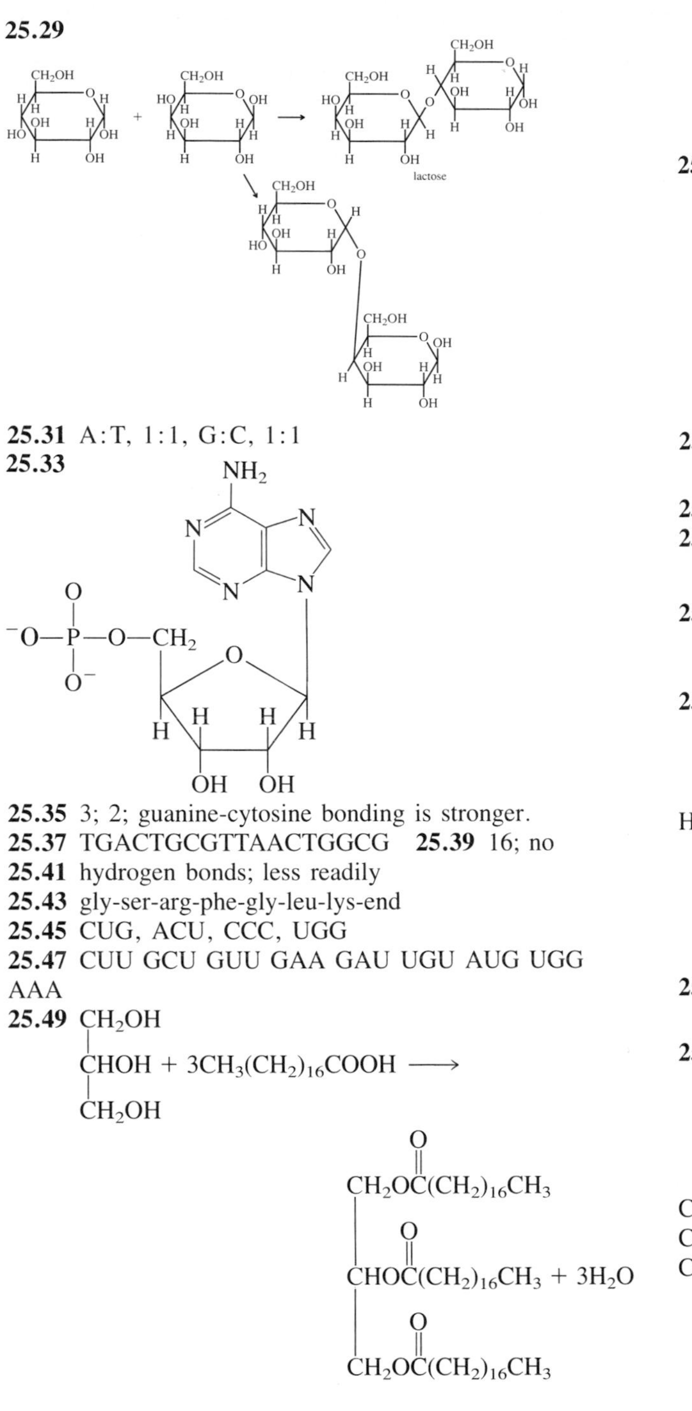

25.29

lactose

25.31 A:T, 1:1, G:C, 1:1

25.33

25.35 3; 2; guanine-cytosine bonding is stronger.

25.37 TGACTGCGTTAACTGGCG **25.39** 16; no

25.41 hydrogen bonds; less readily

25.43 gly-ser-arg-phe-gly-leu-lys-end

25.45 CUG, ACU, CCC, UGG

25.47 CUU GCU GUU GAA GAU UGU AUG UGG AAA

25.49 CH_2OH–$CHOH$–CH_2OH + $3CH_3(CH_2)_{16}COOH$ ⟶ $CH_2OC(O)(CH_2)_{16}CH_3$–$CHOC(O)(CH_2)_{16}CH_3$–$CH_2OC(O)(CH_2)_{16}CH_3$ + $3H_2O$

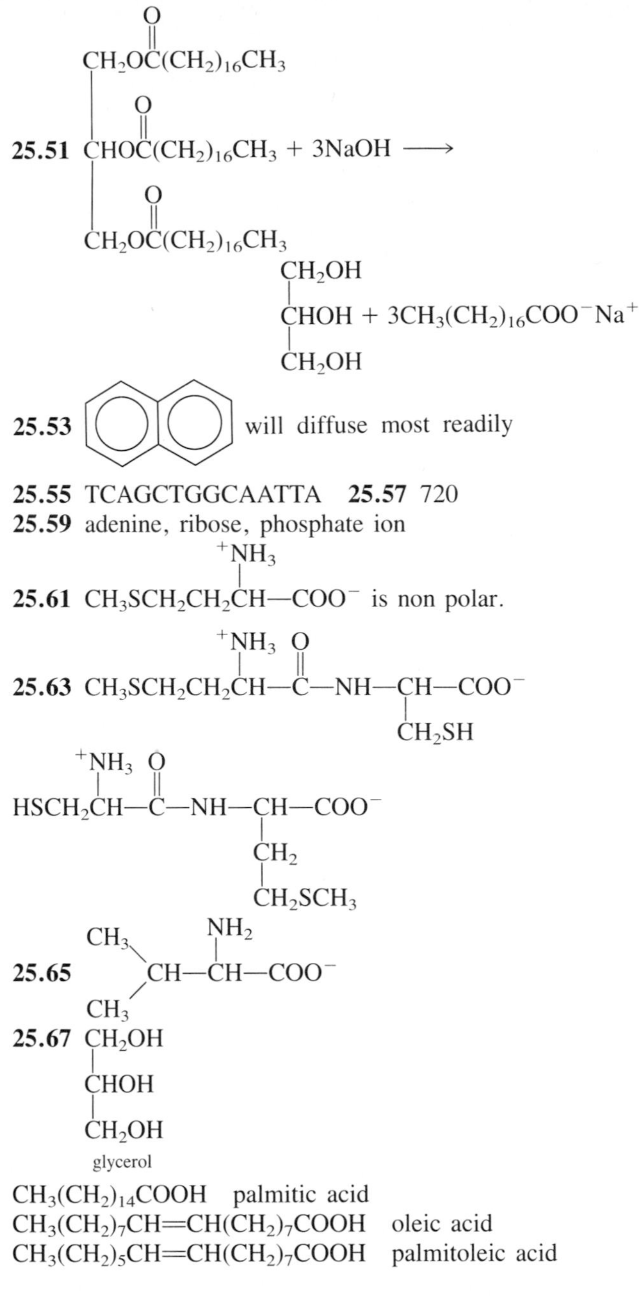

25.51 $CH_2OC(O)(CH_2)_{16}CH_3$–$CHOC(O)(CH_2)_{16}CH_3$–$CH_2OC(O)(CH_2)_{16}CH_3$ + 3NaOH ⟶ CH_2OH–$CHOH$–CH_2OH + $3CH_3(CH_2)_{16}COO^-Na^+$

25.53 will diffuse most readily

25.55 TCAGCTGGCAATTA **25.57** 720

25.59 adenine, ribose, phosphate ion

25.61 $CH_3SCH_2CH_2CH(^+NH_3)$—COO^- is non polar.

25.63 $CH_3SCH_2CH_2CH(^+NH_3)$—C(O)—NH—$CH(CH_2SH)$—COO^-

$HSCH_2CH(^+NH_3)$—C(O)—NH—CH($CH_2CH_2SCH_3$)—COO^-

25.65 $(CH_3)_2CH$—$CH(NH_2)$—COO^-

25.67 CH_2OH–$CHOH$–CH_2OH

glycerol

$CH_3(CH_2)_{14}COOH$ palmitic acid

$CH_3(CH_2)_7CH{=}CH(CH_2)_7COOH$ oleic acid

$CH_3(CH_2)_5CH{=}CH(CH_2)_7COOH$ palmitoleic acid

Glossary

The number given in parentheses at the end of a definition indicates the section (or page) where the term was introduced. In a few cases, several sections are indicated.

Absolute entropy see *standard entropy.*

Absolute temperature scale a temperature scale in which the lowest temperature that can be attained theoretically is zero. (1.4)

Accuracy the closeness of a single measurement to its true value. (1.3)

Acid (Arrhenius definition) any substance that, when dissolved in water, increases the concentration of hydronium ion, H_3O^+ (hydrogen ion, H^+). (3.4 and 13.1)
(Brønsted-Lowry definition) the species (molecule or ion) that donates a proton to another species in a proton-transfer reaction. (3.4 and 13.2)

Acid-base indicator a dye used to distinguish between acidic and basic solutions by means of the color changes it undergoes in these solutions. (3.4)

Acid-base titration curve a plot of the pH of a solution of acid (or base) against the volume of added base (or acid). (16.10)

Acidic oxide (acid anhydride) an oxide that reacts with bases. (8.6)

Acid-ionization (or acid-dissociation) constant (K_a) the equilibrium constant for the ionization of a weak acid. (16.4)

Acid rain rain having a pH lower than that of natural rain, which has a pH of 5.6. (p. 650)

Acid salt a salt that has an acidic hydrogen atom and can undergo neutralization with bases. (3.8)

Actinides the 14 elements following actinium in the periodic table, in which the 5*f* subshell is filling. (p. 923)

Activated complex (transition state) an unstable grouping of reactant molecules that can break up to form products. (14.5)

Activation energy (E_a) the minimum energy of collision required for two molecules to react. (14.5)

Activity of a radioactive source the number of nuclear disintegrations per unit time occurring in a radioactive material. (20.3)

Activity series a table that lists the elements in order of their ease of losing electrons during reactions in aqueous solution. (13.7)

Addition polymer a polymer formed by linking together many molecules by addition reactions. (24.8)

Addition reaction a reaction in which parts of a reagent are added to each carbon atom of a carbon–carbon multiple bond, which becomes a C—C single bond. (24.4)

Adsorption the binding or attraction of molecules to a surface. See *chemisorption* and *physical adsorption.*

Aerosols colloids consisting of liquid droplets or solid particles dispersed throughout a gas. (12.9)

Alcohol a compound obtained by substituting a hydroxyl group (—OH) for an —H atom on an aliphatic hydrocarbon. (24.5)

Aldehyde a compound containing a carbonyl group with at least one H atom attached to it. (24.5)

Aliphatic hydrocarbons hydrocarbons that do not contain benzene rings. (24.1)

Alkali metals the Group IA elements; they are reactive metals. (2.4)

Alkaline dry cell a voltaic cell that is similar to the Leclanché dry cell but uses potassium hydroxide in place of ammonium chloride. (19.7)

Alkaline earth metals the Group IIA elements; they are reactive metals, though less reactive than the alkali metals. (8.2)

Alkanes saturated hydrocarbons that have the general formula C_nH_{2n+2}. (24.1)

Alkenes hydrocarbons that have the general formula C_nH_{2n} and contain a carbon–carbon double bond. (24.2)

Alkyl group an alkane less one hydrogen atom. (24.1)

Alkynes unsaturated hydrocarbons containing a carbon–carbon triple bond. (24.2)

Allotrope one of two or more distinct forms of an element in the same physical state. (6.8)

Alpha emission emission of a $^4_2$He nucleus, or alpha particle, from an unstable nucleus. (20.1)

Aluminosilicate mineral a type of silicate consisting of three-dimensional networks in which some of the SiO_4 tetrahedra are replaced by AlO_4 tetrahedra. (22.1)

Ambidentate ligand a ligand that can bond to a metal atom through one atom or another. (23.5)

Amides compounds derived from the reaction of ammonia, or a primary or secondary amine, with a carboxylic acid. (24.7)

Amide bonds see *peptide bonds.*

Amines compounds that are structurally derived by replacing one or more hydrogen atoms of ammonia by hydrocarbon groups. (24.7)

Amino acids molecules containing a protonated amino group (NH_3^+) and an ionized carboxyl group (COO^-). (25.3)

Amorphous solid a solid that has a disordered structure; it lacks the well-defined arrangement of basic units (atoms, molecules, or ions) found in the crystal. (11.7)

Ampere (A) the base unit of current in the international system. (19.10)

Amphiprotic species a species that can act either as an acid or as a base (that is, it can lose or gain a proton). (13.2)

Amphoteric substance a substance that has both acidic and basic properties. (8.6 and 21.1)

Amphoteric hydroxide a metal hydroxide that reacts with bases as well as with acids. (17.5)

Angstrom (Å) a non-SI unit of length; 1 Å $= 10^{-10}$ m. (1.4)

Angular geometry (bent geometry) a nonlinear geometry in the case of a molecule of three atoms. (10.1)

Angular momentum quantum number (l) also known as the azimuthal quantum number. The quantum number that distinguishes orbitals of given n having different shapes; it can have any integer value from 0 to $n - 1$. (7.9)

Anion a negatively charged ion. (2.5)

Anode the electrode at which oxidation occurs. (19.1)

Antibonding orbitals molecular orbitals that are concentrated in regions other than between two nuclei. (10.5)

Aromatic hydrocarbons hydrocarbons that contain benzene rings or similar structural features. (24.1)

Arrhenius equation the mathematical equation $k = Ae^{-E_a/RT}$, which expresses the dependence of the rate constant on temperature. (14.6)

Association colloid a colloid in which the dispersed phase consists of micelles. (12.9)

Atmosphere (atm) a unit of pressure equal to exactly 760 mmHg; 1 atm = 101.325 kPa (exact). (5.1)

Atom an extremely small particle of matter that retains its identity during chemical reactions. (2.2)

Atomic mass unit (amu) a mass unit equal to exactly one-twelfth the mass of a carbon-12 atom. (2.3)

Atomic number (Z) the number of protons in the nucleus. (2.2 and 7.3)

Atomic orbital a wave function for an electron in an atom; pictured qualitatively by describing the region of space where there is a high probability of finding the electron. (7.9)

Atomic symbol a one-, two-, or three-letter notation used to represent an atom corresponding to a particular element. (2.3)

Atomic theory an explanation of the structure of matter in terms of different combinations of very small particles (atoms). (2.2)

Atomic weight the average atomic mass for a naturally occurring element, expressed in atomic mass units. (2.3)

Aufbau principle see *building-up principle.*

Autoionization see *self-ionization.*

Avogadro's law a law stating that equal volumes of any two gases at the same temperature and pressure contain the same number of molecules. (5.2)

Avogadro's number (N_A) the number of atoms in a 12-g sample of carbon-12, equal to 6.02×10^{23} to three significant figures. (4.2)

Axial direction one of two directions pointing from the center of a trigonal bipyramid along its axis. (10.1)

Azimuthal quantum number see *angular momentum quantum number.*

Band of stability the region in which stable nuclides lie in a plot of number of protons against number of neutrons. (20.1)

Band theory molecular orbital theory of metals. (10.8)

Barometer a device for measuring the pressure of the atmosphere. (5.1)

Base (Arrhenius definition) any substance that, when dissolved in water, increases the concentration of hydroxide ion, OH^-. (3.4 and 13.1) (Brønsted-Lowry definition) The species (molecule or ion) that accepts a proton in a proton-transfer reaction. (3.4 and 13.2)

Base-ionization (or base-dissociation) constant (K_b) the equilibrium constant for the ionization of a weak base. Thus K_b for NH_3 is 1.8×10^{-5}. (16.6)

Base pairing the hydrogen bonding of complementary bases. (25.5)

Basic oxide (basic anhydride) an oxide that reacts with acids. (8.6)

Basic oxygen process a method of making steel by blowing oxygen into the molten iron to oxidize impurities and decrease the amount of carbon present. (p. 736)

Bayer process a process in which purified aluminum oxide is obtained from bauxite. (21.5)

Bent geometry see *angular geometry.*

Beta emission the emission of a high-speed electron from an unstable nucleus. (20.1)

Bidentate ligand a ligand that bonds to a metal atom through two atoms of the ligand. (23.3)

Bimolecular reaction an elementary reaction that involves two reactant molecules. (14.7)

Binary compound a compound composed of only two elements. (2.6)

Binding energy (of a nucleus) the energy needed to break a nucleus into its individual protons and neutrons. (20.6)

Body-centered cubic unit cell a cubic unit cell in which there is a lattice point at the center of the unit cell as well as at the corners. (11.7)

Boiling point the temperature at which the vapor pressure of a liquid equals the atmospheric pressure. (11.2)

Boiling-point elevation a colligative property of a solution equal to the boiling point of the solution minus the boiling point of the pure solvent. (12.6)

Bond distance see *bond length*.

Bond energy the average enthalpy change for the breaking of a bond in a molecule in the gas phase. (9.10)

Bonding orbitals molecular orbitals that are concentrated in regions between nuclei. (10.5)

Bonding pair an electron pair shared between two atoms. (9.4)

Bond length (bond distance) the distance between the nuclei in a bond. (9.9)

Bond order in a Lewis formula, the number of pairs of electrons in a bond. (9.9) In molecular orbital theory, one-half the difference between the number of bonding electrons and the number of antibonding electrons. (10.5)

Boranes binary compounds of boron and hydrogen. (21.4)

Boyle's law a law stating that the volume of a sample of gas at a given temperature varies inversely with the applied pressure. (5.2)

Breeder reactor a nuclear reactor designed to produce more fissionable material than it consumes. (20.7)

Brønsted–Lowry concept a concept of acids and bases in which an acid is the species donating a proton in a proton-transfer reaction, whereas a base is the species accepting a proton in such a reaction. (3.4 and 13.2)

Buffer a solution characterized by the ability to resist changes in pH when limited amounts of acid or base are added to it. (16.9)

Building-up principle (Aufbau principle) a scheme used to reproduce the electron configurations of the ground states of atoms by successively filling subshells with electrons in a specific order (the building-up order). (8.2)

Calorie (cal) a non-SI unit of energy commonly used by chemists originally defined as the amount of energy needed to raise the temperature of 1 g of water by 1°C; now defined as 1 cal = 4.184 J (exact). (6.1)

Calorimeter a device used to measure the heat absorbed or evolved during a physical or chemical change. (6.6)

Carbohydrates polyhydroxy aldehydes or ketones, or substances that yield such compounds when they react with water. (25.4)

Carboxylic acid a compound containing the carboxyl group, —COOH. (24.5)

Catalysis the increase in rate of a reaction as the result of the addition of a catalyst. (14.9)

Catalyst a substance that increases the rate of reaction without being consumed in the overall reaction. (3.1 and p. 548)

Catenation the ability of an atom to bond covalently to like atoms. (22.1)

Cathode the electrode at which reduction occurs. (19.1)

Cathode rays the rays emitted by the cathode (negative electrode) in a gas discharge tube. (7.1)

Cation a positively charged ion. (2.5)

Cation exchange resin a resin (organic polymer) that removes cations from solution and replaces them with hydrogen ions. (p. 494)

Cell reaction the net reaction that occurs in a voltaic cell. (19.1)

Celsius scale the temperature scale in general scientific use. There are exactly 100 units between the freezing point and the normal boiling point of water. (1.4)

Chain reaction, nuclear a self-sustaining series of nuclear fissions caused by the absorption of neutrons released from previous nuclear fissions. (20.7)

Change of state (phase transition) a change of a substance from one state to another. (p. 403)

Charles's law a law stating that the volume occupied by any sample of gas at a constant pressure is directly proportional to the absolute temperature. (5.2)

Chelate a complex formed by polydentate ligands. (23.3)

Chemical bond a strong attractive force that exists between certain atoms in a substance. (p. 312)

Chemical change see *chemical reaction*.

Chemical equation the symbolic representation of a chemical reaction in terms of the chemical formulas. (3.1)

Chemical equilibrium the state reached by a reaction mixture when the rate of forward reaction and that of the reverse reaction have become equal. (15.1)

Chemical formula a notation that uses atomic symbols with numerical subscripts to convey the relative proportions of atoms of the different elements in the substance. (2.5)

Chemical kinetics the study of how reaction rates change under varying conditions and of what molecular events occur during the overall reaction. (p. 548)

Chemical nomenclature the systematic naming of chemical compounds. (2.6)

Chemical property a characteristic of a material involving its chemical change. (2.1)

Chemical reaction (chemical change) a change in which one or more kinds of matter are transformed into a new kind of matter or several new kinds of matter. (2.1) It consists of the rearrangement of the atoms present in the reacting substances to give new chemical combinations present in the substances formed by the reaction. (2.2)

Chemisorption the binding of a species to a surface by chemical bonding forces. (14.9)

Chiral possessing the quality of handedness. A chiral object has a mirror image that is not identical to the object. (23.5)

Chlor–alkali membrane cell a cell for the electrolysis of aqueous sodium chloride in which the anode and the cathode compartments are separated by a special plastic membrane that allows only cations to pass through it. (19.9)

Chlor–alkali mercury cell a cell for the electrolysis of aqueous sodium chloride in which mercury metal is used as the cathode. (19.9)

Chromatography a name given to a group of similar separation techniques that depend on how fast a substance moves, in a stream of gas or liquid, past a stationary phase to which the substance may be slightly attracted. (p. 38)

Clausius–Clapeyron equation an equation that expresses the relation between the vapor pressure P of a liquid and the absolute temperature T: $\log P = -\Delta H_{vap}/2.303RT + B$, where B is a constant. (11.2)

Coagulation the process by which a colloid is made to come out of solution by aggregation. (12.9)

Codon a sequence of three bases in a messenger-RNA molecule that serves as the code for a particular amino acid. (25.5)

Colligative properties properties that depend on the concentration of solute molecules or ions in a solution but not on the chemical identity of the solute. (p. 470)

Collision theory a theory that assumes that, in order for reaction to occur, reactant molecules must collide with an energy greater than some minimum value and with proper orientation. (14.5)

Colloid a dispersion of particles of one substance (the dispersed phase) throughout another substance or solution (the continuous phase). (12.9)

Combination reaction a reaction in which two substances combine to form a third substance. (3.2)

Combustion reaction a reaction of a substance with oxygen, usually with the rapid release of heat to produce a flame. (3.2)

Common-ion effect the shift in an ionic equilibrium caused by the addition of a solute that provides an ion that takes part in the equilibrium. (16.8 and 17.2)

Complementary bases nucleotide bases that form strong hydrogen bonds with one another. (25.5)

Complex (coordination compound) a compound consisting either of complex ions with other ions of opposite charge or of a neutral complex species. (23.3)

Complex ion an ion consisting of a metal atom or ion with Lewis bases attached to it through coordinate covalent bonds. (p. 696 and 23.3)

Compound a substance composed of more than one element chemically combined. (2.1) A type of matter composed of atoms of two or more elements chemically combined in fixed proportions. (2.2)

Concentration a general term referring to the quantity of solute in a standard quantity of solution. (4.9 and 12.4)

Condensation the change of a gas to either the liquid or the solid state. (11.2)

Condensation polymer a polymer formed by linking many molecules together by condensation reactions. (24.8)

Condensation reaction a reaction in which two molecules or ions are joined by the elimination of a small molecule such as water. (22.1 and 24.6)

Conjugate acid in a conjugate acid–base pair, the species that can donate a proton. (13.2)

Conjugate acid–base pair two species in an acid–base reaction, one acid and one base, that differ by the gain or loss of a proton. (13.2)

Conjugate base in a conjugate acid–base pair, the species that can accept a proton. (13.2)

Contact process an industrial method for the manufacture of sulfuric acid. It consists of the oxidation of sulfur dioxide with a solid catalyst (platinum or divanadium pentoxide), followed by the reaction of sulfur trioxide with water. (22.3)

Continuous spectrum a spectrum containing light of all wavelengths. (7.7)

Control rods cylinders composed of substances that absorb neutrons, such as boron or cadmium, and can therefore slow a nuclear chain reaction. (20.7)

Conversion factor a ratio equal to 1 that converts a quantity expressed in one unit to a quantity expressed in another unit. (1.6)

Coordinate covalent bond a bond formed when both electrons of the bond are donated by one atom. (9.4)

Coordination compound a compound consisting of either complex ions with other ions of opposite charge or a neutral complex species. (23.3)

Coordination isomers isomers consisting of complex cations and complex anions that differ in the way the ligands are distributed between the metal atoms. (23.5)

Coordination number in a crystal, the number of nearest-neighbor atoms of an atom. (11.8) In a complex, the total number of bonds the metal atom forms with ligands. (23.3)

Covalent bond a chemical bond formed by the sharing of a pair of electrons between atoms. (p. 323)

Covalent network solid a solid that consists of atoms held together in large networks or chains by covalent bonds. (11.6)

Covalent radii values assigned to atoms in such a way that the sum of covalent radii of atoms A and B predicts an approximate A—B bond length. (9.9)

Critical mass the smallest mass of fissionable material in which a chain reaction can be sustained. (20.7)

Critical pressure the vapor pressure at the critical temperature; it is the minimum pressure that must be applied to a gas at the critical temperature to liquefy it. (11.3)

Critical temperature the temperature above which the liquid state of a substance no longer exists. (11.3)

Crystal a kind of solid having a definite geometrical shape as a result of the regular arrangement of atoms, molecules, or ions that make up the substance. (2.5)

Crystal field splitting (Δ) the difference in energy between the two sets of *d* orbitals on a central metal ion that arises from the interaction of the orbitals with the electric field of the ligands. (23.7)

Crystal field theory a model of the electronic structure of transition-metal complexes that considers how the energies of the *d* orbitals of a metal ion are affected by the electric field of the ligands. (23.7)

Crystal lattice the geometric arrangement of lattice points of a crystal, in which we choose one lattice point at the same location within each of the basic units of the crystal. (11.7)

Crystalline solid a solid composed of one or more crystals; each crystal has a well-defined ordered structure. (11.7)

Crystal systems the seven basic shapes possible for unit cells; a classification of crystals. (11.7)

Cubic close-packed structure (ccp) a crystal structure composed of close-packed atoms (or other units) with the stacking ABCABCABCA ···. (11.8)

Curie (Ci) a unit of activity, equal to 3.700×10^{10} disintegrations per second. (20.3)

Cycloalkanes saturated hydrocarbons in which the carbon atoms form a ring; the general formula is C_nH_{2n}. (24.1)

Cyclotron a type of particle accelerator consisting of two hollow, semicircular metal electrodes, called dees, that are charged by a high-frequency alternating current. The particle is continually accelerated in the region between the dees. (20.2)

Dalton's law of partial pressures a law stating that the sum of the partial pressures of all the different gases in a mixture is equal to the total pressure of the mixture. (5.5)

***d*-Block transition elements** those transition elements with an unfilled *d* subshell in common oxidation states. (8.2 and p. 923)

de Broglie relation the equation $\lambda = h/mv$. (7.8)

Decomposition reaction a reaction in which a single compound reacts to give two or more substances. (3.2)

Degree of ionization the fraction of molecules that react with water to give ions. (16.4)

Delocalized bonding a type of bonding in which a bonding pair of electrons is spread over a number of atoms rather than localized between two. (9.8)

Denitrifying bacteria bacteria that use nitrate ion, NO_3^-, as a source of energy; they convert the ion to gaseous nitrogen. (p. 175)

Density the mass per unit volume of a substance or solution. (1.5)

Deoxyribonucleic acid (DNA) the hereditary constituent of cells; it consists of two polymer strands of deoxyribonucleotide units. (25.5)

Deposition the change of a vapor to a solid. (11.2)

Derived unit see *SI derived unit*.

Desalinate to remove ions from brackish (slightly salty) water or seawater, producing water that is fit to drink (potable). (12.7)

Deuterons nuclei of hydrogen-2 atoms. (20.2)

Dextrorotatory refers to a compound whose solution rotates the plane of polarized light to the right (when we are looking toward the source of light). (23.5)

Diagonal relationship the resemblance of each of the first three elements of the second row of the periodic table to the element located diagonally to the right of it in the third row. (21.1)

Diamagnetic substance a substance that is not attracted by a magnetic field or is very slightly repelled by such a field. This property generally means that the substance has only paired electrons. (8.3)

Diffusion the process whereby a gas spreads out through another gas to occupy the space with uniform partial pressure (5.7)

Dimensional analysis (factor-label method) the method of calculation in which one carries along the units for quantities. (1.6)

Dipole–dipole force an attractive intermolecular force resulting from the tendency of polar molecules to align themselves such that the positive end of one molecule is near the negative end of another. (11.5)

Dipole moment a quantitative measure of the degree of charge separation in a molecule. (10.2)

Dispersion forces see *London forces*.

Displacement reaction (single-replacement reaction) a reaction in which an element reacts with a compound displacing an element from it. (3.2)

Disproportionation a reaction in which a species is both oxidized and reduced. (13.7)

Dissociation constant of a complex (K_d) the reciprocal, or inverse, value of the formation constant. (17.5)

Distillation the process in which a liquid is vaporized and condensed; used to separate substances of different volatilities. (2.1)

Distorted tetrahedral geometry see *seesaw geometry*.

Double bond a covalent bond in which two pairs of electrons are shared by two atoms. (9.4)

Double-replacement reactions see *metathesis reactions*.

Downs cell a commercial electrochemical cell used to obtain sodium metal by the electrolysis of molten sodium chloride. (19.8)

Dry cell see *alkaline dry cell* and *zinc–carbon dry cell.*

Effective nuclear charge the positive charge experienced by an electron from the nucleus, equal to the nuclear charge but reduced by any shielding or screening from any intervening electron distribution. (8.5)

Effusion the process in which a gas flows through a small hole in a container. (5.7)

Electrochemical cell a system consisting of electrodes that dip into an electrolyte and in which a chemical reaction either uses or generates an electric current. (p. 747)

Electrolysis the process of producing a chemical change in an electrolytic cell. (p. 772)

Electrolyte a substance, such as sodium chloride, that dissolves in water to give an electrically conducting solution. (3.3)

Electrolytic cell an electrochemical cell in which an electric current is used to drive a nonspontaneous reaction. (p. 747)

Electromagnetic spectrum the range of frequencies or wavelengths of electromagnetic radiation. (7.5)

Electromotive force (emf) the maximum potential difference between the electrodes of a voltaic cell. (19.3)

Electron a very light particle that carries a unit negative charge and exists in the region around the positively charge nucleus. (2.2 and p. 230)

Electron affinity the energy change for the process of adding an electron to a neutral atom in the gaseous state to form a negative ion. (8.5)

Electron capture the decay of an unstable nucleus by capturing, or picking up, an electron from an inner orbital of an atom. (20.1)

Electron configuration the particular distribution of electrons among available subshells. (8.1)

Electronegativity a measure of the ability of an atom in a molecule to draw bonding electrons to itself. (9.5)

Electron spin quantum number (m_s) see *spin quantum number*.

Electron volt (eV) the quantity of energy that must be imparted to an electron (whose charge is 1.602×10^{-19} C) to accelerate it by one volt potential difference. (20.2)

Element a substance that cannot be decomposed by any chemical reaction into simpler substances. (2.1) A type of matter composed of only one kind of atom, each atom of a given kind having the same properties. (2.2)

Elementary reaction a single molecular event, such as a collision of molecules, resulting in a reaction. (14.7)

Empirical formula (simplest formula) the formula of a substance written with the smallest integer subscripts. (4.5)

Emulsion a colloid consisting of liquid droplets dispersed throughout another liquid. (12.9)

Enantiomers (optical isomers) isomers that are nonsuperimposable mirror images of one another. (23.5)

Endothermic process a chemical reaction or physical change in which heat is absorbed. (6.2)

Energy the potential or capacity to move matter. (6.1)

Energy levels specific energy values in an atom. (7.7)

Enthalpy (H) an extensive property of a substance that can be used to obtain the heat absorbed or evolved in a chemical reaction or physical change at constant pressure. (6.3) It equals the quantity $U + PV$. (18.1)

Enthalpy of formation see *standard enthalpy of formation.*

Enthalpy of reaction (ΔH) the change in enthalpy for a reaction at a given temperature and pressure; it equals the heat of reaction at constant pressure. (6.3)

Entropy (S) a thermodynamic quantity that is a measure of the randomness or disorder in a system. (18.2)

Enzyme a protein that catalyzes a biochemical reaction. (25.1)

Equatorial direction one of the three directions pointing from the center of a trigonal bipyramid to a vertex other than one on the axis. Compare *axial direction.* (10.1)

Equilibrium see *chemical equilibrium.*

Equilibrium constant K_c the value obtained for the equilibrium-constant expression when equilibrium concentrations are substituted. (15.2)

Equilibrium constant K_p an equilibrium constant for a gas reaction, similar to K_c, but in which concentrations of gases are replaced by pressures (in atm). (15.2)

Equilibrium constant K see *thermodynamic equilibrium constant.*

Equilibrium-constant expression an expression obtained for a reaction by multiplying the concentrations of products together, dividing by the concentrations of reactants, and raising each concentration term to a power equal to the coefficient in the chemical equation. (15.2)

Equivalence point the point in a titration at which a stoichiometric amount of reactant has been added. (16.10)

Equivalent (eq) in an acid–base reaction, the quantity of acid that yields 1 mol H^+ or the quantity of base that reacts with 1 mol of H^+. In a redox reaction, the mass of oxidizing or reducing agent that uses or provides one mole of electrons. (13.9)

Equivalent mass the mass of one equivalent. (13.9)

Ester a compound formed from a carboxylic acid, RCOOH, and an alcohol, R′OH. (24.5)

Ether a compound formally obtained by replacing both H atoms of H_2O by hydrocarbon groups R and R′. (24.5)

Exact number a number that arises when we count items or sometimes when we define a unit. (1.3)

Excited state a quantum-mechanical state of an atom or molecule associated with any energy level except the lowest, which is the ground state. (8.2)

Exothermic process a chemical reaction or physical change in which heat is evolved. (6.2)

Experiment an observation of natural phenomena carried out in a controlled manner so that the results can be duplicated and rational conclusions can be obtained. (1.2)

Extensive property a measurable physical property whose magnitude depends on the amount of material. (2.1)

Face-centered cubic unit cell a cubic unit cell in which there are lattice points at the centers of each face of the unit cell, in addition to those at the corners. (11.7)

Factor-label method see *dimensional analysis*.

Faraday constant (*F*) the magnitude of charge on one mole of electrons, equal to 9.65×10^4 C. (19.3)

***f*-Block transition elements (inner-transition elements)** the elements with a partially filled *f* subshell in common oxidation states. (8.2 and p. 923)

Fibrous proteins proteins that form long coils or align themselves in parallel to form long, water-insoluble fibers. (25.3)

First ionization energy (first ionization potential) the minimum energy needed to remove the highest-energy (that is, the outermost) electron from a neutral atom in the gaseous state. (8.5)

First law of thermodynamics a law stating that the change in internal energy of a system, ΔU, equals $q + w$. (18.1)

Flotation a physical method of separating a mineral from the gangue that depends on differences in their wettability. (21.5)

Formal charge (of an atom in a Lewis formula) the hypothetical charge we obtain when we assign electron dots to the different atoms in a prescribed way. (9.6)

Formation constant (stability constant) of a complex (K_f) the equilibrium constant for the formation of a complex ion from the aqueous metal ion and the ligands. (17.5)

Formula unit the group of atoms or ions explicitly symbolized in the formula. (2.5)

Formula weight (FW) the sum of the atomic weights of all atoms in a formula unit of a compound. (4.1)

Fractional (isotopic) abundance the fraction of the total number of atoms that is composed of a particular isotope. (7.4)

Fractional precipitation the technique of separating two or more ions from a solution by adding a reactant that precipitates first one ion, then another, and so forth. (17.3)

Frasch process a process in which underground deposits of solid sulfur are melted in place with superheated water and the molten sulfur is forced upward as a froth, using air under pressure. (22.3)

Free energy (*G*) a thermodynamic quantity defined by the equation $G = H - TS$. (18.4)

Free energy of formation see *standard free energy of formation*.

Freezing the change of a liquid to the solid state. (11.2)

Freezing point the temperature at which a liquid changes to a crystalline solid, or freezes. (11.2)

Freezing-point depression a colligative property of a solution equal to the freezing point of the pure solvent minus the freezing point of the solution. (12.6)

Frequency (ν) the number of wavelengths of a wave that pass a fixed point in one unit of time (usually one second). (7.5)

Frequency factor the symbol *A* in the Arrhenius equation, assumed to be a constant. (14.6)

Fuel cell essentially a battery, but it differs by operating with a continuous supply of energetic reactants or fuel. (19.7)

Fuel rods the cylinders that contain fissionable material for a nuclear reactor. (20.7)

Functional group a reactive portion of a molecule that undergoes predictable reactions. (p. 987)

Fusion see *melting*.

Galvanic cell see *voltaic cell*.

Gamma emission emission from an excited nucleus of a gamma photon, corresponding to radiation with a wavelength of about 10^{-12} m. (20.1)

Gamma photons particles of electromagnetic radiation of short wavelength (0.01 Å, or 10^{-12} m) and high energy per photon. (20.1)

Gangue the impurities in an ore. (21.5)

Gas the form of matter that is an easily compressible fluid; a given quantity of gas will fit into a container of any size and shape. (2.1)

Geiger counter a kind of ionization counter used to count particles emitted by radioactive nuclei. It consists of a metal tube filled with gas, such as argon. (20.3)

Gene a sequence of nucleotides in a DNA molecule that codes for a given protein. (25.5)

Geometric isomers isomers in which the atoms are joined to one another in the same way but that differ because some atoms occupy different relative positions in space. (10.4, 23.5, and 24.2)

Glass electrode a compact electrode used to determine pH by emf measurements. (19.6)

Globular proteins proteins in which long coils fold into compact, roughly spherical shapes. (25.3)

Goldschmidt process a method of preparing a metal by reduction of its oxide with powdered aluminum. (21.4)

Graham's law of effusion a law stating that the rate of effusion of gas molecules from a particular hole is inversely proportional to the square root of the molecular weight of the gas at constant temperature and pressure. (5.7)

Ground state a quantum-mechanical state of an atom or molecule associated with the lowest energy level. States associated with higher energy levels are called excited states. (8.2)

Group (of the periodic table) the elements in any one column of the periodic table. (2.4)

Haber process an industrial process for the preparation of ammonia from nitrogen and hydrogen with a specially prepared catalyst, high temperature, and high pressure. (p. 175 and 22.2)

Half-cell the portion of an electrochemical cell in which a half-reaction takes place. (19.1)

Half-life ($t_{1/2}$) the time it takes for the reactant concentration to decrease by one-half of its initial value in a reaction. (14.4) The time it takes for one-half of the nuclei in a sample to decay. (20.4)

Half-reaction one of two parts of a redox reaction, one of which involves a loss of electrons and the other of which involves a gain of electrons. (13.7)

Hall–Héroult process a commercial method of producing aluminum by electrolysis of aluminum oxide dissolved in molten cryolite, Na_3AlF_6. (21.4)

Halogens the Group VIIA elements; they are reactive nonmetals. (2.4)

Heat the energy that flows into or out of a system because of a difference in temperature between the thermodynamic system and its surroundings. (6.2)

Heat capacity (C) the quantity of heat needed to raise the temperature of a sample of substance one degree Celsius (or one kelvin). (6.6)

Heat of formation see *standard enthalpy of formation*.

Heat of fusion (enthalpy of fusion) the heat needed for the melting of a solid. (11.2)

Heat of reaction the heat absorbed (or evolved) during a chemical reaction; it equals the value of q required to return the chemical system to a given temperature at the completion of the reaction. (6.2)

Heat of solution the heat absorbed (or evolved) when a substance dissolves in a quantity of solvent. (12.3)

Heat of vaporization (enthalpy of vaporization) the heat needed for the vaporization of a liquid. (11.2)

Henderson–Hasselbalch equation an equation for determining the pH of a buffer for different concentrations of conjugate acid and base: $pH = pK_a + \log [base]/[acid]$. (16.9)

Henry's law a law stating that the solubility of a gas is directly proportional to the partial pressure of the gas above the solution. (12.3)

Hess's law of heat summation if a chemical equation can be written as the sum of other equations, the ΔH of this overall equation equals a similar sum of the ΔH's for the other equations. (6.7)

Heterogeneous catalysis the use of a catalyst that exists in a different phase from the reacting species, usually a solid catalyst in contact with a gaseous or liquid solution of reactants. (14.9)

Heterogeneous equilibrium an equilibrium involving reactants and products in more than one phase. (15.3)

Heterogeneous mixture a mixture that consists of physically distinct parts with different properties. (2.1)

Heteronuclear diatomic molecules molecules composed of two different nuclei. (10.6)

Hexagonal close-packed structure (hcp) a crystal structure composed of close-packed atoms (or other units) with the stacking ABABABA ⋯; the structure has a hexagonal unit cell. (11.8)

High-spin complex ion a complex ion in which there is minimum pairing of electrons in the orbitals of the metal atom. (23.6)

Homogeneous catalysis the use of a catalyst in the same phase as the reacting species. (14.9)

Homogeneous equilibrium an equilibrium that involves reactants and products in a single phase. (15.3)

Homogeneous mixture (solution) a mixture that is uniform in its properties throughout given samples. (2.1)

Homologous series a series of compounds in which one compound differs from a preceding one by a $-CH_2-$ group. (24.1)

Homonuclear diatomic molecules molecules composed of two like nuclei. (10.6)

Hoopes process a process for the electrolytic refining of aluminum. (21.5)

Hund's rule a principle stating that the lowest energy arrangement of electrons in a subshell is obtained by putting electrons into separate orbitals of the subshell with parallel spin before pairing electrons. (8.3)

Hybrid orbitals orbitals used to describe bonding that are obtained by taking combinations of atomic orbitals used to describe the isolated atoms. (10.3)

Hydrate isomers isomers of a complex that differ in the placement of water molecules in the complex. (23.5)

Hydrate a compound that contains water molecules weakly bound in its crystals. (2.6)

Hydration the attraction of ions for water molecules. (12.2)

Hydrocarbons compounds containing only carbon and hydrogen. (p. 971)

Hydrogen bonding a weak to moderate attractive force that exists between a hydrogen atom covalently bonded to a very electronegative atom and a lone pair of electrons on another small electronegative atom. It is represented in formulas by a series of dots. (11.5)

Hydrohalic acid an aqueous solution of a hydrogen halide. (22.4)

Hydrologic cycle the natural cycle of water from the oceans to fresh water sources and back to the oceans. (p. 493)

Hydrolysis the reaction of an ion with water to produce the conjugate acid and hydroxide ion or the conjugate base and hydrogen ion. (16.7)

Hydronium ion the H_3O^+ ion; also called the hydrogen ion and written $H^+(aq)$. (3.4 and 13.1)

Hydrophillic colloid a colloid in which there is a strong attraction between the dispersed phase and the continuous phase (water). (12.9)

Hydrophobic colloid a colloid in which there is a lack of attraction between the dispersed phase and the continuous phase (water). (12.9)

Hypothesis a tentative explanation of some regularity of nature. (1.2)

Ideal gas law the equation $PV = nRT$, which combines all of the gas laws. (5.3)

Ideal solution a solution of two or more substances each of which follows Raoult's law. (12.5)

Immiscible fluids fluids that do not mix but form separate layers. (12.1)

Inner-transition elements the two rows of elements at the bottom of the periodic table (2.4); the elements with a partially filled *f* subshell in common oxidation states. (8.2 and p. 923)

Inorganic compounds compounds composed of elements other than carbon. A few simple compounds of carbon, including carbon monoxide, carbon dioxide, carbonates, and cyanides, are generally considered to be inorganic. (2.6)

Intensive property a measurable physical property whose magnitude is independent of the amount of material. (2.1)

Interhalogen a binary compound of one halogen with another. (22.4)

Intermolecular forces the forces of interaction between molecules. (11.5)

Internal energy (U) the sum of the kinetic and the potential energies of the particles making up a system. (6.1 and 18.1)

International System of Units (SI) a particular choice of metric units that was adopted by the General Conference of Weights and Measures in 1960. (1.4)

Ion an electrically charged particle obtained from an atom or a chemically bonded group of atoms by adding or removing electrons. (2.5)

Ion exchange a process in which a water solution is passed through a column of a material that replaces one kind of ion in solution by another kind. (p. 494)

Ionic bond a chemical bond formed by the electrostatic attraction between positive and negative ions. (9.1)

Ionic compound a compound composed of cations and anions. (2.5)

Ionic equation a chemical equation for a reaction involving ions in solution in which soluble substances (ones that dissolve readily) are represented by the formulas of the predominant species in that solution. (3.5)

Ionic radius a measure of the size of the spherical region around the nucleus of an ion within which the electrons are most likely to be found. (9.3)

Ionic solid a solid that consists of cations and anions held together by the electrical attraction of opposite charges (ionic bonds). (11.6)

Ionization energy the energy needed to remove an electron from an atom (or molecule). Often used to mean *first ionization energy* (which see). (8.5)

Ionization isomers isomers of a complex that differ in the anion that is coordinated to the metal atom. (23.5)

Ion product (Q_c) the product of ion concentrations in a solution, each concentration raised to a power equal to the number of ions in the formula of the ionic compound. (17.3)

Ion-product constant for water (K_w) the equilibrium value of the ion product $[H^+][OH^-]$. (16.1)

Ion-selective electrode an electrode whose emf depends on the concentration of a particular ion in solution. (19.6)

Isoelectronic refers to different species having the same number and configuration of electrons. (9.3)

Isomers compounds of the same molecular formula but with different arrangements of the atoms. (10.4)

Isotope dilution a technique designed to determine the quantity of a substance in a mixture or the total volume of solution by adding a known amount of an isotope to it. (20.5)

Isotopes atoms whose nuclei have the same atomic number but different mass numbers. (2.2 and 7.3)

Joule (J) the SI unit of energy; $1\ J = 1\ kg \cdot m^2/s^2$. (6.1)

Kelvin (K) the SI base unit of temperature; a unit on an absolute temperature scale. (1.4)

Ketone a compound containing a carbonyl group with two hydrocarbon groups attached to it. (24.5)

Kilogram (kg) the SI base unit for mass; equal to 2.2 pounds. (1.4)

Kinetic energy the energy associated with an object by virtue of its motion. (6.1)

Kinetic-molecular theory (kinetic theory) the theory that a gas consists of molecules in constant random motion. (p. 162)

Kinetics, chemical see *chemical kinetics*.

Lanthanides the 14 elements following lanthanum in the periodic table, in which the 4*f* subshell is filling. (p. 923)

Laser a source of intense, highly directed beam of monochromatic light; the word *laser* is an acronym meaning *l*ight *a*mplification by *s*timulated *e*mission of *r*adiation. (p. 251)

Lattice energy the change in energy that occurs when an ionic solid is separated into isolated ions in the gas phase. (9.1)

Law a concise statement or mathematical equation about a basic relationship or regularity of nature. (1.2)

Law of combining volumes a relation stating that gases at the same temperature and pressure react with one another in volume ratios of small whole numbers. (5.2)

Law of conservation of energy a law stating that energy may be converted from one form to another but that the total quantity of energy remains constant. (6.1)

Law of conservation of mass a principle stating that mass remains constant during a chemical change (chemical reaction). (1.1)

Law of definite proportions (law of constant composition) a law stating that a pure compound, whatever its source, always contains definite or constant proportions of the elements by mass. (2.1)

Law of mass action a relation stating that the values of the equilibrium-constant expression K_c are constant for a particular reaction at a given temperature, whatever equilibrium concentrations are substituted. (15.2)

Law of multiple proportions a principle stating that when two elements form more than one compound, the masses of one element in these compounds for a fixed mass of the other element are in ratios of small whole numbers. (2.2)

Lead storage cell a voltaic cell that consists of electrodes of lead alloy grids; one electrode is packed with a spongy lead to form the anode, and the other is packed with lead dioxide to form the cathode. (19.7)

Le Chatelier's principle a principle stating that when a system in equilibrium is disturbed by a change of temperature, pressure, or concentration variable, the system shifts in equilibrium composition in a way that tends to counteract this change of variable. (12.3 and 15.7)

LeClanché dry cell see *zinc–carbon dry cell*.

Levorotatory refers to a compound whose solution rotates the plane of polarized light to the left (when we are looking toward the source of light). (23.5)

Lewis acid a species that can form a covalent bond by accepting an electron pair from another species. (13.5)

Lewis base a species that can form a covalent bond by donating an electron pair to another species. (13.5)

Lewis electron-dot formula a formula using dots to represent valence electrons. (9.4)

Lewis electron-dot symbol a symbol in which the electrons in the valence shell of an atom or ion is represented by dots placed around the letter symbol of the element. (9.1)

Ligand a Lewis base attached to a metal atom in a complex. (p. 696 and 23.3)

Limiting reactant (limiting reagent) the reactant that is entirely consumed when a reaction goes to completion. (4.8)

Linear geometry a molecular geometry in which all atoms line up along a straight line. (10.1)

Line spectrum a spectrum showing only certain colors or specific wavelengths of light. (7.7)

Linkage isomers isomers of a complex that differ in the atom of a ligand that is bonded to the metal atom. (23.5)

Lipids biological substances that are soluble in nonpolar organic solvents, such as chloroform and carbon tetrachloride. (25.6)

Liquefaction the process in which a substance that is normally a gas changes to the liquid state. (11.2)

Liquid the form of matter that is a relatively incompressible fluid; a liquid has a fixed volume, but no fixed shape. (2.1)

Liter (L) a unit of volume equal to a cubic decimeter. (1.5)

Lithium–iodine battery a voltaic cell in which the anode is lithium metal and the cathode is an I_2 complex. (19.7)

London forces (dispersion forces) the weak attractive forces between molecules resulting from the small, instantaneous dipoles that occur because of the varying positions of the electrons during their motion about nuclei. (11.5)

Lone pair (nonbonding pair) an electron pair that remains on one atom and is not shared. (9.4)

Low-spin complex ion a complex ion in which there is maximum pairing of electrons in the orbitals of the metal atom. (23.6)

Macromolecules very large molecules having molecular weights that may be several millions of atomic mass units. (p. 1010)

Magic number the number of nuclear particles in a completed shell of protons or neutrons. (20.1)

Magnetic quantum number (m_l) the quantum number that distinguishes orbitals of given n and l—that is, of given energy and shape—but having a different orientation in space; the allowed values are the integers from $-l$ to $+l$. (7.9)

Main-group element (representative element) an element in an A column of the periodic table, in which an outer *s* or *p* subshell is filling. (2.4 and 8.2)

Manometer a device that measures the pressure of a gas or liquid in a sealed vessel. (5.1)

Markownikoff's rule a generalization stating that the major product formed by the addition of an unsymmetrical reagent such as H—Cl, H—Br, or H—OH is the one obtained when the H atom of the reagent adds to the carbon atom of the multiple bond that already has the more hydrogen atoms attached to it. (24.4)

Mass the quantity of matter in a material. (1.1)

Mass defect the total nucleon mass minus the nuclear mass of a nucleus. (20.6)

Mass number (*A*) the total number of protons and neutrons in the nucleus. (2.2 and 7.3)

Mass percentage parts per hundred parts of the total, by mass. (4.3)

Mass percentage of solute the percentage by mass of solute contained in a solution. (12.4)

Mass spectrometer an instrument, such as one based on Thomson's principles, that separates ions by mass-to-charge ratio. (7.4)

Material any particular kind of matter. (2.1)

Matter whatever occupies space and can be perceived by our senses. (1.1)

Maxwell's distribution of molecular speeds a theoretical relationship that predicts the relative number of molecules at various speeds for a sample of gas at a particular temperature. (5.7)

Melting (fusion) the change of a solid to the liquid state. (11.2)

Melting point the temperature at which a crystalline solid changes to a liquid, or melts. (11.2)

Messenger RNA a relatively small RNA molecule that can diffuse about the cell and attach itself to a ribosome, where it serves as a pattern for protein synthesis. (25.5)

Metabolism the process of building up and breaking down organic molecules in cells. (25.1)

Metallic solid a solid that consists of positive cores of atoms held together by a surrounding "sea" of electrons (metallic bonding). (11.6)

Metalloid (semimetal) an element having both metallic and nonmetallic properties. (2.4)

Metal refining in metallurgy, the purification of a metal. (21.5)

Metal a substance or mixture that has a characteristic luster, or shine, is generally a good conductor of heat and electricity, and is malleable and ductile. (2.4)

Metallurgy the scientific study of the production of metals from their ores and of the making of alloys. (21.5)

Metaphosphoric acids acids with the general formula $(HPO_3)_n$. (22.2)

Metastable nucleus a nucleus in an excited state with a lifetime of at least one nanosecond (10^{-9} s). (20.1)

Metathesis reaction (double-replacement reaction) a reaction that appears to involve the exchange of parts of the reactants (when written as a molecular equation). (3.2)

Meter (m) the SI base unit of length. (1.4)

Micelle a colloidal-sized particle formed in water by the association of molecules each of which has a hydrophobic end and a hydrophilic end. (12.9)

Millimeters of mercury (mmHg) a unit of pressure also known as the *torr;* 760 mmHg = 101.325 kPa (exact). (5.1)

Mineral a naturally occurring solid substance or solid solution with definite crystalline form. (21.2)

Miscible fluids fluids that mix with, or dissolve in, each other in all proportions. (12.1)

Mixture a material that can be separated by physical means into two or more substances. (2.1)

Moderator a substance that slows down neutrons in a nuclear fission reactor. (20.7)

Molality the moles of solute per kilogram of solvent. (12.4)

Molar concentration (molarity), *M* the moles of solute dissolved in one liter (cubic decimeter) of solution. (4.9)

Molar gas constant (*R*) the constant of proportionality that relates the molar volume of a gas to T/P. (5.3)

Molar gas volume the volume of one mole of gas. (5.2)

Molarity see *molar concentration*.

Molar mass the mass of one mole of substance. In grams, it is numerically equal to the formula weight in atomic mass units. (4.2)

Mole (mol) the quantity of a given substance that contains as many molecules or formula units as the number of atoms in exactly 12 g of carbon-12. The amount of substance containing Avogadro's number of molecules or formula units. (4.2)

Molecular equation an equation in which the substances are written as though they were molecular substances, even though they may actually exist in solution as ions. (3.5)

Molecular formula a chemical formula that gives the exact number of different atoms of an element in a molecule. (2.5)

Molecular geometry the general shape of a molecule determined by the relative positions of the atomic nuclei. (p. 354)

Molecularity the number of molecules on the reactant side of an elementary reaction. (14.7)

Molecular orbital theory a theory of the electron structure of molecules in terms of molecular orbitals, which may spread over several atoms or the entire molecule. (p. 376)

Molecular solid a solid that consists of atoms or molecules held together by intermolecular forces. (11.6)

Molecular substance a substance that is composed of molecules, all of which are alike. (2.5)

Molecular weight (MW) the sum of the atomic weights of all the atoms in a molecule. (4.1)

Molecule a definite group of atoms that are chemically bonded together—that is, tightly connected by attractive forces. (2.5)

Mole fraction the moles of a component substance divided by the total moles of solution. (5.5 and 12.4)

Monatomic ion an ion formed from a single atom. (2.6)

Monodentate ligand a ligand that bonds to a metal atom through one atom of the ligand. (23.3)

Monomer a compound used to make a polymer (and from which the polymer's repeating unit arises). (24.8)

Monosaccharides simple sugars, each containing three to nine carbon atoms, all but one of which bear a hydroxyl group, the remaining one being a carbonyl group. (25.4)

Nernst equation an equation relating the cell emf, E_{cell}, to its standard emf, E°_{cell}, and the reaction quotient, Q. At 25°C, the equation is $E_{cell} = E^\circ_{cell} - (0.0592/n)\log Q$. (19.6)

Net ionic equation an ionic equation from which spectator ions have been canceled. (3.5)

Neutralization reaction a reaction of an acid and a base that results in an ionic compound (a salt) and possibly water. (3.8)

Neutron a particle found in the nucleus of an atom; it has a mass almost identical to that of the proton but has no electric charge. (2.2 and p. 230)

Neutron activation analysis an analysis of elements in a sample based on the conversion of stable isotopes to radioactive isotopes by bombarding a sample with neutrons. (20.5)

Nickel–cadmium cell a voltaic cell consisting of an anode of cadmium and a cathode of hydrated nickel oxide (approximately NiOOH) on nickel; the electrolyte is potassium hydroxide. (19.7)

Nitrogen cycle the circulation of the element nitrogen in the biosphere, from nitrogen fixation to the release of free nitrogen by denitrifying bacteria. (p. 175)

Nitrogen-fixing bacteria bacteria that can produce nitrogen compounds in the soil from atmospheric nitrogen. (p. 175)

Noble gases (inert gases; rare gases) the Group VIIIA elements; all are gases consisting of uncombined atoms. They are relatively unreactive elements (8.6)

Noble-gas core an inner-shell configuration corresponding to one of the noble gases. (8.2)

Nomenclature see *chemical nomenclature*.

Nonbonding pair an electron pair that remains on one atom and is not shared. (9.4)

Nonelectrolyte a substance, such as sucrose, or table sugar ($C_{11}H_{22}O_{11}$), that dissolves in water to give a nonconducting or very poorly conducting solution. (3.3)

Nonmetal an element that does not exhibit the characteristics of a metal. (2.4)

Nonstoichiometric compound a compound whose composition varies from its idealized formula. (11.7)

Normality the number of equivalents of a substance dissolved in a liter of solution (13.9)

Nuclear bombardment reaction a nuclear reaction in which a nucleus is bombarded, or struck, by another nucleus or by a nuclear particle. (p. 792)

Nuclear equation a symbolic representation of a nuclear reaction. (20.1)

Nuclear fission a nuclear reaction in which a heavy nucleus splits into lighter nuclei and energy is released. (20.6)

Nuclear fission reactor a device that permits a controlled chain reaction of nuclear fissions. (20.7)

Nuclear force a strong force of attraction between nucleons that acts only at very short distances (about 10^{-15} m). (20.1)

Nuclear fusion a nuclear reaction in which light nuclei combine to give a stabler, heavier nucleus plus (possibly) several neutrons, and energy is released. (20.6)

Nucleic acids polynucleotides folded or coiled into specific three-dimensional shapes. (25.5)

Nucleotides the building blocks of nucleic acids. (25.5)

Nucleus the central core of an atom; it has most of the mass of the atom and one or more units of positive charge. (2.2 and p. 230)

Nuclide any particular nucleus characterized by definite atomic number and mass number. (7.3)

Nuclide symbol a symbol for a nuclide, in which the atomic symbol is given as a left subscript and the mass number is given as a left superscript to the symbol of the element. (7.3)

Number of significant figures number of digits reported for the value of a measured or calculated quantity, indicating the precision of the value. (1.3)

Octahedral geometry the geometry of a molecule in which six atoms occupy the vertices of a regular octahedron (a figure with eight faces and six vertices) with the central atom at the center of the octahedron. (10.1)

Octet rule the tendency of atoms in molecules to have eight electrons in their valence shells (two for hydrogen atoms). (9.4)

Oligosaccharides short polymers of two to ten simple sugar units. (25.4)

Optical isomers (enantiomers) isomers that are nonsuperimposable mirror images of one another. (23.5)

Optically active having the ability to rotate the plane of light waves, either as a pure substance or in solution. (23.5)

Orbital diagram a diagram to show how the orbitals of a subshell are occupied by electrons. (8.1)

Ore a rock or mineral from which a metal can be economically produced. (21.3)

Organic compounds compounds that can be thought of as derivatives of compounds of carbon and hydrogen (hydrocarbons). (2.6)

Osmosis the phenomenon of solvent flow through a semipermeable membrane to equalize the solute concentrations on both sides of the membrane. (12.7)

Osmotic pressure a colligative property of a solution equal to the pressure that, when applied to the solution, just stops osmosis. (12.7)

Ostwald process an industrial preparation of nitric acid by the catalytic oxidation of ammonia. (22.2)

Overall order of a reaction the sum of the orders of the species in the rate law. (14.3)

Oxidation the part of an oxidation–reduction reaction in which there is a loss of electrons by a species or an increase in the oxidation number of an atom. (13.7)

Oxidation number (oxidation state) the charge an atom in a substance would have if the pairs of electrons in each bond belonged to the more electronegative atom. (13.6)

Oxidation potential the negative of the standard electrode potential. (19.4)

Oxidation–reduction reaction (redox reaction) a reaction in which electrons are transferred between species or in which atoms change oxidation number. (13.7)

Oxidizing agent a species that oxidizes another species; it itself is reduced. (13.7)

Oxoacid an acid containing hydrogen, oxygen, and another element. (2.6) A substance in which O atoms (and possibly other electronegative atoms) are bonded to a central atom, with one or more H atoms usually bonded to the O atoms. (9.6)

Pairing energy (*P*) the energy required to put two electrons into the same orbital. (23.7)

Paramagnetic substance a substance that is weakly attracted by a magnetic field; this attraction generally results from unpaired electrons. (8.3)

Partial pressure the pressure exerted by a particular gas in a mixture. (5.5)

Particle accelerator device used to accelerate electrons, protons, and alpha particles and other ions to very high speeds. (20.2)

Pascal (Pa) the SI unit of pressure; $1 \text{ Pa} = 1 \text{ kg/(m} \cdot \text{s}^2)$. (5.1)

Pauli exclusion principle a rule stating that no two electrons in an atom can have the same four quantum numbers. It follows from this that an orbital can hold no more than two electrons and can hold two only if they have different spin quantum numbers (8.1)

Peptide (amide) bonds the C—N bonds resulting from a condensation reaction between the carboxyl group of one amino acid and the amino group of a second amino acid. (25.3)

Percentage composition the mass percentages of each element in the compound. (p. 109)

Percentage yield the actual yield (experimentally determined) expressed as a percentage of the theoretical yield (calculated). (4.8)

Period (of the periodic table) the elements in any one horizontal row of the periodic table. (2.4)

Periodic law a law stating that when the elements are arranged by atomic number, their physical and chemical properties vary periodically. (8.5)

Periodic table a tabular arrangement of elements in rows and columns, highlighting the regular repetition of properties of the elements. (2.4)

Peroxides compounds with oxygen in the -1 oxidation state. (22.3)

pH the negative of the logarithm of the molar hydronium-ion concentration. (16.3)

Phase one of several different homogeneous materials present in the portion of the matter under study. (2.1)

Phase diagram a graphical way to summarize the conditions under which the different states of a substance are stable. (11.3)

Phase transition see *change of state*.

Phospholipid a biological substance that resembles a triaglycerol but in which only two fatty acids are present; the third glycerol —OH is bonded to a phosphate group that is bonded in turn to an alcohol. (25.6)

Phospholipid bilayer a part of a biological membrane consisting of two layers of phospholipid molecules. These molecules have hydrophobic and hydrophilic ends, and they aggregate to give this layer structure. The process is similar to the formation of micelles by soap molecules. (25.6)

Photoconductor a material that is normally a poor conductor of electricity but becomes a good conductor when light falls on it. (22.3)

Photoelectric effect the ejection of electrons from the surface of a metal or other material when light shines on it. (7.6)

Photons particles of electromagnetic energy with energy E proportional to the observed frequency of light: $E = h\nu$. (7.6)

Physical adsorption adsorption in which the attraction is provided by weak intermolecular forces. (14.9)

Physical change a change in the form of matter but not in its chemical identity. (2.1)

Physical property a characteristic that can be observed for a material without changing its chemical identity. (2.1)

Pi (π) bond a bond that has an electron distribution above and below the bond axis. (10.4)

Planck's constant (*h*) a physical constant with the value 6.63×10^{34} J · s. It is the proportionality constant relating the frequency of light to the energy of a photon. (7.6)

Plasma an electrically neutral gas made up of ions and electrons. (20.7)

Polar covalent bond a covalent bond in which the bonding electrons spend more time near one atom than near the other. (9.5)

Polyatomic ion an ion consisting of two or more atoms chemically bonded together and carrying a net electric charge. (2.6)

Polydentate ligand a ligand that can bond with two or more atoms to a metal atom. (23.3)

Polymer a chemical species of very high molecular weight that is made up from many repeating units of low molecular weight. (24.8)

Polynucleotide a linear polymer of nucleotide units. (25.5)

Polypeptide a polymer formed by the linking of many amino acids by peptide bonds. (25.3)

Polyphosphoric acids acids that have the general formula $H_{n+2}P_nO_{3n+1}$ and are formed from chains of P—O bonds. (22.2)

Polyprotic acid an acid that yields two or more acidic hydrogens per molecule. (3.8)

Polysaccharides polymers consisting of more than ten simple sugar units. (25.4)

Positron a particle that is similar to an electron and has the same mass but a positive charge. (20.1)

Positron emission emission of a positron from an unstable nucleus. (20.1)

Potential difference the difference in electrical potential (electrical pressure) between two points. (19.3)

Potential energy the energy an object has by virtue of its position in a field of force. (6.1)

Precipitate a solid formed by a reaction in solution. (3.2)

Precision the closeness of the set of values obtained from identical measurements of a quantity. (1.3)

Pressure the force exerted per unit area of surface. (5.1)

Primary alcohol an alcohol in which the hydroxyl group is attached to a carbon atom that is itself bonded to only one other carbon atom. (24.5)

Primary structure (of a protein) the order or sequence of the amino acid units in the protein. (25.3)

Principal quantum number (*n*) the quantum number on which the energy of an electron in an atom principally depends; it can have any positive value: 1, 2, 3,· · ·. (7.9)

Product a substance that results from a chemical reaction. (3.1)

Proteins biological polymers of small molecules called amino acids. (25.3)

Proton a particle found in the nucleus of the atom; it has a positive electric charge equal in magnitude, but opposite in sign, to that of the electron and a mass 1836 times that of the electron. (2.2 and p. 230)

Pseudo-noble-gas core the noble-gas core together with $(n-1)d^{10}$ electrons. (8.2)

Qualitative analysis the determination of the identity of substances present in a mixture. (p. 701)

Quantum (wave) mechanics the branch of physics that mathematically describes the wave properties of submicroscopic particles. (7.8)

Racemic mixture a mixture of equal amounts of optical isomers. (23.5)

Rad The dosage of radiation that deposits 1×10^2 J of energy per kilogram of tissue. (20.3)

Radioactive decay the process in which a nucleus spontaneously disintegrates, giving off radiation. (p. 792)

Radioactive decay constant (*k*) rate constant for radioactive decay. (20.4)

Radioactive decay series a sequence in which one radioactive nucleus decays to a second, which then decays to a third, and so on. (20.1)

Radioactive tracer a radioactive isotope added to a chemical, biological, or physical system to facilitate study of the system. (20.5)

Radioactivity spontaneous radiation from unstable elements. (7.2)

Raoult's law a law stating that the partial pressure, P_A, over a solution equals the vapor pressure of the pure solvent, P_A°, times the mole fraction of the solvent, X_A, in solution: $P_A = P_A^\circ, X_A$. (12.5)

Rate constant a proportionality constant in the relationship between rate and concentrations. (14.3)

Rate-determining step the slowest step in a reaction mechanism. (14.8)

Rate law an equation that relates the rate of a reaction to the concentrations of reactants (and catalyst) raised to various powers. (14.3)

Reactant a starting substance in a chemical reaction. (3.1)

Reaction intermediate a species produced during a reaction that does not appear in the net equation because it reacts in a subsequent step in the mechanism. (14.7)

Reaction mechanism the set of elementary reactions whose overall effect is given by the net chemical equation. (14.7)

Reaction order the exponent of the concentration of a given reactant species in the rate law, as determined experimentally. (14.3)

Reaction quotient (Q_c) an expression that has the same form as the equilibrium-constant expression but whose concentration values are not necessarily those at equilibrium. (15.5)

Reaction rate the increase in molar concentration of product of a reaction per unit time or the decrease in molar concentration of reactant per unit time. (14.1)

Redox reaction see *oxidation–reduction reaction.*

Reducing agent a species that reduces another species; it is itself oxidized. (13.7)

Reduction the part of an oxidation–reduction reaction in which there is a gain of electrons by a species or a decrease of oxidation number of an atom. (13.7)

Reduction potential see *standard electrode potential.*

Reference form the stablest form (physical state and allotrope) of the element under standard thermodynamic conditions. (6.8)

Rem a unit of radiation dosage used to relate various kinds of radiation in terms of biological destruction. It equals the rad times a factor for the type of radiation, called the relative biological effectiveness (RBE): rems = rads × RBE. (20.3)

Representative element see *main-group element.*

Resonance description a representation in which we describe the electron structure of a molecule, having delocalized bonding by writing all possible electron-dot formulas. (9.8)

Reverse osmosis a process in which a solvent, such as water, is forced by a pressure greater than the osmotic pressure to flow through a semipermeable membrane from a concentrated solution to a dilute one. (12.7)

Ribonucleic acid (RNA) a constituent of cells that is used to manufacture proteins from genetic information. It is a polymer of ribonucleotide units. (25.5)

Ribosomal RNA the RNA in a ribosome. (25.5)

Ribosomes tiny cellular particles on which protein synthesis takes place. (25.5)

Roasting the process of heating a mineral in air to obtain the oxide. (21.5)

Root-mean-square (rms) molecular speed a type of average molecular speed; the speed of a molecule that has the average molecular kinetic energy. It equals $\sqrt{3RT/M_m}$, where M_m is the molar mass. (5.7)

Rounding the procedure of dropping nonsignificant digits in a calculation result and adjusting the last digit reported. (1.3)

Salt an ionic compound that is a product of a neutralization reaction. (3.8)

Salt bridge a tube of an electrolyte in a gel that is connected to the two half-cells of a voltaic cell; it allows the flow of ions but prevents the mixing of the different solutions that would allow direct reaction of the cell reactants. (19.1)

Saponification the hydrolysis of an ester in the presence of a base. (24.6)

Saturated hydrocarbon a hydrocarbon in which all carbon atoms are bonded to the maximum number of hydrogen atoms. (24.1)

Saturated solution a solution that is in equilibrium with respect to a given dissolved substance. (12.2)

Scientific method the general process of advancing scientific knowledge through observation, the framing of laws, hypotheses, or theories, and the conducting of more experiments. (1.2)

Scientific notation the representation of a number in the form $A \times 10^n$, where A is a number with a single nonzero digit to the left of the decimal point, and n is an integer, or whole number. (1.3)

Scintillation counter a device that detects nuclear radiations from flashes of light generated in a material by the radiation. (20.3)

Second (s) the SI base unit of time. (1.4)

Second law of thermodynamics a law stating that the total entropy of a system and its surroundings always increases for a spontaneous process. Also, for a spontaneous process at a given temperature, the change in entropy of the system is greater than the heat divided by the absolute temperature. (18.2)

Secondary alcohol an alcohol in which the hydroxyl group is attached to a carbon atom that is itself bonded to two carbon atoms. (24.5)

Secondary structure (of a protein) the relatively simple coiled or parallel arrangement of a protein molecule. (25.3)

Seesaw geometry the geometry of a molecule having four atoms bonded to a central atom, in which two of these outer atoms occupy axial positions of a trigonal bipyramid and the other two occupy equatorial positions. (10.1)

Self-ionization (autoionization) a reaction in which two like molecules react to give ions. (16.1)

Semiconductor a substance that is only slightly conducting at room temperature but becomes a moderately good conductor at higher temperature. When pure semiconducting elements (the metalloids) have certain other elements added to them, they become good conductors at room temperature. (10.8)

Semimetals see *metalloids.*

Shell model of the nucleus a nuclear model in which protons and neutrons exist in levels, or shells, analogous to the shell structure that exists for electrons in an atom. (20.1)

SI see *International System.*

SI base units the SI units of measurement from which all others can be derived. (1.4)

SI derived unit a unit derived by combining SI base units. (1.5)

SI prefix a prefix used in the International System to indicate a power of ten. (1.4)

Sigma (σ) bond a bond that has a cylindrical shape about the bond axis. (10.4)

Significant figures those digits in a measured number (or result of a calculation with measured numbers) that include all certain digits plus a final one having some uncertainty. (1.3)

Silane a hydride of silicon that contains chains of single-bonded silicon atoms. (22.1)

Silicates compounds of silicon and oxygen with various metals. (22.1)

Simple cubic unit cell a cubic unit cell in which lattice points are situated only at the corners of the unit cell. (11.7)

Simplest formula see *empirical formula*.

Single bond a covalent bond in which a single pair of electrons is shared by two atoms. (9.4)

Single-replacement reaction see *displacement reaction*.

Sol a colloid that consists of solid particles dispersed in a liquid. (12.9)

Solid the form of matter characterized by rigidity; a solid is relatively incompressible and has fixed shape and volume. (2.1)

Solubility the amount of a substance that dissolves in a given quantity of solvent (such as water) at a given temperature to give a saturated solution. (12.2)

Solubility-product constant (K_{sp}) the equilibrium constant for the solubility equilibrium of a slightly soluble (or nearly insoluble) ionic compound. (17.1)

Solute in the case of a solution of a gas or solid dissolved in a liquid, the gas or solid; in other cases, the component in smaller amount. (12.1)

Solution see *homogeneous mixture*.

Solvay process an industrial method for obtaining sodium carbonate from sodium chloride, ammonia, and carbon dioxide. (21.2)

Solvent in a solution of a gas or solid in a liquid, the liquid; in other cases, the component in greater amount. (12.1)

Specific heat capacity (specific heat) the quantity of heat required to raise the temperature of one gram of a substance by one degree Celsius (or by one kelvin). (6.6)

Spectator ions ions in an ionic equation that do not take part in the reaction. (3.5)

Spectrochemical series an arrangement of ligands according to the relative magnitudes of the crystal field splittings they induce in the *d* orbitals of a metal ion. (23.7)

Spin quantum number (m_s) the quantum number that refers to the two possible orientations of the spin axis of an electron; possible values are $+\frac{1}{2}$ and $-\frac{1}{2}$. (7.9)

Spontaneous process a physical or chemical change that occurs by itself. (p. 716)

Square pyramidal geometry the geometry of a molecule in which a central atom is at the apex of a pyramid and four other atoms form the square base of the pyramid. (10.1)

Square planar geometry the geometry of a molecule in which a central atom is surrounded by four other atoms arranged in a square and in a plane containing the central atom. (10.1)

Stability constant (of a complex) see *formation constant*.

Standard electrode potential ($E°$) the electrode potential when the concentrations of solutes are 1 *M*, the gas pressures are 1 atm, and the temperature has a specified value—usually 25°C. (19.4)

Standard emf ($E°_{cell}$) the emf of a voltaic cell operating under standard-state conditions (solute concentrations are 1 *M*, gas pressures are 1 atm, and the temperature has a specified value—usually 25°C). (19.4)

Standard enthalpy of formation (standard heat of formation), $\Delta H_f°$ the enthalpy change for the formation of one mole of the substance in its standard state from its elements in their reference forms and in their standard states. (6.8)

Standard entropy (S) the entropy value for the standard state of a species. (18.3)

Standard free energy of formation (ΔG_f) the free-energy change that occurs when one mole of substance is formed from its elements in their stablest states at 1 atm and at a specified temperature (usually 25°C). (18.4)

Standard heat of formation see *standard enthalpy of formation*.

Standard potential diagram a convenient graphical presentation of the standard potentials of an element. (22.4)

Standard state the standard thermodynamic conditions (1 atm and usually 25°C) chosen for substances when we are listing or comparing thermochemical data. (6.8)

Standard temperature and pressure (STP) the reference conditions for gases, chosen by convention to be 0°C and 1 atm. (5.2)

State function a property of a system that depends only on its present state, which is determined by variables such as temperature and pressure, and is independent of any previous history of the system. (6.3 and 18.1)

States of matter the three forms that matter can assume—solid, liquid, and gas. (2.1)

Steam-reforming process an industrial preparation of hydrogen and carbon monoxide mixtures by the reaction of steam and hydrocarbons at high temperature and pressure over a nickel catalyst. (p. 263)

Stereoisomers isomers in which the atoms are bonded to each other in the same order but that differ in the precise arrangement of these atoms in space. (23.5)

Stock system a system of chemical nomenclature in which the charge on a metal atom or oxidation number of an atom is denoted by a Roman numeral in parentheses following the element name. (2.6 and 13.6)

Stoichiometry the calculation of the quantities of reactants and products involved in a chemical reaction. (p. 114)

Stratosphere the region of the atmosphere that lies just above the troposphere and wherein the temperature increases with increasing altitude. (p. 389)

Strong acids an acid that ionizes completely in water. (3.4 and 13.1)

Strong base a base that is present in aqueous solution entirely as ions, one of which is OH^-. (3.4 and 13.1)

Strong electrolyte an electrolyte that exists in solution almost entirely as ions. (3.3)

Structural formula a chemical formula that shows how the atoms are bonded to one another in a molecule. (2.5)

Structural isomers isomers that differ in how the atoms are joined together. (23.5)

Sublimation the change of a solid to the vapor. (11.2)

Substance a kind of matter that cannot be separated into other kinds of matter by any physical process. (2.1)

Substitution reaction a reaction in which part of the reagent molecule is substituted for an H atom on a hydrocarbon or a hydrocarbon group. (24.4)

Substrate the reactant substance whose reaction an enzyme catalyzes. (25.3)

Superoxides binary compounds with oxygen in the $-\frac{1}{2}$ oxidation state; they contain the superoxide ion, O_2^-. (22.3)

Supersaturated solution a solution that contains more dissolved substance than does a saturated solution; the solution is not in equilibrium with the pure substance. (12.2)

Surface tension the energy required to increase the surface area of a liquid by a unit amount. (11.4)

Surroundings everything in the vicinity of a thermodynamic system. (6.2)

Synthesis gas a mixture of CO and H_2 gases used in the industrial preparation of a number of organic compounds, including methanol, CH_3OH. (p. 624)

System (thermodynamic) the substance or mixture of substances under study in which a change occurs. (6.2)

Termolecular reaction an elementary reaction that involves three reactant molecules. (14.7)

Tertiary alcohol an alcohol in which the hydroxyl group is attached to a carbon atom that is itself bonded to three carbon atoms. (24.5)

Tertiary structure (of a protein) the structure associated with the way the protein coil is folded. (25.3)

Tetrahedral geometry the geometry of a molecule in which four atoms bonded to a central atom occupy the vertices of a tetrahedron with the central atom at the center of this tetrahedron. (10.1)

Theoretical yield the maximum amount of product that can be obtained by a reaction from given amounts of reactants. (4.8)

Theory a tested explanation of basic natural phenomena. (1.2)

Thermal equilibrium a state in which heat does not flow between a system and its surroundings because they are both at the same temperature. (6.2)

Thermochemical equation the chemical equation for a reaction (including phase labels) in which the equation is given a molar interpretation and the enthalpy of reaction for these molar amounts is written directly after the equation. (6.4)

Thermodynamic equilibrium constant (*K*) the equilibrium constant in which the concentrations of gases are expressed in partial pressures in atmospheres, whereas the concentrations of solutes in liquid solution are expressed in molarities. (18.6)

Thermodynamics the study of the relationship between heat and other forms of energy involved in a chemical or physical process. (p. 711)

Thermochemistry the study of the quantity of heat absorbed or evolved by chemical reactions. (p. 187)

Third law of thermodynamics a law stating that a substance that is perfectly crystalline at 0 K has an entropy of zero. (18.3)

Three-center bond a bond in which three atoms are held together by two electrons. (21.4)

Titration a procedure for determining the amount of substance *A* by adding a carefully measured volume of solution with known concentration of *B* until the reaction of *A* and *B* is just complete. (4.11)

Torr see *millimeters of mercury*.

Transfer RNA the smallest RNA molecule; it bonds to a particular amino acid, carries it to a ribosome, and then attaches itself (through base pairing) to a messenger-RNA codon. (25.5)

Transition element often used to mean the B columns of elements in the periodic table (the *d*-block transition elements). In general, the term applies to both the *d*-block and the *f*-block transition elements. (2.4 and 8.2)

Transition-state theory a theory that explains the reaction resulting from the collision of two molecules in terms of an activated complex (transition state). (14.5)

Transmutation the changing of one element to another by bombarding the nucleus of the element with nuclear particles or nuclei. (20.2)

Transuranium elements those elements with atomic numbers greater than that of uranium ($Z = 92$), the naturally occurring element of greatest atomic number. (20.2)

Triacylglycerols esters formed from glycerol (a trihydroxy alcohol) and three fatty acids (long-chain carboxylic acids). (25.6)

Trigonal bipyramidal geometry the geometry of a molecule in which five atoms bonded to a central atom occupy the vertices of a trigonal bipyramid (formed by placing two trigonal pyramids base to base) with the central atom at the center of this trigonal bipyramid. (10.1)

Trigonal planar geometry the geometry of a molecule in which a central atom is surrounded by three other atoms arranged in a triangle and in a plane containing the central atom. (10.1)

Trigonal pyramidal geometry the geometry of a molecule in which a central atom is at the apex of a pyramid and three other atoms form the triangular base of the pyramid. (10.1)

Triple bond a covalent bond in which three pairs of electrons are shared by two atoms. (9.4)

Triple point the point on a phase diagram representing the temperature and pressure at which three phases of a substance coexist in equilibrium. (11.3)

Troposphere the lowest region of the atmosphere, wherein the temperature decreases with increasing altitude. (p. 390)

T-shaped geometry the geometry of a molecule in which three atoms are bonded to a central atom to form a T. (10.1)

Tyndall effect the scattering of light by colloidal-sized particles. (12.9)

Uncertainty principle a relation stating that the product of the uncertainty in position and the uncertainty in momentum (mass times speed) of a particle can be no smaller than Planck's constant divided by 4π. (7.8)

Unimolecular reaction an elementary reaction that involves one reactant molecule. (14.7)

Unit a fixed standard of measurement. (1.3)

Unit cell the smallest boxlike unit (each box having faces that are parallelograms) from which we can imagine constructing a crystal by stacking the units in three dimensions. (11.7)

Unsaturated hydrocarbons hydrocarbons that do not contain the maximum number of hydrogen atoms for a given carbon-atom framework. (24.2)

Unsaturated solution a solution that is not in equilibrium with respect to a given dissolved substance and in which more of the substance can dissolve. (12.2)

Valence bond theory an approximate theory to explain the electron pair or covalent bond in terms of quantum mechanics. (10.3)

Valence electron an electron in an atom outside the noble-gas or pseudo-noble-gas core. (8.2)

Valence-shell electron-pair repulsion (VSEPR) model a model for predicting the shapes of molecules and ions in which valence-shell electron pairs are arranged about each atom so that electron pairs are kept as far away from one another as possible, thus minimizing electron-pair repulsions. (10.1)

van der Waals forces a general term for those intermolecular forces that includes dipole–dipole and London forces. (11.5)

van der Waals equation an equation relating P, T, V, and n for nonideal gases. (5.8)

Vapor the gaseous state of any kind of matter that normally exists as a liquid or solid. (2.1)

Vaporization the change of a solid or a liquid to the vapor. (11.2)

Vapor pressure the partial pressure of the vapor over the liquid, measured at equilibrium. (11.2)

Vapor-pressure lowering a colligative property equal to the vapor pressure of the pure solvent minus the vapor pressure of the solution. (12.5)

Viscosity the resistance to flow that is exhibited by all liquids. (11.4)

Volt (V) the SI base unit of potential difference. (19.3)

Voltaic cell (galvanic cell) an electrochemical cell in which a spontaneous reaction generates an electric current. (p. 747)

Water–gas reaction an industrial process in which steam is passed over red-hot coke to give a gaseous mixture of carbon monoxide and hydrogen. (p. 264)

Wavelength (λ) the distance between any two adjacent identical points of a wave. (7.5)

Wave mechanics see *quantum mechanics*.

Weak acid an acid that is only partly ionized as the result of an equilibrium reaction with water. (3.4)

Weak base a base that is only partly ionized as the result of an equilibrium reaction with water. (3.4)

Weak electrolyte an electrolyte that dissolves in water to give an equilibrium between a molecular substance and a small concentration of ions. (3.3)

Work the energy exchange that results when a force F moves an object through a distance d; it equals $F \times d$. (18.1)

Zinc–carbon (Leclanché) dry cell a voltaic cell that has a zinc can as the anode and a graphite (carbon) rod in the center immediately surrounded by a paste of manganese dioxide and carbon black as the cathode. Around this is another paste, this one containing ammonium and zinc chlorides. (19.7)

Zone refining a purification process in which a solid rod of a substance is melted in a small, moving band or zone that carries the impurities out of the rod. (22.1)

Zwitterion an amino acid in the doubly ionized form, in which the carboxyl group has lost an H^+ to give $—COO^-$ and the amino group has gained an H^+ to give $—NH_3^+$. (25.3)

Index

Credits

All photographs by James Scherer unless otherwise noted.

Chapter 1 Opener Charlotte Raymond/Photo Researchers, Inc. **Figure 1.2A** Fisher Scientific **Figure 1.2C** Mettler Instrument Corporation **Figure 1.3** E. R. Degginger **Page 7** © Sidney Harris **Figure 1.8** National Institute of Standards and Technology **Figure 2.10** Digital Instruments, Inc. **Figure 2.12** Finnigan Mat **Figure 2.13 (bottom)** Historical Pictures Service, Chicago **Figure 2.17** © 1985 Martin M. Rotker/Taurus Photos, Inc. **Figure 2.21** Fundamental Photographs, New York **Figure 4.7** Breck P. Kent **Figure 4.15 (left)** Farmland Industries, Inc. **Figure 4.15 (right)** Grant Heilman/Grant Heilman Photography, Inc. **Figure 5.9** M. P. Kahl/Photo Researchers, Inc. **Figure 5.17** Frank Hoffman/Oak Ridge National Laboratory **Figure 5.19** Donna Bise/Photo Researchers, Inc. **Figure 5.24 (bottom)** Vacuum/Atmospheres Company **Figure 6.2** NASA/Johnson Space Center **Figure 6.3** John Running/Stock Boston **Figure 6.12** NASA/Peter Arnold, Inc. **Figure 6.15** Johnson Matthey **Figure 7.3** Cavendish Laboratory/University of Cambridge **Figure 7.16** AP/Wide World Photos **Figure 7.17** Bausch and Lomb **Figure 7.20** Jet Propulsion Laboratory **Figure 7.24** Cambridge Instruments **Figure 7.27** IBM Almaden Research Center **Figure 7.35** NASA/Johnson Space Center **Chapter 8 Opener** Barry L. Runk/Grant Heilman Photography, Inc. **Figure 8.1** Culver Pictures Inc. **Figure 8.4** Paul Shambroom/Photo Researchers, Inc. **Figure 8.17** Barry L. Runk/Grant Heilman Photography, Inc. **Figure 8.19** Yoav/Phototake **Figure 8.21** E. R. Degginger **Figure 8.22** E. R. Degginger **Figure 8.24** NASA/Johnson Space Center **Figure 8.26** Lee Boltin **Chapter 9 Opener** Johnson Matthey **Figure 9.1** Breck P. Kent **Figure 9.15** Nicolet Analytical Instruments **Chapter 10 Opener** Tripos Associates, Inc., St. Louis, MO **Figure 10.20** Yoav/Phototake **Figure 10.32** Argonne National Laboratory **Figure 10.33** Marcello Bertinetti/Photo Researchers, Inc. **Figure 10.36** NASA/Johnson Space Center **Figure 10.37** Breck P. Kent **Figure 10.38** Charles Falco/Photo Researchers, Inc. **Figure 11.14** Biophoto Associates/Photo Researchers, Inc. **Figure 11.37** UCAR Carbon Company Inc., a subsidiary of Union Carbide Corporation **Figure 11.39** From Preston, *Proceedings of the Royal Society,* A, Volume 172, plate 4, figure 5A **Figure 11.42** Siemens Analytical X-Ray Instruments, Inc. **Figure 11.47** NASA/Johnson Space Center **Figure 12.9** Bill Longcore/Photo Researchers, Inc. **Figure 12.11** Physicians & Nurses Manufacturing Corporation **Figure 12.13** Fundamental Photographs, New York **Figure 12.14** Bill Stanton/Rainbow **Figure 12.19** Dan McCoy/Rainbow **Figure 12.25** Polymetrics, Inc. **Figure 12.26** Fundamental Photographs, New York **Figure 13.7** E. R. Degginger **Figure 13.9** National Draeger, Inc. **Figure 13.14** Thomas Eisner **Figure 14.18** General Motors **Figure 15.1** Eastman Kodak Company **Figure 15.5** American Gas Association **Chapter 16 opener** Owen Tomalin/Bruce Coleman Inc. **Figure 16.2** E. R. Degginger **Figure 16.10** Fundamental Photographs, New York **Figure 17.1** Alan Morgan/Peter Arnold, Inc. **Figure 17.2** E. R. Degginger **Figure 17.3** Barry Rokeach/The Image Bank **Figure 17.5** E. R. Degginger **Chapter 18 Opener** E. R. Degginger **Figure 18.1** E. R. Degginger **Figure 18.5** Vance Henry/Taurus Photos, Inc. **Figure 18.7 (left)** Keith Kent/Peter Arnold, Inc. **Figure 18.7 (right)** Wayne State University **Chapter 19 Opener** Spencer Swanger/Tom Stack & Associates **Figure 19.7** Corning Incorporated **Figure 19.12** Delco Remy **Figure 19.13** Rayovac Corporation **Figure 19.15** International Fuel Cells **Figure 19.18 (bottom)** E. R. Degginger **Figure 19.24** Kennecott Minerals Company **Chapter 20 Opener** Edith G. Haun/Stock Boston **Figure 20.1** DuPont, North Billerica, MA **Figure 20.5** Fermilab Visual Media Services **Figure 20.6** Lawrence Berkeley Laboratory **Figure 20.7** Oak Ridge National Laboratory **Figure 20.9** Bicron Corporation, Newbury, OH **Figure 20.11** DuPont, North Billerica, MA **Figure 20.12** NIH/Photo Researchers, Inc. **Figure 20.13** Hank Morgan/Photo Researchers, Inc. **Chapter 21 Opener** Barry H. Frisch/Photo Researchers, Inc. **Figure 21.3** Peter Menzel/Stock Boston **Figure 21.5** DuPont Magazine **Figure 21.7** From *A Field Guide to Rocks and Minerals* by Frederick H. Pough. Copyright © 1953, 1960, 1976 by Frederick H. Pough. Reprinted by permission of Houghton Mifflin Company. **Figure 21.9** The Dow Chemical Company **Figure 21.12** Yoram Lehmann/Peter Arnold, Inc. **Figure 21.13** E. R. Degginger/Bruce Coleman Inc. **Figure 21.14** Roberto Valladares/The Image Bank **Figure 21.17** Hank Morgan/Photo Researchers, Inc. **Figure 21.20** E. R. Degginger **Chapter 22 Opener** Gary Milburn/Tom Stack & Associates **Figure 22.2** General Electric Company Research and Development Center **Figure 22.5 (left)** Tom McHugh/Photo Researchers, Inc. **Figure 22.5 (center)** Yoav/Phototake **Figure 22.5 (right)** Ward's Natural Science

Figure 22.16 Yoav/Phototake **Figure 22.17** Xerox Corporation **Figure 22.20** Vic Cox/Peter Arnold, Inc. **Figure 22.25** Argonne National Laboratory **Chapter 23 Opener** Merrel Wood/The Image Bank **Figure 23.4** E. R. Degginger **Figure 24.9** Amoco Corporation/American Petroleum Institute **Figure 24.13** E. R. Degginger **Chapter 25 Opener** Dan McCoy/Rainbow **Figure 25.3** Computer Graphics Laboratory, University of California, San Francisco **Figure 25.6** Ealing Corporation **Figure 25.7** Carolina Biological

Table of Atomic Weights and Numbers

Names	Symbol	Atomic number	Atomic weight
Actinium	Ac	89	(227)
Aluminum	Al	13	26.981539
Americium	Am	95	(243)
Antimony	Sb	51	121.75
Argon	Ar	18	39.948
Arsenic	As	33	74.92159
Astatine	At	85	(210)
Barium	Ba	56	137.327
Berkelium	Bk	97	(247)
Beryllium	Be	4	9.012182
Bismuth	Bi	83	208.98037
Boron	B	5	10.811
Bromine	Br	35	79.904
Cadmium	Cd	48	112.411
Calcium	Ca	20	40.078
Californium	Cf	98	(251)
Carbon	C	6	12.011
Cerium	Ce	58	140.115
Cesium	Cs	55	132.90543
Chlorine	Cl	17	35.4527
Chromium	Cr	24	51.9961
Cobalt	Co	27	58.93320
Copper	Cu	29	63.546
Curium	Cm	96	(247)
Dysprosium	Dy	66	162.50
Einsteinium	Es	99	(252)
Erbium	Er	68	167.26
Europium	Eu	63	151.965
Fermium	Fm	100	(257)
Fluorine	F	9	18.9984032
Francium	Fr	87	(223)
Gadolinium	Gd	64	157.25
Gallium	Ga	31	69.723
Germanium	Ge	32	72.61
Gold	Au	79	196.96654
Hafnium	Hf	72	178.49
Helium	He	2	4.002602
Holmium	Ho	67	164.93032
Hydrogen	H	1	1.00794
Indium	In	49	114.82
Iodine	I	53	126.90447
Iridium	Ir	77	192.22
Iron	Fe	26	55.847
Krypton	Kr	36	83.80
Lanthanum	La	57	138.9055
Lawrencium	Lr	103	(262)
Lead	Pb	82	207.2
Lithium	Li	3	6.941
Lutetium	Lu	71	174.967
Magnesium	Mg	12	24.3050
Manganese	Mn	25	54.93805
Mendelevium	Md	101	(258)
Mercury	Hg	80	200.59
Molybdenum	Mo	42	95.94
Neodymium	Nd	60	144.24
Neon	Ne	10	20.1797

A value in parentheses is the mass number of the isotope of longest half-life.

Names	Symbol	Atomic number	Atomic weight
Neptunium	Np	93	(237)
Nickel	Ni	28	58.69
Niobium	Nb	41	92.90638
Nitrogen	N	7	14.00674
Nobelium	No	102	(259)
Osmium	Os	76	190.2
Oxygen	O	8	15.9994
Palladium	Pd	46	106.42
Phosphorus	P	15	30.973762
Platinum	Pt	78	195.08
Plutonium	Pu	94	(244)
Polonium	Po	84	(209)
Potassium	K	19	39.0983
Praseodymium	Pr	59	140.90765
Promethium	Pm	61	(145)
Protactinium	Pa	91	(231)
Radium	Ra	88	(226)
Radon	Rn	86	(222)
Rhenium	Re	75	186.207
Rhodium	Rh	45	102.90550
Rubidium	Rb	37	85.4678
Ruthenium	Ru	44	101.07
Samarium	Sm	62	150.36
Scandium	Sc	21	44.955910
Selenium	Se	34	78.96
Silicon	Si	14	28.0855
Silver	Ag	47	107.8682
Sodium	Na	11	22.989768
Strontium	Sr	38	87.62
Sulfur	S	16	32.066
Tantalum	Ta	73	180.9479
Technetium	Tc	43	(98)
Tellurium	Te	52	127.60
Terbium	Tb	65	158.92534
Thallium	Tl	81	204.3833
Thorium	Th	90	232.0381
Thulium	Tm	69	168.93421
Tin	Sn	50	118.710
Titanium	Ti	22	47.88
Tungsten	W	74	183.85
Unnilennium	Une	109	(267)
Unnilhexium	Unh	106	(263)
Unniloctium	Uno	108	(265)
Unnilpentium	Unp	105	(262)
Unnilquadium	Unq	104	(261)
Unnilseptium	Uns	107	(262)
Uranium	U	92	238.0289
Vanadium	V	23	50.9415
Xenon	Xe	54	131.29
Ytterbium	Yb	70	173.04
Yttrium	Y	39	88.90585
Zinc	Zn	30	65.39
Zirconium	Zr	40	91.224

Atomic weights in this table are from the IUPAC report "Atomic Weights of the Elements 1988," *Pure and Applied Chemistry*, Vol. 60, No. 6 (1988), pp. 841–854. (©1988 IUPAC.)

Conversion Factors

1 amu $= 1.6605402 \times 10^{-27}$ kg	1 J $= 1$ kg $\cdot$ m^2/s^2
1 Å $= 10^{-10}$ m	1 cal $= 4.184$ J
1 L $= 1$ dm^3 $= 10^{-3}$ m^3	1 eV/molecule $= 96.485309$ kJ/mol
1 Pa $= 1$ kg/(m $\cdot$ s^2)	1 MeV $= 1.60217733 \times 10^{-13}$ J
1 atm $= 1.01325 \times 10^5$ Pa	
$= 760$ mmHg (torr)	

Physical Constants

Avogadro's number	N_A	$= 6.0221367 \times 10^{23}$/mol
Electronic charge	e	$= 1.6027733 \times 10^{-19}$ C
Electron rest mass	m_e	$= 9.1093897 \times 10^{-31}$ kg
Faraday constant	F	$= 9.6485309 \times 10^4$ C/mol
Molar gas constant	R	$= 0.08205783$ L $\cdot$ atm/(K $\cdot$ mol)
		$= 8.314510$ kPa $\cdot$ dm^3/(K $\cdot$ mol)
		$= 8.314510$ J/(K $\cdot$ mol)
		$= 1.987216$ cal/(K $\cdot$ mol)
Molar volume ideal gas, STP	V_m	$= 0.02241410$ m^3/mol
		$= 22.41410$ L/mol
Neutron rest mass	m_n	$= 1.6749286 \times 10^{-27}$ kg
Planck's constant	h	$= 6.6260755 \times 10^{-34}$ J $\cdot$ s
Proton rest mass	m_p	$= 1.6726231 \times 10^{-27}$ kg
Rydberg constant for H atom	R_H	$= 2.1798741 \times 10^{-18}$ J
Speed of light (in vacuum)	c	$= 2.99792458 \times 10^8$ m/s

Locations of Important Information

Acid-base strengths, relative	p. 509	Hybrid orbitals	p. 369
Acid ionization constants	p. 645, App. E	Ionic radii (main-group)	p. 321
Activity series	p. 519	Ionization energies, H to Ne	p. 296
Base ionization constants	p. 652, App. F	Ions, monatomic (main-group)	p. 50
Boiling-point-elevation constants	p. 480	Ions, polyatomic	p. 51
Bond energies	p. 340	Ions, transition-metal	p. 51
Covalent radii (nonmetals)	p. 338	Molecular geometries	p. 356, 360
Electrode potentials, standard	p. 756, App. I	Nuclear masses	p. 819
Electron affinities (main-group)	p. 297	Oxidation states (main-group)	p. 522, 839
Electron configurations of atoms	p. 280, App. D	Oxidation states (transition-metal)	p. 522
Electronegativities of elements	p. 328	SI base units	p. 12
Enthalpies of formation	p. 208, App. C	SI prefixes	p. 13
Entropies, standard	p. 721, App. C	Solubility product constants	p. 685, App. G
Formation constants of complex ions	p. 696, App. H	Solubility rules	p. 88
Free energies of formation	p. 726, App. C	Vapor pressure of water	p. 161, App. B
Freezing-point-depression constants	p. 480